Quick Reference

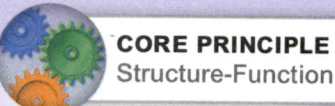
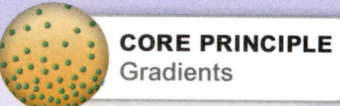
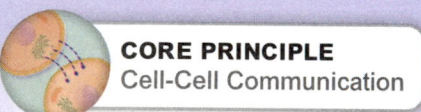
Amerman guides you every step of the way with...

Coaching that clarifies tough concepts

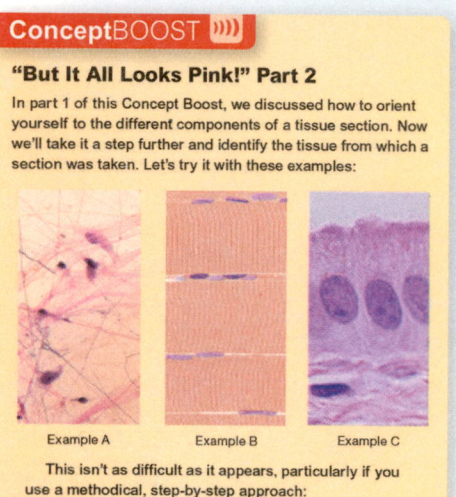

Big Picture Animations that bring **Big Picture Figures** to life and help reinforce key ideas

One-concept-at-a time art with **Big Picture** figures help you focus on key concepts

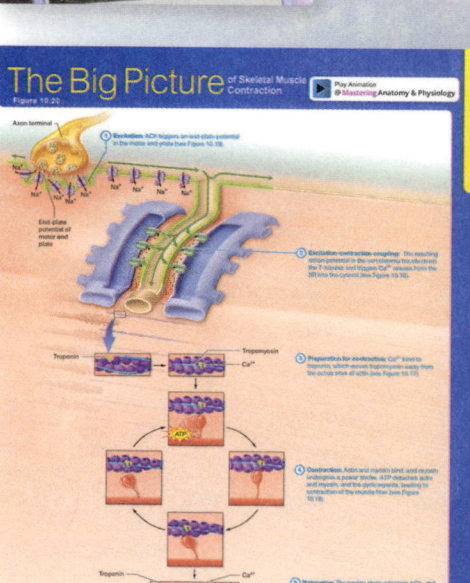

...and easy-to-assign media that maximizes student learning

Assignable in Mastering™ Anatomy & Physiology	Available in Mastering Study Area
Get Ready for A&P Diagnostic Test	—
Get Ready for A&P Learning Styles Assessment	—
Book-specific:	
NEW! Building Vocabulary activities	—
NEW! Flashback questions	—
NEW! Core Principle questions	—
NEW! Apply What You Learned questions	—
EXPANDED! Big Picture Animations	✓
UPDATED! Practice Tests and Quizzes	✓
EXPANDED! Dynamic Study Modules (now customizable)	✓
Concept Boost Video Tutors	✓
Chapter Running Case Studies	✓
General:	
EXPANDED! Interactive Physiology 2.0 Tutorials	✓
NEW! Interactive Physiology Animations	✓
A&P Flix 3D Animations	✓
Clinical Case Studies	✓
PAL Test Bank	**UPDATED!** PAL 3.1 + **NEW!** PAL 3.1 Customizable Flashcards
PhysioEx Activities	✓
Bone and Dissection Videos	✓
Instructor Resources:	
Active-Learning Workbook (in Microsoft Word and PDF) and more	✓
—	Author podcasts and more

A Learner-Centered Approach to the Study of A&P

When Erin Amerman decided to write her own textbook, her mission was clear: to write a textbook that her students could read and learn from; a book that would speak to the way today's diverse students learn and study. With the overwhelmingly positive response to the first edition, it was clear that Amerman's approach resonated with both instructors and students. They singled out three aspects of Amerman's approach that were particularly effective:

One-concept-at-a-time art

Informed by the latest research on cognitive science, Amerman reduces cognitive load by chunking information and visually unpacking art one-concept-at-a-time. She then pulls all the information together in **Big Picture figures** that help students see how the distinct parts work as a whole.

Coaching throughout the book

Amerman begins with a unique **Module 1.1 How to Succeed in Your Anatomy and Physiology Course.** This first module sets the stage for the tone and hand-holding approach found in the rest of the book. **Concept Boosts** provide just-in-time coaching on tough-to-understand concepts. **Core Principles** and **Flashbacks** reinforce recurring principles and anticipate students' knowledge gaps.

Amerman's figures and coaching come to life via author-narrated **Big Picture Animations** and **Concept Boost Video Tutors**. The companion **Active-Learning Workbook** helps students actively read and engage with the chapter as they read it.

Application and Critical Thinking

Amerman provides ample opportunities for students to practice and develop critical thinking skills with **Apply What You Learned** real-world scenario questions that end each module. **Chapter running case studies** and other higher-level application questions are available in Mastering A&P.

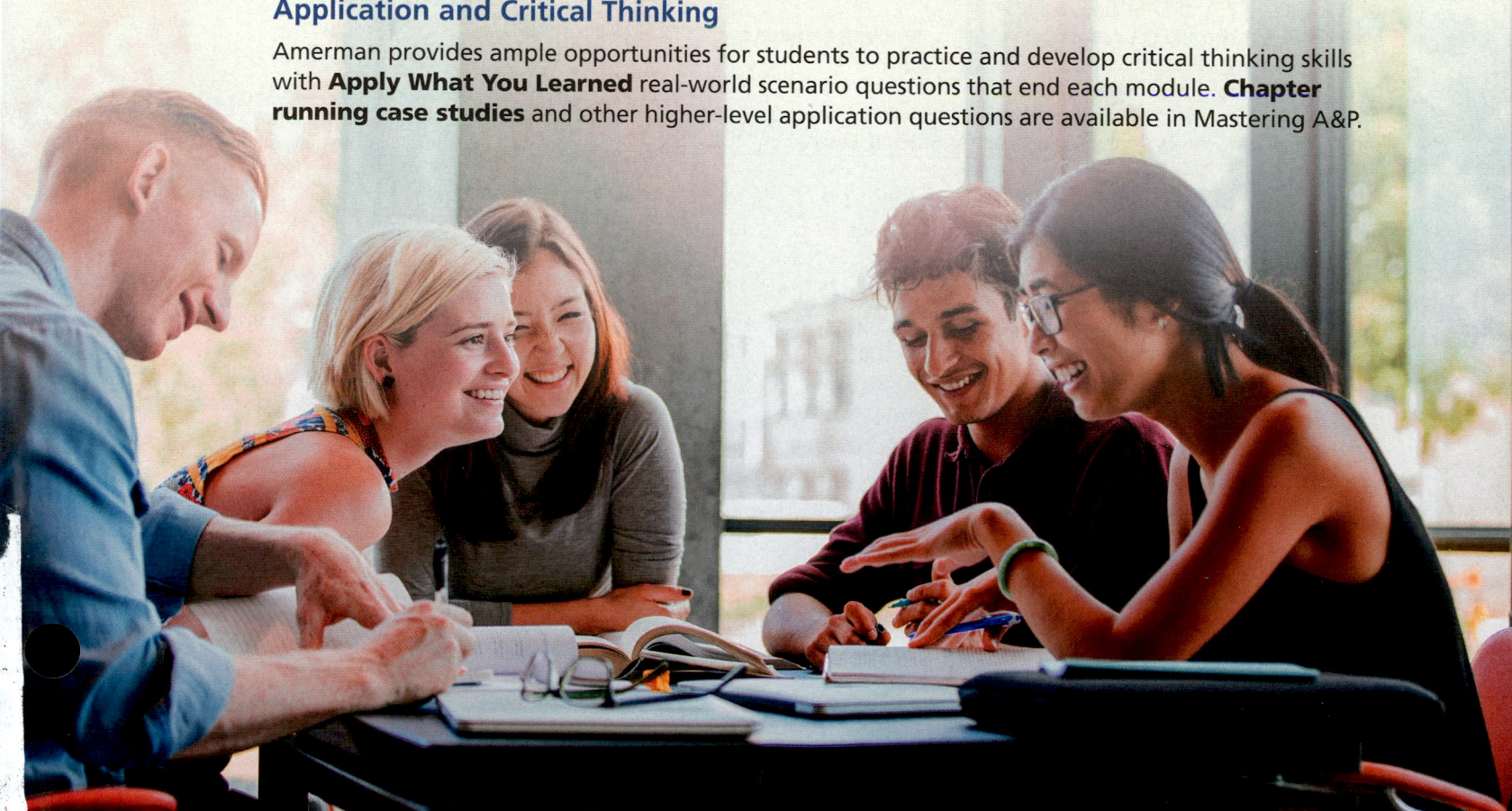

Are your students prepared for the rigor of A&P?

Module 1.1 How to Succeed in A&P

Amerman includes a special section, "How to Succeed in A&P," in Chapter 1, with a discussion about how to manage time, how to take notes, how to study for an A&P exam, and how to use the textbook, the companion *Active-Learning Workbook*, and online tools.

Flashback

Flashback questions encourage students to think about previously learned concepts they will need to apply in order to understand upcoming discussions.

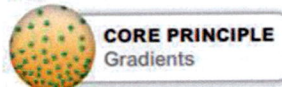

Action Potentials

« FLASHBACK

1. What are negative and positive feedback loops? (pp. 22–24)
2. What takes place during an action potential? (p. 350)

Core Principles

In Chapter 1, Amerman introduces four core principles and then highlights them throughout the textbook to remind students of the overall theme of human anatomy and physiology—homeostasis—and show how the core principles revolve around maintaining it.

Ions moving against their electrochemical gradients move via ATP-consuming pumps. One of the most important pumps in electrophysiology is the **sodium-potassium ion pump, or Na$^+$/K$^+$ ATPase,** which brings two potassium ions into the cytosol as it moves three sodium ions into the extracellular fluid. This pump maintains, and to some extent creates, the vital concentration gradients of sodium and potassium ions that exist across the plasma membrane, an example of the Gradients Core Principle (p. 28). In a neuron, as in a muscle fiber, the concentration of sodium ions is higher in the extracellular fluid than in the cytosol, **CORE PRINCIPLE** Gradients

and the opposite is true for potassium ions—their concentration is higher in the cytosol than in the extracellular fluid.

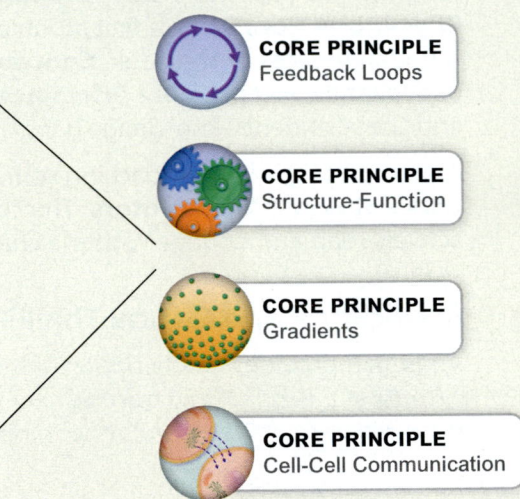

CORE PRINCIPLE Feedback Loops

CORE PRINCIPLE Structure-Function

CORE PRINCIPLE Gradients

CORE PRINCIPLE Cell-Cell Communication

Concept Boosts

In her classroom, Amerman gives her students extra coaching in advance of those tough-to-understand concepts, *right when they need it*, and she has built that same strategy into her textbook. She anticipates where students will need extra help and then provides just-in-time coaching via Concept Boosts. Each Concept Boost focuses on tough-to-understand or tricky concepts.

Concept BOOST)))

How Do Positive Ions Create a Negative Resting Membrane Potential?

Much of the *negative* resting membrane potential is caused by the movement of *positive* ions. But how can positive ions create a negative potential? To understand how this works, let's start with a muscle fiber that has no membrane potential, which means that the charges are distributed equally across the plasma membrane, or sarcolemma. In our diagram here, five positive charges and five negative charges are found on each side of the membrane:

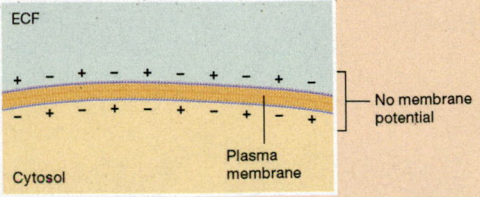

Do your students know how to read a textbook and apply the concepts?

Active-Learning Workbook

This workbook written by Amerman helps engage the kinesthetic learner with labeling, drawing, and build-your-own-summary-table exercises that students can complete as *they read the textbook*. Available in the Study Area of Mastering A&P and as editable Word files in the Instructor Resources in Mastering A&P. The print workbook can also be packaged with the Amerman textbook <u>at no additional cost</u>.

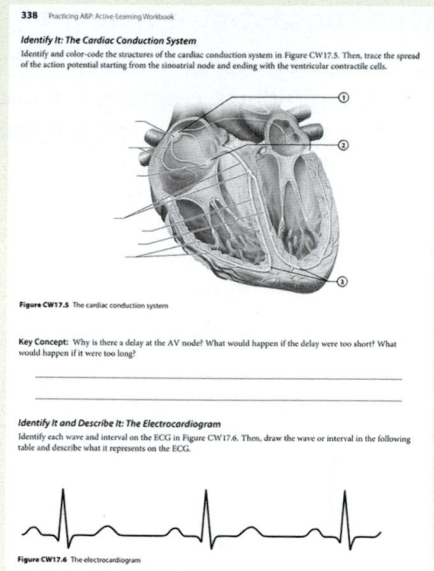

Apply What You Learned

☐ 1. The disease myasthenia gravis (my′-uss-THEE-nee-ah GRAH-viss) results in the destruction of acetylcholine receptors on the motor end plate. Predict the symptoms and effects of this disease.

☐ 2. Predict the effect of improperly functioning troponin that isn't able to bind to tropomyosin.

☐ 3. Researchers discover a genetic mutation that leads to a lack of T-tubules in a person's skeletal muscle fibers. Predict the

Apply What You Learned

Apply What You Learned questions at the end of each module ask students to think critically and apply what they've just learned to a real-world scenario.

Chapter Running Case Studies

Chapter running case studies with assessments challenge students to apply their knowledge of key A&P concepts to a real-world clinical scenario, while allowing instructors to "flip" the classroom and incorporate critical thinking and/or group activities. These cases can be found in the Instructor's Guide and are also assignable in MasteringA&P.

Are your students overwhelmed by the . . .

One-concept-at-a-time art

Drawing from her experience in the classroom and the latest research in cognitive science, Amerman reduces cognitive overload by visually unpacking key information using one-concept-at-a-time art and Big Picture figure visual summaries.

This figure shows the first concept: how excitation occurs at the neuromuscular junction.

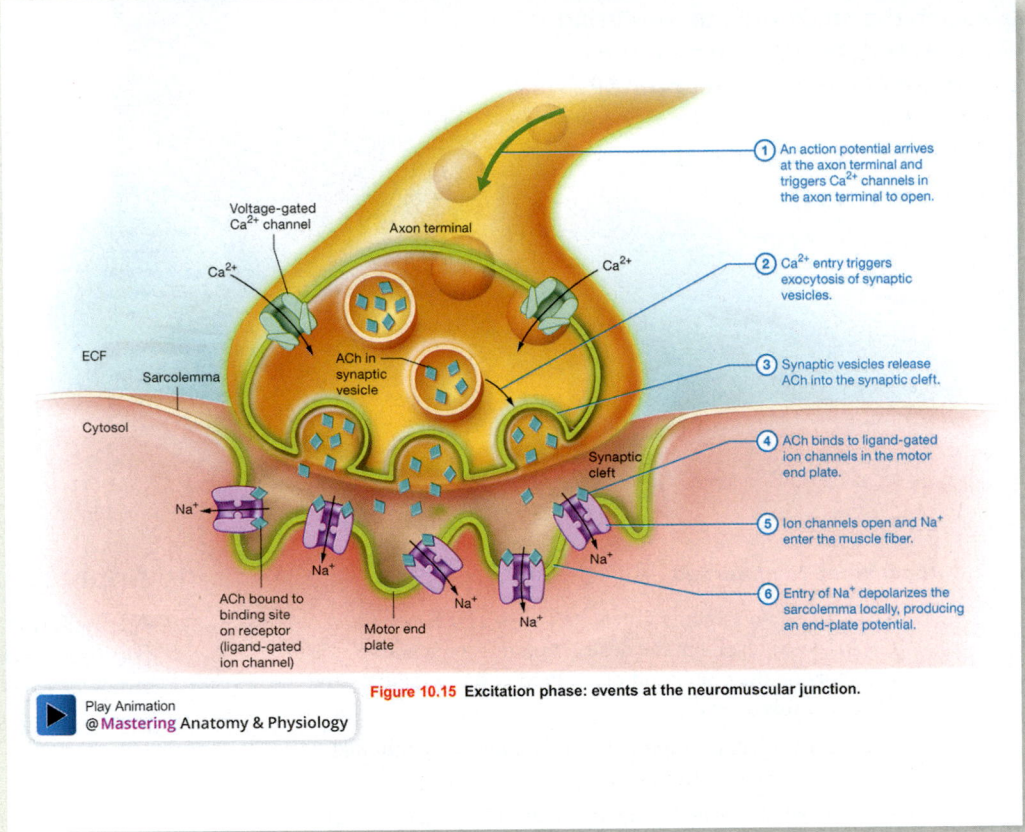

Figure 10.15 Excitation phase: events at the neuromuscular junction.

Play Animation
@ Mastering Anatomy & Physiology

This figure shows the second concept: how excitation-contraction coupling is triggered.

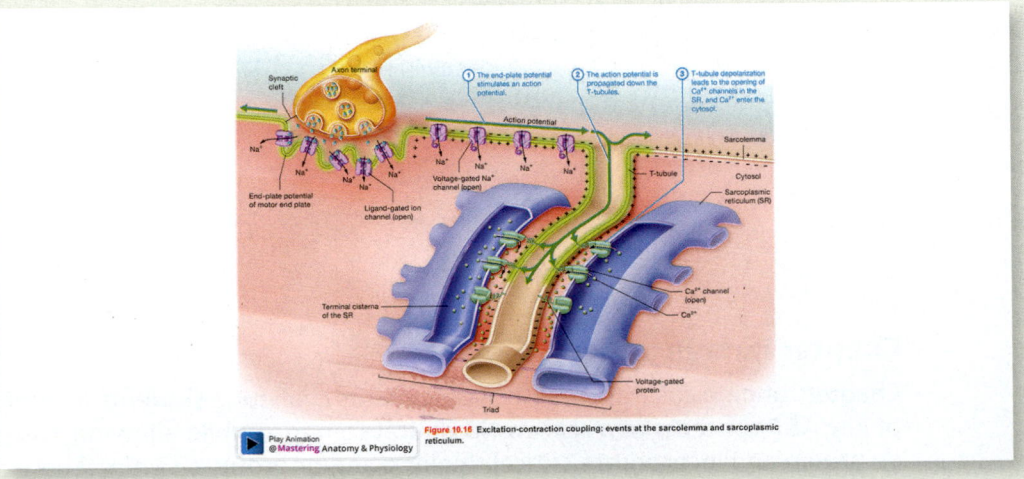

Figure 10.16 Excitation-contraction coupling: events at the sarcolemma and sarcoplasmic reticulum.

Play Animation
@ Mastering Anatomy & Physiology

... amount of information in the course?

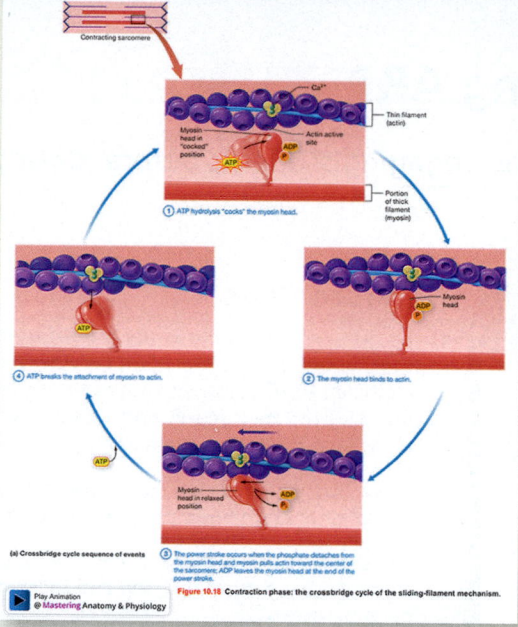

The next figure shows the third concept: the sequence of events of the crossbridge cycle.

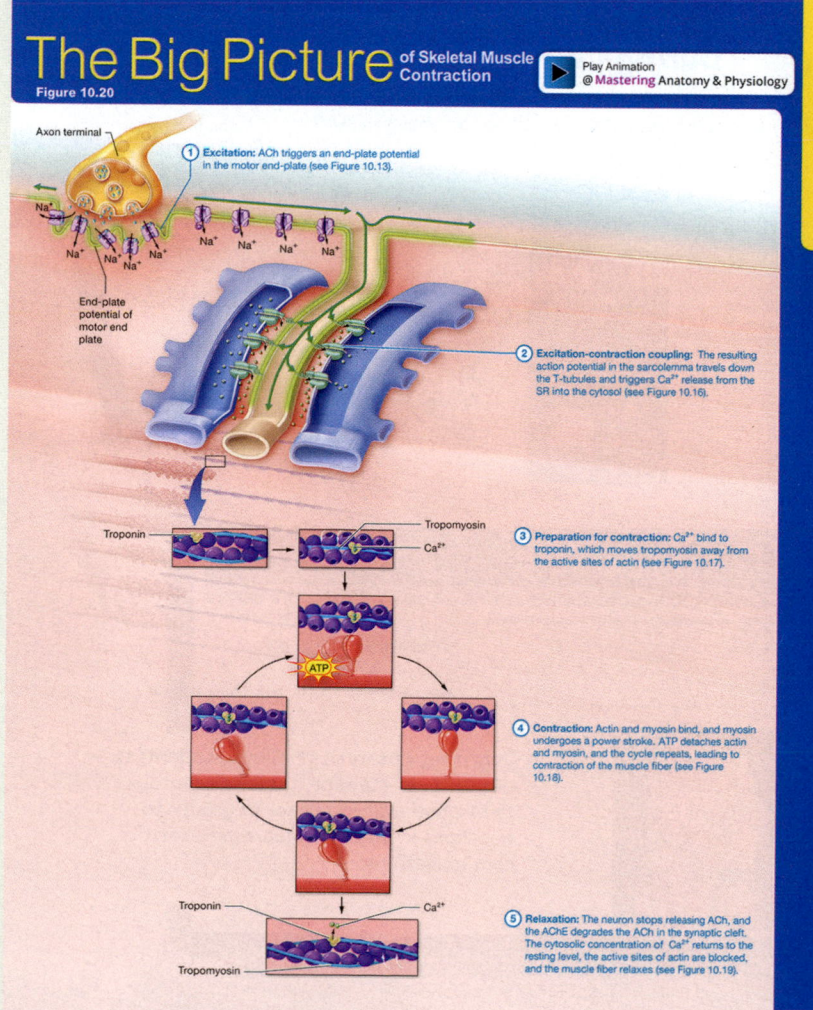

The Big Picture figure summarizes the main steps of skeletal muscle contraction.

Do your students understand A&P concepts . . .

Succeed with Mastering A&P

Mastering A&P improves results by engaging students before, during, and after class.

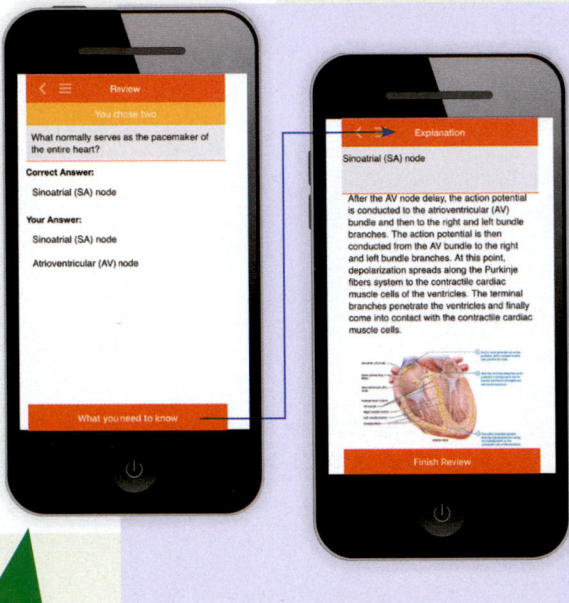

Before Class

EXPANDED! Dynamic Study Modules now include over 1300 questions. The DSMs provide students with multiple sets of questions with extensive feedback so that they can **test, learn, and retest** until they achieve mastery of the textbook material.

NEW! Instructors can now select specific questions to create more customized assignments for their students.

Pre-Class Reading Quizzes help students pinpoint concepts that they understand and concepts that they need to review.

During Class

NEW! Ready-to-Go Teaching Modules provide teaching strategies on tough topics in A&P.

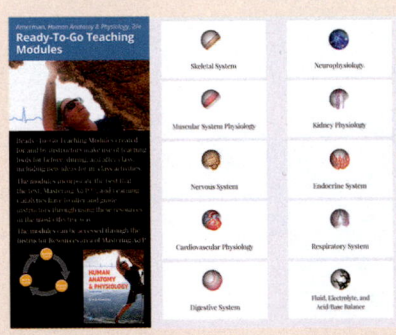

After Class

Hundreds of self-paced tutorials and coaching activities provide students with individualized coaching with specific hints and feedback on the toughest topics in the course.

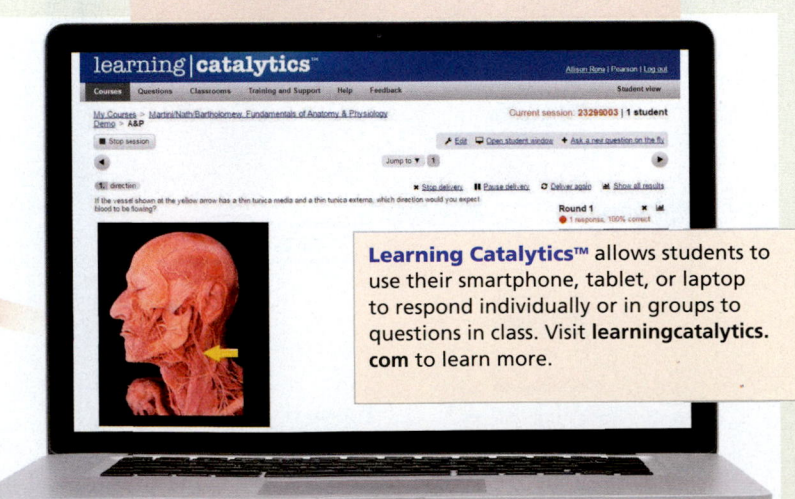

Learning Catalytics™ allows students to use their smartphone, tablet, or laptop to respond individually or in groups to questions in class. Visit **learningcatalytics.com** to learn more.

the first time they encounter them?

Get Ready for A&P Diagnostic Test

allows you and your students to quickly assess at the beginning of the course which foundational concepts students already know coming into the course and which areas will require additional remediation.

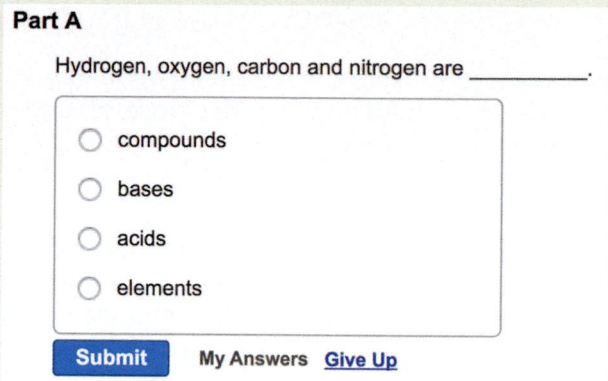

Part A

Hydrogen, oxygen, carbon and nitrogen are _____.

- ○ compounds
- ○ bases
- ○ acids
- ○ elements

Submit   My Answers   **Give Up**

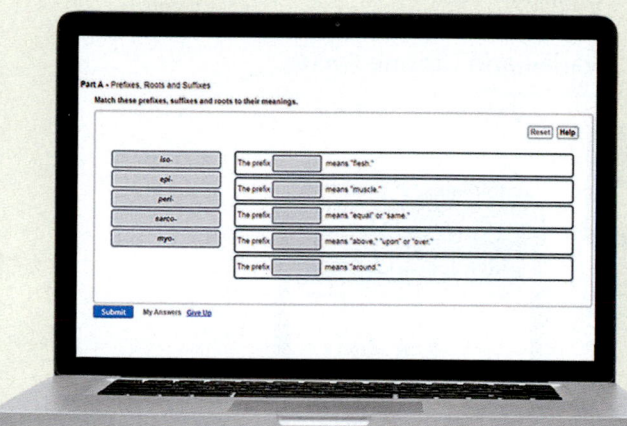

NEW! Building Vocabulary Coaching Activities

Each chapter begins with an assignable activity that gives students practice learning and using word roots in context as they learn new A&P terms.

Concept Boost Video Tutor Coaching

Activities in Mastering A&P feature videos of author, Erin Amerman, teaching directly to students and walking them through select Concept Boost topics that are tricky or tough to understand.

MyReadinessTest for A&P

MyReadinessTest for A&P assesses students' proficiency in study skills and foundational concepts in science and math, and tutors students in core areas where they need additional practice and review before they even set foot in an A&P classroom. Students can get free online access the moment they register for your course. Please contact your Pearson representative for details.

Do your students have tools to succeed in A&P outside of class?

EXPANDED! 5 The Big Picture Animation Coaching Activities (for a total of 10) in Mastering A&P animate key figures from the Amerman textbook, using the same terminology and explanations found in the book, and help students visualize key processes and reinforce the main ideas behind the process. New Big Picture animations include:

- Chapter 16 | The Big Picture of Hormonal Response to Stress
- Chapter 17 | The Big Picture of Blood Flow through the Heart
- Chapter 20 | The Big Picture of the Immune Response to the Common Cold
- Chapter 21 | The Big Picture of Respiration
- Chapter 26 | The Big Picture of Hormonal Regulation of the Ovarian and Uterine Cycles

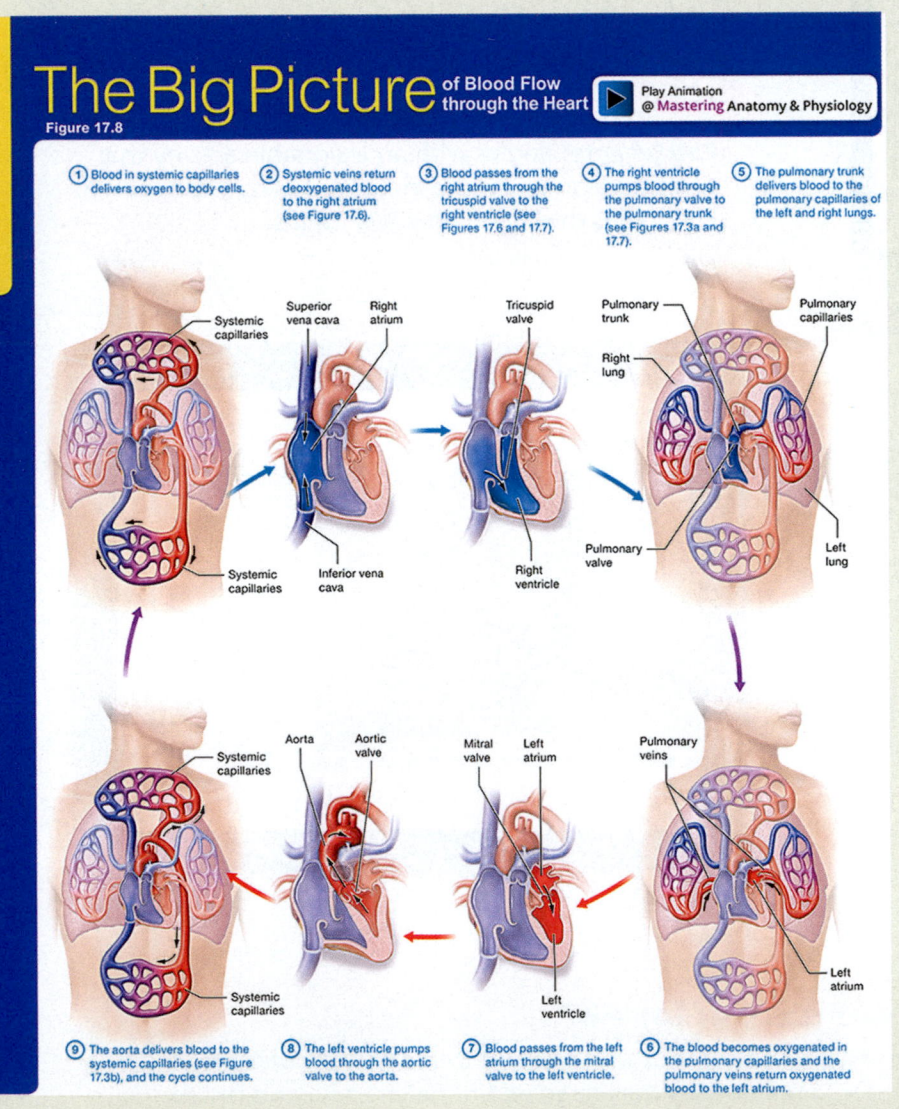

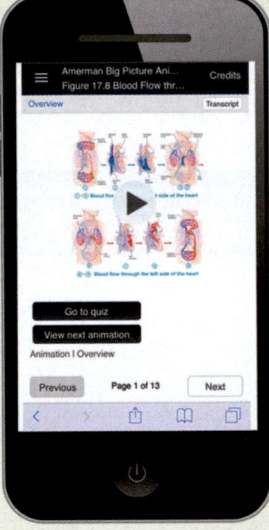

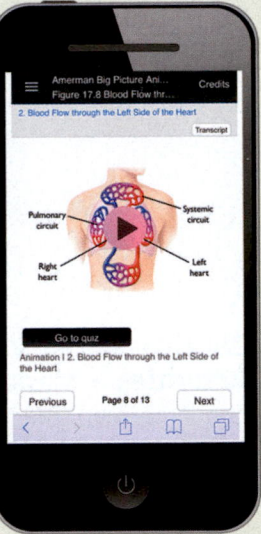

EXPANDED! Interactive Physiology 2.0 Coaching Activities in Mastering A&P help students advance beyond memorization to a genuine understanding of complex physiological processes. Fun, interactive tutorials, games, and quizzes give students additional explanations to help them grasp difficult concepts. IP 2.0 features brand-new graphics, quicker navigation, and more robust interactivity.

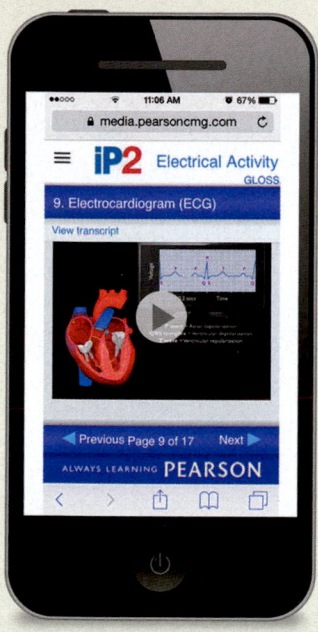

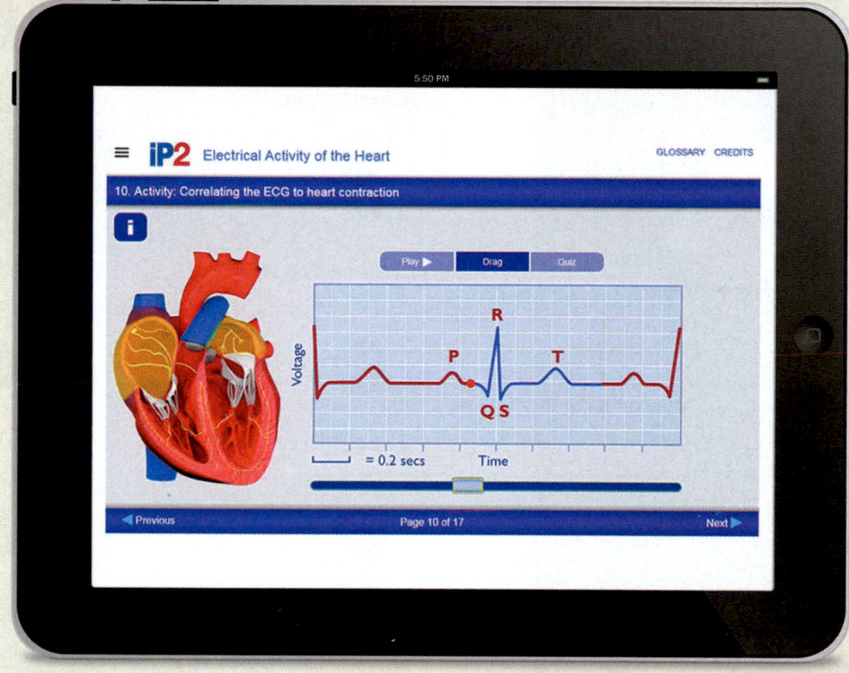

A&P Flix Coaching Activities

These 3D movie-quality animations of key physiological processes include coaching activity assignments that use a variety of question types and levels.

NEW! PAL 3.1 Flashcards

This new student tool in the Mastering A&P Study Area allows students to create a customized, mobile-friendly deck of flashcards and quizzes based on images from PAL. Students generate personalized flashcards by selecting only those structures covered in their course.

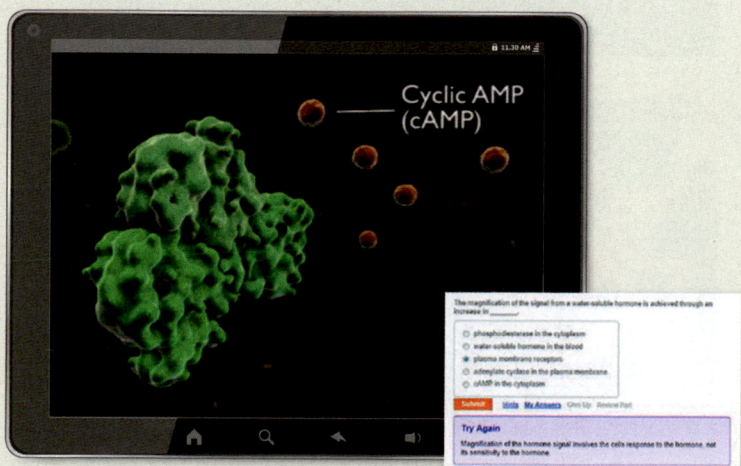

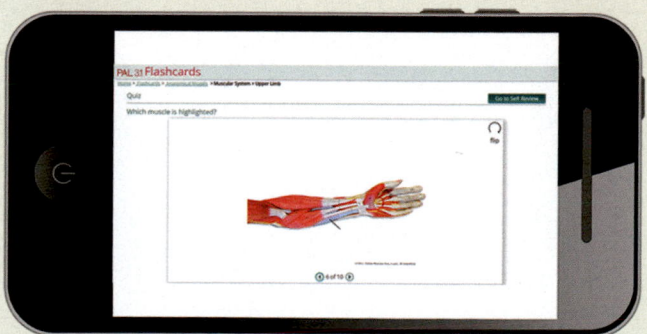

Ready-to-Go Teaching Modules for Instructors

Amerman, *Human Anatomy & Physiology*, 2/e

Ready-To-Go Teaching Modules

Ready-To-Go Teaching Modules created for and by instructors make use of teaching tools for before, during, and after class, including new ideas for in-class activities.

The modules incorporate the best that the text, Mastering A&P ™, and Learning Catalytics have to offer and guide instructors through using these resources in the most effective way.

The modules can be accessed through the Instructor Resources area of Mastering A&P.

HUMAN ANATOMY & PHYSIOLOGY
SECOND EDITION
Erin C. Amerman

- Skeletal System
- Muscular System Physiology
- Nervous System
- Cardiovascular Physiology
- Digestive System
- Neurophysiology
- Kidney Physiology
- Endocrine System
- Respiratory System
- Fluid, Electrolyte, and Acid/Base Balance

Learning Catalytics allows students to use their smartphone, tablet, or laptop to respond to questions in class. Visit learningcatalytics.com to learn more.

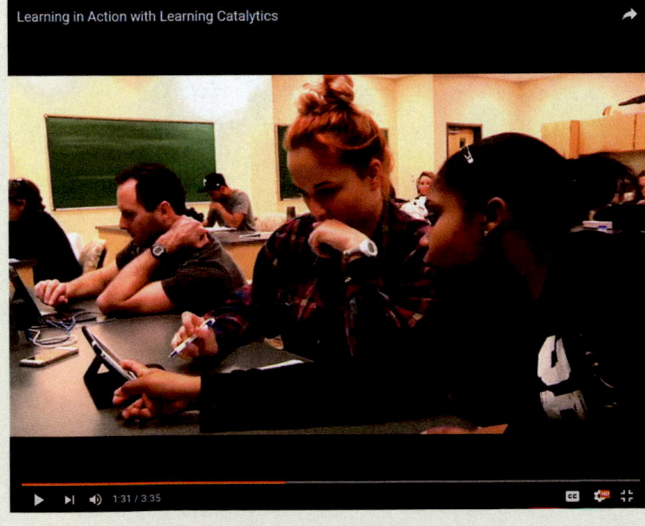

Learning in Action with Learning Catalytics

1:31 / 3:35

Access the Complete Textbook
On or Offline with Pearson eText

NEW! The **Second Edition** of Amerman's text is available in Pearson's fully accessible Pearson eText platform.

NEW! The Pearson eText mobile app offers offline access and can be downloaded for most iOS and Android phones and tablets from the Apple App or Google Play stores.

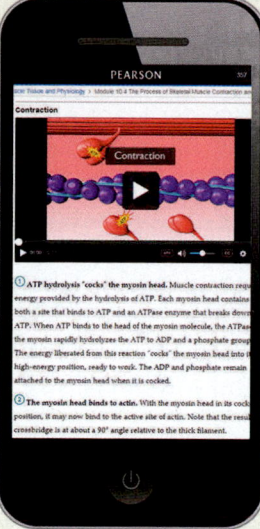

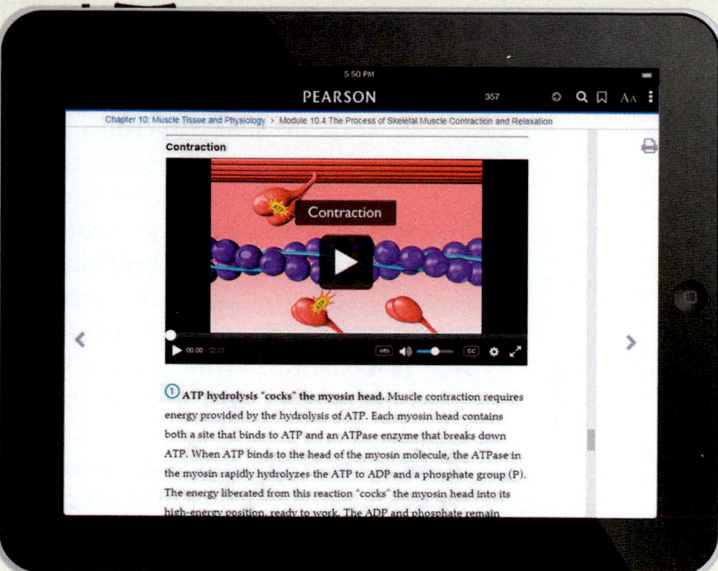

Powerful interactive and customization functions include instructor and student note taking, highlighting, bookmarking, search, and links to glossary terms. The Amerman eText also includes dozens of embedded videos and animations that bring A&P concepts to life.

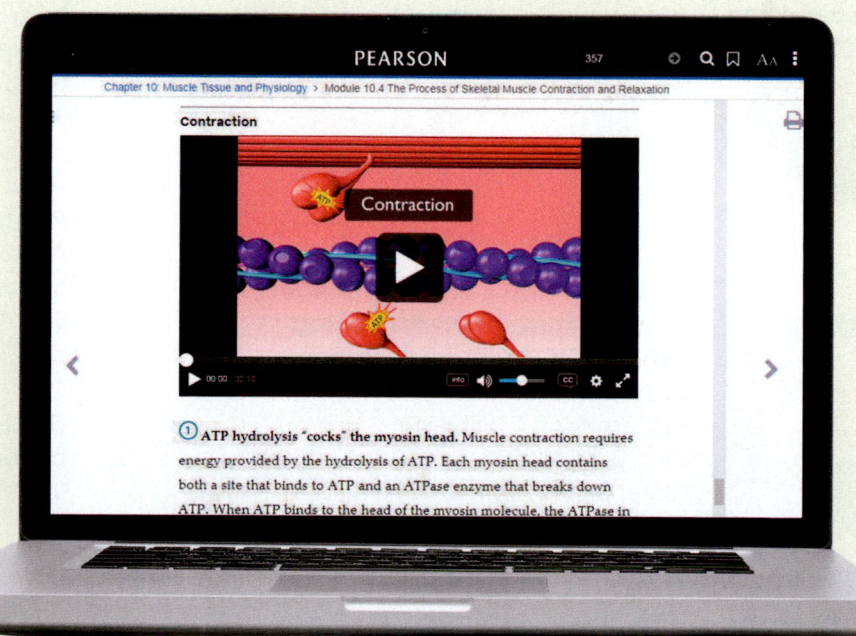

Instructor & Student Supplements

Instructor Resources in Mastering A&P include:

- o **NEW!** **Ready-to-Go Teaching Modules** provides teaching tools for 10 challenging topics in A&P.
- o **PowerPoint Lecture Presentations** for each chapter with lecture notes, editable figures, tables, Core Principle references, and embedded Big Picture Animations, Concept Boost Video Tutors, Interactive Physiology 2.0 Animations and A&P Flix Animations.
- o **All the figures, photos and tables from the text** are available in JPEG and PowerPoint formats in labeled and unlabeled versions, and with customizable labels and leader lines.
- o **Instructor Guide with Chapter Running Case Studies** in Microsoft Word and PDF format and **Active-Learning Workbook** in Microsoft Word and PDF format
- o **Student Lecture Outlines** (with Answers and without Answers) in Microsoft Word and PDF format
- o **Clinical Case Studies with Teaching Strategies** and Worksheets with Answer Keys
- o **Test Bank** questions in TestGen software and in Microsoft Word
- o **Interactive Physiology 2.0** including PowerPoints, Worksheets and Answer Keys and **Interactive Physiology 1.0 Animations**
- o **PAL 3.1 Test Bank** and Additional PAL Instructor Resources and **PhysioEx 9.1 Test Bank** and Instructor Guide

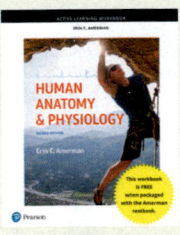

Active-Learning Workbook, 2/e
by Erin C. Amerman
0-13-475750-5 / 978-0-13-475750-6

Instructor Resources DVD with PowerPoint Lecture Outlines 2/e
by Suzanne Pundt
0-13-475758-0 / 978-0-13-475758-2

Printed Test Bank for Human Anatomy & Physiology, 2/e
by Patty Bostwick Taylor
0-13-475756-4 / 978-0-13-475756-8

Human Anatomy & Physiology Laboratory Manual: Making Connections, 2/e
By Catharine C. Whiting
Cat: 0-13-460911-5 / 978-0-13-460911-9
Main: 0-13-474643-0 / 978-0-13-474643-2
Pig: 0-13-474645-7 / 978-0-13-474645-6

Get Ready for A&P, 3/e
by Lori K. Garrett
0-32-181336-7 / 978-0-32-181336-7

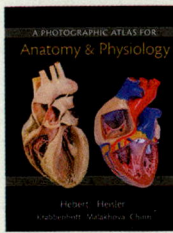

A Photographic Atlas for A&P
by Nora Hebert, Ruth E. Heisler, Jett Chinn, Karen M. Krabbenhoft, Olga Malakhova
0-32-186925-7 / 978-0-32-186925-8

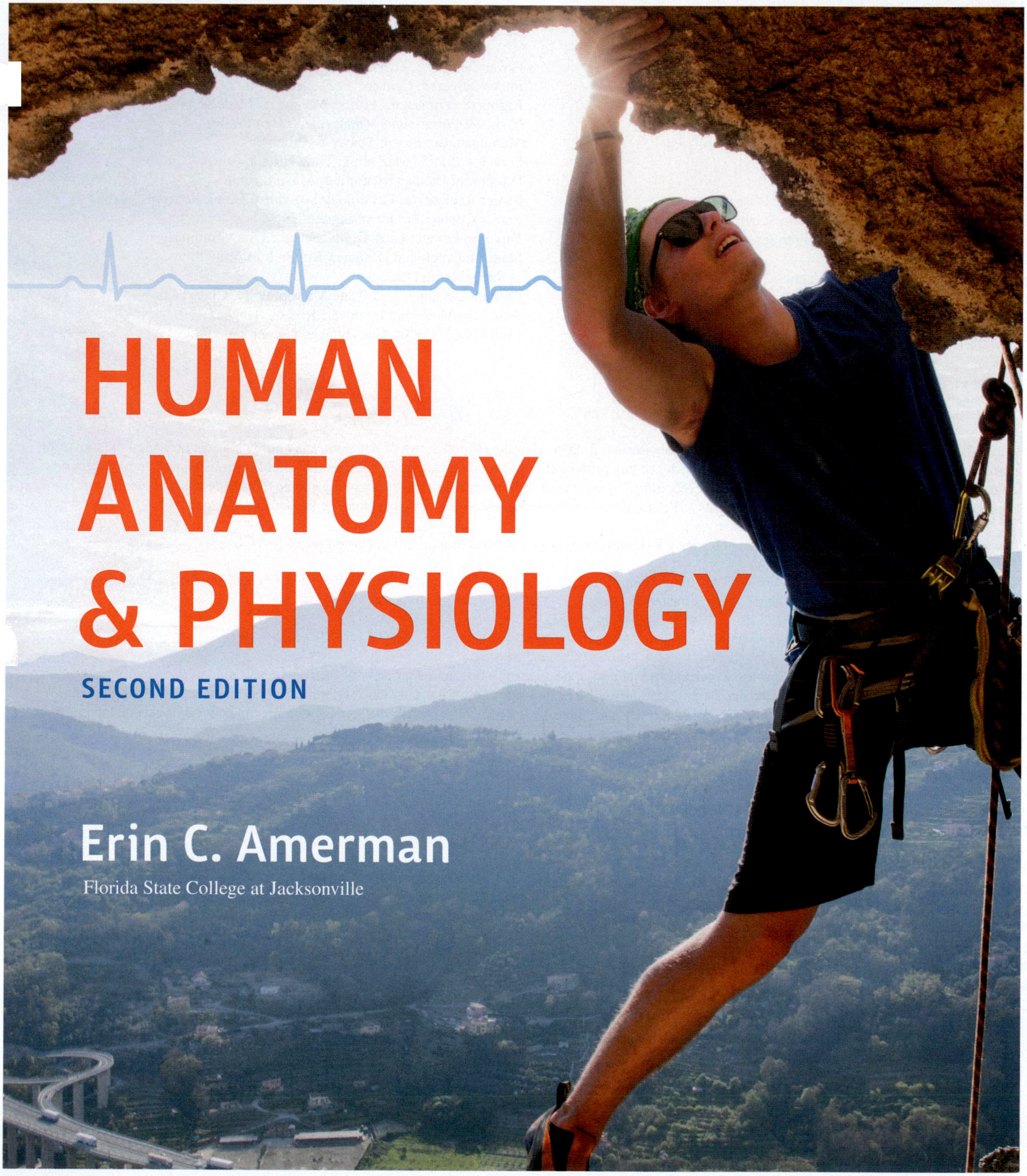

HUMAN ANATOMY & PHYSIOLOGY

SECOND EDITION

Erin C. Amerman
Florida State College at Jacksonville

Pearson

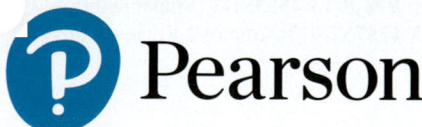

Editor-in-Chief: Serina Beauparlant
Courseware Portfolio Manager: Jennifer McGill Walker
Content Producer: Jessica Picone
Managing Producer: Nancy Tabor
Courseware Director, Content Development: Barbara Yien
Courseware Sr. Analyst: Suzanne Olivier
Courseware Specialist: Laura Southworth
Courseware Editorial Assistant: Dapinder Dosanjh
Associate Rich Media Content Producers: Sarah Shefveland and Libby Reiser
Full-Service Vendor: Cenveo® Publisher Services
Art House: Imagineering
Copyeditor: Joanne E. (Bonnie) Boehme
Compositor: Cenveo® Publisher Services
Design Manager: Mark Ong
Interior Designer: Jeff Puda

Cover Designer: Jeff Puda
Photo Research: Cenveo® Publisher Services
Rights & Permissions Project Manager: Karin Kipp
Rights & Permissions Management: Ben Ferrini
Manufacturing Buyer: Stacey Weinberger
Director of Field Marketing: Tim Galligan
Director of Product Marketing: Allison Rona
Senior Anatomy & Physiology Specialist: Derek Perrigo
Senior Content Producer: Lauren Hill
Director, Production & Digital Studio: Laura Tommasi
Senior Instructional Designer, Big Picture Animations:
    Sarah Young-Dualan
Content Developer, A&P and Microbiology: Cheryl Chi
Associate Mastering Producer: Kristen Sanchez
Cover Photo Credit: GettyImages/Ascent Xmedia

Library of Congress Cataloging-in-Publication Data
Names: Amerman, Erin C., - author.
Title: Human anatomy & physiology / Erin C. Amerman.
Other titles: Human anatomy and physiology
Description: Second edition. I NY, NY : Pearson, [2019] I Includes
    bibliographical references and index.
Identifiers: LCCN 2017039616I ISBN 978-0-13-455351-1 (student edition : alk.
    paper) I ISBN 0-13-455351-9 (student edition : alk. paper) I ISBN
    978-0-13-475752-0 (instructor's review copy : alk. paper) I ISBN 0-13-475752-1
    (instructor's review copy : alk. paper)
Subjects: I MESH: Anatomy I Physiological Phenomena
Classification: LCC QP34.5 I NLM QS 4 I DDC 612--dc23
LC record available at https://lccn.loc.gov/2017039616

1 17

ISBN 10: **0-13-455351-9**; ISBN 13: **978-0-13-455351-1** (Student edition)
ISBN 10: **0-13-475752-1**; ISBN 13: **978-0-13-475752-0** (Instructor's Review Copy)

For Elise, who performs amazing—and sometimes borderline unnatural—feats with the human body.

About the Author

Erin C. Amerman

Dr. Erin Amerman teaches anatomy and physiology at Florida State College at Jacksonville; she has been involved in anatomy and physiology education for more than 17 years as an author and professor. She received a B.S. in cellular and molecular biology from the University of West Florida and a doctorate in podiatric medicine from Des Moines University. She is also the author of the best-selling *Exploring Anatomy and Physiology in the Laboratory* series, now in its third edition, with Morton Publishing Company. Dr. Amerman is deeply committed to helping her students succeed in the A&P course. One of her main goals is to generate curiosity and excitement about the material and its application in their future health careers. A long-time member of the Human Anatomy and Physiology Society (HAPS), she enjoys attending the annual HAPS conferences, especially when they are in locations that have ample hiking opportunities and many different species of snake.

When not writing or teaching, Dr. Amerman enjoys spending time with her family and their menagerie of rescued cats, dogs, cows, snakes, fish, and a turtle. She also tries to not get injured while practicing karate and kobudo and likes learning new languages. A favorite pastime is photographing the local wildlife around her home in rural northern Florida, and she can often be found hiking around the prairie with a camera and snake hook in hand.

About the Media Author

Virginia Irintcheva

Virginia Irintcheva is the author of the Big Picture Animations and interactive figures and assessments in MasteringA&P®. She also served as an editorial advisor and reviewer of key concepts in this book. She is a biology professor at Truckee Meadows Community College in Reno, Nevada. Virginia was previously associate professor of biology at Black Hawk College in Moline, Illinois. She has been teaching courses in anatomy and physiology, medical terminology, and biology for over 12 years. Virginia was born and raised in Bulgaria. She earned a B.S. in biology and Spanish from St. Louis University, under whose auspices she studied in Madrid, Spain, as well as in Missouri. She received her Ph.D. in pharmacological and physiological sciences in 2006 from that school. Her primary research interests were in cell signaling and, specifically, the effects of the coagulation protein thrombin and the family of Ras proteins on cell growth and proliferation. As a teacher, her fundamental goal is to create an environment of effective communication by promoting mutual respect and successful learning. Virginia strives to teach her students not only what to learn but also how to learn and why they are learning.

Outside the classroom, Virginia likes to travel, visit family in Europe, snowboard, hike with her dog, spoil her cat, and ride her horse.

To the Student

Welcome to the fascinating study of the human body! Though you and I might never meet in person, I consider you and every other student who uses this textbook to be "my" student. Just as I want to ensure the success of the students in my classroom, I am similarly invested in your success. For this reason, this book was designed with *you* in mind—every feature, study tool, and media presentation is intended to help you achieve your goals.

This book was written not only for you, but also *about* you. The great thing about human A&P is that no matter what your goals are, it is relevant to your life. Human A&P is you; it's also me, your family, your friends, and indeed every human who ever lived or will live. There's nothing in the study of A&P that is irrelevant or esoteric, because every single detail revolves around you and your life. How many other courses can make that claim?

So dive right in and begin to explore the science of you. I sincerely hope that you enjoy your study of human A&P and find it as fascinating and wondrous as I do.

—Dr. A.
erin.amerman2018@gmail.com

To the Instructor

Why I Wrote This Book

I get the question, "Why did you write this book?" quite regularly. The short answer to this is that writing and teaching are just in my DNA somewhere. For the long answer, we have to look back in time and start with my 5-year-old self.

When I was in kindergarten, I was placed with another kid, Kyle, into a separate group for reading time because we were the only two kids in the class who could already read. It struck my 5-year-old brain as inconceivable that so many of my classmates couldn't read. Reading was so *easy;* anyone could do it! Maybe, I reasoned, they just needed a book to teach them how to read. So I gathered up some construction paper and crayons and got to writing. And thus my first "textbook" was born: *The Bird and Mr. Bear.*

Fast forward a few years to my medical school education. While in medical school, I co-taught a human physiology course, and during my first class I had one of those "aha" moments: Teaching somehow just "felt right." I connected with my students, and they connected with me. This feeling only grew over the next two semesters. But still, I was in medical school, and who would be crazy enough to go through the pain of medical school, graduate, and then not ever practice as a physician?

Well, it turns out that *I* was crazy enough to do just that. I was lucky enough to find a full-time position teaching anatomy and physiology. And while I loved teaching, there were far more challenges than I had anticipated. My students were different from my former classmates. The difference wasn't in intelligence—my students were smart. But, this new generation of students seemed to be ill prepared for the rigors of a college science course. They lacked study skills, they had little to no background in science, and—alarmingly—they couldn't read or understand their textbooks. For these reasons, so many bright, motivated students struggled with the course.

As a teacher, this was the last thing I wanted to see. So I did the same thing I did in kindergarten: grabbed some paper and started writing. First came my own lab exercises, which were followed by lecture outlines and notes. As I wrote, I "Amermanized" the content (a term coined by a student) with concise prose, simple diagrams, stories/analogies, and active-learning exercises. My students' responses were enthusiastic; indeed, many asked if they could return their textbooks and just use my notes instead.

A vision for a new textbook began to form in my mind: one for today's students. It would:

- be written at a level my students could understand and, at the same time, still provide the information they need;
- anticipate where they need help with the science and provide the necessary in-the-moment coaching; and
- reduce cognitive overload and present information—in both text *and* art—in manageable chunks that are more easily digestible.

Eight years later, in 2014, my vision finally became reality with the publication of *Human Anatomy & Physiology.* Today I am thrilled with the very positive response to the first edition and am happy to now be able to offer the second edition. We worked very hard to ensure that it is even stronger in motivating and helping students learn. This is what I have wanted since *The Bird and Mr. Bear*—to help people learn.

Key Features

Many of the key features found in this textbook, the companion workbook, *Active-Learning Workbook,* and media came directly from my experience teaching and working with a range of students and seeing what helps them learn. These features include the following:

- Module 1.1 **How to Succeed in A&P** in Chapter 1 introduces students to core study skills, including how to manage time, how to take notes, and how to study for an A&P exam. I also guide students through how to use the textbook, workbook, and online tools.
- Recurring **Core Principles** icons appear throughout the book and remind students to recall and apply four core principles introduced in Chapter 1: Structure-Function, Feedback Loops, Gradients, and Cell-Cell Communication.
- Over 50 **Concept Boosts** and **Study Boosts** coach students on key A&P concepts that are often difficult or complex. Additional

emphasis is placed on explaining challenging topics, often incorporating familiar analogies and simple illustrations, giving students a boost in fully understanding the content.

- **Concept Boost Video Tutors** walk students through select Concept Boost topics that are particularly tough to understand. **Concept Boost Video Tutors include**:

  ○ Chapter 3: *Understanding Water Movement in Osmosis*
  ○ Chapter 10: *How Do Positive Ions Create a Negative Resting Membrane Potential?*
  ○ Chapter 16: *Understanding the Relationship between Negative Feedback Loops and Thyroid Function*
  ○ Chapter 21: *Making Sense of the Oxygen-Hemoglobin Dissociation Curve*
  ○ Chapter 25: *How Can Respiratory Changes Compensate for Metabolic Acidosis?*

These Video Tutors are assignable in MasteringA&P® and are also available in the Study Area of MasteringA&P.

- **One-concept-at-a-time art** focuses on teaching one concept per figure so that a student can instantly grasp the key idea without being distracted by a sea of details. For key physiology concepts, unique sequence figures unpack information systematically so that each scene contains only the most important information, again making it easier for today's students to focus on key details.
- **In-text simple illustrations** appear as needed to help students visualize concepts being described.
- **Big Picture figures** visually summarize key physiological processes and anatomy concepts, highlighting only what is most important.
- Mobile-ready **Big Picture Animations** with interactive quizzes bring the Big Picture figures to life and help reinforce students' understanding of each step in a key process. These animations are assignable in MasteringA&P and are also available in the Study Area of MasteringA&P.
- **HAPS-based Learning Outcomes** begin each module within a chapter. Additionally, the assessments in MasteringA&P are organized by these Learning Outcomes.
- **Pronunciations** use phonetic sounds (instead of traditional symbols) to help students learn correct pronunciations.
- **Flashback** questions encourage students to think about previously learned concepts they will need to apply in order to understand upcoming discussions.
- **Quick Check** questions appear throughout each module to test students' basic understanding of the material. Answers to Quick Check questions are available in the Study Area of MasteringA&P.
- **Apply What You Learned** questions at the end of each module ask students to think critically and apply what they've just learned to a real-world scenario. Answers to Apply What You Learned questions can be found in Appendix A.
- **A&P in the Real World** features highlight clinical conditions and disorders that illustrate and reinforce key A&P concepts discussed in the chapter.
- **Chapter running case studies** with assessments challenge students to apply their knowledge of key A&P concepts to a real-world clinical scenario, while allowing instructors to "flip" the classroom and incorporate critical thinking and/or group activities. These cases are assignable in MasteringA&P. They can also be found in MasteringA&P both in the Instructor Resources and in the Study Area.
- *Active-Learning Workbook* provides students with a kinesthetic learning modality. It includes labeling, drawing, and build-your-own summary-table exercises that students can complete as they read the textbook. This workbook can be packaged as a print supplement with the Amerman textbook at no additional cost. It is also available in MasteringA&P as an editable Word document and as a downloadable PDF in the Instructor Resources and in the Study Area.

What's New in the Second Edition

- IMPROVED one-concept-at-a-time art and Big Picture figures: The Amerman art program is already touted for visually unpacking concepts, making it easier for students to focus on the key ideas. The second edition of Amerman builds on this one-concept-at-a-time approach by including over 20 new or revised critical figures that conclude with Big Picture visual summaries (e.g., see newly revised The Big Picture of Chemical Synaptic Transmission, The Big Picture of Pulmonary Ventilation, The Big Picture of Tubular Reabsorption and Secretion).
- NEW summary tables help consolidate and summarize key information (e.g., Figure 3.5: Functions of membrane proteins; Figure 10.1: Three types of muscle tissue; Figure 12.9: Structures of the limbic system; Table 20.1: Summary of the First and Second Lines of Defense; Table 21.2: Accessory Muscles of Inspiration and Expiration).
- EXPANDED coaching throughout: Building off the very positive feedback to Amerman's coaching in the first edition, the second edition features several new or revised Concept Boost discussions. The Concept Boosts help demystify foundational principles or concepts that students often find to be stumbling blocks (e.g., new Concept Boosts in Chapter 3: Is Osmosis the Diffusion of Water?; Chapter 10: Connecting the Crossbridge Cycle to the Sliding-Filament Mechanism).
- EXPANDED animations: 5 Big Picture Animations (for a total of 10) animate key figures from the Amerman textbook, using the same visuals, terminology, and explanations found in the book. This helps students visualize key processes and reinforces the main ideas behind the process. These animations are assignable as coaching activities in MasteringA&P and are available in the Study Area of MasteringA&P.

The Big Picture Animations from the first edition included Chapter 3: The Big Picture of Protein Synthesis; Chapter 10: The Big Picture of Skeletal Muscle Contraction; Chapter 11: The Big Picture of Action Potentials and the Big Picture of Chemical Synaptic Transmission; and Chapter 24: The Big Picture of Renal Physiology.

The new Big Picture Animations for the second edition include:

○ Chapter 16: The Big Picture of Hormonal Response to Stress
○ Chapter 17: The Big Picture of Blood Flow through the Heart
○ Chapter 20: The Big Picture of the Immune Response to the Common Cold
○ Chapter 21: The Big Picture of Respiration
○ Chapter 26: The Big Picture of Hormonal Regulation of the Ovarian and Uterine Cycles

● NEW Practice Tests via QR codes at the end of every chapter provide students with quick on-the-go practice on their smartphones, tablets, and computers.

What's New in MasteringA&P

Please see the front of this book for information on the new media and assignments for the second edition of Amerman with MasteringA&P.

Chapter-by-Chapter Changes in the Second Edition

Chapter 1: Introduction

Module 1.1: How to Succeed in Your Anatomy and Physiology Course
● "Learning styles" was changed to "learning modalities."

Module 1.3: The Language of Anatomy and Physiology
● New A&P in the Real World box on Medical Errors has been added.
● In Figure 1.7: Regional terms, part a, the length of the bracket for Cephalic term was fixed.

Module 1.4: Organization of the Human Body
● "Dorsal" was changed to "posterior" and "ventral" to "anterior" when discussing body cavities, including in Figure 1.9: The posterior and anterior body cavities.

Module 1.5: Core Principles in Anatomy and Physiology
● Receptors in Figure 1.13: Comparison of how negative feedback mechanisms control room and body temperature, part b, were changed to those in skin.
● An additional paragraph example of positive versus negative feedback processes was added.
● Figure 1.18: Core principles icons, was updated to combine illustrations and icons with definitions and examples of each core principle.

Chapter 2: The Chemistry of Life

Module 2.2: Matter Combined: Mixtures and Chemical Bonds
● In Figure 2.6: Nonpolar versus polar covalent bonds, the number of electrons in the valence shell of the water molecule has been corrected to be eight.
● In the Concept Boost: Determining the Type of Bonds in a Molecule or Compound, three new small illustrations of the possible atomic and molecular combinations have been added.

Module 2.3: Chemical Reactions
● The text discussing Figure 2.10: Enzyme-substrate interaction, has been updated to cover each step of the process.

Module 2.4: Inorganic Compounds: Water, Acids, Bases, and Salts
● In Figure 2.11: The behavior of hydrophilic and hydrophobic molecules in water, the illustrations have been updated with larger blow-ups that show hydration spheres and label partial charges.

Module 2.5: Organic Compounds: Carbohydrates, Lipids, Proteins, and Nucleotides
● In Figure 2.23: Levels of protein structure, the primary structure (part a) has been redone to show the atomic components of the amino acid molecules, and the secondary structures (part b) have been reworked to show the atoms of the alpha helix and beta-pleated sheet structures. In addition, in part d, the quaternary structures, one chain in each structure now shows the underlying tertiary structure.
● Under ATP, a bit of additional explanation of how ATP is used as energy storage for cellular processes was added.
● In Figure 2.26: Structure of the nucleic acids DNA and RNA, the top adenine in part a has been changed to the correct purple color.

Chapter 3: The Cell

Module 3.2: Structure of the Plasma Membrane
● Figure 3.5: Functions of membrane proteins, has been made into an illustrated table.

Module 3.3: Transport across the Plasma Membrane
● The section on Osmosis has been updated to prevent the over-simplification that this process is the diffusion of water, with the addition of the concept of osmotic pressure. In addition, Figure 3.8: Passive transport: osmosis, has been reworked to clarify the concentrations of solute in all compartments. Also, the Concept Boost has been replaced with a new one—called Is Osmosis the Diffusion of Water? Taking a Closer Look—that includes a new illustration.

Module 3.4: Cytoplasmic Organelles
● In Figure 3.15: The cell and its organelles, labels have been added for additional structures such as those in the nucleus.

Chapter 4: Histology

Module 4.1: Introduction to Tissues
● The first edition (1e) A&P in the Real World box on Diseases of Collagen and Elastic Fibers has been refocused in second edition (2e) on Marfan Syndrome.

Module 4.2: Epithelial Tissues
● In Figure 4.5: Structure of simple epithelia, orientation diagrams have been added to each part showing where in the body the examples were taken from.
● In Figure 4.10: Multicellular exocrine glands, the structures of the compound acinar and compound tubuloacinar glands have been updated and clarified.
● There is now a mention of apocrine secretion under Glandular Epithelia.

Module 4.7 Membranes
- In Figure 4.24: True membranes, in part b on synovial membranes, the orientation diagram has been enlarged and clarified.

Chapter 5: The Integumentary System

Module 5.1: Overview of the Integumentary System
- Figure 5.1: Basic anatomy of the skin, has been enhanced to clarify the structures shown. This enhanced art has been used as orientation diagrams throughout the chapter as appropriate.
- In Figure 5.2: Homeostatic regulation of the body temperature by the integumentary system, part a illustrating the response to rising body temperature has been clarified by showing a brain as the site of thermoreceptors.

Module 5.5: Accessory Structures of the Integument: Hair, Nails, and Glands
- In Figure 5.9: Hair structure, part b has been redone to clarify the structure of the hair bulb.
- In Figure 5.11: Sweat glands and sebaceous glands, a new part has been added showing an illustration of an apocrine sweat gland.

Chapter 6: Bones and Bone Tissue

Module 6.2: Microscopic Structure of Bone Tissue
- In Figure 6.7: Functions of osteoblasts and osteocytes, step 3 has been reworked to show that the "arms" of the osteocytes contact each other through the canaliculi.

Module 6.3: Bone Formation: Ossification
- Figure 6.11: The process of intramembranous ossification, has been reimagined to make the structures more realistic and clearer using a horizontal layout. The steps in the figure and text have been updated to clarify the description as well.
- Figure 6.12: The process of endochondral ossification, has been enhanced to clarify the structures in the process and to make it match the visual presentation in Figure 6.11.

Module 6.4: Bone Growth in Length and Width
- Figure 6.14: Growth at the epiphyseal plate, has been reimagined to show the process more clearly.

Chapter 7: The Skeletal System

Module 7.2: The Skull
- 1e Figure 7.3: Cavities of the skull, and 1e Figure 7.4: Cranial vault and base, have been switched in order to cover all of the cranial cavity before discussing other cavities.
- In Table 7.2: Bones of the skull, view identification labels have been added to all illustrations to keep students oriented.
- In Figure 7.7: Posterior, superior, and inferior views of the skull, in part c the position and extent of the vomer have been clarified.
- In Figure 7.9: Internal view of the skull, the labels have been repositioned to clarify relationships between the bones and their components.

Chapter 8: Articulations

Module 8.1: Overview of Joints
- The first module has been broadened to include an overview of joint function (moved from the 1e chapter introduction) and joint classification.

Module 8.2: Fibrous and Cartilaginous Joints
- The information on fibrous and cartilaginous joints has now been combined into the second module.
- In Figure 8.2: The two types of cartilaginous joints, the epiphyseal plate in part a has been redone to look more realistic.

Module 8.6: Types of Synovial Joints
- In A&P in the Real World, Knee Injuries and the Unhappy Triad, the text and illustration have been reworked to discuss and show the tear in the lateral meniscus.
- Figure 8.15: Anatomical structure of the shoulder joint, has been reimagined. It now has a new part a showing an anterior view of the articular capsule, and a new part c showing a frontal section.

Chapter 9: The Muscular System

Module 9.1: Overview of Skeletal Muscles
- Under Structure of Skeletal Muscles, the subheading Gross Anatomy of a Skeletal Muscle has been added to point out this coverage.
- Figure 9.2: Fascicle pattern and muscle shape, has been rearranged to clarify the relationship between the muscles and the different shapes.
- Under Lever Systems and in Figure 9.5: Lever systems, the text and illustrations have been updated to use some new analogies.
- The 1e Concept Boost on Understanding Lever Systems and Mechanical Advantage has been replaced by a Study Boost on How to Tell the Three Types of Levers Apart that uses simplified illustrations of these concepts.

Module 9.2: Muscles of the Head, Neck, and Vertebral Column
- Figure 9.10: Muscles of chewing, part b, has been replaced with a lateral view showing the pterygoid muscles.
- The Concept Boost on Demystifying Muscle Actions now includes an illustrated example using the muscles of the jaw, so the Boost has been moved up to follow the Muscles of Mastication subsection.
- In Figure 9.13: Muscles of the vertebral column, parts a and b have been combined into a single part, which shows the erector spinae group on the left side of the torso and the transversospinalis group on the right side. Part b is now the cadaver photo.
- The 1e Study Boost: Sorting Out the Erector Spinae has been deleted.

Module 9.4: Muscles of the Pectoral Girdle and Upper Limb
- In Figure 9.18: Muscles that move the scapula, the label Rotator cuff has been added.

Module 9.5: Muscles of the Hip and Lower Limb
- In Figure 9.21: Anterior and medial muscles that move the thigh and leg, parts b and c have been switched.

Chapter 10: Muscle Tissue and Physiology

Module 10.1: Overview of Muscle Tissue
- Figure 10.1: Three types of muscle tissue, has been made into an illustrated table with additional information on the location, structure, and function of each tissue type.

Module 10.2: Structure and Function of Skeletal Muscle Fibers
- The text section called Myofilament Arrangement and the Sarcomere (with 1e Figure 10.7: Structure and bands of the sarcomere, now 2e Figure 10.6) has been moved before Putting It All Together: The Big Picture of Skeletal Muscle Structure. In addition, the sarcomere has now been labeled in 1e Figure 10.6: The Big Picture of Levels of Organization within a Skeletal Muscle (2e Figure 10.8).
- Figure 10.9: The sliding-filament mechanism, has been reworked to show one relaxed sarcomere above one contracted, with all bands indicated in both.

Module 10.3: Skeletal Muscle Fibers as Electrically Excitable Cells
- Several subsections from 1e Chapter 11 (Nervous Tissue) have been moved here in order to present a more complete explanation of action potentials. These subsections, which include Ion Channels and Gradients, Generation of the Resting Membrane Potential, and The Electrochemical Gradient (with the included figures), replace the 1e section called The Na^+/K^+ ATPase Pump and the Sodium and Potassium Ion Concentration Gradients. The Concept Boost on How Do Positive Ions Create a Negative Resting Membrane Potential has also been moved here.
- Calcium ion channels have been added to the depictions of the axon terminal structure and function in 2e Figures 10.14 and 10.15 (1e Figures 10.12 and 10.13).

Module 10.4: The Process of Skeletal Muscle Contraction and Relaxation
- A new Concept Boost on Connecting the Crossbridge Cycle to the Sliding-Filament Mechanism has been added.

Module 10.6: Muscle Tension at the Fiber Level
- The 1e Concept Boost on Understanding How Events at the Myofilaments Produce Tension of a Whole Muscle has been deleted.

Chapter 11: Introduction to the Nervous System and Nervous Tissue

Module 11.1: Overview of the Nervous System
- In Figure 11.3: Summary of the structural and functional divisions of the nervous system, the presentation of the information has been reworked, and illustrations of example organs for the divisions have been added.

Module 11.3: Electrophysiology of Neurons
- A reminder section on Generation of the Resting Membrane Potential has been added, to take into account the sections on this subject that have been moved to Chapter 10.
- 1e Figure 11.11: Measurement of voltage across a plasma membrane, has been added as an orientation diagram to 1e Figure 11.14: Ion movements leading to changes in the membrane potential, forming 2e Figure 11.11 on ion movements.

- The Big Picture of Action Potentials, 1e Figure 11.20, now 2e Figure 11.17, has been enlarged and updated so the depictions of structures are clearer and more 3D.

Module 11.4: Neuronal Synapses
- The section now called Summation of Postsynaptic Potentials and Neural Integration has been moved before the section on Termination of Synaptic Transmission to reflect the chronological order of these processes.
- What was 1e Figure 11.22: The structures of electrical and chemical synapses, now 2e Figure 11.19, has been reworked to focus more on the blow-ups showing the channels and receptors.
- The 1e Concept Boost called How Summation Connects Local Potentials and Action Potentials has been deleted and the information incorporated into the Big Picture of Chemical Synaptic Transmission, 1e Figure 11.26, now 2e Figure 11.25, to illustrate the entire process in one place.
- A new illustrated Concept Boost has been added on Sorting Out the Different Types of Channels and Pumps in the Membrane of a Neuron.
- 1e Figure 11.25: Methods of termination of synaptic transmission, now 2e Figure 11.23, has been reworked to add a blow-up showing events at the synaptic cleft more clearly.

Chapter 12: The Central Nervous System

Module 12.1: Overview of the Central Nervous System
- A new A&P in the Real World box on The Myth of Brain Differences between the Sexes has been added. This box has the new subtitle Pseudoscience Exposed. This subtitle has been added to appropriate A&P boxes throughout the text.

Module 12.2: The Brain
- The section on The Limbic System has been moved after the sections on Basal Nuclei and White Matter.
- A table of information has been added to Figure 12.9: Structure of the Limbic System, to make the information more accessible.

2e Module 12.3: Homeostasis Part I: Role of the Brain in Maintenance of Homeostasis
2e Module 12.4: Higher Mental Functions
- The modules on Homeostasis Part I: Role of the Brain in Maintenance of Homeostasis and Higher Mental Functions have been moved up in the chapter to follow Module 12.2: The Brain, in order to have the discussion of brain functions follow that of brain structures directly.

Chapter 13: The Peripheral Nervous System

Module 13.3 Spinal Nerves
- In Figure 13.4: Structure and function of roots, spinal nerves, and rami, the depiction of the rami communicantes has been reworked to clarify them.
- Under Brachial Plexus, there is a new Concept Boost called Sorting Out the Brachial Plexus, with a new schematic illustration.
- In Figure 13.9: The sacral plexus, text was rearranged to clarify the relationships.

Module 13.4: Sensation Part II: Role of the PNS in Sensation
- Under Sensory Neurons, the text now clarifies when first-, second-, or third-order neurons are being discussed. This information was also added to Figure 13.16: The Big Picture of Detection and Interpretation of Somatic Sensation by the Nervous System.

Module 13.6: Reflex Arcs: Integration of Sensory and Motor Function
- Figure 13.20: The flexion and crossed-extension reflexes, has been reworked to show the reflexes separately.

Chapter 14: The Autonomic Nervous System and Homeostasis

Module 14.1: Overview of the Autonomic Nervous System
- A new Study Boost called Remembering the Difference between Preganglionic and Postganglionic Neurons has been added.

Module 14.2: The Sympathetic Nervous System
- In Figure 14.5: Three possible pathways of sympathetic preganglionic and postganglionic neurons, a third spinal cord cross section and ganglion has been added to clarify the three pathways.
- The subsection called Pharmacology and Sympathetic Nervous System Receptors has been moved to the end of this module.

Module 14.3: The Parasympathetic Nervous System
- In Figure 14.10: The main effects of the parasympathetic nervous system on target cells, the pupil has been adjusted to show that it's constricted.
- A new Concept Boost has been added to the end of this module, called Understanding the Different Effects of the Sympathetic and Parasympathetic Nervous Systems.

Chapter 15: The Special Senses

Module 15.1: Overview of the Special Senses
- The first module has been reworked to include both Comparison of the General and Special Senses, and Sensory Transduction. The 1e Concept Boost on Understanding Transduction has been deleted.

Module 15.2: Olfaction
- In Figure 15.2: Olfactory epithelium and olfactory neurons, labels have been added for the Olfactory nerve and Mitral cell.
- 1e Figure 15.4: The olfactory pathway, has been deleted, as it mostly repeated what was shown in Figure 15.2.

Module 15.4: Anatomy of the Eye
- 1e Figure 15.10: Extrinsic eye muscles, has been deleted, as this information is shown in Figure 9.9.
- Under Layers of the Eyeball, a paragraph on macular degeneration has been added.

Module 15.5: Physiology of Vision
- A new Concept Boost on Understanding Accommodation has been added.

Chapter 16: The Endocrine System

Module 16.1: Overview of the Endocrine System
- A new Figure 16.1 called Overview of Hormone Function has been added.
- The 1e section Types of Chemical Signals is now called Paracrine and Autocrine Signals to reflect the clarified text.
- The second half of the module under the heading Hormones has been reorganized. We now introduce Classes of Hormones first, then a new subsection on Hormone Transport through the Blood. Then we get to Target Cells and Receptors, followed by Mechanisms of Hormone Action. This is followed up with Effects of Hormone Actions, then Hormone Interactions, and a new subsection gathering information on Hormone Half-Life and Elimination. Regulation of Hormone Secretion is its own section following Hormones.
- 1e Figure 16.3 on mechanisms of hormone action has been split into two figures. The first, 2e Figure 16.4, now presents a G-protein second-messenger system; it is titled Mechanism of action of hydrophilic hormones via an adenylate cyclase–cAMP second-messenger system. The second, 2e Figure 16.5, is titled Mechanism of action of hydrophobic hormones via an intracellular receptor mechanism.

Module 16.2: The Hypothalamus and the Pituitary Gland
- All figures in the chapter that depict the anterior and posterior pituitary glands, particularly 2e Figures 16.8 and 16.9 (1e Figures 16.6 and 16.7), now show clarified anatomy of the infundibulum.

Module 16.3: The Thyroid and Parathyroid Glands
- MIT and DIT have been added to the discussion of thyroid hormone synthesis and its depiction in 2e Figure 16.15 (1e Figure 16.13).

Chapter 17: The Cardiovascular System I: The Heart

Module 17.1: Overview of the Heart
- A new part has been added to Figure 17.1: Location and basic anatomy of the heart in the thoracic cavity, showing the mediastinum in a transverse section. What was part c of this figure, on the heart chambers, has now been split off into a separate figure, 2e Figure 17.2.

Module 17.2: Heart Anatomy and Blood Flow Pathway
- In what was 1e Figure 17.3, now 2e Figure 17.4, on the pericardium and layers of the heart wall, a new blow-up has been added to part b showing the heart wall layers in more detail.
- A cadaver photo has been added to 2e Figure 17.6: The internal anatomy of the heart (1e Figure 17.5).
- The Big Picture of Blood Flow through the Heart, 2e Figure 17.8, has been reworked to be shown as a cycle.
- The section on Coronary Circulation has been moved after the sections on heart internal anatomy and blood flow.
- A new A&P in the Real World box on Thoracotomy was added.

Module 17.3: Cardiac Muscle Tissue Anatomy and Electrophysiology

- A new blow-up and a light micrograph have been added to 1e Figure 17.9, now 2e Figure 17.10: Cardiac muscle cells.
- The section on Electrophysiology of Cardiac Muscle Tissue: Pacemaker Cells has been moved before the section on Electrophysiology of Cardiac Muscle Tissue: Contractile Cells.

Chapter 18: The Cardiovascular System II: The Blood Vessels

Module 18.2: Physiology of Blood Flow
- Figure 18.4: Factors that determine blood pressure, has been changed to show vessel radius (rather than diameter) to match the text discussion.
- A new Concept Boost called Taking a Closer Look at Systolic and Diastolic Pressures, with a new illustration, has been added under Systemic Arterial Pressure.

Module 18.3: Maintenance of Blood Pressure
- The text under Nervous System Maintenance of Blood Pressure—and Figure 18.7: Effects of the autonomic nervous system on blood pressure, Figure 18.8: Maintaining homeostasis, and Figure 18.9: Blood pressure maintenance—have all been updated to emphasize that it is not parasympathetic neurons that cause vasodilation, but the autonomic centers in the brainstem inhibiting sympathetic neurons.
- The 1e A&P in the Real World box on Carotid Sinus Massage has been replaced by one on Vasovagal Syncope. A brief discussion of carotid sinus massage has been moved into the text.

Module 18.5: Capillary Pressures and Water Movement
- The discussion of osmotic pressure has been updated to match the current explanation of osmosis in 2e Chapter 3.
- For the same reason, the 1e Concept Boost on Understanding the Pulling Force of Osmotic Pressure has been replaced. In its place there is a Study Boost called Another Way to Think about Hydrostatic and Osmotic Pressure.

Chapter 19: Blood

Module 19.2: Erythrocytes and Oxygen Transport
- The dimensions of a red blood cell have been added both to the text under Erythrocyte Structure and to Figure 19.2: Erythrocyte structure.
- In Figure 19.3: Hemoglobin structure, an orientation diagram of the depiction of hemoglobin from Figure 19.2 has been added, and the heme in oxyhemoglobin has been deleted.

Module 19.4: Platelets
- A new Putting It All Together: The Big Picture of Formed Elements section has been added to the end of this module, with a new two-page Big Picture figure, 2e Figure 19.11, of the same name.

Module 19.5: Hemostasis
- Under Hemostasis Part 3: Coagulation, the newer term contact activation pathway has been added as an alternate for intrinsic pathway, as well as tissue factor pathway as an alter-

nate for extrinsic pathway. These terms have also been added to 2e Figure 19.14: Hemostasis Part 3: Coagulation cascade.

Chapter 20: The Lymphatic System and Immunity

Module 20.1: Structure and Function of the Lymphatic System
- In Figure 20.7: Location and structure of lymph nodes, the lymph node section in part c has been redrawn for clarity and to make it more 3D. In addition, a bit more explanation of these structures has been added to the text under Lymph Nodes.

Module 20.3: Innate Immunity: Internal Defenses
- Under Phagocytes, additional information about dendritic cells has been added.
- Under Complement, the lectin pathway has been added to both the text and to Figure 20.10: Pathways for activation of the complement system.
- At the end of the module, a new Table 20.1 summarizes and compares the first and second lines of defense.

Module 20.4: Adaptive Immunity: Cell-Mediated Immunity
- The new Antigens subsection at the beginning of the module emphasizes coverage of this subject.
- In the Concept Boost called Why Do We Need Both Class I and Class II MHC Molecules?, the CD4 molecule has been added to the illustration of the T_H cell, and the CD8 molecule to the illustration of the T_C cell.
- Figure 20.15: T cell activation, clonal selection, and differentiation, has been updated to show a dendritic cell displaying antigenic fragments to T_H and T_C cells. The clones produced by both of these types of cells are now shown to differentiate into both memory and effector cells. Explanation of the role of dendritic cells in this process has been added to the text section called T Cell Activation, Clonal Selection, and Differentiation.
- In Figure 20.16: Effects of T_H cells, a macrophage is now shown presenting an antigen fragment to and activating the T_H cell.
- In Figure 20.17: Function of T_C cells, a CD8 molecule has been added to each depiction of the T_C cell.

Module 20.5: Adaptive Immunity: Antibody-Mediated Immunity
- Figure 20.19: B cell activation, clonal selection, and differentiation, has been updated with the addition of CD4 molecule to the T_H cell.
- Under Antibody Structure and Classes, class switching is now mentioned.
- The A&P in the Real World box on The Myth of Autism and Vaccines has been given the new subtitle Pseudoscience Exposed.

Module 20.6: Putting It All Together: The Big Picture of the Immune Response
- Figures 20.24, 20.25, and 20.26 are the Big Picture figures of the immune response to the common cold, a bacterial infection, and cancer cells, respectively. All of them have been updated to show the role of the dendritic cell. In addition, CD4 and CD8 molecules have been added as appropriate. The text for each of these has been updated to match these changes.

Chapter 21: The Respiratory System

Module 21.1: Overview of the Respiratory System
- 1e Figure 21.2: The conducting and respiratory zones of the respiratory system, has been deleted, as these concepts are shown in Figure 21.1.

Module 21.2: Anatomy of the Respiratory System
- The illustration of 1e Figure 21.5: Anatomy of the pharynx, (now 2e Figure 21.4) has been extended, using the 1e figure as an orientation diagram to a new larger illustration showing the pharynx with the tonsils and larynx for context.
- In 1e Figure 21.8 (now 2e Figure 21.7): Anatomy of the trachea, a new blow-up of the carina has been added.

Module 21.3: Pulmonary Ventilation
- Under The Process of Pulmonary Ventilation, there are new subsections called Pressure Gradients of Ventilation, which introduces atmospheric, intrapulmonary, and intrapleural pressures, and Mechanics of Inspiration and Expiration, which replaces Volume Changes during Ventilation.
- In the text on Mechanics of Inspiration and Expiration, more discussion of the accessory muscles has been added. In addition, there is a new Table 21.2 on accessory muscles of inspiration and expiration.
- 1e Figures 21.14–21.16 (now 2e Figures 21.13–21.15), on pressure and volume changes in pulmonary ventilation as well as the Big Picture figure, are now presented as cycles, with the addition of step ③ Between inspiration and expiration. What was 1e Figure 21.15 has been moved up to become 2e Figure 21.13, to match the text changes, with the other figures renumbered accordingly. 2e Figure 21.15, The Big Picture of Pulmonary Ventilation, has been extended to a full page, with the addition of the pressures involved.

Module 21.5: Gas Transport through the Blood
- In 1e Figure 21.25 (2e Figure 21.24): Transport of carbon dioxide, the illustrations have been enlarged to make the information more accessible.
- A new Concept Boost called How Does a Buffer Work? has been added.
- The 1e Concept Boost on Relating Ventilation and Blood pH is now a Study Boost.

Module 21.7: Neural Control of Ventilation
- In 1e Figure 21.28, now 2e Figure 21.27: Neural control of the basic pattern of ventilation, the information has been updated to include the respiratory pattern generator (RPG), and to show the nuclei for the glossopharyngeal and vagus nerves in the brainstem and those nerves innervating the lung and heart.
- The RPG is now also shown in 1e Figure 21.29, now 2e Figure 21.28: Role of the central chemoreceptors in regulation of blood pH via regulation of blood pH via the rate of ventilation.
- 1e Figure 21.30: Role of the peripheral chemoreceptors has been deleted. Note that the text section on this is still included, however.
- Under Control of the Rate and Depth of Ventilation, a new subsection called Voluntary Control has been added.

- In 1e Figure 21.31, now 2e Figure 21.29: Control mechanisms of ventilation, a new bottom row has been added to the table showing the pulmonary stretch receptors.

Chapter 22: The Digestive System

Module 22.1: Overview of the Digestive System
- After the section Basic Digestive Functions and Processes, this module has been reorganized. Next there is a new section called Organization of the Digestive System, with subsections on Peritoneal Membranes, Blood and Nerve Supply, and Histology of the Alimentary Canal. Following this is the moved section Regulation of Motility by the Nervous and Endocrine Systems. In the process of this reorganization, Figure 22.2 is now The abdominopelvic cavity, whereas Figure 22.3 is The basic tissue organization of most of the alimentary canal.

Module 22.3: The Stomach
- Under Stomach Mucosa: Gastric Glands, the term enteroendocrine cells has been replaced with diffuse neuroendocrine system (DNES) cells.

Module 22.6: The Pancreas, Liver, and Gallbladder
- In Figure 22.22: Gross anatomy of the liver, the orientation diagram for part b has been replaced with a posteroinferior view.
- The A&P box called Do We Really Need to "Detox"? has been given the new subtitle Pseudoscience Exposed.
- In Figure 22.26: Secretion of bile, the structures shown in the figure have been rearranged for clarity in understanding the process.

Module 22.7: Nutrient Digestion and Absorption
- Under Lipid Digestion, a new Study Boost called An Analogy to Understanding Emulsification has been added.

Chapter 23: Metabolism and Nutrition

Module 23.2: Glucose Catabolism and ATP Synthesis
- The Concept Boost called Why Do We Breathe? has been reworked to separate the discussion into two questions: Why do we inhale oxygen, and why do we exhale the carbon dioxide we produce?
- The Concept Boost now called ATP Yield from Oxidative Catabolism has been reworked to clarify the text and make it more relatable to students.

Module 23.4: Anabolic Pathways
- In Figure 23.14: The Big Picture of Nutrient Anabolism, organ and cell illustrations have been added to appropriate steps to clarify where these processes are occurring.

Module 23.6: The Metabolic Rate and Thermoregulation
- The subtitle Pseudoscience Exposed has been added to the A&P in the Real World box called "Rev" Your Metabolism.
- In Figure 23.18: Maintaining homeostasis: regulation of core body temperature by negative feedback loops, part b, in response to falling body temperature, the fact that the receptors are primarily in the skin rather than the hypothalamus has been clarified.

Module 23.7: Nutrition and Body Mass
- The subtitle Pseudoscience Exposed has been added to the A&P in the Real World box called Vitamin and Mineral Megadoses.

Chapter 24: The Urinary System

Module 24.4: Renal Physiology I: Glomerular Filtration
- In Figure 24.12: Filtration and the filtration membrane, a new blow-up part a has been added that shows a glomerular capillary and podocyte, leading to the existing illustration of a section through the filtration membrane, now part b.
- In Figure 24.13: Net filtration pressure in the glomerular capillaries, a blow-up illustrated table has been added, showing whether the pressures favor or oppose filtration.

Module 24.5: Renal Physiology II: Tubular Reabsorption and Secretion
- Figure 24.19, The Big Picture of Tubular Reabsorption and Secretion, has been reimagined to move the substances being reabsorbed and secreted closer to where these processes are occurring. Blow-up boxes with illustrations of substance movement along the tubes are now shown.

Module 24.6: Renal Physiology III: Regulation of Urine Concentration and Volume
- Part b of Figure 24.21: The countercurrent multiplier in the nephron loop, has moved into the text as part of the Concept Boost on Demystifying the Countercurrent Multiplier.
- Figure 24.22 has been reworked to focus more on the medullary osmotic gradient. It has been retitled Maintenance of the medullary osmotic gradient by the vasa recta and the countercurrent exchanger.

Module 24.10: The Big Picture of Urine Formation, Storage, and Excretion
- There is a new Module 24.10: The Big Picture of Urine Formation, Storage, and Elimination with a new Big Picture figure, Figure 24.28, of the same name.

Chapter 25: Fluid, Electrolyte, and Acid-Base Homeostasis

Module 25.2: Fluid Homeostasis
- Under Movement of Water between Compartments, three new subsections have been added to update the discussion to match the new explanation of osmosis in Chapter 3: The Cell. These subsections are called Hydrostatic Pressure, Osmotic Pressure and Tonicity, and How Hydrostatic Pressure and Osmotic Pressure Influence Water Movement.
- Figure 25.4: Fluid movement between compartments has been updated to go with the new text discussion.

Module 25.4 Acid-Base Homeostasis
- A new A&P in the Real World box called Pseudoscience Exposed: Alkaline Diets was added.

Chapter 26: The Reproductive System

- All figures that show the pituitary glands have been updated in this chapter to match the clarified illustration used in 2e Chapter 16.

Module 26.1: Overview of the Reproductive System and Meiosis
- The 1e Concept Boost called Comparing Mitosis and Meiosis has been made into a regular text section, in order to incorporate Figure 26.2: Comparing meiosis and mitosis, into the discussion.

Module 26.3: Physiology of the Male Reproductive System
- Figure 26.7: Spermatogenesis in the seminiferous tubules, has been changed to clarify the process.

Module 26.5: Physiology of the Female Reproductive System
- In Figure 26.14, the right half of the figure that showed the development of a follicle has been deleted, as this information is shown in Figure 26.15. 2e Figure 26.14 has been retitled The stages of oogenesis.
- The 1e section Hormonal Control of Female Reproduction has been retitled Ovarian Follicles and the Ovarian Cycle to better represent the section's content. The 1e subheading Comparison of Oogenesis and Follicle Development has been deleted, as it was superseded by the new heading; the content of this subsection is still in the text.
- In Figure 26.15: The ovarian cycle, labels for the three phases of the cycle have been added.
- The Uterine Cycle has now been made an overall heading; this section includes the new subheading Phases of the Uterine Cycle as well as the 1e subheading Hormonal Control of the Uterine Cycle. The content of the section is very similar to that in 1e.
- In Figure 26.19: The Big Picture of Hormonal Regulation of the Ovarian and Uterine Cycles, new illustrations of all parts of the ovarian cycle have been added, as well as new explanatory text and labels.

Chapter 27: Development and Heredity

Module 27.2: Pre-embryonic Period: Fertilization through Implantation
- Under Development of Extraembryonic Membranes, explanatory text has been added under the subheading Allantois.

Module 27.4: Fetal Period: Week 9 until Birth
- In Figure 27.10: Development during the fetal period, the part descriptions under the photos have been corrected to identify the conceptus as a fetus.
- In Figure 27.11: Comparison of fetal and newborn cardiovascular systems, the oxygenation of the blood in various parts of the systems has been clarified using the representative colors.

Appendix

NEW Appendix on the Scientific Method

Acknowledgments

Believe it or not, this book you are now holding has been about 12 years in the making—over 9 years for the first edition and nearly 3 for the second edition. When I first started writing it, my daughter wasn't even 2 years old; now she is an eighth grader. But I was certainly not alone on this journey, as a huge number of people were involved in bringing this book to life. Saying a simple "thank you" in the acknowledgments seems so insufficient given the quality and quantity of their contributions, but these thanks are genuine and heartfelt.

I will start with my family because they have gone on this journey with me through both editions. Were it not for the help and understanding of my husband Chris Amerman, my daughter Elise, my mother Cathy Young, and my dear friend David Ferguson, this book would have never been completed. They served as a source of unwavering support, encouragement, and ideas. Elise is also very patient with how much I have to work, and I am so thankful for that. I realize that it isn't easy having a mom who works 7 days a week, 12 hours per day, always chasing another deadline, so thank you for your understanding. I should also thank my dogs for dropping toys in the middle of my laptop, and my cats for never failing to do precisely the least helpful thing possible.

A special thank you must be extended to the brilliant Lourdes Norman-McKay, who has been such an amazing friend and source of support. Chris, Elise, and I have been so fortunate to get to know her and her family. I can't wait to read her textbook, and, hey, I won't just read the ending of the book—I'll definitely start at the beginning and read it all the way through!

Next is the core team of the book, whom I've come to think of as parts of the brain, each performing absolutely vital functions that maintained homeostasis of the whole book. First is Serina Beauparlant, who, as editor-in-chief, is our brainstem. She has tirelessly performed all of those critical behind-the-scenes functions, ranging from wrangling budgets and securing administrative support to running focus groups and analyzing reviewer feedback. It has been Serina's driving force that kept the book alive over these long years. Simply put, without her, there would be no book.

Our team's cerebral hemispheres are our two brilliant developmental editors, Suzanne Olivier and Laura Southworth. As our text development editor, Suzanne is the left cerebral hemisphere. Her ability to logically and patiently approach a chapter from a "big picture" perspective ensures our chapters maintain a consistent narrative flow. It's impossible to overstate her role—not only does Suzanne always manage to find a chapter's sticking points, but she also always proposes solutions to these problems that make the chapter better. The readability, logical flow, and text-art coordination of this book are largely due to Suzanne's efforts.

Laura Southworth, as our art development editor, is the right cerebral hemisphere. Laura not only is a very talented artist but also has an incredible ability to analyze a figure and work magic to make it teach better. This is in part due to her amazing skill for visual-spatial layout (a skill I absolutely lack), which is arguably the most important part of a figure. No matter what we gave her or how rough our ideas or sketches, Laura turned it into gold. This is why "Let's ask Laura" is our mantra when Suzanne and I are working on a chapter. Any time we are perplexed by a figure, Laura unfailingly finds a solution.

The role of team thalamus was played by content producer Jessica Picone, who is new to the team for the second edition. This is a high compliment, as without a functional thalamus, absolutely nothing can get done! Jessica has been a wonderful addition to the team, and I feel very fortunate to be working with her. Like the thalamus, Jessica skillfully manages to monitor, process, and sort absolutely all material for the chapters and supplements for this project. Basically, everything goes through Jessica, and without her ability to juggle it all, we would be lost.

Rounding out the team is Barbara Yien, our cerebellum. Barbara has been involved with this project from the very start, first as a project editor and now as Courseware Director, Content Development. Her even-keeled approach has helped troubleshoot scheduling, budgeting, and our marketing efforts. Whenever we come to a sticking point, we look to Barbara, who always manages to find a way to correct the "motor error" and keep everything balanced and on track.

(Now that I've written this, I'm wondering exactly what part of the brain I represent on the team. The basal nuclei? Maybe the hypothalamus? Hopefully not the pineal gland, as I don't want to make my students sleepy . . .)

Every member of this core "brain" team deserves the highest praise for their skills, dedication, and willingness to persistently climb the mountain that was this book. I am beyond grateful to them for this, and I am also deeply thankful for their friendship. I'd also like to recognize our new editor Jennifer McGill Walker, with whom I can't wait to start working—the third edition is just around the corner!

Assisting the core team was a group of incredibly talented people without whom the book could not have happened: our indomitable marketing team of Allison Rona, Derek Perrigo, Brad Parkins, Maggie Moylan Leen, Tim Galligan, Jessica Moro, Mansour Bethoney, Patrice Jones, and Yez Alayan; design director Mark Ong; copyeditor Bonnie Boehme; the Pearson media team, including Stacy Treco, Laura Tommasi, Caroline Power, Katie Foley, Cheryl Chi, Kristen Sanchez, Patrice Fabel, Sarah Shefveland, and Sarah Young-Dualan; our content production team including Nancy Tabor and Caroline Ayres; our manufacturing buyer Stacey Weinberger; Animated Biomedical Productions; editorial assistant Dapinder Dosanjh; and editorial extern Linh Bui, Grinnell College.

Next I want to thank and acknowledge everyone who contributed to the book, particularly Virginia Irintcheva, who authored the script and storyboards for the book's animations and interactive figures and assessments in MasteringA&P. Everyone who contributed their work devoted a huge amount of time and effort to this project—as I'm sure they will tell you, authoring materials is hard work! I am so grateful that each of them was willing to share his or her talents and play a role in the success of this project. I am also grateful to the Editorial Consultants who provided invaluable feedback on teaching ideas and carefully accuracy-checked pages and to all of the many academic reviewers and focus group attendees who have shared their time, expertise, and ideas with us. I appreciated the detailed feedback from Edwin Griff, University of Cincinnati; Howard Motoike, LaGuardia CC; Laila Nimri, Seminole State College; Paul Nodzak, M.D., University of Cincinnati; immunology researcher Bryan Van Lugt, and Michael Wiley, University of Toronto. Thanks, too to Saeid Baki-Hashemi, Southwest Tennessee Community College for his advice on highlighting the importance of science vs. pseudoscience and for suggesting the addition of an overview of the scientific method (see new Appendix D Scientific Method).

A special "thank you" to Dr. Richard Gonzalez Diaz, Seminole State who shared ideas for now and for the future.

Thanks to Professors Maria Carles and Emily Gonzalez, Northern Essex Community College—Lawrence; Ayanna Alexander-Street, Lehman College; Carlene Tonini-Boutacoff, College of San Mateo and Lori Smith, American River College for giving us an opportunity to hear directly from their students about what types of print and media tools help them learn and what we can do to make these materials even more useful in future editions.

Thanks to Bert Atsma, Union County College for his wonderful update of the Active-Learning Workbook. Thank you as well to Suzi Pundt, University of Texas-Tyler for updating the PPTs and Lecture Outlines and Patty Bostwick Taylor, Florence-Darlington Technical College for updating the Test Bank, Instructor's Guide and other supplemental materials.

Many thanks to the contributors to the assessments in Mastering A&P including Allison Beck, Black Hawk College; Hon-Vu Duong, M.D., Nevada State College; Ken Malachowsky, Florence-Darlington Technical College; Stephanie A. Tacquard, Alvin Community College; and Geraldine Wright, Tidewater Community College.

I'd also like to recognize the superb work on expanded Dynamic Study Modules by Mary Colon, Seminole State College of Florida; Angel Nickens, Northwest Mississippi Community College and Stephen Page, Community College of Baltimore County. Special thanks to Vikash Patel, Nevada State College for his work on the Ready-to-go Teaching Modules.

I would also like to sincerely thank Lauren Harp, Courseware Portfolio Manager. I met Lauren in 2005 when she was the marketing manager for natural sciences. She passed my name along to Serina as a potential author after we had a two-hour-long conversation in my office about what I would like to see in a textbook. Had she not done this, *Human Anatomy & Physiology* likely wouldn't exist.

Finally, none of this would have been possible without the unwavering support of Managing Director of Pearson Science Paul Corey, Editorial Director of Pearson Science Adam Jaworski, and Finance Director of Pearson Science Hogan Nymberg. All have supported this project from the beginning, and it was only because of their continued encouragement and belief in our team that you are holding this book right now. They have my eternal gratitude for allowing us to bring our vision to life.

Editorial Consultants

Emily Allen, *Rowan College of Gloucester County*
Bert Atsma, *Union County College*
Patty Bostwick Taylor, *Florence-Darlington Technical College*
Sheri L. Boyce, *Messiah College*
Robert G. Carroll, *East Caroline University, Brody School of Medicine*
Linda Costanzo, *Virginia Commonwealth University School of Medicine*
Sharon S. Ellerton, *Queensborough Community College*
Jeff E. Engel, *Western Illinois University*
Karen L. Keller, *Frostburg State University*
Naomi Machell, *Delaware County Community College*
Ken Malachowsky, *Florence-Darlington Technical College*
Suzanne Pundt, *University of Texas at Tyler*
Mark Seifert, *Indiana University-Purdue University, Indianapolis, School of Medicine*
Cindy L. Stanfield, *University of South Alabama*
Cynthia Surmacz, *Bloomsburg University*

Second Edition Reviewers

Pius Aboloye, *North Lake College*
Shaheem Abrahams, *Thomas Nelson Community College*
Sandra Acquah, *Montgomery College–Rockville*
Ticiano Alegre, *North Lake College*
Ayanna Alexander-Street, *Lehman College*
Leah Allen, *Montgomery College*
Antoinette Anastasia, *Fairleigh Dickinson University*
John Andreucci, *Seneca College of Applied Art & Technology (Ontario, Canada)*
Meghan Andrikanich, *Lorain County Community College*
Penny P. Antley, *University of Louisiana–Lafayette*
Kanzoni Asabigi, *Wayne County Community College*
Saeid Baki, *Southwest Tennessee Community College*
Donna Balding, *Middle Georgia State University*
Sarah Balizan, *New Mexico State University–Doña Ana Community College*

Amanda Banker, *Southwest Tennessee Community College*
Marcin Baranowski, *Passaic County Community College*
Marilynn Bartels, *Black Hawk College*
Allison L. Beck, *Black Hawk College*
Jerilyn Belle, *Bevill State Community College*
Charles E. Benton Jr., *Madison Area Technical College*
Cathy Bill, *Columbus State Community College*
Evelyn J. Biluk, *Chippewa Valley Technical College–River Fall*
Jennifer Blickwedehl, *Trocaire College*
Patty Bostwick Taylor, *Florence-Darlington Technical College*
Sheri Boyce, *Messiah College*
Tara Breeland-Southam, *Bossier Parish Community College*
Althea M. Brown, *Wayne County Community College*
Julia Brown, *Chippewa Valley Technical College*
Jerry Brunson, *University of Lousiana–Monroe*
Bertha M. Byrd, *Wayne County Community College*
Christie Campbell, *Ozarks Technical Community College*
Susan R. Capasso, *St. Vincents College*
Michelle Carey, *Hutchinson Community College*
Maria Carles, *Northern Essex Community College*
Vlad Chiriac, *Durham College (Ontario, Canada)*
Loraine N. Christie, *Seneca College of Applied Art & Technology (Ontario, Canada)*
Lori D. Coble, *South Dakota School of Mines & Technology*
Elizabeth Collins, *Iowa Central Community College*
Xixuan Collins, *Black Hawk College*
Mary B. Colon, *Seminole State College of Florida*
Matthew Connior, *South Arkansas Community College*
Kelly J. Craig, *Colorado Mesa University*
Kenneth Crane, *Texarkana College*
John Crawford, *Lindenwood University*
Gregory J. Crowther, *University of Washington Bothell*
Judith D'Aleo, *Plymouth State University*
Amy Dawson, *New River Community College*
Carrie Dollar, *St. Clair County Community College*
Barbara Dorsett, *Gadsden State Community College*
Mary L. Dougherty, *Catawba Valley Community College*
Hon-Vu Q. Duong, *Nevada State College*
Curtis Eckerman, *Austin Community College*
Paula K. Edgar, *John Wood Community College*
Ann M. Findley, *University of Louisiana–Monroe*
Julie Fischer, *Wallace Community College*
Teresa G. Fischer, *Indian River State College*
John E. Fishback, *Ozarks Technical Community College*
Robert S. Fitch, *Wenatchee Valley College*
Jodie M. Fleming, *North Carolina Central University*
Lisa Flick, *Monroe Community College*
Christine Foley, *Southwest Texas Junior College*
Eric Forman, *Sauk Valley Community College*
Reza Forough, *Bellevue College*
Polly Foureman, *Chandler-Gilbert Community College*
Mark Garbrecht, *Winona State University*
Linda D. Gaylo, *Mercer County Community College*
Mike Gehner, *Xavier University*
Ellen Genovesi, *Mercer County Community College*
Emily K. Getty, *Ivy Tech Community College–Kokomo*

Diane Gibson, *Hazard Community & Technical College*
Larry E. Gibson, *University of North Georgia*
Sabine Globig, *Hazard Community & Technical College*
Richard Gonzalez Diaz, *Seminole State College of Florida*
Ewa Gorski, *Community College of Baltimore County*
Janelle Green, *Hazard Community & Technical College*
Melissa L. Greene, *Northwest Mississippi Community College–Senatobia*
Edwin R. Griff, *University of Cincinnati*
Geoff Gruenberg, *Baker College–Flint*
Kyle P. Harris, *Temple University, College of Public Health*
Kim Hensley, *Southern West Virginia Community & Technical College*
Kristine Hicks, *University of Central Arkansas*
Austin Hicks, *University of Alabama*
Elizabeth Hodgson, *York College of Pennsylvania*
Jessica C. Hogan, *Central Virginia Community College*
Dale R. Horeth, *Tidewater Community College*
Julie Huggins, *Arkansas State University*
Sandra Hutchinson, *Santa Monica College*
Michael Irowa, *Wayne County Community College*
Joby Jacob, *La Guardia Community College*
Lori Janus-Baxa, *Gateway Technical College*
Naomi Jones, *Bellevue College*
Kebret Kebede, *Nevada State College*
Eric Kenz, *Columbus State Community College*
Peter Kobella, *Owensboro Community & Technical College*
Gopal Krishna, *Moberly Area Community College–Columbia*
Dean Kruse, *Portland Community College*
James A. Landis, *Lakeland Community College*
Shannon Larson, *College of Southern Nevada*
Joyce Ellen Lathrop-Davis, *Community College of Baltimore County*
Steven A. Leadon, *Durham Technical Community College*
Aaron Livingston, *Portland Community College–Southeast*
Alex Lowrey, *University of North Georgia–Gainesville*
Debby Machuca, *Portland Community College*
Ken Malachowsky, *Florence-Darlington Technical College*
Patricia L. Mansfield, *Santa Ana College*
Bruce Maring, *Daytona State College*
Sarah Mattox Holt, *Northwest Mississippi Community College*
John W. McCain, *Owens State Community College–Findlay*
Annie McKinnon, *Howard College*
Larry R. McLean, *Ivy Tech Community College–Lawrenceburg*
Carrie McVean Waring, *Colorado Mesa University*
Laurie S. Meadows, *Roane State Community College*
Jaime Mergliano, *John Tyler Community College*
Glenn Merrick, *Lake Superior College*
Michael Midgley, *Quinnipiac University*
Joseph R. Mikula, *North Central Michigan College*
Sharon Miles, *Itawamba Community College*
Michelle Milner, *Itawamba Community College*
Robert Moldenhauer, *St. Clair County Community College*
Marty Montpetit, *John Tyler Community College*
Erica Morley, *Mesa Community College*
Susan Moss, *Imperial Valley College*
Magdalena Muchlinski, *University of Kentucky*

David Mullaney, *Naugatuck Valley Community College*
Angel Nickens, *Northwest Mississippi Community College*
Paul I. Nodzak, M.D., *University of Cincinnati*
Weston Opitz, *Kansas Wesleyan University*
Stephen H. Page, *Community College of Baltimore County & Townson University*
Ivan Paul, *John Wood Community College*
Emma Phillips, *Blue Ridge Community College*
Jason Pienaar, *University of Alabama*
Christine Priano, *CUNY–Borough of Manhattan Community College*
Candice Pullen, *Central Queensland University (Queensland, Australia)*
Suzanne Pundt, *University of Texas–Tyler*
Kaustubha Qanungo, *Trident Technical College*
Denise Rakestraw, *Itawamba Community College*
James Rayburn, *Jacksonville State University*
Gary Reid, *Trinity Valley Community College*
Nicole Reinke, *University of Sunshine Coast (Queensland, Australia)*
Susan Rohde, *Triton College*
Deborah Rhoden, *Snead State Community College*
Antoninia B. Ries, *Ivy Tech Community College*
Robin Robison, *Northwest Mississippi Community College*
Laurie A. Rocco, *Monroe Community College*
Vanessa Rowan, *Palm Beach Atlantic University*
Hiranya S. Roychowdhury, *New Mexico State University–Doña Ana Community College*
John W. Rumsey, *Indian River State College*
Ali Saleh, *Passaic County Community College*
Methea Sapp, *Spokane Community College*
Michelle M. Scanavino, *Moberly Area Community College*
Michael Schneider, *Ivy Tech Community College of Indiana*
Victoria Schneider, *Montgomery College–Rockville*
Benn Scott, *Louisiana Delta Community College*
Ehsan Siddique, *Broward College*
Scott L. Simerlein, *Purdue University–North Central*
Doug Sizemore, *Bevill State Community College*
Patricia M. Smeltz, *John Tyler Community College*
Pamela S. Smith, *Madisonville Community College*
Gehan Soliman, *Fayetteville Technical Community College*
Allison Stamatis, *Weatherford College*
Lisa Strong, *Northwest Mississippi Community College*
Karla Svedarsky, *Chippewa Valley Technical College*
Stephanie Tacquard, *Alvin Community College*
Candice Thomas, *University of Central Arkansas*
Michael W. Thompson, *Jefferson Community & Technical College*
Sanjay K. Tiwary, *Hinds Community College*
Paula Trilling, *Asheville-Buncombe Technical Community College*
Lisa Tunks, *Broward College*
Albert Urazaev, *Ivy Tech Community College*
Sarah Warrington, *Southwest Tennessee Community College*
Chad Wayne, *University of Houston*
Shay West, *Colorado Mesa University*
Valerie Wheat, *Jefferson Community & Technical College*
Emily C. Whiteley, *Catawba Valley Community College*
Catharine C. Whiting, *University of North Georgia*

Esther Wilczynski, *Trocaire College*
Jeffrey Williams, *Victoria College*
Vanessa L. Williams, *University of Georgia*
Martha T. Wolfe, *Elizabethtown Community & Technical College*
Leon Wooten, *Kilgore College*
Geraldine Wright, *Tidewater Community College*
Imogene Younger, *Southwest Tennessee Community College*
Martin Zahn, *Thomas Nelson Community College*
Gina M. Zainelli, *Gateway Technical College*

Media Reviewers

Big Picture Animations & Concept Boost Video Tutors

Emily Allen, *Rowan College of Gloucester County*
Willie Asobayire, *Essex County Community College*
Vince Austin, *Bluegrass Community & Technical College*
Marianne Baricevic, *Raritan Valley Community College*
Jerry Barton, *Tarrant County College*
C. Audra Bassett-Touchell, *Asheville-Buncombe Technical Community College*
David Bastedo, *San Bernardino Valley College*
Carol Britson, *University of Mississippi*
Jack Brown, *Paris Junior College*
Carolyn Bunde, *Idaho State University*
Susan Burgoon, *Amarillo College*
C. Steven Cahill, *West Kentucky Community & Technical College*
Maria C. Carles, *Northern Essex Community College*
Brendan K. Chastain, *West Kentucky Community & Technical College*
Ken Crane, *Texarkana College*
Kathryn A. Durham, *Lorain County Community College*
Sharon Ellerton, *Queensborough Community College*
Julie Fischer, *Wallace Community College*
Theresa Gillian, *Virginia Tech*
Lauren Gollahon, *Texas Tech University*
Pamela Gregory, *Tyler Junior College*
Kristine Hicks, *University of Central Arkansas*
Mark Hollier, *Georgia State University—Perimeter College–Clarkston*
William F. Huber, *St. Louis Community College*
Julie Huggins, *Arkansas State University*
Tom Jordan, *Pima Community College–NW*
Michelle Klein, *Prince George's Community College*
Adewale Laditan, *Baltimore City Community College*
Jodi Long, *Santa Fe College*
Bruce Maring, *Daytona State University*
Jaime Mergliano, *John Tyler Community College*
Howard Motoike, *LaGuardia Community College*
Maria Oehler, *Florida State College–Jacksonville*
John Patillo, *Middle Georgia State College*
Diane Pelletier, *Green River Community College*
Suzanne Pundt, *University of Texas at Tyler*
Elizabeth Randolph, *Front Range Community College*
Rozanne Redlinski, *Erie Community College*
Ann Riedl, *Front Range Community College*
Michelle M. Scanavino, *Moberly Area Community College*

Sharon Schapel, *Mott Community College*
Joanne Settel, *Baltimore City Community College*
Jason Shaw, *Brigham Young University–Idaho*
Mark Slivkoff, *Collin College*
Lisa Strong, *Northwest Mississippi Community College*
Patricia Visser, *Jackson Community College*
Kathy Warren, *Daytona State College*
Pete Wickley, *Cuyahoga Community College*

Interactive Physiology 2.0
Matthew Abbott, *Des Moines Area Community College*
Emily Allen, *Rowan College at Gloucester County*
Lynne Anderson, *Meridian Community College*
David Babb, *West Hills College Lemoore*
Jerry Barton, *Tarrant County College*
Shawn Bearden, *Idaho State University*
Charles Benton, *Madison Area Technical College*
Gordon Betts, *Tyler Junior College*
Michael Brady, *Columbia Basin College*
Betsy Brantley, *Valencia College*
Carol A. Britson, *University of Mississippi*
Christie Campbell, *Ozarks Technical Community College*
Maria C. Carles, *Northern Essex Community College*
Tamyra Carmona, *Cosumnes River College*
Marien Cendon, *Miami Dade College*
Brendon Chastain, *West Kentucky Community Technical College*
Sam Chen, *Moraine Valley Community College*
Alexander Cheroske, *Mesa Community College*
William M. Clark, *Lone Star College–Kingwood*
Jason Dechant, *University of Pittsburgh*
Smruti Desai, *Lone Star College–Cyfair*
Karen Dougherty, *Hopkinsville Community College*
Sondra Dubowsky, *McLennan Community College*
Kathryn Durham, *Lorain County Community College*
Karen Eastman, *Chattanooga State Community College*
Sharon S. Ellerton, *Queensborough Community College–CUNY*
Paul Emerick, *Monroe Community College*
Colin Everhart, *St. Petersburg Community College*
Brian Feige, *Mott Community College*
Michele Finn, *Monroe Community College*
John E. Fishback, *Ozarks Technical Community College*
Aaron Fried, *Mohawk Valley Community College*
Jane Gavin, *University of South Dakota*
Peter Germroth, *Hillsborough Community College*
Anna Gilletly, *Central New Mexico Community College*
Gary Glaser, *Genesee Community College*
Richard Gonzalez-Diaz, *Seminole State College of Florida*
Abigail Goosie, *Walters State College*
Pattie S. Green, *Tacoma Community College*
Edwin Griff, *University of Cincinnati*
George Hanak, *Pasco Hernando State College*
Mary Beth Hanlin, *Des Moines Area Community College–Boone*
Nora Hebert, *Red Rocks Community College*
Katja Hoehn, *Mount Royal University*
Rodney Holmes, *Waubonsee Community College*
Mark Hubley, *Prince George's Community College*

Carolyn Huffman, *Wichita Area Technical College*
Julie Huggins, *Arkansas State University*
Alexander Ibe, *Weatherford College*
Alexander Imholtz, *Prince George's Community College*
Virginia Irintcheva, *Truckee Meadows Community College*
Thomas Jordan, *Pima Community College*
William M. Karkow, *University of Dubuque*
Suzanne Keller, *Indian Hills Community College*
Michael Kielb, *Eastern Michigan University*
Paul Lea, *Northern Virginia Community College*
Paul Luyster, *Tarrant County College*
Ken Malachowsky, *Florence-Darlington Technical College*
Theresa Martin, *College of San Mateo*
Nicole Mashburn, *Calhoun Community College*
Jennifer Menon, *Johnson County Community College*
Jaime Mergliano, *John Tyler Community College*
Sharon Miles, *Itawamba Community College*
Louise Millis, *North Hennepin Community College*
Justin Moore, *American River College*
Christine Morin, *Prince George's Community College*
Maria Oehler, *Florida State College–Jacksonville*
Betsy Ott, *Tyler Junior College*
Stephen Page, *Community College of Baltimore County & Townson University*
Vikash Patel, *Nevada State College*
Dennis Pearson, *Morton College*
Diane Pelletier, *Green River Community College*
Jessica Petersen, *Pensacola State College*
Jason Pienaar, *University of Alabama*
Becky Pierce, *Delta College*
Gilbert Pitts, *Austin Peay State University*
Renee Prenitzer, *Greenville Technical College*
Fernando Prince, *Laredo Community College*
Suzanne Pundt, *University of Texas at Tyler*
Wendy Rappazzo, *Harford Community College*
Terrence J. Ravine, *University of South Alabama*
Christine S. Rigsby, *Middle Georgia State University*
Cynthia Robison, *Wallace Community College*
Sharon Schapel, *Mott Community College*
Mark Schmidt, *Clark State Community College*
Michael W. Sipala, *Bristol Community College*
Lori Smith, *Amerman River College–Los Rios*
Kerry Smith, *Oakland Community College–Auburn Hills*
Tom Sobat, *Ivy Tech Community College*
Kay Sourbeer, *Tidewater Community College*
Cindy Stanfield, *University of South Alabama*
Laura Steele, *Ivy Tech Community College–Northeast*
George Steer, *Jefferson College of Health Sciences*
Dean Thornton, *South Georgia State College*
Rita Thrasher, *Pensacola State College*
Brenda Tondi, *George Mason University*
Carlene Tonini-Boutacoff, *College of San Mateo*
Sheela Vemu, *Waubonsee Community College*
Khursheed Wankadiya, *Central Piedmont Community College*
Kira L. Wennstrom, *Shoreline Community College*

Shirley Whitescarver, *Bluegrass Community & Technical College–KCTCS*
Darrellyn Williams, *Pulaski Technical College*
Heather Wilson-Ashworth, *Utah Valley University*
Jackie Wright, *South Plains College*

First Edition Reviewers

Joslyn Ahlgren, *University of Florida*
Anitha Akkal, *Joliet Junior College*
Ticiano Alegre, *North Lake College*
Ayanna Alexander-Street, *Lehman College*
Emily Allen, *Rowan College at Gloucester County*
Matt Allen, *Indiana University–Purdue University, Indianapolis, School of Medicine*
Beth Altschafl, *University of Wisconsin–Madison*
Teresa Alvarez, *St. Louis Community College–Forest Park*
Kathy Pace Ames, *Illinois Central College*
Heather Evans Anderson, *Winthrop University*
March Ard, *University of Mississippi Medical Center/Hinds Community College*
Bert Atsma, *Union County College*
Vince Austin, *Bluegrass Community & Technical College*
Stephanie Baiyasi, *Delta College*
Tim Ballard, *University of North Carolina–Wilmington*
Stephen Bambas, *University of South Dakota*
Michelle Baragona, *Northeast Mississippi Community College*
Mary Lou Bareither, *University of Illinois–Chicago*
Marianne Baricevic, *Raritan Valley Community College*
Verona Barr, *Heartland Community College*
David Barton, *Indian Hills Community College*
Jerry Barton, *Tarrant County College–South*
Thomas Bell, *Rutgers University at Camden*
Dena Berg, *Tarrant County College–Northwest*
Tom Betsy, *Bergen County Community College*
Gordon J. Betts, *Tyler Junior College*
Laura Bianco, *Delaware Technical Community College of Wilmington*
Heather Billings, *West Virginia School of Medicine*
Ruth Birch, *St. Louis Community College–Florissant Valley*
Bonnie Blazer-Yost, *Indiana University–Purdue University, Indianapolis*
Rob Blum, *Lehigh Carbon Community College*
Sue Bodine, *University of California, Davis*
Franklyn F. Bolander, *University of South Carolina*
Patty Bostwick Taylor, *Florence-Darlington Technical College*
Sherry Bowen, *Indian River State College*
Sheri Boyce, *Messiah College*
Laura Branagan, *Foothill College*
Eldon Braun, *University of Arizona*
Carol Britson, *University of Mississippi*
David Brown, *East Carolina University, Brody School of Medicine*
Jack Brown, *Paris Junior College*
Kristen Bruzzini, *Maryville University*
Diep Burbridge, *Long Beach City College*
Warren Burggren, *University of North Texas*
Ed Burke, *Truckee Meadows Community College*

Steve Burnett, *Clayton State University*
Rebecca Burton, *Alverno College*
Beth Campbell, *Itawamba Community College*
Jamie Campbell, *Truckee Meadows Community College*
Geralyn Caplan, *Owensboro Community & Technical College*
Tamyra Carmona, *Cosumnes River College*
Steven J. Carlisle, *John Tyler Community College*
Robert Carroll, *East Carolina University, Brody School of Medicine*
Jana Causey, *Pearl River Community College*
Karen Chooljian, *California State University, Fresno*
Robert Clark, *Ozarks Technical Community College*
Pamela Cole, *Shelton State Community College*
Francisco Coro, *Miami Dade College*
Ron Cortright, *East Carolina University, Brody School of Medicine*
Linda Costanzo, *Virginia Commonwealth University*
Ken Crane, *Texarkana College*
Robert Crocker, *SUNY–Farmingdale State College*
James Crowder, *Brookdale Community College*
Michael Cryder, *Riverside Community College*
Paul Currie, *Hazard Community & Technical College*
Judith D'Aleo, *Plymouth State University*
Lynnette Danzl-Tauer, *Rock Valley College*
Mary Dawson, *Kingsborough Community College*
Danielle Desroches, *William Patterson University*
Gary Diffee, *University Wisconsin–Madison*
Josh Drouin, *Lock Haven University*
Joseph D'Silva, *Norfolk State University*
Sondra Dubowsky, *McLennan Community College*
Kathryn Durham, *Lorain County Community College*
Abdeslem El Idrissi, *College of Staten Island*
Sharon Ellerton, *Queensborough Community College*
Kurt Elliott, *Northwest Vista College*
Jeff Engel, *Western Illinois University*
Greg Erianne, *Naugatuck Valley Community College*
Victor Eroschenko, *University of Idaho (retired)*
Martha Eshleman, *University of Arkansas—Pulaski Technical College*
Marirose Ethington, *Genesee Community College*
David Evans, *Pennsylvania College of Technology*
Brian Feige, *Mott Community College*
Michael Ferrari, *University of Michigan, School of Medicine*
Linda Flora, *Delaware County Community College*
Maria Florez, *Lone Star College–CyFair*
Cliff Fontenot, *Southeastern Louisiana University*
Barbara Fritz, *Rochester Community & Technical College*
Larry Frolich, *Miami Dade College–Wolfson*
Van Fronhofer, *Hudson Valley Community College*
Chris Gan, *Highline Community College*
Joseph Gar, *West Kentucky Community & Technical College*
Esther Gardner, *NYU Langone Medical Center*
Lynn Gargan, *Tarrant County College–Northeast*
Lori Garrett, *Parkland College*
Michelle K. Gaston, *Northern Virginia Community College*
Jane Gavin, *University of South Dakota, Sanford School of Medicine*
Michelle Gibson, *Montcalm Community College*

Mike Gilbert, *Fresno City College*

Theresa Gillian, *Virginia Tech*

Lauren Gollahon, *Texas Tech University*

Matthew Gosses, *Owens Community College*

Margaret Grant, *Hudson Valley Community College*

Bruce Gray, *Simmons College*

Melissa Greene, *Northwest Mississippi Community College*

Patricia Halpin, *University of New Hampshire–Manchester*

Chris Harendza, *Montgomery County Community College*

Rebecca Harris, *Pitt Community College*

Nora Hebert, *Red Rocks Community College*

Gary Heisermann, *Salem State University*

DJ Hennager, *Kirkwood Community College*

Kristin Hensley, *Rowan University*

Brent Hill, *University of Central Arkansas*

Karen Hlinka, *West Kentucky Community & Technical College*

Elizabeth Hoffman, *Baker College–Clinton*

Rodney Holmes, *Waubonsee Community College*

Mark Hubley, *Prince George's Community College*

Julie Huggins, *Arkansas State University*

Jason Hunt, *Brigham Young University–Idaho*

Peggy Hunter, *Camosun College*

Jim Hutchins, *Weber State University*

Sandra Hutchinson, *Sinclair Community College*

Alexander Ibe, *Weatherford College*

Virginia Irintcheva, *Truckee Meadows Community College*

Bruce Johnson, *Cornell University*

Corey S. Johnson, *University of North Carolina–Chapel Hill*

Eddie Johnson, *Central Oregon Community College*

Cindy Jones, *Colorado Community College Online*

Margaret Kauffman, *Ohlone College*

Karen Keller, *Frostburg State University*

Suzanne Keller, *Indian Hills Community College*

Will Kleinelp, *Middlesex County College*

Michael Klemsz, *Indiana University–Purdue University, Indianapolis, School of Medicine*

Chad Knight, *Northern Virginia Community College*

Karen Krabbenhoft, *University of Wisconsin-Madison, School of Medicine*

Chris Kule, *Pennsylvania College of Technology*

Edward LaBelle, *Rowan College of Gloucester County*

Jason LaPres, *Lone Star College–North Harris*

Barbara Lax, *Community College of Allegheny*

Steven Leadon, *Durham Technical Community College*

Marian Leal, *Sacred Heart University*

Lisa M. J. Lee, *University of Colorado, School of Medicine*

Peggy LePage, *North Hennepin Community College*

John Lepri, *University of North Carolina–Greensboro*

Michael Levitzky, *Louisiana State University Health Sciences Center*

Robert Logan, *North Shore Community College–Lynn*

Shawn Macauley, *Muskegon Community College*

Naomi Machell, *Delaware County Community College*

Erin MacKenzie, *Howard College*

Ken Malachowsky, *Florence-Darlington Technical College*

Bruce Maring, *Daytona State College–West/Deland*

Patricia Marquardt, *Wayne County Community College*

Karen Martin, *Fulton-Montgomery Community College*

Theresa Martin, *College of San Mateo*

Alice McAfee, *University of Toledo*

Jameson McCann, *Guilford Technical Community College*

Jenny McFarland, *Edmonds Community College*

Cherie McKeever, *Montana State University-Great Falls College of Technology*

Karen McLellan, *Indiana University–Purdue University, Ft. Wayne*

Mark Meade, *Jacksonville State University*

Jaime Mergliano, *John Tyler Community College*

Anthony Mescher, *Indiana University School of Medicine–Bloomington*

Steve Meyer, *Florida State College at Jacksonville*

Justin Moore, *American River College*

Erin Morrey, *Perimeter College–Georgia State University*

Qian Moss, *Des Moines Area Community College*

Susan Moss, *Imperial Valley College*

Howard Motoike, *LaGuardia Community College*

Jen Musa, *Broome Community College*

Barbara Musolf, *Clayton State University*

Cheryl Neudauer, *Minneapolis Community & Technical College*

Chad Newton, *Bakersfield College*

Mary Jane Niles, *University of South Florida*

Maria Oehler, *Florida State College at Jacksonville*

Justicia Opoku-Edusei, *University of Maryland*

David Osborne, *Texas Tech Health Science Center*

Betsy Ott, *Tyler Junior College*

Ellen Ott-Reeves, *Blinn College*

Anthony Paganini, *Michigan State University*

Russ Palmeri, *Asheville-Buncombe Technical Community College*

Michele Paradies, *SUNY–Orange County Community College*

Diane Pelletier, *Green River Community College*

Chris Picken, *SUNY–Suffolk County Community College*

Rebecca Pierce, *Delta Community College*

Melissa Piliang, *Cleveland Clinic*

John Placyk, *University of Texas at Tyler*

Brandon Poe, *Springfield Technical Community College*

Peter Porter, *Moraine Valley Community College*

Frank Powell, *University of California, San Diego*

Renee Prenitzer, *Greenville Technical College*

Cynthia Prentice-Craver, *Chemeketa Community College*

Steven Price, *Virginia Commonwealth University*

Suzanne Pundt, *University of Texas at Tyler*

David Quadagno, *Florida State University*

Saeed Rahmanian, *Roane State Community College*

Scott Rahschulte, *Ivy Tech Community College–Lawrenceburg*

Elizabeth Randolph, *Front Range Community College*

Wendy Rappazzo, *Harford Community College*

Terrence J. Ravine, *University of South Alabama*

Laura Ritt, *Burlington County College*

Dawn Roberts, *Pellissippi State Community College*

Mark Robertson, *Delta College*

Robin Robison, *Northwest Mississippi Community College*

Alex Robling, *Indiana University School of Medicine*

Susan Rohde, *Triton College*

Amanda Rosenzweig, *Delgado Community College*
Kyla Ross, *Georgia State University*
John Rowe, *Florida Gateway College*
Stephen Sarikas, *Lasell College*
Leif Saul, *University of Colorado–Boulder*
Lou Scala, *Passaic County Community College*
Connie Scanga, *University of Pennsylvania School of Nursing*
Sharon Schapel, *Mott Community College*
Steve Schenk, *College of Central Florida-Citrus*
Catherine Scholz, *Gwinnett Technical College*
Mark Seifert, *Indiana University–Purdue University, Indianapolis, School of Medicine*
Donald Shaw, *University of Tennessee at Martin*
Matthew "Doc" Sheehan, *Clinton Community College*
Gidi Shemer, *University of North Carolina–Chapel Hill*
Michael Shipley, *Midwestern State University*
Brian Shmaefsky, *Lone Star College–Kirkwood*
Marilyn Shopper, *Johnson County Community College*
Pam Siergiej, *Roane State Community College*
Lynnda Skidmore, *Wayne County Community College*
Dianne Snyder, *Augusta State University*
Debbie Socci, *Seminole State College of Florida*
Annelle Soponis, *Reading Area Community College*
Ashley Spring, *Brevard Community College*
Maria Squire, *University of Scranton*
Claudia Stanescu, *University of Arizona*
Cindy Stanfield, *University of South Alabama*
George Steer, *Jefferson College of Health Sciences*
Jill Stein, *Essex County College*
Nora Stevens, *Portland Community College–Cascade*
Leo Stouder, *Broward College*
Diana Sturges, *Georgia Southern University*
Carole Subotich, *Rowan College of Gloucester County*
Eric Sun, *Middle Georgia State University*
Cynthia Surmacz, *Bloomsburg University*

Robert Swatski, *Harrisburg Area Community College–York*
Carolyn Szutarski, *Camden County College*
Yong Tang, *Front Range Community College*
George Tanner, *Indiana University School of Medicine*
Shyra Tedesco, *Ivy Tech Community College–Madison*
Alvin Telser, *Northwestern University, Feinberg School of Medicine*
Terry Thompson, *Wor-Wic Community College*
Maureen Tubbiola, *St. Cloud State University*
Katherine "Kate" Van de Wal, *Community College of Baltimore County*
Padmaja Vedartham, *Lone Star College–CyFair*
Heather Walker, *Clemson University*
Michael Walls, *Ivy Tech Community College–Evansville*
Lynn Wandrey, *Mott Community College*
Delon Washo-Krupps, *Arizona State University*
Amy Way, *Lock Haven University*
Chad Wayne, *University of Houston*
Mary Weis, *Collin College*
Lisa Welch, *Weatherford College–Decatur*
Corrie Whisner, *Cornell University*
Shirley Whitescarver, *Bluegrass Community & Technical College*
Sheila Wicks, *Malcolm X College*
Michael Wiley, *University of Toronto*
Samia Williams, *Santa Fe College*
Peggie Williamson, *Central Texas College*
Larry Wilmore, *Lamar State University–Orange*
Colleen Winters, *Towson University*
Diane Wood, *Southeast Missouri State University*
Amber Wyman, *Finger Lakes Community College*
Jim Yount, *Brevard Community College*
Anne Marie Yunker, *Cuyahoga Community College*
Nina Zanetti, *Siena College*
Scott Zimmerman, *Missouri State University*

Brief Contents

Contents

Unit 3 Integration, Control, and Maintenance of Homeostasis

Unit 4 Transport and Immunity

21 The Respiratory System 802

Unit 5 Regulation of the Body's Intake and Output

24 The Urinary System   947

Unit 6 Continuity of Life

1

Introduction to Anatomy and Physiology

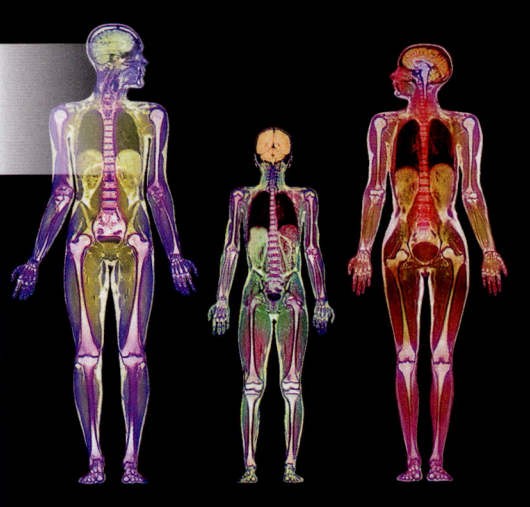

The body has fascinated humankind since ancient times. Many of the earliest ideas about the form and function of the human body were wildly inaccurate. For example, gods and demons were believed to cause most diseases, the heart was perceived as the seat of intelligence, and the brain was viewed as merely a useless mass of tissue. However, through centuries of observation, experimentation, and imagination, we have dramatically expanded our knowledge of the human body. Much of this has resulted from the application of *science,* a way of observing and measuring natural phenomena in order to try to explain them. But so much remains to be learned about the way our bodies are built and how they function. Studying the human body is like a treasure hunt, with fascinating "gems" still to discover.

This chapter introduces you to the world of anatomy and physiology. **Human anatomy** is the study of the structure or form of the human body, whereas **human physiology** (fiz′-ee-AWL-oh-jee) is the study of the body's functions. Although we define the two terms separately, the body's structure and function are closely related, a fact that we explore in depth throughout this text. The common name of your course, anatomy and physiology (A&P), reflects this connection.

The topics we discuss in this chapter form the foundation for the remaining chapters. We begin with a special section on how to succeed in an A&P course. An overview of the field and a discussion of the properties of all living organisms follow. After examining the language of A&P and the basic organization of the human body, the chapter concludes with recurring themes in A&P—those ideas that form the core of the field and the basis of our study in the rest of this book.

MODULE **1.1**

How to Succeed in Your Anatomy and Physiology Course

*For practice applying concepts to a clinical scenario, check out the **Running Case Study** for this chapter @ Mastering Anatomy & Physiology*

Learning Outcomes

1. Describe how to benefit from different learning modalities, read a textbook, budget your time, and study for quizzes and exams.

Photo: Colored magnetic resonance imaging (MRI) whole body scans of a man (left), a woman (right), and a nine-year-old boy.

1

2. Explain how to make the best use of class and laboratory time.

3. Describe how to use this book and its associated materials.

The world of anatomy and physiology often feels like a foreign and intimidating place to new students. And in a way it is—A&P is new territory for most people, and comes with a language all its own. To master the material, you not only need to attend all lecture and laboratory sessions but also must invest a great deal of study time outside class. But what is the best way to study? And how should you make the most of your time in class and in lab? Furthermore, how should you use this book and its associated resources? This introductory module answers these questions and more. Read on, and find out how you can develop the skills required for success in your A&P course.

How to Develop Study Skills

"I don't know why I don't get good grades. I study so much, and yet I still don't do well on the exams. How can I improve my test scores?" Such sentences have been uttered by students everywhere. Many students put in a great deal of effort for little reward. This can be demoralizing, making a person feel like they're just not "smart enough" to excel in tough courses. But this is usually not even remotely the case; instead, the problem centers on simply not knowing how to study and make effective use of time. The next subsections offer help and guidance with these tasks.

Learning Modalities

You might have heard that people have different "learning styles." For example, one person might understand a process just from hearing it described, another might want to see the process diagrammed, and yet another might prefer to manipulate a three-dimensional model. In the past these were called "learning styles," but research has shown that there really is no one way that an individual learns best. So, instead, we now discuss how students learn in terms of **learning modalities,** which refer to the senses that are engaged during the learning process.

There are generally considered to be four main learning modalities:

- **Visual. Visual** learning modalities engage the sense of sight. They include using written material such as notes and textbooks and looking at diagrams, illustrations, and visual multimedia presentations (such as animations).
- **Auditory.** As implied by the name, an **auditory** learning modality engages the sense of hearing. These modalities include listening to lectures or presentations (live or recorded), discussing material with a group, or using auditory memory devices such as mnemonics.
- **Tactile.** The anatomy and physiology laboratory is a great place for **tactile** learning modalities, which engage the sense of touch. Tactile learning modalities include manipulating three-dimensional structures or doing experiments to see how things work.

- **Kinesthetic. Kinesthetic**—or "movement"—learning modalities are covered last because they can be included with any of the previous three. You can color-code your class notes; make textbook reading active by doing workbook pages as you read; draw diagrams; repeat key lecture points aloud or give a lecture to a study group; and draw, write, speak, solve problems, and move during a lab session.

Just from reading these descriptions, you can probably pick out a few things you would prefer to use in studying. But be careful not to limit yourself to just your preferences. In fact, most people learn best when they expand beyond their learning preferences and use different learning modalities as they study. Indeed, research has found that students who use multiple modalities score better on exams than do students who study according to their learning preferences. Stated simply, students who use multiple modalities engage more senses, which enables them to form more connections and retain more information. So branch out, and try a variety of techniques until you find the best combination for you. And be aware that the best combination in one class might not be the best in another class—"learning how to learn" is a process that requires continual adaptation and development throughout your college career.

How to Read a Textbook

Reading your textbook is a must, but what is the best way to go about it? The first mistake many students make is trying to read a textbook as if it were a novel, starting at the first page and reading through it sequentially until the end. However, it's pretty obvious that a textbook isn't a novel, so it doesn't make sense to read it like one. Instead, try an approach known as the SQ3R method, which involves the following steps: *survey, question, read, recite,* and *review.* Let's look at each of these steps in more detail.

- **Survey.** Before reading any chapter, you should do an initial survey, during which you skim the chapter and its figures. Begin by reading the headings and the Learning Outcomes at the beginning of each section. As you skim, take note of the key terms that are in boldface as well as the figures and tables. After finishing the initial survey, read the chapter summary. This initial survey will give an idea of what you will be focusing on and will make you more familiar with the content as you read.
- **Question.** The next step is to form questions about the chapter's content that you should answer as you read. These questions can be fairly broad and simple. For example, if you're reading about the heart, ask the following questions: "What do I know about the heart?" "How is the heart structured?" and "How does the heart's structure relate to its function?" If you find yourself struggling to come up with questions, a good place to start is the Learning Outcomes at the beginning of each section. Take a look back at the Learning Outcomes at the start of this section, and notice how easily each outcome can be turned into a question.
- **Read—actively.** Now you are ready to go on and read the chapter. As you read, it is most important to read *actively,* as active reading—the process of reading while engaging in some

sort of activity with the text—promotes active learning. In other words, to learn most effectively, you must do more than just read: You must also listen, write, discuss, and be engaged in solving problems. A novel is fine to read passively because you are interested in the plot, which keeps you turning pages. You also don't necessarily have to remember the details of a novel—they are more like icing on the cake of the plot. But with a textbook, there is no overarching plot, no matter how interesting you might find the content, and the details are often critical. For this reason, active reading is a must. Reading can be made active in many ways, the first of which is to take notes and make diagrams as you read. In addition, this book has many other resources to help you read actively, which are detailed in a later section of this module.

- **Recite.** Part of reading actively is reciting the material—speaking aloud—as you read through it. This might seem strange, but speaking while reading engages more parts of your brain and helps you retain more of what you've read. It is particularly important in sciences such as A&P, which have a great number of new, often foreign-sounding words. To help, we've provided pronunciations after many new terms.
- **Review.** The final step is reviewing what you have read. This can be done in many ways, including answering quiz questions found in the book, writing summaries, or discussing things aloud with study partners.

Note that you should read your textbook in this manner *before* going to the class at which the topic is discussed. We explain why in a later section.

Managing Your Time

Whatever your approach, one fact remains: Studying takes time, and we don't have a way to add extra hours to the day. Perhaps the number one mistake made by students is not allotting enough time for regular studying every week, which eventually necessitates "cramming" for the exam. It goes (almost) without saying that cramming does not lead to success, in either the short or the long term.

So, how can you avoid the need for cramming? The answer is surprisingly simple: make a schedule and budget your time. Plan out your time at least a week before your exam, and budget several hours each day for studying. Allow more hours than you are likely to need in case something unexpected comes up and disrupts your schedule. You might want to schedule specific activities each day; for example, Monday might be for reviewing notes, Tuesday for meeting with a study group, and Wednesday for studying with a tutor. To maximize study time, think about *all* of your time—you might be able to study, for instance, during lunch or while you are commuting to and from school or a job (**Figure 1.1**).

How to Study for an Exam

Now that you know how to read the book, you are probably wondering how to apply this in studying for quizzes and exams. The answer is, of course, complicated, as no two people are the same. However, the following methods have been found to work for many people:

- **Find out as much as you can about the exam.** The more you know about the exam, the easier it will be to prepare for it. Ask about the format of the exam—are the questions multiple choice, short answer, matching, or a mix of many different types? Will the questions be fact-based or application-based? In addition, your professor may have made old practice exams available for students to use. If you are unable to glean much information, use your first exam in the class for future reference.
- **Take advantage of the resources available to you.** A variety of resources associated with this book can help you study for exams, including the companion workbook, several features within the text, and many online practice questions. (See the section later in this module for details.) The more questions you answer, the better prepared you will be for the exam, because you will find out what you do know and, more importantly, what you *don't* yet know. Your college or university might also have resources for help, including tutors, computer labs, and the A&P lab.

Figure 1.1 Some ways to maximize your study time.

- **Form a study group.** Find a group of classmates at the beginning of the semester and meet regularly to go over difficult concepts, take practice tests, and offer one another support.
- **Use whatever study techniques help you.** As we discussed earlier, students tend to study best when using a variety of different learning modalities. So, try different techniques to help you master the material, such as listening to lectures or podcasts, using coloring and drawing exercises, making flashcards or notes, or watching video tutors or animations. Some materials may prove more useful in preparing for lab quizzes, whereas others may work better for lecture exams.
- **Take care of yourself and manage your stress.** This bit of advice might sound obvious, but it can be easy with a busy schedule to neglect your basic needs, including nutrition, health, and sleep. Multiple studies have shown that neglecting your needs, particularly sleep, interferes with the ability to learn and retain information. Stress can be particularly problematic in the week before an exam, when you might choose to study for an extra hour rather than get that hour of sleep. However, in this case, another hour of sleep might actually be more helpful to your grade than an additional hour of studying.
- **Don't be afraid to ask for help.** We'll let you in on a secret: Your professor genuinely wants you to succeed. Your professor is also likely your best resource and ally in your travels through the world of A&P. But your professor can't read your mind, and can't offer help unless you request it. When you have questions or need help, don't be afraid to ask.

Quick Check

☐ 1. What are some examples of learning modalities?

☐ 2. How should you approach reading a textbook, including this one?

☐ 3. What are some study strategies to improve your chances of success?

How to Make the Best Use of Class and Lab Time

Up to this point, we have been discussing what you do *outside* class to maximize your chances of success. But what you do *in* class is every bit as important to your success. It is doubly important in A&P, because you will likely have a laboratory period as well. The next two subsections discuss how you can make the best use of time in class and in lab.

Come Prepared

Perhaps the most important step you can take to enhance your class and lab experience is to come prepared. You are assigned reading to help you become familiar with the material *before* you come to class or lab. Why is this important? There are several reasons. The first is simple: Attending a lecture without first learning about the material is like visiting a foreign country without knowing one word of its language. It's as difficult to find your way through unfamiliar material as it is to navigate foreign roads whose signs are in a language you don't speak. In other words, it's very easy to wind up lost.

Another important reason to do the assigned reading before class is that your professor prepares lectures with the assumption that you have already read the lecture material. He or she uses the lecture to discuss details of the reading and explain its concepts, not start at the very beginning. If you are unfamiliar with the material in general, you won't benefit from a discussion of its details.

A final reason has to do with the way the brain processes information. Consider this example: A person is driving past a horse field and says to his companion, "Hey, look, my friend told me that there's a zebra in that field." The companion looks at the field blankly and sees only horses. "What zebra?" the companion asks. The driver points and says, "It's right there, in front of you." After a moment of staring, the companion finally realizes that there is, indeed, a zebra standing in the field, surrounded by horses. Why couldn't the companion see the zebra, when it was there all along? The answer is that her brain did not expect a zebra to be present in a horse field, and so her brain simply didn't perceive the zebra's presence. The driver knew the zebra was there and so his brain did perceive its presence. This is a true story—the companion who had trouble seeing the zebra was yours truly, the author of this book.

In the same way that someone might not perceive a zebra that he or she doesn't expect to be there, your brain won't perceive unfamiliar concepts and information in class. If the information in the lecture is foreign to you, your brain simply isn't likely to register it. However, if you read the material before lecture, you will be able to "see the zebra" in your classroom, and do better on your exams and quizzes as a result.

How to Take Good Notes

Many professors present their lectures using PowerPoint®, and many also make these presentations available online for students to read before class. You might think that if you print out your professor's presentation and take it to lecture, your notes for class are as good as completed. This couldn't be further from the truth. Although it's not a bad idea to look over your professor's presentations before class, these are by no means a substitute for class notes.

The best way to take notes during class is to bring along those you took while reading. Leave several lines between each main point in your notes so that you can add as needed while your instructor lectures. Come to class equipped with a highlighter so that you can identify points emphasized by your instructor.

As with other topics we have discussed, note taking should be individualized according to what works best for you and your professor's lecture style. You may also benefit from recording the lectures, if your professor permits it. This can help you fill in any blanks that you might have missed during class.

Quick Check

☐ 4. Why is it important to read the material before you come to class and lab?

☐ 5. What are some strategies for taking good notes in class?

How to Use This Book and Its Associated Materials

The study strategies we have discussed so far are fairly universal and could be applied to any course you are taking, including A&P. Let's now move on to some hints and tips for this course and, in particular, this book and its associated materials. You should know above all that this book was written specifically with *you* in mind. Let's tour the features found in this book and discover how to make the best use of them.

Modules

Each chapter is divided into sections that we call *modules*. A module features definitive starting and stopping points that cover one core concept and its related principles. The text is organized in this manner to divide the material into more manageable chunks and to help you make the most efficient use of limited time. It usually works best to master the material in one module before moving on to the next one. Every module contains several features to help you master the material, which we discuss next.

Learning Outcomes

As you likely noticed at the start of this module, every module begins with Learning Outcomes, or a list of its core concepts and principles that you should come to understand. Students have a tendency to breeze past the Learning Outcomes, but remember that they are there for one main reason: to help you. Use them to your advantage—read them prior to starting the module as part of the "survey" portion of the SQ3R method, and then revisit them after finishing the module for the "review" portion. Take a look at the Learning Outcomes for this module; you'll see they are written in a way that makes it easy to adapt them into a quiz.

Try an experiment for this chapter: Take the end-of-chapter quiz (called Assess What You Learned) after reading the chapter, without first looking at the Learning Outcomes. Then, work with the Learning Outcomes and write down the answers to each one of them. Now, take the quiz a second time to see how much better you can do with the help of the Learning Outcomes.

Concept Boosts and Study Boosts

Without question, certain concepts of A&P are more challenging than others. Such concepts often require some more explanation or another way of looking at them. You'll find this in the form of short sections called Concept Boosts (an example is shown in **Figure 1.2**), which give you a "boost" in understanding a concept. You shouldn't treat these sections as optional; in fact, it's a good idea to give them as much attention as the regular text.

In addition to the Concept Boosts, you'll find Study Boosts in some chapters. Study Boosts give you study tips for particularly challenging material. They often include mnemonic devices to help you memorize anatomical structures or the steps of physiological

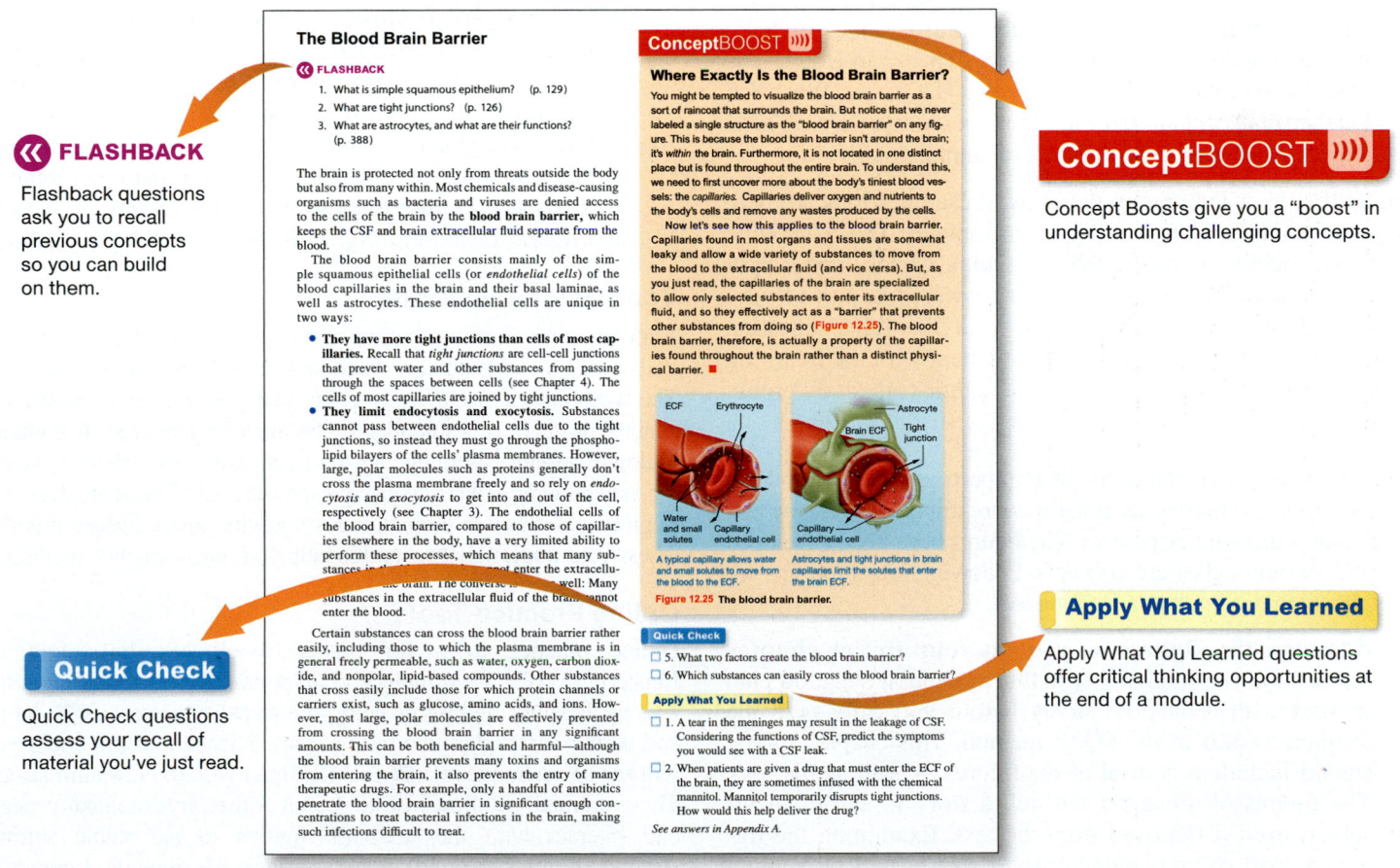

Figure 1.2 Selected features of this textbook.

processes. Like Concept Boosts, Study Boosts are there for your benefit, and it's to your advantage to pay close attention to them.

Questions

You will find many different opportunities for self-assessment throughout each chapter, several examples of which are shown in Figure 1.2. Each type of question serves a different purpose, although all will help you discover how well you met the Learning Outcomes. The following is an overview of the categories of questions you will encounter in this book:

- **Flashback questions.** Many sections start with one or more Flashback questions that ask you to recall material from earlier modules or chapters. Each Flashback question has an associated page number to help you find the concept in the book. Flashbacks help you remember previous terms and concepts so that you can build on them and better understand the current material.

- **Quick Check questions.** Quick Check questions are found after each major section within a module. These are generally simple recall questions so that you can assess how well you remember the basic concepts about which you have just read.

- **Apply What You Learned questions.** A module ends with two or more Apply What You Learned questions. These critical thinking and problem-solving questions ask you to analyze and apply the material you just read. With Apply What You Learned questions, you can test yourself on how well you understand the concepts in the module.

- **Assess What You Learned quiz.** Finally, once you have finished the chapter, you have another chance to assess how well you met the Learning Outcomes. The Assess What You Learned quiz contains three levels of questions, with each level requiring increasingly greater critical thinking skills.

Like the other features in the book, these questions are here for your benefit; use them to your advantage. The in-module questions, including the Flashback, Quick Check, and Apply What You Learned questions, are best answered as part of your active reading process. Experiment with the other question types. You might want to try combining them with the Learning Outcomes to build quizzes for yourself and/or your study group.

Figures

In an A&P course, the art is every bit as important as the text. Indeed, in some cases, the figures are actually *more* important than the text in teaching a structure or process. The figures have been developed to coach you through these concepts. Following are some tips for using the art in this book as a learning tool.

- **Examine the figures as you do your initial chapter survey.** As you read earlier in this module, it's a good plan to start with a chapter survey before you fully read the chapter, as part of the SQ3R method. This chapter survey should include a perusal of the figures as well as the text. The figures of a chapter will tell a story in and of themselves, even if removed from the text. Examining the figures as part of your initial chapter survey will give you a preview of the story told by the art and will make you better prepared to understand its details as you read later.

- **Identify the concept that the figure teaches first.** When you start reading the chapter and delving into the figures, begin by first identifying the exact concept that each figure teaches. The name of this concept can generally be found in the figure title, which you can see in the example shown in **Figure 1.3**. Note that if the concept illustrated in the figure sounds foreign to you, it's a good idea to re-read the text section on the subject so that the terms become more familiar.

- **Break the figure into parts, and understand each part before moving on to the next.** After you have identified what you need to learn from the figure, you can explore its content more deeply. The content of some of the larger figures in this book might look intimidating at first, just in terms of the amount of information they contain. In these cases, it helps to break the figure into smaller, more manageable parts. Often in physiology figures, such as the example in Figure 1.3, this is already done for you in the form of numbered steps with text boxes, which walk you through a process. The blue text explaining these steps represents the author's voice. However, in anatomical figures, it may be helpful to imagine lines that divide the figure into sections. This gives you "boxes" with smaller chunks of information to study.

- **Once you understand each part of the figure, examine it as a whole.** After you have mastered and understood the parts of the figure, you can step back and look at the process or structure as a whole.

- **Combine the figure's content with that of other figures for a more global understanding.** For more complicated anatomical structures or physiology concepts, the content might actually be spread out over several figures. In such cases, follow the same steps outlined previously for each figure. After you have interpreted each one, put them all together so that you can see how all the parts combine to make the whole. Sometimes this is done for you in sections and figures called *Putting It All Together: The Big Picture.*

Companion Workbook

Another active learning feature of this text is your *Active-Learning Workbook*, the companion workbook that comes with your text. The workbook is not a study guide that you use after reading the chapter; instead, it is intended to be used *as* you read. It contains multiple activities: sequencing-the-steps exercises; labeling, coloring, and drawing tasks; and group exercises. All of these are designed to guide you through the reading, to provide many opportunities to assess your understanding, and to help you practice active reading.

Online Practice Tools

Online practice tools for this textbook can be found on MasteringA&P®. They include chapter practice tests, reading quizzes, art-labeling and sequencing-the-steps exercises (⬚ icon), and author-narrated podcasts (🎙 icon). Throughout the book, you will see animation icons (▶), which signal you to view animations. These animations allow you to see an entire physiological process and interact with the textbook figures in an online setting. Additional media found on MasteringA&P include Interactive Physiology® tutorials, PhysioEx™ 9.1 Laboratory Simulations, and Practice Anatomy Lab™ (PAL™) 3.0 virtual anatomy lab study tool.

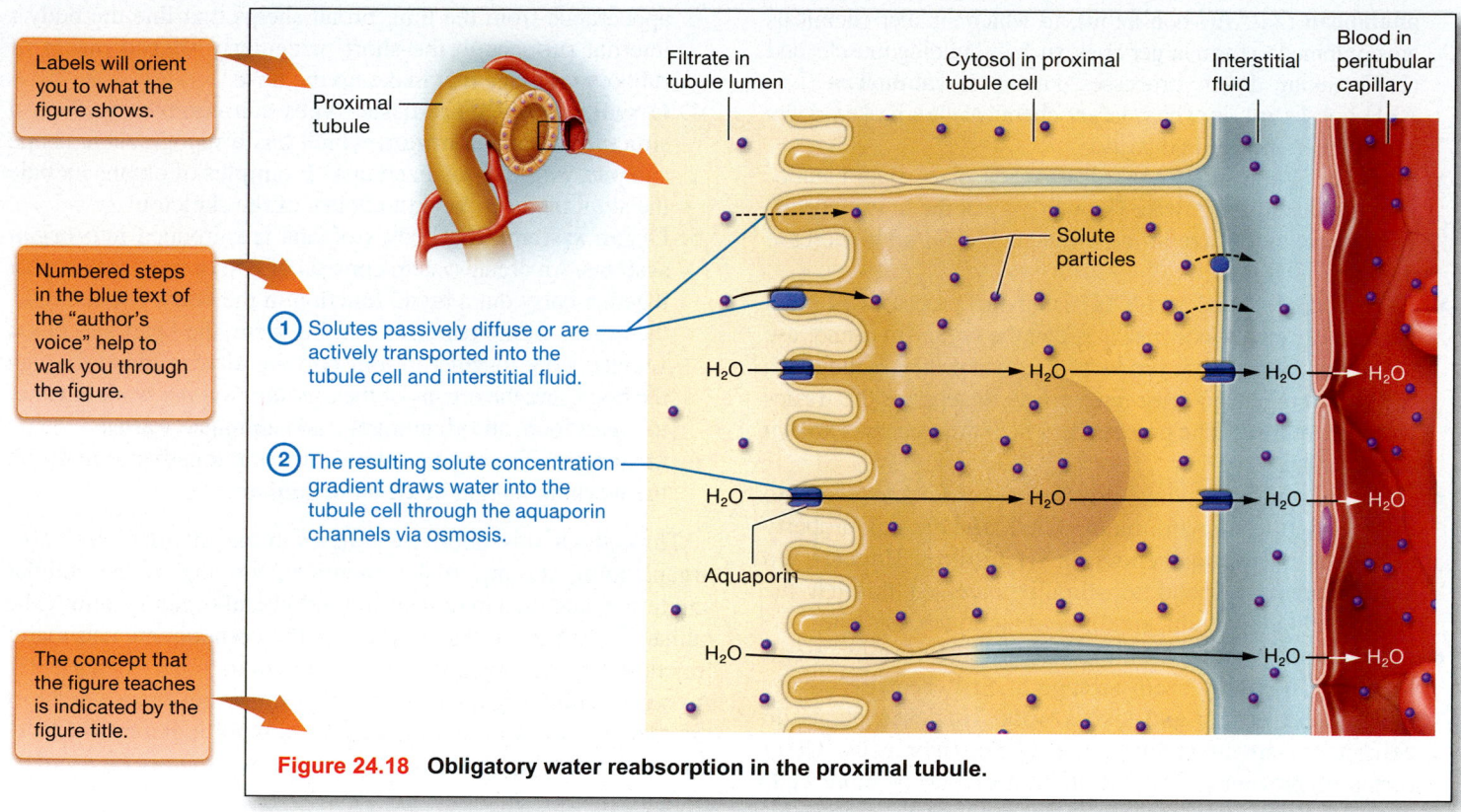

Labels will orient you to what the figure shows.

Proximal tubule

Filtrate in tubule lumen

Cytosol in proximal tubule cell

Interstitial fluid

Blood in peritubular capillary

Numbered steps in the blue text of the "author's voice" help to walk you through the figure.

Solute particles

① Solutes passively diffuse or are actively transported into the tubule cell and interstitial fluid.

② The resulting solute concentration gradient draws water into the tubule cell through the aquaporin channels via osmosis.

Aquaporin

H_2O

The concept that the figure teaches is indicated by the figure title.

Figure 24.18 **Obligatory water reabsorption in the proximal tubule.**

Figure 1.3 **How to approach a physiology figure.**

Quick Check

☐ 6. How can you use the features found in each chapter?

☐ 7. How should you approach the study of figures in this book?

Apply What You Learned

☐ 1. Design a study schedule for yourself up to the first exam.

☐ 2. Determine several study strategies that will help you succeed in your course.

See answers in Appendix A.

MODULE 1.2
Overview of Anatomy and Physiology

Learning Outcomes

1. Describe the characteristics of life and the processes carried out by living organisms.

2. Define the major structural levels of organization in the human body and explain how they relate to one another.

3. Define the types of anatomy and physiology.

4. Describe the organ systems of the human body and their major components.

5. Explain the major functions of each organ system.

Now that you know how to succeed in your A&P course, let's jump right in to our study of the human body. We begin with the characteristics common to all living organisms. Then, discussion turns to the introductory principles of anatomy and physiology, including the "building blocks" of the human body, and the different subfields of A&P.

Before you begin, remember from the previous module that textbooks aren't meant to be read like novels. Use the SQ3R method, and make your reading active by taking notes, making or labeling diagrams, or talking aloud. This will help make the most efficient use of your study time.

Characteristics of Living Organisms

From a biological perspective, we can define living organisms as those sharing the following distinct set of properties:

- **Cellular composition.** The **cell** is the smallest unit that can carry out the functions of life. All living organisms are composed of one or more cells, from single-celled bacteria to complex multicellular organisms such as humans.

- **Metabolism. Chemicals** are substances with unique molecular composition that are used in or produced by chemical processes. Living organisms carry out a wide range of chemical processes known collectively as **metabolism.** Two basic types of metabolic processes are (1) "building" processes, known as anabolism (an-AEH-boh-liz'm), in which smaller chemicals are combined to form larger ones, such as building muscle; and (2) "breaking down" processes, known as **catabolism** (kat-AEH-boh-liz'm), in which larger chemicals are broken down into smaller ones, as in digestion.

- **Growth.** When more anabolism takes place than catabolism, the result is **growth.** Growth may come in two forms: an increase in the size of individual cells and/or an increase in the number of cells.
- **Excretion.** The end result of metabolic processes is often chemicals called *waste products* that the organism cannot use for any purpose. Waste products are toxic if they accumulate, and so the organism must have a way to separate the wastes and remove them. The process by which this occurs is known as **excretion.**
- **Responsiveness.** Living organisms sense and react to changes in their environment called *stimuli;* this property is known as **responsiveness,** or **irritability.** Humans and other animals respond to stimuli perceived through the senses, including sight, smell, hearing, touch, and pain.
- **Movement.** Another key property of life is **movement.** Including this may seem strange at first, because clearly plants don't get up and take a walk. However, plants do exhibit movement inside and between their cells. Other forms of movement include motion of one or more cells within the organism and movement of the organism itself.
- **Reproduction.** The final property common to life is the ability to carry out **reproduction.** Reproduction takes two forms in multicellular organisms: (1) individual cells reproduce within the organism during growth and to replace damaged or old cells, and (2) the organism reproduces to yield offspring similar to itself.

Quick Check

☐ 1. What are the properties common to all living organisms?

Levels of Structural Organization and Body Systems

The body is constructed from a series of progressively larger "building blocks." Each building block is known as a *structural level of organization.* The six levels of organization we'll cover, illustrated in **Figure 1.4,** are as follows:

1. **Chemical.** The smallest level of organization we discuss in the human body is the **chemical level.** Chemicals range in size from tiny *atoms* to complex structures called *molecules,* which are composed of atoms ranging in number from two to thousands. All other levels are made up of combinations of molecules.
2. **Cellular.** Groups of several different types of molecules combine in specific ways to form structures at the **cellular level.** As you will discover in upcoming chapters, the cells in the body vary widely in size, shape, and function.
3. **Tissue.** Groups of similar cells and the material outside them, called the *extracellular matrix,* come together to perform a common function as a **tissue.** Tissues vary in

appearance from the thin, broad sheets that line the body's internal surfaces to the short, irregularly shaped pieces of rubbery cartilage that make up the nose.

4. **Organ.** Two or more tissue types can combine to form a structure called an **organ,** which has a recognizable shape and performs a specialized task. Examples of organs include the skin, the heart, and the bones of the skeleton.
5. **Organ system.** The body's organs are grouped into **organ systems.** An organ system consists of two or more organs that together carry out a broad function in the body. For example, the organs of the *cardiovascular system,* the heart and blood vessels, work together to transport and deliver blood through the body, and the organs of the *digestive system* work together to ingest food, absorb nutrients, and eliminate wastes.
6. **Organism.** The organ systems function together to make up the working human body, an **organism.**

Throughout this book we will examine all these levels of organization, starting at the chemical, moving to the cellular and tissue, and then to organs and individual organ systems. The human body has 11 organ systems; the components and major functions of each organ system are shown in **Figure 1.5** on p. 10. As we examine organs and organ systems, never lose sight of the fact that all of these organ systems work together to ensure the survival of the organism as a whole.

Quick Check

☐ 2. How do the six levels of organization of the human body relate to one another?

☐ 3. What are the 11 organ systems in the body?

Types of Anatomy and Physiology

We can approach the study of anatomy in several ways. In this book, we examine the human body primarily by looking at individual organ systems, which is known as **systemic anatomy.** In another approach, the body can be divided into regions such as the back or head and neck; this is **regional anatomy.** Finally, we can study the surface markings of the body, which is called **surface anatomy.**

Within the broad field of anatomy are numerous subfields for more specialized study. Anatomical subfields are generally classified by the structural level of organization being studied. For example, the field of **gross anatomy** examines structures that can be seen with the unaided eye, such as organs and organ systems. In contrast, the structures studied in **microscopic anatomy** require the aid of a microscope. Topics within the field of microscopic anatomy include **histology** (hiss-TAWL-uh-jee; *histo-* = "tissue," *-logy* = "study of"), the study of tissues, and **cytology** (sy-TAWL-uh-jee; *cyto-* = "cell"), the study of cells.

Physiology also has numerous subfields. Typically, physiological specializations are classified according to the organ or organ system being studied. For example, *neurophysiology* studies the brain and nerves (*neuro-* = "nerve"), and *cardiovascular physiology* studies the heart and blood vessels. Physiologists may also specialize in levels of organization other than the systemic—some

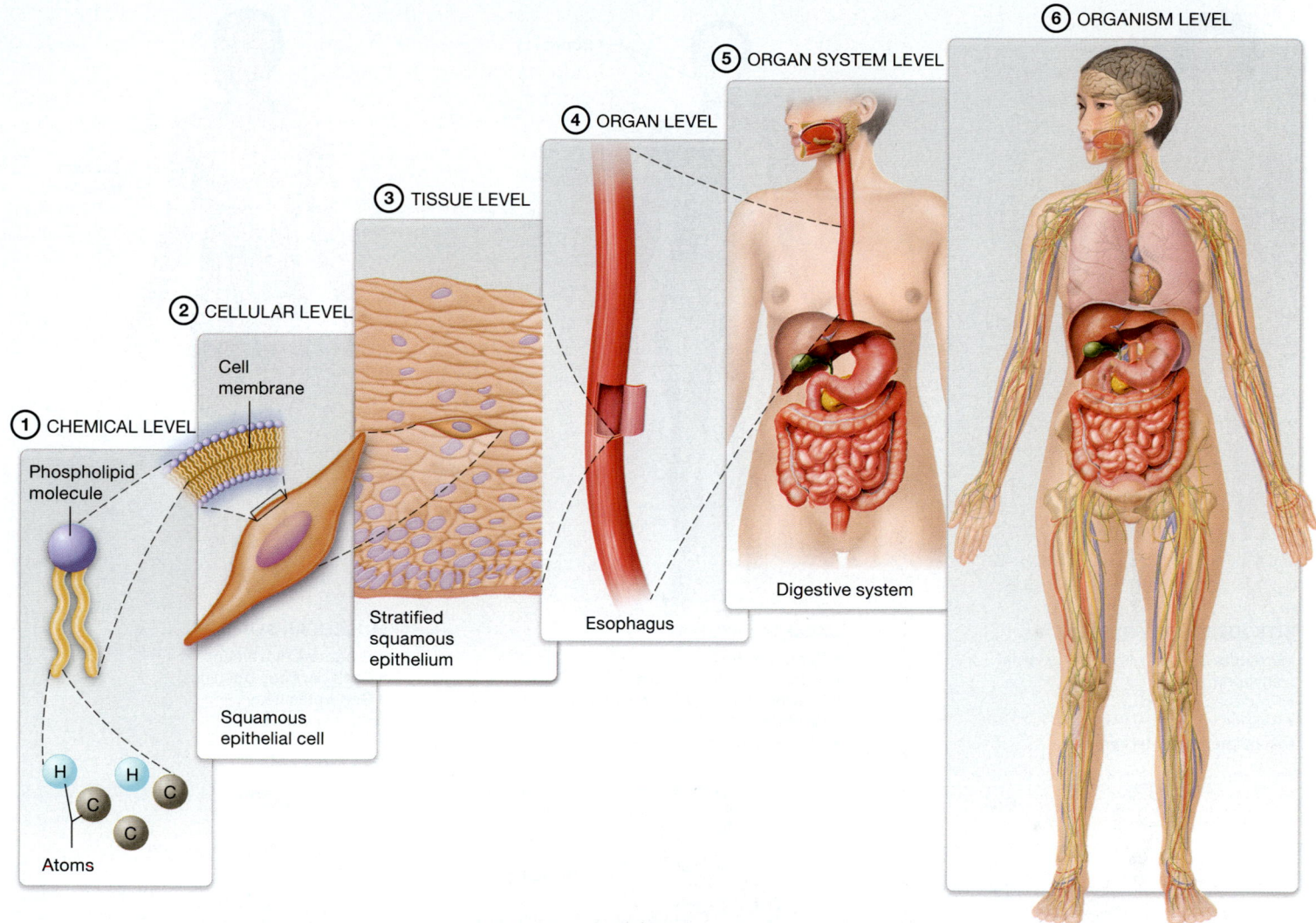

① CHEMICAL LEVEL

Phospholipid molecule

H   H
   C
C   C

Atoms

② CELLULAR LEVEL

Cell membrane

Squamous epithelial cell

③ TISSUE LEVEL

Stratified squamous epithelium

④ ORGAN LEVEL

Esophagus

⑤ ORGAN SYSTEM LEVEL

Digestive system

⑥ ORGANISM LEVEL

Play Video Tutor
@ Mastering Anatomy & Physiology

Figure 1.4 Six structural levels of organization of the human body.

physiologists study the body's chemical and cellular processes, and others study specific tissues or organs.

Quick Check

☐ 4. How do gross anatomy and microscopic anatomy differ?

☐ 5. How are physiological specializations classified?

Apply What You Learned

☐ 1. The condition *hypothyroidism* is characterized by a decrease in the synthesis and secretion of the chemical thyroid hormone from the thyroid gland, an organ in the neck that is part of the endocrine system. Explain how this condition involves all levels of organization in the body.

☐ 2. At first glance, a human and a rose plant seem to have little in common. Explain what these two organisms do actually have in common.

See answers in Appendix A.

MODULE **1.3**
The Language of Anatomy and Physiology

Learning Outcomes

1. Describe a person in anatomical position.

2. Define the major directional anatomical terms.

3. Describe the locations of the major anatomical regions of the body.

4. Describe locations of body structures using regional and directional terminology.

5. Identify the various planes in which a body or body part might be dissected, and describe the appearance of a body sectioned along each of those planes.

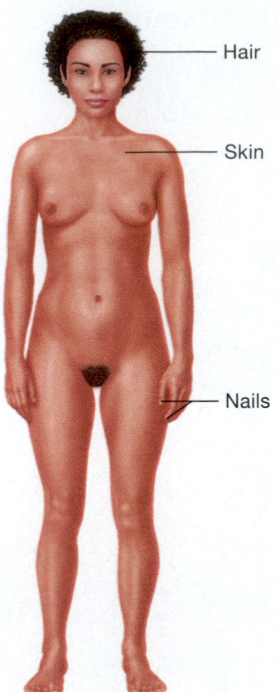

INTEGUMENTARY SYSTEM
- Protects the body from the external environment
- Produces vitamin D
- Retains water
- Regulates body temperature

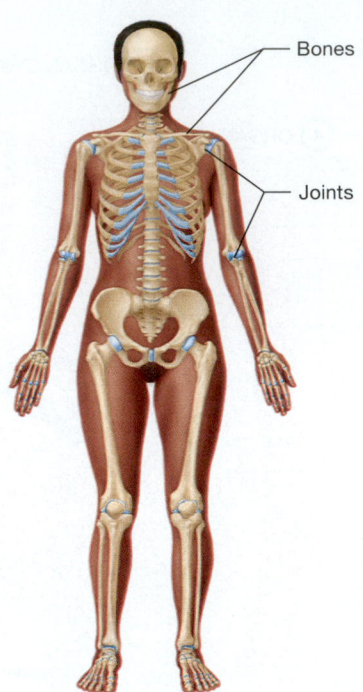

SKELETAL SYSTEM
- Supports the body
- Protects internal organs
- Provides leverage for movement
- Produces blood cells
- Stores calcium salts

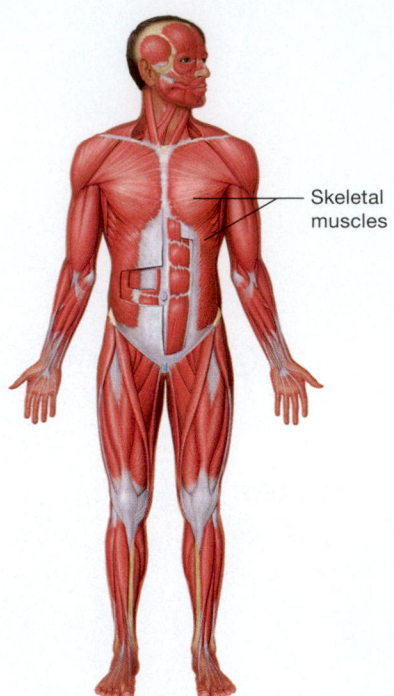

MUSCULAR SYSTEM
- Produces movement
- Controls body openings
- Generates heat

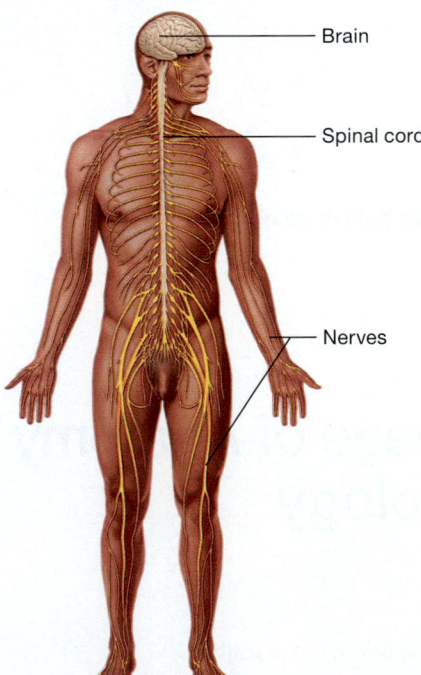

NERVOUS SYSTEM
- Regulates body functions
- Provides for sensation, movement, automatic functions, and higher mental functions via nerve impulses

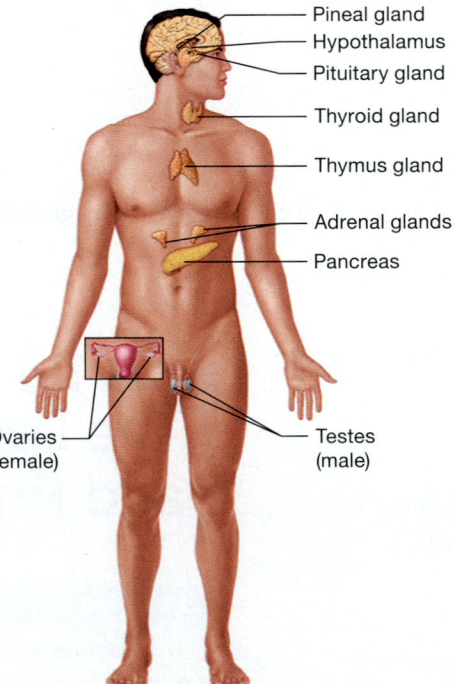

ENDOCRINE SYSTEM
- Regulates body functions
- Regulates the functions of muscles, glands, and other tissues through the secretion of chemicals called hormones

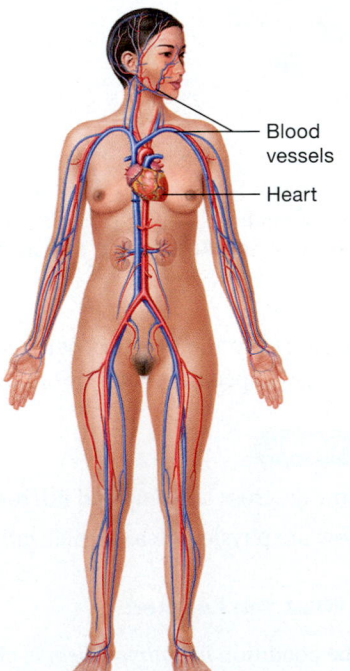

CARDIOVASCULAR SYSTEM
- Pumps and delivers oxygen-poor blood to the lungs and oxygen-rich blood to the tissues
- Removes wastes from the tissues
- Transports cells, nutrients, and other substances

Figure 1.5 The 11 organ systems of the human body.

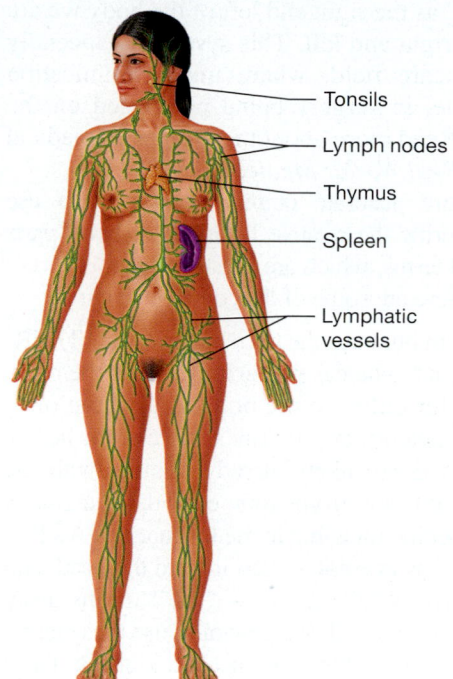

- Tonsils
- Lymph nodes
- Thymus
- Spleen
- Lymphatic vessels

LYMPHATIC SYSTEM
- Returns excess tissue fluid to the cardiovascular system
- Provides immunity (protection against disease)

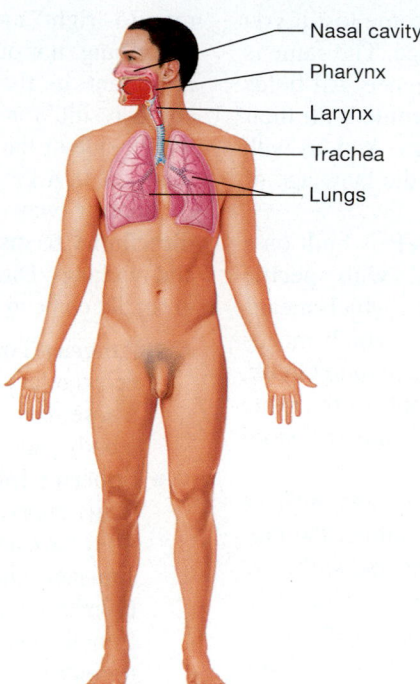

- Nasal cavity
- Pharynx
- Larynx
- Trachea
- Lungs

RESPIRATORY SYSTEM
- Delivers oxygen to the blood
- Removes carbon dioxide from the body
- Maintains the acid-base balance of the blood

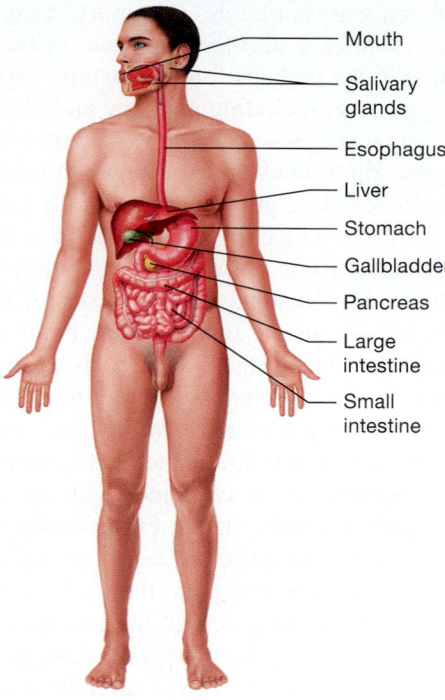

- Mouth
- Salivary glands
- Esophagus
- Liver
- Stomach
- Gallbladder
- Pancreas
- Large intestine
- Small intestine

DIGESTIVE SYSTEM
- Digests food
- Absorbs nutrients into the blood
- Removes food waste
- Maintains fluid, electrolyte, and acid-base balance

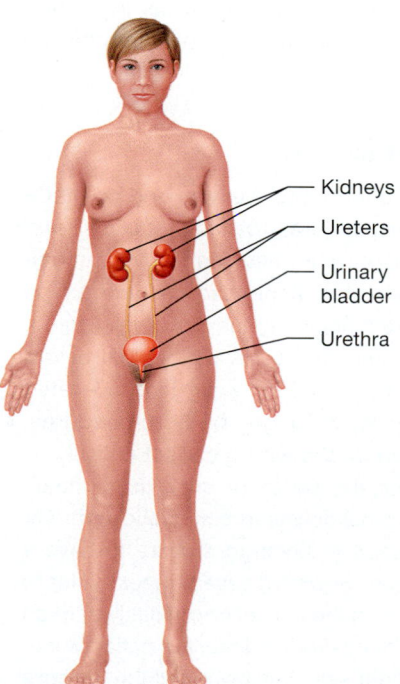

- Kidneys
- Ureters
- Urinary bladder
- Urethra

URINARY SYSTEM
- Removes metabolic wastes from the blood
- Maintains fluid, electrolyte, and acid-base balance
- Stimulates blood cell production

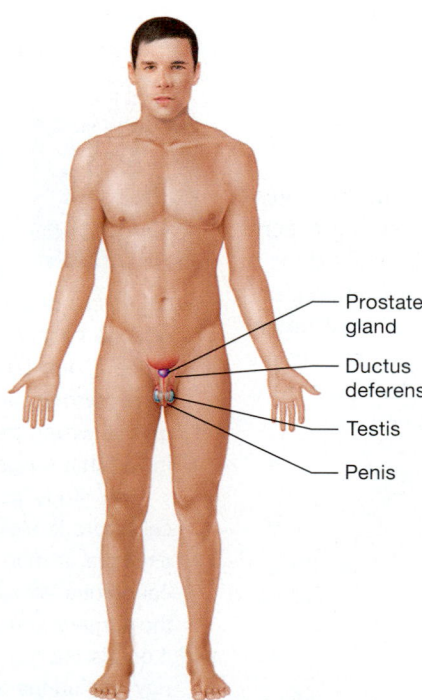

- Prostate gland
- Ductus deferens
- Testis
- Penis

REPRODUCTIVE SYSTEM: MALE
- Produces and transports sperm
- Secretes hormones
- Sexual function

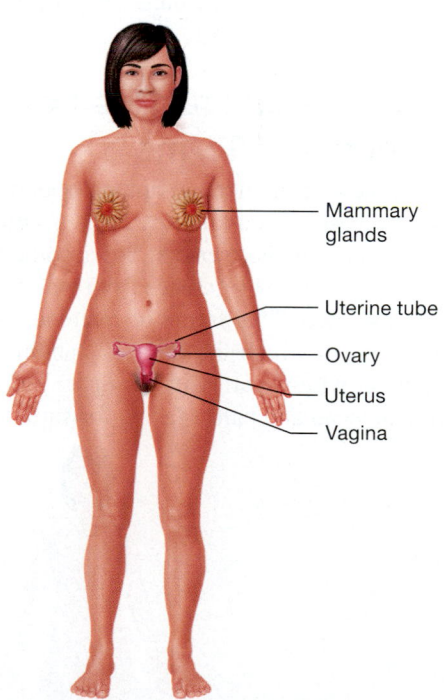

- Mammary glands
- Uterine tube
- Ovary
- Uterus
- Vagina

REPRODUCTIVE SYSTEM: FEMALE
- Produces and transports oocytes (eggs)
- Site of fetal development, fetal nourishment, childbirth, and lactation
- Secretes hormones
- Sexual function

Figure 1.5 **The 11 organ systems of the human body. (*continued*)**

When you visit a foreign country, it's easy to become lost if you don't quickly start learning some of their language. The same is true in the perhaps foreign-seeming world of science. All fields of science, including anatomy and physiology, come with their own language—a specific way of describing things. And, as with your trip to that foreign country, you must learn the language of science before you can hope to fully grasp it.

The language of science, and in particular A&P, is built on a group of *word roots*—core components of words with specific meanings. Examples of word roots include *card-*, which means heart; *encephal-*, which means brain; and *myo-*, which means muscle. Most of these word roots originated in Latin and Greek; this is why there are sometimes more than one root for the same concept. For example, *capit-/crani-* and *cephal-* are the word roots for head in Latin and Greek, respectively.

Word roots are combined with specific prefixes and suffixes to yield scientific terms. For example, we can combine the prefix *an-* ("without"), the word root *encephal-*, and the suffix *-ic* ("condition of") to form the word "anencephalic" (an'-en-seh-FAL-ik), which refers to the condition of lacking a part of the brain. To combine two word roots, add an "o" between them, called a *combining vowel*, as in the term for the heart and blood vessels, *cardiovascular* (*vascul-* = "vessel"). Building anatomical terms from word roots, prefixes, and suffixes is discussed on the inside back cover of the textbook.

This module serves as an introduction to the language of A&P. We'll talk about the accepted terms used by all scientists as well as health and medical professionals to communicate about the human body.

The Anatomical Position and Directional Terms

Accurate communication among scientists and healthcare professionals when discussing the body is crucial in preventing experimental and medical errors. We accomplish this by establishing a common frame of reference from which all body parts and regions are described. This frame of reference is termed the **anatomical position,** in which the body is standing upright and the feet are shoulder width apart, with the upper limbs at the sides of the trunk and the head and palms facing forward, as shown here:

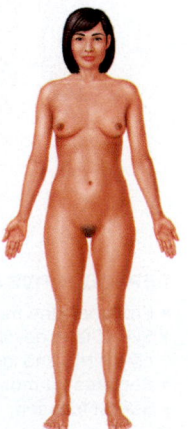

We always refer to the body as if it were in anatomical position, even if it's actually in another position. Note that we also refer to "right" and "left" as the right and left of the body we are describing, not our own right and left. This system is especially important in the healthcare field, where miscommunication could result, for example, in surgery being performed on the wrong side of the body. Read about these types of medical mistakes in *A&P in the Real World: Medical Errors.*

Another way to ensure accurate communication is to use **directional terms** to describe the relative locations of body parts and markings. Directional terms, which are illustrated in Figure 1.6, generally come in pairs. Here are some of the common pairs:

- **Anterior/Posterior.** In humans, the term **anterior** (an-TEER-ee-ur) refers to the front, whereas **posterior** refers to the back. These terms can refer either to the body as a whole or to a body part (e.g., the anterior or posterior surface of the heart).
- **Superior/Inferior.** You are likely already familiar with the terms superior and inferior from your everyday language. They have a more specific meaning in the language of A&P—**superior,** also known as **cranial,** means toward the head, and **inferior** or **caudal** (KAW-d'l; *caud-* = "tail") means away from the head or toward the tail. We generally use these terms to refer to positions only on the head, neck, and trunk; these terms are not applied to the upper and lower limbs.
- **Proximal/Distal.** Instead of using superior and inferior for the limbs, we use the terms proximal and distal. The term **proximal** (PRAWKS-ih-mul) refers to something being closer to the point of origin. You know this from the everyday term *approximate,* which refers to closeness. **Distal** refers to the opposite condition—being farther away or *distant* from the point of origin. In the case of the upper limb,

Medical Errors

Humans are imperfect beings, and so medical mistakes, generally called *medical errors*, are to some extent unavoidable. Most medical errors occur when a patient is dispensed the wrong type or dose of medication, but occasionally they involve surgery. Indeed, one large study found that 0.5% of medical errors were "wrong site" procedures, in which a surgeon operated on the wrong part of the body, or "wrong body" procedures, in which a surgeon operated on the wrong patient entirely.

The study concluded that the cause of such major medical errors is almost always breakdowns in communication. For example, a man was supposed to undergo surgery to have a cancerous left kidney removed. However, several weeks prior to the surgery, someone examining the patient erroneously marked it on his chart as the *right* kidney. And, indeed, the patient's kidney was on the examiner's right side, but it was still the patient's *left* kidney. Due to this error, the surgical team removed the patient's healthy kidney, leaving him only the cancerous one.

Mistakes of this nature have led to surgeons removing the wrong leg, drilling into the wrong side of the skull, removing the wrong ovary or testicle, replacing the wrong hip, and transplanting the wrong lung. Clearly, precise communication is critical to prevent medical errors, both minor and major.

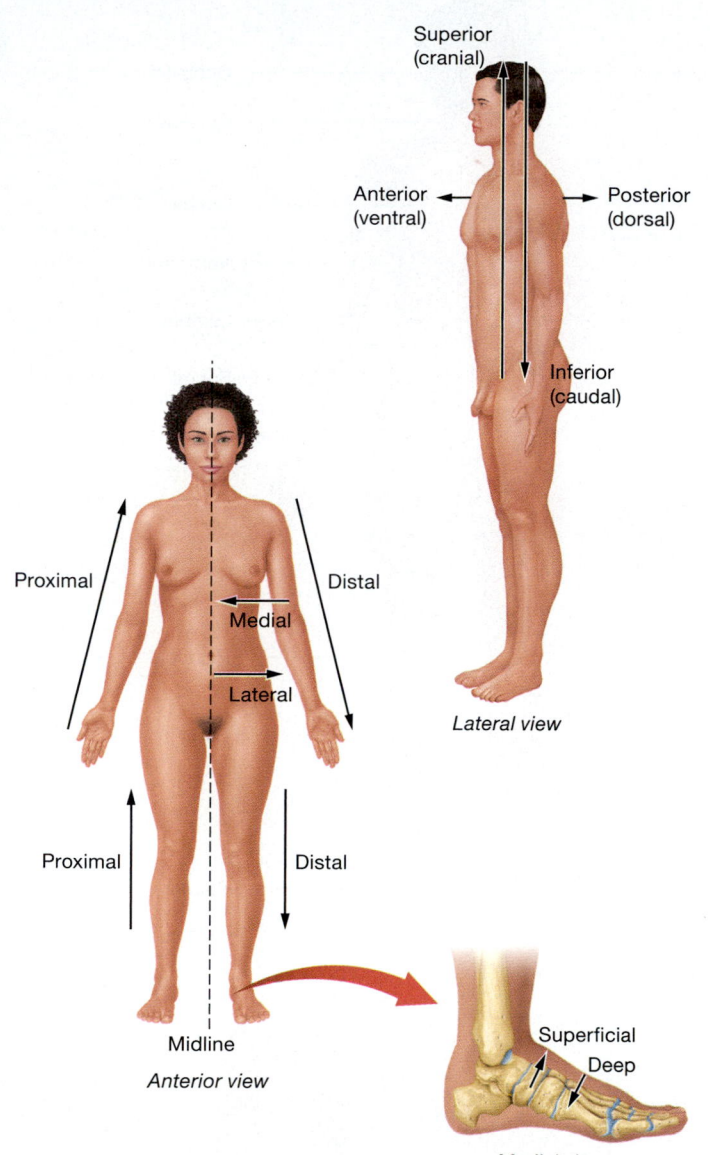

TERM	DEFINITION	EXAMPLES
Anterior (ventral)	Toward the front	• The palms are on the anterior side of the body. • The esophagus is anterior to the spinal cord.
Posterior (dorsal)	Toward the back	• The occipital bone is on the posterior cranium (skull). • The spinal cord is posterior to the esophagus.
Superior (cranial)	Toward the head	• The nose is superior to the mouth. • The neck is superior to the chest.
Inferior (caudal)	Toward the tail	• The nose is inferior to the forehead. • The umbilicus (belly button) is inferior to the chest.
Proximal	Closer to the point of origin (generally the trunk)	• The knee is proximal to the ankle. • The shoulder is proximal to the elbow.
Distal	Farther away from the point of origin (generally the trunk)	• The foot is distal to the hip. • The wrist is distal to the elbow.
Medial	Closer to the midline of the body or a body part; on the inner side of	• The ear is medial to the shoulder. • The index finger is medial to the thumb.
Lateral	Farther away from the midline of the body or a body part; on the outer side of	• The shoulder is lateral to the chest. • The thumb is lateral to the index finger.
Superficial	Closer to the surface	• The skin is superficial to the muscle. • Muscle is superficial to bone.
Deep	Farther below the surface	• Bone is deep to the skin. • Bone is deep to muscle.

Figure 1.6 Directional terms.

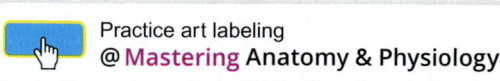
Practice art labeling
@ Mastering Anatomy & Physiology

the point of origin is the shoulder joint; for the lower limb, the point of origin is the hip joint. So structures nearer the shoulder or hip joint are proximal, and those that are farther away are distal.

- **Medial/Lateral.** The body's *midline* is an imaginary line that runs down the middle of the body (see Figure 1.6). The term **medial** refers to a position that is closer to this midline, whereas **lateral** refers to a position that is farther away from the midline. Note that we almost always refer to the midline of the whole body itself, not the midline of an individual body part.

- **Superficial/Deep.** This is another pair of terms that you probably use as part of your everyday language. The term **superficial** refers to structures that are closer to the surface of the body, whereas **deep** refers to those farther below.

Quick Check

☐ 1. What is anatomical position?

☐ 2. Fill in the blanks:

    a. The nose is _____ to the mouth.

    b. The ankle is _____ to the knee.

    c. The ring finger is _____ to the index finger.

    d. The skin is _____ to the bone.

    e. The trachea is _____ to the spine.

    f. The shoulder is _____ to the wrist.

Regional Terms

Broadly, we can divide the body into two regions: the **axial region** (AKS-ee-ul), which includes the head, neck, and trunk, and the **appendicular region** (ap´-en-DIK-yoo-lur), which includes the upper and lower limbs. Notice in **Figure 1.7** that the axial and appendicular

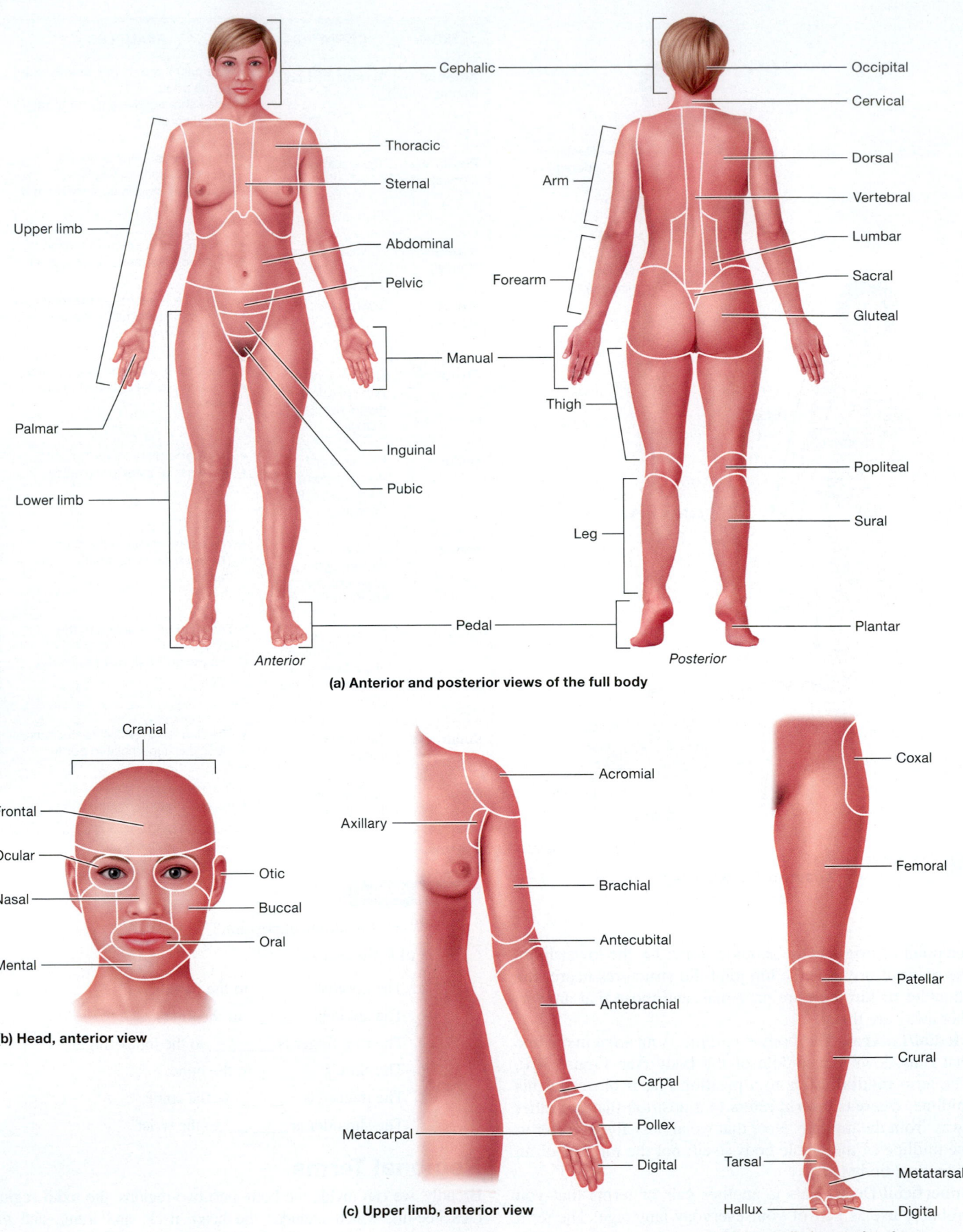

(a) **Anterior and posterior views of the full body**

(b) **Head, anterior view**

(c) **Upper limb, anterior view**

(d) **Lower limb, anterior view**

Figure 1.7 **Regions of the body.**

regions can be divided into several smaller regions. These may be named as nouns, such as *brachium* (BRAY-kee-um; "arm"), *crus* (KROOS; "shin"), and *thorax* ("trunk"), or we can add the suffix *-al* or *-ic* to the terms and turn them into adjectives. To use these adjectives, we must pair them with the word "region," which gives us terms such as the *brachial region, crural region,* and *thoracic region* (thoh-RASS-ik).

patellar region, and deep to the skin and muscle but superficial to the bone. Now we put it all together:

The wound is on the left anteromedial crural region, 6 centimeters proximal to the tarsal region and 10 centimeters distal to the patellar region. The pellet is lodged deep to the skin and muscle but superficial to the bone. ■

A summary of regional terms, including the adjectival forms, is given in **Table 1.1**.

ConceptBOOST)))

Putting Anatomical Terms Together

To put together the directional and regional terms we encountered in the past two sections (see Figures 1.6 and 1.7), let's use these terms to describe the surgical incision shown here:

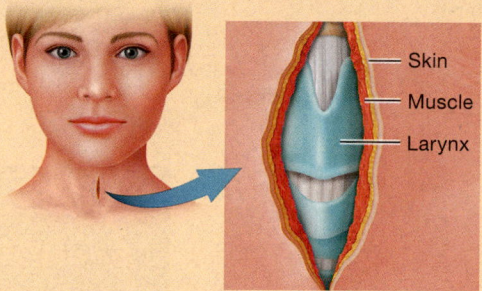

The first thing to do is name the region, which in this case is the cervical region. Now we need to add some descriptive directional terms to tell where, exactly, the incision is located in the cervical region. We can say that the incision is on the anterior side, it is just lateral to the midline, and it begins inferior to the mental region and ends superior to the thoracic region. We could also describe the depth of the incision, which is deep to the skin and muscle but superficial to the underlying larynx. So let's put this all together:

The incision was made on the anterior cervical region just lateral to the midline. It extended vertically from 1 centimeter inferior to the mental region to 2 centimeters superior to the thoracic region, and was deep to the skin and muscle but superficial to the larynx.

Let's do one more example using the following illustration, in which a person has a wound from a pellet gun:

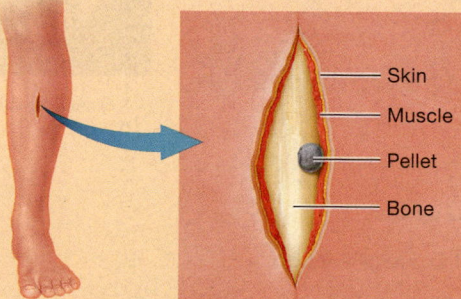

Start with the region, the left crural region, and add descriptive directional terms (don't forget that we need to use proximal and distal here, since we are describing the limbs): anterior, on the medial side, proximal to the tarsal region and distal to the

Table 1.1 Regional Terms

Regions of the Trunk	Pertaining To:
Abdominal	The abdomen
Cervical	The neck
Gluteal	The buttocks
Inguinal	The groin
Lumbar	The lower back
Palmar	The palm
Pelvic	The pelvis
Pubic	The pubis
Sacral	The sacrum
Sternal	The sternum
Thoracic	The chest
Vertebral	The spinal column

Regions of the Head and Face	Pertaining To:
Buccal	The cheek
Cranial	The skull
Cephalic	The head
Frontal	The forehead
Mental	The chin
Nasal	The nose
Occipital	The back of the head
Ocular	The eye
Oral	The mouth
Otic	The ear

Regions of the Upper Limb	Pertaining To:
Acromial	The point of the shoulder
Antebrachial	The forearm
Antecubital	The anterior surface of the elbow
Axillary	The armpit
Brachial	The arm
Carpal	The wrist
Digital	The fingers (or toes)
Manual	The hand
Metacarpal	The metacarpals (bones of the hand)
Pollex	The thumb

Regions of the Lower Limb	Pertaining To:
Coxal	The hip
Crural	The anterior surface of the leg
Femoral	The thigh
Hallux	The great toe
Metatarsal	The metatarsals (bones of the foot)
Patellar	The anterior surface of the knee
Pedal	The foot
Plantar	The sole of the foot
Popliteal	The posterior surface of the knee
Sural	The posterior surface of the leg
Tarsal	The ankle

Quick Check

☐ 3. Fill in the blanks:

   a. The wrist is also known as the _____ region.

   b. The arm is also known as the _____ region.

   c. The armpit is also known as the _____ region.

   d. The neck is also known as the _____ region.

Planes of Section

When studying a body or body part, we can learn only so much by looking at it from the outside. To gain a better understanding of its form and function, we need to look inside. This is accomplished through the use of **planes of section**—standard ways of dividing a body or body part to examine its internal structure. The three primary planes of section, shown in **Figure 1.8**, are as follows:

- **Sagittal plane.** The **sagittal plane** (SAJ-ih-tul) divides the body or body part into right and left sections (see Figure 1.8a). There are two variations of the sagittal plane: (1) the **midsagittal plane,** also known as the **median plane** (not to be confused with medial), which divides the body or body part into equal right and left parts; and (2) the **parasagittal plane** (pehr′-uh-SAJ-ih-tul; *para-* = "near"), which divides the body or body part into unequal right and left sections.
- **Frontal plane.** The **frontal plane,** also known as the **coronal plane,** divides the body or body part into anterior and posterior sections (see Figure 1.8b). The frontal plane is easy to remember because the *frontal* plane gives a *front* section and a back section.
- **Transverse plane.** The third commonly used plane is the **transverse plane,** also known as the **horizontal plane** or a **cross section** (see Figure 1.8c). The transverse plane divides the body or body part into superior and inferior parts. Note that it also divides the appendicular region into proximal and distal parts.

Another type of section that is less standardized is the **oblique plane.** An oblique section is taken at an angle and is useful for examining structures, such as the knee joint, that aren't easy to study with standard planes of section. However, these sections can be difficult to interpret and are used less frequently than the others we have discussed.

Note that planes of section are *not* the same as directional terms. You cannot say, "The bullet is located on the brachial region along the parasagittal plane." You must describe the location of the bullet using directional terms such as medial/lateral, anterior/posterior, and so on. However, as a surgeon you can make a cut along a parasagittal plane of the arm to locate and remove the bullet.

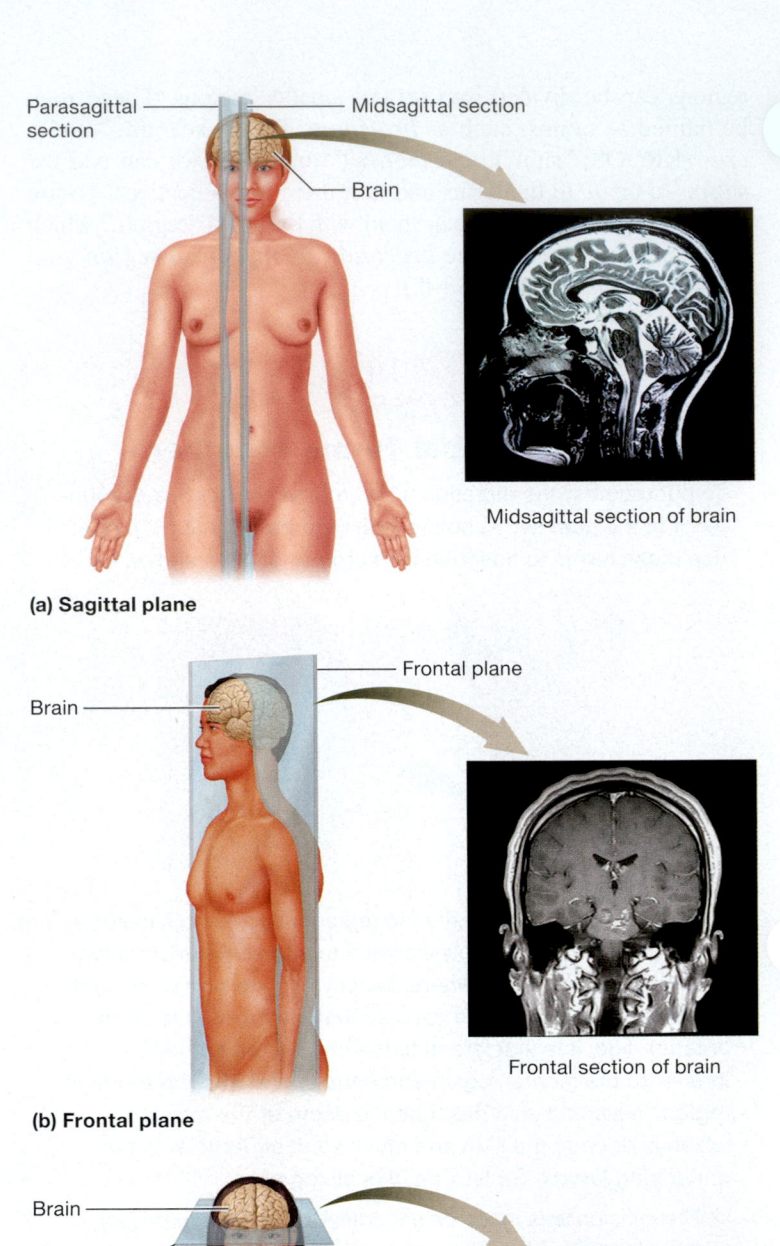

Parasagittal section — Midsagittal section

Brain

Midsagittal section of brain

(a) Sagittal plane

Frontal plane

Brain

Frontal section of brain

(b) Frontal plane

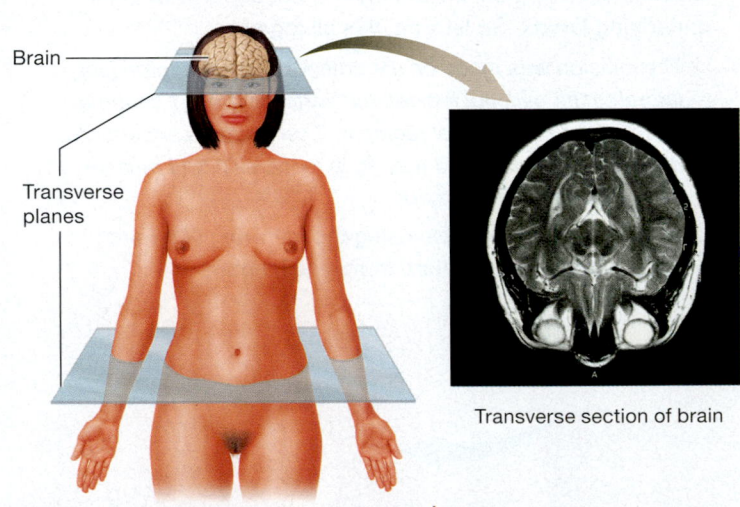

Brain

Transverse planes

Transverse section of brain

(c) Transverse planes

Figure 1.8 Planes of section.

☐ 4. How do the three main planes of section differ?

☐ 1. Locate a mole, scar, tattoo, or other mark on your skin and describe its location using regional and directional terms.

☐ 2. You need to examine the internal anatomy of both lungs at the same time but can make only one cut. Which plane or planes of section would allow you to do this? Explain.

See answers in Appendix A.

The Organization of the Human Body

1. Describe the location of the body cavities, and identify the major organs found in each cavity.

2. Explain how the abdominopelvic cavity is divided into quadrants and regions, and list the major organs located in each quadrant and region.

3. Describe the structure and function of the serous membranes that line the body cavities.

Internally, the axial region of the body—the head, neck, and thorax—is divided into several cavities. A *cavity* is any space within the body. Body cavities both protect internal organs and allow them to move and expand as necessary. The two major cavities are the **posterior body cavity,** which is largely located on the posterior side of the body, and the **anterior body cavity,** which is largely located on the anterior side of the body. This module takes a closer look at these two body cavities, their subdivisions, and their contents.

The Posterior Body Cavity

As you can see in **Figure 1.9a,** two subcavities lie within the posterior body cavity: the **cranial cavity,** which is located within the skull and protects the brain, and the **vertebral or spinal cavity,** which is found within the vertebral column and protects the spinal cord. The two cavities are continuous with each other, which reflects the fact that the brain and spinal cord are actually a single, continuous structure. Both cavities are filled with a fluid called *cerebrospinal fluid* (seh-ree′-broh-SPY-nul; *cerebro-* = "brain"), or *CSF,* which bathes both organs and keeps the brain buoyant within the skull.

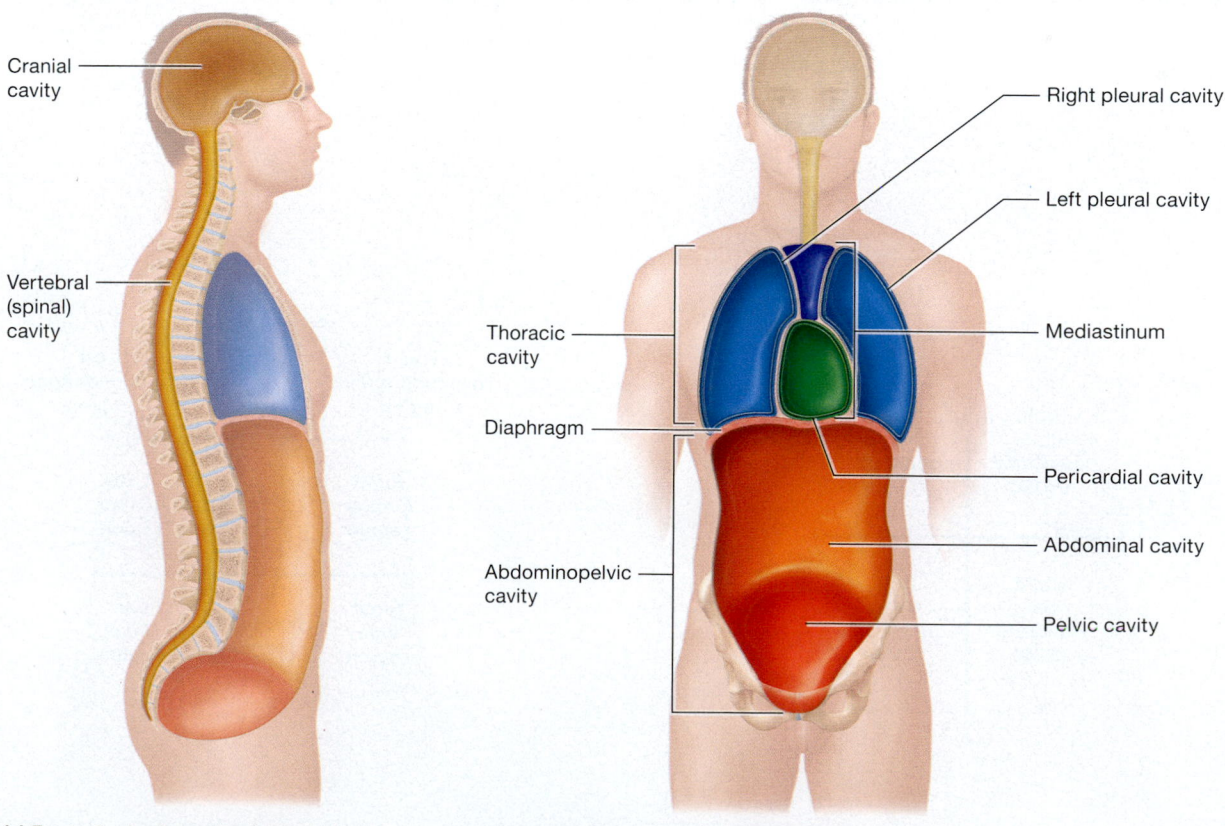

(a) **Posterior body cavity, lateral view**

(b) **Anterior body cavity, anterior view**

Figure 1.9 The anterior and posterior body cavities.

Quick Check

☐ 1. What are the two subcavities of the posterior body cavity?

The Anterior Body Cavity

The anterior body cavity has two main divisions that are separated by a dome-shaped muscle called the **diaphragm** (DY-uh-fram), which functions in breathing (**Figure 1.9b**). The cavity superior to the diaphragm encompasses the area of the thorax, which is why it is referred to as the **thoracic cavity.** The cavity inferior to the diaphragm encompasses the abdomen and pelvis, and so is known as the **abdominopelvic cavity** (ab-dom′-ih-noh-PEL-vik). These cavities are described next, along with the smaller cavities within them. Note that some of these smaller cavities are formed by thin sheets of tissue termed *serous membranes* (SEER-us), and so the discussion concludes with a look at the structure and function of this tissue.

The Thoracic Cavity

If you look at Figure 1.9b, you can see that the thoracic cavity houses three smaller cavities. These cavities are as follows:

- **Pleural cavities.** There are right and left **pleural cavities** (PLOO-rul), each of which surrounds one lung. These cavities are located within serous membranes.
- **Mediastinum.** Between the pleural cavities we find the **mediastinum** (meh′-dee-ah-STY-num; *med-* = "middle").

The mediastinum houses the heart, great blood vessels, trachea, and esophagus. The mediastinum is not located within a serous membrane.

- **Pericardial cavity.** Within the mediastinum we find another cavity within a serous membrane called the **pericardial cavity** (*peri-* = "around"). As its name implies, the pericardial cavity surrounds the heart.

The Abdominopelvic Cavity

There are two divisions within the abdominopelvic cavity: the **abdominal cavity,** which spans the area from the diaphragm to the bony pelvis, and the **pelvic cavity,** which is the area within the bony pelvis. The abdominal cavity contains organs of multiple systems, including the digestive, lymphatic, and urinary systems. Some of these organs are located in another subcavity within serous membranes, called the **peritoneal cavity** (pehr′-ih-tuh-NEE-ul; note that you cannot see the peritoneal cavity in Figure 1.9). We'll revisit the peritoneal cavity shortly. The pelvic cavity also contains the organs of several systems, including the reproductive, digestive, and urinary systems.

Notice in **Figure 1.10a** that we can divide the abdominopelvic cavity into different segments by drawing imaginary lines through its surface. The simplest way to divide the abdominopelvic cavity is to use two lines through the umbilical (belly button) region, one in the transverse plane and one in the midsagittal plane. This gives us four quadrants: the **right upper quadrant (RUQ)**, the **left upper quadrant (LUQ)**, the **right lower quadrant (RLQ),** and the **left lower quadrant (LLQ).**

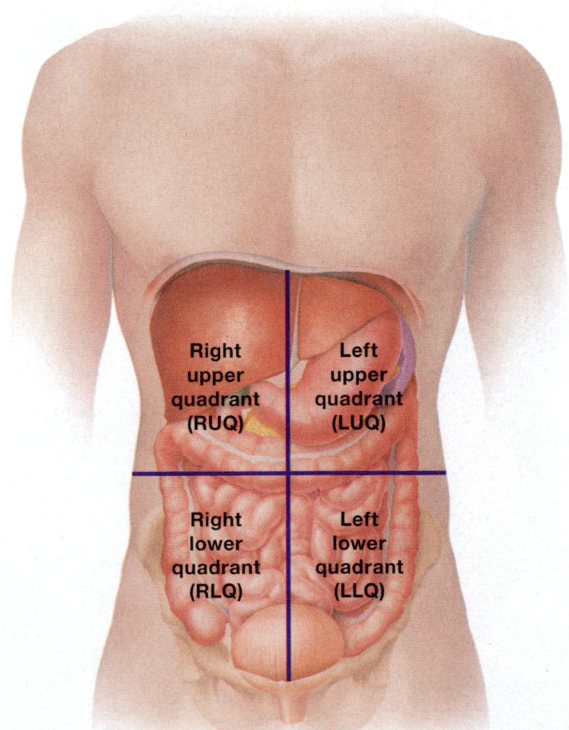

(a) The four abdominopelvic quadrants

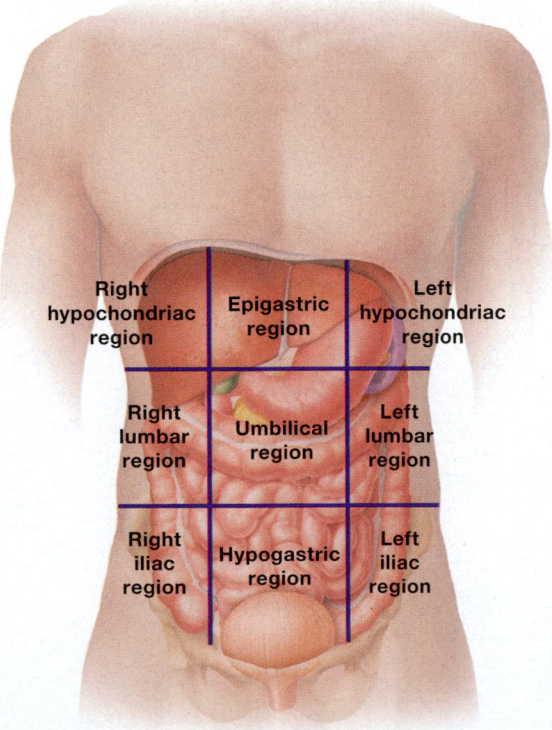

(b) The nine abdominopelvic regions

Figure 1.10 The four quadrants and nine regions of the abdominopelvic cavity.

A second system uses four lines—two in the parasagittal plane and two in the transverse plane—to divide the abdominopelvic cavity into nine regions. These regions, shown in **Figure 1.10b**, are as follows:

- The **right** and **left hypochondriac regions** (hy′-poh-KAWN-dree-ak) are the right and left superior divisions, respectively. They are named for the fact that they lie below the cartilage of the ribs (*hypo-* = "below," *chondro-* = "cartilage").
- The **epigastric region** (ep-ih-GAS-trik) is the region between the right and left hypochondriac regions and superior to the stomach (*epi-* = "above," *gastr-* = "stomach").
- The right and left middle regions are the **right** and **left lumbar regions,** so named because they are located in approximately the same region as the lumbar vertebrae.
- Between the right and left lumbar regions is the **umbilical region,** which contains the umbilicus.
- The right and left inferior regions are the **right** and **left iliac** (IH-lee-ak) or **inguinal regions** (ING-gwih-nul), respectively. Like other regions, these are named for the anatomical regions over which they lie.
- Between the two iliac regions we find the **hypogastric region,** so named because it is inferior to the stomach.

The nine-region system is preferred by anatomists, whereas the four-quadrant system is widely used in the medical field. *A&P in the Real World: Abdominal Pain* discusses the main organs that can be found in each quadrant.

Serous Membranes

As you've just read, certain cavities within the anterior body cavity are formed by thin sheets of tissue called **serous membranes.** When sectioned, serous membranes appear to be two membranes, but they actually consist of a single, continuous layer of tissue that folds over on itself to enclose a single space, much as a bean bag chair is one continuous layer of fabric that encloses stuffing. Within the cavity between the two layers, we find (instead of stuffing) an extremely thin layer of fluid called

Abdominal Pain

Abdominal pain is a common reason for people to seek treatment from a healthcare provider. The cause of such pain can be difficult to diagnose due to the sheer number of structures in the abdominopelvic cavity. However, this process can be made easier by using the four-quadrant system, as potential diagnoses may be narrowed down by the organs located in a specific quadrant. For example, RLQ pain could be caused by problems with the appendix, the right ovary (in females), the first part of the large intestine, or the last portion of the small intestine. Similarly, pain in the LUQ could result from problems with the stomach, the spleen, the pancreas, and parts of the large intestine. Although more information is needed to arrive at a conclusive diagnosis, knowing the quadrant in which the pain is located is a helpful first step in finding the cause of the pain.

serous fluid. This watery, slippery liquid is produced by the cells of the serous membrane. Its primary function is to lubricate organs in the cavity just as oil lubricates the pistons in a car—it prevents friction as an organ moves against adjacent structures.

In the body, serous membranes surround certain organs, including the heart, the lungs, and many abdominal organs. **Figure 1.11** shows how a serous membrane folds around and envelops the heart, forming the pericardial cavity. Part of the membrane comes into contact with the heart itself, and part has contact with the structures surrounding the heart. Although it is all one continuous membrane, we call these two parts of the same membrane different names: The inner layer that contacts the organ is known as its **visceral layer**

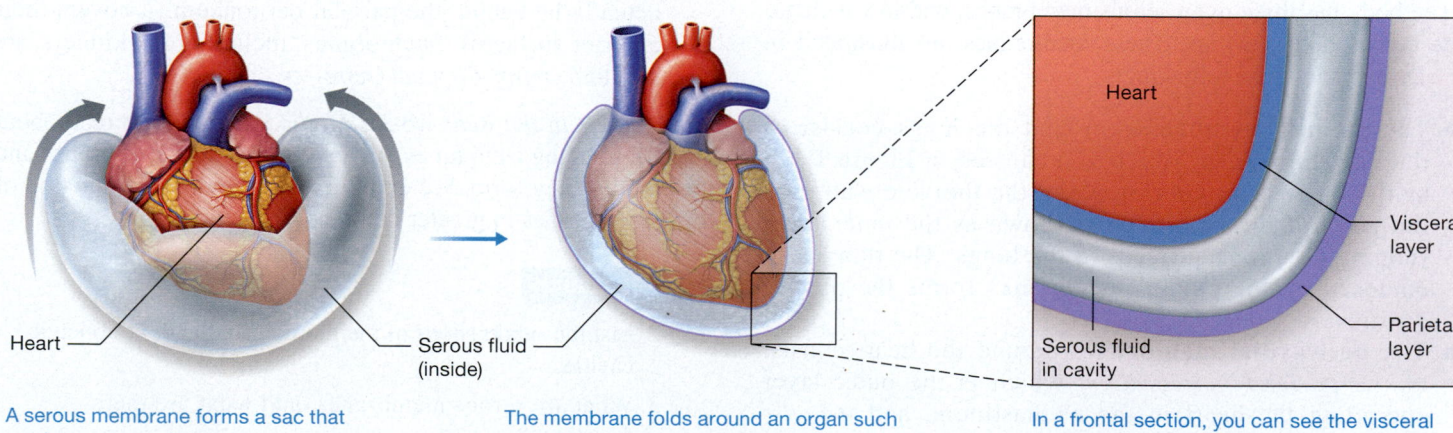

Heart   Serous fluid (inside)

Heart   Visceral layer   Serous fluid in cavity   Parietal layer

A serous membrane forms a sac that produces serous fluid.

The membrane folds around an organ such as the heart, forming the cavity around it (in this case, the pericardial cavity).

In a frontal section, you can see the visceral and parietal layers of the membrane around the heart and the serous fluid between them.

Figure 1.11 How a serous membrane envelops the heart.

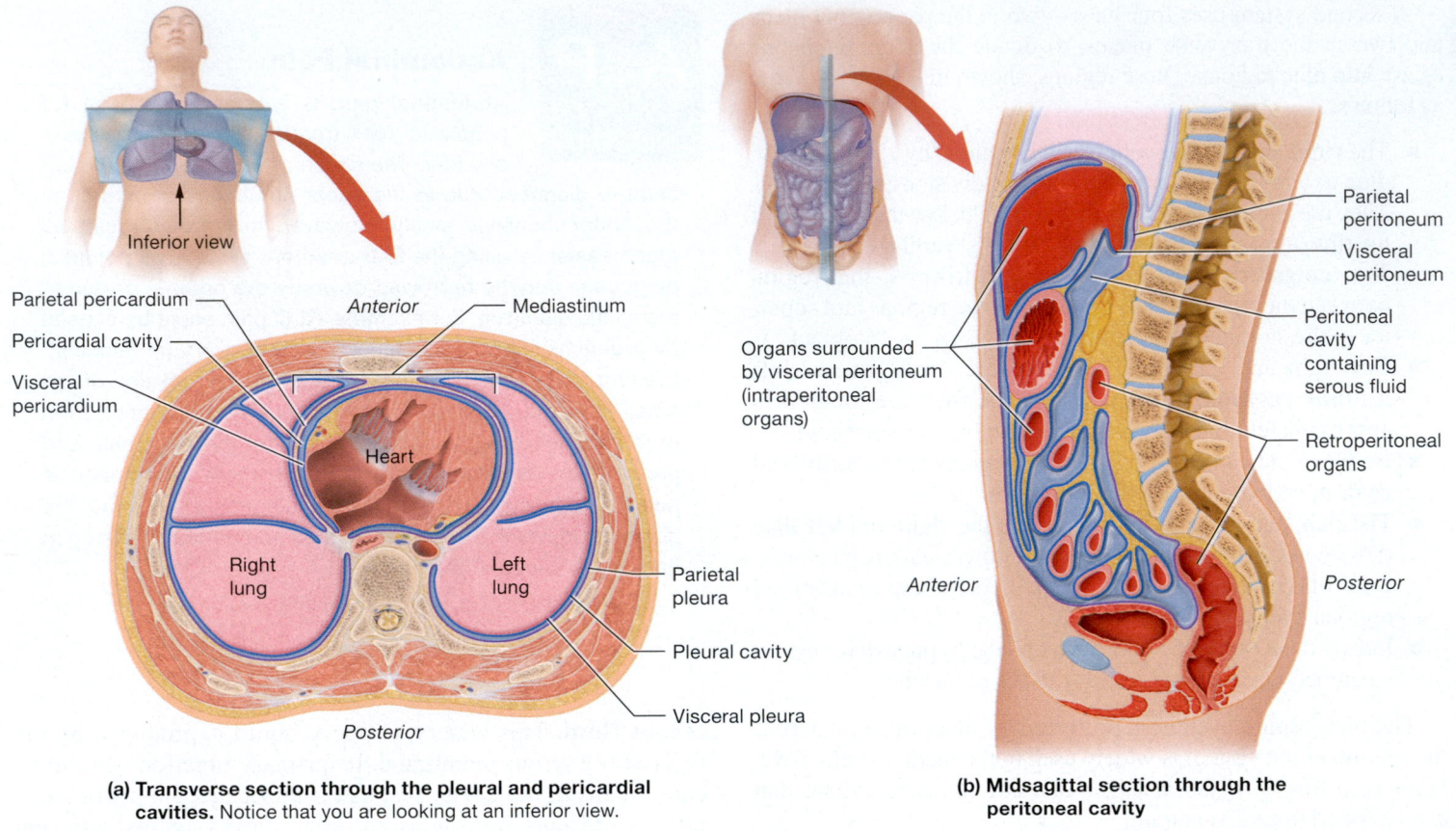

Inferior view

Parietal pericardium
Pericardial cavity
Visceral pericardium

Anterior — Mediastinum

Heart

Right lung
Left lung

Parietal pleura

Pleural cavity

Visceral pleura

Posterior

(a) Transverse section through the pleural and pericardial cavities. Notice that you are looking at an inferior view.

Organs surrounded by visceral peritoneum (intraperitoneal organs)

Parietal peritoneum
Visceral peritoneum

Peritoneal cavity containing serous fluid

Retroperitoneal organs

Anterior

Posterior

(b) Midsagittal section through the peritoneal cavity

Figure 1.12 The serous membranes of the anterior body cavities.

(VISS-er-ul; *viscer-* = "organ"), and the outer layer that attaches to surrounding structures is called the **parietal layer** (puh-RY-eh-tul).

We can envision the relationship between these two layers of membrane by going back to our bean bag chair example. Imagine you are sitting in such a chair. You represent an organ—the part of the chair that is in contact with your body is akin to the visceral layer of a serous membrane, and the part in contact with the floor is like its parietal layer.

The body has three main serous membranes, within which are three serous body cavities. These membranes are illustrated in **Figure 1.12** include the following:

- The **pleural membranes** around the lungs consist of the *parietal pleura,* which you can see in Figure 1.12a as the outer layer curving along the thoracic wall, and the *visceral pleura,* which is shown as the inner layer running along the surfaces of the lungs. The thin space enclosed by the pleural membranes forms the pleural cavities.
- The **pericardial membranes** around the heart consist of the *parietal pericardium,* which is the outer layer encircling the heart in the mediastinum, and the *visceral pericardium,* which is attached to the heart muscle itself (see Figure 1.12a). Between the two we find the pericardial cavity.

- The **peritoneal membranes** around some abdominal organs are the *parietal peritoneum* and the *visceral peritoneum.* The peritoneal cavity is the space between the parietal and visceral peritoneal membranes. As you can see in Figure 1.12b, the peritoneal cavity is much larger than either the pleural cavities or the pericardial cavity and envelops several organs, which are said to be *intraperitoneal* (*intra-* = "inside"). Notice also that several abdominal organs are not located within the peritoneal cavity; instead, these organs actually lie behind the parietal peritoneum (it covers their anterior surfaces). Such organs, including the kidneys, are said to be *retroperitoneal* (*retro-* = "behind").

See *A&P in the Real World: Medical Imaging* to read about modern imaging techniques that can show these membranes and the cavities they form. We explore the structure and function of these membranes in greater detail in later chapters.

Quick Check

☐ 2. List the subdivisions of the thoracic and abdominopelvic cavities.

☐ 3. What are serous membranes, and what are their functions?

☐ 4. Explain how serous membranes form certain anterior body cavities.

Medical Imaging

To look inside patients without resorting to surgery, medical professionals rely on *medical imaging*. These techniques use different forms of radiation to produce images of the body's internal structures.

Many medical images show sections along specific planes so that we can look into different body cavities. Following are several examples of medical images that reveal sections of body planes.

- **X-Ray.** An x-ray uses ionizing radiation to produce an image of internal body structures. The image to the right is a chest x-ray film, which shows the thoracic cavity and all of the smaller cavities contained within it.

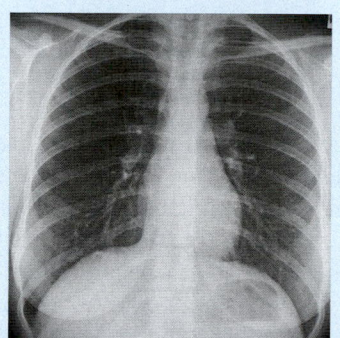

- **Computed tomography scan.** A computed tomography (CT) scan also uses ionizing radiation to gather data, but these data are fed into a computer to produce a three-dimensional image. The image to the right is a CT scan in a transverse section of the abdominopelvic and peritoneal cavities.

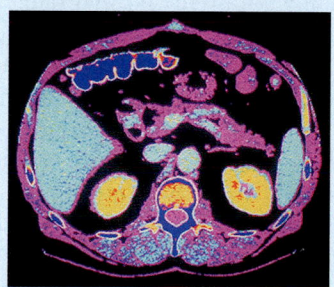

- **Magnetic resonance image.** A magnetic resonance image, or MRI, is produced by placing the body within a magnetic field. As with a CT scan, a computer compiles the data from an MRI scan and produces a three-dimensional image. The following two MRI scans show inside the brain and cranial cavity (left) and inside the vertebral cavity (right), both in a mid-sagittal section.

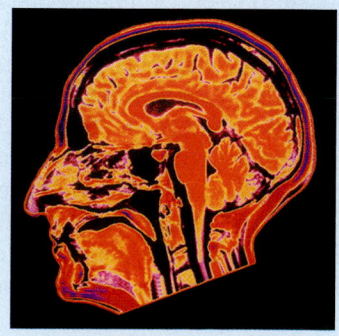 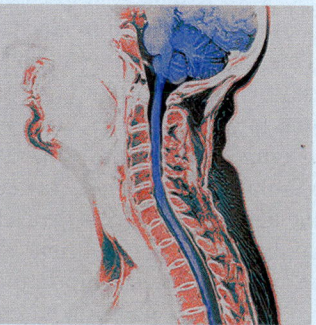

Apply What You Learned

- ☐ 1. In an industrial accident, a piece of shattered equipment enters a man's body near the right anterior pelvis and travels through the organs until it rests inside the left lung. List all of the body cavities and serous membranes through which the fragment passed on its way through this man's body.
- ☐ 2. Certain conditions lead to inadequate serous fluid in the pleural cavities. Predict the effects of these conditions.
- ☐ 3. Inflammation of serous membranes may cause the cavities to fill with excess serous fluid. Predict how this will affect the function of the organs they surround.

See answers in Appendix A.

MODULE 1.5
Core Principles in Anatomy and Physiology

Learning Outcomes

1. Describe the principle of homeostasis.
2. Describe the components of a feedback loop, and explain the function of each component.
3. Compare and contrast negative and positive feedback, and explain why negative feedback is the most commonly used mechanism to maintain homeostasis in the body.
4. Describe how structure and function are related.
5. Define the term gradient, and give examples of the types of gradients that drive processes in the body.
6. Describe how cells communicate with one another, and explain why such communication is necessary in a multicellular organism.

As noted in the introduction, this chapter forms the foundation for the rest of this book, and for your continued studies of the human body. In this module, we complete our foundation with an introduction to a set of basic concepts that we revisit repeatedly in this text. We call these concepts the "core principles" of anatomy and physiology. However, there is one overarching theme in anatomy and physiology that all of these core principles relate to—*homeostasis.* Let's explore this overall theme and the four related principles: (1) feedback loops, (2) the relationship of structure and function, (3) gradients, and (4) cell-cell communication.

Overall Theme: Physiological Processes Operate to Maintain the Body's Homeostasis

In the upcoming chapters, you will discover the incredible number of physiological processes that occur in your body's trillions of cells every second. Yet in spite of this complexity, your body continues to function. This is in large part because of the body's ability to develop and maintain a relatively stable internal environment, a condition called **homeostasis** (hoh'-mee-oh-STAY-sis). Even when your body is subjected to harsh conditions, such as being outside on a cold day without a coat, your internal environment doesn't change very much. This maintenance of homeostasis in the face of changing conditions requires the body to expend an immense amount of energy.

Homeostasis can be likened to a scale that we are using to try to balance a quantity, of, for example, sugar. Adding too much sugar to one side causes the scale to tip to that side. Similarly, removing sugar from one side also causes the scale to lose its balance and tip. This occurs in the body as well—having too little sugar in the blood causes confusion, dizziness, weakness, nausea, and even coma. Conversely, having too much sugar in the blood on a long-term basis damages nerves, blood vessels, the kidneys, and the eyes. Both conditions are disturbances in homeostasis, or *homeostatic imbalances,* which can result in disease or even death if uncorrected. The body must avoid tipping the scale to one side or the other in order to prevent a homeostatic imbalance.

The body's internal environment is the result of a wide range of coordinated processes or *variables,* including the temperature, the chemical composition of blood and other body fluids, and many more. To prevent homeostatic imbalances, most variables in the internal environment are controlled so they stay close to a particular normal value; these variables are known as *regulated variables.*

Quick Check

☐ 1. What is homeostasis, and why is it important?

☐ 2. What is a homeostatic imbalance?

Core Principle One: Feedback Loops Are a Key Mechanism Used to Maintain Homeostasis

A major function of the body's organ systems is to maintain homeostasis to ensure survival. They often do this through control mechanisms called **feedback loops,** in which a change in a regulated variable causes effects that *feed back* and in turn affect that same variable. Feedback loops are made up of a series of events that lead to an effect or *output* of some sort. As the loops continue, this output then influences the events of the loops themselves.

This feedback can be either negative (opposing the initial change and reducing the output) or positive (reinforcing the initial change and increasing the output). These are the two types of feedback loops, examined next.

Negative Feedback Loops

Maintaining homeostasis is based on a relatively simple principle: If one of the body's regulated variables shifts too far from its normal value, then physiological processes increase or decrease their activity to return it to normal. This type of regulation, in which a change in a regulated variable in one direction results in actions that cause changes in the variable in the opposite direction, is called a **negative feedback loop** (or *mechanism*), and this is one of the primary means for maintaining homeostasis in the body.

Each regulated variable has its own **set point,** or normal value. For example, normal body temperature is 37° C (98.6° F), and normal blood pressure is 110/70 mm Hg (millimeters of mercury, a way of measuring pressure). However, in reality, it's not an exact value that is maintained, but rather a **normal range** around that set point, as shown here:

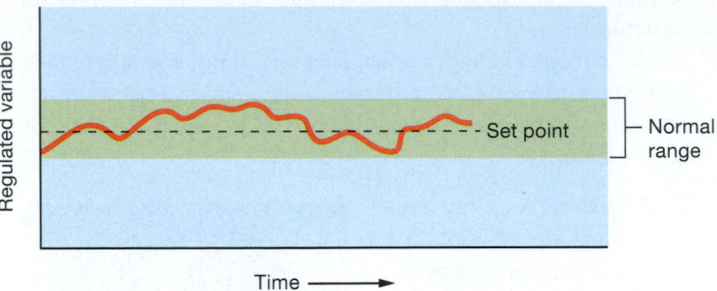

The ranges differ for individual variables, with some having wider ranges and others, narrower ones. Note also that normal ranges for regulated variables can vary between individuals.

A regulated variable has to move outside its normal range in order to trigger a negative feedback mechanism. For example, if your body temperature on a hot day were 37.2° C, your body would probably not initiate a negative feedback loop, as this is usually within the normal range. However, if your body temperature reached 37.4° C, those mechanisms would probably kick in, and your body would start sweating to lose heat and so return your temperature to within the normal range.

Let's see how a negative feedback loop works in general. When a regulated variable is outside its normal range, this information (called a **stimulus**) is picked up by a cellular structure called a **receptor,** or **sensor.** The receptor can then send this stimulus to a **control center** by a variety of mechanisms, usually requiring the use of cells of the nervous or endocrine system. The control center is often cells in the brain or an endocrine organ called a gland. If the control center determines that the variable is out of the normal range, it sends signals to cells or organs, called **effectors.** The effectors cause physiological **responses** that return the variable to the normal range.

As the variable returns to normal, a stimulus for the negative feedback loop no longer exists, and so the control center stops stimulating the effectors. This allows the activity of the effectors to return to normal levels. The activity of the effectors will continue at normal levels until another change in a variable is detected, at which point the negative feedback loop will again be activated.

Negative feedback loops don't occur only in the human body; they can also be found in mechanical devices. A common example of a mechanical negative feedback loop is the control of room temperature by a thermostat (**Figure 1.13a**). Let's say you set the thermostat in your house to 21° C (70° F), the set point. In the winter, cold temperatures outside cause the temperature inside the house to drop below this set point. The receptor, in this case the thermometer in the thermostat, detects the room's current temperature and sends this information via electrical wires to the control center, which is probably a computer chip in the thermostat. The thermostat compares this current value to the set point. When it determines that the room temperature is too far from the set point, it sends commands

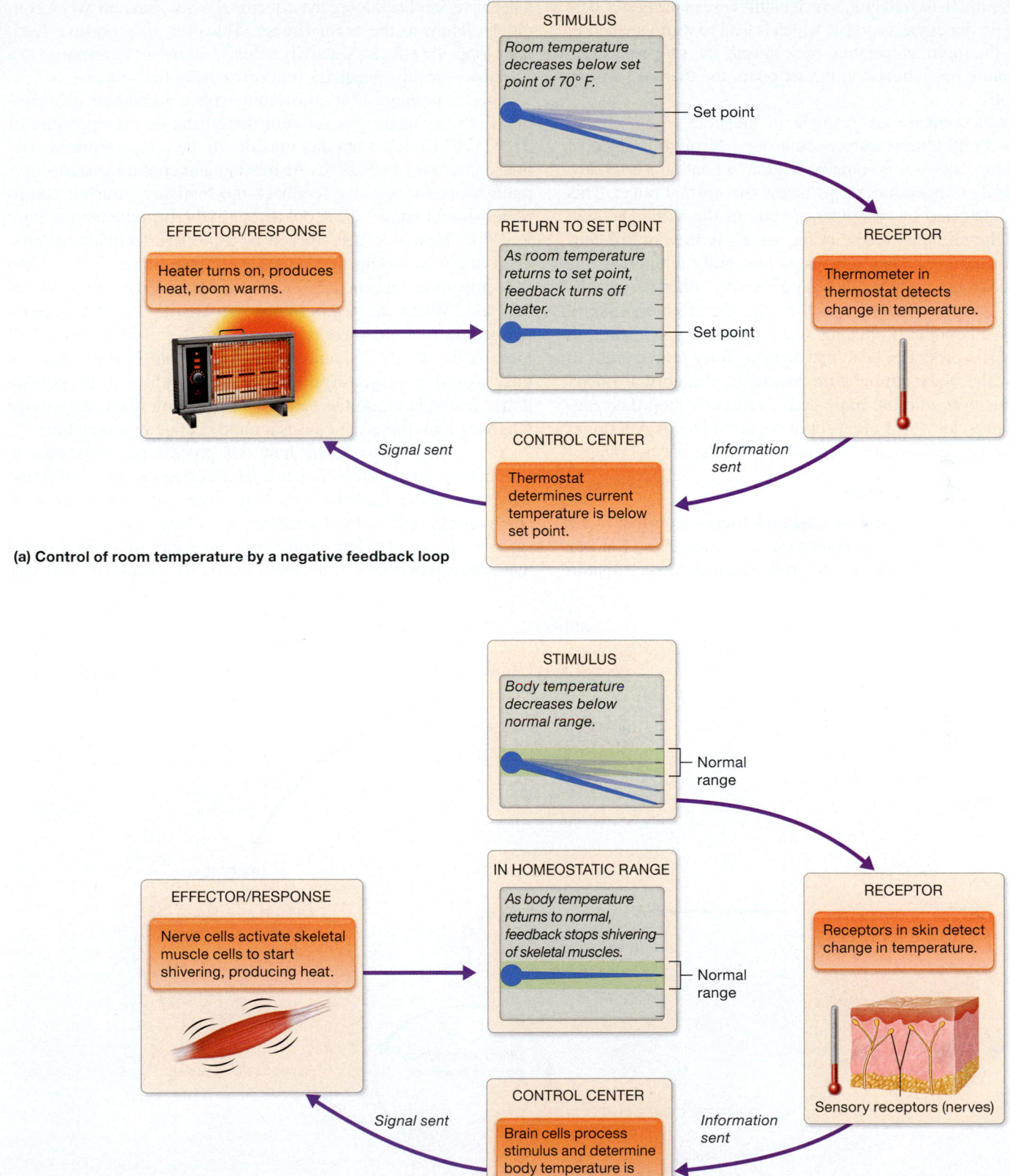

(a) Control of room temperature by a negative feedback loop

(b) Control of body temperature by a negative feedback loop

Figure 1.13 Comparison of how negative feedback mechanisms control room and body temperature.

via wires to turn on the effector, which in this case is the heater. The heater then produces the response, which is heat to warm the house; this moves the room temperature back toward the set point. Once the temperature has returned to the set point, the thermostat turns the heater off.

Now we'll compare an example in the body. **Figure 1.13b** shows how body temperature is controlled through a negative feedback loop. Say you go outside without a coat on a cold day, and your body temperature drops below the normal range. This stimulus is detected by receptors in cells of the skin. The skin cells send the stimulus to the brain, which is the control center. Cells in the brain then determine that body temperature is too far outside the normal range, and activate other nerve cells that send signals to skeletal muscle cells, the effectors. Skeletal muscle cells begin producing small contractions, leading to shivering. This produces heat, causing the body temperature to rise. When the body temperature returns to the normal range, the stimulus ends, and the brain cells eventually stop their signals to the muscles. A different set of negative feedback loops is initiated if body temperature increases above the normal range.

Positive Feedback Loops

Although less common, **positive feedback loops** also occur in the body and contribute to the maintenance of homeostasis. The key difference between negative and positive feedback loops is that in a negative feedback loop, the effector activity shuts off when conditions return to the normal range. However, in a positive feedback loop, the effector's activity actually *increases* in response to a stimulus—positive feedback reinforces the initial stimulus.

This is perhaps best illustrated with a hypothetical example. Let's say again you set your thermostat to a temperature of 21° C (70° F). It's a hot day outside, so the temperature in your house increases to 21.5° C. As heating and cooling systems normally work by negative feedback mechanisms, your air conditioner should come on to cool the air and return the temperature to 21° C. However, if it worked by a positive feedback mechanism, the *heat* would turn on instead. And what's more, when the temperature reached 22° C, the heater's temperature would increase. When the temperature reached 23° C, the heater's temperature would increase again, and it would increase still more at 24° C, 25° C, and so on. Notice that in this loop, not only would it be getting hotter, but it would also be getting hotter faster, because the heater is amplifying itself. A positive feedback loop therefore causes a rapid change in a variable.

You might be wondering how this process can contribute to maintaining homeostasis. Positive feedback loops are often found within a negative feedback loop to produce a quicker response. A common example is the formation of a blood clot (**Figure 1.14**). As you can see, this loop begins with an injury to a blood vessel that causes blood loss. Damage to the blood vessel wall activates

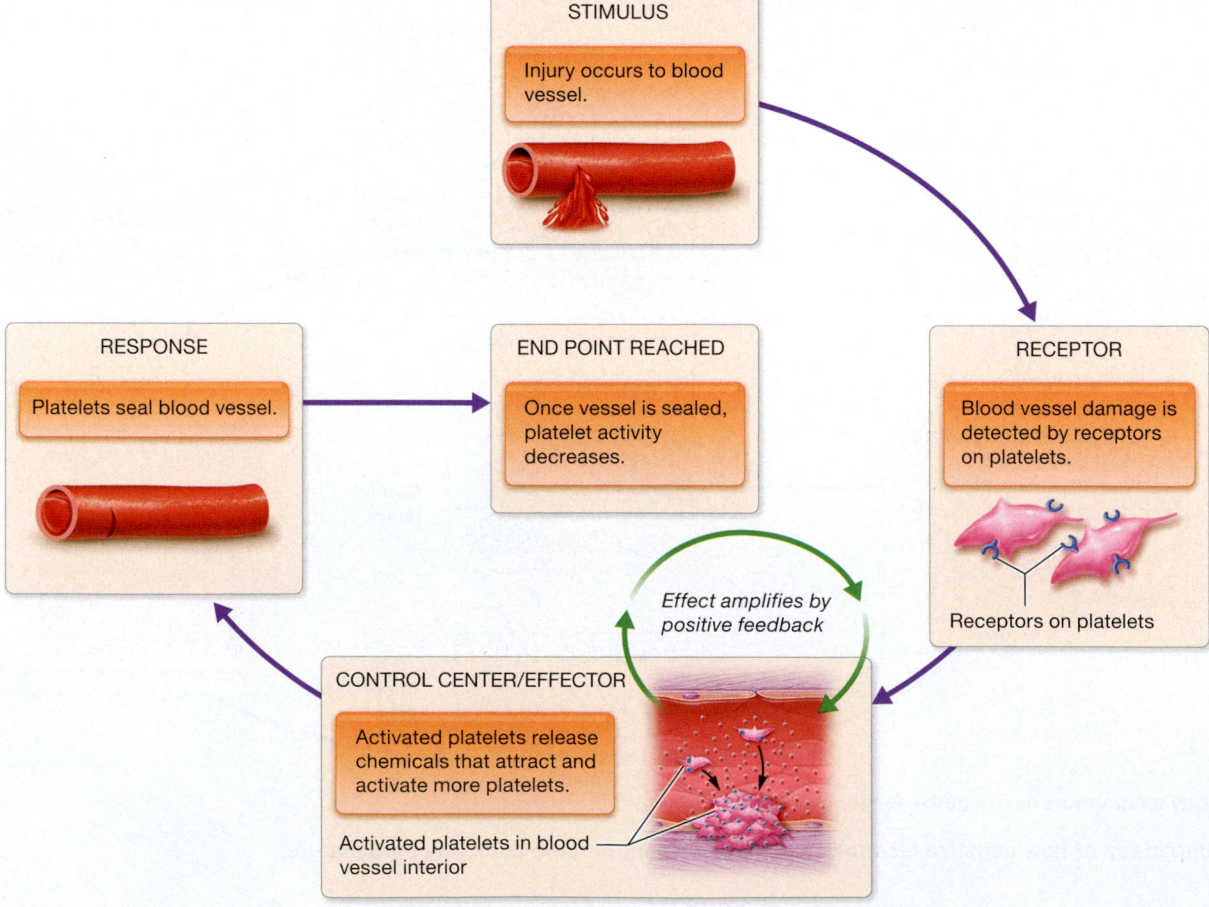

Figure 1.14 Control of blood clotting by a positive feedback mechanism.

receptors on tiny cellular fragments in the blood called *platelets*. The activated platelets act as both control centers and effectors, releasing chemicals that attract and activate more platelets to stick together and plug the wound, and these additional platelets attract and activate more platelets, and so on, amplifying the effect. This continues until the response is complete, when the vessel is sealed by a blood clot and the blood loss stops. In this example, positive feedback occurred when platelets stimulated other platelets to form the clot. The overall effect, however, is negative feedback, as the stimulus—the injury and the bleeding it caused—was stopped.

Although the effects in a positive feedback loop are amplified, they do not continue indefinitely. Every correctly functioning positive feedback loop eventually reaches a point at which positive feedback shuts off in response to an external stimulus (an event that is not part of the loop). In the case of blood clotting, the system shuts down once the torn vessel is sealed, as that removes the stimulus that started the positive feedback loop in the first place. Other positive feedback loops operate in a similar manner and shut down when some end point is reached or some outside event intervenes. Read about the positive feedback loop that causes contractions of the uterus during labor in *A&P in the Real World: Childbirth, Pitocin, and Positive Feedback Loops.*

ConceptBOOST)))

Debunking Some Common Misconceptions about Homeostasis*

As you've read, homeostasis is the overall theme of human anatomy and physiology. With such a broad concept, misconceptions are bound to arise. Here are a few common misconceptions and the truth behind them:

- **Misconception 1: Negative feedback is bad for the body; positive feedback is good.** When most of us hear the term "negative," we think of bad consequences. Likewise, when we hear "positive," good consequences come to mind. But remember that the terms negative feedback and positive feedback refer simply to the fact that the output of the loops goes in opposite directions. Under normal conditions, both negative feedback and positive feedback promote the maintenance of homeostasis.
- **Misconception 2: Maintaining homeostasis means that the body's internal environment is static or unchanging.** Though we talk about the maintenance of "set points" or "normal ranges," it doesn't mean that the body's internal environment is unchanging. Think about what happens in your body when you sit down to eat dinner. The levels of sugar and other nutrients in your blood increase, your water balance changes, the levels of various other chemicals in your blood change, your digestive system undergoes multiple physiological processes to digest the food, and more. That's hardly a static environment, and these sorts of changes are occurring constantly.

- **Misconception 3: Regulatory mechanisms and feedback loops are either "on" or "off," like a switch.** Since the body's internal environment is a dynamic place, feedback loops are constantly engaged in some degree of activity. A good example is blood sugar regulation. The level of sugar in the blood is constantly changing, and the negative feedback loops that regulate blood sugar therefore must always be in operation.
- **Misconception 4: Any physiological variable can be controlled.** A physiological variable can be controlled through feedback loops only if cells with receptors exist to detect changes in a set point. In addition, changes in an unregulated variable can disrupt the homeostasis of a regulated variable. For example, vitamin D is an important chemical involved in calcium ion homeostasis. Sensors exist to detect the level of calcium ions, but not for vitamin D. So if the vitamin D level falls too low, the body has no mechanism to return the level to normal. A lower vitamin D level, in turn, affects the level of calcium ions in the body. ■

* Jenny McFarland, Joel Michael, Mary Pat Wenderoth, Harold Modell, Ann Wright, and William Cliff, "Conceptual Framework and Misconceptions Associated with the Core Principles of Homeostasis," Human Anatomy & Physiology Society workshop (2012).

Quick Check

☐ 3. How do negative feedback loops maintain homeostasis?
☐ 4. How do positive and negative feedback loops differ?

Core Principle Two: Structure and Function Are Related at All Levels of Organization

One of the most basic principles in anatomy and physiology is known as the **principle of complementarity of structure and function.** This principle may be stated more simply as *form follows function:* The form of a structure is always such that it best suits its function. This is easy to see at the organ level of organization—blood vessels are hollow tubes that can transport blood through the body, hard and strong bones support and frame the body, and the hollow and muscular urinary bladder stores and expels urine. However, this principle applies at all levels of organization, even down to the chemical level.

Take a look at **Figure 1.15**, which shows this structure-function relationship in the tissues of the lung. You can see that the lung tissue is extremely thin. This form follows its function of allowing gases such as oxygen and carbon dioxide to cross. If the lung tissue were thick, gases would take too long to cross and our cells would suffocate from lack of oxygen. We will revisit this principle and encounter additional examples in each chapter of this book.

Childbirth, Pitocin, and Positive Feedback Loops

Pregnancy generally lasts about 40 weeks, after which the process of childbirth begins with *labor.* Labor occurs by a positive feedback loop, whereby several stimuli, including the baby's head stretching the cervix, cause the control center in the mother's brain to send signals to the uterus. The uterus responds by producing receptors for the hormone *oxytocin* (awk-sih-TOH-sin), which stimulates uterine contractions. These contractions, in turn, move the baby's head, causing more stretching of the cervix, which stimulates the release of more oxytocin, and the effect continues to be amplified until the baby is born.

Sometimes it becomes necessary to initiate the labor process artificially, which is known as labor induction. Induction is carried out with the help of the drug Pitocin (pih-TOH-sin), the synthetic version of oxytocin. Pitocin stimulates uterine contractions, which then initiate the positive feedback loop that leads to the release of more oxytocin, progressively stronger contractions, and eventual childbirth.

Quick Check

☐ 5. How are structure and function related? To what levels of organization does this relationship apply?

Core Principle Three: Gradients Drive Many Physiological Processes

A **gradient** is present any time more of something exists in one area than another and the two areas are connected. **Figure 1.16** shows three examples of gradients; these same types of gradients exist in the human body. Figure 1.16a shows a *temperature gradient,* in which heat in a room is concentrated closest to a heater and the temperature gradually decreases as we move away from the heater. Figure 1.16b shows a *concentration gradient,* in which a pill is dissolving in a beaker of water. A concentration gradient is present when the amount of a substance changes over a range. In Figure 1.16b, the area around the pill has a high concentration of the medication, and the concentration gradually decreases as we move away from the pill. Finally, Figure 1.16c shows a *pressure gradient,* in which we have a syringe with a plunger pushing down. The contents of the syringe are being compressed and are under high pressure. The air outside the syringe is not as compressed and is therefore under lower pressure. This pressure difference is the pressure gradient.

Gradients form the cornerstone of much of our physiology, as they drive many of our physiological processes. Everything from respiration to nutrient exchange in the blood vessels to the formation of urine is driven by gradients. We explore the roles of different types of gradients in nearly every chapter in which physiology is prominent.

Quick Check

☐ 6. What is a gradient? What are some examples of gradients?

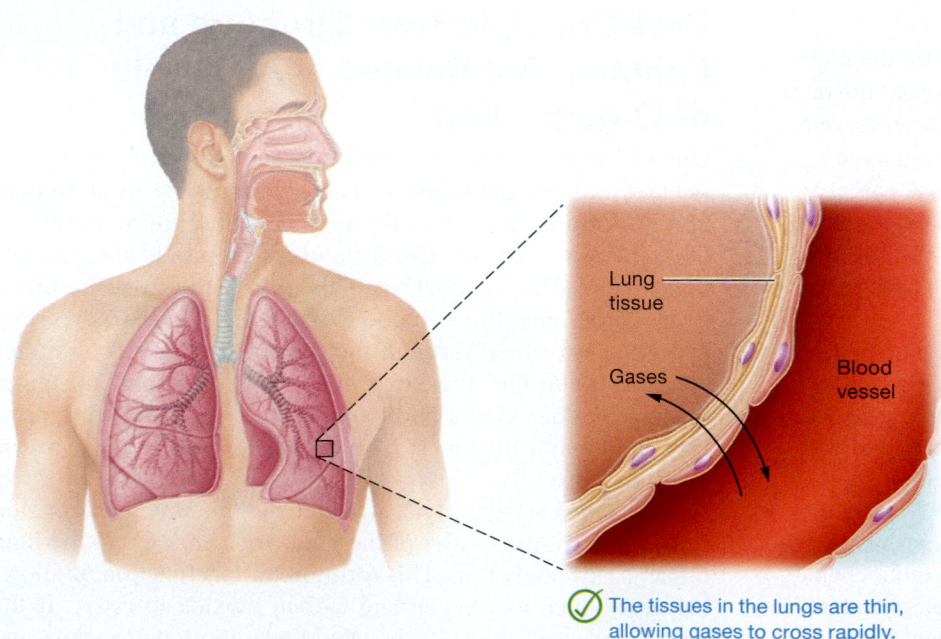
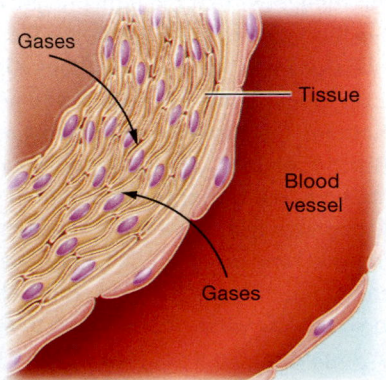

✓ The tissues in the lungs are thin, allowing gases to cross rapidly.

✗ If the tissues were thick, gases would have difficulty crossing rapidly enough to maintain homeostasis.

Figure 1.15 The relationship between structure and function. This figure explores structure and function using the example of what would happen if the tissues in the lung were thick instead of thin.

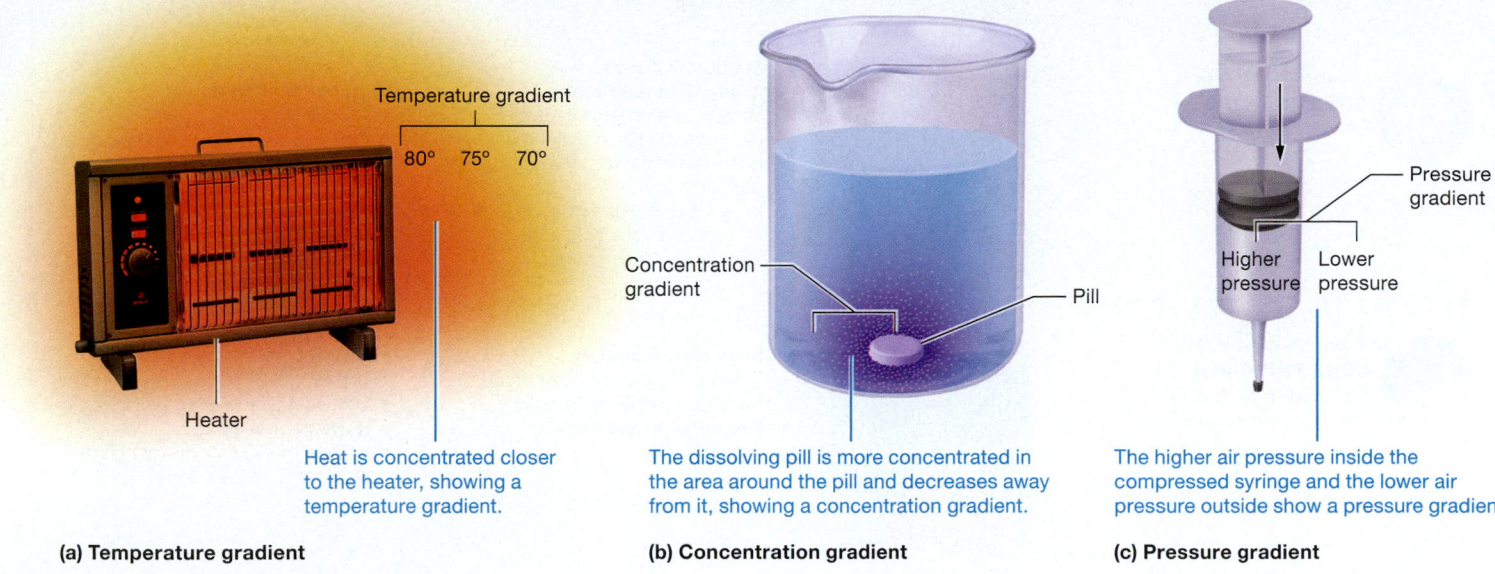

Temperature gradient

80° 75° 70°

Heater

Heat is concentrated closer to the heater, showing a temperature gradient.

(a) Temperature gradient

Concentration gradient

Pill

The dissolving pill is more concentrated in the area around the pill and decreases away from it, showing a concentration gradient.

(b) Concentration gradient

Pressure gradient

Higher pressure   Lower pressure

The higher air pressure inside the compressed syringe and the lower air pressure outside show a pressure gradient

(c) Pressure gradient

Figure 1.16 Examples of gradients.

Core Principle Four: Cell-Cell Communication Is Required to Coordinate Body Functions

Cells in the body must work together in a coordinated fashion to ensure that homeostasis of the entire organism is maintained. Just as you and a friend can coordinate a study date only if you communicate in some way, cells can work together only if they communicate with one another. Such communication generally comes in the form of electrical signals or chemical messengers, in which one cell triggers a response from another cell. Electrical signals are typically transmitted directly between neighboring cells. Chemical messengers, however, may be released from one cell directly onto another cell or into the fluid surrounding another cell, or they may reach another cell after traveling through the blood. In this way, cells can communicate with one another irrespective of their location in the body and the distance between them.

An example of cell-cell communication is shown in **Figure 1.17**, in which a nerve cell is stimulating a muscle cell. The nerve cell releases chemical messengers into the space near the muscle cell; these messengers trigger changes that lead to a muscle contraction. This type of cell-cell communication is the main way in which the cells of the nervous system communicate with other cells of the body. We will see many other types of cell-cell communication throughout this book.

Each of the four core principles is highlighted throughout the book with a core principles icon, which you can see in **Figure 1.18**. Keep on the lookout for these icons to remind yourself of the overall theme of human anatomy and physiology—homeostasis—and the core principles that revolve around maintaining it.

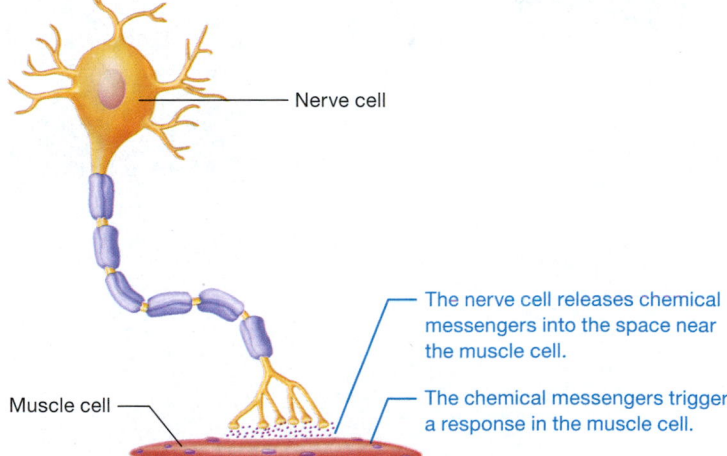

Nerve cell

The nerve cell releases chemical messengers into the space near the muscle cell.

Muscle cell

The chemical messengers trigger a response in the muscle cell.

Figure 1.17 Communication between a nerve cell and a muscle cell.

Quick Check

☐ 7. Why is cell-cell communication important?

☐ 8. What are the two major methods by which cells communicate?

Apply What You Learned

☐ 1. The cells of cancerous tumors undergo changes that cause them to bear little to no resemblance to their original cell type. Predict how this will affect the ability of these cells to perform their functions. (*Hint:* Consider the structure-function relationship.)

☐ 2. Explain what would happen if body temperature increased above its normal range but was regulated by a positive feedback loop.

☐ 3. You plug in an electric air freshener and notice that you can smell the air freshener near the electrical outlet but not on the other side of the room. Explain how this represents a gradient.

See answers in Appendix A.

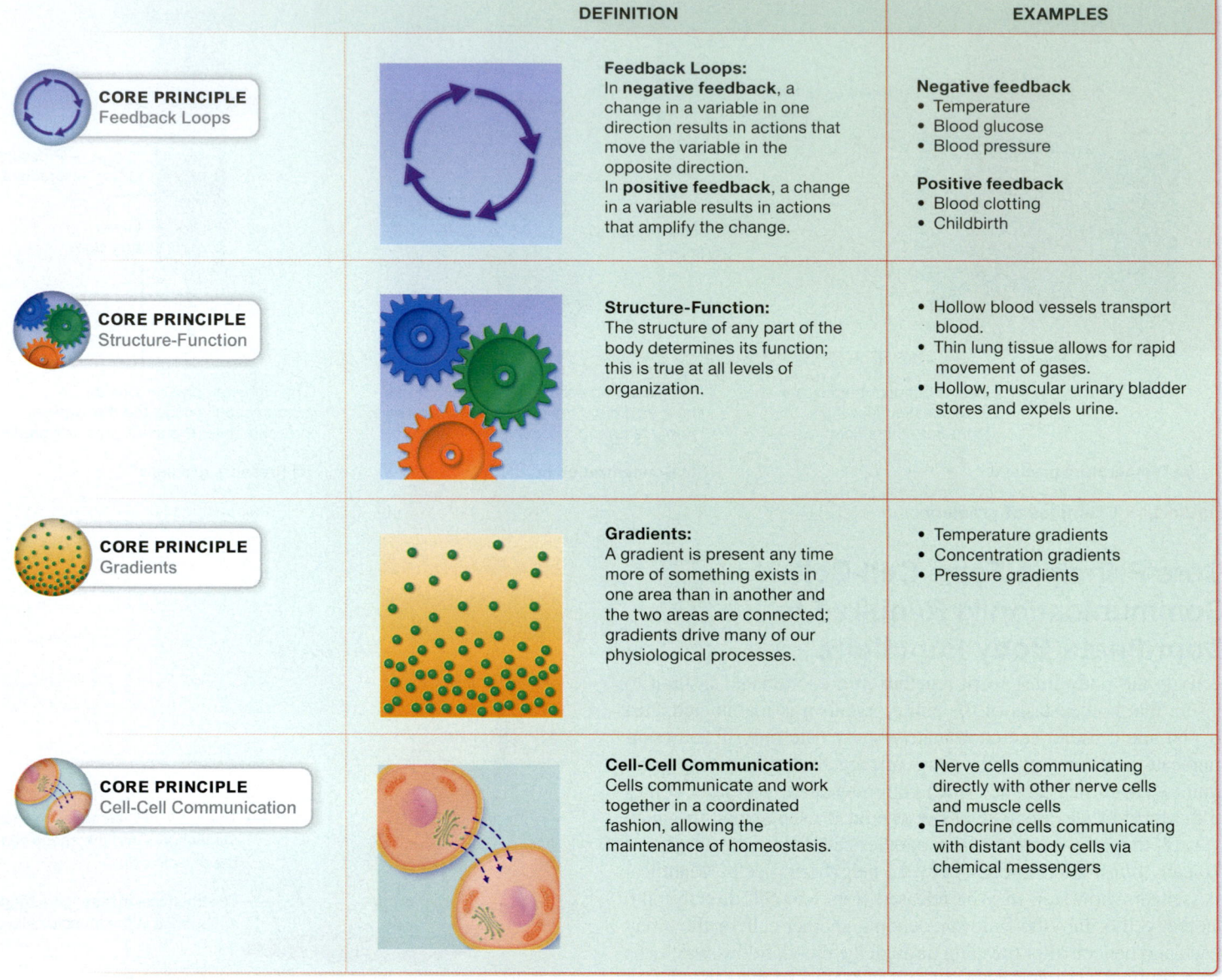

		DEFINITION	EXAMPLES
CORE PRINCIPLE Feedback Loops		**Feedback Loops:** In **negative feedback**, a change in a variable in one direction results in actions that move the variable in the opposite direction. In **positive feedback**, a change in a variable results in actions that amplify the change.	**Negative feedback** • Temperature • Blood glucose • Blood pressure **Positive feedback** • Blood clotting • Childbirth
CORE PRINCIPLE Structure-Function		**Structure-Function:** The structure of any part of the body determines its function; this is true at all levels of organization.	• Hollow blood vessels transport blood. • Thin lung tissue allows for rapid movement of gases. • Hollow, muscular urinary bladder stores and expels urine.
CORE PRINCIPLE Gradients		**Gradients:** A gradient is present any time more of something exists in one area than in another and the two areas are connected; gradients drive many of our physiological processes.	• Temperature gradients • Concentration gradients • Pressure gradients
CORE PRINCIPLE Cell-Cell Communication		**Cell-Cell Communication:** Cells communicate and work together in a coordinated fashion, allowing the maintenance of homeostasis.	• Nerve cells communicating directly with other nerve cells and muscle cells • Endocrine cells communicating with distant body cells via chemical messengers

Figure 1.18 Core principles icons.

Chapter Summary

For everything you need to succeed in this course, go to **Mastering** Anatomy & Physiology. There you will find:

- Practice Tests
- Author Podcasts
- Big Picture Animations
- Concept Boost Video Tutors
- **iP2**™ Interactive Physiology 2.0 Tutorials
- **A&PFlix** A&P Flix 3D Animations
- Active-Learning Workbook

MODULE 1.1
How to Succeed in Your Anatomy and Physiology Course 1–7

- People learn best by using a combination of different **learning modalities.**
- Read a textbook with the SQ3R method: *survey, question, read, recite, review.*
- In studying for an exam, make a schedule, learn about the exam, take advantage of book and school resources, form a study group, use multiple study techniques, and take care of your physical needs.
- You can make the best use of class and laboratory time by coming prepared.
- This book contains several key features to assist you in mastering anatomy and physiology.

MODULE 1.2
Overview of Anatomy and Physiology 7–9

- Human **anatomy** is the study of the form of the human body; human **physiology** is the study of its functions.

- Characteristics common to all forms of life are cellular composition, **metabolism, growth, excretion, responsiveness, movement,** and **reproduction.**

- The six levels of organization of the body are the **chemical, cellular, tissue, organ, organ system,** and **organism** levels. There are 11 major organ systems in the body.

- Subfields of anatomy include **gross anatomy** and **microscopic anatomy.** Microscopic anatomy can be divided into **cytology** and **histology.**

MODULE 1.3
The Language of Anatomy and Physiology 9–17

- The common frame of reference from which all body positions are described is the **anatomical position.**

- Directional terms are used to describe body parts and body markings. The common pairs include: **anterior** and **posterior; superior** and **inferior; proximal** and **distal; medial** and **lateral;** and **superficial** and **deep.**

- The body is divided into two main regions: the **axial region** (the head, neck, and trunk) and the **appendicular region** (the upper and lower limbs). Specific areas of the body are described by **regional terms.**

- The three main planes of section are the **sagittal, frontal,** and **transverse planes.**

MODULE 1.4
The Organization of the Human Body 17–21

- The **posterior body cavity** contains the **cranial** and **vertebral** or **spinal cavities,** which contain *cerebrospinal fluid (CSF).*

- The **anterior body cavity** contains the following divisions:
  - The **thoracic cavity** is the area superior to the diaphragm. Within it we find two **pleural cavities,** the **pericardial cavity,** and the **mediastinum.**
  - The **abdominopelvic cavity** is the area inferior to the diaphragm. Within it we find the **abdominal cavity,** the **pelvic cavity,** and the **peritoneal cavity.**

- The pleural, pericardial, and peritoneal body cavities are formed by double-layered **serous membranes,** continuous sheets of tissue that enclose a single cavity with a thin layer of **serous fluid.**

MODULE 1.5
Core Principles in Anatomy and Physiology 21–28

- Overall Theme: **Homeostasis,** the condition of maintaining a relatively stable internal environment.

- Core Principle One: Feedback Loops.
  - **Negative feedback loops** are the more common type of feedback loop. These loops move in the opposite direction of the stimulus, and the effector's activity decreases when the variable returns to the normal range.
  - **Positive feedback loops** feature an escalating response that amplifies the stimulus.

- Core Principle Two: Structure and Function. Structure and function are related at all levels of organization, a concept known as the **principle of complementarity of structure and function.**

- Core Principle Three: Gradients. We find a **gradient** any time there is more of something in one area than in another. Gradients drive many of our physiological processes.

- Core Principle Four: Cell-Cell Communication. Cells communicate to coordinate their functions.

Assess What You Learned

Scan the QR Code for additional practice test questions

https://goo.gl/bSTG9S

LEVEL 1 Check Your Recall

1. Fill in the blanks: The study of the form of the body is _____; the study of its functions is _____.

2. Mark the following statements as true or false. If a statement is false, correct it to make a true statement.

   a. You should study only according to your learning preferences.
   b. You should budget your time for studying at least two days before an exam.
   c. Textbooks are best read using the survey, question, read, recite, and review method.
   d. It is better to forgo sleep in favor of studying before an exam.

3. Explain why it is important to come to class prepared.

4. Groups of many cells working together to perform a common function are known as a(n):

   a. cell.   c. organ system.
   b. organ.   d. tissue.

5. Which of the following correctly describes the functions of the endocrine system?

   a. Transport of blood through the body and through the lungs
   b. Regulation of body functions through hormone secretion
   c. Regulation of body functions through nerve impulses
   d. Immunity and returning extra tissue fluid to the blood vessels

6. Mark the following statements as true or false. If a statement is false, correct it to make a true statement.

   a. Histology is the division of microscopic anatomy that studies the cellular level of organization.
   b. All living organisms are composed of one or more cells.
   c. Living organisms react to changes in their environment known as stimuli.
   d. The anatomical position features the person facing forward, feet shoulder width apart, and palms facing posteriorly.

7. Match the following terms with the correct definition.

   _____ Anterior   a. Toward the back
   _____ Lateral   b. Closer to the midline
   _____ Proximal   c. Farther away from the point of origin
   _____ Posterior   d. Toward the front
   _____ Inferior   e. Closer to the point of origin
   _____ Distal   f. Toward the head
   _____ Medial   g. Away from the body's midline
   _____ Superior   h. Toward the tail

8. The upper and lower limbs are known broadly as the _____ region.

   a. appendicular   c. crural
   b. axial   d. dorsal

9. The arm is known as the _____ region; the neck is known as the _____ region.

   a. otic; axillary   c. brachial; cervical
   b. antebrachial; dorsal   d. sural; crural

10. A parasagittal section divides the body or body part into:

    a. equal right and left parts.
    b. front and back parts.
    c. superior/proximal and inferior/distal parts.
    d. unequal right and left parts.

11. Fill in the blanks: The two divisions of the posterior body cavity are the _____ and _____ body cavities.

12. Fill in the blanks: The two main divisions of the anterior body cavity are the _____ and _____ body cavities.

13. In which of the following cavities do serous membranes envelop the organs? (Circle all that apply.)

    a. Mediastinum   d. Abdominal cavity
    b. Pericardial cavity   e. Pelvic cavity
    c. Peritoneal cavity   f. Pleural cavity

14. Serous fluid functions in:

    a. providing temperature stability.
    b. lubricating serous membranes as organs move in the cavity.
    c. protecting the organs from mechanical trauma.
    d. decreasing the weight of the organs and keeping them buoyant.

15. Which organs would you expect to find in the left upper quadrant?

    a. Spleen, liver, appendix
    b. Liver, ovary, spleen
    c. Spleen, stomach, pancreas
    d. Gallbladder, liver, large intestine

16. Mark the following statements as true or false. If a statement is false, correct it to make a true statement.

    a. Structure and function are closely related at the organ level only.
    b. Homeostasis is the condition of having a dynamic equilibrium of the internal environment.
    c. Positive feedback loops are triggered by a deviation from the set point of a regulated variable and are shut down when conditions return to the set point.
    d. The effects of negative feedback loops are amplified to create an escalating response.
    e. Cell-cell communication is required to coordinate the activities of the whole body.

17. A gradient exists:

    a. when heat is concentrated in one area of the body.
    b. when more of a substance is located in one place than in another.
    c. when there is higher pressure in one place than in another.
    d. All of the above are correct.
    e. None of the above are correct.

LEVEL 2 Check Your Understanding

1. Examine the structure of the skull, and predict its likely functions based on its form.

2. Use the correct regional and directional terms to describe the location of each of the following organs in the body. You may use Figure 1.7 for reference.

   a. Esophagus
   b. Brain
   c. Urinary bladder (in a female)

LEVEL 3 Apply Your Knowledge

1. Ms. Norman presents to the clinic with right upper quadrant pain. Predict the organs and cavities that may be involved in causing her pain.

2. During a procedure on Ms. Norman's pancreas, a surgeon makes the initial incision in the left anterior hypochondriac region. List all the organs, serous membranes, and body cavities that the surgeon will encounter as she moves through the body to get to the pancreas. (*Hint:* Refer to *A&P in the Real World: Abdominal Pain* on page 19 for help.)

3. Later that same day, the surgeon performs a procedure on Ms. Norman's right kidney. She makes the incision in the right posterior lumbar region. Will she cut through the same serous membrane(s) and cavities as in the previous procedure? Why or why not? How would this change if the incision were made on the anterior lumbar region?

4. The *baroreceptor reflex* causes blood pressure to drop when it rises dangerously high. Predict whether this is a positive or negative feedback loop. Explain your reasoning.

See answers in Appendix A.

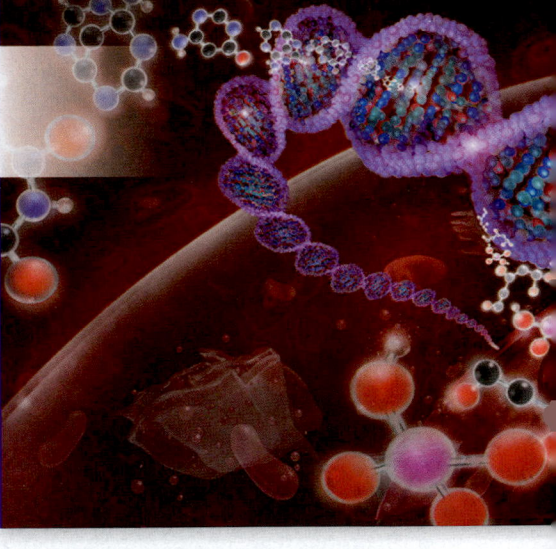

2

The Chemistry of Life

s you begin this chapter, you may wonder why we are starting with chemistry instead of something like bones or digestion or respiration—topics that seem to fit better with the study of the human body. In fact, why study chemistry at all in a human anatomy and physiology course? Well, consider the following examples:

- Bones are strong because they are composed of *minerals* and the *protein collagen.*
- The food we eat is broken down by *chemical reactions* with the help of *acid* and *enzymes.*
- We breathe to take in *oxygen* so our cells can convert the *chemical energy* in food to a form the cells can use.

Even if some of the words in these statements seem foreign to you, one common thread should be clear: We are talking about chemistry (the words in italics) with each statement.

In the first chapter of this book, you learned that the chemical level of organization is the most basic one in the body (look back at Figure 1.4). What this means is that we are made of chemicals, and no modern study of anatomy and physiology could be carried out without discussing them. Chemistry is the foundation not only of our bodies but also for all the concepts in this course.

MODULE 2.1
Atoms and Elements

Learning Outcomes

1. Describe the charge, mass, and relative location of electrons, protons, and neutrons.
2. Compare and contrast the terms atoms and elements.
3. Identify the four major elements found in the human body.
4. Compare and contrast the terms atomic number, mass number, isotope, and radioisotope.
5. Explain how isotopes are produced.

Matter is defined as anything that has mass and occupies space. **Chemistry** is the study of matter and its interactions, and so we begin our exploration of the world of chemistry with matter. In this module, we introduce the components and properties of matter and discuss how they apply to the human body.

For practice applying concepts to a clinical scenario, check out the ***Running Case Study*** *for this chapter @ Mastering Anatomy & Physiology*

Computer-generated image: All the structures that make up our bodies, such as the strand of DNA shown in this image, are composed of chemicals.

Atoms and Atomic Structure

An **atom** is defined as the smallest unit of matter that still retains its original properties. Atoms certainly are tiny—an average human cell has roughly 100 trillion atoms. They are themselves composed of even smaller parts called **subatomic particles.** Let's examine the subatomic particles and how they are organized within an atom.

Subatomic Particles

There are three basic subatomic particles, illustrated in **Figure 2.1**. These particles are as follows:

- **Protons.** The positively charged **protons** (p^+) reside in the central core of the atom, called the **atomic nucleus** (NOO-klee-us).
- **Neutrons.** Notice in Figure 2.1 that protons are not alone in the atomic nucleus—they are joined by a second type of subatomic particle: the **neutrons** (n^0; NOO-trawnz). Neutrons are just slightly larger than protons and, as their name implies, have no charge.
- **Electrons.** Surrounding the atomic nucleus are the tiny, negatively charged **electrons** (e^-).

The number of positively charged protons in an atom equals the number of negatively charged electrons. For this reason, all atoms are electrically neutral, meaning an atom as a whole has no charge. However, the number of protons and the number of neutrons in an atom do not have to be equal.

The bulk of an atom's mass comes from its protons and neutrons. Electrons are so small that they contribute practically no mass to an atom, and 99.95% or more of an atom's mass is in the nucleus. Note, however, that the nucleus makes up only a tiny fraction of an atom's total volume, and most of the atom is simply empty space. As an analogy, think of a football stadium: The entire stadium is the atom, a marble at its center is its nucleus, and the specks of pollen floating around in the stadium are its electrons. Everything except the marble and the pollen grains is just empty space.

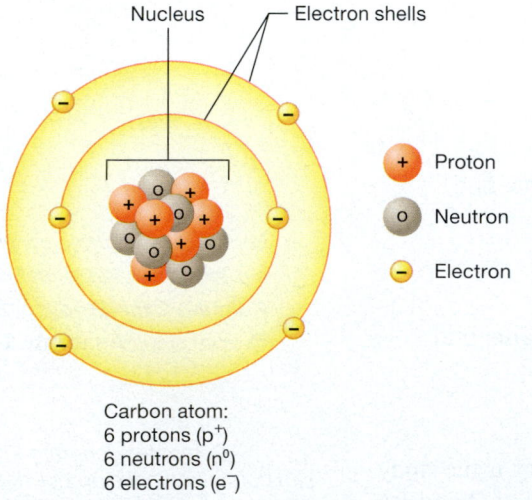

Carbon atom:
6 protons (p^+)
6 neutrons (n^0)
6 electrons (e^-)

Figure 2.1 Structure of a representative atom.

Electron Shells

You may be familiar with the classical atomic model in which the atom looks like a miniature solar system, with the electrons traveling in circular orbits around the nucleus. We often still use that model to draw atoms for the sake of simplicity, but it's not entirely correct. Although we can determine an electron's distance from the nucleus, we can't determine precisely where the electron is going to be found or the path it is going to take. However, we can predict the likelihood, or *probability*, of finding electrons in a given area, and loosely define regions where they are likely to be found. These regions, shown in Figure 2.1, are termed **electron shells.**

Each shell can hold a certain maximum number of electrons:

- the first shell, which is closest to the nucleus, can hold 2 electrons;
- the second shell can hold 8 electrons; and
- the third shell can hold 18 electrons (but is "satisfied" with only 8).

The atom shown in Figure 2.1 has two shells. The number of shells an atom has depends on its number of electrons—the electrons fill the shells closest to the nucleus first. Many atoms have more than three shells, but for simplicity we will restrict our discussion in this chapter to those with three or fewer shells.

Quick Check

☐ 1. What are atoms?

☐ 2. How do the three types of subatomic particles differ?

Elements in the Periodic Table and the Human Body

An **element** is a substance that cannot be broken down into simpler substances by chemical means. Each element is defined by its number of protons, which is called its **atomic number.** For example, every atom of the element oxygen has 8 protons and every atom of the element iron has 26 protons. There are 92 elements that occur naturally.

Elements are listed in order of increasing atomic number in the **periodic table of the elements,** which is shown in abbreviated form in **Figure 2.2**. This way of organizing the elements shows that certain properties repeat in a regular way, or *periodically*. Elements are listed on the periodic table by an abbreviation called a *chemical symbol*. Generally, the chemical symbol is the first letter or two of the element's name, such as C for carbon or Ca for calcium. However, you'll find that some symbols are based on the element's Latin or German name, such as sodium's symbol, Na, which is from the Latin word *natrium*.

Very few elements are found in significant quantities in the human body (see Figure 2.2). In fact, just four *major elements* account for 96% of the body's mass: Oxygen (O) makes

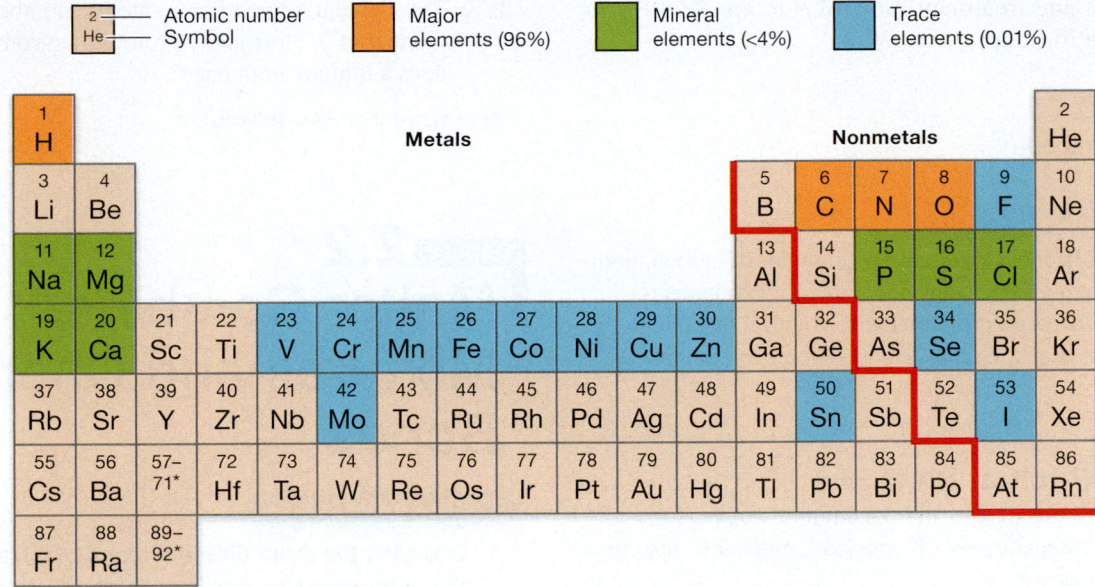

These are the rare earth elements, which are not found in the human body.

Figure 2.2 Elements in the human body and their positions in the periodic table.

up 65%, carbon (C) 18%, hydrogen (H) 10%, and nitrogen (N) 3%. The seven *mineral elements* account for less than 4% of the total mass of the body; these elements are sodium (Na), potassium (K), calcium (Ca), chlorine (Cl), magnesium (Mg), phosphorus (P), and sulfur (S). The 13 *trace elements,* which include iron (Fe), copper (Cu), iodine (I), and zinc (Zn), are found in extremely small amounts in the body.

Notice in Figure 2.2 that a red dividing line that looks like a staircase cuts across the right side of the periodic table. This "staircase" divides the periodic table into two sections: The section to the left contains metals, and the section to the right contains nonmetals. *Metals* are shiny substances that conduct electricity, whereas *nonmetals* exist as gases or brittle solids and are usually poor conductors of electricity. Metals and nonmetals differ in how they interact with other elements, which we discuss in Module 2.2.

Hydrogen is unusual in that it behaves differently than other metals in the periodic table. Although it is placed with other metals in the periodic table, it doesn't experience the right conditions here on Earth to act like a metal. Elemental hydrogen requires extreme pressures and temperatures to exist as a metal; in fact, you would have to go to the core of Jupiter to find conditions extreme enough to produce metallic hydrogen. For this reason, hydrogen on Earth exists as a gas. In addition, it generally acts like a nonmetal when it interacts with other elements.

Isotopes and Radioactivity

Recall that the neutrons in the nucleus of an atom do not contribute to the atom's overall charge or the placement of the element in the periodic table, but they do contribute to the atom's mass. The mass of an atom is represented by a number called the **mass number,** which is equal to the number of neutrons plus the number of protons in the atom. However, atoms can gain or lose neutrons. When this happens, an atom's mass number changes but its atomic number remains the same—in other words, it is still the same element. Atoms with the same atomic numbers but different mass numbers are called **isotopes** (AYE-soh-tohpz; *iso-* = "same"). For example, hydrogen has three isotopes: Regular hydrogen has no neutrons, *deuterium* (doo-TEER-ee-um) has one neutron, and *tritium* (TRIT-ee-um) has two neutrons. The three isotopes of hydrogen are written this way:

$$\begin{array}{ccc} {}^{1}_{1}\text{H} & {}^{2}_{1}\text{H} & {}^{3}_{1}\text{H} \\ \textit{Hydrogen} & \textit{Deuterium} & \textit{Tritium} \end{array}$$

The upper number is hydrogen's mass number, which varies depending on the number of neutrons. The lower number is hydrogen's atomic number, which doesn't change. Another way to represent an isotope is to write the name of the element followed by the isotope's mass number. For example, ${}^{131}_{53}\text{I}$ may also be written as iodine-131.

Certain isotopes have very high energy, which makes them unstable. Such isotopes will release this energy in the form of *radiation*. This energy release, called *radioactive decay,* allows the isotope to assume a more stable form. Isotopes that do this are called **radioisotopes,** and they form the basis for the field of *nuclear medicine,* the use of radioactive materials in medical diagnosis and treatment. See *A&P in the Real World: Nuclear Medicine* for more information.

Quick Check

☐ 5. What are isotopes?

Apply What You Learned

☐ 1. You have collected two equal-size samples of carbon, both of which are pure (i.e., they contain no other elements). However, when you measure them, you find that one has a greater mass than the other. How do you explain this?

Nuclear Medicine

A P in the Real World

Some of the more common applications of radioisotopes in nuclear medicine are as follows:

- **Cancer radiation therapy.** The radiation from certain radioisotopes damages the structural features of cancer cells and interferes with their functions, killing them.
- **Radiotracers.** Many radioisotopes are used as *radiotracers.* Radiotracers are injected into a patient, and the radiation they produce is detected by a camera and analyzed by a computer. The resulting picture shows the size, shape, and activity of certain organs and the cells in the blood. For example, shown here is a colored gamma scan of a healthy human skeleton seen from the front (left) and back (right):

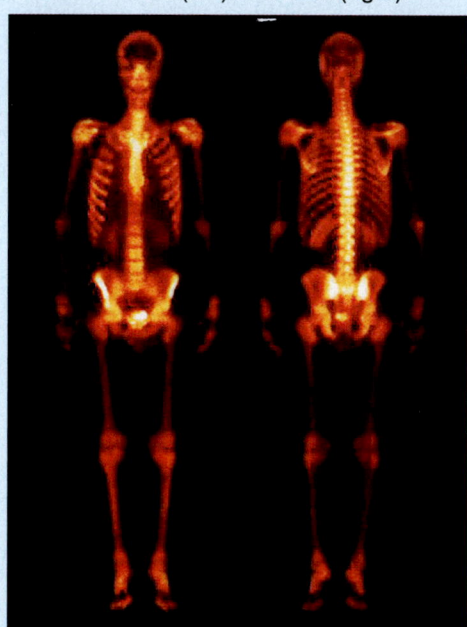

- **Treatment of thyroid disorders.** High doses of the radioisotope iodine-131 are used to treat overactive thyroid and thyroid cancers. The isotope accumulates in the thyroid gland and damages the cells, interfering with their functions and so killing them.

☐ 2. The element lithium has an atomic number of 3 and a mass number of 7. How many protons, neutrons, and electrons does a lithium atom have?

See answers in Appendix A.

MODULE **2.2**

Matter Combined: Mixtures and Chemical Bonds

Learning Outcomes

1. Describe the three different types of mixtures, and explain the difference between a solvent and a solute.
2. Explain how the number of electrons in an atom's valence shell determine its stability and ability to form chemical bonds.
3. Distinguish between the terms molecule, compound, and ion.
4. Explain how ionic, nonpolar covalent, and polar covalent bonds form, and discuss their key differences.
5. Describe how hydrogen bonds form and how they give water the property of surface tension.

Matter may be combined in two ways: either by being physically intermixed or by being chemically combined. When atoms of two or more elements are physically intermixed, producing a **mixture,** the chemical nature of those atoms does not change. The atoms retain their original properties and may be physically separated. In contrast, when atoms of two or more elements are combined by forming *chemical bonds,* the atoms are changed chemically. Two or more atoms combined by chemical bonds form a **molecule** (MAWL-eh-kyool). Molecules have different properties from those of their original atoms, and can be separated only by chemical means. We now look at the ways of combining matter and how they apply to the human body.

Mixtures

There are three basic types of mixtures: *suspensions, colloids,* and *solutions* (**Figure 2.3**). All three are found in the human body, so let's examine each more closely.

Suspensions

A **suspension** generally consists of a liquid mixed with a solid. The solid particles in a suspension are usually visible (making the suspension appear cloudy or opaque); if left alone, these particles tend to settle out of the mixture. An example of a suspension in the human body is blood, in which the blood cells are "suspended" in the plasma, the fluid portion of blood (see

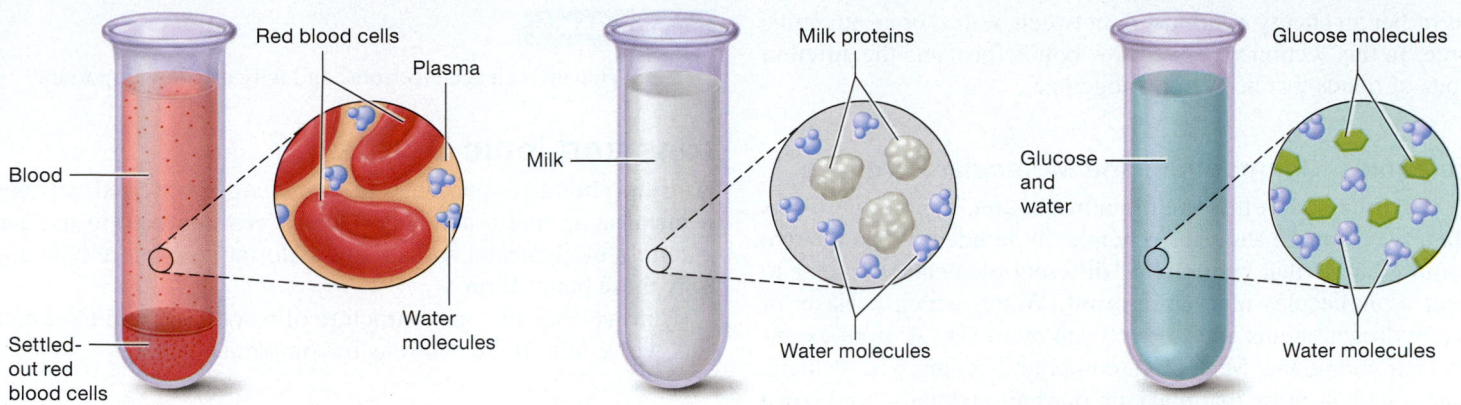

(a) **Suspension:** Particles large and usually visible; settle out.

(b) **Colloid:** Two distinct components; particles small and not visible; do not settle out.

(c) **Solution:** Particles extremely small and not visible; do not settle out; one component dissolves in the other component.

Figure 2.3 **The three types of mixtures.**

Figure 2.3a). If a blood sample is allowed to sit, the blood cells will settle to the bottom of the container. This is one reason why we need our heart to keep our blood moving—when it stops beating, the blood cells stop moving, meaning that they can no longer deliver oxygen to our cells.

Colloids

Like suspensions, **colloids** (KAWL-oydz) consist of two distinct components. Colloids appear opaque; however, the solid particles in a colloid are quite small and are not visible with the naked eye (or even with a microscope). Because of their small size, the particles in a colloid remain dispersed and do not settle out. An example of a colloid in the human body is milk (see Figure 2.3b). Milk contains numerous proteins (PROH-teenz; large compounds that we'll discuss later in this chapter) that are too small to settle out, but large enough to be considered a separate component. Other examples of commonly encountered colloids include gelatin, aerosols, and cytosol, the fluid material inside our cells.

Solutions

A **solution** in a living organism generally consists of a solid, liquid, or gas mixed with a liquid (usually water). A solution is usually translucent because one substance *dissolves* in another. The substance that dissolves is called the **solute** (SAWL-yoot), and the substance in which the solute dissolves is called the **solvent.** For example, in Figure 2.3c, glucose is the solute and water is the solvent. The degree to which a solute dissolves in a solvent is called its *solubility* (sawl'-yoo-BIL-ih-tee). Water is the most important solvent in the human body—it makes up most of our blood plasma and the fluid in and around our cells. Numerous substances dissolve in our plasma and other body fluids, and a chemical's solubility in water is a major factor in determining how it is transported in the body.

The amount of solute present in a solution is called the solution's *concentration*. We can express concentration in several ways; the simplest is by stating the percentage of a solution that is solute. For example, if a solution has 10 grams (g) of salt and 90 milliliters (ml; equal to 90 g) of water, we call it a 10/100 = 10% salt solution; if it has 40 g of salt and 60 ml of water, we call it a 40/100 = 40% salt solution. We often compare solutions by their concentrations, and the solution with the higher concentration of solute is called the *more concentrated* solution. In our example, the 40% salt solution is the more concentrated solution.

The individual solute particles in a solution are extremely small and not individually visible, even with a microscope. Due to their small size, the solute particles will not settle out if the solution is left to stand. However, remember that a solution is still a mixture, and that neither the solvent nor the solute changes chemically when they form a solution. So, if we want to separate the salt from the water in our 40% salt solution, we can do so by simple physical means, such as evaporating the water, which leaves the salt behind.

Quick Check

☐ 1. What is a mixture?

☐ 2. How do the three types of mixtures differ?

Chemical Bonds

Elemental sodium (Na) is a soft metal that reacts violently with water, and elemental chlorine (Cl) is a poisonous green gas. But when you take a small piece of sodium metal and drop it into a beaker containing chlorine gas, something interesting happens: A small explosion occurs, the sodium bursts into flames, and when it finishes burning, a white powder is left behind. The white powder that forms, believe it or not, is something that you sprinkle on your food—sodium chloride (NaCl), or table salt. How exactly did two highly toxic elements become something we can eat? The answer is that when the two elements were combined, each one was altered *chemically* by a **chemical bond** that formed between them. A bond is not a physical structure but rather an energy relationship between atoms, or an attractive force. In this section we cover how bonds form and the different types of bonds that hold atoms together.

Molecules, Compounds, and Molecular Formulas

As we discussed in the module introduction, a molecule forms when two or more atoms are chemically bonded. When the two or more atoms that bond are of different elements, we refer to them as molecules of a **compound.** Water, which consists of two hydrogen atoms and one oxygen atom (H_2O), is an example of a compound. Very large compounds composed of many atoms, such as those that make up our hair and nails, are known as **macromolecules.**

In this discussion, we've seen examples of *molecular formulas,* ways to represent molecules symbolically with letters and numbers. Such formulas show the kinds and numbers of atoms in a molecule. For example, in CH_4, the formula for the compound methane, the use of the element symbols next to each other tells you that this compound is made up of atoms of carbon and hydrogen, and the subscript number only on H tells you that four atoms of hydrogen—but just one atom of carbon—are found in each molecule of the compound. Similarly, in O_2, the subscript 2 indicates that two oxygen atoms are bonded to each other in the molecule. All molecules are formed by joining atoms with chemical bonds, but how exactly are these bonds formed?

Valence Electrons and Chemical Bonding

Chemical bonds are formed when the electrons located in the outermost electron shells of atoms interact. This electron shell is called the **valence shell** (VAY-lents), and the electrons in this shell are called the **valence electrons.** These electrons determine how an atom interacts with other atoms and whether it will form bonds with a specific atom. In general, the atoms that are relevant to physiology follow the **octet rule,** which states that an atom is most stable when it has eight electrons in its valence shell. Atoms that have filled valence shells, such as helium and neon (He and Ne; see the rightmost column of the periodic table), are nonreactive, or *inert.* Atoms that do not meet the octet rule are said to be *reactive*—that is, they are unstable and so will interact with other atoms until they obey the octet rule. Note that an exception to this rule exists for atoms with five or fewer electrons; such atoms are most stable with only their first electron shell filled (because the first electron shell holds only two electrons, this exception is called the *duet rule*).

As you saw with the formation of NaCl, a chemical bond can change an element's properties drastically. This is because a chemical bond alters the atom's valence electrons. The bond allows reactive atoms to obey the octet or duet rule, and the atoms become more stable. One of two things may happen to valence electrons when a chemical bond is formed: (1) The electrons may be transferred from one atom to another, forming an *ionic bond* (AYE-awn-ik); or (2) the electrons may be shared between two or more atoms, forming a *covalent bond.* The upcoming sections examine these two types of chemical bonds.

Quick Check

☐ 3. What are valence electrons, and why are they important?

Ions and Ionic Bonds

An **ionic bond** results when electrons are transferred between a metal atom and a nonmetal atom. Let's continue to use our example of elemental sodium and chlorine to explore how and why these bonds form.

Here you see the basic structure of a sodium atom (Na; note that we've left off the neutrons for simplicity):

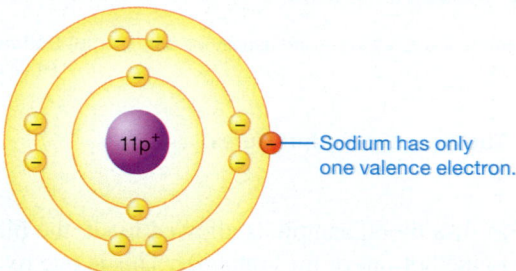

Sodium has only one valence electron.

You can see that a sodium atom's valence shell has only one electron—it does not obey the octet rule, and for this reason, elemental sodium is violently reactive. The sodium can become stable in one of two ways: (1) The sodium atom could give up its lone valence electron, leaving behind its full second shell to become the valence shell; or (2) the sodium atom could steal seven electrons from another atom. The second option would require so much energy that it wouldn't happen naturally. For this reason, we are left with the first option: The sodium atom gives up its lone outer electron.

When this occurs, notice that the sodium still has 11 positively charged protons, but now has only 10 negatively charged electrons. This means that sodium has switched from an electrically neutral atom to a particle with a +1 charge. Such a charged particle is called an **ion** (AYE-awn) or, more specifically, a positive ion called a **cation** (KAT-aye-awn). The positive charge on the sodium ion is represented by adding a superscript plus sign, as Na^+. Note that other metals behave in the same way as sodium—they donate electrons to become cations.

Now let's move on to the chlorine atom (Cl), shown here:

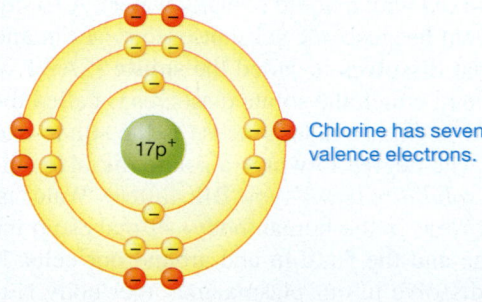

Chlorine has seven valence electrons.

As you can see, the chlorine atom's valence shell has seven electrons—it needs one more electron to obey the octet rule. So, as step ① in **Figure 2.4** shows, the chlorine atom accepts an electron from another atom (in this case sodium), giving it eight electrons in

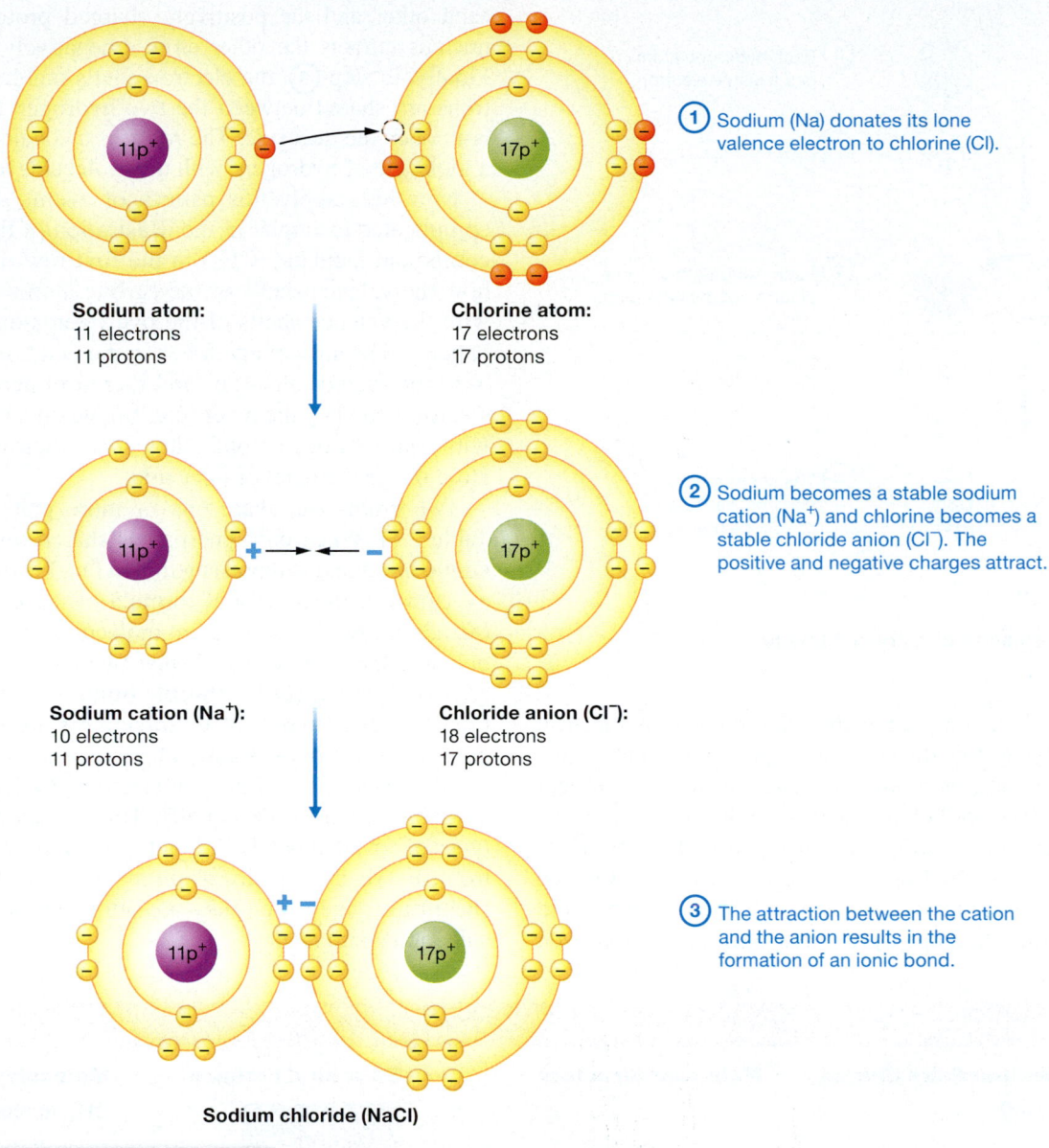

Sodium atom:
11 electrons
11 protons

Chlorine atom:
17 electrons
17 protons

(1) Sodium (Na) donates its lone valence electron to chlorine (Cl).

Sodium cation (Na⁺):
10 electrons
11 protons

Chloride anion (Cl⁻):
18 electrons
17 protons

(2) Sodium becomes a stable sodium cation (Na⁺) and chlorine becomes a stable chloride anion (Cl⁻). The positive and negative charges attract.

(3) The attraction between the cation and the anion results in the formation of an ionic bond.

Sodium chloride (NaCl)

Figure 2.4 **Formation of an ionic bond.**

▶ Play Video Tutor
@ **Mastering** Anatomy & Physiology

its valence shell. The number of protons has not changed, but the extra negative charge turns the chlorine atom into a negative ion, or **anion** (AN-aye-awn), with a −1 charge (note that in this state it is called *chloride* and is represented by adding a superscript minus sign, as Cl⁻). Certain other nonmetals may behave the same way as chlorine, also accepting electrons to become anions.

Now that we have two charged particles, we can get to the actual ionic bond. Stated simply, opposite charges attract. In our example, the positively charged sodium ion and the negatively charged chloride ion attract each other (Figure 2.4, step (2)). This attraction holds, or "bonds," the two ions together, and they form an ionic compound known as a *salt* (Figure 2.4, step (3); salts are discussed in Module 2.4).

Ionic bonds are easy to identify because metals will *only* bond with a nonmetal via an ionic bond. So, if a metal is in the compound, the bond must be ionic. Remember, though, that hydrogen is an exception; it behaves like a nonmetal instead of a metal (unless you are in the core of Jupiter).

Quick Check

☐ 4. What is an ionic bond?
☐ 5. How is an ionic bond formed?

Covalent Bonds

Whereas ionic bonding involves a *transfer* of electrons from a metal to a nonmetal, **covalent bonding** involves *sharing* electrons between two or more nonmetals. The stability created by

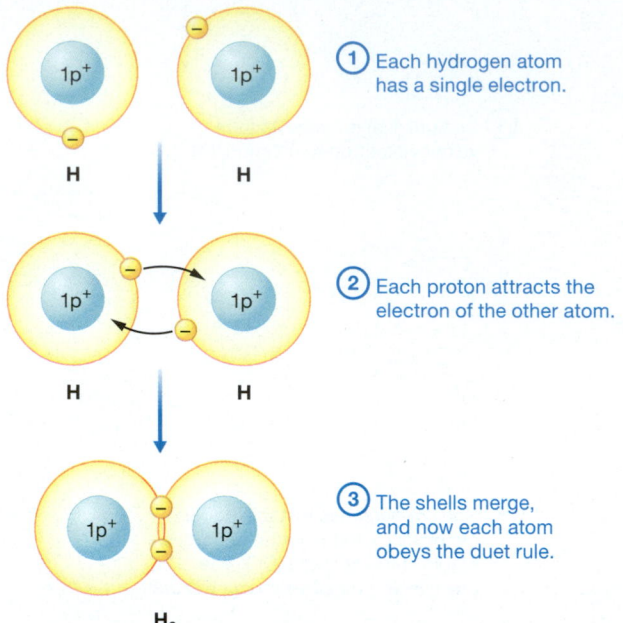

① Each hydrogen atom has a single electron.

② Each proton attracts the electron of the other atom.

③ The shells merge, and now each atom obeys the duet rule.

Figure 2.5 Formation of a covalent bond.

sharing electrons is much greater than that created by electron transfer, which makes this the strongest type of chemical bond.

As with ionic bonds, an atom will share electrons in covalent bonds until its valence shell obeys the octet or duet rule. **Figure 2.5** shows the formation of a simple covalent bond. In step ① of this figure we have two hydrogen atoms, each with one valence electron. To become stable, each atom needs another electron to satisfy the duet rule. In step ②, the hydrogen atoms approach

each other and the positively charged proton in each atom's nucleus attracts the other atom's negatively charged electron. Finally, in step ③, the electron shells merge and the two electrons are shared between the two hydrogen atoms, which now both obey the duet rule. The result is a covalent bond that forms a molecule of hydrogen, with the molecular formula H_2.

Now let's apply this pattern of events to a slightly more complicated example. In the illustration of the molecule of the compound methane (CH_4) in the first row of **Table 2.1**, notice how the valence shells of the carbon atoms come into contact with the valence shells of the hydrogen atoms. Why does this happen? The answer again lies in the octet rule: A carbon atom has four valence electrons and therefore needs four additional electrons to obey the octet rule. So, when a carbon atom bonds with four hydrogen atoms, they share their electrons and each atom obeys the octet or duet rule.

Two atoms can share one or more pairs of electrons (see Table 2.1). When only one pair is shared, the bond is called a **single bond** and is represented by a line between the two atoms, as shown in the structural formula of the compound methane. (*Structural formulas* represent molecules by using element symbols and lines for bonds.) When two electron pairs are shared, as in oxygen gas (O_2), a **double bond** is formed (shown as two lines between atoms). When three pairs are shared, as in nitrogen gas (N_2), a **triple bond** is formed (shown as three lines).

All covalent bonds involve electron sharing, but atoms don't always share electrons equally. This situation gives rise to two types of covalent bonds: A bond is *nonpolar covalent* if the electrons are shared equally or *polar covalent* if the electrons are shared unequally (i.e., they are attracted more to one pole, or side, of a compound).

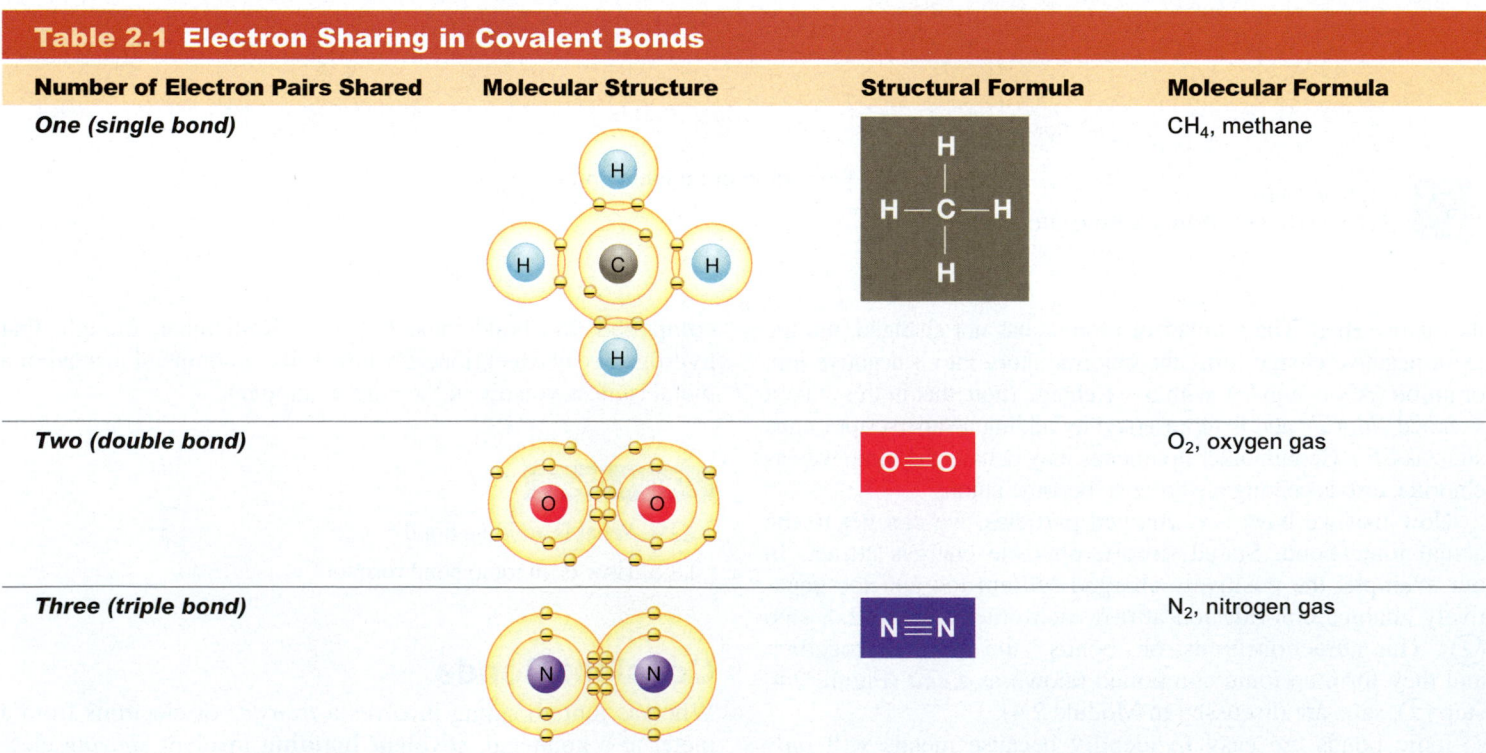

Table 2.1 Electron Sharing in Covalent Bonds			
Number of Electron Pairs Shared	**Molecular Structure**	**Structural Formula**	**Molecular Formula**
One (single bond)			CH_4, methane
Two (double bond)			O_2, oxygen gas
Three (triple bond)			N_2, nitrogen gas

Nonpolar Covalent Bonds

All elements have protons, and those protons can to some extent attract electrons, a property called *electronegativity*. An element's electronegativity generally increases from the bottom left to the upper right of the periodic table—fluorine (F) is the most electronegative element (see Figure 2.2). The more electronegative an element is, the more strongly it attracts electrons. Stated another way, elements with high electronegativity are "greedy" for electrons, and pull them away from less electronegative elements.

When two nonmetals in a molecule have identical or nearly identical electronegativities, they both tug on the electrons with the same force and the electrons are shared equally. This kind of bond is called a **nonpolar covalent bond.** Molecules in which nonpolar bonds predominate are referred to as **nonpolar molecules.** Nonpolar molecules occur in three situations:

- **The atoms sharing the electrons are of the same element.** Identical atoms also have identical electronegativities, and an atom can't pull on electrons any more strongly than its identical twin. You can see an example of a nonpolar covalent molecule in **Figure 2.6a**, which shows H_2. Notice that the yellow electron cloud is evenly distributed around the two atoms. If you look again at Table 2.1, you'll see two other examples: O_2 (a molecule of oxygen) and N_2 (a molecule of nitrogen).

- **The arrangement of the atoms makes one atom unable to pull more strongly than another atom.** Electronegativity isn't the only factor that influences how electrons are shared—the arrangement of the atoms is equally important. Consider, for example, the compound CO_2 (carbon dioxide):

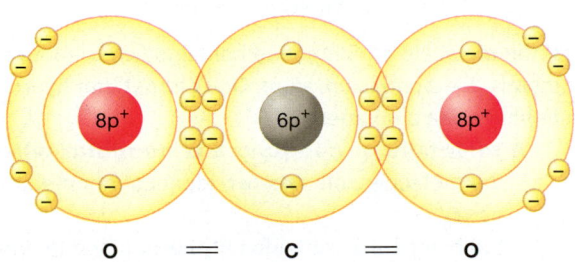

Here the carbon atom is double-bonded to two very electronegative oxygen atoms in two polar double bonds, yet CO_2 is *nonpolar.* This seemingly contradictory fact is due to the arrangement of the oxygen atoms. Notice how the compound is straight, and the oxygen atoms are on either side of the carbon atom. Both oxygen atoms pull the electrons away from the carbon atom in a game of molecular tug-of-war. But the two oxygen atoms are pulling the electrons with equal strength and, as a result, neither "wins."

- **The bond is between carbon and hydrogen.** Technically speaking, carbon has a slightly higher electronegativity than hydrogen. However, this difference is so slight that any bond between carbon and hydrogen atoms is functionally nonpolar.

Note that the electronegativity of most metals is so low that they usually lose electrons (which we just saw in our discussion

(a) Nonpolar covalent bond—H_2 (hydrogen molecule): Electrons spend equal time around the two hydrogen atoms.

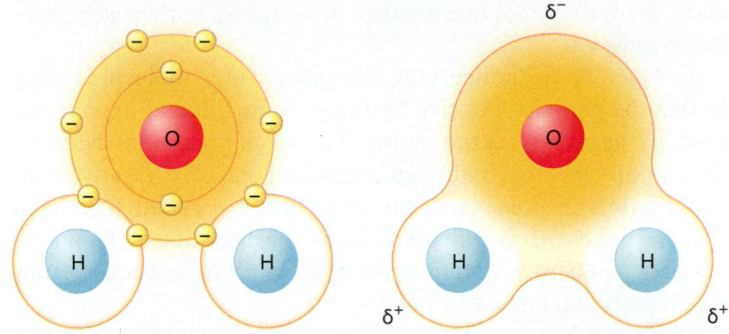

(b) Polar covalent bond—H_2O (water): Electrons spend more time around the more electronegative oxygen atom.

Figure 2.6 **Nonpolar vs. polar covalent bonds.**

of ionic bonds). This is why metals cannot participate in covalent bonds: Electronegative nonmetals tug at the valence electrons of a metal so strongly that they are just pulled away rather than shared.

Polar Covalent Bonds and Hydrogen Bonds

When nonmetals of different electronegativities interact, a tug-of-war between the two nonmetals occurs and the stronger (more electronegative) nonmetal "wins." However, this is only a partial victory, as the winning atom is not electronegative enough to steal the electrons completely. Take a look at **Figure 2.6b**, which shows a molecule of the compound water (H_2O) with its two hydrogen atoms and its single oxygen atom. The oxygen atom is much more electronegative than the two hydrogen atoms, so the oxygen atom attracts the electrons more strongly. This is indicated by the dark yellow color of the electron cloud of the oxygen atom, which means that the electrons spend more time around it. The oxygen atom's stronger electron attraction gives it a partial negative charge (represented by the delta symbol and a superscript minus sign, δ^-). You can see that the electron clouds of the hydrogen atoms are nearly white, which indicates that the electrons spend little time around them. The hydrogen atoms therefore have a partial positive charge (represented by the symbol δ^+).

Molecules of compounds with such partially positive and negative ends have what is known as a **dipole** (DY-pohl; "two sides"), or a **polar covalent bond.** All "polar" really means is that there are two opposite sides, or "poles." In the case of a polar bond, we have a partially positive pole and a partially negative pole. Molecules of compounds that contain primarily polar bonds are known as **polar molecules.**

Keep in mind that a polar molecule as a whole has an equal number of protons and electrons and so is electrically neutral—it is only how the electrons are shared that produces partial negative and partial positive charges. For example, in a water molecule, we have 10 total protons and 10 total electrons, making the molecule electrically neutral as a whole in spite of its partially negative oxygen atom and partially positive hydrogen atoms. This is in contrast to ionic compounds such as NaCl, in which we have two individual charged particles forming a bond based on the attraction of their opposite charges. However, the partial positive and negative charges of polar covalent molecules allow them to attract one another in a similar, but weaker, fashion to ions.

In **Figure 2.7a**, notice how the polar water molecules align so that the partially positive hydrogen atoms line up next to the partially negative oxygen atoms. The weak attractions between the partially positive hydrogen atoms and partially negative nonmetal atoms in polar covalent molecules are called **hydrogen bonds** (indicated in Figure 2.7a by dashed lines). Nonpolar covalent molecules cannot form hydrogen bonds because they lack positive and negative poles.

Hydrogen bonds differ from ionic and covalent bonds in two key ways. First, they are not true bonds because electrons are neither transferred nor shared in a hydrogen bond. Instead, they are simply weak attractions between two polar molecules. Second, hydrogen bonds are generally between *molecules*, whereas covalent and ionic bonds generally form between *atoms* or *ions*.

Hydrogen bonding between water molecules is responsible for one of the key properties of water: *surface tension*. Where air and water meet, the polar water molecules are more strongly attracted to one another than they are to the nonpolar air molecules. This mutual attraction causes the water molecules to cluster together and form hydrogen bonds; each water molecule can form up to four such bonds. The resulting "web" of molecular connections produces **surface tension,** a visible "film" on top of the water. Surface tension is what causes rain to fall in drops, water to form puddles, and a blood sample to form a droplet, as shown in **Figure 2.7b**. Hydrogen bonds play extremely important biological roles. For example, they help to form the structure of our proteins and of our genetic material.

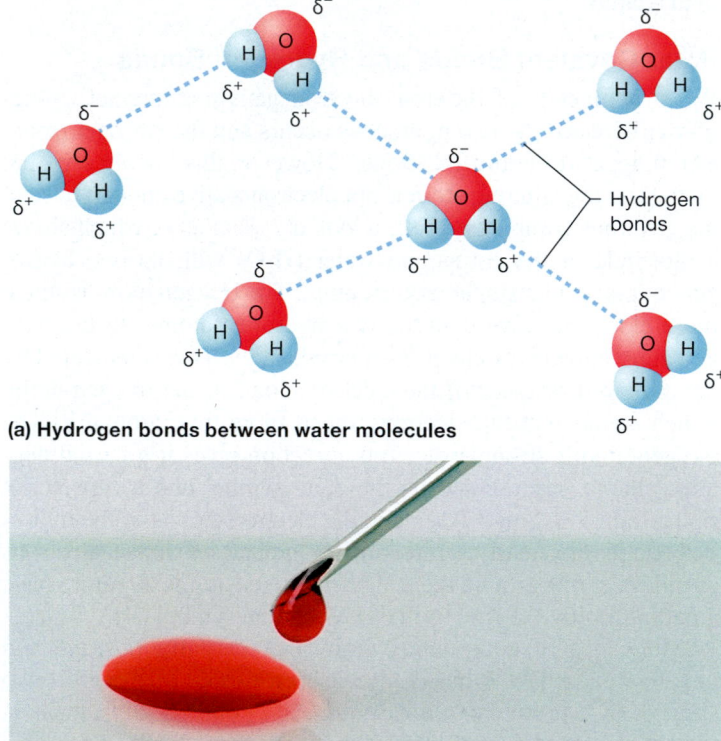

(a) Hydrogen bonds between water molecules

(b) Hydrogen bonds between water molecules create surface tension that causes blood to form droplets.

Figure 2.7 Hydrogen bonding and surface tension between water molecules.

ConceptBOOST)))

Determining the Type of Bonds in a Molecule or Compound

If you're having difficulty determining if a molecule or compound is ionic, polar, or nonpolar, here are some basic "rules" you can keep in mind:

- **A compound is ionic when:** The compound contains both a metal and a nonmetal. Remember that metals won't form covalent bonds, so if you see a metal in the compound, you know immediately that you are looking at an ionic bond.

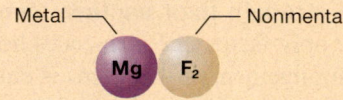

A metal bonded to a nonmetal tells you that it's an ionic compound.

- If the molecule or compound contains two or more nonmetals, the bond is *covalent*. Don't forget that hydrogen behaves like a nonmetal.
  - **A molecule or compound is nonpolar when:**
    - the molecule contains two identical nonmetals, such as O_2;
    - the compound contains only carbon and hydrogen, such as CH_4; and
    - the compound contains primarily carbon and hydrogen, even if it has a couple of strongly electronegative elements such as oxygen (e.g., $C_{16}H_{32}O_2$).

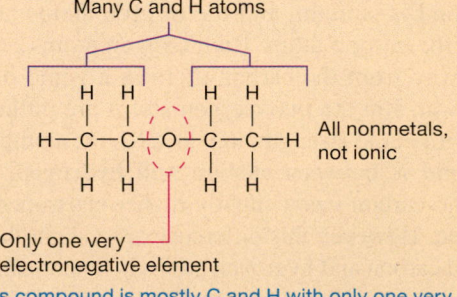

This compound is mostly C and H with only one very electronegative element, and this tells you it's nonpolar.

○ **A compound is polar when:** The molecule contains two nonmetals of significantly different electronegativities. You don't need to memorize electronegativity values to determine this—you can generally go by the rule that hydrogen and carbon have low electronegativities, whereas elements like chlorine, oxygen, nitrogen, and phosphorus have high electronegativities. So, compounds with a proportionally high number of C—O, C—N, P—O, H—N, C—Cl, and H—O bonds are polar (e.g., $C_5H_{10}O_5$, CH_3Cl, H_3PO_4).

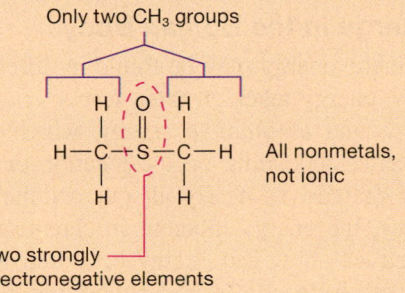

Only two CH_3 groups

All nonmetals, not ionic

Two strongly electronegative elements

This compound has a strongly electronegative S–O group and only two CH_3 groups, and this tells you it's polar.

Remember that these "rules" don't apply to every situation. For example, the arrangement of the atoms plays a role in determining whether a compound is polar or nonpolar. But, in general, they are a good place to start when figuring out if a molecule or compound is ionic, polar, or nonpolar. ■

The types of chemical bonds are summarized in **Table 2.2**.

Quick Check

☐ 6. How do polar and nonpolar covalent bonds differ?

☐ 7. What are hydrogen bonds? Between which types of molecules do hydrogen bonds form?

Apply What You Learned

☐ 1. Which would be more reactive—an atom of fluorine or an atom of neon? Why? (*Hint:* Look at the valence electrons of each element.)

☐ 2. Would a molecule of hydrogen (H_2) form hydrogen bonds? Why or why not?

☐ 3. Explain why the molecule Na_2 does not exist in nature.

See answers in Appendix A.

Chemical Reactions

Learning Outcomes

1. Explain what happens during a chemical reaction.
2. Describe the forms and types of energy, and apply the principles of energy to chemical bonds and endergonic and exergonic reactions.
3. Describe and explain the differences between the three types of chemical reactions.
4. Describe the factors that influence reaction rates.
5. Explain the properties, actions, and importance of enzymes.

In the previous module, we opened our discussion of chemical bonds with the example of combining explosive elemental sodium with poisonous chlorine gas to yield table salt. This is an example of a **chemical reaction,** which happens any time chemical bonds are formed, broken, or rearranged, or when electrons are transferred between two or more atoms (or molecules). Although the reaction of sodium metal with chlorine gas is not one that occurs in the human body, literally quintillions of other chemical reactions are occurring in our bodies at any given time. In this module, we explore the types of reactions that take place in our cells. We also introduce the concept of energy and discuss how it applies to chemical reactions.

Chemical Notation

Before we discuss chemical reactions, we must first address how to represent them. **Chemical notation** is a series of symbols and abbreviations that is used to demonstrate what occurs in a reaction. You have already been introduced to chemical symbols—the letter or letters that represent an element, as well as the subscript numbers in a molecular formula that tell you how many atoms of each element make up a molecule.

Table 2.2 Chemical Bonds		
Type of Bond	**Definition**	**Example**
Ionic Bond	Bond resulting from the transfer of electrons from a metal to a nonmetal	NaCl
Nonpolar Covalent Bond	Bond resulting from the equal sharing of electrons between two nonmetals	H_2
Polar Covalent Bond	Bond resulting from the unequal sharing of electrons between two nonmetals; the electrons spend more time around the more electronegative atom, which creates a dipole	H_2O

The other basic form of chemical notation is the **chemical equation.** Let's use the chemical equation that shows the conversion of carbon dioxide and water into carbonic acid as an example:

$$CO_2 \quad + \quad H_2O \quad \rightleftharpoons \quad H_2CO_3$$
$$\textit{Carbon dioxide} \qquad \textit{Water} \qquad \textit{Carbonic acid}$$

As you can see, this chemical equation has two parts:

- **Reactants.** Both carbon dioxide and water are **reactants**—the substances we are starting with that will undergo a reaction. The plus sign (+) is read as "reacts with." By convention, the reactants are placed on the left-hand side of the equation.
- **Products.** On the right-hand side of the equation, we have the **products**—in this case, carbonic acid—the substance or substances produced in the reaction.

In standard chemical notation, an arrow between the reactants and products signifies that the reactants on the left have interacted to form the products on the right. Notice that in this equation, there are two arrows pointing in opposite directions. This indicates that the reaction is a **reversible reaction:** It can progress in both the forward and the reverse directions—that is, the carbonic acid can also split to form carbon dioxide and water.

Energy and Chemical Reactions

Matter and energy are two of the most basic concepts of the sciences, and the two are inseparable. Broadly, **energy** is defined as the capacity to do work. This means that energy can put matter into motion—it is energy that fuels our chemical reactions, causes the cells in our hearts to contract, drives our digestion and respiration, and allows you to read this book. The next subsections look at the different forms of energy in the human body and how they apply to chemical reactions.

Potential and Kinetic Energy

There are two general types of energy:

- **Potential energy. Potential energy** is energy that is stored, ready to be released and used to do work. The classic example of potential energy is a ball sitting atop a hill. The ball is stationary, but it has the *potential* to roll down the hill.
- **Kinetic energy.** Potential energy becomes **kinetic energy** (*kinet-* = "movement") when it is used. Kinetic energy is energy in motion, which means that work of some sort is being done. When the ball rolls down the hill, its energy has changed from potential to kinetic:

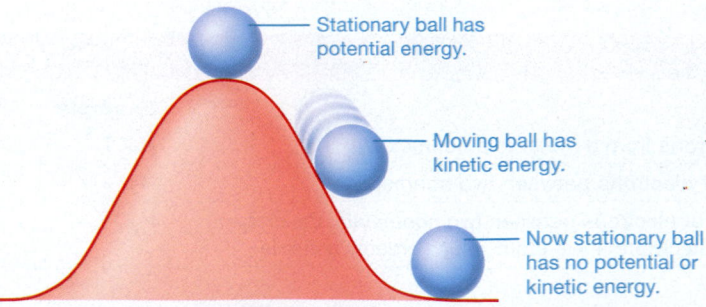

- Stationary ball has potential energy.
- Moving ball has kinetic energy.
- Now stationary ball has no potential or kinetic energy.

All atoms have kinetic energy because atoms are constantly in motion. The faster the atoms move, the higher their kinetic energy.

Note that kinetic energy and potential energy are interconvertible—each may be turned into the other. In our ball and hill analogy, kinetic energy is required to get the ball to the top of the hill. Once the ball is at the top, that kinetic energy has been converted to potential energy. When the ball begins to roll down the hill, its potential energy has been reconverted to kinetic energy. When the ball is at the bottom of the hill and has stopped rolling, however, it has neither potential nor kinetic energy.

Forms of Energy in the Human Body

Potential and kinetic energy may be found in different forms. In the human body, energy takes on three forms: *chemical energy, electrical energy,* and *mechanical energy,* which may each be potential or kinetic, depending on the location or process. The form of energy that drives nearly all our cellular processes is **chemical energy,** the energy inherent in chemical bonds. This might seem abstract—how can energy be stored in a shared or transferred pair of electrons? To understand this, first remember that electrons are always in motion and therefore have kinetic energy. Second, consider that it takes *work* to transfer and share electrons (i.e., protons must attract electrons). We can't do work without energy, so energy must be present in all chemical bonds.

Certain cells in the body also harness **electrical energy.** Electrical energy is generated by the movement of charged particles. For example, the electricity that powers your house is generated by the flow of electrons through metal wiring. In our cells, however, electrical energy results from the flow of ions. This form of energy allows nerve cells to communicate and plays a role in heart and skeletal muscle contractions.

The third form of energy that drives processes in the human body is **mechanical energy.** Mechanical energy is best explained as energy that has been directly transferred from one object to another. For example, when you swing a bat to hit a baseball, you are transferring mechanical energy from your body first to the bat and then to the ball. Similarly, when the tiny proteins in skeletal muscles move past one another and generate a muscle contraction, they transfer their mechanical energy first to the muscle cell, then to the whole muscle, and finally to the body part that moves.

Note that these types of energy are interconvertible (with a few exceptions). For example, electrical energy coming into a blender is converted into mechanical energy that spins its blades. Most of the time, such conversions occur easily but they are rather inefficient, and much of the energy is lost as heat. This is why an appliance or engine feels hot to the touch after it has been running for a while. Heat production is useful in the human body, though, as it helps to maintain our relatively high body temperature. For example, we shiver when our body temperature drops, a process that converts chemical energy into mechanical energy that moves our muscle cells. In this conversion, much of the chemical energy is lost as heat, and the heat raises our body temperature.

Endergonic and Exergonic Reactions

As we have already discussed, energy is inherent in all chemical bonds. This means that some amount of energy must be invested any time a chemical reaction occurs. Some reactions require an input of energy that is *greater* than the energy of the reactants. These reactions, called **endergonic reactions** (en-der-GAHN-ik; *ender-* = "within"), require chemical, electrical, or mechanical energy from another source to proceed. The products of an endergonic reaction contain *more* energy than the reactants because energy was invested in the reaction.

The reverse situation is also true—often the reactants have more energy than is needed for the reaction to proceed; such a reaction is called an **exergonic reaction** (eks-er-GAHN-ik; *exer-* = "without"). In this case, the excess energy stored in the reactants is released, leaving the products of the reaction with *less* energy than the reactants.

A reactant that stores a great deal of energy is often referred to as *unstable*—when it releases this energy in an endergonic reaction, it becomes a more stable product. You have already seen an example of such an exergonic reaction: the one between elemental sodium and chlorine gas. Recall that when the two elements combine, a small explosion occurs. This explosion is due to the release of energy as unstable sodium and chlorine react to form stable NaCl (table salt).

Quick Check

☐ 1. How are kinetic energy and potential energy related?

☐ 2. What are the differences between the three types of energy that drive processes in the human body?

☐ 3. How do endergonic and exergonic reactions differ?

Homeostasis and Types of Chemical Reactions

Before you consider the kinds of reactions commonly carried out in the body, think about what the body must do to maintain *homeostasis,* or a relatively consistent internal environment appropriate to the body's needs (see Chapter 1). First, you need nutrients in the form of food, so you must have a way to break down the food you eat. Then, the energy in the chemical bonds of broken down food must be converted into a form the body can use. Finally, you use that energy to repair or replace damaged cells and for normal growth and maintenance. So we have three fundamental processes that occur in the body: breaking down compounds (e.g., food), converting the energy in food into a usable form, and building new compounds. Each of these processes is carried out by one of three basic types of chemical reactions: (1) *catabolic reactions,* which break chemical bonds; (2) *exchange reactions,* which transfer atoms or electrons between reactants; and (3) *anabolic reactions,* which form new chemical bonds.

Catabolic Reactions

During a **catabolic reaction** (kat-uh-BAWL-ik; *cato-* = "downward"), also called a *decomposition reaction,* larger substances are broken down into smaller ones. These reactions have the general form AB → A + B. Catabolic reactions are required to break down the nutrients in the foods we eat, to break apart and recycle old or damaged cells, to defend the body against invading organisms, and more. Catabolic reactions are generally exergonic because chemical bonds are broken.

Exchange Reactions, Including Oxidation-Reduction Reactions

During an **exchange reaction,** one or more atoms from the reactants are exchanged for another. In this type of reaction, bonds are both broken and formed, as indicated by the general form: AB + CD → AD + BC. A simple example is shown here—the conversion of hydrochloric acid (HCl) and sodium hydroxide (NaOH) to water and sodium chloride.

$$HCl + NaOH \rightarrow H_2O + NaCl$$
Hydrochloric acid   Sodium hydroxide   Water   Sodium chloride

A special kind of exchange reaction is the **oxidation-reduction reaction** (sometimes shortened to *redox reaction*). During an oxidation-reduction reaction, electrons are exchanged instead of atoms. The reactant that loses electrons is said to be *oxidized,* and the reactant that gains electrons is said to be *reduced.* The transferred electrons carry energy with them. Oxidation-reduction reactions are generally exergonic and can release large amounts of energy. In fact, one of the fundamental energy-liberating reactions carried out by our cells is the oxidation of the sugar glucose. This topic is given full coverage in the metabolism chapter (see Chapter 23).

Anabolic Reactions

Our bodies are obviously not made of solitary atoms floating around, but rather of large and often complex compounds. These compounds are built from simpler subunits united by chemical bonds. Reactions that create new chemical bonds are called **anabolic reactions** (an-uh-BAWL-ik; *ana-* = "upward") or **synthesis reactions;** they take the general form A + B → AB. Anabolic reactions in the human body are endergonic, and they are fueled by chemical energy.

Quick Check

☐ 4. What are the differences between the three main types of chemical reactions that take place in the human body?

Reaction Rates and Enzymes

The time it takes for a chemical reaction to finish can vary widely. For example, explosions are chemical reactions that occur in a fraction of a second, but the chemical reactions of metal corrosion (such as rusting) can take years to complete. To understand why reaction rates vary so much, first consider that for a reaction to take place, reactants must come into physical contact with one another. All atoms are in motion, so they sometimes collide. However, the electrons of atoms also repel one another (just as opposite charges attract, like charges repel). So, for a reaction to occur, atoms must collide with enough force to overcome the repulsion of their electrons. The faster these collisions occur, the faster the reaction proceeds.

All chemical reactions require the input of some amount of energy to overcome the repulsion of the atoms' electrons, and to

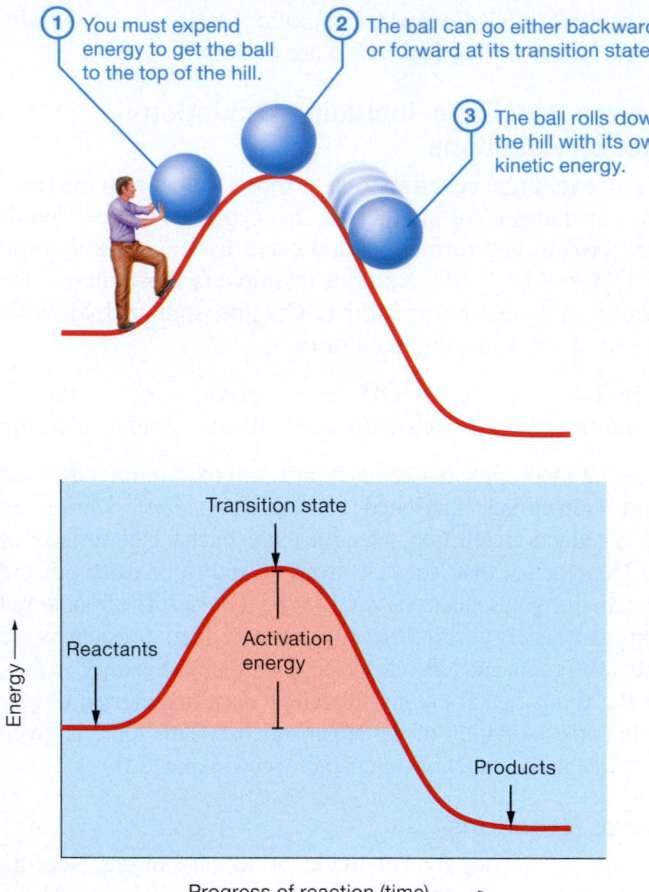

1 You must expend energy to get the ball to the top of the hill.

2 The ball can go either backward or forward at its transition state.

3 The ball rolls down the hill with its own kinetic energy.

Energy →

Transition state

Reactants

Activation energy

Products

Progress of reaction (time) →

Figure 2.8 Activation energy.

allow adequately strong collisions to take place. This energy is called the **activation energy (E_a),** and all reactions, even those that are exergonic, must overcome this energy barrier to proceed. Consider the example in **Figure 2.8**, where you have a ball that you want to get to the other side of a hill. To do this, you must first ① expend energy to roll the ball up the hill. Once the ball is on top of the hill ②, it has an equal probability of rolling backward or forward, and has reached a point called the *transition state*. When the ball has reached this state ③, it can roll down the other side of the hill using its own kinetic energy. This analogy can be applied to a chemical reaction—the activation energy must be supplied so that the reactants reach their transition states (i.e., get to the top of the energy "hill") in order to react and form the products (i.e., roll down the hill).

Several factors increase the rate of a reaction by either reducing the activation energy or increasing the likelihood of strong collisions between atoms. These factors include the following:

- **Concentration.** When the reactant concentration increases, more reactant particles are present, which increases the chances for collisions between the particles.
- **Temperature.** Raising the temperature of the reactants increases the kinetic energy of their atoms, resulting in more forceful and effective collisions. Note that for biological reactions, however, this works only up to a point, because many substances in the human body are destroyed at high temperatures.

- **Properties of the reactants.** Both the size of the particles and the phase (solid, liquid, or gas) of the reactants influence reaction rates. Typically, smaller particles move faster and therefore have more energy than larger particles. Also, reactant particles that are in the gaseous phase have higher kinetic energy than those in the liquid or solid phase. For this reason, reactions with two gases generally proceed slightly faster than those with two liquids, and a great deal faster than those with two solids.
- **Presence or absence of a catalyst.** A *catalyst* (KAT-uh-list) is a substance that increases a reaction rate by lowering the activation energy. Catalysts are neither consumed nor altered in the reaction. Biological catalysts are called **enzymes.**

You may have heard it said that an enzyme "breaks something down," but this is not actually correct. Enzymes are involved in essentially all our catabolic reactions, but they are also involved in all our exchange and anabolic reactions. Furthermore, the enzyme itself doesn't actually chemically alter or "break" the bonds in the reactants. What an enzyme does instead is to increase the rate at which a reaction takes place.

Each cell has roughly 4000 different kinds of enzymes. Most enzymes are named by adding the suffix "-ase" to their root words; for example, the enzyme that catalyzes the reaction breaking down the sugar in milk is called *lactase* (*lact-* = "milk"). Nearly all enzymes are macromolecules called proteins, with the following properties:

- **Enzymes speed up reactions by lowering the activation energy.** Take another look at Figure 2.8, which illustrates our ball and hill analogy. You learned that to get the ball to the other side of the hill, you must first supply the necessary activation energy to move the ball to the hilltop. Notice in **Figure 2.9** what the enzyme does—it lowers the height of the hill. With a smaller hill, less activation energy is required to get the ball to the top. The lowered activation energy also allows the reaction to proceed faster.
- **Enzymes are highly specific for individual substrates and reactions.** Each enzyme contains a unique region, called the **active site,** that binds to a specific substance known as that enzyme's **substrate.** Not only is an enzyme specific for its substrate, but also usually for a single reaction. In this way, one substrate may bind two different enzymes and produce two different products, depending on the conditions in the cell.
- **Enzymes do not alter the chemical reaction.** The nature of a chemical reaction is the same whether or not an enzyme is present—the enzyme does not chemically change the reactants or the products of a reaction.
- **Enzymes are not permanently altered in the reaction.** Although the chemical bonds of the reactants change during a reaction, those of the enzyme do not change. The structure of the enzyme does alter somewhat during the reaction but reverts to its original form when the reaction is complete. The enzyme isn't "used up" in the reaction, and so can go on to bind another substrate and catalyze another reaction.

An enzyme's mechanism of action becomes clear when you look at its interaction with its substrate(s), which is currently

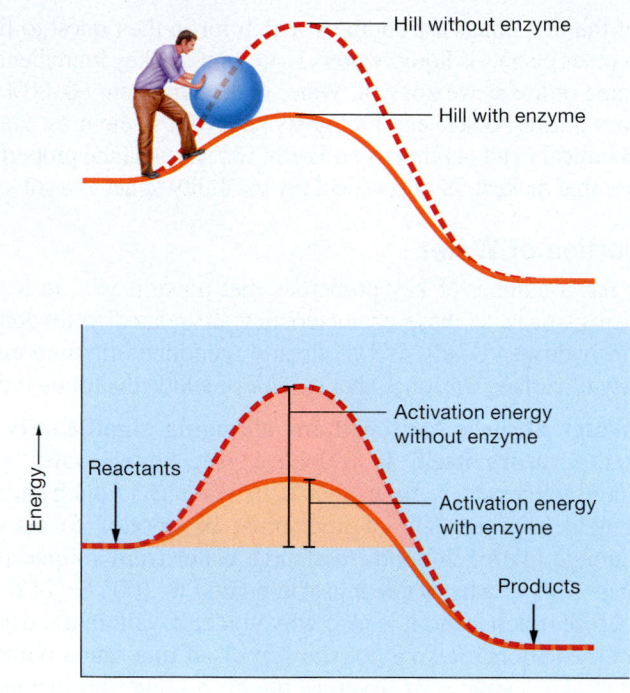

Figure 2.9 The effect of enzymes on activation energy.

▶ Play Animation
@ Mastering Anatomy & Physiology

thought to occur by an "induced fit" mechanism. You can see how this works in **Figure 2.10**. Notice the shape of the substrates and the enzyme in step ①, before the substrates bind the enzyme. Now look to step ②, and see how both the substrates and the enzyme change shape slightly when they bind. As you can see in step ③, the shape change in the substrates is important, as it brings them to their transition states, which significantly reduces the required activation energy. When the product has formed, it dissociates from the enzyme in step ④, and the enzyme returns to its original shape, ready to bind new substrates.

The importance of enzymes in the human body cannot be overemphasized, as they increase the rates of our reactions between 10 million and 100 trillion times. Although many chemical reactions can be carried out in the laboratory without the help of a catalyst, that isn't true for the chemical reactions in the body. Most biological compounds have low kinetic energy and are pretty stable—otherwise, we would risk exploding. The low kinetic energy of biological compounds causes the activation energy for biological reactions to be high. Without enzymes, the chemical reactions required to carry out all our basic processes (to maintain homeostasis) would occur far too slowly for us to remain alive. For this reason, enzymes are absolutely vital to our survival; see *A&P in the Real World: Enzyme Deficiencies* for some examples of diseases that result when enzymes are missing or faulty.

Enzyme Deficiencies

Given the important role of enzymes in the body's homeostasis, deficiencies of certain enzymes can be extremely damaging. The following are some examples of common enzyme deficiencies:

- *Tay-Sachs disease* is a deficiency of the enzyme hexosaminidase-A (hek′-soh-sah-MIN-ih-dayz), which results in the buildup of chemicals called *gangliosides* (GANG-glee-oh-sydz) around the nerve cells in the brain. The disease usually causes death by 3 years of age.
- *Severe combined immunodeficiency syndrome,* or *SCIDS,* may be caused by a deficiency of the enzyme adenosine deaminase (uh-DEN-oh-seen dee-AM-ih-nays). This disease results in a nearly complete absence of the immune system, and affected patients must live in a sterile "bubble."
- *Phenylketonuria* (fee′-nul-kee-tohn-YOOR-ee-uh) is a deficiency of the enzyme phenylalanine hydroxylase (fen′-el-AL-uh-neen hy-DRAWK-syl-ays), which converts the amino acid phenylalanine into a different amino acid called tyrosine. The symptoms of the disease include seizures and intellectual disability (formerly known as mental retardation); however, these symptoms can be prevented with dietary modifications.

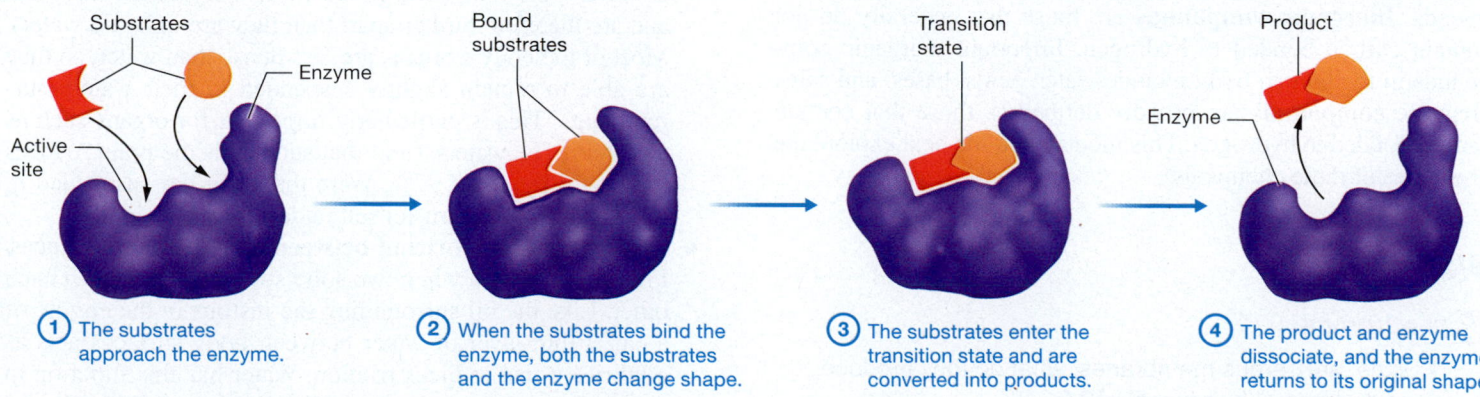

① The substrates approach the enzyme.

② When the substrates bind the enzyme, both the substrates and the enzyme change shape.

③ The substrates enter the transition state and are converted into products.

④ The product and enzyme dissociate, and the enzyme returns to its original shape.

Figure 2.10 Enzyme-substrate interaction.

☐ 5. What factors can influence the rate of a chemical reaction?

☐ 6. What is an enzyme, and what is its function?

Apply What You Learned

☐ 1. Why do you feel hot during and immediately after exercise? (*Hint:* Think about what happens when you convert one form of energy to another.)

☐ 2. Explain why most biological molecules and compounds are stable and nonreactive.

☐ 3. Many naturally occurring poisons function by inhibiting a cell's enzymes. Why might this effect be lethal to the cell?

See answers in Appendix A.

MODULE **2.4**

Inorganic Compounds: Water, Acids, Bases, and Salts

Learning Outcomes

1. Discuss the physiologically important properties of water.

2. Explain why certain molecules and compounds are hydrophilic and why others are hydrophobic.

3. Describe the properties of acids and bases with respect to hydrogen ions.

4. Explain what the pH scale represents, and why a given value is acidic, neutral, or basic.

5. Explain the function of a buffer.

6. Define the terms salt and electrolyte, and give examples of their physiological roles.

In the world of **biochemistry,** the chemistry of life, compounds are grouped into one of two types: inorganic and organic compounds. **Inorganic compounds** are those that generally do not contain carbon bonded to hydrogen. Important inorganic compounds in the human body include water, acids, bases, and salts. **Organic compounds** are broadly defined as those that contain carbon bonded to hydrogen. This module and the next explore the properties of these chemicals.

Water

« FLASHBACK

1. What are serous membranes, what do they produce, and what are their functions? (p. 19)

2. What are hydrogen bonds? (p. 39)

One of the first things astronomers search for in their quest to find life on other planets is liquid **water,** H_2O, as it is a key ingredient in the recipe of life as we know it. Water makes up about 60–80% of the mass of our bodies, and biologists believe it is the most abundant chemical in all organisms on Earth. Let's look at the properties of water that make it unique, especially its ability to act as a solvent.

Properties of Water

Water has a number of key properties that make it vital to living organisms; many of these characteristics are related to its ability to form hydrogen bonds. We've already mentioned the important property of surface tension. Other qualities include the following:

- **Water absorbs heat without changing significantly in temperature itself.** If you have ever boiled water, you have witnessed water's high *heat capacity,* which means that a large amount of heat must be applied to disrupt enough hydrogen bonds to change water from a liquid to a gas. This is why water must be heated to 100° C (212° F) before it boils, and it is also why you can walk in the desert in the summer, take a hot shower, or sit in a sauna without damaging your cells. Even as the air around you increases in temperature, the water in and around your cells doesn't change temperature appreciably.

- **Water carries heat with it when it changes from a liquid to a gas.** If you are outside exercising in the heat of summer, before long you will begin sweating. The sweat will start to evaporate when enough heat is available to disrupt the hydrogen bonds between the water molecules. This heat is provided by your body, which means in essence that the water pulls heat out of your body when it turns into water vapor. The overall effect is that your body cools down. When the humidity in the air is very high, water evaporates much less efficiently, which explains why humid heat often feels hotter than dry heat.

- **Water cushions and protects the body's structures.** The cushioning ability of water has to do with its relatively high *density.* A substance's density relates its mass to its volume. Stated simply, the more of a substance that you can fit into a particular volume, the higher its density. Water's density is high because the hydrogen bonds pull water molecules closely together so that they occupy less space. Such a property is useful because anything less dense than water will float in it to some degree (this explains why ice cubes float in water—the molecules in ice form a crystalline structure and are therefore farther apart than they are in liquid water). Most of the body's organs are less dense than water, so they are able to remain slightly suspended in their watery surroundings. This is particularly important for organs such as the brain. The watery fluid that surrounds the brain reduces its weight by about 97%. Were the brain not suspended in this fluid, it would crush itself under its own weight.

- **Water acts as a lubricant between two adjacent surfaces.** Friction is created when two solid surfaces rub against each other. Like the oil surrounding the pistons in the engine of a car, a thin layer of water between body surfaces acts as a lubricant that reduces friction. Water has this function in the serous membranes surrounding the lungs, the heart, and the abdominal organs.

Water as a Solvent

Water serves as the body's primary solvent. In fact, water is often called the *universal solvent* because so many solutes will dissolve in it fully or at least partially. Water's ability to dissolve substances has to do with both its molecular shape and its polar covalent bonds. Recall that the water molecule is bent—that is, the two hydrogen atoms sit at an angle to each other (see Figure 2.6). Remember also the partial charges on the oxygen and hydrogen atoms that result from its polar covalent bonds.

These two factors combine to make water a polar molecule; this gives it "hands" with which it can "grab" certain solutes and pull them apart. Water molecules can then surround these solutes and keep them apart. Water is only able to dissolve solutes known as **hydrophilic** (hy-droh-FIL-ik; "water loving"). Notice in **Figure 2.11a** and **b** that hydrophilic solutes have fully or partially charged ends. These charged ends attract the oppositely charged ends of water molecules due to the attraction between negative and positive charges. Recall from Module 2.2 that ionic and polar covalent compounds have fully and partially charged ends, respectively, and these compounds are therefore hydrophilic. Solutes that lack charged ends do not have "handles" for water to grab and will not dissolve in water. These solutes are referred to as **hydrophobic** (hy-droh-FOH-bik; "water hating") and include uncharged nonpolar covalent molecules or compounds such as methane, oils, and fats (**Figure 2.11c**).

A solute's solubility in water follows the general rule of thumb "*like dissolves like.*" This is a simple way of saying that partially or fully charged compounds, such as those with polar covalent and ionic bonds, will interact with one another but will not interact with uncharged nonpolar covalent solvents. However, uncharged nonpolar covalent molecules or compounds will not interact with charged compounds but will dissolve readily in nonpolar covalent solvents, such as oil or paint thinner.

Many solutes vital to maintaining homeostasis are hydrophilic, so our bodies are able to carry them in blood plasma and transport them safely and efficiently from one place to another. However, certain hydrophobic substances, such as fats, oxygen, and carbon dioxide, also require transport through the body but have little to no ability to dissolve in water. As we progress through this book, we will see several examples of how our bodies are adapted to meet this challenge.

Quick Check

☐ 1. What four properties of water make it a good solvent for living organisms?

☐ 2. Which molecules or compounds are likely to be hydrophilic? Which are likely to be hydrophobic?

Acids and Bases

The study of acids and bases is really the study of the **hydrogen ion (H$^+$)**. Earlier you learned that hydrogen behaves like a nonmetal most of the time: It forms covalent rather than ionic bonds, and elemental hydrogen exists on Earth as a gas, not a metal. However, when certain hydrogen-containing compounds

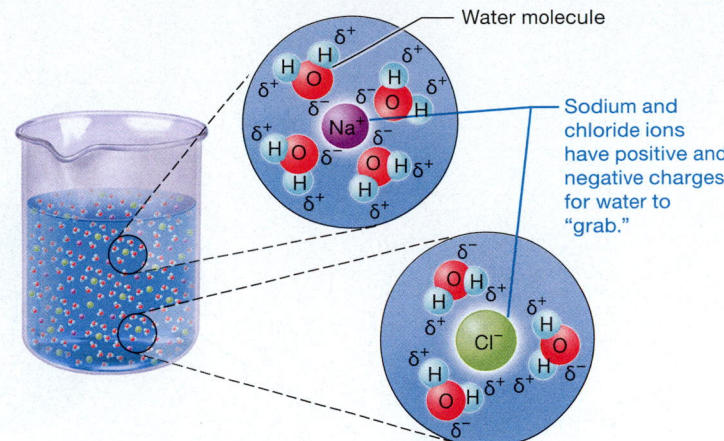

(a) Ionic compounds are hydrophilic.

Water molecule

Sodium and chloride ions have positive and negative charges for water to "grab."

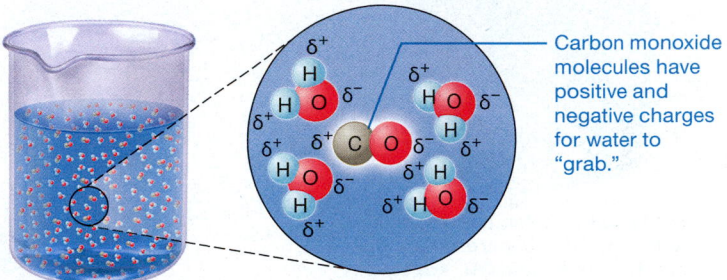

(b) Polar covalent compounds are hydrophilic.

Carbon monoxide molecules have positive and negative charges for water to "grab."

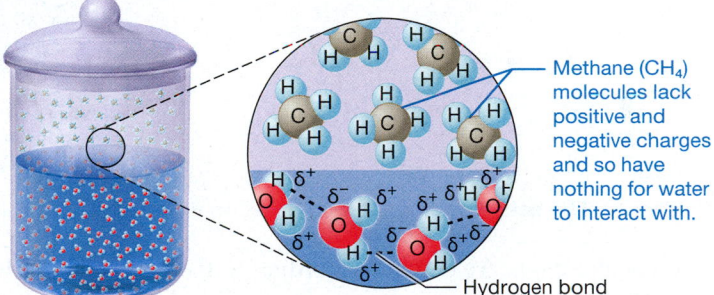

(c) Nonpolar covalent compounds are hydrophobic.

Methane (CH$_4$) molecules lack positive and negative charges and so have nothing for water to interact with.

Hydrogen bond

Figure 2.11 The behavior of hydrophilic and hydrophobic molecules in water.

are placed in water, hydrogen can switch roles and start behaving like a metal—it donates its electron to produce a hydrogen ion. In fact, this even happens to the hydrogen in water to a small extent. As shown in **Figure 2.12a** and the following reaction, a water molecule in a solution may break apart, or *dissociate,* to form a hydrogen ion and an anion called the **hydroxide ion (OH$^-$)**:

$$H_2O \;\rightleftharpoons\; H^+ \;+\; OH^-$$

Water   *Hydrogen*   *Hydroxide*
           ion   *ion*

(Sometimes anions can consist of more than one atom, but they act as single anions.)

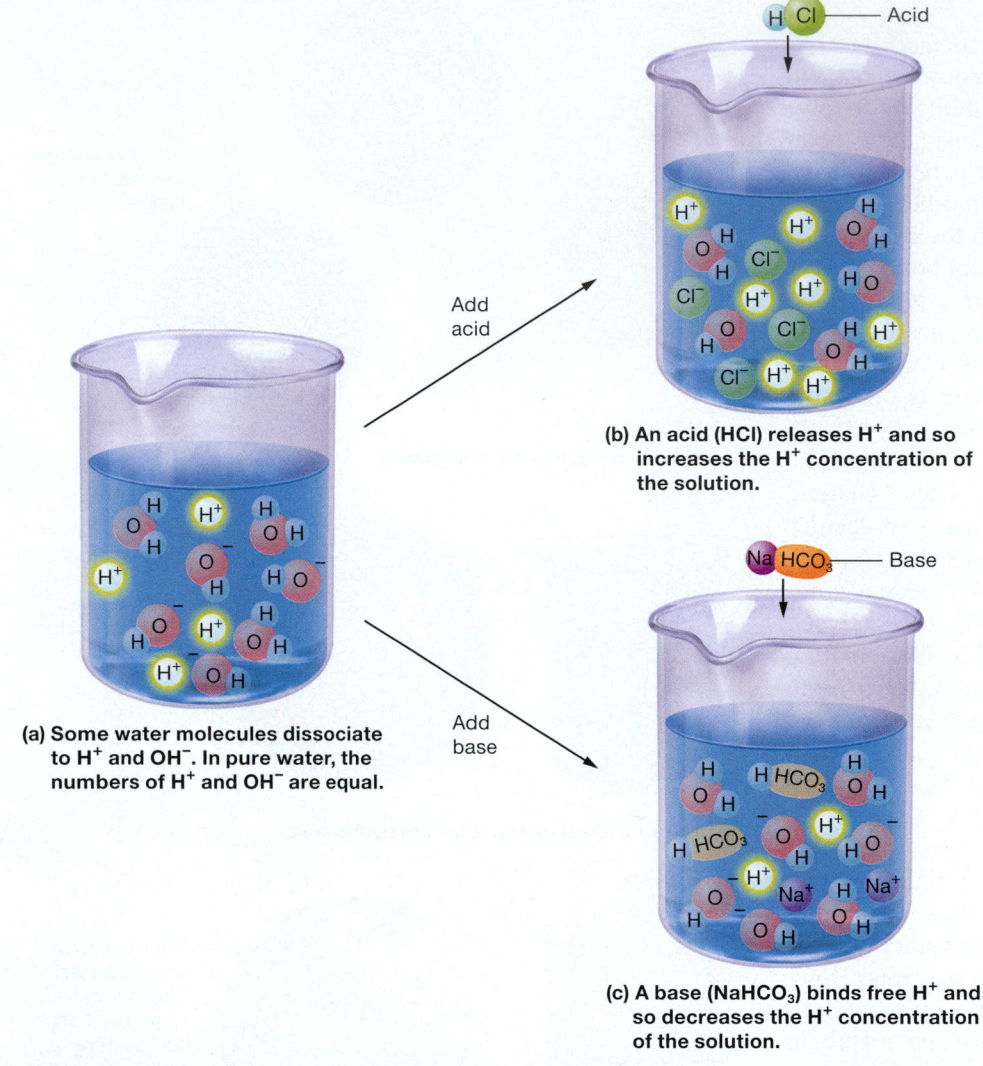

(b) An acid (HCl) releases H⁺ and so increases the H⁺ concentration of the solution.

(a) Some water molecules dissociate to H⁺ and OH⁻. In pure water, the numbers of H⁺ and OH⁻ are equal.

(c) A base (NaHCO₃) binds free H⁺ and so decreases the H⁺ concentration of the solution.

Figure 2.12 The behavior of acids and bases in water.

Acids and bases are defined according to their behavior with respect to hydrogen ions:

- **An acid is a hydrogen ion donor.** As shown in **Figure 2.12b** and the following equation, when you place an **acid** in water, it splits apart into a hydrogen cation (H^+) and an anion. Let's look at what happens with hydrochloric acid (HCl):

$$\textit{In } H_2O$$

HCl	$\rightleftharpoons$	H⁺	+	Cl⁻
Hydrochloric acid		*Hydrogen ion*		*Chloride ion*

In this process, HCl "donates" its hydrogen ion to the solution, and the number of hydrogen ions in the solution increases. After an acid dissociates, the anion that forms can bond with or "accept" the hydrogen ion back if the concentration of hydrogen ions rises again.

Examples of acids found in the human body include HCl in the stomach and H_2CO_3 (carbonic acid) in the blood. Note

that acids can be called *proton donors* rather than hydrogen ion donors. This is because a hydrogen ion has only one subatomic particle—the lone proton in its nucleus.

- **A base is a hydrogen ion acceptor.** When a **base** (also called an **alkali;** AL-kuh-lye) is mixed with water, the number of hydrogen ions in the solution decreases. Notice in **Figure 2.12c** that when the base sodium bicarbonate ($NaHCO_3$) is placed in water, it dissociates into sodium cations (Na^+) and bicarbonate anions (HCO_3^-). The bicarbonate ions then accept the free hydrogen ions in the solution. Most bases used in the chemistry laboratory have −OH (hydroxide) in their formulas. However, the main bases in the human body, such as bicarbonate ions (HCO_3^-), do not.

The pH Scale

Concentrations of acids and bases affect many of the chemical reactions that occur in our bodies. A simple way to represent the hydrogen ion concentration of a solution is via the **pH scale.** The term "pH" has widely, and incorrectly, been reported to mean "potential of hydrogen." Actually, *p* is a mathematical abbreviation for the −log (read as "negative logarithm") of a number, and *H* stands for the concentration of hydrogen ions in the solution. So, the term "pH" means −log[H^+], or the "negative logarithm of the solution's hydrogen ion concentration" (brackets indicate concentration).

The pH scale, shown in **Figure 2.13**, goes from 0 to 14. Notice in the column on the left that the more hydrogen ions present in a solution, the lower its pH. As we move up the scale, we find progressively fewer hydrogen ions in the solution, and the pH increases. However, the column to the right, which represents the number of base ions in the solution, shows the opposite trend. At a lower pH, there are fewer base ions, and as we move up the scale, we have progressively more base ions. Notice now what you see in both columns when the pH is 7—the numbers of hydrogen ions and base ions are equal. At this pH a solution is called **neutral.** Pure water at 25° C has a pH of 7, but few other solutions do—nearly all of them fall into the category of either *acidic* or *basic*.

- **A solution with a pH less than 7 is called acidic.** As you can see in Figure 2.13, **acidic** solutions have more hydrogen ions than base ions. An extremely acidic solution, called a *strong acid,* has a pH near 0, essentially no base ions, and an extremely high concentration of hydrogen

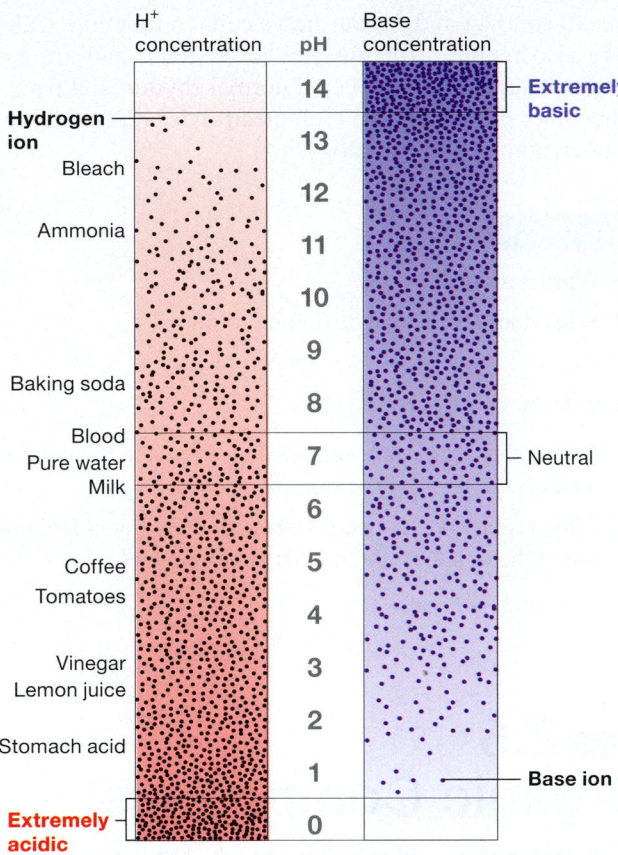

Figure 2.13 The pH scale.

ions. A *weak acid* with a pH just below 7 has only slightly more hydrogen ions than base ions.

- **A solution with a pH greater than 7 is called basic, or alkaline. Basic** solutions have the reverse conditions of acidic solutions—they have fewer hydrogen ions and more base ions. An extremely basic solution, a *strong base,* has a pH near 14, essentially no hydrogen ions, and a very high concentration of base ions. A *weak base* has a pH just above 7 and has only slightly more base ions than hydrogen ions.

The pH of most fluids in the human body hovers just above the neutral mark at a slightly basic pH. For example, the pH of the blood averages about 7.35–7.45, and the pH inside our cells is around 7.2. Examples of other common solutions and their corresponding pH values are also shown in Figure 2.13.

Something important to note about the pH scale is that we are dealing with *base 10* logarithms. What this means is that each single-digit change in the pH number corresponds to a 10-fold change in hydrogen ion concentration. So when the pH decreases from 7 to 6, the hydrogen ion concentration increases 10 times. When the pH decreases from 7 to 5, the hydrogen ion concentration increases 100 times, and it increases 1000 times with a pH decrease from 7 to 4. The same is true in reverse— a pH increase from 7 to 8 represents a 10-fold decrease in the solution's hydrogen ion concentration.

Buffers

The reason that the pH range in our blood and cells is so narrow is that the function of many important compounds in the body, such as enzymes and other proteins, can be disrupted by even slight changes in hydrogen ion concentration. A pH decrease of only about 0.3 reflects a doubling of the hydrogen ion concentration, and as you just learned, a pH decrease of 1 reflects a 10-fold increase in the hydrogen ion concentration. For this reason, it is vital to have mechanisms in place to keep the pH of the body's fluids within their normal ranges.

This task is accomplished by the body's **buffers,** chemical systems that resist changes in pH and prevent large swings in the pH when acid or base is added to a solution. Typically, a buffer consists of a weak acid and its corresponding anion. One of the major buffers in the body is the *carbonic acid–bicarbonate buffer system.* In this system, the weak acid is carbonic acid (H_2CO_3), and its anion is bicarbonate (HCO_3^-), as shown here:

$$H_2CO_3 \quad \rightleftharpoons \quad H^+ \quad + \quad HCO_3^-$$

Carbonic acid   *Hydrogen ion*   *Bicarbonate ion*

This buffer system works in a simple way: When the pH of the blood rises (i.e., the blood becomes too basic), carbonic acid releases hydrogen ions into the blood to offset the increase in pH. When the pH of the blood falls (i.e., the blood becomes too acidic), bicarbonate ions bond to hydrogen ions in the blood, taking them out of solution and offsetting the decrease in pH.

This and the other buffer systems in the body are generally very effective at keeping the blood pH in the normal range. However, these systems can be overwhelmed. One condition that may result is *acidosis* (aeh-sih-DOH-sis), which is caused by an increase in the total body concentration of hydrogen ions and causes the blood pH to fall below 7.35. The other condition is *alkalosis* (al-kah-LOH-sis), which is caused by a loss of hydrogen ions or a gain of basic ions and causes the blood pH to rise above 7.45.

The two main organ systems in the body that work to correct pH imbalances are the respiratory and urinary systems. The specific mechanisms by which these body systems compensate for pH disturbances are discussed fully in the chapters that cover those systems (see Chapters 21 and 24, respectively).

Quick Check

☐ 3. What are acids and bases?

☐ 4. What is the pH scale? Which pH values are considered acidic, basic, and neutral?

☐ 5. What is the effect of a buffer on a solution?

Salts and Electrolytes

⸜⸜ FLASHBACK

1. What type of bond forms between a metal and a nonmetal? (p. 36)

2. Are compounds with this type of bond hydrophilic or hydrophobic? Why? (p. 47)

You are likely familiar with the term "salt" from common table salt, sodium chloride (NaCl). Technically, however, the term **salt** refers to any metal cation and nonmetal anion held together by ionic bonds. As you learned earlier, ionic compounds such as salts dissolve when placed in water. The resulting cations and anions in the solution are called **electrolytes** (eh-LEK-troh-lytz). Electrolytes are named for the fact that they will conduct an electric current in water. This is why you risk electrocution if an electrical appliance comes into contact with tap water with you in it—tap water contains electrolytes, and the electrolytes conduct the electricity from the electrical device into you. All salts dissociate into electrolytes, as do some acids and bases.

Salts play a wide range of vital roles in the body. Solid calcium salts are the major component of bones and teeth. Other salts dissolve into their component electrolytes in the body's fluids. Dissolved sodium and potassium ions are required for

muscles to contract and for our nerve cells to function. Calcium ions are also required for muscles to contract, and are essential for the heart muscle to keep a normal rhythm and pace. We will encounter these salts and their electrolytes repeatedly in our study of anatomy and physiology.

Quick Check

☐ 6. What is a salt?

☐ 7. What does an electrolyte do in a solution?

Apply What You Learned

☐ 1. Explain how the water surrounding a fetus in the womb protects the fetus as it develops.

☐ 2. Why might a person who is severely dehydrated feel hot, even if he or she is not in a hot environment?

See answers in Appendix A.

MODULE **2.5**

Organic Compounds: Carbohydrates, Lipids, Proteins, and Nucleotides

Learning Outcomes

1. Explain the relationship between monomers and polymers, and describe how they are formed and broken down by dehydration synthesis and hydrolysis reactions.

2. Compare and contrast the general molecular structures of carbohydrates, lipids, proteins, and nucleic acids, and identify their monomers and polymers.

3. Identify examples of carbohydrates, lipids, proteins, and nucleic acids, and discuss their functional and structural roles in the human body.

4. Describe the four levels of protein structure, and explain why protein shape is important for protein function.

5. Describe the reaction for ATP hydrolysis, and explain the role of ATP in the cell.

We now begin our exploration of the world of organic compounds. These chemicals are produced by living organisms and so make up much of the human body and other living things. They are generally much larger and more complex than inorganic compounds.

Certain organic compounds, called *hydrocarbons,* contain only hydrogen and carbon. As shown at the top of the next page, they can exist in two forms—chains and rings.

Hydrocarbon chain

Another way to show a hydrocarbon chain; the C atoms are at each point, and the H atoms are not shown.

Hydrocarbon ring

Another way to show a hydrocarbon ring; the C atoms are at each corner, and the H atoms are not shown.

This carbon-hydrogen backbone forms the basis for all our organic compounds, with the addition of groups of atoms called *functional groups.* They have this name because each different group, when bonded to a compound, somehow modifies its function. We'll see examples of functional groups throughout this module.

The organic compounds covered in this chapter are *carbohydrates, proteins, lipids (fats),* and *nucleotides* (including *nucleic acids*). You have likely heard of the nutritional functions of many of these compounds; however, their structural and functional roles in the body extend far beyond this. We explore these roles as well as the structure of each of these compounds in this module.

Monomers and Polymers

Monomers (MAHN-oh-merz; "one part") are single subunits that can be combined to build larger structures. **Polymers** (PAWL-ih-merz; "many parts") are the larger structures that consist of many monomers linked together. Each type of organic compound we discuss here has its own monomer and a corresponding polymer built from those subunits.

Polymers are built by an anabolic reaction called **dehydration synthesis,** in which two monomers are linked by a covalent bond. The products of dehydration synthesis are the polymer compound and a molecule of water; this is where the reaction gets its name. The reverse reaction can also take place by a process called **hydrolysis** (hy-DRAWL-uh-sis; "water splitting"). In this process, a molecule of water is added to the polymer, the atoms in the water molecule are split apart, and the covalent bonds between the monomers are broken. You will encounter several examples of dehydration synthesis and hydrolysis in the upcoming sections.

☐ 1. How do polymers and monomers differ?

Carbohydrates

◀◀ FLASHBACK

1. What types of chemical bonds form between hydrogen and oxygen atoms? (p. 39)

2. Are compounds with primarily these types of bonds hydrophilic or hydrophobic? Why? (p. 47)

The elements that make up the group of organic compounds called **carbohydrates** are easy to figure out just from the name: *carbo-* means "carbon," and *-hydrate* refers to the elements in water—hydrogen and oxygen. Indeed, most carbohydrate monomers are composed of carbon, hydrogen, and oxygen in the ratio of 1C:2H:1O (there are two hydrogen atoms for every carbon and oxygen atom). Because carbohydrates contain several −OH groups (*hydroxyl groups,* an example of a functional group), the compounds on the whole are polar and therefore hydrophilic.

Carbohydrates account for about 1% of the body's mass. They have only a few structural roles in the body—some carbohydrates are scattered throughout the membranes that surround our cells, and others are found in our genetic material. Instead, the main role of carbohydrates in the body is that of fuel. Catabolic and oxidation-reduction reactions liberate the chemical energy stored in the bonds of carbohydrates and use it to drive the synthesis of the nucleotide *adenosine triphosphate,* or *ATP.* ATP is the body's main source of chemical energy, and is discussed shortly.

The monomer of carbohydrates is the so-called *simple sugar,* more formally called the *monosaccharide* (mahn'-oh-SAK-uh-ry'd; *mono-* = "one," *sacchar-* = "sugar"). Carbohydrates are classified according to how many of these subunits are present. Monosaccharides, of course, have just one subunit; *disaccharides* (dy-SAK-uh-rydz; *di-* = "two") have two subunits; and *polysaccharides* (pawl'-ee-SAK-uh-rydz; *poly-* = "many") have many subunits.

Monosaccharides and Disaccharides

Monosaccharides have from three to seven carbon atoms, and all but the smallest generally take on a ring structure. The majority of monosaccharides in the body have either five carbons (the *pentoses*) or six carbons (the *hexoses*). The two primary pentose sugars are *deoxyribose* (dee'-awks-ee-RY-bohs) and *ribose* (RY-bohs; **Figure 2.14a**). Neither pentose is used directly for fuel, but both are found in our genetic material (discussed later in this module).

The main hexose in the body is the sugar **glucose** (GLOO-kohs), shown in **Figure 2.14b**. Glucose is the body's primary source of fuel, and the term "blood sugar" refers to the concentration of glucose dissolved in the blood plasma. Two other hexoses are present to a large extent in our diets: *fructose* (FRUK-tohs), found in fruits and certain vegetables, and *galactose* (guh-LAK-tohs), found in dairy products. Notice in Figure 2.14b that both fructose and galactose have the same molecular formula as glucose: $C_6H_{12}O_6$. However, the arrangement of the atoms is different in all three compounds. Compounds like these with the same molecular formula but different structures are called **isomers** (AYE-soh-merz; "same parts"). In spite of similarities among the three hexoses, however, glucose is the only sugar the body can use directly in its catabolic reactions.

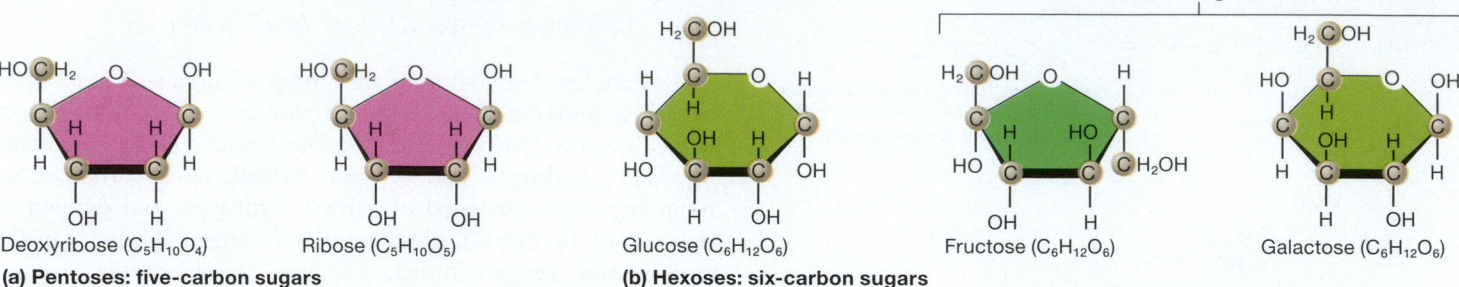

(a) Pentoses: five-carbon sugars   (b) Hexoses: six-carbon sugars

Deoxyribose ($C_5H_{10}O_4$)   Ribose ($C_5H_{10}O_5$)   Glucose ($C_6H_{12}O_6$)   Fructose ($C_6H_{12}O_6$)   Galactose ($C_6H_{12}O_6$)

Figure 2.14 Carbohydrates: structure of monosaccharides.

A **disaccharide** is a compound with two monosaccharides joined by a polar covalent bond. As shown in **Figure 2.15**, disaccharides are formed by dehydration synthesis—a hydrogen atom is removed from fructose and a hydroxyl group (−OH) is removed from glucose, and a molecule of water is formed. Disaccharides can be broken down into monosaccharides by hydrolysis, also shown in Figure 2.15.

The two most prevalent disaccharides are sucrose and lactose. Sucrose is the sweetest of the disaccharides, which is why we use it for sweetening our food and call it "table sugar." It consists of glucose bonded to fructose and comes primarily from sugar cane. The disaccharide lactose, which consists of glucose and galactose, is found in milk and other dairy products. Adults commonly lack the enzyme required to digest lactose, which, as you learned, is lactase, causing the digestive problems associated with *lactose intolerance.*

Polysaccharides

The polymers known as **polysaccharides** are the largest of the carbohydrates. Polysaccharides are composed of long, branching chains of monosaccharides joined by covalent bonds that were formed by dehydration synthesis. Although polysaccharides are composed of polar, hydrophilic monomers, they are not very soluble in water. This is due mostly to their size, as compounds that large are difficult for water molecules to separate from one another, regardless of the types of bonds in the compound.

The structure and low solubility of polysaccharides make them an ideal way for cells to store glucose for later use. Plants store their glucose as a polysaccharide called *starch,* and animals store their glucose as a polysaccharide called **glycogen** (GLY-koh-jen), which is found primarily in the liver and skeletal muscles. As you can see in **Figure 2.16**, glycogen's structure is heavily branched, which allows the glucose in glycogen to be stored in a compact form. This form is a distinct advantage, as glucose molecules are released via enzyme-catalyzed hydrolysis reactions only from the ends of a polysaccharide. The highly branched structure of glycogen gives the enzymes many ends on which to work, so the hydrolysis of glycogen stores can quickly increase the blood glucose concentration when necessary. However, these glycogen stores don't last that long—the supply in the liver and the muscle is exhausted in about 2 hours during strenuous exercise, after which the liver and muscle cells must take in new glucose molecules and "restock" their glycogen supplies.

Some polysaccharides are found attached by covalent bonds to proteins and lipids, yielding compounds called **glycoproteins** and **glycolipids,** respectively. In this form, they help maintain the structural integrity of our cells and play various roles depending on where they are located in the body.

Quick Check

☐ 2. How do monosaccharides, disaccharides, and polysaccharides differ?

☐ 3. How are two monosaccharides linked to form a disaccharide? How is a disaccharide split into two monosaccharides?

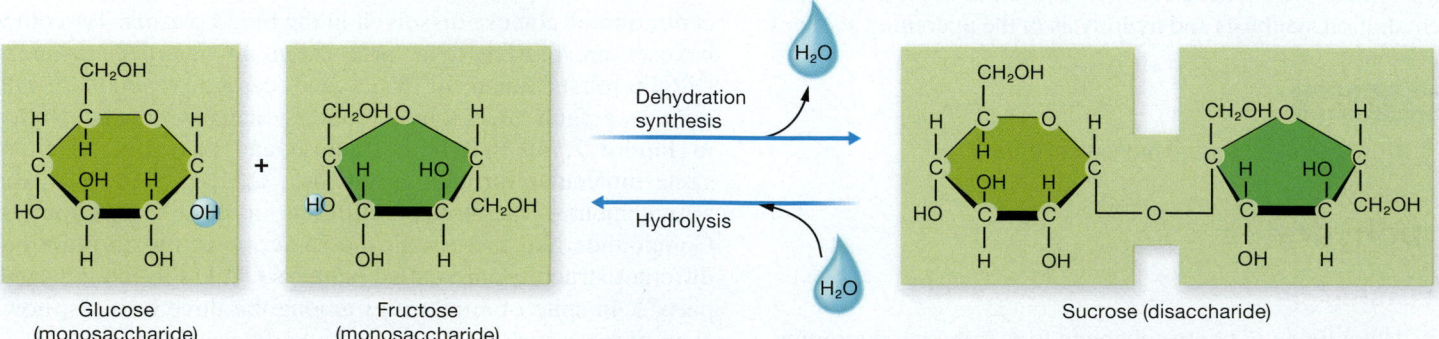

Glucose (monosaccharide)   Fructose (monosaccharide)   Dehydration synthesis   Hydrolysis   Sucrose (disaccharide)

Figure 2.15 Carbohydrates: formation and breakdown of disaccharides.

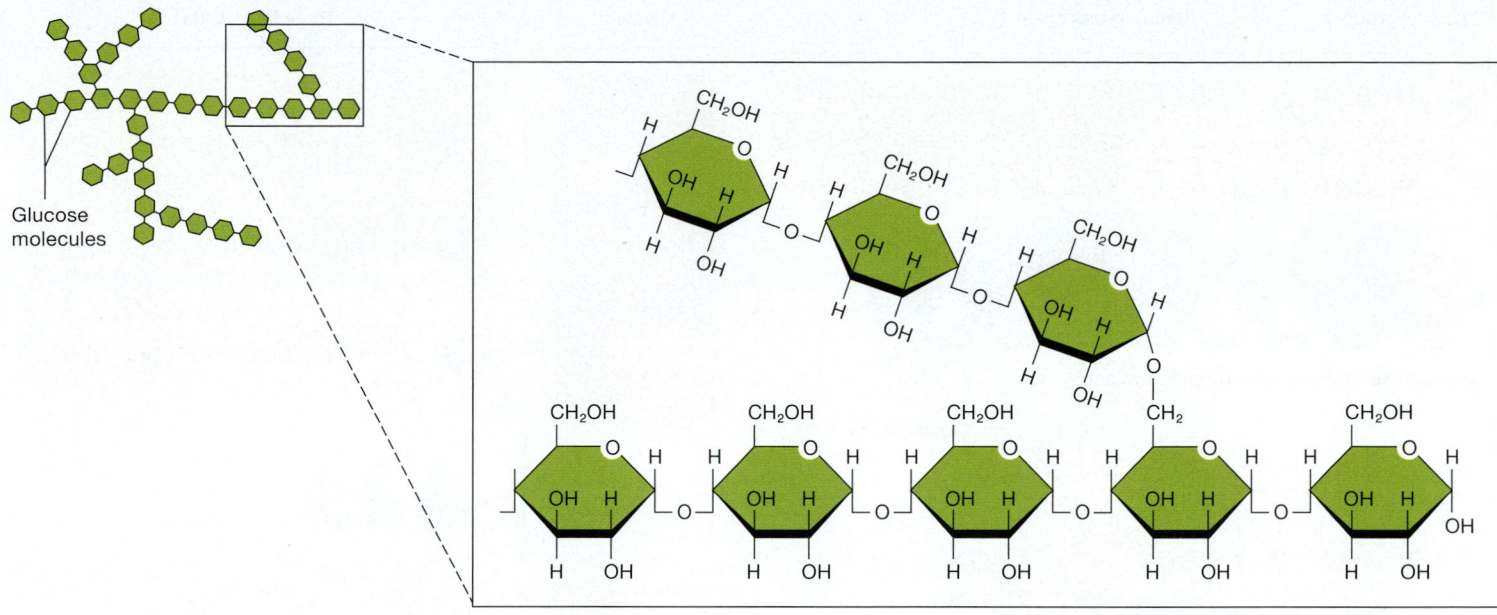

Figure 2.16 Carbohydrates: the polysaccharide glycogen.

Lipids

FLASHBACK

1. What types of chemical bonds form between carbon and hydrogen atoms? (p. 39)

2. Are compounds in which these bonds predominate hydrophilic or hydrophobic? Why? (p. 47)

Like carbohydrates, **lipids** (LIP-idz), which include fats and oils, contain only carbon, hydrogen, and oxygen atoms. However, when you compare the molecular formulas of glucose and a typical lipid, you can see how different they are. Whereas glucose ($C_6H_{12}O_6$) has an equal number of carbon and oxygen atoms and many hydroxyl groups (−OH), the lipid has the molecular formula $C_{15}H_{31}COOH$—carbon and hydrogen are its main components, and oxygen is very much in the minority. The predominance of carbon and hydrogen atoms in the lipid makes it overall very nonpolar and therefore hydrophobic. However, lipids will dissolve in and can act as a solvent for other nonpolar compounds, which are often said to be *lipid-soluble*. Lipids take on many forms in the body, which we examine in the following subsections.

Fatty Acids and Triglycerides

A basic lipid monomer is a compound called a **fatty acid** (**Figure 2.17**). A fatty acid consists of a *carboxylic acid group* (−COOH, another functional group; kar-bawk-SIL-ik) bonded to a hydrocarbon chain that can have 4–20 or more carbon atoms. The hydrocarbon chain may have one or more double bonds between its carbon atoms, which provides a convenient way to classify fatty acids. The three classes are as follows:

- **Saturated fatty acids.** A **saturated fatty acid** has no double bonds between carbon atoms in its hydrocarbon chain, and so its carbon atoms are "saturated" with hydrogen atoms. Saturated fatty acids are found predominantly in animal fats,

and they are mostly solid at room temperature. The fatty acid shown in Figure 2.17a, palmitic acid, is saturated.

- **Monounsaturated fatty acids.** A **monounsaturated fatty acid** has one double bond between two carbons in its hydrocarbon chain, which causes the chain to bend, as you can see in the three-dimensional model in Figure 2.17b. This means that these hydrocarbons can't pack together as tightly. These fats have a lower melting point than saturated fats, so they are generally liquid at room temperature. Plant oils such as olive, grapeseed, and peanut oils have high concentrations of monounsaturated fats. The example shown in the figure is oleic acid (oh-LAY-ik), the predominant fatty acid in olive oil.

- **Polyunsaturated fatty acids.** A **polyunsaturated fatty acid** has two or more double bonds between its carbon atoms. The melting points of polyunsaturated fatty acids are lower than those of both saturated and monounsaturated fatty acids, and they are liquid at room temperature. The example in Figure 2.17c is linoleic acid (lin-oh-LAY-ik).

Our bodies cannot package fatty acids into long chains as they can monosaccharides. Instead, fatty acids are stored by linking three of them via dehydration synthesis to a modified three-carbon sugar called **glycerol** (GLISS-er-ahl). These reactions produce a polymer called a **neutral fat**, or **triglyceride** (try-GLISS-er-ay'd; **Figure 2.18**).

Although all triglycerides share the same basic structure, their properties vary because of the different fatty acids that may be attached to the glycerol. It is partly this variation that determines if a triglyceride is solid (a fat) or liquid (an oil) at room temperature. As you can see in the linoleic and oleic acid structures shown in Figure 2.17, the long hydrocarbon chains of unsaturated fatty acids are kinked. This prevents them from being packed closely together and solidifying. For this reason, a triglyceride that contains one or more unsaturated fatty acids is

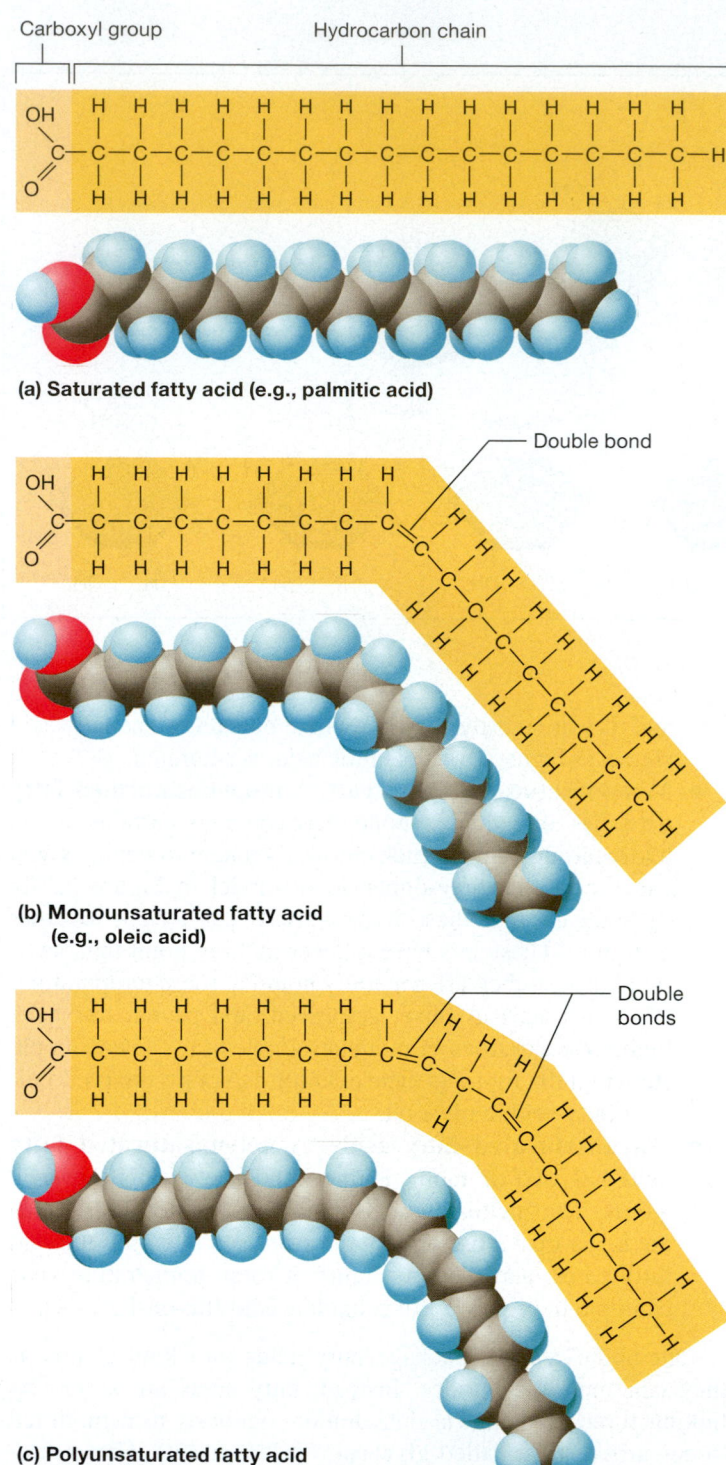

(a) Saturated fatty acid (e.g., palmitic acid)

(b) Monounsaturated fatty acid
(e.g., oleic acid)

(c) Polyunsaturated fatty acid
(e.g., linoleic acid)

Figure 2.17 Lipids: structure of fatty acids.

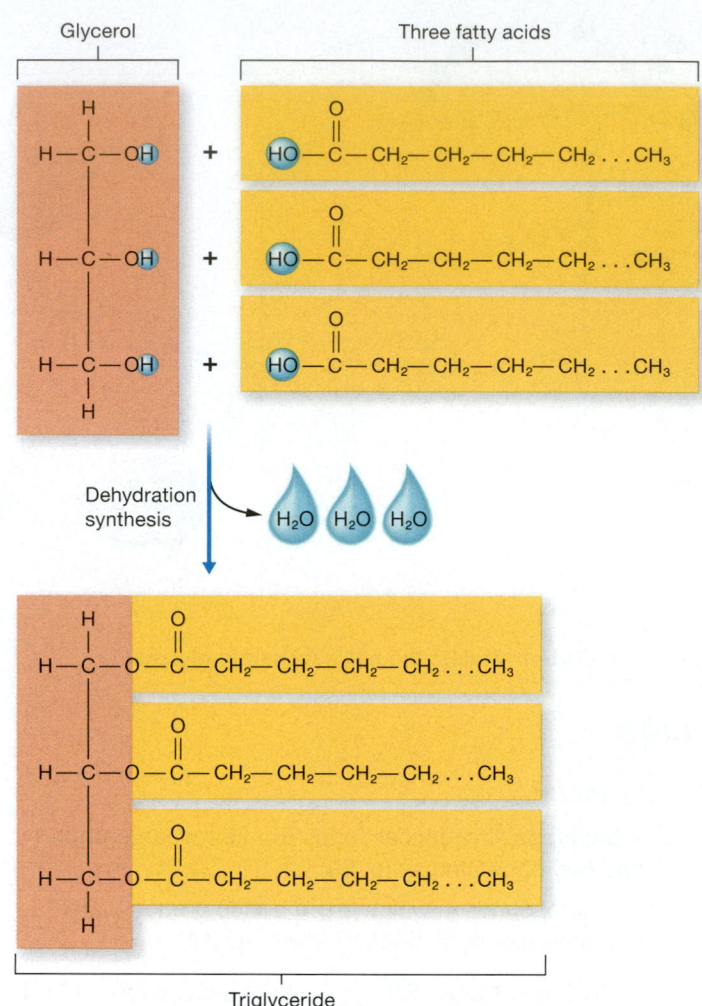

Figure 2.18 Lipids: structure and formation of triglycerides.

likely to be liquid. The long, straight chains of the saturated fatty acids allow them to be packaged together tightly, and triglycerides with three saturated fatty acids are solid. See *A&P in the Real World: The Good, the Bad, and the Ugly of Fatty Acids* for information on how these properties could affect your health.

Triglycerides are stored in fat cells, where they help to insulate our internal organs against physical shock and regulate body temperature. In addition, triglycerides are a ready supply of fuel. When they are released to be catabolized, their fatty acids are removed from the glycerol backbone. These fatty acids are then broken apart into smaller two- and three-carbon subunits that are oxidized, and the chemical energy in their bonds is used to synthesize ATP. Each triglyceride can yield multiple smaller subunits due to its long fatty acid chains, which makes triglycerides a very efficient source of energy.

Phospholipids

Take a look at the **phospholipid** (foss-foh-LIP-id) in **Figure 2.19**. Part of its structure looks familiar, with a glycerol backbone and two fatty acids as in the triglycerides. However, in place of a third fatty acid, we find a structure called a *phosphate group* (FOSS-fayt; another functional group). You already know that the two fatty acids are nonpolar from the previous section. But what about this phosphate group? With its four oxygen atoms and one phosphorus atom, it is clearly polar. In addition, as you can see in Figure 2.19, most phosphate groups are attached to nitrogen-containing groups that are also polar. This leaves us with a two-sided structure: One side, the *phosphate head,* is very polar and hydrophilic, and the other, the *fatty acid tails,* is very nonpolar and hydrophobic.

The Good, the Bad, and the Ugly of Fatty Acids

When it comes to the world of lipids, you can't say fatty acids are all good or all bad. Some fatty acids are good for you, others are bad for you in excess, and still others are downright ugly:

- *The Good: Omega-3 Fats.* Omega-3 fatty acids, found in items such as flaxseed oil and fish oil, are a family of polyunsaturated fatty acids that the human body cannot manufacture, so they must be supplied in the diet. Research has indicated that omega-3 fatty acids have positive effects on cardiovascular health.

- *The Bad: Saturated Fats.* Saturated fatty acids are found in animal fats as well as certain oils such as palm and coconut. Hundreds of scientific studies have demonstrated a link between overconsumption of saturated fats and increased risk for certain diseases, especially heart disease.

- *The Ugly: Trans Fats.* Trans fatty acids are produced by adding hydrogen atoms to certain unsaturated plant oils (producing "partially hydrogenated oils"). There is no dietary requirement for trans fats, and, in fact, there is no safe level of trans fat consumption period, as they significantly increase the risk for heart disease.

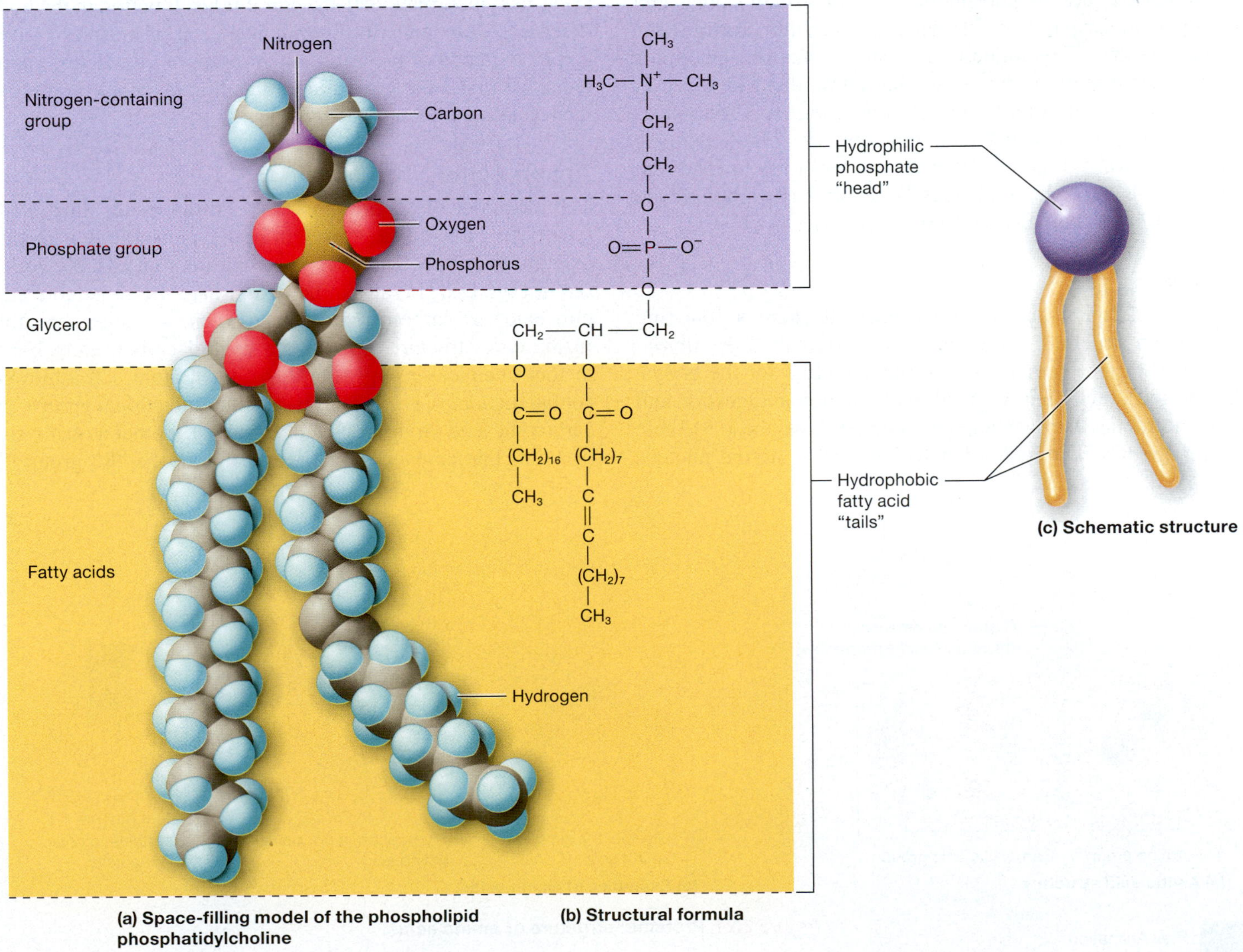

Nitrogen-containing group

Nitrogen

Carbon

Phosphate group

Oxygen

Phosphorus

Glycerol

Fatty acids

Hydrogen

CH_3

$H_3C - N^+ - CH_3$

CH_2

CH_2

O

$O = P - O^-$

O

$CH_2 - CH - CH_2$

O   O

$C = O$ $C = O$

$(CH_2)_{16}$ $(CH_2)_7$

CH_3 C

C

$(CH_2)_7$

CH_3

Hydrophilic phosphate "head"

Hydrophobic fatty acid "tails"

(c) Schematic structure

(a) Space-filling model of the phospholipid phosphatidylcholine

(b) Structural formula

Figure 2.19 Lipids: structure of phospholipids.

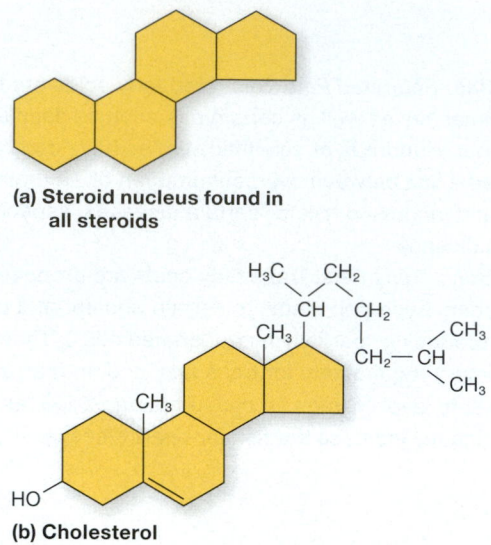

(a) Steroid nucleus found in all steroids

(b) Cholesterol

Figure 2.20 **Lipids: structure of steroids.**

Compounds such as phospholipids, with strongly polar and strongly nonpolar parts, are known as *amphiphilic* (am-fih-FIL-ik; "likes both"). This amphiphilic nature gives phospholipids a number of important properties that allow them to serve as the main structural components of *cell membranes,* the barriers between our cells and their environment. The schematic structure shown in Figure 2.19 is the one we'll use to represent phospholipids throughout this text. We explore phospholipids in more detail in the cell chapter (see Chapter 3).

Steroids

The class of lipids known as **steroids** all share a four-ring hydrocarbon structure, shown in **Figure 2.20**, called the *steroid nucleus.* The steroid *cholesterol* forms the basis for the body's other steroids, including bile acids that function in digestion and the sex hormones estrogen and testosterone (look ahead to Table 2.3 on page 62). The hydrocarbon rings of the steroid nucleus

are nonpolar, and most substances derived from it are nonpolar as well.

Quick Check

☐ 4. Are lipids generally polar or nonpolar?
☐ 5. How do phospholipids and triglycerides differ?
☐ 6. What are steroids?

Proteins

You may not be aware of it, but every time you look at your reflection in the mirror, you are seeing mostly proteins. Proteins make up your hair, nails, and the outer part of your skin; they color your skin, eyes, and hair; they compose the whites of your eyes; they give your bones their structure; and so on. The human body produces from 50,000 to 100,000 different types of proteins, and they constitute about 20% of the body's mass. **Proteins** are macromolecules with numerous functions: They act as enzymes, play vital structural roles, function in the body's defenses, allow cells to communicate, make muscles contract, can be oxidized as fuel, and much more. In this section on proteins, we first look at the monomers that make up proteins and then examine protein structure.

Amino Acids

The monomer of all proteins is the **amino acid,** illustrated in **Figure 2.21**. Each type of protein in the human body is made up of a unique series of the 21 different amino acids. As you can see in Figure 2.21a, all amino acids share the same core structure made of carbon, hydrogen, oxygen, and nitrogen atoms. In the core structure, each has a central carbon atom bonded to four chemical groups: (1) a hydrogen atom, (2) a nitrogen-containing *amino group* ($-NH_2$; another functional group), (3) a carboxylic acid group ($-COOH$), and (4) another atom or group of atoms that is given the generic name of the "R" group ("R"

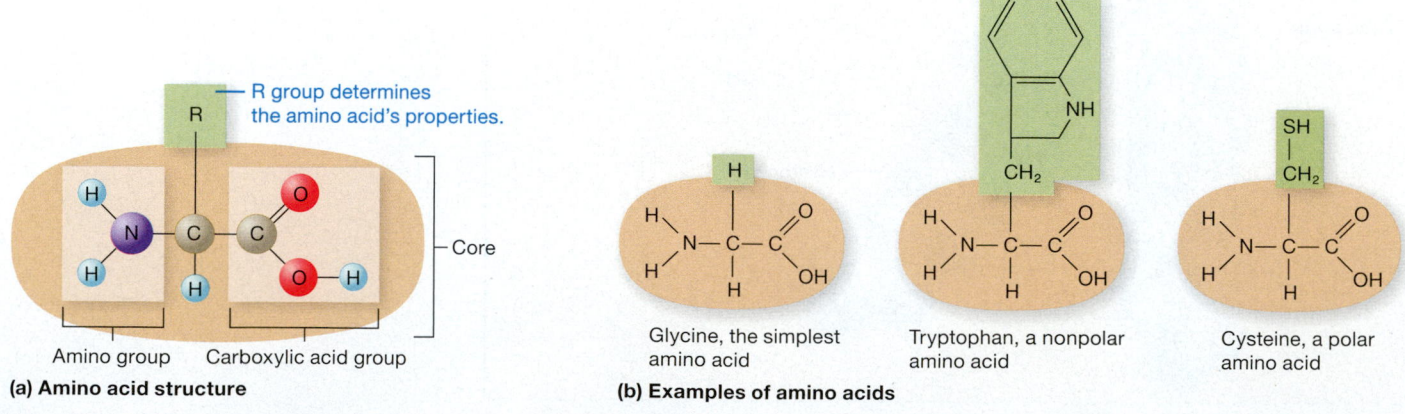

R group determines the amino acid's properties.

Core

Amino group   Carboxylic acid group

(a) Amino acid structure

Glycine, the simplest amino acid

Tryptophan, a nonpolar amino acid

Cysteine, a polar amino acid

(b) Examples of amino acids

Figure 2.21 **Proteins: structure of amino acids.**

▶ Play Animation
@ **Mastering** Anatomy & Physiology

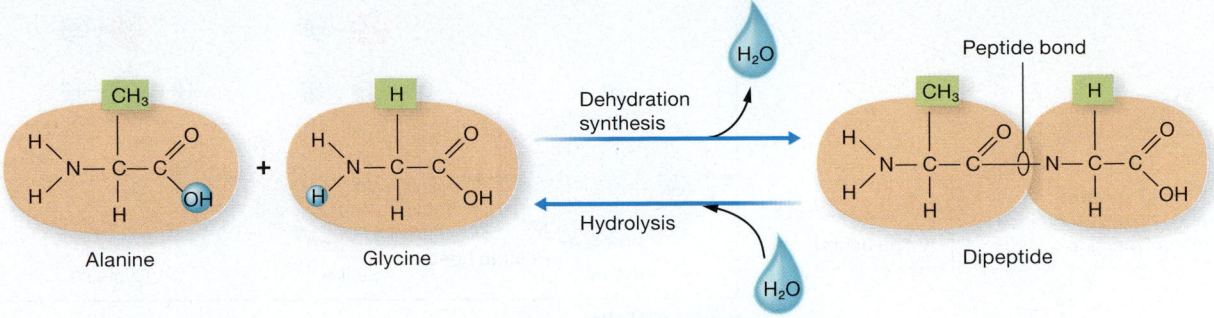

Figure 2.22 Proteins: formation and breakdown of dipeptides.

just stands for "residue"). The name "amino acid" is derived from the *amino* and carboxylic *acid* groups bonded to the central carbon.

Each amino acid is made unique by its R group. Note in Figure 2.21b that the R groups of amino acids vary widely in terms of their complexity and properties, from the hydrogen atom of the simplest amino acid glycine to the large and complicated ring structure of the amino acid tryptophan. R groups also vary in their ability to act as an acid or base and whether they are polar or nonpolar. The behavior of a protein in water is largely determined by the properties of the R groups in its amino acids.

Peptides and Peptide Bonds

Just as we saw earlier with lipids and carbohydrates, amino acid monomers may also be linked by dehydration synthesis to form polymers. Notice in the example in **Figure 2.22** that the amino group of glycine loses a hydrogen atom and the carboxylic acid group (−COOH) of alanine loses a hydroxyl group (−OH), in the process forming a water molecule. The result is the formation of a compound known as a **peptide,** in which two amino acids are joined by a polar covalent bond called a **peptide bond.** In the same way as other organic polymers, peptides may also be degraded by hydrolysis.

Peptides are named for the number of amino acids they contain—*dipeptides* consist of 2, *tripeptides* have 3, and **polypeptides** contain 10 or more amino acids. Proteins consist of one or more polypeptide chains that are folded into a distinct structure. A protein must maintain this structure to be functional.

Protein Structure

Two basic types of proteins may be classified according to their structure: fibrous and globular. The *fibrous proteins* are long protein strands composed of mostly nonpolar amino acids. They are found in structures such as hair, nails, tendons, and bone, and they are tough, durable proteins that give a high degree of strength to a tissue. Fibrous proteins resemble ropes, and indeed they are, in essence, the "ropes" of the body. Just as a rope ties two structures together, fibrous proteins help to connect or "tie" the body's structures together, and allow our tissues to resist stretching and twisting forces. In *globular proteins,* the polypeptide chains assemble into a "globe," or roughly spherical shape.

Most globular proteins are polar and function as enzymes, hormones, and other cell messengers.

The structures of fibrous and globular proteins are often highly complex. Many are even composed of multiple polypeptide chains that interact to form the protein as a whole. Due to this complexity, we divide a protein's structure into four levels (**Figure 2.23**):

1. **Primary structure.** A protein's **primary structure** is simply the amino acid sequence of its polypeptide, which includes all of the amino acids held together by covalent peptide bonds (Figure 2.23a).
2. **Secondary structure.** The first step in organizing a polypeptide into a functional protein is folding one or more segments of the polypeptide over on itself in characteristic ways. These folds are called the polypeptide's **secondary structure.** Two of the most common folding patterns are shown in Figure 2.23b: the **alpha helix,** which resembles a spring, and the **beta-pleated sheet,** which resembles Venetian blinds. Each of these structures is stabilized by hydrogen (H) bonds between the amino acids. Note that a given protein may have several regions of different secondary structures.
3. **Tertiary structure.** To become a functional protein, the polypeptide must fold into its **tertiary structure** (Figure 2.23c), which refers to the final three-dimensional shape that it assumes. The tertiary structure consists of the twists, folds, and coils that form between and around the alpha helices and beta-pleated sheets. This shape is stabilized by hydrogen bonds and ionic interactions that occur between the R groups of the amino acids in the polypeptide chain. Notice that the tertiary structure of the fibrous protein is nearly identical to its secondary structure. This is because most fibrous proteins simply consist of very long alpha helices or beta-pleated sheets.
4. **Quaternary structure.** Many proteins consist of more than one polypeptide chain, and the manner in which these polypeptide chains assemble makes up the **quaternary structure** of the protein (Figure 2.23d). Each polypeptide chain that contributes to the protein's quaternary structure has its own primary, secondary, and tertiary structure.

You learned previously about the Structure-Function Core Principle—the way structures in the body are shaped allows

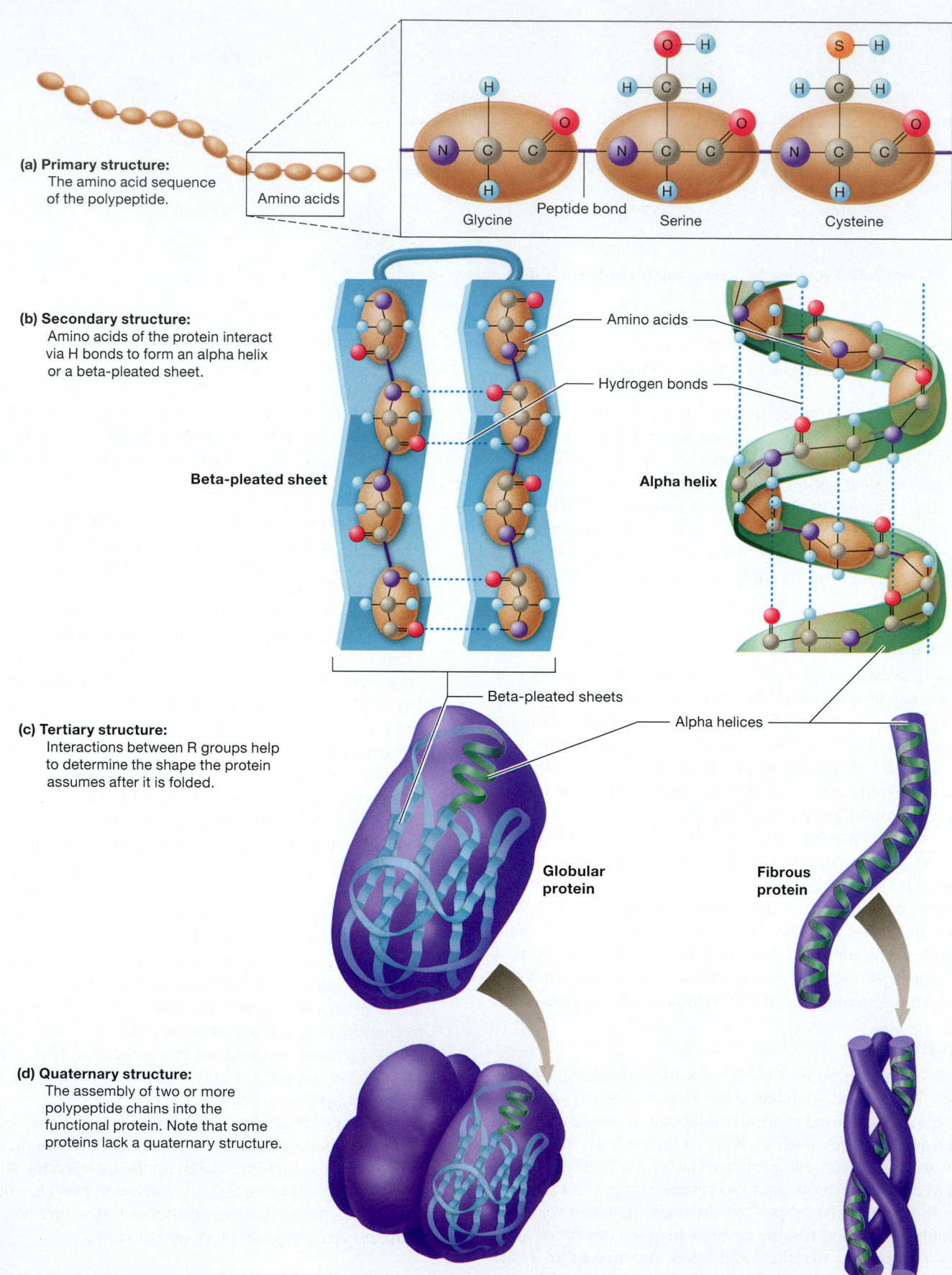

(a) Primary structure:
The amino acid sequence of the polypeptide.

Amino acids

O H — H H — S H
H — C — H   H — C — H
H   H   H   H
N — C — C — O   N — C — C — O   N — C — C — O
H   H   H

Glycine   Peptide bond   Serine   Cysteine

(b) Secondary structure:
Amino acids of the protein interact via H bonds to form an alpha helix or a beta-pleated sheet.

Amino acids

Hydrogen bonds

Beta-pleated sheet

Alpha helix

(c) Tertiary structure:
Interactions between R groups help to determine the shape the protein assumes after it is folded.

Beta-pleated sheets

Alpha helices

Globular protein

Fibrous protein

(d) Quaternary structure:
The assembly of two or more polypeptide chains into the functional protein. Note that some proteins lack a quaternary structure.

Figure 2.23 Levels of protein structure.

Play Animation
@ **Mastering** Anatomy & Physiology

them to best carry out their functions (see page 28). This is true even at the molecular level: A protein's shape determines its

CORE PRINCIPLE
Structure-Function

function. Each polypeptide and protein contains a unique number and sequence of amino acids, and even minor changes to its primary structure can alter all of its subsequent levels of structure and render it nonfunctional. A classic example is found in the disease *sickle cell anemia,* in which the substitution of a single amino acid in the oxygen-carrying protein hemoglobin causes the protein to form chains inside blood cells. The chains of hemoglobin distort the blood cells and cause them to assume the characteristic "sickle" shape, which produces the symptoms of the disease (look ahead to Figure 19.7).

Protein Denaturation

A protein that loses its shape cannot function properly. For example, enzymes are highly specific for their individual substrates. When an enzyme loses its shape, it is no longer able to bind to its substrate and carry out its functions. The process that destroys a protein's shape is called **denaturation.** Many things can denature proteins, including heat, pH changes, and chemicals such as alcohol, as they disrupt hydrogen bonds and ionic interactions within the protein that stabilize its structure. The human body cannot tolerate extensive temperature and pH swings because they denature our proteins.

Quick Check

☐ 7. What four components does each amino acid contain?

☐ 8. How are amino acids linked to form peptides and proteins?

☐ 9. What are the four levels of structural organization of a protein?

Nucleotides and Nucleic Acids

« FLASHBACK

1. What is chemical energy? How is energy stored in chemical bonds? (p. 35 and p. 42)

2. What is an exergonic reaction? (p. 42)

The last class of organic compounds in the body we'll cover is the **nucleotides** (NOO-klee-oh-tydz). The name nucleotide comes from the fact that many of these compounds are found in the central core of the cell, the *nucleus.* Nucleotides are the monomers that form our genetic material, the **nucleic acids** *deoxyribonucleic acid (DNA;* dee-awk-see-RY-boh-noo-klay' -ik) and *ribonucleic acid (RNA).* Unlike the other organic compounds, nucleic acids are not oxidized by the cell for fuel. However, nucleotides do play a key role in the energy systems of the body—the nucleotide adenosine triphosphate (ATP) is our main source of chemical energy that drives nearly all of our cellular processes. In this section, we explore the structure and function of nucleotides in greater depth.

Nucleotide Structure

As you can see in **Figure 2.24a**, a nucleotide is composed of three parts: (1) a nitrogenous base, (2) a five-carbon (pentose) sugar, and (3) a phosphate group. The first part, the nitrogenous **base,** is a nitrogen-containing compound with a hydrocarbon ring structure. (Be careful not to confuse the term "base" here with the bases we discussed earlier in reference to pH.) There are two types of nitrogenous bases: purines and pyrimidines (**Figure 2.24b**). *Purines* (PYOOR-eenz) are double-ringed compounds and include the bases **adenine** and **guanine.** *Pyrimidines* (per-IM-ih-deenz) are single-ringed compounds and include the bases **cytosine, thymine,** and **uracil.** Each nitrogenous base is abbreviated by the first letter of its name, so the five bases are usually written simply as A, G, C, T, and U, respectively.

Attached to the nitrogenous base is the second component of a nucleotide: a pentose sugar, which may be the sugar ribose or deoxyribose (see Figure 2.14a). Covalently bonded to the pentose sugar we find the third component, a group you saw previously in phospholipids: the phosphate group. Nucleotides may contain one, two, or three phosphate groups.

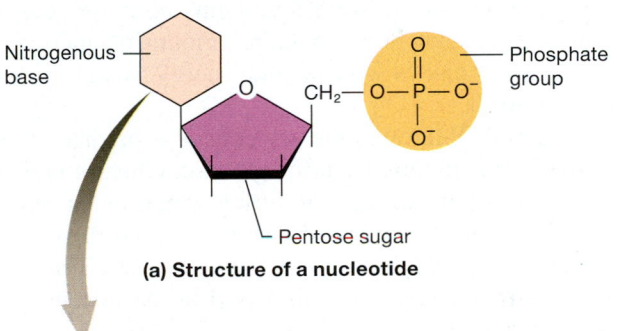

(a) Structure of a nucleotide

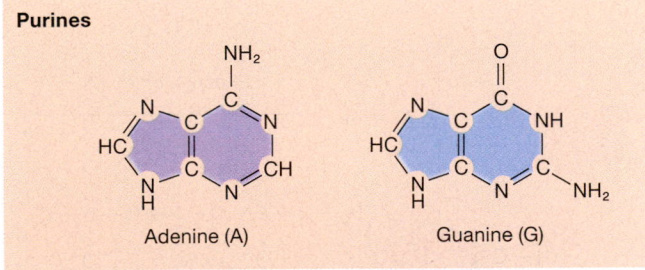

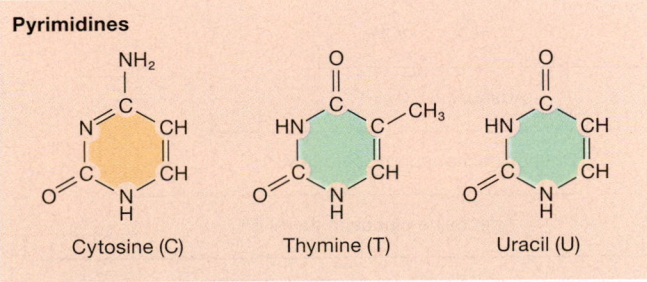

(b) The nitrogenous bases can be either purines or pyrimidines.

Figure 2.24 Structure of nucleotides.

Adenosine Triphosphate (ATP)

As we discussed in Module 2.2, the body's main source of energy is chemical energy, or the energy present in chemical bonds. The primary source of chemical energy in the body comes in the form of a nucleotide called **adenosine triphosphate,** or **ATP** (uh-DEN-uh-seen; **Figure 2.25a**). ATP contains the nitrogenous base adenine attached to ribose—these two together are known as *adenosine*—and three phosphate groups. Adenosine with two phosphates is *adenosine diphosphate (ADP)*, and with only one phosphate is *adenosine monophosphate (AMP)*. Notice that each phosphate group is negatively charged.

ATP is generally synthesized by adding a third phosphate to ADP (**Figure 2.25b**). However, this is a highly endergonic reaction due to the negative charges on the phosphate groups. Just as opposite charges attract, like charges repel each other, so a great deal of energy is required to overcome this repulsive force. When this bond is broken by hydrolysis, resulting in ADP and the inorganic phosphate group (abbreviated P_i), the reaction is highly exergonic because ADP is more stable than ATP. The chemical energy released by the reaction is harnessed by the cell to do work. Note that the bond between the second and third phosphate groups is sometimes referred to as a "high-energy bond," but in truth it contains no more energy than any other molecular bond. A large amount of energy is released when the bond is *broken*, but energy is not stored within the bond itself.

The cell makes ATP through the catabolism of organic compounds, primarily the monosaccharide glucose. Glucose undergoes a series of reactions in the cell during which its chemical bonds are broken, and the chemical energy released drives the cellular processes that add a phosphate group to ADP. The processes by which this happens are discussed in the metabolism chapter (see Chapter 23).

A molecule of ATP lasts maybe 60 seconds in the cell before it is hydrolyzed, as nearly every process in the cell is driven by the energy released from ATP hydrolysis. ATP is not stored in any significant amount by our cells, and our entire supply of ATP is exhausted in only 60–90 seconds. For this reason, a cell must continually replenish its ATP supply or risk failure of essentially all its functions. This becomes critically obvious when a person stops breathing: Oxygen is required for most ATP synthesis, and, in fact, the need for ATP is the main reason we breathe oxygen. When breathing stops, cells can no longer make sufficient ATP, and death occurs within only 3–4 minutes.

The Nucleic Acids: DNA and RNA

The nucleic acids account for only about 2% of the body's mass, but they are some of the most important polymers in the body. The two main nucleic acids are **deoxyribonucleic acid,** or **DNA,** and **ribonucleic acid,** or **RNA.** Together they are responsible for the storage and execution of the genetic code.

Nucleic acids are polymers of linked nucleotides. The nucleotides in a nucleic acid are linked by dehydration synthesis, which joins the phosphate group of one nucleotide to the sugar of the next by a polar covalent bond, as shown here:

Notice in this illustration that the linked sugar and phosphate groups form the nucleic acid's "backbone"; the nitrogenous bases extend out to the side and do not participate in the linkage of the nucleotide strand.

(a) The structure of ATP

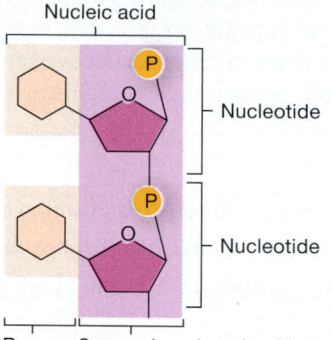

(b) The formation of ATP from ADP requires energy; the reverse reaction liberates energy.

Figure 2.25 Nucleotides: structure and formation of ATP.

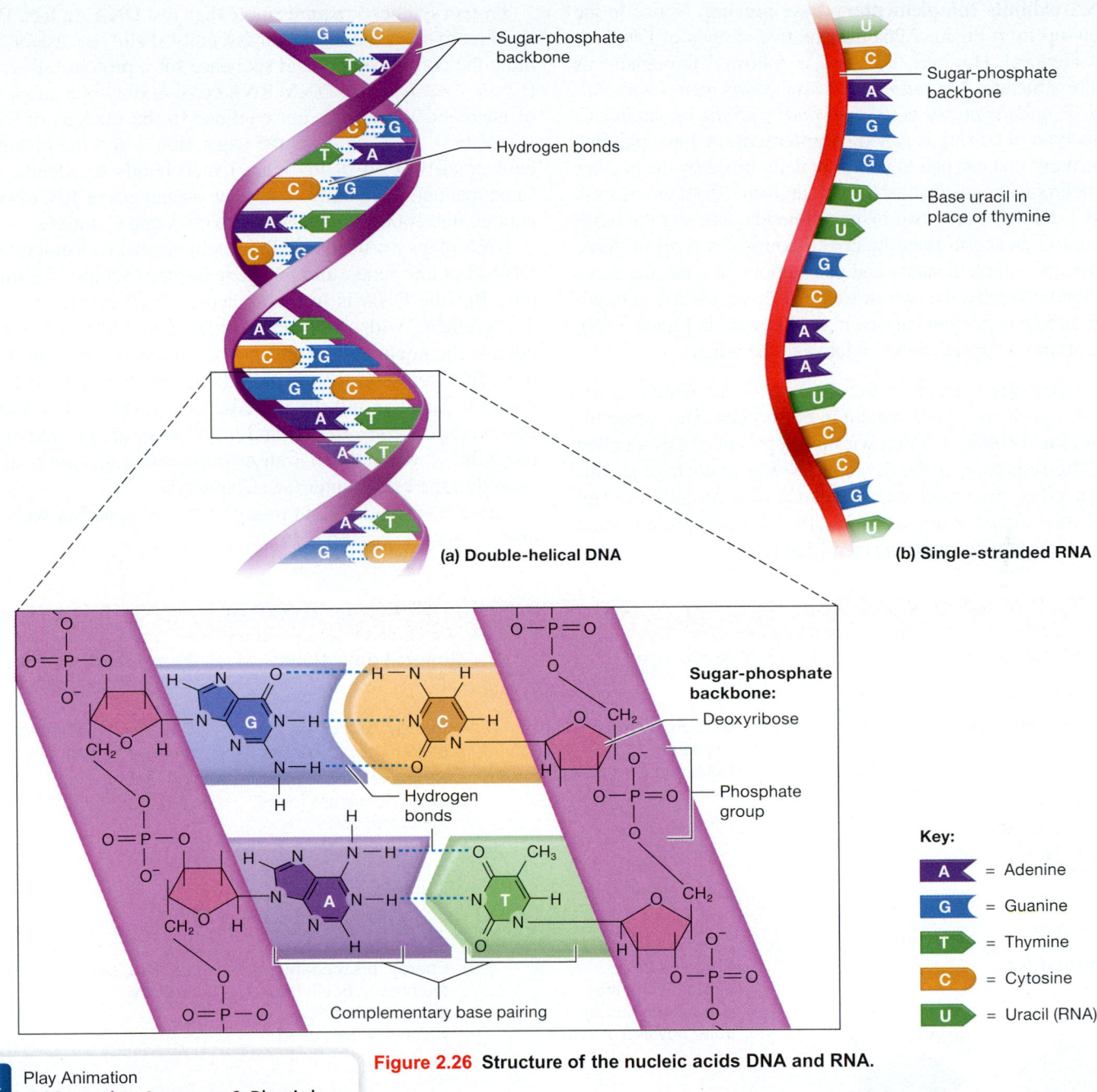

(a) Double-helical DNA

(b) Single-stranded RNA

Sugar-phosphate backbone:
Deoxyribose

Phosphate group

Hydrogen bonds

Complementary base pairing

Key:

A	= Adenine
G	= Guanine
T	= Thymine
C	= Cytosine
U	= Uracil (RNA)

Figure 2.26 Structure of the nucleic acids DNA and RNA.

▶ Play Animation
@ **Mastering** Anatomy & Physiology

Let's examine the role of DNA first. In the previous section, we introduced proteins and their various roles in the body. Recall that proteins are long chains of amino acids joined by covalent peptide bonds, and that each type of protein has a unique amino acid sequence. Each cell produces thousands of different proteins, so how does a cell know which amino acids to join to make a specific protein? The answer lies in the cell's DNA.

DNA is an extremely large polymer found in the nucleus of the cell. As you can see in **Figure 2.26a**, DNA is composed of two long chains or strands of nucleotides that twist around each other to form a **double helix.** Other structural features of DNA include those in the following list.

- **DNA contains the sugar deoxyribose.** Recall from our discussion of monosaccharides that deoxyribose is a five-carbon sugar. It is called *deoxy*ribose because one of its carbon atoms lacks an oxygen-containing hydroxyl group (−OH) and has just a hydrogen atom instead.
- **DNA contains the bases adenine, guanine, cytosine, and thymine.** Only the bases A, G, C, and T are found in DNA.
- **The two strands are held together by hydrogen bonds between the bases.** The nucleotide bases form hydrogen bonds with bases on the other strand. These bonds are the major force that keeps the two strands of DNA together.

- **DNA exhibits complementary base pairing.** Notice in the blow-up from Figure 2.26a that the two strands of DNA are not identical. However, a pattern is followed throughout its entire structure: *The purine A always bonds with the pyrimidine T, and the purine G always bonds with the pyrimidine C.* This type of pairing is called **complementary base pairing.** The bases that can pair together are determined by the number of hydrogen bonds that each base can form. Both the bases A and T can form only two hydrogen bonds, whereas the bases C and G can form three hydrogen bonds. As a result, these bases are complementary, and pair up together. For the bases to bond correctly, the two strands of DNA must run in opposite directions; as you can see in the blow-up in Figure 2.26a, one strand is "upside down" relative to the other.

Within the DNA we find the *genetic code,* which is the "recipe" for each and every protein in our bodies. The recipe for one individual protein is found within a segment of DNA called a *gene.* The sequence of the nucleotides within each gene ultimately specifies the amino acid sequence of a protein; the cell can read this sequence and assemble the encoded protein via a process called *protein synthesis* (see Chapter 3).

Protein synthesis requires more than just DNA. In fact, DNA is not directly involved in protein assembly at all. The task of assembling the correct amino acid sequence for a protein falls to RNA (**Figure 2.26b**). Unlike DNA, RNA consists only of a single strand of nucleotides, and it is not confined to the nucleus of the cell. In addition, RNA contains the sugar ribose and the nitrogenous base uracil instead of thymine. Uracil bonds to adenine in the same manner as thymine, and when a segment of RNA bonds to another nucleotide strand, U pairs with A and C with G.

RNA plays a crucial role in protein synthesis. Remember that DNA contains genes, and one gene has the "recipe" for one protein. But the DNA is in the nucleus, which it can't leave, and the "kitchen," with all of the ingredients to make this protein, is outside the nucleus. This means that to make a protein, the cell must first copy the recipe and take it into the cellular kitchen, where it can then put the ingredients together. These two processes, called *transcription* and *translation,* are carried out with the help of RNA. Both transcription and translation are discussed in the cell chapter (see Chapter 3).

Table 2.3 provides a summary of the information we've covered on organic compounds.

Table 2.3 Organic Compounds

Type of Compound	Structure	General Functions	Examples/Location
Carbohydrates			
Monosaccharides Glucose	Carbon, hydrogen, and oxygen in 1C:2H:1O ratio; can take on a ring form	• Energy: glucose is the main fuel for all cells • Structure: ribose and deoxyribose are found in nucleic acids	Ribose, deoxyribose, glucose, galactose, fructose/ Found in almost all cells
Disaccharides Sucrose	Contain two monosaccharide subunits joined by a polar covalent bond	• Energy: disaccharides are broken down into monosaccharides, which are then used for fuel	Lactose, sucrose, and maltose
Polysaccharides Glycogen	Highly branched polymer with hundreds of monosaccharide subunits	• Energy storage: when needed, stored polysaccharides are broken down into monosaccharides and released to be used as fuel	Glycogen is the main storage polysaccharide/ Stored in liver and skeletal muscle cells

Table 2.3 Organic Compounds (continued)

Type of Compound	Structure	General Functions	Examples/Location
Lipids			
Triglycerides Glycerol Fatty acids	Three fatty acids bonded to a glycerol backbone	• Energy storage: triglycerides are released as needed • Protection and insulation: cushioning and insulation of organs	Multiple combinations of saturated and unsaturated fatty acids are possible/ Stored in adipose (fat) cells
Phospholipids 	Two fatty acids and a phosphate group bonded to a glycerol backbone	• Structure: main component of cell membranes	Phosphatidylcholine/ Found in all membranes
Steroids Testosterone	Four-ringed hydrocarbon with modifications for different compounds	• Regulation: steroid hormones control many physiological processes • Structure: cholesterol forms part of barrier between cells and their environment	Hormones such as testosterone, estrogen, aldosterone, and cortisol; other molecules include cholesterol and bile salts/ Made in glands such as testes and ovaries
Proteins			
Di- and polypeptides Dipeptide	Two or 10–50 amino acids, respectively, joined by a peptide bond or bonds	• Regulation: most function as hormones or chemical messengers in the brain	Substance P, endorphins, glucagon, calcitonin, and insulin

(continued)

Table 2.3 Organic Compounds (continued)

Type of Compound	Structure	General Functions	Examples/Location
Proteins (continued)			
Proteins Globular Fibrous	Long strands of more than 50 amino acids that fold into a characteristic three-dimensional shape	• Structure: primary structural molecule in the body • Movement: involved in cell and muscle movement • Catalysis: function as enzymes • Transport: transport substances through the body • Defense: function in body's immune defenses	Collagen, keratin, amylase, albumin, hemoglobin, and many others/ Found throughout the body
Nucleotides and Nucleic Acids			
ATP	The nitrogenous base adenine bonded to ribose and three phosphate groups	• Energy: main source of chemical energy for the body—drives cellular work	ATP
Nucleic acids DNA RNA	Strings of nucleotides arranged in either a double helix (DNA) or a single strand (RNA)	• Information storage: DNA contains the instructions for building every protein in the body • Information retrieval: RNA carries out the instructions to build proteins	DNA, RNA/ Found in nuclei and cytosol of cells

Quick Check

☐ 10. What are the components and roles of ATP?

☐ 11. How do DNA and RNA differ?

Apply What You Learned

☐ 1. Which compound would be the most soluble in water: glucose, a triglyceride, or a large protein? Explain.

☐ 2. In Module 2.3, you learned that increasing temperature will also increase reaction rates. Why is there a limit on how high the body's temperature can be raised? (*Hint:* What happens to proteins at high temperatures?)

☐ 3. How could a defect in a gene lead to a malfunctioning enzyme?

See answers in Appendix A.

Chapter Summary

MODULE 2.1
Atoms and Elements 31–34

- **Matter** is defined as anything that has mass and takes up space.
- An **atom** is the smallest particle of matter that still retains its properties. It is composed of **subatomic particles: protons, neutrons,** and **electrons.**
- An **element** is a substance that cannot be broken down into simpler substances by chemical means.
- Elements are arranged on the **periodic table** in order of increasing **atomic number.**
- Four major elements are found in the human body: hydrogen, oxygen, carbon, and nitrogen.
- Isotopes are atoms with the same atomic number but different **mass numbers.**

MODULE 2.2
Matter Combined: Mixtures and Chemical Bonds 34–41

- Matter can be combined physically or chemically.
- Physical combinations of matter are called **mixtures.** There are three types of mixtures: **suspensions, colloids,** and **solutions.** A solution contains a **solvent** that dissolves a **solute.** The amount of solute present in a solution is the solution's *concentration.*
- Atoms are united chemically by **chemical bonds** to form **molecules** or **compounds.**
- The electrons in an atom's outermost shell are called the **valence electrons.**

- An **ionic bond** forms when a metal and nonmetal transfer electrons to form **cations** and **anions.** The ionic bond results from the attraction of the positively charged metal cation and the negatively charged nonmetal anion.
- A **covalent bond** forms when two or more atoms share electrons so that each atom obeys the octet rule.
  - **Nonpolar covalent bonds** result from the equal sharing of electrons between two nonmetals.
  - **Polar covalent bonds** result from the unequal sharing of electrons between two nonmetals. The electrons are pulled more strongly by the more electronegative atom, and a **dipole** results.
  - **Hydrogen bonds** form between polar covalent molecules. They result from the weak attractions between the partially positive hydrogen atoms and a fully or partially negative nonmetal.

MODULE 2.3
Chemical Reactions 41–46

- Any time chemical bonds are formed, broken, or rearranged, or electrons are transferred between atoms, a **chemical reaction** has occurred.
- **Energy** is defined as the capacity to do work. There are two types of energy: **potential energy,** which is energy that is stored, and **kinetic energy,** which is energy in motion. Three forms of potential and kinetic energy are found in the human body: **chemical energy, electrical energy,** and **mechanical energy.**
- A reaction that releases energy is an **exergonic reaction;** a reaction that consumes energy is an **endergonic reaction.**
- There are three types of reactions in the human body:
  - **Catabolic** or *decomposition* **reactions** break larger molecules into smaller ones.
  - During an **exchange reaction,** one or more atoms from the reactants are exchanged for another.
  - **Anabolic** or **synthesis reactions** form larger molecules from smaller ones.
- The rate of a reaction is determined by the concentration of the reactants, temperature, size and phase of the reactants, and the presence or absence of a *catalyst.* It is also determined by the **activation energy.**
- **Enzymes** are biological catalysts that increase the speed of a reaction.

MODULE 2.4
Inorganic Compounds: Water, Acids, Bases, and Salts 46–50

- **Water** acts as a solvent for substances with polar covalent and ionic bonds, which are **hydrophilic,** but not for those with nonpolar covalent bonds, which are **hydrophobic.**
- An **acid** is a hydrogen ion donor, and a **base** is a hydrogen ion acceptor.

- The **pH scale** is a logarithmic scale that represents the hydrogen ion concentration of a solution.
  - A pH less than 7 is **acidic;** as the pH decreases, the solution becomes more acidic.
  - A pH of 7 is **neutral.**
  - A pH greater than 7 is **basic** (alkaline); as the pH increases, the solution becomes more basic.
- A **buffer** is a system of chemicals that resist a change in pH.
- A **salt** is a metal cation bonded ionically to a nonmetal anion. Salts are **electrolytes.**

MODULE **2.5**
Organic Compounds: Carbohydrates, Lipids, Proteins, and Nucleotides 50–64

- Each type of organic compound has its own **monomer** and the corresponding **polymer** built from those subunits.
- **Carbohydrates** function primarily as fuels in the body but also play structural roles.
  - **Monosaccharides** are carbohydrate monomers. The most abundant monosaccharides in the body are **glucose,** *fructose, galactose, ribose,* and *deoxyribose.*
  - **Disaccharides** are formed from the union of two monosaccharides by **dehydration synthesis.**
  - **Polysaccharides** consist of many monosaccharides. **Glycogen** is the body's main polysaccharide.
- **Lipids** are nonpolar, hydrophobic molecules composed mostly of carbon and hydrogen.
  - A monomer of fats is the **fatty acid.** Three fatty acids are joined to glycerol form a **triglyceride.**
  - A **phospholipid** contains two fatty acids and a *phosphate group* bonded to glycerol. It is an *amphiphilic* molecule.
  - **Steroids** are a class of lipids based upon a four-ring hydrocarbon *steroid nucleus.*
- **Proteins** function as enzymes, play structural roles, are involved in movement, function in the body's defenses, can be used as fuel, and more.
  - Each protein is composed of a unique sequence of 50 or more **amino acid** monomers joined together into a **polypeptide** by **peptide bonds.**
  - Proteins have four levels of organization to their structure: The **primary, secondary, tertiary,** and possibly **quaternary structures.**
  - Heat, pH changes, and certain chemicals cause a protein to **denature**—permanently lose its shape and therefore its function.
- A **nucleotide** is a monomer composed of a nitrogenous base, a sugar, and a phosphate group.
- There are two types of nitrogenous bases: the *purines,* **adenine** (A) and **guanine** (G), and the *pyrimidines,* **cytosine** (C), **thymine** (T), and **uracil** (U).
- **Adenosine triphosphate (ATP)** is the body's chief source of chemical energy. ATP is synthesized from *adenosine diphosphate (ADP)* and a phosphate group.

- The **nucleic acids** DNA and RNA are polymers of nucleotides linked by dehydration synthesis.
- **Deoxyribonucleic acid,** or **DNA,** is confined to the nucleus of the cell. DNA contains deoxyribose and C, T, A, and G. It has two strands of nucleotides in a **double helix** joined by hydrogen bonds that exhibit **complementary base pairing:** C always bonds with G, and T always bonds with A. DNA contains *genes* that provide the "recipe" or code for every protein in the body.
- **Ribonucleic acid,** or **RNA,** is a single strand of nucleotides that can move between the cell's nucleus and the cytosol. RNA contains ribose and A, C, G, and U. It is a copy of the protein "recipe" from DNA and helps to assemble amino acids into a protein outside the nucleus.

Assess What You Learned

Scan the QR Code for additional practice test questions

https://goo.gl/m0gyfY

LEVEL 1 Check Your Recall

1. Mark the following statements as true or false. If a statement is false, correct it to make a true statement.
   a. The mass number of an atom is the sum of its neutrons and protons.
   b. Protons and neutrons have a positive charge and electrons have a negative charge.
   c. Valence electrons are the electrons in the outermost shell of an atom.
   d. Every element has a characteristic number of protons, which is called the element's mass number.

2. Fill in the blanks: Isotopes are atoms with the same _____ number but different _____ numbers.

3. Which of the following statements correctly describes a solution?
   a. In a solution the solute is chemically dissolved by the solvent.
   b. Solutions involve large particles suspended in another component.
   c. The particles in a solution will settle out if left to sit.
   d. The amount of solute in a solution is expressed as the solution's concentration.

4. Explain the difference between an ionic and a covalent bond.

5. Identify each of the following molecules or compounds as ionic, polar covalent, or nonpolar covalent.
   a. H_2O _____   d. N_2 _____
   b. LiI _____   e. $MgBr_2$ _____
   c. $C_{10}H_{22}$ _____   f. H_2S _____

6. What are hydrogen bonds, and how do they form?

7. Describe what takes place during a chemical reaction.

8. Match the following terms with the correct definition:

_____ Endergonic reaction
_____ Potential energy
_____ Electrical energy
_____ Anabolic reaction
_____ Oxidation-reduction reaction
_____ Chemical energy
_____ Catabolic reaction
_____ Kinetic energy

a. Energy in motion
b. Energy stored in chemical bonds
c. Reaction that consumes energy
d. A decomposition reaction
e. The energy of moving charged particles
f. Stored energy
g. Synthesis reaction
h. Reaction where electrons are transferred between reactants

9. Which of the following would *not* result in an increase in reaction rate?

a. Increase in temperature
b. Particles in the solid phase
c. Presence of a catalyst
d. Smaller particle size

10. An enzyme is a:

a. biological catalyst that increases the concentration of the products.
b. biological catalyst that increases the concentration of the reactants.
c. chemical that is used in a reaction, after which it can no longer be used by the cell.
d. biological catalyst that works by bringing its substrates closer to their transition states.

11. Which of the following is *not* a property of water?

a. Water has a high heat capacity.
b. Water doesn't take heat with it when it evaporates.
c. Water is a polar solvent in which many solutes will dissolve.
d. Water serves as a cushion and lubricant in the body.

12. With respect to their solubility in water, substances with nonpolar covalent bonds are _____, while substances with polar covalent and ionic bonds are _____.

13. Mark the following statements as true or false. If a statement is false, correct it to make a true statement.

a. Acids dissociate in water to give hydrogen ions and an anion.
b. As a solution's pH rises, it becomes more acidic.
c. A buffer is a system of chemicals that resists a change in pH.
d. A solution with a pH of 10 has more hydrogen ions than a solution with a pH of 2.

14. A salt is:

a. a metal cation bonded to a nonmetal anion.
b. a hydrogen ion donor.
c. an electrolyte that conducts electricity in water.
d. Both a and c are correct.
e. Both b and c are correct.

15. Compare and contrast the structures of phospholipids and triglycerides.

16. Mark the following statements as true or false. If a statement is false, correct it to make a true statement.

a. Carbohydrates are the main source of fuel for the human body.
b. The main storage form of glucose in the human body is starch.
c. Lipids contain oxygen and are therefore polar covalent compounds.
d. Proteins are composed of strings of fatty acids linked by glycosidic bonds.

17. Why is it important for a protein to maintain its structure? What is it called when a protein loses its structure?

18. Which of the following is *not* part of a nucleotide?

a. An amino acid
b. A nitrogenous base
c. One or more phosphate groups
d. A sugar

19. Mark the following properties as belonging to either DNA or RNA.

a. Contains the nucleotides A, G, C, and U _____
b. Found only in the nucleus of the cell _____
c. Can move in and out of the cell's nucleus _____
d. Contains two strands of nucleotides linked by hydrogen bonds _____
e. Contains the code for every protein in the body _____
f. Contains the sugar ribose_____

20. Which of the following statements is/are true regarding ATP?

a. It is the body's primary source of chemical energy.
b. Its production is the main reason humans breathe oxygen.
c. It is produced using the energy from the oxidation of molecular fuels like glucose.
d. All of the above statements are true.

LEVEL 2 Check Your Understanding

1. In certain types of radioactive decay, the isotope releases a particle called an alpha particle, which contains two protons and two neutrons. When this happens, is the product still the same element? Why or why not?

2. Considering that water is a main component of the juices in the stomach and intestines, explain why digestion of lipids is more complicated than digestion of carbohydrates and proteins.

3. Explain why monosaccharides are polar and fatty acids are nonpolar even though they both contain the same atoms.

LEVEL 3 Apply Your Knowledge

PART A: Application and Analysis

1. The polysaccharide cellulose is not digestible by humans, as we lack the enzyme *cellulase*, which is required to break it down. Certain dietary supplements contain the enzyme cellulase and claim that being able to break down cellulose will help a person lose weight. But what do you think would happen if we could digest the cellulose we ate?

2. Some claim that the pH of your blood can be affected by eating acidic foods such as citrus. Do you believe this to be true? Explain your answer. (*Hint:* What happens when extra hydrogen ions are added to the blood?)

3. Many drugs and poisons exert their effects by blocking one or more enzymes. How could blocking an enzyme lead to the death of a cell?

PART B: Make the Connection

4. Explain how buffer systems in the body work if the pH of body fluids increases. Is this an example of a negative or a positive feedback loop? Explain. (*Connects to Chapter 1*)

See answers in Appendix A.

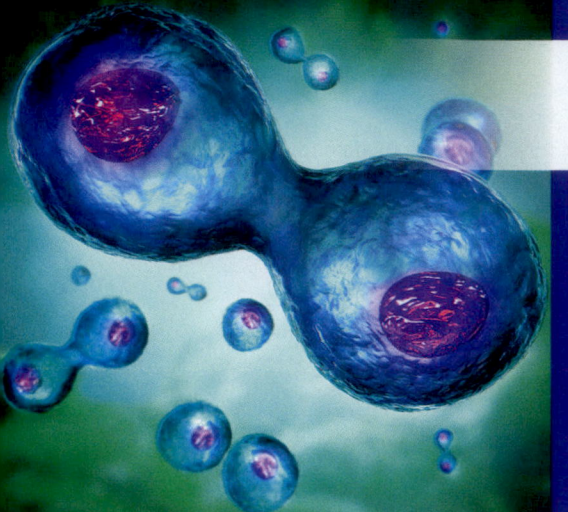

3

The Cell

Cells are the basic units of all living organisms (see Chapter 1). Humans are composed of about 10 trillion cells that work together to maintain homeostasis for the whole organism. In this chapter, we take a tour of a typical human cell, exploring the structure and functions of its various parts, the ways in which it interacts with its environment, and how it reproduces itself.

MODULE **3.1**

Introduction to Cells

Learning Outcomes

1. Identify the three main parts of a cell, and list the general functions of each.
2. Describe the location and components of intracellular and extracellular fluid.
3. Explain how cytoplasm and cytosol are different.
4. Define the term organelle, and describe the basic functions of organelles.

As you learned in Chapter 1, the human body is composed of groups of cells that are specialized to perform specific functions. Yet in spite of their specialization, nearly all cells perform certain basic processes and share the common cellular components that perform them. In this module, we examine these shared cellular structures and functions. We conclude the module with a look at the diverse cell forms found in the human body.

*For practice applying concepts to a clinical scenario, check out the **Running Case Study** for this chapter @Mastering Anatomy & Physiology*

Computer-generated image: The series of events from cell formation to cell division, shown here, is called the cell cycle.

Basic Processes of Cells

Several basic processes are common to all cell types. These processes include cell metabolism, transport of substances, communication, and cell reproduction.

- **Cell metabolism.** The chemical reactions that a cell carries out to maintain life are known collectively as **cell metabolism.** Metabolic reactions include all the different types of reactions that we discussed in the chemistry chapter (see Chapter 2), encompassing:
  - *Anabolic reactions.* Anabolic, or building, reactions produce macromolecules. These reactions include polysaccharide, lipid, protein, and nucleic acid synthesis.
  - *Catabolic reactions.* Cells use catabolic reactions to break down macromolecules into smaller compounds. Some catabolic reactions are required to degrade old and worn-out cellular components, whereas others break down nutrients such as carbohydrates and lipids.

○ *Oxidation-reduction reactions.* An important group of the cell's oxidation-reduction reactions convert the energy in the chemical bonds of nutrients such as glucose into energy the cell can use to fuel the synthesis of *adenosine triphosphate*, or *ATP.*

- **Transport of substances.** A cell must transport compounds it has produced or ingested to a variety of destinations. Some compounds are transported for release from the cell, whereas others are simply transported to a different location within the cell.

- **Communication.** A cell communicates with itself, with its surrounding environment, and with other cells in the body. This communication occurs through a variety of means, including chemical and electrical signals.

- **Cell reproduction.** Many cell types in the body can reproduce themselves through a process called *cell division.* Cell division is necessary for growth and development, as well as for replacement of old and damaged cells.

Quick Check

☐ 1. What general processes are carried out by cells?

Overview of Cell Structure

Most animal cells have three basic components: the plasma membrane, cytoplasm, and nucleus (**Figure 3.1**). We introduce the basic structure and functions of these three components here, and we examine each in greater depth in later modules.

Plasma Membrane

Surrounding each cell is a "fence" called the **plasma membrane.** This membrane serves several important functions, including

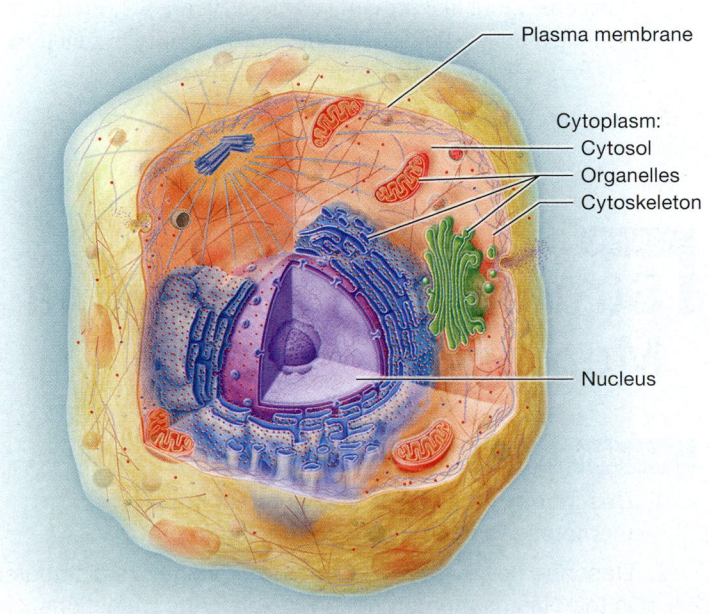

Figure 3.1 **The basic components of a generalized cell.**

structural support, communication with other cells, regulation of transport into and out of the cell, and cell identification. And, of course, a primary role of the plasma membrane is the same as that of any fence—to physically isolate the cell from its surroundings.

But what is the plasma membrane separating the cell from? The body contains two fluid compartments: (1) the *intracellular space,* which is the space within our cells that contains the *cytosol* (SY-toh-sawl), or *intracellular fluid* (*intra* = "within"), and (2) the *extracellular space,* which is the space outside our cells that contains fluid called **extracellular fluid** (*extra* = "outside"), or simply **ECF.** The plasma membrane forms a barrier between the ECF and the cytosol, and keeps the two fluid compartments separate. This allows the cell to control the composition of the cytosol.

Cytoplasm

Inside the plasma membrane is the **cytoplasm.** This consists of both the fluid cytosol and the structures embedded within it—the *organelles* and the *cytoskeleton.*

Cytosol The **cytosol, or intracellular fluid (ICF),** is the fluid portion of the cytoplasm that makes up about half of the cell's total volume. It is a watery gel with many proteins, different forms of RNA, and dissolved solutes.

The cytosol is the site of many critical cellular processes such as protein synthesis. It is also where the series of enzyme-catalyzed reactions known collectively as *glycolytic catabolism,* or *glycolysis* (gly-KAWL-uh-sis; *glyco* = "sugar," *lysis* = "splitting"), take place. As implied by their names, these reactions split a molecule of glucose and in the process generate ATP, the main source of chemical energy for the cell. We will revisit glycolysis in later chapters (see Chapters 10 and 23). The cytosol also functions in storage. In certain cell types, we find clusters of storage bodies called **inclusions** in the cytosol. These inclusions store various materials, such as nutrients from the ECF and proteins made by the cell. For example, cells in the liver and skeletal muscle have inclusions that contain glycogen, the storage form of glucose.

Organelles Suspended in the cytosol are the **organelles** (ohr-gan-ELZ), which are like molecular machines that perform specific functions within the cell. Organelles are vital to a cell because they compartmentalize the cell's functions. Compartmentalization is essential for efficiency—it enables the cell to keep related enzymes and proteins together in one place so that they are not scattered throughout the cell. Compartmentalization is also critical for survival because many organelles contain enzymes and chemicals that would be toxic if they were free in the cytosol. This compartmentalization is achieved for many organelles by enclosure in a membrane. We'll learn more about these organelles later (see Module 3.4).

Cytoskeleton Another component of the cytoplasm is the **cytoskeleton,** which consists of a network of protein filaments. The cytoskeleton supports the cell, helps create and maintain its shape, and holds the organelles in place. The protein filaments also function as molecular "highways" that allow substances and organelles to be transported within the cell. Finally, the cytoskeleton is involved in cellular movement and cell division.

Nucleus

Most cells contain a single roughly spherical structure called the **nucleus** (NOO-klee-us; plural, *nuclei,* NOO-klee-aye). The nucleus is surrounded by a membrane similar to the plasma membrane, known as the *nuclear envelope.* The nucleus contains most of the cell's *deoxyribonucleic acid* (*DNA*) and is the primary location for producing much of the cell's *ribonucleic acid* (*RNA*). The DNA and RNA control cellular functions by coding for and creating proteins.

Quick Check

☐ 2. What three components make up most animal cells, and what are their functions?

☐ 3. Where are intracellular and extracellular fluids located?

Cell Size and Diversity

◀◀ FLASHBACK

1. How are structure and function related? (p. 28)

The cell shown in Figure 3.1 and in later figures in this chapter is a "generic," or *generalized,* animal cell. However, most types of cells do not precisely fit this model. The cells in the human body vary greatly in appearance and size (**Figure 3.2**). The tiny, concave red blood cells that lack nuclei and most organelles measure only 6–8 micrometers (μm; one millionth of a meter), whereas the long, branching nerve cells measure from 1 millimeter (mm) to over a meter. Note that this structural variation is an example of the Structure-Function Core Principle (p. 28). Red blood

CORE PRINCIPLE
Structure-Function

cells carry oxygen through the blood, so they must be small and flexible enough to fit through the smallest blood vessels. Nerve cells, however, contact and interact with many other cells, sometimes in different parts of the body, and so must have long extensions to reach those cells.

The composition of the cytoplasm also varies among cell types, which allows for cells to be *specialized* and so perform functions unique to their cell type. For example, cells of the liver serve critical detoxification functions (the ability to make dangerous chemicals harmless), and so they contain a large number of the organelle that carries out detoxification reactions. Another example is the cells of skeletal muscle, whose cytoplasm is filled with contractile proteins that enable the cells to contract and generate movement. A unique composition of organelles, proteins, and other structures allows each cell type to be specialized. This specialization is vital. Without it, all our organs and tissues would be identical. Cells of the heart would be indistinguishable from those of the bone or kidneys, making these organs unable to carry out their unique functions.

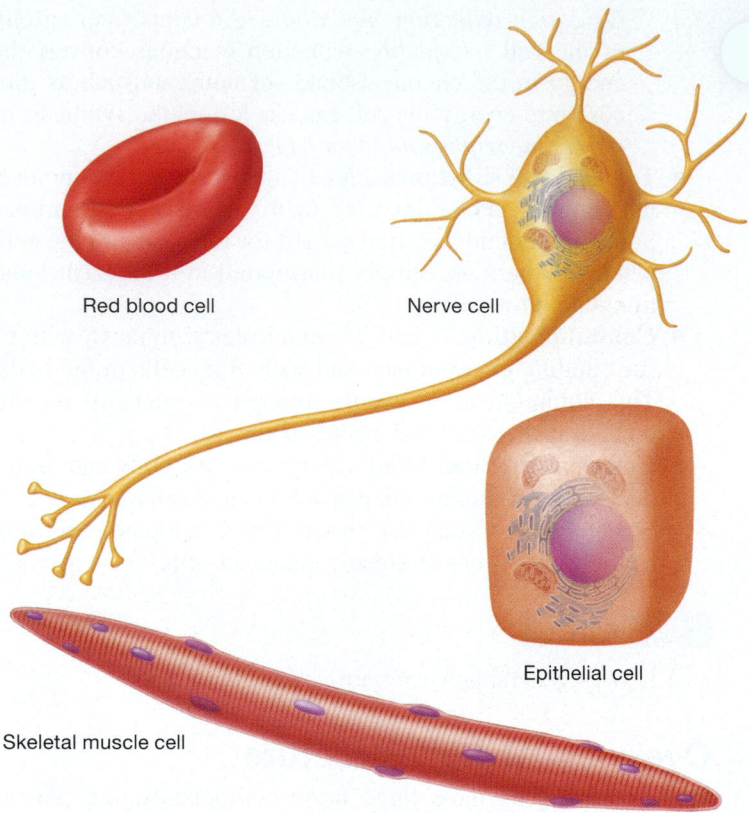

Red blood cell   Nerve cell

Epithelial cell

Skeletal muscle cell

Figure 3.2 Cell diversity. Note that cells are not drawn to same scale.

Apply What You Learned

☐ 1. Bacterial cells lack nuclei and most cytoplasmic organelles. Explain how these cells must differ from our cells in structure and function.

☐ 2. Cancer cells lose their original characteristic features and become more like the cell shown in Figure 3.1. Predict the impact this would have on these cells' ability to function. (*Hint:* Remember that form follows function.)

See answers in Appendix A.

MODULE 3.2
Structure of the Plasma Membrane

Learning Outcomes

1. Describe how lipids are distributed in the plasma membrane, and explain their functions.

2. Describe how carbohydrates and proteins are distributed in the plasma membrane, and explain their functions.

3. Explain the overall structure of the plasma membrane according to the fluid mosaic model.

Let's now begin our exploration of the cell with its outermost component: the plasma membrane. Our main focus in this module is the plasma membrane's structure, but keep in mind that structure and function are tightly correlated. In the next module, we will discuss how the plasma membrane's structure relates to its many functional roles in maintaining homeostasis of the cell and the body as a whole.

The Phospholipid Bilayer

« **FLASHBACK**

1. How do compounds with ionic, polar covalent, and nonpolar covalent bonds differ? (p. 40)
2. How are the terms hydrophilic, hydrophobic, and amphiphilic defined? (p 47 and p. 56)
3. What is a phospholipid? (p. 54)

For the plasma membrane to form an effective barrier between the ECF and cytosol, the chemicals that make up the membrane must have two key properties. First, the chemicals must have parts that can interact with the water in both these fluid compartments. We've discussed that compounds with ionic bonds and compounds with polar covalent bonds are both *hydrophilic,* or able to interact with water (see Chapter 2). Yet ionic compounds wouldn't make good material for our barrier, because they dissociate in water. Obviously, such a barrier would fall apart easily (imagine building your house out of salt—it would dissolve the first time it rained). Compounds with polar covalent bonds, however, are hydrophilic but their individual atoms do not dissociate in water. So, the membrane's hydrophilic regions that interact with water must have polar covalent bonds, whose polar ends can form hydrogen bonds with surrounding water molecules.

Second, the chemicals must also have parts that do not interact with water to keep the water in the ECF and cytosol separated. If the whole plasma membrane were polar and hydrophilic, then water, ions, and many organic compounds could move in and out of our cells freely, and the membrane would not function as a barrier. So, part of the barrier also has to be *hydrophobic*. The major hydrophobic compounds in the body have nonpolar covalent bonds, so the membrane's hydrophobic regions that don't interact with water must contain nonpolar covalent bonds.

As you can see in **Figure 3.3a**, the chemicals that cells use to build the plasma membrane must have both polar and nonpolar parts. Certain chemicals fit this bill precisely: the phospholipids (see Chapter 2). Recall that phospholipids are a special class of lipid that is *amphiphilic*—they have both polar and nonpolar parts. Their phosphate-containing "heads" are relatively large, highly polar regions, and their fatty acid "tails" are large, nonpolar regions.

Notice in **Figure 3.3b** that phospholipids do something interesting when placed on the surface of water: The polar phosphate heads line up so that they face the water, whereas the nonpolar fatty acid tails point away from the water and toward the air (which consists chiefly of nonpolar gases). The result is a single layer, or *monolayer,* of phospholipids. Remember (from Chapter 2) that *like interacts with like*—the polar heads interact with the polar water molecules, and the nonpolar tails interact with the nonpolar gas molecules in air.

However, see what happens when we add another layer of water to the tube in **Figure 3.3c**. The phospholipids rearrange themselves into two layers so that their fatty acid tails still face away from the water molecules, forming a phospholipid "sandwich." The phosphate heads are the "bread" and align on both sides so that they are facing the water molecules. The fatty acid tails are the "cheese" in the middle, and they align so that they face one another and exclude water and polar compounds. This arrangement is known as a **phospholipid bilayer** (*bi* = "two"). Such a bilayer forms the basis of the plasma membrane in every one of our cells. When this bilayer in a cell's membrane

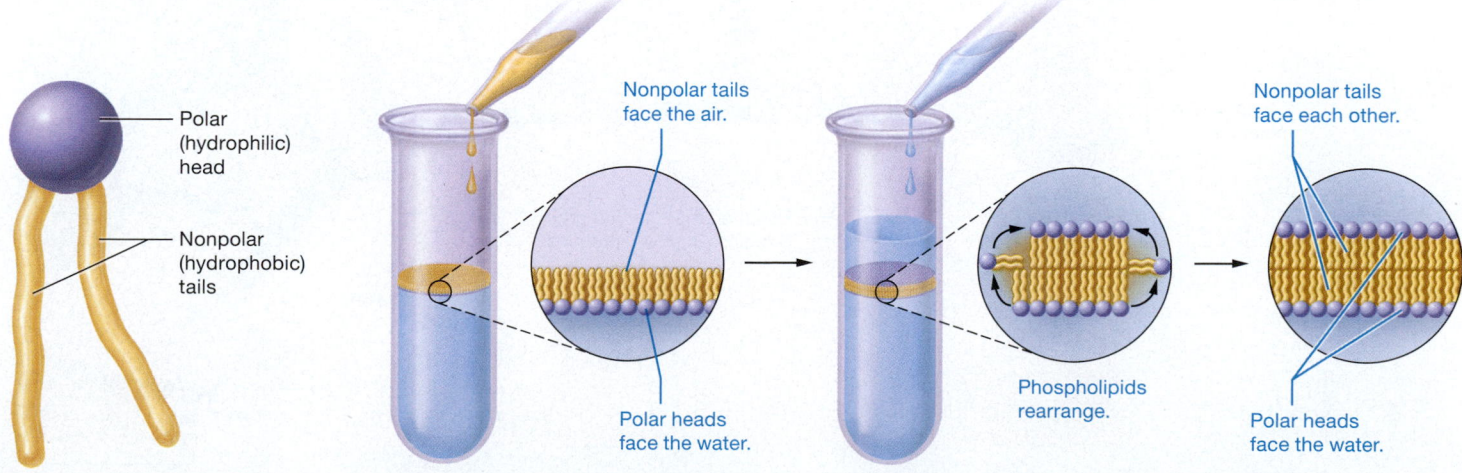

(a) **Schematic structure of a phospholipid molecule**

Polar (hydrophilic) head

Nonpolar (hydrophobic) tails

(b) **A monolayer forms when phospholipids are added to water.**

Nonpolar tails face the air.

Polar heads face the water.

(c) **Phospholipids rearrange into a bilayer when another layer of water is added.**

Nonpolar tails face each other.

Phospholipids rearrange.

Polar heads face the water.

Figure 3.3 **The formation of a phospholipid bilayer.**

is disrupted or torn, the phospholipids behave the same way as those in the tube—they rearrange themselves back into a bilayer, sealing the tear.

Quick Check

☐ 1. How do phospholipids arrange themselves in the plasma membrane? Why?

The Fluid Mosaic Model of the Plasma Membrane

« FLASHBACK

1. What is cholesterol? (p. 56)
2. What are glycoproteins and glycolipids? (p. 52)

To be able to perform all functions required by a living cell, the plasma membrane must contain many components in addition to its phospholipids. Scattered throughout the phospholipid bilayer of the plasma membrane are various proteins, other kinds of lipids, and carbohydrates. All these components together resemble a mosaic, and accordingly, the current model of the plasma membrane is called the **fluid mosaic model** (**Figure 3.4**).

This model defines the plasma membrane as a structure with multiple parts whose arrangement is dynamic, meaning that it changes from moment to moment. Phospholipids in the bilayer shift continually in a lateral, or sideways, direction, and do so extremely rapidly—adjacent phospholipids switch places an average of 10 million times per *second*. Other membrane components also move laterally through this constantly changing "sea" of phospholipids. The ability of phospholipids and other components to move within the plane of the membrane, a property known as *fluidity,* is critical to many plasma membrane functions, such as cell movement, cell reproduction, and transport of substances across the membrane.

Membrane Proteins

Besides phospholipids, the main components of the plasma membrane are **membrane proteins.** These proteins carry out many of the membrane's functions and give different cell types some of their unique properties. Membrane proteins are generally positioned so that their hydrophobic amino acids are in the fatty acid tail portion of the phospholipid bilayer—the middle—and their hydrophilic amino acids are facing the water-filled cytosol and/or the ECF.

There are two basic types of membrane proteins: integral and peripheral (see Figure 3.4). **Integral proteins** typically span the entire width of the membrane; when they do reach both sides of the membrane, they are known as *transmembrane proteins*. **Peripheral proteins** are found on only one side of the membrane. Some integral proteins are anchored in place by the cytoskeleton, but many float freely within the plane of the

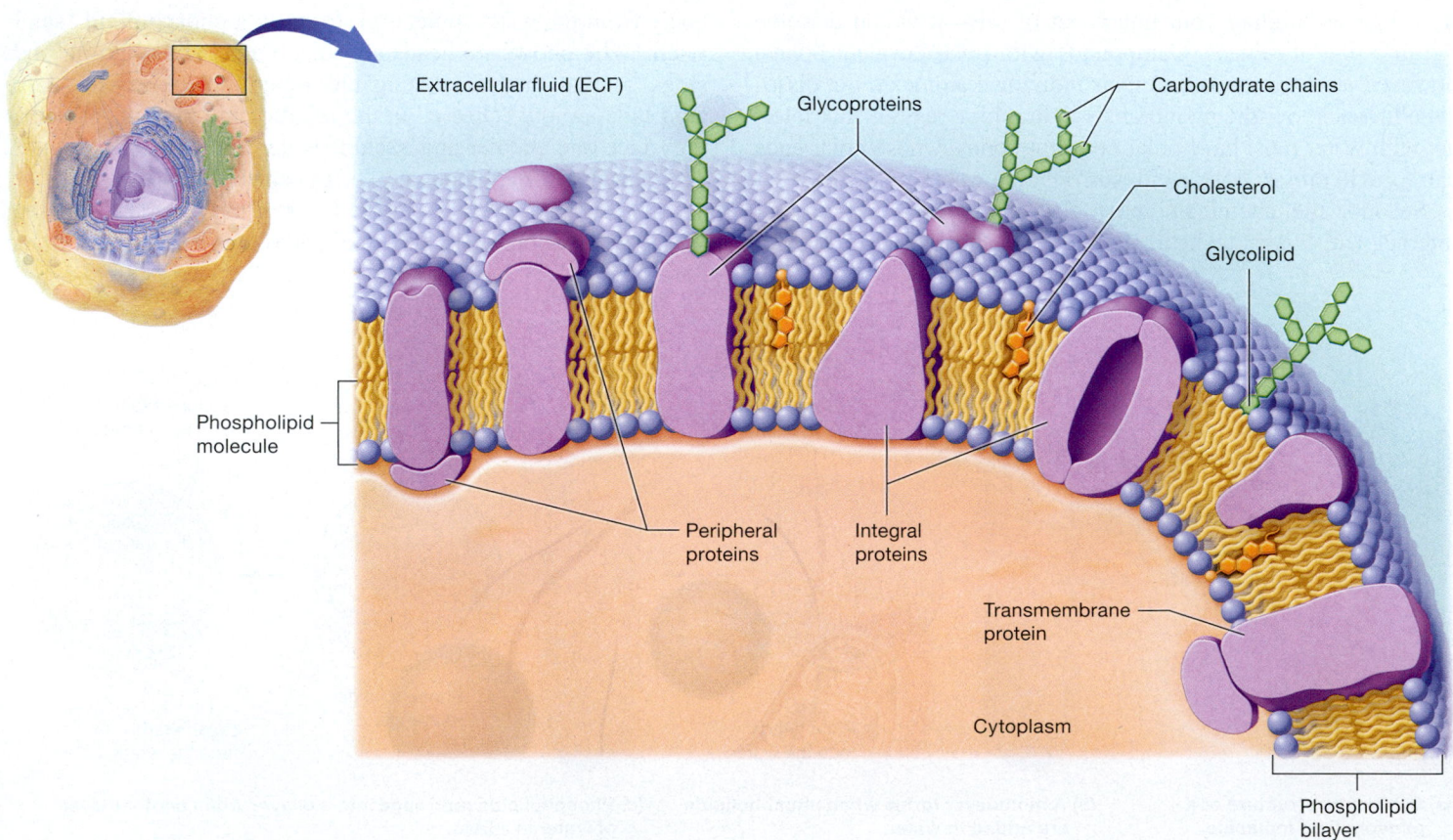

Figure 3.4 The fluid mosaic model of the plasma membrane.

phospholipid bilayer. Peripheral proteins are loosely bound to the membrane surface or to integral proteins, and many can also move horizontally along the membrane surface.

The structures and functions of membrane proteins are illustrated in **Figure 3.5**, and may include the following:

- **Acting as channels.** Many transmembrane proteins serve as **protein channels** that allow certain substances to cross the membrane and pass into or out of the cell.
- **Acting as carriers.** Some integral proteins called **carrier proteins** bind and directly transport substances into and out of the cell. For example, as shown in Figure 3.5, an ion from the ECF binds to a carrier protein, which changes shape and releases the ion into the cytosol.
- **Acting as receptors.** A **receptor** is a membrane protein that binds to a chemical messenger called a **ligand** (LY-gand). When a ligand binds to its receptor, changes are triggered within the cell. For example, notice in Figure 3.5 that a protein channel opens when the ligand binds to its receptor. Receptor-ligand interactions (an

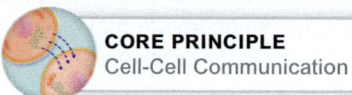 **CORE PRINCIPLE**
Cell-Cell Communication

example of the Cell-Cell Communication Core Principle, p. 28) are responsible for much of the physiology of the muscular, nervous, and endocrine systems, and we discuss them in greater detail in those chapters. See *A&P in the Real World: Drugs and Membrane Receptors* on page 74 to read about how these interactions can be targeted by medications.
- **Acting as enzymes.** As we've discussed, enzymes are proteins that speed up chemical reactions (see Chapter 2). Enzymes that are vital to plasma membrane structure and function are often lodged within the membrane, as are enzymes that catalyze various metabolic reactions. For example, some enzymes in the membrane can catalyze the formation or breakdown of bonds between substrates, as shown in Figure 3.5.
- **Providing structural support.** Membrane proteins may be bound to components of the cytoskeleton and other proteins in both the ECF and the cytosol, which gives the cell shape and helps maintain its structural integrity.
- **Linking adjacent cells.** Certain membrane proteins can attach to membrane proteins in adjacent cells, holding the cells in a tissue together. These "linker" proteins can strengthen a tissue or allow adjacent cells in a tissue to communicate (as we discuss in Chapter 4).

Other Membrane Components

In addition to membrane proteins, several kinds of lipids and carbohydrates form part of the plasma membrane. For example, an important component of the plasma membrane shown in Figure 3.4 is the lipid *cholesterol* (koh-LESS-ter-awl). Cholesterol stabilizes the structure of the plasma membrane when the temperature changes. This is critical because the plasma membrane must maintain a certain state of fluidity in order to function.

FUNCTION	STRUCTURE
Channels: Membrane proteins act as channels through which substances pass to enter or exit the cell.	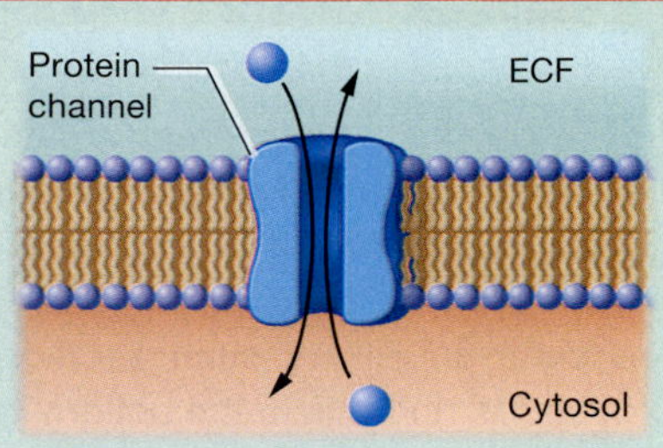
Carriers: Membrane proteins bind and transport substances into or out of the cell.	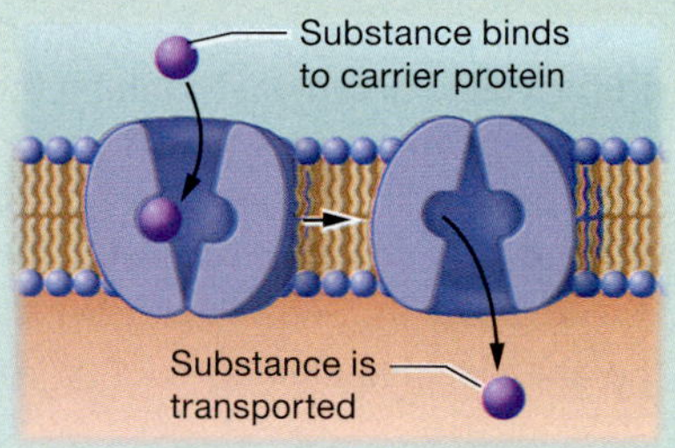
Receptors: Membrane proteins act as receptors, binding to a ligand to trigger a change in the membrane protein or the cell.	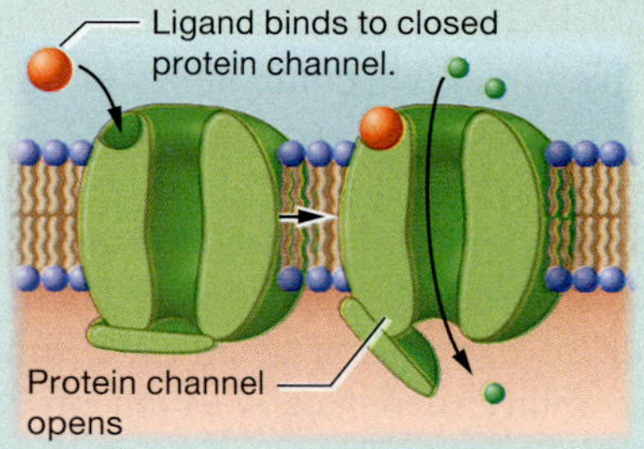
Enzymes: Membrane proteins act as enzymes, catalyzing chemical reactions.	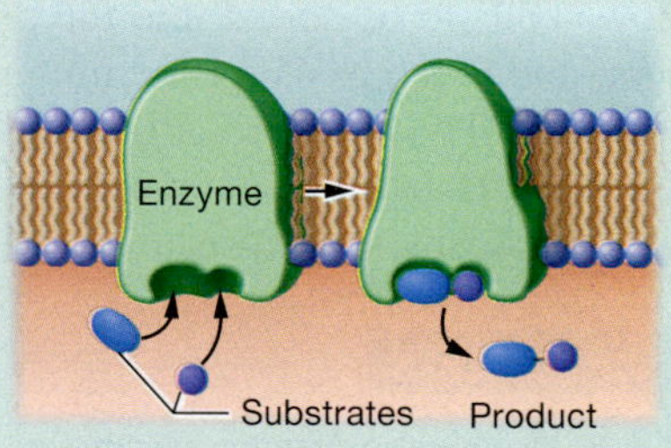
Structural support: Membrane proteins bind other proteins in the ECF and/or cytosol, supporting the cell.	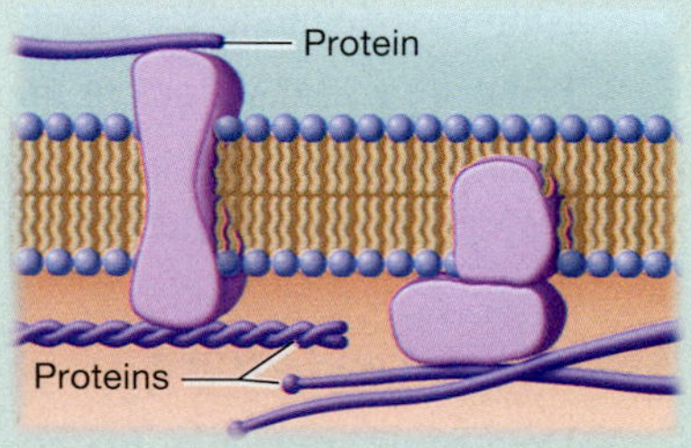
Linking adjacent cells: Membrane proteins link adjacent cells in a tissue together.	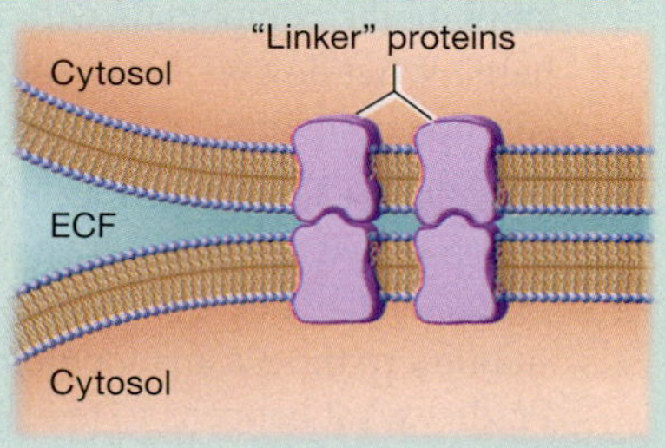

Figure 3.5 Functions of membrane proteins.

The final two components of the plasma membrane shown in Figure 3.4 are *glycolipids* (gly-koh-LIP-idz) and *glycoproteins* (gly-koh-PROH-teenz), which consist of carbohydrate chains (polysaccharides) covalently attached to either membrane lipids or proteins, respectively. These polymers found on the outside of the plasma membrane, function in *cell recognition,* or the ability of our cells to tell one cell type from another. Cell recognition enables a sperm cell to recognize an oocyte (egg), allows cells to be sorted into the proper places during development, and is the basis for the immune system's response to foreign cells. The glycoproteins and glycolipids produced by each of your cells are unique not only to that cell type but also to you as an organism, serving as a sort of identification tag.

Quick Check

☐ 2. How is the plasma membrane described according to the fluid mosaic model?

☐ 3. What are five functions of membrane proteins?

☐ 4. What roles do cholesterol, glycoproteins, and glycolipids play in the plasma membrane?

Apply What You Learned

☐ 1. Detergents are similar to phospholipids in that they are amphiphilic compounds with both polar and nonpolar parts. What do you think would happen to a cell placed in a solution containing a detergent?

☐ 2. The disease Duchenne muscular dystrophy (doo-SHEN DIS-troh-fee) is caused by a defective membrane protein called dystrophin. This protein normally provides structural stability to the plasma membrane of skeletal muscle cells. Predict what happens to the skeletal muscle cells of patients with this disease.

See answers in Appendix A.

MODULE **3.3**

Transport across the Plasma Membrane

Learning Outcomes

1. Describe the energy requirement for and the mechanism by which solute movement occurs in simple and facilitated diffusion.

2. Describe the process of osmosis and the direction of solvent movement.

3. Compare and contrast the effects of hypertonic, isotonic, and hypotonic conditions on cells.

4. Describe the energy requirement for and the mechanism by which solute movement occurs in primary and secondary active transport.

5. Compare and contrast the mechanism by which movement occurs and the types of substances moved for the different types of vesicular transport.

In the previous module, we noted that the plasma membrane acts as a "fence" between the extracellular fluid and the cytosol in our cells. Like other fences, the plasma membrane has "gates" in it. One of the main advantages to having such a membrane is that these gates allow some things—but not others—to pass into or out of the cell. In other words, the membrane is *selective* about what it allows to pass through, or *permeate,* it. This property of being **selectively permeable** is critical to the cell. If the plasma membrane let nothing pass through it, our cells would have no way to exchange oxygen, nutrients, and waste products with the extracellular fluid.

Substances are continuously being exchanged between the cytosol and ECF. Both fluid compartments consist largely of water with dissolved ions and organic compounds of varying size and concentrations. The ECF contains substances the cell needs, including nutrients such as glucose, amino acids, and fatty acids, as well as water, ions, oxygen, vitamins, and other macromolecules (see Chapter 2). Within the cytosol we find substances that must leave the cell, including waste products such as carbon dioxide. We also find substances manufactured by the cell that must enter the ECF to perform their functions. So, both the cell and the body as a whole depend for survival on the transport of substances into and out of the cell.

Substances cross the plasma membrane using several mechanisms. Some transport mechanisms, called **passive transport processes,** require no net expenditure of energy from the cell. Others, known as **active transport processes,** do require the cell to expend energy, usually in the form of ATP. Whether a process is passive or active depends on three main variables: (1) the type of substance crossing the membrane, (2) the membrane's permeability to that substance, and (3) the concentration of that substance both inside and outside the cell. In this module we explore these processes in depth.

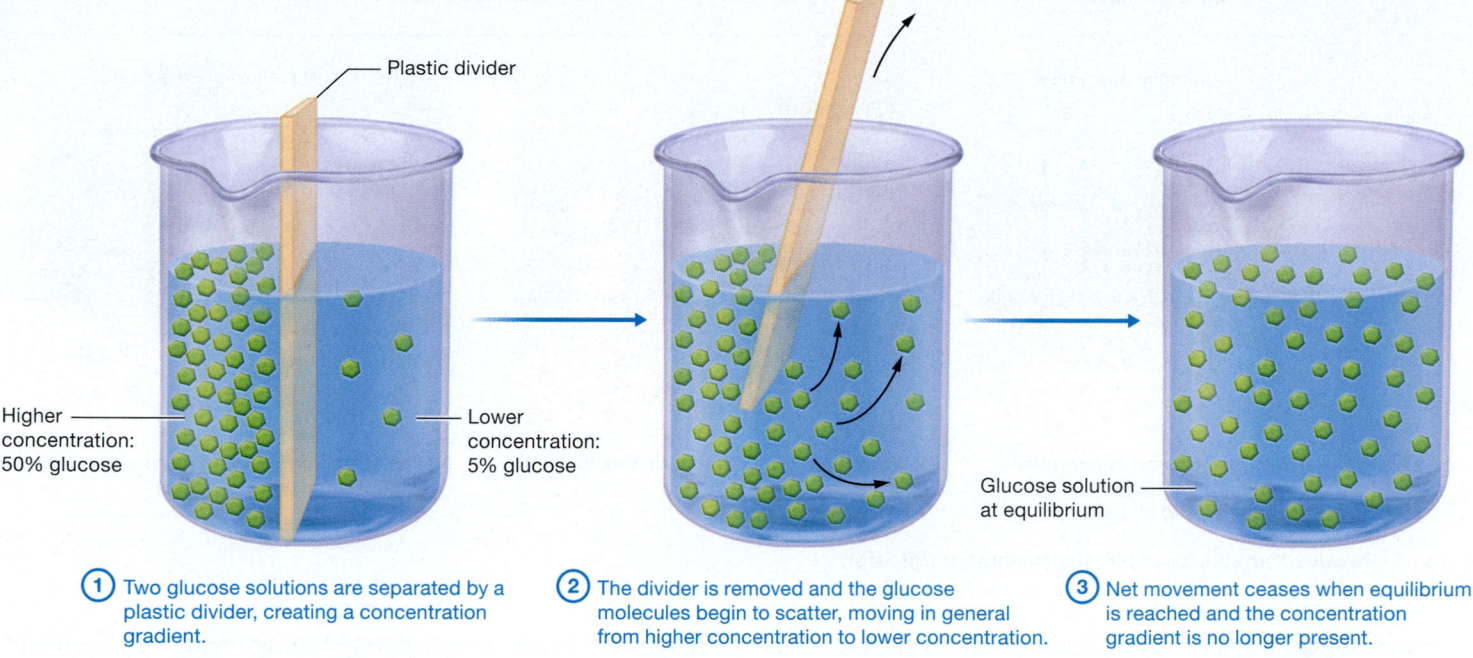

① Two glucose solutions are separated by a plastic divider, creating a concentration gradient.

② The divider is removed and the glucose molecules begin to scatter, moving in general from higher concentration to lower concentration.

③ Net movement ceases when equilibrium is reached and the concentration gradient is no longer present.

Figure 3.6 Diffusion and equilibrium.

Passive Transport Processes

 **FLASHBACK**

1. What is a concentration gradient? (p. 26)
2. What are kinetic energy and potential energy? (p. 42)
3. What are solutes and solvents? (p. 35)

There are two basic types of passive transport: *diffusion* and *osmosis*. We take a look at both in the upcoming subsections.

Diffusion

Diffusion is defined as the movement of *solute* molecules (those that are dissolved) from an area of higher solute concentration to an area of lower solute concentration (**Figure 3.6**). The basic force that drives diffusion is called a **concentration gradient.** Such a gradient involves a difference in the concentration of a substance from one area to another. For example, when you put dye in water, it looks like this:

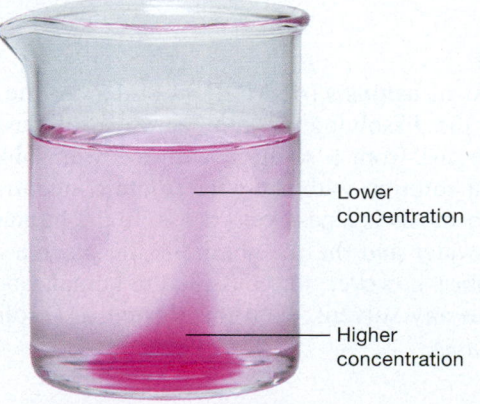

Lower concentration

Higher concentration

Notice that more dye molecules are found on the bottom of the beaker than on the top. When such a concentration difference exists, there is a concentration gradient—an example of the Gradients Core Principle (p. 28).

CORE PRINCIPLE
Gradients

Diffusion allows solutes to move into and out of the cell using the potential energy of a concentration gradient. A solute is said to move *down or "with" its concentration gradient* (from a higher to a lower concentration) during the process of diffusion. Notice in step ① in Figure 3.6 that we have two glucose solutions of different concentrations separated by a plastic divider. When the divider is removed in ②, the glucose molecules move from the side of the container with a higher glucose concentration to the side with a lower glucose concentration. This occurs because the molecules scatter due to their own kinetic energy, which all molecules have as long as thermal energy, or heat, is present.

This movement continues until the glucose concentration is uniform throughout the container and a balanced condition called *equilibrium* is reached, which is shown in ③. No further net movement of the molecules occurs at equilibrium because there is no more potential energy from a gradient to drive it. Note that we use the term *net* movement here, which means that no further change takes place in the solution's concentration. The molecules themselves, however, continue to bounce around even at equilibrium because of their kinetic energy.

The rate of diffusion depends on several factors, including the size and phase (e.g., liquid or gas) of the diffusing particles, the temperature, and the size of the concentration gradient. Generally, smaller particles, particles in the gaseous phase, higher

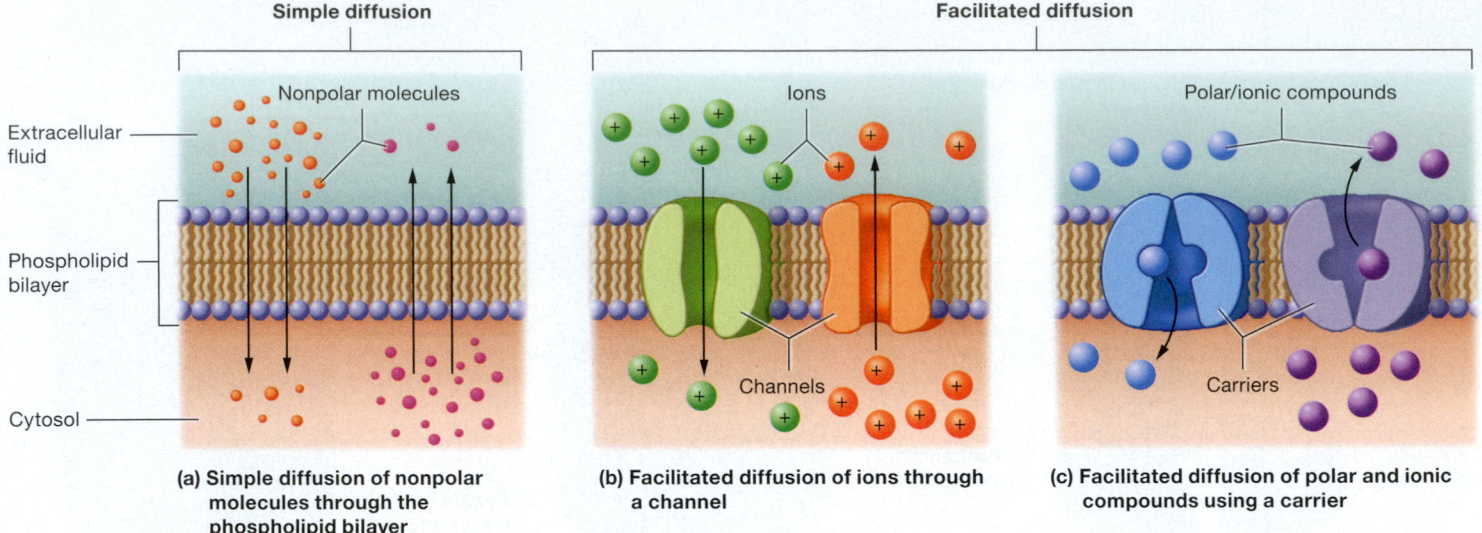

Figure 3.7 Passive transport: simple and facilitated diffusion.

temperatures, and steeper concentration gradients increase the rate of diffusion. All these conditions raise the particles' kinetic energy, which causes them to move more rapidly. Two other factors to consider are the medium through which the particles diffuse and the presence of a barrier. Particles diffuse most rapidly through a gas with no barrier (such as smoke through the air). When a barrier is present, the rate of diffusion depends on the properties of the barrier, such as its thickness, surface area, and permeability to the particles.

The barrier to diffusion into and out of cells is the plasma membrane. Two basic types of diffusion occur through a membrane: simple and facilitated (**Figure 3.7**).

- **Simple diffusion** mostly involves nonpolar solutes (such as hydrocarbons and lipids, and gases such as O_2 and CO_2) that pass straight through the phospholipid bilayer without assistance from a membrane protein.
- **Facilitated diffusion** involves charged or polar solutes (such as ions and glucose) that cross the phospholipid bilayer with the help of a membrane protein.

Simple diffusion is fairly straightforward. Notice in Figure 3.7a that nonpolar molecules such as oxygen, carbon dioxide, and lipids have properties similar to those of the nonpolar fatty acid tail region of the phospholipid bilayer. As a result, such substances can easily cross into or out of the cell with their concentration gradients.

Facilitated diffusion, shown in Figure 3.7b and c, is a bit more complicated. The fatty acid tails of the phospholipid bilayer won't interact with most polar solutes and ions, making it difficult for them to cross the membrane unassisted. For this reason, these solutes usually rely on two types of membrane proteins, called channels and carriers, for their transport into and out of the cell.

In protein channels, the hydrophilic amino acids in the protein are oriented toward the interior of the channel. This arrangement forms a passageway through which charged and polar solutes may pass into or out of the cell during facilitated diffusion, as shown in Figure 3.7b. Most channels are specific for an individual solute. For example, there are specific channels for different ions such as sodium, potassium, and calcium ions.

During facilitated diffusion using carrier proteins, shown in Figure 3.7c, the carriers bind one or two specific solutes and "carry" them into or out of the cell. As the process is currently understood, when a solute binds to its carrier protein, the protein changes shape, allowing the solute to move across the membrane and into or out of the cell. There are three basic types of carrier proteins: (1) a **uniporter,** which transports a single solute; (2) an **antiporter,** which moves two different solutes in opposite directions, one into the cell and one out of the cell; and (3) a **symporter,** which moves two solutes in the same direction.

Don't forget that facilitated diffusion is a type of passive transport, even though it requires the help of membrane proteins. The cell expends no more energy in facilitated diffusion than it does in simple diffusion. Both processes rely on the potential energy of the concentration gradient to drive each solute's movement.

Osmosis

The process of **osmosis** (oz-MOH-sis) refers to the movement of *solvent*, the dissolving medium, across a selectively permeable membrane from a solution with a lower solute concentration to a solution with a higher solute concentration. Like diffusion, osmosis is a passive process. In the human body, the solvent is water and the membrane is the plasma membrane. Do remember, however, that outside the human body osmosis can refer to any solvent, even the solvent in a solution composed of gases.

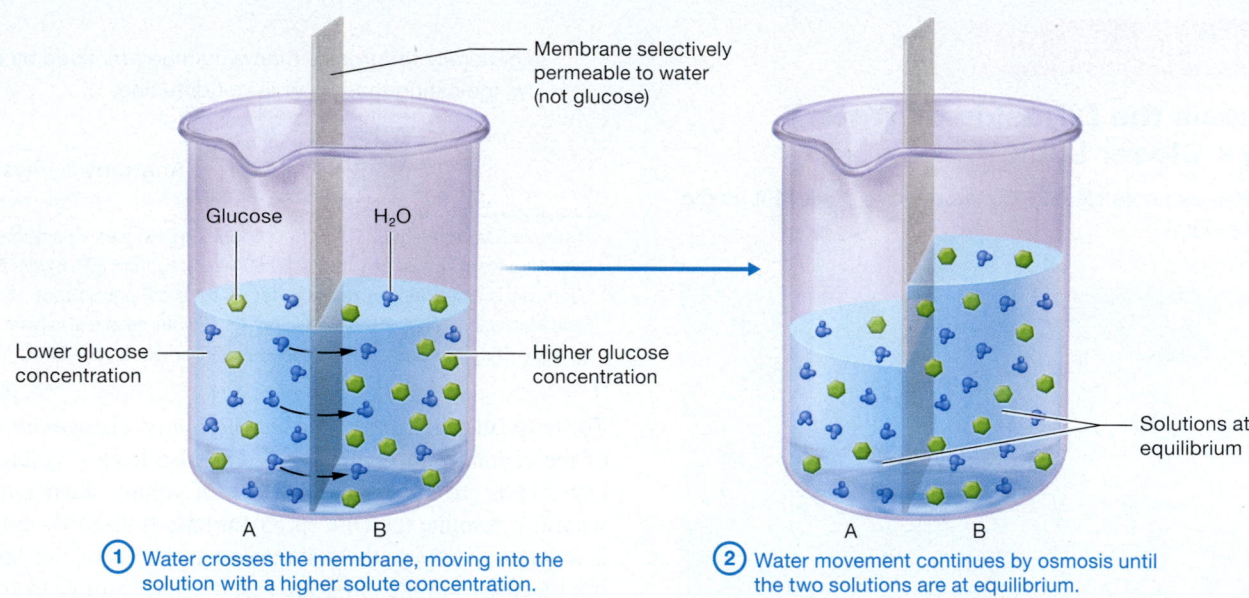

Membrane selectively permeable to water (not glucose)

Glucose   H₂O

Lower glucose concentration

Higher glucose concentration

Solutions at equilibrium

A   B

A   B

① Water crosses the membrane, moving into the solution with a higher solute concentration.

② Water movement continues by osmosis until the two solutions are at equilibrium.

Figure 3.8 Passive transport: osmosis.

Water can cross the plasma membrane in two main ways:

- **Through water channels.** The plasma membrane contains water channels called **aquaporins** that allow water molecules to pass into and out of the cell. Most water that crosses the membrane does so through aquaporins.
- **Between phospholipids in the membrane.** Some water molecules pass directly between the membrane phospholipids. Water molecules are able to pass through the membrane in this manner, in spite of their polarity, mostly because of their small size.

The process of osmosis is illustrated in **Figure 3.8**, where we again have a beaker with two unequal glucose solutions, labeled A and B. This time, however, they are separated by a selectively permeable membrane that allows water—but not glucose—to pass through. As you can see in step ①, the water molecules move through the membrane from the less concentrated glucose solution in side A into the more concentrated solution in side B until the two solutions are in equilibrium in step ②.

Volume Changes in Osmosis You can see an important consequence of osmosis in step of Figure 3.8: Osmosis results in a change in the volume of fluid in each side of our container. As water molecules leave the less concentrated glucose solution in side A, its volume decreases, and as water molecules move into the more concentrated glucose solution in side B, its volume increases. These volume changes have important consequences for our cells, which we discuss shortly.

Osmotic Pressure We said that net diffusion stops when the solute particles are equally distributed and the concentration gradient is extinguished. But what causes net water movement to stop in osmosis? The answer lies in **osmotic pressure,** which

is the pressure that must be applied to a solution to prevent water from moving into it by osmosis.

But osmotic pressure isn't the only force we need to consider with respect to water movement. We also need to consider **hydrostatic pressure,** which is the force that water exerts on the walls of its container. How do osmotic pressure and hydrostatic pressure work in our example in Figure 3.8?

In part ① of Figure 3.8, the hydrostatic pressures in sides A and B are equal because both sides are of equal volume. In part ②, as the water level rises, the hydrostatic pressure in side B gradually starts to increase. If side A in Figure 3.8 were pure water, then water molecules would simply flow into side B until the hydrostatic pressure of side B were equal to its osmotic pressure. At this point, net osmosis would stop.

However, in our figure, both sides have solute, so both sides have an osmotic pressure. The difference between the osmotic pressures in side A and side B is known as the *osmotic gradient*, another example of the Gradients Core Principle (p. 28). In this case, osmosis will stop when the hydrostatic pressure in side B reaches the value of the osmotic gradient. For example, if the osmotic gradient is 10 mmHg (millimeters of mercury, a unit of measure of pressure), net osmosis will stop when the hydrostatic pressure difference between side A and side B is equal to 10 mmHg.

An interesting feature of osmosis is that we can apply a pressure higher than the osmotic pressure to the solution in side B, and water molecules can actually be made to go in the opposite direction, into side A. This is the basis for *reverse osmosis filtration*, which is a common type of filtration system that removes impurities from water.

CORE PRINCIPLE
Gradients

ConceptBOOST

Is Osmosis the Diffusion of Water? Taking a Closer Look

If you look at osmosis closely, it's tempting to think of it as the diffusion of water:

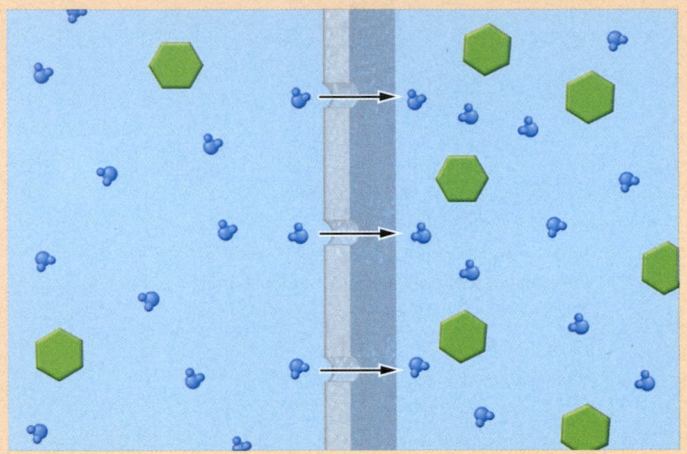

As you can see, it does indeed appear that water molecules are moving from a higher water concentration to a lower water concentration, and this can be a useful way to imagine osmosis for the purpose of remembering it. But is water really "diffusing"? The answer is no. Osmosis and diffusion are two fundamentally different forces of nature for three important reasons:

- *Osmosis requires the presence of a membrane; diffusion does not.* Osmosis will not take place without a selectively permeable membrane. Indeed, if you removed the membrane in Figure 3.8, osmosis wouldn't take place. Diffusion, however, can take place with or without a membrane.
- *Osmosis is reversible; diffusion is not.* Remember that if we apply a pressure greater than the osmotic pressure to a solution, we can reverse the process of osmosis. However, diffusion is not reversible. For example, consider that you have sprayed perfume into a room, and the perfume molecules are evenly distributed in the air. There is no force you could apply to the air to reverse the process of diffusion and return all the perfume molecules to the bottle. Even using the strongest air filter, some molecules would be left behind, and the perfume wouldn't return to its original state, volume, and concentration.
- *Solute movement in diffusion can be predicted by a diffusion law; solvent movement in osmosis cannot.* We can predict how much of a solute diffuses through an area over time, as well as how fast it moves, using an equation known as *Fick's law of diffusion*. If osmosis were the diffusion of water, we should be able to use Fick's law to predict how much water moves across the membrane and how fast it crosses. However, water movement in osmosis

is generally far greater than would be predicted by Fick's law, indicating that water is not diffusing. ■

▶ Play Video Tutor
@ **Mastering** Anatomy & Physiology

*Kramer EM, Myers DR. 2012. Five popular misconceptions about osmosis. American Journal of Physics 80:694–699; Kramer EM, Myers DR. 2013. Osmosis is not driven by water dilution. Trends in Plant Science 18:195–197; and Malińska L, *et al.* 2016. Teaching about water relations in plant cells: An uneasy struggle. CBE – Life Sciences Education 15: ar78,1-78-12.

Tonicity Just as osmosis leads to volume changes in either side of the container in Figure 3.8, it can also lead to volume changes in our cells. To understand the type of volume change that occurs, we must examine the concept of **tonicity** (toh-NISS-ih-tee), which is a way to compare the solute concentrations of two solutions. In our glucose example in Figure 3.8, it's fairly simple to compare the solute concentrations because only water crosses the membrane. However, in the case of our cells, in which we are comparing the cytosol and the ECF, it's somewhat more complicated because some solutes freely cross the plasma membrane. For this reason, tonicity in cells is determined by solutes that are not freely crossing the plasma membrane. To keep things simpler, you might want to think of tonicity in terms of a solution's ability to cause osmosis.

Nearly every cell in the body is surrounded by ECF, which is a water-based solution (see Module 3.1). Ordinarily, as shown in **Figure 3.9a**, the body's ECF is **isotonic** (aye-soh-TOHN-ik; *iso* = "same") to the cytosol of a cell, meaning that the ECF and the cytosol have the same ability to cause osmosis. Under these conditions, water enters and leaves the cell at the same rate, so the cell has no net gain or loss of water over time and its volume remains the same. An example of a solution that is isotonic to human blood is a 0.9% NaCl solution, also called *normal saline*.

However, notice in **Figures 3.9b** and **c** what happens to a cell if the solute concentration of the ECF changes:

- **A hypertonic extracellular solution causes a cell to lose water.** Figure 3.9b shows a red blood cell in a **hypertonic** solution. In this case, the ECF has a *greater* ability to cause osmosis due to the presence of higher numbers of solute particles that do not cross the plasma membrane (*hyper* = "more"). A hypertonic solution drives osmosis, so a cell in a hypertonic ECF loses water as it leaves the cell and enters the ECF. Such water loss makes the cell shrivel, or *crenate* (KREE-nayt), and possibly die. This is why humans cannot drink the salt water of the ocean—it is hypertonic to our cells, and rather than causing our cells to absorb water, it actually causes them to lose water.
- **A hypotonic extracellular solution causes a cell to gain water.** In Figure 3.9c we have a red blood cell immersed in a **hypotonic** solution, which has a *lower* ability to cause osmosis due to fewer solutes (*hypo* = "less"). Now it's the cytosol that has a greater ability to drive osmosis, so water moves from the ECF into the cytosol. This causes the cell to swell and possibly rupture, or *lyse* (LYZ). Pure water is an

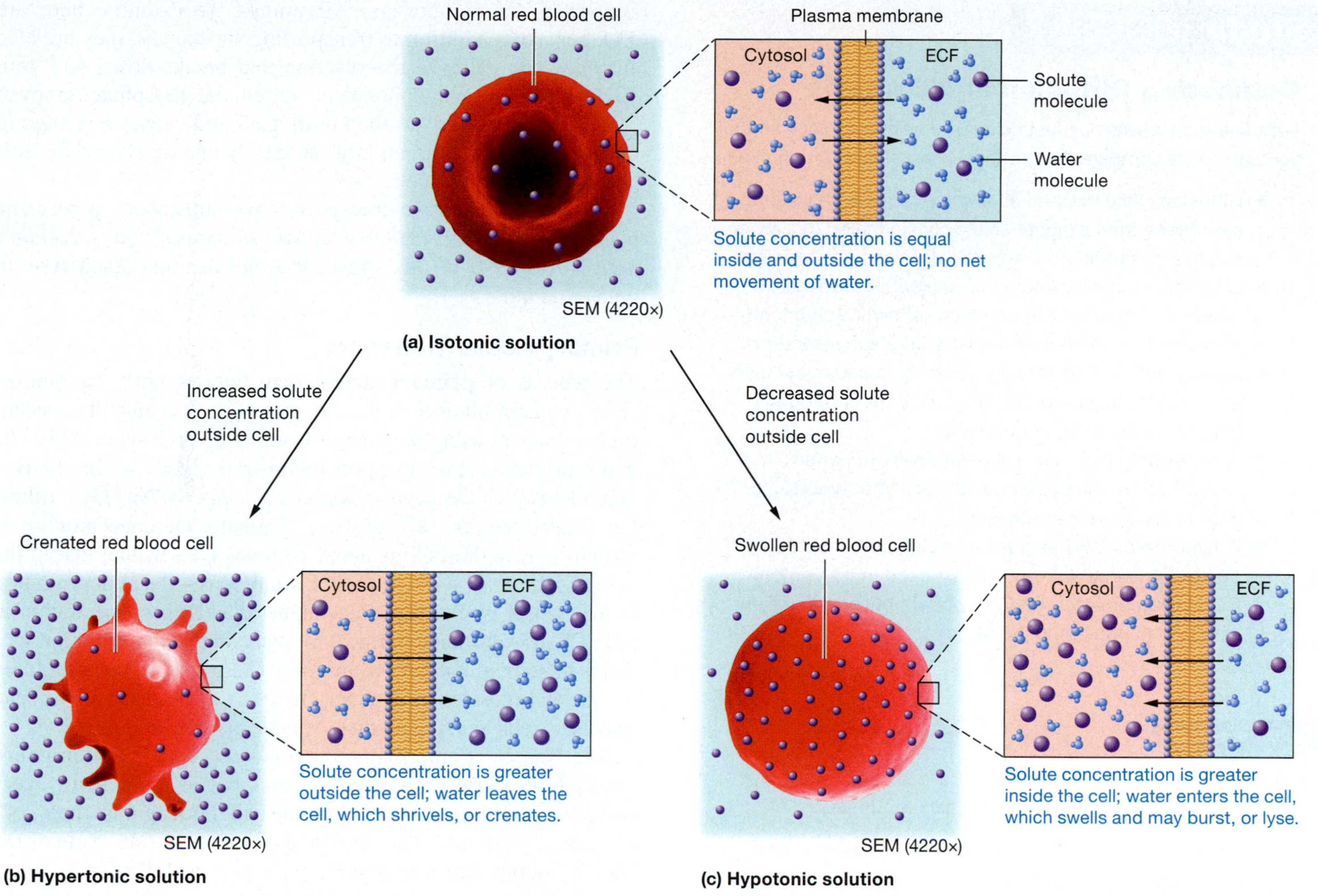

Normal red blood cell

Plasma membrane

Cytosol

ECF

Solute molecule

Water molecule

Solute concentration is equal inside and outside the cell; no net movement of water.

SEM (4220×)

(a) Isotonic solution

Increased solute concentration outside cell

Decreased solute concentration outside cell

Crenated red blood cell

Cytosol

ECF

Solute concentration is greater outside the cell; water leaves the cell, which shrivels, or crenates.

SEM (4220×)

(b) Hypertonic solution

Swollen red blood cell

Cytosol

ECF

Solute concentration is greater inside the cell; water enters the cell, which swells and may burst, or lyse.

SEM (4220×)

(c) Hypotonic solution

Figure 3.9 Tonicity: effects of isotonic, hypertonic, and hypotonic solutions on cell volume.

example of a hypotonic solution. Usually we can drink pure water and other hypotonic solutions because our endocrine and urinary systems work together to ensure that excess water is eliminated in the urine. However, if a person drinks too much water over a short period, these systems may not be able to remove the water quickly enough, and *water poisoning* may result. Water poisoning can cause death from both ion imbalances and cellular swelling. Too little water can cause problems as well, as you can read about in *A&P in the Real World: Dehydration, Sports Drinks, and Water.*

Dehydration, Sports Drinks, and Water

Dehydration, or the loss of cellular water, is a major concern for athletes or any person exercising vigorously. Strenuous exercise results in water and electrolyte loss through sweating. Sweat contains relatively more water than electrolytes (solutes), so the extracellular fluid becomes *hypertonic* to the body's cells. The hypertonic ECF draws water out of the cells by osmosis, and the concentration of solutes such as electrolytes in the cytosol of the cells increases. This can cause major disruptions in the cells' homeostasis.

Sports drinks are a commonly used remedy for mild dehydration. Many popular sports drinks contain mixtures of water, electrolytes, and carbohydrates that are actually hypotonic to human cytosol. When you drink these solutions, they eventually help replenish the water and electrolytes in your ECF and make it mildly *hypotonic* to the cytosol. This condition causes water to move back into the cells by osmosis until the normal concentration of different solutes in their cytosol is restored.

Of course, the most widely available hypotonic solution is plain old water, which generally rehydrates just as well as sports drinks. Care must be taken in cases of severe dehydration, though, because pure water can rehydrate the cells too quickly and result in cellular swelling and possibly water poisoning.

Comparing Diffusion and Osmosis

Let's summarize what we just discussed all in one place so that you can easily compare these critical processes and definitions:

- Diffusion is the movement of *solute* across the plasma membrane from a higher solute concentration to a lower solute concentration, or with the concentration gradient.
- Osmosis is the movement of *solvent* across the plasma membrane from an area of lower solute concentration to an area of higher solute concentration.
- An *isotonic* ECF has the *same* ability to cause osmosis as the cytosol, so no net water movement happens when a cell is in an isotonic ECF.
- A *hypertonic* ECF has a *greater* ability to cause osmosis than the cytosol, and a cell in a hypertonic ECF loses water by osmosis.
- A *hypotonic* ECF has a *lesser* ability to cause osmosis than the cytosol. So, a cell placed in a hypotonic ECF will gain water, as the cytosol now has a greater ability to cause osmosis. ■

Quick Check

Mark each of the following statements as true or false. If a statement is false, correct it to make a true statement.

☐ 1. Diffusion refers to the movement of a solvent with its concentration gradient.

☐ 2. Simple diffusion requires a carrier protein or channel.

☐ 3. Osmosis refers to the movement of solvent from a solution with a higher solute concentration to a solution with a lower solute concentration.

☐ 4. A cell placed in a hypertonic environment will lose water by osmosis and shrivel.

Active Transport via Membrane Proteins

FLASHBACK

1. What is ATP? (p. 60)
2. What is a hydrolysis reaction? (p. 51)

Active transport processes require cells to expend energy in the form of ATP. This use of energy is necessary because, unlike in diffusion, solutes are moved *against* their concentration gradients during active transport, from a lower concentration to a higher one. The concentration gradient does not provide energy to drive the movement of solutes, so energy must be supplied from the hydrolysis of ATP (refer back to Figure 2.25).

Active transport requires carrier proteins called **pumps** in the plasma membrane that bind and transport a solute across the membrane. As we discussed earlier, pumps may be uniporters,

antiporters, or symporters. All pumps we examine here are ATPases, so in addition to transporting substances, they are also enzymes that catalyze the reaction that breaks down ATP into ADP and an inorganic phosphate. Often, the phosphate removed by this reaction binds to the pump itself and causes a change in the shape of the pump, and this shape change transports the solute into or out of the cell.

Two types of active transport move substances across the plasma membrane: *primary active transport* and *secondary active transport*. In this subsection, we examine each type in greater detail.

Primary Active Transport

The process of **primary active transport** is fairly straightforward: A pump binds a solute and transports it against its concentration gradient using the energy from the hydrolysis of ATP. The major primary active transport pump in the body is the antiport pump known as the *sodium-potassium pump,* or Na^+/K^+ **pump** (also called the Na^+/K^+ *ATPase*). Normally the concentration of sodium ions in the ECF is about 10 times greater than that in the cytosol. The reverse is true for potassium ions—their concentration in the cytosol is about 10 times higher than their concentration in the ECF. It is absolutely critical to maintain the concentration gradients of sodium and potassium ions. These gradients are required for skeletal muscles to contract, the heart to beat, nerves to send impulses, cells to maintain their osmotic balance, and more.

The Na^+/K^+ pump maintains this steep concentration gradient by transporting three sodium ions out of the cell for every two potassium ions it moves into the cell, both against their concentration gradients. The sequence of events of one cycle of the Na^+/K^+ pump, illustrated in **Figure 3.10**, is as follows:

1. **The pump binds three sodium ions from the cytosol.** With the pump oriented so that its sodium ion binding sites are facing the cytosol, three sodium ions from the cytosol bind to the pump.
2. **ATP is hydrolyzed, phosphate binds to the pump, and the pump changes shape.** When ATP is hydrolyzed, the released phosphate group binds to the pump, causing it to change shape.
3. **The pump releases the three sodium ions into the ECF and binds two potassium ions.** In response to the shape change, the three sodium ions are released into the ECF. Two potassium ions from the ECF then bind to the pump at their exposed binding sites.
4. **The phosphate detaches and the pump changes back to its original shape.** After this binding, the phosphate group dissociates from the pump, which then returns to its original shape.
5. **The pump releases the two potassium ions into the cytosol.** After the pump changes back to its original shape, it releases the two potassium ions into the cytosol. Three sodium ions can now bind again to their binding sites on the pump, restarting the cycle.

Due to other physiological processes, sodium and potassium ions continue to leak back across the membrane, following their

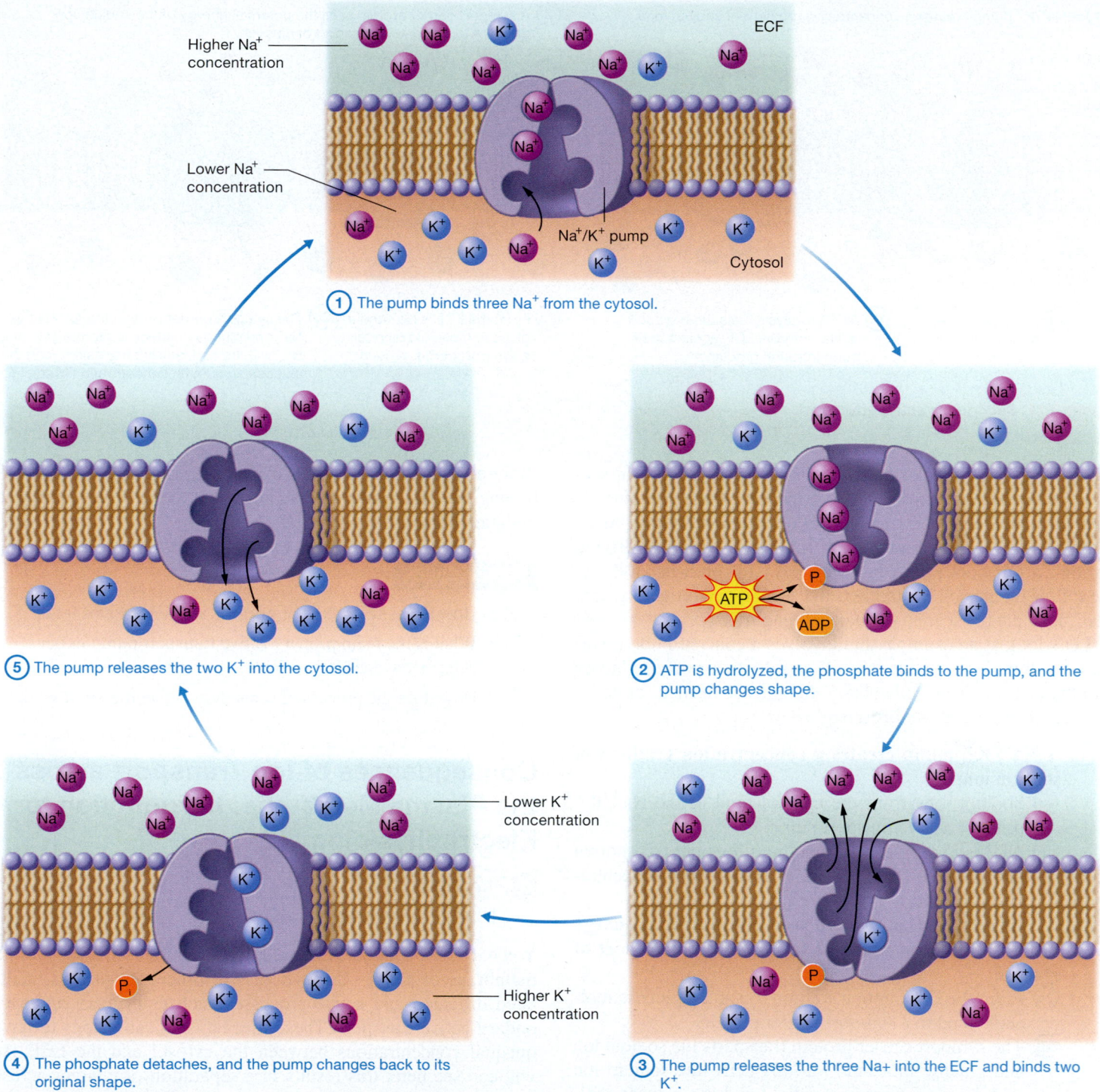

Figure 3.10 Primary active transport by the Na$^+$/K$^+$ pump.

concentration gradients, so Na$^+$/K$^+$ pumps must work continuously. In fact, in some cell populations, as much as 30% of the cell's ATP is used to fuel its Na$^+$/K$^+$ pumps.

Secondary Active Transport

Although primary active transport is fairly uncomplicated, the process of secondary active transport is somewhat more complex. As we just discussed, primary active transport uses ATP directly to fuel a transport pump, but **secondary active transport** uses ATP indirectly. In this type of transport, the cell first uses ATP to create a concentration gradient by pumping one substance across the plasma membrane. Remember that potential energy exists in every concentration gradient, so once the concentration gradient of this first substance is established, the cell then harnesses that potential energy to power the active

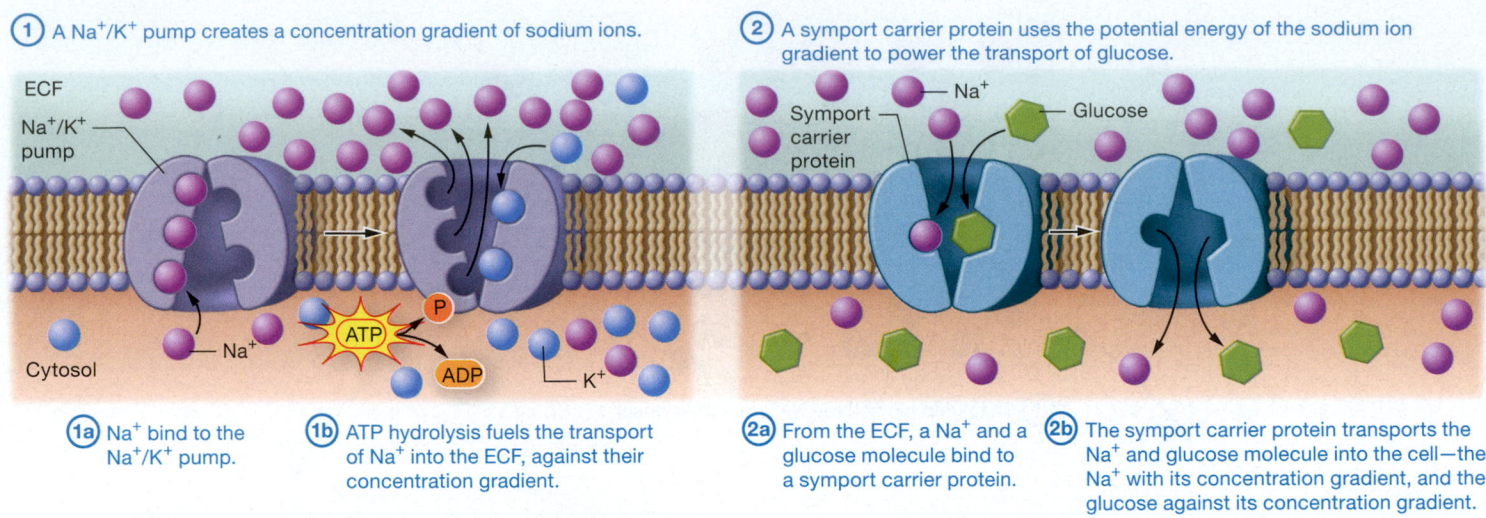

1 A Na⁺/K⁺ pump creates a concentration gradient of sodium ions.

2 A symport carrier protein uses the potential energy of the sodium ion gradient to power the transport of glucose.

1a Na⁺ bind to the Na⁺/K⁺ pump.

1b ATP hydrolysis fuels the transport of Na⁺ into the ECF, against their concentration gradient.

2a From the ECF, a Na⁺ and a glucose molecule bind to a symport carrier protein.

2b The symport carrier protein transports the Na⁺ and glucose molecule into the cell—the Na⁺ with its concentration gradient, and the glucose against its concentration gradient.

Figure 3.11 Secondary active transport.

transport of another substance against its gradient. So, in secondary active transport, the active transport of one substance is coupled with the passive transport of a second substance. As an analogy, think of a random act of kindness—a person driving through a toll booth on a highway pays for both her toll and the toll of the person behind her; the second person gets a "free ride," just like the second substance being transported.

Let's look at an example that involves a concentration gradient of sodium ions outside the cell and the transport of glucose into the cell (one way cells obtain the glucose needed for cellular fuel). The process is illustrated in **Figure 3.11**:

① **A Na⁺/K⁺ pump creates a concentration gradient of sodium ions:**

  ⓐ Sodium ions from the cytosol bind to the Na⁺/K⁺ primary active transport pump.

  ⓑ ATP is hydrolyzed, and sodium ions are transported out of the cell, into the ECF, against their concentration gradient.

② **A symport carrier protein uses the potential energy of the sodium ion gradient to power the transport of glucose:**

  ⓐ From the ECF, both a sodium ion and a glucose molecule bind to a symport carrier protein.

  ⓑ The symport carrier protein transports the sodium ion and glucose molecule into the cell—the sodium ion *with* its concentration gradient and the glucose molecule *against* its gradient.

Many secondary active transport processes involve the steep sodium ion concentration gradient created by the Na⁺/K⁺ pump. This gradient favors the movement of sodium ions back into the cell. In addition to glucose, other substances transported against their concentration gradients using the potential energy of this gradient include amino acids, other sugars, and ions such as chloride and bicarbonate. Secondary active transport is essential for absorption and secretion processes for many cells, particularly in the intestines and kidneys. This key role of the

sodium ion gradient is one of many reasons why properly functioning Na⁺/K⁺ pumps are so important to the body's ability to maintain homeostasis.

Quick Check

  ☐ 5. How does the process of primary active transport work?

  ☐ 6. What is the main primary active transport pump in our cells? How does it function?

  ☐ 7. How does the process of secondary active transport work?

Consequences of Ion Transport across the Plasma Membrane: Introduction to Electrophysiology

《 FLASHBACK

  1. What is electrical energy? (p. 42)

You have read about the different ways that ions cross the plasma membrane of a cell—through passive processes such as facilitated diffusion and active processes such as primary and secondary active transport. These mechanisms cause ions to have unequal concentrations between the cytosol and the ECF. As you can see here, this results in a separation of charges across the plasma membrane: A thin layer of positive charges is present in the ECF, and a thin layer of negative charges is found in the cytosol:

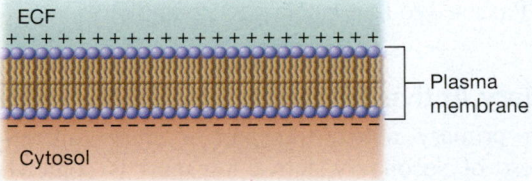

Note that this charge separation is limited to the area on either side of the plasma membrane. On the whole, the cytosol and ECF have the same overall charge.

A separation of charges like the one we see here is known as an **electrical potential.** This name refers to the fact that an electrical potential is a source of potential energy. Why is there potential energy in separated charges? The answer is simple— an electrical potential is really just an *electrical gradient,* similar to the concentration Gradients Core Principle (p. 28) you have

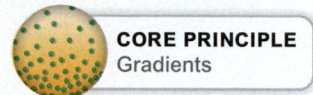

CORE PRINCIPLE
Gradients

seen many times in this chapter. Just as potential energy is found in a concentration gradient, this energy is also found in an electrical gradient. The electrical potential present across the plasma membranes of all cells is known as a **membrane potential,** and the study of these membrane potentials is called **electrophysiology.**

The value of the membrane potential when a cell is at rest, when it is not stimulated or inhibited by any other factor, is called the **resting membrane potential.** The resting membrane potential of most excitable cells, such as muscle and nerve cells, averages about −80 mV (millivolts), which means that their cytosol is about 80 mV more negative than the ECF. In later chapters you'll learn about the basic ability of cells to alter their membrane potential, and how this is critical to the functioning of excitable cells (see Chapters 10 and 11).

Active Transport via Vesicles

Carrier proteins and channels are very effective at helping polar and ionic molecules cross the plasma membrane. However, they have size limitations—polar macromolecules are simply too big to bind to a carrier or fit through a channel. Instead, larger particles are transported into or out of the cell in small sacs called **transport vesicles** (VESS-ih-kulz; *vesic-* = "bubble"). A vesicle is enclosed by a membrane made of a phospholipid bilayer, just like the plasma membrane that encloses the cell. In fact, vesicles form from pinched-off parts of cell membranes, and so they can fuse with the plasma membrane and also with the membranes surrounding the organelles of the cytoplasm. Transport using vesicles is an active process called **vesicular transport** (vess-IK-yoo-lur). Vesicular transport requires the energy from ATP hydrolysis to fuel several steps of the process, including vesicle formation. There are two types of vesicular transport: *endocytosis* (en′-doh-sy-TOH-sis) to bring substances into the cell, and *exocytosis* (eks′-oh-sy-TOH-sis) to move substances out of the cell.

Endocytosis

Extracellular fluid, large molecules, and even sometimes whole cells are taken into the cell by the process of **endocytosis** (*endo-* = "within"). There are two basic types of endocytosis: *phagocytosis* for particles and *pinocytosis* for droplets of fluid.

Phagocytosis During **phagocytosis** (fayg′-oh-sy-TOH-sis; *phago-* = "eat," literally "cell eating"), cells ingest large particles

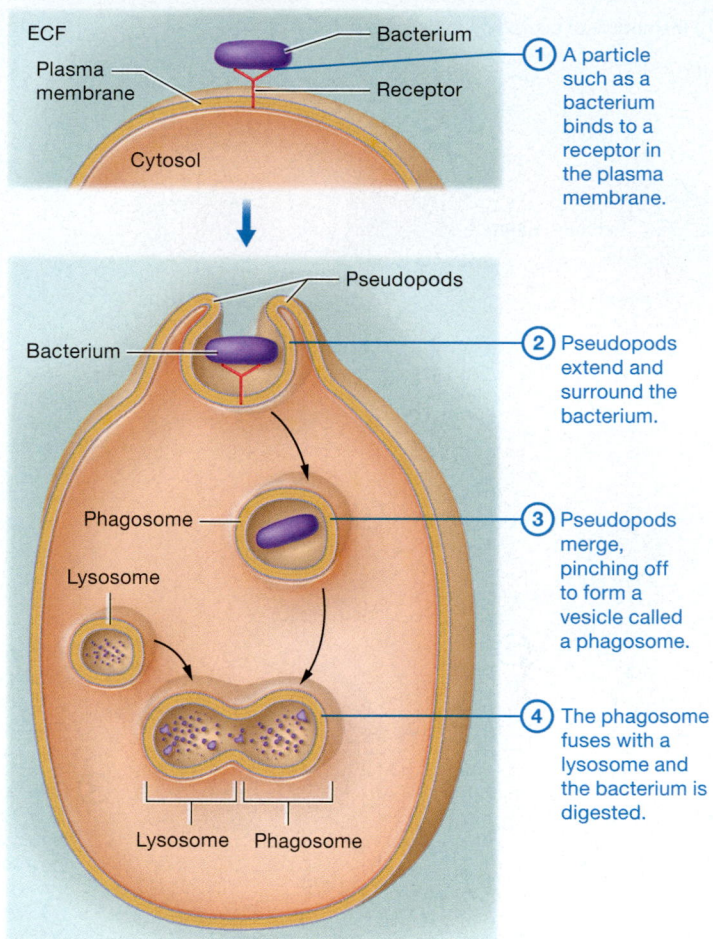

Figure 3.12 Endocytosis: phagocytosis.

such as bacteria, dead body cells, or parts of cells. In humans only certain cells of the immune system called *phagocytes* (FAYG-oh-sytz) carry out phagocytosis.

The process unfolds as follows, shown in **Figure 3.12**:

1. A particle such as a bacterium binds to a receptor in the phagocyte's plasma membrane.
2. This binding triggers the cell to extend bulging "arms" called *pseudopods* (SOO-doh-podz; *pseudo-* = "false," *pod-* = "foot") that surround the bacterium.
3. The pseudopods merge to form a transport vesicle that pinches off from the cell surface into the cytosol, becoming a relatively large structure called a *phagosome* (FAYG-oh-sohm).
4. The phagosome eventually fuses with an organelle called a *lysosome* (LY-soh-sohm) that contains enzymes that digest the bacterium (lysosomes are discussed in Module 3.4).

Pinocytosis **Pinocytosis** (peen′-oh-sy-TOH-sis; *pino-* = "drink," literally "cell drinking"), also called *fluid-phase endocytosis,* is the cellular ingestion of droplets of the ECF. Pinocytosis produces smaller transport vesicles than phagocytosis. These vesicles bring nutrients and other substances dissolved in the ECF into the cell. For this reason, all cells carry out pinocytosis, and it occurs nearly continuously in most cells.

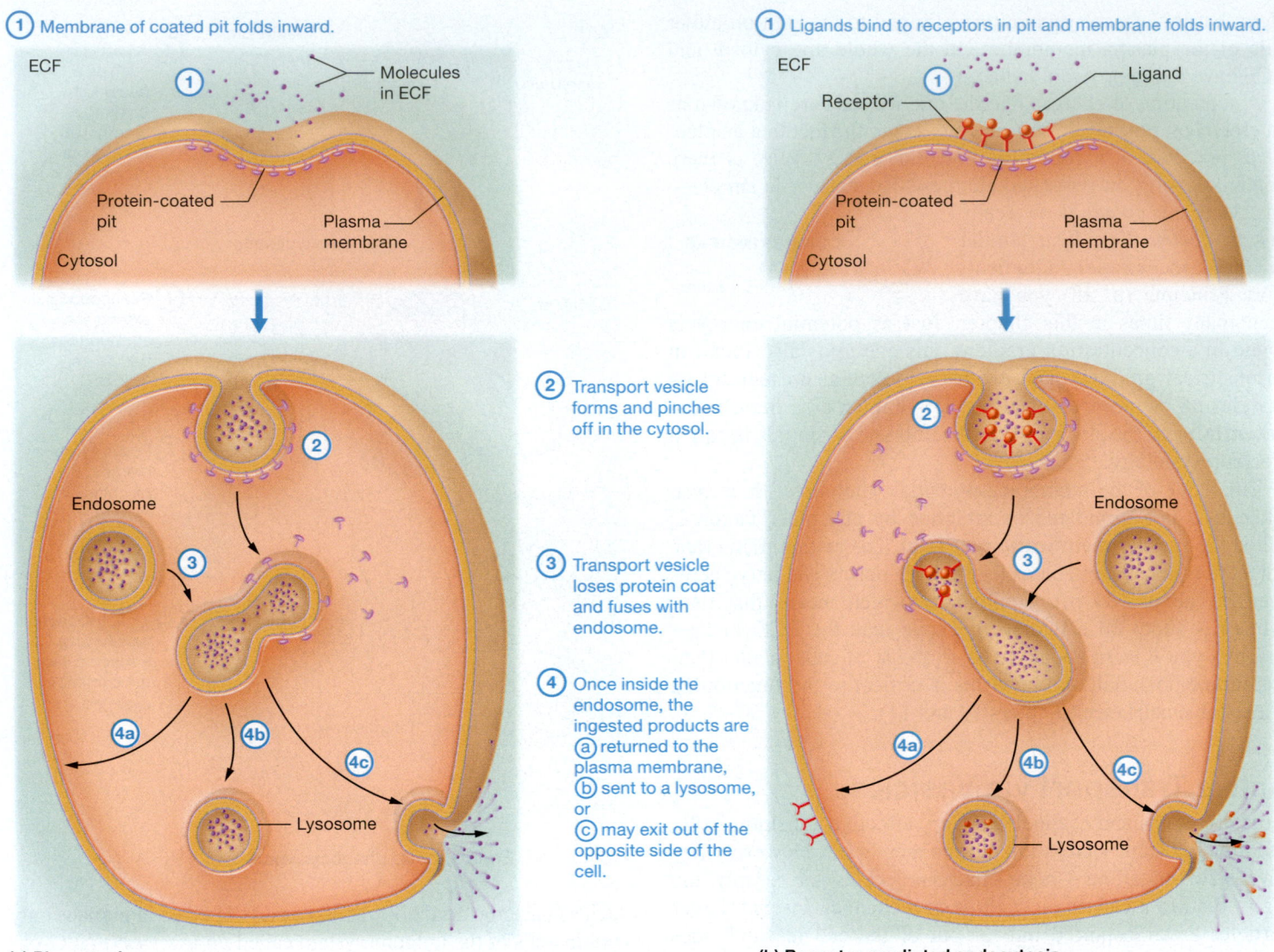

① Membrane of coated pit folds inward.

ECF
① Molecules in ECF
Protein-coated pit
Plasma membrane
Cytosol

② Transport vesicle forms and pinches off in the cytosol.

Endosome
③

④a ④b ④c

Lysosome

③ Transport vesicle loses protein coat and fuses with endosome.

④ Once inside the endosome, the ingested products are
ⓐ returned to the plasma membrane,
ⓑ sent to a lysosome, or
ⓒ may exit out of the opposite side of the cell.

(a) Pinocytosis

① Ligands bind to receptors in pit and membrane folds inward.

ECF
① Ligand
Receptor
Protein-coated pit
Plasma membrane
Cytosol

②

Endosome
③

④a ④b ④c

Lysosome

(b) Receptor-mediated endocytosis

Figure 3.13 Endocytosis: pinocytosis and receptor-mediated endocytosis.

Pinocytosis usually takes place in regions of the plasma membrane known as coated pits, which are small indentations of the membrane that are coated with a protein (**Figure 3.13a**). Soon after these pits form, ① they begin to fold inward, eventually enclosing some extracellular fluid within a transport vesicle. The transport vesicle then ② pinches off in the cytosol, ③ loses its protein coat, and fuses with another type of vesicle called an **endosome** (EN-doh-sohm). In the endosome, the ingested substances are sorted and modified before being sent to their final destinations, which may be one of three places:

④ⓐ Parts of the endosome may return to the plasma membrane, where they are re-inserted.

④ⓑ Other parts may fuse with a lysosome, where their contents are degraded.

④ⓒ Finally, some parts may exit from the other side of the cell in a process called *transcytosis*, which we'll cover soon.

Pinocytosis results in transport vesicles containing a fairly uniform sample of the ECF. However, **receptor-mediated endocytosis**

is a special form of pinocytosis that results in transport vesicles containing a high concentration of a specific substance (**Figure 3.13b**). In this process, ligands such as cholesterol, hormones, and iron bind to unique receptors in a protein-coated pit* in the plasma membrane. Once a certain number of these ligands are bound, the pit folds inward and pinches off to form a transport vesicle. Like the pinocytotic transport vesicle, it then fuses with an endosome and the ingested products may be sorted back to the plasma membrane, sent to lysosomes, or delivered out of the other side of the cell.

Exocytosis

Molecules are released from the cell and components are added to the plasma membrane by the process of **exocytosis** (*exo-* = "outside," literally "out of the cell"), which is functionally the

* Alberts B, et al. Molecular biology of the cell. 6th ed. New York (NY): Garland Science; 2014. p. 730–735.

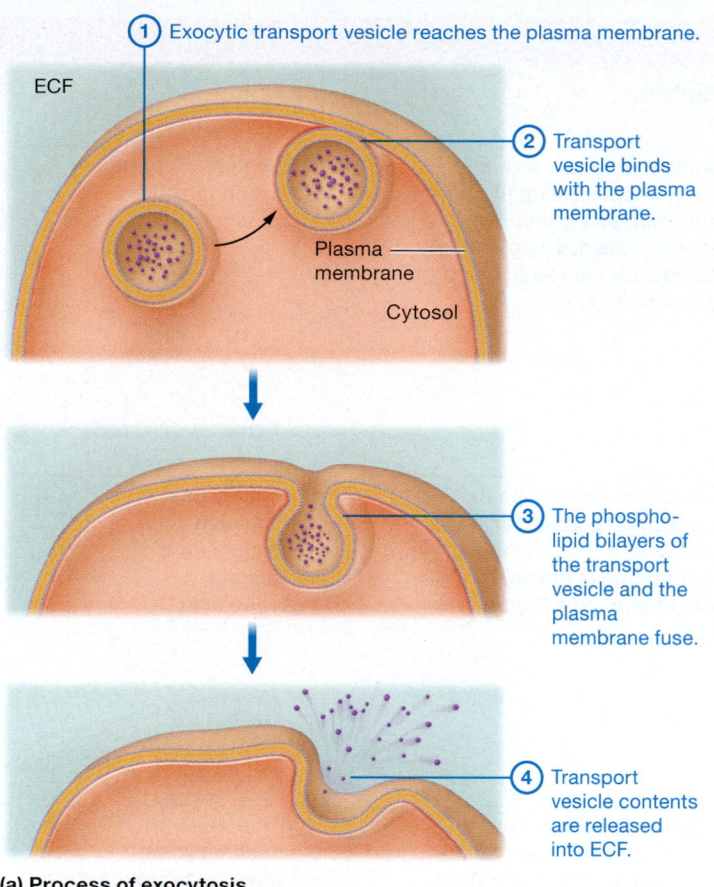

① Exocytic transport vesicle reaches the plasma membrane.

ECF

② Transport vesicle binds with the plasma membrane.

Plasma membrane

Cytosol

③ The phospholipid bilayers of the transport vesicle and the plasma membrane fuse.

④ Transport vesicle contents are released into ECF.

(a) Process of exocytosis

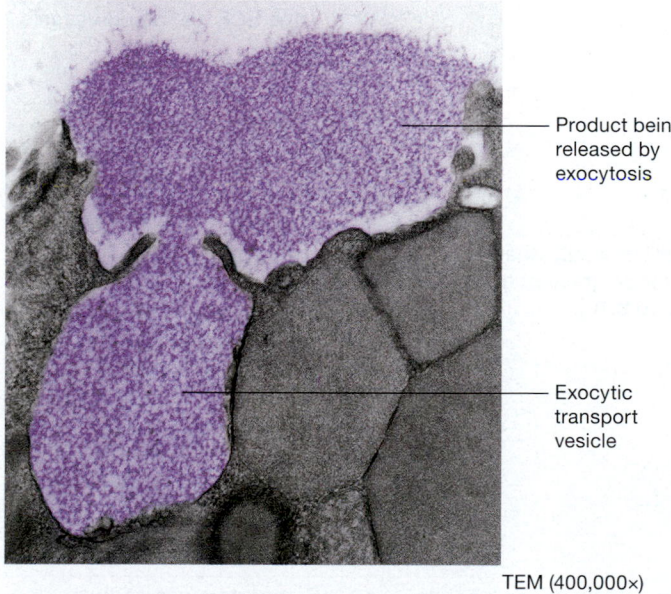

Product being released by exocytosis

Exocytic transport vesicle

TEM (400,000×)

(b) Electron micrograph of exocytosis

Figure 3.14 Exocytosis.

reverse of endocytosis. As illustrated in **Figure 3.14**, particles to be released are packaged into an exocytic transport vesicle. As you can see, ① when the transport vesicle reaches the plasma membrane, ② the transport vesicle binds with the membrane and ③ their phospholipid bilayers fuse. ④ This fusion releases the contents of the transport vesicle into the ECF and incorporates the vesicle membrane components into the plasma membrane. Many molecules are released from the cell in this manner, including components of the ECF, such as glycoproteins and enzymes. In addition, cells that release large volumes of products—including nerve, gland, and digestive cells—release products such as hormones, neurotransmitters (the chemical messengers of nerve cells), and enzymes by a process known as *secretion*. The products of secretion are packaged in specialized vesicles called *secretory vesicles* (SEE-kreh-toh-ree).

You may have noticed that endocytosis removes portions of the plasma membrane; this loss can be quite extensive, depending on the type of cell. For example, certain phagocytes called *macrophages* can ingest the equivalent of their entire membrane in only 30 minutes. However, the process of exocytosis adds pieces of phospholipid bilayer back into the plasma membrane, and portions of endocytic vesicles are recycled and re-inserted into the membrane. Even with endocytosis and exocytosis occurring nearly constantly, the cell never changes size appreciably. This is because the rates of recycling and exocytosis are essentially identical with that of endocytosis. In other words, the rate at which material is added to the plasma membrane equals the rate at which material is removed.

Transcytosis

Some molecules are brought into the cell by endocytosis, transported through the cell, and released from the cell's other side by exocytosis, a process known as **transcytosis.** Transcytosis is particularly important in blood vessels, where large substances exit the blood by passing through cells lining the vessel. Transcytosis also occurs in the kidneys and in the intestinal cells during the absorption of nutrients.

Table 3.1 reviews the different types of transport across the plasma membrane. You can use this table to compare these types of transport and better understand their similarities and key differences.

Quick Check

☐ 8. What are the two types of endocytosis, and how do they differ?

☐ 9. Explain the basic process of exocytosis.

Apply What You Learned

☐ 1. Predict what will happen to a person's red blood cells if he or she is given the following fluids intravenously:

   a. Sterile water (a hypotonic solution)

   b. Normal saline (an isotonic solution)

   c. 10% dextrose in water (a hypertonic solution)

☐ 2. Oleander (*Nerium oleander*) is a common but poisonous ornamental shrub. All parts of the plant contain chemicals that inhibit the Na^+/K^+ pump. Explain why chemicals that have this effect could also disrupt secondary active transport.

See answers in Appendix A.

Table 3.1 Plasma Membrane Transport

Type of Transport	Definition	Example(s)
Passive		
Simple Diffusion 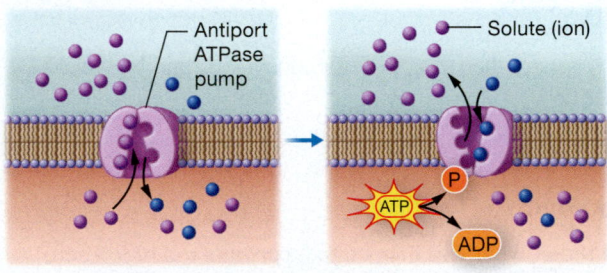	Movement of solute with its concentration gradient through the plasma membrane unaided by a transport protein; energy source is the solute's own kinetic energy (see Figure 3.7a).	• Oxygen • Carbon dioxide • Lipids
Facilitated Diffusion	Movement of solute with its concentration gradient with the help of a carrier or channel protein; energy source is the solute's own kinetic energy (see Figure 3.7b and c).	• Sodium ions • Potassium ions • Calcium ions • Glucose • Amino acids
Osmosis	Movement of solvent (water) from a solution of lower solute concentration to one of higher solute concentration through a selectively permeable membrane (see Figure 3.8).	• Water absorption from the intestinal lining • Water reabsorption from the kidneys • Water movement between the ECF and blood vessels
Active		
Primary Active Transport	Movement of solute against its concentration gradient using ATP (see Figure 3.10).	Na^+/K^+ ATPase antiporter pump removes 3 Na^+ from the cytosol and brings 2 K^+ into the cytosol against the concentration gradient.
Secondary Active Transport 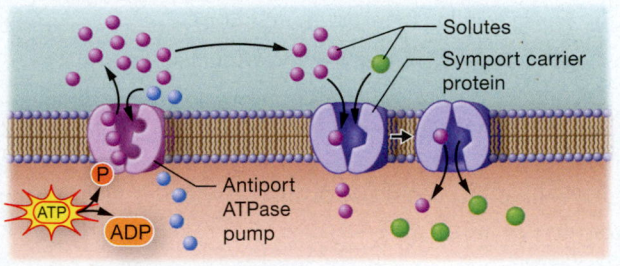	An ATPase pump drives a solute out of (or into) the cell against its concentration gradient. Movement of this solute with its concentration gradient back into the cell is used to power the transport of another solute against its concentration gradient (see Figure 3.11).	Symporters use sodium ion gradient to bring glucose, chloride ions, and bicarbonate ions into the cell.

Simple Diffusion (labels)
ECF — Solute — Plasma membrane — Cytosol

Facilitated Diffusion (labels)
Solute — Plasma membrane — Channel protein — Uniport carrier protein

Osmosis (labels)
Solutes — H_2O molecules — Plasma membrane — Aquaporin channel

Primary Active Transport (labels)
Antiport ATPase pump — Solute (ion) — ATP — ADP — P

Secondary Active Transport (labels)
Solutes — Symport carrier protein — Antiport ATPase pump — ATP — ADP — P

Play A&P Flix Animation
@ Mastering Anatomy & Physiology

Table 3.1 Plasma Membrane Transport *(continued)*

Type of Transport		Definition	Example(s)
Phagocytosis	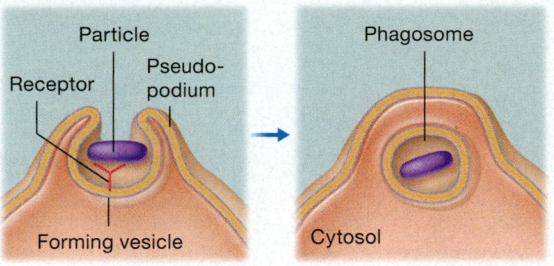	"Cell eating"; bringing large molecules or particles into the cell via a phagosome; ATP required (see Figure 3.12).	Ingestion of bacteria and cell debris by phagocytes
Pinocytosis	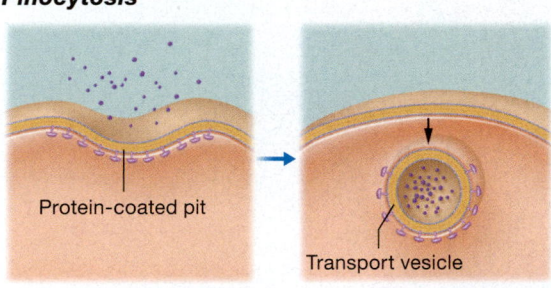	"Cell drinking"; bringing substances in the ECF into the cell via a transport vesicle formed from a protein-coated pit; ATP required (see Figure 3.13a).	Nutrient transport
Receptor-Mediated Endocytosis	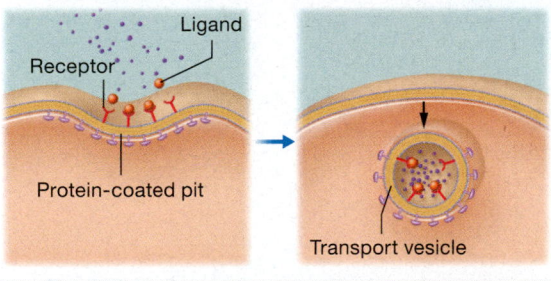	Bringing a specific substance into a transport vesicle using receptors on the plasma membrane; ATP required (see Figure 3.13b).	Cholesterol, iron, and hormone transport
Exocytosis	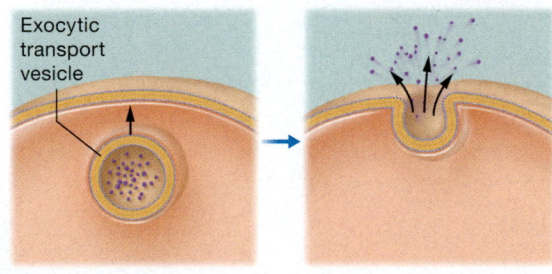	Release of a substance from the cell via an exocytic transport vesicle; ATP required (see Figure 3.14).	• Secretion of hormones, neurotransmitters, and enzymes • Release of proteins and glycoproteins into the ECF • Adding components to the plasma membrane

 Play A&P Flix Animation @ Mastering Anatomy & Physiology

MODULE **3.4**

Cytoplasmic Organelles

Learning Outcomes

1. Describe the structure and function of each type of organelle.
2. Explain how the organelles of the endomembrane system interact.

We now move our discussion inside the cell to the tiny cellular "machines" known as cytoplasmic organelles (**Figure 3.15**). Just as each of our organs has a unique structure that enables it to perform specific functions, so does each of a cell's organelles. Recall from earlier in the chapter that organelles allow the cell's functions to be compartmentalized, which is essential for cellular efficiency (see Module 3.1). To this end, each organelle contains a specific set of enzymes and other proteins unique to its function.

Organelles compartmentalized by membranes are often called *membrane-bounded,* or *membrane-enclosed, organelles.*

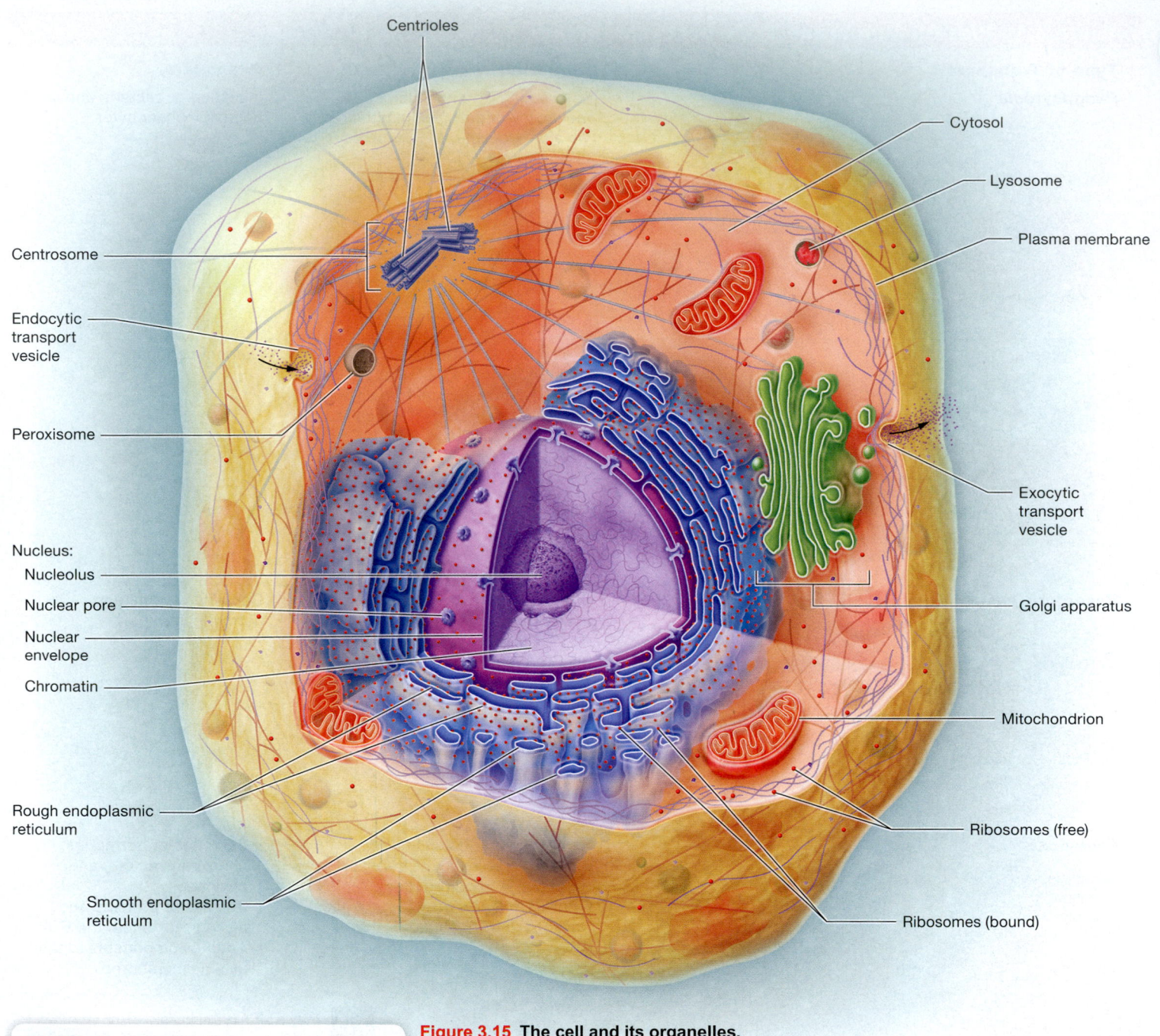

Centrioles

Cytosol

Lysosome

Centrosome

Plasma membrane

Endocytic
transport
vesicle

Peroxisome

Exocytic
transport
vesicle

Nucleus:
    Nucleolus
    Nuclear pore
    Nuclear
    envelope
    Chromatin

Golgi apparatus

Mitochondrion

Rough endoplasmic
reticulum

Ribosomes (free)

Smooth endoplasmic
reticulum

Ribosomes (bound)

Figure 3.15 The cell and its organelles.

Practice art labeling
@ Mastering Anatomy & Physiology

Membrane-enclosed organelles include the *mitochondrion, peroxisome, endoplasmic reticulum, Golgi apparatus,* and *lysosome.* Some organelles are not enclosed by a membrane, including the *ribosome,* covered in this module, and the centrosome, covered in the next module.

Each type of membrane-enclosed organelle has a membrane that is unique in terms of its thickness and composition. In spite of these differences, an organelle's membrane is still structurally similar to the plasma membrane. This similarity allows substances to be packaged in vesicles and transported between the plasma membrane and organelles.

Cells are very small, so we need microscopes to get a good look at them. Microscopes can produce photographs called *micrographs* (*micro-* = "small," *-graph* = "something written or drawn") that you will see throughout this chapter and in the rest of this text. There are two basic types of micrographs, produced with different kinds of microscopes. The first is a *light micrograph* (*LM*), produced with a *light microscope,* which passes a beam of light through specimens to create an image. Light microscopes are used to view relatively large microscopic objects, such as whole cells and tissues. The second is an *electron micrograph,* produced with an *electron microscope,* which uses a beam of electrons instead of light to produce an image. Electron microscopes are used to view considerably smaller objects, such as single organelles. There are two main types of electron micrographs: scanning electron micrographs (SEMs), which are three-dimensional images usually of whole or broken cells, and transmission electron micrographs (TEMs), which are flat images of sectioned cells. Whenever you see a micrograph in this text, you'll see that the type of image is identified, as well as the degree of magnification used.

In this module, we explore the structure and function of many of the cell's remarkable internal components. As you read, don't lose sight of the relationship between an organelle's form and its function. Also bear in mind that no one component of the cell can function independently of the other components. Instead, all must work together to ensure that homeostasis is maintained.

Mitochondria

« **FLASHBACK**

1. How is energy stored in chemical bonds and in ATP? (p. 42 and p. 60)

2. What are oxidation and catabolic reactions? (p. 43)

The organelles known as **mitochondria** (my′-toh-KAHN-dree-uh; singular, *mitochondrion*) are involved in energy production. The main source of energy for cells is *chemical energy,* and the most ready supply of chemical energy for our cells is ATP (see Chapter 2). Although some ATP production occurs via glycolysis in the cytosol, most of our ATP is produced within the mitochondria, which is why they are often called the "power plants"

of the cell. Most cells have hundreds to thousands of mitochondria, and the number that each cell contains is determined by its particular energy needs. For example, cells such as those in the liver, skeletal muscle, and cardiac muscle have a higher demand for ATP (compared to those in the skin, for instance) and therefore require more mitochondria.

Mitochondria are unique among human cell organelles in that each mitochondrion contains its own DNA and the enzymes required for protein synthesis within its matrix (**Figure 3.16**). In addition, each mitochondrion has its own *ribosomes,* which are small, granular organelles that participate in protein synthesis. The vast majority of mitochondrial proteins are encoded by the DNA in the cell's nucleus. The remainder, a mere 14 proteins, are encoded by mitochondrial DNA. In humans, mitochondrial DNA is passed down from the maternal side only, for reasons that we'll discuss in a later chapter (see Chapter 27).

Like most organelles, mitochondria are surrounded by a membrane made up of a phospholipid bilayer and other components. Notice in Figure 3.16 that there are actually *two* mitochondrial membranes: a smooth *outer mitochondrial membrane* and an *inner mitochondrial membrane* that has numerous folds called **cristae** (KRISS-tee), which significantly increase the inner surface area. This double-membrane structure creates two spaces within the mitochondrion: (1) the *intermembrane space,* which is the space between the two membranes, and (2) the *matrix,* which is the innermost space. Each membrane or space contains its own unique enzymes and other proteins, allowing each to carry out specific functions.

The outer mitochondrial membrane has large channels that permit molecules in the cell's cytosol to enter the

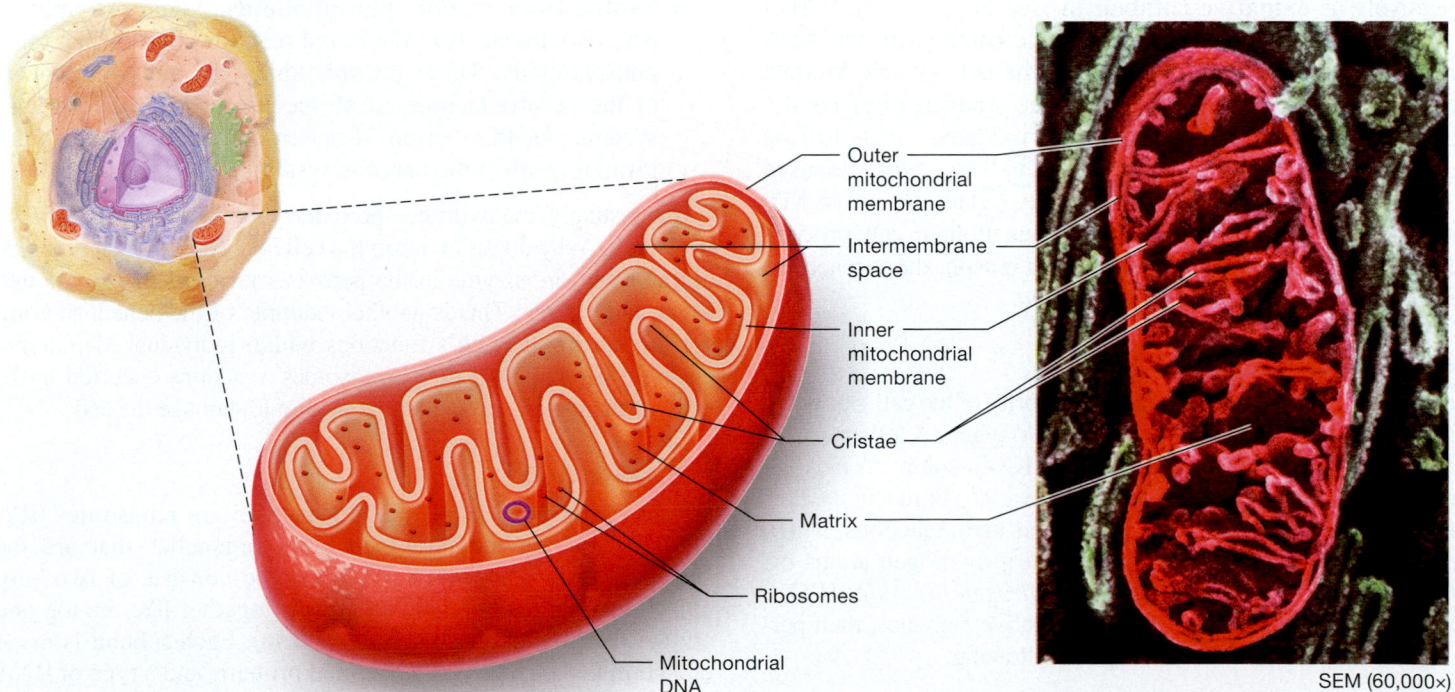

Outer mitochondrial membrane

Intermembrane space

Inner mitochondrial membrane

Cristae

Matrix

Ribosomes

Mitochondrial DNA

SEM (60,000×)

Figure 3.16 **Structure of the mitochondrion.**

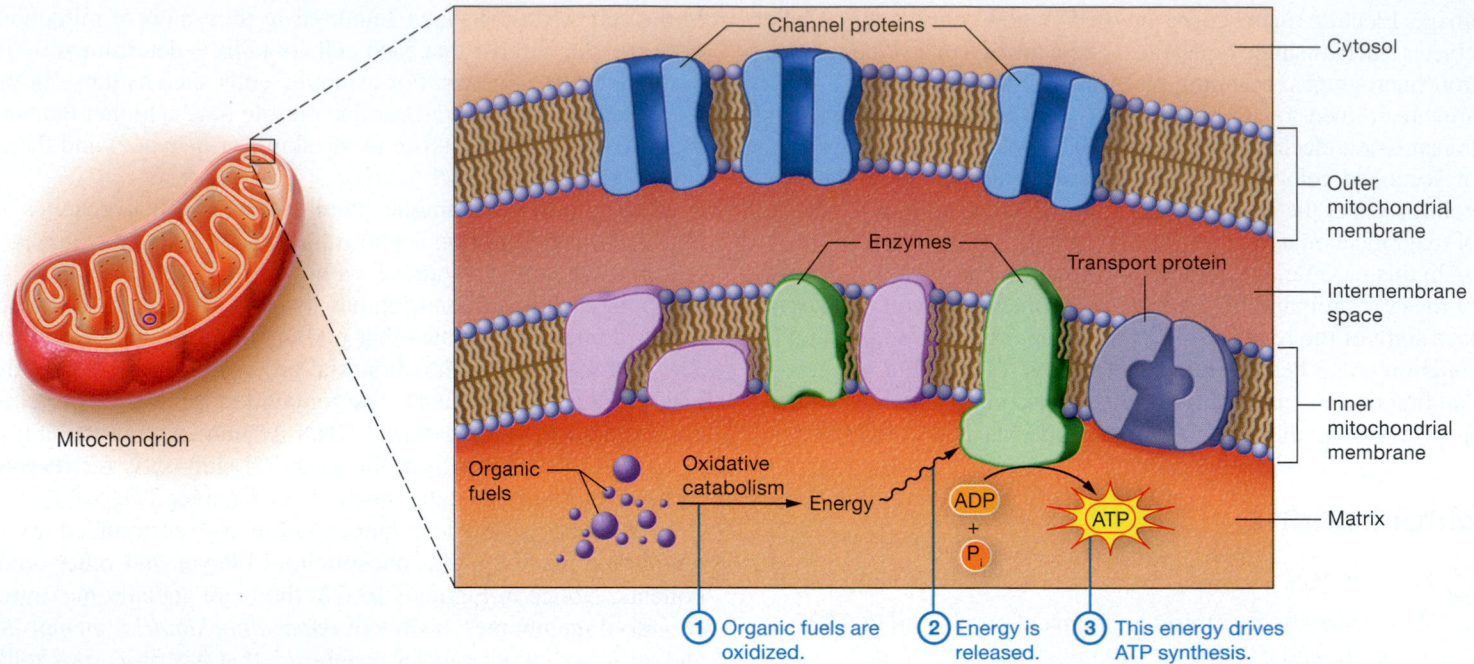

Figure 3.17 Function of the mitochondrion.

intermembrane space (**Figure 3.17**). In contrast, the inner mitochondrial membrane is far more selective. It is essentially impermeable to most solutes, except those for which it has specific transport proteins. As a result, within the matrix we find only certain organic molecules, such as DNA, mitochondrial proteins, and enzymes.

The mitochondrial matrix contains enzymes and proteins that break down organic fuels via a series of reactions known collectively as **oxidative catabolism** (see Figure 3.17). Earlier in the chapter you read about glycolytic catabolism, the ATP-generating reactions that take place in the cytosol (see Module 3.1). During oxidative catabolism, ① the products of glycolytic catabolism and other organic fuels are oxidized (made to lose electrons) by mitochondrial enzymes. ② The energy released by these oxidation reactions is ultimately ③ used to drive ATP synthesis. The reactions of oxidative catabolism can proceed only in the presence of oxygen. For this reason, these reactions are also referred to as *aerobic respiration*.

Peroxisomes

Although mitochondria consume the majority of the cell's oxygen, they are not the only organelle that uses oxygen. Another organelle, called a **peroxisome** (puh-RAWKS-ih-sohm; *soma-* = "body"), also uses oxygen to carry out a variety of reactions. Peroxisomes get their name from one of their main reactions, which uses molecular oxygen to oxidize and strip hydrogen atoms off certain organic molecules to produce *hydrogen peroxide* (H_2O_2). The hydrogen peroxide, along with oxidative enzymes, then performs various functions, including the following:

- **Oxidizing toxic substances.** Hydrogen peroxide oxidizes certain chemicals that would be toxic to the cell, including ethanol (alcohol). The oxidation reactions convert such chemicals

into less toxic compounds, which are then eliminated from the body. This function is particularly important in kidney and liver cells, which perform most detoxification reactions in the body.

- **Breaking down fatty acids.** Many of the catabolic reactions that break down fatty acids into smaller molecules occur in the peroxisomes. The products of these reactions are sent either to the cytosol to participate in anabolic reactions or to the mitochondria to be oxidized to produce ATP.

- **Synthesizing certain phospholipids.** Some enzymes in peroxisomes catalyze the initial reactions that form certain phospholipids. These phospholipids are key components of the plasma membrane of specific cells in the nervous system. For this reason, diseases that affect peroxisomes invariably affect the nervous system.

Interestingly, the hydrogen peroxide formed by peroxisomes is itself toxic. Why doesn't it harm the cell? When hydrogen peroxide accumulates, an enzyme in the peroxisomes turns the excess into water and oxygen. This is another example of the benefit of compartmentalizing the cell's functions within individual membrane-enclosed organelles. If the peroxisomes' reactions occurred in the cytosol, the toxic hydrogen peroxide would damage the cell.

Ribosomes

The tiny, granular particles in Figure 3.15 are **ribosomes** (RY-boh-sohmz), non–membrane-enclosed organelles that are the site of protein synthesis. Each ribosome consists of two subunits, one large and one small, that fit together like the top and bottom of a hamburger bun (**Figure 3.18**). Each subunit is made of a complex assembly of ribosomal proteins and a type of RNA called **ribosomal RNA (rRNA).**

Depending on their location in the cell, ribosomes are grouped into two types: free or bound. Free ribosomes are suspended

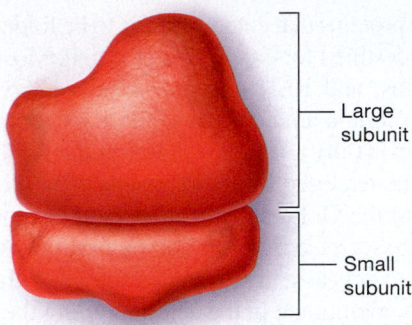

Figure 3.18 Schematic structure of the ribosome.

Large subunit

Small subunit

in the cytosol, where they make proteins that are mainly used in the cytosol itself. Bound ribosomes are associated with the membranes of other cellular structures. All ribosomes start out free, but some will attach themselves to the endoplasmic reticulum or the nuclear envelope. These bound ribosomes typically make proteins that will be exported from the cell, transported to certain organelles such as lysosomes, or inserted into a membrane. Free and bound ribosomes are structurally identical; they differ only in location. We discuss ribosomal structure and function in greater detail later in the chapter (see Module 3.7).

Quick Check

☐ 1. Identify the properties listed in the next column as belonging to mitochondria, peroxisomes, or ribosomes.

a. Detoxify chemicals such as alcohol

b. Can be bound to a membrane or suspended in the cytosol

c. Metabolize fatty acids

d. Make most of the ATP for the cell

e. Synthesize proteins

f. Produce certain phospholipids

The Endomembrane System

As we've discussed, most of the cell's organelles are surrounded by phospholipid bilayer membranes that are very similar to the plasma membrane. Some of these organelles form vesicles that exchange proteins and other molecules. The organelles that transfer molecules in this manner are part of a system called the **endomembrane system,** whose components together synthesize, modify, and package molecules produced by the cell. This system includes the plasma membrane, *endoplasmic reticulum,* the *Golgi apparatus,* and *lysosomes*. Also included in the endomembrane system is the nuclear envelope, which we discuss later in the chapter (see Module 3.6).

Endoplasmic Reticulum

The large, folded membrane shown surrounding the nucleus in **Figure 3.19** is the **endoplasmic reticulum** (en-doh-PLAZ-mik reh-TIK-yoo-lum; *reticulum* = "network"), or simply **ER.** The

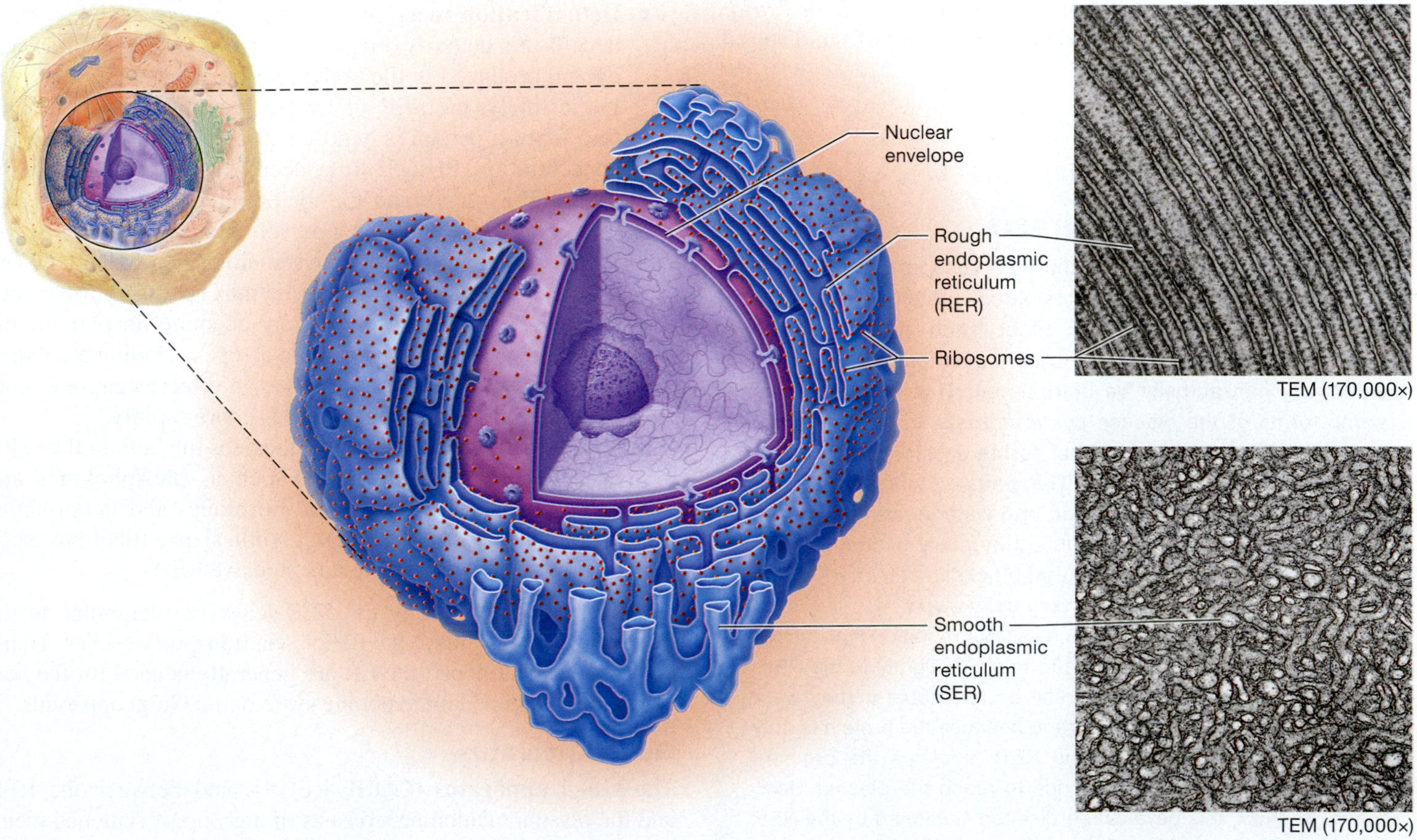

Nuclear envelope

Rough endoplasmic reticulum (RER)

Ribosomes

TEM (170,000×)

Smooth endoplasmic reticulum (SER)

TEM (170,000×)

Figure 3.19 The endoplasmic reticulum.

ER's folded phospholipid bilayer membrane is continuous with the nuclear envelope. Current evidence indicates that the membrane is a single, continuous structure enclosing a fluid-filled space called the *ER lumen*. This space contains enzymes that catalyze a variety of reactions. Figure 3.19 shows two distinct regions of the ER's membrane: the **rough endoplasmic reticulum,** or **RER,** where the membrane is covered in bound ribosomes, and the **smooth endoplasmic reticulum,** or **SER,** which lacks bound ribosomes. Why do you think the RER and the SER differ in structure in this way? You are on the right track if you suspect they have very different functions, which we discuss next.

Rough Endoplasmic Reticulum   Recall that proteins must be folded into their proper shape before they become functional (see Chapter 2). Yet ribosomes, the sites of protein synthesis, only assemble amino acids into a polymer called a *polypeptide* (the protein's primary structure) without a three-dimensional shape. So, ribosomes by themselves do not synthesize fully functional proteins—other processes must modify the proteins so that they can fold into their functional state.

The polypeptides synthesized on bound ribosomes pass through the RER membrane into its lumen, where enzymes catalyze the reactions that fold these polymers into their correct three-dimensional shapes. The RER recognizes proteins that have folded into an incorrect shape and sends them to the cytosol, where they are degraded. In some cases, this can be a surprisingly high percentage—for example, about 90% of certain membrane proteins misfold and for this reason never actually make it to the plasma membrane. See *A&P in the Real World: Cystic Fibrosis* to learn about a surprising result of the disposal of one misfolded protein type.

Many of the proteins that enter the RER to be folded are secretory proteins, those destined for export from the cell. Most such proteins are glycoproteins, and RER enzymes catalyze the reactions that attach sugars to certain amino acids of these glycoproteins. When a secretory protein is fully assembled, it leaves the RER packaged in a small, membrane-enclosed transport vesicle. Transport vesicles are generally sent to the Golgi apparatus for further processing before their contents are secreted from the cell by exocytosis.

The RER also acts as a "membrane factory"—the components for nearly every membrane in the cell, including the plasma membrane, are made there. The ribosomes of the RER synthesize integral and peripheral membrane proteins. The RER also synthesizes some cholesterol and phospholipids and inserts all these components into its membrane. These components are then pinched off the RER membrane and transferred to the plasma membrane or other parts of the endomembrane system via transport vesicles. The RER also makes the membranes for organelles that are not in the endomembrane system. However, the membrane components are transferred to these organelles in other ways.

Smooth Endoplasmic Reticulum   Because the SER is not associated with ribosomes, it plays essentially no role in protein synthesis. The SER performs other vital functions, however, including the following:

- **Calcium ion storage.** In most cells, the SER pumps calcium ions out of the cytosol and stores them for later release. This function is particularly important for muscle cells, which rely on calcium ions for contraction.
- **Detoxification reactions.** The SER is very well developed in cells that detoxify drugs and harmful substances ingested by and produced in the body, particularly cells in the liver. Interestingly, the SER of liver cells can nearly double in size when a person is exposed to certain drugs. When the concentrations of such drugs decrease, the cells remove the excess SER membrane and the organelle returns to its normal size. However, when a person is repeatedly exposed to the drug, the SER membrane remains enlarged and the cell increases its synthesis of the enzymes necessary to degrade the drug. This increased activity accounts in part for the *tolerance* that is seen with some drugs, including alcohol—more drug is required to produce an effect because a more extensive SER detoxifies the drug more rapidly.
- **Lipid synthesis.** The SER synthesizes the bulk of the cell's lipid membrane components, such as phospholipids and cholesterol, as well as steroid hormones and lipoproteins. Cells that perform these lipid-synthesizing functions, such as those of the liver, have an extensive SER.

The lipids produced by the SER leave its membrane in the same way that they leave the RER—via transport vesicles. Transport vesicles that depart the ER are generally headed for the next component of the endomembrane system, the Golgi apparatus.

Golgi Apparatus

The **Golgi apparatus** (GOHL-jee), located between the RER and the plasma membrane, consists of a group of flattened membranous sacs filled with enzymes and other molecules. Within the Golgi apparatus, proteins and lipids produced by the ER are

Cystic Fibrosis

The fact that the RER recognizes and helps dispose of misfolded proteins is usually beneficial, as most such proteins could cause major harm to the cell. However, sometimes the RER's selectivity can actually be detrimental. This is the case in some forms of the disease *cystic fibrosis*, in which certain cells are missing a protein that forms a chloride ion channel in the plasma membrane. This causes deficient chloride ion transport that affects the ion and water secretions of the lungs, digestive system, and integumentary system, resulting in abnormally thick mucus that blocks the airways, digestive enzyme deficiencies, and very salty sweat.

A person with cystic fibrosis has a DNA mutation that causes the chloride ion channel protein to misfold slightly in the RER. The channels would generally still be functional if they were inserted into the plasma membrane because the protein is only slightly misfolded. However, the RER mediates the protein's destruction before it has a chance to reach the plasma membrane. In short, this devastating disease is caused by the RER being "overprotective" regarding protein structure.

further modified, sorted, and packaged for export. Notice in **Figure 3.20** that unlike the sacs of the ER, those of the Golgi apparatus are not continuous, but rather are separated from one another by thin spaces filled with cytosol. The part of the Golgi apparatus nearest the ER receives transport vesicles from the ER, and the part farthest from the ER sends newly formed vesicles off to various destinations. Each sac of the Golgi apparatus contains different enzymes, because each carries out different functions.

As molecules pass via transport vesicles through the sacs of the Golgi apparatus, enzymes catalyze reactions that modify them. In particular, glycoproteins made by the ER are further altered. When product processing in the Golgi apparatus has finished, they are packaged in vesicles bound for different destinations either within or outside the cell. Some secretory vesicles are ready for exocytosis from the plasma membrane, some contain proteins to be inserted into the plasma membrane, and some go to lysosomes, which we discuss next.

Lysosomes

All cells must be able to digest the nutrients they take in, as well as old, worn-out organelles and, in some cases, other cells. These digestive functions take place within organelles known as **lysosomes** (LY-soh-sohmz). Lysosomes are membrane-enclosed sacs that contain water and enzymes called *acid hydrolases.* These hydrolases are synthesized on bound ribosomes, modified in the ER, further modified by the Golgi apparatus, and then delivered to the lysosome in a transport vesicle.

As their name implies, acid hydrolases are enzymes that catalyze hydrolysis reactions (see Chapter 2). They function best in an acidic environment, and the pH within lysosomes is kept low, at around 5.0. Note that enclosing these acidic reactions within a lysosome ensures that other contents of the cell are protected from these digestive enzymes. In addition, if the enzymes happened to leak out, they would also be relatively inactive at the cytosolic pH of 7.2. *A&P in the Real World: Lysosomal Storage Diseases* talks about what happens when these acid hydrolases don't work correctly.

Lysosomal Storage Diseases

A group of diseases collectively called the *lysosomal storage diseases* results from a deficiency of one or more of the acid hydrolases of the lysosomes. This deficiency causes the accumulation of certain undigested molecules within the lysosomes. Symptoms of the disease depend on which specific enzyme(s) is (are) missing. Examples of lysosomal storage diseases include the following:

- *Gaucher's disease* (gow-SHAYZ) is caused by a deficiency in the enzyme that breaks down certain glycolipids, and so these lipids accumulate in cells of the blood, spleen, liver, lungs, bone, and sometimes brain. The most severe form of the disease is generally fatal in infancy or early childhood.
- *Tay-Sachs disease* (TAY saks) results from the deficiency of an enzyme that degrades another specific type of glycolipid. These glycolipids accumulate in the lysosomes of the brain, leading to progressive neural dysfunction and typically death by age 4 or 5 years.
- *Hurler syndrome* is caused by a deficiency in the enzymes that digest certain large polysaccharides. These molecules accumulate in many cells of the body, including those of the heart, liver, and brain. Death can result in childhood from organ damage.
- *Niemann-Pick disease* is caused by a deficiency in an enzyme that is involved in the degradation of a specific lipid. The result is that lipids accumulate in the lysosomes. This disease affects the spleen, liver, brain, and lungs, as well as bone marrow. Its severity varies, and the prognosis is very poor for patients with the severe form, who suffer damage to multiple organs and neural dysfunction.

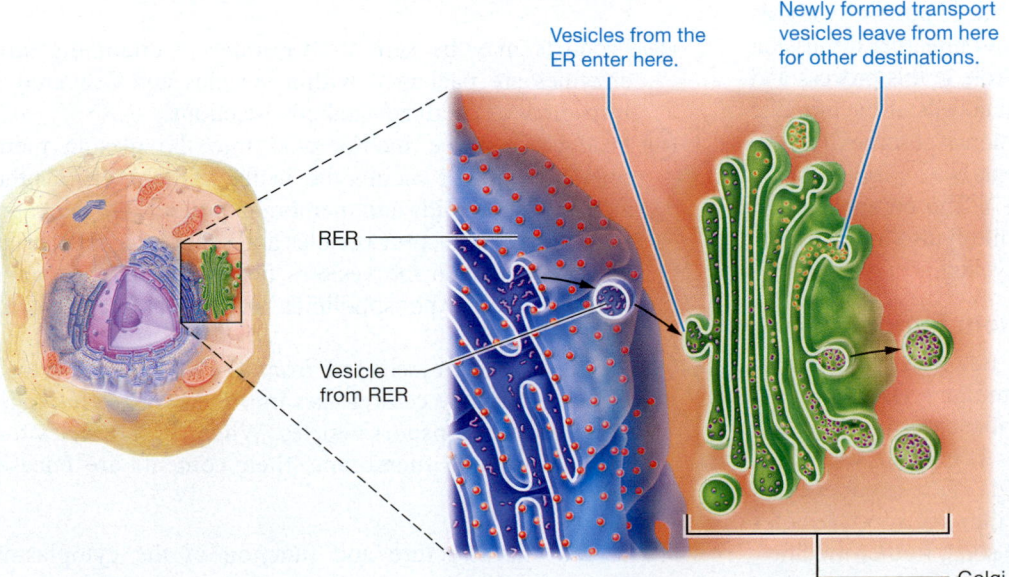

Vesicles from the ER enter here.

Newly formed transport vesicles leave from here for other destinations.

RER

Vesicle from RER

Golgi apparatus

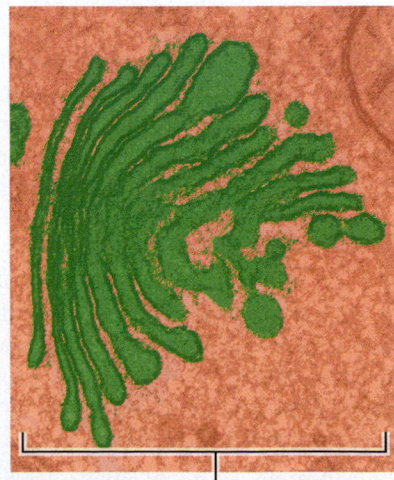

TEM (83,000×)

Figure 3.20 The Golgi apparatus.

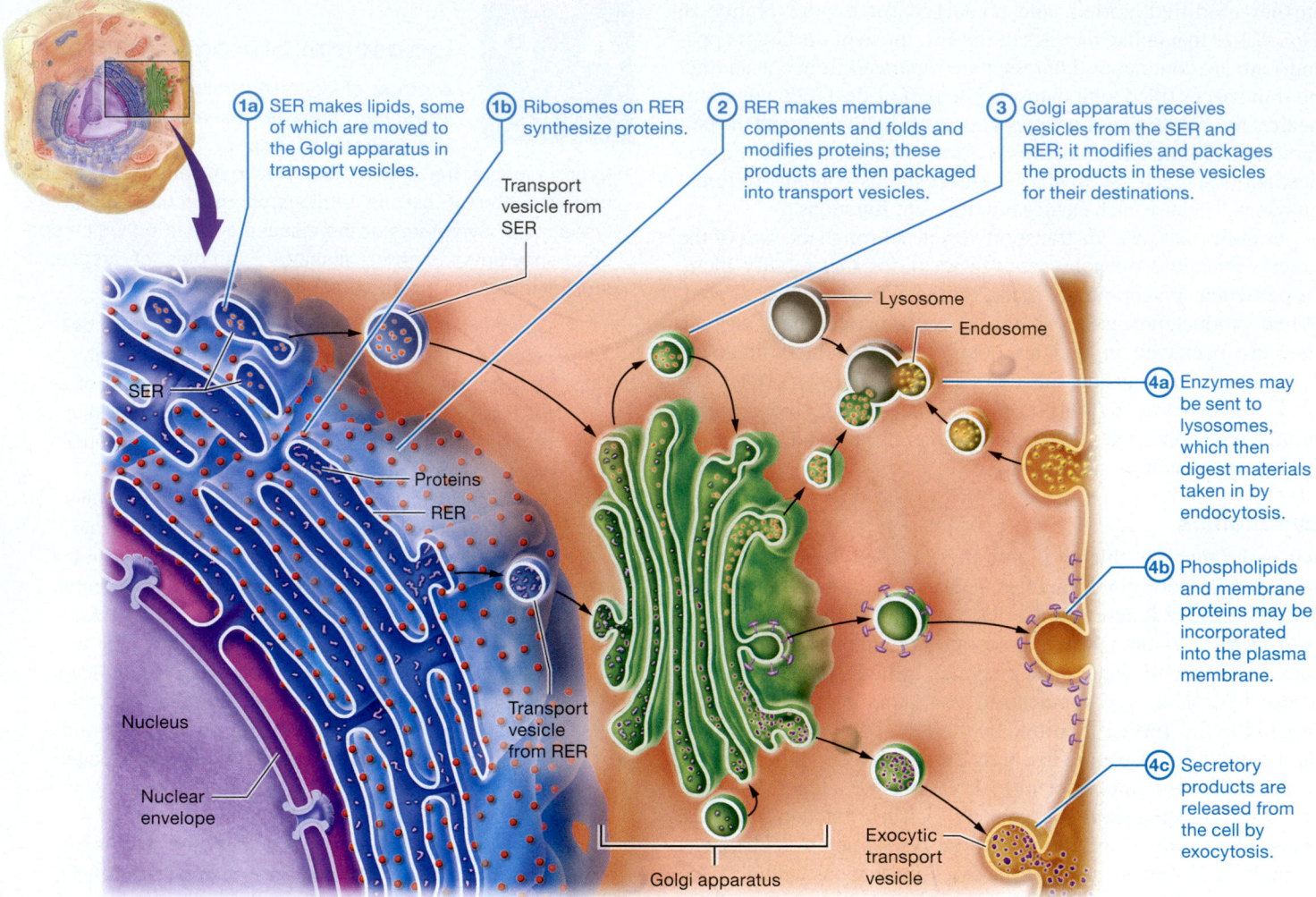

Figure 3.21 Function of the endomembrane system.

Within lysosomes, macromolecules such as proteins, carbohydrates, and fatty acids that are taken in by different types of endocytosis are degraded into smaller units. These subunits are then released into the cytosol for disposal or for use in protein synthesis and various metabolic reactions. Recall from our earlier discussion of phagocytosis that lysosomes play a key role in this process and are critical to the proper functioning of certain cells of the immune system that digest invading bacteria. In addition, lysosomes degrade old, worn-out organelles and cell components via a process called *autophagy* (aw-TAWF-uh-jee; "self-eating"). The components of the digested organelles are then returned to the cytosol to build new cell components, an example of cellular recycling.

Summary of Endomembrane System Function

Figure 3.21 provides an overview of the function of the endomembrane system. We begin with, simultaneously, (1a) the smooth ER making lipids and (1b) bound ribosomes on the rough ER synthesizing proteins. The proteins then enter the RER, where (2) proteins are folded and modified. These products, and the products from the SER, are packed into transport vesicles and sent to the Golgi apparatus, which (3) sorts, further modifies, and packages the products into vesicles.

When the finished products are ready to exit the Golgi apparatus, the vesicles that transport them may take one of three pathways to get to their appropriate destinations:

(4a) Products may be sent to lysosomes. Certain digestive enzymes are packaged within vesicles and delivered to lysosomes to catalyze catabolic reactions.

(4b) Products may be incorporated into the plasma membrane or another membrane in the cell. Newly synthesized phospholipids and membrane proteins are packaged into exocytic transport vesicles and sent to the appropriate membrane. When the vesicles fuse with this membrane, the proteins and phospholipids are incorporated into the membrane.

(4c) Products may be released from the cell by exocytosis. Products that the cell releases into the ECF are packaged into exocytic transport vesicles. When these vesicles fuse with the plasma membrane, their contents are released outside the cell.

To review the structure and function of the cytoplasmic organelles covered in this module, see **Table 3.2**.

Table 3.2 Cytoplasmic Organelles

Organelle	Structure	Function
Mitochondrion 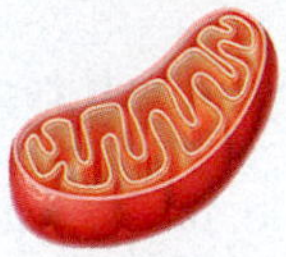	Double membrane; inner membrane folded into cristae; has own DNA and ribosomes (see Figures 3.16 and 3.17).	• Synthesizes the majority of the cell's ATP
Peroxisome 	Membrane-enclosed; similar to large vesicle.	• Detoxifies certain chemicals through oxidation reactions • Metabolizes fatty acids • Synthesizes certain phospholipids
Ribosome 	Two subunits made of proteins and rRNA; not membrane-enclosed (see Figure 3.18).	• Synthesizes proteins
Rough Endoplasmic Reticulum (RER) 	Series of saclike membranes enclosing the ER lumen; surface studded with ribosomes (see Figure 3.19).	• Modifies and folds proteins made by the ribosomes • Manufactures and assembles most components of the plasma membrane
Smooth Endoplasmic Reticulum (SER) 	Series of tubular membranes enclosing the ER lumen; surface does not contain ribosomes (see Figure 3.19).	• Stores calcium ions and synthesizes lipids • Detoxifies certain substances
Golgi Apparatus 	Stack of flattened, membrane-enclosed sacs (see Figure 3.20).	• Sorts, modifies, and packages proteins and other products made by the ER
Lysosome 	Membrane-enclosed structure with digestive enzymes; similar to a large vesicle.	• Digests damaged organelles and products brought into the cell by endocytosis • Recycles damaged organelles

Quick Check

☐ 2. Identify the following properties as belonging to rough ER, smooth ER, the Golgi apparatus, or lysosomes.

a. Synthesizes the majority of the cell's cholesterol

b. Contains digestive enzymes

c. Ensures that proteins fold into the proper shape

d. Packages proteins into secretory vesicles

e. Detoxifies many substances

f. Makes the majority of membrane components

g. Degrades compounds and organisms ingested via endocytosis

h. Contains ribosomes on its surface membranes

☐ 3. To what destinations can products from the Golgi apparatus be sent?

☐ 1. Many toxins and poisons block certain enzymes in the mitochondria. For example, many fruits, including apricots, apples, plums, peaches, and cherries, contain seeds and pits that can cause the formation and release of the poison cyanide if they are ground and ingested. Cyanide binds to an enzyme in the inner mitochondrial membrane critical to energy production. Why would this be fatal for a cell?

☐ 2. What effect do you think deficiencies in lysosomal enzymes would have on phagocytes? Explain.

☐ 3. The deadly poison ricin (RY-sin), found in castor beans, binds to and inhibits free and bound ribosomes. Explain why this effect is lethal.

See answers in Appendix A.

MODULE **3.5**
The Cytoskeleton

Learning Outcomes

1. Describe the structure and function of the three components of the cytoskeleton.

2. Describe the structure and function of centrioles, cilia, and flagella.

3. Explain the role of the cytoskeleton in cellular motion.

The components of the **cytoskeleton** function as a cell's highways. The key difference between highways and the cytoskeleton, however, is that the cytoskeleton is not a static structure—it can rapidly change its form to fit its changing functions and the needs of the cell.

The cytoskeleton is made up of several types of protein filaments organized in different ways. These filaments play a variety of critical roles within the cell, including the following:

- **Giving a cell its characteristic shape and size.** Cells in different parts of the body have different shapes and sizes in accordance with their functions. These differing shapes and sizes are largely a result of the cells' cytoskeletons, which form an internal framework for each cell.

- **Supporting the plasma and nuclear membranes as well as the organelles.** Cytoskeletal filaments maintain the structural integrity and strength of the cells' plasma membranes and the membrane surrounding the nucleus. The cytoskeleton also positions, organizes, and anchors various organelles within the cell.

- **Functioning in movement.** The protein filaments of the cytoskeleton are often associated with *motor proteins.* Motor proteins are like molecular propellers that use the energy derived from ATP hydrolysis to move themselves along the protein filaments. The cytoskeletal filaments and their motor proteins are responsible for moving organelles and vesicles within the cell; splitting the cell during cell division; and, for certain cells, propelling substances past the cell or even propelling the cell itself.

- **Performing specialized functions in different cell types.** The cytoskeleton plays an important role in the specific physiology of certain cell types. For example, the cytoskeleton allows a phagocyte to engulf a bacterium with pseudopodia, enables our muscle cells to contract, and allows our nerve cells to form long "arms" that can reach and communicate with distant targets.

In this module, we examine each of the filaments that make up the cytoskeleton. We also explore the structure and function of extensions from the cell, called *microvilli,* which function in absorption, and *cilia* and *flagella,* which function in movement.

Types of Filaments

The cytoskeleton contains three different types of long protein filaments: *actin filaments, intermediate filaments,* and *microtubules.* Each type of protein filament is composed of smaller protein subunits, which allows the rapid disassembly and reassembly of the filament.

Actin Filaments

As you can see in the left column of **Table 3.3**, **actin filaments,** also called **microfilaments,** are the thinnest members of the cytoskeleton, measuring only about 5–9 nm in diameter. They are composed of two intertwined strands of protein subunits called *actin.* Actin filaments provide structural support, bear tension (stretching forces), and help to maintain the cell's shape. They are particularly common in the outer part of the cell, near the plasma membrane, where they form three-dimensional supportive networks.

Actin filaments also play a role in moving the cell. The motor protein associated with actin filaments, called *myosin,* is capable of moving the entire cell. Many different types of myosin proteins exist, but all contain a globular head that uses energy from ATP hydrolysis to move actin filaments or the myosin protein itself. Myosin-actin filament interactions are responsible for the contraction of muscle cells, for the pinching of the cell in two during cell division, and for changes in the membrane that take place during endocytosis and exocytosis. Additionally, myosin-actin filament interactions allow phagocytes to "crawl" or migrate within our bodies in search of invaders.

Intermediate Filaments

The second type of protein filament in the cytoskeleton is the ropelike **intermediate filament** (shown in the middle column of Table 3.3). These filaments, which measure about 10 nm in diameter, are named "intermediate" because they are intermediate in size between the small actin filaments and the large microtubules. Intermediate filaments are made up of a diverse group of fibrous proteins, including keratin. The long strands are twisted together

Table 3.3 Cytoskeletal Filaments

Property	Actin Filaments	Intermediate Filaments	Microtubules
Location in cell	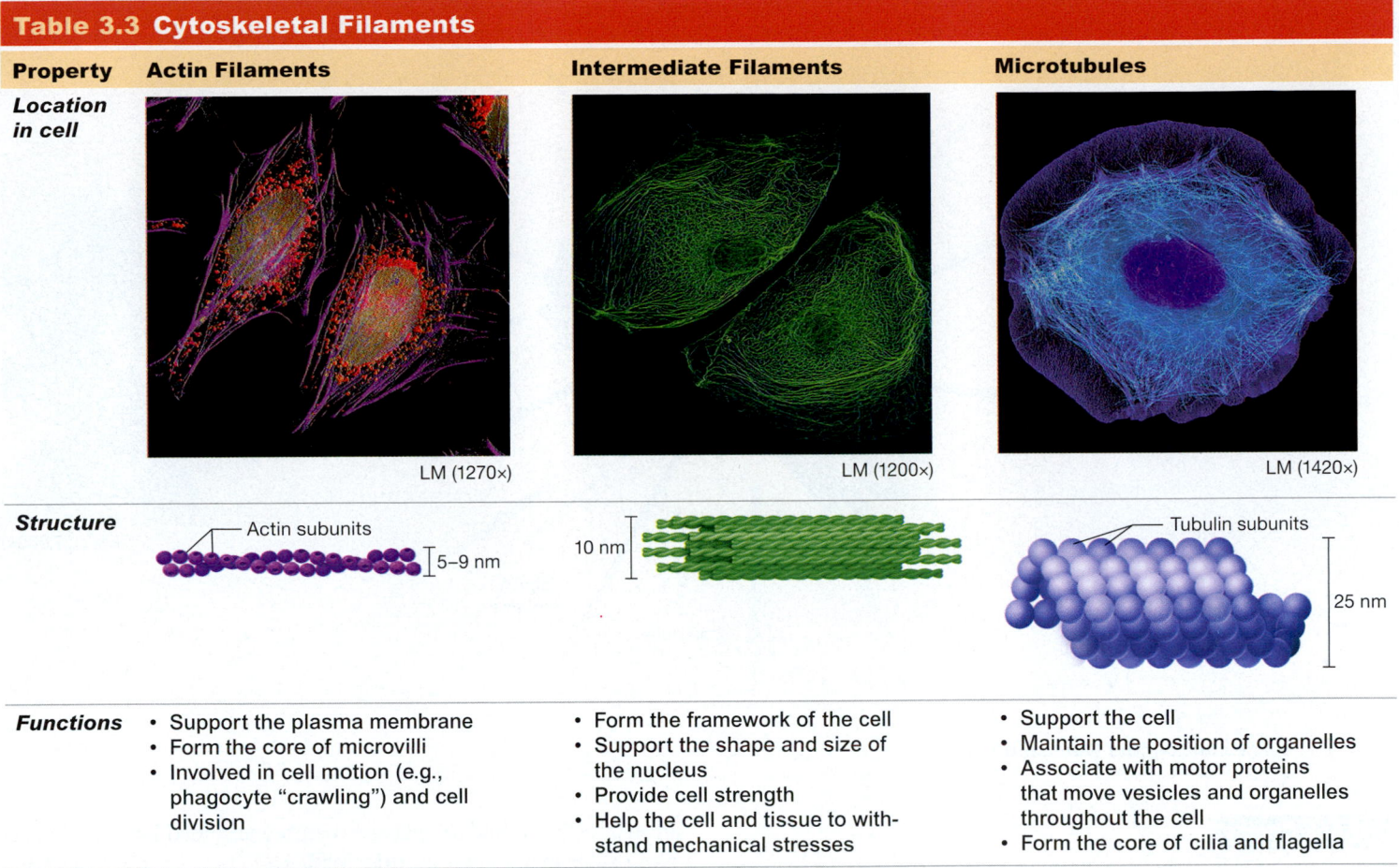 LM (1270×)	LM (1200×)	LM (1420×)
Structure	Actin subunits 5–9 nm	10 nm	Tubulin subunits 25 nm
Functions	• Support the plasma membrane • Form the core of microvilli • Involved in cell motion (e.g., phagocyte "crawling") and cell division	• Form the framework of the cell • Support the shape and size of the nucleus • Provide cell strength • Help the cell and tissue to withstand mechanical stresses	• Support the cell • Maintain the position of organelles • Associate with motor proteins that move vesicles and organelles throughout the cell • Form the core of cilia and flagella

to form miniature "ropes" within the cell. Like a rope, an intermediate filament bends and twists easily but is quite difficult to break. As a result, intermediate filaments are stronger and more difficult to disassemble than other cytoskeletal filaments, and they form more permanent structures within the cell.

The strength and stability of intermediate filaments enable them to have a variety of roles within the cell, which include the following:

- forming a network within the cytoplasm that gives the cell mechanical strength;
- forming a network under the membrane around the nucleus to support its shape and size; and
- uniting adjacent cells in a tissue so that the tissue can withstand mechanical stresses.

Microtubules and Centrioles

The largest members of the cytoskeleton are the **microtubules,** which are hollow rods or "tubes" that measure about 25 nm in diameter (see Table 3.3, right column). Microtubules are composed of protein subunits called *tubulins* that can be rapidly added or removed from a microtubule, allowing it to change size as needed by the cell.

Microtubules have a wide range of functions within the cell. For one, they maintain the internal architecture of the cell,

positioning organelles in their proper places. In fact, if microtubules are disrupted, the endoplasmic reticulum collapses in on itself, and the sacs of the Golgi apparatus disperse in the cytosol.

In addition, microtubules and their associated motor proteins move organelles and vesicles around the cell. The motor proteins called dynein (DYN-een) and kinesin (kih-NEE-sin) move along microtubules, carrying their "cargo," in the same way a train moves along a track. Finally, microtubules form the core of cellular extensions called *cilia* and *flagella*, which we discuss in the next section.

Notice in **Figure 3.22** that microtubules extend outward from a region of the cell called the **centrosome.** The centrosome consists of a gel matrix containing the tubulin proteins needed to build microtubules. When the cell is not dividing, the centrosome acts as a *microtubule-organizing center* for the cell's cytoskeleton. During this time the centrosome is located near the nucleus.

The centrosome also contains a pair of tube-shaped structures called **centrioles.** Centrioles are formed by modified microtubules and other proteins. Notice in the figure that there are 9 groups of microtubules and they are arranged in groups of 3 (for a total of 27). The centrioles play a critical role in cell division, which we explore in a later module (see Module 3.8). Modified centrioles called **basal bodies** are found on the internal surface of the plasma membrane. Both cilia and flagella, which we discuss next, sprout from these basal bodies.

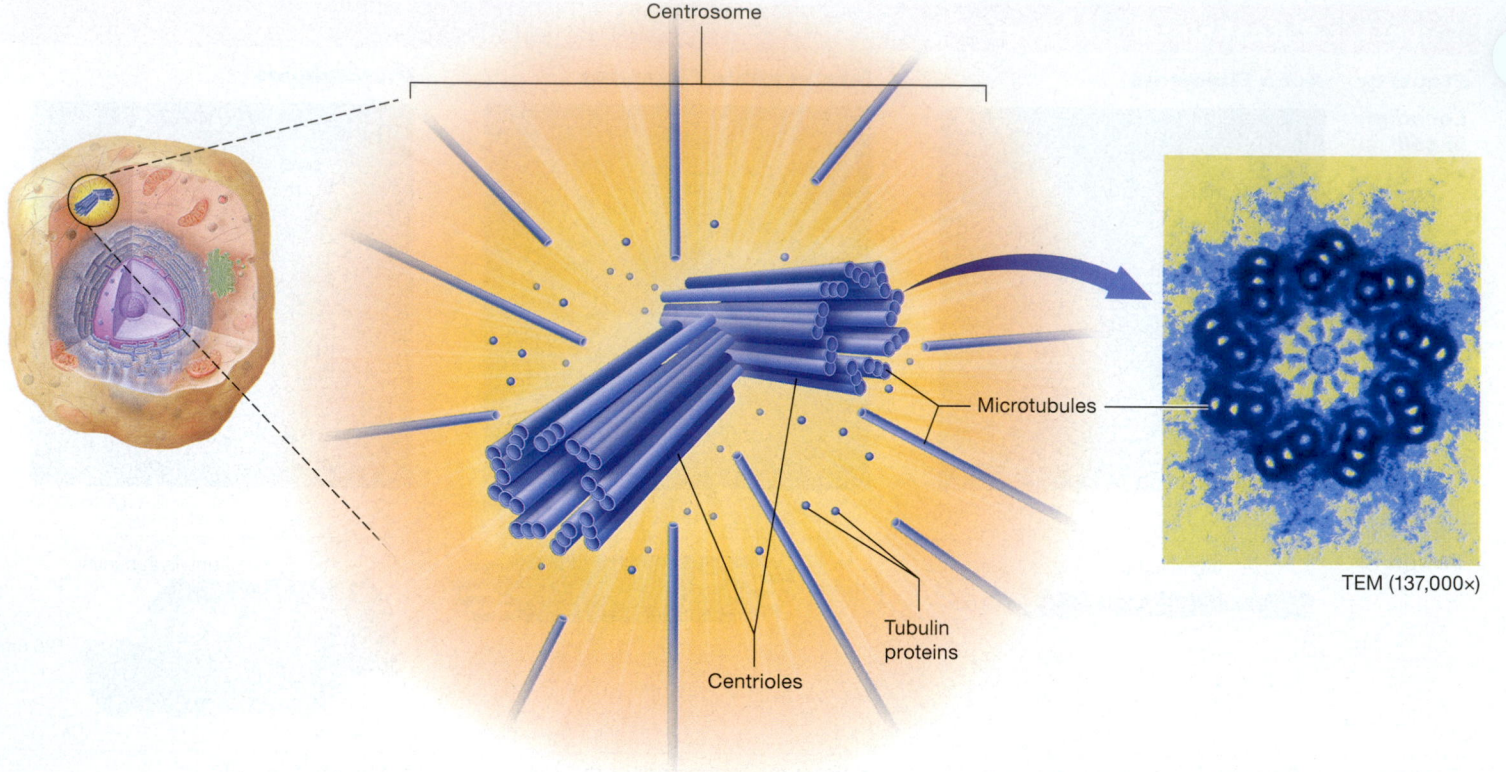

Figure 3.22 **The centrosome with centrioles.**

Quick Check

☐ 1. Determine whether the following statements apply to microtubules, intermediate filaments, or actin filaments.

   a. Thinnest cytoskeletal filaments; are associated with the motor protein myosin

   b. Consist of proteins twisted together into a ropelike form

   c. Allow certain cells to "crawl" and change the shape and size of the cell

   d. Consist of protein subunits surrounding a hollow core

   e. Help a cell to resist mechanical stress

   f. Grow from the cell's centrosome

Cellular Extensions

The cytoskeleton forms the inner framework of cell surface extensions in many cells. These extensions include microvilli, which are folds of the plasma membrane, and cilia and flagella, which are extensions that function in movement. Let's take a closer look at each.

Microvilli

Like all structures in the body, the plasma membrane obeys the rule that its form must follow its function. For this reason, we would expect to see the plasma membrane taking on a slightly different shape in cells of the body where its function is different. This is, in fact, exactly what we see. Cells that are specialized for absorption, such as those in the kidneys and small intestine, have plasma membranes with a large surface area to perform absorption efficiently.

In such cells we find the plasma membrane folded into tiny finger-like extensions known as **microvilli** (**Figure 3.23**; my-kroh-VIL-aye; *villus* = "hair"). On microscopic examination, cells with microvilli resemble a brush, and so these extensions are sometimes referred to collectively as a *brush border*. Microvilli increase the surface area of the plasma membrane up to about 30–40 times (compared to a cell without microvilli), which allows the cell to absorb substances more rapidly, an example of the Structure-Function Core Principle (p. 28). Actin filaments found at the core of microvilli help to maintain their shape.

CORE PRINCIPLE
Structure-Function

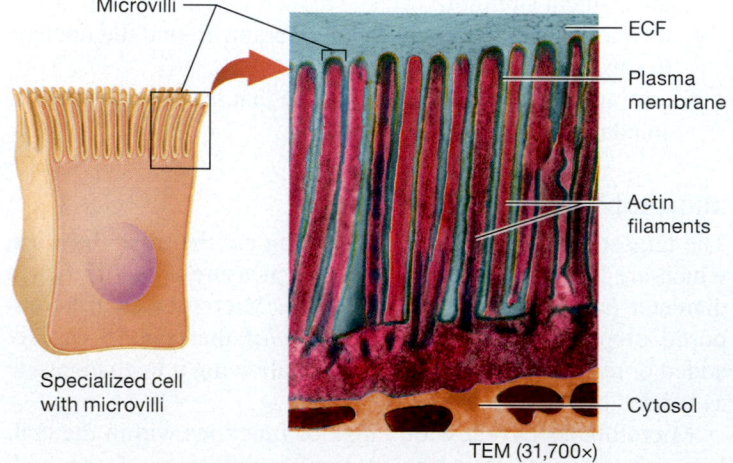

Figure 3.23 **Microvilli.**

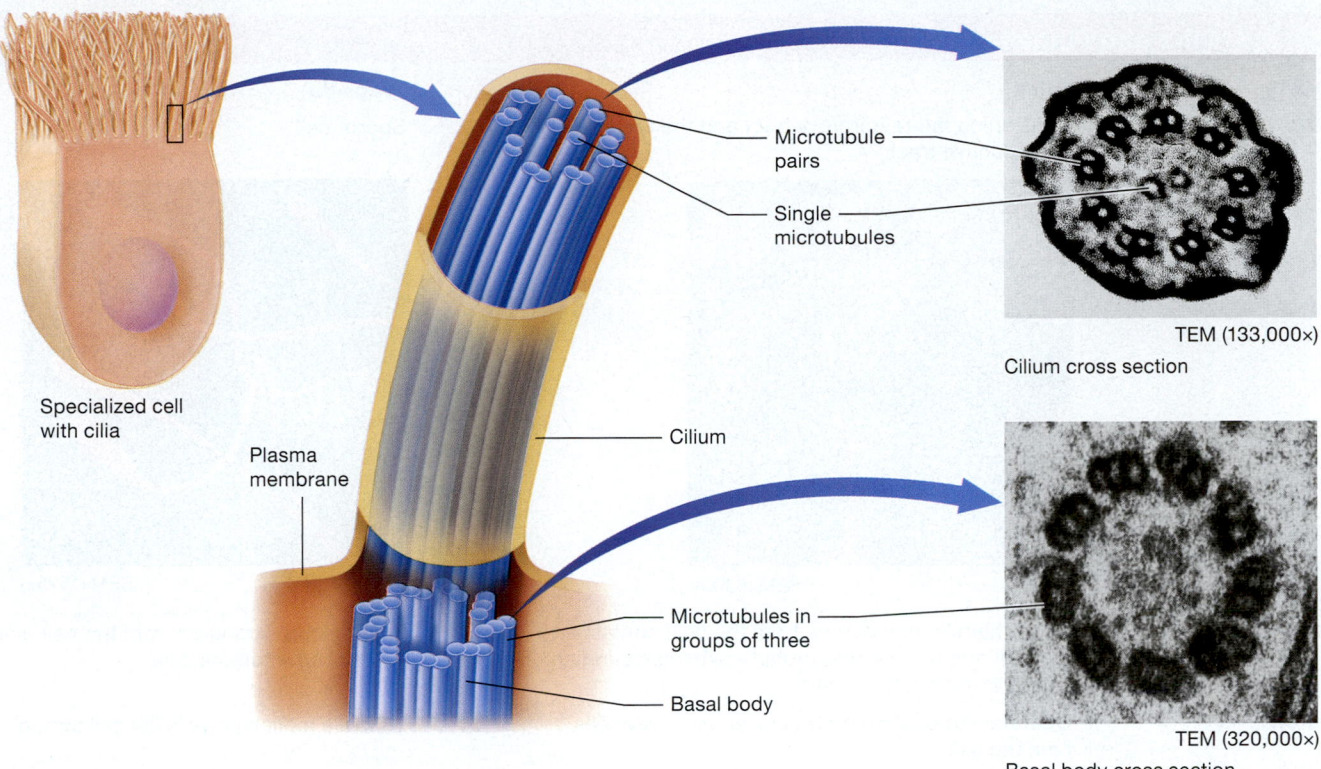

Microtubule pairs

Single microtubules

TEM (133,000×)

Cilium cross section

Specialized cell with cilia

Plasma membrane

Cilium

Microtubules in groups of three

Basal body

TEM (320,000×)

Basal body cross section

Figure 3.24 Structure of cilia and flagella. Although only a cilium is shown here, the flagellum has the same structure.

Cilia and Flagella

Certain cells in the body have extensions from their surfaces called **cilia** (SILL-ee-uh; singular, *cilium*). Although cilia and microvilli may look superficially similar, they actually differ significantly in structure and function. For one, cilia are much larger than microvilli—individual cilia are visible with a light microscope, whereas microvilli appear only as a thickened border along a cell's surface. In addition, they contain different cytoskeletal filaments. Microvilli are supported by actin filaments, whereas cilia contain a core of microtubules associated with motor proteins. This reflects an important functional difference between the two structures: The microtubules and motor proteins enable cilia to be *motile*, or moveable.

Cilia share many structural similarities with another type of motile cellular extension called **flagella** (flah-JEL-uh; singular, *flagellum*). As you can see in **Figure 3.24**, cilia and flagella share the same basic structure, which is similar to that of centrioles. However, instead of the nine groups of three microtubules we saw in centrioles, cilia and flagella are composed of a ring of nine *pairs* of microtubules surrounding two central microtubules. Both cilia and flagella move via the action of motor proteins such as dynein that are associated with their microtubules. The motor proteins crawl along the microtubules, causing the cilium or flagellum to bend.

Although they are similar structurally, cilia and flagella differ in function. Cilia are shorter and move in a coordinated beating motion (**Table 3.4**). They are present in very large numbers on the surfaces of certain cells, where each cilium functions like a tiny broom; these "brooms" beat together to sweep substances over the cell surface. For example, the cells that line our respiratory passages have cilia that sweep out mucus littered with dust and debris we have inhaled. Note on the left side of Table 3.4 that the motion of each cilium resembles the movement of an oar propelling a boat—there is an initial stroke, followed by the cilium curving back to its original position. See *A&P in the Real World: Primary Ciliary Dyskinesia* to read about what happens when these cilia don't function properly.

Primary Ciliary Dyskinesia

The importance of properly functioning cilia is evident in **primary ciliary dyskinesia (PCD),** a rare genetic disorder that results in a defect in one or more of the protein components of cilia and flagella. This condition affects many types of cells, including those lining the respiratory passage, the middle ear, and the uterine tube in females, as well as sperm cells in males.

Ciliary dysfunction in the respiratory passages leads to the buildup of mucus in the lungs, which increases the risk of respiratory infections. Individuals with PCD commonly sustain progressive damage to the lungs due to repeated infections and mucous plugs. In addition, PCD patients may experience repeated ear infections, which can lead to hearing loss. Finally, because their sperm cells lack motility, male patients may be infertile.

Table 3.4 Cilia and Flagella

Property	Cilia	Flagella
Location in body	Cells lining the respiratory tract and the female reproductive tract	Sperm cell
	SEM (5000×)	SEM (1525×)
Structure	Short, hairlike extensions from the cell; contain an internal ring of nine microtubule pairs surrounding a central microtubule core	Single, long extension from the cell; same internal structure as cilia
Function	Coordinated beating motion sweeps substances past the cell.	Whiplike motion propels the cell through liquid.

By contrast, flagella are long extensions that move in a whiplike manner to propel a whole cell (see the right side of Table 3.4). A cell generally has no more than one flagellum, and the only flagellated cell in the human body is the sperm cell in males.

Quick Check

☐ 2. What are microvilli, and what are their primary functions?

☐ 3. How are cilia and flagella similar? How do they differ?

Apply What You Learned

☐ 1. Predict the cellular effects of a disorder that prevents the assembly of microtubules. (*Hint:* Why is it equally important for a microtubule to be easily assembled and disassembled?)

☐ 2. Explain how the disorder in question 1 would affect cilia and flagella.

☐ 3. Explain how the structure of intermediate filaments makes them better suited to provide mechanical strength than either microtubules or actin filaments.

See answers in Appendix A.

MODULE **3.6**

The Nucleus

Learning Outcomes

1. Describe the structure and function of the nucleus.

2. Explain the structure of chromatin and chromosomes.

3. Analyze the interrelationships among chromatin, chromosomes, and sister chromatids.

4. Describe the structure and function of the nucleolus.

As you may recall, the cell consists of multiple components that must function together optimally for the cell to survive. To ensure such function, a cell needs a structure to "govern" each of these components. The structure that serves this purpose in our cells is the **nucleus** (**Figure 3.25**). We discussed earlier that the nucleus houses the cell's **deoxyribonucleic acid (DNA),** and that DNA contains the code, or "plans," for almost every protein in the body (see Chapter 2). The plans, called *genes*, within the DNA are used by different kinds of **ribonucleic acid (RNA)** to build our cells' proteins.

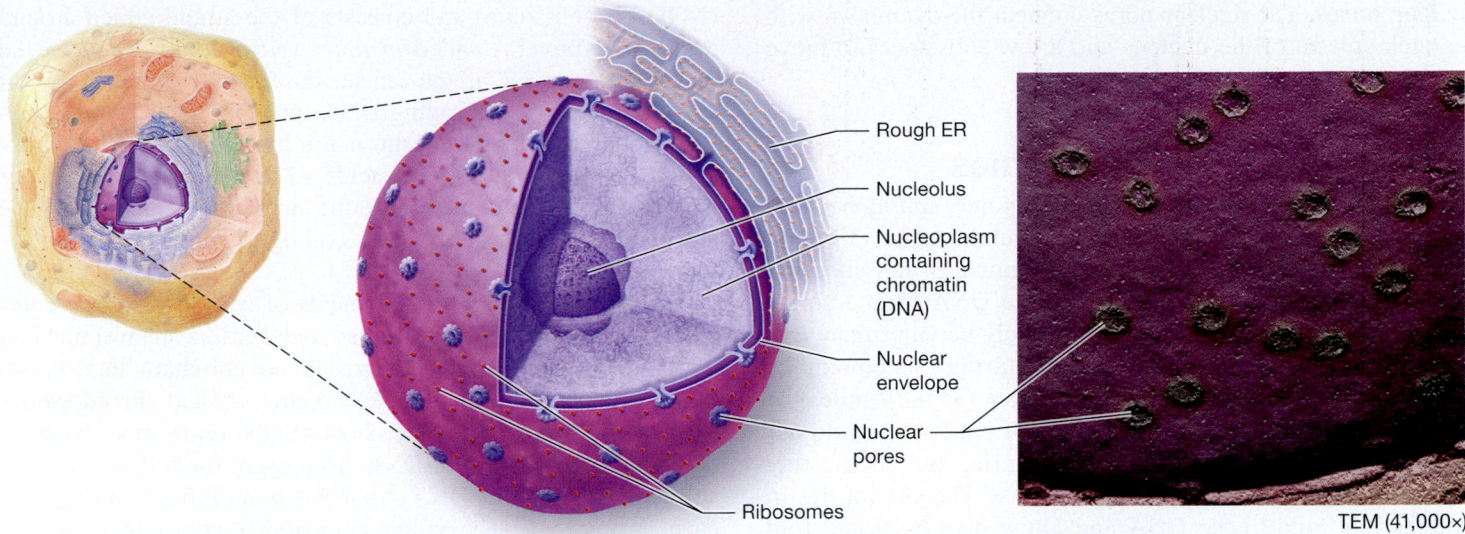

Figure 3.25 The nucleus.

But how does simply holding this code allow the nucleus to coordinate the activities of the cell? Remember that nearly all of a cell's structural components in the cytoskeleton and organelles are proteins, almost all our enzymes are proteins, and proteins are also a vital component of the plasma membrane. The nucleus in large part determines the type of proteins that a cell makes and the rate at which it makes them.

The nucleus consists of three main structures: (1) an enclosing membrane, (2) DNA and its associated proteins, and (3) the *nucleolus* (noo-klee-OH-luss). The enclosing membrane, called the *nuclear envelope,* is pierced by occasional openings known as *nuclear pores.* Both the DNA and the nucleolus are located within a cytosol-like gel called the *nucleoplasm.* Nucleoplasm contains many different components, including water, free-floating nucleotides, enzymes and other proteins, and different types of RNA. The most notable component of the nucleoplasm,

however, is our DNA. Most of the time, DNA is found here in a loosely organized form called *chromatin* (KROH-muh-tin). In this module, we examine the components of the nucleus so that we can address their functions in later modules.

Nuclear Envelope

In **Figure 3.26** you can see that the **nuclear envelope** (or *nuclear membrane*) surrounding the nucleus is composed of a phospholipid bilayer. Like the mitochondrial membrane, the nuclear envelope is a double membrane. The outer membrane is studded with ribosomes and, as mentioned, is continuous with the endoplasmic reticulum. The inner membrane lines the interior of the nucleus and is supported by a network of intermediate filaments called the *nuclear lamina.* The two membranes are joined at large protein complexes that produce the

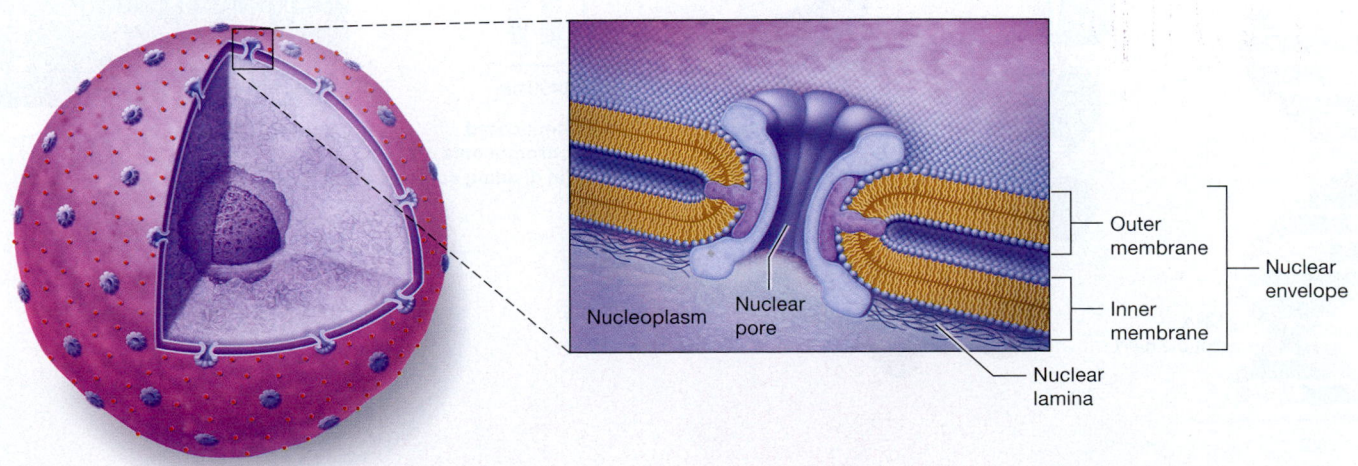

Figure 3.26 The nuclear pore.

nuclear pores. The nuclear pores connect the cytoplasm with the nucleoplasm of the nucleus and allow substances to move between the two areas.

Chromatin and Chromosomes

The task of storing our DNA is a daunting one, simply because of its sheer size: The human genome contains about 3 billion base pairs with about 25,000 genes. A nucleus may measure only about 6 μm, but if you removed the DNA from a single nucleus and stretched it out, it would likely be taller than you, extending about 2 meters. The feat of storing this amount of DNA is equivalent to taking 14 kilometers (about 9 miles) of very thin thread and shoving it into a golf ball! So how do our nuclei manage to squeeze so much material into such a tiny space without it becoming a tangled mess? The answer lies in proteins that bind to the DNA and allow it to twist and fold into an ordered and progressively more compact form.

Chromatin consists of one extremely long DNA strand and its associated proteins. **Figure 3.27a** shows how chromatin is organized and elaborately folded to make it a compact structure. The most basic unit of organization after the DNA strand resembles beads on a string. Each "bead" is called a *nucleosome*

(NOO-klee-oh-sohm) and consists of the strand coiled around a group of proteins called *histones* (HISS-tonz). Between the beads in Figure 3.27a, you can see the "string," which is a short segment of DNA. Packaging DNA into this nucleosome structure reduces the length of chromatin by about one-third. Interactions between the amino acids of the histone proteins of neighboring nucleosomes cause the string of beads to condense and shorten even further, as they coil tightly around one another like a spring.

During the period between rounds of cell division, the chromatin threads are loose and tangled together like an unwound ball of yarn. However, during cell division, the chromatin threads coil tightly and condense into barlike structures called **chromosomes** (KROH-moh-sohmz) that are about 10,000 times more compact than the original strand of DNA. To prepare for cell division, the cell has duplicated its DNA, so at this point it has two copies of each chromosome. The paired and identical copies of a chromosome are called *sister chromatids* and are joined at a region of the chromosome known as the **centromere.** Individual sister chromatids in a dividing cell are visible in the nucleus under a microscope; they look like what you see in **Figure 3.27b**.

Chromosomes vary in size and contain up to about 245 million base pairs each. Most human cells contain two sets of 23

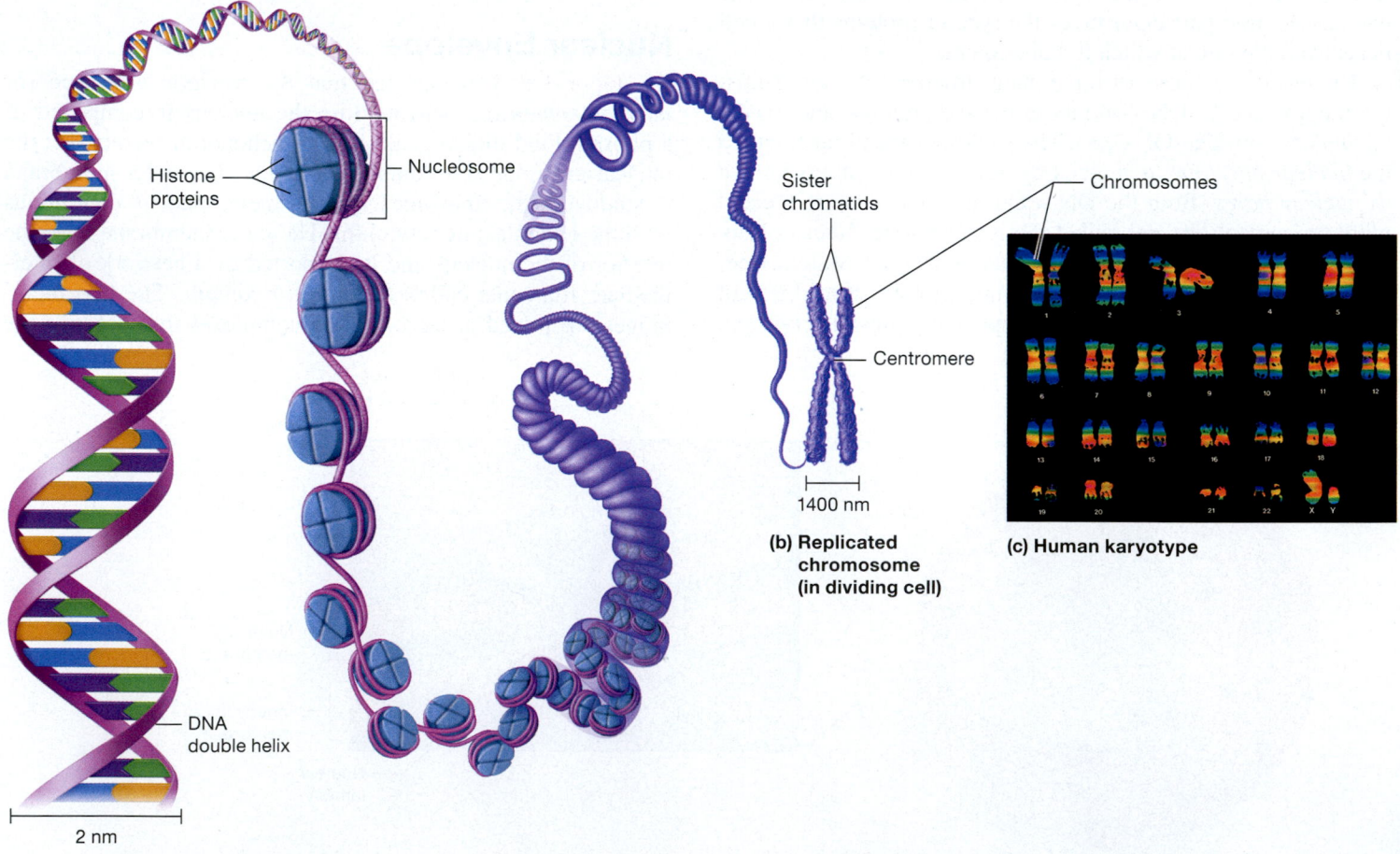

Histone proteins

Nucleosome

DNA double helix

2 nm

(a) The structure of chromatin

Sister chromatids

Centromere

1400 nm

(b) Replicated chromosome (in dividing cell)

Chromosomes

(c) Human karyotype

Figure 3.27 Chromatin and chromosomes.

chromosomes (46 total), with one set of 23 chromosomes inherited from each parent. The maternal and paternal chromosomes are called *homologous chromosomes* (or simply homologs). **Figure 3.27c** shows a *karyotype* (KEHR-ee-oh-ty′p), which contains the full set of 23 pairs of homologous chromosomes. These chromosomes are collected from a dividing cell, photographed, and then organized into their homologous pairs. Note that the pairs of chromosomes differ significantly in size and appearance (see Chapter 27 for more information).

Quick Check

☐ 1. What are the main components of the nucleus? What is its main function?

☐ 2. What is chromatin? How are chromatin and chromosomes related?

Nucleoli

◀◀ FLASHBACK

1. What is the structure of ribonucleic acid (RNA)? (p. 60)

2. What makes up a ribosome? (p. 90)

If you look back at Figure 3.25, you'll see that among the chromatin in the nucleoplasm is a dark-staining structure called the **nucleolus** (cells sometimes have more than one of these *nucleoli*). We can think of a nucleolus as a "ribosome factory"—this is where ribosomes are assembled. It is not enclosed by a membrane, but rather is simply a large aggregate of proteins, DNA, and RNA. As we discussed earlier in this chapter, much of our DNA consists of genes that specify the amino acid sequences of proteins. However, in the nucleolus we find segments of DNA containing genes that don't code for a protein, but instead just specify the nucleotide sequence of an RNA strand. Some of these segments, known as *rRNA genes,* code for ribosomal RNA (rRNA), which is a key component of ribosomes.

Quick Check

☐ 3. What is a nucleolus, and what is its function?

Apply What You Learned

☐ 1. Can a human cell without a nucleus make new ribosomes? Why or why not?

☐ 2. When a biologist is trying to isolate DNA from a cell, he or she must first add enzymes that degrade (break down) proteins. Why is this step necessary?

☐ 3. The nucleolus can change size, depending on how many ribosomes the cell needs to produce. When do you think a cell would need more ribosomes and therefore a larger nucleolus? (*Hint:* What do ribosomes do?)

See answers in Appendix A.

MODULE **3.7**
Protein Synthesis

Learning Outcomes

1. Describe the genetic code, and explain how DNA codes for specific amino acid sequences.

2. Describe the processes of transcription and translation.

3. Explain the roles of rRNA, mRNA, and tRNA and of ribosomes in protein synthesis.

We have already discussed the molecular role of our genetic material: DNA is required to make the proteins within all the cells of our bodies. DNA does this by carrying information that specifies the amino acid sequence for each of the body's proteins and the nucleotide sequence for certain RNA strands. The process of making proteins from DNA via RNA is called **protein synthesis.**

The human *genome*, our complete genetic makeup, contains about 25,000 genes. Over 99% of the coding DNA between individual humans is identical, which makes sense, as critical enzymes and structural proteins aren't going to vary from person to person without causing disease. The remainder of the coding DNA is unique for every individual (except identical twins) and is responsible for each person's unique physical traits.

But how, exactly, do proteins correlate with the vast array of physical traits that we humans possess? Take the simple example of the skin's tendency to either tan or freckle. A gene called the melanocortin-1 receptor gene codes for a membrane protein found in certain skin cells. When this protein is exposed to ultraviolet radiation, it triggers the production of pigment compounds. One variant of the melanocortin-1 receptor gene yields a protein that responds strongly to ultraviolet radiation, resulting in an even tan. A slightly different variation of this same gene yields a protein that responds weakly to ultraviolet radiation, resulting in freckling instead of tanning. The production of a protein from a specific gene is called *gene expression.*

In this module we examine the steps our cells take to build a protein. We begin with an exploration of the rules that govern how DNA specifies the sequence of a protein. We then discuss the two processes that actually make the proteins: transcription and translation. During *transcription,* the code specified by a gene is copied, creating a strand of *messenger RNA* (*mRNA*). In *translation,* ribosomes read the nucleotide sequence of the mRNA strand and synthesize a polypeptide chain containing the correct amino acid sequence. You can picture this process as follows:

Genes and the Genetic Code

◀◀ FLASHBACK

1. What is a nucleotide? (p. 59)

2. What four nucleotides are present in DNA? (p. 61)

Recall from earlier in the chapter that DNA is composed of long chains of nucleotides containing **genes**—segments of DNA, most of which specify the amino acid sequences of proteins. Although there are only four nucleotides in DNA, there are 21 different amino acids. This means that one or even two nucleotides can't code for a single amino acid, because there aren't enough possible combinations. However, there are enough combinations with groups of three nucleotides, and it is these groups (for example, AGT or CCG) that are the "words" that code for amino acids. These words are called **triplets,** and each amino acid is represented by one or more nucleotide triplets.

During transcription, each DNA triplet is transcribed into a complementary three-nucleotide sequence of mRNA called a **codon** (KOH-don). In essence, the codon is a complementary RNA "copy" of the DNA triplet. Recall from an earlier chapter how complementary base pairing works: The nucleotide bases C and G always pair with each other, and A always pairs with T (in DNA) or U (in RNA) (see Chapter 2). Then during translation, each mRNA codon is matched with an amino acid.

The **genetic code** is a list of which amino acid is specified by each possible DNA triplet. **Figure 3.28** shows this code, with each amino acid and the corresponding mRNA codons that specify it. Each three-nucleotide mRNA codon shown in the figure corresponds to the copied DNA triplet (not shown) and represents a single amino acid; often more than one codon codes for a single amino acid. This genetic code is found in all forms of life on Earth.

What happens when something alters the DNA so that its code is changed? The answer can range from having no effect to causing a serious disease such as sickle cell anemia (look ahead to Figure 19.7). Notice that the genetic code in Figure 3.28 is precise, and a difference in as little as a single nucleotide can change the corresponding amino acid entirely. For example, a change from the codon UUU to the codon UCU results in coding for the amino acid serine (Ser) instead of phenylalanine (Phe). Changes to DNA are called **mutations,** and they can be due to mistakes in the copying of DNA or can be induced by agents known as *mutagens*. Common mutagens include ultraviolet light and other forms of radiation, chemicals such as benzene, and infection with certain viruses. DNA mutations are the basis for many diseases, including cancer, which we discuss in the next module.

Now that we have a better understanding of how DNA codes for a protein, we can refine our definition of a gene: A *gene* is a series of nucleotide triplets that specifies the sequence of amino acids in a protein. But this is still not quite the whole story. The average gene is about 8000 nucleotides long, but the average protein contains only about 400 amino acids, which should require only about 1200 nucleotides. The extra parts of the gene are *noncoding*—they are chains of nucleotides that do not specify any amino acids. Much of the noncoding parts of a gene are segments called *introns;* the coding portions of a gene are *exons*. Scientists once thought that introns were "junk DNA," but recent evidence suggests that they play an important role in the evolution of new proteins and that they help to regulate gene expression. Let's now examine how a cell reads the instructions within a gene and uses them to make a protein.

Transcription

 **FLASHBACK**

1. What are the main differences between DNA and RNA? (p. 61)

2. Which nucleotides are complementary to one another and so are able to form hydrogen bonds in both DNA and RNA? (p. 61)

You may recall from earlier in the chapter that most of a cell's DNA resides in the nucleus, and that protein synthesis occurs on ribosomes, which are found outside the nucleus in the cytoplasm. This arrangement presents our cells with a problem: The DNA that codes for the proteins doesn't leave the nucleus, but the information must somehow leave and get into the cytosol so that it can be read there by the ribosomes. Our cells solve this problem by making a copy of the gene, called a *transcript,* which is a strand of RNA that can exit the nucleus and enter the cytosol.

SECOND BASE

	U	C	A	G	
U	UUU ⌉ Phe / UUC ⌋ / UUA ⌉ Leu / UUG ⌋	UCU / UCC / UCA / UCG ⌋ Ser	UAU ⌉ Tyr / UAC ⌋ / UAA Stop / UAG Stop	UGU ⌉ Cys / UGC ⌋ / UGA Stop / UGG Trp	U / C / A / G
C	CUU / CUC / CUA / CUG ⌋ Leu	CCU / CCC / CCA / CCG ⌋ Pro	CAU ⌉ His / CAC ⌋ / CAA ⌉ Gln / CAG ⌋	CGU / CGC / CGA / CGG ⌋ Arg	U / C / A / G
A	AUU ⌉ Ile / AUC / AUA ⌋ / AUG Met or Start	ACU / ACC / ACA / ACG ⌋ Thr	AAU ⌉ Asn / AAC ⌋ / AAA ⌉ Lys / AAG ⌋	AGU ⌉ Ser / AGC ⌋ / AGA ⌉ Arg / AGG ⌋	U / C / A / G
G	GUU / GUC / GUA / GUG ⌋ Val	GCU / GCC / GCA / GCG ⌋ Ala	GAU ⌉ Asp / GAC ⌋ / GAA ⌉ Glu / GAG ⌋	GGU / GGC / GGA / GGG ⌋ Gly	U / C / A / G

(FIRST BASE on left, THIRD BASE on right)

Key:

Abbreviation	Amino acid	Abbreviation	Amino acid
Ala	Alanine	Leu	Leucine
Arg	Arginine	Lys	Lysine
Asn	Asparagine	Met	Methionine
Asp	Aspartic acid	Phe	Phenylalanine
Cys	Cysteine	Pro	Proline
Glu	Glutamic acid	Ser	Serine
Gln	Glutamine	Thr	Threonine
Gly	Glycine	Trp	Tryptophan
His	Histidine	Tyr	Tyrosine
Ile	Isoleucine	Val	Valine

Figure 3.28 The genetic code. This figure shows the mRNA codons that correspond to amino acids.

The transcript is built with the help of an enzyme that serves as the cell's "copy machine," called *RNA polymerase* (remember that the word polymer refers to a compound made of repeating units—see Chapter 2). When RNA polymerase binds to a gene, it brings in nucleotides one at a time that are complementary to the DNA strand and links them to build a polymer of **messenger RNA (mRNA).** As the complementary mRNA strand forms, it joins via temporary hydrogen bonds to the DNA strand.

The process by which the mRNA strand is made is called **transcription,** and it proceeds in three general stages: (1) *initiation*, the beginning of transcription; (2) *elongation*, when nucleotides are linked in a specified order; and (3) *termination*, the end of transcription.

This occurs in the following steps, illustrated in **Figure 3.29**:

1. **Initiation:**

   1a. **Transcription begins when transcription factors bind to the promoter.** The process gets started when proteins called *transcription factors* bind to a DNA segment near the gene, called the *promoter*. The promoter is found on one strand of the DNA, called the *template strand.*

   1b. **RNA polymerase binds to the promoter as well, and a segment of DNA unwinds.** Once certain transcription factors have bound to the promoter, RNA polymerase also binds. At this point, an enzyme called *helicase* (HEE-lih-kayz) catalyzes the unwinding of a segment of the DNA double helix.

2. **Elongation: RNA polymerase builds a complementary mRNA transcript with free nucleotides.** RNA polymerase brings in free nucleotides from the nucleoplasm one by one and catalyzes the formation of hydrogen bonds between the free nucleotides and the nucleotides in the template strand of DNA. Each nucleotide is bound by hydrogen bonds to its complementary base in the DNA. For example, the figure shows the mRNA codon GCU bound to the complementary DNA triplet CGA. As each new nucleotide is brought by the RNA polymerase, this enzyme covalently adds the nucleotide to the end of the growing mRNA transcript. After a section of DNA is copied, the strands re-form the double helix.

3. **Termination: Transcription ends when the end of the gene is reached, and the mRNA transcript is released.** The process continues until the RNA polymerase reaches a nucleotide sequence that signals the end of the gene. At this point enzymes catalyze the release of the mRNA transcript, and the RNA polymerase detaches from the DNA.

If the process of transcription is disrupted, the consequences for the cell, and perhaps the whole body, are severe. *A&P in the Real World: Toxicity of the "Death Cap" Mushroom* talks about one type of poison that can cause such a disruption.

Toxicity of the "Death Cap" Mushroom

What makes some mushrooms poisonous? Consider the "death cap" mushroom (*Amanita phalloides* [fah-LOY-deez]); it and other mushrooms of the same genus (*Amanita*) are responsible for 95% of mushroom-related fatalities worldwide. *A. phalloides* is particularly dangerous because it has a pleasant taste and resembles many species of nontoxic mushrooms. Its main toxin inhibits RNA polymerase, which prevents the formation of new strands of mRNA. This essentially stops protein synthesis and disrupts many of the cell's functions, leading to cell death. No antidote exists for this type of poisoning, although several agents have shown promise as treatments. The liver suffers the most damage from *A. phalloides* poisoning, and patients who survive generally require a liver transplant.

After transcription, the transcript, called *pre-mRNA,* isn't yet ready to be sent out into the cytosol because it must first be modified in several ways. As part of this process, large portions of the pre-mRNA transcript are actually removed before it exits the nucleus. Recall that not all regions of a gene contain actual code, or exons—some of the gene instead consists of introns. The copied introns in the pre-mRNA must be removed and the exons spliced together before the final mRNA transcript is sent to the cytosol:

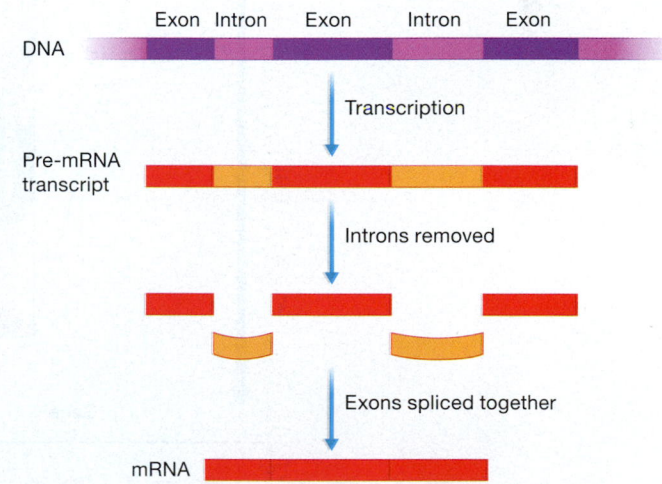

This step is called *RNA processing*. When processing is complete, the mRNA exits the nucleus through a nuclear pore, and enters the cytosol, ready for translation.

Quick Check

☐ 1. How is a codon related to a triplet?

☐ 2. Describe the basic steps of transcription.

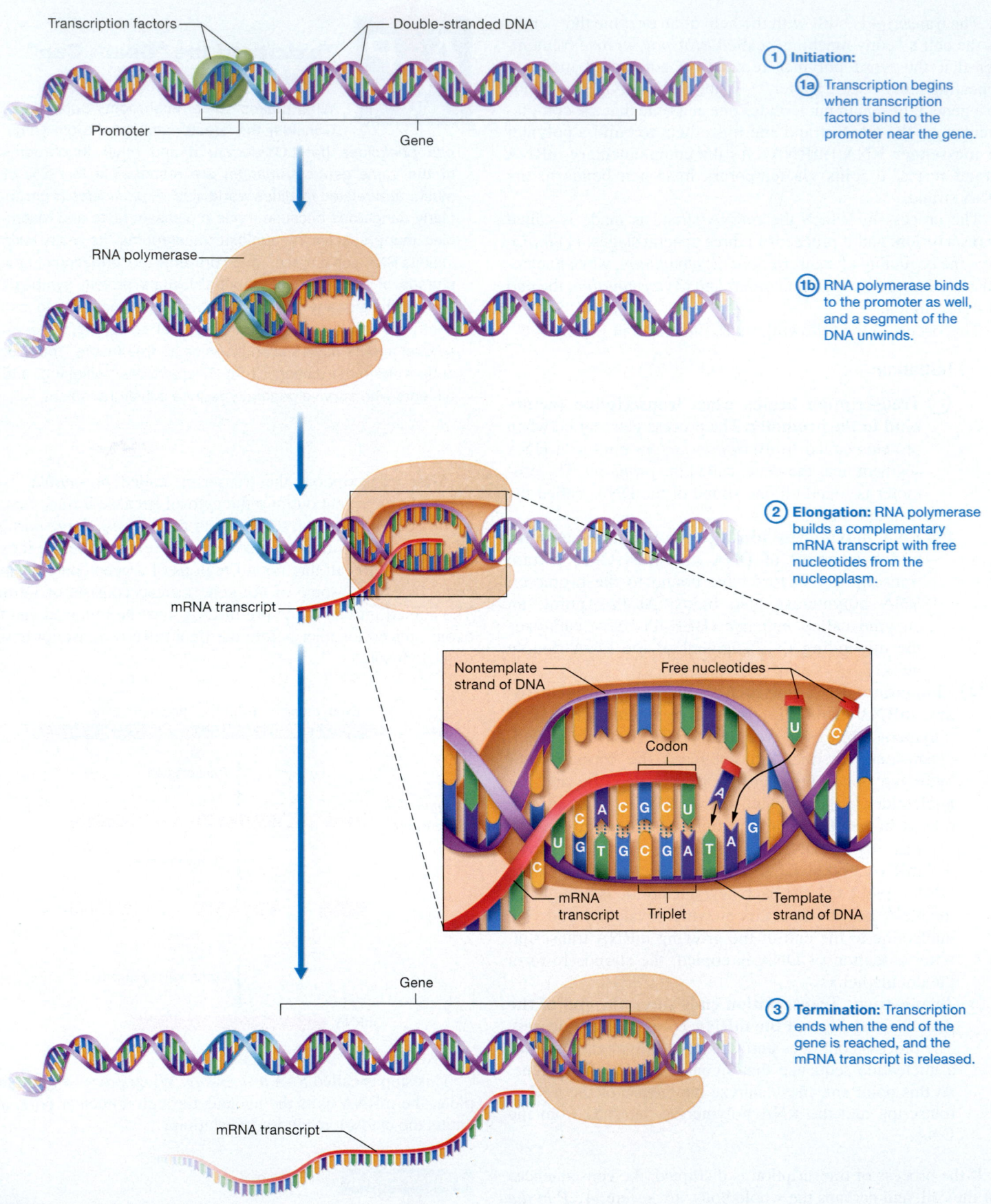

Transcription factors

Double-stranded DNA

Promoter

Gene

1 Initiation:

1a Transcription begins when transcription factors bind to the promoter near the gene.

RNA polymerase

1b RNA polymerase binds to the promoter as well, and a segment of the DNA unwinds.

mRNA transcript

2 Elongation: RNA polymerase builds a complementary mRNA transcript with free nucleotides from the nucleoplasm.

Nontemplate strand of DNA

Free nucleotides

Codon

mRNA transcript

Triplet

Template strand of DNA

Gene

3 Termination: Transcription ends when the end of the gene is reached, and the mRNA transcript is released.

mRNA transcript

Figure 3.29 Transcription.

Translation

« FLASHBACK

1. What are polypeptides and peptide bonds? (p. 57)

At the end of transcription and RNA processing, we have an mRNA strand that contains a copy of the triplet code in the DNA. But this information is still encoded in the language of nucleotides, and our cells need to then *translate* it into the language of amino acids. This occurs by the process that is appropriately called **translation.** Translation takes place with the help of ribosomes and involves another type of RNA that acts as the molecular translator, named **transfer RNA,** or **tRNA** (**Figure 3.30**). Recall from the previous module that tRNA is manufactured within the nucleus from a DNA template. It is called "transfer" RNA because its job is to pick up a specific amino acid from the cytosol and transfer it to the growing polypeptide chain at the ribosome.

Notice in Figure 3.30 that tRNA is a single strand of RNA. This strand forms hydrogen bonds with itself to create the shape of a three-leaf clover. At one end of the tRNA is a region called the **anticodon,** which contains a sequence of three nucleotides that is complementary to a specific mRNA codon. The opposite end of the tRNA carries the specific amino acid that corresponds to the anticodon. At least one type of tRNA specifies each of the 20 amino acids.

Let's see how this process works using the mRNA codon UUC as an example. As you can see by looking back at Figure 3.28, the mRNA codon UUC specifies the amino acid phenylalanine (Phe). By the rules of complementary base pairing, the mRNA codon UUC bonds to the tRNA anticodon AAG, which appears on a tRNA with the amino acid phenylalanine attached to its other end.

(a) Structure of tRNA with bound amino acid

(b) Schematic structure of tRNA (with bound amino acid)

Figure 3.30 Transfer RNA (tRNA).

ConceptBOOST))

Connecting a DNA Triplet to a Particular Amino Acid

It's easy to get the relationships between a *triplet*, a *codon*, and an *anticodon* confused. Here are a couple of quick examples to clarify these relationships:

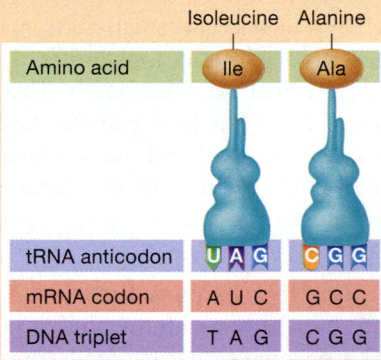

	Isoleucine	Alanine
Amino acid	Ile	Ala
tRNA anticodon	U A G	C G G
mRNA codon	A U C	G C C
DNA triplet	T A G	C G G

Notice in these examples that the nucleotides in the tRNA anticodon are the same as those in the DNA triplet, except that the nucleotide T in DNA is replaced by U in tRNA.

Now, using Figure 3.28 and these examples for reference, figure out the amino acid specified by each of the following DNA triplets (remember that the amino acid correlates to the mRNA codon, not the tRNA anticodon):

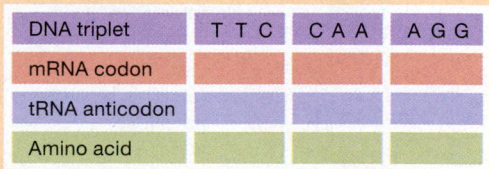

DNA triplet	T T C	C A A	A G G
mRNA codon			
tRNA anticodon			
Amino acid			

Answers: mRNA codon: AAG, GUU, UCC; tRNA anticodon: UUC, CAA, AGG; Amino acid: Lys, Val, Ser

---

As we discussed earlier, protein synthesis takes place with the help of the cell's ribosomes (see Figure 3.18). Each ribosome has three binding sites for tRNA molecules, called the A, P, and E sites. The A (aminoacyl; ah-mee′-noh-AY-sil) site binds an incoming tRNA carrying an amino acid. The P (peptidyl; pep-TID-′l) site is where the tRNA's amino acid is added to the growing protein. The "used" tRNA departs from the E (exit) site of the ribosome; it can pick up another amino acid in the cytosol and be reused.

During translation, the ribosome holds the mRNA transcript in place between its two subunits and facilitates its interaction with tRNAs. Like transcription, the process of translation can be broadly organized into three stages: (1) *initiation,* the beginning of translation; (2) *elongation,* when amino acids are linked together by peptide bonds; and (3) *termination,* when the completed polypeptide is released into either the cytosol or the rough ER. These stages proceed in the following steps, shown in **Figure 3.31**:

1. **Initiation: Translation begins when an initiator tRNA binds the mRNA start codon in the ribosome.** The initiation stage begins when both the mRNA transcript and a tRNA called the *initiator tRNA,* which carries the amino acid methionine (Met), bind to a small ribosomal subunit. With this tRNA in place, the small ribosomal subunit moves down the mRNA transcript looking for a specific nucleotide sequence called the *start codon.* The start codon has a sequence of AUG that base pairs to the initiator tRNA's anticodon, UAC. At this point, the large ribosomal subunit binds to the small subunit, with the initiator tRNA in the P site. Initiation is now complete, and elongation can begin.

2. **Elongation:**

   2a. **Another tRNA binds to the open A site.** The elongation stage begins with the binding of another tRNA to the ribosome's open A site. Note that this tRNA has an anticodon that is complementary to the mRNA codon in the A site.

2b. **The first amino acid is joined to the second amino acid by a peptide bond.** Elongation occurs as the amino acid (or later, growing polypeptide) is transferred from the tRNA in the P site to the amino acid on the tRNA in the A site. These two amino acids are linked covalently by a peptide bond, and both are now attached to the tRNA in the A site.

2c. **The ribosome moves down to the next mRNA codon, the empty tRNA exits, and a new tRNA comes in.** During the next step of elongation, the ribosome moves down to the next mRNA codon. Observe that this move shifts the "empty" tRNA that was in the P site into the E site, from which it exits the ribosome, and the tRNA that was in the A site now moves into the P site. The A site is now vacant, and a new tRNA moves into it. The events from this step make up a cycle called the *elongation cycle,* with the peptide chain gaining an amino acid and the ribosome then shifting to the next mRNA codon each time the cycle is repeated.

3. **Termination: Translation ends when the ribosome reaches the stop codon and the polypeptide is released.** The elongation cycle continues until the ribosome reaches an mRNA codon called the *stop codon* (the stop codon could be UAG, UAA, or UGA; see Figure 3.28). The stop codon doesn't link with a tRNA. Instead, it binds to a protein called the release factor, which releases the completed polypeptide chain from the tRNA in the P site. When this happens, translation is terminated.

At the end of translation we have finished synthesizing a polypeptide. However, such a polypeptide is not a functional protein until it folds into its proper three-dimensional configuration. This folding and other important structural changes take place during *posttranslational modification.* Polypeptides that

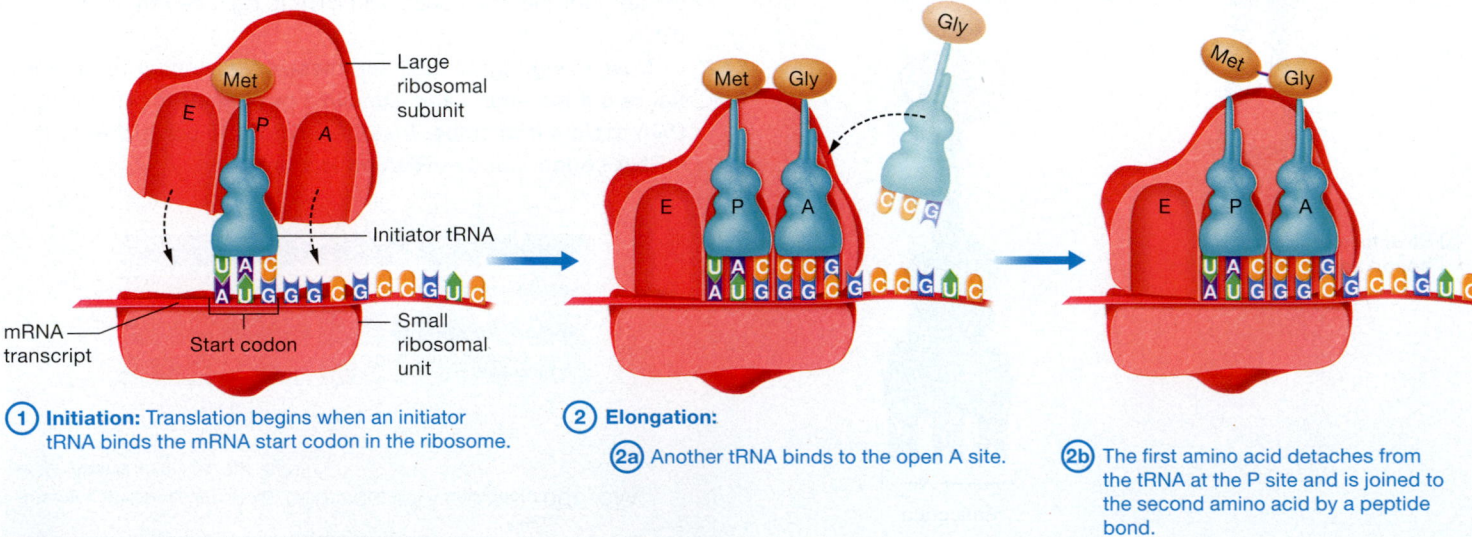

① **Initiation:** Translation begins when an initiator tRNA binds the mRNA start codon in the ribosome.

② **Elongation:**

2a Another tRNA binds to the open A site.

2b The first amino acid detaches from the tRNA at the P site and is joined to the second amino acid by a peptide bond.

Figure 3.31 Translation.

Play Animation
@ **Mastering** Anatomy & Physiology

are destined for the cytosol are synthesized on free ribosomes, and they fold either on their own or with the help of other proteins. Many other polypeptides destined for secretion or insertion into an organelle or membrane require modifications that are performed in the rough endoplasmic reticulum (RER), so they are synthesized on bound ribosomes. When translation is complete, the polypeptide remains in the RER, where it is folded and modified as necessary. The resulting protein is then sent to the Golgi apparatus for final processing, sorting, and packaging.

Quick Check

☐ 3. Explain how tRNA acts as the "translator" of the genetic code.

☐ 4. Describe the basic steps of translation.

☐ 5. Why is posttranslational modification necessary for a protein to be functional?

Putting It All Together: The Big Picture of Protein Synthesis

Figure 3.32 offers a summary of the events of protein synthesis. This process can be summarized in four basic steps: transcription, RNA processing and transit, translation, and posttranslational modification.

Regulation of Gene Expression

The human genome contains about 25,000 genes, but most cells express only a fraction of these genes and synthesize only those proteins required for their specific functions. For example, all cells in the body contain the gene for the protein hormone

insulin, yet only certain cells of the pancreas actually produce insulin. Furthermore, these cells of the pancreas can produce more or less insulin, depending on the conditions in the body. How, then, do cells "know" which proteins to make, and when to make them? The answer is fairly complicated and involves a variety of mechanisms that control the production of a protein.

Many methods of regulating gene expression occur at the initiation of transcription. Specific proteins bind to parts of the gene and initiate transcription, whereas other proteins inhibit its initiation. Regulation of gene expression can also take place after transcription. Some examples of such mechanisms include degradation of mRNA transcripts in the cytosol, inhibition of the start of translation, and, finally, inhibition of posttranslational modification. All these mechanisms and others work together to ensure that a cell produces only the proteins it needs at the appropriate times and in the appropriate amounts to carry out its specific functions.

Quick Check

☐ 6. Why is it important to regulate gene expression?

Apply What You Learned

☐ 1. Your friend has been using a tanning bed and, unbeknownst to him, the radiation from the tanning bed has damaged the DNA in his skin cells. In one particular gene, the triplet TGA has mutated to TGC. Will this change the structure of the protein coded for by this gene? In a separate gene, the triplet GCG has mutated to GGG. Will this change the structure of the protein coded for by this gene? Explain why or why not. (*Hint:* Refer to Figure 3.28 for the genetic code.)

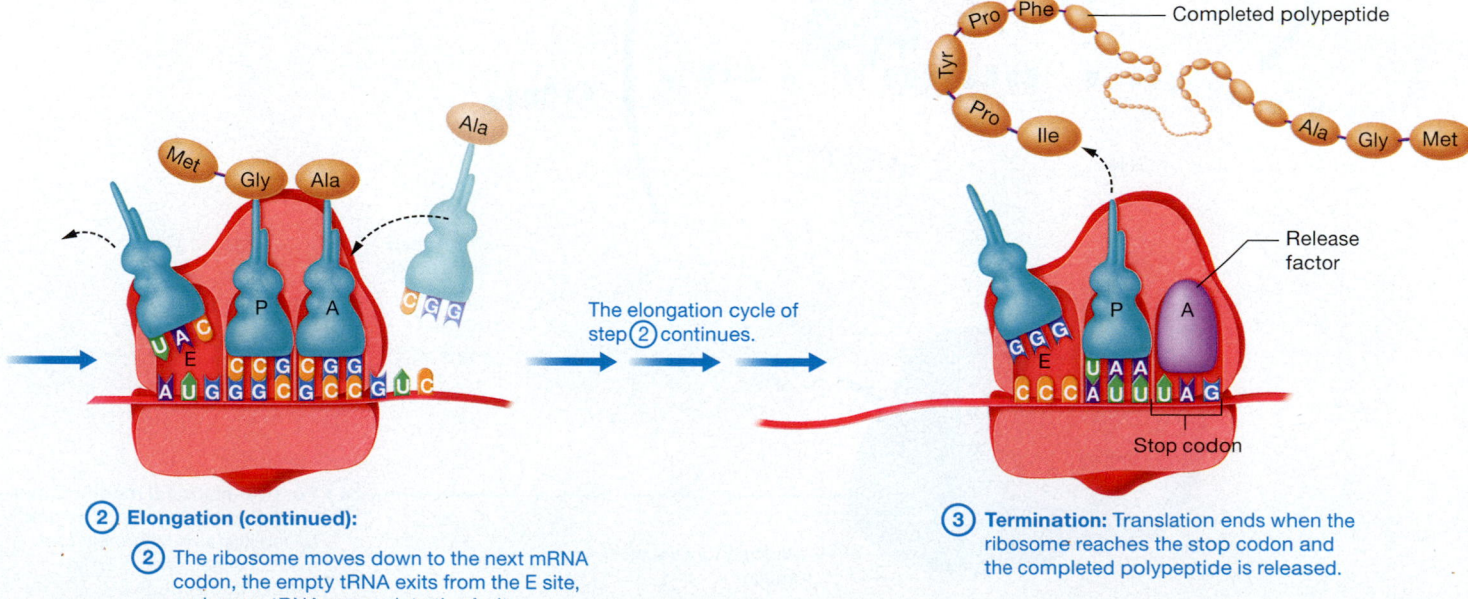

② **Elongation (continued):**

② The ribosome moves down to the next mRNA codon, the empty tRNA exits from the E site, and a new tRNA comes into the A site.

The elongation cycle of step ② continues.

③ **Termination:** Translation ends when the ribosome reaches the stop codon and the completed polypeptide is released.

Completed polypeptide

Release factor

Stop codon

The Big Picture of Protein Synthesis

Figure 3.32

Play Animation
@ Mastering Anatomy & Physiology

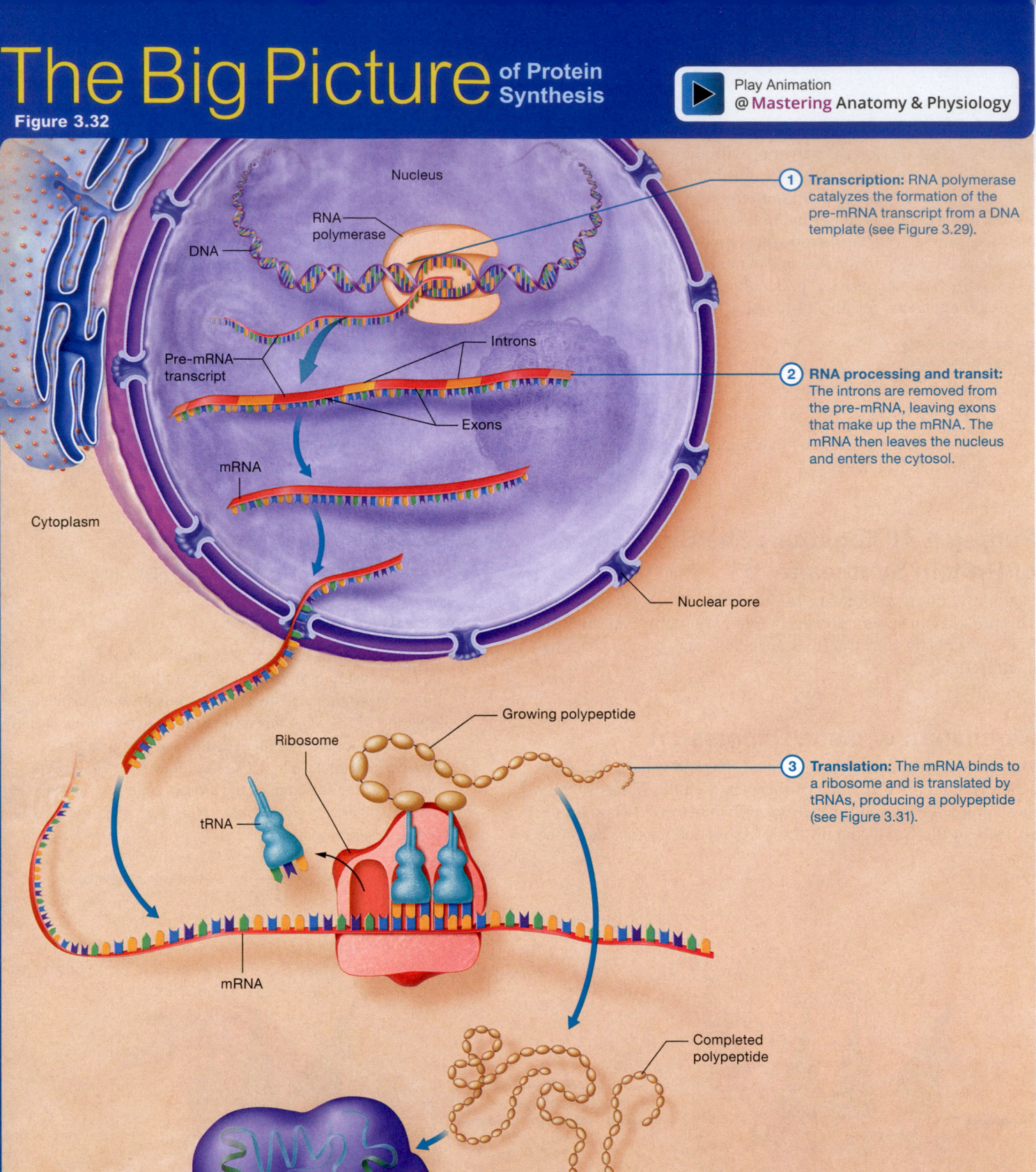

Nucleus

RNA polymerase

DNA

Pre-mRNA transcript

Introns

Exons

mRNA

Cytoplasm

Nuclear pore

Growing polypeptide

Ribosome

tRNA

mRNA

Completed polypeptide

Modified functional protein

1 **Transcription:** RNA polymerase catalyzes the formation of the pre-mRNA transcript from a DNA template (see Figure 3.29).

2 **RNA processing and transit:** The introns are removed from the pre-mRNA, leaving exons that make up the mRNA. The mRNA then leaves the nucleus and enters the cytosol.

3 **Translation:** The mRNA binds to a ribosome and is translated by tRNAs, producing a polypeptide (see Figure 3.31).

4 **Posttranslational modification:** The polypeptide is modified and folded into its final protein form, most of which occurs in the cytosol or the rough ER.

□ 2. You read earlier about the poison ricin, which inhibits the functioning of ribosomes. What part of protein synthesis would be affected by ricin? What effect would this have on a cell's ability to synthesize proteins?

See answers in Appendix A.

See answers in Appendix A.

MODULE **3.8**
The Cell Cycle

Learning Outcomes

1. Describe the events that take place during interphase and their functional significance.
2. For each stage of the cell cycle, describe the events that take place and their functional significance.
3. Distinguish between mitosis and cytokinesis.
4. Describe the process of DNA replication.

The biological principle of the *cell theory* states that cells cannot spontaneously appear, but rather, they must come from the division of cells that already exist. In other words, all forms of life, including humans, are the result of repeated rounds of cell growth and division.

Almost all cells go through a process called the **cell cycle,** which is defined as the ordered series of events from the formation of the cell to its reproduction by cell division. Cell division is vital in a fetus or child, as it is required for growth and development. However, cell division is equally important in adults for tissue repair and renewal, and millions of our cells undergo cell division each *second* throughout our lives. Our cells are not immortal, and they must be replaced as they die; otherwise, we may face major imbalances in our homeostasis. For example, cells in the bone marrow produce millions of oxygen-carrying red blood cells per day. If the cells in the bone marrow cease dividing, which happens in the condition *aplastic anemia,* the body cannot carry oxygen efficiently to its cells, and death may result. In this module we explore the events behind these critical processes of cell growth and division, and touch on the process of cell cycle control and what happens when it is lost.

Phases of the Cell Cycle

« FLASHBACK

1. What are the functions of microtubules? (p. 97)
2. What structure forms and organizes microtubules? (p. 97)

The cell cycle, shown in **Figure 3.33**, is composed of two main phases: *interphase* and *M phase,* or cell division. ① During **interphase**—which includes three subphases known as G_1, S, and G_2—the cell grows and prepares for ② the final M phase, which

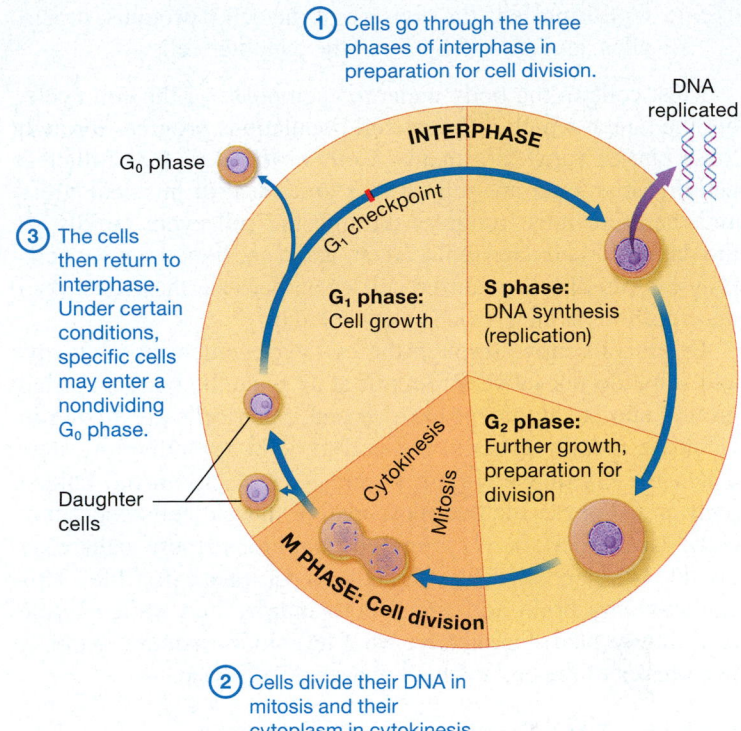

① Cells go through the three phases of interphase in preparation for cell division.

② Cells divide their DNA in mitosis and their cytoplasm in cytokinesis.

③ The cells then return to interphase. Under certain conditions, specific cells may enter a nondividing G_0 phase.

G_0 phase

G_1 checkpoint

INTERPHASE

DNA replicated

S phase: DNA synthesis (replication)

G_1 phase: Cell growth

G_2 phase: Further growth, preparation for division

Cytokinesis

Mitosis

M PHASE: Cell division

Daughter cells

Figure 3.33 The cell cycle.

is the division of the nucleus (*mitosis;* my-TOH-sis) and cytoplasm (*cytokinesis;* sy′-toh-kin-EE-sis; *cyto* = "cell," *kinesis* = "movement"). The cell spends most of its time in the G_1 phase of interphase. During interphase, the cell must duplicate its DNA precisely so that the products of cell division, called **daughter cells,** have an identical genetic makeup. The cell must also grow, synthesize proteins, and make new organelles so that the two daughter cells have adequate amounts of these components in their cytoplasm.

Let's look in more detail at the phases of the cell cycle.

- **Interphase,** the phase of growth and preparation for division, includes the following three subphases:
  ○ **G_1 phase.** During the **G_1 phase,** or first gap phase, the cell is performing its normal metabolic functions as well as growing and carrying out rapid protein synthesis. It uses these proteins to make new organelles and components of the cytoskeleton. (Although "G" stands for "gap," you might think of it as standing for "growth.")
  ○ **S phase.** As you can see in Figure 3.33, the **S phase** is when DNA synthesis takes place. (The "S" stands for "synthesis.")
  ○ **G_2 phase.** The **G_2 phase,** or the second gap phase, is another period of cellular growth during which the proteins necessary for cell division are rapidly produced and the centrioles are duplicated.
- **M phase,** the phase of cell division, has two overlapping parts:
  ○ **Mitosis** is the division of the genetic material between the two daughter cells. It requires a structure called the *mitotic spindle,* which is composed of microtubules organized into *spindle fibers* by the cell's centrosomes.

○ **Cytokinesis** is the division of the cell's proteins, organelles, and cytosol between the daughter cells.

Most cells in the body undergo each phase of the cell cycle, but the rate at which different cell populations progress through these phases varies drastically. Cells of the skin, hair follicles, and digestive tract are subject to a great deal of physical stress and therefore must progress through the cell cycle rapidly so the damaged cells are replaced. In addition, bone marrow cells progress rapidly through the cell cycle because they must produce millions of new blood cells each day.

Other cells move through the cell cycle much more slowly, and some do not move through it at all once the body's normal growth and development have ceased. Such cells are frozen in G_1 phase indefinitely, and are said to be in a nondividing state called the **G_0 phase** (③ in Figure 3.33). Cells in the human body in G_0 phase include most mature muscle cells and nerve cells. Under certain very limited conditions, these cells may divide, but generally they remain in G_0 phase for life. This explains why brain and spinal cord injuries and some muscle tears rarely heal completely. Now let's look in more detail at two phases of the cell cycle, S phase and M phase.

S Phase: DNA Synthesis or Replication

Given the large size of the human genome, it seems like a daunting task to unravel our DNA and make an exact copy. Yet this process, called **DNA synthesis** or **replication,** goes on in millions of cells in the body every second. During DNA synthesis, the chromatin unwinds from the histone proteins, and the entire set of 3.2 billion base pairs is duplicated by a process that builds new DNA strands using each "old" DNA strand as a template. Such replication is called *semiconservative replication* because one of the original strands of DNA is included in each newly formed double helix. The enzyme that catalyzes DNA synthesis is **DNA polymerase.**

Figure 3.34 illustrates DNA synthesis, which proceeds as follows:

① **DNA strands separate as helicase unwinds them.** DNA synthesis starts as the enzyme helicase catalyzes the unwinding of the two strands of DNA.

② **Primase builds RNA primers on the existing DNA strands.** DNA polymerase can add a nucleotide only to an *existing* chain of nucleotides. This means that DNA synthesis cannot start from scratch—a few nucleotides must already be base-paired with the template strand. Our cells have a very simple solution to this problem. An enzyme called *primase* catalyzes the reactions that build a very short segment of RNA, called an *RNA primer,* on the DNA template. RNA primers give the DNA polymerase a place to begin building the new strands of DNA. As this occurs, helicase continues to unwind the DNA double helix.

③ **DNA polymerase catalyzes the addition of nucleotides to the new DNA strands.** DNA polymerase catalyzes the formation of hydrogen bonds between free nucleotides from the nucleoplasm and nucleotides in the DNA templates, determined by the rules of complementary base pairing. Covalent bonds are then formed between adjacent nucleotides, creating the new DNA strand. When DNA polymerase encounters an RNA primer, the primer is removed and replaced with DNA. Note in Figure 3.34 that DNA synthesis progresses in opposite directions along the two strands. This occurs because DNA strands have directionality to them, and DNA polymerase can add nucleotides in only one direction along the template strands.

④ **The end result is two identical double helices, each with one old strand and one newly formed strand of DNA.** This preservation of one original strand is why such replication is called **semiconservative replication.**

When DNA synthesis is complete, the histone proteins reassociate with the DNA and the cell goes into G_2 phase, in which it prepares for cell division.

M Phase: Mitosis and Cytokinesis

Recall that the process of mitosis divides the cell's replicated DNA between two daughter cells. At the beginning of mitosis, we have a cell with 92 chromatids, or 46 replicated chromosomes. At the end, we have two cells with identical sets of 46 chromosomes.

To divide its genetic material, the cell rapidly goes through a series of complex events. Notice in **Figure 3.35a** that during interphase, the nucleus and nucleolus are clearly visible under a light microscope, and the individual chromosomes are not distinguishable. Now look at the first phase of mitosis, called *prophase,* in **Figure 3.35b.** As you can see, the cell undergoes fairly drastic changes in appearance. All these changes ultimately make it easier for the cell to divide.

Division of a cell's genetic material occurs through four stages of the M phase, illustrated in Figure 3.35b:

① **Prophase.** During **prophase,** the chromatin becomes fully compacted, so that individual chromosomes, each composed of two sister chromatids joined at the centromere, are visible. Additionally, the nucleolus breaks down, and a structure called the **mitotic spindle** forms. During this process, the previously duplicated centrosomes, each containing a pair of centrioles, move to opposite poles of the cell and organize the spindle fibers. Spindle fibers from each centriole pair then attach to each sister chromatid at the centromere. To learn about the effects of an inhibited mitotic spindle, see *A&P in the Real World: Spindle Poisons* on page 116. At the end of prophase, the nuclear envelope begins to break apart.

② **Metaphase.** The longest phase of mitosis is **metaphase.** During this phase, spindle fibers from opposite poles of the cell tug the sister chromatids back and forth to eventually line up on the middle, or "equator," of the cell.

③ **Anaphase.** During **anaphase,** the sister chromatids part, and the individual chromosomes, now called *daughter chromosomes,* are pulled to opposite poles of the cell. This movement occurs as the spindle fibers to which they are attached get progressively shorter. By the end of anaphase, each pole of the cell has the complete set of 46 daughter chromosomes. Note that some microtubules are not attached to chromosomes, and these microtubules

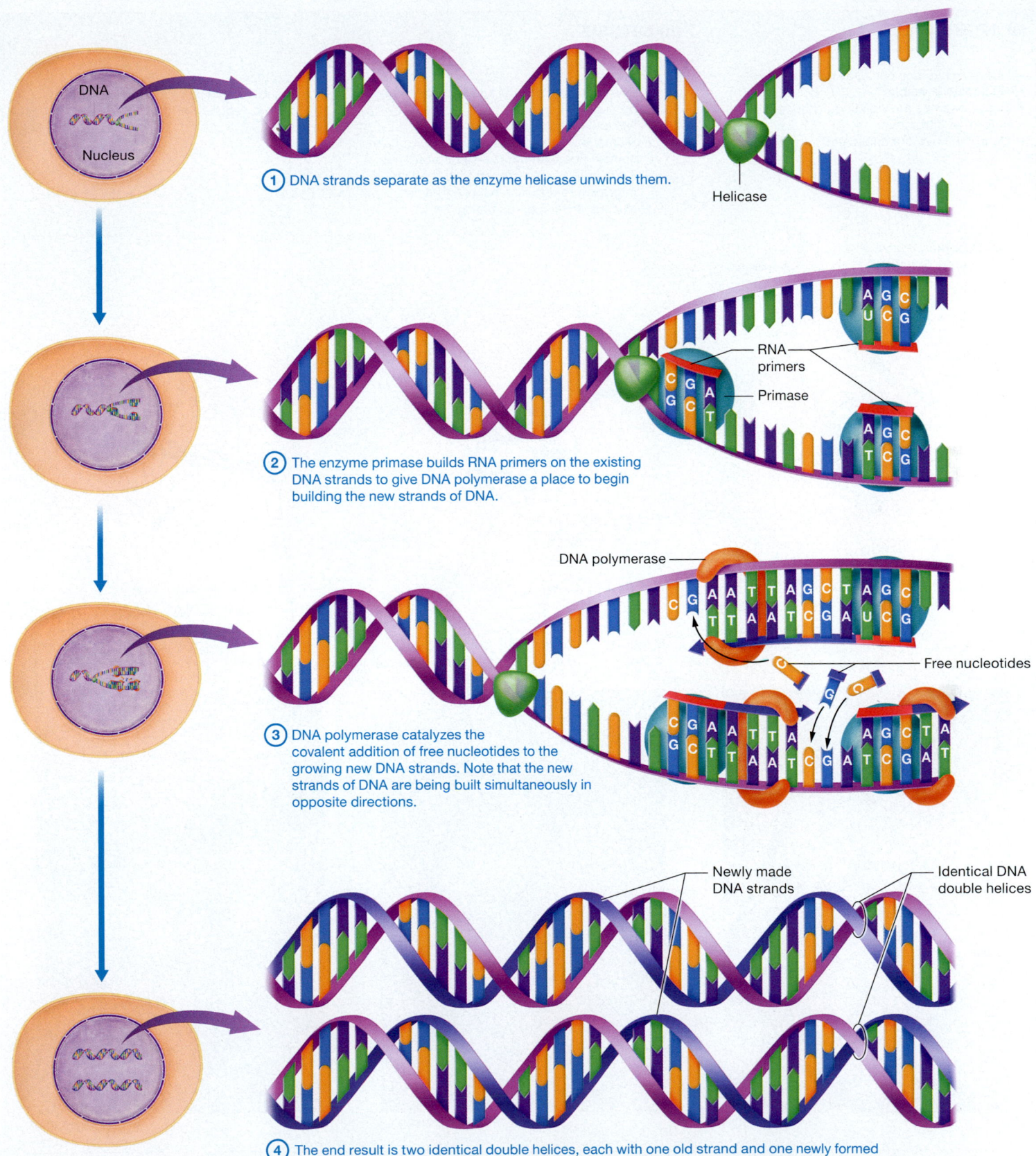

① DNA strands separate as the enzyme helicase unwinds them.

Helicase

② The enzyme primase builds RNA primers on the existing DNA strands to give DNA polymerase a place to begin building the new strands of DNA.

RNA primers

Primase

DNA polymerase

Free nucleotides

③ DNA polymerase catalyzes the covalent addition of free nucleotides to the growing new DNA strands. Note that the new strands of DNA are being built simultaneously in opposite directions.

Newly made DNA strands

Identical DNA double helices

④ The end result is two identical double helices, each with one old strand and one newly formed strand of DNA. For this reason, the whole process is called semiconservative replication.

Figure 3.34 **DNA synthesis.**

(a) INTERPHASE

- Nuclear envelope encloses nucleus.
- Nucleolus is visible.
- Chromosomes are indistinguishable—DNA is in form of chromatin.
- Centriole pairs are duplicated.

(b) MITOSIS

① Prophase:
  - Chromatin condenses so sister chromatids are visible.
  - Nucleolus disperses.
  - Mitotic spindle forms, and in late prophase spindle fibers attach to sister chromatids.
  - Two centriole pairs separate and begin migrating to opposite poles of the cell.
  - Nuclear envelope fragments.

② Metaphase:
  - Spindle fibers pull sister chromatids to align on equator of cell.

Centrosomes (with centriole pairs)   Nucleolus

Nuclei   Chromatin

LM (2900×)

Mitotic spindle   Centriole pair   Chromosomes (sister chromatids)

LM (1430×)

Sister chromatids   Centriole pairs   Spindle fibers

LM (1630×)

Figure 3.35 Interphase, mitosis, and cytokinesis.

Practice art labeling
@ Mastering Anatomy & Physiology

(b) MITOSIS (CONTINUED)

(3) Anaphase:
- Sister chromatids separate as spindle fibers shorten.
- Daughter chromosomes are pulled to opposite poles.
- Cell elongates.
- Cytokinesis begins as organelles and cytosol are divided.

(4) Telophase and Cytokinesis:
- Nuclear envelopes reassemble.
- Nucleoli re-form.
- Chromosomes are no longer distinct—DNA returns to chromatin form.
- Cytokinesis continues as the cleavage furrow forms.
- Daughter cells separate.

(c) END RESULT

Two genetically identical daughter cells that enter interphase

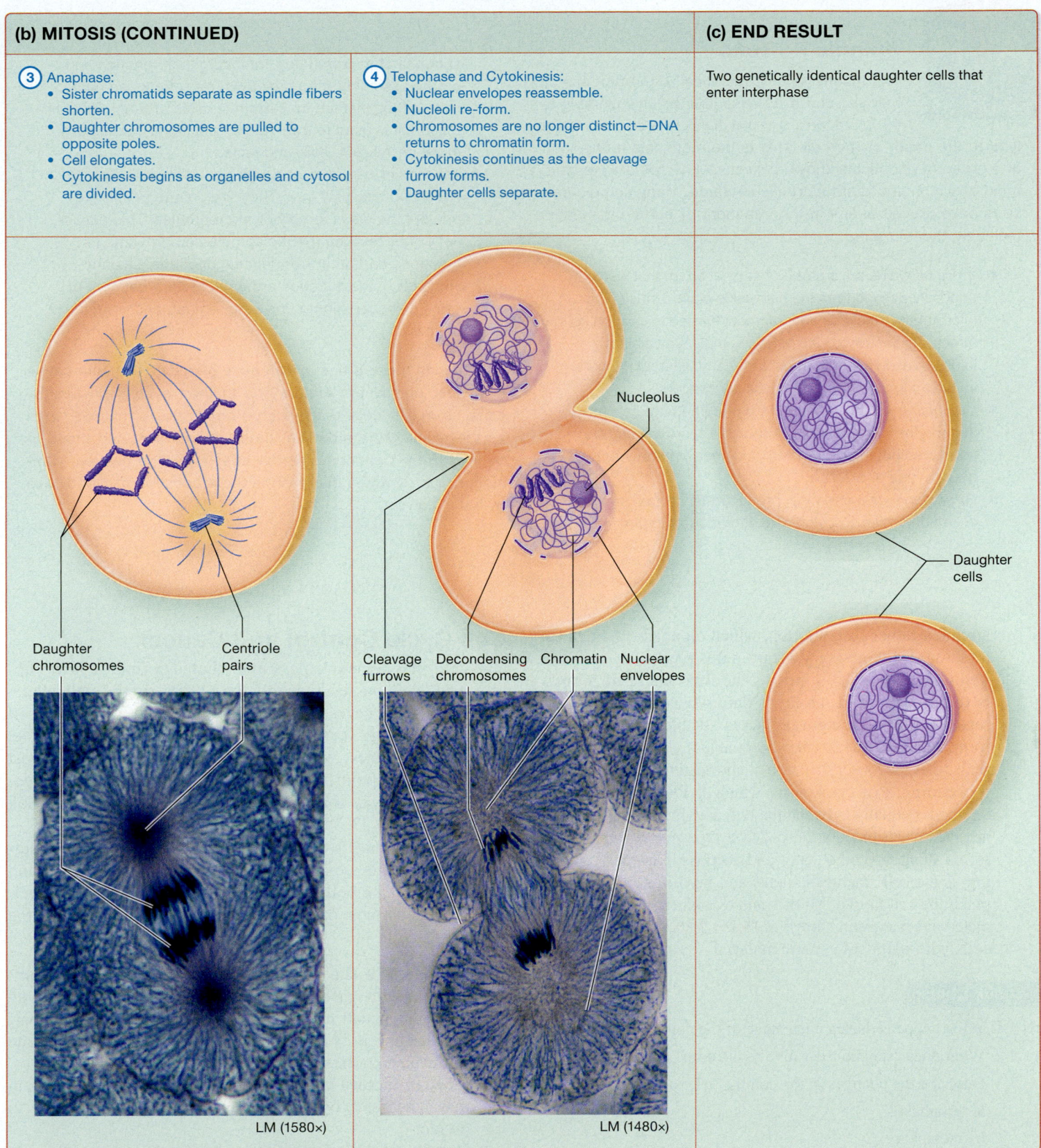

Daughter chromosomes

Centriole pairs

LM (1580×)

Nucleolus

Cleavage furrows

Decondensing chromosomes

Chromatin

Nuclear envelopes

LM (1480×)

Daughter cells

Figure 3.35 **Interphase, mitosis, and cytokinesis.** (*continued*)

Spindle Poisons

The mitotic spindle is critical to the process of mitosis, and if its assembly or disassembly is inhibited, errors in cell division occur that could lead to the death of the cell. This is the mechanism of action of a group of chemicals known collectively as *spindle poisons*, most of which are found naturally in plants. Many spindle poisons have proved useful in the treatment of certain diseases, including the following:

- *Vinca alkaloids* are a group of several drugs that were originally derived from the periwinkle plant. They inhibit microtubule functions and fragment formed microtubules, effectively destroying the mitotic spindle. Vinca alkaloids are used in the treatment of many different cancers, including lung, breast, and testicular cancers.
- *Colchicine* is a drug derived from the autumn crocus. It inhibits the assembly of microtubules, preventing the mitotic spindle from forming. It is used primarily to treat gout, a type of arthritis.
- *Griseofulvin* is a chemical produced by certain strains of mold. Similar to colchicine, it binds to tubulin subunits of microtubules, inhibiting their function and proper assembly. It has relatively few effects on human cells, acting instead on certain species of fungi. For this reason, it is used to treat fungal infections, particularly those of the skin, hair, and nails.
- *Taxanes* are a group of drugs derived from different species of the yew tree. They are unique in that they prevent the *disassembly* of microtubules. This essentially "freezes" the mitotic spindle and prevents its functions and other normal cell functions. Like the vinca alkaloids, taxanes are used in the treatment of certain cancers, including lung, breast, and ovarian cancers.

Although these drugs are all quite effective, they can cause a number of adverse effects due to their actions on healthy cells, especially cells that undergo rapid rates of mitosis, such as those lining the stomach, those of the skin, and those in bone marrow. Common adverse effects include nausea, vomiting, hair loss, and a decrease in the production of blood cells by bone marrow.

---

lengthen rather than shorten, which elongates the cell. In addition, cytokinesis may start in this phase.

④ **Telophase and Cytokinesis.** The final phase of mitosis is **telophase** (TEEL-uh-fayz), during which daughter cells separate. The nuclear envelopes reassemble, the nucleoli reform, and the chromosomes become less visible as the DNA reassumes the looser structure of chromatin. While telophase is occurring, so is cytokinesis, which divides the cytosol and organelles equally between the two daughter cells. The cells are split at an indentation called a *cleavage furrow,* which forms along the cell equator and deepens like a belt tightening as the cell's actin filaments and myosin motor proteins pinch the cell in two. Mitosis and cytokinesis are now complete, and the result, shown in **Figure 3.35c**, is two daughter cells with identical genetic material.

Quick Check

☐ 1. What happens during each stage of the cell cycle?

☐ 2. What does "semiconservative replication" mean?

☐ 3. Describe the changes in the cell that take place during:

    a. prophase.

    b. metaphase.

    c. anaphase.

    d. telophase.

Cell Cycle Control and Cancer

Our cells have a control system that determines how rapidly each cell passes through the stages of the cell cycle. Such control is necessary to balance cell division with cell death. Monitoring occurs at specific points in the cell cycle, called **checkpoints,** which act as stop/go signals. The most important checkpoint, the G_1 *checkpoint,* occurs about three-fourths of the way through G_1, as you can see if you look back at Figure 3.33.

Researchers have worked out many of the extracellular signals that regulate progression through these checkpoints and completion of the cell cycle, but only a few of the intracellular signals. Some of the extracellular signals include the following:

- **Nutrients in the ECF.** Just as your car's engine won't start without gasoline, cells won't proceed all the way through the cell cycle without proper nutrients. If a particular nutrient is lacking in the ECF, the cell stalls at one of the checkpoints.
- **Growth factors in the ECF.** A *growth factor* is a protein that is secreted by one cell and stimulates other cells to divide.
- **Density of cells in a tissue.** In most tissues, cells stop dividing when they reach a certain number or density. This phenomenon is due in part to the fact that if more than a

certain number of cells are in a tissue, the availability of nutrients declines to the point that further cell growth and division are inhibited.

- **Anchorage of cells within the tissue.** In many tissue types, the cells are anchored to the cells around them at the extra-cellular material. If the cell loses its anchorage to the sur-roundings, its growth and division halt.

Cells that cannot pass the checkpoints and cannot be repaired undergo a process of programmed cell death called *apoptosis* (aeh-pop-TOH-sis). This "cellular suicide" will also occur for a variety of other reasons to benefit the body as a whole. For example, during fetal development the hands and feet are initially webbed, but the cells in the "webs" die to separate the fingers and toes.

A cell normally maintains very precise control over the cell cycle via a balancing act between genes whose products inhibit cell growth and division, and genes whose products stimulate cell division. When either of these types of genes is damaged, the DNA may produce a defective protein that results in either insufficient or excessive cellular growth and division. If the rate of cell division in a tissue falls below the rate of cell death, the tissue may progressively deterio-rate. Conversely, if the rate of cell division exceeds the rate of cell death, the tissue may increase in cell number and size. This condition, called *hyperplasia* (hy′-per-PLAY-zee-ah), can strain the surrounding tissues, robbing them of oxygen and nutrients.

When changes in the DNA of a cell cause uncontrolled cell division, the resulting cells may form a growth or mass known as a *tumor.* A benign (beh-NYN) tumor is one that is confined to its original location and does not invade sur-rounding tissues, but it may grow extremely large. A *cancer-ous,* or *malignant,* tumor is one that is made up of *cancer cells.* An example is the tumor shown in **Figure 3.36**, from a patient with renal cell carcinoma (cancer of cells of the kidney). Cancer cells are not inhibited by high cellular den-sity or loss of their anchorage to other cells. In fact, if they have enough nutrients, such cells appear to grow and divide indefinitely. A striking example is a culture of cancerous tumor cells, known as the HeLa cells, that has been grow-ing since the tumor was removed from a patient named Hen-rietta Lacks in 1951. Cancer cells from malignant tumors are able to spread into other tissues, a process called *metastasis* (meh-TASS-tuh-sis), which results in the formation of other

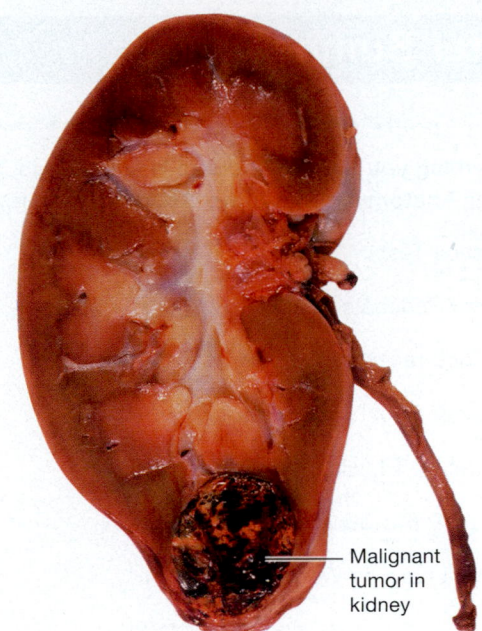

Malignant tumor in kidney

Figure 3.36 Cancerous tumor of kidney cells.

tumors throughout the body. This can cause widespread tissue destruction that, especially if untreated, may lead to death.

Quick Check

☐ 4. What are four external factors that play a role in cell cycle control?

Apply What You Learned

☐ 1. What would happen if a cell completed several rounds of mitosis but did not undergo cytokinesis? (*Hint:* Remember that telophase and cytokinesis are separate processes.)

☐ 2. Both the periwinkle and yew plants produce toxins that interfere with microtubules by binding to them and either inhibiting their assembly (the periwinkle) or preventing their disassembly (the yew). Why would both actions ultimately *prevent mitosis?*

See answers in Appendix A.

Chapter Summary

For everything you need to succeed in this course, go to Mastering Anatomy & Physiology. There you will find:

- 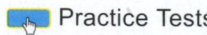 Practice Tests
- Author Podcasts
- Big Picture Animations
- Concept Boost Video Tutors
- iP2™ Interactive Physiology 2.0 Tutorials
- A&PFlix A&P Flix 3D Animations
- Active-Learning Workbook

MODULE 3.1
Introduction to Cells 68–70

- All cells share the same basic functions, including **cell metabolism,** transport of substances through the cell, communication, and cell reproduction.
- Most cells contain three basic components:
  - Cells are surrounded by a **plasma membrane** that separates the body into two compartments: the **extracellular fluid (ECF)** and the **intracellular fluid (ICF),** or **cytosol.**
  - The **cytoplasm** consists of the **cytosol,** organelles, and cytoskeleton.
  - The **nucleus** houses most of the cell's *deoxyribonucleic acid (DNA)* and controls many of the cell's functions.
- Cells vary widely in size and structure, which enables them to better perform specialized functions.

MODULE 3.2
Structure of the Plasma Membrane 70–74

- The plasma membrane is composed primarily of a **phospholipid bilayer** with the hydrophilic phosphate heads facing the water-containing cytosol and ECF, and the hydrophobic fatty acid tails facing one another.
- According to the **fluid mosaic model,** the plasma membrane is a fluid, dynamic structure with multiple components moving laterally within the plane of the phospholipid bilayer.
- **Integral proteins** are embedded in the membrane, with transmembrane proteins spanning its width. **Peripheral proteins** are found on one side of the membrane only. Membrane proteins may function as channels, carriers, enzymes, or receptors. They may also provide structural support to cells and tissues.

MODULE 3.3
Transport across the Plasma Membrane 74–87

- The phospholipid bilayer is **selectively permeable.**
- **Passive transport processes** involve movement of a substance across the membrane with no ATP expenditure. **Active transport processes** involve movement of a substance across the membrane against a gradient and require the use of ATP.
  - **Diffusion** is a type of passive transport during which solutes move down their concentration gradient until equilibrium is reached. Diffusion is driven by a **concentration gradient.** There are two types of diffusion:
    - **Simple diffusion** refers to movement of solutes directly through the phospholipid bilayer.
    - **Facilitated diffusion** refers to movement of polar and ionic solutes across a membrane through a protein channel or with the help of a carrier protein.
- **Osmosis** is a type of passive transport in which the solvent moves across a selectively permeable membrane from a region of lower solute concentration to a region of higher solute concentration.
  - The concept of **tonicity** describes the ability of one solution to cause osmosis relative to another:
    - An **isotonic** ECF has the same ability to cause osmosis as the cytosol and results in no net movement of water by osmosis.
    - A **hypertonic** ECF has a greater ability to cause osmosis than the cytosol and causes water to move out of the cell.
    - A **hypotonic** ECF has a lesser ability than the cytosol to cause osmosis causes water to enter the cell.
- Active transport processes use ATP to move a solute against its concentration gradient.
  - During **primary active transport,** a membrane protein uses ATP to "pump" a solute against its concentration gradient. One of the body's main primary active transport pumps is the Na^+/K^+ **pump.**
  - **Secondary active transport** uses a primary active transport pump to create a concentration gradient, and then uses potential energy from that concentration gradient to pump another solute into or out of the cell against its concentration gradient.
- There is a thin layer of positive charges in the ECF and a thin layer of negative charges in the cytosol. This separation of charges is called a **membrane potential.** The **resting membrane potential** is the membrane potential of a cell at rest. It averages –80 mV.
- **Vesicular transport** is a type of active transport process that allows large substances to enter or exit the cell packaged in membrane-enclosed **transport vesicles.**
  - **Endocytosis** is the process by which large substances enter the cell. The two main types are phagocytosis and pinocytosis:
    - During **phagocytosis,** phagocytes ingest large substances such as bacteria or dead cells.

- During **pinocytosis,** coated pits in the plasma membrane form vesicles containing dissolved substances from the extracellular fluid.
  - During **receptor-mediated endocytosis,** specific ligands bind to their receptors on the plasma membrane and become concentrated in protein-coated pits.
- **Exocytosis** is the process by which large substances are packaged into a vesicle that fuses with the plasma membrane and are released from the cell.

MODULE 3.4
Cytoplasmic Organelles 87–96

- **Organelles** are specialized components within the cell that carry out specific functions. Most organelles are membrane-enclosed.
- **Mitochondria** produce the bulk of the cell's ATP by **oxidative catabolism,** which occurs in the inner mitochondrial membrane.
- **Peroxisomes** oxidize toxic substances and fatty acids and synthesize certain phospholipids.
- **Ribosomes** are small, granular, non–membrane-enclosed organelles with two subunits that synthesize proteins.
- The organelles of the **endomembrane system** interact with one another by packaging and receiving substances in vesicles.
  - The **endoplasmic reticulum** (ER) is a large branching network of tubules and sacs that enclose a single space.
    - **Rough ER** (RER) has ribosomes attached to its surface, and it modifies and folds proteins made by those ribosomes.
    - **Smooth ER** (SER) has no ribosomes on its surface. It stores calcium, synthesizes many lipids, and detoxifies certain substances.
  - The **Golgi apparatus** is a stack of flattened sacs. It modifies and packages products to be secreted from the cell by exocytosis, to become part of the plasma membrane, or to be incorporated into lysosomes.
  - **Lysosomes** contain digestive enzymes that break down substances taken into the cell by endocytosis, as well as worn-out organelles.

MODULE 3.5
The Cytoskeleton 96–100

- The **cytoskeleton** consists of three types of protein filaments made from smaller protein subunits. The cytoskeleton gives the cell its characteristic shape and size, supports the plasma membrane and nucleus, moves substances within the cell, divides the cell, and moves the cell itself.
- **Actin filaments** are small filaments that provide structural support to the cell. Actin filaments associate with the motor protein *myosin.* Together they change the shape and size of the cell during cell movement.
- **Intermediate filaments** are ropelike filaments that provide the cell with mechanical strength and support for its organelles and overall structure.

- **Microtubules,** the largest filaments, support the cell, hold the organelles in their proper places, move organelles within the cell, and form cilia and flagella. They emanate from the **centrosome,** where **centrioles** help to organize their proteins and form microtubules.
- Certain cells feature specialized extensions from their plasma membranes.
  - The plasma membranes of some cells are folded into **microvilli,** which increase the surface area of the membrane.
  - **Cilia** are hairlike projections that beat in unison to sweep substances past the cell.
  - **Flagella** are single projections that propel the cell itself.

MODULE 3.6
The Nucleus 100–103

- The **nucleus** contains most of the cell's **DNA** and is the control center of the cell.
- The nucleus is surrounded by the **nuclear envelope.**
- DNA is found in the nucleoplasm in the form of **chromatin.** During cell division, chromatin condenses into **chromosomes.** Human cells contain 23 pairs of chromosomes, one maternal set and one paternal set.
- **Nucleoli** are regions of the nucleus that make RNA molecules such as ribosomal RNA and assemble ribosomes.

MODULE 3.7
Protein Synthesis 103–111

- **Protein synthesis** is the process by which the cell uses instructions in the DNA to build proteins.
- A **gene** is a segment of DNA that specifies the amino acid sequence of a single protein.
- The **genetic code** is a list of which amino acid is specified by each possible codon.
- Noncoding segments of DNA are called *introns.* Coding segments of the DNA are called *exons.*
- During **transcription,** the cell copies the information in the gene into a **messenger RNA (mRNA)** transcript. The transcription process consists of three stages: *initiation, elongation,* and *termination.*
- Before the pre-mRNA exits the nucleus into the cytosol, it undergoes *RNA processing.*
- During the process of **translation,** the code in the mRNA transcript of a gene is translated into the amino acid sequence of a particular polypeptide. Translation takes place on a ribosome. The "translator" is **transfer RNA (tRNA).**
- Translation consists of three stages:
  - During initiation, an initiator tRNA is brought together with the two ribosomal subunits at the mRNA *start codon.*
  - During elongation, tRNAs bring in specific amino acids that are linked together to form a polypeptide.

○ During termination, the ribosome encounters the mRNA *stop codon* and releases the polypeptide.

• Polypeptides undergo *posttranslational modification*, being modified and folded into functional proteins.

MODULE 3.8
The Cell Cycle 111–117

• The ordered series of events from cell formation to cell division is called the **cell cycle.**

• The cell cycle consists of two main phases, interphase and cell division. **Interphase** includes three subphases: **G₁ phase, S phase,** and **G₂ phase**.

• **M phase** consists of **mitosis,** in which the genetic material divides, and **cytokinesis,** in which the cell divides into two daughter cells.

• During S phase, **DNA synthesis** or **replication** takes place, duplicating a cell's entire set of chromosomes. DNA synthesis is an example of **semiconservative replication,** because each newly formed double helix has one old and one new DNA strand.

• Mitosis divides the cell's replicated DNA between the two **daughter cells.** There are four stages in mitosis: **prophase, metaphase, anaphase,** and **telophase.**

• Cytokinesis occurs simultaneously with anaphase and telophase; it divides the cytoplasm between the daughter cells and pinches the cell in two at the *cleavage furrow.*

• The cell cycle is precisely controlled so that cell formation is balanced with cell death. During the cell cycle, **checkpoints** act as stop/go signals for the cell.

Assess What You Learned

Scan the QR Code for additional practice test questions

https://goo.gl/OliPBy

LEVEL 1 Check Your Recall

1. Which of the following is not a basic function shared by all cells?
   a. Cell metabolism
   b. Communication
   c. Cell reproduction
   d. Cell movement
   e. Transport of substances through the cell

2. Fill in the blanks: The three main components of a cell are the _____, the _____, and the _____.

3. What are the two fluid compartments in the body, and how are they kept separate?

4. Which of the following best describes the arrangement of the main component of the plasma membrane?
   a. A monolayer of phospholipids with the phosphate heads facing the cytosol and the fatty acid tails facing the extracellular fluid
   b. A bilayer of phospholipids with the phosphate heads facing the cytosol and extracellular fluid, and the fatty acid tails facing one another
   c. A bilayer of phospholipids with the phosphate heads of one layer facing the fatty acid tails of the other
   d. A bilayer of phospholipids with a layer of triglycerides sandwiched in the middle

5. Mark the following statements about the plasma membrane as true or false. If a statement is false, correct it to make a true statement.
   a. Integral membrane proteins generally span the width of the plasma membrane, whereas peripheral proteins are found on only one side of the membrane.
   b. Cholesterol provides the plasma membrane with stability in the face of changing ion concentrations.
   c. Membrane cholesterol is vital for cell-cell recognition.
   d. Membrane proteins often function as channels or carriers.
   e. The overall structure of the plasma membrane is a mosaic with the components locked tightly in place.

6. What is the primary difference between active transport processes and passive transport processes?

7. Match the term with its appropriate definition.

   _____Osmosis
   _____Secondary active transport
   _____Exocytosis
   _____Phagocytosis
   _____Simple diffusion
   _____Primary active transport
   _____Pinocytosis
   _____Facilitated diffusion

   a. Type of endocytosis in which a large particle is ingested
   b. Transport across the plasma membrane against the concentration gradient via direct use of energy from ATP
   c. Passive movement of solute across the plasma membrane
   d. Movement of solvent from a solution of lower solute concentration to a solution of higher solute concentration
   e. Passive movement of solute across the plasma membrane via a channel or carrier protein
   f. Type of endocytosis in which ECF is brought into the cell in a protein-coated pit
   g. Release of large substances from the cell through a vesicle
   h. Transport of a substance across the plasma membrane against its concentration gradient using the energy from the "downhill" movement of another substance

8. Fill in the blanks: A hypotonic solution will cause water to move _____ the cell and the cell will _____. A hypertonic solution will cause water to move _____ the cell and the cell will _____.

9. Match the following terms with the correct functions.

_____Peroxisome
_____Ribosome
_____Smooth endoplasmic reticulum
_____Mitochondrion
_____Golgi apparatus
_____Lysosome
_____Rough endoplasmic reticulum
_____Vesicle

a. Modifies and folds proteins into the correct structure
b. Modifies, packages, and sorts proteins
c. Contains digestive enzymes
d. Granular organelle that makes proteins
e. Synthesizes most of a cell's ATP
f. Contains enzymes that oxidize toxins and fatty acids
g. Membrane-enclosed structure used to transport substances through the cell
h. Detoxifies certain chemicals, makes lipids, stores calcium ions

10. Explain how the members of the endomembrane system interact.

11. Mark the following statements about the cytoskeleton as true or false. If a statement is false, correct it to make a true statement.

a. Actin filaments combine with myosin motor proteins to provide the cell with mechanical strength.
b. The cilia found on sperm cells propel the cells through a liquid medium.
c. Microtubules are hollow tubes that align organelles and shuttle them to their proper places in the cell.
d. Intermediate filaments are ropelike structures composed of tubulin proteins.
e. Centrioles are part of the centrosome from which microtubules originate.
f. Endocytosis, exocytosis, muscle cell contraction, and cellular "crawling" are all mediated by intermediate filaments.

12. Our somatic cells' DNA is distributed among 46 _____in the nucleus.

a. nucleosomes
b. chromatids
c. nucleoli
d. chromosomes

13. Explain how and why chromatin is condensed in the nucleus.

14. Which of the following statements correctly describes the function of the nucleolus?

a. Within the nucleolus we find genes for rRNA.
b. Nucleoli manufacture mitochondrial proteins for export to the cytosol.
c. Nucleoli assemble the nuclear envelope.
d. The nucleolus contains the genes that form the Golgi apparatus.

15. Each of the following statements about protein synthesis is false. Correct each to make a true statement.

a. In a gene, each nucleotide specifies one amino acid in a protein sequence.
b. A transcription factor must bind to the promoter region of a gene before the enzyme DNA synthetase is able to bind and begin transcription.

c. The enzyme RNA polymerase builds a strand of transfer RNA, whose codons are complementary to DNA's triplets.
d. Proteins destined for secretion from the cell enter the nucleus after translation, to be folded and modified.
e. During translation, amino acids are delivered by the messenger RNA transcript.

16. Number the following steps of protein synthesis in the order in which they occur, starting with 1 and ending with 9.

a. _____ The stop codon is reached, and the polypeptide is released.
b. _____ The small ribosomal subunit finds the start codon, and the large ribosomal subunit joins.
c. _____ The end of the gene is reached, and the pre-mRNA is released and then edited.
d. _____ The transcription factor binds the promoter.
e. _____ The protein is folded and modified to become functional.
f. _____ RNA polymerase builds the mRNA transcript.
g. _____ mRNA and initiator tRNA bind the small ribosomal subunit.
h. _____ New tRNAs are brought into the A site successively, and the peptide chain of the tRNA in the P site is joined to the amino acid of the tRNA in the A site.
i. _____ mRNA exits the nucleus via a nuclear pore.

17. Which of the following is not a phase of mitosis?

a. Interphase
b. Prophase
c. Anaphase
d. Metaphase

18. Why is regulation of the cell cycle necessary?

19. Mark the following statements about the cell cycle as true or false. If a statement is false, correct it to make a true statement.

a. During the S phase of the cell cycle, the cell stalls until conditions for division are more favorable.
b. During metaphase, the sister chromatids are pulled apart and the chromosomes move to the opposite poles of the cell.
c. The main enzyme that builds the new DNA strands during DNA synthesis is RNA synthetase.
d. The first three stages of the cell cycle are collectively called interphase.

20. Match the following terms with the correct definitions.

_____G_1
_____Metaphase
_____S phase
_____Cytokinesis
_____Telophase
_____M phase
_____G_2
_____Anaphase

a. Division of the cytoplasm
b. Second growth stage of the cell cycle
c. Stage of the cell cycle in which the cell divides
d. Stage of mitosis in which the sister chromatids line up on the cell's equator
e. Stage of the cell cycle in which DNA is replicated
f. Stage of mitosis in which the chromosomes move to opposite poles of the cell
g. Initial growth phase of the cell cycle
h. Stage of mitosis in which the nuclear envelopes reassemble

LEVEL 2 Check Your Understanding

1. Write a single sentence, using no more than 25 words, to summarize each of the following cellular processes:

   a. Diffusion
   b. Osmosis
   c. Primary active transport
   d. Secondary active transport
   e. Transcription
   f. Translation
   g. DNA synthesis
   h. Mitosis

2. Certain diseases are transmitted via mitochondrial DNA. Which cell types do you think would be most affected by such diseases, and why?

3. Explain how the form of each of the following structures is related to its function:

   a. Cilia
   b. Microvilli
   c. Intermediate filaments
   d. Lysosomes
   e. Nuclear envelope

4. Certain types of cancerous lung tumors can secrete hormones normally made by the pancreas, adrenal gland, and hypothalamus. What prevents such secretion from happening in healthy cells?

5. Why do you think the rate of cell division is different for different tissues? Where in the body would you expect to find cells that have a rapid rate of division? Where might you find cells that have a slow rate of division? Explain.

LEVEL 3 Apply Your Knowledge

PART A: Application and Analysis

1. A patient is admitted to the hospital and given intravenous (IV) fluids. Four hours later, the patient complains that his mouth and eyes feel dry. You notice that he displays signs of dehydration, and when you check his IV, you see that he was given the wrong kind of fluids. Were these fluids likely hypotonic, isotonic, or hypertonic? Explain.

2. A popular science fiction program once had an episode that featured an "intron virus" that "turned on" the introns in the genes, causing the synthesis of abnormal proteins. The episode may have been entertaining, but its premise had a large flaw about the nature of introns. What was the flaw?

3. A hypothetical poison prevents transcription factors from binding to the gene for tubulin proteins. What impact would this have on mitosis, and why?

4. What effect would the hypothetical poison of question 3 have on other functions of the cell? Explain.

PART B: Make the Connection

5. The drug *methotrexate* is used to treat several different types of cancer and diseases of the immune system. It works by inhibiting an enzyme in the cell necessary for folic acid synthesis. Without folic acid, the cell cannot make nucleotides. What, specifically, does an enzyme do in the cell? Why would inhibiting this enzyme disrupt folic acid synthesis? What effect would a disruption in folic acid synthesis have on the cell as a whole? (*Hint*: Think about the role that folic acid plays in the cell.) (*Connects to Chapter 2*)

See answers in Appendix A.

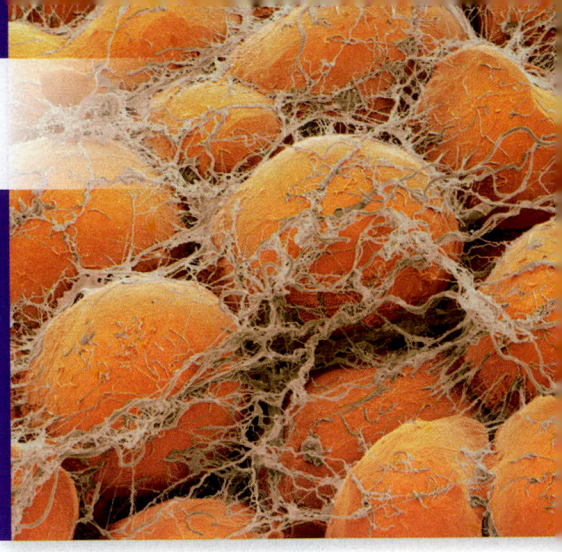

4

Histology

The human body has about 200 different cell types; they often vary dramatically in the genes they express and thus in their structure and functions. Since each cell type is specialized, individual cells cannot perform their functions unless they are in the correct area of the body and surrounded by the appropriate cells. For example, a cell that supports the cells of the brain would do you no good in your big toe, where there are no brain cells to support.

To perform their functions, cells must also be surrounded by the appropriate extracellular materials. For example, a red blood cell carries oxygen through the bloodstream and delivers it to other cells in the body. However, a red blood cell that's in your skin, surrounded by interstitial fluid instead of by blood, would not have access to high amounts of oxygen, so it could not fulfill its role of picking up oxygen. This leads us to a basic principle of human physiology: Cells cannot carry out their functions alone—they require the presence of other cells and other materials in the extracellular fluid (ECF) to maintain homeostasis. For this reason, our cells and ECF are organized into distinct groups called *tissues*. The structure and functions of tissues, in turn, directly determine the structure and functions of organs and organ systems.

Although tissues at first may seem abstract and perhaps less important than the cells or organ systems, a knowledge of tissues is critical not only for the study of anatomy and physiology but also for work in the medical field. One of the most commonly performed medical procedures is a *biopsy*, in which a tissue sample is removed from a patient, prepared, and examined under a microscope. A biopsy is performed on living tissue; however, sometimes tissue examinations are required after death during an *autopsy*. The physician who makes such tissue examinations, a *pathologist*, looks for subtle changes in the shape, size, and characteristics of the cells and components of the extracellular environment. A very slight change in the architecture of a tissue or cell can signal many things, including cancer, infection, diseases of the immune system, and the presence of certain poisons, just to name a few. A pathologist cannot determine if the tissue is abnormal unless he or she has a thorough knowledge of **histology**—the study of the normal structure of tissues. We explore the topic of histology in this chapter.

*For practice applying concepts to a clinical scenario, check out the **Running Case Study** for this chapter @ Mastering Anatomy & Physiology*

Photo: This scanning electron micrograph (SEM) shows fat cells (adipocytes) surrounded by fine protein fibers.

Introduction to Tissues

1. Define the term histology.
2. Explain where tissues fit in the levels of organization of the human body.
3. Compare and contrast the general features of the four major tissue types.
4. Describe the components of the extracellular matrix.
5. Describe the types of junctions that unite cells in a tissue.

After progressing through the first two levels of body organization, chemical and cellular, we have now reached the third level of organization—the tissues. A **tissue** is a group of structurally and functionally related cells and their external environment that together perform common functions. Each tissue type differs in structure and function, yet all tissues share the same two basic components: (1) a discrete population of cells that are related in structure and function, and (2) the surrounding material, called the *extracellular matrix* (*ECM*). Note that the extracellular fluid (ECF) is part of the ECM, along with several other components. In this module, we explore these basic parts of tissues, as well as the molecular "glue" that holds them together. But first, let's look at what types of tissues are found in the body.

Types of Tissues

There are four primary types of tissues, each of which is distinguished by the kinds and number of cells it contains, the amount and composition of ECM present, and the function it performs. The four types are epithelial, connective, muscle, and nervous:

- **Epithelial tissues.** *Epithelial tissues,* also called *epithelia* (ep'-ih-THEE-lee-ah; *epi-* = "upon"), consist of sheets of cells that are tightly packed together with little visible ECM. Epithelia cover and line all body surfaces and cavities. Specialized epithelial cells form *glands,* groups of cells that manufacture and secrete, or release, a product such as sweat, saliva, or a type of chemical messenger called a *hormone.*
- **Connective tissues.** *Connective tissues,* as the name implies, serve as a sort of cellular Velcro® that connects all other tissues in the body to one another. In most connective tissues, the ECM is the most prominent feature, and cells are scattered throughout it. Connective tissues bind, support, protect, and allow the transport of substances through the body.
- **Muscle tissues.** *Muscle tissues* are composed of cells that can contract and generate force, with little ECM between the cells. Muscle tissues are briefly introduced in this chapter, as they are covered fully in their own chapter (see Chapter 10).

- **Nervous tissue.** Within *nervous tissue* we find a unique ECM, as well as cells that can generate, send, and receive messages, and other cells that support this activity. Nervous tissue is covered completely in a later chapter (see Chapter 11).

Organs are made up of two or more of these tissue types working together. For example, the heart as an organ consists of muscle tissue, epithelial tissue, and connective tissue. The brain as an organ consists primarily of nervous tissue, but it also contains some epithelial tissue lining its hollow cavities and is enveloped by three connective tissue coverings.

☐ 1. What are the four types of tissues, and what are their characteristics?

The Extracellular Matrix

1. What are glycoproteins? (p. 74)
2. What are fibrous proteins, and what are their functions? (p. 57)

The **extracellular matrix,** or **ECM,** is composed of the substances surrounding the cells in a tissue (**Figure 4.1**). The ECM has two main components: *ground substance* and *protein fibers.* The proportion of these components varies, so that in different tissues the ECM can be a fluid, a thick gel, or a solid. The ECM is made by the cells of the tissue and performs a variety of functions, some of which are only now becoming clear to scientists. These functions include the following:

- providing the tissue with the strength to resist tensile (stretching) and compressive forces;
- directing cells to their proper places within a tissue;
- regulating the development, mitotic activity, and survival of cells; and
- holding cells in their proper positions.

Let's look at the two components of the ECM in more detail.

Ground Substance

The shapeless, gel-like **ground substance** makes up most of the ECM. Ground substance contains the interstitial or extracellular fluid with water, ions, nutrients, and other solutes, as well as three families of macromolecules: glycosaminoglycans, proteoglycans, and glycoproteins (see Chapter 2). **Glycosaminoglycans** (gly-kohs-uh-mee'-noh-GLY-kanz; **GAGs**) are long, straight polysaccharide chains. Examples of GAGs include the relatively small *chondroitin sulfate* (kahn-DROY-tin) and the enormous *hyaluronic acid* (hy'-al-yoo-RAHN-ik). The negative charges of certain sugars in a GAG attract positively charged ions in the extracellular fluid. In an example of the Gradients Core Principle (p. 28), these ions create a concentration gradient within the ECF

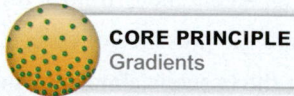
CORE PRINCIPLE
Gradients

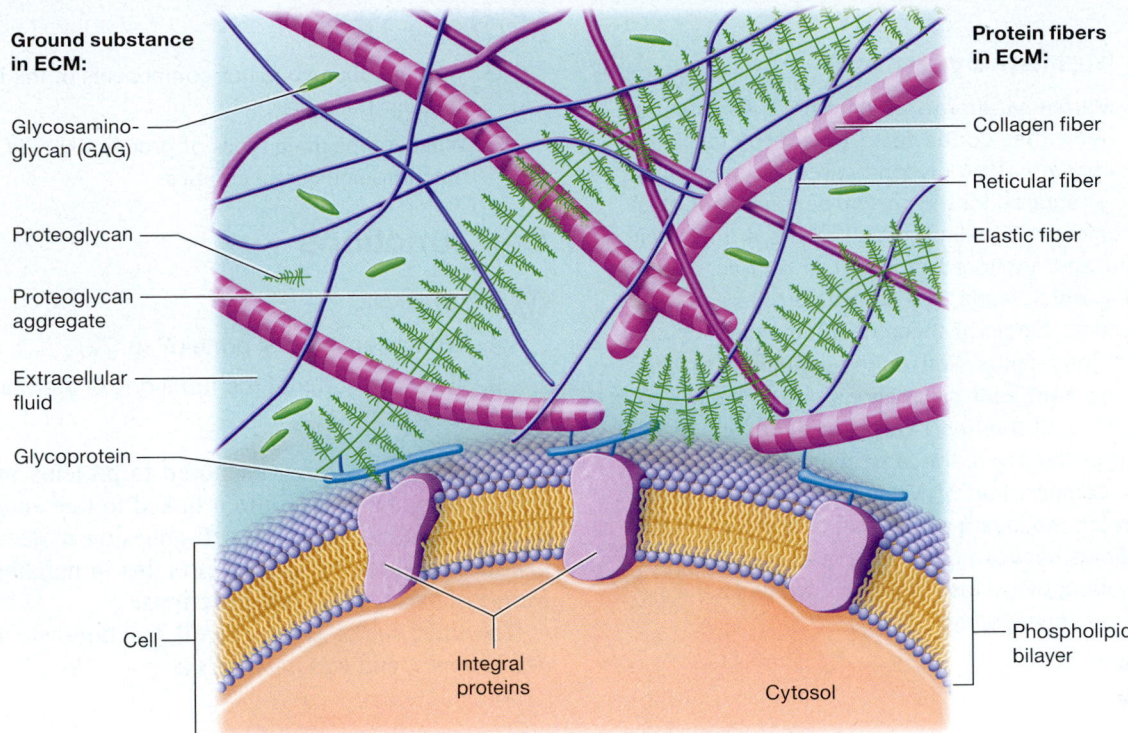

Ground substance in ECM:

Glycosamino-glycan (GAG)

Proteoglycan

Proteoglycan aggregate

Extracellular fluid

Glycoprotein

Cell

Integral proteins

Cytosol

Protein fibers in ECM:

Collagen fiber

Reticular fiber

Elastic fiber

Phospholipid bilayer

Figure 4.1 Extracellular matrix.

that draws water out of cells and blood vessels by osmosis and effectively "traps" it in the ECM, which helps the ECM to resist compression.

Proteoglycans consist of GAGs bound to a protein core; their structure resembles a bottle brush. Thousands of proteoglycans can bind in turn to a very long GAG such as hyaluronic acid, forming huge proteoglycan "aggregates." The size of these proteoglycan aggregates helps make the ECM firmer; tissues that contain more of these aggregates are thus more solid and resistant to compression than are tissues that contain fewer of them. Proteoglycan aggregates also act as a barrier to the diffusion of substances through the ECM, which protects the underlying tissues from invading microorganisms.

The final components of ground substance are **glycoproteins** of different types, together called **cell-adhesion molecules (CAMs).** As their name implies, they are responsible for adhering, or "gluing," the cells both to each other and into their places within the ECM. CAMs bind to cell surface proteins as well as protein fibers and proteoglycans, and help maintain the normal architecture of the tissue.

Protein Fibers

Embedded within the ground substance are numerous large, long structures called **protein fibers.** Protein fibers are composed of multiple fibrous protein subunits that entwine to form a long, rope-like structure with a great deal of tensile strength. Three types of protein fibers are found within the ECM: *collagen, elastic,* and *reticular.*

- **Collagen fibers.** The body makes at least 20 different types of **collagen fibers** out of repeating subunits of the fibrous protein collagen; in fact, collagen proteins make up 20–25% of all protein in the body. Collagen fibers are composed of multiple subunits of a fibrous protein. The structure of collagen proteins, which resembles the entwined pieces of a steel cable, makes them very resistant to tension (pulling and stretching forces) and pressure. Groups of collagen fibers appear white in gross anatomical specimens such as tendons, which attach a muscle to a bone.

- **Elastic fibers.** Unlike collagen fibers, which resist stretching, **elastic fibers** may be stretched to one and a half times their resting length without breaking, a property called *distensibility.* When the stretching force is removed, they return to their original length, a property called *elasticity.* Elastic fibers are composed of a protein known as *elastin* that is surrounded by glycoproteins. Elastin allows the fiber to stretch, whereas the glycoproteins support and organize the elastin. See *A&P in the Real World: Marfan Syndrome* on p. 126 to learn about a disease that affects elastic fibers.

- **Reticular fibers.** The thin **reticular fibers** were once thought to be a separate type of fiber, but they are now known to be a type of collagen fiber. However, they differ in structure, functions, and distribution from the major collagen fibers, so the term "reticular fiber" is still used by many histologists. Reticular fibers are thinner and shorter than regular collagen fibers, and they interweave to form a meshwork or scaffold (*reticul-* = "netlike") that supports the cells and ground substance of many tissues. They also form "webs" in certain organs, such as the spleen, that help these organs trap foreign cells.

Marfan Syndrome

Marfan syndrome results from defects in the gene that codes for a glycoprotein called *fibrillin-I.* This glycoprotein is a component of the ECM that is required for the normal deposition of elastic fibers. With defective fibrillin-I, elastic fibers are not correctly distributed and anchored in the ECM and so cannot function properly, which leads to a number of characteristic signs and symptoms. Some of these characteristics include tall stature with long limbs and fingers, multiple skeletal abnormalities, recurrent joint dislocations due to weak ligaments, abnormalities of the heart valves and the lens of the eye, and dilation of the aorta, the largest artery in the body. The most lethal complication of Marfan syndrome is *aortic dissection,* in which the layers of the wall of the aorta separate and blood flows between them. This leads to aortic rupture, and the ensuing blood loss is nearly always fatal if not caught and treated immediately.

Quick Check

☐ 2. What are the two major components of the ECM, and what are their functions?

☐ 3. What are the three types of protein fibers? Describe their functions and characteristics.

Cell Junctions

‹‹ FLASHBACK

1. What is an integral protein? (p. 72)

2. What are some of the functions of membrane proteins? (p. 73)

Although a cell is often anchored to proteins in the ECM, the cells in a tissue are also often linked to one another. They may be linked in two ways: by cell-adhesion molecules, which we just discussed, or by integral proteins in neighboring cells that form structures called **cell junctions.**

The three major types of cell junctions are *tight junctions, desmosomes,* and *gap junctions* (**Figure 4.2**).

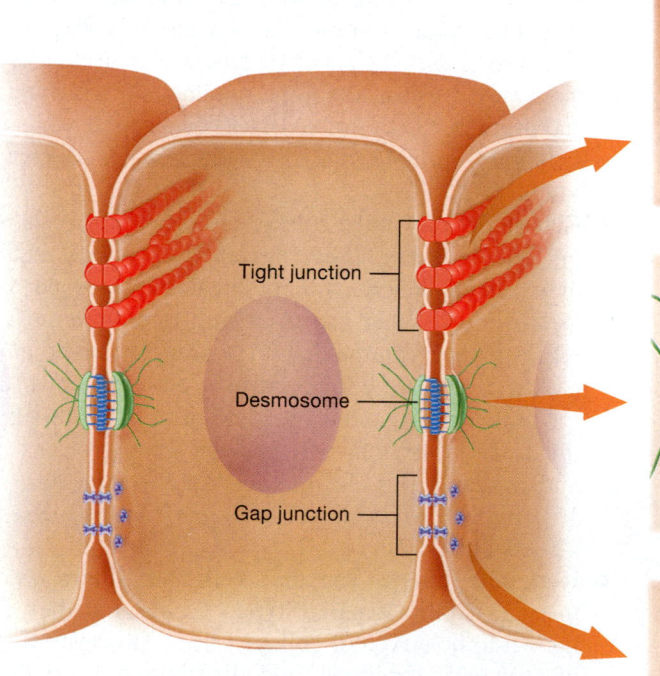

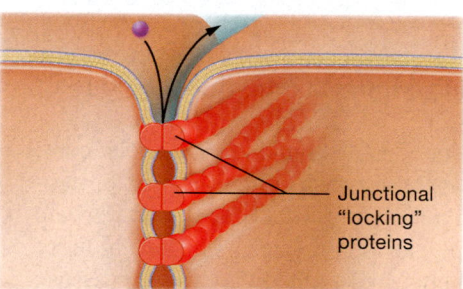

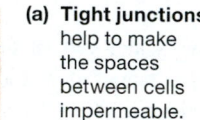

(a) Tight junctions help to make the spaces between cells impermeable.

Junctional "locking" proteins

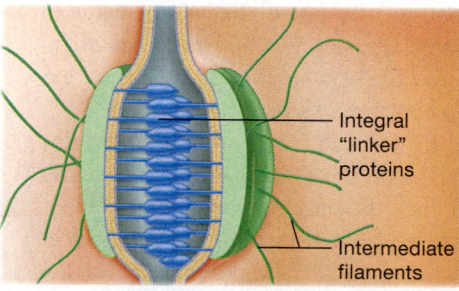

(b) Desmosomes increase the resistance of the tissue to mechanical stress.

Integral "linker" proteins

Intermediate filaments

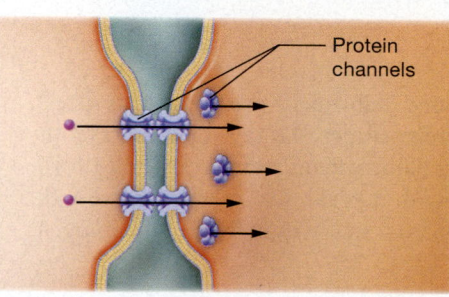

(c) Gap junctions allow small substances to move from one cell to another.

Protein channels

Figure 4.2 Cell junctions.

- **Tight junctions. Tight junctions,** also called *occluding junctions,* hold cells "tightly" together, making the spaces between them impermeable and so preventing macromolecules from passing between adjacent cells (see Figure 4.2a). These junctions are composed of integral proteins in the plasma membranes of adjacent cells. As you can see in the figure, they interweave and lock together, sealing the spaces between adjacent cells, much like the zipper on a plastic freezer bag. Note that some tight junctions are "leaky" and do not form a complete seal, which allows certain substances to pass between the cells. Tight junctions are found in multiple places in the body, such as between the cells of blood vessels, where they prevent substances from leaving the blood.

- **Desmosomes. Desmosomes** (DEZ-moh-sohmz; *desm-* = "band or bond") are also composed of integral proteins that link two cells; however, desmosomes act more like buttons or snaps than like zippers (see Figure 4.2b). And just as rain and snow can pass through the spaces in a snapped coat, materials in the extracellular fluid may pass between cells held together by desmosomes. Desmosomes increase the strength of a tissue by holding the cells together so that mechanical stress is more evenly distributed. Their integral "linker" proteins are attached to intermediate filaments of the cytoskeleton, which reinforces them. Cells that are subject to a great deal of mechanical stress, such as those in the epithelia of the skin, have a large number of desmosomes.

- **Gap junctions.** Notice in Figure 4.2c that gap junctions also link two adjacent cells, but in an entirely different way. **Gap junctions** are small pores in adjacent plasma membranes formed by protein channels, and they allow small substances to pass freely between the cytosol of two cells. Gap junctions illustrate the Cell-Cell Communication Core Principle (p. 28)—they are prominent in cells that can communicate with electrical signals, such as those of cardiac muscle.

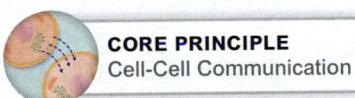

CORE PRINCIPLE
Cell-Cell Communication

Quick Check

☐ 4. What three types of junctions unite cells in a tissue, and what are their functions?

Apply What You Learned

☐ 1. What do you think would happen if the elastic fibers in tissues were distensible but not elastic?

☐ 2. Predict what would happen if the tight junctions in blood vessels were not functional.

☐ 3. Predict the consequences if the desmosomes in the outer layer of the skin were not functional.

See answers in Appendix A.

MODULE 4.2
Epithelial Tissues

Learning Outcomes

1. Classify and identify the different types of epithelial tissues.

2. Describe the location and function of each type of epithelial tissue and correlate that function with structure.

3. Describe and classify the structural and functional properties of exocrine and endocrine glands.

When you look in the mirror, nearly everything you see either is an epithelial tissue or is derived from epithelial tissue. Similarly, if you were able to look inside your hollow organs and body cavities, almost everything viewed there would be an epithelial tissue of some sort. This is because **epithelial tissues,** which are found on every internal and external body surface, primarily act as *barriers* between the body and the external environment, and between our organs and fluid-filled cavities.

The functions of epithelial tissues include the following:

- **Protection.** Epithelial tissues provide a continuous surface that shields the underlying tissues from mechanical and thermal injury. Many epithelial tissues produce substances that help them fulfill this role. For example, the epithelium of the skin produces the hard protein *keratin,* which makes the tissue more resistant to injury. Epithelial cells are subject to damage because of their protective and barrier functions, but they undergo mitosis fairly rapidly, replacing the cells that die.

- **Immune defenses.** Epithelial tissues provide a barrier against invading microorganisms as well. Specialized cells of the immune system are scattered throughout epithelial tissues to protect the underlying tissues.

- **Secretion.** Epithelial cells form glands that produce substances such as oil or hormones. These substances are secreted either through a small tube called a *duct* or into the bloodstream.

- **Transport into other tissues.** Epithelia are selectively permeable barriers, meaning that certain substances can cross them by passive or active transport and so enter other tissues. This function is critical in organs such as the small intestine, in which nutrients must cross the intestinal epithelium to enter the blood.

- **Sensation.** Most epithelia are richly supplied with nerves that detect changes in the internal and external environments. Additionally, specialized epithelial cells form part of the machinery for sensation, as occurs in the olfactory epithelium in the nasal cavity.

In this module, we first examine the components that make up epithelial tissue. Then we discuss the different types of epithelia found in the body, including those that cover and line body surfaces and those that make up glands.

ConceptBOOST 🔊

"But It All Looks Pink!" Part 1

Tissues are generally studied by first taking extremely thin slices of organs called *tissue sections* and mounting them on microscope slides. Each section is then treated with different stains so that components of the cells and ECM are visible. In this module and the ones to follow, we make extensive use of tissue sections to study the various tissue types.

When you're presented with a histological section in this textbook or in the laboratory, you may find yourself unsure of where to begin. The mass of dots and squiggly lines on most sections leads many students to complain, "But it all looks pink!" And, indeed, sometimes it *does* all look pink, due to the chemicals that are used to stain tissue sections. The key to identifying sections is to reduce every tissue to its simplest components: cells and chemicals of the ECM (see Chapter 2). Remember that although tissue sections might look foreign and abstract, they simply show cells and ECM.

Let's take a tour of a typical tissue section so that we can demystify it. We haven't covered the individual tissue types yet, so here we'll just focus on the basics of approaching a tissue section. We will focus on identifying tissues in part 2 of this Concept Boost.

Shown here is a section from the esophagus, the tubular organ that transports food from the mouth to the stomach:

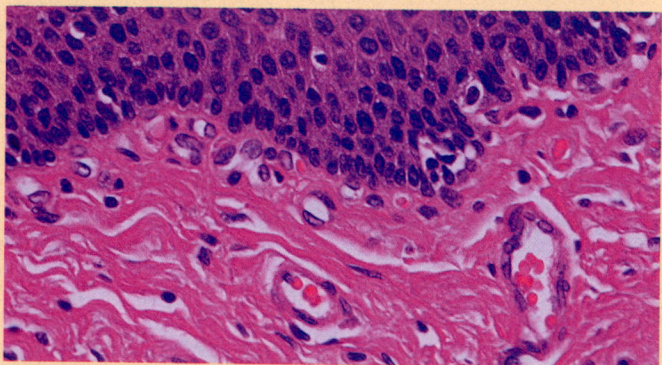

LM (330×)

Let's identify the easiest component first—the cells. Several different types of cells are present in this section, but they share a common, easily identifiable characteristic: They all have dark purple nuclei (due to the stain that binds to their DNA). So we can identify each structure that contains a dark purple nucleus as a cell. This leaves us with the material outside the cells—the ECM. Recall that the ECM consists of ground substance and protein fibers. Ground substance is simple to identify, as it generally either looks clear or has just a slight tinge of color. Protein fibers are also easy to distinguish, as they generally look like wavy or straight lines. In this section, you can see mostly collagen fibers that appear as pink wavy lines. This is the most common appearance of collagen fibers in sections, although sometimes the color varies due to different stains. Occasionally, collagen fibers form bundles that might resemble certain cell types. The easiest way to differentiate the two is to look for nuclei; if the bundles lack nuclei, then they are likely to be collagen fibers.

This next section is from the sublingual gland, a gland that produces saliva under the tongue:

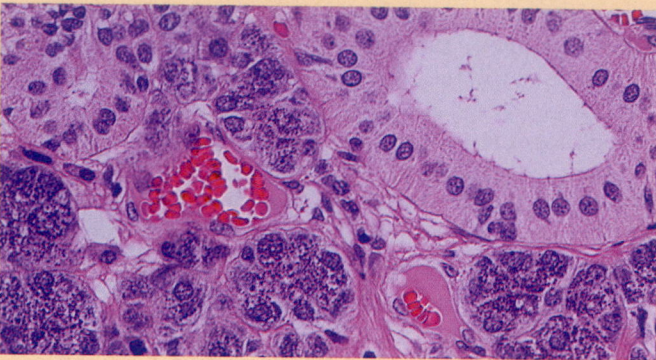

LM (450×)

Go ahead and identify the cells, ground substance, and protein fibers in this section. This section, like many others, contains clusters of small, light red, round discs that lack nuclei. You may not immediately recognize these discs as cells because they lack nuclei, but they are in fact cells called *red blood cells,* or *erythrocytes* (eh-RITH-roh-sytz). Erythrocytes are located in blood vessels, and you are likely to see blood vessels and possibly erythrocytes in many different tissue sections.

This procedure is the most basic way to approach a tissue section and orient yourself to what you are viewing. Once this becomes routine, you can use the cell shapes, the amount of ground substance, and the fiber characteristics to determine the exact type of tissue and the organ from which it was taken. We'll discuss how to do this in part 2 of this Concept Boost. ■

Components and Classification of Epithelia

We know from Module 4.1 that epithelial tissues consist mostly of cells that are closely packed together (**Figure 4.3**). The cells are generally joined by tight junctions and desmosomes that make the sheet of cells fairly impermeable and more resistant to both stresses and mechanical injury.

Notice in Figure 4.3 that epithelial tissue lacks blood vessels—that is, it is *avascular*—so oxygen and nutrients from the blood must diffuse up to the cells in the epithelial tissue from the blood vessels in the tissues that are deep to it, which are nearly always connective tissue. This imposes a limit on the thickness of epithelial tissue. The stacks of epithelial cells can rise only so high, because the oxygen and nutrients are able to diffuse only so far within a certain time. Cells that are too far away from the deeper blood supply will lack sufficient oxygen and nutrients and so will die.

Figure 4.3 also shows that essentially no ECM is present between the cells of epithelial tissue; instead, the ECM is located *beneath* the cells in a structure known as the **basement membrane.** The basement membrane has two components: the **basal lamina** (*basal-* = "bottom," *lamina-* = "layer") and its underlying **reticular lamina.** The basal lamina is the ECM of the epithelial

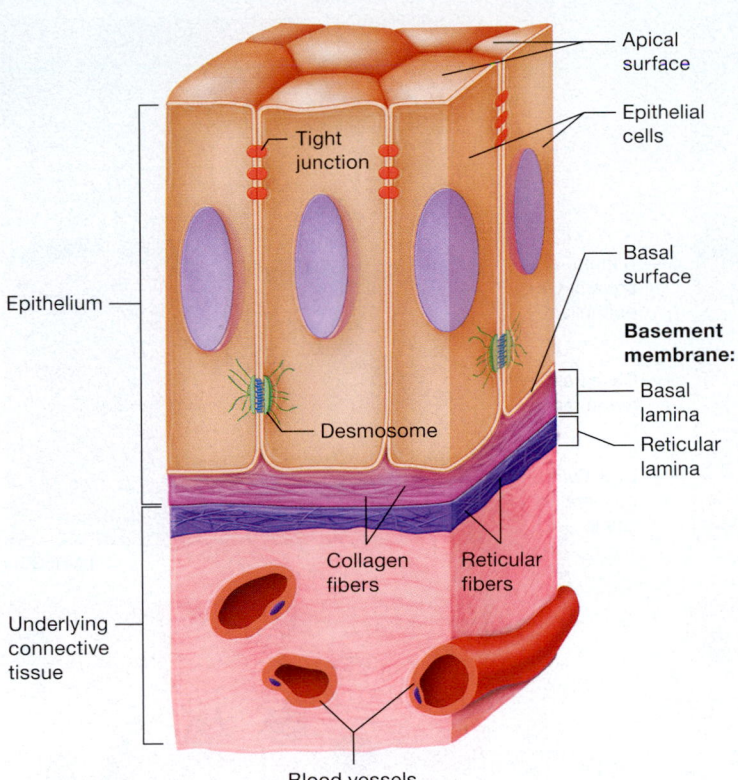

Figure 4.3 **Structure of epithelial tissue.**

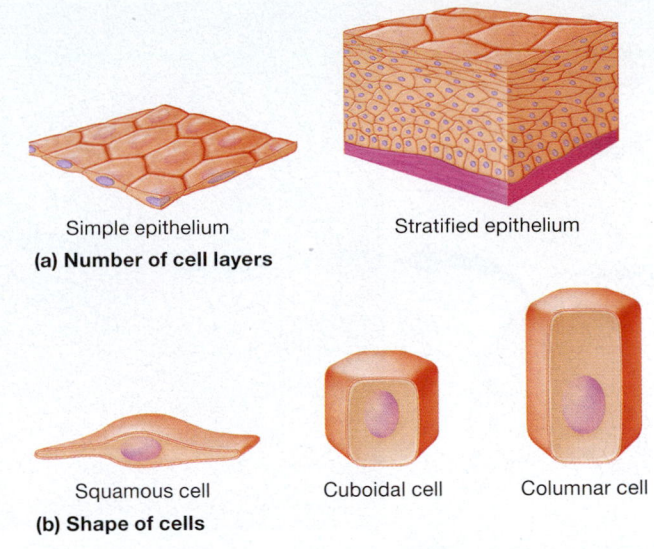

(a) Number of cell layers

Simple epithelium   Stratified epithelium

(b) Shape of cells

Squamous cell   Cuboidal cell   Columnar cell

Figure 4.4 **Classification of epithelial cells.**

by chemicals called *carcinogens,* as discussed in *A&P in the Real World: Carcinogens and Epithelial Tissues* on p. 130.

Quick Check

☐ 1. Why is living epithelial tissue limited to a certain thickness?

☐ 2. Where is the ECM of epithelial tissue located, and what is its function?

☐ 3. What are the three basic shapes of epithelial cells?

Covering and Lining Epithelia

◀◀ FLASHBACK

1. What is diffusion? (p. 75)

2. Which substances cross a plasma membrane by diffusion? (p. 76)

Covering and lining epithelia are found on inner and outer body surfaces. Cells of each shape—squamous, cuboidal, or columnar—are knitted into broad, flat sheets of varying thickness. These sheets are continuous and removable as a whole, so they are often called *membranes* (a term that can include the underlying basement membrane as well). Now let's explore the different classes of epithelia.

Simple Epithelia

As simple epithelia consist of only one layer of cells, and are therefore extremely thin, they are not very resistant to mechanical stresses and thus wouldn't serve very well as outer barriers. However, their thinness makes them perfect for lining hollow organs and for lining surfaces across which substances must diffuse or be transported—a good example of the Structure-Function Core Principle (p. 28).

CORE PRINCIPLE
Structure-Function

Types of Simple Epithelia Simple epithelia can be grouped into four types (**Figure 4.5**):

● **Simple squamous epithelium.** As you can see in Figure 4.5a, **simple squamous epithelium** consists of a single layer of flat

---

tissue and is synthesized by the epithelial cells. It consists mostly of collagen fibers and ground substance. The reticular lamina is manufactured by the connective tissue deep to the epithelial tissue, and consists of reticular fibers and ground substance. Together, the two layers of the basement membrane "glue" the epithelial tissue to the underlying connective tissues. The membrane also anchors underlying blood vessels in place and provides a barrier between the epithelial tissue and the underlying tissue.

Note that the cells of epithelial tissue have one side in contact with the extracellular space (which may be outside the body) and another surface in contact with deeper cells or with the basal lamina. The "free edge" of an epithelial cell or tissue is the **apical surface** (*ap-* = "top"), and the edge attached to deeper cells or the basal lamina is the **basal surface.** The sides are known as the *lateral surfaces.* The properties of epithelial tissues can differ greatly on the apical and basal sides in terms of cell shape, cell functions, and plasma membrane features.

We classify epithelia based on two criteria. The first is the number of cell layers, and the second is the shape of the cells in those layers. Epithelia that consist of only a single layer of cells are **simple epithelia;** those consisting of more than one cell layer are **stratified epithelia** (**Figure 4.4a**). Epithelial cells can take on three basic shapes, shown in **Figure 4.4b**: the flattened **squamous cells** (SKWAY-muss; *squam-* = "scales"), the short **cuboidal cells,** and the tall and elongated **columnar cells.** The number of layers and the shapes of these cells have functional significance, which we discuss in the upcoming sections.

As you might expect, the functions of epithelia are dependent on their structure; when structure changes, function can be disrupted. This is especially evident when their structure is altered

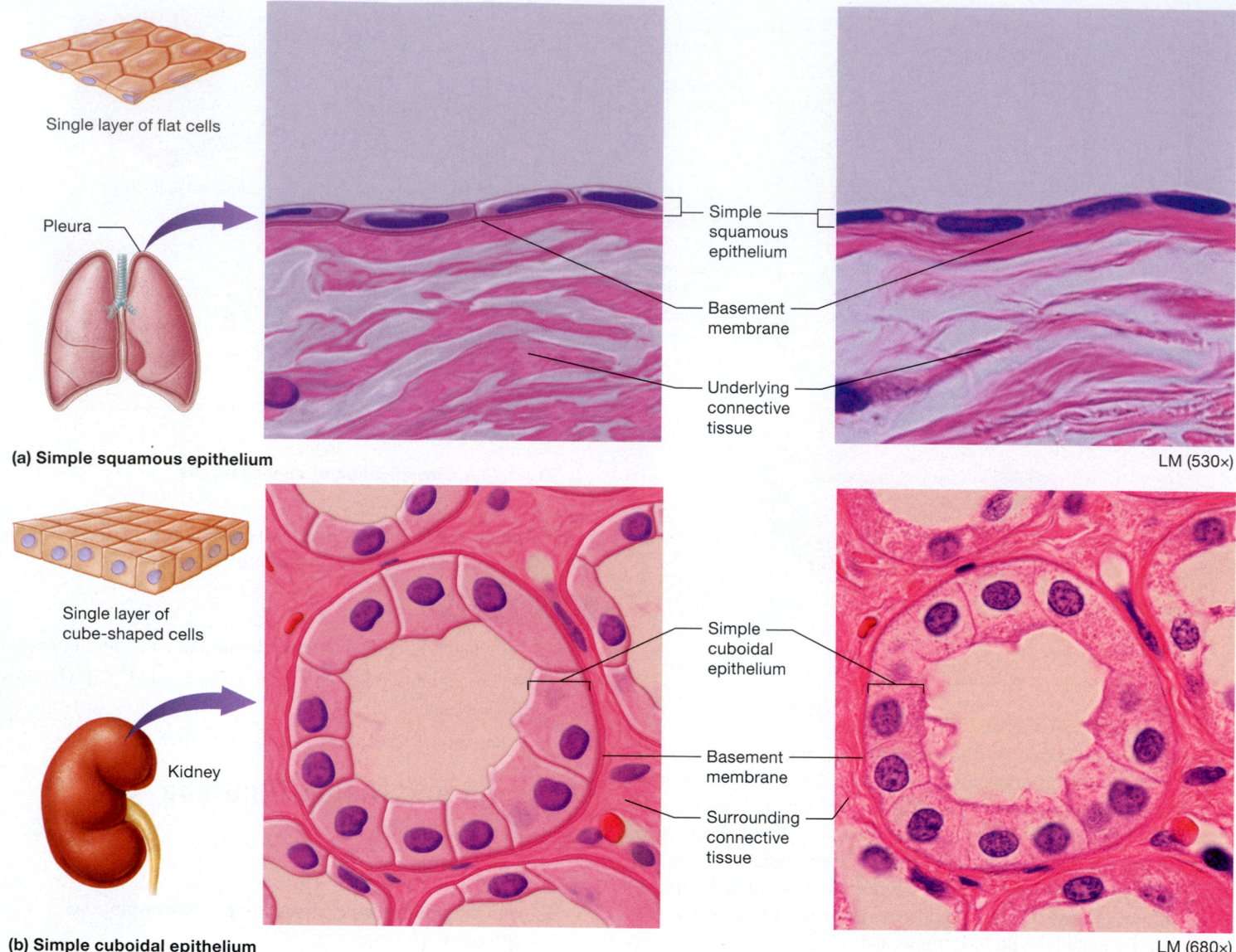

Single layer of flat cells

Pleura

Simple squamous epithelium

Basement membrane

Underlying connective tissue

LM (530×)

(a) Simple squamous epithelium

Single layer of cube-shaped cells

Kidney

Simple cuboidal epithelium

Basement membrane

Surrounding connective tissue

LM (680×)

(b) Simple cuboidal epithelium

Figure 4.5 Structure of simple epithelia.

Carcinogens and Epithelial Tissues

A P in the Real World

Epithelia cover all body surfaces and so are more subject to injuries than most other tissues. One type of injury comes from **carcinogens** (kar-SIN-oh-jenz), which are agents inducing changes in DNA that can lead to cancer. Cancers of epithelial tissues are called **carcinomas** (kar-sih-NOH-mahz); some of the most common carcinomas include the following:

- *lung adenocarcinoma* (ah-den-oh′-kar-sih-NOH-muh), a cancer of the lung;
- *ductal* and *papillary carcinoma* (PAP-ih-lehr-ee), cancers of the breast; and
- *basal cell carcinoma,* a cancer of the skin, shown at right.

The basement membrane provides a barrier that can slow the spread of carcinomas to other tissues. Cancers that have not penetrated the basement membrane, and so have not invaded other tissues, are called "pre-malignant" or *carcinoma in situ* (SEE-too; *in situ* = "in place"). Often cancer cells produce enzymes that degrade components of the basement membrane, which facilitates their spread into neighboring or distant tissues.

Clusters of tumor cells

LM (70×)

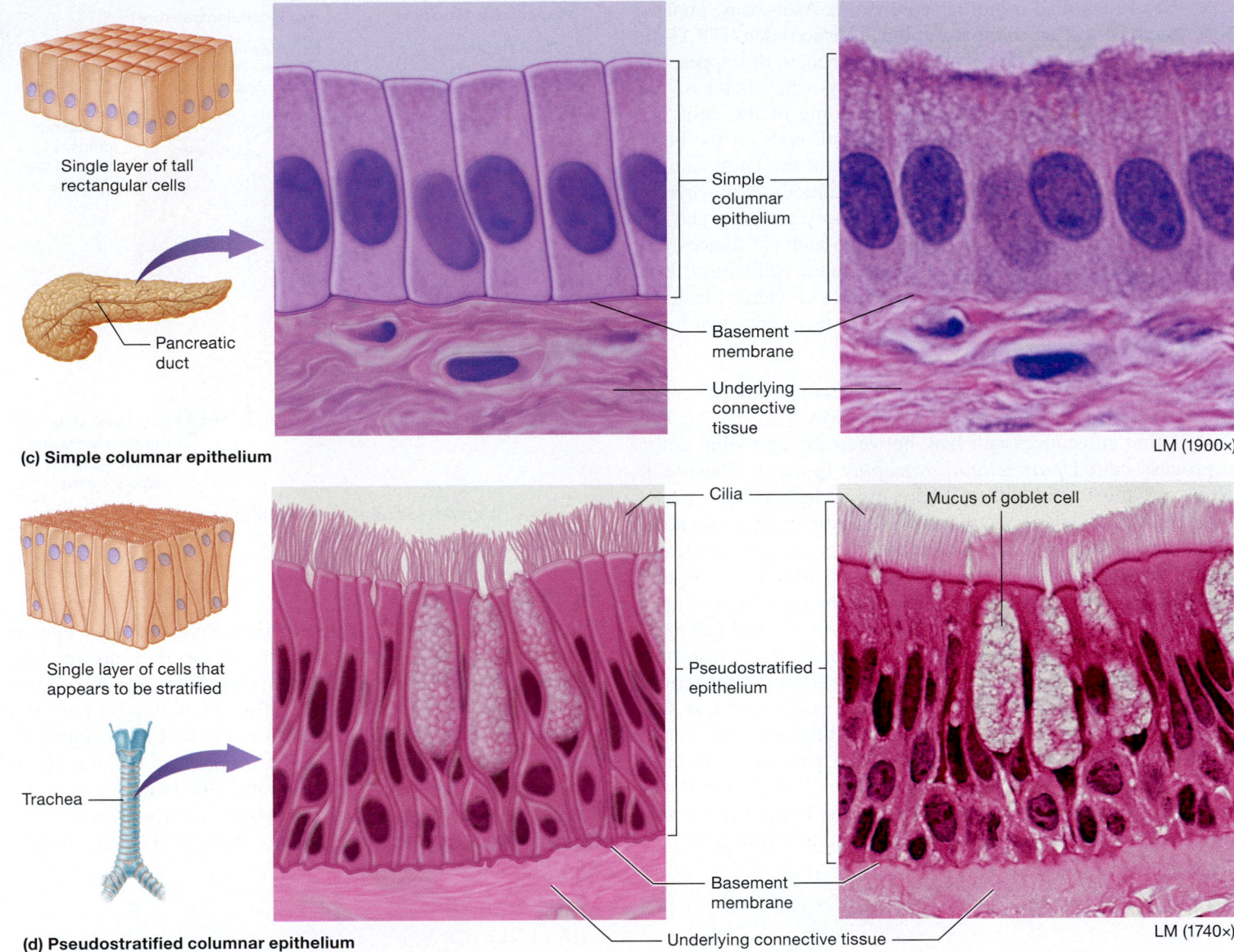

(c) Simple columnar epithelium

Single layer of tall rectangular cells

Pancreatic duct

Simple columnar epithelium

Basement membrane

Underlying connective tissue

LM (1900×)

(d) Pseudostratified columnar epithelium

Single layer of cells that appears to be stratified

Trachea

Cilia

Mucus of goblet cell

Pseudostratified epithelium

Basement membrane

Underlying connective tissue

LM (1740×)

Figure 4.5 (*continued*)

cells. If you look at the apical surfaces of the cells, they resemble fried eggs that fit together like six-sided floor tiles. Simple squamous epithelium is extremely thin, a fact that allows substances such as oxygen, carbon dioxide, fluids, and ions to diffuse across it quickly. We find simple squamous epithelium lining the tiny air sacs of the lungs, forming the outer boundary of serous membranes such as the pleurae (see Chapter 1), forming certain parts of the kidney tubules, and lining blood vessels.

- **Simple cuboidal epithelium. Simple cuboidal epithelium,** shown in Figure 4.5b, is a single layer of roughly cube-shaped cells. When viewed in a section, the cells appear square, with a large, central nucleus. Like simple squamous epithelium, simple cuboidal epithelium is relatively thin, so substances can diffuse across it rapidly. Simple cuboidal epithelium lines structures where rapid diffusion is needed, such as certain kidney tubules and respiratory passages, the ducts of many glands, and the thyroid gland.

- **Simple columnar epithelium.** The cells of **simple columnar epithelium** are tall and appear rectangular in a section, with their nuclei generally in the basal portion of the cell (see Figure 4.5c). The apical plasma membranes of these cells are often folded into *microvilli,* which increases their surface area for absorption (look back at Figure 3.23). Examples of simple columnar epithelia with microvilli are found in the small intestine, the gallbladder, and the kidney tubules. The apical surface of simple columnar epithelial cells also sometimes contains *cilia,* many short cellular extensions that beat in unison to propel something through a hollow organ; examples are found in the uterine tube and certain respiratory passages.

- **Pseudostratified columnar epithelium.** As its name implies, **pseudostratified columnar epithelium** (soo'-doh-STRAT-ih-fy'd; *pseudo-* = "false") is simple epithelium that appears to be stratified. Figure 4.5d shows why—the nuclei of the cells are at different heights, and some of the cells are shorter than others. However, each cell rests on the basal lamina of the basement membrane, and the epithelium is only one cell layer thick. Pseudostratified columnar epithelium is found in the larger respiratory passages and the nasal cavity, where it is ciliated. Although substances can diffuse across pseudostratified columnar epithelium, it is generally too thick for this to take place efficiently. Instead, this relatively thick epithelium performs a protective function in the organs where it is located.

Transport across Simple Epithelia Transport across simple epithelial tissue may occur in two ways (**Figure 4.6**). The first is that substances can leak between the epithelial cells, a process called *paracellular transport* (*para-* = "beside"). However, most epithelial cells are linked by tight junctions, which make the spaces between them relatively impermeable. For this reason, most substances cross epithelia by a second method, *transcellular transport* (*trans-* = "through"). In this process, ① a substance enters the cell through its phospholipid bilayer, ② diffuses through the cytosol, and ③ exits through the other surface of the cell. Transcellular transport operates by the same rules we discussed earlier: Water moves into and out of the cell with an osmotic gradient; some solutes move into and out of the cell by simple diffusion and facilitated diffusion when a concentration gradient is present; other solutes move into and out of the cell by active transport pumps against a concentration gradient; and certain solutes move into and out of the cell by vesicular transport (see Chapter 3).

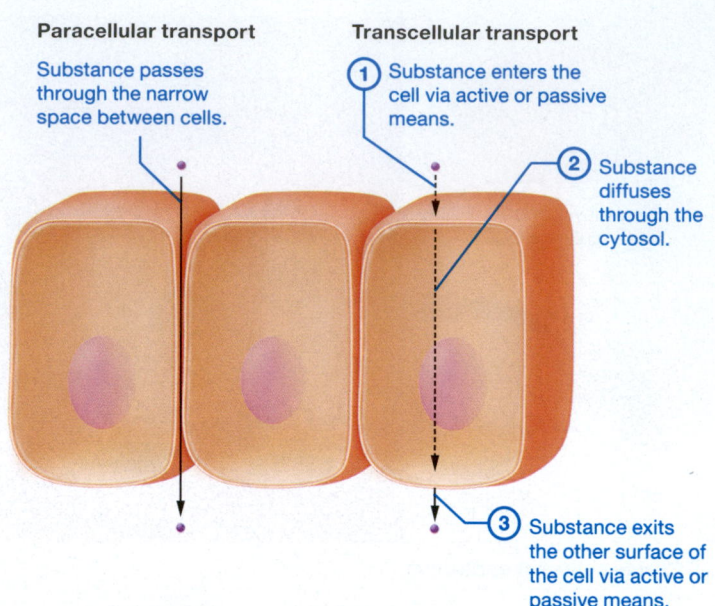

Paracellular transport
Substance passes through the narrow space between cells.

Transcellular transport
① Substance enters the cell via active or passive means.
② Substance diffuses through the cytosol.
③ Substance exits the other surface of the cell via active or passive means.

Figure 4.6 Transport across simple epithelia.

Stratified Epithelia

Stratified epithelia are thicker than simple epithelia due to their additional layers, and for this reason they do not line areas where transcellular transport is required. However, their thickness allows them to be very effective protective barriers, and they are often found in areas subject to a high degree of mechanical stress. As you can see in **Figure 4.7**, the cells in stratified epithelium generally do not maintain the same shape throughout the thickness of the tissue; for simplicity, these epithelia are named according to the shape of the cells in their apical layers.

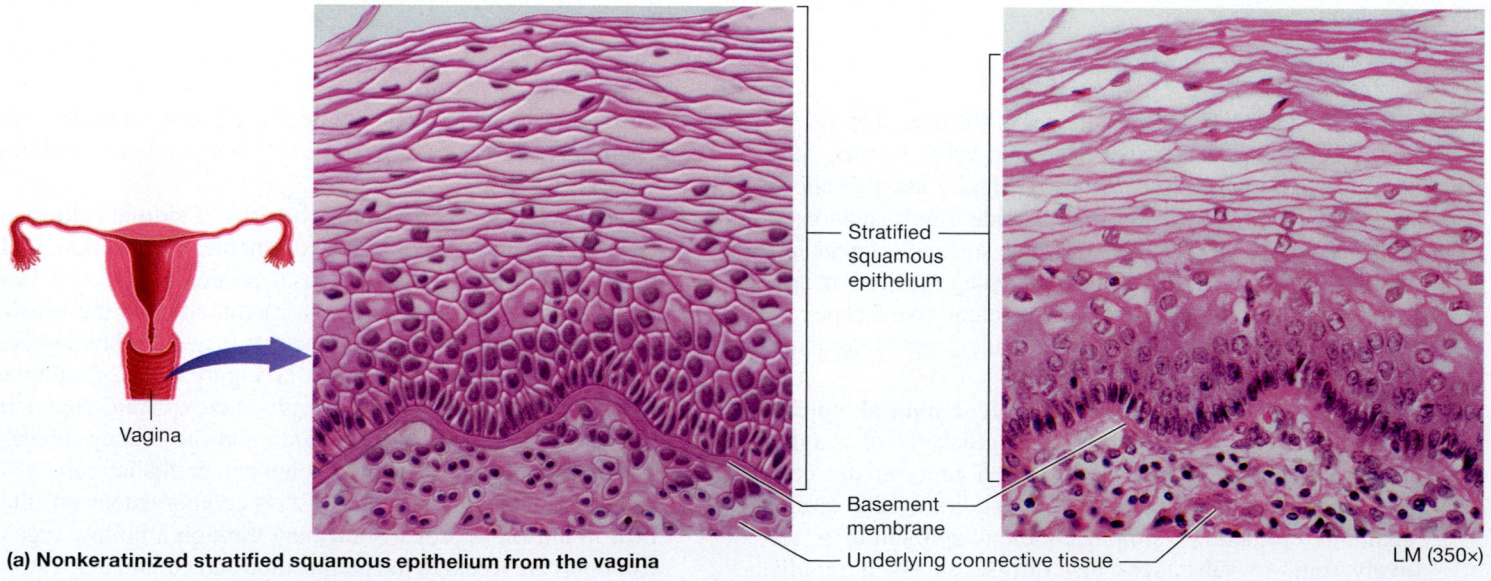

Vagina

Stratified squamous epithelium

Basement membrane

Underlying connective tissue

LM (350×)

(a) Nonkeratinized stratified squamous epithelium from the vagina

Figure 4.7 Structure of stratified epithelia.

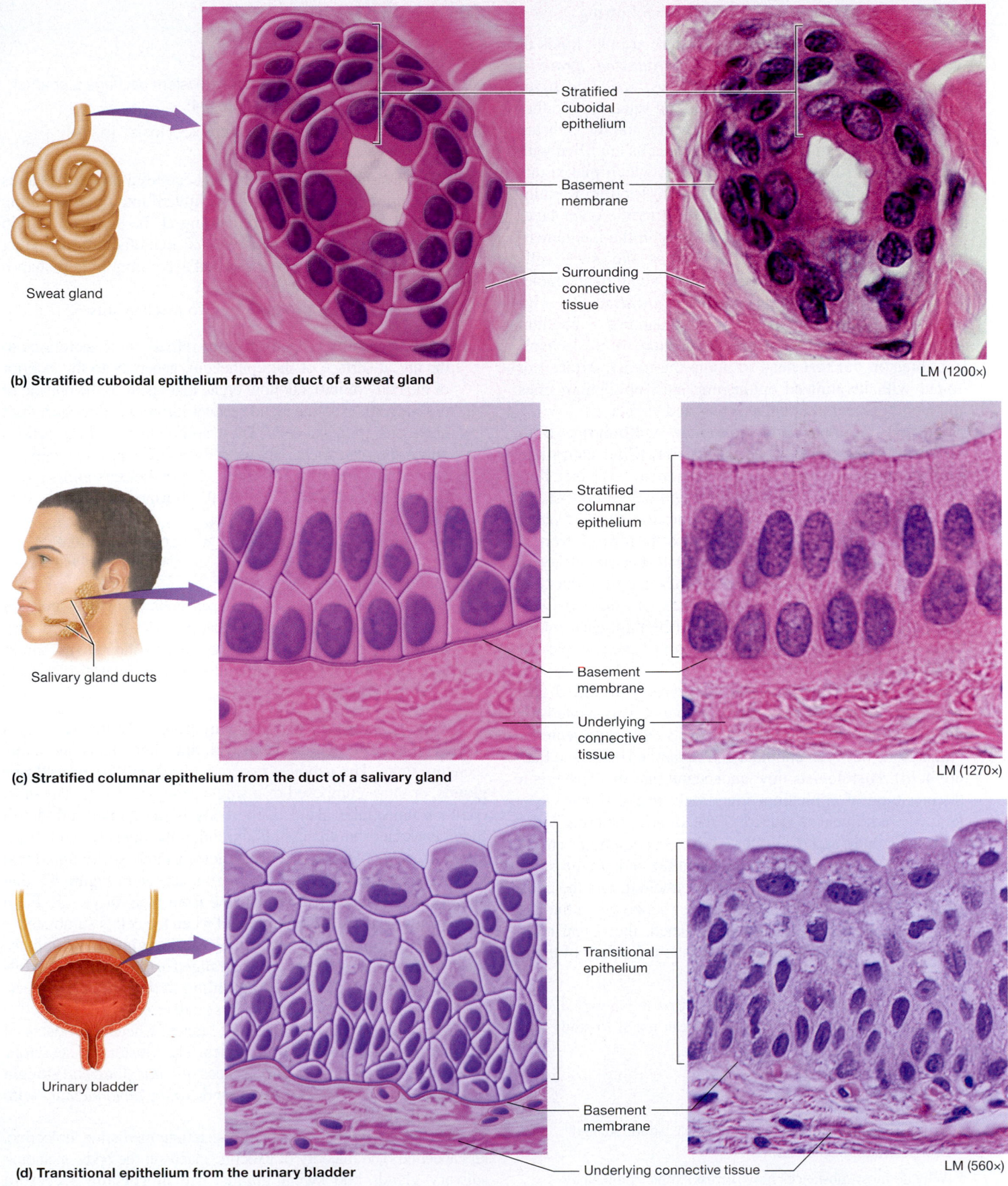

(b) Stratified cuboidal epithelium from the duct of a sweat gland

Sweat gland

Stratified cuboidal epithelium

Basement membrane

Surrounding connective tissue

LM (1200×)

(c) Stratified columnar epithelium from the duct of a salivary gland

Salivary gland ducts

Stratified columnar epithelium

Basement membrane

Underlying connective tissue

LM (1270×)

(d) Transitional epithelium from the urinary bladder

Urinary bladder

Transitional epithelium

Basement membrane

Underlying connective tissue

LM (560×)

Figure 4.7 Structure of stratified epithelia. (*continued*)

The types of stratified epithelia include the following:

- **Stratified squamous epithelium.** There are two kinds of **stratified squamous epithelium:** *keratinized stratified squamous epithelium* and *nonkeratinized stratified squamous epithelium.* In keratinized stratified squamous epithelium, the apical layers of cells lack nuclei, and the cells are no longer living. In addition, the apical cells are filled with the protein **keratin** (*kerato-* = "hard"), which makes this type of epithelium tough and resistant to friction. Keratinized stratified squamous epithelium forms the outer layer of our skin, so it will be discussed fully in the integumentary chapter (see Chapter 5). In contrast, the apical cells of nonkeratinized stratified squamous epithelium are nucleated, the cells are living, and each cell is distinct (see Figure 4.7a). Nonkeratinized stratified squamous epithelium is found in organs that require protection from mechanical abrasion but that need to retain the moist surface not found with keratinized epithelium; such epithelium lines the mouth, throat, esophagus, anus, and vagina.
- **Stratified cuboidal and columnar epithelium.** Both **stratified cuboidal epithelium** and **stratified columnar epithelium** are relatively rare in the human body (see Figure 4.7b and c). Stratified cuboidal epithelium consists of two layers of cuboidal cells, and it lines the ducts of sweat glands. Stratified columnar epithelium is formed from a few layers of cells that are columnar in the apical layers and cuboidal in the basal layers. It makes up the ducts of certain glands, including salivary glands, as well as parts of the male urethra and the conjunctiva (the thin, clear membrane lining the anterior surface of part of the eye and the inner eyelid).
- **Transitional epithelium.** The final type of stratified epithelium is **transitional epithelium,** which was named for the incorrect belief that the shape of its cells represented a "transition" between cuboidal and squamous cells (see Figure 4.7d). Histologists now understand that this tissue is a distinct type of epithelium found only in the urinary system, where it lines the interior of the kidney, the ureters, the urinary bladder, and the urethra. It is now sometimes called *urothelium* for this reason. Notice that the cells in the basal layers of transitional epithelium are cuboidal, and the cells in the apical layers are vaguely dome-shaped when the tissue is relaxed. However, when stretched, the apical cells can become flatter, which contributes to the ability of these urinary organs to stretch.

The types of epithelial tissue are summarized in **Figure 4.8**. This figure is set up in a tabular format so you can use it to study and compare these tissues.

☐ 4. What are the differences between simple, stratified, and pseudostratified epithelium?

☐ 5. Why do most substances move across simple epithelia by transcellular transport?

Glandular Epithelia

« FLASHBACK

1. How does the endomembrane system package a product for secretion from the cell? (p. 94)
2. What is the basic process of exocytosis? (p. 84)

A **gland** is a structure that makes and secretes a product. Glands arise from epithelial tissue that grew inward into the underlying connective tissue rather than remaining at the surface. Certain cells of a gland, called *secretory cells,* manufacture a product and release it. Glands can be classified according to their shapes and how they release their products.

Glands release their products by two mechanisms:

- **Exocrine glands** (EKS-oh-krin) release their secretions to the apical surface of the epithelium, generally to the exterior of the body or into a hollow organ that opens to the outside of the body. An exocrine gland secretes its product through a **duct** lined with epithelial cells. These secretions have *local* actions only—they can affect only the cells in their general vicinity.
- **Endocrine glands** (EN-doh-krin) lack ducts and secrete their products, which are usually hormones, directly into the blood. Hormones are able to reach distant target cells via the bloodstream, so they 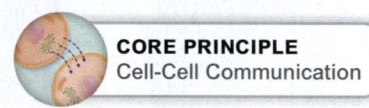 facilitate communication between cells in distant areas of the body (Cell-Cell Communication Core Principle, p. 28). Endocrine glands are discussed fully in their own chapter (see Chapter 16); the remainder of this section discusses exocrine glands.

CORE PRINCIPLE Cell-Cell Communication

Exocrine glands vary in complexity from those that are only a single cell to those that are large with branching ducts and many secretory units. The simplest exocrine glands are the **unicellular glands,** or those composed of a single cell (**Figure 4.9**). The most common unicellular gland in the body is the **goblet cell,** a cell found abundantly in the epithelium lining the digestive and respiratory tracts. Goblet cells secrete **mucus,** a thick, sticky liquid that protects the underlying epithelium. You can see in Figure 4.9 that goblet cells, which derive their name from their shape, are filled with mucus droplets that are released when the cell is stimulated.

Most exocrine glands are **multicellular glands,** which are made of clusters of secretory cells arranged in different ways. Multicellular glands are classified according to the structure of their ducts and the shape of their secretory cell clusters (**Figure 4.10**). **Simple glands** have ducts that don't branch, whereas the ducts of **compound glands** do have branches. The clusters of secretory cells may be arranged into three shapes: *tubular* (long and straight or coiled), *acinar* (AY-sih-nahr; spherical), or *tubuloacinar* (with both tubular and acinar portions).

There are three types of exocrine secretion: merocrine, holocrine, and apocrine. The majority of exocrine glands in the body, including salivary glands and sweat glands, use **merocrine secretion** (MEHR-oh-krin), in which they package their products into secretory

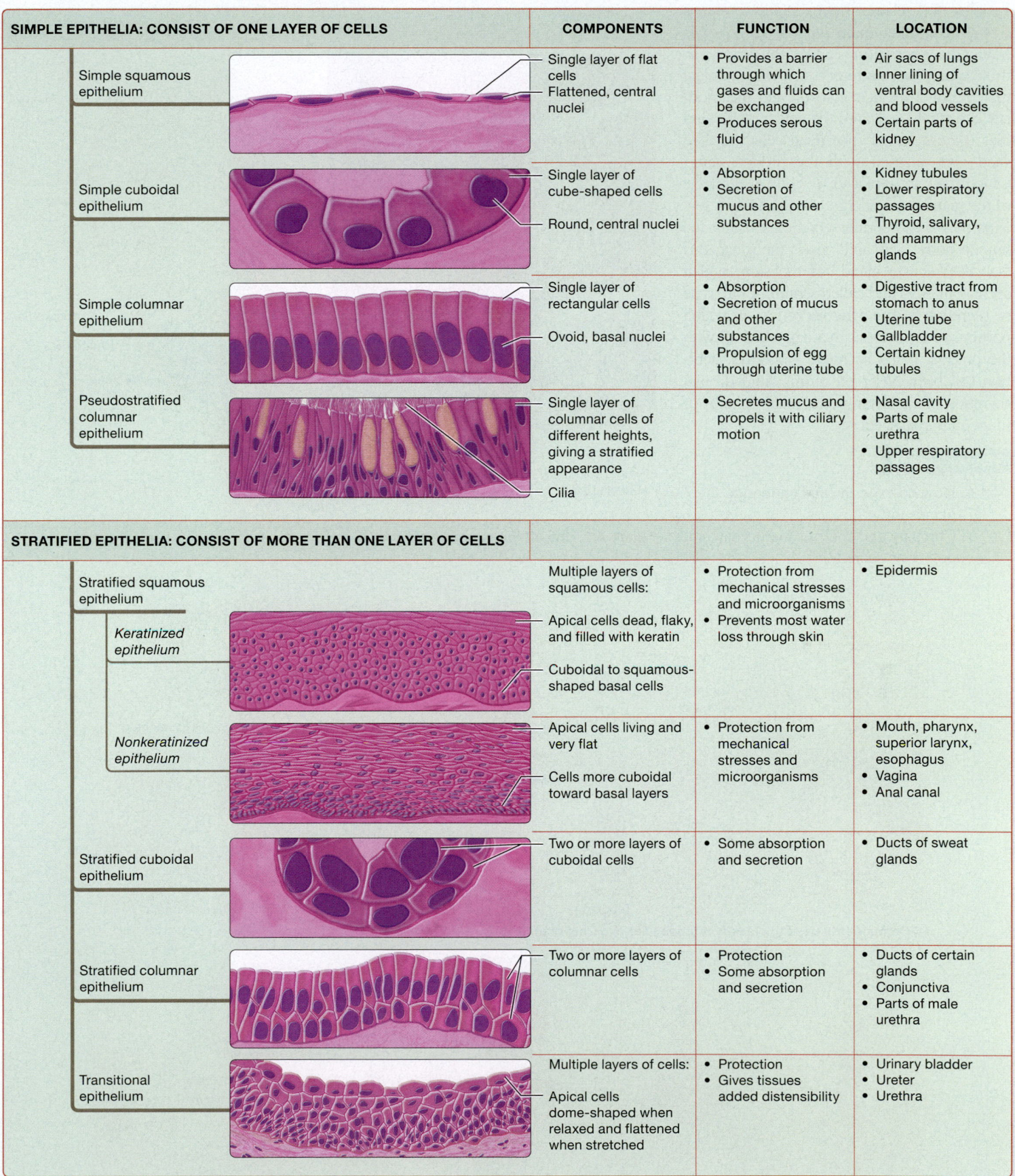

SIMPLE EPITHELIA: CONSIST OF ONE LAYER OF CELLS		COMPONENTS	FUNCTION	LOCATION
Simple squamous epithelium		• Single layer of flat cells • Flattened, central nuclei	• Provides a barrier through which gases and fluids can be exchanged • Produces serous fluid	• Air sacs of lungs • Inner lining of ventral body cavities and blood vessels • Certain parts of kidney
Simple cuboidal epithelium		• Single layer of cube-shaped cells • Round, central nuclei	• Absorption • Secretion of mucus and other substances	• Kidney tubules • Lower respiratory passages • Thyroid, salivary, and mammary glands
Simple columnar epithelium		• Single layer of rectangular cells • Ovoid, basal nuclei	• Absorption • Secretion of mucus and other substances • Propulsion of egg through uterine tube	• Digestive tract from stomach to anus • Uterine tube • Gallbladder • Certain kidney tubules
Pseudostratified columnar epithelium		• Single layer of columnar cells of different heights, giving a stratified appearance • Cilia	• Secretes mucus and propels it with ciliary motion	• Nasal cavity • Parts of male urethra • Upper respiratory passages
STRATIFIED EPITHELIA: CONSIST OF MORE THAN ONE LAYER OF CELLS				
Stratified squamous epithelium		Multiple layers of squamous cells:	• Protection from mechanical stresses and microorganisms • Prevents most water loss through skin	• Epidermis
	Keratinized epithelium	• Apical cells dead, flaky, and filled with keratin • Cuboidal to squamous-shaped basal cells		
	Nonkeratinized epithelium	• Apical cells living and very flat • Cells more cuboidal toward basal layers	• Protection from mechanical stresses and microorganisms	• Mouth, pharynx, superior larynx, esophagus • Vagina • Anal canal
Stratified cuboidal epithelium		• Two or more layers of cuboidal cells	• Some absorption and secretion	• Ducts of sweat glands
Stratified columnar epithelium		• Two or more layers of columnar cells	• Protection • Some absorption and secretion	• Ducts of certain glands • Conjunctiva • Parts of male urethra
Transitional epithelium		Multiple layers of cells: • Apical cells dome-shaped when relaxed and flattened when stretched	• Protection • Gives tissues added distensibility	• Urinary bladder • Ureter • Urethra

Figure 4.8 Summary of epithelial tissues.

View **PAL** Histology
@ **Mastering** Anatomy & Physiology

vesicles for release by exocytosis (**Figure 4.11a**). The other main type of secretion is **holocrine secretion** (HOH-loh-krin). In this type, the secretory cells accumulate their product in their cytosol, and the product is not released until the cell ruptures and dies (**Figure 4.11b**). The dead cells are then shed and released with the product, after which they are replaced by cells at the base of the gland that undergo mitosis. Only a few examples of holocrine secretion exist in the human body; a notable one is the sebaceous (seh-BAY-shuss) glands of the skin, which secrete an oily fluid called sebum. A rare type of secretion is **apocrine secretion,** in which portions of the cytoplasm are pinched off with the product being secreted. Apocrine secretion has been observed during lipid droplet secretion in the lactating mammary glands of many mammal species.

Quick Check

- [ ] 6. How do exocrine and endocrine glands differ?
- [ ] 7. Compare and contrast merocrine and holocrine secretion.

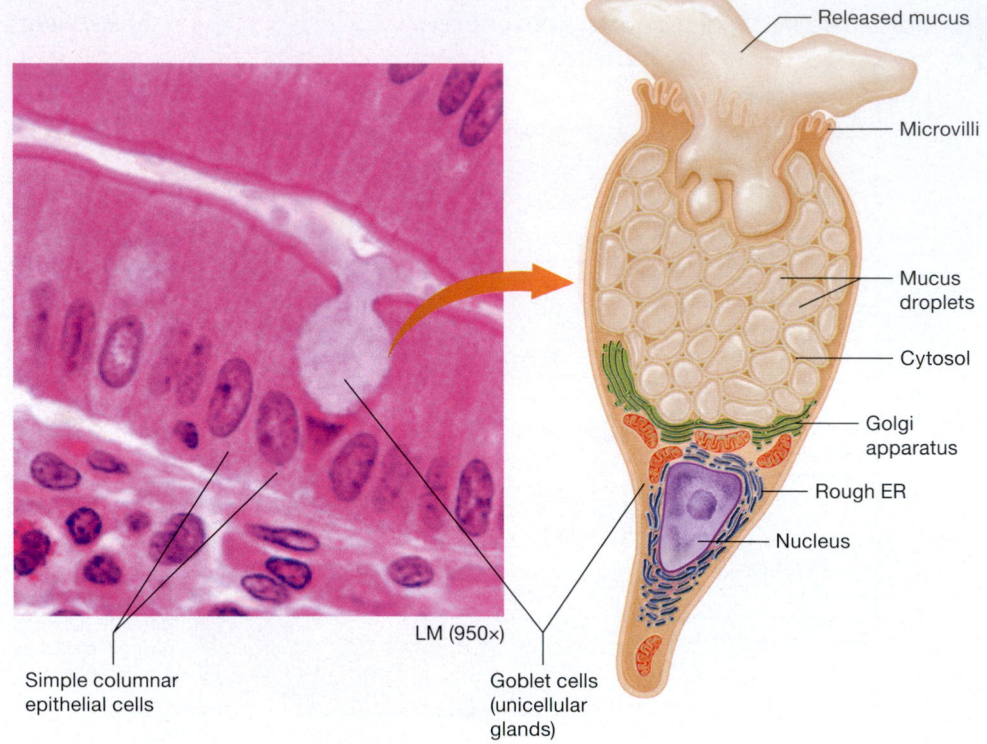

LM (950×)

Simple columnar epithelial cells

Released mucus

Microvilli

Mucus droplets

Cytosol

Golgi apparatus

Rough ER

Nucleus

Goblet cells (unicellular glands)

Figure 4.9 Unicellular exocrine glands.

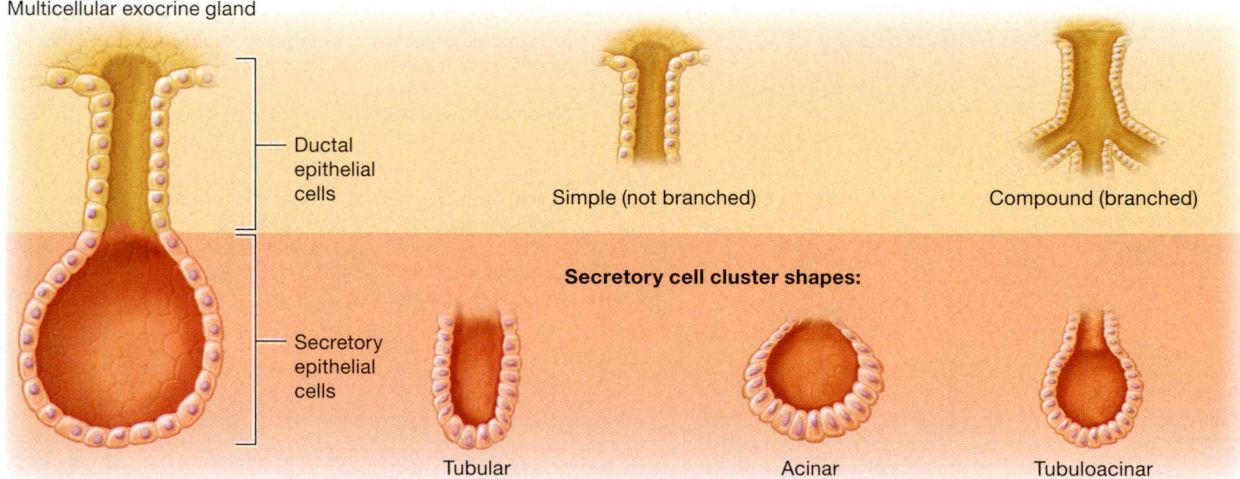

Multicellular exocrine gland

Ductal epithelial cells

Simple (not branched)

Compound (branched)

Secretory cell cluster shapes:

Secretory epithelial cells

Tubular

Acinar

Tubuloacinar

(a) Components used to classify multicellular exocrine glands

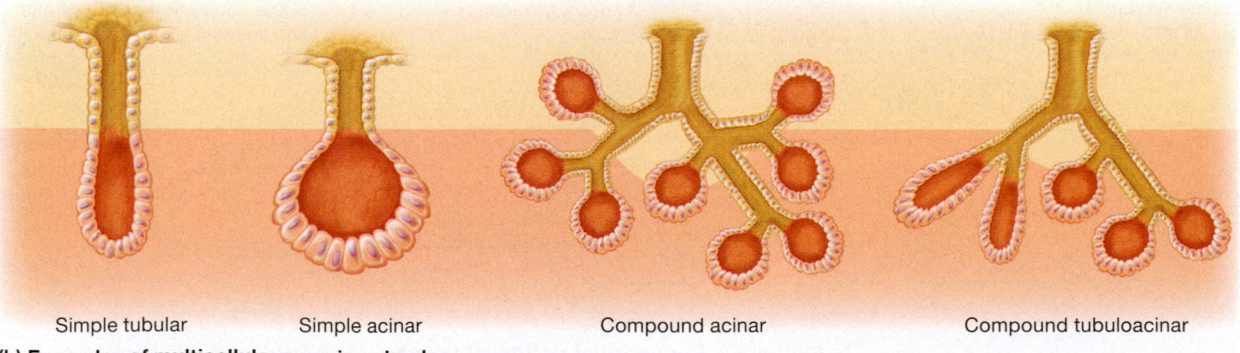

Simple tubular

Simple acinar

Compound acinar

Compound tubuloacinar

(b) Examples of multicellular exocrine glands

Figure 4.10 Multicellular exocrine glands.

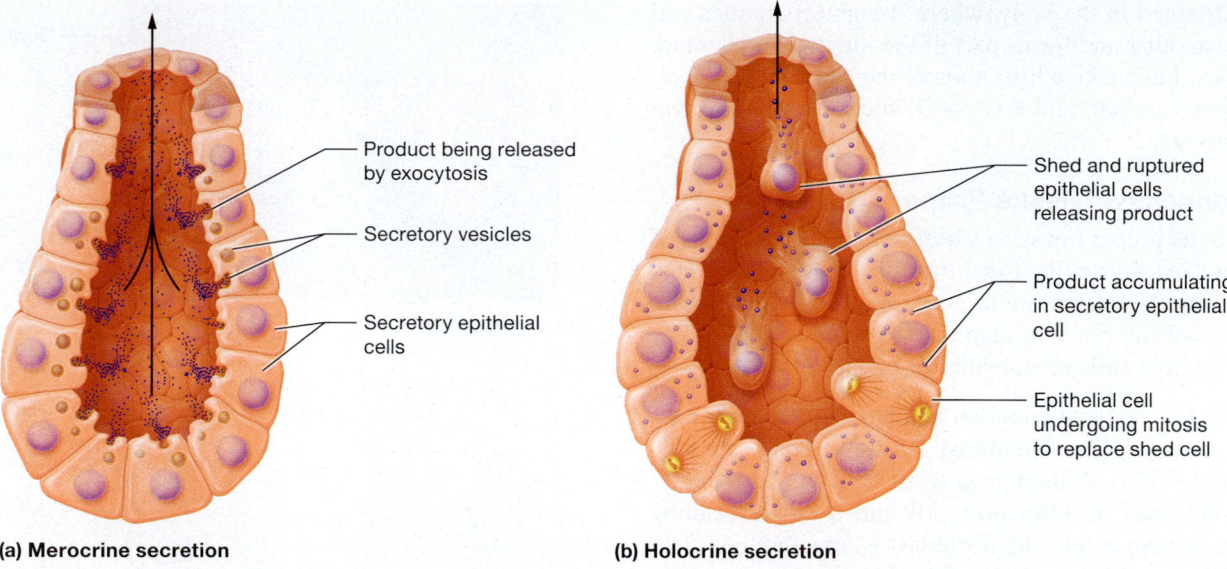

(a) Merocrine secretion

- Product being released by exocytosis
- Secretory vesicles
- Secretory epithelial cells

(b) Holocrine secretion

- Shed and ruptured epithelial cells releasing product
- Product accumulating in secretory epithelial cell
- Epithelial cell undergoing mitosis to replace shed cell

Figure 4.11 Main modes of secretion in exocrine glands.

Apply What You Learned

☐ 1. Certain medical conditions cause the simple squamous epithelium in the lung to thicken. Predict the consequences of this thickening.

☐ 2. As we age, the stratified squamous epithelium of the skin becomes progressively thinner. Explain why this tends to result in a greater number of skin lacerations (cuts) in the elderly.

☐ 3. Explain what might happen if a genetic mutation caused a person's digestive tract to have stratified columnar epithelium instead of simple columnar epithelium.

See answers in Appendix A.

MODULE **4.3**
Connective Tissues

Learning Outcomes

1. Compare and contrast the roles of individual cell and fiber types within connective tissues.

2. Identify the different types of connective tissue, and describe where in the body they are found.

3. Describe the functions of each type of connective tissue, and correlate function with structure for each tissue type.

Connective tissues are a diverse group of tissues with a variety of functions, including the following:

- **Connecting and binding.** As the name implies, connective tissues *connect* structures in the body. In organs, they bind other tissue layers together; they are also found between organs, where they anchor them in place and to one another.

- **Support.** Certain connective tissues such as bone and cartilage support the weight of the body.

- **Protection.** Connective tissues have many protective functions. For example, bone tissue protects certain internal organs, and cartilage and fat tissue provide shock absorption. In addition, many elements of the immune system are found within connective tissues.

- **Transport.** Blood is a fluid connective tissue that is the main transport medium in the body.

Like all tissues, connective tissue is made up of cells and ECM. In this case, the arrangement consists of loosely packed cells surrounded by protein fibers, all of which are embedded in ground substance. Although diverse in function, connective tissues do share a common feature—their ECM plays an extensive role in their functions. We discuss this expanded functional role of the ECM with the individual tissue types.

The classification scheme we'll use identifies two basic classes of connective tissue: *connective tissue proper* and the *specialized connective tissues* (bone, cartilage, and blood). Each class differs in the main types of cells it contains and the components of its ECM. Certain cells are found only in the specialized connective tissues; however, others are more general and are present in many different connective tissues. In this module, we examine the structural and functional properties of both types of connective tissue.

Connective Tissue Proper

« FLASHBACK

1. What are lipids? (p. 53)

2. What is a phagocyte? (p. 83)

First we explore **connective tissue proper,** which is sometimes called *general connective tissue.* Connective tissue proper is widely distributed in the body, where it connects tissues and organs to one another and forms part of the internal architecture of some organs. Let's take a brief tour of the cells that make up connective tissue proper, and then we'll focus on the different types of this tissue.

Cells of Connective Tissue Proper

Connective tissue proper houses a variety of cell types. Some of these cells are *resident* cells, meaning they inhabit the tissue permanently, whereas others are *migrant* cells that can move to different areas of the body in response to various needs. The resident and migrant cells include the following (**Figure 4.12**):

- **Fibroblasts.** The most common resident cell found in connective tissues is the **fibroblast** (*-blast* = "bud, sprout"; see Figure 4.12a). Although in most cell types a "-blast" ending indicates an immature cell and a "-cyte" ending indicates a mature cell, the fibroblast is a mature cell with properties of an immature cell. Fibroblasts produce protein fibers as well as ground substance and other elements of the ECM. Active fibroblasts usually lie close to collagen fibers, as they continually produce collagen proteins.
- **Adipocytes.** Adipocytes (AD-ih-poh-sytz), also called *fat cells,* are another type of resident cell found in many different connective tissues (see Figure 4.12b). The cytoplasm of most adipocytes is dominated by a single large inclusion containing lipids. The nuclei and other organelles of the adipocytes are squashed to the perimeter of the cell, and are often difficult to see in sections.
- **Mast cells.** **Mast cells** are the largest resident cells in connective tissues (see Figure 4.12c). They are cells of the immune system that have cytosolic inclusions (or *granules*) containing chemicals called *inflammatory mediators.* An example of such a mediator is *histamine.* When stimulated, mast cells release the contents of their granules, and inflammation results. *Inflammation* is a protective response that activates the immune system and recruits immune cells to the tissue.
- **Phagocytes.** Recall that **phagocytes** (FAYG-oh-sytz) are cells of the immune system that can ingest foreign substances, microorganisms, and dead and damaged cells by phagocytosis (see Chapter 3 and Figure 4.12d). Two common phagocytes are **macrophages** (MAK-roh-fay-jez), which can be either resident or migrant cells in connective tissues, and *neutrophils* (NOO-troh-filz), which are migrant immune cells.
- **Other immune cells.** Various other cells of the immune system migrate in and out of different connective tissues, depending on the needs of the body at the time.

Types of Connective Tissue Proper

There are four basic types of connective tissue proper, defined by their main components: (1) *loose connective tissue,* (2) *dense connective tissue,* (3) *reticular tissue,* and (4) *adipose tissue.** The next four subsections examine each type.

---

*Gartner, LP. *Textbook of histology*. 4th ed. Philadelphia (PA): Elsevier; 2017. p. 141–147.

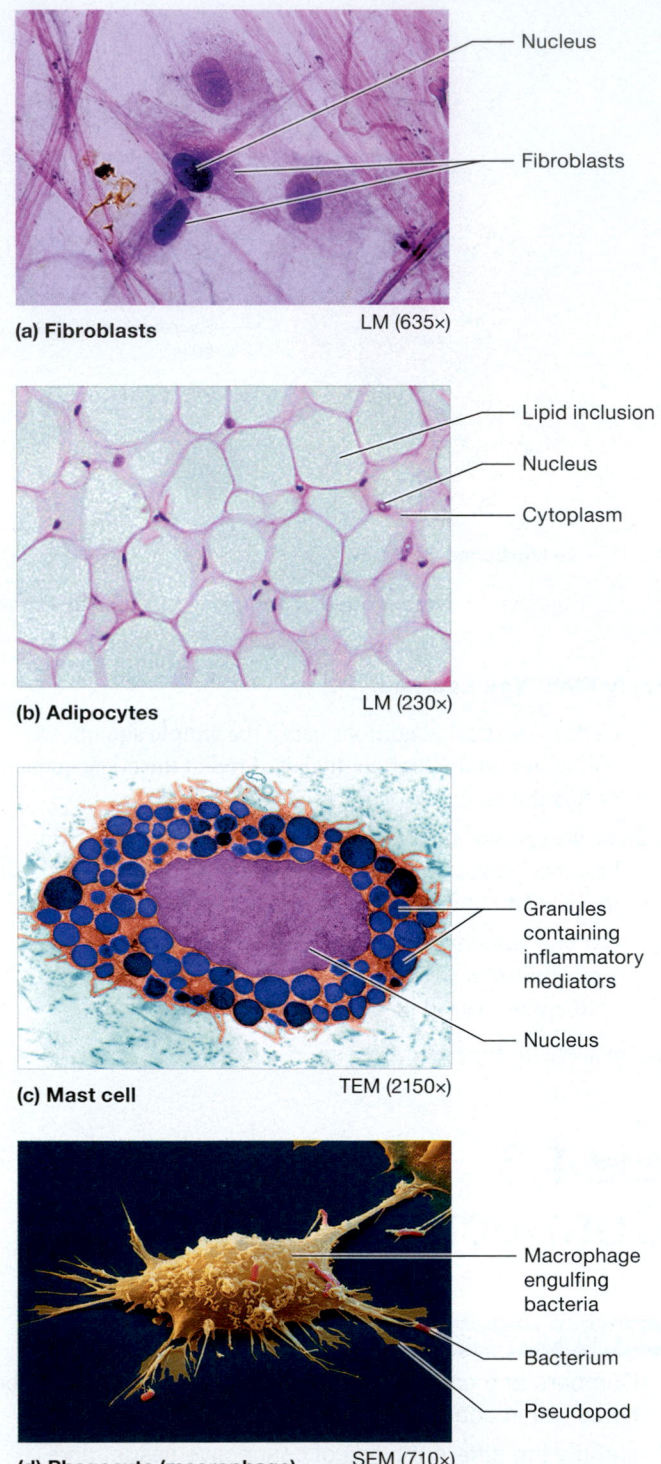

(a) Fibroblasts — LM (635×)
Nucleus · Fibroblasts

(b) Adipocytes — LM (230×)
Lipid inclusion · Nucleus · Cytoplasm

(c) Mast cell — TEM (2150×)
Granules containing inflammatory mediators · Nucleus

(d) Phagocyte (macrophage) — SEM (710×)
Macrophage engulfing bacteria · Bacterium · Pseudopod

Figure 4.12 Cells of connective tissue proper.

Loose Connective Tissue **Loose connective tissue,** also known as **areolar connective tissue** (ah-REE-oh-lur), is composed primarily of ground substance. In addition, all three types of protein fibers, fibroblasts, and other cells such as adipocytes are suspended in the ground substance (**Figure 4.13**). (Reticular fibers require a special stain to be visible, so you can't see them in Figure 4.13.) Loose connective tissue is found deep to the epithelium of the skin,

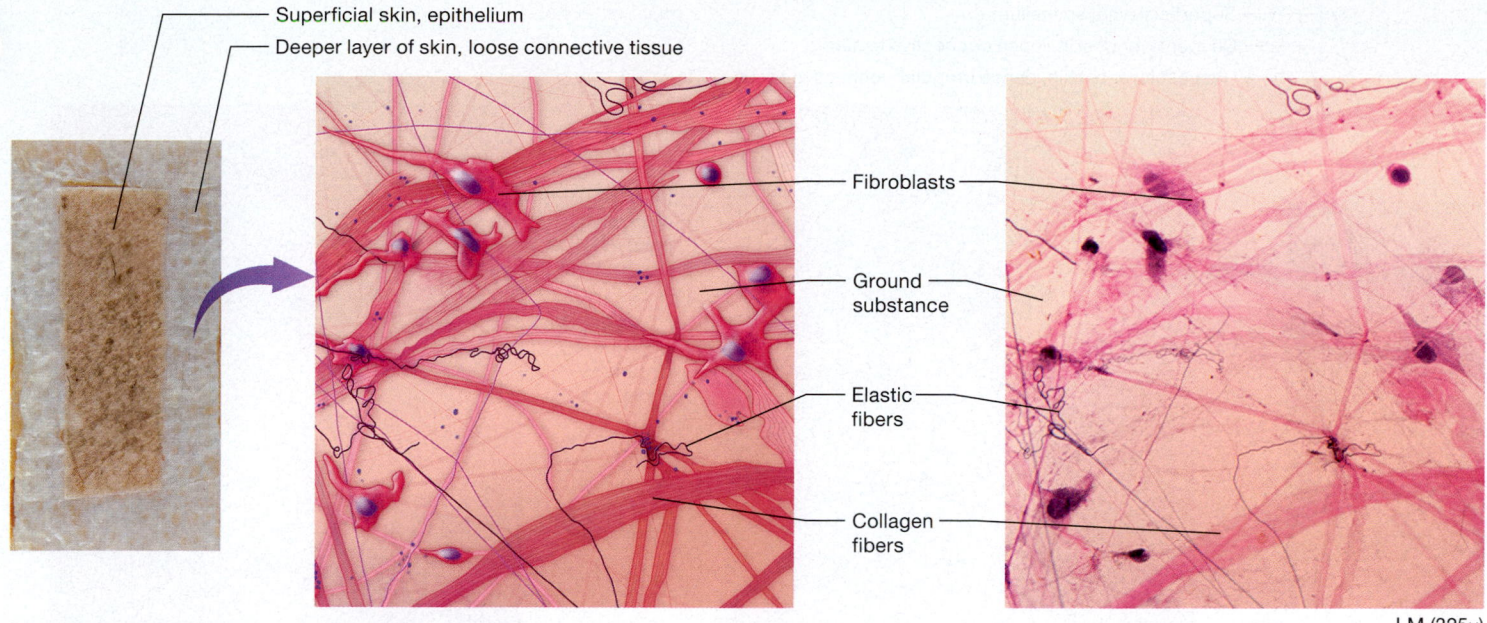

Superficial skin, epithelium
Deeper layer of skin, loose connective tissue

Fibroblasts

Ground substance

Elastic fibers

Collagen fibers

Loose connective tissue from skin

LM (325×)

Figure 4.13　Structure of loose connective tissue.

in the membranes lining the body cavities, and as layers in the walls of hollow organs. It functions in support and contains many blood vessels, whose nutrients and oxygen diffuse up to the superficial epithelial cells. It also houses numerous immune cells that protect against microorganisms invading the epithelium, which reflects the fact that it is the first tissue to come into contact with the epithelium.

Dense Connective Tissue　Whereas ground substance is the primary component of loose connective tissue, **dense connective tissue** is composed primarily of protein fibers. Due to this predominance of fibers, dense connective tissue is often called *fibrous connective tissue.*

Dense connective tissue can be grouped into three classes: irregular, regular collagenous, and regular elastic (**Figure 4.14**).

- **Dense irregular connective tissue.** The predominant fiber type in **dense irregular connective tissue** is the collagen fiber. Notice in Figure 4.14a that the bundles of collagen fibers, which appear as pink clusters due to staining, seem to be arranged haphazardly. This organization makes dense irregular connective tissue quite strong and allows it to resist tension in all three planes. It is found in organs subjected to such tension, such as the dermis (the deep layer of skin), as well as around organs and joints.
- **Dense regular collagenous connective tissue.** As you can see in Figure 4.14b, **dense regular collagenous connective tissue** contains thick collagen fibers arranged parallel to one another to form bundles. The fibers in these bundles are very strong because they are oriented in a single direction, but they resist tension in only one plane. This tissue is therefore found in structures subjected to tension in one direction only, such as tendons, which join muscle to bone, and ligaments, which unite two bones. Note that the

arrangement of the fibers in dense regular collagenous and dense irregular connective tissues is another example of the Structure-Function Core Principle (p. 28).

CORE PRINCIPLE
Structure-Function

- **Dense regular elastic connective tissue. Dense regular elastic connective tissue** (often shortened to simply "elastic tissue") consists mostly of parallel elastic fibers with randomly oriented collagen fibers (see Figure 4.14c). This tissue, which allows certain organs to stretch, is found in the lining of large blood vessels and in certain ligaments, such as those of the spine.

Reticular Tissue　**Reticular tissue** is so named because it is populated by numerous reticular fibers produced by surrounding fibroblasts (often called *reticular cells;* **Figure 4.15**). The thin reticular fibers interweave to form fine networks that support small structures such as blood vessels and lymphatic vessels. Reticular tissue is also found in lymph nodes and the spleen, where reticular fibers form weblike nets that help trap old and foreign cells and house white blood cells. It also forms part of the basement membrane that supports all epithelia and the internal structure of the liver and bone marrow.

Adipose Tissue　**Adipose tissue,** also called *fat tissue,* consists of fat-storing adipocytes and their surrounding fibroblasts and ECM (**Figure 4.16**). Although fibroblasts and an ECM containing collagen fibers are initially the dominant component of adipose tissue, the adipocytes can increase in size to the point that few fibroblasts and little ECM are visible. Adipose tissue has many functions, including insulation, warmth, shock absorption, and protection, and it is also the major energy reserve in the body.

The majority of adipose tissue is *white adipose tissue,* which consists of adipocytes that contain one large lipid inclusion in their cytosol. White adipose tissue is found deep to the skin (*subcutaneous*

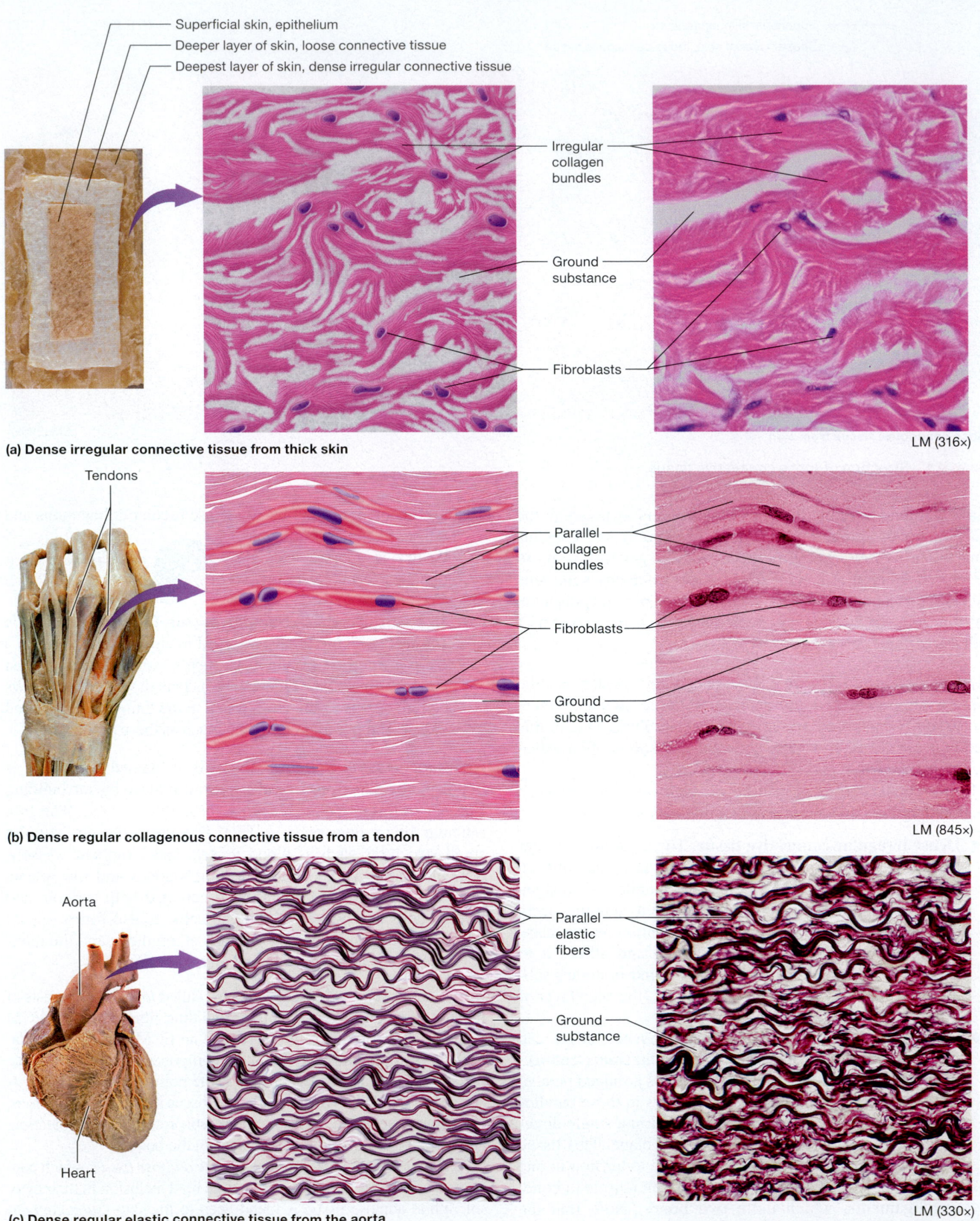

Superficial skin, epithelium
Deeper layer of skin, loose connective tissue
Deepest layer of skin, dense irregular connective tissue

Irregular collagen bundles

Ground substance

Fibroblasts

LM (316×)

(a) Dense irregular connective tissue from thick skin

Tendons

Parallel collagen bundles

Fibroblasts

Ground substance

LM (845×)

(b) Dense regular collagenous connective tissue from a tendon

Aorta

Parallel elastic fibers

Ground substance

Heart

LM (330×)

(c) Dense regular elastic connective tissue from the aorta

Figure 4.14 Structure of dense connective tissue.

fat; sub′-kyoo-TAYN-ee-us) and in characteristic places such as the abdomen, breasts, hips, buttocks, and thighs. It also surrounds certain organs such as the heart and the abdominal organs (*visceral fat;* VISS-er-uhl). As its name suggests, it is white.

A small percentage of adipose tissue is *brown adipose tissue,* which consists of adipocytes that contain multiple lipid inclusions. Such tissue appears brown due to abundant mitochondria in the cells and an extensive blood supply. Infants and young children have brown adipose tissue around the neck and back; however, by adulthood little brown adipose tissue remains. Both types of adipose tissue are used for fuel and insulation. However, the cells of brown adipose tissue can oxidize fatty acids about 20 times as fast as those of white adipose tissue; this additional energy is used to generate heat in cold temperatures.

Obesity is the condition of having excess adipose tissue in proportion to lean body mass. See *A&P in the Real World:*

Adipose Tissue and Obesity on p. 142 for a discussion of the different types of this prevalent condition.

Quick Check

☐ 1. What are the types of cells of connective tissue proper? What are their functions?

☐ 2. What is the primary component in:
   a. loose connective tissue?
   b. dense regular collagenous connective tissue?
   c. dense regular elastic connective tissue?
   d. reticular tissue?
   e. adipose tissue?

☐ 3. What are the primary differences between brown and white adipose tissue?

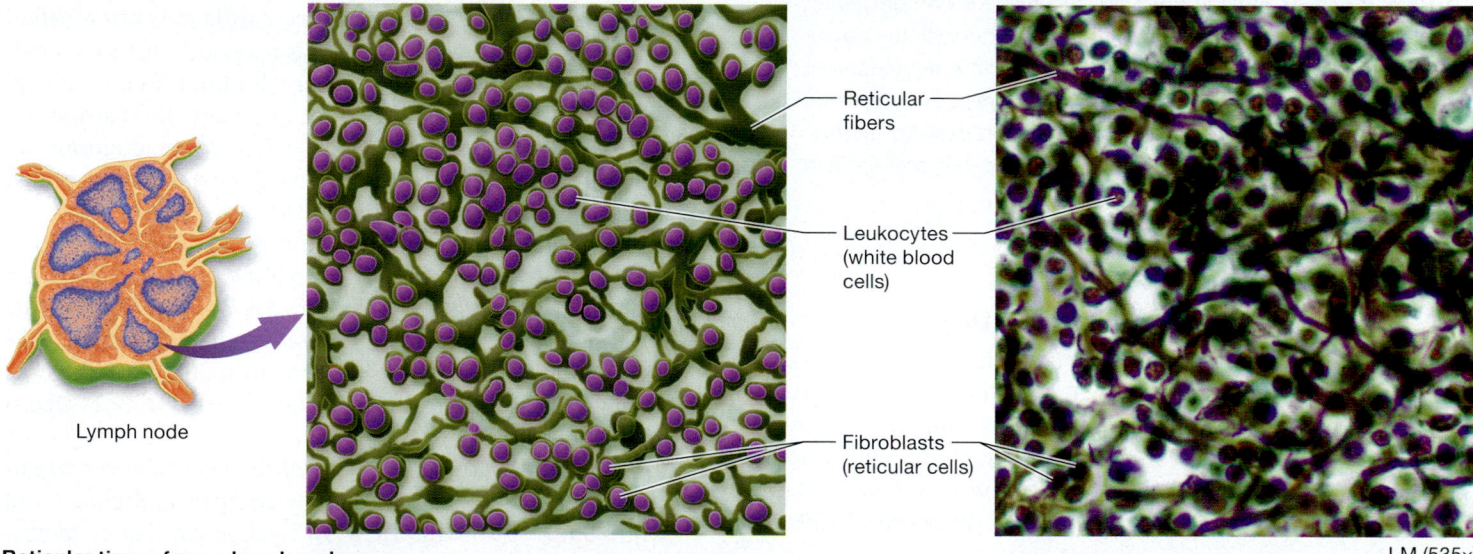

Reticular fibers

Leukocytes (white blood cells)

Fibroblasts (reticular cells)

Lymph node

Reticular tissue from a lymph node

LM (535×)

Figure 4.15 Structure of reticular tissue.

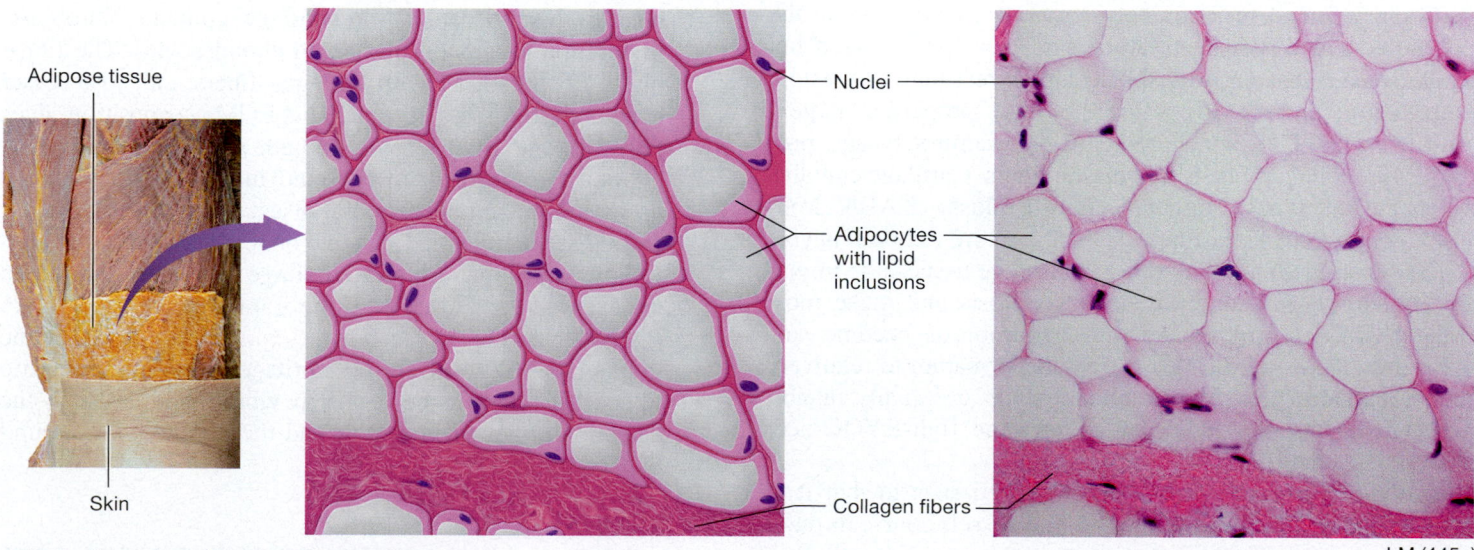

Adipose tissue

Skin

Nuclei

Adipocytes with lipid inclusions

Collagen fibers

Adipose tissue from hypodermis superficial fascia

LM (445×)

Figure 4.16 Structure of adipose tissue.

A&P in the Real World

Adipose Tissue and Obesity

There are two forms of obesity. The first and most common is *hypertrophic obesity,* in which the lipid inclusions in the adipocytes accumulate excess fatty acids and increase in size. Adipocytes may grow to as much as four times their normal size in cases of severe obesity; however, the number of adipocytes remains the same.

The other form is *hypercellular obesity,* in which the actual number of adipocytes increases. Hypercellular obesity, which is generally severe, correlates with the development of obesity in infancy or early childhood. This is because the adipocytes of infants and young children can increase in number, whereas those of adults lack the ability to divide and form new cells.

Both forms of obesity increase the risk for certain health problems; however, they seem to be associated with the development of different disorders. Furthermore, the development of disorders associated with obesity is complex and depends on the distribution of adipose tissue and on genetic factors. For a deeper discussion of this topic, go to the metabolism chapter (see Chapter 23).

Specialized Connective Tissues

The **specialized connective tissues** have more specific functions than connective tissue proper. The three types of specialized connective tissue are *cartilage, bone* or *osseous tissue* (AHS-ee-us), and *blood.* Both bone tissue and blood have chapters dedicated to their discussion (see Chapters 6 and 19, respectively), so we will describe them only briefly here.

Cartilage

Cartilage is found in joints between bones, as well as in the ears, nose, and certain respiratory passages. It is a tough but flexible tissue that absorbs shock and is resistant to tension, compression, and shearing forces. These properties largely result from its ECM, which contains glycosaminoglycans, proteoglycans, collagen fibers, and elastic fibers. Cartilage contains two major cell types: immature **chondroblasts** (KAHN-droh-blasts; *chondro-* = "cartilage") and mature **chondrocytes** (KAHN-droh-sytz). As with other connective tissues, the immature chondroblasts actively divide by mitosis and make most of the ECM of cartilage. As the chondroblasts become surrounded by their own ECM, they gradually mature to relatively inactive chondrocytes. Mature chondrocytes eventually inhabit small cavities in the ECM called **lacunae** (luh-KYOO-nee; *lacuna* = "cavity").

Cartilage is unusual among connective tissues in that it is essentially avascular, as few, if any, blood vessels course through the cartilage itself. The blood supply to this tissue is mostly limited to an outer sheath of dense irregular collagenous connective tissue called the **perichondrium** (pehr'-ih-KAHN-dree-um; *peri-* = "around"). Oxygen and nutrients must diffuse from the blood vessels in the perichondrium and through the cartilage ECM to reach the chondrocytes and chondroblasts. This zone of diffusion is fairly narrow, measuring only a few millimeters, which limits the thickness of living cartilage.

Cartilage can be divided into three classes, based on the composition of their ECM. These three classes are *hyaline cartilage, fibrocartilage,* and *elastic cartilage:*

- **Hyaline cartilage.** The most abundant type of cartilage in the body is **hyaline cartilage** (HY-uh-lin; *hyaline* = "glasslike") (**Figure 4.17a**). The ECM of hyaline cartilage consists of large amounts of ground substance with a fine type of collagen fiber that forms small bundles. This gives hyaline cartilage a uniform, glassy appearance that stains bluish-gray in tissue sections. Hyaline cartilage covers the ends of bones where they form joints (joints are also referred to as *articulations,* and so hyaline cartilage is often called *articular cartilage*). Here it forms a smooth surface so that bones may slide over one another with little friction. It also protects the underlying bone by more equally distributing stresses. Hyaline cartilage is also found where strong yet flexible support is needed, such as at the attachment points of the sternum (breastbone) to the ribs, framing parts of the respiratory tract, and in the nose. Additionally, most of the fetal skeleton begins as hyaline cartilage, which is eventually replaced by bone. See *A&P in the Real World: Osteoarthritis and Glucosamine Supplements* on p. 144 for a discussion of what happens to hyaline cartilage as we age and whether glucosamine (gloo-KOH-suh-meen) helps the resulting problems.

- **Fibrocartilage.** You can see the differences between **fibrocartilage** and hyaline cartilage by comparing Figure 4.17a with **Figure 4.17b**—the ECM of hyaline cartilage is dominated by ground substance, whereas that of fibrocartilage is filled with bundles of collagen fibers, leaving hardly any room for ground substance. We find these collagen fiber bundles here because fibrocartilage contains fibroblasts in addition to chondroblasts and chondrocytes. The fibroblasts fill the ECM with collagen fibers and, to a lesser extent, elastic fibers as well. This ECM composition gives fibrocartilage great tensile strength and a fair amount of elasticity. Fibrocartilage is found in joints called *fibrous joints* and in the intervertebral discs; it also forms articular discs that improve the fit of two bones in a joint.

- **Elastic cartilage.** Elastic cartilage is largely limited to the external ear and parts of the framework of the larynx* (LEHR-inks; the "voice box"; **Figure 4.17c**). As its name implies, the ECM of elastic cartilage is filled with elastic fibers. This allows the tissue to vibrate and assist in the detection of sound in the air and the production of sound by the larynx.

---

*Gartner, LP. *Textbook of histology.* 4th ed. Philadelphia (PA): Elsevier; 2017. p. 400–401.

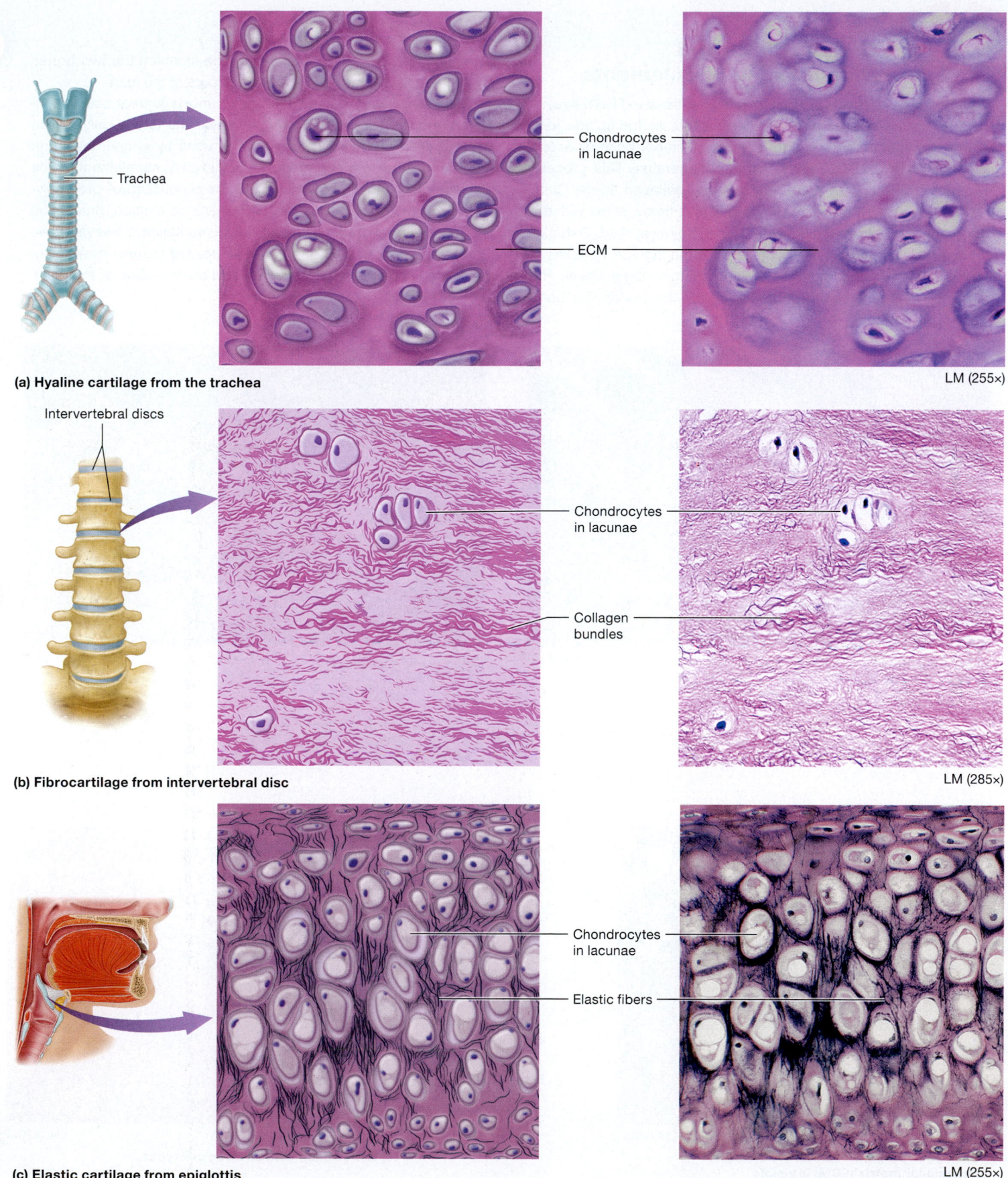

(a) **Hyaline cartilage from the trachea**

Trachea

Chondrocytes in lacunae

ECM

LM (255×)

(b) **Fibrocartilage from intervertebral disc**

Intervertebral discs

Chondrocytes in lacunae

Collagen bundles

LM (285×)

(c) **Elastic cartilage from epiglottis**

Chondrocytes in lacunae

Elastic fibers

LM (255×)

Figure 4.17 Structure of cartilage.

Osteoarthritis and Glucosamine Supplements

Osteoarthritis (ahs-tee-oh-ahr-THRY-tiss) is caused by a variety of factors, including age, joint trauma, genetic disorders, and infection. Osteoarthritis develops as the hyaline cartilage lining joints degenerates. This process leads to destruction of proteoglycans and collagen fibers. Compare the smooth cartilage from a normal joint below at left with the rough, torn cartilage from an arthritic joint below at right. Osteoarthritis makes the cartilage less resistant to mechanical stresses at the joint, which further destroys the remaining collagen fibers. The cartilage may continue to degenerate until the underlying bone is exposed, leading to a painful situation in which the two bones grind on each other any time motion occurs at the joint.

A popular dietary supplement that purports to treat osteoarthritis is the chemical *glucosamine*. Glucosamine is a sugar derived from shellfish and certain fungi that is used by chondroblasts in their synthesis of proteoglycans. Hypothetically, supplementing the diet with glucosamine could increase the production of proteoglycans by chondroblasts and slow cartilage degeneration. Studies on the effectiveness of glucosamine supplementation, however, have yielded mixed results. Further study is needed to determine if glucosamine supplementation can truly stall degeneration of the joints in osteoarthritis.

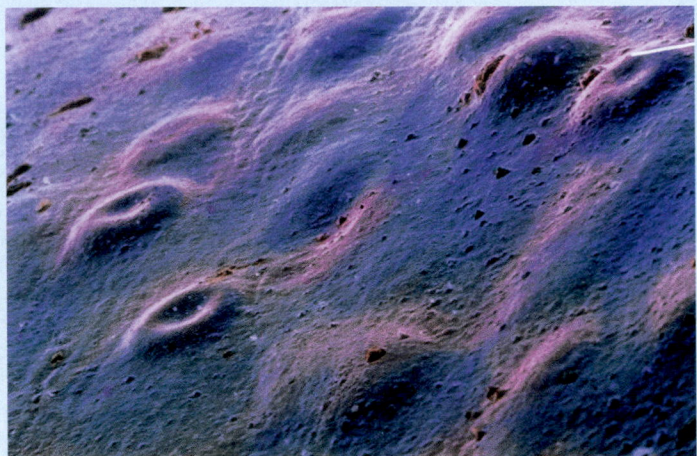

Normal cartilage SEM (65×)

Osteoarthritic cartilage SEM (65×)

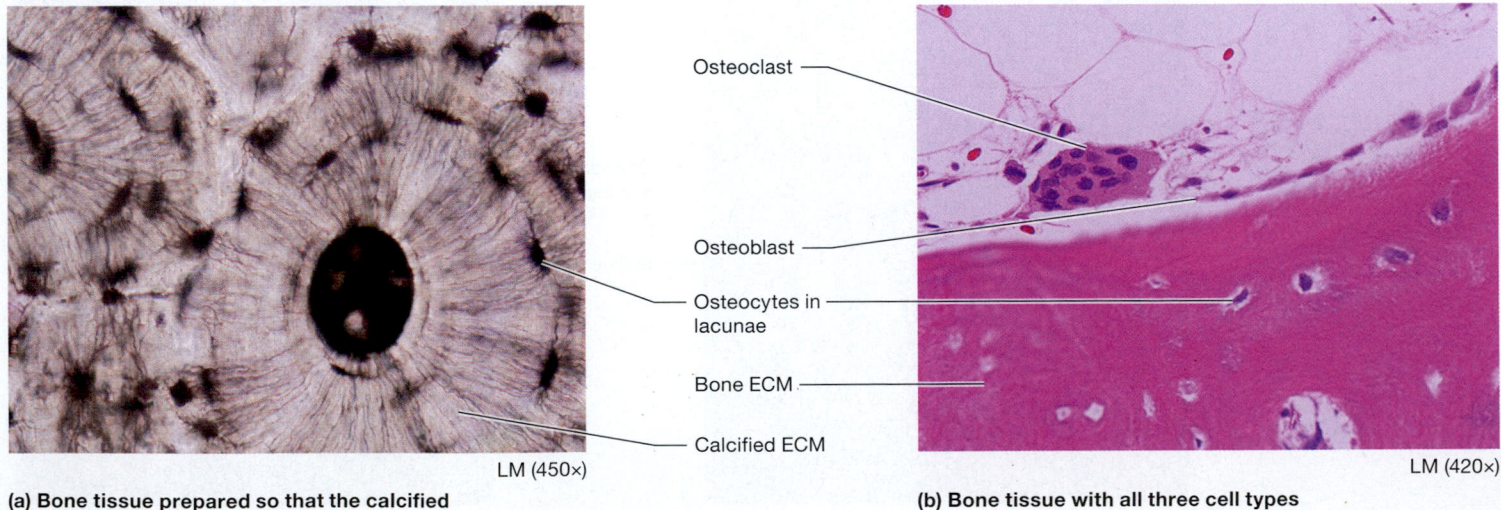

Osteoclast
Osteoblast
Osteocytes in lacunae
Bone ECM
Calcified ECM

LM (450×)

LM (420×)

(a) Bone tissue prepared so that the calcified extracellular matrix (ECM) is visible

(b) Bone tissue with all three cell types

Figure 4.18 Structure of bone.

Bone

Bone tissue, also called **osseous tissue** (*os-, osteo-* = "bone"), serves many functions, including supporting our body; protecting our vital organs; providing a place for attachment of the muscles of voluntary movement; storing calcium salts; and housing the *bone marrow,* the tissue that produces our blood cells and stores fat (**Figure 4.18**). The ECM of bone tissue has two components, organic and inorganic. The organic portion of the ECM accounts for about 35% of the mass of bone and consists of collagen fibers and a type of ground substance called *osteoid* (AHS-tee-oyd). The inorganic portion, which represents the remaining 65% of bone mass, is composed of calcium phosphate crystals, which make bone one of the hardest substances in the body (see Figure 4.18a).

Mature bone tissue contains three cell types: *osteoblasts, osteocytes,* and *osteoclasts* (see Figure 4.18b).

- **Osteoblasts.** The **osteoblasts** (AHS-tee-oh-blasts; *-blast* = "immature cell") are the "bone builders," carrying out the process of *bone deposition.* During this process, they synthesize the organic portion of the ECM, which they secrete by exocytosis, and also produce chemicals required for calcium salts to deposit within the ECM. They are found on the outer surface of the bone, where they are closely associated with a dense irregular collagenous connective tissue covering called the **periosteum** (pehr´-ee-AHS-tee-um).

- **Osteocytes.** Osteoblasts eventually become surrounded by the ECM they secrete. At this point, the osteoblasts reside in lacunae and are called **osteocytes** (AHS-tee-oh-sytz), which are mature bone cells. They produce substances required for bone maintenance, which are released by exocytosis.

- **Osteoclasts.** Whereas osteoblasts are the bone builders, the large and multinucleated **osteoclasts** (AHS-tee-oh-klasts; *-clasia* = "broken") are the bone destroyers. Osteoclasts carry out the process of *bone resorption,* during which they secrete hydrogen ions and enzymes that catalyze reactions to break down the components of the ECM.

Although bone is solid and strong, it is not an inactive tissue. *Bone remodeling,* which includes bone deposition and bone resorption, is constantly occurring in healthy bone. Furthermore, both bone deposition and bone resorption can occur simultaneously in different parts of the bone, depending on the stresses placed on the tissue. Tension on the bone will result in increased osteoblast activity and bone deposition, whereas pressure placed on the bone can result in increased osteoclast activity and bone resorption.

For a more detailed view of bone tissue and the structures in Figure 4.18, see Figure 6.6 (p. 190), which shows the bone cells, and Figure 6.9 (p. 193), which illustrates the structure of bone tissue.

Blood

Blood is unique among the connective tissues in that its ECM is fluid (**Figure 4.19**). The ECM of blood is called **plasma,** and it consists largely of water, dissolved solutes, and proteins. Unlike most connective tissues, plasma does not contain protein fibers; rather, it contains much smaller proteins that serve functions such as transport and blood clotting. Two main types of cells are found in blood. First are the **erythrocytes** (eh-RITH-roh-sytz; *erythro-* = "red"), also called **red blood cells,** which bind and transport oxygen through the body. Second are the **leukocytes** (LOO-koh-sytz; *leuko-* = "white"), also called **white blood cells,** which function in immunity. In addition, blood contains cellular fragments known as **platelets,** which play a role in the blood clotting process.

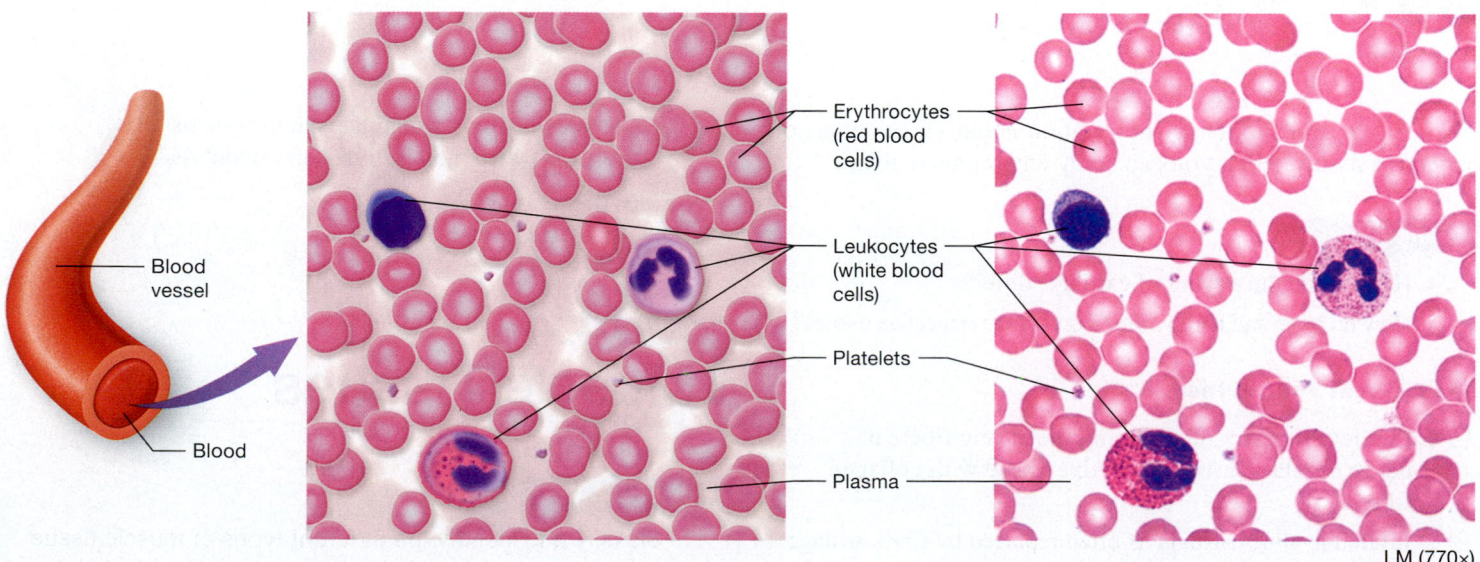

LM (770×)

Figure 4.19 Components of blood.

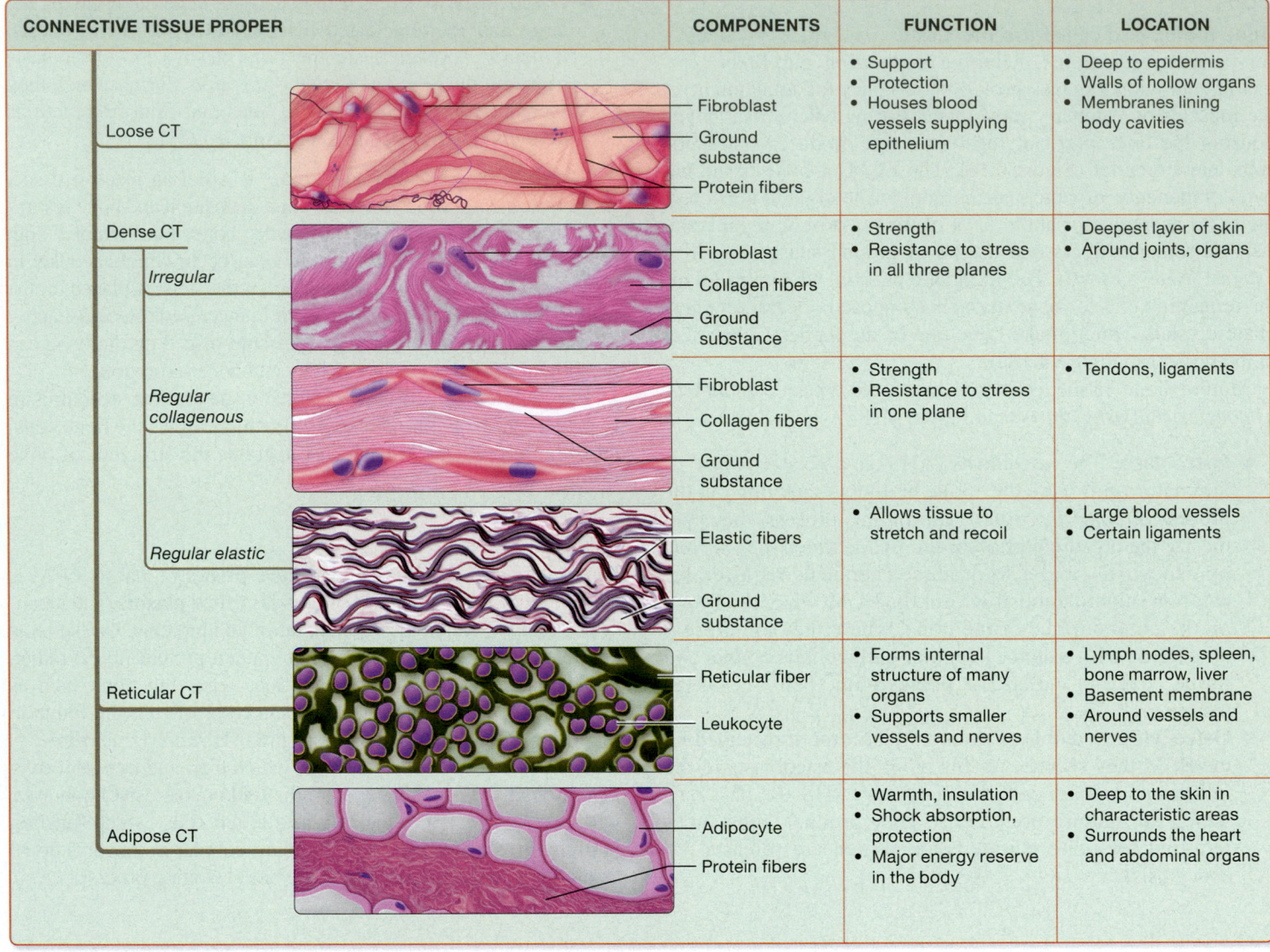

Figure 4.20 Summary of connective tissues.

View PAL™ Histology
@ Mastering Anatomy & Physiology

Figure 4.20 pulls together information about all the types of connective tissue so that you can study and compare them.

Quick Check

☐ 4. How do the three types of cartilage differ?

☐ 5. How do bone and blood differ from other connective tissues?

Apply What You Learned

☐ 1. Predict the effect of replacing the elastic fibers of dense regular elastic connective tissue with collagen fibers.

☐ 2. Injured hyaline cartilage is often replaced by fibrocartilage. Does fibrocartilage provide the same kind of surface for articulation as hyaline cartilage? Why or why not?

☐ 3. What do you think would happen to bone tissue if osteoclasts were more active than osteoblasts?

See answers in Appendix A.

MODULE **4.4**
Muscle Tissues

Learning Outcomes

1. Classify and identify the different types of muscle tissue based on their distinguishing structural characteristics and location in the body.

SPECIALIZED CONNECTIVE TISSUE		COMPONENTS	FUNCTION	LOCATION
Cartilage	Hyaline 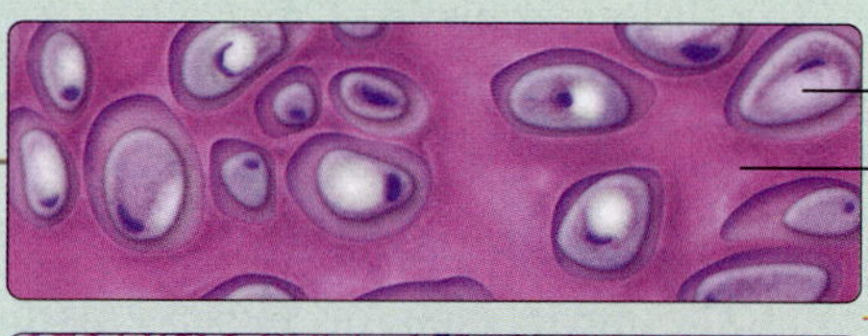	Chondrocyte ECM	• Support • Protection • Resists compression	• Between bones in joints • Between sternum and ribs • Nose • Respiratory tract
	Fibrocartilage	Chondrocyte Collagen fibers ECM	• Support • Protection • Resists compression	• Intervertebral discs
	Elastic	Chondrocyte Elastic fibers ECM	• Involved in producing and detecting sound	• Ears • Epiglottis of larynx
Bone 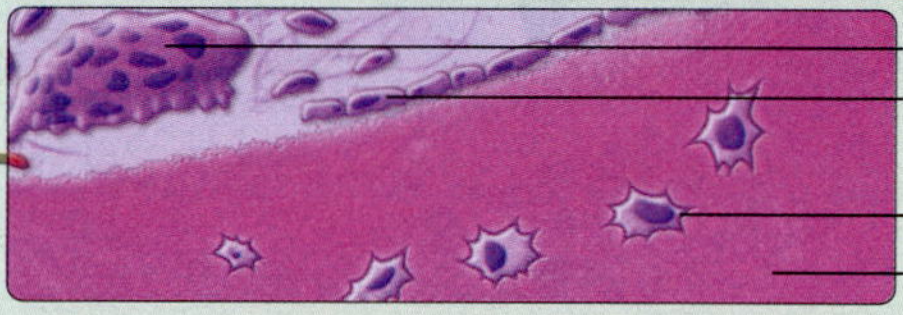		Osteoclast Osteoblast Osteocyte ECM	• Support • Protection • Provides leverage for movement • Stores calcium	• Bones
Blood 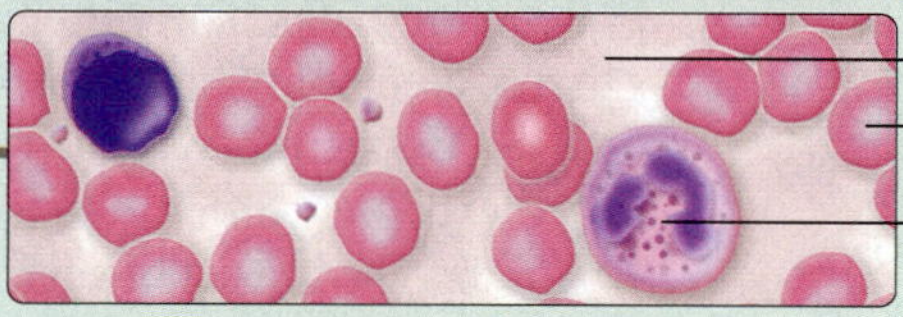		Plasma Erythrocyte Leukocyte	• Transports nutrients, gases, wastes, immune cells	• Within blood vessels and chambers of the heart

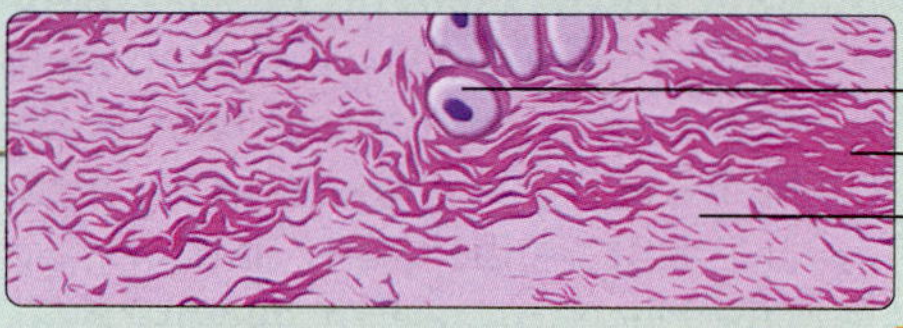

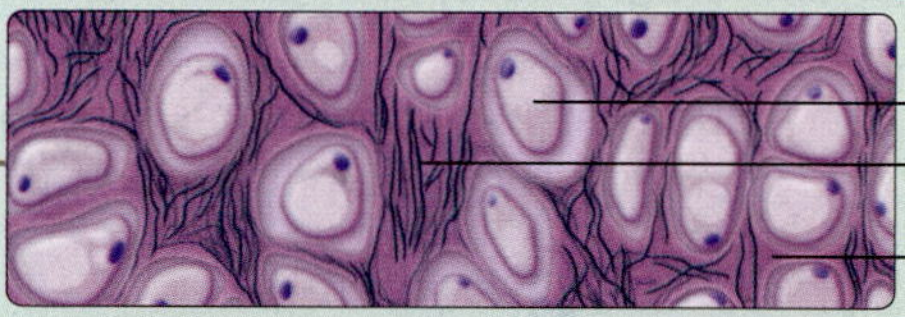

Figure 4.20 Summary of connective tissues. (*continued*)

2. Describe the functions of each type of muscle tissue, and correlate function with structure for each tissue type.

Although many cells have the ability to contract, only the cells of **muscle tissues** are specialized for contraction. The three types of muscle tissues share the common ability to turn the chemical energy of ATP into the mechanical energy of movement. This movement comes in a variety of forms, including walking and breathing, the beating of the heart, and the narrowing of hollow organs. In this module, we discuss first the characteristics of the cells and ECM of muscle tissues, and then the functional and structural differences of the three types of muscle tissue. These topics are only introduced here; they are fully covered in the chapters on muscle tissue and the heart (see Chapters 10 and 17, respectively).

Components of Muscle Tissue

 **FLASHBACK**

1. What are the functions of the cytoskeleton? (p. 69)

The main component of muscle tissues is **muscle cells,** or *myocytes* (MY-oh-sytz; *myo-* = "muscle"). All muscle cells are

excitable cells, meaning they respond to electrical or chemical stimulation. Muscle cells have a very different appearance from that of other cells we have discussed: Their cytoplasm is filled with bundles of proteins called **myofilaments.**

There are two forms of muscle cells, which are classified by the arrangement of their myofilaments:

- **Striated muscle cells.** The myofilaments of certain muscle cells are organized in such a way that there are regions where the myofilaments overlap and regions where they don't. The combination of these regions produces both dark and light areas called "bands" when such cells are viewed under a microscope. These alternating light-dark-light-dark bands are known as **striations** (stry-AY-shunz; *stria-* = "lined" or "streaked"), and muscle cells with this arrangement are called *striated muscle cells.*

- **Smooth muscle cells.** *Smooth muscle cells* have a very different appearance from that of striated muscle cells. Like striated muscle cells, they contain myofilaments; however, these filaments are arranged as irregular bundles scattered throughout the cytoplasm. As a result, no striations are visible, and these cells are called "smooth."

In muscle tissue, a small amount of ECM surrounds each muscle cell and is called the **endomysium** (en′-doh-MY-see-um) or *external lamina*. The endomysium is similar to the basal lamina of epithelial tissue and helps to hold the muscle cells together in the tissue. It blends with connective tissue surrounding muscle cells and so is sometimes referred to as connective tissue.

Quick Check

☐ 1. What are the two forms of muscle cells, and how do these forms differ?

Types of Muscle Tissue

≪ FLASHBACK

1. What is the function of the nucleus? What main process does the DNA in the nucleus direct? (p. 70)
2. What are the functions of gap junctions and tight junctions? (p. 126)

The three kinds of muscle tissue have different structural and functional characteristics (**Figure 4.21**). The first two kinds, *skeletal muscle tissue* and *cardiac muscle tissue,* are composed of striated muscle cells. The third kind, *smooth muscle tissue,* is made up of smooth muscle cells.

Skeletal Muscle Tissue

As its name implies, **skeletal muscle tissue** is found mostly attached to the skeleton, where its contraction produces body movement. Skeletal muscle cells do not contract unless they are first stimulated by the nervous system, and the contraction is generally *voluntary,* or under conscious control. As you can see in Figure 4.21a, skeletal muscle tissue consists of long, thin, striated muscle cells that are arranged parallel to one another. Most skeletal muscle cells are quite long, extending nearly the entire length of the whole muscle. For this reason, they are often called **muscle fibers** (not to be confused with nonliving protein fibers). Skeletal muscle fibers are formed by the fusion of embryonic cells called *myoblasts.* The nucleus of each myoblast is retained in the mature skeletal muscle fiber, causing it to be *multinucleate.* The nuclei are located near the outer edge of the fiber. Having several nuclei is advantageous for skeletal muscle fibers because their size requires nearly constant synthesis of enzymes, structural proteins, and contractile proteins.

Cardiac Muscle Tissue

Cardiac muscle tissue is found only in the heart and is composed of striated muscle cells called **cardiac muscle cells.** Cardiac muscle tissue is *involuntary,* meaning that the brain does not have conscious control over its contraction. Although both skeletal and cardiac muscle tissues contain striated muscle cells, notice in Figure 4.21b that numerous differences between these two cell types can be seen. Cardiac muscle cells are relatively short and thick, as well as branched and often uninucleate, as opposed to the long, cylindrical multinucleated skeletal muscle fibers. The nuclei are located in the centers of the cells. Note also that dark lines separate the cardiac muscle cells, a feature lacking in skeletal muscle tissue. These lines are structures called **intercalated discs** (in-TER-kuh-lay′-t'd; *inter-* = "between"); they contain gap junctions and

modified tight junctions that unite the cardiac muscle cells and permit the heart muscle to contract as a unit. These structures enable communication among the cells (an example of the Cell-Cell Communication Core Principle, p. 28).

CORE PRINCIPLE
Cell-Cell Communication

Smooth Muscle Tissue

The final type of muscle tissue is **smooth muscle tissue** (see Figure 4.21c). As mentioned, it consists of smooth muscle cells, and like cardiac muscle tissue, it is involuntary. Smooth muscle tissue is found in the wall of nearly every hollow organ, as well as the walls of blood vessels, the eyes, the skin, and the ducts of certain glands. Its cells are flattened, with a single ovoid nucleus in the center of the cell. Most smooth muscle cells contain gap junctions in their plasma membranes that link them with other smooth muscle cells.

Quick Check

☐ 2. Compare and contrast the three types of muscle tissue.

Apply What You Learned

☐ 1. When skeletal muscle fibers are damaged and die, they may be replaced with dense irregular collagenous connective tissue, commonly known as scar tissue.
   a. How do these two tissue types differ in structure?
   b. How do they differ in function?
   c. Predict the impact of such an injury to muscle tissue on the function of a whole skeletal muscle.

☐ 2. Predict what might happen to cardiac muscle tissue in which the intercalated discs were not functional.

See answers in Appendix A.

MODULE **4.5**
Nervous Tissue

Learning Outcomes

1. Describe where in the body nervous tissue can be found and its general structural and functional characteristics.
2. Identify and describe the structure and function of neurons and neuroglial cells in nervous tissue.

Nervous tissue makes up the majority of the brain, spinal cord, and nerves. Like all tissues, nervous tissue is composed of cells and their surrounding ECM. The cells are of two main types: *neurons* (NOO-ronz), the cells that send and receive messages, and *neuroglial cells* (noo-roh-GLEE-ul), which perform various supportive functions. The ECM of nervous tissue is quite different from that of the other tissues we have discussed. It contains few protein fibers and instead is mostly made up of ground substance with unique proteoglycans not found in other tissues of the body. In this module, we introduce

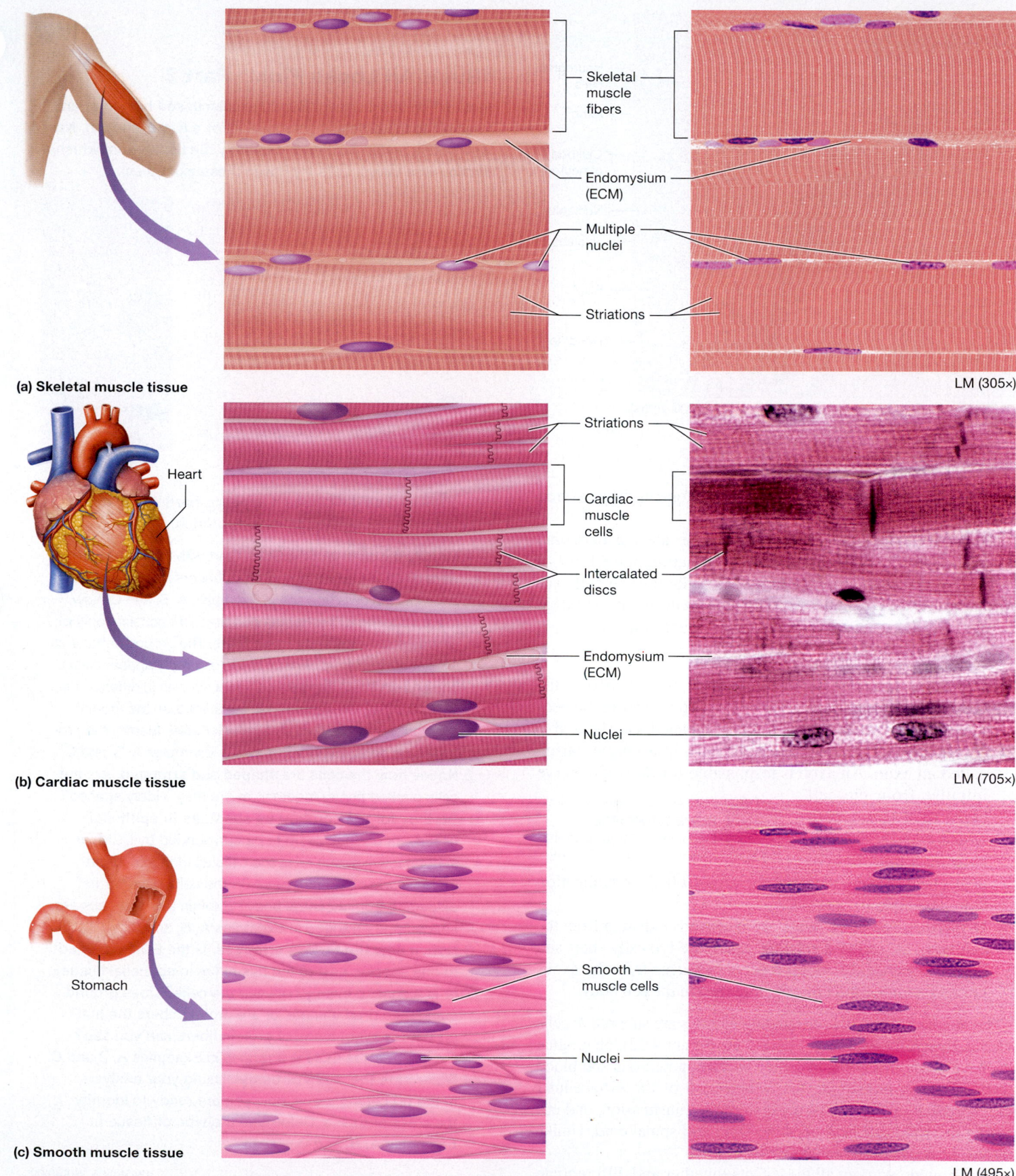

(a) Skeletal muscle tissue

Skeletal muscle fibers

Endomysium (ECM)

Multiple nuclei

Striations

LM (305×)

(b) Cardiac muscle tissue

Heart

Striations

Cardiac muscle cells

Intercalated discs

Endomysium (ECM)

Nuclei

LM (705×)

(c) Smooth muscle tissue

Stomach

Smooth muscle cells

Nuclei

LM (495×)

Figure 4.21 Structure of muscle tissues.

View PAL™ Histology
@ Mastering Anatomy & Physiology

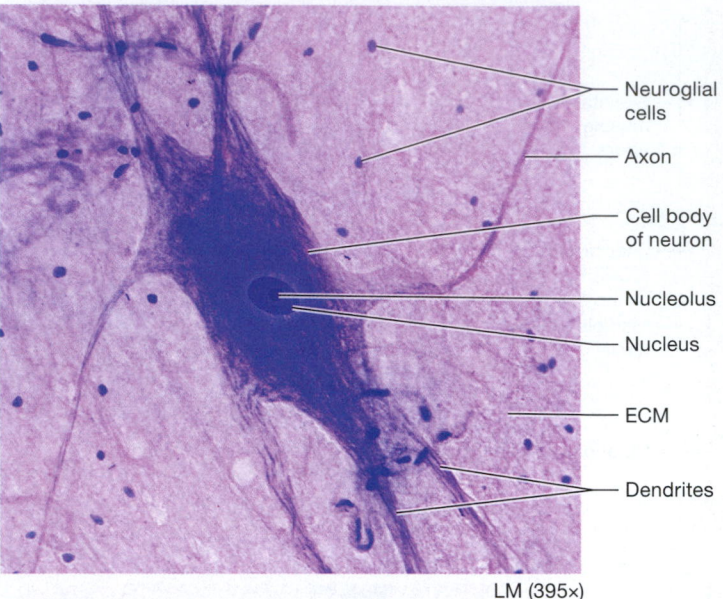

LM (395×)

Figure 4.22 Structure of nervous tissue.

the cells of nervous tissue only briefly, as this topic is fully covered in its own chapter (see Chapter 11).

The most obvious cells of nervous tissue are the **neurons,** which generate, conduct, and receive information in the form of electrical signals called **nerve impulses.** Like muscle cells, all neurons are excitable cells. Most mature neurons are amitotic (ay-my-TAWT-ik), meaning they do not undergo mitosis.

Neurons contain three main parts (**Figure 4.22**):

- **Cell body.** The large, centralized portion of the neuron is the **cell body,** or **soma.** This is the biosynthetic center of the neuron, where the nucleus and most other organelles are housed.
- **Axon.** Extending from the cell body is a single "arm" called an **axon.** An axon is responsible for moving a nerve impulse from the cell body to a target cell, which could be another neuron, a muscle cell, or a gland. Axons illustrate the Cell-Cell Communication Core Principle (p. 28).

 **CORE PRINCIPLE** Cell-Cell Communication

- **Dendrites.** The other "arms" or extensions radiating from the cell body are called **dendrites,** which are typically short and highly branched. Dendrites receive messages from the axons of other neurons and bring the impulses to the cell body.

The much smaller cells surrounding neurons are supportive cells called **neuroglial cells** ("nerve glue"; see Figure 4.22). Neuroglial cells have many functions, including anchoring neurons and blood vessels in place, monitoring the composition of the extracellular fluid, speeding up the rate of nerve impulse transmission, and circulating the fluid that surrounds the brain and spinal cord. Unlike neurons, neuroglial cells can divide by mitosis.

Now, how do you put all these facts together and differentiate one tissue type from another on microscope slides? The following Concept Boost will walk you through this process.

ConceptBOOST

"But It All Looks Pink!" Part 2

In part 1 of this Concept Boost, we discussed how to orient yourself to the different components of a tissue section. Now we'll take it a step further and identify the tissue from which a section was taken. Let's try it with these examples:

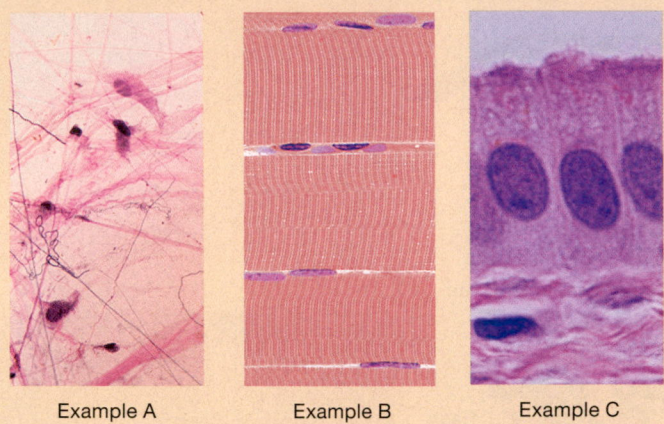

Example A   Example B   Example C

This isn't as difficult as it appears, particularly if you use a methodical, step-by-step approach:

1. **Identify the cells and the ECM.** Start with the basics you were given in the first part of this Concept Boost (on p. 128) and label the cells you see in Examples A, B, and C. Now move on to the ECM. Remember, the ECM consists only of ground substance and protein fibers. The ground substance will generally stain a uniform color (or simply appear clear). The protein fibers can take on various forms in different tissues, but they will generally stain darker than the ground substance, and they will always lack nuclei. Identify the protein fibers and ground substance in Examples A, B, and C.

2. **Notice how the cells are shaped and arranged.** Are the cells packed tightly together, or are they widely spaced? Do they form a continuous sheet, as in epithelial tissue—or do they seem to be surrounded by ECM, as in connective tissue? Are the cells all identical, or are there clearly different types? Do the cells have "arms" extending from a central body? Explain how the cells are shaped and arranged in Examples A, B, and C.

3. **Notice how the ECM is arranged.** Is the ECM confined to one specific part of the tissue (as in epithelial tissue), or is it spaced evenly between the cells? Does ground substance predominate, or are protein fibers the main elements? What types of protein fibers can you see? Explain how the ECM is arranged in Examples A, B and C.

4. **Determine the class of tissue.** Using your analysis in the preceding steps, now you are ready to identify the class of tissue. Determine the type of tissue in Examples A, B, and C. ■

Answers: A: loose connective tissue, B: skeletal muscle tissue, C: simple columnar epithelium

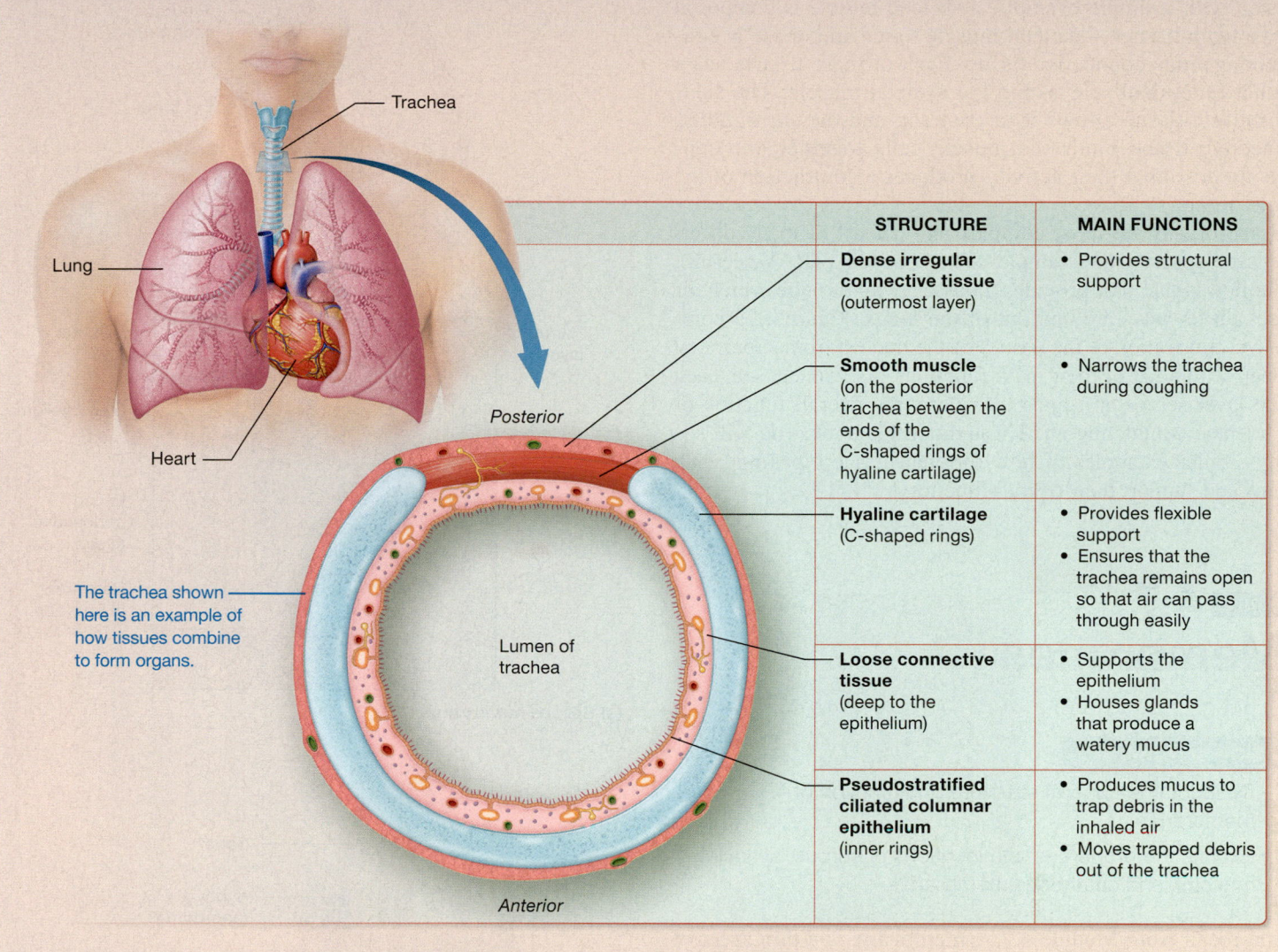

	STRUCTURE	MAIN FUNCTIONS
	Dense irregular connective tissue (outermost layer)	• Provides structural support
	Smooth muscle (on the posterior trachea between the ends of the C-shaped rings of hyaline cartilage)	• Narrows the trachea during coughing
	Hyaline cartilage (C-shaped rings)	• Provides flexible support • Ensures that the trachea remains open so that air can pass through easily
	Loose connective tissue (deep to the epithelium)	• Supports the epithelium • Houses glands that produce a watery mucus
	Pseudostratified ciliated columnar epithelium (inner rings)	• Produces mucus to trap debris in the inhaled air • Moves trapped debris out of the trachea

Trachea

Lung

Heart

Posterior

Lumen of trachea

Anterior

The trachea shown here is an example of how tissues combine to form organs.

Quick Check

☐ 1. What are the two main cell types in nervous tissue, and what are their roles?

☐ 2. What are the three parts of a neuron, and what are their functions?

Apply What You Learned

☐ 1. A tumor consists of cells that divide by mitosis outside the normal controls of the cell cycle. Would you expect tumors of nervous tissue to involve neurons, neuroglial cells, or both? Explain.

☐ 2. A hypothetical poison destroys the axon of a neuron. How will this affect the ability of the neuron to function?

See answers in Appendix A.

MODULE 4.6

Putting It All Together: The Big Picture of Tissues in Organs

Learning Outcomes

1. Describe how tissues work together to form organs.

Now that we have examined each tissue in isolation, let's look at how they function together in the body. As you discovered in the introductory chapter, two or more tissues that combine structurally and functionally form an organ (see Chapter 1; **Figure 4.23**).

To start with a simple example, a skeletal muscle is composed of two main tissues—skeletal muscle tissue and dense irregular collagenous connective tissue. Each of these tissues has a distinct functional role within the skeletal muscle: The skeletal muscle tissue allows it to contract, and the surrounding connective tissue binds the muscle cells together and supports them so that their activity produces a contraction of the whole organ.

Some organs are more complex and consist of many different tissue types. One example is the *trachea* (TRAY-kee-uh), the hollow organ that provides a passageway through which air passes on its way into and out of the lungs. Figure 4.23 combines an illustration of the tissues of the trachea from superficial to deep with a list of their main functions. As you can see, each tissue layer serves an important role in the overall function of the trachea: conducting air. Throughout this book, you will see many similar examples of how different tissue types make up organs and support their overall functions.

MODULE **4.7**
Membranes

Learning Outcomes

1. Describe the general structure and function of membranes.
2. Explain the properties and locations of serous, synovial, mucous, and cutaneous membranes.

If you recall from Module 4.2, a **membrane** is a thin sheet of one or more tissues that lines a body surface or cavity. Most membranes consist of a superficial layer of secretory cells and a layer of connective tissue on which they rest; some also contain a thin layer of smooth muscle. Membranes anchor organs in place, serve as barriers, function in immunity, and secrete various substances. The membranes that fit these structural and functional definitions are considered true membranes. However, other "membrane-like structures" don't necessarily fit these definitions, yet are sometimes called membranes because they perform many of the same functions.

Four general types of membranes are present in the human body: *serous, synovial, mucous,* and *cutaneous.* Of the four, the former two are true membranes and the latter two are membrane-like structures. Each membrane differs in structure, function, and products secreted.

True Membranes

 **FLASHBACK**

1. Where are the pleural, pericardial, and peritoneal body cavities located? (p. 18)

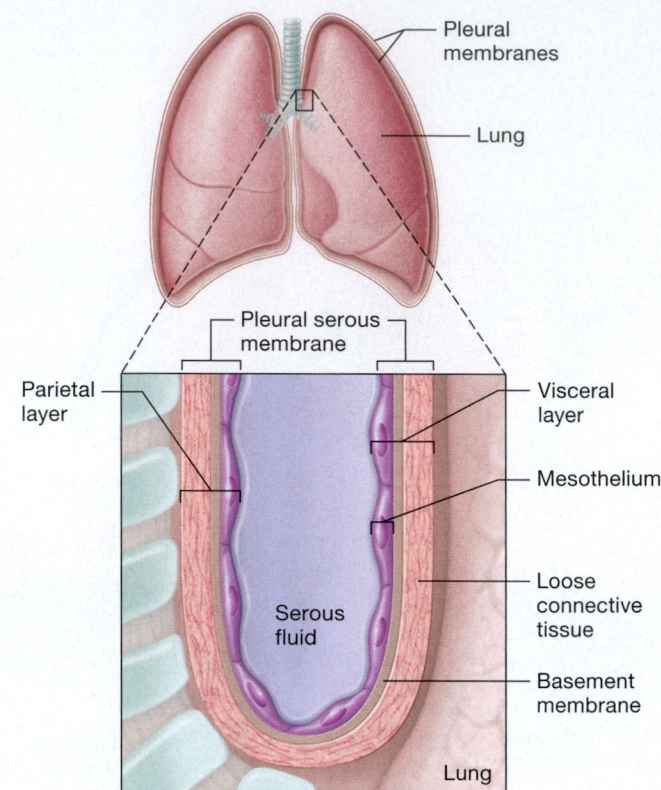

(a) Serous membrane

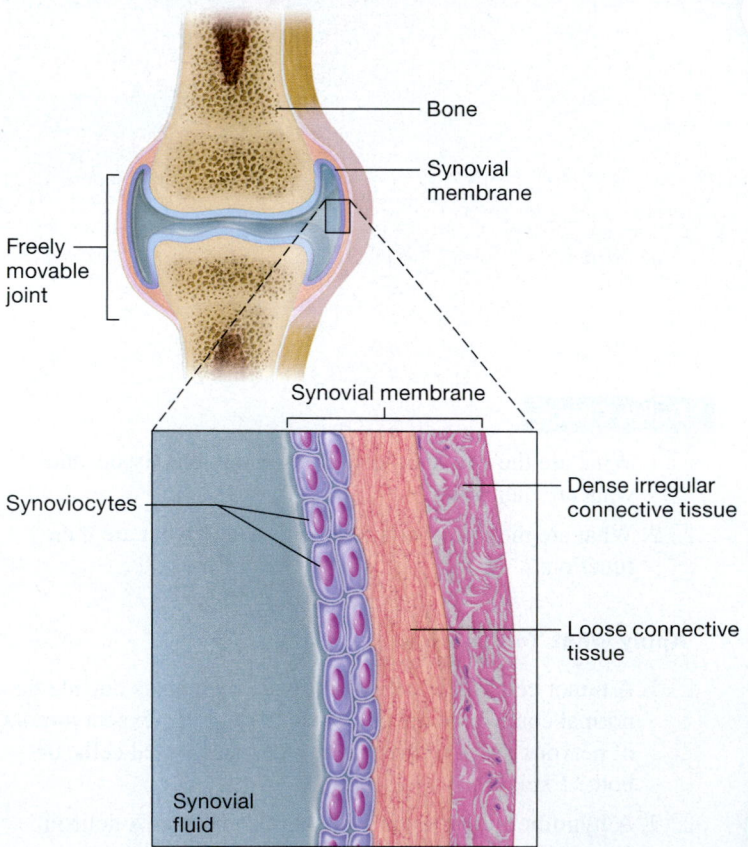

(b) Synovial membrane

Figure 4.24 True membranes.

Both serous and synovial membranes are true membranes that do not open to the outside of the body. We take a closer look at the structure and function of these membranes in this section.

Serous Membranes

Serous membranes (SEE-rus), also known as **serosae** (seh-ROH-see), line the pleural, pericardial, and peritoneal body cavities (**Figure 4.24a** shows the example of the pleural cavity). They consist of a layer of simple squamous epithelium called **mesothelium** (mez´-oh-THEE-lee-um), its basement membrane, and a layer of loose connective tissue. Notice in Figure 4.24a that a serous membrane is a continuous sheet that folds over on itself. This gives it two parts: its outer part, the *parietal layer* (puh-RY-eh-tul), is in contact with the body wall, and its inner part, the *visceral layer,* covers the organs within a cavity. The mesothelial cells produce a thin, watery fluid called **serous fluid,** which fills the very narrow space between the parietal and visceral layers. Here it provides lubrication so that organs can move without friction. When these layers are inflamed, however, this may not provide enough protection (see *A&P in the Real World: Friction Rubs* on p. 154).

Synovial Membranes

Synovial membranes (sih-NOH-vee-ul) line the cavities surrounding freely moveable joints, including the hip, knee, elbow, and shoulder (**Figure 4.24b**). Unlike mucous and serous membranes, synovial membranes do not contain a layer of epithelial cells. Instead, they are made up of two connective tissue layers: The inner layer consists of modified fibroblasts called *synoviocytes* (sih-NOH-vee-oh´-sytz), and the external layer is generally a mixture of loose and dense irregular connective tissue. Synoviocytes secrete a fluid called *synovial fluid,* which is a watery, slippery fluid that lubricates the joint. Synovial membranes are discussed in greater detail with joints, or articulations (see Chapter 8).

Quick Check

☐ 1. Where are serous and synovial membranes located, and what are their functions?

Membrane-like Structures

The body's other two types of membranes, mucous and cutaneous, are actually membrane-like structures. The cutaneous membrane is on the external surface of the body, and mucous membranes line internal body surfaces but open to the outside of the body.

Mucous Membranes

Mucous membranes, also called **mucosae** (myoo-KOH-see), line all body passages as part of the walls of hollow organs that open to the outside of the body, including the respiratory passages, the mouth, the nasal cavity, the digestive tract, and the male and female reproductive tracts (**Figure 4.25a**). Mucosae consist of a layer of epithelium, its basement membrane, a layer of loose connective tissue called the **lamina propria,** and occasionally a thin layer of smooth muscle.

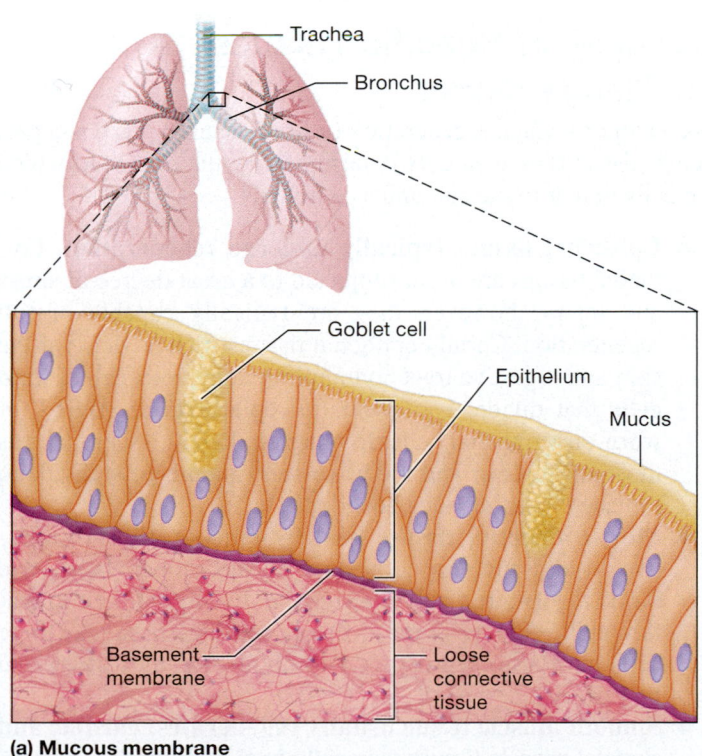

(a) Mucous membrane

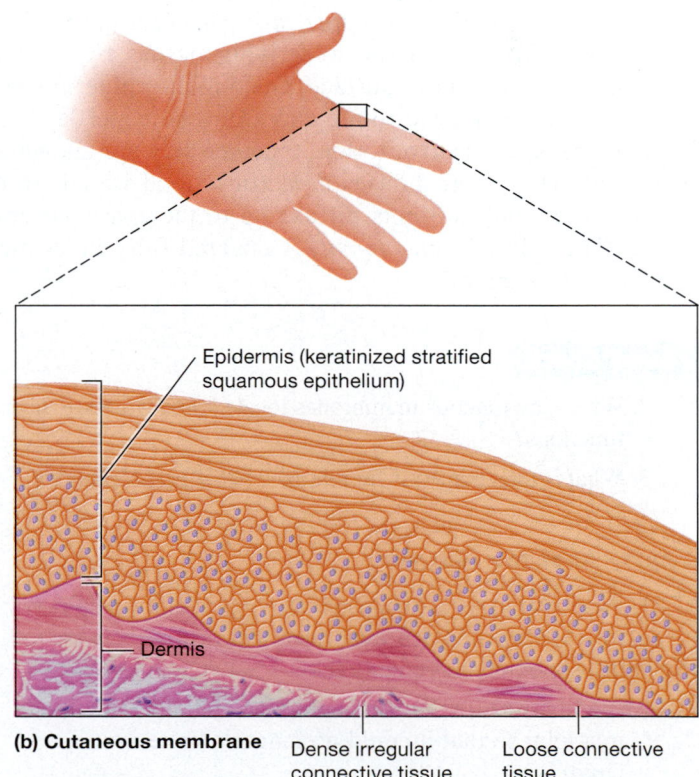

(b) Cutaneous membrane

Figure 4.25 Membrane-like structures.

Friction Rubs

Certain viral and bacterial infections can cause inflammation of the serous membranes of the pleural and pericardial cavities. When these serosae are inflamed, the thin layer of serous fluid produced by the mesothelium is inadequate to reduce friction as the organs move within their cavities. This causes the parietal and visceral layers to rub together as the organs expand and contract, which produces a grating sound called a *friction rub* that can be heard with a stethoscope. Friction rubs are typically quite painful and cause chest pain that worsens with inhalation, body movement, and swallowing. They usually resolve with treatment of the underlying condition.

Not surprisingly, mucosae contain glands with goblet cells that secrete mucus. As we discussed earlier, mucus serves many functions, most of which are protective in nature. For example, in the respiratory tract, mucus traps inhaled debris so that the ciliated epithelial cells can sweep it out. In the urinary bladder and the stomach, mucus provides a protective layer between the acidic contents of these organs and their underlying epithelia.

Cutaneous Membrane

The term **cutaneous membrane** (kyoo-TAYN-ee-us) refers to the **skin,** which is the largest organ of the body (**Figure 4.25b**). Skin consists of an outer layer of keratinized stratified squamous epithelium called the *epidermis*. The keratinized cells of the epidermis protect the underlying tissues by providing a hard and continuous surface. Deep to the epidermis is a layer of loose connective tissue and an even deeper layer of dense irregular connective tissue; together they are known as the *dermis*. The dermis houses numerous blood vessels from which oxygen and nutrients diffuse up to the avascular epidermis. The skin (or *integument*) is covered fully in its own chapter (see Chapter 5).

Quick Check

☐ 2. Where are mucous membranes located, and what are their functions?

☐ 3. What is the cutaneous membrane? What is its function?

Apply What You Learned

☐ 1. Predict what would happen if the organs of the respiratory tract and the stomach were lined with serous membranes instead of mucous membranes.

☐ 2. Predict the effect of damaging the synovial membrane in a joint so that it could no longer secrete synovial fluid.

See answers in Appendix A.

MODULE **4.8**

Tissue Repair

Learning Outcomes

1. Describe how injuries affect epithelial, connective, muscular, and nervous tissues.
2. Describe the process of regeneration.
3. Explain the process of fibrosis.

When cells are damaged and die as a result of normal wear and tear or from injury, the body must remove the dead cells and fill in the remaining gap in order to maintain homeostasis. This process of wound healing is called **tissue repair,** and it occurs differently in different tissues. Certain tissues are capable of **regeneration,** during which the damaged or dead cells are replaced with cells of the same type (**Figure 4.26a**). When regeneration is complete, the function of the tissue is in general completely restored. Other tissues are incapable of full regeneration, and a different process called **fibrosis** (fy-BROH-sis) ensues (**Figure 4.26b**). During fibrosis, fibroblasts divide by mitosis and produce collagen to fill in the defect left by the injury, and the tissue does not regain its ability to function normally. The end result of fibrosis is the formation of **scar tissue,** which is a type of dense irregular connective tissue. Let's look at the factors that affect whether a tissue goes through regeneration or fibrosis.

Capacity of Specific Tissues for Tissue Repair

The extent to which regeneration or fibrosis takes place in a particular tissue type is largely determined by the degree to which the cells in that tissue can undergo mitosis.

- **Epithelial tissues typically undergo regeneration.** Epithelial tissues are often subjected to a great degree of stress and injury; however, they are typically capable of full regeneration. Certain epithelial tissues such as those of the skin and digestive tract contain immature cells called **stem cells** that divide to continually replace dead, injured, or worn-out epithelial cells. In other epithelial tissues, such as those of the liver and blood vessels, mature cells divide to replace those that have been lost.
- **Most connective tissues heal by regeneration.** Connective tissues such as connective tissue proper, bone, and blood regenerate easily through the division of their immature cells. An exception to this is cartilage tissue, as these cells have a limited capacity to divide and thus often heal by fibrosis.
- **Smooth muscle tissue usually regenerates; cardiac and skeletal muscle tissues generally heal by fibrosis.** The cells of smooth muscle tissue largely retain their ability to undergo mitosis and so heal by regeneration. However, mature skeletal

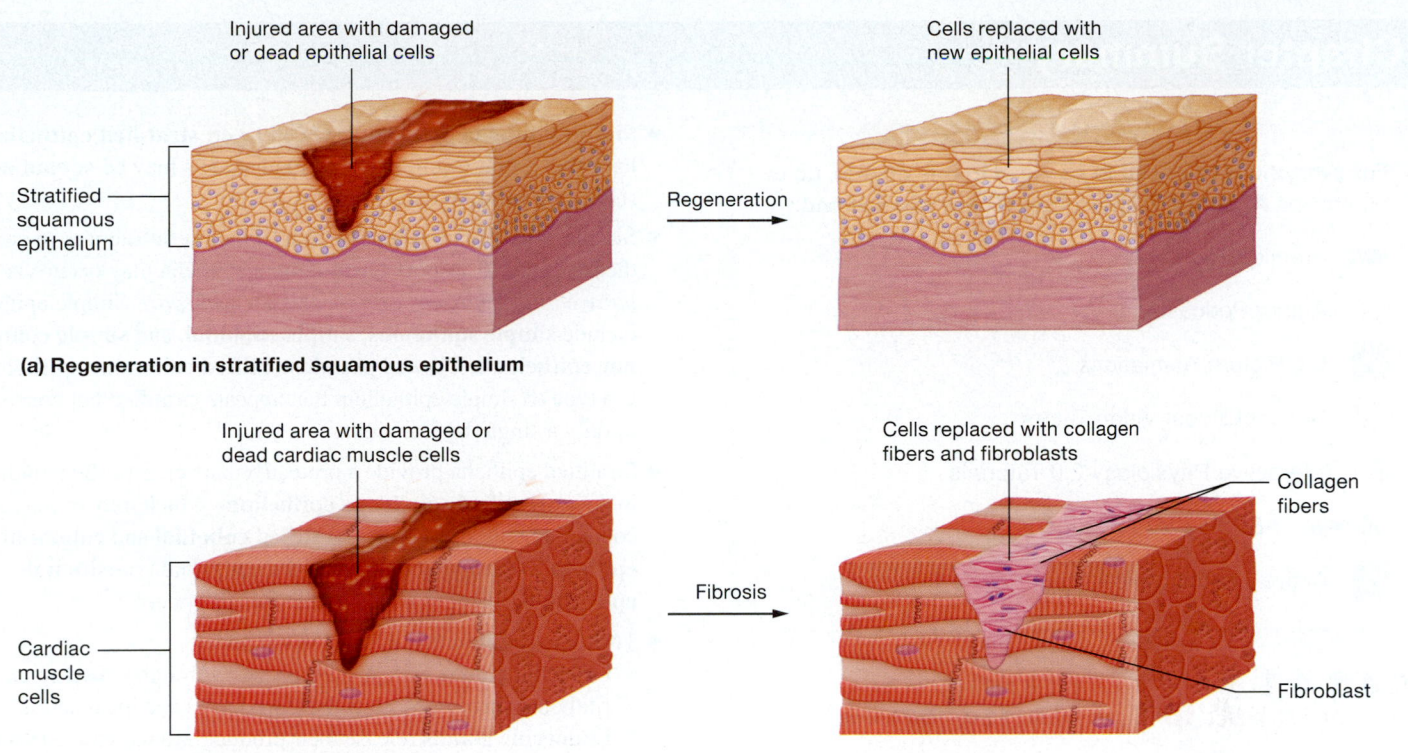

Injured area with damaged or dead epithelial cells

Cells replaced with new epithelial cells

Stratified squamous epithelium

Regeneration

(a) Regeneration in stratified squamous epithelium

Injured area with damaged or dead cardiac muscle cells

Cells replaced with collagen fibers and fibroblasts

Collagen fibers

Cardiac muscle cells

Fibrosis

Fibroblast

(b) Fibrosis in cardiac muscle tissue

Figure 4.26 Tissue repair by regeneration or fibrosis.

muscle fibers and cardiac muscle cells cannot undergo mitosis due to their large size and complicated cellular architecture. In skeletal muscle tissue, a population of cells called *satellite cells* can divide and become skeletal muscle fibers, allowing a limited degree of regeneration. However, cardiac muscle tissue lacks satellite cells and heals only by fibrosis.

● **Neurons of nervous tissue generally do not regenerate.** Although neuroglial cells retain the ability to divide, neurons are generally unable to undergo mitosis. In general, damaged neurons in the brain and spinal cord are replaced by neuroglial cells that divide to produce a scar. However, if the cell body of a neuron is intact and only the axon is damaged, there is some chance that the axon will regenerate, depending on the location and nature of the damage. This generally applies only to axons outside the brain and spinal cord.

Quick Check

☐ 1. How do regeneration and fibrosis differ?

☐ 2. Which tissues generally heal by regeneration? Which generally heal by fibrosis?

Other Factors Affecting Tissue Repair

In addition to the ability of cells in a tissue to undergo mitosis, two other factors—nutrition and blood supply—are also important determinants of tissue repair. Because tissue

repair involves production of proteins such as collagen, protein (amino acid) intake must be adequate for repair to occur. Another dietary requirement is sufficient intake of vitamin C, which is needed by the fibroblasts to produce collagen. In addition to dietary factors, the blood supply to the damaged area must be taken into consideration, as blood brings in oxygen, nutrients, and cells of the immune system that are needed for tissue repair. This is why people suffering from diseases of the arteries often have nonhealing wounds, even in tissues capable of regeneration. Details of the process of tissue repair are covered in the immune system chapter (see Chapter 20).

Quick Check

☐ 3. Which factors influence the ability of a tissue to heal?

Apply What You Learned

☐ 1. During a heart attack, the blood flow to cardiac muscle tissue is blocked, and cardiac muscle cells in the affected area generally die. Why will cardiac function rarely return to normal after this takes place?

☐ 2. Why do you think that homeless patients often have problems with wound healing?

See answers in Appendix A.

Chapter Summary

For everything you need to succeed in this course, go to Mastering Anatomy & Physiology. There you will find:

 Practice Tests

 Author Podcasts

 Big Picture Animations

 Concept Boost Video Tutors

 iP2™ Interactive Physiology 2.0 Tutorials

 A&PFlix A&P Flix 3D Animations

 Active-Learning Workbook

MODULE 4.1
Introduction to Tissues 124–127

- **Histology** is the study of the normal structure of tissues.
- Every **tissue** is composed of cells and the surrounding **extracellular matrix (ECM).**
- There are four classes of tissues: **epithelial, connective, muscle,** and **nervous tissue.**
- The ECM provides a tissue with strength, regulates cell activity, and anchors cells in place. It is composed of **ground substance** and **protein fibers.** There are three types of protein fibers: **collagen, elastic,** and **reticular fibers.**
- Cells in a tissue are joined by **cell junctions,** which include the following:
  - **Tight junctions** make the space between adjacent cells impermeable to macromolecules.
  - **Desmosomes** anchor neighboring cells together to increase the strength of a tissue with respect to mechanical stress.
  - **Gap junctions** are small pores in the plasma membranes of two adjacent cells that allow small substances to pass from the cytosol of one cell to that of another.

MODULE 4.2
Epithelial Tissues 127–137

- Epithelial tissues act as barriers between the body and the external environment and between our organs and fluid-filled cavities. They function in protection, immune defenses, secretion, transport, and sensation.
- Epithelial cells are packed closely together and joined by tight junctions and desmosomes.
- Epithelia are avascular, and their ECM is mostly confined to the **basal lamina** of the **basement membrane,** which anchors them to the underlying tissues.

- **Simple epithelia** have only one cell layer; **stratified epithelia** have two or more cell layers. Epithelial cells may be **squamous** (flat), **cuboidal,** or **columnar.**
- Simple epithelia are generally thin and allow substances to cross them rapidly. Transport across simple epithelia may occur via *paracellular transport* or *transcellular transport.* Simple epithelia include **simple squamous, simple cuboidal,** and **simple columnar epithelium.** Ciliated **pseudostratified columnar epithelium** is a type of simple epithelium that appears stratified but consists of only a single layer of cells.
- Stratified epithelia provide a protective barrier. Stratified epithelia include **stratified squamous epithelium,** which may or may not contain a keratinized layer; **stratified cuboidal** and **columnar epithelia,** which are both rare in the body; and **transitional epithelium,** which is found in the urinary system.
- Two types of **glands** are found in the body:
  - **Exocrine glands** release their product through a **duct** to an epithelial surface. Most of their products have local actions.
  - **Endocrine glands** release their product into the bloodstream, and their products can have actions on distant target cells in the body.
- Exocrine glands may be **unicellular** or **multicellular.** Multicellular glands are of varying shape and complexity.
- Exocrine glands release their products in one of three ways: **merocrine secretion,** in which the product is released by exocytosis; **holocrine secretion,** in which the cell ruptures; and **apocrine secretion,** in which a portion of the cytoplasm is pinched off and released with the product.

MODULE 4.3
Connective Tissues 137–146

- Functions of connective tissues include connecting and binding, support, protection, and transport.
- There are two basic types of connective tissue: **connective tissue proper** and **specialized connective tissue.**
- Cells of connective tissue proper may be resident cells or migrant cells. These cells include **fibroblasts, adipocytes, mast cells,** and **phagocytes.**
- There are four types of connective tissue proper:
  - **Loose (areolar) connective tissue** has all three fiber types and has ground substance as its primary element. It is part of the skin and lines body cavities and hollow organs.
  - **Dense connective tissue** features an ECM composed primarily of protein fibers. **Dense irregular connective tissue** contains bundles of collagen fibers arranged at various angles to one another; **dense regular collagenous connective tissue** contains parallel bundles of collagen fibers; and **dense regular elastic connective tissue** contains parallel bundles of elastic fibers.

○ **Reticular tissue** contains numerous reticular fibers in its ECM that form weblike networks in the spleen and lymph nodes.

○ **Adipose tissue** contains adipocytes as its primary element.

● The specialized connective tissues are cartilage, bone, and blood.

○ **Cartilage** is a tough but flexible tissue that absorbs shock and is resistant to tension and compression. There are three types of cartilage:

  • **hyaline cartilage,** which is found on the ends of bones in joints, in the nose, in certain respiratory passages, and where the ribs meet the sternum;

  • **fibrocartilage,** which contains many bundles of collagen fibers in its ECM and is found in certain joints and the intervertebral discs; and

  • **elastic cartilage,** which contains elastic fibers in its ECM and is found in the ear and in parts of the larynx.

○ **Bone,** or **osseous tissue,** contains collagen fibers, ground substance, and calcium phosphate crystals. **Osteoblasts** make the organic component of the ECM and **osteoclasts** break down bone.

○ **Blood** is a fluid connective tissue. Its ECM is **plasma** and the main cell types are **erythrocytes** and **leukocytes.**

MODULE **4.4**
Muscle Tissues 146–148

● The cells of muscle tissue are specialized for contraction by turning the chemical energy of ATP into the mechanical energy of movement.

● **Muscle cells** are **excitable cells** that respond to electrical or chemical stimulation. They may be *striated* or *smooth.*

● There are three types of muscle tissue:

○ **Skeletal muscle tissue** is voluntary and consists of long, striated, multinucleated **muscle fibers.**

○ **Cardiac muscle tissue** is involuntary and found in the heart. Its cells are striated, short, wide, and branched, and are united by **intercalated discs.**

○ **Smooth muscle tissue** consists of uninucleate smooth muscle cells and is found in the eye, lining most hollow organs, in the ducts of certain glands, and in the skin.

MODULE **4.5**
Nervous Tissue 148–151

● Nervous tissue makes up most the brain, spinal cord, and nerves.

● **Neurons** are excitable cells that send and receive messages in the form of **nerve impulses** to other neurons, muscle cells, and/ or glands. They contain a central **cell body,** a single **axon** that carries nerve impulses away from the cell body, and one or more **dendrites** that bring impulses from other neurons into the cell body.

● **Neuroglial cells** are the supporting cells of nervous tissue.

MODULE **4.6**
Putting It All Together: The Big Picture of Tissues in Organs 151–152

● Organs are made up of several different types of tissues that function together.

MODULE **4.7**
Membranes 152–154

● **Membranes** are thin sheets of one or more tissues that anchor organs in place, serve as barriers, function in immunity, and secrete substances. There are also membrane-like structures that have similar functions.

● **Serous membranes** are true membranes that secrete **serous fluid** and form the pleural, pericardial, and peritoneal body cavities.

● **Synovial membranes** are true membranes that secrete *synovial fluid* and line the cavities surrounding freely moveable joints.

● **Mucous membranes** are membrane-like structures that secrete **mucus** and line body passages and hollow organs that open to the outside of the body.

● The **cutaneous membrane** is the **skin,** a membrane-like structure, and it consists of the superficial *epidermis* and the deeper *dermis.*

MODULE **4.8**
Tissue Repair 154–155

● During **regeneration,** damaged cells are replaced with cells of the same type. During **fibrosis,** fibroblasts fill in the defect with dense irregular connective tissue, forming **scar tissue.**

● Epithelial, most connective, and smooth muscle tissues typically undergo regeneration. Nervous tissue generally does not regenerate.

● Cartilage, skeletal muscle, and cardiac muscle tissues generally heal by fibrosis.

Assess What You Learned

Scan the QR Code for additional practice test questions

https://goo.gl/ExcLeJ

LEVEL 1 Check Your Recall

1. Explain how connective tissues differ from epithelial tissues in structure and function.

2. State whether each of the following describes epithelial, connective, muscle, or nervous tissue.

   a. _____ ECM is often the primary element.
   b. _____ Consists of excitable cells that are specialized for contraction.
   c. _____ Sheets of tightly packed cells with little ECM.
   d. _____ Makes up the majority of the brain and spinal cord.
   e. _____ Cells may be smooth or striated.
   f. _____ Binds, connects, supports, and transports substances throughout the body.

3. Describe the roles of each of the following components of the ECM:

   a. Collagen fibers
   b. Glycosaminoglycans
   c. Reticular fibers
   d. Proteoglycans
   e. Glycoproteins (cell-adhesion molecules)
   f. Elastic fibers

4. Fill in the blanks: Tight junctions make the spaces between cells _____, whereas desmosomes increase the _____ of a tissue.

5. Mark the following statements as true or false. If a statement is false, correct it to make a true statement.

   a. Epithelial tissues are classified by cell shape and the number of cell layers.
   b. Epithelial tissues function in protection, immune defenses, secretion, transport, and sensation.
   c. Epithelial tissue is highly vascular.
   d. A goblet cell is a unicellular exocrine gland that secretes mucus.
   e. Pseudostratified epithelium appears to be simple epithelium but is actually stratified.
   f. Stratified epithelia are specialized to allow substances to cross their cells rapidly.

6. Match each type of epithelium with its correct location in the body.

   _____ Simple squamous   a. Skin
   _____ Pseudostratified columnar   b. Urinary bladder
   _____ Keratinized stratified   c. Air sacs of the lungs,
   _____ Simple columnar   squamous blood vessels
   _____ Transitional   d. Kidney tubules, thyroid gland
   _____ Simple cuboidal   e. Respiratory passages, nasal
                                         cavity
                                      f. Digestive tract

7. Compare and contrast the following pairs of terms:

   a. Endocrine gland and exocrine gland
   b. Unicellular gland and multicellular gland
   c. Holocrine secretion and merocrine secretion

8. Which of the following best describes the position of a tissue in the levels of organization of the human body?

   a. Tissues are the most fundamental level of organization.
   b. Tissues are between cells and organs in the levels of organization.
   c. Tissues are the most complex level of organization.
   d. Tissues are between organs and systems in the levels of organization.

9. Mark the following statements as true or false. If a statement is false, correct it to make a true statement.

   a. Fibroblasts store lipids in a large inclusion in their cytoplasm.
   b. Loose connective tissue features protein fibers as its primary component.
   c. Adipose tissue functions in insulation, warmth, protection, and shock absorption, and is the major energy reserve in the body.
   d. Fibrocartilage provides a smooth surface on which bones may articulate with little friction.
   e. The ECM of bone tissue consists exclusively of calcium phosphate crystals.

10. Match the following types of connective tissue with their correct location(s).

    _____ Hyaline cartilage   a. Dermis, surrounding
    _____ Elastic cartilage   joints, organs
    _____ Smooth muscle   b. Most hollow organs,
    _____ Dense irregular   eye, skin
          connective tissue   c. Large arteries, certain
    _____ Fibrocartilage   ligaments
    _____ Cardiac muscle   d. Ear and epiglottis
    _____ Dense regular elastic tissue e. Lymph nodes, spleen, liver
    _____ Reticular connective tissue   f. Freely moveable joints
                                        g. Intervertebral discs, knee
                                           joint
                                        h. Heart

11. Which of the following statements about muscle tissue is *false?*

    a. Smooth muscle tissue is involuntary and the cells are nonstriated.
    b. Cardiac muscle tissue is involuntary and the cells are striated.
    c. Skeletal muscle fibers are joined by intercalated discs.
    d. Skeletal muscle fibers are long and cylindrical striated cells.

12. Fill in the blanks: _____ are the cells of nervous tissue that send and receive messages, and _____ are the supporting cells of nervous tissue. A(n) _____ carries a nerve impulse toward a neuron cell body, and a(n) _____ carries a nerve impulse away from a neuron cell body.

13. Each of the following statements is false. Correct each to make a true statement.

    a. Mucous membranes are composed of mesothelium and the underlying loose connective tissue.
    b. Mesothelial cells secrete mucus.
    c. Synoviocytes are the secretory cells of mucosae.
    d. The cutaneous membrane is composed of simple squamous epithelium, loose connective tissue, and dense irregular connective tissue.
    e. Serous membranes line all hollow organs that open to the outside of the body.

14. Which tissues undergo regeneration? Which tissues undergo fibrosis? Why?

LEVEL 2 Check Your Understanding

1. If you were to cut through epithelial tissue without penetrating the basement membrane, would you expect bleeding to occur? Why or why not?

2. Vitamin C is required for synthesis of collagen. Predict the effect that a vitamin C deficiency (a disease called *scurvy*) would have on bone tissue, dense regular collagenous connective tissue, dense irregular connective tissue, and cartilage. What symptoms would you expect to see from this disease?

3. Predict what would happen if the mucous membranes of the body stopped secreting mucus or if they secreted excess mucus.

LEVEL 3 Apply Your Knowledge

PART A: Application and Analysis

1. The disease pemphigus vulgaris involves a patient's own immune system attacking the desmosomes between the epithelial cells of the skin. What changes would you expect to see with this disease?

2. In the disease scleroderma, excessive collagen, glycosaminoglycans, and glycoproteins are produced and deposited in tissues throughout the body. Predict the effects of this disease.

3. In the disease pulmonary fibrosis, elastic fibers of the lung are destroyed and replaced with collagen fibers. Predict the effect this would have on breathing.

4. Imagine that a disease turns the simple epithelia of the lungs, kidney tubules, and intestines into keratinized stratified squamous epithelia. What effect would this change in form have on the functions of these tissues?

PART B: Make the Connection

5. Epithelial cells of the kidneys have pumps that drive the transcellular transport of sodium ions.

   a. Is this epithelium likely to be simple or stratified? Why?
   b. The movement of sodium ions drives the transcellular transport of water. Explain why water follows sodium. (*Connects to Chapter 3*)

6. Explain why it would be difficult for a mature multinucleate cell such as a skeletal muscle fiber to divide by mitosis. (*Connects to Chapter 3*)

7. Predict which organelles are likely to be abundant in cells such as fibroblasts that actively produce and secrete proteins. (*Connects to Chapter 3*)

See answers in Appendix A.

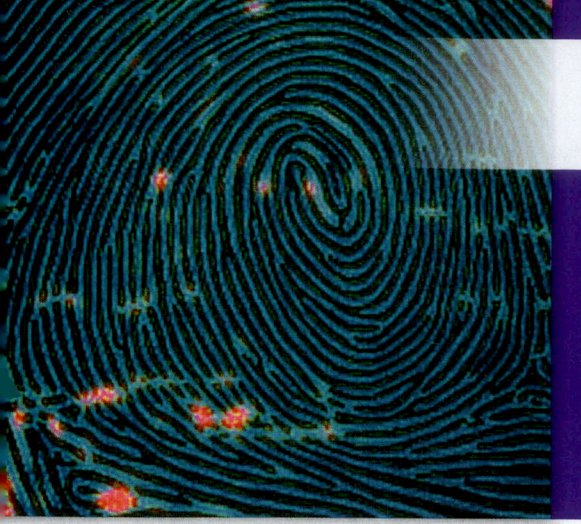

5

The Integumentary System

Perhaps more than any other system, the **integumentary system** (in-teg´-yoo-MEN-tuh-ree; *integu-* = "covering")—consisting of the **skin** and its accessory structures, such as hair, glands, and nails, collectively called the **integument**—is associated with how we identify and express ourselves and how others perceive us. This is evidenced by our spending habits. How much money have you spent in the past year on shampoo, soap, hair and nail care, deodorant, lotions, and similar items?

In the previous chapters, we examined the three simplest levels of organization—the chemical, cellular, and tissue levels. Now we change our focus to the organ and system levels of organization. We first take an introductory tour of the integumentary system, then examine the anatomy of the skin and its accessory structures, as well as the functions of the integumentary system in maintaining homeostasis. We conclude with a look at common conditions and diseases that affect the skin.

MODULE **5.1**

Overview of the Integumentary System

Learning Outcomes

1. Describe the basic structure of the skin.
2. Describe the basic functions carried out by the components of the integumentary system.

The skin doesn't get much respect as an organ, but it's actually the largest organ in the body and makes up approximately 10–15% of our total body weight. Indeed, this remarkable organ has a complex structure and performs numerous functions that are vital to maintaining homeostasis.

Skin Structure

« FLASHBACK

1. What is the cutaneous membrane? Which tissues make up the cutaneous membrane? (p. 154)
2. What is adipose tissue? (p. 139)

For practice applying concepts to a clinical scenario, check out the **Running Case Study** *for this chapter* @ Mastering Anatomy & Physiology

Photo: Tiny sweat pores open along epidermal ridges and sweat leaves a thin film called a fingerprint when we touch most surfaces.

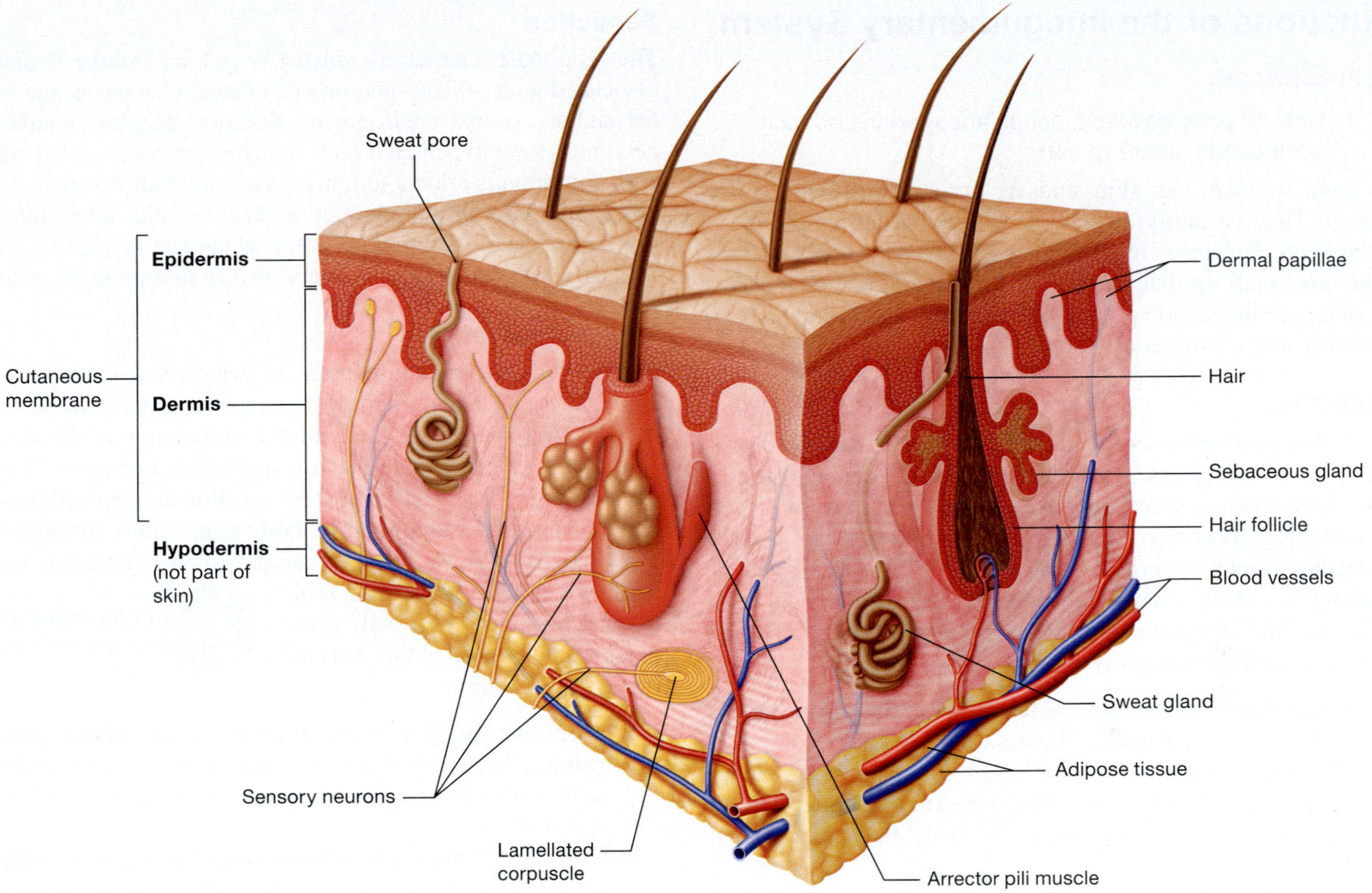

Sweat pore

Epidermis

Cutaneous membrane

Dermis

Hypodermis (not part of skin)

Sensory neurons

Lamellated corpuscle

Dermal papillae

Hair

Sebaceous gland

Hair follicle

Blood vessels

Sweat gland

Adipose tissue

Arrector pili muscle

Figure 5.1 Basic anatomy of the skin.

Practice art labeling
@Mastering Anatomy & Physiology

The skin, also known as the **cutaneous membrane,** has two main components (**Figure 5.1**):

- The superficial **epidermis** (eh-pih-DER-miss; *epi-* = "on top," *derm-* = "skin") consists of keratinized stratified squamous epithelium that rests on top of a basement membrane.
- The deep **dermis** consists of loose connective tissue and dense irregular connective tissue.

Embedded within the epidermis and dermis we find the skin's accessory structures: *hair, nails*, sweat-producing *sweat glands,* and oil-producing *sebaceous glands* (Figure 5.1 does not show nails). The skin also has *sensory neurons* and their associated receptors, such as *lamellated corpuscles*. In addition, we find small bands of smooth muscle called *arrector pili muscles* (uh-REK-tohr PIL-ee) that attach to hairs in the dermis.

Like all epithelia, the epidermis is avascular, or lacks a blood supply. For this reason, oxygen and nutrients follow their concentration gradients and diffuse from blood vessels located in the dermis below. This is an example of the Gradients Core Principle

CORE PRINCIPLE
Gradients

(p. 28). The need for diffusion limits the thickness of the epidermis in which we can find living cells. Indeed, nearly 50% of the cells of the epidermis are too far away from the blood supply in the dermis to sustain life, so the superficial epidermis consists entirely of dead cells.

Deep to the dermis we find the **hypodermis** (hy-poh-DER-miss; *hypo-* = "below"), which is also known as the **superficial fascia** (FASH-uh) or **subcutaneous tissue** (sub′-kyoo-TAY-nee-us). The hypodermis is not actually part of the skin, but it anchors the skin to the muscle and bone deep to it. The hypodermis consists of loose connective tissue that is richly supplied with blood vessels and varying amounts of adipose tissue. The number of blood vessels makes the hypodermis a good place to administer medicines via *subcutaneous injections* using a hypodermic needle. The hypodermis can also affect the appearance of the skin, as discussed in *A&P in the Real World: Cellulite* on p. 164.

Quick Check

☐ 1. What are the major structures of the skin, and what are their main properties?

Functions of the Integumentary System

◀◀ FLASHBACK

1. How do polar covalent, nonpolar covalent, and ionic compounds differ? (p. 39)

It's easy to take our skin and its accessory structures for granted. They certainly do not *seem* as important as the brain or the heart. But in reality, the integumentary system performs numerous functions that are critical to maintaining homeostasis. These functions include protection, sensation, thermoregulation, excretion, and synthesis of vitamin D.

Protection

One of the most obvious functions of the integument is to protect underlying tissues from damage. Tissue damage comes from three main sources: mechanical (physical) trauma, such as stretching, pressure, or abrasion; disease-causing microorganisms known as *pathogens* (*patho-* = "disease," *-gen* = "causing"); and environmental hazards, such as chemicals and radiation. The integument provides protection from all three sources in the following ways:

- **Protection from mechanical trauma.** Recall from the tissues chapter that the superficial aspect of the skin is composed of stratified squamous, keratinized epithelium (see Chapter 4). This structure provides a continuous, durable, yet flexible surface that protects the body from mechanical trauma.
- **Protection from pathogens.** The continuous structure of the skin also protects against invasion by pathogens. This function is evident any time the skin is compromised. The risk of infection rises dramatically when the skin is cut, which is why you often cover the laceration with a bandage. In addition, the skin contains cells of the immune system that destroy pathogens before they can invade deeper tissues. The glands of the skin also protect the body by secreting a variety of antimicrobial substances. For example, the secretions of sebaceous glands give the surface of the skin a slightly acidic pH, which deters the growth of many pathogens. This slightly acidic pH is known as the skin's *acid mantle*.
- **Protection from the environment.** The integument protects against multiple threats in the environment. For example, the skin absorbs *ultraviolet (UV) radiation* from the sun before it can damage the underlying tissues. The skin also secretes lipid-based chemicals that have nonpolar covalent bonds. These chemicals repel compounds with polar covalent bonds, such as water, and those with ionic bonds, such as salts, which keeps them from both leaving and entering the body via the skin. This is crucial to maintaining water and electrolyte homeostasis in the body by preventing the exit or entrance of excess water and electrolytes. Be aware, however, that these lipid-based substances do allow other nonpolar covalent substances to cross the skin, which we discuss in a later module.

Sensation

The skin houses numerous *sensory receptors,* cellular structures associated with sensory neurons that detect changes in the internal and/or external environment. Sensory receptors enable the nervous system to perceive such changes, a process called *sensation*. Sensation is critical to homeostasis, as it allows us to detect potentially harmful stimuli such as heat or cold. They allow us to perceive painful stimuli, as well, which can help avoid tissue damage (extreme heat and cold can also be interpreted as pain).

Thermoregulation

The body's internal temperature is determined largely by the many reactions that are part of its metabolism and muscle activity. Unless it is extreme, the external temperature of the air outside the body has little effect on internal temperature. This is due in large part to the process of **thermoregulation**—the maintenance of a stable internal body temperature through negative feedback loops—in an example of the Feedback Loops Core Principle (p. 28). Let's look at what happens when body temperature rises above the normal range (**Figure 5.2a**):

CORE PRINCIPLE
Feedback Loops

- **Stimulus: Body temperature increases above normal range.** Body temperature may rise due to normal conditions, such as exercise and hot weather, or abnormal conditions, such as fever.
- **Receptor: Thermoreceptors detect increase in temperature.** Receptors in the brain called thermoreceptors detect temperature changes in the internal body fluids.
- **Control center: Thermoregulatory center in brain receives signals from receptors.** The thermoreceptors send signals to a part of the brain called the hypothalamus, which acts as the body's "thermostat," or thermoregulatory center.
- **Effector/response: The control center stimulates effectors to respond with cooling mechanisms.** The hypothalamus triggers accessory organs of the skin known as sweat glands to release a watery fluid called sweat. Recall from the chemistry chapter that water takes a great deal of heat with it when it evaporates, so evaporating sweat takes heat away from the body and produces a cooling effect (see Chapter 2). The hypothalamus also adjusts blood flow through the dermis: When body temperature rises, blood vessels widen, or dilate, causing a large volume of blood to fill the dermis. This radiates heat from the blood into the environment, cooling the body.
- **Homeostasis and negative feedback: Body temperature returns to normal range, and negative feedback stops cooling mechanisms.** Sweating and vasodilation help return body temperature to normal. The thermoreceptors in the brain detect this, and negative feedback causes the thermoregulatory center to decrease its output to blood vessels and sweat glands.

Figure 5.2b shows what happens in the opposite situation, when body temperature decreases, generally due to cold environmental conditions. In this feedback loop, the parts are mostly the same, except the thermoreceptors that detect the lower

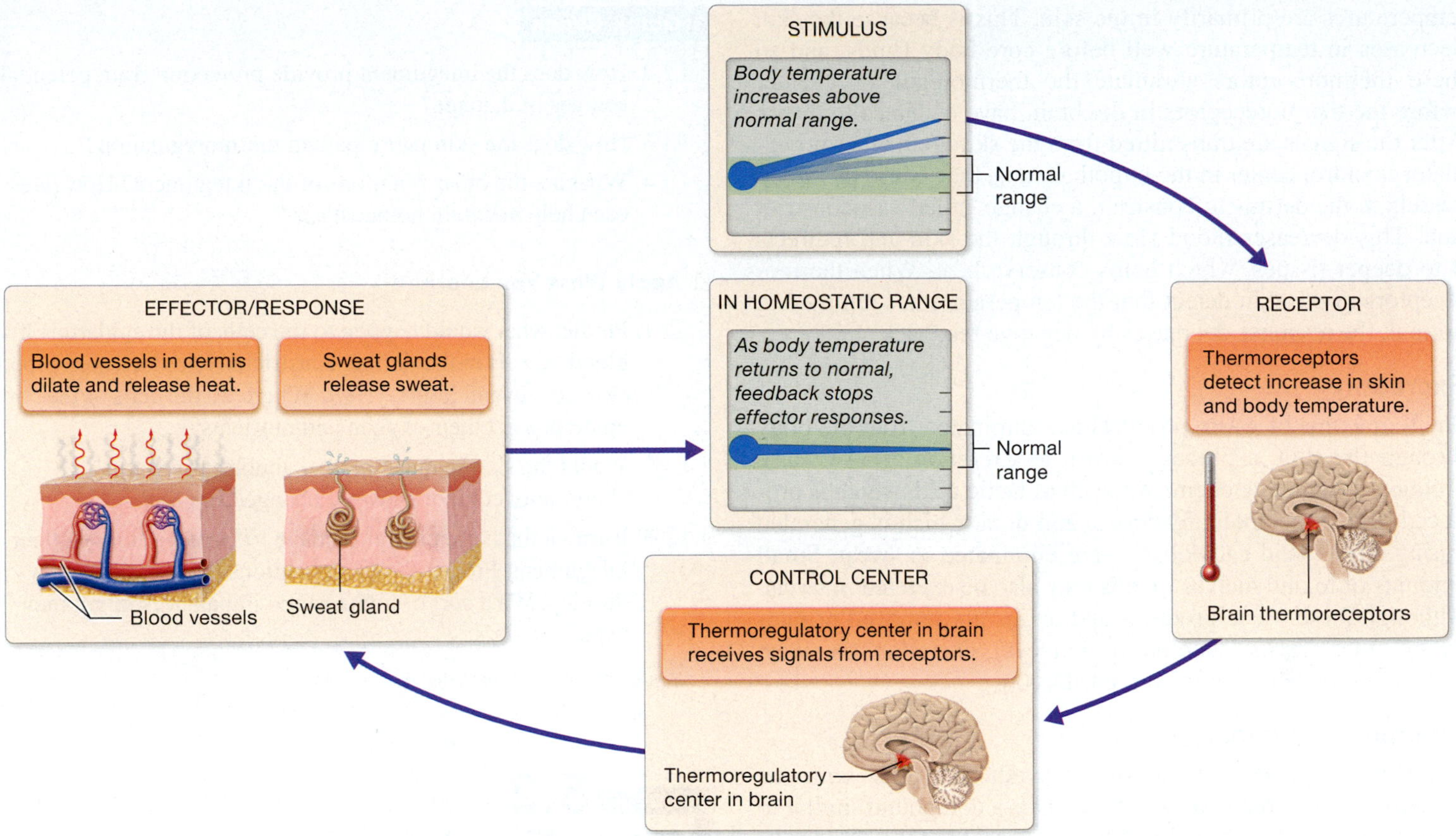

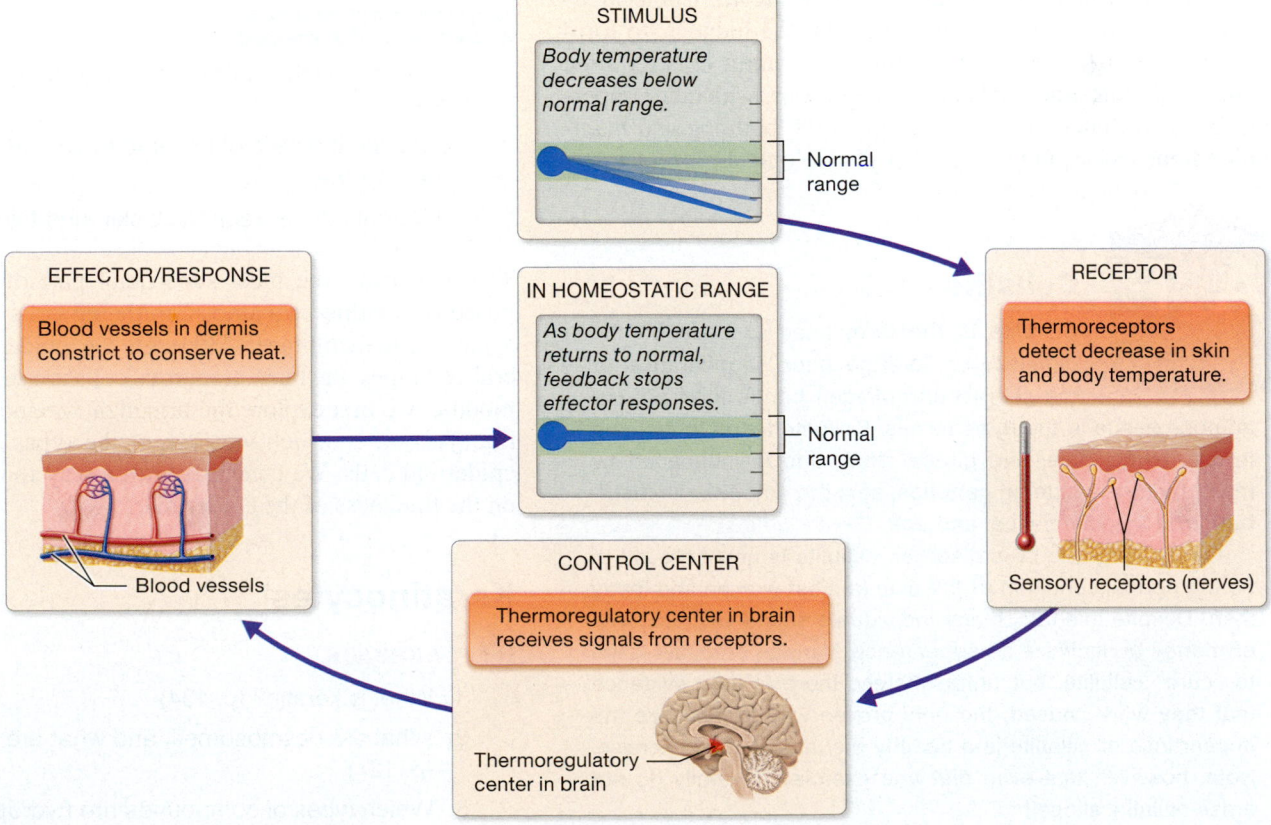

(a) Response of the integument to rising body temperature

(b) Response of the integument to falling body temperature

Figure 5.2 **Homeostatic regulation of body temperature by the integumentary system.**

temperatures are primarily in the skin. This is because the skin decreases in temperature well before core body fluids, and so these thermoreceptors stimulate the thermoregulatory center before the thermoreceptors in the brain have a chance to do so. After the signals are transmitted from the skin to the thermoregulatory control center in the hypothalamus, they cause the blood vessels in the dermis to constrict, a change called vasoconstriction. This decreases blood flow through the skin and redirects it to deeper tissues, which helps conserve heat. When thermoreceptors in the skin detect that the temperature has returned to normal, the response decreases by negative feedback.

Excretion

Small amounts of waste products are eliminated from the body through the skin, a process known as *excretion*. These waste products—including chemicals such as lactic acid, which is produced during metabolic reactions, and urea, which is generated during amino acid catabolism—are eliminated in sweat. Small amounts of toxins such as metals may also be excreted in sweat. Although these waste products and toxins are removed primarily by other organs (such as the kidneys), excretion via sweat makes a small but significant contribution.

Vitamin D Synthesis

The skin also plays a vital role in synthesizing the hormone **vitamin D.** The precursor of vitamin D is a compound similar to cholesterol found in the cells of the deeper part of the epidermis. When exposed to certain wavelengths of ultraviolet radiation, this compound is converted to the inactive *cholecalciferol* (vitamin D_3), which then enters the bloodstream. Cholecalciferol is modified first by the liver, then by the kidneys, to form *calcitriol,* the active form of vitamin D. Vitamin D is required for the small intestine to absorb calcium ions, which are important for nerve function, muscle contraction, building and maintaining bone tissue, and many other physiological processes.

A&P in the Real World

Cellulite

Cellulite is the term used to describe the dimpled, or "orange peel," appearance of the skin when collagen bands form around adipose tissue in the hypodermis. Cellulite tends to develop in the thighs, hips, and gluteal area, and is influenced by many factors, including genetics, sex, the amount and distribution of adipose tissue, and age.

Once thought to be a disorder, cellulite is now believed to be the normal condition of the skin in most women and many men. Despite this fact, many individuals spend vast amounts of money to minimize its appearance. Various products claim to "cure" cellulite, but unfortunately, there is little evidence that they work. Indeed, the only proven way to minimize the appearance of cellulite is a healthy diet and regular exercise. Note, however, that even diet and exercise generally do not erase cellulite altogether.

Apply What You Learned

☐ 1. Predict what would happen to the cells of the epidermis if blood vessels in the dermis constricted and decreased blood flow for several hours. (*Hint:* Where do the cells of the epidermis get their oxygen and nutrients?)

☐ 2. Would the skin in question feel abnormally warm, abnormally cold, or show no change? Explain.

☐ 3. Burn victims often suffer damage to large portions of their integument. Predict the complications these patients may develop. What sort of homeostatic imbalances might they experience?

See answers in Appendix A.

MODULE 5.2
The Epidermis

Learning Outcomes

1. Explain how the cells of the epidermis are arranged into layers.

2. Describe the cells of the epidermis and the life cycle of a keratinocyte.

3. Differentiate between thick skin and thin skin.

The epidermis, the most superficial part of the skin, is composed of stratified squamous epithelial cells that rest on top of a basement membrane. Although the epidermis contains several cell types, the most numerous cells are *keratinocytes*. In this module, we first explore the organization and life cycle of keratinocytes, after which we look at the types and roles of other epidermal cells. We conclude with a way to classify skin based on the thickness of the epidermis.

Keratinocytes

≪ FLASHBACK

1. What is keratin? (p. 134)

2. What are desmosomes, and what are their functions? (p. 127)

3. Which types of compounds are hydrophobic and which are hydrophilic? (p. 47)

Approximately 95% of cells in the epidermis are **keratinocytes** (kehr′-ah-TIN-oh-sytz). Two structural features of keratinocytes make the epidermis stronger and less susceptible to mechanical trauma, demonstrating the Structure-Function Core Principle (p. 28). First, as their name implies, keratinocytes are cells that manufacture keratin (KEHR-uh-tin; *kerato-* = "horn," so called because it makes up the horns of other animals). Recall that **keratin** is a fibrous protein that makes a tissue tougher and more resistant to mechanical stress (see Chapter 4). Second, keratinocytes are linked by numerous *desmosomes,* intercellular junctions that hold cells together.

CORE PRINCIPLE
Structure-Function

Keratinocytes are arranged into layers, or *strata,* which we examine in the next few subsections.

Layers of the Epidermis

Figure 5.3 illustrates the five structurally distinct layers of the epidermis. From deep to superficial, they are the stratum basale, stratum spinosum, stratum granulosum, stratum lucidum, and stratum corneum.

Stratum Basale The deepest layer of the epidermis is the thin **stratum basale** (STRAH-tum buh-SAY-leh; *basal* = "bottom"), also known as the *basal layer.* The stratum basale consists primarily of a single layer of stem cells that rests on the basement membrane and appears slightly cuboidal or columnar rather than squamous. The keratinocytes in the stratum basale are closest to the blood supply in the dermis, and so are the most metabolically active keratinocytes of the epidermis. These cells actively produce the precursor to vitamin D when ultraviolet radiation from the sun interacts with specific cholesterol-like steroids in the cells.

The nearby blood supply also allows these keratinocytes to be mitotically active (engaged in cell division), which accounts for the alternate name of the stratum basale: *stratum germinativum* (jer′-min-uh-TYV-um; "germinating" or "generating layer"). All cells in the epidermis come from the mitotic activity of cells in the stratum basale or the layer above it, the stratum spinosum.

Stratum Spinosum Immediately superficial to the stratum basale we find what is normally the thickest stratum of the epidermis: the **stratum spinosum** (spin-OH-sum; *spino-* = "prickly").

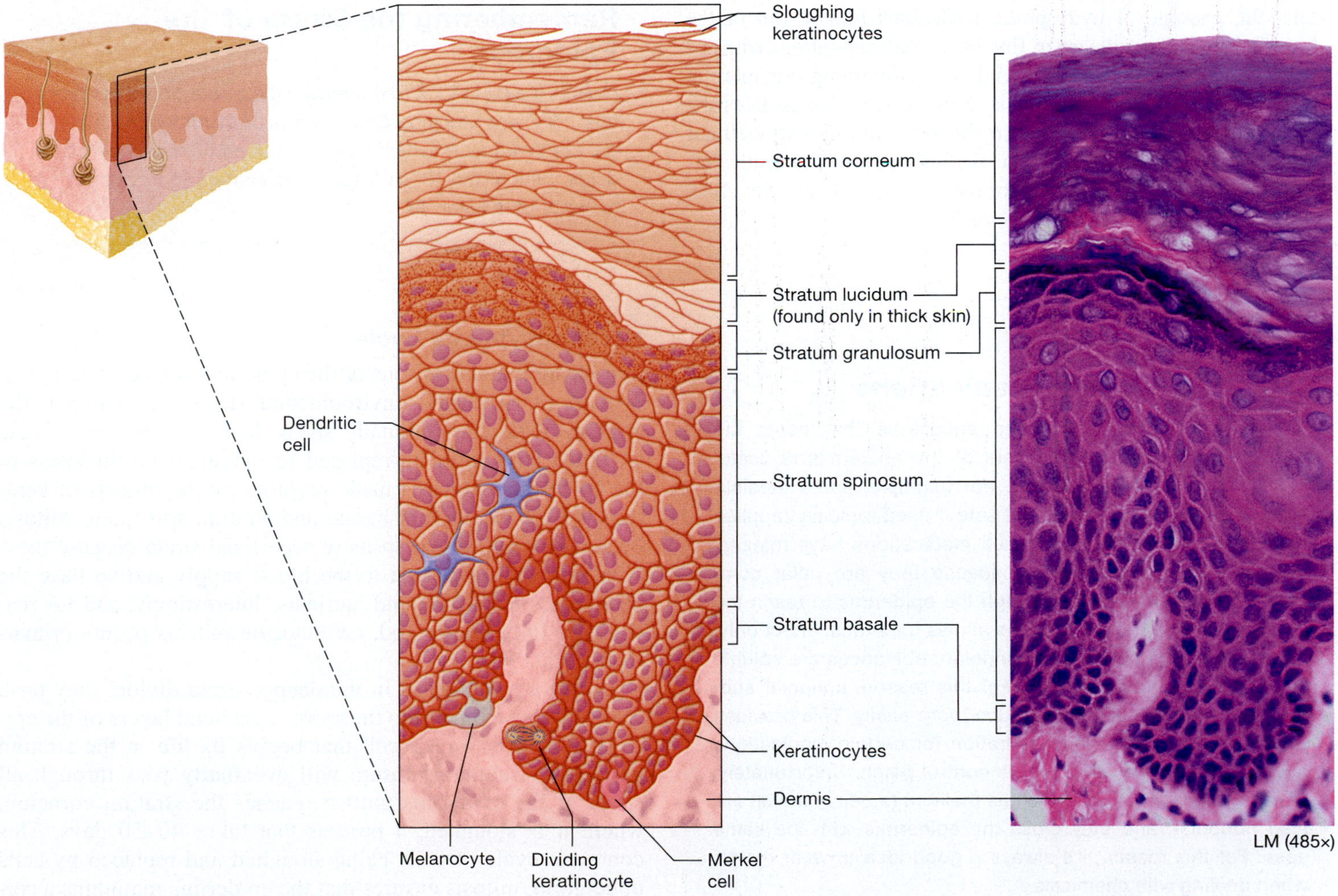

Sloughing keratinocytes

Stratum corneum

Stratum lucidum (found only in thick skin)

Stratum granulosum

Dendritic cell

Stratum spinosum

Stratum basale

Keratinocytes

Dermis

Melanocyte

Dividing keratinocyte

Merkel cell

LM (485×)

Figure 5.3 Structure of the epidermis.

The stratum spinosum, or *prickle cell layer*, is named for the fact that its cells appear spiky due to bundles of cytoskeletal filaments in the periphery of the cells that attach to desmosomes (this appearance is mostly an artifact of preparing them for a microscope slide; in living tissue these cells are not noticeably spiky). Like the cells of the stratum basale, those of the stratum spinosum are near the blood supply in the dermis and so are metabolically and mitotically active; for this reason, they are sometimes classed as part of the stratum germinativum. They also help produce vitamin D.

Stratum Granulosum The middle stratum of keratinocytes, the **stratum granulosum** (gran-yoo-LOH-sum), consists of three to five rows of cells. The stratum granulosum, or *granular layer*, is named for the prominent cytoplasmic granules in its cells. At least two types of granules are present in the cells: One type contains keratin bundles, and the other a lipid-based substance.

Keratinocytes release this lipid-based substance by exocytosis. When released, this substance coats keratinocytes in the stratum granulosum and the more superficial strata of the epidermis. Recall that lipids are hydrophobic, meaning they do not interact with hydrophilic, or water-soluble, substances (see Chapter 2). So this lipid coating acts as a water barrier and prevents the passage of hydrophilic molecules into or out of the skin. As we discussed earlier, this lipid coating—along with the multiple layers of cells—is crucial for maintaining our internal fluid and electrolyte homeostasis. However, it also isolates the cells from water and nutrients in the surrounding extracellular fluid, which promptly kills them. To find out how the lipid coating of these cells can be useful for medication delivery, see *A&P in the Real World: Topical Medications.*

Stratum Lucidum Superficial to the stratum granulosum we find a narrow, clear layer of dead keratinocytes called the **stratum lucidum** (LOO-sih-dum; *lucid* = "clear"), also known as the *clear layer*. This stratum is found only in thick skin, which we discuss shortly.

Stratum Corneum The most superficial stratum of the epidermis is the **stratum corneum** (KOHR-nee-um; *cornu* = "horn"), or *cornified layer*, which consists of several layers of dead, flattened keratinocytes. These cells have lost their normal organelles and are little more than bundles of keratin with thickened plasma membranes. Eventually, the most superficial cells of the stratum corneum also lose their desmosomes, which allows them to be sloughed (shed) or exfoliated mechanically. Every day you shed about a million dead keratinocytes, which amounts to roughly eight pounds of cells over the course of a year. Where do all these dead cells go? Just look around your house for the answer—most of them are visible as dust, which is then ingested by microscopic dust mites.

Topical Medications

The lipid-based substance that coats the superficial layers of the epidermis is useful therapeutically. For example, some medications are toxic if swallowed, but safe if used topically (applied to the surface of the skin). Such medications have minimal risk of systemic absorption because they are polar compounds that cannot pass through the epidermis to reach the blood vessels in the dermis. This allows for a local effect only.

However, hydrophobic or nonpolar substances are soluble in other nonpolar substances. For this reason, nonpolar substances cross the epidermis much more easily. This provides a convenient route of administration for certain medications such as the hormones in a birth control patch. Unfortunately, many poisons and toxins such as thallium (a heavy metal) are also nonpolar, and they cross the epidermis with the same ease. For this reason, it's always a good idea to wear gloves when dealing with chemicals.

Keratinocyte Life Cycle

The location and functions of the epidermis subject it to a great deal of physical and environmental stress. As a result, the stratum corneum continually sheds dead keratinocytes. These sloughed cells must be replaced to maintain the thickness of the epidermis, which is made possible by the mitosis of keratinocytes in the stratum basale and stratum spinosum. Mitosis occurs here rather than in more superficial strata because these keratinocytes are closest to the blood supply and so have the best access to oxygen and nutrients. Interestingly, and for reasons not fully understood, keratinocyte mitosis occurs primarily at night.

As the keratinocytes in the deeper strata divide, they push the cells above them into the more superficial layers of the epidermis. In this way, a cell that begins its life in the stratum basale or stratum spinosum will eventually pass through all strata of the epidermis until it reaches the stratum corneum, where it is sloughed, a process that takes 40–50 days. This continual cycle of cells being sloughed and replaced by cells undergoing mitosis ensures that the epidermis maintains a certain thickness.

ConceptBOOST))))

Understanding Epidermal Growth

As you've just read, a keratinocyte begins in the stratum basale or spinosum, and eventually ends up in the stratum corneum. But how does the cell travel from one stratum to another? The answer is that the cell doesn't actually travel—it is pushed. The original cell in the stratum basale is a stem cell that undergoes mitosis to produce two daughter cells. One daughter cell remains in the stratum basale while the other daughter cell is pushed up through the epidermal layers. In this way the dividing stem cells push the products of mitosis up through the epidermal strata.

To see how this works, suppose for simplicity's sake that the skin has one row of cells in each epidermal stratum, as shown here:

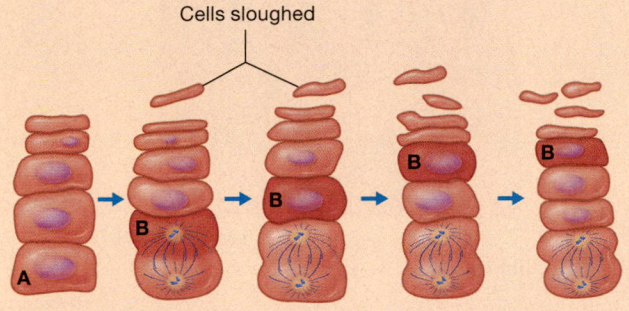

Cells sloughed

We start with a stem cell in the stratum basale, marked cell A. This cell undergoes mitosis, and one of its two daughter cells (cell B in the diagram) is now in the stratum spinosum. The other daughter cell of cell A divides again, producing two more daughter cells, one of which pushes cell B into the stratum granulosum.

Cell B is now quite far from its blood supply and becomes coated with the lipid-based substance, which causes it to die. Stem cells in the stratum basale continue to divide, pushing cell B even farther away from the blood supply, into the stratum lucidum and then into the stratum corneum. Cell B is now a dead cell filled with keratin, and will eventually be sloughed off the skin surface. ■

Quick Check

☐ 1. What are the five strata of the epidermis? How does each stratum differ from the others?

☐ 2. How does a keratinocyte that begins its life in the deepest stratum of the epidermis eventually end up on the epidermal surface?

Other Cells of the Epidermis

◀◀ FLASHBACK

1. What is a phagocyte? (p. 83)

In addition to keratinocytes, three other cell types are scattered throughout the epidermis (see Figure 5.3, p. 165):

- **Dendritic (Langerhans) cells.** In the stratum spinosum we find **dendritic cells,** also known as **Langerhans cells.** Dendritic cells are phagocytes of the immune system that protect the skin and deeper tissues from pathogens.
- **Merkel cells.** Oval cells called **Merkel cells** are scattered throughout the stratum basale. They are sensory receptors associated with small neurons in the dermis, and together they function to detect light touch and differentiate shapes and textures. Merkel cells are especially numerous in body regions specialized for touch, such as the fingertips, lips, and at the base of hairs.
- **Melanocytes. Melanocytes** are cells located in the stratum basale that produce the protein **melanin** (MEL-uh-nin; *melan-* = "black"). Melanin is a skin pigment ranging from orange-red to brown-black. We discuss melanin and skin pigmentation in Module 5.4.

Quick Check

☐ 3. What are the three other types of cells found in the epidermis? What are their functions?

Thick and Thin Skin

As with all structures in the body, the form of the epidermis in various parts of the body differs to match its function in accordance with the Structure-Function Core Principle (p. 28). Certain body locations are subject to a great degree of mechanical stress, including the palms of the hands, the palmar surfaces of the fingers, the soles of the feet, and the plantar surfaces of the toes. In these locations, we find **thick skin,** a type of epidermis about as thick as a paper towel. (Note that this measurement refers to the thickness of the epidermis only, not the skin as a whole.) Thick skin contains all five epidermal strata and a very thick stratum corneum (**Figure 5.4a**). This type of skin lacks hair follicles but contains numerous sweat glands.

CORE PRINCIPLE
Structure-Function

The remaining parts of the body are subject to less mechanical stress, and here we find **thin skin** (**Figure 5.4b**), which is about as thick as a sheet of printer paper. Thin skin lacks the stratum lucidum and so has only four epidermal strata. In addition, its stratum corneum and other strata are much thinner (have fewer layers) than those in thick skin. Thin skin has abundant hairs, sebaceous glands, and sweat glands.

In both thick and thin skin, the stratum corneum may develop additional layers when subjected to repeated pressure. This forms a **callus** (KAL-us), which is sometimes painful due to the added pressure placed on tissues deep to the callus. A callus may be treated by shaving off the excess layers of stratum corneum, but unless the source of the pressure is removed, the callus generally returns.

Quick Check

☐ 4. How do thin and thick skin differ?

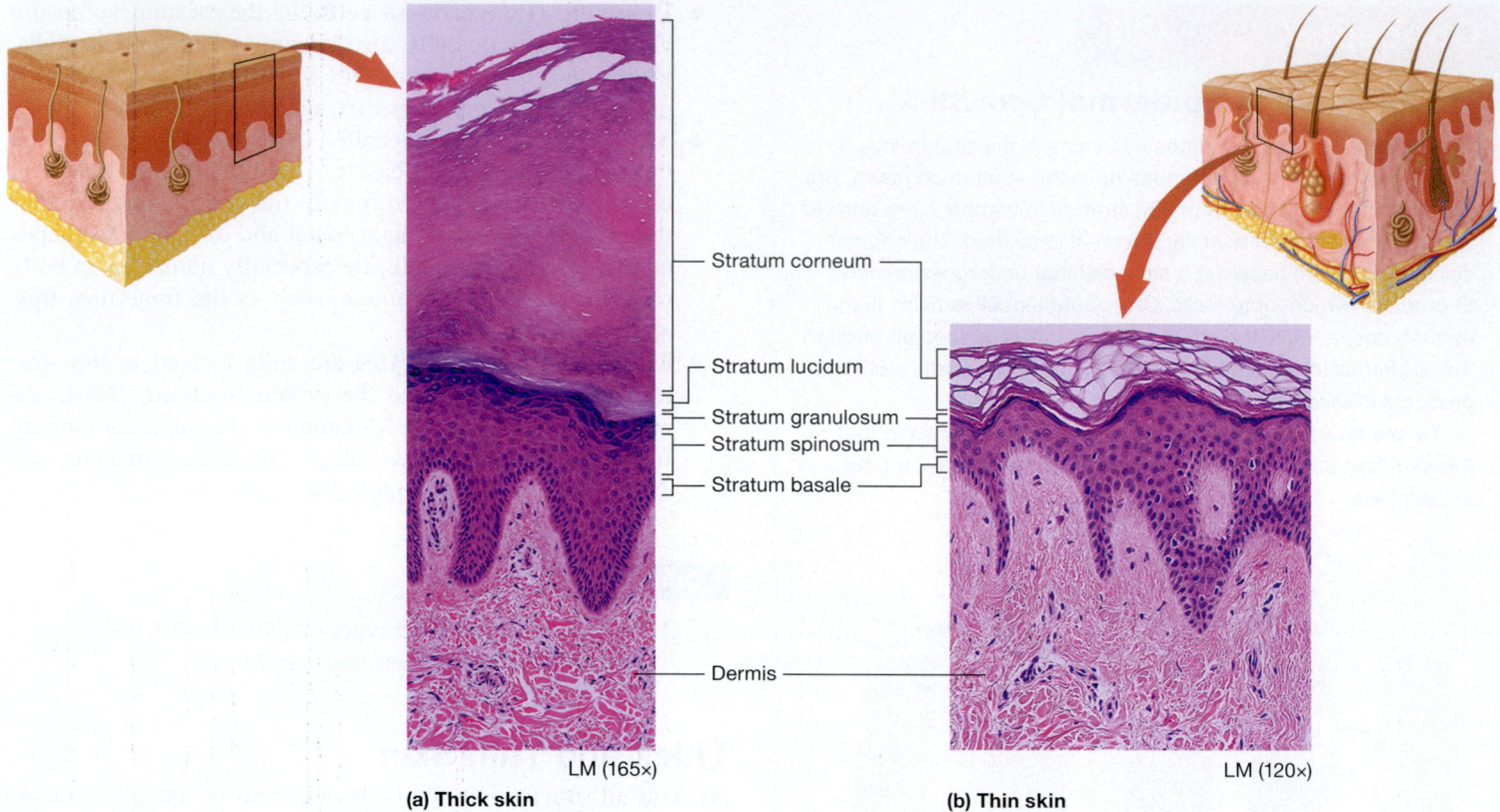

Stratum corneum

Stratum lucidum

Stratum granulosum
Stratum spinosum

Stratum basale

Dermis

LM (165×)

(a) Thick skin

LM (120×)

(b) Thin skin

Figure 5.4 Thick and thin skin.

☐ 1. Paper cuts often involve only the epidermis. You get a paper cut and notice that it doesn't bleed, but you can feel it. Explain why you can feel it but the cut isn't bleeding.

☐ 2. A hypothetical drug causes blood vessels to grow from the dermis into the superficial stratum granulosum of the epidermis. Would this affect the cells in the most superficial epidermal layers? Why or why not?

See answers in Appendix A.

MODULE 5.3
The Dermis

Learning Outcomes

1. Describe the layers and basic structure and components of the dermis.

2. Explain the functions of the dermal papillae.

3. Explain how skin markings such as epidermal ridges are formed.

Deep to the epidermis we find the highly vascular dermis. The dermis serves many functions: It houses the blood supply of the epidermis, contains sensory receptors, and anchors the epidermis in place. As with the epidermis, we can divide the dermis into distinct layers. However, the two layers of the dermis are not simply layers of cells but, rather, two different types of connective tissue.

Papillary Layer

« FLASHBACK

1. What are fibroblasts and ground substance? (p. 124 and p. 138)

2. Define loose connective tissue. (p. 138)

The most superficial layer of the dermis is the thin **papillary layer** (PAP-ih-lehr-ee), which accounts for about 20% of the depth of the dermis (**Figure 5.5**). The papillary layer consists of loose connective tissue. Recall that loose connective tissue is primarily ground substance with collagen, elastic, and reticular fibers (see Chapter 4). Within the papillary layer are the usual cells that we find in connective tissue, including fibroblasts and phagocytes. At the dermis-epidermis junction, special collagen fibers extend up into the basement membrane to anchor the epidermis in place. Repetitive trauma, such as rubbing from a pair of ill-fitting shoes, can disrupt these collagen fibers and separate the epidermis from the dermis. This usually results in a fluid-filled pocket called a *blister.*

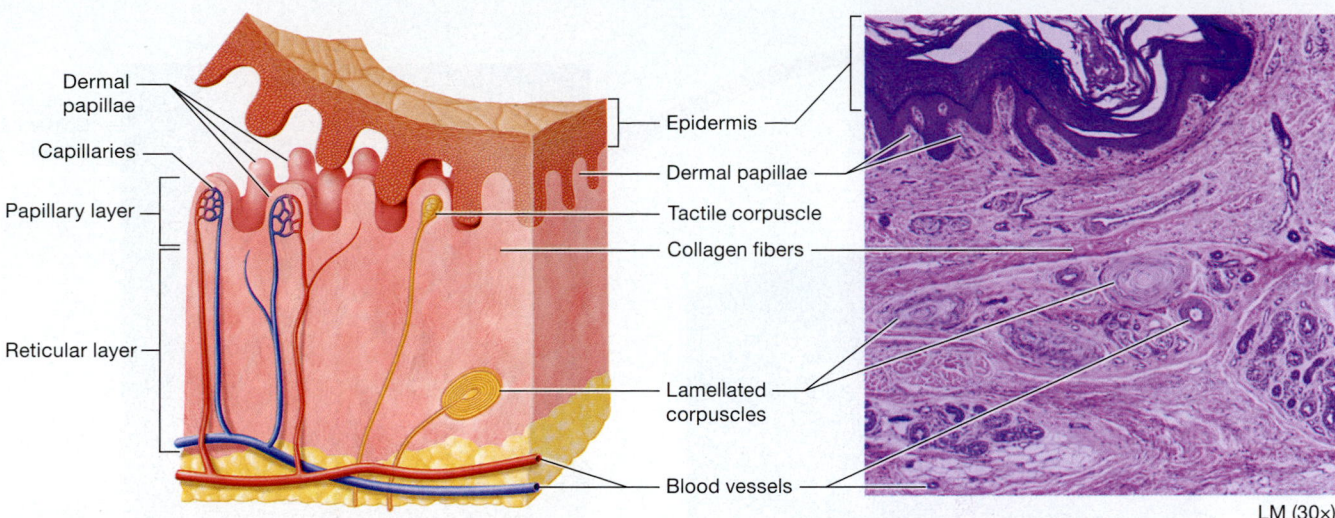

Dermal papillae
Capillaries
Papillary layer
Reticular layer

Epidermis
Dermal papilla
Tactile corpuscle
Collagen fibers
Lamellated corpuscles
Blood vessels

LM (30×)

Figure 5.5 Structure of the dermis.

The surface of the papillary layer where it contacts the epidermis is folded into projections referred to as **dermal papillae** (DER-mul pah-PILL-ee). The dermal papillae house tiny blood vessels called *capillaries* that are arranged into loops. These loops extend up into the superficial part of the dermal papillae, where oxygen and nutrients diffuse into the extracellular fluid of the dermis and up into the cells of the avascular epidermis.

The dermal papillae also house sensory receptors called **tactile (Meissner) corpuscles** (MYS-ner KOHR-pus-ulz), which respond to light touch stimuli that help us distinguish the shape and texture of different objects. Tactile corpuscles are more numerous in areas of the body that have sensation as a primary function, such as the fingertips, lips, face, and external genitalia.

Quick Check

☐ 1. Which type of tissue makes up the papillary layer of the dermis?

☐ 2. What are the functions of the dermal papillae?

Reticular Layer

« FLASHBACK

1. What is dense irregular connective tissue, and what are its properties? (p. 139)

2. What is smooth muscle tissue, and what are its properties? (p. 148)

The deeper and thicker **reticular layer** is composed mainly of dense irregular connective tissue. Note that the reticular layer is continuous with both the papillary layer and the hypodermis, which makes it difficult to distinguish a clear boundary between them. Recall that dense irregular connective tissue consists largely of irregularly arranged bundles of

collagen fibers (see Chapter 4). Many more fibers are found in the reticular layer than in the papillary layer, and the fibers of the reticular layer are bundled together. These collagen fibers strengthen the dermis and help prevent traumatic injuries from reaching deeper tissues.

We also find elastic fibers in the reticular layer of the dermis, which enable the dermis to return to its original shape and size after being stretched. Note that if stretching is extreme, such as that occurring with pregnancy or significant weight gain, the elastic fibers may tear, forming *striae* (STRY-ee), or stretch marks, that are generally permanent. The reticular layer also contains ground substance rich in proteoglycans, as well as cells such as fibroblasts. The proteoglycans draw water into the ground substance, which hydrates the skin and makes it appear firm. In addition, small pockets of adipose tissue are scattered throughout the reticular layer, although most of the adipose tissue is confined to the hypodermis.

Embedded within the reticular layer we find blood vessels and some accessory structures, including sweat glands, hairs, sebaceous glands, and a number of sensory receptors. One prominent sensory receptor is the **lamellated**, or **Pacinian, corpuscle** (puh-SIN-ee-in; see Figure 5.5). Lamellated corpuscles are named for their layered appearance (*lamell-* = "layer") and resemble cut onions when viewed in cross section. These receptors respond mainly to changes in deep pressure and vibration applied to the skin.

Quick Check

☐ 3. Which type of tissue makes up the reticular layer of the dermis?

☐ 4. What other structures are located in the dermis?

Skin Markings

Interactions between the dermis and the epidermis are visible as small lines in the epidermis known as *skin markings*. Some of the most obvious skin markings are seen in thick skin on the

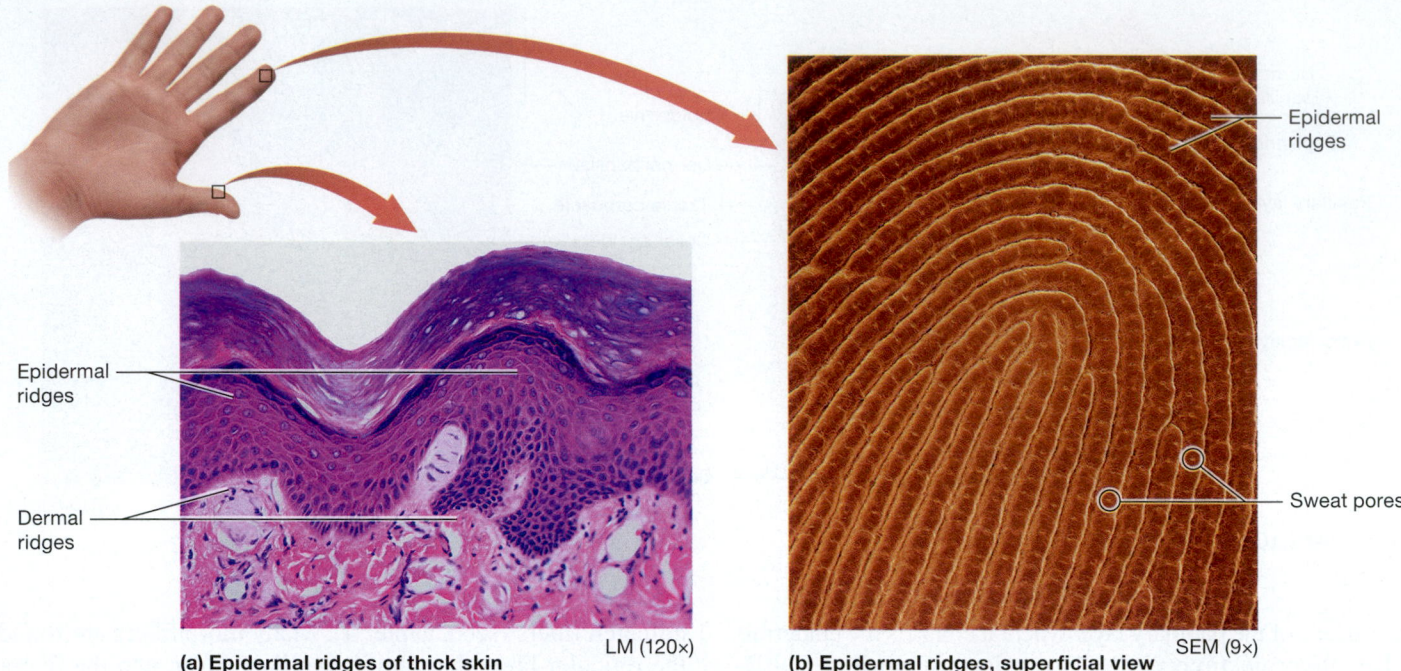

(a) Epidermal ridges of thick skin   LM (120×)

(b) Epidermal ridges, superficial view   SEM (9×)

Figure 5.6 Epidermal ridges and fingerprint patterns.

palmar surfaces of the hands and fingers and the plantar surfaces of the feet and toes. Here the dermal papillae are extremely prominent, and thick collagen fibers arrange them into *dermal ridges* (**Figure 5.6**). This structure indents the overlying epidermis and produces **epidermal ridges** that enhance the gripping ability of the hands and feet.

Epidermal ridges occur in characteristic patterns such as a *loop, arch,* and *whorl* that are genetically determined and unique to each individual. Notice in Figure 5.6 that tiny *sweat pores,* where sweat exits from sweat glands, open along the ridges and leave a thin film called a **fingerprint** when we touch most surfaces. (The palms, toes, and soles of the feet also leave characteristic prints.)

The reticular layer is also responsible for certain skin markings. Throughout the dermis, gaps between the bundles of collagen indent the epidermis, forming lines called **tension,** or **cleavage, lines.** Tension lines normally run in a circular pattern in the neck and trunk, but longitudinally in the head and upper and lower limbs. **Figure 5.7** shows why they are important clinically: If a skin wound is parallel to tension lines, its edges tend to stay closed and heal better with less risk of scarring. However, a wound that runs perpendicular to tension lines tends to gape and is more likely to form a scar. This is why surgeons plan incisions to be as parallel as possible to tension lines.

The final type of skin marking also is a product of the reticular layer. In certain locations—for example, around joints—the reticular layer is anchored tightly to deeper structures. This creates deep creases known as **flexure lines.** Flexure lines are pronounced in areas such as the palms and on the posterior

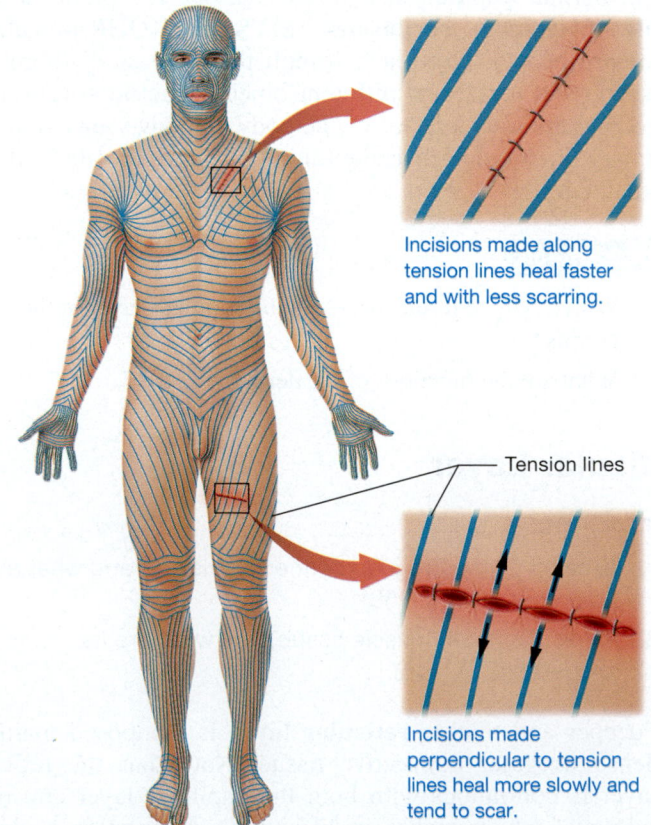

Incisions made along tension lines heal faster and with less scarring.

Tension lines

Incisions made perpendicular to tension lines heal more slowly and tend to scar.

Figure 5.7 Importance of tension lines for surgical incisions.

surface of the fingers and toes. Another product of the reticular layer of the dermis is explored in *A&P in the Real World: Skin Wrinkles.*

See answers in Appendix A.

Quick Check

☐ 5. How does the papillary layer of the dermis affect the appearance of the epidermis?

☐ 6. What causes tension lines and flexure lines? How do they differ?

Apply What You Learned

☐ 1. While woodworking, your friend Lily cuts her forearm with a saw. You examine the wound and see large quantities of adipose tissue at its base. Does her wound extend down to the epidermis, dermis, or hypodermis? How do you know?

☐ 2. Lily wants to know if the wound is likely to form a scar. The wound is located on her left anterior forearm and cuts straight across the width of her arm. What do you tell her, and why?

Skin Wrinkles

A hallmark of aging is the formation of *skin wrinkles,* which many individuals consider the least desirable type of skin marking. Wrinkles are due to an age-related decrease in collagen fibers, elastic fibers, proteoglycans, and adipose tissue in the dermis. This reduces the skin's firmness, hydration, and recoil ability after stretching. Age-related wrinkles tend to be deeper in areas of repetitive muscle movement, such as in the forehead and around the eyes and mouth. Exposure to UV radiation (both from natural sunlight and tanning beds) and cigarette smoking accelerate the formation of wrinkles.

The appearance of wrinkles can be minimized with products and procedures such as the following:

- *BOTOX* contains a bacterial toxin that temporarily paralyzes facial muscles for 4–6 months, causing the skin over them to appear smoother.
- *Fillers* contain adipose tissue, collagen, and/or proteoglycans that are injected into wrinkles to temporarily "fill" them.
- *Peels* use lasers, chemicals, or abrasion to remove varying amounts of the epidermis and superficial dermis to cause the formation of new, hopefully firmer skin.
- *Topical creams* contain a variety of ingredients that claim to reduce the appearance of wrinkles. These creams, particularly nonprescription products, have little or no effect on wrinkles. Such creams are classified as cosmetics rather than drugs, which means they're not required to undergo testing for safety and efficacy.

Of course, the simplest—and least expensive—way to minimize the appearance of wrinkles is to delay and reduce their formation in the first place: avoid the sun, use sunscreen, stay well hydrated, and don't smoke cigarettes.

MODULE
5.4
Skin Pigmentation

Learning Outcomes

1. Explain how melanin is produced and its role in the integument.

2. Describe the other pigments that contribute to skin color.

3. Explain how skin coloration may indicate pathology.

Skin color is one of the most readily identifiable physical features in humans—so much so that it is used as a basis for individuals to form groups and for one individual or group to discriminate against others. But from a physiological standpoint, human skin color is little more than the amount of a pigment called *melanin* produced in the epidermis. In this module, we examine the cells that manufacture melanin, its functions, and how it influences skin color. We also discuss two other pigments that impact skin color in minor ways: *hemoglobin* and *carotene.* Finally, we explore how changes in skin color can serve as diagnostic clues to various health conditions.

Melanin

◀◀ FLASHBACK

1. What are amino acids? (p. 56)

2. What is DNA, and what are DNA mutations? (p. 60 and p. 104)

The primary skin pigment is **melanin,** which, as you read earlier, is produced by melanocytes in the stratum basale. Melanin ranges in color from orange-red to black. It is composed of two molecules of the amino acid tyrosine joined by a series of reactions catalyzed by the enzyme *tyrosinase.* These reactions occur in vesicles within the melanocyte known as **melanosomes** (meh-LAN-oh-sohmz), shown in step ① of **Figure 5.8.**

You can see that a melanocyte somewhat resembles an octopus, with numerous "arms" that reach out and contact or even pierce the plasma membranes of neighboring keratinocytes. Notice in step ② of the figure that within the melanocytes, melanosomes migrate to the tips of these arms, where they are released into or near a neighboring keratinocyte. The

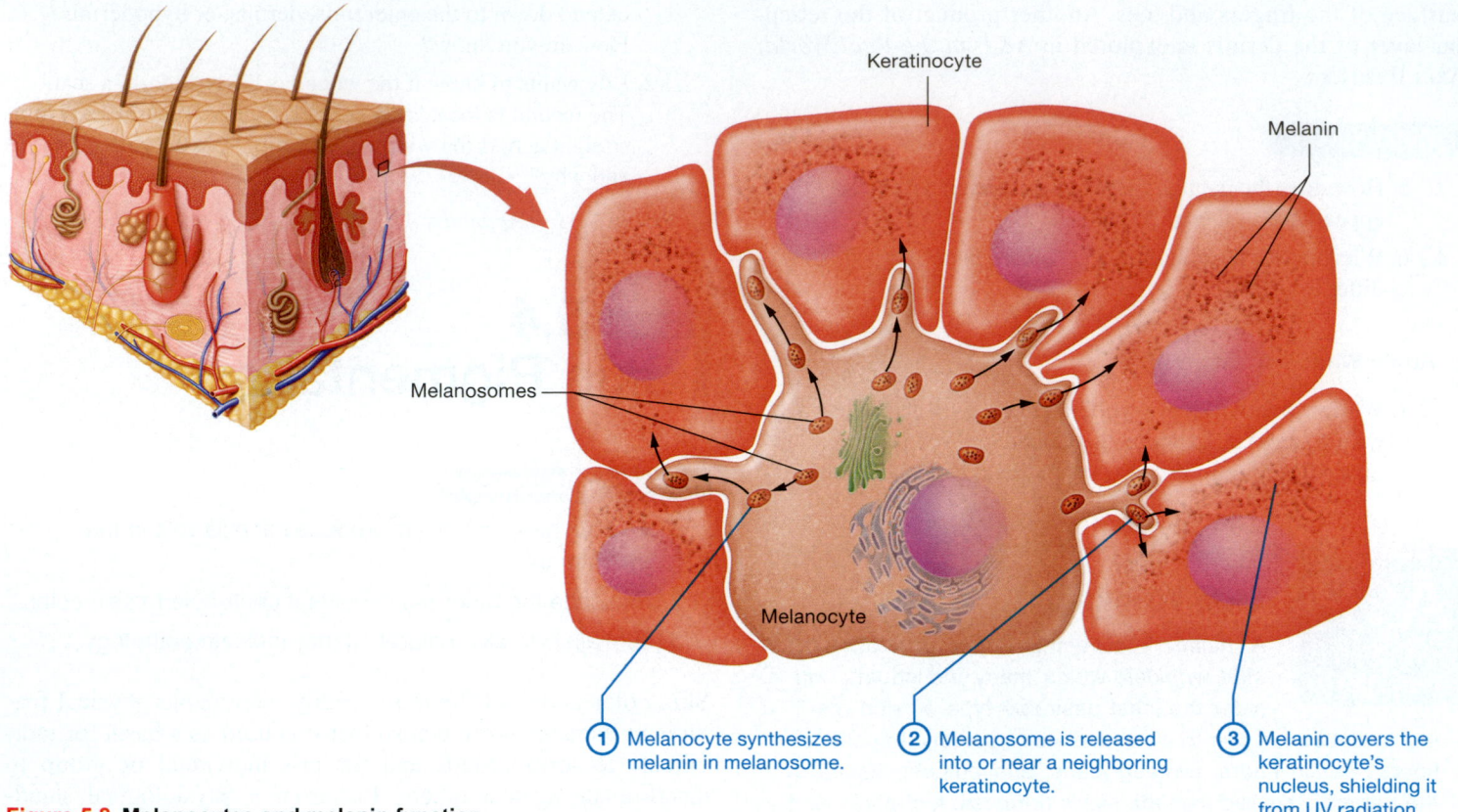

Keratinocyte

Melanin

Melanosomes

Melanocyte

1 Melanocyte synthesizes melanin in melanosome.

2 Melanosome is released into or near a neighboring keratinocyte.

3 Melanin covers the keratinocyte's nucleus, shielding it from UV radiation.

Figure 5.8 Melanocytes and melanin function.

keratinocyte then takes in the melanosome by phagocytosis, after which the melanin is transported to the region superficial to the nucleus, where it shields the DNA of the keratinocyte like an umbrella, shown in step . The pigment degrades after a few days, so melanocytes must continually produce new melanosomes to maintain a specific skin color.

Melanin synthesis increases upon exposure to UV radiation from the sun or sunlamps. This leads to *tanning,* or darkening of skin pigmentation. UV radiation has both immediate and delayed effects on skin pigmentation. Its immediate action is to oxidize the melanin already formed within the keratinocytes, which causes the melanin to rapidly darken. UV radiation also damages the DNA of the melanocytes, which increases the production of additional melanin. These secondary effects become visible about 72 hours after exposure but last longer than melanin oxidation. To find out if there is any such thing as a "healthy tan," see *A&P in the Real World Pseudoscience Exposed: Tanning and a "Healthy Tan."*

One of the functions of melanin is to protect the DNA of keratinocytes from mutations induced by UV radiation. Recall from the cell chapter that such mutations may form cancerous cells (see Chapter 3). Note, however, that melanin does not provide complete protection from DNA damage—it absorbs some UV radiation, but not all of it, especially if the radiation is intense. For this reason, people of all skin pigmentations can develop skin cancer (we discuss skin cancer in Module 5.6). Melanin also does not provide much protection from burns due to UV radiation, so individuals of all skin colors can sunburn.

A second function of melanin is to decrease the synthesis of vitamin D in response to UV radiation. This may seem counterintuitive—after all, why would the body want to produce *less*

A&P in the Real World

Pseudoscience Exposed: Tanning and a "Healthy Tan"

Skin tanning is big business—in fact, a $5-billion-a-year business in the United States alone. In the past decade, the number of tanning salons has soared from about 10,000 to 50,000, with most of these salons promoting the notion of a "healthy tan." But is there such a thing as a healthy tan?

Two types of UV rays reach Earth's surface: UVA and UVB rays. UVB rays are associated with producing sunburn, whereas UVA rays are linked with tanning. This has led tanning salons to claim that UVA rays are safe and will not damage skin. But the mechanism of increased melanin production is the same for both types of rays—oxidative damage to existing melanin and DNA damage to melanocytes that can lead to skin cancer. Not only does UVA radiation do just as much DNA damage as UVB radiation, but UVA actually ages the skin at a much faster rate, as well.

So can you get a healthy tan? The answer is no—any amount of tanning damages melanocytes and other elements of the skin, ages the skin prematurely, and increases the risk for skin cancer.

vitamin D? The answer is that vitamin D levels must be kept within a specific range, as excess amounts can lead to calcium ion imbalance and kidney failure.

In fact, the role of melanin in vitamin D synthesis may explain the evolution of different skin pigmentation in different parts of the world. It was advantageous for individuals living in regions exposed to high amounts of UV radiation, such as Africa, to have darker skin to prevent excess vitamin D production. Conversely, it was advantageous for people in areas with less UV radiation, such as northern Europe, to have lighter skin so they could synthesize enough vitamin D.

As you have probably noticed, skin pigment isn't evenly distributed throughout the body. The unequal distribution of melanocytes largely accounts for this. For example, fewer melanocytes are located on the palms of the hands and the insides of the arms and thighs. This distribution typically gives these areas a lighter color compared to parts of the body with more melanocytes, such as the face. Pigment is also sometimes concentrated in one spot. This may produce a **freckle,** which is due to localized higher melanin production, or a **mole** (also known as a **nevus** [NEE-vus]), which results from a local proliferation of melanocytes.

Interestingly, the number of melanocytes is virtually identical among all individuals, irrespective of skin color. What, then, causes the wide variety of skin colors in humans? This spectrum of skin tones is due to differences in the amount of tyrosinase activity and the type of melanin produced. This is evident from *albinism,* a condition in which melanocytes fail to manufacture tyrosinase. Albinism results in lack of skin pigmentation and a greatly increased risk of keratinocyte DNA damage from UV radiation.

Quick Check

☐ 1. How is melanin produced, and how does it interact with keratinocytes?

☐ 2. What are the functions of melanin?

Other Pigments That Affect Skin Color: Carotene and Hemoglobin

❰❰ FLASHBACK

1. What is an oxidation reaction? (p. 43)

Although melanin is the major skin pigment, two other minor contributors to skin color are *carotene* and *hemoglobin.* Normally, humans ingest the yellow-orange pigment **carotene** (KEHR-uh-teen) in their diet, from egg yolks and yellow and orange vegetables such as carrots. Carotene is a lipid-soluble compound that tends to accumulate in the stratum corneum. This imparts a slight yellow-orange color to the skin that is particularly visible in thick skin, where the stratum corneum has the most layers of cells.

Hemoglobin (HEE-moh-gloh-bin) is an iron-containing protein found within our red blood cells that binds oxygen and transports it through the body. The interaction of hemoglobin

with oxygen gives blood its characteristic red color. How does a blood protein affect the color of skin? The answer is that it doesn't affect the actual color, but, rather, the apparent skin color. Recall that the dermis is highly vascular, so a great deal of the body's blood resides in the dermis at any given time. In fair-skinned individuals, the epidermis is partly translucent, so we are able to look through the epidermis to see the blood in the vessels of the dermis. Under normal conditions, this gives the skin a faint pinkish hue. In regions where the stratum corneum is thinner, such as the lips, the pink is generally darker.

Quick Check

☐ 3. What is carotene, and what color does it give to the skin?

☐ 4. How does hemoglobin affect skin color?

Skin Color as a Diagnostic Tool

The skin color imparted by hemoglobin may change based on the amount of blood flowing through the dermis. The skin becomes brighter red when blood flow through the dermis increases, such as when the body is trying to release heat during exercise, a color change called **erythema** (ehr-uh-THEE-muh; *eryth-* = "red"). We also see erythema when the body is injured by trauma (e.g., breaking a bone) or infection, as blood flow increases to the injured area.

Conversely, the pinkish hue becomes faint and nearly disappears when blood flow through the dermis decreases, such as when the body is trying to conserve heat. In extreme cases, so little blood is flowing through the skin that collagen in the dermis is visible, which gives the skin a white color termed **pallor** (PAL-ur). Pallor is especially visible in lighter-pigmented skin.

The color resulting from hemoglobin also depends on how much oxygen is bound to it. When high amounts are bound, the interaction between oxygen and hemoglobin causes it to assume a brighter red color. This effect of oxygen on the color of hemoglobin is particularly obvious if a person is exposed to the poison *cyanide,* which prevents oxygen from leaving hemoglobin and entering the tissues. This causes the cheeks of cyanide poisoning victims have a characteristic "cherry" color due to the bright red color of the highly oxygen-bound hemoglobin.

However, when extremely low amounts of oxygen are bound to hemoglobin, it turns a darker red-purple color. This causes the skin to take on a faint bluish color known as **cyanosis** (sy-uh-NOH-sis; *cyan-* = "blue"). Skin may appear cyanotic when an individual has breathing difficulties, when hemoglobin and red blood cell levels are too low, or when hemoglobin is unable to bind to oxygen. Cyanosis is a sign that the person is in trouble and needs immediate assistance.

Quick Check

☐ 5. How can the oxygen content of the blood affect the color of skin?

☐ 6. What is cyanosis, and what can it tell us about a person's health? Be specific.

☐ 1. *Vitiligo* (vit-ih-LY-goh) is a condition characterized by death or dysfunction of scattered groups of melanocytes in the skin. Predict how vitiligo may cause the skin to appear.

☐ 2. Max and Breanna go to the beach, and Max tells Breanna she does not need sunscreen because she is African American and individuals with dark skin cannot get sunburns or skin cancer. What should Breanna tell him in response?

☐ 3. A hypothetical poison prevents oxygen from binding to hemoglobin. Predict how this would affect skin color in a fair-skinned person.

See answers in Appendix A.

MODULE **5.5**

Accessory Structures of the Integument: Hair, Nails, and Glands

Learning Outcomes

1. Describe the structure and function of hair and nails.
2. Explain the process by which hair and nails grow.
3. Summarize the structural properties of sweat and sebaceous glands.
4. Explain the composition and function of sweat and sebum.

Let's now turn our attention to the **accessory structures** or **appendages** of the integument: *hair, nails,* and *glands.* Each of these structures derives primarily from a single tissue type—epithelium—and assists the integument in performing its overall functions. First we explore the structure and function of hair and nails, which are both derived from the epidermis. Then, we focus on the two types of glands in the skin: *sweat glands* and *sebaceous glands.*

Hair

« FLASHBACK

1. What type of tissue makes up the epidermis? (p. 161)

Like most mammals, humans have **hair,** or **pili**—small, filamentous structures that project from the surface of the skin over the entire body except the regions with thick skin, the lips, and parts of the external genitalia. The hair of most animals assists with thermoregulation, but human hair over the majority of the body is too sparse to play any significant role in this task. However, it does help the integument perform its protective functions. For example, hairs around the eyes prevent the entrance of foreign objects, hairs inside the nose keep us from inhaling insects and harmful particles, and hairs on the head protect the body from UV radiation and mechanical trauma. Hairs also participate in sensation because they are associated with small sensory neurons to detect changes in the environment.

Hair is derived from the epidermis, and so consists of squamous keratinized epithelial cells. However, in its structure and the manner in which it grows, hair differs from the epidermis. In the next three subsections, we look more closely at the structure of hair, the process by which it grows, and the variations in hair texture and color among individuals.

Hair Structure

A hair is composed of two main parts: the **shaft,** or the portion projecting from the skin's surface, and the **root,** the portion embedded in the dermis (**Figure 5.9**). Both the shaft and the root are composed of columns of keratinized epithelial cells, but these cells are at different stages of development in each location. The cells in the shaft have completed their keratinization process and are all dead. However, many cells in the root are still living and undergoing keratinization. A small population of keratinocytes called the **matrix** at the base of the root actively undergoes mitosis.

The root is embedded in the **hair follicle.** Notice in Figure 5.9 that the hair follicle is simply an infolding of the epidermis known as the **epithelial root sheath,** which extends down into the dermis or hypodermis. This epithelial root sheath has two components: The outer component anchors the follicle to the dermis, and the inner component is anchored tightly to the hair root. In fact, when a hair is pulled out, this inner root sheath is usually visible as a shiny, clear covering around the base of the hair.

Surrounding the epithelial root sheath is a *dermal root sheath,* which consists of connective tissue that supports the follicle and separates it from the dermis. Tiny bands of smooth muscle called **arrector pili** muscles attach to the dermal root sheath on one end and the dermal papillary layer on the other end. When an arrector pili muscle contracts, the hair stands up, which is known as **piloerection,** and the skin where the muscle attaches to the papillary layer indents slightly. This happens when we are cold or frightened and gives the skin a dimpled appearance we call "goosebumps."

Near its bottom, the root enlarges to form the **hair bulb.** At the base of the hair bulb, the epithelial root sheath folds inward to allow a projection of the dermis known as the **hair papilla** to abut the cells of the hair matrix. The hair papilla houses numerous capillaries that provide oxygen and nutrients to the dividing cells.

On a transverse section (see Figure 5.9), we can see that a strand of hair has three regions:

- The inner *medulla* is present only in thick hairs, such as those on your head, and is the soft core that consists of keratinocytes with *soft keratin,* the same type of keratin found in the epidermis.
- The middle *cortex* is a highly structured and organized region with several layers of keratinocytes that contain *hard keratin.* These keratinocytes do not flake the way keratinocytes containing soft keratin do. This is advantageous

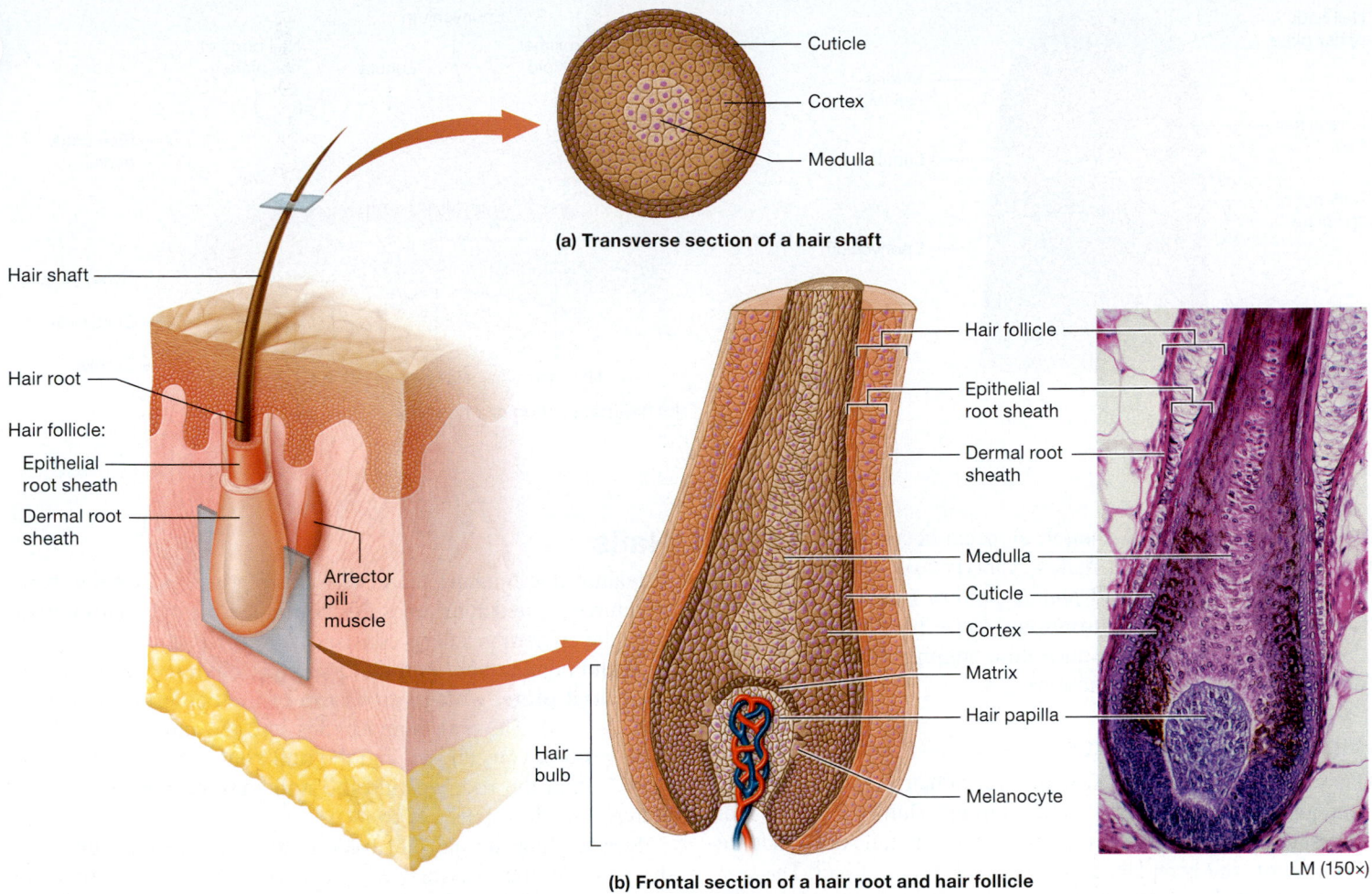

(a) Transverse section of a hair shaft

(b) Frontal section of a hair root and hair follicle

LM (150×)

Figure 5.9 Hair structure.

because if the cells of the cortex flaked off like the superficial layers of the epidermis, the hair strands would quickly whittle down and break.

- The outer *cuticle,* the most superficial layer, consists of a single layer of keratinocytes arranged in an overlapping manner, much like shingles on a roof. Like the cells of the cortex, those of the cuticle contain hard keratin, enabling the cuticle to compact the underlying layers and giving it mechanical strength to resist abrasion. Over time, the cuticle can become worn, particularly near the ends of a hair strand, which allows the cortex and medulla to fray, resulting in "split ends."

Hair Growth

Hair grows at different rates for different individuals, but averages about 1–1.5 cm per month. This growth is not continuous but occurs via a cycle with two main phases: the *growth stage* and the *resting stage.*

During the growth stage, mitosis occurs within the cells of the matrix in the root. As the cells divide, those above them are pushed farther away from the blood supply, keratinize, and die, in a similar manner to what takes place in the epidermis. The

growth stage lasts anywhere from about a month in the eyelashes to as long as 6 years in the scalp.

During the resting stage, the cells in the matrix cease dividing and start to die. The follicle shortens somewhat and the hair is pushed up near the surface of the follicle, which remains dormant for one to several months. At this point the hair may fall out on its own, or a new hair may push it out when the growth stage resumes.

The duration of the growth and resting stages determines both the rate at which a hair grows and its maximal length. For example, a scalp hair will typically be in the growth stage for about six years. If the rate of mitosis in the matrix is rapid during the growth stage, this hair could grow to be nearly a meter long during that 6-year time span. Even if the rate of mitosis is slow, hair length can reach 250–500 cm. In contrast, the growth stage of an eyelash is only about 30 days. This limits the length of an upper eyelash to about 11 mm and a lower eyelash to about 7 mm, again depending on the rate of mitosis in the matrix.

Hair grows most rapidly from puberty to the fifth decade of life, after which its growth slows, a factor influenced by both environment and genetics. Gradually, hairs begin to fall out faster than they grow, which leads to thinning hair with possible

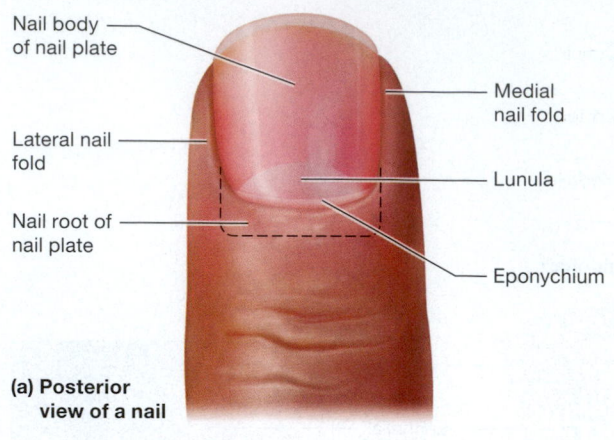

(a) **Posterior view of a nail**

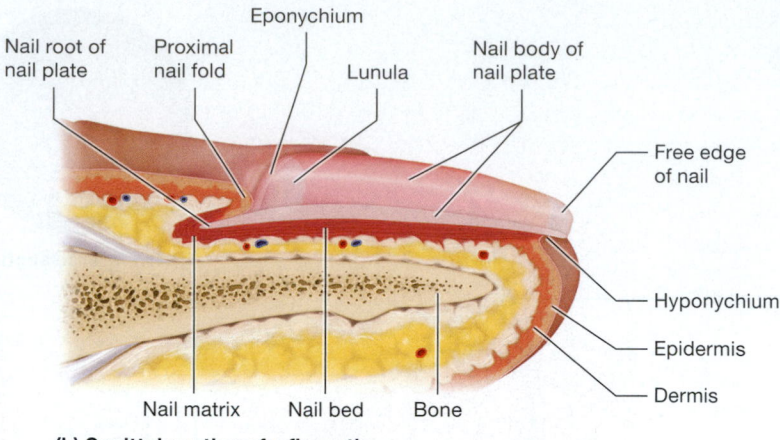

(b) **Sagittal section of a fingertip**

Figure 5.10 Nail structure.

alopecia, or baldness. Although alopecia occurs in both sexes, it is more pronounced in males. Subtypes of alopecia include *male pattern baldness* and *female pattern baldness,* in which hormones, particularly testosterone, cause hair follicles to die. The remaining follicles produce thin, nonpigmented *vellus hair* (discussed in the next subsection).

Hair Pigment and Texture

Hair color and texture vary with the type of hair. A fetus has a thin, nonpigmented type of hair called **lanugo** (luh-NOO-goh) that covers nearly the entire body. Lanugo typically falls out about the time of birth and is replaced by one of two types of hair. The first is **terminal hair,** which is thicker, coarser, and pigmented. We find terminal hair on the scalp and around the eyes. The other type is **vellus hair** (VELL-us), which is thinner and nonpigmented, and is found over the remainder of the body. After puberty, much of the vellus hair on the body is replaced with terminal hair. This is determined by sex—terminal hair replaces about 90% of the vellus hair in males, but only about 35% in females.

Like skin color, hair color is largely determined by the pigment melanin. Melanocytes located in the matrix manufacture melanin that interacts with keratinocytes just as it does in the epidermis. The melanocytes in blonde hair produce very little melanin, whereas those in darker hair produce progressively more melanin. Red hair has a special reddish pigment that contains iron. As we age, the melanocytes produce less melanin, and the growing hair gradually loses pigment, turning gray or white. The myth of a person's hair turning white overnight from fear is technically impossible—even if all melanocytes stopped producing pigment simultaneously, the existing hair would already have color and would need to grow out before the hair appeared white.

Quick Check

☐ 1. How do the hair shaft and hair root differ?

☐ 2. How does a hair grow in length?

☐ 3. What influences hair color and texture?

Nails

A feature that humans share with other primates is **nails,** hard structures located at the ends of our digits. Like hair, nails derive from the epidermis and so consist of stratified squamous epithelium and contain hard keratin. The most obvious part of a nail is the **nail plate,** which sits on top of the epidermal **nail bed** (**Figure 5.10**). The nail plate divides into the **nail body,** which is the portion you can see, and the **nail root,** which is under the skin. Within the root we find the **nail matrix,** which contains actively dividing cells.

Surrounding the nail are folded areas of skin that reinforce the nail. On the proximal edge covering the root we find the **proximal nail fold.** The most distal part of the proximal nail fold, called the **eponychium** (ep-oh-NIK-ee-um) or *cuticle,* consists only of its stratum corneum. On the medial and lateral edges of the nail are the **medial** and **lateral nail folds,** respectively. The distal end of the nail is attached to the underlying nail bed by an accumulation of stratum corneum referred to as the **hyponychium.**

Nail growth occurs at the nail matrix, where actively dividing cells push those above them distally. As with a hair, the keratinocytes making up a nail die and keratinize as they are pushed farther from the blood supply. Fingernails grow an average of 0.5 mm per week, whereas toenails grow more slowly. Contrary to a popular urban legend, nails do not continue to grow after death. Instead, the tissue around the nails shrinks, which makes the nails appear longer.

Unlike hair, nails contain no melanocytes and so are generally translucent. The translucent nature of the nails makes them a good indicator of health: Nail beds are pinkish in a well-oxygenated individual, but bluish (cyanotic) in a poorly oxygenated individual. Certain nails contain a slightly more opaque region near the proximal nail fold, called the **lunula** (LOON-yoo-luh), named for its half-moon shape. The lunula is on top of the distal part of the nail matrix where it is thick and the cells have accumulated keratin.

The primary function of nails is to safeguard the underlying tissue. Their hardness protects the distal fingers and toes from trauma. Nails also act as tools, enabling more precise gripping than with fingertips alone.

One of the most common nail conditions is an *ingrown nail.* Ingrown nails most often occur in the toenails, and are caused by factors such as poorly fitting shoes that pinch the nails and cause them to penetrate the tissue lateral or medial to the nail. This results in inflammation, pain, softening of the nail, and often infection. An ingrown nail is managed by removing part or all of the nail, and occasionally the nail matrix is treated with chemicals to kill its cells and prevent regrowth of the nail.

Quick Check

☐ 4. Define the following terms: nail bed, nail plate, nail matrix, and proximal nail fold.

☐ 5. How does nail growth occur?

Glands

《 FLASHBACK

1. What are exocrine glands? (p. 134)
2. How do holocrine and merocrine glandular secretion differ? (p. 134)

The skin has two basic kinds of glands: **sweat (sudoriferous) glands** (soo′-doh-RIF-er-us), which form *sweat,* and **sebaceous glands** (suh-BAY-shus), which produce *sebum* or oil. Both types of glands are derived from epithelial cells in the epidermis but are located in the dermis. Let's look at the properties of sweat and sebaceous glands and their products.

Sweat Glands

The body contains four types of sweat glands: eccrine, apocrine, ceruminous, and mammary. These glands have different structures and release different products. However, in all four types the secretory cells of the gland release their product by exocytosis, a type of secretion called *merocrine secretion.*

Eccrine Sweat Glands **Eccrine sweat glands** (EK-rin), the most prevalent type of sweat gland, are simple coiled tubular glands situated in the dermis (**Figure 5.11a**). Sweat exits the gland through a duct that passes from the dermis into the epidermis, and leaves the duct through a tiny **sweat pore** in the epidermis (note that these are not the same "pores" created by the infolding of hair follicles).

The sweat released by eccrine sweat glands is 99% water, with small amounts of solutes—including electrolytes such as sodium, potassium, and chloride ions—and waste products such as lactic acid. Sweat also contains antimicrobial compounds and has a slightly acidic pH, both of which deter the growth of pathogens. Although sweat functions in excretion and protection, its primary function is thermoregulation. Recall that when sweat evaporates it takes heat with it, reducing body temperature.

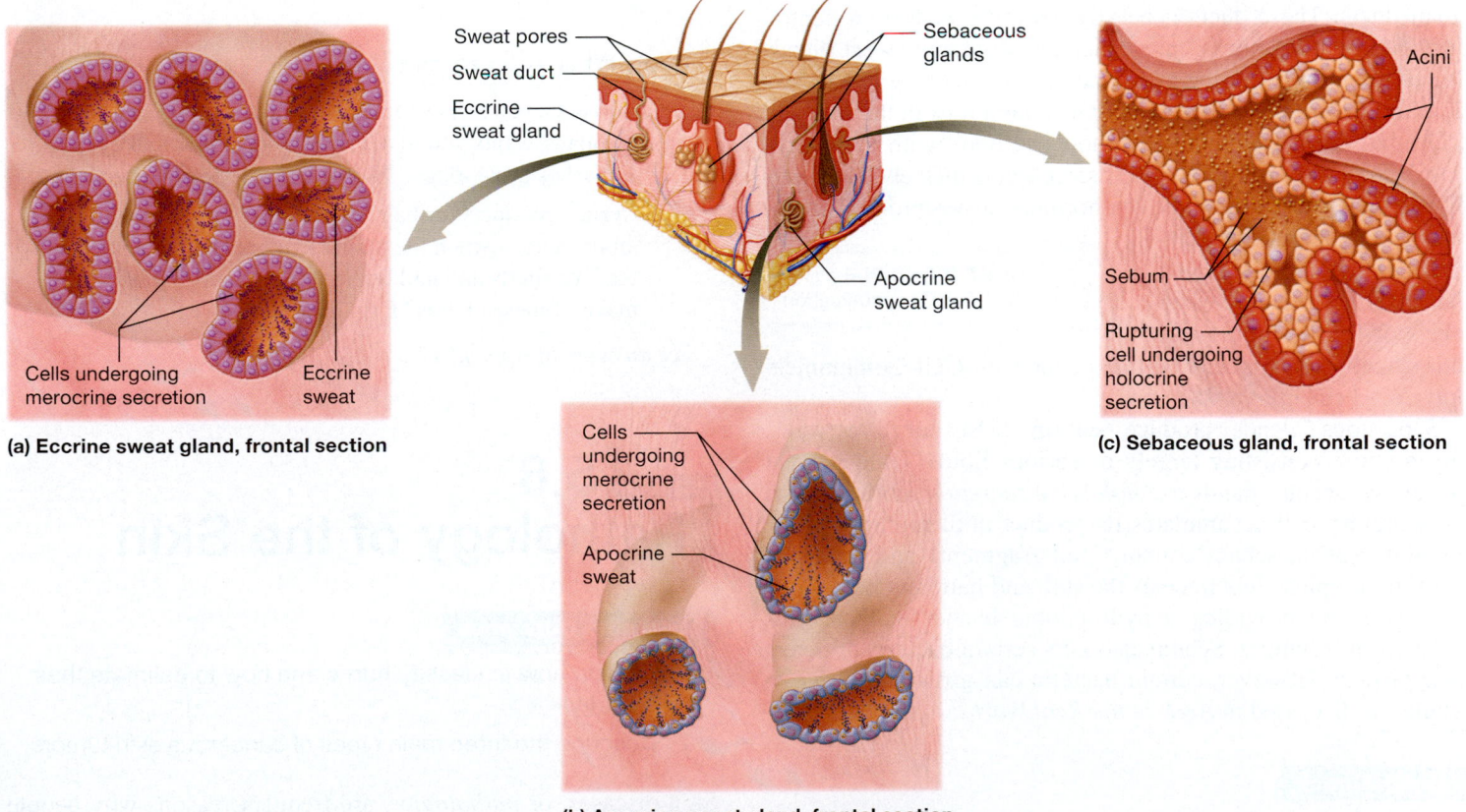

(a) **Eccrine sweat gland, frontal section**

(b) **Apocrine sweat gland, frontal section**

(c) **Sebaceous gland, frontal section**

Figure 5.11 Sweat glands and sebaceous glands.

Apocrine Sweat Glands We find the large **apocrine glands** (AP-oh-krin) only in certain locations, including the axillae (armpits), the anal area, and the areolae (the darkened areas around the nipples). Note in **Figure 5.11b** that unlike eccrine sweat glands, apocrine glands release their product into a hair follicle rather than through a pore. This product is a thick sweat rich in proteins. Although sweat itself is odorless, after being secreted it is metabolized by bacteria in the skin, which produces a characteristic odor. Secretion from apocrine glands is influenced by sex hormones, and does not begin until puberty.

Ceruminous Glands The modified apocrine glands known as **ceruminous glands** (seh-ROO-min-us) are located in the ear, where they secrete a thick product called **cerumen** (seh-ROO-min) into hair follicles. Cerumen, better known as ear wax, lines the canal leading into the ear, where it lubricates the tympanic membrane and traps incoming particles before they reach the membrane.

Mammary Glands **Mammary glands** are highly specialized sweat glands that produce a modified sweat called **milk.** Milk contains proteins, lipids, sugars, and a variety of other substances to nourish a newborn infant. The chapter on the reproductive system discusses the structure and function of mammary glands (see Chapter 26).

Sebaceous Glands

As you can see in **Figure 5.11c**, sebaceous glands are branched glands with clusters of secretory cells called *acini* that surround small ducts. These ducts generally converge to form a central duct that empties into a hair follicle, as apocrine sweat glands do, although in some locations the products are secreted via a small pore. Sebaceous glands are most numerous in the face and scalp, but they are found throughout the body with the exception of the palms and soles. Their secretion is influenced by hormones, particularly the male sex hormone testosterone. For this reason, secretion from sebaceous glands increases after the onset of puberty, when levels of sex hormones rise

CORE PRINCIPLE
Cell-Cell Communication

dramatically. This is an example of the Cell-Cell Communication Core Principle (p. 28).

Sebaceous glands produce **sebum** (SEE-bum), a waxy, oily mixture consisting largely of various lipids. Unlike sweat glands, sebaceous glands use *holocrine secretion,* during which the secretory cell accumulates its product until rupture occurs. For this reason, sebum contains cell fragments and debris in addition to lipids. Sebum coats the skin and hair, and helps keep skin moist by providing a hydrophobic barrier that prevents water from leaving it. Sebum also kills certain bacteria or deters their growth. However, certain bacteria can grow and thrive in sebum, as discussed in *A&P in the Real World: Acne.*

Quick Check

☐ 6. What are eccrine sweat glands, and what are the functions of sweat?

Acne

One of the most common skin problems to plague adolescents is *acne vulgaris,* or simply *acne.* Indeed, 96% of teenagers and young adults experience some degree of acne.

It is caused by an accumulation of sebum and dead cells within sebaceous glands, which produces a *comedone* (KOHM-eh-dohn), or blackhead. Occasionally, these become infected by the bacterium *Propionibacterium acnes,* which results in inflammation and the formation of a *pustule,* or pimple. Although many individuals experience only mild acne, in some people it may become severe and lead to permanent scarring.

Male sex hormones such as testosterone strongly contribute to the development of acne. For this reason, acne tends to be more pronounced in males who are entering puberty, which is when levels of these hormones rise. (Note that females produce small amounts of these hormones, as well.) Acne generally decreases and may disappear by age 20–25, but in some individuals it persists much longer.

☐ 7. What are the other three types of sweat glands, and how do they differ from one another?

☐ 8. How do sebaceous glands and sebum differ from sweat glands and sweat?

Apply What You Learned

☐ 1. You are a researcher for a cosmetics company, and the company wants you to devise a product that makes eyelashes grow longer. How could this be achieved?

☐ 2. Certain products for hair and nails contain proteins and amino acids, which the manufacturers claim are needed by the hair shafts and nail bodies for strength. What do you make of these claims? Explain your answer.

See answers in Appendix A.

MODULE **5.6**
Pathology of the Skin

Learning Outcomes

1. Explain how to classify burns and how to estimate their severity.

2. Describe the three main types of cancerous skin tumors.

Skin diseases, or *pathologies,* are frequent reasons why people seek medical treatment. One common type of skin pathology is a **wound,** defined as any disruption in the skin's integrity. Wounds come in various forms, such as *lacerations* (cuts), *burns,* and *skin cancers.* They may involve the epidermis, the dermis, and occasionally deeper tissues.

Treatment of wounds depends on their type and severity. For example, lacerations may be closed with a few *sutures* or stitches, whereas severe burns may require extensive surgical repair. In this module, we first examine burns, including how to classify and treat them. We finish by discussing skin cancer, a pathology that is extremely common and potentially deadly.

Burns

A **burn** is a skin wound caused by agents such as heat, extreme cold, electricity, chemicals, and radiation. Burns are grouped into three classes according to the extent and depth of tissue damage (**Figure 5.12**):

- **First-degree burns.** The most minor burns are **first-degree burns** (or *superficial burns*), in which only the epidermis is damaged. Erythema and minor pain are typically present, but no blisters or permanent damage. First-degree burns generally require no treatment. An example is a minor sunburn.
- **Second-degree burns. Second-degree burns** (also called *partial thickness burns*) involve the epidermis and part or all of the dermis, and result in significant pain, blistering, and possibly scarring. Second-degree burns generally require medical intervention.

- **Third-degree burns.** The most damaging burns are **third-degree burns** (or *full thickness burns*), which involve the epidermis, dermis, hypodermis, and possibly deeper tissues including muscle and bone. Interestingly, third-degree burns are not generally painful at first because the nerves are destroyed too. These burns typically result in major tissue damage and significant scarring with loss of hair follicles and diminished or absent keratin production. Patients often have problems with dehydration due to massive fluid loss from the swelling that accompanies such burns, and are also at great risk for infection. Treatment of third-degree burns depends on the location and extent of the burn, but patients may require extensive *skin grafting,* in which skin is removed from another part of the body and transplanted onto the burn wound. These patients often face a prolonged and difficult recovery.

We can also describe burns by estimating the percentage of body surface area they affect. **Figure 5.13** illustrates the **rule of nines,** which divides the body into 11 areas, each representing 9% of the total body surface area (plus the genital area, which represents 1%). So a burn involving the entire right lower limb covers 18% of the body, and a burn involving the entire right upper limb and anterior trunk covers 27% of the body. Such estimates are important because they predict the complications that can arise from a burn. For example, the total body surface area affected by a burn indicates the patient's level of dehydration and the amount of fluids the patient should receive. The total body surface area of the burn helps determine prognosis,

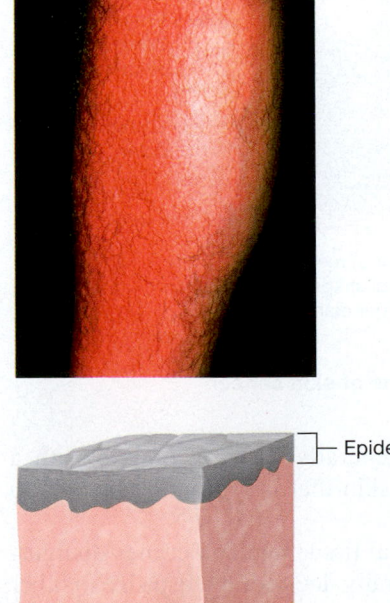
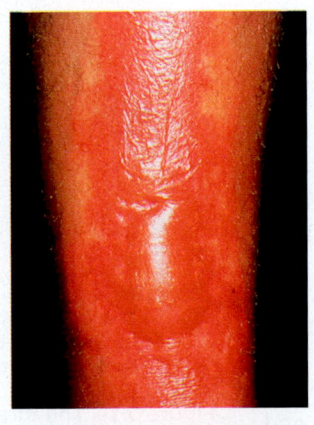
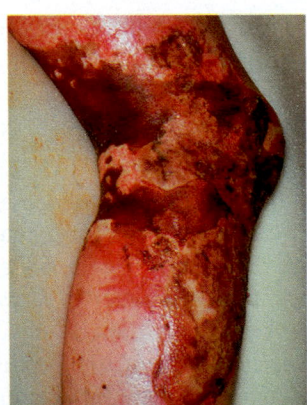
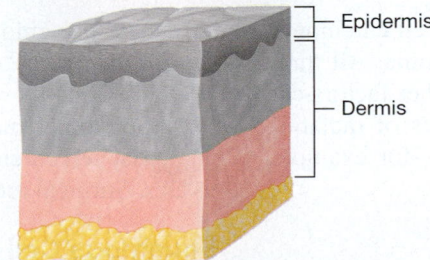
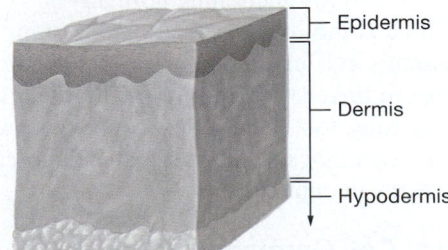

(a) First-degree burn involves epidermis only.

(b) Second-degree burn involves epidermis and part or all of the dermis.

(c) Third-degree burn involves epidermis, dermis, hypodermis, and possibly deeper tissues.

Figure 5.12 The three classes of burns.

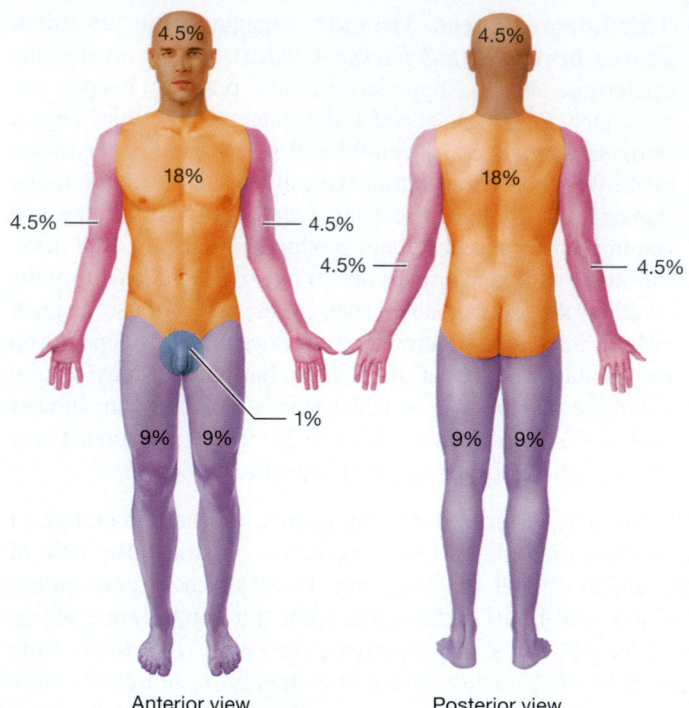

Figure 5.13 Rule of nines: estimating the extent of a burn.

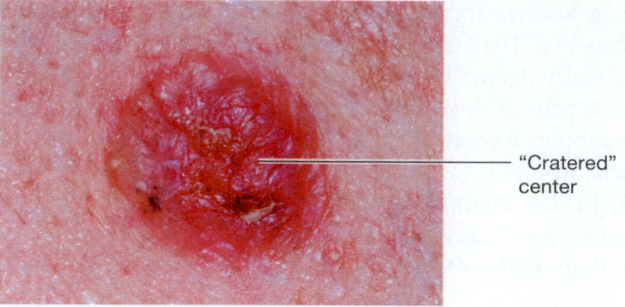

(a) Basal cell carcinoma: cancer of keratinocytes in stratum basale; generally forms a nodule with a "cratered" center

"Cratered" center

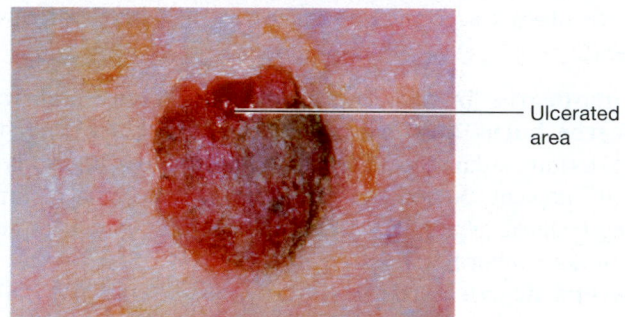

(b) Squamous cell carcinoma: cancer of keratinocytes in stratum spinosum; forms plaques that bleed or ulcerate

Ulcerated area

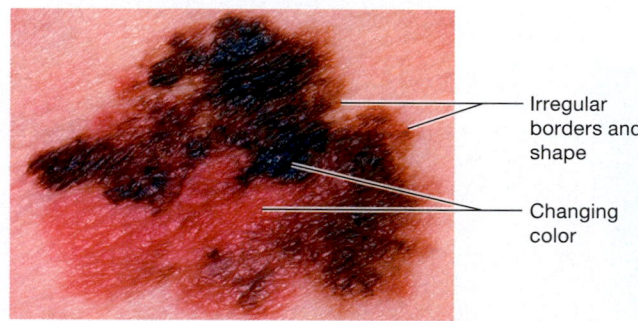

(c) Malignant melanoma: cancer of melanocytes; characterized by asymmetrical shape, irregular borders, blue-black color, larger diameter, and evolving nature

Irregular borders and shape

Changing color

Figure 5.14 The three main forms of skin cancer.

although this is also influenced by factors such as age and other health problems.

Quick Check

☐ 1. What are the three classes of burns, and how do they differ?
☐ 2. What is the clinical importance of the rule of nines?

Skin Cancer

Cancer, one of the most common diseases in the world, claims millions of lives every year. The root cause of cancer is mutations in the DNA that induce a cell to lose control over its cell cycle. This results in the formation of a *tumor,* a cluster of undifferentiated cells. Cancerous tumors are characterized by their ability to *metastasize,* or spread to other tissues through the cardiovascular or lymphatic systems. Tumors then grow in new areas, where the tumor cells alter the structure of a tissue and prevent it from functioning normally.

Three common cancers affect the skin: basal cell carcinoma, squamous cell carcinoma, and malignant melanoma. All three types are linked to exposure to UV radiation. Other factors can play a role, too, such as exposure to other forms of radiation and to *carcinogens* (agents that induce cancer)—for example, the toxins in cigarettes.

Basal Cell Carcinoma

The most common skin cancer—indeed, the most common of all cancers—is **basal cell carcinoma** (kar-sih-NOH-muh; *-oma* = "tumor") (**Figure 5.14a**). As implied by its name, basal cell carcinoma arises from keratinocytes in the stratum

basale of the epidermis. It generally forms a nodule with a central ulcerated "crater" on skin that is regularly exposed to UV radiation.

Although destructive to local tissue and sometimes disfiguring, basal cell carcinoma generally does not metastasize to other tissues. Surgical removal usually resolves it completely.

Squamous Cell Carcinoma

The second most common skin cancer is **squamous cell carcinoma,** a cancer of the keratinocytes of the stratum spinosum (**Figure 5.14b**). It tends to form scaly plaques that bleed or ulcerate, most commonly on the head and neck. Squamous cell carcinoma is more likely to metastasize than basal cell carcinoma but typically can still be managed with surgery.

Malignant Melanoma

The most dangerous skin cancer is **malignant melanoma** (mel-uh-NOH-muh), a cancer of melanocytes (**Figure 5.14c**). Melanomas can be distinguished from other skin cancers and from normal moles using the ABCDE rule:

- **A**symmetrical shape (the two sides do not match)
- **B**order irregularity
- **C**olor, usually blue-black or a variety of colors
- **D**iameter generally larger than 6 mm (the size of a pencil eraser)
- **E**volving (changing) shape and size

Early detection of melanoma is critical because it has a much greater tendency to metastasize than basal or squamous cell carcinoma. This is because the "arms" of the cancerous melanocytes extend down into the dermis and access dermal blood vessels, enabling the cells to spread to other tissues via the bloodstream. Like basal and squamous cell carcinoma, malignant melanoma is treated with surgical excision. Treatment may also include chemotherapy and radiation therapy, although their effectiveness is limited. Prognosis depends on the size of the tumor, the depth to which it extends into the dermis, and whether it has metastasized to other tissues.

Quick Check

☐ 3. What is cancer?

☐ 4. How do the three types of skin cancer differ?

Apply What You Learned

☐ 1. Your friend just spent a day at the beach but forgot to use sunscreen. That evening she complains of a great deal of pain and you notice that her skin is red and beginning to form large blisters. What type of burn has she suffered? What other potential long-term damage has she done to her skin?

☐ 2. A hypothetical chemical is found to be carcinogenic when applied to cells in a tissue culture. Chemical analysis shows that it is a water-soluble compound (hydrophilic). What type of cancer, if any, might this chemical cause when applied to the skin? Explain your reasoning. (*Hint:* Do hydrophilic substances pass through the epidermis? Why or why not?)

See answers in Appendix A.

Chapter Summary

For everything you need to succeed in this course, go to **Mastering** Anatomy & Physiology. There you will find:

- Practice Tests
- Author Podcasts
- Big Picture Animations
-))) Concept Boost Video Tutors
- **iP2**™ Interactive Physiology 2.0 Tutorials
- **A&PFlix** A&P Flix 3D Animations
- Active-Learning Workbook

MODULE **5.1**
Overview of the Integumentary System 160–164

- The skin consists of the superficial **epidermis** and the deeper **dermis,** which rests on the **hypodermis.**
- The integument serves many functions: protection, sensation, **thermoregulation,** excretion, and **vitamin D** synthesis.

MODULE **5.2**
The Epidermis 164–168

- Most of the cells in the epidermis are **keratinocytes,** which manufacture **keratin.**
- **Dendritic cells** are phagocytes in the **stratum spinosum. Merkel cells** are sensory receptors in the **stratum basale.**
- Thick skin has all five epidermal strata, and all five strata are thicker, particularly the **stratum corneum.** Thin skin lacks the **stratum lucidum,** and its four strata are thinner.

MODULE **5.3**
The Dermis 168–171

- The dermis consists of the **papillary layer** and the **reticular layer.**
- Folds of the papillary layer include **dermal papillae** and **epidermal ridges.**
- Gaps between collagen fibers in the reticular layer cause **tension,** or **cleavage, lines.**

MODULE **5.4**
Skin Pigmentation 171–174

- The primary pigment that influences skin color is **melanin.** Other substances that affect skin color are **carotene** and **hemoglobin.**

- **Hair,** or **pili,** are composed of dead, keratinized keratinocytes that perform protective functions and sense the environment.
- **Nails** are located on the ends of digits and are derived from the epidermis.
- **Sweat (sudoriferous) glands** reside in the dermis and secrete sweat by merocrine secretion. The most common sweat gland is the **eccrine sweat gland.**
- Other sweat glands include **apocrine glands, ceruminous glands,** and **mammary glands.**
- **Sebaceous glands** secrete **sebum** into a hair follicle via holocrine secretion.

- Skin pathologies may include a **wound,** a **burn,** or **cancer.**

Assess What You Learned

Scan the QR Code for additional practice test questions

https://goo.gl/06Hh6e

LEVEL 1 Check Your Recall

1. Explain why the skin is an organ.

2. Which of the following correctly describes the structure of the skin?

   a. It consists of the superficial epidermis, middle dermis, and deep hypodermis.
   b. It consists of the superficial epidermis and deep dermis.
   c. It consists of the superficial dermis and deep epidermis.
   d. It consists of the superficial dermis and deep hypodermis.

3. Which of the following is *not* a function of the integument?

   a. Protection from mechanical trauma
   b. Thermoregulation
   c. Protection from acid-base imbalances
   d. Vitamin D synthesis

4. Explain what happens to dermal blood vessels when heat needs to be conserved (i.e., due to a cold environment).

5. Number the strata of thick skin epidermis from deepest (1) to most superficial (5).

   _____ Stratum spinosum
   _____ Stratum corneum
   _____ Stratum basale
   _____ Stratum lucidum
   _____ Stratum granulosum

6. Keratinocytes in the superficial strata of the epidermis die because:

   a. they are too far away from the blood supply in the dermis.
   b. they are surrounded by a lipid-based substance that makes them more permeable to water.
   c. They do not die.
   d. No keratinocytes in the epidermis are alive.

7. Mark the following statements as true or false. If a statement is false, correct it to make a true statement.

   a. Melanocytes account for the bulk of the epidermis.
   b. Keratinocytes begin life in the stratum corneum and gradually are pushed into the stratum basale.
   c. Dendritic cells are phagocytes of the immune system that protect the skin and deeper tissues from invasion by pathogens.
   d. Merkel cells produce the pigment melanin.

8. Which of the following statements is *false*?

   a. Thin skin lacks the stratum lucidum, whereas thick skin has all five epidermal layers.
   b. Thick skin is located on the palms, the palmar surfaces of the fingers, the soles of the feet, and the plantar surface of the toes, whereas thin skin is located everywhere else.
   c. Thick skin has numerous hairs, whereas thin skin lacks hairs.
   d. Thin skin has a thin stratum corneum and the other layers are thinner than what we find in thick skin.

9. What are the functions of the dermal papillae?

10. Epidermal ridges are created by:

    a. the epidermal papillae.
    b. mounds of papillary dermis arranged into dermal ridges.
    c. gaps between collagen bundles in the reticular layer.
    d. tight binding of the reticular layer to deeper structures.

11. Mark the following statements as true or false. If a statement is false, correct it to make a true statement.

    a. The primary skin pigment is melanin, which is derived from the amino acid tyrosine.
    b. Melanin is produced by melanocytes and covers the nuclei of neighboring dendritic cells.
    c. Carotene is a brown-black pigment that accumulates in the stratum corneum.
    d. Increased amounts of blood flowing through the dermis lead to pallor in the skin.
    e. The pigment hemoglobin in red blood cells gives skin a pinkish hue when it binds high amounts of oxygen.

12. Which of the following is not a function of body hair in humans?

    a. Protection from the environment
    b. Sensation
    c. Protection from mechanical trauma
    d. Regulation of blood flow

13. Fill in the blanks: The portion of the hair that projects from the surface of the skin is the _____, and the portion within the dermis is the _____, which is embedded in a(n) _____. The portion of the hair that contains cells that undergo mitosis during the growth stage of the hair is known as the _____.

14. Nail growth occurs when:

   a. cells in the nail plate undergo mitosis.
   b. cells in the nail matrix undergo mitosis.
   c. cells in the eponychium undergo mitosis.
   d. cells in the medial and lateral nail folds undergo mitosis.

15. Fill in the blanks: Sebaceous glands secrete by _____ secretion; sweat glands secrete by _____ secretion.

16. Match each type of gland with its correct properties.

   _____ Eccrine sweat gland
   _____ Sebaceous gland
   _____ Ceruminous gland
   _____ Mammary gland
   _____ Apocrine sweat gland

   a. Branched gland that secretes sebum into a hair follicle
   b. Gland that secretes a protein-rich sweat into a hair follicle
   c. Simple coiled tubular gland found over most of the body that secretes sweat through a sweat pore
   d. Gland that secretes ear wax
   e. Modified sweat gland that produces milk

17. How do sweat and sebum differ?

18. Which type of burn involves the epidermis and all or part of the dermis?

   a. First-degree burn
   b. Second-degree burn
   c. Third-degree burn
   d. Fourth-degree burn

19. The type of skin tumor that involves the keratinocytes of the stratum spinosum is:

   a. basal cell carcinoma.
   b. squamous cell carcinoma.
   c. malignant keratocytoma.
   d. malignant melanoma.

LEVEL 2 Check Your Understanding

1. Why don't you bleed when you cut your nails? Why don't you bleed when a hair is pulled?

2. Manufacturers of shampoos and conditioners often claim their products contain vitamins that are necessary to keep hair shafts healthy. Do hair shafts need vitamins? (*Hint:* Vitamins are required for cells to carry out certain reactions.) Why or why not? How valid are these claims?

3. The hair and nails are sometimes called *accessory organs*. Are these structures technically organs? Why or why not?

LEVEL 3 Apply Your Knowledge

PART A: Application and Analysis

1. You are working in the emergency department when paramedics rush in with an unconscious patient. You notice that the patient is wearing athletic clothing and the skin on his face and elsewhere on his body is bright red. The paramedics tell you that he was picked up after collapsing during a bike race. What does the color of the patient's skin tell you about the probable cause of his illness? Explain.

2. After Ramon's skin came into contact with a poison ivy plant in biology lab, he developed a painful, itchy rash. However, after the skin of his colleague Cathy came into contact with snake venom in lab, she developed no skin irritation. Poison ivy plants contain lipid-soluble oils, whereas snake venoms contain mostly water-soluble peptides. Explain why Ramon developed a rash, whereas Cathy did not.

3. Nguyen comes to your clinic with a mole that has recently changed in appearance. You examine the mole and note that its borders are irregular, it has a deep blue-black color, and the color is unevenly distributed throughout the mole.

   a. What is your immediate concern? Why?
   b. Nguyen tells you that she has used a tanning booth once per week for the past several years, and that the tanning salon advertises it as safe. What do you tell her about the tanning salon's claim? How does any UV exposure affect keratinocytes and melanocytes? Explain.

PART B: Make the Connection

4. What would happen to the skin if the oil produced by sebaceous glands was instead a polar covalent compound? (*Connects to Chapter 2*)

5. Many antiaging skin creams contain collagen and hyaluronic acid. Manufacturers claim that the collagen and hyaluronic acid applied to the surface of the skin will be absorbed into the dermis, where they will be incorporated into dermal tissue.

   a. What are collagen and hyaluronic acid, and what are their functions? (*Connects to Chapter 4*)
   b. Are these substances polar or nonpolar? (*Connects to Chapter 2*)
   c. Predict whether or not these substances are likely to be absorbed by the epidermis.
   d. Predict the effectiveness of the creams.

6. Would a mild second-degree burn be likely to heal by regeneration or fibrosis? Would the same type of healing take place in a third-degree burn that involved muscle tissue? Why or why not? (*Connects to Chapter 4*)

See answers in Appendix A.

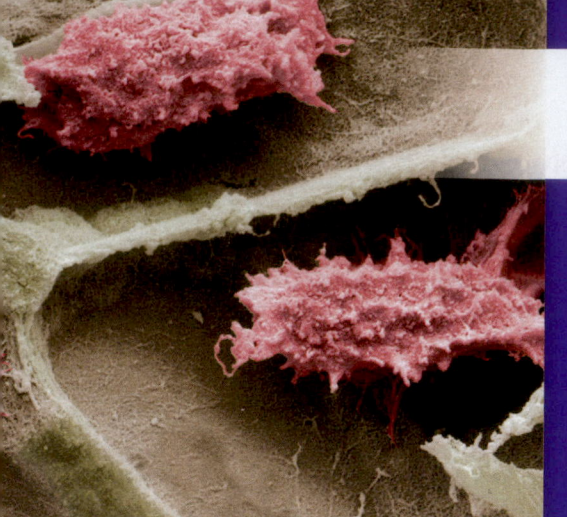

6

Bones and Bone Tissue

Have you ever seen a slug creep along the ground? Well, that's how we would move without our bones. We're now starting our tour of the **skeletal system,** which includes the bones, joints, and other supporting tissues. As you learned earlier, an *organ system* is defined as a collection of organs with related functions, and organs are composed of two or more tissues (see Chapter 1). At first glance, the main organs of the skeletal system—the bones—may seem to be only one type of tissue. But bones contain several tissue types, including *bone tissue,* which is more formally called *osseous tissue* (AHS-ee-us); dense regular and irregular collagenous connective tissues; and a tissue called *bone marrow.* These different tissues make each bone in the skeletal system an organ.

People are said to have "healthy bones" when they have properly functioning cells and adequate quantities and quality of extracellular matrix. When a bone has both of these, it is able to perform its functions much more effectively. When it lacks these components, bone disease nearly always results. In this chapter we explore the gross and microscopic structure of these organs, the functions of the skeletal system, and the relationship of "healthy bones" to the body's overall homeostasis. We discuss individual bones and joints in the skeletal and articulations chapters (see Chapters 7 and 8, respectively).

MODULE 6.1
Introduction to Bones as Organs

For practice applying concepts to a clinical scenario, check out the **Running Case Study** for this chapter @ Mastering Anatomy & Physiology

Learning Outcomes

1. Describe the functions of the skeletal system.
2. Describe how bones are classified by shape.
3. Describe the gross structure of long, short, flat, irregular, and sesamoid bones.
4. Explain the differences between red and yellow bone marrow.

Before we dive into the microscopic anatomy of bone and associated tissues, let's take a look at the structure and functions of bones as organs. The upcoming sections introduce you to the overall functions of bones and the skeletal system, the different bone shapes, and the general structure of bones.

Photo: Colored scanning electron micrograph showing osteoclasts in bone lacunae.

Functions of the Skeletal System

◀◀ FLASHBACK

1. Define the terms mineral, electrolyte, and buffer. (p. 33 and p. 49)

2. What are adipocytes, and what do they store? (p. 138)

Some functions of the skeletal system are fairly obvious. For example, it probably comes as no surprise that the skeleton provides support for the body. But the skeleton is more than simply a scaffold for other body parts—a number of its functions are vital to the maintenance of homeostasis for the body as a whole. These roles, shown in **Figure 6.1**, include the following:

- **Protection.** Certain bones, including the skull, sternum, ribs, and pelvis, protect their underlying organs.

Bone is one of the hardest substances in the body and it provides a good strong "shell" for organs such as the brain, the heart, and the lungs, an example of the Structure-Function Core Principle (p. 28). Bone also protects the sense organs, such as the eyes and the organs of hearing.

- **Mineral storage and acid-base homeostasis.** Bone is the most important storehouse in the body for minerals such as calcium, phosphorus, and magnesium salts. These minerals, also found in the blood as electrolytes, acids, and bases, are vital to the maintenance of electrolyte and acid-base balance in the body.

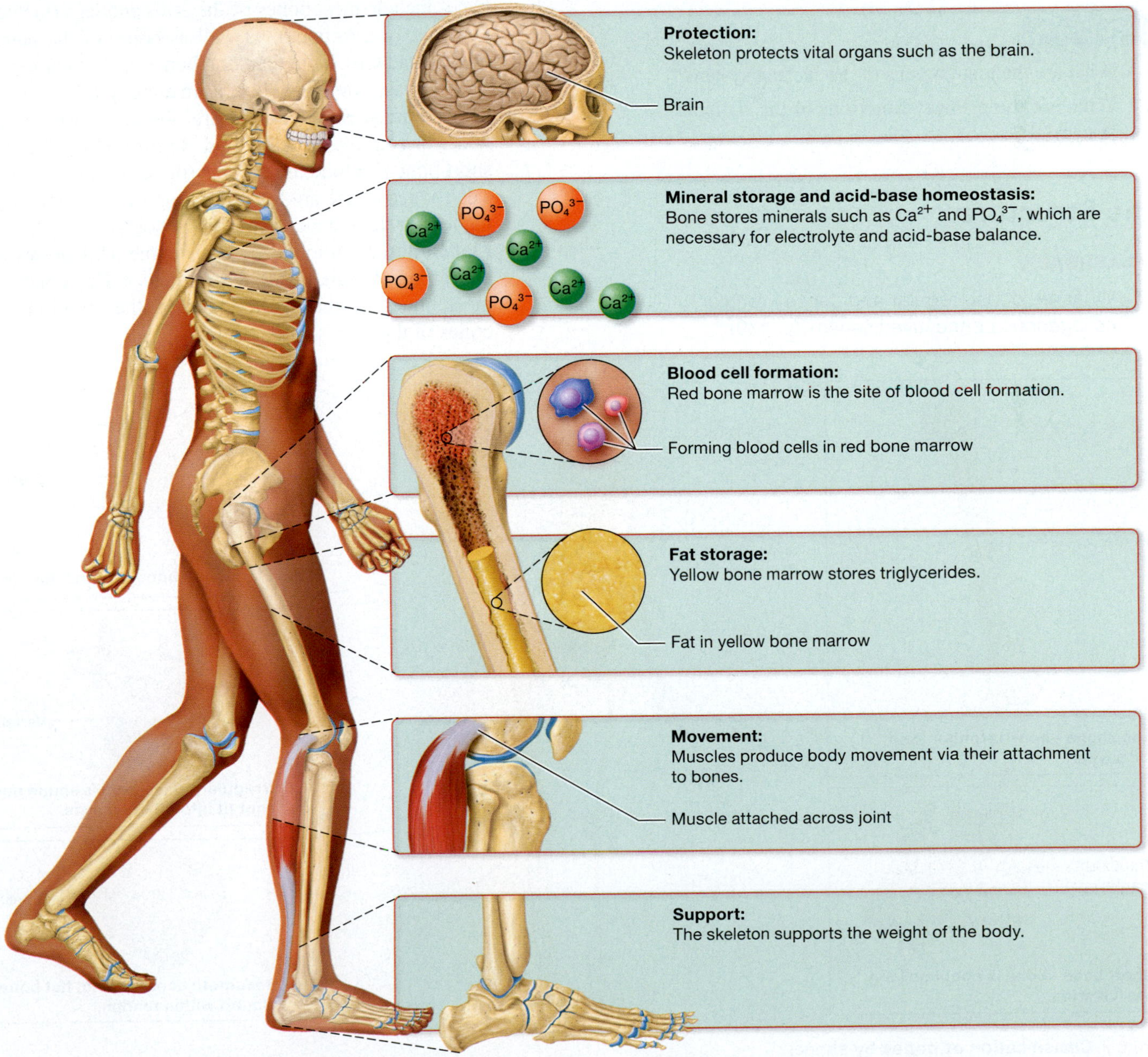

Protection:
Skeleton protects vital organs such as the brain.

- Brain

Mineral storage and acid-base homeostasis:
Bone stores minerals such as Ca^{2+} and PO_4^{3-}, which are necessary for electrolyte and acid-base balance.

Blood cell formation:
Red bone marrow is the site of blood cell formation.

- Forming blood cells in red bone marrow

Fat storage:
Yellow bone marrow stores triglycerides.

- Fat in yellow bone marrow

Movement:
Muscles produce body movement via their attachment to bones.

- Muscle attached across joint

Support:
The skeleton supports the weight of the body.

Figure 6.1 Functions of the skeletal system.

- **Blood cell formation.** Bones house *red bone marrow,* which is a special form of connective tissue. In this tissue, the process of **hematopoiesis** (heh'-mah-toh-poy-EE-sis; *hemato-* = "blood," *poiesis-* = "to make"), or formation of blood cells, takes place. Hematopoiesis is examined in detail in the blood chapter (see Chapter 19).
- **Fat storage.** As you can see in Figure 6.1, not all bone marrow is red. Some, called *yellow bone marrow,* contains adipocytes with stored triglycerides. Fatty acids from the breakdown of these triglycerides can be released and used as fuel by cells if necessary.
- **Movement.** Bones serve as sites of attachment for most skeletal muscles. When muscles contract, they pull on the bones, which generates movement around a joint.
- **Support.** The skeleton supports the weight of the body and provides its structural framework.

> **Quick Check**
>
> ☐ 1. What are the main organs of the skeletal system?
> ☐ 2. What are the primary functions of the skeletal system?

Bone Structure

《 FLASHBACK

1. What are dense regular and dense irregular collagenous connective tissues? (p. 139)

2. What are collagen fibers? (p. 125)

3. What is hyaline cartilage? (p. 142)

Looking at all the bones in a skeleton, you can see that they show great diversity in appearance. However, even with all this variation, we can define five structural bone classes, and each of our 206 bones fits into one of these classes. The criterion used to categorize a bone is its general shape. In this section we examine bone shapes, then discuss their structure in more detail.

Classification of Bones by Shape

The five general bone shapes, shown in **Figure 6.2**, are long, short, flat, irregular, and sesamoid. Let's take a closer look at what defines each shape.

- **Long bones. Long bones** are named for the fact that they are longer than they are wide (Figure 6.2a). Examples of long bones include most bones of the arms and legs (including the humerus, or arm bone), as well as bones of the hands, feet, fingers, and toes. Note that long bones are not named for their size—in fact, some long bones are quite small.
- **Short bones. Short bones** are about as long as they are wide, or roughly cube-shaped (Figure 6.2b). Examples of short bones include the carpals (bones of the wrist) and tarsals (bones of the ankle). Like long bones, short bones are named for their shape rather than their size.
- **Flat bones.** As implied by their name, **flat bones** are thin and broad (Figure 6.2c). Examples of flat bones include most bones of the skull, the ribs, the sternum, and the bones of the pelvis.

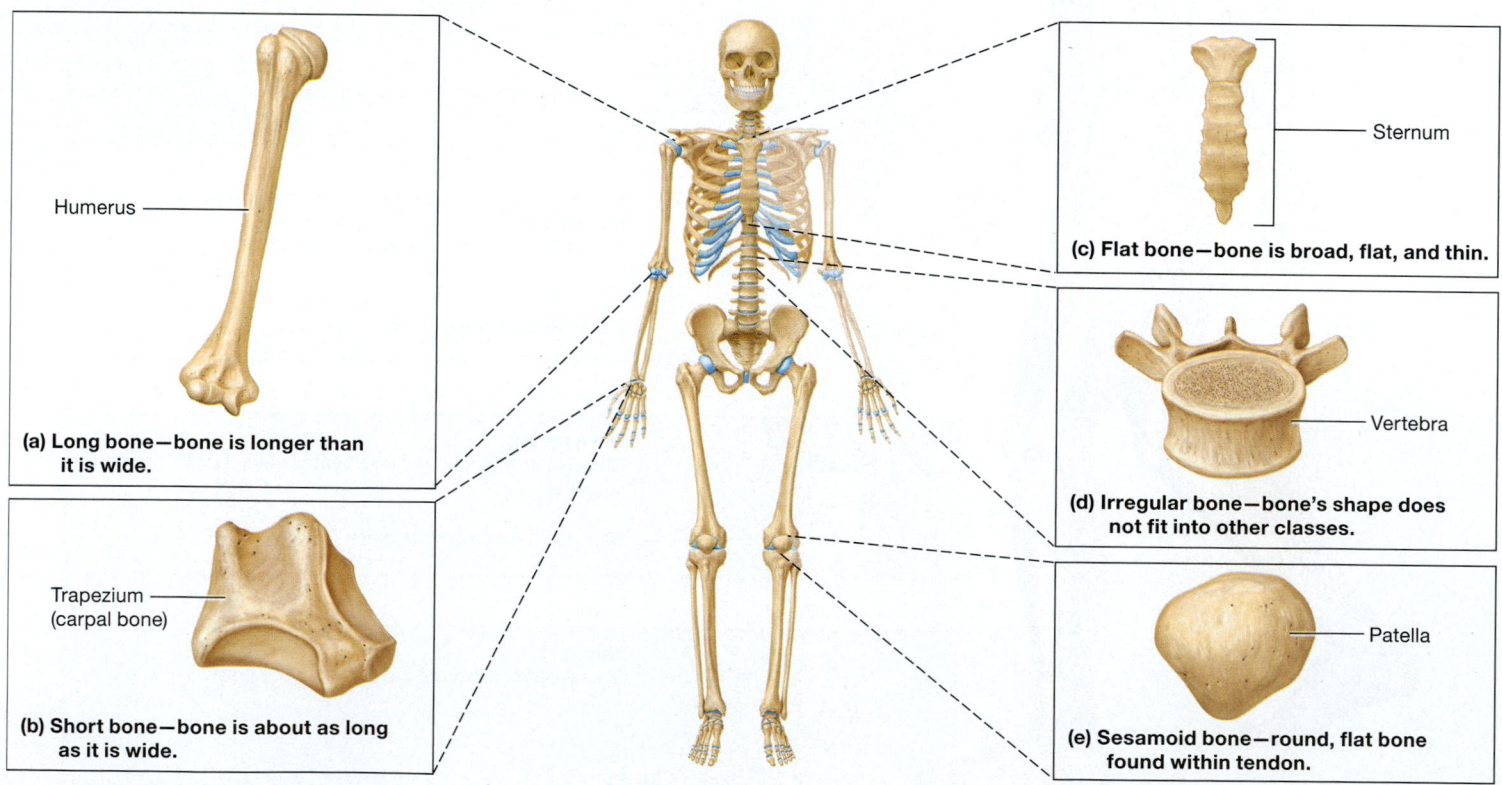

Humerus

(a) Long bone—bone is longer than it is wide.

Trapezium (carpal bone)

(b) Short bone—bone is about as long as it is wide.

Sternum

(c) Flat bone—bone is broad, flat, and thin.

Vertebra

(d) Irregular bone—bone's shape does not fit into other classes.

Patella

(e) Sesamoid bone—round, flat bone found within tendon.

Figure 6.2 Classification of bones by shape.

- **Irregular bones.** The **irregular bones** do not fit into the other classes due to their irregular shapes (Figure 6.2d). Examples of irregular bones include the vertebrae and certain skull bones.
- **Sesamoid bones.** Sesamoid bones (SEH-suh-moyd; "sesame-shaped") are small, relatively flat, and oval-shaped bones located within tendons (Figure 6.2e). They give the tendon a mechanical advantage, providing better leverage for muscles, and also reduce wear and tear on the tendon. An example of a sesamoid bone is the patella (kneecap).

Structure of a Long Bone

When you look at a long bone's surface, you'll see that most of its surfaces are covered with a membrane, the **periosteum** (pehr'-ee-AHS-tee-um; *peri-* = "around," *oste-* = "bone") (**Figure 6.3a**). The periosteum is composed of dense irregular

collagenous connective tissue that is richly supplied with blood vessels and nerves. As Figure 6.3a shows, the periosteum is firmly attached to the underlying bone by collagen fibers called **perforating fibers.** These fibers penetrate deeply into the bone matrix, securing the periosteum in place. The innermost surface of the periosteum contains different types of bone cells, as we'll see in the next module.

Each long bone has a middle shaft and two rounded ends, features unique to long bones. The shaft of a long bone is called its **diaphysis** (dy-AF-uh-sis; *dia-* = "across"). At the ends of the bones we find the enlarged, rounded **epiphyses** (eh-PIF-uh-seez; singular, *epiphysis; epi-* = "upon"). Notice in Figure 6.3a that the epiphyses are covered with a thin layer of hyaline cartilage, also known as **articular cartilage,** which allows bones to rub together with reduced friction at joints.

Inside a long bone (**Figure 6.3b**) the diaphysis surrounds a hollow cavity called the **medullary cavity** (MED-yoo-lehr-ee), also known as the *marrow cavity (medullary* = "marrow"). As its name implies, this is where much of the marrow of a long bone is housed. The marrow can be either yellow or red, depending on the bone and the person's age.

As you can see in Figure 6.3b, bone has two obviously different textures. The hard, dense outer bone is **compact bone.** The structure of compact bone enables it to resist the majority of stresses placed on it, which are linear compression and twisting forces. The inner, honeycomb-like bone is **spongy bone,** or *cancellous bone* (KAN-sel-us). Spongy bone forms a framework of bony struts that allows it to resist forces in many directions and provides a place for the bone marrow to reside. The diaphyses of long bones consist of a very thick layer of compact bone with only scant inner spongy bone surrounding the medullary cavity. As with the diaphysis, the outer parts of the epiphyses are compact bone; however, the interior of the epiphyses consists of abundant spongy bone. Both compact and spongy bone will be discussed further in Module 6.2.

The bony struts of spongy bone, and indeed all inner surfaces of bone, are lined with a membrane called the **endosteum** (en-DAHS-tee-um; *endo-* = "within"). Like the periosteum, the endosteum contains different types of bone cells that help maintain bone homeostasis. However, the endosteum is thinner and lacks the fibrous outer layer that we see with the periosteum.

One final thing to note in Figure 6.3b is the presence of lines running across the proximal and distal ends of the bone between the epiphysis and the diaphysis. These lines, called *epiphyseal lines* (eh-PIF-ih-see-uhl), are remnants of a structure termed the *epiphyseal plate.* The epiphyseal plate, also known as the *growth plate,* is a line of hyaline cartilage from which a long bone grows in length in children and adolescents. Bone growth and the role of the epiphyseal plate are discussed in Module 6.4.

Structure of Short, Flat, Irregular, and Sesamoid Bones

Short, flat, irregular, and sesamoid bones do not have diaphyses or epiphyses. As a result, they contain no medullary cavity, epiphyseal lines, or epiphyseal plates. But as you can see

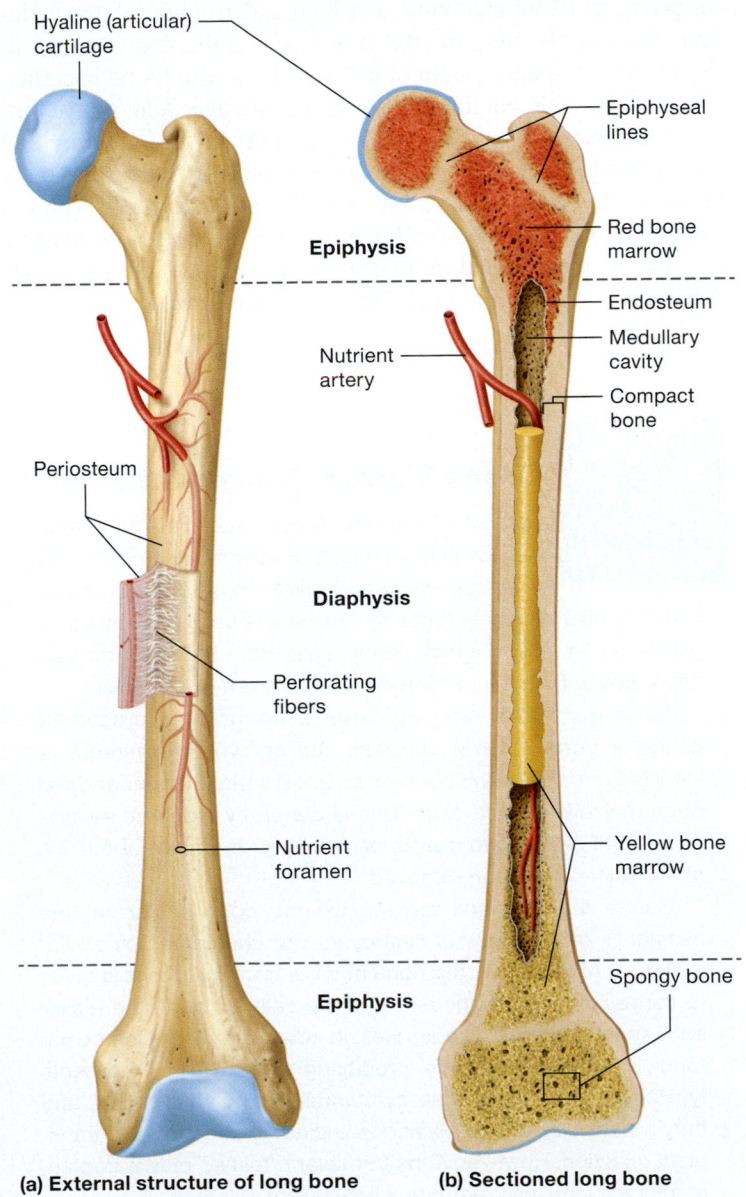

(a) External structure of long bone

Hyaline (articular) cartilage

Epiphysis

Periosteum

Nutrient artery

Diaphysis

Perforating fibers

Nutrient foramen

Epiphysis

(b) Sectioned long bone

Epiphyseal lines

Red bone marrow

Endosteum

Medullary cavity

Compact bone

Yellow bone marrow

Spongy bone

Figure 6.3 Structure of long bones.

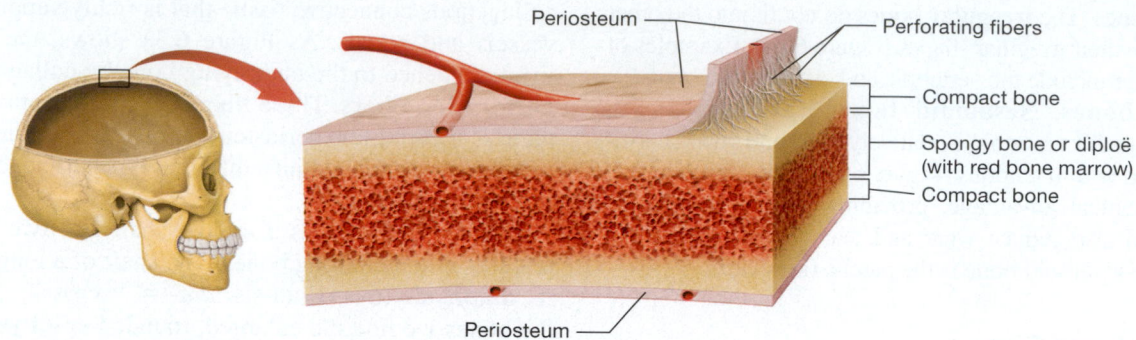

Figure 6.4 Structure of short, flat, irregular, and sesamoid bones.

in **Figure 6.4**, they do share some similarities with long bones. Externally these bones are covered with a periosteum attached by perforating fibers and well supplied with blood vessels and nerves. In addition, they have an outer layer of compact bone that surrounds the inner spongy bone.

Notice that their internal structure is fairly simple. A section through one of these bones resembles a "sandwich" made up of two thin layers of compact bone and a middle layer of spongy bone housing bone marrow. In flat bones, the spongy bone is called the **diploë** (dip-LOH-ee; *diploë* = "fold"). The interior of some flat and irregular bones of the skull contains hollow, air-filled spaces called *sinuses* that make the bones lighter.

Blood and Nerve Supply to Bone

Bones are well supplied with blood vessels (look again at Figures 6.3 and 6.4), as evidenced by the extensive bleeding that occurs whenever a bone is injured. Bones are also supplied with many sensory fibers, which is why such an injury hurts so acutely. The blood supply to short, flat, irregular, and sesamoid bones comes largely from vessels in the periosteum that penetrate the bone. Long bones also get a significant portion (about one-third) of their blood supply from the periosteum, which supplies primarily the compact bone. However, their greatest source of blood comes from one or two *nutrient arteries* that enter the bone via a small hole in the diaphysis called the **nutrient foramen** (fohr-AY-men; see Figure 6.3b). The nutrient artery passes through compact bone into the inner medullary cavity, where it supplies the internal structures of the long bone. The epiphyses receive some of their blood supply from the nutrient artery, but most of it comes from small blood vessels that enter and exit via numerous small holes in their compact bone.

Red and Yellow Bone Marrow

As we discussed earlier, there are two types of bone marrow with different functions. **Red bone marrow** consists of a network of reticular fibers supporting islands of blood-forming, or *hematopoietic,* cells. **Yellow bone marrow,** which stores triglycerides, consists mostly of blood vessels and adipocytes. In infants and young children, most bone marrow is red because their rapid rate of growth requires a constant supply of new blood cells. At about age 5, yellow bone marrow begins to replace some of the red bone marrow.

By the time we reach adulthood, most bone marrow in the body is yellow. Red marrow remains only in the bones of the pelvis, the proximal femur and humerus (thigh and arm bones), the vertebrae, the ribs, the sternum, the clavicles, and the scapulae (shoulder blades). A small amount of red marrow is also found in certain bones of the skull, but this degenerates as we age. Adults have limited red bone marrow because we are not actively growing and do not need hematopoiesis to occur as rapidly as it does in young children. However, some yellow marrow can be replaced with red marrow if increased blood cell production is needed. Bone marrow can also be transplanted to treat some diseases of the blood—see *A&P in the Real World: Bone Marrow Transplantation.*

A&P in the Real World   Bone Marrow Transplantation

Diseases of the blood such as leukemia, sickle-cell anemia, and aplastic anemia feature improperly functioning hematopoietic cells. One potential treatment for such diseases is *bone marrow transplantation,* in which hematopoietic cells from red bone marrow are removed from a matching donor and given to a recipient.

Bone marrow is removed from a donor by a procedure called a *bone marrow harvest,* during which a needle is inserted into the pelvic bone of an anesthetized donor and red bone marrow is withdrawn. This is generally repeated several times until one to two quarts of red bone marrow—about 2% of the total—has been removed.

Before the recipient can receive the donor's marrow, the recipient's own marrow is destroyed with chemotherapy and/or radiation. At this point, the bone marrow from the donor is given to the recipient intravenously, and the cells travel to the recipient's spongy bone. If all goes well, in about 2–4 weeks the hematopoietic tissue will begin producing blood cells. Recipients typically experience flu-like symptoms during this period, and they run the risk of complications, including infection and transplant rejection. However, if the transplant "takes," or is accepted, many recipients can return to a healthy life.

☐ 3. Look at Figure 6.2 and classify the following bones as long, short, flat, or irregular.

    a. Phalanges (fingers)

    b. Ribs

    c. Carpals

    d. Pelvis

    e. Humerus

☐ 4. Where are compact bone and spongy bone located, and what are their functions?

☐ 5. What are the two types of bone marrow, and what are their functions?

☐ 1. Predict how the characteristics of a person's bones would change if compact bone were located on the inside and spongy bone on the outside.

☐ 2. Sometimes a bone is injured at the epiphyseal plate. Predict the long-term effect this might have on the injured bone.

☐ 3. Predict the potential consequences of damaging the nutrient artery in a bone injury. How would this affect the ability of the bone to heal?

See answers in Appendix A.

MODULE **6.2**

Microscopic Structure of Bone Tissue

Learning Outcomes

1. Describe the inorganic and organic components of the extracellular matrix of bone tissue.

2. Explain the functions of the three main cell types in bone tissue.

3. Describe the microscopic structure of compact bone and the components of the osteon.

4. Describe the microscopic structure of spongy bone.

The primary tissue type found in bones is **bone tissue,** formally known as **osseous tissue.** We introduced this tissue earlier as one of the specialized connective tissues (see Chapter 4). Bone shares many properties with the other connective tissues. For example, like most connective tissues, bone consists mainly of extracellular matrix with cells scattered throughout. However, as we discuss in this module, the extracellular matrix and cell types in bone are unique. Let's now explore these components.

The Extracellular Matrix of Bone

« FLASHBACK

1. Compare organic and inorganic substances. (p. 46 and p. 50)

2. What are protein fibers and ground substance? (pp. 124–125)

The extracellular matrix (ECM) of bone tissue, sometimes simply called the *bone matrix,* is not seen in any other tissue. About 65% of the total weight of bone consists of an **inorganic matrix** composed of minerals. The remaining 35% of total bone weight consists of an **organic matrix,** which contains many of the usual ECM "ingredients," including collagen fibers.

Although we generally think that it's the inorganic matrix that makes bone strong, the organic matrix contributes equally to its strength. Note in **Figure 6.5a** what happens if we destroy

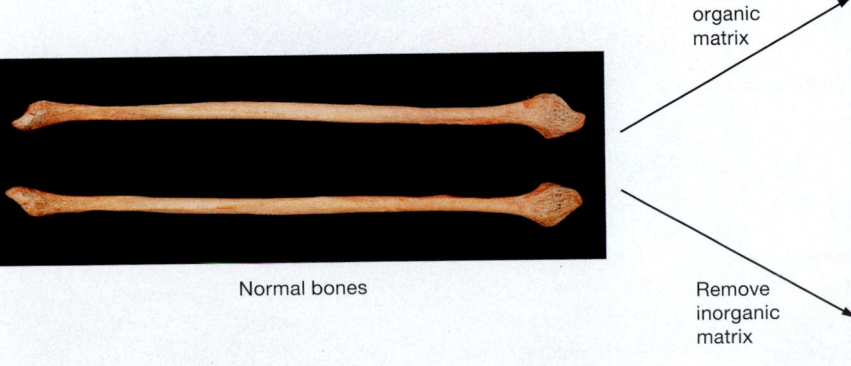

Normal bones

Remove organic matrix →

(a) **Bone without its organic matrix (collagen) is brittle and shatters easily.**

Remove inorganic matrix →

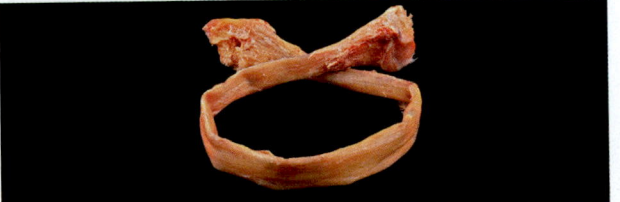

(b) **Bone without its inorganic matrix (minerals) cannot resist compression.**

Figure 6.5 The importance of bone matrices.

the organic matrix of bone and leave the inorganic matrix intact—the bone becomes brittle and shatters easily. Similarly, in **Figure 6.5b**, if we dissolve the inorganic matrix and leave the organic matrix behind, the bone becomes flexible and unable to resist compression. So, clearly, both components are required to maintain bone strength.

Inorganic Matrix

The predominant ingredient of the inorganic matrix is calcium salts (in fact, bone stores about 85% of the total calcium ions in the body), with a good amount of phosphorus as well. Most of the calcium and phosphorus salts exist as part of a large mineral called **hydroxyapatite crystals** $[Ca_{10}(PO_4)_6(OH)_2]$ (hydrawk′-see-AP-uh-ty′t). This mineral makes bone one of the hardest substances in the body and gives bone its strength and the ability to resist compression, which allows it to perform its functions of support and protection. This is another example of the Structure-Function Core Principle (p. 28). Other ingredients of the inorganic matrix include bicarbonate (HCO_3^-), potassium, magnesium, and sodium salts.

CORE PRINCIPLE
Structure-Function

Organic Matrix

The organic matrix of bone, known as **osteoid** (AHS-tee-oyd), consists of protein fibers, proteoglycans, glycosaminoglycans, glycoproteins, and bone-specific proteins such as *osteocalcin* (see Chapters 2 and 4). The predominant protein fibers in bone are collagen fibers, which form cross-links with one another and help bone to resist torsion (twisting) and tensile (pulling or stretching) forces. They also align with hydroxyapatite crystals, significantly enhancing the hardness of bone. Collagen fibers are one of the most important components of the bone ECM in terms of strength; when these fibers are inadequate or defective, bone weakens and any minor twisting force may break it.

The other components of organic matrix perform support functions. Recall that large compounds such as glycosaminoglycans and proteoglycans in the ECM draw water out of the blood vessels and cells by osmosis (see Chapter 2). This process traps water within the ECM, which helps the tissue to resist compression. Recall also that most glycoproteins in the ECM act like molecular "glue" that binds the different components together. The glycoproteins in bone perform this same function by binding to both hydroxyapatite and bone cells.

Quick Check

☐ 1. What are the two components of bone ECM? How do these components differ?

Bone Cells

《 FLASHBACK

1. What are osteoblasts and osteocytes? (p. 145)
2. What is an osteoclast? (p. 145)

Bone is a dynamic tissue, and new bone is continually being formed as older bone is broken down. The dynamic nature of bone is due to the actions of three cell types that we discussed in the tissues chapter (see Chapter 4): osteoblasts, osteocytes, and osteoclasts (**Figure 6.6**). *Osteoblasts* build bone and mature into *osteocytes,* which help to maintain bone. In contrast, the multinucleated *osteoclasts* break down bone. Let's look in more detail at the location, primary functions, and structure of each of these cell types.

- **Osteoblasts.** Osteoblasts (AHS-tee-oh-blasts; *-blast* = "immature cell") are cuboidal to columnar cells found in the inner periosteum and endosteum that build bone (remember the mnemonic "osteoBlasts Build Bone"). As shown in **Figure 6.7 ①**, osteoblasts are derived from flattened cells called **osteogenic cells,** which differentiate into osteoblasts when stimulated by certain chemical signals. Osteoblasts perform the process of *bone deposition,* during which they secrete the organic matrix and aid in formation of the inorganic matrix.
- **Osteocytes.** Osteoblasts become surrounded and eventually trapped by secreted bone matrix in a small cavity known as a *lacuna* (luh-KOO-nuh; *lacun-* = "cavity"; **Figure 6.7 ②**). At this point, these cells are called **osteocytes** (AHS-tee-oh-sytz). Osteocytes secrete chemicals that are required for maintaining the ECM (**Figure 6.7 ③**). They

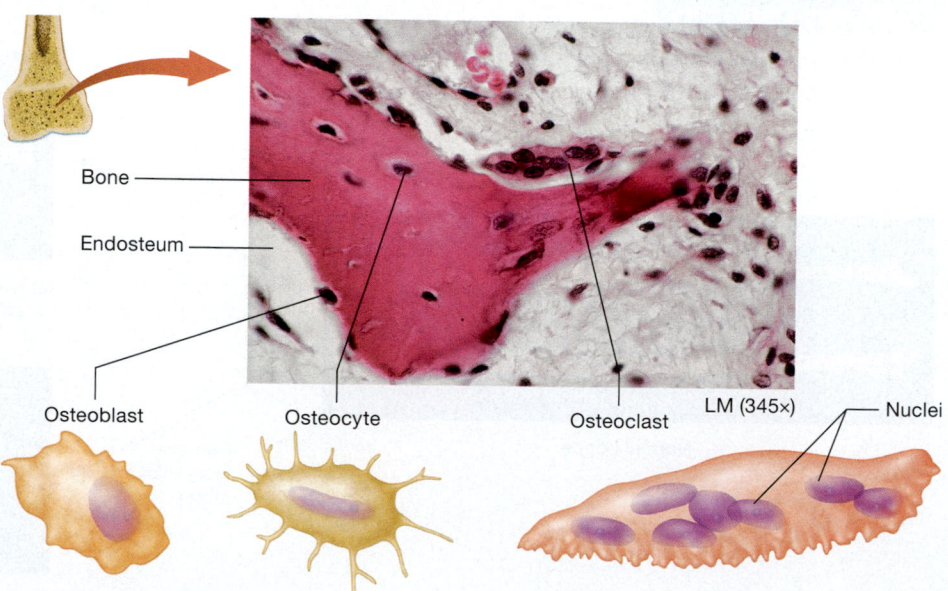

Bone

Endosteum

Osteoblast

Osteocyte

Osteoclast

LM (345×)

Nuclei

Figure 6.6 Types of bone cells.

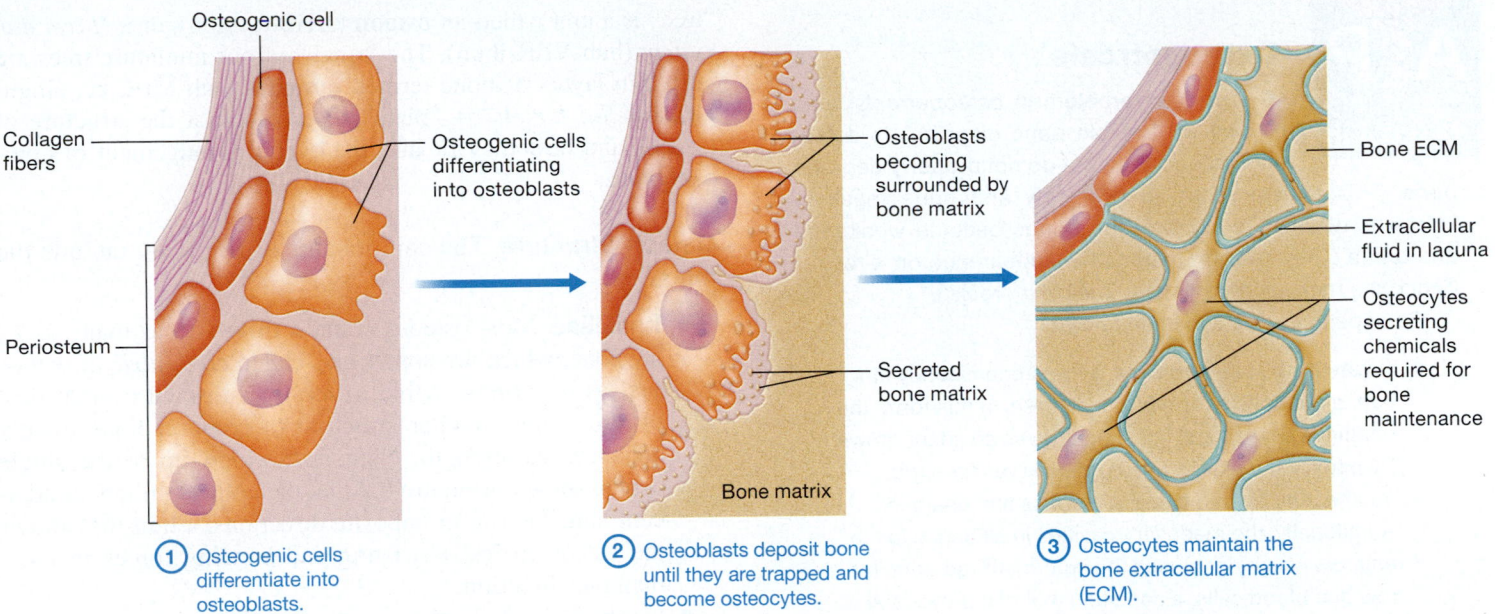

1 Osteogenic cells differentiate into osteoblasts.

2 Osteoblasts deposit bone until they are trapped and become osteocytes.

3 Osteocytes maintain the bone extracellular matrix (ECM).

Figure 6.7 Functions of osteoblasts and osteocytes.

also appear to recruit osteoblasts to build up areas of the bone under tension.

- **Osteoclasts. Figure 6.8** shows that **osteoclasts** (AHS-tee-oh-klasts; *-clasia* = "broken") look quite different from osteoblasts and osteocytes, as they somewhat resemble jellyfish. Osteoclasts are large, multinucleated cells derived from the fusion of cells formed in the bone marrow. They reside in shallow depressions on the internal or external surfaces of bone. Osteoclasts are responsible for the process of *bone resorption,* during which they break down the bone ECM. They accomplish bone

resorption by secreting hydrogen ions and enzymes from a region of the cell called the ruffled border. The hydrogen ions create an acidic environment that dissolves the inorganic matrix, and the enzymes catalyze reactions that break down the organic matrix. Notice that the liberated minerals, amino acids, and sugars then enter the osteoclast, after which they are eventually delivered to the blood for reuse in the body or excretion as waste. See *A&P in the Real World: Osteopetrosis* on p. 192 for an explanation of what happens when osteoclasts cannot perform their functions.

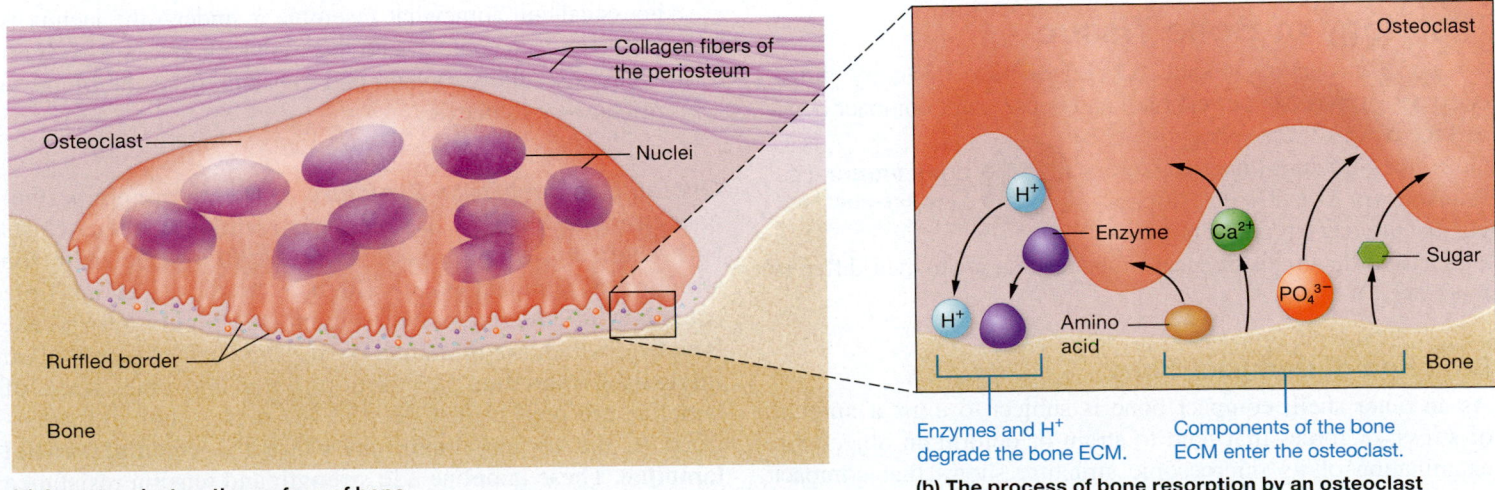

(a) **An osteoclast on the surface of bone**

(b) **The process of bone resorption by an osteoclast**

Figure 6.8 Function of osteoclasts.

Osteopetrosis

The primary problem in *osteopetrosis*, also known as "marble bone disease," is defective osteoclasts that do not properly degrade bone. This causes bone to overgrow and bone mass to increase, while paradoxically the bones become weakened and brittle. They take on a stonelike appearance on x-rays—hence the name of the disease (*petra* = "rock").

Osteopetrosis has two main forms:

- **Infantile osteopetrosis** is the predominantly inherited, more severe form of the disease. In this form, the openings of the skull fail to enlarge as an infant grows. Eventually the skull may trap the nerves for sight and hearing and so cause blindness and deafness. Additionally, the medullary cavities in all bones fail to enlarge, which decreases the amount of red bone marrow and blood cells, a condition that can prove fatal.
- **Adult osteopetrosis** is also inherited but develops during adolescence or adulthood. The symptoms may include bone pain, recurrent fractures, nerve trapping, and joint pain.

Infantile osteopetrosis requires treatment to prevent blindness, deafness, and death. Treatments include drugs to stimulate both osteoclasts and bone marrow. Currently, adult osteopetrosis is only treated symptomatically.

The processes of bone deposition and bone resorption are discussed further with bone remodeling in Module 6.5.

Quick Check

☐ 2. Compare and contrast the locations and functions of osteoblasts, osteocytes, and osteoclasts.

Histology of Bone Tissue

As we discussed in Module 6.1, bones have two very different types of bone tissue: the hard and dense outer compact bone, and the porous inner spongy bone. Given the Structure-Function Core Principle (p. 28), it's not surprising that these two types of bone tissues have different histological structures relating to their different functions.

CORE PRINCIPLE
Structure-Function

Structure of Compact Bone

As an outer shell, compact bone is subject to a great amount of stress, or forces that tend to strain or deform an object. An examination of its microscopic structure shows that compact bone is built in a way that enables it to withstand these stresses quite well. Notice in **Figure 6.9** that the structure of compact bone resembles a forest of small, tightly packed trees. Each

"tree" is a unit called an **osteon** (AHS-tee-ahn), or a *Haversian system* (huh-VER-jhun). The "rings" of our miniature trees are very thin layers of bone termed **lamellae** (luh-MEL-ee; singular, *lamella; lamell-* = "plate"). Let's look at the structure of osteons and how they fit into the overall arrangement of compact bone.

Osteon Structure   The components of an osteon include the following:

- **Lamellae.** Most osteons contain from 4 to as many as 20 lamellae, which are sometimes called *concentric lamellae.* Just as tree rings enable a tree to withstand a great deal of stress, the lamellar structure of compact bone greatly enhances its strength. Note in the blow-up of the single osteon shown in Figure 6.9 that the collagen fibers of adjacent lamellae run in opposite directions, a trait that allows the osteon to resist twisting and bending forces in more than one direction.
- **Central canal.** In the center of each osteon is a passage called the **central canal,** or *Haversian canal.* The central canal contains blood vessels and nerves that supply the cells of the osteon. This canal, like all internal surfaces of bone, is lined by endosteum.
- **Lacunae.** Recall that osteoblasts eventually are surrounded by the ECM they secrete to become osteocytes. These cells reside in the remaining **lacunae,** small cavities that are filled with extracellular fluid and located between lamellae. About 20,000–30,000 osteocytes and lacunae are found in each cubic millimeter of bone.
- **Canaliculi.** Lacunae are connected to one another by tiny canals called **canaliculi** (kan´-al-IK-yoo-lee; "little canals"). Osteocytes have long, thin "arms," cytoplasmic extensions that extend through the canaliculi to contact the arms of other osteocytes. At these places of contact are gap junctions that allow small substances to pass from cell to cell. Note in the illustration and photomicrograph of a sectioned osteon in Figure 6.9 that a lacuna and its surrounding canaliculi somewhat resemble a spider—the lacuna is the body of the spider, and the canaliculi are its legs. Canaliculi also allow oxygen and nutrients from the blood to reach every osteocyte.

Overall Compact Bone Structure   Osteons are not permanent structures, as they are regularly resorbed by osteoclasts and rebuilt by osteoblasts. This process is reflected in the structure of compact bone, which is composed of multiple osteons with lamellae between them (see Figure 6.9). These lamellae, which are remnants of resorbed osteons, are called **interstitial lamellae.** In addition, outer and inner rings of lamellae are present just deep to the periosteum and superficial to the spongy bone; these are the **circumferential lamellae.** These lamellae add strength and tension resistance to the bone as a whole.

The central canals of neighboring osteons are connected by a second type of canal called a **perforating canal** (also known

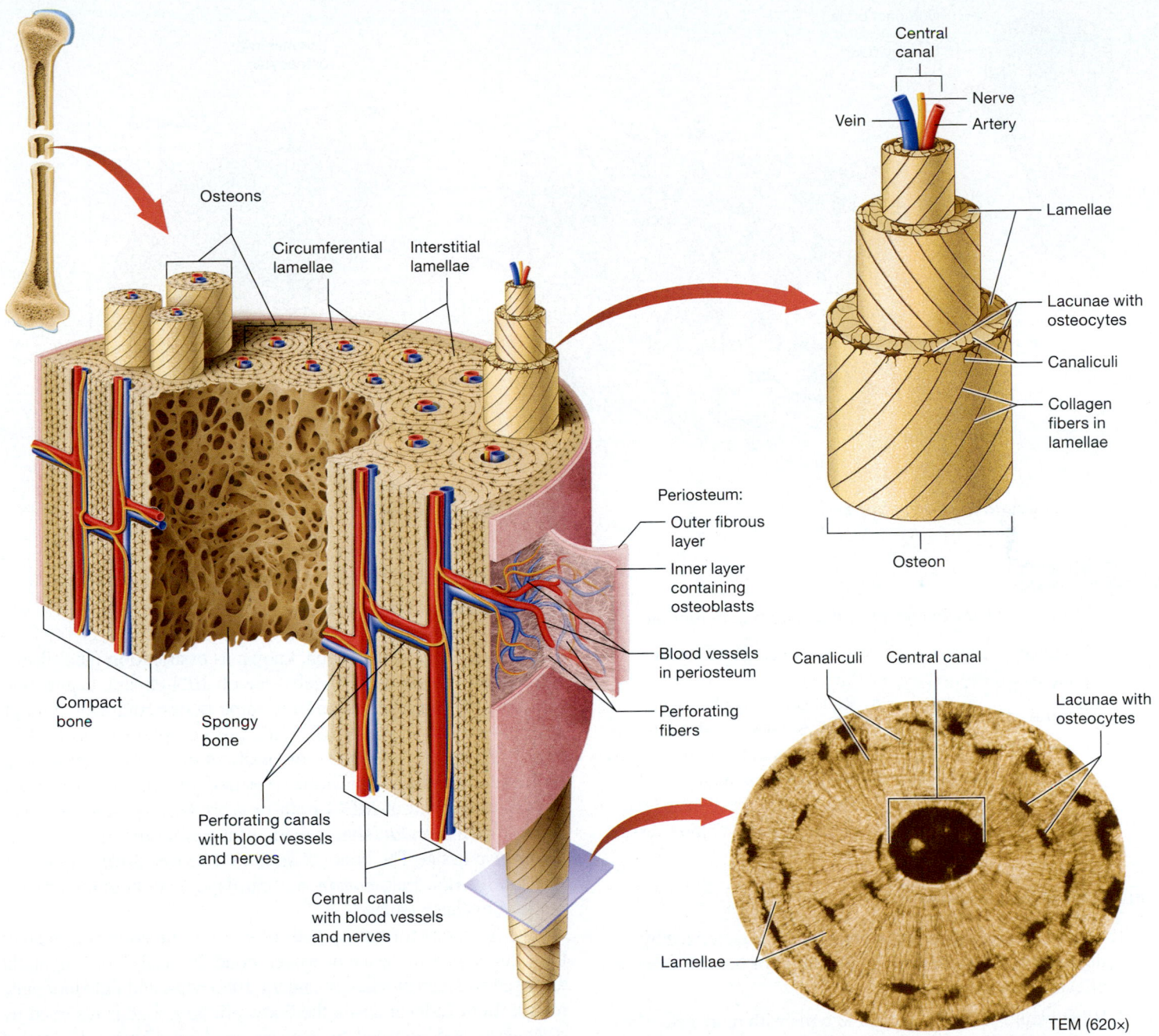

Figure 6.9 Structure of compact bone.

as a *Volkmann canal;* FOLK-men). Observe in Figure 6.9 that perforating canals run perpendicular to the osteons. They carry blood vessels from the periosteum that merge with vessels in the central canals.

Structure of Spongy Bone

Spongy bone is usually not the weight-bearing part of a bone, so it does not have a dense structure like that of compact bone. Instead, spongy bone resists forces from many directions and forms a protective framework for the bone marrow. These functions are performed by branching "ribs" of bone called

trabeculae (truh-BEK-yoo-lee; *trab-* = "beam"), which project into the marrow cavity (**Figure 6.10**). Trabeculae are covered with endosteum and usually do not contain osteons. However, they do contain concentric lamellae, within which we find canaliculi and lacunae housing osteocytes. No central or perforating canals are present within trabeculae, and the cells obtain their oxygen and nutrients from the blood vessels in the bone marrow.

Read *A&P in the Real World: Osteoporosis and Healthy Bone Tissue* on p. 197 to see how both compact and spongy bone can deteriorate over time.

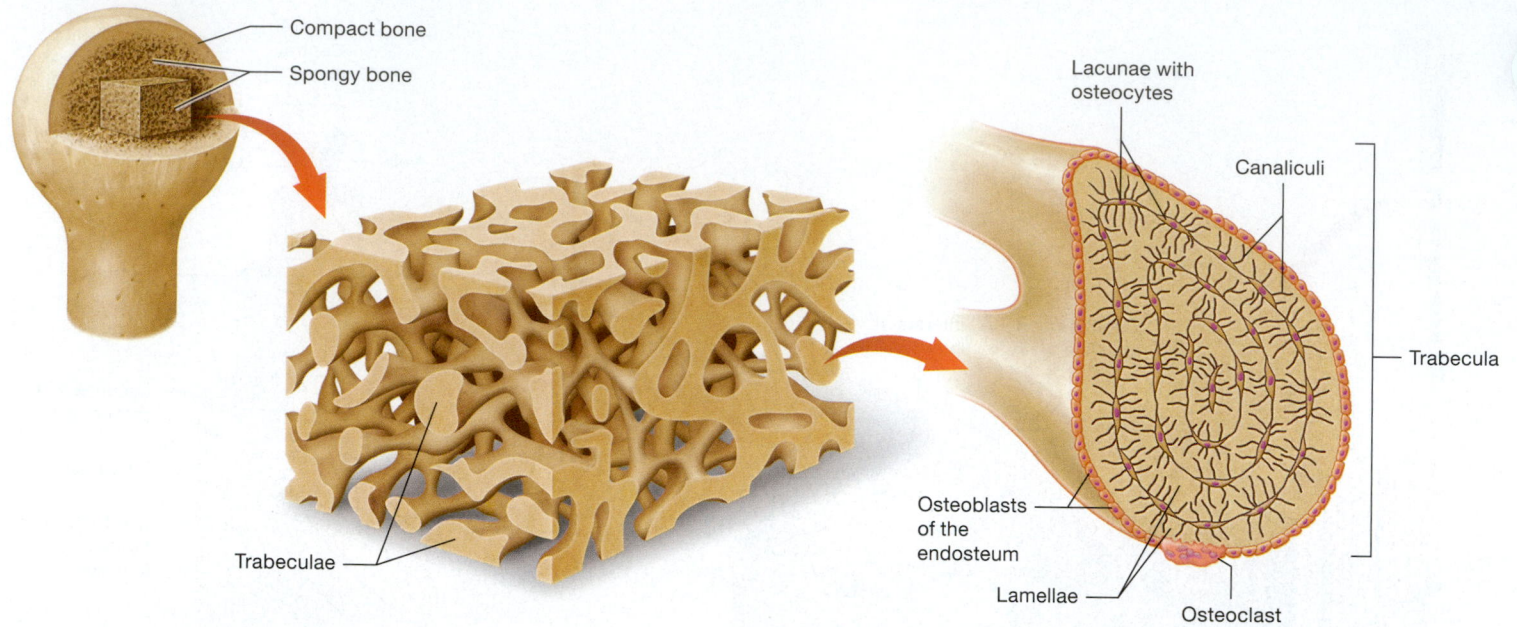

Figure 6.10 **Structure of spongy bone.**

Quick Check

☐ 3.   Match the following terms with the correct definition.

____ Lacunae
____ Lamellae
____ Central canal
____ Canaliculi
____ Perforating canal

a. Tiny canals that connect lacunae
b. Runs perpendicular to the osteons, contains blood vessels
c. Cavities that house osteocytes
d. Rings of bone matrix
e. Runs down the center of the osteon, contains blood vessels

Apply What You Learned

☐ 1. The disease *osteogenesis imperfecta* is characterized by defective collagen in the organic matrix of bone. Predict the effects of such a disease.

☐ 2. What would probably happen to bone with more osteoclast than osteoblast activity?

☐ 3. How does the structure of compact bone follow its function?

See answers in Appendix A.

MODULE 6.3
Bone Formation: Ossification

Learning Outcomes

1. Explain the differences between primary and secondary bone.

2. Describe the process of intramembranous ossification.

3. Describe the process of endochondral ossification.

The process of bone formation, known as **ossification** (ahs′-ih-fih-KAY-shun) or **osteogenesis** (ahs′-tee-oh-JEN-eh-sis), begins during the embryonic period and for some bones continues through childhood. The majority of bones are completely ossified by age 7. There are two types of ossification—intramembranous and endochondral. The bones formed by *intramembranous ossification* (in′-trah-MEM-brah-nus) are built on starting material known as a *model* that is made of a membrane of embryonic connective tissue. The bones formed by *endochondral ossification* (en-doh-KAHN-drul; *chondr-* = "cartilage") are built on a model made of hyaline cartilage.

The first bone formed by both types of ossification is immature bone called *primary bone* or woven bone. It consists of irregularly arranged collagen bundles, abundant osteocytes, and little inorganic matrix. In most locations in the body, primary bone is resorbed by osteoclasts and replaced by mature *secondary bone,* also called lamellar bone. As its alternative name implies, secondary bone has fully formed lamellae with regularly arranged collagen bundles that are parallel to one another, which makes it much stronger than primary bone. In addition, secondary bone contains a higher percentage of inorganic matrix, which contributes to its strength.

In this module, we examine in greater depth each type of ossification and the bones in which they occur. Note, however, that bones formed by either type of ossification are histologically identical once they mature.

Intramembranous Ossification

 **FLASHBACK**

1. Which bones are flat bones? (p. 186)

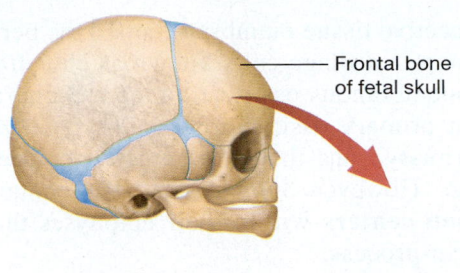

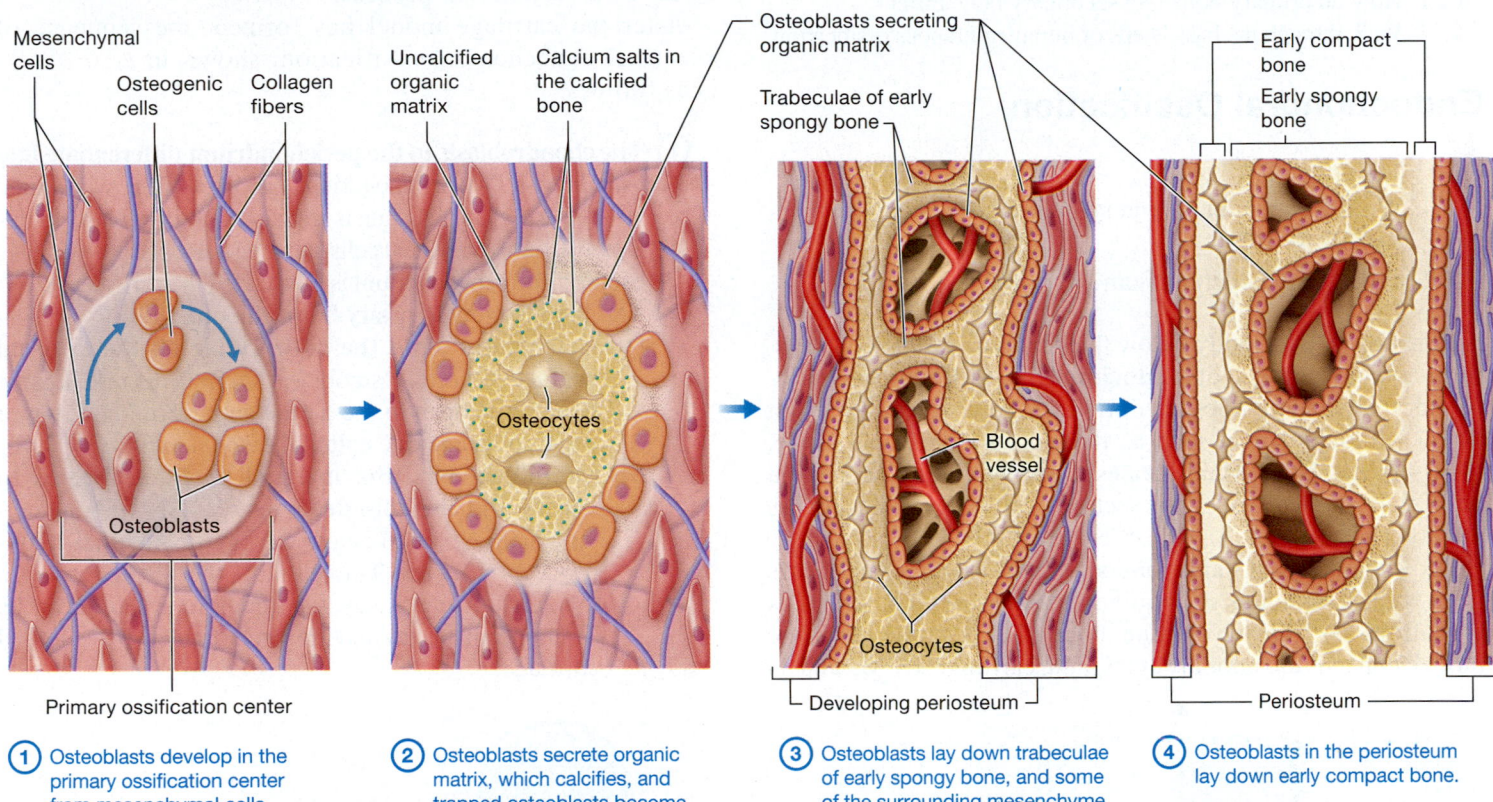

Figure 6.11 **The process of intramembranous ossification.**

① Osteoblasts develop in the primary ossification center from mesenchymal cells.

② Osteoblasts secrete organic matrix, which calcifies, and trapped osteoblasts become osteocytes.

③ Osteoblasts lay down trabeculae of early spongy bone, and some of the surrounding mesenchyme differentiates into the periosteum.

④ Osteoblasts in the periosteum lay down early compact bone.

Many flat bones, including the skull and the clavicles, form during fetal development by the process of **intramembranous ossification.** Bones formed in this way are primary bone, which is eventually resorbed and replaced with secondary bone. This type of ossification occurs within a mesenchymal membrane (mez-en-KY-mal). Recall that a *membrane* is simply a flat sheet of one or more tissues (see Chapter 4). In this case, the membrane is composed of a sheet of embryonic connective tissue called mesenchyme that is rich with blood vessels and mesenchymal cells.

As you consider the events of intramembranous ossification, remember the structure of flat bones—they are essentially a "sandwich" with two layers of compact bone as the "bread" and the spongy bone as the "meat." During intramembranous ossification, the inner spongy bone forms before the outer compact bone, beginning at a place called the **primary ossification center.**

The events of intramembranous ossification, shown in **Figure 6.11**, are as follows:

① **Osteoblasts develop in the primary ossification center from mesenchymal cells.** At the primary ossification center, mesenchymal cells differentiate first into osteogenic cells and then into osteoblasts.

② **Osteoblasts secrete organic matrix, which calcifies, and trapped osteoblasts become osteocytes.** The newly formed osteoblasts secrete the organic matrix of bone. In a few days' time, calcium salts and other components of the inorganic matrix are deposited in the primary ossification center, a process called *calcification,* and the early bone hardens. Trapped osteoblasts then become osteocytes.

③ **Osteoblasts lay down trabeculae of early spongy bone, and some of the surrounding mesenchyme differentiates into the periosteum.** Osteoblasts continue to lay down new bone, forming the trabeculae of the early spongy bone. Over time, the trabeculae enlarge and merge, forming larger trabeculae. About this time, some of the mesenchyme surrounding the developing bone differentiates into the periosteum. Some of the vascular tissue in the forming spongy bone will become bone marrow.

④ **Osteoblasts in the periosteum lay down early compact bone.** Osteoblasts within the periosteum continue to secrete organic bone matrix. This matrix becomes more heavily calcified than the deeper spongy bone trabeculae, and its structure is remodeled to become the immature compact bone.

Larger bones have more than one primary ossification center, and so the resulting "pieces" of bone fuse together to form one complete bone. This process is often incomplete at birth. For example, the *fontanels,* or "soft spots," in the skull of a newborn represent areas of incomplete fusion.

Quick Check

☐ 1. How do primary bone and secondary bone differ?
☐ 2. Walk through the basic steps of intramembranous ossification.

Endochondral Ossification

« FLASHBACK

1. What are the mature and immature cells of cartilage called? (p. 142)
2. What is the perichondrium? (p. 142)

All the bones in the body below the head, except the clavicles, form by the process of **endochondral ossification.** Endochondral ossification begins during the fetal period for most bones, although some, such as those in the wrist and ankle, ossify much later. Many bones complete this ossification process by about age 7. In this section, we'll follow this process in a long bone.

As implied by its name, endochondral ossification occurs from within a model of hyaline cartilage, which serves as a scaffold for the developing bone. The hyaline cartilage model consists of chondrocytes and cartilage ECM, and is surrounded by a connective tissue membrane called the perichondrium and immature cartilage cells known as chondroblasts. As with intramembranous ossification, endochondral ossification begins at primary ossification centers, and the bone laid down is primary bone that is eventually replaced with secondary bone. However, long bones also contain **secondary ossification centers** within their epiphyses that ossify by a very similar process.

After the cartilage model has formed, the sequence of events of endochondral ossification, shown in **Figure 6.12,** is as follows:

① **The chondroblasts in the perichondrium differentiate into osteoblasts.** The perichondrium becomes filled with blood vessels, and chemical signals trigger its chondroblasts to differentiate into osteogenic cells and then osteoblasts. When this happens, the perichondrium is now called the periosteum.

② **The bone begins to ossify from the outside:**

 ②ᵃ **Osteoblasts build the bone collar on the external surface of bone.** Osteoblasts begin to secrete organic bone ECM deep to the periosteum, forming a ring of early compact bone called the *bone collar.*

 ②ᵇ **Simultaneously, the internal cartilage begins to calcify and the chondrocytes die.** As the bone collar calcifies, ECM surrounding the internal chondrocytes calcifies. This cuts off the blood supply to the chondrocytes, causing them to eventually die. Their death leaves cavities surrounded by calcified cartilage.

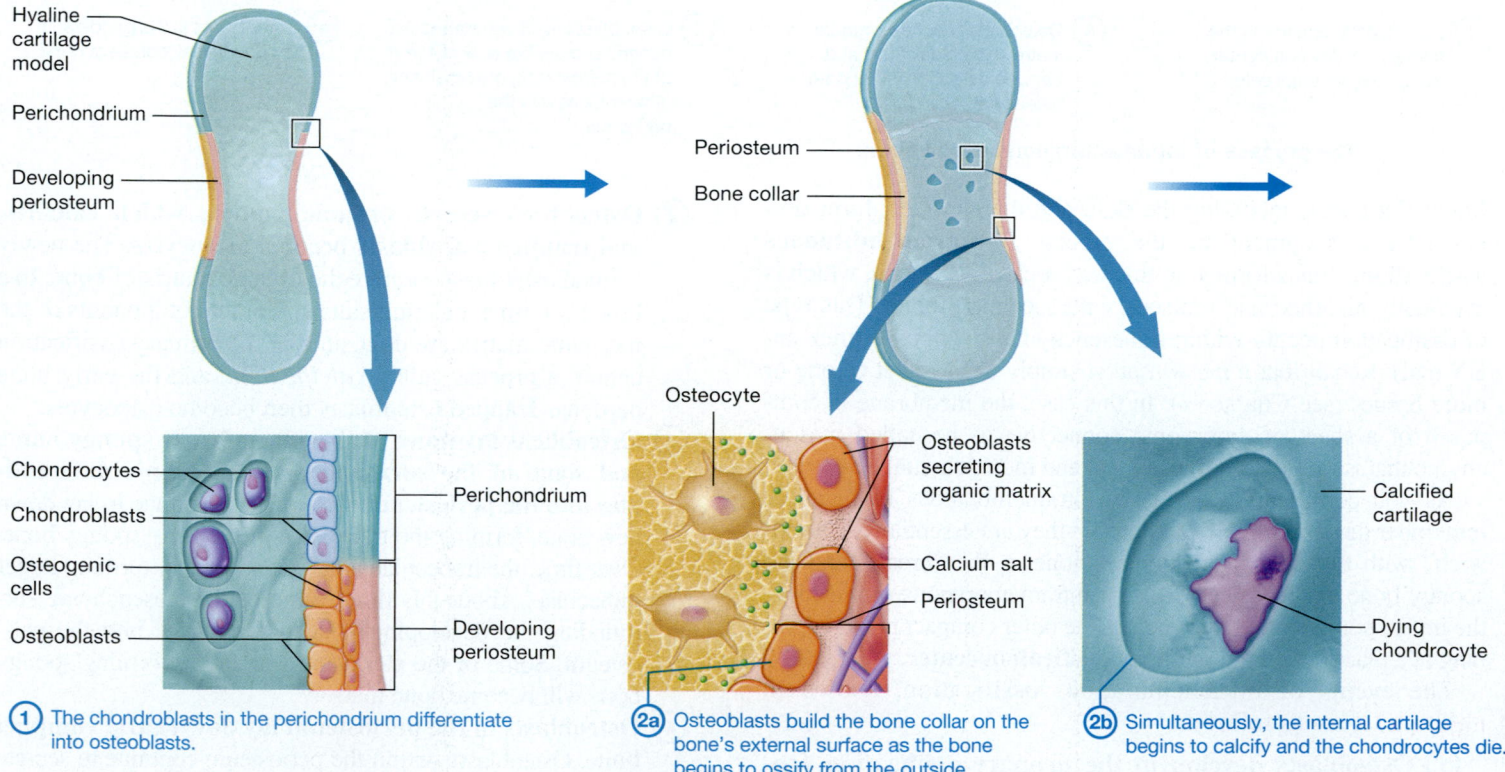

Hyaline cartilage model
Perichondrium
Developing periosteum

Periosteum
Bone collar

Osteocyte

Chondrocytes
Chondroblasts
Osteogenic cells
Osteoblasts
Perichondrium
Developing periosteum

① The chondroblasts in the perichondrium differentiate into osteoblasts.

Osteoblasts secreting organic matrix
Calcium salt
Periosteum

②ᵃ Osteoblasts build the bone collar on the bone's external surface as the bone begins to ossify from the outside.

Calcified cartilage
Dying chondrocyte

②ᵇ Simultaneously, the internal cartilage begins to calcify and the chondrocytes die.

Figure 6.12 The process of endochondral ossification.

Osteoporosis and Healthy Bone Tissue

The bone disease **osteoporosis** is due to inadequate inorganic matrix in the ECM. You can see here the differences between healthy and osteoporotic bone:

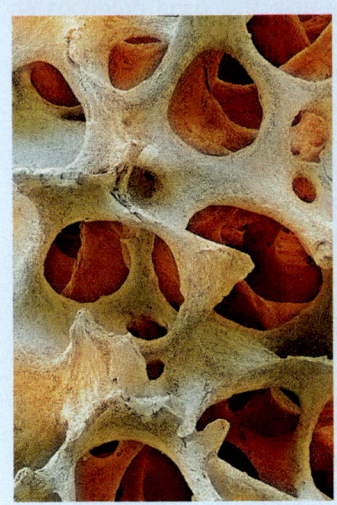

SEM (15×)

Normal bone in vertebra

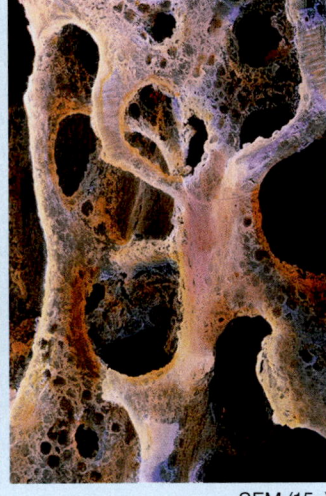

SEM (15×)

Osteoporotic bone in vertebra

Osteoporosis has a variety of causes, including dietary factors (such as calcium ion and vitamin D deficiency); female sex; advanced age; lack of exercise; hormonal factors (such as lack of protective estrogen in postmenopausal women); genetic factors; and diseases of the skin, digestive system, and urinary system. Many of these causes are preventable, and a variety of methods are used to keep osteoporosis from developing, including:

- ensuring adequate dietary calcium salt intake;
- engaging in weight-bearing exercise; and
- replacing estrogen in women, if appropriate.

Osteoporosis makes bone brittle and so dramatically raises the risk of *fractures,* and osteoporotic bone heals more slowly when it is fractured. Although a fractured bone may be little more than a painful inconvenience for a younger person, in an elderly person with osteoporosis, a fracture such as a broken hip can lead to serious complications.

Osteoporosis makes bone brittle and so is treated by preventing further bone loss and by increasing bone mass. All of the preventative methods just described can treat osteoporosis to some extent. Other treatments include drugs that inhibit osteoclasts or stimulate osteoblasts.

③ **In the primary ossification center, osteoblasts replace the calcified cartilage with early spongy bone; the secondary ossification centers and medullary cavity develop.**

Osteoclasts etch a hole in the bone collar that allows a group of blood vessels and bone cells to enter the primary ossification center. Here, osteoblasts replace the calcified cartilage

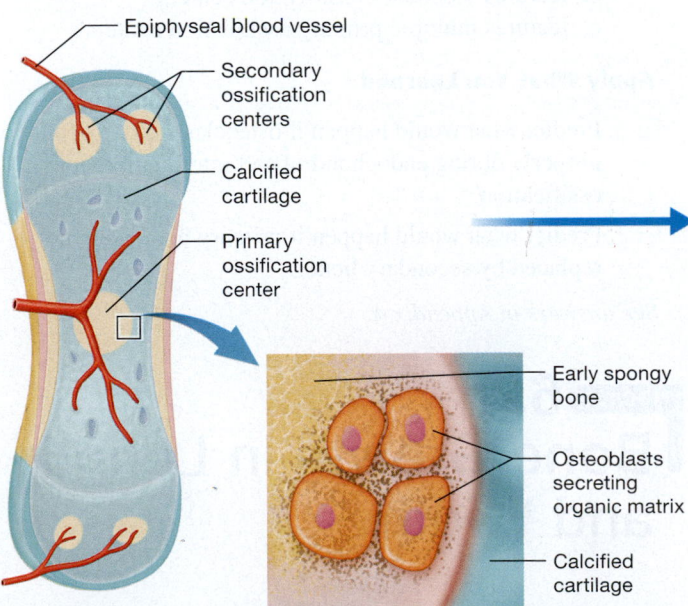

Epiphyseal blood vessel

Secondary ossification centers

Calcified cartilage

Primary ossification center

Early spongy bone

Osteoblasts secreting organic matrix

Calcified cartilage

Articular cartilage

Spongy bone

Epiphyseal plate

Compact bone

Medullary cavity

Osteoclasts enlarging medullary cavity

Osteocytes

Medullary cavity

③ In the primary ossification center, osteoblasts replace the calcified cartilage with early spongy bone; the secondary ossification centers and medullary cavity develop.

④ As the medullary cavity enlarges, the remaining cartilage is replaced by bone; the epiphyses finish ossifying.

Figure 6.12 **The process of endochondral ossification. (*continued*)**

with early spongy bone while other osteoblasts continue to increase the size of the bone collar. As this happens, the cavities enlarge and combine, forming the medullary cavity. Also, secondary ossification centers develop and the epiphyses begin to ossify.

④ **As the medullary cavity enlarges, the remaining cartilage is replaced by bone; the epiphyses finish ossifying.** As ossification continues, the calcified cartilage is replaced with bone. Osteoclasts degrade much of the newly formed spongy bone, which increases the size of the medullary cavity. This space becomes filled with bone marrow. Cartilage remains in only two locations: the epiphyseal plates and the articular cartilage. Cartilage in the epiphyseal plate will eventually be replaced with bone when growth in length ceases; the articular cartilage, however, remains for life.

Although we discussed long bones in this example, a similar process occurs in all bones that form by endochondral ossification. This is evident in the following radiograph of a child's hand, which shows the hyaline cartilage persisting in both the long bones (the metacarpals and phalanges) and the short bones (the carpals):

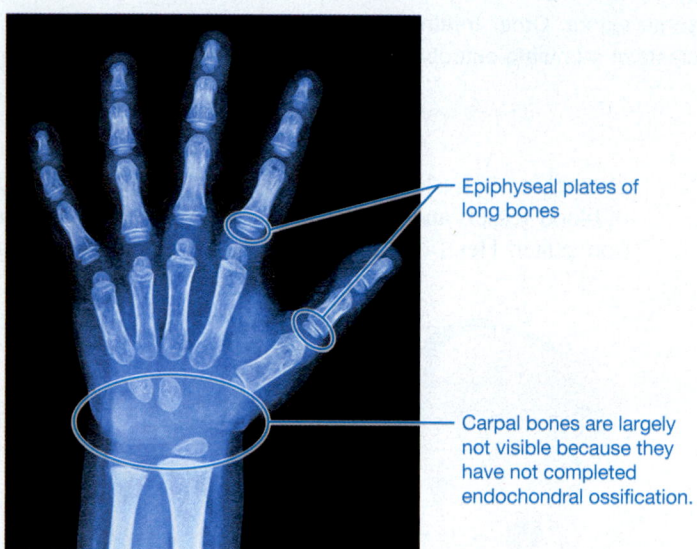

Epiphyseal plates of long bones

Carpal bones are largely not visible because they have not completed endochondral ossification.

Hand of a young child

Notice that both the epiphyseal plates of the long bones and the unfinished carpal bones appear dark on the radiograph, which reflects that both are made of hyaline cartilage.

Certain inherited diseases affect bones formed by endochondral ossification. See *A&P in the Real World: Achondroplasia* to read about one such condition.

Quick Check

☐ 3. What are primary and secondary bone, and how do they differ?

☐ 4. For each type of ossification:
   a. Does spongy bone or compact bone form first?
   b. What is the model for ossification?
   c. Which bones form by each type?

A&P in the Real World

Achondroplasia

The disease *achondroplasia* (ay'-kahn-droh-PLAY-jhah) is the most common cause of dwarfism. It results from a defect in a gene that is either inherited from an affected parent or, more commonly, caused by a new mutation. The defective gene produces an abnormal growth factor receptor on cartilage. In normal development, this growth factor receptor *negatively regulates* growth, meaning that growth is inhibited when it is activated. The mutated gene causes it to be overly active, which decreases endochondral ossification, cell division in the epiphyseal plate, and normal development of articular cartilage.

The abnormal cartilage development of achondroplasia (*achondroplasia* = "without cartilage development") leads to characteristic skeletal features. For example, individuals with achondroplasia are invariably short in stature, averaging around 4 feet in height, with shortened limbs, a disproportionately long trunk, and characteristic facial features. In addition, abnormal spinal curvatures, chest wall deformities, and narrowing of the vertebral cavity are common.

Over the long term, individuals with achondroplasia may develop problems such as joint disorders, respiratory difficulties, and spinal cord compression. However, these symptoms may be managed with medications, and, if necessary, surgery to correct structural deformities.

☐ 5. Which type of ossification:
   a. involves a bone collar?
   b. features secondary ossification centers?
   c. features multiple primary ossification centers?

Apply What You Learned

☐ 1. Predict what would happen if osteoclasts did not function properly during endochondral and intramembranous ossification.

☐ 2. Predict what would happen if primary bone were not replaced by secondary bone.

See answers in Appendix A.

MODULE **6.4**
Bone Growth in Length and Width

Learning Outcomes

1. Describe how long bones grow in length.
2. Compare longitudinal and appositional bone growth.

3. Describe the hormones that play a role in bone growth.

One of the most obvious changes we see in children and young adults is that they grow taller each year. Not only do a child's bones grow in length, they also increase in width. Bone growth is, in many ways, a continuation of the ossification processes that we discussed in the previous module. In this module, we follow the process of bone growth in childhood and throughout life. We also discuss how hormones control bone growth.

Growth in Length

◀◀ FLASHBACK

1. Where is the epiphysis? (p. 187)
2. What is the epiphyseal plate? (p. 187)

Long bones lengthen by a process known as **longitudinal growth.** Interestingly, longitudinal growth occurs not from the division of osteocytes or other bone cells, but from the

division of chondrocytes in the **epiphyseal plate.** Recall from the previous module that the epiphyseal plate is composed of hyaline cartilage that does not ossify during endochondral ossification.

As shown in **Figure 6.13**, the epiphyseal plate contains five different zones of cells, each having a distinct appearance. The zone closest to the epiphysis is the *zone of reserve cartilage,* and it contains cells that are not directly involved in bone growth but that can be called upon to divide if needed. The region just "above" (closer to the diaphysis) the zone of reserve cartilage is the *zone of proliferation,* and it has actively dividing chondrocytes in lacunae. The next region toward the diaphysis is the *zone of hypertrophy and maturation,* and it contains mature chondrocytes. Above that is the *zone of calcification;* it contains dead chondrocytes, some of which are calcified. And finally, the *zone of ossification* has calcified chondrocytes and osteoblasts that build bone.

The last four of these zones are actively involved in longitudinal growth. Most mitotic activity in the epiphyseal plate occurs in the zone of proliferation. As cells divide in this zone, the cells "above" them progressively become part of the next zones.

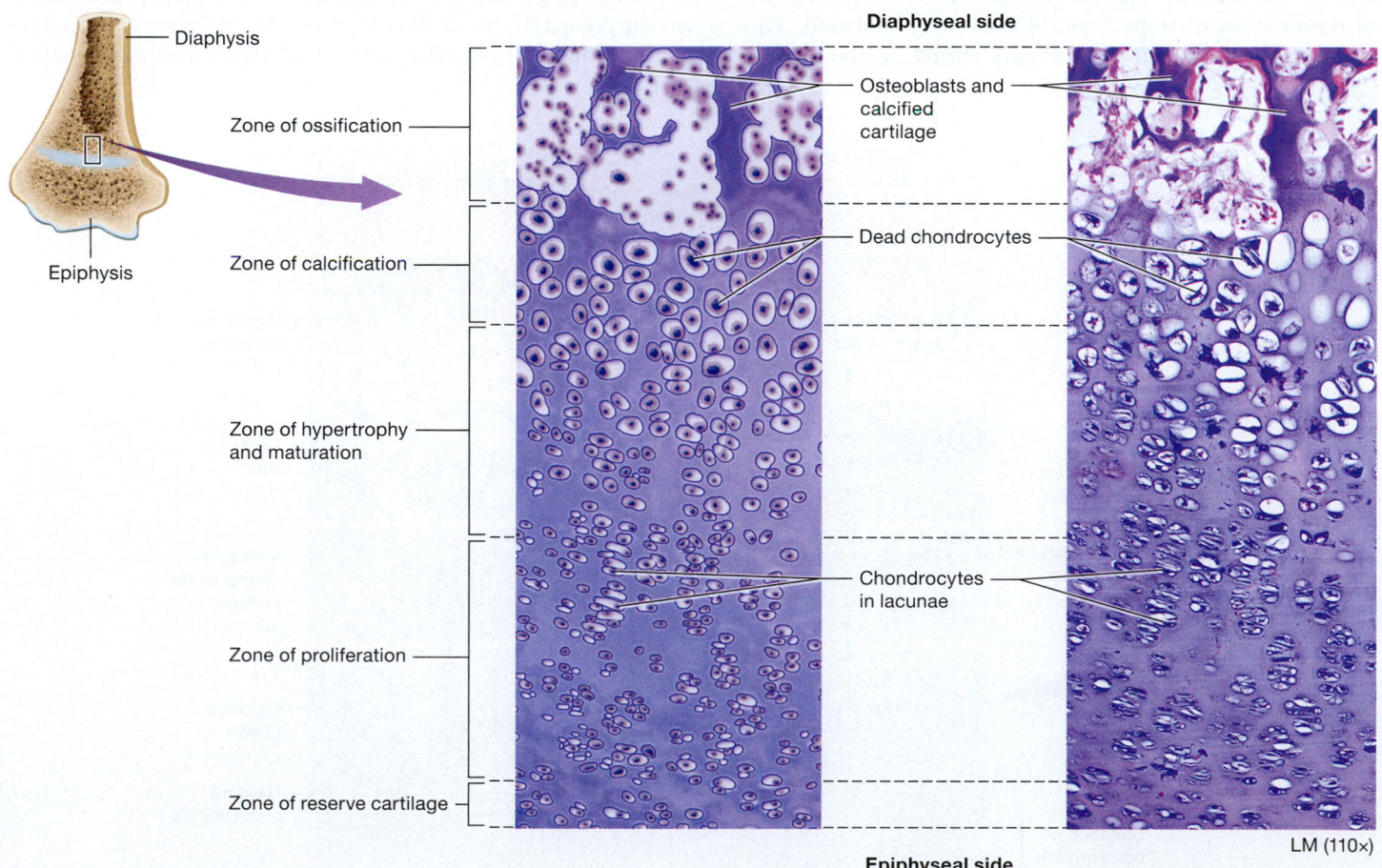

Figure 6.13 Structure of the epiphyseal plate.

Let's look closely at the process of longitudinal growth, which proceeds by the following sequence (**Figure 6.14**):

① **Chondrocytes divide in the zone of proliferation.** The stacked chondrocytes in the zone of proliferation divide, and the cells above them become part of the next zone.

② **Chondrocytes that reach the next zone enlarge and mature.** Chondrocytes that reach the zone of hypertrophy and maturation enlarge and cease dividing. Notice that the lacunae surrounding the chondrocytes are larger in this zone.

③ **Chondrocytes die and their matrix calcifies.** Chondrocytes then reach the zone of calcification. Here, chondrocytes are quite far from the blood supply, so they die and their ECM accumulates calcium salt deposits.

④ **Calcified cartilage is replaced with bone.** In the final zone, the zone of ossification, osteoblasts invade the calcified cartilage and begin to lay down bone on top of it. Eventually, the calcified cartilage/bone is resorbed by osteoclasts and completely replaced with bone. Notice that the new bone is added to the diaphysis.

Longitudinal growth continues at the epiphyseal plate as long as mitosis is happening in the zone of proliferation. However, at about 12–15 years of age, the rate of mitosis slows, but the rate of ossification in steps 3 and 4 continues unabated. This causes the epiphyseal plate to gradually shrink, as the zone of proliferation is progressively overtaken by the zones of calcification and ossification. Eventually, at about age 18–21, the zone of proliferation completely ossifies; at this point, the epiphyseal plate is said to be "closed." Recall from Module 6.1 that a closed epiphyseal plate leaves behind a calcified remnant called the **epiphyseal line.** When just the epiphyseal line remains, the long bone can no longer grow in length.

Quick Check

☐ 1. What tissue type makes up the epiphyseal plate?

☐ 2. How does bone grow in length from the epiphyseal plate?

Growth in Width

Although only long bones undergo longitudinal growth from the epiphyseal plate, all bones grow in width. This growth occurs through a process known as **appositional growth,** during which osteoblasts between the periosteum and the bone surface lay down new bone. Appositional bone growth does not initially result in the formation of new osteons; instead, new circumferential lamellae are formed. As these new lamellae are added, the deeper circumferential lamellae are either removed or incorporated into osteons.

In actively growing bones, appositional growth primarily thickens the compact bone of the diaphysis. As this occurs, osteoclasts in the medullary cavity digest the inner circumferential lamellae,

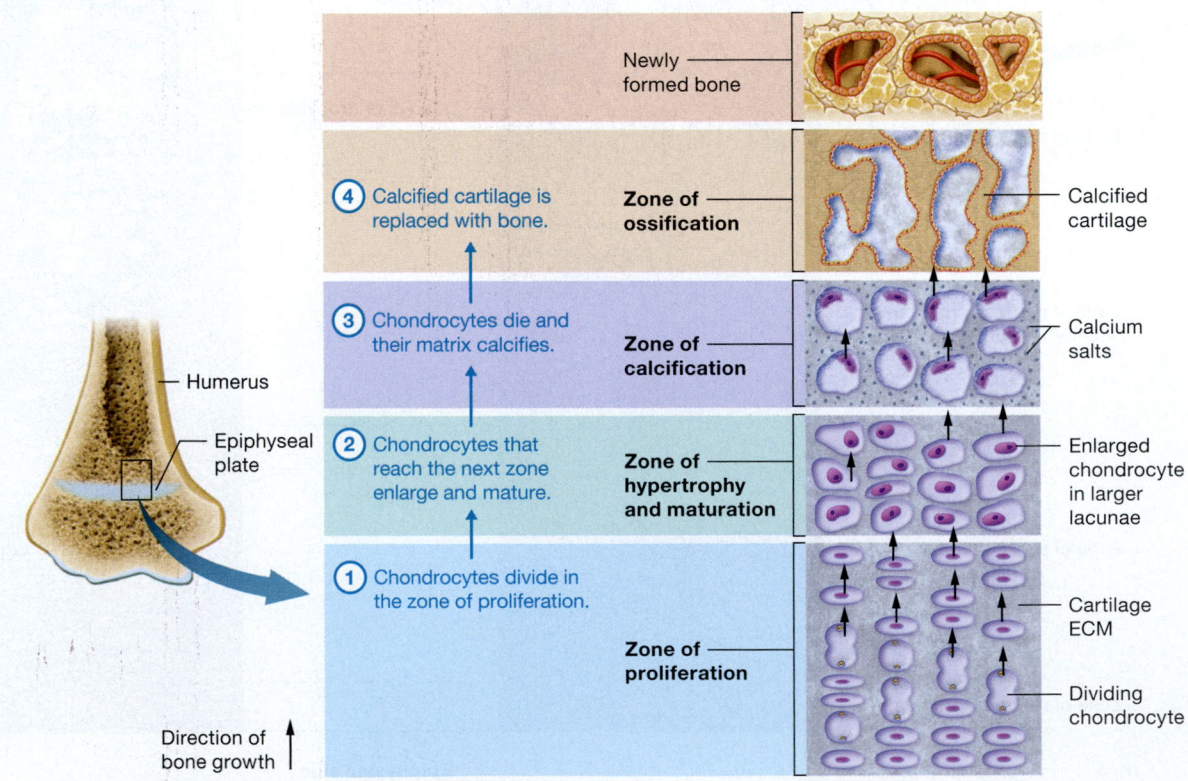

Figure 6.14 Growth at the epiphyseal plate.

so that as the bones increase in width, their medullary cavities enlarge as well. Bone growth in width may continue after bone growth in length ceases, depending on factors such as hormones, the forces to which the bone is subjected, and diet. These factors are discussed in Module 6.5.

> **Quick Check**
>
> ☐ 3. How is appositional growth different from longitudinal growth?

The Role of Hormones in Bone Growth

◀◀ FLASHBACK

> 1. What is an endocrine gland? (p. 134)

Although the processes of longitudinal and appositional bone growth are the same for everyone, clearly some of us grow more than others. This is because multiple factors play a role in determining how much cell division occurs at the epiphyseal plate and how long this process continues. One of the main factors that influence bone growth is a group of chemicals called *hormones,* which are secreted by the cells of endocrine glands into the blood, another example of the Cell-Cell Communication Core Principle (p. 28). An important hormone is *growth hormone,* produced by the anterior pituitary, an endocrine gland below the brain. Growth hormone is secreted throughout life but is produced in the highest amounts during infancy and childhood. It enhances protein synthesis and cell division in nearly all tissues, and it has several effects on bone tissue. These effects include the following:

 **CORE PRINCIPLE** Cell-Cell Communication

- an increase in the rate of mitosis of chondrocytes in the epiphyseal plate, promoting longitudinal growth;
- an increase in the activity of osteogenic cells, including their activity in the zone of ossification; and
- direct stimulation of osteoblasts in the periosteum, triggering appositional growth.

Take a look at *A&P in the Real World: Gigantism and Acromegaly* for a discussion of what happens when a person has too much growth hormone.

A second hormone with a pronounced effect on bone growth is the male sex hormone *testosterone.* Testosterone increases appositional bone growth, causing bones in males to become much thicker and have greater calcium salt deposits than bones in females. Testosterone also increases the rate of mitosis at the epiphyseal plate, and so the large increases in testosterone that occur during the teenage years are accompanied by "growth spurts." Although testosterone increases longitudinal bone growth, it also accelerates the closure of the epiphyseal plates. This is why males castrated at a young age are often taller than unaltered adult males.

> ### A&P in the Real World
>
> ## Gigantism and Acromegaly
>
> The presence of excess growth hormone can result in one of two conditions, depending on when in life it develops. If excess growth hormone is secreted in childhood before closure of the epiphyseal plates, a condition called *gigantism* develops. People with gigantism are very tall due to excessive longitudinal and appositional bone growth.
>
> If excessive growth hormone secretion occurs after closure of the epiphyseal plates, a different condition known as *acromegaly* (ak′-roh-MEG-uh-lee) results. Acromegaly does not cause an increase in height; however, it does cause enlargement of bone, cartilage, and soft tissue. For this reason, we see enlargement of certain parts of the body, such as the skull and the bones of the face, the tongue, and the hands and feet. In addition to significant disfigurement, acromegaly can cause heart and kidney malfunction and is associated with the development of diabetes mellitus.
>
> Both gigantism and acromegaly are generally caused by a tumor that secretes growth hormone. Removal of the tumor is often an effective treatment for both conditions.

The female sex hormone *estrogen* also plays a role in bone growth. Like testosterone, estrogen increases the rate of longitudinal bone growth and inhibits osteoclasts. When the estrogen level increases during the teenage years, a "growth spurt" similar to that in males begins in females. Like testosterone, estrogen also accelerates closure of the epiphyseal plate, although it has a much more potent effect on epiphyseal plate closure than does testosterone. This is partly the reason why women are generally shorter in stature than men.

> **Quick Check**
>
> ☐ 4. How does growth hormone affect bone growth?
>
> ☐ 5. How is bone growth affected by the sex hormones estrogen and testosterone?

> **Apply What You Learned**
>
> ☐ 1. Malnutrition in children often leads to a decreased rate of mitosis in different cell populations. What effect would this have on growth at the epiphyseal plate? Explain.
>
> ☐ 2. How can excessive growth hormone increase the size of bones even after the epiphyseal plates have closed?
>
> ☐ 3. A young girl develops a tumor that secretes excess estrogen. Predict the effect of this excessive secretion on her bone growth and final height.

See answers in Appendix A.

MODULE **6.5**
Bone Remodeling and Repair

Learning Outcomes

1. Describe the processes of bone resorption and bone deposition.
2. Describe the physical, hormonal, and dietary factors that influence bone remodeling.
3. Explain the role of calcitonin, parathyroid hormone, and vitamin D in bone remodeling and calcium ion homeostasis.
4. Describe the general process of bone repair.

Even when bones finish growing in length, bone tissue is far from static. Bone undergoes a continual process of formation and loss called **bone remodeling.** During this process, new bone is formed by **bone deposition,** and old bone is destroyed by **bone resorption.** It may seem odd that our cells make all this new bone only to eventually destroy it. But this cycle of bone formation and loss is carried out for a number of reasons, including maintenance of calcium ion homeostasis; bone repair; replacement of primary bone with secondary bone; replacement of older, brittle bone with newer bone; and bone adaptation to tension and stresses. In this module, we explore the basic process of bone remodeling, the factors that influence it, and how bone is repaired when it's damaged.

Bone Remodeling

《 FLASHBACK

1. Where and how is vitamin D produced? (p. 164)

In Module 6.2, we introduced the cells responsible for bone remodeling: osteoblasts and osteoclasts. In healthy bone, the two processes of formation and loss occur simultaneously, with bone continually being built by osteoblasts and broken down by osteoclasts:

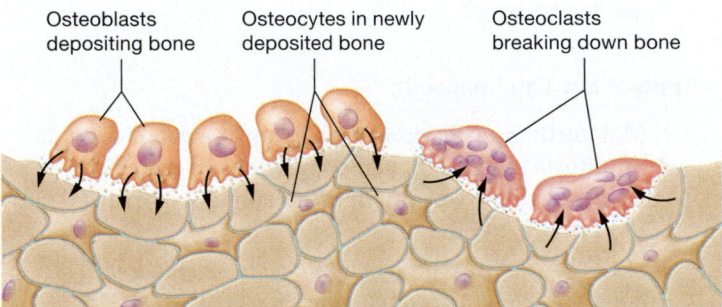

Osteoblasts depositing bone   Osteocytes in newly deposited bone   Osteoclasts breaking down bone

Bone deposition   Bone resorption

Now we turn to how these cells carry out their functions. The next subsections examine how bone remodeling takes place and the factors that influence it. We then focus on how bone remodeling affects calcium ion homeostasis.

Process of Bone Remodeling

Let's look in more depth at the processes of bone deposition and resorption.

- **Bone deposition.** Bone deposition is carried out by osteoblasts in the periosteum and endosteum. Osteoblasts make the components of the organic matrix, as well as facilitate the formation of inorganic matrix. They secrete certain proteoglycans and glycoproteins that bind to calcium ions, and it appears that they also secrete vesicles containing calcium ions, ATP, and enzymes. These vesicles bind to collagen fibers, and their calcium ions eventually crystallize, which ruptures the vesicle and begins the overall process of calcification.

- **Bone resorption.** During bone resorption, osteoclasts secrete hydrogen ions (H^+) from their ruffled borders onto the bone ECM. The hydrogen ions make the pH more acidic, which breaks down the pH-sensitive hydroxyapatite crystals in the inorganic matrix. The released calcium ions and other minerals may then be reused in the body. To degrade organic matrix, osteoclasts secrete enzymes to catalyze reactions that break down proteoglycans, glycosaminoglycans, and glycoproteins. The breakdown products are taken into the osteoclasts for possible reuse.

In childhood, deposition far outweighs resorption, which makes sense, since this is a period of active growth. However, when the epiphyseal plates close and longitudinal growth ceases, bone deposition approximately equals bone resorption—in healthy young adults, total bone mass remains constant. In addition, osteons are continually being rebuilt in response to many factors, including the signals that maintain calcium ion homeostasis.

Bone Remodeling in Response to Tension and Stress

One of the primary influences on bone remodeling is the stress placed on bones because bone deposition occurs in proportion to it. Stated simply, the heavier the load a bone must carry, the more bone tissue is deposited in that bone. This load doesn't necessarily refer to *weight* per se, but rather to the amount of *compression.* Compression is the act of squeezing or pressing together; this force results when the bones are, for example, pressed between the body's weight and the ground. So an athlete who exercises and trains extensively with weights places intense compression forces on his or her bones, and as a result deposits more bone tissue, leading to higher bone mass. Conversely, a sedentary person who engages in few weight-bearing activities has decreased osteoblast activity but normal osteoclast activity. Overall, this results in net bone resorption, which decreases total bone mass.

Tension and pressure placed on the bone also affect bone remodeling. Tension is a stretching force, whereas pressure is application of a continuous downward force. Where tension is placed, osteoblasts are stimulated and bone deposition occurs. However, where continuous pressure is placed, osteoclasts are stimulated and bone resorption occurs. These principles allow an orthodontist to straighten a person's teeth with braces.

Other Factors Influencing Bone Remodeling

Numerous other factors influence bone deposition and resorption, including a variety of hormones. The male hormone testosterone promotes bone deposition, and the female hormone estrogen depresses osteoclast activity. Age also plays an important role in bone remodeling, because levels of hormones such as growth hormone decline with advancing age, which decreases protein synthesis in bone.

Nutrient intake also influences bone remodeling, including:

- **Calcium ion intake.** Obviously, if calcium ion intake is too low, bone deposition will not take place. This explains why people are encouraged to include calcium ion–rich foods in their diets.
- **Vitamin D intake.** Recall that vitamin D is a steroid synthesized by the body in response to exposure of the skin to UV light (see Chapter 5). Vitamin D acts mainly on the intestines, where it promotes calcium ion absorption, and on the kidneys, where it prevents loss of calcium ions to the urine. The net effect of vitamin D on bone is to increase bone deposition and bone mass. However, not everyone gets enough sun exposure to produce as much as the body needs. In these cases, supplementation of the diet with vitamin D may be required.
- **Vitamin C intake.** Remember that vitamin C is required for the synthesis of collagen proteins (see Chapter 4). If vitamin C intake is inadequate, collagen synthesis in bone decreases.
- **Vitamin K intake.** There is good evidence that vitamin K aids in the production of calcium ion–binding glycoproteins by osteoblasts. Inadequate vitamin K intake may decrease bone deposition.
- **Protein intake.** An adequate dietary intake of protein is necessary for osteoblasts to synthesize the collagen fibers needed for the organic matrix of bone.

Bone Remodeling and Calcium Ion Homeostasis

As we discussed in Module 6.1, bone stores most of the body's calcium ions. Calcium ions are required for a number of critical processes, including contraction of the heart and skeletal muscles, transmission of nerve impulses, and blood clotting. If the calcium ion concentration in the blood drops too low or rises too high, major disruptions in homeostasis, up to and including death, can occur.

One main negative feedback loop works to maintain calcium ion homeostasis (**Figure 6.15**), in an example of the Feedback

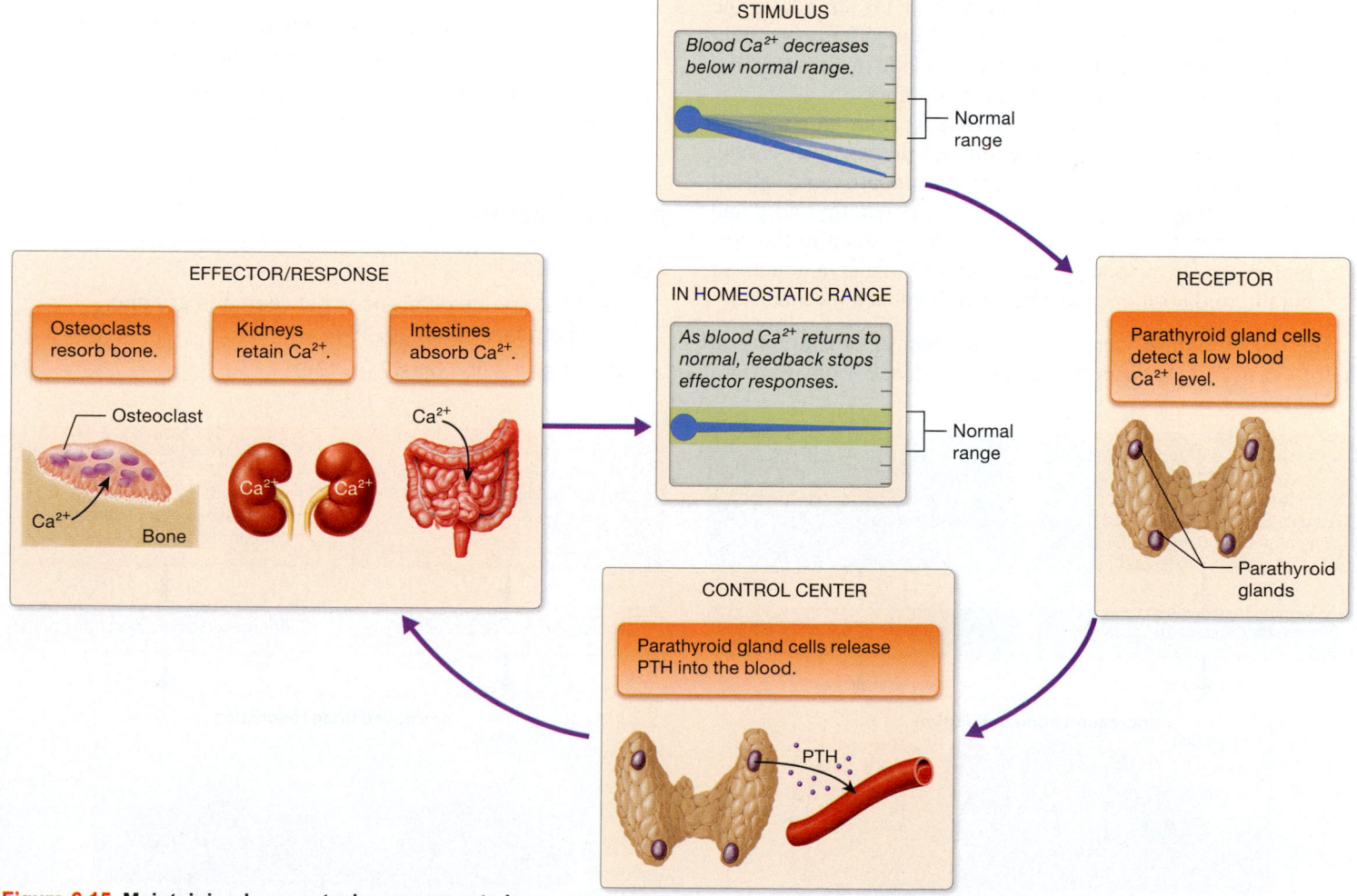

Figure 6.15 Maintaining homeostasis: response to low blood calcium ion level by a negative feedback loop.

Loops Core Principle (p. 28). Like all negative feedback loops, it begins with deviation from a set point—in this case, a decreased concentration of calcium ions in the blood. The response of the feedback loop involves a hormone called parathyroid hormone (PTH). The loop proceeds as follows:

CORE PRINCIPLE
Feedback Loops

- **Stimulus: Blood calcium ion level decreases below normal range.**
- **Receptor: Parathyroid gland cells detect a low blood calcium ion level.** The cells of the parathyroid gland act as receptors that detect a lower than normal concentration of calcium ions in the blood.
- **Control center: Parathyroid gland cells release parathyroid hormone (PTH) into the blood.** The cells of the parathyroid gland also act as control centers. When they are stimulated, they release PTH into the bloodstream.
- **Effector/response: Parathyroid hormone stimulates effects that increase the blood calcium ion level.** PTH leads to increased osteoclast activity, which causes the breakdown of bone and the subsequent release of calcium ions into the blood. Note that this actually weakens the bones, but prevents much more serious imbalances in calcium ion homeostasis. In addition, PTH stimulates the intestines to absorb calcium ions and the kidneys to retain calcium ions. Each of these actions works to increase the blood calcium ion level to the normal range.
- **Homeostasis and negative feedback: The calcium ion concentration returns to the normal homeostatic range, and negative feedback decreases parathyroid gland cell secretion of PTH.** The cells of the parathyroid gland detect when the calcium ion level has returned to the normal range. In response, they secrete smaller amounts of parathyroid hormone via a negative feedback mechanism.

The response to an increase in the number of calcium ions in the blood triggers a different negative feedback loop. The first response is a decline in parathyroid hormone secretion. Next, secretion of the hormone **calcitonin** (kal-sih-TOH-nin) is triggered; this hormone is produced by the cells of the thyroid gland, located in the anterior neck. Calcitonin has essentially the opposite effect of PTH, as it decreases both the activity of osteoclasts and the formation of new osteoclasts. This increases bone deposition, which pulls calcium ions out of the blood and deposits them into the bone. Note that calcitonin seems to be most critical during times of active bone remodeling, as occurs during longitudinal bone growth. In adults, calcitonin doesn't appear to be as potent a regulator of blood calcium ion concentration as is PTH.

Vitamin D also plays an important role in calcium ion homeostasis. If vitamin D is absent, few calcium ions are absorbed from the intestines, regardless of how much is taken in via the diet. Inadequate vitamin D intake in children leads to the condition called *rickets,* characterized by bone pain, bony deformities, increased risk of fractures, and muscle weakness.

The factors that influence bone remodeling are summarized in **Figure 6.16**.

Quick Check

☐ 1. How do bone resorption and bone deposition differ?

☐ 2. Explain how compression, continuous pressure on a bone, and dietary factors influence bone remodeling.

☐ 3. What role does parathyroid hormone play in calcium ion homeostasis?

Bone Repair

« FLASHBACK

1. What are the differences between primary and secondary bone? (p. 194)
2. What is a fibroblast? (p. 138)

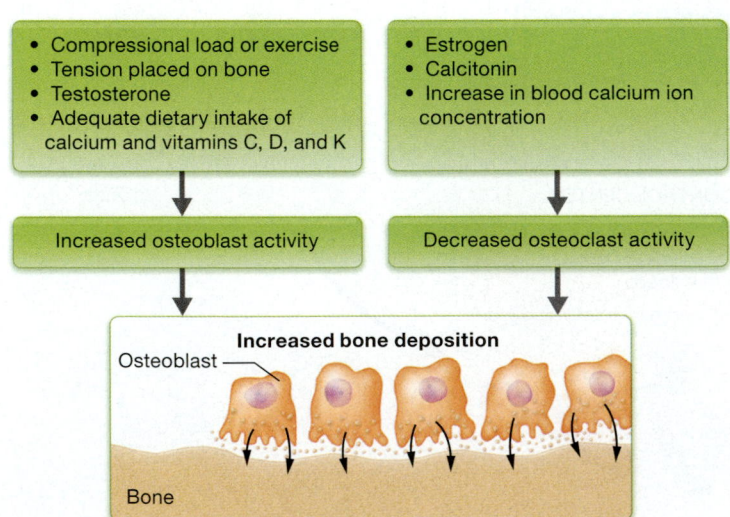

Figure 6.16 Factors that influence bone remodeling.

Given the intricate and complex structure of bone, it's not surprising that many different components are damaged or destroyed when a bone is injured, including the bone ECM, bone cells, the periosteum, the endosteum, and blood vessels and nerves. The most dramatic bone injury is a **fracture,** commonly referred to as a "broken bone." Fractures are grouped into many different classes, some of which are shown in **Table 6.1**. The most basic way to classify fractures is simple or compound.

Table 6.1 Types of Fractures

Fracture Type	Description	Fracture Type	Description
Spiral	Fracture resulting from twisting forces applied to the bone	Compression	Fracture in which the bone is crushed under the weight it is meant to support; common in the elderly and those with reduced bone mass
Comminuted [kom-ih-NOOT-'d]	Fracture in which the bone is shattered into multiple fragments; difficult to repair	Avulsion [ah-VUL-shun]	Fracture in which a tendon or ligament pulls off a fragment of bone; often seen in ankle fractures
Greenstick	Fracture in which the bone breaks on one side but only bends on the other side, similar to the break observed when a young ("green") twig is bent; common in children, whose bones are more flexible	Epiphyseal plate	Fracture that involves at least part of the epiphyseal plate; occurs only in children and young adults; may interfere with growth

Simple fractures (also called closed fractures) are those in which the skin and tissue around the fracture remain intact, whereas *compound fractures* (also called open fractures) involve damage around the fracture. Beyond this basic classification, fractures are also classified according to the type of trauma inflicted on the bone. For example, a *spiral fracture* results when twisting forces are applied to a bone, and a *compression fracture* occurs when a bone is crushed under the weight it is meant to support.

The general process of fracture healing occurs via the following steps (**Figure 6.17**):

① **A hematoma fills the gap between the bone fragments.** When a bone is damaged, its blood vessels rupture and bleed into the injured site, forming a **hematoma** (hee-muh-TOH-muh; also known as a *blood clot*), a mass of blood cells and proteins that resembles grape jelly. The hematoma cuts off the blood supply to the damaged area, and the bone cells surrounding the area die.

② **Fibroblasts and chondroblasts infiltrate the hematoma and a soft callus forms.** Fibroblasts and small blood vessels in the periosteum invade the hematoma in the first week after the fracture. The fibroblasts secrete collagen fibers and form dense irregular collagenous connective tissue that starts to bridge the gap between the bone fragments. At the same time, some of the osteogenic cells from the endosteum become chondroblasts that secrete hyaline cartilage. This mixture of hyaline cartilage and collagenous connective tissue is called a **soft callus** (KAL-us; *callus* = "hard skin").

③ **Osteoblasts build a bone callus.** Osteoblasts within the periosteum begin laying down a collar of primary bone called a **bone callus** (or *hard callus*). Initially, only part of the bone callus is actually bone tissue; the remainder is the soft callus. This process continues over the next several weeks until the entire callus is made of primary bone that forms a bridge between the bone fragments. As we discussed earlier, primary bone is not as strong as secondary bone; however, notice in Figure 6.17 that the bone

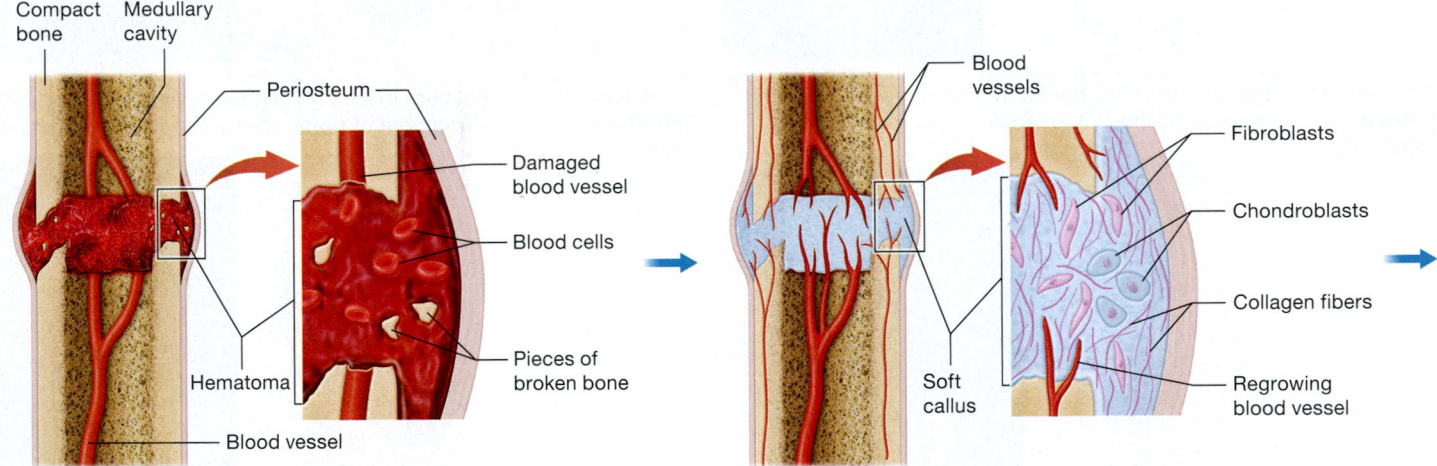

① A hematoma fills the gap between the bone fragments.

② Fibroblasts and chondroblasts infiltrate the hematoma, and a soft callus forms.

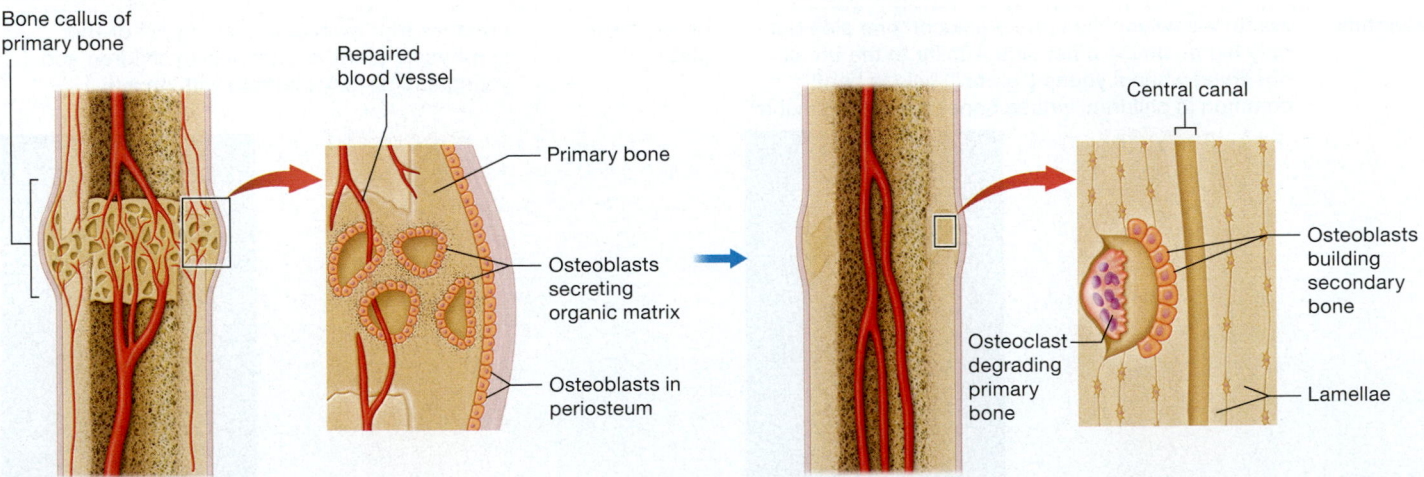

③ Osteoblasts build a bone callus.

④ The bone callus is remodeled and primary bone is replaced with secondary bone.

Figure 6.17 The process of fracture repair.

callus extends beyond the border of the bone, which helps to strengthen it.

④ **The bone callus is remodeled and primary bone is replaced with secondary bone.** Over the next several months, the bone callus is remodeled and the primary bone is resorbed and replaced with secondary bone. Eventually, the bone callus is fully resorbed and the bone regains its previous structure and strength. However, the bone callus often remains visible for months or longer following full healing of the injury.

The primary treatment for fractures is stabilization of the fracture, followed by immobilization. Many fractures may be stabilized through *closed reduction,* in which the bone ends are brought into contact by simply manipulating the body part. More severe fractures may require *open reduction,* in which the fracture is fixated surgically with, for example, plates, wires, and screws. After the fracture is stabilized, it is immobilized for a period of around 6 weeks while the repair process occurs.

The process of fracture repair is imperfect and doesn't always proceed as we would hope. For example, if the fracture is inadequately stabilized or immobilized, or if the blood supply to the bone is poor, the ends of the bone may not fuse and grow together. In this case, the bone ends are united only by scar tissue or cartilage. This situation generally requires surgical removal of the scar tissue and refixation of the bone.

Quick Check

☐ 4. Walk through the basic steps of fracture repair.

Apply What You Learned

☐ 1. Why do you think that astronauts are often faced with decreased bone mass after returning from periods of prolonged weightlessness?

☐ 2. Predict the effect that a parathyroid hormone–secreting tumor would have on bone tissue. What would happen to the concentration of calcium ions in the blood with such a condition?

See Answers in Appendix A.

Chapter Summary

For everything you need to succeed in this course, go to Mastering Anatomy & Physiology. There you will find:

- 🖱 Practice Tests
- 🎙 Author Podcasts
- ▶ Big Picture Animations
-))) Concept Boost Video Tutors
- **iP2™** Interactive Physiology 2.0 Tutorials
- **A&PFlix** A&P Flix 3D Animations
- ✏ Active-Learning Workbook

MODULE **6.1**
Introduction to Bones as Organs 184–189

- Bones are the main organs of the **skeletal system,** which also includes joints and other supporting structures.
- Functions of the skeletal system include protection, mineral storage, maintenance of acid-base homeostasis, blood cell formation, fat storage, movement, and support.
- There are five different shapes of bones: **long bones, short bones, flat bones, irregular bones,** and **sesamoid bones.**
- Bones are covered with the **periosteum,** which sits on top of hard, outer **compact** bone. The inner, honeycomb-like bone is **spongy bone.**
- A long bone has a **diaphysis,** or shaft, and two **epiphyses,** the ends of the long bone.

- Short, flat, irregular, and sesamoid bones consist of two thin layers of compact bone with spongy bone in the middle.
- **Red bone marrow** contains hematopoietic tissue. **Yellow bone marrow** houses blood vessels and adipocytes.

MODULE **6.2**
Microscopic Structure of Bone Tissue 189–194

- Bone has a unique extracellular matrix (ECM):
  - The **inorganic matrix** is composed primarily of calcium and phosphorus salts existing as **hydroxyapatite crystals.**
  - The **organic matrix** is called **osteoid** and consists of collagen fibers and other ECM components.
- Three main types of cells are found in bone:
  - **Osteoblasts** are responsible for **bone deposition.**
  - **Osteocytes** are mature osteoblasts that have become surrounded by the ECM they have secreted.
  - **Osteoclasts** are responsible for **bone resorption.**
- Compact bone is composed primarily of subunits called **osteons** that consist of rings of bone ECM called **lamellae.**
- Spongy bone forms a framework for bone marrow with its branching **trabeculae** that are not composed of osteons.

MODULE **6.3**
Bone Formation: Ossification 194–198

- Bones form by the process of **ossification** or **osteogenesis.**
- The first bone formed is *primary bone,* which is replaced by much stronger *secondary bone.*

- **Intramembranous ossification** occurs from within a *membrane* composed of embryonic mesenchyme. Spongy bone is formed first at the **primary ossification center,** after which the periosteum differentiates, and finally compact bone forms.

- **Endochondral ossification** takes place within a hyaline cartilage model. Ossification begins with early compact bone in the *bone collar,* after which the inner early spongy bone forms.

MODULE 6.4
Bone Growth in Length and Width 198–201

- **Longitudinal growth** occurs at the **epiphyseal plate** in long bones, which consists of five zones of cells.

- When the epiphyseal plate fully ossifies, no further longitudinal bone growth is possible.

- Short, flat, irregular, and sesamoid bones grow from their ossification centers by either endochondral or intramembranous ossification.

- Bones grow in width by **appositional growth,** during which osteoblasts in the periosteum secrete new circumferential lamellae.

- Three *hormones* exert a significant effect on bone growth: *growth hormone, testosterone,* and *estrogen.*

MODULE 6.5
Bone Remodeling and Repair 202–207

- **Bone remodeling** is a combination of the continual processes of **bone deposition,** carried out by osteoblasts, and **bone resorption,** carried out by osteoclasts.

- Bone remodeling is influenced by several variables:
  - Tension placed on a bone triggers osteoblasts and bone deposition. Osteoclasts are stimulated when pressure is placed on a bone, which leads to bone resorption.
  - Hormones such as vitamin D, testosterone, and estrogen generally lead to net bone deposition.
  - Dietary factors such as adequate intake of calcium ions, vitamin D, vitamin C, vitamin K, and protein all influence bone remodeling.
  - Parathyroid hormone stimulates osteoclasts, calcium ion absorption in the intestines, and calcium ion retention by the kidneys. These effects increase the calcium ion concentration in the blood.

- **Fractures** can be classified as *simple* (or *closed*) or *compound* (or *open*).

- A bone fracture is healed first by chondroblasts that secrete cartilage, forming the **soft callus.** Osteoblasts later replace the soft callus with a **bone callus** made of primary bone.

Assess What You Learned

Scan the QR Code for additional practice test questions

https://goo.gl/glBckb

LEVEL 1 Check Your Recall

1. Which of the following is *not* a function of the skeletal system?
   a. Primary storage site in the body for the minerals sodium and potassium
   b. Location of the red bone marrow, which produces red blood cells
   c. Storage of triglycerides in yellow bone marrow
   d. Support and protection of the body and vital organs
   e. Functions in movement as the site of attachment for skeletal muscles

2. Match the following terms with the correct definition.

   _____ Long bone
   _____ Epiphysis
   _____ Nutrient artery
   _____ Flat bone
   _____ Short bone
   _____ Periosteum
   _____ Diaphysis
   _____ Medullary cavity

   a. Main source of blood to the medullary cavity
   b. Membrane surrounding the bone
   c. Bone that is about as wide as it is long
   d. Shaft of a long bone
   e. Bone that is longer than it is wide
   f. Canal running down the center of the diaphysis
   g. End of a long bone
   h. Bone that is broad and thin

3. Explain the differences between red bone marrow and yellow bone marrow.

4. Mark the following statements about bone tissue as true or false. If a statement is false, correct it to make a true statement.
   a. The primary mineral in the inorganic matrix is hydroxyapatite.
   b. The inorganic matrix of bone is solely responsible for the strength of bone tissue.
   c. Collagen fibers are one of the predominant parts of the inorganic matrix.
   d. The collagen fibers of bone help it to resist torsion and tension.
   e. Osteoblasts are responsible for bone resorption, and osteoclasts are responsible for bone deposition.
   f. Osteocytes are mature and less active osteoblasts that have become surrounded by bone ECM.

5. Fill in the blanks: The subunit of compact bone is the _____. It consists of rings of bone matrix called _____ that surround a structure called the _____ that contains blood vessels and nerves. Other structures called _____ also contain blood vessels and nerves. Osteocytes are housed in _____ and communicate via _____.

6. The branching pieces of bone in spongy bone are called:
   a. lamellae.
   b. lacunae.
   c. osteoclasts.
   d. trabeculae.

7. Which bones form via intramembranous ossification?

 a. Irregular bones
 b. Certain flat bones
 c. Long bones
 d. Short bones
 e. More than one of the above

8. Of the following statements, identify those that are properties of intramembranous ossification, endochondral ossification, or both.

 a. The bone is formed via a hyaline cartilage model.
 b. Bone tissue forms from ossification centers.
 c. Bone forms from within a mesenchyme membrane.
 d. The early spongy bone is formed, after which the early compact bone develops.
 e. The original primary bone is replaced with secondary bone.
 f. A bone collar forms, followed by the early spongy bone.

9. What is the difference between a primary and secondary ossification center in a long bone?

10. The part of the epiphysis that does not ossify during a person's lifetime is the:

 a. articular surface.
 b. secondary ossification center.
 c. diaphyseal notch.
 d. nutrient foramen.

11. Long bones grow in length from the:

 a. diaphyseal line.
 b. epiphyseal line.
 c. epiphyseal plate.
 d. medullary cavity.

12. Correctly order the following steps of bone growth in length, by placing a 1 by the first step, a 2 by the second step, and so on.

 _____ Calcified cartilage is replaced with bone in the zone of ossification.
 _____ Chondrocytes in the zone of proliferation divide by mitosis.
 _____ Chondrocytes enter the zone of calcification and die as their matrix calcifies.
 _____ Chondrocytes enlarge and cease dividing.

13. Explain the effect that the following hormones have on growth of bone tissue:

 a. Growth hormone
 b. Testosterone
 c. Estrogen

14. Fill in the blanks. Bone deposition is carried out by _____, which secrete the _____ of the ECM by exocytosis. They facilitate the formation of the _____ of the ECM by secreting _____ filled with calcium ions, enzymes, and ATP. Bone resorption is carried out by _____, which secrete _____ and _____ from their ruffled border.

15. Mark the following statements as true or false. If a statement is false, correct it to make a true statement.

 a. Bone resorption is triggered by pressure placed on the bone.
 b. Bone deposition is triggered by tension placed on the bone.
 c. The greater the load the bone must carry, the more bone that is resorbed by osteoclasts.
 d. Calcitonin is released in response to an elevated calcium ion concentration in the blood.

 e. Parathyroid hormone increases the blood calcium ion concentration by increasing the activity of osteoblasts.
 f. Vitamin D is required for calcium ion absorption from the intestines and retention in the kidneys.

16. Which of the following influences bone remodeling?

 a. Vitamins such as vitamins D, C, and K
 b. Intake of calcium ions
 c. Hormones such as estrogen and testosterone
 d. All of the above

17. Correctly order the following steps of fracture repair, placing a 1 by the first step, a 2 by the second step, and so on.

 _____ Osteoblasts in the periosteum lay down a bone callus of primary bone.
 _____ Damaged blood vessels bleed and fill the gap between the bone fragments with a hematoma.
 _____ The bone callus is remodeled and replaced with secondary bone.
 _____ Fibroblasts, chondroblasts, and blood vessels enter the clot, and the soft callus begins to bridge the gap between bone fragments.

LEVEL 2 Check Your Understanding

1. Explain why a person who is wheelchair-bound or bed-bound often has very low bone mass, especially in his or her legs.

2. How could diseases of the kidney, skin, and/or intestines cause diseases of the bone?

3. How could a disease that affects primarily cartilage impact bone growth and bone formation?

LEVEL 3 Apply Your Knowledge

PART A: Application and Analysis

1. It used to be common practice in surgical fracture repair to strip and remove the periosteum. Now, however, surgeons take great care to avoid damaging the periosteum. Explain why, and describe what would likely happen to bone healing with the periosteum missing.

2. Explain why young men who take anabolic steroids (which mimic the actions of testosterone) often display stunted growth and have an overall shorter stature.

3. Lucy Dupre is a 2-year-old girl living in northern Canada. You notice that her growth seems abnormally slow and she has exhibited signs of weakened bones, including fractures. Her parents admit they have not supplemented her diet with vitamin D, as they feel that any supplementation is "unnatural" and "not organic." Explain why Lucy is having problems with her bones and bone growth.

PART B: Make the Connection

4. Fouz Akkad is a 6-year-old girl who has been diagnosed with a rare genetic defect in which her lysosomes are unable to maintain an acidic pH. What are lysosomes? Why do lysosomes require an acidic pH? How will this affect the ability of certain bone cells to function? What consequences will this disease have on Fouz's bones? (*Connects to Chapter 3*)

See answers in Appendix A.

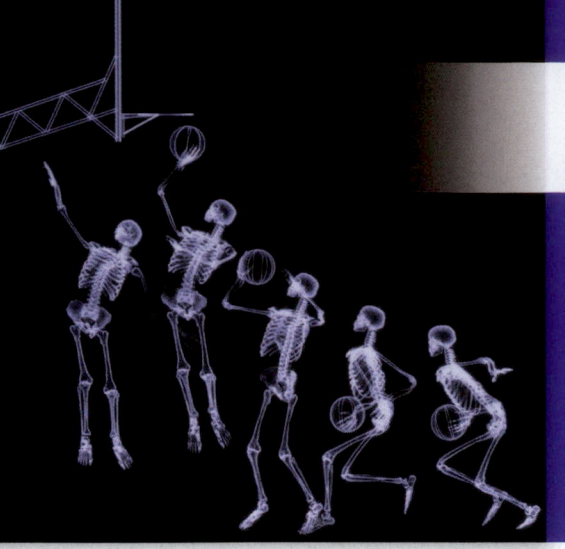

7

The Skeletal System

I n the previous chapter, we introduced bones and bone tissue. Now we look at the bones and cartilages that make up the **skeletal system,** or *skeleton*. Knowledge of the skeletal system will help build a foundation for later chapters. For example, many muscles, blood vessels, and nerves share names with the bones in their anatomical area, which can help you remember their location. In addition, the lobes of the brain are named for bones of the skull.

In progressing through the chapter, you'll realize there are a lot of bones and bone markings to remember. To help you, a number of memory devices are collected into two Study Boosts, where you can find them easily when you study. These Study Boosts are on pages 236 and 254.

MODULE **7.1**

Overview of the Skeletal System

Learning Outcomes

1. Compare and contrast the two structural divisions of the skeleton.
2. Identify and describe bone markings, and explain their function.
3. Identify the types of cartilage tissues found in the skeletal system.

On average, the human body has about 206 bones that together make up the **skeleton.** Before getting into the structure of individual bones, let's take an introductory tour of the skeleton. We start with the basic components of the skeleton and its cartilages, and then move on to the types of markings found on bones.

*For practice applying concepts to a clinical scenario, check out the **Running Case Study** for this chapter @ Mastering Anatomy & Physiology*

Computer-generated image: The skeletal system consists of bones and skeletal cartilages that function together in movement, support, and protection.

Structure of the Skeleton and Skeletal Cartilages

 **FLASHBACK**

1. What are the properties of cartilage? (p. 142)
2. How do hyaline cartilage and fibrocartilage differ? (p. 142)

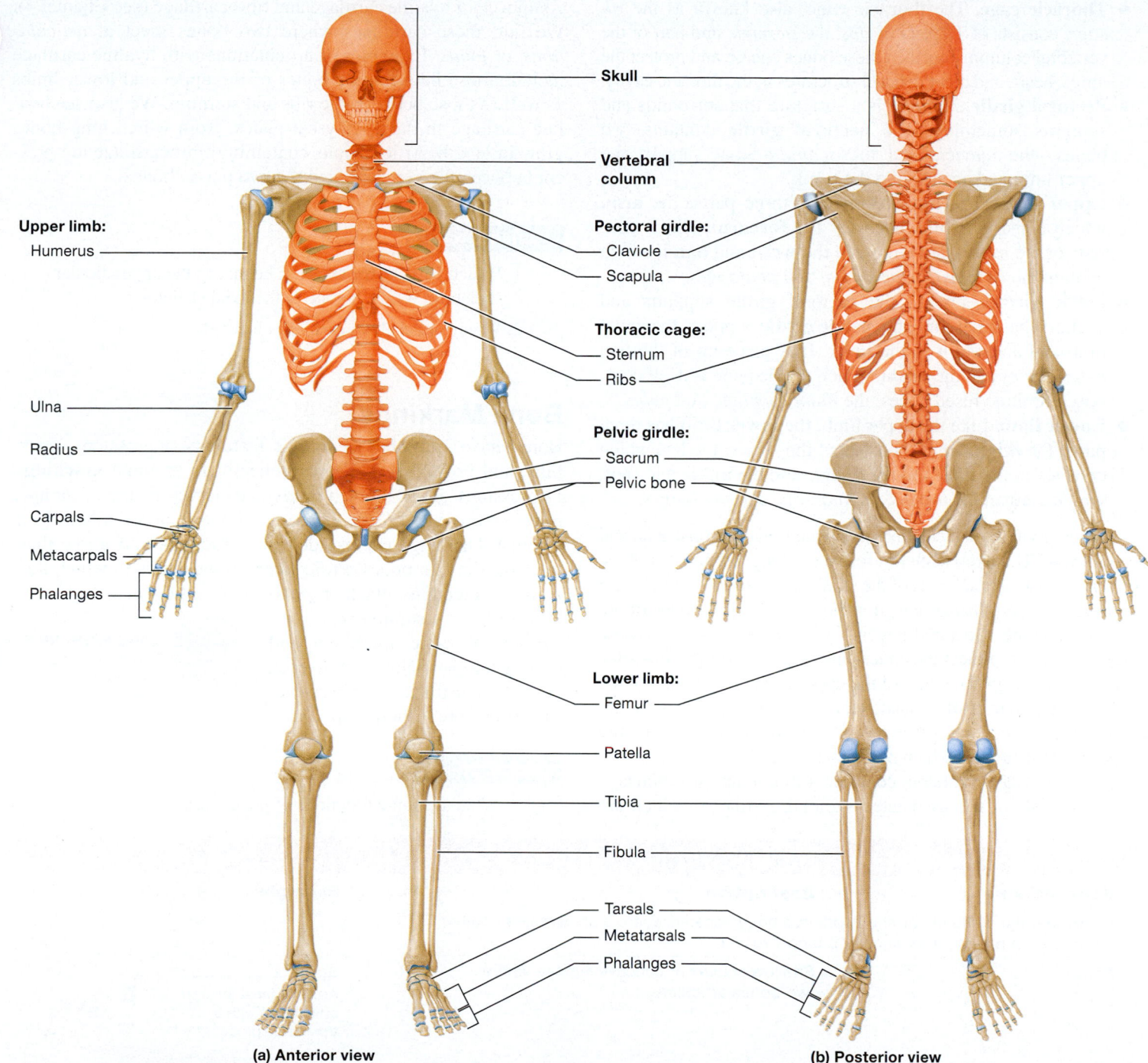

Skull

Vertebral column

Upper limb:
Humerus

Ulna

Radius

Carpals

Metacarpals

Phalanges

Pectoral girdle:
Clavicle
Scapula

Thoracic cage:
Sternum
Ribs

Pelvic girdle:
Sacrum
Pelvic bone

Lower limb:
Femur

Patella

Tibia

Fibula

Tarsals
Metatarsals
Phalanges

(a) Anterior view

(b) Posterior view

Figure 7.1 **Divisions of the skeletal system.** The axial skeleton is shaded orange; the appendicular skeleton is shaded tan.

The skeleton consists of groups of bones that work together to perform common functions. The basic parts of the skeleton include (**Figure 7.1**):

- **Skull.** The *skull* is the skeleton's most complex structure. It has a total of 22 bones: 8 *cranial bones,* which encase the brain, and 14 *facial bones,* which form the framework for the face.

- **Vertebral column.** Inferior to the skull we find the **vertebral column,** which in an adult consists of approximately 33 bones, called *vertebrae,* stacked on top of one another. The first 24 bones are individual vertebrae, which encase the spinal cord. The two inferior bones, the *sacrum* and the *coccyx,* are made up of fused vertebrae.

- **Thoracic cage.** The **thoracic cage,** also known as the *rib cage,* consists of 12 pairs of *ribs,* the *sternum,* and part of the vertebral column. Together, these bones encase and protect the lungs, heart, and other delicate structures in the thoracic cavity.
- **Pectoral girdle.** A *girdle* is a structure that surrounds and supports something. The **pectoral girdle** contains two bones—the *clavicle* and the *scapula*—that support the upper limb and anchor it to the trunk.
- **Upper limb.** The **upper limb** has three parts: the **arm,** which consists of the *humerus;* the **forearm,** which consists of the *radius* and *ulna;* and the **wrist** and **hand,** which contain the *carpals, metacarpals,* and *phalanges.*
- **Pelvic girdle.** Just as the pectoral girdle supports and anchors the upper limb, the **pelvic girdle** supports the lower limb and anchors it to the trunk. It is made up of the two *pelvic bones* and the sacrum. Each pelvic bone is itself composed of three fused bones: the *ilium, ischium,* and *pubis.*
- **Lower limb.** Like the upper limb, the **lower limb** has three parts: the **thigh,** which consists of the *femur;* the **leg,** which consists of the *tibia* and the *fibula;* and the **ankle** and **foot,** which contain the *tarsals, metatarsals,* and *phalanges.*

We can group the skeleton into two structural divisions: axial and appendicular. The **axial skeleton** forms the longitudinal axis of the body. It consists of the bones of the skull, the vertebral column, and the thoracic cage (shaded orange in Figure 7.1). The bones of the axial skeleton are structured largely for protection, as they encase body cavities and protect the underlying organs. The **appendicular skeleton** consists of the bones of the upper and lower limbs and the pectoral and pelvic girdles (shaded tan in Figure 7.1). These bones are structured largely for motion and act primarily as levers and supportive structures to which muscles attach.

Recall that the skeleton contains more than just bones— it also consists of the associated skeletal cartilages, which are composed of hyaline cartilage and fibrocartilage (see Chapter 4). We find these cartilages where two bones meet at *articulations,* or *joints.* Examples of articulations with hyaline cartilage include those between the bones of the upper and lower limbs as well as those between the ribs and sternum. We also see hyaline cartilage in the epiphyseal plates, from which long bones grow in length. Articulations containing fibrocartilage are present between the vertebrae and the two pelvic bones.

Quick Check

☐ 1. Which parts of the skeleton belong to the appendicular skeleton? Which belong to the axial skeleton?

☐ 2. Where are skeletal cartilages located?

Bone Markings

Bones have a number of surface features—depressions, openings, and projections—known collectively as **bone markings.** Depressions allow blood vessels and nerves to travel along a bone, or provide a place where two bones can articulate (form a joint). Openings enclose delicate structures and allow them to travel through bones. Projections provide sites to which ligaments and tendons attach or where bones articulate.

Table 7.1 summarizes categories of bone markings and gives examples. Notice how each type demonstrates the Structure-Function Core Principle (p. 28).

CORE PRINCIPLE
Structure-Function

Quick Check

☐ 3. What are some functions of bone markings?

Table 7.1 Bone Markings

Bone Marking	Description	Example
Depressions: clefts of varying depth in a bone; located where a bone meets another structure, such as another bone or a blood vessel.		
Facet	Shallow convex or concave surface where two bones articulate	Rib: Articular facet for articulation with a transverse process
Fossa (plural, fossae)	Indentation in a bone into which another structure fits	Humerus: Distal portion with olecranon fossa
Fovea	Shallow pit	Femur: Fovea capitis

Table 7.1 Bone Markings (*continued*)

Bone Marking	Description	Example
Groove (or sulcus)	Long indentation along which a narrow structure travels	**Rib:** Costal groove

Openings: holes that allow blood vessels and nerves to travel through a bone; permit access to the middle and inner ear; encase delicate structures and protect them from trauma.

Bone Marking	Description	Example
Canal (or meatus)	Tunnel through a bone	**Temporal bone:** External acoustic meatus
Fissure	Narrow slit in a bone or between adjacent parts of bones	**Sphenoid bone:** Superior orbital fissure
Foramen (plural, *foramina*)	Hole in a bone	**Frontal bone:** Supraorbital foramen

Projections: bony extensions of varying shapes and sizes; some provide locations for attachment of muscles, tendons, and ligaments; some fit into depressions of other bones to stabilize joints.

Bone Marking	Description	Example
Condyle	Rounded end of a bone that articulates with another bone	**Mandible:** Mandibular condyle
Crest	Ridge or projection	**Ilium:** Iliac crest
Head	Round projection from a bone's epiphysis	**Humerus:** Head
Tubercle and tuberosity	Small, rounded bony projection; a tuberosity is a large tubercle	**Humerus:** Deltoid tuberosity
Epicondyle	Small projection usually proximal to a condyle	**Humerus:** Medial epicondyle
Process	Prominent bony projection	**Scapula:** Coracoid process
Spine	Sharp process	**Scapula:** Spine

(*continued*)

Table 7.1 Bone Markings (continued)

Bone Marking	Description	Example
Protuberance	Outgrowth from a bone	Occipital bone: External occipital protuberance
Trochanter	Large projection found only on the femur	Femur: Greater trochanter
Line	Long, narrow ridge	Femur: Linea aspera

Apply What You Learned

☐ 1. Why might a fractured bone of the axial skeleton be more damaging to overall homeostasis than a fractured bone of the appendicular skeleton?

☐ 2. Many bone abnormalities are characterized by changes in the normal structure of bone markings. Defective openings that are smaller or larger than normal can be particularly problematic. Why do you think this is true?

See answers in Appendix A.

MODULE **7.2**

The Skull

Learning Outcomes

1. Describe the location, structural features, and functions of each of the cranial bones and the main sutures that unite them.

2. Describe the location, structural features, and functions of each of the facial bones and the hyoid bone.

3. Explain the structural features and functions of the orbit, nasal cavity, and paranasal sinuses.

4. Compare and contrast the skull of a fetus or infant with that of an adult.

We begin our tour of the skeletal system with its most complex bony structure, the **skull,** which generally has 22 bones. Note that some sources consider the six tiny bones in the ear, the *auditory*

ossicles, to be skull bones (for a total of 28); however, we cover the auditory ossicles in the special senses chapter (see Chapter 15). In this module, we examine only the 22 main skull bones and their major markings, as well as considering the unique features of the fetal and infant skull. But before we focus on individual bones, let's look at the overall structure of the skull.

Overview of Skull Structure

« FLASHBACK

1. Where is the cranial cavity located, and what is housed within this cavity? (p. 17)

The skull consists of two groups of bones: (1) the **cranial bones,** which are known collectively as the **cranium** and encase the brain, and (2) the **facial bones,** which form the framework for the face (**Figure 7.2**). Both groups also contribute to the bony structures that house the special sensory organs. There are eight cranial bones: four single bones, which are the *frontal, occipital, ethmoid, and sphenoid bones,* and two paired bones, which are the *temporal* and *parietal bones.* The 14 facial bones include the paired *maxillary, zygomatic, nasal, lacrimal, palatine, and inferior nasal conchal bones,* as well as the unpaired *mandible* and *vomer.* With the exception of the mandible (the lower jaw bone), all skull bones are united in adults by immoveable joints called **sutures.**

Together, the eight cranial bones form the large **cranial cavity,** which surrounds the brain (**Figure 7.3**). The cranial cavity consists of a superior portion, called the **cranial vault,** or **calvaria,** and an inferior portion, the **cranial base.** Internally, the cranial base is divided into three indentations, or *fossae,* into which the brain fits snugly: the *anterior, middle,* and *posterior cranial fossae.*

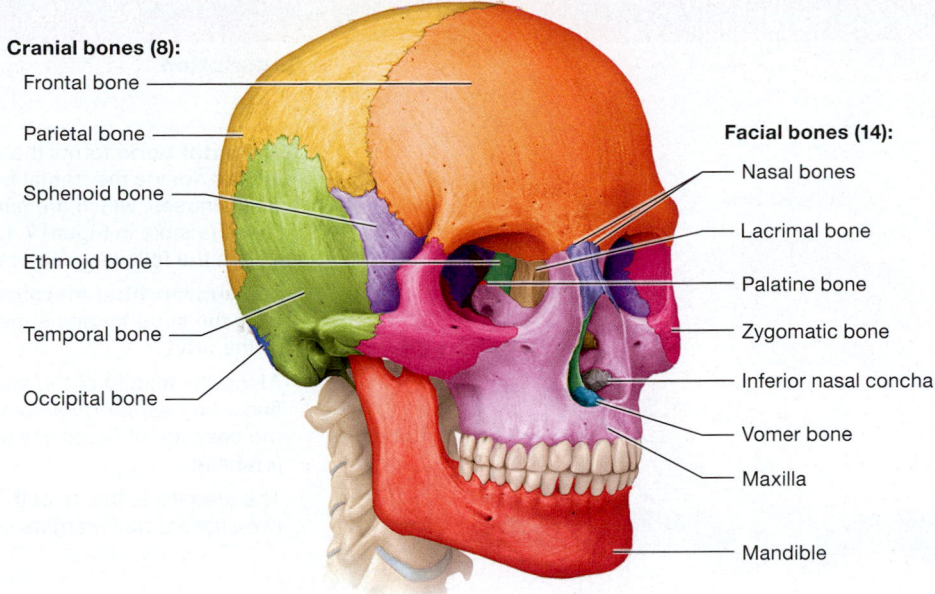

Cranial bones (8):
- Frontal bone
- Parietal bone
- Sphenoid bone
- Ethmoid bone
- Temporal bone
- Occipital bone

Facial bones (14):
- Nasal bones
- Lacrimal bone
- Palatine bone
- Zygomatic bone
- Inferior nasal concha
- Vomer bone
- Maxilla
- Mandible

Explore PAL™ Cadaver
@ Mastering Anatomy & Physiology

Figure 7.2 Basic structure of the skull: anterolateral view of the cranial and facial bones.

Smaller cavities in the skull house other, often delicate, structures (**Figure 7.4**):

- The *orbits* contain the eyeballs.
- The *nasal cavity* houses the sensory receptors for smell.
- The *oral cavity* surrounds the teeth and tongue.
- Other small cavities house the sense organs for hearing and balance (not visible in Figure 7.4).

In addition, many bones contain air-filled, membrane-lined spaces known as *sinuses*. In particular, four bones around the nasal cavity have large sinuses called *paranasal sinuses*.

To help with your studying, we have put the structural and functional features of the individual cranial and facial bones, as well as illustrations of the individual bones, into **Table 7.2**. When you've become acquainted with each bone and its characteristics, move on to **Figures 7.5–7.9**. These figures present several views and sections of the complete skull, and Concept Boost: Understanding How Skull Bones Relate to One Another shows how the individual bones fit together.

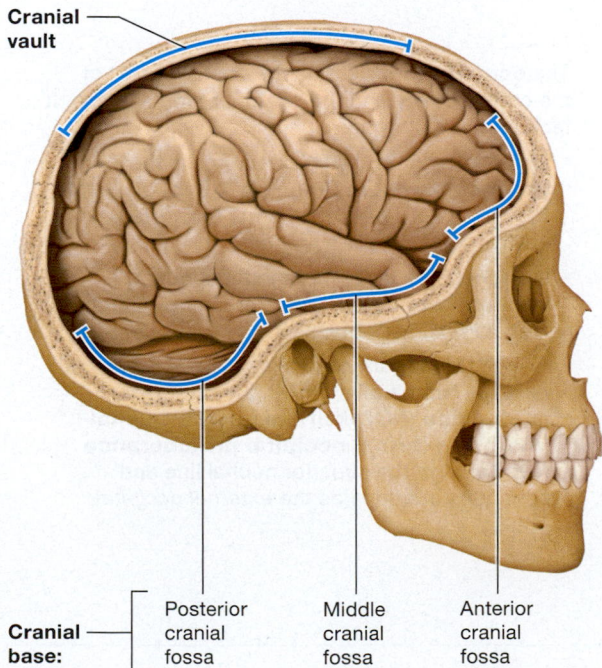

Cranial vault

Cranial base:
- Posterior cranial fossa
- Middle cranial fossa
- Anterior cranial fossa

Figure 7.3 Cranial vault and base of the cranial cavity.

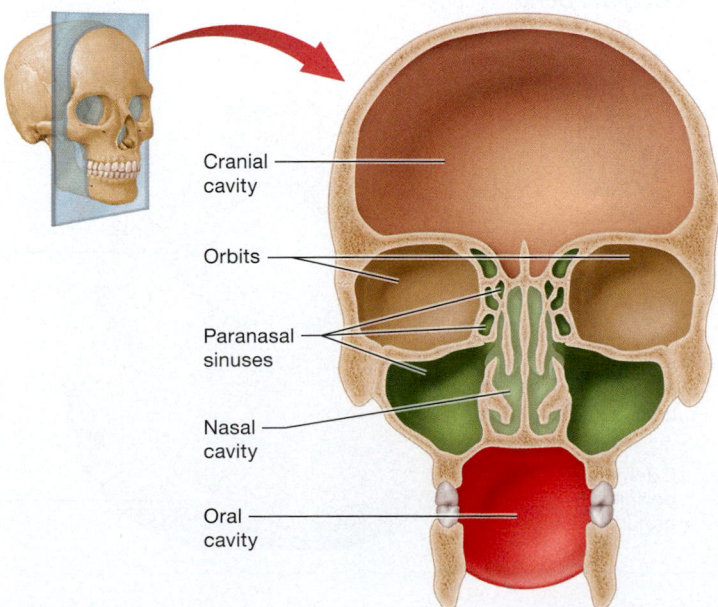

- Cranial cavity
- Orbits
- Paranasal sinuses
- Nasal cavity
- Oral cavity

Figure 7.4 Cavities of the skull.

Table 7.2 Bones of the Skull

Bone	Description

Cranial Bones

Frontal bone

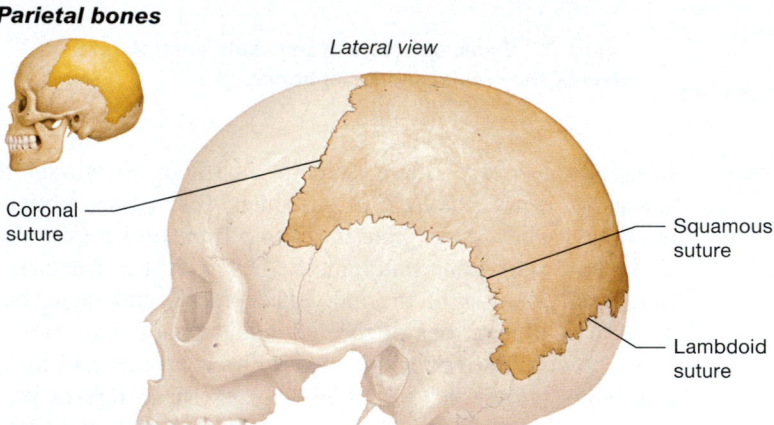

Anterior view

Frontal squama — Glabella — Supraorbital margin — Supraorbital foramen — Superior wall of orbit

The **frontal bone** forms the *frontal squama,* or forehead. Inside the frontal bone, we find the frontal sinuses, which are part of the paranasal sinuses (visible in Figure 7.13). The frontal bone also has the following features:

- The **supraorbital margins** are sharp ridges that form the superior and superomedial boundaries of the orbit.
- Along the middle of the supraorbital margin, we find a tiny *supraorbital foramen,* which allows the passage of blood vessels and nerves to the forehead.
- The *glabella* is the smooth region between the two supraorbital margins.

Parietal bones

Lateral view

Coronal suture — Squamous suture — Lambdoid suture

The two **parietal bones** form the superior wall and part of the lateral wall of the cranial vault. They meet at the **sagittal suture** and articulate with several cranial bones at other sutures :

- together they meet the frontal bone at the **coronal suture** (*coron-* = "crown");
- each parietal bone meets a temporal bone at a **squamous suture;** and
- together they meet the occipital bone at the **lambdoid suture.**

Occipital bone

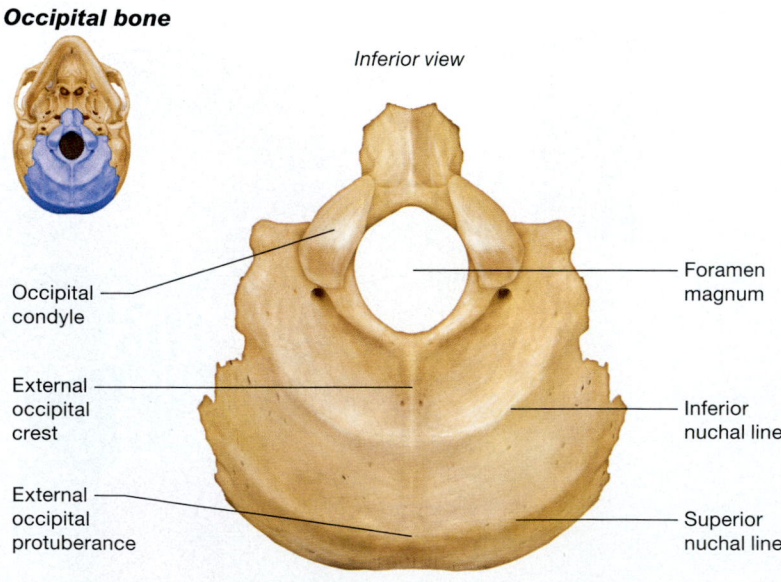

Inferior view

Occipital condyle — Foramen magnum — External occipital crest — Inferior nuchal line — External occipital protuberance — Superior nuchal line

The **occipital bone** forms the posterior part of the cranial cavity and the posterior cranial fossa. It features the following:

- On its anterior and inferior surface is the **foramen magnum** ("big hole"), through which the spinal cord passes to enter the vertebral cavity.
- On either side of the foramen magnum are two **occipital condyles,** which articulate with the first cervical vertebra.
- The posterior surface of the occipital bone features two parallel ridges, the **superior nuchal line** (NOO-kul) and **inferior nuchal line. The external occipital protuberance** runs through the superior nuchal line and continues inferiorly as the *external occipital crest.*

Table 7.2 **Bones of the Skull** (*continued*)

Bone	Description

Temporal bones

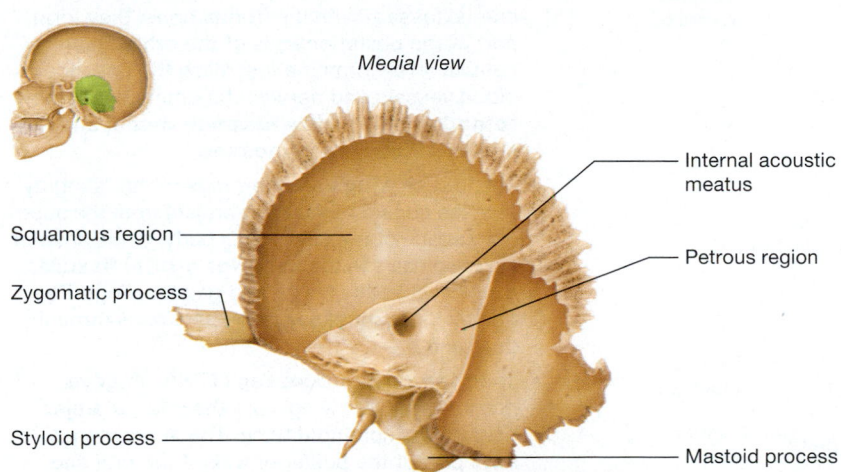

Lateral view

- Squamous region
- Mandibular fossa
- Zygomatic process
- External acoustic meatus
- Tympanic region
- Mastoid region
- Styloid process
- Mastoid process

Medial view

- Squamous region
- Internal acoustic meatus
- Zygomatic process
- Petrous region
- Styloid process
- Mastoid process

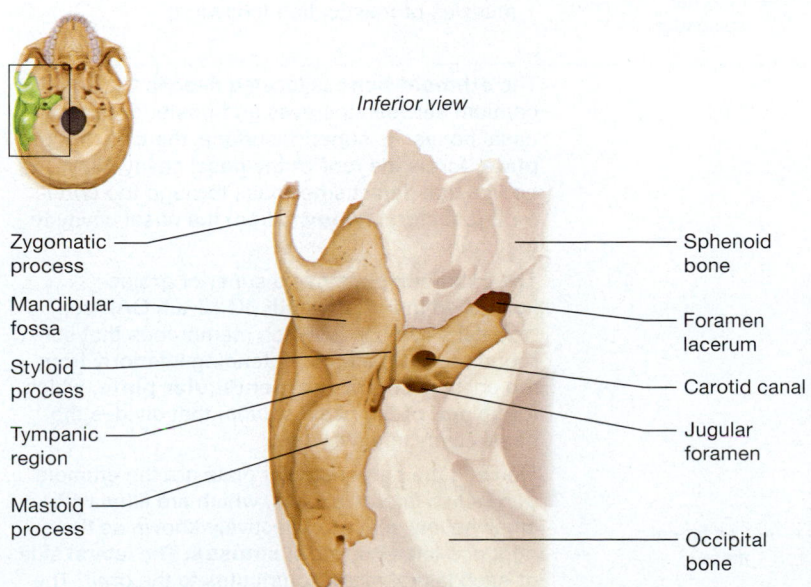

Inferior view

- Zygomatic process
- Sphenoid bone
- Mandibular fossa
- Foramen lacerum
- Styloid process
- Carotid canal
- Tympanic region
- Jugular foramen
- Mastoid process
- Occipital bone

The two **temporal bones** form the lateral walls of the cranium. Each temporal bone has four regions:

- The **squamous region** (*squam-* = "flat") is the bone's broad, flat surface. It features the **zygomatic process,** which forms part of the zygomatic arch, and the **mandibular fossa,** which articulates with the mandible.

- The **tympanic region** houses the **external acoustic meatus,** which is the entry into the canal that leads to the middle ear. Inferiorly, it features the **styloid process** (*stylus* = "needle"), a needle-like spur of bone.

- The **mastoid region** contains the thick projection called the **mastoid process,** which is posterior and lateral to the styloid process. The mastoid process is filled with tiny sinuses called **mastoid air cells.**

- The **petrous region** (*petr-* = "rock") is located on the internal or medial surface of the temporal bone, where it forms part of the middle cranial fossa. It contains four important features (see the inferior view for the second, third, and fourth):

  - the **internal acoustic meatus,** a canal leading from the inner ear;

  - the **jugular foramen,** a posterior opening where the temporal bone meets the occipital bone, through which the *internal jugular vein* and nerves pass*;*

  - the **carotid canal,** through which the *internal carotid artery* passes as it enters the cranial cavity; and

  - the **foramen lacerum** (LASS-er-um)**,** which allows the passage of small blood vessels and nerves.

(*continued*)

Table 7.2 Bones of the Skull (*continued*)

Bone	Description

Sphenoid bone

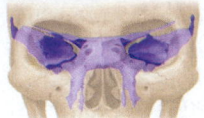

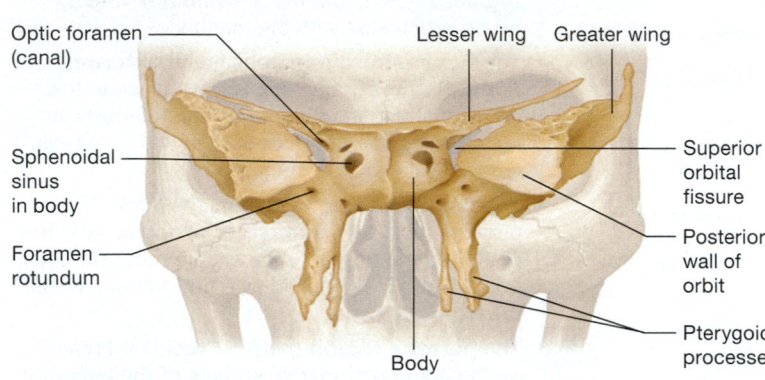

Anterior, deep view

Optic foramen (canal) — Lesser wing — Greater wing
Sphenoidal sinus in body
Foramen rotundum
Superior orbital fissure
Posterior wall of orbit
Body — Pterygoid processes

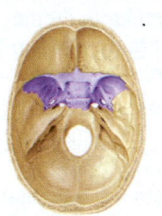

Superior, internal view

Greater wing — Sella turcica — Lesser wing
Foramen rotundum
Foramen ovale
Foramen spinosum

The **sphenoid bone** (SFEE-noyd) is a deep cranial bone that is unique in that it articulates with every other cranial bone. It forms part of the anterior and middle cranial fossae. From an anterior view (upper figure), it resembles a bat; from a superior view (lower figure), it resembles a stingray. It has four main components:

- The central **body** is largely hollow, containing air-filled **sphenoidal sinuses.** On the anterior surface of the sphenoid's body, we find the two **optic foramina** leading to two **optic canals,** through which the nerves that transmit vision pass. On its superior surface, a saddle-shaped depression called the **sella turcica** (SELL-uh TER-sih-kuh; "Turk's saddle") houses the *pituitary gland.*

- The **greater wings** are the wings of the "bat" that extend laterally from the body. With the temporal bone, they form part of the middle cranial fossae. With the frontal bone, they form part of the posterior walls of the orbit. They contain three foramina that allow the passage of blood vessels and nerves: the anterior **foramen rotundum,** the middle **foramen ovale,** and the posterior **foramen spinosum.**

- The **lesser wings** are the wings of the "stingray" (see the superior view) that project from the superior surface of the sphenoid's body. In the anterior view, we can see the **superior orbital fissure,** a slit between the lesser and greater wings. The nerves that control eye movement pass through this fissure.

- The **pterygoid processes** (TEHR-uh-goyd; *ptery-* = "fin" or "wing") are the inferior projections of the sphenoid bone. These processes form part of the posterior wall of the oral and nasal cavities, and are the site of attachment for muscles of mastication (chewing).

Ethmoid bone

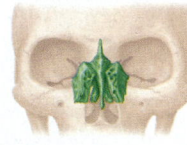

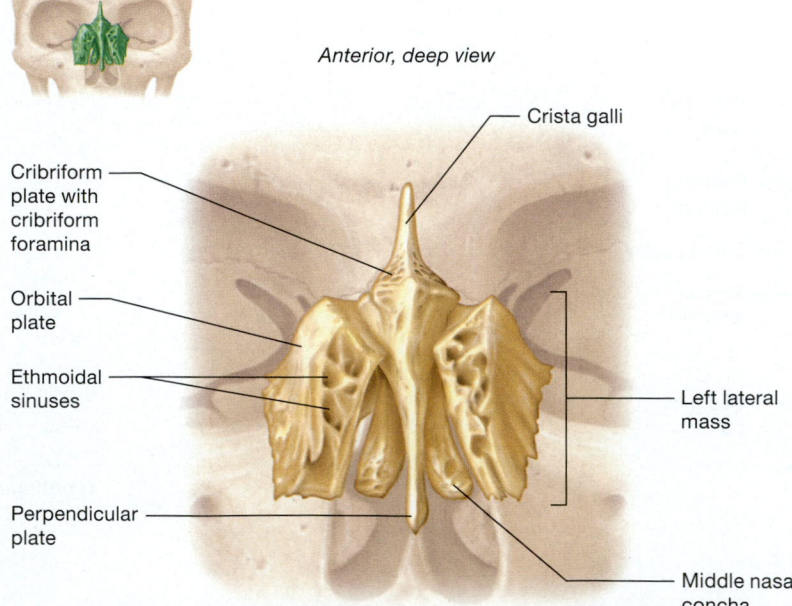

Anterior, deep view

Crista galli
Cribriform plate with cribriform foramina
Orbital plate
Ethmoidal sinuses
Left lateral mass
Perpendicular plate
Middle nasal concha

The **ethmoid bone** is located deep in the anterior cranium between the eyes and posterior to the nasal bones. Its superior surface, the **cribriform plate,** forms the roof of the nasal cavity. Tiny nerves that detect smell pass through the *cribriform (olfactory) foramina* from the nasal cavity to the brain.

The cribriform plate has a superior projection called the **crista galli** (KRIS-tuh GAL-ee; "rooster's comb"), to which membranes that surround the brain attach. Extending inferiorly from the crista galli is the **perpendicular plate,** which forms part of the *nasal septum* that divides the nasal cavity into two sides.

Flanking the perpendicular plate are the ethmoid bone's two *lateral masses,* which are filled with small hollow spaces collectively known as the right and left **ethmoidal sinuses.** The lateral side of each lateral mass contributes to the orbit. The medial side of each lateral mass features two projections into the nasal cavity called the **superior** and **middle nasal conchae** (KAHN-kee; singular, *concha*), so named because their curved shape resembles a conch shell (*concha* = "shell").

Table 7.2 Bones of the Skull (*continued*)

Bone	Description

Facial Bones

Nasal, lacrimal, and zygomatic bones

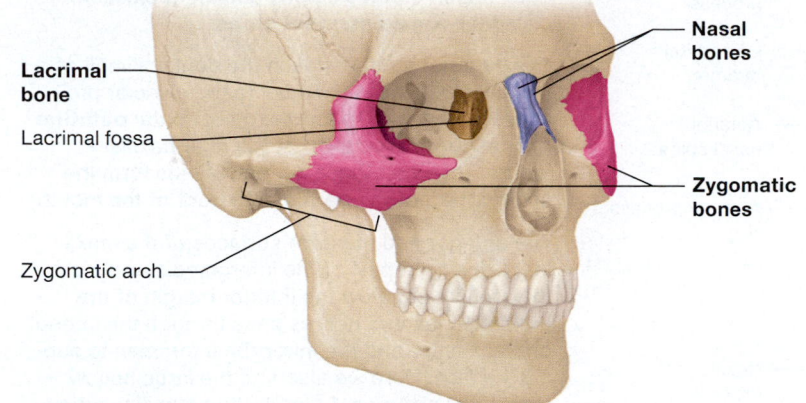

The two **nasal bones** form the bridge of the nose. They articulate with the hyaline cartilages that form most of the framework of the nose.

The two **lacrimal bones** (LAK-rih-mul) are the smallest and most delicate facial bones. They are found in the medial wall of the orbit, and contain a depression called the **lacrimal fossa** through which tears drain.

The two **zygomatic bones,** along with the zygomatic processes of the temporal bones and maxillae, form the bulk of the cheekbone, or *zygomatic arch.* They also form the lateral wall and part of the inferior border of the orbit.

Palatine bones

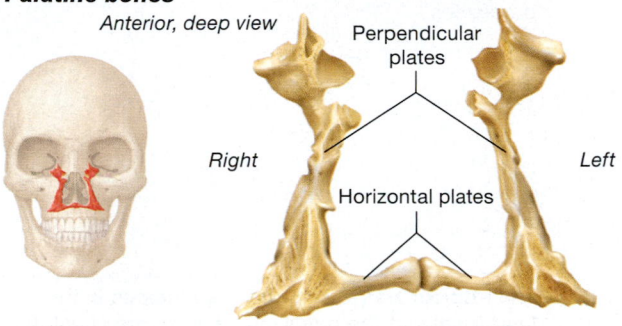

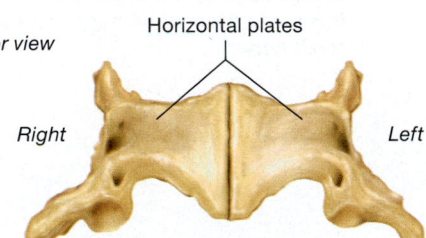

The two L-shaped **palatine bones** (pal-uh-TYN) are located in the posterior nasal cavity between the maxillae and the pterygoid processes of the sphenoid. Each bone is composed of two *plates.* The smaller *horizontal plates* form the posterior part of the hard palate. The larger *perpendicular plates* are vertical projections that form part of the lateral walls of the nasal cavity and a tiny piece of the orbit.

Mandible

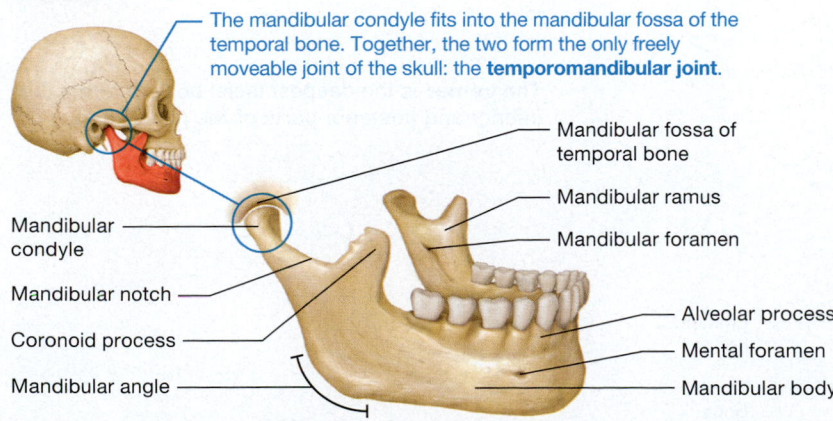

The **mandible** is the inferior jaw bone and the only moveable bone of the adult skull. It consists of a central **mandibular body,** and the right and left **mandibular rami** (RAY-mee; *ramus* = "branch").

The mandibular body, which forms the chin, meets the two rami at the right and left **mandibular angles.** Within the mandibular body on each side is a small **mental foramen** (*menta* = "chin"). Also in the mandibular body are the inferior teeth, housed in deep sockets called *dental alveoli* within a ridge known as the *alveolar process* (al-vee-OH-lur; *alve-* = "hollow").

Each mandibular ramus is topped by the U-shaped **mandibular notch,** inferior to which we find the **mandibular foramen.** On either side of the mandibular notch are two processes:

- on the anterior side is the **coronoid process,** which serves as the attachment site for a major muscle of mastication (chewing); and
- on the posterior side is the **condylar process,** at the top of which is the **mandibular condyle.** The mandibular condyle articulates with the mandibular fossa of the temporal bone. Together, they form the *temporomandibular joint.*

(*continued*)

Table 7.2 Bones of the Skull (continued)

Bone	Description
Maxillae 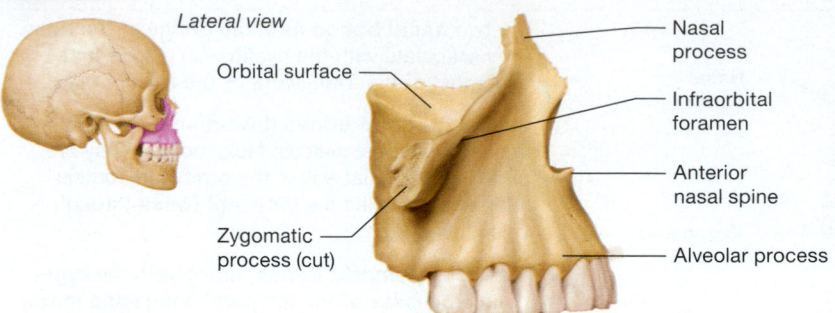 *Lateral view* — Nasal process, Orbital surface, Infraorbital foramen, Anterior nasal spine, Zygomatic process (cut), Alveolar process *Medial view, midsaggital section* — Maxillary sinus, Nasal process, Anterior nasal spine, Palatine process, Alveolar process *Note: The inferior suture, intermaxillary orbital fissure, and maxillary sinuses are not visible in these views. (See Figures 7.5, 7.7, and 7.13.)*	The two **maxillary bones** or **maxillae** (mak-SILL-ee; *maxil-* = "jaw") are the superior jaw bones. Their fusion creates a bony midline projection called the **anterior nasal spine**. The superior teeth reside in the dental alveoli of the alveolar process. Medial to the alveolar processes, the maxillae form two horizontal **palatine processes,** which meet at the *intermaxillary suture.* Together, these two processes form the anterior part of the hard palate (roof of the mouth). The anterior and superior surfaces of the maxillary bone form part of the inferomedial walls of the orbit. Just below the inferior margin of the orbit, arteries and nerves pass through the inferior orbital fissure and the infraorbital foramen to supply the face. Here we also find the large hollow **maxillary sinuses.** Laterally, the maxillae articulate with the zygomatic bones via their *zygomatic processes.*
Inferior nasal conchae 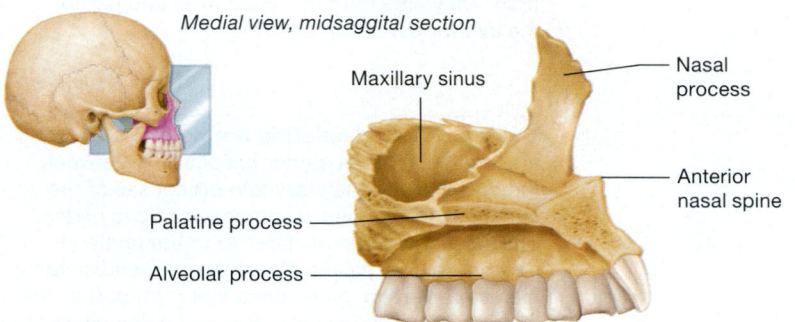 *Medial view, parasagittal section* — Frontal bone, Ethmoid bone, Nasal bone, Sphenoid bone, Middle nasal concha, Inferior nasal concha, Maxilla, Palatine bone	The **inferior nasal conchae** are located in the lateral walls of the nasal cavity. They are situated inferior to the middle nasal conchae of the ethmoid bone.
Vomer bone 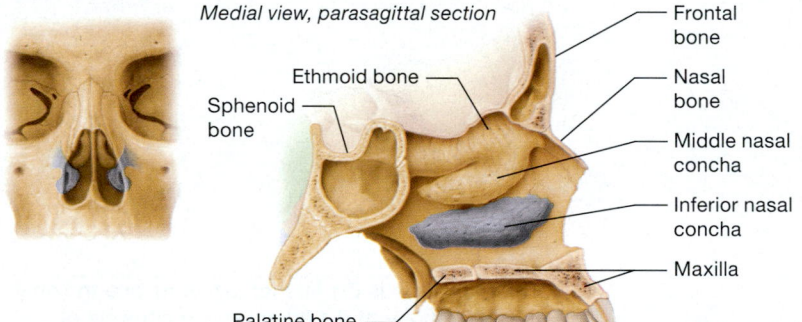 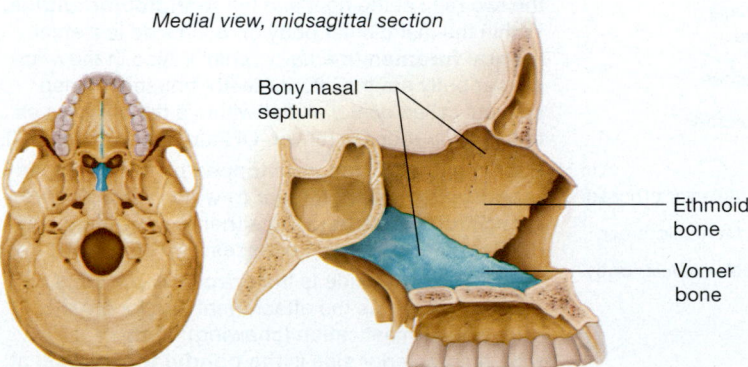 *Medial view, midsagittal section* — Bony nasal septum, Ethmoid bone, Vomer bone	The **vomer** is the deepest facial bone. It forms the inferior and posterior parts of the *nasal septum.*

The first view of the whole skull is the anterior view. While all facial bones are visible here, some are visible only in the orbit (a tiny piece of palatine bone) or the nasal cavity (the vomer and inferior nasal conchae). Certain cranial bones are visible here as well, including the frontal bone and portions of the sphenoid and ethmoid bones.

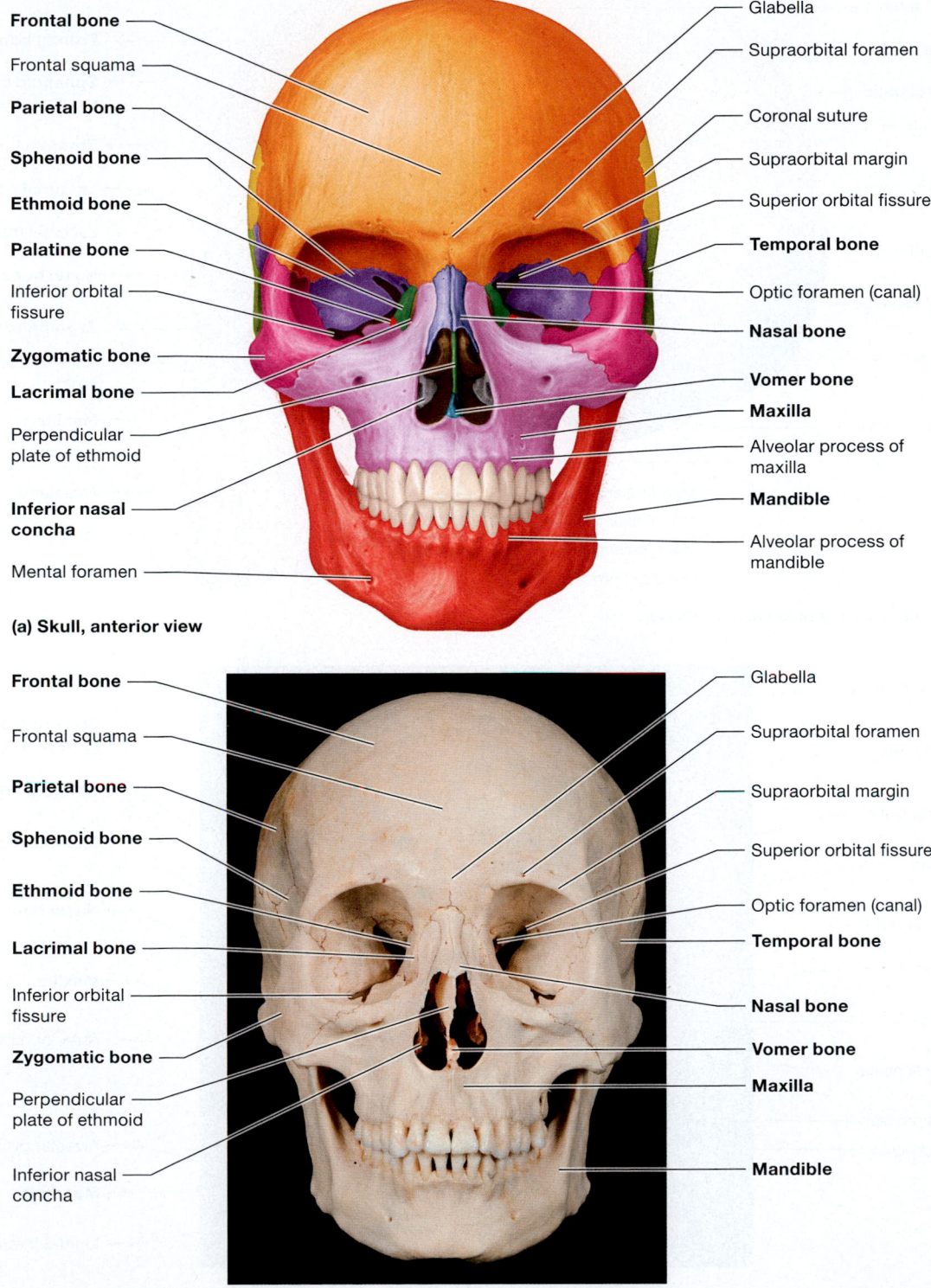

Frontal bone

Frontal squama

Parietal bone

Sphenoid bone

Ethmoid bone

Palatine bone

Inferior orbital fissure

Zygomatic bone

Lacrimal bone

Perpendicular plate of ethmoid

Inferior nasal concha

Mental foramen

Glabella

Supraorbital foramen

Coronal suture

Supraorbital margin

Superior orbital fissure

Temporal bone

Optic foramen (canal)

Nasal bone

Vomer bone

Maxilla

Alveolar process of maxilla

Mandible

Alveolar process of mandible

(a) Skull, anterior view

Frontal bone

Frontal squama

Parietal bone

Sphenoid bone

Ethmoid bone

Lacrimal bone

Inferior orbital fissure

Zygomatic bone

Perpendicular plate of ethmoid

Inferior nasal concha

Glabella

Supraorbital foramen

Supraorbital margin

Superior orbital fissure

Optic foramen (canal)

Temporal bone

Nasal bone

Vomer bone

Maxilla

Mandible

(b) Photo of skull, anterior view

Figure 7.5 Anterior view of the skull.

Notice that in the lateral view, we can see all of the cranial bones and many of the facial bones. This is also the best view to see the bones that contribute to the zygomatic arch and how the temporal bone and mandible articulate.

The sphenoid bone articulates with the temporal, parietal, and frontal bones, forming a "weak spot" called the temple. This is a dangerous location for a head injury because bones fracture more easily along sutures.

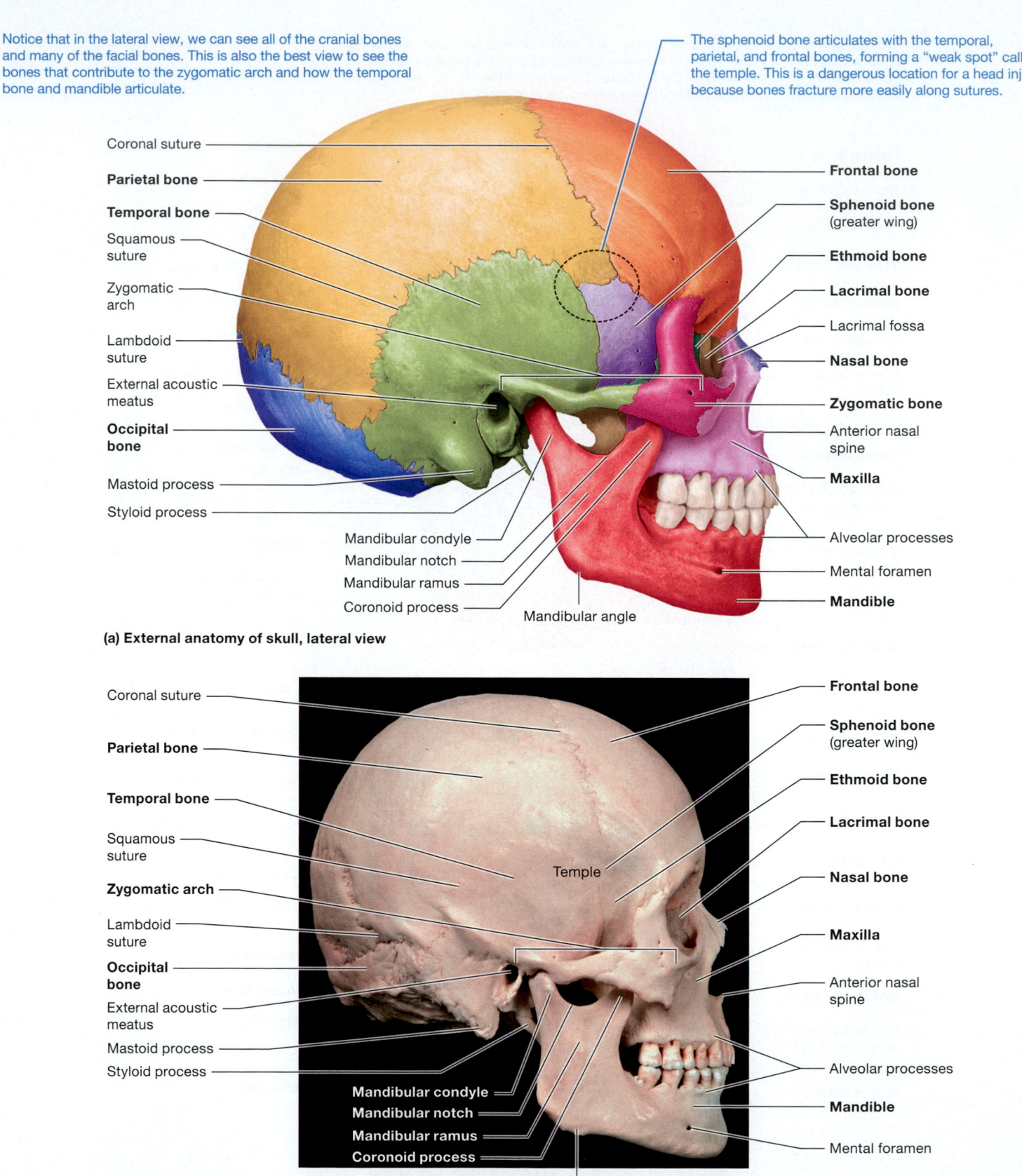

Coronal suture

Parietal bone

Temporal bone

Squamous suture

Zygomatic arch

Lambdoid suture

External acoustic meatus

Occipital bone

Mastoid process

Styloid process

Mandibular condyle

Mandibular notch

Mandibular ramus

Coronoid process

Mandibular angle

Frontal bone

Sphenoid bone (greater wing)

Ethmoid bone

Lacrimal bone

Lacrimal fossa

Nasal bone

Zygomatic bone

Anterior nasal spine

Maxilla

Alveolar processes

Mental foramen

Mandible

(a) External anatomy of skull, lateral view

Coronal suture

Parietal bone

Temporal bone

Squamous suture

Zygomatic arch

Lambdoid suture

Occipital bone

External acoustic meatus

Mastoid process

Styloid process

Mandibular condyle

Mandibular notch

Mandibular ramus

Coronoid process

Mandibular angle

Frontal bone

Sphenoid bone (greater wing)

Ethmoid bone

Lacrimal bone

Nasal bone

Maxilla

Anterior nasal spine

Alveolar processes

Mandible

Mental foramen

Temple

(b) Photo of skull, lateral view

Figure 7.6 Lateral view of the skull.

Here we can see the occipital, parietal, and parts of the temporal bones, as well as the posterior sides of the mandible and maxillae. This is a good view in which to see the surface markings of the occipital bone and the sagittal, squamous, and lambdoid sutures.

In this view, we can also see several sutures, particularly where the parietal bones meet the frontal bone at the coronal suture.

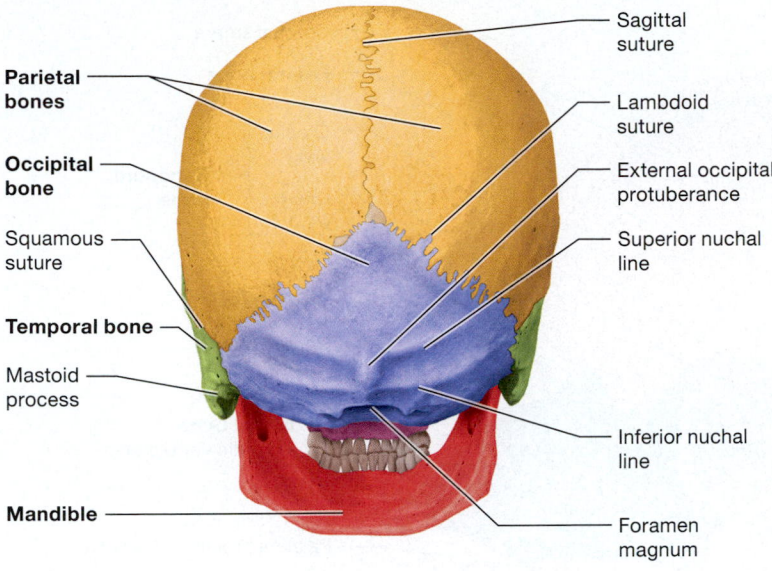

(a) Posterior view

(b) Superior view

For this inferior view, imagine holding the skull (without its mandible) and flipping it upside down so you are looking at its base. This is a good view to appreciate the structure of the hard palate. The two maxillae meet at the intermaxillary suture and articulate with the palatine bones posteriorly. We can also see the sphenoid bone's pterygoid processes and laterally its greater wings.

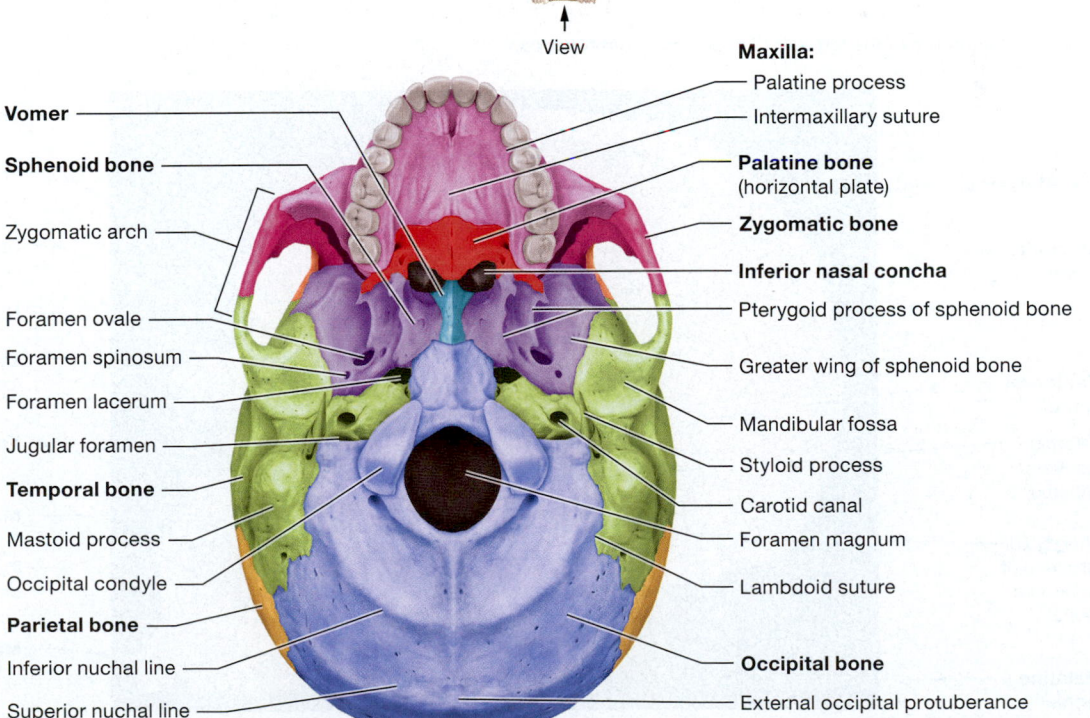

(c) Inferior view

Figure 7.7 Posterior, superior, and inferior views of the skull.

This midsagittal section is the best view to appreciate the structure of the nasal septum. Its superior portion is composed of the perpendicular plate of the ethmoid bone and the inferior portion of the vomer bone. We can also see two paranasal sinuses—the frontal sinus and sphenoid sinus.

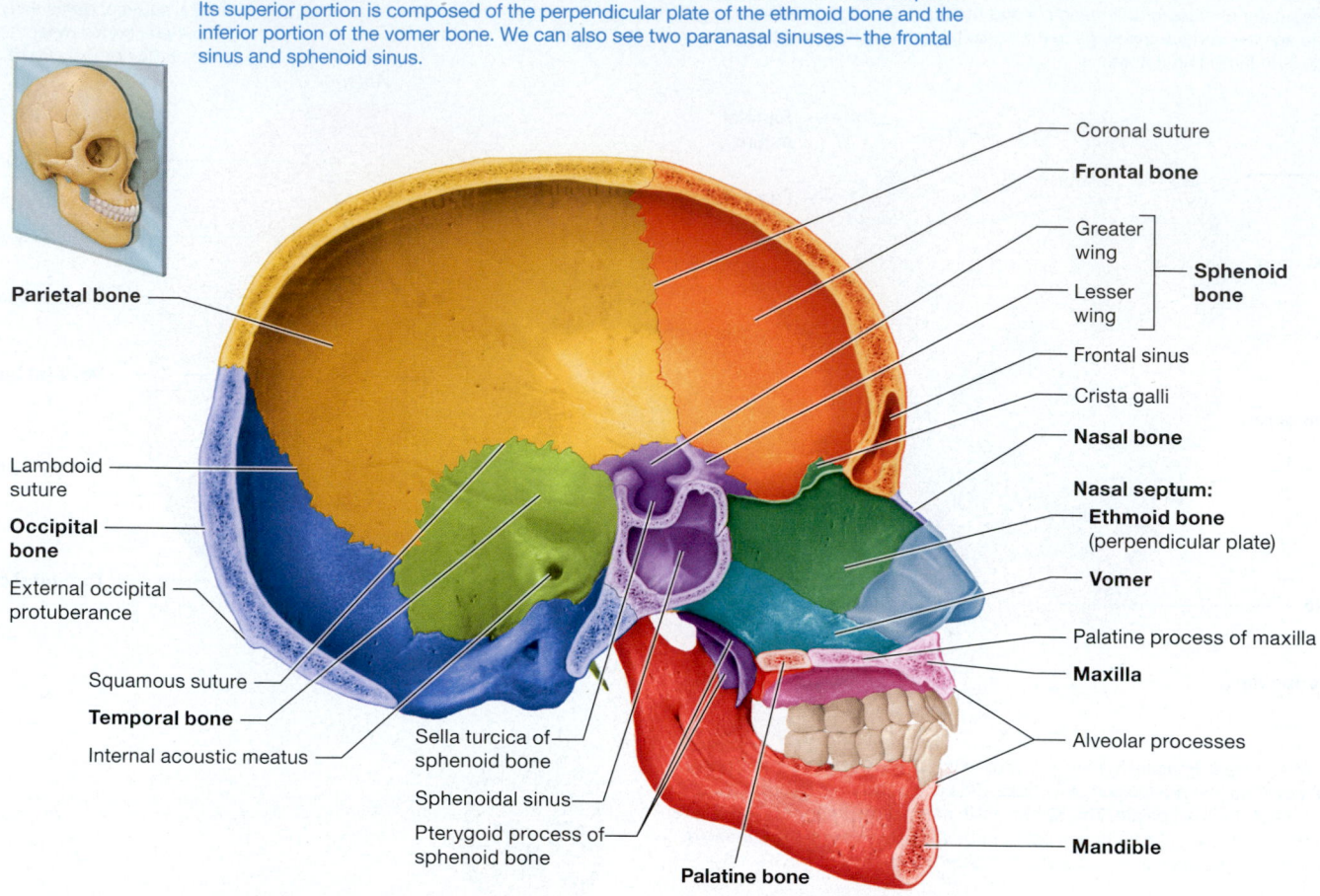

Coronal suture

Frontal bone

Greater wing

Lesser wing

} **Sphenoid bone**

Frontal sinus

Crista galli

Nasal bone

Nasal septum:

Ethmoid bone (perpendicular plate)

Vomer

Palatine process of maxilla

Maxilla

Alveolar processes

Mandible

Parietal bone

Lambdoid suture

Occipital bone

External occipital protuberance

Squamous suture

Temporal bone

Internal acoustic meatus

Sella turcica of sphenoid bone

Sphenoidal sinus

Pterygoid process of sphenoid bone

Palatine bone

(a) Internal anatomy of the left half of skull, midsagittal section

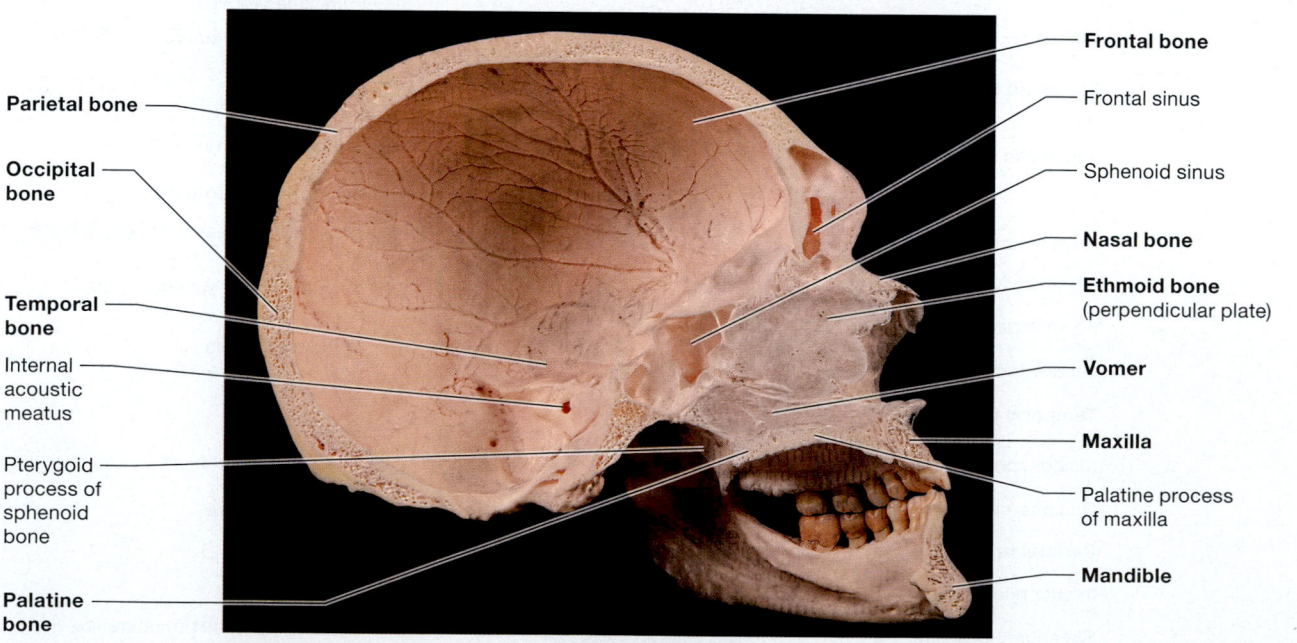

Parietal bone

Occipital bone

Temporal bone

Internal acoustic meatus

Pterygoid process of sphenoid bone

Palatine bone

Frontal bone

Frontal sinus

Sphenoid sinus

Nasal bone

Ethmoid bone (perpendicular plate)

Vomer

Maxilla

Palatine process of maxilla

Mandible

(b) Photo of the skull, midsagittal section

Play Bone Video
@ **Mastering** Anatomy & Physiology

Figure 7.8 **Medial view of midsagittal section of the skull.**

Imagine you are holding a skull, looking down at its superior surface. Now imagine removing the top of the skull and the brain and looking inside. This is the best view to appreciate the way the cranial fossae cradle the brain. Also note the cribriform plate of the ethmoid bone, the lesser and greater wings and sella turcica of the sphenoid bone, and the petrous part of the temporal bone.

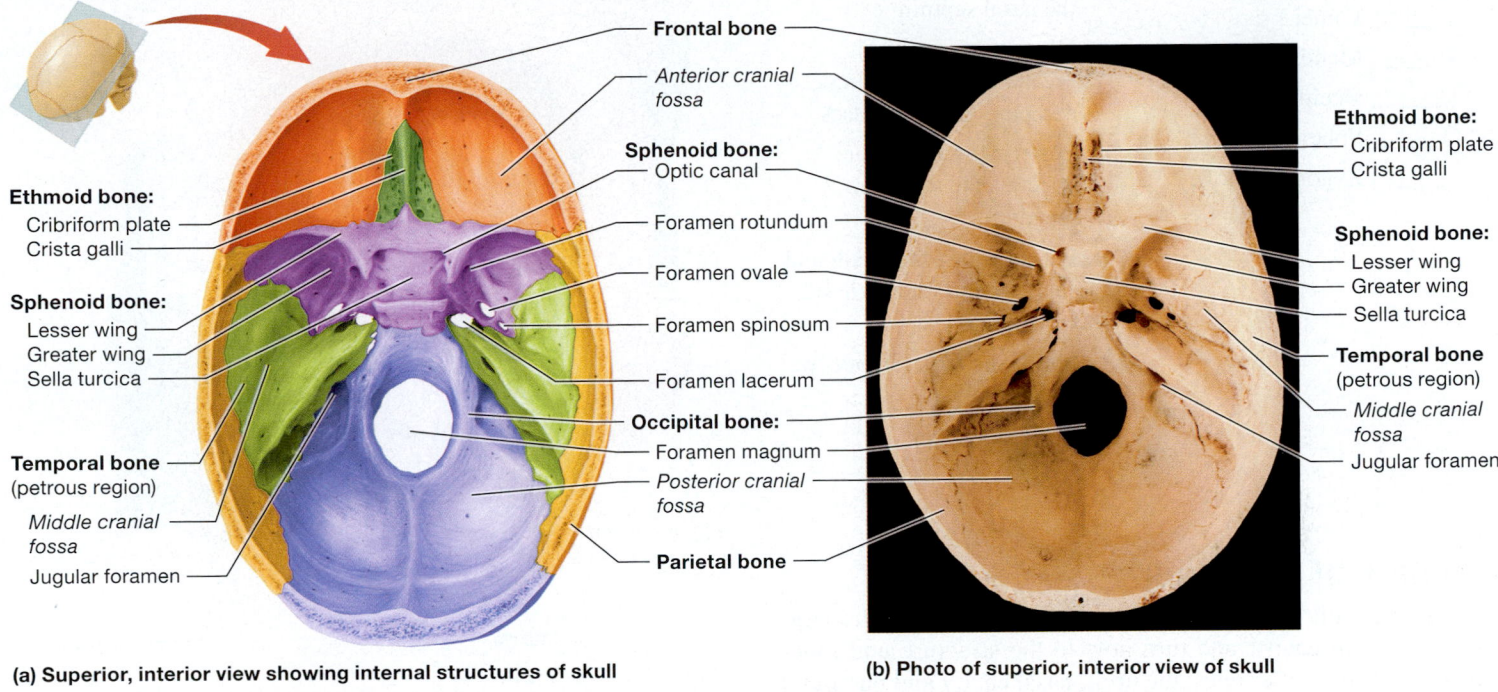

(a) Superior, interior view showing internal structures of skull

(b) Photo of superior, interior view of skull

Figure 7.9 Internal view of the skull.

ConceptBOOST)))

Understanding How Skull Bones Relate to One Another

It can be hard to appreciate the skull bones in terms of their structure and how they relate to one another. For this reason, we present **Figure 7.10**, which shows a disarticulated, or "exploded," skull. The bones are separated from one another so that you can see each individual bone clearly, as well as how they fit together. Think of the skull as a three-dimensional jigsaw puzzle with the pieces taken apart and separated. This is the best way to appreciate the structure of complex bones like the ethmoid and sphenoid, and see how they relate to other skull bones. ■

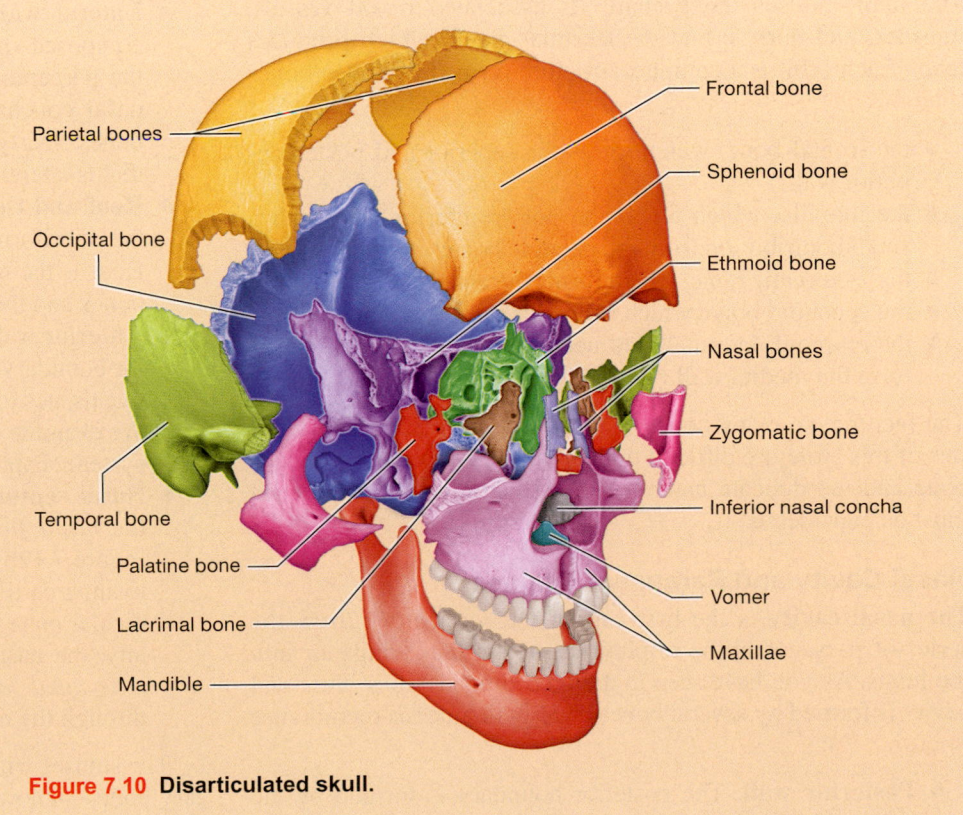

Figure 7.10 Disarticulated skull.

☐ 1. Match each bone with the correct description from the right column.

_____ Ethmoid bone

_____ Vomer

_____ Mandible

_____ Occipital bone

_____ Sphenoid bone

_____ Zygomatic bone

a. Forms the inferior part of the nasal septum

b. Composed of a body, greater and lesser wings, and pterygoid processes

c. Posterior cranial bone; contains the foramen magnum

d. Features the crista galli and superior and middle nasal conchae

e. Forms the anterior parts of the cheek

f. Lower jaw bone

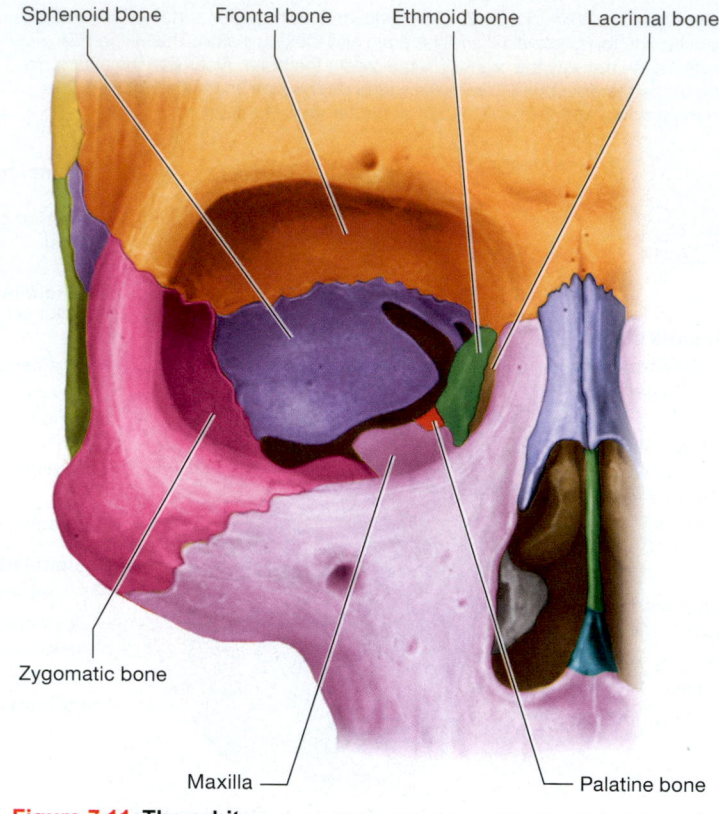

Figure 7.11 **The orbit.**

Cavities of the Skull

The skull is full of cavities. We've already discussed the largest, the cranial cavity, and turn now to the structure and function of several smaller ones: the orbit, nasal cavity and paranasal sinuses, and oral cavity.

Orbit

The **orbit** houses the eyeball; its associated blood vessels, muscles, and nerves; and the lacrimal gland, which produces tears. Each orbit is a complex structure formed by seven bones (**Figure 7.11**):

- the frontal bone, which forms the superior and posterosuperior walls;
- the maxilla, which forms the posteroinferior wall with a small contribution from the palatine bone;
- the zygomatic bone, which forms the anterolateral wall;
- the sphenoid bone, which forms the posterior wall; and
- the ethmoid, lacrimal, and palatine bones, which together form the medial wall.

The fact that so many bones contribute to the orbit makes injuries of this structure difficult to treat, as the contribution of each bone and the delicate nature of the structures within the orbit must be considered.

Nasal Cavity and Paranasal Sinuses

The **nasal cavity** is the first part of the *respiratory tract,* the series of passages in the respiratory system that brings air into the lungs. As you have seen in the preceding sections, the nasal cavity is formed by several bones lined with mucous membranes (**Figure 7.12**):

- **Posterior wall.** The posterior boundary is formed by the sphenoid body and pterygoid processes.

- **Lateral walls.** The lateral walls of the nasal cavity are composed of several bones, including the ethmoid bone, the perpendicular plate of the palatine bones, the inferior nasal conchae, and the maxilla. You can only see these bones clearly in a parasagittal section of the skull (a section that is lateral to the nasal septum; see Figure 7.12a).
- **Roof and floor.** The roof of the nasal cavity is formed by the cribriform plate of the ethmoid bone. The nasal cavity's floor is the hard palate, which is composed of the palatine bones and the palatine processes of the maxillae.
- **Anterior wall.** The nasal bones and maxillae form the anterior boundary of the nasal cavity. The cartilage and connective tissues of the nose attach to the margins of these bones. We examine these cartilages more closely in the respiratory system chapter (see Chapter 21).
- **Nasal septum.** The two sides of the nasal cavity, called the *nasal fossae,* are divided by the **nasal septum** (see Figure 7.12b). Anteriorly, the nasal septum is composed of hyaline cartilage. Posteriorly, it is made up of the perpendicular plate of the ethmoid bone and the vomer. Occasionally, the nasal septum is shifted to one side, resulting in a *deviated septum,* which may make it difficult to breathe through the nose.

The sinuses within the frontal, sphenoid, ethmoid, and maxillary bones—known collectively as the **paranasal sinuses**—are located around the nasal cavity and connect to it through small

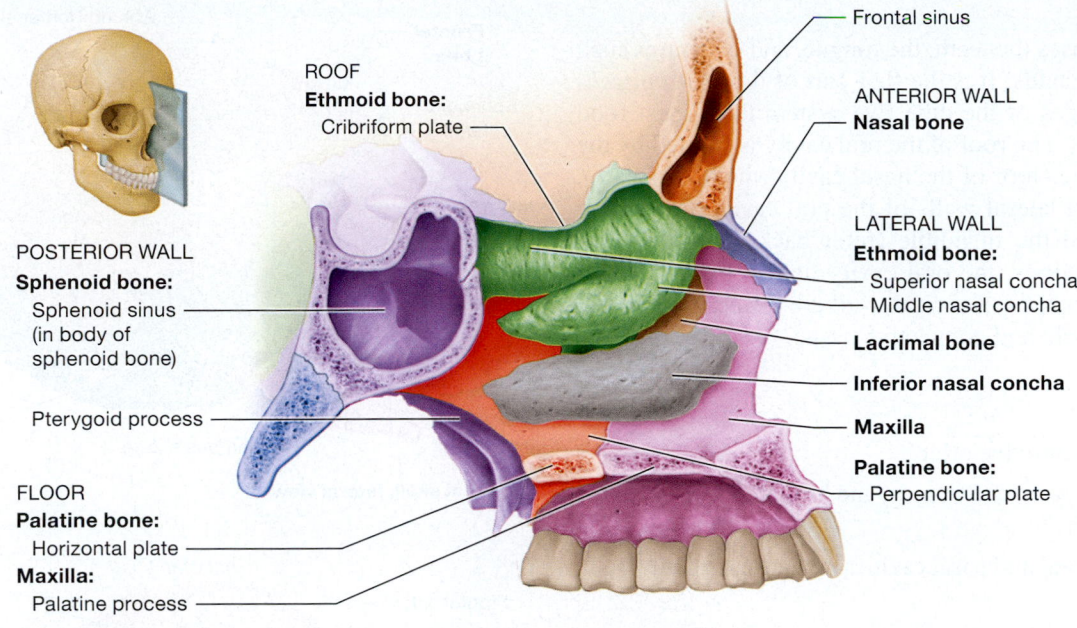

ROOF
Ethmoid bone:
Cribriform plate

Frontal sinus

ANTERIOR WALL
Nasal bone

POSTERIOR WALL
Sphenoid bone:
Sphenoid sinus
(in body of
sphenoid bone)

Pterygoid process

LATERAL WALL
Ethmoid bone:
Superior nasal concha
Middle nasal concha
Lacrimal bone
Inferior nasal concha
Maxilla
Palatine bone:
Perpendicular plate

FLOOR
Palatine bone:
Horizontal plate
Maxilla:
Palatine process

(a) Parasagittal section through nasal cavity (nasal septum removed)

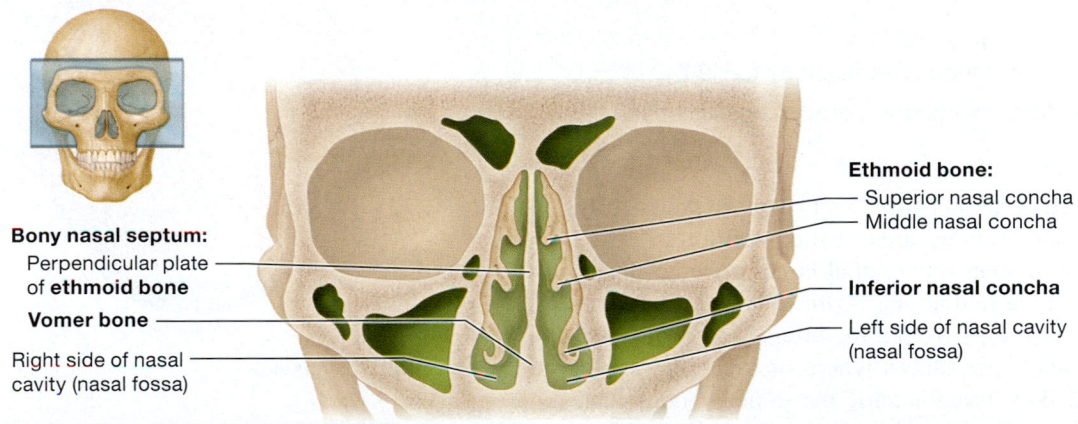

Ethmoid bone:
Superior nasal concha
Middle nasal concha

Inferior nasal concha

Left side of nasal cavity
(nasal fossa)

Bony nasal septum:
Perpendicular plate
of **ethmoid bone**
Vomer bone

Right side of nasal
cavity (nasal fossa)

(b) Anterior view of nasal cavity

Figure 7.12 Nasal cavity.

bony openings (**Figure 7.13**). Like the nasal cavity, they are lined with mucous membranes. Air flowing through the nasal cavity passes through the openings into the sinuses, where the mucous membranes filter, warm, and humidify the air. The paranasal sinuses also lighten the skull considerably and enhance voice resonance.

The connection of the sinuses to the nasal cavity has the unfortunate consequence of making the transmission of bacteria and viruses between the two structures quite easy. When infectious agents enter the paranasal sinuses, a condition called *sinusitis* results. The swollen mucous membranes produce excess mucus and block the drainage of mucus into the nasal cavity. Excess mucus leads to the common symptoms of pressure and a sinus headache. You can use this to your advantage, though, to help you remember the location of the sinuses: Sinus headaches occur above the eyes (frontal sinuses), between the eyes (ethmoidal sinuses), behind the eyes (sphenoidal sinuses), and below the eyes (maxillary sinuses).

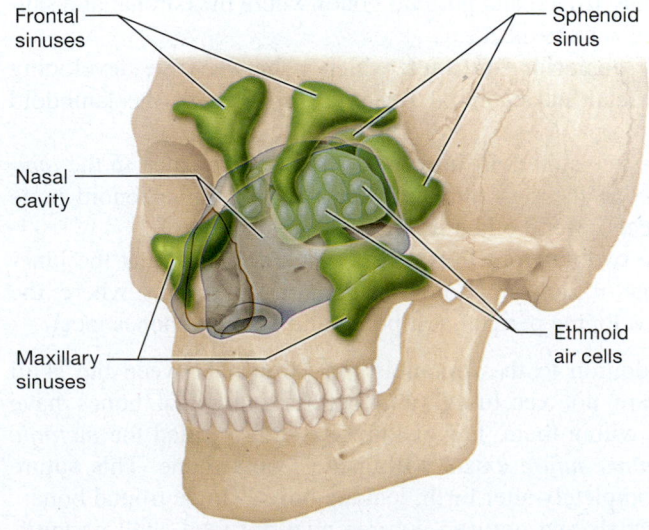

Frontal
sinuses

Sphenoid
sinus

Nasal
cavity

Maxillary
sinuses

Ethmoid
air cells

Figure 7.13 Paranasal sinuses.

Oral Cavity

The **oral cavity** houses the teeth, the tongue, and structures such as certain salivary glands. It is the first part of the *gastrointestinal tract,* the passages of the digestive system that digest food and absorb nutrients. The roof of the oral cavity is formed by the same structure as the floor of the nasal cavity (the hard palate), and the anterior and lateral walls of the oral cavity are formed by the maxillae and the mandible (refer back to Figure 7.8). Unlike the other cavities, the oral cavity has no bony floor or posterior wall. Instead, the floor and wall are formed by soft tissues, including muscle and connective tissue.

Quick Check

☐ 2. Which bones form the orbit?

☐ 3. What are the paranasal sinuses, and how are they related to the nasal cavity?

☐ 4. How are the oral and nasal cavities related structurally?

Fetal Skull

« FLASHBACK

1. What is intramembranous ossification? (p. 194)
2. Which bones form by intramembranous ossification? (p. 195)

You may have heard that infants have "soft spots" in their skulls. These soft spots, membranous areas called **fontanels,** result from the ossification process that cranial bones undergo. In the previous chapter, you learned about *intramembranous ossification,* in which bone develops from a membrane of fibrous connective tissue. Fontanels are places where ossification has not yet completed, and they remain until the cranial bones have ossified when a child is 18–24 months old. The skull of a fetus or infant has several fontanels, including (**Figure 7.14**):

- the **anterior fontanel,** which is located between the developing frontal and parietal bones where the coronal and sagittal sutures meet;
- the **posterior fontanel,** which is between the developing parietal and occipital bones at the apex of the lambdoid suture;
- the two **sphenoid fontanels,** which are located in the *temple* on the right and left sides where the sphenoid bone meets several other cranial bones; and
- the two **mastoid fontanels,** which are located at the junction of the lambdoid and squamous sutures where the developing parietal, temporal, and occipital bones meet.

In addition to the fontanels, the sutures between the skull bones are not yet fused, and certain individual bones have sutures within them. For example, a suture called the *metopic* (or *frontal) suture* exists within the frontal bone. This suture fuses completely after birth, leaving only a single frontal bone.

Fontanels and unfused sutures give the fetal skull a significant amount of flexibility. This flexibility ensures that an

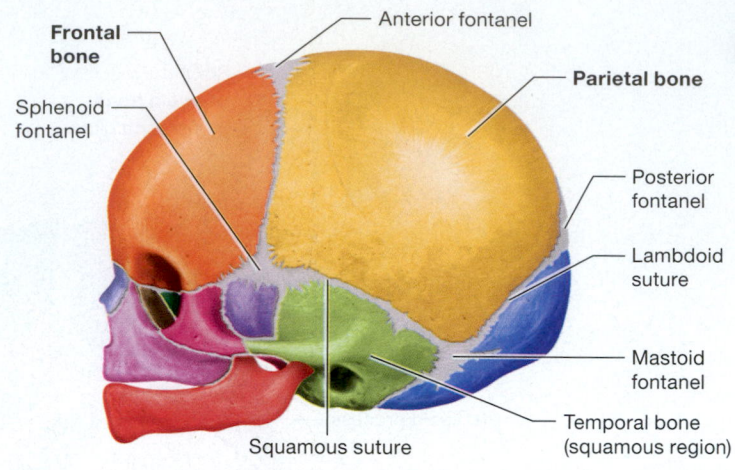

(a) Fetal skull, lateral view

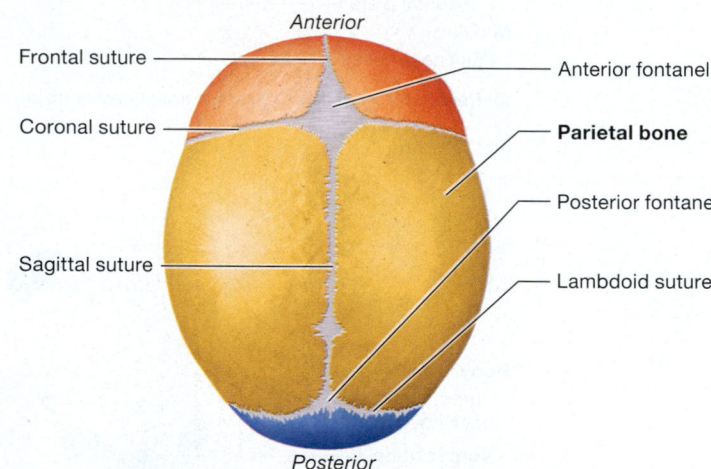

(b) Fetal skull, superior view

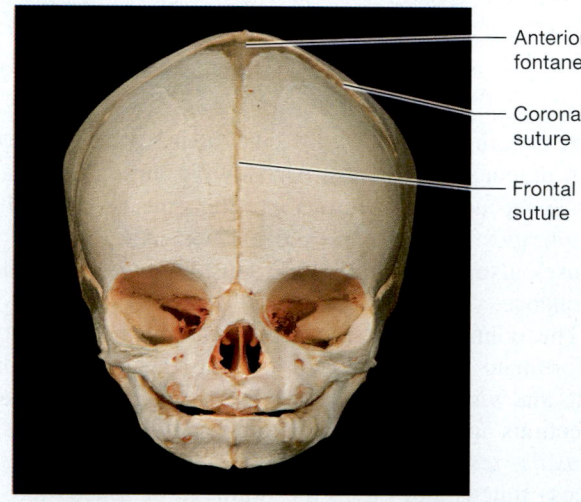

(c) Photo of fetal skull, anterior view

Figure 7.14 Fetal skull.

infant's skull is able to fit through the mother's vaginal canal during birth. A common consequence of this flexibility is that the heads of many infants are

CORE PRINCIPLE
Structure-Function

misshapen after a vaginal birth. Fortunately, their heads generally return to a normal shape in a few days. The fontanels also allow the skull to enlarge as an infant experiences normal brain growth, another example of the Structure-Function Core Principle (p. 28).

In a forensic setting, the degree to which skull sutures have fused can give important information about an individual's age. *A&P in the Real World: Forensic Skull Anatomy* examines sutures and other identifying features of the skull.

Quick Check

☐ 5. What are fontanels, and why are they important in the fetal skull?

☐ 6. Where are the six main fontanels located?

Hyoid Bone

The **hyoid bone** is a small, C-shaped bone in the superior neck (**Figure 7.15**). It is actually not a skull bone, but we discuss it with the skull bones because of its close proximity to them. However, it does not articulate with any skull bones or, indeed, with any bones at all. Instead, the hyoid bone is suspended in the superior neck by muscles and ligaments that attach it to the styloid processes of the temporal bone and to the larynx. It serves

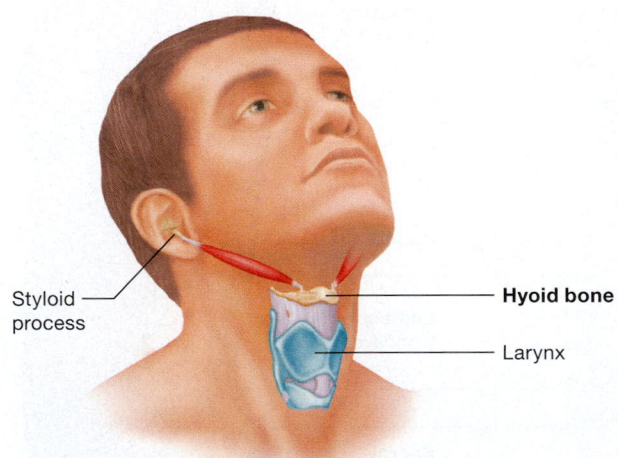

(a) Position of the hyoid bone

Styloid process

Hyoid bone

Larynx

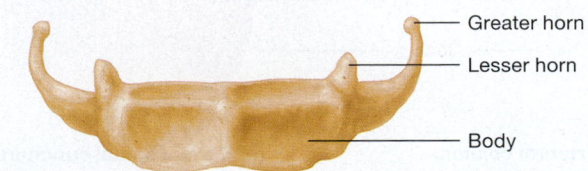

Greater horn

Lesser horn

Body

(b) Hyoid bone, anterior view

Figure 7.15 Structure of the hyoid bone.

as the attachment point for numerous muscles, including those involved in swallowing and speech. It is also an important bone in forensics, as a broken hyoid bone can indicate that a crime victim was strangled.

Forensic Skull Anatomy

Forensic investigators often must identify human remains with little to go on except bones. Fortunately, bones, particularly the skull, can give investigators many clues. One of the most basic traits that can be identified from a skull is the individual's sex. The two skulls shown here are from a female (left) and a male (right):

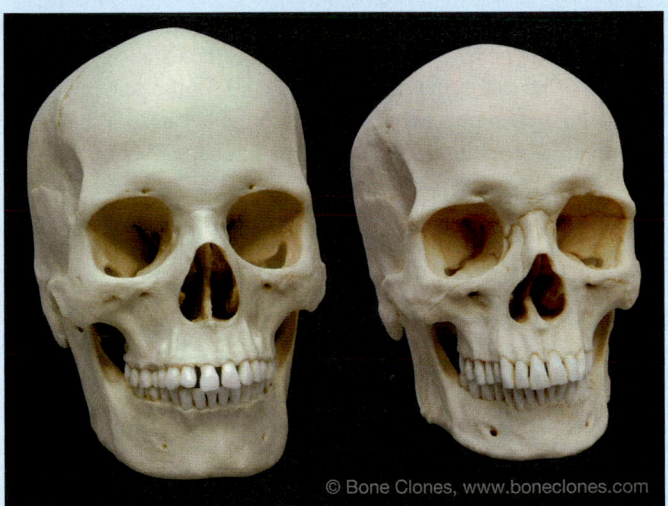

© Bone Clones, www.boneclones.com

Female skull   Male skull

There are several obvious differences between the male and female skulls:

- The forehead of the male skull is sloped and the female's is straight.
- The male's supraorbital ridge is more prominent.
- The mandibular angle of the male skull is closer to 90 degrees, and that of the female skull is generally greater than 90 degrees.
- The mastoid process of the male skull is larger and more prominent.

These features and several others are factored into sex identification.

In addition, the size of the skull, the appearance of the sutures, and the teeth can help approximate an individual's age. Other skull features provide clues to the individual's ethnic heritage. Once these factors have been analyzed, a forensic artist can develop a sketch of what the person looked like in life.

☐ 7. What is unique about the hyoid bone?

☐ 1. Ms. Mayer has been brought to the emergency department after sustaining a head injury during a cycling collision.

   a. During the physical exam, you note that a clear fluid is leaking from her nose; you recognize this as cerebrospinal fluid, the fluid surrounding the brain. Which bone has likely been fractured? Why would this cause cerebrospinal fluid to leak from the nasal cavity? (*Hint:* Which bone forms part of the cranial and nasal cavities?)

   b. Ms. Mayer is reporting vision problems. An x-ray of her skull shows a fracture of the lesser wing of the sphenoid bone. Why could this interfere with her vision?

☐ 2. Ms. Midna presents to your clinic with a complaint of pain in her face. She points to two areas, between her eyes and inferior to her eyes over her cheeks, as the locations of greatest pain. She states that she has had a cold for the past few days and has been unable to breathe through her nose. What is causing her pain? How and why is it related to her cold?

See answers in Appendix A.

MODULE **7.3**
The Vertebral Column and Thoracic Cage

1. Describe the curvatures of the vertebral column.
2. Compare and contrast the three classes of vertebrae, the sacrum, and the coccyx.
3. Explain the structure and function of intervertebral discs.
4. Describe the location, structural features, and functions of the bones of the thoracic cage.

The previous module focused on the superiormost part of the axial skeleton, the skull. Now, our attention shifts to the rest of the axial skeleton: the bones and skeletal cartilages of the vertebral column and thoracic cage.

Overview of the Vertebral Column

The **vertebral column,** or **spine,** consists of, on average, 33 bones called **vertebrae** (**Figure 7.16a**). We classify vertebrae on the basis of their structure and location:

- 7 **cervical vertebrae** located in the neck;
- 12 **thoracic vertebrae** that articulate with the ribs;

- 5 **lumbar vertebrae** in the lower back;
- 5 fused **sacral vertebrae** (collectively called the *sacrum*) that articulate with the pelvic bones; and
- 3–5 fused **coccygeal vertebrae** (collectively called the *coccyx*) at the most inferior end of the vertebral column.

We refer to each individual vertebra with a letter abbreviation, which signifies its type, and a subscript number that indicates its position, from superior to inferior, in the vertebral column. For example, C_4 is the 4th cervical vertebra and T_{11} is the 11th thoracic vertebra.

The lateral view (**Figure 7.16b**) reveals a space between most of the moveable (unfused) vertebrae; this is called the *intervertebral foramen.* The spinal cord is visible through these foramina, where spinal nerve roots emerge.

Spinal Curvatures

In a newborn infant, the vertebral column is C-shaped, reflecting its position within the mother's uterus as the fetus develops. As an infant grows, the vertebral column changes shape to develop a series of *curvatures* that result in a

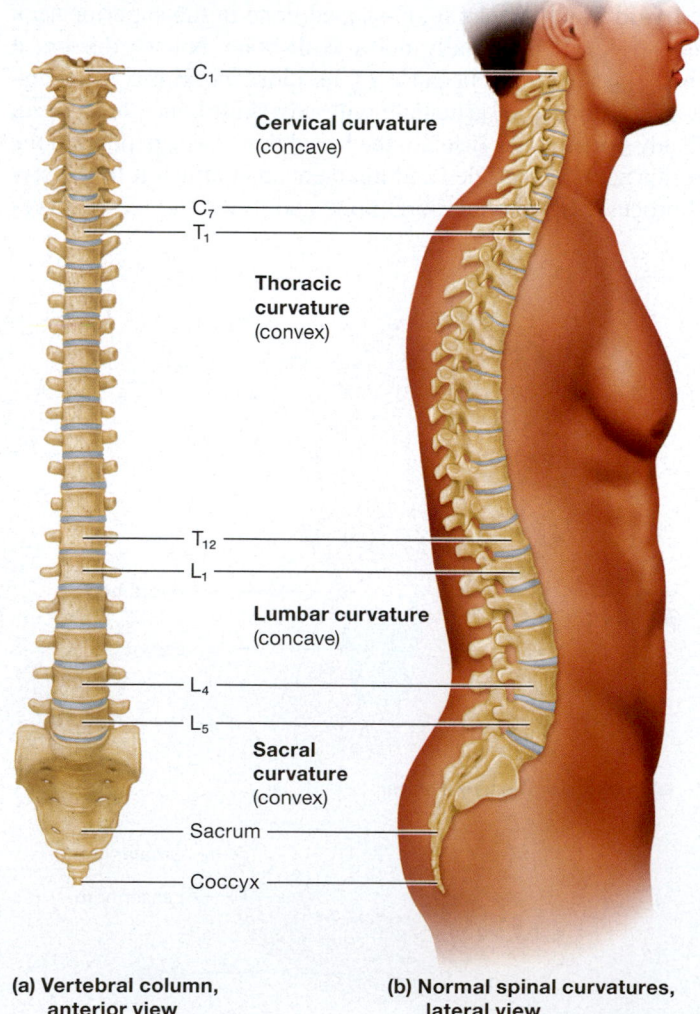

C₁

Cervical curvature (concave)

C₇
T₁

Thoracic curvature (convex)

T₁₂
L₁

Lumbar curvature (concave)

L₄
L₅

Sacral curvature (convex)

Sacrum

Coccyx

(a) Vertebral column, anterior view

(b) Normal spinal curvatures, lateral view

Figure 7.16 The vertebral column and normal spinal curvatures.

vaguely S-shaped vertebral column (see Figure 7.16b). The concave **cervical curvature,** which extends from about C_2 to T_2, develops as an infant begins to lift the head and crawl. As the child ages and begins to walk, the **lumbar curvature** develops, which is a concave curvature extending from about T_{12} to L_5.

The remaining two curvatures are the **thoracic curvature,** from T_2 to T_{12}, and the **sacral curvature,** from the lumbosacral junction to the coccyx. Both are convex curves that are present in fetal life as part of the fetus's original C-shaped vertebral column. For this reason, the thoracic and sacral curvatures are known as *primary curvatures.* The cervical and lumbar curvatures, in contrast, are called *secondary curvatures* because they develop after the fetal period.

The development of these secondary curvatures is critical to our species' ability to walk upright. The cervical curvature

 **CORE PRINCIPLE** Structure-Function

allows us to hold our heads up, and the lumbar curvature shifts the weight of the body onto the sacrum, which lends the balance and support needed to walk on two legs. Here we have another example of the Structure-Function Core Principle (p. 28).

Abnormal Spinal Curvatures

The spinal curvatures discussed so far are normal anatomical features of an adult vertebral column. However, abnormal curvatures may develop. Three conditions are marked by such curvatures:

- **Scoliosis.** Scoliosis (skoh-lee-OH-sis; *scolios* = "bending") is characterized by lateral curvatures in the vertebral column that give it a C or S shape when viewed from the posterior or anterior side (**Figure 7.17a**). Scoliosis may be *congenital* (caused by deformities present at birth), *neuromuscular* (caused by abnormalities of or trauma to the nerves and muscles around the vertebral column), or *idiopathic* (of unknown cause). Idiopathic cases are the most common. Mild scoliosis may cause no symptoms and require no treatment. Severe cases, however, can put pressure on the heart and lungs and so require treatment, which may include back braces, physical therapy, or surgery.

- **Lordosis.** Lordosis (lohr-DOH-sis; *lordos* = "bent forward"), commonly known as "swayback," is characterized by exaggerated cervical and lumbar curvatures (**Figure 7.17b**). Some degree of lordosis, particularly of the lumbar curvature, is normal in young children but generally diminishes with

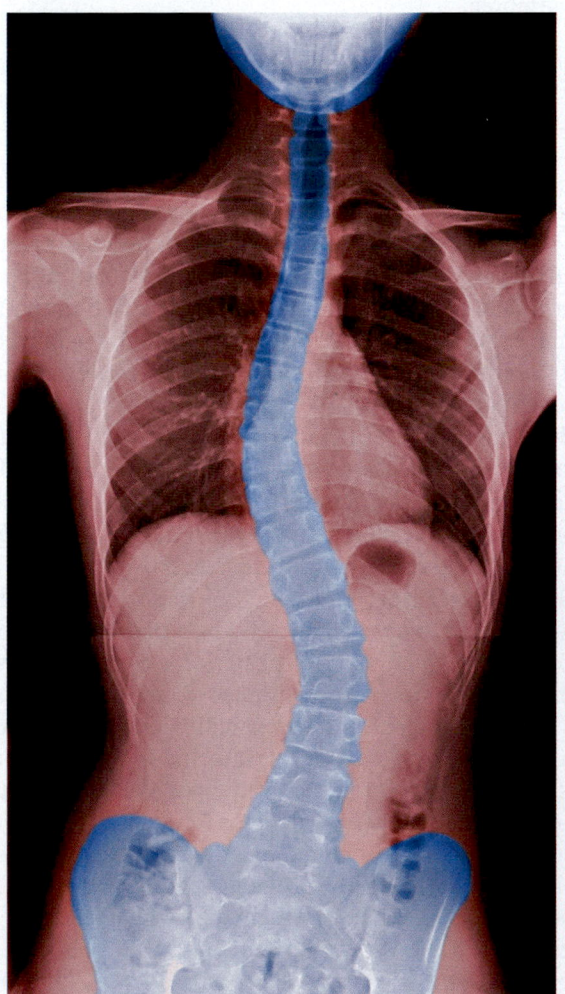

(a) Scoliosis, posteroanterior radiograph

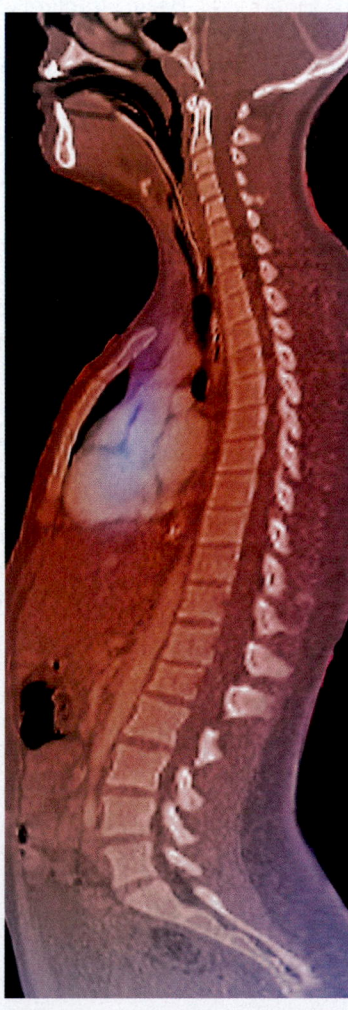

(b) Lordosis, lateral view, CT scan

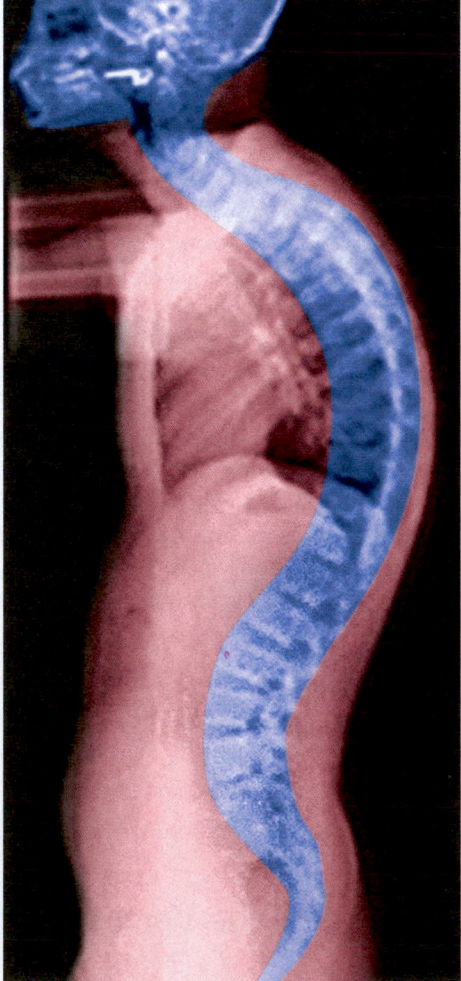

(c) Kyphosis, lateral radiograph

Figure 7.17 Three examples of abnormal spinal curvatures.

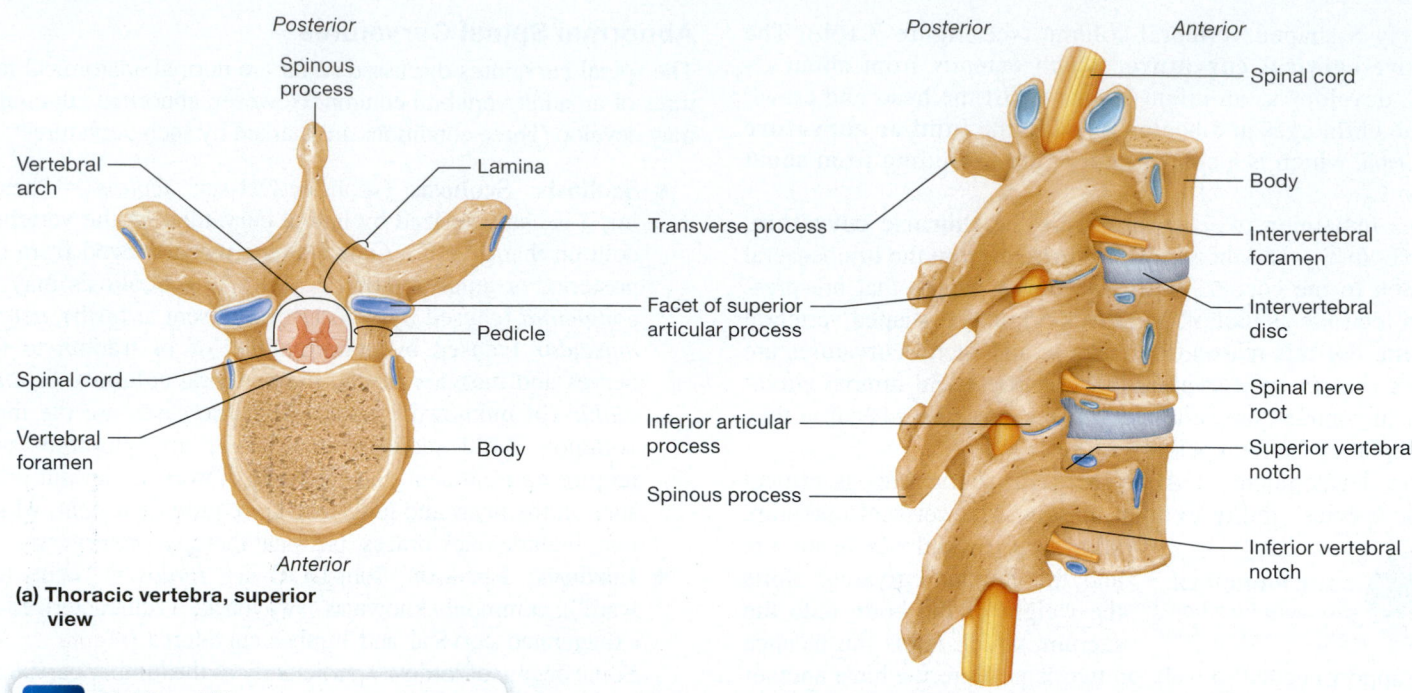

(a) Thoracic vertebra, superior view

Posterior

- Spinous process
- Vertebral arch
- Lamina
- Transverse process
- Facet of superior articular process
- Pedicle
- Spinal cord
- Vertebral foramen
- Body

Anterior

(b) Posterolateral view of three articulated thoracic vertebrae

Posterior — Anterior

- Spinal cord
- Body
- Intervertebral foramen
- Intervertebral disc
- Spinal nerve root
- Superior vertebral notch
- Inferior vertebral notch
- Inferior articular process
- Spinous process
- Transverse process

Play Bone Video @ Mastering Anatomy & Physiology

Figure 7.18 Basic structure of vertebrae.

age. Abnormal lordosis, sometimes called *hyperlordosis,* is commonly seen in adults bearing extra abdominal weight, such as pregnant women or overweight people. It may also have skeletal and neuromuscular causes. For example, dancers often develop lordosis due to their training, which can lead to muscle imbalances. Irrespective of the cause, lordosis puts extra stress on the lumbar vertebrae and can lead to lower back pain. It is generally treated with weight loss (if the person is overweight) and physical therapy.

- **Kyphosis.** In **kyphosis** (ky-FOH-sis; *kyphos* = "hump"), the thoracic curvature is exaggerated, giving a "hunchback" appearance (**Figure 7.17c**). It is caused by joint conditions such as arthritis, bone conditions such as osteoporosis and vertebral fractures, and developmental abnormalities of the skeleton. Mild kyphosis generally requires no treatment, but severe cases can be debilitating, leading to heart and lung dysfunction, nerve compression, and significant pain. Such severe cases generally require surgical correction.

Quick Check

☐ 1. How many cervical, thoracic, lumbar, sacral, and coccygeal vertebrae are normally present in the vertebral column?

☐ 2. What are the normal spinal curvatures?

☐ 3. Compare scoliosis, lordosis, and kyphosis.

Structure of the Vertebrae

« FLASHBACK

1. Where is the vertebral cavity? (p. 17)
2. Which structure resides in the vertebral cavity? (p. 17)

All vertebrae share common features, reflecting their shared function of protecting the spinal cord and supporting the head and trunk. **Figure 7.18a** shows a typical thoracic vertebra:

- **Body.** The most anterior structure (the part of the vertebra facing the thoracic cavity) is the blocklike **body,** also called the **centrum.** The body is the vertebra's primary weight-bearing structure. Vertebral discs cushion adjacent vertebral bodies and help absorb the shock of running and jumping. Fractures of the vertebral body can be painful and debilitating (see *A&P in the Real World: Vertebral Compression Fractures* on p. 235).

- **Vertebral foramen.** Just posterior to the body is the large **vertebral foramen,** through which the spinal cord and its associated tissues pass. The vertebral foramina of the entire vertebral column form the *vertebral cavity,* or *vertebral canal* (which you read about in Chapter 1).

- **Pedicles and laminae.** The vertebral foramen is bordered anteriorly by the body, laterally by two **pedicles** that project from the body, and most posteriorly by the **laminae.** Together, the pedicles and laminae form the **vertebral arch.** The inferior side of the pedicles curves up, creating a deep **inferior vertebral notch,** and the superior side curves down slightly, creating a **superior vertebral notch.** The inferior and superior notches of successive vertebrae stacked on top of one another form the intervertebral foramina that you saw in the previous section (**Figure 7.18b**). The pedicles merge into the laminae, which enclose the posterior part of the vertebral foramen.

- **Superior and inferior articular processes.** At the junctions of the pedicles and laminae we find two small processes, the **superior** and **inferior articular processes,** on the superior and inferior sides, respectively. The surfaces of both processes contain smooth regions known as *facets* that are covered with hyaline cartilage. The superior articular facets of one vertebra form joints with the inferior articular facets of the vertebra above it.

- **Transverse processes.** Projecting from the lateral sides of the vertebral arch are two **transverse processes.** These processes serve as attachment sites for muscles.
- **Spinous process.** The posterior projection from the vertebral arch is the **spinous process.** Like the transverse processes, the spinous processes are a site for muscle attachment. The spinous processes are palpable along your back as your "spine." Notice in Figure 7.18b how the spinous processes help protect the posterior spinal cord, which would otherwise be fairly vulnerable.

Although most vertebrae share these general features, their specific shape and size differ among cervical, thoracic, and lumbar vertebrae. In addition, cervical and thoracic vertebrae contain structures unique to those regions. The upcoming subsections focus on these differences.

Cervical Vertebrae

The seven cervical vertebrae are the smallest ones (**Figure 7.19**). Their most easily identifiable features are holes in their transverse processes, called **transverse foramina.** The transverse foramina allow the passage of the *vertebral arteries* and *vertebral veins.*

Vertebrae C_3–C_7 generally have oval bodies, large triangular vertebral foramina, and short straight spinous processes that are forked like the tongue of a snake (see Figure 7.19a). However, C_7 has a longer spinous process that is not forked. This long process is easily palpable and is used clinically as a landmark, which has led to the common name for C_7: *vertebra prominens,* or "prominent vertebra." The first and second cervical vertebrae, known as the *atlas* and the *axis,* respectively, are notably different from C_3–C_7.

C_1: *Atlas* The first cervical vertebra (C_1) is the **atlas** (see Figure 7.19b). Its name stems from the fact that it holds the head, just as Atlas, the Titan from Greek mythology, supported the world. The atlas is immediately recognizable by its unusual shape and unique features. It has no vertebral body or spinous

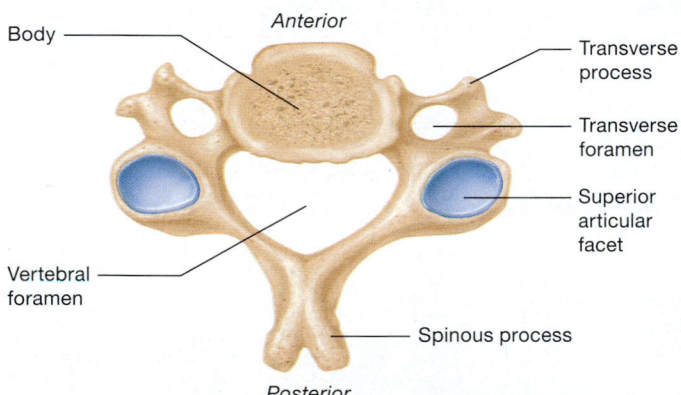

(a) Typical cervical vertebra (C_5)

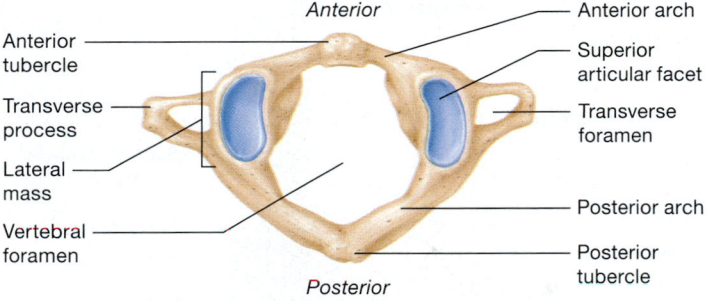

(b) Atlas (C_1), superior view

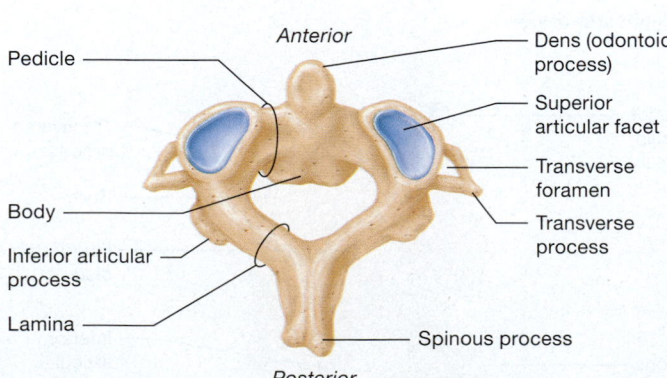

(c) Axis (C_2), superior view

Figure 7.19 Cervical vertebrae.

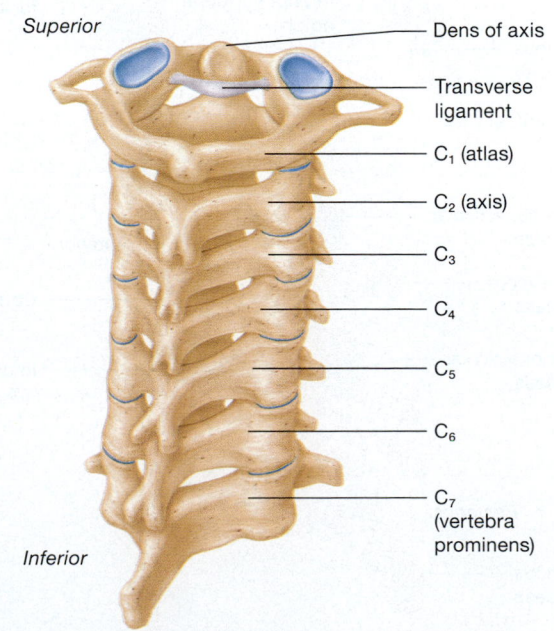

(d) Posterior view of articulated cervical vertebrae

process, and instead has two *anterior arches* that meet at the *anterior tubercle,* and two *posterior arches* that meet at the *posterior tubercle*. The anterior and posterior arches enclose a large, teardrop-shaped vertebral foramen. Where the anterior and posterior arches meet, we find the *lateral masses* that contain the

superior and inferior articular facets. Its superior articular facets articulate with the occipital condyles at the *atlanto-occipital joint,* the articulation that allows you to nod your head "yes." The inferior facets articulate with the second cervical vertebra.

C₂: Axis The second cervical vertebra (C_2) is the **axis** (see Figure 7.19c). Its most notable feature is a superior projection from the body known as the **dens** or **odontoid process.** Both names derive from the fact that the process is shaped like a tooth (*den/dont* = "tooth"). Figure 7.19d shows how the dens fits up into the vertebral foramen of the atlas where the body of the

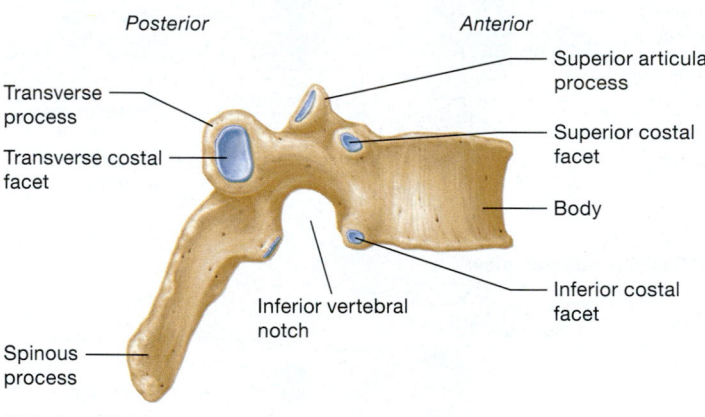

(a) Superior view

(b) Lateral view

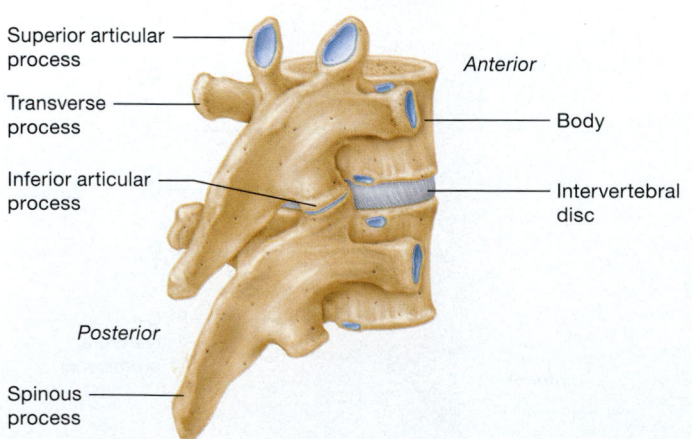

(c) Two articulated thoracic vertebrae, superior view

Figure 7.20 **Thoracic vertebrae.**

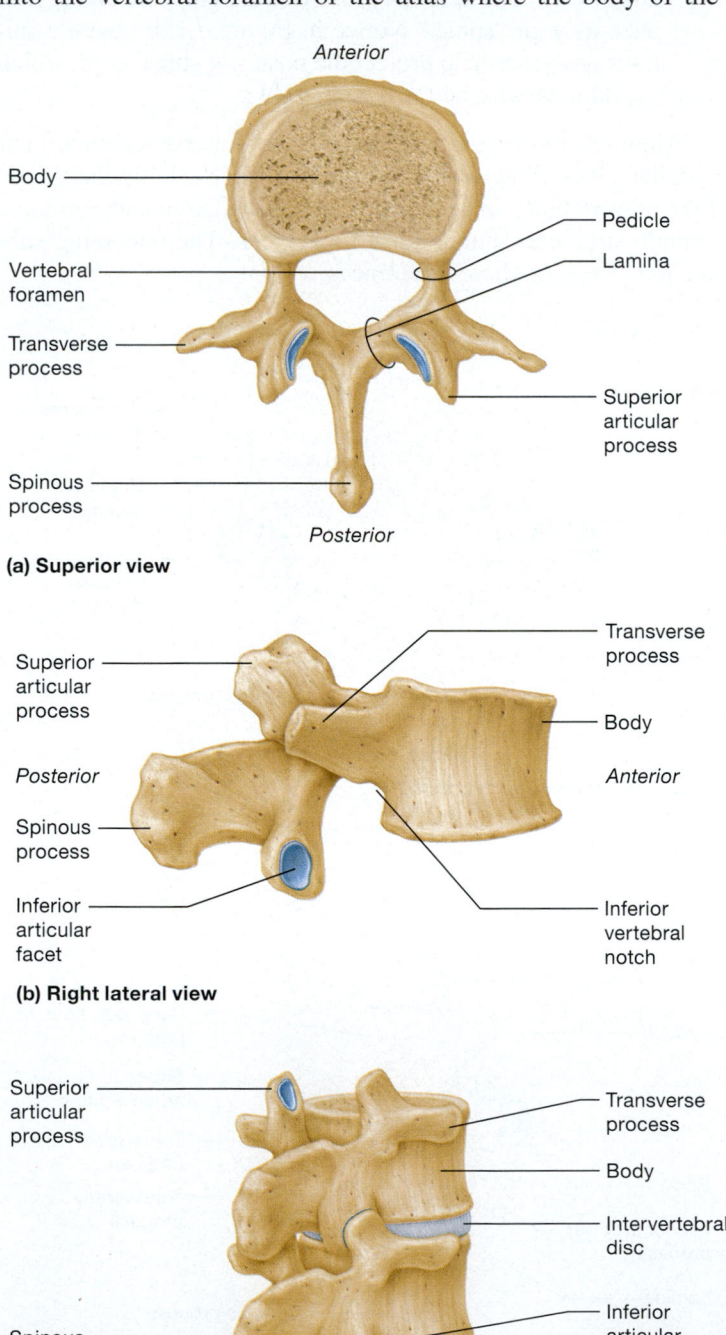

(a) Superior view

(b) Right lateral view

(c) Posterolateral view

Figure 7.21 **Lumbar vertebrae.**

atlas would be located, and is held snugly in place by the *transverse ligament*. Together, dens and atlas form the *atlantoaxial joint*. The way the dens fits into the atlas allows this joint to perform its function of rotational motion of the neck, such as shaking your head to indicate "no." Here we see another example of the Structure-Function Core Principle (p. 28).

CORE PRINCIPLE
Structure-Function

Thoracic Vertebrae

The 12 thoracic vertebrae are larger than the cervical vertebrae and grow progressively larger as we move down the vertebral column. Thoracic vertebrae are identifiable by their heart-shaped bodies, circular vertebral foramina, and long spinous processes that point inferiorly (**Figure 7.20**). On the lateral sides of the bodies of thoracic vertebrae, we find two small facets: the **superior costal facet** and **inferior costal facet** (*cost* = "rib"; note that the bodies of T_{10}–T_{12} have only a single costal facet on each side). As implied by their name, the costal facets provide the points of articulation for the ribs. Two additional costal facets are located on the transverse processes of T_1–T_{10}, the **transverse costal facets,** which articulate with another portion of the ribs, to be discussed shortly.

Lumbar Vertebrae

The five lumbar vertebrae are the largest and heaviest vertebrae, which reflects their primary function: bearing the weight of the torso. Their large kidney-shaped bodies, thick spinous processes that point posteriorly, and vertebral foramina shaped like a flattened triangle make them easy to recognize (**Figure 7.21**). In addition, lumbar pedicles and laminae are thicker and shorter than in other regions of the vertebral column.

Table 7.3 summarizes and compares the structure of cervical, thoracic, and lumbar vertebrae.

Vertebral Compression Fractures

A *vertebral compression fracture* is defined as a fracture of the vertebral body that reduces the height of the vertebra by more than 20%. Most often these fractures result from trauma or from diseases that destroy bone, such as osteoporosis and bone tumors. Compression fractures are quite common.

There are two types of compression fractures: wedge and burst. *Wedge fractures* involve only the anterior portion of the vertebra. These are most commonly seen in elderly women with kyphosis and osteoporosis, and can be induced by very minor trauma, such as sneezing or rolling over in bed. *Burst fractures* involve the entire vertebral body and are generally due to severe trauma. The symptoms of a wedge fracture vary but ordinarily involve pain that may radiate to other locations. The symptoms of a burst fracture are usually more severe, as bone fragments can damage the spinal cord and nerves.

Treatment depends on many factors, including the type and severity of the fracture and the overall health of the patient. Many patients respond well to conservative treatment, such as bed rest, bracing, and physical therapy, whereas others require surgery.

Table 7.3 Comparison of Cervical, Thoracic, and Lumbar Vertebrae

Characteristic	Cervical Vertebrae	Thoracic Vertebrae	Lumbar Vertebrae
Body shape and size	Small and oval; C_1 lacks a body; C_2 has the dens on the superior surface of its body	Larger and heart-shaped; contain costal facets	Largest and kidney-shaped
Vertebral foramen shape	Triangular	Circular	Flattened triangular
Transverse processes	Contain transverse foramina	Long; contain articular facets for ribs	Short with no facets or foramina
Spinous processes	Most are fork-shaped; C_1 lacks a spinous process	Long; point inferiorly	Thick; point posteriorly
Appearance (superior view)	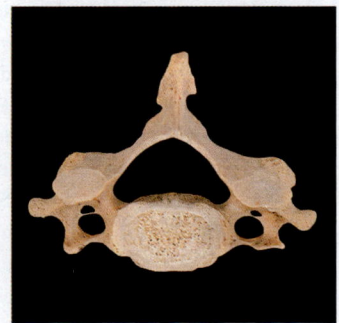	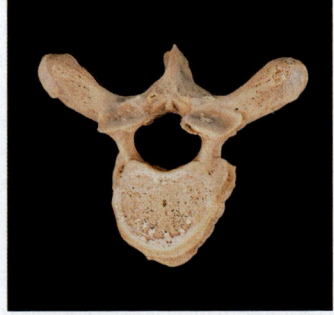	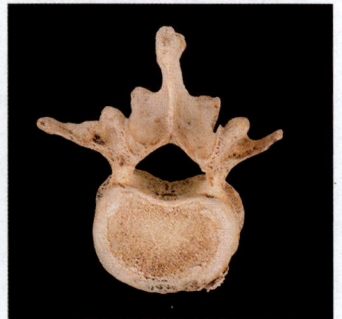

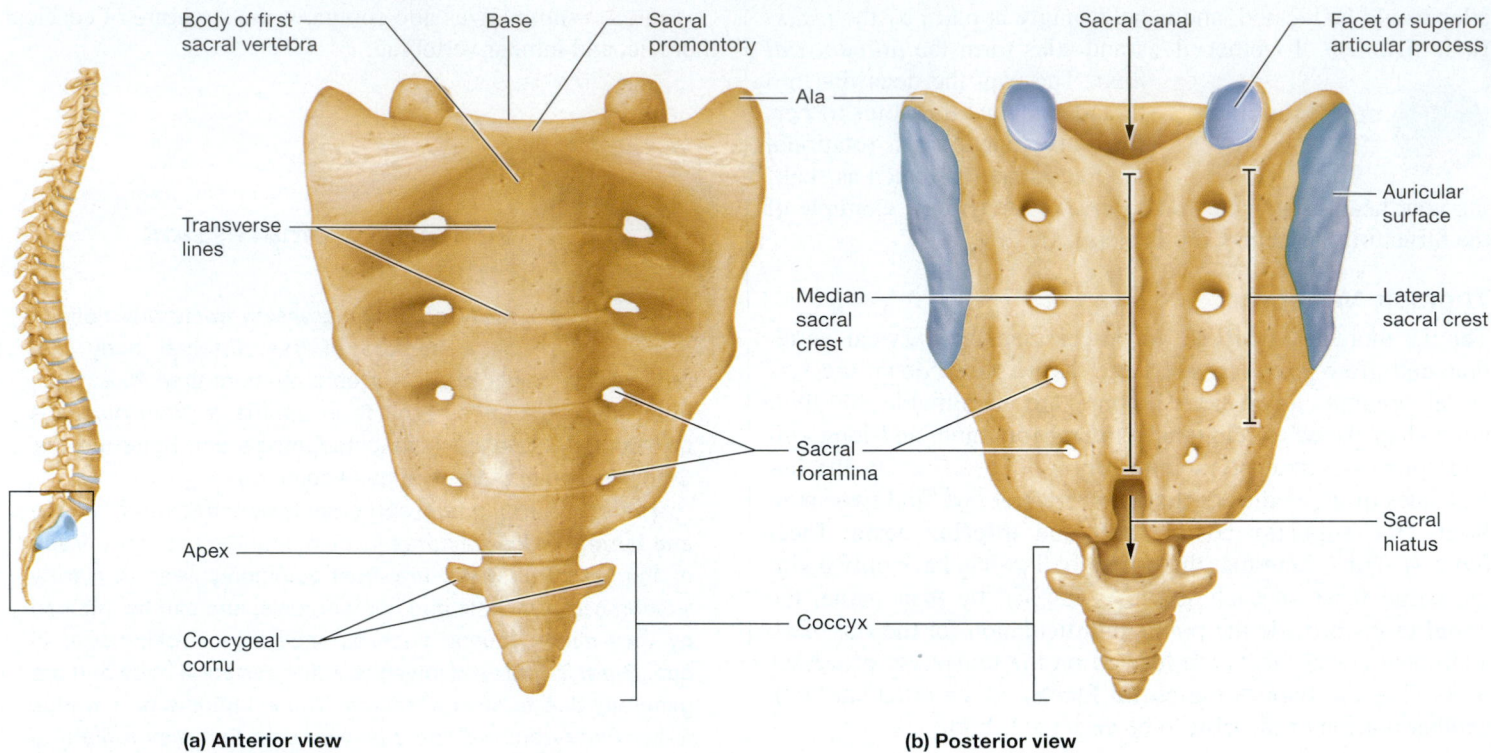

(a) Anterior view   (b) Posterior view

Figure 7.22 The sacrum and coccyx.

Sacrum and Coccyx

The five sacral vertebrae (S_1–S_5) form the posterior wall of the pelvic cavity. They begin as individual vertebrae but generally fuse by ages 20–25 to form the thick, triangular sacrum (**Figure 7.22**). Recall that the sacrum is curved as part of the concave sacral curvature. Its superior surface, the flattened **base,** articulates with the fifth lumbar vertebra at its superior articular processes. The inferior surface, which articulates with the coccyx, is called the **apex.** The sacrum's two lateral surfaces, the **auricular surfaces,** articulate with the two pelvic bones.

The anterior sacrum (see Figure 7.22a) shows the fused bodies of the sacral vertebrae, with *transverse lines* demarcating where the vertebrae fused. At the anterior margin of the base is a bony projection called the **sacral promontory.** Lateral to the sacral promontory are two relatively smooth regions, the **alae** (AY-lee; singular, *ala*).

On the posterior surface of the sacrum (see Figure 7.22b) is the opening to the **sacral canal,** a continuation of the vertebral canal, which contains nerve roots from the spinal cord as well as surrounding connective tissue membranes. The posterior boundary of the sacral canal is formed by the remnants of the sacral spinous processes, a ridge of bone called the **median sacral crest.** Flanking either side of the median sacral crest are four pairs of holes, the **sacral foramina,** through which exit nerves. Lateral to the sacral foramina are the **lateral sacral crests;** these are remnants of the transverse processes of the sacral vertebrae. Near the apex of the sacrum is the **sacral hiatus,** which is the termination of the sacral canal.

The final region of the vertebral column, the coccyx, is generally composed of four vertebrae that begin to fuse at about age 25, although it can contain anywhere from three to five vertebrae. Like

the sacrum, the coccyx has transverse ridges where the vertebrae fused. At its superior end, it has two hornlike projections, each called a **coccygeal cornu.** In many adults, some mobility exists between the coccyx and sacrum, although this generally diminishes with age.

StudyBOOST

Remembering Skull Bones and Vertebrae

Are you feeling overwhelmed by the names of all these bones? Here are some tricks to help you remember the skull bones and vertebrae.

- **PEST OF 6.** This identifies the six cranial bones: Parietal, Ethmoid, Sphenoid, Temporal, Occipital, Frontal.
- **Virgil Is Now Making My Pet Zebra Laugh.** This silly sentence will help you remember the facial bones: Vomer, Inferior nasal conchae, Nasal, Mandible, Maxillae, Palatine, Zygomatic, Lacrimal.
- **Every Student Fancies Learning Zillions More Parts.** This optimistic statement will help you with the bones of the orbit: Ethmoid, Sphenoid, Frontal, Lacrimal, Zygomatic, Maxilla, Palatine.
- **For Easier Sinus Memorization.** As you can tell, this trick is for remembering the paranasal sinuses: Frontal, Ethmoidal, Sphenoidal, Maxillary.
- **Breakfast at 7, lunch at 12, dinner at 5.** This will help you remember the number of each type of vertebrae: 7 cervical, 12 thoracic, and 5 lumbar.

To remember the different bones, you may also use visual analogies that compare an anatomical structure to a more familiar object. Here are a few examples:

- The sphenoid bone looks like a bat when viewed from the anterior side (it's the bone with "wings").
- The ethmoid bone is positioned somewhat like an iceberg in the skull, with only its tip (the crista galli) visible in the anterior view (refer to Table 7.2).
- A thoracic vertebra looks like the head of a giraffe when viewed from the posterior side (think "thoRAcic giRAffe"):

 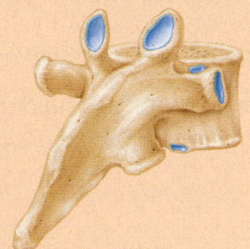

Thoracic giraffe

- A lumbar vertebra, in contrast, resembles the head of a "lumbering" moose:

 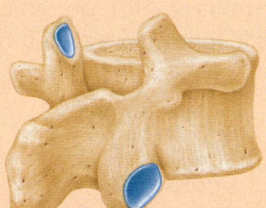

Lumbering moose

You aren't necessarily going to use everything on this list. Choose whatever tricks work for you and use them. If you find that a memory aid doesn't stick after repeating it a couple of times, move on. Search for a new one (the Internet has many), make up your own, or simply don't use one at all. Be careful not to spend your energy memorizing mnemonics that don't work for you—these aids should make studying easier, not harder. ■

Quick Check

☐ 4. How do the atlas and axis differ from other vertebrae?

☐ 5. Identify each of the following characteristics as belonging to cervical, thoracic, or lumbar vertebrae; the sacrum; or the coccyx.

_____ Have thick, blocklike bodies with thick spinous processes

_____ Makes up the inferiormost portion of the vertebral column

_____ Contain transverse foramina

_____ Consists of five fused vertebrae

_____ Have downward-pointing spinous processes and costal facets

Intervertebral Discs

« FLASHBACK

1. What are the properties of fibrocartilage? (p. 142)

An **intervertebral disc** is a fibrocartilage pad between two vertebrae that absorbs shock, binds the vertebral column together, and helps support the weight of the body. There are 23 intervertebral discs; the first one is between C_2 and C_3, and the final one is between L_5 and S_1 (the sacrum).

A disc has two main components (**Figure 7.23**). The soft, inner, jelly-like substance called the **nucleus pulposus** (NOO-klee-us pull-POH-sus) is a resilient shock absorber. It is surrounded by an outer ring of fibrocartilage, the **anulus fibrosus** (AN-yoo-lus fy-BROH-sus; *anulus* = "ring"), which holds both the nucleus pulposus in place and the vertebrae together. *A&P in the Real World: Herniated Disc* discusses what happens when the anulus fibrosus tears.

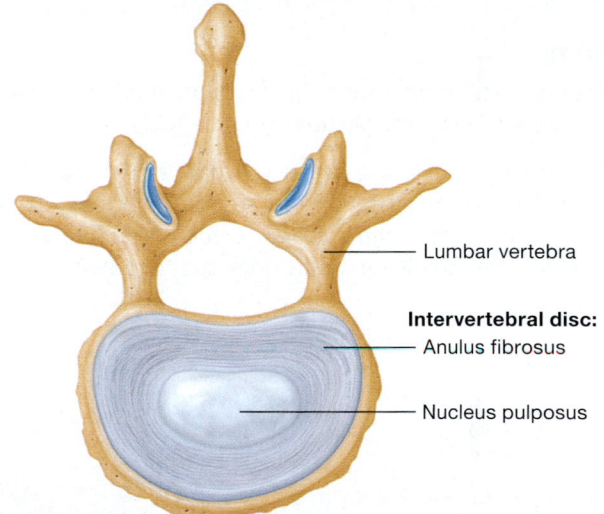

Lumbar vertebra

Intervertebral disc:
— Anulus fibrosus

— Nucleus pulposus

Figure 7.23 Structure of an intervertebral disc, superior view.

Herniated Disc

A tear in the anulus fibrosus can allow the nucleus pulposus to protrude, a condition known as a **herniated disc** (commonly called a *slipped disc*). Tears generally result from trauma, injuries secondary to improper lifting, or longer-term damage due to repeated motions such as lifting and squatting. The anulus fibrosus generally tears on its posterolateral side because of the arrangement of ligaments around the vertebral column. The bulging nucleus pulposus compresses nerve roots as they emerge from the spinal cord.

Nerve root compression can cause significant pain that may radiate to other parts of the body, as well as numbness, tingling, and muscle weakness. Treatments include injection of anti-inflammatory steroid medications, physical therapy, and surgery to repair the tear and relieve nerve compression.

In addition to the intervertebral discs, a number of ligaments and muscles support the vertebral column (see the muscular system chapter).

Quick Check

☐ 6. Describe the structure of an intervertebral disc.

The Thoracic Cage

❮❮ FLASHBACK

1. Where is the thoracic cavity? (p. 18)
2. Which organs does the thoracic cavity enclose? (p. 18)

The **thoracic cage** consists of the *sternum,* the 12 pairs of *ribs,* and the thoracic vertebrae. These bones form the peripheral boundary of the thoracic cavity, and they shield the heart, lungs, and great blood vessels.

Sternum

The flattened **sternum** forms the anterior, median part of the thoracic cage. It has three portions (**Figure 7.24**):

- **Manubrium.** The sternum's top part is the **manubrium** (muh-NOO-bree-um). On its superior surface, it has a notch called the *suprasternal notch.* Just lateral to the suprasternal notch on both sides are the two *clavicular*

notches, which articulate with the clavicles of the pectoral girdles. Inferior to the clavicular notches we find the sites of attachment for the cartilages of the first ribs.

- **Body.** The middle **body** is the largest part of the sternum. The cartilages of the second ribs attach where the manubrium and the body meet, a location called the **sternal angle.** The lateral edges of the body have several notches where the cartilages of the third through seventh ribs attach.

- **Xiphoid process.** The most inferior portion of the sternum is the pointed **xiphoid process** (ZY-foyd; *xiph-* = "sword"), which is the site of attachment for certain abdominal muscles. Composed of hyaline cartilage initially, it is generally fully ossified by about age 40.

To read about an important clinical implication of sternal anatomy, see *A&P in the Real World: The Sternum and CPR.*

Ribs

In both men and women, the **rib cage** consists of 12 pairs of ribs and the anterior **costal cartilages** (*cost-* = "rib"). The spaces between successive ribs are the **intercostal spaces** (see Figure 7.24). Recall from earlier in the chapter that the costal cartilages are composed of mostly hyaline cartilage, which gives the rib cage some flexibility during movement and breathing. Each rib has a posterior attachment to a thoracic vertebra, from which it curves around anteriorly in a C shape.

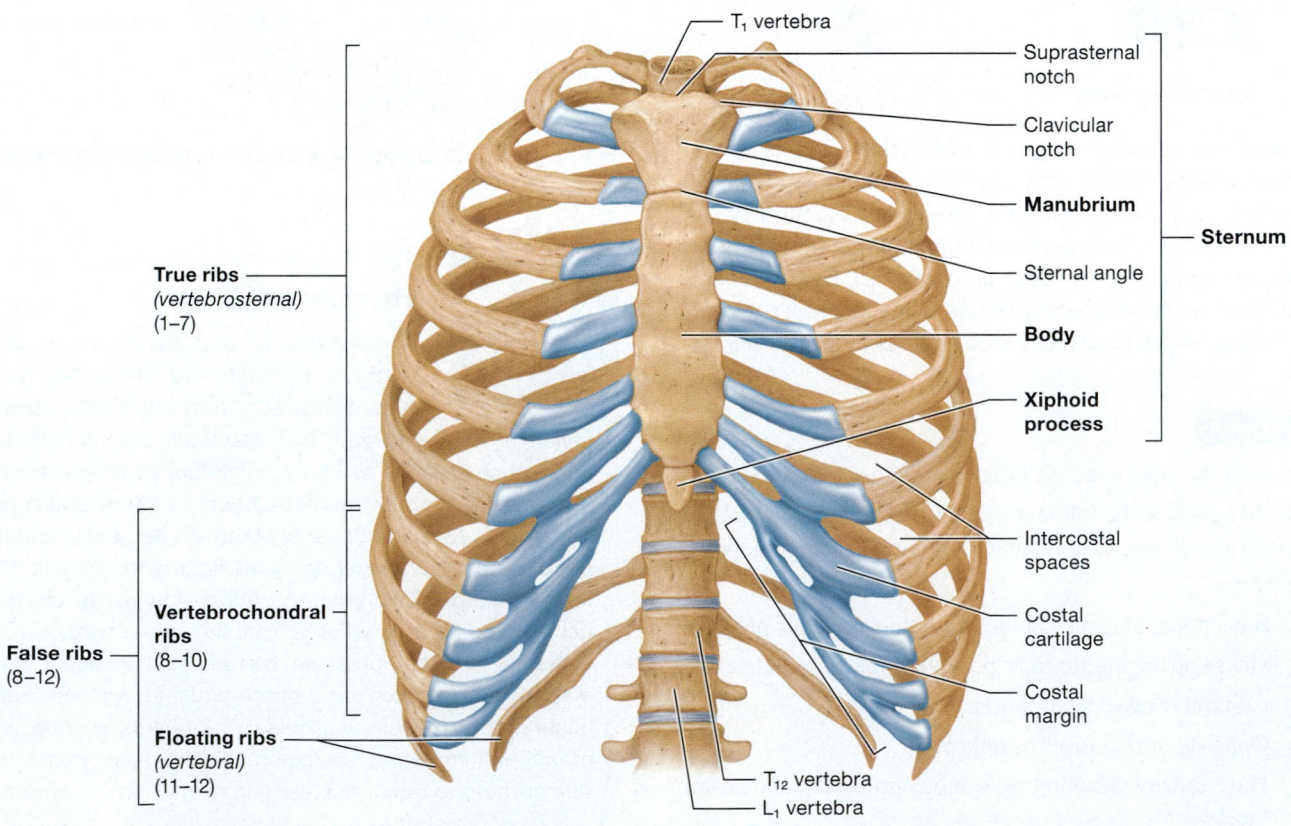

Figure 7.24 The thoracic cage, anterior view.

We group ribs into two classes based on their anterior attachment:

- Ribs 1–7 are called **true ribs** or *vertebrosternal ribs* because they attach to the sternum via their own costal cartilage.
- Ribs 8–12 are called **false ribs** because they do not attach to the sternum directly. There are two types of false ribs. The first, consisting of ribs 8–10, are the *vertebrochondral ribs,* which have costal cartilages that attach to the cartilage of the seventh rib. Their costal cartilages form a prominent rim called the **costal margin.** The second type, consisting of ribs 11 and 12, are referred to as **floating ribs,** or *vertebral ribs,* because they lack any attachment to the sternum.

Figure 7.25 shows the structure of a typical rib. Each rib has a rounded **head** where it articulates with the body of a thoracic vertebra. The heads of most ribs have two articular facets: an *inferior articular facet* that articulates with the body of the thoracic vertebra of the same number as the rib, and a *superior articular facet* that articulates with the body of the vertebra superior to it. For example, the third rib articulates with T$_3$ at its inferior articular facet and T$_2$ at its superior articular facet.

Lateral to the rib's head is its **neck,** which runs along the transverse process of the vertebra. The neck is followed by a projection of bone called the **tubercle** (TOO-bur-kuhl), which articulates with the transverse process of the vertebra. The rib then curves anteriorly at its **angle** to become its **shaft,** or **body.** The concave internal surface of the shaft contains a groove called the *costal groove,* along which blood vessels and nerves travel. The ends of the shafts for ribs 1–10 are square and attach to their costal cartilage; the ends of the floating ribs are also capped with a small layer of hyaline cartilage in spite of their lack of sternal attachment.

Ribs 2–9 contain all of the features just discussed, but ribs 1 and 10–12 have slight variations in their structure. For instance, each of these ribs attaches to the body of only one thoracic vertebra, and so their heads lack separate superior and inferior articular facets. In addition, the floating ribs have no articulation with the transverse processes of the vertebrae and so lack tubercles.

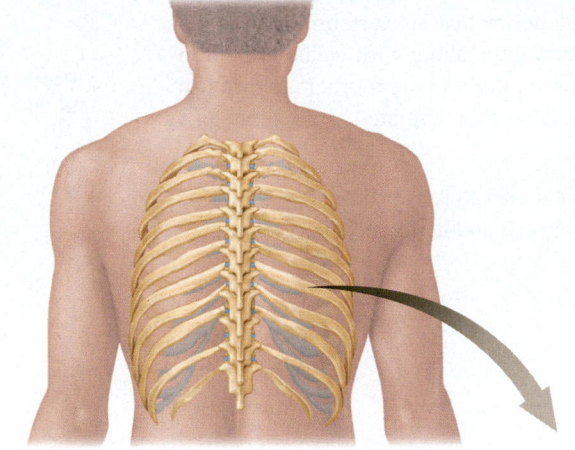

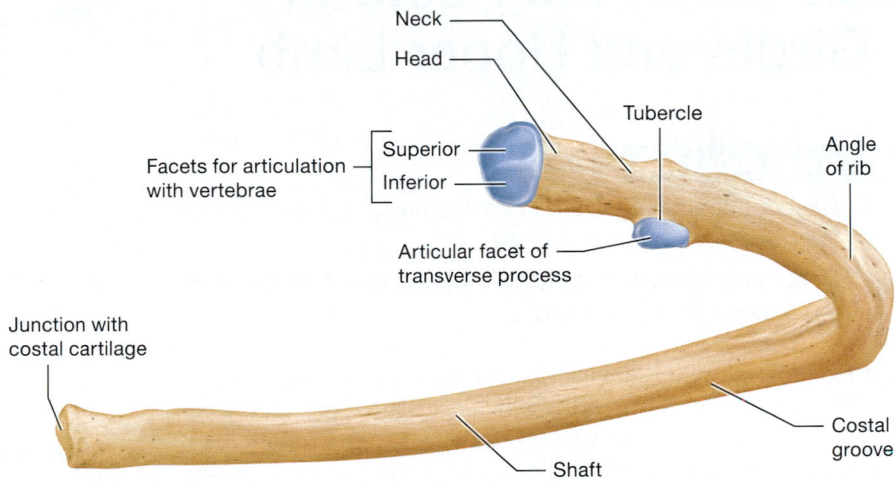

Figure 7.25 **Structure of a typical rib.**

Neck

Head

Tubercle

Angle of rib

Facets for articulation with vertebrae — Superior / Inferior

Articular facet of transverse process

Junction with costal cartilage

Costal groove

Shaft

Right rib, posterior view

Quick Check

☐ 7. What are the three components of the sternum?

☐ 8. How do true, false, and floating ribs differ?

Apply What You Learned

☐ 1. Ms. Cho has fractured the transverse process of her fourth thoracic vertebra. She reports back pain, as well as pain with movement such as breathing. What else was likely fractured with her vertebra (be specific)? Explain. Why does this cause pain when she breathes?

The Sternum and CPR

Cardiopulmonary resuscitation (CPR) is a lifesaving technique administered when an individual's heart has stopped producing functional contractions and/or breathing has stopped. CPR may restore circulation to the body by the application of repeated compressions to the chest over the sternum. Correct placement of the hands on the sternum is critical—they must be positioned in the center of the body of the sternum. This way, the rescuer avoids compressing the xiphoid process, as compressions applied directly to the xiphoid process can cause it to break off and damage underlying organs, particularly the liver. Note though that even when performed correctly, CPR is often a traumatic process. It can break the sternal body and ribs, especially in elderly individuals with low bone mass.

☐ 2. Tran is a 12-year-old patient who is experiencing back pain. Upon examination, you notice that his vertebral column has a pronounced lateral curvature, along with weakness in the muscles on one side, particularly in the thoracic region. An MRI scan of his back shows two herniated discs between T_8 and T_9 and T_9 and T_{10}.

   a. What abnormal spinal curvature is present?
   b. Are the muscle weakness and herniated discs related? Explain.

See answers in Appendix A.

MODULE **7.4**

Bones of the Pectoral Girdle and Upper Limb

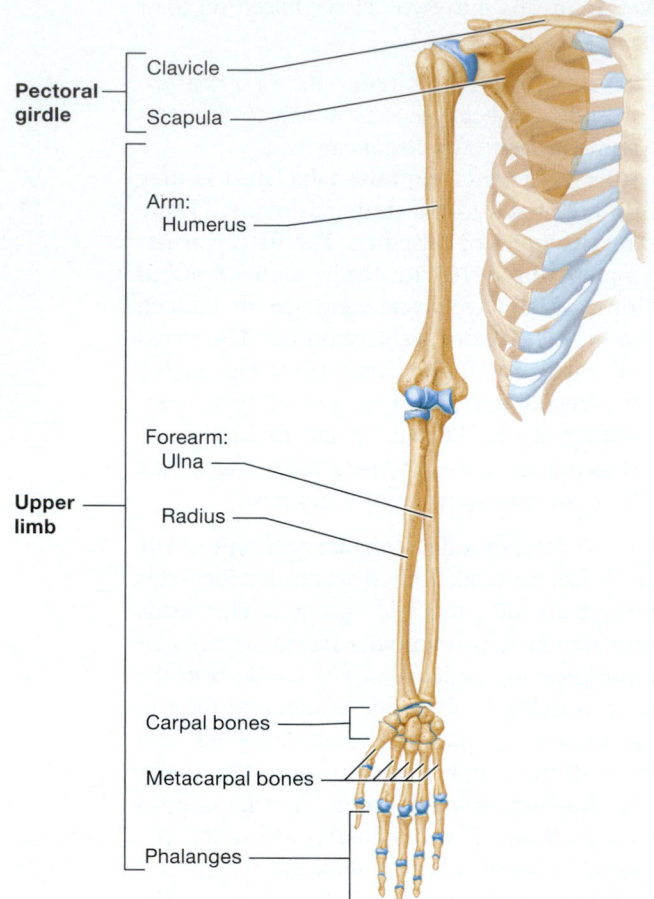

Figure 7.26 Overview of the bones of the pectoral girdle and upper limb.

Learning Outcomes

1. Describe the location, structural features, and functions of the bones of the pectoral girdle.
2. Describe the location, structural features, and functions of the bones of the upper limb.

We move on now to the appendicular skeleton. We start in this module with the pectoral girdle, which consists of two bones, the *clavicle* and *scapula,* that support the upper limb (**Figure 7.26**). We then move on to the upper limb, which consists of 30 bones, including the *humerus, ulna, radius, carpal bones, metacarpal bones,* and *phalanges.*

The Pectoral Girdle

The two bones of the pectoral girdle—the *clavicle* and the *scapula*—reside along the superior and posterior thoracic cage (**Figure 7.27**). Together they support the upper limb, particularly the humerus, and are the sites of attachment of numerous muscles.

Clavicle

The **clavicle** has two distinct ends that are usually easily palpable through the skin. From an anterior view the clavicle appears straight (see Figure 7.27a), but from a superior or inferior view it is actually S-shaped (see Figures 7.27b and c). Its medial **sternal end** articulates with the manubrium at the sternum, forming the *sternoclavicular joint.* Note that this joint is the only place where the pectoral girdle articulates with the axial skeleton. The lateral **acromial end** articulates with a process of the scapula called the *acromion* to form the *acromioclavicular joint* (discussed in the next section). Near the acromial end we find the **conoid tubercle,** which is the site of ligament attachment.

The structure of the clavicle enables it to function like a brace in a building that supports two opposing beams (an example of the Structure-Function Core Principle, p. 28). It sits between the shoulder and thoracic cage and braces the upper limb so that  it rests laterally to the trunk. Its role in supporting the upper limb is evident in clavicular fractures, in which the arm falls anteriorly and medially. Most clavicular fractures result from direct trauma or from falling onto an outstretched arm.

Scapula

The triangular **scapula** (SKAP-yoo-luh; plural, *scapulae*) sits on the posterosuperior rib cage, extending from approximately ribs 2 to 7 (**Figure 7.28**). Its largest portion is the *body,* which has three borders: the **medial, lateral,** and **superior borders.** The apices of the scapular triangle are called *angles:* the *superior, inferior,* and *lateral angles.*

The anterior surface of the scapula (see Figure 7.28a) features a hook-shaped projection called the **coracoid process** (KOHR-uh-koyd; *corac-* = "crow"; be careful not to confuse this with the cor*onoid* processes of the mandible and ulna). Inferior to the coracoid process is a broad indentation, the *subscapular fossa,* to which a muscle called the *subscapularis muscle* attaches.

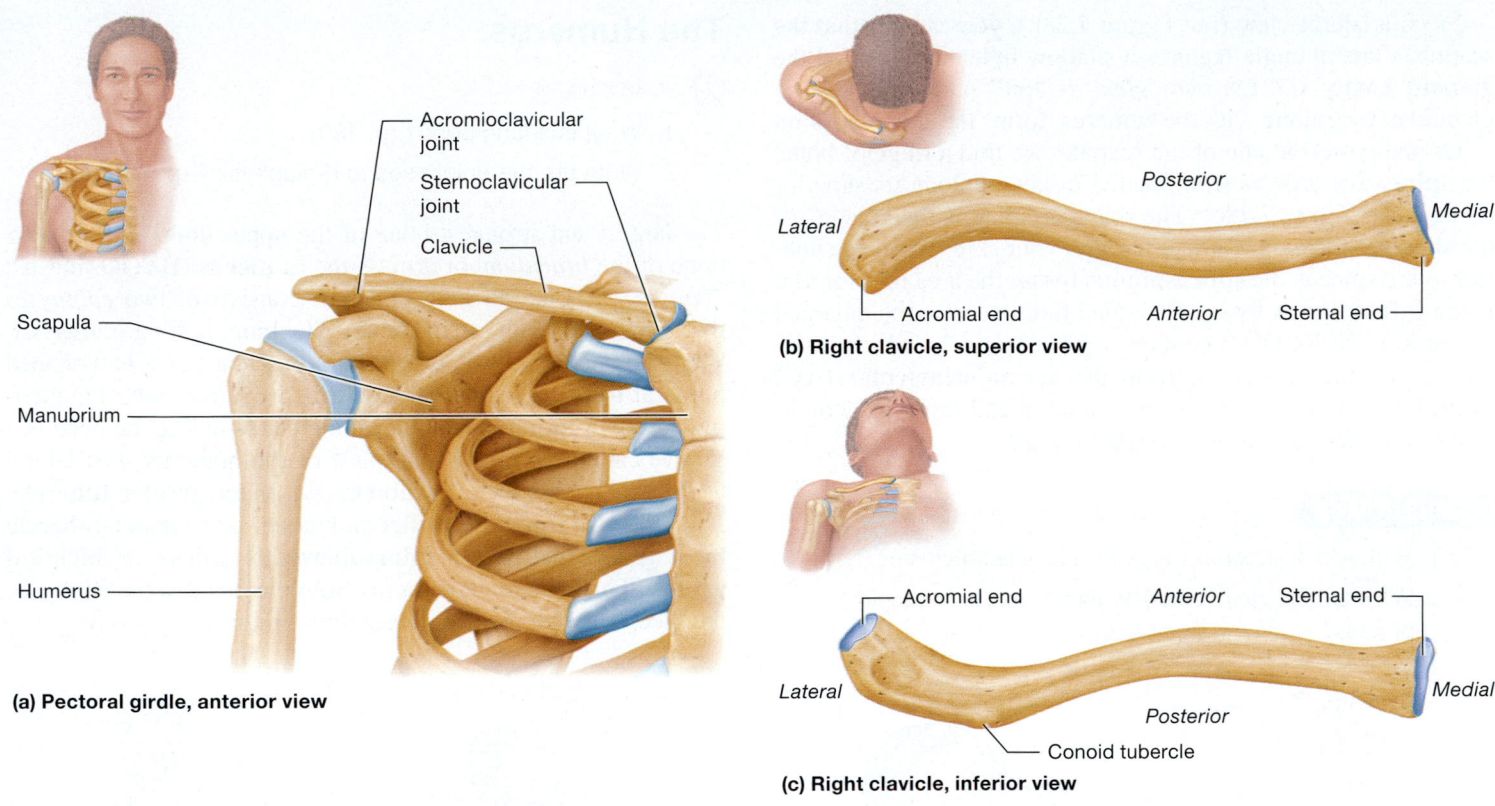

(a) Pectoral girdle, anterior view

(b) Right clavicle, superior view

(c) Right clavicle, inferior view

Figure 7.27 **The pectoral girdle.**

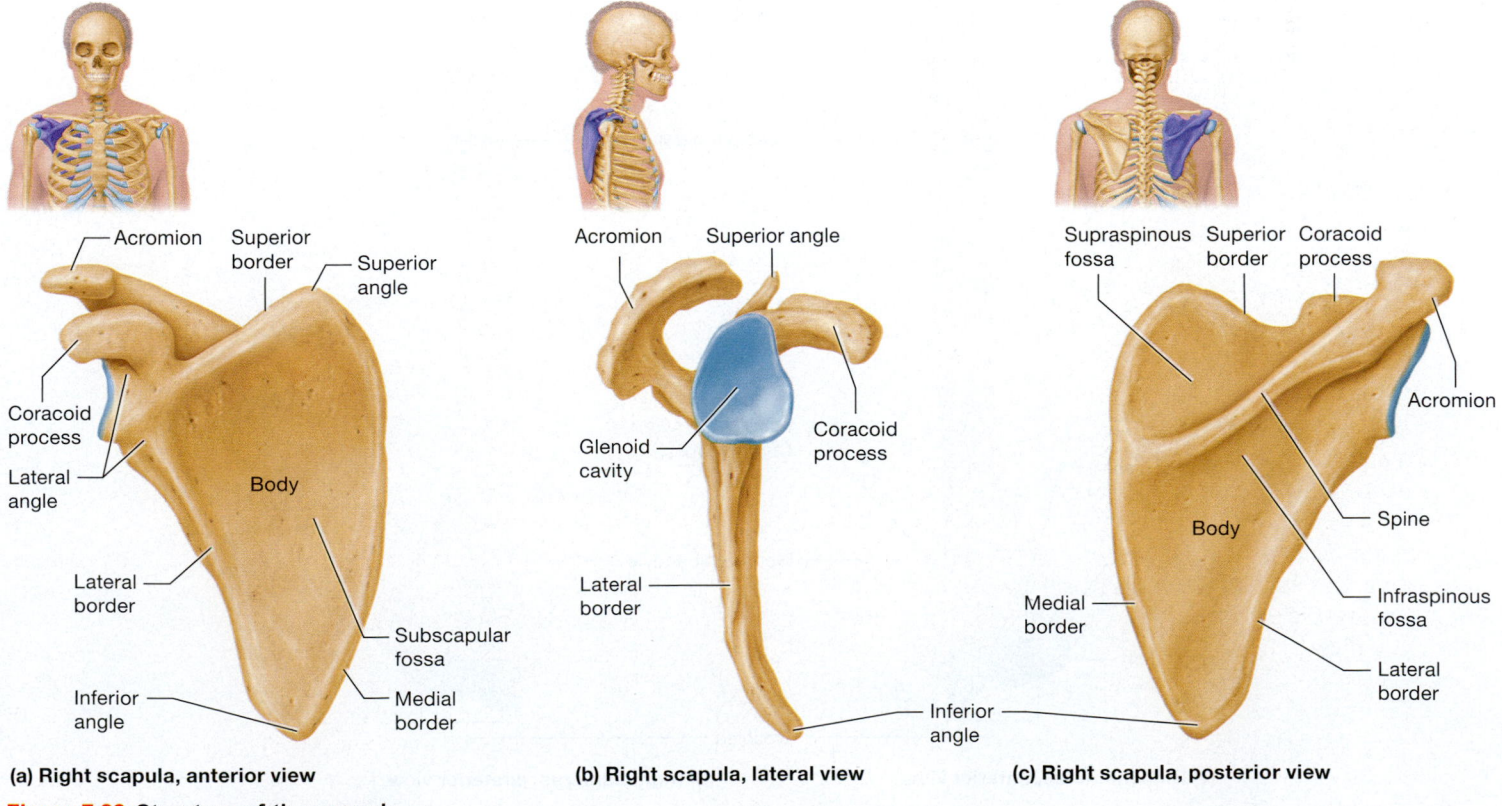

(a) Right scapula, anterior view

(b) Right scapula, lateral view

(c) Right scapula, posterior view

Figure 7.28 **Structure of the scapula.**

From a lateral view (see Figure 7.28b), you can see that the scapula's lateral angle features a shallow indentation called the **glenoid cavity** (GLEN-oyd; *glen-* = "pit" or "cavity"). The glenoid cavity, along with the humerus, forms the shoulder joint.

On the posterior side of the scapula, we find a ridge of bone, the **spine,** that crosses from medial to lateral along its superior border (see Figure 7.28c). The scapular spine is the "blade" of the shoulder that you can feel through your skin. The area superior to the spine is the **supraspinous fossa;** the area inferior to it is the **infraspinous fossa.** The spine terminates in the enlarged **acromion** (uh-KROH-mee-ahn; *acr-* = "point"), which articulates with the clavicle to form the **acromioclavicular (AC) joint.** Injuries to the AC joint are common and result in a condition often referred to as a *separated shoulder.*

Quick Check

☐ 1. With which structures does the clavicle articulate?

☐ 2. What are the glenoid cavity, acromion, and coracoid process?

The Humerus

« FLASHBACK

1. What is a long bone? (p. 186)
2. What are the epiphyses and diaphysis? (p. 187)

The largest and strongest bone of the upper limb, and the only bone of the *brachium,* or arm, is the **humerus** (HYOO-mur-us; **Figure 7.29**). Like all long bones, it consists of two *epiphyses* that articulate with other bones flanking a long *diaphysis.* The proximal epiphysis of the humerus features a ball-shaped humeral **head** on its medial side that articulates with the glenoid cavity to form the shoulder joint. Surrounding the head is a groove called the *anatomical neck* of the humerus. Just lateral to the neck we find a projection known as the **greater tubercle.** It is separated from the smaller and more medial **lesser tubercle** by a groove called the **intertubercular sulcus** or **bicipital groove.** This second name stems from the fact that the tendon of the biceps brachii muscle passes through this bony groove.

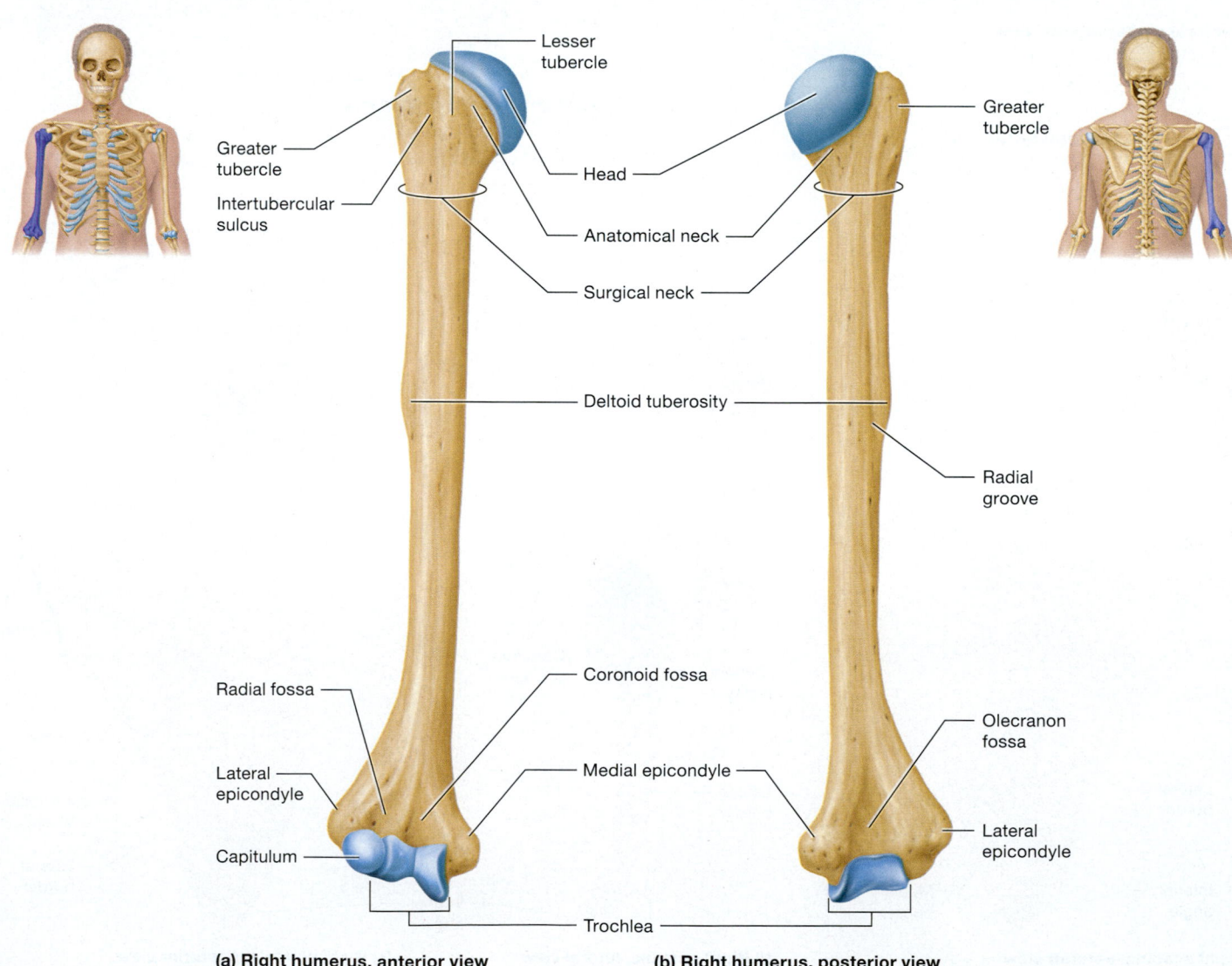

(a) Right humerus, anterior view   (b) Right humerus, posterior view

Figure 7.29 The humerus.

The junction between the proximal epiphysis and the diaphysis is a region called the **surgical neck,** so named because it is a frequent site of fractures (which may require surgical repair). The humeral diaphysis is fairly featureless compared to the two epiphyses. It contains only one major projection in about the middle of the bone on the lateral side, which is known as the **deltoid tuberosity.** As its name implies, this provides the site of attachment for the *deltoid muscle.* The only other feature of note on the diaphysis is the **radial groove,** located on the posterior side of the bone, along which travels the nerve that gives this groove its name.

The distal epiphysis has two flared ends, the **medial epicondyle** and **lateral epicondyle,** which are sites of muscle attachment. The remaining features are involved in the humeral articulation with the ulna and radius at the elbow joint. On the anterior side, we find two rounded knobs: the lateral **capitulum** (kah-PIT-yoo-lum; *capit-* = "head"), named for its spherical shape and resemblance to a head, and the medial **trochlea** (TROH-klee-uh; *troch-* = "wheel"), also named for its shape, which resembles a wheel or spool of thread. Just proximal to the capitulum and trochlea are two small indentations: the

lateral **radial fossa** and the medial **coronoid fossa.** On the posterior side of the distal epiphysis, we see the continuation of the trochlea and a deep indentation called the **olecranon fossa** (oh-LEK-ruh-nahn).

Quick Check

☐ 3. With which structures does the humerus articulate at its proximal and distal epiphyses?

☐ 4. Describe the structure and location of the following: the humeral head, capitulum, trochlea, deltoid tuberosity, and greater tubercle.

Bones of the Forearm: The Radius and Ulna

The forearm, or *antebrachium,* consists of two bones: the lateral **radius** (RAY-dee-us) and the medial **ulna,** held together by a fibrous structure called the **interosseous membrane** (**Figure 7.30**).

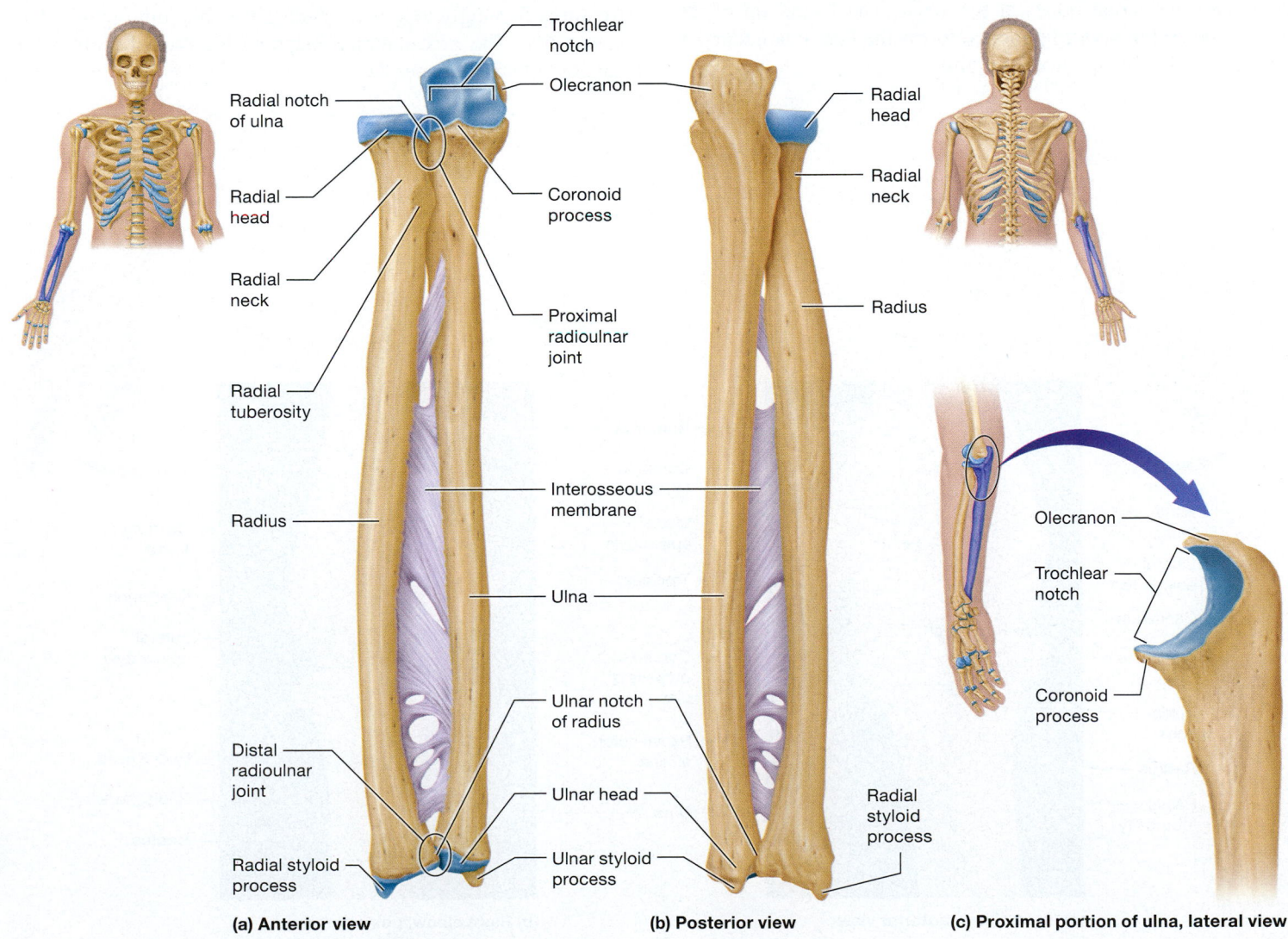

(a) Anterior view (b) Posterior view (c) Proximal portion of ulna, lateral view

Figure 7.30 The bones of the forearm: the radius and ulna.

The interosseous membrane distributes the force borne by the bones more equally, reducing the load on each bone. These two bones articulate with the humerus at their proximal ends and with the carpal bones at their distal ends. In addition, they also articulate with each other at two joints called the *radioulnar joints.*

Radius

Notice in Figures 7.30a and b that the radius is narrow proximally and becomes progressively broader as we move distally. Its narrow proximal epiphysis is called the **radial head,** which is a round, flattened structure. It articulates with the capitulum of the humerus to form part of the elbow joint and with the ulna to form the proximal radioulnar joint. Just distal to the radial head is the *radial neck,* which ends in the **radial tuberosity.** This process sits on the medial side of the bone and is the site of attachment of the biceps brachii muscle.

At the widened distal epiphysis of the radius, we find an indentation called the **ulnar notch,** which is the site of the distal radioulnar joint. The flattened end of this epiphysis articulates with the carpal bones at the wrist. The lateral tip of the radius, the **radial styloid process,** forms the lateral boundary of the wrist and helps stabilize the joint.

Ulna

Unlike the radius, the ulna is wide proximally and narrow distally. Its wide proximal epiphysis has several prominent features. The first is a U-shaped notch called the **trochlear notch,** into which the trochlea of the humerus fits (see Figure 7.30c). On the posterior side of the trochlear notch is a projection, the **olecranon,** which is the knob of the elbow. The anterior lip of the trochlear notch is a projection known as the **coronoid process,** which fits into the coronoid fossa of the humerus. Just lateral to the coronoid process is a smooth area called the **radial notch of the ulna.** This area articulates with the radial head. Note that these notches are named for the bone with which they articulate, not the bone on which they are located.

The narrow distal epiphysis of the ulna contains the **ulnar head.** This is another difference between the radius and ulna: The ulnar head is at the distal epiphysis, and the radial head is at the proximal epiphysis. The medial side of the ulnar head has a small **styloid process.** The styloid processes of both the radius and the ulna are palpable through the skin.

The proximal radius and ulna fit together with the distal humerus to form the elbow joint. **Figure 7.31** shows how the structures fit snugly together, which gives this joint a great deal of stability. The articulations chapter discusses the elbow in more detail (see Chapter 8).

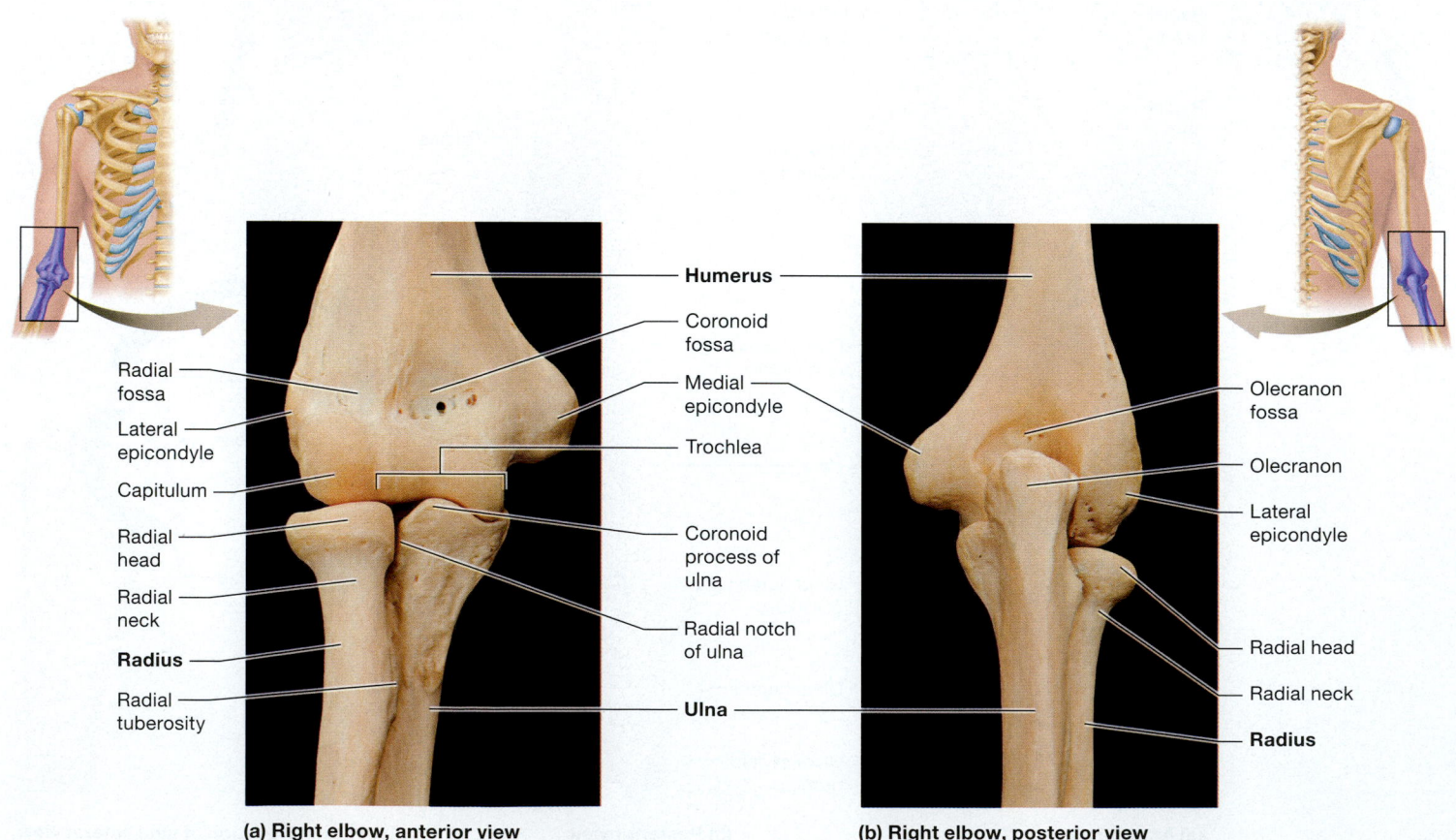

(a) Right elbow, anterior view

- Humerus
- Coronoid fossa
- Medial epicondyle
- Trochlea
- Coronoid process of ulna
- Radial notch of ulna
- Ulna
- Radial fossa
- Lateral epicondyle
- Capitulum
- Radial head
- Radial neck
- **Radius**
- Radial tuberosity

(b) Right elbow, posterior view

- Humerus
- Olecranon fossa
- Olecranon
- Lateral epicondyle
- Radial head
- Radial neck
- **Radius**

Figure 7.31 The elbow joint.

Quick Check

☐ 5. How do the radius and ulna differ in their shape and features?

☐ 6. Which parts of the radius and ulna articulate with the humerus?

☐ 7. With what other bones do the radius and ulna articulate?

Bones of the Wrist: Carpals

« FLASHBACK

1. What is a short bone? (p. 187)

2. How does a short bone differ from a long bone? (pp. 187–188)

The wrist, or *carpus,* consists of eight short bones—known collectively as the **carpals**—arranged in two rows containing four bones each (**Figure 7.32**). The four proximal carpal bones are the following, from lateral to medial (see the Study Boost on p. 236 for help with remembering their names):

- the boat-shaped **scaphoid** (SKAF-oyd; *scaph-* = "boat"), which articulates with the radius;
- the slightly moon-shaped **lunate,** which articulates primarily with the radius but also contacts the ulna;
- the triangular **triquetrum** (try-KWEE-trum), which articulates with the ulna; and
- the small, round **pisiform** (PY-zih-form; "pea-shaped"), which articulates with the anterior surface of the triquetrum (note that it is visible in the anterior view only).

The distal carpal bones, from lateral to medial, are as follows:

- the **trapezium,** which articulates proximally with the scaphoid;
- the **trapezoid,** which also articulates primarily with the scaphoid proximally;
- the rounded **capitate,** which articulates proximally with the scaphoid and lunate; and
- the **hamate,** which articulates proximally with the triquetrum and is named for its anterior hooklike projection (*ham-* = "hook").

The carpal bones, along with the radius and ulna, are often injured during falls. To learn more, see *A&P in the Real World: Wrist Fractures.*

Quick Check

☐ 8. List the proximal and distal carpal bones from lateral to medial.

Bones of the Hand and Fingers: Metacarpals and Phalanges

The hand, or *manus,* has five long bones, the **metacarpals** (numbered I–V from lateral to medial), that articulate with the distal carpal bones proximally and the bones of the fingers distally. Each metacarpal consists of three parts: The proximal epiphysis is the **base,** the diaphysis is the **body,** and the distal epiphysis is the **head.** You can see the heads of your metacarpals when you

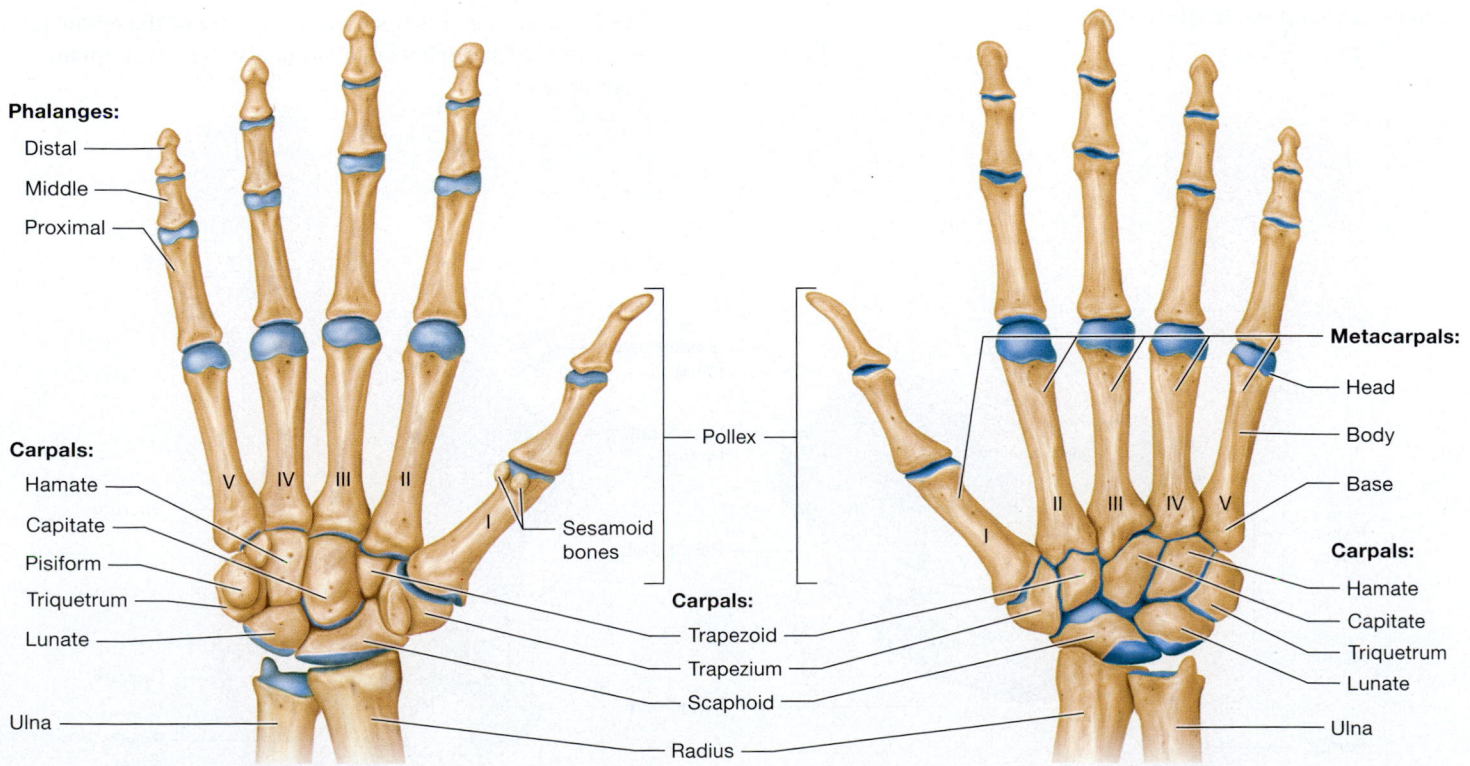

(a) Right wrist and hand, anterior view

(b) Right wrist and hand, posterior view

Figure 7.32 The hand and wrist.

Wrist Fractures

The wrist is the most frequently injured region of the upper limb, usually from direct trauma or from an attempt to stop a fall with an outstretched arm. Fractures commonly involve the distal radius and ulna, and may also involve the carpal bones, particularly the scaphoid. The fracture shown at right is a *Colles fracture* (KOL-eez) of the distal radius.

Wrist fractures occur most commonly in children and in adults aged 60–69. The incidence of wrist fractures declines with advancing age (and the

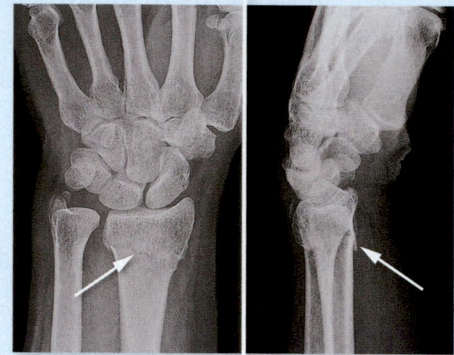

incidence of other fractures, such as hip fractures, increases). This change in incidence likely occurs because people fail to catch themselves and land on their hip instead.

Like most fractures, those of the wrist generally present with pain and swelling over the injured area. A common misconception is that an individual with a fractured bone will be unable to move the affected part. In fact, this is only the case with severe fractures in which a bone is dislocated. Generally, a person with a wrist fracture will have slightly less ability to move the wrist due to pain and swelling, but it is possible for motion to be totally normal (a fact to which this author can attest personally).

Treatment of wrist fractures varies according to the severity of the injury and which bone is injured. Some cases are treated with simple casting, whereas others require surgical intervention. Many fractures could be successfully prevented by teaching individuals how to fall in a way that distributes the force across their trunk rather than concentrating it on an outstretched arm.

make a fist, as they form your "knuckles." Fractured metacarpal heads are called *boxer's fractures*.

The bones of the fingers consist of 14 total **phalanges** (fuh-LAN-jeez; singular, *phalanx*). Each finger has three bones: a **proximal, middle** (or **intermediate**), and **distal phalanx** (FAY-langks). The exception is the thumb, or **pollex,** which has only a proximal and a distal phalanx. Like the metacarpals, the phalanges have a base, body, and head.

Quick Check

☐ 9. How many metacarpals and phalanges are in the hand and fingers?

☐ 10. What are the three parts of a metacarpal and a phalanx?

Apply What You Learned

☐ 1. Predict what might happen to the stability of the elbow joint if the trochlear notch were abnormally shallow. Explain your answer.

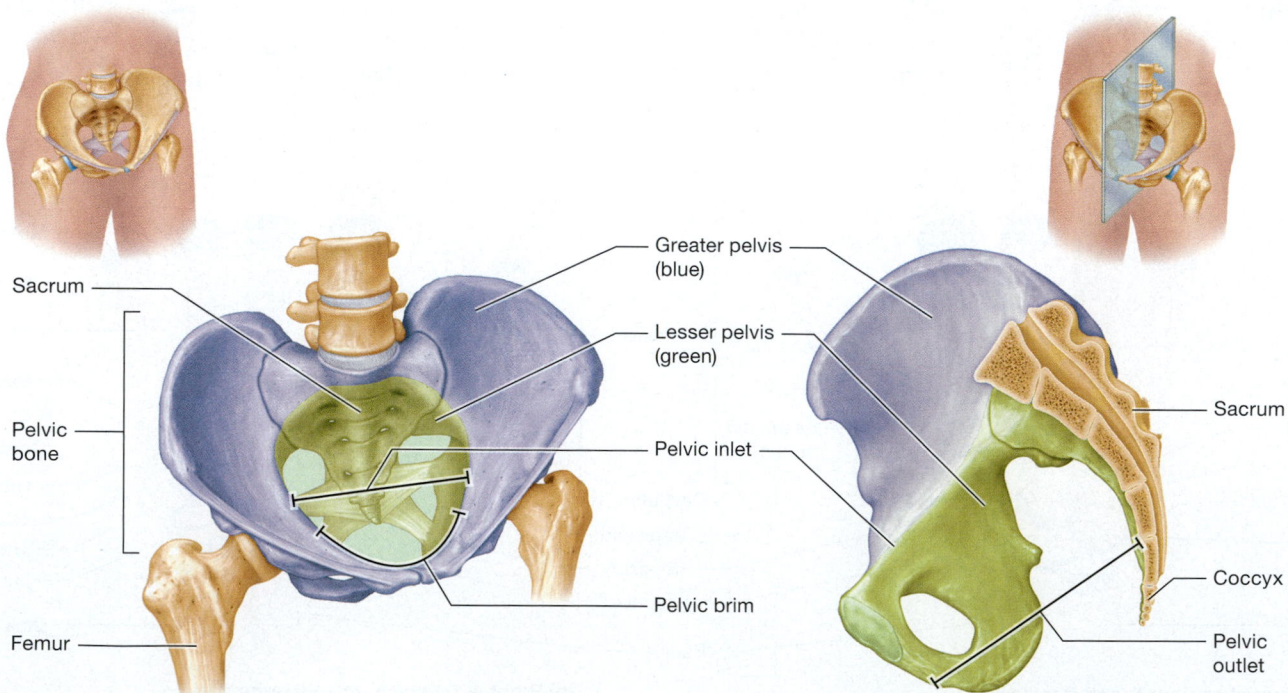

(a) Anterolateral view of pelvis

Sacrum

Pelvic bone

Femur

Greater pelvis (blue)

Lesser pelvis (green)

Pelvic inlet

Pelvic brim

(b) Midsagittal section of pelvis

Sacrum

Coccyx

Pelvic outlet

Figure 7.33 The pelvis. The greater (false) pelvis is shaded blue; the lesser (true) pelvis is shaded green.

☐ 2. Mr. Heller presents to the emergency department following a bicycle crash during which he landed on his shoulder. During the physical exam, you notice that his upper limb is hanging in a more medial position than normal and he can hold it in a lateral position only with assistance. What has likely happened? Explain.

See answers in Appendix A.

MODULE
7.5
Bones of the Pelvic Girdle and Lower Limb

Learning Outcomes

1. Describe the location, structural features, and functions of the bones of the pelvic girdle.

2. Compare and contrast the adult male and female pelvic bones.

3. Describe the location, structural features, and functions of the bones of the thigh, the leg, the ankle, and the foot.

This module explores the lower part of the appendicular skeleton: the pelvic girdle and the lower limb. We begin the module with the **pelvic girdle,** which consists of the sacrum and **pelvic bones** (also known as the *hip bones, os coxae,* or *coxal bones*). We next turn to the lower limb, which has three parts: the thigh, which has the *femur;* the leg, which is made up of the *tibia* and *fibula;* and the foot

and ankle, which consist of the *tarsals, metatarsals,* and *phalanges.* As you read this module, note the similarities and differences between these bones and those of the pectoral girdle and upper limb.

The Pelvis and Bones of the Pelvic Girdle

« FLASHBACK

1. Where is the pelvic cavity located? (p. 18)

2. Which organs are located within the pelvic cavity? (p. 18)

The two pelvic bones, sacrum, and coccyx together form the bowl-shaped **pelvis** (**Figure 7.33**), which establishes the boundaries for the pelvic cavity that houses organs of the digestive, urinary, and reproductive systems. The pelvic girdle and sacrum form an oval opening called the **pelvic inlet.** The bony ridge surrounding the inlet is the **pelvic brim.** As you can see in Figure 7.33a, the pelvic brim defines the boundaries between the **greater** or *false* **pelvis,** which is the area superior to the pelvic brim, and the **lesser** or *true* **pelvis,** which is inferior to the pelvic brim. At the inferior boundary of the lesser pelvis, we find the **pelvic outlet** (see Figure 7.33b).

Each pelvic bone is composed of three bones that fuse during childhood: the *ilium, ischium,* and *pubis* (**Figure 7.34**). All three bones contribute to the deep socket on the lateral side of the pelvic bone, referred to as the **acetabulum** (aeh´-suh-TAB-yoo-lum), which, along with the femur, forms the hip joint. In addition, two bones, the ischium and pubis, contribute to the large opening in the anterior pelvic bone, called the **obturator foramen** (AHB-tuh-ray-tur), through which nerves and blood vessels pass. Let's now look at the individual bones and their features.

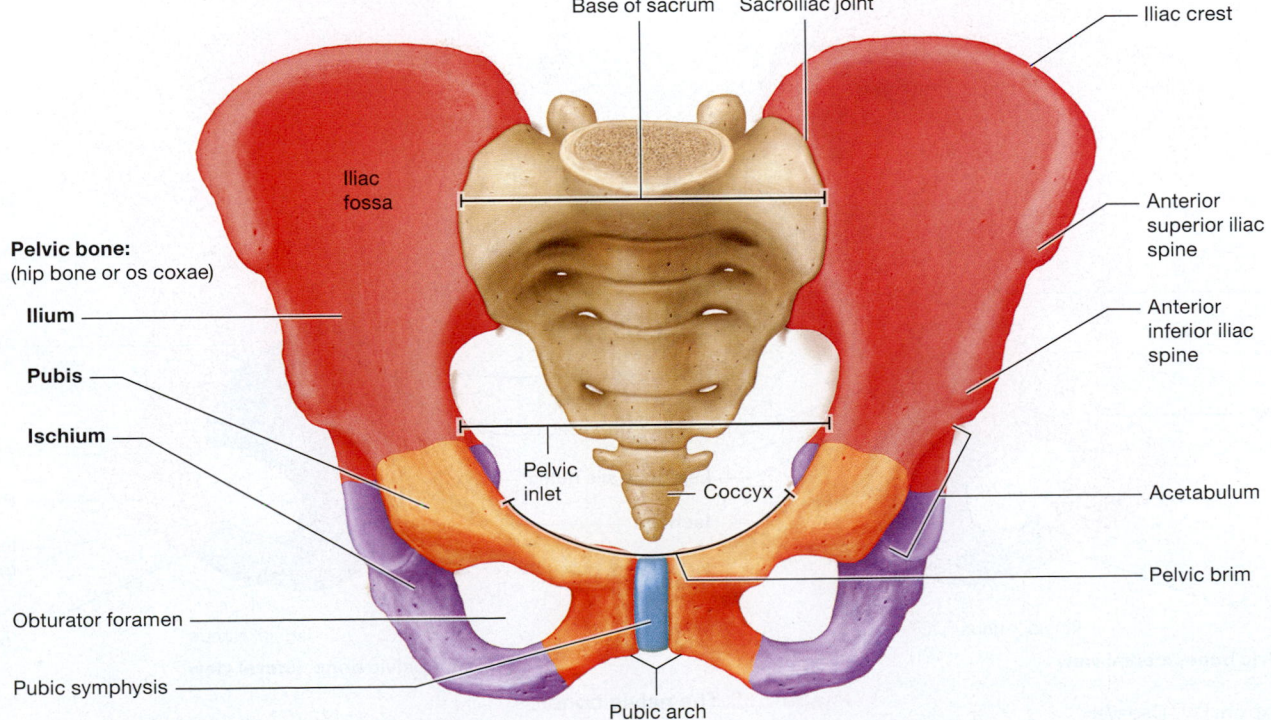

Figure 7.34 The pelvis and pelvic bones, anterior view. Each pelvic bone (also called the hip bone) consists of three fused bones: the ilium, ischium, and pubis.

Ilium

The **ilium** (ILL-ee-um; plural, *ilia*) forms the top part of the pelvic bone. Its superior portion is the flared **ala** (AY-luh), named for its resemblance to a wing; its inferior portion is the **body** of the ilium (**Figure 7.35**). The superior boundary of the ala is the **iliac crest,** which you feel every time you put your hands on your hips. At its anterior end, the iliac crest terminates in a bony projection known as the **anterior superior iliac spine** (ASIS), which can be palpated in some people on the anterior pelvis. Just inferior to the ASIS is another, smaller ridge called the *anterior inferior iliac spine.*

Notice in the Figure 7.35a that the iliac crest terminates posteriorly as the **posterior superior iliac spine,** which has a smaller projection inferior to it, the *posterior inferior iliac spine.* The ilium indents sharply at the posterior inferior iliac spine to form a deep groove called the **greater sciatic notch** (sy-AT-ik), through which the large sciatic nerve passes. Also looking at the ilium's medial surface, notice the **iliac fossa,** which is the large, smooth, anterior part, and the *auricular surface,* which is the smaller, rough, posterior part. Finally, on this side, we can see the **arcuate line,** which marks where the ilium forms part of the pelvic brim. It runs from the auricular surface along the body of the ilium.

The smooth lateral surface of the ilium (see Figure 7.35b) is known as its *gluteal surface* (GLOO-tee-uhl). It is marked by three lines: the **posterior, anterior,** and **inferior gluteal lines,** to which the gluteal muscles attach.

Ischium

The posteroinferior portion of the pelvic bone is formed by the **ischium** (ISS-kee-um; see Figure 7.35). It consists of two components that together approximate a C shape: the posterior **ischial body,** which, along with the ilium and pubis, forms part of the acetabulum, and the anterior **ischial ramus,** which, along with the pubis, forms part of the obturator foramen.

Projecting posteriorly and medially from the ischial body is the **ischial spine,** to which a ligament from the sacrum attaches. Just inferior to the ischial spine is the **lesser sciatic notch.** Like the greater sciatic notch, the lesser notch provides a passageway for blood vessels and nerves.

The most prominent feature of the ischium is the thick **ischial tuberosity,** located on its posteroinferior side. Together, the ischial tuberosities bear our weight when we sit. When someone is said to have a "bony butt," it simply means that the individual's ischial tuberosities are prominent.

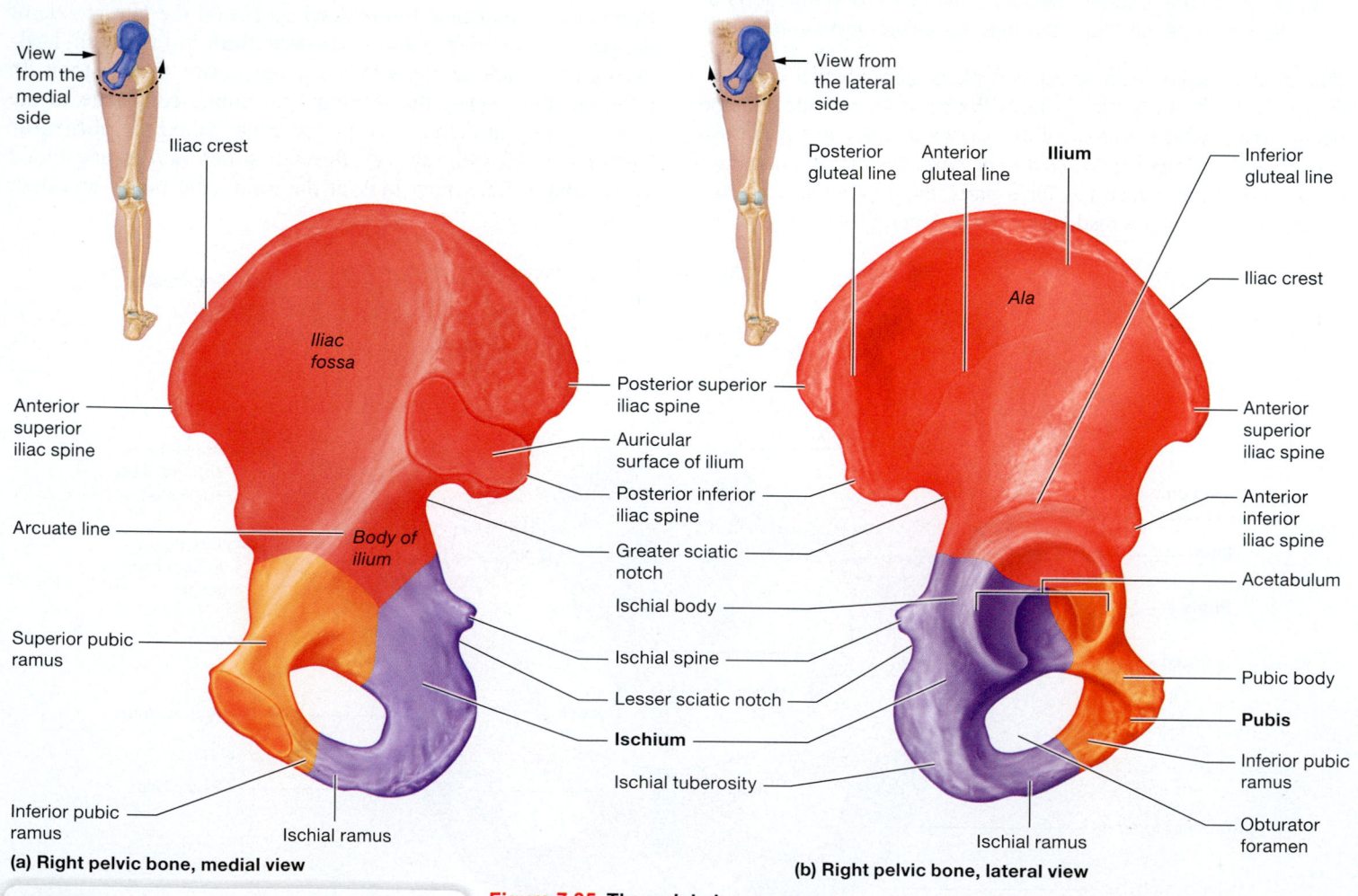

(a) Right pelvic bone, medial view

(b) Right pelvic bone, lateral view

Figure 7.35 The pelvic bone.

Explore PAL™ Cadaver
@ Mastering Anatomy & Physiology

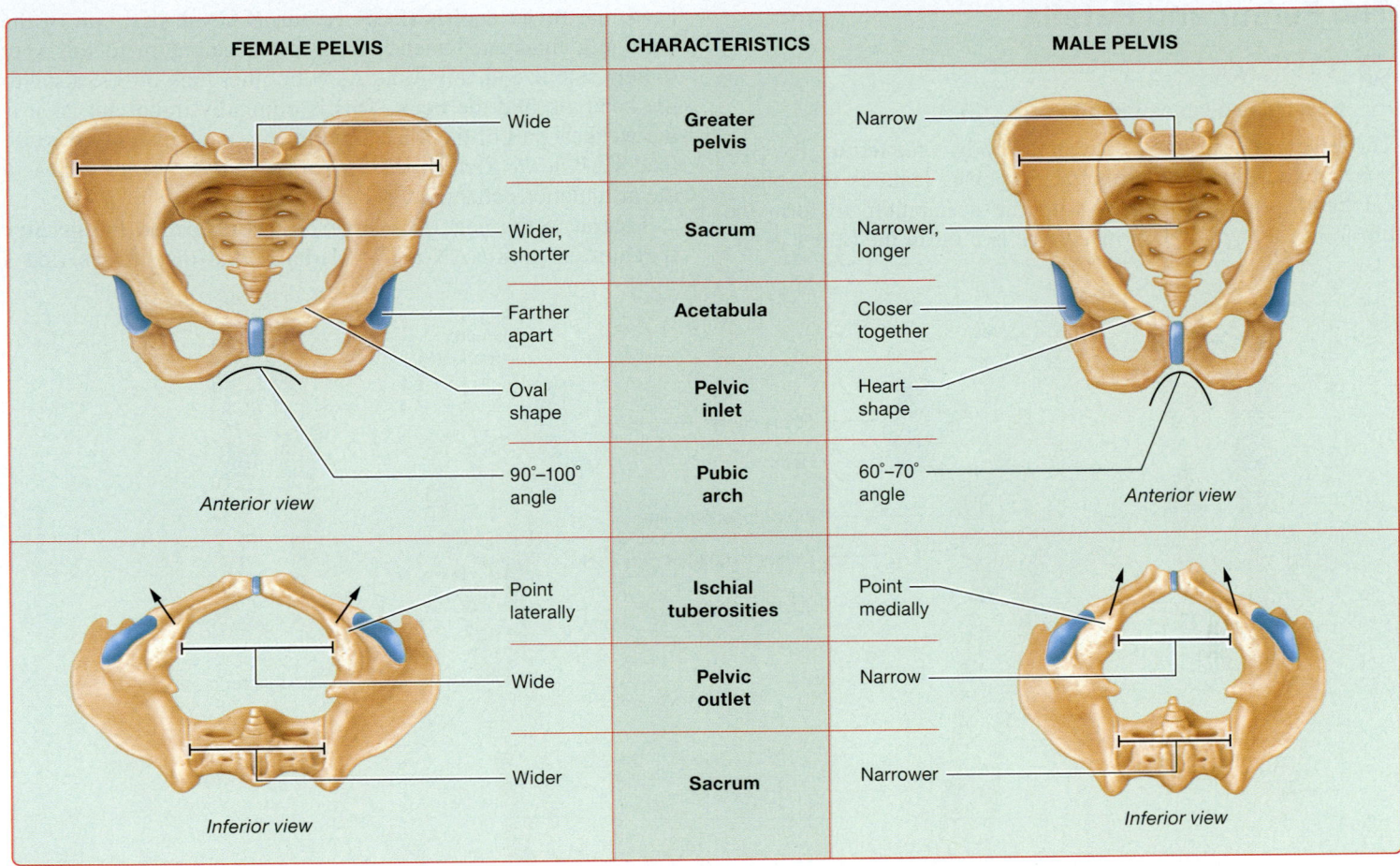

FEMALE PELVIS	CHARACTERISTICS	MALE PELVIS
Wide	**Greater pelvis**	Narrow
Wider, shorter	**Sacrum**	Narrower, longer
Farther apart	**Acetabula**	Closer together
Oval shape	**Pelvic inlet**	Heart shape
90°–100° angle *Anterior view*	**Pubic arch**	60°–70° angle *Anterior view*
Point laterally	**Ischial tuberosities**	Point medially
Wide	**Pelvic outlet**	Narrow
Wider *Inferior view*	**Sacrum**	Narrower *Inferior view*

Figure 7.36 Differences between the female and male pelves.

Pubis

The smallest component of the pelvic bone is the anterior **pubis** (PYOO-biss), also called the *pubic bone* (see Figure 7.35). It consists of three parts that together approximate a C shape: the **superior pubic ramus,** the **pubic body,** and the **inferior pubic ramus.** The superior pubic ramus contributes to the acetabulum, and both rami form parts of the boundary of the obturator foramen. The two pubic bodies meet at a joint called the **pubic symphysis** (SIM-fih-sis), where they are united by a pad of fibrocartilage. The angle where the two bodies meet is the *pubic arch;* this pelvic feature is important in determining sex, as discussed next.

Female and Male Pelves

Apart from the skull, the other skeletal structure that can be reliably used to determine sex is the pelvis (plural, *pelves;* **Figure 7.36**). The female pelvis is structured for pregnancy and childbirth and, as a result, is generally wider and shallower than the male pelvis. We can therefore examine a number of features to determine sex, including the following:

- **Shape of greater pelvis.** The greater pelvis is wider in females, with the anterior superior iliac spines farther apart, and with flared iliac crests.

- **Coccyx and sacrum.** The female sacrum tends to be wider and shorter than the male sacrum. In addition, the female coccyx is generally situated more posteriorly and is more moveable than the male coccyx. A woman with a coccyx situated anteriorly could suffer coccygeal fractures during childbirth.
- **Pelvic inlet and outlet.** The female pelvic inlet is usually wider and oval, whereas the male pelvic inlet is narrow and vaguely heart-shaped. The female pelvic outlet is also typically wider than that of the male.
- **Acetabula.** Female acetabula are generally farther apart and pointed more anteriorly than those of males. This affects gait patterns, giving many women a "swaying" walk.
- **Pubic arch.** The pubic arch in females tends to have an angle between 90° and 100°, whereas the male pubic arch has a narrower angle of 60°–70°. If an investigator has only a single pelvic bone, the angle of the pubic arch can still be estimated by holding the bone against a mirror and measuring the angle in the reflection.
- **Ischial tuberosities.** The female ischial tuberosities tend to point laterally, whereas those of the male point medially. This gives the female buttocks a more rounded appearance.

Also, the female pelvis is generally lighter and less robust than the male pelvis due to the male's greater muscle mass and weight.

The Femur and Patella

« **FLASHBACK**

1. What is a sesamoid bone? (p. 187)

The largest and strongest bone in the body is the **femur,** the only bone of the thigh. Its proximal epiphysis features the spherical **head,** which articulates with the acetabulum to form the hip joint (**Figure 7.37**). Notice there is a pit in the center of the head, the **fovea capitis** (FOH-vee-uh KAP-uh-tiss; "pit of the head"); a ligament attaches from the acetabulum to this spot to help stabilize the hip joint. As with other long bones, distal to the head we find the **neck.** This is clinically important, as it is the weakest part of the bone and the area most likely to fracture (see *A&P in the Real World: Hip Joint Replacement Surgery* in the articulations chapter, Chapter 8).

Lateral to the neck is a large projection called the **greater trochanter** (troh-KAN-tuhr). Medially and distally we find a

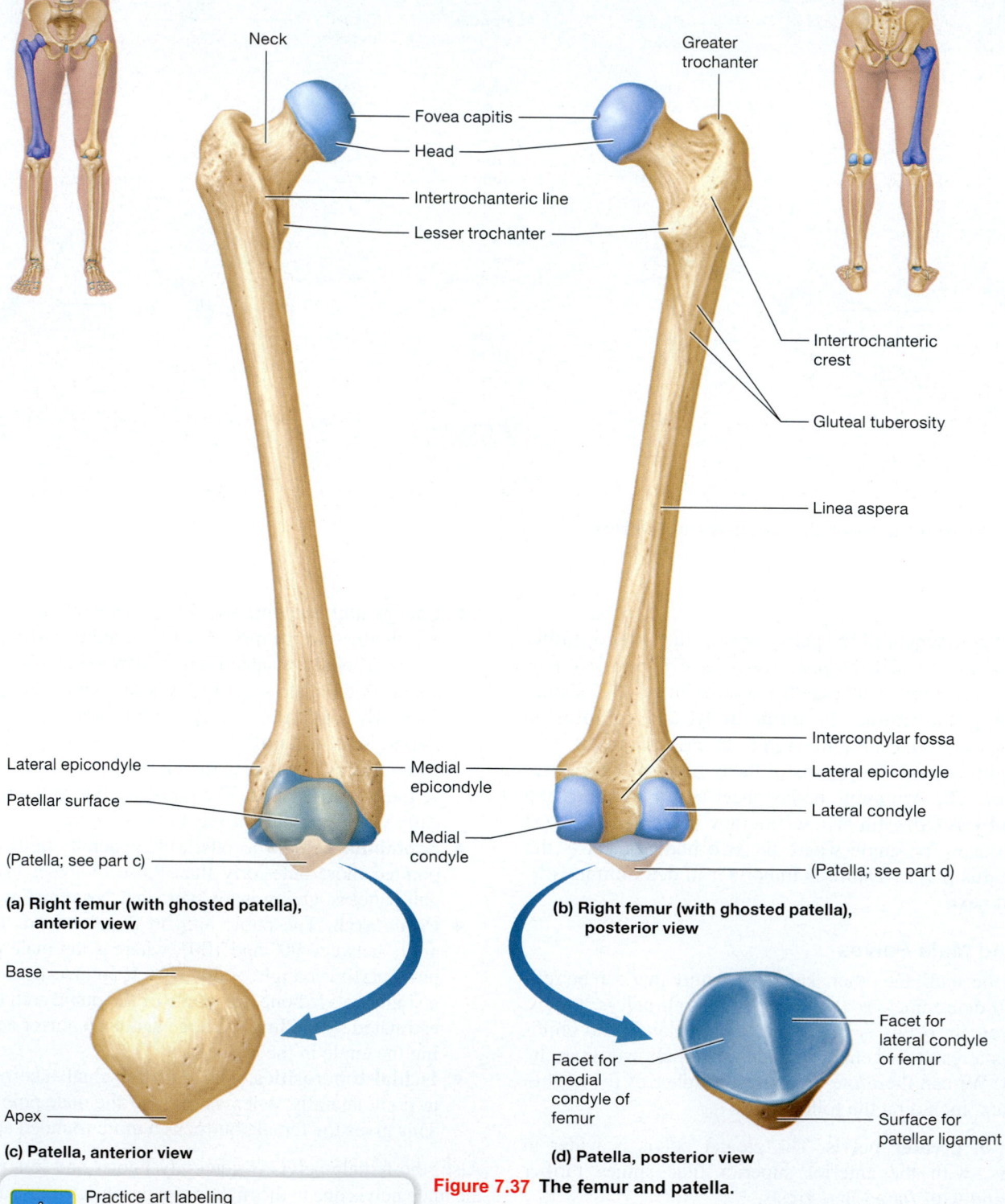

Neck

Fovea capitis

Head

Intertrochanteric line

Lesser trochanter

Greater trochanter

Intertrochanteric crest

Gluteal tuberosity

Linea aspera

Lateral epicondyle

Patellar surface

(Patella; see part c)

Medial epicondyle

Medial condyle

Intercondylar fossa

Lateral epicondyle

Lateral condyle

(Patella; see part d)

(a) Right femur (with ghosted patella), anterior view

(b) Right femur (with ghosted patella), posterior view

Base

Apex

(c) Patella, anterior view

Facet for medial condyle of femur

Facet for lateral condyle of femur

Surface for patellar ligament

(d) Patella, posterior view

Figure 7.37 The femur and patella.

Practice art labeling
@ Mastering Anatomy & Physiology

similar projection, the **lesser trochanter.** The two are connected by a bony ridge on the anterior side, the **intertrochanteric line,** which continues on the posterior surface of the bone as the **intertrochanteric crest.**

The femoral diaphysis, also known as the *shaft,* is fairly smooth and featureless on the anterior side. On the posterior side, a prominent line called the **linea aspera** (LIN-ee-uh ASP-er-uh; "rough line") runs down the shaft. Near the distal epiphysis, it splits into two lines that lead to two projections, the **medial epicondyle** and **lateral epicondyle,** which are the widest points of the femur. From the epicondyles, the femur tapers into the **medial condyle** and **lateral condyle,** which articulate with the tibia to form the knee joint. On the posterior side (see Figure 7.37b) is the **intercondylar fossa,** which is an indentation between the two condyles. Anteriorly, the space between the condyles is smooth; this area is known as the *patellar surface.*

The bone that articulates with the femur's patellar surface is the triangular **patella** (see Figure 7.37c). The patella is a sesamoid bone located within the common tendon of the *quadriceps*

femoris muscle group, a group of four muscles in the anterior thigh. The *patellar ligament* is a continuation of this tendon that inserts into the tibia and secures the patella over the anterior knee (see the next section). The proximal end of the patella is the *base,* and its pointed distal end is the *apex.* Even though the patella is not technically a thigh bone, we include it here because it articulates with the femur.

Quick Check

☐ 1. With which bones does the femur articulate? Be specific.

☐ 2. Which parts of the femur form these articulations?

☐ 3. What type of bone is the patella, and where is it located?

Bones of the Leg: The Tibia and Fibula

Two bones make up the leg: the medial **tibia** (TIB-ee-uh) and the lateral **fibula** (FIB-yoo-luh; **Figure 7.38**). (Be careful with these names; many a student has missed a test question by

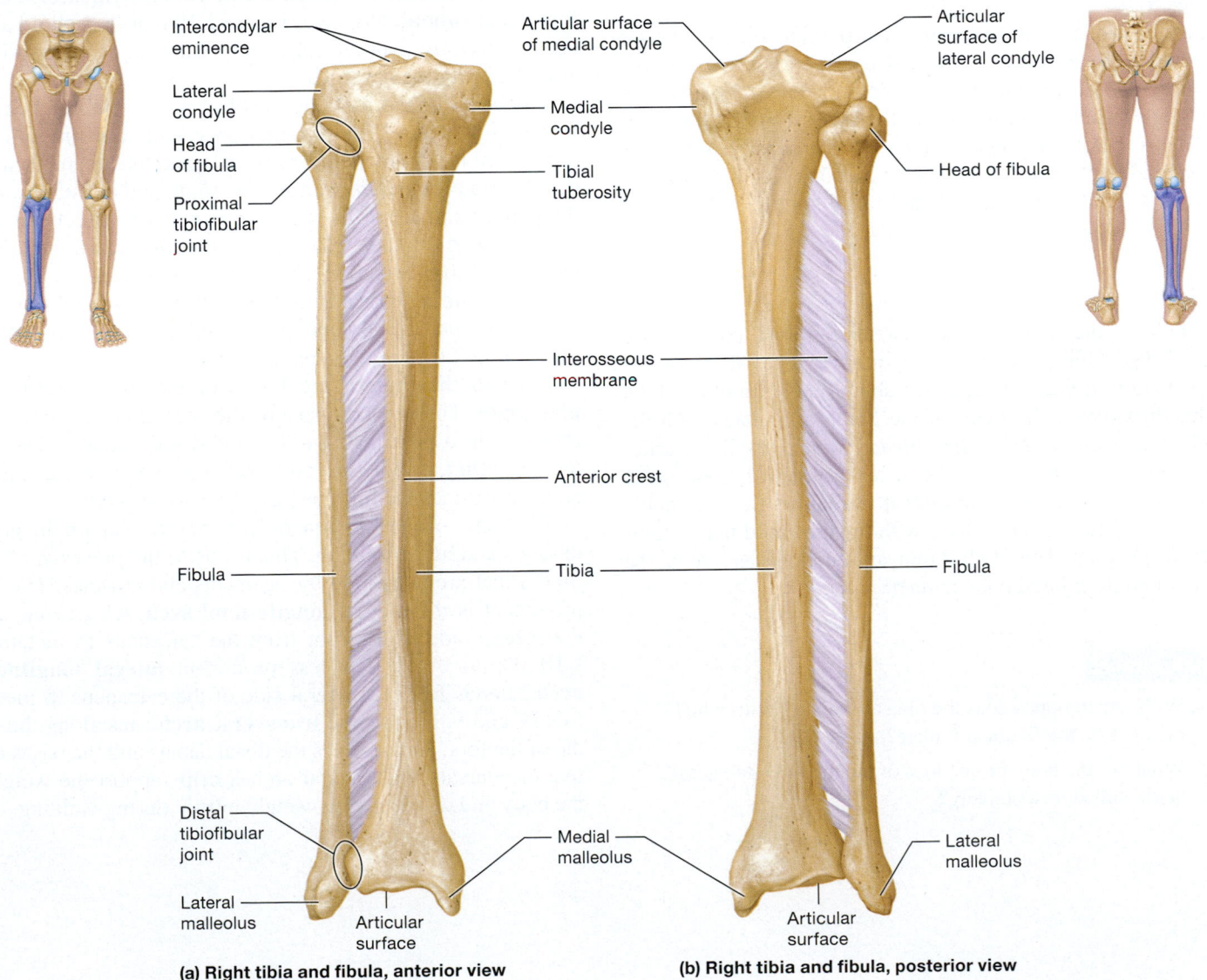

(a) Right tibia and fibula, anterior view

(b) Right tibia and fibula, posterior view

Figure 7.38 The tibia and fibula.

writing "tibula" and "fibia.") Like the radius and ulna, the tibia and fibula are united by an interosseous membrane. They articulate with each other at both the distal and the proximal *tibiofibular joints*. However, unlike the radioulnar joints, the tibiofibular joints allow very little movement, which makes the lower limb more stable.

Tibia

The larger, medial tibia is the second strongest bone in the body, which reflects its primary function as the main weight-bearing bone of the leg. At its proximal end, it features two concave depressions called the **medial condyle** and **lateral condyle,** which articulate with the femoral condyles to form the knee joint. The two condyles are separated by a ridge, the **intercondylar eminence.** As you'll see in the next chapter, the intercondylar eminence is an important site for ligament attachment in the knee joint. Just distal to the tibial condyles is a rough projection from the tibia's anterior surface, known as the **tibial tuberosity,** which is where the patellar ligament attaches to the tibia.

The tibial diaphysis features a sharp ridge on its anterior side, called the **anterior crest** and commonly known as the *shin.* Near the distal epiphysis, the anterior crest curves to the medial side, where it terminates in the **medial malleolus** (mal-lee-OH-lus; plural, *malleoli*), which you can easily palpate as your medial "ankle bone." Lateral to this is a flat articular surface where the tibia articulates with a tarsal bone called the *talus* to form the ankle joint.

Fibula

The fibula is the much thinner, lateral bone of the leg (see Figure 7.38). Often incorrectly reported to bear none of the body's weight, it does, in fact, bear about one-sixth of our total weight. Proximally, the **head** of the fibula articulates with the lateral tibia at the *proximal tibiofibular joint.* Distally, it articulates with the tibia again at the *distal tibiofibular joint,* after which it expands to form the **lateral malleolus** (lateral "ankle bone"). This structure articulates with the talus and helps stabilize the ankle joint. The fibula's thin structure causes it to be the most commonly injured bone in *ankle fractures.*

Quick Check

☐ 4. With which bones does the tibia articulate? With which bones does the fibula articulate?

☐ 5. What are the bony projections of the medial and lateral ankle called, respectively?

Bones of the Ankle and Foot: The Tarsals, Metatarsals, and Phalanges

« FLASHBACK

1. Where is the plantar surface of the foot located? (p. 14)

The structure of the ankle parallels that of the wrist in many ways. The ankle consists of seven short bones, the **tarsals,** that connect the leg with the foot (**Figure 7.39**). However, unlike the arrangement of the wrist, only a single tarsal bone articulates with the leg: the dome-shaped **talus** (TAYL-us), which forms the ankle joint with the tibia and fibula. The talus rests on top of the largest tarsal bone, the **calcaneus** (kal-KAYN-ee-us), which makes up the heel of the foot. The posterior side of the calcaneus attaches to a large tendon called the *calcaneal (Achilles) tendon.* Distally, the talus articulates with a bone called the **navicular,** named for its resemblance to the shape of a boat. Together, the talus, calcaneus, and navicular are the proximal tarsal bones.

The distal four tarsal bones, from medial to lateral, include the **medial cuneiform, intermediate cuneiform, lateral cuneiform,** and **cuboid.** Notice in Figure 7.39 that the cuboid articulates posteriorly with the calcaneus, whereas the cuneiforms all articulate with the navicular.

The bones of the foot and toes largely parallel the hand and fingers in structure. The five foot bones are called **metatarsals** (numbered I–V from medial to lateral), and like the metacarpals, they have a proximal *base,* a middle *shaft,* and a distal *head.* On the plantar surface of the first metatarsal we usually find two small sesamoid bones (often just called the *sesamoids*). These bones, and the tendon in which they are located, can become inflamed with repetitive activity, resulting in *sesamoiditis.* This condition is particularly common in athletes playing on artificial turf, giving it the common name "turf toe."

As with the fingers, the bones of the toes consist of 14 **phalanges.** The second through fifth toes generally have three bones each: a proximal, middle, and distal phalanx. The great toe, or **hallux,** usually has only two, a proximal and distal phalanx (sometimes the fifth toe has only two, as well).

The bones of the foot generally do not rest flat on the ground during standing or walking. This is due to the presence of three *arches* that are supported by ligaments and muscles. The most prominent is the **medial longitudinal arch,** which runs along the medial side of the foot from the calcaneus to metatarsals I–III (**Figure 7.40**). The less prominent **lateral longitudinal arch** extends from the lateral side of the calcaneus to metatarsals IV and V. Finally, the **transverse arch** runs along the middle of the foot, and involves the distal tarsals and the bases of all five metatarsals. The plantar arches help support the weight of the body and distribute this weight evenly during walking.

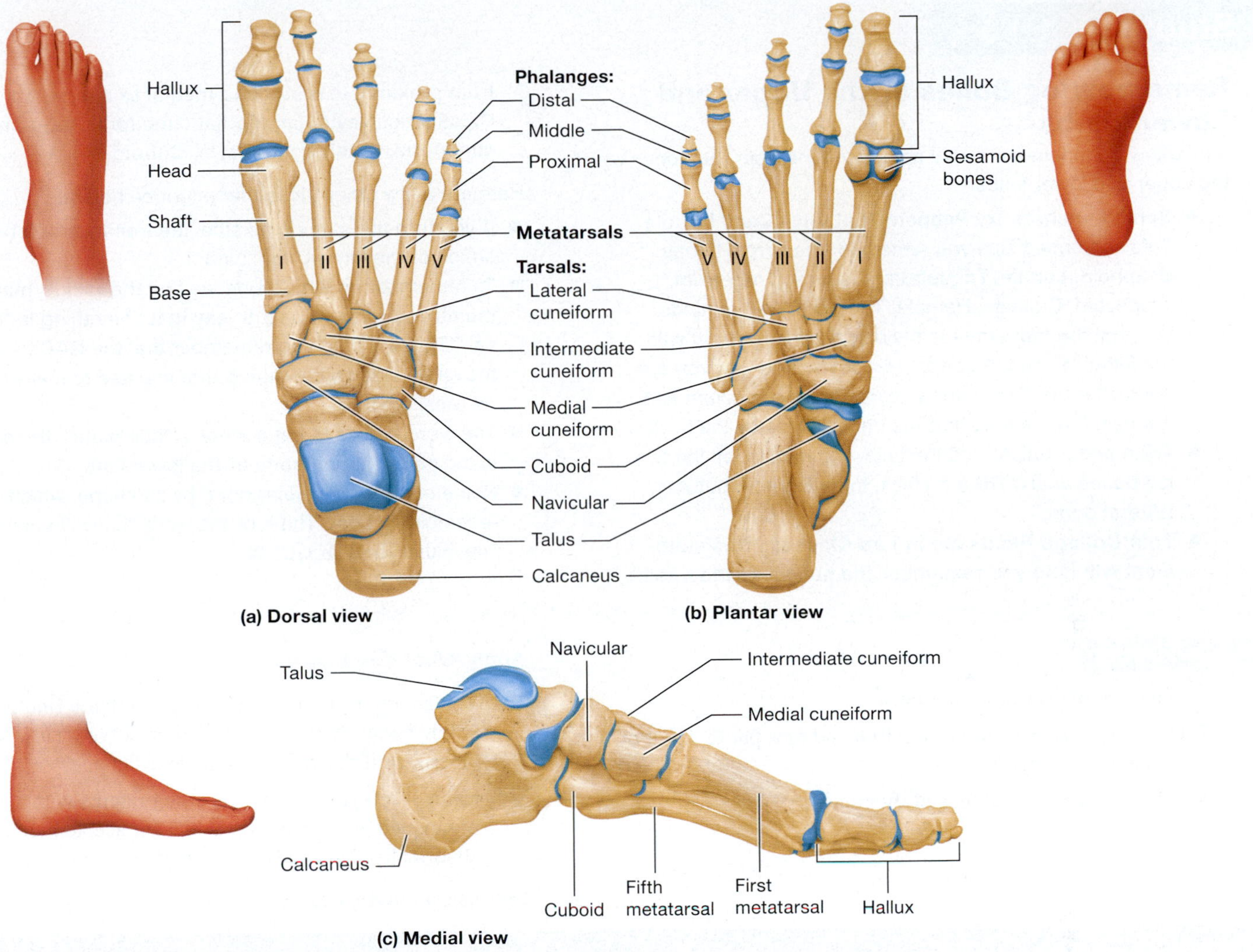

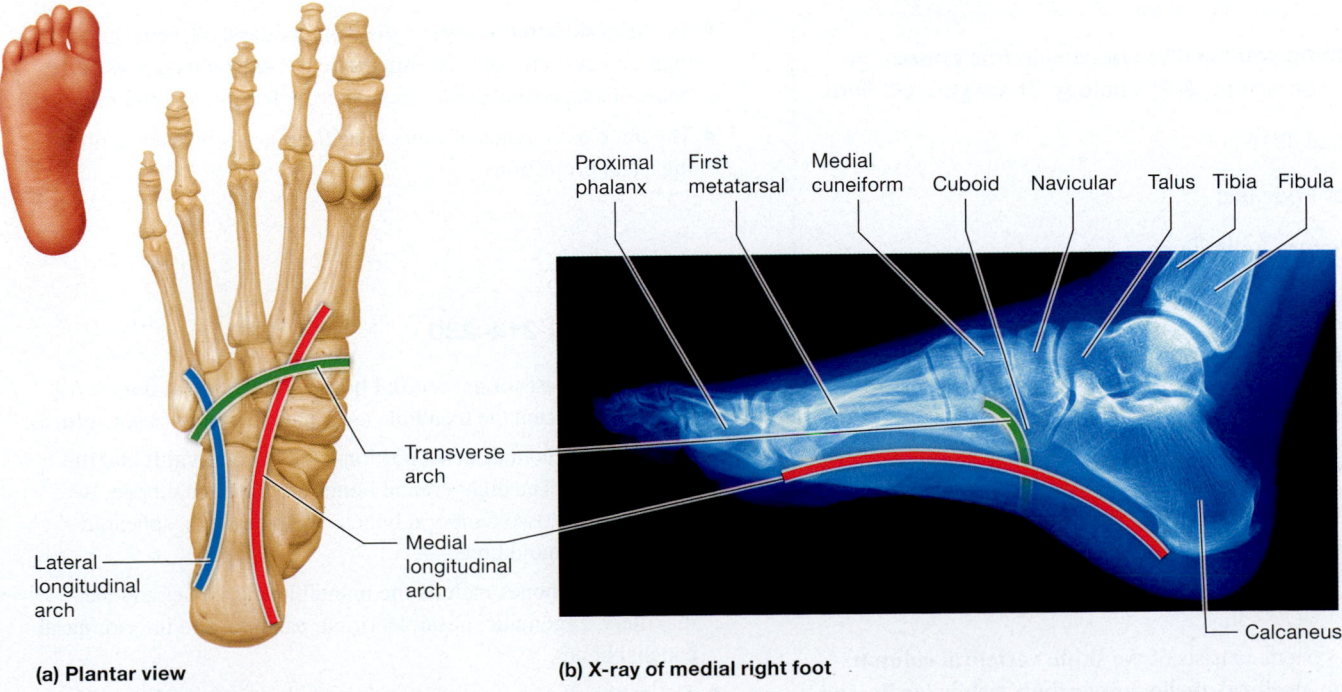

Phalanges:
- Distal
- Middle
- Proximal

Hallux

Head

Shaft

Base

Metatarsals

I II III IV V

Tarsals:
- Lateral cuneiform
- Intermediate cuneiform
- Medial cuneiform
- Cuboid
- Navicular
- Talus
- Calcaneus

Hallux

Sesamoid bones

V IV III II I

(a) Dorsal view

(b) Plantar view

Navicular

Intermediate cuneiform

Talus

Medial cuneiform

Calcaneus

Cuboid

Fifth metatarsal

First metatarsal

Hallux

(c) Medial view

Figure 7.39 The ankle and foot.

Transverse arch

Lateral longitudinal arch

Medial longitudinal arch

Proximal phalanx

First metatarsal

Medial cuneiform

Cuboid

Navicular

Talus

Tibia

Fibula

Calcaneus

(a) Plantar view

(b) X-ray of medial right foot

Figure 7.40 The three arches of the foot.

StudyBOOST ⬆

Remembering Bones of the Upper and Lower Limbs

The following mnemonics may help you remember the bones of the upper and lower limbs:

- **Some Lunchers Try Peppers That They Can't Handle.** This sentence helps you remember the carpal bones: Scaphoid, Lunate, Triquetrum, Pisiform, Trapezium, Trapezoid, Capitate, Hamate. The word "try" reminds you that the *tri*quetrum is the carpal bone starting with the letter "T" on the bottom row. If you find that you are confusing the other two "T" carpals, the trapezium and trapezoid, remember that the trapeziUM is by the thUMb.
- **TIBia and FibuLA.** Use the following trick to tell the two leg bones apart: **TIB**ia = **T**hick, **I**nner **B**one; Fibu**LA** = **LA**teral bone.
- **This College Needs Me In Lab Classes.** This statement will help you remember the names of the tarsals from proximal to distal and medial to lateral: Talus, Calcaneus, Navicular, Medial cuneiform, Intermediate cuneiform, Lateral cuneiform, Cuboid.

Here are a few other tricks that might be helpful:

- If you turn the ulna on its side, the trochlear notch is the shape of the letter U for "ulna."
- To remember on which side we find the radius, make a thumb's up gesture as if to say that something is "radical." This will help you remember that the RADius is on the same side as the thumb that you use to make the RADical gesture.
- The word *patella* in Latin means "little plate"; the patella is the plate-shaped bone of the lower limb.
- Remember the navicular bone by its shape, which resembles a boat. Think of the word "navy" to connect "navicular" and "boat." ■

Quick Check

- ☐ 6. What are the seven tarsal bones?
- ☐ 7. How does the structure of the foot and toes parallel that of the hand and fingers?
- ☐ 8. What are the three arches of the foot?

Apply What You Learned

- ☐ 1. You are on an archaeological dig, and you find two pelvic bones that are quite well preserved. How will you determine if the pelves are male or female?
- ☐ 2. Callie is a 6-year-old patient who presents with a complaint of hip pain. An x-ray reveals that her acetabulum is abnormally shallow. How might this explain her symptoms?

See answers in Appendix A.

Chapter Summary

For everything you need to succeed in this course, go to Mastering Anatomy & Physiology. There you will find:

- Practice Tests
- Author Podcasts
- Big Picture Animations
- Concept Boost Video Tutors
- **iP2™** Interactive Physiology 2.0 Tutorials
- **A&PFlix** A&P Flix 3D Animations
- Active-Learning Workbook

MODULE 7.1
Overview of the Skeletal System 210–214

- The skeletal system consists of the **skull, vertebral column, thoracic cage, pectoral girdle, upper limb, pelvic girdle,** and **lower limb,** as well as the skeletal cartilages.

- The **axial skeleton** consists of the bones of the skull, vertebral column, and thoracic cage. The **appendicular skeleton** consists of the bones of the pectoral girdle, upper limb, pelvic girdle, and lower limb.

- The three basic types of **bone markings** are depressions, openings, and projections.

MODULE 7.2
The Skull 214–230

- The skull consists of the **cranial bones** and the **facial bones.** All skull bones except the mandible are joined by immoveable **sutures.**

- The **cranium** contains two portions: the **cranial vault** and the **cranial base.** The eight cranial bones are the frontal bone, two parietal bones, two temporal bones, occipital bone, sphenoid bone, and ethmoid bone.

- The 14 facial bones include the mandible and vomer and the maxillary, zygomatic, nasal, lacrimal, palatine, and inferior nasal conchal bones.

- The **orbit** is formed from contributions by the frontal, maxillary, zygomatic, sphenoid, ethmoid, lacrimal, and palatine bones.

- The **nasal cavity** is formed by the ethmoid, vomer, nasal, palatine, sphenoid, and maxillary bones.
- The **paranasal sinuses** are air-filled cavities in the frontal, maxillary, sphenoid, and ethmoid bones that warm, filter, and humidify inspired air.
- The **oral cavity** is formed by the maxillary, palatine, and mandibular bones.
- The fetal skull largely forms by intramembranous ossification, leading to the presence of **fontanels.**
- The **hyoid bone** is suspended in the neck by muscles.

MODULE 7.3
The Vertebral Column and Thoracic Cage 230–240

- The **vertebral column** consists of 33 **vertebrae** and the ligaments and cartilages that join them. There are 7 **cervical vertebrae,** 12 **thoracic vertebrae,** 5 **lumbar vertebrae,** 5 fused **sacral vertebrae,** and 3-5 fused **coccygeal vertebrae.**
- There are four normal spinal curvatures: the concave **cervical** and **lumbar curvatures** and the convex **thoracic** and **sacral curvatures.**
- Vertebrae share common features, including a central **body,** a **vertebral foramen** through which the spinal cord passes, lateral **transverse processes,** and a posterior **spinous process.**
- Cervical vertebrae are located in the neck and have **transverse foramina** in their transverse processes. C_1 is known as the **atlas,** and C_2 is the **axis.**
- Thoracic vertebrae have **superior** and **inferior costal facets** where they articulate with the ribs.
- Lumbar vertebrae are the thickest and heaviest vertebrae.
- The fused vertebrae of the sacrum feature **sacral foramina** through which nerves pass.
- An **intervertebral disc** is a fibrocartilage pad between two vertebrae.
- The **thoracic cage** consists of the **sternum,** 12 pairs of **ribs,** and the thoracic vertebrae.

MODULE 7.4
Bones of the Pectoral Girdle and Upper Limb 240–247

- The pectoral girdle consists of the **clavicle** and the **scapula.**
  - The clavicle articulates with the manubrium of the sternum and the acromion of the scapula.
  - The scapula rests on the posterosuperior rib cage. Notable features include the posterior **acromion** and the lateral **glenoid cavity.**
- The only bone of the arm is the **humerus.** Proximally its head articulates with the glenoid cavity, and distally the **capitulum** and **trochlea** articulate with the radius and ulna, respectively.
- The bones of the forearm are the medial **ulna** and the lateral **radius.**

- The wrist consists of 7 **carpals,** the hand consists of 5 **metacarpals,** and the fingers consist of 14 **phalanges.**

MODULE 7.5
Bones of the Pelvic Girdle and Lower Limb 247–254

- The **pelvic girdle** consists of the two **pelvic bones** that articulate posteriorly with the sacrum. Each pelvic bone consists of the superior **ilium,** the posteroinferior **ischium,** and the anterior **pubis,** which together form the **acetabulum.**
- The only bone of the thigh is the **femur.** Proximally the **head** articulates with the acetabulum; distally the **medial** and **lateral condyles** articulate with the tibia.
- The **patella** is a sesamoid bone that resides within the tendon of the quadriceps femoris muscle group.
- The **tibia** is the medial leg bone. The **fibula** is the lateral leg bone.
- The bones of the foot and ankle are composed of 7 **tarsals,** 5 **metatarsals,** and 14 **phalanges.** The bones of the foot form three arches: the **medial** and **lateral longitudinal arches** and the **transverse arch.**

Assess What You Learned

Scan the QR Code for additional practice test questions

https://goo.gl/TBYWvO

LEVEL 1 Check Your Recall

1. Which of the following are considered parts of the axial skeleton? (Circle all that apply.)

   a. Pectoral girdle   d. Vertebral column
   b. Lower limb   e. Pelvic girdle
   c. Skull   f. Thoracic cage

2. _____ is the anatomical name for a hole in a bone.

   a. Fossa   c. Condyle
   b. Foramen   d. Tubercle

3. Fill in the blanks: The two parietal bones are united at the _____ suture; they meet the frontal bone at the _____ suture, the temporal bones at the _____ sutures, and the occipital bone at the _____ suture.

4. Mark the following statements as true or false. If a statement is false, correct it to make a true statement.

   a. The four paranasal sinuses are the frontal, parietal, sphenoidal, and mandibular sinuses.
   b. The cribriform plate is a component of the ethmoid bone.
   c. The sella turcica of the sphenoid bone houses the pituitary gland.
   d. The styloid process of the temporal bone is a thick, posterior projection.
   e. The most conspicuous feature of the temporal bone is the foramen magnum.

5. The only moveable bone in the adult skull is the:

a. maxilla.
b. lacrimal bone.
c. mandible.
d. frontal bone.

6. The structure(s) that divide the nasal cavity into right and left sides is/are the:

a. nasal bones.
b. perpendicular plate of the ethmoid bone.
c. vomer.
d. Both a and b are correct.
e. Both b and c are correct.

7. The "soft spots" in an infant's skull are known as:

a. fontanels.
b. sutures.
c. metopic joints.
d. hyaline cartilage.

8. Mark the following statements as true or false. If a statement is false, correct it to make a true statement.

a. The thoracic and sacral curvatures are the vertebral column's concave curvatures.
b. A vertebral disc is composed of an inner anulus fibrosus and an outer nucleus pulposus.
c. The posterior projection from a vertebra is the spinous process.
d. The sacral, coccygeal, and cervical vertebrae are fused in an adult.

9. Transverse foramina are a characteristic of which kind of vertebra?

a. Thoracic
b. Lumbar
c. Sacral
d. Cervical

10. Fill in the blanks: The inferior portion of the sternum is the _____. The superior portion of the sternum is the _____, and it articulates with the _____ and the first rib.

11. How do true, false, and floating ribs differ from one another?

12. Which of the following portions of the scapula articulates with the clavicle?

a. Coracoid process
b. Acromion
c. Spine
d. Glenoid cavity

13. Fill in the blanks: The only bone of the arm is the _____. The forearm consists of the medial _____ and the lateral _____.

14. The "elbow bone" is called the:

a. trochlea.
b. capitulum.
c. olecranon.
d. deltoid tuberosity.

15. Which of the following is not a proximal carpal bone?

a. Hamate
b. Pisiform
c. Scaphoid
d. Lunate

16. Mark the following statements as true or false. If a statement is false, correct it to make a true statement.

a. The obturator foramen articulates with the head of the femur at the hip joint.
b. The superior border of the pelvic bone is the iliac crest.
c. The weight of the body in the sitting position is supported by the ischial tuberosities.
d. The two pubic bones articulate at the acetabulum.
e. The pelvic brim is the boundary between the greater and lesser pelvis.

17. The most lateral projection of the proximal epiphysis of the femur is the:

a. lesser trochanter.
b. gluteal tuberosity.
c. greater trochanter.
d. femoral neck.

18. Fill in the blanks: The bones of the leg are the medial _____ and the lateral _____. The sesamoid bone that articulates with the distal femur is the _____.

19. The heel bone is more properly known as the:

a. talus.
b. navicular.
c. cuboid.
d. calcaneus.

20. The arch(es) of the foot are the:

a. transverse arch.
b. medial longitudinal arch.
c. lateral longitudinal arch.
d. Both a and b are correct.
e. All of the above are correct.

LEVEL 2 Check Your Understanding

1. How do the atlas (C_1) and the axis (C_2) differ from other cervical vertebrae? How does this difference enable them to perform their functions?

2. Explain how abnormal bone structure could affect muscle function.

3. What structures form the knee and elbow joints? Of the two joints, which do you think would be more stable? Why?

LEVEL 3 Apply Your Knowledge

PART A: Application & Analysis

1. A *deviated septum* results when the nasal septum is shifted to one side or the other. What bones might be involved in this condition? Why might this make breathing difficult?

2. Mrs. Dent presents to the clinic with back pain. During the exam, you notice that she has severe kyphosis, and you suspect a vertebral fracture. What specific part of her vertebra is likely to be fractured, considering her deformity? Explain.

3. You arrive on the scene where a person without a pulse was found. Someone on the scene performed CPR, but the individual unfortunately could not be revived. On postmortem examination, it is discovered that several ribs and the xiphoid process were fractured. What likely caused these fractures?

PART B: Make the Connection

4. Predict where each of the following structures is located, based on your knowledge of skeletal anatomy and anatomical terms from Chapter 1 (your answers should be along the lines of "lateral crural region" or "posterior cervical region").

a. Frontal lobe of brain
b. Suprahyoid muscle
c. Ulnar artery
d. Tibial nerve
e. Intercostal muscle
f. External iliac artery

See answers in Appendix A.

8

Articulations

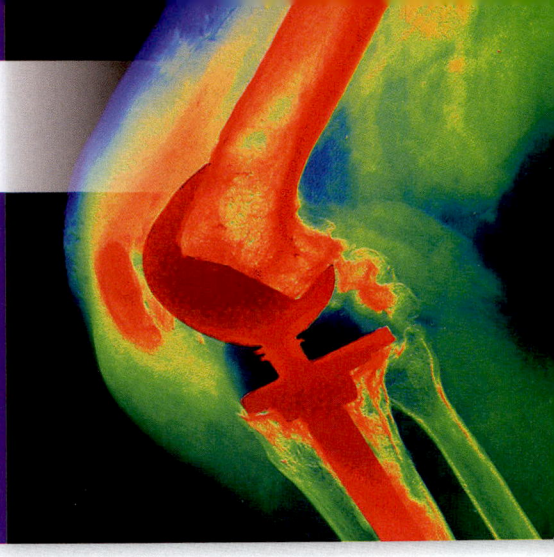

The human skeleton contains over 200 bones having a variety of functions. Almost all these bones contact other bones at **articulations,** or **joints,** which form a critical part of the skeletal system. In this chapter, we examine the structural and functional classes of joints, the different types of motion they allow, and the gross anatomy of selected joints. As you read, notice how one of the core principles you have seen throughout this textbook—the structure-function relationship—applies to each class of articulation that we study.

MODULE **8.1**

Overview of Joints

Learning Outcomes

1. Describe the basic functions of joints.
2. Describe how joints are classified both structurally and functionally.

Before we examine specific types of joints in the body, we introduce the functional and structural classes of joints. But first, let's take a look at the basic functions of joints.

Functions of Joints

Obviously, joints connect—or *articulate*—two bones. Indeed, two bones that have been separated from their articulation, and so are no longer connected, are said to be *disarticulated*. But how, specifically, do these connections help the skeletal system perform its functions? By connecting two bones, joints perform three important functions:

- **Joints enable movement.** The link between bones can allow movement when surrounding muscles and tendons exert the necessary amount of force across the joint.
- **Joints provide stability.** Some joints allow limited or no movement. Such joints are very stable, which is critical for structures such as the skull that protect underlying structures.
- **Joints allow long bones to lengthen.** Recall that the *epiphyseal plate* is the location from which long bones grow in length during skeletal development (see Chapter 6). Epiphyseal plates are actually temporary joints, which we discuss in a later module.

Quick Check

☐ 1. What are the main functions of joints?

*For practice applying concepts to a clinical scenario, check out the **Running Case Study** for this chapter @ Mastering Anatomy & Physiology*

Photo: Colored x-ray of a prosthetic knee (dark red) in a patient with osteoarthritis. The knee replacement attaches to the femur and tibia and has a flexible hinge joint.

Classes of Joints

« FLASHBACK

1. What is dense regular collagenous connective tissue? (p. 139)
2. What is cartilage? (p. 142)

Joints can be classified in two ways: (1) according to the amount of motion they allow, which is their functional classification; or (2) according to their anatomical features, which is their structural classification. As you might expect, the functional and structural classifications relate and overlap in significant ways. We examine both in the next two sections.

Functional Classification

As we just discussed, the functions of most joints revolve around stabilizing two bones where they meet and possibly allowing movement between the bones. A joint's functional classification, then, is based on the amount of movement, and so the degree of stability, provided by a joint. Given these criteria, we have three functional classes:

- A **synarthrosis** (sin-ahr-THROH-sis) does not allow any movement between articulating bones and provides the greatest stability.
- An **amphiarthrosis** (am′-fee-ahr-THROH-sis) allows only a small amount of movement between articulating bones and provides a significant amount of stability, but less than a synarthrosis.
- A **diarthrosis** (dy-ahr-THROH-sis) is freely moveable, allowing a wide variety of specific movements, and provides the least amount of stability.

Notice that the more movement allowed by a joint, the less stability it offers. This tradeoff of motion at the expense of stability is a theme that we examine repeatedly throughout this chapter.

Structural Classification

Joints can also be grouped on the basis of structure. Two anatomical features are considered: the type of connective tissue that links bones, and the presence or absence of a space between them. These features give us three structural classes:

- **Fibrous joints** are united by the short collagen fibers of dense regular collagenous connective tissue. No joint space is present in fibrous joints, and so functionally they are synarthroses or amphiarthroses.
- **Cartilaginous joints** have cartilage between the articulating bones. Like fibrous joints, there is no joint space, which makes cartilaginous joints functionally synarthroses or amphiarthroses.
- **Synovial joints** (sih-NOH-vee-uhl) are the only class to have a joint space, or *cavity,* filled with fluid between articulating bones. The joint cavity makes synovial joints diarthroses, and gives them the greatest range of motion of any joint class.

Although structural classifications are the primary focus of this chapter, keep the structure-function relationship in mind. We cannot really separate the structural and functional aspects of a joint.

Quick Check

- ☐ 2. How are joints classified functionally?
- ☐ 3. How are joints classified structurally?

Apply What You Learned

- ☐ 1. Do you think the cranial bones are joined by synarthroses, amphiarthroses, or diarthroses? Explain.
- ☐ 2. A surgical procedure called a *fusion* takes two bones and joins them, making them functionally one bone. What functional class of joint does this resemble? What properties would the fused joint have?

See answers in Appendix A.

MODULE **8.2**
Fibrous and Cartilaginous Joints

Learning Outcomes

1. Compare and contrast the three subclasses of fibrous joints.
2. Give examples of fibrous joints, and describe how they function.
3. Compare and contrast the two subclasses of cartilaginous joints.
4. Give examples of cartilaginous joints, and describe their function.

Now that we've introduced fibrous and cartilaginous joints, let's take a closer look at them. In this module, we explore the structure and function of the different types of fibrous and cartilaginous joints. Several examples of these joints, such as the epiphyseal plate and intervertebral joints, are structures that you have studied in previous chapters.

Fibrous Joints

« FLASHBACK

1. What are the major sutures joining the cranial bones? (p. 216)
2. What are dental alveoli? (p. 219 and p. 220)

In the tissues chapter, you learned about dense regular collagenous connective tissue. Recall that it contains fibers of the tough

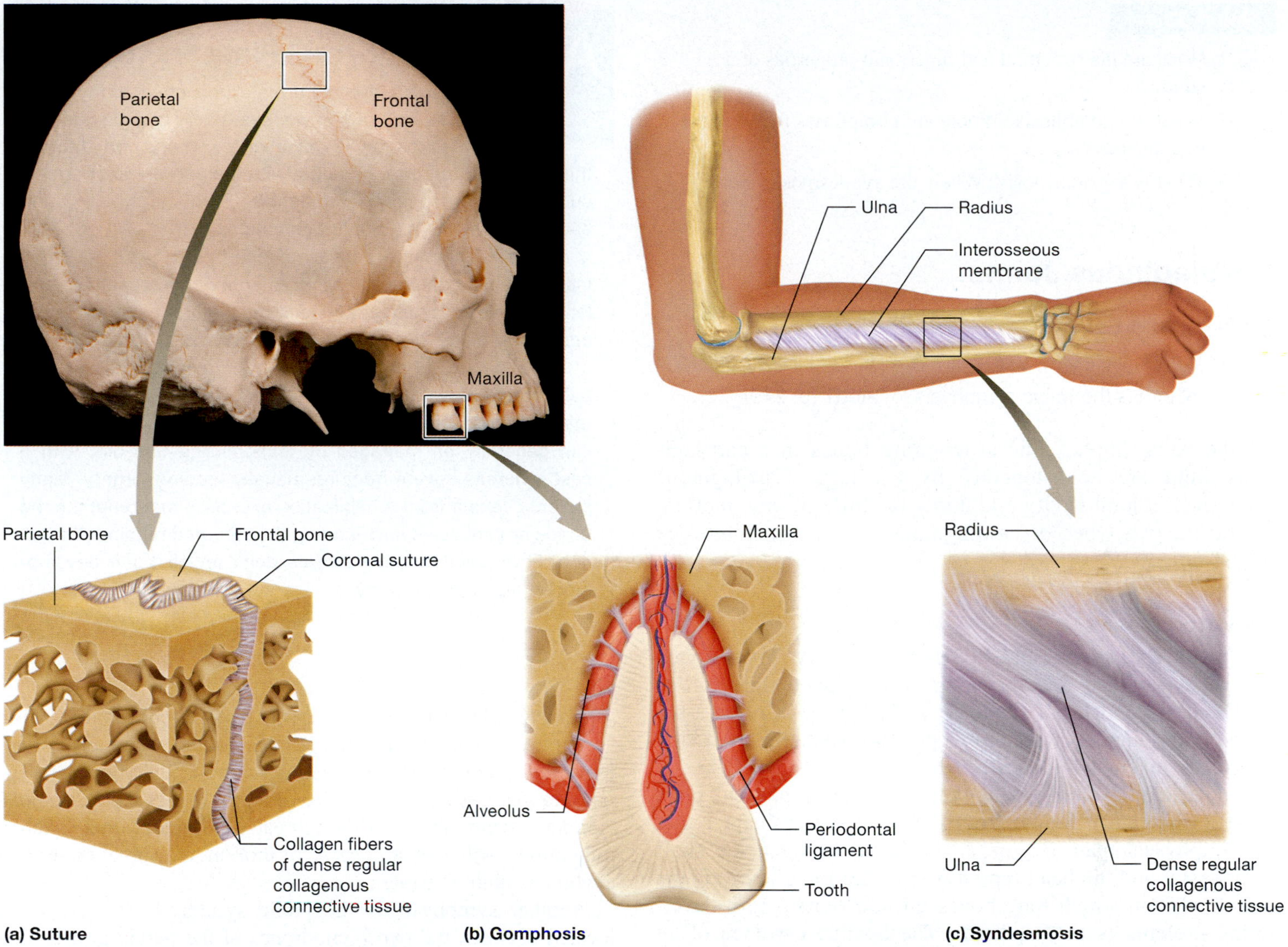

(a) Suture

Parietal bone — Frontal bone — Coronal suture

Collagen fibers of dense regular collagenous connective tissue

(b) Gomphosis

Maxilla

Alveolus — Periodontal ligament — Tooth

(c) Syndesmosis

Radius — Ulna — Dense regular collagenous connective tissue

Ulna — Radius — Interosseous membrane

Parietal bone — Frontal bone — Maxilla

Figure 8.1 The three types of fibrous joints.

protein collagen (see Chapter 4). Collagen fibers found in fibrous joints lend stability to these joints but permit little, if any, motion. There are three types of fibrous joints—*sutures, gomphoses,* and *syndesmoses*—which are discussed in the following subsections.

Sutures

Recall that a **suture** is a joint between bones that make up the skull (**Figure 8.1a**). The edges of skull bones have tiny finger-like projections that interweave somewhat like a closed zipper. When skull development is complete, these interweaved edges are held together by very short collagen fibers that are part of dense regular collagenous connective tissue. This makes sutures very stable synarthroses and well suited for reinforcing the bones that protect the brain, an example of the Structure-Function Core Principle (p. 28).

CORE PRINCIPLE
Structure-Function

Gomphoses

A **gomphosis** (gahm-FOH-sis) is a fibrous joint between a tooth and its corresponding alveolus in the mandible or maxilla

(**Figure 8.1b**). As you might expect, a gomphosis is a synarthrosis, as its primary function is to provide stability and hold the tooth in place. Each individual tooth is firmly attached to the bone by a group of connective tissue fibers collectively called the *periodontal ligament.* The articulations between the teeth and the maxilla or mandible are the only gomphoses in the human body.

Syndesmoses

The third type of fibrous joint is the **syndesmosis** (sin-dez-MOH-sis), in which the articulating bones are joined by a long membrane—the *interosseous membrane* or *ligament*—composed of dense regular collagenous connective tissue. Syndesmoses are amphiarthroses because the interosseous membrane provides a limited amount of movement, allowing the articulating bones to pivot around one another. Examples of syndesmoses are found between the radius and ulna in the forearm (**Figure 8.1c**) and between the fibula and tibia in the leg.

Quick Check

☐ 1. What are the structural and functional properties of a suture?

☐ 2. What is a gomphosis? Where are gomphoses found in the human body?

☐ 3. What is a syndesmosis? Where are syndesmoses located?

Cartilaginous Joints

≪ FLASHBACK

1. What is the epiphyseal plate? What is its function? (p. 199)

2. Where is the pubic symphysis located? (p. 249)

As the name implies, the articulating bones in a cartilaginous joint are held together by cartilage. Cartilaginous joints lack a joint cavity and allow for little, if any, motion. There are two types of cartilaginous joints: *synchondroses* and *symphyses.*

Synchondroses

A **synchondrosis** (sin-kahn-DROH-sis) consists of bones united by hyaline cartilage. Recall from the tissues chapter that *hyaline cartilage* is a tough but flexible connective tissue often associated with joints. Synchondroses permit essentially no motion and so are functionally synarthroses.

One example of a synchondrosis is the epiphyseal plate (eh-PIF-ih-see-ul). Recall from the bone tissue chapter that the epiphyseal plate (**Figure 8.2a**), found between the shaft (diaphysis) and the head (epiphysis) of the long bones, is the structure from which long bones grow in length in a developing skeleton (see Chapter 6). The hyaline cartilage of an epiphyseal plate is replaced with bone when bone growth is complete. As you might expect, motion occurring at an epiphyseal plate could disrupt its structure, which would in turn disturb its function and possibly the development of the skeletal system (see *A&P in the Real World: Epiphyseal Plate Fractures*).

Synchondroses that persist into adulthood are found in the rib cage. The first sternocostal joint, where the first rib connects to the manubrium of the sternum, stabilizes the rib cage (**Figure 8.2b**). Additional synchondroses, the costochondral joints, are found where the ribs attach to their costal cartilages.

Symphyses

A **symphysis** (SIM-fih-sis) is a joint in which the bones are united by a tough fibrocartilage pad. These joints permit a small amount of motion and so are amphiarthroses. In an example of the Structure-Function Core Principle (p. 28), a symphysis is best suited for regions of the skeleton that must resist compression and tension while still allowing a small amount of motion.

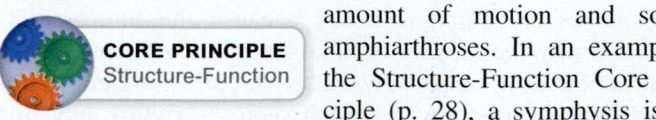

CORE PRINCIPLE
Structure-Function

Epiphyseal Plate Fractures

The epiphyseal plate in a child's long bone is one of the weakest parts of a developing skeleton. Even a minor injury can fracture this delicate structure, possibly with lifelong consequences, such as differences in limb length, limb deformities, and early-onset arthritis (joint inflammation).

The most common causes of epiphyseal plate fractures include recreational activities, accidents, and competitive athletics. Injuries may affect any joint but occur most often at the epiphyseal plates of the bones of the forearm.

Symptoms of epiphyseal plate fractures include swelling, pain, and redness over the injured joint. Treatment depends on the severity of the fracture. Minor epiphyseal plate fractures can generally be managed by immobilizing the joint with a cast, whereas severe fractures usually require surgery. Many patients benefit from rehabilitation exercises to strengthen the bones and muscles surrounding the joint and regain full function. Fortunately, most fractures don't impair bone development if managed properly.

An example of a symphysis is the intervertebral joint (**Figure 8.2c**); these joints are located between adjacent vertebral bodies of the spinal column. Intervertebral joints have fibrocartilage pads, or *discs,* which protect the vertebral bodies from compression and absorb shock. Although each individual intervertebral joint allows only a small degree of motion, the spinal column as a whole is quite flexible.

Another symphysis is the pubic symphysis (**Figure 8.2d**), located between the two pubic bones of the pelvic girdle. This region of the skeleton is under enormous tension, stress, and compression, especially when the body is moving. Normally, this joint limits motion; however, in women during childbirth, it allows increased motion, creating a larger opening through the pelvic outlet where the head of the newborn emerges.

Quick Check

☐ 4. What are the features of a synchondrosis?

☐ 5. What are the general features of a symphysis, and what are two examples of a symphysis?

Apply What You Learned

☐ 1. Certain diseases cause the production of collagen that is weaker than normal. Predict how such diseases could affect the structure and function of gomphoses.

☐ 2. What might happen if sutures were fused at birth?

☐ 3. Explain how increased motion at a synchondrosis would impair the function of the joint.

See answers in Appendix A.

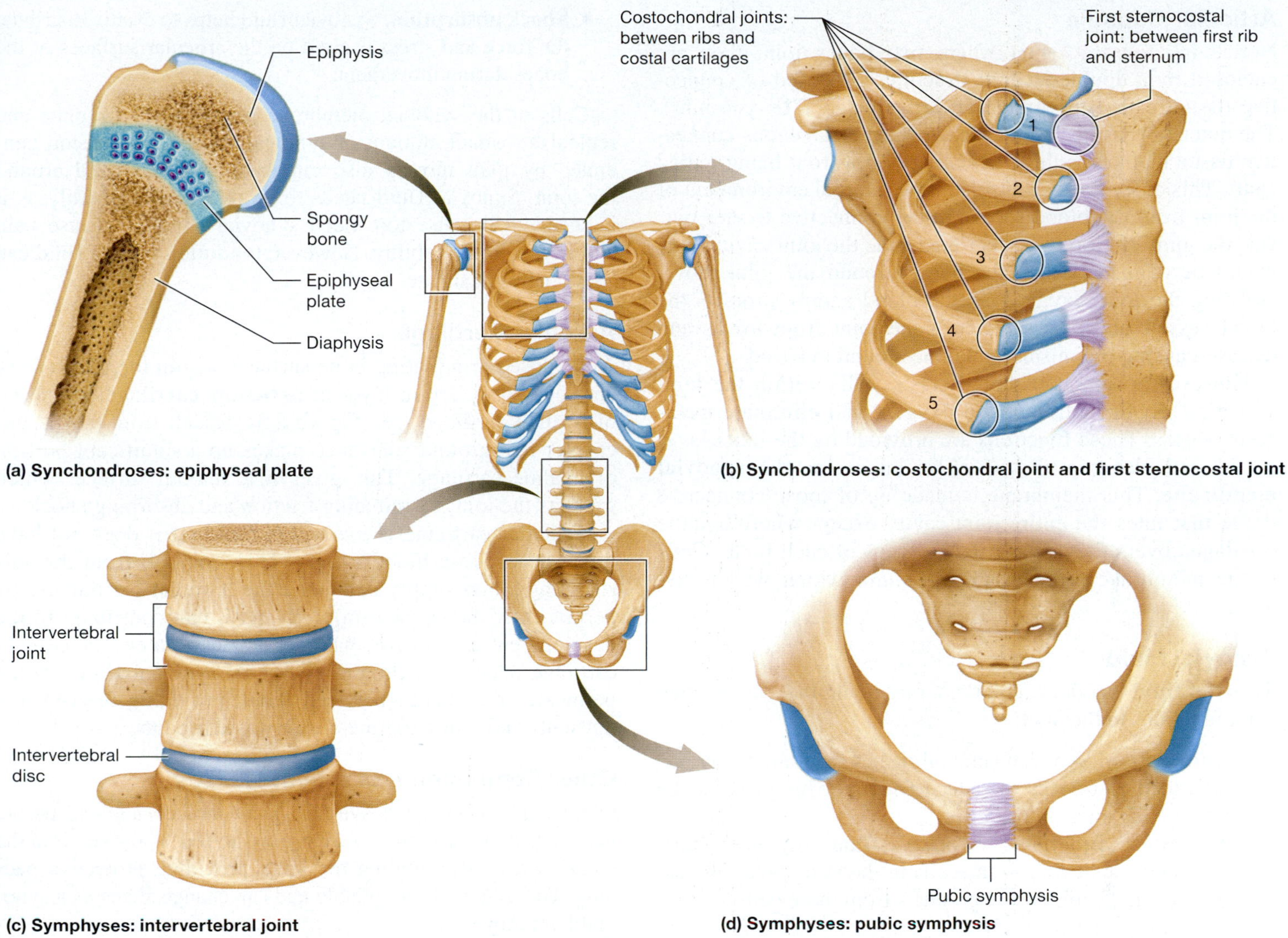

Epiphysis
Spongy bone
Epiphyseal plate
Diaphysis
(a) Synchondroses: epiphyseal plate
Costochondral joints: between ribs and costal cartilages
First sternocostal joint: between first rib and sternum
1
2
3
4
5
(b) Synchondroses: costochondral joint and first sternocostal joint
Intervertebral joint
Intervertebral disc
(c) Symphyses: intervertebral joint
Pubic symphysis
(d) Symphyses: pubic symphysis

Figure 8.2 **The two types of cartilaginous joints.**

MODULE **8.3**

Structure of Synovial Joints

Learning Outcomes

1. Identify the structural components of a synovial joint.
2. Compare and contrast synovial joints with fibrous and cartilaginous joints.

Most joints in the human body are synovial joints. Recall that these are functionally diarthroses (freely moveable joints), which gives them many structural differences from fibrous and cartilaginous joints. We examine these structural features next.

The Joint Cavity

FLASHBACK

1. What are the properties of hyaline cartilage? (p. 142)
2. How do hyaline cartilage and fibrocartilage differ? (p. 142)

One of the key differences between synovial joints and other types of joints is the presence of a space—the **joint cavity,** or **synovial cavity**—between the two articulating bones. The joint cavity is characterized by three unique features: an *articular capsule, synovial fluid,* and an *articular cartilage.*

Articular Capsule

Notice in **Figure 8.3** that where two bones join, they are enclosed by a double-layered structure composed of connective tissue, called the **articular capsule** (ahr-TIK-yoo-luhr). The outer fibrous layer is dense irregular collagenous connective tissue that keeps the articulating bones from being pulled apart. This outer layer also isolates the internal environment of the joint from the blood supply in the connective tissues outside the joint. If blood vessels were inside the joint cavity, they would be damaged when the bones around the joint move. Isolating the joint cavity from the blood supply protects the blood vessels and also safeguards the joint from toxins and disease-causing organisms sometimes found in blood.

However, to maintain homeostasis, cells within the joint still must obtain nutrients and oxygen, and eliminate metabolic wastes. These functions are provided by the inner layer of the articular capsule, which is known as the **synovial membrane.** This membrane is made up of loose connective tissue that lines the entire joint cavity except where hyaline cartilage covers the articulating surfaces of each bone. Cells in the membrane secrete *synovial fluid,* which we discuss next.

Synovial Fluid

Synovial fluid is a thick, colorless, oily liquid that serves three main functions in the joint:

- **Lubrication.** Synovial fluid lubricates the joint cavity and articulating surfaces to reduce friction. This protects the articulating ends of the bones.
- **Metabolic functions.** Synovial fluid supplies nutrients such as glucose to the cells in the joint cavity. It also removes metabolic waste products from these cells.

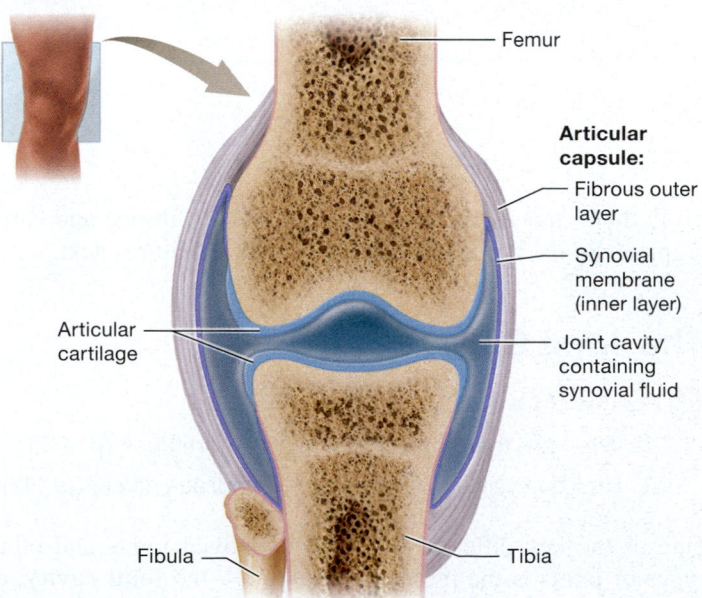

Figure 8.3 Structure of a typical synovial joint.

Labels: Femur; Articular capsule: Fibrous outer layer; Synovial membrane (inner layer); Joint cavity containing synovial fluid; Articular cartilage; Fibula; Tibia

- **Shock absorption.** Synovial fluid helps to evenly distribute the force and stress exerted on the articular surfaces of the bones during movement.

Cells of the synovial membrane continuously circulate and replenish a small amount of synovial fluid. Compression generated by joint motion also circulates synovial fluid around the joint. Synovial fluid needs to be removed as quickly as it is secreted, because too much synovial fluid can cause pain and impair joint mobility. However, too little synovial fluid can result in joint damage.

Articular Cartilage

All exposed articulating bone surfaces within the joint cavity are covered by a thin layer of **articular cartilage** composed of hyaline cartilage (see Figure 8.3). Recall from the tissues chapter that ground substance makes up a significant portion of hyaline cartilage. This gives it a smooth surface, which protects the joint by reducing friction and absorbing shock.

Articular cartilage is *avascular,* meaning it does not have direct access to a blood supply. This isolation from the surrounding blood supply can cause the articular cartilage to be permanently damaged if injured, and it can contribute to the development of *arthritis,* which we discuss later. The cells in cartilage depend on the circulation of synovial fluid, which permeates the spaces between them, providing oxygen and nutrients and removing metabolic waste products.

Other Components of a Synovial Joint

Other components of a synovial joint include adipose tissue, nerves, and blood vessels. Adipose tissue (fat) is packed into the empty spaces surrounding the joint, providing protective padding. These fat pads are pliable and can change shape as a synovial joint moves.

Synovial joints are surrounded by a copious supply of blood vessels and nerves found in the supportive connective tissues. Blood vessels deliver nutrients to, and remove waste products from, the articular capsule, with the aid of synovial membrane cells and synovial fluid. Neurons in nerves transmit painful stimuli to the brain to prevent damage to the joint. They also transmit stimuli relating to the position of the joint, which helps with body movement and postural adjustments.

Quick Check

☐ 1. What are the layers of the articular capsule, and what are their functions?

☐ 2. What functions are served by synovial fluid?

☐ 3. Which type of cartilage makes up articular cartilage, and what is its function?

Stabilizing and Supportive Structures

Synovial joints allow more mobility than other types of joints, but they are also the least stable. These joints are continually stressed as part of their normal function. Although the articular capsule provides some structural support, synovial joints must also rely on several structures outside the joint itself for stability and support. These include *ligaments, tendons, bursae,* and *tendon sheaths* (**Figure 8.4**).

Ligaments

A **ligament** is a strand of dense regular collagenous connective tissue that connects one bone to another to strengthen and reinforce the joint. Two types of ligaments are present around synovial joints. *Intrinsic ligaments* (not visible in Figure 8.4) are thickened regions of the articular capsule; they are called "intrinsic" because they are found within the articular capsule. *Extrinsic ligaments* are not part of the articular capsule. They may be inside or outside the joint cavity. For instance, the coracoacromial ligament (see Figure 8.4) is an extrinsic ligament that stabilizes the shoulder joint.

Tendons

A **tendon**—a structural component of a skeletal muscle composed of dense regular collagenous connective tissue—connects the muscle to a bone or another structure. Tendons typically cross over or around a joint so that when a muscle contracts, the force generated is transmitted across the joint. This force results in motion as the tendon pulls on the bone to which it attaches. Tendons often stabilize joints, particularly joints that have multiple tendons crossing them. For instance, the tendon of the long head of the biceps brachii muscle crosses the shoulder joint, stabilizing the head of the humerus in the joint cavity (see Figure 8.4).

The muscles to which the tendons attach also help stabilize joints. Muscles constantly generate small contractions, collectively known as *muscle tone,* that pull the tendons taut around or across the joint. Muscle tone keeps the ends of the articulating bones near one another.

Bursae and Tendon Sheaths

A **bursa** (plural, *bursae* [BER-see]) is a synovial fluid–filled structure resembling a limp water balloon. Like the articular capsule, it is a fibrous structure lined with a synovial membrane. Bursae may be attached to the articular capsule or completely separate. Generally, bursae are found in regions of high stress where bones, tendons, muscles, and skin interact in a small space. Here they minimize friction between all the moving parts of a synovial joint. What happens to bursae when excessive stress is placed on them? See *A&P in the Real World: Bursitis.*

Bursitis

Bursitis refers to inflammation of a bursa. Bursitis can result from a single traumatic event such as a fall, repetitive movements like pitching a softball, or an inflammatory disease such as arthritis. The most common sites of bursitis are the shoulder, elbow, hip, and knee.

Clinical features of bursitis include pain both at rest and with motion of the affected joint. The joint may feel tender, swollen, and warm.

Treatment is aimed primarily at reducing pain and swelling. Rest, ice, compression of the injured area, and anti-inflammatory medications are beneficial in the early stages of the injury. Anti-inflammatory steroid medications may be injected directly into the bursa itself. Fluid can also be removed from the bursa to relieve swelling. If left untreated, bursitis can become chronically painful and increasingly difficult to treat.

Tendon sheaths are long bursae that surround some tendons in high-stress regions of the human body. Tendon sheaths protect long tendons as they course over and around synovial joints. For example, the stabilizing force that the biceps brachii tendon exerts on the shoulder joint is further enhanced by its surrounding tendon sheath.

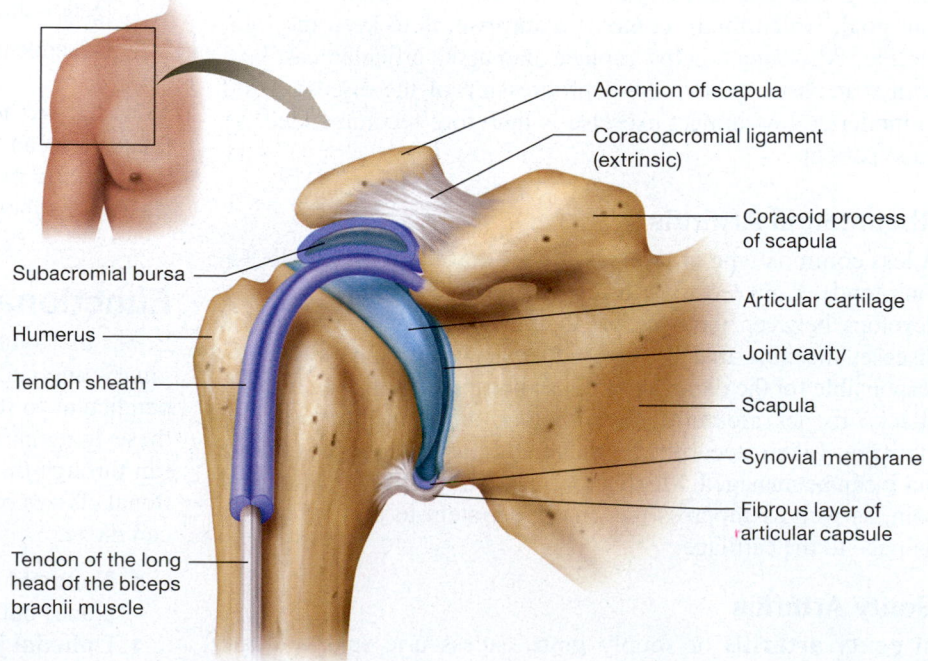

Figure 8.4 Supportive structures of a synovial joint.

Acromion of scapula

Coracoacromial ligament (extrinsic)

Coracoid process of scapula

Subacromial bursa

Articular cartilage

Humerus

Joint cavity

Tendon sheath

Scapula

Synovial membrane

Fibrous layer of articular capsule

Tendon of the long head of the biceps brachii muscle

☐ 4. Why do synovial joints require stabilization?

☐ 5. What are the basic structural and functional properties of ligaments, tendons, and bursae?

Arthritis

Arthritis is defined as the inflammation of one or more joints, resulting in pain and decreased range of motion. It results from the breakdown of articular cartilage, which leads to damage of the underlying bone. Three common forms of arthritis are osteoarthritis, rheumatoid arthritis, and gouty arthritis. Each type results in similar symptoms, but the mechanism by which the cartilage is damaged differs.

Osteoarthritis

The most common type, **osteoarthritis** (ahs′-tee-oh-ahr-THRY-tiss), is the arthritis we generally associate with wear and tear. Although younger people may develop the condition due to accidents, competitive athletics, and recreational injuries, most individuals with osteoarthritis are over 65. Obesity, female sex, and advancing age increase the risk of developing osteoarthritis. Any joint can be affected, but the back, hands, knees, and shoulders are the most common sites.

Common features of osteoarthritis include pain and stiffness, as well as decreased joint mobility. Surrounding muscles, tendons, tendon sheaths, and bursae can also become painful and swollen.

Although osteoarthritis is not reversible, it can be managed. The first goal is to minimize pain, primarily through pain-relieving and anti-inflammatory medications. Another important goal, which might come as a surprise, is to keep the joint mobile. Movement helps replace damaged articular cartilage with scar tissue, which slows progression of the disease. Mild to moderate low-impact exercise is therefore recommended for most patients.

Rheumatoid Arthritis

A less common type of arthritis is **rheumatoid arthritis** (ROO-muh-toyd). Unlike osteoarthritis, rheumatoid arthritis generally develops between the ages of 30 and 50. It is an autoimmune disease, meaning that the individual's own immune system is responsible for the damage. In this condition, the immune system attacks tissues around synovial joints, damaging the articular cartilage. Like osteoarthritis, rheumatoid arthritis is not curable but it can be managed. Medications can reduce inflammation and pain, as well as suppress the immune system to prevent further damage to the cartilage.

Gouty Arthritis

In **gouty arthritis** or simply **gout,** excess uric acid, a waste product, crystallizes and forms deposits in the connective tissue surrounding a joint. Any joint can be affected, but it most commonly involves the first metatarsophalangeal joint (great toe).

The inflammatory reaction to the crystals damages the articular cartilage, leading to gouty arthritis. As with other forms of arthritis, medications that reduce inflammation and pain can help. In addition, patients may take medications to lower the blood concentration of uric acid.

☐ 6. What is osteoarthritis, and in what age group might you expect the disease to develop? How does osteoarthritis differ from rheumatoid and gouty arthritis?

Apply What You Learned

☐ 1. Certain conditions cause *ligamentous laxity,* which means that the ligaments are very loose. Predict how this would affect synovial joints. Would it have the same effect on cartilaginous and fibrous joints? Why or why not?

☐ 2. Explain why joint movement is often painful when a tendon—which is outside the joint—is injured.

See answers in Appendix A.

MODULE 8.4
Function of Synovial Joints

Learning Outcomes

1. Define the functional classes of synovial joints.
2. Describe and demonstrate the movements of synovial joints.

Now that we have discussed the structural features of synovial joints, we can start to appreciate how those features translate into specific movements. This module examines the functional classes of synovial joints and the motions that occur at them.

Functional Classes of Synovial Joints

Bones in a synovial joint travel through a plane around an imaginary line called an *axis.* Each plane has an axis that is perpendicular to its flat surface. Motion of a joint is described by these imaginary flat surfaces and the perpendicular lines that run through them. We can group synovial joints into four functional classes based on the number of axes around which a bone can move:

- **Nonaxial joints** allow motion to occur in one or more planes but do not move around an axis.
- **Uniaxial joints** allow motion around only one axis.
- **Biaxial joints** allow motion around two axes.
- **Multiaxial** or **triaxial joints** allow motion around three axes.

ConceptBOOST)))

Understanding Axes of Motion

Axes of motion might seem complex at first, so let's break down their functions with simple examples. The elbow joint has only one axis (axis 1 in the figure at right) that acts like a hinge and allows motion in one plane perpendicular to the axis. This uniaxial joint only allows the forearm and hand to move toward the shoulder or to make the opposite movement away from the shoulder.

Now let's look at an example of a biaxial joint, which is capable of movement around two separate axes. The metacarpophalangeal joints are a group of biaxial joints found between the proximal phalanges and the metacarpals in the hand (the "knuckles"), as shown in the figures below. The metacarpophalangeal joints can move around axis 1, allowing the proximal phalanges to move toward and away from the palm of the hand in the same way that the elbow joint moves. The metacarpophalangeal joints can also move around axis 2, allowing the fingers to be squeezed together or fanned out, as you can see in the next part of the figure.

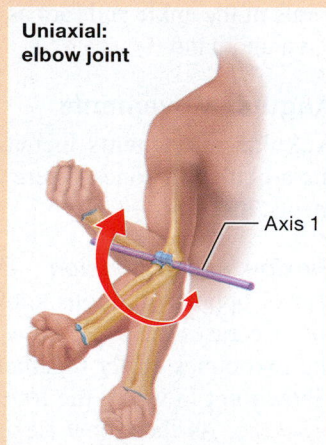

Uniaxial: elbow joint

Axis 1

A third axis allows an additional motion that our uniaxial and biaxial joints are not able to perform. The shoulder is an example of a multiaxial joint. As the final figure shows, your humerus can move forward and backward within the shoulder joint around axis 1 (as when you swing your arms back and forth while walking). The humerus can also move away from and toward your body around axis 2 (as when you do jumping jacks). Finally, it can rotate, or move in a circular fashion, around axis 3 (as when you throw a Frisbee). ■

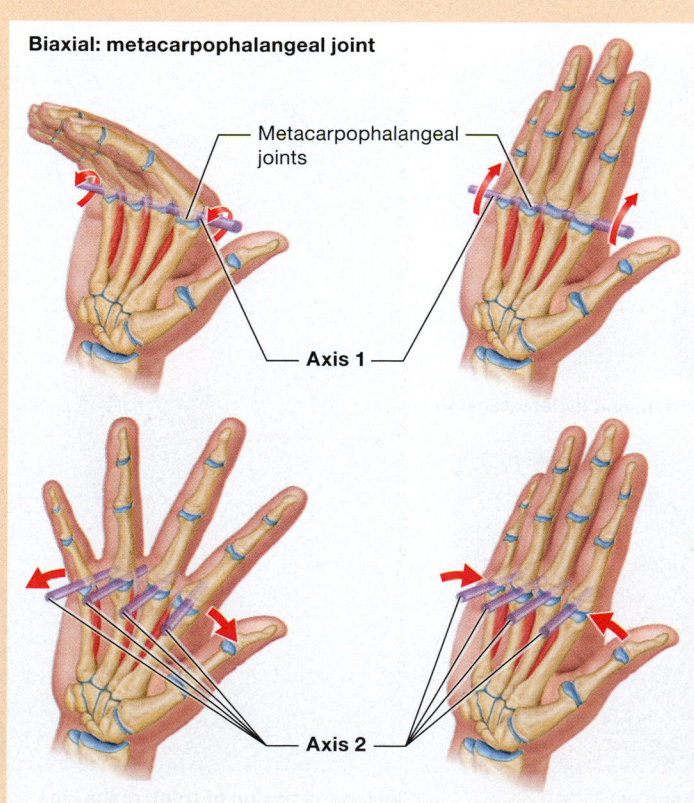

Biaxial: metacarpophalangeal joint

Metacarpophalangeal joints

Axis 1

Axis 2

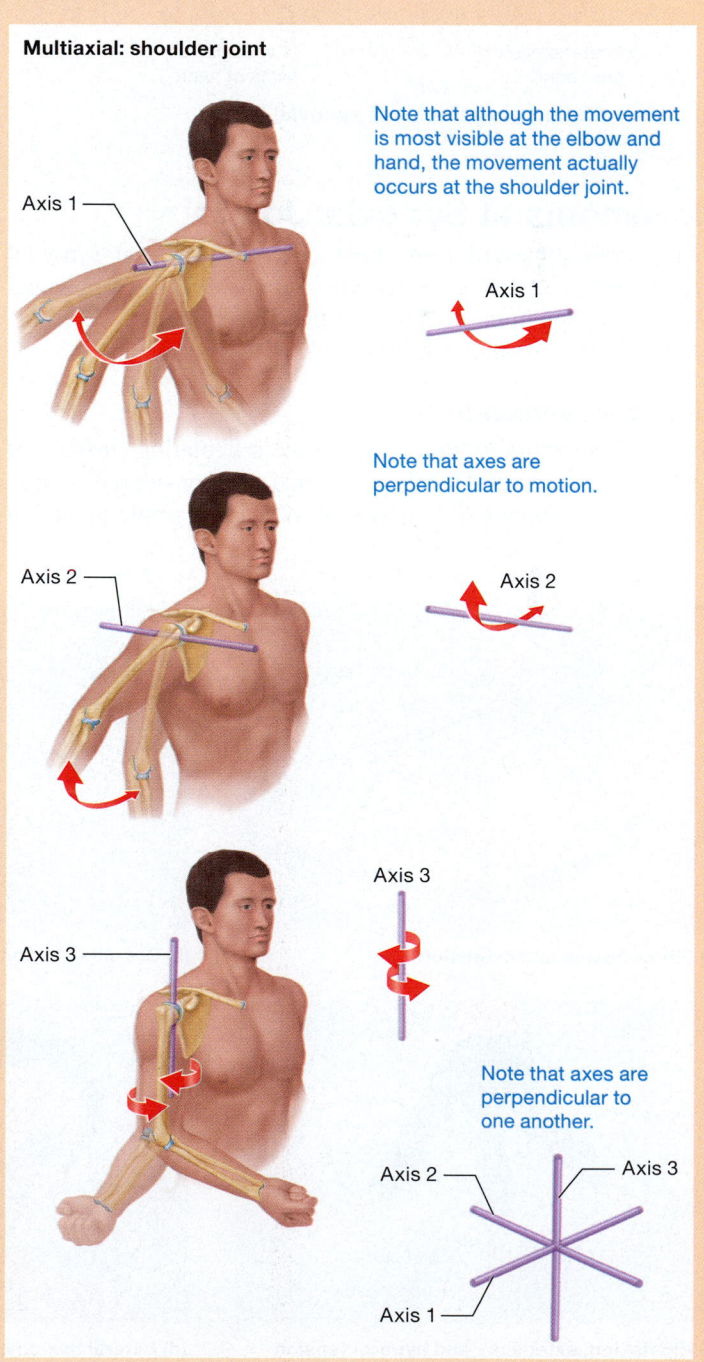

Multiaxial: shoulder joint

Axis 1

Note that although the movement is most visible at the elbow and hand, the movement actually occurs at the shoulder joint.

Axis 1

Axis 2

Note that axes are perpendicular to motion.

Axis 2

Axis 3

Axis 3

Note that axes are perpendicular to one another.

Axis 2   Axis 3

Axis 1

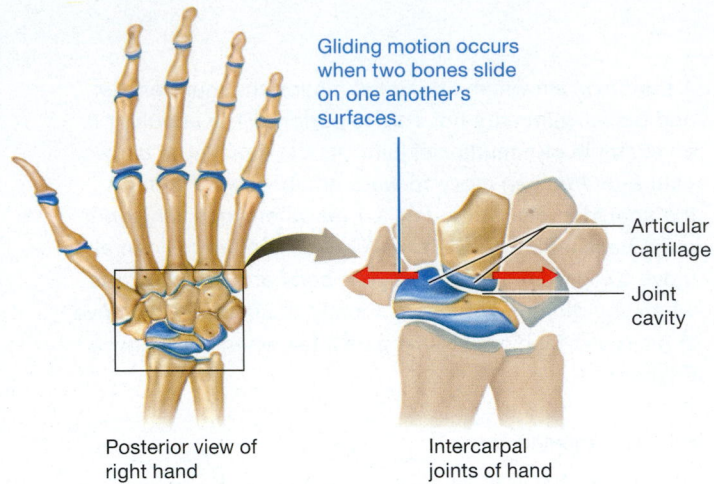

Gliding motion occurs when two bones slide on one another's surfaces.

Articular cartilage

Joint cavity

Posterior view of right hand

Intercarpal joints of hand

Figure 8.5 Gliding movements of synovial joints.

Movements at Synovial Joints

Four general types of movement can take place at synovial joints: *gliding, angular, rotational,* and *special.* Each synovial joint can engage in one or more of these movements. We examine each in the upcoming subsections.

Gliding Movements

Gliding is a sliding motion between the articulating surfaces of the bones in a joint. These movements are considered nonaxial because the bones slide past each other in a single plane but not around an axis. The direction of movement depends on the shape of the articulating surfaces at the joint as well as the limitations of the surrounding supportive connective tissues. For instance, we find gliding movements between the intertarsal joints of the ankle and foot as well as the intercarpal joints of the wrist and hand (**Figure 8.5**).

Angular Movements

Angular movements increase or decrease the angle between the articulating bones. There are several types of angular movements, discussed next.

Flexion and Extension   The first angular movements are a pair of opposite motions known as *flexion* and *extension.* **Flexion** decreases the angle between articulating bones by bringing the two bones closer together. An example is the action of the elbow joint in which the forearm is pulled toward the arm (**Figure 8.6a**). As the elbow flexes, the angle between the arm and forearm decreases. Flexion also occurs at the shoulder and the hip (**Figure 8.6b** and **c**). Flexion of these joints moves the arm and thigh anteriorly, in front of the body.

Lateral flexion is sideways movement, such as when you tilt your head to one side or the other toward the shoulder (**Figure 8.6d**). Bending the trunk sideways at the hips in either direction is also lateral flexion (**Figure 8.6e**). In each of these examples, the angle decreases between the head and neck or the trunk and hips.

Extension, the opposite of flexion, increases the angle between articulating bones. For instance, in Figure 8.6a, the

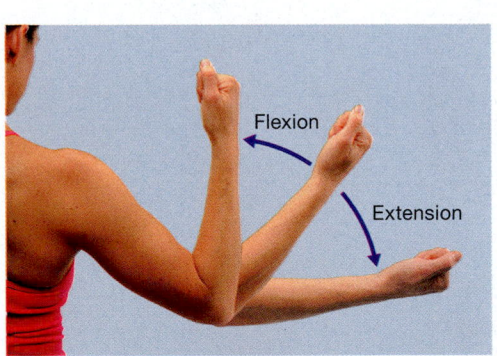

Flexion

Extension

(a) Elbow flexion and extension

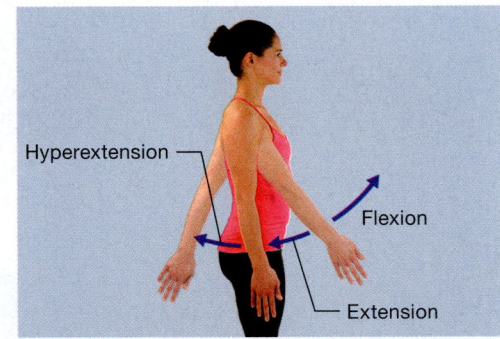

Hyperextension

Flexion

Extension

(b) Shoulder flexion, extension, and hyperextension

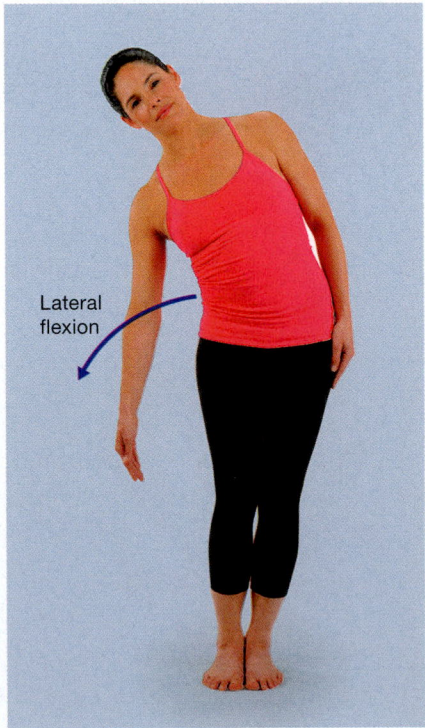

Lateral flexion

Hyperextension

Flexion

Extension

(c) Hip flexion, extension, and hyperextension

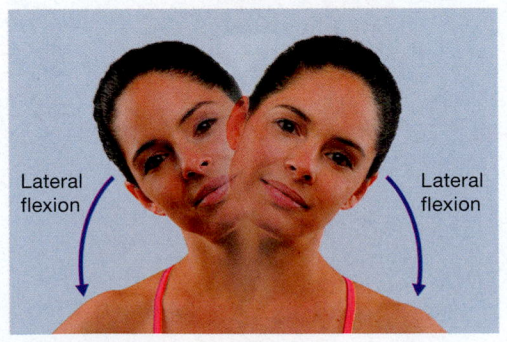

Lateral flexion

Lateral flexion

(d) Lateral flexion of head and neck

(e) Lateral flexion of trunk at the hips

Figure 8.6 Angular movements: flexion and extension of synovial joints.

elbow is flexed with the forearm close to the arm. If we extend the elbow, the arm and forearm move away from each other, increasing the angle. Extension of the shoulder and hip is motion in a posterior direction, moving the hand or foot behind the rest of the body (see Figure 8.6b and c).

A motion related to extension is **hyperextension,** or extension beyond the anatomical position of the joint (see Figure 8.6b and c). For instance, you hyperextend your shoulder when you prepare to pitch a softball as your arm moves behind the trunk of your body beyond anatomical position of the arm at the shoulder joint. You hyperextend your hip as it moves behind your trunk and beyond anatomical position while you are running.

Abduction and Adduction Another pair of angular movements is abduction and adduction. **Abduction** is the motion of a body part away from the midline of the body or another reference point (such as the midline of the hand or foot). The first movement of a jumping jack exercise—spreading the legs at the hips and raising the arms at the shoulders—is an example of abduction (**Figure 8.7a**). Note that the arms and legs are moving away from the midline of the body.

Another example of abduction is fanning out the fingers and toes (**Figure 8.7b**). Note that the reference point has changed from the midline of the body to the midline of the hand or foot, respectively. Again, the fingers and toes are moving away from the midline of the reference point.

Adduction, the opposite of abduction, is the motion of a body part toward the midline of the body or some other reference point. Completing the second movement of the jumping jack adducts the arms and legs by bringing them back toward the body (**Figure 8.7c**). Similarly, if we bring our fingers and toes back toward the reference point, we are adducting these joints (**Figure 8.7d**).

Be careful to remember anatomical position, especially with abduction and adduction of the wrist. Recall that in anatomical position the palm of the hand faces anteriorly. If you move the wrist away from the trunk of the body or the midline of the hand, by definition this is abduction. The opposite motion, movement of the wrist toward the midline of the body, is adduction.

Circumduction In **circumduction,** a freely moveable distal bone moves around a stationary proximal bone in a cone-shaped motion. This is a complex movement that is the sum total of flexion-extension and abduction-adduction, best seen at the hip and shoulder joints. If you draw a circle on the wall by moving your shoulder only, that is circumduction (**Figure 8.7e**).

Rotation

Rotation is a nonangular, pivoting motion, in which one bone rotates or twists on an imaginary line running down its middle, known as its *longitudinal axis*. Shaking the head to indicate

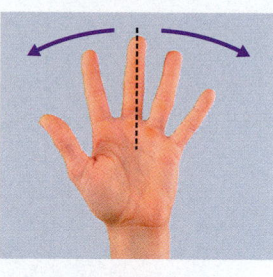
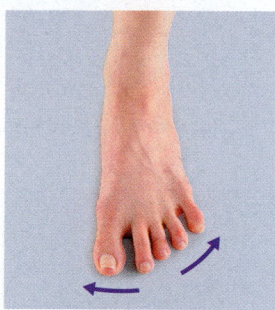

(a) Abduction of shoulders and hips

(b) Abduction of fingers and toes

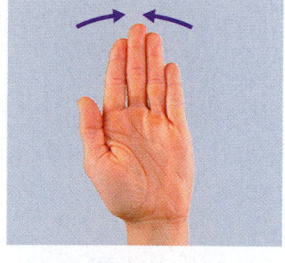
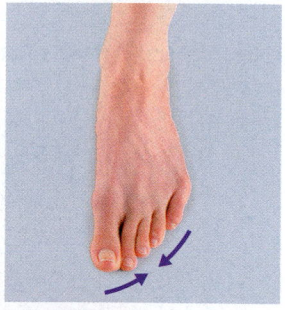

(c) Adduction of shoulders and hips

(d) Adduction of fingers and toes

(e) Circumduction of shoulder

Figure 8.7 Angular movements: abduction, adduction, and circumduction of synovial joints.

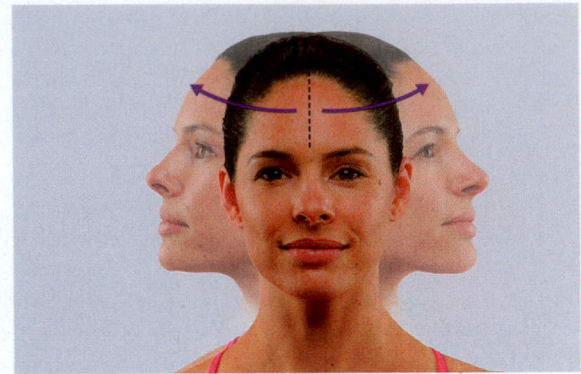

(a) Rotation of atlantoaxial joint in neck

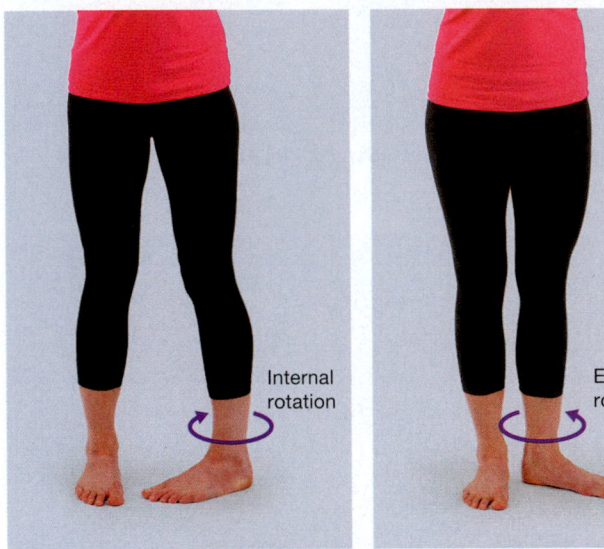

Internal
rotation

External
rotation

(b) Internal and external rotation of hip

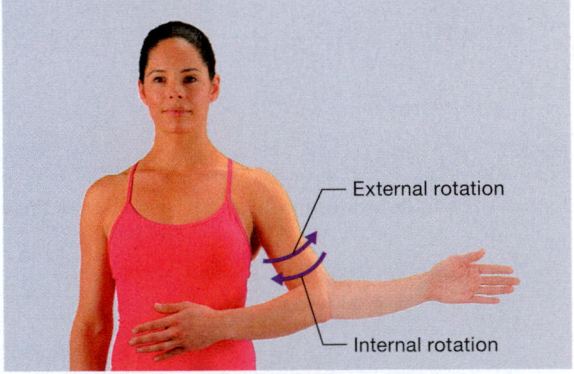

External rotation

Internal rotation

(c) Internal and external rotation of shoulder

Figure 8.8 Rotational movements: internal and external rotation of synovial joints.

"no" is a rotational movement of the atlantoaxial joint in the neck (**Figure 8.8a**).

Multiaxial joints such as the hip and shoulder are also able to rotate. *Internal (medial) rotation* of the hip is best appreciated by using the foot as a reference point. When the hip is internally rotated, the toes point medially toward the midline of the body

(**Figure 8.8b**). In the opposite movement, *external (lateral) rotation,* the toes point away from the midline of the body. In both motions, the femur rotates along its longitudinal axis, while the head of the femur articulates with the acetabulum at the hip joint.

Internal and external rotation of the shoulder is best appreciated with the elbow in a flexed position as if you are holding a coffee mug. If you swing the mug away from your body, the shoulder rotates externally (**Figure 8.8c**). Swinging the mug inward as you raise it to your mouth would be internal rotation. Similar to the femur, the humerus rotates along its longitudinal axis while the head of the humerus articulates with the scapula in the shoulder joint.

Be careful not to confuse rotation with circumduction. Although both involve circular movements, they are entirely separate classes of motion. In rotation, the bone turns around itself. In circumduction, the bone can move in a circle around something else:

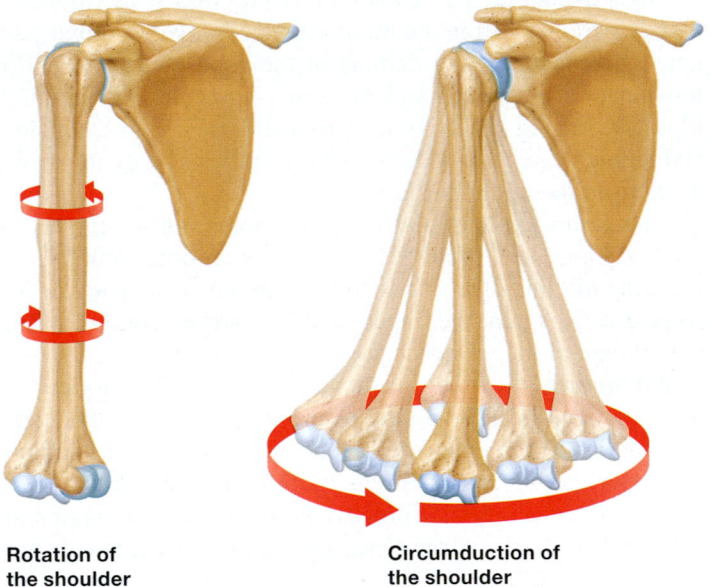

**Rotation of
the shoulder**

**Circumduction of
the shoulder**

Special Movements

Special movements include those that are not well described by the previous categories or that pertain to only one or a few synovial joints. The following subsections explain the different kinds of special movements.

Opposition and Reposition   **Opposition** occurs only at the thumb or first carpometacarpal joint, and involves movement of the thumb across the palmar surface of the hand (which is why we say humans have *opposable thumbs*; **Figure 8.9a**). You can perform this motion by using your thumb to touch the tips of your other four fingers. Opposition is important for grasping objects. **Reposition** is the return of the thumb to its anatomical position (**Figure 8.9b**).

Depression and Elevation   **Depression** is the movement of a body part in an inferior direction. For instance, opening the mouth depresses the mandible, which moves inferiorly relative to the maxilla (**Figure 8.9c**). **Elevation** is the opposite of depression in which, for example, the mandible is pulled up toward the maxilla in a superior direction (**Figure 8.9d**). These two movements are combined in mastication (chewing).

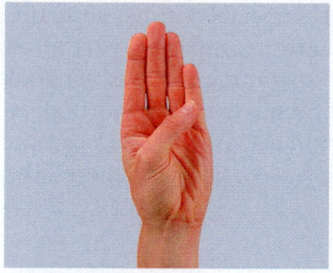

(a) Opposition of thumb

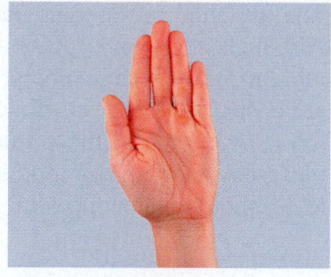

(b) Reposition of thumb

(c) Depression of mandible

(d) Elevation of mandible

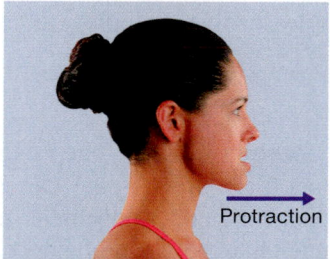

(e) Protraction of mandible

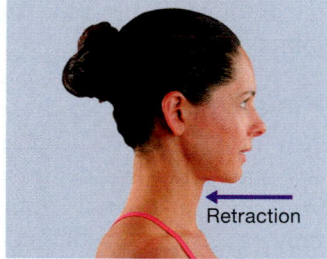

(f) Retraction of mandible

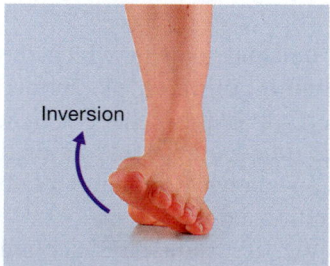

(g) Inversion of foot

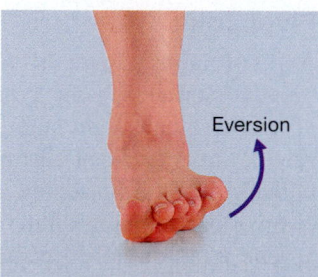

(h) Eversion of foot

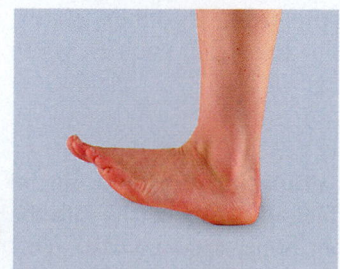

(i) Dorsiflexion of foot

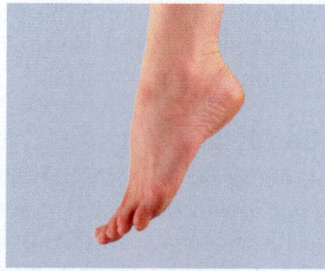

(j) Plantarflexion of foot

Figure 8.9 Special movements of synovial joints.

Protraction and Retraction **Protraction** moves a body part in the anterior direction. Moving the mandible forward so the inferior teeth stick out is an example of protraction (**Figure 8.9e**). The mandible moves anteriorly to the maxilla during this motion.

Retraction is the opposite motion, in which the body part moves posteriorly. The mandible, once protracted, must be retracted to return the mandible to its normal position (**Figure 8.9f**). Protraction and retraction are also important in mastication (chewing).

Inversion and Eversion **Inversion** is a rotational motion of the foot in which the plantar surface (the sole of the foot) rotates medially toward the midline of the body (**Figure 8.9g**). **Eversion** (ee-VER-zhuhn) is the opposite motion, in which the plantar surface of the foot rotates laterally away from the midline of the body (**Figure 8.9h**). This combination of motions is important for walking.

Dorsiflexion and Plantarflexion Dorsiflexion and plantarflexion are a combination of movements that involve the foot and ankle. In **dorsiflexion** (dohr-see-FLEK-shuhn), the angle between the foot and the tibia decreases (**Figure 8.9i**). In other words, the toes are pulled up toward the head.

In **plantarflexion** (plant-uhr-FLEK-shuhn), the opposite motion, the angle between the foot and the tibia increases—the toes point toward the ground (**Figure 8.9j**). As with inversion and eversion, this pair of opposing movements is instrumental in walking.

Supination and Pronation **Supination** (soo-pih-NAY-shuhn) and **pronation** (proh-NAY-shuhn) are rotational movements that occur at the proximal radioulnar joint of the forearm and at a combination of the ankle and tarsal joints. Your forearm is supinated when your palm faces anteriorly with the thumb pointing laterally, as in anatomical position (**Figure 8.10a**). Starting with the hand in a supinated position, you pronate the forearm by turning the palmar surface medially until it faces posteriorly with the thumb pointing toward the body (**Figure 8.10b**).

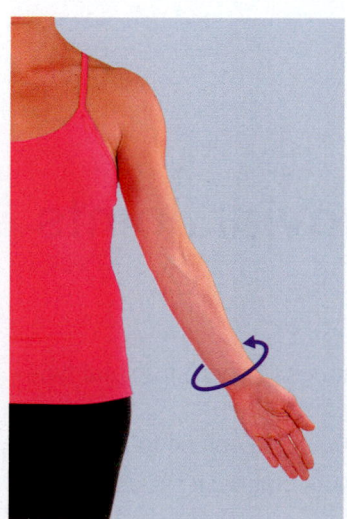

(a) Supination of forearm

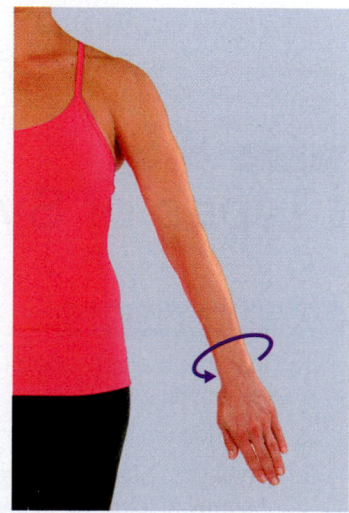

(b) Pronation of forearm

Figure 8.10 Special movements: supination and pronation.

StudyBOOST ⬆

Keeping Synovial Joint Movements Straight

Remember the difference between supination and pronation with this tip: You hold a cup of *soup* when your hand is *supinated*, and you *pour* it out when your hand *pronates*.

If you find abduction and adduction confusing, remember this tip: With *abduc*tion, you *abduct* (take away) the part from the body. With *add*uction, you *add* the part back to the body. ■

Quick Check

☐ 1. Compare and contrast: gliding and angular motions, circumduction and rotation, abduction and adduction.

Range of Motion

The **range of motion** of a joint is the amount of movement it is capable of under normal circumstances. For example, when you move your knee joint from a relaxed state to full flexion, and then return the joint to its fully extended state, that is the range of motion of the knee. Nonaxial joints (such as the intercarpal joints) tend to have the smallest range of motion, whereas multiaxial joints (such as the shoulder) tend to have the greatest.

Apply What You Learned

☐ 1. In the condition known as *drop foot,* the foot and ankle are unable to dorsiflex. Predict the consequences of this condition for walking.

☐ 2. Marfan syndrome can affect connective tissues around joints and make ligaments loose. Predict how this might affect the range of motion of a biaxial joint.

See answers in Appendix A.

MODULE 8.5
Types of Synovial Joints

Learning Outcomes

1. Describe the anatomical features of each structural type of synovial joint.

2. Describe where each structural type can be found.

3. Predict the kinds of movements that each structural type of synovial joint will allow.

4. Compare and contrast the structural features of the knee and elbow and of the shoulder and hip.

The function of a synovial joint depends on its structure. Although all synovial joints are diarthroses, some joints allow more movement than others. In the upcoming sections, we look at the structural classes of synovial joints, from those that allow the least motion to those that allow the most. We then explore the gross anatomy of four specific synovial joints: the elbow, knee, shoulder, and hip joints.

Structural Classes of Synovial Joints

The amount of motion a joint allows is largely determined by the shape of the surfaces of its two articulating bones and the number of axes around which the bones move. On the basis of these criteria, we can identify six structural classes: plane, hinge, pivot, condylar, saddle, and ball-and-socket joints.

Plane Joint

A **plane joint,** the simplest and least mobile synovial joint, features two bones whose flat surfaces sit next to each other (**Figure 8.11a**). Plane joints are nonaxial, as they allow only side to side "gliding" movements (hence their alternate name, *gliding joints*). The intercarpal joints of the wrist and intertarsal joints of the ankle are examples of these simple, stable joints.

Hinge Joint

At a **hinge joint** (**Figure 8.11b**), a convex surface of one bone fits into a concave depression of another bone. As the name implies, the motion of a hinge joint is restricted to movement in only one plane (uniaxial motion), much like the hinge of a door. Examples include the elbow, knee, and interphalangeal joints of the fingers and toes.

Pivot Joint

A **pivot joint** (**Figure 8.11c**) is a uniaxial joint in which the rounded surface of one bone fits into a groove on the surface of another bone. Joint stability at a pivot joint is enhanced by a ringlike ligament that surrounds the rounded bone and holds it in the groove of the second bone. This allows the bone to rotate or pivot on its longitudinal axis within this ring.

The atlantoaxial joint between the first (atlas) and second (axis) cervical vertebrae is an example of a pivot joint. The rounded dens process of the axis articulates with the atlas, and is surrounded and reinforced by the transverse ligament. When muscles move the bones at this joint, it rotates, resulting in the side to side movement of the head, such as when motioning "no."

Condylar Joint

A **condylar** (KAHN-duh-luhr), or **ellipsoid, joint** is a biaxial joint in which the oval, convex surface of one bone fits into the shallow concave surface of the other bone (**Figure 8.11d**). This alignment allows both flexion and extension, and abduction and adduction. Examples of condylar joints include the metacarpophalangeal joints (the "knuckles") of the fingers. Notice that

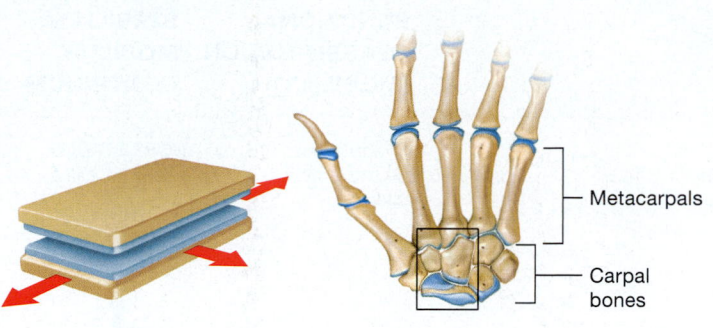

(a) Plane joint, nonaxial: intercarpal joint

Metacarpals

Carpal bones

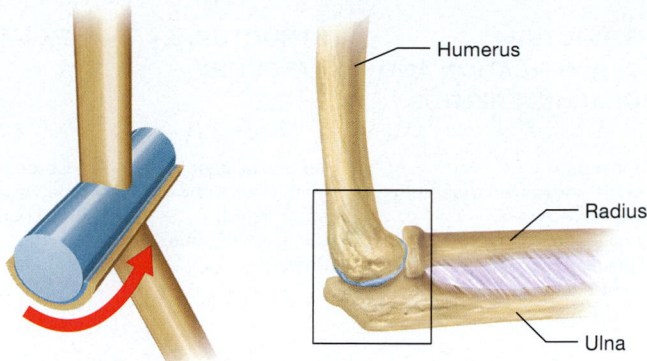

Humerus

Radius

Ulna

(b) Hinge joint, uniaxial: elbow joint

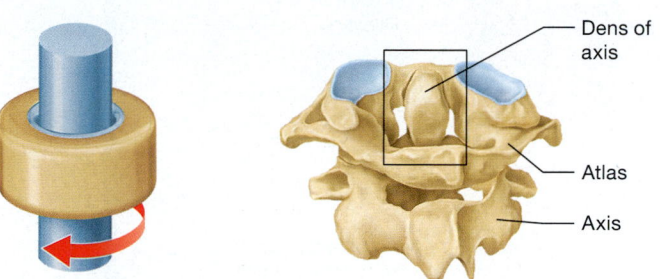

Dens of axis

Atlas

Axis

(c) Pivot joint, uniaxial: atlantoaxial joint

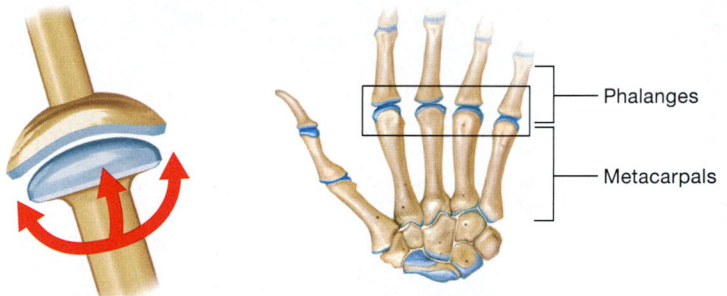

Phalanges

Metacarpals

(d) Condylar joint, biaxial: metacarpophalangeal joint

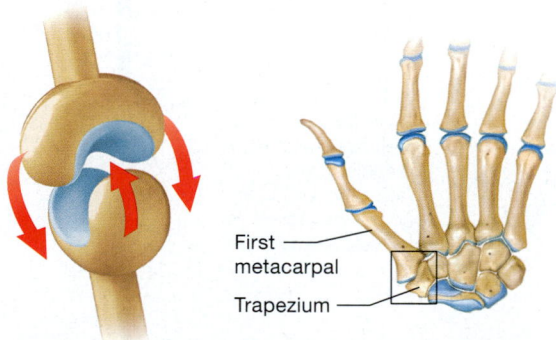

First metacarpal

Trapezium

(e) Saddle joint, biaxial: carpometacarpal joint of thumb

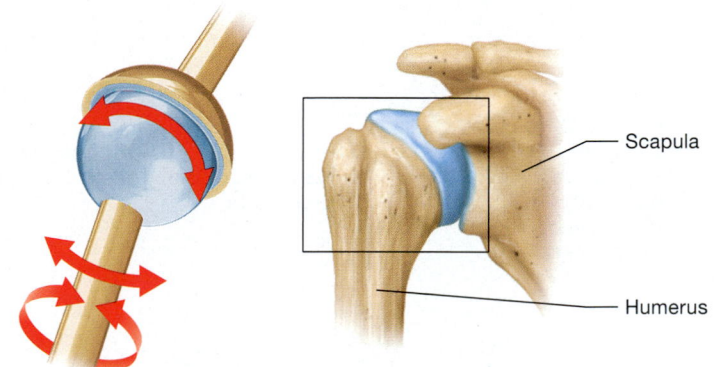

Scapula

Humerus

(f) Ball-and-socket joint, multiaxial: shoulder joint

Figure 8.11 **The six types of synovial joints and the motion allowed at each.**

you can move these joints to bunch your fingers into a fist or spread them out like a fan. Other examples are found in certain parts of the wrist joint.

Saddle Joint

The articulating surfaces of a **saddle joint** resemble a horseback riding saddle (**Figure 8.11e**). The surface of each articulating bone has both convex and concave regions that complement each other. This allows for a good fit, much like putting one saddle on the ground and then stacking a second one on top, while also permitting a large amount of motion.

A saddle joint is biaxial but allows greater motion than condylar joints. The carpometacarpal joint between the first metacarpal of the thumb and the trapezium is an example. Move your fingers and thumb and note how much more motion is available to the thumb compared to the fingers. This arrangement is crucial in allowing the fingers and thumb to work together for opposition and reposition.

Ball-and-Socket Joint

A **ball-and-socket joint** is a multiaxial joint in which the articulating surface of one bone is ball-shaped or spherical and fits into a cup or socket formed by the articulating surface of the other bone (**Figure 8.11f**). Both the shoulder and hip joints are ball-and-socket joints and will be discussed later in this module.

The Big Picture of Joint Classifications and Stability versus Mobility

Figure 8.12

STRUCTURAL CLASSIFICATION AND CHARACTERISTICS	STRUCTURAL CATEGORY	EXAMPLE		FUNCTIONAL CLASSIFICATION (MOBILITY)	STABILITY/ MOBILITY CONTINUUM
Fibrous Bone edges held together by dense regular collagenous connective tissue; no joint cavity; no articular capsule (see Figure 8.1)	Suture: skull bones held together by short fibrous connective tissue fibers	Coronal suture between the frontal and parietal bones 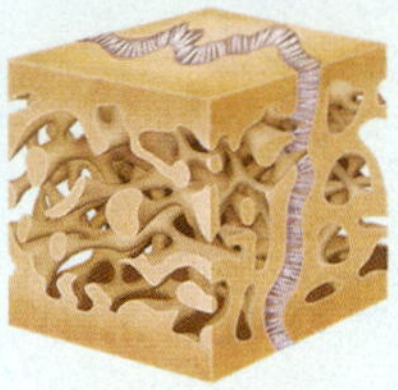		Synarthrosis: no movement allowed	*MOST STABLE, LEAST MOBILE*
	Gomphosis: tooth within bony cavity held by periodontal ligament	Tooth in the mandible or maxilla 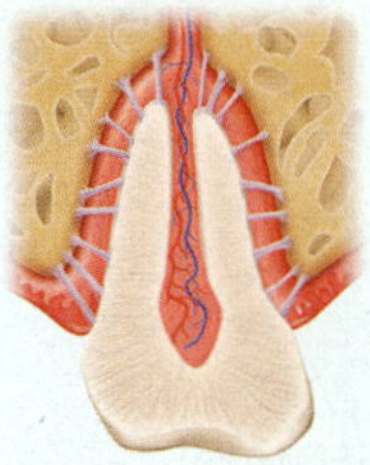		Synarthrosis: no movement allowed	
	Syndesmosis: dense regular collagenous connective tissue between two long bones	Interosseous membrane between radius and ulna and between tibia and fibula 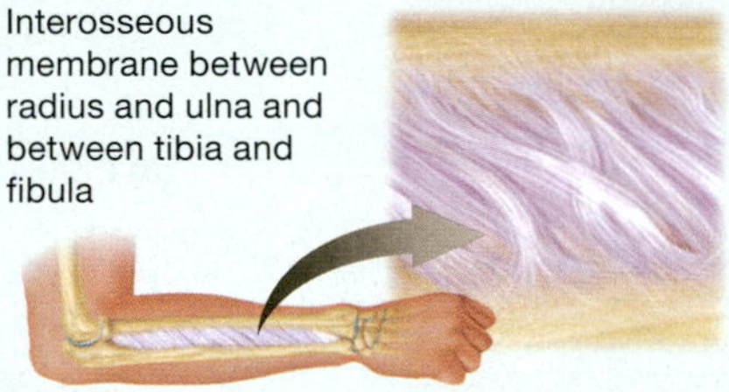		Amphiarthrosis: some movement allowed	
Cartilaginous Wedge of cartilage located between articulating bones; no joint cavity; no articular capsule (see Figure 8.2)	Synchondrosis: hyaline cartilage plate between bones	Epiphyseal plates; sternocostal joint 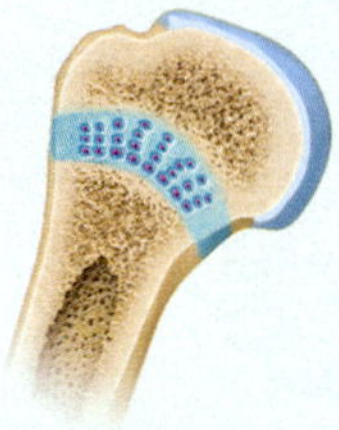		Synarthrosis (epiphyseal plate) or amphiarthrosis (sternocostal joint): no or little movement allowed, respectively	*LESS STABLE, SOMEWHAT MOBILE*
	Symphysis: fibrocartilage pad between the hyaline cartilage on the two articulating bones	Pubic symphysis between two pubic bones 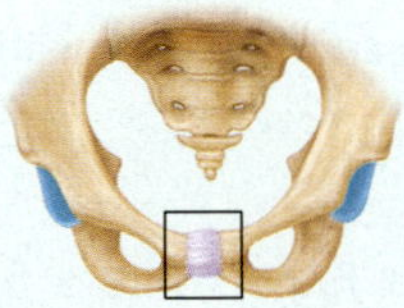		Amphiarthrosis: some movement allowed	

Ball-and-socket joints allow movement in all three axes of motion, which allows the most freedom of movement. Remember, though, that this mobility comes with a stability tradeoff. Ball-and-socket joints are rather unstable, especially the shoulder joint, which is very susceptible to dislocation.

Putting It All Together: The Big Picture of Joint Classifications and Stability versus Mobility

Throughout this chapter, we have discussed two important functions of joints: enabling movement and providing stability. As

STRUCTURAL CLASSIFICATION AND CHARACTERISTICS	STRUCTURAL CATEGORY	EXAMPLE		FUNCTIONAL CLASSIFICATION (MOBILITY)	STABILITY/ MOBILITY CONTINUUM
Synovial Layer of hyaline cartilage covers surface of each articulating bone; separated by a fluid-filled joint cavity; articular capsule present (see Figure 8.3)	Plane: nonaxial; flat bone surfaces glide across one another (see Figures 8.5 and 8.11a)	Intercarpal and intertarsal joints		Diarthrosis: freely moveable	*LESS STABLE, MORE MOBILE*
	Hinge: uniaxial; convex surface of one bone fits into a concave surface of another bone to allow angular motion along axis (see Figure 8.11b)	Elbow and knee joints		Diarthrosis: freely moveable	
	Pivot: uniaxial; projection of one bone fits into another bone; the second bone rotates around the first (see Figure 8.11c)	Atlantoaxial joint		Diarthrosis: freely moveable	
	Condylar: biaxial; convex oval surface of one bone articulates with concave oval surface of another bone (see Figure 8.11d)	Second through fifth metacarpophalangeal joints		Diarthrosis: freely moveable	
	Saddle: biaxial; saddle-shaped surface of one bone fits into a depression of another bone (see Figure 8.11e)	Articulation between the first metacarpal and the trapezium (thumb joint)		Diarthrosis: freely moveable	
	Ball-and-socket: multiaxial; round head of one bone fits into a cup-shaped surface of another bone (see Figure 8.11f)	Shoulder and hip joints		Diarthrosis: freely moveable	*LEAST STABLE, MOST MOBILE*

you have seen, a joint can be mobile or stable, but generally not both at the same time. As joints allow more movement, they become increasingly unstable. The most mobile joints in the body are also the most easily injured. **Figure 8.12** summarizes the classes of joints and where they fall on this continuum of joint stability versus mobility.

Quick Check

☐ 1. Compare the six structural classes of synovial joints.

☐ 2. Which type of synovial joint has the greatest range of motion? Which type has the least?

Specific Hinge Joints: The Elbow and the Knee

« FLASHBACK

1. What are the functions of the trochlea and trochlear notch of the humerus and ulna, respectively? (pp. 243–244)

2. Where are the femoral and tibial condyles located? (pp. 251–252)

Now that we have discussed the general properties of synovial joints, let's focus on specific examples. First, we examine the elbow and the knee, and highlight their key similarities and differences. Both are hinge joints whose primary motions are flexion and extension. We concentrate on these two hinge joints because they are commonly injured due to their structural complexity and relative instability.

The Elbow

The **elbow joint** is actually composed of two separate articulations (**Figure 8.13a**). The larger articulation is the *humeroulnar joint* (hyoo´-muhr-oh-UHL-nuhr) between the trochlea of the humerus and the trochlear notch of the ulna. The smaller articulation is the *humeroradial joint* (hyoo´-muhr-oh-RAY-dee-uhl) between the capitulum of the humerus and the head of the radius.

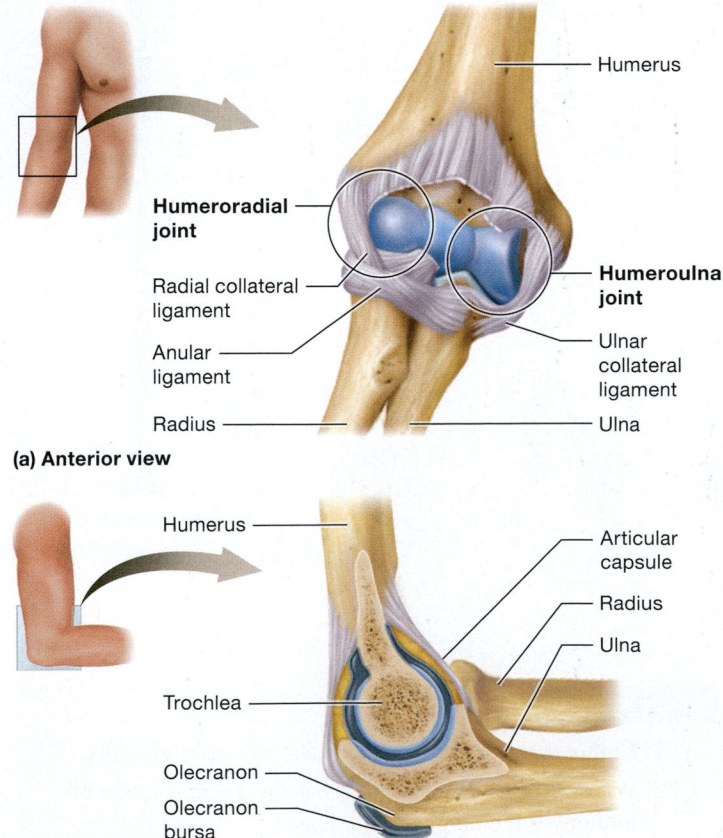

(a) Anterior view

Humerus

Humeroradial joint

Radial collateral ligament

Anular ligament

Radius

Humeroulnar joint

Ulnar collateral ligament

Ulna

Humerus

Articular capsule

Radius

Ulna

Trochlea

Olecranon

Olecranon bursa

(b) Sagittal section

Figure 8.13 Anatomical structure of the elbow joint.

Three strong extrinsic ligaments reinforce and support the articular capsule, which strengthens the elbow joint (**Figure 8.13b**). The **radial collateral ligament** (or lateral collateral ligament) supports the lateral side of the joint, and the **ulnar collateral ligament** (or medial collateral ligament) supports the medial side. An accessory ligament, the **anular ligament** (AN-yoo-luhr), binds the head of the radius to the neck of the ulna to stabilize the radial head in the elbow joint.

The elbow is more stable than the knee: Its durable articular capsule encloses both of the articulations between the humerus, ulna, and radius. Notice how the articulating surfaces of the humeroulnar joint fit together snugly, further stabilizing the joint.

The Knee

The **knee joint** is the largest diarthrosis in the human body. This complex joint is functionally a hinge joint that allows flexion and extension, but when the knee is flexed, it also allows some degree of rotation and lateral gliding. Like the elbow, the knee is made up of two distinct articulations (**Figure 8.14**): the *tibiofemoral joint* (tib´-ee-oh-FEM-ohr-uhl) between the femoral and tibial condyles, and the *patellofemoral joint* (puh-tel´-oh-FEM-ohr-uhl) between the patella and the patellar surface of the femur.

The articular capsule encloses all but the anterior surface of the knee joint, where it is covered by the patella. The patella is surrounded by the tendon of the anterior thigh muscles, collectively known as the *quadriceps femoris muscle group* (KWAHD-rih-seps FEM-oh-ris; see Figure 8.14a and b). Distal to the patella, this structure is considered a ligament because it now connects bone to bone. The patellar ligament continues distally to its destination on the anterior surface of the tibia. This structural arrangement leaves the knee joint without a continuous articular capsule.

The knee joint is inherently unstable due to its structure and the many forces placed on it. However, several anatomical structures enhance its stability. For instance, a pair of C-shaped fibrocartilage pads—the **medial meniscus** and **lateral meniscus** (men-ISS-kuhs; plural, *menisci*)—sit on the tibial condyles. The menisci provide shock absorption and cushioning between the articular cartilages of the femoral and tibial condyles. They also improve the fit between the two bones, which further stabilizes the joint (see Figure 8.14c).

Two extrinsic *collateral ligaments* on either side of the knee joint tighten when the leg is extended to provide additional stabilization during weight bearing. The **tibial collateral ligament** (or medial collateral ligament) connects the femur with the tibia and attaches to the medial meniscus. This ligament prevents the tibia from moving too far laterally on the femur (hyperabduction of the knee). The **fibular collateral ligament** (or lateral collateral ligament) links the femur to the fibula and prevents the tibia from moving too far medially on the femur (hyperadduction). Notice in Figures 8.14a and b that the fibular collateral ligament does not attach to the lateral meniscus.

We find two other stability-promoting extrinsic ligaments within the joint cavity of the knee: the two *cruciate ligaments* (KROO-shee-iht; *cruci* = "X-shaped"). These ligaments, named

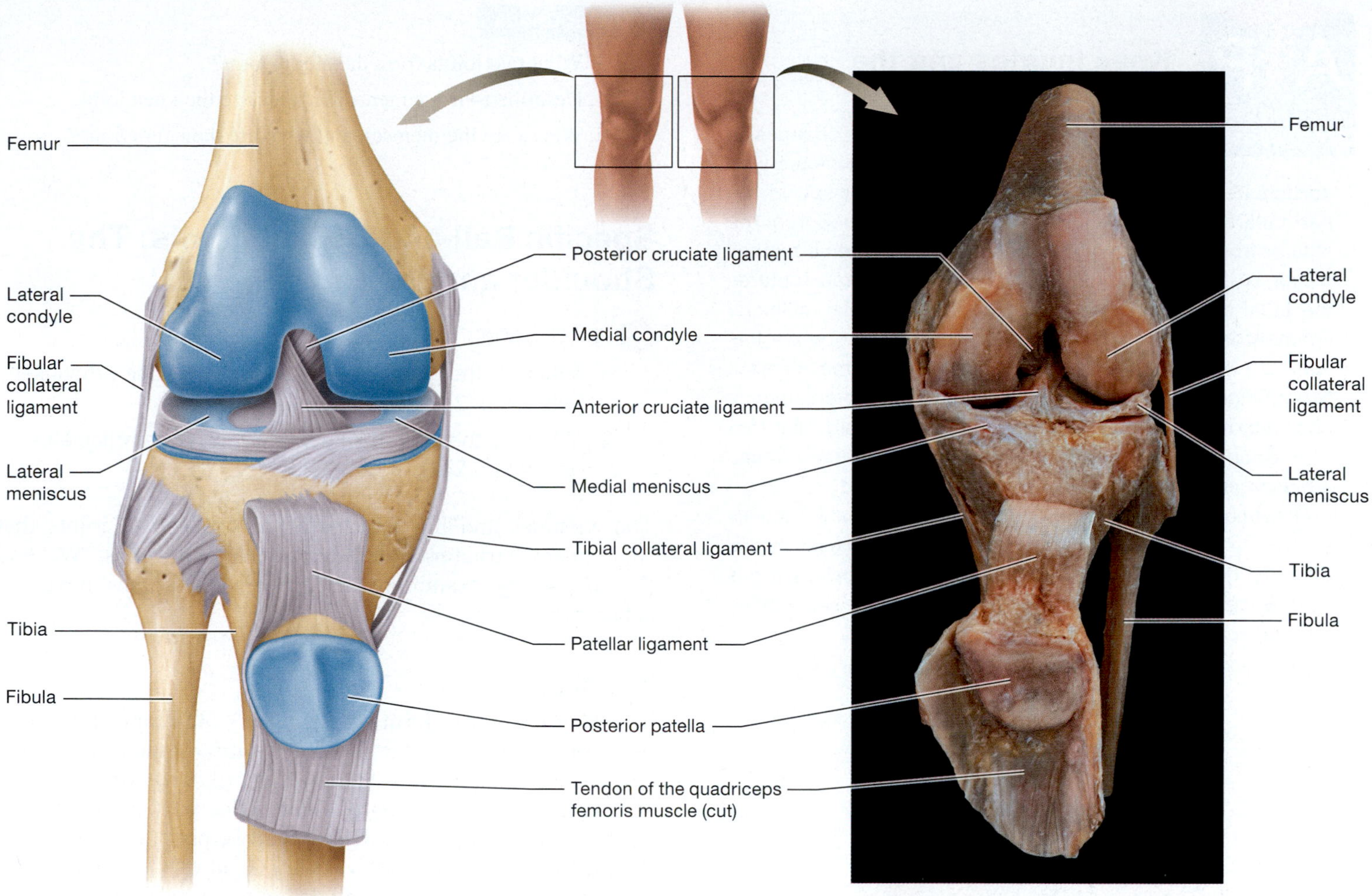

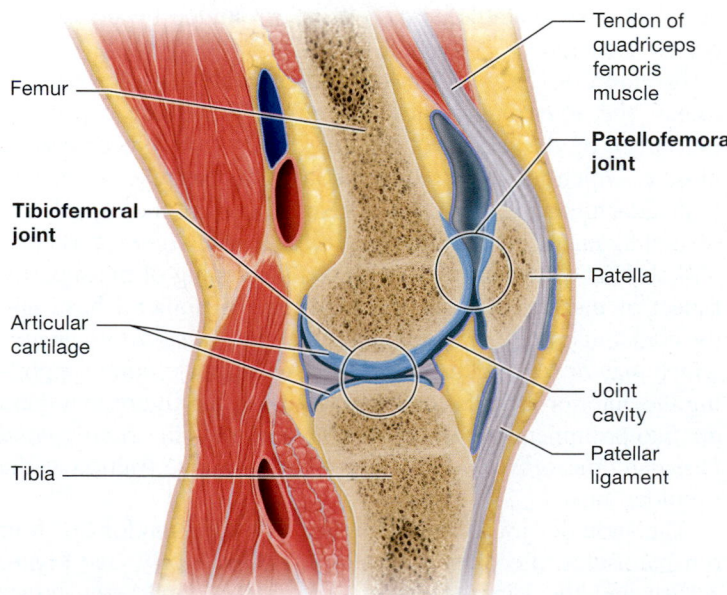

Femur

Lateral condyle

Fibular collateral ligament

Lateral meniscus

Tibia

Fibula

Posterior cruciate ligament

Medial condyle

Anterior cruciate ligament

Medial meniscus

Tibial collateral ligament

Patellar ligament

Posterior patella

Tendon of the quadriceps femoris muscle (cut)

Femur

Lateral condyle

Fibular collateral ligament

Lateral meniscus

Tibia

Fibula

(a) Anterior view, right knee

(b) Cadaver photo, anterior view, left knee

Femur

Tibiofemoral joint

Articular cartilage

Tibia

Tendon of quadriceps femoris muscle

Patellofemoral joint

Patella

Joint cavity

Patellar ligament

(c) Sagittal section of knee

Figure 8.14 Anatomical structure of the knee joint.

for their tibial attachments, limit motion of the knee joint in the anterior and posterior directions. The **anterior cruciate ligament** (ACL) runs from an anterior insertion site on the tibia to the posterior aspect of the femur. When the knee joint is extended, the ACL tightens to prevent hyperextension and to prevent the tibia from moving too far anteriorly on the femur. The **posterior cruciate ligament** (PCL) travels from a posterior position on the tibia to the anterior femur. When the PCL tightens during knee flexion, it prevents the tibia from displacing itself posteriorly (i.e., sliding backward) from the femur.

An important feature of this synovial joint is its ability to "lock out." When an individual is weight-bearing, the knee joints can be locked in extension for long periods of time without causing undue muscle fatigue. When fully extended, the tibia rotates laterally, which tightens the ACL and squeezes the menisci between the articular surfaces of the tibial and femoral condyles. All of this locks the knees while the individual is in a standing position.

The instability of the knee joint makes it one of the most commonly injured joints in the body. See *A&P in the Real World: Knee Injuries and the Unhappy Triad* on p. 276 to read about one type of knee injury.

Knee Injuries and the Unhappy Triad

Any activity that involves quick changes in direction can injure the knee. Athletes who participate in contact sports such as football or soccer are particularly at risk, especially if the knee is struck from the side or from behind. A lateral blow—such as when a football player is illegally blocked below the knees—often ruptures the tibial collateral ligament. Recall that the tibial collateral ligament is attached to the medial meniscus, so when this ligament ruptures, the force is transmitted to the menisci. Surprisingly, this often results in tearing of the *lateral* meniscus, although the medial meniscus occasionally tears as well. Such a force generally also ruptures the anterior cruciate ligament, forming the so-called unhappy triad.

Treatment of unhappy triad injuries almost invariably requires surgery to repair the damaged anterior cruciate ligament and often the damaged meniscus. After surgery, physical therapy is used to improve range of motion and strengthen the surrounding muscles to reduce the risk of re-injury.

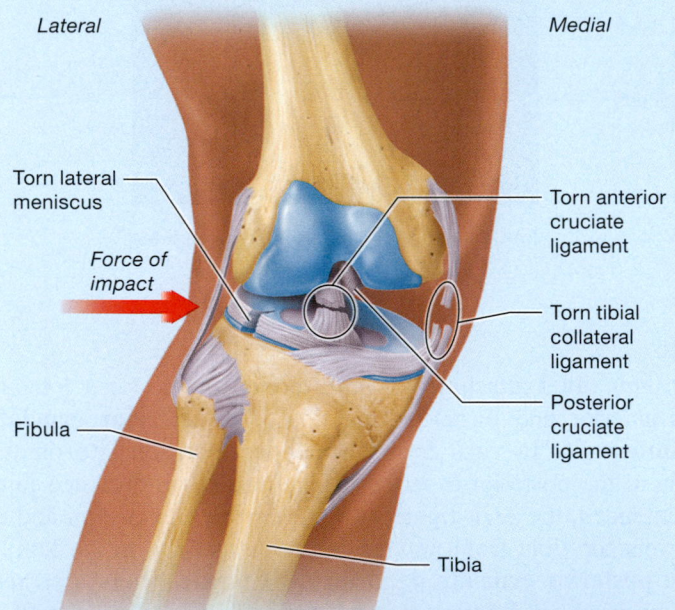

Lateral *Medial*

Torn lateral meniscus
Force of impact
Fibula
Torn anterior cruciate ligament
Torn tibial collateral ligament
Posterior cruciate ligament
Tibia

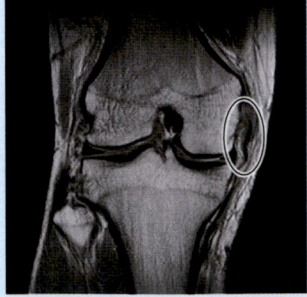

Anterior MRI of torn tibial collateral ligament

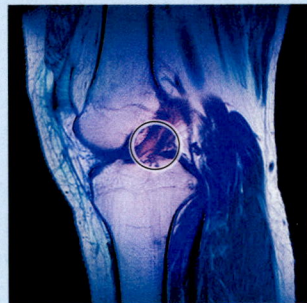

Sagittal MRI of torn anterior cruciate ligament

Quick Check

☐ 3. What two joints form the elbow joint?

☐ 4. Describe four ligaments that stabilize the knee joint.

☐ 5. Where are the menisci, and what are their functions?

Specific Ball-and-Socket Joints: The Shoulder and the Hip

◀◀ FLASHBACK

1. Where is the glenoid cavity, and which bone articulates with it? (p. 242)

2. Where is the acetabulum, and which bone articulates with it? (p. 247)

The shoulder and hip joints are ball-and-socket joints that share many structural and functional similarities. We end the chapter by examining these two joints in the next two subsections.

The Shoulder

The **glenohumeral joint** (glen'-oh-HYOO-muhr-uhl), more commonly known as the **shoulder joint,** is made up of the ball-shaped humeral head and the *glenoid cavity* on the lateral scapula (**Figure 8.15**). The shoulder is the most freely moving joint in the human body. This is due in part to the fact that the glenoid cavity is shallow, so the head of the humerus fits in it somewhat like a golf ball fits in a tee, allowing for freer motion. However, this wide range of motion makes the shoulder joint very unstable, and so it is surrounded by many supportive structures.

In Figure 8.15a, you can see the articular capsule that surrounds the anterior part of the shoulder joint. The capsule is reinforced by the tendon of the long head of the biceps brachii muscle, which passes through the articular capsule on its way to its attachment to the scapula. Also supporting the capsule are several ligaments, including the *coracohumeral ligament* (kohr'-uh-koh-HYOO-muhr-uhl), which is a thickening of the superior aspect of the articular capsule between the humeral head and the coracoid process. In addition, three *glenohumeral ligaments,* which may or may not be present, are sometimes found supporting the anterior articular capsule. Notice in the figure that there are two prominent bursae around the capsule—the *subacromial bursa* and the *subscapular bursa*—that minimize friction as the shoulder moves.

The shoulder joint is also reinforced by the tendons of four other muscles: the subscapularis muscle anteriorly (see Figure 8.15b) and the supraspinatus, infraspinatus, and teres minor muscles posteriorly (see Figure 8.15c). Collectively, the tendons of these four muscles are called the **rotator cuff,** which surrounds the shoulder joint and provides much of its strength and stability. Rotator cuff tears are common injuries, as we discuss in the muscular system chapter (see Chapter 9). Deep to

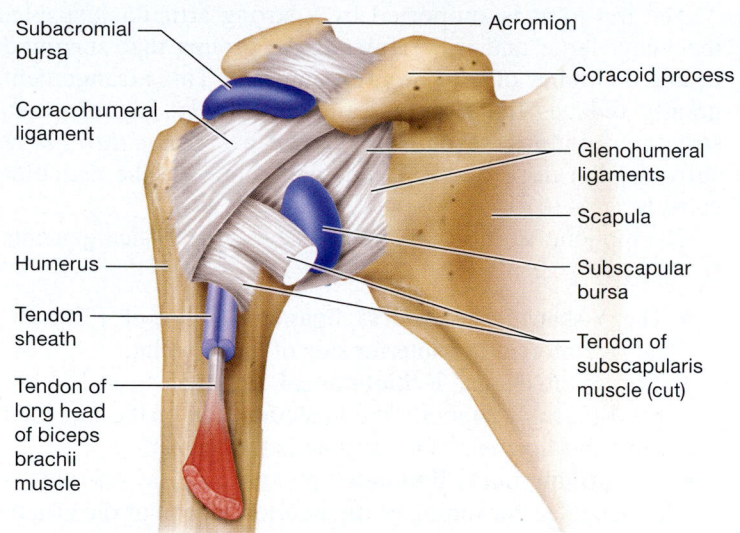

Subacromial bursa
Coracohumeral ligament
Humerus
Tendon sheath
Tendon of long head of biceps brachii muscle
Acromion
Coracoid process
Glenohumeral ligaments
Scapula
Subscapular bursa
Tendon of subscapularis muscle (cut)

(a) Anterior view of the articular capsule

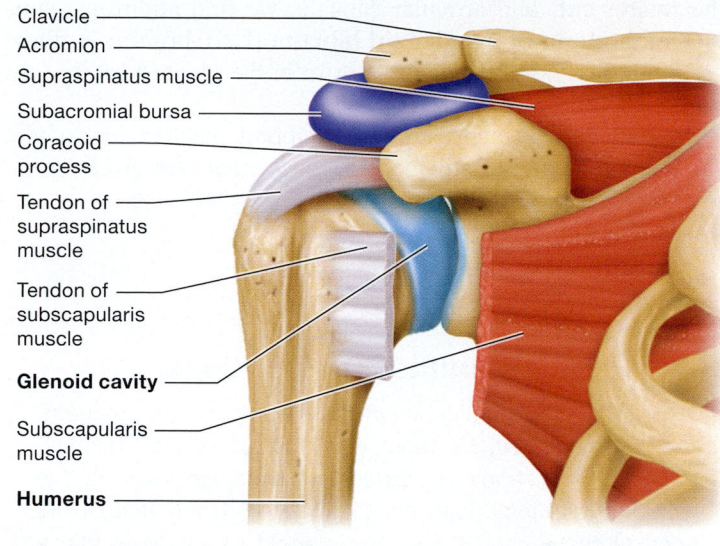

Clavicle
Acromion
Supraspinatus muscle
Subacromial bursa
Coracoid process
Tendon of supraspinatus muscle
Tendon of subscapularis muscle
Glenoid cavity
Subscapularis muscle
Humerus

(b) Anterior view

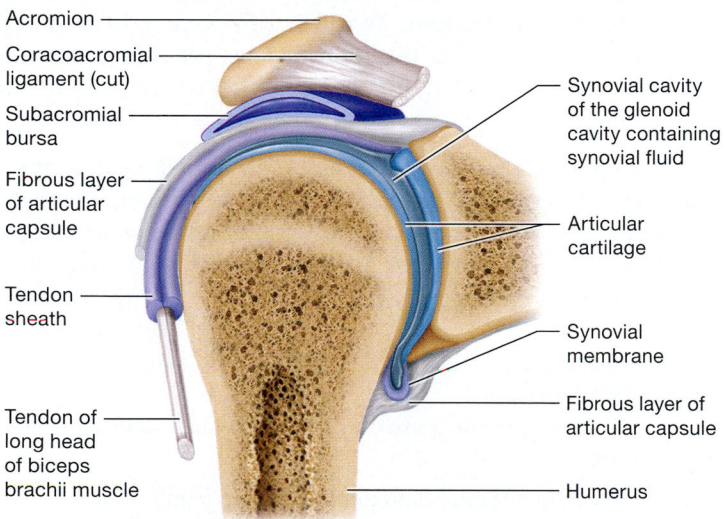

Acromion
Coracoacromial ligament (cut)
Subacromial bursa
Fibrous layer of articular capsule
Tendon sheath
Tendon of long head of biceps brachii muscle
Synovial cavity of the glenoid cavity containing synovial fluid
Articular cartilage
Synovial membrane
Fibrous layer of articular capsule
Humerus

(c) Frontal section

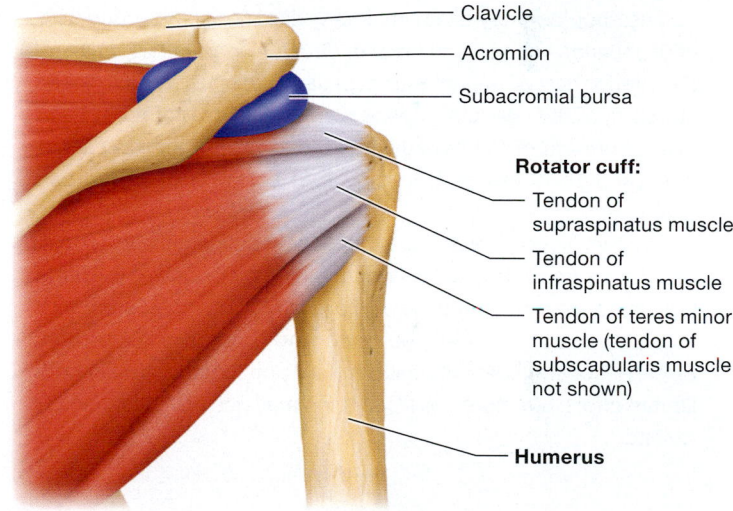

Clavicle
Acromion
Subacromial bursa
Rotator cuff:
Tendon of supraspinatus muscle
Tendon of infraspinatus muscle
Tendon of teres minor muscle (tendon of subscapularis muscle not shown)
Humerus

(d) Posterior view

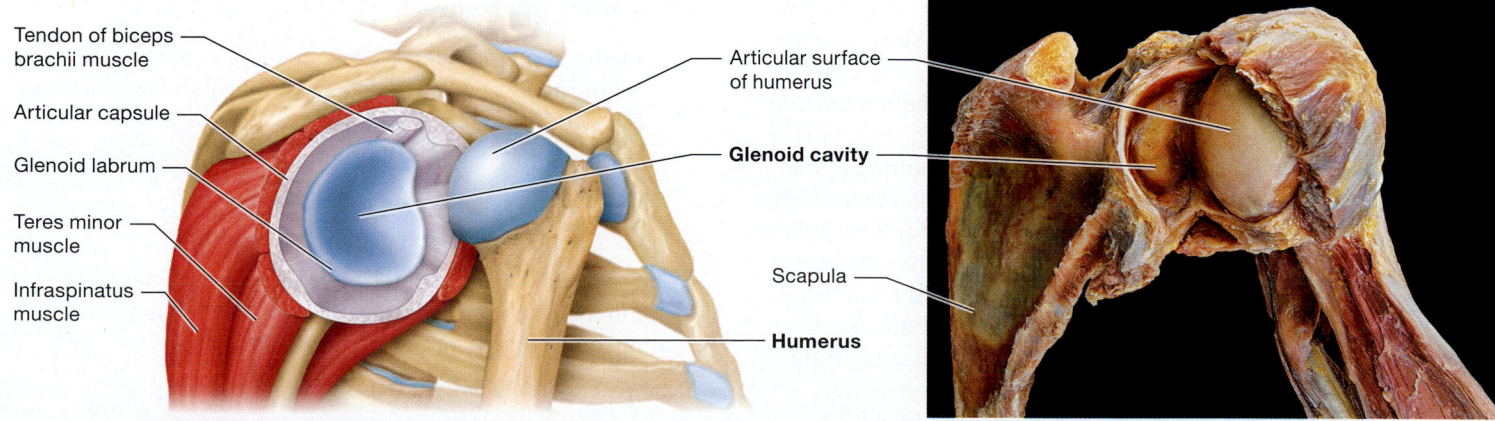

Tendon of biceps brachii muscle
Articular capsule
Glenoid labrum
Teres minor muscle
Infraspinatus muscle
Articular surface of humerus
Glenoid cavity
Scapula
Humerus

(e) Lateral view with head of humerus removed from glenoid cavity

(f) Cadaver photo, lateral view with head of humerus removed from glenoid cavity

Figure 8.15 Anatomical structure of the shoulder joint.

the rotator cuff and articular capsule, we find additional support in the form of the **glenoid labrum** (LAY-bruhm), a fibrocartilaginous ring that sits on the rim of the glenoid cavity (see Figures 8.15d and e).

In spite of the many structural support mechanisms of the shoulder joint, it can still become *dislocated* (see *A&P in the Real World: Shoulder Dislocations*).

Shoulder Dislocations

Shoulder injuries are very common, accounting for more than half of all joint dislocations. A *dislocated shoulder* involves the glenohumeral joint, with the traumatic displacement of the head of the humerus from the glenoid cavity. Note that a *separated shoulder* actually involves the acromioclavicular joint, which is not a component of the shoulder.

The shoulder may dislocate in any direction, but around 90% of dislocations occur in an anterior direction, particularly through the inferior part of the anterior capsule. This is largely due to the structure of the capsule—notice in Figure 8.15a that the antero-inferior portion of the capsule has the fewest supporting structures and so is weakest, leading it to be the most likely part to tear.

Contact sport athletes are especially susceptible to shoulder dislocations, but anyone can suffer them under the right circumstances. Common causes of dislocation injuries are falls in which one lands with the hand and forearm outstretched. Some minor shoulder dislocations can "pop" back into place with limited effort, but more severe injuries may need to be surgically repaired.

The Hip

The **hip joint,** also known as the **coxal joint** (KAHK-suhl), is the articulation between the acetabulum and the ball-shaped head of the femur. Recall from the skeletal system chapter that the femur fits snugly into the acetabulum (see Chapter 7). A fibrocartilaginous ring called the *acetabular labrum* further strengthens the fit between the bones (**Figure 8.16a**).

Although like the shoulder joint, the hip is a multiaxial ball-and-socket joint with a wide range of motion, it is much more

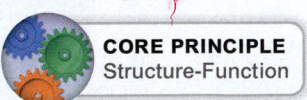

CORE PRINCIPLE Structure-Function

stable than the shoulder joint. This is due to its role in weight bearing and locomotion, an example of the Structure-Function Core Principle (p. 28). Its increased stability comes with some reduction in mobility, as you might expect.

The hip joint is supported by a strong articular capsule, the many large and powerful muscle groups that surround it, and a number of reinforcing ligaments. This arrangement greatly reduces the risk of joint dislocation. Ligamentous structures within the capsule known as *retinacular fibers* surround the neck of the femur and reinforce the articular capsule.

The hip joint is also supported by several extrinsic ligaments (**Figure 8.16b–d**):

- The Y-shaped **iliofemoral ligament** (il′-ee-oh-FEM-oh-ruhl) reinforces the anterior side of the hip joint.
- The spiral-shaped **ischiofemoral ligament** (iss′-kee-oh-FEM-oh-ruhl) supports the posterior side of the hip joint (note this ligament is not visible in the figure).
- The **pubofemoral ligament** (pyoo′-boh-FEM-oh-ruhl) is a triangular thickening of the inferior portion of the articular capsule.
- The **ligament of the head of the femur** is a small ligament that links the center of the head of the femur with the acetabulum. While it offers very little structural support to the hip joint, it does provide a pathway for a small artery that supplies blood to the head of the femur.

The hip joint's role in locomotion puts a great deal of stress on it, and it frequently suffers damage. One possible treatment for a damaged hip is replacement with an artificial prosthesis, which you can read about in *A&P in the Real World: Hip Joint Replacement Surgery* on p. 280.

Quick Check

☐ 6. What features increase the stability of the shoulder joint?

☐ 7. Which bones articulate to form the hip joint?

☐ 8. Why is the shoulder joint less stable than the hip joint?

Apply What You Learned

☐ 1. Predict the consequences that an injury to each of the following ligaments would have for the function and range of motion of the knee.
   a. Anterior cruciate ligament
   b. Tibial collateral ligament
   c. Posterior cruciate ligament

☐ 2. Predict how the functions of the human hand would be different if the carpometacarpal joint of the thumb were a uniaxial hinge joint instead of a saddle joint.

See answers in Appendix A.

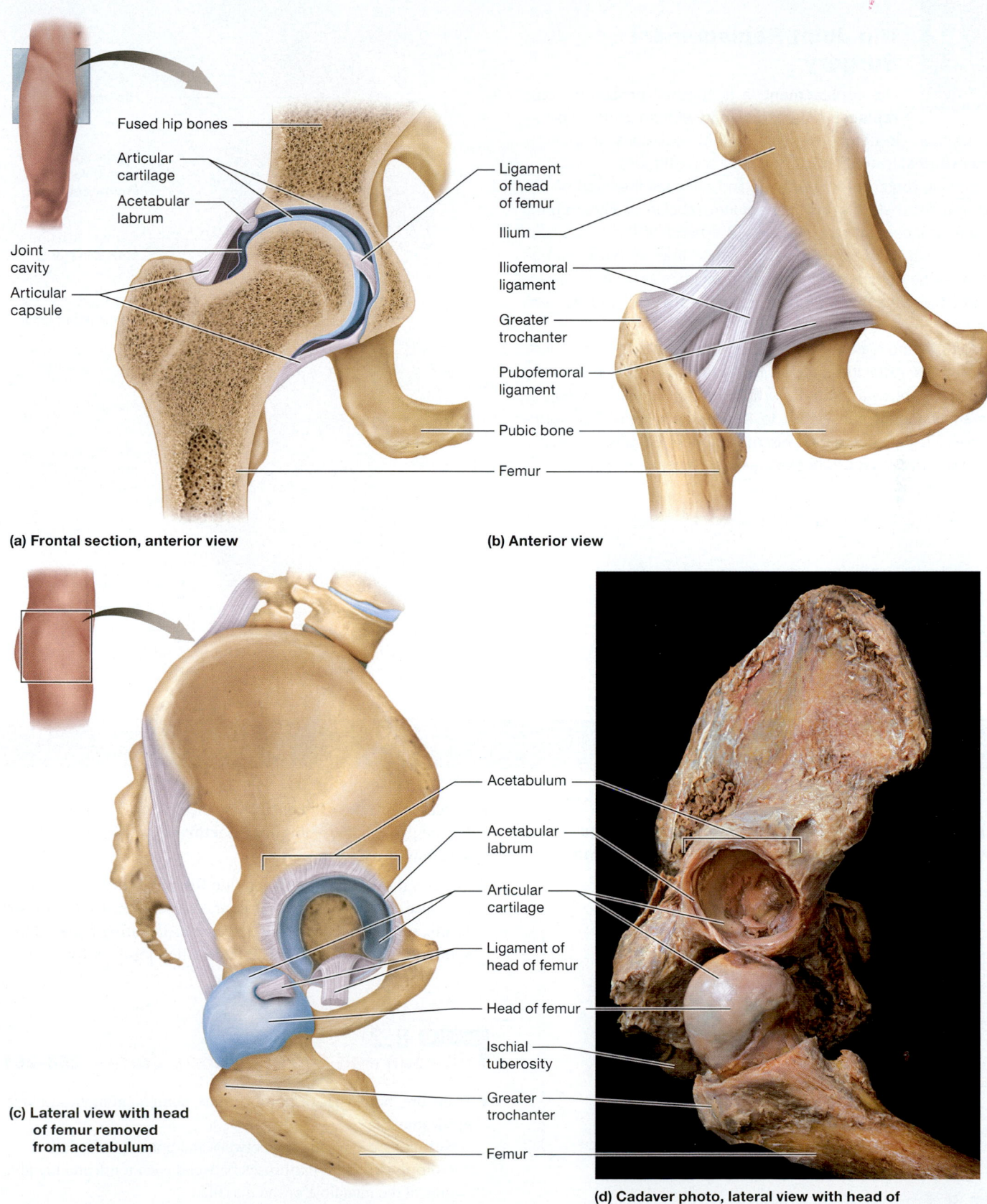

(a) Frontal section, anterior view

Fused hip bones

Articular cartilage

Acetabular labrum

Joint cavity

Articular capsule

Ligament of head of femur

Ilium

Iliofemoral ligament

Greater trochanter

Pubofemoral ligament

Pubic bone

Femur

(b) Anterior view

(c) Lateral view with head of femur removed from acetabulum

Acetabulum

Acetabular labrum

Articular cartilage

Ligament of head of femur

Head of femur

Ischial tuberosity

Greater trochanter

Femur

(d) Cadaver photo, lateral view with head of femur removed from acetabulum

Figure 8.16 Anatomical structure of the hip joint.

Hip Joint Replacement Surgery

Hip replacement is a surgical procedure that replaces a damaged joint with an artificial prosthetic device. Common reasons for hip replacement include severe arthritis, trauma, fractures, and bone tumors.

A total hip replacement removes and replaces the head of the femur and reconstructs the acetabulum (shown in illustration). A partial replacement removes only the head of the femur and replaces it with a prosthetic device, while leaving the acetabulum intact. The choice of replacement depends on many factors, including the type of injury and the patient's age and general health.

Surgical complications are rare but may include blood clots and infections. After the surgery, patients generally begin physical therapy right away, often starting the day of the procedure, and many patients can return to normal daily activities within 2–8 weeks. The prognosis for hip replacement surgery is usually quite good, with most prostheses lasting 15 years or more.

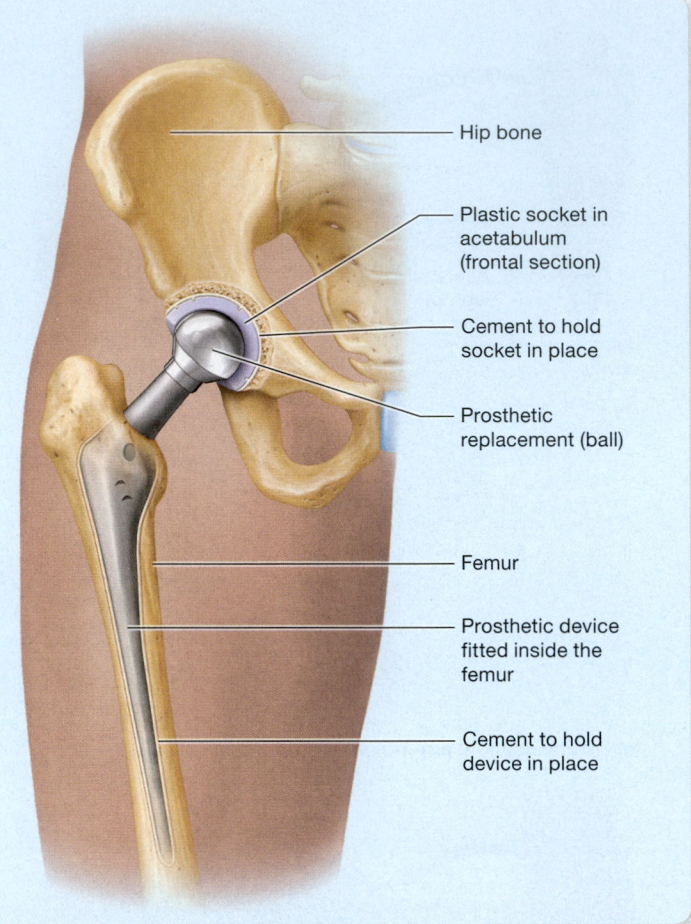

- Hip bone
- Plastic socket in acetabulum (frontal section)
- Cement to hold socket in place
- Prosthetic replacement (ball)
- Femur
- Prosthetic device fitted inside the femur
- Cement to hold device in place

Chapter Summary

For everything you need to succeed in this course, go to Mastering Anatomy & Physiology. There you will find:

- Practice Tests
- Author Podcasts
- Big Picture Animations
- Concept Boost Video Tutors
- **iP2** Interactive Physiology 2.0 Tutorials
- **A&PFlix** A&P Flix 3D Animations
- Active-Learning Workbook

MODULE 8.1
Overview of Joints 257–258

- Functions of joints include enabling movement, providing stability, and allowing for the growth of long bones.

- Functional classes of joints include the immoveable **synarthrosis,** the partially moveable **amphiarthrosis,** and the freely moveable **diarthrosis.**

- Structural classes of joints include **fibrous joints,** held together by dense regular collagenous connective tissue; **cartilaginous joints,** which have cartilage between articulating bones; and **synovial joints,** which have a fluid-filled joint cavity.

MODULE 8.2
Fibrous and Cartilaginous Joints 258–261

- There are three types of fibrous joints: sutures, gomphoses, and syndesmoses.
  - **Sutures** are synarthroses between the bones of the skull.
  - **Gomphoses** are synarthroses between each tooth and the alveolus in the mandible or the maxilla.
  - **Syndesmoses** are amphiarthroses in which the articulating bones are joined by an interosseous membrane.

- There are two types of cartilaginous joints: synchrondroses and symphyses.

○ **Synchondroses** are synarthroses in which the bones are held together by hyaline cartilage.

○ **Symphyses** are amphiarthroses in which the bones are connected by a fibrocartilaginous pad.

MODULE 8.3
Structure of Synovial Joints 261–264

- The bones of a synovial joint are separated by a **joint cavity**, or **synovial cavity.** They are held together by an **articular capsule,** which is lined by a **synovial membrane.**

- The ends of the bones in a synovial joint are covered with a layer of **articular cartilage.**

- Synovial fluid is secreted by the synovial membrane. It lubricates the joint, absorbs shock, and performs metabolic functions.

- **Ligaments, tendons, bursae,** and **tendon sheaths** all provide the synovial joint with additional support.

MODULE 8.4
Function of Synovial Joints 264–270

- Synovial joints may be classified based on the number of axes of motion around which they move as **nonaxial, uniaxial, biaxial,** or **multiaxial.**

- **Gliding** is simple back and forth or up and down motion between articulating surfaces.

- **Angular movements** are those in which the angle between articulating bones changes. They include **flexion, extension, abduction, adduction,** and **circumduction.**

- **Rotation** is a nonangular motion in which a bone pivots or twists along its long axis.

- *Special movements* include **opposition** and **reposition, depression** and **elevation, protraction** and **retraction, inversion** and **eversion, dorsiflexion** and **plantarflexion,** and **supination** and **pronation.**

MODULE 8.5
Types of Synovial Joints 270–280

- Synovial joints may be classified as **plane, hinge, pivot, condylar, saddle,** or **ball-and-socket joints.** Plane joints are nonaxial and the least mobile, and ball-and-socket joints and multiaxial are the most mobile.

- The **elbow** is a hinge joint made up of two separate joints enclosed in a single articular capsule.

- The **knee** is a hinge joint made up of two articulations and is the largest diarthrosis in the body.

- The **shoulder joint** (**glenohumeral joint**) is the most mobile and least stable joint in the body.

- The **hip joint** is a ball-and-socket articulation between the head of the femur and the acetabulum.

Assess What You Learned

Scan the QR Code for additional practice test questions

https://goo.gl/ZkpKW2

LEVEL 1 Check Your Recall

1. Which of the following is *not* a function of articulations?

 a. Movement
 b. Blood cell formation
 c. Stability
 d. Providing growth in length for long bones

2. Which functional joint class includes freely moveable joints?

 a. Amphiarthroses
 b. Synarthroses
 c. Diarthroses
 d. All functional joint classes are freely moveable.

3. Identify each of the following joints as synovial, fibrous, or cartilaginous.

 a. Pubic symphysis _____
 b. Elbow joint _____
 c. Epiphyseal plate _____
 d. Frontal suture _____
 e. Gomphosis _____

4. In general, when mobility of a joint _____, its stability _____.

 a. increases; increases
 b. decreases; decreases
 c. increases; remains unchanged
 d. increases; decreases

5. Mark the following statements as true or false. If a statement is false, correct it to make a true statement.

 a. Fibrous joints are united by collagen fibers.
 b. A syndesmosis is a type of cartilaginous joint.
 c. Cartilaginous joints are synarthroses.
 d. The joint between the two pubic bones and the intervertebral joints are examples of symphyses.

6. Fill in the blanks: The articulating ends of bones of synovial joints are covered in _____. The remaining internal surfaces of the joint are lined by the _____, which produces synovial fluid. The entire joint is encased by the _____, which is composed of dense irregular collagenous connective tissue.

7. What is/are the function(s) of synovial fluid?

8. Which of the following correctly describes the function of a ligament?

   a. It connects a muscle to its attachment point, such as a bone.
   b. It is a fluid-filled structure that minimizes friction during movement.
   c. It connects two bones to each other in a joint.
   d. It surrounds a tendon and protects it in high-stress areas.

9. Bone movement at a joint is described around an invisible line known as a/an:

   a. synarthrosis.
   b. axis.
   c. sagittal plane.
   d. amphiarthrosis.

10. Match the following terms with the correct definition from the right column.

    _____ Plane joint
    _____ Saddle joint
    _____ Ball-and-socket joint
    _____ Condylar joint
    _____ Pivot joint
    _____ Hinge joint

    a. Uniaxial joint in which the rounded articular surface of one bone fits into a groove of another bone and is held in place by a ligamentous ring
    b. Multiaxial joint in which the spherical articular surface of one bone fits into a cup or socket of another bone
    c. Nonaxial joint in which two flat surfaces glide over each other
    d. Uniaxial joint in which the convex articular surface of one bone fits into a concave articular depression
    e. Biaxial joint in which the oval, convex articulating surface of one bone fits into the shallow depression of another bone
    f. Biaxial joint in which each articular surface has both convex and concave regions

11. Define each of the following movements or movement pairs.

    a. Flexion and extension
    b. Adduction and abduction
    c. Rotation
    d. Circumduction
    e. Dorsiflexion and plantarflexion
    f. Elevation and depression

12. Mark the following statements as true or false. If a statement is false, correct it to make a true statement.

    a. The knee and the elbow are multiaxial joints.
    b. The elbow joint consists of two separate articulations.
    c. The patella is encased within the tendon of the quadriceps femoris muscle group.
    d. The shoulder joint is stabilized by the medial and lateral menisci.
    e. The hip joint is less stable than the shoulder joint, but it allows more motion.

13. Which of the following best describes the function of the anterior cruciate ligament?

    a. It prevents posterior displacement of the tibia on the femur.
    b. It prevents the tibia from moving too far laterally on the femur.
    c. It prevents anterior displacement of the tibia on the femur and hyperextension.
    d. It improves the fit between the femur and the tibia.

14. The structure that stabilizes the shoulder joint is known as the:

    a. rotator cuff.
    b. radial collateral ligament.
    c. posterior cruciate ligament.
    d. fibular collateral ligament.

15. Which factors contribute to the stability of the hip joint?

LEVEL 2 Check Your Understanding

1. Explain how the structure of each of the following joint types follows its function.

   a. Fibrous joint
   b. Cartilaginous joint
   c. Synovial joint

2. The primary action of the biceps brachii muscle of the anterior arm is to flex the forearm at the elbow. However, when this muscle is inflamed, pain is felt in the shoulder. Explain this finding.

3. Some individuals have an abnormally small and shallow glenoid labrum. How would this affect the stability of the shoulder joint?

LEVEL 3 Apply Your Knowledge

PART A: Application and Analysis

1. Some health practitioners claim that the cranial bones are moveable and that they are able to move these bones to treat a variety of conditions. Is this likely to be true in an adult? Why or why not?

2. Lauren has hurt her knee playing soccer. She explains that during the match, someone tackled her, hitting her on the lateral side of her knee. You notice during your examination that she has excessive range of motion in her knee; specifically, you are able to hyperextend her knee and anteriorly displace her tibia on her femur. Her tibia also displaces laterally on her femur. What has likely happened to her knee? Explain.

PART B: Make the Connection

3. When articular cartilage is damaged, often fibrocartilage forms instead of new hyaline cartilage. Does fibrocartilage have the same properties as hyaline cartilage? Is it likely to provide the same type of surface as hyaline cartilage? Explain. (*Connects to Chapter 4*)

See answers in Appendix A.

The Muscular System

9

Eighty-eight keys on a piano, struck a few at a time in the right combinations, can produce beautifully complex music. Over 600 muscles in the human body, stimulated a few at a time in the right combinations, can produce amazingly complex movements. But even apparently simple actions require a highly coordinated series of muscle contractions. Speaking, walking, tying your shoes, and holding and reading this book—all these motions demand the coordinated teamwork of multiple skeletal muscles. We explore the anatomy of the skeletal muscles in this chapter, and how they work together to produce these types of movements.

MODULE **9.1**
Overview of Skeletal Muscles

Learning Outcomes

1. Explain how the name of a muscle can help identify its action, appearance, or location.
2. Summarize the major functions of skeletal muscles.
3. Define the terms agonist, antagonist, synergist, and fixator.
4. Differentiate among the three classes of levers in terms of their structure and function, and give examples of each class in the human body.

Movement is obviously a key function of skeletal muscles. As with all organs, a muscle's structure and function are intimately related, so its shape, attachments, and the way that it crosses joints all affect how it produces movement. For this reason, in this module we first look at the basic microscopic and gross structure of a skeletal muscle. We then consider how this information is used to name many of the body's muscles.

The second half of the module explores how muscle structure relates to function and how groups of muscles work together. In the final section, we consider the lever systems formed by our skeleton and the mechanical advantages they confer to skeletal muscles as they perform the work of movement.

Structure of Skeletal Muscles

Like all organs, skeletal muscles are composed of multiple types of tissues, including skeletal muscle, connective, and nervous tissues. Here we take a brief look at the structure of skeletal muscles as organs before we dig deeper into the functions of the muscular system.

For practice applying concepts to a clinical scenario, check out the **Running Case Study** *for this chapter @ Mastering Anatomy & Physiology*

Computer-generated image: Most movements at joints require teamwork by several muscles, each playing a unique role.

283

Gross Anatomy of a Skeletal Muscle

Let's build up a skeletal muscle from the cell to the organ level. At the microscopic level, skeletal muscle cells are called *skeletal muscle fibers* due to their long, thin shape (**Figure 9.1**). Each individual skeletal muscle fiber is surrounded by a thin layer of extracellular matrix called **endomysium** (en'-doh-MY-see-um). Between 10 and 100 muscle fibers are bundled together into a group known as a **fascicle** (FASS-ih-kul), which is surrounded by a connective tissue sheath called the **perimysium** (pehr'-ee-MY-see-um). All the fascicles in a muscle are surrounded by another layer of connective tissue, the **epimysium** (ep'-ee-MY-see-um). The epimysium is continuous with the most superficial connective tissue sheath, known as the **fascia** (FASH-uh), which separates individual muscles from one another. These interconnected connective tissues merge and taper down to become **tendons** or other structures that attach the muscle to the bone or other part that the muscle moves.

A skeletal muscle is richly supplied with blood vessels, which bring the muscle oxygen and nutrients and remove waste products. A muscle also has an extensive nerve supply. Recall from the histology chapter that skeletal muscle 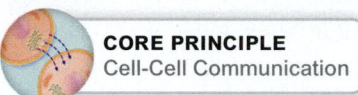 **CORE PRINCIPLE** Cell-Cell Communication is *voluntary,* meaning it is under conscious control and so must be stimulated by the nervous system in order to contract (see Chapter 4). For this reason, a close relationship exists between neurons and skeletal muscle fibers. This is another example of the Cell-Cell Communication Core Principle (p. 28).

Sometimes muscle fibers contract without relaxing, causing painful muscle knots (see *A&P in the Real World: Muscle Knots*).

A&P in the Real World

Muscle Knots

A muscle knot, technically known as a *myofascial trigger point* (my'-oh-FASH-ee-uhl), is defined as a discrete place within the fascia surrounding a muscle that is overly irritable or susceptible to inflammation. When the trigger point becomes inflamed, the muscle fibers around it engage in a sustained contraction, which results in pain. You can usually feel this as a nodule or "knot" in the muscle, which is the source of its common name.

Trigger points may be activated in a variety of situations. In athletes, repetitive exercises that stress muscles, such as running, can cause localized microscopic tears in the fascia that set up trigger points. In addition, improper training or a sudden increase in activity can activate trigger points in both athletes and nonathletes. Trauma, stress, certain disease states, and trapped nerves may also contribute to trigger point development.

Treatments for trigger points are generally aimed at relaxing the muscle fibers in and around the trigger point. This can be accomplished with massage, anti-inflammatory medications, and muscle relaxants. Most treatment plans also involve passive stretching of the affected muscle fibers and fascia to prevent the trigger point from becoming re-inflamed.

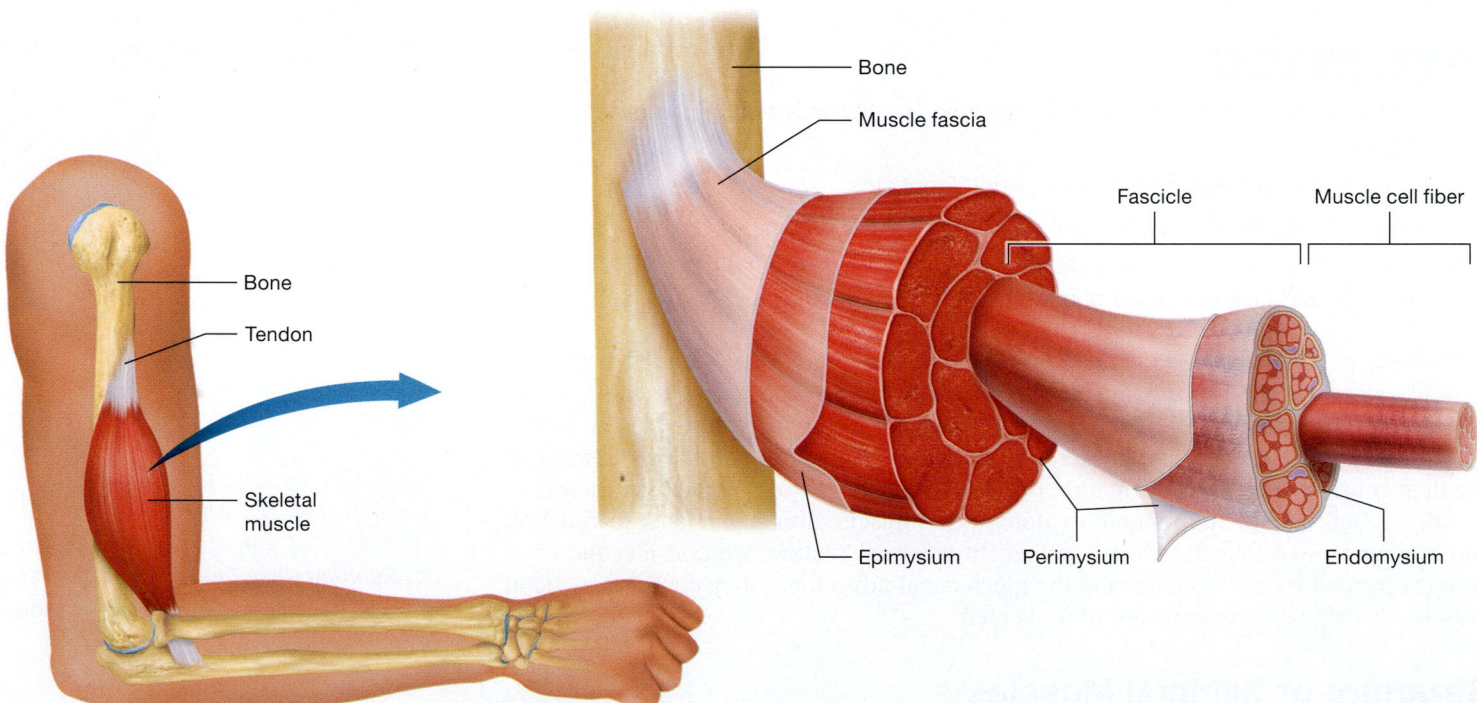

(a) Position of a representative skeletal muscle (brachialis)

(b) Structure of a skeletal muscle

Figure 9.1 Position and structure of a skeletal muscle.

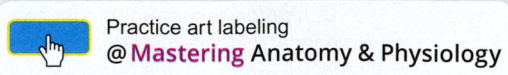
Practice art labeling
@ Mastering Anatomy & Physiology

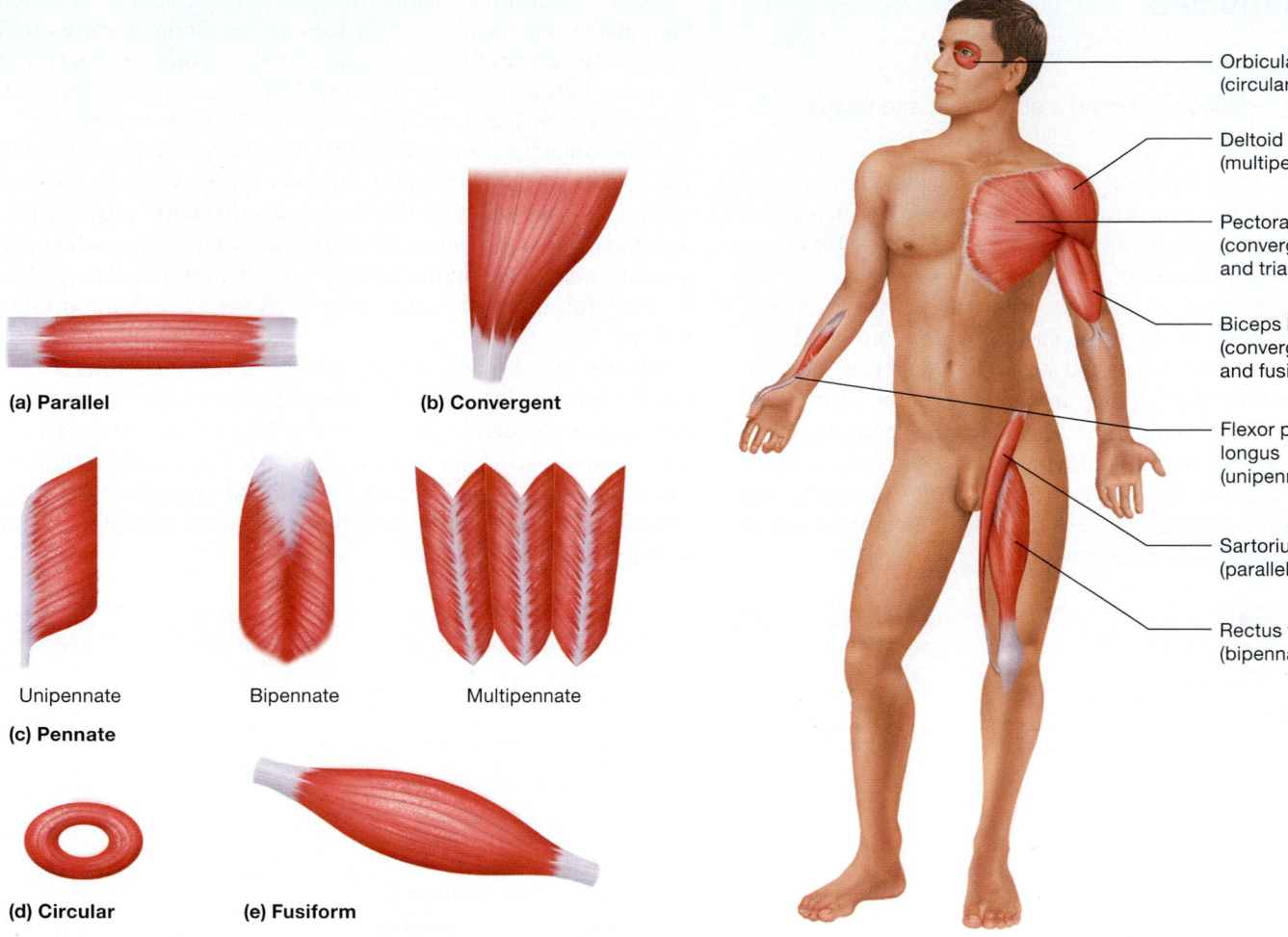

Figure 9.2 Fascicle pattern and muscle shape.

Fascicle Patterns and Muscle Shapes

As structure affects function, it may come as no surprise that the pattern in which fascicles are arranged influences not only the appearance but also the function of a skeletal muscle (**Figure 9.2**). There are six main fascicle patterns: *parallel, convergent, pennate, circular, spiral,* and *fusiform.*

- A **parallel** muscle has evenly spaced fascicles attaching to a tendon that is about the same width as the muscle (see Figure 9.2a). This orientation often produces a straplike muscle, an example of which is the sartorius muscle in the thigh.
- A **convergent** muscle is broad at one end and uniformly tapers to a single tendon (see Figure 9.2b). Triangular muscles, such as the pectoralis major muscle in the chest, usually have a convergent fascicle arrangement.
- A **pennate** (PEN-et) muscle has fibers and fascicles that attach to the tendon at an angle in such a way that the muscle resembles a feather (see Figure 9.2c). The pennate pattern has three basic variations. A **unipennate** muscle has a single tendon, and its fascicles feather out at an angle from only one side of it (an example is the flexor pollicis longus muscle of the forearm). A **bipennate** muscle also has a single tendon, but its fascicles angle out from both sides of its tendon (e.g.,

the rectus femoris muscle of the thigh). A **multipennate** muscle is usually composed of several tendons, and the way the fascicles are attached makes the muscle look like several feathers joined together (e.g., the deltoid muscle of the shoulder).

- As its name implies, a **circular** muscle encircles a structure, such as the opening of the eye, to close or constrict it when it contracts (e.g., the orbicularis oculi muscle; see Figure 9.2d). Circular muscles are often referred to as **sphincters** (SFINGK-terz).
- A **spiral** muscle may wrap around a bone or have the twisted appearance of a towel wrung out to dry (e.g., the supinator muscle in the forearm).
- A **fusiform** (FYOOZ-ih-form) muscle is thicker in its **belly,** or middle region, and tapered at its ends (e.g., the biceps brachii muscle; see Figure 9.2e).

Quick Check

□ 1. What is a fascicle, and how does it relate to the perimysium?

□ 2. How do the terms pennate, parallel, and circular relate to fascicle orientation?

Naming Muscles

« FLASHBACK

1. What are flexion, extension, abduction, and adduction? (p. 266)

Many muscles are named for their shape, appearance, or position. Others are named for structural considerations, such as the number of points of proximal attachment, or *heads*. Still others are named for their size or location.

Muscle names based on size use words such as major, minor, longus, brevis (brief or short), or vastus (vast or broad). Those based on location use directional terms like superior, inferior, medial, and lateral, often slightly modified based on their Greek or Latin roots. For example, the vastus lateralis muscle of the thigh is a broad muscle on the lateral side of the femur. Many other muscle names use the regional anatomical terms you learned in earlier chapters. An example is the pectoralis minor muscle—a small muscle in the chest.

Some muscles have names that reveal the structures to which they attach. For instance, let's look at the sternocleidomastoid muscle. The name of this neck muscle sounds complex, but within its name you can learn that it is attached to the sternum (*sterno-*), clavicle (*cleido-*), and mastoid process of the temporal bone.

Other muscles are named for their functions. The terms *flexion* and *extension* from the articulations chapter apply to naming muscles as **flexors** (FLEK-sohrz) and **extensors**, respectively. Similarly, a **levator** (leh-VAY-ter) muscle elevates, an **adductor** adducts, and an **abductor** abducts a body part. For this reason, it is helpful to review joint movements before tackling muscle actions.

Finally, some muscles and their locations relate to common vocabulary. For example, the flexor digitorum profundus muscle (proh-FUN-dus) in the forearm is located deep to the flexor digitorum superficialis muscle (soo'-per-fish-ee-AL-iss). *Profundus* means "deep," just as a profound statement has deep meaning. Table 9.1 summarizes these and other helpful terms for understanding muscle names.

Table 9.1 Common Terms in Muscle Anatomy

Term and Meaning	Example
Muscle Size	
Brevis—short	Fibularis brevis muscle
Longus—long	Adductor longus muscle
Vastus—wide/large	Vastus lateralis muscle
Muscle Location	
Anterior—toward the front	Tibialis anterior muscle
External—toward the outside	External intercostal muscle
Infra—below	Infraspinatus muscle
Intercostal—between the ribs	Internal intercostal muscle
Internal—toward the inside	Internal oblique muscle
Posterior—toward the back	Tibialis posterior muscle
Profundus—deep	Flexor digitorum profundus muscle
Superficialis—nearer the surface	Flexor digitorum superficialis muscle
Supra—above	Supraspinatus muscle
Muscle Action	
Abductor—pulls away from the midline	Abductor pollicis longus muscle
Adductor—pulls toward the midline	Adductor magnus muscle
Depressor—pulls down	Depressor labii inferioris muscle
Erector—holds erect or straight	Erector spinae muscle
Extensor—increases the angle between bones	Extensor digitorum longus muscle
Flexor—decreases the angle between bones	Flexor digitorum longus muscle
Levator—raises a body part	Levator scapulae muscle
Pronator—turns palm posteriorly	Pronator teres muscle
Supinator—turns palm anteriorly	Supinator muscle
Body Region	
Abdominis—abdominal area	Rectus abdominis muscle
Brachii—arm area	Biceps brachii muscle
Capitis—head area	Semispinalis capitis muscle

☐ 3. The adductor magnus muscle is a muscle in the thigh. What can you deduce about its function and appearance just from its name?

Functions of Skeletal Muscles

《 FLASHBACK

1. What are some functions of bone markings? (p. 212)
2. What are tendons and ligaments? (p. 263)

The basic function of skeletal muscles is to contract in order to generate a force called *muscle tension*. The ability to generate muscle tension allows skeletal muscles to produce body movements, or **actions,** as they are called in muscle physiology. Tension production also enables muscles to perform another task—generate heat.

Any process that converts chemical energy to mechanical energy is inefficient and results in energy loss, usually as heat. In muscle contraction, the chemical energy of ATP is converted to the mechanical energy of movement, a process that generates a great deal of heat as a waste product. In fact, the heat generated from muscle contraction is one reason why we are able to maintain a stable body temperature. This is why we shiver when we are cold—the extra heat helps raise the body temperature back to normal levels. We discuss heat generation further in the metabolism chapter (see Chapter 23).

This chapter focuses on muscle actions that produce movement. Before we explore the individual muscles and their actions, let's examine the general functions of skeletal muscles. The upcoming subsections take a look at how functional groups of muscles work together, how muscles cause movement on the body parts on which they act, and how they work with bones to form lever systems.

Table 9.1 Common Terms in Muscle Anatomy (*continued*)

Term and Meaning	Example
Body Region (*continued*)	
Carpi—wrist area	Extensor carpi ulnaris muscle
Cervicis—neck area	Splenius cervicis muscle
Digitorum/Digiti—related to fingers/toes	Extensor digitorum muscle
Femoris—femur or thigh	Biceps femoris muscle
Gluteal—buttocks	Gluteus maximus muscle
Hallucis—great toe	Abductor hallucis muscle
Oculi—eye area	Orbicularis oculi mucle
Oris—mouth area	Orbicularis oris muscle
Pectoralis—chest area	Pectoralis minor muscle
Pollicis—thumb	Flexor pollicis brevis muscle
Muscle Fiber Orientation	
Oblique—at an angle	External oblique muscle
Orbicular—circular	Orbicularis oculi muscle
Rectus—straight	Rectus femoris muscle
Transversus—across/transverse	Transversus abdominis muscle
Muscle Heads	
Biceps—two heads	Biceps brachii muscle
Quadriceps—four heads	Quadriceps femoris muscle group
Triceps—three heads	Triceps brachii muscle
Muscle Shape	
Deltoid—triangular (as in the Greek letter delta)	Deltoid muscle
Maximus—largest	Gluteus maximus muscle
Minimus/Minimi—smallest	Gluteus minimus muscle
Minor—small	Pectoralis minor muscle
Quadratus—shaped like a rectangle	Pronator quadratus muscle
Rhomboid—shaped like a rhombus	Rhomboid major muscle
Serratus—serrated or jagged	Serratus anterior muscle
Trapezius—shaped like a trapezoid	Trapezius muscle

Functional Groups of Muscles

In the chapter opener, we mentioned that teamwork is needed for most muscle actions. A car needs different components such as motors, brakes, and steering mechanisms to move in a controlled manner. So it isn't surprising that most actions at joints involve several muscles, each with a particular job to do. These muscles are the *agonists*, *antagonists*, *synergists*, and *fixators* (**Figure 9.3**).

The **agonist** provides most of the force required for a given movement. For this reason, we also call an agonist the **prime mover**. An agonist is typically easy to identify as one of the larger muscles spanning the joint to be moved. In Figure 9.3, the agonist to lift a glass of water is the brachialis muscle. Although it may seem as though the brachialis muscle could do all the work of lifting the glass, it would be jerky and imprecise without the help of other team members.

The **antagonist** usually lies on the opposite side of a joint from its agonist partner, and tends to oppose and slow the action. Many of our slower, more graceful movements result from highly controlled action by antagonists. Gymnastics, ballet, and yoga all involve carefully balancing the force of agonists and antagonists. Even lifting a glass of water requires an antagonist: the triceps brachii muscle.

Synergists (SIN-er-jists) are muscles that work together with the agonist. But synergists are more than just "secondary movers"—they help guide the movement and ensure it is smooth. In Figure 9.3, the biceps brachii muscle acts as the synergist to

the brachialis muscle that helps guide the glass to the mouth. In addition, some synergists make a movement more efficient by stabilizing a joint, a function that overlaps with fixators.

Fixators are muscles that hold a bone in place, an anchoring function that makes movement more efficient and reduces the risk of injury. You can easily visualize the role of a fixator if you imagine trying to push open a door while wearing roller blades. If you're not careful, you are just as likely to push yourself backward as you push the door forward. But if you lock the wheels of the roller blades in place, all of your force goes forward into opening the door. In our example of lifting the glass of water, several scapular muscles, such as the supraspinatus muscle, act as fixators of the shoulder joint. Without the actions of these fixators, the movement wouldn't be smooth and water would splash out of the glass.

Muscle Origin and Insertion

A skeletal muscle's action depends on the bones or parts to which it is attached and the joints that it crosses. We describe muscle actions in terms of the muscle's *origin* and *insertion*. A muscle's **origin** is the part to which it attaches that is generally more fixed, whereas its **insertion** is usually the part that the muscle moves. **Figure 9.4** illustrates these attachments for the biceps brachii muscle before contraction (see Figure 9.4a) and after contraction (see Figure 9.4b). At one end, this muscle attaches to the scapula, which remains relatively stationary. The other end attaches to the radius, which is the bone that moves.

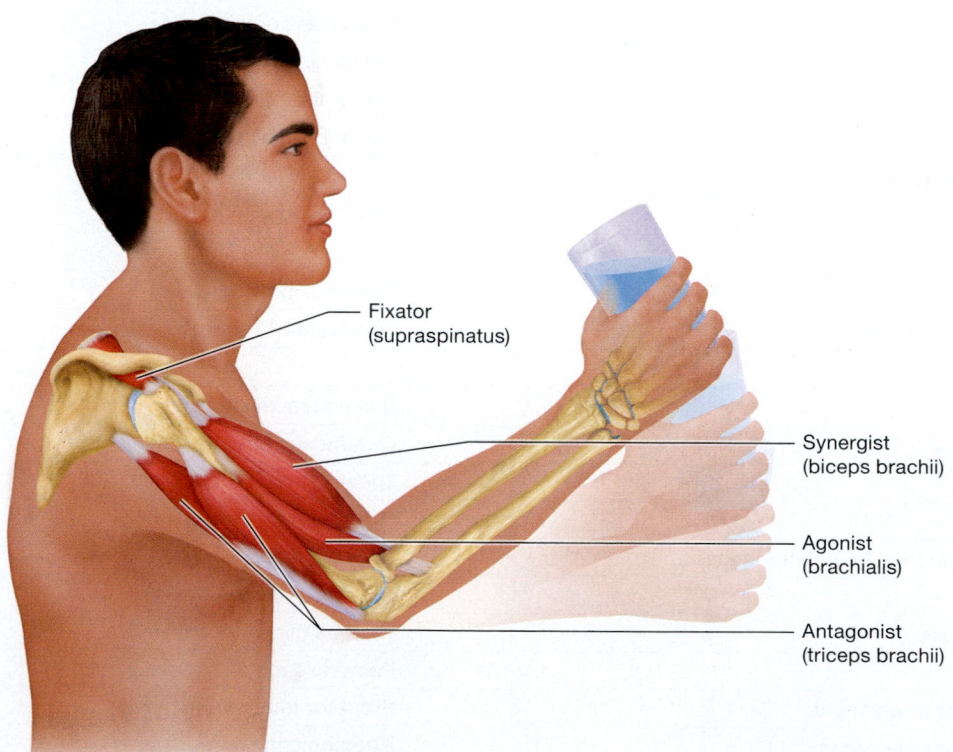

Figure 9.3 Functional groups of muscles.

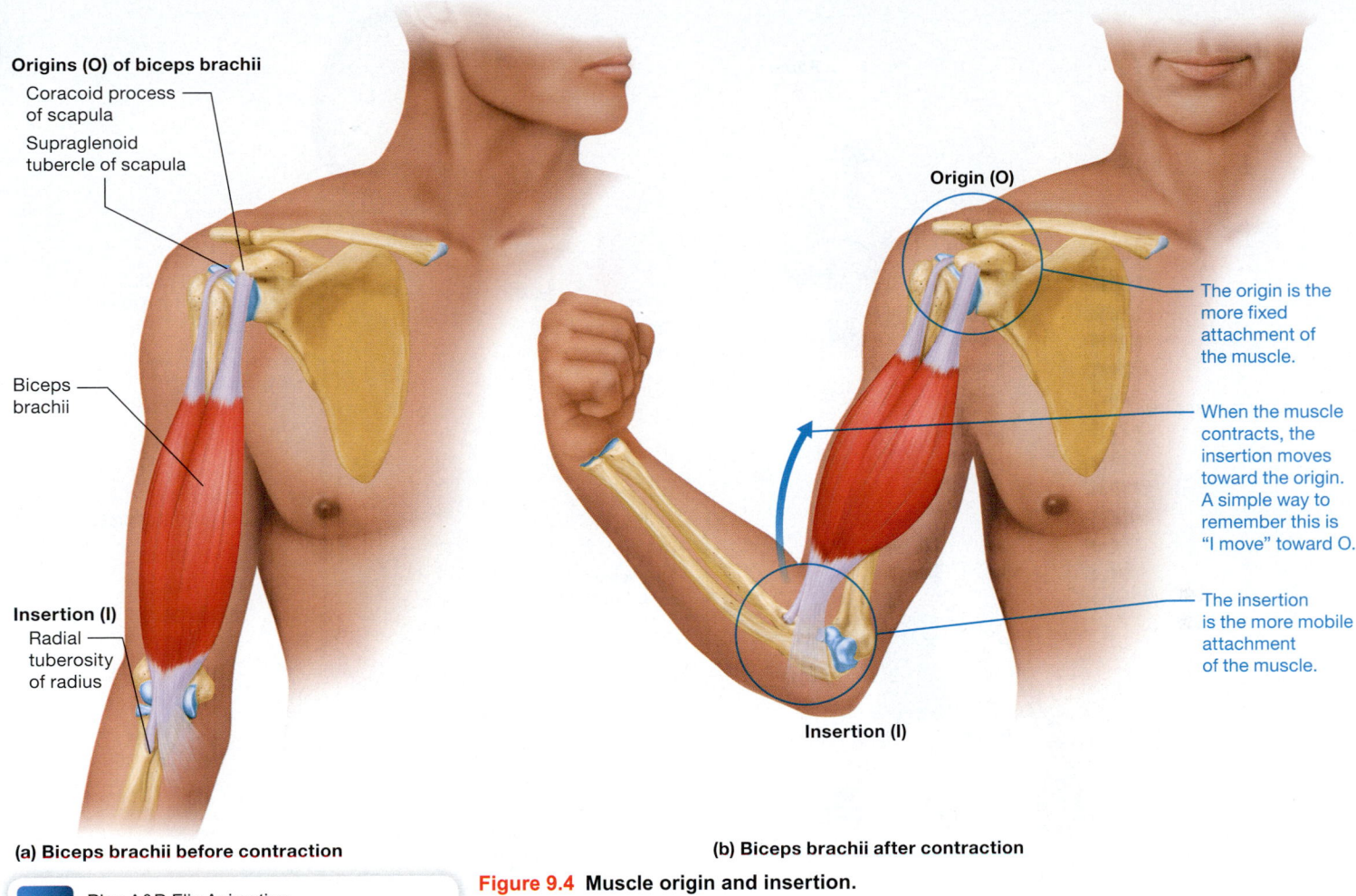

Origins (O) of biceps brachii

Coracoid process
of scapula

Supraglenoid
tubercle of scapula

Biceps
brachii

Insertion (I)
Radial
tuberosity
of radius

(a) Biceps brachii before contraction

Origin (O)

The origin is the
more fixed
attachment of
the muscle.

When the muscle
contracts, the
insertion moves
toward the origin.
A simple way to
remember this is
"I move" toward O.

The insertion
is the more mobile
attachment
of the muscle.

Insertion (I)

(b) Biceps brachii after contraction

Figure 9.4 Muscle origin and insertion.

▶ Play A&P Flix Animation
@**Mastering** Anatomy & Physiology

This is why we call the scapula the origin of the biceps brachii muscle, and the radius its insertion.

This stationary versus moving bone rule does have exceptions for muscles that act on more than one joint, because the origin and insertion can sometimes switch. This depends largely on which fixators are active. For example, the iliopsoas muscle attaches to the pelvis and lower vertebral column at one end and to the femur at the other end. The femur is considered the insertion of the iliopsoas muscle, since it is the more commonly moved body part. However, when we flex the trunk at the hip (bend forward), the origin and insertion switch—the femur is fixed, and so it becomes the origin, and the pelvis and vertebrae are moved, becoming the insertion.

Lever Systems Used in Body Movements

You have almost certainly used many levers in your lifetime, as shovels, scissors, and even brooms are all examples of common lever systems. A **lever system** has four components: (1) the lever itself; (2) the *load,* or object, you are trying to move; (3) the *force* applied to the lever to move the load; and (4) the **fulcrum,** or hinge point, around which the lever moves. In moving the body, the bones are the levers, the load is the weight of

the body part being moved, the force is the tension generated by muscle contractions, and the fulcrum is the joint at which the movement occurs.

There are three types of levers: first-, second-, and third-class (**Figure 9.5**). In a **first-class lever,** the load to be moved is on one side of the lever, the fulcrum is in the middle, and the force is applied to the other side of the lever. An example of a first-class lever, shown in Figure 9.5a, is a simple seesaw. In the body we can compare this to the atlanto-occipital joint. Notice that the fulcrum is the joint itself in the middle, the load is the weight of the head, and the force is the muscle contraction of the posterior neck muscles that lift the head (pull it back).

A **second-class lever** is one in which the fulcrum is located on one end of the lever, the load to be moved is in the middle, and the force to move the load is applied on the other end of the lever. An example of a simple second-class lever is a hand truck, as you can see in Figure 9.5b. Second-class lever systems are uncommon in the body, but one example is the motion of rising up on your toes. In this example, the metatarsophalangeal joints are the fulcrums and the weight of the body in the middle is the load. The force is provided by contraction of the posterior leg

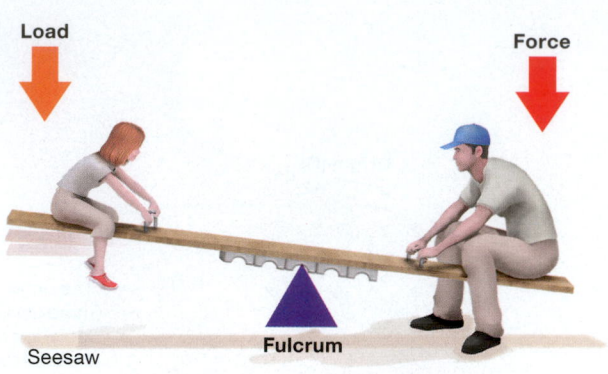

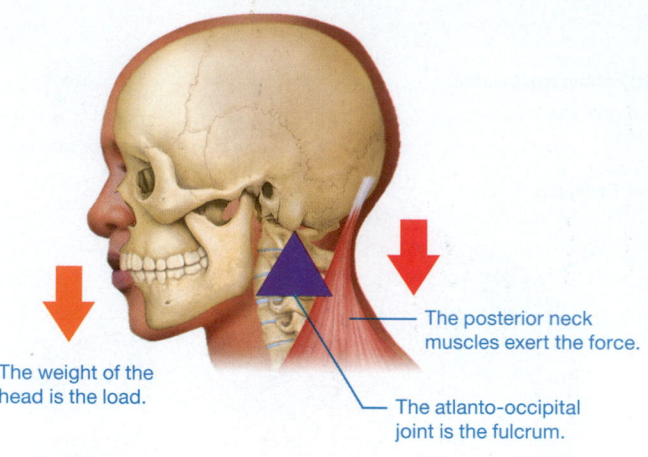

Load

Force

Seesaw

Fulcrum

(a) First-class lever

The weight of the head is the load.

The posterior neck muscles exert the force.

The atlanto-occipital joint is the fulcrum.

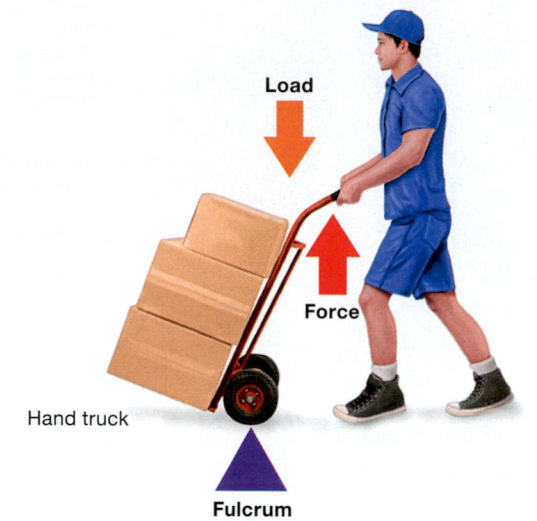

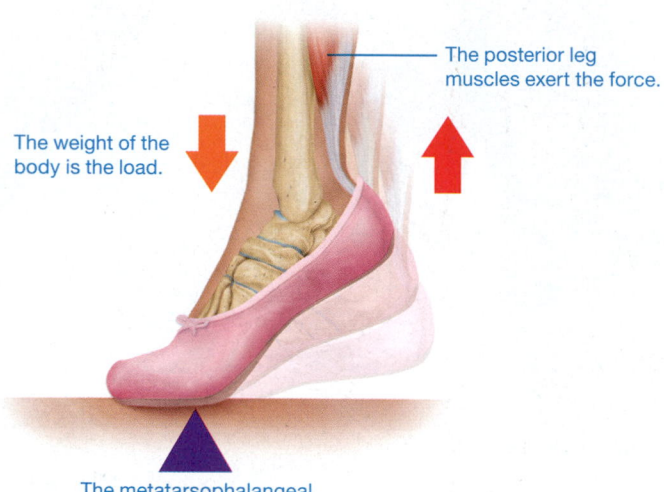

Load

Force

Hand truck

Fulcrum

(b) Second-class lever

The weight of the body is the load.

The posterior leg muscles exert the force.

The metatarsophalangeal joints are the fulcrum.

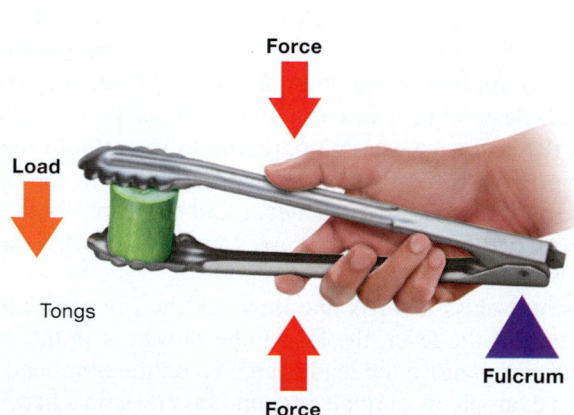

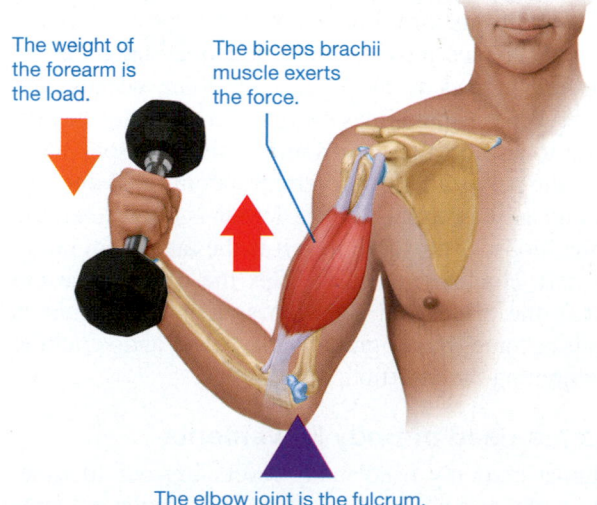

Force

Load

Tongs

Fulcrum

Force

(c) Third-class lever

Figure 9.5 Lever systems.

The weight of the forearm is the load.

The biceps brachii muscle exerts the force.

The elbow joint is the fulcrum.

muscles—as they contract, they lift the body's weight up onto the toes.

In a **third-class lever** system, the load to be moved is on one end of the lever, the force moving the load is applied in the middle, and the fulcrum is at the other end of the lever. An everyday example of a third-class lever is a pair of tongs, shown in Figure 9.5c. Third-class lever systems are very common in the body, as illustrated with a muscle to which you've already been introduced—the biceps brachii muscle. In the figure, note that the fulcrum is at the elbow joint and the load is the weight of the forearm and hand. The force that moves the forearm and hand is the contraction of the biceps brachii muscle. As you can see, this muscle inserts between the elbow (the fulcrum) and the forearm and hand (the load).

The concepts of *mechanical advantage* and *mechanical disadvantage* are important in understanding how lever systems work. In a lever system that works at a mechanical advantage, a relatively small force can move a large load. For a lever system to have a mechanical advantage, the fulcrum must be located closer to the load being moved and farther from the force that is moving the load:

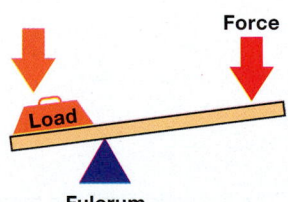

Mechanical advantage: Placing the fulcrum close to the load allows a large load to be moved with a smaller force.

If you look at Figure 9.5b, you can see that second-class lever systems have a definite mechanical advantage. This makes sense that the body would need a mechanical advantage when standing on the toes, as supporting the body's entire weight would otherwise require a much greater force.

In a lever system that works at a mechanical disadvantage, the fulcrum is located farther from the load being moved and closer to the force moving the load:

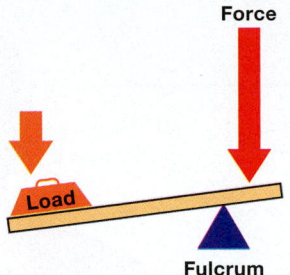

Mechanical disadvantage: Placing the fulcrum farther from the load requires a greater force to move the load.

Being at a mechanical disadvantage means that more force has to be applied in order to move the load—in other words, more work for less reward. Third-class levers, the most common type in the human body, work with a mechanical disadvantage. However, they sacrifice force for advantages in speed of movement and distance through which the load is moved. You may be able to move a lot of weight when you push up onto your toes with a second-class lever, but consider how much faster you can flex your elbow with the third-class lever controlled by your biceps brachii muscle. In addition, the range of motion of the elbow joint is much greater.

Quick Check

☐ 4. What is the role of an agonist in a muscle contraction? What is the role of an antagonist?

☐ 5. What are a muscle's origin and insertion, and how does this relate to the muscle's action?

☐ 6. How do first-, second-, and third-class lever systems differ?

Studying Muscles

In the sections that follow, we examine the muscles of the human body, categorized as:

- Muscles of the head, neck, and vertebral column
- Muscles of the trunk and pelvic floor
- Muscles of the pectoral girdle and upper limb
- Muscles of the hip and lower limb

Before reading on, you may find it useful to review the anterior and posterior superficial muscles of the human body in **Figures 9.6** and **9.7**. That way, some of the muscles in each section will be familiar to you.

As you go through the rest of this chapter, you'll note that each section has three parts: text description, figures, and tables. The text description talks you through the muscles of a section and cites examples of major muscles in each region. These major muscles are also illustrated in the figures of each section. For detailed information on each muscle, you can turn to the tables. The information in tables includes the muscle's origin, insertion, action(s), and the nerves that innervate it (stimulate its activity). Many tables also contain a concept figure to demonstrate the muscle's actions;

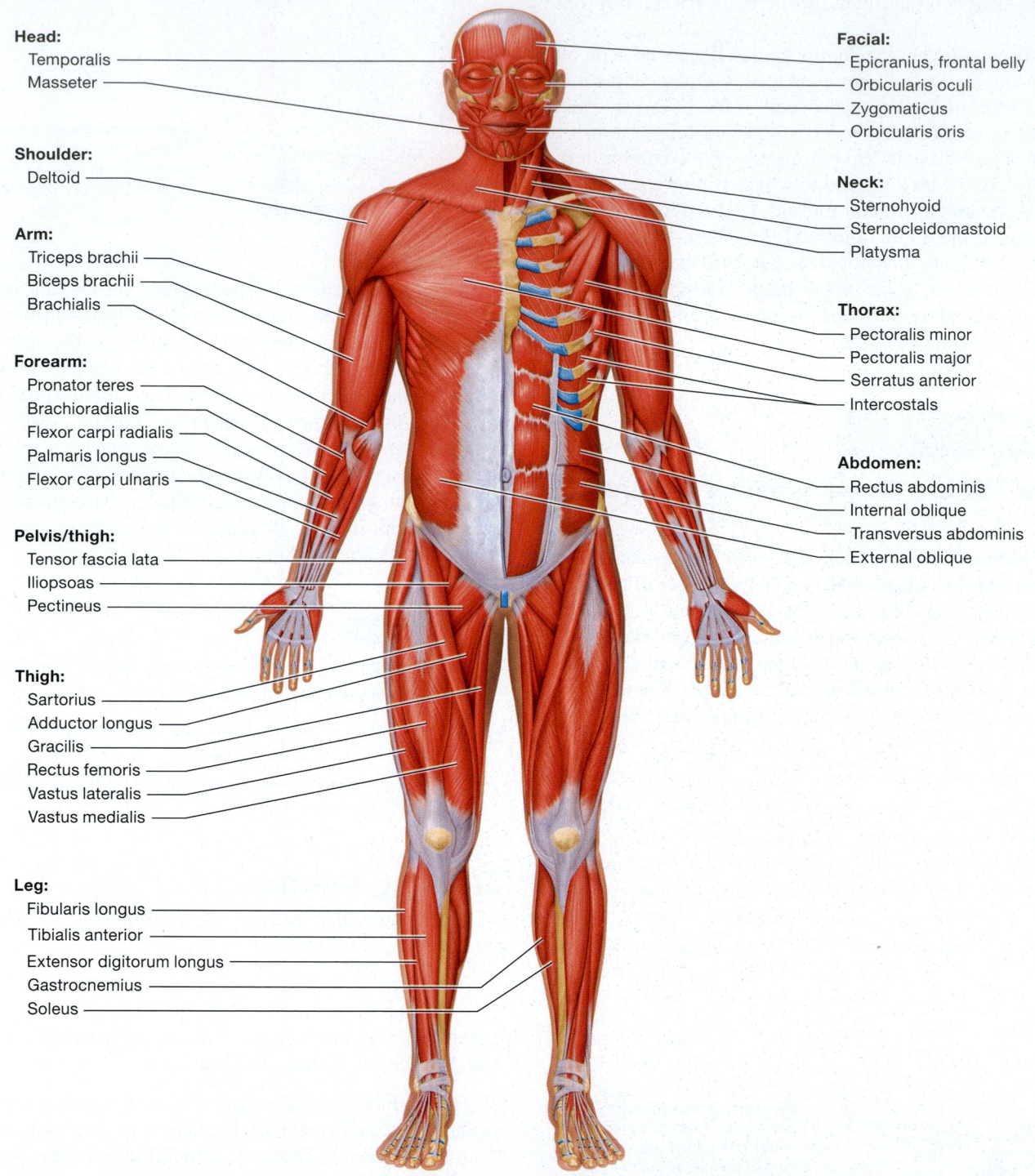

Head:
- Temporalis
- Masseter

Shoulder:
- Deltoid

Arm:
- Triceps brachii
- Biceps brachii
- Brachialis

Forearm:
- Pronator teres
- Brachioradialis
- Flexor carpi radialis
- Palmaris longus
- Flexor carpi ulnaris

Pelvis/thigh:
- Tensor fascia lata
- Iliopsoas
- Pectineus

Thigh:
- Sartorius
- Adductor longus
- Gracilis
- Rectus femoris
- Vastus lateralis
- Vastus medialis

Leg:
- Fibularis longus
- Tibialis anterior
- Extensor digitorum longus
- Gastrocnemius
- Soleus

Facial:
- Epicranius, frontal belly
- Orbicularis oculi
- Zygomaticus
- Orbicularis oris

Neck:
- Sternohyoid
- Sternocleidomastoid
- Platysma

Thorax:
- Pectoralis minor
- Pectoralis major
- Serratus anterior
- Intercostals

Abdomen:
- Rectus abdominis
- Internal oblique
- Transversus abdominis
- External oblique

Figure 9.6 Superficial muscles: anterior view. The left abdominal and thoracic surfaces have been partially dissected to show somewhat deeper muscles.

Practice art labeling
@ **Mastering** Anatomy & Physiology

note that the colors of the written actions and/or directions of action in the Action(s) column match the colors of directional arrow(s) in the concept figures. A good study approach might be to read the text description and view the art first, and then study the tables to learn about the muscles assigned by your instructor.

Apply What You Learned

☐ 1. Predict the name that might be assigned to a muscle that attaches to the sternum and the hyoid bone.

☐ 2. The temporalis muscle is involved in *mastication*, or chewing. It has attachments to the temporal bone and the

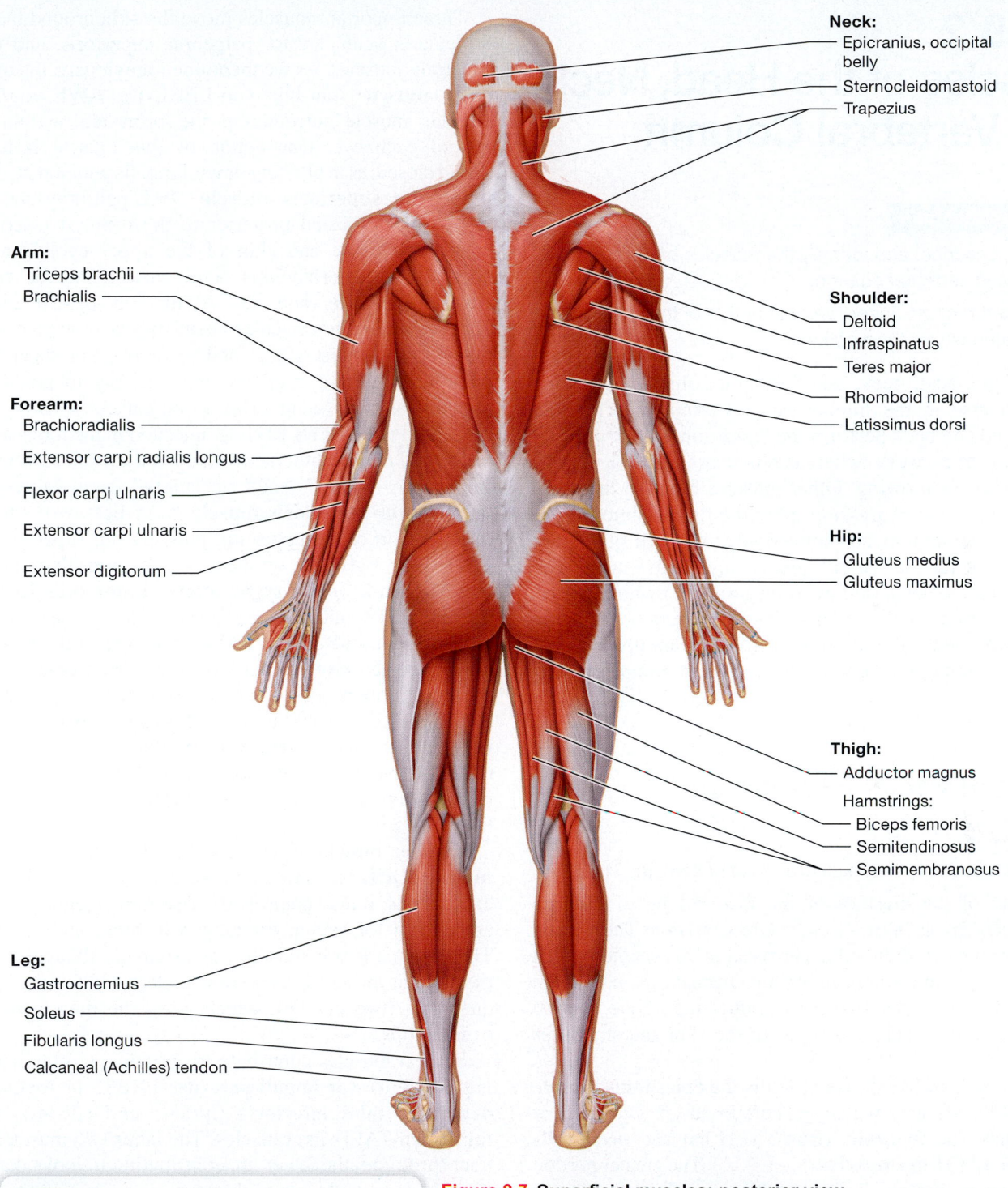

Neck:
Epicranius, occipital belly
Sternocleidomastoid
Trapezius

Arm:
Triceps brachii
Brachialis

Shoulder:
Deltoid
Infraspinatus
Teres major
Rhomboid major
Latissimus dorsi

Forearm:
Brachioradialis
Extensor carpi radialis longus
Flexor carpi ulnaris
Extensor carpi ulnaris
Extensor digitorum

Hip:
Gluteus medius
Gluteus maximus

Thigh:
Adductor magnus
Hamstrings:
Biceps femoris
Semitendinosus
Semimembranosus

Leg:
Gastrocnemius
Soleus
Fibularis longus
Calcaneal (Achilles) tendon

Figure 9.7 Superficial muscles: posterior view.

Practice art labeling
@ Mastering Anatomy & Physiology

mandible. Which bone serves as the origin, and which is the insertion? Explain your answer.

3. You are trying to pry a large rock out of the soil using a lever. You have wedged one end of the lever underneath the rock (the load), and have placed another rock underneath the lever to act as the fulcrum. You have placed your fulcrum very close to the load that you want to move. What kind of lever system are you using to dislodge the rock? Is your work being done at a mechanical advantage or disadvantage? Explain your answer.

See answers in Appendix A.

Muscles of the Head, Neck, and Vertebral Column

Learning Outcomes

1. Name, describe, and identify the muscles of the head, neck, and vertebral column.
2. Identify the origin, insertion, and action of these muscles, and demonstrate their actions.

Muscles of the head, neck, and vertebral column are among the most diverse in the human body. Certain muscles that move the head and neck perform the "traditional" task of moving body parts to do work, whereas others are part of a coordinated reflex for swallowing. Other muscles found in the head, the muscles of facial expression, play the more subtle role of enabling us to communicate nonverbally, and still others are responsible for the extremely precise eye movements that allow us to track objects (and read lines of text) without moving our heads. And finally, the muscles of the vertebral column not only move the trunk but allow us to maintain upright posture. In this module, we explore these diverse groups of skeletal muscles.

Muscles of Facial Expression

« **FLASHBACK**

1. Which bones of the skull are facial bones? (p. 214)

The majority of the muscles of the face belong to a group known collectively as *muscles of facial expression*. These muscles differ from most of the others covered in this chapter in that they do not cause movement at a joint. Instead, most of them insert into skin or connective tissue rather than bone, and so their actions produce shape changes to the skin and structures of the face.

A large muscle of facial expression is the **epicranius muscle** (eh′-pih-KRAY-nee-us), which is composed of two distinct muscle bellies: the **frontalis** (frun-TAEH-lis) and **occipitalis** (awk′-sip-ih-TAEH-lis) **muscles** (Figure 9.8). The greater portion of the epicranius muscle is not actually muscle tissue, but a sheet of connective tissue called the **epicranial aponeurosis** (ap′-oh-noo-ROH-sis) that connects the two muscle bellies. The most important actions of this muscle are to elevate the eyebrows and skin of the forehead into horizontal wrinkles, as we do when looking surprised. This action makes the frontalis muscle one of the common injection sites for the cosmetic product BOTOX® (a registered trademark of Allergan), which paralyzes affected muscles with the intent of reducing fine lines and wrinkles. Paralysis of the frontalis muscle is the reason why some people may have "frozen forehead" after receiving BOTOX injections.

Three important muscles move the skin around the eyes: the orbicularis oculi, levator palpebrae superioris, and corrugator supercilii muscles. As we mentioned previously, the **orbicularis oculi muscle** (ohr-bik′-yoo-LEHR-iss AWK-yoo-lye) is a circular muscle surrounding the orbit and within the eyelids of each eye. One action of this muscle is to pull the eyelid closed, as in blinking or winking. Its antagonist, the **levator palpebrae superioris muscle** (PAL-peh-bray soo-peer′-ee-OHR-iss), is located posterior to the orbit. It inserts into the connective tissue and skin of the upper eyelid and pulls it open. Both the orbicularis oculi muscle and the **corrugator supercilii muscle** (kohr-uh-GAY-tur soo′-per-SIL-ee-aye) produce squinting: The orbicularis oculi muscle draws the skin around the eye like a purse string, and the corrugator supercilii muscle pulls the eyebrows inferiorly and medially to produce vertical wrinkles, as in frowning. (This action makes the corrugator supercilii muscle a target for BOTOX injection in the forehead as well.)

Several muscles interact with the mouth and surrounding area to produce "happy," "sad," or "angry" facial expressions. The **levator labii superioris muscle** (LAY-bee-aye), which inserts into the skin of the upper lip, pulls the lip superiorly in folds, to produce a grimace or sneer. The **zygomaticus major** (zy′-goh-MAT-ih-kus) and **zygomaticus minor muscles** insert into the skin and connective tissue around the corners of the mouth and so assist with smiling. Also attaching to the corners of the mouth are the **risorius muscles** (ry-ZOHR-ee-us). They originate from connective tissue anterior to the ear and so course almost directly across the face. This causes them to pull the corners of the mouth laterally to produce a closed-mouth smile or smirk. Note that the risorius muscles develop differently in many individuals, which partly explains why we have distinctive smiles.

Another muscle around the mouth is the **orbicularis oris muscle** (OHR-iss). Although this muscle is involved in facial expressions, it also controls the fine movements of the lips that are critical for eating, drinking, whistling, and proper speech. The orbicularis oris muscle was originally thought to be a simple circular muscle, but it is now understood to have four curved quadrants (two per lip), which gives the muscle finer control over the lips.

Several muscles contribute to frowning expressions, including the **depressor anguli oris** (dee-PRESS-ur ANG-gyoo-lee), **depressor labii inferioris** (in-feer′-ee-OHR-iss), and **mentalis** (men-TAEH-lis) **muscles.** The latter two muscles also protract (protrude) the lower lip, contributing to looks that might be described as pouting or doubting.

The final two muscles in this section have more subtle effects on facial expression. The **buccinator muscle** (BUK-sih-nay-tur) is the main muscle of the cheek (remember that the *buccal region* is the region over the cheek). This muscle pulls the cheeks inward, as we do when sucking, chewing food, or whistling. The **platysma muscle** (plah-TIZ-mah) is a broad, flat, sheetlike muscle found in the superficial neck. It helps produce that open-mouthed, "jaw-dropping" look of horror, and also tightens the skin of the neck.

Table 9.2 summarizes the details of these muscles.

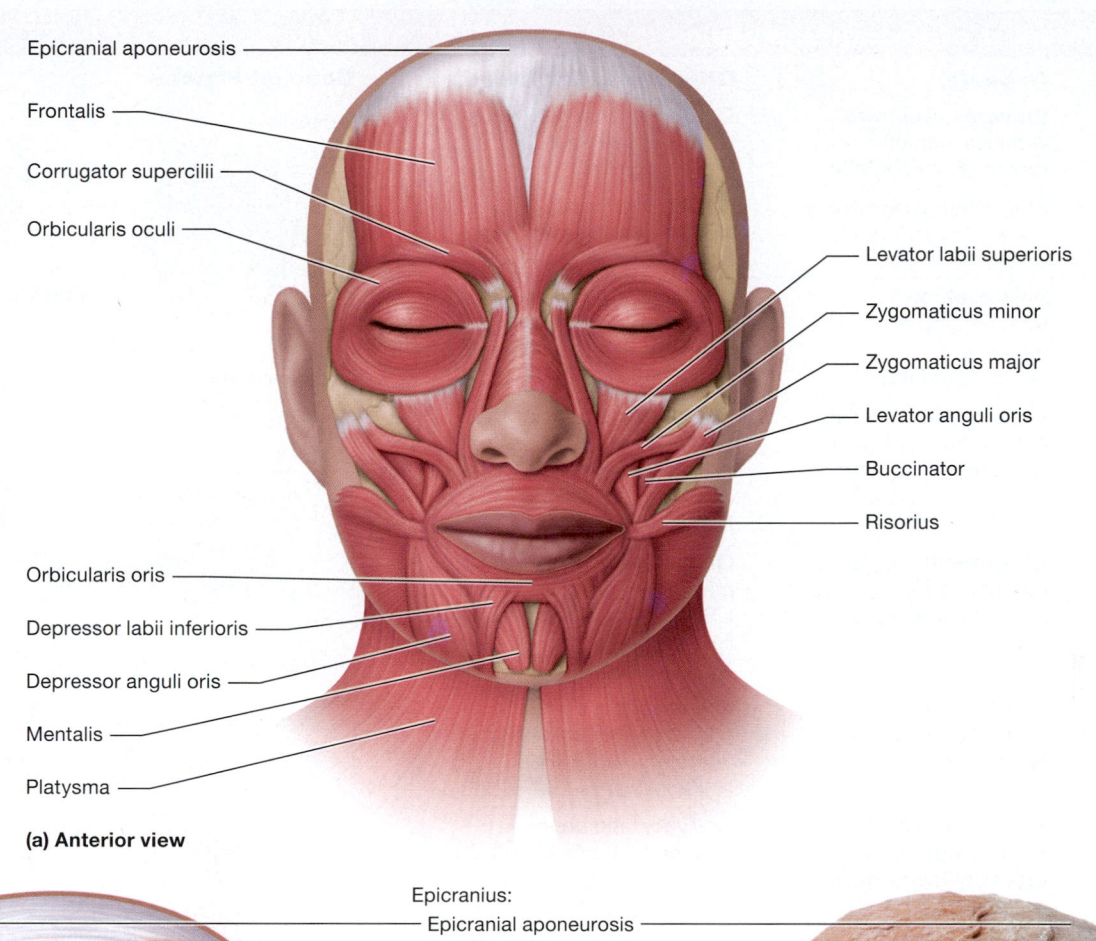

Epicranial aponeurosis

Frontalis

Corrugator supercilii

Orbicularis oculi

Levator labii superioris

Zygomaticus minor

Zygomaticus major

Levator anguli oris

Buccinator

Risorius

Orbicularis oris

Depressor labii inferioris

Depressor anguli oris

Mentalis

Platysma

(a) Anterior view

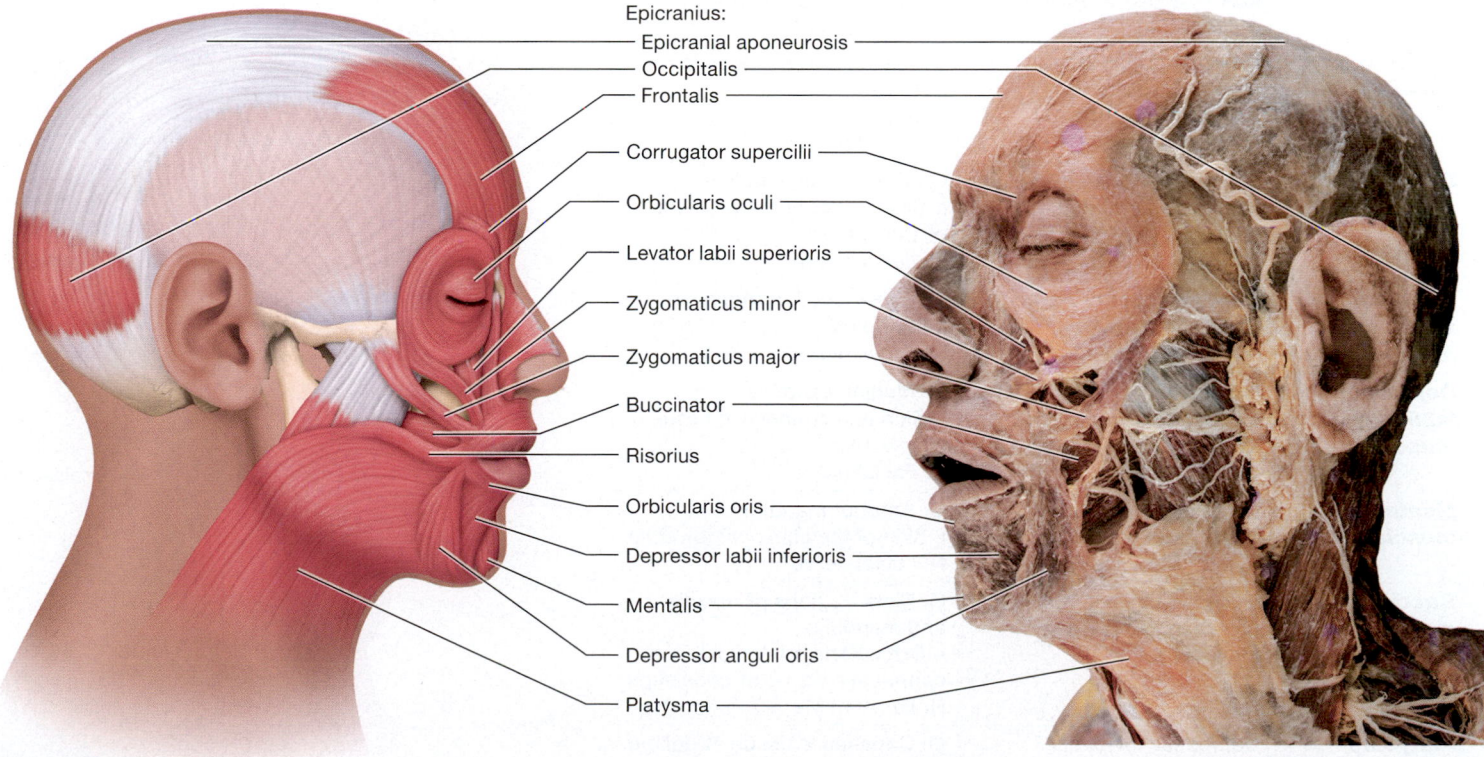

Epicranius:
Epicranial aponeurosis
Occipitalis
Frontalis

Corrugator supercilii

Orbicularis oculi

Levator labii superioris

Zygomaticus minor

Zygomaticus major

Buccinator

Risorius

Orbicularis oris

Depressor labii inferioris

Mentalis

Depressor anguli oris

Platysma

(b) Lateral view

Figure 9.8 Muscles of facial expression.

Explore **PAL**™ Cadaver
@ **Mastering** Anatomy & Physiology

Table 9.2 Muscles of Facial Expression

Muscle	Action(s)	Origin/Insertion/Nerve(s)	Concept Figures
Frontalis muscle	**Elevates** eyebrows; wrinkles skin of forehead horizontally	O: Epicranial aponeurosis I: Skin of eyebrows N: Facial nerve	
Occipitalis muscle	Pulls scalp **posteriorly**	O: Occipital bone I: Epicranial aponeurosis N: Facial nerve	
Corrugator supercilii muscle	Pulls eyebrows **inferiorly** and medially (as in squinting)	O: Medial supraorbital margin of frontal bone I: Skin of medial eyebrows N: Facial nerve	
Orbicularis oculi muscle	**Closes** eye; pulls skin around the eyes (as in blinking and squinting)	O: Orbital portions of the frontal bone and maxilla I: Skin of the orbital area and eyelids N: Facial nerve	
Levator labii superioris muscle	**Elevates** the upper lip; everts and furrows upper lip (as in sneering)	O: Zygomatic bone and upper maxilla near orbit I: Skin and muscle of the upper lip N: Facial nerve	
Zygomaticus minor muscle	**Elevates** lateral portion of the upper lip (as in smiling)	O: Zygomatic bone I: Skin and muscle of the lateral upper lip N: Facial nerve	
Zygomaticus major muscle	Pulls the angle of the mouth **superiorly** and **laterally** (as in smiling or laughing)	O: Zygomatic bone I: Lateral muscle fibers of corner/angle of mouth N: Facial nerve	
Risorius muscle	Pulls the angle of the mouth **laterally** to make a closed-mouth smile or smirk	O: Connective tissue anterior to the ear I: Modiolus* N: Facial nerve	
Orbicularis oris muscle	**Closes** and **protrudes** lips (as in pursing the lips)	O: Maxilla and mandible I: Skin and connective tissue of the lips N: Facial nerve	
Depressor anguli oris muscle	Draws corners of the mouth **inferiorly** (unhappy face)	O: Lower body of mandible I: Modiolus* N: Facial nerve	
Depressor labii inferioris muscle	**Protrudes** lower lip (sad or pouting expressions)	O: Medial mandible I: Skin and connective tissue of lower lip N: Facial nerve	
Mentalis muscle	**Protrudes** the lower lip and chin for drinking and "doubtful" expression	O: Anterior mandible I: Skin of the chin near lower lip N: Facial nerve	
Buccinator muscle	**Pulls** the cheeks in, producing sucking movements	O: Molar regions of maxilla and mandible I: Orbicularis oris muscle and connective tissue of cheek/lips N: Facial nerve	
Platysma muscle	Depresses lower lip and opens mouth by **depressing** the mandible; tenses skin of the neck	O: Connective tissue of deltoid and pectoralis major muscles I: Mandible; skin and connective tissue inferior to mouth N: Facial nerve	

Frontalis

Occipitalis

Corrugator supercilii

Orbicularis oculi

Levator labii superioris

Zygomaticus minor — Zygomaticus major

Risorius

Orbicularis oris

Depressor anguli oris

Depressor labii inferioris — Mentalis

Buccinator

Platysma

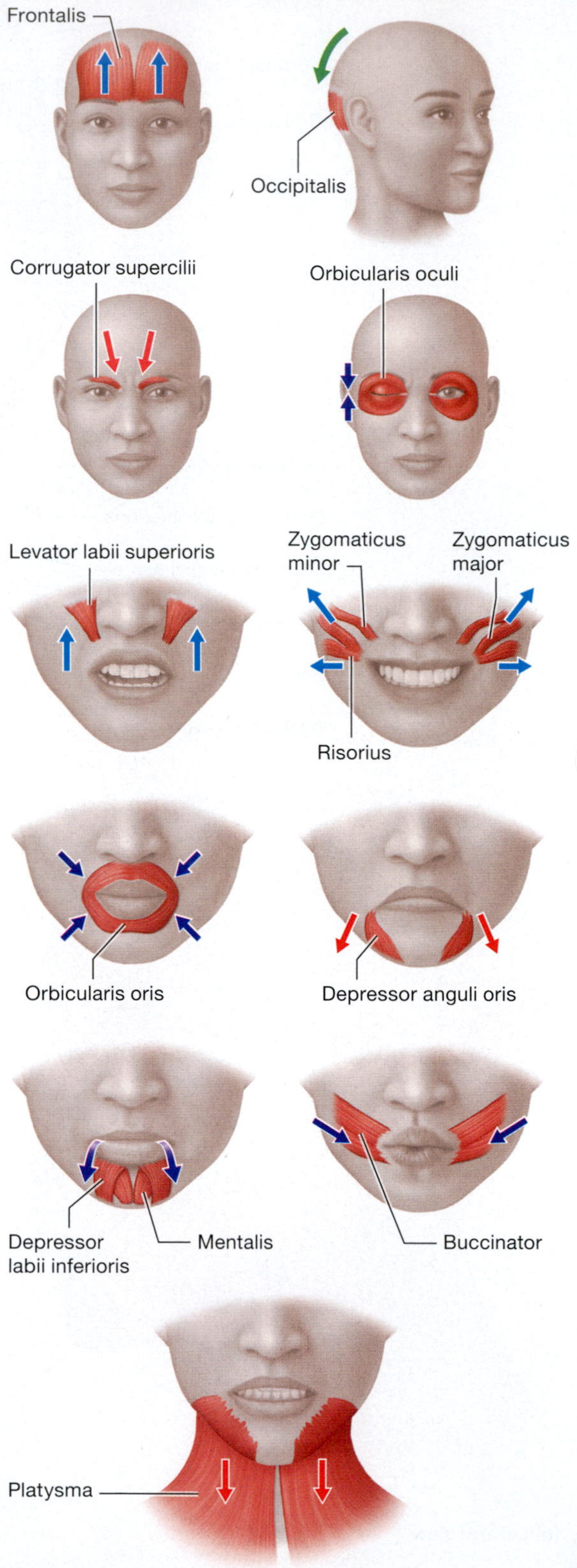

Note: Colors of actions and/or directions of action in Action(s) column match colors of directional arrow(s) in Concept Figures.

*The modiolus (moh-DY-oh-lus) is a structure that mixes muscle and connective tissue at the corners of the mouth.

Quick Check

☐ 1. How do the muscles of facial expression differ from most other skeletal muscles?

☐ 2. Which muscles are involved in opening and closing the eye?

☐ 3. What are three muscles involved in producing happy facial expressions?

☐ 4. Which muscles are involved in producing frowning or pouting facial expressions?

Extrinsic Eye Muscles

《 FLASHBACK

1. What do the directional terms superior, inferior, medial, and lateral mean? (p. 12)

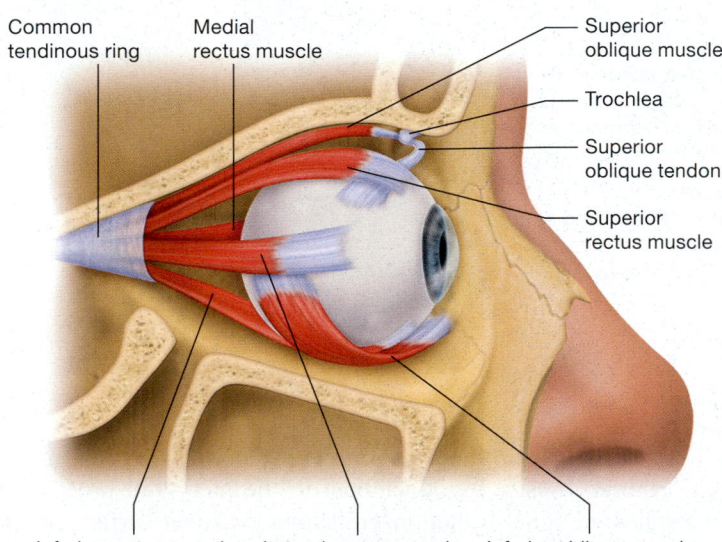

(a) Lateral view of the right eye

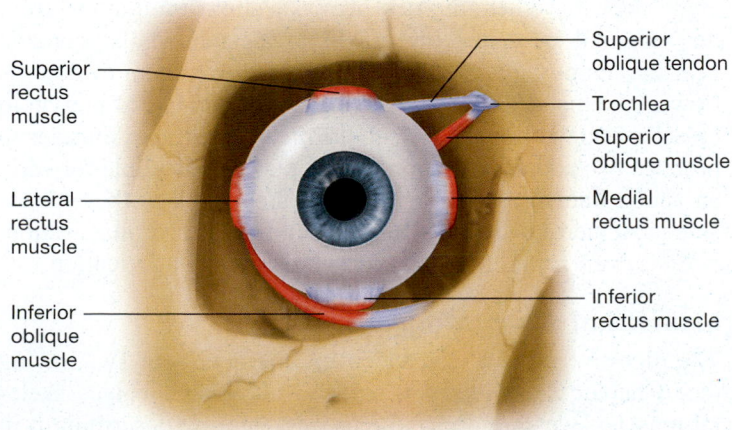

(b) Anterior view of the right eye

Figure 9.9 Extrinsic eye muscles.

Six skeletal muscles, known as **extrinsic eye muscles,** move each eye. Four of our six extrinsic eye muscles are *rectus* muscles, named for their straight-line placement around the eyeball ("rectus" means "straight"; **Figure 9.9**). As a group, these muscles originate from a circular ring of dense collagenous connective tissue called the *common tendinous ring* at the posterior orbit. From here the **superior rectus**, **inferior rectus, medial rectus,** and **lateral rectus muscles** diverge to insert on the corresponding surface of the eyeball for which they are named.

These names tell you not only the location of these muscles, but also some of their main actions. The medial and lateral rectus muscles are fairly straightforward: They move the eye medially (toward the nose) and laterally (toward the temple), respectively. However, the superior and inferior rectus muscles have more than one action. The superior rectus muscle moves the eye superiorly and the inferior rectus muscle moves the eye inferiorly, as their names would imply. But both muscles also move the eye medially.

The two other extrinsic eye muscles, the superior and inferior oblique muscles, attach to the eyeball at an angle, which allows them to pull the eyeball in a direction opposite to their names. The **superior oblique muscle** originates near the common tendinous ring at the posterior orbit. On the way to its insertion, its tendon passes through a fibrocartilaginous structure called the *trochlea* (TROHK-lee-uh; "pulley") near the anterior orbit, which redirects it at an angle to its insertion onto the superolateral eyeball. So, unlike the superior rectus, which pulls the eye superiorly and medially, the trochlea causes the superior oblique muscle to rotate the eye *inferiorly* and laterally.

Whereas the other five extrinsic eye muscles originate from the posterior orbit, the **inferior oblique muscle** originates from the anterior orbit, specifically the anterior maxilla. From the maxilla, the inferior oblique muscle crosses under the eyeball to insert between the inferior and lateral rectus muscles. When it contracts, it rotates the eyeball *superiorly* and laterally.

See **Table 9.3** on p. 298 for more information about the extrinsic eye muscles.

Quick Check

☐ 5. Which extrinsic eye muscles pull the eyeball medially? Which pull it laterally?

☐ 6. Which extrinsic eye muscles pull the eyeball superiorly? Which pull it inferiorly?

Muscles of the Head and Neck

《 FLASHBACK

1. Where are the pterygoid processes of the sphenoid bone located? (p. 218)

2. What is the hyoid bone? (p. 229)

Many of the muscles of the head and neck go relatively unnoticed in their day-to-day functions. For instance, you likely don't think about the skeletal muscles that maintain your posture or allow

Table 9.3 Extrinsic Eye and Orbit Muscles

Muscle(s)	Action(s)	Origin/Insertion/Nerve(s)	Concept Figures
Superior rectus muscle	**Elevates** (rotates upward) anterior eye, as in looking up, and turns eye **medially**	O: Common tendinous ring of posterior orbit I: Superior surface of eye N: Oculomotor nerve	
Inferior rectus muscle	**Depresses** (rotates downward) anterior eye, as in looking down, and turns eye **medially**	O: Common tendinous ring of posterior orbit I: Inferior surface of eye N: Oculomotor nerve	
Medial rectus muscle	Pulls anterior eye **medially**, toward the nose	O: Common tendinous ring of posterior orbit I: Medial surface of eye N: Oculomotor nerve	
Lateral rectus muscle	Pulls anterior eye **laterally**, as when looking toward the temple	O: Common tendinous ring of posterior orbit I: Lateral surface of eye N: Abducens nerve	
Superior oblique muscle	Pulls (and helps rotate) anterior eye **inferiorly** and **laterally**, as when left eye rolls to left and down	O: The posterior orbit I: Superior/lateral surface of eye after passing through the trochlea N: Trochlear nerve	
Inferior oblique muscle	Pulls (and helps rotate) anterior eye **superiorly** and **laterally**, as when left eye rolls to left and up	O: Maxilla just inside the anterior portion of the orbit I: Inferior/lateral surface of eye N: Oculomotor nerve	
Levator palpebrae superioris muscle	**Elevates** (opens) upper eyelid	O: Posterior orbit I: Connective tissue and skin of upper eyelid N: Oculomotor nerve	

you to swallow. We examine these muscles, and the important roles they play, in the next subsections. First, however, we look at the muscles involved in chewing, also known as mastication.

Muscles of Mastication

Have you ever heard that a person can hold an alligator's jaws closed with just their bare hands? It's at least partly true, with the right grip, a fair amount of hand strength, and some luck. This is because alligators have more muscles associated with clamping their jaws shut than opening them. Similarly, most of our powerful jaw muscles involve *elevation* of the mandible, or closing the jaw.

Two major muscles elevate the mandible: the masseter and temporalis muscles. The **masseter muscle** (MASS-uh-tur) is a thick, bandlike muscle that originates from the zygomatic arch and inserts into the lateral surface of the mandible (**Figure 9.10a**). The **temporalis muscle** (tem-pohr-AEH-lis) is a fan-shaped, convergent muscle that originates from the flat portion of the temporal bone and inserts into the coronoid process of the mandible. Together, these two muscles provide much of the force needed for elevating the mandible to masticate.

Two additional sets of mastication muscles attach to the mandible (**Figure 9.10b**). The **medial pterygoid muscle** (TEHR-ih-goyd) originates from the posterior mouth and pharynx (throat) mostly by attachment to the pterygoid process of the sphenoid bone. It then inserts along the inner surface of the inferior mandible, which allows it to elevate the mandible as a synergist to the masseter and temporalis muscles. Notice in the figure that the **lateral pterygoid muscles** attach at a different angle than the other muscles of mastication, originating from the sphenoid bone and inserting into the mandibular condyle near the temporomandibular joint. This allows them to depress (lower) the mandible rather than elevate it. The lateral pterygoid muscles also work together with the medial pterygoid muscles to produce two additional actions: They pull the mandible forward, an action known as *protraction*, and they produce the side-to-side movements necessary for grinding food during mastication. See **Table 9.4** for more details on the muscles of mastication.

Muscles of Swallowing

The digestive system chapter covers the details of swallowing (see Chapter 22). However, since swallowing involves skeletal muscles, we summarize the process here by describing it in two stages of muscular action: (1) muscles that push food to the posterior oral cavity, and (2) muscles that initiate swallowing

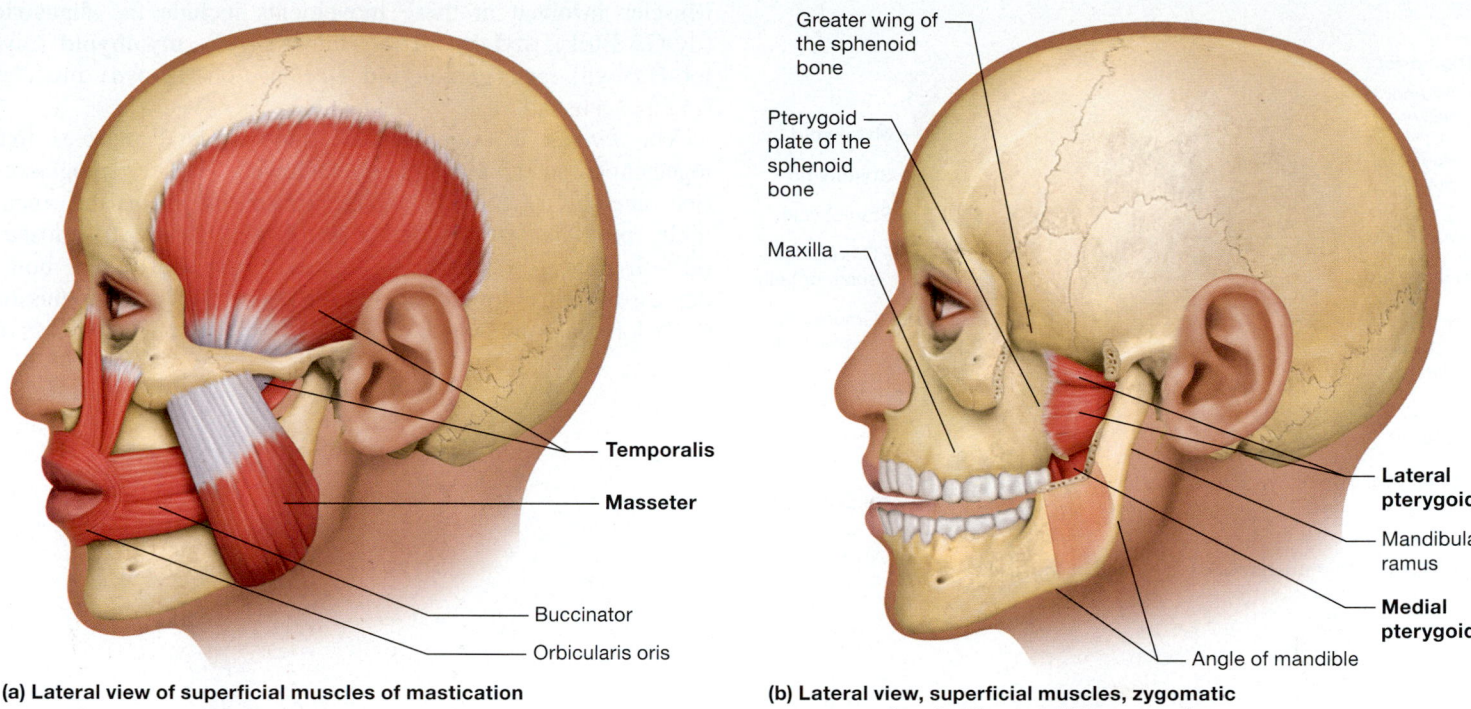

(a) Lateral view of superficial muscles of mastication

- Temporalis
- Masseter
- Buccinator
- Orbicularis oris

- Greater wing of the sphenoid bone
- Pterygoid plate of the sphenoid bone
- Maxilla
- **Lateral pterygoid**
- Mandibular ramus
- **Medial pterygoid**
- Angle of mandible

(b) Lateral view, superficial muscles, zygomatic process, and part of the mandible removed.

Play A&P Flix Animation
@ Mastering Anatomy & Physiology

Figure 9.10 Muscles of mastication. Labels in **bold** indicate muscles involved in this action.

Table 9.4 Muscles of Mastication

Muscle	Action(s)	Origin/Insertion/Nerve(s)	Concept Figures
Masseter muscle	**Elevates** the mandible	O: Zygomatic arch I: Angle and lateral surface of ramus of mandible N: Mandibular nerve (branch of the trigeminal nerve)	Temporalis / Masseter
Temporalis muscle	**Elevates** and **retracts** the mandible	O: Lateral surface of skull I: Coronoid process of the mandible N: Mandibular nerve	
Medial pterygoid muscle	**Elevates** and **protracts** the mandible; assists in lateral movements to grind food	O: Pterygoid plate of the sphenoid, small portion of the maxilla I: Medial angle and ramus of the mandible N: Mandibular nerve	Lateral pterygoid
Lateral pterygoid muscle	**Protracts** and **depresses** the mandible; lateral movements to grind food	O: Pterygoid plate and greater wing of the sphenoid I: Condyle and neck of mandible N: Mandibular nerve	Medial pterygoid

in the **pharynx** (FEHR-inks), or throat, and push food into the *esophagus* (eh-SAWF-uh-gus), the passageway that connects the pharynx to the stomach.

Pushing Food to the Posterior Oral Cavity   Just before we begin to swallow, muscles of mastication elevate the mandible.

At the same time, muscles originating from the temporal bone and mandible and inserting into the hyoid bone contract, an action that elevates the hyoid bone. When this bone is elevated, the tongue and floor of the mouth rise in preparation for swallowing, and food is pushed posteriorly toward the pharynx.

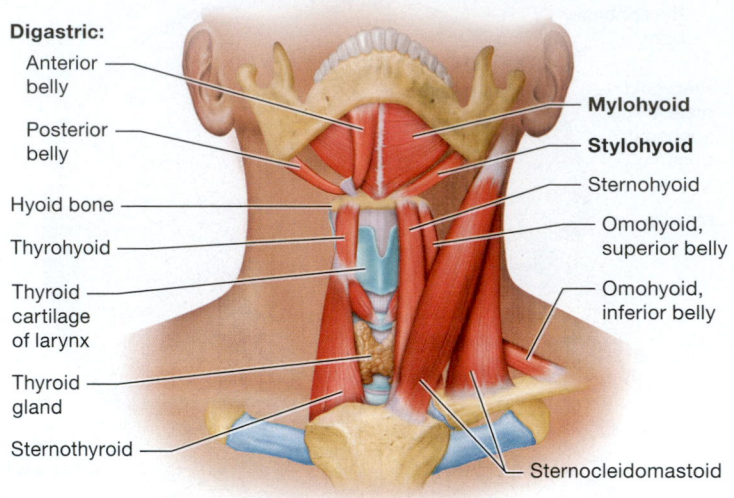

Digastric:
- Anterior belly
- Posterior belly

Hyoid bone
Thyrohyoid
Thyroid cartilage of larynx
Thyroid gland
Sternothyroid

Mylohyoid
Stylohyoid
Sternohyoid
Omohyoid, superior belly
Omohyoid, inferior belly
Sternocleidomastoid

(a) Muscles that push food to the back of the oral cavity

Muscles involved in these movements include the **digastric** (dy-GAS-trik), **stylohyoid** (sty-loh-HY-oyd), **mylohyoid** (my-loh-HY-oyd), and **geniohyoid** (jee′-nee-oh-HY-oyd) **muscles** (**Figure 9.11a** and **b**).

The *tongue* is composed of three skeletal muscles that together play an important role in swallowing: (1) The **genioglossus** (jee′-nee-oh-GLOSS-us), which originates from the mandible, protrudes (sticks out) the tongue; (2) the **hyoglossus** (HY-oh-gloss-us), which originates from the hyoid bone, depresses (pulls down) the tongue; and (3) the **styloglossus** (STY-loh-gloss-us), which originates from the styloid process of

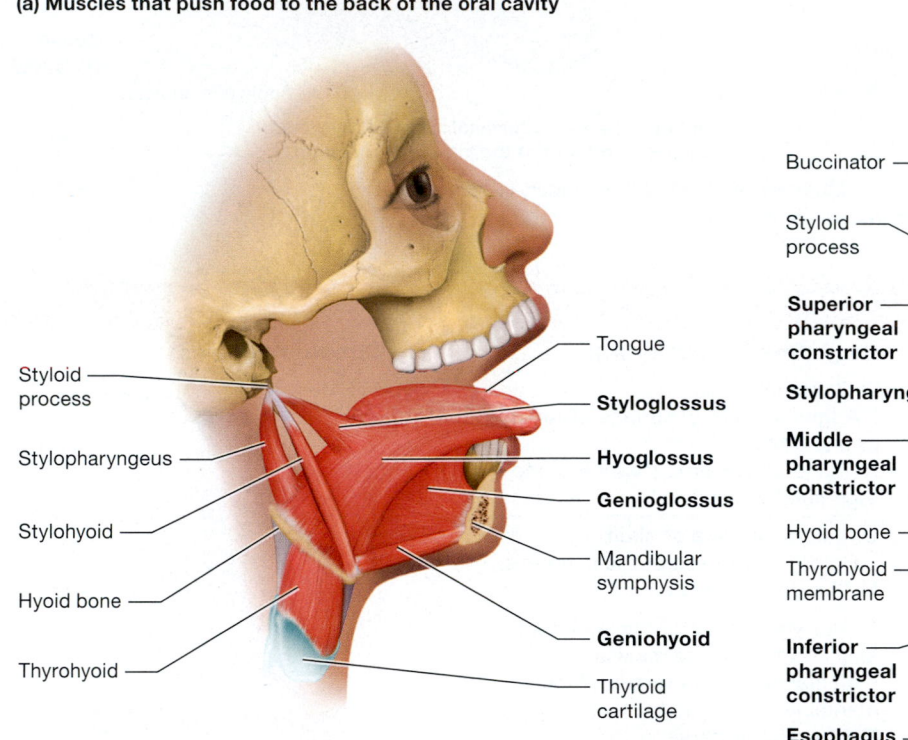

Styloid process
Stylopharyngeus
Stylohyoid
Hyoid bone
Thyrohyoid

Tongue
Styloglossus
Hyoglossus
Genioglossus
Mandibular symphysis
Geniohyoid
Thyroid cartilage

(b) Muscles that change the shape of the oral cavity and manipulate food

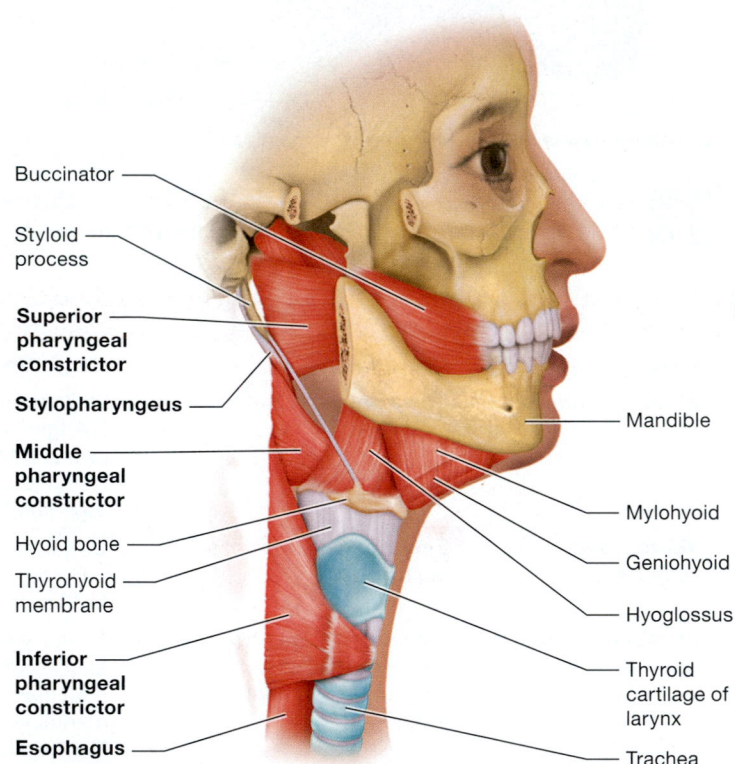

Buccinator
Styloid process
Superior pharyngeal constrictor
Stylopharyngeus
Middle pharyngeal constrictor
Hyoid bone
Thyrohyoid membrane
Inferior pharyngeal constrictor
Esophagus

Mandible
Mylohyoid
Geniohyoid
Hyoglossus
Thyroid cartilage of larynx
Trachea

(c) Muscles that are involved in swallowing

Figure 9.11 Muscles of swallowing. Labels in **bold** indicate muscles involved in the phases of swallowing.

Table 9.5 Muscles of the Tongue

Muscle(s)	Action(s)	Origin/Insertion/Nerve(s)	Concept Figure
Genioglossus muscle	**Protrudes** tongue	O: Internal/anterior mandible (near symphysis) I: Majority of muscle blends into the tongue; body of hyoid bone N: Hypoglossal nerve	
Hyoglossus muscle	**Depresses** tongue	O: Hyoid bone (greater horn) I: Posterolateral surface of the tongue N: Hypoglossal nerve	Styloglossus
Styloglossus muscle	**Retracts** and **elevates** tongue	O: Styloid process of temporal bone I: Lateral/posterior border of tongue N: Hypoglossal nerve	Hyoglossus Genioglossus

the temporal bone, elevates and retracts (pulls up and back) the tongue (see Figure 9.11b). Together the tongue muscles manipulate food within the oral cavity during mastication, which helps to break it down and mix it with saliva. As swallowing is initiated, these muscles contract and move food posteriorly toward the pharynx. **Table 9.5** summarizes the muscles of the tongue.

Pushing Food into the Esophagus The pharynx is briefly elevated to receive food as it enters. You can feel this yourself by placing your hand over your hyoid bone as you swallow—see how it moves superiorly. However, the pharynx and hyoid bone are quickly depressed by the contraction of the next group of muscles. This action pushes food farther down into the pharynx while also preventing us from inhaling the food (having it "go down the wrong pipe"). These muscles insert into either the hyoid bone or a piece of cartilage in the anterior neck called the *thyroid cartilage*, which is reflected in their names. Muscles in this group include the **sternohyoid** (ster-noh-HY-oyd), **sternothyroid** (ster-noh-THY-royd), **omohyoid** (oh-moh-HY-oyd), and **thyrohyoid** (thy-roh-HY-oyd) **muscles.** Finally, the **pharyngeal constrictor muscles** (fah-RIN-jee-uhl) rapidly contract in sequence from superior to inferior to push the food into the esophagus (**Figure 9.11c**). **Table 9.6** on p. 302 summarizes the muscles of swallowing.

ConceptBOOST)))

Demystifying Muscle Actions

The number of muscle actions that we discuss in this chapter might seem daunting, so you may be wondering at this point how you can possibly memorize all of them. Well, the key to success here is not to try to *memorize* them. Instead, learn how to figure out the muscles' actions by looking at their origins and insertions.

Here's how this works. First, remember that a muscle's origin is usually the anchoring point, the part that doesn't move, and the insertion is the part that moves. Second, remember that the muscle usually moves the insertion *toward* the origin. With this in mind you can use an extremely simple technique we'll call the "fingertip method": Place the base of your palm over the origin, and your fingertips of the same hand over the insertion. Then pull your fingertips in like you're pulling on the insertion to mimic the muscle's contraction. In this way, you can imagine how the muscle pulls on the insertion when it contracts.

Let's try it with a muscle you probably didn't know about before reading this chapter: the medial pterygoid muscle. Take a look at the illustration of the fingertip method showing the hand placement over the muscle's origin and insertion, and then the movement of the fingertips "pulling" on the mandible.

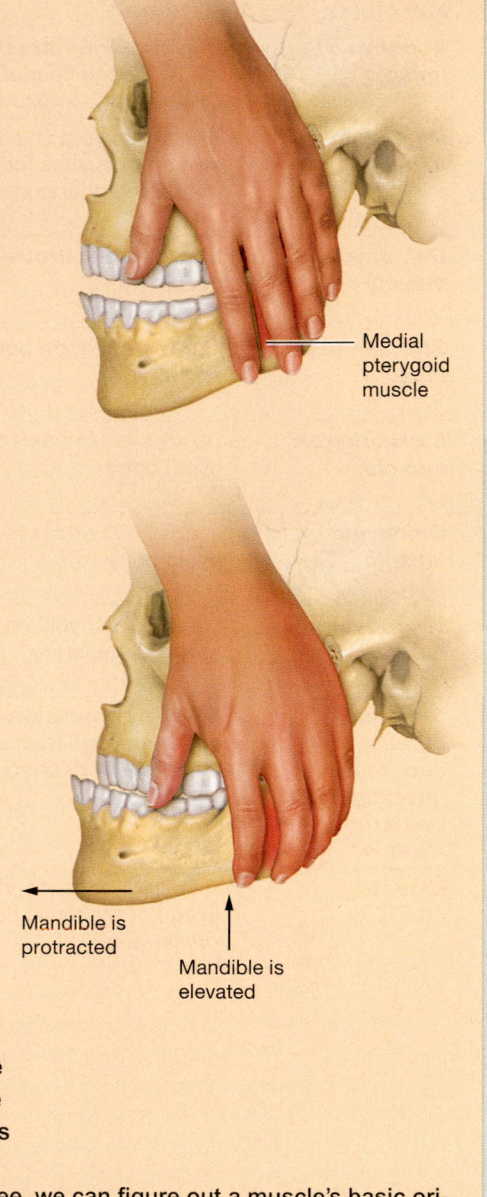

You can see that the fingers would "pull" the mandible up (elevation) and forward (protraction). From this you can easily figure out that the actions of the medial pterygoid muscle are elevation and protraction of the mandible.

One nice thing about this method is that it works even if you can't remember a muscle's exact origin and insertion, because usually you can see which part is less moveable, and so figure out the origin. In our example of the medial pterygoid muscle, clearly the sphenoid bone is not going to move, so that has to be the origin. The mandible is the only moveable bone to which the muscle attaches, so that has to be the insertion.

So, as you can see, we can figure out a muscle's basic origin, insertion, and actions with no really heavy memorization required. All that's needed is some logic, observation, and a bit of imagination. ■

Medial pterygoid muscle

Mandible is protracted

Mandible is elevated

Muscles That Move the Head and Neck

Although we have discussed many muscles *found* in the head and neck region, few of them so far play any role in actually moving the head or neck. One of the main muscles that does play this role is one we mentioned earlier, the **sternocleidomastoid muscle** (ster-noh-KLY-doh-mass′-toyd). This muscle originates from the sternum and clavicle, and crosses diagonally over the lateral neck to insert into the mastoid process of

Table 9.6 Muscles of Swallowing

Muscle(s)	Action(s)	Origin/Insertion/Nerve(s)	Concept Figures
Stylohyoid muscle	**Elevates** and **retracts** the hyoid and floor of the mouth during swallowing	O: Styloid process of temporal bone I: Hyoid bone N: Facial nerve	
Mylohyoid muscle	**Elevates** hyoid and floor of mouth, pushes food toward the pharynx	O: Medial portion of the mandible I: Hyoid bone and median connective tissue raphe (seam) N: Mylohyoid nerve	
Geniohyoid muscle	**Elevates** and **protracts** hyoid	O: Inner surface of mandibular symphysis I: Hyoid bone N: Branch of the cervical plexus*	
Sternohyoid muscle	**Depresses** hyoid bone and larynx	O: Posterior manubrium of sternum I: Inferior portion of hyoid bone N: Branch of the cervical plexus	
Sternothyroid muscle	**Depresses** larynx and hyoid bone	O: Posterior manubrium of sternum I: Thyroid cartilage of the larynx N: Branch of the cervical plexus	
Omohyoid muscle	**Depresses** and **retracts** hyoid bone	O: Superior border of scapula I: Inferior portion of hyoid bone N: Branch of the cervical plexus	
Thyrohyoid muscle	**Depresses** hyoid bone; may elevate larynx	O: Thyroid cartilage of the larynx I: Hyoid bone N: Branch of the cervical plexus*	
Superior, middle, and inferior pharyngeal constrictor muscles	Stepwise **contraction** squeezes food from top to bottom of pharynx into esophagus	O: Superior—sphenoid Middle—hyoid Inferior—larynx I: Posterior portion of pharynx N: Branches of vagus nerve	
Digastric muscle	**Depresses** the mandible; fixator of the hyoid during swallowing	O: Anterior belly: anterior inner margin of the mandible; posterior belly: into mastoid notch, medial to the mastoid process I: Connective tissue of the hyoid bone N: Mylohyoid nerve (anterior belly) and facial nerve (posterior belly)	

*Cervical nerve fibers are distributed to this muscle via the hypoglossal nerve.

the temporal bone (**Figure 9.12**). This diagonal pathway allows each sternocleidomastoid muscle acting individually to rotate the head toward the opposite shoulder. When acting together, the sternocleidomastoid muscles flex the head (as in bowing it).

Another muscle group that moves the head are the anterior, middle, and posterior **scalene muscles** (SKAY-leen). The scalene muscles originate from the transverse processes of cervical vertebrae and insert into the first two ribs. When each group contracts individually, the head is pulled laterally (i.e., the ear is pulled toward the shoulder). However, in this muscle group, the origin and insertion can switch. When fixators hold the head and neck in place and the muscle groups on both sides contract together, the ribs are elevated instead. This movement plays an important part in deep breathing.

On the posterior head and neck we find the diamond-shaped **trapezius muscle** (trah-PEE-zee-us). This muscle has a long

origin that extends from the 12th thoracic vertebra all the way to the occipital bone. It is divided into three parts; the superior part that attaches to the occipital bone extends the head (we discuss its other actions in Module 9.4). Deep to the trapezius muscle we find the **splenius capitis muscle** (SPLEN-ee-us KAP-ih-tus) and **splenius cervicis muscle** (SER-vis-us). These muscles originate along the cervical and thoracic vertebrae and are named for their insertions: The splenius capitis muscle inserts into the head (*capit-* = "head") at the mastoid process, and the splenius cervicis muscle inserts into the neck along the transverse processes of cervical vertebrae. When contracting on both sides, these muscles extend the head and neck. When they contract on only one side, they rotate the head and neck to the same side.

See **Table 9.7** for more details about these muscles.

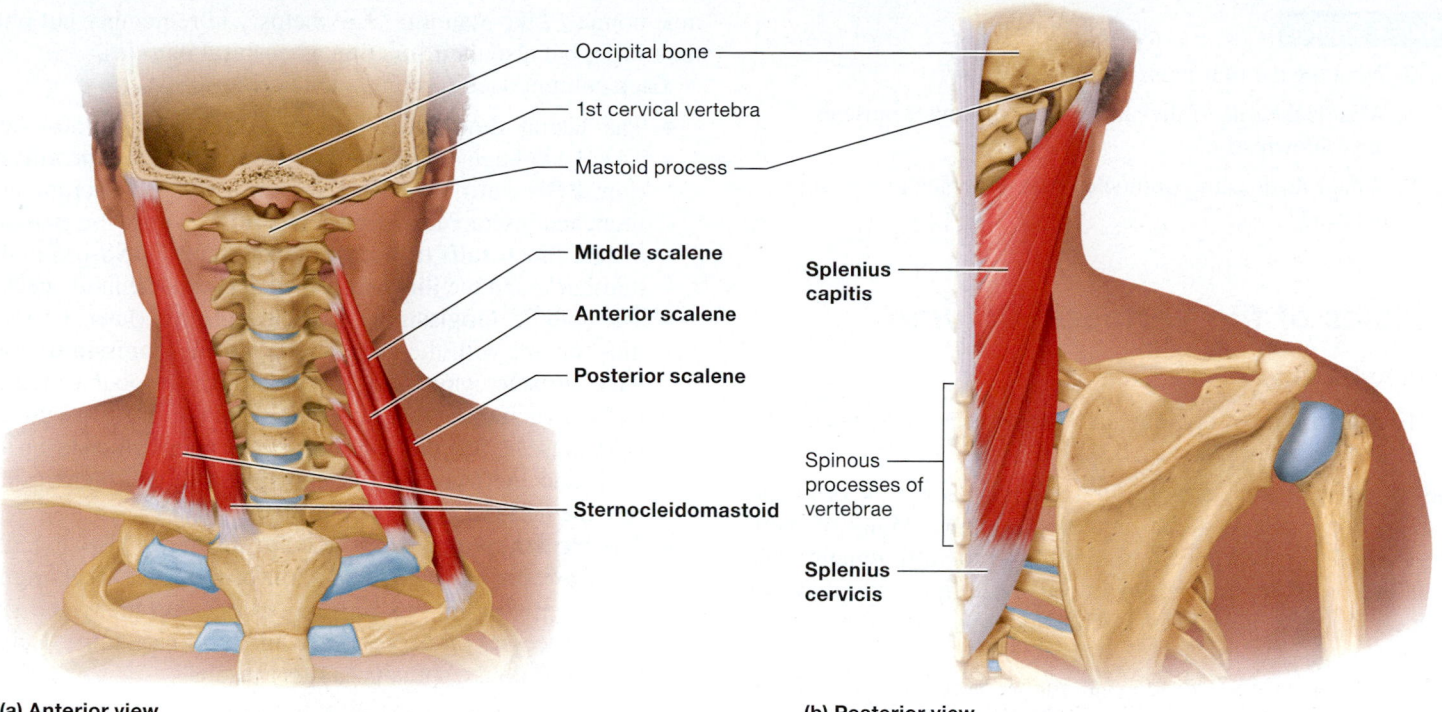

(a) Anterior view

Occipital bone
1st cervical vertebra
Mastoid process
Middle scalene
Anterior scalene
Posterior scalene
Sternocleidomastoid

(b) Posterior view

Splenius capitis
Spinous processes of vertebrae
Splenius cervicis

Figure 9.12 Muscles that move the head and neck. Labels in **bold** indicate muscles involved in these actions.

Table 9.7 Muscles That Move the Head and Neck

Muscle(s)	Action(s)	Origin/Insertion/Nerve(s)	Concept Figures
Sternocleidomastoid muscle	Together: **flex** head; individually: flex and **rotate** the head toward opposite side; accessory muscles of inspiration	O: Manubrium of sternum, medial portion of clavicle I: Mastoid process of temporal bone N: Accessory nerve	
Scalene muscles	Move head **laterally** when contracted individually; **elevate** rib cage with vertebral column fixed; accessory muscles of inspiration	O: Tubercles of transverse processes of C2–C7 I: Laterally on the first two ribs N: Spinal nerve branches of C3–C8	
*Trapezius muscle (superior section)**	**Extends** the head (raises head from "bowed" position)	O: External occipital protuberance and cervical vertebrae I: Lateral portion of clavicle N: Accessory nerve	
Splenius capitis muscle	**Extends** the head; with other muscles, **rotates** the head to the same side as the muscle that is contracting	O: Spinous processes of cervical vertebrae and connective tissue of the posterior neck (the ligamentum nuchae) I: Mastoid process and occipital bone N: Branches of C2–C3 posterior rami (posterior branches of spinal nerves)	
Splenius cervicis muscle	**Extends** the head and neck; **rotates** the upper cervical vertebrae to the same side as the muscle that is contracting	O: Spinous processes of T3–T6 I: Transverse processes of C1–C4 N: Branches of the lower cervical posterior rami	

*The trapezius muscle is covered in more detail in Module 9.4.

☐ 7. What are the four muscles of mastication?

☐ 8. What is the role of the pharyngeal constrictor muscles in swallowing?

☐ 9. What role does the sternocleidomastoid play in moving the head?

Muscles of the Vertebral Column

« FLASHBACK

1. Where are the spinous and transverse processes of vertebrae located? (p. 233)

The muscles of the vertebral column are those that originate from and/or insert into one or more vertebrae. Many of them are *postural muscles*, meaning that they generally remain contracted to some degree to help us remain upright. You may not think about these muscles and their functions very often, unless you injure one of them.

One of the largest muscles of the vertebral column is the **erector spinae muscle group** (eh-REK-tohr SPY-nee; **Figure 9.13**). It is made up of nine muscles, which are divided into three vertical groups, or *columns,* arranged along each side of the vertebral column: the lateral *iliocostalis muscle*, the middle *longissimus muscle*, and the medial *spinalis muscle*. To remember the order of these muscles from lateral to medial, keep in mind the

mnemonic: *I Like Standing*. This helps you remember not only their order but also their function as postural muscles.

Each column itself has three divisions:

- The lateral **iliocostalis muscle column** (il′-ee-oh-kaws-TAEH-lis) begins with the **iliocostalis lumborum muscle** (lum-BOH-rum), which originates from the sacrum and ilium and inserts into the posterior ribs. Its next two muscles are the **iliocostalis thoracis muscle** (thoh-RASS-iss) in the mid-back, and the **iliocostalis cervicis muscle** in the neck.
- The middle **longissimus muscle column** (lawn-JISS-ih-mus) begins with the large and powerful **longissimus thoracis muscle,** which originates from the lumbar vertebrae and inserts into thoracic vertebrae and their nearby ribs. Its next muscles are the **longissimus cervicis muscle** in the neck and the **longissimus capitis muscle,** which inserts into the mastoid process of the head.
- The medial **spinalis muscle column** (spy-NAEH-lis) also has **thoracis, cervicis,** and **capitis muscles.** The origins and insertions for this muscle column run along the spinous processes of the vertebrae and terminate at the occipital bone with the spinalis capitis muscle.

In addition to being one of the most important sets of postural muscles in the body, most of the erector spinae muscles are powerful extensors of the vertebral column. Some also help rotate or laterally flex the vertebral column when contracted on one side only. The capitis and cervicis muscles also play minor roles in extending (or rotating) the head and neck.

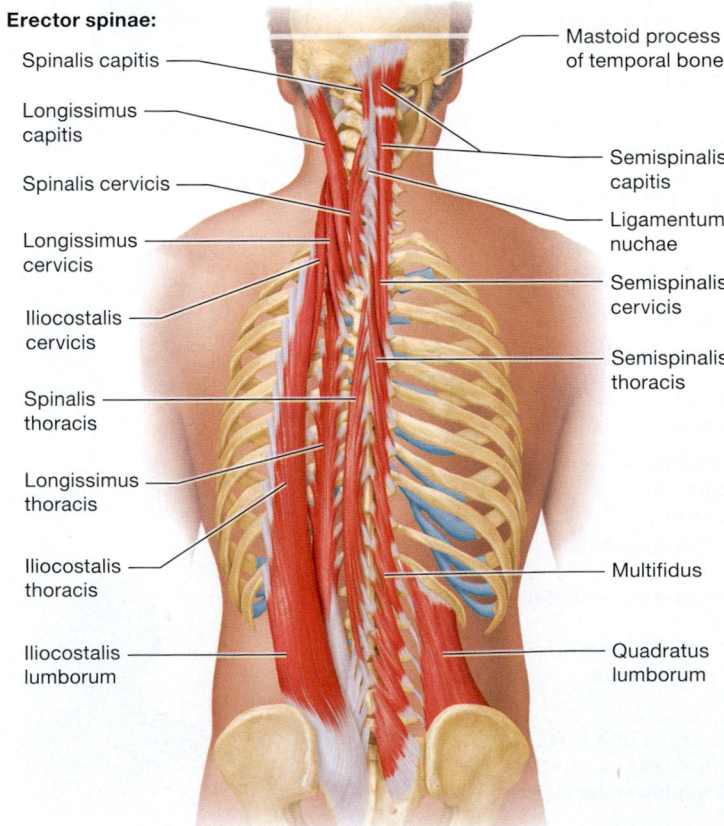

(a) The deep muscles of the back

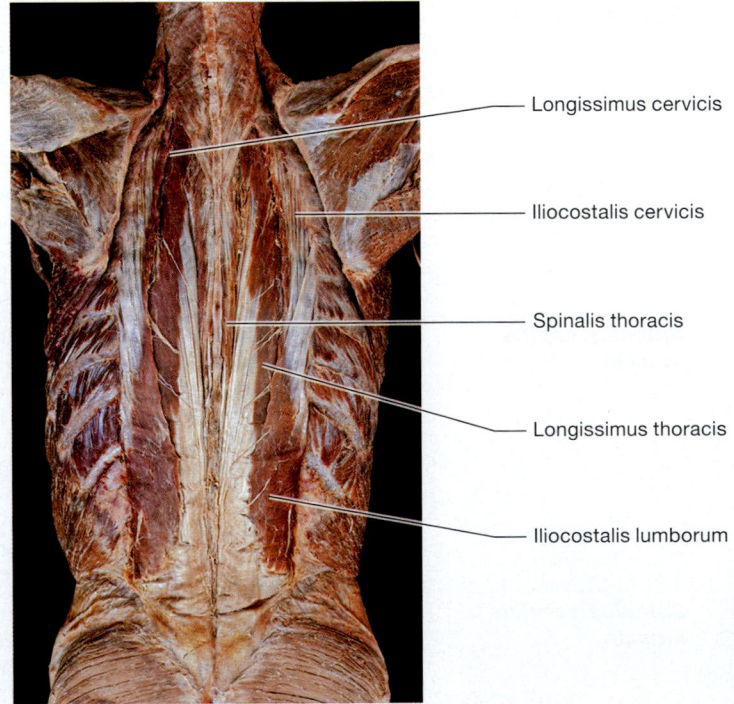

(b) Cadaver photo

Figure 9.13 Muscles of the vertebral column.

Another vertebral muscle group is the *transversospinal group*. As the group's name implies, these muscles are located along the vertebral column between the transverse and spinous processes of vertebrae. This group is composed of the **semispinalis, multifidus** (mul-TIF-ih-dus), and **rotatores** (roh-tah-TOHR-eez) **muscles.** Collectively, their actions are similar to those of the erector spinae muscle group: They extend the head, neck, and vertebral column (see Figure 9.13a). Another muscle of the vertebral column that helps extend and laterally flex the vertebral column is the **quadratus lumborum muscle** (kwad-RAY-tus) in the inferior back.

See **Table 9.8** for details of these muscles as well as the others described in this section.

Table 9.8 Muscles of the Vertebral Column

Muscle(s)	Action(s)	Origin/Insertion/Nerve(s)	Concept Figures
Erector Spinae Muscle Group			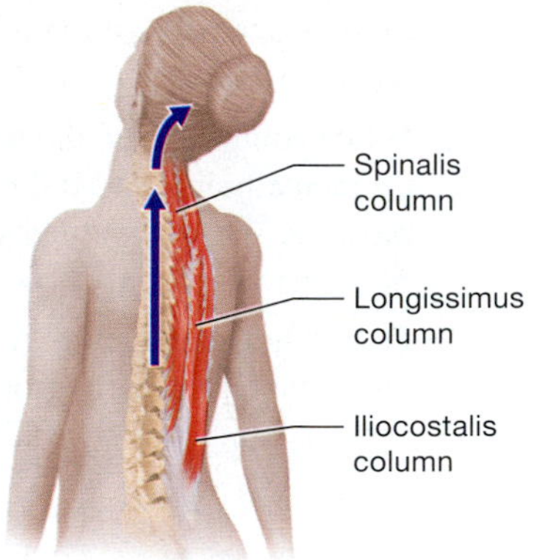
Spinalis muscle column Capitis Cervicis Thoracis	**Extend** the vertebral column (particularly the thoracis); capitis and cervicis may play a minor role in **extending** the head	O: Spinous processes of superior lumbar, thoracic, and inferior cervical vertebrae I: Spinous processes of thoracic and cervical vertebrae and skull N: Posterior rami of cervical, thoracic, and lumbar spinal nerves	
Longissimus muscle column Capitis Cervicis Thoracis (largest and most powerful component of the erector spinae)	**Extend** the vertebral column (particularly the thoracis) and maintain posture; capitis and cervicis may play a minor role in **extending** the head and rotating it to the same side; **laterally flex** the vertebral column when contracted on one side only	O: Transverse processes of vertebrae—usually several vertebrae inferior to insertions I: Mastoid process (capitis); transverse processes of vertebrae; medial posterior portions of ribs in the thoracic area N: Posterior rami of cervical, thoracic, and lumbar spinal nerves	
Iliocostalis muscle column Cervicis Thoracis Lumborum	**Extend** the vertebral column (particularly the thoracis and lumborum); maintain **posture**; **laterally flex** the vertebral column when contracted on one side only	O: Posterior surfaces of ribs; portions of sacrum and ilium I: Transverse processes of inferior cervical vertebrae; posterior surfaces of ribs N: Posterior rami of cervical, thoracic, and lumbar spinal nerves	
Transversospinal Group			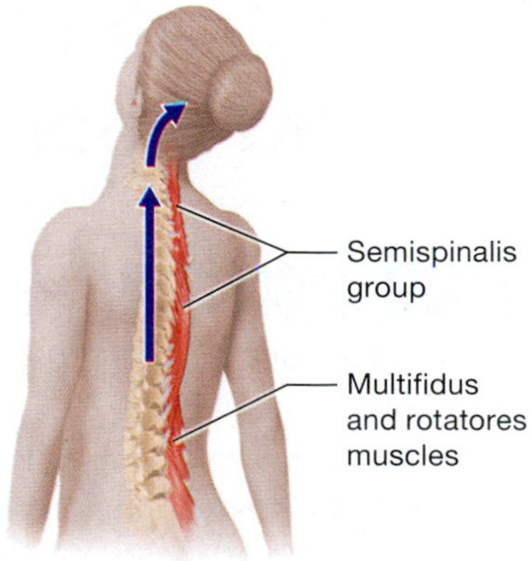
Semispinalis capitis, cervicis, and thoracis muscles	**Extend** the vertebral column; capitis and cervicis **extend** the head and rotate the head to the opposite side	O: Transverse and articular processes of C4–T6 I: Spinous processes (several vertebrae superior to origins for cervicis and thoracis) and skull N: Posterior rami of cervical and thoracic nerves	
Multifidus muscles	Synergist of other muscles of vertebral column movement and assists in maintaining **posture**	O: In its inferior half, sacrum and transverse processes of lumbar vertebrae; in its superior half, transverse processes of thoracic vertebrae and articular processes of inferior cervical vertebrae I: Spinous processes of vertebrae from L5 to C2 N: Posterior rami of cervical and thoracic nerves	
Rotatores muscles	**Extend** the vertebral column	O: Transverse process of one vertebra I: Spinous process of next one or two vertebrae superior N: Posterior rami of cervical, thoracic, and lumbar nerves	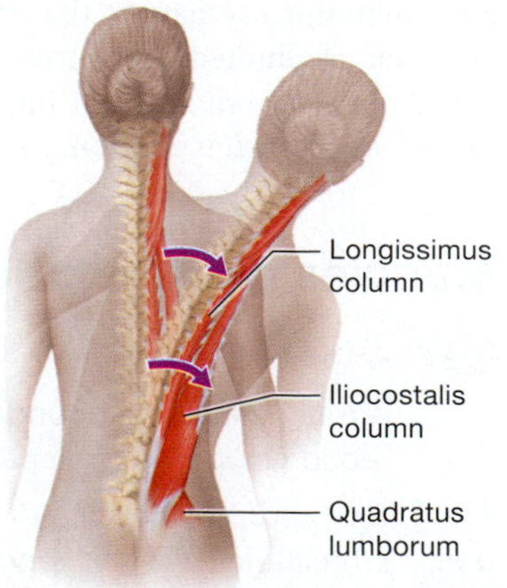
Other Spinal Extensors			
Quadratus lumborum muscle	**Extend** the vertebral column; maintain **posture**; **laterally flex** the vertebral column when contracted on one side only	O: Iliac crest and connective tissue of lumbar region I: Rib 12 and transverse processes of lumbar vertebrae N: Anterior rami of T12 and lumbar spinal nerves	

Concept Figure labels: Spinalis column; Longissimus column; Iliocostalis column; Semispinalis group; Multifidus and rotatores muscles; Longissimus column; Iliocostalis column; Quadratus lumborum

☐ 10. What are the three main muscles of the erector spinae muscle group, and what is the main function of the group?

☐ 11. What additional roles besides supporting posture are played by the muscles of the vertebral column?

☐ 1. Mr. Dawson is planning to visit his cosmetic surgeon for BOTOX injections to remove wrinkles around his mouth and in his forehead. This treatment involves temporarily paralyzing the frontalis muscle, corrugator supercilii muscle, and parts of the orbicularis oris muscle. Afterward, what movements will he be unable to perform?

☐ 2. While administering the treatment, the physician accidentally injects Mr. Dawson's buccinator and masseter muscles. What movements will this affect? Will it affect more than facial expression? Explain.

☐ 3. Predict how contracture (abnormal shortening) of the erector spinae muscle group on the left side of the vertebral column would affect childhood growth and posture.

See answers in Appendix A.

MODULE 9.3

Muscles of the Trunk and Pelvic Floor

1. Identify and describe the muscles of the trunk and pelvic floor.

2. Identify the origin, insertion, and action of these muscles, and demonstrate their actions.

We have already discussed some muscles of the trunk, such as those of the vertebral column. Others will be discussed in Module 9.4, as a number of muscles that move the arm are also anatomically part of the trunk. This section explores muscles that are limited to the trunk by both structure and function. We also discuss muscles of the pelvis and the pelvic floor, also known as the *pelvic diaphragm*.

Muscles of Ventilation

≪ FLASHBACK

1. What defines the boundaries of the thoracic and abdominopelvic body cavities? (p. 18)

The lungs have no muscles of their own, so to cause *ventilation*, or breathing, we rely on the actions of skeletal muscles.

The main muscle of ventilation is the dome-shaped **diaphragm muscle** (DY-uh-fram), which separates the thoracic and abdominopelvic body cavities (**Figure 9.14a**). The diaphragm muscle has a broad origin, extending around the thoracic cavity from the xiphoid process of the sternum and the lower ribs and costal cartilages anteriorly to the lumbar vertebrae posteriorly. In addition, it has an unusual insertion—it inserts into itself, into a central tendon. Due to this insertion, when the diaphragm contracts, it changes from dome-shaped to flat. As it pulls down and flattens, it pulls the thoracic cavity and lungs with it, which increases the size and so the volume of the lungs. This ultimately causes air to rush into the lungs as we *inspire*, or inhale. For this reason, the diaphragm is often called the main *muscle of inspiration* (see Chapter 21).

Other muscles that function in ventilation are those located between the ribs: the **external** and **internal intercostal muscles** (**Figure 9.14b**). Both sets of muscles originate from the rib superior to them and insert into the rib inferior to them. However, notice in the figure that the orientation of their fibers differs, which allows them to have different actions. With other parts of the rib cage fixed, the external intercostal muscles raise and spread the ribs, an action that helps the diaphragm in increasing the volume of the thoracic cavity and lungs. This assists with inspiration, so the external intercostal muscles are also considered muscles of inspiration. The internal intercostal muscles have the opposite action, which we'll get to momentarily.

When the diaphragm muscle relaxes, it moves superiorly and returns to its dome shape. This pushes the diaphragm up into the thoracic cavity, decreasing the volume of the thoracic cavity and lungs, an action that causes us to *expire*, or exhale. Take note that expiration is passive—it doesn't require the contraction of any muscles. However, when expiration is forced, the internal intercostal muscles come into play and depress the rib cage. This further decreases the lungs' volume and increases the amount of air expired.

During forced inspiration and expiration, in addition to the work of the diaphragm and external intercostal muscles, *accessory muscles of ventilation* are also active. Two muscles discussed in a previous section, the sternocleidomastoid and scalene muscles, can elevate the rib cage, assisting in forced inspiration. Other muscles of the trunk that we cover in later sections also function as accessory muscles of ventilation. **Table 9.9** summarizes the muscles of ventilation.

☐ 1. What is the main muscle of inspiration, and where is it located?

☐ 2. What roles do the external and internal intercostal muscles play in ventilation?

Abdominal Muscles

If you've ever heard of someone trying to get "six pack abs" at the gym, those "abs" are part of the **rectus abdominis muscle** (ab-DAWM-ih-nis), the superficial muscle of the anterior

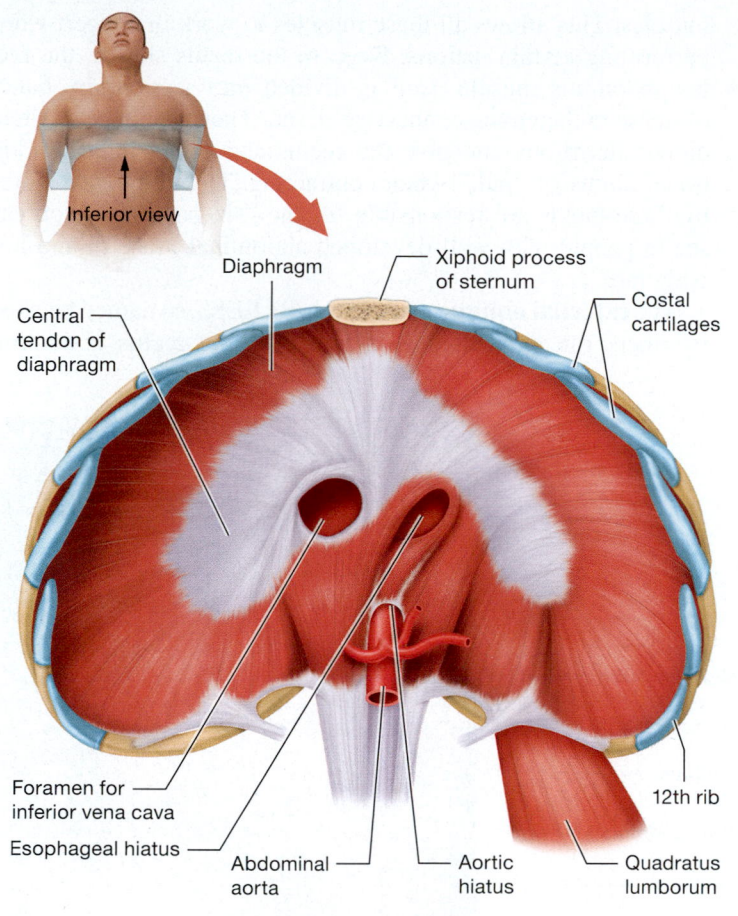

Inferior view

Diaphragm

Xiphoid process of sternum

Central tendon of diaphragm

Costal cartilages

Foramen for inferior vena cava

Esophageal hiatus

Abdominal aorta

Aortic hiatus

12th rib

Quadratus lumborum

(a) Inferior surface of diaphragm as viewed from the abdominal cavity

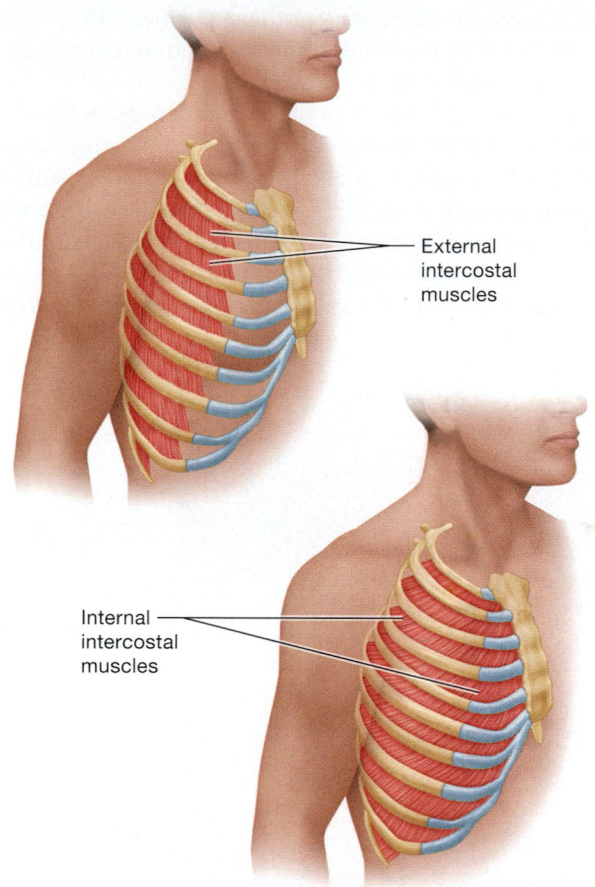

External intercostal muscles

Internal intercostal muscles

(b) The external and internal intercostal muscles

Figure 9.14 Muscles of ventilation. Notice that part (a) shows an inferior view of this transverse section of the trunk.

Table 9.9 Muscles of Ventilation

Muscle(s)	Action	Origin/Insertion/Nerve(s)	Concept Figures
Diaphragm muscle	**Flattens**, lowering the "floor" of the thoracic cavity and increasing its volume, ultimately causing inspiration (inhalation)	O: Xiphoid process of sternum, lower ribs and costal cartilages, lumbar vertebrae I: Central tendon of diaphragm N: Phrenic nerve	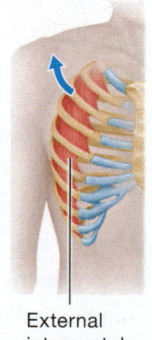 Diaphragm
External intercostal muscles	**Elevate** the rib cage, spreading the ribs, assisting inspiration	O: Lower edge of rib superior to its insertion I: Upper edge of rib inferior to its origin N: Intercostal nerves	
Internal intercostal muscles	**Depress** the rib cage, pulling ribs closer together, assisting forced expiration	O: Upper edge of rib inferior to its insertion I: Lower edge of rib superior to its origin, deep to the insertion of the external intercostals N: Intercostal nerves	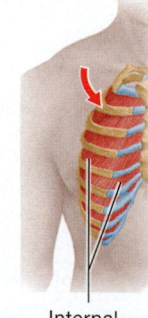 External intercostal muscles Internal intercostal muscles

abdomen (**Figure 9.15**). It originates from the superior pubic bone and inserts into the costal cartilages, which allows it to be a powerful flexor of the trunk (think of bending over as in taking a bow or doing a crunch or sit-up).

The structure and function of the rectus abdominis muscle are influenced by dense collagenous connective tissue in and around it. For one, it is enclosed by a structure called the **rectus sheath,** which is formed by fibers from the other abdominal muscles. A line of connective tissue runs along the midline of the rectus sheath, known as the **linea alba** (LIN-ee-uh AHL-buh), which forms a sort of central tendon for all the abdominal

muscles. This allows all these muscles to work in concert when performing certain actions. Deep to the rectus sheath, the rectus abdominis muscle itself is divided into sections by bands of dense collagenous connective tissue. These bands act as tendinous insertions and give the rectus abdominis muscle additional points of "pull" when contracting. The divisions created by these bands are responsible for the "six pack abs" you can see in people with well-developed abdominal muscles and low body fat.

The **external oblique muscle** (oh-BLEEK), so named because its fibers run at an angle, is lateral to the rectus abdominis

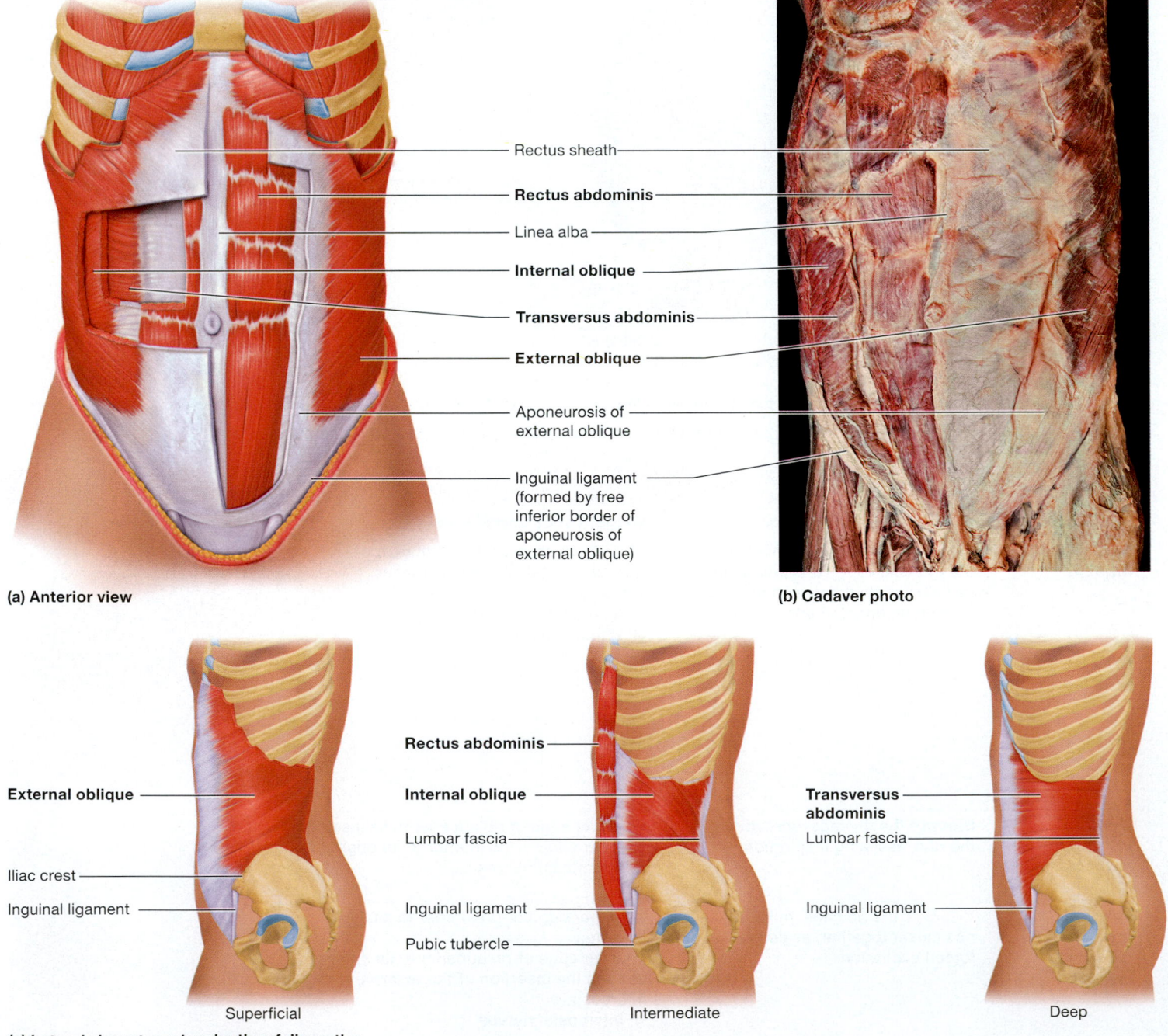

Rectus sheath

Rectus abdominis

Linea alba

Internal oblique

Transversus abdominis

External oblique

Aponeurosis of external oblique

Inguinal ligament (formed by free inferior border of aponeurosis of external oblique)

(a) Anterior view

(b) Cadaver photo

External oblique

Iliac crest

Inguinal ligament

Superficial

Rectus abdominis

Internal oblique

Lumbar fascia

Inguinal ligament

Pubic tubercle

Intermediate

Transversus abdominis

Lumbar fascia

Inguinal ligament

Deep

(c) Lateral view at varying depths of dissection

Figure 9.15 Abdominal muscles. Labels in **bold** indicate muscles involved in abdominal movement.

muscle (see Figure 9.15). It originates from the inferior ribs and inserts into the iliac crest, which allows it to rotate the trunk and flex it laterally. Deep to this muscle is the **internal oblique muscle,** whose fibers run opposite to those of the external oblique muscle. However, in spite of their differently oriented fibers, both muscles have the same actions.

The deepest muscle of the abdomen is the **transversus abdominis muscle** (tranz-VER-sus). Notice in the figure that its fibers run horizontally across the abdomen like a belt. For this reason, its main action is to compress the abdominal cavity, which raises the intra-abdominal pressure. Indeed, the other three abdominal muscles, contracting in concert with the transversus abdominus muscle, can raise intra-abdominal pressure quite significantly. This has several functions, including facilitating urination, defecation, and childbirth.

Increasing intra-abdominal pressure also allows these muscles to act as accessory muscles of expiration. An increase in intra-abdominal pressure pushes on the diaphragm, which raises pressure in the thoracic cavity. This facilitates forced expiration, such as during a cough. This is why your abdominal muscles are often sore after you have been coughing for several days.

Table 9.10 provides more information about the abdominal muscles.

Quick Check

☐ 3. Which muscles are involved in trunk flexion?

☐ 4. Which muscles compress the abdominal cavity, and what purpose does this serve?

Muscles of the Pelvic Diaphragm, Urogenital Diaphragm, and Perineum

« FLASHBACK

1. What three bones make up the pelvis? (p. 247)

2. Which organs are found in the pelvic body cavity? (p. 18)

Table 9.10 Abdominal Muscles

Muscle(s)	Action(s)	Origin/Insertion/Nerve(s)	Concept Figures
Rectus abdominis muscle	**Flexes** the trunk; **compresses** abdominal cavity	O: Superior aspect of pubic bones I: Costal cartilages of inferior ribs N: Branches of anterior rami of the lower six to seven thoracic nerves	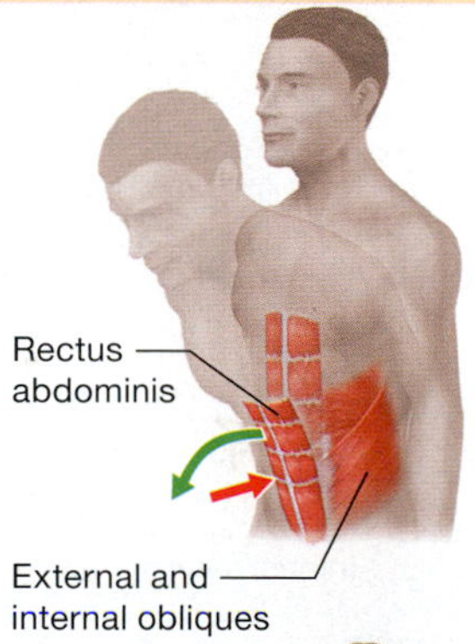 Rectus abdominis — External and internal obliques
External oblique muscle	**Flexes** and **laterally flexes** the trunk; **compresses** abdominal cavity	O: Inferior eight ribs I: Iliac crest, pubic tubercle, and linea alba N: Branches of the lower six thoracic nerves	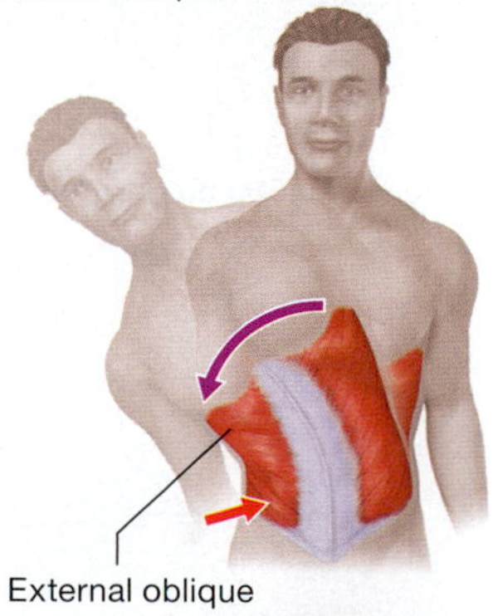
Internal oblique muscle	**Flexes** and **laterally flexes** the trunk; **compresses** abdominal cavity	O: Iliac crest and inguinal ligament I: Inferior three to four ribs, linea alba, and pubic bone N: Branches of the lower six thoracic nerves and first lumbar nerve	External oblique
Transversus abdominis muscle	**Compresses** abdominal cavity	O: Inferior six costal cartilages, thoracolumbar fascia, iliac crest, and inguinal ligament I: Linea alba and pubic bone N: Branches of the lower six thoracic nerves and first lumbar nerve	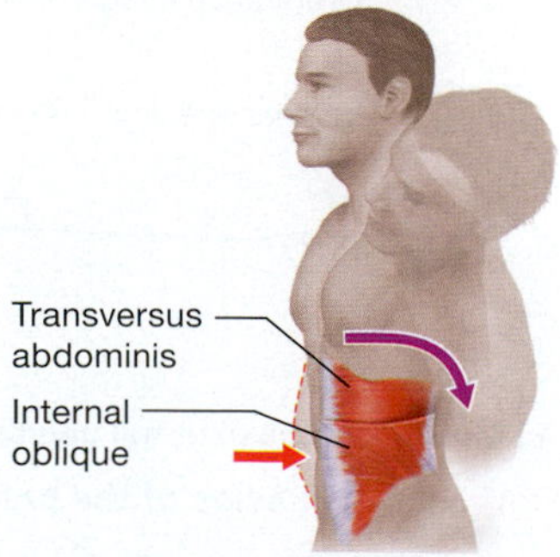 Transversus abdominis — Internal oblique

The *pelvic floor*, also known as the **pelvic diaphragm,** is the broad, flat sheet that essentially holds all your organs inside your pelvis. It is formed by a muscle called the **levator ani muscle** (AY-nee), which is divided into three smaller muscles: the anterior **pubococcygeus** (pyoo′-boh-kok-SIJ-ee-us), the middle **iliococcygeus** (il′-ee-oh-kok-SIJ-ee-us), and the posterior **ischiococcygeus** (isk′-ee-oh-kok-SIJ-ee-us) **muscles** (**Figure 9.16a**). These muscles originate from the interior of the pelvic bones and insert into the coccyx and a connective tissue structure that extends from the rectum to the coccyx, called the *anococcygeal ligament* (ah′-noh-kok-SIJ-ee-uhl). You can see in Figure 9.16a how the anal canal, urethra, and vagina in the

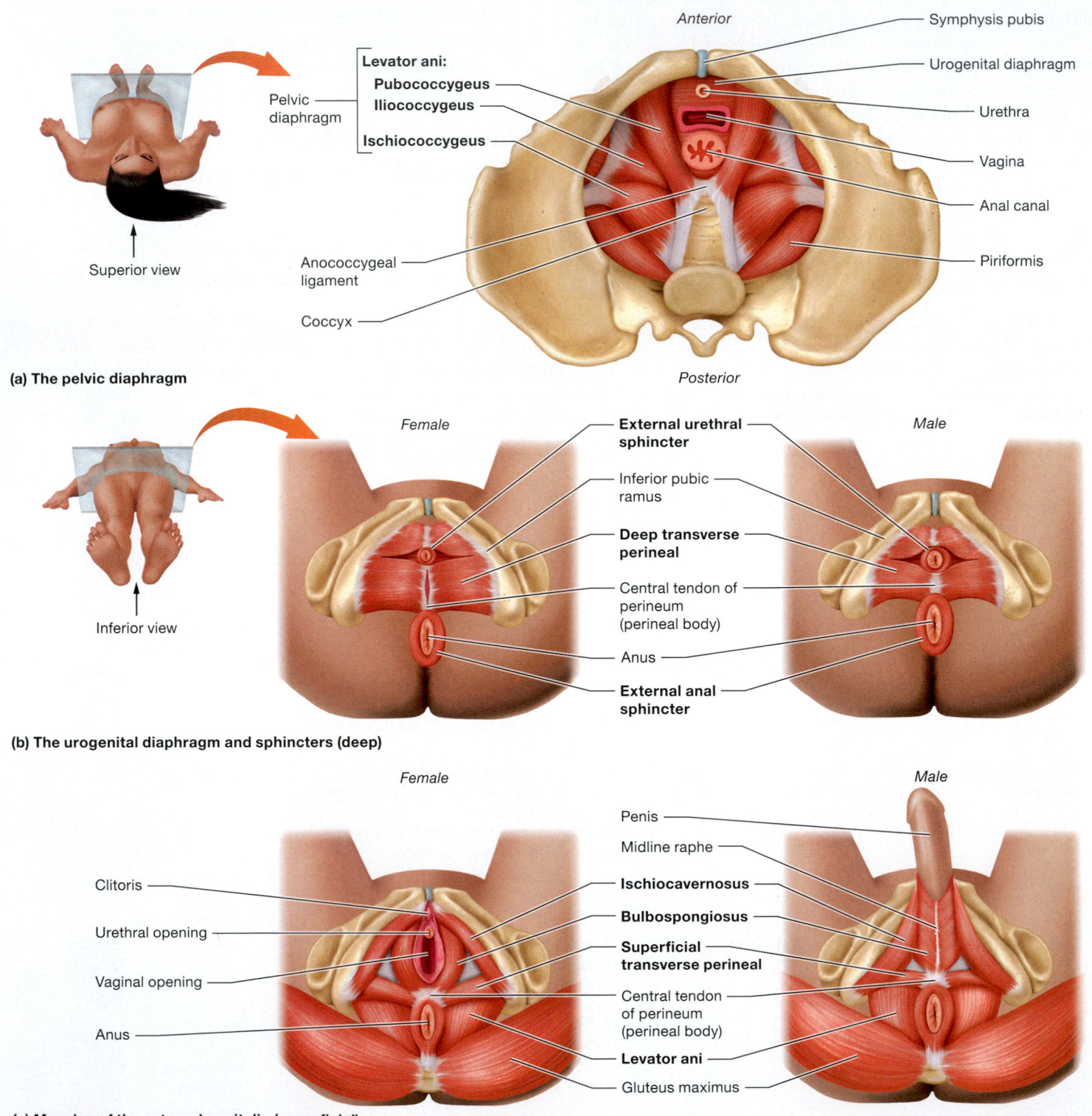

(a) The pelvic diaphragm

(b) The urogenital diaphragm and sphincters (deep)

(c) Muscles of the external genitalia (superficial)

Figure 9.16 Muscles of the pelvic floor and perineum. Labels in **bold** indicate muscles that act in each region of the pelvic floor and peritoneum.

female pass through the levator ani muscle. This muscle plays an important role in maintaining urinary and bowel continence. Indeed, problems with this muscle can result in lack of bladder control, or *urinary incontinence* (see *A&P in the Real World: Urinary Incontinence and Kegel Exercises* on p. 312).

Inferior to the pelvic diaphragm is a diamond-shaped region called the *perineum* (pehr-ih-NEE-um); it lies between the proximal thighs and contains the external genitalia (the penis and testes in the male and the vulva in the female) and the opening of the anus. In the deeper portion of the perineum, shown in **Figure 9.16b**, is the **urogenital diaphragm,** through which the urethra passes in the male, and the urethra and vagina pass in the female. The urogenital diaphragm contains three structures: (1) the **external urethral sphincter** (yoo-REE-thruhl), a small circular muscle that surrounds the urethra; (2) a circular muscle that surrounds the vagina in the female; and (3) the broad, flat **deep transverse perineal muscle** (pehr-ih-NEE-uhl), which mainly supports the genitourinary organs.

Posterior to the urogenital diaphragm is another circular muscle, the **external anal sphincter,** which surrounds the terminal part of the anal canal, the *anus*. It attaches to the tip of the coccyx and a band of connective tissue running down the middle of the perineum, called the *perineal body,* or *central perineal tendon*. The external urethral and anal sphincters provide voluntary control of urination and defecation, respectively.

The superficial structures of the perineum are shown in **Figure 9.16c.** Notice that the **superficial transverse perineal muscles** attach to the underside of the urogenital diaphragm and stabilize the perineal body. Anterior to the superficial transverse perineal muscle are two perineal muscles associated with the external genitalia: the **bulbospongiosus** (bul-boh-spun´-jee-OH-sus) and **ischiocavernosus** (iss´-kee-oh-kav-er-NOH-sus) **muscles.** These thin, delicate muscles help make the external genitalia erect during sexual excitement. In males, the bulbospongiosus muscle also helps expel semen from the urethra.

See **Table 9.11** for more details about the muscles of the pelvic diaphragm, urogenital diaphragm, and perineum.

Quick Check

☐ 5. What are the functions of the levator ani muscles?

☐ 6. Which muscles are involved in conscious control of urination and defecation?

Apply What You Learned

☐ 1. Bruised ribs are generally caused by a traumatic blow to the chest. The most common symptom of bruised ribs is pain with each inspiration (inhalation). Why do you think the action of inspiration causes pain to injured ribs?

Table 9.11 Muscles of the Pelvic Diaphragm, Urogenital Diaphragm, and Perineum

Muscle(s)	Action(s)	Origin/Insertion/Nerve(s)
Pelvic Floor (or Pelvic Diaphragm)		
Levator ani muscle group Pubococcygeus muscle Iliococcygeus muscle	Supports pelvic floor, anal canal, and genitourinary organs	O: Interior of pelvic bones I: Coccyx and anococcygeal ligament N: Branches of S3 and S4
Ischiococcygeus muscle	Supports pelvic floor	O: Ischial spine I: Lateral surface of coccyx and lower sacrum N: Branches of S3 and S4
Urogenital Diaphragm		
Deep transverse perineal muscle	Supports pelvic organs; fixes central tendon	O: Ischiopubic rami I: Perineal body N: Pudendal nerve
External urethral sphincter	Compresses urethra (encircles urethra in both sexes and, in females, also vagina)	N: Pudendal nerve
Perineum		
External anal sphincter	Closes anal opening	O: Coccyx and perineal body I: Connective tissues around the anal canal N: Pudendal nerve
Bulbospongiosus muscle	Expels semen and urine in males; assists erection of penis in males and clitoris in females; constricts opening of vagina in females	O: Perineal body I: Corpus spongiosum of the penis; clitoris N: Pudendal nerve
Ischiocavernosus muscle	Maintains erection of the penis or clitoris	O: Ischiopubic rami I: Corpus cavernosum of the penis; clitoris N: Pudendal nerve
Superficial transverse perineal muscle	Assists with stabilizing the perineal body	O: Ischial tuberosity I: Perineal body N: Pudendal nerve

☐ 2. Normal, quiet expiration isn't generally painful to bruised ribs. However, forced expiration, as occurs when one coughs or sneezes, is usually quite painful. Why is the action of forced expiration, unlike quiet expiration, painful to bruised ribs?

See answers in Appendix A.

Urinary Incontinence and Kegel Exercises

Weak pelvic diaphragm muscles are a likely suspect in many cases of urinary incontinence. This is particularly true in women who have undergone vaginal childbirth, which often causes stretching and trauma to those muscles. Other factors that may weaken the muscles of the pelvic diaphragm include older age, weight gain, changing hormone levels after menopause, and other conditions that affect structures in the pelvic cavity.

Pelvic floor exercises, also called *Kegel exercises* (KEG-el; for Dr. Arnold Kegel), help strengthen the levator ani muscle and treat this type of urinary incontinence. The goal is to engage in repeated rounds of contraction and relaxation of the levator ani muscle, often by starting and stopping the urine stream during urination. The exercises usually need to be performed several times per day for a few months to have a noticeable effect.

Pelvic floor exercises are also recommended for pregnant women to strengthen the levator ani muscle in the hope of minimizing pelvic tearing and trauma during childbirth. Strengthening this muscle also guards against a condition known as pelvic organ *prolapse*, in which women's pelvic organs, such as the uterus, slip out of place. Pelvic floor exercises may also be recommended for men suffering from certain types of incontinence or sexual dysfunction.

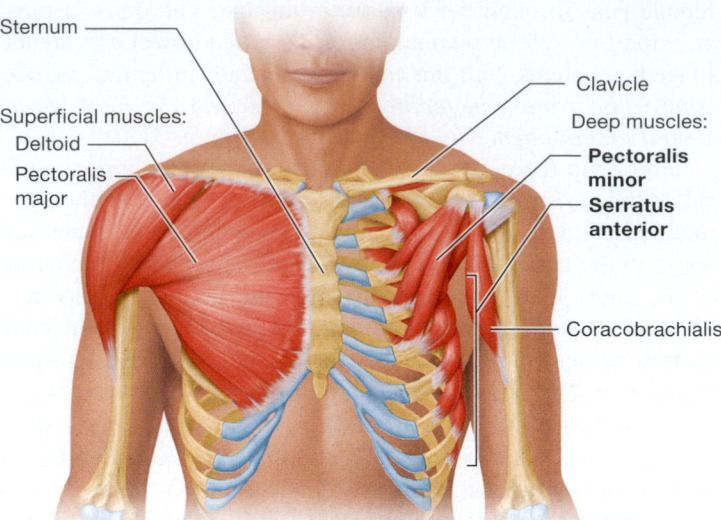

(a) Anterior view: superficial muscles shown on left; deep muscles shown on right, with superficial removed

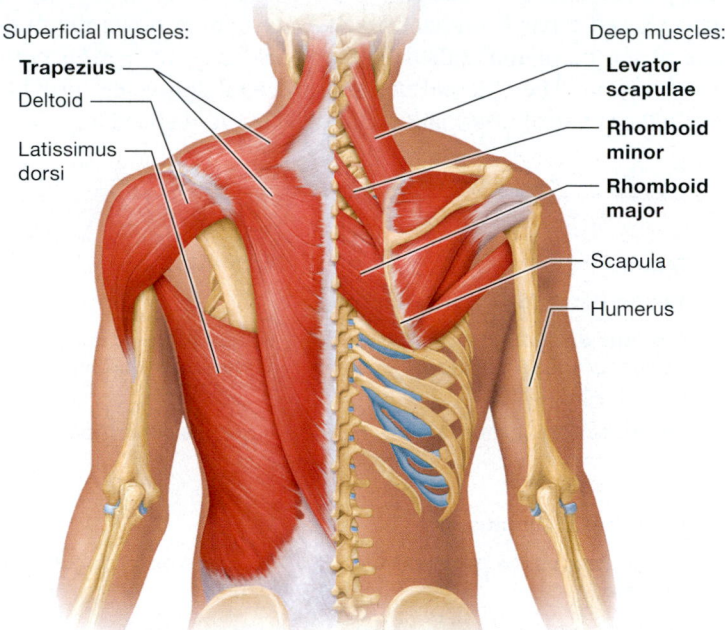

(b) Posterior view: superficial muscles shown on left; deep muscles shown on right, with superficial removed

Figure 9.17 Muscles that move the scapula. Labels in **bold** indicate muscles involved in these actions.

Play A&P Flix Animation
@ Mastering Anatomy & Physiology

MODULE 9.4

Muscles of the Pectoral Girdle and Upper Limb

Learning Outcomes

1. Identify and describe the muscles that move the pectoral girdle and upper limb.
2. Identify the origin, insertion, and action of these muscles, and demonstrate their actions.

Recall from our study of the skeletal system that the upper limb—the arm, forearm, wrist, and hand—attaches to the axial skeleton by way of the pectoral girdle. But what attaches the pectoral girdle to the axial skeleton? The answer is largely skeletal muscles. Not only that, but the main connectors of the upper limb to the pectoral girdle are also skeletal muscles (along with a few ligaments). We explore these muscles, their attachments, and their actions in this module.

Muscles That Move the Scapula at the Pectoral Girdle

 **FLASHBACK**

1. Define elevation and depression. (p. 268)
2. Describe protraction and retraction. (p. 269)

The muscles that move the scapula maneuver the pectoral girdle and stabilize it for coordinated, powerful movements such as rowing or throwing a punch. One of the main muscles involved in these actions is the **serratus anterior muscle** (seh-RAY-tus). Its name refers to its many pointed origins from the ribs, which make it look somewhat like the edge of a serrated knife (**Figure 9.17a**). The muscle wraps around the contour of the rib cage to insert along the medial border of the scapula. This insertion allows the serratus anterior muscle to protract the scapula, as in when you are pushing something in front of you. In fact, the serratus anterior muscle is such a powerful protractor of the scapula that it's nicknamed the "boxer's muscle," because it is key in the motion necessary to throw a punch. The serratus anterior muscle also rotates the scapula superiorly, as in when you are lifting something above your head.

Another muscle that assists with scapular protraction is the small **pectoralis minor muscle** (pek-toh-RAL-iss), which is located deep to the much larger *pectoralis major muscle* (which we cover later with muscles that move the arm). The pectoralis minor muscle originates from the third through fifth ribs and inserts into the coracoid process of the scapula. This insertion allows it to assist with scapular protraction, but causes it to also depress the scapula (pull it down). Interestingly, the origins and insertions of both the serratus anterior and pectoralis minor muscles can switch if needed, and both can elevate the rib cage to act as accessory muscles of inspiration.

A muscle that acts on the scapula from the posterior side is the trapezius muscle (**Figure 9.17b**), which you already met in the section on muscles that move the head. Recall that the trapezius muscle has three parts, and each part has a different origin and insertion and so a different primary action. Its superior part originates from the occipital bone and cervical vertebrae and inserts into the lateral clavicle. This allows it to elevate the scapula, a motion you perform when shrugging your shoulders. The middle part of the trapezius muscle originates from the first three thoracic vertebrae and inserts into the scapular acromion. This fairly straight pathway across the back causes it to retract the scapula—pull it back—when it contracts. The trapezius muscle's inferior part originates from the inferior thoracic vertebrae and inserts into the scapular spine. This angle of insertion causes it to depress the scapula when it contracts. In addition, the inferior part can work together with the superior part and the serratus anterior muscle to rotate the scapula superiorly.

Several other muscles work with the trapezius muscle to move the scapula. For example, the **levator scapulae muscle** (SKAP-yoo-lee) works with the superior part of the trapezius muscle. The origin of the levator scapulae muscle is superior to the scapula along the transverse processes of the first four cervical vertebrae, and it inserts into the scapula's superior angle. Following this origin and insertion, it makes sense that this muscle elevates the scapula, as its name would imply. Another is a pair of muscles called the **rhomboid major** (RAHM-boyd) and **rhomboid minor muscles,** which work with the middle portion of the trapezius muscle. The rhomboid muscles have their origin on the spinous processes of the seventh cervical vertebra through the fifth thoracic vertebra, and they insert into the medial border of the scapula. This allows them to retract the scapula.

For more information on muscles that move the scapula, see **Table 9.12**.

Table 9.12 Muscles That Move the Scapula

Muscle(s)	Action(s)	Origin/Insertion/Nerve(s)	Concept Figures
Levator scapulae muscle	**Elevates** scapula	O: Transverse processes of C1–C4 I: Superior angle of scapula N: Dorsal scapular nerve	
Rhomboid major and rhomboid minor muscles	**Retracts** scapula	O: Spinous processes of: T2–T5 (major); C7–T1 (minor) I: Medial border of scapula N: Dorsal scapular nerve	
Trapezius muscle (superior section)	**Elevates** the scapula; **rotates** scapula superiorly in collaboration with inferior portion	O: External occipital protuberance and cervical vertebrae I: Lateral clavicle N: Accessory nerve	
Trapezius muscle (all sections)	**Retracts** (via middle portion) and **depresses** (via inferior section) the scapula; **rotates** scapula superiorly and **elevates** scapula (via superior section)	O: Spinous processes of thoracic vertebrae I: Spine and acromion of scapula N: Accessory nerve	
Pectoralis minor muscle	**Protracts** and **depresses** scapula	O: Anterior part of ribs 3–5 I: Coracoid process N: Medial pectoral nerve	
Serratus anterior muscle	Prime mover of scapula in **protraction**; assists in superior **rotation** of scapula	O: First nine ribs I: Medial border of scapula N: Long thoracic nerve	

Note: Some of these actions vary depending on which fibers are stimulated.

Quick Check

☐ 1. What are the anterior muscles that act on the scapula, and what are their actions?

☐ 2. Which posterior trunk muscles act on the scapula? What are their main actions?

Muscles That Move the Arm at the Shoulder Joint

⟪ FLASHBACK

1. What is a multiaxial joint? (p. 264)
2. What is rotation? (p. 267)

Recall from the articulations chapter that the shoulder joint has the greatest range of motion in the body (see Chapter 8). It is a *multiaxial joint* that permits movement in all axes of motion, so it can undergo flexion, extension, abduction, adduction, and rotation. Accordingly, it has several muscles that provide these movements; however, it also has many muscles that stabilize the joint in the face of this extensive range of motion.

We'll start with the **pectoralis major muscle,** the large muscle of the chest area (**Figure 9.18a**). Note in the figure its broad origin from the clavicle, sternum, and superior costal cartilages that then tapers to a tendon that inserts into the proximal part of the humerus. This origin, insertion, and shape allow the pectoralis major muscle to be a powerful agonist of arm flexion: the action you take when you swing to hit a volleyball with your forearms. This muscle also adducts the arm at the shoulder (pulls it toward the midline) and internally rotates it. The pectoralis

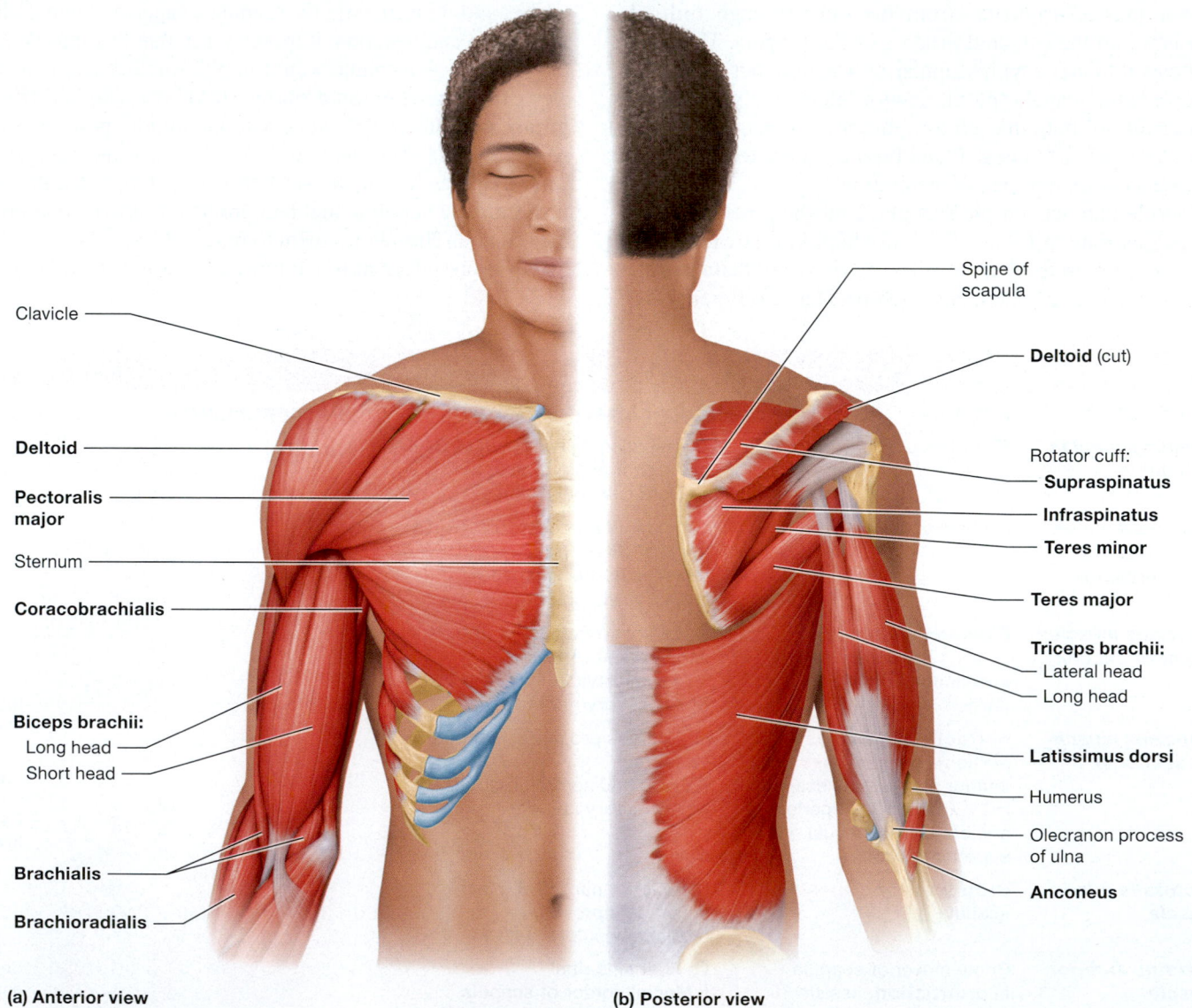

(a) Anterior view

(b) Posterior view

Figure 9.18 Muscles that move the arm and forearm. Labels in **bold** indicate muscles involved in these actions.

major muscle is assisted in flexion and adduction of the humerus by the small, cylindrical **coracobrachialis muscle** (kohr´-uh-koh-bray-kee-AL-iss), which originates on the coracoid process and inserts midway down the diaphysis of the humerus.

Next is the **deltoid muscle,** the thick, rounded mass of the shoulder. You can see in the figure how all fibers of the deltoid muscle insert via a central tendon into the deltoid tuberosity of the humerus. However, this muscle has three very different origins, and for this reason it has three different actions: (1) Its anterior fibers, which originate from the lateral clavicle, assist the pectoralis major muscle with arm flexion; (2) its central portion, which originates from the acromion, is the agonist of arm abduction, such as the upward motion of the arm when you're doing jumping jacks; and (3) its posterior fibers, which originate from the scapular spine, assist with arm extension (pulling the arm back). On a clinical note, the deltoid muscle is a popular site for intramuscular ("into a muscle") injections because of its bulk and easy access.

Turning to the posterior side we find the large, triangular **latissimus dorsi muscle** (lah-TISS-ih-muss DOHR-see) in the middle and inferior back (**Figure 9.18b**). The muscle has a broad origin from the iliac crest, the inferior thoracic vertebrae, and all of the lumbar vertebrae, and its fibers converge to a tendon that inserts into the proximal humerus. This gives it an ideal placement to extend, adduct, and internally rotate the arm. The latissimus dorsi muscle is assisted in its actions by the **teres major muscle** (TEHR-eez), which originates from the posteroinferior scapula and inserts on the humerus close to the latissimus dorsi tendon.

The next muscles that act on the shoulder joint belong to a group collectively known as the **rotator cuff muscles.** There are four muscles in this group: (1) the **teres minor muscle,** which is superior to the teres major muscle; (2) the **supraspinatus muscle** (soo´-prah-spy-NAY-tus), which is superior to the scapular spine; (3) the **infraspinatus muscle** (in´-frah-spy-NAY-tus), which is inferior to the scapular spine; and (4) the **subscapularis muscle** (sub-skap´-yoo-LAY-riss), which is on the anterior surface of the scapula (and for this reason is not visible in Figure 9.18b). The tendons of the rotator cuff muscles attach firmly to the shoulder joint capsule and strengthen it. This helps to hold the head of the humerus in place in the glenoid cavity so that other muscles can move the arm, which stabilizes the shoulder joint during forceful movement. Damage to the rotator cuff is one of the most common shoulder injuries, as described in *A&P in the Real World: Rotator Cuff Injuries.*

Table 9.13 on p. 316 lists the other actions of the rotator cuff muscles. See this table also for more information on the muscles that move the arm.

Quick Check

☐ 3. What are two muscles that abduct the arm?

☐ 4. What are two muscles that adduct the arm?

A&P in the Real World

Rotator Cuff Injuries

The rotator cuff muscles are critical for stabilizing the shoulder joint when executing forceful overhead arm movements, so it's not surprising that activities involving repetitive types of these motions can damage the rotator cuff muscles or their tendons. These activities range from throwing a fastball to working in construction and spending a lot of time with your hands over your head. Such repetitive actions, particularly if they are carried out with improper form, cause microscopic tears in the rotator cuff tendons that cause inflammation. Over time, this can lead to partial or complete tears in one of more of the tendons.

The shape of the scapular acromion also plays a role in the development of rotator cuff tears, as the rotator cuff tendons, particularly the supraspinatus tendon, pass under the acromion. If the acromion is hooked or curved, or has a bone spur, the rotator cuff is more likely to tear. Interestingly, chronic inflammation of the rotator cuff can lead to deformation of the acromion, which produces further inflammation of the rotator cuff, in a sort of positive feedback cycle.

Symptoms of rotator cuff injuries include pain, weakness, and decreased range of motion in the shoulder joint, especially with flexion and abduction. Treatment depends on the severity of the tear but may include medications to decrease inflammation, rest from the activities that caused the injury, and physical therapy to improve range of motion and strength of the surrounding muscles. Certain injuries require surgical intervention to repair the tear and possibly remove part of the acromion.

Muscles That Move the Forearm and Hand

« FLASHBACK

1. What are supination and pronation? (p. 269)

A lot of muscles move the forearm and hand. Fortunately, most of their names are based on either their location or their action. Recall from Table 9.1 that *brachii* = "arm," *carpi* = "wrist," *superficialis* = "nearer the surface," *profundus* = "deep," *brevis* = "short," *digiti* = "fingers and toes," and *minimi* = "smallest." You can put these terms together to decipher a relatively complicated-sounding muscle. For example, the flexor digiti minimi brevis simply translates to a short muscle that flexes the smallest finger (the fifth digit). Make sure to refer back to this table when needed as we examine the muscles in this section.

Table 9.13 Muscles That Move the Arm

Muscle(s)	Action(s)	Origin/Insertion/Nerve(s)	Concept Figures
Pectoralis major muscle	**Flexes** and **adducts** the arm; **rotates** arm medially	O: Medial clavicle, sternum, costal cartilages 1–7 I: Greater tubercle and lateral lip of intertubercular sulcus of humerus N: Medial and lateral pectoral nerves	
Latissimus dorsi muscle	**Adducts** and **extends** the arm; **rotates** arm medially	O: Iliac crest, spinous processes of lower thoracic and all lumbar vertebrae I: Floor of intertubercular sulcus of the humerus N: Thoracodorsal nerve	
Deltoid muscle	**Abducts** the arm; secondarily **flexes** and **extends** arm	O: Acromion and spine of scapula; lateral clavicle I: Deltoid tuberosity of humerus N: Axillary nerve	
Teres major muscle	**Adducts**, **extends**, and **rotates** arm medially	O: Posterior, inferior portion of scapula I: Medial lip of intertubercular sulcus of the humerus N: Lower subscapular nerve	
Coracobrachialis muscle	**Flexes** and **adducts** the arm	O: Coracoid process I: Medial diaphysis of humerus N: Musculocutaneous nerve	
Rotator Cuff Muscles			
Supraspinatus muscle	Assists **abduction**; holds the humerus and stabilizes shoulder joint	O: Supraspinous fossa of scapula I: Greater tubercle of humerus N: Suprascapular nerve	
Infraspinatus muscle	Laterally **rotates** the humerus; stabilizes shoulder joint	O: Infraspinous fossa of scapula I: Greater tubercle of humerus N: Suprascapular nerve	
Teres minor muscle	Laterally **rotates** the humerus; stabilizes shoulder joint	O: Posterior, lateral border of scapula I: Greater tubercle of humerus N: Axillary nerve	
Subscapularis muscle	**Adducts** and **rotates** the humerus medially; stabilizes shoulder joint	O: Subscapular fossa of scapula I: Lesser tubercle of humerus N: Subscapular nerves	

Muscles That Move the Forearm at the Elbow Joint

Most of the muscles we think of as "arm" muscles actually move the forearm at the elbow joint. The largest muscle on the anterior arm is the two-headed **biceps brachii muscle** (BY-seps BRAYK-ee-aye), which you met earlier (see Figure 9.18a). Recall that its two heads originate from the scapula and its common tendon inserts into the radial tuberosity of the proximal radius. This allows the biceps brachii muscle to flex the forearm, which is the action we tend to associate with this muscle. However, its most powerful action actually is to supinate the forearm, meaning that it turns the forearm anteriorly.

The smaller, deeper **brachialis muscle** (bray-kee-AL-iss) is actually the agonist of elbow flexion. It originates from the distal half of the humerus and inserts into the proximal ulna, which gives it better leverage than the biceps brachii muscle to flex the forearm. The brachialis muscle is assisted by a muscle located on the lateral forearm called the **brachioradialis muscle** (bray'-kee-oh-ray-dee-AL-iss), which originates on the distal humerus and inserts into the distal radius.

Turning to the posterior side of the arm (see Figure 9.18b) we find the large, three-headed **triceps brachii muscle** (TRY-seps). All three heads of this muscle insert via a common tendon into the olecranon process of the ulna (the medial head is deep to the lateral and long heads and so is not visible in Figure 9.18b). However, one of the heads has a different origin and so a different action. Both the lateral and medial heads originate only from the humerus, and so only extend the forearm. But the long head originates from the scapula, so it can play a supporting role in adducting the arm and stabilizing the shoulder as well as extending the forearm. The triceps brachii muscle is assisted in forearm extension by the small, triangular *anconeus muscle* (ang-KOH-nee-us).

See **Table 9.14** for more information.

Muscles That Move the Hand and Fingers

Many of the muscles that move the hand can be divided into two broad groups: flexors, located on the anterior and medial forearm, and extensors, located on the posterior and lateral forearm. We show these muscles in **Figure 9.19** and list them in **Table 9.15**. Let's look at the flexors first. Notice in Figure 9.19a that we divide the muscles into three layers—superficial, intermediate, and deep—based on their closeness to the skin. In the

superficial layer, we find three hand flexors that share a common tendon that originates from the medial epicondyle of the humerus: the **flexor carpi radialis** (KAR-pee ray-dee-AEH-lis), **palmaris longus** (pahl-MEHR-iss LONG-gus), and **flexor carpi ulnaris** (uhl-NAY-ris) **muscles.** Also in this superficial layer is a muscle that pronates (turns posteriorly) the forearm, called the **pronator teres muscle.**

In the intermediate layer, we find a large flexor of the fingers called the **flexor digitorum superficialis muscle** (dij-ih-TOH-rum). There are additional flexors of the fingers and thumb in the deep layer, including the **flexor digitorum profundus** and **flexor pollicis longus** (POHL-uh-sis) **muscles.** In this layer is also another muscle of pronation, the **pronator quadratus muscle,** and its antagonist, the **supinator muscle** (SOO-pih-nay-tuhr).

Now let's turn to the extensors in Figure 9.19b, which we divide into superficial and deep layers. Several of these muscles are hand extensors, including the **extensor carpi radialis longus, extensor carpi radialis brevis,** and **extensor carpi ulnaris muscles.** Others are extensors of the fingers and thumb, including the **extensor digitorum** and **extensor digiti minimi** (DIJ-ih-ty MIN-ih-my) **muscles.** In the deep layer we find more finger and thumb extensors: the **extensor pollicis longus, extensor pollicis brevis,** and **extensor indicis** (IN-dih-kis) **muscles.** In this layer we also find a muscle that abducts the thumb, the **abductor pollicis longus muscle.**

The muscles shown in **Figure 9.20** and listed in **Table 9.16** are known as *intrinsic hand muscles,* as they originate from within the hand. As in Table 9.15, the groupings and names suggest their function and location. For example, the **abductor pollicis brevis muscle** is a short muscle that abducts the thumb, and the **flexor**

Table 9.14 Muscles That Move the Forearm at the Elbow Joint

Muscle(s)	Action(s)	Origin/Insertion/Nerve(s)	Concept Figure
Biceps brachii muscle	**Supinates** the forearm; **flexes** the forearm	O: Supraglenoid tubercle (long head) and coracoid process (short head) I: Radial tuberosity N: Musculocutaneous nerve	
Brachialis muscle	**Flexes** the forearm (prime mover)	O: Distal half of the diaphysis of the anterior humerus I: Coronoid process of ulna N: Musculocutaneous nerve	
Brachioradialis muscle	**Flexes** the forearm (synergist)	O: Ridge proximal to the lateral condyle of the distal humerus I: Distal radius proximal to the styloid process N: Radial nerve	
Triceps brachii muscle	**Extends** the forearm	O: Infraglenoid tubercle (long head); posterior, proximal diaphysis (lateral head); most of posterior diaphysis (medial head) I: Olecranon process N: Radial nerve	
Anconeus muscle	Assists **extension** of the forearm	O: Lateral epicondyle of humerus I: Olecranon process N: Radial nerve	

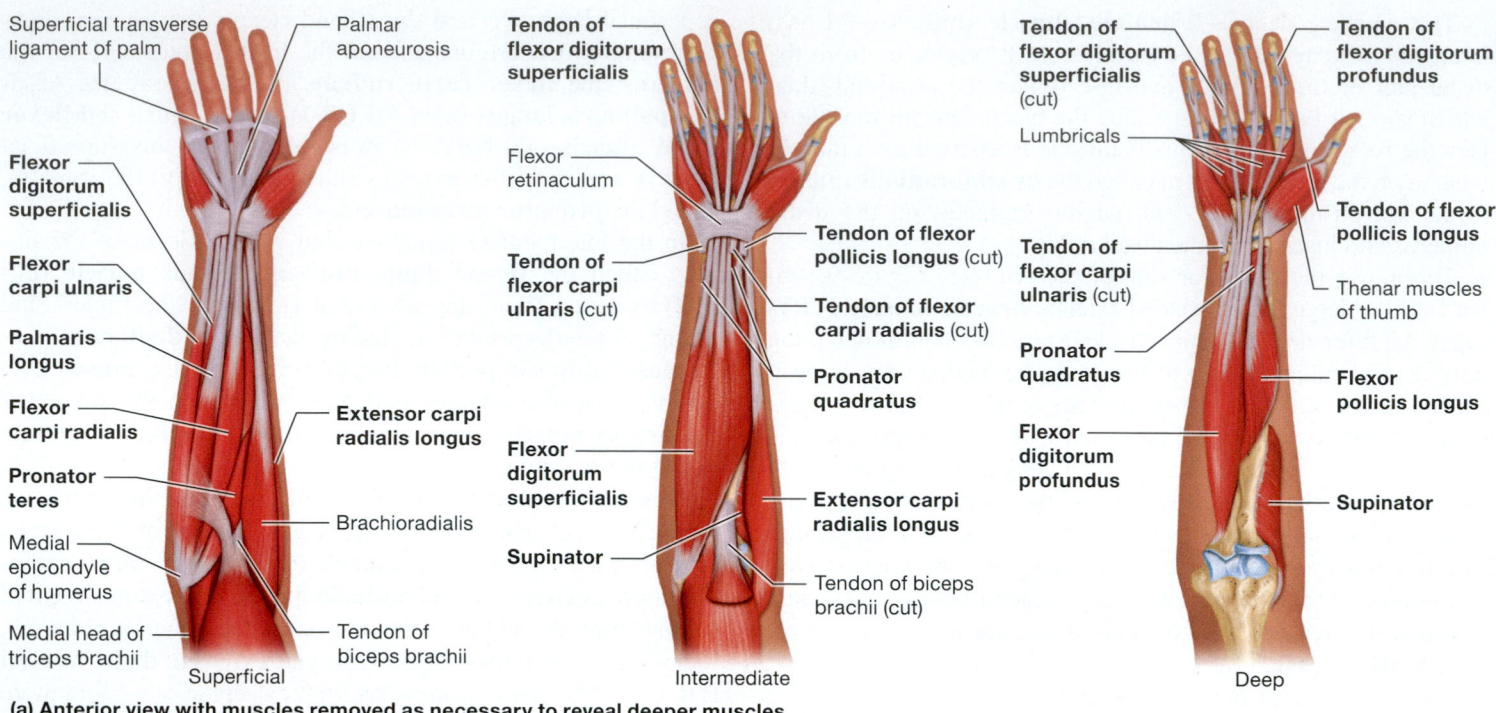

Superficial transverse ligament of palm — Palmar aponeurosis

Flexor digitorum superficialis

Flexor carpi ulnaris

Palmaris longus

Flexor carpi radialis — Extensor carpi radialis longus

Pronator teres

Medial epicondyle of humerus — Brachioradialis

Medial head of triceps brachii — Tendon of biceps brachii

Superficial

Tendon of flexor digitorum superficialis

Flexor retinaculum

Tendon of flexor carpi ulnaris (cut) — Tendon of flexor pollicis longus (cut)

Tendon of flexor carpi radialis (cut)

Pronator quadratus

Flexor digitorum superficialis

Supinator — Extensor carpi radialis longus

Tendon of biceps brachii (cut)

Intermediate

Tendon of flexor digitorum superficialis (cut) — Tendon of flexor digitorum profundus

Lumbricals

Tendon of flexor carpi ulnaris (cut) — Tendon of flexor pollicis longus

Tendon of flexor carpi ulnaris (cut) — Thenar muscles of thumb

Pronator quadratus — Flexor pollicis longus

Flexor digitorum profundus — Supinator

Deep

(a) Anterior view with muscles removed as necessary to reveal deeper muscles

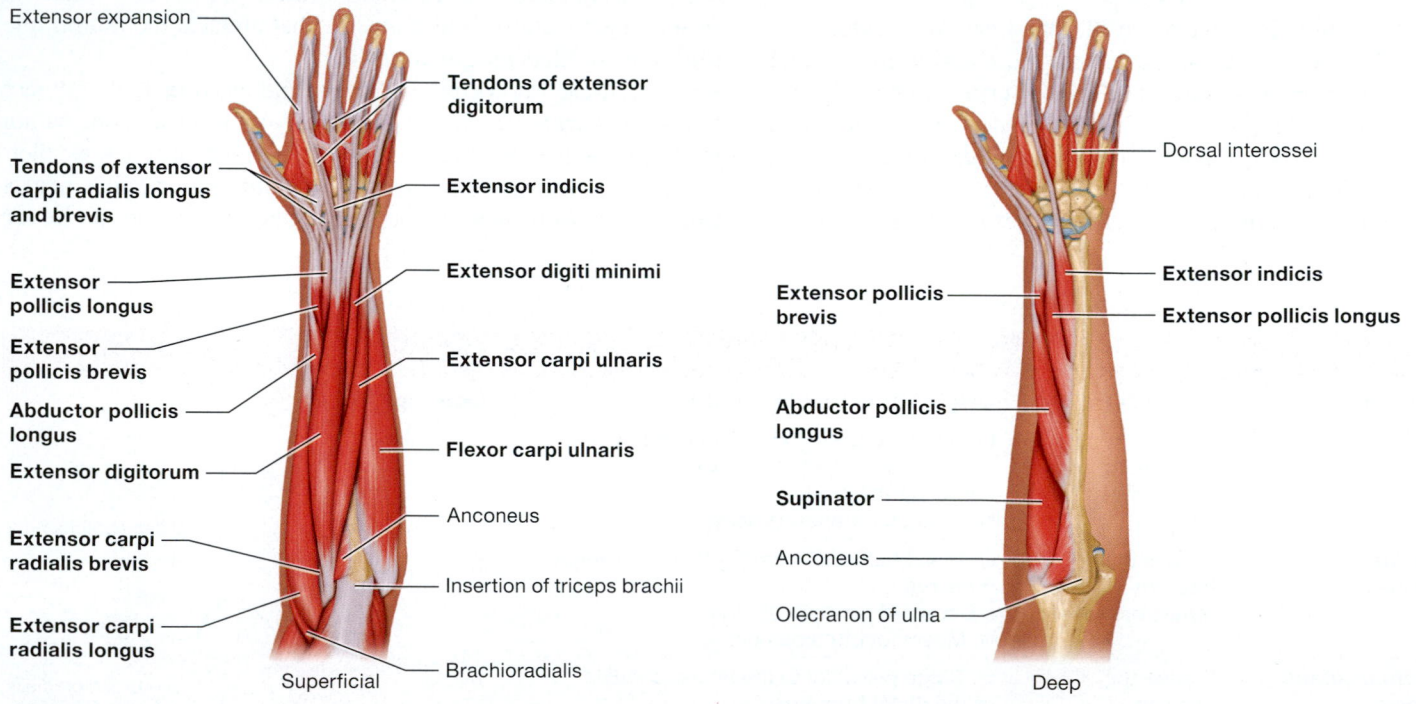

Extensor expansion — **Tendons of extensor digitorum**

Tendons of extensor carpi radialis longus and brevis — **Extensor indicis**

Extensor pollicis longus — **Extensor digiti minimi**

Extensor pollicis brevis

Abductor pollicis longus — **Extensor carpi ulnaris**

Extensor digitorum — **Flexor carpi ulnaris**

Anconeus

Extensor carpi radialis brevis — Insertion of triceps brachii

Extensor carpi radialis longus — Brachioradialis

Superficial

Dorsal interossei

Extensor pollicis brevis — **Extensor indicis**

Extensor pollicis longus

Abductor pollicis longus

Supinator

Anconeus

Olecranon of ulna

Deep

(b) Posterior view with muscles removed as necessary to reveal deeper muscles

Figure 9.19 Muscles that move the hand. Labels in **bold** indicate muscles and tendons involved in these actions.

pollicis brevis muscle is a short muscle that flexes the thumb (see Figure 9.20a and b). These muscles are part of the *thenar muscles* of the thumb. The other intrinsic hand muscles, including the **lumbricals** (LUM-brih-kulz) and **dorsal** and **palmar interossei muscles** (in-ter-AWSS-ee-aye; "between bones"), move the fingers and are illustrated in Figures 9.20c, d, and e.

Quick Check

☐ 5. What are the main muscles that flex the forearm at the elbow joint?

☐ 6. What are the main muscles that extend the hand? Which muscles flex the hand?

Table 9.15 Muscles That Move the Hand

Muscle(s)	Action(s)	Origin/Insertion/Nerve(s)
Pronator teres muscle	Pronates forearm (palm posterior)	O: Medial epicondyle of humerus, coronoid process of ulna I: Lateral surface of radius N: Median nerve
Flexor carpi radialis muscle	Flexes hand (bends hand toward anterior forearm); abducts hand	O: Medial epicondyle of humerus I: Second and third metacarpals N: Median nerve
Palmaris longus muscle	Flexes hand; tenses dense connective tissue in palm of hand	O: Medial epicondyle of humerus I: Flexor retinaculum and surrounding connective tissue in palm of hand N: Median nerve
Flexor carpi ulnaris muscle	Flexes hand; adducts hand	O: Medial epicondyle of humerus, medial olecranon process of ulna I: Pisiform, hamate, fifth metacarpal N: Ulnar nerve
Flexor digitorum superficialis muscle	Flexes fingers; flexes hand	O: Medial epicondyle of humerus, coronoid process of ulna, anterior proximal radius I: Middle phalanges of fingers 2–5 N: Median nerve
Flexor pollicis longus muscle	Flexes thumb	O: Anterior diaphysis of radius I: Distal phalanx of thumb N: Median nerve
Flexor digitorum profundus muscle	Flexes fingers; flexes hand	O: Medial/anterior ulna and coronoid process I: Distal phalanges of fingers 2–5 N: Median and ulnar nerve branches
Pronator quadratus muscle	Pronates forearm	O: Anterior distal ulna I: Anterior distal radius N: Median nerve
Extensor carpi radialis longus muscle	Extends hand; abducts hand	O: Lateral lower diaphysis of humerus I: Second metacarpal N: Radial nerve
Extensor carpi radialis brevis muscle	Extends hand; abducts hand	O: Lateral epicondyle of humerus I: Third metacarpal N: Deep radial nerve
Extensor digitorum muscle	Extends fingers; extends hand	O: Lateral epicondyle of humerus I: Posterior phalanges of fingers 2–5 N: Posterior interosseous nerve
Extensor digiti minimi muscle	Extends fifth finger	O: Lateral epicondyle of humerus I: Phalanges of finger 5 N: Posterior interosseous nerve
Extensor carpi ulnaris muscle	Extends hand; adducts hand	O: Lateral epicondyle of humerus, lateral proximal ulna I: Fifth metacarpal N: Posterior interosseous nerve
Supinator muscle	Supinates forearm (palm anterior)	O: Lateral epicondyle of humerus, area of the radial notch of ulna I: Anterior/lateral area of proximal third of radius, distal to tuberosity N: Deep radial nerve
Abductor pollicis longus muscle	Abducts and extends thumb	O: Posterior diaphysis of radius and ulna I: First metacarpal N: Posterior interosseous nerve
Extensor pollicis longus muscle	Extends thumb ("thumbs up" motion)	O: Posterior diaphysis of ulna I: Distal phalanx of thumb N: Posterior interosseous nerve
Extensor pollicis brevis muscle	Extends thumb	O: Lower posterior diaphysis of radius I: Proximal phalanx of thumb N: Posterior interosseous nerve
Extensor indicis muscle	Extends index finger	O: Lower posterior diaphysis of ulna I: Phalanges of finger 2 N: Posterior interosseous nerve

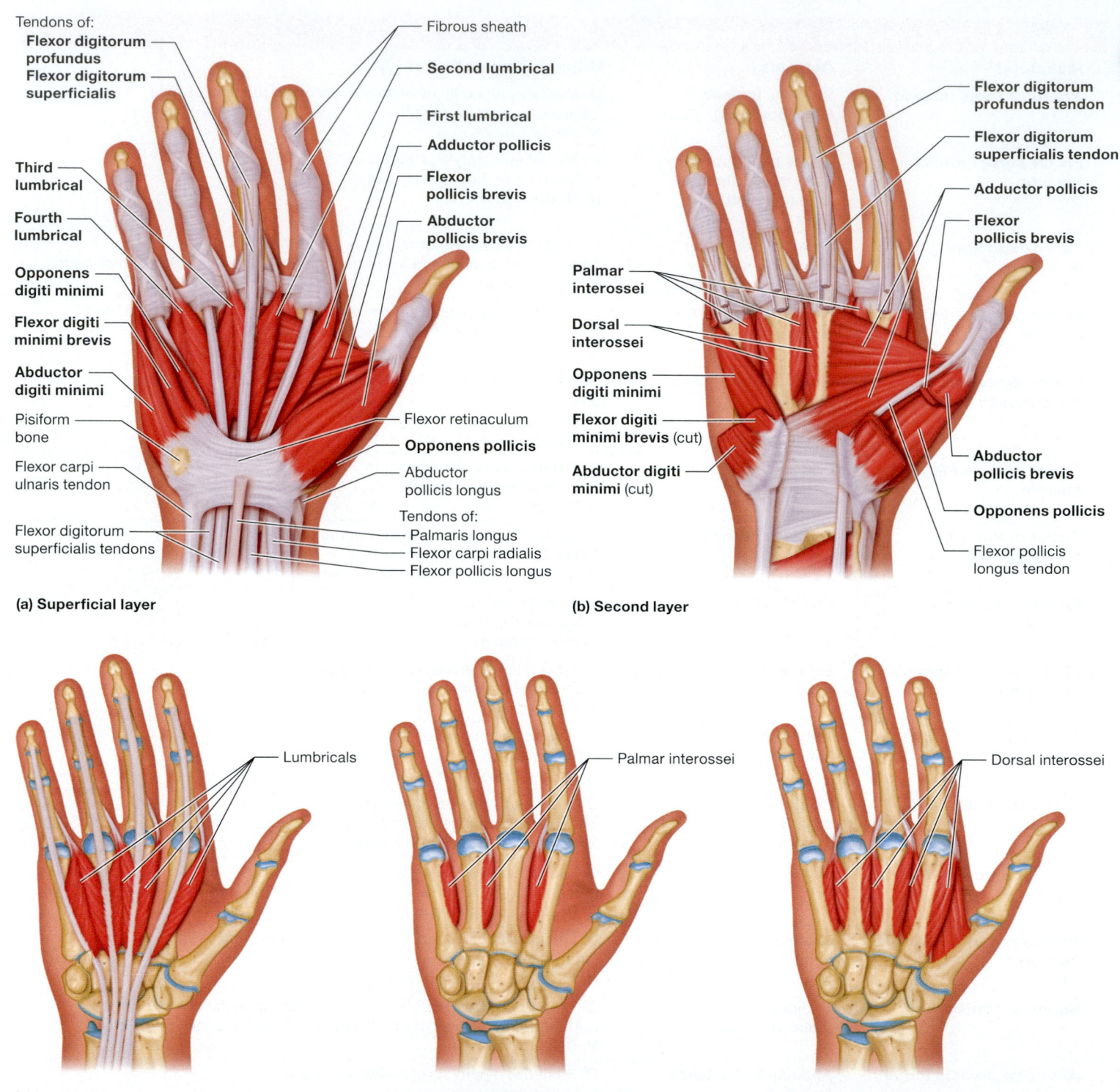

(a) Superficial layer

Tendons of:
Flexor digitorum profundus
Flexor digitorum superficialis

Fibrous sheath
Second lumbrical
First lumbrical
Adductor pollicis
Flexor pollicis brevis
Abductor pollicis brevis

Third lumbrical
Fourth lumbrical
Opponens digiti minimi
Flexor digiti minimi brevis
Abductor digiti minimi
Pisiform bone
Flexor carpi ulnaris tendon
Flexor digitorum superficialis tendons

Flexor retinaculum
Opponens pollicis
Abductor pollicis longus
Tendons of:
Palmaris longus
Flexor carpi radialis
Flexor pollicis longus

(b) Second layer

Flexor digitorum profundus tendon
Flexor digitorum superficialis tendon
Adductor pollicis
Flexor pollicis brevis

Palmar interossei
Dorsal interossei
Opponens digiti minimi
Flexor digiti minimi brevis (cut)
Abductor digiti minimi (cut)

Abductor pollicis brevis
Opponens pollicis
Flexor pollicis longus tendon

(c) Lumbricals

Lumbricals

(d) Palmar interossei

Palmar interossei

(e) Dorsal interossei

Dorsal interossei

Figure 9.20 **Muscles of the hand that move the fingers.** These muscles are shown in a palmar view of the right hand. Labels in **bold** indicate muscles and tendons involved in these actions.

Apply What You Learned

☐ 1. A potential consequence of shaving the acromion to repair the rotator cuff is detachment of part of the origin of the deltoid muscle. This leads to weakening of the deltoid muscle, which can be treated with physical therapy. What general types of exercises do you think would be helpful in strengthening this muscle? Explain your reasoning.

☐ 2. A common activity many athletes perform in training is climbing a rope with only the hands. What are the main muscles used to climb a rope in this manner? (*Hint:* Break it down into two steps: reaching up and grabbing the rope first, then pulling yourself up second.)

See answers in Appendix A.

Table 9.16 Muscles of the Hand That Move the Fingers

Muscle(s)	Action(s)	Origin/Insertion/Nerve(s)
*Abductor pollicis brevis** muscle	Abducts thumb	O: Flexor retinaculum, scaphoid and trapezium bones I: Lateral side of proximal phalanx of thumb N: Median nerve
*Flexor pollicis brevis** muscle	Flexes thumb	O: Flexor retinaculum, trapezium bone I: Lateral proximal phalanx of thumb N: Median nerve
*Opponens pollicis** muscle	Opposes thumb	O: Flexor retinaculum, trapezium bone I: First metacarpal N: Median nerve
Adductor pollicis muscle	Adducts thumb	O: Capitate bone and metacarpals 2–3 I: Proximal phalanx of thumb N: Ulnar nerve
Abductor digiti minimi muscle	Abducts fifth digit	O: Pisiform bone I: Proximal phalanx of finger 5 N: Ulnar nerve
Flexor digiti minimi brevis muscle	Flexes fifth digit	O: Flexor retinaculum, hamate bone I: Proximal phalanx of finger 5 N: Ulnar nerve
Opponens digiti minimi muscle	Opposes fifth digit to thumb	O: Flexor retinaculum, hamate bone I: Fifth metacarpal N: Ulnar nerve
Lumbrical muscle	Flex fingers at metacarpophalangeal joints; extend fingers at interphalangeal joints	O: Tendons of the flexor digitorum profundus muscle (in the palm) I: Proximal phalanx of fingers 2–5 N: Median nerve and ulnar nerve
Palmar interossei muscles	Adduct fingers (pull extended fingers together)	O: Adductor side of metacarpals 2, 4, and 5 I: Proximal phalanx of fingers 2, 4, and 5 N: Ulnar nerve
Dorsal interossei muscles	Abduct fingers (spread them)	O: Interior sides of metacarpals I: First phalanx of fingers 2, 3, and 4 N: Ulnar nerve

* Part of the thenar muscles of the thumb.

MODULE 9.5

Muscles of the Hip and Lower Limb

Learning Outcomes

1. Identify and describe the muscles that move the hip and lower limb.
2. Identify the origin, insertion, and action of these muscles, and demonstrate their actions.

As you learned in the chapter on the skeleton, the pelvic girdle attaches each lower limb—the thigh, leg, ankle, and foot—to the axial skeleton (see Chapter 7). However, unlike the pectoral girdle, the pelvic girdle has a firm bony attachment to the axial skeleton via the sacrum. This is partly due to the role of the pelvic girdle and lower limbs in supporting the weight of the body, an example of the Structure-Function Core Principle (p. 28). This "frees up" muscles that would otherwise be required to fix and stabilize the pelvis, as many do for the scapula. For this reason, nearly all of the muscles in the pelvis and thigh are also involved in moving the thigh and leg. We examine these muscles, as well as the muscles that move the ankle and foot, in this module.

CORE PRINCIPLE
Structure-Function

Muscles of the Hip, Thigh, Knee, and Leg

 **FLASHBACK**

1. Where are the lesser trochanter and linea aspera located? (p. 250)
2. Where are the iliac crest and the anterior superior iliac spine? (p. 248)

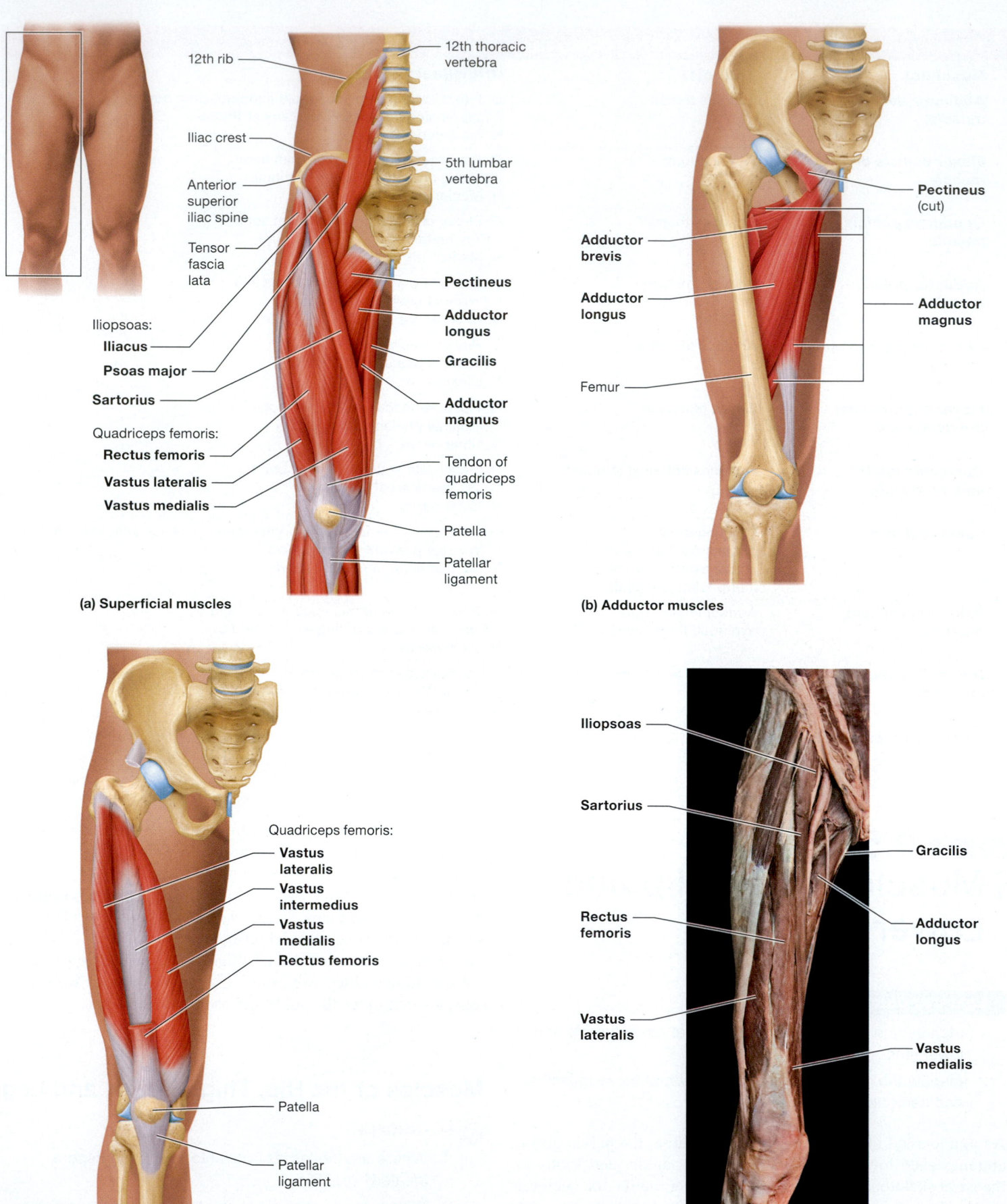

12th rib

12th thoracic vertebra

Iliac crest

Anterior superior iliac spine

5th lumbar vertebra

Tensor fascia lata

Pectineus

Iliopsoas:

Iliacus

Psoas major

Adductor longus

Sartorius

Gracilis

Quadriceps femoris:

Adductor magnus

Rectus femoris

Vastus lateralis

Tendon of quadriceps femoris

Vastus medialis

Patella

Patellar ligament

(a) Superficial muscles

Pectineus (cut)

Adductor brevis

Adductor longus

Adductor magnus

Femur

(b) Adductor muscles

Quadriceps femoris:

Vastus lateralis

Vastus intermedius

Vastus medialis

Rectus femoris

Patella

Patellar ligament

(c) Deeper muscles under the rectus femoris

Iliopsoas

Sartorius

Gracilis

Rectus femoris

Adductor longus

Vastus lateralis

Vastus medialis

(d) Cadaver photo

Figure 9.21 Anterior and medial muscles that move the thigh and leg. Labels in **bold** indicate muscles involved in these actions.

Let's start with muscles that flex the thigh, shown in **Figure 9.21**. One of the most powerful thigh flexors is the **iliopsoas muscle** (il′-ee-oh-SOH-az; see Figure 9.21a). This deep muscle is formed by the union of two muscles, the **iliacus** (il-ee-AK-us) and **psoas major** (SOH-us) **muscles,** which insert via a common tendon into the lesser trochanter of the femur. The iliacus muscle originates from the inner surface of the ilium, whereas the psoas major muscle has a much more superior origin, from the lateral sides of the lumbar vertebrae. This more superior origin allows the origin and insertion of the psoas major muscle to switch when the hip joint is fixed, in which case it can flex the trunk. The iliopsoas muscle is assisted in its job of thigh flexion by a small muscle that originates from the pubic bone, called the **pectineus muscle** (pek-TIN-ee-us).

Other muscles that move the thigh are located on its medial surface. First is a superficial straplike muscle in the medial thigh called the **gracilis muscle** (grass-IL-us) that originates from the pelvis and inserts into the proximal tibia (see Figure 9.21a and d). It adducts the thigh—moves it medially—along with three muscles known collectively as the *adductor group*: the **adductor magnus, adductor longus,** and **adductor brevis muscles** (see Figure 9.21b). These muscles originate from the pubis and/or ischium and insert into the linea aspera of the femoral diaphysis, which makes them very powerful adductors. The gracilis and adductor group muscles are assisted in adduction by the pectineus.

Let's now move on to muscles that move the leg, although as we'll see, many of these muscles move the thigh as well. First is the **sartorius muscle** (sar-TOHR-ee-us), a long muscle that crosses the thigh from its origin on the anterior superior iliac spine to its insertion on the medial tibial condyle (see Figure 9.21a). This unusual pathway allows the sartorius muscle to carry out four actions: flexion, abduction, and lateral rotation of the thigh; and flexion of the leg. You perform all four motions when you sit "criss-crossed" (or "cross-legged") or when you lift up your foot to check the bottom of your shoe.

Next is the **quadriceps femoris muscle group** (KWAH-drih-seps FEM-oh-ris) of the anterior thigh, which consists of four muscles (*quad-* = "four"; see Figure 9.20c and d). All muscles of this group converge into a common tendon called the *quadriceps tendon*, which wraps around the patella (kneecap). Distal to the patella, the tendon becomes the *patellar ligament* and inserts into the tibial tuberosity, an insertion that allows these muscles to extend the leg at the knee joint.

The centrally located, superficial muscle of the quadriceps femoris muscle group is the **rectus femoris muscle.** This muscle is the only one to originate from the hip, an origin that gives it the additional action of thigh flexion. The other three muscles are part of a group known as the *vastus group*: the **vastus lateralis** (VAST-us lah-ter-AHL-iss), **vastus intermedius** (in-ter-MEE-dee-us), and **vastus medialis** (mee-dee-AHL-iss) **muscles.** Their names suggest their relative location—lateral, "in between," and medial, respectively—along the diaphysis of the femur. All muscles of the vastus group originate from the femur and so only cross the knee joint. For this reason, their only action is extension of the leg at the knee joint.

See **Table 9.17** on p. 324 for more information on anterior and medial muscles.

Now let's move on to the posterior muscles (**Figure 9.22**). The **gluteal muscle group** located on the posterior aspect of the pelvis consists of three separate muscles. First is the large **gluteus maximus muscle** (GLOO-tee-us MAX-ih-mus; see Figure 9.22a and b), which has a broad origin ranging from the coccyx and sacrum to the ilium, and inserts into the gluteal tuberosity of the femur. The shape and insertion of the gluteus maximus muscle make it a powerful extensor of the thigh. This muscle also abducts and laterally rotates the thigh; all three motions combined are used when you extend the thigh forcibly, as in climbing stairs. Several small muscles—the **piriformis** (peer-ih-FORM-iss), **obturator internus, gemelli** (jeh-MEH-lee), **quadratus femoris,** and **obturator externus muscles** (see Figure 9.22c)—also help laterally rotate the thigh.

The two smaller gluteal muscles, the **gluteus medius** and **gluteus minimus muscles** (see Figure 9.22c), insert onto the greater trochanter of the femur, which makes them play little to no part in thigh extension. However, both muscles have an important role in walking. During the swing phase of walking, when the moving lower limb swings forward toward the next step, gravity would tend to cause the pelvis to sag on the side of the moving limb because it's unsupported. However, the gluteus medius and minimus muscles abduct and medially rotate the thigh and pelvis of the moving limb, which pulls it up and actually elevates it above the level of the supported lower limb (the limb still on the ground).

Moving distally we find the **hamstring muscle group**—the medial and superficial **semitendinosus muscle** (sem′-ee-ten-dih-NOH-sus), the medial and deep **semimembranosus muscle** (sem′-ee-mem-brah-NOH-sus), and the lateral **biceps femoris muscle** (see Figure 9.22a). The hamstring muscle group originates from the ischial tuberosity and inserts into the fibula and/or tibia. As they cross both the hip and the knee, they act on both joints: They extend the thigh and flex the leg.

See **Table 9.18** on p. 326 for more details regarding posterior muscles that move the thigh and leg.

Quick Check

☐ 1. What are the major muscles of thigh flexion and extension?

☐ 2. What are the main muscles of leg flexion and extension?

Muscles of the Ankle, Foot, and Toes

《 FLASHBACK

1. What are the bones of the leg? With which tarsal bone do they articulate? (p. 251)

2. What occurs during the motions dorsiflexion, plantarflexion, inversion, and eversion? (p. 269)

There are a surprising number of intricate, muscle-controlled movements in the feet and toes that are involved in an activity as seemingly simple as walking. As the activities get more

Table 9.17 Muscles That Move the Thigh and Leg: Anterior and Medial Muscles

Muscle(s)	Action(s)	Origin/Insertion/Nerve(s)	Concept Figures
*Iliacus muscle**	**Flexes** the thigh	O: Iliac fossa I: Lesser trochanter of the femur N: Femoral nerve	
*Psoas major muscle**	**Flexes** the thigh; **flexes** and **laterally flexes** the vertebral column	O: Lateral sides and transverse processes of T12 and lumbar vertebrae I: Lesser trochanter of the femur N: L1–L3	
Sartorius muscle	**Flexes** the thigh and **flexes** the leg; **abducts** and laterally **rotates** the thigh	O: Anterior superior iliac spine I: Proximal portion of the medial condyle of the tibia N: Femoral nerve	
Tensor fascia lata muscle	**Abducts** the thigh; **flexes** the thigh; **rotates** the thigh medially; tightens ("tenses") the fascia lata, the deep fascia of the lateral thigh	O: Iliac crest and anterior superior iliac spine I: Iliotibial tract, the thickened portion of the fascia lata, which itself inserts into the proximal, lateral tibia N: Superior gluteal nerve	
Adductor magnus muscle	**Adducts** the thigh; **rotates** the thigh medially	O: Ischial tuberosity, ischial ramus, inferior ramus of pubis I: Linea aspera and adductor tubercle of femur N: Obturator and sciatic nerves	
Adductor longus muscle	**Adducts** the thigh; **flexes** the thigh; **rotates** the thigh medially	O: Body of pubis I: Linea aspera N: Obturator nerve	
Adductor brevis muscle	**Adducts** the thigh; **flexes** the thigh; **rotates** the thigh medially	O: Inferior ramus of the pubis I: Proximal linea aspera N: Obturator nerve	
Pectineus muscle	**Adducts** the thigh; **flexes** the thigh; **rotates** the thigh medially	O: Superior ramus of pubis I: Posterior femur between the lesser trochanter and linea aspera N: Femoral nerve	
Gracilis muscle	**Adducts** the thigh; **flexes** the leg; **rotates** the thigh medially	O: Body and inferior ramus of pubis; ischial ramus I: Medial surface of proximal tibia N: Obturator nerve	
Quadriceps Femoris Muscle Group			
Rectus femoris muscle	**Extends** the leg; **flexes** the thigh	O: Anterior inferior iliac spine, superior margin of acetabulum I: Patella and tibial tuberosity N: Femoral nerve	
Vastus lateralis muscle	**Extends** the leg; stabilizes patella	O: Greater trochanter, intertrochanteric line, proximal half of the linea aspera I: Patella and tibial tuberosity N: Femoral nerve	
Vastus medialis muscle	**Extends** the leg; stabilizes patella	O: Linea aspera, intertrochanteric line I: Patella and tibial tuberosity N: Femoral nerve	
Vastus intermedius muscle	**Extends** the leg	O: Anterior and lateral portions of proximal two-thirds of the diaphysis of femur I: Patella and tibial tuberosity N: Femoral nerve	

*Part of the iliopsoas muscle that shares a common insertion tendon.

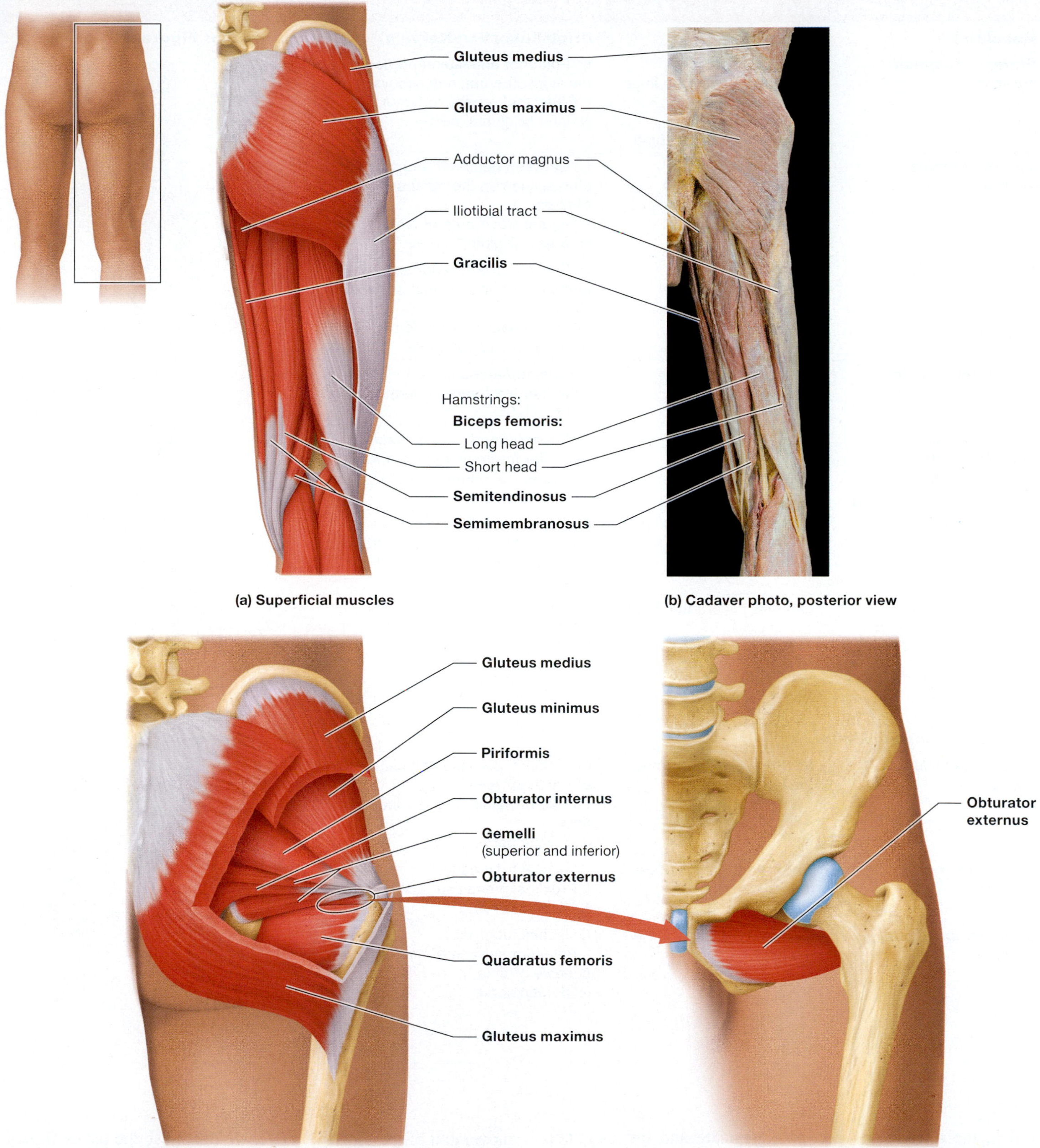

Gluteus medius

Gluteus maximus

Adductor magnus

Iliotibial tract

Gracilis

Hamstrings:
Biceps femoris:
Long head
Short head
Semitendinosus
Semimembranosus

(a) Superficial muscles

(b) Cadaver photo, posterior view

Gluteus medius

Gluteus minimus

Piriformis

Obturator internus

Gemelli
(superior and inferior)

Obturator externus

Quadratus femoris

Gluteus maximus

Obturator
externus

(c) Muscles under the gluteus maximus and gluteus medius

(d) The obturator externus muscle, anterior view

Explore PAL™ Cadaver
@ **Mastering** Anatomy & Physiology

Figure 9.22 Posterior muscles that move the thigh and leg. Labels in **bold** indicate muscles involved in these actions.

Table 9.18 Muscles That Move the Thigh and Leg: Posterior Muscles

Muscle(s)	Action(s)	Origin/Insertion/Nerve(s)	Concept Figures
Gluteus maximus muscle	**Extends** the thigh (especially when the thigh is in flexed position); laterally **rotates** the thigh; **abducts** the thigh	O: Posterior and lateral portions of the ilium, sacrum, and coccyx I: Gluteal tuberosity of femur N: Inferior gluteal nerve	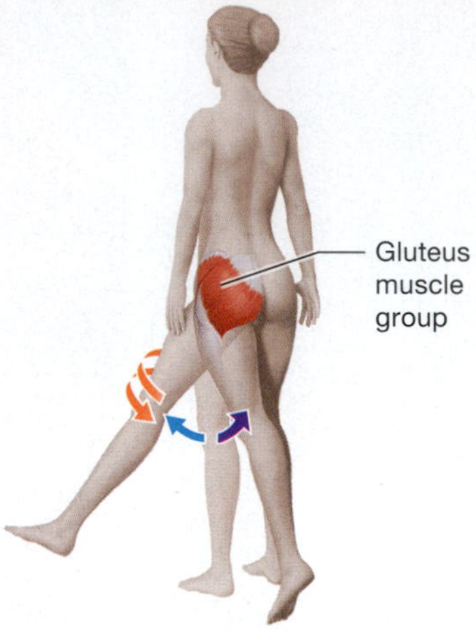
Gluteus medius muscle	**Abducts** the thigh; medially **rotates** the thigh; stabilizes pelvis while walking	O: Between posterior and anterior gluteal lines on the outer surface of ilium I: Greater trochanter of femur N: Superior gluteal nerve	
Gluteus minimus muscle	**Abducts** the thigh; medially **rotates** the thigh; stabilizes pelvis while walking	O: Between the anterior and inferior gluteal lines on the outer surface of ilium I: Greater trochanter of femur N: Superior gluteal nerve	
Piriformis muscle	**Abducts** the thigh; laterally **rotates** the thigh	O: Anterior/lateral sacrum I: Greater trochanter of femur N: L5, S1–S2	
Obturator externus muscle	Laterally **rotates** the thigh	O: Edges of and outer surface of obturator foramen and membrane I: Greater trochanter of femur N: Obturator nerve	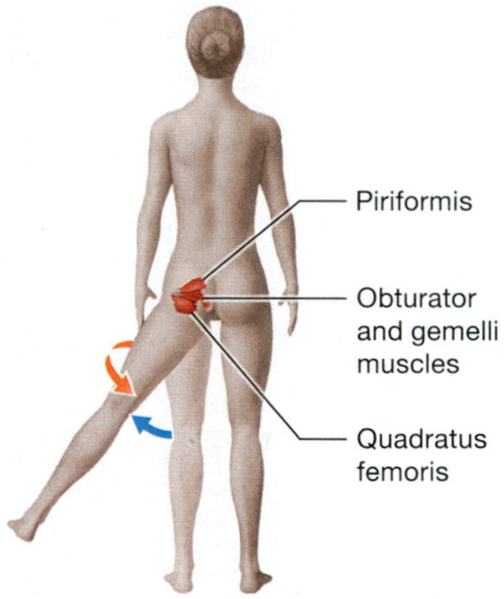
Obturator internus muscle	Laterally **rotates** the thigh	O: Edges of and inner surface of obturator foramen and membrane I: Greater trochanter of femur N: L5, S$_1$	
Gemelli (superior gemellus and inferior gemellus muscles)	Laterally **rotate** the thigh	O: Ischial spine and tuberosity I: Greater trochanter of femur N: L5, S1	
Quadratus femoris muscle	Laterally **rotates** the thigh	O: Ischial tuberosity I: Intertrochanteric crest of femur N: L5, S1	
Hamstring Muscle Group			
Biceps femoris muscle	**Extends** the thigh; **flexes** the leg	O: Ischial tuberosity and distal half of posterior femur I: Head of fibula; lateral condyle of tibia N: Sciatic nerve	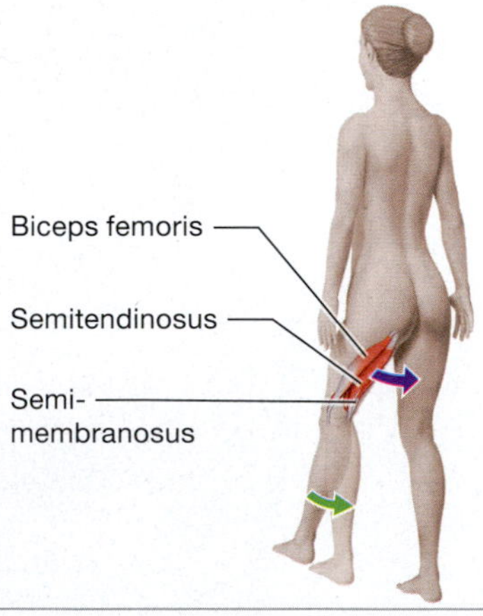
Semitendinosus muscle	**Extends** the thigh; **flexes** the leg	O: Ischial tuberosity I: Proximal medial surface of tibia N: Sciatic nerve	
Semimembranosus muscle	**Extends** the thigh; **flexes** the leg	O: Ischial tuberosity I: Posterior surface of medial condyle of tibia N: Sciatic nerve	

Gluteus muscle group

Piriformis

Obturator and gemelli muscles

Quadratus femoris

Biceps femoris

Semitendinosus

Semimembranosus

complex, the movements get more intricate and involve finer muscle control. Let's take a look at some of the muscles involved in moving the feet and toes.

Muscles That Move the Foot

We'll start with the anterior muscles in the leg, which are shown in **Figure 9.23**. First remember that when we talk about

movement of the foot and ankle, we don't use the terms flexion and extension. Instead, we use the terms *dorsiflexion,* in which the toes are brought toward the leg, and *plantarflexion,* in which the toes are pointed away from the leg. The muscles of the anterior leg, including the **tibialis anterior** (tib-ee-AL-iss) and **extensor digitorum longus muscles,** are the main dorsiflexors of the foot. The extensor digitorum longus muscle inserts

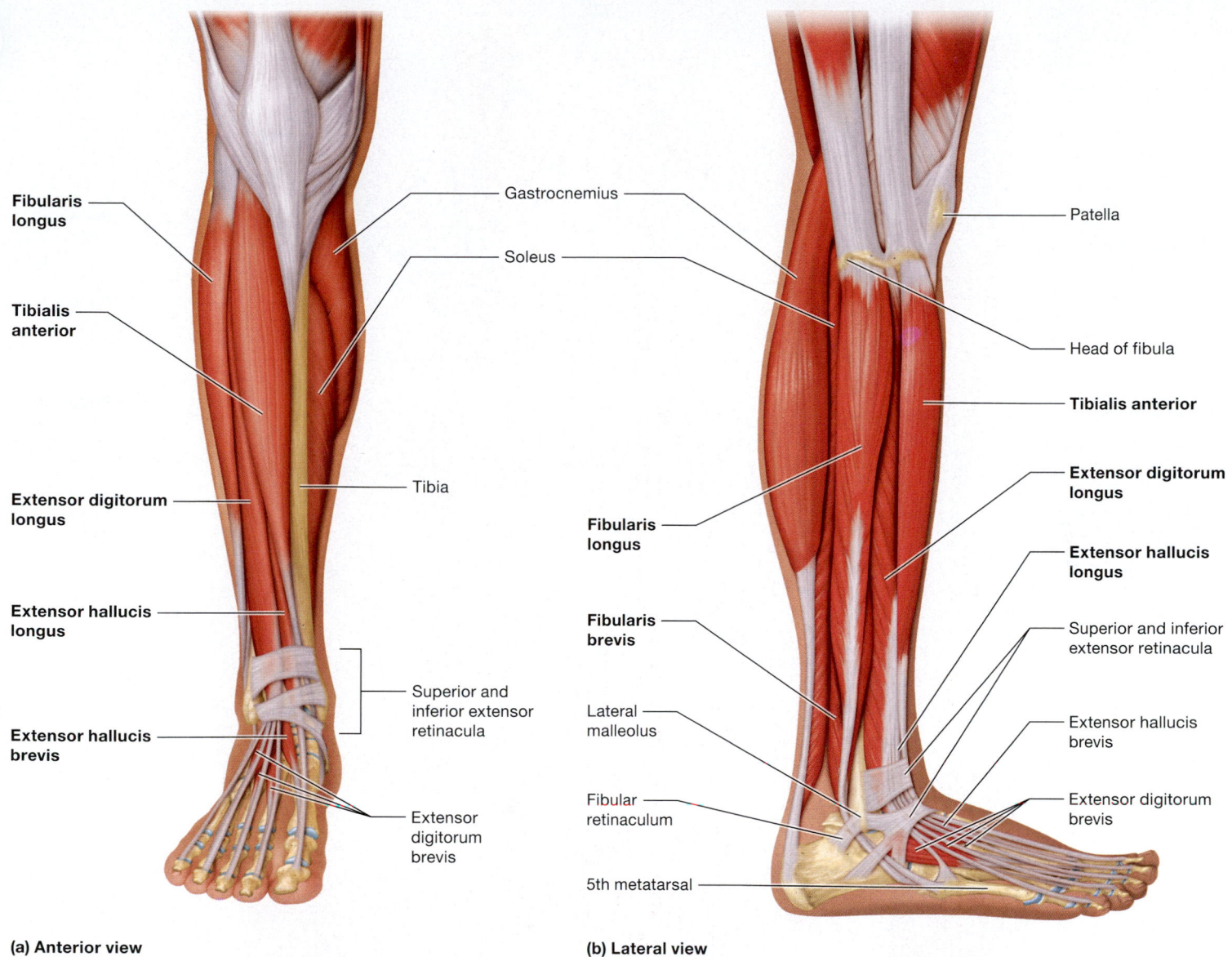

Fibularis longus

Gastrocnemius

Patella

Tibialis anterior

Soleus

Head of fibula

Tibialis anterior

Tibia

Extensor digitorum longus

Extensor digitorum longus

Fibularis longus

Extensor hallucis longus

Extensor hallucis longus

Fibularis brevis

Superior and inferior extensor retinacula

Superior and inferior extensor retinacula

Extensor hallucis brevis

Lateral malleolus

Extensor hallucis brevis

Fibular retinaculum

Extensor digitorum brevis

Extensor digitorum brevis

5th metatarsal

(a) Anterior view

(b) Lateral view

Figure 9.23 Anterior and lateral muscles that move the foot and toes. Labels in **bold** indicate muscles involved in these actions.

into the phalanges of the second through fifth digits, so it also extends these digits, as its name implies. Another anterior muscle, the **extensor hallucis longus muscle** (hal-OO-sis) assists dorsiflexion by extending the hallux (great toe).

In the lateral view of the leg in Figure 9.23b, we find two muscles that run along the fibula. Accordingly, they are called the **fibularis longus** (fib-yoo-LAH-ris) and **fibularis brevis muscles.** These muscles insert into the lateral side of the foot, which enables them to pull the plantar surface of the foot outward, a movement known as *eversion.* The tibialis anterior muscle acts as an antagonist to these muscles, as it functions in *inversion* of the foot (turning the plantar surface inward).

Turning now to the posterior leg in **Figure 9.24**, you can see that far more muscles are involved in plantarflexion than in dorsiflexion—so many that we divided the figure into three layers in order to show all the muscles. The largest posterior muscle and most

powerful plantarflexor of the foot is the *triceps surae muscle group* (SOO-ree), which consists of the superficial two-headed **gastrocnemius muscle** (gas′-trok-NEE-mee-us) and the deeper **soleus muscle.** Notice that the gastrocnemius muscle originates from the femur, and so has a minor action of assisting with leg flexion. The soleus muscle originates from the tibia and fibula and so has no action on the leg (see Figure 9.24b). Both muscles converge to form a common tendon that inserts into the posterior calcaneus called the **calcaneal tendon.** The calcaneal tendon is subject to a great deal of tension, and as such, is prone to injuries. Read more about this topic in *A&P in the Real World: Calcaneal Tendon Injuries.*

Other muscles that plantarflex the foot include the **plantaris, tibialis posterior,** fibularis longus, and fibularis brevis muscles. Note that the tibialis posterior muscle assists the tibialis anterior muscle with inversion of the foot. See **Table 9.19** for more information on muscles that move the foot.

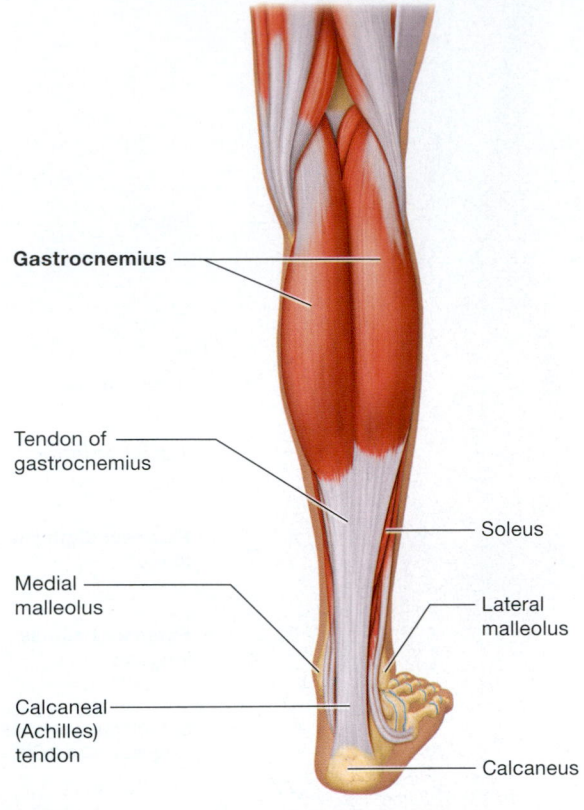

Gastrocnemius

Tendon of
gastrocnemius

Soleus

Medial
malleolus

Lateral
malleolus

Calcaneal
(Achilles)
tendon

Calcaneus

(a) Superficial muscles

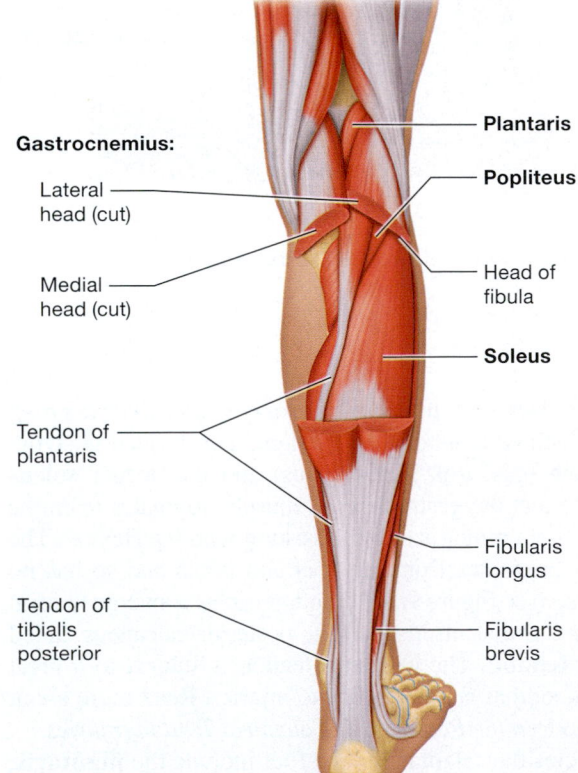

Gastrocnemius:

Lateral
head (cut)

Medial
head (cut)

Plantaris

Popliteus

Head of
fibula

Soleus

Tendon of
plantaris

Tendon of
tibialis
posterior

Fibularis
longus

Fibularis
brevis

(b) Muscles under the gastrocnemius

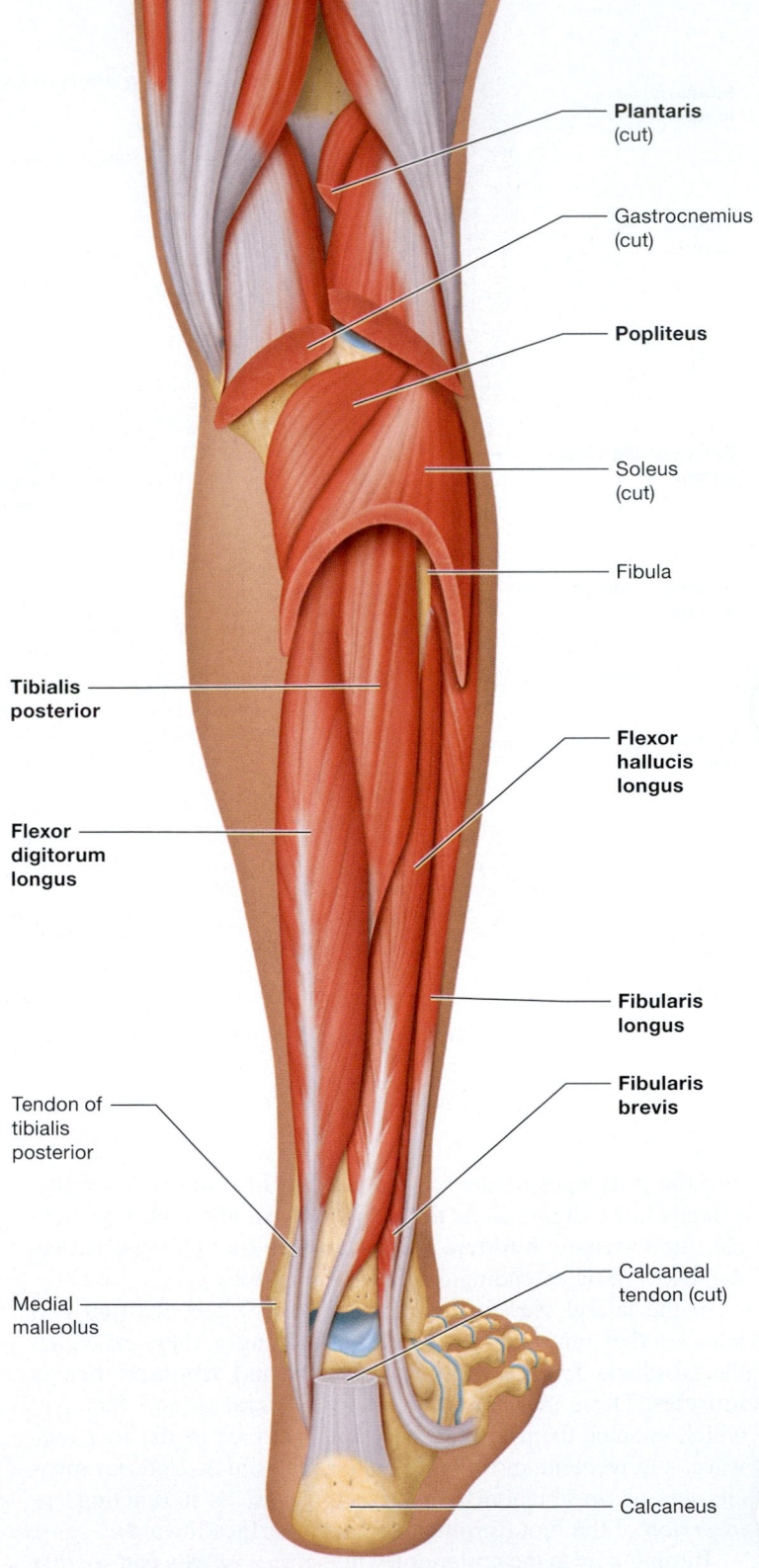

Plantaris
(cut)

Gastrocnemius
(cut)

Popliteus

Soleus
(cut)

Fibula

**Tibialis
posterior**

**Flexor
hallucis
longus**

**Flexor
digitorum
longus**

**Fibularis
longus**

**Fibularis
brevis**

Tendon of
tibialis
posterior

Medial
malleolus

Calcaneal
tendon (cut)

Calcaneus

(c) Deeper muscles under the gastrocnemius and soleus

Figure 9.24 **Right leg: posterior muscles that move the ankle and toes.** Labels in **bold** indicate muscles involved in these actions.

Calcaneal Tendon Injuries

Recall from our discussion on levers that the gastrocnemius and soleus muscles form a second-class lever. This type of system generates a lot of force, which makes sense because these two muscles are lifting the entire weight of the body. The high forces generated by the second-class lever put a great deal of strain on the calcaneal tendon. Indeed, injuries to the calcaneal tendon are relatively common, especially in sports such as football, basketball, and gymnastics. The tendon may also be injured by overuse or trauma—stepping in a hole is a common culprit.

Symptoms of calcaneal tendon injuries depend on their severity. An individual with tendonitis or a partial tear will likely have pain and inflammation over the area that worsens with activity. An individual with a complete tear often reports feeling a "pop" along with severe pain. Such patients will be almost completely unable to plantarflex the affected foot. This is assessed clinically with the *Simmonds test*, during which a patient lies face down and the gastrocnemius muscle is squeezed. If the calcaneal tendon is intact, the foot will plantarflex. If it has ruptured, the foot will not move.

Treatment for a ruptured tendon may involve nonsurgical options, such as wearing a walking boot while the tendon heals, or surgical repair of the torn tendon. Both treatment protocols involve physical therapy to increase the range of motion of the ankle and gently stretch and strengthen the tendon as it heals.

Table 9.19 Muscles That Move the Foot and Toes

Muscle(s)	Action(s)	Origin/Insertion/Nerve(s)	Concept Figures
Tibialis anterior muscle	**Dorsiflexes** the foot; **inverts** the foot	O: Lateral condyle and proximal diaphysis of the tibia I: Medial cuneiform bone and first metatarsal N: Deep fibular nerve	
Extensor digitorum longus muscle	**Extends** the toes; **dorsiflexes** the foot	O: Lateral condyle of the tibia and proximal portion of the fibula I: Phalanges and connective tissues of toes 2–5 N: Deep fibular nerve	
Extensor hallucis longus muscle	**Extends** the hallux (great toe); **dorsiflexes** the foot	O: Diaphysis of the fibula and interosseous membrane I: Distal phalanx of hallux N: Deep fibular nerve	
Fibularis longus muscle	**Everts** the foot; **plantarflexes** the foot	O: Proximal lateral fibula I: Medial cuneiform bone and first metatarsal N: Superficial fibular nerve	
Fibularis brevis muscle	**Everts** the foot; **plantarflexes** the foot	O: Distal fibula I: Fifth metatarsal N: Superficial fibular nerve	
*Gastrocnemius muscle**	**Plantarflexes** the foot; flexes leg	O: Medial and lateral condyles of the femur I: Posterior calcaneus N: Tibial nerve	
*Soleus muscle**	**Plantarflexes** foot	O: Head of the fibula, proximal tibia, and interosseous membrane I: Posterior calcaneus N: Tibial nerve	
Plantaris muscle	**Plantarflexes** foot	O: Posterior femur near lateral condyle I: Calcaneus and calcaneal tendon N: Tibial nerve	
Tibialis posterior muscle	**Inverts** foot; **plantarflexes** foot; stabilizes foot	O: Proximal tibia, fibula, and interosseous membrane I: Metatarsals 2–4 N: Tibial nerve	
Popliteus muscle	**Flexes** leg; rotates leg	O: Lateral condyle of the femur I: Posterior surface of proximal tibia N: Tibial nerve	

Concept figure labels: Tibialis anterior, Extensor muscles, Fibularis muscles, Plantaris, Gastrocnemius and soleus; Tibialis anterior, Tibialis posterior, Fibularis muscles; Popliteus, Extensor muscles

*Part of the triceps surae muscle that shares a common insertion tendon, the calcaneal (Achilles) tendon.

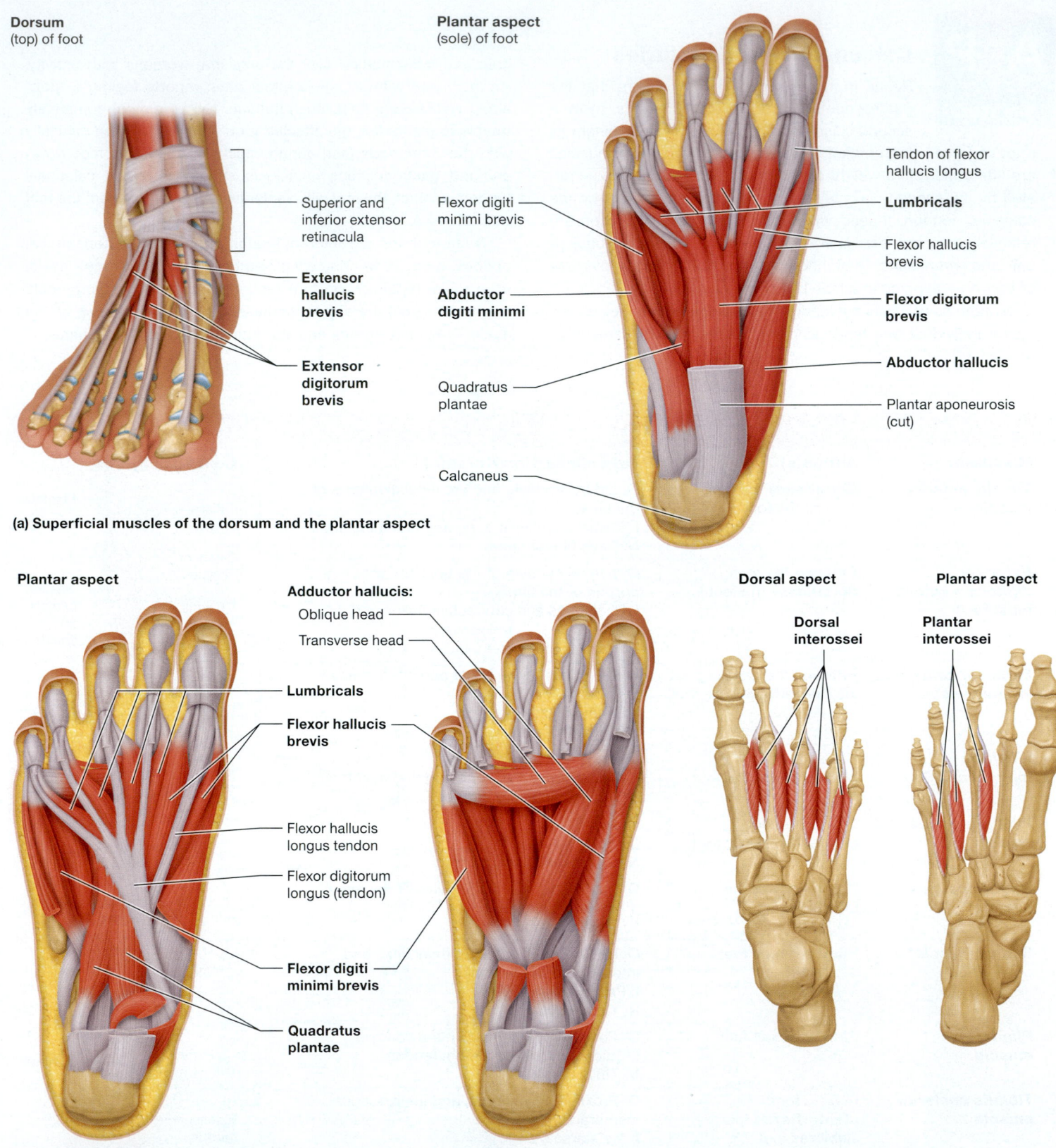

Dorsum
(top) of foot

Superior and inferior extensor retinacula

Extensor hallucis brevis

Extensor digitorum brevis

Plantar aspect
(sole) of foot

Flexor digiti minimi brevis

Abductor digiti minimi

Quadratus plantae

Calcaneus

Tendon of flexor hallucis longus

Lumbricals

Flexor hallucis brevis

Flexor digitorum brevis

Abductor hallucis

Plantar aponeurosis (cut)

(a) Superficial muscles of the dorsum and the plantar aspect

Plantar aspect

Lumbricals

Flexor hallucis brevis

Flexor hallucis longus tendon

Flexor digitorum longus (tendon)

Flexor digiti minimi brevis

Quadratus plantae

Adductor hallucis:
Oblique head
Transverse head

(b) Deeper muscles of the plantar aspect

Dorsal aspect

Dorsal interossei

Plantar aspect

Plantar interossei

(c) Muscles between the toes

Figure 9.25 Muscles that move the toes. Labels in **bold** indicate muscles involved in these actions.

Muscles That Move the Toes

The first muscles we'll examine are actually in the leg (see Figure 9.24c): the **flexor hallucis longus muscle,** which flexes the hallux, and the **flexor digitorum longus muscle,** which flexes the second through the fifth digits. These muscles originate from the posterior fibula and tibia, respectively, and so they also play a role in plantarflexing the foot.

The rest of the muscles that move the toes are *intrinsic foot muscles* and are shown in **Figure 9.25**. Note that many of the intrinsic foot muscles have similar names to the intrinsic hand muscles you read about earlier; both groups of muscles are named for their actions. For example, the **flexor digitorum brevis muscle** is a small muscle that flexes the second through fourth digits. These muscles support the arches of the foot and allow it to adapt to changing terrain as we walk. They can also be trained to perform quite skilled movements such as drawing, as shown by individuals with upper limb disabilities.

See **Table 9.20** for more details about muscles that move the toes.

Table 9.20 Muscles That Move the Toes		
Muscle(s)	**Action(s)**	**Origin/Insertion/Nerve(s)**
Flexor digitorum longus muscle	Flexes toes; stabilizes foot	O: Posterior tibia I: Distal phalanges of toes 2–5 N: Tibial nerve
Flexor hallucis longus muscle	Flexes hallux	O: Posterior fibula and interosseous membrane I: Distal phalanx of hallux N: Tibial nerve
Extensor digitorum brevis muscle	Extends toes	O: Superior lateral surface of calcaneus I: Dorsal surfaces of toes 2–4 N: Deep fibular nerve
Extensor hallucis brevis muscle	Extends hallux	O: Superior lateral surface of calcaneus I: Dorsal surface of proximal phalanx of hallux N: Deep fibular nerve
Flexor digitorum brevis muscle	Flexes the toes	O: Calcaneal tuberosity I: Middle phalanx of toes 2–5 N: Medial plantar nerve
Abductor digiti minimi muscle	Flexes and abducts fifth toe	O: Calcaneal tuberosity I: Lateral side of the base of proximal phalanx of toe 5 N: Lateral plantar nerve
Abductor hallucis muscle	Abducts hallux	O: Calcaneal tuberosity I: Medial side of the base of proximal phalanx of hallux N: Medial plantar nerve
Quadratus plantae muscle	Flexes toes	O: Lateral surface of the calcaneus I: Tendon of flexor digitorum longus muscle N: Lateral plantar nerve
Lumbrical muscles	Flex toes close to ball of foot; extend toes at phalanges	O: Tendons of the flexor digitorum longus muscle I: Connective tissues of toes 2–5 N: Lateral and medial plantar nerves
Adductor hallucis muscle	Adducts hallux; stabilizes the metatarsals and arch of the foot	O: Metatarsals and surrounding connective tissues, including the tendon of the fibularis longus muscle I: Base of proximal phalanx of lateral hallux N: Lateral plantar nerve
Flexor hallucis brevis muscle	Flexes hallux	O: Cuboid bone, lateral cuneiform bone, and tendon branches of other muscles (e.g., the tibialis posterior muscle) I: Base of hallux and tendons of other muscles (e.g., the adductor hallucis muscle) N: Medial plantar nerve
Flexor digiti minimi brevis muscle	Flexes fifth toe	O: Fifth metatarsal and tendon of fibularis longus muscle I: Base of proximal phalanx of toe 5 N: Lateral plantar nerve
Plantar interossei muscles	Adduct and flex toes 3–5	O: Metatarsals 3–5 I: Medial base of proximal phalanges of toes 3–5 N: Lateral plantar nerve
Dorsal interossei muscles	Abduct and flex toes 2–4	O: Metatarsals 2–5 I: Base of proximal phalanges of toes 2–4 N: Lateral plantar nerve

The Big Picture of Muscle Movement

Figure 9.26

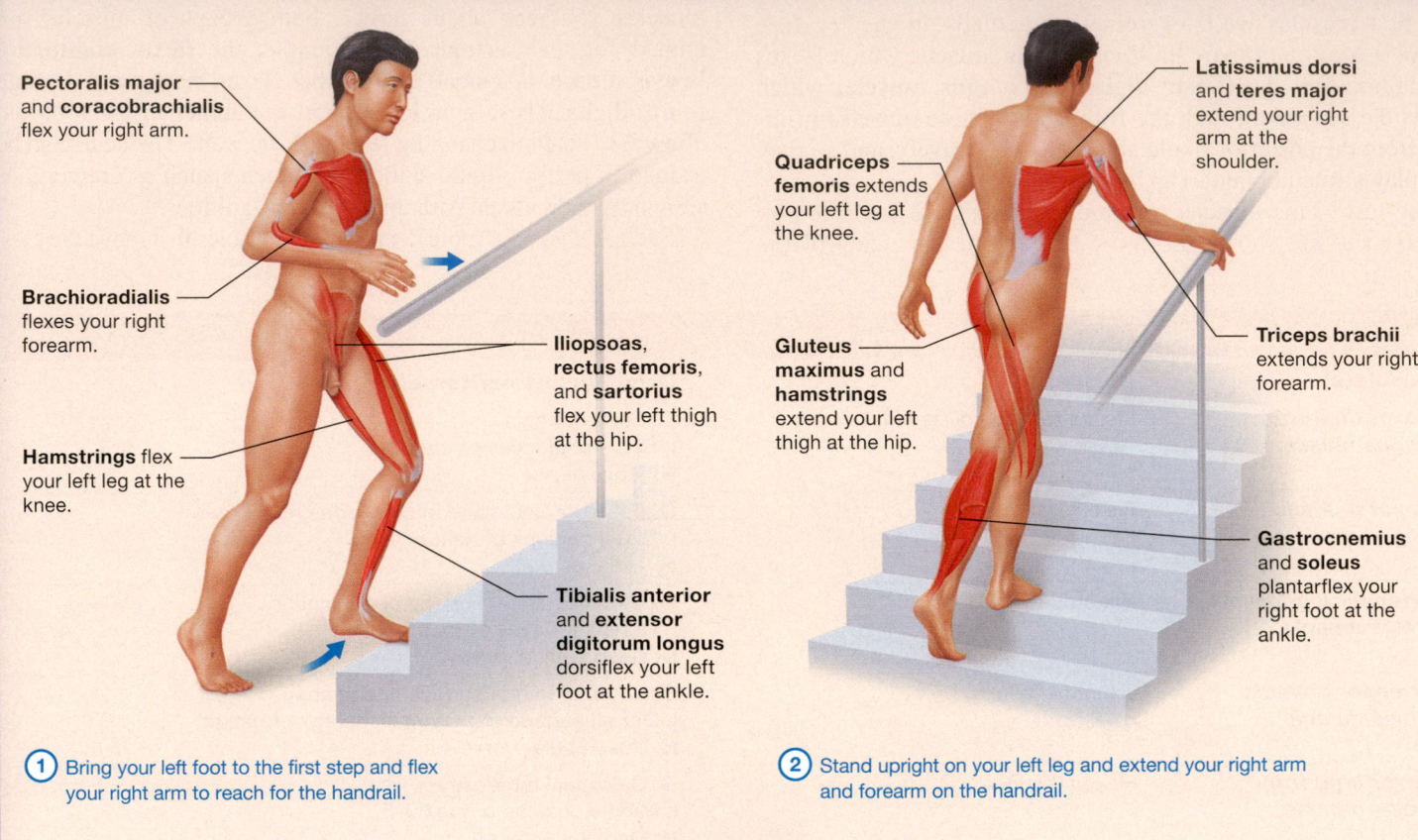

Pectoralis major and **coracobrachialis** flex your right arm.

Brachioradialis flexes your right forearm.

Hamstrings flex your left leg at the knee.

Iliopsoas, **rectus femoris,** and **sartorius** flex your left thigh at the hip.

Tibialis anterior and **extensor digitorum longus** dorsiflex your left foot at the ankle.

① Bring your left foot to the first step and flex your right arm to reach for the handrail.

Latissimus dorsi and **teres major** extend your right arm at the shoulder.

Quadriceps femoris extends your left leg at the knee.

Gluteus maximus and **hamstrings** extend your left thigh at the hip.

Triceps brachii extends your right forearm.

Gastrocnemius and **soleus** plantarflex your right foot at the ankle.

② Stand upright on your left leg and extend your right arm and forearm on the handrail.

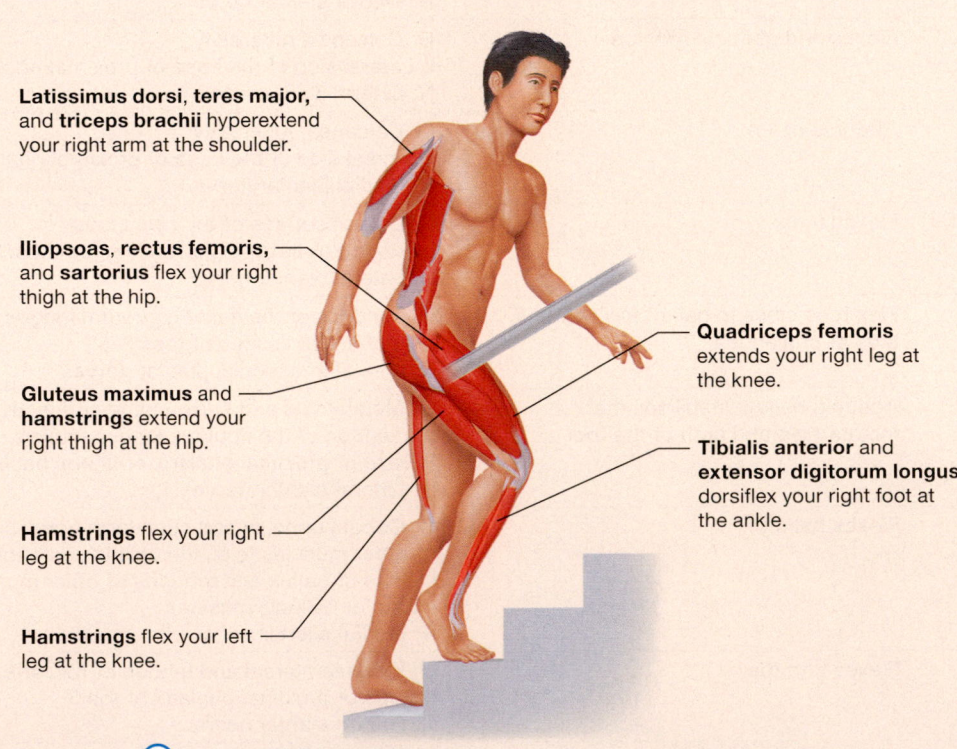

Latissimus dorsi, teres major, and **triceps brachii** hyperextend your right arm at the shoulder.

Iliopsoas, **rectus femoris,** and **sartorius** flex your right thigh at the hip.

Gluteus maximus and **hamstrings** extend your right thigh at the hip.

Hamstrings flex your right leg at the knee.

Hamstrings flex your left leg at the knee.

Quadriceps femoris extends your right leg at the knee.

Tibialis anterior and **extensor digitorum longus** dorsiflex your right foot at the ankle.

③ Bring your right leg to the next step and extend it. Take the weight off your left leg and hyperextend your right arm to swing it forward again, reaching for the handrail.

☐ 3. What are the major muscles of dorsiflexion?

☐ 4. What are the major muscles of plantarflexion?

Apply What You Learned

☐ 1. Ms. Sadler presents with an injury that occurred during a surgical procedure that damaged her superior gluteal nerve, which has effectively paralyzed her gluteus medius and gluteus minimus muscles. What movements will she be unable to perform?

☐ 2. How will paralysis of the gluteus medius and gluteus minimus muscles affect Ms. Sadler when she walks? Be specific.

☐ 3. Sydney is a gymnast who is having muscle pain in her left leg when she runs. You notice that she has pain when her foot is plantarflexed. However, she notes no pain when her knee is flexed or extended, but she does report pain when her foot is everted. Which muscle or muscles are potentially strained? Explain.

See answers in Appendix A.

MODULE 9.6
Putting It All Together: The Big Picture of Muscle Movement

Learning Outcomes

1. Identify the locations of the major skeletal muscles, and demonstrate their actions.

Throughout this chapter, we have demonstrated that muscles must work together in a coordinated fashion to accomplish even the simplest movements. We finish the chapter by presenting a common activity—climbing stairs—and how muscles work together to accomplish it.

Figure 9.26 breaks this process down. Note that for simplicity, the figure focuses on the main muscles involved in each action. The complete process is more complicated and involves more muscles than can be shown here.

Chapter Summary

For everything you need to succeed in this course, go to **Mastering** Anatomy & Physiology. There you will find:

- Practice Tests
- Author Podcasts
- Big Picture Animations
- Concept Boost Video Tutors
- **iP2** Interactive Physiology 2.0 Tutorials
- **A&PFlix** A&P Flix 3D Animations
- Active-Learning Workbook

MODULE 9.1
Overview of Skeletal Muscles 283–293

- A muscle fiber is surrounded by an extracellular matrix called **endomysium.** A **fascicle** is a bundle of many muscle fibers enclosed by the **perimysium.** The whole muscle consists of many fascicles wrapped by the **epimysium.**

- The pattern of fascicle arrangement greatly contributes to the appearance and function of a skeletal muscle. Muscles may have a parallel, convergent, pennate, circular, or spiral orientation.

- Some muscles are named for their shape, appearance, size, position, or other structural considerations, such as number of heads (e.g., triceps).

- Skeletal muscles commonly move at joints and move the bones to which they are attached, which is known as the muscle's **action.**

- Other functions of skeletal muscle include facial expression, breathing, and generating heat to regulate body temperature.

- Muscles often work in coordinated groups, with each muscle having a specific role.

- Points of attachment of a muscle are its more stationary **origin** and more mobile **insertion.**

- Bones and muscles work together as **lever systems.**
  - A **first-class lever** places the **fulcrum** between the point of force and the load to be moved.
  - A **second-class lever** places the load to be moved between the force and the fulcrum.
  - A **third-class lever** places the force between the load to be moved and the fulcrum.

MODULE 9.2
Muscles of the Head, Neck, and Vertebral Column 294–306

- The muscles of facial expression include the following:
  - The **frontalis, occipitalis,** and **corrugator supercilii muscles** move the forehead and eyebrows.
  - The circular **orbicularis oculi muscle** closes the eye.

○ The **zygomaticus major, zygomaticus minor, levator labii superioris, risorius,** and **orbicularis oris muscles** produce expressions of smiling, grimacing, grinning, and sneering.

○ The **depressor anguli oris, depressor labii inferioris,** and **mentalis muscles** contribute to sad and "doubtful" expressions.

○ The **buccinator muscle** pulls the cheeks interiorly, producing sucking motions.

● The six extrinsic eye muscles move the eyeball. They include the **superior, inferior, medial,** and **lateral rectus muscles,** and the **superior** and **inferior oblique muscles**.

● Muscles of mastication include the **masseter, temporalis, medial pterygoid,** and **lateral pterygoid muscles.**

● The muscles of swallowing work in the following manner:

○ The **genioglossus, hyoglossus,** and **styloglossus muscles** move the tongue to manipulate food during mastication and push the food into the pharynx.

○ The **stylohyoid, mylohyoid, geniohyoid,** and **digastric muscles** elevate the hyoid bone and nudge the food posteriorly.

○ The **sternohyoid, sternothyroid, omohyoid,** and **thyrohyoid muscles** depress the hyoid, larynx, and pharynx as the food enters the pharynx.

○ The **pharyngeal constrictor muscles** push the food into the esophagus.

● Muscles that move the head include the **sternocleidomastoid** and **scalene muscles,** which rotate and flex the head; the **trapezius muscle,** which extends the head; and the **splenius capitis** and **splenius cervicis muscles,** which extend and rotate the head.

● The muscles of the vertebral column include the following:

○ The **erector spinae muscle group** is composed of three columns of muscles, each of which is made up of three individual muscles: the lateral **iliocostalis muscle,** the middle **longissimus muscle,** and the medial **spinalis muscle.** The erector spinae muscle group maintains posture and extends and laterally flexes the vertebral column.

○ The *transversospinal muscle group* includes the **semispinalis, multifidus,** and **rotatores muscles,** which support posture and vertebral column extension.

○ The **quadratus lumborum** extends the vertebral column.

MODULE **9.3**
Muscles of the Trunk and Pelvic Floor 306–312

● Muscles of inspiration include the **diaphragm** and **external intercostal muscles.** The **internal intercostal muscles** cause air to leave the lungs during forced expiration.

● The **rectus abdominis, external oblique,** and **internal oblique muscles** of the abdomen flex the trunk and compress the abdominal cavity. The **transversus abdominis muscle** compresses the abdominal cavity.

● Muscles of the pelvic floor include the following:

○ The pelvic diaphragm is formed by the **levator ani muscle** group, which forms the floor of the pelvis and works like sphincters around the urethra, vagina, and anal canal.

○ The **external urethral** and **anal sphincters** allow voluntary control of urination and defecation.

○ The **deep transverse perineal** and **superficial transverse perineal muscles** support the pelvic organs.

○ The **bulbospongiosus** and **ischiocavernosus muscles** support erection of the penis and clitoris.

MODULE **9.4**
Muscles of the Pectoral Girdle and Upper Limb 312–321

● The muscles of the pectoral girdle include the following:

○ The **trapezius muscle** elevates, retracts, depresses, and rotates the scapula superiorly.

○ The **levator scapulae muscle** elevates the scapula.

○ The **rhomboid major** and **rhomboid minor muscles** retract the scapula and rotate it inferiorly.

○ The **serratus anterior muscle** protracts the scapula and rotates it superiorly.

○ The **pectoralis minor muscle** protracts and depresses the scapula.

● The muscles that move the arm include the following:

○ The **pectoralis major** and **coracobrachialis muscles** flex, adduct, and medially rotate the arm.

○ The **latissimus dorsi** and **teres major muscles** adduct, medially rotate, and extend the arm.

○ The **deltoid muscle** is the agonist of arm abduction.

○ The four **rotator cuff muscles** stabilize the shoulder joint. They include the **teres minor, infraspinatus, supraspinatus,** and **subscapularis muscles.**

● Muscles that flex the arm at the elbow joint include the **biceps brachii, brachialis,** and **brachioradialis muscles.**

● Muscles that extend the arm at the elbow joint include the **triceps brachii** and the *anconeus* muscles.

● The **pronator teres** and **pronator quadratus muscles** pronate the forearm, whereas the **supinator** and biceps brachii muscles supinate the forearm.

● Flexor muscles flex the hand, thumb, or fingers, whereas extensors extend the hand, thumb, or fingers.

● Numerous small muscles move the fingers, and are usually named for their action and placement.

MODULE **9.5**
Muscles of the Hip and Lower Limb 321–333

● The **iliopsoas, sartorius,** and **rectus femoris muscles** flex the thigh.

- The **adductor muscle group, pectineus muscle,** and **gracilis muscle** adduct and medially rotate the thigh.

- The **quadriceps femoris muscle group** extends the leg at the knee.

- The **gluteus maximus muscle** extends, abducts, and laterally rotates the thigh during climbing. The **gluteus medius** and **gluteus minimus muscles** abduct and medially rotate the thigh during walking.

- The muscles of the **hamstring muscle group** (the **biceps femoris, semitendinosus,** and **semimembranosus muscles**) extend the thigh and flex the leg.

- The main dorsiflexors of the foot are the **tibialis anterior** and **extensor digitorum longus muscles.**

- The **fibularis longus** and **fibularis brevis muscles** evert the foot.

- The main plantarflexors of the foot are the **gastrocnemius** and **soleus muscles.**

- Numerous small muscles move the toes, and are usually named for their action and placement.

MODULE 9.6
Putting It All Together: The Big Picture of Muscle Movement 332–333

- An example of the big picture of muscle movement is shown in Figure 9.26.

Assess What You Learned

Scan the QR Code for additional practice test questions

https://goo.gl/Qstgky

LEVEL 1 Check Your Recall

1. Which type of muscle fascicle pattern has an appearance similar to a feather?

   a. Fusiform
   b. Triangular
   c. Pennate
   d. Parallel

2. If "adductor" is part of the name of a muscle, what does that tell you about the muscle?

   a. It is a wide muscle.
   b. It is a long muscle.
   c. It raises a body part.
   d. It pulls a body part toward the midline.

3. Match the term with its description:

   _____ Brevis   a. Related to the fingers/toes
   _____ Digitorum   b. Straight
   _____ Hallucis   c. Short
   _____ Rectus   d. Decreases the angle between
   _____ Flexor   bones
   _____ Pronator   e. Turns palm down
                        f. Related to the hallux
                           (great toe)

4. The action of the biceps brachii muscle on the hinge joint of the elbow is an example of which kind of lever system?

   a. First class   c. Third class
   b. Second class   d. Fourth class

5. Which function is being fulfilled by a muscle that holds a bone steady during movement?

   a. Antagonist   c. Supinator
   b. Synergist   d. Fixator

6. Mark the following statements as true or false. If the statement is false, correct it to make a true statement.

   a. The main action of the zygomaticus major and minor muscles is to pull the corners of the mouth up to produce smiling.
   b. The orbicularis oris muscle is located around the eye.
   c. The temporalis and occipitalis muscles are attached to each other by the epicranial aponeurosis.
   d. The buccinator muscle pulls the cheeks in, producing sucking movements.

7. Which eye muscle passes through the trochlea and turns the eye inferiorly and laterally?

   a. Lateral rectus muscle
   b. Inferior rectus muscle
   c. Inferior oblique muscle
   d. Superior oblique muscle

8. The thick, bandlike muscle of mastication that covers much of the ramus of the mandible is the:

   a. masseter muscle.
   b. temporalis muscle.
   c. lateral pterygoid muscle.
   d. risorius muscle.

9. Which of the erector spinae muscles is most lateral to the vertebral column?

   a. Spinalis muscle
   b. Iliocostalis muscle
   c. Semispinalis muscle
   d. Longissimus muscle

10. Which of the following muscle groups is considered muscles of inspiration?

    a. External intercostal muscles
    b. Internal intercostal muscles
    c. External oblique muscles
    d. Internal oblique muscles

11. Match the muscle with its main action:

_____ Sternocleidomastoid muscle
_____ Transversus abdominis muscle
_____ Internal oblique muscle
_____ Rectus abdominis muscle
_____ Splenius capitis muscle
_____ Quadratus lumborum muscle

a. Laterally flexes the trunk
b. Flexes the trunk
c. Compresses the abdominal cavity
d. Extends the vertebral column
e. Flexes the head
f. Extends the head

12. Which of the following is not a muscle of the rotator cuff?

a. Teres major muscle
b. Teres minor muscle
c. Subscapularis muscle
d. Infraspinatus muscle
e. Supraspinatus muscle

13. Mark the following statements as true or false. If the statement is false, correct it to make a true statement.

a. The pectoralis major and coracobrachialis muscles are antagonists.
b. The latissimus dorsi muscle is a major abductor of the arm.
c. The rhomboid major and minor muscles retract the scapula.
d. The deltoid muscle is a main abductor of the forearm.

14. Match the muscle with its main action:

_____ Biceps brachii muscle
_____ Pronator teres muscle
_____ Extensor digitorum muscle
_____ Brachialis muscle
_____ Triceps brachii muscle
_____ Flexor carpi radialis muscle

a. Extends fingers and hand
b. Supinates forearm
c. Flexes hand
d. Extends forearm
e. Pronates forearm
f. Flexes forearm

15. Which of the following muscles flexes the thigh, but also laterally flexes the vertebral column when its origin and insertion switch?

a. Sartorius muscle
b. Psoas major muscle
c. Iliacus muscle
d. Rectus abdominis muscle

16. Match the muscle with its main action:

_____ Vastus medius muscle
_____ Gracilis muscle
_____ Gluteus medius muscle
_____ Gastrocnemius muscle
_____ Biceps femoris muscle
_____ Tibialis anterior muscle

a. Adducts thigh
b. Extends thigh and flexes leg
c. Plantarflexes foot
d. Abducts thigh
e. Dorsiflexes foot
f. Extends leg

LEVEL 2 Check Your Understanding

1. Predict the location and function of the flexor digitorum profundus muscle based only on its name.

2. Explain how the diaphragm muscle is able to increase the size of the thoracic cavity even though it doesn't insert into a bone.

3. Why do you think that the muscles of the erector spinae and transversospinal muscle groups remain in a state of continual contraction when we are upright?

4. Why do the three parts of the trapezius muscle have three different actions?

5. Explain why strengthening the levator ani muscle may help improve bladder and bowel continence (voluntary control over urination and defecation).

6. The rectus femoris muscle originates from the anterior superior iliac spine and inserts into the tibial tuberosity. Predict its main actions with only this information.

LEVEL 3 Apply Your Knowledge

PART A: Application and Analysis

1. Mr. Bell presents with the inability to move certain muscles on one side of his face. You ask him to make various facial expressions and find that on his right side he is unable to purse his lips, pull in his cheeks, elevate his upper lip, and smirk. What muscles is Mr. Bell unable to contract?

2. Ms. Cho presents with muscle pain in the area around her anterior neck and superior chest. She explains that she has had a respiratory infection over the past 2 weeks that made it hard for her to breathe. What is likely causing her muscle soreness? (*Hint:* Remember that the origin and insertion of some muscles can switch and that one likely has to breathe more deeply with a respiratory infection.)

3. Elise is a competitive gymnast who strained muscles in her left lower limb doing a tumbling pass. She has pain with extension of her leg and inversion of her foot. Which muscles did she potentially strain?

4. Chris is training for his black belt in karate and is working on developing a stronger punch. Which muscles do you recommend that he strengthen to improve his punch? Explain.

PART B: Make the Connection

5. Ms. Hendrix suffered a severe hip fracture that required hip replacement surgery. After an extended recovery period, she is undergoing physical therapy to regain strength and mobility.

a. Which bone is involved in a hip fracture, and what part of the bone is likely to be fractured?
b. Which muscles were likely affected by the hip replacement surgery, and to which parts of the bone do they attach?
c. Which actions could Ms. Hendrix perform to strengthen these muscles? (*Connects to Chapter 7*)

See answers in Appendix A.

10

Muscle Tissue and Physiology

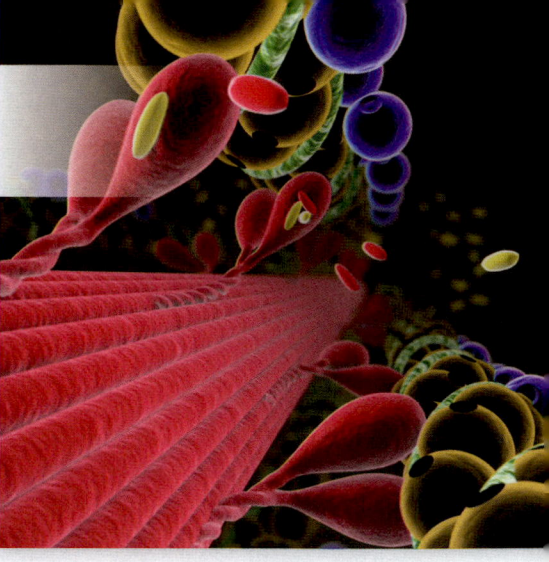

How many times did you move in the last minute? Did you sit down, maybe take out your earbuds, open your book, and follow these words with your eyes? Every one of your movements is the result of a process that happens over and over again in millions of your muscle cells. These cells make up the tissues and organs of the **muscular system**—the skeletal muscles (see Chapter 9). To understand how movements happen, we need to focus on skeletal muscles at the cellular and tissue levels of organization.

Although this chapter deals mostly with skeletal muscle tissue, in the last module we also discuss smooth muscle tissue and provide an overview of cardiac muscle tissue. (You will find complete coverage of cardiac muscle in the heart chapter—see Chapter 17.) Let's begin with a look at what all types of muscle tissue have in common.

MODULE 10.1
Overview of Muscle Tissue

Learning Outcomes

1. Describe the major functions of muscle tissue.
2. Name and describe the structural elements and properties common to all types of muscle cells.
3. Compare and contrast the characteristics of skeletal, cardiac, and smooth muscle tissue.

This module introduces you to the basic structure and properties common to all types of muscle tissue. As we know, there are three types of muscle tissue: skeletal, cardiac, and smooth (see Chapter 4). Although each type has unique features, they all have the same basic function: generating a force called **muscle tension.** At the organ level, this force often creates movement, but it also maintains posture, stabilizes joints, generates heat, and regulates the flow of materials through hollow organs.

Muscle tissue consists of **muscle cells,** sometimes called *myocytes* (MY-oh-sytz; *myo-* = "muscle"), and the surrounding extracellular matrix, the **endomysium.** (The endomysium blends with surrounding connective tissue and so is often referred to as connective tissue.) The endomysium holds the muscle cells together within muscle tissue and transmits tension generated by muscle cells to neighboring cells.

As a result of their common function, muscle cells share many structural properties. However, the three types of muscle tissue also have notable differences. In this module,

For practice applying concepts to a clinical scenario, check out the **Running Case Study** for this chapter @ Mastering Anatomy & Physiology

Computer-generated image: Myosin heads are shown binding to actin subunit in a key step of skeletal muscle contraction.

we examine both the similarities of and differences between the types of muscle tissue.

Types of Muscle Tissue

There are two basic forms of muscle cells: *striated,* which feature alternating light and dark bands called **striations** (stry-AY-shunz) and *smooth,* which lack striations (see Chapter 4). As you can see in **Figure 10.1a**, **skeletal muscle tissue** is made up of long, multinucleated (containing more than one nucleus) muscle cells that are arranged parallel to one another. Some skeletal muscle cells are quite long, extending nearly the entire length of the muscle. In keeping with their shape, these cells are often called **muscle fibers** (the terms cell and fiber are often used interchangeably). Skeletal muscle fibers are mostly found attached by connective tissue to the skeleton, where their contraction can produce the movement of a body part. However, this contraction occurs only when muscle fibers are stimulated by the nervous system. Although we're not always conscious of the contractions of skeletal muscle fibers, they can be controlled *voluntarily,* or by conscious thought.

Although **cardiac muscle tissue,** like skeletal muscle tissue, consists of striated muscle cells, **Figure 10.1b** reveals numerous differences between these two types of tissue. For example, cardiac muscle cells are shorter and wider, are branched, and generally have only a single nucleus (although some have two nuclei). Notice also what look like dark, wavy "lines" that join the cardiac muscle cells. These "lines" represent *intercalated discs* (in-TER-kuh-lay′-t'd; *inter-* = "between"), which are specialized structures that contain gap junctions and modified tight junctions. Intercalated discs unite cardiac muscle cells and permit them to coordinate contraction so that the heart contracts as a unit. Cardiac muscle tissue is found only in the heart and is *involuntary,* meaning that the brain does not have conscious control over its contraction.

As its name implies, **smooth muscle tissue** consists of smooth muscle cells, which are long and flattened with two pointed ends and which have a single, centrally located, oval

TYPE OF MUSCLE TISSUE	STRUCTURE	LOCATION	VOLUNTARY/ INVOLUNTARY	FUNCTION
(a) Skeletal 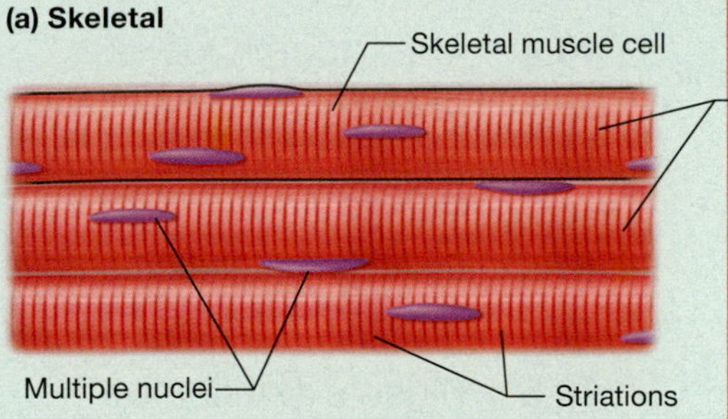	Long, cylindrical striated muscle fibers; cells are multinucleated.	Mostly attached to skeleton	Voluntary	• Produces movement of the body
(b) Cardiac 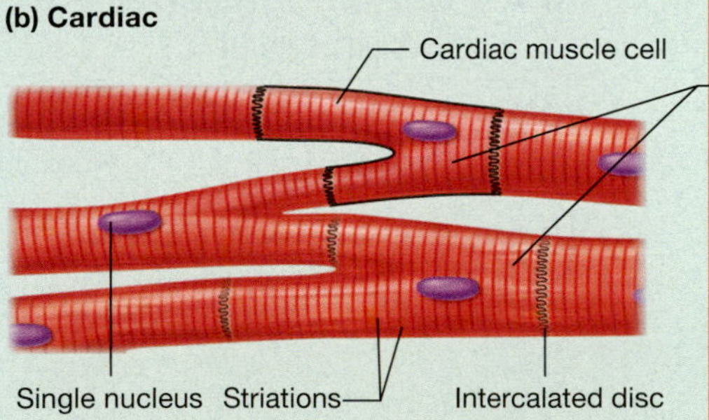	Short, wide, branching striated cardiac muscle cells with intercalated discs; cells have a single nucleus or two nuclei.	Heart	Involuntary	• Produces beating of the heart
(c) Smooth 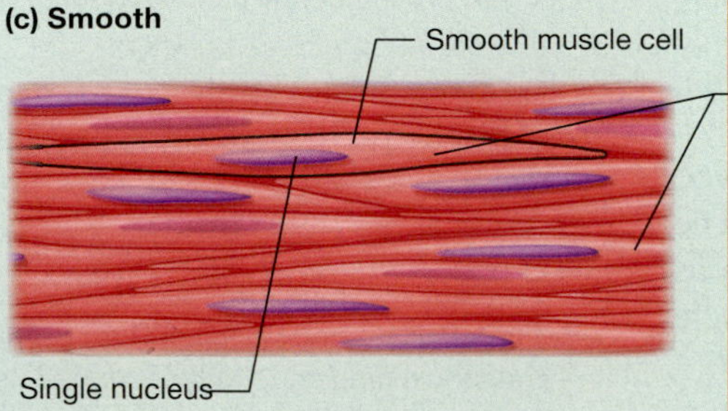	Thin, smooth muscle cells, generally joined by gap junctions; cells have a single nucleus.	Walls of hollow organs, as well as in the skin, and the eyes	Involuntary	• Changes diameter of hollow organs • Causes hairs to stand erect • Adjusts the shape of the lens and the size of the pupil of the eye

Figure 10.1 Three types of muscle tissue.

nucleus (**Figure 10.1c**). These cells line nearly every hollow organ, and are found as well in the eyes, the skin, and the ducts of certain glands. Many smooth muscle cells are linked to one another by gap junctions in their plasma membranes. Like cardiac muscle tissue, smooth muscle tissue is involuntary.

Properties of Muscle Cells

What properties allow all types of muscle cells to contract and produce muscle tension? Imagine muscle cells as mechanical devices somewhat like your car—in the same way your car converts the chemical energy of gasoline into the mechanical energy of movement, a muscle cell converts the chemical energy of ATP into the mechanical energy of muscle tension. All muscle cells share a set of common properties that enable them to generate tension:

- **Contractility.** *Contractility* is, of course, the ability of cells to contract. You might think that "contracting" means "shortening," but the term contraction actually refers to the ability of proteins within muscle cells to draw together. As you will discover in later modules, a muscle cell does not necessarily shorten when it contracts.
- **Excitability.** Muscle cells are **excitable,** or responsive, in the presence of various stimuli; these might include chemical signals from the nervous or endocrine systems, mechanical stretch signals, or local electrical signals. Such stimuli generate electrical changes across the plasma membrane of the muscle cell.
- **Conductivity.** When a muscle cell is excited, the electrical changes across the plasma membrane do not stay in one place. Instead, they are rapidly *conducted* along the entire

length of the plasma membrane, similar to how an electrical impulse is conducted through a copper wire.
- **Distensibility.** Most cells will rupture when stretched, but muscle cells are **distensible**—they can be stretched up to three times their resting length without damage.
- **Elasticity.** Muscle cells can also return to their original shape after being stretched, a property called **elasticity.** Often, elasticity is mistaken for stretch, but distensibility, not elasticity, refers to stretch.

Structure of Muscle Cells

《 FLASHBACK

1. What is the basic structure of the plasma membrane? (p. 70)
2. What is the smooth endoplasmic reticulum, and what is its main function? (p. 91)

The right side of **Figure 10.2** illustrates a generic muscle cell. Note that it has most of the same organelles as the generalized cell pictured to its left, including nuclei and mitochondria. However, muscle cells also have many structural differences, so we use different terminology when describing them. For example, the term *cytoplasm* is replaced with **sarcoplasm** (SAR-koh-plazm; *sarco-* = "flesh"), and *plasma membrane* is replaced with **sarcolemma** (sar-koh-LEM-ah; *-lemma* = "husk or shell"). Sarcoplasm, like cytoplasm, contains cytosol and all organelles in the muscle cell. The sarcolemma is composed of a phospholipid bilayer with multiple specialized integral and peripheral proteins.

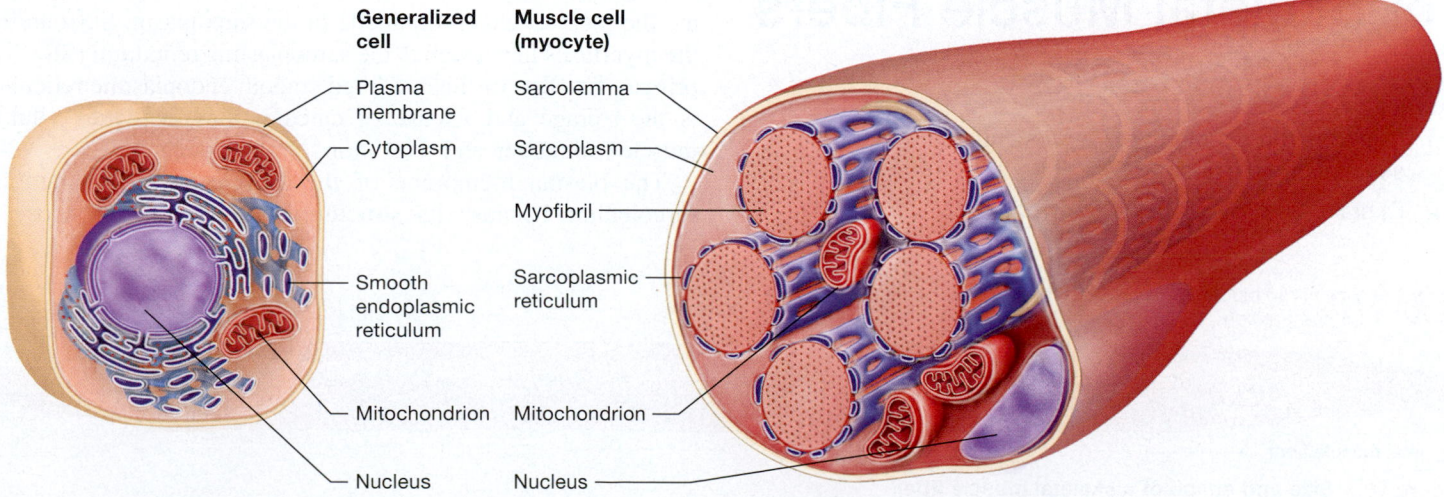

Generalized cell | Muscle cell (myocyte)

Plasma membrane — Sarcolemma

Cytoplasm — Sarcoplasm

Myofibril

Smooth endoplasmic reticulum — Sarcoplasmic reticulum

Mitochondrion — Mitochondrion

Nucleus — Nucleus

Figure 10.2 A generalized cell (left) compared with a generic muscle cell (right).

Muscle cells also have unique structures that allow them to serve their functions. You can see in Figure 10.2 that the sarcoplasm contains cylindrical organelles called **myofibrils** (my-oh-FY-brilz). All three types of muscle cells contain myofibrils, although they are arranged differently in smooth muscle cells than in skeletal and cardiac cells. Myofibrils are essentially bundles of specialized proteins, including those involved in muscle contraction. A myofibril measures about 1 μm in diameter, or about $\frac{1}{100}$ the thickness of a human hair. Each muscle cell has hundreds to thousands of myofibrils—they make up about 50–80% of its volume. Other organelles such as mitochondria are found packed between the myofibrils.

Another unique structure common to all muscle cells is an organelle called the **sarcoplasmic reticulum** (sar-koh-PLAZ-mik reh-TIK-yoo-lum), or simply **SR.** The SR is a modified smooth endoplasmic reticulum that forms a weblike network surrounding each myofibril. The structure of the SR varies in the three types of muscle tissue, as we'll discuss in later modules.

Quick Check

☐ 5. What basic structures are unique to muscle cells?

Apply What You Learned

☐ 1. Suppose that a new type of muscle tissue is discovered and, like the other types, it generates tension. Would you expect this new tissue type to have the same basic structure as skeletal, smooth, and cardiac muscle tissues? Why or why not?

☐ 2. Predict what might happen to a muscle cell that is distensible but not elastic. What if the reverse were true—a muscle cell that is elastic but not distensible?

See answers in Appendix A.

MODULE 10.2
Structure and Function of Skeletal Muscle Fibers

Learning Outcomes

1. Describe the structural properties and components of a skeletal muscle fiber.
2. Explain the organization of a myofibril.

3. Describe the structure and components of thick, thin, and elastic filaments.
4. Name and describe the function of each of the contractile, regulatory, and structural protein components of a sarcomere.
5. Explain the sliding-filament mechanism of muscle contraction.

In this module, we explore the structure of skeletal muscle fibers and the proteins found within them. Our discussion also covers how the arrangement of these proteins within a myofibril correlates with the mechanism of muscle contraction.

Structure of the Skeletal Muscle Fiber

Skeletal muscle tissue consists of many skeletal muscle fibers and the endomysium surrounding each of them. These fibers are quite different from most of the cells we have studied up to this point. Some differences include the following:

- **Cell shape and size.** **Figure 10.3** shows that skeletal muscle fibers are shaped like thin cylinders and are much longer than most cells. The longest fibers, found in the thigh muscles, are about 100 μm in diameter and 30 cm (almost a foot) long. (Compare this with a red blood cell, which is only about 7.5 μm in diameter.)
- **Striations.** Notice also in Figure 10.3 that skeletal muscle fibers appear striated, or striped, when viewed through a microscope, a characteristic we've mentioned before.
- **Multiple nuclei.** Each skeletal muscle fiber contains multiple sausage-shaped nuclei on the inner surface of its sarcolemma. As you may recall, this feature is a result of the embryonic origin of the fiber, as each muscle fiber arises from the fusion of multiple embryonic cells called *myoblasts* (see Chapter 4).

In addition to these differences, skeletal muscle fibers have unique features that enable them to perform their functions. As you know, a skeletal muscle fiber, like all types of muscle fibers, is surrounded by its sarcolemma (**Figure 10.4**). Within the boundary of the sarcolemma, we find the cylindrical myofibrils, which are made up primarily of proteins involved in contraction. Myofibrils are the most abundant organelle in the sarcoplasm. Surrounding the myofibrils like a web is the sarcoplasmic reticulum (SR). The primary function of this modified smooth endoplasmic reticulum is the storage and release of calcium ions, activities vital to muscle contraction and relaxation.

The plasma membrane of the muscle fiber is called the sarcolemma because its structure is different from those of

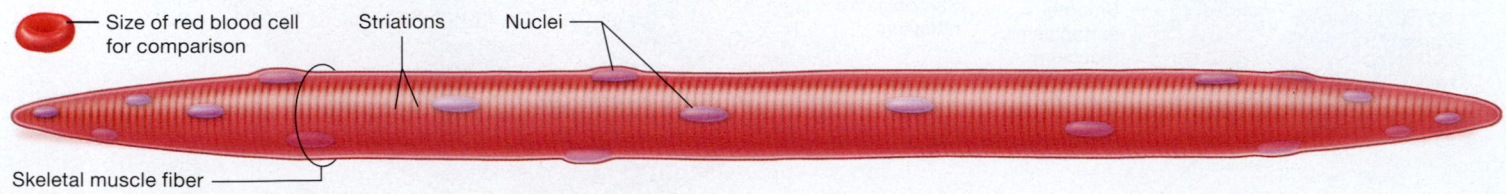

Size of red blood cell for comparison   Striations   Nuclei

Skeletal muscle fiber

Figure 10.3 Size and shape of a skeletal muscle fiber.

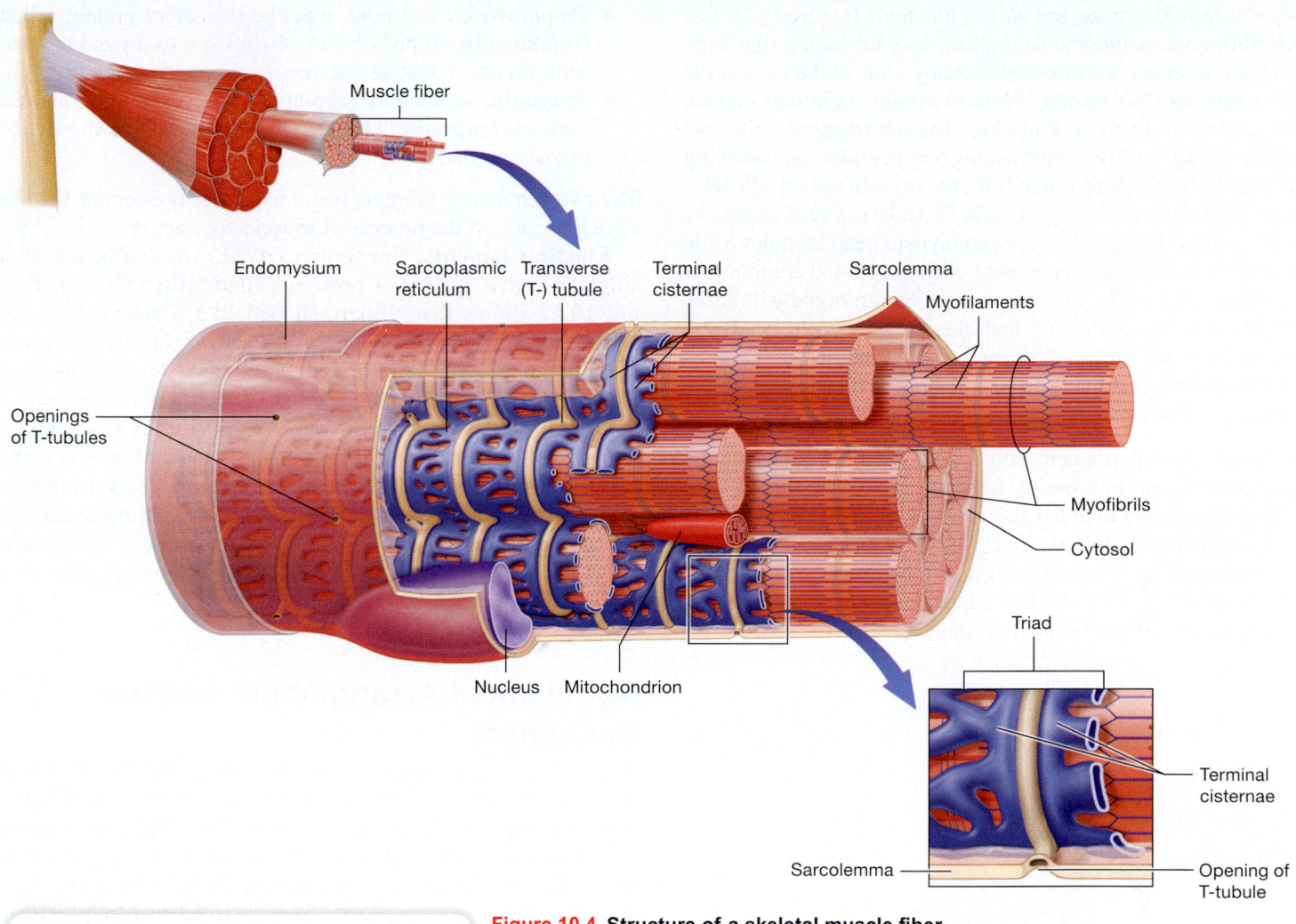

Figure 10.4 Structure of a skeletal muscle fiber.

Practice art labeling
@ Mastering Anatomy & Physiology

generalized cells. You can see in Figure 10.4 that the sarcolemma isn't confined to the exterior of the cell. Rather, it forms inward extensions called **transverse tubules** (or **T-tubules**) that dive into the muscle fiber and surround each myofibril, forming a tunnel-like network within the muscle fiber. These tunnels are continuous with the exterior of the cell and so are filled with extracellular fluid. Flanking each side of a T-tubule are enlarged portions of the SR called **terminal cisternae** (sis-TER-nee; "end sacs"). The combination of a T-tubule and the two terminal cisternae on either side is known as a **triad** (see the blow-up in Figure 10.4). This association is important in muscle contraction, as we'll discuss in Module 10.4.

Quick Check

☐ 1. How does the structure of a skeletal muscle fiber differ from the structure of other cells?

☐ 2. How are the terminal cisternae related to the sarcoplasmic reticulum?

☐ 3. What is a triad?

Structure of the Myofibril

« FLASHBACK

1. What are actin filaments? (p. 96)

As you can see in Figure 10.4, a myofibril is composed of hundreds to thousands of protein bundles called **myofilaments.** These consist of one or more of the following types of proteins:

- **contractile proteins,** which produce tension;
- **regulatory proteins,** which control when the muscle fiber can contract; and
- **structural proteins,** which hold the myofilaments in their proper places and ensure the structural stability of the myofibril and the muscle fiber.

See *A&P in the Real World: Duchenne Muscular Dystrophy* on p. 346 to read about what happens when the connections between the proteins of the muscle fiber break down.

There are three types of myofilaments, each with characteristic contractile, regulatory, and/or structural proteins: *thick filaments, thin filaments,* and *elastic filaments* (**Figure 10.5**). The **thick filaments,** as their name implies, have the largest diameter. Each thick filament is composed of many units of the contractile protein **myosin** (MY-oh-sin). Note in Figure 10.5a that myosin looks somewhat like two golf clubs twisted together, with two globular "heads" and two intertwining polypeptide chains making up a "tail." The heads protrude from the myosin tail on a "neck." The neck of each myosin protein is flexible where it meets the tail at a point called the *hinge.* Each myosin head includes a site that binds to a thin filament, among other functional components we'll discuss later. The myosin proteins are arranged within the thick filament in such a way that clusters of myosin heads are found at each end, with only myosin tails found in the middle.

A **thin filament** is made up of both contractile and regulatory proteins (see Figure 10.5b), as follows:

- **Actin.** A thin filament consists of many subunits of the contractile protein **actin** (AK-tin). This bead-shaped protein has an area, called the **active site,** that can bind to a myosin head. Multiple actin subunits string together like beads on a necklace to form the largest part of the thin filament. This actin "string" appears as two intertwining strands in the functional thin filament.

- **Tropomyosin.** The long, ropelike regulatory protein called **tropomyosin** (trohp'-oh-MY-oh-sin) spirals around the two actin strands so that, at rest, it covers the active sites on actin.

- **Troponin.** A second regulatory protein is the smaller, globular **troponin** (TROH-poh-nin) that holds the tropomyosin in place.

The two regulatory proteins tropomyosin and troponin help to switch on and off the process of muscle contraction.

Elastic filaments, the thinnest type, are composed of a single massive structural protein called **titin** (TY-tin). Figure 10.5c shows that titin is shaped like a spring that can uncoil when stretched and recoil to its original shape when the stretching force is removed. Titin runs through the core of the thick filament, which helps to stabilize the myofibril structurally (you can't see titin through the thick filaments in the figures because it is inside them). Elastic filaments serve several purposes in addition to holding the thick filaments in place. Some of these functions include resisting excessive stretching and providing elasticity to the muscle fiber—that is, helping it to "spring" back to its original length after it is stretched.

Myofilament Arrangement and the Sarcomere

In **Figure 10.6** you can see how the arrangement of thin and thick myofilaments forms a pattern within the myofibril—there are regions where we find only thin filaments, regions where the thin and thick filaments overlap, and regions that have only thick filaments. This pattern forms alternating light and dark bands, or striations. As explained in Module 10.6, this region with overlapping filaments, referred to as the *zone of overlap,* is where tension is generated during a muscle contraction.

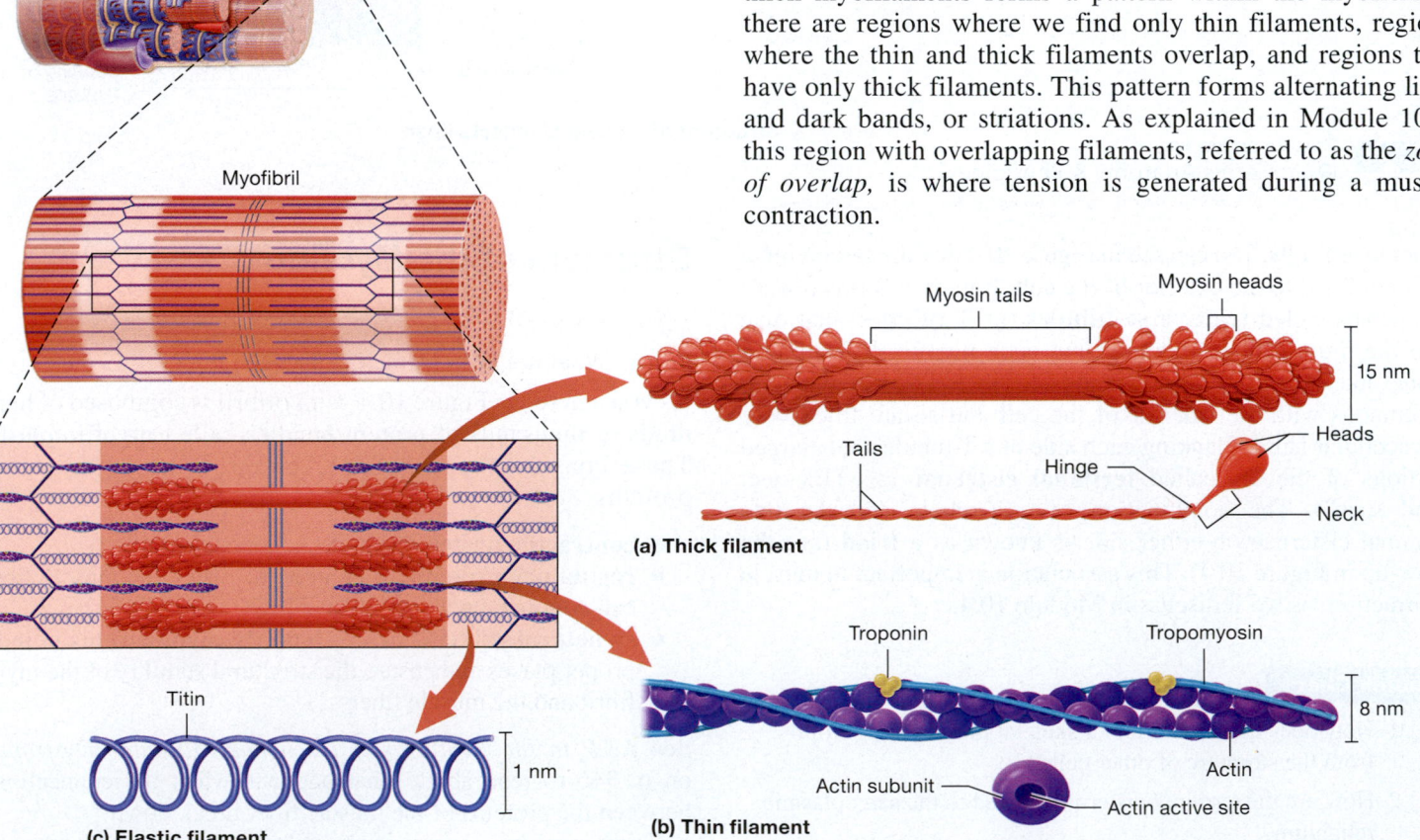

(a) **Thick filament**

(b) **Thin filament**

(c) **Elastic filament**

Figure 10.5 Structure of myofilaments.

The dark and light areas of the striations are named as follows (see Figure 10.6):

- **I bands.** The **I band** is the light region of a striation. It appears lighter because it contains only thin filaments, which allow more light to pass through them.
- **A bands.** The **A band** is the dark region of a striation, which contains thick filaments. Thick filaments block more light than thin filaments, making the A band appear darker in micrographs. In Figure 10.6 we find both thick and thin filaments in the outer portion of the A band, and only thick filaments in the middle of the A band. This middle area, which is slightly lighter than the rest of the A band, is called the **H zone.**

You can also see that both the A and I bands in Figure 10.6 have a dark line running down their middle. The dark line in the middle of the A bands is called the **M line.** The M line consists of structural proteins that hold the thick filaments in place and serve as an anchoring point for the elastic filaments. The dark line in the I bands is known as a **Z-disc.** The Z-disc is composed of structural proteins that have many functions: They anchor the thin filaments in place and to one another, they are an attachment point for elastic filaments, and they attach myofibrils to one another across the whole diameter of the muscle fiber.

All this may sound at first like alphabet soup, but these details are vital: It is this arrangement of myofilaments within the myofibrils that allows muscle to contract. The functional unit of contraction—where muscle tension is produced—is the **sarcomere** (SAR-koh-meer; "muscle part"). A sarcomere is the section of a myofibril that extends from one Z-disc to the next Z-disc. Each sarcomere, then, includes a full A band and half of two I bands.

Figure 10.7 shows the three-dimensional arrangement of the sarcomere in a myofibril. Notice in the cross sections through the different regions of the sarcomere that the A band consists of both thin and thick filaments, and each thick filament is surrounded by six thin filaments. However, the I band contains only thin filaments, and the H zone only thick filaments.

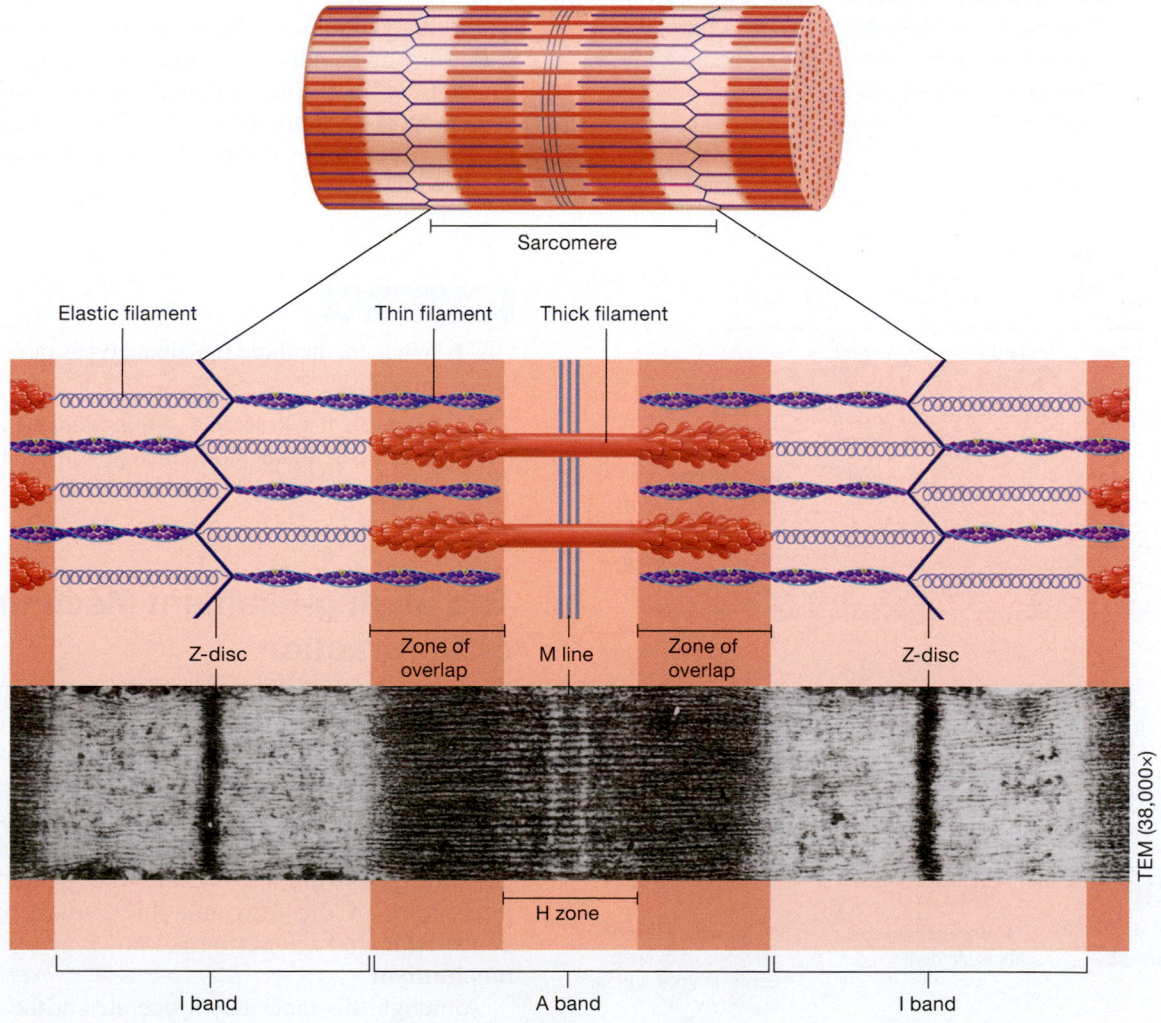

Figure 10.6 Structure and bands of the sarcomere.

StudyBOOST ⬆

Remembering the Bands of the Sarcomere

The following are some simple descriptions and mnemonics to help you remember the basic organization of the sarcomere.

Structure	Description	Mnemonic
A band	The dark band that contains both thick and thin filaments.	A is the d**A**rk band.
H zone	The middle portion of the A band in which only thick filaments are found.	"Ha!" because H is in the A band.
M line	Middle line of the A band; contains structural proteins that hold the thick filaments in place.	Think of the "M" as standing for "middle." Also remember that the "Myosin tails point toward the M line."
I band	The light band that contains only thin filaments.	I is the l**I**ght band.
Z-disc	Line bisecting the I band. Both the thin and elastic filaments anchor to the Z-discs, which also attach myofibrils to one another.	Identify the Z-disc by its "Z" shape (see Figure 10.7).

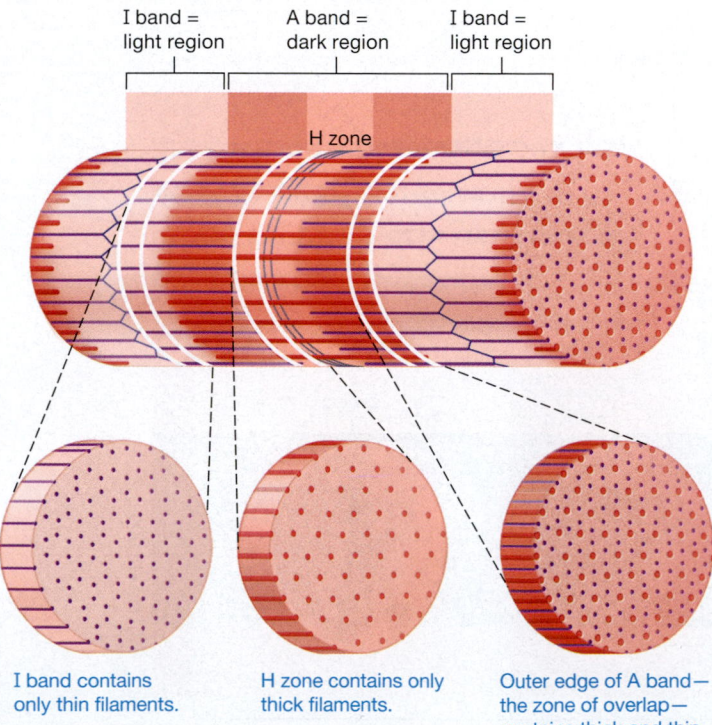

I band = light region
A band = dark region
I band = light region

H zone

I band contains only thin filaments.

H zone contains only thick filaments.

Outer edge of A band—the zone of overlap—contains thick and thin filaments.

Figure 10.7 Three-dimensional structure of the sarcomere.

Quick Check

☐ 4. How does the arrangement of myofilaments produce the characteristic striations of skeletal muscle fibers?

☐ 5. Describe the structure of a sarcomere. What is its function?

Putting It All Together: The Big Picture of Skeletal Muscle Structure

Figure 10.8 combines the gross anatomy of the skeletal muscle from Chapter 9 with the microscopic anatomy of skeletal muscle fibers from this section to give you a "big picture" view of the organizational levels of a skeletal muscle. Remember that skeletal muscles are enclosed by a layer of thick connective tissue called *fascia*, which anchors them to the surrounding tissues and holds groups of muscles together. Deep to the fascia is another layer of connective tissue, the *epimysium*, which surrounds the whole muscle. The epimysium blends with a deeper layer of connective tissue, the *perimysium*, to form tendons, which bind the muscle to its attaching structure. The perimysium surrounds individual *fascicles*, or groups of muscle fibers. The muscle fiber is then of course made up of myofibrils, which are composed of myofilaments. The arrangement of myofilaments within the myofibrils creates sarcomeres, the functional units of contraction.

Quick Check

☐ 6. What are the three functional types of proteins found in a myofibril?

☐ 7. Describe the structures of thin filaments, thick filaments, and elastic filaments. Which proteins make up these filaments?

The Sliding-Filament Mechanism of Contraction

At this point you know that muscle contraction occurs at the level of the sarcomere, but how, precisely, does contraction occur? Physiologists previously believed that the myofilaments themselves shortened. However, after detailed study of skeletal muscle, the currently accepted model is that the thin filaments instead *slide past* the thick filaments, generating tension throughout the whole sarcomere. This interaction of the thin and thick filaments during muscle contraction and relaxation is known as the **sliding-filament mechanism.**

Although this mechanism operates at the microscopic level, it's easy to picture how it works on a gross level. Take your

The Big Picture of Levels of Organization within a Skeletal Muscle

Figure 10.8

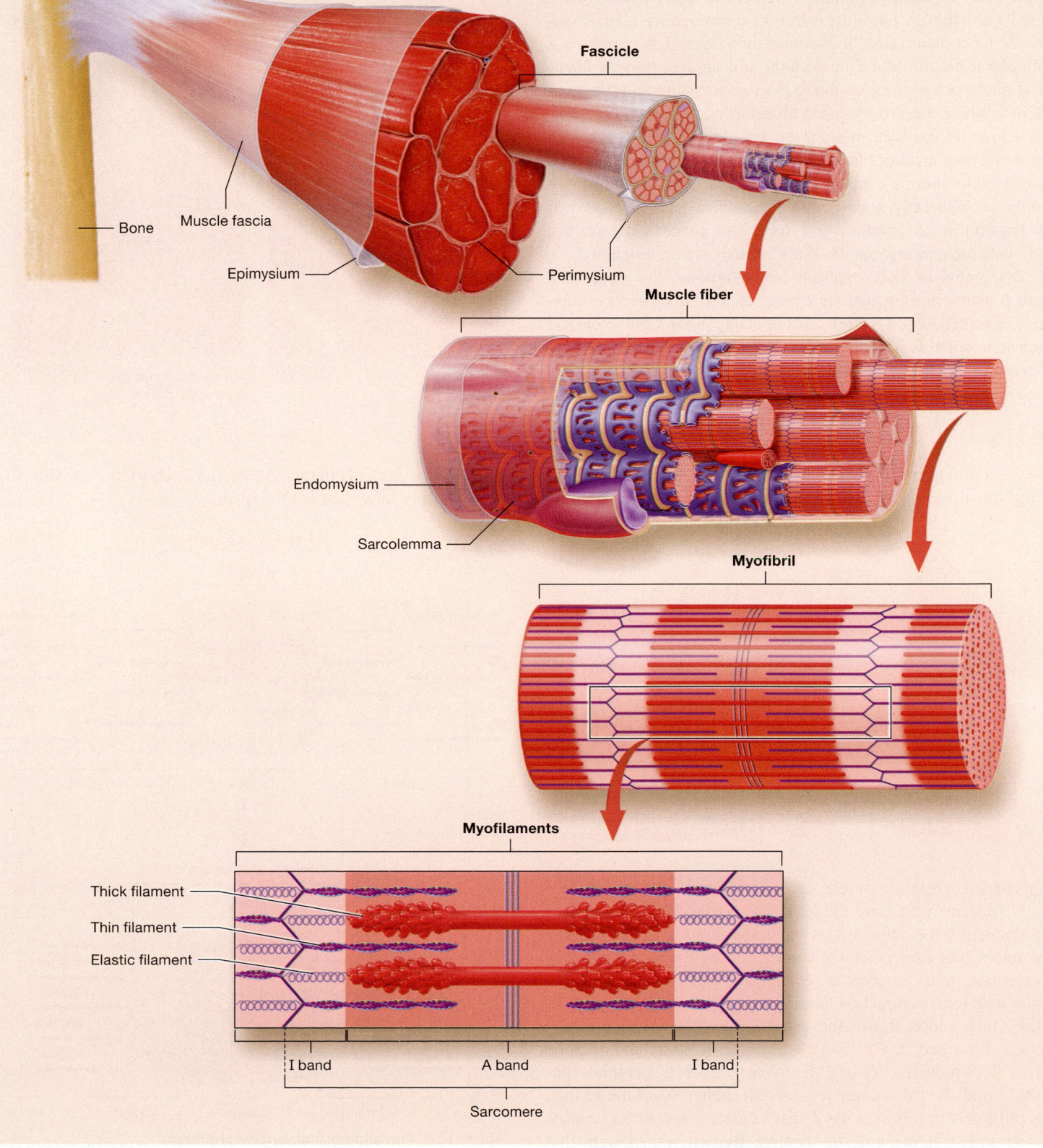

Tendon

Muscle

Fascicle

Bone

Muscle fascia

Epimysium

Perimysium

Muscle fiber

Endomysium

Sarcolemma

Myofibril

Myofilaments

Thick filament

Thin filament

Elastic filament

I band

A band

I band

Sarcomere

Duchenne Muscular Dystrophy

Duchenne muscular dystrophy (doo-SHEN), or **DMD,** is a degenerative muscular disease occurring almost exclusively in boys. It is caused by a defective gene for the protein **dystrophin** (diss-TROH-fin) that is carried on the X chromosome (see Chapter 27 for a discussion of X-linked disorders). Dystrophin is a structural protein found in skeletal and cardiac muscle fibers that anchors the sarcolemma to the endomysium and the myofibrils; this is the arrangement found in normal muscle tissue, as shown in the light micrograph on the left. In the absence of functional dystrophin, the sarcolemma breaks down and the muscle fiber is destroyed. As muscle fibers die, they are replaced with fatty and fibrous connective tissue, as shown in the light micrograph on the right. This process causes the whole muscle to enlarge while it's actually losing function.

Symptoms of DMD typically arise between the ages of 2 and 6 years, and include weakness of the proximal limb muscles, a waddling gait pattern, and difficulty in performing certain activities such as climbing stairs. Generally by 12 years of age, patients with DMD are wheelchair-bound, and death from cardiac and respiratory failure occurs by about age 20. Researchers have made a lot of progress toward understanding the mechanism of this disease. Although DMD remains incurable, current research that involves modifying the mutated gene so that the muscle cells can produce functional dystrophin appears promising.

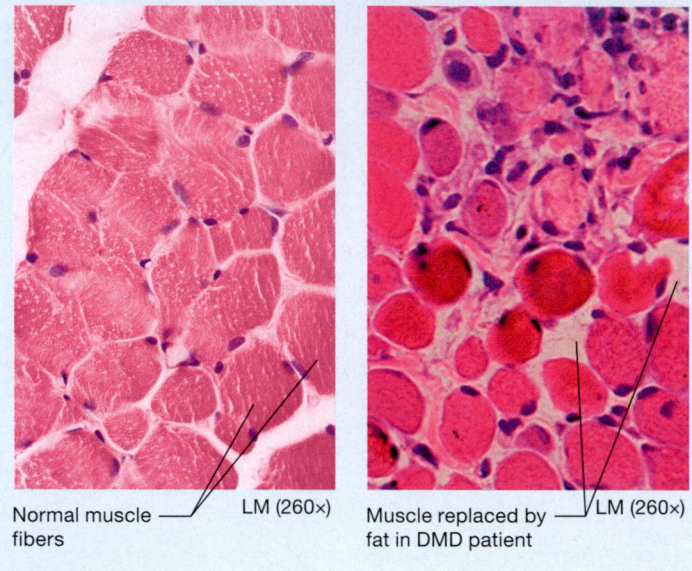

Normal muscle fibers — LM (260×)   Muscle replaced by — LM (260×) fat in DMD patient

---

hands and turn them so your fingers are facing one another, as in the top illustration here:

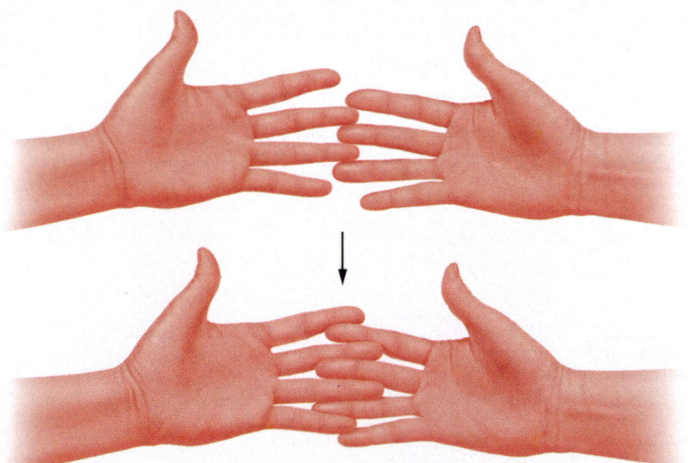

Your hands represent a single, large sarcomere, and where your fingers overlap represents the zone of overlap of the thick and thin filaments; the inner edges of your palms are the Z-discs. Now move your fingers slowly together about 1 mm at a time. As you can see, your "sarcomere" (your hands) grows progressively shorter with each contraction, ending up as the bottom illustration.

Let's take a look at how the sarcomeres change during a contraction. First see in **Figure 10.9** how the I bands and H zone narrow. This happens because the myosin heads of the thick filaments "grab" the thin filaments and pull them toward the M line. This pulling action brings the Z-discs closer together and causes the sarcomere as a whole to shorten. Remember, though, that none of the filaments themselves actually shorten—the thin filaments simply move toward the M line.

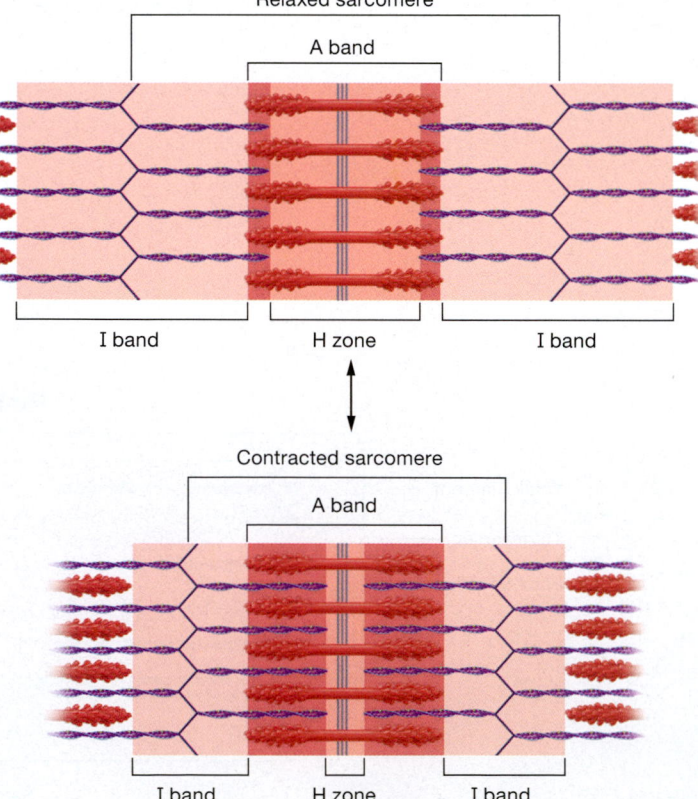

Figure 10.9 The sliding-filament mechanism

The size of the A band, however, remains unchanged, as you can see in Figure 10.9. The A band doesn't change in size because the myosin heads are actually doing the pulling. Think about a rope that you're climbing—you get progressively closer to the top, but the rope itself never changes in size.

Sarcomeres are arranged end to end within each myofibril. As many sarcomeres simultaneously contract, the whole muscle fiber contracts. We discuss details of the sliding-filament mechanism in Module 10.4.

Quick Check

☐ 8. What is the basic mechanism of muscle contraction?

Apply What You Learned

☐ 1. A hypothetical poison denatures the protein titin, making it straight instead of spring-shaped. Predict the effect of this change on muscle tissue.

☐ 2. Predict what might happen if the thin and thick filaments were unable to move relative to one another within a sarcomere.

See answers in Appendix A.

MODULE 10.3
Skeletal Muscle Fibers as Electrically Excitable Cells

Learning Outcomes

1. Contrast the relative concentrations of sodium and potassium ions inside and outside a cell.

2. Differentiate between a concentration gradient and an electrochemical gradient.

3. Describe how the resting membrane potential is generated.

4. Describe the sequence of events of a skeletal muscle fiber action potential.

Although skeletal muscle contraction takes place at the level of the sarcomere, the events leading to contraction actually begin with electrical changes that occur across the sarcolemma. For this reason, we need to discuss the electrical properties of the sarcolemma before we get into the mechanism of contraction. The branch of physiology that studies electrical changes across plasma membranes, and the accompanying physiological processes, is called **electrophysiology.** Many of our body's processes, including skeletal muscle contraction, involve electrophysiology. This module presents the principles of electrophysiology and examines how they apply to muscle fibers. To begin, we'll review a concept from the cell chapter: *membrane potentials* (see Chapter 3).

Membrane Potentials in Our Cells

《 FLASHBACK

1. What are potential energy and kinetic energy? (p. 42)

2. What is the resting membrane potential? How are charges separated across the plasma membranes of cells? (p. 82)

Earlier in this text we discussed a characteristic common to all cells, including electrically excitable cells: an unequal distribution of electrically charged ions near the cells' plasma membranes (see Chapter 3). Specifically, there is a thin layer of negative ions in the cytosol and a thin layer of positive ions in the extracellular fluid. Away from the plasma membrane, positive and negative ions are present in equal numbers. This means that the cytosol and extracellular fluid are always electrically neutral.

The separation of electrical charges across the plasma membrane really just repeats the Gradients Core Principle (p. 28). This separation of charges is called an *electrical gradient,* where we have unequal numbers of positive and negative ions separated by a barrier. As with all gradients, an electrical gradient represents a source of potential energy. When the barrier separating the ions is removed, they follow their gradients, creating a flow of electrical charges, and the potential energy becomes kinetic energy. For this reason, we can call the electrical gradient an **electrical potential,** because we are referring to its potential energy.

CORE PRINCIPLE
Gradients

A difference in electrical potential between two points is called a **voltage;** the size of the voltage, measured in *volts,* indicates the size of the potential. For example, the potential difference present between an electrical outlet in your wall and an appliance such as a toaster is relatively large, measuring about 110 volts. In comparison, the potential difference across the two sides of a plasma membrane is small, and is measured in a unit called a *millivolt (mV),* or 1/1000 of a volt.

The potential difference across a cell's plasma membrane is called the **membrane potential** (see Chapter 3). The voltage across the membrane may be measured with a voltmeter, as shown here:

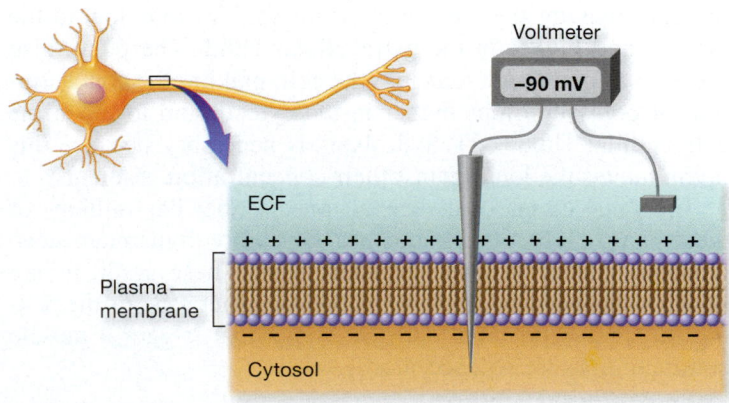

Notice that as you measure from outside to inside the cell with a voltmeter, the voltage becomes more negative. This negative voltage is present when the cell is at rest (not being stimulated), and for this reason is called the **resting membrane potential.** The cell in this state is said to be **polarized,** which simply means that the voltage difference across the plasma membrane of the cell is not at 0 mV, but rather measures to either the positive or the negative side (or *pole*) of zero. Different cell types have different resting membrane potentials. As we discuss shortly, changes in this potential are responsible for the electrical events of a muscle fiber. First, however, we look at how this negative resting membrane potential is generated.

Ion Channels and Gradients

Two types of ions are involved in generating the resting membrane potential: sodium ions and potassium ions. However, these ions cannot pass through the hydrophobic portion of the phospholipid bilayer of the plasma membrane because they are charged particles. For this reason, their movement across the plasma membrane is dependent on specific membrane proteins such as protein channels. As we discussed in the cell chapter, different types of channels function in passive and active transport processes (see Chapter 3).

There are two main classes of channels, *leak* and *gated*.

- **Leak channels** are always open, so they continually allow ions to follow their gradients into or out of the cell.
- **Gated channels** are closed at rest, and open only in response to certain stimuli. Some gated channels, called **ligand-gated channels,** open in response to a certain chemical binding to the channel (or to an associated receptor). Other channels, called **voltage-gated channels,** open or close in response to changes in voltage across the membrane.

Ions can pass through these channels only when a gradient is present to drive their passive diffusion.

To move ions against a gradient requires a different type of membrane protein called an active transport pump. The most important active transport pump for sodium and potassium ions is the **sodium-potassium ion pump,** or **Na⁺/K⁺ ATPase.** Recall from the cell chapter that this pump maintains concentration gradients of sodium and potassium ions across the sarcolemma, using the energy from ATP hydrolysis (**Figure 10.10**). Specifically, the pump moves three sodium ions out of the cell, making the concentration of sodium ions low in the cytosol and higher in the extracellular fluid. The pump also moves two potassium ions into the cell, making the concentration of potassium ions higher in the cytosol and lower in the extracellular fluid. ATP hydrolysis is necessary because this pump moves the ions against their concentration gradients.

The sarcolemma of a skeletal muscle fiber has millions of these pumps, which work together constantly to maintain steep gradients of sodium and potassium ions. These gradients are critical because they are involved in the generation of the resting membrane potential and the events that trigger a muscle contraction, as we will see shortly.

Electrochemical Gradients

The resting membrane potential in our cells has important implications for ion transport. Several times in this book, you have seen how solutes move across membranes by diffusion

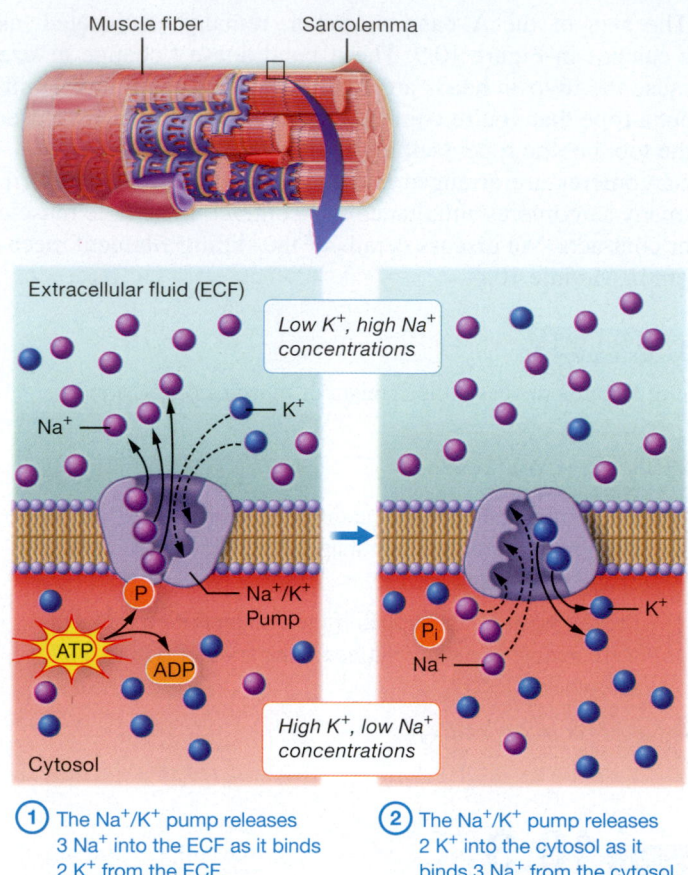

1 The Na⁺/K⁺ pump releases 3 Na⁺ into the ECF as it binds 2 K⁺ from the ECF.

2 The Na⁺/K⁺ pump releases 2 K⁺ into the cytosol as it binds 3 Na⁺ from the cytosol.

Figure 10.10 Ion gradients maintained by the Na⁺/K⁺ pump.

according to their concentration gradient. In fact, the concentration gradient is the main factor that determines the movement of uncharged solutes such as carbon dioxide, glucose, and oxygen. But the story for ions is more complicated because they are also affected by electrical gradients, as charge also exerts a force on ions. For this reason, diffusion of an ion across the plasma membrane is determined by both its concentration gradient and its electrical gradient. Combined, these two forces are called the **electrochemical gradient.**

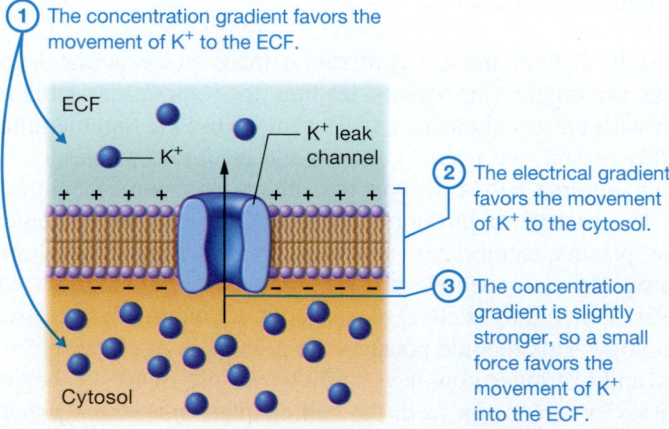

1 The concentration gradient favors the movement of K⁺ to the ECF.

2 The electrical gradient favors the movement of K⁺ to the cytosol.

3 The concentration gradient is slightly stronger, so a small force favors the movement of K⁺ into the ECF.

Figure 10.11 The electrochemical gradient for potassium ions.

As an example, consider a potassium ion in the cytosol of a muscle fiber (**Figure 10.11**). You have already seen that ① the concentration gradient for potassium ions favors their diffusion into the extracellular fluid. But now let's add the force of the electrical gradient. The −90 mV resting potential means that the cytosol is negatively charged relative to the extracellular fluid. As you know, opposite charges attract, so the positively charged potassium ion is attracted to the negatively charged cytosol. ② This electrical gradient then favors the movement of potassium ions in the opposite direction, into the cytosol. The overall electrochemical gradient is the sum of these two forces—one drawing potassium ions into the cytosol and one drawing them into the extracellular fluid. If these two forces were equal, no net movement of potassium ions would occur.

However, ③ the concentration gradient for potassium ions is stronger than the electrical gradient by a small amount. For this reason, the net electrochemical gradient is a small force that draws potassium ions into the extracellular fluid. The small size of the electrochemical gradient for potassium ions in a muscle fiber at rest helps to ensure that the cell doesn't lose too many potassium ions to the extracellular fluid through leak channels.

When we look at sodium ions, however, a different picture emerges. You already know that the concentration gradient favors the movement of sodium ions into the cytosol. The electrical gradient also favors their movement into the cytosol, as the positively charged sodium ions are attracted to its negative charges. This creates a strong electrochemical gradient for sodium ions that draws them into the cytosol.

Generation of the Resting Membrane Potential

Now let's talk about how the resting membrane potential is generated. Imagine a muscle fiber that isn't polarized—its membrane potential has been temporarily changed to 0 mV. What happens to return the membrane to its resting state of −90 mV? To understand this, we need to discuss two key points:

- Ion electrochemical gradients favor diffusion of potassium ions out of the cell and sodium ions into the cell; and
- Potassium ions diffuse through leak channels more easily than do sodium ions.

As you know, the electrochemical gradients of potassium ions in the cytosol favor their diffusion through leak channels out of the cell. Similarly, the electrochemical gradients of sodium ions in the extracellular fluid favor their diffusion through leak channels into the cell.

To get to the resting state, the membrane potential must become more negative, so the cell must *lose* more positive charges than it gains. This means that more potassium ions must leave the cell than sodium ions enter. So why does this happen? It occurs because of the second factor: Potassium ions flow through leak channels more easily than do sodium ions.

As these two factors work together, the cytosol loses more positive charges than it gains. This net loss causes the membrane

potential to become more negative, until the value of the resting membrane potential is reached.

ConceptBOOST 🔊

How Do Positive Ions Create a Negative Resting Membrane Potential?

Much of the *negative* resting membrane potential is caused by the movement of *positive* ions. But how can positive ions create a negative potential? To understand how this works, let's start with a muscle fiber that has no membrane potential, which means that the charges are distributed equally across the plasma membrane, or sarcolemma. In our diagram here, five positive charges and five negative charges are found on each side of the membrane:

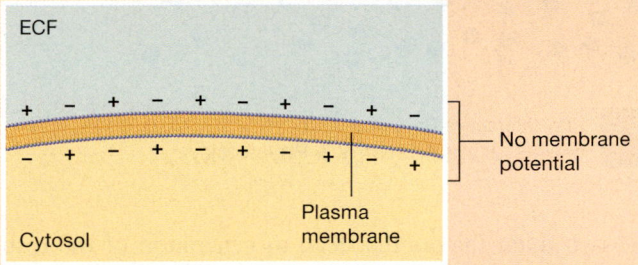

Now, imagine that a potassium ion diffuses out of the cytosol with its gradient through a leak channel:

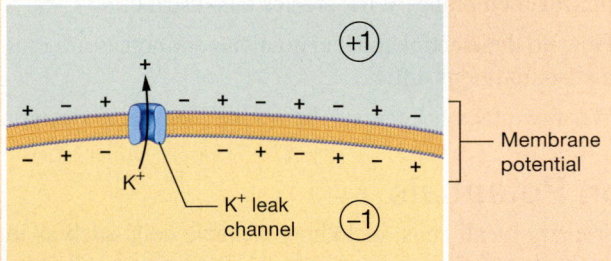

We now find six positive charges outside the membrane and four positive charges inside. This makes the overall charge in the cytosol −1 and in the extracellular fluid +1—a membrane potential has been created. Now imagine that many thousands of potassium ions exit through leak channels. You can see how this would cause the membrane potential to become progressively more negative. ■

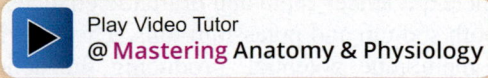

Play Video Tutor
@ **Mastering** Anatomy & Physiology

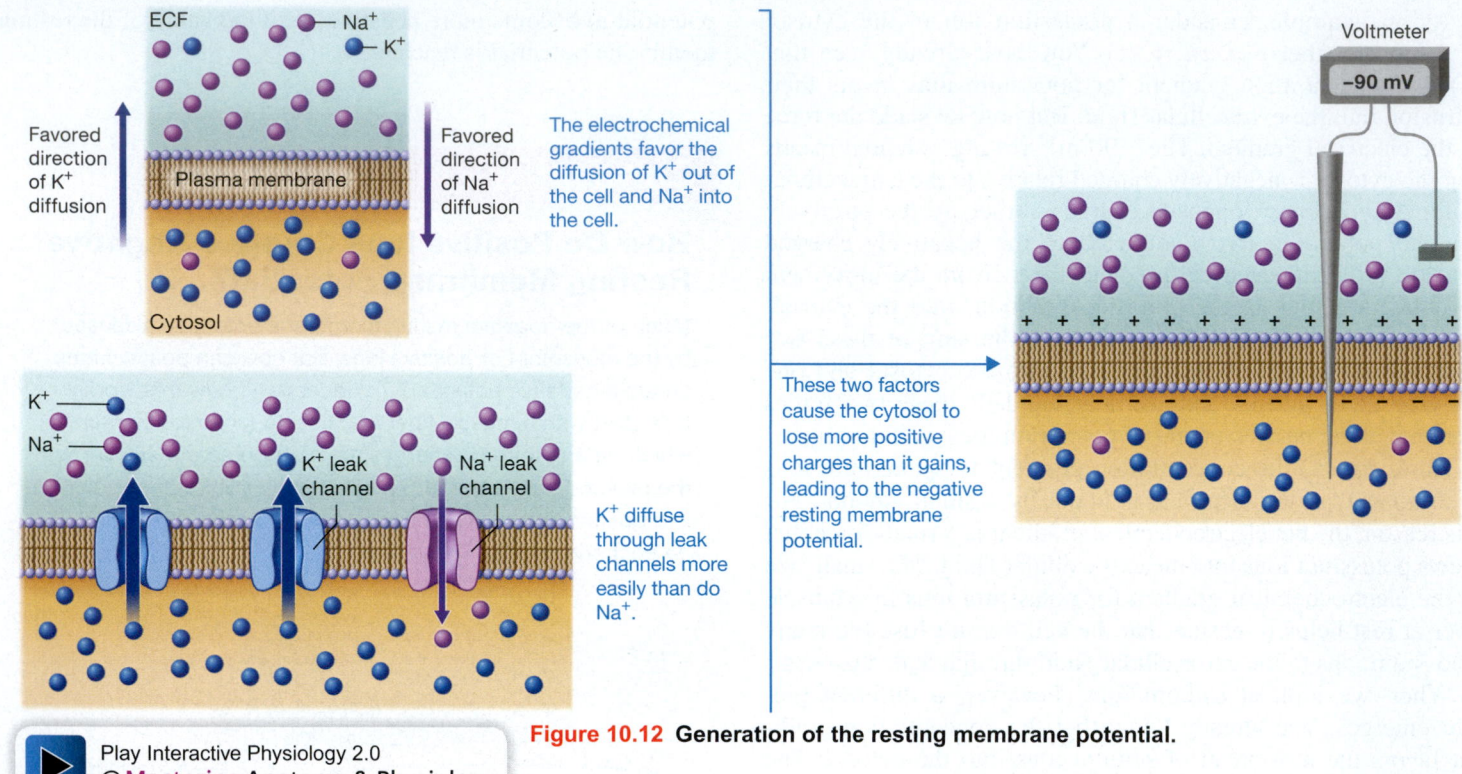

Figure 10.12 Generation of the resting membrane potential.

Play Interactive Physiology 2.0
@ Mastering Anatomy & Physiology

The two main factors that lead to generation of the resting membrane potential are illustrated in **Figure 10.12**.

Quick Check

☐ 1. What is the resting membrane potential?

☐ 2. How are sodium and potassium ions distributed across the plasma membrane? What creates this distribution?

☐ 3. How do the electrochemical gradients for potassium ions and sodium ions differ?

☐ 4. What two factors generate the resting membrane potential?

Action Potentials

A key property of all cells, including excitable cells such as muscle fibers, is that their resting membrane potential can be altered to become more negative or more positive. One such change is the **action potential,** which is a quick, temporary change in the membrane potential in a single region of the plasma membrane. During an action potential, the membrane potential goes from its negative resting value to a more positive value and then back to its negative resting potential again. All of this happens extremely rapidly, within a few milliseconds—much faster than the muscle contraction itself.

Action potentials are generated by the opening and closing of gated sodium and potassium ion channels in the plasma membrane in response to a stimulus. When these channels open (or close) the membrane experiences rapid and dramatic changes in the movement of both sodium and potassium ions. This activity in turn alters the membrane potential, producing the action potential.

Let's look at a single region of the sarcolemma to see how an action potential is generated in a muscle cell. We begin with the membrane at the resting membrane potential of about −90 mV and the voltage-gated channels closed (**Figure 10.13**). Then, in response to a stimulus, the action potential occurs and proceeds through two basic stages, *depolarization* and *repolarization:*

① **Depolarization stage: In response to a stimulus, voltage-gated sodium ion channels open and sodium ions enter the cell, making the membrane potential less negative.** When a stimulus reaches the sarcolemma, the **depolarization stage** begins, during which voltage-gated sodium ion channels open. As these channels open, sodium ions follow their electrochemical gradient and rush into the cell. The entry of these positively charged sodium ions makes the membrane potential become less negative, or less polarized (hence the term *de*polarization). The membrane potential rises toward zero and rapidly reaches a peak of about +30 mV. Notice that the membrane potential is reversed now—the outside of the cell is negative, and the inside is positive.

② **Repolarization stage: Sodium ion channels close while voltage-gated potassium ion channels open and potassium ions leave the cell, making the membrane potential more negative again.** During the **repolarization stage,** the sarcolemma returns to its negative resting membrane potential, meaning it goes back to a polarized state (hence the term *re*polarization). Repolarization occurs when voltage-gated sodium ion channels close and voltage-gated potassium ion channels open simultaneously. As potassium ion channels

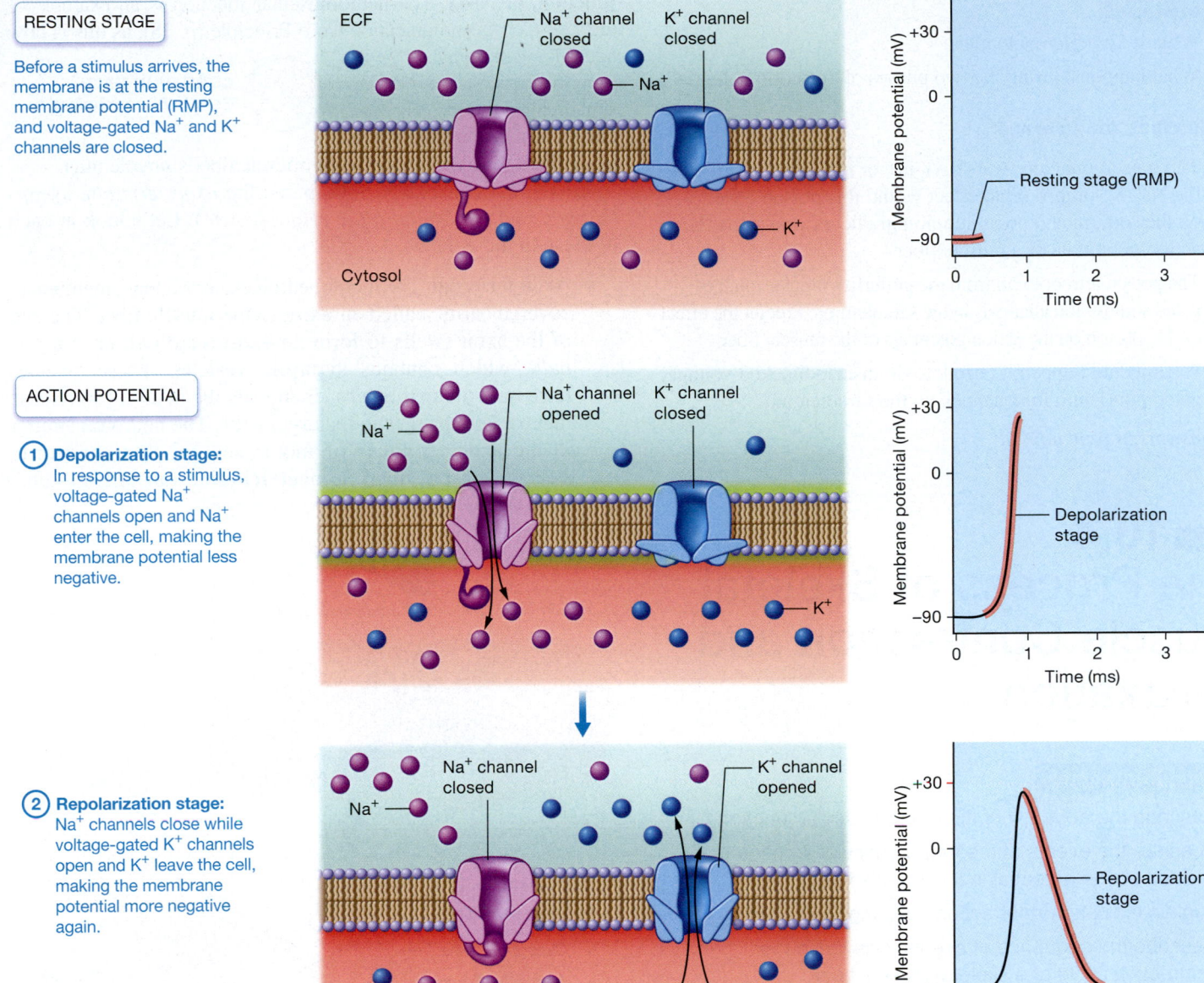

RESTING STAGE

Before a stimulus arrives, the membrane is at the resting membrane potential (RMP), and voltage-gated Na$^+$ and K$^+$ channels are closed.

ACTION POTENTIAL

① **Depolarization stage:** In response to a stimulus, voltage-gated Na$^+$ channels open and Na$^+$ enter the cell, making the membrane potential less negative.

② **Repolarization stage:** Na$^+$ channels close while voltage-gated K$^+$ channels open and K$^+$ leave the cell, making the membrane potential more negative again.

Figure 10.13 Stages of an action potential.

open, potassium ions follow their electrochemical gradient and rush out of the cell. As the cell loses positively charged potassium ions, the membrane potential becomes more negative. When the sarcolemma returns to the resting potential of around −90 mV, the potassium ion channels close. This marks the end of the action potential.

Note that the total intracellular and extracellular concentrations of sodium and potassium ions change very little during an action potential. This is because only a few ions have to cross the sarcolemma to cause large changes in its membrane potential. Any minor shifts in ion concentration are corrected by the action of the Na$^+$/K$^+$ pump.

After this detailed discussion of how action potentials occur, let's step back and consider the big picture of *why*

action potentials happen: long-distance signaling. Up to this point, we have been discussing action potentials as if they were single events in one area of the sarcolemma. However, recall that a property of muscle fibers is conductivity, which means that electrical changes across the sarcolemma are not isolated events. Action potentials don't stay in one place; rather, they are conducted, or **propagated,** throughout the entire sarcolemma like ripples in a pond. The process happens incredibly fast and results in depolarization of the entire sarcolemma, including the T-tubules. This rapid propagation enables a single stimulus in one region of the sarcolemma to have nearly instant, far-reaching impact. As we discuss in the next module, this is needed so that the muscle fiber can contract as a unit.

Quick Check

☐ 5. What is an action potential?

☐ 6. What happens during the two phases of an action potential?

Apply What You Learned

☐ 1. The poison ouabain (wah-BAY-in), or *arrow poison*, blocks the Na⁺/K⁺ pump. What effect would this poison have on the sodium and potassium ion gradients, and so on the action potentials of a muscle fiber?

☐ 2. The poison tetrodotoxin from the pufferfish blocks voltage-gated sodium ion channels in the sarcolemma. Predict the effect of this poison on the action potentials of the muscle fiber.

☐ 3. What would happen if tetrodotoxin instead blocked voltage-gated potassium ion channels in the sarcolemma?

See answers in Appendix A.

MODULE 10.4

The Process of Skeletal Muscle Contraction and Relaxation

Learning Outcomes

1. Describe the anatomy of the neuromuscular junction.

2. Describe the events at the neuromuscular junction that elicit an action potential in the muscle fiber.

3. Explain excitation-contraction coupling.

4. Describe the sequence of events involved in the contraction cycle of a skeletal muscle fiber.

5. Explain the process of skeletal muscle fiber relaxation.

Muscles contract when skeletal muscle fibers are stimulated by the nervous system, which initiates the sliding-filament mechanism. This module explores the sliding-filament mechanism of contraction in detail, along with the events that lead to muscle relaxation. First, however, we need to understand the relationship between the muscular system and the nervous system.

The Neuromuscular Junction

« FLASHBACK

1. What is a vesicle? (p. 83)

2. What is a neuron, and what are its components? (p. 148)

All skeletal muscle fibers are **innervated,** meaning they are connected to a neuron. A single neuron called a **motor neuron** communicates with many muscle fibers; each connection is known as a **synapse** (SIN-aps; *syn-* = "clasp together"). The synapse of a motor neuron with a muscle fiber is known as the **neuromuscular**

junction, or **NMJ.** The neuromuscular junction is an example of the Cell-Cell Communication Core Principle (p. 28), as this is how neurons communicate with muscle fibers. At the NMJ, a signal called a *nerve impulse,* or neuronal action potential, is transmitted from the neuron to the muscle fiber's sarcolemma.

CORE PRINCIPLE
Cell-Cell Communication

Each NMJ consists of three parts: the *axon terminal, synaptic cleft,* and *motor end plate* (**Figure 10.14**). Let's look at each of these in more detail:

● **Axon terminal.** The motor neuron extends a long, membrane-covered "arm" called an **axon** to the muscle fiber. The end of the axon swells to form an **axon terminal,** or *synaptic bulb,* which contains **synaptic vesicles.** These vesicles have **neurotransmitters,** chemicals the neuron can release that trigger changes in its target cells. The neurotransmitter in the axon terminals of motor neurons that connect to skeletal muscle fibers is **acetylcholine** (ah-SEE-til-koh′-leen; **ACh**).

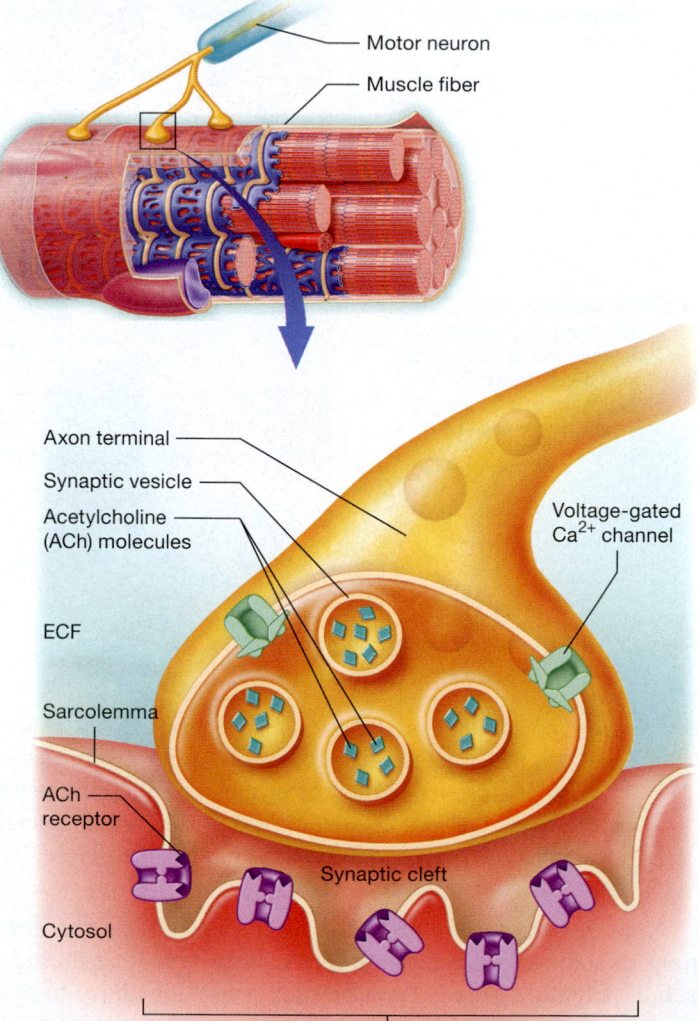

Figure 10.14 Structures of the neuromuscular junction.

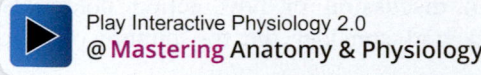

Play Interactive Physiology 2.0
@ **Mastering** Anatomy & Physiology

- **Synaptic cleft.** The **synaptic cleft** is the narrow space between the axon terminal and the muscle fiber into which ACh is released. It is filled with collagen fibers and an extracellular gel that anchors the neuron in place. It also contains enzymes that break down ACh.
- **Motor end plate.** The **motor end plate** is a specialized region of the sarcolemma whose folded surface contains many receptors for ACh. Recall that a receptor is a protein within the plasma membrane that binds to a specific ligand. These receptors are actually ligand-gated ion channels; ACh is the ligand.

Take note that when a neuron stimulates a muscle cell at the NMJ, an electrical signal (the neuronal action potential) is first converted to a chemical signal (the ACh molecules). The chemical signal is then converted back into an electrical signal—the muscle fiber action potential. This is a pattern we see repeatedly throughout physiology.

Quick Check

☐ 1. What are the three components of the neuromuscular junction?

☐ 2. What is the function of each part of the neuromuscular junction?

Skeletal Muscle Contraction

It's now time to talk about how ion channels, action potentials, and the neuromuscular junction work together to cause a muscle contraction. We know that skeletal muscle fibers must be stimulated to contract by the nervous system, but what this means more specifically is

that the sarcolemma of a muscle fiber must be stimulated by ACh from a motor neuron to have an action potential. This is a process known as *excitation*.

Once the muscle fiber is excited by such stimulation, a process called *excitation-contraction coupling* conveys this excitation to the parts of the fiber that produce the contraction, the myofilaments. It's at this point that the sliding-filament mechanism occurs and the sarcomere contracts. Contractions in many sarcomeres produce contraction of the whole muscle. So the overall process of muscle contraction can be broken down into three parts: the excitation phase, excitation-contraction coupling, and the contraction phase.

Excitation Phase

The **excitation phase** involves transmission of a signal from the motor neuron to the sarcolemma of a muscle fiber. This phase, which occurs at the neuromuscular junction, proceeds with the following steps (**Figure 10.15**):

① **An action potential arrives at the axon terminal and triggers Ca^{2+} channels in the axon terminal to open.** An action potential from the brain or spinal cord reaches the axon terminal of the motor neuron.

② **Calcium ion entry triggers exocytosis of synaptic vesicles.** The action potential causes voltage-gated calcium ion channels in the membrane of the axon terminal to open,

Voltage-gated Ca^{2+} channel

Axon terminal

Ca^{2+}

Ca^{2+}

ECF

Sarcolemma

ACh in synaptic vesicle

Cytosol

Na^+

Na^+

ACh bound to binding site on receptor (ligand-gated ion channel)

Motor end plate

Synaptic cleft

Na^+

Na^+

Na^+

① An action potential arrives at the axon terminal and triggers Ca^{2+} channels in the axon terminal to open.

② Ca^{2+} entry triggers exocytosis of synaptic vesicles.

③ Synaptic vesicles release ACh into the synaptic cleft.

④ ACh binds to ligand-gated ion channels in the motor end plate.

⑤ Ion channels open and Na^+ enter the muscle fiber.

⑥ Entry of Na^+ depolarizes the sarcolemma locally, producing an end-plate potential.

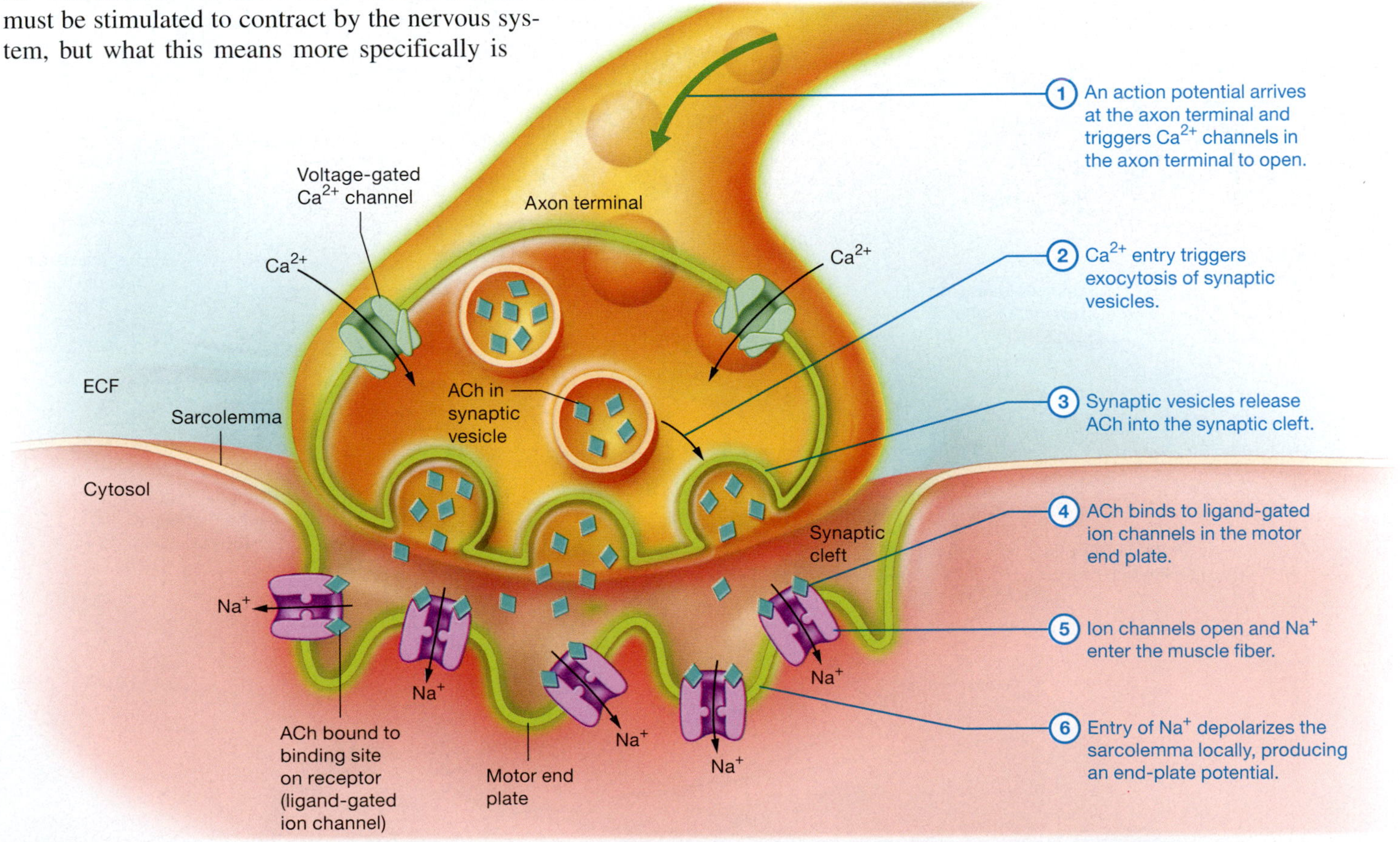

Figure 10.15 Excitation phase: events at the neuromuscular junction.

Play Animation
@**Mastering** Anatomy & Physiology

and calcium ions follow their electrochemical gradient and enter the cytosol of the axon terminal. This entry of calcium ions triggers exocytosis of the synaptic vesicles.

③ **Synaptic vesicles release acetylcholine into the synaptic cleft.** These vesicles release the neurotransmitter ACh into the synaptic cleft between the axon terminal and the motor end plate of the muscle fiber.

④ **Acetylcholine binds to ligand-gated ion channels in the motor end plate.** ACh diffuses across the cleft until it reaches and binds to ACh receptors on the motor end plate. These receptors are ligand-gated ion channels in the sarcolemma.

⑤ **Ion channels open and sodium ions enter the muscle fiber.** When ACh binds, the ligand-gated ion channels of the ACh receptors open and allow sodium ions to follow their electrochemical gradient and enter the muscle fiber.

⑥ **Entry of sodium ions depolarizes the sarcolemma locally, producing an end-plate potential.** As sodium ions enter the fiber, a small area of the sarcolemma depolarizes, producing an effect called an **end-plate potential.** This

potential is simply a local depolarization in the area of the motor end plate.

Multiple end-plate potentials must be generated to produce a functional contraction of a muscle fiber. However, the ACh released from the synaptic vesicles is degraded and inactivated almost immediately by an enzyme in the synaptic cleft called **acetylcholinesterase** (ah-SEET-'l-kohl-in-ess′-teh-rayz; **AChE**). For this reason, the neuron must continue to fire action potentials and release new ACh in order to stimulate repeated end-plate potentials in the muscle fiber.

Excitation-Contraction Coupling

Up to this point, excitation of the sarcolemma and contraction of the myofilaments might seem like two separate processes. However, you've read that a muscle fiber will not contract without first being excited, so these two processes must be connected, or "coupled" together. This connection occurs by a series of events known collectively as **excitation-contraction coupling** (**Figure 10.16**). In this phase, the end-plate potential leads to an action potential in the sarcolemma, which in turn triggers events that result in contraction.

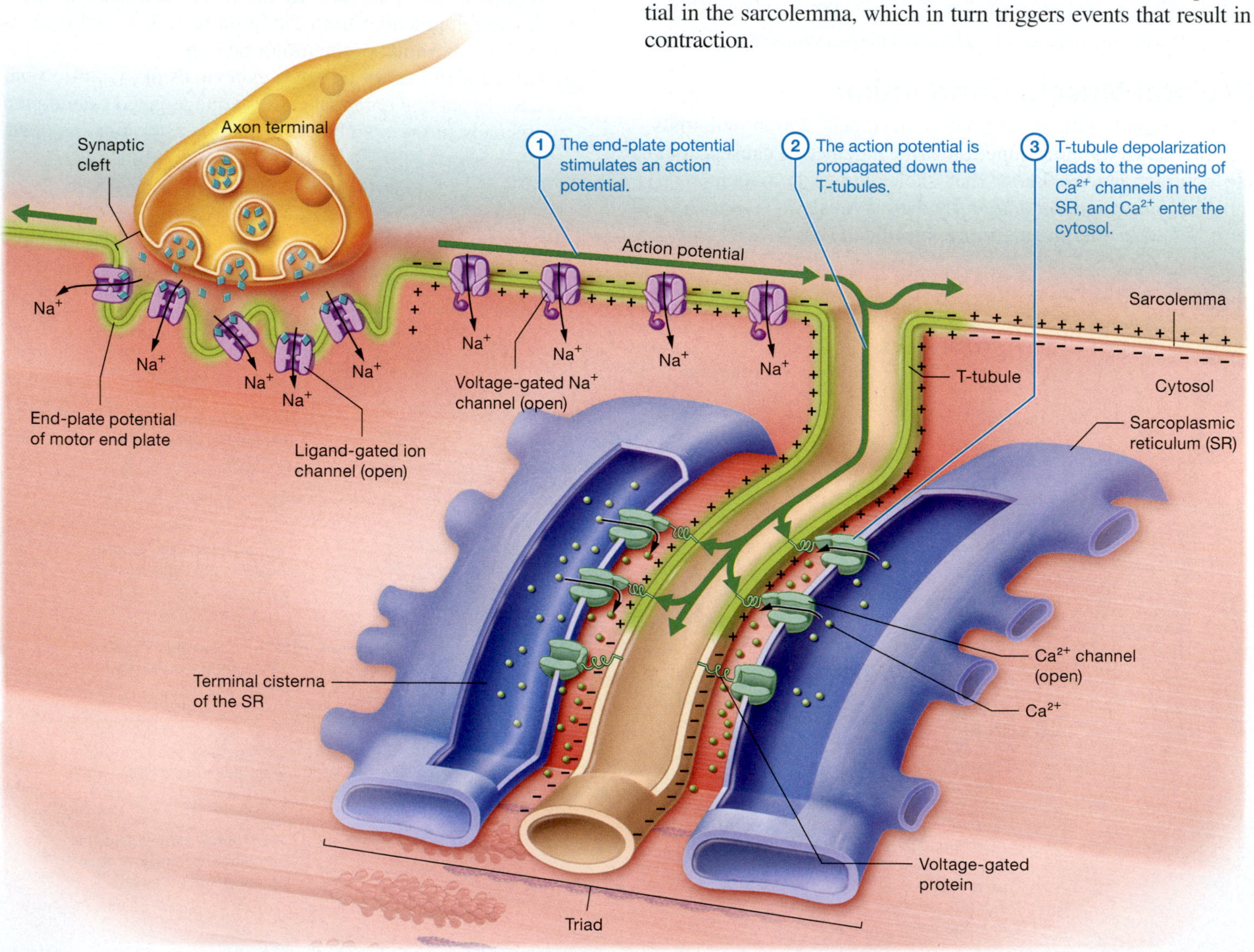

① The end-plate potential stimulates an action potential.

② The action potential is propagated down the T-tubules.

③ T-tubule depolarization leads to the opening of Ca^{2+} channels in the SR, and Ca^{2+} enter the cytosol.

Axon terminal

Synaptic cleft

Action potential

Sarcolemma

Na^+

End-plate potential of motor end plate

Ligand-gated ion channel (open)

Voltage-gated Na^+ channel (open)

Na^+

T-tubule

Cytosol

Sarcoplasmic reticulum (SR)

Terminal cisterna of the SR

Ca^{2+} channel (open)

Ca^{2+}

Voltage-gated protein

Triad

Figure 10.16 Excitation-contraction coupling: events at the sarcolemma and sarcoplasmic reticulum.

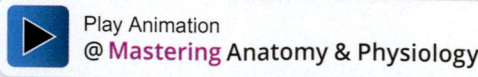

Play Animation
@ **Mastering** Anatomy & Physiology

The steps of excitation-contraction coupling are as follows:

① **The end-plate potential stimulates an action potential.** The end-plate potential spreads to the areas of the sarcolemma adjacent to the motor end plate. The change in voltage created by the end-plate potential opens enough voltage-gated sodium ion channels in the sarcolemma to stimulate an action potential.

② **The action potential is propagated down the T-tubules.** The action potential propagates like a wave along the sarcolemma, as depolarization of one area of the membrane triggers the next few voltage-gated sodium ion channels to open. This process continues like a chain reaction along the sarcolemma, diving into the fiber via the T-tubules.

③ **T-tubule depolarization leads to the opening of calcium ion channels in the sarcoplasmic reticulum, and calcium ions enter the cytosol.** Remember that on either side of a T-tubule is a terminal cisterna of the SR. The terminal cisternae are directly linked to the T-tubules by voltage-gated proteins. When the T-tubules depolarize, these proteins "twist" open calcium ion channels in the terminal cisternae. In essence, these proteins open the "floodgates" of the SR, releasing many calcium ions into the cytosol.

As calcium ions flood the cytosol, the sarcolemma is repolarizing.

The events described here involve only a single action potential. However, a skeletal muscle contraction in the body requires this process to occur repeatedly. So, new action potentials are generated over and over, and the concentration of calcium ions in the cytosol increases with each one. Now let's see how large numbers of calcium ions in the cytosol lead to the contraction phase.

Contraction Phase: The Crossbridge Cycle of the Sliding-Filament Mechanism

We've said that myofilaments slide past one another during the **contraction phase.** Recall that the myosin heads of the thick filaments attach to the actin of the thin filaments at their active sites and pull them toward the middle of the sarcomere. But before the contractile proteins can do their work, the regulatory proteins must let them.

As you can see in **Figure 10.17a**, in a resting muscle fiber the regulatory protein tropomyosin curls around actin, blocking its active sites. Calcium ions initiate the mechanism that moves the attached troponin and tropomyosin away from the active sites. At rest, the concentration of calcium ions in the cytosol is very low. However, as we just saw, once excitation-contraction

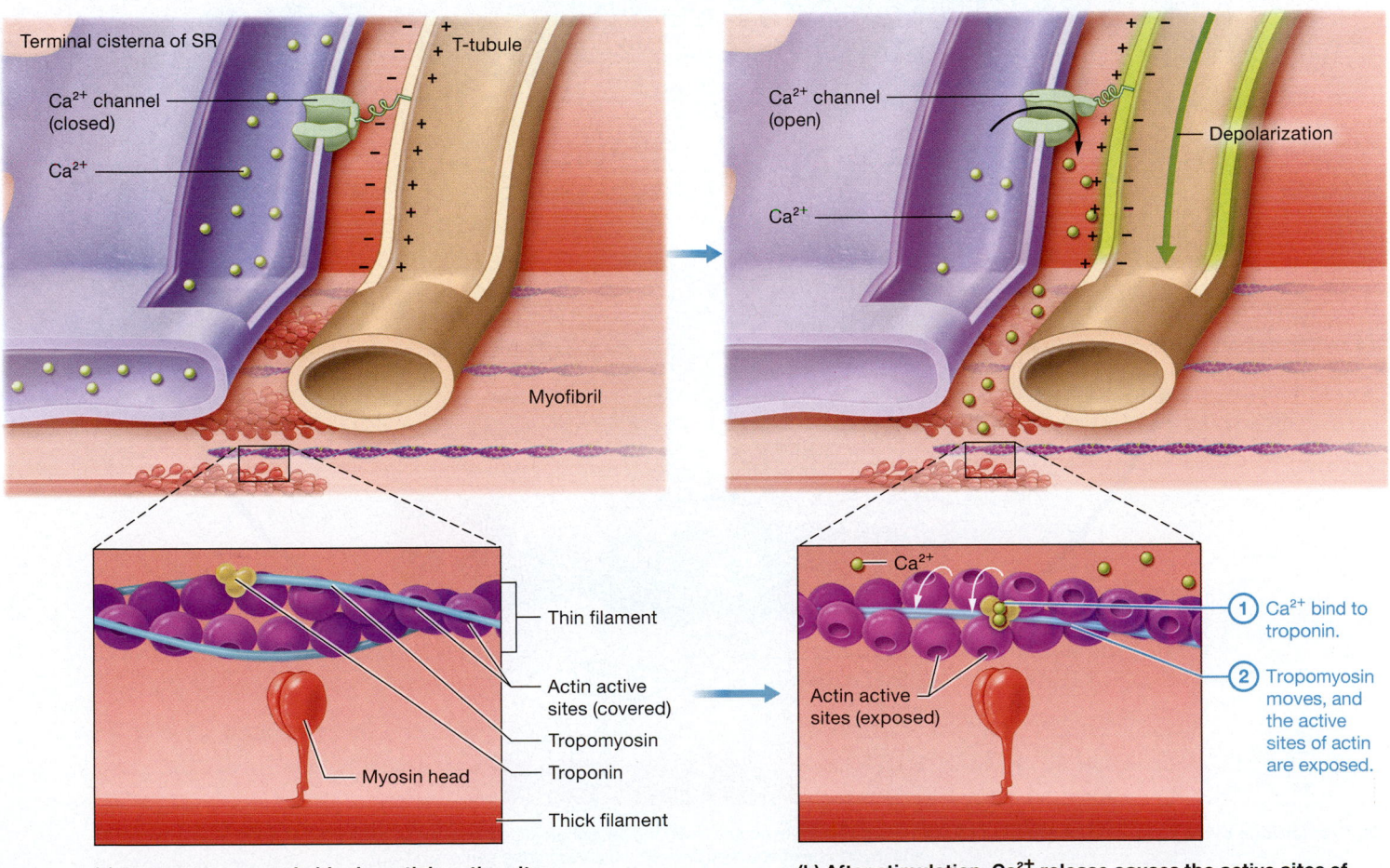

(a) At rest, tropomyosin blocks actin's active sites.

(b) After stimulation, Ca²⁺ release causes the active sites of actin to be exposed.

Figure 10.17 Preparation for contraction: regulatory events at the myofibril.

▶ Play Animation
@ **Mastering** Anatomy & Physiology

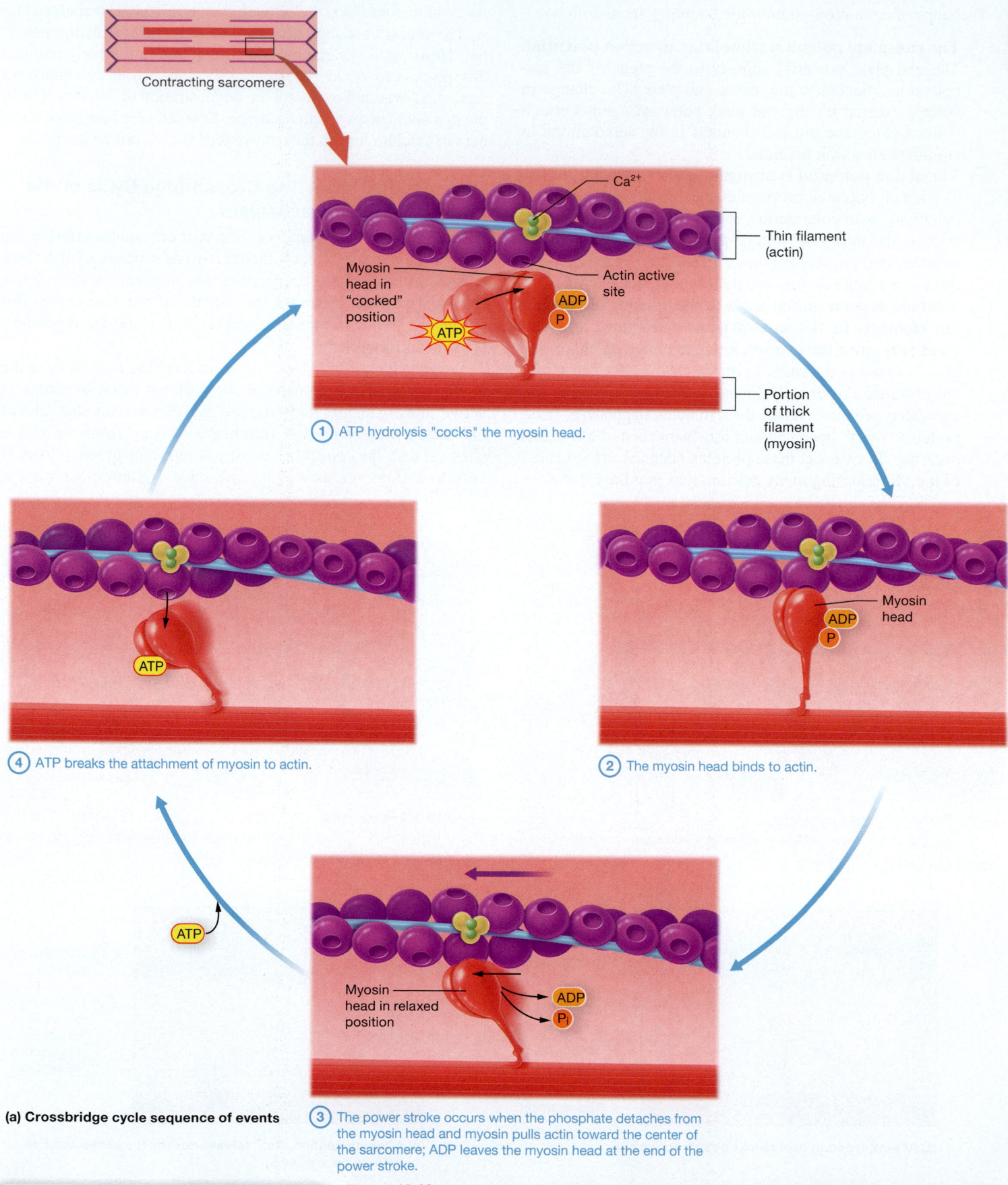

Contracting sarcomere

Ca²⁺

Thin filament (actin)

Myosin head in "cocked" position

Actin active site

ADP

P

ATP

Portion of thick filament (myosin)

① ATP hydrolysis "cocks" the myosin head.

Myosin head

ADP

P

② The myosin head binds to actin.

ATP

④ ATP breaks the attachment of myosin to actin.

ATP

Myosin head in relaxed position

ADP

Pᵢ

③ The power stroke occurs when the phosphate detaches from the myosin head and myosin pulls actin toward the center of the sarcomere; ADP leaves the myosin head at the end of the power stroke.

(a) Crossbridge cycle sequence of events

Figure 10.18 Contraction phase: the crossbridge cycle of the sliding-filament mechanism.

▶ Play Animation
@ **Mastering** Anatomy & Physiology

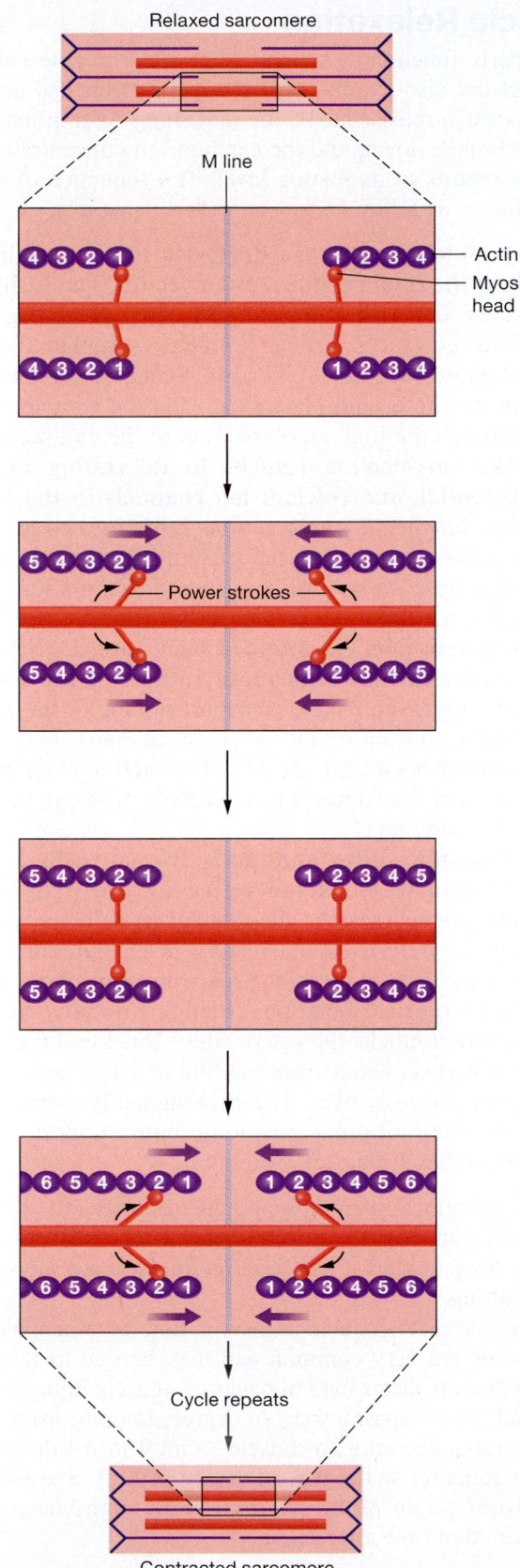

(b) Crossbridge cycle repeats, pulling actin toward the center of the sarcomere.

Figure 10.18 Contraction phase: the crossbridge cycle of the sliding-filament mechanism. (*continued*)

coupling has occurred, the concentration of calcium ions in the cytosol increases dramatically. After stimulation, in preparation for muscle contraction (**Figure 10.17b**):

1. **Calcium ions bind to troponin.** The regulatory protein troponin has three subunits: One binds calcium ions, one binds actin, and the other tropomyosin. In this step, calcium ions bind to the appropriate troponin subunit.

2. **Tropomyosin moves, and the active sites of actin are exposed.** The binding of calcium ions causes troponin to shift its position, allowing tropomyosin to move away from the active sites.

When the active sites of actin are exposed, the myosin heads are able to bind tightly to them. A myosin head bound to an actin subunit is known as a **crossbridge.** When this binding occurs, a **crossbridge cycle** is initiated. A muscle **contraction** is simply a succession of crossbridge cycles and the resulting production of tension.

The sequence of events of the crossbridge cycle, illustrated in **Figure 10.18a**, is as follows:

1. **ATP hydrolysis "cocks" the myosin head.** Muscle contraction requires energy provided by the hydrolysis of ATP. Each myosin head contains both a site that binds to ATP and an ATPase enzyme that catalyzes the breakdown of ATP. When ATP binds to the myosin head, the ATPase in the myosin rapidly catalyzes the hydrolysis of ATP to ADP and a phosphate group (P). The energy liberated from this reaction "cocks" the myosin head into its high-energy position, ready to work. The ADP and phosphate remain attached to the myosin head when it is cocked.

2. **The myosin head binds to actin.** With the myosin head in its cocked position, it may now bind to the active site of actin. Note that the resulting crossbridge is at about a 90° angle relative to the thick filament.

3. **The power stroke occurs when the phosphate detaches from the myosin head and myosin pulls actin toward the center of the sarcomere; ADP leaves the myosin head at the end of the power stroke.** The phosphate (inorganic phosphate, represented as P_i) detaches from the myosin head as it pivots on its hinge and moves from its cocked, high-energy position to its relaxed, low-energy position. As the myosin pivots, it pulls the actin toward the center of the sarcomere. This action is known as the **power stroke.** Notice that the myosin crossbridge is now at about a 45° angle relative to the thick filament. At the end of the power stroke, ADP dissociates from the myosin head.

4. **ATP breaks the attachment of myosin to actin.** Another ATP binds to the myosin head, which breaks its attachment to actin.

The contraction cycle is then repeated, as shown in **Figure 10.18b**. ATP is hydrolyzed, the myosin head is recocked, it binds to the next actin subunit, and the power stroke repeats. You can see in the figure how myosin first attaches to the first actin, then the second, and so on. For an average contraction, this process will repeat about 20–40 times for each myosin head in each sarcomere of the muscle fiber.

ConceptBOOST))))

Linking a Crossbridge Cycle to the Sliding-Filament Mechanism

In Module 10.2, we introduced you to the sliding-filament mechanism, and said that the thick and thin filaments slide past one another during a muscle contraction. But in our discussion of the crossbridge cycle, we have been focused on the very small movement produced by the interaction between actin and myosin, so you may be wondering how this connects to sliding filaments. Well, remember that during these cycles, the myosin head grabs onto a series of actin subunits in the thin filament, pulling the filament progressively closer to the M line of the sarcomere.

As the process repeats, the thin filament slides past the thick filament, edging progressively closer to the M line of the sarcomere. This is the "sliding" part of the sliding-filament mechanism—the thin filament slides past the thick filament as the myosin heads pull it toward the M line. As this occurs, the zone of overlap becomes larger and the sarcomere becomes smaller as tension is generated (look back at Figure 10.9).

You may also be wondering why the thin filaments don't simply slide backward when ATP breaks the actin-myosin attachment in preparation for the next crossbridge cycle. To picture this process, let's use the analogy of sailors pulling on a rope to bring up an anchor from the sea floor. The rope represents the thin filament, the anchor the Z-disc, and the hands of the sailors the myosin heads. At any given time, the hands of some sailors are holding the rope and pulling, like myosin heads attached to actin and pulling the thin filament with power strokes.

In addition, at any given time, some sailors are repositioning their hands farther up the rope and also allowing their arms to rest for a second. In this way, they work together to continue to pull the rope without allowing it to slide backward. This is also true of the myosin heads—while some are pulling the thin filament toward the M line, others are detached and being reactivated ("recocked") by ATP. This causes the thin filament to slide in only one direction: toward the M line. ∎

Quick Check

☐ 3. How does excitation from a neuron trigger muscle fiber contraction?

☐ 4. How are excitation and contraction coupled?

☐ 5. What are the steps of the crossbridge cycle?

Muscle Relaxation

A properly functioning muscle fiber must be able not only to contract but also to relax. **Muscle relaxation** has three components: ACh release stops, the remaining ACh in the synaptic cleft is broken down, and the calcium ion concentration in the cytosol returns to its resting level. The sequence of events in relaxation is as follows (**Figure 10.19**):

① **Acetylcholinesterase degrades the remaining ACh, and the final repolarization occurs.** The AChE breaks down the ACh still in the synaptic cleft, degrading it into substances that can no longer stimulate the muscle. Without such stimulation, the ligand-gated ion channels in the motor end plate close, and the sarcolemma goes through the final repolarization of the contraction.

② **The sarcolemma returns to its resting membrane potential, and calcium ion channels in the SR close.** The sarcolemma returns to its resting state once repolarization is complete. The calcium ion channels in the SR then close, so no further release of calcium ions from the SR takes place.

③ **Calcium ions are pumped back into the SR, returning the calcium ion concentration in the cytosol to its resting level.** Active transport pumps in the SR membrane consume ATP to pump calcium ions from the cytosol back into the SR. This activity decreases the calcium ion concentration of the cytosol, returning it to its resting level.

④ **Troponin shifts and pulls tropomyosin back into position to block the active sites of actin, and the muscle relaxes.** As the number of calcium ions in the cytosol returns to its resting level, calcium ions dissociate from troponin. This causes troponin to return to its original position, pushing tropomyosin back to where it blocks the active sites. This blocking prohibits the myosin heads from binding to actin, and the muscle contraction is over. The myofilaments then slide back into their original positions, with support from titin and other structural proteins.

The calcium ion pumps and AChE work all the time to allow relaxation of a muscle fiber. This makes it possible for the fiber to start a new contraction when a new stimulus comes along.

A muscle that is unable to relax is said to be in **spasm.** Muscle spasms are very common and may be due to factors such as dehydration, electrolyte imbalances, muscle injury, or muscle overload. These spasms vary in degree, ranging from a "tight" muscle (also known as a muscle "knot") to a full spasm that impairs function and is typically very painful. See *A&P in the Real World: Rigor Mortis* to discover why muscles can't relax for a period of time after death.

Quick Check

☐ 6. What are the roles of AChE and calcium ions in muscle relaxation?

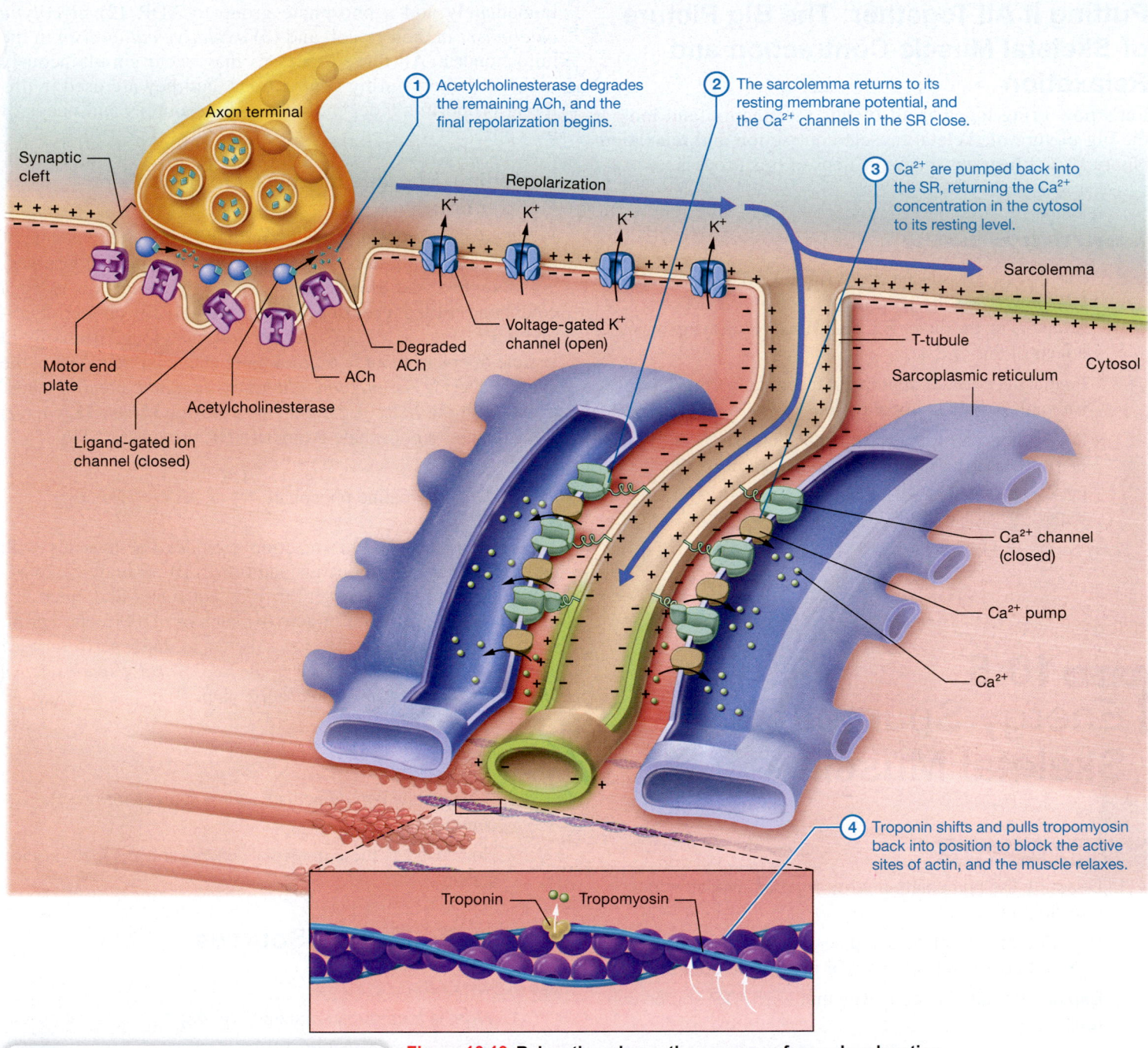

① Acetylcholinesterase degrades the remaining ACh, and the final repolarization begins.

② The sarcolemma returns to its resting membrane potential, and the Ca^{2+} channels in the SR close.

③ Ca^{2+} are pumped back into the SR, returning the Ca^{2+} concentration in the cytosol to its resting level.

④ Troponin shifts and pulls tropomyosin back into position to block the active sites of actin, and the muscle relaxes.

Figure 10.19 Relaxation phase: the process of muscle relaxation.

Play Animation
@ Mastering Anatomy & Physiology

Rigor Mortis

After death, muscle spasms occur over the entire body, a condition known as *rigor mortis* (*rigor* = "rigidity," *mortis* = "death"). Rigor mortis is the progressive stiffening of skeletal muscles that begins about 3–4 hours after death, as the pumps that drive calcium ions back into the SR no longer have ATP to fuel their activity. As a result, calcium ions remain in the cytosol, where they bind with troponin and initiate a muscle contraction. The muscle fibers are unable to relax without ATP, so the myosin heads can't detach from actin. This causes the muscles to remain in spasm until the proteins in the myofilaments begin to degenerate, about 48–72 hours after death.

Putting It All Together: The Big Picture of Skeletal Muscle Contraction and Relaxation

Let's now bring together all of this information and look at the big picture of skeletal muscle contraction and relaxation. **Figure 10.20** illustrates the main steps of this process.

Apply What You Learned

☐ 1. The disease myasthenia gravis (my'-uss-THEE-nee-ah GRAH-viss) results in the destruction of acetylcholine receptors on the motor end plate. Predict the symptoms and effects of this disease.

☐ 2. Predict the effect of improperly functioning troponin that isn't able to bind to tropomyosin.

☐ 3. Researchers discover a genetic mutation that leads to a lack of T-tubules in a person's skeletal muscle fibers. Predict the possible effects of such a mutation. (*Hint:* How would it impact the structure and function of a triad?)

See answers in Appendix A.

MODULE **10.5**

Energy Sources for Skeletal Muscle

Learning Outcomes

1. Describe the immediate energy sources available to muscle fibers.

2. Describe the glycolytic and oxidative mechanisms that muscle fibers use to obtain ATP for muscle contraction.

3. Explain the duration of activity that each ATP source can fuel.

Muscle fibers perform many functions that require energy. From Module 10.4, we know that energy in the form of ATP is required for multiple aspects of skeletal muscle contraction. For example, cells need ATP to run the Na^+/K^+ pumps that maintain the ion gradients involved in the action potential of skeletal muscle fibers. ATP is also necessary for both contraction *and* relaxation of the muscle fiber.

With so much ATP required for muscle function, you might expect muscle fibers to contain a great deal of stored ATP. Surprisingly, they store only a few seconds' worth of ATP in their cytosol. Muscle fibers, then, must have a way to quickly regenerate ATP. Regeneration may be accomplished through three processes: (**1**) reactions in the cytosol that

immediately add a phosphate group to ADP, (**2**) *glycolytic catabolism* in the cytosol, and (**3**) *oxidative catabolism* in the mitochondria. All three processes may occur simultaneously in muscle fibers during contractions, but they are used in different proportions, depending on the resources and needs of the cells.

Immediate Sources of Energy for Muscle Contraction

When contraction begins, the main immediate energy source of the muscle fiber is stored ATP. This ATP is rapidly consumed, but is regenerated almost immediately by a reaction using a compound called **creatine phosphate** (KREE-uh-teen FOS-fayt; **CP**; **Figure 10.21a**). Creatine phosphate, found primarily in muscle fibers, is about 5–6 times more abundant than ATP in the cytosol. During the *creatine phosphate reaction*, creatine phosphate, with the help of the enzyme **creatine kinase** (KY-nayz; **CK**), donates a phosphate group to ADP, producing ATP:

$$ADP + Creatine\ phosphate \underset{}{\overset{CK}{\rightleftharpoons}} ATP + Creatine$$

ATP produced by this reaction provides the muscles with enough energy for about an additional 10 seconds of maximal muscle activity. Though this may not sound like much, it's enough for a short burst of activity, such as a 100-meter sprint. Does more creatine make for a better athletic performance? To find out, take a look at *A&P in the Real World, Pseudoscience Exposed: Creatine Supplementation* on p. 363.

Quick Check

☐ 1. What are the two immediate energy sources for muscle contraction?

☐ 2. How long can these immediate energy sources fuel muscle activity?

Glycolytic Energy Sources

« FLASHBACK

1. What is a catabolic reaction? (p. 43)

2. What is glycolysis? (p. 69)

When immediate energy sources are depleted, muscle fibers turn to glycolysis, also known as **glycolytic** or **anaerobic catabolism,** to make ATP. As you may recall, *glycolysis* is a series of reactions that takes place in the cytosol of all cells, including muscle fibers (see Chapter 3). During this process, glucose is broken down to produce two ATP per molecule of glucose (left side of **Figure 10.21b**).

A muscle fiber has two potential sources of glucose for glycolysis: glucose that enters the fiber from the bloodstream and a storage form of glucose called *glycogen* (GLY-koh-jen). Glycogen granules are found in the cytosol of both muscle fibers and liver cells. Muscle fibers that depend primarily on

The Big Picture of Skeletal Muscle Contraction

Figure 10.20

Play Animation @ Mastering Anatomy & Physiology

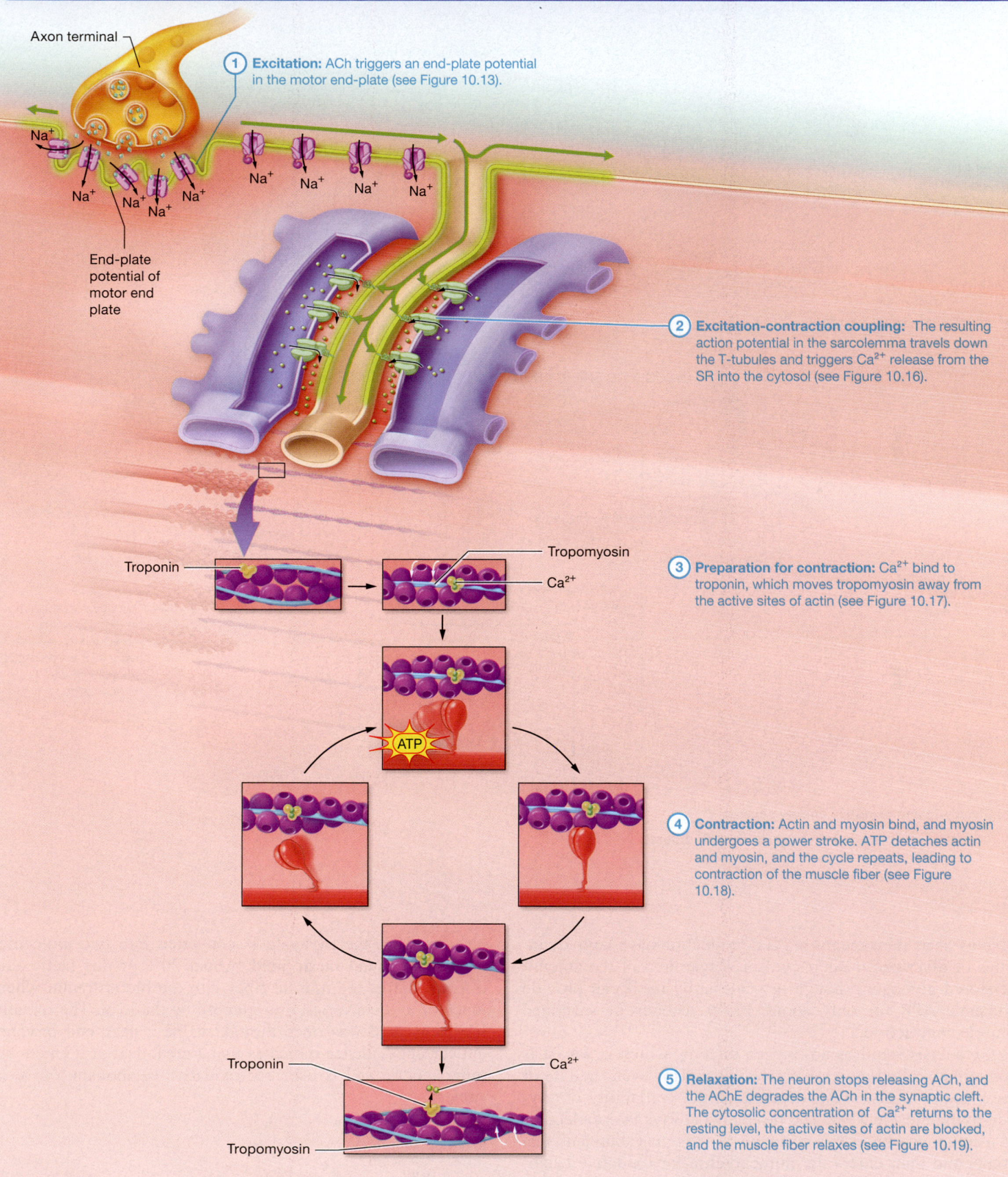

Axon terminal

1 Excitation: ACh triggers an end-plate potential in the motor end-plate (see Figure 10.13).

Na^+

End-plate potential of motor end plate

2 Excitation-contraction coupling: The resulting action potential in the sarcolemma travels down the T-tubules and triggers Ca^{2+} release from the SR into the cytosol (see Figure 10.16).

Troponin

Tropomyosin

Ca^{2+}

3 Preparation for contraction: Ca^{2+} bind to troponin, which moves tropomyosin away from the active sites of actin (see Figure 10.17).

ATP

4 Contraction: Actin and myosin bind, and myosin undergoes a power stroke. ATP detaches actin and myosin, and the cycle repeats, leading to contraction of the muscle fiber (see Figure 10.18).

Troponin

Ca^{2+}

Tropomyosin

5 Relaxation: The neuron stops releasing ACh, and the AChE degrades the ACh in the synaptic cleft. The cytosolic concentration of Ca^{2+} returns to the resting level, the active sites of actin are blocked, and the muscle fiber relaxes (see Figure 10.19).

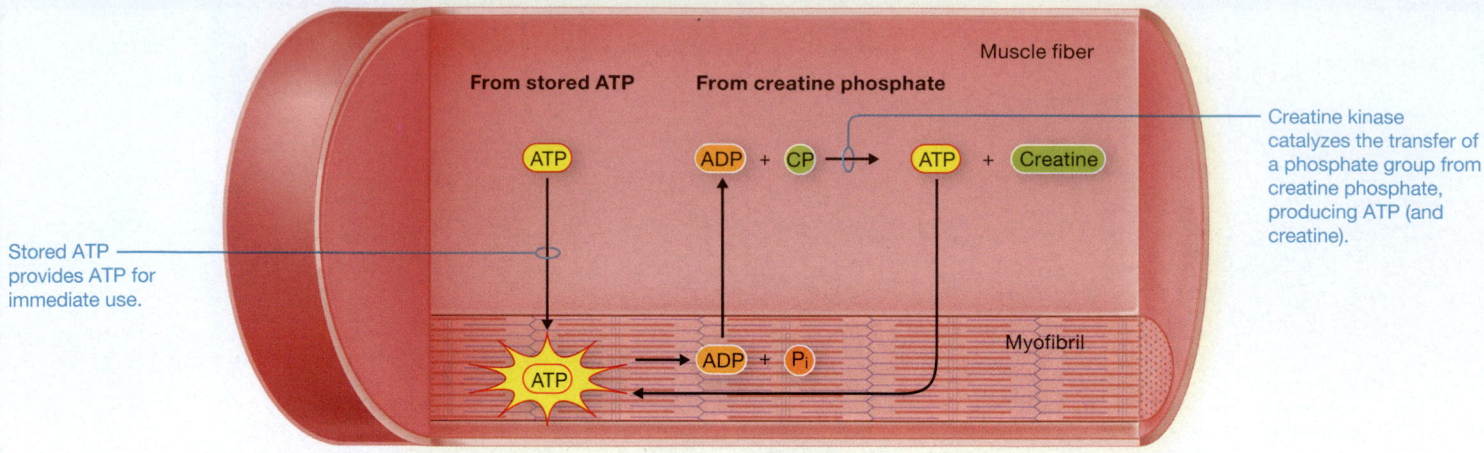

(a) Immediate energy sources

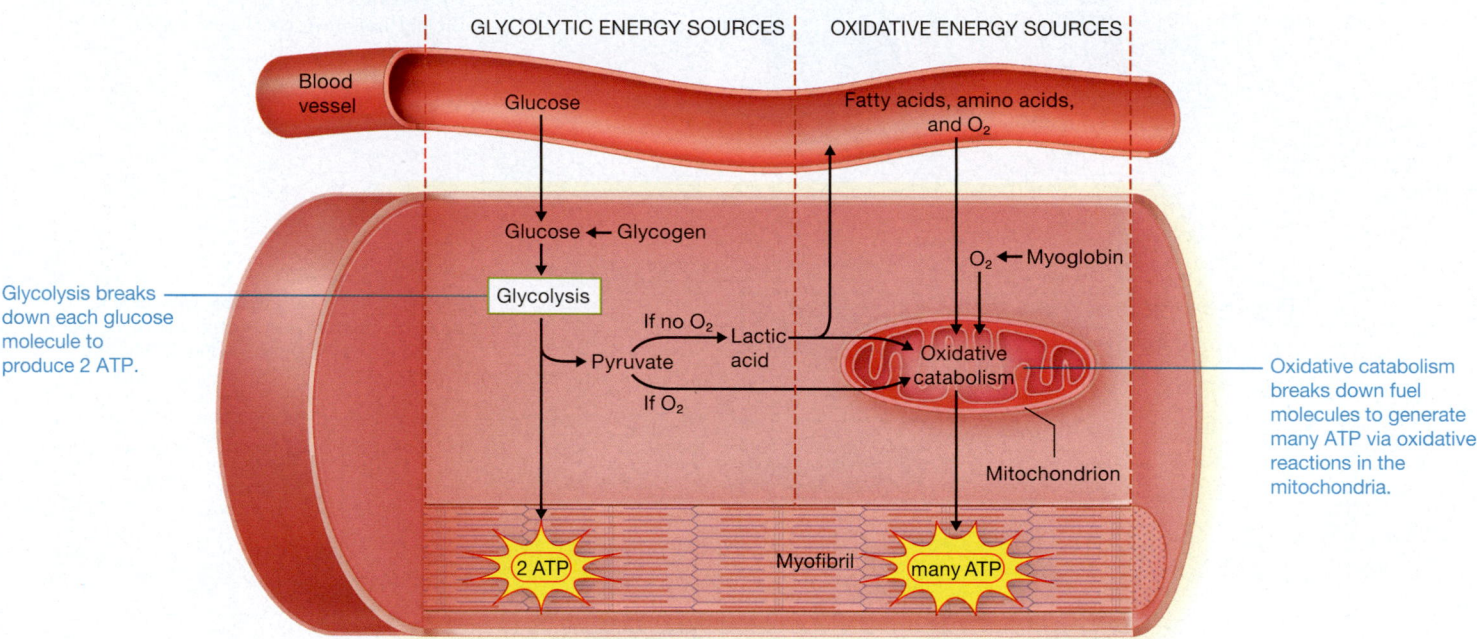

(b) Glycolytic and oxidative energy sources

Figure 10.21 Sources of energy for muscle fibers.

glycolysis as their means of ATP production have large quantities of glycogen in their cytosol. Regardless of the amount of stored glycogen, though, glycolysis by itself can provide adequate ATP for only about 30–40 seconds of sustained muscle contraction.

Glycolysis requires no oxygen directly, which is why it is sometimes called *anaerobic catabolism*. However, the fate of the product of glycolysis depends on the availability of oxygen to the muscle fiber. The product is always two molecules of a compound known as *pyruvate*. If oxygen is abundant, this compound then enters the mitochondria for oxidative catabolism, which at that point will be occurring simultaneously with glycolysis as long as glucose is available. If oxygen is

not abundant, the pyruvate is converted into two molecules of the compound **lactic acid.** About 20% of this lactic acid diffuses out of the muscle fiber into the bloodstream, where much of it is converted into glucose by the liver. The remaining lactic acid was once thought to be a dead-end product, but current evidence indicates that a great deal of it enters the mitochondria and is used for oxidative catabolism, our next subject.

Quick Check

☐ 3. What is glycolysis?

☐ 4. How long can glycolytic catabolism fuel muscle activity?

Pseudoscience Exposed: Creatine Supplementation

Dietary supplement manufacturers market creatine supplements as a way to improve muscle strength and performance. The evidence to support these claims, however, is mixed. Research has demonstrated that supplementation with creatine does mildly improve performance for activities that require short bursts of muscle activity. However, the effects of creatine on endurance-type activities are minimal to nonexistent. And creatine may actually be detrimental to athletes in certain sports, because supplementation causes weight gain from water retention.

One of the most pressing questions of creatine supplementation isn't about its effectiveness, but rather its safety. Most of the recent data suggest that supplementation within the recommended dose range is safe in the short term. But long-term studies remain to be done. In addition, massive doses of creatine can have negative health implications such as kidney damage. Beyond the possible health issues, huge doses are a waste of money. Skeletal muscle fibers have a maximal storage capacity for creatine, and any amount ingested in excess of this limit is simply excreted in the urine.

One final important thing to consider with creatine is it's a dietary supplement, so it is regulated as a food, not as a drug. Dietary supplements don't need be approved by the Food and Drug Administration (FDA), nor do they have to be proven to be safe or effective in order to be sold. Quality control is left up to individual manufacturers, and there is no requirement that the active ingredient actually be present in the bottle you are purchasing (e.g., you could be purchasing an herbal drug in which the active ingredient is found in the roots, but the pills contain only ground-up leaves). So, when you buy creatine powder, are you actually getting what's on the label?

Oxidative Energy Sources

‹‹ FLASHBACK

1. What is an oxidation-reduction reaction? (p. 43)
2. Where does oxidative catabolism take place, and what does it yield? (p. 90)

Combined, immediate and anaerobic glycolytic energy sources are adequate for short bursts of activity, such as running a 400-meter race. However, longer-lasting muscle activity, such as running a 5K or 10K race, requires a muscle fiber to use mostly **oxidative** or **aerobic catabolism** to generate ATP (right side of Figure 10.21). Recall that the reactions of

oxidative catabolism occur in the mitochondria (see Chapter 3). During this process, electrons are removed from carbon-based compounds, and the energy liberated is then used to fuel the synthesis of ATP.

Oxidative catabolism produces more ATP than does glycolysis; the amount of ATP depends on the type of fuel used by the muscle fiber. Muscle fibers can use multiple fuels for oxidative catabolism, including the products of glycolysis, as well as fatty acids and amino acids. Glucose is generally the preferred fuel for muscle fibers—the one they use first. As glucose becomes less available, muscle fibers will catabolize fatty acids and amino acids if necessary.

The final step in this process is the transfer of electrons to a molecule of oxygen, which is why this type of metabolism is also called *aerobic catabolism*. In fact, oxidative catabolism, which takes place in nearly all cells, is the main reason we need oxygen for survival. To meet this requirement, muscle fibers must have a plentiful supply of oxygen during activity. Much of this oxygen diffuses into the cytosol from the bloodstream. The rest is bound to the oxygen-binding protein **myoglobin** (MY-oh-glohb-in), which is similar to hemoglobin, the oxygen-carrying protein in the blood. Myoglobin, which is found in the cytosol, binds to oxygen that has diffused into the muscle fiber from the extracellular fluid, and releases it as the available oxygen is depleted by mitochondria performing oxidative catabolism.

Oxidative catabolism can provide ATP for hours, as long as oxygen and fuels are available. After only 1 minute of skeletal muscle activity, oxidative catabolism is the predominant source of ATP for almost all muscle fibers. After several minutes, nearly 100% of the ATP is produced by this aerobic process. This is why exercise that lasts continuously for at least several minutes is often called "aerobics."

(This module provides only an overview of the three main sources of ATP for skeletal muscle fibers. You'll find full coverage of this topic in the metabolism chapter—see Chapter 23.)

Quick Check

☐ 5. How do oxidative catabolism and glycolytic catabolism differ?
☐ 6. How long can oxidative catabolism fuel muscle activity?
☐ 7. Why is oxidative catabolism referred to as "aerobic"?

Apply What You Learned

☐ 1. Would extra creatine phosphate likely enhance performance of activities lasting several minutes to an hour? Explain.
☐ 2. Explain why engaging in 30–60 minutes of "aerobics" is promoted as an activity that "burns fat." Would you get this same "fat burning" from several 20-second sprints? Explain your answer in biochemical terms.

See answers in Appendix A.

Muscle Tension at the Fiber Level

Learning Outcomes

1. Describe the stages of a twitch contraction, and explain how a twitch is affected by the frequency of stimulation.
2. Relate tension production to the length of a sarcomere.
3. Compare and contrast the anatomical and metabolic characteristics of type I and type II muscle fibers.

As we discussed earlier, a basic principle of muscle physiology is that muscle tissue contracts to produce tension. The smallest muscle contraction, known as a **muscle twitch,** is the response of a muscle fiber to a single action potential in a motor neuron. Twitches are laboratory phenomena—they don't occur in whole muscles in the body. In spite of this fact, twitches are useful in experiments to study different types of contractions.

Every twitch produces some degree of tension, which translates to some degree of force. Twitches can create vastly different amounts of tension, both within an individual muscle fiber and within an entire muscle. In this module, we explore the specifics of the production and strength of a twitch contraction, as well as the relationship between the length of a sarcomere and the amount of tension that can be produced. We then examine the different classes of muscle fibers and the relative speeds with which they proceed through a twitch contraction. Finally, we conclude by considering how tension is transmitted from the muscle fibers to a whole muscle.

Twitch Contraction

A muscle twitch is produced in a laboratory by direct electrical stimulation of a muscle fiber. An instrument records the twitch and generates a recording known as a **myogram** (**Figure 10.22**). Notice in this myogram that the muscle twitch has three distinct phases—the latent, contraction, and relaxation periods:

1. **Latent period.** The latent period is the 1- to 2-ms (millisecond) time that it takes for the action potential to spread through the sarcolemma. It begins with the start of the action potential, and by the end, the action potential has spread past the T-tubules and triggered the release of calcium ions from the terminal cisternae of the SR. These ions then bind to troponin, and tropomyosin moves away from the active sites of actin. The myofibril is now ready to enter a crossbridge cycle. Note that the sarcolemma completes the repolarization phase of the action potential at the end of this period.

2. **Contraction period.** This period is marked by a rapid increase in tension as crossbridge cycles occur repeatedly.

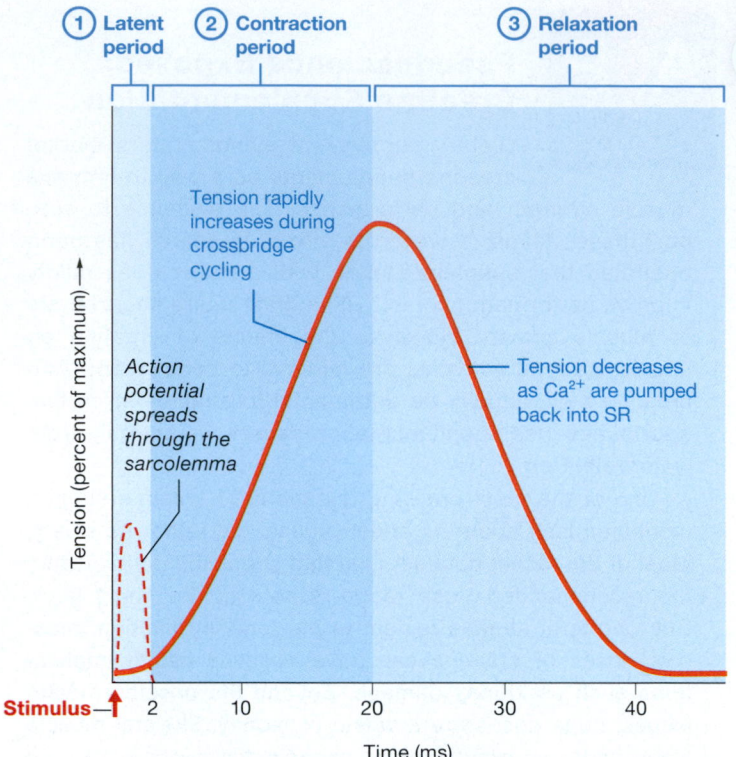

① **Latent period**   ② **Contraction period**   ③ **Relaxation period**

Tension rapidly increases during crossbridge cycling

Action potential spreads through the sarcolemma

Tension decreases as Ca^{2+} are pumped back into SR

Tension (percent of maximum) →

Stimulus— 2   10   20   30   40

Time (ms)

Figure 10.22 Myogram of a twitch contraction.

The amount of tension produced during the contraction phase, and the duration of this phase, depend on the type of muscle fiber, which we discuss shortly.

3. **Relaxation period.** During the relaxation period, tension decreases due to the decreasing calcium ion concentration in the cytosol. It takes the pumps in the SR from 10 to 100 ms to pump calcium ions from the cytosol back into the SR. Then, as tropomyosin once again blocks the active sites of actin, the muscle fiber relaxes.

Between the start of the latent period and the start of the contraction period, there is an interval of about 5 ms during which the muscle fiber cannot respond to another stimulus. This period, called the *refractory period,* is a property of all excitable cells. Cardiac muscle and smooth muscle have refractory periods as long as their contractions, so the cells must fully relax before they can contract a second time. However, skeletal muscle fibers have a much shorter refractory period, allowing them to maintain a sustained contraction phase (more on this shortly).

Figure 10.22 shows an average duration for contraction and relaxation periods. However, these times vary greatly among skeletal muscle fibers, and each may last from about 7 to 100 ms, depending largely on the type of muscle fiber involved and the level of activity of the ATPase on the myosin proteins during a crossbridge cycle. The tension produced during a twitch also varies considerably with several factors, including the timing and frequency of stimulation, the length of the fiber at rest, and the type of muscle fiber. These factors are examined in the next sections.

☐ 1. What is a twitch contraction?

☐ 2. What are the phases of a twitch contraction?

Tension Production and the Timing and Frequency of Stimulation

In general, repeated stimulation of a muscle fiber by a motor neuron results in twitches with progressively greater tension. This happens because the pumps in the SR membranes have inadequate time to pump all of the released calcium ions back into the SR before the fiber is restimulated, and so the concentration of calcium ions in the cytosol increases with each stimulus.

This phenomenon is known as **wave summation,** so named because the waves of contraction add together. The amount of tension produced during wave summation depends on the frequency of stimulation by the motor neuron. Depending on the number of times per second the muscle fiber is stimulated, we end up with one of two states, unfused or fused *tetanus* (**Figure 10.23**):

- **Unfused tetanus.** If the fiber is stimulated about 50 times per second, it can only partially relax between contractions. A sustained contraction called **unfused (or incomplete) tetanus** results (Figure 10.23a). During unfused tetanus, the tension pulsates, decreasing slightly and then increasing a bit more with each successive twitch until a level of maximal tension is reached.
- **Fused tetanus.** If the fiber is stimulated at a higher rate of 80–100 times per second, the muscle fiber does not have time to relax between stimuli because the calcium ion concentration in the cytosol remains high. The availability of calcium ions allows more and more crossbridges to form, contributing to the increase in tension. This increase results in a condition called **fused (or complete) tetanus,** in which the tension remains constant at a maximal level (Figure 10.23b). Note that fused tetanus is possible only because of the extremely short refractory period of the skeletal muscle fiber. This period is so short that the fiber does not have to relax between stimuli, and so can respond to a new stimulus before the last twitch is over, producing a sustained contraction.

Both unfused and fused tetanus generate more tension than a single twitch contraction. Tension is highest during fused tetanus, as the muscle fiber produces a constant tension several times greater than that created during a single twitch.

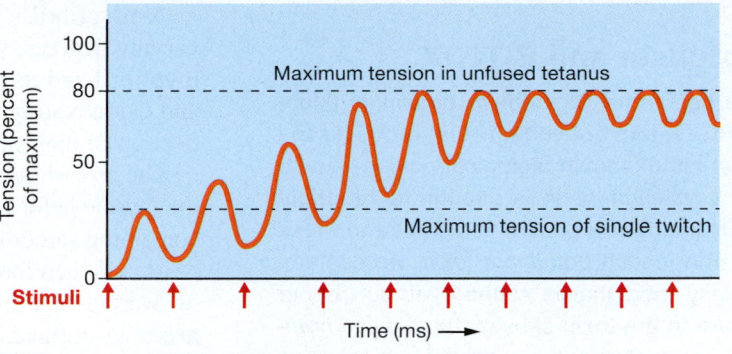

(a) Wave summation: unfused tetanus

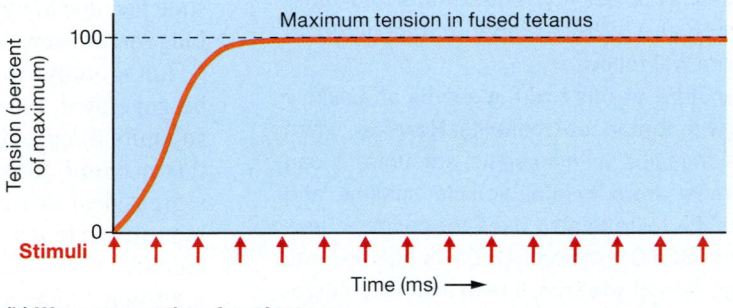

(b) Wave summation: fused tetanus

Figure 10.23 Wave summation: unfused and fused tetanus.

Take care not to confuse the states of fused and unfused tetanus with the disease tetanus. The disease is caused by the bacterium *Clostridium tetani* (TET-an-ee), which releases a toxin that enters the nervous system and excessively stimulates skeletal muscles. The result is simultaneous contraction of muscle groups, known as a *tetanic spasm.* The disease tetanus is almost completely preventable with vaccination, but it remains a serious health threat for unvaccinated individuals, particularly in countries where people have little access to healthcare.

A related bacterium, *Clostridium botulinum,* produces a toxin that can actually be used to relieve muscle spasms. Read more about this bacterium in *A&P in the Real World: Botulism and BOTOX* on p. 366.

☐ 3. How does the timing of a stimulus impact the tension produced by the resulting muscle contraction?

☐ 4. How do fused and unfused tetanus differ?

The Length-Tension Relationship

The second factor that determines the amount of tension produced by a twitch contraction is the number of crossbridges that can form within each sarcomere of the muscle fiber. The number of crossbridges depends on the length of the sarcomere prior to contraction, a principle known as the **length-tension relationship.** The *optimal length* of the sarcomere is the length of the muscle fiber at which the most crossbridges can form, allowing the

Botulism and BOTOX

The bacterium *Clostridium botulinum* (kloh-STRID-ee-um bot-yoo-LIN-um) produces the most lethal known biological poison—as little as one gram of crystalline toxin is enough to kill about one million adults. Humans are generally exposed to these bacteria and their toxin, called *botulinum toxin,* via contaminated food, especially food that is canned without proper sterilization. Exposure to the toxin causes the disease *botulism* (BOT-yoo-lizm), in which botulinum toxin binds to motor neurons of the NMJ and blocks the release of acetylcholine from synaptic vesicles. This activity renders affected muscles unable to contract, and without proper treatment, death from respiratory failure will follow.

Given the deadly nature of this toxin, it seems an unlikely candidate for a drug with therapeutic benefits. However, when injected into certain muscles in minuscule amounts, it can provide temporary relief from painful muscle spasms and migraine headaches. The toxin is also used for cosmetic purposes under the name BOTOX. When BOTOX is injected into the muscles under the skin of the face, it relaxes them, temporarily reducing the appearance of fine lines and wrinkles.

A myofibril's elastic filaments enable it to be stretched to varying degrees, which results in different possible lengths for a myofibril and its sarcomeres. In Figure 10.24, sarcomeres A, B, and C are resting at different lengths, as shown by the degree of overlap of their thin and thick filaments, or the **zone of overlap.**

The size of the zone of overlap is largely due to the position of the muscle before contraction. Figure 10.24 shows that the muscle containing sarcomere A is shortened by flexion of the hand at the wrist, and therefore all the sarcomeres are shorter, with large zones of overlap. The muscle that sarcomere B is part of, however, is nearer its optimal length, as the hand is in a neutral, extended position. As a result, all the sarcomeres are longer and have smaller zones of overlap. Finally, the muscle containing sarcomere C is stretched due to hyperextension of the hand at the wrist, leading to long sarcomeres with very small zones of overlap.

Let's apply this concept to muscle tension: For tension to be generated, the myosin heads must be able to grab the actin subunits to pull them toward the center of the sarcomere. With this in mind, look at sarcomere A. Although sarcomere A has a great deal of overlap, this is actually an obstacle to producing muscle tension, because there is little room for the thin filaments to be pulled together before they get to the M line and the thick filaments run into the Z-discs. As you can see from the graph line in Figure 10.24, this results in decreased tension production. To demonstrate this yourself, flex your hand at the wrist and try to grip your pen tightly in your fist. Notice how difficult it is to maintain a tight grip—this is due to the limited amount of tension that the shortened sarcomeres in your forearm can generate.

fiber to generate almost 100% of the tension that is possible to produce (**Figure 10.24**).

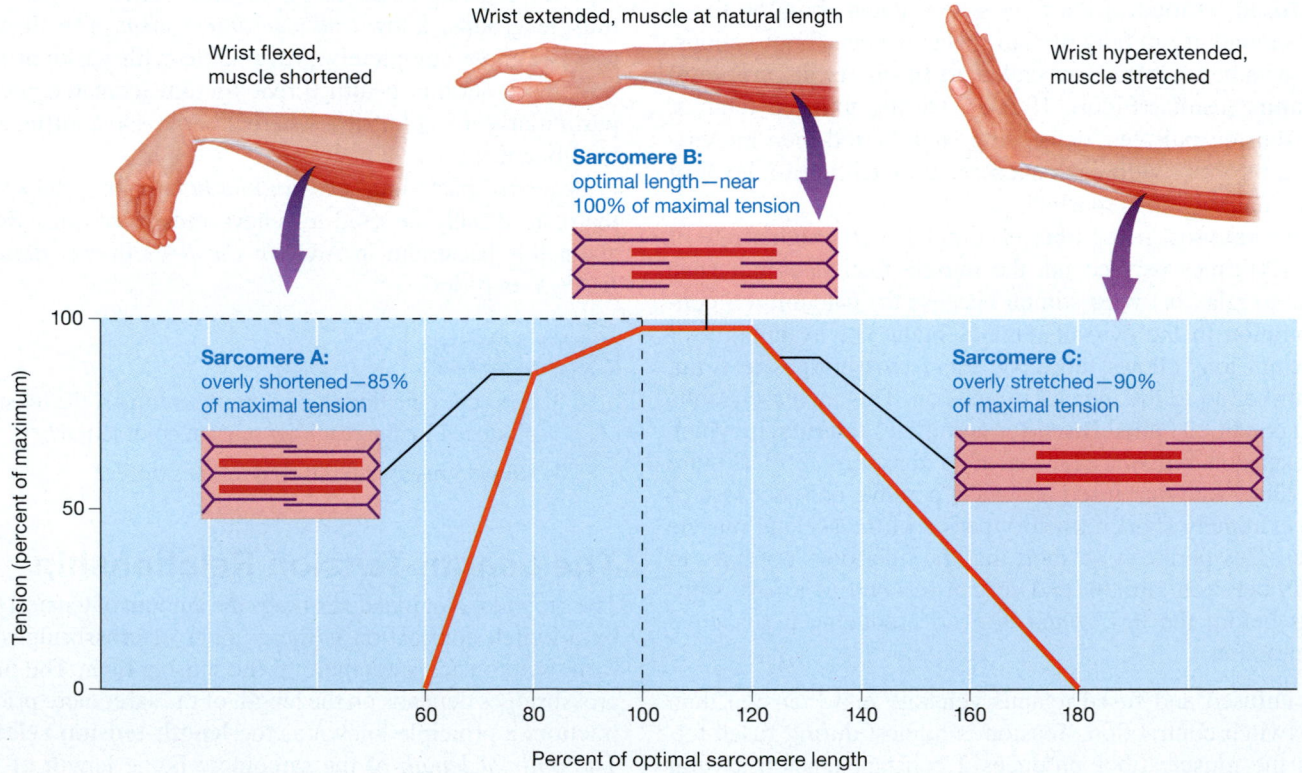

Figure 10.24 The length-tension relationship.

Now consider sarcomere C. You may think that sarcomere C could produce more tension than sarcomere A because it has a small zone of overlap, but actually the two sarcomeres produce about the same amount of tension. Why? Because the zone of overlap in sarcomere C is so short that the myosin heads are unable to get much of a "grip" on the actin subunits, so less tension results from this contraction. Try the experiment with your pen again, but this time hyperextend your hand at the wrist. The results are about the same—your forearm muscles are still unable to generate enough tension to grip the pen very tightly.

From our discussion so far, it follows that less tension will be produced in a sarcomere that is either overly shortened or overly stretched at rest. Maximal tension production, therefore, occurs in a narrow range of resting sarcomere lengths, near the optimal length of the sarcomere. We find this situation with sarcomere B—it has enough room to shorten and plenty of actin subunits for the myosin heads to grab. The graph line in Figure 10.24 shows that sarcomere B produces close to 100% of the maximal tension of the muscle fiber, whereas sarcomeres A and C are able to produce only about 85% and 90% of this tension, respectively.

Quick Check

☐ 5. At what length will a sarcomere be able to generate the greatest amount of tension? Why?

Classes of Skeletal Muscle Fibers

Up to this point in the chapter, we've been describing the twitch contraction of a generic skeletal muscle fiber. However, skeletal muscle fibers are neither structurally nor functionally identical, and so some proceed through a twitch contraction more rapidly than others. How fast this process happens depends mostly on the level of myosin ATPase activity in the fiber, which may be "fast" or "slow." Recall that myosin ATPase catalyzes the hydrolysis of ATP so that a power stroke can occur. Muscle fibers with high myosin ATPase activity are called **fast-twitch fibers** because they proceed rapidly through crossbridge cycles. Fast-twitch fibers are found in muscles that must move body parts rapidly, such as those that move the eyeballs. Fibers with low myosin ATPase activity are called **slow-twitch fibers** because they proceed through crossbridge cycles more slowly. Slow-twitch fibers are found in muscles that require slow, sustained contractions, such as the postural muscles of the back.

The speed of a twitch contraction is combined with its predominant energy source (oxidative versus glycolytic catabolism) to give us two main classes of skeletal muscle fibers, types I and II:

- **Type I fibers** are slow-twitch fibers that are small to intermediate in diameter. Type I fibers contract more slowly and less forcefully than other fibers, but they can maintain extended periods of contraction. This ability requires the continual generation of large quantities of ATP via oxidative catabolism; for this reason, type I fibers are also called *slow oxidative* fibers. To support oxidative catabolism, these fibers have large quantities of myoglobin, many mitochondria, and a well-developed blood supply. The high myoglobin content of type I muscle fibers makes them red, so they are sometimes known as "red muscle."

- **Type II fibers** are fast-twitch fibers that are often larger in diameter and contract more rapidly than type I fibers, but they are quickly fatigued. Type II fibers rely more heavily on glycolytic energy sources, and they have less myoglobin, fewer mitochondria, and a less extensive blood supply than type I fibers. Due to their low myoglobin content, type II fibers are lighter in color than type I fibers, and so these muscle fibers are sometimes called "white muscle." (This visible difference between white muscle and red muscle explains the "white meat" and "dark meat" of chicken.) Type II fibers have two subtypes: types IIa (also known as *fast oxidative glycolytic,* or FOG) and IIx (*fast glycolytic,* or FG). These subtypes have progressively faster, stronger twitches and rely increasingly on glycolytic energy sources, with type IIx fibers having extremely fast, powerful twitches and using glycolytic catabolism almost exclusively.

Most muscles contain all fiber classes (**Figure 10.25**), each of which is stimulated under different conditions. To tie it together, imagine this: You are a baseball player at a game and you are sitting in the dugout, using primarily type I fibers in your back and abdominal muscles to remain sitting upright. When it's your turn at bat, you get up and jog to the plate, using primarily type IIa fibers in your leg muscles. When the pitch comes toward you, you swing your bat and hit the ball, using type IIx fibers in

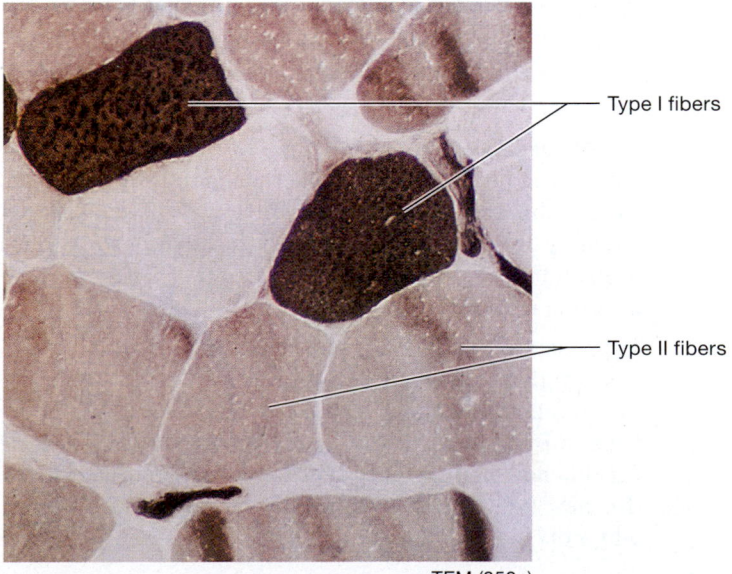

Type I fibers

Type II fibers

TEM (850×)

Figure 10.25 Comparison of differently stained type I and type II muscle fibers.

Table 10.1 Three Classes of Skeletal Muscle Fibers

Class	I	IIa	IIx
Fiber Property	Slow Oxidative	Fast Oxidative Glycolytic	Fast Glycolytic
Primary type of catabolism	Oxidative	Oxidative and glycolytic	Glycolytic
Blood supply	Extensive	Less extensive	Limited
Number of mitochondria	Many	Intermediate	Few
Amount of myoglobin	Large	Intermediate	Little
Amount of glycogen	Little	Intermediate	Large
Myosin ATPase activity	Low	High	Highest
Fatigability	Low	Intermediate	High
Diameter of fiber	Small to intermediate	Large	Intermediate*
Color of muscle	Red	Light red	Light pink to white
Example of activities	Standing, sitting	Walking, writing	Heavy lifting, sprinting

*Although other fibers are usually reported to be the largest fiber type, these data are for non-human animals. Studies have found that type IIa fibers are the largest fiber type in nontrained human skeletal muscle.

your arm muscles. The classes of muscle fibers are summarized in **Table 10.1**.

Quick Check

☐ 6. How do type I and type II muscle fibers differ?

Apply What You Learned

☐ 1. When you lift a heavy box, your muscles need to generate a stronger than average contraction. What will happen to the timing and frequency of nerve stimulation of your muscles, and why?

☐ 2. Your friend sits with her forearm flexed 45° at the elbow. You place a weight in her hand, and she notes that she has difficulty lifting it further. However, when she attempts to lift the weight with her forearm in a relaxed or neutral position, she has little trouble. Explain these differing results.

☐ 3. Suppose a new type of muscle fiber is discovered that has a high concentration of myoglobin, large numbers of mitochondria, and a well-developed blood supply. What do you think would be the primary energy source for this new muscle fiber type? Explain your answer. Is this new type likely to be resistant to fatigue? Why or why not?

See answers in Appendix A.

MODULE 10.7

Muscle Tension at the Organ Level

Learning Outcomes

1. Describe the structure and function of a motor unit.
2. Explain how muscle tone is produced.
3. Compare and contrast the three types of contractions.

We've already examined the production of tension via a twitch contraction by a single muscle fiber. However, in a living organism, we don't find twitches of single muscle fibers. Rather, a twitch is produced by multiple muscle fibers that are innervated by the same motor neuron, a structure known as a *motor unit*. In this module, we explore the structure of a motor unit and how motor unit activation controls tension of the entire muscle. We then look at three types of skeletal muscle contractions and the amount of tension that each contraction can generate.

Motor Units

As a motor neuron approaches a muscle, its axon branches out and innervates multiple muscle fibers. A single motor neuron along with the muscle fibers it innervates is called a

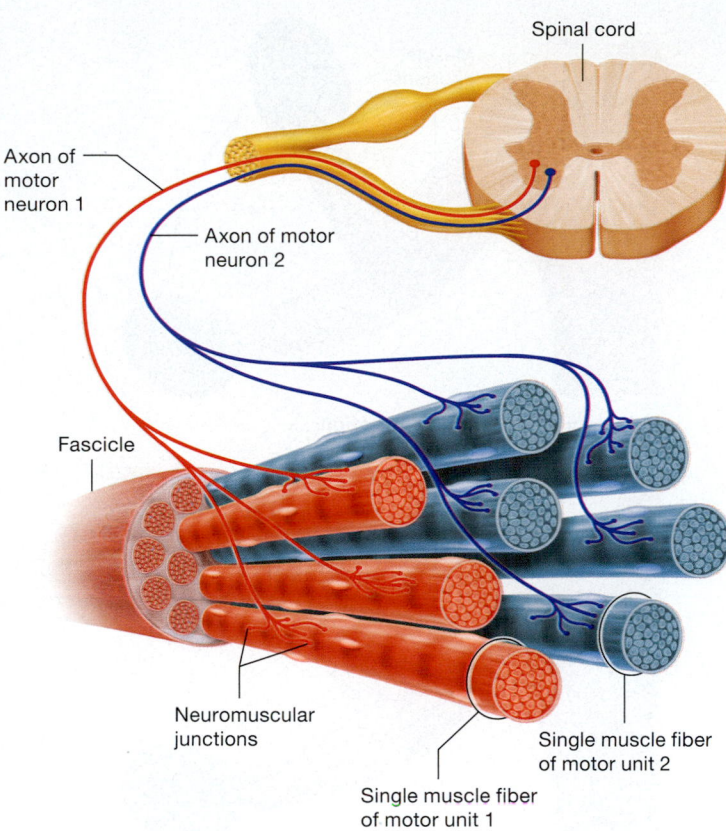

Spinal cord

Axon of motor neuron 1

Axon of motor neuron 2

Fascicle

Neuromuscular junctions

Single muscle fiber of motor unit 1

Single muscle fiber of motor unit 2

Figure 10.26 The motor unit. Two motor units (colored red and blue) are shown here.

motor unit (**Figure 10.26**). When the motor neuron fires an action potential, all of the muscle fibers within its motor unit respond and produce about the same amount of tension. Note that this applies only to a motor unit, not to an entire muscle.

An average motor unit consists of about 150 muscle fibers, but this number can vary widely with the degree of motor control needed for the muscle— another example of the Structure-Function Core Principle (p. 28). Muscles that require fine control,  **CORE PRINCIPLE** Structure-Function such as those around the larynx (voice box) or in the fingers, will have multiple small motor units, often containing as few as 10 muscle fibers per motor unit. This gives the nervous system more precise control over the amount and rate of tension produced. Conversely, large, powerful muscles, such as the postural muscles of the back and the gastrocnemius in the calf, can have 2000–3000 muscle fibers in each motor unit.

Motor units contain one class of muscle fiber, either type I or type II. Those with type I fibers are called **slow motor units,** and those with type II fibers are called **fast motor units.**

Now let's see how motor units function in the body.

Recruitment

We know that an increased frequency in the stimulation of a muscle fiber affects muscle tension. However, when thinking about the tension generated by whole muscles, we need to consider another factor: the number of motor units that are stimulated. Initiation of a contraction activates a small number of motor units. As greater force is needed, more motor units are activated, a phenomenon called **recruitment.** Slow motor units are typically activated first, followed by fast motor units if additional tension is needed. For example, if you were lifting this textbook off your desk, you would likely use your slow motor units for the most part. However, if you were to lift a stack of textbooks from the floor, your nervous system would first activate the slow motor units and then the fast motor units.

Muscle Tone

If you reach down right now and squeeze your calf muscle, you will probably discover that it feels somewhat taut despite being relaxed. Even when a muscle is at rest, it still has some degree of tension. This small amount of tension produces what is known as **muscle tone.** This tone is due to the involuntary activation of motor units by the brain and spinal cord. The nervous system alternates which motor units it activates, so that some can rest while others contract. Although the contractions that produce muscle tone are not strong enough to cause movement, they do serve important functions, including maintaining erect posture, stabilizing joints, generating heat, and ensuring that the muscle is ready to respond if movement is initiated.

If the tone in a skeletal muscle is abnormally low, generally because of a nervous system disorder, a condition known as **hypotonia** results. A hypotonic muscle looks flattened and feels soft and loose rather than firm. Often, the joints around the affected muscles are hyperextended, as the muscles are no longer able to generate enough tension to stabilize them. Alternately, a muscle may have abnormally high tone, a condition called **hypertonia.** A hypertonic muscle often feels rock hard, and it may shorten so much that it causes painful joint contractures (see Chapter 8). Normal hypertonia occurs during *shivering,* which takes place in response to a drop in body temperature, as more motor units are activated to generate additional heat.

Quick Check

☐ 1. What is a motor unit?

☐ 2. Explain the process of recruitment.

☐ 3. How is muscle tone produced?

Types of Muscle Contractions

Besides the number of motor units that are activated, another factor influencing the amount of tension a muscle generates involves how the length of the muscle changes during contraction. Although you may equate muscle contraction with muscle shortening, a muscle shortens only in a specific type of contraction. Some muscle contractions result in no change in fiber length, and others actually occur during muscle fiber *lengthening.*

On the basis of muscle length during a contraction, we can classify contractions into three types: *isotonic concentric, isotonic eccentric,* and *isometric contractions.* An isotonic contraction (*iso-* = "same," *tono-* = "tension") is one that produces enough tension to initially move a load, such as a weight, and then maintains that same level of tension throughout the contraction. During an isotonic contraction, the tension stays the same, but the muscle changes length. An isometric contraction (*-metr* = "measure"), however, results when the external load is equal to the force generated by the muscle. During an isometric contraction, the muscle remains the same length, but the tension changes throughout the contraction.

Let's look more closely at each type of contraction (**Figure 10.27**):

- **Isotonic concentric contractions.** You are probably familiar with the type of muscle contraction in which the muscle shortens; this is known as an **isotonic concentric contraction,** or a *miometric contraction* (my-oh-MET-rik; *mio-* = "shorter"). During an isotonic concentric contraction, the force generated by the muscle is greater than that of the external load. Let's consider the example of holding a 10-lb weight in your hand and lifting it with your forearm (see Figure 10.27a). As you begin to pick up the weight (which is the external load), the motor units in your forearm flexors generate force to flex your forearm at the elbow. When they generate force of more than 10 lb, the muscle shortens, and an isotonic concentric contraction results.
- **Isotonic eccentric contractions.** The second type of contraction, an **isotonic eccentric contraction** or a *pliometric contraction* (ply-oh-MET-rik; *plio-* = "longer"), results in muscle lengthening. It occurs when the force generated by the muscle is less than that of the external load. As a result, the muscle lengthens as tension is produced. A muscle is able to lengthen while it's contracting because the elastic filaments in its myofibrils allow it to stretch considerably. Consider the example of the 10-lb weight that you have just lifted. As you slowly set the weight down (see Figure 10.27b), the force produced by your muscle becomes less than the load of the weight. However, your motor units are still generating tension, even though the sarcomeres are stretching and lengthening.
- **Isometric contractions.** The third type of contraction, in which the length of the muscle doesn't change, is the **isometric contraction** (aye-soh-MET-rik). This time, lift the 10-lb weight by abducting your arm at the shoulder rather than flexing your forearm (see Figure 10.27c). The initial lifting involved an isotonic concentric contraction; however, now your arm is stationary in an abducted position and your

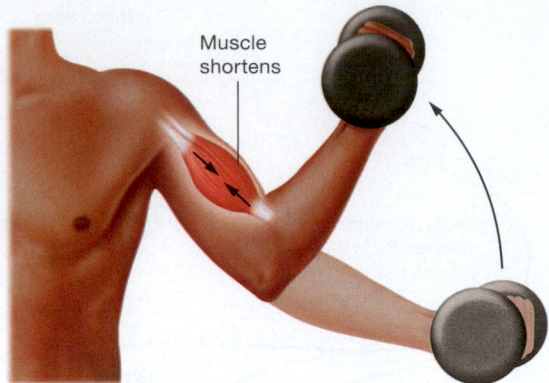

(a) Isotonic concentric contraction

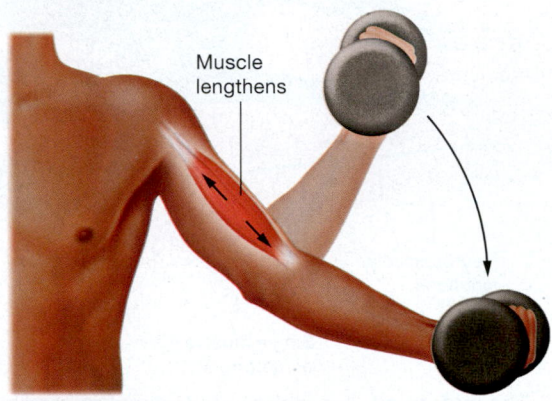

(b) Isotonic eccentric contraction

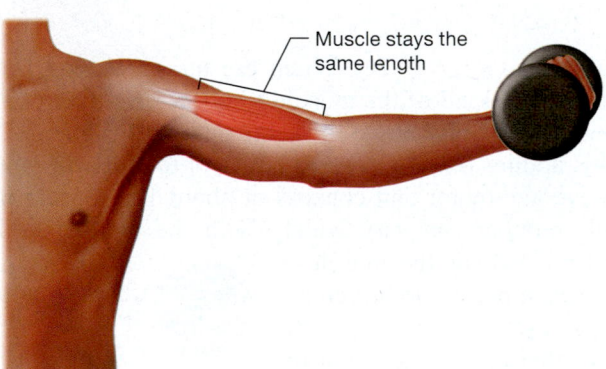

(c) Isometric contraction

Figure 10.27 The three types of muscle contractions.

forearm is stationary in a supinated position. So, the muscle is neither shortening nor lengthening, even though it is still producing 10 lb of force to hold the weight in the air.

These types of contractions can be induced in the laboratory in individual muscle fibers. But as we discussed earlier, in the body these contractions occur at the organ level through activation of motor units.

Of the three types of contractions, isotonic eccentric contractions require the greatest amount of tension, and therefore produce the greatest amount of force. As a result, activities that require a large number of isotonic eccentric contractions (e.g., running downhill) have a greater tendency to produce exercise-induced injuries than do those using more isotonic concentric or isometric contractions (e.g., running on a flat surface). To understand a common type of muscle injury, see *A&P in the Real World: Delayed-Onset Muscle Soreness.*

A&P in the Real World

Delayed-Onset Muscle Soreness

Whether we've tried a new workout at the gym or had an overly ambitious day of snowboarding, all of us have probably experienced muscle soreness at one time or another. For years physiologists thought that this phenomenon, called **delayed-onset muscle soreness** (or **DOMS**), was due to the lactic acid produced during glycolysis. However, research studies within the past 15 years have shown that lactic acid accumulation is not to blame. Instead, studies have suggested another culprit—minor structural damage, in particular that caused by isotonic eccentric muscle contractions. The force produced by these contractions can result in minor damage to the myofilaments, particularly in muscles not trained for specific activities. Recent data indicate that the resulting pain may also be due to remodeling of the myofibrils, which is an adaptive response to the stress placed on the muscle.

Oddly, the most consistently effective treatment for DOMS is more exercise. Other treatment modalities, including massage, topical therapies, acupuncture, and oral medications, have shown little benefit. The effectiveness of exercise in treating DOMS, however, is temporary. Once the exercise has ceased, the pain returns until the muscle is sufficiently conditioned through training.

Quick Check

☐ 4. How do isotonic concentric, isotonic eccentric, and isometric contractions differ?

Apply What You Learned

☐ 1. Using biochemical terms, explain why lifting a stack of textbooks over and over again would likely make you feel fatigued more quickly than repeatedly lifting a single textbook. (*Hint:* What types of motor units are activated when more tension is needed?)

☐ 2. A fitness trainer advises her client to lower a weight slowly and in a controlled manner to reduce the risk of injury. Which type of muscle contraction is being used when the weight is lowered? Explain why the trainer advises her client to train in this manner.

See answers in Appendix A.

Skeletal Muscle Performance

Learning Outcomes

1. Describe the effects of physical conditioning on skeletal muscle tissue, and compare endurance and resistance training.

2. Explain the factors that contribute to muscular fatigue.

3. Summarize the events that occur during the recovery period.

Exercise has both short-term and long-term effects on the structure and performance of skeletal muscle tissue. For example, in the short term the body's metabolic rate increases during exercise as the skeletal muscle fibers try to keep up with the demand for ATP. In the process, waste products such as heat and carbon dioxide are generated, which the body must deal with in order to recover. Over the long term, consistent exercise yields far more than just waste products—it results in changes to the architecture of the muscles themselves. In this module we examine how the body responds to exercise, factors that contribute to muscular fatigue, and how the body recovers from exercise.

Changes Caused by Physical Training

Physical training refers to the repetitive use of skeletal muscles that leads to changes in the structure and biochemistry of the muscle fibers. The changes that occur in the muscle are governed by the **principle of myoplasticity,** which is just another way to say  **CORE PRINCIPLE** Structure-Function

that a muscle will alter its structure to follow its function—an example of the Structure-Function Core Principle (p. 28). For example, when a person engages in little physical activity and is "out of shape," the muscle fibers will change structurally and biochemically to meet this lesser demand. Conversely, when someone physically trains to "get in shape," his or her muscle fibers change again to meet this increased demand.

Note that the great majority of mature skeletal muscle fiber nuclei are *amitotic,* meaning that they generally do not undergo mitosis under normal conditions. Therefore, these changes are within the muscle fibers themselves, not in the number of muscle fibers. (A small population of unspecialized *satellite cells,* however, does retain the ability to divide. These cells can help repair injured skeletal muscle.) The precise structural and biochemical changes depend on the type of training chosen—endurance training and resistance training are the two basic types.

Endurance Training

Endurance training is defined as training with a large increase in the frequency of motor unit activation and a moderate

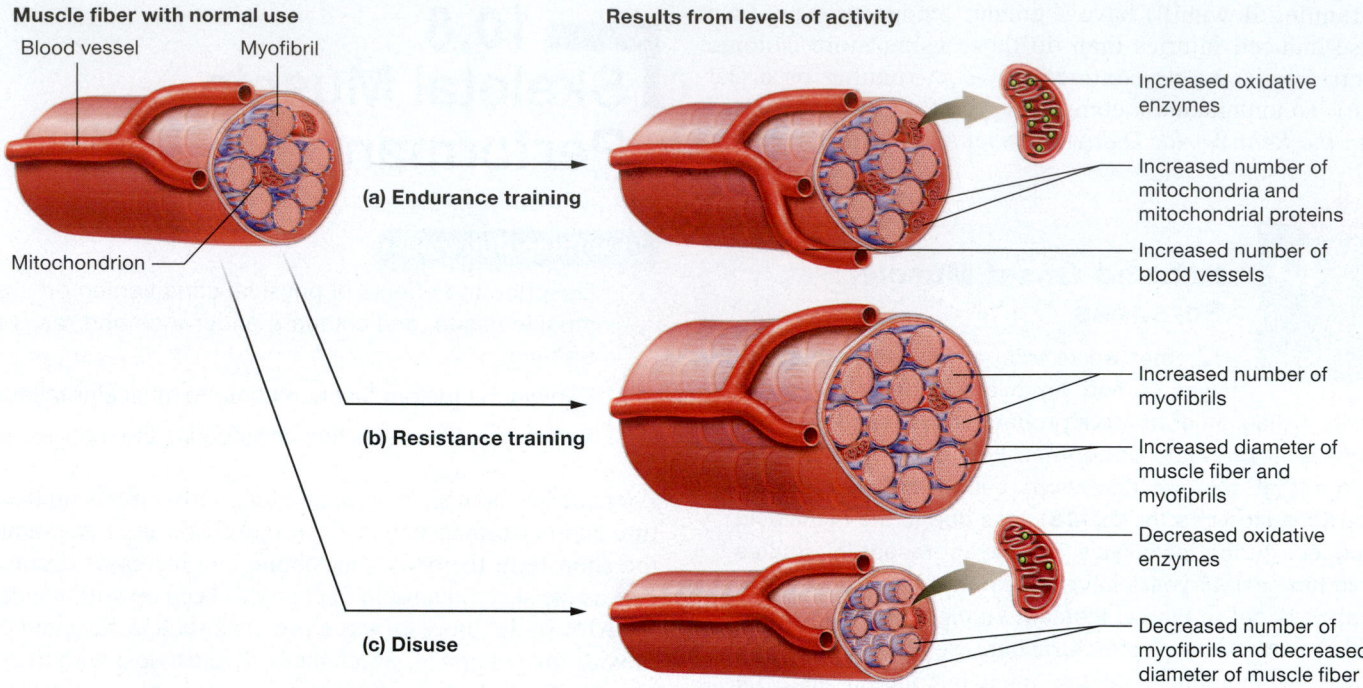

Muscle fiber with normal use

Blood vessel   Myofibril

Mitochondrion

Results from levels of activity

(a) **Endurance training**

Increased oxidative enzymes

Increased number of mitochondria and mitochondrial proteins

Increased number of blood vessels

(b) **Resistance training**

Increased number of myofibrils

Increased diameter of muscle fiber and myofibrils

(c) **Disuse**

Decreased oxidative enzymes

Decreased number of myofibrils and decreased diameter of muscle fiber

Figure 10.28 Adaptive changes of muscle fibers due to training and disuse.

increase in force—in other words, more repetitions with lighter weight. Endurance training involves activities such as cycling, jogging, cross-country skiing, and distance swimming. **Figure 10.28a** shows that after endurance training the muscle fiber has increased amounts of oxidative enzymes, more mitochondria and mitochondrial proteins, and a greater number of blood vessels. These adaptations enhance the oxidative capabilities of the muscle dramatically, resulting in a better ability to use fatty acids and other fuels to make ATP, and so an increased resistance to fatigue. These changes occur even in type II fibers, but are more dramatic in type I fibers.

Resistance Training

Resistance training, also called *strength training,* features a moderate increase in the frequency of motor unit activation and a large increase in force production—in other words, fewer repetitions with heavier weight. Resistance training typically involves the use of either free weights or a resistance-exercise machine. Whereas many of the changes due to endurance training are primarily biochemical, several changes that result from resistance training are in large part anatomical. In **Figure 10.28b** you can see that both the number of myofibrils and the diameter of the muscle fibers increase, a change called **hypertrophy** (hy-PER-troh-fee). With hypertrophy comes a *decreased* proportion of mitochondrial proteins and a lower blood supply. However, this decrease is a function of the fiber enlarging rather than actually losing mitochondria or blood vessels. As you might expect, resistance training can decrease the capacity for endurance. For this reason, a balanced program combining both resistance and endurance training is recommended for most people.

Disuse

Regardless of the type of training, the old adage "Use it or lose it" holds true. In response to physical inactivity, the diameter of the muscle fiber decreases due to loss of myofibrils (**Figure 10.28c**), a condition called **atrophy** (AEH-troh-fee). In addition, the amount of oxidative enzymes decreases, and the fiber has a lower capacity for oxidative catabolism. The result is a decline in both strength and endurance. Muscle atrophy is a particular problem for people who are bedridden or have lost the use of their limbs.

Quick Check

☐ 1. What changes in the muscle fiber accompany endurance training, resistance training, and disuse?

Muscular Fatigue

If you were to pick up a 10-lb weight and hold it out in front of you in an isometric contraction, your muscles would probably begin to feel weak after about a minute (or sooner, for some people). This weakness is called **muscular fatigue,** defined as the inability to maintain a given level of intensity of a particular exercise. Though the idea of fatigue seems straightforward, it actually results from a combination of several different factors, which may include the following:

- **Depletion of key metabolites.** Certain metabolites, such as creatine phosphate, glycogen, and blood glucose, are depleted by intense activity, rendering the muscle fiber less able to replenish the ATP consumed by the activity.

- **Decreased availability of oxygen to muscle fibers.** Exercise increases the oxygen requirement of muscle fibers because of the need for more ATP. The amount of oxygen bound to myoglobin may also be depleted by intense exercise, and the amount of oxygen taken in by the lungs may be inadequate to replace it. The muscle fiber must then rely more heavily on the less efficient process of glycolysis to generate ATP. Keep in mind that we are referring only to the amount of oxygen available to muscle fibers relative to what they need; the actual amount of oxygen in the body does not decrease significantly.
- **Accumulation of certain chemicals.** Many chemicals that may contribute to fatigue are produced by muscle fibers. For example, calcium ions accumulate in the mitochondria, where they interfere with the mitochondrial ability to carry out oxidative catabolism. Also, as ATP is split, phosphate ions and ADP accumulate in the cytosol and interfere with excitation-contraction coupling.
- **Environmental conditions.** Severe environmental conditions, particularly extreme heat, disrupt the body's homeostasis, leading to more rapid muscular fatigue. Sweating in response to excessive heat may cause electrolyte disturbances that can interfere with the production of action potentials in a muscle fiber.

Many other causes of muscular fatigue exist, some of which don't involve the muscular system directly. For example, psychological factors play a role in fatigue. The mechanism behind psychological input is poorly understood, but some athletes seem able to minimize distressing stimuli such as pain or breathlessness and push their physical performance to the limits, whereas others are sidelined by these stimuli.

As you can see, the many potential causes of muscular fatigue make a precise cause difficult to pin down. Further complicating the scenario is the fact that all the body's physiological functions are interrelated (as you've seen many times in this textbook already). Thus, a decline of something as simple as one enzyme system may unleash a cascade of events that hampers the ability of the entire organism to maintain maximal output during exercise.

Quick Check

☐ 2. How is muscular fatigue defined?

☐ 3. What factors influence muscular fatigue?

Excess Postexercise Oxygen Consumption and the Recovery Period

During physical activity, whether or not you reach the fatigued state, you've likely noticed that your rate and depth of breathing, or *ventilation*, increase. You've also probably observed that these changes persist for a time, even after you've finished exercising. Have you ever wondered why? The fact is that your body needs time to return to the pre-exercise state. This time is called the **recovery period,** and the persisting increased rate of ventilation is referred to as **excess postexercise oxygen consumption (EPOC).** EPOC was formerly called the *oxygen debt* or *oxygen deficit,* but these terms are misnomers, as there is no debt or deficit of oxygen that must be "repaid."

For many years, physiologists thought that EPOC was due to an accumulation of lactic acid in skeletal muscle fibers and in the bloodstream. However, as discussed earlier, current evidence indicates that lactic acid is oxidized by the mitochondria to generate ATP. So physiologists now think that EPOC results from the responses necessary to correct the disturbances to homeostasis that were brought on by exercise. To return to homeostasis, the body must accomplish several goals, which include dissipating heat, restoring ion concentrations, and correcting blood pH. Let's see why each of these is important:

- **Heat dissipation.** Energy is lost as heat during both glycolytic and oxidative catabolism. Exercise increases the rate of catabolic processes, so it also raises the amount of heat generated. The body must lose this heat and return to the pre-exercise temperature level to maintain temperature homeostasis.
- **Restoration of intracellular and extracellular ion concentrations.** During muscle contractions, the ATP-consuming pumps in the sarcolemma and SR must work harder to maintain the normal concentrations of calcium and sodium ions in the cytosol and of potassium ions in the extracellular fluid.
- **Correction of blood pH.** Certain products of metabolism, including lactic acid and carbon dioxide, can cause the pH of the blood to decrease (which means acidity increases) from its normal value of 7.35–7.45. The body must return the pH of the blood to the pre-exercise level in order to maintain homeostasis.

So, how does increasing the rate and depth of ventilation after exercise help to accomplish these goals? Most obvious is that a relationship exists between the rate of ventilation and the level of carbon dioxide in the blood: As the rate and depth of ventilation increase, more carbon dioxide is exhaled, helping to return the pH of the body's fluids to normal. However, the relationship may be less obvious for temperature and electrolyte homeostasis. To make it easier to understand, consider facts you already know—the Na^+/K^+ pump restores ion gradients, pumps move calcium ions back into the sarcoplasmic reticulum, and sweating produces cooling. These three mechanisms have one thing in common: They all consume ATP. Since oxidative catabolism—the most efficient way to generate ATP—requires oxygen, it follows that the body's cells will need "extra" oxygen after exercise.

Quick Check

☐ 4. What conditions does excess postexercise oxygen consumption help to correct?

Apply What You Learned

☐ 1. Predict the appearance of the muscle fibers of an athlete who trains to run marathons.

☐ 2. You have finished exercising with a friend, who says that he is breathing rapidly because he must have "used up all of his oxygen." What do you tell him?

See answers in Appendix A.

10.9
Smooth and Cardiac Muscle

Learning Outcomes

1. Describe the structure, location in the body, and functions of smooth and cardiac muscle tissue.
2. Describe the contraction process of smooth muscle fibers, and contrast it with skeletal muscle fiber contraction.

Although smooth and cardiac muscle tissues share many structural and functional properties with skeletal muscle tissue, they also have notable differences. One such difference is that both smooth and cardiac muscle tissues are involuntary. In this module we examine these differences in greater detail. (We emphasize smooth muscle here, as the structure and function of cardiac muscle tissue are covered in detail in the heart chapter—see Chapter 17.)

Smooth Muscle

◀◀ FLASHBACK

1. What are intermediate filaments? (p. 96)
2. What are the primary structural and functional differences between smooth muscle tissue and skeletal muscle tissue? (p. 148)
3. What is the function of a gap junction? (p. 127)

Smooth muscle tissue is widely distributed in the body. Much of it is found lining hollow organs, but it is also present elsewhere, such as in the arrector pili muscles in the dermis and the iris of the eye. Some functions of smooth muscle include the following:

- **Peristalsis.** In many organs, such as the small intestine, the smooth muscle tissue is arranged into two differently oriented layers of fibers, the circular and longitudinal layers (**Figure 10.29a**). These two layers alternately contract and relax, producing rhythmic waves called **peristalsis** (pehr-ih-STAL-sis; "constricting around"). Peristalsis propels materials through hollow organs of the digestive, urinary, and reproductive systems.
- **Formation of sphincters.** In both the digestive and urinary systems, smooth muscle forms rings called **sphincters** (SFINGK-terz). Sphincters are usually contracted but relax periodically to allow substances to pass through them.
- **Regulation of flow.** Smooth muscle controls the flow of materials through certain hollow organs by changing the diameter of the passages. For example, the smooth muscle

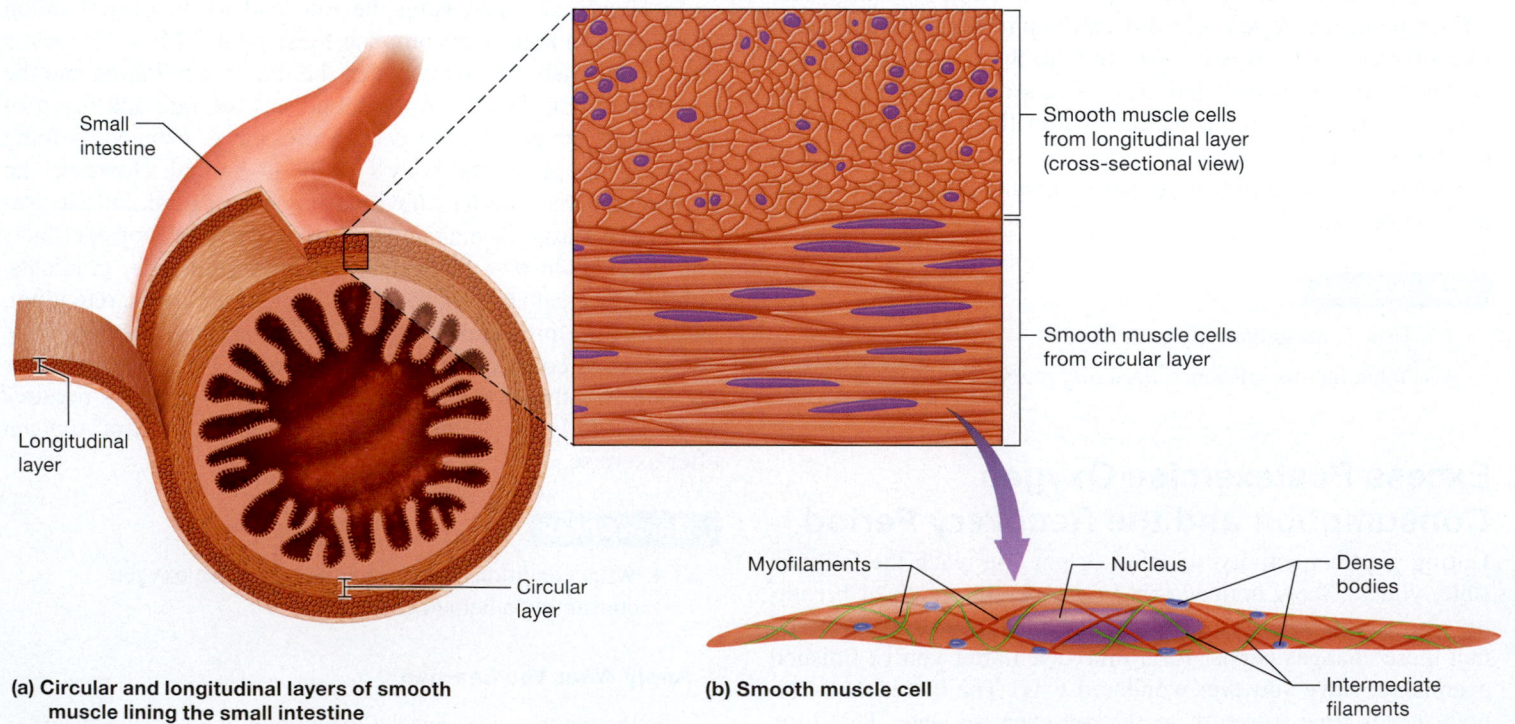

(a) Circular and longitudinal layers of smooth muscle lining the small intestine

(b) Smooth muscle cell

Figure 10.29 Structure of smooth muscle tissue and cells.

found in the walls of all but the smallest blood vessels controls both blood pressure and blood flow to organs and tissues (see Chapter 18). In addition, the smooth muscle in the walls of the respiratory tract regulates air flow to the smaller airway passages (see Chapter 21).

The next section discusses the structure and contractile mechanism of smooth muscle cells. As you read, note the significant overlap between smooth and skeletal muscle physiology.

Structure of Smooth Muscle Cells

Smooth muscle cells share the same contractile proteins found in skeletal muscle fibers: myosin and actin. However, the arrangement of these contractile proteins differs greatly from that in skeletal muscle. To begin with, smooth muscle cells lack striations and sarcomeres, which gives them their characteristic "smooth" appearance (**Figure 10.29b**). In addition, the actin filaments are arranged obliquely in the sarcoplasm and are anchored to proteins called **dense bodies.** Some dense bodies are associated with the sarcolemma, where they attach to the dense bodies of other smooth muscle cells and transmit tension from one cell to the next. Other dense bodies are found in the sarcoplasm, where they are bound to scaffold-like intermediate filaments that connect the dense bodies to one another. Several thin filaments radiate from each dense body, and these thin filaments surround one thick filament. The ratio of thin to thick filaments is higher in smooth muscle than in skeletal muscle, with more thin filaments for every thick filament.

Differences are also found in the structure and composition of thin and thick filaments in smooth muscle cells. Both types of filaments are longer than those in skeletal muscle fibers, and the thin filaments lack troponin. You can see in **Figure 10.30a** that another difference is found in the myosin heads, which have opposite-facing hinges—the heads on one side hinge in one

direction, whereas those on the other side hinge in the opposite direction. Notice also that a thick filament contains myosin heads along its entire length.

Beyond contractile proteins, there are three other important structural differences between skeletal muscle fibers and smooth muscle cells:

- Smooth muscle cells lack motor end plates.
- The sarcoplasmic reticulum is much less extensive.
- T-tubules are absent.

The importance of these differences will become apparent in the next subsection.

Smooth Muscle Contraction and Relaxation

Smooth muscle and skeletal muscle share some structural features and also have some of the same functional features. For example, the thick and thin filaments slide past one another, calcium ions are required for a contraction to occur, and ATP is consumed in the process. However, just as we find significant structural variations between smooth and skeletal muscle, we also find significant functional variations. It is these structural differences that allow smooth muscle to carry out specialized functions.

Earlier in the chapter, you read that only stimulation by a motor neuron can excite a skeletal muscle fiber. However, many different stimuli may elicit a smooth muscle cell contraction, including mechanical (stretch), hormonal, and nervous stimuli. In addition, some smooth muscle cells are able to depolarize and contract spontaneously. These cells, known as **pacemaker cells,** have unstable membrane potentials that cause them to spontaneously depolarize in a rhythmic fashion. Their depolarizations cause the firing of an action potential that spreads through the surrounding muscle cells and initiates a wave of contraction. Pacemaker cells of this type are responsible for the waves of contraction that move through the stomach and intestines.

The triggering event for a smooth muscle contraction is the same as that for skeletal muscle—calcium ions flooding the cytosol. However, calcium ions come from two sources in a smooth muscle cell: They are released from the SR and they enter from the extracellular fluid. Calcium ions also play a different role in smooth muscle contraction because these cells lack troponin. Instead, when calcium ions flood the cytosol, the following series of events occurs:

1. Calcium ions bind a protein in the cytosol called **calmodulin** (kal-MOD-yoo-lin; **Cam**).
2. The calcium ion–Cam complex activates an enzyme associated with myosin called **myosin light-chain kinase (MLCK).**
3. MLCK causes the activation of myosin ATPase.
4. Crossbridge cycles then ensue.

As the crossbridge cycles repeat, the myosin heads pull actin along the length of the thick filament, and the muscle cell changes shape from long and thin to fat and globular, which you can see in **Figure 10.30b**. Smooth muscle cells can contract up to

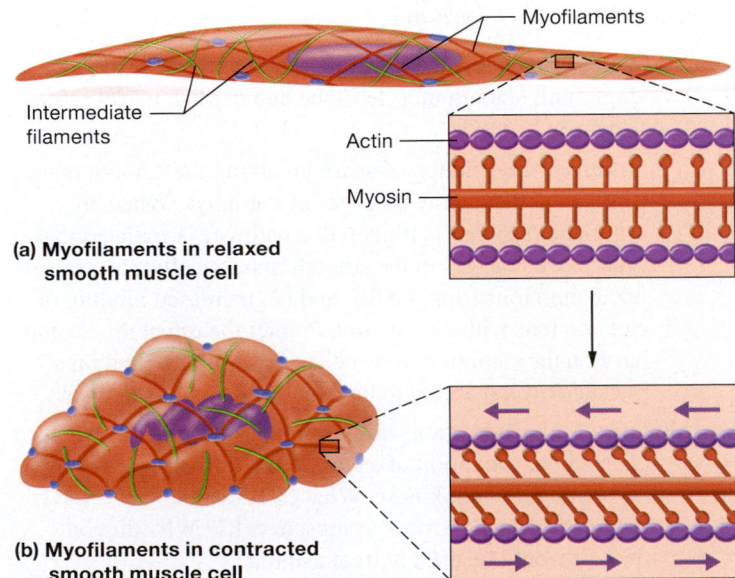

(a) Myofilaments in relaxed smooth muscle cell

Myofilaments

Intermediate filaments

Actin

Myosin

(b) Myofilaments in contracted smooth muscle cell

Figure 10.30 Contraction of smooth muscle cells.

80% of their resting length (in contrast, skeletal muscle fibers can contract a maximum of 30–40% of their resting length).

Although smooth muscle is often functionally required to maintain sustained contractions, it consumes surprisingly little ATP. Most smooth muscle cell contractions consume as little as $\frac{1}{100}$ of the ATP used by skeletal muscle fibers, a fact that physiologists think is due to low myosin ATPase activity. In spite of this low energy consumption, smooth muscle cells are still able to generate significant force, because the percentage of time that the crossbridges are attached is much greater than in skeletal muscle fibers. Anyone who has gone through childbirth has felt the strength of smooth muscle contractions.

Relaxation of smooth muscle begins with removal of calcium ions from the cytosol. As this concentration declines, calcium ions dissociate from calmodulin, and MLCK is inactivated. At this point, the muscle cell either relaxes or enters the *latch state.* During the latch state, the muscle cell maintains tension while consuming very little ATP. The latch state is very important in terms of the ability of smooth muscle to maintain a state of energy-efficient sustained contraction, as is found in sphincters, which must stay contracted to remain closed.

Types of Smooth Muscle

Smooth muscle tissue varies in structure and organization with each organ of the body in which it is found. However, from all this variation, two general types emerge:

- **Single-unit** (or **unitary**) **smooth muscle,** also called **visceral smooth muscle** (VISS-er-ul), is the predominant type of smooth muscle in the body and is found in nearly all hollow organs, including the uterus. Single-unit smooth muscle consists of hundreds to thousands of muscle cells whose plasma membranes are linked electrically via gap junctions. Action potentials spread rapidly through the cells via the gap junctions, causing the muscle cells to contract in a coordinated wave as a single unit.
- **Multi-unit smooth muscle** is the second, and rarer, of the two types. It is found in such locations as the muscles in the eye and the arrector pili muscles in the dermis. Multi-unit smooth muscle consists of individual muscle cells whose plasma membranes are not joined by gap junctions. This characteristic allows each cell to contract independently of the others, permitting precise control of contractions. As with skeletal muscle, the amount of tension produced by multi-unit smooth muscle varies with the number of muscle cells activated.

Whereas multi-unit smooth muscle responds primarily to nerve stimulation, single-unit smooth muscle is able to respond to multiple types of stimulation. Keep in mind, however, that not all single-unit smooth muscle is the same. Though it shares the property of contracting as a single unit, its structure and function vary in different organs of the body.

Quick Check

☐ 1. What are some functions of smooth muscle tissue?

☐ 2. How does contraction differ in smooth muscle cells and skeletal muscle fibers?

☐ 3. How are single-unit and multi-unit smooth muscle different?

Cardiac Muscle

Cardiac muscle cells more closely resemble skeletal muscle fibers than smooth muscle cells, as we saw in Figure 10.1. Like skeletal muscle fibers, cardiac muscle cells are striated and consist of sarcomeres. They also have both T-tubules and extensive networks of sarcoplasmic reticulum. However, notable structural differences between cardiac muscle cells and skeletal muscle fibers do exist. Cardiac muscle cells are typically shorter, branched cells with one or two nuclei and abundant myoglobin. Mitochondria account for about 30% of their cytoplasmic volume. In addition, cardiac muscle cells are connected by **intercalated discs** that contain desmosomes and gap junctions (see Chapter 4). These discs join cardiac muscle cells to one another physically and electrically, which permits the heart to contract as a unit.

Unlike skeletal muscle fibers, cardiac muscle cells do not require stimulation from the nervous system to generate action potentials, because their electrical activity is coordinated by a small population of pacemaker cells. Like those in smooth muscle, these cells rhythmically and spontaneously generate action potentials that trigger the remaining cardiac muscle cells to have action potentials as well. This quality allows cardiac muscle tissue to be **autorhythmic**—it sets its own rhythm. In this way, cardiac muscle tissue is similar to single-unit smooth muscle in that all the cells depolarize and contract with every contraction cycle.

Quick Check

☐ 4. What are the basic structural features of cardiac muscle cells?

Apply What You Learned

☐ 1. Would you expect to find motor units and recruitment in single-unit smooth muscle tissue and cardiac muscle tissue? Why or why not?

☐ 2. Asthma is a respiratory disease involving the smooth muscle cells lining the airway passages of the lungs. When an asthmatic response is triggered, a pathway is initiated that causes two changes in the smooth muscle cells: (1) a release of calcium ions from the SR, and (2) increased binding of calcium ions with calmodulin. Predict the effect this would have on the smooth muscle cells of the airways and on a person's overall ability to breathe. Explain your answer.

☐ 3. One of the medications used to treat asthma works by causing the formation of a compound that inactivates myosin light-chain kinase. What effect would this medication have on smooth muscle cells? Why do you think it would be used to treat asthma?

See Answers in Appendix A.

Chapter Summary

For everything you need to succeed in this course, go to **Mastering** Anatomy & Physiology. There you will find:

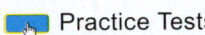 Practice Tests

Author Podcasts

Big Picture Animations

Concept Boost Video Tutors

iP2 Interactive Physiology 2.0 Tutorials

A&PFlix A&P Flix 3D Animations

Active-Learning Workbook

MODULE 10.1
Overview of Muscle Tissue 337–340

- The three types of muscle tissue are skeletal, cardiac, and smooth muscle tissue.
- All muscle cells turn chemical energy into mechanical energy by contracting to generate **muscle tension.**
- All types of muscle tissue consist of **muscle cells** and the surrounding extracellular matrix, called the **endomysium.**
- All types of muscle tissue share common properties, including *contractility,* **excitability, conductivity, distensibility,** and **elasticity.**
- The plasma membrane of a muscle cell is the **sarcolemma,** and the cytoplasm is the **sarcoplasm.** The sarcoplasm contains **myofibrils** and the **sarcoplasmic reticulum (SR).**

MODULE 10.2
Structure and Function of Skeletal Muscle Fibers 340–347

- Skeletal muscle fibers are very long cells that consist of striated, multinucleated voluntary muscle cells.
- The sarcolemma has inward extensions called **T-tubules** that surround myofibrils. The Ca^{2+}-storing SR swells where it meets T-tubules to form **terminal cisternae.**
- **Myofibrils** are composed of **myofilaments.** Myofilaments are composed of **contractile proteins, regulatory proteins,** and/or **structural proteins.**
- There are three types of myofilaments:
  - **Thick filaments** are composed of **myosin,** a contractile protein.

  - **Thin filaments** are composed of contractile **actin** proteins and the smaller regulatory proteins **troponin** and **tropomyosin.**
  - **Elastic filaments** are composed of the structural protein **titin.**
- The **striations** of skeletal muscle tissue are due to the arrangement of myofilaments:
  - The **I bands** are the light regions of striations where only thin filaments are found. In the middle of each I band is the **Z-disc.**
  - The **A bands** are the dark regions of striations where the thick and thin filaments overlap. The central region of the A band is the **H zone.** It is bisected by the **M line.**
- The **sarcomere** is the functional unit of contraction. It is defined as the area from one Z-disc to the next.
- The **sliding-filament mechanism** is the currently accepted model of muscle contraction. In this process, the thick and thin filaments slide past one another.

MODULE 10.3
Skeletal Muscle Fibers as Electrically Excitable Cells 347–352

- A separation of charges occurs across the sarcolemma, which is an electrical gradient or **electrical potential. Membrane potentials** are electrical potentials across plasma membranes.
- An unstimulated muscle fiber shows a decrease in voltage across the membrane, called the **resting membrane potential.** Muscle fibers at rest have a resting membrane potential of about –90 mV.
- Ions move across the sarcolemma via channels. Two types of channels are **leak** and **gated channels.**
- The concentration of Na^+ is higher in the extracellular fluid, and the concentration of K^+ is higher in the cytosol. The **Na^+/K^+ pump** maintains this gradient.
- The negative resting membrane potential is due to the loss of K^+ through leak channels and the actions of the Na^+/K^+ pumps.
- Ion movement is driven by the forces of the concentration gradient and the electrical gradient. The sum of these two forces is the **electrochemical gradient.**
- An **action potential** is a temporary, quick reversal in the membrane potential. It consists of: (1) **depolarization,** in which Na^+ enter the fiber, causing the membrane potential to become more positive; and (2) **repolarization,** in which K^+ exit the fiber, returning it to its resting state.

MODULE 10.4
The Process of Skeletal Muscle Contraction and Relaxation 352–360

- All skeletal muscle fibers are innervated by **motor neurons.** The neuron meets the fiber at the **neuromuscular junction,** or **NMJ.**

The NMJ consists of the **axon terminal,** the **synaptic cleft,** and the **motor end plate.**

- Skeletal muscle contraction can be divided into three parts: excitation, excitation-contraction coupling, and contraction.

- The **excitation phase** begins as the axon terminal releases ACh into the synaptic cleft, which binds ACh receptors on the motor end plate. The result is an **end-plate potential.**

- In **excitation-contraction coupling,** the end-plate potential triggers an action potential in the sarcolemma, which propagates down the T-tubules. This triggers the opening of Ca^{2+} channels in the SR, and Ca^{2+} flood the cytosol.

- In preparation for contraction, the Ca^{2+} in the cytosol bind troponin, which allows tropomyosin to move away from the active sites of actin.

- The **contraction phase** then begins, and the **crossbridge cycle** occurs as ATP hydrolysis "cocks" the myosin head and it binds to actin. When myosin undergoes **a power stroke,** it pulls actin toward the center of the sarcomere.

- **Muscle relaxation** has two components: (1) The ACh in the synaptic cleft is broken down, and (2) the Ca^{2+} concentration in the cytosol returns to the resting level.

MODULE 10.5
Energy Sources for Skeletal Muscle 360–364

- Immediate energy sources for muscle fibers include stored ATP within the cytosol and **creatine phosphate.**

- **Glycolytic catabolism** is an anaerobic process during which glucose in the cytosol is split and ATP is produced.

- **Oxidative catabolism** is an aerobic process that takes place in the mitochondria, during which the products of glycolysis, fatty acids, and amino acids are oxidized to generate ATP.

MODULE 10.6
Muscle Tension at the Fiber Level 364–368

- A single contraction-relaxation cycle of a muscle fiber is called a **twitch.** It consists of the **latent period,** the **contraction period,** and the **relaxation period.** Muscle fibers may be classified as **fast-twitch** or **slow-twitch** fibers.

- The amount of tension generated by the contraction of a muscle fiber depends on the frequency of stimulation by the motor neuron and the resulting concentration of Ca^{2+}.
  - If a muscle fiber is stimulated before the relaxation period is completed, **unfused tetanus** results.
  - If a muscle fiber is stimulated 80–100 times per second before the relaxation period begins, **fused tetanus** results.

- Per the **length-tension relationship,** a contraction can produce a maximal amount of tension at the optimal length of a sarcomere.

- **Type I** muscle fibers are slow-twitch fibers that use primarily oxidative catabolism. **Type II** muscle fibers are fast-twitch fibers that use primarily glycolytic catabolism.

MODULE 10.7
Muscle Tension at the Organ Level 368–371

- A **motor unit** consists of a single motor neuron and the muscle fibers that it innervates.
  - **Recruitment** occurs when more motor units are activated for more forceful contractions.
  - **Muscle tone** is produced by small, involuntary contractions of alternating motor units.

- There are three types of muscle contractions: an **isotonic concentric** (shortening) **contraction,** an **isotonic eccentric** (lengthening) **contraction,** and an **isometric** (unchanging length) **contraction.**

MODULE 10.8
Skeletal Muscle Performance 371–374

- **Endurance** and **resistance** training result in changes in the structure and biochemistry of skeletal muscle fibers.

- Disuse leads to **atrophy** of the muscle fibers.

- **Muscular fatigue** is caused by depletion of key metabolites, inadequacy of oxygen delivery to muscle fibers, accumulation of certain metabolites, and environmental conditions.

- **Excess postexercise oxygen consumption (EPOC)** occurs after exercise to correct the homeostatic imbalances that were caused by exercise.

MODULE 10.9
Smooth and Cardiac Muscle 374–376

- The functions of **smooth muscle** tissue include peristalsis, forming sphincters, and regulating the flow of material through hollow organs.

- Smooth muscle cells are uninucleate; lack T-tubules, striations, and sarcomeres; and have a less extensive SR than do skeletal muscle fibers.

- Smooth muscle contraction is stimulated by the autonomic nervous system, stretch, hormones, and the activity of pacemaker cells.

- Smooth muscle cell contraction is activated by Ca^{2+} binding to **calmodulin.** This activates **myosin light-chain kinase (MLCK),** which initiates crossbridge cycles.

- Smooth muscle cell relaxation involves removal of Ca^{2+} from the cytosol.

- There are two types of smooth muscle tissue:
  - **Single-unit smooth muscle** cells contract together as a single unit.
  - **Multi-unit smooth muscle** cells that can contract independently of one another.

- Cells of **cardiac muscle** are joined physically and electrically by **intercalated discs.** Their electrical activity is coordinated by **pacemaker cells.**

Assess What You Learned

Scan the QR Code for additional practice test questions  https://goo.gl/GJyrs6

LEVEL 1 Check Your Recall

1. Mark the following statements as true or false. If a statement is false, correct it to make a true statement.

    a. A property of all muscle cells is elasticity, which means that the tissue is able to stretch.

    b. The common function of all types of muscle tissue is to generate tension.

    c. The plasma membrane of a muscle cell is called the sarcoplasmic reticulum.

    d. Muscle cells are contractile, conductive, distensible cells.

2. How does a skeletal muscle fiber differ structurally from typical cells?

3. Thick filaments are composed of the protein

    a. myosin.

    b. actin.

    c. troponin.

    d. tropomyosin.

    e. Choices b, c, and d are correct.

4. Match the following terms with the correct definition.

    _____ Z-disc
    _____ Sarcomere
    _____ A band
    _____ H zone
    _____ I band
    _____ M line

    a. The dark band containing the entire length of the thick filament
    b. The band of proteins in the middle of the H zone
    c. The boundary between sarcomeres
    d. The functional unit of contraction
    e. The middle region of the A band containing only thick filaments
    f. The light band containing only thin filaments

5. What is the basic mechanism of contraction at the level of myofilaments?

6. Mark the following statements as true or false. If a statement is false, correct it to make a true statement.

    a. The resting membrane potential refers to the voltage difference across the membranes of excitable cells at rest.

    b. The concentration of Na^+ is highest in the cytosol, and the concentration of K^+ is highest in the extracellular fluid.

    c. The Na^+/K^+ pumps and gated channels maintain the Na^+ and K^+ gradients necessary for action potentials to occur.

    d. A depolarization is a change in membrane potential that makes the potential less negative.

7. Describe the three components of the neuromuscular junction.

8. Order the following events of excitation and excitation-contraction coupling. Put 1 by the first event, 2 by the second, and so on.

    _____ The motor end plate generates an end-plate potential.
    _____ The action potential spreads along the T-tubules, SR Ca^{2+} channels are pulled open, and Ca^{2+} flood the cytosol.
    _____ Acetylcholine binds to receptors on the motor end plate, and ligand-gated ion channels open.
    _____ Ca^{2+} bind troponin, which allows tropomyosin to move away from the actin active site, initiating a contraction cycle.
    _____ The action potential propagates through the sarcolemma and dives deeply into the cell along the T-tubules.

9. Which of the following statements accurately describes the role of ATP in a muscle contraction?

    a. ATP is directly responsible for the power stroke.

    b. ATP moves troponin and tropomyosin away from actin.

    c. ATP breaks the actin/myosin attachment and "cocks" the myosin head.

    d. ATP causes the myofilaments to shorten.

10. A muscle fiber relaxes when:

    a. the concentration of Ca^{2+} in the cytosol returns to resting levels.

    b. the supply of ATP is exhausted.

    c. Ca^{2+} flood the cytosol.

    d. acetylcholine is released from the axon terminal and the sarcolemma depolarizes.

11. Which of the following energy sources would provide the majority of the ATP for a person running a 26-mile marathon?

    a. Stored ATP

    b. Glycolytic catabolism

    c. Oxidative catabolism

    d. Creatine phosphate

12. Mark the following statements as true or false. If a statement is false, correct it to make a true statement.

    a. Muscle fibers generate more tension if the starting length of their sarcomeres is very short.

    b. Stimulation by a motor neuron before a muscle fiber has fully relaxed results in a condition called wave summation.

    c. Muscles that require a great deal of precise control will have large motor units.

    d. A muscle fiber changes length during isotonic concentric and isotonic eccentric contractions.

13. Which of the following types of muscle fibers have low myosin ATPase activity and are classified as slow-twitch?

    a. Type I fibers

    b. Type IIa fibers

    c. Type IIx fibers

    d. Type III fibers

14. Muscle tone is:

    a. the result of voluntary shortening of the muscle.

    b. the result of a small amount of involuntary activation of motor units by the nervous system.

    c. abnormal—a person's muscles should be relaxed normally.

    d. present only where there is damage to the nervous system.

15. Fill in the blanks: Resistance-type activities will likely rely on _____ energy sources, whereas endurance activities will probably rely on _____ energy sources.

16. Which of the following is *not* likely to result from endurance training alone?

    a. Increase in oxidative enzymes in the muscle fiber
    b. Increased numbers of mitochondria
    c. Hypertrophy of the muscle fibers
    d. Increase in blood supply to the muscle fibers

17. Which of the following factors is/are responsible for muscular fatigue? (Circle all that apply.)

    a. Accumulation of chemicals, including calcium and phosphate ions
    b. Increased blood flow to the muscle
    c. Decreased availability of oxygen
    d. Psychological and environmental factors
    e. Depletion of key metabolic fuels, such as creatine phosphate

18. What is thought to cause excess postexercise oxygen consumption?

19. List some of the functions of smooth muscle tissue.

20. Which of the following best describes single-unit smooth muscle tissue?

    a. The fibers function individually.
    b. It is found in organs that require precise control of contraction.
    c. It contains gap junctions that couple the fibers electrically.
    d. The amount of tension produced varies with the number of muscle cells recruited.

21. Mark the following statements as true for smooth muscle tissue, cardiac muscle tissue, and/or skeletal muscle tissue.

    a. _____ Actin attaches to dense bodies.
    b. _____ Cells are joined by intercalated discs.
    c. _____ The thick and thin filaments are arranged into sarcomeres.
    d. _____ The thick filaments contain myosin heads along their entire length.
    e. _____ The cells depolarize and contract as a unit.
    f. _____ Ca^{2+} binding to troponin is the initiating event of contraction.
    g. _____ Ca^{2+} binding to calmodulin is the initiating event of contraction.
    h. _____ The sarcolemma has a distinct motor end plate.

LEVEL 2 Check Your Understanding

1. Would you expect to find larger motor units in the postural muscles of the back or the muscles of the hand? Explain your answer.

2. A hypothetical poison blocks K^+ leak channels. How would this affect the resting membrane potential of skeletal muscle fibers? Explain your reasoning.

3. The drug *neostigmine* blocks the activity of the enzyme acetylcholinesterase in the synaptic cleft. Predict the effects of this drug.

4. Explain why cardiac muscle cells and some smooth muscle cells will continue to contract even when their nerve supply has been removed.

LEVEL 3 Apply Your Knowledge

PART A: Application and Analysis

1. The poison *curare* (kyoo-RAH-ray) blocks the binding of acetylcholine to its receptors at the neuromuscular junction. What effects would you predict from such a poison? Can you think of any useful applications for it? Why might an overdose of it be lethal?

2. Some athletes will consume only protein for several days before a competition, which reduces the amount of glycogen in both the muscle fibers and the liver. What effect would this have on their ability to perform activities that require short, powerful bursts of activity? How would it affect their ability to perform endurance activities?

3. Ms. Sanchez was in a motorcycle accident in which she lost the use of her right upper limb muscles due to significant nerve damage. However, when an electrode is inserted into her muscles, they are able to contract. Explain specifically why nerve damage caused her to lose the use of her muscles. Why can they still respond to stimulation from an electrode?

4. Mr. Nasheed has cerebral palsy and suffers severe skeletal muscle spasms as a result of his condition. He is prescribed the drug *dantrolene,* which prevents the release of Ca^{2+} from the SR. Explain how this will treat his muscle spasms.

PART B: Make the Connection

5. Jesse is a 2-year-old boy who presents with difficulty in walking and poor control of movements. When the doctor examines Jesse, she notices that when his muscles contract, they are very slow to relax and remain contracted well after the movement has been performed. She sends a sample of his tissue for genetic analysis, and the lab reports a genetic defect that causes the pumps in the SR to operate much more slowly than normal. How does a defect in DNA lead to a malfunctioning protein? How does this finding explain Jesse's symptoms? (*Connects to Chapter 3*)

6. Paola is a 3-year-old girl with a disease that reduces the ability of her mitochondria to generate ATP. Explain the specific effects of this disease on the ability of Paola's muscles to function properly. What other tissues and organs are likely to be especially affected by her disease, and why? (*Connects to Chapter 3*)

See answers in Appendix A.

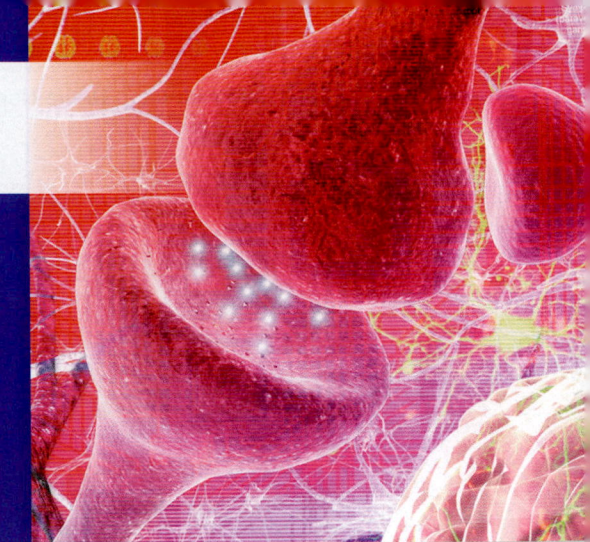

11

Introduction to the Nervous System and Nervous Tissue

You can't turn on the television or radio, much less go online, without seeing something to remind you of the nervous system. From advertisements for medications to treat depression and other psychiatric conditions to stories about people's battles with illegal drugs, information about the nervous system is everywhere in our popular culture. And there is good reason for this—the **nervous system** controls our perception and experience of the world. In addition, it directs voluntary movement, and is the seat of our consciousness, personality, and learning and memory. Along with the endocrine system, the nervous system regulates many aspects of homeostasis, including respiratory rate, blood pressure, body temperature, the sleep/wake cycle, and blood pH.

In this chapter we introduce the multitasking nervous system and its basic functions and divisions. We then examine the structure and physiology of the main tissue type of the nervous system: nervous tissue. As you read, notice that many of the principles you discovered in the muscle tissue chapter (see Chapter 10) apply here as well.

MODULE 11.1
Overview of the Nervous System

Learning Outcomes

1. Describe the major functions of the nervous system.
2. Describe the structures and basic functions of each organ of the central and peripheral nervous systems.
3. Explain the major differences between the two functional divisions of the peripheral nervous system.

In this module we introduce the organs of the nervous system and how they fit within anatomical and functional divisions. These organs and their classifications are covered in more detail in later chapters (see Chapters 12, 13, and 14).

Structural Divisions of the Nervous System

⟪ FLASHBACK

1. Define neuron, neuroglial cell, and axon. (p. 150)
2. Where is the foramen magnum located, and what is the main nervous system structure that passes through it? (p. 216)
3. What are vertebral foramina? (p. 232)

For practice applying concepts to a clinical scenario, check out the **Running Case Study** *for this chapter* @ Mastering *Anatomy & Physiology*

Computer-generated image: A synapse between two nerve cells is shown.

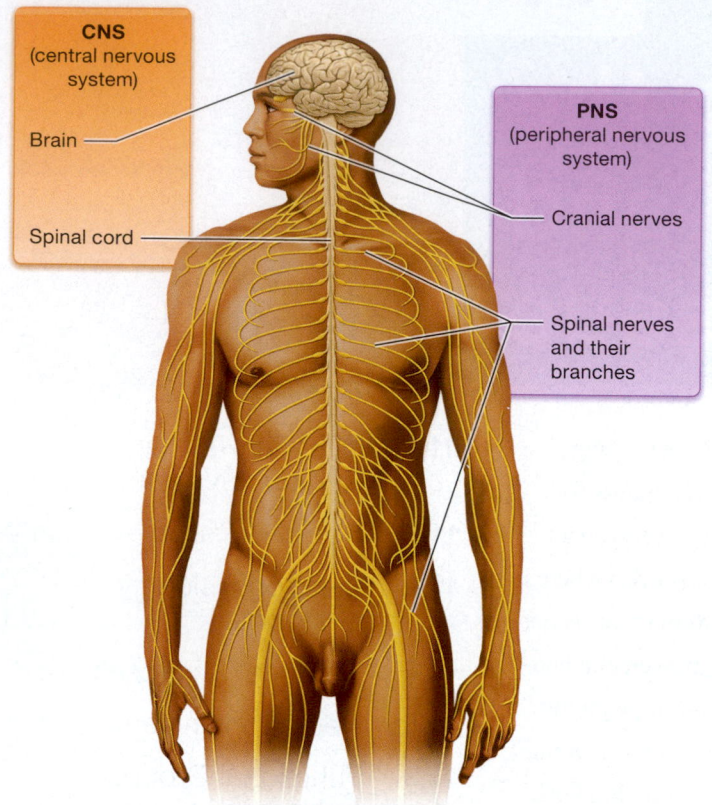

Figure 11.1 Structure of the nervous system.

The nervous system can be divided into two structural divisions: the **central nervous system (CNS)** and the **peripheral nervous system (PNS).** The CNS is made up of the *brain* and *spinal cord,* whereas *nerves* make up the PNS (**Figure 11.1**). Let's look at each of these divisions more closely.

The Central Nervous System

The organ of the central nervous system that is likely most familiar to you, yet still holds the greatest mysteries for physiologists, is the **brain.** Enclosed completely by the skull, the brain is composed primarily of nervous tissue. This remarkable organ consists of about 100 billion cells called *neurons* (NOOR-onz), or *nerve cells,* that enable everything from the regulation of breathing and the processing of algebra to performing in the creative arts. The cells that make up nervous tissue are discussed in Module 11.2.

At the foramen magnum, the brain merges with the other organ of the central nervous system: the **spinal cord.** The spinal cord passes through the vertebral foramen of the first cervical vertebra and continues inferiorly to the first or second lumbar vertebra (see Chapter 7). It contains fewer cells than the brain, with only about 100 million neurons. The spinal cord enables the brain to communicate with most parts of the body below the head and neck; it is also able to carry out certain functions on its own (which are discussed in later chapters).

The Peripheral Nervous System

The peripheral nervous system is made up of the most numerous organs of the nervous system, the **nerves,** which carry signals to and from the central nervous system. A nerve consists of a bundle of long neuron "arms" known as *axons* that are packaged together with blood vessels and surrounded by connective tissue sheaths. Nerves are classified according to their origin or destination: Those originating from or traveling to the brain are called *cranial nerves,* and those originating from or traveling to the spinal cord are called *spinal nerves* (see Figure 11.1). There are 12 pairs of cranial nerves and 31 pairs of spinal nerves. The PNS has separate functional divisions, which we discuss next.

> **Quick Check**
>
> ☐ 1. What are the organs of the CNS?
> ☐ 2. What are the organs of the PNS?

Functional Divisions of the Nervous System

As the nervous system performs its many tasks, millions of processes may be occurring simultaneously. However, all these tasks or functions generally belong to one of three types: sensory, integrative, or motor. **Sensory functions** involve gathering information about the internal and external environments of the body. **Integrative functions** analyze and interpret the detected sensory stimuli and determine an appropriate response. **Motor functions** are the actions performed in response to integration. An example of these functions is illustrated in **Figure 11.2**.

Sensory input is gathered by the **sensory,** or **afferent, division** (AF-er-ent; "carrying toward") of the PNS. Integration is performed entirely by the CNS, mostly by the brain. Motor output is performed by the **motor,** or **efferent, division** (EE-fer-ent; "carrying away") of the PNS (remember this with the mnemonic "The *M*otor *E*fferent division moves ME"). Let's look more at these three functional divisions:

- **PNS sensory division.** Sensory stimuli are first detected by structures of the PNS called **sensory receptors.** The form of these receptors is diverse—they range from small tips of neurons found in the skin that sense temperature to complex receptors within muscles that sense muscle stretch. Depending on the location of the sensory receptors, the PNS sensory division may be further classified as follows:
  - The **somatic sensory division** (*soma-* = "body") consists of neurons that carry signals from skeletal muscles, bones, joints, and skin. This division also includes sensory neurons that transmit signals from the organs of vision, hearing, taste, smell, and balance (see Chapter 15). Sometimes these particular neurons are referred to as the *special sensory division.*
  - The **visceral sensory division** consists of neurons that transmit signals from viscera (organs) such as the heart, lungs, stomach, intestines, kidneys, and urinary bladder.

Sensory input from both divisions is carried from sensory receptors to the spinal cord and/or the brain by cranial and spinal nerves of the PNS.

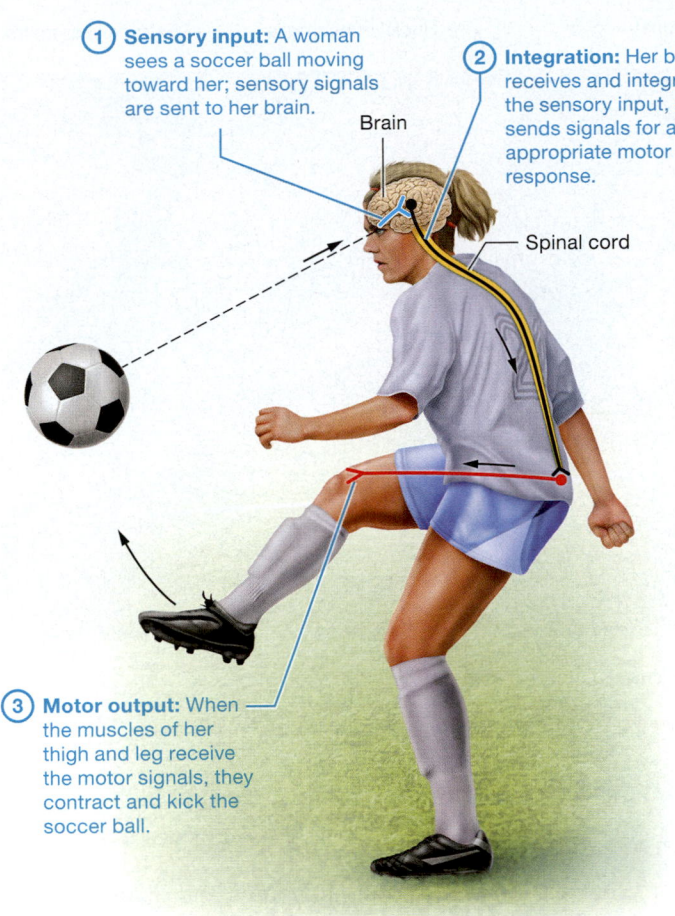

① **Sensory input:** A woman sees a soccer ball moving toward her; sensory signals are sent to her brain.

② **Integration:** Her brain receives and integrates the sensory input, and sends signals for an appropriate motor response.

Brain

Spinal cord

③ **Motor output:** When the muscles of her thigh and leg receive the motor signals, they contract and kick the soccer ball.

Figure 11.2 Functions of the nervous system.

sensory input, it responds by disregarding about 99% of such integrated data, a process that happens subconsciously. For example, you are likely unaware of any jewelry you're wearing or the hum of the air conditioner, because these stimuli are filtered out as unimportant. However, that small percentage of sensory stimuli to which the CNS does respond generally leads to a motor response.

- **PNS motor division.** The PNS motor division consists of motor neurons that carry out the motor functions of the nervous system. Motor output traveling from the brain and spinal cord via cranial and spinal nerves of the PNS may be used to control the contraction of muscle cells or secretion from a gland. Organs that carry out the effects of the nervous system are often called **effectors.** Like the sensory division, the motor division may be further classified based on the organs that the neurons contact:
  - ○ The **somatic motor division** consists of neurons that transmit signals to skeletal muscles. Skeletal muscle tissue is under conscious control, so this division is sometimes referred to as the *voluntary motor division.*
  - ○ The *visceral motor division,* better known as the **autonomic nervous system** (aw-toh-NAHM-ik; **ANS**), consists of neurons that carry signals primarily to thoracic and abdominal viscera. The ANS regulates secretion from certain glands, the contraction of smooth muscle, and the contraction of cardiac muscle in the heart. These functions are not generally under voluntary control, so the ANS is sometimes called the *involuntary motor division.* The ANS, which is very important for maintaining homeostasis of the internal environment, is discussed in its own chapter (see Chapter 14).

- **CNS.** The neurons of the CNS put together the many different types of sensory input, or integrate them, to form a more complete picture that can then elicit response if necessary. Interestingly, once the CNS integrates

Although the divisions of the nervous system are classified separately, both functionally and anatomically, remember that all functions of the nervous system rely on these divisions working together smoothly—no division operates independently. **Figure 11.3** summarizes the divisions, organs, and functions of the nervous system.

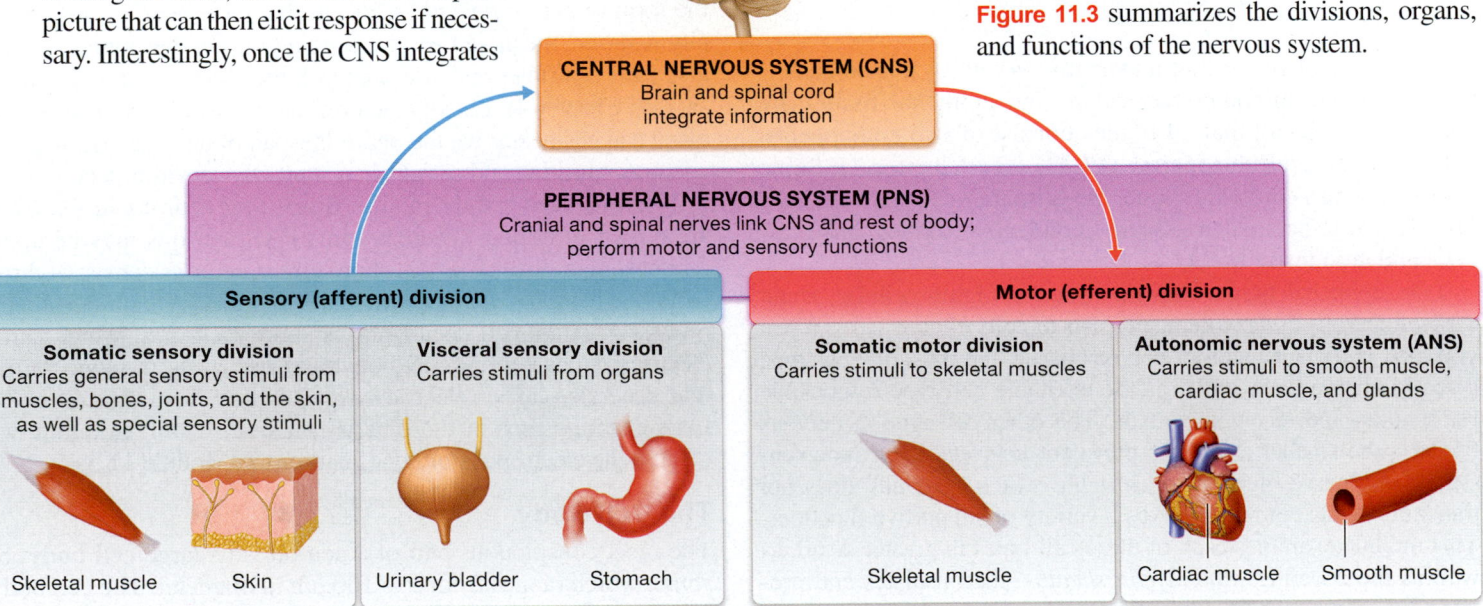

CENTRAL NERVOUS SYSTEM (CNS)
Brain and spinal cord integrate information

PERIPHERAL NERVOUS SYSTEM (PNS)
Cranial and spinal nerves link CNS and rest of body; perform motor and sensory functions

Sensory (afferent) division		**Motor (efferent) division**	
Somatic sensory division Carries general sensory stimuli from muscles, bones, joints, and the skin, as well as special sensory stimuli	**Visceral sensory division** Carries stimuli from organs	**Somatic motor division** Carries stimuli to skeletal muscles	**Autonomic nervous system (ANS)** Carries stimuli to smooth muscle, cardiac muscle, and glands
Skeletal muscle Skin	Urinary bladder Stomach	Skeletal muscle	Cardiac muscle Smooth muscle

Figure 11.3 Summary of the structural and functional divisions of the nervous system.

☐ 3. Describe the sensory, integrative, and motor functions of the nervous system.

☐ 4. What are the differences between the somatic and visceral sensory divisions of the PNS?

☐ 5. How does the somatic motor division of the PNS differ from the ANS?

Apply What You Learned

☐ 1. Imagine you have just picked up a cup of coffee. List all the sensory, integrative, and motor functions that your nervous system is performing as you do so.

☐ 2. Injuries may damage the nerves of any motor or sensory division of the PNS. In which PNS subdivision would a nerve injury be most threatening to survival? Explain.

See answers in Appendix A.

MODULE **11.2**
Nervous Tissue

Learning Outcomes

1. Describe the structure and function of each component of the neuron.
2. Describe the structure and function of each type of neuron.
3. Describe how the structure of each type of neuron supports its function.
4. Describe the structure and function of the four types of CNS neuroglial cells and the two types of PNS neuroglial cells.
5. Explain how the structure of each neuroglial cell supports its function.

The majority of tissue that makes up nervous system organs is nervous tissue, although connective and epithelial tissues are also present. Recall that all tissues consist of two components: cells and extracellular matrix (ECM) (see Chapter 4). Some tissues, such as epithelial tissue, are primarily cellular with very little ECM. Others, such as many connective tissues, have few cells and are mostly ECM.

Like epithelial tissue, nervous tissue is highly cellular; about 80% of nervous tissue volume consists of cells (**Figure 11.4**). When you look at a micrograph of nervous tissue, the most obvious type of cell is the *neuron*, which is the excitable cell type responsible for sending and receiving signals. The other cell type in nervous tissue is the smaller and more prevalent *neuroglial cell* (noo-roh-GLEE-ul; "nerve glue"), or *neuroglia*, which generally does not transmit signals but rather serves a variety of supportive functions. This module examines each of these cell types in greater detail, as well as the covering—the *myelin sheath*—that insulates and protects certain neurons. We also discuss how some of these cells can be regenerated if they are damaged.

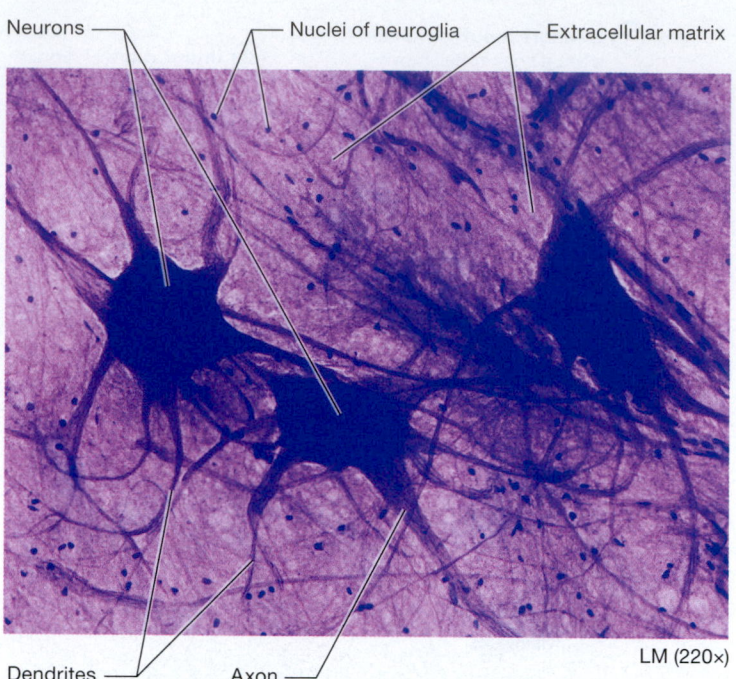

Neurons — Nuclei of neuroglia — Extracellular matrix

Dendrites — Axon — LM (220×)

Figure 11.4 Nervous tissue.

Neurons

« FLASHBACK

1. What are the functions of ribosomes and rough ER? (pp. 90–92)
2. What are the three components of a neuron? (p. 150)
3. Do neurons undergo mitosis? (p. 150)

The billions of neurons in nervous tissue are directly responsible for its sensory, integrative, and motor functions. **Neurons** are the excitable cell type responsible for sending and receiving signals in the form of *action potentials*. Recall that most neurons are generally *amitotic*, meaning that at a certain point in development, they lose their centrioles and after that lack the ability to undergo mitosis (see Chapter 4). Luckily, neurons are very long-lived cells, and some can easily survive the entire lifespan of an organism if given adequate nutrition and oxygen in a supportive environment.

Neurons vary greatly in size. Some tiny neurons in the CNS are only 1 mm long, whereas some PNS neurons may be up to 1 m or longer. As **Figure 11.5** shows, most neurons consist of three parts: the central *cell body*, where the majority of the biosynthetic processes of the cell occur; one or more *dendrites*, which carry electrical signals to the cell body; and one *axon*, the long "arm" that generally carries electrical signals away from the cell body. Let's examine each of these parts in greater detail. Note that we discuss the electrophysiology of neurons in Module 11.3.

The Cell Body

The most conspicuous part of a neuron is its large **cell body,** or *soma,* which ranges from 5 to 100 μm in diameter. The cell body is the most metabolically active part of the neuron, because it is responsible for maintaining the sometimes huge cytoplasmic

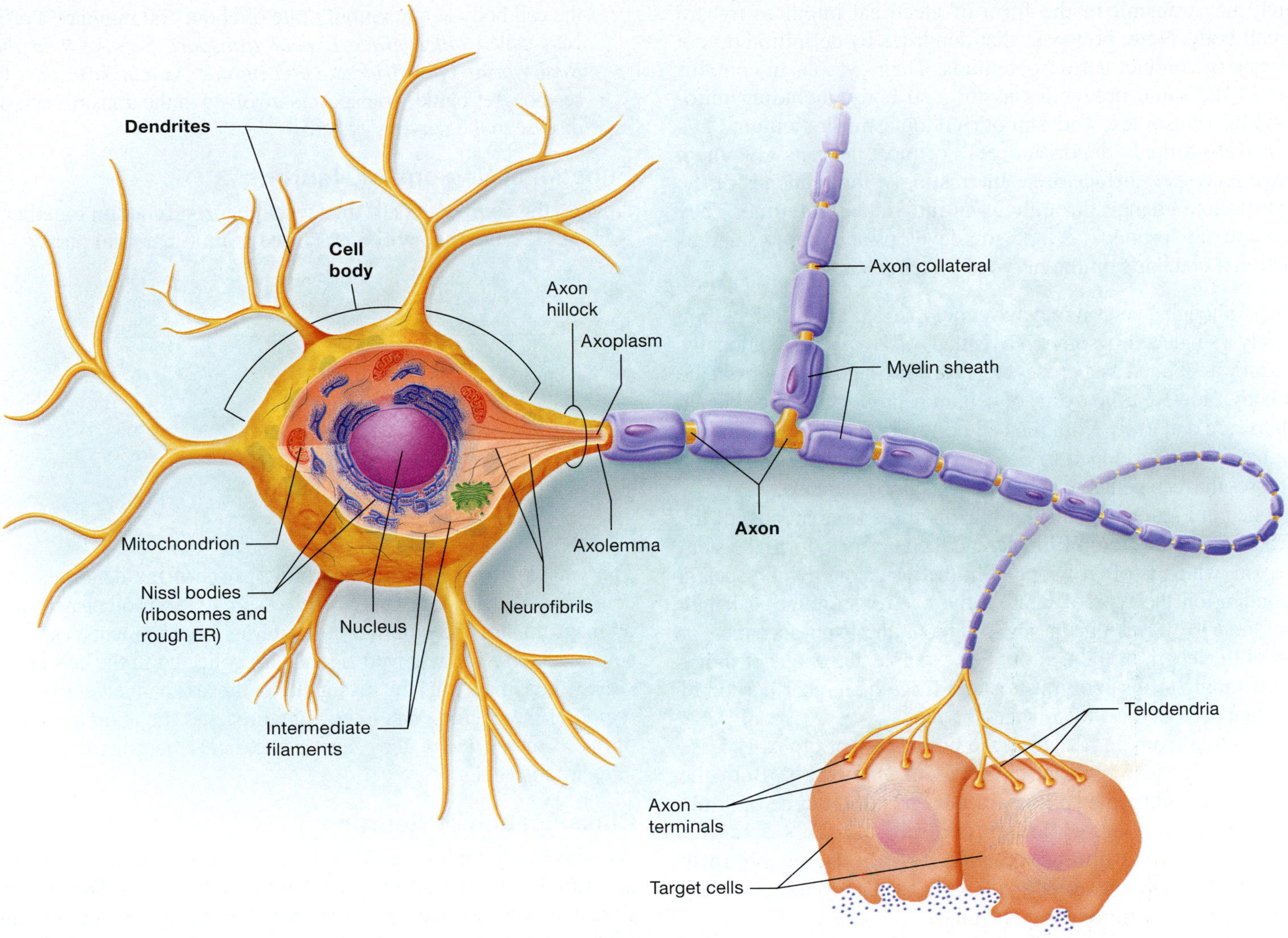

Dendrites

Cell
body

Axon
hillock

Axoplasm

Axon collateral

Myelin sheath

Axon

Axolemma

Mitochondrion

Nissl bodies
(ribosomes and
rough ER)

Nucleus

Neurofibrils

Intermediate
filaments

Telodendria

Axon
terminals

Target cells

Figure 11.5 Neuron structure.

Practice art labeling
@**Mastering** Anatomy & Physiology

volume of the neuron and also for manufacturing all the proteins the neuron needs. This high level of biosynthetic activity is reflected in the composition of the organelles within its cytoplasm:

- Free ribosomes and rough endoplasmic reticulum (RER) are found in abundance, reflecting the commitment of the cell body to protein synthesis. Note that the association of ribosomes and RER forms what appears under a microscope as dark-staining clusters called *Nissl bodies;* these are represented in Figure 11.5.
- Other organelles involved in protein synthesis, including the Golgi apparatus and one or more prominent nucleoli, are present.
- Mitochondria are found in large numbers, indicating the high metabolic demands of the neuron.

Additionally, the cytoplasm of the cell body includes lysosomes, smooth ER, and other organelles found in most cells.

The characteristic shape of the cell body is maintained by another component of the cytoplasm—the neuronal cytoskeleton, which is composed largely of intermediate filaments. These filaments bundle together to form larger structures called **neurofibrils,** which extend out into the dendrites and axon of the neuron (see Figure 11.5). The cytoskeleton also contains microtubules that provide structural support and a means for transporting chemicals between the cell body and the axon.

Processes: Dendrites and Axons

Extending from all neuron cell bodies are long "arms," or extensions, that are called *processes*. These processes allow the neuron to communicate with other cells. Most neurons have one or more dendrites and one axon.

Dendrites Dendrites (DEN-drytz; *dendr-* = "branch or tree") are typically short, highly forked processes that resemble the branches of a tree limb. They receive input from other neurons, which they transmit in the form of electrical impulses toward the cell body. Note, however, that dendrites by definition do not generate or conduct action potentials. Their cytoplasm contains most of the same organelles as the cell body, including mitochondria, ribosomes, and smooth endoplasmic reticulum. The extensively forked "dendritic trees" of most neurons give them a huge receptive surface area. Interestingly, the branches of the dendritic tree change throughout an individual's lifetime: They grow and are "pruned" as a person develops and matures and as functional demands on the nervous system change.

Axon Although a neuron may have multiple dendrites, each neuron has only a single **axon,** sometimes called a *nerve fiber.* Traditionally, an axon was defined as a process that carried a signal away from the cell body. However, the axons of certain neurons can carry a signal both toward and away from the cell body. For this reason, new criteria have been developed to define an axon: They are considered processes that can generate and conduct action potentials.

Notice in Figure 11.5 that each axon arises from an area of the cell body called the **axon hillock,** and then tapers to form the slender axon, which is often wrapped in the insulating myelin sheath. Depending on the type of neuron, the axon may range in length from short to very long; in some neurons the axon accounts for most of the length of the neuron. For example, the axons of motor neurons going to the foot must extend from the lumbar portion of the spinal cord all the way down to the foot.

Extending from some axons are branches that typically arise at right angles to the axon, called **axon collaterals.** Both the axon and its collaterals split near their ends to produce multiple fine branches known as **telodendria** (tee´-loh-DEN-dree-ah). The telodendria terminate in **axon terminals,** or **synaptic knobs,** that communicate with a target cell. Each axon generally splits into 1000 or more axon terminals.

The plasma membrane that envelops the axon is called the **axolemma** (aks-oh-LEM-ah), and its cytoplasm is known as **axoplasm.** Although dendrites have most of the same organelles as the cell body, axons do not. Axons contain mitochondria, abundant intermediate filaments, vesicles, and lysosomes; however, they do not contain protein-making organelles such as ribosomes or Golgi apparatus. The composition of the axoplasm is dynamic, as substances move both toward and away from the cell body along the axon's length.

Substances may travel through the axoplasm using one of two types of transport, which are together termed *axonal transport* or *flow:*

- **Slow axonal transport.** Substances within the axoplasm, such as cytoskeletal proteins and other types of proteins, move by *slow axonal transport.* These substances move only away from the cell body and do so at a rate of about 1–3 mm/day.
- **Fast axonal transport.** Vesicles and membrane-enclosed organelles use *fast axonal transport* to travel much more rapidly through the axon. This type of transport relies on motor proteins in the axoplasm that consume ATP to move components along microtubules. Components may move either toward the cell body at a maximum rate of about 200 mm/day, a process called *retrograde axonal transport,* or away from

the cell body at a maximum rate of about 400 mm/day, a process called *anterograde axonal transport.* See *A&P in the Real World: Poliovirus and Retrograde Axonal Transport* to see how retrograde transport is involved in the transmission of disease to the nervous system.

Functional Regions of Neurons

How do the components of the typical neuron function together? As you see here, the neuron has three main functional parts:

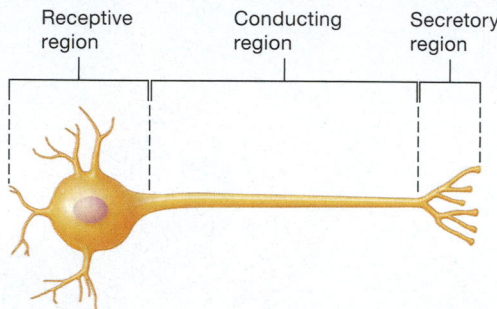

The *receptive region* of the neuron consists of the dendrites and cell body. The dendrites may receive signals from other neurons, or may monitor the external and internal environments via sensory receptors. The received signals are collected in the cell body, which then may transmit a signal to the axon, the *conducting region* of the neuron. When the signal reaches the axon terminals of the *secretory region,* they secrete chemicals that trigger changes in their target cells.

Classification of Neurons

As with many topics we've covered, neurons can be classified according to both their structure and their function. These classification schemes overlap—certain functional groups of neurons often have the same structural features, another example of how "form follows function" in the body.

Structural Classification Neurons vary widely in shape, with the greatest structural variation seen in the number and form of

Poliovirus and Retrograde Axonal Transport

Poliomyelitis (poh´-lee-oh-my-eh-LY-tus), the disease caused by *poliovirus,* is an infection that can affect the central nervous system (CNS), particularly the spinal cord, and may result in deformity and paralysis. No cure for polio is available, but it can be easily prevented by vaccination.

The virus is thought to gain access to the CNS by entering muscle fibers and then passing into motor neurons at the neuromuscular junction. The virus then travels the length of the axon, using retrograde axonal transport, until it eventually enters the spinal cord. Other toxins and viruses such as herpes simplex virus, rabies virus, and tetanus toxin also have the ability to invade the nervous system via retrograde axonal transport.

the processes extending from the cell body. On this basis, neurons are classed structurally into three groups:

- **Multipolar neurons.** Over 99% of neurons in the human body fall into the group known as **multipolar neurons.** These neurons have a single axon and typically multiple highly branched dendrites. This group of neurons has the widest variability in terms of shape and size.
- **Bipolar neurons.** A **bipolar neuron** has only two processes: one axon and one dendrite. In humans the majority of bipolar neurons are sensory neurons, located in places such as the retina of the eye and the olfactory epithelium of the nasal cavity.
- **Pseudounipolar neurons.** Pseudounipolar neurons (soo'-doh-yoo-nih-POH-lar; formerly referred to as *unipolar neurons*) begin developmentally as bipolar neurons, but their two processes fuse to give rise to a single axon. As the axon extends from the cell body, it splits into two processes: one that brings stimuli from sensory receptors to the cell body, called the *peripheral process* or *axon,* and one that travels to the spinal cord away from the cell body, called the *central process* or *axon.* The pseudounipolar neurons are sensory neurons that detect stimuli such as touch, pressure, and pain.

Functional Classification Functionally, neurons are grouped into three classes based on the direction in which they transmit signals. The three classes are as follows, in order of information flow:

1. **Sensory,** or **afferent, neurons** carry signals *toward* the central nervous system. The sensory receptors of these neurons detect stimuli, and the electrical changes are transmitted to their cell bodies in the PNS, then down their axons to the brain or spinal cord. They are generally pseudounipolar or bipolar in structure because they receive stimuli from only one area. Sensory neurons detect the internal and external environments (such as from the skin and viscera) and facilitate motor coordination (such as in joints and muscles).

2. **Interneurons,** also called *association neurons,* relay messages *within* the CNS, primarily between sensory and motor neurons, and are the location of most information processing. The vast majority of neurons are interneurons. Multipolar in structure, interneurons generally communicate with many other neurons (for example, one Purkinje cell [per-KIN-jee] of the cerebellum can receive as many as 150,000 contacts from other neurons).

3. **Motor,** or **efferent, neurons** carry stimuli *away from* their cell bodies in the CNS to muscles and glands. Most motor neurons are multipolar, as motor tasks are generally complicated and require input from many other neurons.

The classification systems of neurons are summarized in **Table 11.1.** Note that Table 11.1 includes three different examples of multipolar neurons—one from the spinal cord (spinal motor neuron), one from the hippocampus of the brain (pyramidal cell), and another from the cerebellum of the brain (Purkinje cell).

Groups of Neuron Cell Bodies and Axons

In the CNS and PNS, cell bodies of neurons are typically found within clusters, most of which are in the CNS, where they are called **nuclei.** Within the PNS, clusters of cell bodies are called **ganglia** (GANG-glee-ah; singular, *ganglion; gangli-* = "knot"). In addition, axons tend to be bundled together in the CNS and the PNS. In the CNS, these bundles are referred to as *tracts,* and in the PNS, as *nerves.*

Table 11.1 Neuron Classification				
Structural Class	**Multipolar Neurons**		**Bipolar Neurons**	**Pseudounipolar Neurons**
Structural Features	One axon with two or more dendrites; typically have highly branched dendritic tree		One axon and one dendrite	Single short process that splits into two axons (no dendrites)
	Spinal motor neuron / Pyramidal cell / Purkinje cell		Special sensory neuron	General sensory neuron
Typical Functional Class	Motor (efferent) neurons, interneurons		Sensory (afferent) neurons	Sensory (afferent) neurons
Location	Most neurons in the CNS, motor neurons in the PNS		Special sense organs in the PNS, such as the retina and olfactory epithelium	Sensory neurons in the PNS associated with touch, pain, and vibration sensations

Quick Check

☐ 1. What are the functions of the cell body, dendrites, and axon?

☐ 2. What are the structural differences between multipolar, bipolar, and pseudounipolar neurons?

☐ 3. What are the functional differences between sensory neurons, interneurons, and motor neurons?

Neuroglia

◀◀ FLASHBACK

1. Why do nonpolar, lipid-based substances diffuse easily across cell membranes, but polar compounds do not? (p. 76)

2. What are tight junctions and gap junctions? How does their form follow their function? (p. 127)

Neuroglia (noo-roh-GLEE-uh), or **neuroglial cells,** were named for the early scientific idea that these cells "glued together" the neurons, as the word root *glia* means "glue." However, we now recognize that neuroglia also serve many more functions. Examples include maintaining the environment around neurons, protecting them, and assisting in their proper functioning. Unlike the mostly amitotic neurons, neuroglia retain their ability to divide, and they fill in gaps left when neurons die.

Six different types of neuroglia can be found in the nervous system, four in the CNS and two in the PNS. Like all cells we've

covered, the form of each type of neuroglial cell is specialized for its function, another example of the Structure-Function Core Principle (p. 28). Keep this in mind as we examine the six types of cells.

CORE PRINCIPLE
Structure-Function

Neuroglia in the CNS

Neuroglia are about 10 times more abundant in the CNS than neurons, and they make up about half the mass of the brain. Within the CNS we find four types of neuroglia: *astrocytes, oligodendrocytes, microglia,* and *ependymal cells* (**Figure 11.6**).

Astrocytes The star-shaped **astrocytes** (ASS-troh-syt'z; "star cells") are the most numerous and the largest of the neuroglia in the CNS. Note in Figure 11.6 that each astrocyte has a central portion and numerous processes, all of which terminate in structures called *end-feet*. This anatomical feature equips astrocytes to perform multiple functions, including the following:

- **Anchoring neurons and blood vessels in place.** Astrocytes help form the three-dimensional structure of the brain by using their end-feet to anchor neurons and blood vessels in place. In addition, astrocytes may facilitate the transport of nutrients and gases from the blood vessels to neurons.
- **Regulating the extracellular environment of the brain.** Astrocytes are connected by gap junctions that allow them to communicate with one another about the local extracellular environment within the brain. Via this communication they

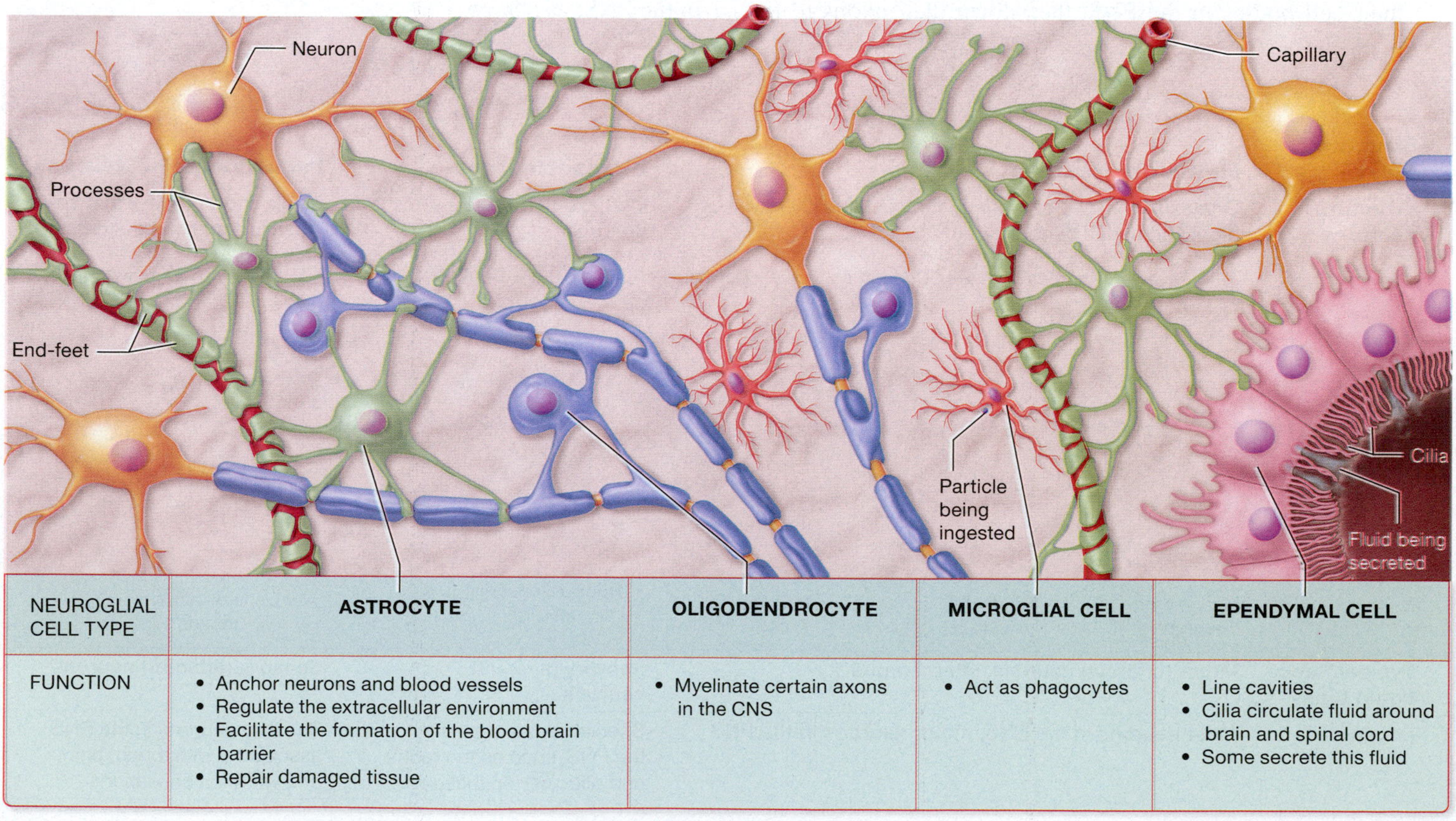

Neuron

Capillary

Processes

End-feet

Particle being ingested

Cilia

Fluid being secreted

NEUROGLIAL CELL TYPE	ASTROCYTE	OLIGODENDROCYTE	MICROGLIAL CELL	EPENDYMAL CELL
FUNCTION	• Anchor neurons and blood vessels • Regulate the extracellular environment • Facilitate the formation of the blood brain barrier • Repair damaged tissue	• Myelinate certain axons in the CNS	• Act as phagocytes	• Line cavities • Cilia circulate fluid around brain and spinal cord • Some secrete this fluid

Figure 11.6 Neuroglial cells of the CNS.

can act as a "clean-up crew," removing excess extracellular potassium ions as well as chemicals known as *neurotransmitters*. Although neurons use neurotransmitters to send signals, their extracellular accumulation can lead to toxicity.

- **Assisting in the formation of the blood brain barrier.** Astrocytes facilitate the formation of a protective structure called the *blood brain barrier* by ensheathing capillaries and inducing their cells to form tight junctions. These tight junctions prevent most proteins and polar compounds from leaving the capillaries and entering the brain extracellular fluid (ECF). The only substances that can cross these capillaries easily are those that are nonpolar and lipid-soluble and/or those for which special transporters exist. The double barrier separates the blood from the brain ECF, which ensures selective transport of substances between the two fluids. The blood brain barrier is discussed fully in the CNS chapter (see Chapter 12).
- **Repairing damaged brain tissue.** When brain injury occurs, astrocytes are triggered to divide rapidly. Although this growth stabilizes the damaged tissue, it may also impede complete healing. Recent research has demonstrated that excess astrocyte activity actually inhibits the regrowth of neurons, leading to more permanent defects.

Astrocytes are critical to normal functioning of the nervous system, so when they undergo rapid, uncontrolled cell division, the results can be devastating. Find out more about this in *A&P in the Real World: Gliomas and Astrocytomas* on p. 390.

Oligodendrocytes Like astrocytes, **oligodendrocytes** (oh-lig′-oh-DEN-droh-syt′z; *oligo-* = "few") also have radiating processes, but they are fewer in number and smaller than those of astrocytes. The flattened ends of some of these processes wrap around part of the axons of certain neurons. These wrapped processes form concentric layers of plasma membrane that are collectively called **myelin** (MY-eh-lin). Repeating segments of myelin along the length of an axon form the myelin sheath. As you can see in Figure 11.6, each oligodendrocyte has several of these processes that wrap around multiple axons. We consider the formation of the myelin sheath and its functional significance later in this module.

Microglia The least numerous neuroglial cells are the small and branching **microglia** (my-KROH-glee-ah). Although many functions of microglia are still under investigation, we do know that they are activated by injury within the brain and become wandering phagocytes—cells that "clean up" the environment in the brain. When activated, microglia ingest disease-causing organisms, dead neurons, and other cellular debris. They also secrete chemicals that stimulate inflammation.

Ependymal Cells The neuroglia known as **ependymal cells** (eh-PEN-dih-mal) are ciliated cells with a variety of functions. One of their main functions is circulating *cerebrospinal fluid,* which is the fluid in the cavities of the brain and spinal cord. Certain ependymal cells also play a role in the formation of this fluid, and others are thought to monitor its composition.

Neuroglia in the PNS

In the PNS the two types of neuroglia are *Schwann cells* and *satellite cells* (**Figure 11.7**). Like those in the CNS, the neuroglia of the PNS serve supportive and protective functions, with, once again, their form specialized for their function.

Schwann Cells Just as some axons of the CNS are covered with a myelin sheath, axons of the PNS can also be myelinated. However, the myelin sheath of the PNS is created by a different type of neuroglial cell: the sausage-shaped **Schwann cells** (see Figure 11.7, left). As we will see, unmyelinated axons are also encased in Schwann cells. Additionally, Schwann cells play a vital role in repair of damaged axons in the PNS.

Satellite Cells **Satellite cells** are flat cells that surround the cell bodies of neurons in the PNS (see Figure 11.7, right). The most poorly understood of the neuroglia, they appear to enclose and support the cell bodies, and have intertwined processes that link them with other parts of the neuron, other satellite cells, and also neighboring Schwann cells. They also appear to regulate the extracellular environment around the neuronal cell body, a function analogous to that of astrocytes in the CNS.

Quick Check

- ☐ 4. What are the functions of astrocytes?
- ☐ 5. What are the functions of microglia?
- ☐ 6. Which neuroglial cell forms and circulates the fluid surrounding the brain and spinal cord?

The Myelin Sheath

《 FLASHBACK

1. What is the difference between polar and nonpolar covalent bonds? (p. 39)
2. What are the differences between hydrophobic and hydrophilic compounds? (p. 47)
3. Are lipids polar covalent or nonpolar covalent compounds? Are they hydrophilic or hydrophobic? (p. 53)

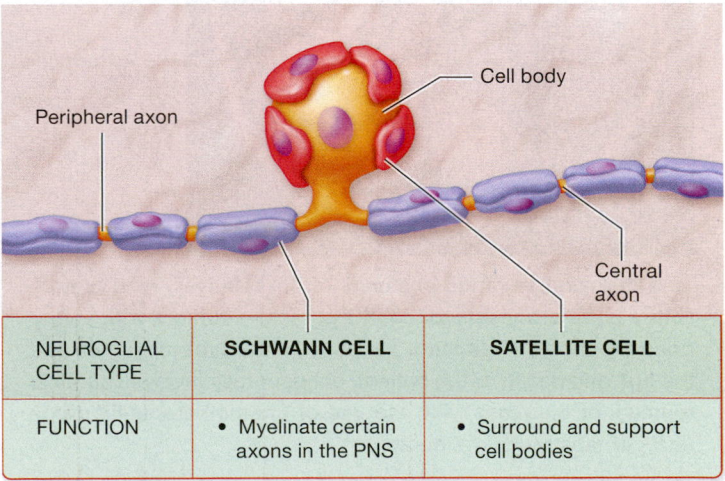

NEUROGLIAL CELL TYPE	SCHWANN CELL	SATELLITE CELL
FUNCTION	• Myelinate certain axons in the PNS	• Surround and support cell bodies

Figure 11.7 Neuroglial cells of the PNS.

As we discussed, Schwann cells and oligodendrocytes wrap themselves around the axons of certain neurons to create a structure known as the **myelin sheath** (Figure 11.8). Myelin is composed of repeating layers of the plasma membrane of the neuroglial cell, so it has the same substances as any plasma membrane: phospholipids, other lipids such as cholesterol, and proteins. In addition, myelin contains lipids unique to Schwann cells and oligodendrocytes.

Myelin plays an important role in the electrophysiology of many neurons. Recall that in the body, the flow of charged particles, or ions, creates an *electric current*. In *unmyelinated axons*, the electric current "leaks" out of the axon and has to be continually regenerated. However, as you learned in the muscle tissue chapter, ions do not pass easily through the hydrophobic portion of the phospholipid bilayer (see Chapter 10). For this reason, the high lipid content of myelin makes it an excellent insulator, akin to the rubber tubing around a copper wire. The overall effect of this insulation is to increase the speed of conduction of action potentials: *Myelinated axons* conduct action potentials about 15–150 times faster than unmyelinated axons. This is a good example of the Structure-Function Core Principle (p. 28).

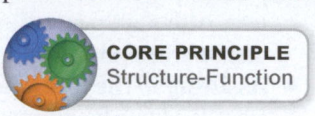

CORE PRINCIPLE Structure-Function

Myelination (my′eh-lin-AY-shun) is the process of myelin sheath formation. During this process in the PNS, a Schwann cell wraps itself outward away from the axon in successively tighter bands, forming a myelin sheath up to 100 layers thick (see Figure 11.8a). The basic process is similar for an oligodendrocyte in the CNS. However, in the CNS the arms of an oligodendrocyte wrap inward toward the axon—the opposite direction from the Schwann cells (see Figure 11.8b).

Many other differences can be found between myelination in the PNS and CNS, including the following:

- **Presence or absence of a neurolemma.** Note in Figure 11.8a that we find the nucleus and the bulk of the Schwann cell's cytoplasm and organelles on the outer surface of a myelinated axon, a structure called the **neurolemma** (noor-oh-LEM-ah). No outer neurolemma is found in the CNS because the nucleus and cytoplasm of the oligodendrocyte remain in a centralized location (see Figure 11.8b).
- **Number of axons myelinated by a single glial cell.** Each oligodendrocyte may send out multiple processes to envelop parts of several axons. However, Schwann cells can encircle only a portion of a single axon.
- **Timing of myelination.** The timing of myelination is also different within the CNS and the PNS. In the PNS myelination begins during the early fetal period, whereas myelination in the CNS, particularly in the brain, begins much later. Very little myelin is present in the brain of the newborn (which is why babies and toddlers need adequate fat in their diets).

In both the CNS and the PNS, axons are generally much longer than a single oligodendrocyte or Schwann cell, so more than one cell is needed to myelinate the entire axon. The segments of an axon that are covered by neuroglia are called **internodes,** and they range from 0.15 to 1.5 mm in length. Between each internode is a gap about 1 μm wide, called a **node of Ranvier** (rahn-vee-AY), or *myelin sheath gap*, where no myelin is found. Also unmyelinated is a short region from the axon hillock to the first neuroglial cell; this is known as the *initial segment*.

Short axons in both the CNS and the PNS are nearly always unmyelinated. However, in the PNS, even axons that lack a myelin sheath associate with Schwann cells (Figure 11.9). Take note, though, that the Schwann cells do not wrap themselves around these axons. Instead, they enclose them much like a hot dog in a bun. A single Schwann cell can envelop multiple axons in this manner.

In the CNS you can actually see which regions of the brain and spinal cord contain myelinated axons and which do not. In sections of both the spinal cord and the brain, regions of darker- and lighter-colored tissue (look ahead to Figure 12.2) can be noted.

Gliomas and Astrocytomas

Tumors that originate within the brain are called *primary brain tumors* and most such tumors are *gliomas* (glee-OH-mahz). A glioma is caused by an abnormally high rate of cellular division in neuroglial cells. Some conditions are known to predispose patients to gliomas, such as exposure to ionizing radiation and certain diseases like the genetic condition *neurofibromatosis,* but the cause of the tumor usually goes undiscovered. Tumors of astrocytes, called *astrocytomas* (ass′-troh-sy-TOH-mahz), are the most common type of glioma. An example of an astrocytoma is shown here:

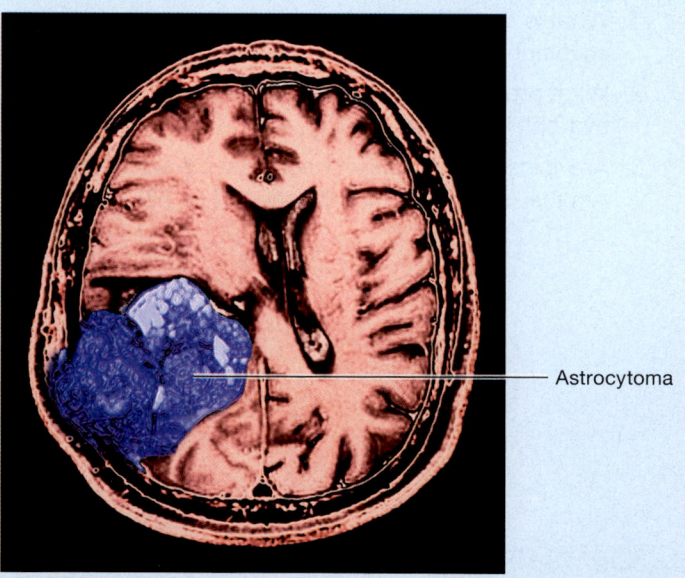

Astrocytoma

Astrocytomas range in severity from relatively mild tumors with a good prognosis to highly aggressive tumors with a very poor prognosis. Treatment varies with the type of tumor and the age and health of the patient, but generally involves surgical removal of the mass, with the use of chemotherapeutic drugs and perhaps radiation therapy.

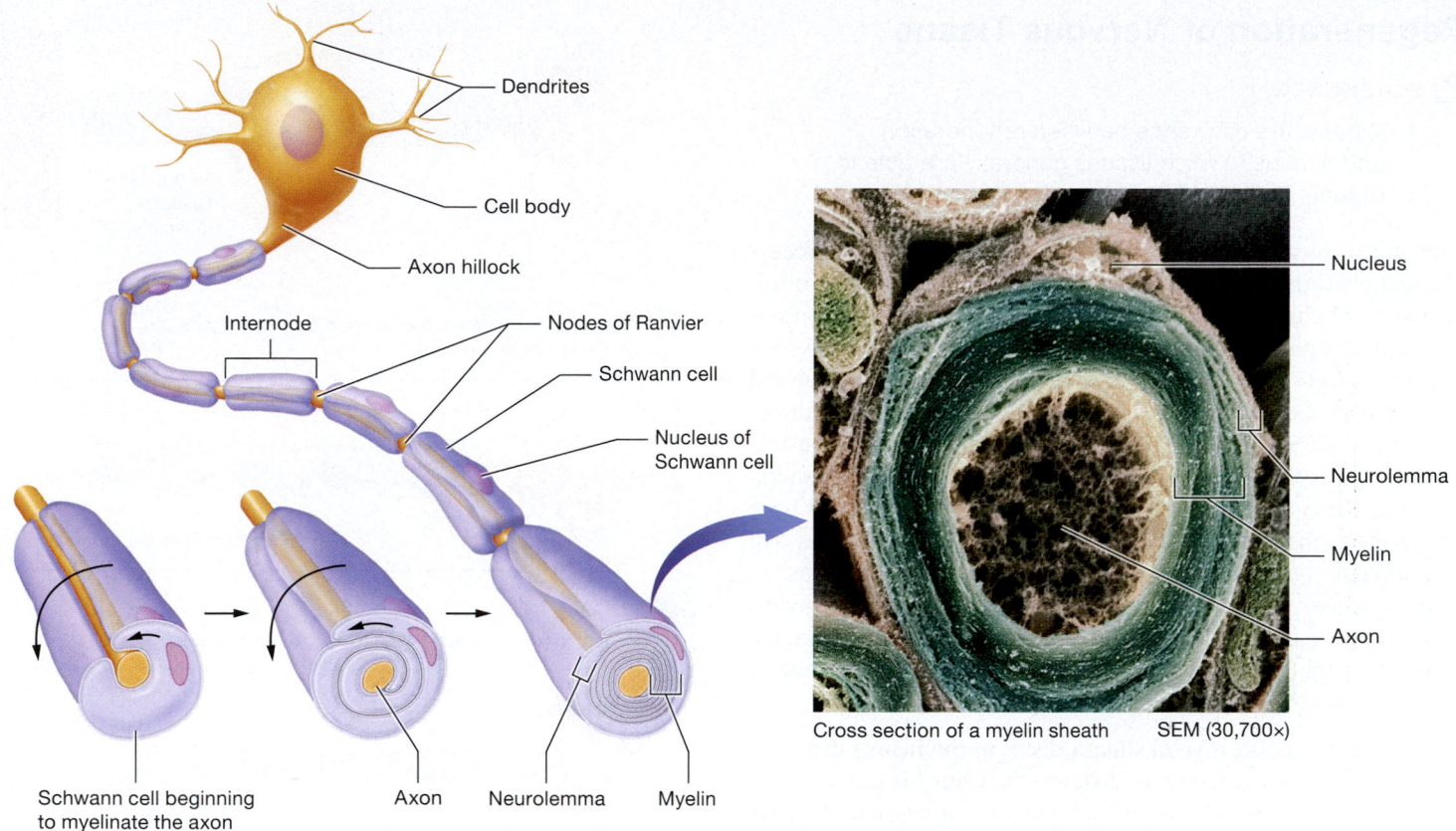

(a) **The myelin sheath and myelination in the PNS**

Cross section of a myelin sheath   SEM (30,700×)

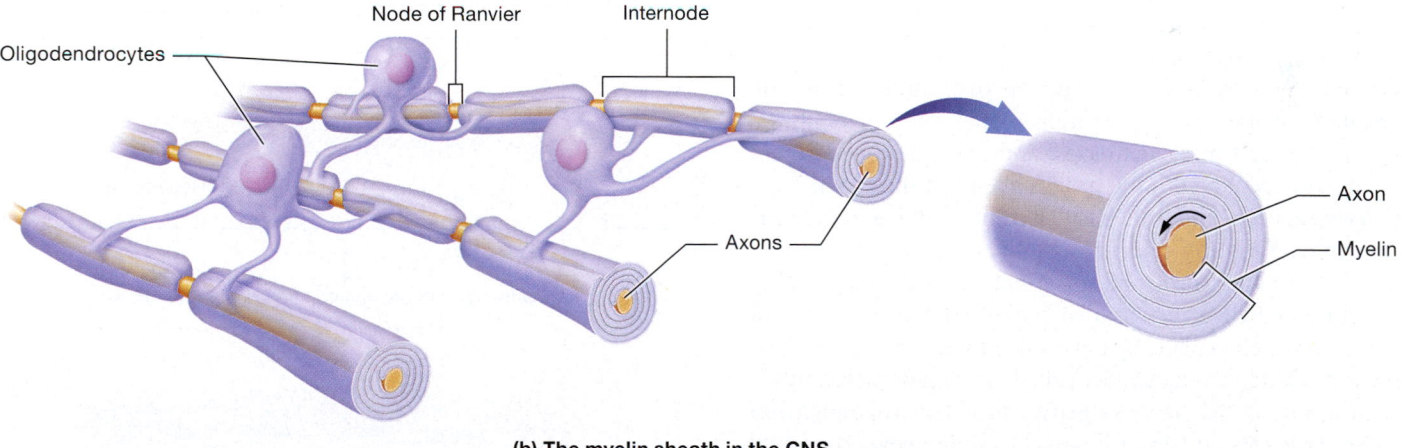

(b) **The myelin sheath in the CNS**

Figure 11.8 The myelin sheath in the PNS and CNS.

This color difference reflects the distribution of the myelin sheath. The lighter-colored areas, or **white matter,** are composed of myelinated axons. The darker-colored areas, or **gray matter,** are made up primarily of cell bodies and dendrites, which are never myelinated, as well as small unmyelinated axons.

Quick Check

☐ 7. What is the function of the myelin sheath?

☐ 8. How does the myelin sheath differ in the CNS and the PNS?

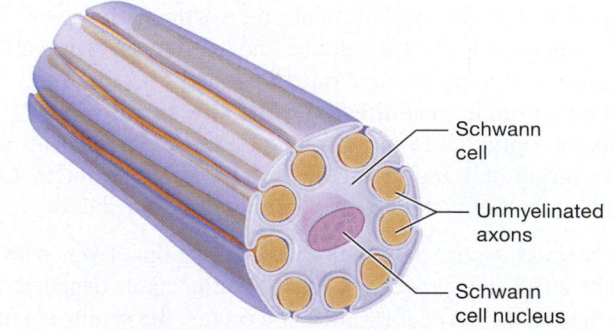

Figure 11.9 Unmyelinated peripheral axons and Schwann cells.

Regeneration of Nervous Tissue

« FLASHBACK

1. What is the difference between regeneration and fibrosis? Which tissues generally are able to regenerate? (p. 154)

Human nervous tissue has a fairly limited capacity for the process of **regeneration,** or replacement of damaged tissue with the original tissue. Damaged axons and dendrites in the CNS almost never regenerate, a phenomenon apparently due to several factors. For example, oligodendrocytes may inhibit the process of neuronal growth, and chemicals called *growth factors* that trigger mitosis are largely absent in the CNS. In addition, the growth of astrocytes creates space-filling scar tissue that also prohibits regeneration. For these reasons, injuries to the brain or spinal cord have largely permanent effects. However, in some circumstances lost function may be regained by retraining the remaining neurons.

In contrast, neural tissue in the PNS is capable of regeneration to some extent, but only if the cell body remains intact. When a peripheral axon is damaged, the following sequence of events repairs the damaged neuron (**Figure 11.10**):

1. **The axon and myelin sheath distal to the injury degenerate.** The damaged axon distal to the injury is cut off from the cell body where all the protein-synthesis machinery is housed. It therefore has no way to repair itself, and so this part of the axon, along with its myelin sheath, begins to degenerate. This occurs via a process called *Wallerian degeneration* (vah-LEHR-ee-an), in which phagocytes digest the cellular debris.

2. **Growth processes form from the proximal end of the axon.** As Wallerian degeneration occurs, protein synthesis within the cell body increases, and several small *growth processes* sprout from the proximal end of the axon.

3. **Schwann cells and the basal lamina form a regeneration tube.** Schwann cells near the site of the injury begin to proliferate along the length of a collagen-rich surrounding structure known as the basal lamina (also called the *external lamina*), which is made by connective tissue cells around the neuron. This forms a cylinder called the **regeneration tube.**

4. **A single growth process grows into the regeneration tube.** Note in step ② of Figure 11.10 that several growth processes form; however, only one will make it into the regeneration tube. In the tube, Schwann cells secrete growth factors that stimulate regrowth of the axon. The regeneration tube then guides the axon to grow toward its target cell at an average rate of about 1.5–3 mm/day.

5. **The axon is reconnected with the target cell.** If the axon continues to grow, it most likely will meet up with its target cell and re-establish its synaptic contacts. Over time, the Schwann cells re-form the myelin sheath.

This process occurs only under ideal conditions. Even with the cell body intact, the process often stalls after axon degeneration, and the neuron dies. And if regeneration occurs, the results are often imperfect. Occasionally, the axon will contact the wrong target cell, or contact between the cells will not be re-established.

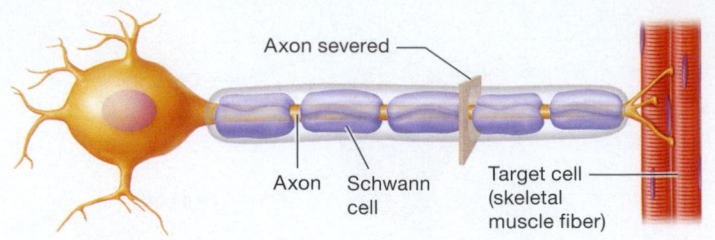

① Axon and myelin sheath distal to the injury degenerate (Wallerian degeneration).

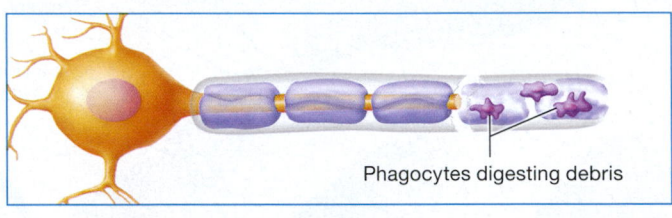

② Growth processes form from the proximal end of the axon.

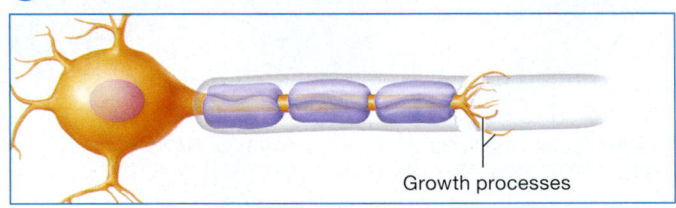

③ Schwann cells and the basal lamina form a regeneration tube.

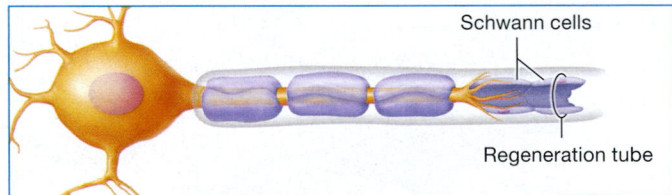

④ A single growth process grows into the regeneration tube.

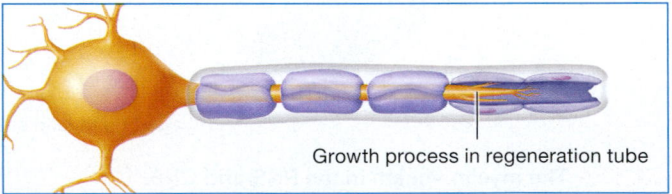

⑤ The axon is reconnected with the target cell.

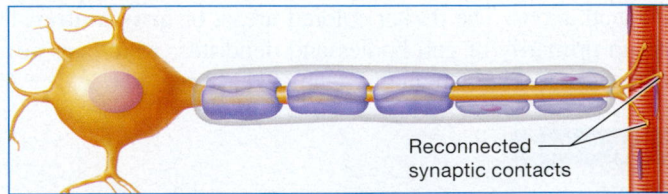

Figure 11.10 Repair of axon damage in the PNS.

Chapter 10). In this module we re-examine these principles in terms of the electrophysiology of neurons, and how they allow the nervous system to perform virtually all its functions.

Principles of Electrophysiology

◀◀ FLASHBACK

1. What is the resting membrane potential? (pp. 347–348)
2. Is the concentration of sodium ions greater in the cytosol or in the extracellular fluid? How about the concentration of potassium ions? What maintains these two gradients? (p. 348)
3. What is an electrochemical gradient? (pp. 348–349)

In the muscle tissue chapter you were introduced to the concepts of *electrophysiology*—the study of electrical changes across the plasma membrane and the accompanying physiological processes (see Chapter 10). Although discussion in that chapter revolved around the electrophysiology of the muscle fiber, the same basic principles apply to the electrophysiology of neurons. Like muscle fibers, electrical changes across the plasma membrane of neurons rely on the presence of *ion channels* in the membrane and a *resting membrane potential*. So, before we move on, let's review these important concepts.

The Resting Membrane Potential

Recall that there is a separation of charges across the plasma membrane—there is a thin layer of negative charges in the cytosol lining the inside of the membrane and a thin layer of positive charges in the extracellular fluid lining the outside of the membrane, as you can see here:

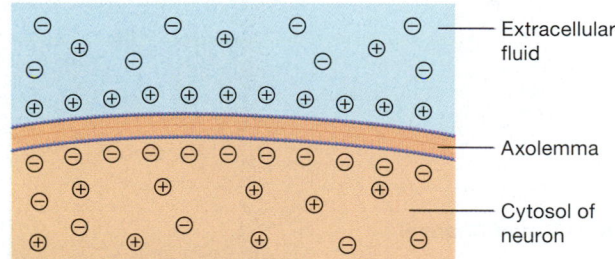

As you learned, this separation of charges, called a *voltage,* is a type of gradient referred to as an *electrical gradient.* The electrical gradient across the plasma membrane is known as a **membrane potential,** named for the fact that, like any gradient, an electrical gradient is a source of potential energy for the cell.

When you measure the membrane potential of a neuron at rest, or the **resting membrane potential,** it measures about –70 mV. Recall that the resting membrane potential is negative because the cell constantly loses small numbers of positively charged potassium ions. The potassium ions are able to exit the cell due to the presence of proteins in the plasma membrane known as *leak channels,* which are always open. This continual loss of positive charges makes the cytosol on the inside of the membrane negative with respect to the extracellular fluid. You may have noticed that the resting membrane potential for a neuron is less negative at –70 mV than that of a skeletal muscle fiber at –90 mV. This is largely due to the number of potassium ion leak channels in the skeletal muscle fiber

MODULE **11.3**
Electrophysiology of Neurons

Learning Outcomes

1. Describe the voltage-gated ion channels that are essential for the development of the action potential.
2. Interpret a graph of an action potential, and describe the depolarization, repolarization, and hyperpolarization phases of an action potential.
3. Explain the physiological basis of the absolute and relative refractory periods.
4. Compare and contrast continuous and saltatory conduction.
5. Explain how axon diameter and myelination affect conduction speed.

Neurons share two key properties with skeletal muscle fibers (see Chapter 10). First, all neurons are *excitable* (responsive) in the presence of various stimuli, including chemical signals, local electrical signals, and mechanical deformation. These stimuli generate electrical changes across the plasma membrane of the neuron. Another property is *conductivity,* which means that electrical changes across the plasma membrane don't stay in one place. Instead, they are rapidly conducted along the entire length of the membrane, similar to how an electrical impulse is conducted through a copper wire.

The electrical changes across a neuron's plasma membrane come in two forms: (1) *local potentials,* which travel only short distances, and (2) *action potentials,* which travel the entire length of an axon. Both types of potentials rely on the same principles of electrophysiology that we discussed with muscle tissue (see membrane—the skeletal muscle fiber loses more positive charges and so has a more negative resting membrane potential.

A cell at its resting membrane potential is **polarized.** The word polarized simply means that the membrane potential is at the negative or positive side (think of the negative and positive "poles" of a magnet). The closer the membrane potential comes to 0 mV, the less polarized the membrane potential becomes. As we discuss shortly, changes of this sort are responsible for the electrical events of a neuron.

Ion Channels and Gradients

As in muscle fibers, ions must cross the plasma membrane of neurons through protein channels or pumps because charged particles cannot pass through the hydrophobic portion of the phospholipid bilayer. Ions moving with their electrochemical gradients by diffusion pass through channels. As we have discussed, there are two main classes of channels: the leak channels that we just mentioned, and gated channels, which open or close in response to a specific stimulus. There are three types of gated channels:

- **Ligand-gated channels** open in response to a certain chemical, called a **ligand,** binding to the channel or to a receptor associated with the channel.
- **Voltage-gated channels** open or close in response to changes in the cell's membrane potential.
- **Mechanically gated channels** open or close in response to mechanical stimulation such as stretch, pressure, and vibration.

Table 11.2 reviews the different types of channels involved in the electrophysiology of neurons.

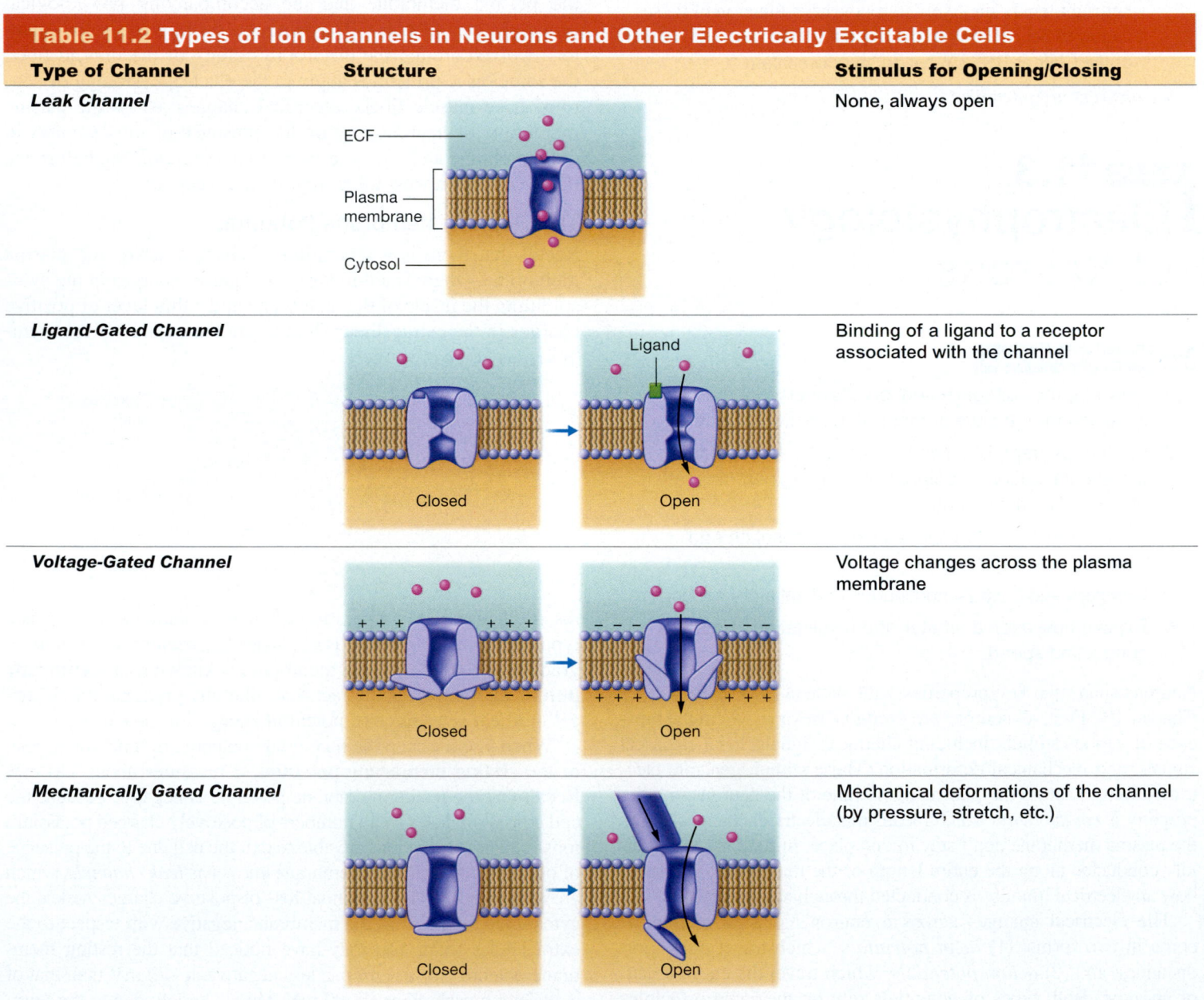

Table 11.2 Types of Ion Channels in Neurons and Other Electrically Excitable Cells		
Type of Channel	**Structure**	**Stimulus for Opening/Closing**
Leak Channel	ECF / Plasma membrane / Cytosol	None, always open
Ligand-Gated Channel	Closed → Ligand Open	Binding of a ligand to a receptor associated with the channel
Voltage-Gated Channel	Closed → Open	Voltage changes across the plasma membrane
Mechanically Gated Channel	Closed → Open	Mechanical deformations of the channel (by pressure, stretch, etc.)

Ions moving against their electrochemical gradients move via ATP-consuming pumps. One of the most important pumps in electrophysiology is the **sodium-potassium ion pump,** or **Na$^+$/K$^+$ ATPase,** which brings two potassium ions into the cytosol as it moves three sodium ions into the extracellular fluid. This pump maintains, and to some extent creates, the vital concentration gradients of sodium and potassium ions that exist across the plasma membrane, an example of the Gradients Core Principle (p. 28). In a neuron, as in a muscle fiber, the concentration of sodium ions is higher in the extracellular fluid than in the cytosol, and the opposite is true for potassium ions—their concentration is higher in the cytosol than in the extracellular fluid.

CORE PRINCIPLE
Gradients

Changes in the Membrane Potential: Ion Movements

Now let's connect the two concepts we have been discussing: ion channels and gradients plus the resting membrane potential. Consider the three facts in the following list.

- A cell starts with a negative resting membrane potential, meaning it is negative with respect to the extracellular fluid.
- There is an unequal distribution of ions across the plasma membrane. We have discussed the sodium and potassium ion gradients, but there are also gradients for other ions such as chloride and calcium. These gradients are maintained by gated channels and pumps such as the Na$^+$/K$^+$ pump.
- When the gated channels for a specific ion are triggered to open, those ions will follow their electrochemical gradient into or out of the cell.

When we think about this, it follows that we can alter the cell's membrane potential by opening gated channels and causing ions to flow into or out of the cell. Take a look at **Figure 11.11** to see how this works. In Figure 11.11a we have the membrane at rest, or not being stimulated, and its gated ion channel is closed. In Figure 11.11b, a ligand binds a ligand-gated cation channel, and cations (such as sodium ions) follow their electrochemical gradient and enter the cell. The influx of positive charges makes the membrane potential less negative, a change called **depolarization.** During this process, the cell becomes less polarized as its

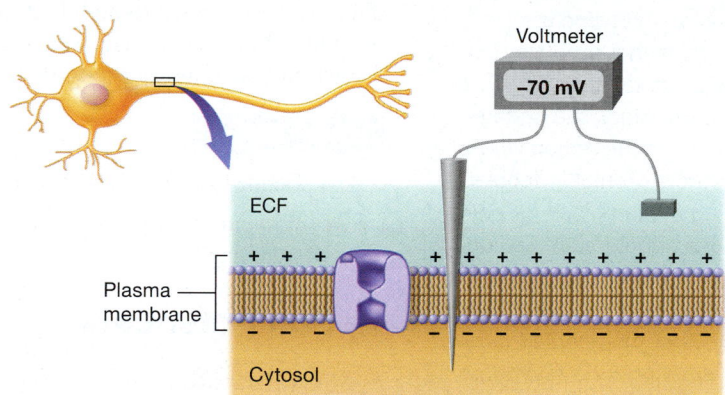

(a) The membrane at its negative resting membrane potential, before stimulation.

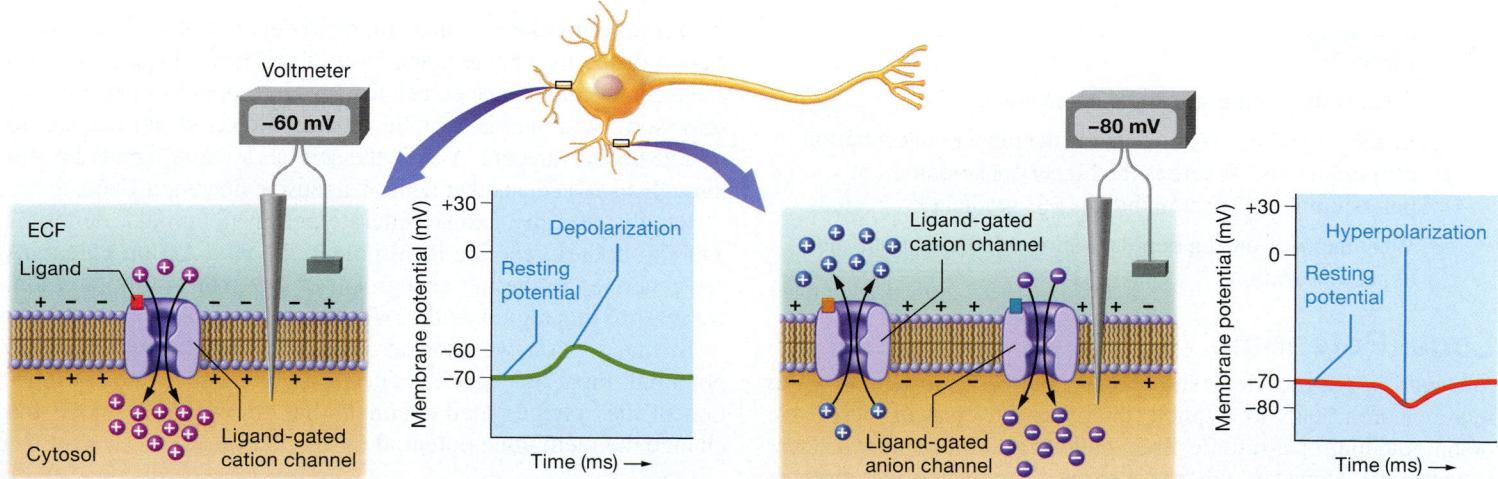

(b) Depolarization: Gain of positive charges makes the inside of the cell less negative, causing depolarization.

(c) Hyperpolarization: Loss of positive charges (or gain of negative charges) makes the inside of the cell more negative, causing hyperpolarization.

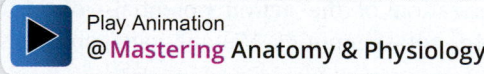

Play Animation
@ **Mastering** Anatomy & Physiology

Figure 11.11 Ion movements leading to changes in the membrane potential. The changes shown here are local potentials.

membrane potential approaches 0 mV. (Find out what happens when depolarization is blocked in *A&P in the Real World: Local Anesthetic Drugs.*) When a cell returns to its resting membrane potential, **repolarization** has occurred.

In Figure 11.11c, a ligand binds to a cation channel (such as a potassium ion channel) for which the electrochemical gradient is reversed, allowing cations to flow out of the cell into the extracellular fluid. As the cell loses positive charges, the membrane potential becomes more negative than it is at rest, a change termed **hyperpolarization**. Note that hyperpolarization may also result from the opening of channels for anions, such as chloride ions, which would allow these negatively charged ions to flow into the cell.

Both depolarization and hyperpolarization are seen in neurons. In the upcoming sections, we see how this applies to nervous system physiology and the ability of the neuron to send signals.

Local Anesthetic Drugs

Local anesthetics such as *lidocaine* are commonly administered agents that produce temporary numbness in an area, usually for a surgical or dental procedure. These drugs block the voltage-gated sodium ion channels of the neurons in the region where they are injected. Blocking these channels prohibits depolarization, and so action potentials relaying the impulses for pain are not transmitted to the CNS. Because agents such as lidocaine block sodium ion channels nonselectively, they will also affect the sodium ion channels of muscles in the area. This causes temporary weakness or paralysis of the affected muscle. As a result, you may be prone to drooling for a few hours after leaving the dentist's office.

Quick Check

☐ 1. What is the resting membrane potential?

☐ 2. In and around the axon, where is the higher concentration of sodium ions? Where is the higher concentration of potassium ions? What maintains this gradient?

☐ 3. What happens during depolarization, repolarization, and hyperpolarization?

Local Potentials

You read in the muscle tissue chapter that each stimulus from a motor neuron leads to a quick, temporary reversal in the membrane potential of a muscle fiber, called an *action potential* (see Chapter 10). However, when a neuron is stimulated just once, a full action potential rarely results. Instead, a small, local change in the membrane potential of the neuron, called a **local potential,** is produced (see Figure 11.11).

A local potential may have one of two effects:

- It may cause a depolarization, in which positive charges enter the cytosol and make the membrane potential less negative (e.g., a change from −70 to −60 mV).
- Alternatively, it may cause a hyperpolarization, in which either positive charges exit or negative charges enter the cytosol to make the membrane potential more negative (e.g., a change from −70 to −80 mV).

Local potentials are sometimes called *graded potentials* because they vary greatly in size—some produce a larger change in membrane potential than others. The degree of change in the membrane potential during a local potential depends on multiple factors, including the length of stimulation, number of ion channels that open, and type(s) of ion channels that open. Another feature of local potentials is that they are reversible; when the stimulus that caused the ion channels to open stops, the neuron quickly returns to its resting potential. Local potentials are also *decremental* in nature: The changes in membrane potential they produce are small, and the current generated is lost across the membrane over the distance of a few millimeters. Consequently, local potentials cannot send signals over great distances, and are useful for short-distance signaling only (which is why they're called *local* potentials). However, even though they occur only over short distances, we will see in the next section that local potentials are vital triggers for action potentials, our long-distance signals.

Quick Check

☐ 4. Define local potential. Why is it also called a graded potential?

☐ 5. Why are local potentials useful only for short-distance signaling?

Action Potentials

◄◄ FLASHBACK

1. What are negative and positive feedback loops? (pp. 22–24)

2. What takes place during an action potential? (p. 350)

An **action potential** is a uniform, rapid depolarization and repolarization of the membrane potential of a cell (see Chapter 10). This change in the membrane potential causes a response—or *action*—of some sort. For a muscle fiber, the change initiates events that lead to muscle fiber contraction. Within the nervous system, signals are sent through an axon to another neuron, a muscle fiber, or a gland.

Recall that only axons generate action potentials; dendrites and cell bodies generate local potentials only. Action potentials are generated in the initial segment of the axon; for this reason, we refer to this region as the *trigger zone*.

In this section we look at what happens during an action potential. First, however, we need to delve deeper into the function of the voltage-gated channels that allow ions to move and change the membrane potential of the neuron.

States of Voltage-Gated Channels

Two types of voltage-gated channels are involved in the depolarization and repolarization of the action potential—one for sodium ions and one for potassium ions. Voltage-gated channels

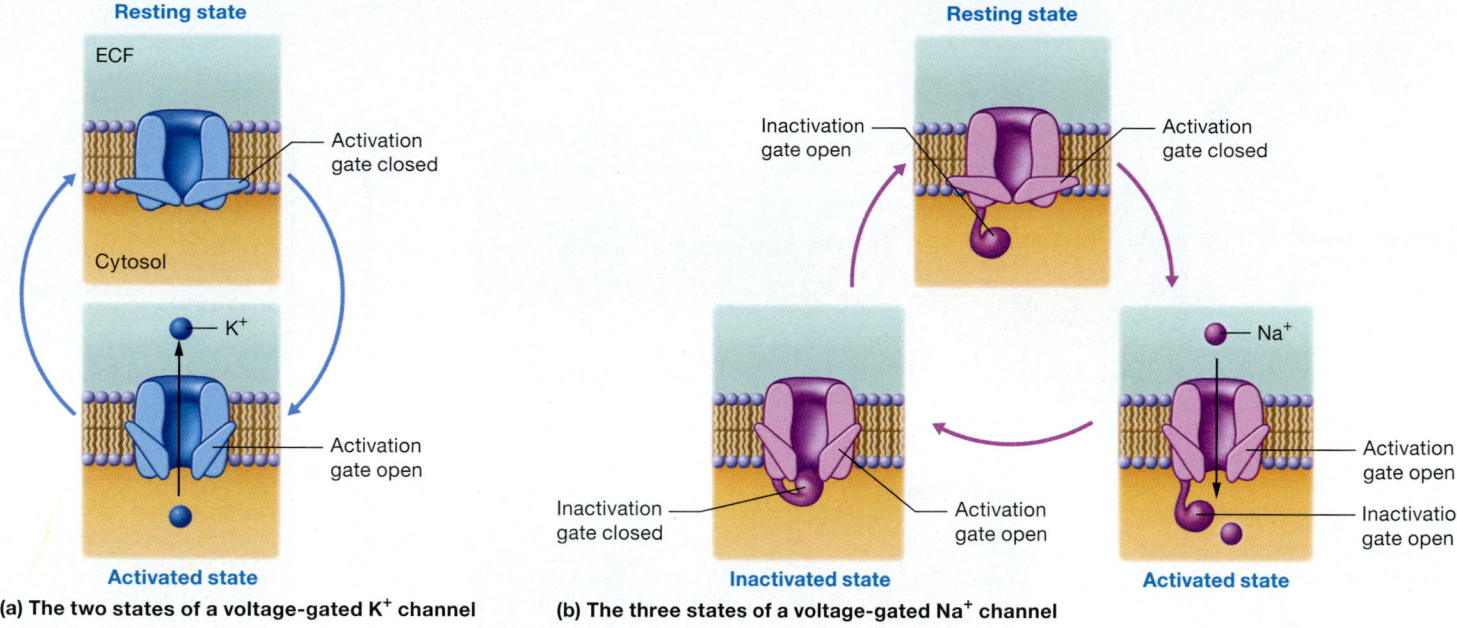

(a) The two states of a voltage-gated K⁺ channel

(b) The three states of a voltage-gated Na⁺ channel

Figure 11.12 States of voltage-gated channels.

Play Interactive Physiology 2.0
@ Mastering Anatomy & Physiology

are found most abundantly in the axolemma of the neuron, which is why only axons have action potentials.

The structures of voltage-gated potassium and sodium ion channels are depicted in **Figure 11.12**. Notice in Figure 11.12a that the voltage-gated potassium ion channel has two possible states: resting and activated. In the resting state, the channel is closed. In the activated state, the channel is open and allows potassium ions to cross the axolemma.

The voltage-gated sodium ion channel shown in Figure 11.12b is more complicated. It has two gates: an *activation gate* and an *inactivation gate*. This means a sodium ion channel has three potential "states":

- **Resting state: Inactivation gate opened, activation gate closed.** During the resting state the neuron is not being stimulated, and the activation gate is closed and the inactivation gate is open. No sodium ions cross the membrane when the channel is in the resting state.
- **Activated state: Both activation and inactivation gates opened.** When an action potential is initiated, the voltage change opens the activation gates and the channel is in its activated state. The channel in the activated state allows sodium ions to cross the axolemma.
- **Inactivated state: Inactivation gate closed, activation gate opened.** When the inactivation gate closes, the channel is in its inactivated state. The channel in this state no longer allows sodium ions to pass through. Notice that during this state, the activation gate remains open. When the action potential is finished, the channel returns to the resting state.

Events of an Action Potential

Let's examine the sequence of events of an action potential in a section of axon, illustrated in **Figure 11.13** on p. 398. The entire

sequence takes just a few milliseconds. Neuronal action potentials have three general phases: the depolarization phase, the repolarization phase, and the hyperpolarization phase. During the **depolarization phase,** the membrane potential rises toward zero and then becomes briefly positive. The membrane potential returns to a negative value during the **repolarization phase,** and then becomes temporarily more negative than resting during the **hyperpolarization phase.** Each phase occurs because of the selective opening and closing of specific voltage-gated ion channels. Note that before the action potential, when the membrane is at rest, both the sodium and the potassium ion channels are in the resting state.

The action potential proceeds as follows:

1. **A local potential depolarizes the axolemma of the trigger zone to threshold.** The action potential begins when the voltage-gated sodium ion channels in the axolemma of the trigger zone enter the activated (open) state (see Figure 11.12b). However, these voltage-gated channels will become activated only if the membrane is already depolarized to a level known as **threshold,** usually −55 mV. The source of this depolarization is generally local potentials that arrive from the cell body.

2. **Voltage-gated sodium ion channels activate, sodium ions enter, and the axon section depolarizes.** When threshold is reached, the sodium ion channels in the trigger zone are activated (open) and sodium ions rush into the neuron with their electrochemical gradient. As the membrane potential becomes more positive, more voltage-gated sodium ion channels are activated. This cycle continues, and the more the axon depolarizes, the more voltage-gated sodium ion channels are activated. This influx of positive charges causes rapid depolarization to about +30 mV. You may recognize this as an example

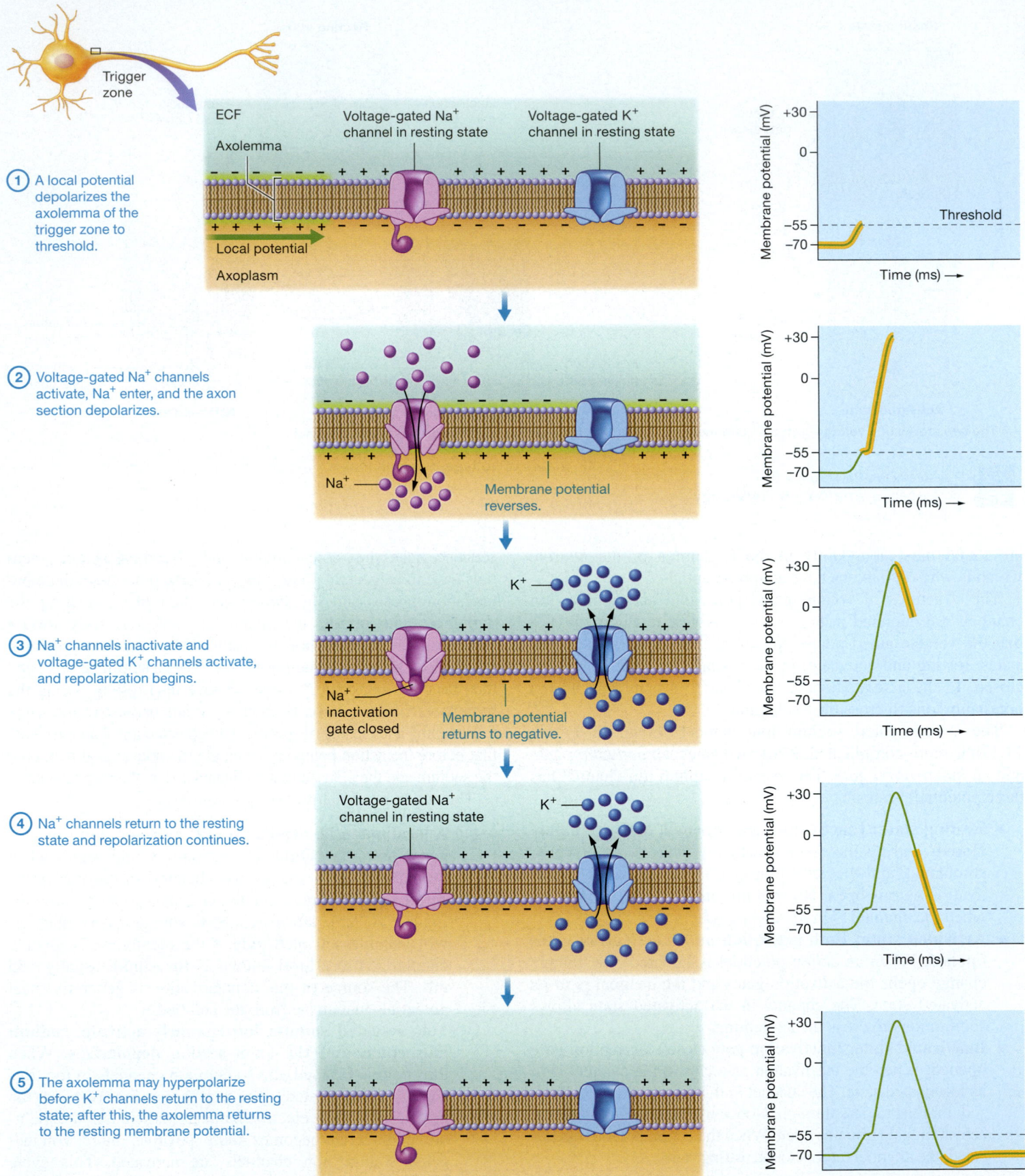

Trigger
zone

ECF
Axolemma
Voltage-gated Na⁺
channel in resting state
Voltage-gated K⁺
channel in resting state

1 A local potential
depolarizes the
axolemma of the
trigger zone to
threshold.

Local potential
Axoplasm

Membrane potential (mV)
+30
0
−55   Threshold
−70
Time (ms)

2 Voltage-gated Na⁺ channels
activate, Na⁺ enter, and the axon
section depolarizes.

Na⁺
Membrane potential
reverses.

Membrane potential (mV)
+30
0
−55
−70
Time (ms)

3 Na⁺ channels inactivate and
voltage-gated K⁺ channels activate,
and repolarization begins.

K⁺

Na⁺
inactivation
gate closed
Membrane potential
returns to negative.

Membrane potential (mV)
+30
0
−55
−70
Time (ms)

4 Na⁺ channels return to the resting
state and repolarization continues.

Voltage-gated Na⁺
channel in resting state
K⁺

Membrane potential (mV)
+30
0
−55
−70
Time (ms)

5 The axolemma may hyperpolarize
before K⁺ channels return to the resting
state; after this, the axolemma returns
to the resting membrane potential.

Membrane potential (mV)
+30
0
−55
−70
Time (ms)

Figure 11.13 Events of an action potential.

▶ Play Animation
@ **Mastering** Anatomy & Physiology

of a *positive* feedback loop—the initial input (activation of sodium ion channels and depolarization) amplifies the output (more sodium ion channels are activated and the axolemma depolarizes further), an example of the Feedback Loops Core Principle (p. 28).

CORE PRINCIPLE
Feedback Loops

③ **Sodium ion channels inactivate and voltage-gated potassium ion channels activate, and repolarization begins.** When the axolemma is fully depolarized (about +30 mV), the inactivation gates of the voltage-gated sodium ion channels close, and sodium ions stop entering the axon. As this occurs, voltage-gated potassium ion channels slowly open and potassium ions flow out of the axon along their electrochemical gradient, causing the axolemma of the trigger zone to lose positive charges and so to begin repolarization.

④ **Sodium ion channels return to the resting state and repolarization continues.** As potassium ions exit the axon and repolarization continues, the activation gates of the sodium ion channels close and the inactivation gates open, returning the sodium ion channels to their resting state.

⑤ **The axolemma may hyperpolarize before potassium ion channels return to the resting state; after this, the axolemma returns to the resting membrane potential.** In many axons, the outflow of potassium ions continues until the membrane potential of the axolemma hyperpolarizes, possibly becoming as negative as –90 mV. The axolemma hyperpolarizes because the gates of the potassium ion channels are slow to close, allowing additional potassium ions to leak out of the cell. Hyperpolarization finishes as the voltage-gated potassium ion channels return to their resting state. After the action potential, the potassium leak channels and Na^+/K^+ pumps re-establish the resting membrane potential.

Throughout the preceding sequence of events of a single action potential, very little change occurs in the intracellular or extracellular concentration of sodium or potassium ions, and therefore the gradient isn't too disturbed. However, with repetitive action potentials, the gradient will eventually deplete, and the neuron relies on the Na^+/K^+ pumps in the axolemma to restore it.

Quick Check

☐ 6. What takes place during the depolarization phase of an action potential? How is it an example of a positive feedback loop?

☐ 7. What must be reached in order for voltage-gated sodium ion channels to open?

☐ 8. What takes place during the repolarization and hyperpolarization phases of an action potential?

The Refractory Period

Neurons are limited in how often they can fire action potentials. For a brief time after a neuron has produced an action potential, the membrane cannot be stimulated to fire another one. This time is called the **refractory period** (**Figure 11.14**). The refractory period may be divided into two phases: the *absolute refractory period* and the *relative refractory period*.

During the **absolute refractory period,** no additional stimulus, no matter how strong, is able to produce an additional action potential. Notice in Figure 11.14 that this period coincides with the voltage-gated sodium ion channels being in their activated and inactivated states; sodium ion channels may not be activated until they return to their resting states with their activation gates closed and their inactivation gates open.

Immediately following the absolute refractory period is the **relative refractory period,** during which only a strong stimulus will produce an action potential. The relative refractory period is marked by a return of voltage-gated sodium ion channels to their resting state while some potassium ion channels remain activated. It's difficult to depolarize the membrane to threshold during this period because the potassium ion channels are activated and the membrane is repolarizing or even hyperpolarizing. However, if a greater than normal stimulus is applied, the membrane may depolarize to threshold, and the axon may fire off another action potential.

The absolute and relative refractory periods limit the frequency of action potential production. In addition, the relative refractory period ensures that stronger stimuli trigger more frequent action potentials.

Quick Check

☐ 9. What are the absolute and relative refractory periods?

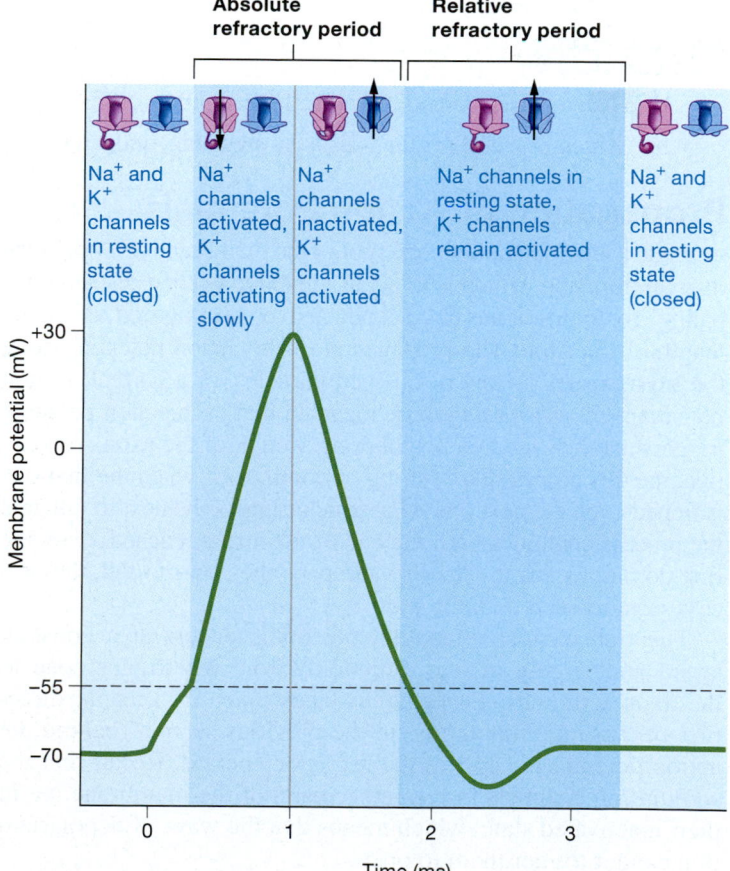

Figure 11.14 Refractory periods of an action potential.

Local and Action Potentials Compared

Now that we have discussed both local potentials and action potentials, let's highlight their differences. You discovered earlier that local potentials are graded, and so produce changes in membrane potential of varying degree; however, each action potential will cause a maximum depolarization of the same amount, to about +30 mV. This is due to a phenomenon called the **all-or-none principle.** Simply put, this principle refers to an event, in this case an action potential, that either happens completely or doesn't happen at all. If a neuron does not depolarize to threshold, an action potential does not occur. If the neuron does depolarize to threshold, the result is an action potential of a characteristic strength. The size of the action potential is not determined by the strength, frequency, or length of the stimulus, and therefore is not graded like a local potential.

The all-or-none principle leads us to a second difference between local potentials and action potentials: their reversibility. Recall that a local potential is reversible; once the stimulus stops, the ion channels close and the resting membrane potential is restored. However, a key feature of an action potential is that when one occurs, it is irreversible—once threshold is reached, it cannot be stopped and will proceed to completion.

Finally, a third important difference between local potentials and action potentials is the distance the signal must travel. Whereas local potentials are decremental and decrease over short distances, action potentials are *nondecremental;* that is, their strength does not diminish. This property of action potentials is key, as otherwise, signals could not be sent over long distances in the nervous system.

Quick Check

☐ 10. How do local potentials and action potentials differ?

☐ 11. Which is useful for long-distance signaling, and why?

Propagation of Action Potentials

A single action potential in one spot of the membrane can't perform its main function, which is to act as a method of long-distance signaling. To do this, it has to be conducted, or **propagated,** down the length of the axon. The propagation of the action potential along the axon creates a flow of charged particles, or a current. Action potentials are *self-propagating,* meaning that each action potential triggers another one in a neighboring section of the axon. You can imagine this process like a string of dominoes—when the first one is tipped over, the next one falls, which triggers the next to fall, and the process continues until the end of the line is reached. Only the first domino needs the "push," and once they start to fall, the process sustains itself until the end.

The transmission of action potentials occurs at a constant speed and largely in one direction—from the trigger zone to the axon terminals. Propagation takes place in a single direction because the membrane in the previous section (behind the action potential) is still in the refractory period. Recall that the sodium ion channels in refractory parts of the membrane are in their inactivated state, which means that the wave of depolarization cannot trigger them to open.

Let's take a look at how propagation occurs and the factors that influence the speed of action potential conduction.

Events of Propagation

The action potential is propagated along the axon by the following sequence of events, shown in **Figure 11.15**:

① **The axolemma depolarizes to threshold due to local potentials.** The axolemma of the trigger zone is depolarized to threshold by local potentials from the dendrites, cell body, or axon.

② **As sodium ion channels activate, an action potential is triggered and spreads positive charges down the axon.** Voltage-gated sodium ion channels are activated (open) and an action potential occurs. When this happens, positive charges flow down the axon through the axoplasm.

③ **The next section of the axolemma depolarizes to threshold and fires an action potential as the previous section of the axolemma repolarizes.** As the depolarizing current reaches the next section of the axolemma, it depolarizes that section to threshold. The voltage-gated sodium ion channels in that part of the axolemma are activated, and that section then generates an action potential. The current then flows down to the next section of the axon. Note that the section of the axolemma that had an action potential in step ② is repolarizing (its potassium ion channels are activated and potassium ions are exiting the axoplasm) and is in its refractory period, so any current that flows backward can't trigger an action potential.

④ **The current continues to move down the axon, and the process repeats.** The current flowing into the next section of the axolemma causes it to depolarize to threshold, activating voltage-gated sodium ion channels and producing an action potential there.

The process then repeats down the length of the axon until it reaches the axon terminals.

Conduction Speed

The rate at which propagation occurs is called **conduction speed,** and it determines how rapidly signaling can occur within the nervous system. Conduction speed is influenced by two main factors: the diameter of the axon and the presence or absence of a myelin sheath. The diameter of the axon affects the conduction of current through the axon because larger axons have lower resistance to conduction, and therefore current flows through them more easily.

The second determinant of conduction speed is the presence or absence of a myelin sheath. Two types of conduction can take place in an axon: **saltatory conduction,** in which the myelin sheath is present, and **continuous conduction,** in which it is absent (**Figure 11.16** on p. 402). Recall from our discussion in Module 11.2 that myelin is an excellent insulator of electrical charge. So, the flow of current is far more efficient in a myelinated axon, which causes saltatory conduction to be significantly faster than continuous conduction.

Figure 11.16 compares conduction along a myelinated axon and an unmyelinated axon, and depicts the effect of the myelin sheath on conduction speed. You can see that on the myelinated axon, the nodes of Ranvier are the only segments that must be depolarized to threshold. When the node, rich in voltage-gated

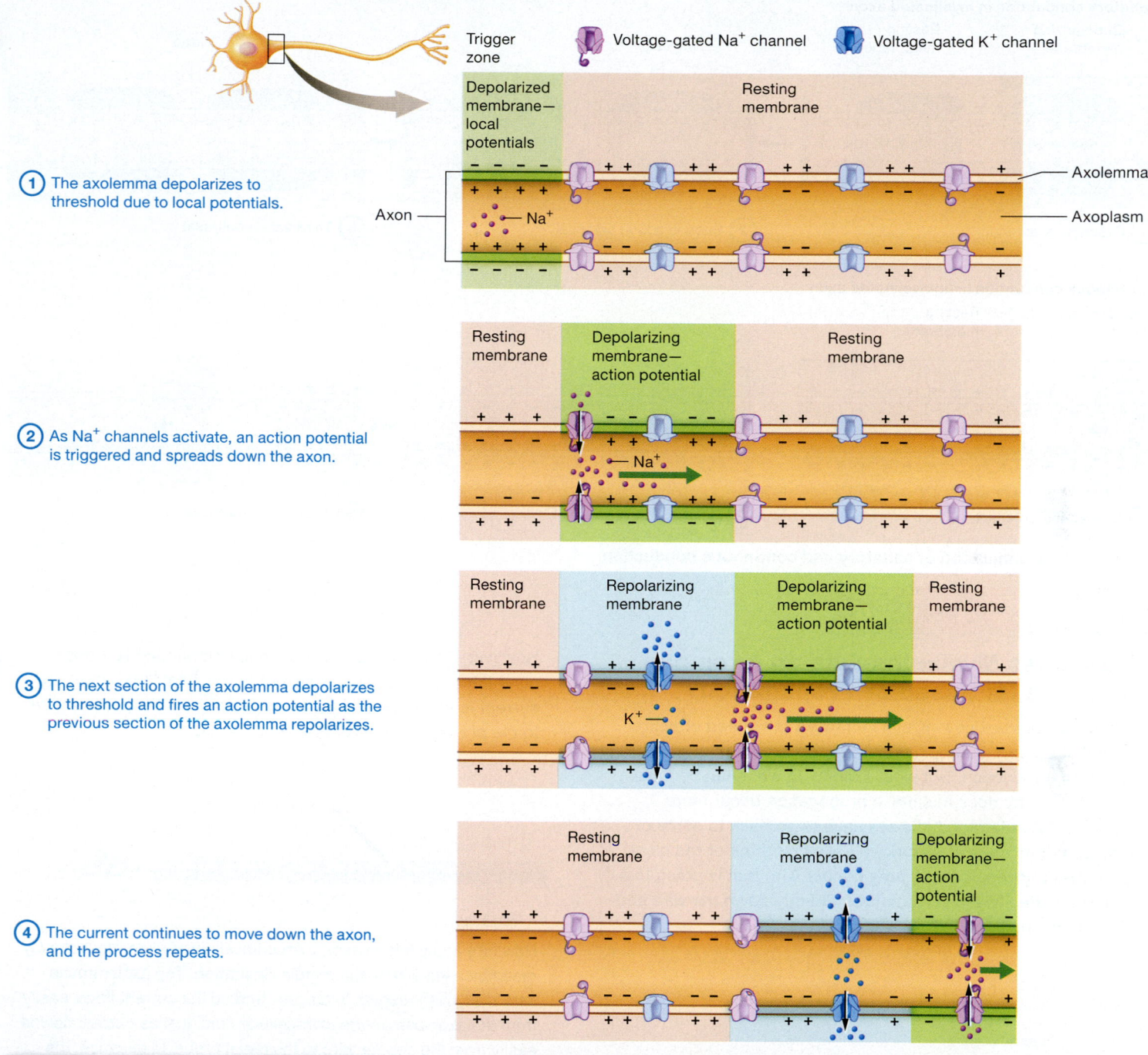

Trigger zone

🟣 Voltage-gated Na⁺ channel 🔵 Voltage-gated K⁺ channel

① The axolemma depolarizes to threshold due to local potentials.

② As Na⁺ channels activate, an action potential is triggered and spreads down the axon.

③ The next section of the axolemma depolarizes to threshold and fires an action potential as the previous section of the axolemma repolarizes.

④ The current continues to move down the axon, and the process repeats.

Figure 11.15 Propagation of an action potential.

Play Animation
@ **Mastering** Anatomy & Physiology

sodium ion channels, is depolarized to threshold, an action potential is triggered. This action potential generates a current that flows passively and efficiently with little loss of charge through the next myelinated segment, or internode (see Figure 11.8 for a review of anatomical terminology). When the current reaches the next node of Ranvier, another action potential is generated. This cycle is repeated down the length of the axon, and the current "jumps" from one node to the next (in fact, "saltatory" comes from the Latin word *saltare,* which means "leaping").

Compare these "leaping" action potentials with the much slower and more gradual continuous conduction seen in the unmyelinated axon. Figure 11.16 shows that in continuous conduction, absence of the myelin sheath means that each section of the axolemma must be depolarized to threshold. This is why conduction of this sort is called *continuous*—action potentials must be generated in a continuous sequence along the entire axolemma for the current to spread down the length of the axon. The flow of current in a myelinated axon is much faster than the process of triggering action potentials in each part of an unmyelinated axon. See *A&P in the Real World: Multiple Sclerosis* on p. 403 for information on how loss of the myelin sheath affects conduction speed.

Saltatory conduction in myelinated axon

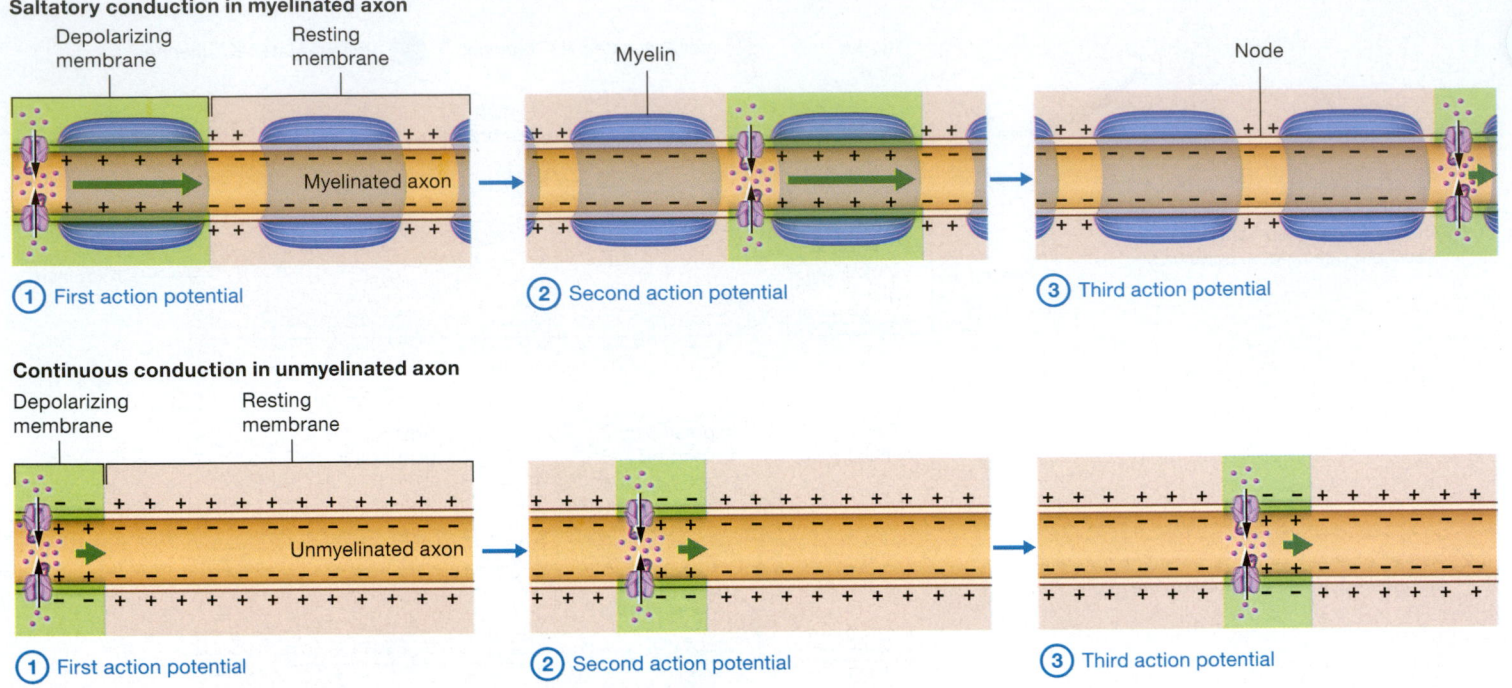

Figure 11.16 **Comparison of saltatory and continuous conduction.**

ConceptBOOST)))

How Does Myelin Insulate an Axon and Increase Its Speed of Propagation?

We just discussed how myelin increases the speed of action potential propagation by "insulating" the axon and enabling saltatory conduction. But what exactly do we mean by "insulated," and why does this make propagation occur more quickly? To understand this, we have to go back to some of the basic principles of electricity. Let's first think of the electric current flowing down a bare copper wire that is going to a light bulb. Ideally, the current flows directly down the wire and illuminates the light bulb, as shown here:

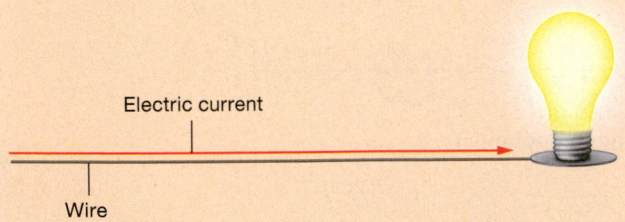

But, if you touch the wire with a metal probe, as depicted in the next illustration, most of the current might instead flow down the probe, a situation known as a *short circuit*:

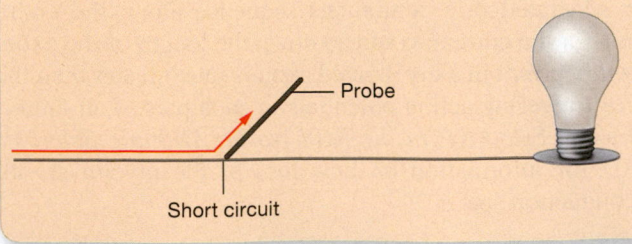

However, if the wire is encased in a material that is a poor conductor of electricity, as shown in this final illustration, the current is *insulated* and unable to move from the copper wire to the probe. This prevents a short circuit.

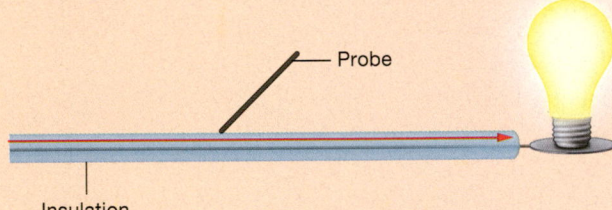

Now let's apply this to axons. An unmyelinated axon most closely resembles the wire in the middle illustration. The axolemma is very leaky with respect to current, and so the current flows easily from the axoplasm to the extracellular fluid, just as current flowed easily from the copper wire to the metal probe. In essence, the current in an unmyelinated axon short circuits, and so has to be continuously regenerated down the length of the axon.

A myelinated axon more closely resembles the wire in the final illustration. Myelin is a very good insulator because it is a poor conductor of electricity, and so it prevents current from leaking out through the axolemma. This means that the signal decreases very little in strength as it travels through an internode. The action potential can then propagate through the internode without having to be regenerated. It is not until the current reaches the unmyelinated node of Ranvier that it starts to dissipate and the action potential must be regenerated. This is why the action potentials of saltatory conduction can "leap" from node to node and why this type of propagation is so much faster than continuous conduction. ■

Multiple Sclerosis

in the Real World

Multiple sclerosis (MS) is a disease in which certain cells of the immune system attack the myelin sheath around axons of the CNS. This disease is therefore a type of autoimmune disorder (one in which the patient's own immune system attacks a certain part of the body). In most cases MS causes a progressive loss of the myelin sheath, which in turn produces loss of current from the neurons. Symptoms of the disease result from gradual slowing of action potential propagation as saltatory conduction becomes less efficient.

The exact symptoms of MS depend on the regions of the CNS that are affected. Over time most patients exhibit changes in sensation (e.g., numbness), alterations in behavior and cognitive abilities, and motor dysfunction including paralysis, all of which may cause significant disability.

Classification of Axons by Conduction Speed

Axons are often classified according to conduction speed. The two primary defining features in this classification are the diameter of the axon and the presence or absence of the myelin sheath. The three main classes are as follows:

- **Type A fibers** are the largest-diameter axons (5–20 μm), all of which are myelinated. These characteristics give them the fastest conduction speed, with a maximum speed of about 120 m/sec (about 250 mi/h). Type A fibers are found in parts of the body with which the CNS must communicate extremely rapidly, such as certain sensory axons from joints and muscle fibers, as well as motor axons to skeletal muscles.
- **Type B fibers** are intermediate in diameter (2–3 μm) and most are myelinated. They typically have a maximum conduction speed of about 15 m/sec (about 32 mi/h), and include certain efferent fibers of the autonomic nervous system and certain sensory axons coming from organs.
- **Type C fibers** are the smallest fibers (0.5–1.5 μm) and are unmyelinated. Their conduction speed is the slowest, conducting action potentials only at about 0.5–2 m/sec (about 1–5 mi/h). Type C fibers include other efferent fibers of the autonomic nervous system and certain sensory axons that transmit pain, temperature, and certain pressure sensations.

Quick Check

- ☐ 12. How is an action potential propagated down an axon?
- ☐ 13. How do saltatory conduction and continuous conduction differ? Which is faster, and why?

Putting It All Together: The Big Picture of Action Potentials

After all this talk of action potentials, you may still find some of these concepts rather abstract. Before moving on to the next module, make sure you think about the big picture of how and why the action potential occurs. The *how* of the action potential is shown in

Figure 11.17. The *why* of the action potential is long-distance signaling. As you will see in the upcoming module, the arrival of the action potential at the axon terminal is what allows the neuron to communicate with its target cells.

Apply What You Learned

- ☐ 1. Predict the effect of the poison ouabain (wah-BAY-in), which blocks Na^+/K^+ pumps, on the neuronal action potential. (*Hint:* What would happen to the sodium and potassium ion gradients?)
- ☐ 2. What do you think would happen to the neuronal action potential if the concentration of sodium ions in the extracellular fluid decreased significantly, to the point of reversing the gradient?
- ☐ 3. Sometimes when you pull your dinner out of the microwave you have to hold your fingertips to the food for a second or two before you can tell if it is hot or cold. Explain why this happens. (*Hint:* What type of fiber transmits temperature stimuli?)

See answers in Appendix A.

MODULE 11.4
Neuronal Synapses

Learning Outcomes

1. Compare and contrast electrical and chemical synapses.
2. Describe the structures that make up a chemical synapse.
3. Discuss the relationship between a neurotransmitter and its receptor.
4. Describe the events of chemical synaptic transmission in chronological order.
5. Define excitatory postsynaptic potential (EPSP) and inhibitory postsynaptic potential (IPSP), and interpret graphs of an EPSP and an IPSP.
6. Explain temporal and spatial summation of synaptic potentials.

Up to this point, we have discussed how signals are generated and propagated *within* a neuron. Remember, though, that neurons must communicate with other cells, including other neurons, to carry out their functions—an example of the Cell-Cell Communication Core Principle (p. 28). So now in this module, we turn to how signals are transmitted *between* neurons at loca-

CORE PRINCIPLE
Cell-Cell Communication

tions called *synapses*. Recall from the muscle tissue chapter that a **synapse** (SIN-aps; *syn-* = "to clasp or join") is where a neuron meets its target cell (see Chapter 10). The discussion in that chapter revolved around a specific type of synapse—the neuromuscular junction. Here we explore the synapses that occur between two neurons, or *neuronal synapses*. These synapses may be of two types, *electrical* and *chemical*. We examine both types in this module, although we focus on chemical synapses.

The Big Picture of Action Potentials

Figure 11.17

Play Animation
@ Mastering Anatomy & Physiology

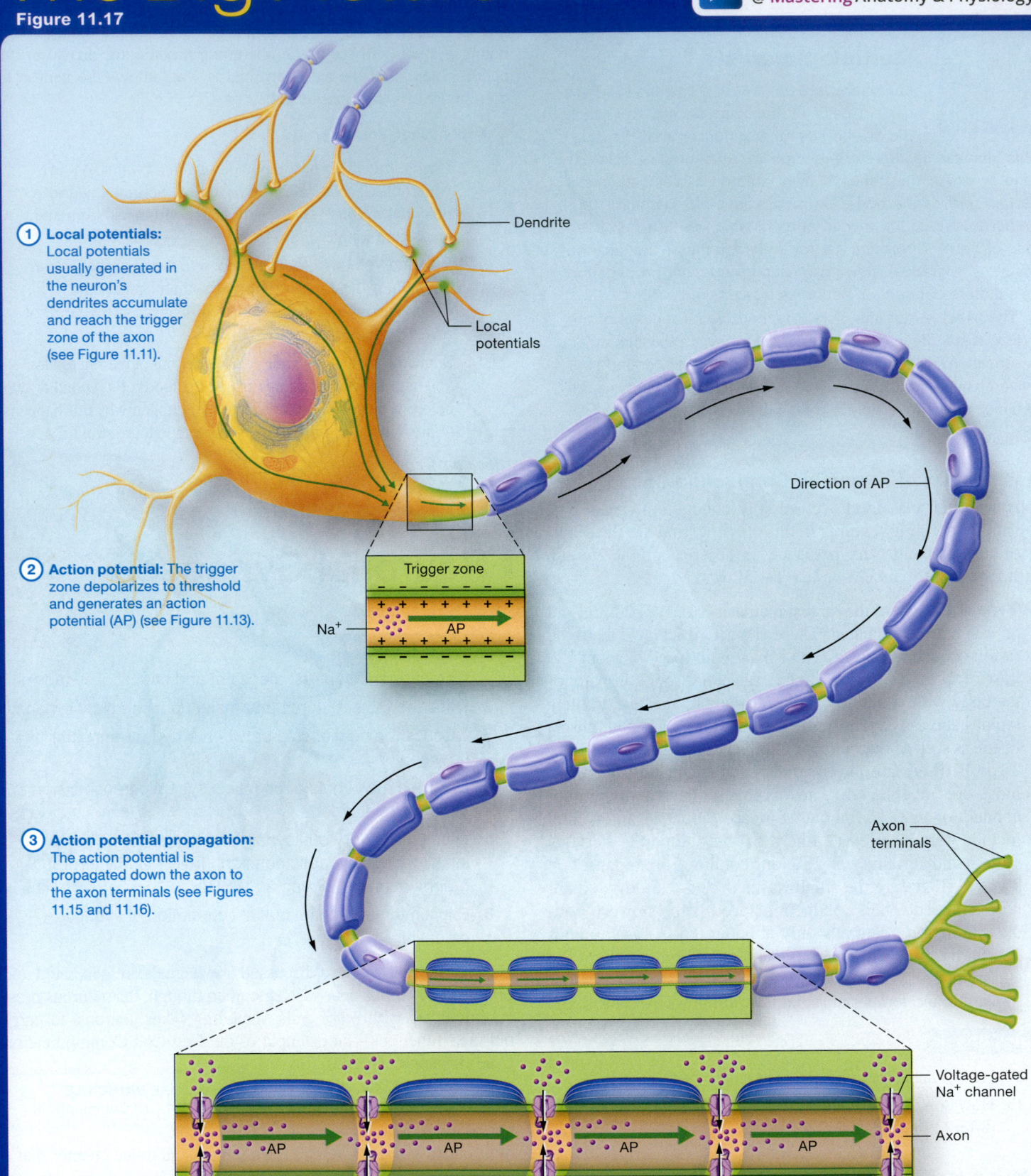

1 Local potentials: Local potentials usually generated in the neuron's dendrites accumulate and reach the trigger zone of the axon (see Figure 11.11).

Dendrite

Local potentials

2 Action potential: The trigger zone depolarizes to threshold and generates an action potential (AP) (see Figure 11.13).

Trigger zone

Na⁺

AP

Direction of AP

3 Action potential propagation: The action potential is propagated down the axon to the axon terminals (see Figures 11.15 and 11.16).

Axon terminals

Voltage-gated Na⁺ channel

Axon

AP

Na⁺

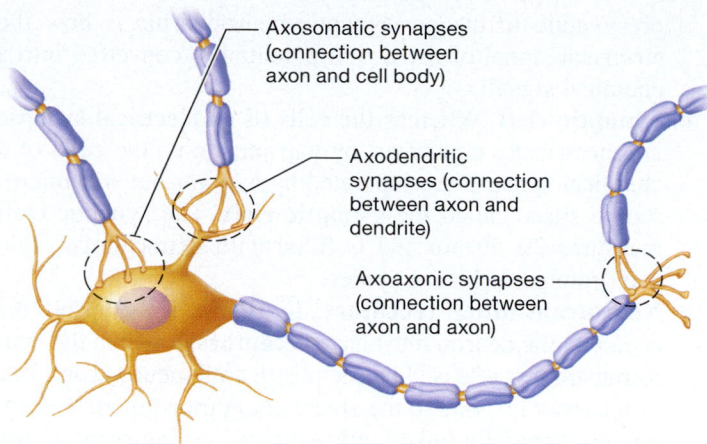

Figure 11.18 Structural types of synapses.

The transfer of chemical or electrical signals between neurons at a synapse is called **synaptic transmission,** and it is the fundamental process for most functions of the nervous system. Synaptic transmission allows voluntary movement, cognition, sensation, and emotion, as well as countless other processes. Each neuron has an enormous number of synapses. Recall from Module 11.2 that each axon generally splits into 1000 or more axon terminals, and each terminal meets up with another axon, dendrite, or cell body. An average presynaptic neuron, then, generally forms synapses with about 1000 postsynaptic neurons. A postsynaptic neuron can receive input from even more synapses—an average neuron can have as many as 10,000 synaptic connections from different presynaptic neurons.

Quick Check

☐ 1. What are three locations where presynaptic axons connect with a postsynaptic neuron?

☐ 2. Define synaptic transmission.

Overview of Neuronal Synapses

Neuronal synapses generally occur between an axon and another part of a neuron; they may occur between an axon and a dendrite, an axon and a cell body, and an axon and another axon. These types are called *axodendritic, axosomatic,* and *axoaxonic synapses,* respectively (**Figure 11.18**).

Regardless of the type of synapse, we use certain terms to describe the neurons sending and receiving the message:

- **Presynaptic neuron.** The presynaptic neuron is the neuron that is sending the message from its axon terminal.
- **Postsynaptic neuron.** The postsynaptic neuron is the neuron that is receiving the message from its dendrite, cell body, or axon.

Electrical Synapses

An **electrical synapse** (**Figure 11.19a**) occurs between cells that are electrically coupled via gap junctions. In these synapses, the axolemmas of the two neurons are nearly touching (they are separated by only about 3.5 nm) and the gap junctions contain precisely aligned channels that form pores through which ions and other small substances may travel. This allows the electric current to flow directly from the axoplasm of one neuron to that of the next.

This arrangement creates two unique features of electrical synapses:

- **Synaptic transmission is bidirectional.** In an electrical synapse, transmission is usually bidirectional, which means that either neuron may act as the presynaptic or the

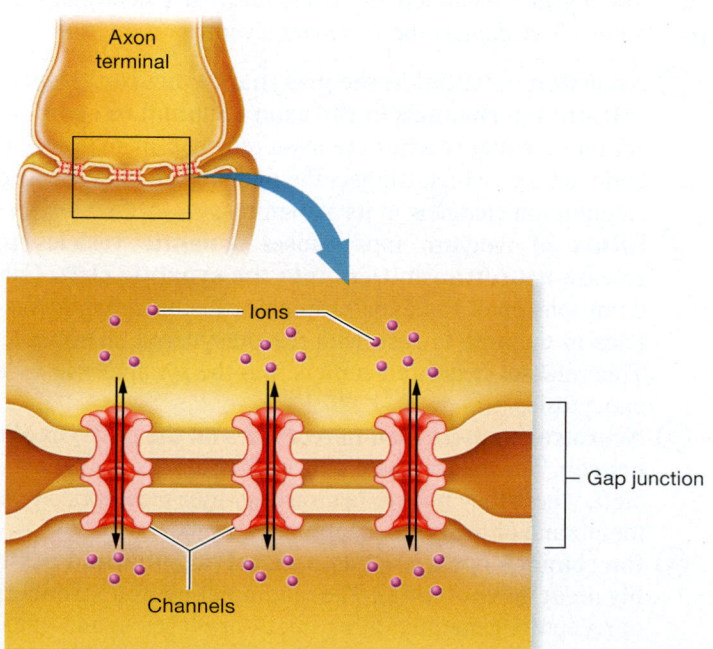

(a) Electrical synapse

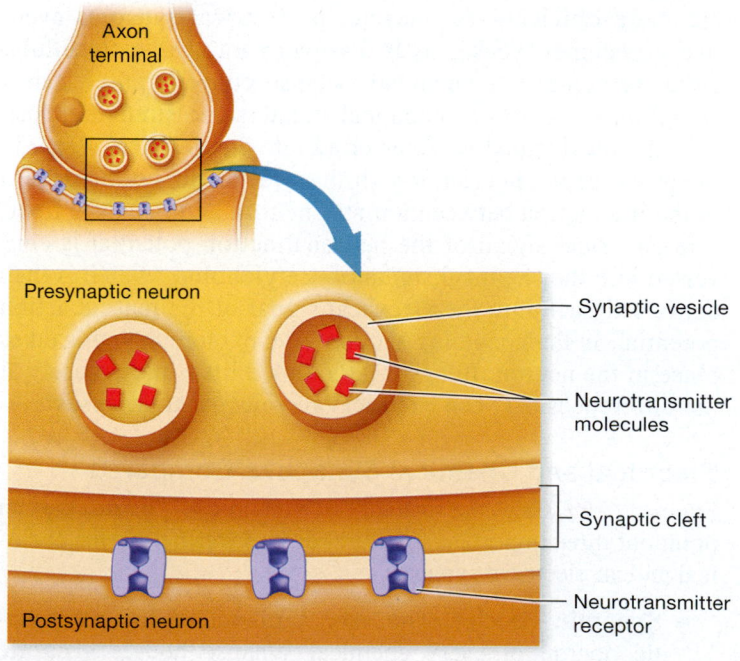

(b) Chemical synapse

Figure 11.19 The structures of electrical and chemical synapses.

postsynaptic neuron and that current may flow in either direction between the two cells.

- **Synaptic transmission is nearly instantaneous.** The delay between depolarization of the presynaptic neuron and change in potential of the postsynaptic neuron is less than 0.1 ms (millisecond), which is extraordinarily fast (we will see that transmission at most chemical synapses requires from one to a few milliseconds).

These features of electrical synapses allow the activity of a group of cells to be synchronized—when stimulated, the cells will produce action potentials in unison. Electrical synapses are found primarily in areas of the brain that are responsible for programmed, automatic behaviors such as breathing. They are also present in developing nervous tissue in the embryo and fetus and are thought to assist in the development of the brain. In addition, electrical synapses are found outside the nervous system in locations such as cardiac and visceral smooth muscle, where they allow those tissues to engage in coordinated muscle activity.

Quick Check

☐ 3. What are the two main features of an electrical synapse?

Chemical Synapses

« FLASHBACK

1. What is a synaptic vesicle? (p. 352)
2. What is a neurotransmitter? (p. 352)

The vast majority of synapses in the nervous system are **chemical synapses.** These synapses are more common because they are more efficient—the current in electrical synapses eventually becomes weaker as it dissipates into the extracellular fluid. In contrast, a chemical synapse converts an electrical signal into a controlled chemical signal, so no strength is lost. The chemical signal is reconverted into an electrical signal in the postsynaptic neuron. Recall that we saw this same pattern in the interaction between a motor neuron and a muscle fiber: The electrical signal of the neuronal action potential is converted into the chemical signal of acetylcholine, which is then converted back into the electrical signal of the muscle action potential. In the upcoming sections, we explore how this takes place in the neuron. But first let's look a little more closely at the differences between chemical and electrical synapses.

Electrical and Chemical Synapses Compared

Figure 11.19b shows the structure of a chemical synapse. We can point out three important structural differences between a chemical and an electrical synapse:

- **Synaptic vesicles.** The axon terminal of the presynaptic neuron of every chemical synapse houses **synaptic vesicles.** These vesicles contain chemical messengers called **neurotransmitters** that transmit signals from the

presynaptic to the postsynaptic neuron. This is how the electrical signal of the action potential is converted into a chemical signal.

- **Synaptic cleft.** Whereas the cells of an electrical synapse are electrically connected by gap junctions, the cells of a chemical synapse are separated by a larger but still microscopic space called the **synaptic cleft.** The synaptic cleft measures 20–50 nm and is filled with extracellular fluid and proteins such as enzymes.

- **Neurotransmitter receptors.** In chemical synapses, the postsynaptic neuron must have **receptors** to which the neurotransmitters released by the presynaptic neuron can bind, or it cannot respond to the signal being transmitted. Receptors are generally linked either directly or indirectly to ion channels. This is how the chemical signal is converted back into an electrical signal.

These three features of chemical synapses cause them to transmit signals more slowly than do electrical synapses. In fact, there is about a 0.5-ms gap between the arrival of the action potential at the axon terminal and the effects on the postsynaptic neuron's membrane, known as **synaptic delay.** Also, chemical synapses are unidirectional—the message can be sent only by the presynaptic neuron. However, these three structural differences also allow something not permitted by the structure of the electrical synapse: The signal can vary in size. If more neurotransmitters are released, then the presynaptic neuron has a greater effect on the postsynaptic neuron. The signal in an electrical synapse, by contrast, will always be the same size. In addition, the effect that the presynaptic neuron triggers can vary with different neurotransmitters and receptors.

Events at a Chemical Synapse

Let's now dig into what actually takes place at a neuronal synapse. **Figure 11.20** depicts the following events:

1. **An action potential in the presynaptic neuron triggers calcium ion channels in the axon terminal to open.** An action potential reaches the axon terminal of the presynaptic neuron, which triggers the opening of voltage-gated calcium ion channels in its axolemma.

2. **Influx of calcium ions causes synaptic vesicles to release neurotransmitters into the synaptic cleft.** Calcium ions enter the axon terminal, causing synaptic vesicles in the area to fuse with the presynaptic membrane. This releases neurotransmitters into the synaptic cleft via exocytosis.

3. **Neurotransmitters bind to receptors on the postsynaptic neuron.** The neurotransmitters diffuse across the synaptic cleft, where they bind to neurotransmitter receptors on the membrane of the postsynaptic neuron.

4. **Ion channels open, leading to a local potential and possibly an action potential.** The binding of neurotransmitters to receptors generally either opens or closes ligand-gated ion channels in the postsynaptic membrane, resulting in a local potential. Such local potentials may or may not lead to an action potential in the postsynaptic neuron.

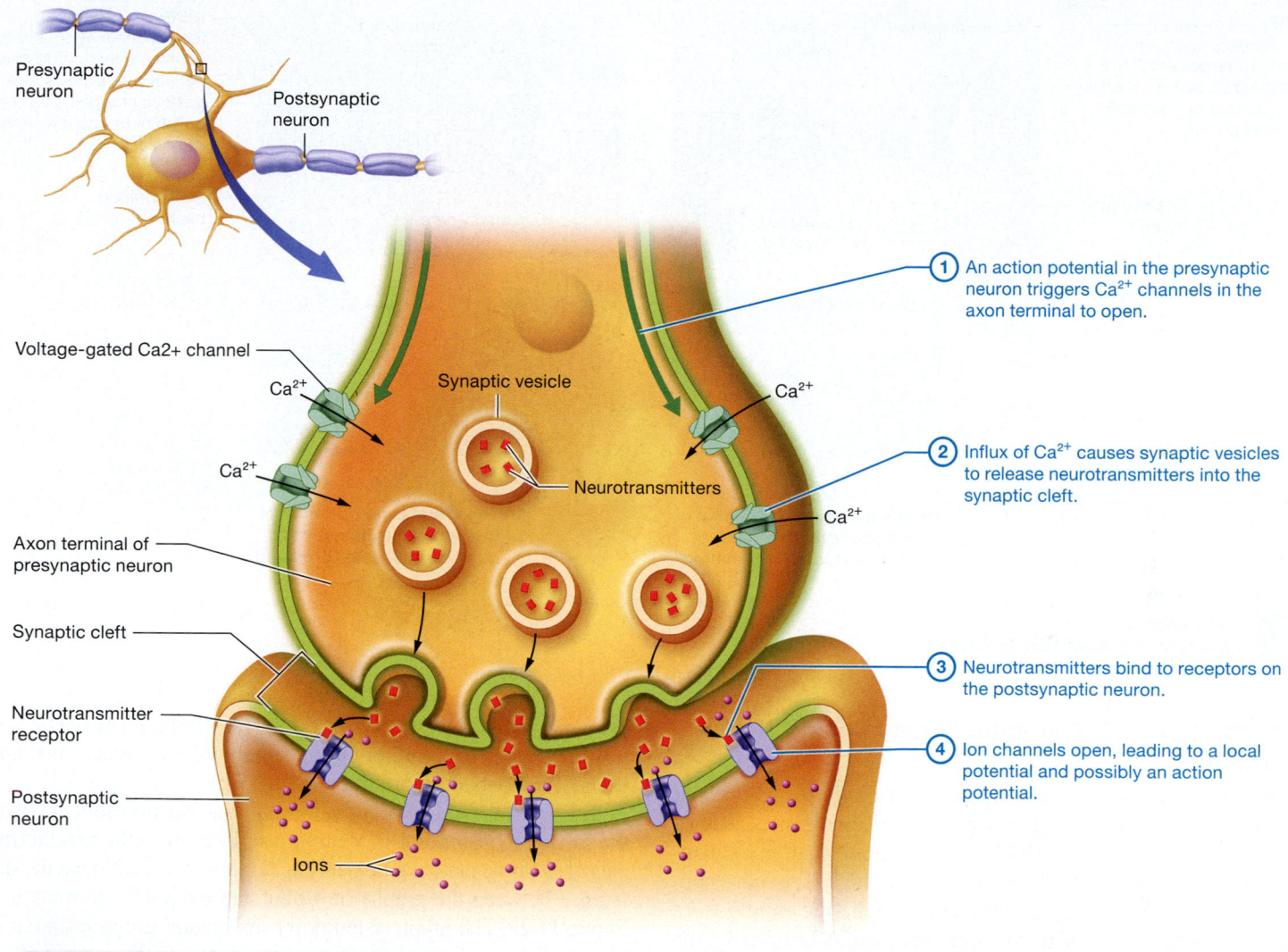

Presynaptic neuron

Postsynaptic neuron

Voltage-gated Ca2+ channel

Ca^{2+}

Synaptic vesicle

Ca^{2+}

Ca^{2+}

Neurotransmitters

Ca^{2+}

Axon terminal of presynaptic neuron

Synaptic cleft

Neurotransmitter receptor

Postsynaptic neuron

Ions

1 An action potential in the presynaptic neuron triggers Ca^{2+} channels in the axon terminal to open.

2 Influx of Ca^{2+} causes synaptic vesicles to release neurotransmitters into the synaptic cleft.

3 Neurotransmitters bind to receptors on the postsynaptic neuron.

4 Ion channels open, leading to a local potential and possibly an action potential.

Figure 11.20 Events at a chemical synapse: synaptic transmission.

▶ Play Animation
@ Mastering Anatomy & Physiology

Postsynaptic Potentials

Local potentials in the membrane of the postsynaptic neuron, which are called **postsynaptic potentials,** can move the membrane potential at the trigger zone either closer to or farther away from threshold. Therefore, depending on which channels are opened, one of two events may occur:

- **The membrane potential of the postsynaptic neuron moves closer to threshold.** A small, local depolarization called an **excitatory postsynaptic potential (EPSP)** may occur, which brings the membrane potential at the trigger zone closer to threshold (**Figure 11.21a**). If the membrane potential reaches threshold, an action potential is generated.
- **The membrane potential of the postsynaptic neuron moves away from threshold.** Alternatively, a small, local hyperpolarization known as an **inhibitory postsynaptic potential (IPSP)** may occur, moving the membrane potential at the trigger zone farther away from threshold (**Figure 11.21b**). This tends to inhibit an action potential from firing.

EPSPs typically result when ligand-gated channels such as sodium or calcium ion channels open, and these positively charged ions enter the postsynaptic neuron (see Figure 11.21a). A single EPSP produces only a very small, local potential across the membrane. However, each successive EPSP makes the membrane more depolarized, and so makes the trigger zone more likely to reach threshold and fire an action potential.

Note in Figure 11.21b that an IPSP can be produced in two ways. First, ligand-gated potassium ion channels can open, which causes the cytosol to lose positive charges and so makes the membrane potential become more negative. Second, ligand-gated chloride ion channels can open. Chloride ions are more abundant in the extracellular fluid than in the neuron, so when chloride ion channels open, these anions enter the neuron and make the membrane potential more negative. The opening of either type of channel will yield the same result: The membrane potential at the trigger zone moves farther away from threshold, and an action potential becomes less likely.

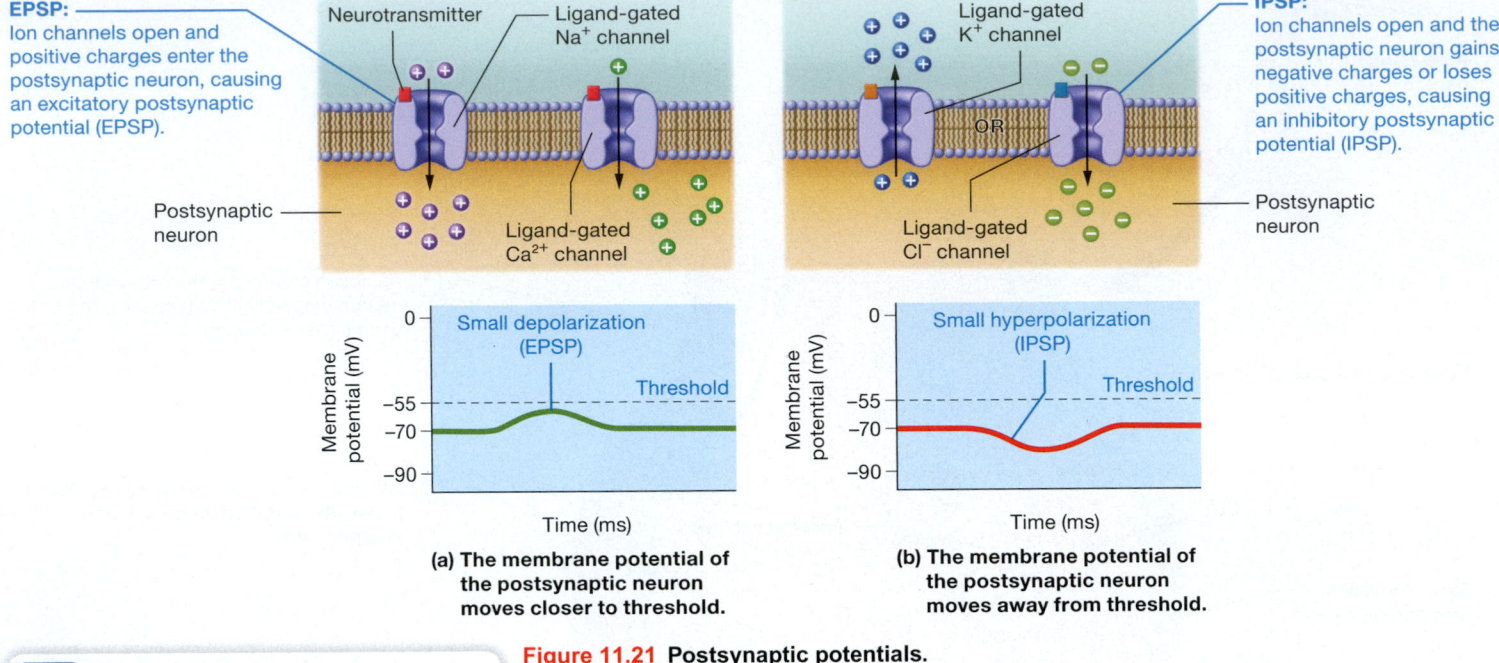

Figure 11.21 Postsynaptic potentials.

▶ Play Animation
@ Mastering Anatomy & Physiology

Summation of Postsynaptic Potentials and Neural Integration

A neuron very rarely receives input from a single source; rather, it receives input from multiple presynaptic neurons, each of which causes an EPSP or IPSP. To complicate matters, synaptic transmission in the CNS occurs continuously for most neurons—they are constantly bombarded by synaptic inputs from hundreds to thousands of presynaptic neurons. Additionally, the input from each presynaptic neuron may be different: The input may be excitatory or inhibitory, and the strength and location of each input may vary.

All this input from presynaptic neurons combines to have one cumulative effect on the postsynaptic neuron. The process by which this occurs is known as **neural integration.** Recall that in Module 11.1 you read about the integrative functions of the nervous system. Put simply, these integrative functions refer to this process of putting together all the excitatory and inhibitory stimuli that determine whether a neuron will or won't fire an action potential.

As we have discussed, to fire an action potential, the trigger zone of an axon must be depolarized to threshold. But a single EPSP produces only a small, local depolarization that is often quite far away from the trigger zone. The small size of the EPSP and the distance from the trigger zone mean that a single EPSP is insufficient to trigger an action potential. However, the depolarizations of many EPSPs can be added together to produce a much greater overall effect. This phenomenon of adding the input from several postsynaptic potentials to affect the membrane potential at the trigger zone is known as **summation.**

There are two types of summation, *temporal* and *spatial.* The first, **temporal summation,** occurs when neurotransmitters are released repeatedly from the axon terminal of a single presynaptic neuron (**Figure 11.22a**). The EPSPs must occur rapidly in succession for temporal summation to occur, because each EPSP lasts no more than about 15 msec.

The second type of summation, **spatial summation,** involves the simultaneous release of neurotransmitters from the axon terminals of multiple presynaptic neurons (**Figure 11.22b**). Notice the difference between the graphs of spatial and temporal summation in Figure 11.22. The graph of temporal summation shows a staircase-like rise in membrane potential as the postsynaptic neuron is hit with successive bursts of neurotransmitters. In contrast, the membrane potential in the graph of spatial summation shows a smooth rise as large quantities of neurotransmitters are released at once.

Spatial summation can combine with temporal summation. When several presynaptic neurons fire together and trigger EPSPs in the postsynaptic neuron, spatial summation occurs and the membrane potential at the trigger zone approaches threshold. The closer the membrane potential gets to threshold, the more likely it becomes that the next EPSP will trigger an action potential due to temporal summation, even if the stimulus is smaller.

Although we have discussed summation of EPSPs, IPSPs can summate both temporally and spatially as well. With summation of IPSPs, the postsynaptic neuron becomes less and less likely to fire an action potential. Additionally, IPSPs and EPSPs can summate. The overall result of this will depend on the individual strength of the IPSP and EPSP—if the IPSP is stronger, the membrane potential will hyperpolarize slightly, and if the EPSP is stronger, the membrane potential will depolarize slightly.

Termination of Synaptic Transmission

Now we have reached the final step of synaptic transmission—termination. But why terminate synaptic transmission? The answer

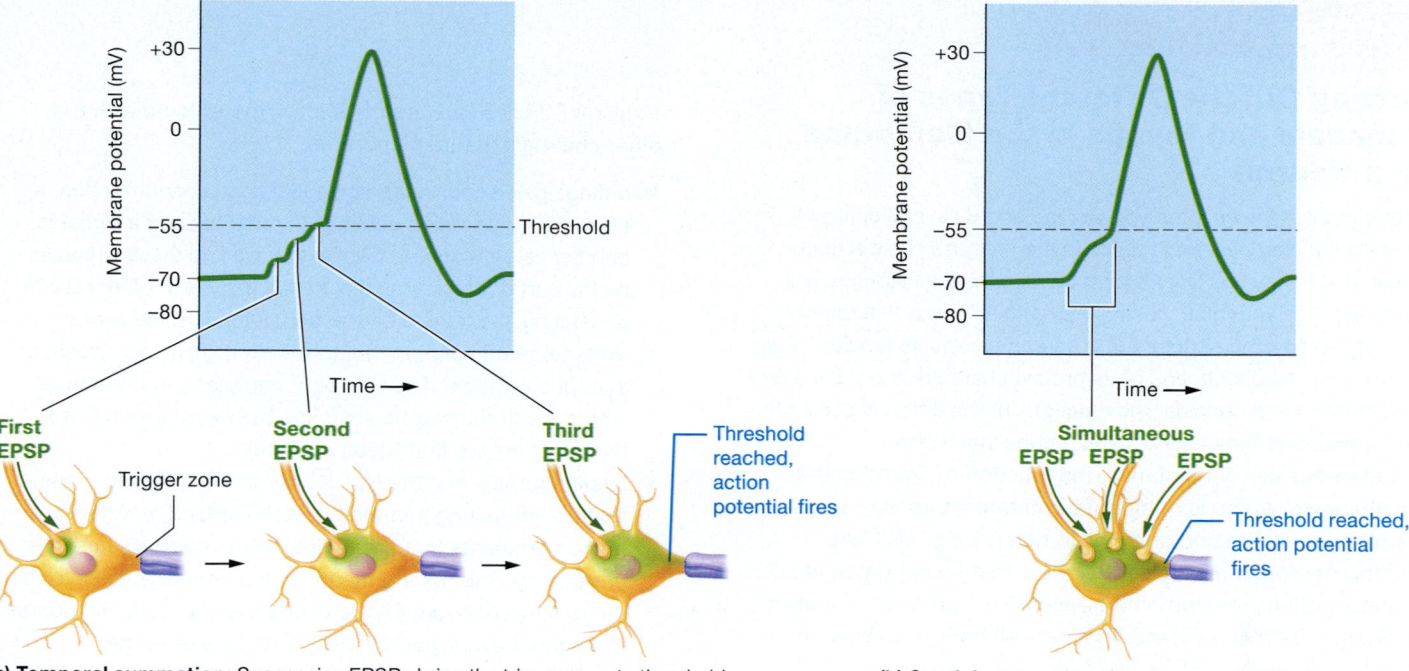

(a) **Temporal summation:** Successive EPSPs bring the trigger zone to threshold.

(b) **Spatial summation:** Simultaneous EPSPs from multiple neurons bring the trigger zone to threshold.

Figure 11.22 **Temporal and spatial summation of excitatory postsynaptic potentials (EPSPs).**

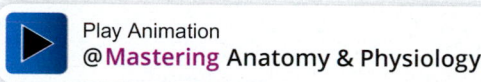
Play Animation
@**Mastering** Anatomy & Physiology

is simple: After presynaptic neurons have generated a specific response in the postsynaptic neuron, the response cannot be initiated again until the postsynaptic neuron stops being stimulated. In other words, it has to be "turned off" in order to be "turned on" again. The neurons involved in breathing provide a simple example. When we need to inhale, specific neurons are stimulated to trigger our respiratory muscles to contract. Once we have taken a breath, our nervous system needs to stop stimulating these neurons or we will continue to inhale. This is accomplished by stopping synaptic transmission.

The messenger of synaptic transmission is the neurotransmitter released by the presynaptic neuron. So, synaptic transmission may be terminated by ending the effects of the neurotransmitter. In general, this happens in three ways (**Figure 11.23**):

- **Diffusion and absorption.** Some neurotransmitters simply diffuse away from the synaptic cleft through the extracellular matrix, where they diffuse through the plasma membrane of a neuron or astrocyte and are then returned to the presynaptic neuron.
- **Degradation in the synaptic cleft.** Certain neurotransmitters are broken down by reactions catalyzed by enzymes that reside in the synaptic cleft. The components of the destroyed neurotransmitter are often then taken back up by the presynaptic neuron and resynthesized into the original neurotransmitter.
- **Reuptake into the presynaptic neuron.** Some neurotransmitters are removed by a process called *reuptake*, in which they are transported back into the presynaptic neuron by proteins in its axolemma. Depending on their

type, these neurotransmitters may be repackaged into synaptic vesicles or degraded in reactions catalyzed by enzymes.

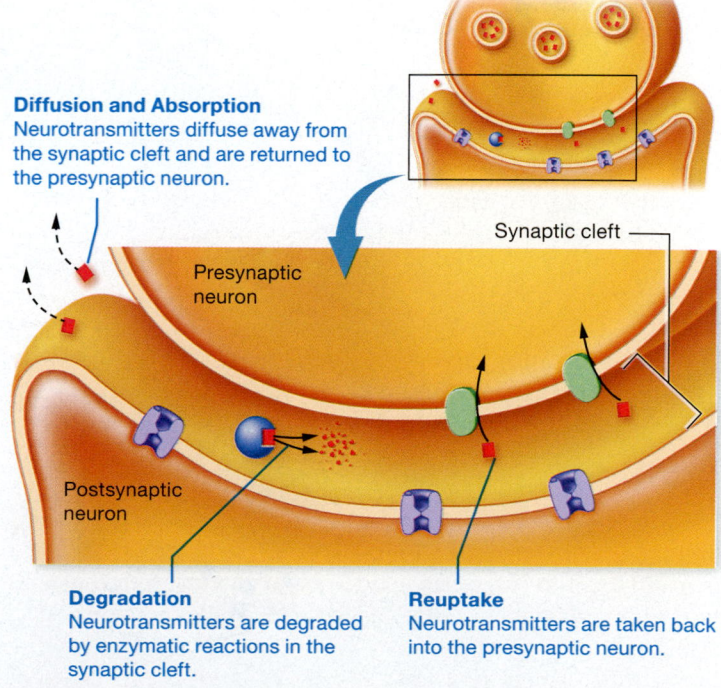

Diffusion and Absorption
Neurotransmitters diffuse away from the synaptic cleft and are returned to the presynaptic neuron.

Synaptic cleft

Presynaptic neuron

Postsynaptic neuron

Degradation
Neurotransmitters are degraded by enzymatic reactions in the synaptic cleft.

Reuptake
Neurotransmitters are taken back into the presynaptic neuron.

Figure 11.23 **Methods of termination of synaptic transmission.**

Play Animation
@**Mastering** Anatomy & Physiology

ConceptBOOST))))

Sorting Out the Different Types of Channels and Pumps in the Membrane of a Neuron

At this point in the chapter, you've read about several different types of ion channels and pumps in the neuron's plasma membrane, and it may feel like it's hard to keep them all straight. But it's easier than you think, because as with all things in anatomy and physiology, the structure of a neuron follows its function. For this reason, the distribution of its protein channels and pumps is predictable if you consider the functions of the different parts of a neuron and of the channels and pumps themselves.

Let's consider, for instance, the function of ligand-gated ion channels, which is to bind neurotransmitters from another neuron—or, stated another way, to *receive* signals from another neuron. It makes sense, then, that ligand-gated ion channels will be located on the *receptive* regions of a neuron, meaning primarily the dendrites and cell body. As shown in

Figure 11.24, we can use the same type of logic with the other channels in the membrane:

- Voltage-gated sodium and potassium ion channels open or close during an action potential in order to send a signal to another cell. It therefore follows that we find these channels on the part of the neuron that sends signals to other cells by generating and transmitting action potentials: the axon.
- Voltage-gated calcium ion channels trigger exocytosis of synaptic vesicles. There is only one place in the neuron where we find synaptic vesicles, the axon terminal, so that is where we find these channels.
- Leak channels and the Na⁺/K⁺ pump are involved in generating the resting membrane potential and maintaining the ion gradients that are critical to the neuron's electrophysiology. The resting membrane potential applies to the entire neuron, so we find leak channels and Na⁺/K⁺ pumps throughout every part of the neuron's membrane. ■

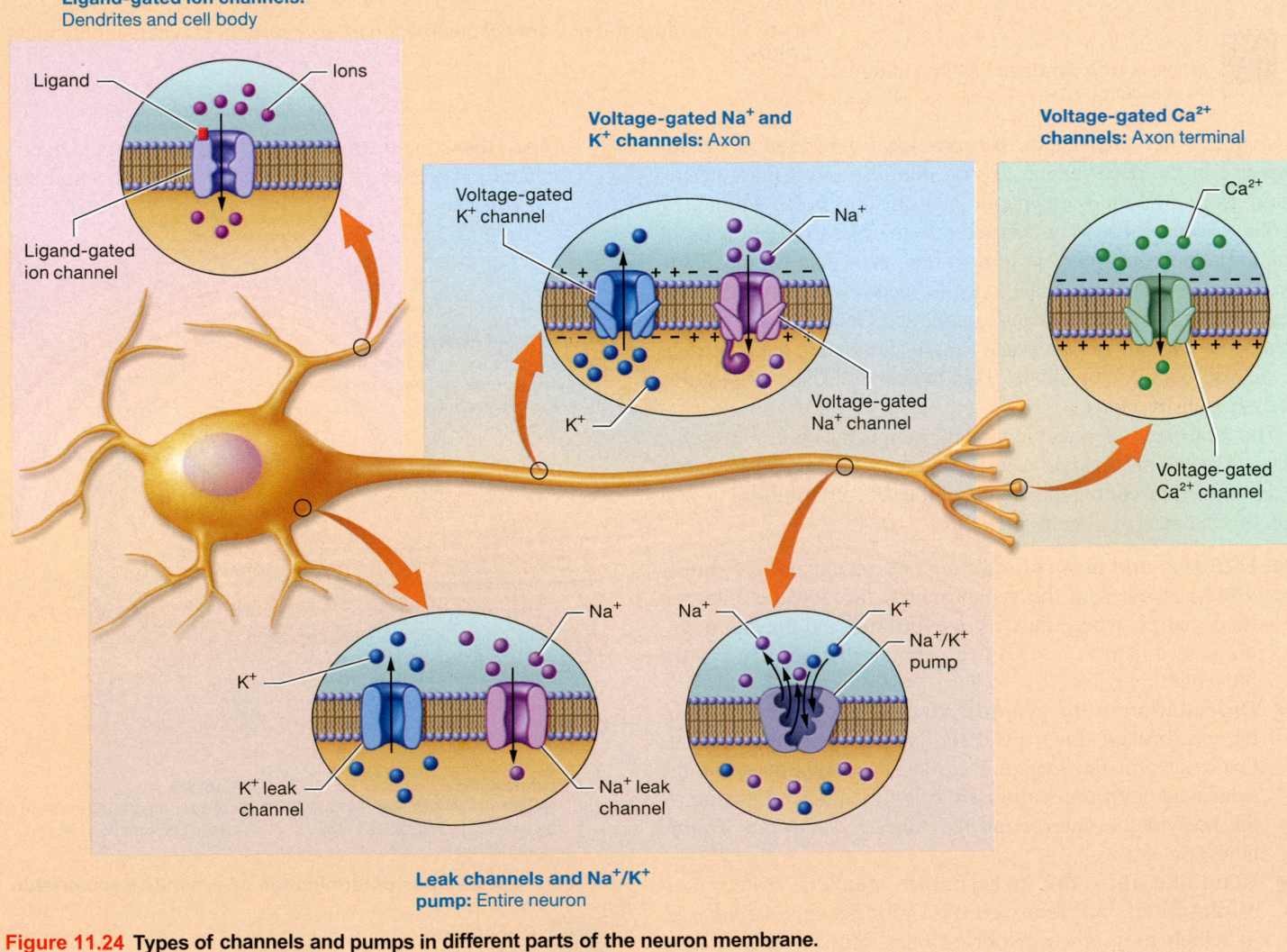

Figure 11.24 **Types of channels and pumps in different parts of the neuron membrane.**

When the neurotransmitter molecules have been removed from the synaptic cleft, synaptic transmission is complete. See *A&P in the Real World: Arthropod Venom* to learn what happens when continued synaptic transmission causes postsynaptic neurons to be overstimulated.

Arthropod Venom

Within the United States live several species of venomous arthropods, including spiders and scorpions. Many of their venoms affect neuronal synapses and are therefore called *neurotoxins.*

One of the most notorious venomous spiders is the female black widow spider (*Latrodectus mactans*), which produces a neurotoxic venom. When a person is injected with venom, or *envenomated,* the toxin interacts with their presynaptic neurons, causing massive release of neurotransmitters and repetitive stimulation of postsynaptic neurons.

The United States is also home to about 40 species of scorpions. The only species with a potentially lethal neurotoxin is the bark scorpion. The venom of the bark scorpion prohibits inactivation of voltage-gated sodium ion channels in the postsynaptic neuron, causing the neuron to continue to fire action potentials.

Although the toxins work in different ways, they produce similar effects because both lead to overstimulation of postsynaptic neurons. Common symptoms include muscle hyperexcitability, muscle cramps, sweating, nausea and vomiting, and difficulty breathing. Treatment and prognosis for both types of bites depend on the amount of venom received and the availability of medical care. Severe cases may require an *antivenin* to block the effects of the toxin. Fortunately, most envenomations are mild, and death is exceedingly rare.

Quick Check

☐ 4. How do the two types of postsynaptic potentials differ?

☐ 5. What is the difference between temporal and spatial summation?

☐ 6. How is synaptic transmission terminated?

Putting It All Together: The Big Picture of Chemical Synaptic Transmission

At this point we've discussed the particulars of synaptic transmission at a chemical synapse: how the action potential triggers the release of neurotransmitters from the presynaptic neuron, and how these neurotransmitters induce EPSPs or IPSPs in the postsynaptic neuron. We then looked at the link between local potentials and action potentials: Local potentials can summate and depolarize the trigger zone to threshold, triggering an action potential in the postsynaptic neuron. Finally, we explored how synaptic transmission is ended. Now we can summarize the whole process, as shown in **Figure 11.25**.

Apply What You Learned

☐ 1. Predict how a poison that blocks voltage-gated calcium ion channels in the axon terminal would affect synaptic transmission.

☐ 2. A new drug opens ligand-gated calcium ion channels in the membrane of the postsynaptic neuron. Would this produce an EPSP or an IPSP? Would this make an action potential more or less likely to occur? Why?

☐ 3. Explain how you could increase the likelihood that a certain neuron will reach threshold and have an action potential. (*Hint:* Think about the different types of summation.)

See answers in Appendix A.

MODULE 11.5

Neurotransmitters

Learning Outcomes

1. Explain how a single neurotransmitter may be excitatory at one synapse and inhibitory at another.

2. Describe the structural and functional properties of the major classes of neurotransmitters.

3. Describe the most common excitatory and inhibitory neurotransmitters in the CNS.

The search for new neurotransmitters is still going on in laboratories around the world every day. The precise number of neurotransmitters operating in the human nervous system is not yet known but is well above 100, and they are diverse in structure and function. In this module we explore the properties and effects of this diverse group of chemicals and their receptors.

Neurotransmitter Receptors

Nearly all neurotransmitters induce postsynaptic potentials by binding to their receptors in the postsynaptic membrane. The type of receptor to which a neurotransmitter binds determines the postsynaptic response. Two types of neurotransmitter

The Big Picture of Chemical Synaptic Transmission

Figure 11.25

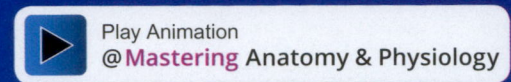
Play Animation
@Mastering Anatomy & Physiology

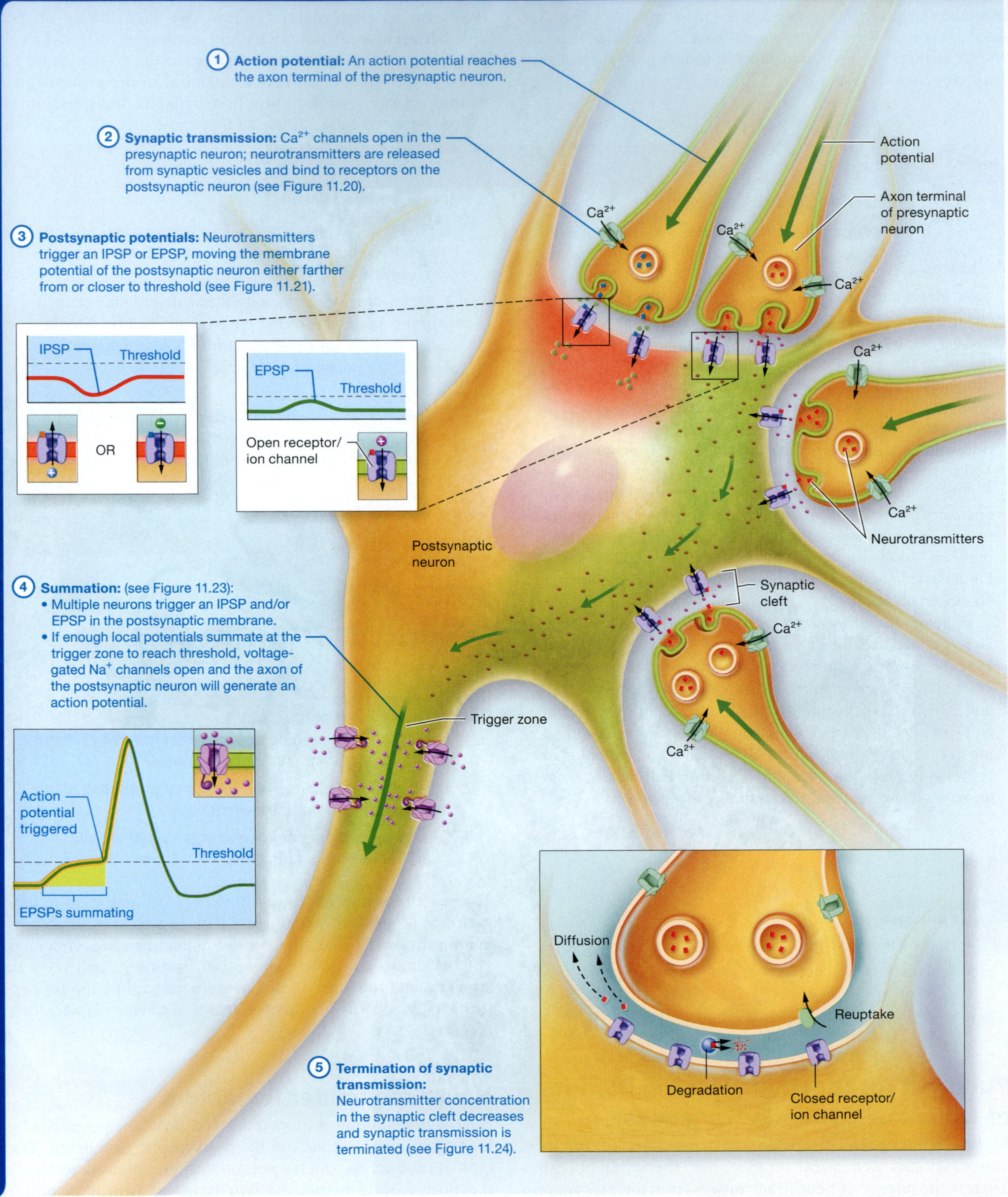

1. Action potential: An action potential reaches the axon terminal of the presynaptic neuron.

2. Synaptic transmission: Ca²⁺ channels open in the presynaptic neuron; neurotransmitters are released from synaptic vesicles and bind to receptors on the postsynaptic neuron (see Figure 11.20).

3. Postsynaptic potentials: Neurotransmitters trigger an IPSP or EPSP, moving the membrane potential of the postsynaptic neuron either farther from or closer to threshold (see Figure 11.21).

IPSP — Threshold

OR

EPSP — Threshold

Open receptor/ ion channel

4. Summation: (see Figure 11.23):
- Multiple neurons trigger an IPSP and/or EPSP in the postsynaptic membrane.
- If enough local potentials summate at the trigger zone to reach threshold, voltage-gated Na⁺ channels open and the axon of the postsynaptic neuron will generate an action potential.

Action potential triggered

Threshold

EPSPs summating

Action potential

Axon terminal of presynaptic neuron

Ca²⁺

Neurotransmitters

Postsynaptic neuron

Synaptic cleft

Ca²⁺

Trigger zone

5. Termination of synaptic transmission: Neurotransmitter concentration in the synaptic cleft decreases and synaptic transmission is terminated (see Figure 11.24).

Diffusion

Degradation

Reuptake

Closed receptor/ ion channel

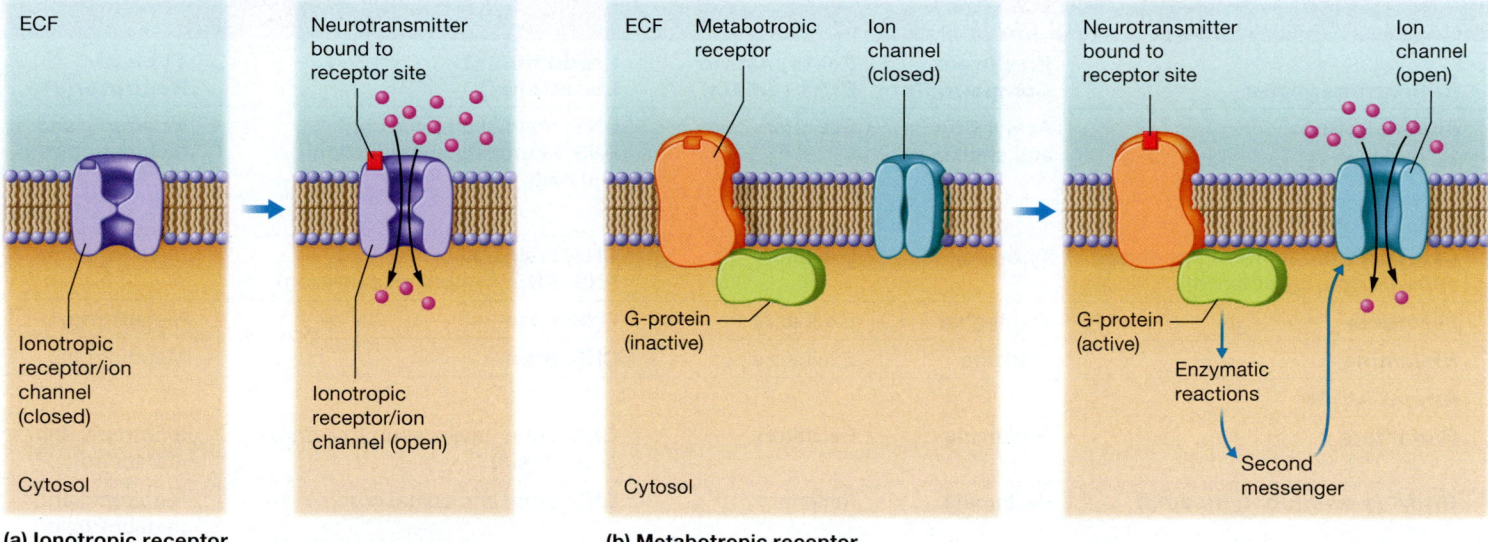

(a) **Ionotropic receptor**

(b) **Metabotropic receptor**

Figure 11.26 Types of neurotransmitter receptors.

receptors have been identified: ionotropic and metabotropic (**Figure 11.26**):

- *Ionotropic receptors* (aye-AHN-oh-troh′-pik) are simply receptors that are part of ligand-gated ion channels. They are called ionotropic because they directly control the movement of ions into or out of the neuron when bound by a neurotransmitter. Neurotransmitters that bind ionotropic receptors have very rapid but short-lived effects on the membrane potential of the postsynaptic neuron.

- *Metabotropic receptors* (meh-TAB-oh-troh′-pik) are receptors within the plasma membrane that are connected to a separate ion channel in some fashion. They are called metabotropic because they are directly connected to metabolic processes that begin when they are bound by neurotransmitters. Most are connected through a group of intracellular enzymes called **G-proteins.** When the neurotransmitter molecule binds to the receptor, it activates one or more G-proteins and begins a cascade of enzyme-catalyzed reactions. The end result of the cascade is the formation of a compound inside the postsynaptic neuron called a **second messenger** (in this system the neurotransmitter molecule is considered the "first messenger"). The second messenger then opens or closes a ligand-gated ion channel in the plasma membrane of the postsynaptic neuron.

Metabotropic receptors elicit much slower changes in the membrane potential of the postsynaptic neuron, but the effects are typically longer-lasting and more varied than those of ionotropic receptors. An example of a common second messenger is **cyclic adenosine monophosphate** (or **cAMP**), which is derived from ATP. In the neuron, cAMP has multiple functions, including binding a group of enzymes that catalyze reactions adding phosphate groups to ligand-gated ion channels, triggering them to open or close. Second messengers are covered more fully in the endocrine chapter (see Chapter 16).

□ 1. What are the two classes of neurotransmitter receptors?

□ 2. What are G-proteins and second messengers?

Major Neurotransmitters

« FLASHBACK

1. What are amino acids and peptides? (pp. 56–57)

2. What are ATP and adenosine? (p. 60)

Regardless of the type of receptor a neurotransmitter binds, that binding leads to either EPSPs or IPSPs. Neurotransmitters that induce EPSPs in the postsynaptic neuron are said to have **excitatory** effects; those that induce IPSPs have **inhibitory** effects. Most neurotransmitters can have both excitatory and inhibitory effects, depending on which postsynaptic neuron receptors they bind. In fact, a single neurotransmitter can have several receptor types. This makes a purely functional classification of neurotransmitters difficult. For this reason, the major neurotransmitters operating within the nervous system are usually classified into four groups by their chemical structures, which we will now explore.

For quick reference, **Table 11.3** summarizes the location, function, and effects of selected neurotransmitters.

Acetylcholine

The best-studied, and one of the most widely used neurotransmitters by the nervous system overall, is the small-molecule neurotransmitter **acetylcholine (ACh)** (ah-seet′l-KOH-leen). Synapses that use ACh, called **cholinergic synapses,** are located at the neuromuscular junction, within the brain and spinal cord, and within the autonomic nervous system (ANS). Its effects are

Table 11.3 Major Neurotransmitters

Neurotransmitter	Precursor Compound(s)	Postsynaptic Effect (Main)	Predominant Location(s)	Type of Receptor(s)
Acetylcholine	Acetyl-CoA and choline	Excitatory	CNS: brain and spinal cord PNS: neuromuscular junction and ANS	Ionotropic and metabotropic
Biogenic Amines				
Catecholamines (norepinephrine, epinephrine, dopamine)	Tyrosine	Excitatory	CNS: brain and spinal cord PNS: ANS (sympathetic division)	Metabotropic
Serotonin	Tryptophan	Excitatory	CNS: brain	Metabotropic
Histamine	Histidine	Excitatory	CNS: brain	Metabotropic
Amino Acids				
Glutamate	Glutamine	Excitatory	CNS: brain (major neurotransmitter of the brain)	Ionotropic and metabotropic
GABA (γ-aminobutyric acid)	Glutamate	Inhibitory	CNS: brain and spinal cord	Ionotropic and metabotropic
Glycine	Serine	Inhibitory	CNS: brain and spinal cord (most common inhibitory neurotransmitter in the spinal cord)	Ionotropic
Neuropeptides				
Substance P	Amino acids	Excitatory and inhibitory	CNS: brain and spinal cord (major neurotransmitter for pain perception) PNS: enteric nervous system (neurons in the digestive tract)	Metabotropic
Opioids (enkephalin, α-endorphin, dynorphin-A)	Amino acids	Excitatory and inhibitory	CNS: brain and spinal cord (major neurotransmitters for pain control)	Metabotropic
Neuropeptide Y	—	Excitatory and inhibitory	CNS: brain PNS: ANS	Metabotropic

largely excitatory; however, it does exhibit inhibitory effects at some PNS synapses.

ACh is synthesized from the precursors choline and acetyl-CoA (an acetic acid molecule bound to coenzyme A) and then packaged into synaptic vesicles. Once ACh is released from the synaptic vesicles, its activity is rapidly terminated by an enzyme in the synaptic cleft known as **acetylcholinesterase (AChE;** ah-seet'l'-koh-leh-NESS-ter-ayz). AChE degrades ACh back into acetic acid and choline. The presynaptic neuron then takes the choline back up, to be used in the synthesis of new ACh molecules.

The Biogenic Amines

The **biogenic amines,** also called the *monoamines,* are a class of neurotransmitters synthesized from amino acids. Most biogenic amines are widely used by the CNS and the PNS, and have diverse functions including maintenance of homeostasis and cognition (thinking). The biogenic amines are implicated in a wide variety of psychiatric disorders and are often the targets of drug therapy for these disorders.

Three of the biogenic amines form a subgroup called the **catecholamines** (kat'-eh-KOHL-ah-meenz), all of which are synthesized from the amino acid tyrosine and share a similar chemical structure. Though many of their synapses are excitatory, like most neurotransmitters, catecholamines can cause inhibition as well. The three catecholamines are as follows:

- **Norepinephrine.** *Norepinephrine* (nor'-ep-ih-NEF-rin; also called *noradrenalin*) is widely used by the ANS, where it influences functions such as heart rate, blood pressure, and digestion. Neurons that secrete norepinephrine in the CNS are largely confined to the brainstem, where they work to regulate the sleep/wake cycle, attention, and feeding behaviors.

- **Epinephrine.** *Epinephrine* (also called *adrenalin*) is also used by the ANS, where it has the same effects as norepinephrine. However, it is more widely used as a hormone by the endocrine system (see Chapter 16 for details).

- **Dopamine.** *Dopamine,* used extensively in the CNS, has a variety of functions. It helps to coordinate movement, and is also involved in emotion and motivation. The receptor for dopamine in the brain is a target for certain illegal drugs, such as cocaine and amphetamine, and is likely responsible for the behavioral changes seen with addiction to these drugs.

Another biogenic amine is *serotonin* (sehr-oh-TOH-nin), which is synthesized from the amino acid tryptophan. Most neurons that use serotonin are found in the brainstem, and their axons project to multiple places in the brain. Serotonin is thought to be one of the major neurotransmitters involved in mood regulation (likely along with norepinephrine), and it is a common target in the treatment of depression. Additionally, serotonin acts to affect emotions, attention and other cognitive functions, motor behaviors, feeding behaviors, and daily rhythms.

The final biogenic amine we'll discuss is *histamine* (HISS-tah-meen), which is synthesized from the amino acid histidine. Histamine is involved in a large number of processes in the CNS, including regulation of arousal and attention. In addition, outside the nervous system, histamine is an important mediator of allergic responses. Drugs called *antihistamines* block histamine receptors outside the nervous system to alleviate allergy symptoms, but most also block histamine receptors in the CNS. As histamine plays a part in arousal, blocking its actions often leads to the common side effect of drowsiness seen with these drugs.

Amino Acid Neurotransmitters

There are three major **amino acid neurotransmitters:** glutamate; glycine; and γ-aminobutyric acid, or GABA. *Glutamate* is the most important excitatory neurotransmitter in the brain—most neurons in the brain are thought to have at least one type of receptor for it. There are both ionotropic and metabotropic receptors for glutamate; both generally lead to the opening of sodium or calcium ion channels and the production of EPSPs.

Glycine and *GABA* are the two major inhibitory neurotransmitters of the nervous system. Both induce IPSPs in the postsynaptic neurons, primarily by opening ligand-gated chloride ion channels and hyperpolarizing the axolemma. GABA is the most important inhibitory neurotransmitter in the brain, whereas glycine is the most widely used inhibitory neurotransmitter in the spinal cord.

Neuropeptides

The **neuropeptides** are a group of neurotransmitters that have a wide variety of effects within the nervous system. They must be synthesized in the cell body, as axons lack the organelles for protein synthesis. Multiple neuropeptides have been identified, and a few are described next.

- **Substance P.** Substance P was the first identified neuropeptide (its name comes from the fact that it was extracted from brain and gut *powder*). It is released from type C sensory afferent fibers that transmit stimuli about pain and temperature (leading many students to use the mnemonic that the "P" stands for "pain"). It is also released by neurons in the brain, spinal cord, and gut.
- **Opioids.** The opioids (OH-pee-oydz) make up a family of more than 20 neuropeptides that includes three classes: the *endorphins,* the *enkephalins,* and the *dynorphins.* All share

the same property of eliciting pain relief (called *analgesia* [an′-al-JEE-zee-ah]), and all are nervous system depressants. They also appear to be involved in sexual attraction and aggressive or submissive behaviors.
- **Neuropeptide Y.** Neuropeptide Y is a large neuropeptide with 36 amino acids. It appears to function in feeding behaviors, and may mediate hunger or feeling "full."

Quick Check

☐ 3. How do neurotransmitters excite a postsynaptic neuron? How do they inhibit a postsynaptic neuron?

☐ 4. Which neurotransmitters have largely excitatory effects?

☐ 5. Which neurotransmitters have largely inhibitory effects?

Neuromodulation

Neurons also release neurotransmitters that don't directly excite or inhibit a postsynaptic neuron. Instead, these neurotransmitters alter, or *modulate*, synaptic transmission in some way, a process called **neuromodulation.** Accordingly, neurotransmitters functioning in this manner are known as **neuromodulators.**

Neuromodulators bind to metabotropic receptors in the membrane of the presynaptic or postsynaptic neuron, where they generally initiate a "slow" cascade of events that can last from several minutes to several days. These actions can cause several effects, including increasing or decreasing neurotransmitter release from a presynaptic neuron, affecting the sensitivity of the postsynaptic membrane, and altering gene transcription. Neuromodulator effects can be wide-ranging, as they are not limited to a single synapse—they can diffuse through the extracellular fluid and interact with multiple neurons.

It was previously thought that only certain chemicals released by neurons could function as neuromodulators. However, we now understand that essentially any neurotransmitter can act as a neuromodulator, and some may have neuromodulation as a primary function. In addition, certain hormones, known as *neurohormones*, modulate the activity of neurons.

Read about how medications can modulate synaptic transmission in *A&P in the Real World: Psychiatric Disorders and Treatments* on p. 416.

Quick Check

☐ 6. What happens during neuromodulation?

Apply What You Learned

☐ 1. Toxins from the cone snail block glutamate receptors in the postsynaptic membrane. What will be the effect of these toxins?

☐ 2. Predict the effects of the poison strychnine, which blocks glycine receptors on postsynaptic neurons of the CNS.

☐ 3. What would happen to synaptic transmission if you blocked the degradation and/or reuptake of excitatory neurotransmitters in the synaptic cleft? What if the neurotransmitters were inhibitory?

See answers in Appendix A.

AP in the Real World

Psychiatric Disorders and Treatments

Psychiatric disorders, those that affect the thought processes of the brain, are generally treated by modulating synaptic transmission to change how neurons communicate with one another. Much of the science of *psychopharmacology* (the study of drugs that affect higher brain functions) targets either action potential generation or some aspect of neurotransmitter physiology. Examples of disorders and drug actions include the following:

- **Schizophrenia.** The disease schizophrenia (skit'-zoh-FREEN-ee-ah) is characterized by repetitive *psychotic episodes*—periods during which a person is unable to appropriately test his or her beliefs and perceptions against reality. Schizophrenia is thought to result from excessive release of dopamine, so pharmacological management of the disorder primarily involves blocking postsynaptic dopamine receptors.

- **Depressive disorders.** The depressive disorders are marked by disturbances in mood and are thought to be caused by a deficiency in synaptic transmission of serotonin, norepinephrine, and/or dopamine. Pharmacological treatment makes use of drugs that prolong the lifespan of these neurotransmitters, particularly serotonin, in the synaptic cleft. The most widely used antidepressants are *selective serotonin reuptake inhibitors* (SSRIs), which block only the serotonin transporter, preventing the reuptake of serotonin by the presynaptic neuron.

- **Anxiety disorders.** The hallmark of anxiety disorders is an exaggerated and inappropriate fear response, believed to stem from abnormalities in norepinephrine, serotonin, and GABA transmission. Drugs used to treat anxiety disorders may include the antidepressants already discussed, drugs that enhance GABA activity, and other drugs that modulate norepinephrine transmission.

- **Bipolar disorder.** Bipolar disorder is a group of disorders characterized by episodes of abnormally elevated mood (called *mania*) followed by episodes of depression. Many treatments for bipolar disorder involve decreasing the ease with which axons generate action potentials, generally by blocking sodium ion channels in the axolemma.

Functional Groups of Neurons

Learning Outcomes

1. Define a neuronal pool, and explain its purpose.
2. Compare and contrast the two main types of neural circuits in the central nervous system.

So far, we have mostly discussed the behavior of individual neurons—how action potentials are generated and conducted, how an action potential leads to synaptic transmission, and how neurotransmitters affect the postsynaptic neuron. Now we explore the behavior of *groups* of neurons. Neurons don't typically operate as discrete entities but instead form networks called *neuronal pools* that perform a common function. Neuronal pools are organized into functional groups known as *neural circuits* (or *neural networks*). In this module, we examine neuronal pools and neural circuits, and discuss how groups of neurons work together to carry out the many activities of the nervous system.

Neuronal Pools

Neuronal pools are groups of interneurons within the CNS. These pools typically are a tangled mat of neuroglial cells, dendrites, and axons in the brain, whereas their cell bodies may lie in other parts of the CNS. The type of information that can be processed by a pool is defined by the synaptic connections of that pool. The connections between pools allow for complex mental activity such as planned movement, cognition, and personality.

Each neuronal pool begins with one or more neurons called *input neurons* that initiate the series of signals. The input neuron branches repeatedly to serve multiple neurons in the pool; however, it may have different effects on different neurons. For some neurons, it may generate EPSPs that trigger an action potential, and for others, it may simply bring the trigger zone closer to threshold. This difference is determined by the number of contacts the input neuron makes with the postsynaptic neuron.

A small neuronal pool with one input neuron and its postsynaptic neurons is illustrated in **Figure 11.27**. You can see that the postsynaptic neurons in the center (surrounded by green) have the highest number of synaptic contacts with the input neuron. Because of these connections, spatial summation is possible and the firing of the input neuron is likely to generate adequate EPSPs to trigger an action potential.

Notice, however, that the neurons in the light orange area on either side have fewer synaptic contacts with the input neuron. As a result, the input neuron acting alone will not be able to bring the trigger zones of these neurons to threshold and elicit action potentials. However, it can help another input neuron trigger action potentials.

Until now we've been discussing only excitatory input; however, remember that inhibitory synapses occur as well. The

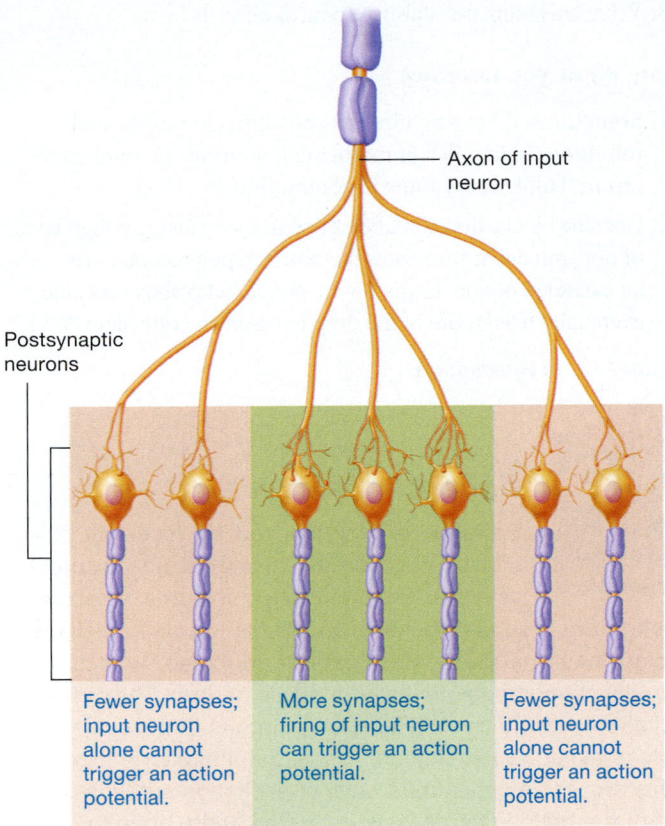

Postsynaptic neurons

Axon of input neuron

| Fewer synapses; input neuron alone cannot trigger an action potential. | More synapses; firing of input neuron can trigger an action potential. | Fewer synapses; input neuron alone cannot trigger an action potential. |

Figure 11.27 A neuronal pool.

degrees of inhibition also correlate strongly with the number of synaptic contacts. Action potentials are effectively prevented in the postsynaptic neurons that receive the greatest number of IPSPs from the input neuron.

Neural Circuits

As you know, form follows function, so the functional characteristics of a neuronal pool are determined largely by its pattern of structural organization. The patterns of synaptic connection between neuronal pools are called **neural circuits.** Each neuronal pool in a circuit receives input from other pools, and then produces output that travels to additional pools. How the pools are connected determines the function of the circuits.

There are two basic types of neural circuits, diverging and converging:

- **Diverging circuit.** As shown in **Figure 11.28a**, a **diverging circuit** begins with one axon of an input neuron branching to make contacts with multiple postsynaptic neurons. The axons of these postsynaptic neurons then branch to contact more neurons, which in turn make contact with yet more neurons, and so on. So, when a signal is transmitted down the circuit's pathway, an increasing number of neurons are excited. Diverging circuits are critical because they allow a single neuron to communicate with multiple parts of the brain and/or body. Notice in Figure 11.28a, left, that some diverging circuits are *amplifying circuits,* in which the signal passes through a progressively greater number of neurons. We start with one neuron, which branches to excite two, the two then excite four, and so on. Some diverging circuits (Figure 11.28a, right) split into multiple tracts, each of which goes in a different direction. This type of circuit is characteristic of those transmitting incoming sensory stimuli, which is sent from neurons in the spinal cord to different neuronal pools in the brain for processing.
- **Converging circuit.** A converging circuit (**Figure 11.28b**) is essentially the opposite of a diverging circuit. In converging circuits, axon terminals from multiple input neurons converge onto a single postsynaptic neuron, allowing

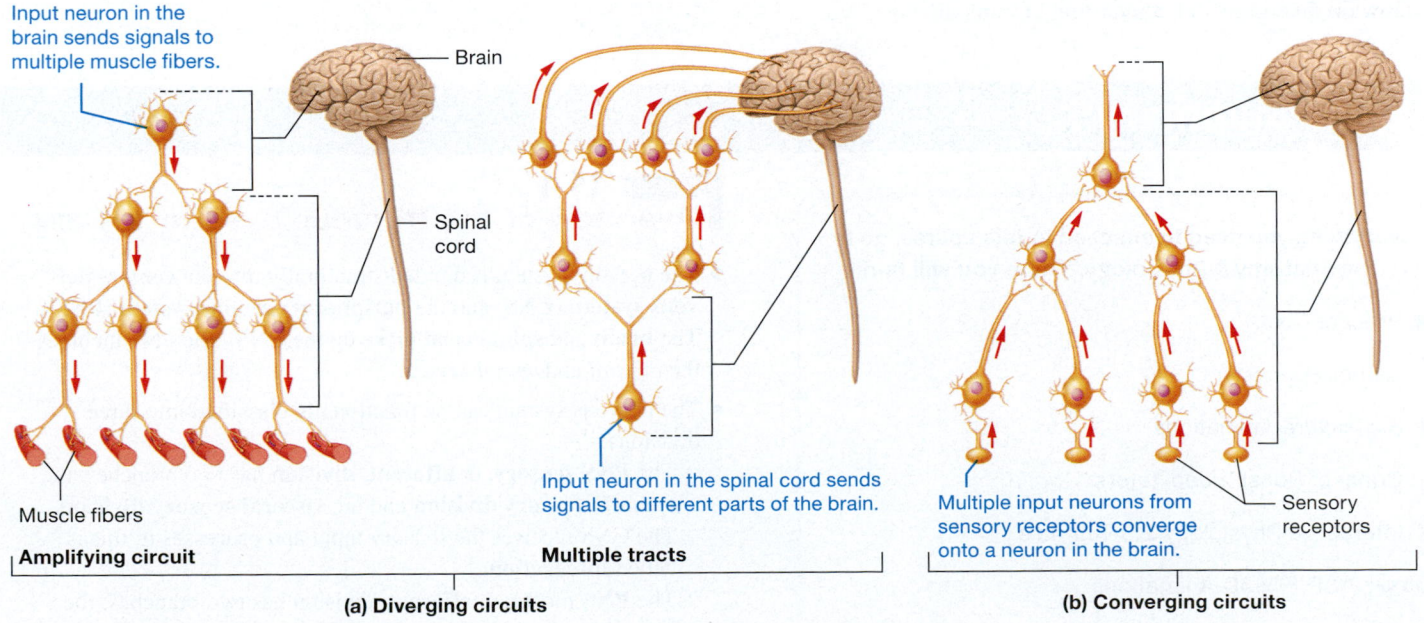

Input neuron in the brain sends signals to multiple muscle fibers.

Brain

Spinal cord

Muscle fibers

Amplifying circuit

Input neuron in the spinal cord sends signals to different parts of the brain.

Multiple tracts

(a) Diverging circuits

Multiple input neurons from sensory receptors converge onto a neuron in the brain.

Sensory receptors

(b) Converging circuits

Figure 11.28 Types of neural circuits.

for spatial summation of synapses. Converging circuits are important for control of skeletal muscle movement—the interneurons in the spinal cord receive input from neurons in different regions of the brain, which then converges to synapse on the motor efferent neurons that stimulate skeletal muscle contraction. Converging circuits also allow the nervous system to respond to the sensory stimuli that it collects and processes; an example is shown in Figure 11.28b.

Given that every part of the brain contacts virtually every other part of the brain via some neural circuit, disorganized electrical activity can be disastrous (see *A&P in the Real World: Epileptic Seizures*). To prevent electrical activity in the brain from becoming chaotic and overly excitatory, the CNS has two basic mechanisms to stabilize neural circuits. The first is simply inhibitory circuits. Most neural circuits have an intrinsic negative feedback mechanism mediated by neuromodulators that inhibits either the input neurons or the postsynaptic neurons of their pools if they become overly excited. Additionally, some neuronal pools consist largely of neurons that release neuromodulators that control the activity of other neural circuits.

The second mechanism for stabilizing neural circuits is a property of synapses called *synaptic fatigue.* Fatigue refers to the fact that synaptic transmission becomes progressively weaker with prolonged and intense excitation. This is in part due to a gradual decrease in the sensitivity of the postsynaptic neurons in the circuit to neurotransmitters when they are overused. Further, over the long term, the number of neurotransmitter receptors in the plasma membrane of the postsynaptic neuron actually decreases when high levels of neurotransmitters are present for extended periods. This "downregulation" of postsynaptic receptors is thought to be why people develop a tolerance to certain medications that modulate neurotransmitter release.

Quick Check

☐ 1. Why are neurons organized into neuronal pools?

☐ 2. How do diverging and converging circuits differ?

☐ 3. What mechanisms stabilize neural circuits?

Apply What You Learned

☐ 1. Sometimes diverging circuits split into excitatory and inhibitory paths. When might such a circuit be required? (*Hint:* Think about muscle contraction.)

☐ 2. Cocaine blocks the reuptake of dopamine, causing a high level of dopamine that stimulates the postsynaptic receptors for an extended period. Explain why people who abuse cocaine eventually need more of the drug to reach an equivalent "high."

See answers in Appendix A.

A&P in the Real World

Epileptic Seizures

Epilepsy is characterized by recurrent episodes of abnormal, disorganized electrical activity in the brain called *seizures*. A seizure results from a sudden burst of excitatory electrical activity within a neuronal pool, which may be triggered by instability in the membrane potential of a single neuron. The excess excitation overwhelms the inhibitory circuits that would normally prevent overexcitation. Once the inhibitory mechanisms are lost, a continuous wave of excitation spreads over part of the brain (a *partial seizure*) or the entire brain (a *generalized seizure*). When the brain is inundated with signals of this nature, no meaningful signals can be transmitted. This can lead to a wide variety of symptoms, ranging from mild sensory disturbances to loss of consciousness to characteristic jerking movements from uncontrolled muscle activity.

Although a number of medications will stop seizures, they generally end on their own due to synaptic fatigue. So, most therapy for epilepsy is aimed at preventing seizures and allowing the inhibitory circuits to function optimally.

Chapter Summary

For everything you need to succeed in this course, go to **Mastering** Anatomy & Physiology. There you will find:

- Practice Tests
- Author Podcasts
- Big Picture Animations
- Concept Boost Video Tutors
- **iP2** Interactive Physiology 2.0 Tutorials
- *A&PFlix* A&P Flix 3D Animations
- Active-Learning Workbook

MODULE 11.1
Overview of the Nervous System 381–384

- The nervous system is divided structurally into the **central nervous system (CNS)** and the **peripheral nervous system (PNS)**. The **brain** and **spinal cord** make up the CNS. The PNS includes the *cranial* and *spinal nerves.*

- The nervous system can be functionally classified into three divisions:
  - The **PNS sensory,** or **afferent, division** has two branches: the **somatic sensory division** and the **visceral sensory division.**
  - The CNS receives the sensory input and processes it; this is called **integration.**
  - The **PNS motor,** or **efferent,** division has two branches: the **somatic motor division** and the **autonomic nervous system (ANS).**

MODULE **11.2**
Nervous Tissue 384–393

- Nervous tissue consists of **neurons** and **neuroglial cells.**
- Neurons are excitable cells that send, propagate, and receive **action potentials.** They consist of three parts: the **cell body,** one or more receptive **dendrites,** and a single **axon.**
- Neurons are classified both structurally and functionally.
  - Structural classes include **multipolar neurons, bipolar neurons,** and **pseudounipolar neurons.**
  - Functional classes include **sensory,** or **afferent, neurons; interneurons;** and **motor,** or **efferent, neurons.**
- The **neuroglia** in the CNS include **astrocytes,** which anchor neurons and blood vessels in place; **oligodendrocytes,** which form the **myelin sheath; microglia,** which are phagocytes; and **ependymal cells,** which produce and circulate cerebrospinal fluid.
- The neuroglia of the PNS include **Schwann cells,** which form the myelin sheath, and **satellite cells,** which surround cell bodies of neurons in the PNS.
- Oligodendrocytes in the CNS and Schwann cells in the PNS wrap around the axon up to 100 times to form the myelin sheath. This covering significantly speeds up conduction of an action potential through the axon.
- Unmyelinated axons in the PNS are embedded in Schwann cells.
- Axons of the PNS may be regenerated if the cell body remains intact. When **regeneration** occurs, an axonal *growth process* is guided toward its target cell by a **regeneration tube** made of Schwann cells and the basal lamina.

MODULE **11.3**
Electrophysiology of Neurons 393–403

- A separation of charges occurs across the membrane of neurons. An unstimulated neuron shows a decrease in voltage across the membrane, called the **resting membrane potential.** Neurons at rest are **polarized** with a resting membrane potential of about –70 mV.
- Ions move across the axolemma via channels. Two types of channels are **leak** (always open) and **gated channels.**
- Two important ion gradients are those of Na^+ and K^+: The concentration of Na^+ is higher in the extracellular fluid, and the concentration of K^+ is higher in the cytosol.
- A **local potential** is a small, local change in the membrane potential of a neuron.
  - A local potential may either **depolarize** the neuron, making it less negative, or **hyperpolarize** the neuron, making it more negative.
  - Local potentials are *graded,* reversible, *decremental* with distance, and useful for short-distance signaling only.
- Voltage-gated K^+ channels have two states, resting and activated, whereas voltage-gated Na^+ channels have three: resting, activated, and inactivated.
- An **action potential** is a rapid depolarization and repolarization of the membrane potential of the cell.
  - During the **depolarization phase** Na^+ flood the axon, and the membrane potential rises toward a positive value.

- During the **repolarization phase** K^+ flow out of the axon, returning the axon to its negative resting membrane potential. Many neurons hyperpolarize after repolarization completes.
- Action potentials are *nondecremental,* they obey the **all-or-none principle,** they are irreversible, and they are long-distance signals.
- The **refractory period** is the span of time during which it is difficult or impossible to elicit another action potential.
- Action potentials are **propagated** along an axon via the flow of current.
- The speed of action potential propagation depends on the diameter of the axon (larger axons conduct more rapidly) and on the presence or absence of a myelin sheath. Conduction may occur in two ways:
  - **Saltatory conduction** occurs rapidly because the current is insulated as it flows through each internode and action potentials are generated only at nodes of Ranvier.
  - **Continuous conduction** occurs much more slowly, as each consecutive region of the membrane must be depolarized to threshold and generate an action potential.

MODULE **11.4**
Neuronal Synapses 403–411

- A **synapse** is the location where a neuron meets its target cell. The transfer of stimuli between neurons at a synapse is called **synaptic transmission.**
- **Electrical synapses** occur between neurons whose axolemmas are electrically joined via gap junctions. Stimulus transfer at an electrical synapse is nearly instantaneous and bidirectional.
- The majority of synapses in the nervous system are **chemical synapses,** which rely on **neurotransmitters** to send signals. Chemical synapses are slower than electrical synapses and are unidirectional.
- The events at a chemical synapse start with an action potential reaching the axon terminal of the **presynaptic neuron,** which triggers exocytosis of neurotransmitters stored in synaptic vesicles. The neurotransmitters bind to receptors on the membrane of the **postsynaptic neuron** and cause a local **postsynaptic potential.**
- One of two things may happen during a postsynaptic potential:
  - The postsynaptic neuron may be depolarized by an **excitatory postsynaptic potential (EPSP).**
  - The postsynaptic neuron may be hyperpolarized by an **inhibitory postsynaptic potential (IPSP).**
- **Neural integration** is the process of putting together all the excitatory and inhibitory stimuli that determine whether a neuron will or won't fire an action potential.
  - **Summation** is the phenomenon that combines local postsynaptic potentials.
  - In **temporal summation,** a single presynaptic neuron fires at a rapid pace. In **spatial summation,** multiple presynaptic neurons fire simultaneously.
- The effects of synaptic transmission are terminated by removal of the neurotransmitters from the synaptic cleft.

MODULE **11.5**
Neurotransmitters 411–416

- Neurotransmitters produce their effects by influencing the opening or closing of ion channels in the axolemma of the postsynaptic neuron.

- There are two types of neurotransmitter receptors: (1) *ionotropic receptors* and (2) *metabotropic receptors.*

- The effects of a neurotransmitter are described as **excitatory** if they generally induce EPSPs and **inhibitory** if they generally induce IPSPs. Many neurotransmitters are capable of generating both EPSPs and IPSPs.

- The major neurotransmitters include the following:
  - **Acetylcholine** is mostly excitatory, and is degraded by **acetylcholinesterase.**
  - The **biogenic amines** include the **catecholamines** (*norepinephrine, dopamine,* and *epinephrine*), *serotonin,* and *histamine.*
  - The **amino acid neurotransmitters** include *glutamate, glycine,* and *γ-aminobutyric acid* (*GABA*). Glutamate is the major excitatory neurotransmitter in the brain. Both GABA and glycine are major inhibitory neurotransmitters in the CNS.

MODULE 11.6
Functional Groups of Neurons 416–418

- Interneurons are organized into **neuronal pools** that enable specialization within the CNS and so higher mental activity.

- An *input neuron* is the presynaptic neuron that initiates the series of signals in a neuronal pool.

- The pattern of connection between neuronal pools is called a **neural circuit.** There are two main types of neural circuits:
  - A **diverging circuit** begins with one or more input neurons that contact an increasing number of postsynaptic neurons.
  - A **converging circuit** is one in which the signals from multiple neurons converge onto one or more final postsynaptic neurons.

Assess What You Learned

Scan the QR Code for additional practice test questions

https://goo.gl/ucxE5f

LEVEL 1 Check Your Recall

1. Which of the following statements about the general functions of the nervous system is *false?*

   a. The three primary functions of the nervous system include sensory, integrative, and motor functions.
   b. The integrative functions of the nervous system are its processing functions.
   c. Sensory stimuli are transmitted on sensory efferent fibers to a sensory receptor.
   d. Motor functions are carried out by fibers that carry signals to an effector.

2. Regulation of heart rate, blood pressure, and digestive functions is carried out by the:

   a. somatic motor division of the peripheral nervous system.
   b. central nervous system.
   c. visceral sensory division of the peripheral nervous system.
   d. autonomic nervous system.

3. Match each type of neuroglial cell with its correct function.

   _____ Schwann cells
   _____ Ependymal cells
   _____ Microglial cells
   _____ Oligodendrocytes
   _____ Satellite cells
   _____ Astrocytes

   a. Phagocytic cells of the CNS
   b. Surround the cell bodies of neurons in the PNS
   c. Create the myelin sheath in the PNS
   d. Anchor neurons and blood vessels, maintain extracellular environment around neurons, assist in repair of damaged brain tissue
   e. Create the myelin sheath in the CNS
   f. Ciliated cells in the CNS that produce and circulate the fluid around the brain and spinal cord

4. Mark the following statements as true or false. If the statement is false, correct it to make it a true statement.

   a. Aggregates of Golgi apparatus and lysosomes form dark-staining Nissl bodies within the cell body.
   b. The axon contains a high density of ribosomes, rough endoplasmic reticulum, and Golgi apparatus.
   c. Axons arise from the axon hillock.
   d. Substances can move toward or away from the cell body through the axon via fast axonal transport.

5. An axon is *best* defined as a process that:

   a. transmits signals only toward the cell body.
   b. can generate action potentials.
   c. transmits signals only away from the cell body.
   d. cannot generate action potentials.

6. Fill in the blanks: The myelinated segment of an axon that is covered by a glial cell is called a/an _____; the gaps between glial cells where the axolemma is exposed are called _____.

7. Fill in the blanks: The _____ is the period of time during which it is impossible to stimulate a neuron to have an action potential, whereas the _____ is the period of time during which a larger-than-normal stimulus is required to elicit an action potential.

8. Which of the following statements best describes saltatory conduction?

   a. Every section of the axolemma must be depolarized and triggered to generate an action potential.
   b. The internodes must generate action potentials.
   c. The dendrites and cell bodies propagate EPSPs toward the trigger zone.
   d. Only the nodes of Ranvier must generate action potentials.

9. Identify the following as properties of electrical synapses (ES), chemical synapses (CS), or both (B).

   a. _____ The plasma membranes of presynaptic and postsynaptic neurons are joined by gap junctions.
   b. _____ Transmission is unidirectional and delayed.
   c. _____ A presynaptic neuron and a postsynaptic neuron are involved.
   d. _____ The use of neurotransmitters packaged into synaptic vesicles is required.
   e. _____ Transmission is nearly instantaneous and bidirectional.

10. The trigger for exocytosis of synaptic vesicles from the presynaptic neuron is:

   a. arrival of an action potential at the axon terminal and influx of calcium ions.
   b. summation of IPSPs at the presynaptic neuron.
   c. binding of neurotransmitters to the axon hillock.
   d. influx of Na$^+$ into the postsynaptic neuron.

11. Match the following neurotransmitters with their correct description.

   _____ GABA
   _____ Dopamine
   _____ Substance P
   _____ Acetylcholine
   _____ Glutamate
   _____ Endorphins
   _____ Norepinephrine

   a. Neuropeptide involved in transmission of pain
   b. Neurotransmitter released at the neuromuscular junction
   c. Major excitatory neurotransmitter in the brain
   d. Major inhibitory neurotransmitter in the brain
   e. Neuropeptide involved in relief of pain
   f. Catecholamine involved in the autonomic nervous system
   g. Catecholamine involved in movement and behavior

12. Which of the following is *not* a method by which the effects of neurotransmitters are terminated?

   a. Reuptake into the presynaptic neuron
   b. Diffusion away from the synaptic cleft and uptake by glial cells
   c. Movement back to the cell body by retrograde axonal transport
   d. Degradation by enzymes in the synaptic cleft

13. A _____ is characterized by multiple input neurons synapsing on one postsynaptic neuron.

   a. diverging circuit
   b. discharge zone
   c. facilitation zone
   d. converging circuit

14. Mark the following statements as true or false. If a statement is false, correct it to make a true statement.

   a. The resting membrane potential refers to the voltage difference across the membranes of excitable cells at rest.
   b. The concentration of Na$^+$ is highest in the cytosol, and the concentration of K$^+$ is highest in the extracellular fluid.
   c. The Na$^+$/K$^+$ pumps and gated channels maintain the Na$^+$ and K$^+$ gradients necessary for action potentials to occur.
   d. A depolarization is a change in membrane potential that makes the potential less negative.
   e. A local potential is a change in membrane potential that conducts the long-distance signals of the nervous system.

15. Sequence the following list of events of a neuronal action potential by placing 1 next to the first event, 2 next to the second event, and so on.

   a. _____ The activation gates of voltage-gated Na$^+$ channels open, Na$^+$ flood the cytoplasm, and depolarization occurs.
   b. _____ K$^+$ continue to flow out of the axon until the membrane is hyperpolarized.
   c. _____ Local potentials depolarize the membrane to threshold.
   d. _____ The inactivation gates of voltage-gated Na$^+$ channels close as voltage-gated K$^+$ channels open, K$^+$ begin to exit the axon, and repolarization begins.
   e. _____ Repolarization continues and Na$^+$ channels return to resting.

16. Mark the following statements as true or false. If a statement if false, correct it to make a true statement.

   a. An excitatory postsynaptic potential is caused by K$^+$ or Cl$^-$ channels opening in the membrane of the postsynaptic neuron.
   b. Postsynaptic potentials may summate by spatial summation in which multiple neurons fire onto a single postsynaptic neuron.
   c. An inhibitory postsynaptic potential causes the membrane potential of the postsynaptic neuron to approach threshold.
   d. Spatial summation can combine two EPSPs, two IPSPs, or an EPSP and an IPSP.

LEVEL 2 Check Your Understanding

1. A drug that blocks Na$^+$ channels in neurons does so not only in the axon but also in the dendrites and cell body. What overall effect would this have on action potential generation?

2. What would happen if the drug blocked K$^+$ channels instead?

3. Why must a cell body be intact for an axon to regenerate? (*Hint*: What structure is housed only in the cell body, and what are its functions?)

4. Explain how an action potential is propagated down an axon in continuous conduction. Why is saltatory conduction faster than continuous conduction?

LEVEL 3 Apply Your Knowledge

PART A: Application and Analysis

1. The drug neostigmine blocks the actions of acetylcholinesterase in the synaptic cleft. What effect would this have on synaptic transmission? What effects might you expect to see as a result of this drug?

2. During a surgical procedure, an anesthesiologist administers to the patient an inhaled anesthetic agent that opens Cl$^-$ channels in the postsynaptic membranes of neurons in the brain. Explain why this would put the patient "to sleep" for the duration of the surgical procedure.

3. Albert accidentally ingests the poison *tetrodotoxin* from the pufferfish, which you know blocks voltage-gated Na$^+$ channels. Predict the symptoms Albert will experience from this poisoning.

4. Albert, the patient in question 3, takes the drug lithium, which reduces the permeability of the neuronal axolemma to Na$^+$ (that is, it allows fewer Na$^+$ to enter the axon). Predict the effect this would normally have on his neuronal action potentials. Do you think this drug would be beneficial or harmful, considering his condition?

PART B: Make the Connection

5. Predict the effect that tetrodotoxin would have on Albert's muscle fiber action potentials (see question 3). Would it affect end-plate potentials at the motor end plate? Why or why not? (*Connects to Chapter 10*)

6. Explain what would happen if depolarization of the trigger zone led to a negative feedback loop instead of a positive one. (*Connects to Chapter 1*)

See answers in Appendix A.

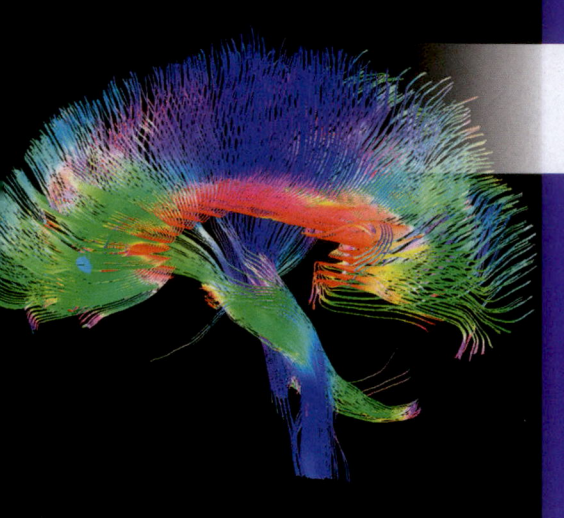

12

The Central Nervous System

For practice applying concepts to a clinical scenario, check out the **Running Case Study** *for this chapter* @ Mastering Anatomy & Physiology

Photo: *Colored three-dimensional magnetic resonance imaging (MRI) scan showing the white matter pathways of the brain, lateral view.*

We throw around the idea of "the mind" a lot in our day-to-day language—we have things on our minds, we give someone a piece of our minds, we change our minds, and sometimes we even lose our minds. But what *is* the mind, exactly, and where is it located? Scientists, philosophers, and theologians have spent centuries trying to answer these questions. The current working definition for the *mind* is the collection of thoughts, emotions, intellect, and imagination that is unique to each individual. The idea that the mind resides within and can be explained by the processes of the body has been met with much resistance; in fact, the mind was believed to be completely separate from the body for millennia, and this belief is still held by some. However, abundant scientific evidence shows that the mind does indeed reside in the body and that normal physiological processes can account for it. The most dramatic evidence comes from disease processes or injuries to regions of the brain that lead to changes in the functions of the mind.

In this chapter, we explore the structure and functions of the remarkable organs of the **central nervous system (CNS):** the *brain* and the *spinal cord.* After an overview, we take a look at the brain's anatomy. Next, we move on to some of the brain's processes, such as how the brain works to maintain homeostasis and how it performs functions relating to the mind. Then we cover the structure and function of the spinal cord, including its roles in movement and transmitting sensory stimuli. Finally, we look at the "big picture" of how the different parts of the brain and spinal cord work together to transmit and interpret sensation and to control movement.

Note, however, that some of the body's critical processes—sensation, movement, and maintenance of homeostasis—rely on components in both the CNS and the peripheral nervous system (PNS). For this reason, in the next two chapters we discuss how the PNS contributes to movement, sensation, and maintenance of homeostasis (see Chapters 13 and 14). You will notice in all three chapters, though, that we can't ever really discuss the functions of the CNS or the PNS exclusively, as the two are able to function only as an integrated whole.

Overview of the Central Nervous System

Learning Outcomes

1. Describe the structure and function of each major area of the brain.
2. Describe the five developmental regions of the brain, and identify the major areas of the adult brain that arise from each region.

Before we get into the specifics of the CNS, an introductory tour of its components is in order. This module first examines the basic functions carried out by the CNS. Next it presents the general structure of the brain and spinal cord. It concludes with an overview of the embryonic development of the CNS.

Overview of CNS Functions

« FLASHBACK

1. How do motor, sensory, and integrative functions differ? (p. 382)

The functions of the nervous system can be broken down into three categories: *sensory functions,* or detection of sensations within and outside the body; *integrative functions,* or "decision-making" processes (see Chapter 11); and *motor functions,* or stimulation of a muscle cell contraction or a gland secretion. Both sensory and motor functions are performed by the PNS, whereas integrative functions are carried out exclusively by the CNS. Integrative functions encompass a wide variety of processes, including maintenance of homeostasis; higher mental functions such as use of language and learning; interpretation of sensory stimuli; and planning and monitoring movement.

As you saw in Figure 11.2 (see p. 383), the brain receives and processes input from a number of sources, including sensory neurons of the PNS, interneurons from other parts of the brain, and interneurons from the spinal cord. Once the information has been processed, the brain makes a "decision" on how to respond to the input and sends this output to other interneurons of the CNS or to a muscle or gland via motor neurons of the PNS.

Quick Check

☐ 1. What types of functions are performed by the CNS?

Basic Structure of the Brain and Spinal Cord

« FLASHBACK

1. Where are the cranial and vertebral cavities located? (p. 17)

2. Which types of cells make up nervous tissue? (p. 148)
3. What are the differences between gray matter and white matter? What names are given to structural groups of gray matter and white matter in the CNS? (p. 387 and p. 391)

The brain and spinal cord are anatomically a single, continuous structure. The **brain** is a soft, whitish-gray organ that resides in the cranial cavity. It is composed primarily of nervous tissue, although it contains epithelial and connective tissue as well (see Chapter 11). It has internal cavities called *ventricles* (VEN-trih-kulz) that are filled with a protective *cerebrospinal fluid.* An adult brain weighs between 1250 and 1450 g and tends to be slightly larger in males than in females (all male organs tend to be larger than those of females). However, the size of the brain has little to do with intelligence, and the ratio of brain weight to body weight of females generally is equal to that of males. But is there such a thing as a "male brain" or a "female brain"? To find out, read *A&P in the Real World, Pseudoscience Exposed: The Myth of Brain Differences between the Sexes.*

The brain is richly supplied with blood vessels. Indeed, during rest about 20% of total blood flow in the body goes to the brain. This reflects its need for large amounts of oxygen, glucose, and other nutrients. For a view of blood supply to the brain, please refer to the blood vessels chapter (see Figure 18.17).

The brain consists of four divisions: the cerebrum, diencephalon, cerebellum, and brainstem (**Figure 12.1**). Each division is distinct in terms of the type of input it receives and where it sends its output. Let's look at each division in more detail.

- **Cerebrum.** The *cerebrum* (seh-REE-brum; Latin *cerebr-* = "brain") is the enlarged superior portion of the brain. It is composed of two halves known as the right and left **cerebral hemispheres.** The cerebrum is responsible for our higher mental functions, including learning, memory, personality, cognition (thinking), language, and conscience, and it also plays major roles in sensation and movement.
- **Diencephalon.** Buried beneath the cerebral hemispheres is the central core of the brain: the *diencephalon* (dy'-en-SEF-ah-lahn; Greek *encephal-* = "brain"). The diencephalon consists of four distinct structural parts and is responsible for processing, integrating, and relaying information to different parts of the brain; maintaining homeostasis of various physiological variables; regulation of movement; and biological rhythms.
- **Cerebellum.** The posterior and inferior portion of the brain is the *cerebellum* (sehr-uh-BEL-um; "little brain"). Like the cerebrum, the cerebellum is composed of right and left hemispheres. The cerebellum figures importantly in the planning and coordination of movement, particularly for complex activities such as playing an instrument or a sport.
- **Brainstem.** The oldest part of the brain from an evolutionary standpoint, the *brainstem* connects the brain and spinal cord. Its functions include control of basic involuntary processes such as the rate and depth of breathing, mediating certain reflexes, monitoring movement, and integrating and relaying information to other parts of the nervous system.

The **spinal cord** is a long, tubular organ encased within and protected by the vertebral cavity. It begins at the foramen

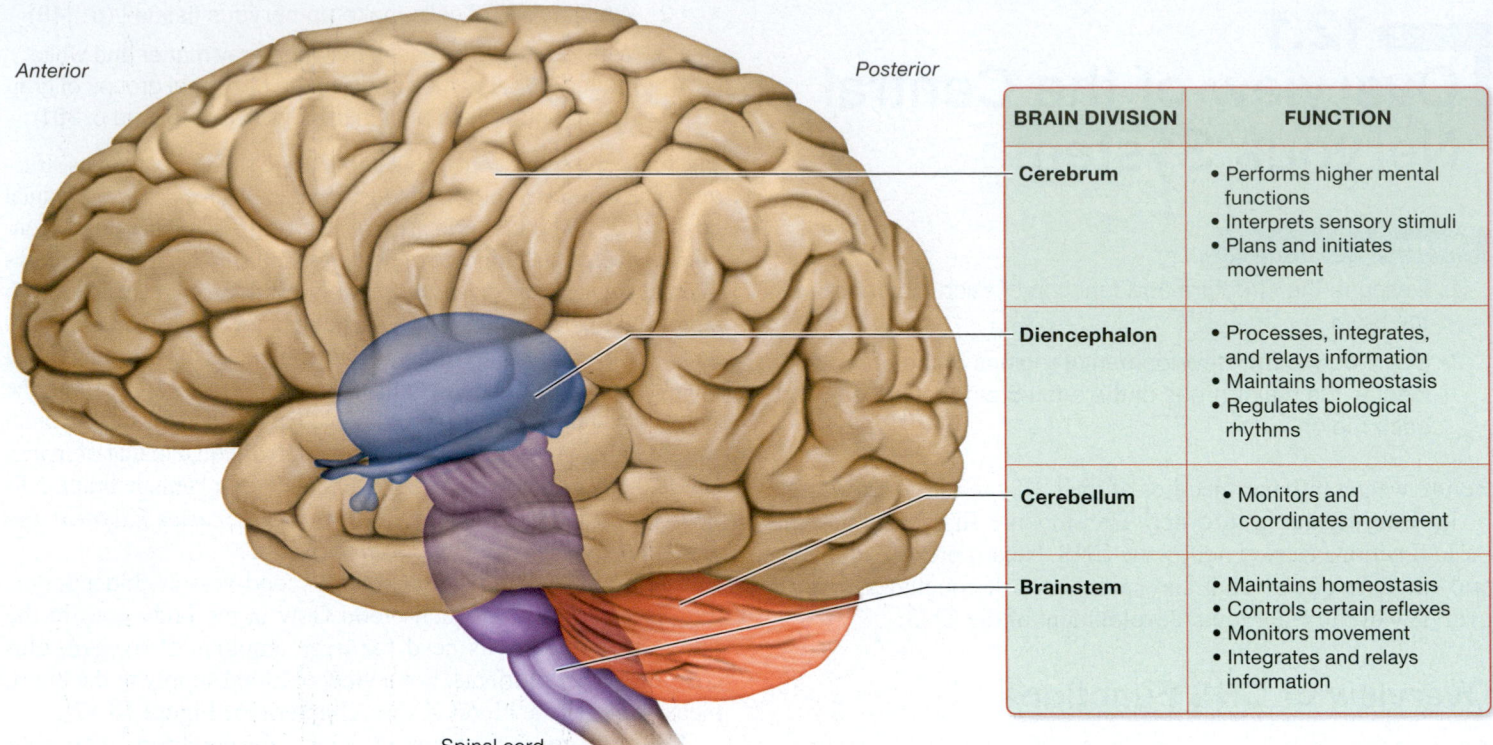

Anterior   Posterior

BRAIN DIVISION	FUNCTION
Cerebrum	• Performs higher mental functions • Interprets sensory stimuli • Plans and initiates movement
Diencephalon	• Processes, integrates, and relays information • Maintains homeostasis • Regulates biological rhythms
Cerebellum	• Monitors and coordinates movement
Brainstem	• Maintains homeostasis • Controls certain reflexes • Monitors movement • Integrates and relays information

Spinal cord

Figure 12.1 Divisions of the brain (lateral view).

magnum of the skull, where it blends with the inferior part of the brainstem, and ends approximately between the first and second lumbar vertebrae. It measures about 43–46 cm (17–18 inches) in length and is quite thin in diameter, ranging from about 0.65–1.25 cm (0.25–0.5 inch). Like the brain, the spinal cord has an internal cavity filled with cerebrospinal fluid; its cavity, the *central canal,* is continuous with the ventricles of the brain.

Both the brain and the spinal cord consist of **white matter,** which contains myelinated axons, and **gray matter,** which is made up of neuron cell bodies, dendrites, and unmyelinated axons (**Figure 12.2**). Within the brain, the gray matter is located in the

Pseudoscience Exposed: The Myth of Brain Differences between the Sexes

Have you ever heard the adage "Men are from Mars, women are from Venus"? Or seen the female brain represented as pink and the male brain represented as blue? Without a doubt, humans are sexually dimorphic creatures, meaning that there are differences in physical characteristics between males and females, and this dimorphism extends far beyond the reproductive system to essentially every system in the body. But does it extend to the brain? Is there such a thing as "brain sex"?

For centuries, scientists have been trying to prove the existence of a "male brain" and a "female brain," much as scientists a century ago tried to prove that people of different skin colors had different brains. And, indeed, researchers have found differences between the brains of men and women. For example, men, on average, have higher visual-spatial intelligence, whereas women, on average, have higher verbal-emotional intelligence. However, evidence strongly suggests that this is due to a phenomenon called *neuroplasticity,*

in which a person's socialization influences the development of neural networks. For example, engaging in sports or building activities encourages the development of visual-spatial skills, and participating in imaginative play fosters verbal-emotional skills. How do we know that neuroplasticity is at play? Because when we test very young children, the sex differences in visual-spatial and verbal-emotional intelligence become statistically insignificant.

Other data used to support the hypothesis of brain sex include MRI studies appearing to show vastly dissimilar "brain maps" in males and females. Yet when these maps are corrected for the size discrepancies in male and female brains, those differences virtually disappear. In addition, it has long been held that a part of the brain called the hippocampus is larger in females than in males, accounting for women's higher "emotional intelligence." However, a recent analysis of over 80 studies found no significant size difference between the male and female hippocampus.

In spite of these findings, the media frequently still publish articles about how men are from Mars and women are from Venus. Unfortunately, the myth of brain sex doesn't appear to be going away any time soon.

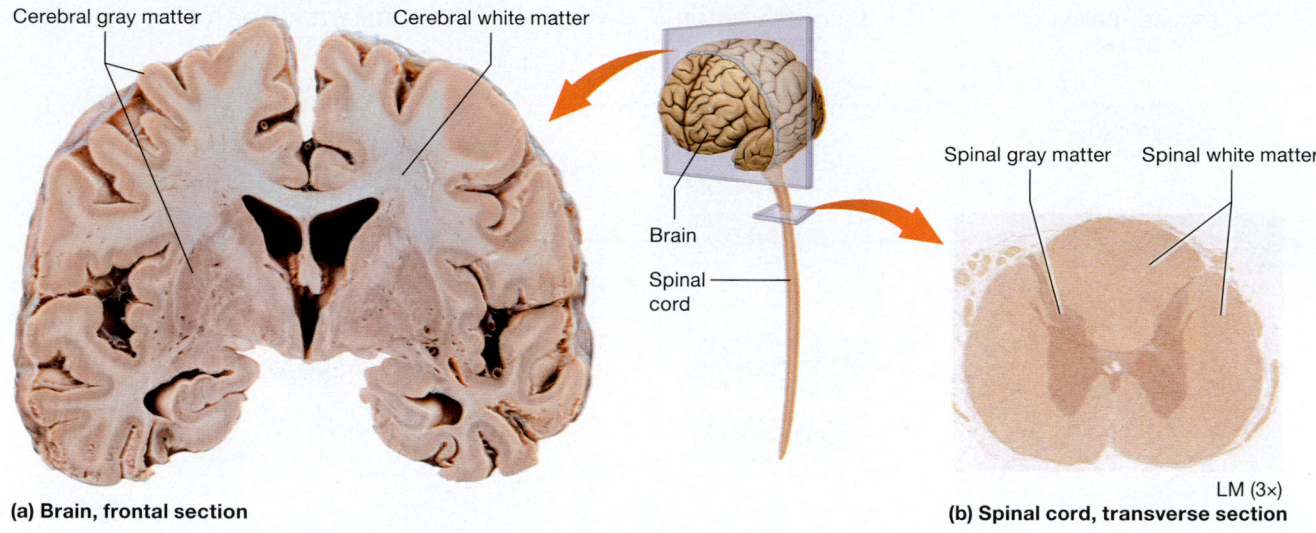

Cerebral gray matter   Cerebral white matter

Brain

Spinal cord

Spinal gray matter   Spinal white matter

LM (3×)

(a) Brain, frontal section   **(b) Spinal cord, transverse section**

Figure 12.2 White and gray matter in the CNS.

cerebrum's outer few millimeters and is also scattered throughout its deeper portions; the remainder of the cerebrum is white matter (see Figure 12.2a). Each lobe of the cerebrum contains bundles of white matter, called **tracts,** that receive input from and send output to the different clusters of cell bodies and dendrites in the cerebral gray matter, called **nuclei.** This communication connects the different parts of the brain and spinal cord. The myelin sheath on the axons that form these tracts enables nearly instantaneous communication between parts of the CNS.

The spinal cord consists internally of gray matter containing nuclei that process stimuli, as well as tracts of white matter that shuttle these stimuli to and from the brain. As you can see in Figure 12.2b, the spinal gray matter is surrounded by white matter and in cross section has the shape of a butterfly. Note that the organization of gray and white matter in the brain and in the spinal cord are reversed: The spinal white matter is superficial, whereas the cerebral white matter is deep.

Quick Check

☐ 2. What are the divisions of the brain?

☐ 3. Where is the spinal cord located? Where does it begin and end?

Overview of CNS Development

Although the CNS structure is complicated, it can be useful to think of it as just a tube with an enlarged end. We can see how the brain and spinal cord end up with this structure by looking at how they develop in the embryo (**Figure 12.3**). Nervous tissue develops from a hollow structure called the **neural tube,** which is completely formed by the fourth week of development. Just after the neural tube forms, its caudal end (*cauda-* = "tail") goes on to become the spinal cord, and the cranial end rapidly enlarges and forms three saclike structures called **primary brain vesicles:** the *forebrain, midbrain,* and *hindbrain.* These primary brain vesicles

continue to enlarge and give rise to five **secondary brain vesicles** by about the fifth week of development.

The secondary vesicles eventually become the four divisions of the mature brain:

● The forebrain expands into two secondary brain vesicles: the *telencephalon* and the *diencephalon.* The two lobes of the telencephalon enlarge to become the cerebral hemispheres, and the diencephalon, as you might predict, becomes the diencephalon of the mature brain.

● The midbrain expands into a secondary brain vesicle known as the *mesencephalon.* This term means "midbrain," and indeed the mesencephalon goes on to become a portion of the brainstem called the *midbrain.*

● The hindbrain develops into two secondary brain vesicles: the *metencephalon* and *myelencephalon.* Both vesicles develop into the remainder of the brainstem, and part of the metencephalon becomes the cerebellum.

Note that the cavity in the hollow neural tube becomes the ventricles in the brain and the central canal in the spinal cord.

Quick Check

☐ 4. What is the neural tube?

☐ 5. What does each of the secondary brain vesicles become in the mature brain?

Apply What You Learned

☐ 1. In terms of survival, to which of the four divisions of the brain would an injury be most damaging? Explain.

☐ 2. In which of the four divisions of the brain might an injury cause changes in personality? Explain.

See answers in Appendix A.

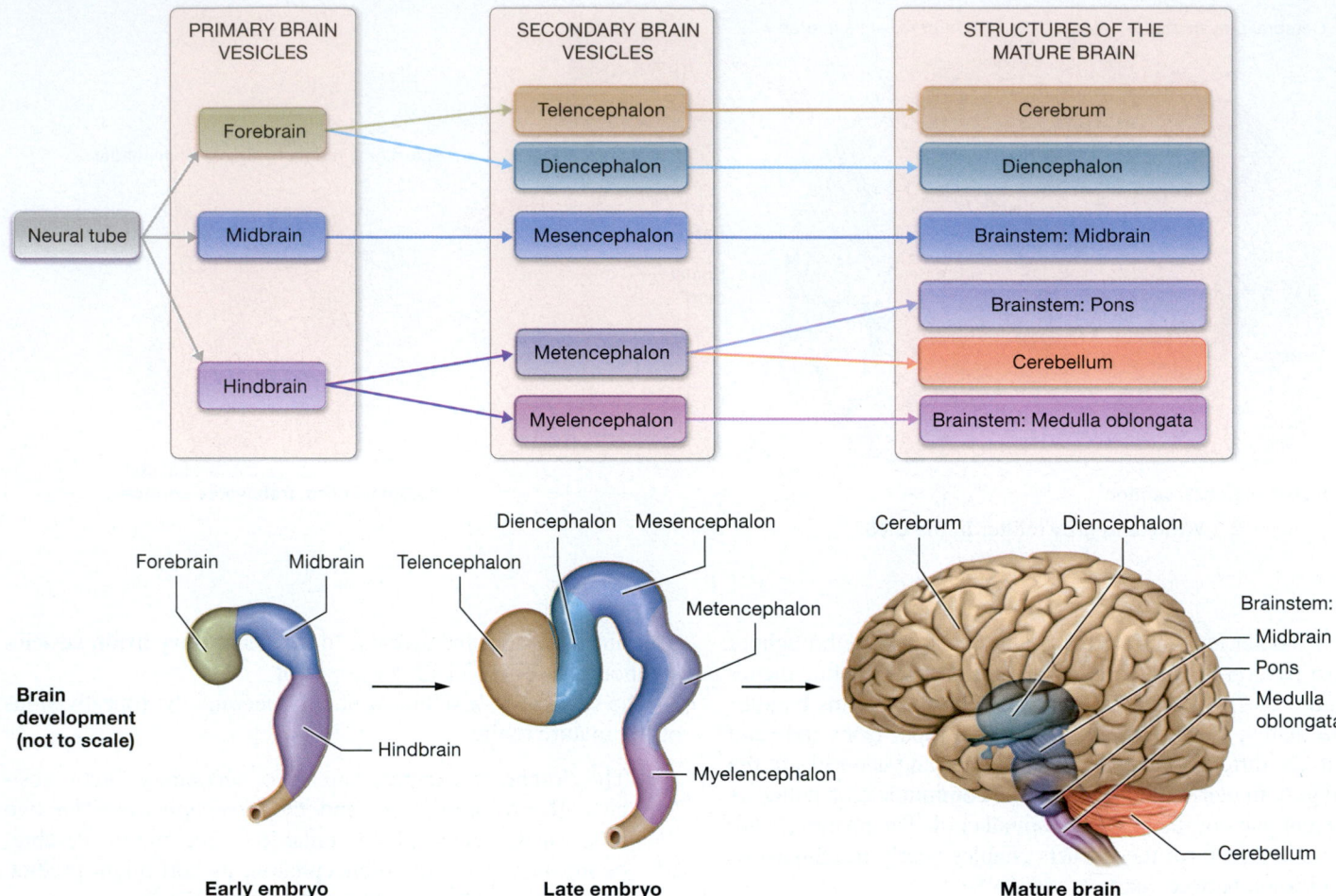

Figure 12.3 Development of the brain.

12.2
The Brain

Learning Outcomes

1. Describe and identify the five lobes of the cerebral hemisphere, and explain how motor and sensory functions are distributed among the lobes.

2. Describe the structure, components, and general functions of the regions of the diencephalon, cerebellum, and brainstem.

3. Describe the location and functions of the limbic system and the reticular formation.

We use only 10% of our brains, right? Wrong—this is a myth; we use 100% of our brains. This error has been widely spread by the media, by psychics looking to explain their purported abilities, and by those wishing to sell you books and courses that claim to teach you how to "tap into the potential" of the 90% of your brain you supposedly don't use. Although it is widely believed, this myth doesn't even really make much sense when

you think about it: Why would the body invest so much of its blood flow, nutrients, and energy in something that was 90% nonfunctional? Although we don't precisely know the function of every region of the brain, it is fairly safe to say that each part is doing *something* important.

Now we explore the four divisions of the brain that were introduced in the previous module. We touch briefly on the function of each component of the divisions, with full discussions coming in subsequent modules.

The Cerebrum

As the previous module indicated, the **cerebrum** is that part of the brain responsible for our higher mental functions, which is why deep thinkers are sometimes called "cerebral." Its large, paired cerebral hemispheres somewhat resemble a shelled walnut (**Figure 12.4**). The superficial parts of the cerebrum are separated by both shallow and deep grooves. The shallow grooves are called **sulci** (SUL-kee; singular, *sulcus*), and the deep grooves are called **fissures.** The two hemispheres are separated from one another by a long fissure called the **longitudinal fissure.** Each hemisphere contains multiple sulci, and between the sulci are elevated ridges known as **gyri** (JY-ree; singular, *gyrus*). In an example of the Structure-Function

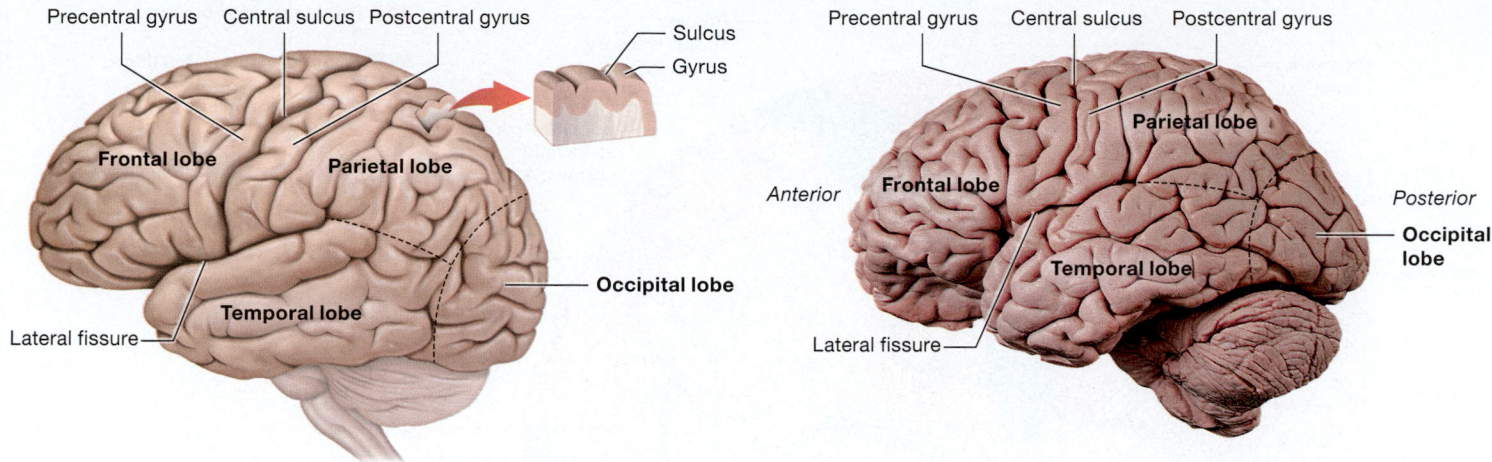

(a) Lateral view (left hemisphere)

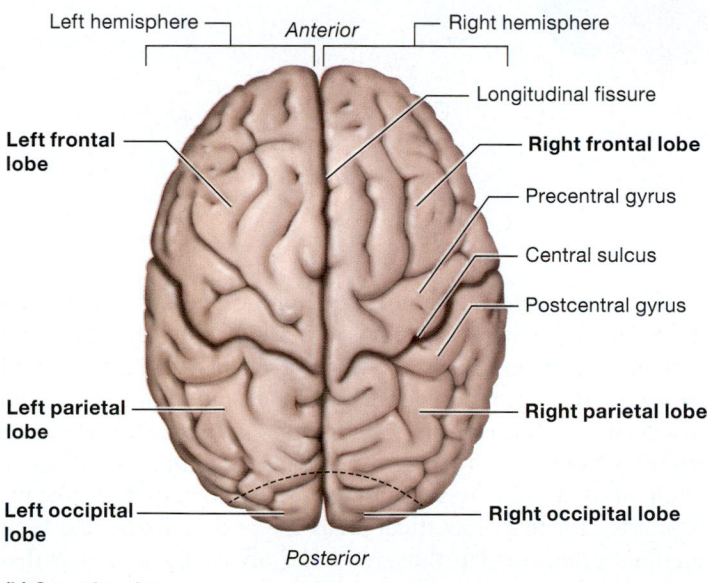

(b) Superior view

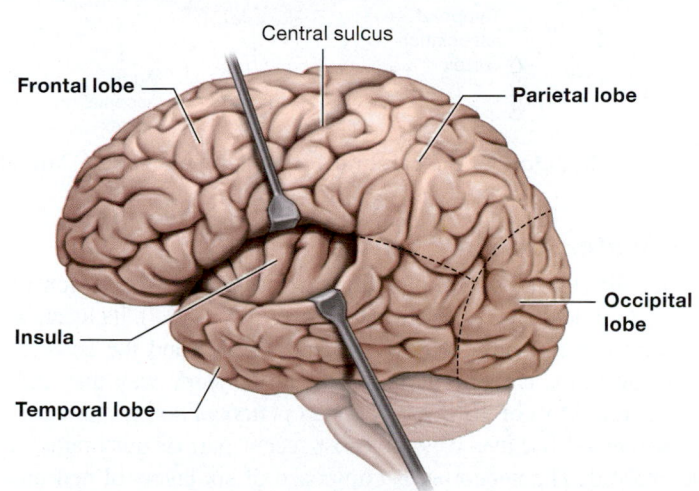

(c) Lateral view (frontal, parietal, and temporal lobes pulled back)

Figure 12.4 Structure of the cerebrum.

Core Principle (p. 28), the gyri and sulci increase the surface area of the brain and make it a much more compact structure; if our brains were smooth, we  **CORE PRINCIPLE** Structure-Function would require much larger skulls to hold them. Although the gyri and sulci can vary in location and size from one brain to the next, certain gyri and sulci are found in almost all brains and thus are named.

The cerebral hemispheres house two ventricles, the right and left *lateral ventricles.* They also contain a number of structures that carry out specialized CNS functions. The upcoming subsections provide a closer look at these structures and their roles.

Lobes of the Cerebral Hemispheres

The cerebral hemispheres are divided into five lobes (rounded subdivisions) that contain groups of neurons performing specific tasks. The names and boundaries of the five lobes are detailed in the following list.

- **Frontal lobes.** The anteriormost lobe is the **frontal lobe.** Note in Figure 12.4a and b that its posterior boundary is called the **central sulcus.** Just anterior to the central sulcus is the **precentral gyrus.**
- **Parietal lobes.** Posterior to the frontal lobes are the paired **parietal lobes** (pah-RY-eh-tal). The major gyrus of each of these lobes is the **postcentral gyrus,** which sits just posterior to the central sulcus.
- **Temporal lobes.** The paired **temporal lobes** are on the lateral surfaces of the hemisphere, separated from the frontal and parietal lobes by the **lateral fissure.**
- **Occipital lobes.** The posterior lobe of each cerebral hemisphere is the **occipital lobe.** It is separated from the parietal lobe by the *parieto-occipital sulcus* (look ahead to Figure 12.12).
- **Insulas.** As you can see in Figure 12.4c, the paired **insulas** (IN-syoo-lahz) are visible only when you pry the frontal, parietal, and temporal lobes apart at the lateral fissure (remember this with the mnemonic "the insula is *insula*ted by the other lobes").

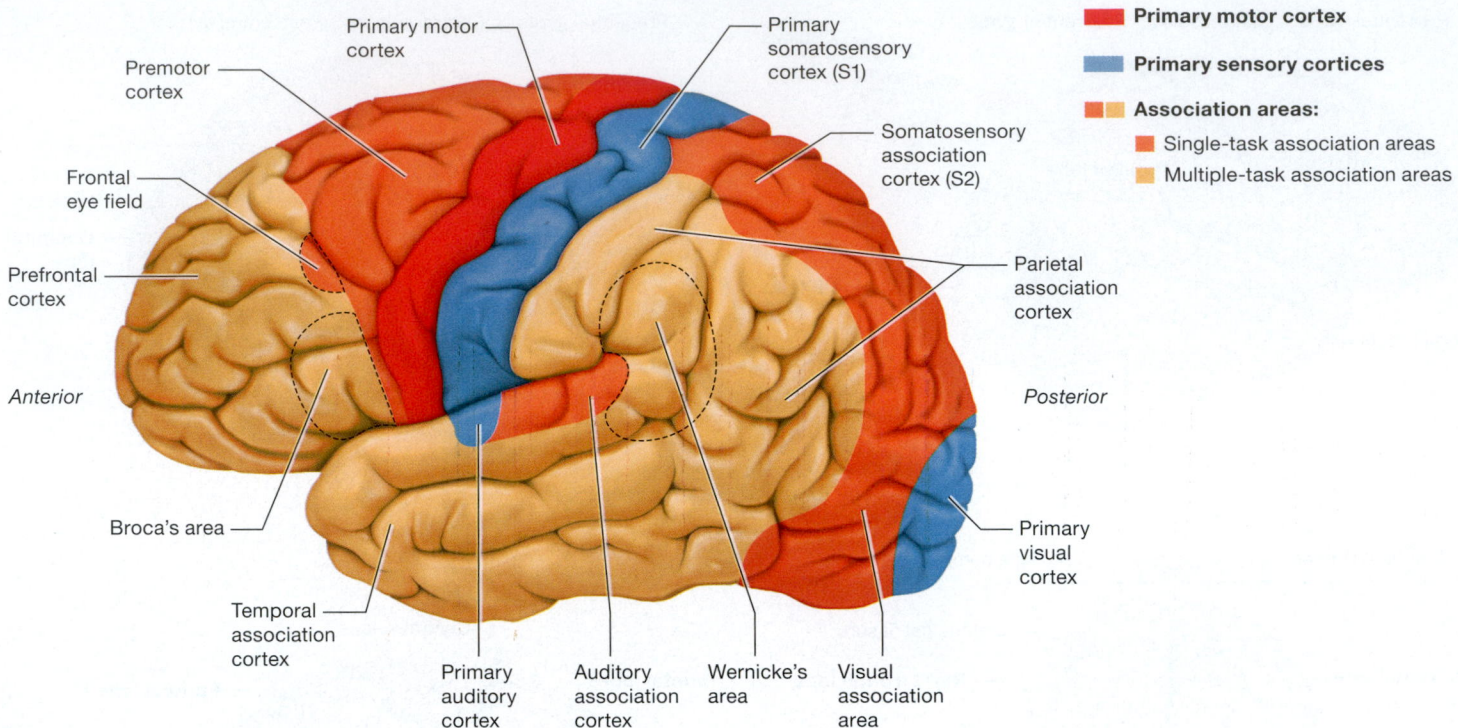

Figure 12.5 Structure of the cerebral cortex (left hemisphere, lateral view).

Gray Matter: Cerebral Cortex

Functionally, the most complex part of the brain is the region of gray matter known as the **cerebral cortex** (**Figure 12.5**). Its location is reflected in its name; "cortex" means "bark," and the cerebral cortex covers the cerebral hemispheres in the same way that bark covers a tree. Most of the cerebral cortex is **neocortex,** or "new cortex," so named because it is the most recent part of our brains to have evolved. The neocortex is composed of six layers of neurons and neuroglia and varies in thickness from about 1.5 to 4.5 mm. The many convolutions of the gyri give the neocortex quite a large surface area—about 0.25 square meter (2.5 square feet).

The functions of the neocortex revolve around *conscious* processes (those in which we are aware of ourselves and our environment), such as planning movement, interpreting incoming sensory stimuli, and complex higher functions like performing long division. In other words, it allows us to become aware of—and respond to—our surroundings. Other components of the cerebrum and the remainder of the brain function largely in *subconscious* processes, that is, those carried out below the level of conscious thought. Such processes are often called "subcortical" because they do not directly involve the cerebral cortex. Many of these subconscious processes, such as control of blood pressure and body temperature, are vital, arguably even more so than our ability to perform long division.

As you can see in Figure 12.5, the neocortex has three generally recognized areas distributed among the five lobes: the *primary motor cortex,* which plans and executes movement; the *primary sensory cortices* (the plural of cortex), which are the first to receive and process sensory input; and the **association areas,** which integrate different types of stimuli. Some association areas integrate only one type of information ("single-task"),

and others integrate several different types of stimuli ("multiple-task"). The following subsections will briefly introduce the areas of the neocortex responsible for motor, sensory, and higher-level functions. (Be sure not to confuse sensory and motor areas with sensory and motor neurons—all neurons in the cortex are interneurons.) Note that although we discuss each area as a discrete region, it is a vast oversimplification to do so. Realistically, boundaries for the neocortical areas are at best fuzzy and their functions often overlap those of other areas. Each area is discussed more fully later in the chapter and in the special senses chapter (see Chapter 15).

Motor Cortices and Upper Motor Neurons Most of the motor areas are located in the frontal lobe; these areas contain interneurons called **upper motor neurons.** Movement is consciously planned by upper motor neurons of the **primary motor cortex,** located in the precentral gyrus of the frontal lobe. The upper motor neurons of each cerebral hemisphere control the motor activity of the *opposite* side of the body. The movements are executed by neurons of the PNS called *lower motor neurons* that directly contact skeletal muscles.

Movement requires input from several motor association areas. For example, the large **premotor cortex,** located anterior to the primary motor cortex in the frontal lobe, is involved in the planning, guidance, coordination, and execution of movement. Another motor association area consists of the paired **frontal eye fields,** one on each side of the frontal lobe anterior to the premotor cortex. These fields are required for back-and-forth eye movements, such as those relating to reading.

Sensory Cortices Each of the five classic senses—touch, vision, hearing, smell, and taste—is represented by both a primary and an association area in the cerebral cortex. In addition, the sense of *equilibrium,* which gives us information about balance, also has primary and association areas. These areas, most of which are shown in Figure 12.5, are found in all lobes except the frontal lobe (although some overlap does occur) and include the following:

- **Somatosensory areas.** Two main areas of the cerebral cortex deal with the *somatic* senses (*soma-* = "body"), which are those pertaining to temperature or touch (mechanical deformation of a tissue), including vibration, pressure, stretch, and joint position. The first is the **primary somatosensory cortex** (also called **S1**), located in the postcentral gyrus of the parietal lobe. The second is the **somatosensory association cortex** (also called **S2**), located posterior to S1, also in the parietal lobe.
- **Visual areas.** Humans have a large amount of cortical area dedicated to processing visual input, including all of the occipital lobes and part of the temporal lobes. This reflects the importance of vision in humans and other primates. The **primary visual cortex,** which lies at the posterior end of the occipital lobe, is the first area to receive visual input. It feeds this information into surrounding **visual association areas,** which process color, object movement, and depth.
- **Auditory areas.** The **primary auditory cortex,** located in the superior temporal lobe, is the first to receive auditory stimuli. From here, stimuli are sent to the adjacent **auditory association cortex** as well as other association areas.
- **Other sensory areas.** Taste stimuli are processed by the poorly localized **gustatory cortex,** which appears to be in the insula and part of the parietal lobe. Stimuli pertaining to equilibrium and positional sense are processed by **vestibular areas,** which are located in the parietal and temporal lobes. The sense

of smell is processed by the **olfactory cortex,** which consists of several areas in and around the medial temporal lobes (see Chapter 15). (These areas are not shown in Figure 12.5.)

Multiple-Task Association Areas   The association areas that allow for higher-level functions are some of the most fascinating parts of the brain. Some of the best-characterized of these association areas include the following:

- **Language areas.** Two cortical areas are directly involved in language: **Broca's area,** a premotor area for speech sounds, located in the anterolateral frontal lobe, and **Wernicke's area** (VER-nih-keez)**,** or the integrative speech area, located in the temporal and parietal lobes. Broca's area is responsible for the ability to *produce* language, whereas Wernicke's area is responsible for the ability to *understand* language.
- **Prefrontal cortex.** As you can see in Figure 12.5, the **prefrontal cortex** is located in and occupies most of the frontal lobe. It communicates with many areas of the brain, including parts of the diencephalon, other cerebral gray matter, and the association cortices of other cerebral lobes. The prefrontal cortex has many functions, among which are modulating behavior, personality, learning, memory, and an individual's psychological state.
- **Parietal and temporal association cortices.** The *parietal and temporal association cortices* occupy much of their respective lobes. These areas perform a variety of tasks, including integration of sensory stimuli, language, maintaining attention, recognition, and spatial awareness.

The major motor, sensory, and association areas of the cerebral cortex are summarized in **Table 12.1**.

Table 12.1 Motor, Sensory, and Association Areas of the Cerebral Cortex

Area Name	Type of Cortex	Location	Function
Primary motor cortex	Motor	Precentral gyrus of frontal lobe	Plans and executes movement
Premotor cortex	Association	Widespread throughout lateral and medial frontal lobe	Plans and executes complex movement
Frontal eye fields	Association	Anterior to the premotor cortex	Back-and-forth eye movements
Primary somatosensory cortex (S1)	Sensory	Postcentral gyrus of parietal lobe	Interprets incoming somatic sensory stimuli
Somatosensory association cortex (S2)	Association	Posterior to the primary somatosensory cortex in the parietal lobe	Integrates somatic sensory stimuli
Primary visual cortex	Sensory	Posterior occipital lobe	Interprets and processes visual stimuli
Primary auditory cortex	Sensory	Superior temporal lobe	Processes auditory stimuli
Gustatory cortex	Sensory	Insula, parietal lobe	Processes taste stimuli
Vestibular areas	Sensory	Several in both parietal and temporal lobes	Processes stimuli relating to equilibrium and balance
Olfactory cortex	Sensory	Limbic lobe, medial temporal lobe	Processes smell stimuli
Broca's area	Association	Superolateral frontal lobe	Language production
Wernicke's area	Association	Superolateral temporal lobe	Language comprehension
Prefrontal cortex	Association	Anterior frontal lobe	Planning, personality, higher cognitive functions
Parietal association cortex	Association	Widespread in the parietal lobe	Spatial awareness and attention
Temporal association cortex	Association	Widespread in the temporal lobe	Recognition and associations

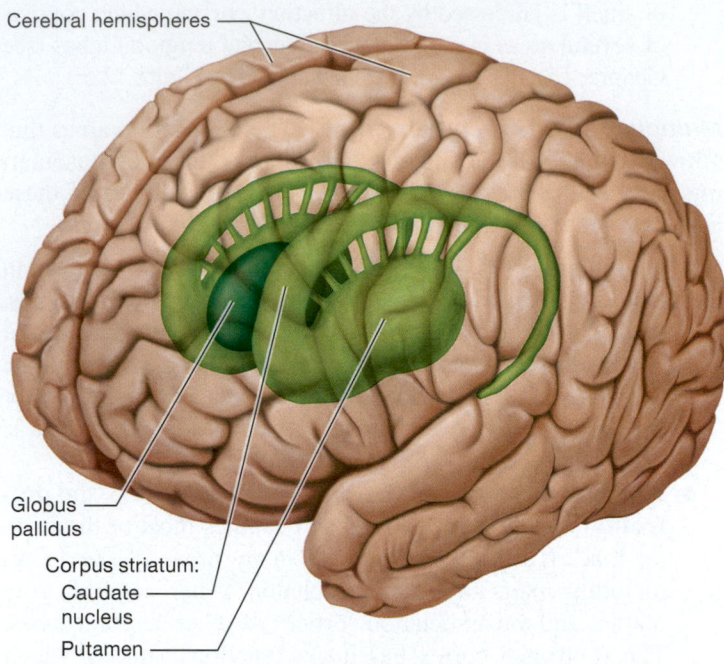

Cerebral hemispheres

Globus
pallidus

Corpus striatum:

Caudate
nucleus

Putamen

Figure 12.6 Structure of the basal nuclei (anterolateral view).

Gray Matter: Basal Nuclei

Buried within the cerebral hemispheres on either side of the diencephalon lie clusters of neuron cell bodies collectively called the **basal nuclei** (**Figure 12.6**). (Note that these nuclei have traditionally been referred to as *basal ganglia;* however, this is a misnomer, as ganglia are collections of cell bodies in the PNS.) The basal nuclei are separated from the gray matter of the diencephalon by an area of white matter known as the *internal capsule.*

The basal nuclei include the following pairs:

- **Caudate nucleus.** As you can see in Figure 12.6, the **caudate nuclei** (KOW-dayt) are C-shaped rings of gray matter that

sit lateral to the lateral ventricles in each hemisphere. The caudate nucleus is named for its long, slender "tail" that curls around and points anteriorly.

- **Putamen.** The **putamen** (poo-TAY-men; "pod") lies posterior and inferior to the caudate nucleus and is connected to it via small bridges of gray matter. The caudate nucleus and putamen are similar structurally and functionally, and are thus sometimes referred to as a single structure called the *corpus striatum.*
- **Globus pallidus.** The **globus pallidus** ("pale globe"), which sits medial to the putamen, is named for the fact that it contains more myelinated fibers than the other basal nuclei and is thus paler.

The basal nuclei have a vital role in inhibiting involuntary movement and initiating voluntary movement. When movement is not occurring, the globus pallidus prevents upper motor neurons from causing spontaneous, inappropriate muscle contractions. To begin movement, the caudate nucleus and putamen inhibit the globus pallidus, which allows the upper motor neurons to initiate the desired motion.

Although the basal nuclei are primarily associated with movement, disorders involving the basal nuclei also cause problems with behavior, cognition, and perception. This demonstrates that these nuclei have multiple functional roles in the brain.

Cerebral White Matter

The white matter in the cerebrum can be classified as one of three types—commissural, projection, or association fibers—depending on its function (**Figure 12.7**):

- **Commissural fibers.** The **commissural fibers** (kahm-ih-SHUR-ul; *commis-* = "united") connect the right and left cerebral hemispheres. Of the four groups of commissural fibers, the largest is the **corpus callosum,** which sits in the middle of the brain at the base of the longitudinal fissure.

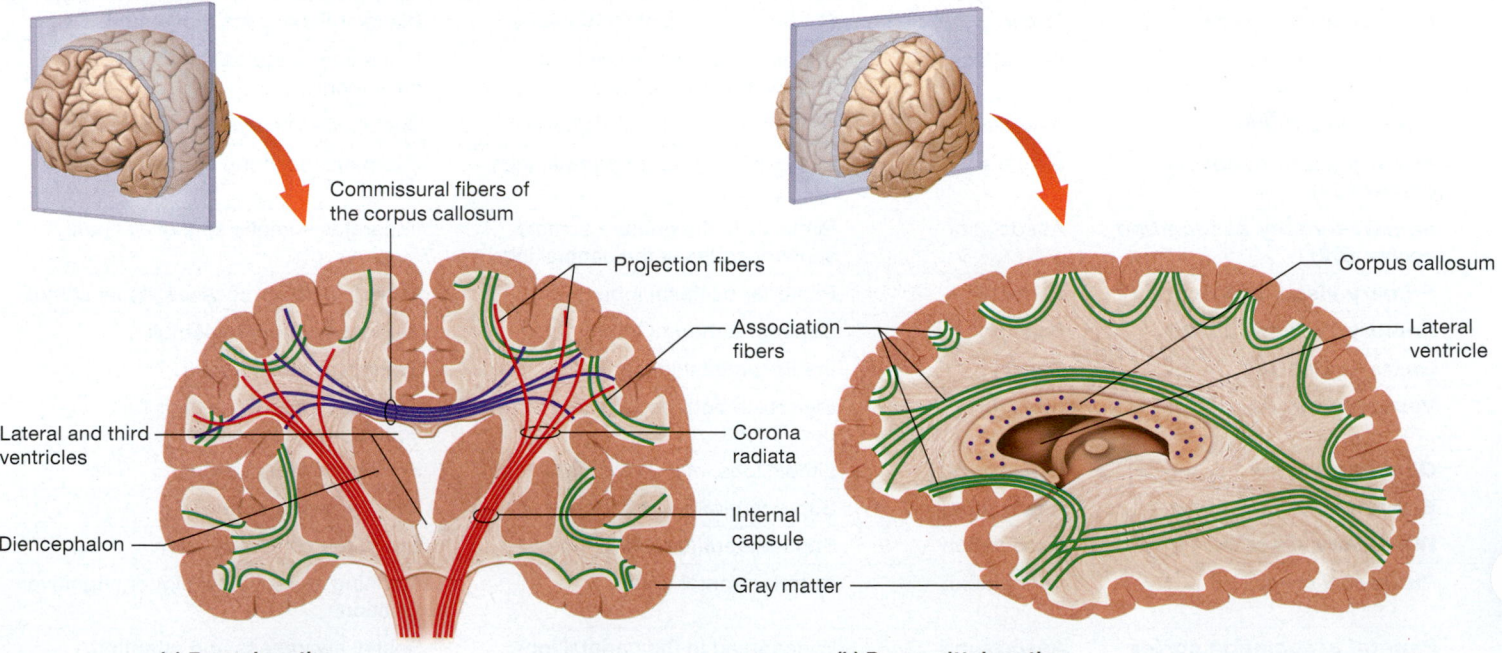

Commissural fibers of
the corpus callosum

Projection fibers

Association
fibers

Corpus callosum

Lateral
ventricle

Lateral and third
ventricles

Diencephalon

Corona
radiata

Internal
capsule

Gray matter

(a) Frontal section

(b) Parasagittal section

Figure 12.7 Structure of cerebral white matter.

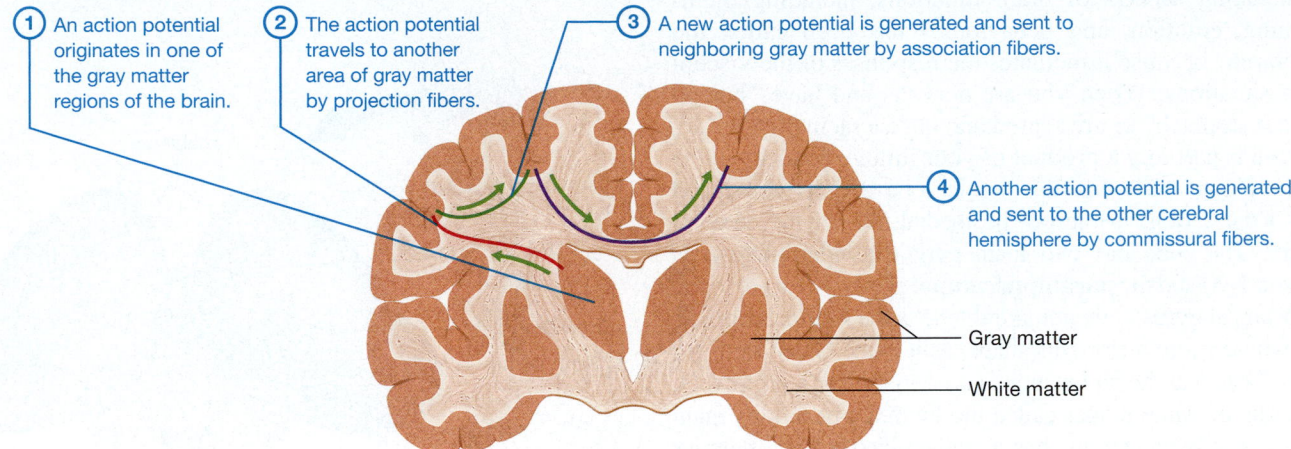

① An action potential originates in one of the gray matter regions of the brain.

② The action potential travels to another area of gray matter by projection fibers.

③ A new action potential is generated and sent to neighboring gray matter by association fibers.

④ Another action potential is generated and sent to the other cerebral hemisphere by commissural fibers.

Gray matter

White matter

Figure 12.8 A possible pathway for conduction of an action potential in the brain.

- **Projection fibers.** The **projection fibers** connect the cerebral cortex of one hemisphere with other areas of the same hemisphere, as well as with other parts of the brain and the spinal cord. Note in Figure 12.7 that the cerebral projection fibers in each hemisphere form a radiating pattern called the *corona radiata;* these fibers condense around the diencephalon on the right and left sides to form two V-shaped bands known as the **internal capsules.**
- **Association fibers.** The **association fibers** are also restricted to a single hemisphere and connect the gray matter of cortical gyri with one another.

A possible pathway for information transfer by the conduction of an action potential from one region of the brain to another is shown in **Figure 12.8**. Note that ① an action potential originates in gray matter and ② is then sent to another area of gray matter by projection fibers. This area of gray matter in turn ③ sends a new action potential to neighboring gray matter by association

fibers, and ④ another action potential is generated and sent to the other cerebral hemisphere by commissural fibers. Some nuclei send impulses to only a single nucleus; however, many communicate with multiple places, and their tracts of white matter resemble dizzying three-dimensional highway systems.

Structures of the Limbic System

Groups of gray matter that share the same function and the white matter that connects them are sometimes referred to as *functional brain systems*. An important functional brain system is the group of structures known collectively as the **limbic system.** This system includes a region of the medial cerebrum sometimes referred to as the *limbic lobe,* the *hippocampus* (hip-poh-KAM-pus), a collection of gray matter called the *amygdala* (uh-MIG-duh-luh), and the pathways that connect these regions of gray matter with one another and with other parts of the brain (**Figure 12.9**). The limbic system is found only in mammals and contains some of the most ancient parts of the mammalian brain from an evolutionary perspective. It participates in a number of the

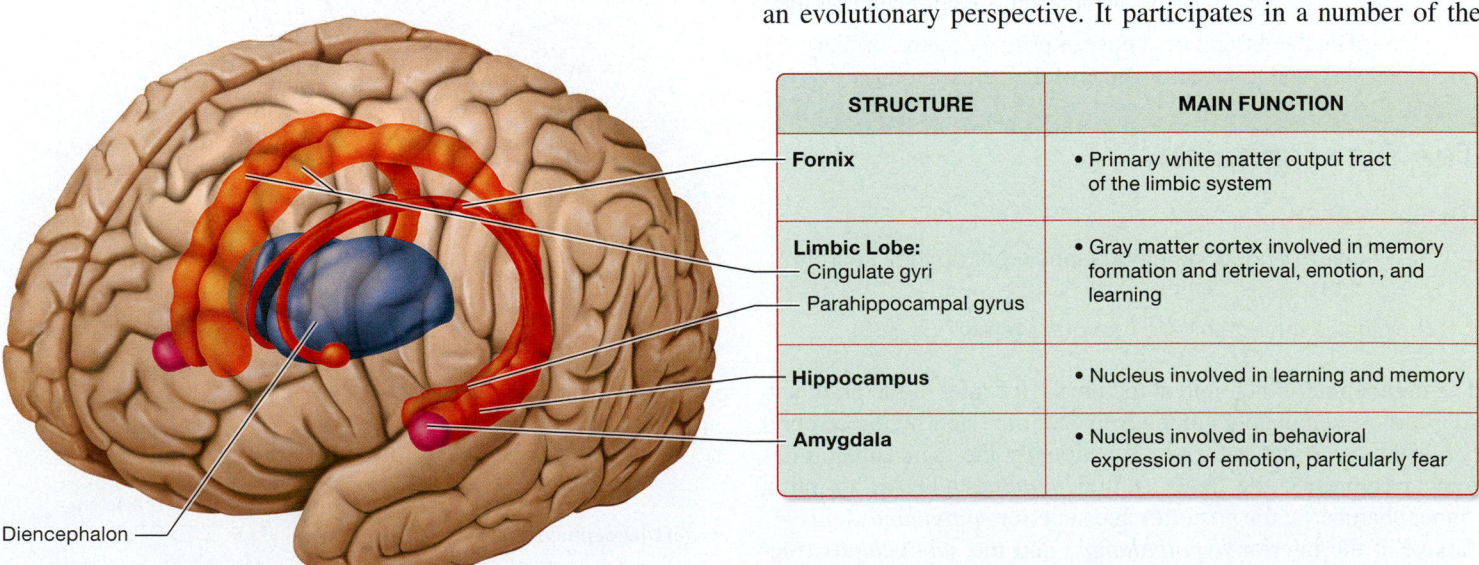

Diencephalon

STRUCTURE	MAIN FUNCTION
Fornix	• Primary white matter output tract of the limbic system
Limbic Lobe: Cingulate gyri Parahippocampal gyrus	• Gray matter cortex involved in memory formation and retrieval, emotion, and learning
Hippocampus	• Nucleus involved in learning and memory
Amygdala	• Nucleus involved in behavioral expression of emotion, particularly fear

Figure 12.9 Structures of the limbic system (anterolateral view).

more fascinating aspects of brain functions, including memory, learning, emotion, and behavior. It has been called the "visceral brain" because it mediates the responses of the viscera to certain situations. When you are nervous and have "butterflies in your stomach" or are scared and have a racing heart rate, that reaction is partially a product of your limbic system.

Note in Figure 12.9 that the limbic lobe and its associated structures form a ring visible on the medial side of the cerebral hemisphere. This lobe has two main gyri: the **cingulate gyrus** (SING-gyoo-layt) and the **parahippocampal gyrus.** Adjacent to the parahippocampal gyrus in the temporal lobe is the **hippocampus,** a structure whose name reflects its shape, as it is the Greek term for "seahorse." Note that the hippocampus is connected to a prominent, C-shaped ring of white matter called the **fornix,** which is its main output tract. The hippocampus has a well-studied role in memory and learning, which we find out more about in Module 12.4.

Anterior to the hippocampus is the **amygdala,** a structure whose shape is also reflected in its name, which is the Greek word for "almond." The amygdala functions in the behavioral expression of emotion, particularly fear.

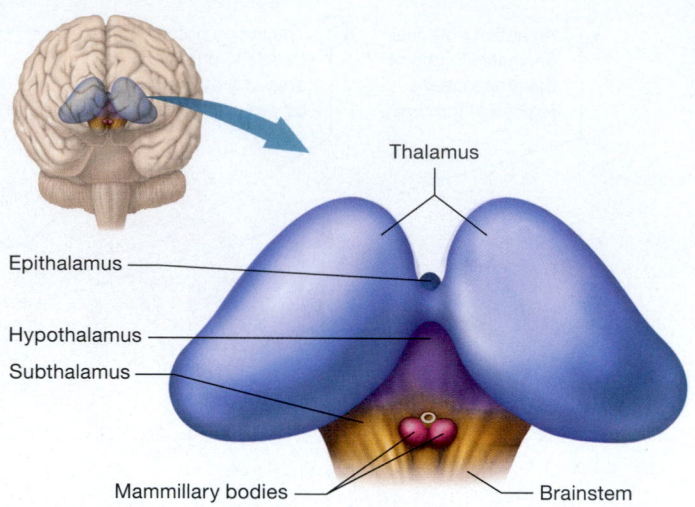

(a) Diencephalon, anterior view

Labels: Thalamus, Epithalamus, Hypothalamus, Subthalamus, Mammillary bodies, Brainstem

Quick Check

☐ 1. Match the following terms with the correct definition/description:

_____ Insula
_____ Occipital lobe
_____ Cerebral cortex
_____ Cerebral white matter
_____ Parietal lobe
_____ Basal nuclei

a. Outer few millimeters of gray matter of the cerebral hemispheres
b. Cerebral lobe deep to the frontal, parietal, and temporal lobes
c. Clusters of gray matter within the cerebrum that control movement
d. Posterior cerebral lobe
e. Groups of myelinated axons that connect regions of the cerebrum and other parts of the brain
f. Middle and superior cerebral lobe

☐ 2. Explain the differences between primary motor, primary sensory, and association areas of the cerebral cortex.

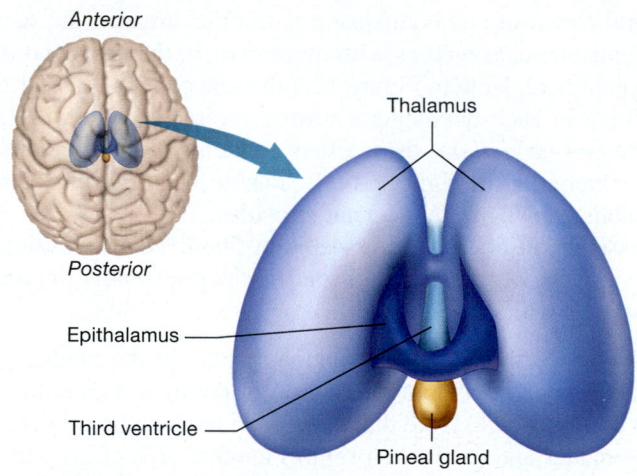

(b) Diencephalon, superior view

Labels: Anterior, Posterior, Thalamus, Epithalamus, Third ventricle, Pineal gland

The Diencephalon

◀◀ FLASHBACK

1. What are endocrine glands, and what do they secrete? (p. 134)
2. What is the autonomic nervous system? (p. 383)

The **diencephalon** is found at the physical center of the brain and so is almost completely hidden from external view by the cerebral hemispheres. Structurally and functionally the diencephalon has four components: the large, central *thalamus* (THAL-ah-muss; "inner chamber"); the posterior and superior *epithalamus* (epi- = "above"); the inferior *hypothalamus;* and the *subthalamus* (**Figure 12.10**). Each component consists of various nuclei that receive specific input and send output to other parts of the brain.

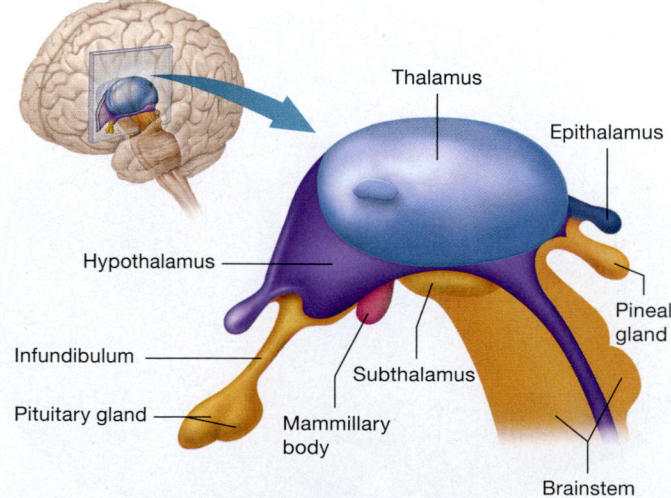

(c) Diencephalon, midsagittal section

Labels: Thalamus, Epithalamus, Hypothalamus, Infundibulum, Pituitary gland, Mammillary body, Subthalamus, Pineal gland, Brainstem

Figure 12.10 Structure of the diencephalon.

Thalamus

As you can see in Figure 12.10a and b, the **thalamus** consists of two large, egg-shaped masses of gray matter that together make up about 80% of the diencephalon. Between these two masses lies a cavity known as the *third ventricle*. Internally, each half of the thalamus has three main groups of nuclei separated by two thin layers of white matter.

Thalamic nuclei receive input from many sources, including the cerebral cortex, the cerebellum, the basal nuclei, structures of the limbic system, and the sensory system (except for the sense of smell), and their main output travels to the cerebral cortex. The thalamus is literally the "main entrance" into the cerebral cortex—nearly all stimuli destined for the cerebral cortex must first pass through the thalamus. This allows the thalamus to control which information reaches the cerebral cortex and where the information is sent, which means that ultimately the thalamus regulates cortical activity. For this reason, the thalamus plays a role in nearly all processes occurring in the brain, including initiation and regulation of movement, integration of sensory stimuli, emotions, memory, arousal, consciousness, and the level of responsiveness and excitability of the cerebral cortex.

Hypothalamus

The **hypothalamus** is a collection of nuclei that sits anterior and inferior to the thalamus. It is much smaller than the thalamus, weighing only about 4 g and making up less than 1% of the mass of the brain. Don't let its small size fool you, though, as the hypothalamic nuclei perform several functions that are vital to our survival. Some of these include regulating much of the *autonomic nervous system,* the sleep/wake cycle, thirst and hunger, and body temperature.

In addition to these roles, the hypothalamus has a close anatomical and functional connection to an endocrine organ called the **pituitary gland** (pih-TOO-ih-tehr′-ee). Note in Figure 12.10c that the pituitary gland attaches to the inferior hypothalamus by an extension referred to as the **infundibulum,** and hypothalamic tissue makes up the posterior portion of this gland. The hypothalamus secretes a variety of hormones that affect the pituitary gland—hypothalamic releasing or inhibiting hormones increase or decrease the secretion of other hormones from the pituitary gland, respectively. The pituitary gland, in turn, secretes hormones that influence secretion from other endocrine glands throughout the body. The hypothalamus also produces two additional hormones that do not act on the pituitary gland—*antidiuretic hormone* (*ADH*), which influences water balance in the body, and *oxytocin* (awk-sih-TOH-sin), which in women stimulates contraction of the uterus during childbirth and in both sexes may promote emotional bonding (see Chapter 16).

The hypothalamus receives input from a variety of sources, including the cerebral cortex and basal nuclei. It also connects with the limbic system via two small projections called the **mammillary bodies.** These structures receive input from the hippocampus, with which they play a role in regulating memory and behavior. Input also arrives at the hypothalamic nuclei from sources outside the nervous system, including the endocrine system, receptors that detect changes in body temperature, and receptors that detect the blood's concentration.

Epithalamus and Subthalamus

As implied by their names, the epithalamus lies superior to the thalamus and the subthalamus lies inferior. The major part of the epithalamus, located at the posterosuperior portion of the diencephalon, is an endocrine organ called the **pineal gland** (PIN-ee-ul). Named for its resemblance to a pine cone, the pineal gland secretes the hormone **melatonin** (mel-uh-TOH-nin), which helps to regulate the sleep/wake cycle. The largest nucleus of the subthalamus is functionally connected to the basal nuclei and works with them to control movement.

Quick Check

☐ 3. Which component of the diencephalon performs each of the following functions?

 a. Controls body temperature, thirst, and hunger

 b. Integrates emotion, memory, and sensory stimuli and sends them to association areas of the cerebral cortex

 c. Produces the hormone melatonin

 d. Works with the basal nuclei to monitor and control movement

 e. Controls the ANS and parts of the endocrine system

 f. Determines which stimuli reach the cerebral cortex

The Cerebellum

The **cerebellum** makes up the posterior and inferior portion of the brain. It functions with the cerebral cortex, basal nuclei, brainstem, and spinal cord to coordinate ongoing movement to reduce what is known as *motor error*. It is composed of two **cerebellar hemispheres** connected by a structure called the **vermis** (VER-miss; "worm") (**Figure 12.11a**). The cerebellar surface contains ridges known as **folia** ("leaves"), which are separated by shallow sulci; this arrangement increases the cerebellar surface area in the same way that gyri and sulci increase the surface area of the cerebral cortex. The cerebellum is divided into three different lobes: the **anterior lobe,** the **posterior lobe,** and the small **flocculonodular lobe** (flok′-yoo-loh-NOD-yoo-lar; **Figure 12.11b**).

As you can see in **Figure 12.11c** and **d**, the cerebellar interior is arranged in the same way as the cerebrum. It has an outer layer of gray matter—the *cerebellar cortex*—and inner white matter in which we find clusters of gray matter called the *deep cerebellar nuclei*. The cerebellar cortex is so extensively folded that the white matter resembles tree branches and so is named the **arbor vitae** (AHR-bohr VEE-tay; "tree of life"). The cerebellar white matter converges into three large tracts of white matter known as **cerebellar peduncles** (PEH-dung-k′lz) that connect the cerebellum to the brainstem. These form the only route by which stimuli flow into and out of the cerebellum.

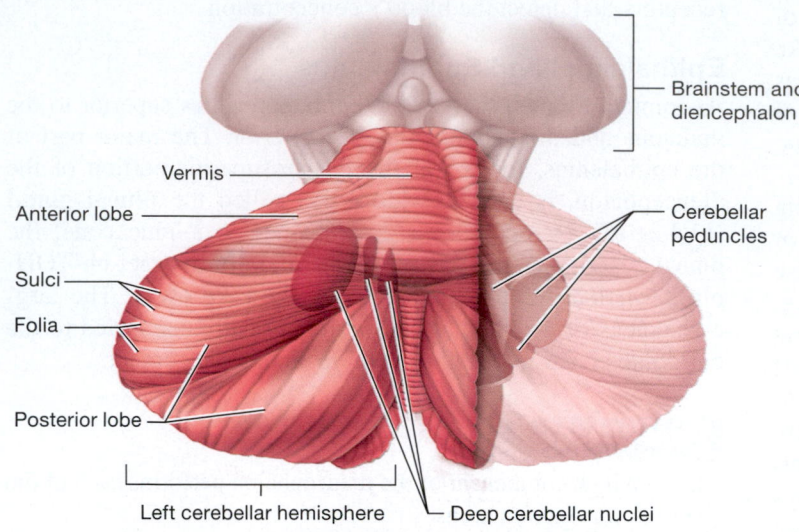

Brainstem and diencephalon

Vermis

Anterior lobe

Sulci

Folia

Posterior lobe

Cerebellar peduncles

Left cerebellar hemisphere — Deep cerebellar nuclei

(a) Cerebellum, posterior view

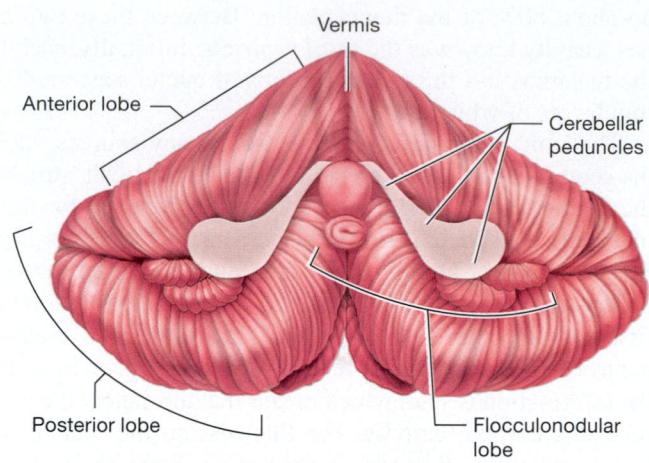

Vermis

Anterior lobe

Cerebellar peduncles

Posterior lobe

Flocculonodular lobe

(b) Cerebellum, anterior view

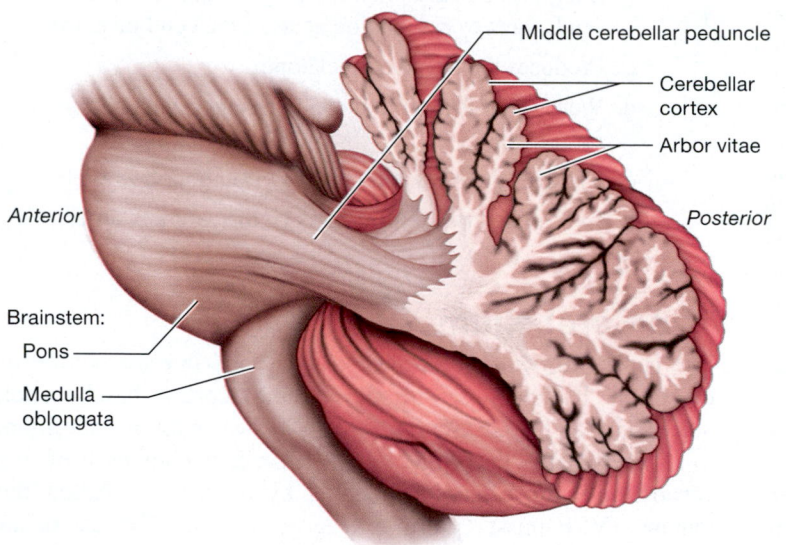

Middle cerebellar peduncle

Cerebellar cortex

Arbor vitae

Anterior

Posterior

Brainstem:

Pons

Medulla oblongata

(c) Cerebellum, parasagittal section

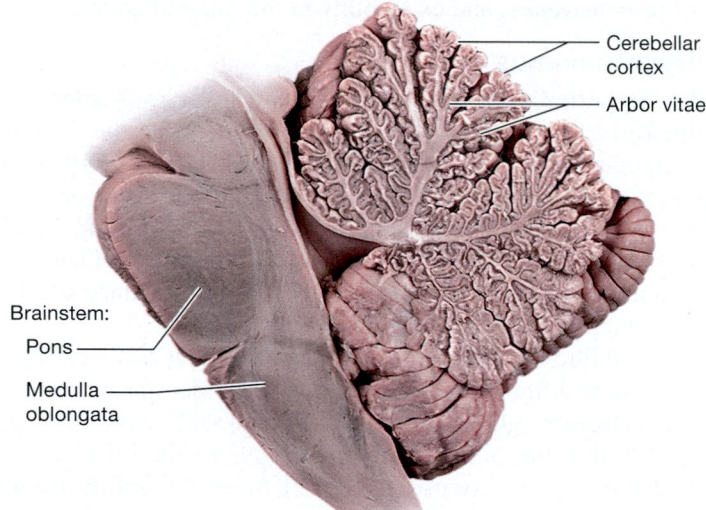

Cerebellar cortex

Arbor vitae

Brainstem:

Pons

Medulla oblongata

(d) Photo of cerebellum, midsagittal section

Figure 12.11 **Structure of the cerebellum.**

The role of the cerebellum in movement (as well as its inter- actions with the cerebral cortex, basal nuclei, brainstem, and spinal cord) is discussed fully in Module 12.6.

Quick Check

☐ 4. Describe the basic anatomical arrangement of the cerebellum.

☐ 5. What is the primary function of the cerebellum?

The Brainstem

One of the most evolutionarily ancient components of the brain is the **brainstem.** In terms of our immediate survival, it is perhaps the most important part of the brain, as its nuclei

control many basic functions, including maintenance of heart rate and breathing rhythm. This is why head injuries involv- ing brainstem damage have poor prognoses (prospects of recovery)—even relatively small lesions of the brainstem can lead to death. The brainstem also mediates numerous *reflexes,* which are programmed, automatic responses to stimuli (see Chapter 13), and functions in movement, sensation, and main- taining alertness.

In the midsagittal section of the brain shown in **Figure 12.12** note that the brainstem is located inferior to the diencephalon; it is also anterior to the cerebellum and superior to the spinal cord. It extends to the level of the foramen magnum of the occipital bone and contains the *fourth ventricle.*

The brainstem has three subdivisions: the superior *mid- brain,* the middle *pons,* and the inferior *medulla oblongata*

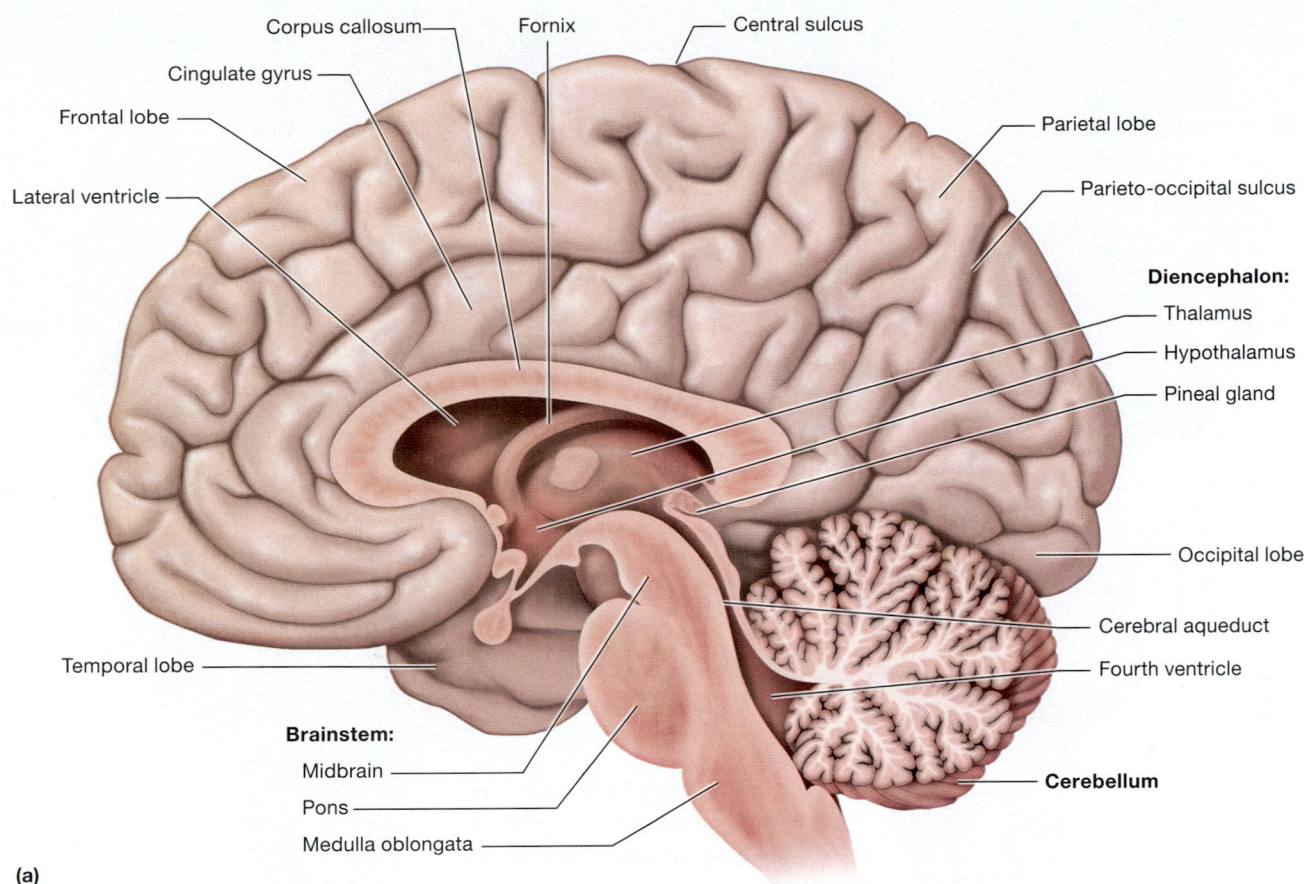

(a)

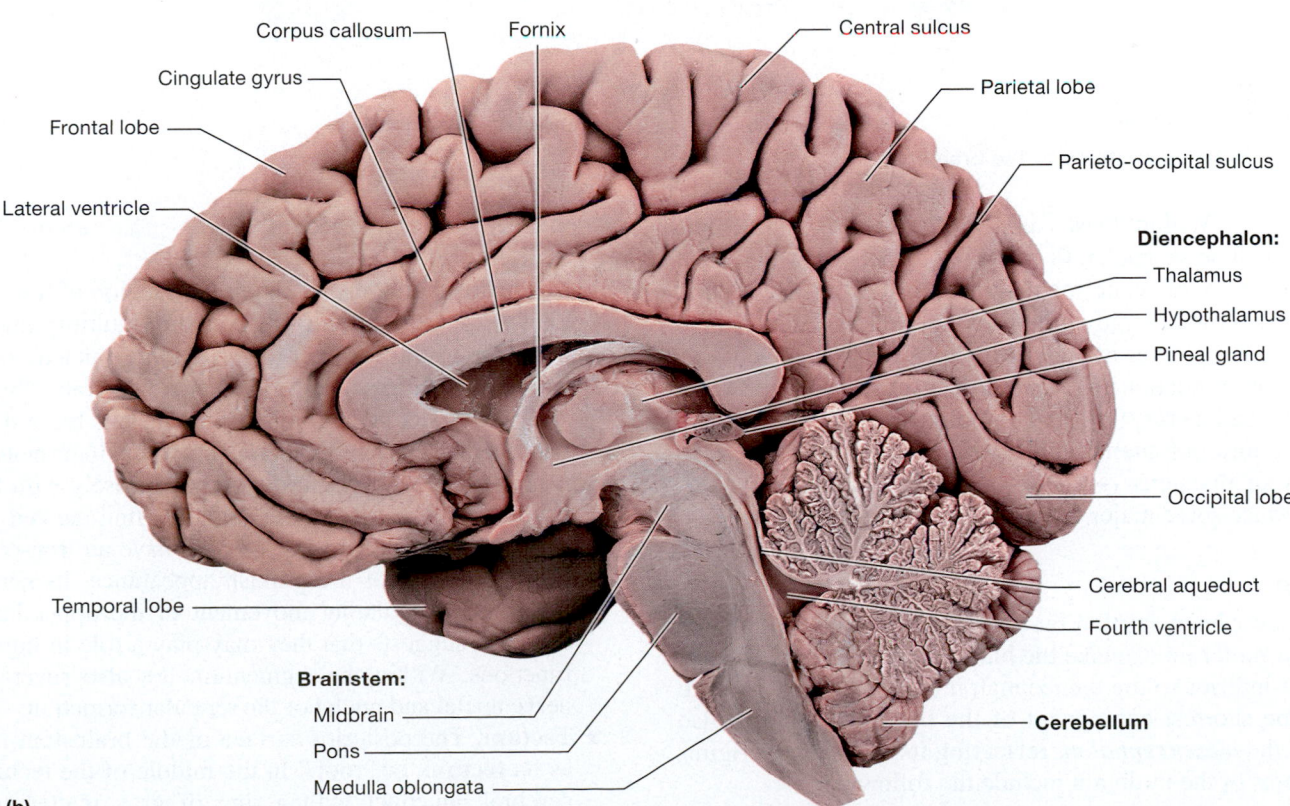

(b)

Figure 12.12 Midsagittal section of the brain showing the brainstem. Part (a) is an illustration, and part (b) is a photo showing the same view.

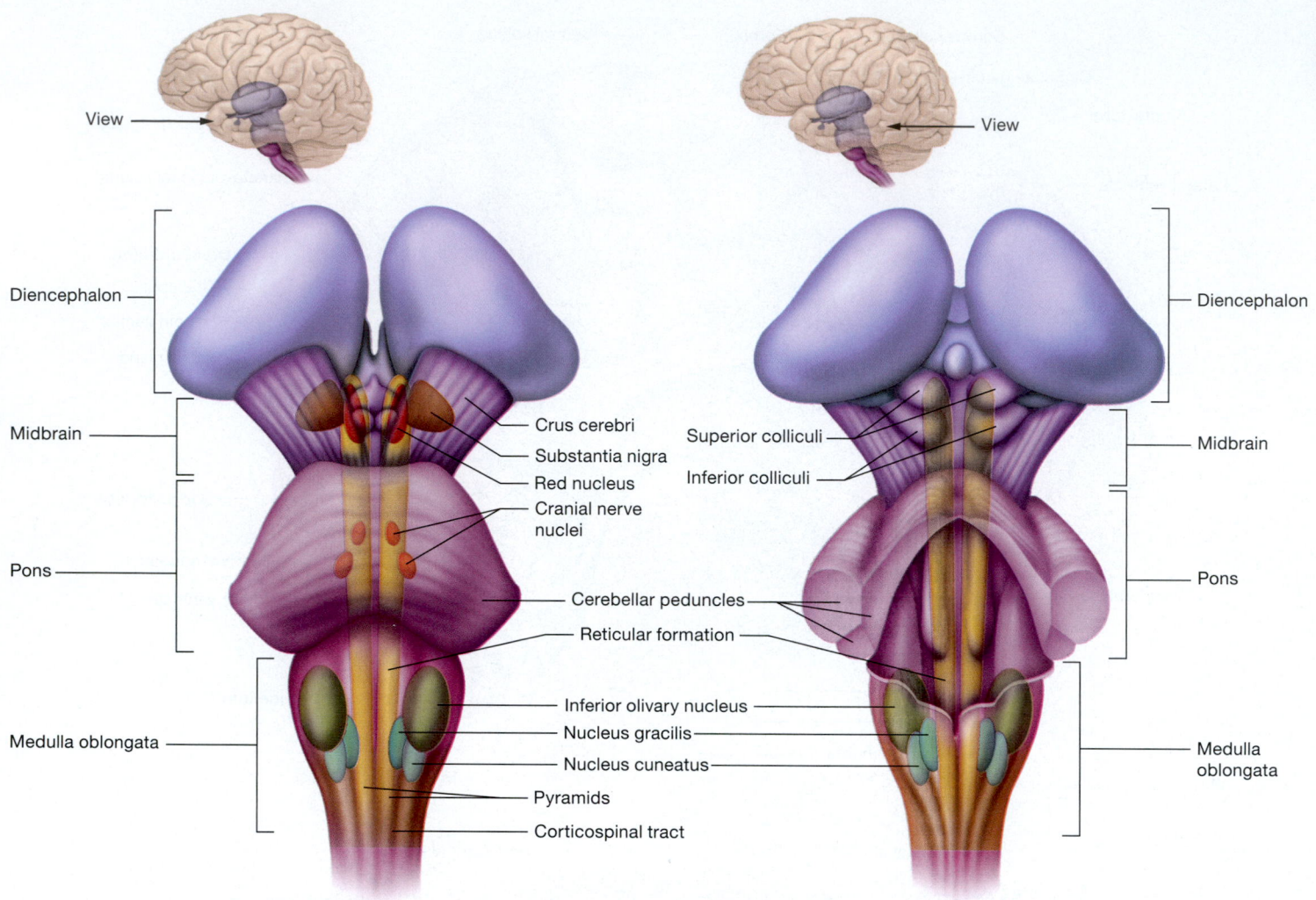

(a) Brainstem, anterior view

(b) Brainstem, posterior view

Figure 12.13 External anatomy of the brainstem.

(**Figure 12.13**). Within these subdivisions we find structures such as cranial nerve nuclei, fibers of the cerebellum and related nuclei, and tracts of white matter between the spinal cord and brain. We also find a large group of connected nuclei called the *reticular formation* that has a multitude of functions, including regulation of ventilation (breathing), blood pressure, the sleep/wake cycle, pain perception, and consciousness.

Now we turn our attention to the basic anatomy of the three brainstem subdivisions (**Figure 12.14**). The following subsections introduce some major functions of each subdivision.

Midbrain

If you've ever been startled by a loud noise or an unexpected sight, your midbrain can take the blame for it. The **midbrain** is found just inferior to the diencephalon (see Figure 12.14a and b). It is the shortest component of the brainstem, and is also known as the *mesencephalon,* reflecting its embryonic origin.

Structures of the midbrain include the following:

- **Crus cerebri.** On the anterior surface of the brainstem are descending tracts of white matter originating from

the cerebrum, referred to as the *crus cerebri* (KROOS SER-eh-bry).

- **Midbrain tegmentum.** The middle region of the midbrain, known as the **tegmentum,** contains multiple nuclei. Just posterior to the crus cerebri is a large nucleus called the **substantia nigra** (sub-STAN-chah NY-grah; "black substance"). As its name implies, this nucleus has a dark color due to a pigment similar to melanin found in its neurons. The neurons of the substantia nigra work closely with the basal nuclei to control movement. Next we find the **red nucleus,** an area so named because its cells have an iron-containing pigment that lends it a pinkish appearance. Its neurons are involved in regulating movement of the upper limbs, and evidence suggests that they may play a role in higher brain functions. Within the tegmentum are also several cranial nerve nuclei and nuclei of the reticular formation.

- **Tectum.** The posterior surface of the brainstem is known as its **tectum,** or "roof." In the middle of the tectum is the cerebral aqueduct with a ring of gray matter surrounding it. Protruding from its posterior surface are two pairs of rounded projections called the **superior** and **inferior**

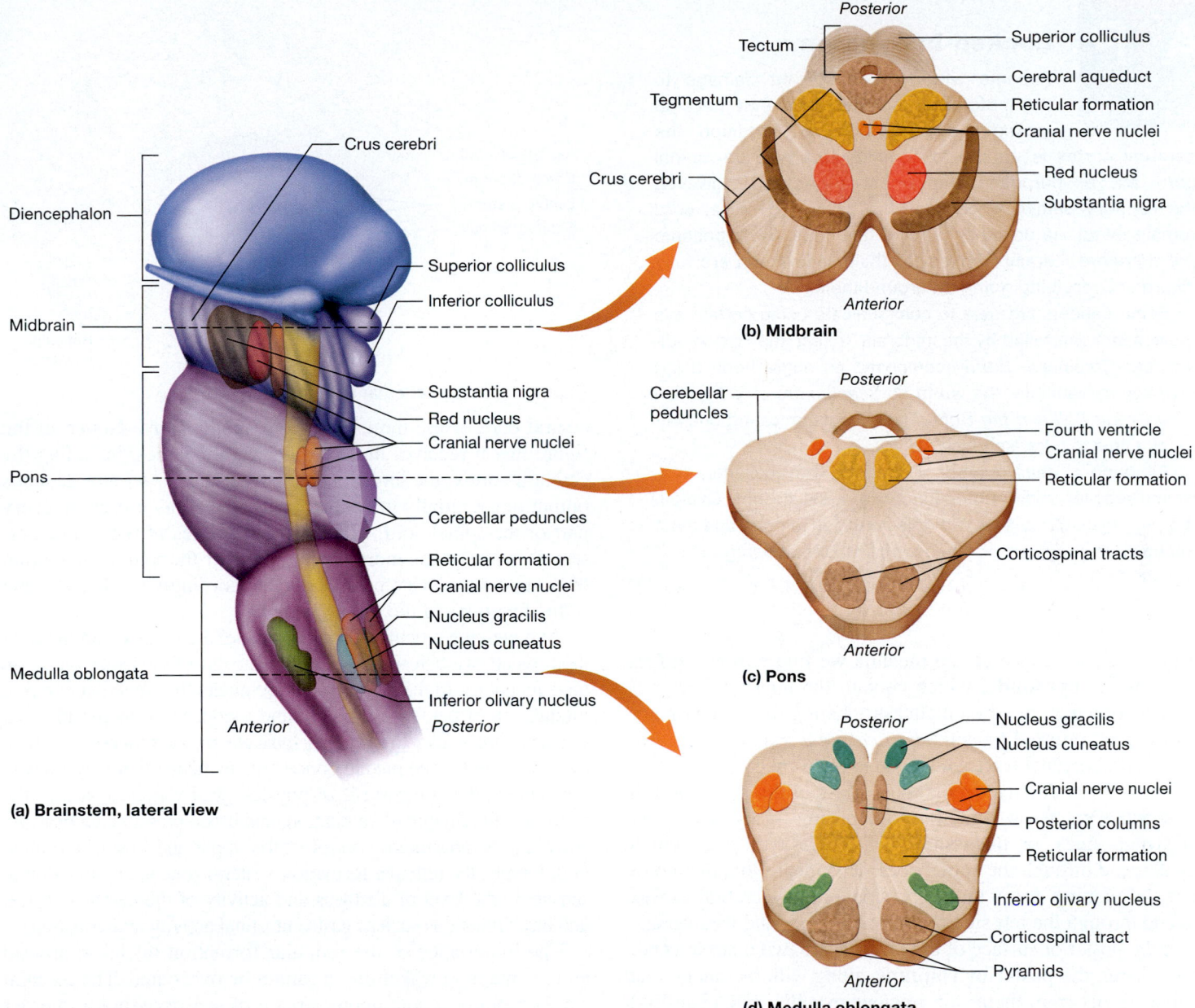

Figure 12.14 Internal anatomy of brainstem divisions.

colliculi (koh-LIH-kyoo-lye; singular, *colliculus*). These structures are involved with visual and auditory functions, respectively, and project to the thalamus.

The nuclei of the midbrain have diverse functions, including roles in movement, sensation, and certain reflexes, such as the previously mentioned startle reflex. These functions will be discussed in later modules and in the peripheral nervous system chapter with the cranial nerves (see Chapter 13).

Pons

Inferior to the midbrain is the **pons,** whose anterior surface is rounded and prominent (see Figure 12.14c). The anterior pons contains the descending motor tracts from the crus cerebri, some of which pass through the pons on their way to the spinal cord;

others enter the cerebellum via the middle cerebellar peduncle. (See *A&P in the Real World: Locked-In Syndrome* on p. 438 to discover what can happen if these motor tracts are damaged.) Posterior to these tracts is an area that contains reticular formation nuclei and cranial nerve nuclei. Surrounding the pontine tegmentum are the middle cerebellar peduncles.

The pontine nuclei have diverse roles. Certain nuclei help regulate movements, including those required for breathing. Others, particularly the cranial nerve nuclei, participate in reflexes, and some are involved in complex functions such as sleep and arousal.

Medulla Oblongata

The inferiormost portion of the brainstem is the **medulla oblongata** (or simply *medulla*), which blends with the spinal cord after it passes through the foramen magnum (see Figure 12.14d).

Locked-In Syndrome

Patients who have sustained damage to the motor tracts of the pons may develop *locked-in syndrome.* In this condition, the cerebral cortex is unable to communicate with the spinal cord, and no purposeful movement is possible. However, the sensory pathways between the brain and spinal cord remain intact, as do other cortical functions. Such patients are therefore literally "locked in" their bodies and are fully aware of everything going on around them.

Some patients are able to communicate using certain eye movements controlled by the midbrain. In fact, the French editor Jean-Dominique Bauby composed an entire book using only eye movements—he wrote *Le Scaphandre et le Papillon* (*The Diving Bell and the Butterfly*) by communicating one letter at a time to a transcriptionist.

Although a small number of patients with locked-in syndrome recover some motor functions, most do not. Overall, the prognosis of patients with this condition is poor, and most succumb to infections or other complications of paralysis.

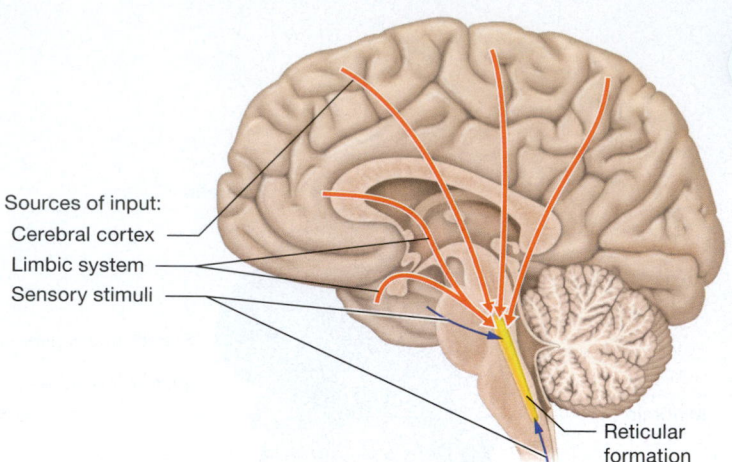

Sources of input:
Cerebral cortex
Limbic system
Sensory stimuli

Reticular formation

Figure 12.15 **The reticular formation.**

On the anterior surface of the medulla we find elevated ridges known as the **pyramids,** which contain the upper motor neuron fibers of a tract of white matter originating from the primary motor cortex, referred to as the *corticospinal tract* (for this reason the corticospinal tract is also sometimes called the *pyramidal tract*). Within the pyramids most fibers of the right and left corticospinal tracts do something interesting—they cross over and switch sides, or **decussate** (deh-KUSS-ayt), after which they descend through the spinal cord. This means that the motor fibers originating from the right side of the cerebral cortex descend through the left side of the spinal cord, and vice versa.

On the posterior surface of the medulla are two other tracts of white matter, the **posterior columns,** along with the nuclei that receive signals from them: the **nucleus gracilis** (GRAS-ih-liss) and **nucleus cuneatus** (kyoo-nee-AH-tus). The posterior columns carry sensory stimuli from the spinal cord. Like most fibers of the corticospinal tracts, some also decussate in the medulla, so sensory fibers from the right side of the spinal cord are processed by the left side of the brain, and vice versa.

The medulla also contains cranial nerve nuclei and nuclei of the reticular formation. In addition, it contains a protuberance lateral to each pyramid called the **olive.** This structure contains a nucleus called the *inferior olivary nucleus* that receives sensory fibers from the spinal cord and passes them to the cerebellum.

Reticular Formation

Many of the functions that physiologists formerly assigned to specific brainstem subdivisions are now understood to instead be performed by a complex region of the brain known as the reticular formation. As you can see in **Figure 12.15**, the **reticular formation** is a collection of more than 100 nuclei that forms the central core of the three brainstem subdivisions. Notice in the figure that it receives input from several sources, including the cerebral cortex, the limbic system, and sensory stimuli. It sends output to essentially the entire brain as well as just about every part of the spinal cord. Like many structures of the limbic system, it is one of the more ancient parts of the mammalian brain from an evolutionary perspective. Accordingly, it shares many of the functions of the limbic system.

Although the reticular formation nuclei are diverse and many of them blend, we can loosely divide them into four zones based on their functions. In the central zone are nuclei that function in sleep, modulation of pain transmission, and mood. Surrounding the central zone are nuclei involved in intricate motor functions such as eye movements. The medial zone contains nuclei that play roles in maintaining the homeostasis of physiological variables such as the heart rate, the rhythm of ventilation, and blood pressure; arousal and wakefulness; and motor control of the upper and lower limb muscles. Finally, the reticular formation's lateral zone nuclei function in sensation, the level of alertness and activity of the cerebral cortex, and basic functions such as gastrointestinal activity and urination.

The importance of the reticular formation nuclei in arousal and maintaining wakefulness cannot be overstated. The cerebral cortex requires a continuous stream of sensory stimuli in order to maintain a state of consciousness. This stream of sensory stimuli is provided by reticular formation nuclei of the medial zone. These nuclei send their output largely to the thalamus, which in turn relays it throughout cerebral cortex. This system of fibers from the reticular formation to the cerebral cortex is known as the **reticular activating system (RAS)**, so named because it arouses the cortex from sleep and keeps it "activated" (awake). Blocking the RAS, as occurs during certain types of anesthesia, results in unconsciousness.

Loud, sudden stimuli tend to be stronger triggers to activate the RAS, whereas a lack of sensory stimuli can significantly decrease its activity. This is why you likely use a loud, unpleasant sound for your alarm clock in the morning, and why it's much more difficult to fall asleep at a concert than in a dark, quiet room. This also explains why you might have dozed off in class once or twice when bored—although that obviously would never happen in your A&P course.

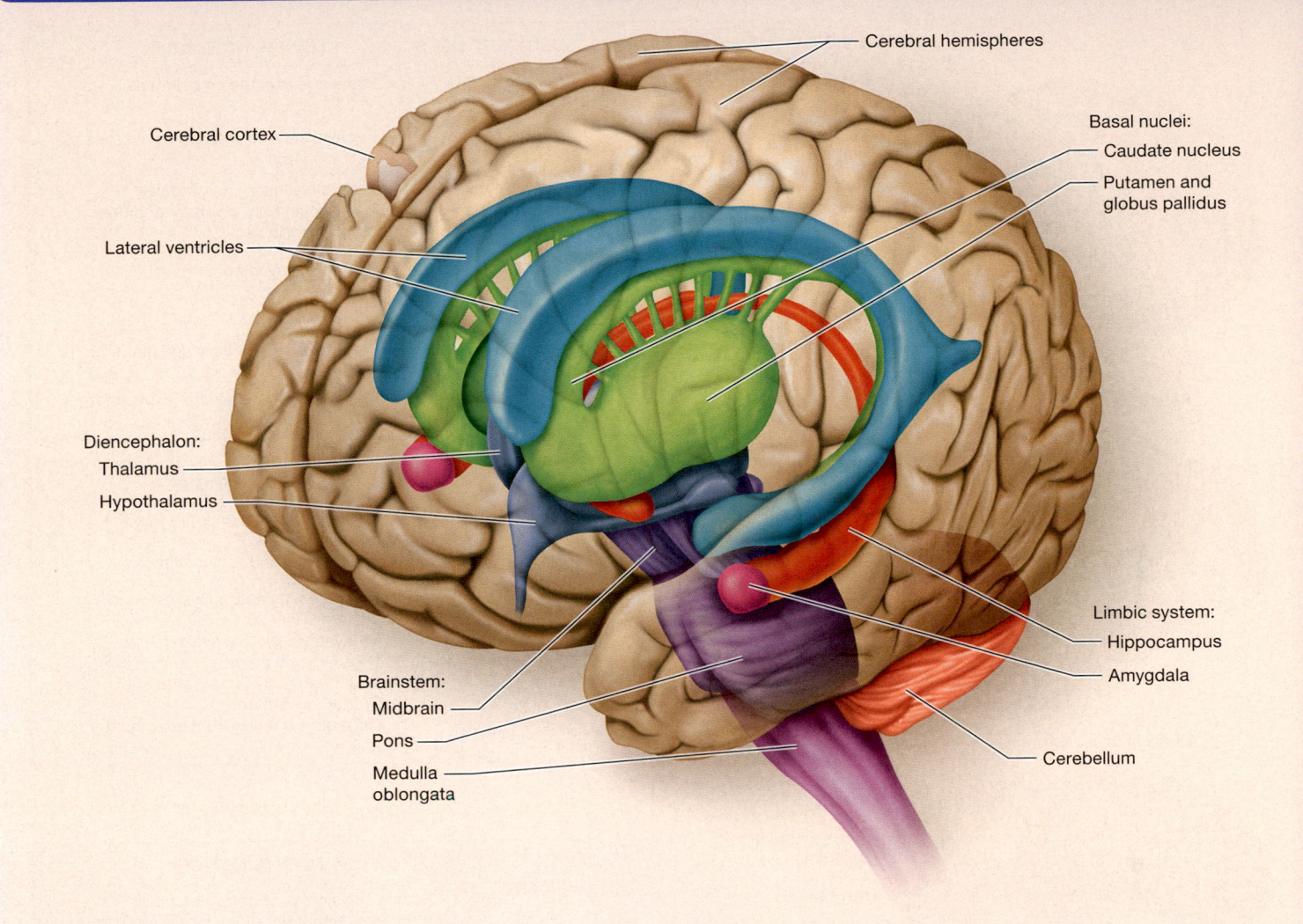

Cerebral hemispheres

Cerebral cortex

Basal nuclei:
Caudate nucleus
Putamen and globus pallidus

Lateral ventricles

Diencephalon:
Thalamus
Hypothalamus

Limbic system:
Hippocampus
Amygdala

Brainstem:
Midbrain
Pons
Medulla oblongata

Cerebellum

☐ 6. What are some functions of the brainstem?

☐ 7. What happens to certain motor tracts in the medulla oblongata?

☐ 8. What are the general functions of the reticular formation? Where are the reticular formation nuclei located?

Putting It All Together: The Big Picture of Major Brain Structures and Their Functions

A big picture overview of the brain and its many structures is shown in **Figure 12.16** for your review. As in previous figures, the cerebrum is presented in a "see-through" fashion so that you can appreciate the location of the brain's deeper structures. When you know the structures fairly well, take a look at **Figure 12.17**, which summarizes the relationships between those structures and connects them with their functions.

☐ 1. A person who is in a vegetative state has no activity in his or her cerebrum. Why can a person in this condition continue living for a certain period? Can such a person feel pain or initiate voluntary movements? Explain.

☐ 2. Predict the effects of damage to the basal nuclei or the cerebellum.

☐ 3. The condition *lissencephaly* (liss′-en-SEF-uh-lee; "smooth brain") is characterized by a lack of gyri and sulci in the cerebral cortex, which gives the brain hemispheres a smooth appearance. Predict the effects of such a condition.

☐ 4. Ms. Greer sustained major head trauma in an automobile accident. She is unresponsive to sensory stimuli, and scans of her cerebral cortex demonstrate no cortical activity. During a surgical procedure to relieve pressure on her brain, it is discovered that she has sustained major damage to her thalamus. How does this damage explain her symptoms?

See answers in Appendix A.

The Big Picture of Major Brain Structures and Their Functions

Figure 12.17

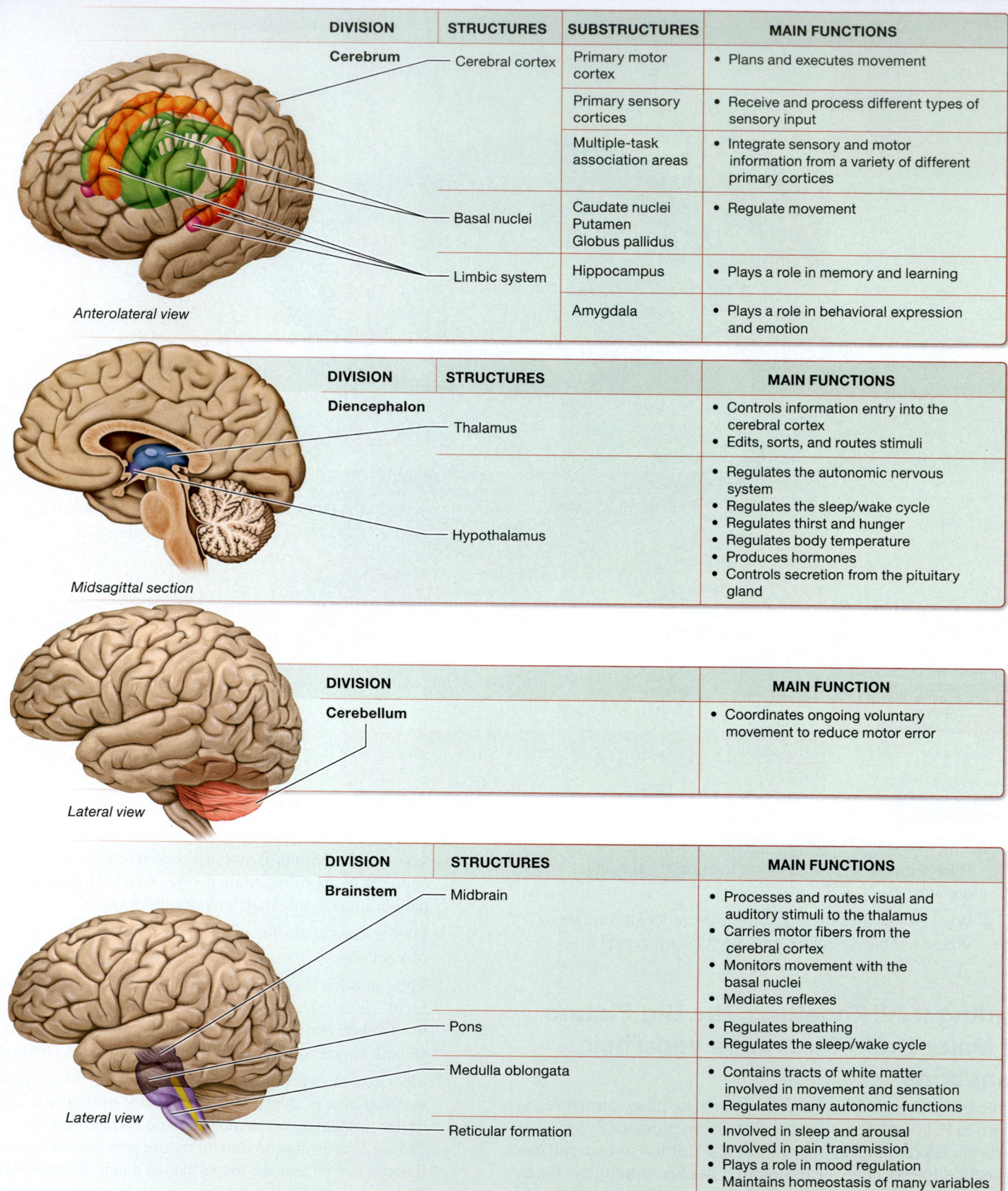

Anterolateral view

DIVISION	STRUCTURES	SUBSTRUCTURES	MAIN FUNCTIONS
Cerebrum	Cerebral cortex	Primary motor cortex	• Plans and executes movement
		Primary sensory cortices	• Receive and process different types of sensory input
		Multiple-task association areas	• Integrate sensory and motor information from a variety of different primary cortices
	Basal nuclei	Caudate nuclei Putamen Globus pallidus	• Regulate movement
	Limbic system	Hippocampus	• Plays a role in memory and learning
		Amygdala	• Plays a role in behavioral expression and emotion

Midsagittal section

DIVISION	STRUCTURES	MAIN FUNCTIONS
Diencephalon	Thalamus	• Controls information entry into the cerebral cortex • Edits, sorts, and routes stimuli
	Hypothalamus	• Regulates the autonomic nervous system • Regulates the sleep/wake cycle • Regulates thirst and hunger • Regulates body temperature • Produces hormones • Controls secretion from the pituitary gland

Lateral view

DIVISION	MAIN FUNCTION
Cerebellum	• Coordinates ongoing voluntary movement to reduce motor error

Lateral view

DIVISION	STRUCTURES	MAIN FUNCTIONS
Brainstem	Midbrain	• Processes and routes visual and auditory stimuli to the thalamus • Carries motor fibers from the cerebral cortex • Monitors movement with the basal nuclei • Mediates reflexes
	Pons	• Regulates breathing • Regulates the sleep/wake cycle
	Medulla oblongata	• Contains tracts of white matter involved in movement and sensation • Regulates many autonomic functions
	Reticular formation	• Involved in sleep and arousal • Involved in pain transmission • Plays a role in mood regulation • Maintains homeostasis of many variables

MODULE **12.3**

Homeostasis Part I: Role of the Brain in Maintenance of Homeostasis

Learning Outcomes

1. Describe the differences between the endocrine system and the nervous system in terms of how they maintain homeostasis.
2. Provide specific examples demonstrating how the brain responds to maintain homeostasis of regulated physiological variables in the body.

If you recall, *homeostasis* is defined as the maintenance of a relatively stable internal environment in the face of ever-changing conditions (see Chapter 1). The body maintains homeostasis of different physiological variables, including fluid, electrolyte, and acid-base balance; the blood pressure; the concentration of glucose and other nutrients in the blood; biological rhythms; and body temperature. The two main systems in the body dedicated to maintaining homeostasis are the nervous and endocrine systems. Although they work together to accomplish this function, they do so in different ways. The endocrine system secretes hormones into the blood, which regulate the functions of other cells, whereas the nervous system sends action potentials that excite or inhibit target cells. The actions of the endocrine system are typically slow and take several hours to several days to have an effect, whereas those of the nervous system are generally immediate.

Two structures of the brain are concerned directly with maintenance of homeostasis: certain nuclei of the brainstem reticular formation and the hypothalamus. In Module 12.2, we discussed how the reticular formation and hypothalamus control the functions of many internal organs as well as aspects of behavior. The reticular formation and hypothalamus also have numerous connections with each other that enable them to coordinate many of their functions.

In this module, we focus on the role of the brain in maintaining homeostasis in the body. Our discussion of homeostasis continues in the chapters on the autonomic nervous system and the endocrine system (see Chapters 14 and 16, respectively).

Homeostasis of Vital Functions

The homeostasis of vital functions such as heart pumping, blood pressure, and digestion is largely maintained by the **autonomic nervous system (ANS)**. Recall from the nervous tissue chapter that the ANS is a branch of the PNS that regulates the function of the body's viscera (see Chapter 11). Although the ANS is

classified as a division of the PNS, it is controlled by structures of the CNS.

The main "boss" of the ANS is the hypothalamus, which receives sensory input from the viscera, the components of the limbic system, and the cerebral cortex. This allows the hypothalamus to adjust ANS output to preserve homeostasis in response to both normal physiological changes and emotional changes. The hypothalamus accomplishes this task largely by relaying instructions to nuclei in the reticular formation of the pons and medulla, including the following:

- **Vasopressor center.** The neurons of the **vasopressor center** (VAYZ-oh-press-er; *vaso-* = "vessel") are located in nuclei of the anterolateral pons and medulla. When stimulated by the hypothalamus, this center increases the rate and force of heart pumping and causes blood vessels to constrict (narrow). All three of these effects raise blood pressure.
- **Vasodepressor center.** The **vasodepressor center** is located in nuclei inferior and medial to those of the vasopressor center. Its effects on the heart and blood vessels are the opposite of those of the vasopressor center—it decreases the rate and force of contractions of the heart and dilates (opens) the blood vessels. All three of these effects decrease blood pressure.
- **Other centers.** Many nuclei in the reticular formation participate in regulation of digestive processes and control of urination. Much of this activity is discussed in the ANS chapter (see Chapter 14).

Ventilation is one of the few vital functions not under direct control of the ANS. Instead, the rate and depth of ventilation are regulated by reticular formation nuclei that are mostly in the pons. Several factors influence the rate at which these neurons fire, including input from the cerebral cortex, the limbic system, the hypothalamus, certain sensory receptors, and other brainstem nuclei. We discuss control of ventilation more completely in the respiratory system chapter (see Chapter 21).

Quick Check

☐ 1. Which two body systems coordinate the maintenance of homeostasis? How do they differ in the ways they accomplish this task?

☐ 2. Which branch of the PNS controls most of the body's vital functions? What structure is the "boss" of this branch?

Body Temperature and Feeding

The hypothalamus regulates both body temperature and feeding behaviors. It acts as the body's thermostat and sets a normal range of temperatures (around 36.5–37.2° C, or 97.8–99° F). It receives input from temperature-sensitive neurons located in several places, including the skin and areas deeper in the body, and from neurons in the hypothalamus itself. When the body temperature increases above the normal range, a negative feedback loop initiates, and neurons in certain hypothalamic nuclei trigger changes such as sweating that cool the body, an example of the

Feedback Loops Core Principle (p. 28). Conversely, when body temperature decreases below the normal range, another negative feedback loop sets in motion mechanisms to conserve body heat, such as shivering. When the set point for body temperature is temporarily set higher than normal, a fever ensues; see *A&P in the Real World: Fever* for more information on this topic.

CORE PRINCIPLE
Feedback Loops

The hypothalamus also regulates feeding, which in turn helps maintain levels of glucose and other nutrients in the bloodstream. Stimulation of specific hypothalamic nuclei induces hunger and feeding behaviors, which indirectly works to preserve homeostasis of the concentration of glucose in the blood. These effects are believed to be partly due to neurotransmitters called *orexins* (*orexis* = "appetite"), which are secreted in high amounts by hypothalamic neurons during periods of fasting and appear to induce the urge to eat. Stimulating different hypothalamic nuclei seems to inhibit feeding; however, the mechanisms controlling food intake are complex and involve hormones and several nuclei in the hypothalamus as well as other parts of the brain. Interestingly, the same area of the hypothalamus that mediates hunger also mediates aggression and rage, which perhaps factors into "hanger," or the anger some people feel when hungry. The orexins have also been found to play an important role in regulating the sleep/wake cycle, which is our next topic.

Quick Check

☐ 3. How does the hypothalamus regulate hunger?

☐ 4. What role does the hypothalamus play in temperature homeostasis?

A&P in the Real World

Fever

At some point, everyone has had a *fever*—an elevation of normal body temperature that can accompany a variety of infectious and noninfectious conditions. Fevers are caused by chemicals called *pyrogens* secreted by cells of the immune system and by certain bacteria that interact with the temperature-controlling nuclei of the hypothalamus. Pyrogens increase the hypothalamic set point to a higher temperature, which makes the hypothalamus initiate a negative feedback loop to raise body temperature. This produces the familiar and unpleasant symptoms of shivering and muscle aches due to increased muscle tone. The hypothalamus also causes the blood vessels serving the skin to constrict, which produces the feeling of coldness that accompanies a fever.

Treatment of a fever is not always necessary, as it is often more important to address the underlying cause of the fever. To treat a fever, agents called *antipyretics,* such as acetaminophen and aspirin, are generally administered. Antipyretics work by blocking the formation of pyrogens, which permits the hypothalamus to reset back to the normal range.

Sleep and Wakefulness

Sleep is one of the most fundamental, yet most mysterious, processes carried out by humans and most other animals. **Sleep** is defined as a reversible and normal suspension of consciousness. We have uncovered a great deal about how we sleep, how sleep is regulated, and brain activity during sleep, but we are still unsure *why* most animals sleep. Physiologically, sleep seems to serve an energy restoration function, allowing the brain to replenish its glycogen supply; however, if there are psychological reasons why we sleep, they remain elusive. Nonetheless, it is clearly required for our survival; sleep deprivation may cause imbalances in temperature homeostasis, weight loss, a decrease in cognitive abilities, hallucinations, and even death.

To feel rested, most adults require 7–8 hours of sleep per night. The requirement for infants is much higher, at about 17 hours per day, but this requirement decreases with increasing age. As we grow older, we tend to sleep for shorter periods and less deeply for reasons that are at present unclear. The following discussion will examine what is known about this mysterious state and some of the possible reasons we spend so much time in it.

Circadian Rhythms and the "Biological Clock"

Among the most basic questions about sleep are those that revolve around *when* and *how* we sleep. Human sleep generally follows a **circadian rhythm** (sir-KAY-dee-an; *circa-* = "around," *dia-* = "day"), meaning that we spend a period of the 24-hour cycle awake and another period of the cycle asleep. This rhythm is controlled by our own hypothalamic "biological clock," which causes changes in the level of wakefulness in response to day and night cycles. As shown in **Figure 12.18**, the clock brings about sleep in the following fashion:

① **Neurons from the eye signal the suprachiasmatic nucleus (SCN) that the light level is decreasing.** Receptors in the eye that detect the decreasing light level carry this information to a region of the hypothalamus called the *suprachiasmatic nucleus* (soop′-rah-ky-az-MAT-ik; *SCN*). The SCN is often referred to as the body's "master clock."

② **The SCN stimulates the ventrolateral preoptic nucleus.** The SCN sends this stimulus to another hypothalamic nucleus, the *ventrolateral preoptic nucleus.* The SCN also sends signals to the pineal gland that stimulate the secretion of melatonin.

③ **The ventrolateral preoptic nucleus decreases the activity of the reticular formation.** When excited, the ventrolateral preoptic nucleus sends signals that decrease the activity of the nuclei of the reticular activating system. Recall that these nuclei stimulate the thalamus, which in turn stimulates the cerebral cortex, during wakefulness. Melatonin also depresses the activity of this part of the reticular formation.

④ **The decreased activity of the reticular formation "disconnects" the thalamus from the cerebral cortex, in turn decreasing the level of consciousness.** A decrease

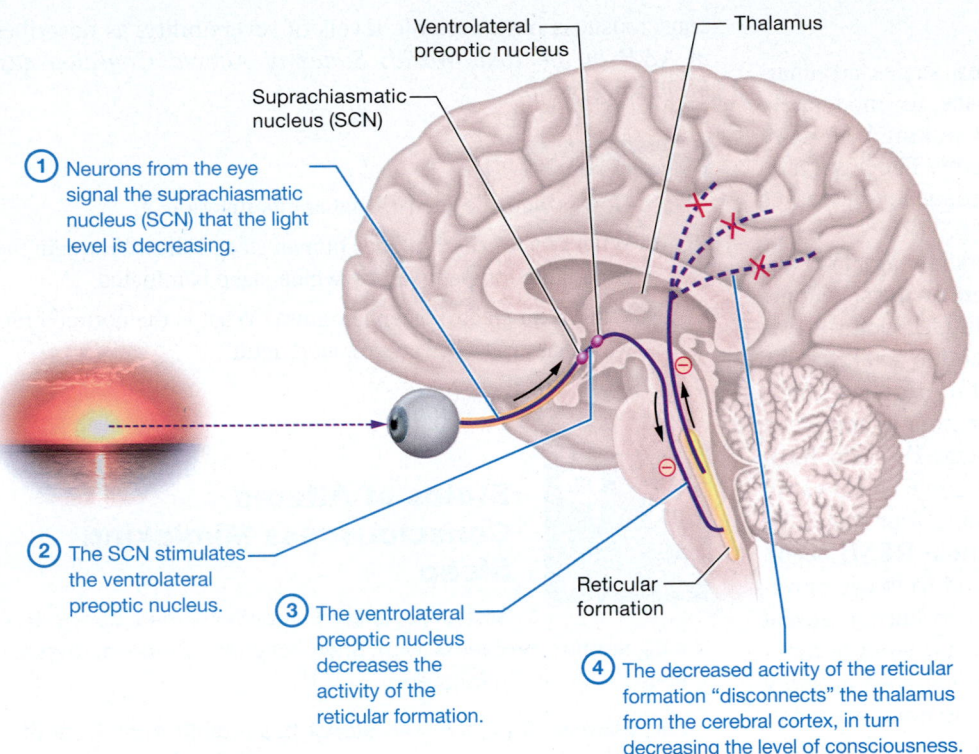

① Neurons from the eye signal the suprachiasmatic nucleus (SCN) that the light level is decreasing.

② The SCN stimulates the ventrolateral preoptic nucleus.

③ The ventrolateral preoptic nucleus decreases the activity of the reticular formation.

④ The decreased activity of the reticular formation "disconnects" the thalamus from the cerebral cortex, in turn decreasing the level of consciousness.

Figure 12.18 The process of falling asleep.

in the activity of multiple nuclei of the reticular formation lowers the rate at which the thalamus transmits external stimuli to the cerebral cortex. This in effect "disconnects" the cortex from the outside world (represented by red Xs in Figure 12.18) and reduces the level of consciousness.

Arousal from sleep is mediated by a different group of hypothalamic neurons (not those regulating satiety) that secrete the neurotransmitter orexin, in an example of the Cell-Cell Communication Core

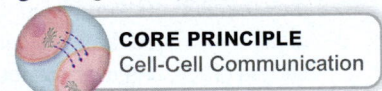
CORE PRINCIPLE
Cell-Cell Communication

Principle (p. 28). Orexin release increases before waking and helps maintain a state of wakefulness throughout the day. This effect is thought to be due to stimulation of the reticular activating system nuclei by these orexin-secreting neurons. When the reticular formation neurons are active, they cause the thalamus to begin transmitting external sensory stimuli to the cerebral cortex. Histamine is an important excitatory neurotransmitter in this system, which is why *antihistamines,* drugs that block histamine receptors, often produce drowsiness as an adverse effect. Interestingly, an orexin deficiency is believed to be the culprit behind *narcolepsy,* a disease in which an individual sleeps poorly at night, leading to very strong urges to fall asleep during the day.

Brain Waves and Stages of Sleep

Research has demonstrated that in one night we progress through several different cycles of sleep, each having many stages. We can monitor these stages with an **electroencephalogram,** or **EEG,** a test that measures the electrical activity of the brain via electrodes attached to the skin. The tracings of electrical activity, called **brain waves,** show characteristic patterns during wakefulness and the different stages of sleep. Waves differ in their height, also known as *amplitude,* and in the speed at which they are generated, or their *frequency.* As you can see in **Figure 12.19,** when we are awake and engaged in mental activity, the brain waves are low-amplitude and high-frequency, a pattern called **beta waves.** When we fall asleep and proceed through the stages of sleep, the amplitude of the waves progressively increases and their frequency progressively decreases.

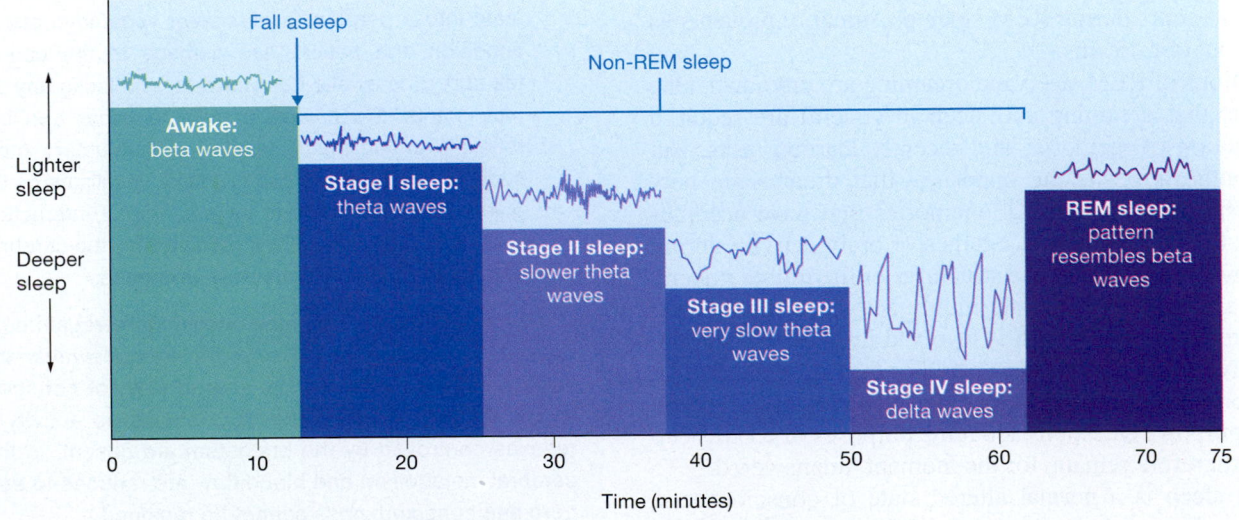

Figure 12.19 Stages of wakefulness and sleep as shown by EEG patterns.

The stages of sleep are as follows:

- **Stages I–III sleep.** The first three sleep stages are characterized by drowsiness that progresses to moderately deep sleep and a slowing of the normal waking beta wave rhythm into what is known as **theta waves.** The amplitude of the theta waves increases and their frequency decreases as we move from stage I to stage III sleep.
- **Stage IV sleep.** The deepest sleep occurs in stage IV. During this stage, the brain waves, called **delta waves,** become lower in frequency and higher in amplitude. Delta waves are also seen during general anesthesia and in awake adults with significant brain damage. The entire sequence from stage I to stage IV sleep takes less than an hour to complete.
- **REM sleep.** As a person emerges from stage IV sleep, he or she enters a puzzling portion of the sleep cycle called **rapid eye movement (REM) sleep** (the preceding four stages are referred to as **non–rapid eye movement [non-REM] sleep**). REM sleep is named for the rapid, back-and-forth eye movements that occur during this stage of sleep but are absent from other stages. REM sleep is when most *dreams* occur—curious hallucination-like episodes that often have a highly emotional content but make little sense. The brain wave patterns during REM sleep resemble the beta waves of an awake adult, although clearly the two brain states are not equivalent. REM sleep generally lasts only about 10–15 minutes before cycling back into stage I sleep; however, the REM sleep phase generally gets longer with each sleep cycle.

Keep in mind that an individual won't necessarily progress through each stage in order. For example, one may transition from stage IV sleep into REM, stage I, or stage II sleep.

The contrast between the activity of the brain and the body during non-REM and that during REM sleep extends beyond eye movements. During stages I–IV of the sleep cycle, the body is able to move ("tossing and turning" occurs during these stages), and blood pressure, heart rate, respiratory rate, body temperature, and metabolic rate decrease. During REM sleep, however, blood pressure, heart rate, and metabolic rate increase, and most of the major muscle groups are paralyzed. The muscle paralysis that occurs during REM sleep presumably prevents us from acting out our dreams.

The functions of REM sleep and dreaming are unknown. One hypothesis is that dreaming and sleep in general are required for consolidation of memories and recently learned tasks. But another hypothesis posits the opposite—that dreams are necessary to dispose of "unwanted" memories that have accumulated during the waking period. Studies of brain activity during dreaming have indicated that the structures of the limbic system, which play a role in emotion, are highly active during dreams, but the prefrontal cortex, which is involved in logic and reason, is barely active at all. This perhaps explains the highly emotional and often nonsensical content of dreams but sheds little light on their actual purpose. Questions about the purposes of REM sleep and dreams therefore remain, for the moment, unanswered.

Although sleep is a normal altered state of consciousness, a key feature is that it is *reversible.* Other altered states of consciousness have variable levels of reversibility, as described in *A&P in the Real World: States of Altered Consciousness Mimicking Sleep.*

☐ 5. How is sleep defined? What are its functions?

☐ 6. What type of rhythm does human sleep follow? Explain the general neural pathway by which sleep is initiated.

☐ 7. What is an electroencephalogram? What is the normal brain wave pattern for an awake, alert adult?

A&P in the Real World

States of Altered Consciousness Mimicking Sleep

States of altered consciousness can indicate serious problems with brain function. Some of these states include the following:

- **Stupor.** A person with *stupor* has a diminished level of cortical activity, although he or she is often arousable to strong or painful stimuli. Common causes of stupor include infections, mental illnesses, and conditions of the brain such as tumors.
- **Coma.** *Coma* is defined as unarousable unconsciousness in which there are no purposeful responses to any stimuli, even pain. The underlying defect in the coma state is damage to the reticular formation or a related component that prohibits normal arousal of the cerebral cortex. EEGs demonstrate a slow, fixed pattern with no observable sleep/wake cycles. Brainstem reflexes, however, remain intact, and the patient may exhibit certain involuntary movements. Coma may result from head trauma, various neurological conditions, certain drugs, and disturbances in acid-base homeostasis.
- **Persistent vegetative state.** Sometimes patients with head trauma or neurological injury will move from the coma state into a perplexing *persistent vegetative state.* In this condition, the patient has damage to the cerebral cortex and so is awake but unaware, and lacks any voluntary movements. Sleep/wake cycles do occur and brainstem reflexes remain intact, leading to involuntary movements such as head turning and gruntlike vocalizations that caregivers often misinterpret as meaningful interaction. However, these actions are not mediated by the cerebral cortex and so do not imply conscious awareness.

People in these states may occasionally regain consciousness, depending on the cause of the state. However, this is not the case in the most extreme altered state of consciousness, **brain death,** in which the EEG shows no activity and the reflexes controlled by the brainstem are absent. In this state, cerebral metabolism and blood flow are reduced to essentially zero and consciousness cannot be regained.

Apply What You Learned

☐ 1. You are about to take your anatomy and physiology exam and you notice that your heart is beating rapidly and your rate of breathing has increased. What parts of the brain are responsible for these changes, and how do they accomplish them?

☐ 2. Predict the effects of a tumor that secretes excess orexins.

☐ 3. Explain why sleepwalking and sleeptalking can occur only during non-REM sleep.

See answers in Appendix A.

MODULE 12.4
Higher Mental Functions

Learning Outcomes

1. Describe the areas of the cortex responsible for cognition and language.

2. Discuss the concept of cerebral hemispheric specialization.

3. Describe the parts of the brain involved in storage of long-term memory, and discuss possible mechanisms of memory consolidation.

Our current understanding of biology, evolution, and physiology leaves little doubt that higher mental functions, such as morality, language, and learning, are a product of normal electrochemical processes occurring in the human brain. In this module, we explore these complex mental functions and the remarkable physiological processes behind them.

Cognition and Language

≪ FLASHBACK

1. Where are Wernicke's area and Broca's area located? What are their functions? (p. 429)

The majority of the cerebral cortex consists of association areas that perform the diverse group of tasks collectively known as **cognition.** Our cognitive functions include processing and responding to complex external stimuli, recognizing related stimuli, processing internal stimuli, and planning appropriate responses to stimuli. Our cognitive processes are responsible for social and moral behavior, "intelligence," thoughts, problem-solving skills, language, and personality. The upcoming subsections focus on those parts of the brain responsible for cognitive functions and how language is produced.

Localization of Cognitive Function

Aspects of cognition have been localized to different multiple-task association areas in the frontal, parietal, and temporal lobes. These areas and their functions include those listed next.

- **Parietal association cortex.** This area of cortex is responsible for spatial awareness and attention. It allows us to focus on distinct aspects of a specific object, such as the direction in which it is moving, its color, or its shape. This part of the brain also has visual-spatial functions and enables us to recognize the position of an object in space. A curious consequence of damage to this area is called *neglect;* in this condition, patients may not acknowledge the existence of an entire half of their visual field or even an entire half of their body in spite of intact sensory pathways.

- **Temporal association cortex.** This area of cortex is primarily responsible for recognizing stimuli, especially complex stimuli such as faces. Patients with damage to this part of the brain suffer from *agnosias* (ag-NOH-see-ahz)—the awareness of a stimulus but the inability to identify it.

- **Prefrontal cortex.** The prefrontal cortex is the largest and most complex of the association cortices. It performs most cognitive functions that make up a person's "character" or "personality." It receives information from the temporal and parietal association cortices as well as from multiple other sensory and motor cortices. This allows for the integration of complex sensory and motor information that creates an awareness of "self" in the world, which leads to planning and execution of behaviors appropriate for given circumstances. For example, our prefrontal cortex enables us to recognize that it is appropriate to laugh aloud when having lunch with friends, but inappropriate to do the same during a funeral. The prefrontal cortex also allows us to select "morally appropriate" behaviors, even those not necessarily to our immediate advantage. Additionally, it plays a role in governance of emotions, which are discussed later in this module. Damage to the prefrontal cortex can be devastating and may produce dramatic changes in personality, impulsive behavior with little restraint, difficulty in making plans, and disordered thought processes. Brain disorders such as *dementia* (deh-MEN-shah) can also cause loss of cognitive and intellectual functions. The consequences of different types of dementias for mental faculties are discussed in *A&P in the Real World: Dementia.*

Cerebral Lateralization

Many of the cognitive functions we just discussed are unequally represented in the right and left hemispheres, a phenomenon called **cerebral lateralization.** The idea that certain cognitive functions are lateralized has led to the unfortunate misconception by the public that one hemisphere is "dominant" over the other, and that people can be "right-brained" or "left-brained." What cerebral lateralization actually represents is a division of labor of sorts between the hemispheres—it is how our brains make the most use out of a limited amount of space.

The functions that appear to be lateralized include the following (although lateralization is not absolute):

- *emotional functions,* with the left frontal cortex mostly being responsible for "positive" emotions such as happiness and the right frontal cortex for "negative" emotions such as anger and fear;

Dementia

Some of the most devastating diseases of the brain are those collectively referred to as dementias. Dementia has many causes, but all patients with dementia exhibit a progressive loss of recent memory, degeneration of cognitive functions, and changes in personality. In effect, a person with dementia literally "loses his or her mind." Common forms of dementia, listed in descending order from the most frequently occurring, include the following:

- **Alzheimer's disease.** *Alzheimer's disease* (ALZ-hy-merz; AD), the most common form of dementia, affects about 30 million people worldwide. Some characteristic findings in brain tissue of those with AD are neurofibrillary tangles (tangled aggregates of proteins in the neurons), senile plaques (extracellular deposits of a specific type of protein around neurons), and degeneration of cortical neurons and synaptic connections. These changes are found throughout the brain, but are especially numerous in the cortical association areas and the hippocampus. The earliest signs of the disease typically involve recent memory loss and forgetfulness. As the disease progresses, impairment of attention, language skills, critical thinking, and visual-spatial abilities, as well as changes in personality, becomes apparent. Eventually, motor skills, sensory perception, and long-term memory are also affected.
- **Vascular dementia.** Vascular dementia is a group of disorders that share the common feature of a disruption in blood flow to parts of the brain, particularly the cerebral cortex. The signs and symptoms of vascular dementia are highly variable because of the range of brain regions affected. There is significant overlap of vascular dementia with AD, and in fact patients often have both forms, known as *mixed dementia.*
- **Lewy body dementia.** Lewy body dementia is characterized by cytoplasmic inclusions called *Lewy bodies* in neurons in certain parts of the brain. Unlike other forms of dementia, Lewy body dementia features fluctuations in cognitive functions, a less profound impact on memory formation, hallucinations and delusions, and sleep disorders. In addition, patients with Lewy body dementia often have the motor features of Parkinson's disease.
- **Pick's disease.** In Pick's disease, the cerebral cortex of the frontal and temporal lobes progressively degenerates, resulting in severe atrophy due to loss of neurons, as well as excess neuroglial cells. Pick's disease is unusual in that it strikes a younger population, with onset typically around age 55–65. Patients tend to deteriorate rapidly, and because the degeneration affects the prefrontal cortex, dramatic personality changes and behavioral problems are common.

No proven methods exist for preventing the many forms of dementia, nor are any of the forms curable. Some drugs may slow the progression of AD in certain patients; however, they do not reverse the changes that have already taken place. These drugs have not shown effectiveness for any other forms of dementia. Dementias have profound and devastating effects on a person's quality of life; research into the causes, prevention, and treatment of these disorders is therefore intensive and ongoing.

- *attention,* which is lateralized to the right parietal cortex;
- *facial recognition,* which tends to be performed by the right temporal cortex; and
- *language-related recognition,* including the ability to identify an object with its proper name, which is lateralized to the left temporal cortex.

Other language functions, which we discuss next, appear to be strongly lateralized.

Language

Language is arguably one of the most important cognitive functions of the brain in terms of the evolution of the human species. The ability to assign a word as a symbolic representation of an object or concept allows us to communicate with one another through speaking, writing, and/or signing. This ability has made possible many of our most important cultural, technological, and social achievements. Indeed, it's difficult for most of us to even imagine life without language, as it is so integral to our day-to-day lives.

In the context of neurology, **language** refers to the ability to comprehend and produce words through speaking, writing, and/or signing, and to assign and recognize the symbolic meaning of a word correctly. But language is far more complex than simple words; it involves syntax (the arrangement of words and phrases to create sentences), grammar, and context, as well. In addition, correct understanding of a statement also relies on cues such as tone, volume, and the speaker's body language. The phrase "oh, great" could have a number of meanings based on the way it's said and in what context, including excitement, dread, apathy, sarcasm, or joy.

Many parts of the brain are required for us to communicate successfully through speech, writing, and/or sign language (**Figure 12.20**). However, most of these areas are concerned with the motor and sensory aspects of communication rather than with language itself. For example, a blind or deaf person does not lose the capacity for language, nor does someone with an inability to speak. Instead, language appears to depend on two association areas. The first is Broca's area in the frontal lobe, which is responsible for the *production* of language, including the planning and ordering of words with proper grammar and syntax. The second is Wernicke's area in the temporal lobe, which is responsible for *understanding* language and linking a word with its correct symbolic meaning. The functions of these two areas become readily apparent when one area is damaged,

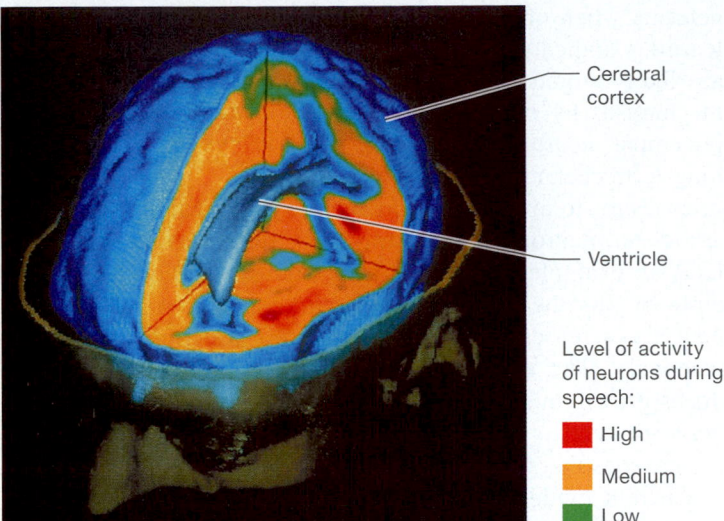

Figure 12.20 **Functional neuroimaging (functional magnetic resonance imaging, fMRI) of the brain during speech.**

which results in a language deficit called an **aphasia** (ah-FAY-zhee-ah). You can find out more about these deficits in *A&P in the Real World: Aphasias.*

As we noted in the previous section, language functions are heavily lateralized, with most language functions residing in the left cerebral hemisphere in about 97% of people. However, the right hemisphere does have some important language functions: those relating to the emotional nature of language.

Aphasias

The two major types of aphasia (language deficits) are Broca's and Wernicke's aphasias.

- **Broca's aphasia** is characterized by the inability to produce speech, because the patient is unable to correctly plan and order the grammar and syntax of a statement. Speech in a patient with Broca's aphasia is not fluent; words and phrases tend to be repeated; and grammar, syntax, and the structure of individual words are disordered. Broca's aphasia tends to be frustrating for patients because language comprehension remains intact—they understand what is being said to them but are unable to organize a reply.
- **Wernicke's aphasia** is characterized by the inability to understand language. Patients with this form of aphasia can speak fluently with adequate grammar and syntax, but the words make little to no sense, as words and meanings are not correctly linked. In Wernicke's aphasia, language comprehension is not intact, so patients cannot fully understand what is being said to them and cannot realize that what they're saying makes no sense.

Damage to language areas of the right hemisphere results in the loss of emotional and tonal components of language (imagine not being able to detect whether someone intends sarcasm or humor); so, both hemispheres are required for normal language function.

Quick Check

☐ 1. What is cognition? Which part of the brain is responsible for cognitive processes?

☐ 2. What is cerebral lateralization? Which functions are known to be lateralized?

☐ 3. Define language in the context of neurology. What are the functions of Broca's and Wernicke's areas?

Learning and Memory

When we *learn,* our nervous systems acquire new information that is observable as some sort of behavioral change, whether that is providing the correct answer on an examination or riding a bicycle. What we learn generally is put into our **memory,** where it is encoded and stored in our neural circuitry and is retrievable at will. There are two basic types of memory: **declarative (fact) memory,** which is defined as the memory of things that are readily available to consciousness that could in principle be expressed aloud (hence the term "declarative"); and **nondeclarative memory** (also called *procedural,* or *skills, memory*), which tends to include skills and associations that are largely subconscious. Examples of declarative memory are a phone number, a quote, or the pathway of the corticospinal tracts. Nondeclarative memory can be exemplified by how to enter the phone number on a phone, how to move your mouth to speak the quote, and how to read this chapter.

Both declarative memory and nondeclarative memory can be classified by the length of time for which they are stored. These classes include the following:

- **Immediate memory.** Your immediate memory is stored for only a few seconds. This may not seem significant, but it is crucial for carrying out normal conversation, reading, and daily tasks, such as remembering the areas of the floor you just swept.
- **Short-term (working) memory.** Your short-term memory is stored for several minutes and allows you to remember and manipulate information with a general behavioral goal in mind. Short-term memory comes into play when you have lost your keys and are searching for them—without such memory, you wouldn't be able to remember where you had already looked, making an effective search impossible.
- **Long-term memory.** Your long-term memory is a more permanent form of storage for days, weeks, or even your entire lifetime.

Some of the information in immediate and working memory is transferred into long-term memory, a process called

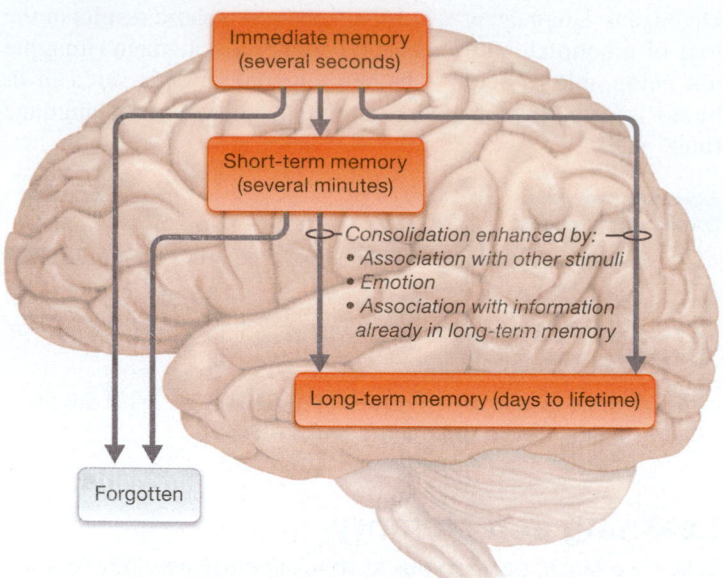

Figure 12.21 Pathways for consolidation of memories.

consolidation (**Figure 12.21**). However, much of the information is simply forgotten. Humans have a very limited ability to consolidate relatively meaningless data into long-term memory; however, if these data are presented in a form that is associated with strong emotion, another stimulus, or something one has already learned, consolidation occurs much more easily. Consider, for example, the method by which children are taught the alphabet—by song. By setting the string of 26 letters to music, one can associate a letter with a musical note. Try right now to recite the alphabet backward. Although you know the sequence of letters, you likely had a hard time doing this, as you have no notes with which to associate it.

The process of consolidation is notoriously flawed. This same use of association to form memories can lead individuals to make false associations that become encoded in memory. Additionally, strong emotions and previous perceptions can color the memory significantly. For instance, people who believe they have seen a "flying saucer" will often dramatically exaggerate the speed and size of the object and their relative peril during the incident. Such exaggeration doesn't make these people liars—they genuinely remember the event occurring that way. However, even those memories about which people feel most confident can be false.

The mechanism by which the physical embodiment of a memory, called an *engram,* is formed differs for declarative and nondeclarative memory. The following subsections discuss both mechanisms.

Formation and Storage of Declarative Memory

Formation of new declarative memories appears to require involvement of the hippocampus, which, as mentioned, is a component of the limbic system. Brain imaging studies have shown dramatically increased activity in the hippocampal

neurons when one is studying information with the intent of learning, indicating that immediate and short-term memories are likely stored here. The mechanism by which hippocampal neurons encode long-term declarative memories seems to involve an increase in synaptic activity between associated neurons, called **long-term potentiation (LTP)**. In LTP, an example of the Cell-Cell Communication Core Principle (p. 28), the strength of synaptic connection between two neurons increases when the two neurons are stimulated simultaneously. The resulting increase in synaptic activity produces growth in dendrites, greater release of neurotransmitters between the two neurons, and induction of protein synthesis in the neurons that maintain this high degree of synaptic activity.

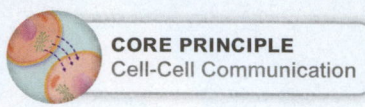

CORE PRINCIPLE
Cell-Cell Communication

Current evidence indicates that LTP in the hippocampus is required for new declarative memories to form, but it is clear from patients with damage to the hippocampus that long-term declarative memories aren't *stored* here. Instead, memories appear to be stored in different areas of the cerebral cortex that correlate with their function. So, for example, the image of a cat is stored in the visual association area, the sound of the cat in the auditory association area, and the feel of a cat's fur in the somatosensory association area. Retrieval of these memories seems to be mediated by pathways that involve the hippocampus and the prefrontal cortex, although researchers are still working to understand these pathways.

Formation and Storage of Nondeclarative Memory

Damage to the hippocampus and associated structures does not affect the ability to form new nondeclarative memories. Instead, it appears from studies of patients with certain brain diseases that the motor cortices, the cerebellum, and the basal nuclei are required for the formation and storage of nondeclarative memories. The exact locations and the mechanism involved are not clear; however, many researchers think a form of LTP comes into play, similar to what happens in the formation of declarative memories.

Quick Check

☐ 4. Explain the differences between declarative memory and nondeclarative memory.

☐ 5. How do immediate, short-term, and long-term memory differ?

Emotion

What we commonly think of as *emotion* is actually a complex combination of three separate phenomena:

- visceral motor responses, such as blushing when one is embarrassed or a racing heart when one is afraid;
- somatic motor responses, such as smiling, laughing, frowning, and crying; and

- highly subjective "feelings" that respond to and are integrated with sensory and/or cognitive stimuli, such as feeling sad when remembering a lost pet or feeling tense when watching a suspenseful movie.

As an example, let's discuss the different parts of the brain responsible for the visceral and somatic motor responses that accompany a feeling of excitement. The hypothalamus plays a role in both the visceral and somatic motor components. It sends instructions to nuclei of the reticular formation, which in turn activate autonomic neurons that innervate the viscera; this causes the heart rate and blood pressure to increase. Both the hypothalamus and the limbic lobe influence the somatic motor response by sending instructions to other nuclei in the reticular formation. These nuclei relay the instructions to lower motor neurons, which causes us to open our eyes widely and smile.

The third component of emotion—the highly subjective "feelings" that we experience—is the most complex. A key player in this aspect of emotion is the amygdala. The amygdala receives sensory stimuli from a number of sources, including the brainstem, the thalamus, the cerebral cortex, and the basal nuclei. These connections allow the amygdala to analyze the emotional significance of sensory stimuli and create associations between different sensory stimuli, a type of "emotional learning." For example, a young child learns to associate a caregiver's cheerful tone of voice with happiness and a loud, harsh tone of voice with anger and fear. The amygdala also has connections with basal nuclei circuits, which appear to select behaviors that maximize rewards and minimize punishments.

Still, none of these circuits or connections explains why we "feel" sad or happy or angry. Neither do they explain why the same stimulus can make one person happy and another person sad. This enigmatic aspect of emotion is not understood on a cellular level, but it is assumed to involve the amygdala's projections to the prefrontal cortex. Given the importance of emotions in the experiences of all human beings, this is likely to be an area of intensive research for years to come.

Quick Check

☐ 6. What three phenomena constitute emotion, and which components of the nervous system mediate them?

Apply What You Learned

☐ 1. Mr. Jacobs suffers widespread damage to his right cerebral hemisphere. His wife believes he will be fine because his left cerebral hemisphere is undamaged. Is his wife correct? Explain. What deficits, if any, can Mr. Jacobs expect to face?

☐ 2. Ms. Marcos undergoes treatment for a seizure disorder that involves removal of the right and left hippocampus. Will this affect her declarative memory, nondeclarative

memory, or both? Will this remove her old memories, impair the formation of new memories, or both?

☐ 3. The patient in question 2, Ms. Marcos, also has her right and left amygdala removed. What effect is this likely to have?

See answers in Appendix A.

MODULE **12.5**
Protection of the Brain

Learning Outcomes

1. Describe the functions of cerebrospinal fluid as well as the details of its production, its circulation within the CNS, and its ultimate reabsorption into the bloodstream.

2. Describe the structural basis for and the importance of the blood brain barrier.

3. Identify and describe the cranial meninges, and explain their functional relationship to the brain.

The brain isn't a very durable organ—it is rather gelatinous and soft, and its neurons can be damaged by even slight pressure. Given the importance of the structures of the brain and their fragility, it should come as no surprise the body offers substantial protection for the brain. As you already know, it is housed within the cranial cavity, and the thick skull provides strong protection from external forces such as trauma, movement, and temperature changes.

Inside the skull, three features further shelter the brain from external and internal forces:

- the *cranial meninges,* which are a set of three membranes that surround the brain;
- the *cerebrospinal fluid* (*CSF*), a protective fluid that bathes the brain and fills its cavities; and
- the *blood brain barrier,* which prevents many substances in the blood from gaining access to the cells of the brain.

This module explains the structure and function of each of these features.

The Cranial Meninges

⟪ FLASHBACK

1. What is dense irregular connective tissue, and what are its properties? (p. 139)

2. What is the periosteum? (p. 145)

The three **cranial meninges** (meh-NIN-jeez; singular, *meninx*) are protective membranes made primarily of dense irregular

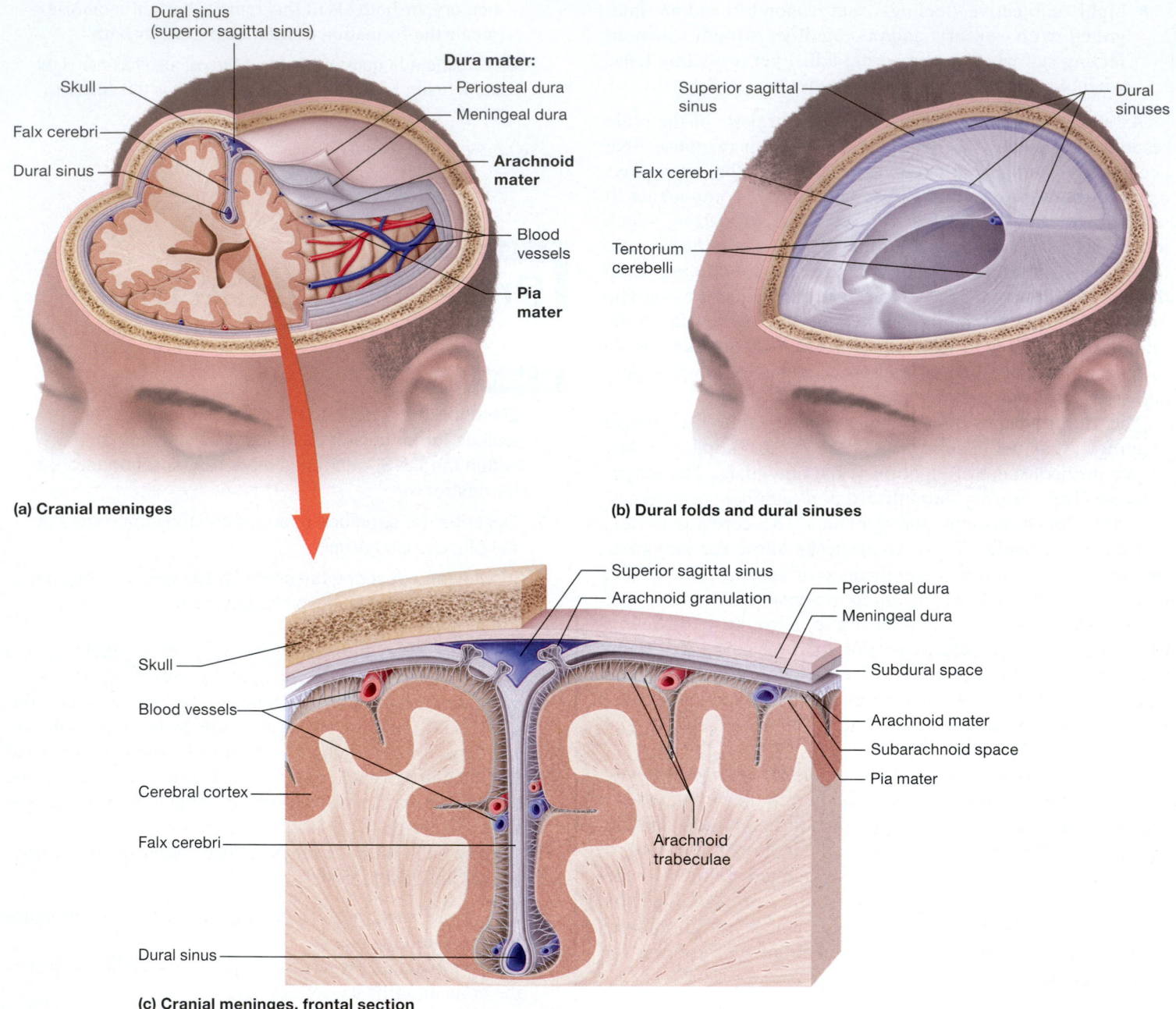

(a) Cranial meninges

(b) Dural folds and dural sinuses

(c) Cranial meninges, frontal section

Figure 12.22 Structure of the cranial meninges and dural sinuses.

collagenous connective tissue (**Figure 12.22**). The outermost meninx, known as the **dura mater** (DOO-rah MAH-ter; "tough womb"), or simply *dura,* is the thickest and toughest of these membranes. Between the dura and the cranial bones is a potential space called the *epidural space.* (This is referred to as "potential" space because normally the dura is bound tightly to the bone and nothing is found between them except blood vessels.) Deep to the dura is another very narrow space, called the **subdural space,** that houses a thin layer of serous fluid and certain veins that drain the brain.

The middle meninx is the thinner **arachnoid mater** (ah-RAK-noyd), or simply *arachnoid,* and the thinnest and innermost

meninx is the **pia mater** (PEE-ah; "gentle mother"), or *pia.* The arachnoid and pia are separated by another narrow, fluid-filled space called the **subarachnoid space.** The subarachnoid space contains CSF and the major blood vessels of the brain. *A&P in the Real World: Infectious Meningitis* discusses what can happen when viruses or bacteria invade this space.

Each of the potential spaces around and between the meninges may become filled with blood; such conditions are called *epidural, subdural,* and *subarachnoid hematomas* (hee-mah-TOH-mahz), respectively, and tend to result from trauma or from the rupture of a weakened blood vessel. These serious conditions are potentially life-threatening, as they increase

Infectious Meningitis

A&P in the Real World

Infectious meningitis (meh-nin-JY-tis) is a potentially life-threatening infection of the meninges in the subarachnoid space. The infection causes inflammation of the meninges and leads to the classic signs of headache, lethargy, stiff neck, and fever. Infectious meningitis is diagnosed by examining a sample of CSF for infectious agents and white blood cells (cells of the immune system).

Certain viruses and bacteria are the most common agents of infectious meningitis. Differentiating between the two types of pathogens to ensure a correct diagnosis and proper treatment is critical—viral meningitis is generally mild and resolves in 1–2 weeks, whereas bacterial meningitis can rapidly progress to brain involvement and death, so aggressive antibiotic treatment is necessary. Fortunately, some of the most common forms of bacterial meningitis are largely preventable with safe and effective vaccines.

intracranial pressure, or pressure on the brain. Treatment of a hematoma in a meningeal space generally consists of drilling a hole through the skull and implanting a device that allows the blood to drain from the space, which relieves pressure on the brain.

Now let's examine each of the three cranial meninges.

Dura Mater

As implied by its name, the dura mater is a tough, leathery membrane with abundant collagen fibers and very few elastic fibers. As shown in Figure 12.22a, it has two layers (from superficial to deep):

1. **Periosteal dura.** The outer *periosteal dura* is attached to the inner surface of the bones of the cranial cavity; it functions as the periosteum of those bones and has an extensive blood supply that resides in the epidural space.
2. **Meningeal dura.** The *meningeal dura* is the inner, avascular layer that lies superficial to the arachnoid mater.

Throughout most of the dura these two layers are fused, giving rise to a single, thick, inelastic membrane. However, in certain locations the two layers separate and form cavities called **dural sinuses.** The dural sinuses are venous channels that drain CSF and deoxygenated blood from the brain's many veins. In Figure 12.22 the dural sinuses are blue, reflecting the fact that they are filled with deoxygenated blood.

Dural sinuses may also be found where the meningeal dura folds over on itself and dives between certain structures of the brain (see Figure 12.22b). These locations include those in the following list.

- **Falx cerebri.** The **falx cerebri** (SEHR-eh-bry) is the partition between the right and left cerebral hemispheres, and lies in the longitudinal cerebral fissure. A large dural sinus called the **superior sagittal sinus** is situated superior to the falx cerebri.
- **Tentorium cerebelli.** The **tentorium cerebelli** (ten-TOH-ree-um seh-reh-BEL-aye) is the partition between the cerebellum and the occipital lobe of the cerebrum. It is so named because it resembles a "tent" covering the cerebellum.
- **Falx cerebelli.** The small **falx cerebelli** (not visible in Figure 12.22b) separates the right and left cerebellar hemispheres.

Arachnoid Mater

The middle meninx, the arachnoid mater, is named for its resemblance to a spider web ("arachnoid" means "spiderlike"). Like the dura, the arachnoid is composed of dense irregular collagenous connective tissue; however, it is thinner and somewhat more elastic. If you look at Figure 12.22c, you can see that the arachnoid has inward extensions called *arachnoid trabeculae.* These trabeculae, made up of bundles of collagen fibers and fibroblasts, anchor the arachnoid mater to the deeper pia mater. Take note also that small bundles of the arachnoid project up through the meningeal dura into the dural sinuses. These *arachnoid granulations* (also known as *arachnoid villi*) play an important role in returning CSF to the bloodstream.

Pia Mater

The innermost and very thin pia mater is the only meninx that physically touches the brain, and it closely follows each contour of the brain and dives into its sulci and fissures (see Figure 12.22). The pia mater is permeable to substances in the brain extracellular fluid and CSF. This allows substances to move between the two fluids and helps to balance the concentration of different solutes in them.

Quick Check

☐ 1. What are the three meninges, from superficial to deep?

☐ 2. What are the three spaces (potential and actual) around and between the meninges? Which of these spaces is filled with cerebrospinal fluid?

The Ventricles and Cerebrospinal Fluid

≪ FLASHBACK

1. What properties of water make it effective for protecting structures of the body? (p. 46)

2. Which type of neuroglial cell lines the cavities of the brain and spinal cord? (p. 389)

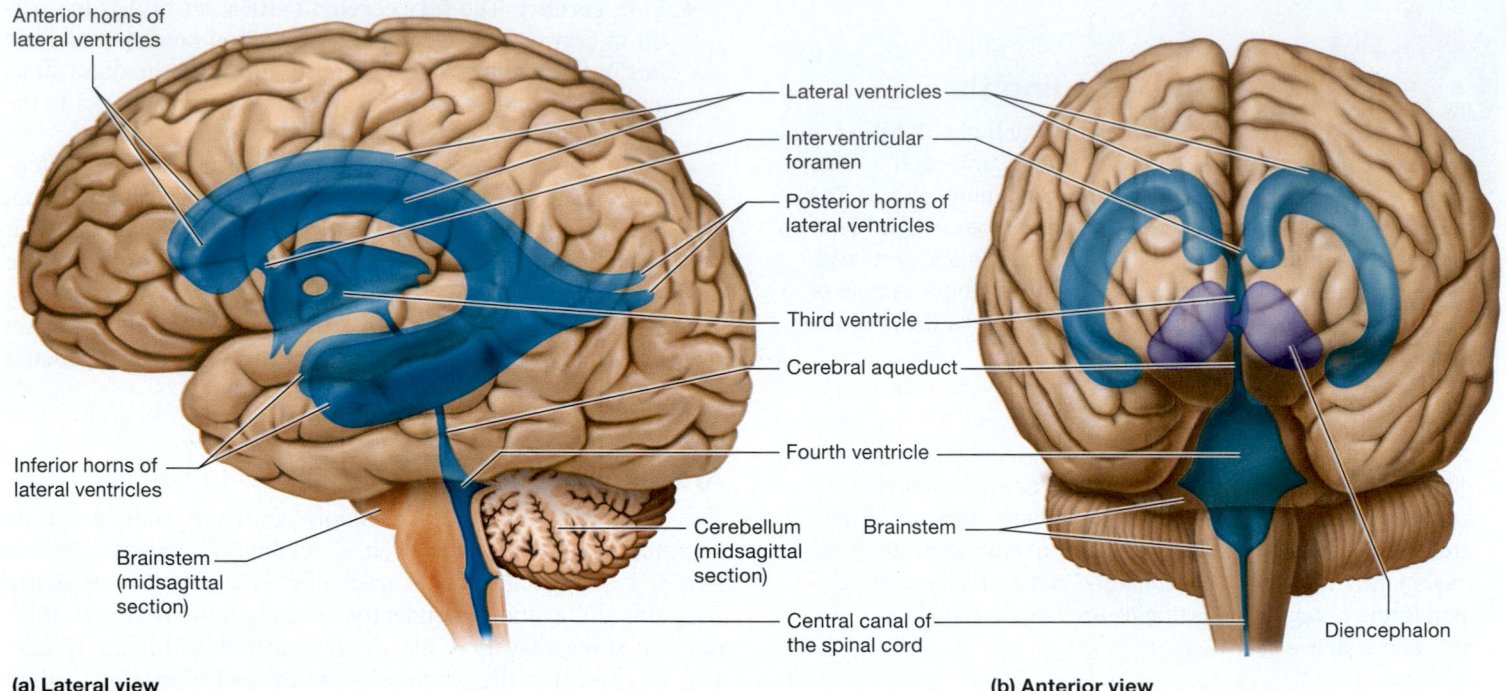

(a) Lateral view

Anterior horns of lateral ventricles

Lateral ventricles

Interventricular foramen

Posterior horns of lateral ventricles

Third ventricle

Cerebral aqueduct

Fourth ventricle

Inferior horns of lateral ventricles

Brainstem (midsagittal section)

Cerebellum (midsagittal section)

Brainstem

Central canal of the spinal cord

(b) Anterior view

Diencephalon

Figure 12.23 Ventricles of the brain.

As we discussed earlier, the brain has cavities called **ventricles** (**Figure 12.23**). The ventricles are continuous with one another and with the cavity in the spinal cord, called the *central canal.* Both the ventricles and the central canal are lined with ependymal cells and filled with cushioning cerebrospinal fluid.

Recall from Module 12.1 that four ventricles are found within the brain, as follows:

- **Right and left lateral ventricles.** The cavities within the right and left cerebral hemispheres are called the right and left **lateral ventricles;** these are considered the first and second ventricles. In Figure 12.23, from the anterior view, the lateral ventricles resemble ram's horns; from the lateral view, they are horseshoe-shaped and have three regions— the *anterior horn,* the *inferior horn,* and the *posterior horn.*
- **Third ventricle.** The narrow **third ventricle** is housed between the two lobes of the diencephalon. It is continuous with the lateral ventricles via an opening that connects all three called the **interventricular foramen.**
- **Fourth ventricle.** The third ventricle drains into the **fourth ventricle** via the **cerebral aqueduct,** a small canal that passes through the midbrain of the brainstem. The fourth ventricle is situated between the pons and the cerebellum and is continuous inferiorly with the central canal of the spinal cord. Figure 12.23a shows that this ventricle contains several posterior openings that allow CSF in the ventricles to flow into the subarachnoid space around the brain and the spinal cord.

Cerebrospinal fluid (CSF) is a clear, colorless liquid similar in composition to plasma (the fluid portion of blood). CSF protects the brain by cushioning it and maintaining a constant temperature within the cranial cavity, removing wastes, and

increasing the buoyancy of the brain. This buoyancy is particularly important, as without adequate CSF, the brain can literally crush itself under its own weight.

Most CSF is formed within each of the four ventricles by structures called **choroid plexuses** (KOHR-oyd). The choroid plexuses are located where blood capillaries come into direct contact with the ependymal cells (which also produce some CSF) lining the ventricles. The capillaries within the choroid plexuses are *fenestrated*—their cells have small gaps between them and within their membranes (*fenestra-* = "window") that allow fluid and electrolytes to escape the blood and enter the extracellular fluid.

The general pathway for the formation, circulation, and reabsorption of CSF is shown in **Figure 12.24**, and consists of the following steps:

1. **Fluid and electrolytes leak out of the capillaries of the choroid plexuses.** Note that the fluid and electrolytes are leaking out toward the ventricles, *not* the extracellular fluid of the brain.

2. **Ependymal cells secrete CSF into the ventricles.** The fluid and electrolytes that have escaped the capillaries enter the ependymal cells. The ependymal cells then secrete CSF into the ventricles by exocytosis.

3. **CSF circulates through and around the brain and spinal cord in the subarachnoid space.** The CSF circulates through the ventricles with the help of ciliated ependymal cells. Some CSF then flows into the subarachnoid space around the spinal cord.

4. **Some of the CSF is reabsorbed into the blood in the dural sinuses via the arachnoid granulations.** As the CSF flows through the subarachnoid space, some of it is

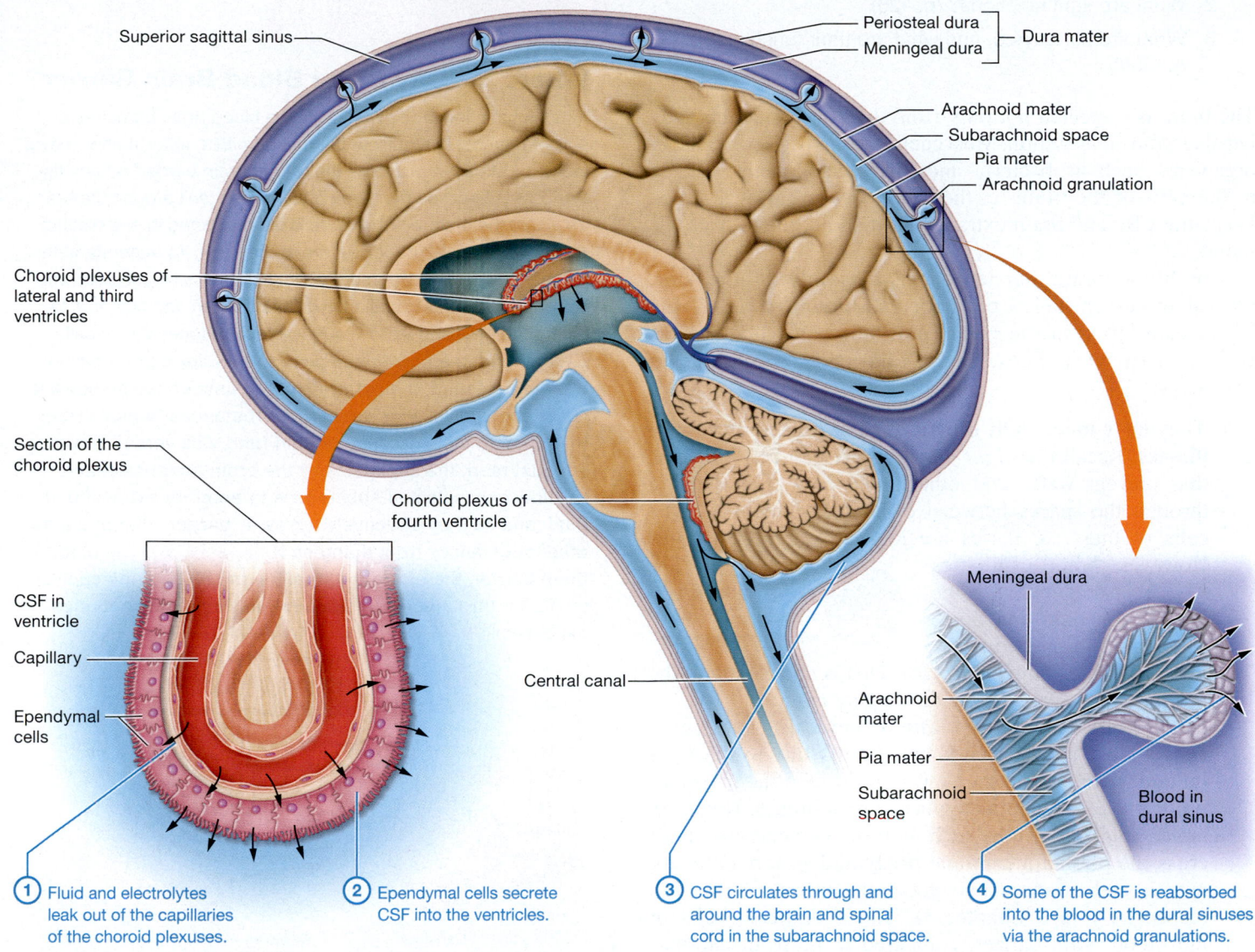

Figure 12.24 Formation and flow of cerebrospinal fluid (CSF).

reabsorbed through the arachnoid granulations. The reabsorbed CSF ends up in the dural sinuses, where it mixes with venous blood.

At any given time about 150 ml (roughly two-thirds of a cup) of CSF circulates through and around the brain and spinal cord; approximately 50 ml is in the ventricles, the subarachnoid space, and around the spinal cord. However, about 750–800 ml of CSF is produced per day, which means that the CNS must drain old CSF as the choroid plexuses secrete new CSF. This occurs constantly, and in fact roughly 50% of the total CSF is completely replaced every 5–6 hours.

Hydrocephalus (hy′-droh-SEF-ah-luss; "water on the head") results if too much CSF is formed by the choroid plexuses, if the flow of CSF is blocked, or if not enough CSF is reabsorbed by the arachnoid granulations. If left untreated, hydrocephalus can increase intracranial pressure and lead to brain damage.

Quick Check

☐ 3. Where are the four ventricles of the brain located?

☐ 4. What do the ventricles contain, and why?

The Blood Brain Barrier

« FLASHBACK

1. What is simple squamous epithelium? (p. 129)

2. What are tight junctions? (p. 126)

3. What are astrocytes, and what are their functions? (p. 388)

The brain is protected not only from threats outside the body but also from many within. Most chemicals and disease-causing organisms such as bacteria and viruses are denied access to the cells of the brain by the **blood brain barrier,** which keeps the CSF and brain extracellular fluid separate from the blood.

The blood brain barrier consists mainly of the simple squamous epithelial cells (or *endothelial cells*) of the blood capillaries in the brain and their basal laminae, as well as astrocytes. These endothelial cells are unique in two ways:

- **They have more tight junctions than cells of most capillaries.** Recall that *tight junctions* are cell-cell junctions that prevent water and other substances from passing through the spaces between cells (see Chapter 4). The cells of most capillaries are joined by tight junctions. However, brain capillaries have an exceptionally large number of tight junctions, an example of the Structure- **CORE PRINCIPLE** Structure-Function
Function Core Principle (p. 28). This is due mostly to the effects of astrocytes on the developing brain.
- **They limit endocytosis and exocytosis.** Substances cannot pass between endothelial cells due to the tight junctions, so instead they must go through the phospholipid bilayers of the cells' plasma membranes. However, large, polar molecules such as proteins generally don't cross the plasma membrane freely and so rely on *endocytosis* and *exocytosis* to get into and out of the cell, respectively (see Chapter 3). The endothelial cells of the blood brain barrier, compared to those of capillaries elsewhere in the body, have a very limited ability to perform these processes, which means that many substances in the blood simply cannot enter the extracellular fluid of the brain. The converse is true as well: Many substances in the extracellular fluid of the brain cannot enter the blood.

Certain substances can cross the blood brain barrier rather easily, including those to which the plasma membrane is in general freely permeable, such as water, oxygen, carbon dioxide, and nonpolar, lipid-based compounds. Other substances that cross easily include those for which protein channels or carriers exist, such as glucose, amino acids, and ions. However, most large, polar molecules are effectively prevented from crossing the blood brain barrier in any significant amounts. This can be both beneficial and harmful—although the blood brain barrier prevents many toxins and organisms from entering the brain, it also prevents the entry of many therapeutic drugs. For example, only a handful of antibiotics penetrate the blood brain barrier in significant enough concentrations to treat bacterial infections in the brain, making such infections difficult to treat.

Concept BOOST))))

Where Exactly Is the Blood Brain Barrier?

You might be tempted to visualize the blood brain barrier as a sort of raincoat that surrounds the brain. But notice that we never labeled a single structure as the "blood brain barrier" on any figure. This is because the blood brain barrier isn't around the brain; it's *within* the brain. Furthermore, it is not located in one distinct place but is found throughout the entire brain. To understand this, we need to first uncover more about the body's tiniest blood vessels: the *capillaries.* Capillaries deliver oxygen and nutrients to the body's cells and remove any wastes produced by the cells.

Now let's see how this applies to the blood brain barrier. Capillaries found in most organs and tissues are somewhat leaky and allow a wide variety of substances to move from the blood to the extracellular fluid (and vice versa). But, as you just read, the capillaries of the brain are specialized to allow only selected substances to enter its extracellular fluid, and so they effectively act as a "barrier" that prevents other substances from doing so (**Figure 12.25**). The blood brain barrier, therefore, is actually a property of the capillaries found throughout the brain rather than a distinct physical barrier. ■

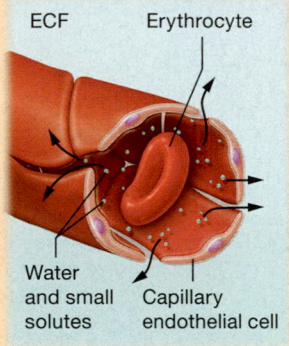
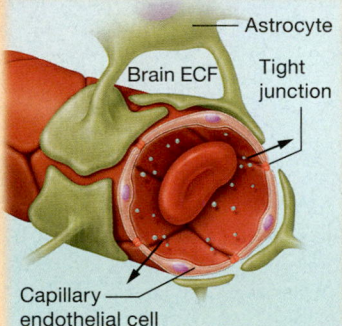

A typical capillary allows water and small solutes to move from the blood to the ECF.

Astrocytes and tight junctions in brain capillaries limit the solutes that enter the brain ECF.

Figure 12.25 The blood brain barrier.

Quick Check

☐ 5. What two factors create the blood brain barrier?

☐ 6. Which substances can easily cross the blood brain barrier?

Apply What You Learned

☐ 1. A tear in the meninges may result in the leakage of CSF. Considering the functions of CSF, predict the symptoms you would see with a CSF leak.

☐ 2. When patients are given a drug that must enter the ECF of the brain, they are sometimes infused with the chemical mannitol. Mannitol temporarily disrupts tight junctions. How would this help deliver the drug?

See answers in Appendix A.

The Spinal Cord

Learning Outcomes

1. Describe the gross anatomy and location of the spinal cord.
2. Identify and describe the anatomical features seen in a cross-sectional view of the spinal cord.
3. Identify and describe the spinal meninges and the spaces between and around them.
4. Describe the differences between ascending and descending tracts in the spinal cord.

Although the spinal cord is less complex structurally than the brain, there is no denying its importance; the dramatic consequences of spinal cord injuries attest to this fact. The **spinal cord** is composed primarily of nervous tissue and is responsible for both relaying and processing stimuli:

- **Relay station.** The spinal cord acts as a *relay station* and an intermediate point between the body and the brain; in fact, it is the only means by which the brain can communicate with most of the body below the head and neck. It receives outgoing stimuli from the brain and sends them to the rest of the body; it also receives incoming stimuli from the body and sends them to the brain.
- **Processing station.** Although the spinal cord cannot perform the higher functions characteristic of the brain, it does do some integration and processing. In fact, certain activities called *spinal reflexes* can be carried out by the spinal cord alone, without influence from the brain (see Chapter 13).

This module explores the structure, organization, and functions of the nervous tissue in the spinal cord. First, however, let's examine how the spinal cord is protected.

Protection of the Spinal Cord

The meninges of the brain pass through the foramen magnum and continue covering the spinal cord. These sheaths continue even after termination of the spinal cord and envelop the nerves at its base. As with the brain, three **spinal meninges** are present: the *dura,* the *arachnoid,* and the *pia mater* (**Figure 12.26a**). The spinal meninges are similar in structure to those of the brain except for the dura mater, which lacks a periosteal layer and consists only of a meningeal layer. Also, the spinal pia plays a role that the cranial pia does not—it helps to anchor the spinal cord in the vertebral cavity. Thin pieces of the spinal pia called **denticulate ligaments** extend outward through the arachnoid and attach to the spinal dura.

We find three actual or potential spaces around or between the spinal meninges (**Figure 12.26b**), as detailed in the following list (from superficial to deep).

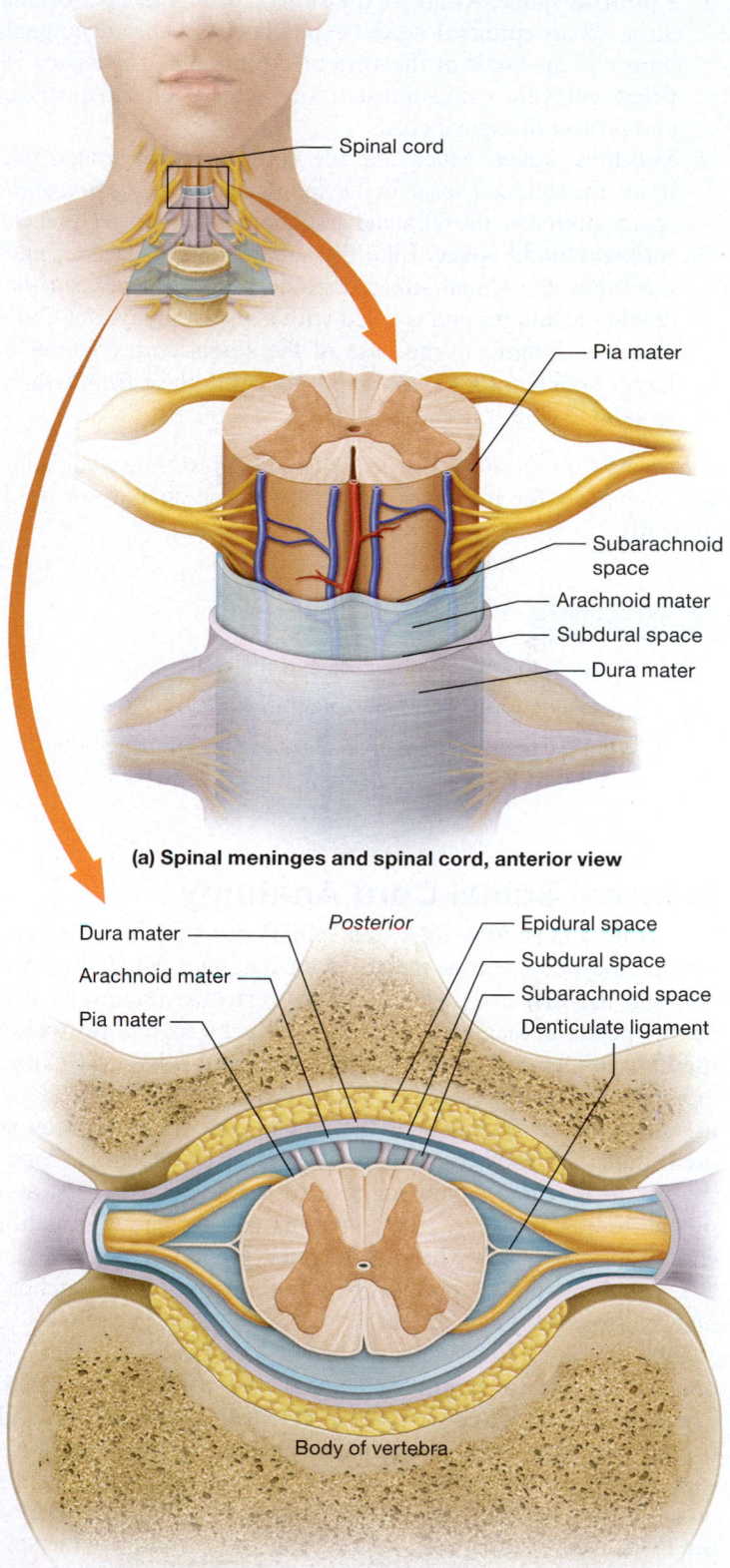

Spinal cord

Pia mater

Subarachnoid space

Arachnoid mater

Subdural space

Dura mater

(a) Spinal meninges and spinal cord, anterior view

Dura mater — *Posterior* — Epidural space

Arachnoid mater — Subdural space

Pia mater — Subarachnoid space

Denticulate ligament

Body of vertebra

Anterior

(b) Spinal meninges and spinal cord, transverse section

Figure 12.26 Structure of the spinal meninges.

1. **Epidural space.** As noted, the spinal cord lacks a periosteal dura, so an **epidural space** exists between the meningeal dura and the walls of the vertebral foramina. This space is filled with veins and adipose tissue, which helps to cushion and protect the spinal cord.

2. **Subdural space.** Much like the epidural space around the brain, the subdural space in the spinal cord is only a *potential* space; normally, the dura and arachnoid adhere to each other.

3. **Subarachnoid space.** Like the subarachnoid space around the brain, the spinal subarachnoid space lies between the arachnoid and pia and is filled with a very thin layer of CSF. The area inferior to the base of the spinal cord contains a larger volume of CSF, making it a useful place from which to sample CSF if needed.

See *A&P in the Real World: Epidural Anesthesia and Lumbar Punctures* for information on how these spaces are used clinically.

Quick Check

☐ 1. What are the two primary roles of the spinal cord?

☐ 2. List and describe the three spinal meninges.

☐ 3. What are the three meningeal spaces around the spinal cord? Which are actual and which are potential spaces?

External Spinal Cord Anatomy

As you read in Module 12.1, the spinal cord starts as an extension of the brainstem at the foramen magnum and terminates between the first and second lumbar vertebrae (**Figure 12.27**). On the posterior side of the spinal cord is the narrow **posterior median sulcus,** and on its anterior side is a wider slit called the **anterior median fissure,** both of which are present along its entire length. The end of the spinal cord forms the **conus medullaris** (KOHN-us med-yoo-LEHR-us), which is cone-shaped, as its name would suggest. At the level of the first and second lumbar vertebrae, the spinal pia gathers into a very thin structure known as the **filum terminale** (FY-lum ter-mee-NAL-ay), which continues through the vertebral cavity and anchors into the first coccygeal vertebra.

Note in Figure 12.27a that the spinal cord has two "bulges"—one in the cervical region and another in the lumbar region. These bulges are known as the **cervical enlargement** and **lumbar enlargement,** respectively. The **nerve roots** that fuse to form the spinal nerves serving the upper and lower limbs attach to these enlargements. **Spinal nerves,** which are part of the PNS, carry sensory and motor impulses to and from the spinal cord. Projecting off either side of the spinal cord between the vertebrae are *posterior* and *anterior nerve roots.* Roots of spinal nerves extend inferiorly from the conus medullaris and fill the remainder of the vertebral cavity. This bundle of spinal nerve roots is collectively called the **cauda equina** (KOW-dah eh-KWY-nah), so named because it resembles a horse's tail (*equin-* = "horse").

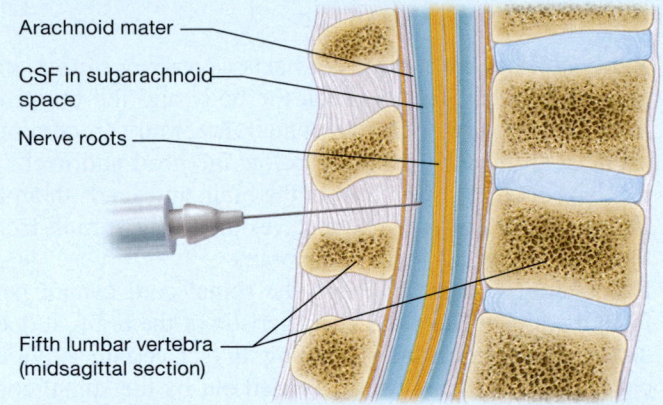
Quick Check

☐ 4. Where are the two bulges of the spinal cord?

☐ 5. What is the cauda equina?

Internal Spinal Cord Anatomy

Recall from Module 12.1 that the butterfly-shaped spinal gray matter is surrounded by tracts of white matter (see Figures 12.2b and 12.27b). A **central canal** filled with CSF runs down the middle of the cord. The two "butterfly wings" are connected by a thin strip of gray matter, the **gray commissure** (KAHM-ih-shoor), which surrounds the central canal of the spinal cord. You can tell the anterior and posterior sides of

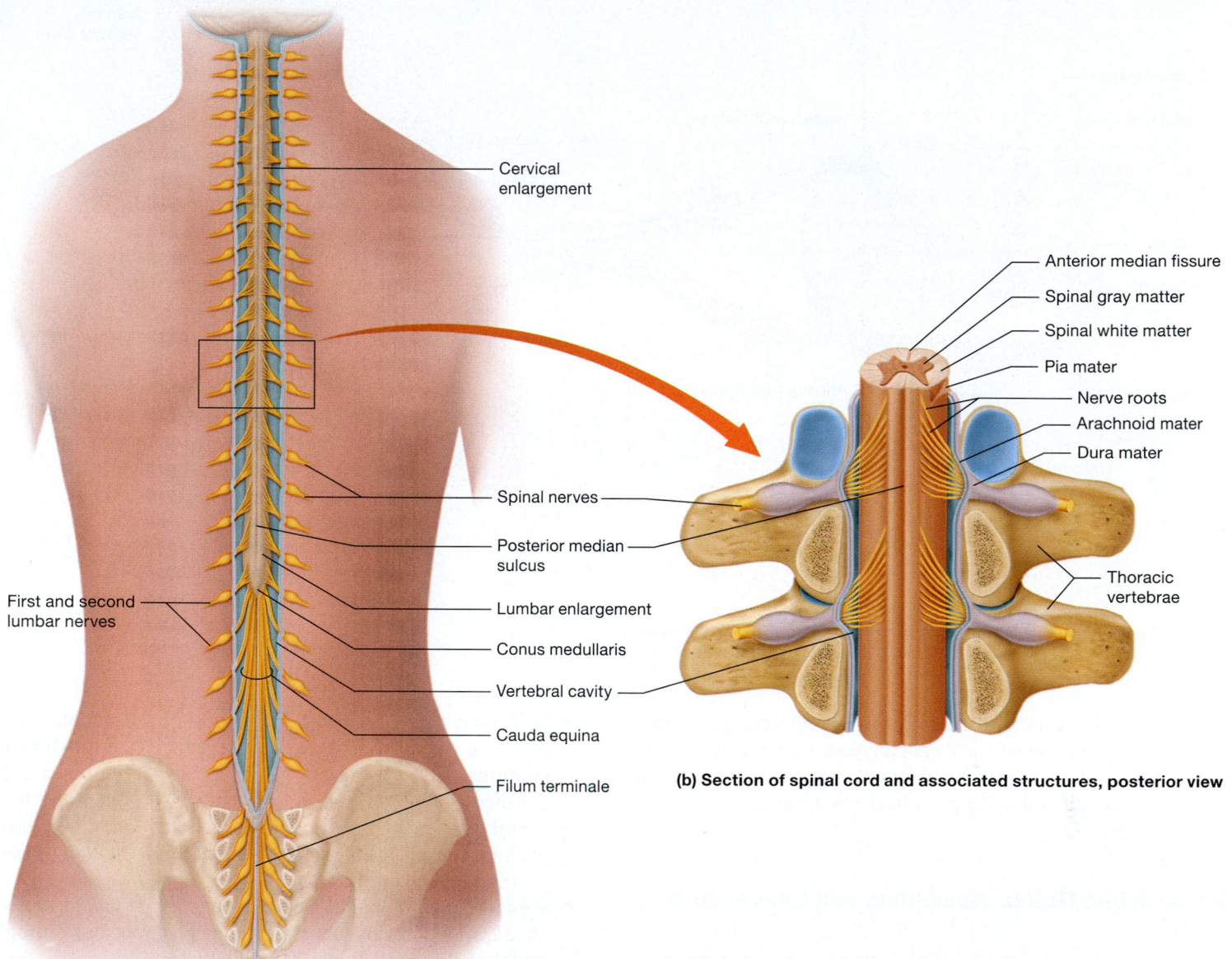

Cervical enlargement

Anterior median fissure

Spinal gray matter

Spinal white matter

Pia mater

Nerve roots

Arachnoid mater

Dura mater

Spinal nerves

Posterior median sulcus

Thoracic vertebrae

First and second lumbar nerves

Lumbar enlargement

Conus medullaris

Vertebral cavity

Cauda equina

Filum terminale

(b) Section of spinal cord and associated structures, posterior view

(a) Spinal cord, posterior view

Figure 12.27 External structure of the spinal cord.

the spinal cord apart by the different shapes of the wings: Those on the anterior side are broader, whereas those on the posterior side are thinner and extend nearly to the outer surface of the spinal cord.

Spinal Gray Matter

Three distinct regions of gray matter are found within the spinal cord, each of which houses neurons with specific related functions (**Figure 12.28**). The three regions include the following:

- **Anterior horn.** The **anterior horn** (or *ventral horn*) is the area of gray matter that makes up the anterior wing

of the butterfly. The neurons of the anterior horn are concerned with somatic motor functions (those of the skeletal muscles).

- **Posterior horn.** The longer and thinner **posterior horn** (or *dorsal horn*) makes up the posterior wing of the butterfly. It contains the cell bodies of neurons that are involved in processing both somatic and visceral incoming sensory stimuli.

- **Lateral horn.** The **lateral horn** is present only from the first thoracic vertebra to the lumbar portion of the cord. It contains the cell bodies of neurons responsible for motor control of the viscera via the autonomic nervous system (discussed in Chapter 14).

As you can see in Figure 12.28, extending to and from the horns are anterior (motor) and posterior (sensory) nerve roots. Notice that the posterior root has an enlarged area just lateral to the cord—this is the *posterior* or *dorsal root ganglion*,

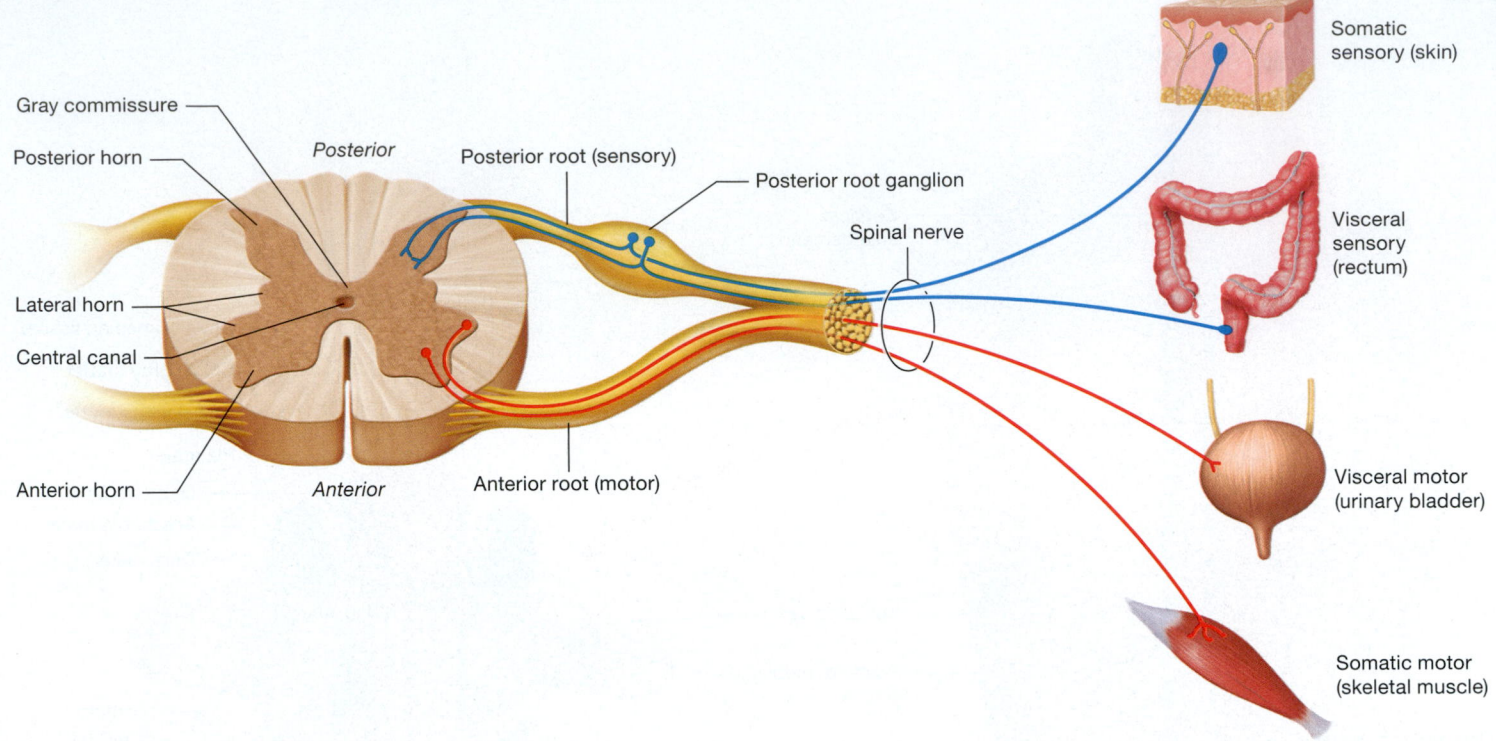

Figure 12.28 Overview of internal spinal cord structure and function.

where the cell bodies of these neurons are housed. These two roots form spinal nerves that contain both motor and sensory axons. The nerves carry motor and sensory stimuli to and from, respectively, different parts of the body, such as the muscles and skin.

Spinal White Matter: Ascending and Descending Tracts

One of the main functions of the spinal cord is to act as a relay station. This function is carried out by its white matter, which contains the axons of neurons that travel to and from the brain. The spinal white matter, like the gray matter, is organized into regions. Each general region of spinal white matter is called a **funiculus** (foo-NIK-yoo-luss; "long rope"), and three funiculi lie on each side of the spinal cord: the **posterior funiculus**, the **lateral funiculus**, and the **anterior funiculus**. The white matter within each funiculus is further organized into tracts, also known as *columns*. The ascending and descending tracts bring stimuli to and from a specific part of the brain, respectively. Like the rest of the spinal cord, the tracts are bilaterally symmetrical, meaning that the right and left sides of the cord contain the same ascending and descending tracts that serve the right and left sides of the body.

Ascending (Sensory) Tracts Several ascending tracts carry various kinds of sensory stimuli (**Figure 12.29a**). The major tracts include those in the following list.

- **Posterior columns.** Within the posterior funiculus we find two tracts that together are called the **posterior columns.** The two tracts of the posterior columns are the medial *fasciculus gracilis* (fah-SIK-yoo-las) and the lateral *fasciculus cuneatus*. These tracts carry somatosensory stimuli such as touch and *proprioception* (joint position) to the brain.
- **Spinocerebellar tracts.** Several tracts in the lateral funiculi transmit information about joint position and muscle stretch to the cerebellum, including the anterior and posterior *spinocerebellar tracts*.
- **Anterolateral system.** Another set of ascending tracts is the *anterolateral system,* which includes the *spinothalamic tracts*. These tracts travel in the anterior and lateral funiculi; they bring pain and temperature stimuli to the brain.

Descending (Motor) Tracts The main descending spinal tracts are shown in **Figure 12.29b**. The largest descending tracts are the **corticospinal tracts,** which control the skeletal muscles below the head and neck. Recall that the corticospinal tracts originate primarily from the motor areas of the cerebral cortex. The axons descend first as part of the internal capsule, then through the brainstem, where the majority of them decussate. As they travel through the lateral funiculi of the spinal cord, they bring motor stimuli to the appropriate places in the anterior horn. Other, smaller descending tracts are also shown in Figure 12.29b.

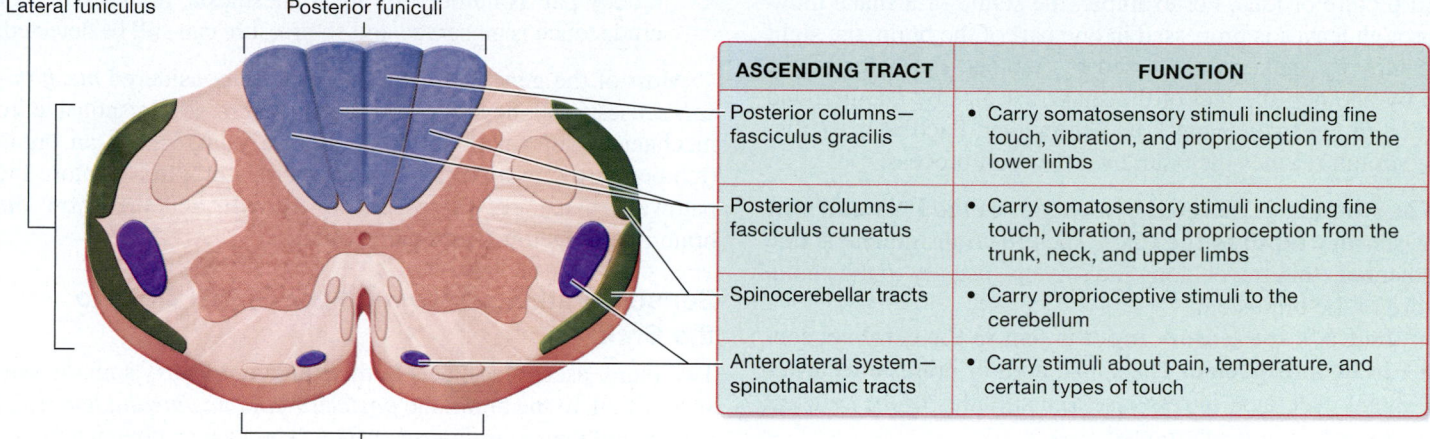

(a) Ascending tracts (sensory)

ASCENDING TRACT	FUNCTION
Posterior columns—fasciculus gracilis	• Carry somatosensory stimuli including fine touch, vibration, and proprioception from the lower limbs
Posterior columns—fasciculus cuneatus	• Carry somatosensory stimuli including fine touch, vibration, and proprioception from the trunk, neck, and upper limbs
Spinocerebellar tracts	• Carry proprioceptive stimuli to the cerebellum
Anterolateral system—spinothalamic tracts	• Carry stimuli about pain, temperature, and certain types of touch

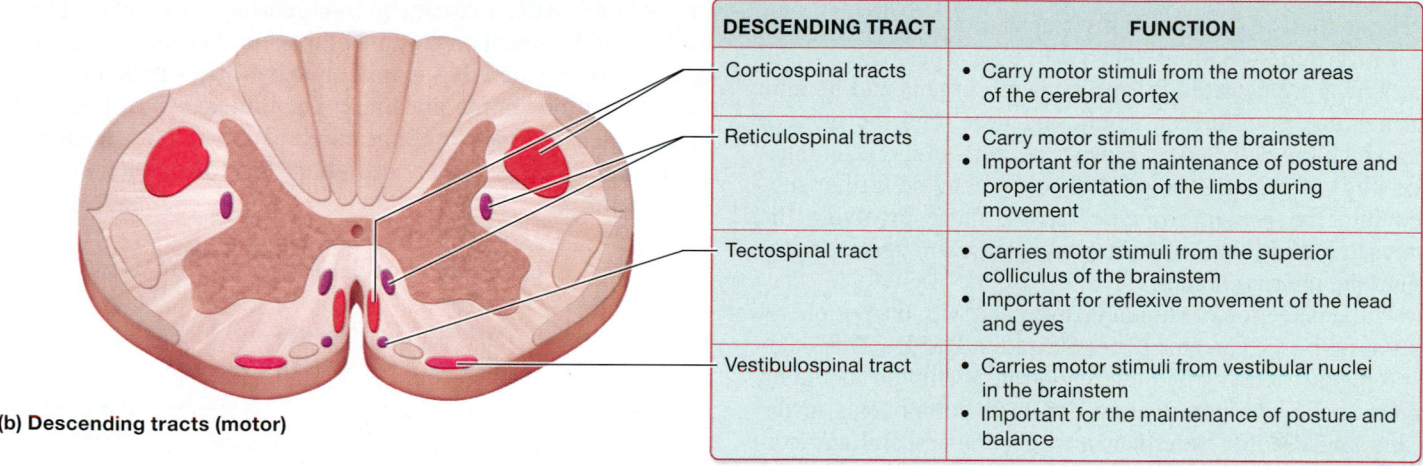

(b) Descending tracts (motor)

DESCENDING TRACT	FUNCTION
Corticospinal tracts	• Carry motor stimuli from the motor areas of the cerebral cortex
Reticulospinal tracts	• Carry motor stimuli from the brainstem • Important for the maintenance of posture and proper orientation of the limbs during movement
Tectospinal tract	• Carries motor stimuli from the superior colliculus of the brainstem • Important for reflexive movement of the head and eyes
Vestibulospinal tract	• Carries motor stimuli from vestibular nuclei in the brainstem • Important for the maintenance of posture and balance

Figure 12.29 Ascending and descending tracts of the spinal cord.

Quick Check

☐ 6. What are the functions of the neurons in the three horns of spinal gray matter?

☐ 7. What are the major ascending tracts of spinal white matter? What are their functions?

☐ 8. What are the major descending tracts of spinal white matter? What are their functions?

Apply What You Learned

☐ 1. Why are no dural sinuses present in the spinal cord?

☐ 2. Mandy is a 13-year-old patient with suspected bacterial meningitis. A resident performs a lumbar puncture to collect a sample of CSF and sends it to the lab. However, when the lab technician examines the sample, she finds only blood and adipocytes (fat cells), and no CSF. What has likely happened?

See answers in Appendix A.

MODULE 12.7

Sensation Part I: Role of the CNS in Sensation

Learning Outcomes

1. Describe the roles of the central and peripheral nervous systems in processing sensory stimuli.

2. Describe the locations and functions of first-, second-, and third-order neurons in a sensory pathway.

3. Explain the ways in which special sensory stimuli are processed by the CNS.

We humans have an amazing ability not only to perceive different **sensory stimuli** (things that cause our senses to respond) but also to assemble multiple sensory stimuli into a single mental picture or idea. For example, the sound of a snake moving through leaves is processed in one part of the brain, the sight of the moving snake is processed by another, the color of the snake by another, the feel of the snake's scales by another, and the smell of the snake's musk by yet another. Each of these disparate stimuli reaches the brain by a two-part process:

1. **The stimulus is detected by neurons in the PNS and sent as sensory input to the CNS.** These neurons initiate action potentials that travel along their axons, usually to the spinal cord or the brainstem.
2. **In the CNS, the sensory input is sent to the cerebral cortex to be interpreted.** The spinal cord or brainstem carries a signal generated in response to a stimulus to the cerebral cortex, where it is interpreted.

When the CNS has received all the different sensory inputs, it integrates them into a single **perception** (a conscious awareness of the sensation). In this case, we very quickly perceive a moving snake so that we can decide if we need to move, too.

We can group sensations into two basic types: the *special senses,* which are detected by special sense organs and include vision, hearing, equilibrium, smell, and taste, and the *general senses,* which are detected by sensory neurons in the skin, muscles, or walls of organs. The general senses may be further subdivided into the *general somatic senses,* those involving the skin, muscles, and joints, and the *general visceral senses,* those involving the internal organs.

The initial detection of stimuli is discussed with the peripheral nervous system (see Chapter 13 for Sensation Part II). This module explores the delivery and interpretation of stimuli, including the central pathways of sensation and their components in the brain and spinal cord. Note that most general visceral and certain general somatic sensory inputs are delivered to parts of the brain such as the cerebellum and the brainstem; these pathways are discussed in a later chapter (see Chapter 14). The following discussion covers only those sensory inputs that are consciously interpreted by the cerebral cortex to produce sensations.

General Somatic Senses

≪ FLASHBACK

1. What are the components of a pseudounipolar neuron, and what are their functions? (p. 387)

Recall that the *general somatic senses* pertain to touch, stretch, joint position, pain, and temperature. Two types of touch stimuli are delivered by different pathways to the appropriate part of the cerebral cortex:

- **Tactile senses** pertain to fine or *discriminative touch,* and include vibration, two-point discrimination (see Chapter 5), and light touch. The tactile senses allow you to discriminate between different shapes and textures without visual input; your tactile senses are at work when you can tell with just your fingers whether the coin in your pocket is a dime or a nickel.
- **Nondiscriminative touch,** also called *crude touch,* lacks the fine spatial resolution of the tactile senses. Often when

a body part is numbed by local anesthesia, the sensation of crude touch is preserved and so pressure can still be detected.

Most of the general somatic senses are considered *mechanical* senses—the neurons that detect them are responsive to mechanical deformations in the skin, a joint, and/or an organ (temperature sensation is the exception). Let's first explore the pathways that carry these sensory stimuli, and then how the brain interprets touch and pain stimuli.

Sensory Pathways through the Spinal Cord to the Brain

Two major ascending tracts in the spinal cord carry somatic sensory stimuli to the brain: the *posterior columns/medial lemniscal system* and the *anterolateral system.* The basic pathway consists of an initial pseudounipolar **first-order neuron,** the sensory neuron that detects the initial stimulus in the PNS. The central process of the first-order pseudounipolar neuron then synapses on a **second-order neuron,** an interneuron located in the posterior horn of the spinal cord or the brainstem. The axons of second-order neurons generally synapse on **third-order neurons,** which are interneurons in the thalamus. Then the axons of third-order neurons deliver impulses to the cerebral cortex. Here is the basic pathway, as well as its relationship to the motor functions of the spinal cord:

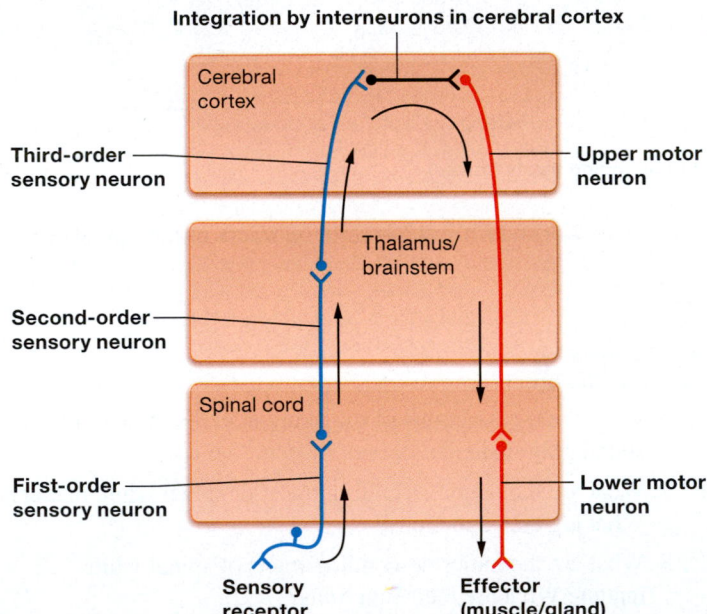

These sensory pathways warrant a closer look.

Posterior Columns/Medial Lemniscal System The axons of neurons that transmit tactile discriminative touch and proprioceptive (joint position) stimuli make up tracts known as the **posterior columns/medial lemniscal system.** This system, shown in **Figure 12.30,** begins with a stimulus from the first-order neuron that enters the spinal cord on its posterior side. Axons from these neurons ascend through the spinal cord in the posterior columns. As mentioned, two tracts are located within

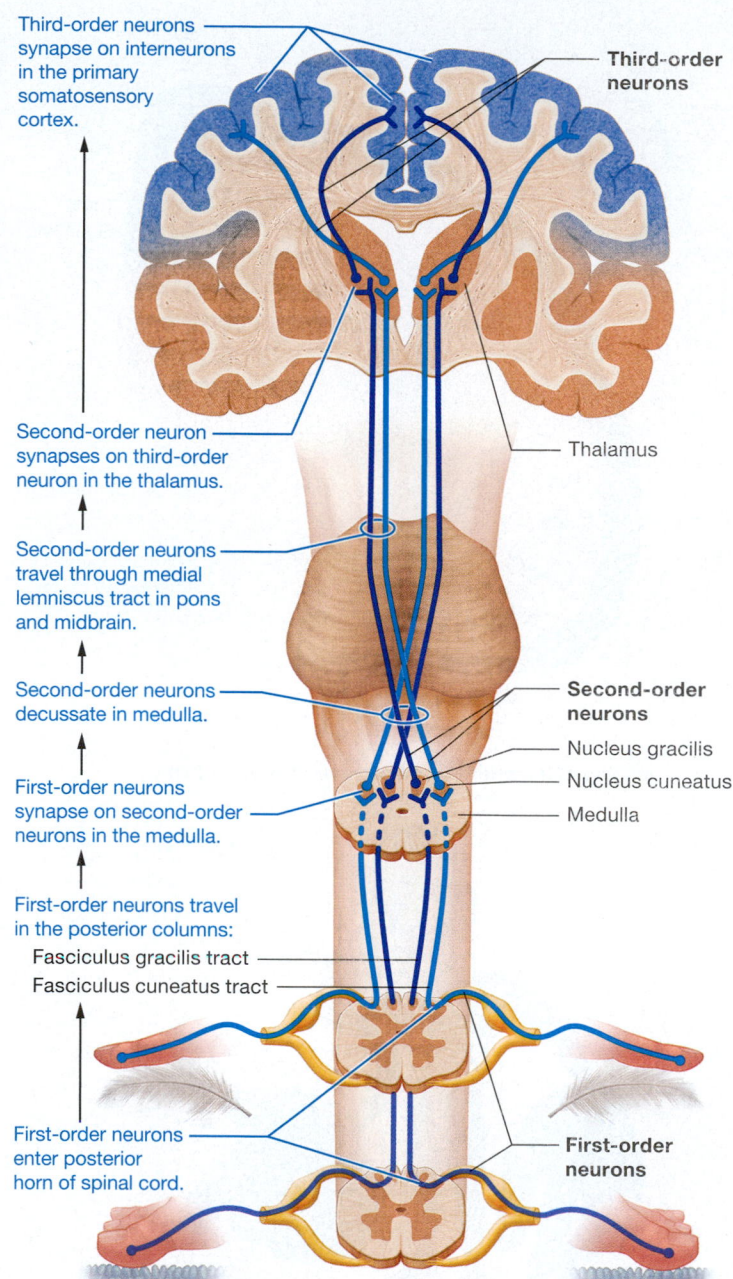

Figure 12.30 Ascending (sensory) pathways: the posterior columns/ medial lemniscal system in the right and left sides of the body.

Third-order neurons synapse on interneurons in the primary somatosensory cortex.

Third-order neurons

Second-order neuron synapses on third-order neuron in the thalamus.

Thalamus

Second-order neurons travel through medial lemniscus tract in pons and midbrain.

Second-order neurons decussate in medulla.

Second-order neurons

Nucleus gracilis

Nucleus cuneatus

First-order neurons synapse on second-order neurons in the medulla.

Medulla

First-order neurons travel in the posterior columns:

Fasciculus gracilis tract

Fasciculus cuneatus tract

First-order neurons enter posterior horn of spinal cord.

First-order neurons

respectively. The axons of these neurons then decussate and form tracts called the **medial lemniscus.** The fibers of the medial lemniscus ascend through the pons and midbrain until they reach the thalamus. In the thalamus, the neurons synapse on third-order neurons, which ascend to the cerebral cortex for integration.

Anterolateral System As with the tactile senses and proprioception, the fibers that carry the stimuli of pain, temperature, and nondiscriminative touch enter the posterior spinal cord. However, their pathways are in the anterolateral spinal cord, and so this is referred to as the **anterolateral system.** In **Figure 12.31** you can see a difference between the posterior columns and the

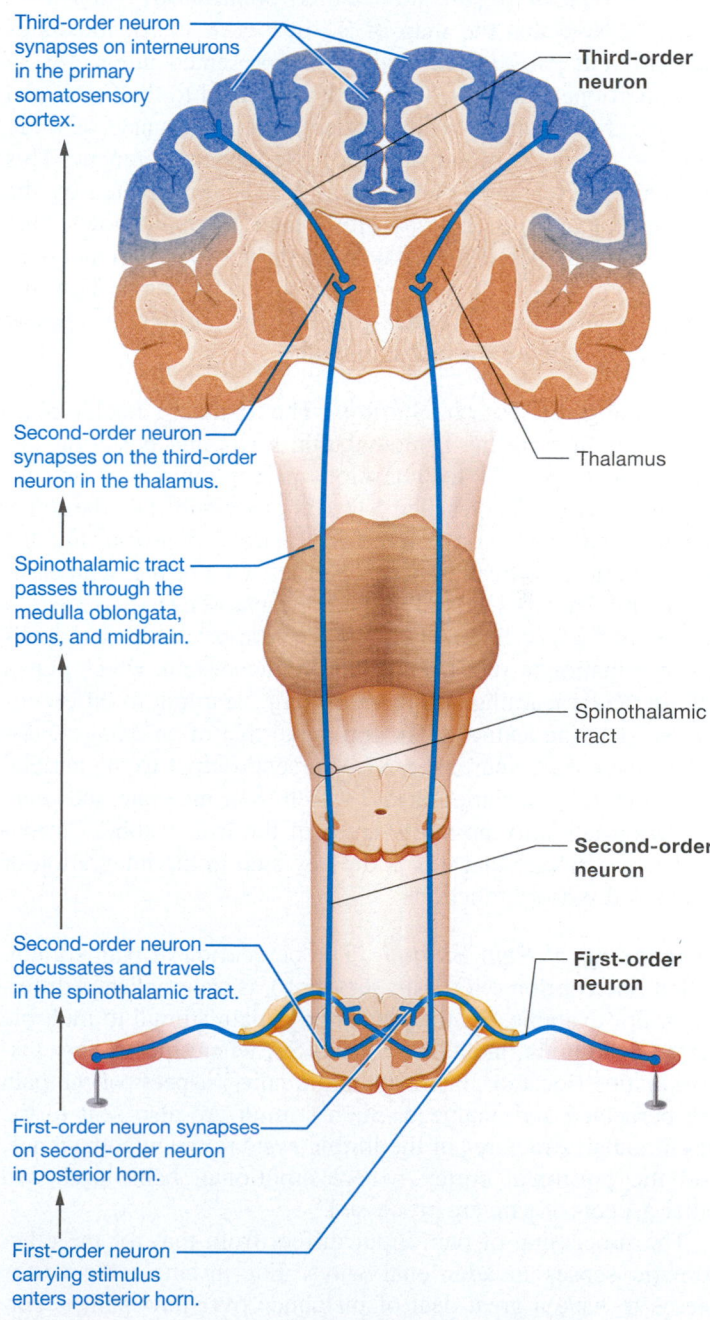

Figure 12.31 Ascending (sensory) pathways: the right and left spinothalamic tracts (part of the anterolateral system).

Third-order neuron synapses on interneurons in the primary somatosensory cortex.

Third-order neuron

Thalamus

Second-order neuron synapses on the third-order neuron in the thalamus.

Spinothalamic tract passes through the medulla oblongata, pons, and midbrain.

Spinothalamic tract

Second-order neuron

Second-order neuron decussates and travels in the spinothalamic tract.

First-order neuron

First-order neuron synapses on second-order neuron in posterior horn.

First-order neuron carrying stimulus enters posterior horn.

the posterior columns: (1) the medial **fasciculus gracilis,** which contains neurons that transmit stimuli from the lower limbs; and (2) the lateral **fasciculus cuneatus,** which contains neurons that transmit stimuli from the trunk, neck, and the upper limbs. (If you have trouble remembering which tract is which, remember that the *gracilis* muscle is in the lower limb, and the fasciculus *gracilis* transmits stimuli from the lower limb.) Figure 12.30 shows that both tracts of the posterior columns ascend on the same side of the body as the original stimulus.

When the axons of the fasciculus gracilis and fasciculus cuneatus neurons enter the medulla, they synapse on the second-order neurons of the *nucleus gracilis* and *nucleus cuneatus,*

anterolateral system: The first-order neurons synapse on the second-order neurons in the posterior horn, after which they decussate.

Several tracts lie within the anterolateral system, the largest of which are the right and left **spinothalamic tracts.** As the name implies, they send signals through the spinal cord to nuclei in the thalamus. Third-order neurons from the thalamus then transmit the stimuli to the cerebral cortex.

Role of the Cerebral Cortex in Sensation

The thalamus relays most incoming information to the primary somatosensory cortex, or S1, in the postcentral gyrus (see Figure 12.5). Each part of the body is represented by a specific region of S1, a type of organization called **somatotopy** (-*topos* = "place"). Note that the map of S1 in **Figure 12.32a** illustrates that different parts of the body are represented unequally—a disproportionate amount of space is dedicated to the hands and the face, for example, which reflects the importance of these parts of the body in sensing the external environment. This unequal representation of body parts in S1 is depicted by the *sensory homunculus* (literally "little man"), which shows what we would look like if the body were proportionate to the areas of S1 (**Figure 12.32b**). We now turn our attention to how the primary somatosensory cortex processes two different types of sensory input, those for touch and pain.

Processing of Touch Stimuli   The thalamic nuclei relay touch stimuli from the spinothalamic tracts and posterior columns primarily to S1 for conscious perception. Once the sensory information has reached S1, it is processed (i.e., the touch is perceived) and axons of the S1 neurons then transmit the information to cortical association areas. One of the main areas to which S1 axons send output is the somatosensory association cortex, or S2 (see Figure 12.5). S2 in turn processes and sends the information to structures of the limbic system, which play a role in tactile learning and memory (e.g., learning to differentiate between the texture of an apple and that of an orange without visual cues). Neurons in S1 also send output to the parietal and temporal association areas, which then integrate and relay this information to the motor areas of the frontal lobe. Current evidence indicates that this is the key step in the integration of motor and sensory functions.

Processing of Pain Stimuli   The perception of pain stimuli, called **nociception** (NOH-sih-sep-shun), is accomplished differently. For example, the thalamus relays pain stimuli to multiple parts of the brain, including S1 and S2, where the sensory discriminative (location, intensity, and quality) aspects of the pain are perceived and analyzed. Such stimuli are also sent to the basal nuclei, structures of the limbic system, the hypothalamus, and the prefrontal cortex, where emotional, behavioral, and other aspects of pain are processed.

The processing of pain input differs from that for the other somatic senses in additional ways. For instance, the cortex seems to have a great deal of influence over how pain is perceived. This is evident in a phenomenon called the *placebo effect,* in which a patient is given a *placebo* (plah-SEE-boh)—a

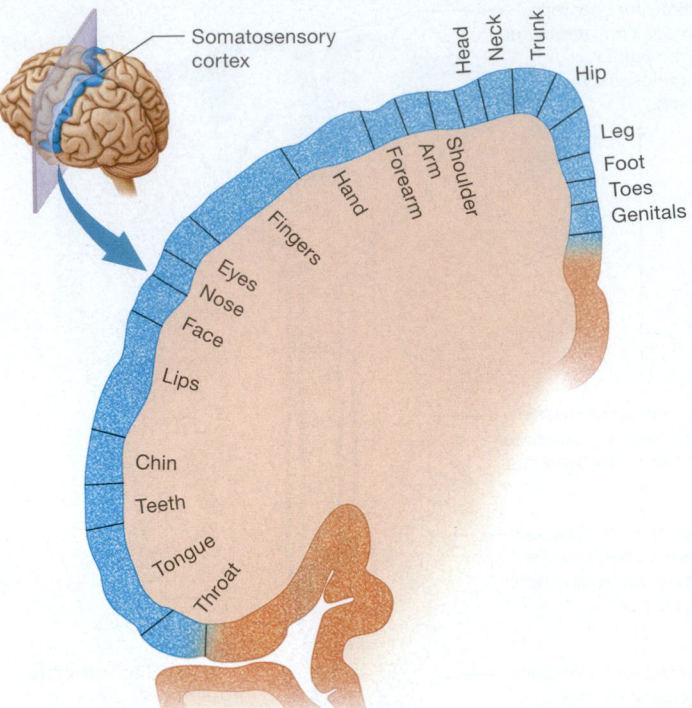

(a) Map of the primary somatosensory cortex (S1)

(b) The sensory homunculus

Figure 12.32 Representations of the primary somatosensory cortex.

"dummy treatment" with no actual therapeutic value. The use of a placebo as an *analgesic* (an-ahl-JEE-zik; pain reliever) can be surprisingly effective; in most studies, up to 30% of patients report relief from pain when a placebo is administered, but this number has been reported to be as high as 75%. This does not mean that the patient is "crazy" or that the pain is imagined; instead, it reflects the very real ability of the brain to modulate the perception of pain.

The explanation for the placebo effect appears to involve a descending pathway that originates largely in S1, the amygdala, and a region of the midbrain known as the *periaqueductal gray matter* (so named because it surrounds the cerebral aqueduct). The neurons of the periaqueductal gray matter release *endorphins,* neurotransmitters that render the neurons of the posterior horn in the spinal cord less sensitive to pain input, an example of the Cell-Cell Communication Core Principle (p. 28). The stimulus causing the pain (e.g., a broken bone) is still present at the same intensity, but the CNS neurons *perceive* it as being less intense or even absent. Painkillers such as morphine work by binding to the same receptors as endorphins. Researchers don't yet know why some people can modulate pain perception in this way more effectively than others. How pain is processed can have other unusual consequences, one of which we explore in *A&P in the Real World: Phantom Limb Pain.*

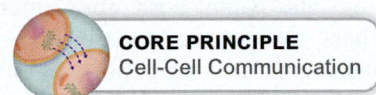

CORE PRINCIPLE
Cell-Cell Communication

Quick Check

☐ 1. Where are the posterior columns and their two tracts? What type of stimuli do they carry?

☐ 2. What is the anterolateral system? Which stimuli are carried via this system?

☐ 3. How are touch and pain processed by the cerebral cortex?

Introduction to the Special Senses

The *special senses* include vision, hearing (audition), taste (gustation), smell (olfaction), and balance (vestibular sensation). Like the general somatic senses, each special sense involves neurons that detect a stimulus and bring it to the CNS for processing and integration. The details of the special senses are

A&P in the Real World

Phantom Limb Pain

The role of the CNS in sensory perception becomes quite apparent in the phenomenon known as *phantom limb.* A phantom limb occurs after a patient has had a limb, a digit, or even a breast amputated; it is characterized by the perception that the body part is still present and functional even in the absence of sensory input. A small percentage of patients with phantom limbs also develop *phantom pain*—the sensation of burning, tingling, or even severe pain in the missing part. Phantom limb pain is extremely difficult to treat due to the complicated way in which the CNS processes pain.

The existence of phantom limbs and phantom pain supports the idea that S1 has a "map" of the body that exists independently of the PNS. The map is the reason why S1 continues to perceive the limb as present even when no stimuli from the limb are being sent to the brain. Over time, the map generally will rearrange itself so that the body is represented accurately, and the phantom limb and pain sensations decrease.

discussed in a later chapter (see Chapter 15); in this chapter, we introduce their basic pathways within the CNS.

We can note many similarities in the ways general somatic and special sensory stimuli are processed. As with the somatic senses, the cerebral cortex interprets the majority of special sensory stimuli, and the thalamus is the "gateway" for the entry of most special sensory stimuli into the cerebral cortex (olfaction is an exception). The usual pathway for processing each type of sensory stimulus is as follows:

- **Visual stimuli.** Most visual stimuli are delivered directly to the thalamus, which relays them to the primary visual cortex in the occipital lobe. The primary visual cortex processes these stimuli and perceives an object's depth and color, as well as detecting rapidly changing stimuli (which occurs when you are watching a TV program, for example). It also sends information to association areas in the temporal and parietal lobes, which are crucial for object recognition and spatial awareness, respectively.

- **Auditory stimuli.** Stimuli from sound waves are delivered first to nuclei in the brainstem. Some of these stimuli are processed in those nuclei; the remainder are routed to the thalamus and then to the primary auditory cortex in the superior temporal lobe. The neurons in this part of the cerebral cortex are specialized for processing sounds with complex tempos, such as speech and music. Information from the primary auditory cortex is relayed to many association areas, including Wernicke's area for language comprehension.

- **Gustatory stimuli.** Stimuli pertaining to taste are delivered to a nucleus in the medulla and relayed to the thalamus. From the thalamus, gustatory stimuli are sent to gustatory cortices in both the insula and the parietal lobes that analyze the different components of the taste. These cortices also communicate with the hypothalamus and limbic system, which presumably influence taste preferences and food-seeking behaviors.

- **Olfactory stimuli.** As stated earlier, olfactory stimuli actually enter the cerebral cortex of the limbic system for initial processing, without going through the thalamus. From the limbic lobe, these stimuli are relayed to several places, including the thalamus (which then transmits them to the prefrontal cortex), the hypothalamus, and other components of the limbic system. This feature allows olfactory stimuli to influence many behaviors, including those related to feeding, emotion, and cognition.

- **Vestibular stimuli.** The processing of vestibular stimuli involves multiple brainstem nuclei, the cerebellum, descending pathways through the spinal cord, and pathways through the thalamus to the cerebral cortex.

Quick Check

☐ 4. How is the processing of olfactory stimuli different from that of other special sensory stimuli?

Apply What You Learned

☐ 1. Would you expect the fingertips or the back to be represented by a greater area of the primary somatosensory cortex? Explain.

☐ 2. A person with a spinal cord injury is experiencing loss of pain sensation in his left leg and loss of tactile sensation in his right leg. On which side of the spinal cord is the injury located? Explain.

☐ 3. The term painkiller is used for drugs that bind to receptors for endorphins in the CNS. How do these drugs relieve pain? Do they actually treat the cause of the pain? (*Hint:* Where do they bind?)

See answers in Appendix A.

MODULE 12.8
Movement Part I: Role of the CNS in Voluntary Movement

Learning Outcomes

1. Describe the locations and functions of the upper and lower motor neurons in a motor pathway.

2. Explain the roles of the cerebral cortex, basal nuclei, and cerebellum in movement.

3. Describe the overall pathway from the decision to move to the execution and monitoring of a motor program.

4. Explain how decussation occurs in sensory and motor pathways, and predict how brain and spinal cord injuries affect these pathways.

We don't often stop to consider the complex neurological processes behind even the simplest movements, or the rapidity with which these occur. For example, we don't have to think about which muscles to activate in standing up or how to remain balanced as we walk across the room, because our nervous system does this automatically once a motor skill is learned. Planning and coordination of voluntary movement are carried out within the CNS, and involve the motor areas of the cerebral cortex, the basal nuclei, the cerebellum, and the spinal cord. Each of these components is required for motion to occur smoothly and rapidly, and if one is disrupted, the consequences can be dramatic.

Three types of neurons are directly involved in eliciting a muscle contraction. The first are the upper motor neurons, described in Module 12.2, whose cell bodies reside in the motor areas of the cerebral cortex (some upper motor neurons

are located in the brainstem, as well). Their axons descend through the cerebral white matter, and as they enter the brainstem and spinal cord, they interact with local interneurons. These interneurons pass on the messages of the upper motor neurons to their neighbors, the lower motor neurons. The cell bodies of these lower motor neurons lie in the anterior horn of the spinal gray matter, and their axons exit the spinal cord and innervate the skeletal muscles. Lower motor neurons are part of the PNS.

This module discusses the pathway from the upper motor neurons to the lower motor neurons leaving the spinal cord. Then we explore how movement is achieved, including the roles of the upper motor neurons and how the cerebellum and basal nuclei modulate their activity. In the PNS chapter, we trace the pathway from the lower motor neurons to the muscles (see Chapter 13).

Motor Pathways from the Brain through the Spinal Cord

Recall from Module 12.6 that the axons from the cortical motor areas unite to form several tracts of white matter. Among the largest tracts are the right and left **corticospinal tracts,** which, as noted, control the muscles below the head and neck via the lower motor neurons of spinal nerves, and the **corticonuclear tracts,** which control the muscles of the head and neck via the lower motor neurons of the cranial nerves. Let's trace the pathway taken by motor stimuli in each of these tracts.

Corticospinal Tracts

As you can see in **Figure 12.33**, the axons that form the corticospinal tracts originate from the cell bodies of upper motor neurons of the primary motor cortex and premotor cortex. These axons combine and descend first through the corona radiata and internal capsule until they reach the brainstem, where they go through the midbrain and pons. Recall from Module 12.6 that at the level of the medullary pyramids most of the fibers of the corticospinal tracts decussate. For this reason, the neurons on the right side of the brain control the left side of the body, and vice versa. This is why patients with damage to certain parts of the brain show loss of function of muscles on the *opposite* side of the body from their injury.

The fibers that decussate travel in the lateral funiculi of the spinal cord and are known as the right and left lateral corticospinal tracts. As shown in Figure 12.33, when the fibers of the lateral corticospinal tracts reach the anterior horn of spinal gray matter, they synapse on local interneurons. Note that over half of the corticospinal fibers terminate at the cervical levels of the spinal cord to control motor functions of the upper limbs. This location reflects the importance of precise motor control of our arms and hands.

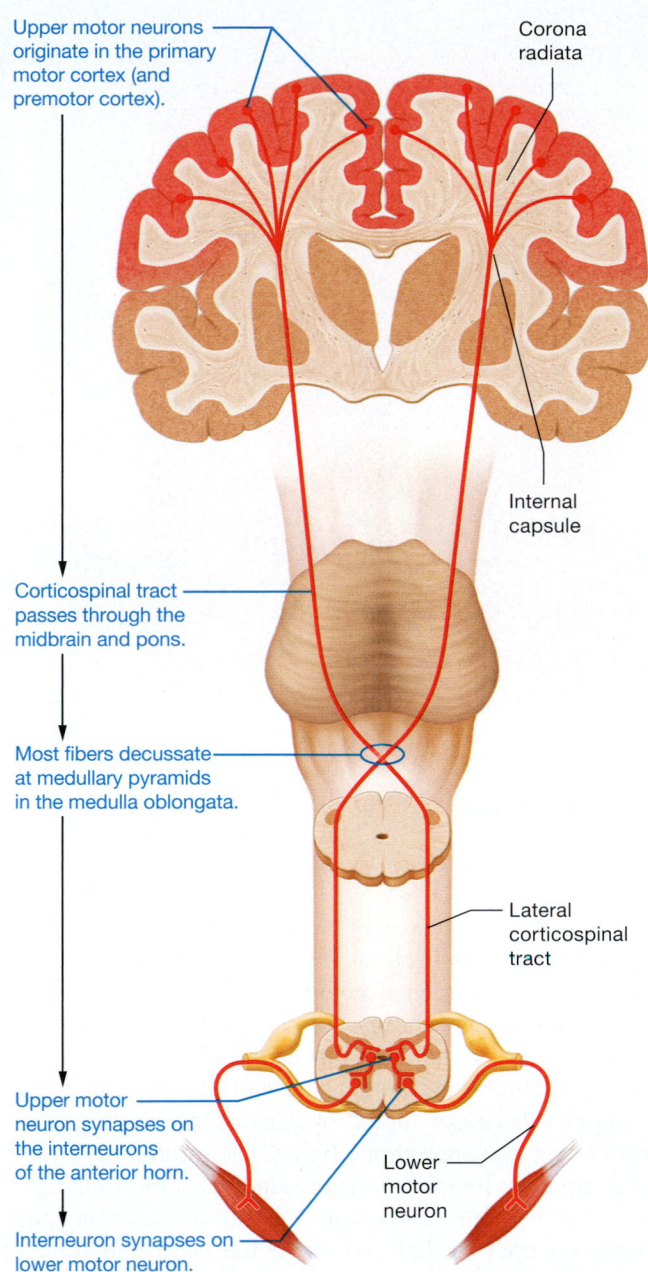

Upper motor neurons originate in the primary motor cortex (and premotor cortex).

Corona radiata

Internal capsule

Corticospinal tract passes through the midbrain and pons.

Most fibers decussate at medullary pyramids in the medulla oblongata.

Lateral corticospinal tract

Upper motor neuron synapses on the interneurons of the anterior horn.

Lower motor neuron

Interneuron synapses on lower motor neuron.

Figure 12.33 Descending (motor) pathways: the right and left lateral corticospinal tracts.

About 10–15% of the motor fibers from the cerebrum, however, do not decussate and so remain on the same side. These fibers travel through the anterior funiculi of the spinal white matter and are therefore called the right and left *anterior corticospinal tracts.*

Corticonuclear Tracts

The pathway for the corticonuclear tracts (formerly called the *corticobulbar tracts*) is similar to that for the corticospinal tracts. Like the corticospinal tracts, they originate from the cell bodies of upper motor neurons. The corticospinal and corticonuclear tracts travel together through the corona radiata and internal capsule to the brainstem. However, the fibers of the corticonuclear tracts do not decussate and do not enter the spinal cord. Rather, they synapse on interneurons that communicate with cranial nerve nuclei at various levels of the brainstem, whose lower motor neurons innervate muscles of the head and neck.

Quick Check

☐ 1. What is the main difference between the corticospinal and corticonuclear tracts?

☐ 2. Where do the fibers of the corticospinal tracts decussate?

Role of the Brain in Voluntary Movement

◀◀ FLASHBACK

1. What are multipolar neurons? What are the functions of their dendrites? (p. 387)

2. What happens at excitatory and inhibitory synapses? (p. 407)

The pathways just discussed tell only part of the story of how we achieve movement. Realistically, even a motion as simple as pushing open a door requires the simultaneous firing of a dizzying number of neurons as part of a selected group of actions called a **motor program.** The motor program for pushing open a door includes the muscles you will need to open it, the amount of force that will be required, the muscles that will be used to maintain your balance as you push it open, and more. Execution of this motor program requires the firing of neurons in motor association areas; the firing of upper motor neurons; and input from the basal nuclei, the cerebellum, the spinal cord, association areas such as the prefrontal cortex, and various sensory areas. In addition, the firing of lower motor neurons in the PNS is necessary. If simply pushing open a door demands this much activity, imagine what is required to execute the motor program for playing an instrument, painting a portrait, or catching a football.

Role of the Cerebral Cortex

The upper motor neurons of the primary motor cortex and the premotor and motor association areas plan and initiate voluntary movement by selecting an appropriate motor program. This coordinates the sequence of skilled movements, such as those used to ride a bicycle or send a text message. Upper motor neurons are also located in certain nuclei of the brainstem, where they work to maintain posture, balance, and body position, particularly during locomotion. These brainstem upper motor neurons also produce motor responses to sensory stimuli, such as turning your head and eyes toward the source of a loud noise.

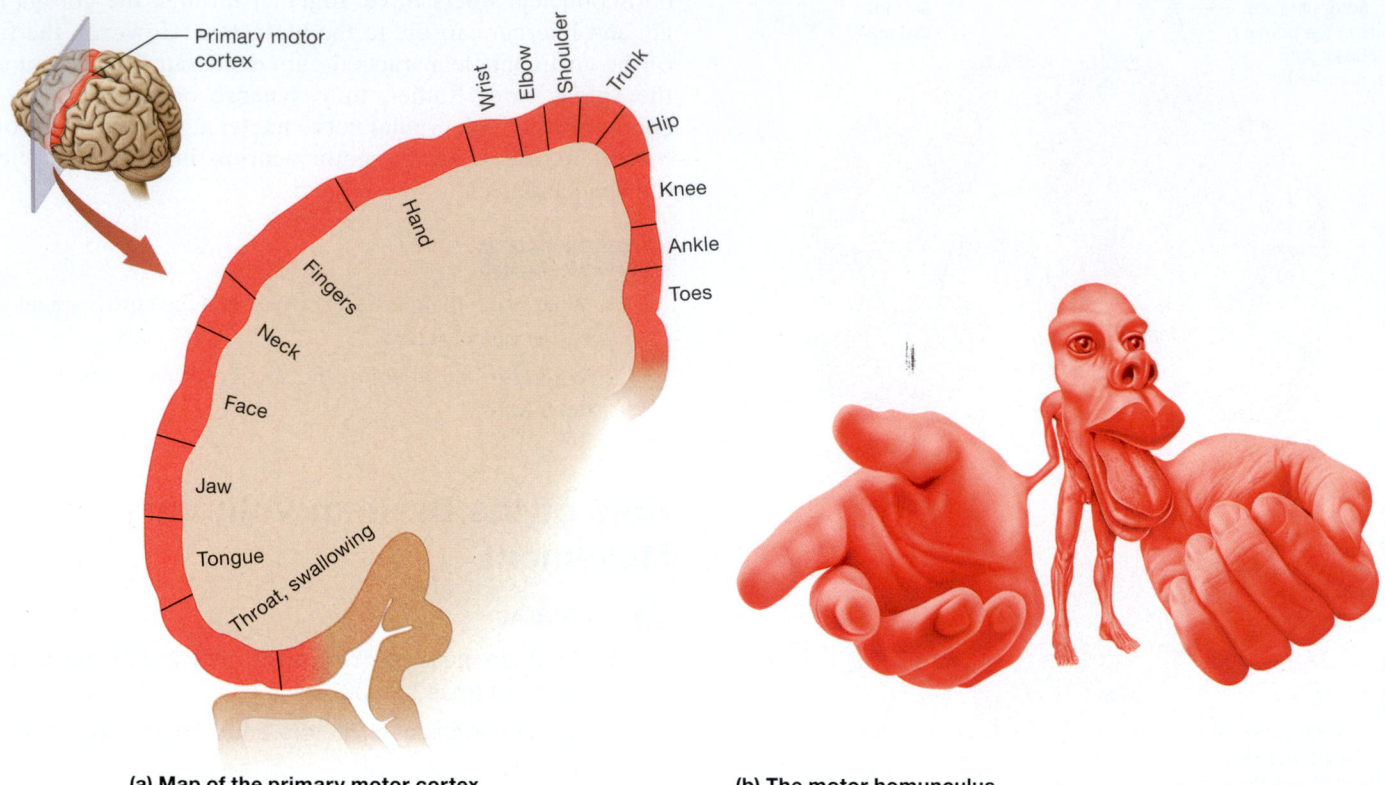

(a) Map of the primary motor cortex

(b) The motor homunculus

Figure 12.34 **Representations of the primary motor cortex.**

In **Figure 12.34** you can see that the upper motor neuron map of the primary motor cortex and the *motor homunculus* very much resemble the somatosensory map and sensory homunculus discussed earlier (see Figure 12.32). Like the primary somatosensory cortex, the primary motor cortex is organized somatotopically, and certain areas have disproportionately more cortical area devoted to them, especially the lips, tongue, and hands. This indicates the importance of vocalization and manual dexterity to human survival.

The upper motor neurons in the cortical motor areas don't act alone to deliver the commands for movement to the lower motor neurons. In fact, movement initiated by the cerebral cortex alone is jerky, uncoordinated, and difficult to control. Smooth, fluid motion requires input from the basal nuclei and the cerebellum.

Role of the Basal Nuclei

In Module 12.2 you read about the three collections of cell bodies that make up the basal nuclei: the caudate nucleus, the globus pallidus, and the putamen. The substantia nigra of the midbrain works very closely with the basal nuclei and is often grouped with them because of their shared functions. Unlike the upper motor neurons that modify the activity of lower motor neurons (via spinal interneurons), the basal nuclei modify the activity of upper motor neurons to produce voluntary movements and inhibit involuntary, unintentional ones.

Complicated interconnections form a circuit between the basal nuclei and other brain structures. Let's consider how this circuit functions in the absence of voluntary movement. As shown in **Figure 12.35a**, the neurons of the globus pallidus fire continuously to inhibit the motor nuclei of the thalamus. This, in turn, inhibits the upper motor neurons of the cortical motor areas from firing at inappropriate times. The importance of this function is obvious: Imagine you are standing in line at the grocery store. You certainly don't want the muscles of your limbs to suddenly contract so that you strike the people in front of or behind you, nor do you want to spontaneously begin barking like a dog. This part of the basal nuclei circuit helps prevent this; therefore, a critical function of the basal nuclei is to *inhibit inappropriate movements.*

Now consider how voluntary movement is begun by this same circuit (**Figure 12.35b**). Notice that the neurons of the caudate nucleus and the putamen receive input from the cerebral cortex, with which they form excitatory synapses. The caudate nucleus and putamen then send action potentials to the globus pallidus that inhibit its neurons from firing. This inhibitory effect is enhanced by the substantia nigra, which increases the output of the caudate nucleus and putamen. The overall effect is that input from these two nuclei inhibits the inhibitor (i.e., the globus pallidus). This removes the block on the thalamus, allowing its motor nuclei to stimulate the upper motor neurons of the cerebral cortex, which enables motion to begin. Thus, a second critical function of the basal nuclei is to *initiate voluntary motion.*

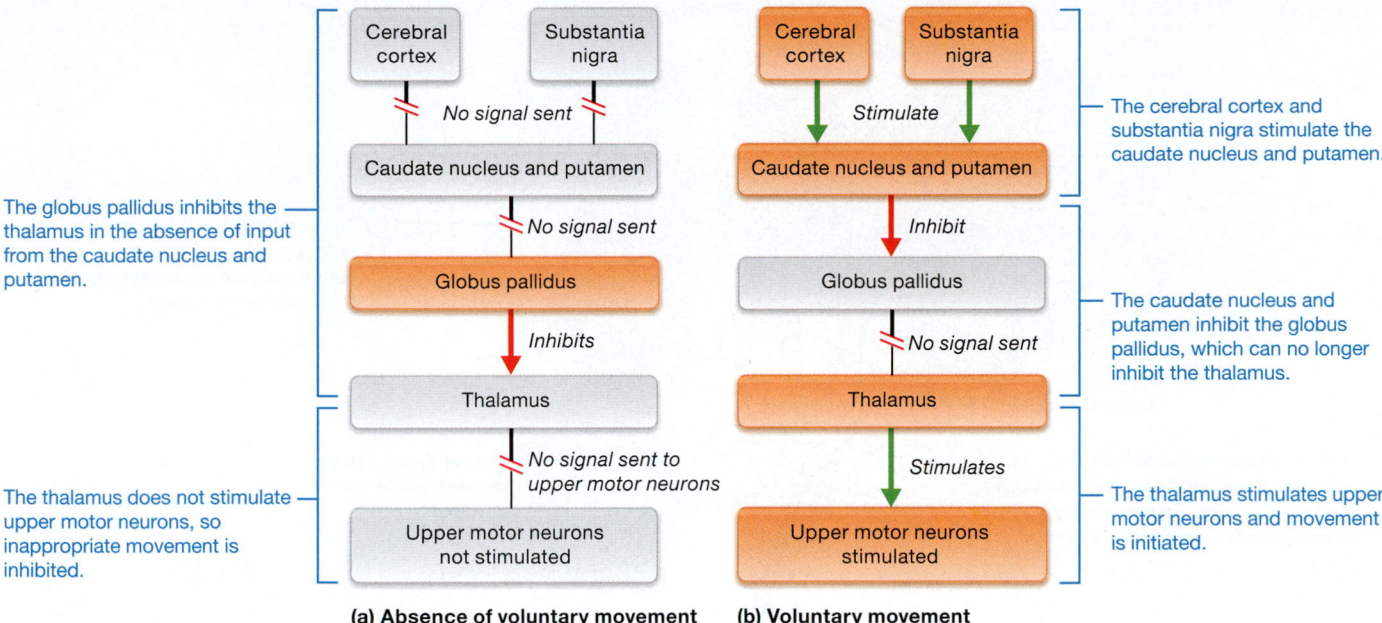

Figure 12.35 Role of the basal nuclei in voluntary movement.

The final player in the basal nuclei drama is the substantia nigra. The neurons of the substantia nigra are *dopaminergic neurons* because they release the neurotransmitter *dopamine.* These dopaminergic neurons project to the caudate nucleus and putamen, and when dopamine binds their receptors, their inhibitory output to the globus pallidus is enhanced. The overall effect is an increase in the eventual activation of the upper motor neurons that will stimulate the desired motor activity. For this reason, the input of the substantia nigra is vital for initiating movement.

Damage to any component of the basal nuclei system results in a **movement disorder.** There are a vast number of movement disorders, but they come in two main forms. The first causes an inability to initiate voluntary movement, making simple activities such as walking or talking difficult. The second causes an inability to inhibit inappropriate, involuntary movements. Some such movements are mild and may take the form of throat clearing or blinking; however, others may be severe enough to cause disability. One of the more common movement disorders, Parkinson's disease, is discussed in *A&P in the Real World: Parkinson's Disease.*

Parkinson's Disease

One of the most common movement disorders is *Parkinson's disease.* This disease is called a *hypokinetic* ("low movement") movement disorder, which means that movement is difficult to initiate and, once started, difficult to terminate. This leads to characteristic symptoms such as minimal facial expression, a shuffling gait with no arm swing, a resting tremor (uncontrollable shaking movements when at rest), difficulty in swallowing, and rigid movements of the arms and legs.

Degeneration of dopamine-secreting neurons of the substantia nigra causes Parkinson's disease. The degeneration tends to progress slowly over a 10- to 20-year period. Its underlying cause is unknown, although genetic mechanisms are suspected in approximately 10% of cases. Parkinson's disease is generally treated by medications that increase the level of dopamine in the brain to compensate for the decreased level.

Role of the Cerebellum

The cerebellum monitors *ongoing* movement and integrates information about the contraction and relaxation of the muscles; the position of the joints; and the direction, force, and type of movement that is going to occur next. Once this information has been integrated, the cerebellum then determines the *motor error*—the difference between the intended movement and the actual movement that is taking place. It then influences different parts of the brain to minimize this error.

Interestingly, the correction of motor error can occur over both the short and the long term by the process of **motor learning.** In essence, the corrections for motor error are added to the motor program over time. The more you perform a certain action, the more corrections for motor error that are added to the program, and the more fluid and error-free the motion becomes. In this case, practice really does make perfect, or at least mostly error-free.

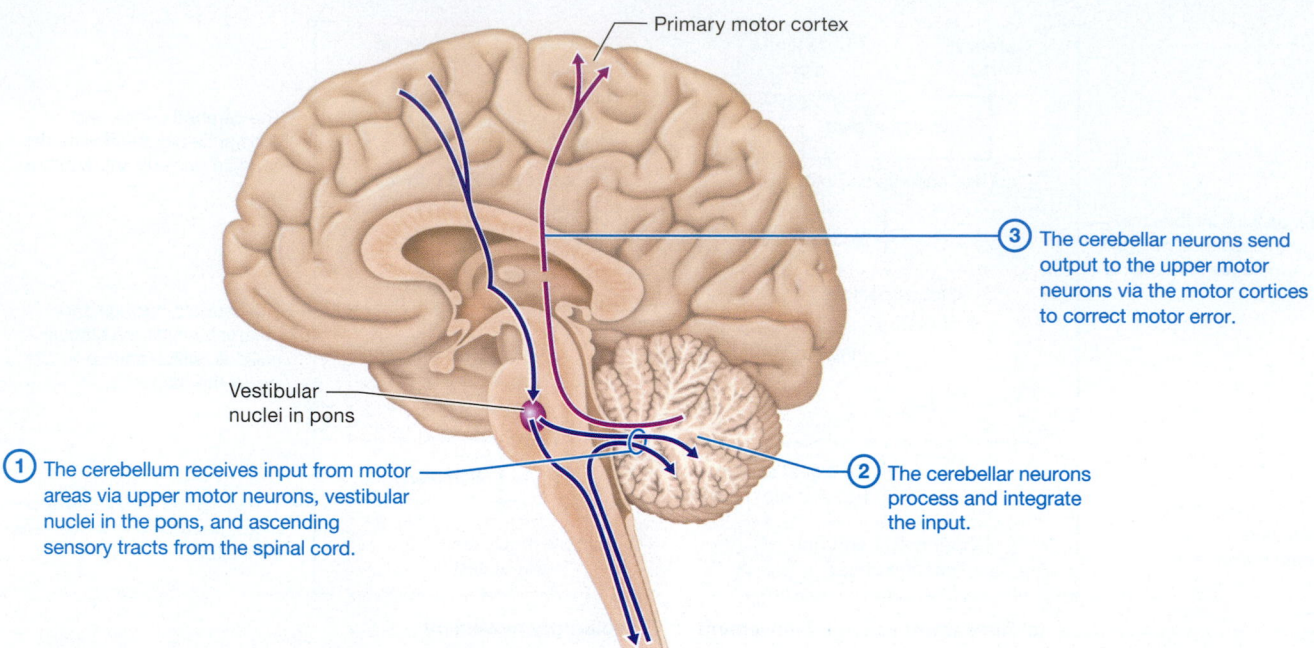

Figure 12.36 Role of the cerebellum in voluntary movement.

As step ① in **Figure 12.36** shows, the cerebellum receives input from three sources simultaneously:

- motor areas of the cerebral cortex via upper motor neurons,
- vestibular (balance) nuclei of the pons, and
- ascending sensory tracts from the spinal cord.

The motor areas provide information on intended movement, and the vestibular nuclei and ascending tracts supply data on the actual movement performed. Note that in step ②, cerebellar neurons process and integrate the input from these sources. Then, in step ③, cerebellar neurons send output to correct motor error, primarily to the upper motor neurons via the premotor cortex and the primary motor cortex. Like the basal nuclei, the cerebellum affects movement by modifying the activity of upper motor neurons; the cerebellum does not have direct connections with lower motor neurons.

As you would expect, damage to the cerebellum makes fluid, well-coordinated movement nearly impossible—instead, movements are jerky and inaccurate. This condition, called **cerebellar ataxia** (ah-TAKS-ee-ah; *a-* = "without," *taxis* = "order"), typically features a wide, staggering gait; difficulty with rapid, alternating motions; overreaching or underreaching for an object; and/or an *intention tremor* (shaking that occurs when attempting to move a body part with accuracy or precision, such as reaching for a glass). Keep in mind that intention tremors may also be present in individuals without a history of cerebellar injury. In such cases, it's called *essential tremor,* and it appears to result from degeneration of certain cerebellar neurons. Although the precise cause of essential tremor is unknown, a genetic link appears to be present.

Quick Check

☐ 3. Where do upper motor neurons reside, and what are their functions?

☐ 4. What are the two parts of the basal nuclei circuit?

☐ 5. What is the overall function of the cerebellum?

Putting It All Together: The Big Picture of CNS Control of Voluntary Movement

As you've seen, multiple parts of the CNS work together to execute and monitor a motor program. The big picture view of this role and all the different components is shown in **Figure 12.37**.

Quick Check

☐ 6. Trace the overall voluntary movement pathway from deciding to move to monitoring of the ongoing movement.

Apply What You Learned

☐ 1. Predict the effects of degeneration of the globus pallidus.

☐ 2. How would degeneration of the globus pallidus differ from degeneration of the caudate nucleus, putamen, and substantia nigra? Explain.

☐ 3. A potential effect of chronic alcohol abuse is the development of cerebellar damage. Predict the symptoms you would likely see in a person with such a condition.

☐ 4. Ms. Nazari presents to the emergency department with loss of muscle function on the right side of her body. The physician suspects she has suffered a stroke (a blood clot in the brain). On which side of her brain did the stroke likely occur? Explain.

See answers in Appendix A.

The Big Picture of CNS Control of Voluntary Movement

Figure 12.37

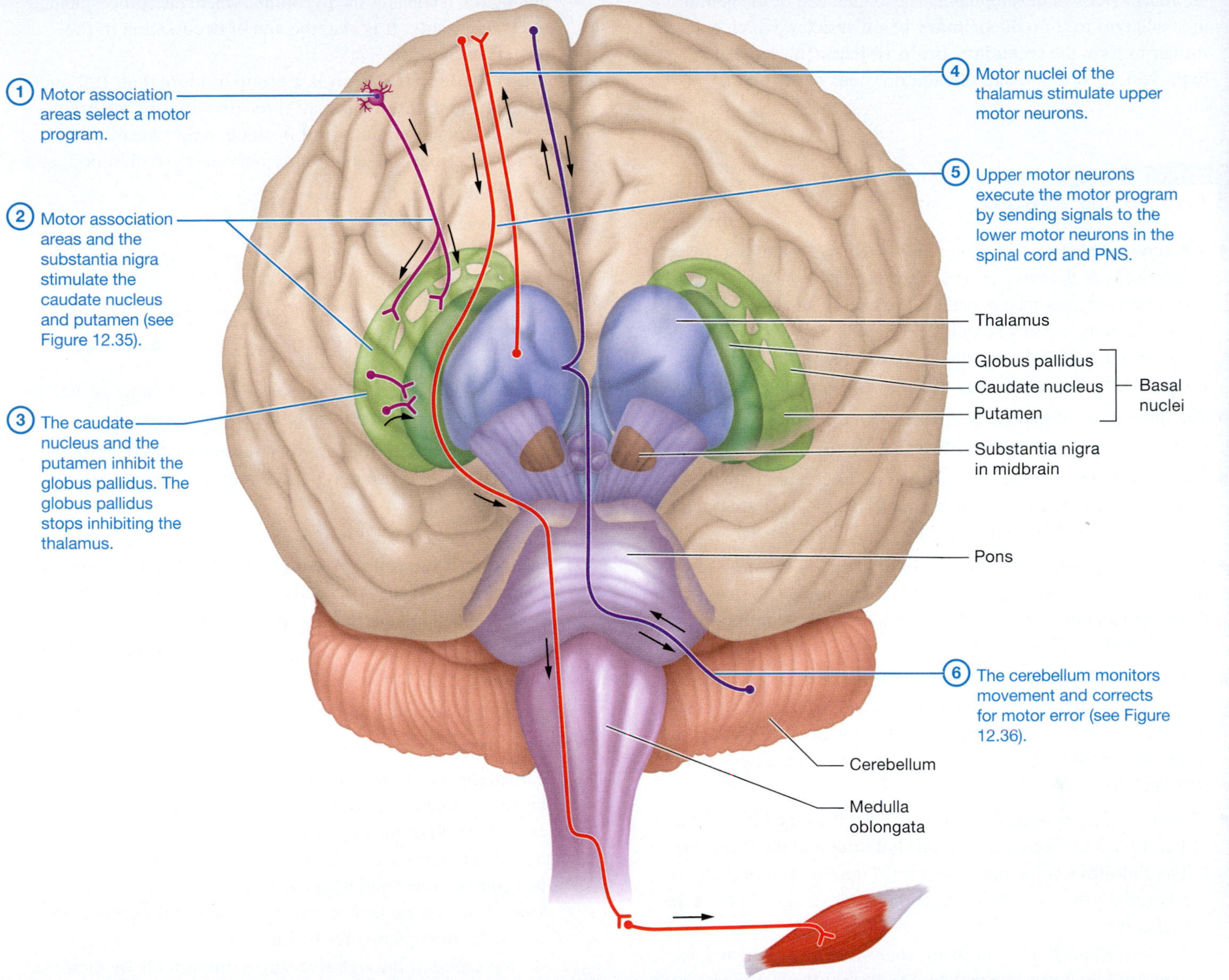

1 Motor association areas select a motor program.
2 Motor association areas and the substantia nigra stimulate the caudate nucleus and putamen (see Figure 12.35).
3 The caudate nucleus and the putamen inhibit the globus pallidus. The globus pallidus stops inhibiting the thalamus.
4 Motor nuclei of the thalamus stimulate upper motor neurons.
5 Upper motor neurons execute the motor program by sending signals to the lower motor neurons in the spinal cord and PNS.
Thalamus
Globus pallidus
Caudate nucleus
Putamen
Basal nuclei
Substantia nigra in midbrain
Pons
6 The cerebellum monitors movement and corrects for motor error (see Figure 12.36).
Cerebellum
Medulla oblongata

Chapter Summary

For everything you need to succeed in this course, go to Mastering Anatomy & Physiology. There you will find:

- Practice Tests
- Author Podcasts
- Big Picture Animations
- Concept Boost Video Tutors
- iP2™ Interactive Physiology 2.0 Tutorials
- A&P Flix 3D Animations
- Active-Learning Workbook

MODULE 12.1

Overview of the Central Nervous System 423–426

- The **central nervous system** performs the integrative nervous system functions.
- The **brain** is an organ that consists mostly of nervous tissue. It is housed within the cranial cavity.
- The four divisions of the brain are the **cerebrum,** the **diencephalon,** the **cerebellum,** and the **brainstem.**
- The **spinal cord** is nervous tissue that forms an extension of the inferior brainstem.
- The CNS contains both gray matter (nuclei) and white matter (myelinated axons).

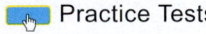

- The CNS develops from tissue that folds into the **neural tube** by the fourth week of development. The cranial end of the neural tube enlarges to form the **primary brain vesicles,** which enlarge further to form the **secondary brain vesicles.** The secondary brain vesicles develop into the four divisions of the brain.

- The **cerebral cortex** is the cerebrum's outer covering of gray matter. The majority of the cerebral cortex is **neocortex,** which functions in conscious processes such as planning of movement and interpretation of incoming sensory stimuli. The cerebral cortex has five lobes: the anterior **frontal lobe,** the middle **parietal lobe,** the posterior **occipital lobe,** the lateral **temporal lobe,** and the deep **insula.**

- The three types of neocortex are the **primary motor cortex,** the *primary sensory cortices,* and **association areas.**

- The **basal nuclei** are a group of nuclei in the cerebrum that function in movement. The three basal nuclei are the **caudate nucleus,** the **putamen,** and the **globus pallidus.**

- The cerebral white matter may be of three types: **Commissural fibers** connect areas of the right and left hemispheres, **projection fibers** connect the outer cerebrum with other parts of the CNS, and **association fibers** connect gyri.

- Structures of the **limbic system** include the gyri of the *limbic lobe,* the **hippocampus,** the **amygdala,** and the tracts that connect these structures with one another and with other parts of the brain. The limbic system functions in memory, learning, emotion, and behavior.

- The diencephalon is the central core of the brain and is composed of four divisions, including the **subthalamus** and the following:
  ○ The **thalamus** is the main "gateway" into the cerebral cortex, controlling the flow of information both to and from the cerebral cortex.
  ○ The **hypothalamus** controls functions relating to the maintenance of homeostasis; regulates the sleep/wake cycle; regulates the autonomic nervous system; functions in the endocrine system; and plays a role in emotion, behavior, and memory.
  ○ The *epithalamus* contains the **pineal gland,** which secretes the hormone **melatonin.**

- The cerebellum plays a role in the coordination of movement. It is located inferior to the occipital lobe and consists of three lobes and two hemispheres, which are connected by the **vermis.** It has an outer layer of gray matter, the *cerebellar cortex,* and inner white matter, the **arbor vitae.**

- The brainstem controls many basic involuntary processes as well as numerous reflexes, and participates in movement, sensation, and complex functions such as sleep. It houses many important nuclei and tracts of white matter.
  ○ The **midbrain** is the superior portion of the brainstem. It functions in movement, sensation, and reflexes such as the startle reflex.
  ○ The **pons** is the middle portion of the brainstem. The pontine nuclei play a role in movement, breathing, reflexes, and complex functions such as sleep and arousal.

○ The inferiormost portion of the brainstem is the **medulla oblongata.** It contains the **pyramids,** where the corticospinal tracts **decussate.** It is also the site of decussation of the **posterior columns.**

○ The **reticular formation** is a group of more than 100 nuclei that are found throughout all three divisions of the brainstem. Its nuclei are involved in sleep, pain transmission, motor functions, the autonomic nervous system, mood, and sensation.

- The two systems in the body dedicated to maintaining homeostasis are the endocrine system and the nervous system.

- The two structures of the nervous system most concerned with homeostasis are the hypothalamus and the reticular formation.

- The homeostasis of vital functions is largely maintained by the **autonomic nervous system (ANS).** The hypothalamus is the main "boss" of the ANS and adjusts its output by relaying instructions to the nuclei of the reticular formation.

- Respiration is regulated by neurons in the medullary reticular formation.

- The hypothalamus regulates body temperature and feeding behaviors.

- **Sleep** is the normal and reversible suspension of consciousness.
  ○ Human sleep follows a **circadian rhythm** set by the hypothalamus. When light levels decrease, the *suprachiasmatic nucleus (SCN)* of the hypothalamus decreases the activity of the reticular formation, which "disconnects" the thalamus from the cerebral cortex and triggers sleep.
  ○ Arousal from sleep is mediated by *orexins* that stimulate the neurons of the reticular formation.
  ○ Sleep progresses through five stages, discernible by their patterns of brain activity measured on an **electroencephalogram (EEG).**
  ○ Stages I, II, and III are light to moderately deep sleep stages and stage IV is deep sleep.

- **Rapid eye movement,** or **REM, sleep** is characterized by rapid movements of the eyes.

- **Cognition** consists of processes that include social and moral behavior, intelligence, thoughts, language, problem-solving skills, and personality. These functions are carried out by the association areas, and many are lateralized to one hemisphere.

- Most aspects of **language** are lateralized to the left cerebral hemisphere. **Broca's area** and **Wernicke's area** control the ability to produce and understand language, respectively.

- **Declarative (fact) memory** is memory of things readily available to consciousness; **nondeclarative memory** includes skills and associations that are largely subconscious.
  - Immediate memory is stored for only seconds, short-term (working) memory is stored for several minutes, and long-term memory is stored for days or longer.
  - Moving the information in immediate and working memory into long-term memory is called **consolidation.**
  - Declarative memories may be formed by **long-term potentiation (LTP)** that occurs within the hippocampus. Nondeclarative memories are formed by systems that do not rely on the hippocampus.
- Emotion involves responses mediated by the hypothalamus and the limbic system, particularly the amygdala.

MODULE 12.5
Protection of the Brain 449–454

- The three cranial meninges include the thick outer **dura mater,** the middle **arachnoid mater,** and the thin inner **pia mater.** Between the dura and the arachnoid is the **subdural space;** between the arachnoid and the pia is the **subarachnoid space.**
- The cavities within the brain are the **ventricles.** The ventricles contain **cerebrospinal fluid (CSF)** and are continuous with the central canal of the spinal cord.
- Cerebrospinal fluid increases the buoyancy of the brain, cushions it, and maintains a constant temperature. It is formed in the ventricles by a cluster of capillaries and ependymal cells called the **choroid plexus,** and is reabsorbed into the blood through the *arachnoid granulations.*
- The **blood brain barrier** is a barrier between the blood in the capillaries of the brain and the extracellular fluid of nervous tissue.

MODULE 12.6
The Spinal Cord 455–459

- The spinal cord has two main roles: It acts as a relay station to connect the brain with the rest of the body, and it carries out some processing and integration.
- The spinal cord is protected by its **spinal meninges.** There is a fat-filled **epidural space** between the dura and the vertebral periosteum.
- The spinal cord begins at the foramen magnum and terminates as the **conus medullaris** between the first and second lumbar vertebrae.
- **Spinal nerves** project off each side of the spinal cord. The terminal bundle of spinal nerves is the **cauda equina.**
- Spinal gray matter is surrounded by an outer sheath of spinal white matter. The spinal gray matter is divided into three regions known as horns:
  - The **anterior horn** houses the cell bodies of lower motor neurons.

- The **posterior horn** contains the cell bodies of neurons that process and transmit sensory stimuli.
- The **lateral horn** contains cell bodies of the ANS.
- There are three regions of white matter called funiculi: the **posterior funiculus,** the **lateral funiculus,** and the **anterior funiculus.** Each funiculus contains ascending and/or descending tracts.

MODULE 12.7
Sensation Part I: Role of the CNS in Sensation 459–464

- Stimuli are processed in two steps: (1) The stimulus is detected by neurons of the PNS, and (2) the stimulus is interpreted by the CNS.
- The *general somatic senses* include the **tactile senses, nondiscriminative touch,** pain, stretch, proprioception (joint position), and temperature. Stimuli relating to these senses are delivered to the brain via two major pathways:
  - The **posterior columns/medial lemniscal pathway** carries stimuli concerning the tactile senses and proprioception. These tracts decussate in the brainstem. Sensory information is relayed by the thalamus to the primary somatosensory cortex and cortical association areas for interpretation.
  - The **anterolateral system** carries pain, temperature, and nondiscriminative touch sensations. The stimuli ascend through a number of tracts, including the **spinothalamic tracts,** which decussate in the spinal cord, with the thalamus as the ultimate destination. The thalamus relays temperature and nondiscriminative touch stimuli to the primary somatosensory cortex, as well as pain stimuli to many areas of the brain.
- Each part of the body is represented by a specific region of the primary somatosensory cortex.
- Visual, auditory, and gustatory stimuli are relayed to the thalamus and then to a primary sensory cortex for interpretation.
- Olfactory stimuli are first processed by the cortex of the limbic system, after which they are relayed to the thalamus, hypothalamus, and other components of the limbic system.
- Vestibular stimuli are processed by multiple brainstem nuclei, the cerebellum, and areas within the cerebral cortex.

MODULE 12.8
Movement Part I: Role of the CNS in Voluntary Movement 464–469

- **Upper motor neurons** reside mostly in the motor areas of the cerebral cortex. They stimulate local interneurons in the brainstem or spinal cord. The interneurons pass the message to *lower motor neurons* of the PNS, which innervate muscle fibers.
- Axons from the cortical motor areas unite to form the **corticospinal tracts,** which control the muscles below the head and neck,

and the **corticonuclear tracts,** which control muscles of the head and neck.

○ The corticospinal tracts descend through the cerebral white matter into the brainstem, where the tracts decussate in the medullary pyramids. Most of the fibers then travel through the lateral funiculus of the spinal cord as the lateral corticospinal tracts.

○ The corticonuclear tracts are similar to the corticospinal tracts except their axons terminate in the brainstem.

- Every motion requires the selection of a **motor program** by the upper motor neurons of the primary motor cortex in the cerebral cortex, which plan and execute voluntary motion.

- Each part of the body is represented by a specific region of the primary motor cortex.

- The basal nuclei and substantia nigra inhibit inappropriate movements and initiate motion.

- The cerebellum regulates ongoing movement and modifies the activity of the upper motor neurons to reduce *motor error*. The correction of motor error can occur over the long term as the process of **motor learning.**

Assess What You Learned

Scan the QR Code for additional practice test questions

https://goo.gl/XlK2Z5

LEVEL 1 Check Your Recall

1. The central nervous system is responsible for:

a. integrative functions.
b. sensory functions.
c. motor functions.
d. Both b and c are correct.

2. Mark the following statements about the brain as true or false. If a statement is false, correct it to make a true statement.

a. Humans use only 10% of their brains.
b. The four main components of the brain are the cerebrum, the diencephalon, the cerebellum, and the brainstem.
c. The right and left lateral ventricles are the largest of the ventricles in the brain and are located in the diencephalon.
d. The cerebrum is responsible for our basic, involuntary functions and reflexes.

3. Which of the following is *not* one of the basal nuclei?

a. Caudate nucleus
b. Globus pallidus
c. Cerebellum
d. Putamen

4. Which statement about cerebral white matter is *false*?

a. Commissural fibers connect the right and left cerebral hemispheres.
b. Projection fibers connect the cerebral cortex of one hemisphere with structures in the other hemisphere.
c. The corpus callosum is the largest bundle of white matter in the brain.
d. Association fibers connect the gyri of the cerebral cortex with one another.

5. Mark the following statements about the cerebral cortex as true or false. If a statement is false, correct it to make a true statement.

a. The neocortex is the "newest" component of the cerebral cortex from an evolutionary perspective.
b. The cerebral cortex is composed of white matter.
c. The primary visual cortex is located in the occipital lobe.
d. The prefrontal cortex is located in the frontal lobe and is concerned with movement.

6. The central sulcus separates the:

a. parietal and temporal lobes.
b. parietal and occipital lobes.
c. frontal and temporal lobes.
d. frontal and parietal lobes.

7. Match the term on the left with its correct description in the column on the right.

_____ Amygdala
_____ Thalamus
_____ Hippocampus
_____ Midbrain
_____ Medulla oblongata
_____ Hypothalamus
_____ Cerebellum
_____ Pons

a. Serves as the main "entryway" into the cerebral cortex
b. Contains the pyramids where the corticospinal tracts decussate
c. Portion of the limbic system involved in emotion
d. Regulates homeostasis, the autonomic nervous system, the endocrine system, and the sleep/wake cycle
e. Consists of three lobes and coordinates movement
f. Portion of the limbic system; involved in memory
g. Middle portion of brainstem; plays a role in movement, sleep, and arousal
h. First component of the brainstem; participates in reflexes, sensation, and movement

8. Which statement about the cranial meninges is *true*?

a. The subdural space is between the dura and the arachnoid mater.
b. The arachnoid mater closely follows the contours of the cerebral gyri and sulci.
c. The pia mater is the tough outer meninx.
d. The subarachnoid space is between the dura and the arachnoid mater.

9. What structure forms cerebrospinal fluid? How does this occur?

10. Explain the importance of the blood brain barrier, and list the structures that form it.

11. Mark the following statements about the spinal cord as true or false. If a statement is false, correct it to make a true statement.

    a. The spinal cord functions as a relay and processing station.

    b. The spinal cord extends from the foramen magnum to the level between the first and second lumbar vertebrae.

    c. The epidural space around the spinal cord is only a potential space.

    d. The posterior horn of spinal gray matter contains the cell bodies of motor neurons.

    e. The corticospinal tracts are the main sensory tracts in the spinal cord.

12. Fill in the blanks: The tracts of the posterior columns decussate in the _____, whereas the tracts of the anterolateral system decussate in the _____.

13. Which parts of the body have the greatest amount of space dedicated to them in the primary somatosensory cortex? Why?

14. Which of the following statements is *false*?

    a. The spinothalamic tracts are part of the anterolateral system.

    b. Pain, temperature, and crude touch stimuli are carried by the anterolateral system.

    c. Descending pathways from the brain and spinal cord can make the spinal cord less receptive to pain stimuli.

    d. The thalamus serves as the "gateway" for entry of all special sensory stimuli into the cerebral cortex, with the exception of audition (hearing).

15. Fill in the blanks: The cell bodies of upper motor neurons reside in the _____ and function to _____, whereas the cell bodies of lower motor neurons reside in the _____ and function to _____.

16. Label the following components of the corticospinal tracts with numbers 1 through 6, with 1 being the origin of the tracts and 6 their destination.

    _____Medullary pyramids where most fibers decussate
    _____Anterior horn of the spinal gray matter
    _____Midbrain and pons
    _____Upper motor neurons in the primary motor and premotor cortices
    _____Corona radiata and internal capsule
    _____Lateral funiculus of the spinal cord

17. Mark the following statements on the role of the brain in movement as true or false. If a statement is false, correct it to make a true statement.

    a. The dopaminergic neurons of the substantia nigra enhance the actions of the caudate nucleus and putamen.

    b. The cerebellum monitors the initiation of movement but does not monitor ongoing movements.

    c. The basal nuclei inhibit inappropriate movements and are required for the initiation of movement.

    d. The correction of motor error by the cerebellum can occur over the long term by motor learning.

18. Which of the following somatic sensations is *not* carried by the posterior columns?

    a. Tactile senses   c. Proprioception
    b. Temperature   d. Stretch

19. Fill in the blanks: The two components of the CNS that are responsible for maintenance of homeostasis are the _____ and the _____.

20. Which of the following statements is *false*?

    a. The suprachiasmatic nucleus is often called the body's "master clock."

    b. Melatonin and the ventrolateral preoptic nucleus increase the activity of the reticular formation.

    c. A decrease in the activity of the reticular formation disconnects the thalamus from the cerebral cortex and decreases the level of consciousness.

    d. Orexins are neurotransmitters that stimulate the reticular formation.

21. Match the term on the left with its correct description from the column on the right.

    _____REM sleep
    _____Stage I sleep
    _____Delta waves
    _____Beta waves
    _____Theta waves
    _____Stage IV sleep

    a. The EEG pattern seen in the deepest stage of sleep
    b. The deepest stage of sleep
    c. The EEG pattern seen in an alert, awake adult
    d. Drowsiness or the lightest stage of sleep
    e. The EEG pattern seen in lighter stages of sleep
    f. The stage of sleep during which most dreaming takes place

22. The part of the brain responsible for the production of language is known as:

    a. Wernicke's area.
    b. the prefrontal cortex.
    c. the auditory association area.
    d. Broca's area.

23. Fill in the blanks: Declarative memories are formed by the process of _____ that takes place in the _____, and are stored in the _____.

24. Which of the following is *not* a component of emotion?

    a. Visceral motor responses mediated by the hypothalamus

    b. Somatic motor responses mediated by the limbic system and the hypothalamus

    c. Somatic motor responses mediated by the upper motor neurons of the corticospinal tracts

    d. Subjective feelings mediated by the amygdala and the cerebral cortex

LEVEL 2 Check Your Understanding

1. Huntington's disease is characterized by a loss of normal inhibition mediated by the basal nuclei. Predict the symptoms you are likely to see with this disease.

2. How could you tell the difference between an injury that damaged the cerebellum and one that damaged the basal nuclei?

3. Why do injuries to the hippocampus interfere with the formation of new declarative memories but do not impact memories already encoded in long-term memory? Why do such injuries not affect the formation of nondeclarative memories?

LEVEL 3 Apply Your Knowledge

PART A: Application and Analysis

1. Ms. Norris is brought to the emergency department with injuries to the posterior of her head resulting from a motor vehicle accident. Which lobes of the cerebrum likely sustained injuries? Predict the main signs and symptoms you will see resulting from the injury to these cerebral lobes.

2. On further examination, Ms. Norris is shown to be in a state of unarousable unconsciousness, or coma. Which area or system of the brain is likely damaged? Explain.

3. In a diving accident, Arlene damages the left side of her upper thoracic spinal cord. On which side of the body will she lose: muscle control, tactile sensation, and pain sensation? Explain.

4. A new diet wonder drug is designed to block the release of orexins. How might this cause weight loss? Predict the potential adverse effects that might come from blocking orexin release.

PART B: Make the Connection

5. An athlete engaging in strenuous activity sometimes experiences a "runner's high," which is characterized by release of endorphins from the midbrain periaqueductal gray matter. What effect is this likely to have on the athlete's perception of pain and ability to resist muscle fatigue? Will the release of endorphins stop the source of the athlete's pain? (*Connects to Chapter 11*)

See answers in Appendix A.

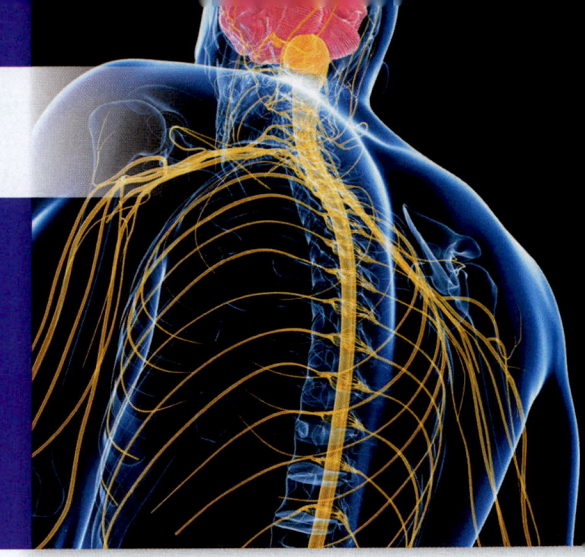

The Peripheral Nervous System

Has anyone ever snuck up right behind you and yelled your name? You jumped a foot in the air, right? Well, that was an automatic response that depended on your nervous system receiving input and responding to it, both functions that rely on your peripheral nervous system. In the previous chapter, we explored the *central nervous system* (*CNS*) and examined its role in movement, sensory perception, and the maintenance of homeostasis (see Chapter 12). Now we turn to the second part of the nervous system story: the *peripheral nervous system,* or *PNS.* Although we discuss the PNS and CNS in three separate chapters (continuing in the autonomic nervous system chapter—see Chapter 14), remember that the central and peripheral systems are continuous with one another anatomically and functionally. In fact, the cell bodies of many neurons in the PNS actually reside in the CNS, and the two systems cannot function independently of each other.

In this chapter, we examine the anatomy of the PNS, as well as its role in sensation and movement. A "big picture" view then ties together the roles of the CNS and PNS, showing how the nervous system as a whole controls these critical life functions. Finally, we wrap up the chapter with a discussion of the processes that cause the automatic movements called *reflexes.*

MODULE **13.1**

Overview of the Peripheral Nervous System

Learning Outcomes

1. Explain the differences between the sensory and motor divisions of the peripheral nervous system.
2. Differentiate between the somatic motor and visceral motor (autonomic) divisions of the nervous system.
3. Describe the structure of a peripheral nerve, and explain the differences between spinal nerves and cranial nerves.

The **peripheral nervous system (PNS)** links the CNS with the body and the external environment. The PNS does this by first detecting sensory stimuli and delivering

*For practice applying concepts to a clinical scenario, check out the **Running Case Study** for this chapter @ Mastering Anatomy & Physiology*

Computer-generated image: Thirty-one pairs of spinal nerves carry sensory and motor signals throughout the body.

them to the CNS as sensory input. The CNS processes the input and then transmits the impulses through the PNS to *effectors*—muscle cells and glands—for motor output. This module explains the basics of the PNS, including its functional divisions and the structure of its main organs, the *peripheral nerves*. It concludes with an overview of how the PNS functions and how those activities are integrated with the CNS.

Divisions of the PNS

❮❮ FLASHBACK

1. How do motor (efferent) and sensory (afferent) neurons differ in structure and function? (p. 382)

2. What is a ganglion? (p. 387)

Recall that the PNS is classified functionally into two main divisions: the *sensory* (or *afferent*) *division* and the *motor* (or

efferent) *division* (see Chapter 11). Each division is further categorized anatomically—by the types of structures that are innervated—into *somatic* (body) and *visceral* (internal organ) branches (**Figure 13.1**).

Sensory Division

As its name implies, the **sensory division** of the PNS consists of sensory (or afferent) neurons that detect various sensory stimuli and bring them to the CNS (see the left side of Figure 13.1). The sensory division has two anatomical subdivisions:

- The **somatic sensory division** contains neurons that detect sensory stimuli from the skin and structures of the musculoskeletal system. Its neurons respond to stimuli of the *general senses* that arise external to the body, such as touch, temperature, and pain, as well as those originating from within the body, such as muscle stretch and the concentration of different chemicals in the body's fluids. The

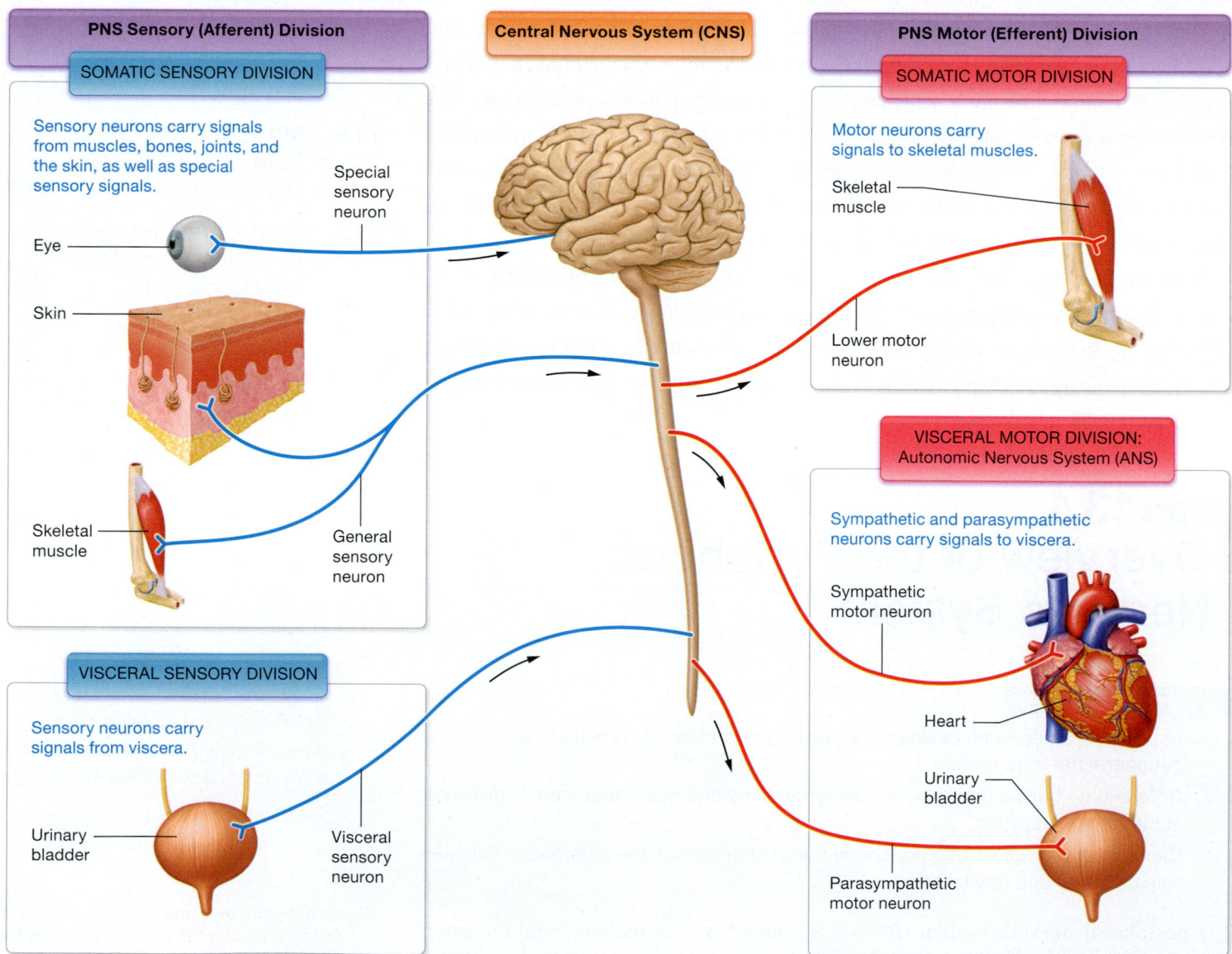

Figure 13.1 The organization of the peripheral nervous system.

somatic sensory division also contains special sensory neurons that are responsible for detecting stimuli of the *special senses* (vision, hearing, equilibrium, taste, and smell).

- The **visceral sensory division** contains neurons that relay sensory stimuli from the organs of the abdominopelvic and thoracic cavities. Its neurons detect internal changes such as blood pressure and the degree to which organs such as the urinary bladder are stretched.

Modules 13.2 and 13.4 focus on the somatic sensory division. The special somatic sensory neurons are discussed in the special senses chapter (see Chapter 15).

Motor Division

The **motor division** of the PNS consists of motor neurons that carry out the motor functions of the nervous system (see the right side of Figure 13.1). Like the sensory division, it may be further classified—based on the organs that the neurons contact—into the *somatic motor division* and the *visceral motor division,* or *autonomic nervous system.*

Somatic Motor Division   The **somatic motor division** is responsible for the body's *voluntary* motor functions. It is made up of *lower motor neurons* (or *somatic motor neurons*) that directly contact skeletal muscle fibers and trigger a contraction when stimulated by *upper motor neurons* in the CNS. We'll see in Module 13.6 that the activities of the somatic sensory and somatic motor divisions are closely linked.

Visceral Motor Division: The Autonomic Nervous System
As we discussed in the chapters on nervous tissue and the CNS (see Chapters 11 and 12, respectively), the **visceral motor division,** or **autonomic nervous system (ANS),** is responsible for maintaining the homeostasis of many physiological variables through its control of the body's *involuntary* motor functions. Its neurons innervate cardiac muscle cells, smooth muscle cells, and the secretory cells of glands.

The ANS has two divisions: the **sympathetic nervous system** and the **parasympathetic nervous system.** The sympathetic nervous system is usually described as the "fight or flight" division of the ANS—this name reflects its role in preparing the body for emergency situations that would involve fighting off an attacker or fleeing from danger. Its role is broader than this, however, as it maintains homeostasis when the body is engaged in any type of physical work and mediates the body's visceral responses to emotion. The parasympathetic nervous system is often described as the "rest and digest" division of the ANS, which reflects its role in digestion and in maintaining the body's homeostasis at rest. The ANS is discussed at greater length in its own chapter (see Chapter 14).

Quick Check

- ☐ 1. What two subclasses make up the sensory division of the PNS, and how do they differ?
- ☐ 2. What is a lower motor neuron? How are upper and lower motor neurons different?
- ☐ 3. In what ways do the somatic and visceral motor divisions of the PNS differ?

Overview of Peripheral Nerves and Associated Ganglia

❮❮ FLASHBACK

1. What are the anterior and posterior horns of the spinal cord, and what types of neuronal cell bodies are found in them? (p. 457)

2. Where are the spinal nerves and cranial nerves located? (p. 382)

The main organs of the PNS are its **peripheral nerves** (this term is generally shortened to just *nerves*), which consist of the axons of many neurons bound together by a common connective tissue sheath (**Figure 13.2**). The many nerves of the PNS contact, or **innervate** (IN-er-vayt), the majority of structures in the body. Most nerves are **mixed nerves,** meaning they contain both sensory and motor neurons. This is why a damaged nerve affects both sensation and movement to some degree. There are, however, **sensory nerves** that contain only sensory neurons and **motor nerves** that contain mostly motor neurons. All motor nerves contain a small population of sensory neurons that carry stimuli pertaining to muscle stretch and tension, so no nerve is a pure motor nerve.

Remember from the nervous tissue chapter that the two types of nerves are classified according to their location (see Chapter 11):

- **Spinal nerves.** The **spinal nerves** originate from the spinal cord and mainly innervate structures inferior to the head and neck. As you can see in Figure 13.2a, two groups of axons connect the PNS with the spinal cord's gray matter—an **anterior root** containing the axons of motor neurons exiting the anterior horn, and a **posterior root** containing the axons of sensory neurons entering the posterior horn. Note that the posterior root features an enlarged area that houses the cell bodies of sensory neurons just lateral to the spinal cord. A collection of cell bodies in the PNS is called a *ganglion,* and accordingly, this swollen area in the posterior root is the *posterior root ganglion* (or dorsal root ganglion). Lateral to the posterior root ganglion, the anterior and posterior roots fuse to form a spinal nerve. Each spinal nerve contains both sensory and motor neurons, so all are mixed nerves. There are 31 pairs of spinal nerves, which are discussed fully in Module 13.3.
- **Cranial nerves.** The **cranial nerves** attach to the brain and mainly innervate structures of the head and neck. Unlike spinal nerves, cranial nerves are not formed from the fusion of motor and sensory nerve roots, and so some cranial nerves are purely sensory, others are mixed, and others are predominantly motor. Module 13.2 covers the 12 pairs of cranial nerves.

Figure 13.2b and c show how the motor and sensory axons making up a nerve are held together. The nerve as a whole is wrapped by a connective tissue sheath called the **epineurium** (ep′-ih-NOOR-ee-um). Within the nerve, axons are bundled into smaller groups known as **fascicles** (FASS-ih-kulz). Fascicles are in turn bound by another connective tissue sheath, the **perineurium** (pehr′-ih-NOOR-ee-um). Nestled between the fascicles are blood vessels that supply the axons with oxygen and nutrients. Each axon

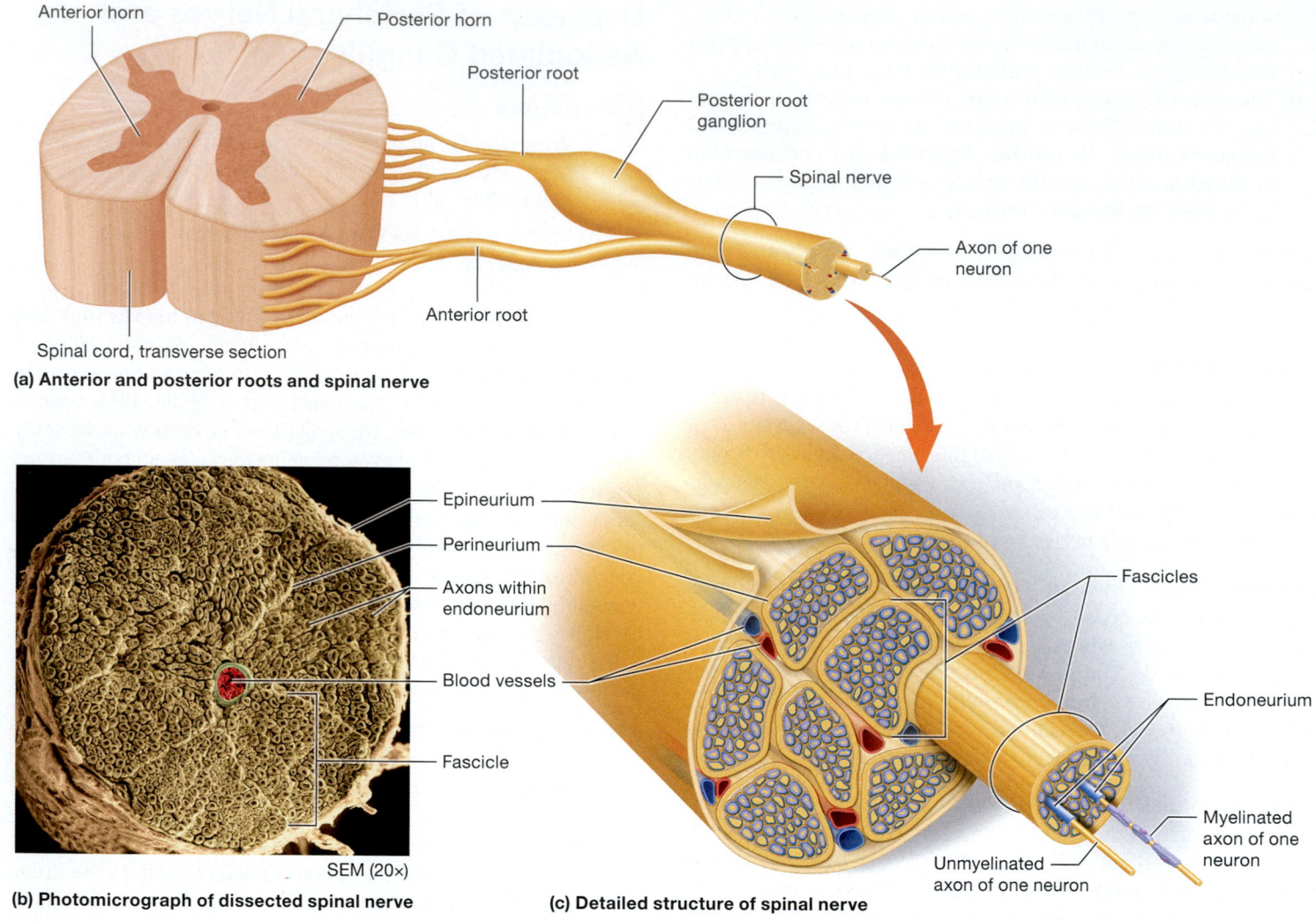

Anterior horn
Posterior horn
Posterior root
Posterior root ganglion
Spinal nerve
Axon of one neuron
Anterior root
Spinal cord, transverse section

(a) Anterior and posterior roots and spinal nerve

Epineurium
Perineurium
Axons within endoneurium
Blood vessels
Fascicle

SEM (20×)

(b) Photomicrograph of dissected spinal nerve

Fascicles
Endoneurium
Myelinated axon of one neuron
Unmyelinated axon of one neuron

(c) Detailed structure of spinal nerve

Figure 13.2 The structure of roots and spinal nerves.

within a fascicle is surrounded by its own connective tissue sheath, the **endoneurium** (en′-doh-NOOR-ee-um) (note that although the figure shows a spinal nerve, the arrangement within spinal and cranial nerves is the same). Notice how the arrangement of a nerve closely resembles that within a skeletal muscle, which we discussed in the muscle tissue chapter (see Chapter 10).

> **Quick Check**
>
> ☐ 4. How do cranial and spinal nerves differ?
> ☐ 5. Which structures unite to form a spinal nerve?
> ☐ 6. What structures are found in a peripheral nerve?

Functional Overview of the PNS

Let's look at how the PNS functions and how those functions are integrated with the CNS. We start with the sensory arm of the PNS. Sensory neurons detect stimuli at structures known as *sensory receptors*. The detected stimuli are transmitted along the sensory neurons via spinal or cranial nerves to sensory neurons

of the CNS, which transmit the impulses to the cerebral cortex for interpretation and integration. Depending on the nature of the sensory stimuli, an appropriate motor response is initiated. For example, if you hear the microwave beep, your CNS interprets this input to mean that your dinner is ready, and initiates a motor response so you can walk to the microwave and remove, and presumably eat, your dinner.

The motor response is initiated by commands from the motor areas of the brain to the upper motor neurons. These impulses travel to the spinal cord, where they synapse on local interneurons and then lower motor neurons of the PNS. The lower motor neurons then carry the impulses to the appropriate muscle fibers via specific cranial or spinal nerves, where they trigger their contraction. Later modules will delve into the specifics of these processes.

> **Quick Check**
>
> ☐ 7. How are sensations detected in the PNS and delivered to the CNS?
> ☐ 8. How are motor impulses transmitted from the CNS to the PNS?

Apply What You Learned

☐ 1. How do a nerve and a neuron differ? Which would most affect motor or sensory function: damage to a neuron or damage to a nerve? Explain.

☐ 2. Explain how the arrangement of axons within a nerve resembles the arrangement of muscle fibers within a skeletal muscle.

☐ 3. To which division of the PNS would damage be most life-threatening? Explain.

See answers in Appendix A.

MODULE **13.2**
The Cranial Nerves

Learning Outcomes

1. Identify the cranial nerves by name and number.
2. Describe the specific functions of each pair of cranial nerves, and classify each pair as sensory, motor, or mixed nerves.
3. Describe the locations of selected cranial nerve nuclei and the ganglia associated with the cranial nerves.

As we discussed in Module 13.1, the cranial nerves transmit impulses to and from the brain. Each pair of cranial nerves is identified in two ways: First, a Roman numeral reflects its attachment site in the brain, and second, a name reflects its function.

Recall that certain cranial nerves contain only sensory axons, others contain both motor and sensory axons, and the remainder contain mostly motor axons. Generally speaking, the olfactory (I), optic (II), and vestibulocochlear (VIII) nerves are sensory nerves; the oculomotor (III), trochlear (IV), abducens (VI), accessory (XI), and hypoglossal (XII) nerves are motor nerves; and the trigeminal (V), facial (VII), glossopharyngeal (IX), and vagus (X) nerves are mixed nerves. The order, names, and function of the 12 pairs of cranial nerves are shown in **Figure 13.3**.

In this module, we discuss the structure, location, and functions of all 12 pairs of cranial nerves. Although it is traditional to cover the cranial nerves sequentially in ascending order, here we categorize them functionally as sensory, motor, or mixed nerves.

StudyBOOST ⬆

Remembering the Cranial Nerves

It's easy to feel overwhelmed when you're learning the cranial nerves. You have to remember not only what all 12 pairs do and where they are located, but also two names for each nerve! A way to make this easier, however, is to use a mnemonic to recall the order and names of the cranial nerves. A popular mnemonic is the following:

Oh	(I, Olfactory)
Once	(II, Optic)
One	(III, Oculomotor)
Takes	(IV, Trochlear)
The	(V, Trigeminal)
Anatomy	(VI, Abducens)
Final	(VII, Facial)
Very	(VIII, Vestibulocochlear)
Good	(IX, Glossopharyngeal)
Vacations	(X, Vagus)
Are	(XI, Accessory)
Happening	(XII, Hypoglossal)

If you have trouble remembering which Roman numeral is assigned to the optic nerve and which to the olfactory nerve, just remember that you have *one* nose (I, olfactory) and *two* eyes (II, optic).

A second mnemonic can help you recall the cranial nerves by their main function—sensory (S), motor (M), or both (B):

Some	(I, Olfactory—Sensory)
Say	(II, Optic—Sensory)
Money	(III, Oculomotor—Motor)
Matters	(IV, Trochlear—Motor)
But	(V, Trigeminal—Both)
My	(VI, Abducens—Motor)
Brother	(VII, Facial—Both)
Says	(VIII, Vestibulocochlear—Sensory)
Big	(IX, Glossopharyngeal—Both)
Brains	(X, Vagus—Both)
Matter	(XI, Accessory—Motor)
More	(XII, Hypoglossal—Motor)

A final hint that can help you remember the functions of the cranial nerves is to look closely at their names and connect them with word roots (see the back of the book). For example, cranial nerve III is the oculomotor nerve. If you break this word into its two components, you get *oculo-*, which means "eye," and *-motor,* which means "movement." So you can deduce just from this nerve's name one of its main functions, which is to move the eye. ■

The Sensory Cranial Nerves

« FLASHBACK

1. What is the limbic system, and what are its functions? (p. 431)
2. Define the term decussate. (p. 438)

There are three cranial nerves that contain the axons of only sensory neurons: the olfactory (I), optic (II), and vestibulocochlear (VIII) nerves. Each of these nerves is involved in one of the special senses. To make these nerves easier to study, we have arranged the illustrations and text into **Table 13.1**.

Quick Check

☐ 1. What are the Roman numeral and function(s) of the following nerves?

    a. Olfactory nerve   c. Vestibulocochlear nerve

    b. Optic nerve

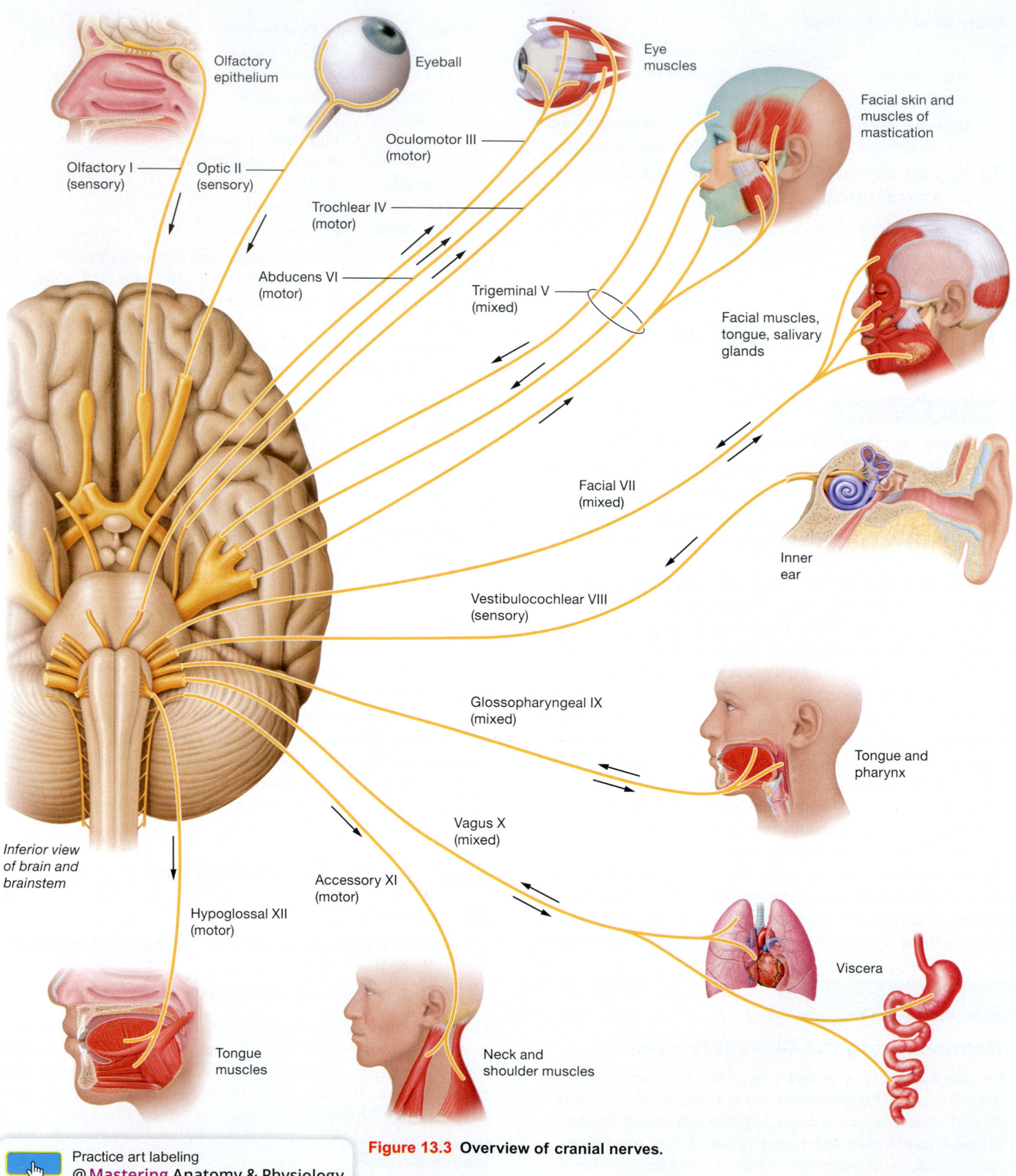

Olfactory epithelium

Eyeball

Eye muscles

Olfactory I (sensory)

Optic II (sensory)

Oculomotor III (motor)

Facial skin and muscles of mastication

Trochlear IV (motor)

Abducens VI (motor)

Trigeminal V (mixed)

Facial muscles, tongue, salivary glands

Facial VII (mixed)

Inner ear

Vestibulocochlear VIII (sensory)

Glossopharyngeal IX (mixed)

Tongue and pharynx

Inferior view of brain and brainstem

Vagus X (mixed)

Accessory XI (motor)

Hypoglossal XII (motor)

Viscera

Tongue muscles

Neck and shoulder muscles

Figure 13.3 Overview of cranial nerves.

Practice art labeling
@ Mastering Anatomy & Physiology

Table 13.1 The Sensory Cranial Nerves

Nerve	Origin, Course, and Destination	Function
Olfactory nerve (I) (ohl-FAK-toh-ree) 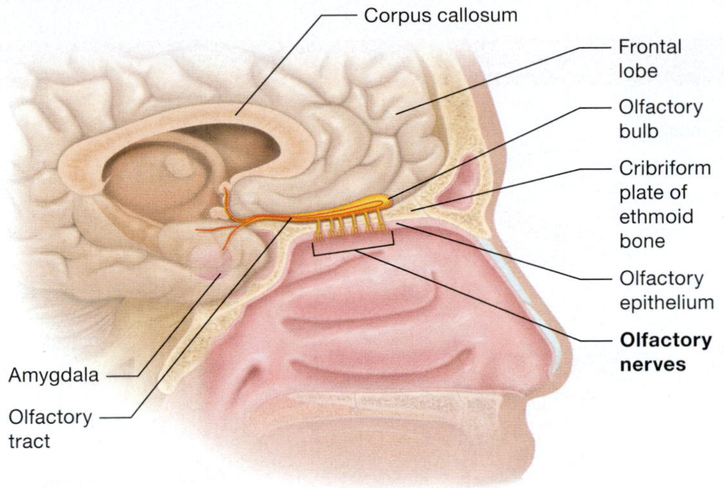 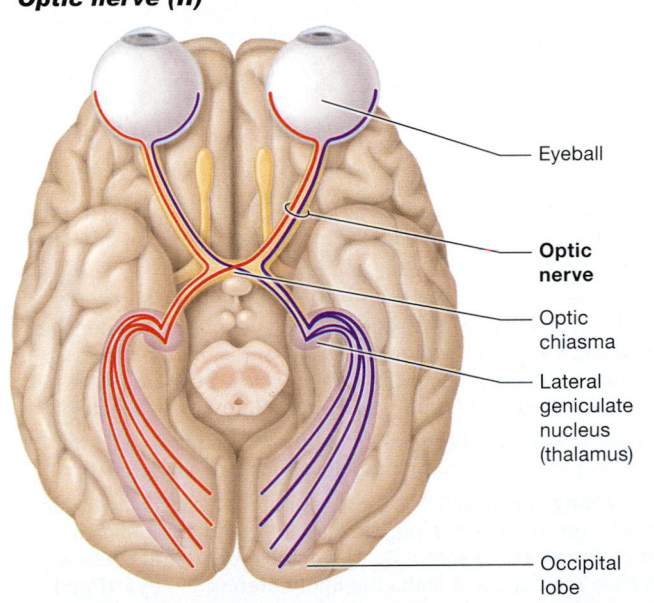	*Origin:* Originates from the unmyelinated axons of neurons whose cell bodies are located in the olfactory epithelium in the roof of the nasal cavity. *Course:* The axons form bundles that penetrate the holes in the cribriform plate of the ethmoid bone, and end in a bulbous structure known as the *olfactory bulb* (shown to the left). This bulb sits on the inferior surface of the brain's frontal lobe and continues posteriorly as the *olfactory tract*. *Destination:* Terminates in the medial temporal lobe and structures of the limbic system, including the limbic cortex, the hippocampus, and the amygdala.	The nerve for *olfaction* (*olfact-* = "smell"), or the sense of smell. Its nerve endings contain chemoreceptors that depolarize in response to chemicals in the air that we breathe. These stimuli are interpreted in the brain by the primary olfactory cortex.
Optic nerve (II)	*Origin:* Originates from myelinated axons of neurons in the posterior eye. *Course:* As shown to the left, the two optic nerves meet and form an "X" called the **optic chiasma** (ky-AZ-mah), where some of the axons decussate (switch sides). *Destination:* Axons from the optic chiasma are destined for structures such as the lateral geniculate nucleus of the thalamus, the midbrain, and ultimately the primary visual cortex in the occipital lobe.	The nerve for vision (*opt-* = "vision"). It transmits visual stimuli in the form of action potentials triggered when light hits the eye's photoreceptors. These stimuli are processed in the brain by the primary visual cortex.
Vestibulocochlear nerve (VIII) (ves-tib'-yoo-loh-KOHK-lee-ur) 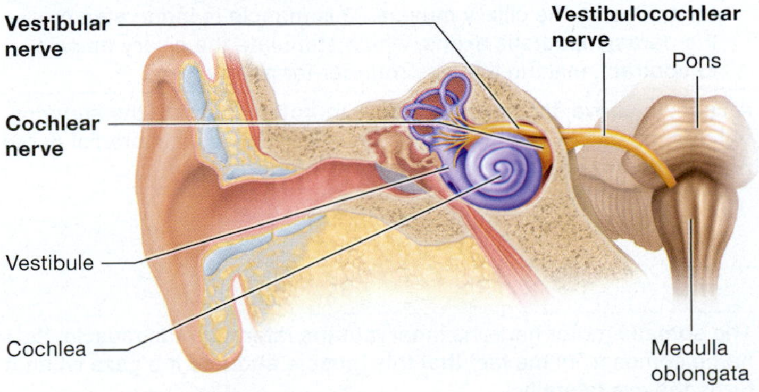	*Origin:* Actually two separate nerves, the vestibular and cochlear, which originate in the inner ear and share a common epineurium. *Courses:* Fibers from the vestibular and cochlear nerves fuse shortly after they leave the temporal bone to become the vestibulocochlear nerve. *Destination of vestibular nerve:* Axons terminate in the cerebellum and in nuclei in the medulla oblongata. *Destination of cochlear nerve:* Axons travel to the medulla oblongata and terminate in auditory cochlear areas.	*Vestibular nerve:* Its neurons depolarize in response to body position and so are concerned with balance and equilibrium. *Cochlear nerve:* Its neurons depolarize in response to sound waves and are responsible for *audition* (aw-DIH-shun; *aud-* = "hear"), or the sense of hearing.

Labels (Olfactory nerve figure): Corpus callosum, Frontal lobe, Olfactory bulb, Cribriform plate of ethmoid bone, Olfactory epithelium, **Olfactory nerves**, Amygdala, Olfactory tract

Labels (Optic nerve figure): Eyeball, **Optic nerve**, Optic chiasma, Lateral geniculate nucleus (thalamus), Occipital lobe

Labels (Vestibulocochlear nerve figure): Vestibular nerve, **Vestibulocochlear nerve**, Pons, Cochlear nerve, Vestibule, Cochlea, Medulla oblongata

The Motor Cranial Nerves

The motor cranial nerves—the oculomotor (III), trochlear (IV), abducens (VI), accessory (XI), and hypoglossal (XII) nerves—also technically contain axons of proprioceptive sensory neurons. In spite of these sensory axons, they are still viewed as motor nerves because the main function of their sensory neurons is to allow the brain to monitor the contraction of the muscles they innervate. The locations and functions of the motor cranial nerves are covered in **Table 13.2**.

Table 13.2 The Motor Cranial Nerves

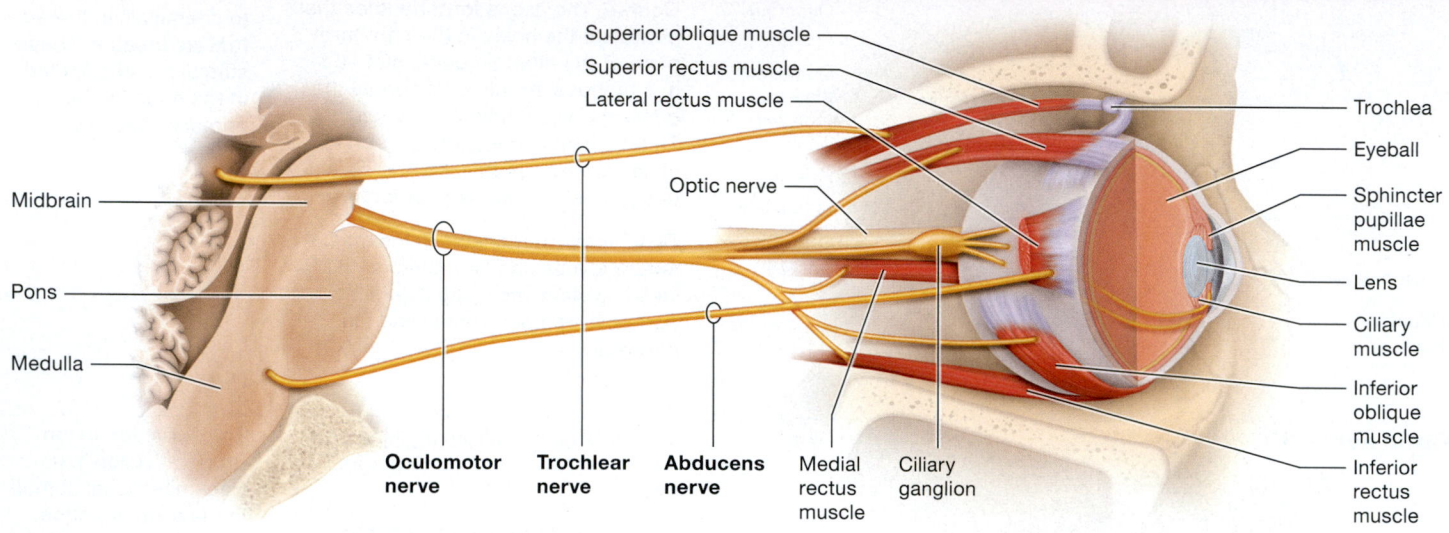

Nerve	Origin and Destination	Function
Oculomotor nerve (III) (awk´-yoo-loh-MOH-ter)	*Origin:* Arises from the superior and lateral portion of the midbrain. *Destination:* Extrinsic eye muscles and smooth muscles of the eye (*oculo-* = "eye")	Contains axons of both somatic motor neurons and visceral motor neurons of the parasympathetic nervous system, and has the following four primary functions: • **Moving the eyeball.** As shown above, the somatic motor axons innervate four of the six extrinsic eye muscles that move the eyeball: the medial rectus, superior rectus, inferior rectus, and inferior oblique muscles. These muscles move our eyes medially, superiorly, inferiorly, and superolaterally, respectively. • **Opening the eye.** Other somatic motor axons innervate the levator palpebrae superioris muscle, which opens the eyelid. • **Constricting the pupil.** The parasympathetic axons innervate a muscle surrounding the pupil called the *sphincter pupillae muscle.* These axons stimulate this muscle to contract, which constricts the pupil and limits the light entering the eye. (Pupil dilation is mediated by a different nerve.) • **Changing the lens shape.** The thickness of the lens (which focuses light on the photoreceptors in the posterior eyeball) is controlled by the ciliary muscle. This muscle is innervated by the parasympathetic axons, which stimulate the ciliary muscle to contract, making the lens rounder for near vision.
Trochlear nerve (IV) (TROHK-lee-ur)	*Origin:* Originates from the inferior portion of the midbrain. Named for the cartilaginous trochlea through which the tendon of the superior oblique muscle passes. *Destination:* Superior oblique muscle of the eye	As shown above, the somatic motor neurons innervate the superior oblique muscle (a common mnemonic is "SO$_4$," as it is cranial nerve IV), which moves the eye medially and inferiorly.
Abducens nerve (VI) (ab-DOO-senz)	*Origin:* The cell bodies of the abducens nerve are located in the pons. *Destination:* Lateral rectus muscle of the eye	The somatic motor neurons innervate the lateral rectus muscle. Its name comes from the fact that this muscle abducts the gaze when it turns the eye laterally.

Table 13.2 The Motor Cranial Nerves (*continued*)

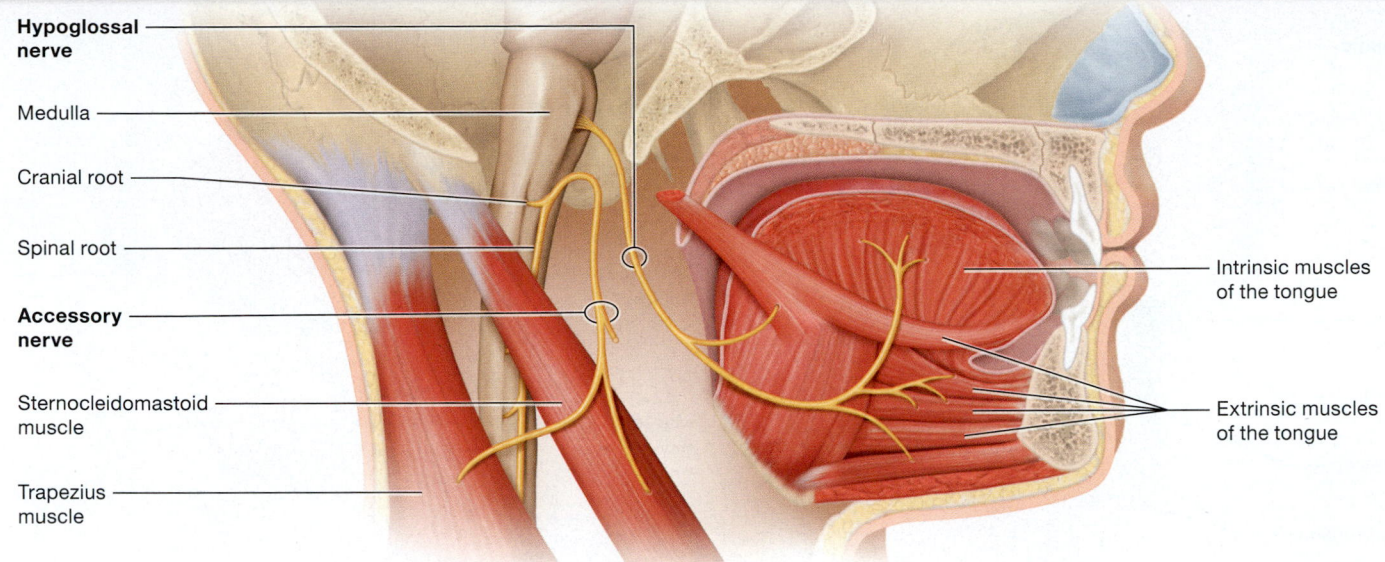

Nerve	Origin and Destination	Function
Accessory nerve (XI)	*Origin:* Unique in that its origin has both a cranial component from the medulla and a spinal component from the cervical spinal cord. The somatic motor neurons of the spinal component travel superiorly and enter the cranial cavity through the foramen magnum, after which they merge with the somatic motor neurons of the cranial component. As shown above, the two travel together for only a short distance, after which they diverge. *Cranial Destination:* The cranial component accompanies the vagus nerve (covered in Table 13.3) and innervates certain muscles of the larynx. *Spinal Destination:* The spinal component turns inferiorly and exits the cranial cavity to innervate the trapezius and sternocleidomastoid muscles of the neck and shoulders.	The cranial component of the accessory nerve innervates certain muscles of speech, whereas the spinal component innervates muscles that move the head and shoulder.
Hypoglossal nerve (XII)	*Origin:* Arises from the inferiormost part of the medulla (see above). *Destination:* As its name suggests, (*hypo-* = "below," *glosso-* = "tongue"), its destination lies inferior to the tongue. It innervates most of the intrinsic and extrinsic muscles of the tongue.	This is a motor nerve that innervates the muscles of the tongue—it plays no role in taste sensation.

Quick Check

☐ 2. What are the Roman numerals and main function(s) of the following nerves?

    a. Oculomotor nerve

    b. Trochlear nerve

    c. Abducens nerve

    d. Accessory nerve

    e. Hypoglossal nerve

The Mixed Cranial Nerves

As you may expect, the mixed cranial nerves—the trigeminal (V), facial (VII), glossopharyngeal (IX), and vagus (X) nerves—are generally large and have a fairly wide distribution. All but one (the trigeminal nerve) contain somatic sensory, somatic motor, and parasympathetic neurons wrapped up in the same nerve. The mixed cranial nerves have multiple and diverse functions due to their size and the different types of neurons they house. Their structure and functions are described in **Table 13.3**.

Table 13.3 The Mixed Cranial Nerves

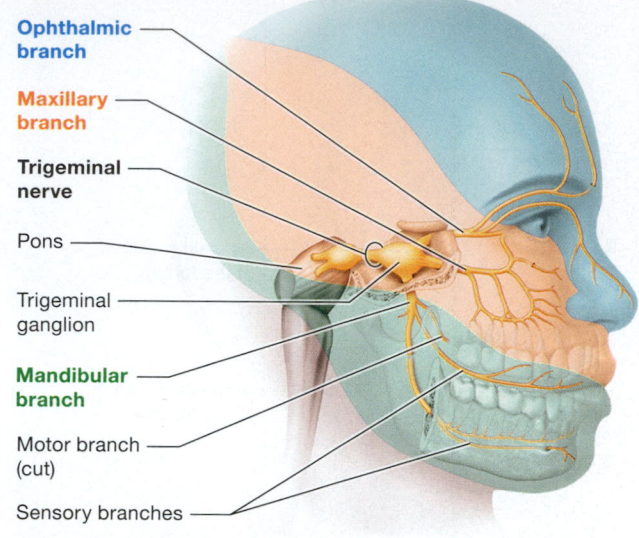

Trigeminal nerve: sensory function

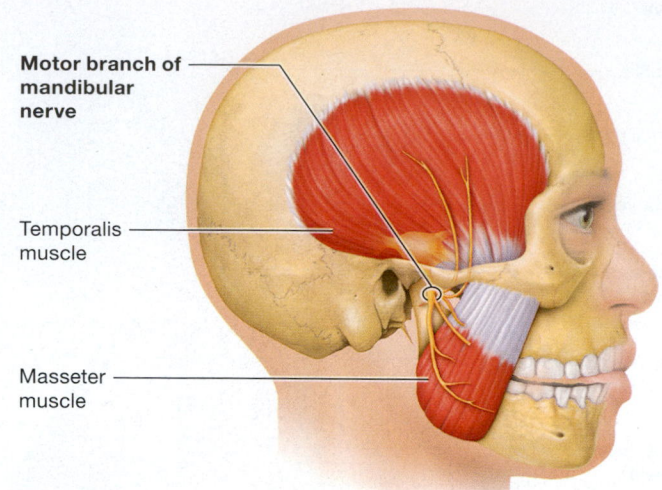

Mandibular nerve: motor function

Nerve	Origin, Course, and Destination	Sensory Function	Motor Function
Trigeminal (V) (try-JEM-ih-nul)	*Origin:* The motor portion originates at the midbrain and pons junction, and the sensory portion from sensory receptors around the face. *Course:* A short distance from its origin is the large, bulbous **trigeminal ganglion,** which houses the cell bodies of its sensory neurons. Anterior to this ganglion, it splits into three branches (*tri-* = "three"): the *ophthalmic* (awf-THAL-mik), *maxillary,* and *mandibular* nerves. *Destinations:* All three branches have sensory fibers that terminate in the primary somatosensory cortex. The mandibular nerve has motor fibers that terminate at the muscles of mastication.	The sensory root detects facial sensation, including stimuli from the oral and nasal cavities. *Ophthalmic nerve:* Its somatic sensory axons supply the area shaded blue in the illustration above, which includes the skin over much of the scalp, the forehead, around the eyes, and over the anterior nose; they also supply the nasal mucosa and structures of the eye. *Maxillary nerve:* As shown above, its somatic sensory axons supply the area shaded orange, which includes the skin over the middle of the face. *Mandibular nerve:* The sensory axons of the inferior mandibular nerve supply the area shaded green above, which includes the skin of the chin and the lateral part of the face.	*Ophthalmic nerve:* No motor function. *Maxillary nerve:* No motor function. *Mandibular nerve:* Shown above, its motor axons supply the masseter and temporalis muscles, which elevate the mandible (close the jaw) during mastication (chewing) and swallowing.
Facial nerve (VII)	*Origin:* As shown in the illustrations on the next page, the motor portion (or *motor root*) originates in nuclei in the pons and medulla, and the sensory portion (or *sensory root*) from the tongue, external ear, palate, and nasal cavity. *Course:* Several ganglia house cell bodies of the facial nerve's sensory root; the largest is called the **geniculate ganglion** (jen-IK-yoo-lit).	The sensory root provides the following: • taste sensation from chemoreceptors in specialized receptor cells in the mucosa of the anterior two-thirds of the tongue; and • somatic sensation from the external ear, palate, and nasal cavity.	The somatic motor neurons of the five branches of the motor root supply the muscles of facial expression and other facial muscles.

Table 13.3   The Mixed Cranial Nerves (*continued*)

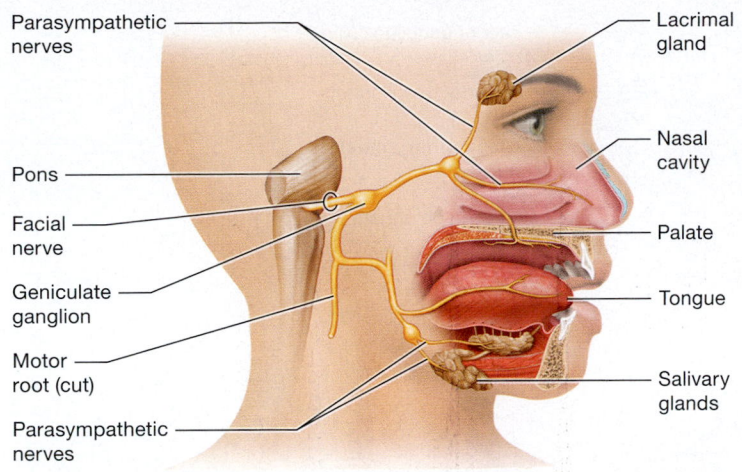

Facial nerve: mixed sensory and visceral motor roots

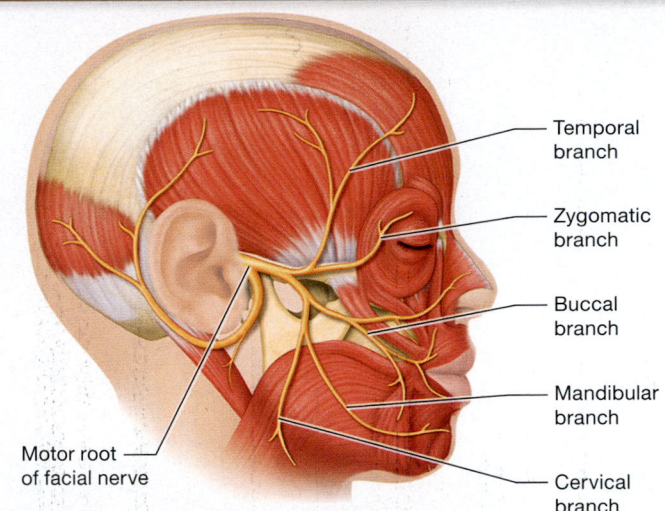

Facial nerve: motor function

Nerve	Origin, Course, and Destination	Sensory Function	Motor Function
Facial nerve (VII) (continued)	*Sensory Destination:* Various somatosensory areas of the cerebral cortex. *Motor Destination:* As shown above, these neurons innervate the muscles of facial expression as well as several other facial and neck muscles. The motor root splits into five branches: the *temporal, zygomatic, buccal, mandibular,* and *cervical nerves* (a common mnemonic is "**T**o **Z**anzibar **B**y **M**otor **C**ar").		In addition, the parasympathetic neurons trigger secretion from certain salivary glands in the mouth, the lacrimal (tear) glands over the eye, and the nasal mucous glands.
Glossopharyngeal nerve (IX) (gloss´-oh-fah-RIN-jee-ul)	*Origin:* The motor neurons originate in nuclei in the medulla; the sensory neurons originate in the tongue, pharynx, around the ear, and in blood vessels of the neck. *Course:* The cell bodies of its sensory neurons are located in two ganglia: the *superior ganglion* and the *inferior ganglion.* *Destination:* Its location and the structures it innervates are reflected in its name—recall that *glosso-* means "tongue" and *pharynx* means "throat."	The sensory portion of this nerve detects sensation as follows: • The chemoreceptors of the posterior one-third of the tongue are associated with special sensory axons of this nerve. The cell bodies of these neurons are in the inferior ganglion. • A small branch of this nerve contains somatic sensory neurons that innervate the external ear alongside the facial nerve. It also contains visceral sensory neurons that provide sensation for the posterior pharynx and surrounding structures. In addition, certain visceral neurons detect changes in blood pressure via receptors in the carotid artery of the neck.	The motor branches of this nerve supply a muscle around the pharynx that is responsible for swallowing movements. Also, parasympathetic neurons trigger salivation from the salivary gland called the *parotid gland.* You can feel this stimulation and subsequent salivation in your cheeks when you eat very salty or acidic foods such as pickles or lemons.

(continued)

Table 13.3   The Mixed Cranial Nerves (*continued*)

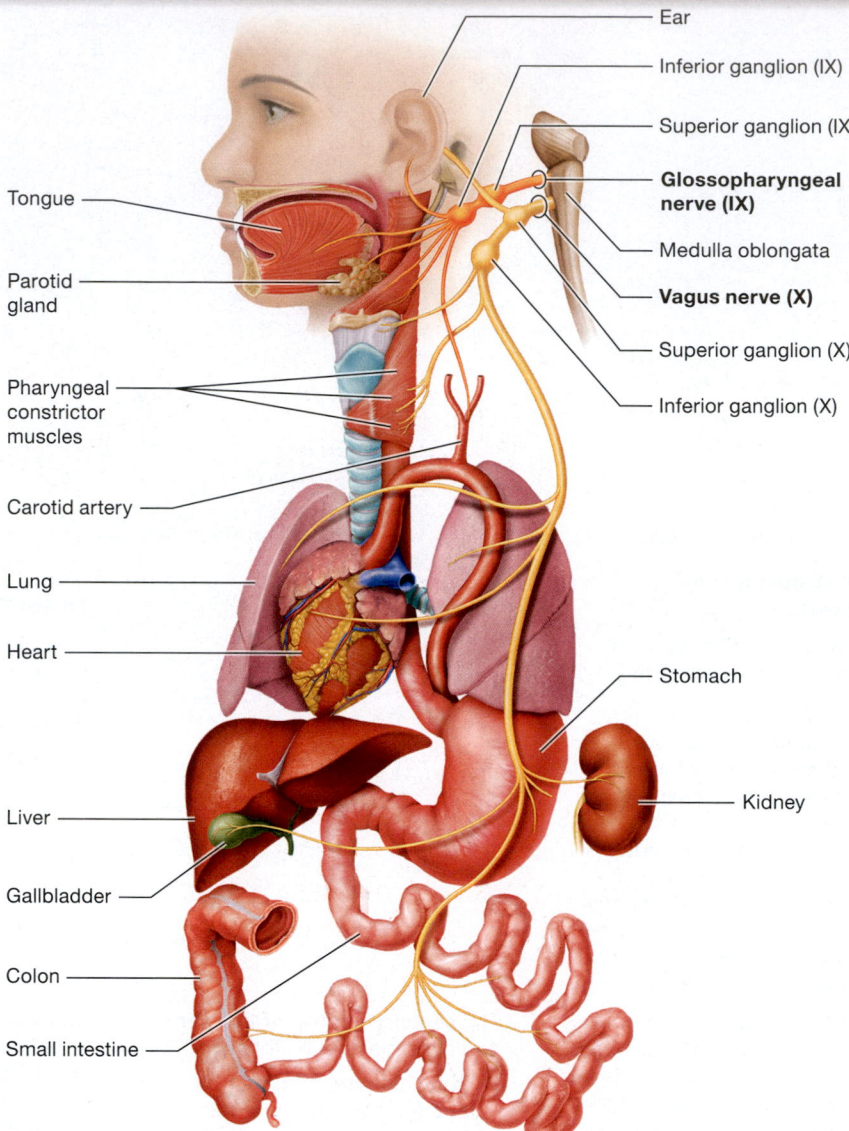

Illustration is not to scale; orange is used simply to distinguish the glossopharyngeal nerve.

Nerve	Origin, Course, and Destination	Sensory Function	Motor Function
Vagus nerve (X) (VAY-guss)	*Origin:* The motor neurons originate from the brain at the medulla. The sensory neurons originate around the tongue, pharynx, skin of the ear, and certain blood vessels of the neck. *Course:* As shown above, cell bodies of its sensory neurons are housed in the *superior ganglion* and the *inferior ganglion.* *Destination:* It is the main parasympathetic nerve, and the most widely distributed nerve in the body (see above). It innervates both the throat and anterior neck, and nearly all the thoracic and abdominal viscera.	The somatic sensory neurons serve the skin around the ear. The special sensory neurons convey taste sensation from the pharynx, and the visceral sensory neurons detect sensation in the mucous membranes of the pharynx. This nerve also contains visceral sensory neurons whose chemoreceptors detect the blood CO_2 concentration.	The somatic motor fibers supply the muscles surrounding the pharynx and larynx (the "voice box") that contract during speaking and swallowing. We continue our discussion of the visceral innervation of this nerve in the ANS chapter (see Chapter 14).

Having multiple functions means that these mixed nerves are subject to many problems when they are damaged. See *A&P in the Real World: Trigeminal Neuralgia* and *A&P in the Real World: Bell's Palsy* to read about two such conditions.

Trigeminal Neuralgia

Trigeminal neuralgia (noo-RAL-jah) is a chronic pain syndrome that involves one or more branches of the trigeminal nerve. Patients with trigeminal neuralgia suffer brief attacks of intense pain that last from a few seconds to 2 minutes. The pain is typically unilateral and may occur several times per day. Certain stimuli are known to trigger attacks, such as chewing and light touch or vibratory stimuli to the face (even a light breeze has been said to trigger an attack for certain patients).

In spite of the recurrent painful episodes, neurological examinations of patients with trigeminal neuralgia are normal, and the cause of the disease is unknown. Pain medications are typically ineffective, and treatment is instead aimed at reducing the aberrant transmission through the nerve, often by severing it.

Quick Check

☐ 3. What are the Roman numerals and main function(s) of the following nerves?

    a. Trigeminal nerve

    b. Facial nerve

    c. Glossopharyngeal nerve

    d. Vagus nerve

☐ 4. List the 12 pairs of cranial nerves in ascending order by both their name and number.

☐ 5. Which cranial nerves are sensory, which are motor, and which are mixed nerves?

Apply What You Learned

☐ 1. Often when the brain is damaged, the cranial nerves are damaged as well. A simple physical exam can be used to test clinically for the function of each cranial nerve. Damage to which cranial nerve or nerves would lead to the following findings? Explain your answer.

    a. Complete loss of taste sensation

    b. Inability to move the tongue

    c. Inability to move the eyes in any direction

    d. Loss of balance and equilibrium

    e. Inability to close the jaw

Bell's Palsy

A common problem associated with the facial nerve is *Bell's palsy,* in which the facial nerve's motor root is impaired by a virus, tumor, trauma, or an unknown cause. Patients with Bell's palsy have weakness or complete paralysis of the muscles of facial expression on the affected side. The unaffected side, however, appears normal. The paralysis leads to problems with blinking, closing the eye, and making general facial expressions such as smiling. In addition, other structures innervated by the facial nerve may be affected, including the lacrimal gland, salivary glands, and taste sensation to the anterior two-thirds of the tongue.

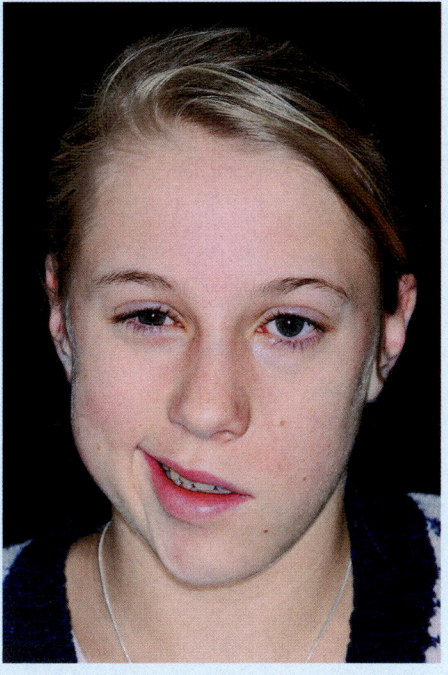

Typically, the individual experiences the onset of symptoms rapidly, within about 72 hours. Treatment may include anti-inflammatory medication, antiviral medication, physical therapy, and surgery. However, even without treatment, many individuals recover function of the paralyzed muscles in about 3–8 weeks.

☐ 2. Drugs that stimulate the parasympathetic nervous system often lead to the adverse effect of excessive drooling. Which cranial nerve(s) is/are involved in this effect? What would happen if the activity of these nerves was instead inhibited?

See answers in Appendix A.

MODULE 13.3
The Spinal Nerves

Learning Outcomes

1. Discuss the relationships between structures of the spinal nerves: root, nerve, ramus, plexus, and ganglion.

2. Identify and describe the four spinal nerve plexuses, and give examples of nerves that emerge from each.

3. Describe the functions of the major spinal nerves.

Now let's turn our attention to the structure and function of the 31 pairs of spinal nerves. First, we focus on the structure of

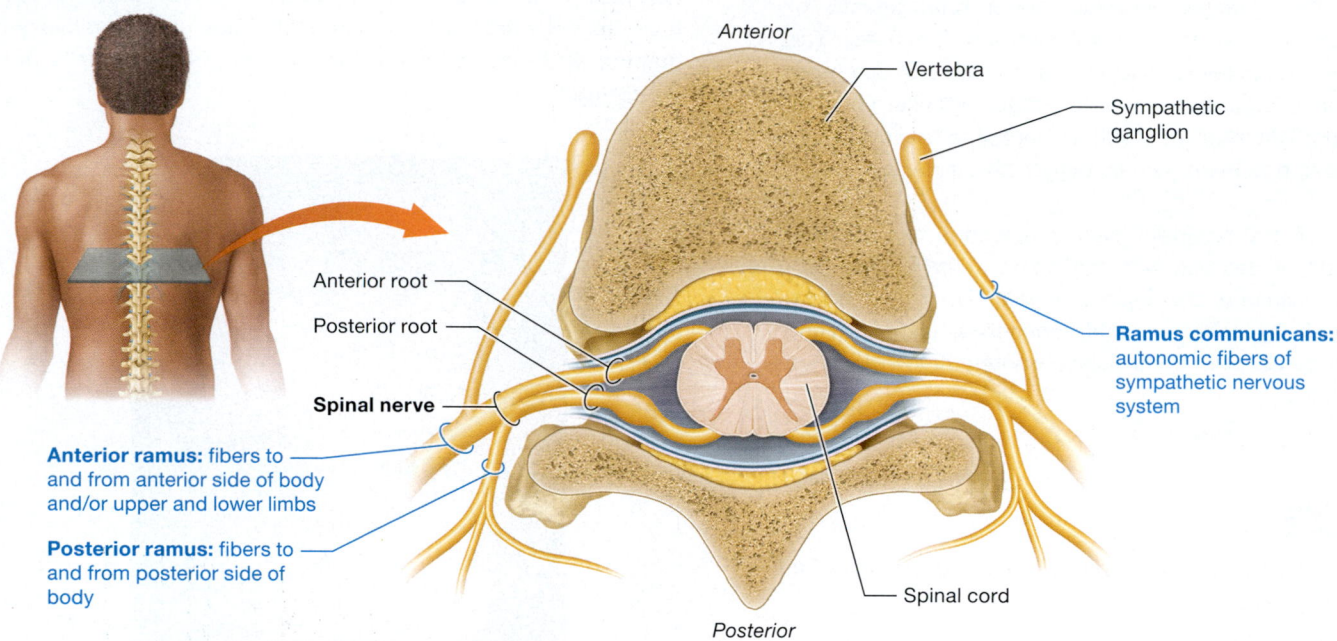

Anterior

Vertebra

Sympathetic ganglion

Anterior root

Posterior root

Ramus communicans: autonomic fibers of sympathetic nervous system

Spinal nerve

Anterior ramus: fibers to and from anterior side of body and/or upper and lower limbs

Posterior ramus: fibers to and from posterior side of body

Spinal cord

Posterior

(a) Structure of anterior and posterior rami of spinal nerves

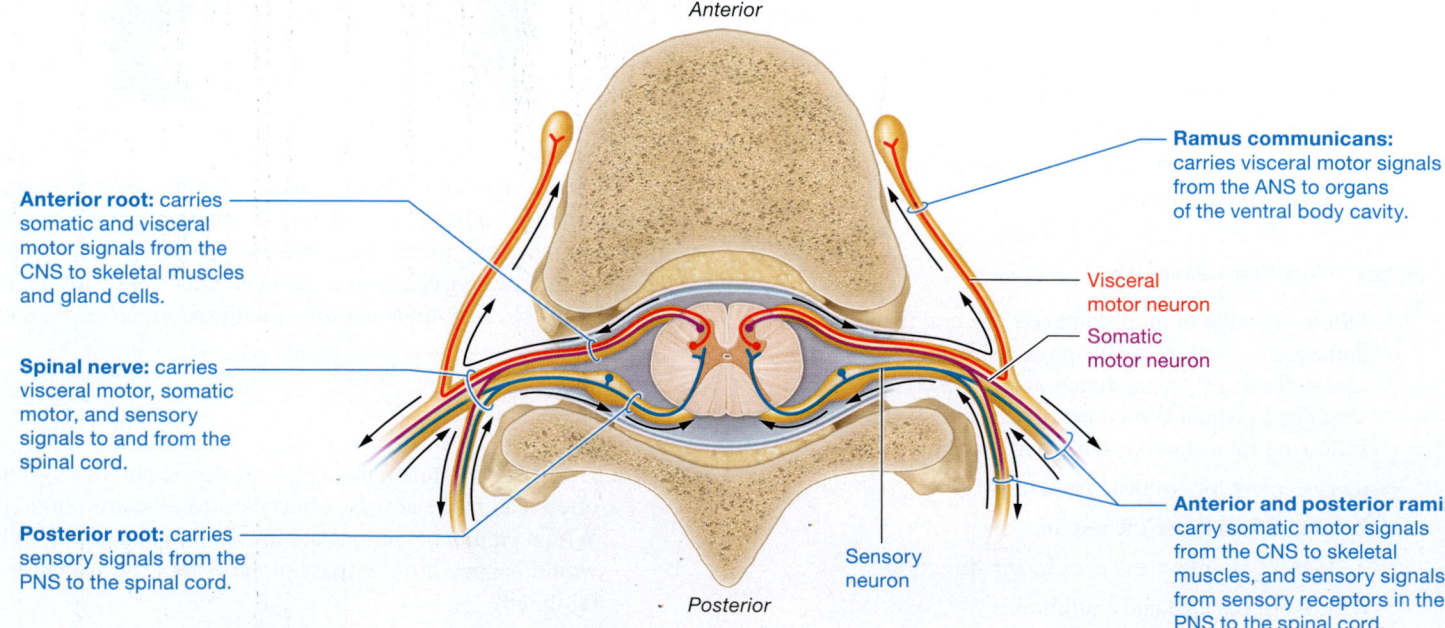

Anterior

Anterior root: carries somatic and visceral motor signals from the CNS to skeletal muscles and gland cells.

Spinal nerve: carries visceral motor, somatic motor, and sensory signals to and from the spinal cord.

Posterior root: carries sensory signals from the PNS to the spinal cord.

Ramus communicans: carries visceral motor signals from the ANS to organs of the ventral body cavity.

Visceral motor neuron

Somatic motor neuron

Sensory neuron

Anterior and posterior rami: carry somatic motor signals from the CNS to skeletal muscles, and sensory signals from sensory receptors in the PNS to the spinal cord.

Posterior

(b) Function of roots, spinal nerve, and rami

Figure 13.4 Structure and function of roots, spinal nerves, and rami.

spinal nerves and how most of them cluster to form networks, or *plexuses*. Then we examine each plexus and the main nerves that branch from it.

Structure of Spinal Nerves and Spinal Nerve Plexuses

As we have discussed, all spinal nerves are mixed because they are formed from the fusion of an anterior and posterior root (look back at Figure 13.2). The posterior root transmits sensory stimuli from the PNS to the CNS. Its partner, the anterior root, transmits somatic and visceral motor signals from the CNS to the PNS. When these roots come together in a spinal nerve, the nerve contains both sensory and motor axons.

As you can see in **Figure 13.4a**, the spinal nerve itself is actually quite short, because about 1–2 cm after it forms and leaves the vertebral cavity, it splits into two nerves: (1) a **posterior ramus** (RAY-muss; plural, *rami*; *ramus* = "branch"), which travels to the posterior side of the body; and (2) an **anterior ramus,** which travels to the anterior side of the body and/or the upper and lower limbs. Each ramus typically branches multiple times along its course through the body and ultimately supplies different skeletal muscles and regions of the skin. Both the anterior and posterior rami are mixed nerves, as they contain sensory and somatic motor axons (**Figure 13.4b**).

Note that another small branch stems from the anterior ramus. These small branches are called *rami communicantes* (singular, *ramus communicans*), and they contain visceral motor or autonomic neurons of the sympathetic nervous system (see Figure 13.4). Unlike the other branches of the anterior and posterior rami, the rami communicantes contain visceral motor axons only and so are not mixed nerves.

The 31 pairs of spinal nerves consist of 8 pairs of cervical spinal nerves (C_1–C_8), 12 pairs of thoracic spinal nerves (T_1–T_{12}), 5 pairs each of lumbar and sacral spinal nerves (L_1–L_5 and S_1–S_5), and 1 tiny pair of coccygeal nerves (Co_1) (**Figure 13.5**). The anterior rami of the cervical, lumbar, and sacral spinal nerves come together and merge to form complicated networks of nerves called **nerve plexuses** (PLEK-suss-ez). The axons of each spinal nerve cross over one another to enter different plexus branches. For this reason, the muscles supplied by a single branch of a nerve plexus are often served by two or more different spinal nerves. This works to our advantage, as it means that injury to one spinal nerve does not completely cut off motor or sensory innervation to that body part.

Next, we explore the anatomy of the nerve plexuses and the main structures supplied by the larger nerves. We also examine the nerves that arise from the anterior rami of the thoracic nerves, which do not form plexuses.

Quick Check

☐ 1. Which structures form a spinal nerve? Why are all spinal nerves mixed?

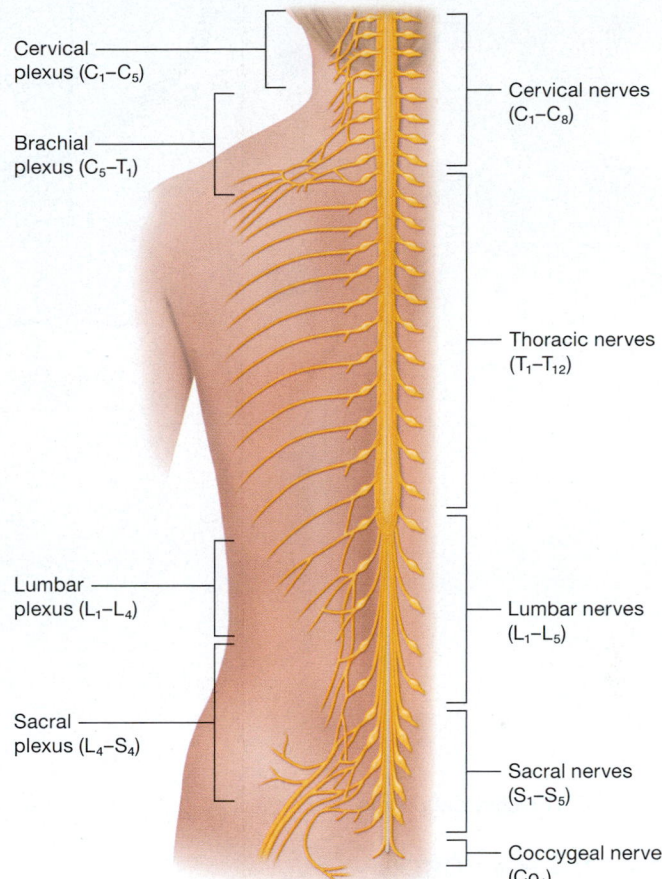

Figure 13.5 Overview of spinal nerves.

Labels in figure:
Cervical plexus (C_1–C_5)
Brachial plexus (C_5–T_1)
Lumbar plexus (L_1–L_4)
Sacral plexus (L_4–S_4)
Cervical nerves (C_1–C_8)
Thoracic nerves (T_1–T_{12})
Lumbar nerves (L_1–L_5)
Sacral nerves (S_1–S_5)
Coccygeal nerve (Co_1)

☐ 2. What are the anterior and posterior rami, and what do they supply?

Cervical Plexuses

The right and left **cervical plexuses** are located deep in the neck lateral to the first through the fourth cervical vertebrae. These plexuses consist largely of the anterior rami of C_1–C_4, although they also receive small contributions from C_5 and the hypoglossal nerve (cranial nerve XII). As you can see in **Figure 13.6**, the nerves of each cervical plexus contain cutaneous branches that supply the skin of the neck and portions of the head, chest, and shoulders. They also contain motor branches that supply muscles of the neck, including the infrahyoid and scalene muscles. The major named motor branch is the **phrenic nerve** (FREN-ik), which contains axons from C_3 to C_5, although its main contributor is C_4. The phrenic nerve supplies the diaphragm muscle and is the main nerve that drives ventilation (a good mnemonic to remember the roots of the phrenic nerve is "3, 4, 5 to stay alive"). An interesting application of phrenic nerve anatomy is discussed in *A&P in the Real World: A Hiccups Cure That Really Works* on p. 494.

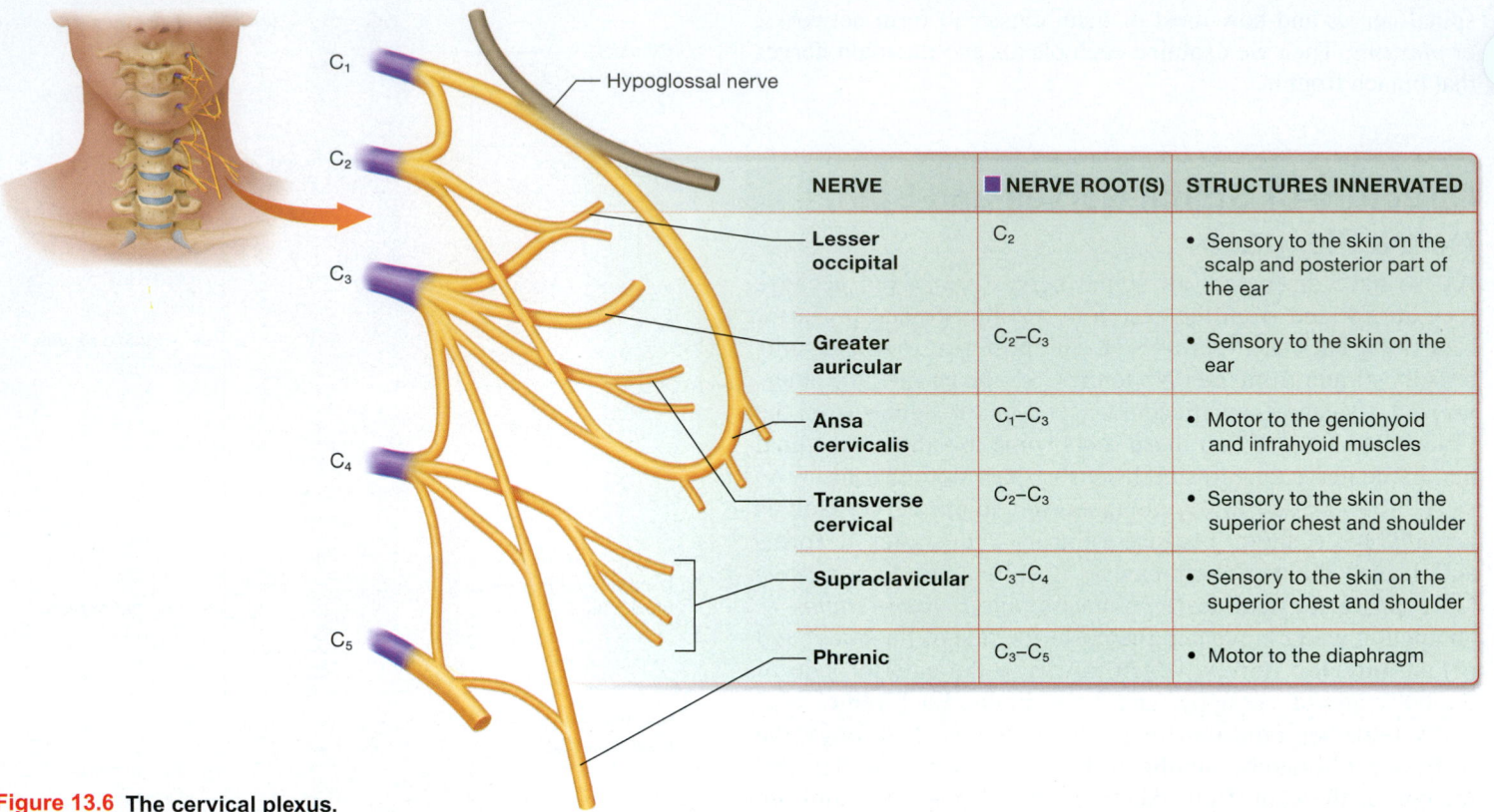

NERVE	■ NERVE ROOT(S)	STRUCTURES INNERVATED
Lesser occipital	C_2	• Sensory to the skin on the scalp and posterior part of the ear
Greater auricular	C_2–C_3	• Sensory to the skin on the ear
Ansa cervicalis	C_1–C_3	• Motor to the geniohyoid and infrahyoid muscles
Transverse cervical	C_2–C_3	• Sensory to the skin on the superior chest and shoulder
Supraclavicular	C_3–C_4	• Sensory to the skin on the superior chest and shoulder
Phrenic	C_3–C_5	• Motor to the diaphragm

Figure 13.6 The cervical plexus.

Brachial Plexuses

As implied by their name, the **brachial plexuses** provide motor and sensory innervation to the upper limb. You can see in the following lateral view that they originate from the anterior rami of spinal nerves C_5–T_1; the first structures formed in the brachial plexus are its large **trunks.** Typically, C_5 and C_6 unite to form the large *superior trunk,* C_7 forms the *middle trunk,* and C_8 and T_1 combine to form the *inferior trunk.*

This next step is where things get tricky: Each trunk splits into an *anterior division* and a *posterior division* that become the **cords** of the plexus. The anterior division of the inferior trunk forms the *medial cord,* which descends in the medial arm. The anterior divisions of the superior and middle trunks combine to form the *lateral cord,* which descends in the lateral arm. The posterior divisions of each trunk unite to form the *posterior cord,* which lies in the posterior arm; you can see it here:

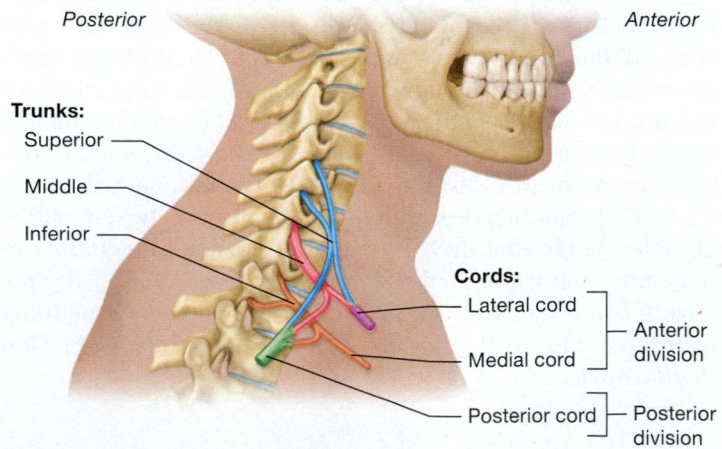

Figure 13.7a shows the entire brachial plexus in an anterior view. From different points along the trunks and cords of the brachial plexus, various nerves originate and travel to different muscles and skin regions in the arm, forearm, and hand. The five major nerves are as follows (**Figure 13.7b**):

- **Axillary nerve.** The **axillary nerve** (AKS-ih-lehr-ee) is a branch of the posterior cord of the brachial plexus. True to its name, the axillary nerve serves structures near the axilla, including the deltoid and teres minor muscles and the skin over the deltoid region.

- **Radial nerve.** As the posterior cord descends in the posterior arm, it becomes the **radial nerve.** In the posterior arm, this nerve supplies the triceps brachii muscle, after which it supplies most of the muscles in the forearm that extend the hand. A small branch continues down to the hand, where it supplies the skin over the posterior thumb, the second and third digits, and the lateral half of the fourth digit.

- **Musculocutaneous nerve.** The continuation of the lateral cord of the brachial plexus is the **musculocutaneous nerve** (muss'-kyoo-loh-kyoo-TAYN-ee-us). This nerve supplies most of the muscles that flex the forearm, including the biceps brachii muscle, and the skin over the lateral forearm.

- **Median nerve.** If you look back at Figure 13.7a, you can see that the lateral and medial cords unite to form the **median nerve,** so named because it runs approximately down the middle of the arm and forearm. In the forearm this nerve innervates the muscles in the forearm that flex the hand and the digits, and in the hand it supplies some of the intrinsic hand muscles. It also provides sensory innervation to the skin over the anterior thumb, the second and third digits, and the lateral half of the fourth digit. As the median nerve

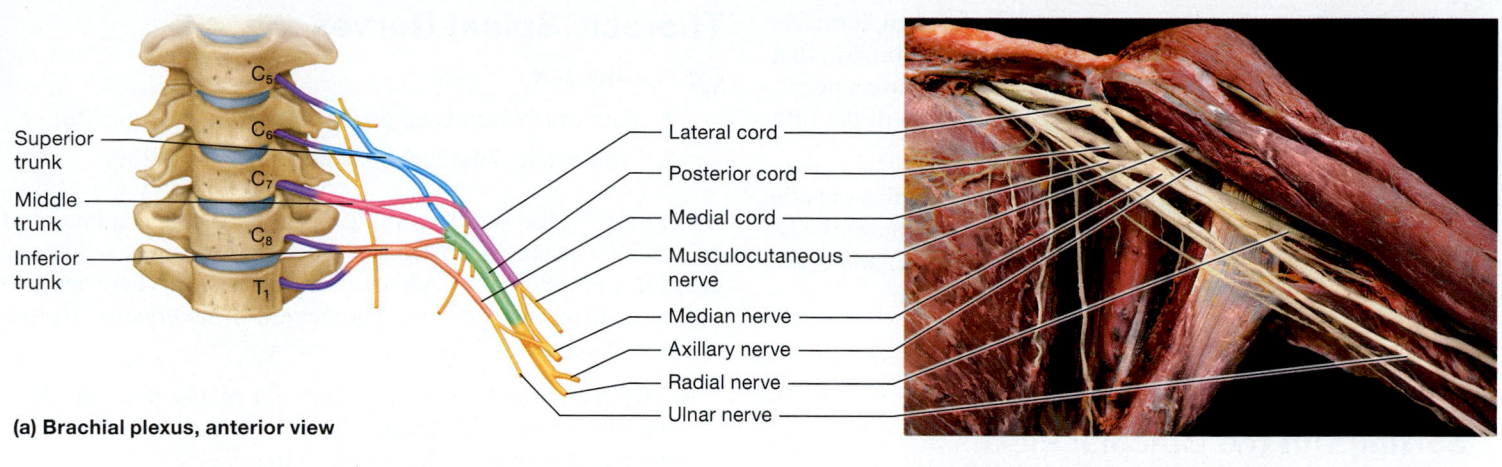

(a) Brachial plexus, anterior view

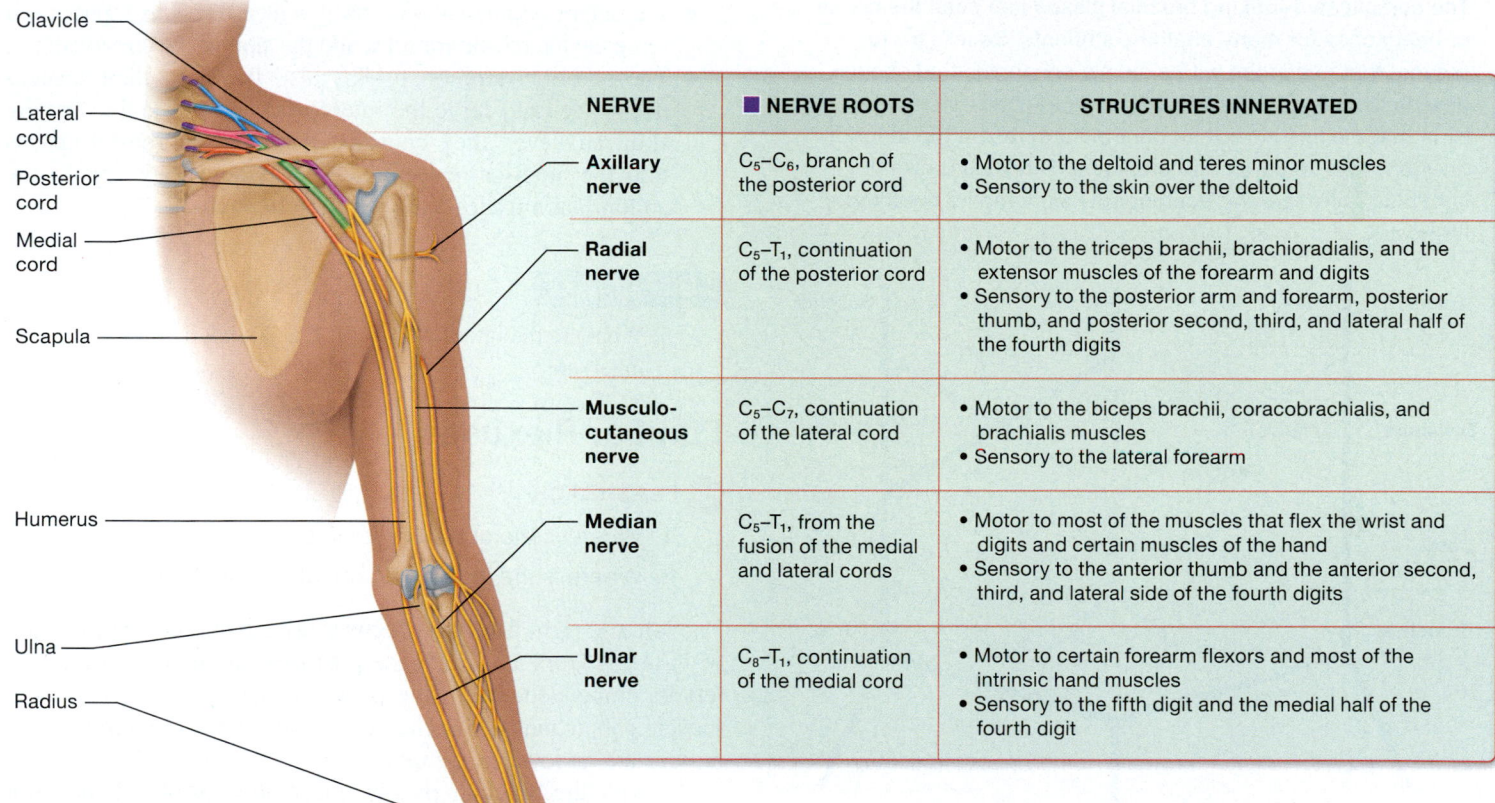

NERVE	■ NERVE ROOTS	STRUCTURES INNERVATED
Axillary nerve	C_5–C_6, branch of the posterior cord	• Motor to the deltoid and teres minor muscles • Sensory to the skin over the deltoid
Radial nerve	C_5–T_1, continuation of the posterior cord	• Motor to the triceps brachii, brachioradialis, and the extensor muscles of the forearm and digits • Sensory to the posterior arm and forearm, posterior thumb, and posterior second, third, and lateral half of the fourth digits
Musculo-cutaneous nerve	C_5–C_7, continuation of the lateral cord	• Motor to the biceps brachii, coracobrachialis, and brachialis muscles • Sensory to the lateral forearm
Median nerve	C_5–T_1, from the fusion of the medial and lateral cords	• Motor to most of the muscles that flex the wrist and digits and certain muscles of the hand • Sensory to the anterior thumb and the anterior second, third, and lateral side of the fourth digits
Ulnar nerve	C_8–T_1, continuation of the medial cord	• Motor to certain forearm flexors and most of the intrinsic hand muscles • Sensory to the fifth digit and the medial half of the fourth digit

(b) Nerves of brachial plexus, anterior view

Figure 13.7 The brachial plexus.

Explore PAL™ Cadaver
@ Mastering Anatomy & Physiology

enters the hand, it passes under a band of connective tissue that you encountered in the muscular system chapter, called the *flexor retinaculum* (see Figure 9.19). This anatomical arrangement unfortunately sometimes causes the median nerve to become inflamed and "trapped" under the retinaculum, resulting in *carpal tunnel syndrome.* This painful condition is generally treated by making a small incision in the retinaculum and releasing the pressure on the nerve.

● **Ulnar nerve.** The **ulnar nerve** is the continuation of the medial cord of the brachial plexus. It begins in the posterior arm, then passes over the medial epicondyle of the humerus to enter the forearm. At this point, the ulnar nerve is very superficial, a fact of which you may become acutely aware if you bang your elbow on a hard surface. The painful, tingling, electrical sensation that results from hitting your "funny bone" (although it is decidedly unfunny) is the effect of the slight contusion this nerve receives with such an injury. Once the ulnar nerve enters the forearm, it travels along the ulna and supplies the muscles in the forearm that flex the hand but that are not supplied by the median nerve, most of the intrinsic hand muscles, and the skin of the fifth digit and the medial side of the fourth digit.

Smaller nerves of the brachial plexus serve structures outside the arm, including the pectoralis major and minor muscles, the serratus anterior muscle, the latissimus dorsi muscle, and some of the rotator cuff muscles.

ConceptBOOST)))

Sorting Out the Brachial Plexus

The complicated-looking brachial plexus has been the cause of headaches for many anatomy students. So let's try to prevent that headache with a simple schematic that shows how the trunks, divisions, cords, and branches are related. Note that the lines drawn in dark blue represent the anterior divisions and its cords and branches. The lines drawn in light blue represent the posterior division and its cord and branches. ■

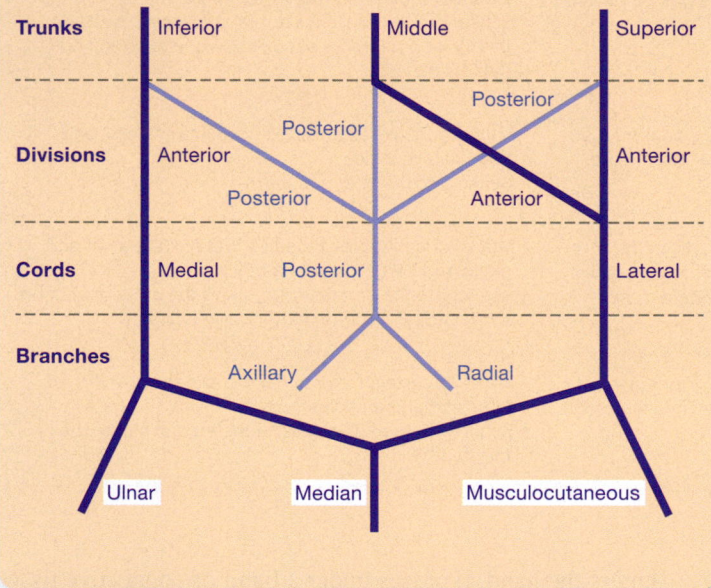

Quick Check

- ☐ 3. What are the key structures supplied by each of the following nerves?
  - a. Phrenic nerve
  - b. Ulnar nerve
  - c. Musculocutaneous nerve
  - d. Median nerve
  - e. Radial nerve
- ☐ 4. Differentiate between the trunks and cords of the brachial plexus.

Thoracic Spinal Nerves

《 FLASHBACK

1. With which vertebrae do the ribs articulate? (p. 238)
2. Where is an intercostal space located? (p. 238)

Except for T_1, the nerves coming from the thoracic spinal cord do not form plexuses. Instead, each posterior ramus serves the deep back muscles, and each anterior ramus travels between two ribs as an **intercostal nerve.** The pattern of innervation for the intercostal nerves is as follows:

- The intercostal branch of the anterior ramus of T_1 travels in the first intercostal space, where it innervates the intercostal muscles and the skin of the axilla.
- The anterior rami of T_2–T_6 travel in their respective spaces and serve the intercostal muscles and the skin of the chest wall.
- The anterior rami of T_7–T_{12} also travel in their respective spaces and serve the intercostal muscles and overlying skin. However, they continue from the intercostal spaces into the anterior abdominal wall, where they supply the abdominal muscles and the overlying skin.

Quick Check

- ☐ 5. What are the intercostal nerves, and what do they innervate?

Lumbar Plexuses

《 FLASHBACK

1. Where is the obturator foramen located? (p. 247)
2. Where is the inguinal ligament located? (p. 308)

The left and right **lumbar plexuses** arise from the anterior rami of L_1–L_4 (**Figure 13.8a**). These plexuses lie anterior to the vertebrae, embedded within the posterior part of the psoas major muscle. Their many branches primarily serve structures of the pelvis and the lower extremity.

As with the brachial plexus, most nerve roots of the lumbar plexus also separate into anterior and posterior divisions; **Figure 13.8b** shows the nerves of the anterior division. The largest nerve of this division is the **obturator nerve** (AHB-too-ray-ter). This nerve travels through the pelvis and enters the thigh by passing through the obturator foramen. Its branches serve the adductor muscles in the thigh, the hip joint, and the skin over the medial part of the thigh.

The largest nerve from the posterior division, and of the lumbar plexus as a whole, is the **femoral nerve.** It passes from the psoas major muscle through the pelvis and under the inguinal ligament to enter the thigh. The branches of the femoral nerve supply the muscles in the anterior thigh that extend the knee, including the quadriceps femoris muscle group, the skin over the anterior and medial thigh and leg, and the knee joint.

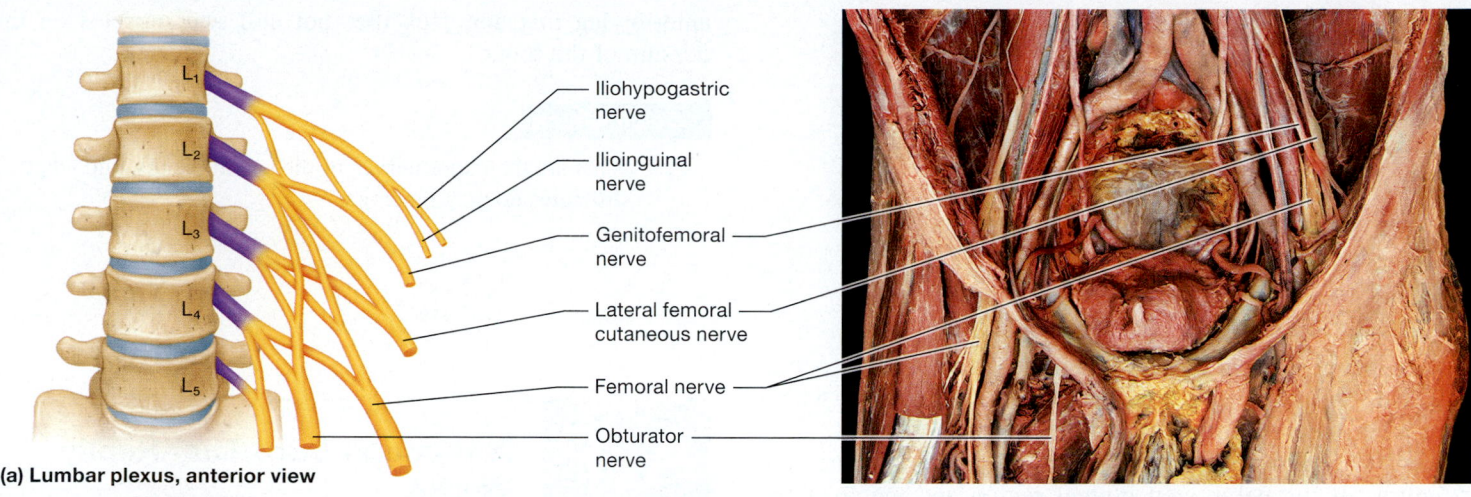

(a) Lumbar plexus, anterior view

Iliohypogastric nerve
Ilioinguinal nerve
Genitofemoral nerve
Lateral femoral cutaneous nerve
Femoral nerve
Obturator nerve

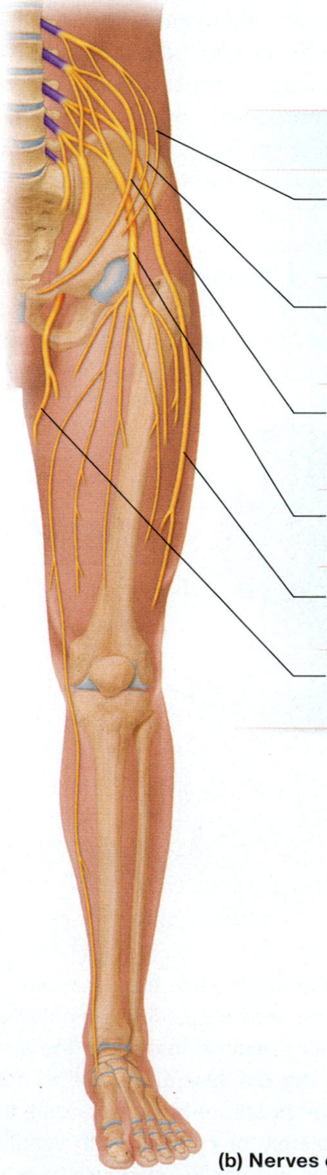

NERVE	■ NERVE ROOTS	STRUCTURES INNERVATED
Iliohypogastric nerve	L_1	• Motor to the transversus abdominis and internal oblique muscles • Sensory to the skin over the lateral gluteal region and the suprapubic region
Ilioinguinal nerve	L_1	• Motor to the transversus abdominis and internal oblique muscles • Sensory to the medial thigh, scrotum (males), and labia majora (females)
Genitofemoral nerve	L_1–L_2	• Motor to the cremaster muscle (males) • Sensory to the anteromedial thigh, scrotum (males), and labia majora (females)
Femoral nerve	L_2–L_4, posterior division	• Motor to the quadriceps femoris, iliopsoas, and sartorius muscles • Sensory to the anterior thigh, medial thigh, medial leg, and foot
Lateral femoral cutaneous nerve	L_2–L_3	• Sensory to the anterolateral thigh
Obturator nerve	L_2–L_4, anterior division	• Motor to the thigh adductors and gracilis muscle • Sensory to the superomedial thigh

(b) Nerves of lumbar plexus, anterior view

Figure 13.8 The lumbar plexus.

Quick Check

☐ 6. What are the key structures supplied by each of the following nerves?

    a. Femoral nerve

    b. Lateral femoral cutaneous nerve

    c. Obturator nerve

Sacral Plexuses

Inferior to the lumbar plexuses against the posterior pelvic wall we find the right and left **sacral plexuses,** which form from the anterior rami of spinal nerves L_4–S_4 (**Figure 13.9a**). The nerves from the sacral plexuses innervate structures of the pelvis and gluteal region and much of the lower extremity. Like the nerves of the brachial and lumbar plexuses, those of the sacral plexus separate into anterior and posterior divisions.

The largest nerve of the sacral plexus—indeed, the largest and longest nerve in the body—is the **sciatic nerve** (sy-AT-ik; *sciatic* = "of the hip"). Unlike the nerves in other plexuses, the sciatic nerve actually contains axons from both the anterior and posterior divisions that travel together and share an epineurium. The sciatic nerve travels from the pelvis through the greater sciatic notch and enters the thigh by passing between the greater trochanter and the ischial tuberosity. Within the posterior thigh, it serves the hip joint before it splits into its two main branches, typically near the lower third of the thigh: the **tibial nerve** and the **common fibular nerve.**

As you can see in **Figure 13.9b**, the tibial nerve is the larger branch—it contains the axons from the anterior division of the sacral plexus. In the thigh its branches supply muscles that extend the thigh and flex the leg, including almost all the muscles of the hamstring group (note that sometimes the sciatic nerve also supplies this muscle group before it splits into the tibial and common fibular nerves). As implied by its name, it then descends through the leg alongside the tibia, where it innervates part of the knee joint, the ankle joint, and the muscles of the leg that plantarflex the foot, such as the gastrocnemius muscle. One of the tibial nerve's early branches, at about the level of the gastrocnemius, is the *sural nerve* (SOO-rul; *sural* = "calf"), which supplies the posterior and lateral skin of the distal leg and part of the foot. At the level of the medial malleolus, the tibial nerve divides into its terminal branches, which supply the muscles and skin of the foot.

The smaller common fibular nerve, also known as the *common peroneal nerve* (pehr-OH-nee-ul; *peron-* = "fibula"), contains axons from the posterior division of the sacral plexus. This nerve descends along the lateral leg, where it supplies part of the knee joint and the skin of the anterior and distal leg. It terminates by dividing into superficial and deep branches. The superficial branch supplies the skin on the dorsum of the foot and the muscles in the lateral leg that evert the foot, whereas the deep branch supplies the muscles in the

anterior leg that dorsiflex the foot and two muscles on the dorsum of the foot.

Quick Check

☐ 7. What are the major nerves of the sacral plexus, and what structures do they innervate?

A Hiccups Cure That Really Works

The common annoyance known as *hiccups* is actually caused by spasms of the diaphragm muscle that cause a forceful inhalation of air. Numerous remedies are purported to cure hiccups, from holding your breath to having someone sneak up on you and scare you, but none have ever been shown to work reliably. However, there is one way to call a halt to many cases of hiccups, and it involves the phrenic nerve. Here's how it works: Find the approximate area of the third through fifth cervical vertebrae, which is roughly in the middle of the neck. Place your fingers about 1 centimeter lateral to the vertebral column on both sides of the neck, illustrated here:

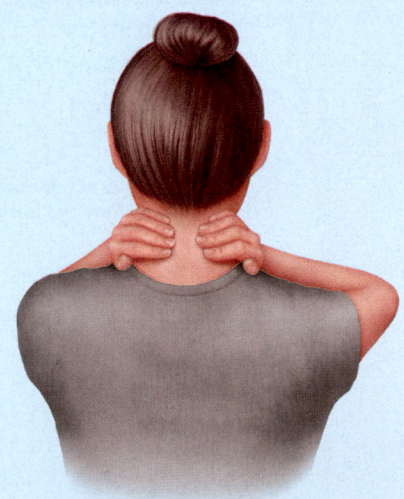

Apply firm pressure to the muscles of the neck that overlie the phrenic nerve until the hiccups stop, which should happen in about 5–10 seconds. This pressure interrupts the aberrant impulses that are causing the diaphragm muscle to contract inappropriately (although the pressure is not adequate to stop the nerve from firing completely or interfere with ventilation). Try this trick on your friends and family—they will think you're a genius when you cure their hiccups.

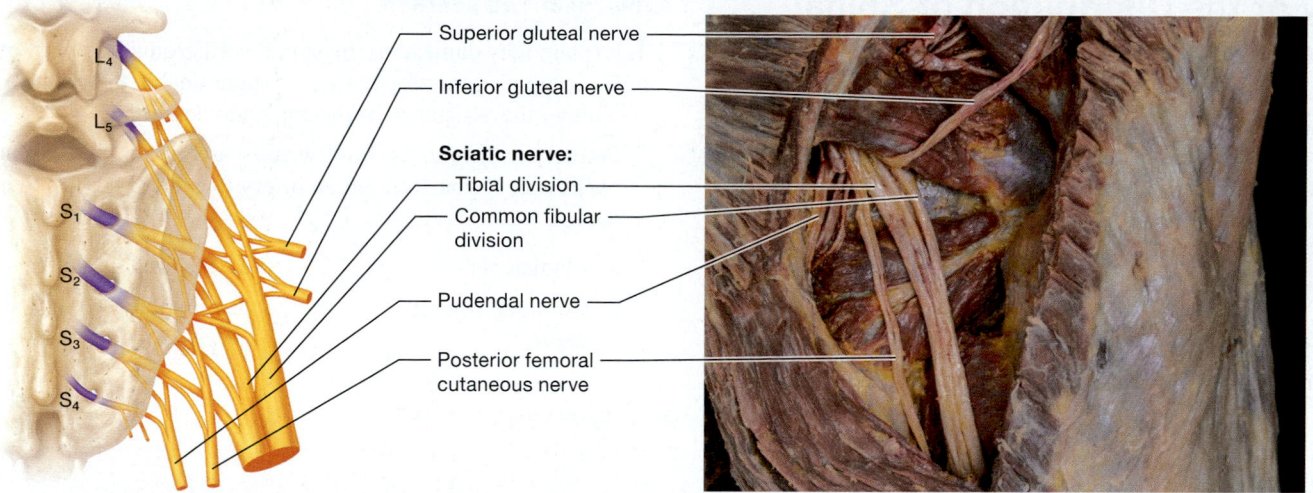

(a) Sacral plexus, posterior view

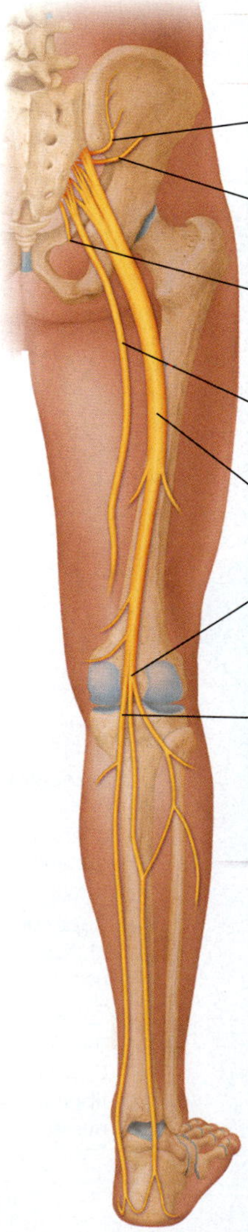

NERVE	▪ NERVE ROOTS	STRUCTURES INNERVATED
Superior gluteal nerve	L_4–S_1	• Motor to the gluteus medius, gluteus minimus, and tensor fasciae latae muscles
Inferior gluteal nerve	L_5–S_2	• Motor to the gluteus maximus muscle
Pudendal nerve	L_4–S_3	• Motor to the muscles of the pelvic floor, the external anal sphincter, and the external urethral sphincter • Sensory to the skin of the external genitalia
Posterior femoral cutaneous nerve	S_2–S_4	• Sensory to the skin of the posterior thigh
Sciatic nerve	S_1–S_3	• Sensory to the hip joint
Common fibular nerve	L_4–S_2, terminal branch of sciatic nerve	• Motor to the lateral leg muscles (superficial branch), the anterior leg muscles, and two foot muscles (deep branch) • Sensory to the knee joint, the skin of the anterior and distal leg, and the dorsum of foot
Tibial nerve	L_4–S_3, terminal branch of sciatic nerve	• Motor to the hamstring muscles, posterior leg muscles, and plantar foot muscles • Sensory to the knee joint, ankle joint, skin of the posterior and lateral leg (via the sural nerve), and skin of the plantar surface of the foot

(b) Nerves of sacral plexus, posterior view

Figure 13.9 The sacral plexus.

Summary of the Distribution of Spinal Nerve Branches

After looking at all the plexuses, let's take a moment to think about how they function together. **Figure 13.10a** summarizes the cutaneous distribution of the spinal plexuses, indicating the areas of the skin from which these nerves carry sensory stimuli. **Figure 13.10b** illustrates the motor distribution of these plexuses, showing to which groups of muscles these nerves carry motor signals. Note that the nerves are coded with the same color in both parts of the figure to help you more easily connect the nerves' sensory and motor distribution (for example, the cutaneous and motor distribution of the median nerve are both coded in yellow).

☐ 1. Explain why damage to the spinal cord around the level of C_4 or higher generally leaves a patient unable to breathe without the assistance of mechanical ventilation.

☐ 2. Predict the symptoms (both sensory and motor) you might experience if you suffered an injury to each of the following nerves:

a. Median nerve   c. Femoral nerve

b. Common fibular nerve   d. Sciatic nerve

    e. Ulnar nerve

See answers in Appendix A.

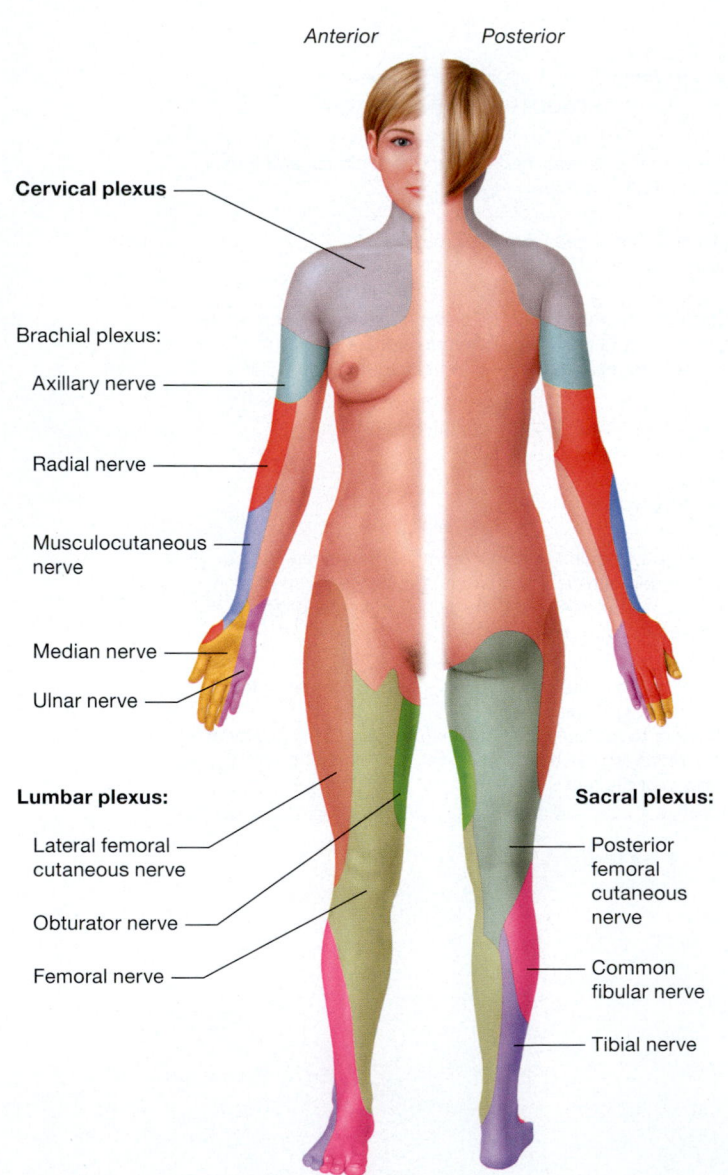

(a) Cutaneous distribution of spinal nerve plexuses

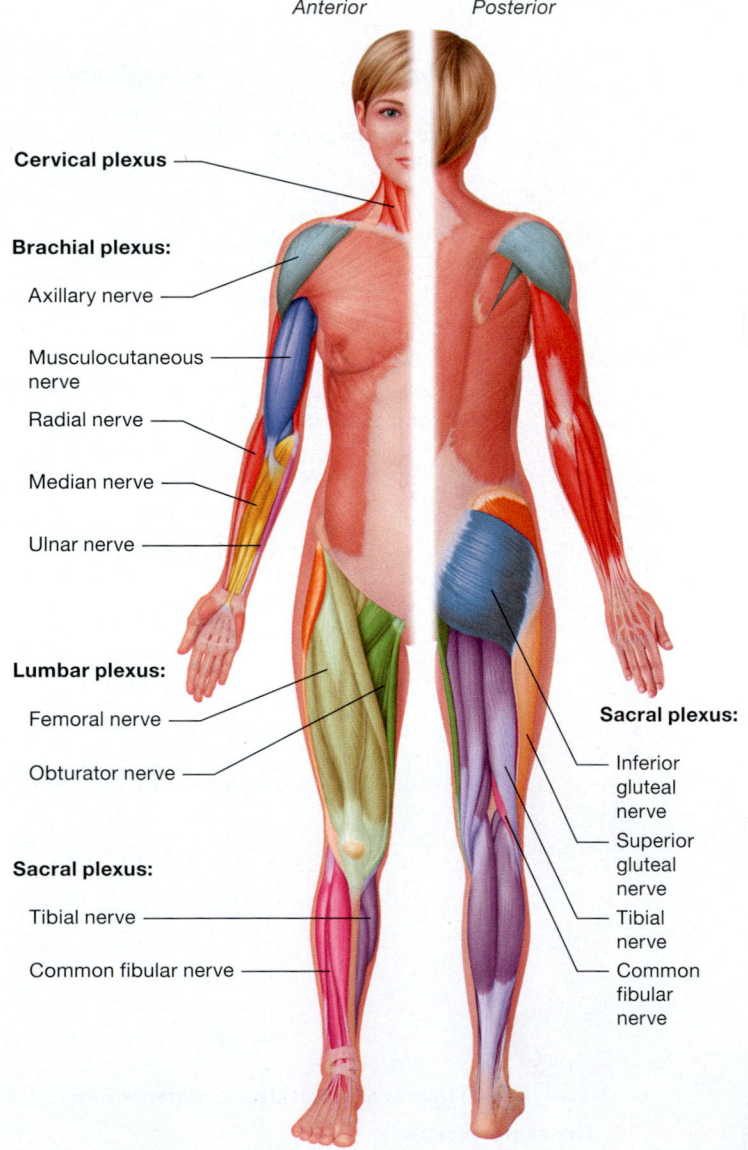

(b) Motor distribution of spinal nerve plexuses

Figure 13.10 **The distribution of spinal nerve branches.**

Sensation Part II: Role of the PNS in Sensation

Learning Outcomes

1. Compare and contrast the structure and functions of exteroceptors, interoceptors, and proprioceptors.

2. Describe the location, structure, and function of nociceptors, thermoreceptors, mechanoreceptors, chemoreceptors, and photoreceptors.

3. Explain how sensory transduction takes place at a sensory receptor.

4. Describe the pathway that a sensation takes from its detection in the PNS to its delivery to the CNS.

In the process of sensation, stimuli are initially detected by sensory neurons, which then transmit impulses to the CNS. The basic pathway of information flow is as follows:

1. Stimulus is detected by sensory receptors of the PNS →
2. transmitted by PNS sensory neurons to the CNS →
3. integrated and interpreted by CNS neurons.

We discussed the role of the CNS in sensation in the CNS chapter: the pathways through which sensory stimuli travel in the spinal cord and brain and how these stimuli are processed and interpreted (part of step 2 and step 3; see Chapter 12). Now we focus on the role of the PNS in sensation: the initial detection and transmission of sensory stimuli by sensory neurons (step 1 and part of step 2).

This module examines the cellular structures that detect sensations: sensory neurons. These neurons are located throughout the body but are especially numerous in the skin, skeletal muscles, and special sense organs. First we look at the portion of a sensory neuron that actually detects stimuli, known as *sensory receptors;* from there the focus shifts to the structure and function of entire sensory neurons. The module concludes with an explanation of how stimuli are delivered to the spinal cord and brain, and the overall pathway that a stimulus takes from detection in the PNS to interpretation by the CNS. The detection and transmission of special sensory stimuli are covered in the special senses chapter (see Chapter 15).

Sensory Receptors

« FLASHBACK

1. What are mechanically gated and voltage-gated ion channels? (p. 394)

2. What is depolarization? What is an action potential? (p. 350)

3. Where are tactile corpuscles, lamellated corpuscles, and Merkel cell fibers located? (p. 167 and p. 169)

The initial portion of the peripheral process belonging to a somatic sensory neuron features multiple small *nerve endings*. It is here that a stimulus is detected and the process of **sensory transduction**—the conversion of a stimulus into an electrical signal—takes place. Sensory transduction begins at a region of the nerve ending called a **sensory receptor,** which can be of several types. Some sensory receptors are surrounded by specialized supportive cells; these receptors are known as **encapsulated nerve endings.** Others are **free nerve endings** that lack specialized supportive cells and are "naked" (these were formerly called *free dendritic endings,* but they are part of an axon, not a dendrite, so this name is misleading).

Sensory Transduction

Figure 13.11 illustrates the basic mechanism behind sensory transduction. The axolemma of a nerve ending contains many gated ion channels that respond to various stimuli (such as the mechanically gated ion channels shown in this figure). In response to the stimulus, one or more types of ion channel open or close, changing the flow of ions across the membrane.

The basic steps in sensory transduction are as follows: ① Before any stimulus arrives, the ion channels in the axolemma of the somatic sensory neuron are closed. The process of transduction begins with a stimulus, which in our example is pressure on a finger from a probe. ② When a stimulus such as pressure is applied, mechanically gated sodium ion channels open. Sodium ions enter the axoplasm, generating a temporary depolarization referred to as a **receptor potential**. ③ If enough sodium ions enter that the membrane potential reaches threshold, voltage-gated sodium ion channels open. This triggers an action potential, which will be propagated along the axon to the spinal cord.

Different types of sensory receptors respond to stimuli with different speed, intensity, and duration. Some receptors respond rapidly and with high intensity but stop sending the stimuli after a certain period, a phenomenon known as *adaptation*. These **rapidly adapting receptors** are important for detecting the initiation of stimuli, but they ignore ongoing stimuli. Rapidly adapting receptors are the reason why you can walk around your home in search of your sunglasses only to find out that they were on top of your head the whole time. Other receptors, called **slowly adapting receptors,** respond to stimuli with constant action potentials that do not diminish with time. The dull, throbbing pain you feel for a week after spraining your ankle is the work of slowly adapting receptors.

Classification of Sensory Receptors

Sensory receptors can be broadly classified according to the location of the stimuli they detect. **Exteroceptors** (ek´-ster-oh-SEP-terz)

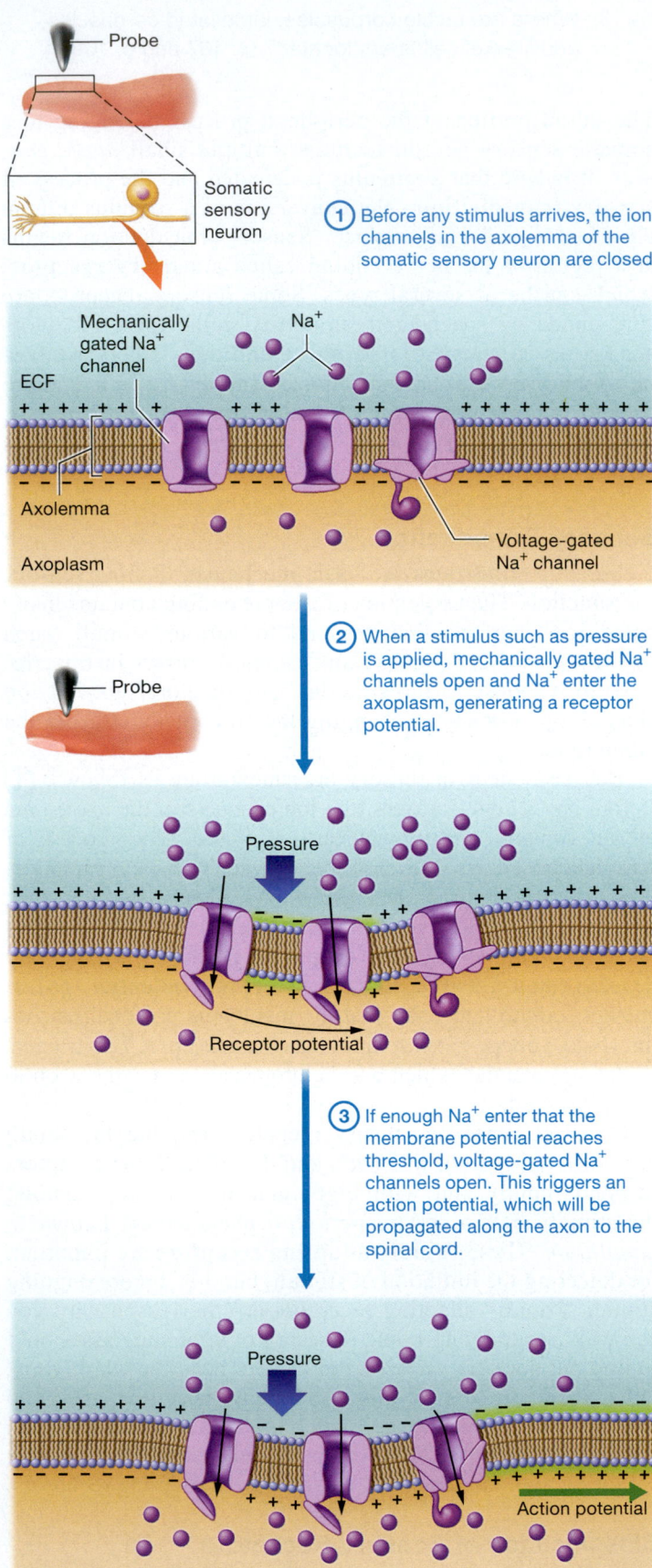

① Before any stimulus arrives, the ion channels in the axolemma of the somatic sensory neuron are closed.

② When a stimulus such as pressure is applied, mechanically gated Na⁺ channels open and Na⁺ enter the axoplasm, generating a receptor potential.

③ If enough Na⁺ enter that the membrane potential reaches threshold, voltage-gated Na⁺ channels open. This triggers an action potential, which will be propagated along the axon to the spinal cord.

Figure 13.11 Sensory transduction.

are typically close to the surface of the body. They detect stimuli originating outside the body, including an object's texture, temperature, and color; chemical odors in the air; and the level of light. **Interoceptors** (in′-ter-oh-SEP-terz) lie generally within the body's interior, where they detect stimuli originating inside the body, including blood pressure, the stretch of an organ such as a skeletal muscle or the urinary bladder, the concentration of certain chemicals in body fluids, and body temperature.

A more detailed way to classify sensory receptors is by the type of stimuli that causes them to depolarize and generate a receptor potential. This classification system gives us five types of receptors, which are the following:

- **Mechanoreceptor. Mechanoreceptors** (mek′-ah-noh-ree-SEP-terz) are encapsulated exteroceptors or interoceptors found in the skin, the musculoskeletal system, and many different organs. They depolarize in response to anything that *mechanically* deforms the tissue, including external stimuli such as light touch and vibration and internal stimuli such as stretch and pressure. The mechanism behind their sensory transduction is found within their specialized ion channels, called *mechanically gated ion channels,* which are shown in Figure 13.11.
- **Thermoreceptor.** As you can probably guess by their name, **thermoreceptors** are exteroceptors that respond to *thermal* stimuli, depolarizing in response to temperature changes. Most thermoreceptors are slowly adapting receptors. Separate thermoreceptors detect hot and cold stimuli.
- **Chemoreceptor.** A **chemoreceptor** (kee′-moh-ree-SEP-ter) is an exteroceptor or interoceptor that depolarizes in response to certain chemicals in body fluids or in the air. Chemicals that are specific for the receptor bind and trigger ion channels to open, which generates a receptor potential and perhaps an action potential. Internal chemoreceptors detect the hydrogen ion concentration, the level of carbon dioxide, and the level of oxygen in the body's fluids. External chemoreceptors are responsible for the special senses of smell and taste.
- **Photoreceptor. Photoreceptors,** found only in the eye, are special sensory exteroceptors whose membrane potentials change in response to light.
- **Nociceptor.** As you learned in the CNS chapter, *nociception* is how you detect noxious stimuli, and *pain* is how you perceive and interpret these stimuli (see Chapter 12). The receptors that depolarize in response to these stimuli are accordingly called **nociceptors** (noh-sih-SEP-terz) and are generally exteroceptors. Like thermoreceptors, nociceptors are slowly adapting receptors.

Although nociception is not a primary function of mechanoreceptors, thermoreceptors, chemoreceptors, and photoreceptors, each type of receptor can transmit stimuli that the brain can perceive as painful. Common examples include a bright light shining in your eyes and toxic chemicals with a pungent odor—neither stimulus will cause depolarization of nociceptors, but they are nonetheless perceived as painful.

See *A&P in the Real World: Capsaicin* to find out why hot peppers feel as though they burn your mouth.

This chapter primarily deals with mechanoreceptors, so we discuss these in more detail next. We also briefly address the types of thermoreceptors.

Types of Mechanoreceptors   There are six classes of mechanoreceptors in the body. These receptors, ordered from superficial to deep, are as follows (**Figure 13.12**):

- **Merkel cell fibers.** Recall from the integumentary chapter that **Merkel cell fibers,** also called *tactile cell fibers,* consist of a nerve ending surrounded by a capsule of Merkel, or tactile, cells (see Chapter 5). We find these slowly adapting receptors in the floors of the epidermal ridges, where they are most numerous in the skin of the hands, especially the fingertips. Action potentials appear to stem from mechanically gated ion channels in the nerve ending. Merkel cell fibers have the finest spatial resolution of any of the skin mechanoreceptors. For this reason, they primarily detect discriminative touch stimuli such as form and texture.

- **Tactile corpuscles. Tactile corpuscles,** also known as *Meissner corpuscles* (MYS-ner KOHR-pus-ulz), are found in the dermal papillae, projections of the dermis into the epidermis (see Chapter 5). These rapidly adapting receptors are more numerous than Merkel cell fibers. Like Merkel cell fibers, tactile corpuscles transmit discriminative touch stimuli, although their resolution is not as fine.

- **Ruffini endings.** The spindle-shaped **Ruffini endings,** also called as *bulbous corpuscles,* are located in the dermis and the hypodermis, as well as in ligaments. They are slowly adapting receptors that respond to stretch and movement; they do not transmit discriminative touch stimuli.

- **Lamellated corpuscles. Lamellated corpuscles,** formerly called *Pacinian corpuscles* (pah-SIN-ee-an), are named for their layered, onion-like appearance and are located deep

Capsaicin

Anyone who has ever eaten hot peppers knows they can make your mouth feel as though it's on fire. This happens because of a chemical in peppers called *capsaicin* (kap-SAY-sin). Capsaicin opens specific ligand-gated ion channels in nociceptors and triggers action potentials, which causes the CNS to perceive the chemical as painful. Mice that lack this receptor due to genetic engineering can ingest capsaicin as if it were water.

Interestingly, repeated application of capsaicin to nociceptors seems to desensitize them. This makes the nociceptors less likely to generate receptor potentials in response to painful stimuli. For this reason, capsaicin may be applied as a topical cream to relieve the pain of peripheral nerve disorders called *neuropathies;* the viral infection shingles, caused by the chickenpox virus; and other conditions. Remember that capsaicin doesn't do anything to treat the actual cause of pain; it simply makes the nociceptors less able to relay the painful stimuli to the CNS.

within the dermis. The layered capsule enables these receptors to perform their function—the layers act somewhat like a filter that allows only high-frequency vibratory stimuli and deep pressure to activate them. This is an example of the Structure-Function Core Principle (p. 28). Lamellated corpuscles are rapidly adapting and, like Ruffini endings, do not transmit discriminative touch stimuli.

CORE PRINCIPLE
Structure-Function

- **Hair follicle receptors.** Recall that **hair follicle receptors** are free nerve endings wrapped around the base of a hair follicle in the dermis or hypodermis (see Chapter 5). These receptors are not found in thick skin, the type of skin

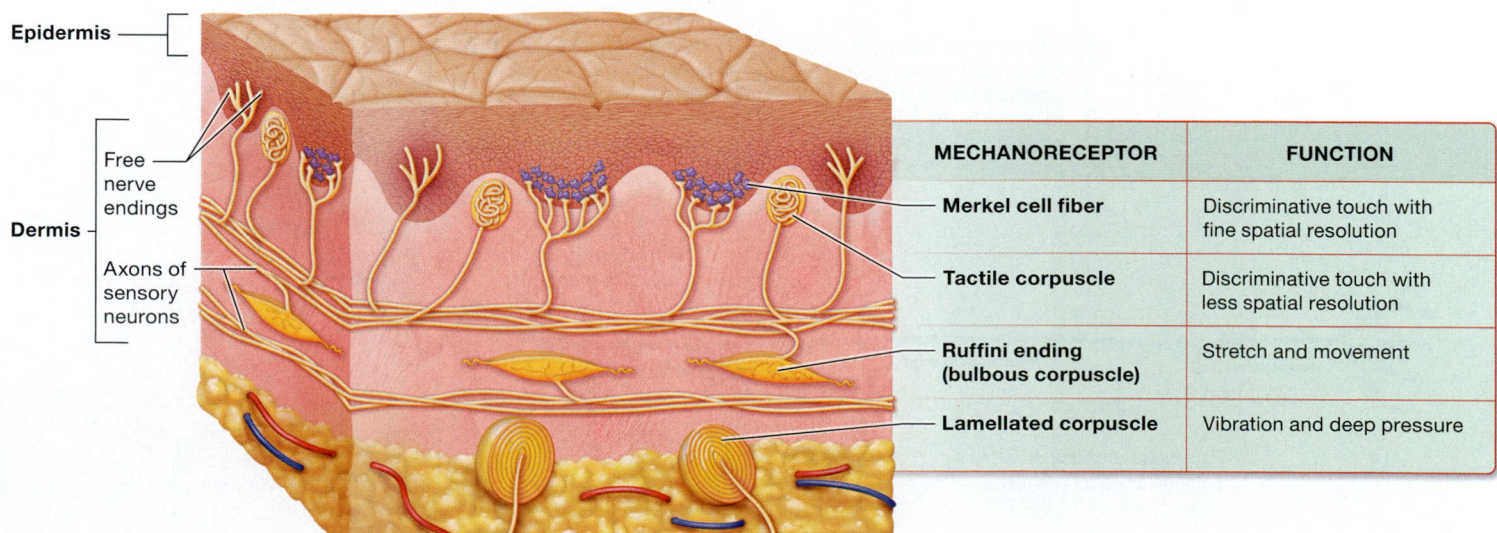

MECHANORECEPTOR	FUNCTION
Merkel cell fiber	Discriminative touch with fine spatial resolution
Tactile corpuscle	Discriminative touch with less spatial resolution
Ruffini ending (bulbous corpuscle)	Stretch and movement
Lamellated corpuscle	Vibration and deep pressure

Figure 13.12 Mechanoreceptors in the skin.

shown in Figure 13.12. Hair follicle receptors respond to stimuli that cause the hair to bend, such as an insect landing on your arm.

- **Proprioceptors.** We find **proprioceptors** in the musculoskeletal system, where they detect the movement and position of a joint or body part (they are not found in the skin, so they are not included in Figure 13.12). These receptors are integral to the body's ability to sense its position in space and to monitor ongoing movement. In other words, they are critical to the integration of sensory and motor functions.

Types of Thermoreceptors   Thermoreceptors are generally small knobs on the ends of free nerve endings in the skin. There are two types of thermoreceptors: so-called cold receptors that respond to temperatures between 10° C and 40° C (50–104° F) and hot receptors that respond to warmer temperatures between 32° C and 48° C (about 90–118° F). Cold receptors are located in the superficial dermis, whereas hot receptors lie in the deep portion of the dermis. Temperatures outside the ranges of hot and cold receptors are detected by nociceptors, which is why very hot or very cold objects feel painful to the touch.

Quick Check

☐ 1. What is sensory transduction?
☐ 2. What role do sensory receptors play in sensory transduction?

Sensory Neurons

⟪ FLASHBACK

1. What is a pseudounipolar first-order neuron? (p. 387)
2. What is an axon? (p. 386)
3. What two factors determine the speed of action potential conduction in an axon? (p. 400)

This section examines the structure and classification of somatic sensory neurons (those involved in the general senses). It also discusses the area of skin served by a somatic sensory neuron, known as the *receptive field*.

Structure of Sensory Neurons

A typical somatic sensory neuron is shown in **Figure 13.13**. Recall that we refer to these as *first-order neurons* because they are the first neurons to detect and transmit sensory stimuli along the way to the primary somatosensory cortex in the CNS (see Chapter 12). As you can see, first-order somatic sensory neurons are *pseudounipolar neurons* with three main components—a cell body and two axons:

- The cell body of the neuron is located in a **posterior root ganglion** (or *dorsal root ganglion*) found lateral to the spinal cord. The cell bodies of cranial nerves are situated in **cranial nerve ganglia** present in the head and neck.
- The *peripheral process* of the neuron is a long axon. At one end, it splits into nerve endings; associated with each nerve ending is a sensory receptor. At the other end, the peripheral process terminates near the neuron's cell body.
- The *central process* exits the cell body and travels through the posterior root of the spinal cord to enter the posterior horn (or into the brainstem for cranial nerves).

You can see how information flow works in a first-order somatic sensory neuron in Figure 13.13. In step ①, the peripheral process transmits an action potential from the sensory receptor to the neuron's other axon, the central process. In step ②, the central process transmits an action potential from the peripheral process to the posterior horn, eventually synapsing on a *second-order neuron* in the spinal cord or brainstem. An action potential propagated down the peripheral process does not generally reach the cell body. Instead, the stimulus is usually transmitted to the central process in the area where the peripheral and central processes come into contact near the cell body at the axon "stem."

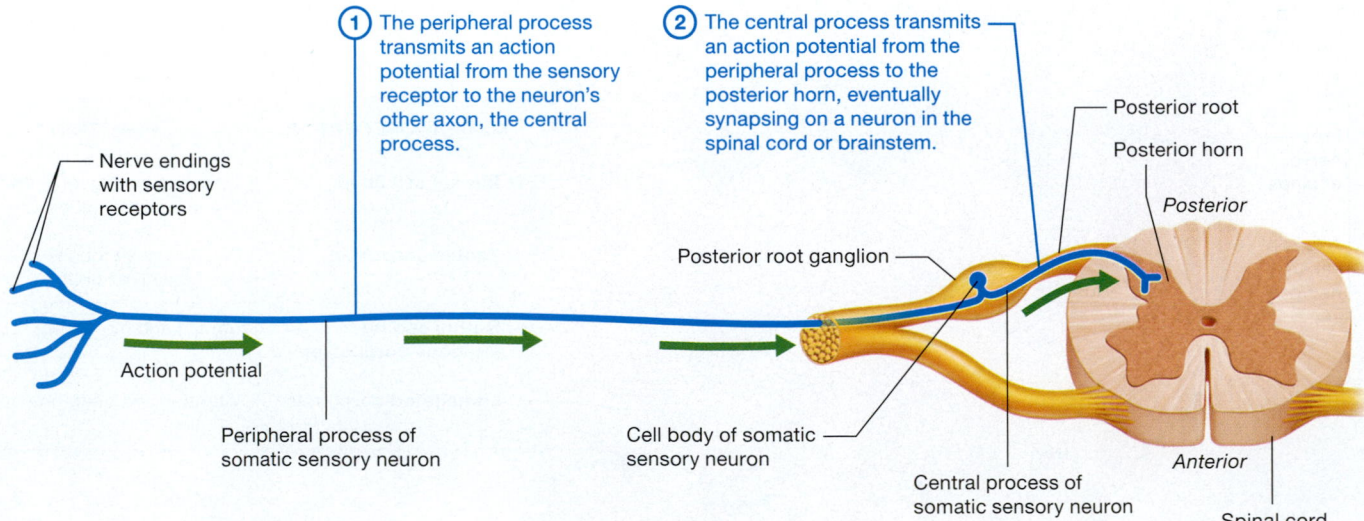

① The peripheral process transmits an action potential from the sensory receptor to the neuron's other axon, the central process.

② The central process transmits an action potential from the peripheral process to the posterior horn, eventually synapsing on a neuron in the spinal cord or brainstem.

Nerve endings with sensory receptors

Action potential

Peripheral process of somatic sensory neuron

Posterior root ganglion

Cell body of somatic sensory neuron

Central process of somatic sensory neuron

Posterior root

Posterior horn

Posterior

Anterior

Spinal cord

Figure 13.13 Somatic sensory neuron structure and function.

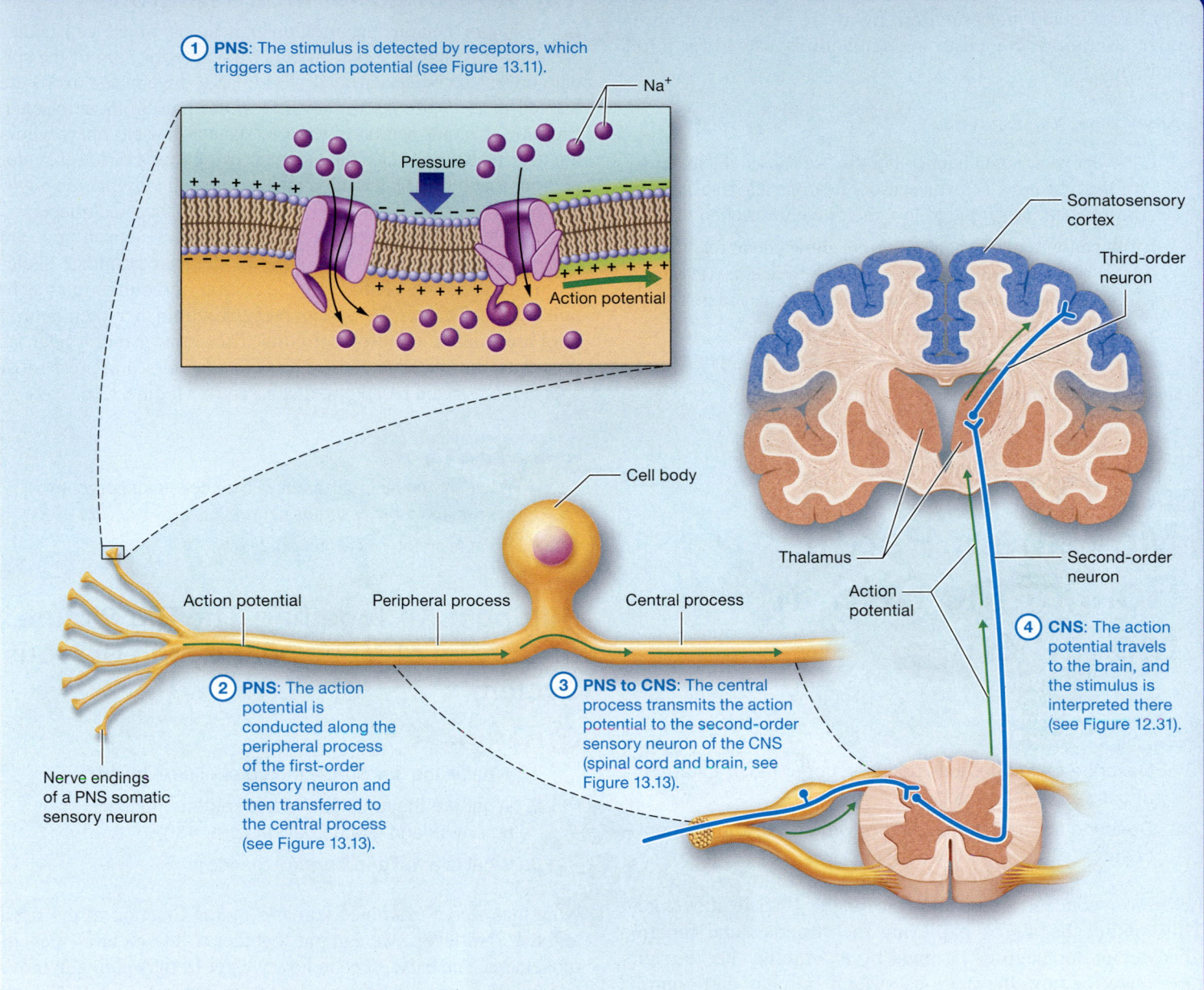

① **PNS:** The stimulus is detected by receptors, which triggers an action potential (see Figure 13.11).

Na⁺

Pressure

Action potential

Somatosensory cortex

Third-order neuron

Cell body

Thalamus

Second-order neuron

Action potential

Action potential

Peripheral process

Central process

② **PNS:** The action potential is conducted along the peripheral process of the first-order sensory neuron and then transferred to the central process (see Figure 13.13).

③ **PNS to CNS:** The central process transmits the action potential to the second-order sensory neuron of the CNS (spinal cord and brain, see Figure 13.13).

④ **CNS:** The action potential travels to the brain, and the stimulus is interpreted there (see Figure 12.31).

Nerve endings of a PNS somatic sensory neuron

---

body) along dermatomes. No known treatments cure shingles, but it may be prevented effectively with a vaccine.

Quick Check

☐ 6. What is a dermatome?

☐ 7. Why is visceral pain often perceived as cutaneous pain?

Putting It All Together: The Big Picture of the Detection and Perception of Somatic Sensation by the Nervous System

◀◀ **FLASHBACK**

1. What is the role of the thalamus in sensation? (p. 433)

2. What are the different ascending tracts in the spinal cord, and which types of stimuli are carried on each tract? (pp. 460–461)

The overall pathway for the detection and perception of somatic sensory stimuli is shown in **Figure 13.16**. This figure uses the example of a pressure or nondiscriminative touch stimulus. The stimulus is conducted to the first-order sensory neuron's central process, which enters the brainstem or the spinal cord and transmits the action potential to a second-order sensory neuron in the CNS. Recall that the pathway a stimulus takes once it enters the CNS depends on the type of stimulus (see Chapter 12). In the case of a pressure stimulus, it will synapse on a third-order sensory neuron in the thalamus and be

503

relayed to the primary and secondary somatosensory cortices for processing and interpretation. Be aware that many proprioceptive stimuli are routed to the cerebellum instead for initial processing.

☐ 1. Predict what might happen if a disease caused the axons of the first-order somatic sensory neurons that transmit proprioceptive stimuli to lose their myelin sheaths.

☐ 2. Which types of mechanoreceptors allow one to read Braille (raised dots that enable a blind person to read text)? (*Hint:* What receptors allow for fine, discriminative touch?)

☐ 3. Why is it advantageous to our survival that our thermoreceptors and nociceptors are slowly adapting?

See answers in Appendix A.

MODULE 13.5

Movement Part II: Role of the PNS in Movement

Learning Outcomes

1. Describe the differences between upper motor neurons and lower motor neurons.
2. Describe the overall "big picture" view of how movement occurs.

Let's now concentrate on the role of the PNS in movement. This module begins by exploring the structure and function of lower motor neurons. It ends by examining the "big picture" view of how the nervous system initiates and controls movement.

From CNS to PNS: Motor Output

As we discussed in the muscle tissue chapter, there is a close relationship between the muscular and nervous systems (see Chapter 10). Skeletal muscle fibers are voluntary and contract only when stimulated to do so by a somatic motor neuron. In the CNS chapter, you explored how the CNS initiates movement—**upper motor neurons** in the primary motor cortex of the cerebrum make the "decision" to move and initiate movement (with the help of the basal nuclei; see Chapter 12). However, the upper motor neurons do not contact skeletal muscle fibers, and so by themselves cannot stimulate a muscle contraction. Instead, the messages from upper motor neurons are relayed to **lower motor neurons,** which release acetylcholine onto the muscle fiber and initiate a muscle contraction.

The Role of Lower Motor Neurons

Lower motor neurons are multipolar neurons whose cell bodies are located within the CNS (in either the anterior horn of the spinal cord or the brainstem) and whose large, myelinated axons are located in the PNS. As we covered in the muscle tissue chapter, each lower motor neuron innervates skeletal muscle fibers within a single skeletal muscle (see Chapter 10). Groups of lower motor neurons that innervate the same muscle, called *motor neuron pools,* are located together in the anterior horn in rod-shaped clusters.

Most lower motor neurons within a motor neuron pool are the large α-**motor neurons.** α-Motor neurons stimulate skeletal muscle fibers to contract by the excitation-contraction mechanism (see Chapter 10). Also present within a motor neuron pool are smaller γ-**motor neurons.** These lower motor neurons innervate muscle fibers called *intrafusal fibers* that are part of specialized stretch receptors, discussed in Module 13.6.

Quick Check

☐ 1. What are the main differences between an upper motor neuron and a lower motor neuron?

☐ 2. What is a motor neuron pool?

Putting It All Together: The Big Picture of Control of Movement by the Nervous System

« FLASHBACK

1. What is the function of the cerebellum? (p. 467)
2. What are the functions of the caudate nucleus, putamen, and globus pallidus? (p. 430)
3. What is a motor program? (p. 465)

Now that we've examined somatic motor function at the CNS and the PNS levels, we can put together a "big picture" view of movement. The basic steps in how we get from deciding to move to actually making the movement are shown in **Figure 13.17**.

Quick Check

☐ 3. What is the general sequence of events for movement, and which parts of the CNS and PNS are involved?

☐ 1. Predict the effects of damage to each of the following motor neurons:

 a. Upper motor neurons of the premotor cortex

 b. Upper motor neurons of the primary motor cortex

 c. Lower motor neurons

☐ 2. Predict the effect of damage to proprioceptors and sensory neurons that communicate with the cerebellum.

See answers in Appendix A.

The Big Picture of Control of Movement by the Nervous System

Figure 13.17

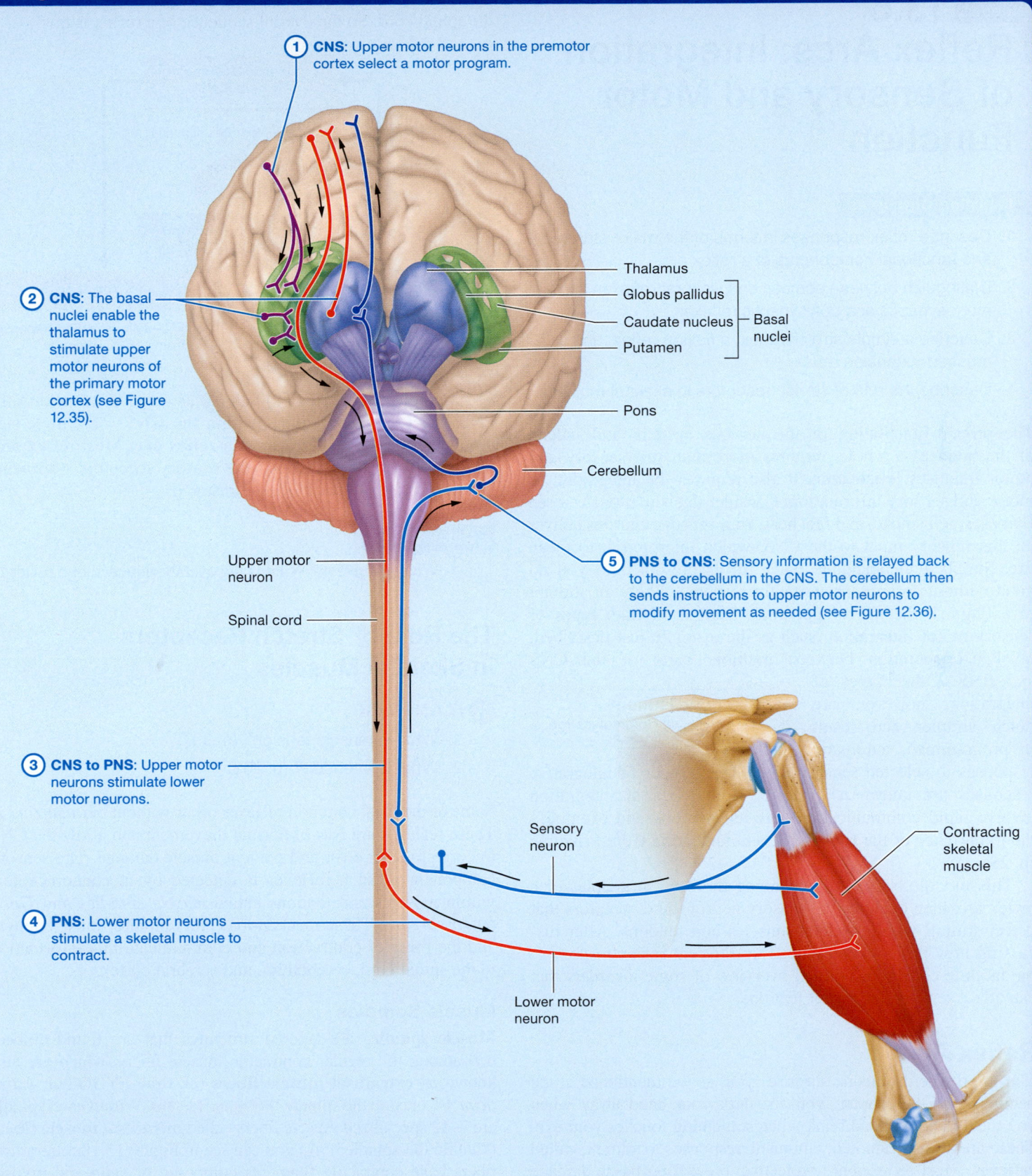

1 **CNS:** Upper motor neurons in the premotor cortex select a motor program.

2 **CNS:** The basal nuclei enable the thalamus to stimulate upper motor neurons of the primary motor cortex (see Figure 12.35).

Thalamus

Globus pallidus

Caudate nucleus — Basal nuclei

Putamen

Pons

Cerebellum

Upper motor neuron

Spinal cord

5 **PNS to CNS:** Sensory information is relayed back to the cerebellum in the CNS. The cerebellum then sends instructions to upper motor neurons to modify movement as needed (see Figure 12.36).

3 **CNS to PNS:** Upper motor neurons stimulate lower motor neurons.

Sensory neuron

Contracting skeletal muscle

4 **PNS:** Lower motor neurons stimulate a skeletal muscle to contract.

Lower motor neuron

MODULE 13.6

Reflex Arcs: Integration of Sensory and Motor Function

Learning Outcomes

1. Describe reflex responses in terms of the major structural and functional components of a reflex arc.
2. Distinguish between somatic and visceral reflexes and monosynaptic and polysynaptic reflexes.
3. Describe a simple stretch reflex, a flexion reflex, and a crossed-extension reflex.
4. Describe the role of stretch receptors in skeletal muscles.

The proper functioning of the nervous system, and indeed of the body as a whole, requires more than just sensory and motor impulses in isolation. It also requires the integration of motor and sensory information. Consider this situation: You see flames, smell smoke, and feel heat. Your sensory neurons deliver all this sensory input to the CNS, which interprets it to mean fire. Simply understanding that a fire is present does you no good without an appropriate motor response, likely an attempt to extinguish the fire or get away from it. Certain types of sensory-motor integration, such as the example just described, involve cooperation between multiple parts of the CNS and PNS. Other types of integration, however, are much simpler and result in programmed, automatic

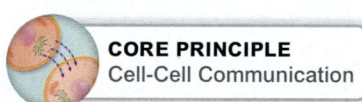

CORE PRINCIPLE
Cell-Cell Communication

responses to selected sensory input. These types of integrative responses are known as *reflexes*. Both types of integration involve rapid communication between neurons and exemplify the importance of the Cell-Cell Communication Core Principle (p. 28).

This module begins by explaining the basic principles of a reflex arc, then explores the sensory neurons and receptors that detect stimuli within skeletal muscles and tendons. After discussing how these receptors lead to different reflex responses, the module concludes with an overview of some disorders that can affect sensory and motor neurons.

Reflex Arcs

You jump when someone surprises you, as we mentioned at the beginning of the chapter; you also jerk your hand away when you touch a hot pot and blink when something touches your eye. These are all programmed, automatic responses to stimuli, called **reflexes.** Most reflexes are protective, preventing tissue damage in some way.

Notice that each of these reflexes begins with a sensory stimulus and finishes with a rapid motor response. Between the sensory stimulus and the motor response, neural integration

takes place within the CNS, typically within the brainstem or the spinal cord. So the overall sequence of a reflex is this:

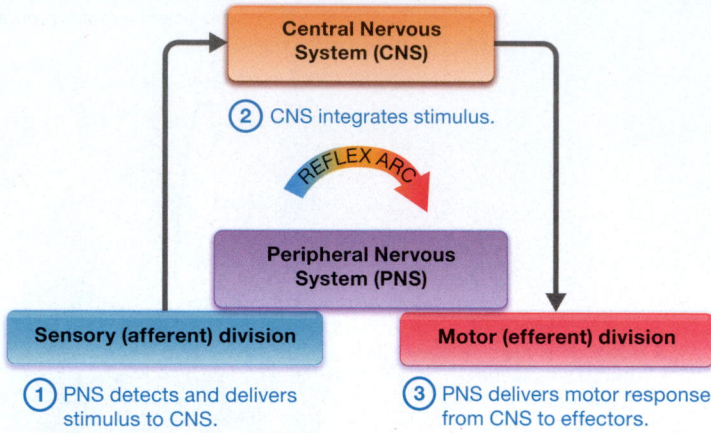

As you can see, steps ① and ③ are carried out within the PNS and ② is carried out within the CNS. This three-step sequence of events is known as a **reflex arc.** Most reflex arcs are negative feedback loops—the sensory receptors stop firing when the original conditions are restored.

Quick Check

☐ 1. What is a reflex? What is the overall sequence of a reflex?

The Role of Stretch Receptors in Skeletal Muscles

⟪ FLASHBACK

1. What is muscle tension? (p. 337)
2. What is a tendon? (p. 263)

Some of the most common reflexes occur without our realizing it. These reflexes are part of normal movement and allow the CNS to correct motor error and prevent muscle damage. The sensory component of such reflexes is detected by mechanoreceptors within muscles and tendons called *muscle spindles* and *Golgi tendon organs.* These mechanoreceptors monitor muscle length and the force of contraction and communicate this information to the spinal cord, cerebellum, and cerebral cortex.

Muscle Spindles

Muscle spindles are tapered structures that are found embedded among the regular contractile muscle fibers, which are also known as **extrafusal muscle fibers** (ek-strah-FYOO-zul; *extrafusal* = "outside the spindle") (**Figure 13.18a**). Within each spindle are 2–12 specialized muscle fibers called **intrafusal muscle fibers** ("within the spindle"). As you can see in Figure 13.18a, intrafusal fibers have contractile filaments composed of actin and myosin at their poles and a central area where contractile filaments are absent. These contractile poles are innervated by γ-motor neurons.

Two classes of sensory neurons innervate the intrafusal fibers. Both types of neurons contain mechanically gated ion channels

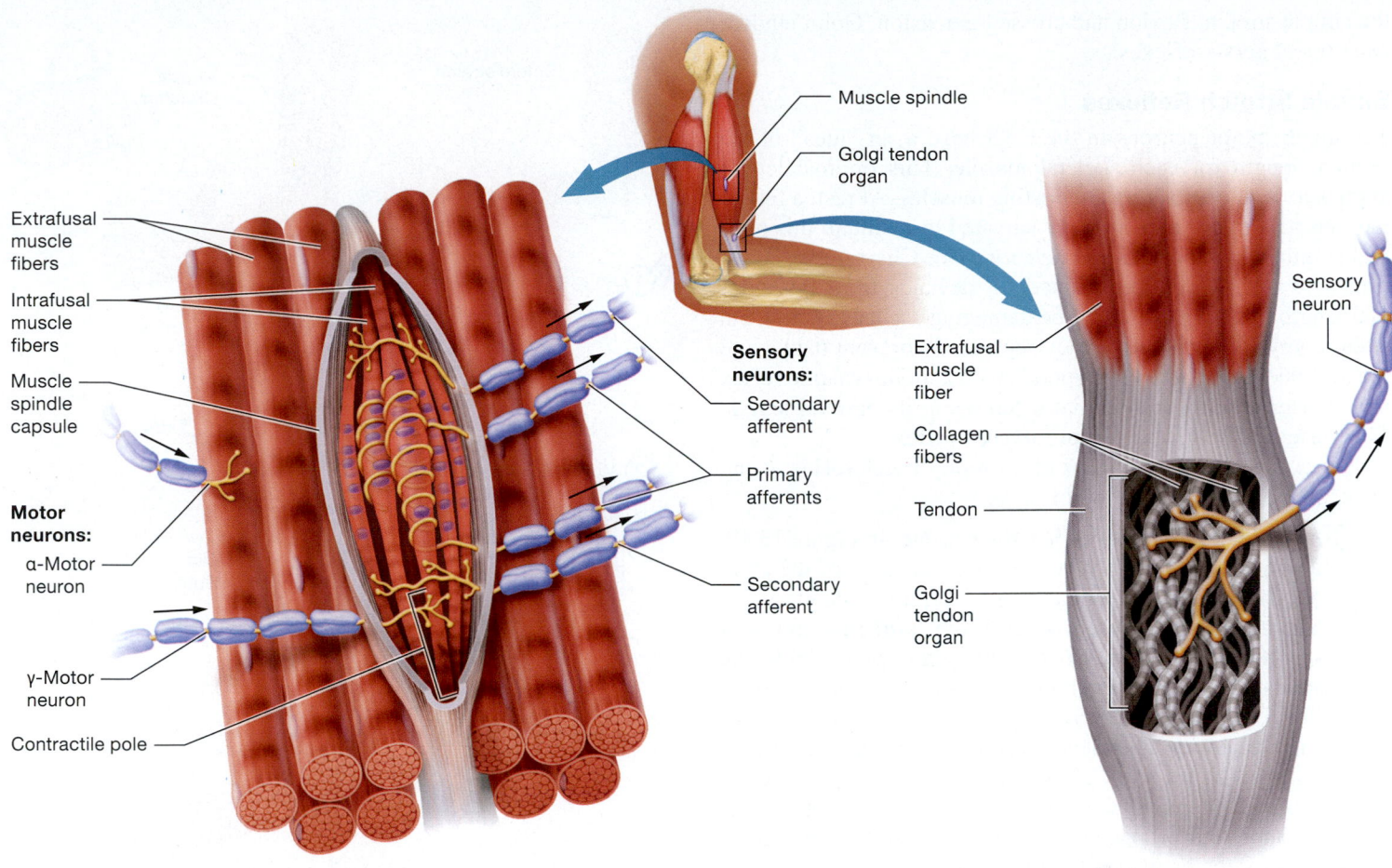

(a) Muscle spindle—stretch, length, and position

(b) Golgi tendon organ—force

Figure 13.18 **Muscle spindles and Golgi tendon organs.**

that open when the intrafusal fiber is stretched. The first type of neuron, known as a *primary afferent,* responds to stretch when it is first initiated. The second type, known as a *secondary afferent,* responds to both the static length of a muscle and the position of a limb.

Muscle groups differ in their number of muscle spindles. Those groups that produce fine movements, such as those of the hand and the extrinsic eye muscles, have large numbers of muscle spindles. This allows for precise control of muscle contractions. Conversely, muscle groups that produce coarse movements, such as the postural muscles of the back, have relatively few muscle spindles.

Golgi Tendon Organs

Golgi tendon organs are mechanoreceptors within tendons near the muscle-tendon junction; they monitor the tension generated by a muscle contraction. Note in **Figure 13.18b** that a Golgi tendon organ consists of an encapsulated bundle of collagen fibers attached to about 20 extrafusal muscle fibers. Each Golgi tendon organ contains a single somatic sensory axon whose endings are wrapped around its enclosed collagen fibers. The rate at which these neurons fire depends on the amount of muscle tension generated with each contraction—the greater the tension, the more rapidly they fire. This feedback provides the CNS with information about the force being generated by each muscle contraction.

Quick Check

☐ 2. How do intrafusal and extrafusal muscle fibers differ?

☐ 3. What are the functions of primary and secondary afferent fibers?

☐ 4. How do Golgi tendon organs and muscle spindles differ?

Types of Reflexes

Reflexes can be classified by at least two criteria. For the first criterion, we determine the number of synapses that occur between the neurons involved in the arc. The simplest reflex arcs involve only a single synapse within the spinal cord between the sensory and motor neurons; this is called a **monosynaptic reflex.** A more complicated type of reflex arc, the **polysynaptic reflex,** involves multiple synapses.

The second way to classify reflex arcs is according to the type of organ in which the reflex takes place. *Visceral reflexes* involve neurons that innervate our internal organs and are largely connected with the autonomic nervous system. We discuss some of these reflexes in the ANS chapter (see Chapter 14), and others are discussed in specific chapters throughout the book. The reflexes covered in the upcoming subsections are primarily *somatic reflexes,* which involve somatic sensory and motor neurons. We'll look closely at four examples of somatic reflexes:

the simple stretch, flexion and crossed-extension, Golgi tendon, and cranial nerve reflexes.

Simple Stretch Reflexes

The upper motor neurons in the CNS have a set "idea" of the optimal length for each skeletal muscle. This optimal length applies to both contracting and resting muscles. At rest, a few of the muscle's motor units remain activated to maintain this optimal length, which produces *muscle tone* (see Chapter 10). When a muscle is stretched (lengthened), it deviates from this optimal length in a way that could be damaging—muscle fibers can stretch only so much before they rupture. To prevent damage to the muscle fibers, the body responds with a monosynaptic reflex that shortens the muscle so that it returns to its optimal length. This reflex is known as a **simple stretch reflex.**

Figure 13.19 shows the steps of a simple stretch reflex, using the example of getting a glass of juice:

① **An external force stretches the muscle.** In Figure 13.19, the muscle is stretched by the external force of the extra weight that results from adding the liquid to the glass.

② **Muscle spindles detect the stretch, and primary and secondary afferents transmit an action potential to the spinal cord.** The stretching force stretches the intrafusal fibers in the muscle spindles, which opens the mechanically gated ion channels in their primary and secondary afferents. If the stretch is strong enough to depolarize the neuronal membranes to threshold, an action potential is initiated and propagated along their axons to the spinal cord (or the brainstem for cranial nerves).

③ **In the spinal cord, sensory afferents synapse on α-motor neurons and trigger an action potential.** The sensory afferents synapse on α-motor neurons in the spinal cord; note that upper motor neurons are not directly involved in this process. This causes the α-motor neurons to generate action potentials.

④ **The α-motor neurons stimulate the muscle to contract, and it returns to its optimal length.** The action potentials are propagated along the axons of the α-motor neurons, and when they arrive at the neuromuscular junction, they trigger the muscle to contract. In Figure 13.19, the biceps brachii muscle is stimulated to contract, and the forearm is returned to its original position.

Despite being classified as a monosynaptic reflex, this reflex arc has more going on than the stimulation of a single synapse. The sensory afferents also synapse with interneurons in the spinal cord that inhibit the antagonist muscles (in Figure 13.19, the triceps brachii muscle is the antagonist muscle that would be inhibited). This permits the agonist muscle to contract with no interference from the antagonist muscle groups.

The simple stretch reflex is easily elicited by tapping on certain tendons. Perhaps the best-known example is the **patellar,** or **knee-jerk, reflex.** The patellar reflex is elicited by using a blunt object such as a reflex hammer to tap the patellar tendon. This stretches the quadriceps femoris muscle group of the thigh, producing a quick "knee jerk" with extension of the leg. Although you likely associate this reflex with trips to the doctor's office, it is critical in keeping you upright when you're standing. When

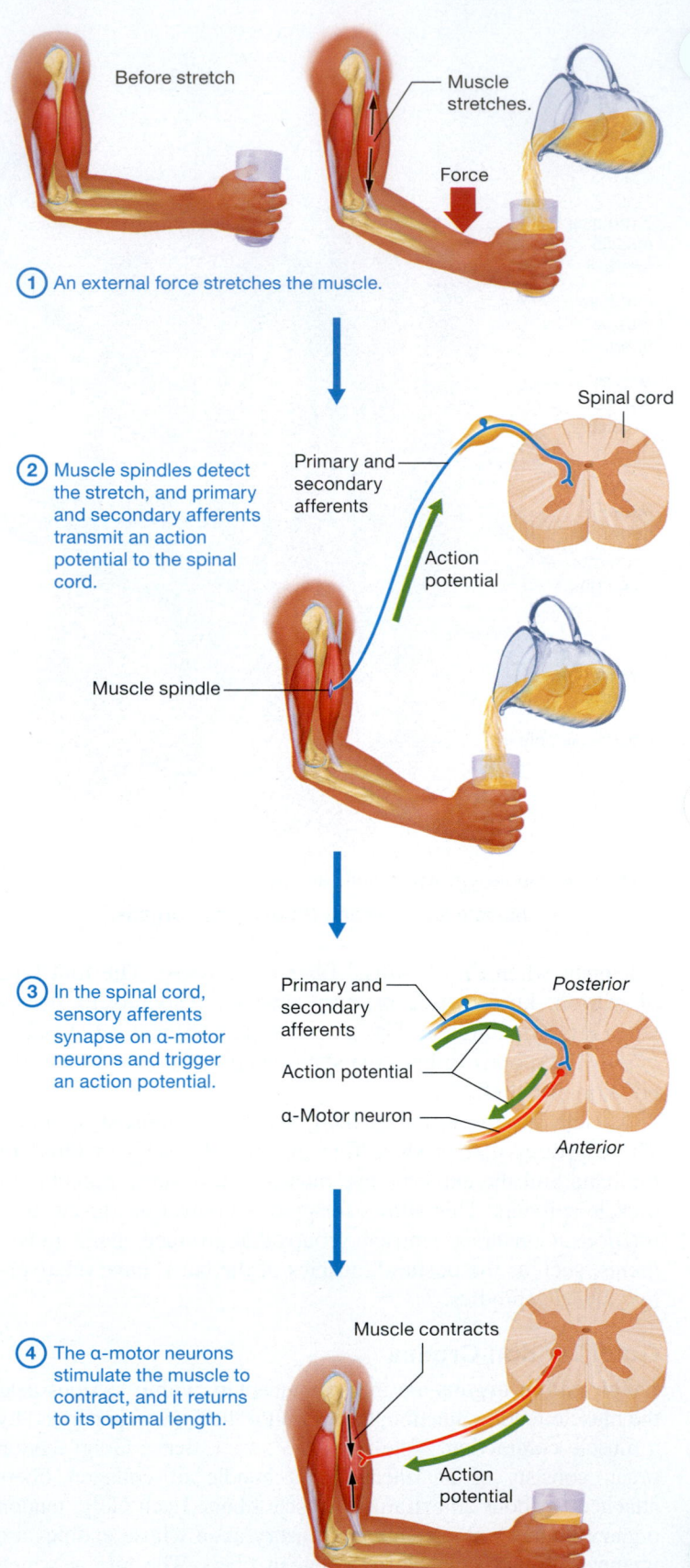

① An external force stretches the muscle.

② Muscle spindles detect the stretch, and primary and secondary afferents transmit an action potential to the spinal cord.

③ In the spinal cord, sensory afferents synapse on α-motor neurons and trigger an action potential.

④ The α-motor neurons stimulate the muscle to contract, and it returns to its optimal length.

Figure 13.19 A simple stretch reflex.

you're standing upright, any slight buckling of the knee joint will also stretch the muscles of the quadriceps femoris group. The muscle spindles detect this stretch and initiate the reflex, which results in these muscles firing to straighten the lower limb and prevent the knee joint from buckling further.

There are many examples of similar simple stretch reflexes that may be elicited with a tap. For example, the **jaw-jerk reflex,** which involves the trigeminal nerve, is elicited by tapping the chin. This slight stretch leads to a reflexive contraction of the masseter and temporalis muscles. Reflex contractions may also be generated by tapping the triceps brachii tendon and the calcaneal (Achilles) tendon. But you can see these reflexes in action in an even easier way: Simply stand up and bend over to touch your toes. When doing so, feel your posterior lower limb muscles as they become progressively tighter the more you try to stretch them. This tightening occurs because they are contracting in response to the stretch.

Flexion (Withdrawal) and Crossed-Extension Spinal Reflexes

When you touch a very hot object or step on a tack, you almost immediately pull your hand or foot away from the source of the pain. These are examples of the **flexion,** or **withdrawal, reflex.** As you can see in **Figure 13.20a,** the flexion reflex involves rapidly conducting nociceptive afferents and multiple synapses in the spinal cord, which makes it a polysynaptic reflex. ① When stimulated, the nociceptive afferents transmit the stimulus to interneurons in the spinal cord. These interneurons then synapse on to α-motor neurons. ②ₐ The α-motor neurons generate an action potential and stimulate contraction of muscles that flex the limb receiving the painful stimulus. This withdraws the affected limb from the stimulus.

A second reflex that occurs simultaneously with the flexion reflex is the **crossed-extension reflex,** which triggers extension of the opposite limb to help preserve balance. Note in **Figure 13.20b** that the spinal interneurons can synapse on two different populations of motor neurons: some that supply the flexor muscles of the affected limb, and some that supply the extensor muscles of the opposite limb. For this reason, when the flexion reflex is stimulated, step ②ₐ, flexion of the affected limb, and step ②ᵦ, extension of the opposite limb, occur simultaneously.

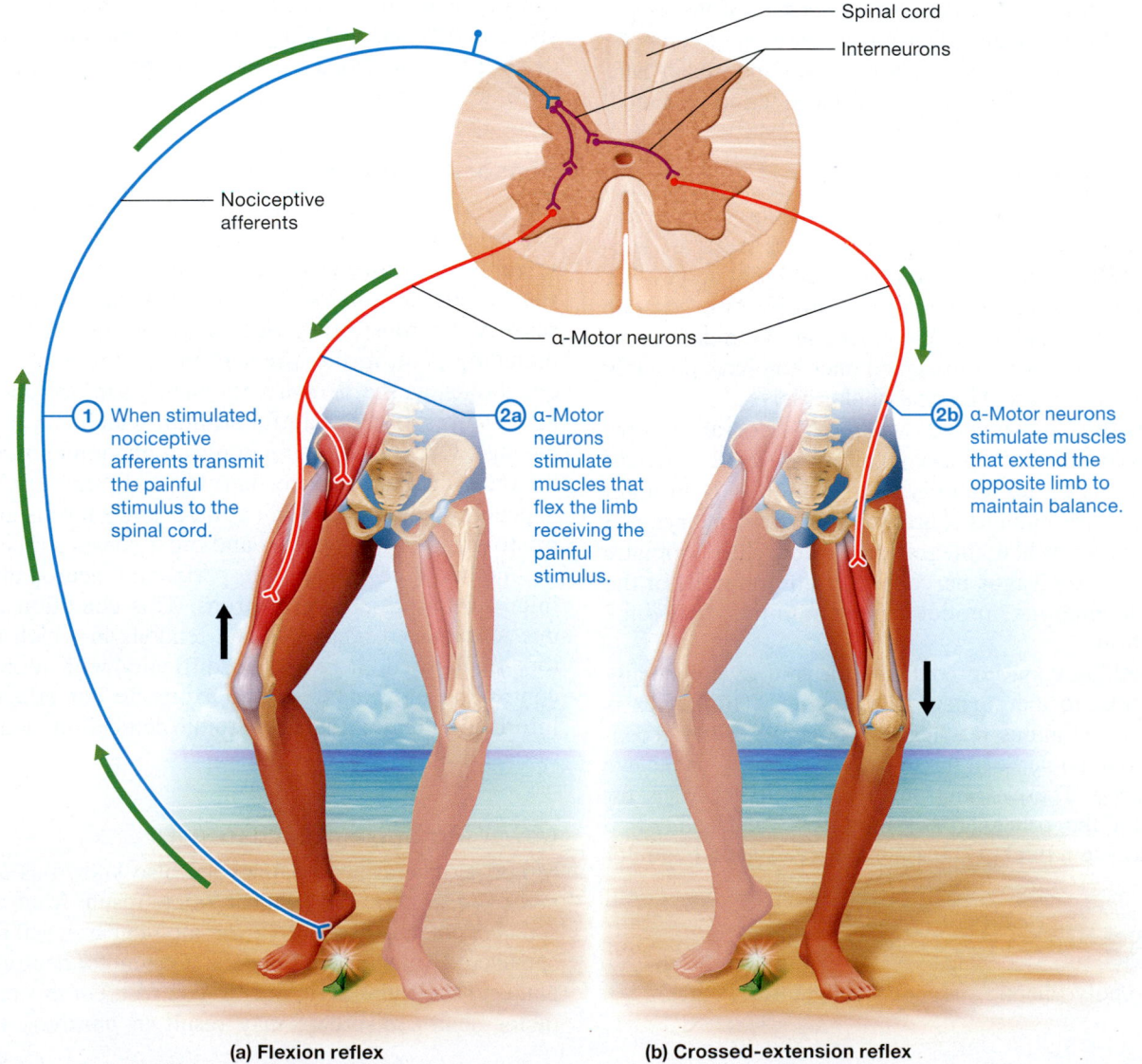

(a) Flexion reflex

(b) Crossed-extension reflex

Figure 13.20 **The flexion and crossed-extension reflexes.**

In our example, muscles that flex the right lower limb contract at the same time as muscles that extend the left lower limb. Together, these two motions allow for withdrawal from the painful stimulus (the broken glass) while providing balance and postural support. Without the crossed-extension reflex and flexion reflex happening simultaneously, we would likely lose our balance and fall when withdrawing our limb from the painful stimulus.

Golgi Tendon Reflex

A polysynaptic reflex that also protects muscles and tendons from damage is the **Golgi tendon reflex.** However, the effect of this reflex is opposite that of the simple stretch reflex—it causes muscle *relaxation.* As its name implies, this reflex involves Golgi tendon organs. When tension in the muscle and tendon increases dramatically, the Golgi tendon organs signal the spinal cord and cerebellum. This leads to inhibition of the motor neurons supplying the contracting agonist muscles and simultaneous activation of antagonist muscles. As a result, the agonist muscle relaxes and the antagonist muscles contract, protecting the muscle and tendon from potentially damaging forces.

The Golgi tendon reflex is part of the reason why you should perform certain weight-lifting exercises with a spotter—if the weight is too heavy, the Golgi tendon reflex might cause you to drop the weight. Although the reflex would prevent muscle and tendon damage, you could drop the weight on yourself and be seriously injured.

Cranial Nerve Reflexes

Several polysynaptic reflex arcs involve the cranial nerves. Recall that some structures of the head and the neck are supplied with motor and sensory neurons from different nerves. For this reason, these cranial nerve reflexes involve two separate nerves—one afferent and one efferent. Among the more important of these cranial nerve reflexes, and ones you have probably experienced, are the gag and corneal blink reflexes.

The **gag reflex** is triggered when the visceral sensory nerve endings of the glossopharyngeal nerve in the posterior throat are stimulated unilaterally. This stimulus is brought back to a medullary nucleus shared by the glossopharyngeal and vagus nerves. When integration is complete, somatic motor neurons of the vagus nerve trigger contractions of the muscles of the pharynx, producing the familiar "gagging" sound and action.

When something comes in contact with your eye, generally you blink, thanks to the **corneal blink reflex.** This reflex is triggered when a stimulus reaches the somatic sensory receptors of the trigeminal nerve in the thin outer covering of the eye called the *cornea.* These stimuli are returned to the pons for integration, and the orbicularis oculi muscle is triggered to contract, producing a blink, via the facial nerve's somatic motor neurons.

Quick Check

☐ 5. How do polysynaptic and monosynaptic reflex arcs differ?

☐ 6. Describe the reflex arcs of the following reflexes:

    a. Simple stretch reflex

    b. Flexion reflex

    c. Gag reflex

    d. Corneal blink reflex

☐ 7. How are the flexion and crossed-extension reflexes related?

Sensory and Motor Neuron Disorders

As you have seen in the preceding sections, homeostasis requires the proper functioning of both sensory and motor neurons. For this reason, the effects of damage to either type of neuron can lead to problems. Disorders that affect sensory and motor neurons of the PNS are collectively called *peripheral neuropathies* (noor-AW-puh-theez; *patho-* = "disease"). One of the most common causes of peripheral neuropathy is injury to a spinal or cranial nerve from trauma, systemic illness such as diabetes mellitus (see Chapter 16), or certain drugs and poisons such as arsenic. The upcoming subsections direct our attention to damaged sensory and motor neurons and their effect on homeostasis. The effects of damage to lower motor neurons (a type of peripheral neuropathy) and to upper motor neurons of the CNS are compared, as well.

Sensory Neuron Disorders

The symptoms and severity of sensory peripheral neuropathy depend on which spinal or cranial nerve is injured. Damage to visceral sensory neurons can produce a variety of symptoms, such as urinary and/or fecal incontinence. Injury to somatic sensory neurons often decreases or eliminates sensation, including pain, in the affected part of the body. This can be dangerous, as the person with such an injury doesn't recognize and respond to painful stimuli, which can lead to tissue damage. Occasionally, the opposite phenomenon results, and the individual experiences burning and "shooting" pain. This is generally due to inflammation of the neurons and inappropriate firing of nociceptors and other sensory receptors.

Another type of sensory peripheral neuropathy involves injury to proprioceptive neurons. This condition is generally due to damaged ligaments and tendons in which proprioceptors are housed. It results in difficulty with monitoring and controlling movement, as the axons do not effectively relay information on joint position to the cerebellum and other parts of the CNS.

Lower Motor Neuron Disorders

Motor peripheral neuropathies are also known as *lower motor neuron disorders.* They most often result from injury to a spinal or cranial nerve or to lower motor neuron cell bodies in the spinal cord. Such injuries prevent an α-motor neuron from stimulating a skeletal muscle fiber to contract. For this reason, lower motor neuron disorders may result in *paralysis* (inability to

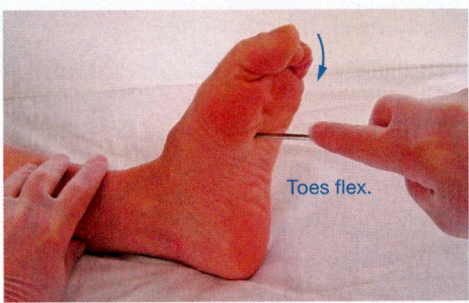

Toes flex.

(a) Plantar reflex—normal response

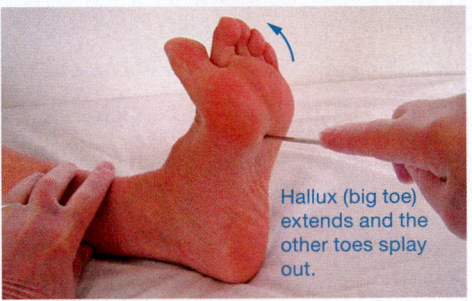

Hallux (big toe) extends and the other toes splay out.

(b) Positive Babinski sign—present in adults with upper motor neuron disorders

Figure 13.21 The Babinski sign.

move a given muscle) or *paresis* (weakness of a given muscle). The damaged α-motor neuron is also unable to respond to feedback from muscle spindles. This causes a reduction or absence of stretch reflexes and loss of muscle tone.

Upper Motor Neuron Disorders

Lower motor neuron disorders have very different symptoms from those of *upper motor neuron disorders* (note that upper motor neuron disorders are not peripheral neuropathies, as they impact neurons of the CNS). These conditions can result from damage or disease anywhere along the pathways from the motor cortices to the spinal cord. The body's initial response to upper motor neuron damage is *spinal shock,* characterized by paralysis. Spinal shock is believed to result from the "shock" experienced by spinal cord circuits when input from the upper motor neurons is removed.

For reasons that are not well understood, after a few days spinal shock wears off and in its place *spasticity* often develops. Spasticity is characterized by an increase in stretch reflexes, an increase in muscle tone, and a phenomenon called *clonus,* or the alternating contraction and relaxation of a stretched muscle. Spasticity is likely due to a loss of normal inhibition mediated by upper motor neurons. In addition to spasticity, the *Babinski sign* (bah-BIN-skee) develops. This sign is elicited by stroking the bottom of the foot, which in a healthy adult will cause flexion of the toes, a response known as the *plantar reflex* (**Figure 13.21a**). However, a patient with an upper motor neuron disorder will extend the hallux (first toe) and splay out the other toes (**Figure 13.21b**). A positive Babinski sign is often present in infants up to 18 months old and does not signify pathology, but the same response in an adult is always considered abnormal.

Some diseases affect both the upper and lower motor neurons. For one example, see *A&P in the Real World: Amyotrophic Lateral Sclerosis.*

Quick Check

☐ 8. What are some potential effects of sensory peripheral neuropathy?

☐ 9. How do upper and lower motor neuron disorders differ?

Apply What You Learned

☐ 1. An experimental drug blocks the body's muscle spindles from detecting stretch. Predict the effects of such a drug on movement, posture, and balance.

☐ 2. Mr. Pratchett has damage to his facial nerve, but not his trigeminal nerve. What would happen to his blink reflex? Explain.

☐ 3. Explain why a lower motor neuron disorder leads to absent or weakened simple stretch reflexes but an upper motor neuron disorder does not.

See answers in Appendix A.

Amyotrophic Lateral Sclerosis

The devastating disease *amyotrophic lateral sclerosis* (ay´-my-oh-TROH-fik; ALS), also known as *Lou Gehrig's disease,* involves degeneration of the cell bodies of α-motor neurons in the anterior horn of the spinal cord as well as upper motor neurons in the cerebral cortex. The cause of the degeneration is unknown at present, and many factors likely play a role.

The most common early feature of the disease is muscle weakness, particularly in the distal muscles of the limbs and the muscles of the hands. Over time the muscle weakness spreads to other muscle groups, and upper motor neuron symptoms develop as well. Death from ALS typically results within 5 years of the disease's onset. In most forms of ALS, cognitive functions are spared, so the patient is fully aware of the effects and complications of the disease. Although intensive research efforts are ongoing, at this time there is no cure or treatment that prevents the disease's progression.

Chapter Summary

For everything you need to succeed in this course, go to Mastering Anatomy & Physiology. There you will find:

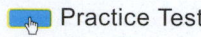 Practice Tests

Author Podcasts

Big Picture Animations

Concept Boost Video Tutors

iP2 Interactive Physiology 2.0 Tutorials

A&PFlix A&P Flix 3D Animations

Active-Learning Workbook

MODULE 13.1
Overview of the Peripheral Nervous System 475–479

- The **peripheral nervous system (PNS)** performs the sensory and motor functions of the nervous system and links the CNS with the rest of the body and its external environment.

- The PNS is classified functionally into the **sensory (afferent) division** and the **motor (efferent) division.**
  ○ The sensory division detects sensory stimuli and delivers them to the CNS. The **somatic sensory division** detects general and special sensory stimuli.
  ○ The **visceral sensory division** detects internal sensory stimuli from within the thoracic and abdominopelvic organs.
  ○ The **somatic motor division** is responsible for the body's voluntary motor functions.
  ○ The **visceral motor division,** or **autonomic nervous system (ANS),** is responsible for the body's involuntary motor functions.

- The **peripheral nerves** are the main organs of the PNS. A nerve consists of the axons of many neurons, multiple connective tissue sheaths, and blood vessels.

- The two types of nerves are based on their location:
  ○ The 12 pairs of **cranial nerves** bring impulses to and from the brain and mostly innervate structures of the head and neck.
  ○ The 31 pairs of **spinal nerves** bring impulses to and from the spinal cord and mostly innervate structures below the head and neck.

MODULE 13.2
The Cranial Nerves 479–487

- Each of the 12 pairs of cranial nerves is given a specific descriptive name and a Roman numeral that reflects the pair's position on the brain.

- The sensory cranial nerves are the following:
  ○ **Cranial nerve I,** or the **olfactory nerve,** is the nerve for the sense of smell. It originates from the neurons of the olfactory epithelium in the roof of the nasal cavity.
  ○ **Cranial nerve II,** or the **optic nerve,** is the nerve for vision. It arises from neurons in the posterior eye.
  ○ **Cranial nerve VIII,** or the **vestibulocochlear nerve,** detects the position of the body and detects sound waves.

- The motor cranial nerves are the following:
  ○ **Cranial nerve III,** or the **oculomotor nerve,** innervates four of six extrinsic eye muscles and the levator palpebrae superioris muscle, constricts the pupil, and changes the shape of the lens for better near vision.
  ○ **Cranial nerve IV,** or the **trochlear nerve,** innervates one of six extrinsic eye muscles.
  ○ **Cranial nerve VI,** or the **abducens nerve,** innervates one of six extrinsic eye muscles.
  ○ **Cranial nerve XI,** or the **accessory nerve,** innervates certain muscles of the larynx and the trapezius and sternocleidomastoid muscles.
  ○ **Cranial nerve XII,** or the **hypoglossal nerve,** innervates the intrinsic and extrinsic muscles of the tongue.

- The mixed cranial nerves are the following:
  ○ **Cranial nerve V,** or the **trigeminal nerve,** supplies much of the skin of the face and the muscles of mastication.
  ○ The neurons of **cranial nerve VII,** or the **facial nerve,** innervate the muscles of facial expression; trigger secretion from the lacrimal, salivary, and nasal mucous glands; provide taste sensation; and provide some somatic sensation.
  ○ **Cranial nerve IX,** or the **glossopharyngeal nerve,** supplies taste sensation; provides some somatic sensation; detects changes in blood pressure; supplies muscles involved in swallowing; and triggers secretion from certain salivary glands.
  ○ **Cranial nerve X,** or the **vagus nerve,** provides some somatic and visceral sensation; provides taste sensation to the pharynx; detects the carbon dioxide concentration in the blood; and supplies muscles of swallowing. It supplies parasympathetic innervation to most of the thoracic and abdominal viscera.

MODULE 13.3
The Spinal Nerves 488–496

- Spinal nerves form from the fusion of an **anterior root** and a **posterior root.**

- About 1–2 cm after a spinal nerve forms, it splits into a **posterior ramus** and an **anterior ramus.**

- Anterior roots of the spinal nerves carry only motor signals, and posterior roots only sensory signals.

- There are 31 pairs of spinal nerves. The anterior rami of the cervical, lumbar, and sacral spinal nerves merge to form **nerve plexuses.**

- The **cervical plexuses** consist of the anterior rami of C₁–C₄; they innervate the skin of the neck and portions of the head, chest, and shoulders. Their motor branches innervate certain muscles of the neck, and the **phrenic nerve** innervates the diaphragm muscle.

- The **brachial plexuses** stem from C₅–T₁ and innervate the skin and muscles of the upper limb. The nerves of the brachial plexus include the **axillary nerve,** the **radial nerve,** the **musculocutaneous nerve,** the **median nerve,** and the **ulnar nerve.**

- The anterior rami of the thoracic spinal nerves form the **intercostal nerves.**

- The **lumbar plexuses** stem from the anterior rami of L₁–L₄ and innervate structures of the pelvis and the lower limb. Their nerves include the **obturator nerve** and the **femoral nerve.**

- The **sacral plexuses** stem from the anterior rami of spinal nerves L₄–S₄ and innervate structures of the pelvis, the gluteal region, and much of the lower limb. The largest nerve in the body is the **sciatic nerve,** which splits into the **tibial nerve** and the **common fibular nerve.**

MODULE 13.4
Sensation Part II: Role of the PNS in Sensation 497–504

- **Sensory transduction,** the conversion of a stimulus into an electrical signal, takes place at a **sensory receptor.** During sensory transduction, a **receptor potential** is generated, which may trigger an action potential.

- **Interoceptors** detect stimuli originating within the body. **Exteroceptors** detect stimuli originating outside the body.

- **Mechanoreceptors** detect stimuli that mechanically deform the tissue; **thermoreceptors** detect thermal stimuli; **chemoreceptors** detect the presence or concentration of chemicals; **photoreceptors** detect light; and **nociceptors** detect painful stimuli.

- Sensory, or afferent, neurons detect and transmit sensory stimuli.

- Somatic sensory neurons are first-order pseudounipolar neurons.

- Large-diameter and heavily myelinated sensory neurons transmit impulses the fastest.

- The *receptive field* of a neuron is the area served by that neuron. The size of a receptive field may be measured by assessing the **two-point discrimination threshold.**

- The skin can be divided into **dermatomes** that are innervated by a single cranial or spinal nerve. Visceral pain is often perceived as cutaneous pain along the dermatome for the nerve that supplies the organ. This is called **referred pain.**

MODULE 13.5
Movement Part II: Role of the PNS in Movement 504–505

- The CNS initiates movement when **upper motor neurons** in the motor cortices make the "decision" to move. **Lower motor neurons** contact the muscle fiber and initiate a muscle contraction.

- Lower motor neurons are multipolar neurons whose cell bodies are located in the anterior horn of the spinal cord or in the brainstem. Groups of lower motor neurons that innervate the same muscle are located together in the spinal cord in *motor neuron pools.*

MODULE 13.6
Reflex Arcs: Integration of Sensory and Motor Functions 506–511

- A **reflex** is a programmed, automatic motor response to a specific sensory input. The series of events that culminates in a reflex response is a **reflex arc.** Reflexes may be *visceral* or *somatic,* and **monosynaptic** or **polysynaptic.**

- **Muscle spindles** and **Golgi tendon organs** detect the sensory component of reflexes that monitor and control movement.
  ○ Muscle spindles contain **intrafusal muscle fibers,** which detect stretch.
  ○ Golgi tendon organs detect the tension generated by a muscle contraction.

- A **simple stretch reflex** is a monosynaptic reflex that returns a muscle to its optimal length after being stretched.

- The Golgi tendon reflex causes the muscle to lengthen in response to increased muscle tension.

- The **flexion,** or **withdrawal, reflex** stimulates contraction of flexor muscles to withdraw the affected limb from a painful stimulus. The **crossed-extension reflex** occurs simultaneously and causes contraction of the extensor muscles of the opposite limb.

- The cranial nerves are responsible for the **gag reflex** and the **corneal blink reflex.**

- Disorders of nerves of the PNS are *peripheral neuropathies.*

Assess What You Learned

Scan the QR Code for additional practice test questions

https://goo.gl/6lwuW3

LEVEL 1 Check Your Recall

1. Mark the following statements as true or false. If a statement is false, correct it to make a true statement.

   a. The somatic sensory division of the PNS detects sensory stimuli from the organs in the thoracic and abdominopelvic cavities.

   b. The somatic motor division of the PNS consists of lower motor neurons that directly innervate skeletal muscle fibers.

   c. The visceral motor division is also known as the autonomic nervous system and maintains homeostasis of many physiological variables.

   d. The term nerve is the equivalent of the term neuron.

   e. There are 31 pairs of spinal nerves and 12 pairs of cranial nerves.

2. Fill in the blanks: A spinal nerve divides into a(n) _____ that serves the anterior side of the body and the limbs and a(n) _____ that serves the posterior side of the body.

3. Define each of the following terms in your own words, using 20 or fewer words.

   a. Peripheral nerve
   b. Nerve plexus
   c. Posterior root ganglion

4. First, write the Roman numeral that corresponds to each named cranial nerve (after the abbreviation CN). Second, match the cranial nerve with its correct function from the column on the right.

   CN _____
   _____ Vestibulocochlear nerve

   CN _____
   _____ Trigeminal nerve

   CN _____
   _____ Hypoglossal nerve

   CN _____
   _____ Abducens nerve

   CN _____
   _____ Vagus nerve

   CN _____
   _____ Olfactory nerve

   CN _____
   _____ Accessory nerve

   CN _____
   _____ Oculomotor nerve

   CN _____
   _____ Facial nerve

   CN _____
   _____ Optic nerve

   CN _____
   _____ Glossopharyngeal nerve

   CN _____
   _____ Trochlear nerve

   a. Motor to the lateral rectus muscle
   b. Motor to the muscles of facial expression; lacrimation; salivation; taste to the anterior two-thirds of the tongue
   c. Sense of smell
   d. Motor to the muscles for swallowing; salivation; taste to the posterior one-third of the tongue; somatic sensation from the throat
   e. Senses of hearing and equilibrium
   f. Motor to the superior oblique muscle
   g. Motor to the tongue
   h. Motor to the sternocleidomastoid and trapezius muscles
   i. Sense of vision
   j. Motor to muscles of swallowing and speaking; parasympathetic innervation to thoracic and abdominal viscera; sense of taste from the throat
   k. Sensory to the face; motor to the muscles of mastication
   l. Motor to four of six extrinsic eye muscles; constricts the pupil; changes the shape of the lens; opens the eyelid

5. Which cranial nerves are sensory only, primarily motor, and mixed?

6. Match the following nerves with the structures they innervate.

   _____ Phrenic nerve
   _____ Median nerve
   _____ Femoral nerve
   _____ Tibial nerve
   _____ Radial nerve
   _____ Intercostal nerves
   _____ Common fibular nerve
   _____ Musculocutaneous nerve

   a. Motor to the triceps brachii muscle and muscles in the forearm that extend the hand; sensory from the posterior hand
   b. Motor to the muscles in the anterior arm that flex the forearm; sensory from skin over the lateral forearm
   c. Motor to the muscles in the anterior and lateral leg that evert and dorsiflex the foot; sensory from the skin of the anteroinferior leg
   d. Motor to the diaphragm muscle
   e. Motor to the muscles in the anterior thigh extend the knee; sensory from the skin over the anterior thigh and leg
   f. Motor to the hamstring muscles that extend the thigh and flex the leg, muscles of the leg that plantarflex the foot, and muscles of the foot; sensory from the skin over the posterior and lateral leg and foot
   g. Motor to the muscles between the ribs and the abdominal muscles; sensory from the skin over the abdomen
   h. Motor to the muscles in the forearm that flex the hand, certain intrinsic hand muscles; sensory from the skin of the anterior hand

7. First-order somatic sensory neurons are _____ neurons whose cell bodies are located in the _____.

   a. multipolar, posterior horn
   b. pseudounipolar, posterior root ganglion
   c. bipolar, anterior horn
   d. pseudounipolar, posterior horn

8. A receptor potential:

   a. always leads to an action potential.
   b. never leads to an action potential.
   c. causes hyperpolarization of the neuron.
   d. leads to an action potential if the stimulus is strong enough.

9. Why is visceral pain often perceived as cutaneous pain?

10. Merkel cell fibers, tactile corpuscles, Ruffini endings, and lamellated corpuscles are all types of:

    a. nociceptors.   c. photoreceptors.
    b. mechanoreceptors.   d. chemoreceptors.

11. Place the following sequence of events for the detection of somatic sensation in the proper order. Place a 1 by the first event, a 2 by the second event, and so on.

    a. _____ The central process transmits the action potential to a second-order sensory neuron in the CNS.

    b. _____ The action potential is transferred to the central process in the posterior root ganglion.

    c. _____ The stimulus triggers an action potential.

    d. _____ The signal is transferred to other CNS sensory neurons for eventual perception and interpretation.

    e. _____ The action potential is propagated along the peripheral process of the neuron.

12. How do upper and lower motor neurons differ?

13. List and describe the basic steps involved in producing movement, beginning with the upper motor neurons in the cerebral cortex.

14. The lower motor neurons that innervate contractile skeletal muscle fibers are called:

    a. α-motor neurons.

    b. β-motor neurons.

    c. upper motor neurons.

    d. γ-motor neurons.

15. Fill in the blanks: _____ detect the degree to which a muscle is stretched, whereas _____ detect the force of a muscle contraction.

16. Which of the following is the correct order of events of a reflex arc?

    a. Stimulus detection and delivery → motor response → integration in the CNS

    b. Motor response → stimulus detection and delivery → integration in the CNS

    c. Stimulus detection and delivery → integration in the CNS → delivery of motor response

    d. Integration in the CNS → motor response → stimulus detection and delivery

17. Mark the following statements as true or false. If a statement is false, correct it to make a true statement.

    a. The simple stretch reflex is a monosynaptic reflex with only a single synapse in the spinal cord between the sensory and motor neurons.

    b. The Golgi tendon organs detect stretch in a simple stretch reflex.

    c. A flexion reflex is a monosynaptic reflex with only one synapse in the spinal cord.

    d. The crossed-extension reflex occurs simultaneously with the simple stretch reflex.

18. Which two cranial nerves mediate the gag reflex?

LEVEL 2 Check Your Understanding

1. Devise a series of physical examination tests that assess the function of each of the cranial nerves.

2. Predict what might happen if nociceptors were rapidly adapting instead of slowly adapting receptors. Would this change be beneficial or potentially harmful?

3. Explain why you lose both motor and sensory function of a part of your body when a spinal nerve is numbed with anesthetic agents.

LEVEL 3 Apply Your Knowledge

PART A: Application and Analysis

1. Complaining of muscle weakness, Delia goes to the doctor. Devise a series of physical examination tests that would help to determine if her problem involves upper or lower motor neurons.

2. Jason presents for evaluation after a severe shoulder injury during which his entire brachial plexus suffered damage. What effects would you expect Jason to have from this injury?

3. When Mr. Williams goes to the emergency department with pain in the area along the midline of the diaphragm, he worries he is having a heart attack. Could the pain be related to his heart? Could it be related to any other organ(s)? Explain.

4. Maria is a 3-year-old who has been diagnosed with CIPA, or congenital insensitivity to pain with anhidrosis. This disease results from a genetic mutation that causes essentially all general sensory neurons to not function properly. What types of sensations will Maria be unable to detect (be specific)? Predict what problems she might face from her condition.

PART B: Make the Connection

5. Another feature of CIPA is anhidrosis, or the inability to sweat. Explain why Maria's inability to produce sweat could potentially be life-threatening. (*Connects to Chapter 5*)

See answers in Appendix A.

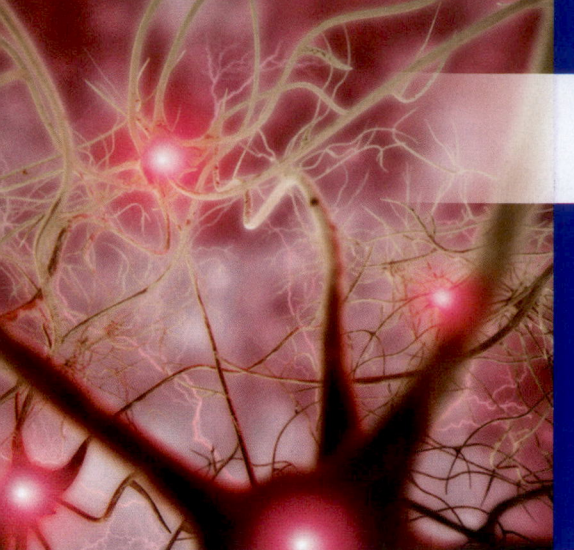

14

The Autonomic Nervous System and Homeostasis

Have you ever had to slam on the brakes of your car unexpectedly to avoid hitting another car? Do you remember how shaky you felt right after that, maybe also sweaty and out of breath? Well, that feeling, along with your ability to hit the brakes so fast, was courtesy of the neurotransmitter norepinephrine, produced by neurons of the largely involuntary arm of the peripheral nervous system (PNS)—the *visceral motor division,* or **autonomic nervous system (ANS).**

We've already discussed the roles of the central nervous system (CNS) in maintaining homeostasis (see Chapter 12). Recall that homeostasis is mostly maintained by the hypothalamus and the reticular formation of the brainstem. Much of this is mediated by the neurons of the ANS. Recall that there are two branches of the ANS, the *sympathetic* and *parasympathetic nervous systems*; look back at Figure 13.1 to see again how the PNS divisions are related.

As in any other part of the nervous system, the neurons of the ANS communicate with one another at synapses, and the neurons' effects are mediated by neurotransmitters that bind to receptors on target cells. The sympathetic and parasympathetic nervous systems are covered in separate modules, but bear in mind that the two systems work together constantly to maintain the delicate balance that is homeostasis. We conclude the chapter by examining how the ANS and the nervous system as a whole maintain homeostasis.

MODULE 14.1

Overview of the Autonomic Nervous System

*For practice applying concepts to a clinical scenario, check out the **Running Case Study** for this chapter @ Mastering Anatomy & Physiology*

Computer-generated image: The autonomic nervous system is composed of the sympathetic and parasympathetic nervous systems, which generally have opposite effects on a body function or organ.

Learning Outcomes

1. Describe the structural and functional details of the sensory and motor (autonomic) components of visceral reflex arcs.
2. Distinguish between the target cells of the somatic and autonomic nervous systems.
3. Contrast the cellular components of the somatic and autonomic motor pathways.
4. Discuss the physiological roles of each division of the autonomic nervous system.

Before delving into the intricacies of the ANS, let's first take an introductory tour of its properties. This module begins with an overview of ANS functions and then compares its structures and roles with those of the somatic motor division of the PNS. The discussion then moves on to differences between the two divisions of the ANS, the sympathetic and parasympathetic nervous systems.

Functions of the ANS and Visceral Reflex Arcs

« FLASHBACK

1. What is a reflex arc? (p. 506)
2. What is a ganglion? (p. 387)

The autonomic nervous system oversees such vital functions as heart rate, blood pressure, and digestive and urinary processes. The ANS is named for the fact that it can operate autonomously, without conscious control. It performs these functions via a series of **visceral reflex arcs.** As you have learned, a reflex arc is a series of events in which a sensory stimulus leads to a predictable motor response (see Chapter 13).

Notice in **Figure 14.1** that in the visceral reflex arcs of the ANS, (1) sensory signals from the viscera and skin are sent by afferent sensory neurons to the brain or spinal cord. (2) The stimuli are then integrated by the CNS. (3) Next, motor impulses from the CNS are sent out via efferent motor neurons in cranial and spinal nerves. These nerves usually lead to ganglia in the PNS, called **autonomic ganglia.** (4) Finally, the autonomic ganglia send the impulses via other efferent motor neurons to various target organs, where they trigger a motor response in the target cells.

CORE PRINCIPLE
Cell-Cell Communication

Note that visceral reflex arcs provide another example of the Cell-Cell Communication Core Principle (p. 28).

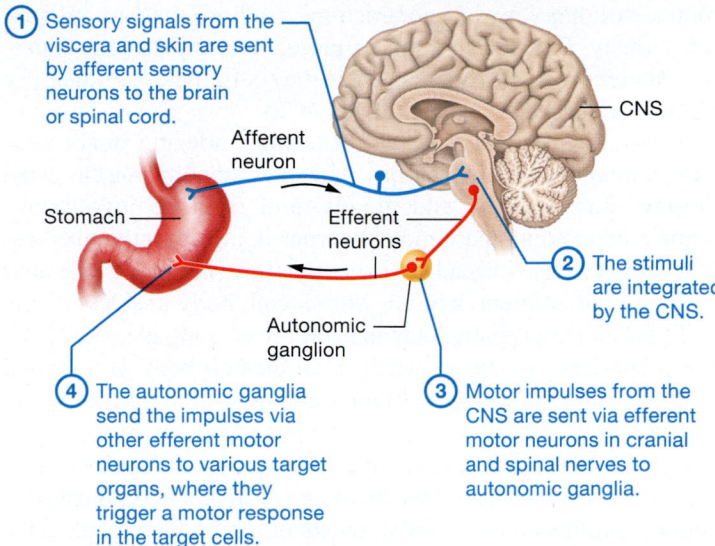

① Sensory signals from the viscera and skin are sent by afferent sensory neurons to the brain or spinal cord.

Afferent neuron

Stomach

Efferent neurons

CNS

Autonomic ganglion

② The stimuli are integrated by the CNS.

④ The autonomic ganglia send the impulses via other efferent motor neurons to various target organs, where they trigger a motor response in the target cells.

③ Motor impulses from the CNS are sent via efferent motor neurons in cranial and spinal nerves to autonomic ganglia.

Figure 14.1 Visceral reflex arcs.

Quick Check

☐ 1. What are the basic functions of the ANS?
☐ 2. What are autonomic ganglia?

Comparison of Somatic and Autonomic Nervous Systems

« FLASHBACK

1. What is the somatic motor division of the PNS? (p. 477)
2. What is a somatic motor neuron? (p. 477)

In the PNS chapter, we talked about the somatic motor division and the somatic motor neurons that innervate skeletal muscle fibers (see Chapter 13; **Figure 14.2a**). These neurons largely

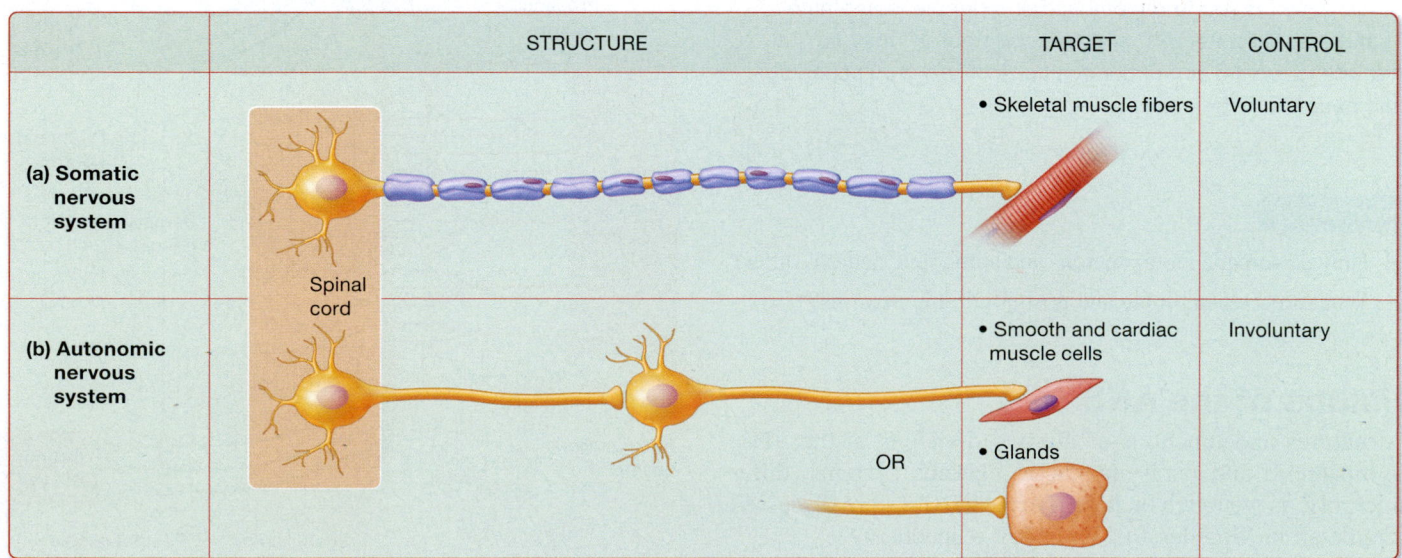

	STRUCTURE		TARGET	CONTROL
(a) Somatic nervous system	Spinal cord		• Skeletal muscle fibers	Voluntary
(b) Autonomic nervous system		OR	• Smooth and cardiac muscle cells	Involuntary
			• Glands	

Figure 14.2 Comparison of the somatic and autonomic nervous systems.

control voluntary muscle contractions, or those that are initiated consciously. Recall that somatic motor neurons directly innervate skeletal muscle fibers, which they stimulate by releasing the neurotransmitter acetylcholine (ACh).

In **Figure 14.2b** you can see that, unlike somatic motor neurons, autonomic motor neurons innervate smooth muscle cells, cardiac muscle cells, and glands, and produce involuntary actions. In addition, these motor neurons do not directly innervate their target cells. Instead, two neurons are involved. The first is the initial efferent neuron whose cell body resides in the CNS, called the **preganglionic neuron** (pree´-gang-glee-AHN-ik). The preganglionic neuron synapses on the cell body of a second neuron, called the **postganglionic neuron,** within an autonomic ganglion in the PNS. The axon of the postganglionic neuron then synapses on the target cell, in which it triggers a change by releasing various neurotransmitters, including ACh and norepinephrine. These neurotransmitters may stimulate or inhibit their target cells.

nervous system, the preganglionic axons are often short and the postganglionic axons are generally long. The opposite is true for the axons of the parasympathetic nervous system:

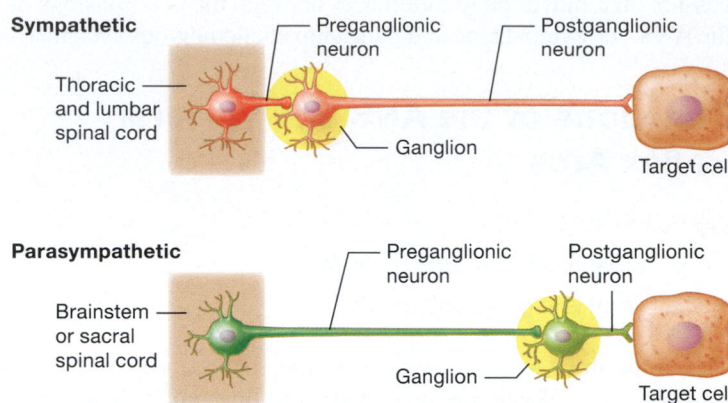

StudyBOOST ⬆

Remembering the Difference between Preganglionic and Postganglionic Neurons

Admittedly, the terms "preganglionic" and "postganglionic" sound very similar and possibly confusing. But if you break down the terms and think about their parts, they make a lot of sense. Remember that a ganglion is simply a cluster of neuronal cell bodies. Recall also that the prefixes *pre-* and *post-* mean "before" and "after," respectively. So, a *pre*ganglionic neuron is simply the neuron that comes *before* the ganglion. Similarly, the *post*ganglionic neuron is the one whose cell body is located within the ganglion, and whose axon comes *after* the ganglion.

If you still find yourself getting confused with the terminology, try to associate different words with "pre-" and "post-" that are easy for you to remember. For example, something as simple as "first neuron" and "second neuron" may be adequate to remind you of the positions of pre- and postganglionic neurons in the pathway. ■

Quick Check

☐ 3. How do somatic motor neurons and autonomic neurons differ?

☐ 4. What are preganglionic and postganglionic neurons?

Divisions of the ANS

The structures and functions of the two divisions of the ANS, the sympathetic and parasympathetic nervous systems, differ considerably, as we touch on here. These divisions are discussed in more detail in Modules 14.2 and 14.3, respectively.

To start, we can note a major difference between the two systems in the arrangement of their neurons: In the sympathetic

The Sympathetic Nervous System

As you can see on the left side of **Figure 14.3**, in the sympathetic nervous system, the cell bodies of the preganglionic neurons originate in the thoracic and upper lumbar spinal cord. For this reason, the sympathetic nervous system is often called the **thoracolumbar division** (thor´-aeh-koh-LUM-bar) of the ANS. These neurons synapse first in sympathetic ganglia, which are generally located near the spinal cord (with a few exceptions). The postganglionic neurons innervate their target cells.

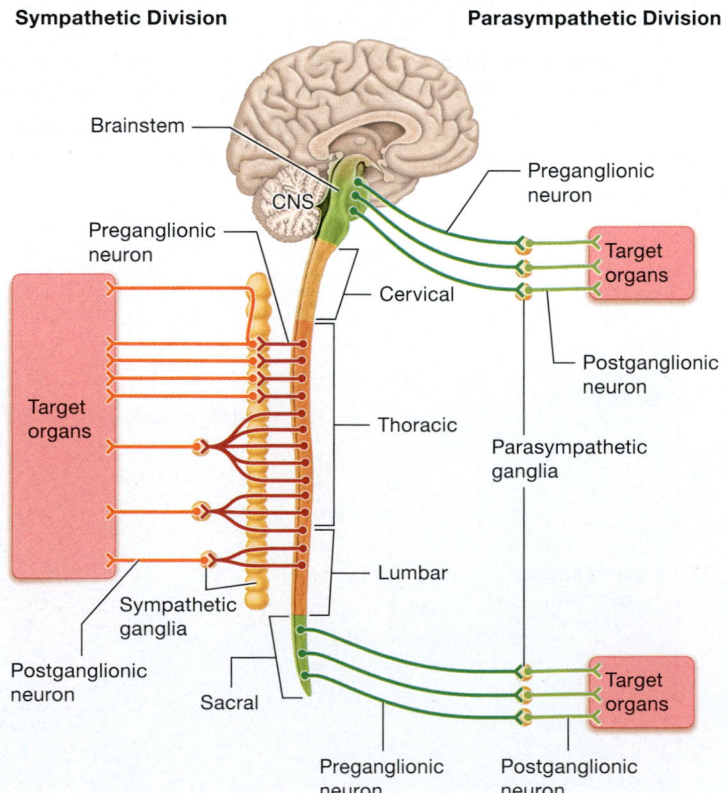

Figure 14.3 Overview of the structure of ANS divisions.

The sympathetic nervous system is usually described as the "fight or flight" division of the ANS—this reflects its role in preparing the body for emergency situations in which one would need to fight off an attacker or flee from danger. The sympathetic nervous system's role is broader than this, however, as it maintains homeostasis when the body is engaged in any type of physical work and mediates the body's visceral responses to emotion.

The Parasympathetic Nervous System

As the right side of Figure 14.3 shows, the cell bodies of the preganglionic parasympathetic neurons are located within the nuclei of several cranial nerves in the brainstem and in the sacral region of the spinal cord. For this reason, the parasympathetic nervous system is often called the **craniosacral division** (kray′-nee-oh-SAY-krul) of the ANS. The cranial nerves innervate structures of the head and neck, the thoracic viscera, and most abdominal viscera. The sacral nerves innervate structures within the pelvic cavity. As in the sympathetic division, the preganglionic neurons synapse first in parasympathetic ganglia, and the postganglionic neurons then innervate the target organs. However, parasympathetic ganglia are typically located near or within the target organs, as opposed to near the spinal cord.

The parasympathetic nervous system is often described as the "rest and digest" division of the ANS. This reflects its role in digestion and in maintaining the body's homeostasis when at rest.

The Balance between the Parasympathetic and Sympathetic Nervous Systems

The actions of the parasympathetic nervous system directly antagonize those of the sympathetic nervous system. For example, parasympathetic neurons decrease the rate of the heart's contractions, whereas sympathetic neurons increase their rate. Together, these two divisions maintain a delicate balance to ensure that homeostasis is maintained at all times.

Quick Check

- ☐ 5. What are the two divisions of the ANS?
- ☐ 6. How are the two divisions of the ANS alike, and how do they differ?

Apply What You Learned

- ☐ 1. Certain types of peripheral neuropathy (diseases of PNS neurons) affect somatic motor neurons, whereas others affect autonomic motor neurons. Predict the effects of autonomic neuropathy. How would autonomic neuropathy differ from somatic motor neuron neuropathy?
- ☐ 2. Mrs. Black is taking a medication for high blood pressure that blocks the effects of the sympathetic nervous system. What is likely to happen if she is faced with an emergency situation? Explain.

See answers in Appendix A.

MODULE **14.2**

The Sympathetic Nervous System

Learning Outcomes

1. Explain how the sympathetic nervous system maintains homeostasis.
2. Describe the anatomy of the sympathetic nervous system.
3. Describe the neurotransmitters and neurotransmitter receptors of the sympathetic nervous system.
4. Explain the effects of the sympathetic nervous system on the cells of its target organs.

As you know, the **sympathetic nervous system** is the division of the ANS that prepares the body for emergency situations. But the "fight or flight" story grossly oversimplifies this system, as it is also active, for example, when you experience emotions (particularly strong ones, such as fear, excitement, rage, and embarrassment) and even during minor physical activities. In fact, even the act of standing up requires the sympathetic nervous system; if its neurons don't temporarily initiate changes in your heart rate and blood vessel diameter, your blood pressure can drop quite low due to the effect of gravity on your blood (a condition called *orthostatic hypotension*), and cause you to fall.

This module explores the structure and function of the neurons and ganglia of the sympathetic nervous system and explains how this system, when dominant, maintains homeostasis. The interactions between this system and the parasympathetic nervous system are discussed in Module 14.4.

Gross and Microscopic Anatomy of the Sympathetic Nervous System

 **FLASHBACK**

1. What are the three structural components of a chemical synapse? (p. 406)

The organization of the sympathetic nervous system and its preganglionic and postganglionic neurons is shown in **Figure 14.4**. Notice that most cell bodies of the postganglionic neurons are found in a series along the vertebral column. Their chain-like appearance has given them the name **sympathetic chain ganglia.** These ganglia extend beyond the thoracic and lumbar spinal cord, from the *superior cervical ganglion* down to the *inferior sacral ganglion.*

Preganglionic neurons begin in the lateral horns of the thoracic and lumbar spinal cord. The axons of preganglionic neurons exit the spinal cord with the axons of lower motor neurons via the anterior root. They then travel with the spinal nerve and

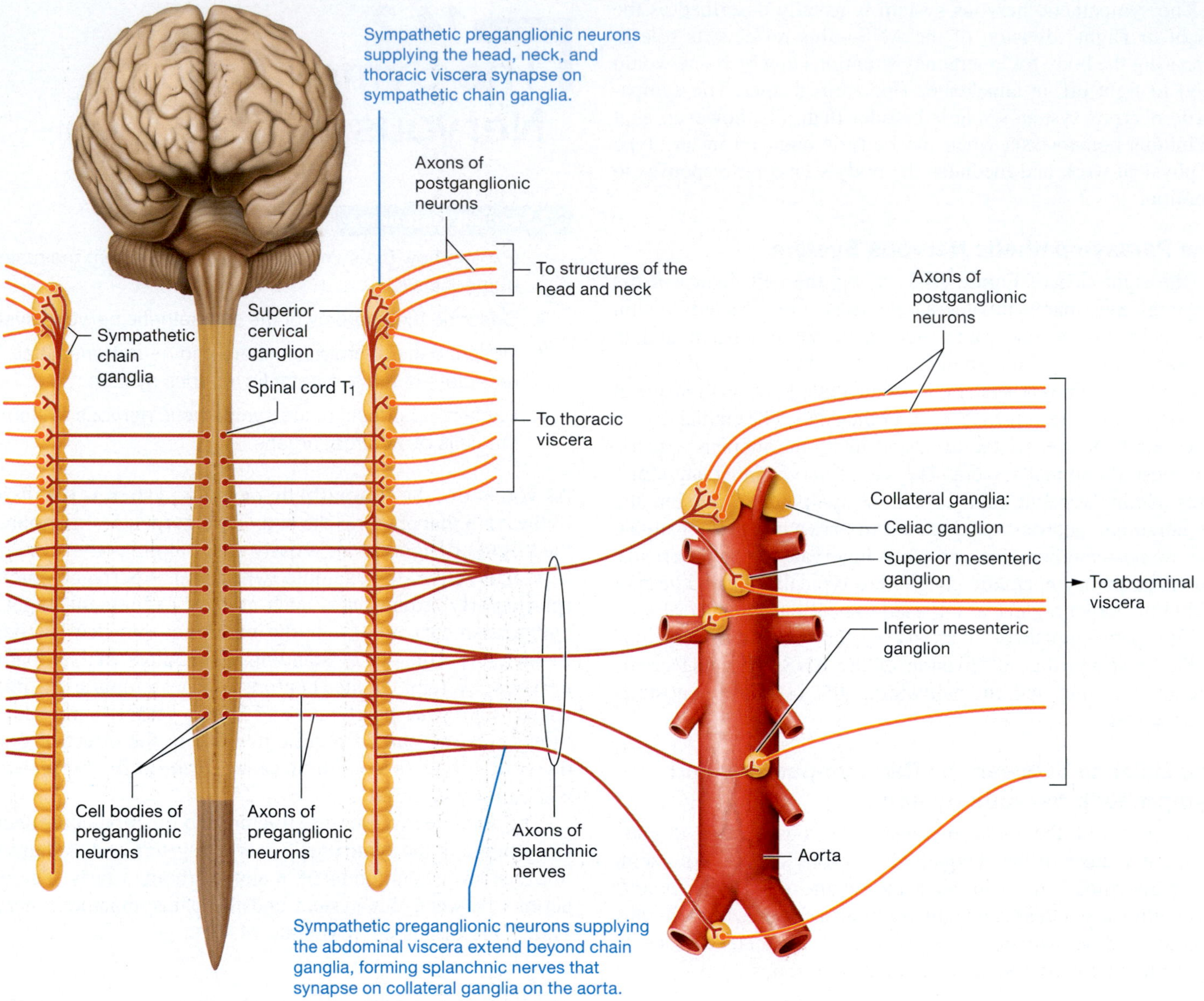

Sympathetic preganglionic neurons supplying the head, neck, and thoracic viscera synapse on sympathetic chain ganglia.

Axons of postganglionic neurons

To structures of the head and neck

Sympathetic chain ganglia

Superior cervical ganglion

Spinal cord T₁

To thoracic viscera

Axons of postganglionic neurons

Collateral ganglia:
Celiac ganglion
Superior mesenteric ganglion
Inferior mesenteric ganglion

To abdominal viscera

Cell bodies of preganglionic neurons

Axons of preganglionic neurons

Axons of splanchnic nerves

Aorta

Sympathetic preganglionic neurons supplying the abdominal viscera extend beyond chain ganglia, forming splanchnic nerves that synapse on collateral ganglia on the aorta.

Figure 14.4 Organization of the sympathetic nervous system.

the anterior ramus for a short distance before branching off to form small nerves called **white rami communicantes** (kuh-myoo´- nih-KAN-teez; singular, *communicans;* called *white* because they are myelinated), as shown here:

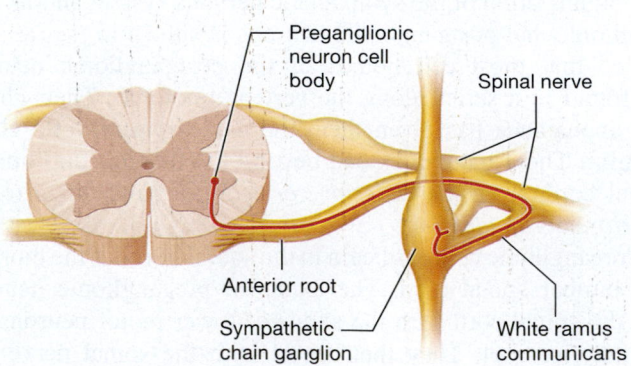

Preganglionic neuron cell body

Spinal nerve

Anterior root

Sympathetic chain ganglion

White ramus communicans

The axons in the white rami communicantes then enter the sympathetic chain ganglia that house the cell bodies of the post-ganglionic sympathetic neurons.

Some preganglionic neurons pass through the sympathetic chain ganglia without synapsing. Instead, they may synapse on cell bodies in different chain ganglia, or on **collateral ganglia,** which are located near the target organ. Many collateral ganglia are located near the aorta, and for this reason are called *preaor-tic ganglia.* Other collateral ganglia are located near the organs of the abdominopelvic cavity. The preganglionic axons that synapse here are part of nerves called the **splanchnic nerves** (SPLANK-nik; *splanchn-* = "viscera") that synapse on ganglia including the *celiac ganglion,* the *superior mesenteric ganglion,* and the *inferior mesenteric ganglion.*

So, as shown in **Figure 14.5,** a preganglionic neuron may synapse with a postganglionic neuron in one of three ways.

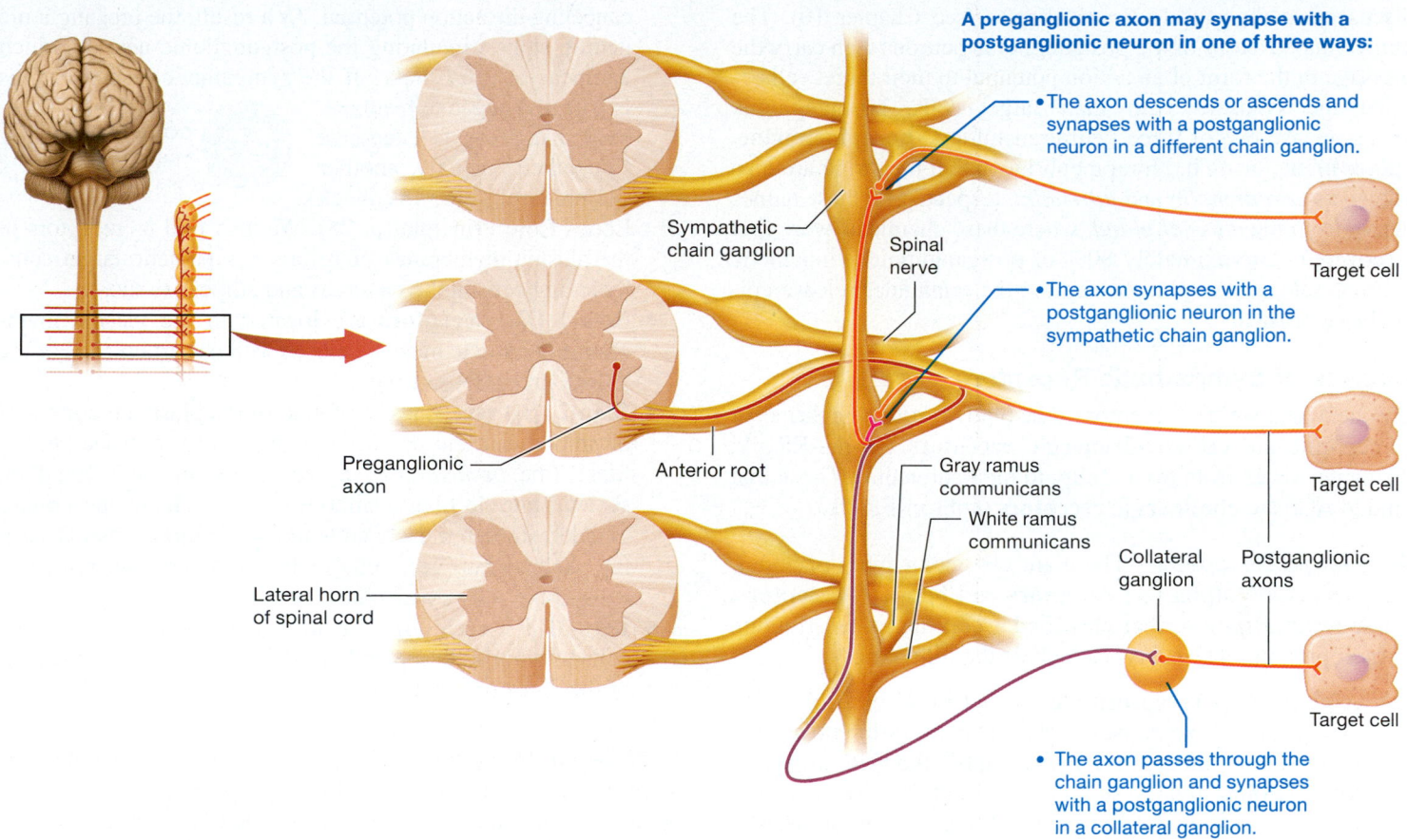

A preganglionic axon may synapse with a postganglionic neuron in one of three ways:

- The axon descends or ascends and synapses with a postganglionic neuron in a different chain ganglion.

- The axon synapses with a postganglionic neuron in the sympathetic chain ganglion.

- The axon passes through the chain ganglion and synapses with a postganglionic neuron in a collateral ganglion.

Sympathetic chain ganglion

Spinal nerve

Target cell

Preganglionic axon

Anterior root

Gray ramus communicans

White ramus communicans

Collateral ganglion

Postganglionic axons

Target cell

Lateral horn of spinal cord

Target cell

Figure 14.5 Three possible pathways of sympathetic preganglionic and postganglionic neurons.

- the axon descends or ascends, and synapses with a postganglionic neuron in a different chain ganglion; or
- the axon synapses with a postganglionic neuron in the sympathetic chain ganglion;
- the axon passes through the chain ganglion and synapses with a postganglionic neuron in a collateral ganglion.

The postganglionic neuron then innervates its target cells. Notice in Figure 14.5 that postganglionic axons that take the first two pathways exit the ganglia as small **gray rami communicantes** (called *gray* because they are unmyelinated), which rejoin and travel with spinal nerves to reach their target cells. In contrast, postganglionic axons that take the third pathway travel to their target cells directly. The largest population of postganglionic axons that take this third path are the splanchnic nerves.

Quick Check

☐ 1. Where are the cell bodies of sympathetic preganglionic neurons located? Where are their axons located, and what do the axons contact?

☐ 2. Where are the cell bodies of sympathetic postganglionic neurons located? Where are their axons located, and what do the axons contact?

Sympathetic Neurotransmitters and Receptors

FLASHBACK

1. What are the functions of the neurotransmitters norepinephrine, epinephrine, and acetylcholine? (p. 413)

2. What is a neurotransmitter receptor? What happens when a neurotransmitter binds to a receptor? (p. 411)

Recall that neurons interact with other neurons and target cells through the release of chemicals called *neurotransmitters* (see Chapter 11). A neurotransmitter interacts with a target cell by binding to a protein-based receptor for that neurotransmitter embedded in the plasma membrane of the target cell. The following subsections discuss the neurotransmitters released by sympathetic neurons and the receptors to which these neurotransmitters bind on their target cells.

Classes of Sympathetic Neurotransmitters

The sympathetic preganglionic axon communicates with the postganglionic neuron by way of an excitatory synapse using the neurotransmitter **acetylcholine (ACh).** This is the same neurotransmitter released by somatic motor neurons, as we discussed in the muscle tissue chapter (see Chapter 10). The unmyelinated axons of the postganglionic neurons then carry the message in the form of an action potential to their target cells.

At the synapses with their target cells, postganglionic axons release one of three neurotransmitters: **norepinephrine, epinephrine,** or ACh. Norepinephrine and epinephrine are also known as *noradrenalin* and *adrenalin,* respectively; these names derive from the *adrenal gland,* where these chemicals were first discovered. Approximately 80% of postganglionic sympathetic neurons release norepinephrine, and the remainder release epinephrine or ACh.

Classes of Sympathetic Receptors

The neurotransmitter receptors that bind to norepinephrine or epinephrine are called **adrenergic receptors** (ad-ren-ER-jik; the name comes from their ability to bind adrenalin). Those that bind to ACh are **cholinergic receptors** (kohl-in-ER-jik).

Adrenergic Receptors   There are two major types of adrenergic receptors: **alpha (α) receptors** and **beta (β) receptors.** These receptors are further classified according to the structure of their proteins. The subtypes include the following:

- **Alpha-1 (α_1) receptors** are located in the plasma membranes of smooth muscle cells of many different organs, including the blood vessels that supply the skin, organs of the gastrointestinal system, and the kidneys. These receptors are also found on arrector pili muscles in the dermis, the uterus (during pregnancy), and certain organs of the genitourinary system.
- **Alpha-2 (α_2) receptors** are different from the other adrenergic receptors—we find most of them in the membrane of preganglionic sympathetic neurons rather than in peripheral target cells. Recall that normally an action potential in a preganglionic neuron leads to ACh release, which stimulates the postganglionic neuron (**Figure 14.6a**). However, notice in **Figure 14.6b** that when norepinephrine binds to α_2 receptors, the axon terminal is hyperpolarized, slowing or even

canceling the action potential. As a result, the preganglionic neuron stops stimulating the postganglionic neuron, which dampens or even shuts off the sympathetic response. This is part of a negative feedback loop that prevents excessive sympathetic activity, another example of the Feedback Loops Core Principle (p. 28). We also find α_2 receptors in the plasma membranes of certain sympathetic target cells, including cells in the pancreas and adipose tissue.

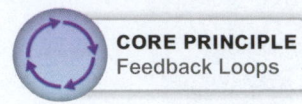

CORE PRINCIPLE
Feedback Loops

- **Beta-1 (β_1) receptors** are located in the plasma membranes of cardiac muscle cells as well as certain cells of the kidney and adipose tissue.
- **Beta-2 (β_2) receptors** are found in the plasma membranes of smooth muscle cells lining the airway passages in the lungs (the *bronchioles*). β_2 receptors are also found on skeletal muscle fibers, smooth muscle cells of the urinary bladder, smooth muscle cells lining the blood vessels serving skeletal muscles, cells of the liver and pancreas, and cells of the salivary glands.
- **Beta-3 (β_3) receptors** are located predominantly on the cells of adipose tissue and smooth muscle cells in the wall of the digestive tract.

Cholinergic Receptors   There are two types of cholinergic receptors: the **muscarinic receptor** (muss-kuh-RIN-ik; so named because it binds the poison muscarine) and the **nicotinic receptor** (nik-oh-TIN-ik; so named because it binds the poison found in tobacco, nicotine). Muscarinic receptors are found on sweat glands in the skin. Nicotinic receptors are located in the membranes of all postganglionic neurons and the cells of the adrenal medulla (which consists of modified postganglionic neurons).

Quick Check

☐ 3. Which neurotransmitter(s) is (are) released by sympathetic preganglionic neurons? Which are released by sympathetic postganglionic neurons?

☐ 4. What types of receptors do these neurotransmitters bind?

Effects of the Sympathetic Nervous System on Target Cells

« FLASHBACK

1. What are the differences between smooth, skeletal, and cardiac muscle tissue? (p. 148)

2. What are endocrine and exocrine glands? (p. 134)

The effects of the sympathetic nervous system on its target cells are shown in **Figure 14.7** on p. 524. They are easy to understand if you consider the system's most basic function: to ensure survival and maintain homeostasis during times of physical or emotional stress. Think about what happens to your body when you're running a race—your heart beats faster and harder, you begin to

Preganglionic neurons

α_2 receptors

NE bound to α_2 receptor

ACh

Postganglionic neuron

Postganglionic neuron

When α_2 receptors are not bound to NE, the preganglionic neuron releases ACh, exciting the postganglionic neuron.

When α_2 receptors bind NE, the preganglionic neuron hyperpolarizes, decreasing ACh release and postganglionic neuron activity.

(a) α_2 receptor with no NE

(b) α_2 receptor bound to NE

Figure 14.6 The effect of α_2 receptors on preganglionic neurons. Note that NE stands for norepinephrine.

sweat, and your blood pressure increases, among many other changes. Much the same happens, although to a lesser degree, when you're nervous about a big exam or excited about an upcoming vacation. Each of these situations triggers a release of norepinephrine, epinephrine, and ACh from pre- and postganglionic sympathetic neurons, which in turn triggers changes in the cells of the target organs and tissues.

In the upcoming subsections, we'll examine specific effects of the sympathetic nervous system on the body's cells. Since most adrenergic synapses use norepinephrine, we will refer primarily to this neurotransmitter, although a small percentage of postganglionic neurons release epinephrine.

Effects on Cardiac Muscle Cells

As you already know, when you are nervous or exercising, your heart rate and blood pressure rise. This is due to the effect of norepinephrine binding to β_1 receptors on cardiac muscle cells, which opens ion channels that increase both the rate and force of contractions. This heightens not only the amount of blood delivered to the tissues but also blood pressure, which maintains homeostasis during periods of increased activity.

Effects on Smooth Muscle Cells

The smooth muscle cells of many organs also have receptors for norepinephrine. When norepinephrine binds, it mediates the following changes:

- **Constriction of blood vessels serving the digestive, urinary, and integumentary systems.** Norepinephrine binds to α_1 receptors on smooth muscle cells of blood vessels serving the organs of the digestive, urinary, and integumentary systems and causes them to contract. This narrows the blood vessels, an action called *vasoconstriction* (vay'-zoh-kun-STRIK-shun; *vaso-* = "vessel"). Vasoconstriction decreases the blood flow to those organs and diverts blood to tissues that are temporarily "more important," such as skeletal and cardiac muscle. Decreasing blood flow to the urinary and digestive organs slows the rate of urine formation and digestion, respectively (if you are running from a bear, digesting the lunch you just ate can wait). This wide-scale vasoconstriction also increases the overall blood pressure. Reduced blood flow to the skin explains why people with light-colored skin "turn white as a sheet" when they are scared—as the blood vessels serving the skin constrict, the pinkish color of the blood under the skin is less visible and the skin appears paler.
- **Dilation of the bronchioles.** During times of stress more oxygen needs to be delivered to the body's cells, especially the skeletal muscle cells, to keep up with their metabolic demands. The sympathetic nervous system accomplishes this through norepinephrine's effect on the smooth muscle cells lining the smaller airway passages (the *bronchioles*). When norepinephrine binds to the β_2 receptors on these cells, they relax. This action, *bronchodilation* (brong'-koh-dy-LAY-shun), increases the diameter of the bronchioles and so the amount of oxygen that can be inhaled

with each breath. Although the sympathetic nervous system does increase oxygen intake, note that this system has no effect on the *rate* of ventilation. The rate does tend to increase during exercise and emergency situations, but this is due to a mechanism independent of the sympathetic nervous system (see Chapter 21).

- **Dilation of blood vessels serving skeletal and cardiac muscle.** Although norepinephrine constricts most of the body's blood vessels, when it binds to β_2 receptors on the smooth muscle cells of blood vessels serving skeletal and cardiac muscle, it causes them to relax. This opens the blood vessels, an action called *vasodilation* (vay'-zoh-dy-LAY-shun), which increases the blood flow and the delivery of oxygen and nutrients to these cells. If this vasodilation occurred alone, the blood pressure would decrease. However, it occurs simultaneously with the vasoconstriction of nearly every other blood vessel in the body, so the effect of skeletal and cardiac muscle blood vessel vasodilation on blood pressure is minimal.
- **Contraction of urinary and digestive sphincters.** The binding of norepinephrine to the smooth muscle cells of the sphincters of the urinary and digestive systems causes them to contract. This makes emptying the bowel and bladder more difficult during times of exercise, stress, or emergency.
- **Relaxation of the smooth muscle of the digestive tract.** Although stimulation by the sympathetic nervous system causes the smooth muscle of the digestive tract sphincters to contract, when norepinephrine binds to β_3 receptors it leads to relaxation of the remainder of the smooth muscle in the digestive tract. This decreases the movement of food through the digestive system and slows digestive processes overall.
- **Dilation of the pupils.** The dilator pupillae muscles are innervated by sympathetic nervous system neurons. When norepinephrine binds to the α_1 receptors on these smooth muscle cells, it causes them to contract, which dilates the pupils and allows more light to enter the eyes.
- **Constriction of blood vessels serving most exocrine glands.** The sympathetic nervous system almost universally decreases secretion from exocrine glands—apart from sweat glands—via norepinephrine binding to β_2 receptors. This affects most exocrine cells of the digestive system, including the salivary glands around your mouth and the cells that secrete digestive products. This effect on salivary glands is why your mouth is often dry when you're nervous. The decreased secretion is generally due to reduced blood flow to these glands rather than a direct effect on the gland itself.

Effects on the Cellular Metabolic Rate

During times of sympathetic nervous system activation, nearly all cells, and in particular skeletal muscle cells, consume dramatically more ATP as their metabolic rates increase. To ensure that skeletal muscle cells have a steady supply of fuels with

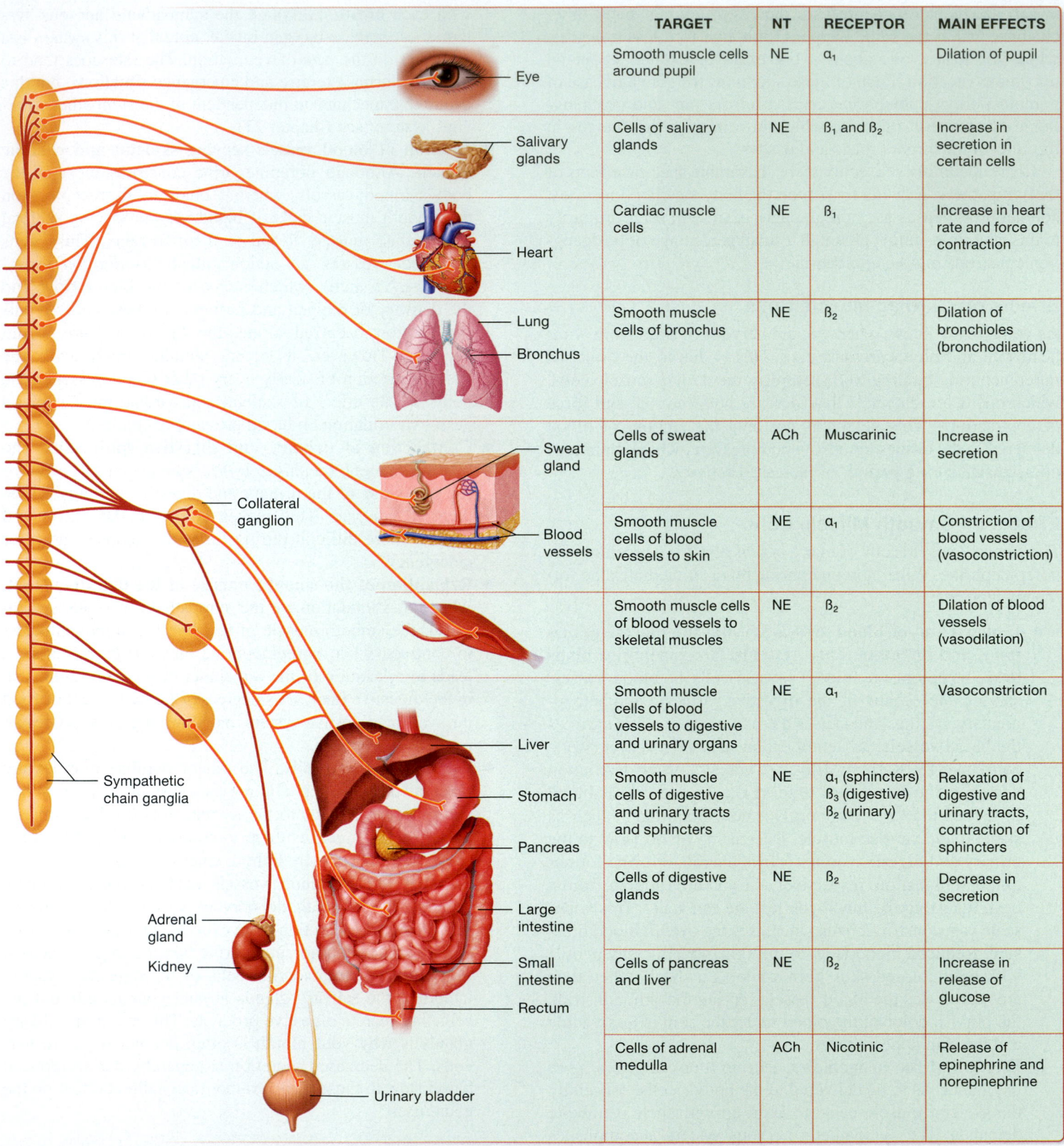

	TARGET	NT	RECEPTOR	MAIN EFFECTS
Eye	Smooth muscle cells around pupil	NE	α_1	Dilation of pupil
Salivary glands	Cells of salivary glands	NE	β_1 and β_2	Increase in secretion in certain cells
Heart	Cardiac muscle cells	NE	β_1	Increase in heart rate and force of contraction
Lung / Bronchus	Smooth muscle cells of bronchus	NE	β_2	Dilation of bronchioles (bronchodilation)
Sweat gland	Cells of sweat glands	ACh	Muscarinic	Increase in secretion
Blood vessels	Smooth muscle cells of blood vessels to skin	NE	α_1	Constriction of blood vessels (vasoconstriction)
	Smooth muscle cells of blood vessels to skeletal muscles	NE	β_2	Dilation of blood vessels (vasodilation)
Liver	Smooth muscle cells of blood vessels to digestive and urinary organs	NE	α_1	Vasoconstriction
Stomach	Smooth muscle cells of digestive and urinary tracts and sphincters	NE	α_1 (sphincters) β_3 (digestive) β_2 (urinary)	Relaxation of digestive and urinary tracts, contraction of sphincters
Pancreas	Cells of digestive glands	NE	β_2	Decrease in secretion
Large intestine / Small intestine / Rectum	Cells of pancreas and liver	NE	β_2	Increase in release of glucose
Urinary bladder	Cells of adrenal medulla	ACh	Nicotinic	Release of epinephrine and norepinephrine

Labels in figure: Eye, Salivary glands, Heart, Lung, Bronchus, Sweat gland, Blood vessels, Collateral ganglion, Sympathetic chain ganglia, Liver, Stomach, Pancreas, Large intestine, Small intestine, Rectum, Adrenal gland, Kidney, Urinary bladder

Figure 14.7 The main effects of the sympathetic nervous system on target cells. Note that NT stands for neurotransmitter, NE for norepinephrine, and ACh for acetylcholine.

which to make ATP, norepinephrine brings about the following three main effects: (1) It binds to β_3 receptors on adipocytes and triggers the breakdown of lipids, which releases free fatty acids into the bloodstream; (2) it binds to β_2 receptors on cells of the liver and triggers the release of glucose from glycogen and also the synthesis of glucose from other precursors; and (3) it binds to β_2 receptors on cells of the pancreas and triggers the release of a hormone called *glucagon* that increases the concentration of glucose in the blood.

Be careful here not to confuse the metabolic rate with digestion. Digestion is the breakdown of foods into nutrients that can be absorbed into the blood, and these processes *decrease* when sympathetic activity increases. The metabolic rate refers to the rate at which cells consume ATP, and this *increases* when sympathetic activity rises.

Given all the effects of the sympathetic nervous system on the metabolic rate, it's not surprising that sympathetic receptors are purported targets for weight loss supplements. See *A&P in the Real World, Pseudoscience Exposed: The Sympathetic Nervous System and Weight Loss Supplements* for more information.

Effects on Secretion from Sweat Glands

When you're nervous or exercising, you sweat. This useful (but often annoying) attempt to help the body maintain temperature

Pseudoscience Exposed: The Sympathetic Nervous System and Weight Loss Supplements

As you just read, one effect of the sympathetic nervous system is to increase the metabolic rate, or raise the rate at which ATP is produced and consumed. This has led dietary supplement manufacturers to create products intended to result in weight loss by capitalizing on this effect. These products are generally sold as topical creams that the user rubs over "problem areas" with excess adipose tissue.

One chemical found in many of these creams is *yohimbine* (yoh-HIM-been), which is the active ingredient derived from a plant called *yohimbe* (yoh-HIM-bay). Manufacturers of the creams claim that yohimbine binds to β_3 receptors on adipocytes and triggers the breakdown of lipids. This is, at best, a misleading statement. Yohimbine actually blocks α_1 receptors in blood vessels and α_2 receptors in the spinal cord. This has a dual effect: It causes vasodilation while also increasing the activity of sympathetic neurons. The higher sympathetic activity can, briefly, increase the metabolic rate. However, it can also dangerously elevate the heart rate; cause seizures, high blood pressure, and kidney failure; and lead to insomnia and panic attacks.

Fortunately, yohimbine is not actually absorbed through the epidermis in any significant amount, so it never reaches the blood vessels in the dermis and hypodermis. This limits the amount of harm it can do to your body, although the supplement can still do significant harm to your wallet.

homeostasis is due to postganglionic sympathetic neurons that release ACh onto the cells of sweat glands in the skin. ACh binds to these cells' muscarinic receptors and increases their secretions. This occurs as part of a negative feedback loop that controls body temperature, which reminds us of the Feedback Loops Core Principle (p. 28).

CORE PRINCIPLE
Feedback Loops

Effects on Cells of the Adrenal Medulla

The **adrenal medulla** (uh-DREE-nul muh-DOOL-uh) is the internal part of the adrenal glands, which sit atop each kidney. It's unique in that it is contacted directly by preganglionic sympathetic neurons. This is because the adrenal medulla is not like the remainder of the adrenal gland—instead of glandular epithelium, it actually consists of modified sympathetic postganglionic neurons. In other words, each adrenal medulla is functionally a ganglion.

As shown in **Figure 14.8**, ① when the preganglionic neuron releases ACh, ② it binds to the nicotinic receptors of adrenal medulla cells. ③ ACh then stimulates these cells to release additional epinephrine and norepinephrine into the bloodstream. Although norepinephrine and epinephrine are usually considered neurotransmitters, they can both also function as *hormones*, long-distance chemical messengers. This means that the adrenal medulla effectively acts as an interface between the sympathetic nervous system and the endocrine system (see Chapter 16).

The ratio of chemicals the adrenal medulla produces is generally opposite that of the ANS, with about 80% epinephrine and 20% norepinephrine. This additional epinephrine and norepinephrine is useful in many ways. For one, it prolongs the duration of the sympathetic nervous system's effects. The effects of neurotransmitters released by postganglionic neurons last only a few seconds, whereas in the blood the effects of these

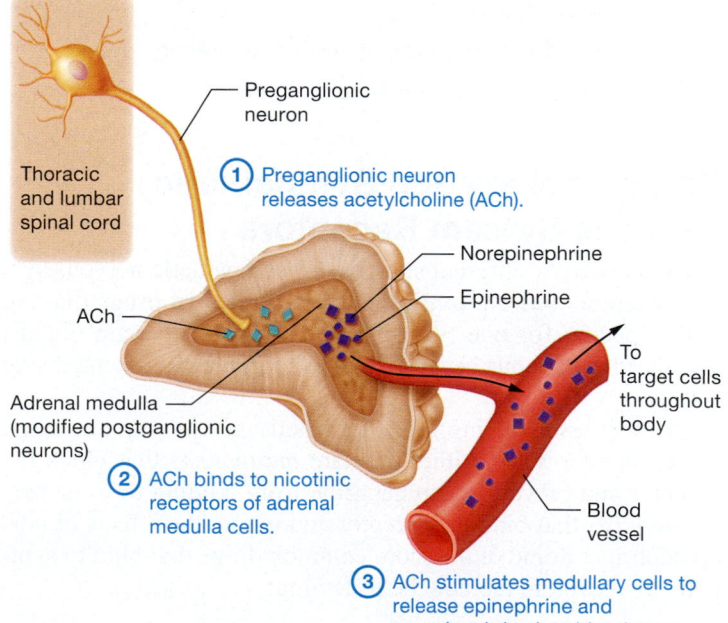

Figure 14.8 Sympathetic nervous system stimulation of the adrenal medulla.

chemicals continue for several minutes. A second benefit is that the epinephrine and norepinephrine in the blood can reach cells that are not innervated by sympathetic neurons. This allows the sympathetic nervous system to affect the entire body without innervating every body cell. Finally, the adrenal medulla acts as a "backup" for the sympathetic nervous system—even if the pathways connecting this system with many of its target organs are disrupted, it can still affect these organs indirectly via the adrenal medulla.

Effects on Other Cells

The sympathetic nervous system influences many of the body's other processes via its effects on target cells. For instance, this system enhances mental alertness, increasing neuron activity in the association areas of the cerebral cortex. Stimulation by this system also increases the blood's tendency to clot, which can be helpful if a person is injured during a "fight" or "flight" situation. In addition, the sympathetic nervous system can temporarily elevate the tension generated by skeletal muscle cells during a muscle contraction, which is why some people have been known to perform unusual feats of strength under the influence of an "adrenaline (epinephrine) rush." Furthermore, sympathetic neurons trigger contraction of the arrector pili muscles, which produces "goose bumps." Finally, these neurons also cause ejaculation of semen via effects on smooth muscle cells of the male reproductive ducts.

Quick Check

☐ 5. Describe the effects of the sympathetic nervous system on each of the following tissues.

    a. Cardiac muscle

    b. Blood vessels serving skeletal muscles

    c. Sweat glands

    d. Blood vessels serving digestive and urinary organs

    e. Smooth muscle lining the airways

Pharmacology and Sympathetic Nervous System Receptors

The existence of different subtypes of sympathetic nervous system receptors has allowed researchers to design drugs that are fairly specific for one type of receptor and so certain organs. This is advantageous because it helps to minimize potential side effects.

Most drugs targeting the sympathetic nervous system work in one of two ways: Either they are *antagonists* that block the receptor and prevent norepinephrine from binding to it, or they are *agonists* that bind the receptor and mimic the effects of norepinephrine. Some of the more common drugs that bind to sympathetic receptors include the following:

- α_1 *Blockers* (*antagonists*) bind to α_1 receptors, particularly those on the smooth muscle cells lining blood vessels. They block the action of norepinephrine and prevent the

blood vessels from constricting, an effect that lowers blood pressure and is useful in treating hypertension (high blood pressure). Certain α_1 blockers also cause relaxation of the smooth muscle in the prostate gland; these are used to treat benign prostatic hyperplasia (see Chapter 26).

- α_2 *Agonists* bind to the presynaptic α_2 receptors and activate them, which decreases the output of both preganglionic and postganglionic sympathetic neurons. These drugs may be used in the treatment of hypertension and other conditions such as opiate withdrawal.

- β *Blockers* are antagonists that bind to β_1 receptors on the heart and decrease its rate and force of contraction. These drugs are widely used in the treatment of hypertension and other diseases of the cardiovascular system.

- β_2 *Agonists* bind to β_2 receptors on the smooth muscle of the bronchioles and cause bronchodilation. These drugs are commonly used to treat asthma.

Quick Check

☐ 6. What are agonists and antagonists?

Apply What You Learned

☐ 1. A nurse injects a patient with a drug that selectively binds to and activates α_2 adrenergic receptors. What will happen to the sympathetic nervous system response? Why? What useful applications might such a drug have?

☐ 2. If there were a drug that selectively blocked nicotinic receptors on sympathetic postganglionic neurons, the effects observed in a patient would be similar to those of the drug administered in question 1. Explain why.

☐ 3. Mr. Nguyen is involved in an accident that destroys the sympathetic ganglia housing the neurons that innervate his heart. However, you notice that his heart rate and blood pressure increase in response to exercise. What is causing this increase? (*Hint:* Remember that epinephrine and norepinephrine are hormones as well as neurotransmitters.)

See answers in Appendix A.

MODULE **14.3**

The Parasympathetic Nervous System

Learning Outcomes

1. Identify the role of the parasympathetic nervous system, and explain how it maintains homeostasis.

2. Describe the anatomy of the parasympathetic nervous system.

3. Describe the neurotransmitters and neurotransmitter receptors of the parasympathetic nervous system.

4. Describe the effects of the parasympathetic nervous system on its target cells.

As we discussed, the **parasympathetic nervous system** is often characterized as the "rest and digest" division of the ANS. This relates to its role in the body's maintenance functions, such as digestion and urine formation, that are typically carried out during periods of rest. Like the sympathetic nervous system, this system is rarely completely silent, although it is in general considered subordinate to the sympathetic nervous system. Let's take a look at the structure and function of the parasympathetic nervous system.

Gross and Microscopic Anatomy of the Parasympathetic Nervous System

The cranial nerves that house parasympathetic preganglionic neurons are the oculomotor (III), facial (VII), glossopharyngeal (IX), and vagus (X) nerves. The sacral nerves that house parasympathetic preganglionic neurons include S2–S4. **Figure 14.9** shows the organization of this division.

As with preganglionic sympathetic neurons, the axons of preganglionic parasympathetic neurons synapse with postganglionic neurons located within ganglia. The ganglia, known as *terminal ganglia,* are typically found near the postganglionic neurons' target cells, and the postganglionic axons are in general fairly short.

Parasympathetic Cranial Nerves

Although four cranial nerves carry parasympathetic fibers, the main parasympathetic nerves are the two vagus nerves (CN X), which together provide about 90% of parasympathetic innervation to the body. Branches of the vagus nerve contribute to several groups of nerves, or *plexuses,* that innervate specific organs. For example, the *cardiac plexus* innervates the heart; the *pulmonary plexus,* the lungs and airway passages; and the *esophageal plexus,* the esophagus. These preganglionic fibers typically synapse on terminal ganglia in the walls of the organ being innervated.

Other cranial nerves also carry parasympathetic fibers. The parasympathetic preganglionic neurons of the oculomotor nerves (CN III) synapse on terminal ganglia called the **ciliary ganglia** (SIL-ee-ehr-ee). Several terminal ganglia, including the **submandibular ganglia** and the **pterygopalatine ganglia** (tehr′-uh-goh-PAL-uh-tyn), house the cell bodies of sensory neurons and are the sites where preganglionic parasympathetic neurons of the facial nerves (CN VII) synapse on postganglionic parasympathetic neurons. The preganglionic parasympathetic neurons of the glossopharyngeal nerves (CN IX) synapse on postganglionic parasympathetic neurons in small terminal ganglia called the *otic ganglia* (OH-tik).

Parasympathetic Sacral Neurons

The cranial nerve parasympathetic fibers supply most viscera of the thoracic and abdominopelvic cavities. The remaining organs, including the last segment of the large intestine,

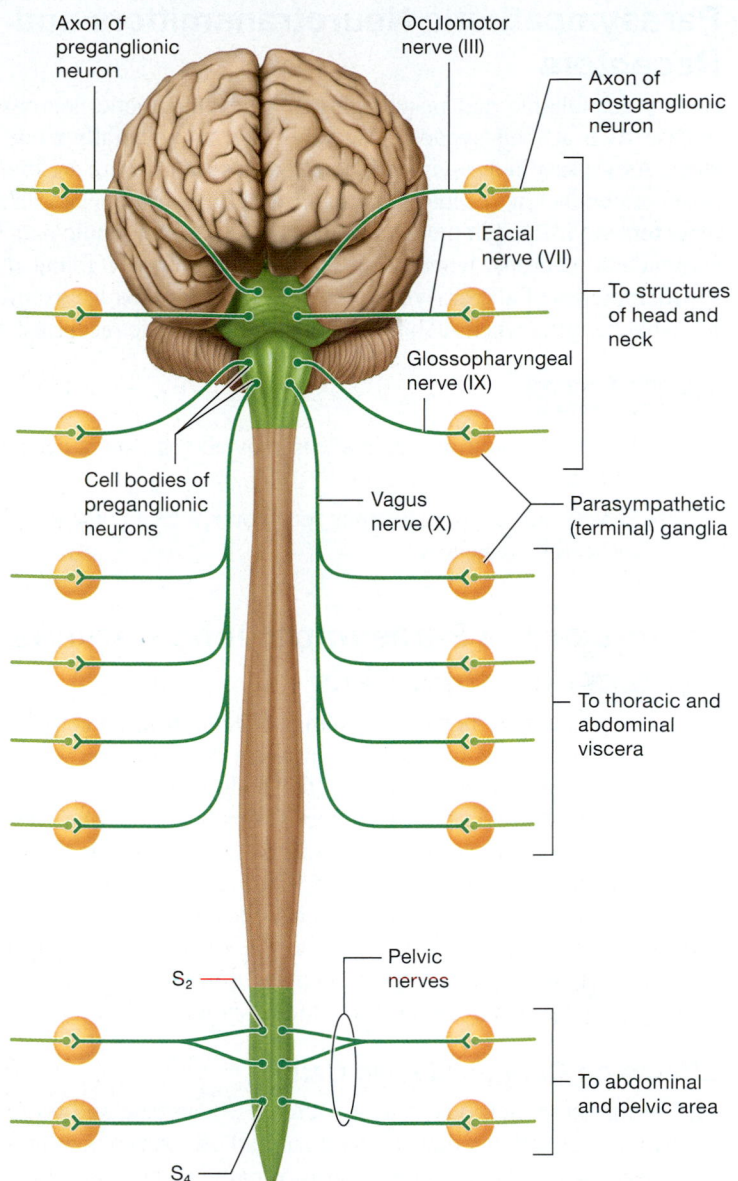

Figure 14.9 Organization of the parasympathetic nervous system.

the urinary bladder, and the reproductive organs, are supplied by parasympathetic sacral neurons. Branches from the sacral spinal cord form the **pelvic splanchnic nerves,** which in turn form plexuses in the pelvic floor. Some preganglionic neurons synapse in terminal ganglia in these plexuses, but most synapse in terminal ganglia in the walls of the organs.

Quick Check

☐ 1. What is the basic function of the parasympathetic nervous system?

☐ 2. Where are the preganglionic parasympathetic cell bodies located? Where are the postganglionic parasympathetic cell bodies located?

☐ 3. How does the arrangement of neurons in the sympathetic nervous system differ from that in the parasympathetic nervous system?

Parasympathetic Neurotransmitters and Receptors

Both preganglionic and postganglionic parasympathetic neurons release ACh at their synapses, and the effect is generally excitatory. As we saw in the sympathetic nervous system, there are two types of cholinergic receptors: nicotinic and muscarinic. Nicotinic receptors are located in the membranes of all postganglionic parasympathetic neurons, whereas muscarinic receptors are found in the membranes of all parasympathetic target cells. Next we examine what happens when ACh binds to these muscarinic receptors.

Quick Check

☐ 4. Which neurotransmitter is released by all parasympathetic neurons?

☐ 5. Which cells contain nicotinic receptors? Which contain muscarinic receptors?

Effects of the Parasympathetic Nervous System on Target Cells

The parasympathetic nervous system's effects on its target cells are easily understood if you keep in mind its most basic function: to maintain homeostasis when the body is at rest. Consider what happens after you eat dinner and sit down to read your anatomy and physiology textbook—your heart rate slows, your blood pressure decreases, you digest your food, and your eyes adjust for near vision in order to read the book. Each effect results from a decrease in sympathetic activity and the release of ACh from pre- and postganglionic parasympathetic neurons. The main effects of the parasympathetic nervous system are shown in **Figure 14.10**.

Effects on Cardiac Muscle Cells

The sympathetic nervous system increases heart rate and blood pressure, so it follows that the parasympathetic nervous system decreases them. Preganglionic parasympathetic neurons travel to the heart via the vagus nerve (CN X) and stimulate postganglionic neurons. These neurons in turn act in the heart to reduce its rate of contraction, which lowers blood pressure.

Effects on Smooth Muscle Cells

Postganglionic parasympathetic neurons innervate the smooth muscle cells in many different organs, and trigger the following effects:

- **Constriction of the pupil.** Fibers of the oculomotor nerve in the ciliary ganglion innervate the sphincter pupillae muscle, which constricts the pupil and so allows less light into the eye.
- **Accommodation of the lens for near vision.** Fibers of the oculomotor nerve also innervate the smooth muscle cells of the ciliary muscle. When this muscle contracts, the lens becomes rounder, allowing accommodation for near vision.
- **Constriction of the bronchioles.** When vagal parasympathetic neurons stimulate contraction of the smooth muscle cells lining the airways, the airways narrow, a response called *bronchoconstriction* (brong′-koh-kun-STRIK-shun). This effect is mild, except with very strong stimulation.

- **Contraction of the smooth muscle lining the digestive tract.** Parasympathetic neurons of the vagus nerve also trigger contraction of the smooth muscle cells lining the digestive tract. This produces rhythmic contractions known as *peristalsis* that help to propel ingested food from one part of the digestive tract to the next.
- **Relaxation of digestive and urinary sphincters.** Parasympathetic neurons from the vagus and pelvic splanchnic nerves stimulate relaxation of the smooth muscle cells of the urinary and digestive sphincters. These actions promote urination, as well as the movement of digesting food from one area of the digestive tract to the next, ultimately resulting in defecation.
- **Engorgement of the penis or clitoris.** In men, the blood vessels in the penis receive parasympathetic innervation, as do the clitoral blood vessels in women. When stimulated by parasympathetic pelvic splanchnic neurons, the smooth muscle cells in these vessels relax, which leads to vasodilation. This engorges the penis or clitoris with blood.

Unlike the sympathetic nervous system, the parasympathetic nervous system innervates virtually no blood vessels except in specific areas of the body such as the penis. Yet many blood vessels do dilate when the parasympathetic nervous system is active, particularly those in the digestive and urinary systems. However, this dilation is due to a decrease in *sympathetic* nervous system activity and removal of epinephrine from the bloodstream rather than any direct action by parasympathetic neurons.

Effects on Glandular Epithelial Cells

Recall that the sympathetic nervous system promotes sweating while decreasing secretion from other exocrine glands. So it probably won't be surprising that the parasympathetic nervous system has little to no effect on sweat glands while increasing secretion from other exocrine glands. For example, parasympathetic neurons of the facial nerve stimulate tear production from the lacrimal (tear) glands and mucus production from glands in the nasal mucosa. Fibers from the both the facial and glossopharyngeal nerves stimulate secretion of saliva from the salivary glands. Finally, fibers from the vagus nerve stimulate secretion of enzymes and other products from the cells of the digestive tract.

Effects on Other Cells

Unlike the sympathetic nervous system, the parasympathetic nervous system has no direct effect on the metabolic rate, mental alertness, the force generated by skeletal muscle contractions, blood clotting, adipocytes, or most endocrine secretions. Instead, each of these factors returns to a normal, or "resting," state during times of parasympathetic activity simply because the sympathetic nervous system is no longer dominant. This return to the resting state is an important time for the body to store glucose and other fuels in preparation for the next round of sympathetic activity.

Some drugs block cholinergic synapses, and so many also block the effects of the parasympathetic nervous system as a side effect. For more information, see *A&P in the Real World: Side Effects of Anticholinergic Drugs* on p. 531.

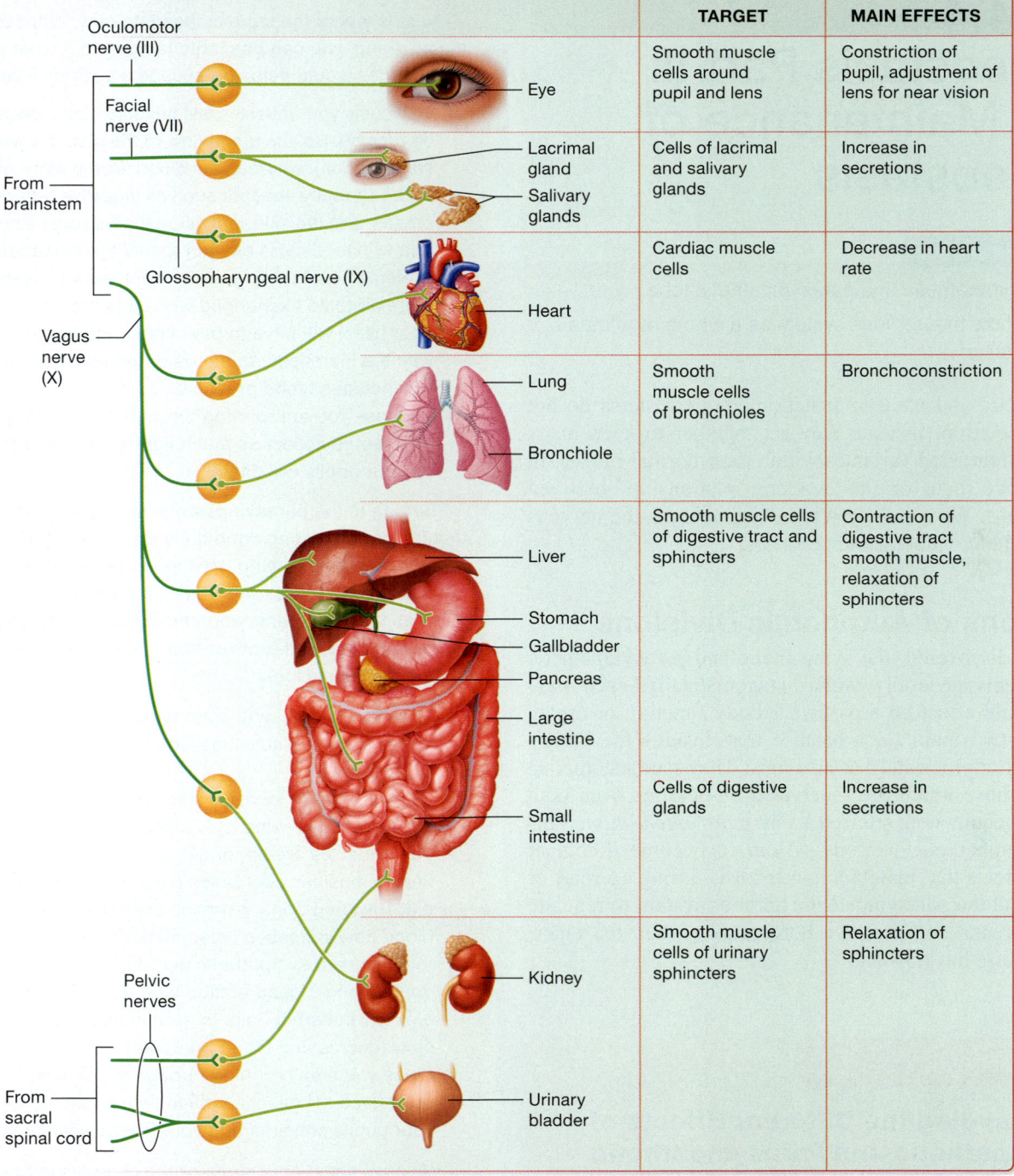

	TARGET	MAIN EFFECTS
Eye	Smooth muscle cells around pupil and lens	Constriction of pupil, adjustment of lens for near vision
Lacrimal gland / Salivary glands	Cells of lacrimal and salivary glands	Increase in secretions
Heart	Cardiac muscle cells	Decrease in heart rate
Lung / Bronchiole	Smooth muscle cells of bronchioles	Bronchoconstriction
Liver / Stomach / Gallbladder / Pancreas / Large intestine	Smooth muscle cells of digestive tract and sphincters	Contraction of digestive tract smooth muscle, relaxation of sphincters
Small intestine	Cells of digestive glands	Increase in secretions
Kidney	Smooth muscle cells of urinary sphincters	Relaxation of sphincters
Urinary bladder		

Figure 14.10 The main effects of the parasympathetic nervous system on target cells. All these effects are stimulated by ACh at nicotinic receptors on postganglionic neurons and muscarinic receptors in target organs.

Quick Check

☐ 6. How does the parasympathetic nervous system affect blood pressure?

☐ 7. Summarize the effects of the parasympathetic nervous system on the smooth muscle cells of the following structures.

   a. Pupil   d. Digestive tract

   b. Lens   e. Urinary and digestive sphincters

   c. Blood vessels serving the penis

Apply What You Learned

☐ 1. The drug *atropine,* derived from the poisonous plant deadly nightshade, blocks muscarinic receptors. Predict the effects of administering this drug.

☐ 2. Drugs that stimulate ACh release often have adverse effects, such as drooling, poor distance vision, and diarrhea. Explain these effects, considering the functions of the parasympathetic nervous system.

See answers in Appendix A.

MODULE 14.4
Homeostasis Part II: PNS Maintenance of Homeostasis

Learning Outcomes

1. Define sympathetic and parasympathetic tone.
2. Explain how the nervous system as a whole regulates homeostasis.

The sympathetic and parasympathetic nervous systems do not function in isolation. Instead, they act together to keep many of the body's regulated variables within their normal ranges. In this module, we focus on the "working relationship" between the two systems. We conclude with a view of how the nervous system as a whole maintains homeostasis.

Interactions of Autonomic Divisions

As you have discovered, the sympathetic and parasympathetic nervous systems generally work antagonistically, each having the opposite effect on a particular body function or organ. Together the two maintain a balance that ensures the body's needs are met appropriately at all times. They can accomplish this because most organs are innervated by neurons from both systems, a phenomenon referred to as *dual innervation*. This allows the sympathetic nervous system to become dominant and trigger effects that maintain homeostasis during exercise or emergency, and the parasympathetic nervous system to regulate these same organs and preserve homeostasis when the emergency or exercise has finished.

ConceptBOOST))))

Understanding the Different Effects of the Sympathetic and Parasympathetic Nervous Systems

The similar names of the sympathetic and parasympathetic nervous systems can make it easy to confuse the two divisions and their functions. But this issue can be addressed by using a couple of simple mnemonics and a bit of logic. For the sympathetic nervous system, remember that it promotes survival in an emergency situation. With that in mind, try the following mnemonic: You are walking through the woods when suddenly a swarm of bees starts chasing you. Your *sympathetic* nervous system is *sympathetic* to your situation, and so is going to initiate changes in your body that will make it easier to escape the bees.

Next is where the logic comes into play. Without memorizing anything, you can use logic to figure out what physiological changes would help you escape a swarm of bees:

- Obviously your muscles will be more active because you will be running. So, the blood flow to the muscles will increase (vasodilation), whereas the blood flow to other organs that aren't currently needed, such as digestive organs, urinary organs, and the skin, will decrease (vasoconstriction).
- You will need more oxygen to fuel the demands of your skeletal muscles, so your bronchioles will dilate to allow more air to be exchanged with each breath.
- Your heart will have to beat faster and harder to keep up with the increased demands of the skeletal muscles. This will increase blood pressure.
- Because you are running through the woods, your eyes will want to collect as much light as possible to see better, so your pupils will dilate.

Turning to the parasympathetic nervous system, remember that it restores resting conditions after a burst of sympathetic activity. With that in mind, try the following mnemonic: You have escaped the swarm of bees, but not without receiving a few stings. Your *para*sympathetic nervous system will help you re*pair*, rest, and recover from your escape. Here again comes the simple logic:

- To repair tissues, you need to take in nutrients, which requires digestive activities, so all digestive processes need to increase.
- When we take in food, we also take in liquids, which results in the formation of urine, so urinary processes increase as well.
- Your muscles are no longer actively moving, so blood flow to muscles decreases (vasoconstriction), but your digestive and urinary organs are actively working, so their blood flow increases (vasodilation). Note this is due only to decreased sympathetic activity.
- You no longer need additional oxygen, so your bronchioles will constrict back to normal and your heart rate will slow (decreasing blood pressure).
- Since you aren't running through the woods, your eyes don't need to collect as much light or see as far away, so your pupils constrict and your lenses adjust for near vision.

We have listed only some of the changes mediated by the sympathetic and parasympathetic nervous systems here, but if you look over Figures 14.7 and 14.10, you'll see that their other effects also make sense if you think about them in these settings (fleeing to escape a swarm of bees or recovering from the escape). As is often the case, understanding a concept rather than simply memorizing it proves much more useful in the end. ■

See **Figure 14.11** for a summary and comparison of the structural and functional features of the sympathetic and parasympathetic nervous systems.

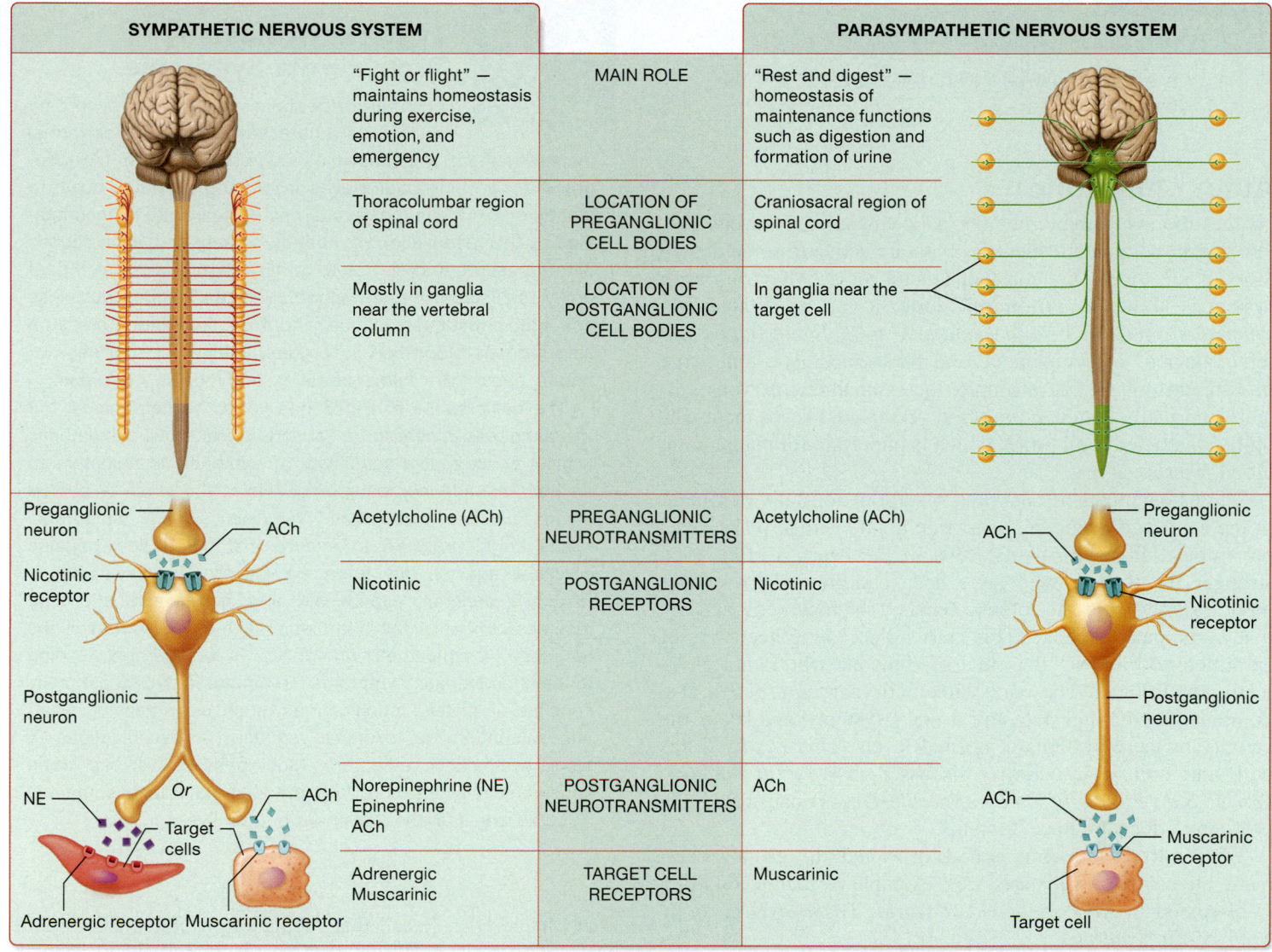

SYMPATHETIC NERVOUS SYSTEM			PARASYMPATHETIC NERVOUS SYSTEM
"Fight or flight" — maintains homeostasis during exercise, emotion, and emergency	MAIN ROLE	"Rest and digest" — homeostasis of maintenance functions such as digestion and formation of urine	
Thoracolumbar region of spinal cord	LOCATION OF PREGANGLIONIC CELL BODIES	Craniosacral region of spinal cord	
Mostly in ganglia near the vertebral column	LOCATION OF POSTGANGLIONIC CELL BODIES	In ganglia near the target cell	
Acetylcholine (ACh)	PREGANGLIONIC NEUROTRANSMITTERS	Acetylcholine (ACh)	
Nicotinic	POSTGANGLIONIC RECEPTORS	Nicotinic	
Norepinephrine (NE) Epinephrine ACh	POSTGANGLIONIC NEUROTRANSMITTERS	ACh	
Adrenergic Muscarinic	TARGET CELL RECEPTORS	Muscarinic	

Figure 14.11 Comparison of structures and effects of the sympathetic and parasympathetic nervous systems.

A P in the Real World

Side Effects of Anticholinergic Drugs

Many drugs block either ACh release or ACh receptors as an unintended side effect. Such drugs are prescribed for many different conditions, such as allergies, respiratory diseases, and gastrointestinal disorders. These *anticholinergic drugs* have a wide range of potential effects. Some of the more common unwanted effects are the following:

- *Urinary retention.* Anticholinergics block the relaxing effect of the parasympathetic nervous system on the urinary sphincters, and make passing urine more difficult.
- *Constipation.* Anticholinergics also block parasympathetic effects on the smooth muscle of the digestive tract. This causes digested food to move more slowly through the tract and can lead to constipation.
- *Dry mouth.* One of the most common anticholinergic effects is a dry mouth, caused by a decrease in the parasympathetic nervous system's ability to stimulate the secretion of saliva.
- *Blurred vision.* By blocking the actions of ACh on the smooth muscle of the eye, anticholinergics have two effects. First, they prevent constriction of the pupil, leading to pupillary dilation. Second, they prevent contraction of the ciliary muscle, preventing accommodation for near vision. Both effects can cause blurred vision.

Quick Check

☐ 1. What is dual innervation?

☐ 2. How do the sympathetic and parasympathetic nervous systems interact?

Autonomic Tone

Neither the sympathetic nor the parasympathetic nervous system is ever completely silent, as both are active to some degree most of the time. This constant amount of activity from each system is known as **autonomic tone;** it can be divided into *sympathetic tone* and *parasympathetic tone.* Interestingly, different degrees of sympathetic and parasympathetic tone exist in different organs. For example, the sympathetic nervous system is normally dominant in blood vessels and keeps them partially constricted at all times, which is important for maintaining blood pressure at rest.

The parasympathetic nervous system is normally dominant in the heart and keeps the heart rate at an average of 72 beats per minute. Parasympathetic tone in the heart is often even stronger in athletes, who have been conditioned to recover more quickly from the intense bursts of sympathetic activity that accompany exercise. This causes the resting heart rate of an athlete to be lower than that of someone who is not physically conditioned. The parasympathetic nervous system also dominates in the digestive and urinary systems, and keeps the activity of these systems at normal levels. This is why drugs with anticholinergic actions (discussed in *A&P in the Real World: Side Effects of Anticholinergic Drugs*) can have such profound effects on these systems.

When normal autonomic tone is disrupted, this can have dramatic effects on homeostasis. One example of such a condition is discussed in *A&P in the Real World: Postural Orthostatic Tachycardia Syndrome.*

Quick Check

☐ 3. What is autonomic tone?

Summary of Nervous System Control of Homeostasis

FLASHBACK

1. Where is the hypothalamus located? Which regulated physiological variables does it control? (p. 433)
2. Where is the reticular formation located? Which regulated physiological variables does it control? (p. 438)
3. What are the primary functions of the cerebral cortex and the amygdala? (p. 428 and p. 432)

Maintenance of homeostasis is the body's most essential function, and the two divisions of the ANS figure importantly in how the nervous system overall fulfills this role (**Figure 14.12**). Now that we know more about the ANS, we can revisit what you learned in the CNS section on homeostasis of vital functions

A&P in the Real World Postural Orthostatic Tachycardia Syndrome

Postural orthostatic tachycardia syndrome (POTS) is characterized by an abnormal increase in heart rate (known as *tachycardia;* tak´-ih-KAR-dee-uh) when an individual moves from lying or sitting down to standing up. This inappropriate rise of heart rate is accompanied by the vasodilation of most blood vessels, which causes blood pressure to drop due to gravity. Symptoms, which generally result from the low blood pressure, include dizziness and lightheadedness, fatigue, and thirst. Low blood pressure also reduces blood flow to organs, leading to shortness of breath, chest pain, cold extremities, and muscle weakness.

The basic cause of POTS has yet to be determined, but the symptoms appear to be caused by excessive sympathetic activity or excessive sensitivity of sympathetic receptors to epinephrine and norepinephrine. Normally, when a person stands up, the sympathetic nervous system temporarily raises blood pressure to ensure that blood flow remains constant against the force of gravity. The sympathetic response generally tapers off, and parasympathetic tone resumes control of the heart rate. With POTS, however, the response to sympathetic stimulation is exaggerated, leading to the characteristic symptoms. Treatment for POTS generally consists of dietary modifications, such as increasing water and salt intake; an exercise regimen; and medications to block sympathetic receptors. Most patients will see some improvement in the condition gradually, although some will struggle with it for the remainder of their lives.

(see Chapter 12). Recall that regulated physiological variables are largely controlled centrally by the hypothalamus and the brainstem reticular formation. Many actions of these structures are mediated peripherally via the sympathetic and parasympathetic nervous systems.

Note in Figure 14.12 that many signals sent from the hypothalamus are directed toward areas of the reticular formation called **autonomic centers.** These centers contain neurons that control the activity of preganglionic sympathetic and parasympathetic neurons. Although the hypothalamus does exert some degree of control over these centers, the reticular formation can function even if the circuits connecting it with the hypothalamus are severed. This indicates that the reticular formation is capable of controlling many of our most critical autonomic functions, such as heart rate and blood pressure, on its own.

Figure 14.12 also shows that both the hypothalamus and reticular formation receive input from higher centers in the brain, including the amygdala and multiple areas of the cerebral cortex. This in part explains why emotion has such a profound effect on our visceral functions. A state of excitement, mediated by the cerebral cortex and amygdala, will be passed on to the hypothalamus, reticular formation, and finally to the ANS, after which we notice an increased heart rate, elevated blood pressure, dilated pupils, and other signs of sympathetic activity.

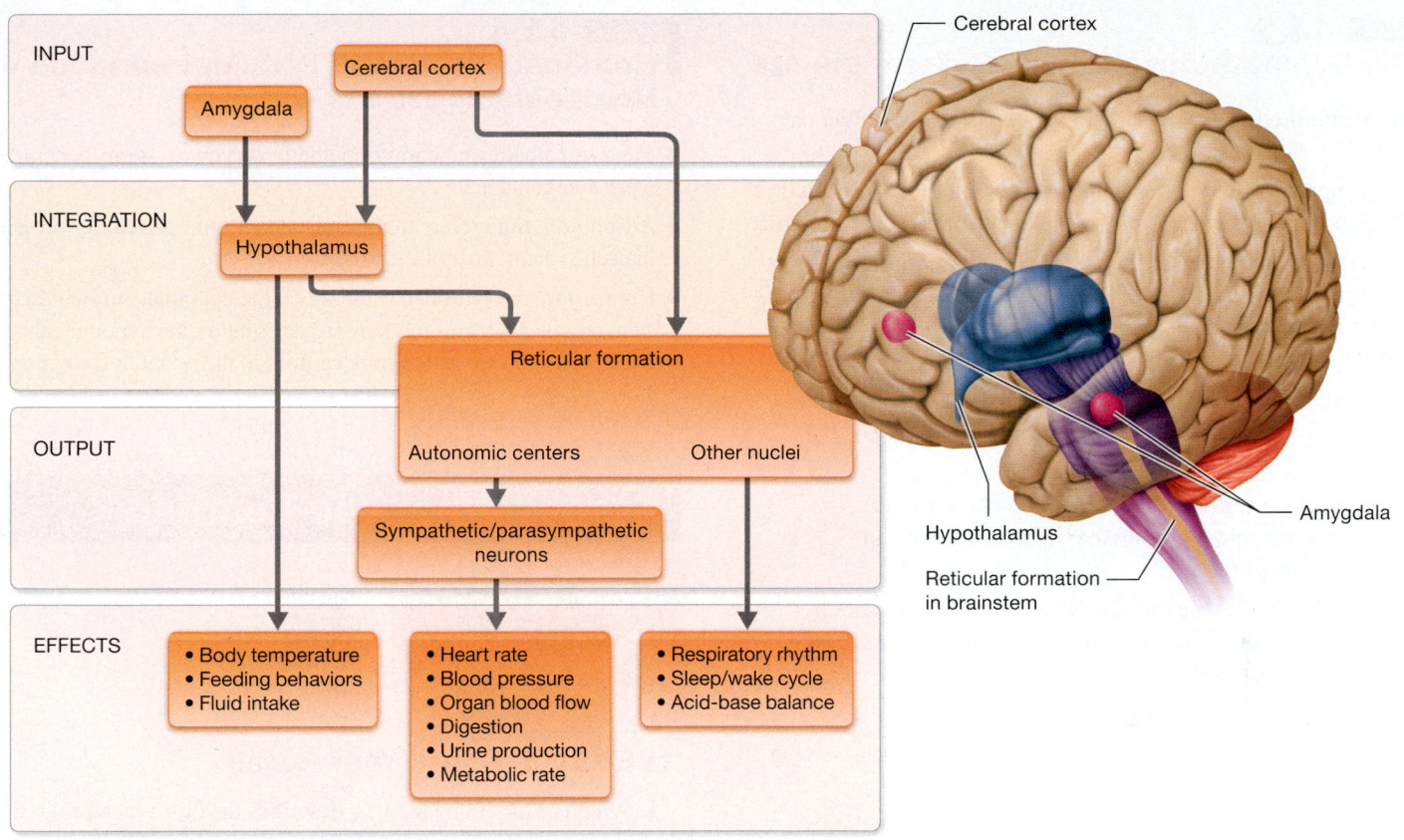

Figure 14.12 Summary of nervous system control of homeostasis.

Although the nervous system plays an important role in maintaining homeostasis, it is only part of the story. We explore the other part of it when we discuss the endocrine system (in Chapter 16).

☐ 4. How does the CNS mediate control over the ANS?

☐ 1. Explain why the nervous systems of individuals with brainstem injuries generally have difficulty controlling their autonomic functions.

☐ 2. Predict what would happen if parasympathetic tone were dominant in the airways.

See answers in Appendix A.

Chapter Summary

For everything you need to succeed in this course, go to Mastering Anatomy & Physiology. There you will find:

- Practice Tests
- Author Podcasts
- ▶ Big Picture Animations
-))) Concept Boost Video Tutors
- **iP2**™ Interactive Physiology 2.0 Tutorials
- **A&PFlix** A&P Flix 3D Animations
- Active-Learning Workbook

MODULE 14.1
Overview of the Autonomic Nervous System 516–519

- The autonomic nervous system (ANS) performs its functions via a series of **visceral reflex arcs.**
- Autonomic motor neurons consist of **preganglionic neurons** that synapse on **autonomic ganglia,** and then **postganglionic neurons** that contact the target cells.
- The autonomic nervous system is made up of the sympathetic and parasympathetic nervous systems.

MODULE **14.2**
The Sympathetic Nervous System 519–526

- The **sympathetic nervous system** is often called the "fight or flight" division of the ANS.
- The sympathetic nervous system is called the **thoracolumbar division.** Most ganglia are part of the sympathetic chain ganglia.
- Preganglionic sympathetic axons release **acetylcholine (ACh)** at their synapses with postganglionic neurons. Postganglionic axons release **norepinephrine, epinephrine,** or ACh at their synapses with target cells.
- **Adrenergic receptors** bind to epinephrine and norepinephrine. There are five classes of adrenergic receptors: $\alpha_1, \alpha_2, \beta_1, \beta_2,$ and β_3.
- **Cholinergic receptors** bind ACh. The two types of cholinergic receptors are **muscarinic receptors** and **nicotinic receptors.**
- The sympathetic nervous system increases blood pressure by raising the rate and force of contraction of the heart and also by causing *vasoconstriction*.
- Sympathetic neurons cause *bronchodilation*.
- Norepinephrine causes *vasodilation* of the blood vessels serving skeletal muscles.
- The adrenal medulla is made up of modified sympathetic postganglionic neurons. When stimulated, the adrenal medulla releases epinephrine and norepinephrine into the blood.
- Sympathetic activity decreases urinary and digestive functions and the secretion from most glands, dilates the pupil, increases the levels of metabolic fuels in the blood, and increases the secretion from sweat glands.

MODULE **14.3**
The Parasympathetic Nervous System
526–529

- The **parasympathetic nervous system** is the "rest and digest" division of the ANS that performs the body's maintenance functions such as digestion and urine formation.
- The parasympathetic nervous system is called the **craniosacral division.** Preganglionic axons synapse with the cell bodies of postganglionic neurons in ganglia near the target cells.
- Both preganglionic and postganglionic parasympathetic neurons release ACh at their synapses.
- The parasympathetic nervous system lowers blood pressure by decreasing the rate of contraction of the heart.
- Parasympathetic activity causes constriction of the pupil, accommodation of the lens for near vision, *bronchoconstriction,* contraction of digestive tract smooth muscle, relaxation of digestive and urinary sphincters, engorgement of the penis or clitoris, and an increase in secretion from most glands except sweat glands.

MODULE **14.4**
Homeostasis Part II: PNS Maintenance of Homeostasis 530–533

- The sympathetic and parasympathetic nervous systems generally work antagonistically.
- **Autonomic tone** refers to the constant amount of activity present in each system most of the time.
- Homeostasis is controlled centrally by the hypothalamus and brainstem reticular formation. Many signals sent by the hypothalamus are directed toward **autonomic centers** in the reticular formation.

Assess What You Learned

Scan the QR Code for additional practice test questions

https://goo.gl/T7xtvQ

LEVEL 1   Check Your Recall

1. Which of the following best describes the basic function of the autonomic nervous system?

   a. Controls the somatic nervous system
   b. Controls automatic functions to maintain homeostasis
   c. Detects somatic sensory information
   d. All of the above

2. Fill in the blanks: The sympathetic nervous system is also known as the _____ division because the cell bodies of its preganglionic neurons are located in the _____.

3. Sympathetic preganglionic neurons synapse:

   a. directly on target cells.
   b. on skeletal muscle fibers.
   c. on sympathetic chain or collateral ganglia.
   d. on vagal ganglia.

4. Differentiate between adrenergic and cholinergic receptors.

5. Which neurotransmitter(s) is/are used by sympathetic postganglionic neurons?

   a. Epinephrine   d. Both a and b
   b. Norepinephrine   e. All of the above
   c. Acetylcholine

6. Explain what happens when sympathetic neurons stimulate the adrenal medulla. What purpose does this serve?

7. Which of the following actions would you expect when sympathetic neurons release norepinephrine onto β_1 receptors?

   a. Decreased sweat production
   b. Constriction of blood vessels serving the digestive, urinary, and integumentary systems
   c. Adjustment of the shape of the lens
   d. Increase in the rate and force of contraction of the heart.

8. Mark each of the following as an effect of the sympathetic nervous system or the parasympathetic nervous system.

   a. _____ Bronchodilation
   b. _____ Bronchoconstriction
   c. _____ Constriction of pupil
   d. _____ Vasoconstriction of blood vessels serving organs of digestive, urinary, and integumentary systems
   e. _____ Increased metabolic rate
   f. _____ Dilation of pupil
   g. _____ Contraction of smooth muscle of digestive tract

9. Mark the following statements as true or false. If a statement is false, correct it to make a true statement.

   a. The parasympathetic nervous system generally decreases the secretion from digestive glands.
   b. Sympathetic stimulation increases sweat secretion.
   c. The parasympathetic nervous system releases acetylcholine onto all its synapses.
   d. Sympathetic tone controls the resting rate of the heart.

10. Fill in the blanks: _____ receptors are located on parasympathetic postganglionic neurons, and _____ receptors are located on parasympathetic target cells.

11. Parasympathetic ganglia are typically:

    a. located along the spinal cord.
    b. located near their target cells.
    c. located within the central nervous system.
    d. Parasympathetic neurons do not synapse in ganglia.

12. Central nervous system control over the ANS is mediated by:

    a. the reticular formation.   c. the thalamus.
    b. the hypothalamus.   d. both a and b.

LEVEL 2 Check Your Understanding

1. Using 20 or fewer words, define each of the following terms in your own words.

   a. Sympathetic nervous system
   b. Parasympathetic nervous system

2. You are running a race to the top of a mountain. Explain all the changes your sympathetic nervous system will initiate to maintain homeostasis as you run the race.

3. Describe all the changes initiated by the parasympathetic nervous system that will take place when you finish the race. How will these changes maintain homeostasis?

LEVEL 3 Apply Your Knowledge

PART A: Application and Analysis

1. Which cranial nerves would be affected by a drug that stimulates the parasympathetic nervous system? Predict potential adverse effects that one might experience.

2. A patient, Dr. Young, has both asthma and high blood pressure. Her physician prescribed the drug propranolol to treat her hypertension; this drug blocks all types of β-adrenergic receptors. She also takes the drug albuterol for asthma, which activates β_2 receptors on bronchial smooth muscle. Will the pairing of these two drugs cause problems for Dr. Young? Explain.

3. Mr. Chevalier has been diagnosed with Horner syndrome, which is caused by dysfunction of the sympathetic neurons in the superior cervical ganglion that innervate structures of the head, face, and neck. What symptoms is Mr. Chevalier likely to face due to his disease? Explain.

PART B: Make the Connection

4. Many chemical warfare agents, such as the poisonous gas *sarin,* block the enzyme acetylcholinesterase in the synaptic cleft. What effects will this poison have on muscle contraction? (*Connects to Chapter 10*) What effects will it have on the autonomic nervous system? What symptoms would you expect to see from this poison?

See answers in Appendix A.

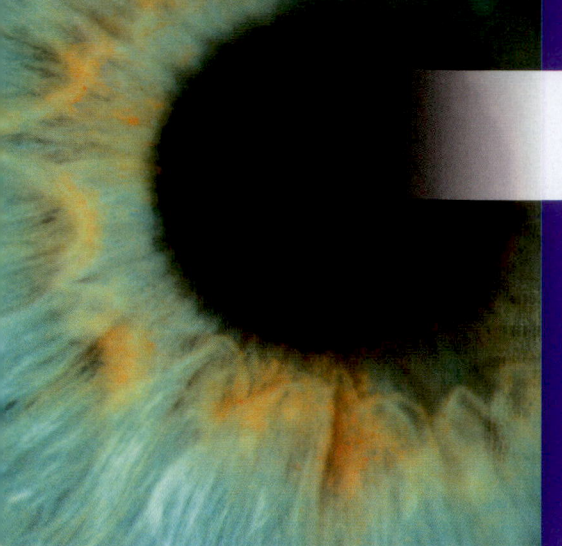

15
The Special Senses

*For practice applying concepts to a clinical scenario, check out the **Running Case Study** for this chapter @ Mastering Anatomy & Physiology*

Photo: The eye is critical to vision, one of the five special senses; vision is the perception of light reflected by various objects.

536

I f a tree falls in the forest when no one is around, does it still make a sound? In studying the special senses, we can distinguish between physical stimuli—the sound waves produced by the falling tree—and the receptors that are specialized to detect them. This chapter examines the role of the special senses in detecting stimuli (such as sounds, aromas, and light), *transducing* (converting) them into action potentials, and interpreting them. But first, let's see how the special senses are similar to, and different from, the general senses.

MODULE **15.1**
Overview of the Special Senses

Learning Outcomes

1. Describe the basic process of sensory transduction.
2. Compare and contrast the general and special senses.

The **special senses** are so named because they convey specific stimuli from specialized sensory organs in discrete locations of the head. You can probably list four of the special senses quite easily—smell (*olfaction*), taste (*gustation*), *vision*, and hearing (*audition*). However, can you name the fifth special sense? It is *vestibular sensation*, the detection of head movement and position that helps maintain equilibrium, or balance.

In this module, we first compare the special senses with general senses that you examined in previous chapters. We also review how a sensory stimulus is converted into an action potential by the process of *sensory transduction*.

Comparison of the General and Special Senses

 **FLASHBACK**

1. What are the receptors for the general senses? (p. 460)

You have already been introduced to sensation—the *general senses*—in the chapters on the central and peripheral nervous systems. So before we dig deeper into the special senses, let's take a look at how they differ from the general senses (**Figure 15.1**):

- **Stimuli detected.** The general senses involve the detection of stimuli such as touch, pain, and temperature. As we have discussed, each special sense organ detects very specific stimuli—light, sound waves, head movements, and the chemicals that produce tastes and smells.

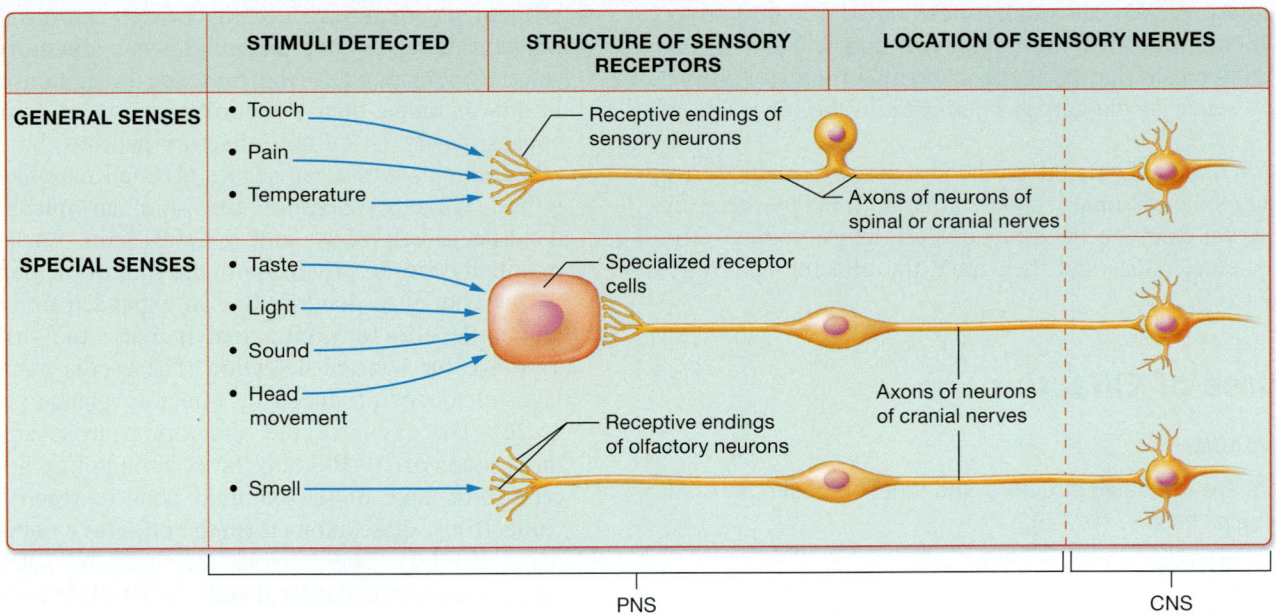

	STIMULI DETECTED	STRUCTURE OF SENSORY RECEPTORS	LOCATION OF SENSORY NERVES
GENERAL SENSES	• Touch • Pain • Temperature	Receptive endings of sensory neurons	Axons of neurons of spinal or cranial nerves
SPECIAL SENSES	• Taste • Light • Sound • Head movement • Smell	Specialized receptor cells Receptive endings of olfactory neurons	Axons of neurons of cranial nerves

PNS   CNS

Figure 15.1 Comparison of general and special senses.

- **Structure of sensory receptors.** In general sensation, stimuli are detected at the receptive endings of sensory neurons. These receptors have relatively simple structures and are located throughout the body in both the skin and the internal organs. In contrast, the receptors of most special senses are not neurons but instead are specialized cells that detect special sensory stimuli. The sense organs convert these stimuli into electrical signals that are passed on to neurons via neurotransmitters. Olfaction is the only special sense in which the receptor cells are neurons.
- **Location of sensory nerves.** General sensory stimuli are transmitted to the CNS by the axons of neurons located in both spinal and cranial nerves. However, the special sense organs are all housed in the head, so only the axons of neurons of cranial nerves convey special sensory stimuli.

Quick Check

□ 1. What are the three major differences between the general and special senses?

Sensory Transduction

In the peripheral nervous system chapter, you read that **transduction** occurs when a physical or chemical stimulus is converted into an action potential that can be interpreted by the brain (see Chapter 13). Neurons involved in the general senses have specialized receptive endings to detect touch, temperature, or pain. The presence of these stimuli alters the resting membrane potential of the sensory neurons to produce an action potential that reaches the CNS.

The special senses also detect stimuli—such as light, chemicals, or sounds—in the environment and transduce them into action potentials. These action potentials also are propagated through the axons of peripheral neurons to the CNS.

For both general and most special sensory stimuli, the stimuli are first processed by sensory nuclei and transmitted to the thalamus and primary cortex areas for awareness and identification. Sensory signals are then passed to association areas for further interpretation and integration.

Quick Check

□ 2. What is sensory transduction?

Apply What You Learned

□ 1. Stimuli such as light or sound are not detected directly by neurons. Explain why this is true. (*Hint:* Focus on the differences between the general and special senses.)

See answer in Appendix A.

MODULE 15.2
Olfaction

Learning Outcomes

1. Describe and identify the location of olfactory receptors.
2. Explain how odorants activate olfactory receptors.
3. Describe the path of action potentials from the olfactory receptors to various parts of the brain.

Our olfactory system allows us to detect the presence of **odorants** (OH-dohr-entz), or chemicals, in the air. Once detected, neurons of the olfactory system transduce the chemical signals of the odorants into electrical signals that our brain can interpret. We can smell an estimated 400,000 odor-ants, of which about 80% are unpleasant odors. This speaks to the protective role that olfaction plays in animals, warning us to the presence of dangerous chemicals in the air, such as smoke.

In addition to identifying odors, the nervous system allows us to experience the emotional responses that accompany them. In this module, we examine the structures and physiological pro-cesses that detect odors and then trace the olfactory pathway through the CNS.

Structures of Olfaction

◀◀ FLASHBACK

1. What are the basic structure and function of simple epithelial tissue? (p. 129)
2. What are bipolar neurons? (p. 387)

Olfaction (ohl-FAK-shun) begins in the **olfactory epithelium,** a small patch of specialized epithelium located in the superior region of each nasal cavity (**Figure 15.2a**). The olfactory epithelium is yellowish-brown and about the size of your thumbnail. As described next, it contains three types of cells (**Figure 15.2b and c**).

- **Olfactory neurons** are modified bipolar neurons that detect odorants. Olfactory neurons are known as **chemoreceptors,** since they respond to the presence of certain chemicals in the air rather than to neurotransmitters. Humans have approximately 10 million olfactory neurons, but dogs and other animals with keen senses of smell may have several billion. Olfactory neurons consist of an apical modified dendrite, a cell body, and a basal axon. Each neuron's modified dendrite projects into the roof of the nasal cavity. At the end, each dendrite has an expanded tip containing nonmotile **olfactory cilia** that increase the surface area available for odorant detection. These cilia extend into a layer of mucus produced by olfactory glands (see Figure 15.2c). The axons of the olfactory neurons are bundled into groups of 10–100 and travel through tiny holes in the cribriform plate of the ethmoid bone to reach the CNS. Collectively, these axons form the **olfactory nerve** (cranial nerve [CN] I). These axons terminate by synapsing on other neurons called **mitral cells** (MY-trul) in the **olfactory bulb,** a structure located in the brain just superior to the ethmoid bone and inferior to the frontal lobe of the brain. Axons leave the olfactory bulb in the **olfactory tract** (see Figure 15.2a and b) and transmit olfactory stimuli to other parts of the brain.

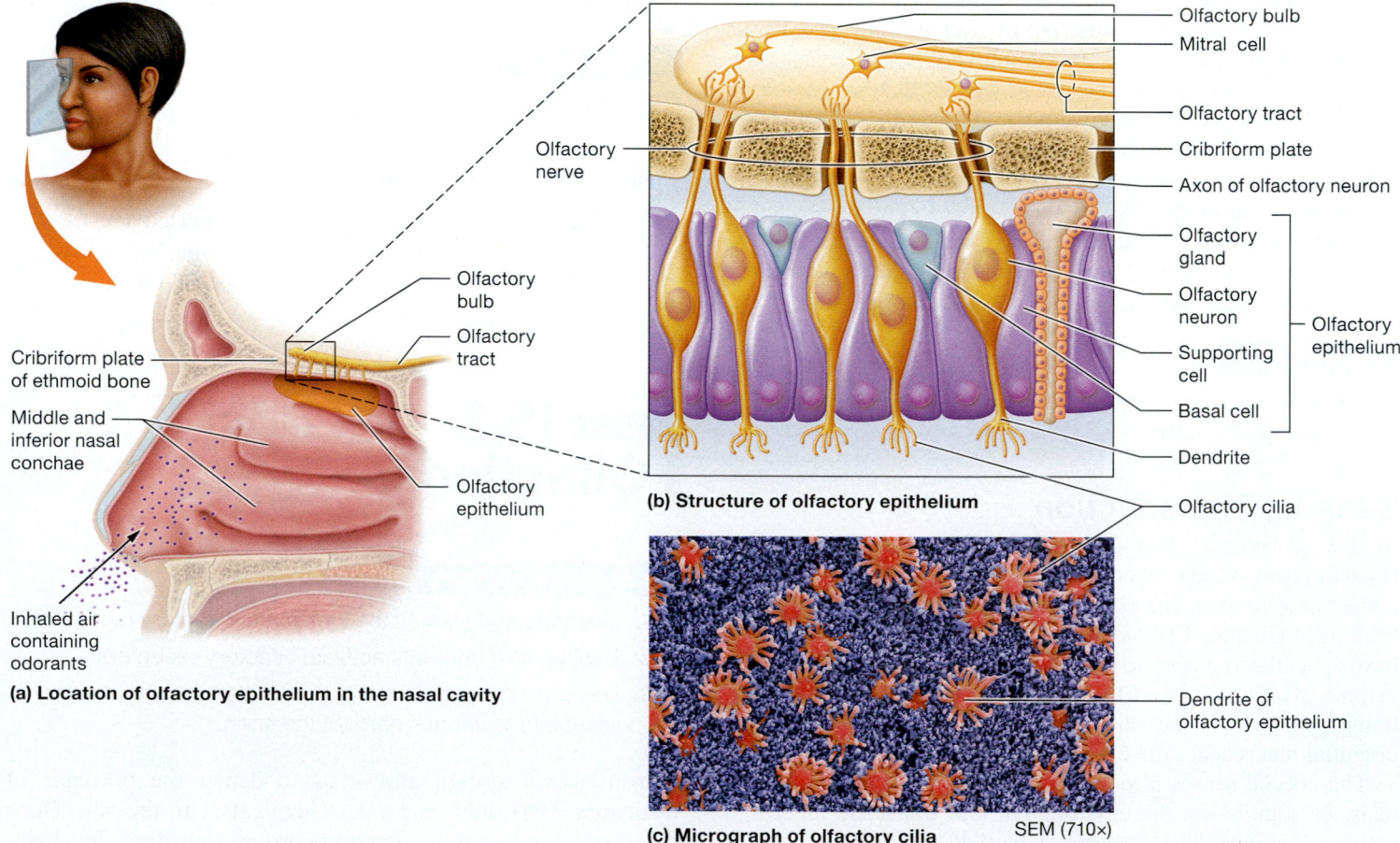

(a) Location of olfactory epithelium in the nasal cavity

(b) Structure of olfactory epithelium

(c) Micrograph of olfactory cilia   SEM (710×)

Figure 15.2 Olfactory epithelium and olfactory neurons.

- **Basal cells** are stem cells that develop into olfactory neurons. An olfactory neuron has a lifespan of only 30–60 days due to its highly exposed location; the basal cells ensure the continual replacement of olfactory neurons.
- **Supporting cells** are columnar cells that support and surround the olfactory neurons. They contain a pigment that gives the olfactory epithelium its color.

Quick Check

☐ 1. Where is the olfactory epithelium located?
☐ 2. What types of cells make up the olfactory epithelium?
☐ 3. Describe the structure of an olfactory neuron.

Physiology of Olfaction

FLASHBACK

1. What is a local potential? How does it differ from an action potential? (p. 396)
2. What is the function of the limbic system? (p. 431)

Odorants are carried into the nasal cavity by the air, but few of them reach the olfactory epithelium during normal breathing. If you detect a faint odor of something, your typical response is to sniff or inhale more deeply. This draws in more air at a faster rate and forces it across the olfactory epithelium. When odorants come into contact with the olfactory epithelium, they are detected by olfactory neurons, which begin a series of action potentials that are transmitted to specific parts of the brain, allowing you to identify the scent. In the next subsections, we discuss how these processes take place.

Activation of Olfactory Receptors

When odorants reach the olfactory epithelium, most of them bind to **odorant-binding proteins,** which shuttle them to the

receptors on the olfactory cilia. When the odorants reach their receptors, the following events occur (**Figure 15.3**):

① **Binding of an odorant to its receptor activates a G-protein.** The activated G-protein then detaches from the receptor.

② **The activated G-protein triggers the enzyme adenylate cyclase to convert ATP into cyclic AMP (cAMP).**

③ **cAMP opens ion channels that allow sodium and calcium ions to enter the cell.** The entry of these ions into the cell creates a local potential that depolarizes the membrane of the olfactory neuron. If enough local depolarizations summate so that threshold is reached at the axon hillock, an action potential is triggered.

Although humans can detect about 400,000 different odorant substances, we only have about 350 genes coding for receptor proteins. However, each can bind more than one type of odorant, and we identify a particular smell by the combination of receptors that are stimulated. Just as you can play thousands of different chords using the 88 keys on a piano, the olfactory neurons can code for thousands of odors using combinations of a few hundred receptors.

Not all chemical odorants bind to olfactory neurons. Some substances, such as ammonia or chlorine, bind to branches of the trigeminal nerve in the epithelium lining the nasal cavity and produce a burning sensation. If you've ever gotten a large whiff of ammoniated window cleaner or very strong mustard, you know how these substances can irritate your nasal cavity.

The Olfactory Pathway

For us to detect and identify an odor, olfactory stimuli must be transmitted to the CNS and delivered to various regions of the brain:

① **The axons of olfactory neurons (making up the olfactory nerve) carry olfactory stimuli to the olfactory bulb in the CNS.** The axons of the olfactory neurons terminate

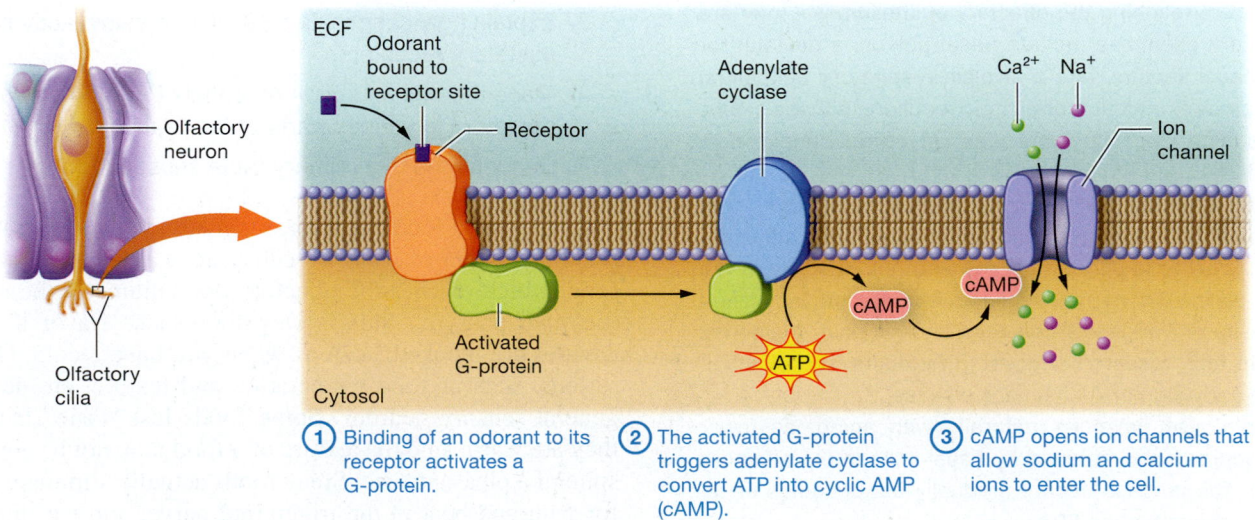

① Binding of an odorant to its receptor activates a G-protein.

② The activated G-protein triggers adenylate cyclase to convert ATP into cyclic AMP (cAMP).

③ cAMP opens ion channels that allow sodium and calcium ions to enter the cell.

Figure 15.3 Transduction of olfaction in an olfactory neuron.

in the olfactory bulb just inferior to the frontal lobe. The axon terminals synapse with the dendrites of mitral cells and trigger local depolarizations in their dendrites; many olfactory neurons with the same receptor type end on the same mitral cell. If enough local depolarizations are triggered, the mitral cells will have an action potential.

② **An olfactory stimulus travels from the olfactory bulb to the primary olfactory cortex in the temporal lobe.** If an action potential is initiated, it travels along the axons of the mitral cells, which form the bulk of the olfactory tract. The stimuli are delivered to the **primary olfactory cortex,** located in the inferomedial temporal lobe near the optic chiasma. This is the only sensory pathway that has no synapse in the thalamus. The primary olfactory cortex is responsible for awareness and identification of an odor. Interestingly, it is also a common location for seizures, which are often preceded by the sensation of an unpleasant odor.

Disruptions of this pathway can cause significant dysfunction, as explained in *A&P in the Real World: Anosmia.*

Neurons from the olfactory cortex then contact other neurons in the amygdala, hippocampus, hypothalamus, and other parts of the limbic system. This evokes emotional responses to odors as well as visceral reactions such as salivation, sexual arousal, and nausea. Connections also exist between the primary olfactory cortex and the cortex in the base of the frontal lobe, which are thought to play a part in integrating olfaction and taste.

Apply What You Learned

☐ 1. You are working in a lab when you find a bottle of unlabeled chemicals. You open the bottle and your nasal passages immediately begin to feel as though they are burning. Your lab partner wonders what sort of "smell" produces burning, as in what type of olfactory receptors lead to this sensation. What do you tell your partner?

☐ 2. Which structure(s) in the olfactory pathway would be most affected by a large tumor in the inferior frontal lobe? How would this affect olfaction?

See answers in Appendix A.

Anosmia

Although the thought of losing your sense of smell might not be as frightening as that of losing your vision or hearing, **anosmia** (an-AWZ-mee-uh; the lack of olfaction) can be serious. People with anosmia may not be able to detect smoke or the odor of spoiled food. They may also suffer from malnutrition—olfaction contributes significantly to the sense of taste, so food loses much of its appeal.

Many conditions result in anosmia or **hyposmia** (hy-PAWZ-mee-uh; reduced olfactory sensitivity). Anything that blocks air from reaching the olfactory epithelium will interfere with olfaction; examples include nasal polyps, a deviated or fractured nasal septum, and a swollen respiratory epithelium from a respiratory infection or allergic reaction. In these cases, anosmia is often temporary, and olfaction returns once air flow is restored.

However, anosmia can be permanent. Head injuries that damage the neural pathways at any point can disrupt olfaction; the olfactory nerve is particularly vulnerable to fracture of the cribriform plate. The neural degeneration caused by Alzheimer's disease or Parkinson's disease often impairs the sense of smell. In fact, research suggests that decreased olfaction may be an early sign of Alzheimer's disease.

Hyposmia also develops naturally with aging, as fewer olfactory neurons are replaced by basal cells in the olfactory epithelium. The process occurs gradually but becomes especially noticeable after age 60.

MODULE **15.3**
Gustation

Learning Outcomes

1. Describe the location and structure of taste buds.
2. Explain how chemicals dissolved in saliva activate gustatory receptors.
3. Trace the path of action potentials from the gustatory receptors to various parts of the brain.
4. Describe the five primary taste sensations.

The flavor of a food encompasses many types of sensation. Surprisingly, much of what we call taste is actually the smell of a food, which reaches the olfactory epithelium via the nasal cavities and pharynx. This is why food lacks flavor if the nasal cavities are blocked, such as when you have a cold. Other sensations, such as food temperature and texture, are detected by general sensory neurons. Some foods just "taste" better when they are warm, and the texture of a food may not be appealing in spite of a pleasant taste. Spicy foods actually stimulate receptors for pain and heat in the trigeminal nerve, and we interpret the sensation as "spicy," or "hot."

However, a large part of our experience in eating a food does depend on the sense of **gustation** (gus-TAY-shun), or *taste.* Like olfaction, gustation involves chemoreceptors that respond to chemicals. Taste begins with stimulation of specialized receptor cells in **taste buds,** which are small clusters of receptor cells and supporting cells scattered across the tongue and parts of the oral cavity. Each taste bud is contacted by sensory neurons that transmit stimuli to the central nervous system. This module explores taste sensation, from the detection of chemicals on the tongue to the processing of neuronal signals in various regions of the brain.

Structures of Gustation: Taste Buds

Notice that your tongue is covered with rounded projections called **papillae** (puh-PILL-ee; singular, *papilla*). Papillae are classified into four groups based on shape (**Figure 15.4a**):

- **Vallate** (or **circumvallate**) **papillae** are the largest and are dome-shaped; each contains hundreds of taste buds (**Figure 15.4b**).
- **Fungiform papillae** (FUN-jih-form) are mushroom-shaped, and each contains only a few taste buds.
- **Foliate papillae** (FOH-lee-ayt) are ridges on the sides of the tongue and contain taste buds only in childhood.
- **Filiform papillae** (FILL-ih-form) are long, thin cylinders scattered across the tongue. They do not contain any taste buds but have sensory nerve endings that detect the texture and temperature of food.

Taste buds are typically located on the lateral surfaces of the papillae and contain three types of cells (**Figure 15.4c**):

- **Gustatory (taste) cells** (GUS-tuh-tohr-ee) are specialized epithelial cells with microvilli that contain receptors detecting different tastes. The microvilli are found on the apical surface of the cell and project into a small pocket on the surface of the papilla, called the **taste pore.** The basal end of each gustatory cell forms a synapse with the receptive endings of a sensory neuron that carries taste stimuli into the CNS via the facial (CN VII), glossopharyngeal (CN IX), or vagus (CN X) nerve. A typical taste bud has approximately 100 gustatory cells.
- **Basal cells** are stem cells that differentiate into new gustatory cells, which have a lifespan of only 10–14 days.
- **Supporting cells** surround the gustatory cells. They provide physical support but have no role in taste sensation.

Taste sensation tends to decline as people age because the number of basal cells decreases and so the gustatory cells are not replaced when they degenerate. We are typically born with approximately 10,000 taste buds but lose half of them by adulthood, and the rate of decline is greatest after age 50.

Quick Check

☐ 1. Where are taste buds located?

☐ 2. How does a gustatory cell stimulate a sensory neuron?

☐ 3. What are the roles of basal and supporting cells?

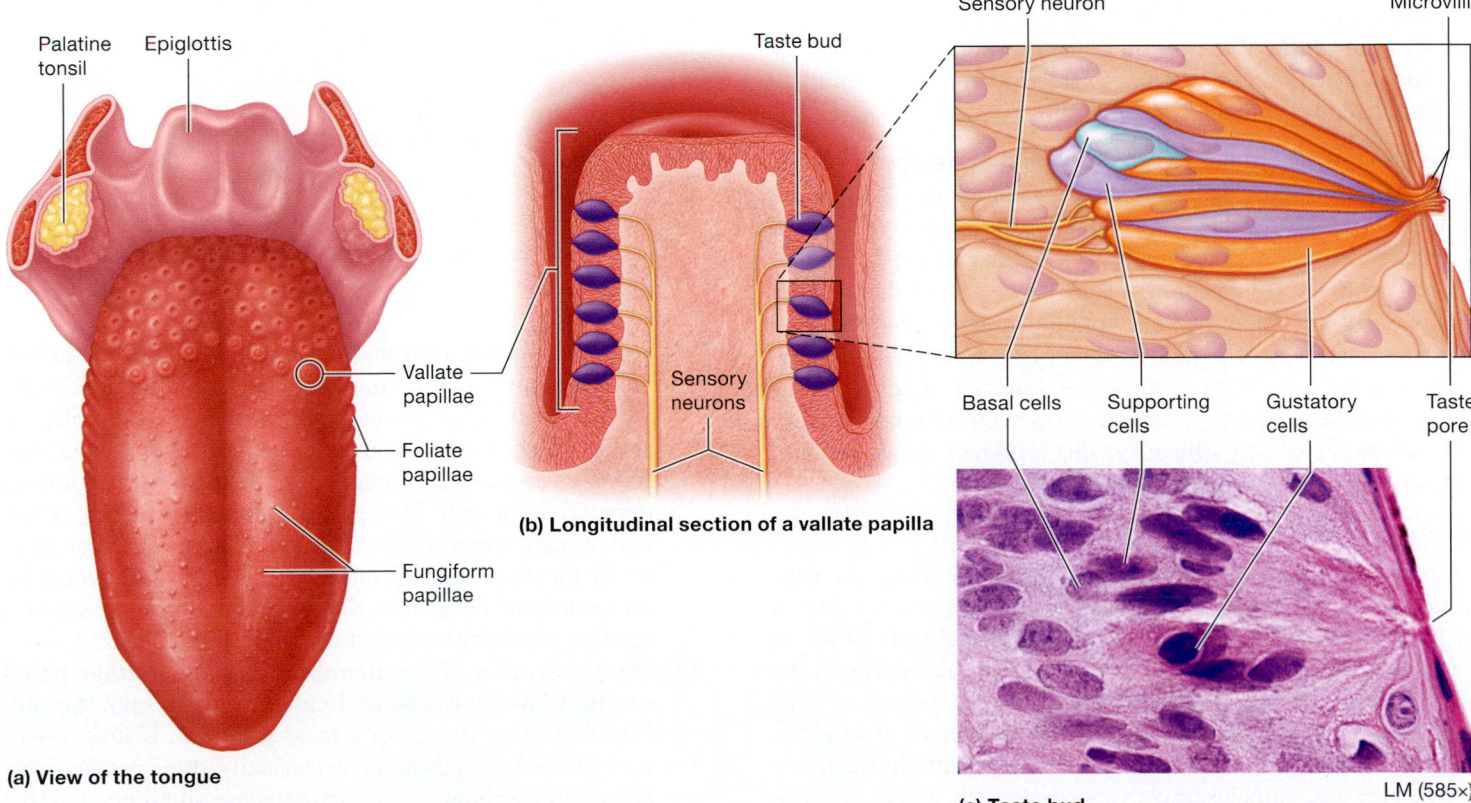

(a) View of the tongue

(b) Longitudinal section of a vallate papilla

(c) Taste bud

LM (585×)

Figure 15.4 Anatomy of the tongue and taste buds. Note that this illustration doesn't include filiform papillae, as they do not contain taste buds.

Physiology of Gustation

Like olfaction, the sense of taste involves transducing chemicals—in this case, in the food we eat—into electrical signals. This transduction from chemical to electrical signal takes place in the gustatory cells, and the action potential is then transmitted to the CNS by one of three cranial nerves. To understand this process, we first need to look at the chemicals we perceive as tastes.

Taste Sensations

The sense of taste relies on detecting various chemicals, which are typically grouped into five classes:

- **Sweet** tastes are mostly due to simple sugars such as glucose and fructose. Ethylene glycol (found in antifreeze) and lead (in lead paint) can also taste sweet, which is why young children may ingest these substances, leading to accidental poisonings.
- **Sour** tastes are produced by hydrogen ions, as found in the citric acid of lemon juice.
- **Salty** foods taste salty due to the presence of metal ions such as sodium and potassium ions.
- **Bitter** flavors are imparted by many different compounds, including alkaloids such as those found in coffee, many plant poisons, and rancid foods. Humans are most sensitive to bitter tastes, which serves a protective function, as a large number of bitter substances are toxic.
- **Umami** (oo-MAWM-ee; the Japanese word for "savory" or "delicious"), a taste often associated with meat or broth, is produced by glutamate or other amino acids.

We detect taste stimuli when they bind to receptors on the microvilli of gustatory cells. The taste receptors are classified by the substance to which they respond, and only one type of receptor is typically found on a gustatory cell. Contrary to popular belief, different tastes are not localized to separate regions of the tongue.

Obviously, the foods we eat contain more flavors than the five mentioned here. The unique flavor of a food lies in its particular combination of taste chemicals and the unique combination of taste receptors it activates (as well as other sensory components, especially olfactory). So a pretzel tastes different from a potato chip, even though both are salty.

Activation of Taste Receptors

The first step in activating a taste receptor is helping the substance to actually reach the taste buds in the crevices between the papillae. This is accomplished with the help of *saliva*, in which most substances will dissolve. You can demonstrate the importance of saliva to taste by completely drying part of your tongue and then touching it with a piece of pretzel or candy—you'll notice that the taste of the food is greatly diminished.

When the taste stimulus reaches a gustatory cell, it must be transduced into an electrical signal. This occurs through three steps, beginning with the depolarization of the gustatory cell (**Figure 15.5**).

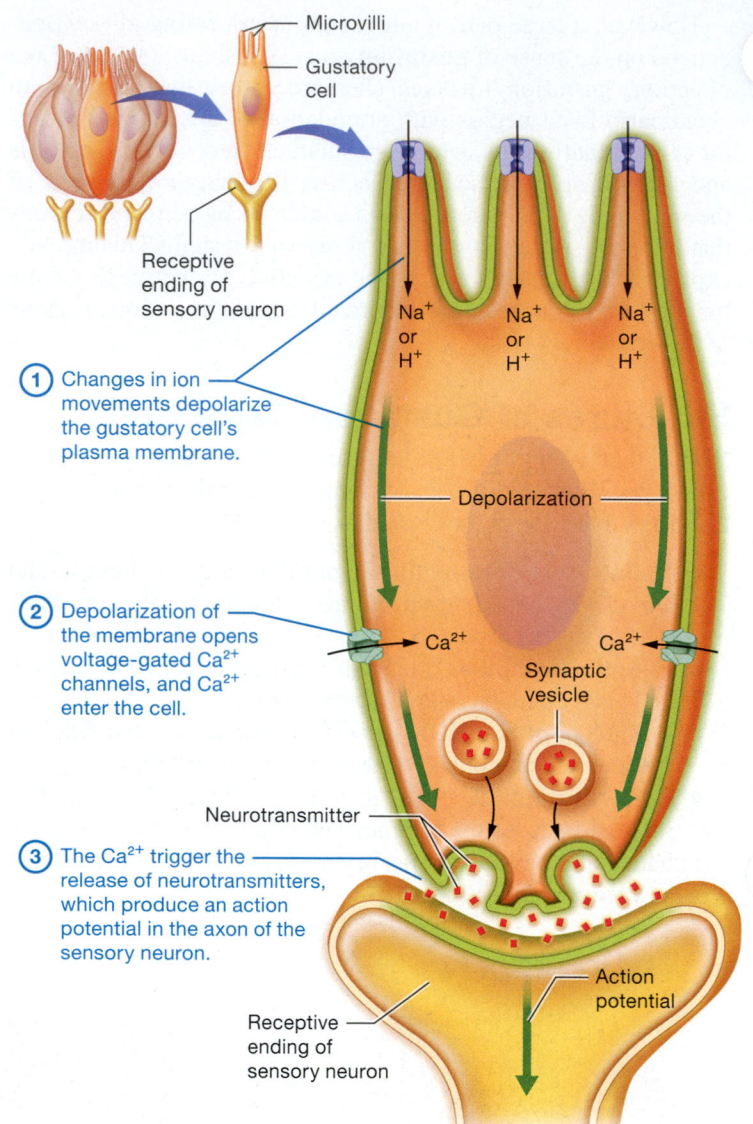

Figure 15.5 Transduction of salty and sour tastes in a gustatory cell.

① **Changes in ion movements depolarize the gustatory cell's plasma membrane.** The specific ion movements vary depending on the particular taste. For example, the sodium ions in salty foods and the hydrogen ions in sour foods travel directly through their ion channel receptors to depolarize the cell. However, receptors for bitter, sweet, and umami compounds activate a G-protein that indirectly closes potassium ion leak channels. The decreased movement of potassium ions out of the cell produces a gradual depolarization (not shown in Figure 15.5).

② **Depolarization of the membrane opens voltage-gated calcium ion channels, and calcium ions enter the cell.** Depolarization results in a local potential. Unlike a neuron, however, a gustatory cell usually does not generate an action potential. Gustatory cells are small enough that the local potential can quickly spread across their membrane, making an action potential unnecessary.

③ **The calcium ions trigger the release of neurotransmitters, which produce an action potential in the axon of the sensory neuron.** Gustatory cells release a variety of neurotransmitters onto their associated sensory axons, including serotonin, glutamate, acetylcholine, and GABA. In the next section, we find out where these sensory neurons transmit the taste stimuli in the CNS.

Quick Check

☐ 4. What types of chemicals produce each of the five taste sensations?

☐ 5. How does a gustatory cell transduce a chemical taste into an action potential?

The Gustatory Pathway

Taste stimuli are transmitted to the CNS via sensory neurons, where they are sent to various parts of the brain for processing (**Figure 15.6**):

① **Axons of the facial, glossopharyngeal, and vagus nerves carry taste stimuli from the tongue into the CNS.** The facial nerve transmits taste stimuli from the anterior two-thirds of the tongue, and the glossopharyngeal nerve transmits taste stimuli from the posterior third of the tongue. The vagus nerve has a minor role in taste,

innervating taste buds that are scattered on the base of the tongue and the posterior palate.

② **Axons of these three nerves terminate in the solitary nucleus in the medulla oblongata by synapsing on central sensory neurons.** The *solitary nucleus* is a cluster of cell bodies located in a column in the medulla oblongata where the axons transmitting taste stimuli synapse. Some neurons in the solitary nucleus connect with parasympathetic neurons of the facial and glossopharyngeal nerves to control reflexive production of saliva.

③ **Axons from the solitary nucleus synapse on neurons in the thalamus, which then send the taste signals to the primary gustatory cortex in the parietal lobe.** The primary gustatory cortex integrates incoming stimuli, leading to awareness and identification of a particular taste.

From the primary gustatory cortex, taste stimuli are relayed to the insula, where integration is believed to occur. They are also relayed to the inferior part of the frontal lobe, where they are integrated with visual and olfactory stimuli to form a complete "picture" of what you are eating, and to the limbic system, which produces emotional reactions to taste (both pleasant and unpleasant).

Quick Check

☐ 6. Which cranial nerves transmit taste sensation to the CNS?

☐ 7. Which part of the brain is responsible for identifying a particular taste?

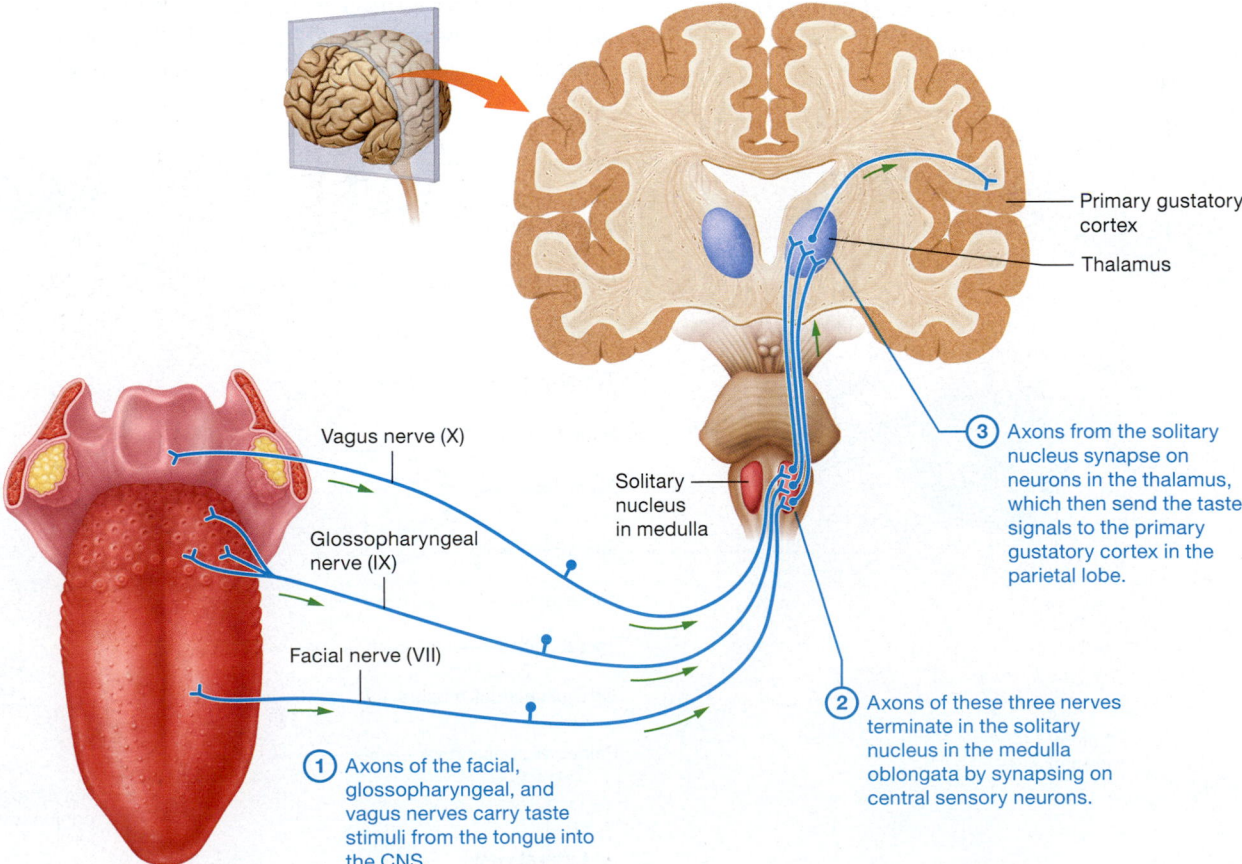

Figure 15.6 The gustatory pathway.

☐ 1. Explain why a tumor that destroys saliva-producing cells can interfere with taste sensation.

☐ 2. Mr. Finn has had a stroke that damaged the trigeminal nerve but not the facial, glossopharyngeal, or vagus nerve. Would he still be able to taste the difference between hot peppers and French fries? Explain your answer.

☐ 3. You ordered coleslaw at a new restaurant but unfortunately suffered repeated bouts of vomiting for the next 12 hours. Three years later, your stomach turns at the very thought of eating coleslaw. Explain this reaction.

See answers in Appendix A.

MODULE 15.4
Anatomy of the Eye

Learning Outcomes

1. Discuss the structure and functions of the accessory structures of the eye.
2. Describe the innervation and actions of the extrinsic eye muscles.
3. Identify and describe the three layers of the eyeball.
4. Describe the structure of the retina.

The eye consists of two main parts: the *eyeball* itself, and the surrounding *accessory structures* that support, protect, or move the eyeball. In this module, we examine the anatomy of the eyeball and its accessory structures. In the following module, we apply this anatomy to the study of vision.

Accessory Structures of the Eye

FLASHBACK

1. What is the function of sebaceous glands? (p. 178)
2. What is a motor unit? (p. 368)

The accessory structures of the eye include the *eyelids, eyebrows* and *eyelashes, conjunctiva, lacrimal apparatus,* and *extrinsic eye muscles.* These structures have important roles in protecting the eyeball from injury and in moving the eyeball.

Eyelids

The **eyelids,** or **palpebrae** (pal-PEE-bree), are two thin folds that cover the anterior part of the orbit (**Figure 15.7a**). They prevent foreign objects from entering the eye and distribute tears across the surface of the eye during blinking. Each eyelid is stiffened by a **tarsal plate,** a thin piece of dense regular collagenous connective tissue (**Figure 15.7b**). Modified sebaceous glands called **tarsal glands** are located within the tarsal plate that secrete oil to prevent the eyelids from sticking together. The superior and inferior palpebrae meet at the edges of the orbit at the medial and lateral **commissures** (KAHM-ih-shur; also known as the medial and lateral canthi). The **lacrimal caruncle** (kar-UN-kul)

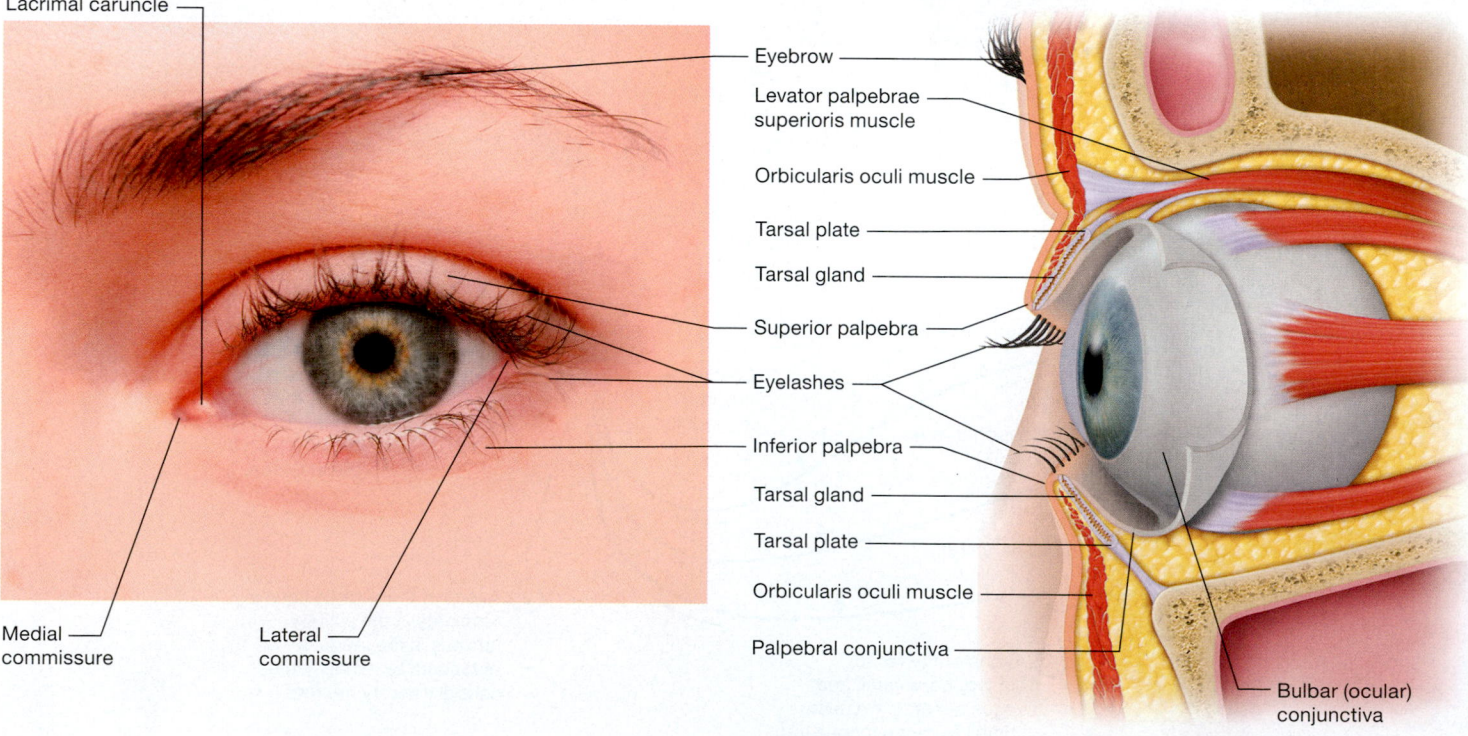

Lacrimal caruncle

Medial commissure

Lateral commissure

Eyebrow

Levator palpebrae superioris muscle

Orbicularis oculi muscle

Tarsal plate

Tarsal gland

Superior palpebra

Eyelashes

Inferior palpebra

Tarsal gland

Tarsal plate

Orbicularis oculi muscle

Palpebral conjunctiva

Bulbar (ocular) conjunctiva

(a) Anterior view

(b) Lateral view

Figure 15.7 Accessory structures of the eye.

is a fleshy structure found at the medial commissure; it contains sebaceous glands that secrete a whitish lubricating substance.

Two skeletal muscles move the eyelids. Both the superior and inferior palpebrae contain portions of the orbicularis oculi muscle, which allows you to tightly close them. The levator palpebrae superioris muscle inserts in the superior palpebra and elevates it.

Eyebrows and Eyelashes

The eyebrows and eyelashes serve a protective function (see Figure 15.7). Hairs along the ridge of the brow form the **eyebrows,** which prevent perspiration from running into the eyes. They also reduce glare in bright light and are important in forming facial expressions. **Eyelashes** are stiff hairs located in the edges of the superior and inferior palpebrae. They are richly innervated by nerve endings so that the slightest touch triggers the blink reflex, which reduces the risk of eye injury.

Conjunctiva

The **conjunctiva** (kon-junk-TY-vah) is a thin epithelial membrane that lines the posterior surfaces of the eyelids (the **palpebral conjunctiva**) and then turns back on itself as the **bulbar** (or **ocular) conjunctiva** to cover the anterior surface of the white part of the eye (see Figure 15.7b). The conjunctiva is translucent, which permits you to see tiny blood vessels beneath it. Inflammation of the conjunctiva is called *conjunctivitis*; when caused by a viral or bacterial infection, conjunctivitis is known as "pink eye," a highly contagious condition.

Lacrimal Apparatus

The **lacrimal apparatus** (LAK-rih-mul) produces tears and drains them from the eye. The **lacrimal gland** is located in the superolateral region of the orbit, just posterior to the conjunctiva (**Figure 15.8a**). When stimulated by autonomic neurons, tiny ducts release tears and mucus into the conjunctival sac to lubricate and wash away dust and debris.

Blinking sweeps the tears medially and inferiorly across the surface of the eye, where they drain into a series of passages that lead to the nasal cavity. Tears first enter the **lacrimal puncta** (singular, *punctum*), which are tiny holes in the medial edge of each eyelid. The puncta are continuous with the **lacrimal canaliculi** (also called *canals*), small ducts that empty into the **lacrimal sac,** located in a small depression in the lacrimal bone. The lacrimal sac drains into the **nasolacrimal duct,** which travels through the lacrimal and maxillary bones to reach the inferior nasal meatus (just inferior to the inferior concha). You can see how narrow the ducts of the lacrimal apparatus are in **Figure 15.8b**, which shows an x-ray of the ducts with dye injected into them. If tear production increases in response to debris, allergens, or emotions, you end up with a runny nose, and excess tears may spill over onto your cheek. The connection between the nasal cavity and the eye can be a two-way street— pathogens can travel from the nose to the eye, resulting in conjunctivitis or a painful infection of the nasolacrimal duct. It also explains, from an anatomical perspective, why some people can actually eject liquids like water or milk from their eyes.

Extrinsic Eye Muscles

Our vision would be more limited if our eyes were fixed in a single position, because we'd have to move our heads every time we wanted to focus on something new or follow a moving object (imagine watching a tennis match). Fortunately, we can move our eyes thanks to the **extrinsic eye muscles,** which

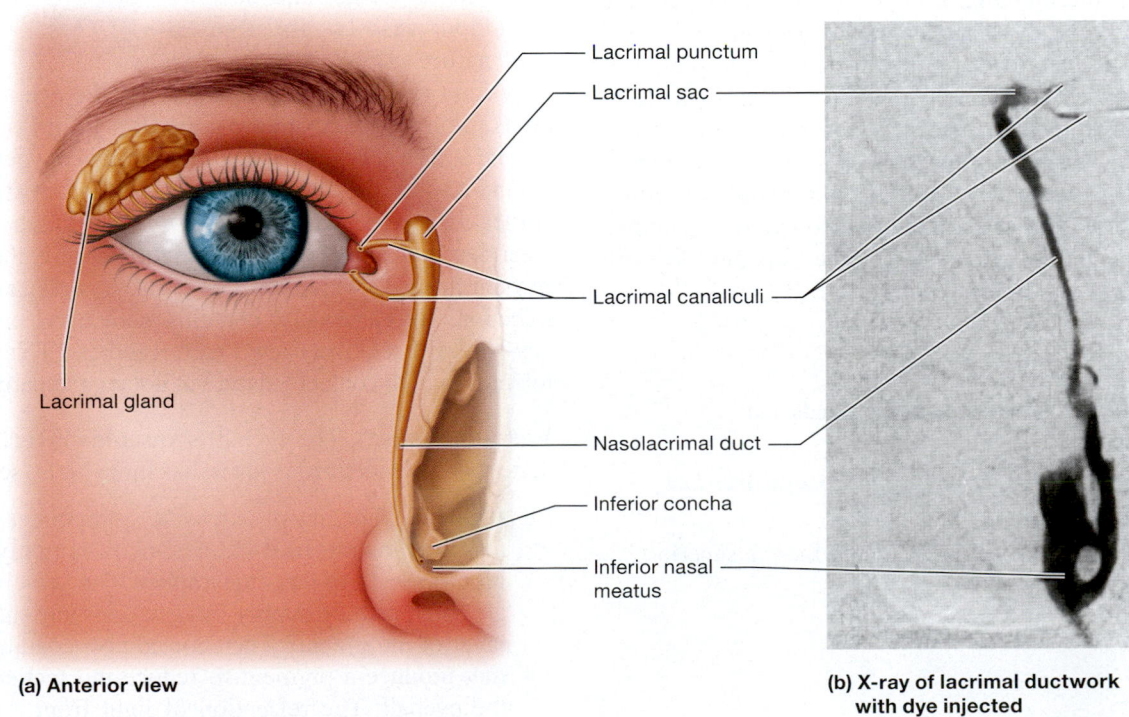

(a) Anterior view

Lacrimal punctum
Lacrimal sac
Lacrimal canaliculi
Nasolacrimal duct
Inferior concha
Inferior nasal meatus
Lacrimal gland

(b) X-ray of lacrimal ductwork with dye injected

Figure 15.8 The lacrimal apparatus.

are the six skeletal muscles that insert into the outer layer of each eyeball. As you saw previously in Figure 9.9, each extrinsic muscle originates from a common tendinous ring on the posterior wall of the orbit. The individual muscles are as follows:

- The **superior and inferior rectus muscles** insert into the superior and inferior portions of the eyeball and move the eyeball superiorly and inferiorly, respectively. However, their origin lies slightly medial to the eyeball, so both muscles are able to move it slightly medially, as well.
- The **medial and lateral rectus muscles** insert into the medial and lateral aspects of the eyeball and move it medially and laterally, respectively.
- The **superior oblique muscle** travels from the posterior orbit along the medial wall and through a fibrous loop called the **trochlea** (TROHK-lee-ah) before inserting on the superolateral part of the eyeball. It depresses the eye and draws it laterally. When acting together with the inferior rectus muscle, it moves the eye directly inferiorly.
- The **inferior oblique muscle** sweeps from the medial floor of the orbit to insert on the inferolateral part of the eyeball. The inferior oblique elevates the eye and moves it laterally.

You can also look back at Table 9.3 for a summary of extrinsic eye muscle movements.

The extrinsic eye muscles must produce very small, precise movements, so they are among the most highly innervated skeletal muscles in the body. Some of the motor units consist of only two or three muscle fibers (compared to a thousand in muscles of the leg). Three of the 12 cranial nerves—the oculomotor (CN III), trochlear (CN IV), and abducens (CN VI)—innervate the extrinsic eye muscles.

Injury to any of these cranial nerves can weaken or paralyze the eye muscles. If the eyes cannot move together, they send slightly different images to the brain, which is interpreted as blurred or double vision, a condition known as *diplopia* (dih-PLOH-pee-uh). If diplopia is present at birth, the child is said to have "lazy eye," or **strabismus** (struh-BIZ-mus). Strabismus can sometimes be corrected by covering the good eye with a patch, thereby forcing the use of the affected eye to strengthen the weak muscles. If not corrected in time, the muscles will atrophy and diplopia will become permanent.

Quick Check

☐ 1. What are the functions of the tarsal glands and conjunctivae?

☐ 2. Trace the path of tears as they travel from the lacrimal gland to the nasal cavity.

☐ 3. Which two extrinsic eye muscles move the eye superiorly?

The Eyeball

 **FLASHBACK**

1. What is the function of collagen fibers? (p. 125)

2. What are the effects of the sympathetic and parasympathetic nervous systems on the pupil of the eye? (p. 523 and p. 530)

The eyeball is a hollow sphere approximately 2.5 cm (1 inch) in diameter that rests in the anterior portion of the orbit. Each eyeball occupies only the anterior one-third of the orbit; the remainder of the orbit is filled with adipose tissue that cushions the eyeball and holds it in position. The eyeball consists of an outer wall that encloses several inner chambers and the *lens*, which focuses light as it enters the eye. The next subsections examine the structure of this complex organ.

Layers of the Eyeball

Like many organs we've discussed, the wall of the eyeball consists of three tissue layers or tunics: the *fibrous, vascular,* and *neural layers*. These layers perform diverse functions, which include maintaining the shape of the eyeball and transducing light into action potentials.

Fibrous Layer The eyeball's outermost layer is the **fibrous layer.** It has two parts: the sclera (SKLEHR-ah) and the cornea (KOHR-nee-ah) (**Figure 15.9**). The **sclera,** the "white" part of the eye, covers approximately five-sixths of its surface. The sclera is white due to its numerous collagen fibers, which allow it to resist deformation from external and internal forces and maintain its shape. These fibers are arranged irregularly, which is why the sclera is opaque.

The sclera is continuous with the translucent **cornea,** which covers the anterior one-sixth of the eyeball. The cornea is translucent due to its orderly parallel arrangement of collagen fibers, relative lack of water in its tissue, and absence of blood vessels. Its translucent nature allows it to play an important role in admitting light into the eye and focusing it, an example of the Structure-Function Core Principle (p. 28).

CORE PRINCIPLE
Structure-Function

The cornea is avascular, so it must obtain oxygen from the fluid behind it as well as from the air. Contact lenses sit directly on the cornea and are designed to allow oxygen to reach it. Extended-wear lenses can be worn overnight because they allow more oxygen through, which reduces corneal swelling and the risk of infection. The cornea also contains many nerve endings that convey pain, but relatively few touch receptors. This is why we can tolerate smooth contact lenses, but not jagged particles of dust.

Vascular Layer The eyeball's **vascular layer,** located immediately deep to its fibrous layer, has three parts (see Figure 15.9):

1. **Choroid.** The most extensive component of the vascular layer, the **choroid** (KOHR-oyd), is richly supplied with blood vessels, as its name implies. The vessels of the choroid are vital, as they supply oxygen and nutrients to the neural layer of the eyeball. The choroid also contains melanocytes that produce a pigment to reduce the scattering of light in the eyeball. The reflection of light from a camera's flash off the choroid is responsible for producing the "red eye"

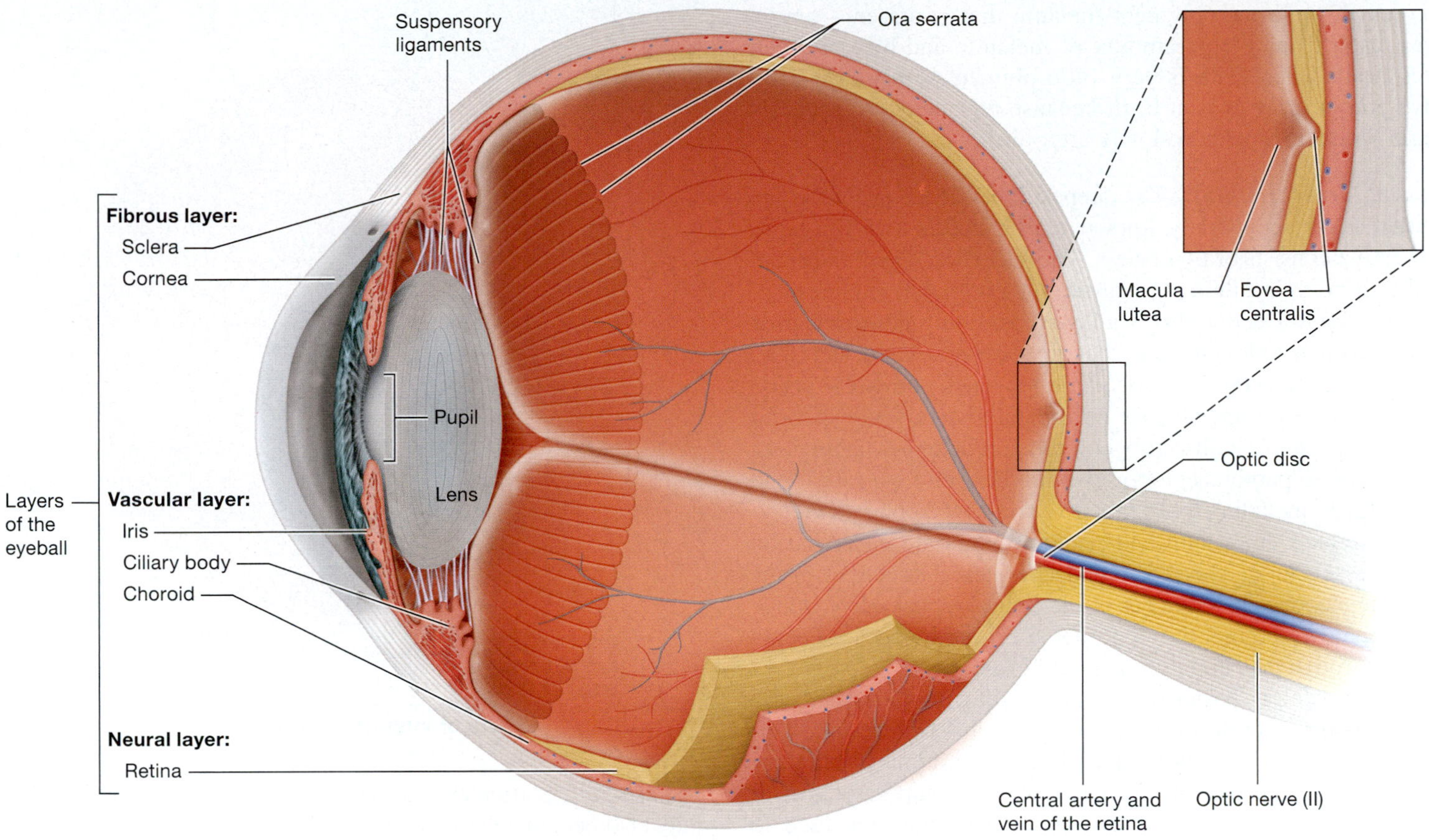

Suspensory
ligaments

Ora serrata

Fibrous layer:

Sclera

Cornea

Macula
lutea

Fovea
centralis

Pupil

Optic disc

Lens

**Layers
of the
eyeball**

Vascular layer:

Iris

Ciliary body

Choroid

Neural layer:

Retina

Central artery and
vein of the retina

Optic nerve (II)

Figure 15.9 Midsagittal section of internal structures of the eye.

effect that has ruined many family photos (the red is primarily due to the large amount of blood in the choroid).

2. **Ciliary body.** The choroid is continuous anteriorly with the **ciliary body** (SIL-ee-ehr-ee), which contains a ring of smooth muscle that surrounds the lens. You can see where the two meet at the scalloped edge of the ciliary body, which is known as the *ora serrata.* Fine threads called **suspensory ligaments** connect the ciliary body to the lens; contraction and relaxation of the smooth muscle change the shape of the lens to focus light on the retina.

3. **Iris.** The **iris** is the extension of the vascular layer just anterior to the ciliary body. It is the visibly pigmented portion of the eye that surrounds the circular **pupil,** an opening through which light enters the eye. The iris contains two smooth muscles that control how much light passes through the pupil (**Figure 15.10**). First is the **pupillary sphincter muscle** (PYOO-pih-lehr-ee), which constricts the pupil to reduce the amount of light entering the eye in response to stimulation from the parasympathetic nervous system. Second is the **pupillary dilator muscle,** which has fibers arranged around the pupil in a radial pattern like the spokes around a bicycle tire. When these fibers are stimulated by the sympathetic nervous system, they dilate the pupil to allow more light into the eye. If you've ever had

your "eyes dilated" during an eye exam, you have experienced the effect of blocking parasympathetic activity, which allows sympathetic activity to dilate the pupils without any opposition.

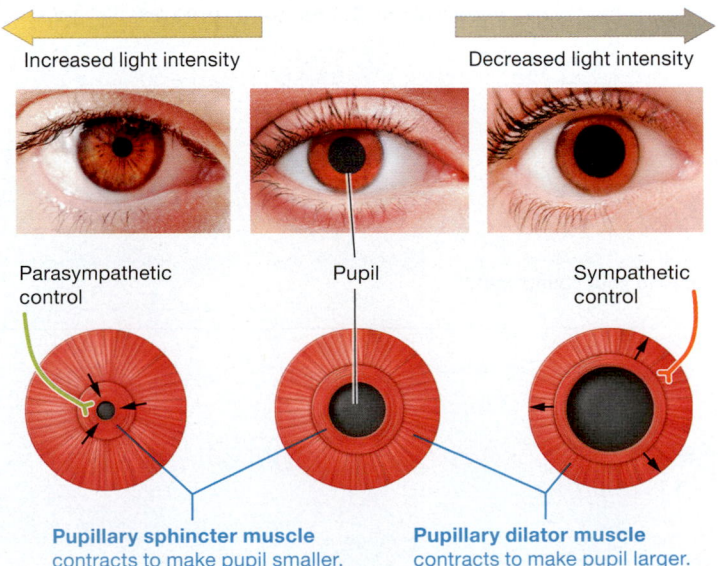

Increased light intensity

Decreased light intensity

Parasympathetic
control

Pupil

Sympathetic
control

Pupillary sphincter muscle
contracts to make pupil smaller.

Pupillary dilator muscle
contracts to make pupil larger.

Figure 15.10 Constriction and dilation of the pupil.

The color of the iris in humans is determined largely by the amount of the brown pigment melanin that is present—brown eyes result from a large amount of melanin, and blue or green eyes result when there is very little pigment. Infants typically have blue or gray eyes at birth because melanocytes in the iris begin producing melanin slowly after birth.

Neural Layer (Retina) The deepest layer of the eyeball is the **neural layer,** commonly known as the **retina** (RET-ih-nuh). The retina consists of two layers. The superficial layer (adjacent to the choroid) is a thin, pigmented epithelium that reduces scattering of light. Deep to this is a layer containing **photoreceptor cells** known as rods and cones that are specialized to detect light. *Rods* are photoreceptors responsible for black and white vision in low light levels and also for peripheral vision, whereas *cones* are responsible for high-acuity color vision in higher light levels. In addition to photoreceptors, several other cell types are found in this layer, including those that give rise to the optic nerve.

The retina's epithelium and photoreceptors are nourished by blood vessels in the choroid. The deeper neural structures of the retina are supplied by branches of the central retinal artery, which enters the eye through the optic nerve. These blood vessels are visible through the pupil during an eye exam (**Figure 15.11**).

At the central posterior retina we find a yellowish area known as the **macula lutea** (MAK-yoo-luh LOO-tee-uh; "yellow spot"), which contains a high concentration of photoreceptors (note that the macula lutea does not appear yellow in Figure 15.11 due to the red eye effect from the light used to view the retina). In the center of the macula lutea is an indented area called the **fovea centralis** (FOH-vee-uh; *fovea* = "pit") in which we find cones packed tightly together. This high density of cones makes the fovea centralis the "high-definition" area of the retina—it allows for extremely detailed vision as we focus on a particular object.

The importance of the macula lutea to high-acuity vision becomes apparent when the cells of this area suffer damage, as occurs with the condition *macular degeneration*. As you might expect, macular degeneration leads to a progressive loss of

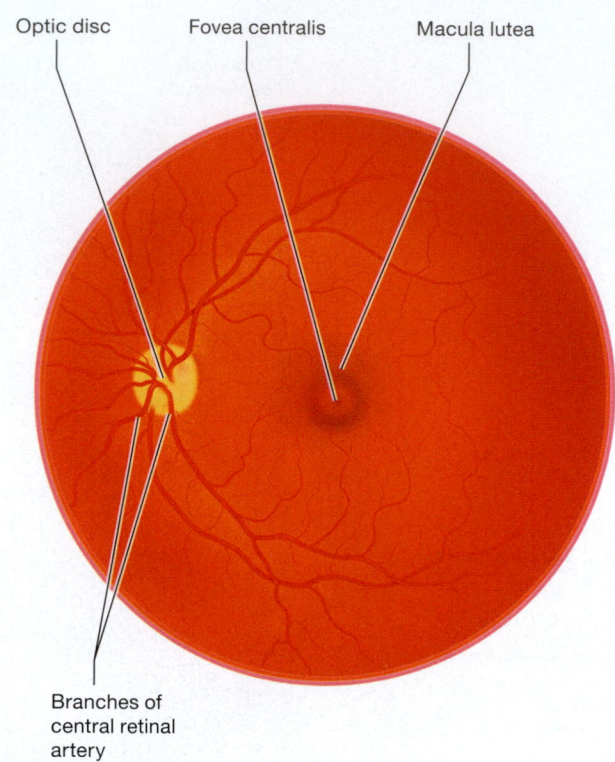

Optic disc   Fovea centralis   Macula lutea

Branches of central retinal artery

Figure 15.11 Photo of interior of the eye.

visual acuity, particularly in the center of the visual field. Vision may also become distorted, and individuals often have difficulty with color perception.

The axons of the optic nerve gather at the **optic disc,** a portion of the retina that does not contain any photoreceptors. For this reason, this area does not capture any visual image and is called the *blind spot*. We don't typically notice our blind spot because the brain is adept at "filling in" the missing information based on the images next to it. However, you can find your blind spot by using the test in **Figure 15.12**.

The two layers of the retina are usually in contact with each other but are not physically linked by desmosomes or tight

(a) Find your "blind spot."

① To test your left eye, cover your right eye and focus on the X. Slowly move your book toward (or away from) you until the dot disappears.

② To test your right eye, cover your left eye and focus on the dot. Move the book until the X disappears.

(b) Fill in your "blind spot."

Figure 15.12 Demonstration of the blind spot and "filling in."

① To test your left eye, cover your right eye and focus on the X. Slowly move your book toward (or away from) you until the pencil appears intact.

② When the gap in the pencil hits your blind spot, areas in your brain fill in the missing information so the pencil appears intact.

③ To test your right eye, turn your book upside down, cover your left eye, and repeat the process.

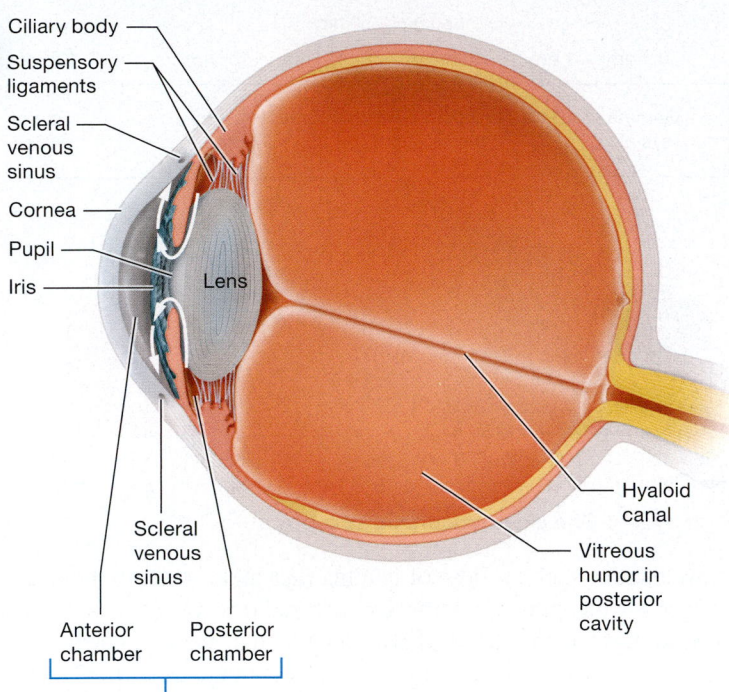

Anterior and posterior chambers form the anterior cavity and are filled with aqueous humor. The arrows show the flow of aqueous humor before it drains into the scleral venous sinus.

Figure 15.13 The lens, cavities, and chambers of the eye.

junctions. For this reason, certain conditions—such as trauma, diabetes, or an abnormally shaped eyeball—can pull the inner layer away from the pigmented epithelium, a condition commonly referred to as a *detached retina*. Since this isolates the photoreceptors from their blood supply, permanent loss of vision can occur if it is not corrected within several days.

Lens

The **lens** is a transparent, flexible, rounded disc located posterior to the iris and pupil that focuses light on the retina from objects near the eye. It is typically maintained in its roundest shape by elastic fibers found in its capsule, unless external forces flatten it. As you can see in **Figure 15.13**, it is surrounded by the ciliary body and connected to it by the suspensory ligaments. The lens contains elongated cells known as lens fibers that lack nuclei and are tightly packed to make them transparent. To discover how a clouded lens impairs vision, see *A&P in the Real World: Cataracts.*

Cavities and Chambers of the Eye

Notice in Figure 15.13 that the eyeball is divided into anterior and posterior cavities by the lens and ciliary body. The large **posterior cavity,** located posterior to the lens, is filled with a gelatinous material called the **vitreous humor** (VIT-ree-us). Vitreous humor presses the retina against the choroid and helps maintain the shape of the eyeball. It consists mostly of water, with about 1% of its volume composed of collagen fibers, hyaluronic acid, and a few cells on the periphery. Some people experience "floaters," which are shadows cast by cellular debris trapped in the vitreous humor. The number of floaters increases with age, as the vitreous humor tends to shrink but is not replaced. Also note

Cataracts

A transparent lens is essential for vision; if light cannot pass through the lens, vision is impaired even if functional photoreceptors are present in the retina. Indeed, a clouded lens, or **cataract** (KAT-uh-rakt), is one of the most frequent causes of blindness. Trauma, exposure to UV radiation, and diseases such as diabetes can promote cataracts. However, the most common cause is simply aging, as the lens fibers tend to progressively darken and become less organized, turning the lens milky white. A cataract cannot be reversed, so the usual treatment is to replace the entire lens surgically with a synthetic lens. This restores vision, although glasses or contact lenses may be required for minor adjustments.

in the figure that the small *hyaloid canal* passes through the vitreous humor from the lens to the optic nerve. After birth, it contains fluid that assists in changing the shape of the lens.

The **anterior cavity** is found anterior to the lens and ciliary body. It consists of two smaller sub cavities: the **posterior chamber,** between the lens and the iris, and the **anterior chamber,** between the iris and the cornea. These two chambers are both filled with watery **aqueous humor** (AY-kwee-us), secreted by the ciliary body. Aqueous humor flows from the posterior chamber through the pupil into the anterior chamber. To prevent buildup, the fluid must then drain at the same rate from the anterior chamber into a blood vessel that circles the anterior edge of the iris, called the **scleral venous sinus.** What happens when the aqueous humor doesn't drain properly? For the answer, see *A&P in the Real World: Glaucoma.*

Glaucoma

In **glaucoma** (glaw-KOH-muh), the aqueous humor cannot drain and fluid builds up in the anterior and posterior chambers. This excess fluid raises the pressure inside the eyeball, or *intraocular pressure*. This elevated pressure pushes on the vitreous humor, which in turn compresses and damages the retina and the optic nerve as it leaves the eye. Glaucoma, which is the second leading cause of blindness, can result from eye infections, certain medications, or congenital defects in the scleral venous sinus. However, in most cases the cause is unknown.

Most people who have glaucoma do not have any symptoms other than gradual loss of vision, which can occur so slowly that it may not be detected until the disease is advanced. Unfortunately, lost vision cannot be restored, but progression can be stopped or slowed with medications that either improve drainage or reduce the amount of aqueous humor produced. If medications fail to control intraocular pressure, surgical procedures may improve drainage.

☐ 4. In which layer of the eyeball are the photoreceptors found?

☐ 5. What are the functions of each component of the vascular layer of the eyeball?

☐ 6. What are the two fluids found in the eyeball? How do they differ in location, composition, and function?

Apply What You Learned

☐ 1. Explain why it is not uncommon to develop an eye infection after having a cold.

☐ 2. When a patient is asked to look to the left, her right eye moves to the left, but her left eye does not. Which extrinsic eye muscle is not functioning properly?

☐ 3. In this module you read about macular degeneration. Would you expect the same type of vision loss if a person experienced degeneration of the rods? Why or why not? Explain.

See answers in Appendix A.

MODULE **15.5**
Physiology of Vision

Learning Outcomes

1. Describe how light activates photoreceptors.

2. Explain how the optical system of the eye creates an image on the retina.

3. Compare the functions of rods and cones in vision.

4. Trace the path of light as it passes through the eye to the retina and the path of action potentials from the retina to various parts of the brain.

5. Explain the processes of light and dark adaptation.

The structures described in the previous module work together to produce **vision,** which is the perception of light reflected by various objects. However, vision is more than simply detecting light. Using our eyes and the visual pathways in the CNS, we can determine the size, shape, and color of an object (such as a round, purple Frisbee), as well as the rate and direction of movement (it's rapidly flying toward your head), and the distance between us and the object (it's close enough that you'd better duck). This module explores how the structures of the visual system provide these capabilities.

Principles of Light

In the chemistry chapter, we discussed different forms of *energy,* defined as the capacity to do work (see Chapter 2). One form of energy is **electromagnetic radiation,** which has different wavelengths, measured in nanometers (nm). These range from short

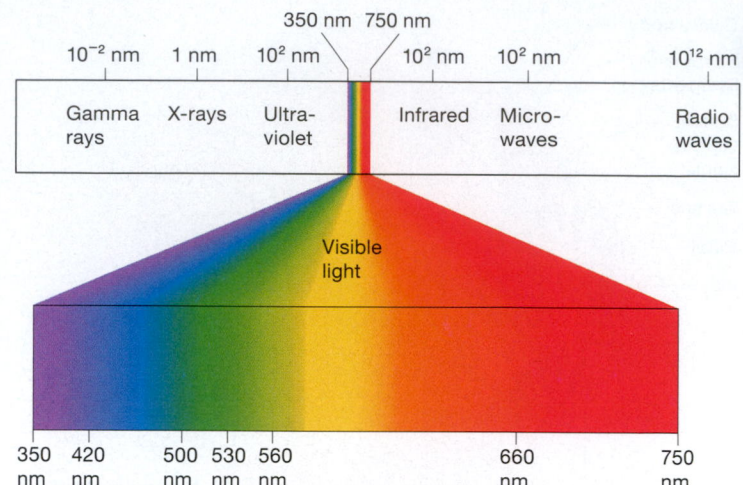

Figure 15.14 **The electromagnetic spectrum.**

wavelengths, such as those of gamma rays and x-rays, to the much longer wavelengths of microwaves and radio waves (**Figure 15.14**). **Visible light** is the range of wavelengths that we can detect with our eyes, and we perceive segments of that range as particular colors. For example, we see shorter wavelengths of visible light (about 420 nm) as blue and violet, and longer wavelengths (about 660 nm) as red. Regardless of wavelength, the basic unit of light is the **photon** (FOH-tawn), which stimulates the photoreceptors in the retina.

Rays of light bend, or **refract,** when they pass through a translucent object. The **refractive index** of an object measures the amount of refraction it provides. Air has a refractive index of 1.0, meaning that light passing through air is not significantly bent (**Figure 15.15a**). If something has a refractive index greater than that of air, such as the water in **Figure 15.15b**, light is refracted. You have seen refraction if you have ever put a spoon into a glass of water. The portion of the spoon sticking out of the water appears straight, but the portion under water appears bent. This is due to the water having a higher refractive index than air, and so the light rays entering the water refract.

The degree to which light refracts also depends on the angle at which it strikes the surface: The greater the angle, the greater the refraction. This has particular consequences for curved surfaces, as they are more angled at the edges. For this reason, light hitting the edge of a curved surface is bent more than light hitting its middle. As you can see here, a *convex lens* has a surface

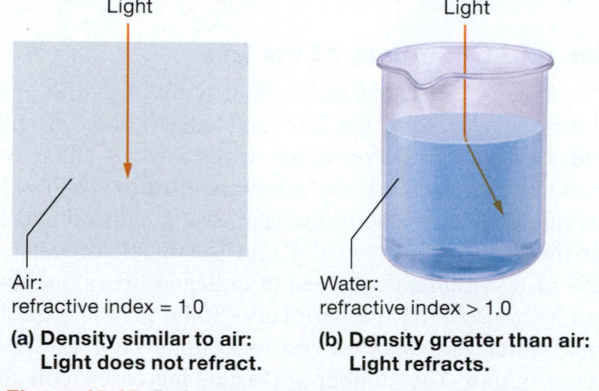

(a) Density similar to air:
   Light does not refract.

(b) Density greater than air:
   Light refracts.

Figure 15.15 **Refraction of light.**

that bulges outward, which causes the light hitting the edges of the lens to bend inward, or **converge.**

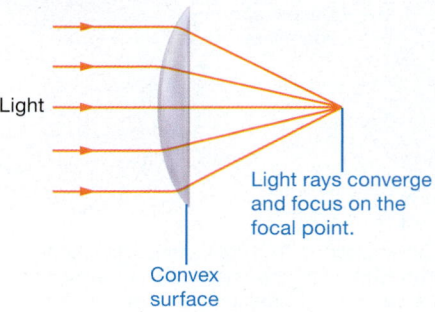

Light rays converge and focus on the focal point.

Convex surface

When the rays converge on one point, known as the *focal point,* they are said to be *focused.* A *concave lens* has a surface that "caves in." Notice that it causes the light hitting the edges of the lens to bend outward, or **diverge,** so they do not focus.

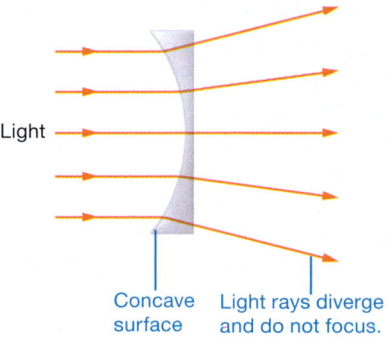

Concave surface   Light rays diverge and do not focus.

Quick Check

☐ 1. What is refraction?

☐ 2. What is the difference between convergence and divergence of light?

Focusing Light on the Retina

To provide clear vision, rays of light must focus on the retina, not behind or in front of it. Unfortunately, when light strikes a nontransparent object (such as an apple), the light rays are reflected and scatter away from the object in a diverging pattern. Notice in **Figure 15.16a** that light rays reflecting close to the object are still diverging sharply, but when they have reflected a distance of about 20 feet from the object, they are nearly parallel. In either case, the rays must be refracted to converge and focus on your retina. The cornea and lens provide this refraction.

The greatest degree of refraction—about two-thirds of the eye's refractive power—occurs as light passes from air through the cornea, which has a refractive index similar to that of water. (This explains why your vision underwater is blurry—the light passing from the water to the cornea does not refract much.) Some additional refraction occurs as light passes through the lens, but the refractive index of the lens is similar to that of the aqueous humor, so the lens cannot provide as much refraction as the cornea. For this reason, the lens typically provides for fine refractive adjustments.

As we just discussed, rays of light from distant objects (such as if the apple were on a tree) are nearly parallel when they reach the eye, so they need little refraction to be focused on the retina. The cornea provides most of the necessary refraction, and the lens in its normal flattened shape provides the rest (**Figure 15.16b**). When the eye is relaxed and focusing on a distant object, this is known as the **emmetropic** (em-eh-TROH-pik) state of the eye (or *emmetropia*).

However, diverging rays from nearby objects (such as the apple after you pick it) require a greater degree of refraction, far beyond what the cornea can provide. The lens increases total refraction by changing its shape from flattened to round; this ability of the lens to "round up" is called **accommodation.** When the lens is rounded, its curved surface refracts light to a greater extent and can focus it on the retina (**Figure 15.16c**).

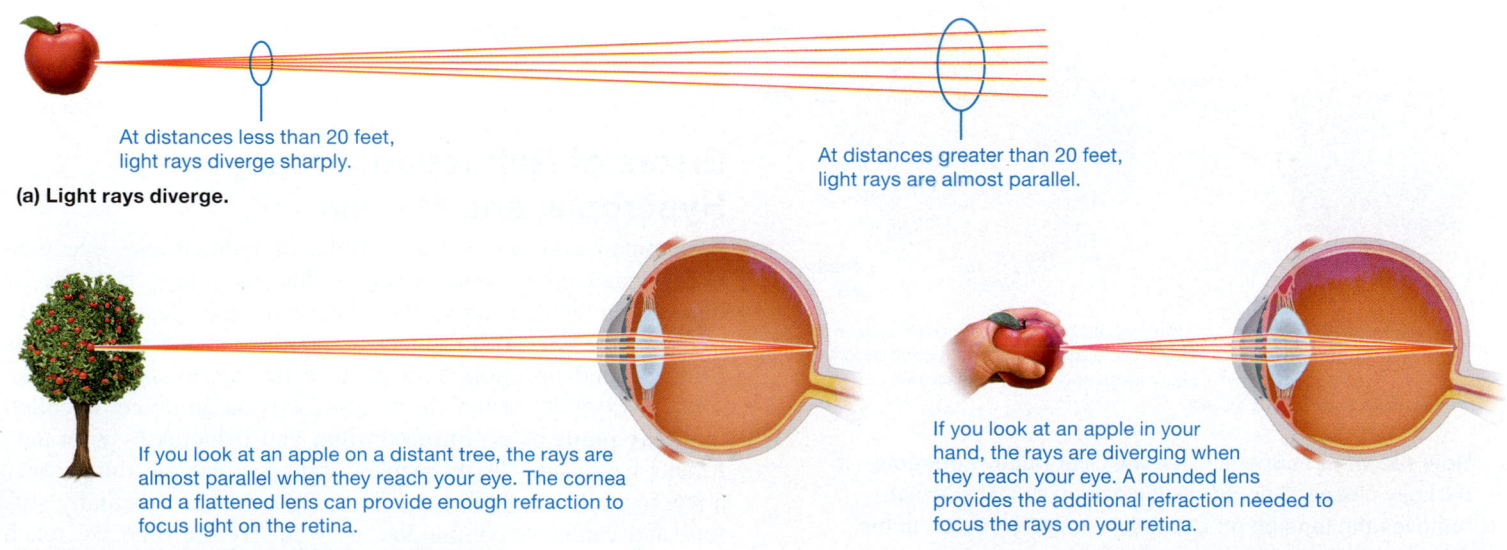

At distances less than 20 feet, light rays diverge sharply.

At distances greater than 20 feet, light rays are almost parallel.

(a) Light rays diverge.

If you look at an apple on a distant tree, the rays are almost parallel when they reach your eye. The cornea and a flattened lens can provide enough refraction to focus light on the retina.

If you look at an apple in your hand, the rays are diverging when they reach your eye. A rounded lens provides the additional refraction needed to focus the rays on your retina.

(b) Emmetropia: The lens flattens for distant vision.

(c) Accommodation: The lens "rounds up" for near vision.

Figure 15.16 **Refraction by the lens of the eye.**

How does the lens change its shape? The answer lies in the sphincter-like smooth muscle found in the ciliary body that surrounds the lens. When viewing distant objects, the eye is in a state of emmetropia and the smooth muscle of the ciliary body relaxes, moving the ciliary body away from the lens. This movement increases tension in the suspensory ligaments, flattens the lens, and reduces refraction.

When accommodation occurs in the viewing of nearby objects, the ciliary muscle contracts and the ciliary body moves closer to the lens. This decreases tension on the suspensory ligaments, which reduces their pull on the lens. Without this tension, the elastic fibers of the lens make it "round up" into a nearly spherical shape. This change in shape makes the lens more convex, which increases refraction and focuses light from nearby objects on the retina.

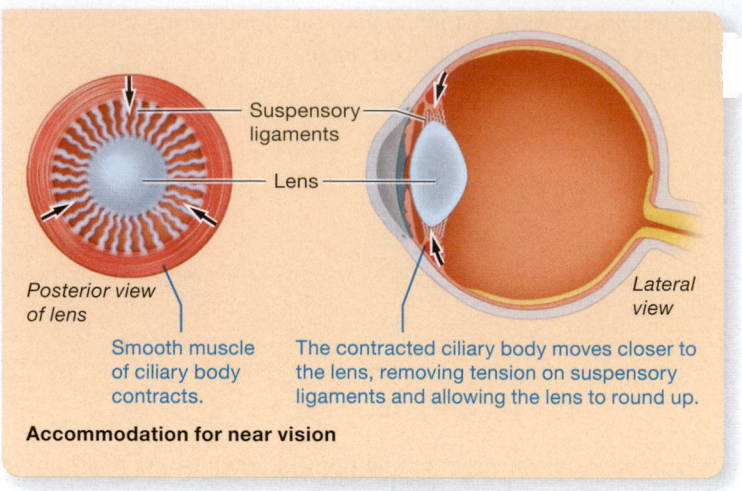

Smooth muscle of ciliary body contracts.

The contracted ciliary body moves closer to the lens, removing tension on suspensory ligaments and allowing the lens to round up.

Posterior view of lens

Lateral view

Suspensory ligaments

Lens

Accommodation for near vision

ConceptBOOST)))

Understanding Accommodation

Movement of the ciliary body seems counterintuitive, because the suspensory ligaments tighten when the muscle *relaxes*, and generally we associate this kind of tension with muscle *contraction*. To understand how this works, we need to consider how a sphincter functions. Recall that sphincters are circular muscles that usually control the flow of materials from one part of the body to another. For most sphincters, when they are relaxed, they are open and so allow materials to flow, and when they are contracted, they are closed and so prevent the flow of materials.

   Considering this, let's look at the ciliary body. Notice here that when the ciliary body is in its relaxed state, it is open, which places it farther away from the lens. Its distance from the lens puts tension on the suspensory ligaments and so flattens the lens.

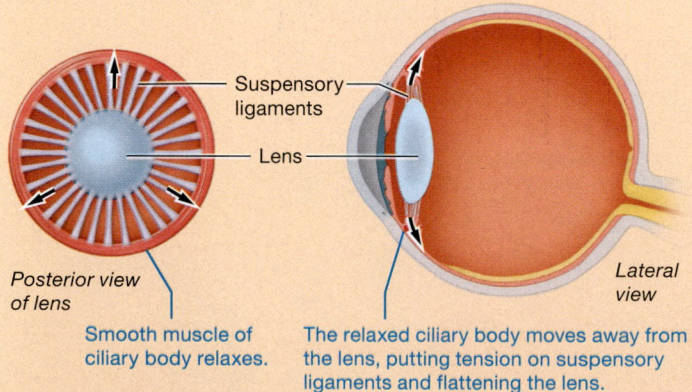

Posterior view of lens

Lateral view

Suspensory ligaments

Lens

Smooth muscle of ciliary body relaxes.

The relaxed ciliary body moves away from the lens, putting tension on suspensory ligaments and flattening the lens.

Emmetropia for distant vision

Now see what happens when the ciliary body contracts—it partially closes. This moves it closer to the lens, which removes the tension on the suspensory ligaments. In the absence of this tension, the lens becomes rounder. ■

Accommodation isn't the only event necessary to focus on near objects. Two other processes must also occur:

- **Pupillary constriction.** Light rays that enter the edge of the lens cannot be refracted enough and therefore produce blurry images on the retina. To prevent this, the pupils constrict slightly to cover the edge of the lens. Dilation of the pupils during an eye exam prevents this constriction, which is why you can't focus well on close objects after the exam until your pupils return to normal.
- **Convergence.** To focus near objects on the fovea of each eye, the eyeballs must move medially. This movement is called **convergence.**

You can easily view these processes by watching a friend's eyes as he or she shifts focus from a far object to a near object.

Quick Check

☐ 3. What is accommodation?

☐ 4. How do the ciliary body and lens focus light from a near object on the retina?

☐ 5. What is the near point of accommodation, and how does it change as we age?

Errors of Refraction: Presbyopia, Hyperopia, and Myopia

Accommodation serves the normal, or emmetropic, eye well under many circumstances, but it has its limits. For one, it depends on the distance of the object from the eye and the flexibility of the lens. If an object is too close to the eye, the lens cannot round up enough to provide the necessary refraction. The closest point at which we can focus on an object is called the **near point of accommodation,** and it increases with age. Infants have very flexible lenses and can focus on things only a few inches from their eyes. As we age, the lens gradually stiffens and cannot accommodate as well. By the time we reach our mid-40s, the near point is approximately 10–20 inches, and most people begin to have difficulty with reading or activities requiring close vision. This condition, **presbyopia** (prez′-bee-OH-pee-uh), is usually corrected with reading glasses or bifocals (glasses with one set of lenses for distance vision and another set of lenses for close vision). Note that many glasses now have *progressive lenses,* in which there is a smooth transition between the distance and close lenses to correct for visual deficits at multiple focal lengths.

Accommodation by the lens also depends on having the correct eye shape and proper curvature of the cornea. In emmetropia, the length of the eyeball is normal in the anterior-posterior direction and the lens can focus light on the retina. However, in **hyperopia** (hy′-per-OH-pee-uh; farsightedness), the eyeball is too short (**Figure 15.17a**) or the cornea is too flat. The lens cannot round up enough to adequately bend the light, and the image focuses behind the retina, causing blurry images when viewing objects up close. Refraction is increased by using a lens with convex surfaces that provide additional convergence, which causes the light to focus on the retina.

The opposite occurs in **myopia** (my-OH-pee-uh; nearsightedness): The distance between the cornea and the retina is too long (**Figure 15.17b**), or the cornea curves too much. In this case, the lens cannot flatten enough and bends the light too much, which focuses the light in front of the retina. This blurs the image when viewing distant objects. Myopia is corrected with a concave lens that diverges light before it strikes the lens of the eye. The lens of the eye can then provide enough refraction to focus light on the retina.

Astigmatism (uh-STIG-muh-tizm) occurs when the curvature of the lens or the cornea is irregular and the rays of light are not evenly refracted. It results in blurred vision at all distances,

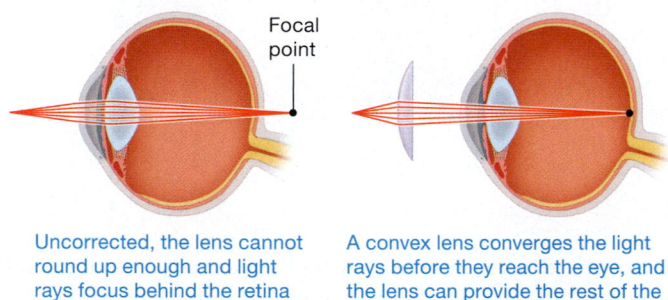

Uncorrected, the lens cannot round up enough and light rays focus behind the retina when viewing near objects.

A convex lens converges the light rays before they reach the eye, and the lens can provide the rest of the needed refraction.

(a) Hyperopia (farsightedness): Eyeball is too short.

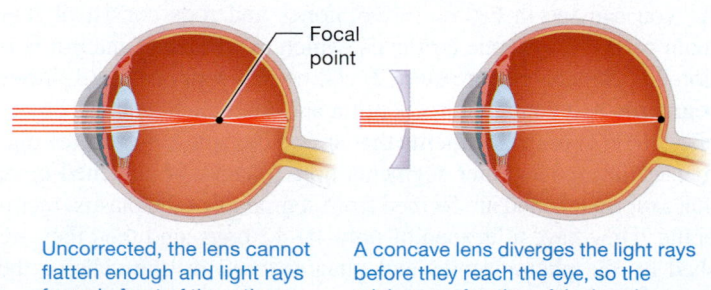

Uncorrected, the lens cannot flatten enough and light rays focus in front of the retina when viewing distant objects.

A concave lens diverges the light rays before they reach the eye, so the minimum refraction of the lens is enough to focus the rays on the retina.

(b) Myopia (nearsightedness): Eyeball is too long.

Figure 15.17 Errors of refraction.

and can be treated with corrective lenses that adjust for the specific abnormal corneal or lens curvatures. If the astigmatism involves only the corneal surface, it may also be treated with a surgical procedure known as *LASIK,* or *laser-assisted in situ keratomileusis.* During LASIK, a laser is used to remodel the cornea and smooth out the irregularities in its shape. If the astigmatism is due to deeper problems such as abnormal lens curvatures, a special form of LASIK may be used that actually induces asymmetry in the corneal surface, which compensates for the deeper irregularities. Note that LASIK is also a popular procedure for the correction of myopia and hyperopia. The laser reshapes the cornea in such a way that it adjusts the amount of refraction so that light may be focused on the focal point in the retina.

Quick Check

☐ 6. How do myopia and hyperopia differ?

Photoreceptors and the Retina

« FLASHBACK

1. How do depolarization and hyperpolarization differ? (p. 397)
2. What is an inhibitory postsynaptic potential (IPSP)? (p. 407)

As described earlier, the retina consists of two layers, an outer pigmented epithelium and an inner layer that contains photoreceptors and other cell types arranged in layers. Adjacent to the pigmented layer we find the two types of photoreceptors, cones and rods. These cells synapse with neurons called **bipolar cells,** which communicate with **retinal ganglion cells,** the most anterior cells in the retina (**Figure 15.18**). The axons of the retinal ganglion cells course across the retinal surface and form the optic nerve (CN II). Two other cell types are also involved in image processing. **Horizontal cells** form connections between photoreceptors and bipolar cells and modulate transmission to enhance visual contrast. **Amacrine cells** (AM-uh-kreen) are located among the dendrites of the retinal ganglion cells, where they respond to changes in light intensity and moving objects.

As noted, the two types of photoreceptors are cones and rods. **Cones** contain pigments that allow us to perceive color and they produce high-acuity (sharp) vision. They function best in bright light and cannot respond at all in the dark. **Rods** produce low-acuity vision, cannot detect color, and are unable to function in bright light. However, they are extremely sensitive in dim light. Indeed, our ability to see at night is almost entirely due to rods. The differences between rods and cones explains why objects appear fuzzy and in grayscale at night. The cones are nonfunctional, so only the rods can produce the images you see, resulting in low-acuity, grayscale images.

The rods and cones are arranged in the retina in an opposite fashion. Each retina has about 3 million cones, which are highly

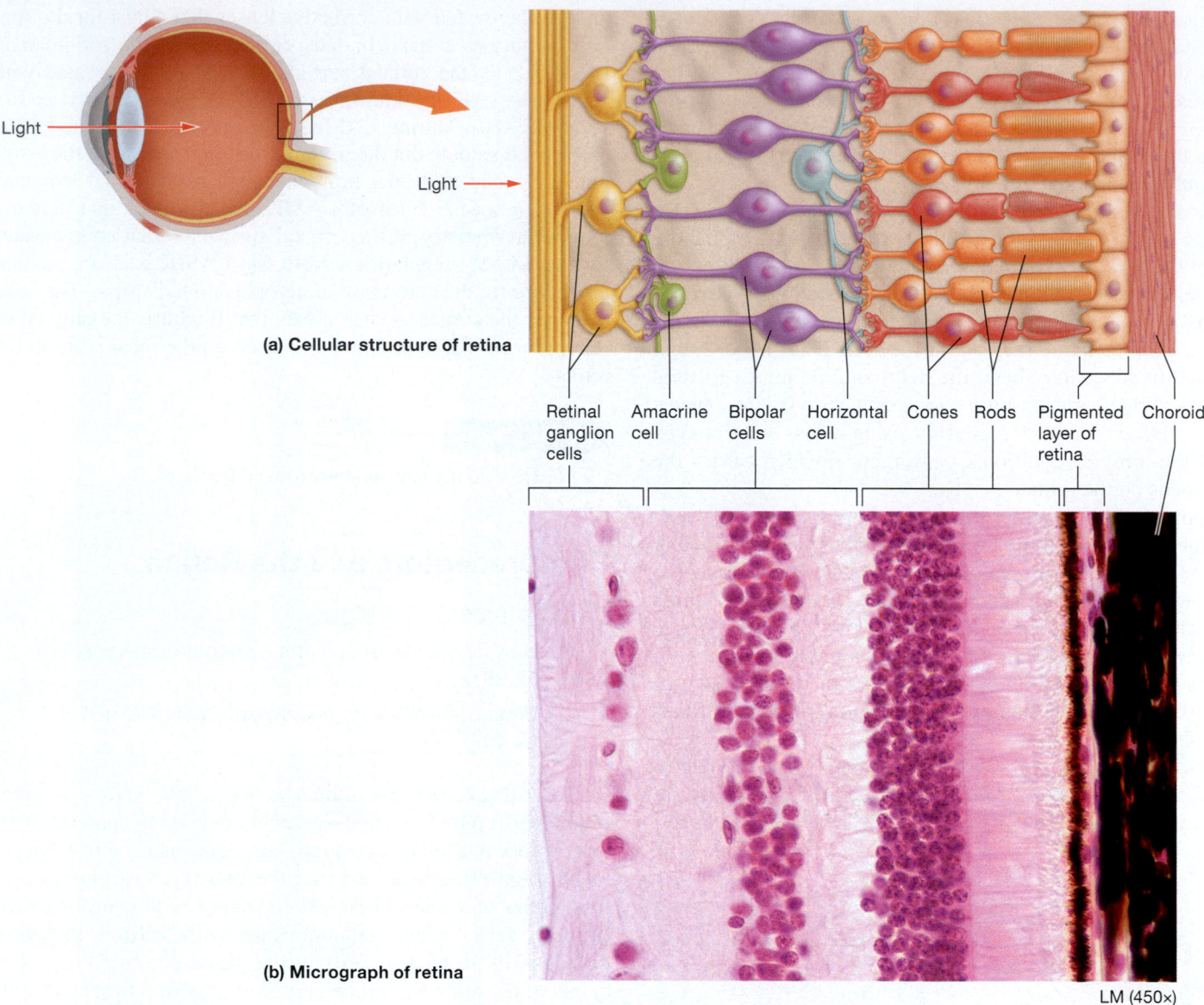

(a) Cellular structure of retina

Light

Retinal ganglion cells | Amacrine cell | Bipolar cells | Horizontal cell | Cones | Rods | Pigmented layer of retina | Choroid

(b) Micrograph of retina

LM (450×)

Figure 15.18 Layers of the retina.

concentrated in the fovea centralis. As you move outward away from the fovea, the concentration of cones decreases, until, at the very edge of the retina, no cones are found at all. Conversely, there are 100 million rods in each retina, but they are more concentrated away from the fovea. Their concentration decreases as we move toward the fovea, and there are no rods in the fovea itself. The concentration of cones in the fovea allows us to focus on an object. However, the concentration of rods in the outer retina makes them important for peripheral vision, or detecting images at the edge of our vision.

This arrangement of cones and rods explains why, at night, you can see an object if you look just to the side of it but it disappears when you look directly at it. Looking directly at the object focuses it on the fovea, where there are only cones, but cones do not function in dim light. Looking to the side of the object focuses it on the retina adjacent to the fovea, where rods are present.

Now let's discuss how cones and rods transduce light into action potentials that can be interpreted by the brain.

Structure of Photoreceptors

As you can see in **Figure 15.19a**, cones and rods consist of four main parts: (1) synaptic terminals, which are the most anterior portion that contact bipolar cells; (2) cell bodies with nuclei; (3) **inner segments** that contain mitochondria and the typical cellular organelles; and (4) **outer segments** that are on the side closer to the pigmented layer. The outer segments house stacks of flattened *discs* that absorb light and are formed from segments of the plasma membrane. They have a lifespan of only 10–12 days, and then they are shed by the photoreceptor and phagocytosed by the cells of the pigmented layer.

Rods have an outer segment that is shaped like a cylinder. It contains as many as 1000 discs that fold and bud off from the plasma membrane and float in the cytosol. Each disc

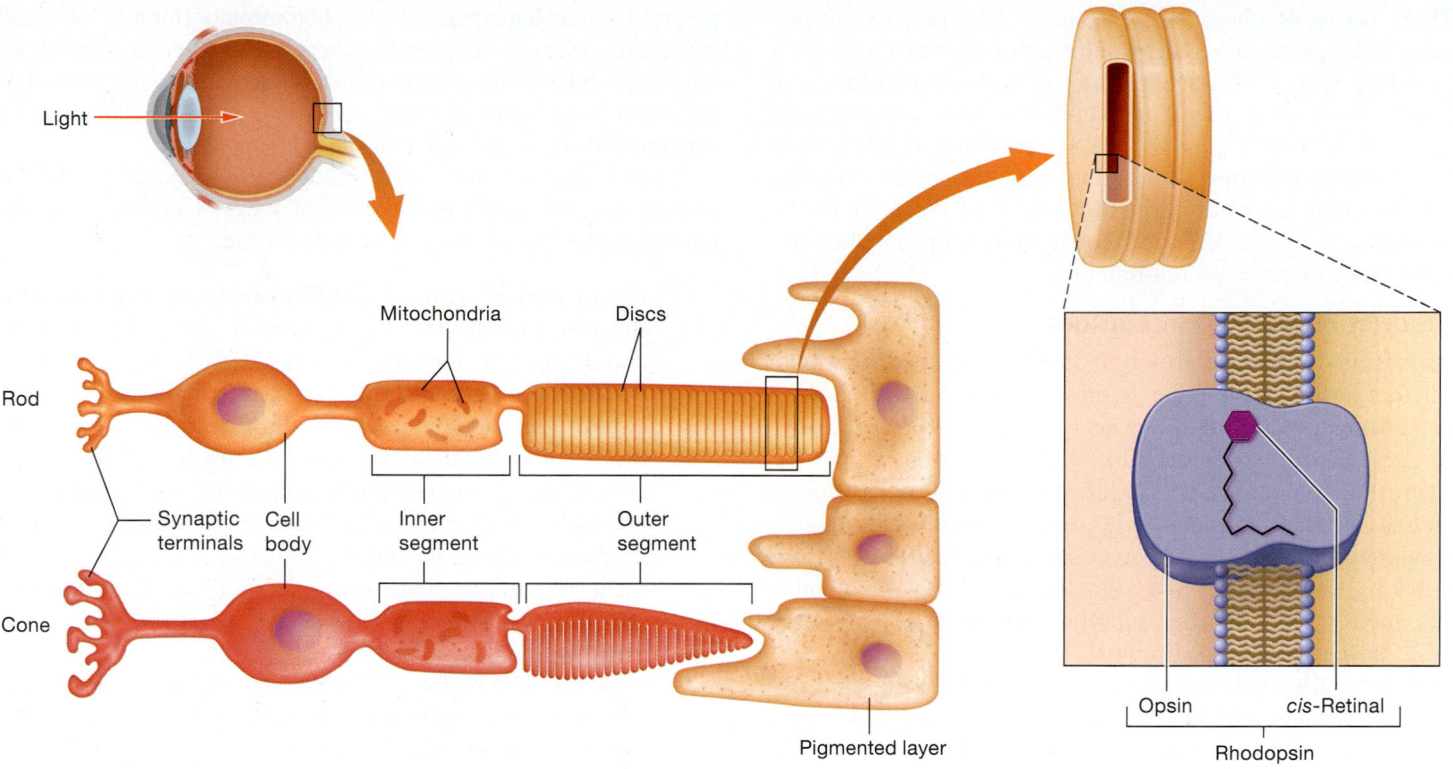

(a) Structure of a rod and a cone

(b) Rhodopsin in the disc membrane of a rod

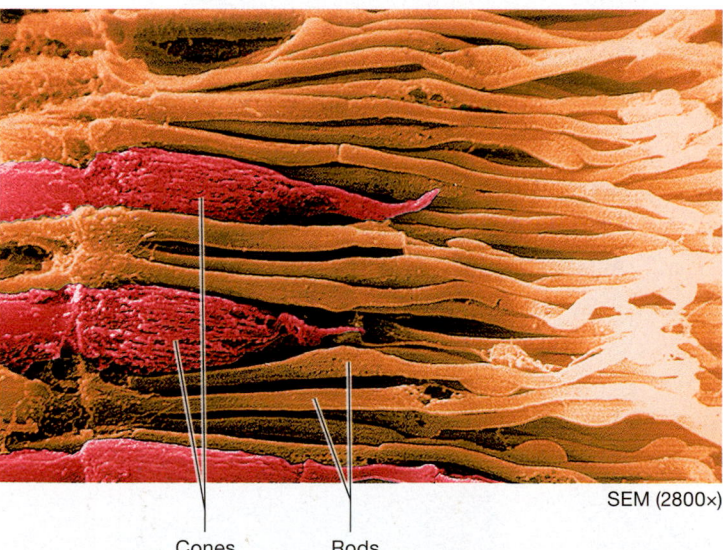

(c) Micrograph of rods and cones in a human retina

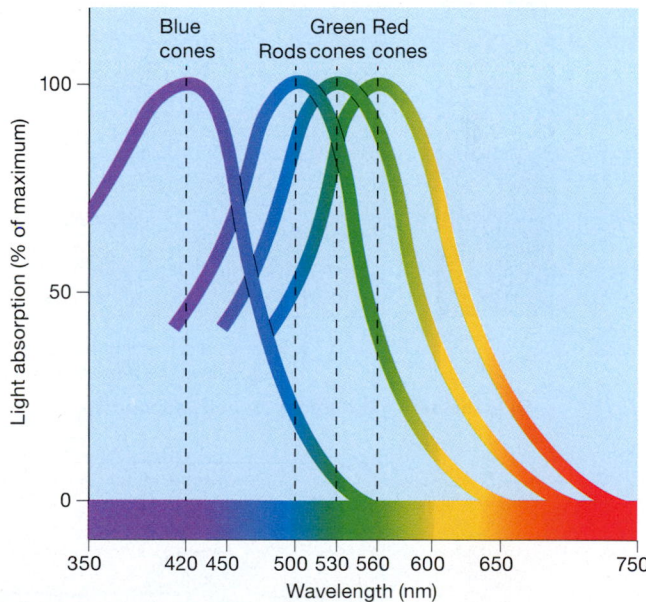

(d) Absorption of various wavelengths of light by cones and rods

Figure 15.19 Cones and rods.

contains the pigment **rhodopsin** (roh-DAWP-sin) in its membrane (**Figure 15.19b**). Rhodopsin has two components: the protein **opsin** and the pigment **retinal** (RET-ih-nal), which is derived from vitamin A. In the dark, the retinal is in a bent configuration known as *cis*-retinal. All rods contain this pigment and do not distinguish between different wavelengths (colors) of light.

Like rods, cones are named because of the shape of their outer segment, which contains discs composed of infoldings of the plasma membrane (Figure 15.19a and **Figure 15.19c**). The discs contain the pigment **iodopsin,** which is made of two components: retinal and the protein **photopsin** (foh-TAWP-sin). Photopsin is similar to opsin but has a slightly altered structure that allows it to absorb different wavelengths of light.

Three forms of photopsin respond to wavelengths we perceive as blue, green, or red. The color pigments react to slightly overlapping ranges of wavelengths, so that our perception of each color is generated by stimulation of varying combinations of cone populations (**Figure 15.19d**). For example, red light stimulates primarily red cones, but orange light stimulates red cones and some green cones as well. Read *A&P in the Real World: Color Blindness* on p. 562 to discover what happens when one or more types of cones do not function.

Transduction in Photoreceptors

Transduction of light into an action potential begins when a photon strikes a disc in the outer segment of a rod or cone. Much more is known regarding the events in rods, so we'll focus on the process there, although it is likely similar in cones.

There is a fundamental difference between transduction in the retina and that in other special senses. Recall that olfactory neurons and gustatory cells detect smell and taste stimuli and respond by depolarizing and signaling another cell; in other words, the olfactory neurons and gustatory cells are "off" until they are turned "on" by a stimulus.

In the visual system, the process is reversed. In the dark (the absence of a stimulus), the photoreceptor cells are depolarized ("on") and continually release neurotransmitters onto adjacent neurons. The presence of light hyperpolarizes the photoreceptor (turns it "off") and reduces the release of neurotransmitters, which alters the activity of other cells in the retina and sends that information to the brain. The end result is the same: The nervous system detects a change in the environment and transduces it into an action potential.

What happens when light strikes a photoreceptor cell? To answer this, we first need to look at what is occurring in the photoreceptor cell in the dark (**Figure 15.20a**):

(1) **Opsin and *cis*-retinal combine to form rhodopsin in the disc membrane.** The G-protein complex transducin and the enzyme phosphodiesterase (PDE) are inactive.

(2) **Sodium ions enter the outer segment of the photoreceptor and depolarize it.** Sodium ion channels in the plasma membrane of the outer segment are opened by the second messenger **cyclic guanosine monophosphate (cGMP),** which is bound to them. There is a concentration gradient favoring the movement of sodium ions into the cell, so sodium ions flow into the cell (an example 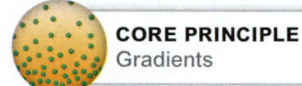 of the Gradients Core Principle, p. 28). This depolarizes the cell from its resting membrane potential of –70 mV to a membrane potential of –40 mV.

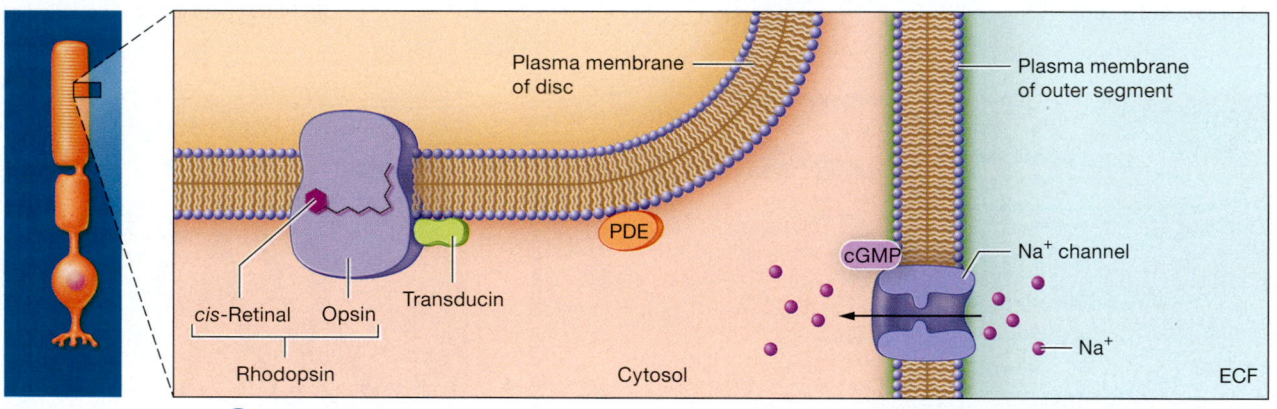

(1) Opsin and *cis*-retinal combine to form rhodopsin in the disc membrane.

(2) Na⁺ enter the outer segment of the photoreceptor and depolarize it.

(a) In the dark, photoreceptor cells depolarize.

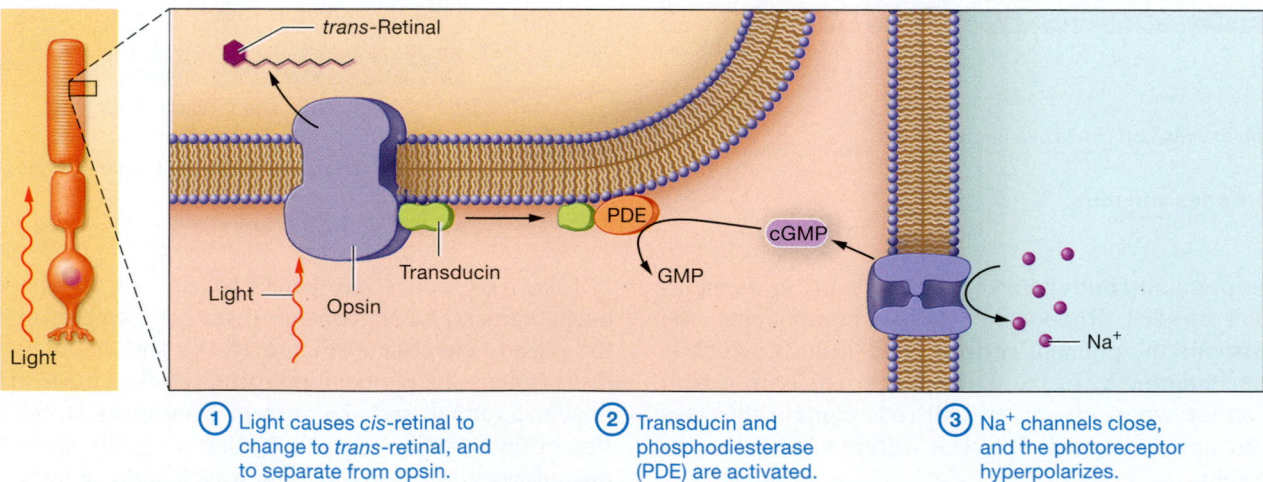

(1) Light causes *cis*-retinal to change to *trans*-retinal, and to separate from opsin.

(2) Transducin and phosphodiesterase (PDE) are activated.

(3) Na⁺ channels close, and the photoreceptor hyperpolarizes.

Figure 15.20 Transduction of light in a photoreceptor cell.

As shown in **Figure 15.20b**, when light strikes the photoreceptor:

① **Light causes *cis*-retinal to change to *trans*-retinal, and to separate from opsin.** When a photon hits the outer segment of the rod, it is absorbed by the rhodopsin embedded in the disc membrane. This causes the retinal to change from a bent shape (*cis*-retinal) to a straightened form (*trans*-retinal). The *trans*-retinal doesn't fit into opsin, which causes the two to split apart.

② **Transducin and phosphodiesterase (PDE) are activated.** The unbound opsin changes shape, which triggers a series of chemical reactions, beginning with the activation of transducin, which then activates PDE.

③ **Sodium ion channels close, and the photoreceptor hyperpolarizes.** When activated by transducin, PDE catalyzes the conversion of cGMP to GMP, leaving less cGMP available to bind to sodium ion channels. Without cGMP, the channels close and reduce the influx of sodium ions. The membrane potential becomes more negative than it was at rest, and is now hyperpolarized.

When *trans*-retinal dissociates from opsin, rhodopsin is said to be **bleached** and cannot respond to light until it reassembles. Reassembly occurs via a complicated process in which the *trans*-retinal leaves the photoreceptor and travels to the epithelial cells of the pigmented layer. Here it is converted back to *cis*-retinal in a process that uses ATP and, after transport back to the photoreceptor, combines with opsin to make rhodopsin. For this reason, the pigmented layer plays an essential part in the function of photoreceptors.

Dark and Light Adaptation

Suppose you leave a sunny parking lot to enter a dark theater, or you are relaxing in a dark room when someone suddenly turns on the light. In both cases, although you may feel blinded at first, your eyes gradually adjust in processes known as **dark adaptation** and **light adaptation**, respectively. These adjustments to different light intensities partially depend on pupil size. The pupils constrict in bright light and dilate in dim light, but this occurs almost immediately and does not explain why it can take several minutes for your eyes to adjust to bright light and much longer to completely adjust to darkness. Instead, the adjustment is explained by changes that occur in the photoreceptor cells.

The adjustment to darkness depends on the rods, as they are the photoreceptors that function in dim light. But in bright light, all the rhodopsin in the rods is bleached and the rods are unable to function, and vision depends on the cones. When light is suddenly reduced, the cones can no longer function and rods are slow to regenerate enough rhodopsin to function. It can take as long as 40 minutes for dark adaptation to occur and for rods to be completely functional.

Bleaching also has an important function in light adaptation. In the dark, rods are fully functional and the retina is extremely sensitive to light. When bright light hits the retina, it immediately bleaches the rods and cones, resulting in a blinding glare. The rods become nonfunctional; however, the cones can regenerate functional pigments faster and are able to respond within a few minutes as their sensitivity decreases.

Image Processing by the Retina

Let's now look at how depolarization and hyperpolarization of the photoreceptors affect neighboring bipolar cells and retinal ganglion cells. In the dark, the following steps occur (**Figure 15.21a**):

① **The photoreceptor depolarizes and releases glutamate onto bipolar cells.** Note here that photoreceptor depolarizations are local potentials. Action potentials in these cells are not necessary, as the cells are short enough for a local potential to spread over the entire membrane without dying out. As the depolarization spreads, the neurotransmitter glutamate is released onto bipolar cells.

② **Glutamate inhibits the bipolar cell and reduces its release of neurotransmitters.** In most bipolar cells, glutamate produces an inhibitory postsynaptic potential (IPSP). Some bipolar cells depolarize in response to glutamate, but they will not be discussed here.

③ **The retinal ganglion cell does not produce an action potential.** The bipolar cell does not stimulate the retinal ganglion cell, so no signal is sent through the optic nerve to the brain.

Figure 15.21b illustrates what happens when light strikes the photoreceptors:

① **Light hyperpolarizes the photoreceptor, and it stops releasing glutamate.** Hyperpolarizing the photoreceptor inhibits it, preventing it from releasing glutamate onto bipolar cells.

② **The bipolar cell is freed from inhibition and depolarizes, releasing neurotransmitters onto the retinal ganglion cell.** In bipolar cells, a local potential is enough to trigger the release of an excitatory neurotransmitter (the specific neurotransmitter is not yet known).

③ **The retinal ganglion cell produces action potentials that are sent to the brain via the optic nerve.** When stimulated by the bipolar cells, the axons of the retinal ganglion cells generate action potentials. These axons collectively form the innermost layer of the retina, and they exit the eyeball together as the optic nerve (CN II).

The pathways in Figure 15.21 are simplified to show what happens to each type of cell. The retina has more than 105 million photoreceptors, but only 1 million retinal ganglion cells, so most retinal ganglion cells receive input from multiple photoreceptors. However, vision is sharpest when a ganglion cell receives input from only a single photoreceptor. This is the case in the fovea, where one cone may synapse with only two bipolar cells, which in turn contact only one retinal ganglion cell. For this reason, sharp vision is possible here, because when a specific ganglion cell is stimulated, the brain knows exactly which tiny part of the retina (and the visual field) has been stimulated.

Toward the edges of the retina, however, hundreds or thousands of photoreceptors contact a small number of bipolar cells, which then synapse with one retinal ganglion cell. The retinal ganglion

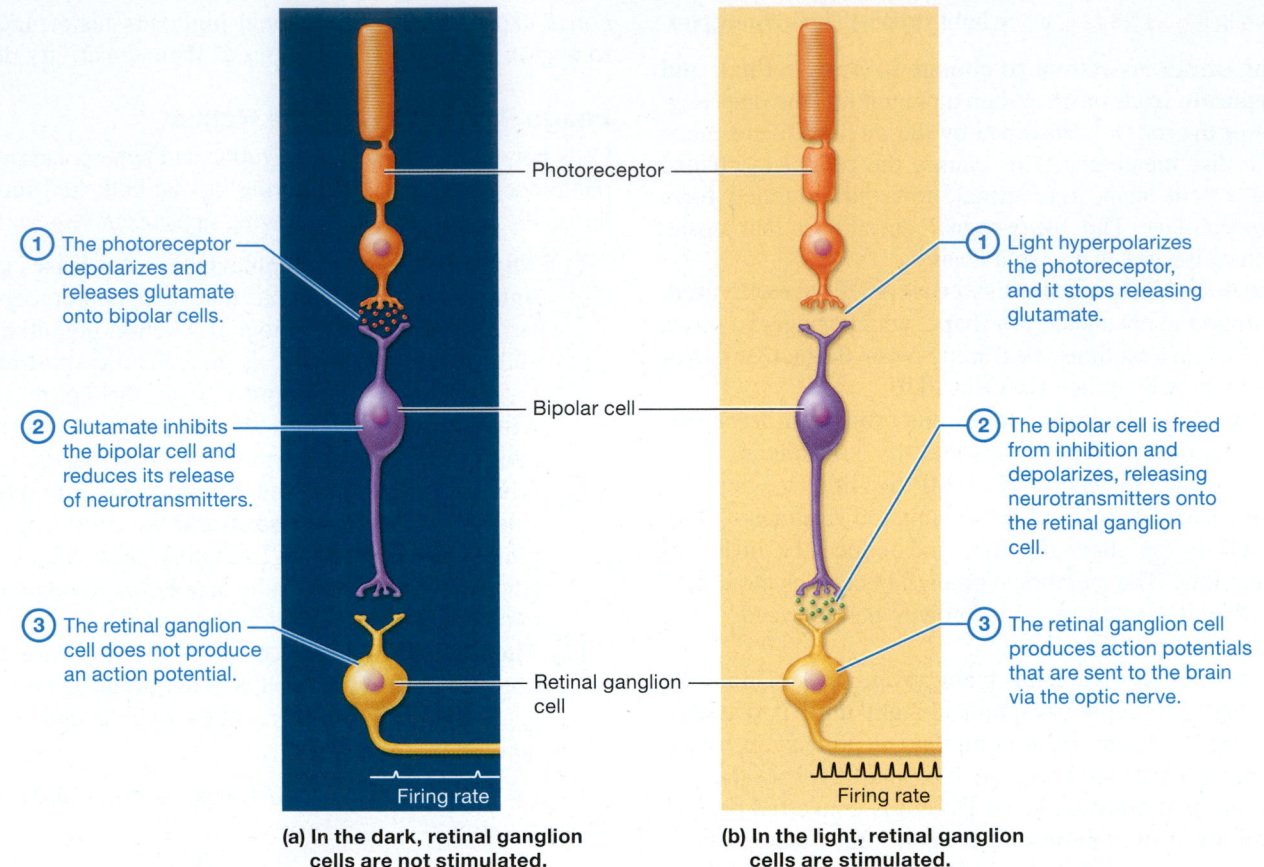

(a) In the dark, retinal ganglion cells are not stimulated.

① The photoreceptor depolarizes and releases glutamate onto bipolar cells.

② Glutamate inhibits the bipolar cell and reduces its release of neurotransmitters.

③ The retinal ganglion cell does not produce an action potential.

Photoreceptor

Bipolar cell

Retinal ganglion cell

Firing rate

(b) In the light, retinal ganglion cells are stimulated.

① Light hyperpolarizes the photoreceptor, and it stops releasing glutamate.

② The bipolar cell is freed from inhibition and depolarizes, releasing neurotransmitters onto the retinal ganglion cell.

③ The retinal ganglion cell produces action potentials that are sent to the brain via the optic nerve.

Firing rate

Figure 15.21 Image processing in the retina.

cell cannot determine exactly which one of the photoreceptors has been stimulated, so the brain does not receive precise visual information. So, peripheral vision is less detailed than central vision, which explains why you can clearly read the menu at a restaurant but not immediately recognize a friend who walks up beside you.

Quick Check

☐ 7. How are rods different from cones?

☐ 8. Why are photoreceptors depolarized in the dark?

☐ 9. How do rods hyperpolarize when light strikes them? How does this affect the firing rate of a ganglion cell?

The Visual Pathway

« FLASHBACK

1. Where is the occipital lobe, and what is the function of the primary visual cortex? (pp. 427–428)

2. Where is the optic chiasma? (p. 481)

Detection of light and image formation in the retina is only part of the story of how we see. The information from the retina must be transferred to the visual areas of the brain for conscious awareness and interpretation. This involves many structures in the brain, which we now examine.

Structures of the Visual Pathway

To understand how visual information travels through the brain, you must understand the concept of a **visual field.** The visual field of each eye is what you see with that eye alone. However, the term also refers to what you see with both eyes open, where it is divided into right and left visual fields with the focal point (what your gaze is focused on) in the center.

In **Figure 15.22**, the right visual field is shaded orange and the left visual field is shaded purple. Notice that the center of the visual field is viewed by both eyes (see the dark orange and dark purple sections), which is known as **binocular vision.** However, the edges of the visual field are seen only by the eye on the same side, which is known as **monocular vision.** The visual pathway is as follows:

① **The retina of each eye detects visual stimuli from portions of the right and left visual fields.** Images in the left visual field—the portion shaded in purple in Figure 15.22—can be seen by the left eye's *nasal retina* (nearest the nose) and the right eye's *temporal retina* (nearest the temple). Conversely, images in the right visual field—shaded orange in the figure—can be seen by the right eye's nasal retina and the left eye's temporal retina.

② **Some visual stimuli cross at the optic chiasma so that all stimuli from the right visual field are processed by the left hemisphere, and stimuli from the left visual field by the right hemisphere.** All axons of the retinal ganglion cells gather at the optic disc and leave

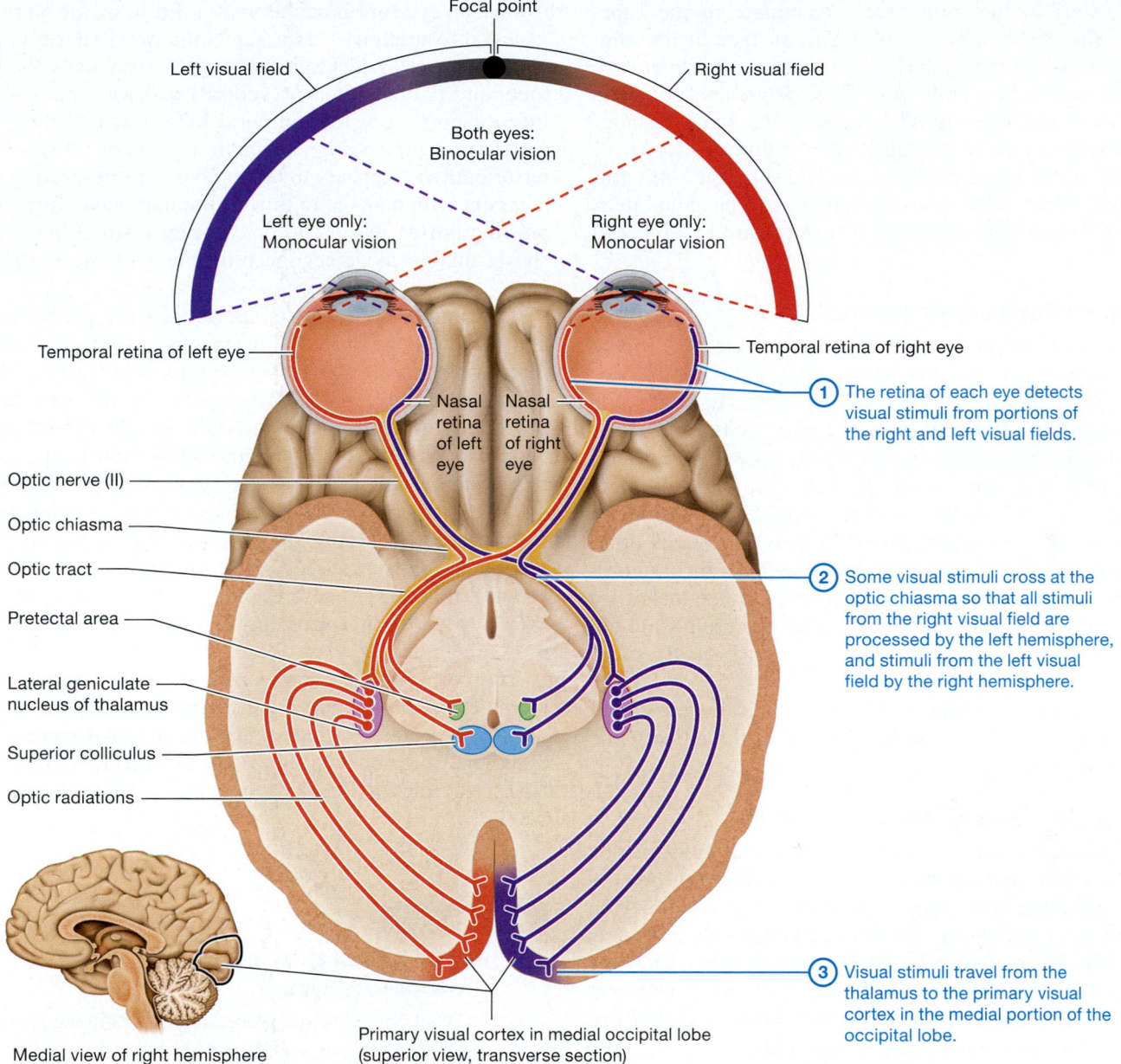

Figure 15.22 The visual pathway.

① The retina of each eye detects visual stimuli from portions of the right and left visual fields.

② Some visual stimuli cross at the optic chiasma so that all stimuli from the right visual field are processed by the left hemisphere, and stimuli from the left visual field by the right hemisphere.

③ Visual stimuli travel from the thalamus to the primary visual cortex in the medial portion of the occipital lobe.

the posterior wall of the eye as the optic nerve. They exit the orbit and enter the skull through a foramen in the sphenoid bone, then meet at the midline in the X-shaped **optic chiasma** (KY-az-muh) inferior to the hypothalamus. In the optic chiasma, axons from each nasal retina cross to the opposite side of the brain. However, axons from each temporal retina remain on the same side of the brain. Notice what this does—it delivers all stimuli from the left visual field to the left hemisphere, and all stimuli from the right visual field to the right hemisphere.

After passing through the optic chiasma, the retinal ganglion axons are called the **optic tracts.** Most axons in the optic tracts terminate in the **lateral geniculate nucleus** (jeh-NIK-yoo-lit) of the thalamus. However, about 10% terminate in other locations such as the *superior colliculus* or the *pretectal area*, both in the midbrain.

③ **Visual stimuli travel from the thalamus to the primary visual cortex in the medial portion of the occipital lobe.** The axons of the lateral geniculate neurons are grouped as bundles called *optic radiations* that fan out deep to the temporal and parietal lobes. Note that when the optic radiations reach the **primary visual cortex** of the occipital lobe, each half of the visual field is now represented entirely in the opposite half of the cerebral hemisphere. At this point, we become consciously aware of visual stimuli, and basic analysis of objects begins in terms of shape, color, movement, and location. From the primary visual cortex, information is relayed to other parts of the occipital lobe and to portions of the temporal and parietal lobes for more interpretation and integration with other senses.

Some axons in the optic tracts terminate in the superior colliculus, located on the posterior surface of the midbrain. Neurons in the superior colliculus coordinate eye movements when you shift your gaze from one object to another. Each eye must move precisely the same distance in the same direction at the same time in order to focus the new object on the fovea of each eye. The superior colliculus also communicates with neurons that control head and neck movement so that you can quickly turn toward an object of interest.

Consensual Pupillary Response

Have you ever watched what happens to one eye if you shine a bright light into the other eye? Both pupils constrict, not just the one exposed to the bright light. This is the **consensual pupillary response,** and it occurs because of the neural connections between the eyes. Instead of continuing to the thalamus, some axons of each optic tract terminate in both sides of the midbrain in the *pretectal area,* which you can see in Figure 15.22. Pretectal neurons contact other neurons that send their axons back through the oculomotor nerve to the pupillary muscles in both eyes. Since the signal about bright light hitting the retina in one eye travels to both sides of the brain, the reflexive response constricts the pupil in both eyes.

The consensual pupillary response is clinically important. If it does not occur, this indicates damage to the retina, optic nerve, or brainstem.

Stereoscopic Vision (Depth Perception)

Our eyes face anteriorly, so much of the visual field for one eye overlaps with that for the other eye, resulting in binocular vision. Each eye sees many of the same objects, but from a slightly different viewpoint. The brain compares these images and uses the differences to generate **stereoscopic vision** (or **depth perception**)—in other words, the distance of an object from the eye. The loss of vision in one eye destroys the ability to compare images, and stereoscopic vision is lost.

Image Processing by the Brain

The act of seeing involves much more than simple detection and conscious awareness of visual stimuli, and as just described, areas of the brain involved in processing visual information are extensive. Retinal axons project to the lateral geniculate nucleus and the primary visual cortex so that the image from the retina is "mapped" onto these areas. However, the image is distorted: Parts of it falling on the fovea of the retina take up much more area than parts from peripheral areas. This occurs because the increased numbers of photoreceptors in the fovea provide enhanced sensitivity, much like general sensation from your fingertips occupies a disproportionately large area of cortex compared to sensations from your calf.

In addition to conscious awareness of visual stimuli, the primary visual cortex also provides initial processing of location, movement, shape, and color, and combines images from each eye for binocular vision. From the primary cortex, visual information is then sent into two general pathways: (1) the posterior (dorsal) pathway that involves the parietal lobe, and (2) the anterior (ventral) pathway that ends in the inferior portion of the temporal lobe. Each pathway is specialized for processing particular aspects of vision. The posterior pathway appears to be involved in interpreting motion; a person with damage in this region may have difficulty with pouring coffee into a cup or crossing a street because he or she is unable to detect the movement of the coffee or the cars.

The anterior pathway is specialized for processing color and the forms of objects. For example, some neurons in the inferior temporal lobe respond more vigorously to faces as opposed to furniture or skyscrapers. People suffering from *prosopagnosia* (pros'-oh-pag-NOH-see-uh) cannot recognize people (even close family members) by their faces; they must rely on other cues such as the sound of their voices. Damage to the anterior pathway can also result in complete loss of color perception despite the presence of functional cones; this rare condition is called *achromatopsia* (ah-kroh'-mah-TAWP-see-uh) and can be more unpleasant than you might expect. Would you find it appetizing to put black ketchup on your gray French fries?

The posterior and anterior pathways are not the end of the story. The neurons here form myriad connections with other parts of the brain so that we can form visual memories, read aloud the words on a page or name an object, direct the extrinsic eye muscles, and make behavioral decisions based on what we see.

Quick Check

☐ 10. What happens at the optic chiasma? What is the consequence of this?

☐ 11. What type of visual processing occurs in the posterior and anterior pathways?

Putting It All Together: The Big Picture of Vision

Now let's take a look at all the processes involved in vision. Imagine you are sitting on a beach, watching a friend surf the waves and tumble off his surfboard. **Figure 15.23** summarizes the process by which you detect, transduce, and analyze these visual stimuli.

Apply What You Learned

☐ 1. If your vision is normal and you put on your hyperopic friend's glasses, is the focal point in front of your retina or behind it? (*Hint:* How does a lens that corrects hyperopia affect rays of light?)

The Big Picture of Vision

Figure 15.23

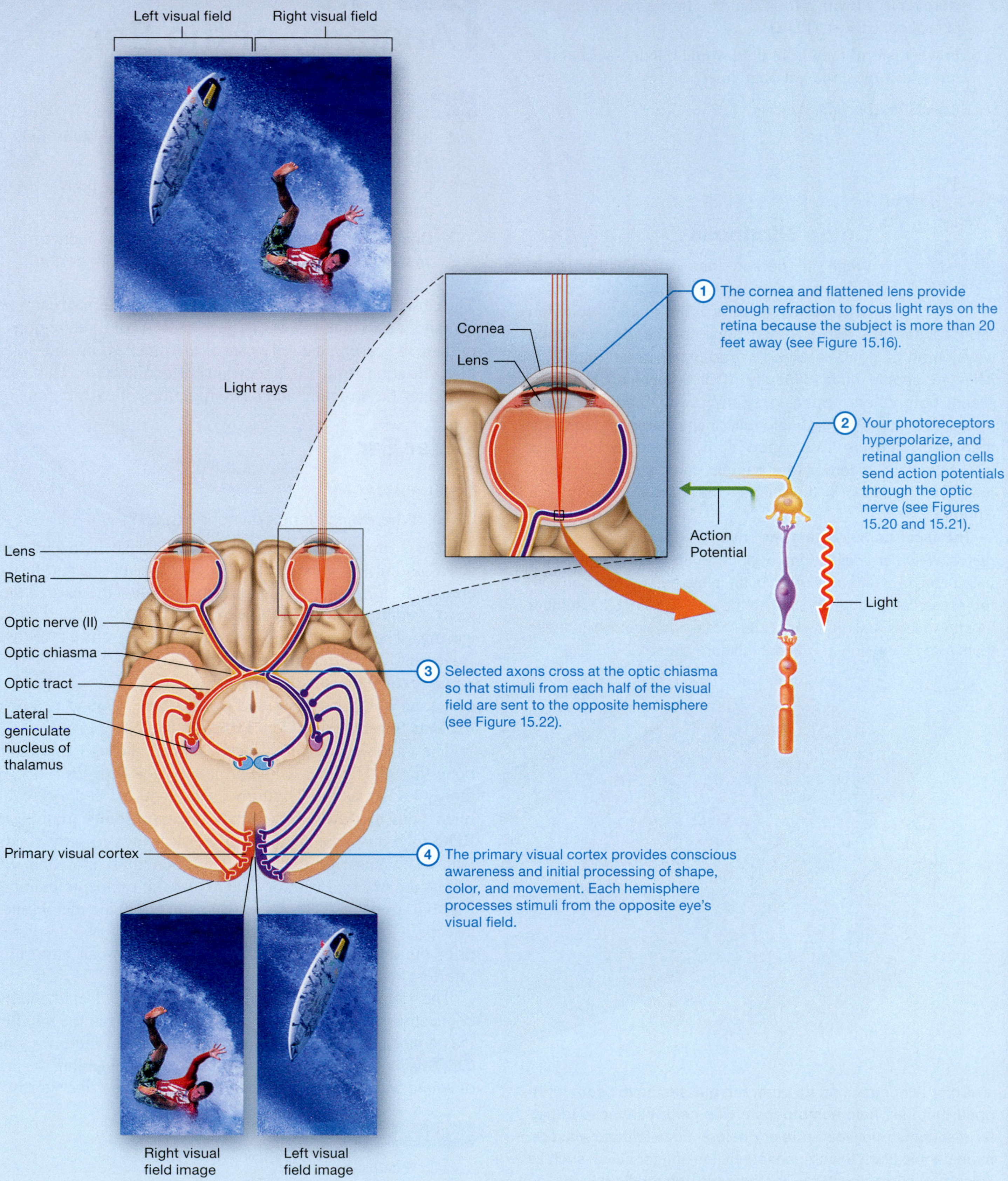

Left visual field   Right visual field

Light rays

Cornea

Lens

Lens

Retina

Optic nerve (II)

Optic chiasma

Optic tract

Lateral geniculate nucleus of thalamus

Action Potential

Light

Primary visual cortex

Right visual field image

Left visual field image

1 The cornea and flattened lens provide enough refraction to focus light rays on the retina because the subject is more than 20 feet away (see Figure 15.16).

2 Your photoreceptors hyperpolarize, and retinal ganglion cells send action potentials through the optic nerve (see Figures 15.20 and 15.21).

3 Selected axons cross at the optic chiasma so that stimuli from each half of the visual field are sent to the opposite hemisphere (see Figure 15.22).

4 The primary visual cortex provides conscious awareness and initial processing of shape, color, and movement. Each hemisphere processes stimuli from the opposite eye's visual field.

☐ 2. Explain what would happen in a rod if the cGMP were replaced with a form that was insensitive to phosphodiesterase (PDE).

☐ 3. In what part of the visual field would blindness occur if a tumor disrupted the left optic tract?

See answers in Appendix A.

Color Blindness

When a person has a defective gene for one or more cone pigments, **color blindness** results. The most common form causes the red or green cone to absorb the wrong wavelength of light, resulting in red-green color blindness. Affected people have difficulty in distinguishing red from green; both colors appear grayish-brown. About 8–10% of males have some form of color blindness, compared to fewer than 1% of females. This pattern shows that color blindness is a *sex-linked disorder;* the genetic causes of such disorders are discussed in the chapter that covers heredity (see Chapter 27).

The most common test for color blindness uses *Ishihara plates,* which are collections of dots with a number embedded using dots of a different color. In the plate below, a person with red-green color blindness will not be able to detect the number 74 and will see a 21 instead. Note that reliable color blindness tests require looking at original Ishihara plates.

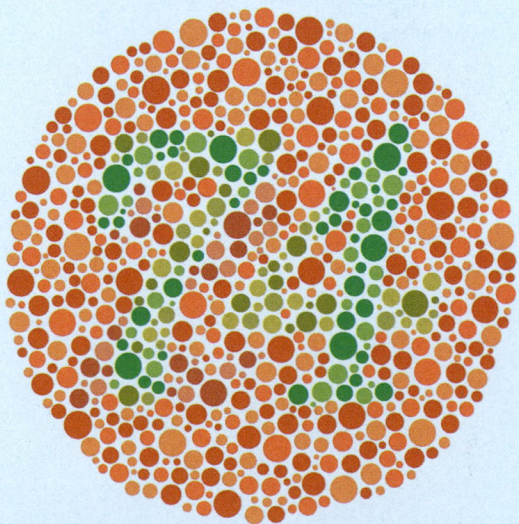

Color blindness has no cure, but glasses have been developed that may help some people with color blindness to better distinguish between primary colors. Even without glasses, most people learn to compensate by looking for clues, such as color intensity or location—for instance, the red light is the top light of a traffic signal.

MODULE **15.6**
Anatomy of the Ear

Learning Outcomes

1. Describe the structure and function of the outer and middle ear.
2. Discuss the role of the pharyngotympanic tube in draining and equalizing pressure in the middle ear.
3. Describe the structure and function of the cochlea, vestibule, and semicircular canals in the inner ear.

The ear is divided into three regions: outer (external) ear, middle ear, and inner (internal) ear (**Figure 15.24**). All three regions participate in hearing, but only part of the inner ear provides vestibular sensation, which is important for equilibrium. This module examines the structure and basic functions of each region.

Outer Ear

« FLASHBACK

1. Where is the temporal bone? (p. 217)

The most obvious part of the outer ear is the shell-shaped **auricle** (OHR-ih-kul), or *pinna*. It largely consists of elastic cartilage covered with skin, except for the fleshy *lobule*—the part of the ear most commonly pierced for earrings. The auricle funnels sound waves into the **external auditory canal** (**external acoustic meatus**),  a task at which it excels due to its shape, a good example of the Structure-Function Core Principle (p. 28).

The canal is a slightly curved tunnel through the temporal bone that ends at the *tympanic membrane* (tim-PAN-ik), commonly known as the "eardrum." The external auditory canal is lined with modified sweat glands, **ceruminous glands,** (seh-ROO-meh-nus) that secrete yellowish-brown to gray **cerumen** (seh-ROO-men), or *ear wax*. Cerumen lubricates and waterproofs the external auditory canal and the tympanic membrane; it also traps debris before it reaches the tympanic membrane and sweeps it out of the auditory canal. In most people, cerumen dries out and exits the canal, but in others it can build up and interfere with hearing.

The **tympanic membrane** is a thin sheet of epithelium and connective tissue that separates the outer ear from the middle ear. It is cone-shaped, with its tip, or apex, pointing into the middle ear. This membrane plays an important role in enabling the energy of sound waves to reach the inner ear, as we describe shortly.

Quick Check

☐ 1. What is the auricle?

☐ 2. What is the purpose of cerumen?

☐ 3. Where is the tympanic membrane located?

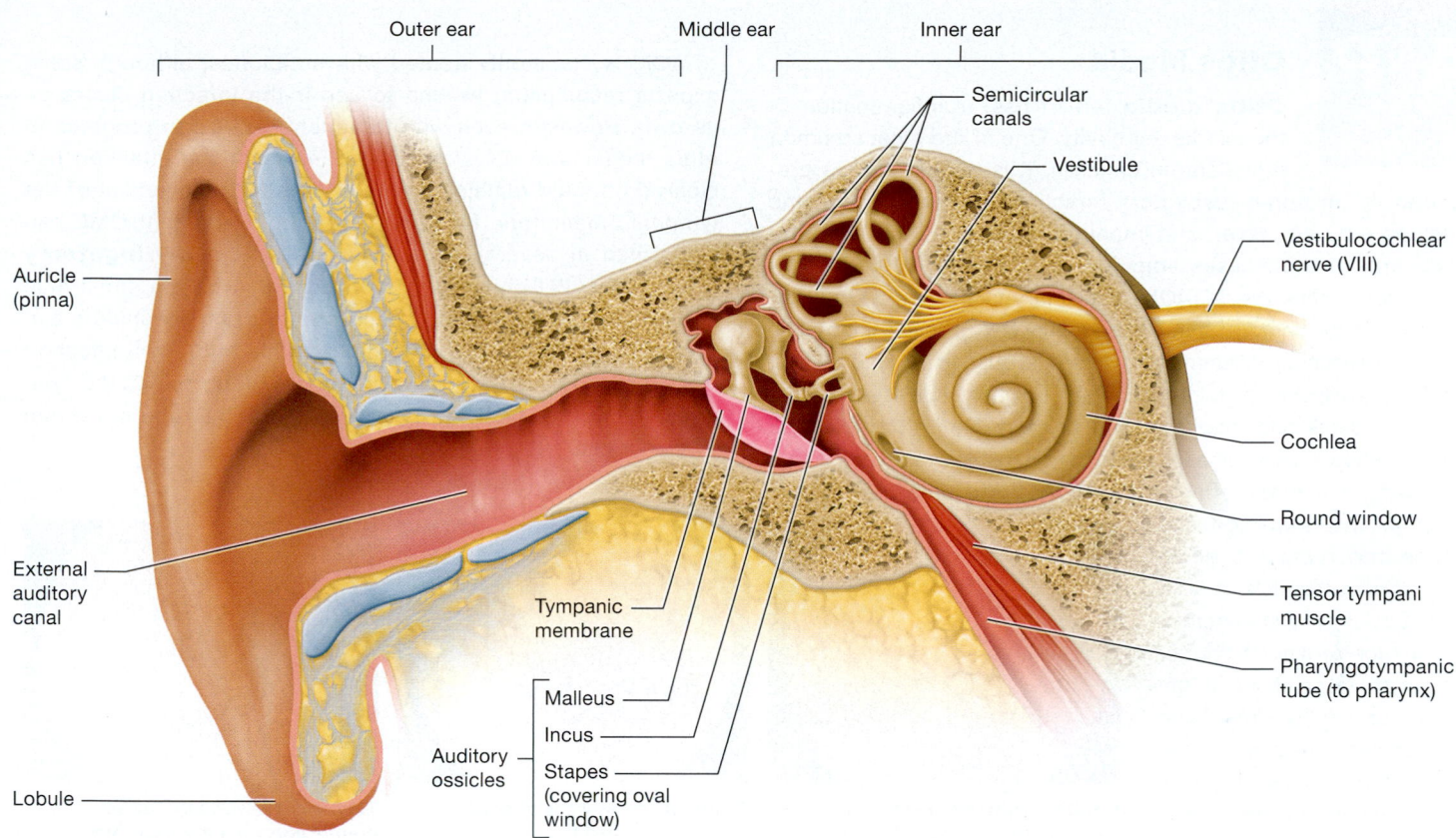

Figure 15.24 Regions of the ear.

Middle Ear

The middle ear is a hollow, air-filled chamber in the temporal bone lined with a mucous membrane. It is bounded by the tympanic membrane laterally and the wall of the inner ear medially. The middle ear and nasopharynx (the part of the throat posterior to the nasal cavity) are connected by a passage known as the **pharyngotympanic tube** (fah-ring'-goh-tim-PAN-ik), or *auditory tube*.

The pharyngotympanic tube equalizes air pressure on both sides of the tympanic membrane. Normally closed and flattened, it opens briefly when we yawn or swallow. You no doubt have had firsthand experience with this when your ears "pop" during an airplane takeoff or landing. As the elevation changes, air pressure in the external auditory canal changes relative to that in the middle ear. Any pressure difference causes the tympanic membrane to stretch and bulge inward or outward. Excess tension reduces the ability of the membrane to vibrate, interfering with hearing and resulting in a feeling of pressure in the ear. Yawning or swallowing opens the pharyngotympanic tube and equalizes pressure in the middle and outer ears. The pharyngotympanic tube can also allow pathogens in the respiratory tract to enter the middle ear, which is examined in *A&P in the Real World: Otitis Media.*

The middle ear contains the three **auditory ossicles** (AWS-ih-kulz), which are the smallest bones in the body (all three would fit on your fingernail). They are named for their shape: The

malleus (MAL-ee-us) is shaped like a hammer, the **incus** (ING-kus) like an anvil, and the **stapes** (STAY-peez) like a stirrup. The ossicles are connected by synovial joints and together form a bridge that extends from the middle ear to the inner ear. Notice in Figure 15.24 how the malleus is connected to the tip of the cone-shaped tympanic membrane, and the stapes is attached to an oval membrane in the medial wall called the **oval window.** The oval window is the boundary between the air-filled middle ear and the fluid-filled inner ear. The ossicles have an important part in amplifying and converting sound waves in the air into fluid movement. We'll soon see how this leads to action potentials.

Two very small skeletal muscles attach to the ossicles. The **tensor tympani muscle** (TEN-sohr TIM-pah-nee) has its origin on the walls of the pharyngotympanic tube and inserts on the malleus to pull the tympanic membrane medially and tense it. The **stapedius muscle** (stuh-PEE-dee-us) (not shown in Figure 15.24) arises from the posterior wall of the middle ear and inserts on the stapes; it reduces movement of the ossicles as a unit. These muscles are part of a reflex that prevents damage from loud sounds at low frequencies by reducing the movements of the tympanic membrane and ossicles. Since sound travels faster than the muscles can contract, they will not prevent damage from sudden short sounds like gunshots, but they can be protective for longer-lasting sounds, such as a nearby train. The muscles also contract just before we speak and while we chew so that these sounds are not uncomfortably loud.

Otitis Media

Otitis media (oh-TY-tiss) is inflammation of the middle ear cavity. One of the most common infections in children, it is almost always preceded by an upper respiratory infection. Common symptoms include ear pain, fever, and impaired hearing, as well as irritability and, in some cases, vomiting.

In *acute otitis media* (*AOM*), the pharyngotympanic tube acts as a route for pathogens in the nasopharynx to reach the middle ear. The resulting inflammation causes a buildup of fluid or pus, and the tympanic membrane bulges painfully into the external auditory canal and interferes with hearing. The left and middle photos shown here compare a healthy tympanic membrane with a tympanic membrane affected by AOM. If the pressure is great enough, the tympanic membrane may rupture. Children are more susceptible to AOM because their pharyngotympanic tubes are much shorter and more horizontal than are those of adults. They also suffer more frequent upper respiratory infections because their immune systems are less developed.

AOM is commonly treated with antibiotics, although some experts recommend waiting to see if the infection clears on its own. However, even with treatment, AOM can progress to *otitis media with effusion* (*OME*), in which the infection has cleared but fluid remains and interferes with movement of the tympanic membrane for normal hearing. Persistent OME can be treated in several ways. One option is a **myringotomy** (meer'-in-GAWT-oh-mee), a procedure in which a small hole is pierced in the tympanic membrane to drain the middle ear. A small tube may be inserted into the hole to prevent infection from recurring (see the right photo shown here). As the tympanic membrane heals, it naturally pushes the tube out within a year or two.

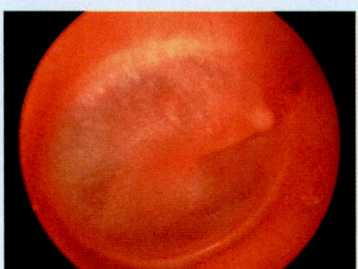

Healthy tympanic membrane viewed through the external auditory canal

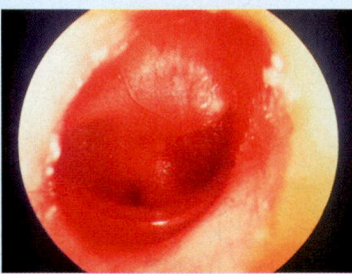

Tympanic membrane in acute otitis media (AOM)

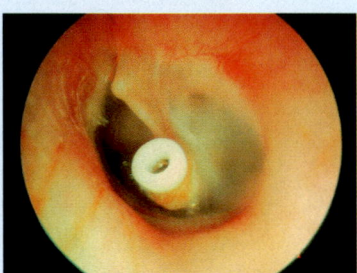

Tube inserted into tympanic membrane via a myringotomy

Quick Check

☐ 4. What are the three auditory ossicles, and what are their functions?

☐ 5. What is the role of the pharyngotympanic tube?

Inner Ear

« FLASHBACK

1. What are the functions of the vestibulocochlear nerve? (p. 481)

2. What is the concentration of sodium ions relative to that of potassium ions in cytosol versus extracellular fluid? (p. 348)

The inner ear is a cavity within the temporal bone that is divided into three regions (**Figure 15.25**). The *cochlea* houses structures involved in hearing, and the *vestibule* and *semicircular canals* contain structures that detect head movement and position (vestibular sensation) (see Module 15.8).

Bony and Membranous Labyrinths

A labyrinth (LAB-uh-rinth) is a network of winding passages. The inner ear is sometimes called the **bony labyrinth** because of its mazelike series of tunnels. A membrane referred to as the **membranous labyrinth** lines the inner walls of the bony labyrinth. Within the membranous labyrinth, we find **endolymph,** a fluid with a higher concentration of potassium ions than sodium ions, similar to cytosol. Between the walls of the bony labyrinth and the membranous labyrinth is **perilymph,** a fluid that has a higher concentration of sodium ions than potassium ions, which resembles extracellular fluid. Endolymph and perilymph play an important part in transducing sound waves and head movements into action potentials.

Vestibule

The **vestibule** is located medial to the middle ear, and its wall contains the membranous oval window to which the stapes is connected (see Figure 15.25). The vestibule houses two portions of the membranous labyrinth, the **utricle** (YOO-trih-kul) and the **saccule** (SAK-yool). These chambers contain receptor cells that detect head tilting and linear movement. The receptor cells transmit the stimuli to neurons that form the vestibular portion of the *vestibulocochlear nerve* (CN VIII; ves-tib'-yoo-loh-KOHK-lee-ur).

Semicircular Canals

The utricle is continuous with three tubes of the membranous labyrinth called **semicircular ducts,** which are enclosed in the

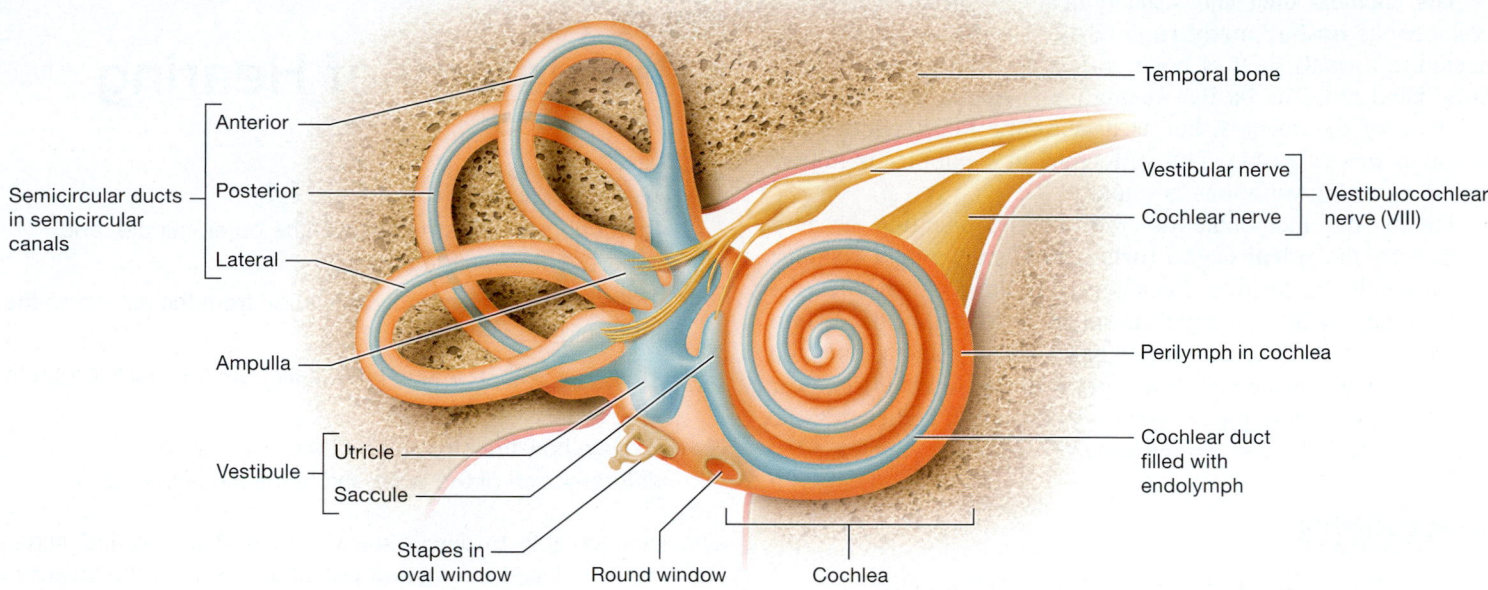

Figure 15.25 The membranous labyrinth. Structures containing perilymph are shaded orange, and structures containing endolymph are shaded blue. This color coding is also used in Figures 15.26 and 15.28.

bony **semicircular canals.** Note in Figure 15.25 that the anterior, posterior, and lateral semicircular ducts are oriented at right angles to one another so they can detect rotational movement of the head in any plane. At the base of each duct is a swollen bulb, the **ampulla** (am-PEWL-uh), which contains receptor cells in contact with neurons. As in the vestibule, the receptor cells are innervated by the vestibular portion of the vestibulocochlear nerve.

Cochlea

The bony **cochlea** (KOHK-lee-uh) features a membrane-covered opening called the **round window,** which separates the middle and inner ear. Within the cochlea, we find a spiral-shaped portion of the membranous labyrinth known as the **cochlear duct,** or **scala media** (SKAY-luh), which is connected to the saccule and filled with endolymph. You can see how the cochlear duct spirals about two and a half times around a screw-shaped core of bone called the **modiolus** (muh-DY-oh-lus; labeled in **Figure 15.26**) and then ends at the tip of the cochlea.

Two other chambers spiral through the cochlea, both filled with perilymph. Superior to the cochlear duct is the **scala vestibuli** (ves-TIB-yoo-lye; see Figure 15.26), which is continuous with the vestibule near the oval window. The inferior chamber is the **scala tympani,** which ends in the vestibule near the round window. Together, the scala tympani and scala vestibuli form a "sandwich" around the cochlear duct, and are continuous with each other at the tip of the cochlea.

A sheet of epithelium called the **vestibular epithelium** forms the boundary between the cochlear duct and scala vestibuli. This membrane separates the endolymph and perilymph and serves as a diffusion barrier between the two, maintaining the concentrations of ions in the fluids.

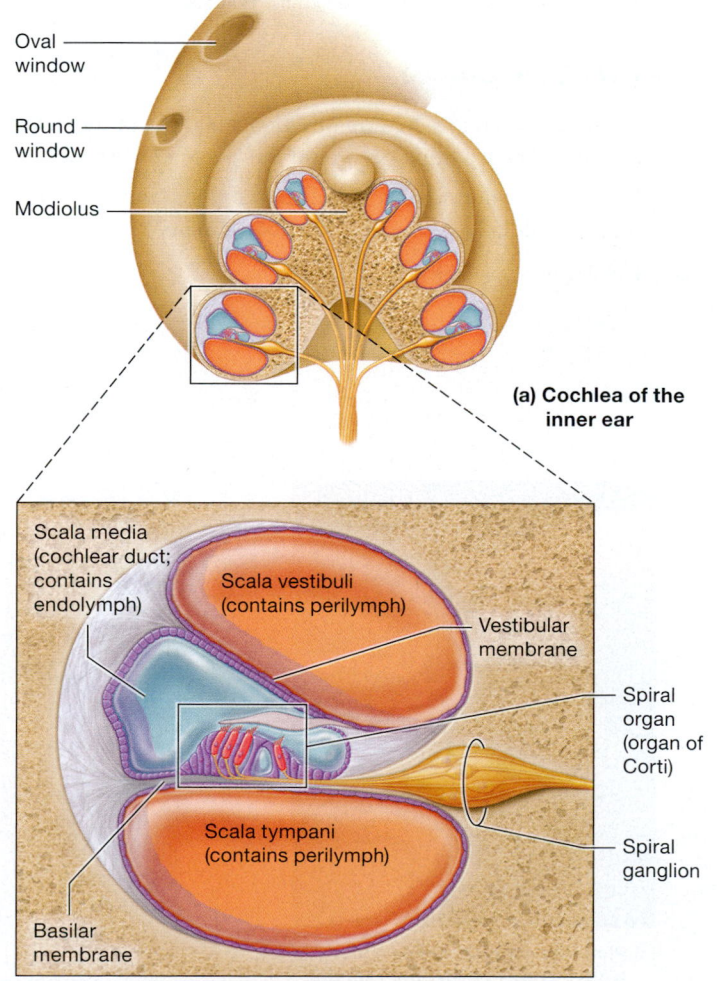

(a) Cochlea of the inner ear

(b) Spiral organ and ducts of the cochlea

Figure 15.26 Structure of the cochlea.

The cochlear duct and scala tympani are separated by the collagenous **basilar membrane** (BAZ-ih-lur), a structure connected to a small shelf of bone projecting from the wall of the bony labyrinth. The basilar membrane is narrow and stiff near the base of the cochlea, but as it nears the tip, it widens and becomes more flexible. This difference in flexibility is important in detecting variations in sound.

The basilar membrane deserves special attention because it supports the **spiral organ** (**organ of Corti**), which contains receptor cells for hearing. Located within the cochlear duct, the spiral organ is where energy from sound waves is transduced into action potentials. The action potentials are propagated into the CNS by axons in the cochlear portion of the vestibulocochlear nerve, also known as the *cochlear nerve;* the cell bodies of these neurons form the *spiral ganglion* at the base of the cochlea.

Quick Check

☐ 6. What are the functions of the utricle, saccule, and semicircular ducts?

☐ 7. How are the scala tympani, scala media, and scala vestibuli arranged with respect to one another?

☐ 8. What is the spiral organ, and where is it located?

Apply What You Learned

☐ 1. Suppose someone's ceruminous glands did not produce any cerumen. What symptoms would this person experience?

☐ 2. How would your hearing be affected if your tensor tympani muscle suddenly contracted and did not relax?

See answers in Appendix A.

MODULE **15.7**
Physiology of Hearing

Learning Outcomes

1. Describe how the structures of the outer, middle, and inner ear function in hearing.

2. Trace the path of sound conduction from the auricle to the fluids of the inner ear.

3. Trace the path of action potentials from the spiral organ to various parts of the brain.

4. Explain how the structures of the ear enable differentiation of the pitch and loudness of sounds.

Our ears are able to detect sound waves. These sound waves must be transduced into action potentials for us to be aware of and interpret the sound. As a starting point for understanding this process of **hearing,** or *audition,* let's first look at how sound waves are produced and travel through the air.

Principles of Sound

Sound waves are nothing more than the displacement of air molecules in response to an object moving. When you pluck a string on an instrument, the string vibrates, or moves back and forth (**Figure 15.27a**). As it moves in one direction, it compresses molecules of air and creates an area of slightly increased air pressure. When the string moves in the opposite direction, it causes the air pressure to drop slightly.

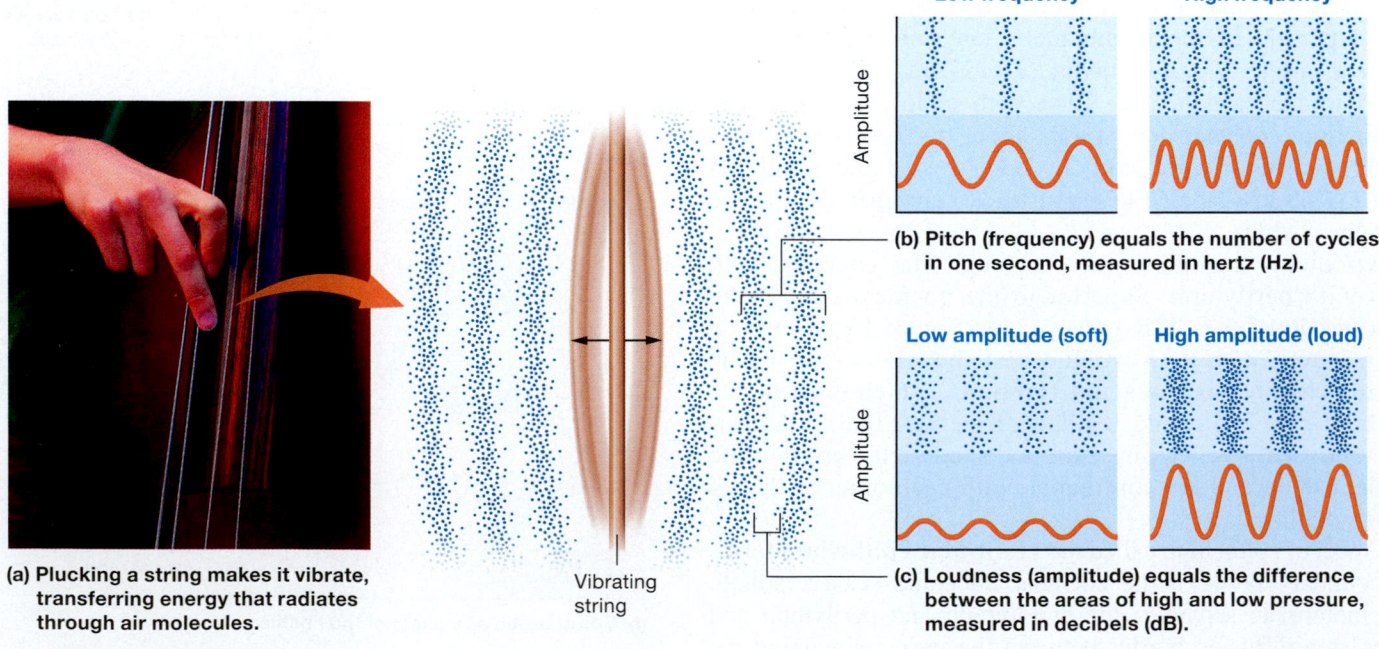

(a) Plucking a string makes it vibrate, transferring energy that radiates through air molecules.

Vibrating string

(b) Pitch (frequency) equals the number of cycles in one second, measured in hertz (Hz).

(c) Loudness (amplitude) equals the difference between the areas of high and low pressure, measured in decibels (dB).

Figure 15.27 Sound waves.

Although sound waves are often drawn as if the air molecules radiate away from the source, this is not the case. The molecules barely move from their original position; instead, they bump into adjacent molecules and transfer their energy, which causes those molecules to collide with others (much like the chain reaction in a row of dominos). The energy is transmitted in waves that radiate away from the vibrating object like ripples spreading across a pond, except that sound waves spread in three dimensions.

We often describe sound in terms of pitch and loudness. Every sound has a certain *pitch:* We hear the low-pitched rumble of thunder and the high-pitched yipping of a poodle frightened by the storm. Pitch, or *frequency,* is determined by how many times the object vibrates back and forth during a certain period (**Figure 15.27b**). Pitch is measured in hertz (Hz), which is the number of cycles or vibrations in one second. Our ears can typically detect sounds between 20 Hz and 20,000 Hz (20 kHz). The hearing range for dogs is much higher and explains why they respond to a seemingly "silent" dog whistle. We can feel the vibrations of sounds between 4 and 16 Hz, even though we can't hear them. Elephants and some whales communicate by producing sounds at frequencies well below 20 Hz.

Loudness, or the *amplitude* of a sound wave, is determined by the difference between areas of high and low air pressure (**Figure 15.27c**). If you gently pluck a string, it moves a limited distance as it vibrates, and the difference between the waves of high and low air pressure is also limited. Plucking the string with more force produces a louder sound because the string vibrates over a greater distance and displaces more air as it moves. Loudness is measured in decibels (dB); the rustling of leaves is barely audible at 10 dB, whereas the roar of a jet engine at 120 dB is painful and damaging to the ears.

Quick Check

☐ 1. What physical properties determine the pitch and loudness of a sound?

☐ 2. What is the range (in hertz) of human hearing?

Transmission of Sound to the Inner Ear

What happens when sound waves reach the ear? The auricle funnels sound waves into the auditory canal, after which the following steps occur (**Figure 15.28a**):

① **When sound waves strike the tympanic membrane, movement of the connected auditory ossicles causes the oval window to vibrate at the same frequency.** Although the oval window vibrates at the same frequency

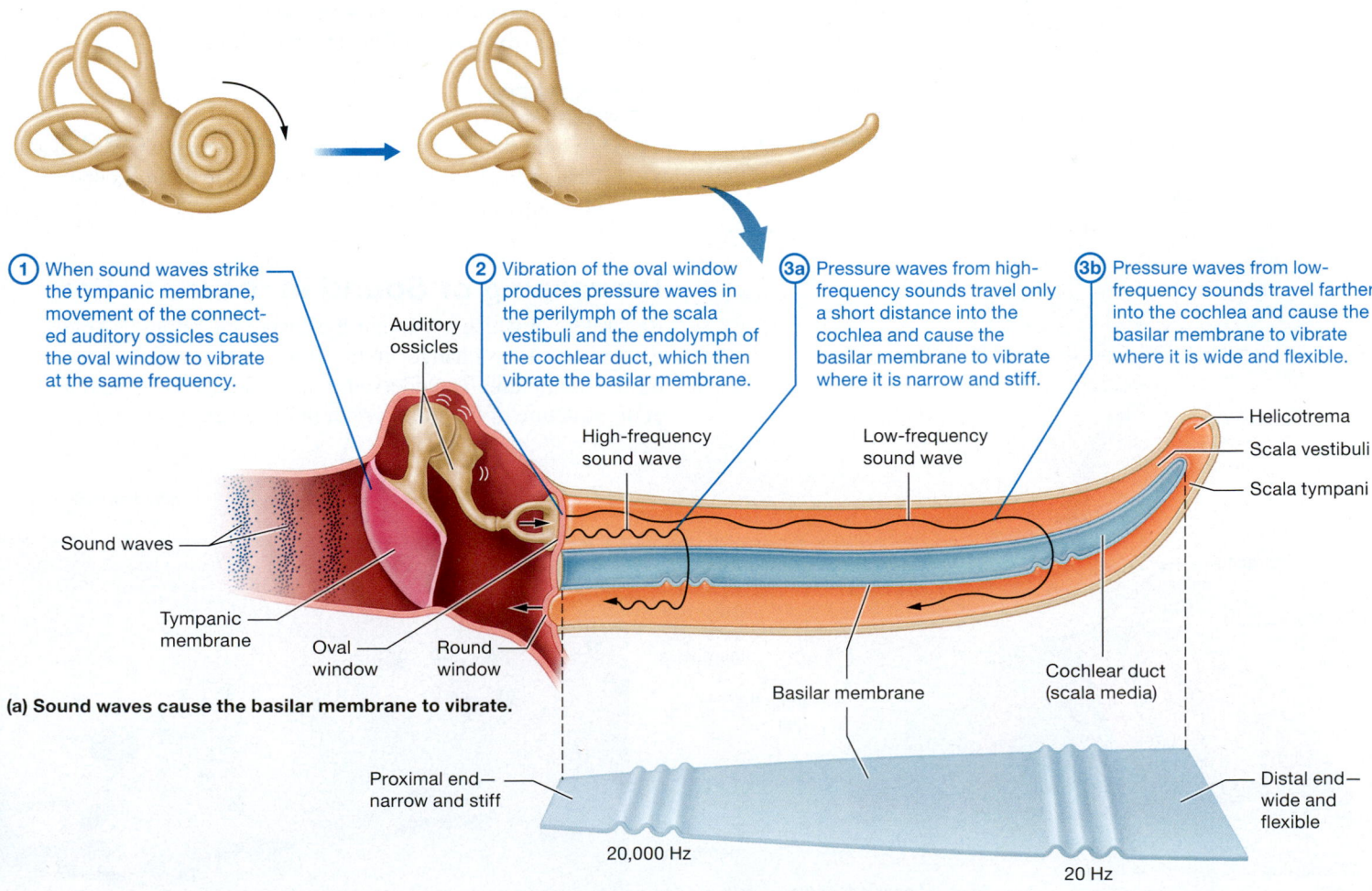

① When sound waves strike the tympanic membrane, movement of the connected auditory ossicles causes the oval window to vibrate at the same frequency.

② Vibration of the oval window produces pressure waves in the perilymph of the scala vestibuli and the endolymph of the cochlear duct, which then vibrate the basilar membrane.

③a Pressure waves from high-frequency sounds travel only a short distance into the cochlea and cause the basilar membrane to vibrate where it is narrow and stiff.

③b Pressure waves from low-frequency sounds travel farther into the cochlea and cause the basilar membrane to vibrate where it is wide and flexible.

Auditory ossicles

High-frequency sound wave

Low-frequency sound wave

Helicotrema

Scala vestibuli

Scala tympani

Sound waves

Tympanic membrane

Oval window

Round window

Basilar membrane

Cochlear duct (scala media)

(a) Sound waves cause the basilar membrane to vibrate.

Proximal end— narrow and stiff

Distal end— wide and flexible

20,000 Hz

20 Hz

(b) The shape of the basilar membrane varies.

Figure 15.28 Vibration of the basilar membrane.

as the tympanic membrane, the ossicles transmit the force from the larger tympanic membrane to the smaller oval window. This concentrates the force in a smaller area and so increases it significantly. The ossicles also act as a lever system to amplify the vibration of the oval window.

② **Vibration of the oval window produces pressure waves in the perilymph of the scala vestibuli and the endolymph of the cochlear duct, which then vibrate the basilar membrane.** More force is needed to generate motion in fluid than in air (think about the effort required to move your cupped hand through water as opposed to air). This is why the force needs to be concentrated and amplified on the oval window—the extra force is necessary to move the endolymph and perilymph.

③ₐ **Pressure waves from high-frequency sounds travel only a short distance into the cochlea and cause the basilar membrane to vibrate where it is narrow and stiff.** The pressure waves generated at the oval window travel through the perilymph of the vestibule to the scala vestibuli of the cochlea. High-frequency vibrations (such as those produced by squealing tires) take a shortcut through the cochlear duct (scala media) to reach the scala tympani. As they pass through the cochlear duct to the scala tympani, they cause the endolymph and basilar membrane to vibrate. These vibrations occur near the proximal end of the basilar membrane, close to the base of the cochlea, where the basilar membrane is narrow and stiff (**Figure 15.28b**). Notice in the figure that when waves reach the vestibule at the end of the scala tympani, they cause the round window to bulge outward into the middle ear. The movement of the round

window is essential, as without it the perilymph would not move. Fluid cannot be compressed; if it is pushed from one area, it must move into another area (like squeezing a water balloon at one end forces it to bulge at the other end). In the same way, as the oval window is pushed into the inner ear, the round window bulges outward, and vice versa.

③b **Pressure waves from low-frequency sounds travel farther into the cochlea and cause the basilar membrane to vibrate where it is wide and flexible.** Low-frequency vibrations (like those of the rumble of an engine) travel farther into the cochlea, near the distal end of the basilar membrane, where the basilar membrane is wider and more flexible. For very low-frequency sounds that we cannot hear, the waves travel all the way to the tip of the cochlea, where the scala vestibuli connects with the scala tympani at an opening called the **helicotrema** (hel'-ih-koh-TREE-muh).

Note that whereas pitch is determined by which *area* of the basilar membrane vibrates, loudness is determined by *how much* the basilar membrane vibrates at that area. A loud noise produces sound waves with more energy, which in turn cause greater movement of the tympanic membrane, middle ear ossicles, and the perilymph and endolymph. Increased movement of these fluids produces greater vibration of the basilar membrane and stronger stimulation of nearby receptor cells.

Quick Check

☐ 3. Why is the force exerted on the oval window greater than that exerted on the tympanic membrane during sound transmission?

Processing of Sound in the Inner Ear

Regardless of which portion of the basilar membrane vibrates, the movement causes changes in the spiral organ, which rests on the basilar membrane. To understand these changes, we need to look at the structure of the spiral organ in more detail (**Figure 15.29**).

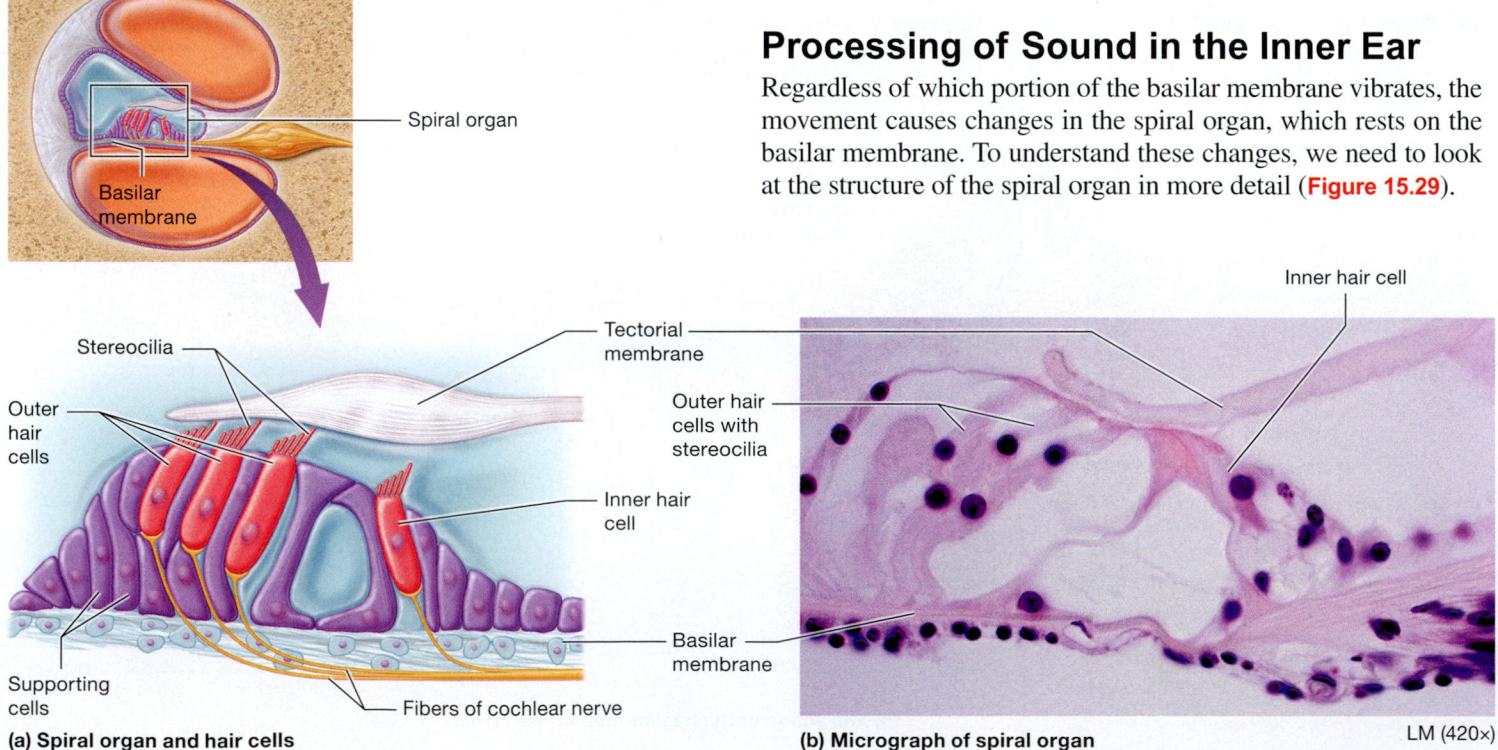

Spiral organ

Basilar membrane

Stereocilia

Outer hair cells

Supporting cells

Fibers of cochlear nerve

Tectorial membrane

Outer hair cells with stereocilia

Inner hair cell

Basilar membrane

Inner hair cell

(a) Spiral organ and hair cells

(b) Micrograph of spiral organ

LM (420×)

Figure 15.29 Structure of the spiral organ.

The spiral organ contains receptor cells called **hair cells.** They are grouped into a single row of **inner hair cells** and three rows of **outer hair cells.** The 3500 inner hair cells are primarily responsible for detecting sound. Each hair cell has microvilli of varying lengths known as **stereocilia** (stehr´-ee-oh-SILL-ee-uh; singular, *stereocilium*). They project from the cell into the endolymph of the cochlear duct, arranged from tallest to shortest. The stereocilia themselves are stiff but "bend" by flexing at the junction with the cell body (like a joystick).

What makes the stereocilia bend? A stiff membrane, called the tectorial membrane (tek-TOHR-ee-ul), sits on top of the hair cells so that they are sandwiched between the basilar and tectorial membranes. The stereocilia on the outer hair cells contact the tectorial membrane, but the stereocilia in the inner hair cells do not. When the basilar membrane vibrates up and down in response to sound waves, the hair cells move with it toward and away from the tectorial membrane.

Each stereocilium is connected to its neighbor by an elastic filament called a *tip link* (**Figure 15.30a**). When the tallest stereocilium bends in one direction, the other stereocilia are pulled in the same direction.

The sequence of events that takes place when sound is transmitted to the inner ear is shown in **Figure 15.30b** and proceeds as follows:

① **The basilar membrane moves up toward the tectorial membrane, bending the stereocilia toward the tallest stereocilium.** As the basilar membrane moves toward the tectorial membrane, the stereocilia on the outer hair cells push up against it and bend toward the tallest stereocilium. Stereocilia of inner hair cells also bend due to movement of the endolymph between the two membranes. When the basilar membrane moves away from the tectorial membrane, the stereocilia bend in the opposite direction.

② **Bending the stereocilia opens potassium ion channels that depolarize the hair cell.** As the stereocilia bend toward the tallest one, the tip links stretch and pull open a potassium ion channel in the tip of each stereocilium. In typical neurons, opening potassium ion channels would hyperpolarize the cell, but the endolymph contains such a high concentration of potassium ions that the gradient is reversed, causing them to flow into the hair cells. This leads to depolarization and a local potential. This is an example of the Gradients Core Principle (p. 28).

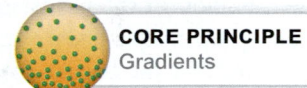

CORE PRINCIPLE
Gradients

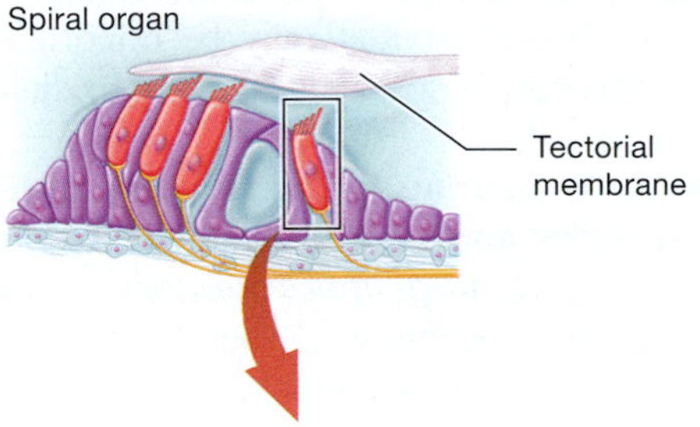

Spiral organ

Tectorial membrane

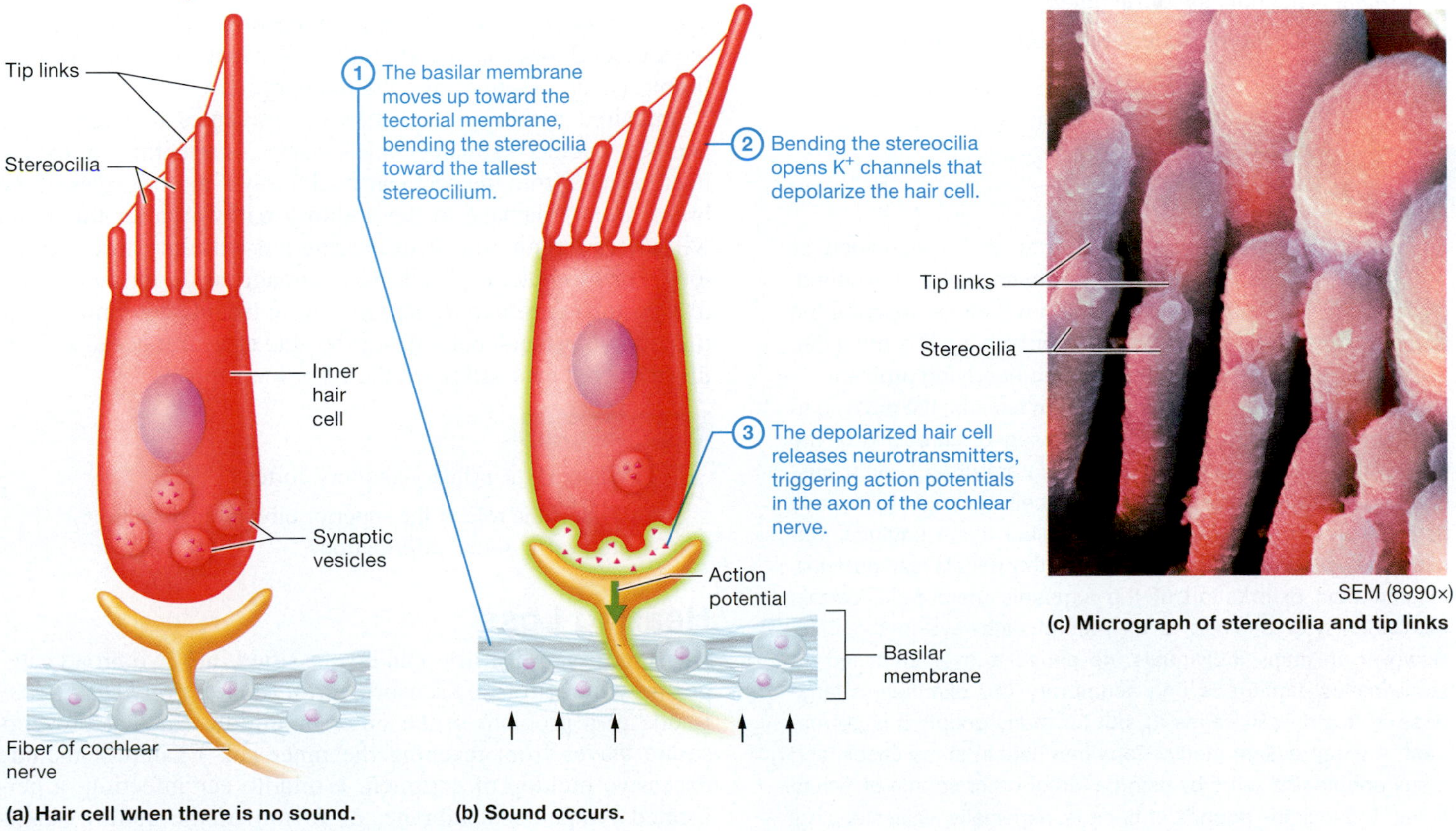

Tip links

Stereocilia

Inner hair cell

Synaptic vesicles

Fiber of cochlear nerve

(a) Hair cell when there is no sound.

① The basilar membrane moves up toward the tectorial membrane, bending the stereocilia toward the tallest stereocilium.

② Bending the stereocilia opens K⁺ channels that depolarize the hair cell.

③ The depolarized hair cell releases neurotransmitters, triggering action potentials in the axon of the cochlear nerve.

Action potential

Basilar membrane

(b) Sound occurs.

Tip links

Stereocilia

SEM (8990×)

(c) Micrograph of stereocilia and tip links

Figure 15.30 Transduction of sound in hair cells.

③ **The depolarized hair cell releases neurotransmitters, triggering action potentials in the axon of the cochlear nerve.** The neurotransmitter is likely glutamate, and it triggers an action potential in the cochlear nerve neuron that innervates the hair cell.

When the basilar membrane moves back, away from the tectorial membrane, the stereocilia bend in the opposite direction. The tip links are no longer stretched and the potassium ion channels close. The hair cells hyperpolarize, which stops the release of neurotransmitters.

The inner hair cells perform most of the "work" of detecting sounds. Indeed, about 90% of auditory neurons contact the inner hair cells, and multiple neurons contact each cell. In contrast, the outer hair cells are poorly innervated—a single neuron contacts multiple cells. However, in spite of their lesser innervation, destruction of the outer hair cells (for example, by certain drugs such as the anticancer drug cisplatin) can result in nearly complete deafness. This is because the outer hair cells have an essential role in increasing the sensitivity of the inner hair cells and amplifying the vibration of the basilar membrane. The outer hair cells have motor proteins that allow them to generate forces that enhance the movement of the basilar membrane as much as a hundred-fold. To see how damage to hair cells or other structures of the ear can interfere with hearing, read *A&P in the Real World: Tinnitus.*

Quick Check

☐ 4. How does the cochlea distinguish different frequencies?

☐ 5. What is the function of tip links?

☐ 6. What is the role of the tectorial membrane in sound detection?

Tinnitus

Tinnitus (tin-AYE-tus) is the sensation of hearing noise in the absence of actual sound. People who suffer from tinnitus describe the noise as ringing, buzzing, whistling, or hissing. It is not a disease itself, but often a symptom of an underlying problem.

Almost one in five adults suffers from tinnitus; the most common cause is hearing loss due to damaged stereocilia of the cochlear hair cells. Damaged stereocilia are often bent or misshapen, so they depolarize the hair cells enough to release some neurotransmitters onto the cochlear nerve endings. Tinnitus may also result from fusion of the middle ear ossicles, inflammation or irritation of the tympanic membrane, certain medications, or damage to the neural pathways in the CNS. However, in many individuals, no cause is ever identified. In some cases, tinnitus is only temporary (for example, a large dose of aspirin can cause it), but for many people it is permanent. Antidepressant medications can help in some cases, and many people find relief by using a fan or other source of "white noise" to block the sounds of tinnitus, especially while sleeping.

The Auditory Pathway

After the hair cells transduce the energy from sound waves into action potentials, these signals travel into the CNS along the pathway shown in **Figure 15.31**.

① **Action potentials propagate through axons of the cochlear portion of the vestibulocochlear nerve to the cochlear nuclei at the medulla-pons junction.** Axons of the cochlear nerve synapse in the cochlear nuclei, which are located at the junction of the medulla oblongata and pons.

② **Axons from the cochlear nuclei synapse on the superior olivary nucleus in the pons.** Neurons in the **superior olivary nucleus** compare information from both ears to determine the location of the sound.

③ **Auditory stimuli are then sent to the inferior colliculus of the midbrain.** The stimuli are coordinated with head and trunk movement to produce the *startle reflex*. This reflex, a response to unexpected sounds, causes you to turn your head and body toward the sound to investigate it.

④ **The auditory stimuli are relayed to the medial geniculate nucleus of the thalamus.** The role of the thalamus in human hearing is not clear, but it may be important for processing combinations of sounds according to their frequency and time interval.

⑤ **The thalamus stimulates neurons of the primary auditory cortex in the superior portion of the temporal lobe.** As with other sensory cortices, the **primary auditory cortex** is where our first conscious awareness of sound begins, as well as analysis of its location, pitch, and loudness.

The primary auditory cortex has connections with other parts of the temporal lobe that are specialized for language, and with the limbic system for emotions and memory.

Impaired hearing in just one ear indicates damage to the cochlear nuclei, vestibulocochlear nerve, or structures in the ear itself. Axons from the cochlear nuclei travel to both sides of the brain, so any damage to the auditory pathway after this point will affect certain aspects of hearing but will not cause hearing loss. For example, a person with damage to the primary auditory cortex in one hemisphere may have trouble determining the location of a sound, but will still be able to hear it because stimuli from both ears still reach the other hemisphere.

Quick Check

☐ 7. Where is the primary auditory cortex?

☐ 8. What is the role of the superior olivary nucleus in processing sound information?

Hearing Loss

Hearing loss or deafness can be classified into two broad categories: conduction and sensorineural. *Conduction hearing loss* is due to a problem in the outer or middle ear that prevents sound waves from reaching the inner ear. Examples include excessive buildup of cerumen, a middle ear infection, a perforated tympanic membrane, or fusion of the auditory ossicles. Vibrations can reach the fluid of the inner ear through the

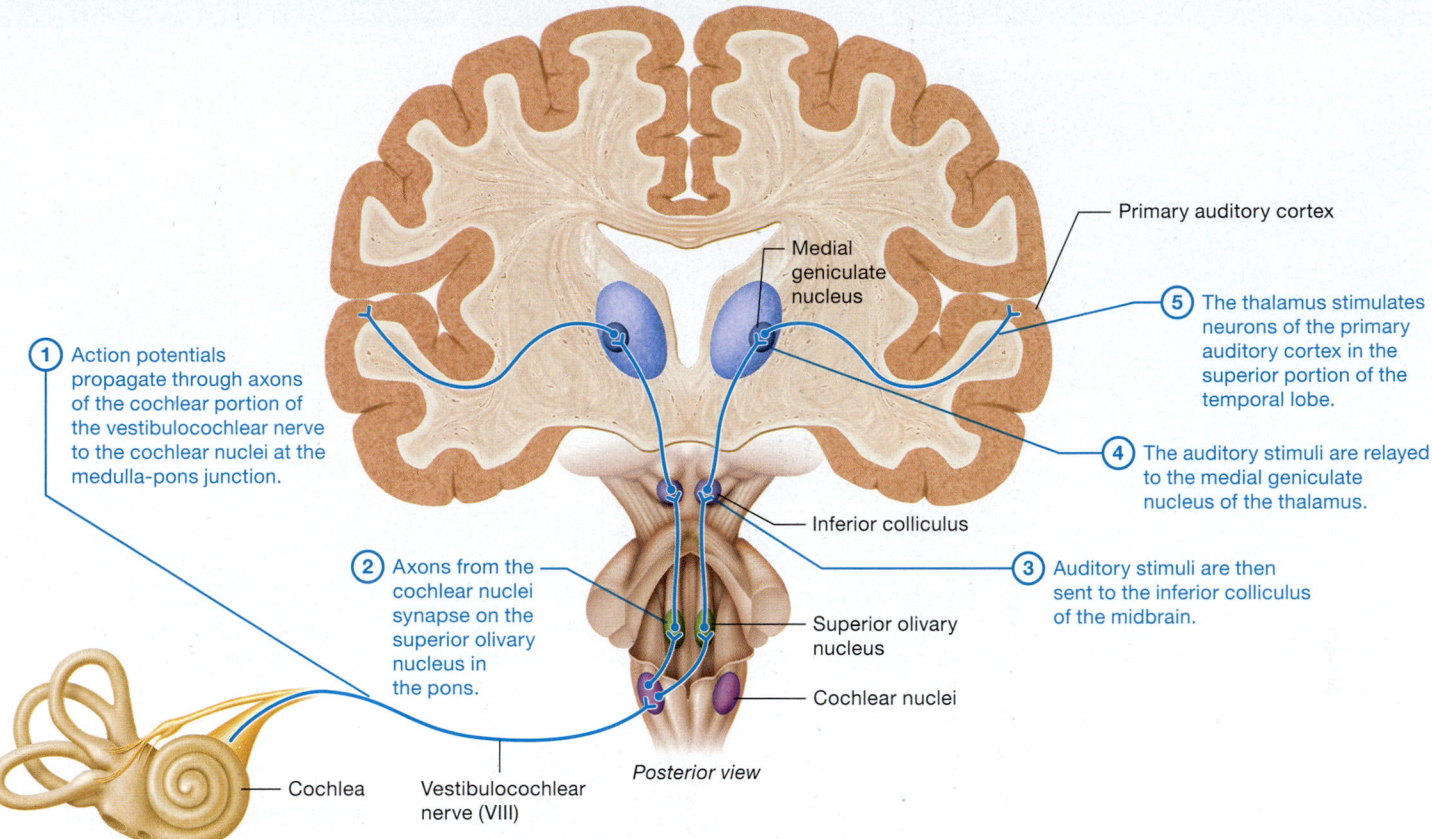

1. Action potentials propagate through axons of the cochlear portion of the vestibulocochlear nerve to the cochlear nuclei at the medulla-pons junction.

Medial geniculate nucleus

Primary auditory cortex

5. The thalamus stimulates neurons of the primary auditory cortex in the superior portion of the temporal lobe.

4. The auditory stimuli are relayed to the medial geniculate nucleus of the thalamus.

Inferior colliculus

2. Axons from the cochlear nuclei synapse on the superior olivary nucleus in the pons.

3. Auditory stimuli are then sent to the inferior colliculus of the midbrain.

Superior olivary nucleus

Cochlear nuclei

Posterior view

Cochlea

Vestibulocochlear nerve (VIII)

Figure 15.31 **The auditory pathway.**

temporal bone (bone conduction), but with much less accuracy than conduction through air. Many forms of conduction hearing loss are temporary and can be corrected, for example, by removing excess cerumen or repairing a perforated tympanic membrane.

Sensorineural hearing loss refers to a defect in the cochlea or any of the neural pathways in the cochlear nerve or CNS. There are actually two types of sensorineural hearing loss, and they differ significantly in terms of cause and treatment:

- *Sensory hearing loss* occurs when action potentials cannot be generated in the cochlea, usually due to dysfunction of the hair cells. Exposure to loud sounds or certain medications may damage hair cells. We also lose stereocilia as we age, especially those in the proximal part of the cochlea that are responsible for detecting high-frequency sounds. Hearing aids can help mild to moderate cases of sensory hearing loss. A cochlear implant may also be an option (see *A&P in the Real World: Cochlear Implants*).
- *Neural hearing loss* occurs when the action potentials fail to propagate through the cochlear branch of the vestibulocochlear nerve or CNS pathways. Typical causes include strokes or tumors such as an acoustic neuroma, a benign tumor outside the brainstem at the junction of the medulla and pons. Cochlear implants are not effective for neural hearing loss.

Cochlear Implants

A cochlear implant involves surgically implanting electrodes that bypass damaged hair cells and directly stimulate the cochlear nerve. Although implants do not fully restore hearing, they enable a person to perceive enough sound to understand speech or hear warnings such as smoke alarms.

An implant consists of several components. First, an external microphone, usually worn just behind the ear, sends sound waves to a processor, which converts the waves to electronic signals. Then, an external transmitter sends the signals to an internal receiver implanted beneath the skin. Finally, the receiver relays the signals to a long, thin bundle of electrodes that are threaded through the cochlea, where they stimulate the nerve endings of the vestibulocochlear nerve.

The success of a cochlear implant varies. For children who are born deaf or suffer hearing loss shortly after birth, implants are usually inserted before age 3, during the period of maximal language and speech acquisition. These children often learn to speak as expected. However, if a deaf child receives an implant after age 9, it is unlikely that regular speech will develop. Children and adults who lose their hearing after acquiring speech and language often do well with cochlear implants and no longer have to solely use sign language or lip reading. Regardless of age, extensive training is required to use the implant successfully.

The Big Picture of Hearing

Figure 15.32

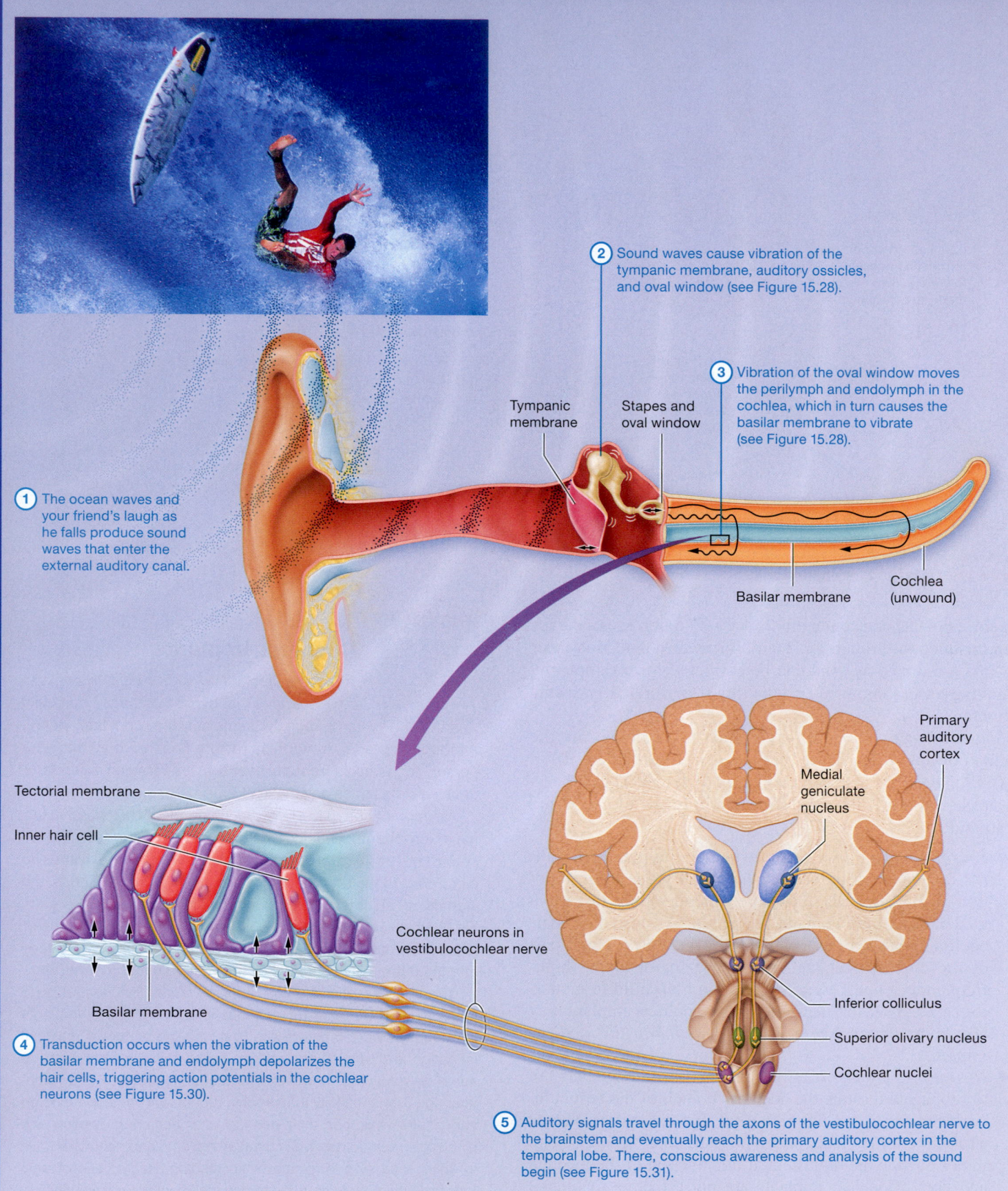

2 Sound waves cause vibration of the tympanic membrane, auditory ossicles, and oval window (see Figure 15.28).

3 Vibration of the oval window moves the perilymph and endolymph in the cochlea, which in turn causes the basilar membrane to vibrate (see Figure 15.28).

Tympanic membrane

Stapes and oval window

1 The ocean waves and your friend's laugh as he falls produce sound waves that enter the external auditory canal.

Basilar membrane

Cochlea (unwound)

Tectorial membrane

Inner hair cell

Basilar membrane

Cochlear neurons in vestibulocochlear nerve

Primary auditory cortex

Medial geniculate nucleus

Inferior colliculus

Superior olivary nucleus

Cochlear nuclei

4 Transduction occurs when the vibration of the basilar membrane and endolymph depolarizes the hair cells, triggering action potentials in the cochlear neurons (see Figure 15.30).

5 Auditory signals travel through the axons of the vestibulocochlear nerve to the brainstem and eventually reach the primary auditory cortex in the temporal lobe. There, conscious awareness and analysis of the sound begin (see Figure 15.31).

Putting It All Together: The Big Picture of Hearing

Let's now look at the overall, "big picture" process of hearing, shown in **Figure 15.32**. We return to our earlier example: You are at the beach and your friend is surfing. Your ears detect many auditory stimuli as you listen to ocean waves on the sand, people talking, and your friend laughing as he falls off his surfboard. All these sounds travel through the air as waves that enter your auditory canals, after which they are transduced into action potentials that travel to the CNS for interpretation.

> **Apply What You Learned**
>
> ☐ 1. A sound produces substantial vibration in the basilar membrane at the base of the cochlea. Describe the sound in terms of pitch and loudness.
>
> ☐ 2. Eva has been diagnosed with sensorineural deafness. She wonders if she might be helped by a bone-anchored hearing aid, a device that is surgically implanted in the temporal bone so that the bone may be used to amplify sound. Will this device be beneficial for her type of deafness? Why or why not?

See answers in Appendix A.

MODULE

15.8
Vestibular Sensation

> **Learning Outcomes**
>
> 1. Distinguish between static and dynamic equilibrium.
> 2. Describe the structure of the maculae, and explain their function in static equilibrium.
> 3. Describe the structure of the crista ampullaris, and explain its function in dynamic equilibrium.

How many times have you suddenly tripped and flailed your arms wildly to keep from falling? Each time it happened, you probably didn't consider the rapid responses of your inner ear, nervous system, and skeletal muscles that kept you on your feet. This is a classic example of how we think about our sense of equilibrium—that is, we don't think about it much at all.

The sense of equilibrium depends on input from three sources: (1) vision; (2) the proprioceptors in muscles and joints that provide feedback about body position; and (3) the **vestibular system** in the inner ear (vestibule and semicircular canals), which provides information about head position and movement, known as **vestibular sensation.** The inner ear contains organs that detect two types of equilibrium: static and dynamic equilibrium. The meaning of each term can be gleaned from its name: *static* means stationary or unmoving, and *dynamic* means changing or moving. **Static equilibrium** therefore refers to maintaining balance when the head and body are not moving but the head is tilted. **Dynamic equilibrium,** however, refers to maintaining balance when the head or body is moving. This module examines how the structures of the inner ear detect head position and movement, as well as maintain static and dynamic equilibrium.

Utricle and Saccule: Static Equilibrium and Linear Acceleration

« FLASHBACK

> 1. What is proprioception? (p. 458)

Static equilibrium is monitored by the utricle and the saccule, the portions of the membranous labyrinth located in the vestibule. Each chamber has a structure known as a **macula** (plural, *maculae*; **Figure 15.33a**) in the epithelium lining its wall, which contains the receptor cells for head position and movement. The receptor cells are hair cells similar to those in the cochlea, except that each cell contains stereocilia and one true cilium, called a **kinocilium** (ky´-noh-SIL-ee-um), which is taller than the stereocilia (**Figure 15.33b**).

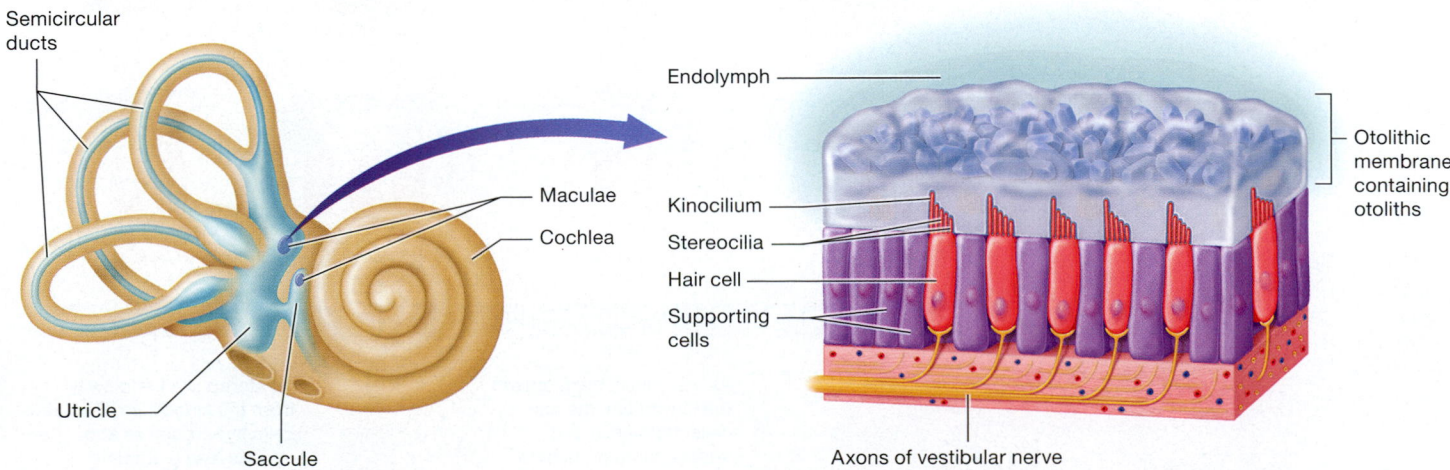

(a) Location of the utricle and saccule

Semicircular ducts
Maculae
Cochlea
Utricle
Saccule

(b) Relationship of hair cells to the otolithic membrane in a macula

Endolymph
Kinocilium
Stereocilia
Hair cell
Supporting cells
Otolithic membrane containing otoliths
Axons of vestibular nerve

Figure 15.33 Maculae of the utricle and saccule.

The stereocilia and kinocilium are embedded in a gelatinous mass referred to as the **otolithic membrane** (oh-toh-LITH-ik), which is suspended in the endolymph of the chamber. The otolithic membrane gets its name because it contains **otoliths** ("ear stones"), which are crystals of calcium carbonate that increase its density. The hair cells release the neurotransmitter glutamate continuously onto dendrites of the bipolar neurons that form the vestibular portion of the vestibulocochlear nerve. As you may guess, bending the stereocilia alters the release of glutamate, which changes the activity of the vestibular neurons.

The hair cells respond differently in the utricle and saccule because they are oriented differently. In the utricle, the stereocilia are oriented vertically. When your head is upright and stationary, the stereocilia remain vertical, but they bend when the head tilts to either side or to the front or back. However, in the saccule, the stereocilia are oriented horizontally. When your head is upright and stationary, they remain horizontal, but they bend in response to up-and-down movements (such as when you sit up straight from a slouched position).

Figure 15.34 shows how moving the stereocilia alters their function. When the head tilts, gravity pulls on the otolithic membrane, which bends the stereocilia. If the stereocilia bend toward the kinocilium, the hair cells have a local potential and depolarize. They then release more glutamate, increasing the number of action potentials produced in the axons of the vestibular nerve. If the stereocilia bend away from the kinocilium, the hair cells hyperpolarize and they release less glutamate, reducing the number of action potentials in the axons of the vestibular nerve.

The maculae are paired, so that tilting your head in one direction depolarizes hair cells in one ear and hyperpolarizes hair cells in the other ear. This increases action potentials in one vestibular nerve and decreases action potentials in the other. The brain interprets the combination of signals as the tilting of your head in a particular direction.

The utricle and saccule are also important in detecting linear acceleration, a form of dynamic equilibrium, such as riding in a car or elevator. As movement begins, the hair cells move with the rest of the body, but the endolymph and otolithic membrane do not move immediately because of a physical property known as *inertia*. Essentially, inertia is matter's resistance to change unless acted upon by a force—matter at rest stays at rest, and matter in motion remains in motion in a straight line. Due to inertia, the endolymph and otolithic membrane "lag behind" the hair cells. This lag bends the stereocilia, which either depolarizes or hyperpolarizes the hair cells and changes the activity in the vestibular nerve. When the movement stops, the hair cells stop with the body, but the endolymph and the otolithic membrane continue moving. This bends the stereocilia in the opposite direction, and the hair cells respond by altering glutamate release.

Take note that the maculae are sensitive only to changes in the *rate* of movement, not movement itself. If acceleration stops, the endolymph and otolithic membrane eventually "catch up" to the hair cells and the stereocilia no longer bend. Activity in the vestibular nerve returns to its baseline level, which your brain perceives as no movement. You can easily experience this the next time you are a passenger in a car. When you close your eyes, you can sense the acceleration of the car as the driver pulls away from a stop sign. However, once the driver stops accelerating and maintains a steady speed, it can be difficult to tell if the car is moving unless you open your eyes.

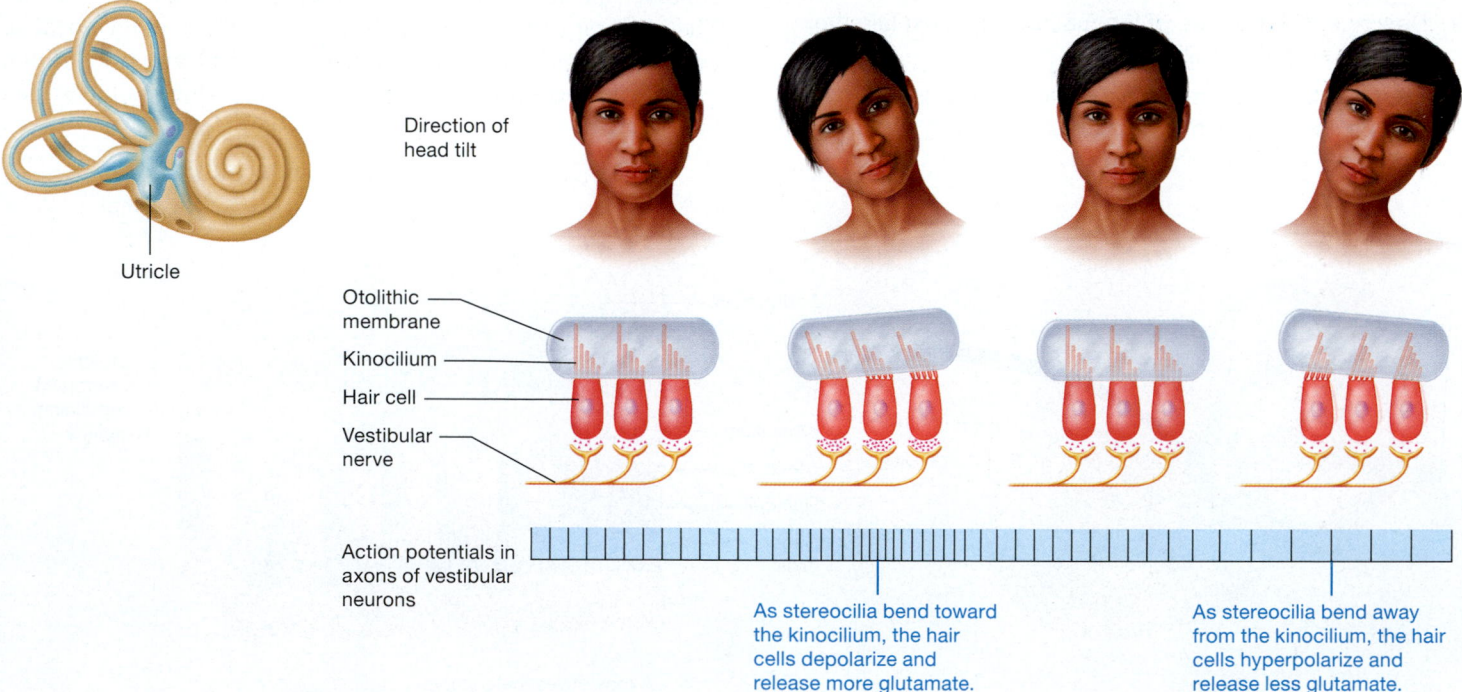

Direction of head tilt

Utricle

Otolithic membrane

Kinocilium

Hair cell

Vestibular nerve

Action potentials in axons of vestibular neurons

As stereocilia bend toward the kinocilium, the hair cells depolarize and release more glutamate.

As stereocilia bend away from the kinocilium, the hair cells hyperpolarize and release less glutamate.

Figure 15.34 Head tilting and activity of the maculae.

To see what can happen when your CNS receives contradictory signals about movement, see *A&P in the Real World: Motion Sickness.*

hair cells in the direction opposite to that when the car accelerated, and the activity of the vestibular neurons changes accordingly. In other words, if nerve activity increased when you first started moving, it will now decrease. The brain interprets this as slowing down, or decelerating. ■

ConceptBOOST)))

How Inertia Influences Movement of the Otolithic Membrane and Endolymph

If you are having difficulty grasping how inertia influences the movement of the otolithic membrane and endolymph, here's an analogy that may help. Imagine you are stopped at a red light while riding in the front passenger seat of a friend's car. You have a cup of coffee (without a lid) in your hand. The light turns green and your friend stomps on the gas pedal. Why are you now wearing your coffee on the front of your shirt? The answer is inertia. The car, your body, and the cup all move together as one unit. However, the coffee (a fluid) moves independently. It also has inertia, so it does not move immediately with the cup. Instead, it "lags behind" the cup, as shown in the photo, and ends up all over your shirt.

The coffee is equivalent to the endolymph and otolithic membrane in the vestibule. Its movement lags behind that of the hair cells in the wall of the utricle and saccule (the "cup"), which bends the stereocilia. Depending on the direction in which they bend, neuron activity either increases or decreases, which the brain interprets as the body speeding up, or accelerating.

Now, let's turn back to the coffee cup. Imagine that after driving at a constant speed, your friend slams on the brakes to avoid hitting another car. Where is your coffee now? It's on the floor of the car, which is also explained by inertia. While traveling at a constant speed, the coffee "catches up" and moves with the cup. However, when the car, your body, and the cup suddenly stop (decelerate), inertia causes the coffee to keep moving forward, out of the cup.

In the same way, when your hair cells and the rest of your body come to a stop, your endolymph and otolithic membrane continue to move forward. This bends the

Quick Check

☐ 1. How do static and dynamic equilibrium differ?

☐ 2. What role do the utricle and saccule play in static equilibrium?

☐ 3. What role do they play in dynamic equilibrium?

Semicircular Ducts: Rotational Equilibrium

Another type of dynamic equilibrium is *rotational equilibrium,* which allows you to maintain your balance during rotational movements such as turning your head or spinning in a chair. This type of movement is detected by hair cells located in the anterior, posterior, and lateral semicircular ducts of the membranous labyrinth. The three ducts, which contain endolymph, are oriented to detect rotation in all three planes.

Motion Sickness

Motion sickness refers to symptoms such as nausea, dizziness, and sometimes vomiting, which can occur during particular types of movement, such as riding in a car, boat, or roller coaster. Motion sickness occurs when the brain receives mismatched sensory information from the eyes and the vestibular system. For instance, when you are on a boat and it rocks on the water, your vestibular system reports that you are moving. However, your visual system indicates that you are not moving relative to the boat. Similar effects occur when motion is seen but not felt, as occurs in panoramic movie theaters that simulate movement for the viewer. In such cases, the brain cannot integrate the contradictory information. Through pathways not well understood, this causes the familiar symptoms of motion sickness.

Certain antihistamines or a scopolamine patch can help deal with motion sickness by relieving nausea. It may also help to sit in the front seat of a car or look at the horizon while on a boat, both of which reduce the conflicting information about movement reported by the visual system.

The semicircular ducts are connected to the utricle; near this connection, each duct has a swelling, the **ampulla** (plural, *ampullae;* **Figure 15.35a**). Within each ampulla we find a **crista ampullaris** (KRIS-tuh am-pyoo-LEHR-iss), a cluster of hair cells and supporting cells. The hair cells have stereocilia and one kinocilium that are embedded in a gelatinous mass called the **cupula** (KUH-pyoo-luh).

The structure of the crista ampullaris is similar to that of the maculae, so it's not surprising that it functions very much like the maculae—another example of the Structure-Function Core Principle (p. 28). For example, when you turn your head, the endolymph lags behind and pushes on the cupula, as shown in **Figure 15.35b**. This bends the stereocilia, which either increases or decreases glutamate release from the hair cells and causes a corresponding change in the activity of neurons in the vestibular nerve. Your brain interprets the change in neuron activity as the

CORE PRINCIPLE
Structure-Function

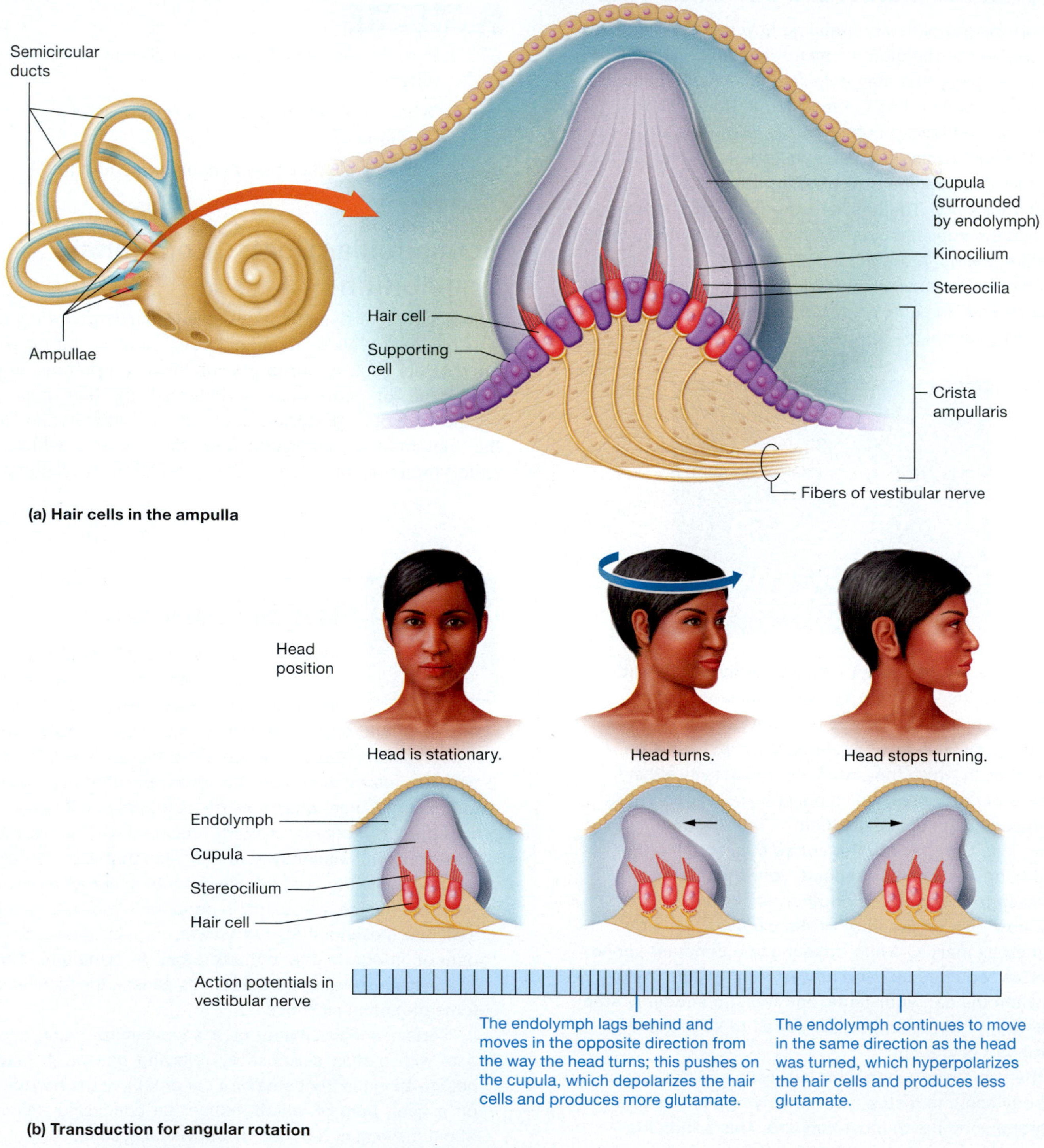

(a) Hair cells in the ampulla

Semicircular ducts

Ampullae

Cupula (surrounded by endolymph)

Kinocilium

Stereocilia

Hair cell

Supporting cell

Crista ampullaris

Fibers of vestibular nerve

Head position

Head is stationary.

Head turns.

Head stops turning.

Endolymph

Cupula

Stereocilium

Hair cell

Action potentials in vestibular nerve

The endolymph lags behind and moves in the opposite direction from the way the head turns; this pushes on the cupula, which depolarizes the hair cells and produces more glutamate.

The endolymph continues to move in the same direction as the head was turned, which hyperpolarizes the hair cells and produces less glutamate.

(b) Transduction for angular rotation

Figure 15.35 Angular movement and the semicircular canals.

start of head rotation (acceleration). The endolymph "catches up" quickly during continued rotation. When you stop turning your head, the endolymph continues to move and bends the cupula in the direction opposite to that when you started turning your head. This has the opposite effect on glutamate release and the activity of the neurons of the vestibular nerve, and your brain interprets the change to mean that head rotation has stopped.

Quick Check

☐ 4. How do the hair cells in the crista ampullaris detect rotation of the head?

The Vestibular Sensation Pathway

Detecting head movement and position is only part of the battle to maintain equilibrium; the brain must receive and analyze the stimuli and then coordinate a response. The responses can be grouped into three categories: (1) muscle movement to produce changes in posture (for example, flailing your arms to avoid falling when you trip), (2) cognitive awareness of head position and/or movement, and (3) compensatory eye motions for some types of head movement.

To produce these responses, the vestibular stimuli need to reach various regions of the central nervous system, which happens by the pathway described next and shown in **Figure 15.36**.

① **Action potentials propagate to the vestibular nuclei located at the medulla-pons junction.** The action potentials generated by the vestibular neurons are propagated along their axons, and they synapse on the vestibular nuclei in the brainstem.

② **The vestibular nuclei relay the signals simultaneously to the following areas of the CNS:**

②ₐ **The thalamus and then the inferior parietal lobe, for conscious awareness of head position and movement.** The inferior parietal lobe integrates the stimuli so you are aware of the direction in which your head is moving (for example, knowing that your head is turning to the right and not the left).

②b **The cranial nerve nuclei, to coordinate eye movement in response to head movement.** Axons from the vestibular nuclei terminate in the oculomotor, trochlear, and abducens nuclei, which give rise to the cranial nerves that control eye movement. One example of coordinated eye movement is the *vestibulo-ocular reflex,* which enables us to keep our eyes fixed on an object while our head is moving.

②c **The cerebellum and spinal cord, to coordinate muscle movement that maintains balance in response to head movement.** For example, tilting your head back to look at the sky requires contracting muscles in your torso and lower limbs so you don't fall over backward.

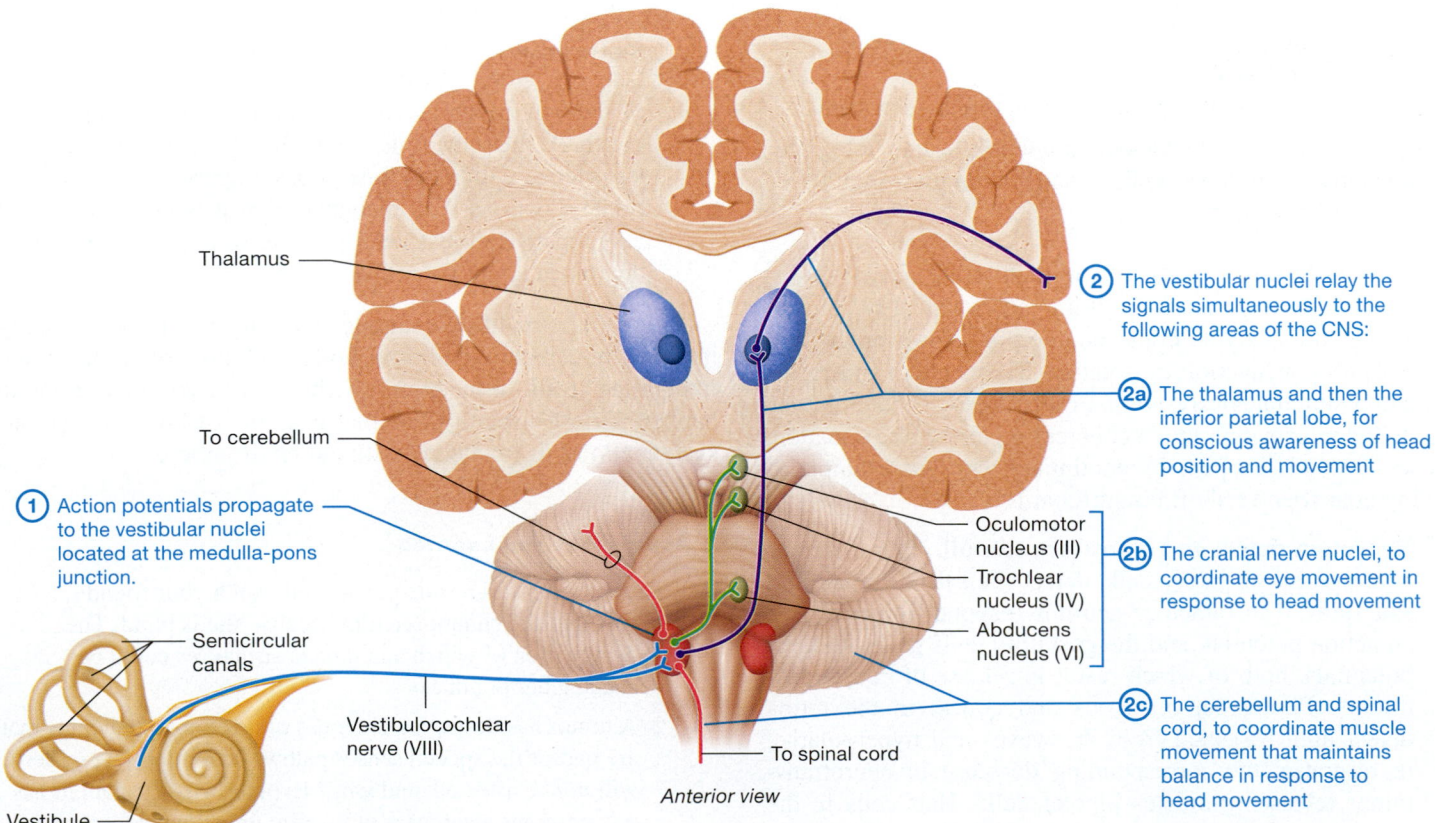

Figure 15.36 The vestibular sensation pathway.

☐ 5. What are the three primary areas of the central nervous system that receive input from the vestibular nuclei?

Apply What You Learned

☐ 1. You feel dizzy and lose your balance whenever you tilt your head back to look at the sky. In which part of the inner ear is dysfunction likely causing your symptoms?

☐ 2. You fix your gaze on your book as you turn your head to the right. In which direction does the vestibulo-ocular reflex cause your eyes to move? Which cranial nerve nuclei receive input from the vestibular nuclei to produce the vestibulo-ocular reflex?

☐ 3. Your grandmother suffers from dizziness and loses her balance easily. She asks you why her doctor recommended that she visit an ear, nose, and throat specialist. What would you tell her?

See answers in Appendix A.

MODULE **15.9**

How the Special Senses Work Together

Learning Outcomes

1. Summarize the pathways for each of the special senses.
2. Describe how the frontal lobe and limbic system integrate the signals from the special senses into a meaningful picture of a situation.

In this chapter, you have learned how each of the special senses detects a stimulus and transduces it into an electrical signal—an action potential—that your brain can interpret. However, the special senses do not function in isolation. They collect and report information simultaneously, so that other areas of the brain can integrate the information into a coherent picture of the situation. Let's see how this happens by continuing with our example of watching your friend fall off his surfboard (**Figure 15.37**):

① **Receptors detect and transduce stimuli.** Your olfactory neurons and gustatory cells detect the salt in the air and salt spray. The olfactory neurons respond by generating an action potential, and the gustatory cells generate local potentials, both of which result in release of neurotransmitters. Photoreceptors (rods and cones) in the retina detect light reflected from the waves and hyperpolarize in response; the corresponding decrease in neurotransmitter release stimulates bipolar cells. Hair cells in the cochlea respond to the energy of sound waves and release neurotransmitters. Hair cells in the vestibule detect the position and movement of your head and also release neurotransmitters.

② **Neurons of cranial nerves transmit action potentials to the CNS.** Olfactory stimuli are transmitted to the olfactory bulb, which leads to the olfactory tract. The brainstem receives taste, auditory, and vestibular stimuli from the facial, glossopharyngeal, and vagus nerves (taste) and the vestibulocochlear nerve. Visual stimuli are transmitted to the thalamus by the optic nerve.

③ **Neurons synapse in the thalamus (except those for olfaction).** With the exception of olfaction, all the special senses pathways synapse in the thalamus for processing and analysis at a subconscious level. Neurons in the thalamus then project to the primary sensory cortex for each sensation. Olfaction travels through the olfactory tract directly to the primary olfactory cortex.

④ **Awareness occurs in the primary sensory cortices.** The primary olfactory cortex and the primary gustatory cortex allow you to notice the aroma of salty air and the taste of ocean spray on your lips. The primary visual cortex enables you to see the large wave behind your friend and his movement as he falls off his surfboard. The primary auditory cortex detects and analyzes the sounds of the waves and your friend's laughter. Several regions of the parietal lobe provide awareness that you are sitting upright on your beach towel.

⑤ **The frontal lobe and limbic system integrate the special senses.** Each primary cortical area has significant connections with the frontal lobe and limbic system that allow you to convert all these pieces of information into a meaningful picture. For example, the frontal lobe integrates the taste and smell of the salty air with the sight and sound of the waves, which enables you to know you are at the beach. At the same time, the limbic system integrates your knowledge with memories of past experiences and appropriate emotions. You recognize that it is your friend, not a stranger, who just went into the water, and you realize that his laughter means he is safe.

From this simplified example, you can see the complexity involved in detecting and processing each type of stimulus and integrating it into a meaningful whole. And yet, this seems to happen almost instantly—another example of how our nervous system allows us to interact with our environment.

Apply What You Learned

☐ 1. Your sister sits beside you as you watch your friend surf, but she cannot see him because she is blind. The dysfunction of which anatomical structures could be causing her blindness?

☐ 2. A tumor has destroyed the areas of a patient's thalamus that are part of the special senses pathways. Which special sense will *not* be affected, and why? Explain why the patient has no conscious awareness of the remaining special senses.

See answers in Appendix A.

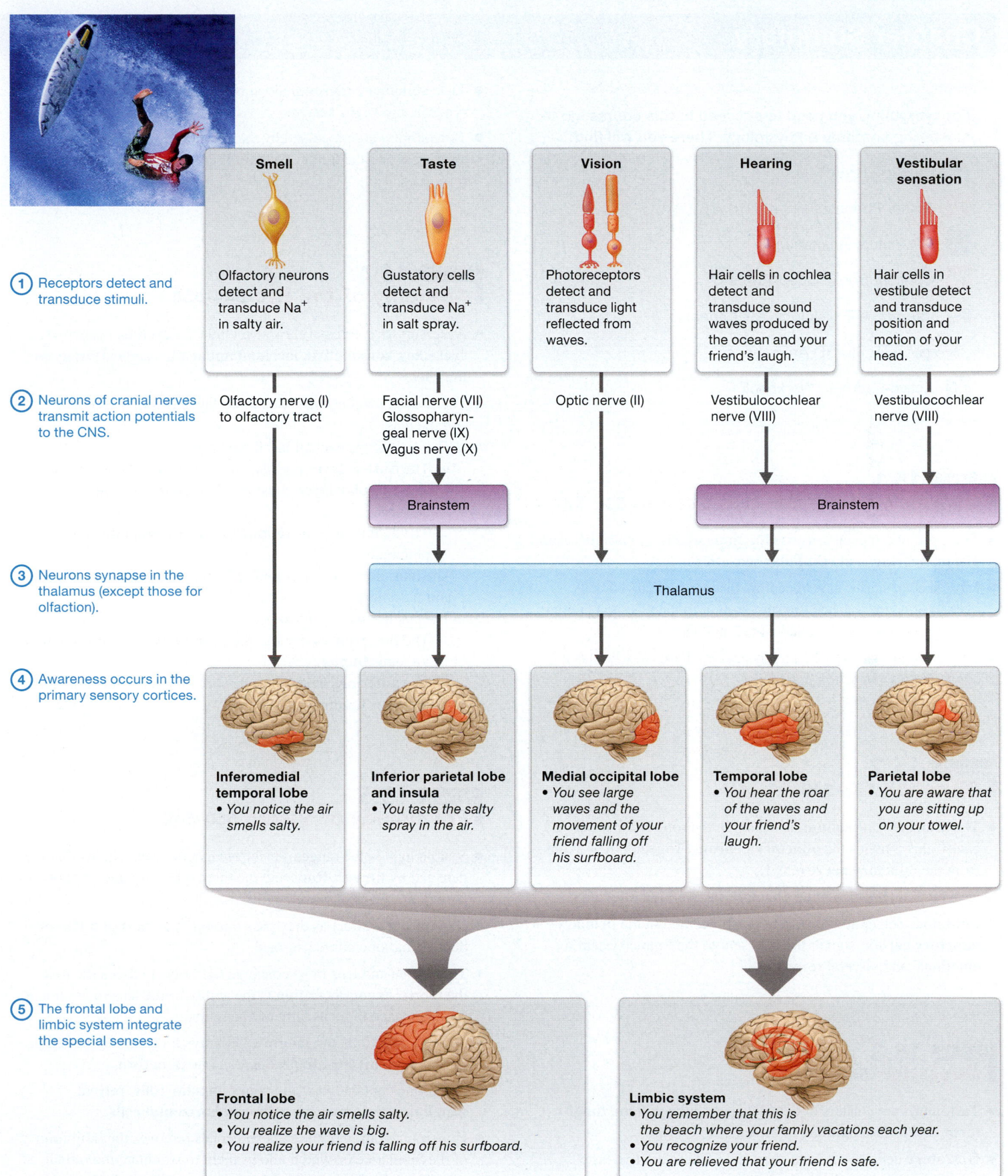

① Receptors detect and transduce stimuli.

② Neurons of cranial nerves transmit action potentials to the CNS.

③ Neurons synapse in the thalamus (except those for olfaction).

④ Awareness occurs in the primary sensory cortices.

⑤ The frontal lobe and limbic system integrate the special senses.

Smell
Olfactory neurons detect and transduce Na$^+$ in salty air.

Taste
Gustatory cells detect and transduce Na$^+$ in salt spray.

Vision
Photoreceptors detect and transduce light reflected from waves.

Hearing
Hair cells in cochlea detect and transduce sound waves produced by the ocean and your friend's laugh.

Vestibular sensation
Hair cells in vestibule detect and transduce position and motion of your head.

Olfactory nerve (I) to olfactory tract

Facial nerve (VII)
Glossopharyn-geal nerve (IX)
Vagus nerve (X)

Optic nerve (II)

Vestibulocochlear nerve (VIII)

Vestibulocochlear nerve (VIII)

Brainstem

Brainstem

Thalamus

Inferomedial temporal lobe
• *You notice the air smells salty.*

Inferior parietal lobe and insula
• *You taste the salty spray in the air.*

Medial occipital lobe
• *You see large waves and the movement of your friend falling off his surfboard.*

Temporal lobe
• *You hear the roar of the waves and your friend's laugh.*

Parietal lobe
• *You are aware that you are sitting up on your towel.*

Frontal lobe
• *You notice the air smells salty.*
• *You realize the wave is big.*
• *You realize your friend is falling off his surfboard.*

Limbic system
• *You remember that this is the beach where your family vacations each year.*
• *You recognize your friend.*
• *You are relieved that your friend is safe.*

Figure 15.37 How the special senses work together.

Chapter Summary

For everything you need to succeed in this course, go to **Mastering** Anatomy & Physiology. There you will find:

- Practice Tests
- Author Podcasts
- Big Picture Animations
- Concept Boost Video Tutors
- **iP2™** Interactive Physiology 2.0 Tutorials
- **A&PFlix** A&P Flix 3D Animations
- Active-Learning Workbook

MODULE 15.1
Overview of the Special Senses 536–537

- There are five special senses: **olfaction** (smell), **gustation** (taste), **vision, hearing,** and **vestibular sensation.**
- Most special senses have receptors that are specialized cells instead of neurons. They transmit stimuli to the CNS via cranial nerves only instead of cranial and spinal nerves.
- **Transduction** occurs when a physical or chemical stimulus is converted into an action potential.

MODULE 15.2
Olfaction 537–540

- The **olfactory epithelium** is located in the roof of the nasal cavity and contains the **olfactory neurons.** Together their axons form the **olfactory nerve** (CN I).
- Receptors on the cilia of an olfactory neuron bind **odorants.**
- Conscious perception of olfaction is processed in the **primary olfactory cortex;** signals are also sent to the limbic system for emotional and visceral responses.

MODULE 15.3
Gustation 540–544

- **Taste buds** are clusters of cells located on the **vallate, fungiform,** and **foliate papillae** of the tongue.
- **Gustatory cells** detect taste chemicals. There are five taste sensations: **sweet, salty, sour, bitter,** and **umami** (savory). Each gustatory cell responds to only one taste.

- Taste stimuli are transmitted via neurons of the facial, glossopharyngeal, and vagus nerves.
- Taste stimuli are processed by the parietal lobe for conscious awareness and the limbic system for emotional and visceral responses.

MODULE 15.4
Anatomy of the Eye 544–550

- Accessory structures of the eye include the **eyelids, eyebrows, eyelashes, conjunctiva, lacrimal apparatus,** and **extrinsic eye muscles.**
- The eyeball itself consists of a three-layered wall, two cavities, and the lens.
  - The wall of the eyeball has three layers:
    - The **fibrous layer** consists of the **sclera** and the **cornea.**
    - The **vascular layer** consists of the **choroid, ciliary body,** and **iris.**
    - The neural layer, or **retina,** transduces light into action potentials.
  - The **lens** alters the refraction of light so we can focus on near and far objects.
  - The eye has two cavities:
    - The **posterior cavity** lies posterior to the lens and contains **vitreous humor.**
    - The **anterior cavity** lies anterior to the lens and contains **aqueous humor.**

MODULE 15.5
Physiology of Vision 550–562

- Visible light is the range of electromagnetic wavelengths that are detected by the eye. **Rods** and the three types of **cones** respond to particular ranges of wavelengths.
- Rays of light refract as they pass through materials of different density and/or a curved surface.
- The cornea and lens must converge light rays to focus them on the retina. **Accommodation** is the ability of the lens to assume a rounder shape to focus light on the retina.
- Several conditions can interfere with rays focusing on the retina: **presbyopia, myopia, hyperopia,** and **astigmatism.**
- The retina contains rods and cones, **bipolar cells, retinal ganglion cells, amacrine cells,** and **horizontal cells.**
- Hyperpolarization of the rods and cones removes the inhibition of the bipolar cells, which allows them to stimulate the retinal ganglion cells. The axons of the retinal ganglion cells form the optic nerve.

- The optic nerves meet at the **optic chiasma,** where axons from the nasal retinas cross to the opposite side. All axons from the opposite visual field then travel in the **optic tract** to the **lateral geniculate nucleus.**

- Conscious interpretation of vision occurs in the primary visual cortex, the temporal lobe, and parts of the parietal lobe.

MODULE 15.6
Anatomy of the Ear 562–566

- The ear is divided into three regions: the outer, middle, and inner ear.

- The **outer ear** consists of the **auricle,** which funnels sound into the **external auditory canal.**

- The **middle ear** is a cavity in the temporal bone. The **tympanic membrane** is on its lateral wall. The **pharyngotympanic tube** connects the middle ear and superior pharynx. The three **auditory ossicles** (the **malleus, incus,** and **stapes**) connect the tympanic membrane to the inner ear.

- The **inner ear** is a series of tunnels in the temporal bone. It is lined with the **membranous labyrinth,** which contains **endolymph;** the **perilymph** separates the membranous labyrinth from the walls of the inner ear. The inner ear is divided into three regions:
  - The **vestibule** contains the **utricle** and the **saccule,** which detect head position and linear acceleration.
  - The **semicircular ducts** are in the **semicircular canals** and detect rotational movement of the head.
  - The **cochlea** contains structures for hearing, including the **cochlear duct,** the **scala vestibuli,** and the **scala tympani.**

MODULE 15.7
Physiology of Hearing 566–573

- Sound waves are alternating areas of compressed and less dense air molecules that are generated by the vibration of an object.

- To hear a sound, the sound waves must reach the inner ear and be transduced into action potentials that are sent through the neurons of the vestibulocochlear nerve to the CNS.
  - Sound waves move through the auditory canal and trigger vibration of the tympanic membrane, auditory ossicles, and **oval window.**
  - Vibrations in the inner ear move the **basilar membrane** on which the **spiral organ** sits. The **hair cells** move with the basilar membrane and release neurotransmitters onto neurons of the cochlear nerve, triggering action potentials that travel to the cochlear nuclei in the brainstem.
  - Auditory stimuli are processed in the **primary auditory cortex** in the temporal lobe for conscious perception. From there, the signals travel to other parts of the temporal and parietal lobes, as well as the limbic system.

MODULE 15.8
Vestibular Sensation 573–578

- The utricle and saccule detect head tilting and linear acceleration, which displace the otolithic membrane, bending the **stereocilia** and **kinocilium.** Bending toward the kinocilium depolarizes the hair cell, and bending in the opposite direction hyperpolarizes it.

- The semicircular ducts detect rotational head movements. Rotation of the head pushes the endolymph against the cupula, bending the stereocilia and kinocilium. This depolarizes or hyperpolarizes the hair cells, depending on the direction of rotation.

- **Vestibular sensations** are processed in the spinal cord and cerebellum to coordinate muscle movement and maintain balance; in the inferior parietal lobe; and in cranial nerve nuclei for eye movement.

MODULE 15.9
How the Special Senses Work Together 578–579

- Specialized receptor cells—olfactory neurons, gustatory cells, photoreceptors, and hair cells—detect the stimuli for each special sense, which are transmitted to the CNS via cranial nerves.

- Except for olfaction, the basic pathway special sensory stimuli take is as follows: brainstem → thalamus → primary sensory cortex. Olfactory stimuli pass from the olfactory tracts to olfactory cortices.

- The frontal lobe integrates the information from the special senses into a coherent picture, and the limbic system provides appropriate emotional and behavioral responses.

Assess What You Learned

Scan the QR Code for additional practice test questions

https://goo.gl/WTPxTZ

LEVEL 1 Check Your Recall

1. Match the cell type with the correct stimulus.

_____ Rod or cone	a. Head movement
_____ Hair cell in cochlea	b. Odorant
_____ Gustatory cell	c. Photon
_____ Olfactory neuron	d. Taste substance
_____ Hair cell in vestibule	e. Sound wave

2. The axons of the olfactory nerve terminate in the:

a. olfactory epithelium.
b. olfactory bulb.
c. olfactory tract.
d. primary olfactory cortex.

3. Fill in the blanks: In an olfactory neuron, the binding of a(n) _____ to its membrane receptor triggers a(n) _____ potential in the axons of the _____ nerve.

4. The primary olfactory cortex is located in the:

   a. frontal lobe.
   b. occipital lobe.
   c. parietal lobe.
   d. temporal lobe.

5. Which of the following statements is *true* regarding gustatory cells?

   a. They have microvilli that project into the taste pore.
   b. There are only 10–20 gustatory cells in a typical taste bud.
   c. They have a lifespan of approximately 6 months.
   d. Some form synapses with neurons that give rise to the trigeminal nerve.

6. Match the taste with the chemical substance that produces it.

   _____ Sweet   a. Many alkaloids
   _____ Sour   b. Sucrose
   _____ Salty   c. Glutamate
   _____ Bitter   d. Sodium ions
   _____ Umami   e. Hydrogen ions

7. Which of the following cranial nerves is *not* involved in the gustatory sense?

   a. Vagus nerve (X)
   b. Hypoglossal nerve (XII)
   c. Facial nerve (VII)
   d. Glossopharyngeal nerve (IX)

8. Tears normally flow from the lacrimal sac into the:

   a. lacrimal gland.
   b. lacrimal canaliculi.
   c. lacrimal puncta.
   d. nasolacrimal duct.

9. Which cells in the retina are depolarized in darkness?

   a. Rods and cones
   b. Bipolar cells
   c. Retinal ganglion cells
   d. All of the above
   e. Only rods

10. Why can you see an object better in dim light by looking to the side of it instead of directly at it?

11. Each of the following statements is false. Correct each statement to make it true.

   a. Photons are absorbed by rhodopsin in retinal ganglion cells.
   b. Rods and cones are found in the fovea centralis.
   c. In the dark, rods and cones produce action potentials.

12. The axons from the nasal retina in the left eye terminate in the:

   a. right lateral geniculate nucleus.
   b. left lateral geniculate nucleus.
   c. right medial occipital lobe.
   d. left medial occipital lobe.

13. Mark the following statements as true or false. If a statement is false, correct it to make a true statement.

   a. The incus is connected to the tympanic membrane.
   b. The stapes is attached to the oval window.
   c. The auditory canal is separated from the middle ear by the round window.
   d. The cochlear duct is filled with perilymph.
   e. The semicircular ducts are connected to the utricle, and the cochlear duct is continuous with the saccule.
   f. The spiral organ is located in the scala tympani.

14. Explain how sounds of different frequencies are detected in the cochlea.

15. How do the hair cells of the crista ampullaris detect rotation of the head?

16. The macula in the utricle detects:

   a. rotation of the head to the right.
   b. very low-frequency sound waves that we can feel but not hear.
   c. tilting of the head to one side.
   d. linear acceleration of the head in a vertical plane.

17. Fill in the blanks: The spiral organ sits on the _____ membrane, and the stereocilia of the outer hair cells contact the _____ membrane.

18. True or false: Hair cells in the spiral organ will depolarize or hyperpolarize, depending on the direction in which the stereocilia are bent.

19. Which of the following is *not* part of the auditory pathway in the CNS?

   a. Superior temporal lobe
   b. Superior colliculus
   c. Medial geniculate nucleus
   d. Nuclei in the pons
   e. Cochlear nuclei

20. Fill in the blanks: In the ampulla of a semicircular duct, the hair cells are located in the _____ _____ and their stereocilia and kinocilium are embedded in the _____.

21. Stimuli from the inner ear regarding head movement and position are sent to all of the following except the:

   a. vestibular nuclei.
   b. oculomotor nucleus.
   c. parietal lobe.
   d. cerebellum.
   e. trigeminal nucleus.

LEVEL 2 Check Your Understanding

1. Explain what would happen to your sense of smell if there were no basal cells present in the olfactory epithelium.

2. In which direction would you be unable to move your right eye if your right abducens nerve were damaged?

3. If a patient suffers visual impairment only in one eye, why must the damage be located in the visual pathway prior to the optic chiasma?

4. Following a stroke, a patient lost vision in the left visual field. Name several structures in the brain that could have been damaged by the stroke to have caused this vision loss. Be specific as to right or left side.

5. Why do you have the sensation that you are still spinning immediately after stopping?

LEVEL 3 Apply Your Knowledge

PART A: Application and Analysis

1. Mr. Spencer suffers loss of taste sensation. Tests reveal that his taste buds and salivary glands are functional. What other structures might be impaired that would explain his loss of taste?

2. Your friend tells you that she has just been diagnosed with myopia, and she wonders if staring at her computer screen too much might have caused her problem. What do you tell her?

3. If severe congestion from a cold prevented your pharyngotympanic tube from opening, what could happen to your tympanic membrane? Explain your answer.

4. Suppose the round window in your middle ear became very stiff. What effect would this have on hearing, and why?

PART B: Make the Connection

5. Mrs. Flores is a 45-year-old female who suffered a fracture of the ethmoid bone as the result of a car accident. Several days later, she complained that she couldn't taste her food very well. How is her injury related to the loss of taste sensation? *(Connects to Chapter 7)*

6. Your 60-year-old patient, Mr. Guster, has Bell's palsy, which causes dysfunction of those portions of the facial nerve that control muscles of facial expression. He is surprised when you tell him to tape the eyelids of his right eye closed at night to prevent dryness. Why might Bell's palsy cause dryness of the eye on the affected side? *(Connects to Chapter 9)*

See answers in Appendix A.

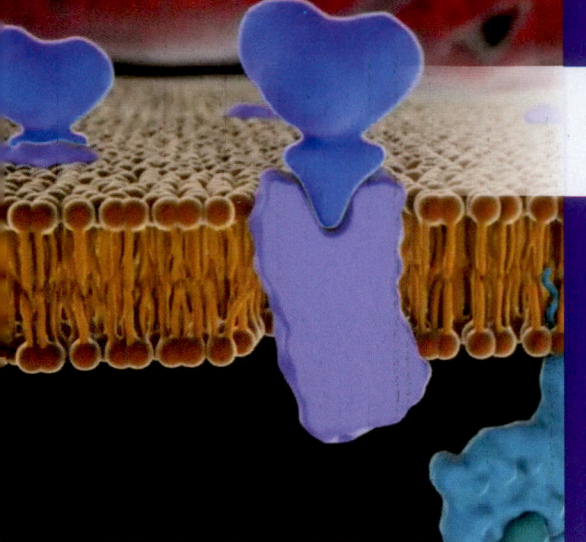

16 The Endocrine System

For practice applying concepts to a clinical scenario, check out the **Running Case Study** for this chapter @ Mastering Anatomy & Physiology

Computer-generated image: A hormone molecule binds to its receptor in the plasma membrane of its target cell.

Throughout this book, we have seen that a cell's activities can be altered in the face of changing conditions to maintain homeostasis. For example, when the concentration of calcium ions in the blood decreases, osteoclasts become more active and resorb more bone to release calcium ions. But how does a cell "know" how and when to alter its activities? The answer lies in the body's two major *regulatory systems,* meaning those that regulate the functions of other cells. These two systems are the nervous system, which we explored in previous chapters, and the endocrine system, which we turn to in this chapter. Organs of the **endocrine system** work by synthesizing and secreting chemical messengers called **hormones** into the blood. Hormones, in turn, interact with specific cells known as *target cells.* Tissues in which a hormone's target cells reside are referred to as *target tissues.* Target cells have specific *receptors* to which the hormones can bind, leading to changes in the cells' functions. Many of these functions revolve around maintaining fluid, electrolyte, and acid-base homeostasis; promoting growth; regulating metabolic reactions; and responding to stressors.

In this chapter, we explore the organs and cells of the endocrine system, the hormones they produce, and how these hormones maintain the homeostasis of various physiological variables. Along the way, we examine the impact of endocrine system disorders and the often dramatic effect they can have on the body's functions. The chapter concludes with three examples of endocrine control of physiological variables, providing a "big picture" look at how hormones work together and thereby maintain overall homeostasis.

MODULE 16.1
Overview of the Endocrine System

Learning Outcomes

1. Compare and contrast how the endocrine and nervous systems control body functions.
2. Describe the major structures and functions of the endocrine system.
3. Explain the different types of chemical signaling used by the body.
4. Describe the major chemical classes of hormones found in the human body, and compare and contrast the types of receptors to which each class of hormone binds.

5. Describe several types of stimuli that control production and secretion of hormones, including the roles of negative and positive feedback.

We start our study of the endocrine system with an overview of its basic properties. This module first examines the differences between the ways in which the nervous and endocrine systems maintain homeostasis. We continue with a look at the types of chemical signals used in the body, the anatomy of the endocrine organs, and finally the main properties of the endocrine system's chemical messengers: hormones.

Comparison of the Endocrine and Nervous Systems

‹‹ FLASHBACK

1. How does a neuron communicate with its target cell? (p. 406)

Together, the endocrine and nervous systems complement each other to maintain the homeostasis of nearly every regulated physiological variable. The cells of both systems use chemicals to communicate with other cells, an example of the Cell-Cell Communication Core Principle (p. 28). However, the two systems differ significantly in how they regulate body functions. As you discovered in previous chapters, the

CORE PRINCIPLE
Cell-Cell Communication

nervous system operates through a series of neurons that directly affect their target cells through the release of neurotransmitters (see Chapter 11). The effects of the nervous system are almost immediate but are short-lasting unless the stimulation is repetitive.

The cells of the endocrine system, however, do not come into close contact with their target cells. Instead, as you can see in **Figure 16.1**, ① hormones are first secreted into the interstitial fluid, after which they enter the blood ("endo" means "within," as in "within the blood"). ② They do this by diffusing through or between the cells of tiny blood vessels called *blood capillaries,* which are located in networks called *capillary beds.* When blood exits the capillaries, it transports the hormones to the heart through larger blood vessels called *veins.* ③ After leaving the heart, the blood transports the hormones to the rest of the body through *arteries.* Eventually the arteries branch out repeatedly into smaller vessels and finally capillary beds. ④ Once in the capillaries, the hormones will diffuse out of the blood and into the interstitial fluid to reach the receptors on their target cells.

Due to their different mechanisms of action, hormones can require seconds to several hours or days to elicit their effects, but these effects are generally longer-lasting than those of the nervous system. Note also that the endocrine and nervous systems can work together to increase each other's effects.

Quick Check

☐ 1. How do the endocrine and nervous systems differ in the ways they maintain homeostasis? How are they similar?

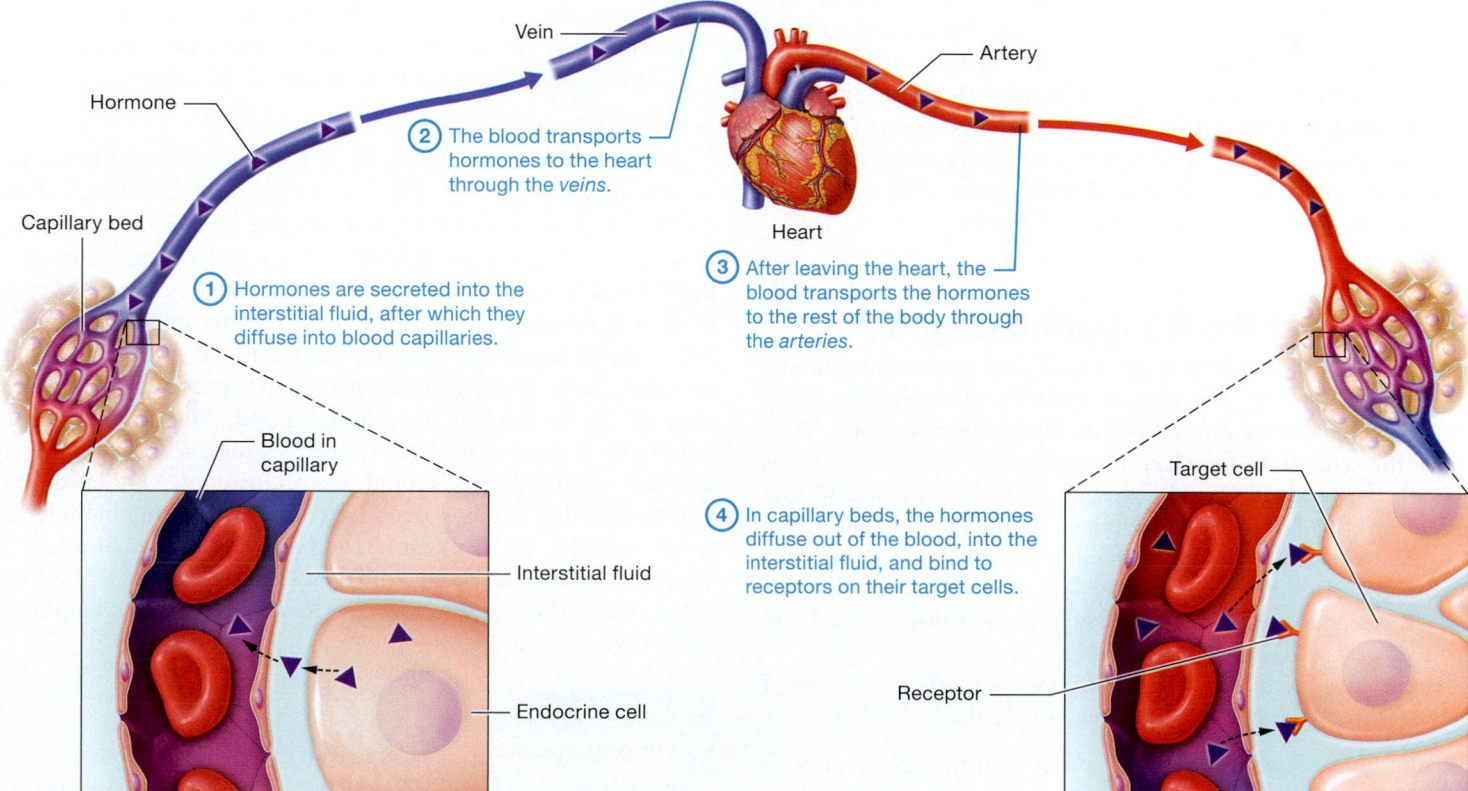

Figure 16.1 Overview of hormone secretion and distribution by the blood.

Vein

Artery

Hormone

② The blood transports hormones to the heart through the *veins.*

Capillary bed

Heart

① Hormones are secreted into the interstitial fluid, after which they diffuse into blood capillaries.

③ After leaving the heart, the blood transports the hormones to the rest of the body through the *arteries.*

Blood in capillary

Target cell

④ In capillary beds, the hormones diffuse out of the blood, into the interstitial fluid, and bind to receptors on their target cells.

Interstitial fluid

Endocrine cell

Receptor

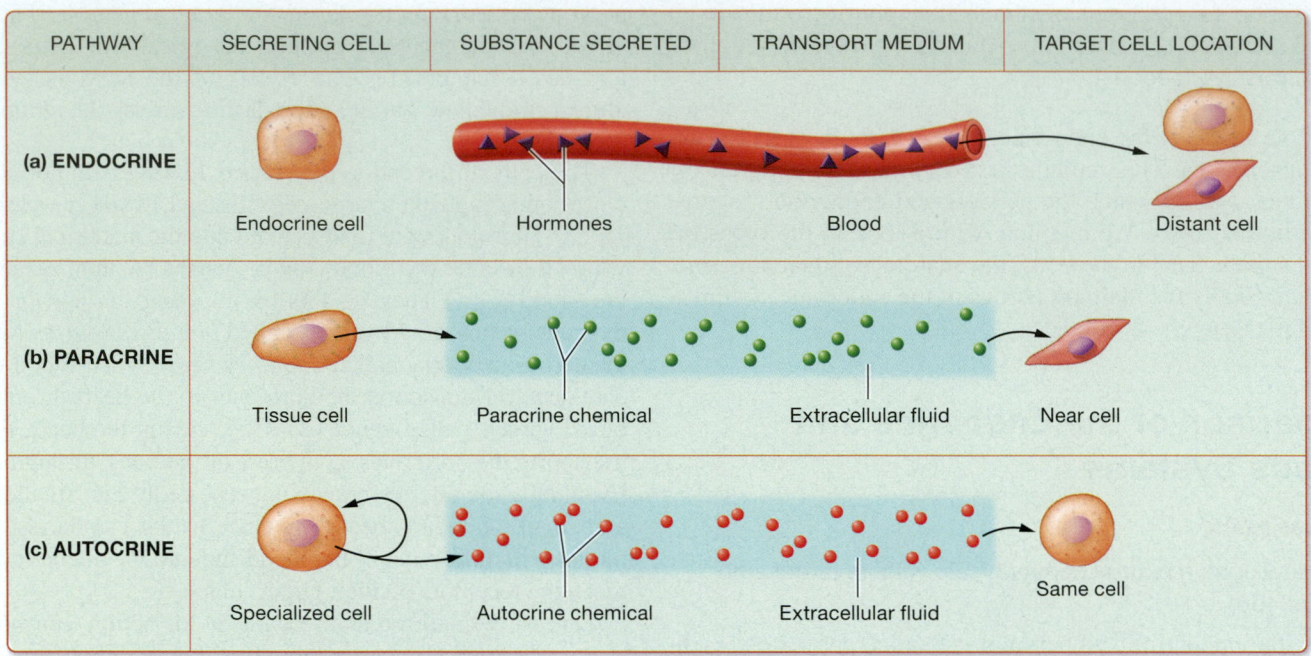

PATHWAY	SECRETING CELL	SUBSTANCE SECRETED	TRANSPORT MEDIUM	TARGET CELL LOCATION
(a) ENDOCRINE	Endocrine cell	Hormones	Blood	Distant cell
(b) PARACRINE	Tissue cell	Paracrine chemical	Extracellular fluid	Near cell
(c) AUTOCRINE	Specialized cell	Autocrine chemical	Extracellular fluid	Same cell

Figure 16.2 Three basic signaling pathways.

Paracrine and Autocrine Signals

Hormones that are secreted into the blood to affect distant targets are known as *classic endocrine signals* (**Figure 16.2a**). However, hormones can also be secreted into the interstitial fluid and affect local cells without actually entering the blood. There are two types of local signals: *paracrine* (PEHR-uh-krin) and *autocrine* (AW-toh-krin). **Paracrine** signals consist of a chemical secreted by cells into the extracellular fluid to influence nearby cells (**Figure 16.2b**). In **autocrine** signals, a chemical secreted by cells into the ECF elicits effects from the same cell (**Figure 16.2c**). Although we primarily consider classic endocrine signals in this chapter, all three types of signals are an important part of the body's regulatory system.

Overview of the Endocrine Organs

The organs of the endocrine system, called **endocrine glands,** are shown in **Figure 16.3**. Notice that they are a diverse group of organs, located in various places throughout the body. They share the common function of regulating other cell types through the production and secretion of hormones. Recall that endocrine glands consist of ductless glandular epithelial cells that secrete their hormones into the interstitial fluid for transport by the bloodstream (see Chapter 4), in contrast to *exocrine glands,* which secrete their products into ducts that lead to body surfaces or cavities (see Chapter 22).

There are generally considered to be seven primary organs of the endocrine system. These organs include the following:

- the *anterior pituitary gland,* situated in the sphenoid bone of the skull;
- the *thyroid gland,* located in the anterior neck;
- three to five small *parathyroid glands,* found on the posterior thyroid gland;
- the paired *adrenal cortices,* located on the superior surface of the kidneys;
- the *endocrine pancreas,* found in the left side of the abdominal cavity mostly posterior to the stomach;
- the *thymus,* located in the superior mediastinum; and
- the paired *ovaries* or *testes,* the former in the pelvic cavity in women and the latter suspended below the pelvic cavity in men.

In addition to these glands, there are several organs known as *secondary endocrine glands* that are not considered part of the endocrine system. These organs produce hormones but belong to other systems (note that the testes and ovaries are often considered secondary endocrine organs, as they are also a part of the reproductive system). Many secondary endocrine organs are **neuroendocrine organs,** including the *hypothalamus* and the *pineal gland* (pin-EE-ul) in the brain and the *adrenal medulla* in the core of the adrenal gland. Anatomically these structures consist of nervous tissue, yet they secrete chemicals that act as hormones, called *neurohormones.* Other organs and tissues that act as secondary endocrine organs include the heart, kidneys, small intestine, and adipose tissue. Sometimes cancer cells can act as secondary endocrine tissues. Find out more about this in *A&P in the Real World: Paraneoplastic Syndrome.*

Quick Check

☐ 2. What are the primary organs of the endocrine system?

☐ 3. What are examples of secondary endocrine organs? Which organs are considered neuroendocrine?

Paraneoplastic Syndrome

The heart and kidneys aren't the only secondary endocrine tissues able to produce hormones—many types of cancer cells can, too. The signs and symptoms that accompany hormone secretion from cancer cells are collectively termed *paraneoplastic syndrome.* Many different types of cancer cells can cause paraneoplastic syndrome, but the most common are cells of lung and gastrointestinal cancers.

The effects of paraneoplastic syndrome depend on the type of hormone secreted by the cancer cells; common findings include imbalances in fluid, calcium ion, and sodium ion homeostasis. The symptoms of paraneoplastic syndrome often precede other symptoms of the cancer, and investigating them may lead to an earlier diagnosis.

Hormones

◀◀ FLASHBACK

1. Are peptides and steroids generally hydrophilic or hydrophobic? (p. 56)
2. What is gene expression? (p. 103)
3. What is a second-messenger system? (p. 413)

Hormones are chemicals that regulate some functions of other cells. In a sense, they represent the "middle managers" of the body—they "tell" other cells what to do by stimulating some sort of change in those cells. Before we discuss individual hormones, let's look at their general properties. The upcoming subsections focus on the different types of hormones, how they travel through the blood, and how they interact with their target cells to produce certain effects. We also explore the mechanisms by which hormone secretion is regulated, how hormones interact with one another, and how they are eventually eliminated.

Classes of Hormones

There are two basic classes of hormones, grouped according to their chemical structure: *amino acid–based* and *steroid.* **Amino acid–based hormones,** including those produced by the anterior pituitary gland, pancreas, thymus, thyroid, and parathyroid glands, consist of one or more amino acids (look back at Figure 2.21). They vary in size from single amino acids (amine hormones) to several amino acids (peptide hormones) to complete proteins (protein hormones). Most amino acid–based hormones have a structure that makes them hydrophilic, so they can freely interact with water molecules. Note, however, that thyroid hormone is an exception. It is derived from the hydrophobic amino acid tyrosine, and so does not freely interact with water.

Steroid hormones are cholesterol derivatives, with a core of hydrocarbon rings. Endocrine glands that produce steroid hormones include the adrenal cortices, testes, and ovaries.

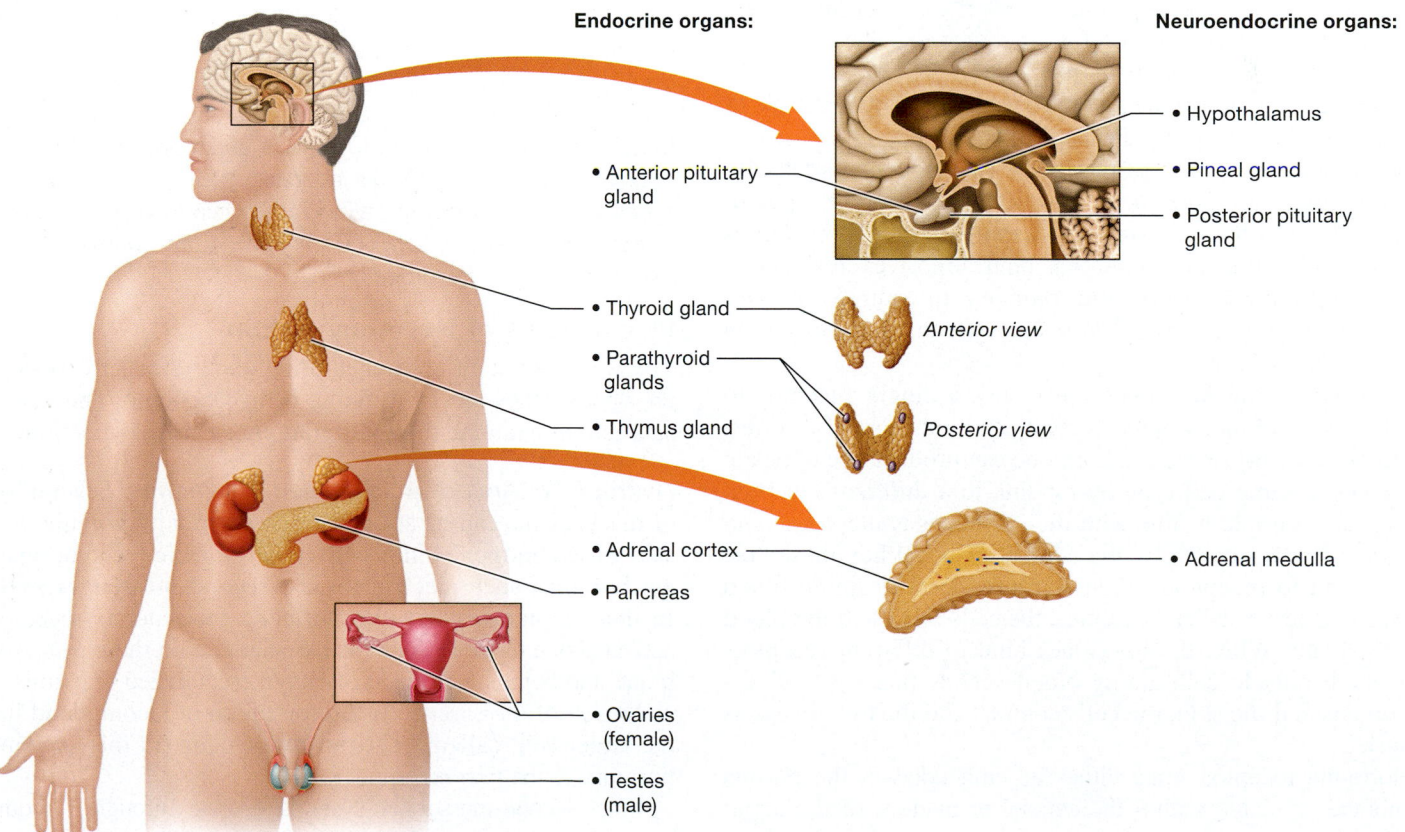

Figure 16.3 **Overview of the endocrine organs.**

Like cholesterol, all steroid hormones are hydrophobic. These hormones are also sometimes referred to as *lipid-soluble* because they are able to mix with fats and can be stored in adipose tissue.

Hormone Transport through the Blood

Hormones may travel in the blood either freely or bound to proteins in the plasma. **Free hormones** are generally small, amino acid–based hormones that are hydrophilic and able to freely associate with water. Their hydrophilic nature allows them to travel freely through the water-based plasma of blood. However, **bound hormones,** which form complexes with binding proteins in the plasma, are usually hydrophobic and so do not associate with the water molecules in plasma. Note though that a few notable hydrophilic hormones, such as *growth hormone* produced by the anterior pituitary gland, are also protein-bound.

Protein binding serves several important functions. First, it allows hydrophobic hormones to be transported safely through the watery environment of blood. Second, it gives the body a reservoir of hormones that can be released when the need arises, which helps prevent the concentration of free hormone in the plasma from experiencing large fluctuations. Finally, it extends the lifespan of a hormone in the blood, which we discuss later in this module in the section on *half-life.*

Target Cells and Receptors

As we discussed earlier, hormones can interact only with cells— their **target cells**—that contain proteins, referred to as **receptors,** to which the hormones can bind. Receptors have three-dimensional shapes that are highly specific for their hormones, and can bind to hormones present at extremely low concentrations, an example of the Structure-Function Core Principle (p. 28).

CORE PRINCIPLE
Structure-Function

Some hormones bind primarily to one type of target cell, whereas others are able to bind to multiple different types of target cells in different tissues. For example, the hypothalamus produces *releasing hormones* that bind almost exclusively to specific cell types in the anterior pituitary. In contrast, the thyroid gland produces *thyroid hormone*, which binds to nearly every cell in the body.

Similarly, some hormones have only a single receptor to which they can bind, whereas others can bind to several different receptors. Surprisingly, a hormone can produce the opposite effect in the same cell type by binding to a different receptor. We see an example of this with the hormones *epinephrine* and *norepinephrine* secreted by the adrenal gland. When these hormones bind to receptors on smooth muscle cells lining blood vessels that serve skeletal muscles, the cells relax and the blood vessels dilate. When the hormones bind to different receptors on smooth muscle cells lining blood vessels that serve digestive organs and the skin, the cells contract and the blood vessels constrict.

Hormone receptors may either be embedded in the plasma membrane or reside within the cytosol or nucleus of the target cell. The location of the receptor that a hormone binds depends on whether it is hydrophilic or hydrophobic. Hydrophilic hormones

cannot readily cross the phospholipid bilayer of the plasma membrane to enter target cells because they are repelled by the fatty acid tails of the phospholipid bilayer. In contrast, as shown here, hydrophobic hormones can cross membranes because of the attraction between the hormone and the membrane's fatty acid tails:

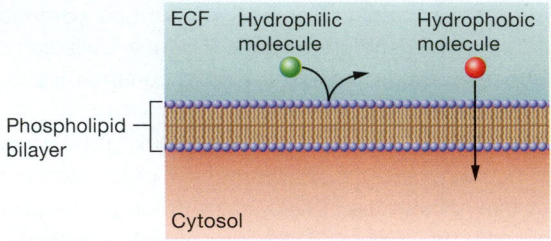

Due to these differences, hydrophilic hormones interact with receptors embedded in the target cell's plasma membrane, with the binding site exposed on the cell surface. The receptor is often associated with another membrane protein that produces a response inside the cell. These other proteins can be ion channels, enzymes, or peripheral proteins. Hydrophobic hormones, however, can bind to receptors found within the plasma membrane, in the cytosol, or in the nucleus.

The number of receptors in the target cell varies with the body's needs. Certain conditions will cause the cell to produce more hormone receptors. For example, when the level of a particular hormone in the blood declines, target cells will make more receptors for the hormone, thereby increasing the cells' sensitivity to the hormone, a process called **upregulation.** Note that in some cases upregulation can also occur in response to a temporary *increase* in the circulating hormone level, which enhances the hormone's effect on the target cells. However, prolonged exposure to a high level of a given hormone causes the opposite effect—target cells decrease the number of receptors specific for that hormone, which is known as **downregulation.** These mechanisms allow cells to have tight control over how they interact with hormones.

Mechanisms of Hormone Action

The ways in which hormones bring about their effects depend on whether they are hydrophilic or hydrophobic. The next two subsections examine these different mechanisms of action.

Hydrophilic Amino Acid–Based Hormones Several types of mechanisms can be set in motion when a hydrophilic amino acid–based hormone binds with its cell surface receptor on a target cell. One such mechanism is a **second-messenger system.** In these systems, when the hormone binds to its receptor, it activates one or more proteins associated with the plasma membrane and begins a cascade of enzyme-catalyzed reactions. The end result of the cascade is the formation of a compound inside the target cell, called a **second messenger** (in this system the hormone is the *first messenger*).

Most second-messenger systems operate through a group of proteins associated with receptors called **G-proteins.** When a hormone binds to a receptor linked to a G-protein, the G-protein changes shape and dissociates into two subunits. One of the subunits then interacts with another protein, generally an enzyme. The G-protein subunit may activate or inhibit the enzyme, depending on which G-protein has been activated. If the enzyme is activated, it catalyzes the formation of a second messenger, which then initiates a series of changes in the cell that lead to some change in its activity.

One key feature of second-messenger systems is that they allow for *signal amplification*. This refers to the fact that a single hormone molecule can bind a receptor and lead to the formation of hundreds of second-messenger molecules. The second messengers in turn can activate or inhibit thousands of other chemicals downstream that then change events in the cell.

A well-studied example of a G-protein second-messenger system is the *adenylate cyclase–cAMP system*. In this signal cascade, the G-protein activates an enzyme called **adenylate cyclase,** which then catalyzes the formation of the second messenger **cyclic adenosine monophosphate,** or **cAMP.** The basic steps of an adenylate cyclase–cAMP system, shown in **Figure 16.4,** are as follows:

① **The hydrophilic hormone (first messenger) binds to its receptor in the plasma membrane, causing the G-protein to split into two subunits.** Notice that the target cell's receptor is linked to a G-protein. As the hormone binds, the G-protein changes shape and dissociates into two subunits.

② **The G-protein subunit activates adenylate cyclase.** One of the G-protein subunits leaves the receptor to interact with adenylate cyclase, which it binds and activates.

③ **Adenylate cyclase catalyzes the formation of cAMP, the second messenger.** Activated adenylate cyclase forms cAMP from ATP in the cytosol of the cell.

④ **cAMP activates protein kinase A.** Many of cAMP's effects are due to the activation of the enzyme **protein kinase A.** Notice the signal amplification—cAMP activates multiple protein kinase A enzymes. *Kinases* are enzymes that catalyze the transfer of a phosphate group from ATP to another substance, a process called *phosphorylation.*

⑤ **Protein kinase A phosphorylates specific proteins.** Activated protein kinase A phosphorylates specific amino acids on proteins. Phosphorylation changes the activity or shape of a protein and so can have a wide range of effects within a cell. Note the further signal amplification, as each protein kinase A enzyme catalyzes the phosphorylation of multiple proteins.

Note that the adenylate cyclase–cAMP system is just one of many second-messenger systems that involve G-proteins and protein kinases. Other such systems function in a similar manner. For example, some G-proteins activate the enzyme *phospholipase C,* which produces the second messengers *inositol trisphosphate (IP3)* and *diacylglycerol (DAG).*

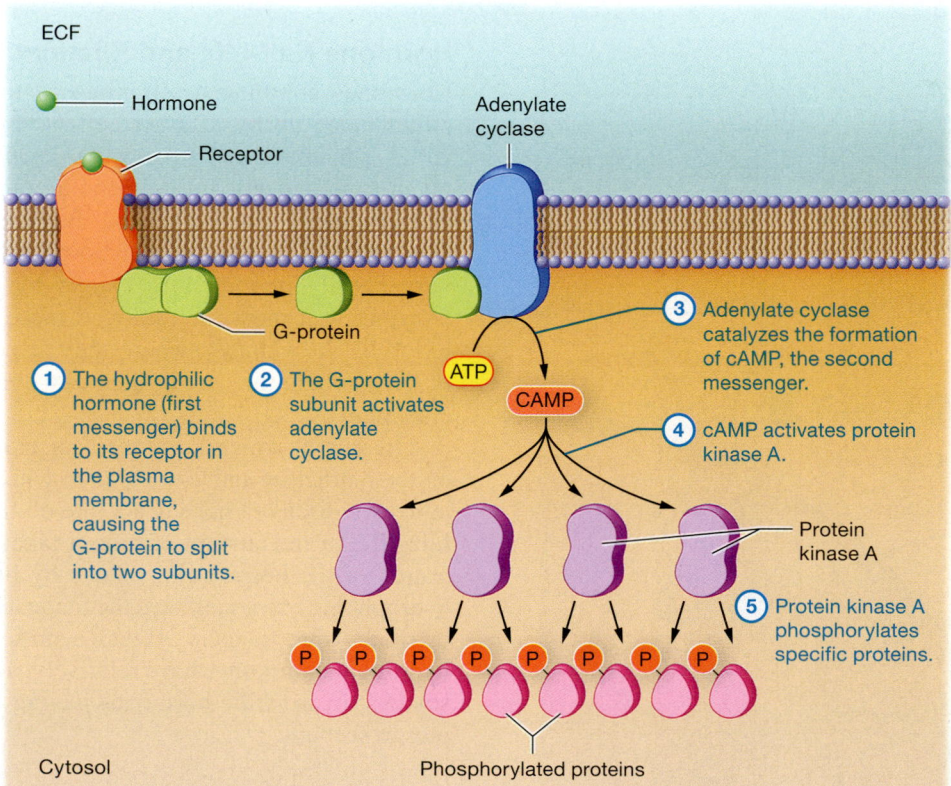

Figure 16.4 **A hydrophilic amino acid–based hormone mechanism of action via second-messenger system.** This example shows the adenylate cyclase-cAMP – second-messenger system.

Play A&P Flix Animation
@ Mastering Anatomy & Physiology

G-protein–mediated second-messenger systems are only one possible way that hydrophilic hormones interact with their target cells. Other hydrophilic hormones bind to receptors that also function as ion channels. Binding of the hormone to such receptors may cause the channel to open or close. In addition, some hydrophilic hormones can open or close ion channels indirectly by binding to receptors linked to G-proteins.

Hydrophobic Hormones: Steroids and Thyroid Hormone
As discussed earlier, hydrophobic hormones, including steroids and thyroid hormone, are able to diffuse across the plasma membrane and enter the cytosol of target cells. You can see in **Figure 16.5** that this enables them to interact with intracellular receptors that may be located in the cytosol or the nucleus. When a hydrophobic hormone binds to an intracellular receptor, it forms a *hormone-receptor complex*. This complex then binds to specific regions of the DNA known as **hormone-response elements,** an action that changes the rate of synthesis of one or more proteins.

For a long time, interaction with DNA was thought to be the only mechanism of action of hydrophobic hormones; however, current evidence indicates they can exert many other effects not involving DNA. These effects are carried out via binding plasma membrane receptors and/or certain cytosolic receptors and producing the responses we described for hydrophilic hormones.

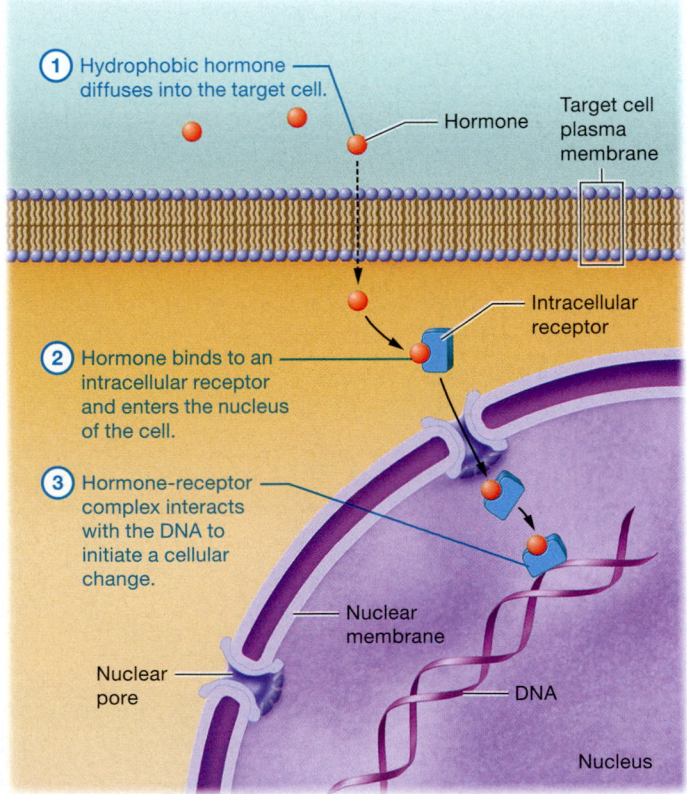

① Hydrophobic hormone diffuses into the target cell.

Hormone

Target cell plasma membrane

② Hormone binds to an intracellular receptor and enters the nucleus of the cell.

Intracellular receptor

③ Hormone-receptor complex interacts with the DNA to initiate a cellular change.

Nuclear membrane

Nuclear pore

DNA

Nucleus

Figure 16.5 Mechanism of action of hydrophobic hormones (steroids and thyroid hormones) via binding an intracellular receptor.

Effects of Hormone Actions

Activation of either cell surface or intracellular receptors triggers different mechanisms inside the cell that lead to cellular effects. The many effects of hormone action may include:

- stimulating secretion from an endocrine or exocrine cell;
- activating or inhibiting enzymes;
- stimulating or inhibiting mitosis and/or meiosis;
- opening or closing ion channels in the cell's plasma membrane and/or altering its membrane potential; and
- activating or inhibiting transcription of genes that code for RNA or proteins (also known as gene expression).

Hormone Interactions

The maintenance of homeostasis requires not just individual hormones, but the interaction of multiple hormones. Many hormones have *complementary* actions, meaning they each affect different target cells to accomplish a common goal. For example, hormones of the adrenal gland and pancreas work together to maintain homeostasis during exercise. Other hormones can act on the same target cell to exert the same effect. Such hormones are known as **synergists** (SIN-er-jists), and they have more pronounced effects when acting together than when acting alone. Sometimes, however, hormones that act on the same target cells have opposite effects. These hormones are called **antagonists.**

Hormone Half-Life and Elimination

Hormones continue to circulate in the blood until they are either taken up by a target cell or broken down and deactivated. The amount of a particular hormone in the blood at any given time depends on both the amount of hormone secreted from the endocrine gland and the rate at which the hormone is removed from the blood. Some hormones are inactivated very rapidly in their target cells by reactions catalyzed by enzymes, but most are removed more slowly from the circulation by the kidneys (where they are eliminated in the urine) or liver (where they are broken down in reactions catalyzed by enzymes).

The rate at which hormones are eliminated largely depends on their structure and the degree to which they are bound to proteins. We discuss a hormone's rate of elimination in terms of its **half-life,** or the amount of time it takes for the plasma concentration of the hormone to reduce by half. Generally speaking, hydrophobic hormones have the longest half-lives—often a week or more—due to their extensive protein binding. In contrast, hydrophilic hormones have relatively shorter half-lives, with the smallest hydrophilic hormones having half-lives of a few minutes to seconds.

Quick Check

☐ 4. What are the two major classes of hormones, and how do they interact with their target cells?

☐ 5. How do synergistic and antagonistic hormones differ?

□ 6. What is the half-life of a hormone, and how is it related to that hormone's elimination?

Regulation of Hormone Secretion

« FLASHBACK

1. What are negative feedback loops, and how do they function? (p. 22)

Hormone secretion can be initiated by hormonal, humoral, or neural stimuli (**Figure 16.6**). Let's look at each type of stimulus in more detail:

- **Hormonal stimuli.** Some endocrine cells increase or decrease their secretion in response to other hormones. For example, the hypothalamus regulates secretion from the anterior pituitary gland through secretion of *releasing* and *inhibiting hormones*. In Figure 16.6a, we use the example of two hypothalamic hormones: growth hormone–releasing hormone and somatostatin, which stimulate and inhibit secretion of growth hormone from the anterior pituitary gland, respectively.

- **Humoral stimuli.** Many endocrine cells respond to the concentration of a certain ion or compound in the blood or extracellular fluid, such as glucose or calcium ions. These stimuli are known as *humoral stimuli,* a term reflecting the historical name for body fluids, *humors.* The example in Figure 16.6b shows a cell of the pancreas releasing the

hormone insulin in response to an elevated level of glucose in the blood.

- **Neural stimuli.** Finally, some cells respond to signals from the nervous system, as shown in Figure 16.6c. The best example of this is the adrenal medulla, which is stimulated by sympathetic neurons to release the neurohormones epinephrine and norepinephrine.

Hormone secretion is generally regulated as part of a *negative feedback loop*—recall the Feedback Loops Core Principle (p. 28). Negative feedback loops of the endocrine system proceed by the following general steps (**Figure 16.7**; note that we are using an example in which the endocrine cell acts as the receptor and control center).

CORE PRINCIPLE
Feedback Loops

- **Stimulus: A regulated physiological variable deviates from its normal range.** The change can be an increase or decrease from the normal range; in this figure, we are showing a decrease.
- **Receptor: Receptors on endocrine cells detect the deviation of the variable.** The receptors of the endocrine gland cells continuously monitor regulated variables in the blood. Such variables include glucose and calcium ion concentrations and body temperature.
- **Control center: The stimulated control center (often the same endocrine cell) increases or decreases its**

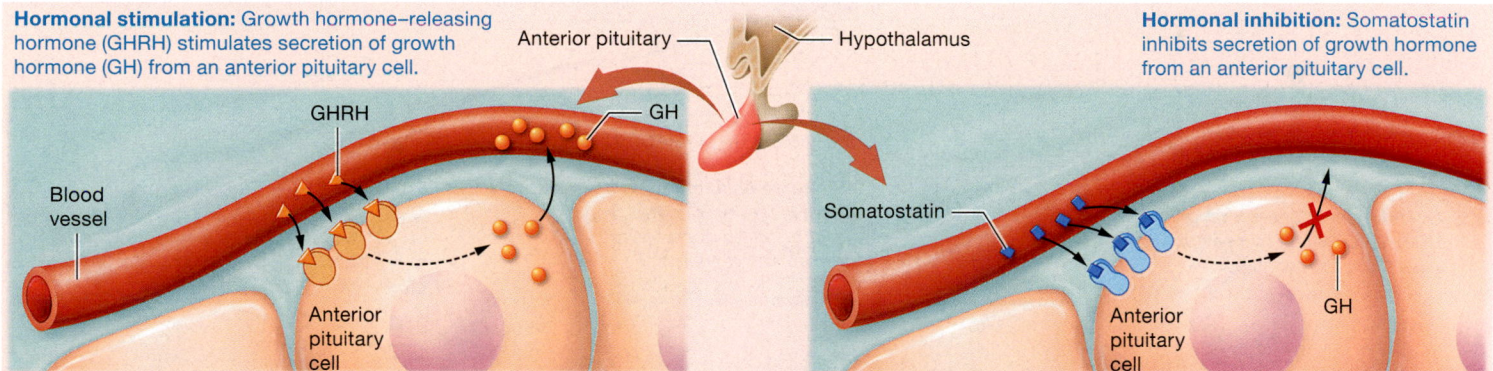

(a) Hormonal stimuli

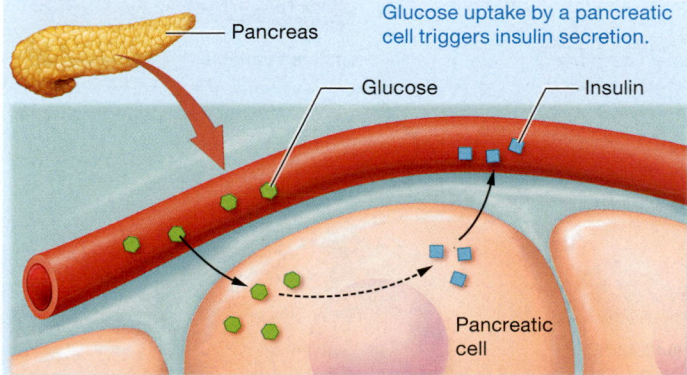

(b) Humoral stimulus

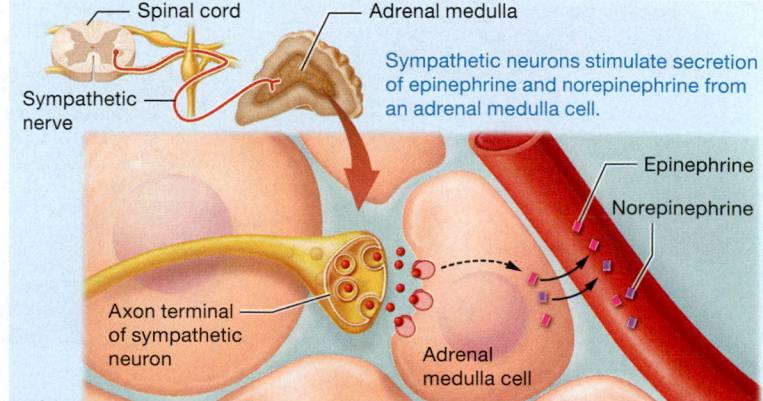

(c) Neural stimulus

Figure 16.6 Types of stimuli for hormone secretion.

secretion of a particular hormone. In the endocrine system, the endocrine cell that detects a change in the variable from the normal range is generally also the control center. When the control center (the endocrine cell) is stimulated, it changes its secretion of the appropriate hormone.

- **Effector/response: The hormone triggers a response in its target cells that moves conditions toward the normal range.** The hormone enters the blood and reaches its target tissue or tissues, where it triggers changes that return the variable to its normal range.
- **Homeostatic range: As the variable returns to its normal range, feedback to the control center decreases the effector response.** When homeostatic conditions have been restored, the control center returns hormone secretion to the normal range by either decreasing or increasing it. This in turn stops stimulation to the target cells, decreasing or ending the response.

As the upcoming modules discuss, many negative feedback loops in the endocrine system are more complex than this simple example, and many involve several levels of control from multiple endocrine glands. There are also rare instances of hormones regulated by positive feedback mechanisms.

Quick Check

☐ 7. What are the three types of stimuli that initiate hormone secretion? How do they perform this function?

☐ 8. How is hormone secretion generally regulated?

Apply What You Learned

☐ 1. A newly discovered hormone X is determined to be structurally similar to cholesterol. Predict how this new hormone interacts with its target cells.

☐ 2. Hormone X appears to regulate the blood concentration of chemical Y via a negative feedback loop.

  a. Predict what happens to the secretion of hormone X when the blood concentration of chemical Y *decreases*.

  b. What effect will this likely have on the blood concentration of chemical Y?

  c. Predict what happens to the secretion of hormone X when the blood concentration of chemical Y *increases*.

See answers in Appendix A.

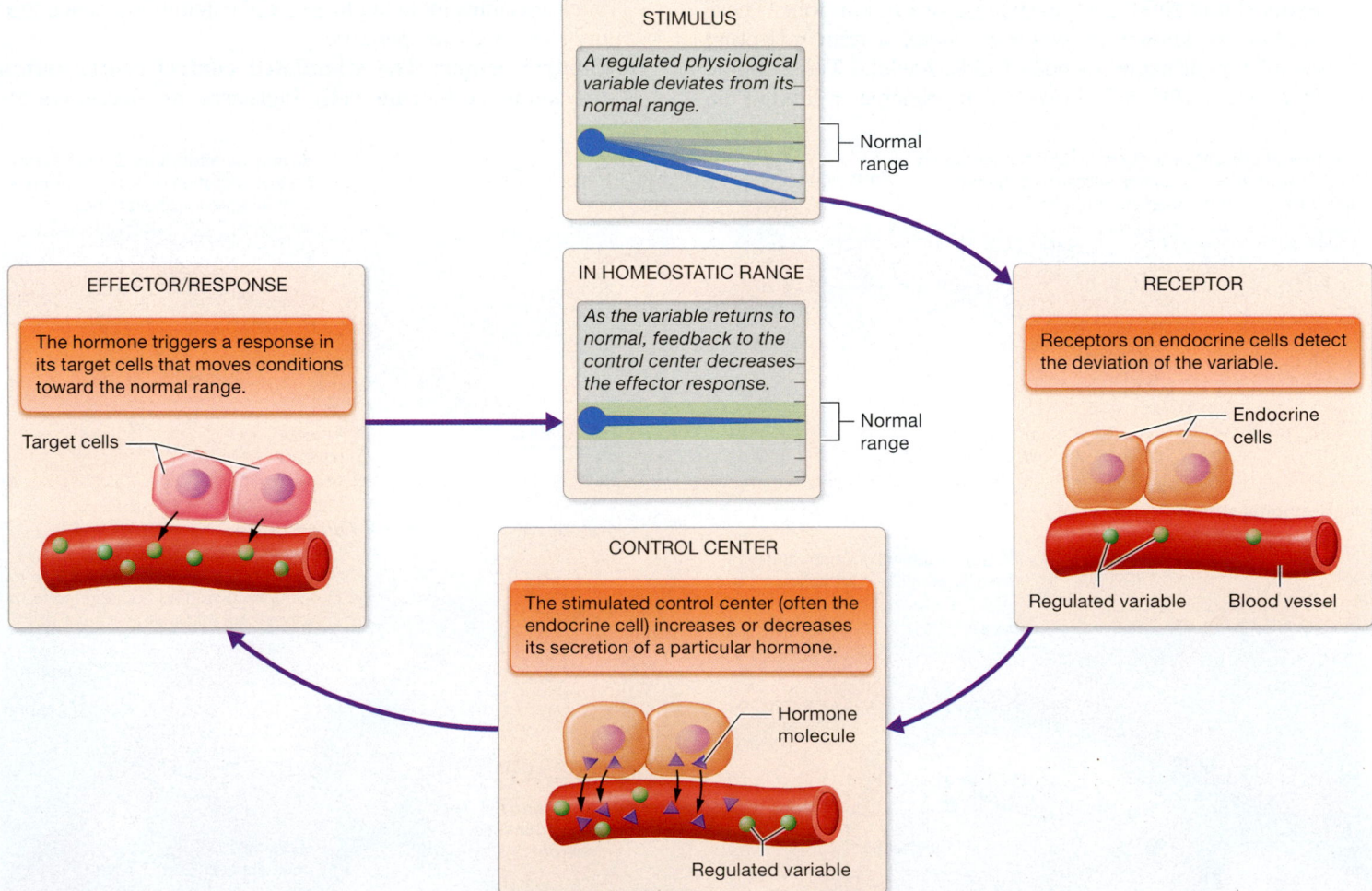

Figure 16.7 Maintaining homeostasis: regulation of hormone secretion by negative feedback loops.

The Hypothalamus and the Pituitary Gland

1. Describe the anatomical and functional relationships between the hypothalamus and the anterior and posterior pituitary glands.

2. Explain how tropic hormones affect the secretion of other hormones.

3. Describe the stimulus for release, the target tissue, and the functional effect of each hormone released from the posterior pituitary.

4. Describe the stimulus for release, the target tissue, and the functional effect of each hormone produced and secreted by the anterior pituitary.

5. Explain the negative feedback loops that regulate the production and release of anterior pituitary hormones.

Recall that the hypothalamus of the brain is a small collection of nuclei with many important physiological functions, including regulation of hunger, thirst, fluid balance, body temperature, the sleep/wake cycle, and certain reproductive functions (see Chapter 12). The hypothalamus works together with the pituitary gland, with which it has a close anatomical and physiological relationship. Several hypothalamic and anterior pituitary hormones directly control certain body functions, whereas others are *tropic hormones,* meaning they control hormone secretion from other endocrine glands. Be careful not to confuse a tropic hormone with the similar-sounding *trophic hormone* (TROH-fik; *troph-* = "change"), which induces growth in its target cell. Note that some hormones have both tropic and trophic effects on their target cells.

Some of the primary functions of the hypothalamus are performed directly by hypothalamic neurons whose axon terminals are found in the posterior pituitary gland, whereas others are accomplished by hormones targeted to the anterior pituitary gland. In this module we examine this relationship in discussing the cells, hormones, and target tissues of the hypothalamus and the anterior and posterior pituitary glands. We revisit these structures in later modules to more closely explore their roles in regulating other endocrine organs.

Structure of the Hypothalamus and Pituitary Gland

« FLASHBACK

1. What is the sella turcica, and where is it located? (p. 218)

As you can see in **Figure 16.8a**, the **hypothalamus** is the small anteroinferior portion of the diencephalon of the brain. It is connected to the **pituitary gland** (also called the *hypophysis;* hy-PAWF-eh-sis) by a stalk called the **infundibulum** (in´-fun-DIB-yoo-lum). The pituitary gland is a small organ, about the size of a bean, that sits in the sella turcica of the sphenoid bone.

As **Figure 16.8b** shows, the pituitary gland is composed of two structurally and functionally distinct components: the **anterior pituitary gland,** or **adenohypophysis** (ad´-eh-noh-hy-PAWF-eh-sis), and the **posterior pituitary gland,** or **neurohypophysis** (noo´-roh-hy-PAWF-eh-sis). The differences in the two components are inherent in their names: The *adeno-* in adenohypophysis means "gland," which reflects the fact that the anterior pituitary is a true gland composed of hormone-secreting glandular epithelium. The *neuro-* in neurohypophysis refers to the fact that the posterior pituitary is actually made up of nervous tissue.

The hypothalamus and pituitary gland have a specialized blood supply that allows them to deliver their hormones to the target cells. Capillaries merge in the hypothalamus to form larger blood vessels, the *portal veins,* that travel through the

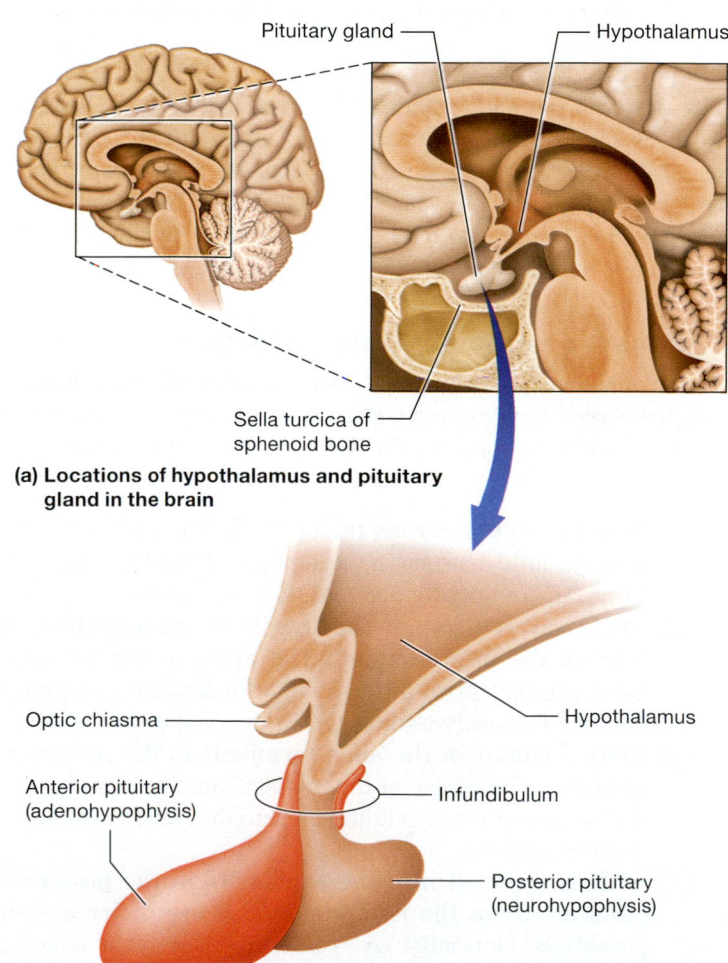

Pituitary gland — Hypothalamus

Sella turcica of sphenoid bone

(a) Locations of hypothalamus and pituitary gland in the brain

Optic chiasma — Hypothalamus

Anterior pituitary (adenohypophysis)

Infundibulum

Posterior pituitary (neurohypophysis)

(b) Structure of hypothalamus, and anterior and posterior pituitary glands

Figure 16.8 Structure of the hypothalamus and pituitary gland.

infundibulum. These veins lead to a second group of capillaries in the anterior pituitary gland. Systems like this, in which capillaries are drained by veins that lead to another set of capillaries, are called *portal systems*. Accordingly, this group of vessels is called the **hypothalamic-hypophyseal portal system** (hy-paw'-fih-SEE-ul; look ahead to Figure 16.9).

□ 1. How do the anterior pituitary and posterior pituitary differ structurally and functionally?

□ 2. What is the hypothalamic-hypophyseal portal system?

Hormones of the Hypothalamus and Posterior Pituitary

« FLASHBACK

1. What name is given to groups of neuronal cell bodies in the central nervous system? (p. 387)

2. What are the primary functions of the hypothalamus? (p. 433)

The posterior pituitary makes no hormones of its own. Instead, it stores and releases two neurohormones produced by clusters of cell bodies in the hypothalamus, known as the *supraoptic* and *paraventricular nuclei*. These hormones are *antidiuretic hormone* (an'-ty-dy-yoo-RET-ik), and *oxytocin* (awk-see-TOH-sin). Let's look more closely at these hormones and their roles.

Antidiuretic Hormone: Water Retention

The hypothalamic hormone that controls water balance is known as **antidiuretic hormone (ADH),** or **vasopressin.** As you can see in **Figure 16.9a**, the process for ADH synthesis and release is as follows:

① **Hypothalamic neurons make ADH.** The cell bodies of hypothalamic neurons in the supraoptic and paraventricular nuclei continually make ADH in low amounts.

② **ADH travels through the hypothalamic-hypophyseal tract in the infundibulum.** The axons of the hypothalamic neurons travel through the infundibulum as a group known collectively as the *hypothalamic-hypophyseal tract.*

③ **ADH is stored in the axon terminals in the posterior pituitary.** The axons of the hypothalamic neurons terminate in the posterior pituitary, where the ADH is stored in synaptic vesicles.

④ **ADH is secreted into the capillaries in the posterior pituitary when the hypothalamic neurons fire action potentials.** Normally, an axon terminal forms a synapse with another cell. However, in the posterior pituitary, the axon terminal simply releases hormones by exocytosis into the interstitial fluid, and they diffuse into the local capillaries.

The primary function of ADH is to increase the amount of water retained by the kidneys. This function is apparent from its name: A *diuretic* (dy-yoo-RET-ik) is a chemical that increases urine production by the kidneys, which decreases the amount of water in the body. As you may guess, an *anti*diuretic chemical has the opposite effect—it decreases urine production by increasing the amount of water retained by the body.

So that you can better understand the effect of ADH on the kidneys, let's briefly look at the kidneys' microscopic structure. The fluid that eventually becomes urine is located within small *tubules* in the kidney. The relationship between the tubules, interstitial fluid, and blood is shown here:

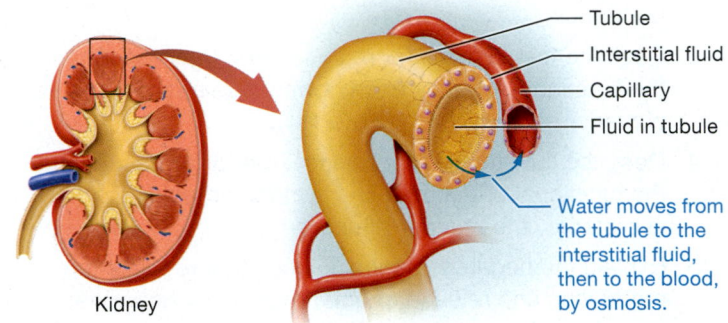

Kidney

- Tubule
- Interstitial fluid
- Capillary
- Fluid in tubule

Water moves from the tubule to the interstitial fluid, then to the blood, by osmosis.

ADH increases the amount of water in the body by causing the insertion of water channels called *aquaporins* (ah-kwah-POHR-inz) into the plasma membranes of the cells forming the kidney tubules. Water can enter these cells only via aquaporins, and aquaporins are inserted only in the presence of ADH. Aquaporins allow water in the tubules to enter the cytosol of the kidney cells by osmosis, reducing the amount of water remaining in the kidney tubules. From the cytosol, water moves into the interstitial fluid and then into the blood in capillaries, again by osmosis. So the net effect of ADH is to return water to the blood that would have otherwise been eliminated from the body as urine.

Cells of the hypothalamus contain *osmoreceptors* that monitor changes in the solute concentration of the blood. These osmoreceptors respond to increasing solute concentration by stimulating ADH release from the posterior pituitary gland. This leads to water retention, which decreases the solute concentration of the blood (keep in mind that the increased amount of water dilutes the blood, which causes the relative concentration of solutes to decrease). Conversely, these same receptors respond to abnormally low solute concentrations by inhibiting ADH release, which leads to water loss and an increased solute concentration of the blood.

An abnormal lack of ADH secretion or activity results in a disease called **diabetes insipidus** (dy-uh-BEET-eez in-SIP-ih-dus). Signs and symptoms include extreme thirst, dehydration, and a very high solute concentration of the blood because the body is unable to conserve most of the water that is consumed. Most forms of diabetes insipidus can be treated by administration of synthetic ADH. By the way, don't confuse diabetes insipidus with *diabetes mellitus* (MELL-ih-tus),

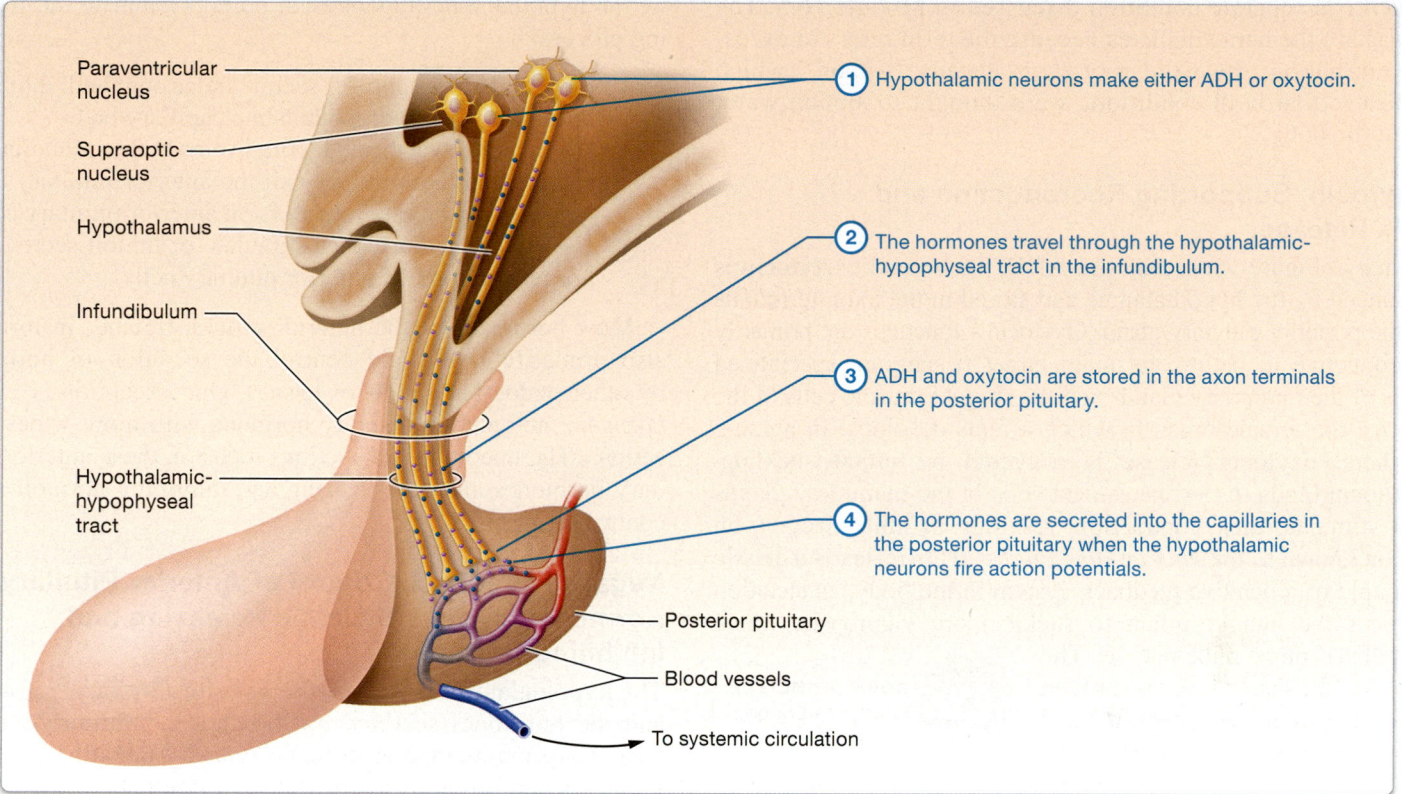

Paraventricular nucleus

Supraoptic nucleus

Hypothalamus

Infundibulum

Hypothalamic-hypophyseal tract

① Hypothalamic neurons make either ADH or oxytocin.

② The hormones travel through the hypothalamic-hypophyseal tract in the infundibulum.

③ ADH and oxytocin are stored in the axon terminals in the posterior pituitary.

④ The hormones are secreted into the capillaries in the posterior pituitary when the hypothalamic neurons fire action potentials.

Posterior pituitary

Blood vessels

To systemic circulation

(a) Relationship between the hypothalamus and posterior pituitary

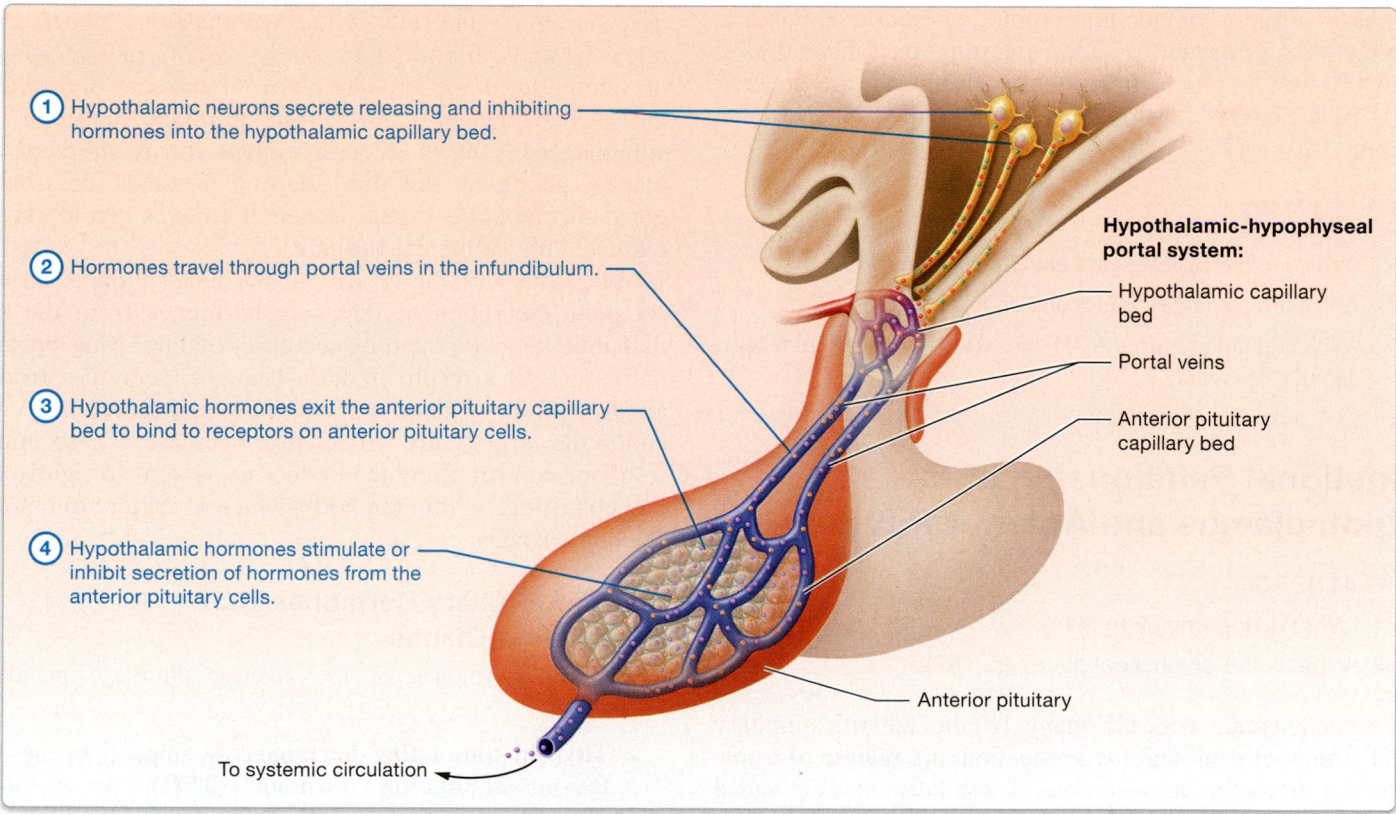

① Hypothalamic neurons secrete releasing and inhibiting hormones into the hypothalamic capillary bed.

② Hormones travel through portal veins in the infundibulum.

③ Hypothalamic hormones exit the anterior pituitary capillary bed to bind to receptors on anterior pituitary cells.

④ Hypothalamic hormones stimulate or inhibit secretion of hormones from the anterior pituitary cells.

Hypothalamic-hypophyseal portal system:

Hypothalamic capillary bed

Portal veins

Anterior pituitary capillary bed

Anterior pituitary

To systemic circulation

(b) Relationship between the hypothalamus and anterior pituitary

Figure 16.9 **Functional relationships between the hypothalamus and pituitary gland.**

which is a separate condition discussed in Module 16.5. The two share the name diabetes because the term refers to excessive urination; the word root *diab-* means "siphon," which indicates that both conditions were thought to siphon water from the body.

Oxytocin: Supporting Reproduction and Milk Release

Notice in Figure 16.9a that, like ADH, the hormone **oxytocin** is produced by the hypothalamus and stored in the axon terminals of the posterior pituitary gland. Oxytocin's functions are primarily reproductive in nature. Its main target tissues are specialized cells of the mammary glands and the smooth muscle cells of the *uterus,* the female organ in which a fetus develops. In nursing mothers, oxytocin release is triggered by infant suckling. Oxytocin binds to the specialized cells in the mammary glands that stimulate their contraction, resulting in milk ejection, an action known as the *milk let-down reflex.* This reflex is a classic example of a positive feedback system in the body, as lactation induces the hungry infant to suckle more vigorously, which stimulates more milk release. The positive feedback loop is broken when the infant becomes full and so no longer suckles. Note that the

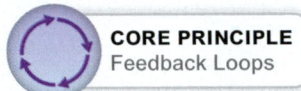

CORE PRINCIPLE
Feedback Loops

milk let-down reflex is another example of the Feedback Loops Core Principle (p. 28).

Oxytocin may have effects on the brain as well. For example, the possible role of oxytocin in promoting emotional bonding in both sexes is currently under investigation. These topics, including how oxytocin functions in childbirth, are covered more fully in the development and heredity chapter (see Chapter 27).

Quick Check

☐ 3. What are the target tissues and effects of ADH?

☐ 4. What are the target tissues and effects of oxytocin?

☐ 5. Which gland produces ADH and oxytocin, and from where are they secreted?

Functional Relationship of the Hypothalamus and Anterior Pituitary

« FLASHBACK

1. What is dopamine? (p. 414)

2. What is the epiphyseal plate? (p. 187)

The hypothalamus controls many of the anterior pituitary gland's functions through the production and release of tropic hormones that affect its secretions. These hormones are called **releasing** and **inhibiting hormones** because they either stimulate or inhibit the release of hormones from the anterior pituitary. As shown in **Figure 16.9b**, the hypothalamic releasing and inhibiting hormones are delivered to the anterior pituitary

by the hypothalamic-hypophyseal portal system in the following process:

① Hypothalamic neurons secrete releasing and inhibiting hormones into the hypothalamic capillary bed.

② Hormones travel through portal veins in the infundibulum.

③ Hypothalamic hormones exit the anterior pituitary capillary bed and bind to receptors on anterior pituitary cells.

④ Hypothalamic hormones stimulate or inhibit secretion of hormones from the anterior pituitary cells.

Most hormones of the anterior pituitary gland, in turn, are also primarily tropic and control the secretion of hormones by other endocrine glands or tissues. One exception is *growth hormone,* an anterior pituitary hormone with more widespread actions. The upcoming subsections focus on these anterior pituitary hormones and the releasing and inhibiting hormones that control them.

Negative Feedback Control of Anterior Pituitary Hormones by Hypothalamic Releasing and Inhibiting Hormones

The hypothalamic hormones, the anterior pituitary hormones, and the hormones secreted by the anterior pituitary's target tissues are maintained at normal levels by negative feedback loops with multiple levels, or tiers, of control (**Figure 16.10**). Each tier includes a stimulus, receptor, control center, and response/effector. The first tier of feedback control involves the neuroendocrine cells of the hypothalamus, which secrete releasing and inhibiting hormones. These hormones stimulate or inhibit hormone production from the anterior pituitary, the second tier of control. The pituitary hormones then stimulate cells of their target organs, many of them other glands, acting as the third tier of feedback control. The glands of the target organs secrete hormones that affect some regulated physiological variable.

When the variable returns to the normal range around a set point, secretion of releasing hormones from the hypothalamus decreases and/or secretion of inhibiting hormones increases. As a result of these changes, secretion from the anterior pituitary gland will decrease, and secretion from the endocrine target cells will in turn decrease. These multiple levels of control allow the endocrine system to tightly regulate conditions within the body, which is critical to maintaining homeostasis.

Anterior Pituitary Hormones That Affect Other Glands

The tropic hormones of the anterior pituitary include the following:

● **Thyroid-stimulating hormone.** As implied by its name, **thyroid-stimulating hormone (TSH),** or *thyrotropin,* stimulates development of the thyroid gland and its secretion of *thyroid hormones.* TSH is produced and secreted by cells called *thyrotrophs* (THY-roh-trohfs). Its secretion is triggered by a hypothalamic releasing hormone known as

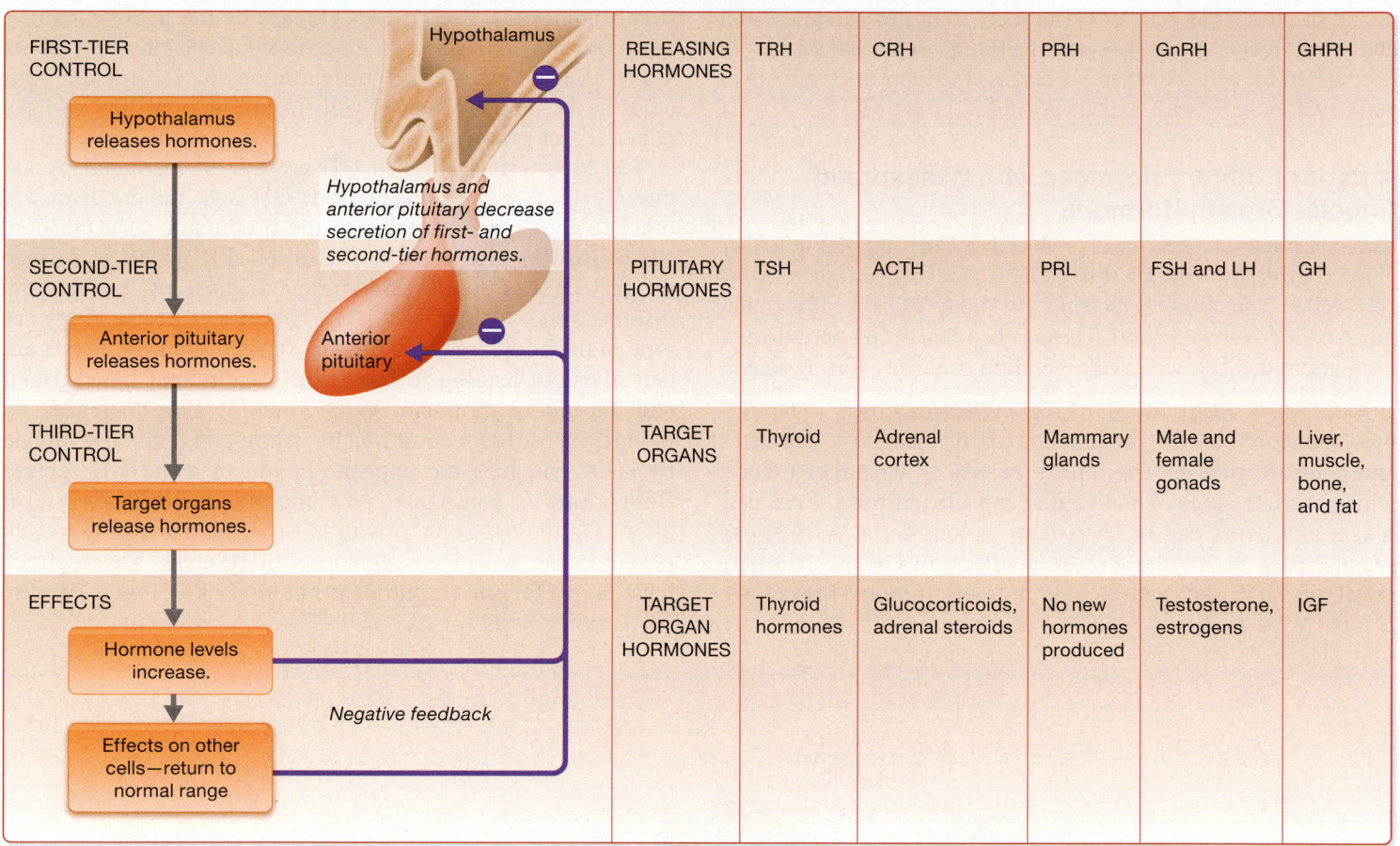

FIRST-TIER CONTROL	Hypothalamus	RELEASING HORMONES	TRH	CRH	PRH	GnRH	GHRH
Hypothalamus releases hormones.	*Hypothalamus and anterior pituitary decrease secretion of first- and second-tier hormones.*						
SECOND-TIER CONTROL		PITUITARY HORMONES	TSH	ACTH	PRL	FSH and LH	GH
Anterior pituitary releases hormones.	Anterior pituitary						
THIRD-TIER CONTROL		TARGET ORGANS	Thyroid	Adrenal cortex	Mammary glands	Male and female gonads	Liver, muscle, bone, and fat
Target organs release hormones.							
EFFECTS		TARGET ORGAN HORMONES	Thyroid hormones	Glucocorticoids, adrenal steroids	No new hormones produced	Testosterone, estrogens	IGF
Hormone levels increase.							
Effects on other cells—return to normal range	*Negative feedback*						

Figure 16.10 **Multi-tiered negative feedback control of hormones of the hypothalamus, anterior pituitary gland, and target organs.**

thyrotropin-releasing hormone (TRH). Secretion of TSH is inhibited by the hypothalamic hormone **somatostatin** (soh′-mah-toh-STAH-tin; *-statin* = "inhibiting or blocking").

- **Adrenocorticotropic hormone. Adrenocorticotropic hormone (ACTH),** or *corticotropin,* is produced by cells called *corticotrophs.* It stimulates the development of the adrenal glands and their synthesis of various steroid hormones. ACTH secretion is stimulated by a hypothalamic releasing hormone known as **corticotropin-releasing hormone (CRH).**

- **Prolactin.** The hormone **prolactin (PRL;** *lact-* = "milk") is produced by cells called *lactotrophs.* It has several target tissues in both males and females where its effects are tropic, but its primary target tissue is mammary gland cells, where its effects are non-tropic. Here it is secreted generally in women who have given birth, and it stimulates growth of the mammary glands, the initiation of milk production after childbirth, and the maintenance of milk production for the duration of breastfeeding. The main stimulus for prolactin release is suckling of the nipple by an infant. Normally, prolactin secretion is inhibited by the hypothalamic inhibiting hormone **prolactin-inhibiting factor,** now known to be dopamine. Suckling by the infant decreases the release of dopamine, which facilitates prolactin release.

- **Luteinizing hormone. Luteinizing hormone (LH;** LOO-tee-in-aye′-zing) is one of two hormones known as **gonadotropins** (goh-nad′-uh-TROHP-inz) secreted by cells called *gonadotrophs.* The gonadotropins are so named because they stimulate the male and female *gonads,* the testes and ovaries, respectively. LH secretion is stimulated by the hypothalamic releasing hormone **gonadotropin-releasing hormone (GnRH).** In males, LH stimulates production of the male hormone *testosterone* by cells of the testes. In females, it stimulates production of female hormones called *estrogens* and *progesterone.* It also triggers the release of an oocyte (OH-oh-syt; an immature egg cell) in the process of ovulation.

- **Follicle-stimulating hormone.** The other gonadotropin produced by gonadotrophs is **follicle-stimulating hormone (FSH).** Like LH, FSH secretion is stimulated by GnRH from the hypothalamus. In males, FSH stimulates cells of the testes to produce chemicals that bind and concentrate testosterone. In females, FSH works with LH to trigger the production of estrogens. Its name derives from the fact that it also triggers the maturation of *ovarian follicles,* which house the developing oocytes.

We discuss TSH and ACTH in more detail in Modules 16.3 and 16.4, respectively. LH and FSH are explored further in Module 16.6, and also when we discuss the reproductive system and development (see Chapters 26 and 27).

Anterior Pituitary Hormone with Widespread Effects: Growth Hormone

Another important hormone produced by the anterior pituitary gland is **growth hormone (GH).** GH is produced and secreted by cells called *somatotrophs* (*somato-* = "body"; GH itself is sometimes called *somatotropin*). These cells release GH periodically throughout the day, with peak secretion occurring during sleep.

Effects and Regulation of Growth Hormone The primary function of GH is inherent in its name—it regulates and controls growth. The effects of GH on its target tissues, which include skeletal and cardiac muscle, adipose, liver, cartilage, and bone, can be short-term or long-term. As you can see in **Figure 16.11a**, the short-term effects of GH are primarily metabolic in nature and include promotion of fat breakdown

(lipolysis), production of new glucose by the liver (*gluconeogenesis*), and inhibition of glucose uptake by muscle fibers. These effects increase the concentrations of glucose and fatty acids in the blood, allowing cells to use them as fuel and raw materials for growth.

The long-term effects of GH are not all mediated by GH directly (**Figure 16.11b**). Instead, GH acts on the liver and other target tissues to promote the production of a hormone called **insulin-like growth factor (IGF);** its name reflects the fact that its actions are similar to those of the hormone *insulin,* which is discussed later. IGF affects nearly every cell type in the body, where it triggers rapid protein synthesis and cell division, leading to increased longitudinal bone growth and muscle development in children. It also decreases the blood glucose concentration by stimulating glucose uptake by cells, which is the opposite action of acute GH release. Both hormones continue to play important parts in adults even after longitudinal bone growth has ceased, promoting muscle development as well as regulating body mass. These effects have led to the unscrupulous promotion of GH as a "health"

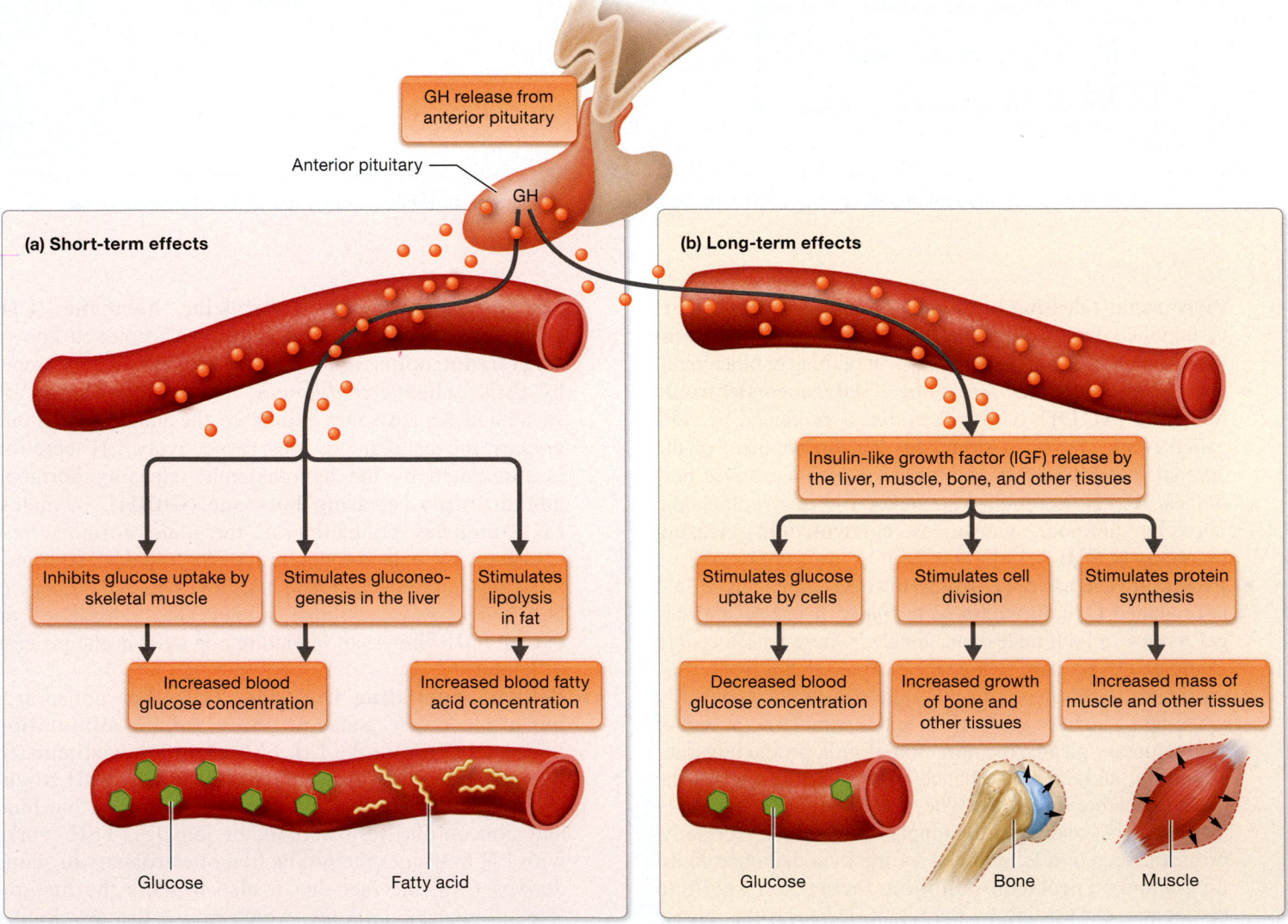

Figure 16.11 Effects of growth hormone (GH).

supplement, which is explored in *A&P in the Real World, Pseudoscience Exposed: Human Growth Hormone and the "Fountain of Youth."*

GH release is regulated by two hypothalamic hormones, as shown in **Figure 16.12**. **Growth hormone–releasing hormone (GHRH)** stimulates the release of GH, and its secretion increases during exercise, fasting, and stress, and after the ingestion of a protein-rich meal. GH release, like that of TSH, is inhibited by the hypothalamic hormone somatostatin, which is also known as *growth hormone–inhibiting hormone.*

Growth Hormone Disorders Imbalances in GH secretion can result in various disruptions in growth. Hypersecretion of GH, which is generally caused by a GH-secreting pituitary tumor, produces different effects depending on the age of the individual. If GH hypersecretion occurs *before* closure of the epiphyseal plates, **gigantism** results. Individuals with gigantism are unusually tall, generally more than 7 feet. Not only does the skeleton increase in size, but all other tissues, including the heart and other organs, enlarge as well. If the condition is detected early enough, it can be managed by removal of the tumor. However, if it is left untreated, affected individuals generally succumb to heart failure at an early age.

If GH hypersecretion occurs *after* closure of the epiphyseal plates, the result is a different condition, **acromegaly** (aeh´-kroh-MEG-uh-lee; "large extremities"). Longitudinal growth can no longer take place, but tissues do increase in girth, particularly those of the head, face, hands, and feet, as well as organs such as the heart and liver. This condition results in progressively distorted facial features, thickening of the tongue and skin, and enlargement of the hands and feet. Like gigantism, acromegaly can, if left untreated, lead to disruption of the functions of various organ systems, up to and including headaches, high blood pressure, and heart failure.

Hyposecretion of GH prior to closure of the epiphyseal plates results in **pituitary dwarfism.** Those with this condition are generally short in stature, but their limbs and trunk are proportional. A similar form of dwarfism results when IGF secretion is abnormally low, although in these patients GH secretion is normal to elevated.

Table 16.1 on p. 600 gives a summary of the hypothalamic hormones stored in and released from the posterior pituitary. It also provides details on the regulatory hormones of the hypothalamus and anterior pituitary.

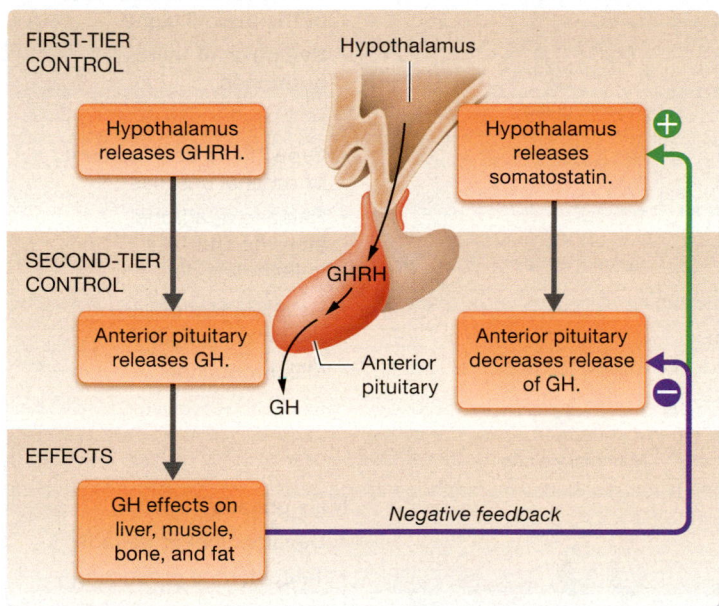

FIRST-TIER CONTROL

Hypothalamus

Hypothalamus releases GHRH.

Hypothalamus releases somatostatin.

SECOND-TIER CONTROL

GHRH

Anterior pituitary releases GH.

Anterior pituitary

Anterior pituitary decreases release of GH.

GH

EFFECTS

GH effects on liver, muscle, bone, and fat

Negative feedback

Figure 16.12 **The regulation of growth hormone (GH) release.**

Quick Check

☐ 6. How does the hypothalamus control the secretion of hormones from the anterior pituitary gland?

☐ 7. What are the tropic hormones of the anterior pituitary gland? What glands do they affect?

☐ 8. Describe the target tissues and effects of growth hormone and IGF.

A P in the Real World

Pseudoscience Exposed: Human Growth Hormone and the "Fountain of Youth"

In your e-mail inbox you may have received spam advertising the wonders of synthetic human growth hormone, or hGH. Proponents claim that hGH promotes fat loss and muscle growth, and is literally a "fountain of youth." It has become a popular supplement with body builders, those seeking weight loss, and individuals looking to stave off the inevitable effects of aging.

As with most such claims, there is a grain of truth to these assertions. GH does indeed promote fat breakdown and protein synthesis. In addition, adult patients with GH deficiency who are treated with hGH do lose fat and gain muscle. However,

serious problems can also arise with hGH supplementation. Excess GH has many potentially harmful effects, including elevated blood glucose level (hyperglycemia) and growth of organs such as the tongue and the bones of the hands, face, and feet.

However, most hGH supplements fortunately come with few risks, because they are sold as oral preparations to be ingested. Why does this minimize risk? Well, GH is a protein hormone, so it is destroyed by acid and enzymatic reactions in the stomach. The only way to effectively administer the drug is via injection, so the oral preparations amount to little more than expensive feces. Someone taking such supplements should still be careful, however, because they could still contain ingredients that are untested for safety or efficacy.

Table 16.1 Hormones of the Hypothalamus and Pituitary Gland

Hormone	Stimulus for Release	Inhibitor(s) of Release	Target Tissue(s)	Effects
Hypothalamic Hormones (stored in the posterior pituitary)				
Antidiuretic hormone (ADH)	Increased solute concentration of the blood	Decreased solute concentration of the blood	Kidneys Brain	• Water reabsorption from the kidney tubules • Increases blood volume
Oxytocin	• Stretching of the uterus • Infant suckling at the nipple	Lack of appropriate stimuli	Uterus Mammary gland	• Uterine contractions • Milk let-down reflex
Anterior Pituitary Hormones				
Thyroid-stimulating hormone (TSH)	• Thyrotropin-releasing hormone (TRH) from the hypothalamus • Exposure to cold • Stress	• Increased levels of thyroid hormones • Somatostatin from the hypothalamus	Thyroid gland	• Growth and development of the thyroid gland • Synthesis of thyroid hormones
Adrenocortico-tropic hormone (ACTH)	• Corticotropin-releasing hormone (CRH) from the hypothalamus • Stress	• Increased level of cortisol • Increased level of aldosterone	Adrenal cortex	• Growth and development of adrenal cortices • Release of adrenal steroids and catecholamines
Prolactin	• Infant suckling at the nipple • Prolactin-releasing hormone (PRH) from the hypothalamus	Prolactin-inhibiting factor (dopamine) from the hypothalamus	Mammary gland	• Development of mammary glands • Milk production
Luteinizing hormone (LH)	Gonadotropin-releasing hormone (GnRH) from the hypothalamus	Increased levels of testosterone (males) and estrogens and progesterone (females)	Male gonads Female gonads	Male gonads: • Development of gonads • Testosterone production Female gonads: • Development of gonads • Production of estrogens and progesterone • Ovulation
Follicle-stimulating hormone (FSH)	Gonadotropin-releasing hormone (GnRH) from the hypothalamus	Increased levels of testosterone (males) and estrogens (females)	Male gonads Female gonads	• Production of factors that bind and concentrate testosterone • Production of estrogens; maturation of ovarian follicles
Growth hormone (GH)	• Growth hormone–releasing hormone (GHRH) from the hypothalamus • Stress/exercise • Ingestion of protein • Fasting	Somatostatin from the hypothalamus	Liver Adipose tissue Muscle tissue Bone and cartilage	• Gluconeogenesis • Fat breakdown (lipolysis) • Protein breakdown • Production of insulin-like growth factor (IGF), which stimulates cell division and protein synthesis

☐ 1. A hypothetical poison destroys the hypothalamic cells that produce TRH. Predict the effects of this poison on hormone secretion by the anterior pituitary and thyroid glands.

☐ 2. *Syndrome of inappropriate ADH secretion,* or *SIADH,* is characterized by an abnormally increased secretion of ADH from the hypothalamus. Predict the effects of this condition.

☐ 3. A tumor increases the secretion of GHRH. Predict the effects of this tumor on secretion of GH and IGF. How would these effects differ in a child versus an adult?

See answers in Appendix A.

MODULE 16.3
The Thyroid and Parathyroid Glands

Learning Outcomes

1. Describe the gross and microscopic anatomy of the thyroid and parathyroid glands.

2. Identify and describe the types of cells within the thyroid gland that produce thyroid hormone and calcitonin.

3. Describe the stimulus for release, the target tissue, and the effects of thyroid hormones.

4. Explain how negative feedback loops regulate the production of thyroid hormones.

5. Describe the stimulus for release, the target tissue, and the effects of parathyroid hormone.

As you can see in **Figure 16.13a**, the *thyroid gland* is found in the anterior neck, just superficial to the larynx. The thyroid gland secretes two types of hormones: *thyroid hormones,* which regulate growth and metabolism, and *calcitonin* (kal-sih-TOH-nin), which helps to regulate calcium ion homeostasis. Embedded in the posterior surface of the thyroid gland we find several small glands, the *parathyroid glands* (*para-* = "next to"; refer back to Figure 16.3). These glands secrete a hormone called *parathyroid hormone,* which is a major player in maintaining the level of calcium ions in the ECF within the normal range.

This module begins by describing the structure of the thyroid gland and parathyroid glands. Next we explore the thyroid hormones and their many physiological roles. We then turn our attention to parathyroid hormone and calcitonin, and the module concludes by discussing hormonal maintenance of calcium ion homeostasis.

Structure of the Thyroid and Parathyroid Glands

The butterfly-shaped **thyroid gland** consists of right and left lobes connected by a small band called the **isthmus** (ISS-muss). Microscopically, the thyroid gland is composed of multiple spheres known as **thyroid follicles.** Note in **Figure 16.13b** that follicles are bounded by a layer of simple cuboidal epithelial cells referred to as *follicle cells,* which produce thyroid hormones. The interior of a follicle is filled with **colloid** (KAWL-oyd), a protein-rich, gelatinous material where the precursor of thyroid hormones is stored. Colloid also contains a high concentration of iodine atoms, which are required for thyroid hormone synthesis.

Between the thyroid follicles lie clusters of **parafollicular cells.** These cells are larger than follicle cells, and they have no

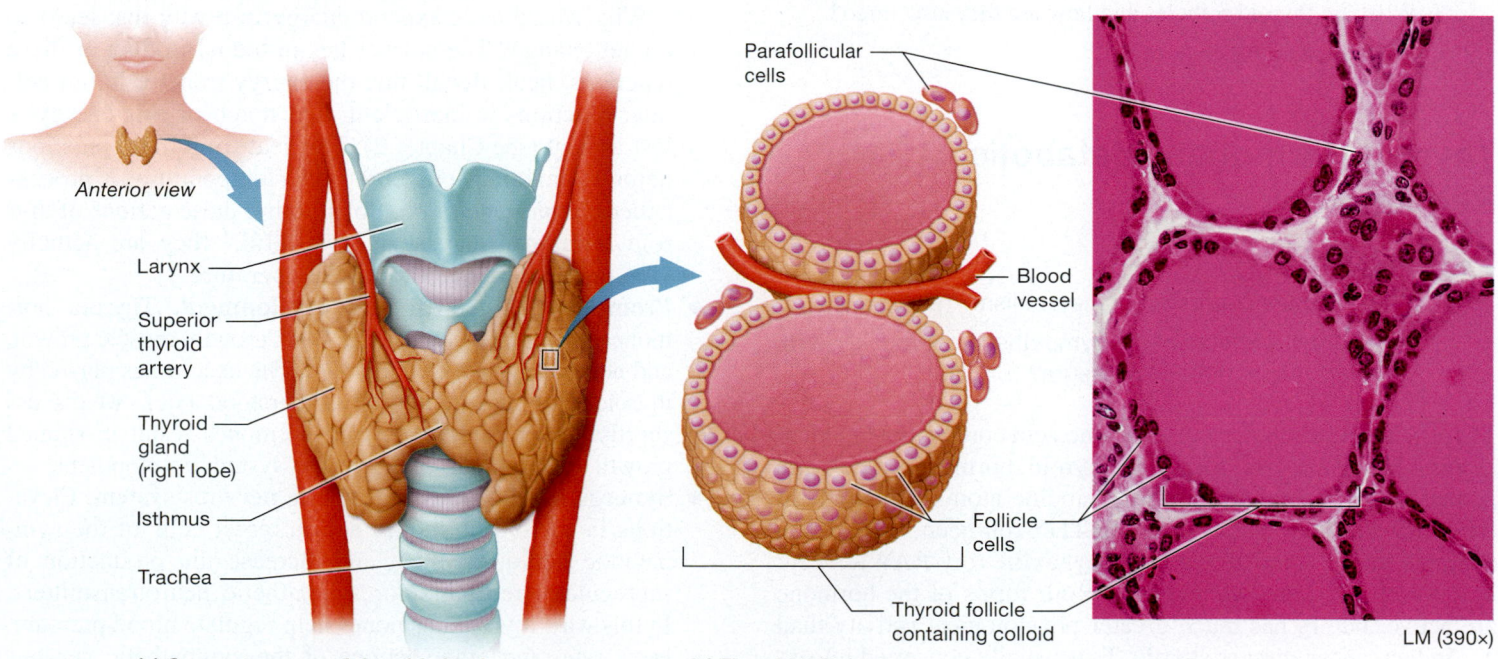

Anterior view

Larynx

Superior thyroid artery

Thyroid gland (right lobe)

Isthmus

Trachea

Parafollicular cells

Blood vessel

Follicle cells

Thyroid follicle containing colloid

LM (390×)

(a) Gross structure of thyroid gland   (b) Thyroid follicles

Figure 16.13 Anatomy and histology of the thyroid gland.

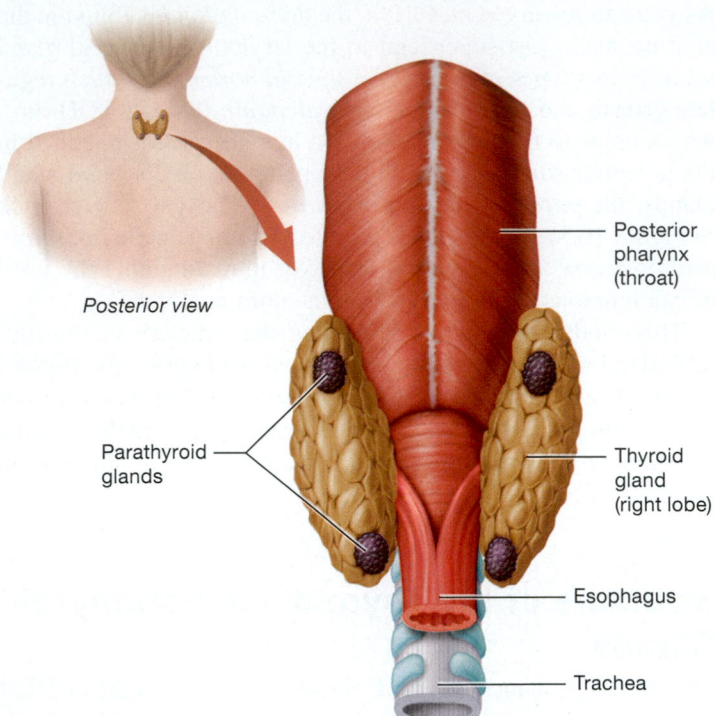

Posterior view

Posterior pharynx (throat)

Parathyroid glands

Thyroid gland (right lobe)

Esophagus

Trachea

Figure 16.14 Anatomy of the parathyroid glands.

role in thyroid hormone production. Instead, they produce the hormone calcitonin, which we discuss later in this module.

The **parathyroid glands** are small glands embedded in the posterior surface of the thyroid gland (**Figure 16.14**). Typically, there are three to five parathyroid glands, with four being most common. The hormone-secreting cells of the parathyroid gland, which produce parathyroid hormone, are called **chief cells.**

1. What are thyroid follicles and how are they structured?
2. What are parafollicular cells?

Thyroid Hormones: Metabolic Regulators

⟪ FLASHBACK

1. Why is heat a byproduct of metabolism? (p. 42)
2. What are the effects of the sympathetic nervous system on heart rate and blood pressure? (p. 524)

Thyroid hormones consist of an amino acid core bound to iodine atoms. There are two forms of thyroid hormone, which are named according to the number of iodine atoms they contain: **triiodothyronine** (try′-aye-oh-doh-THY-roh-neen; **T_3**), which contains three iodine atoms, and **thyroxine** (thy-RAWKS-een; **T_4**), which has four iodine atoms. Both forms of the hormone are active, but T_3 has much greater physiological activity than T_4. In fact, as we discuss shortly, T_4 is usually converted into T_3 in target cells.

Although T_3 and T_4 are amino acid–based hormones, their amino acids are nonpolar and hydrophobic, so they do not bind to cell surface receptor proteins. Instead, they diffuse into target cells and then bind to intracellular receptors in the nucleus. Here, like steroid hormones, they either activate or inhibit the transcription of certain genes.

This section first covers the functions of thyroid hormones. Then we see how they are produced, how their production is regulated, and what happens when this production is imbalanced.

Effects of Thyroid Hormones

Nearly every cell in the body contains receptors for thyroid hormones, so their effects are widespread. We can broadly divide these effects into three main categories: (1) regulation of the metabolic rate and thermoregulation, (2) promotion of growth and development, and (3) synergism with the sympathetic nervous system.

- **Regulation of the metabolic rate and thermoregulation.** Thyroid hormones set the *basal metabolic rate*—the amount of energy required by the body to carry out all its reactions when at rest. The hormones accomplish this in several ways. For one, they increase the synthesis of the ATP-requiring Na^+/K^+ pumps, which raises the amount of ATP the cell consumes. Thyroid hormones also act on a number of different cell types to trigger other energy-requiring reactions. For example, they act on cells of the liver and trigger gluconeogenesis. They also stimulate reactions that break down proteins in skeletal muscle and fats in adipose tissue in order to provide these compounds for gluconeogenesis. Finally, thyroid hormones initiate energy-requiring reactions in these same tissues that *build* proteins and fats. So, thyroid hormones increase the rate at which cells carry out both catabolic (breaking down) and anabolic (building up) reactions, leading to essentially no net change in the cell's overall composition.

  Why would cells expend energy in a way that leads to no net change? The answer lies in the byproduct of these reactions: heat. Recall that the energy transfer in our cellular reactions is inefficient, and much of this energy is lost as heat (see Chapter 2). The heat from these reactions helps to maintain our core body temperature, a process called *thermoregulation*. So although these actions of thyroid hormones may seem "wasteful," they are actually vital to maintaining our body temperature.

- **Promotion of growth and development.** Thyroid hormones are required for normal bone growth, muscle growth, and nervous system development. The exact roles played by thyroid hormones are not fully understood, but as we discuss shortly, deficiencies in thyroid hormones result in stunted growth and problems with nervous system development.

- **Synergism with the sympathetic nervous system.** Elevations in thyroid hormones act on target cells of the sympathetic nervous system and increase the production of (upregulate) receptors for sympathetic neurotransmitters. In this way, thyroid hormones help regulate blood pressure, heart rate, and other actions of the sympathetic nervous system without directly causing these effects.

Thyroid Hormone Production

The synthesis of thyroid hormones is a multistep process that occurs both in the colloid and in the follicle cells. The process is illustrated in **Figure 16.15** and proceeds as follows:

1. **Iodide ions and thyroglobulin are secreted into the colloid.** The synthesis of thyroid hormones begins with the accumulation of iodide ions (I⁻) in the colloid. Iodide ions are ingested in the diet, after which they are absorbed into the blood. In the capillaries of the thyroid gland, the iodide ions diffuse into the interstitial fluid and are actively pumped into the follicle cells. Finally, the ions enter the colloid via transport proteins in the plasma membranes of the follicle cells. In addition, follicle cells secrete the protein **thyroglobulin** (thy′-roh-GLOB-yoo-lin) into the colloid by exocytosis. Thyroglobulin is a very large protein that contains about 100 residues of the amino acid *tyrosine*. Thyroid hormones are made from these tyrosine residues.

2. **Iodide ions are converted to iodine atoms that attach to thyroglobulin.** An enzymatic reaction converts iodide ions to iodine atoms (I^0; the superscript zero indicates that the atom has no charge). Iodine atoms then move into the colloid and attach to selected tyrosine residues on the thyroglobulin, forming *iodinated thyroglobulin*. Some of the tyrosine residues receive only a single iodine atom, forming *monoiodothyronine*, or *MIT*, whereas others receive two iodine atoms, forming *diiodothyronine*, or *DIT*.

3. **Iodinated thyroglobulin enters the follicle cell by endocytosis, and T_3 and T_4 are cleaved from the molecule in reactions catalyzed by lysosomal enzymes.** The iodinated thyroglobulin is taken up by the follicle cells via endocytosis. The endocytic vesicles fuse with lysosomes, where enzymatic reactions cleave the MIT and DIT from thyroglobulin. If an MIT molecule is combined with a DIT molecule, the result is T_3, and if two DIT molecules are combined, the result is T_4. About 90% of the hormone produced is T_4, and the remainder is T_3.

4. **T_3 and T_4 are released into the blood.** T_3 and T_4 are released into the blood, where the vast majority of both bind to plasma proteins. The unbound hormones are known as *free* T_3 and T_4. Only free T_3 and T_4 have biological activity.

The thyroid gland produces T_4 in much greater amounts than T_3, yet T_3 is the far more active hormone. Why would the thyroid gland produce more of the less active hormone? The answer is that many peripheral tissues can convert T_4 to T_3, so having large amounts of T_4 in the blood provides a ready supply of potential T_3. This is important because T_3 has a short half-life of about 24 hours, whereas T_4 has a much longer half-life of about 7 days.

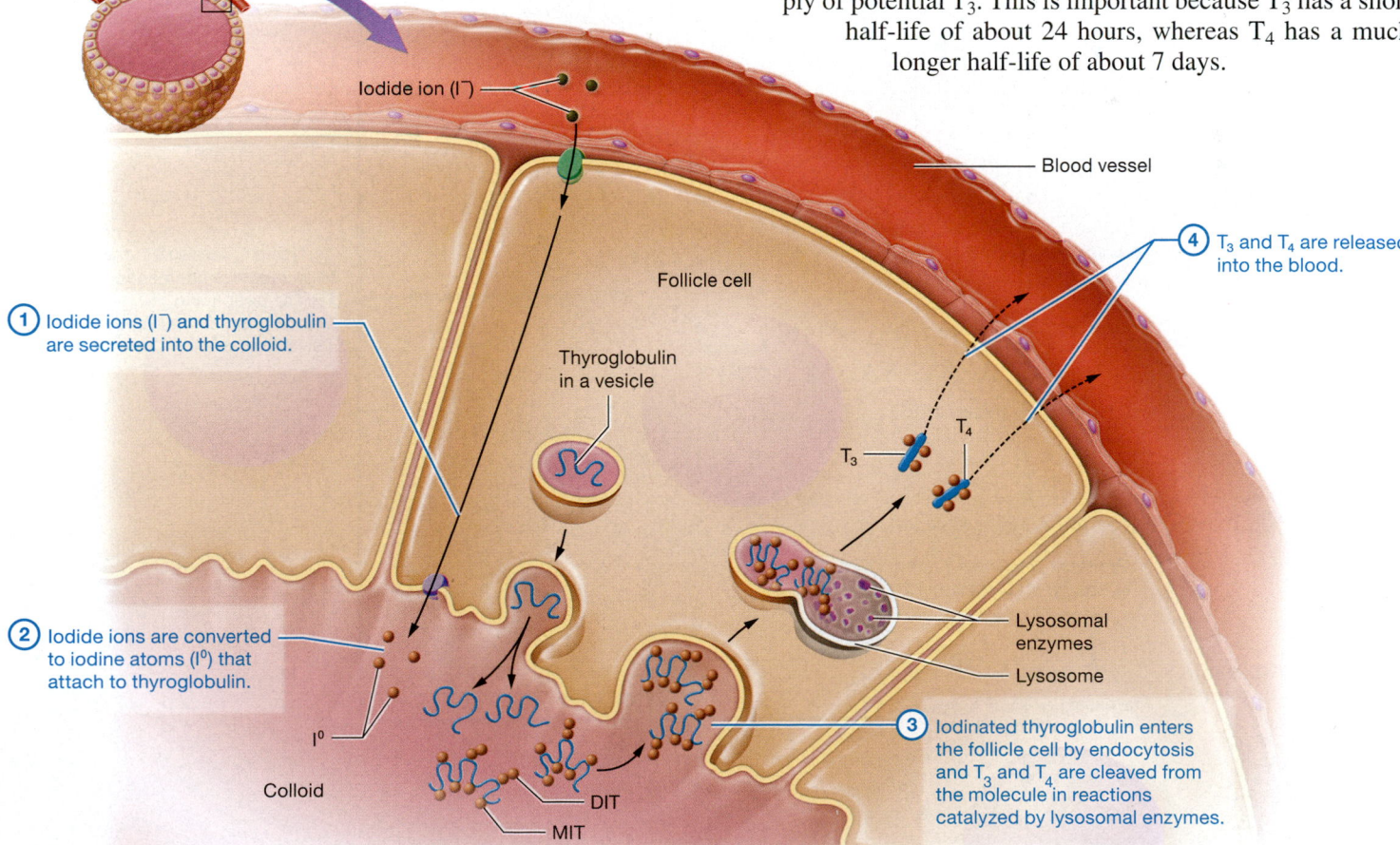

Iodide ion (I⁻)

Blood vessel

Follicle cell

④ T_3 and T_4 are released into the blood.

① Iodide ions (I⁻) and thyroglobulin are secreted into the colloid.

Thyroglobulin in a vesicle

T_4

T_3

② Iodide ions are converted to iodine atoms (I^0) that attach to thyroglobulin.

Lysosomal enzymes

Lysosome

I^0

Colloid

DIT

MIT

③ Iodinated thyroglobulin enters the follicle cell by endocytosis and T_3 and T_4 are cleaved from the molecule in reactions catalyzed by lysosomal enzymes.

Figure 16.15 Production of thyroid hormones.

Regulation of Thyroid Hormone Production

As with most hormones, production of T_3 and T_4 is regulated by a negative feedback loop. However, note in **Figure 16.16** that this negative feedback loop has multiple tiers of control. It involves both thyrotropin-releasing hormone (TRH) from the hypothalamus as the first tier of control and thyroid-stimulating hormone (TSH) from the anterior pituitary as the second tier. Hypothalamic neurons secrete TRH at a fairly constant rate; TRH then travels to the anterior pituitary via the hypothalamic-hypophyseal portal system and stimulates production and release of TSH. TSH stimulates multiple responses from the thyroid gland, including:

- production of T_3 and T_4 by follicle cells,
- secretion of T_3 and T_4 from follicle cells into the blood, and
- growth and development of the thyroid gland.

Production and secretion of TRH and TSH increase when levels of free T_3 and T_4 fall or when the body is exposed to cold temperatures. Secretion of both TRH and TSH is inhibited by rising levels of free T_3 and T_4. As mentioned earlier, the hypothalamic hormone somatostatin also inhibits TSH secretion.

Thyroid Disorders

Due to the systemic effects of thyroid hormones, serious imbalances in homeostasis can accompany their over- or underproduction. Overproduction of thyroid hormones is known as **hyperthyroidism;** underproduction is called **hypothyroidism.**

The most common cause of hyperthyroidism is **Graves' disease,** which results from the immune system producing abnormal proteins that mimic the actions of TSH on the thyroid gland. The abnormal proteins stimulate the thyroid, and the levels of both T_3 and T_4 in the blood increase. The common signs and

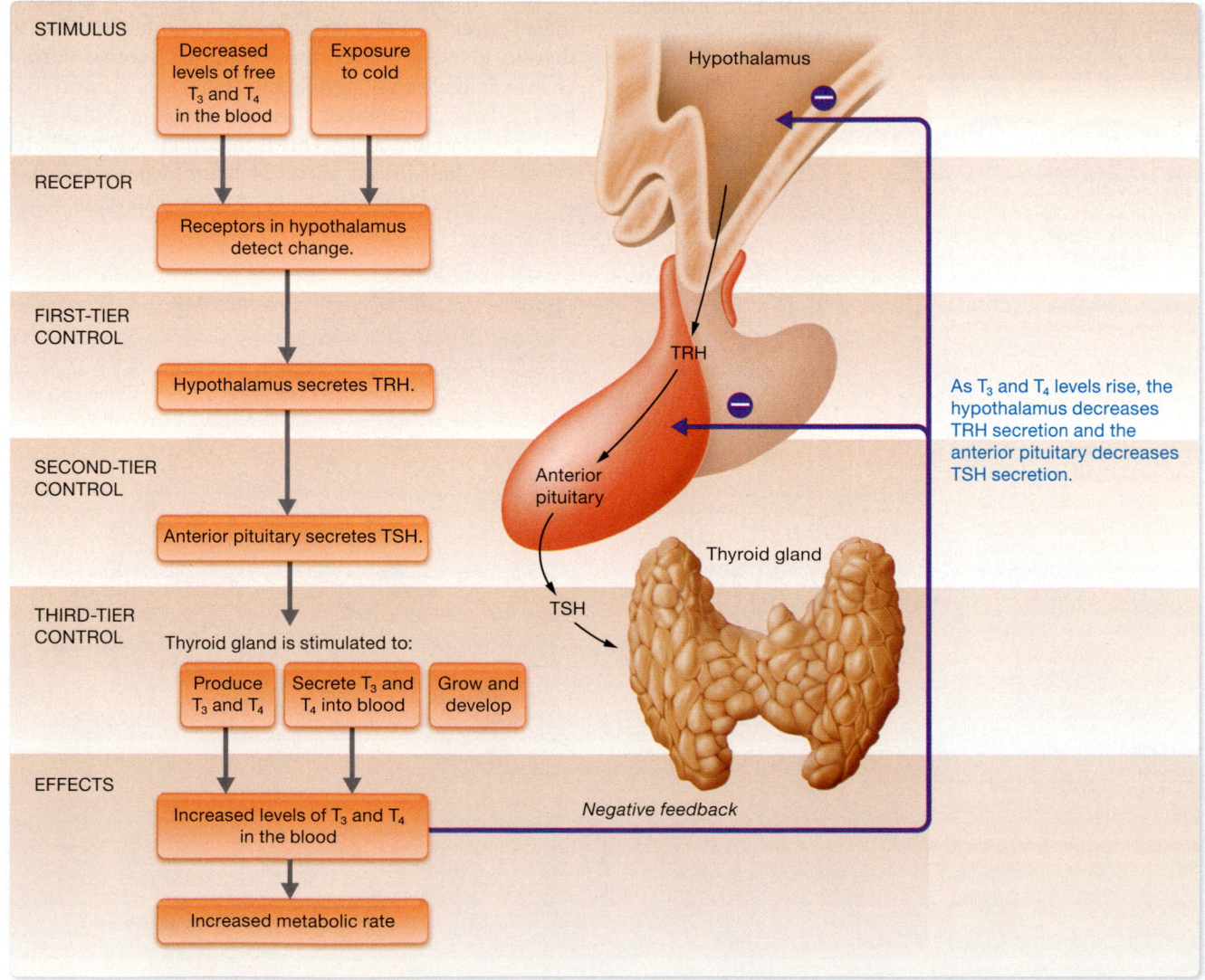

Figure 16.16 **Maintaining homeostasis: regulation of thyroid hormone production by a negative feedback loop.**

symptoms of Graves' disease result from excessive T$_3$ and T$_4$ activity, and include:

- weight loss due to an elevated metabolic rate,
- heat intolerance due to excessive heat production, and
- disruptions in heart rhythm and increase in blood pressure due to the synergism of thyroid hormones with the sympathetic nervous system.

Another sign of Graves' disease is *exophthalmos* (eks-off-THAL-mohs). Exophthalmos is characterized by bulging of the eyeballs caused by weakened extrinsic eye muscles and water retention in the tissues around the orbits. A final sign of hyperthyroidism is enlargement of the thyroid gland, known as a **goiter** (GOY-t'r), shown in **Figure 16.17**. Thyroid enlargement results from the abnormal TSH-like proteins secreted in Graves' disease, which stimulate growth of the thyroid gland.

Hyperthyroidism may be treated with drugs that decrease the production of T$_3$ and T$_4$, but such drugs must be taken for the rest of a patient's life and are not always a good permanent solution. Permanent treatments involve the surgical removal or nonsurgical destruction of the thyroid gland. The gland is destroyed nonsurgically with radioactive iodine, which is selectively taken up by the follicle cells. The radiation produced as the iodine decays kills the follicle cells.

The two most common causes of hypothyroidism are iodine deficiency and destruction of the thyroid gland by the immune system. The latter condition is known as *Hashimoto thyroiditis*. Iodine-deficiency hypothyroidism is relatively uncommon in the developed world, because most salt is supplemented with iodine. Also, people who regularly consume food from the ocean may receive their iodine naturally. However, in some developing countries, iodine deficiency remains relatively common. Both causes of hypothyroidism result in decreased T$_3$ and T$_4$ production and concurrent increases in TRH and TSH production. This can be explained by the negative feedback loops that attempt to increase T$_3$ and T$_4$ production from the thyroid gland.

Most signs and symptoms of hypothyroidism are opposite those of hyperthyroidism and include:

- weight gain due to a decreased metabolic rate,
- cold intolerance due to decreased heat production, and
- slow heart rate and low blood pressure due to decreased synergism with the sympathetic nervous system.

Surprisingly, many patients with hypothyroidism present with one of the same signs found in hyperthyroidism—a goiter. A goiter develops in Hashimoto thyroiditis because the increased level of TSH promotes thyroid gland growth. In iodine-deficiency

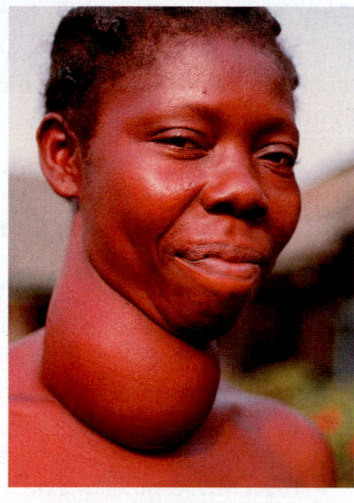

Figure 16.17   Disorder of thyroid hormone secretion: goiter.

hypothyroidism, the elevated TSH level contributes to the goiter, but the goiter also results from the accumulation of thyroglobulin in the colloid of thyroid follicles. For this reason, goiters seen with iodine-deficiency hypothyroidism can become quite large.

The treatment for hypothyroidism depends on the cause. Iodine deficiency typically resolves with iodine supplementation. Hashimoto thyroiditis has no cure, but can be easily treated with thyroid hormone supplementation for the remainder of the patient's life.

A special variant of hypothyroidism, called **congenital hypothyroidism** (formerly known as *cretinism*), develops when an infant is born with inadequate thyroid function. If untreated, it results in delayed physical and nervous system development and eventual intellectual disability. Congenital hypothyroidism is completely avoidable with early diagnosis and treatment. For this reason, infants are screened for hypothyroidism at birth.

ConceptBOOST)))

Understanding the Relationship between Negative Feedback Loops and Thyroid Function

As we've discussed, TSH secretion *increases* with hypothyroidism and *decreases* with hyperthyroidism. This may seem counterintuitive, so let's look at why this happens. You've read that the production of T$_3$ and T$_4$ is regulated by a negative feedback loop. This means that when T$_3$ and T$_4$ levels in the blood drop, the receptors in the cells of the hypothalamus detect the change and secrete more TRH, which stimulates the anterior pituitary to secrete more TSH. Under normal conditions, the elevated TSH will stimulate the thyroid gland to produce more T$_3$ and T$_4$. When T$_3$ and T$_4$ levels rise, receptors in both the hypothalamus and anterior pituitary gland detect this change, and TRH and TSH levels decrease through negative feedback.

However, in hypothyroidism, the thyroid gland is unable to produce more T$_3$ and T$_4$. The hypothalamus and anterior pituitary don't "know" that the thyroid gland isn't functioning; all their receptors sense is that T$_3$ and T$_4$ levels are low, so the cells continue to secrete more TRH and TSH in an attempt to stimulate the thyroid. This leads to the characteristic elevated TSH and decreased T$_3$ and T$_4$ levels seen with hypothyroidism.

This situation works in reverse with hyperthyroidism. Under normal conditions, the cells of the hypothalamus and anterior pituitary gland decrease production of TRH and TSH when levels of T$_3$ and T$_4$ rise. In the case of hyperthyroidism, levels of T$_3$ and T$_4$ are extremely elevated. The hypothalamus detects these abnormally high levels of T$_3$ and T$_4$, and "decides" that there is no further need to stimulate the thyroid gland. For this reason, production and secretion of TRH and TSH fall to nearly zero. This leads to the characteristic decreased TSH with elevated T$_3$ and T$_4$ levels seen in hyperthyroidism. ■

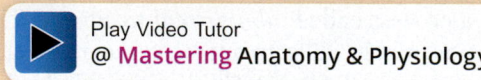

Play Video Tutor
@ **Mastering** Anatomy & Physiology

☐ 3. What are the main functions of thyroid hormones?

☐ 4. How are thyroid hormones produced? How is this production regulated?

☐ 5. What homeostatic imbalances may accompany hyperthyroidism and hypothyroidism?

Parathyroid Hormone and Calcitonin: Calcium Ion Homeostasis

《 FLASHBACK

1. What are osteoblasts and osteoclasts, and what are their functions? (p. 190)

2. What is the function of vitamin D? (p. 203)

As we discussed in the introduction to this module, calcitonin and parathyroid hormone are produced by the thyroid and parathyroid glands, respectively. We now take a closer look at the functions and target tissues of these hormones, as well as their roles in maintaining calcium ion concentration in the blood within a specific range.

Parathyroid Hormone

The chief cells of the parathyroid glands secrete the peptide hormone **parathyroid hormone (PTH),** which is a major factor in maintaining the blood calcium ion concentration in the normal range. PTH is secreted in response to a declining blood calcium ion concentration, a condition known as *hypocalcemia.* Its actions raise the calcium ion level of the blood by triggering the following effects:

- **Increasing release of calcium ions from bone by stimulating osteoclasts.** PTH induces changes in osteoblasts that enhance osteoclast activity. Recall that osteoclasts are "bone-destroying" cells—they release calcium ions into the blood by breaking down the extracellular matrix of bone (see Chapter 6). As the osteoclasts degrade bone, calcium ions are released into the blood.
- **Increasing absorption of dietary calcium ions by the small intestine.** The effects of PTH on the small intestine are indirect and involve the hormone vitamin D, which is produced by the skin in response to certain wavelengths of ultraviolet light (see Chapter 5). However, the skin produces an *inactive* form of vitamin D. PTH acts on the kidneys to stimulate the conversion of vitamin D to its active form, vitamin D3, or **calcitriol** (kal-sih-TRY-awl). In this form, vitamin D3 increases the number of calcium ions absorbed from the small intestine.
- **Increasing reabsorption of calcium ions from the fluid in the kidneys.** In the process of filtering blood, the kidneys remove large amounts of water and solutes from the blood, including calcium ions. The cells of the kidney tubules then reclaim most of the water and solutes through a process called *tubular reabsorption.* PTH acts on the kidney tubule cells to increase the number of calcium ions reclaimed from the fluid in the tubules before it becomes urine and is lost to the body.

Regulation of Calcium Ion Homeostasis

The negative feedback loop involving PTH that maintains the blood calcium ion concentration in its normal range proceeds as follows (**Figure 16.18**).

- **Stimulus: The blood calcium ion level decreases below the normal range.**
- **Receptor: Chief cells in the parathyroid gland detect a low blood calcium ion level.** A low calcium ion concentration in the blood leads to a decreased calcium ion concentration in the ECF around the chief cells. This is then detected by their cell surface receptors.
- **Control center: The chief cells increase PTH secretion.** The chief cells, which also act as the control center, increase their production and secretion of PTH into the blood.
- **Effector/response: The effects of PTH on target cells increase the blood calcium ion concentration.** PTH affects the functions of its target cells in bone (osteoclasts), the kidneys, and the small intestine. In addition, it stimulates the conversion of inactive vitamin D to the active form, calcitriol. All these effects increase the concentration of calcium ions in the blood.
- **Homeostatic range and negative feedback: As the blood calcium ion level returns to the normal range, negative feedback to chief cells decreases PTH secretion.** The calcium ion concentration in the blood returns to the normal range. Receptors on the chief cells detect this change, and negative feedback decreases their production and secretion of PTH. This in turn decreases stimulation to the target cells, and the response is reduced.

Calcitonin

Hypercalcemia occurs when the calcium ion level in the blood increases above normal. In response to hypercalcemia, the hormone **calcitonin** is released. Calcitonin, which is produced by the thyroid gland's parafollicular cells, decreases the blood calcium ion concentration. Calcitonin's primary actions involve inhibiting the activity of the osteoclasts in bone. This leaves the effects of the osteoblasts, the "bone-building" cells, unopposed. Osteoblasts deposit calcium ions into the bone, which reduces the blood concentration of calcium ions.

Calcitonin's effects have been difficult to characterize because they are fairly transient—the effects of a high calcitonin level appear to last only a few hours. Its activity seems to be most important during times of active bone turnover, such as during bone growth. At present, calcitonin is not believed to be a major regulator of calcium ion homeostasis, and might instead play a more important role in bone homeostasis.

Both calcitonin and PTH can be used to treat the bone disease *osteoporosis* (ahs´-tee-oh-por-OH-sis). See *A&P in the Real World: Calcitonin, Parathyroid Hormone, and Osteoporosis* on p. 608 for more information.

The hormones of the thyroid and parathyroid glands are summarized in **Table 16.2.**

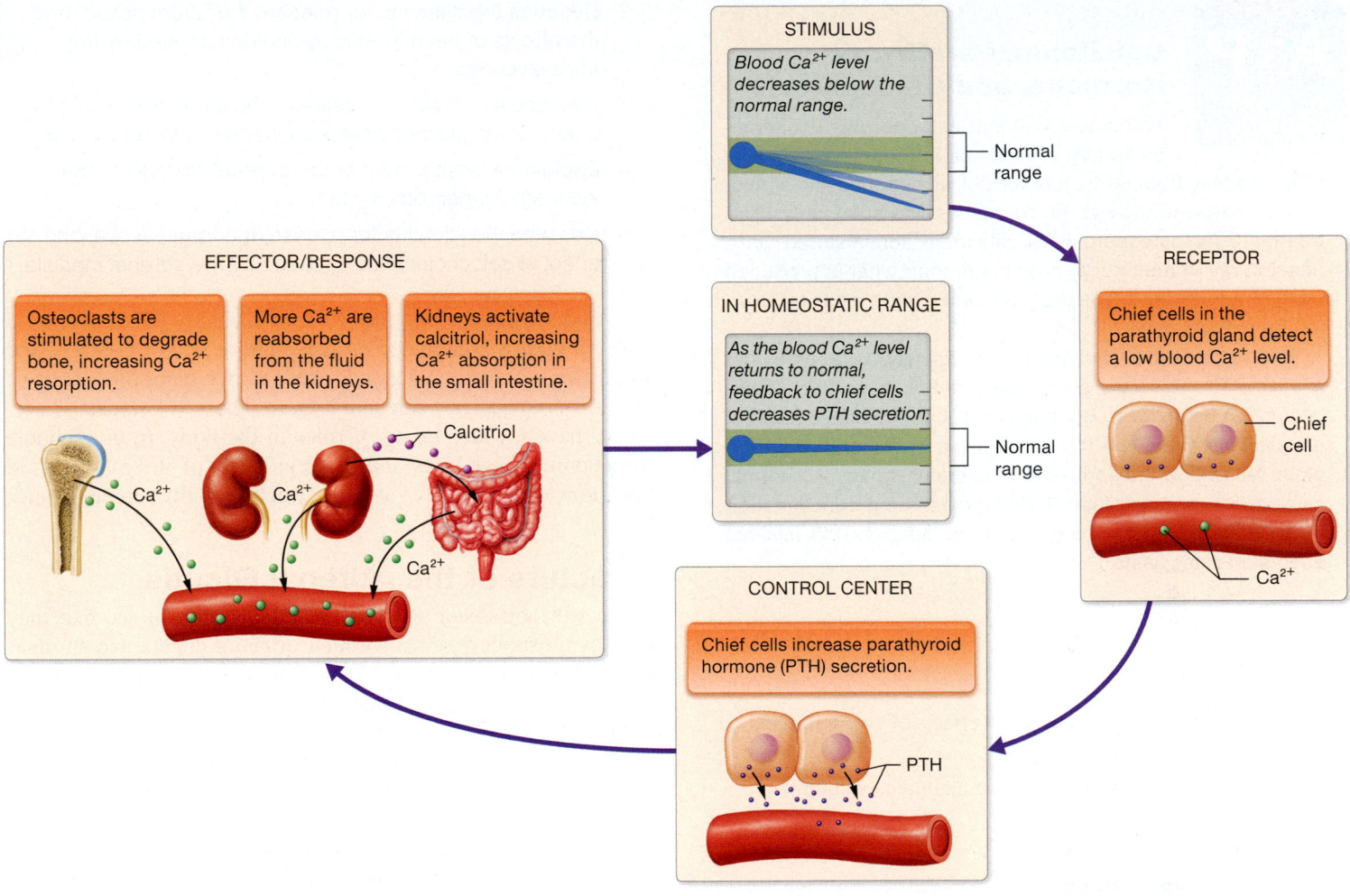

Figure 16.18 Maintaining homeostasis: regulation of blood calcium ion concentration by a negative feedback loop.

Table 16.2 Hormones of the Thyroid and Parathyroid Glands

Cell Type	Hormone(s)	Stimulus for Secretion	Inhibitor(s) of Secretion	Target Tissue(s)	Effects
Thyroid Gland					
Follicle cells	Triiodothyronine (T_3) and thyroxine (T_4)	TSH from the anterior pituitary	Increased levels of T_3 and T_4 inhibit TRH and TSH.	Nearly every cell in the body	• Set the basal metabolic rate • Thermoregulation • Growth and development • Synergism with the SNS
Parafollicular cells	Calcitonin	Increased concentration of calcium ions in the blood	Decreased concentration of calcium ions in the blood	Osteoclasts	Inhibits osteoclast activity under certain conditions, which transiently decreases the blood calcium ion concentration.
Parathyroid Glands					
Chief cells	Parathyroid hormone (PTH)	Decreased concentration of calcium ions in the blood	Increased concentration of calcium ions in the blood	Bone, kidneys, intestines	• Increases calcium ion reabsorption from the fluid in the kidneys • Increases calcium ion absorption from the contents of the small intestine • Indirectly increases osteoclast activity • All effects lead to an increased blood calcium ion concentration.

Calcitonin, Parathyroid Hormone, and Osteoporosis

You may recall that osteoporosis is characterized by a decrease in the mineral density of bones (see Chapter 6). It results in weak bones and significantly increases the risk for fractures. One commonly used treatment for osteoporosis is calcitonin administered via a nasal spray in much higher amounts than what is produced by the thyroid gland. The drug increases bone mass by slowing bone breakdown.

Another common treatment, surprisingly, is a drug that mimics PTH. This may seem counterintuitive, as PTH stimulates bone breakdown. But it appears that PTH has this effect on osteoclasts only if the calcium ion concentration in the blood falls below the normal level. In the presence of a normal calcium ion concentration, PTH increases the concentration through its effects on the gut and the kidneys, with minimal effects on osteoclasts.

Quick Check

☐ 6. What are the target tissues and effects of calcitonin and parathyroid hormone?

☐ 7. How does parathyroid hormone maintain calcium ion homeostasis?

Apply What You Learned

☐ 1. Certain dietary supplements marketed for weight loss contain thyroid hormone. How would these supplements cause weight loss? Would these products generally be safe?

☐ 2. Mr. Scully, a patient with hypothyroidism, has undergone blood tests to check his thyroid function. The lab technician reports that his TSH level is abnormally high, but forgets to record the levels of T_3 and T_4. Based on the TSH level, predict whether Mr. Scully's T_3 and T_4 levels will be elevated, normal, or decreased. Explain your answer.

☐ 3. Tumors of the parathyroid gland often secrete excessive parathyroid hormone. Predict the effects of such a tumor.

See answers in Appendix A.

MODULE **16.4**
The Adrenal Glands

Learning Outcomes

1. Describe the gross and microscopic anatomy of the cortex and medulla of the adrenal gland.

2. Describe the stimulus for release, the target tissue, and the effects of the mineralocorticoids secreted by the adrenal cortex.

3. Describe the stimulus for release, the target tissue, and the effects of the glucocorticoids secreted by the adrenal cortex.

4. Explain the relationship of the adrenal medulla to the sympathetic nervous system.

5. Describe the stimulus for release, the target tissue, and the effect of catecholamines secreted by the adrenal medulla.

As you discovered in Module 16.1, the roughly pyramid-shaped **adrenal glands** are located on the superior end of the right and left kidneys. The adrenal glands produce two types of hormones: steroid hormones and catecholamines. These two types of hormones have a wide variety of roles in the body. In this module we examine these roles and the regulation of these hormones. First, however, let's delve into the structure of the adrenal glands.

Structure of the Adrenal Glands

When we look inside the adrenal glands we can see that they have two distinct regions: an outer **adrenal cortex** and an inner **adrenal medulla** (**Figure 16.19a**). Whereas the adrenal cortex is a typical endocrine gland that functions in much the same way as the other such organs we've discussed, the adrenal medulla is a neuroendocrine organ that secretes neurohormones.

As you can see in **Figure 16.19b**, the adrenal cortex has three distinct zones: the outer **zona glomerulosa** (glahm-ehr´-yoo-LOH-suh), the middle **zona fasciculata** (fah-sik´-yoo-LAH-tuh), and the inner **zona reticularis** (reh-tik´-yoo-LEHR-us). The zona glomerulosa consists of densely packed cells, whereas the middle zona fasciculata consists of cells stacked on top of one another in columns. The cells of the thin zona reticularis are arranged more loosely in clusters.

Hormones of the Adrenal Cortex

◀◀ FLASHBACK

1. Is the concentration of sodium ions higher in the cytosol of our cells or the extracellular fluid? What about potassium ions? (p. 80)

Each zone of the adrenal cortex produces steroid hormones derived from cholesterol. The zona glomerulosa cells produce hormones called *mineralocorticoids* (min´-er-aeh-loh-KOHR-tih-koydz). The cells of the zona fasciculata produce and secrete *glucocorticoids* (gloo´-koh-KOHR-tih-koydz). Small amounts of glucocorticoids are also produced and secreted by the cells of the zona reticularis, which in addition make *androgenic steroids*.

Production of these steroids by the adrenal cortex is partially regulated by a multi-tiered negative feedback system referred to as the *hypothalamic-pituitary-adrenocortical,* or *HPA, axis.* Let's examine the primary hormones of the zones of the adrenal cortex and how their production and secretion are regulated by the HPA axis and other factors.

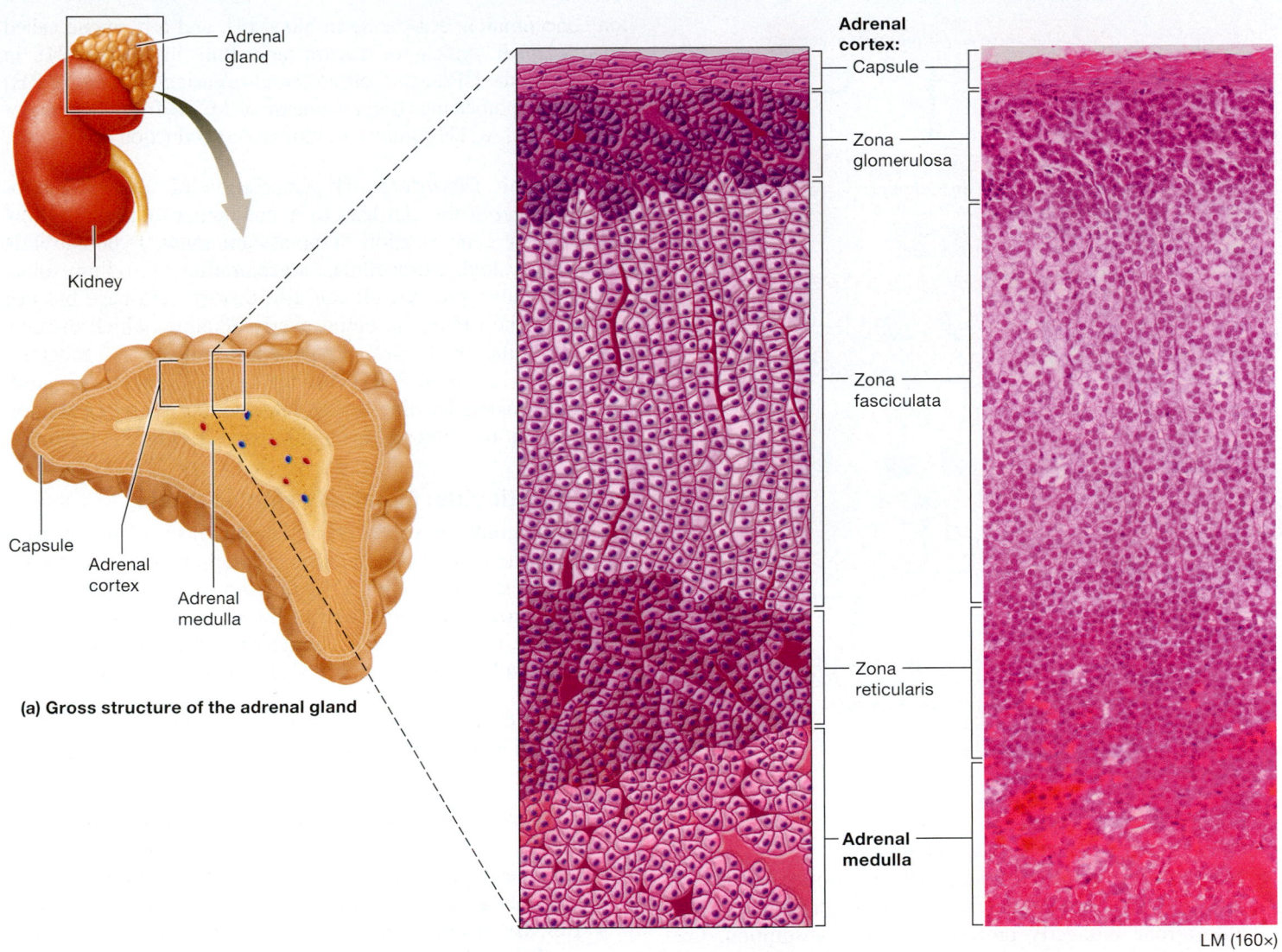

(a) Gross structure of the adrenal gland

(b) Histology of the adrenal gland: illustration (left) and light micrograph (right)

Figure 16.19 Anatomy and histology of the adrenal gland.

Mineralocorticoids: Fluid and Electrolyte Homeostasis

Mineralocorticoids regulate the concentration of certain *minerals* in the body (such as sodium and potassium ions). The main mineralocorticoid is **aldosterone** (al-DAWS-ter-ohn). Let's examine the functions of aldosterone and how it is regulated.

Effects of Aldosterone   Notice in **Figure 16.20** that aldosterone has several functions:

- **Maintaining the concentrations of extracellular sodium and potassium ions within their normal ranges.** As we discussed in earlier chapters, the gradients of sodium and potassium ions across plasma membranes are critical to the functioning of muscle cells and neurons (see Chapters 10 and 11). Specifically, such excitable cells require high extracellular and low intracellular concentrations of sodium ions, and low extracellular and high intracellular

concentrations of potassium ions. These gradients are maintained largely by the actions of the sodium-potassium ion pumps in the cells' plasma membranes. The concentrations of these ions in the ECF are also partially maintained by aldosterone acting on the cells of the kidney tubules. Here, it stimulates transcription of more sodium-potassium ion pumps, sodium ion channels, and sodium/chloride/potassium ion cotransporters. These actions have two net effects: (1) Sodium and chloride ions are transported from the fluid in the kidney tubules to the extracellular fluid for return to the blood, and (2) potassium ions are transported from the extracellular fluid into the fluid in the kidney tubules, where they will be excreted in the urine.

- **Regulating extracellular fluid volume.** Aldosterone's effects on sodium and chloride ions in the tubule cells create a more concentrated interstitial fluid. This causes water to move by osmosis from the fluid in those tubules to the interstitial fluid and blood. Note that this is an indirect

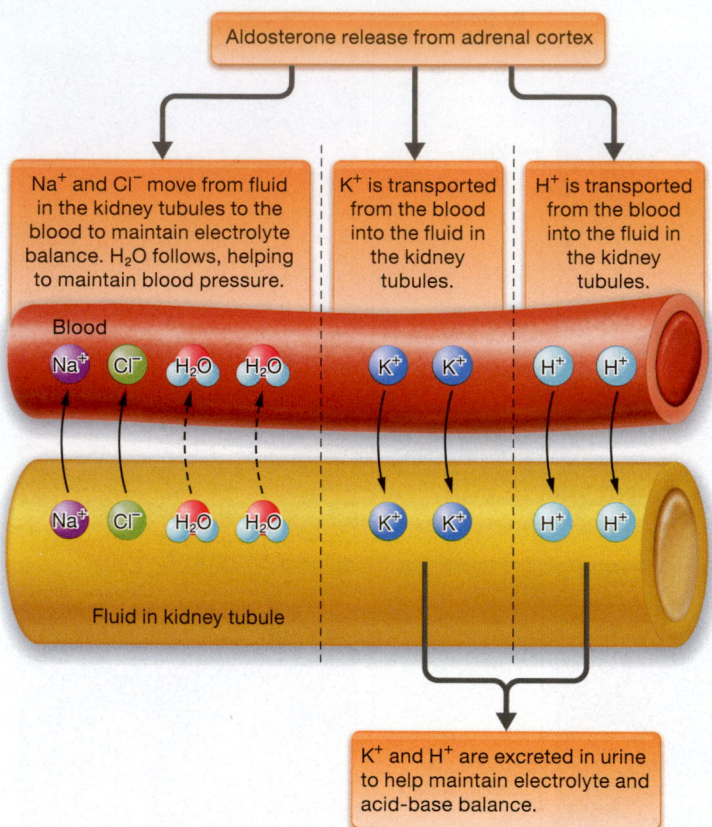

Figure 16.20 **Effects of aldosterone.**

effect—aldosterone has no direct actions on water molecules or water channels in the tubule cells.

- **Maintaining blood pressure.** Blood volume is one of the key factors that determine blood pressure. When aldosterone indirectly promotes water reabsorption, this maintains the extracellular fluid and blood volume, which in turn maintains blood pressure. Aldosterone's effects on blood pressure are a key part of a larger system involving the kidneys, lungs, and liver, called the *renin-angiotensin-aldosterone system* (*RAAS*). The overall physiological functions of this system are to increase blood pressure, and to preserve blood flow to the heart, brain, and kidneys. We further discuss this system with the kidneys as other endocrine organs in Module 16.6, and fully with the urinary system (see Chapter 24).
- **Maintaining acid-base homeostasis.** The pH of the blood is slightly alkaline, at about 7.35–7.45. Aldosterone helps to maintain this normal range by activating hydrogen ion pumps in certain cells of the kidney tubules. These pumps transport hydrogen ions from the extracellular fluid into the fluid of the tubules, after which they are excreted in the urine. This lowers the hydrogen ion concentration of the blood, preserving its slightly alkaline pH.

Regulation of Aldosterone Synthesis Multiple factors regulate aldosterone synthesis and secretion. The major factors that stimulate aldosterone release include an elevated blood potassium

ion concentration, a decrease in blood pH, and a hormone called *angiotensin-II* (which we discuss more fully in Chapter 24). In addition, in the HPA axis, corticosteroid-releasing hormone (CRH) from the hypothalamus triggers release of ACTH from the anterior pituitary, and ACTH stimulates aldosterone production and release.

Aldosterone Disorders Hypersecretion of aldosterone, or *hyperaldosteronism,* can lead to hypokalemia (abnormally low extracellular concentration of potassium ions), hypernatremia (abnormally high extracellular concentration of sodium ions), and high blood pressure. It can also disrupt acid-base balance and make the pH of the blood overly alkaline, which disrupts many cellular processes. Hyperaldosteronism is generally *idiopathic* (id'-ee-oh-PATH-ik), meaning of unknown cause, but is occasionally due to an aldosterone-secreting tumor. Hyposecretion of aldosterone is discussed later in this module.

Glucocorticoids: Metabolic Homeostasis

Glucocorticoids are produced in the zona fasciculata and zona reticularis, and their main function is to help mediate the body's response to stress. As implied by their name, they help respond to stress through the regulation of blood glucose, among other effects. The most potent glucocorticoid in the body is the hormone **cortisol** (KOHR-tih-zohl), also known as **hydrocortisone.**

Effects of Cortisol Cortisol's primary target tissues are liver, muscle, and adipose, where it stimulates the following processes (**Figure 16.21**):

- **Gluconeogenesis in the liver.** Cortisol binds to genes in liver cells (hepatocytes) and stimulates them to synthesize enzymes that convert amino acids and fats into glucose by gluconeogenesis. This increases the blood glucose level.
- **Release of amino acids from muscle tissue.** Cortisol acts on skeletal muscle to induce the breakdown of muscle proteins into amino acids. These amino acids enter the bloodstream, where they are then available to the liver for use in gluconeogenesis.
- **Release of fatty acids from adipose tissue.** The fatty acids stored in adipose tissue can be used by cells as an alternative fuel to glucose and by the liver in gluconeogenesis. For this reason, cortisol acts on adipocytes and triggers lipolysis, or fat breakdown, releasing fatty acids into the bloodstream.

Cortisol's role in the body extends beyond its influence on blood glucose concentration. In fact, most tissues in the body have receptors for cortisol, so its effects are wide-ranging. One of its most pronounced effects is on the immune system, where cortisol acts as a potent anti-inflammatory agent by decreasing the levels of certain leukocytes. The strength of cortisol as an anti-inflammatory agent led to the development of synthetic cortisol derivatives called *corticosteroids,* which are used to treat a variety of conditions in which inflammation is a key component. Cortisol is also used to suppress the immune response; for example, it is given to patients with *autoimmune disorders,* in which a patient's own immune system attacks normal body

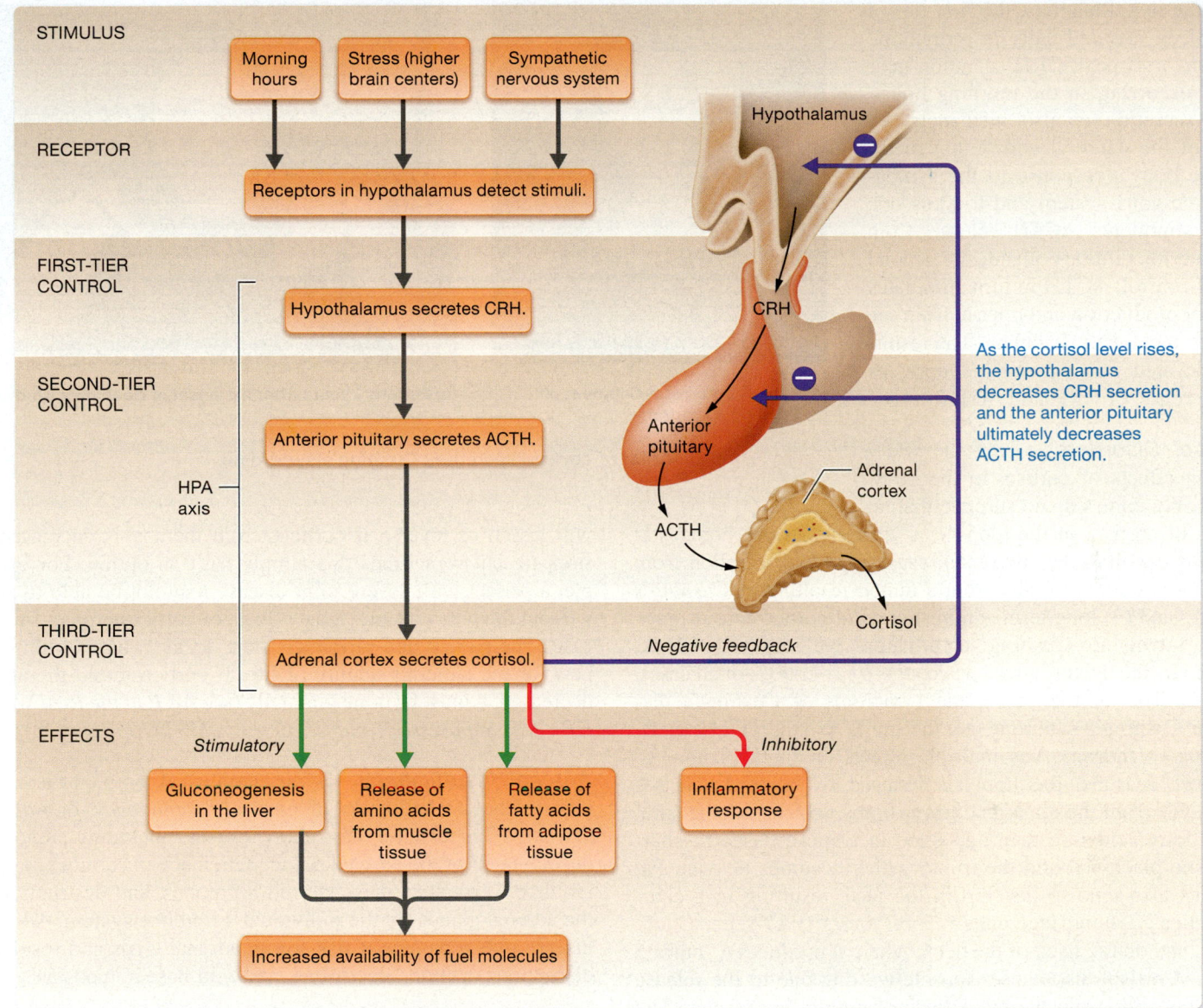

Figure 16.21 Regulation of cortisol production by a negative feedback loop.

cells, and to organ transplant recipients to help prevent rejection of the new organ.

Cortisol's actions make it an important component of the **stress response,** which is the series of changes that maintains homeostasis when the body is faced with a *stressor* (which can be anything from exercise or emotional excitement to being chased by an alligator). Indeed, cortisol is sometimes called the "stress hormone," which reflects this role. To understand this relationship, step back and think about what would happen if you were subjected to the physical and emotional stress of being chased by an alligator:

- You are running from the alligator, which means your skeletal muscle cells will need additional glucose to fuel their activity.
- Your muscle activity will quickly use up the supply of glycogen, so these cells will need other fuel sources, such as fatty acids.

- If you fall and break your toe while running, you will be unable to stop and nurse your injury. Instead, you need to continue running until you are safe.

How does cortisol help with each of these points? First, cortisol triggers breakdown of fats and proteins for gluconeogenesis, which increases the glucose level in the blood. Second, the fatty acids released from adipose tissue provide additional fuels for continued muscle activity. Finally, the anti-inflammatory actions of cortisol prevent inflammation from the injury to your toe, which enables you to continue running. These effects make cortisol an extremely important hormone in the stress response, which we examine more fully in Module 16.7.

Regulation of Cortisol Synthesis Cortisol synthesis is primarily controlled by the HPA axis (see Figure 16.21). CRH from the hypothalamus, the first tier of control, is secreted daily in a rhythmic fashion, with peak CRH secretion generally occurring in the morning hours. CRH secretion is also influenced by parts of the cerebral cortex that mediate the body's response to the sympathetic nervous system and to stressors. CRH stimulates ACTH release from the anterior pituitary gland, the second tier of control. ACTH in turn stimulates cortisol production and release from the cells of the zona fasciculata. The resulting elevated cortisol level suppresses the secretion of CRH and ACTH.

(a) Patient before development of Cushing's syndrome

(b) Patient 3 years after the onset of Cushing's syndrome

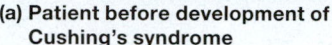

Figure 16.22 **Disorder of cortisol secretion: Cushing's syndrome.**

Cortisol Disorders Given the wide-ranging effects of cortisol in the body, it probably comes as no surprise that an excess of cortisol in the blood can seriously disrupt homeostasis. This condition has two main causes: (1) oversecretion from the adrenal cortex, usually from a tumor, resulting in *Cushing's disease;* and (2) long-term administration of corticosteroids, producing **iatrogenic Cushing's syndrome** (aye′-aeh-troh-JEN-ik; "caused by the doctor"), which accounts for over 90% of all cases.

Regardless of the cause, the symptoms of Cushing's disease or Cushing's syndrome are the same. As you might expect, Cushing's syndrome has multiple effects on metabolism. For unknown reasons, the lipolysis induced by cortisol releases fatty acids from the upper and lower limbs, which become slim. These fatty acids are then deposited in adipose tissue in characteristic places around the trunk, which becomes heavier. Fat deposits also tend to develop in the face, resulting in a characteristically round face called *moon facies* (FAY-seez; **Figure 16.22**), and on the back of the neck, where it produces a "buffalo hump." Cortisol also causes muscle wasting due to the release of amino acids from the breakdown of protein in muscles. The released fats and amino acids are used to make an abnormally high amount of glucose, which raises the blood glucose level, a condition known as *hyperglycemia* (hy′-per-gly-SEE-mee-uh). This and the other metabolic effects of cortisol tend to increase appetite, which leads to weight gain.

Excess cortisol has effects on other body systems, as well. For example, it has slight mineralocorticoid effects on the kidneys, resulting in greater renal retention of sodium ions (and water by osmosis). Over time, this excess water increases the volume of blood in the circulation, which raises blood pressure. In addition, cortisol's effects on leukocytes lead to overall suppression of the immune response. As a result, patients with Cushing's syndrome have a higher risk of developing infections. Finally, the excess cortisol decreases osteoblast activity and interferes with the absorption of dietary calcium ions by the small intestine. Both effects lead to bone loss, which can eventually cause osteoporosis.

Cushing's disease is generally treated by removing the ACTH- or cortisol-secreting tumor. The treatment for iatrogenic Cushing's syndrome is somewhat more complex. Many symptoms will lessen or reverse if corticosteroid therapy is decreased or stopped, but sometimes this simply isn't an option. For example, a patient with severe joint disease may not be able to walk without the anti-inflammatory effects of corticosteroids. In such cases, the patient and physician must decide what is ultimately best for the patient's quality of life. If corticosteroid therapy is decreased, it must be done gradually (see *A&P in the Real World: HPA Axis Suppression and Corticosteroid Therapy* on p. 614).

Adrenal Insufficiency Hyposecretion of both cortisol and aldosterone are hallmarks of *adrenal insufficiency,* or *Addison disease.* This disease has many causes, including abnormal development of the adrenal gland, deficiency in certain enzymes required to produce these steroid hormones, and destruction of the adrenal glands by the individual's immune system. Addison disease puts a patient at risk for an adrenal crisis and results in disruptions in fluid, electrolyte, and acid-base homeostasis.

Adrenal Production of Androgenic Steroids: Sex Hormones

The **androgenic steroids** (*andro-* = "male") are sex hormones that affect the reproductive organs, or gonads, and multiple other tissues. The majority of androgenic steroids are produced by the gonads, but they are also synthesized in small amounts in both males and females by the adrenal cortex, largely as byproducts of the pathway that yields cortisol. Androgenic steroids made by the adrenal cortex have the same general effects as those made by the gonads, and can be converted in the circulation to the androgen testosterone or the female hormone estrogen. We discuss the target tissues and effects of androgens in Module 16.6.

Quick Check

☐ 1. What are the three zones of the adrenal cortex, and which hormones are produced by each zone?

☐ 2. What are the target tissues and effects of aldosterone?

☐ 3. What are the target tissues and effects of cortisol?

Hormones of the Adrenal Medulla: Messengers of the Sympathetic Nervous System

« FLASHBACK

1. What are pre- and postganglionic neurons? (p. 518)
2. Which neurotransmitters are used by the neurons of the sympathetic nervous system? (p. 521)

The adrenal medulla consists of neuroendocrine cells called **chromaffin cells** (KROH-maf-in), which are derived from nervous tissue rather than glandular tissue. They are stimulated by acetylcholine released by preganglionic neurons of the sympathetic nervous system. Recall that a sympathetic preganglionic neuron exits the spinal cord and triggers a postganglionic neuron to release neurotransmitters onto a target cell (see Chapter 14). Chromaffin cells are very similar to these postganglionic neurons, except they secrete their products into the bloodstream (which is why these products are considered neurohormones).

As with sympathetic postganglionic neurons, chromaffin cells release chemicals called *catecholamines* (kat'-eh-KOHL-uh-meenz), chiefly **epinephrine** and **norepinephrine.** Chromaffin cells produce and secrete primarily epinephrine (the reverse is true for sympathetic neurons, for which the primary neurotransmitter is norepinephrine). Hormonal epinephrine and norepinephrine have almost the same effects on target cells

as the neurotransmitters of the sympathetic nervous system. Specifically, they mediate the body's immediate response to a stressor by:

- increasing the rate and force of heart contraction;
- dilating the bronchioles (the airway passages in the lungs);
- constricting the blood vessels supplying the skin, digestive organs, and urinary organs (which increases blood pressure);
- dilating the blood vessels supplying skeletal muscles;
- dilating the pupils; and
- decreasing digestive and urinary functions.

As we discussed previously, the additional epinephrine and norepinephrine secreted by the adrenal medulla prolong the duration of the sympathetic response (see Chapter 14). Additionally, the epinephrine and norepinephrine from the adrenal medulla can reach cells that are not innervated by the sympathetic nervous system. In this way, the sympathetic nervous system can affect the entire body without innervating every individual cell.

Secretion from the adrenal medulla is regulated not only by the sympathetic nervous system but also by the hormones of the HPA axis. ACTH can directly stimulate epinephrine synthesis by chromaffin cells, and cortisol increases synthesis of enzymes required to produce epinephrine. The overall result is a coordinated, integrated response between the endocrine and nervous systems that gives us our best chance of survival when faced with a stressor.

The hormones of the adrenal cortex and adrenal medulla are summarized in **Table 16.3**.

Table 16.3 Hormones of the Adrenal Gland

Hormone	Stimulus for Release	Inhibitor(s) of Release	Target Tissue(s)	Effects
Adrenocortical Hormones				
Aldosterone (mineralocorticoids)	• Angiotensin-II • Elevated potassium ion concentration • ACTH • Elevated hydrogen ion concentration	• Increased level of aldosterone • Decreased potassium ion concentration • Decreased hydrogen ion concentration • Increased blood pressure	Tubules of the kidneys	• Increases sodium ion retention directly and water retention indirectly • Increases potassium ion loss in the urine • Increases hydrogen ion loss in the urine • Regulates blood pressure
Cortisol (glucocorticoids)	ACTH	Increased level of cortisol	• Liver • Muscle • Adipose • White blood cells	• Increases gluconeogenesis in the liver • Increases protein breakdown in muscle • Increases lipolysis in adipose tissue • Inhibits the inflammatory response
Androgens (androgenic steroids)	ACTH	Poorly understood	Multiple target tissues, including: • Organs of the reproductive tract • Brain • Bone • Skeletal muscle	• Can be converted to testosterone in the circulation • Likely responsible for development of female pubic hair and libido
Adrenal Medullary Hormones				
Catecholamines (epinephrine and norepinephrine)	• Stimulation from preganglionic sympathetic neurons • ACTH	Lack of sympathetic nervous system stimulation	Nearly every cell in the body	• Increase rate and force of heart contractions • Dilate bronchioles • Constrict blood vessels to the digestive and urinary organs and the skin • Increase the metabolic rate; dilate the pupils

HPA Axis Suppression and Corticosteroid Therapy

When discontinuing corticosteroid therapy, the dose of corticosteroids must be tapered off gradually over a period of weeks or months. This is necessary because of the HPA axis feedback loops that regulate cortisol production and secretion. When a patient receives high doses of corticosteroids for an extended period of time, the negative feedback loops essentially halt production of CRH and ACTH via negative feedback, a phenomenon known as *HPA axis suppression.* This makes a patient's body unable to produce adequate amounts of cortisol for several weeks or even months after discontinuation of corticosteroid therapy.

The most serious potential consequence of HPA axis suppression is an *adrenal crisis.* Cortisol is absolutely vital to survival when a patient is subjected to stress such as surgery, trauma, or illness. If a patient is unable to produce an adequate amount of cortisol in the face of such a stressor, symptoms such as low blood glucose level, low blood pressure, fainting, vomiting, and convulsions may develop. Untreated, an adrenal crisis can easily be fatal. However, it can be prevented by slowly reducing the dose of corticosteroids to give a patient's HPA axis adequate time to resume normal levels of CRH, ACTH, and cortisol production.

Quick Check

☐ 4. What two hormones are produced by the adrenal medulla? What are their target tissues and their effects?

☐ 5. What is the relationship between the adrenal medulla and the sympathetic nervous system?

Apply What You Learned

☐ 1. A patient has a tumor of the anterior pituitary gland that is found to secrete ACTH. Predict the effects this tumor will have on CRH and cortisol levels in the patient's blood. Explain your answers.

☐ 2. Certain drugs used to treat high blood pressure work by blocking the effects of aldosterone on the kidneys. Predict the effects these drugs would have on the:

    a. sodium and potassium ion concentrations in the blood.

    b. pH of the blood.

    c. solute concentration of the blood.

☐ 3. A *pheochromocytoma* (fee´-oh-kroh-moh-sy-TOHM-uh) is a tumor of the adrenal medulla that secretes large amounts of epinephrine and norepinephrine. Predict the effects of this kind of tumor.

See answers in Appendix A.

MODULE **16.5**

The Endocrine Pancreas

Learning Outcomes

1. Describe the structure of the endocrine pancreas and its hormone-secreting cells.

2. Describe the stimulus for release, the target tissue, and the effect of glucagon.

3. Describe the stimulus for release, the target tissue, and the effect of insulin.

4. Explain how insulin and glucagon work together to maintain the blood glucose level within the normal range.

5. Describe the causes, symptoms, and treatments for the two types of diabetes mellitus.

We've already discussed three hormones that have an impact on blood glucose: thyroid hormone, growth hormone, and cortisol. Although their effects are important, they are not the primary hormones that maintain the blood glucose level within its normal range. Instead, this task falls to two hormones produced by the endocrine pancreas: *insulin* and *glucagon.* This module examines the pancreas as well as these hormones and their roles in the maintenance of blood glucose homeostasis.

Structure of the Pancreas

The **pancreas** is a club-shaped organ located in the abdominal cavity, mostly posterior to the stomach. It consists of three major parts: the rounded *head,* the middle *body,* and the thin *tail* (**Figure 16.23a**). The pancreas is unusual in that it is both an endocrine and an exocrine gland. Notice in **Figure 16.23b** that there are two groups of cells within the pancreas: (1) small, round "islands" of endocrine cells called **pancreatic islets,** and (2) exocrine cells called *acinar cells* (AY-sih-nahr). Clusters of acinar cells are located around small ducts, into which they secrete enzymes and other products that are delivered into the digestive tract (see Chapter 22).

Pancreatic islets secrete hormones into the bloodstream. These islets contain three main cell types:

- **alpha (α) cells,** which secrete the peptide hormone **glucagon;**
- **beta (β) cells,** which secrete the protein hormone **insulin;** and
- **delta (δ) cells,** which secrete the peptide hormone somatostatin (note that this is identical to the somatostatin produced by the hypothalamus).

This module focuses on the alpha and beta cells.

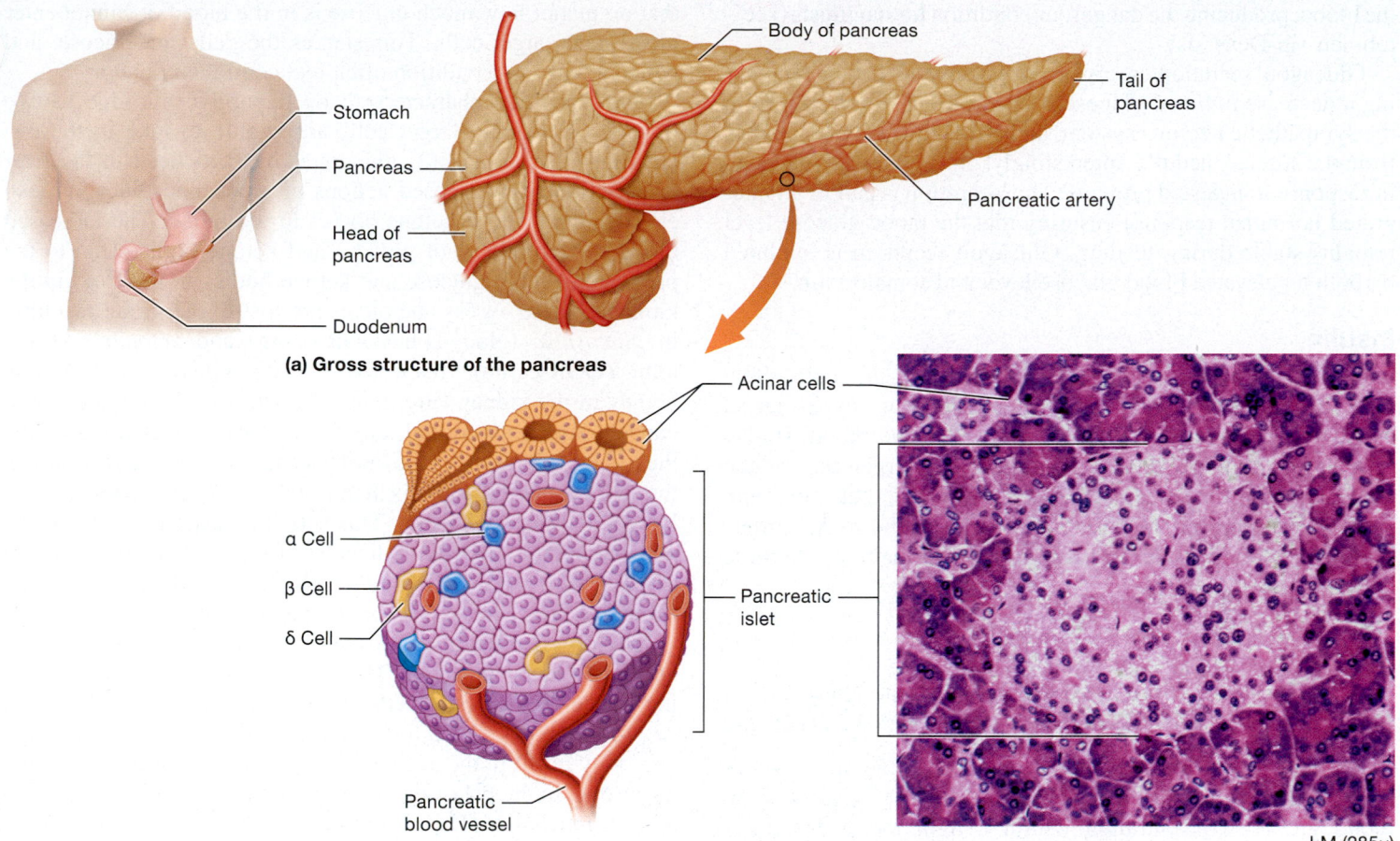

(a) Gross structure of the pancreas

Body of pancreas

Tail of pancreas

Pancreatic artery

Stomach

Pancreas

Head of pancreas

Duodenum

Acinar cells

α Cell

β Cell

δ Cell

Pancreatic islet

Pancreatic blood vessel

LM (285×)

(b) Histology of pancreatic islet and acinar cells: illustration (left) and light micrograph (right)

Figure 16.23 **Anatomy and histology of the pancreas.**

Hormones of the Endocrine Pancreas: Glucose Homeostasis

FLASHBACK

1. What is glycogen, and where is it stored? (p. 52)

As we noted in the introduction to this module, glucagon and insulin regulate the concentration of glucose in the blood. Let's now take a look at their specific roles. For a summary of this information, look ahead to Table 16.4 on p. 621.

Glucagon

Glucagon's major targets are the cells of the liver, muscle tissue, and adipose tissue. In these tissues, it promotes reactions that increase the levels of glucose and other metabolic fuels in the blood. These reactions include the following:

- the breakdown of glycogen into glucose, a process called *glycogenolysis;*
- the formation of new glucose by gluconeogenesis in the liver;
- protein breakdown in muscle tissue to release amino acids for gluconeogenesis;
- the release of fats from adipose tissue for gluconeogenesis and for an additional source of fuel; and
- the formation in the liver of fuel substances called *ketone bodies* (KEE-tohn) from fatty acids.

Ketone bodies are formed when the products of fatty acid metabolism are combined to yield four-carbon ketone compounds. The liver cannot itself catabolize ketone bodies in any appreciable amount, and so they are released into the bloodstream and delivered to other cells. Some tissues, such as those of the heart and skeletal muscle, are able to take up and then oxidize ketone bodies. Other tissues, such as the brain, are not as readily able to oxidize ketone bodies, although under starvation conditions, the brain can adapt and generate some of its ATP from ketone bodies.

During extreme calorie restriction or starvation, glucagon stimulates rapid ketone body production. However, the cells that can use ketone bodies as fuel are able to use only so many of them, and as a result, they accumulate in the blood. If too many ketone bodies accumulate, they can lower the pH of the blood, producing the dangerous condition **ketoacidosis** (kee′-toh-aeh-sih-DOH-sis).

Glucagon secretion is triggered by several factors, including a decrease in blood glucose concentration, stimulation from the sympathetic nervous system, and circulating catecholamines from the adrenal medulla. Interestingly, glucagon is also secreted in response to ingested proteins. This secretion is part of an integrated hormonal response ensuring that the blood glucose level remains stable during feeding. Glucagon secretion is inhibited by both an elevated blood glucose level and somatostatin.

Insulin

Insulin is the primary antagonist of glucagon; its main effect is to promote the uptake of ingested nutrients by its target cells, which lowers the level of glucose in the blood. It also promotes storage of these nutrients by its target cells if the nutrients aren't immediately needed by the cell for fuel. Insulin's primary target tissues are those of the liver, cardiac muscle, skeletal muscle, and certain parts of the brain. In these cells, insulin stimulates:

- uptake of lipids, amino acids, and glucose;
- synthesis of glycogen in the liver;
- synthesis of fat from lipids and carbohydrates; and
- promotion of *satiety* (suh-TY-eh-tee; the feeling of fullness).

Through these effects, insulin replenishes the fuel sources that were depleted during fasting, and helps to maintain the blood glucose concentration within a tight range. Insulin is absolutely required for its target cells to take up glucose (note, however, that the liver is an exception). In the absence of insulin, these cells can be bathed in glucose yet essentially starve.

Insulin secretion decreases during fasting and increases during feeding. Its secretion from the β cells is stimulated directly by a rising blood glucose concentration, and is inhibited by somatostatin and the sympathetic nervous system.

When the insulin level rises too high, the blood glucose level drops, a condition called **hypoglycemia** (hy′-poh-gly-SEE-mee-uh). Symptoms of hypoglycemia include weakness, dizziness, rapid breathing, nausea, and sweating. Severe hypoglycemia can lead to confusion, hallucinations, seizures, coma, and even death. These symptoms generally arise because the brain has inadequate glucose to fuel its metabolic processes. Hypoglycemia is most often caused by overadministration of insulin or other agents that lower the blood glucose level, but it may also result from prolonged starvation, insulin-secreting tumors, alcohol, poisons, infection, and genetic metabolic conditions.

Conversely, insufficient insulin secretion or decreased insulin sensitivity cause the blood glucose level to become elevated, a condition called **hyperglycemia.** The two most common causes of chronic hyperglycemia are diabetes mellitus types 1 and 2.

Type 1 Diabetes Mellitus The disease **type 1 diabetes mellitus,** or *insulin-dependent diabetes mellitus,* affects about 5–10% of people with diabetes mellitus in the United States. It is caused by the destruction of the insulin-producing β cells of the pancreas by the immune system. The lack of insulin means

that no matter how much glucose is in the blood, it cannot enter most of its target cells. This starves the cells for glucose, and patients with this condition often feel continually hungry.

The disease is characterized by hyperglycemia due to two factors: (1) Certain target cells are unable to take in the circulating glucose, and (2) glucose is overproduced in the liver because of the unopposed actions of glucagon. Glucagon also elevates the level of ketone bodies in the blood. The increased blood concentrations of glucose and ketone bodies lead to the presence of more glucose and ketone bodies in the fluid in the kidneys. These excess chemicals are lost to the urine, resulting in *glucosuria* (gloo′-koh-SOOR-ee-uh) and *ketonuria* (kee′-tohn-YOOR-ee-uh), respectively. The glucose and ketone bodies in the kidney fluid make the fluid in the kidneys more concentrated than normal, and so less water than usual leaves the kidney tubules by osmosis. This causes more water to stay in the kidney tubules, and it is then lost to the urine, a phenomenon known as *osmotic diuresis*. This activity causes two other common symptoms of type 1 diabetes: frequent urination, known as *polyuria* (pawl′-ee-YOOR-ee-uh); and dehydration causing excessive thirst, called *polydipsia* (pawl′-ee-DIP-see-uh).

The only treatment at present for type 1 diabetes is insulin administered by injection. The disease requires strict monitoring of diet and blood glucose level to ensure the level doesn't rise too high or fall too low. Failure to closely monitor the disease may result in the accumulation of ketones in the blood and a decrease in the pH of the blood, a condition referred to as *diabetic ketoacidosis*. This dangerous condition can lead to coma and death from nervous system shutdown.

The chronic hyperglycemia that occurs in poorly controlled type 1 diabetes has wide-ranging effects on the body. Hyperglycemia damages blood vessels, particularly those in the heart and lower limbs. This results in decreased circulation to these tissues, which increases the risk of heart attack, nonhealing wounds, and amputation. Hyperglycemia also damages peripheral nerves, again particularly in the lower limbs. This leads to *peripheral neuropathy,* which produces numbness, tingling, and burning pain in the affected areas. Other tissues affected by hyperglycemia include the lens of the eye and the capillaries of the retina and the kidneys, possibly resulting in blindness and kidney failure.

Type 2 Diabetes Mellitus **Type 2 diabetes mellitus,** also known as *non–insulin-dependent diabetes mellitus,* is the much more common form of diabetes, affecting up to 95% of people with diabetes mellitus in the United States. This form of diabetes generally develops in adults, and so it was formerly referred to as *adult-onset diabetes mellitus.* Patients with type II diabetes generally have functional β cells capable of producing insulin. However, the β cells do not respond to normal increases in blood glucose, and other target cells are less responsive to insulin, an effect called *insulin resistance.* This results in hyperglycemia and the accompanying characteristic signs and symptoms, such as glucosuria, polyuria, and polydipsia. However, unlike individuals with type 1 diabetes, those with type 2 diabetes generally produce enough insulin to keep ketoacidosis from developing. The development of type 2 diabetes is strongly associated with heredity and obesity.

The complications of poorly controlled type 2 diabetes are the same as those of type 1 diabetes—damage to the blood vessels, peripheral nerves, retina and lens, and kidneys. Patients with type 2 diabetes seem to be at higher risk for these complications, probably because the signs and symptoms of hyperglycemia in these patients are often vague, leading to a delay in diagnosis and treatment. By contrast, the signs and symptoms of type 1 diabetes are more immediately evident because these individuals produce no insulin, and so diagnosis and treatment are likely to be earlier.

Nearly all patients with type 2 diabetes are advised to make lifestyle changes, including weight loss, healthy diet, and regular exercise. In fact, some patients can manage the disease with lifestyle modifications alone. However, many require additional treatment. One of the more common treatments is a class of drugs, called *oral hypoglycemics,* that lower the circulating blood glucose level. Occasionally, patients with type 2 diabetes require insulin therapy to maintain a stable blood glucose concentration, but this is not a mainstay of treatment for most patients.

Quick Check

☐ 1. What are the main target tissues of glucagon? What are the stimuli for secretion of glucagon?

☐ 2. What are the main target tissues of insulin?

☐ 3. What are the signs and symptoms of the two types of diabetes mellitus? How do the two types differ?

Blood Glucose Regulation

As you have read, insulin and glucagon act as antagonists in a negative feedback loop that regulates the blood glucose concentration. But this is somewhat more complex than most feedback loops, because both insulin and glucagon are secreted during feeding to maintain blood glucose homeostasis. The regulation of blood glucose level is illustrated in **Figure 16.24**. When the blood glucose level increases, the following feedback loop is initiated (see Figure 16.24a):

- **Stimulus: Blood glucose level increases above its normal range.** The main reason for an increase in blood glucose concentration is feeding, although it can also rise in response to hormones such as cortisol.
- **Receptor: Beta cells of the pancreas detect the increased blood glucose concentration.** Like α cells, β cells act as sensors that detect the increase in blood glucose level.
- **Control center: Beta cells increase insulin secretion.** The β cells also act as the control center, regulating the secretion of insulin. The same stimulus reduces glucagon secretion from α cells as well.
- **Effector/response: Insulin decreases the blood glucose level by increasing glucose uptake by cells and storage of glucose, amino acids, and fats.** Insulin triggers the uptake of glucose by muscle and fat cells as well as the storage of glucose, amino acids, and fatty acids. These effects are balanced by a slight increase in glucagon secretion in

response to proteins in food. This prevents hypoglycemia if the food ingested contains limited carbohydrates.
- **Homeostatic range and negative feedback: As the blood glucose level returns to its normal range, negative feedback to β cells decreases insulin secretion.** The falling blood glucose concentration is detected by β cells. These cells then slow insulin secretion, decreasing the responses through negative feedback so that the blood glucose level never falls very far below the normal range.

If the blood glucose level decreases below the normal range, as during fasting, a different loop is initiated (see Figure 16.24b):

- **Stimulus: Blood glucose level decreases below its normal range.** The blood glucose concentration can decrease below the normal range due to fasting or an increased use of injected insulin in diabetic individuals.
- **Receptor: Alpha cells of the pancreas detect the decreased blood glucose concentration.** The α cells act as the sensors and detect the falling blood glucose level. (Note that α cells also detect the presence of ingested protein, which we discuss in Module 16.7.)
- **Control center: Alpha cells increase glucagon secretion.** The α cells also act as the control center, increasing glucagon secretion. At the same time, insulin secretion by β cells falls.
- **Effector/response: Glucagon increases the blood glucose level by triggering breakdown of glycogen into glucose and formation of new glucose.** Glucagon triggers glycogenolysis and gluconeogenesis, raising the blood glucose level. In addition, it increases the availability of other fuels such as ketone bodies, free amino acids, and fatty acids.
- **Homeostatic range and negative feedback: As the blood glucose level returns to its normal range, negative feedback to α cells decreases glucagon secretion.** When the blood glucose concentration returns to its normal range, it is detected by the α cells. They, in turn, decrease glucagon secretion, which decreases the responses through negative feedback.

Quick Check

☐ 4. How do glucagon and insulin work together to regulate the level of blood glucose?

Apply What You Learned

☐ 1. Predict the effects of a pancreatic tumor that secretes glucagon. What other disease would this condition resemble? Explain.

☐ 2. Predict the effects of a pancreatic tumor that secretes insulin. What signs and symptoms would you expect to see from such a condition?

☐ 3. Explain why drugs that activate or suppress the sympathetic nervous system have effects on the blood glucose concentration.

See answers in Appendix A.

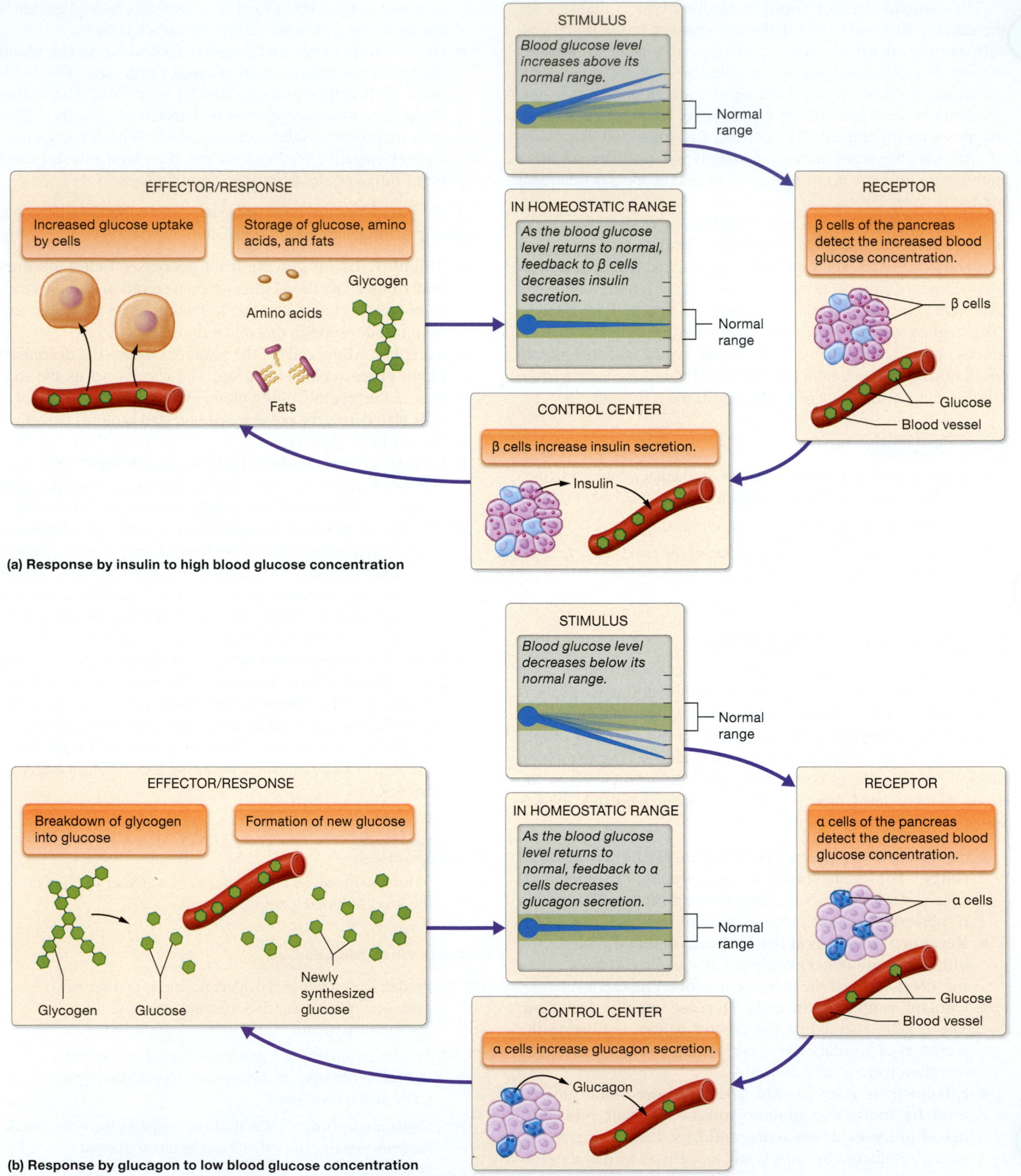

(a) Response by insulin to high blood glucose concentration

(b) Response by glucagon to low blood glucose concentration

Figure 16.24 Maintaining homeostasis: regulation of blood glucose concentration by negative feedback loops.

Other Endocrine Glands and Hormone-Secreting Tissues

Learning Outcomes

1. Describe the stimulus for release, the target tissue, and the effects of the hormones produced by the pineal and thymus glands.

2. Describe the stimulus for release, the target tissue, and the effects of the hormones produced by the gonads, adipose tissue, the heart, and the kidneys.

We still have three primary endocrine organs to discuss: the thymus gland and the gonads (the testes and the ovaries). In addition, as we mentioned in Module 16.1, many secondary endocrine organs and tissues secrete hormones, including those of nervous tissue, connective tissue, muscle tissue, and glandular epithelium. This module covers the thymus gland and gonads, as well as secondary organs and tissues, their hormones, and their effects.

The Thymus: Thymosin and Thymopoietin

The **thymus** is an irregularly shaped organ found in the mediastinum (look back at Figure 16.2). It is the site of maturation of a class of white blood cells (leukocytes) called *T lymphocytes* (LIM-foh-sytz) that are involved in the immune response. It also secretes the hormones **thymosin** and **thymopoietin** (thy′-moh-POY-eh-tin), which function mainly as paracrine signals that assist in T lymphocyte maturation. The thymus gland is larger and far more active in infants and children, who have developing immune systems (look ahead to Figure 20.9). In adults, the thymus gland shrinks (or *atrophies*) and eventually is replaced mostly with adipose and other connective tissues.

The Gonads: Sex Hormones

The **testes** and **ovaries** are the primary male and female reproductive organs, or **gonads,** respectively. The gonads are responsible for the production of *gametes* (GAM-eetz)—sperm in males and ova in females. In addition to these functions, the gonads produce the sex steroid hormones that are responsible for gamete production and have multiple other functions throughout the body. We now take a closer look at the hormones produced by these organs.

The Testes: Testosterone

The testes produce the steroid hormone **testosterone** (tes-TAHS-ter-ohn). Although testosterone is considered the "male hormone," females do produce small amounts of it in the adrenal cortex and ovaries. Testosterone production and secretion in the male are regulated by a multi-tiered negative feedback loop involving the hypothalamus and anterior pituitary. Recall that GnRH from the hypothalamus triggers secretion of both LH and FSH from the anterior pituitary. LH stimulates testosterone synthesis, and FSH stimulates production of a protein that binds testosterone and concentrates it in the testes. Through negative feedback, a high blood level of testosterone decreases GnRH secretion and so decreases LH and FSH secretion.

Like all steroid hormones, testosterone is hydrophobic and so easily crosses the plasma membranes of its target cells and binds to intracellular receptors that influence the expression of various genes. Testosterone has two basic types of effects on its target cells: (1) *anabolic effects,* which include those that stimulate bone growth and increase muscle mass; and (2) *androgenic effects* ("male generating"), which encompass those involving the development of male secondary sex characteristics, such as a deeper voice, more prominent facial hair, and increased bone and muscle mass. (These characteristics are called "secondary" because the affected organs are not part of the reproductive system.)

The Ovaries: Estrogens and Progesterone

Cells of the ovary produce the female steroid hormones known as **estrogens.** In parallel with testosterone, estrogens are considered "female" hormones but are produced in smaller amounts in males. As with testosterone, the secretion of estrogens is regulated by negative feedback loops involving GnRH from the hypothalamus and LH and FSH from the anterior pituitary. Estrogens stimulate the development of female secondary sex characteristics such as the breasts; regulate the menstrual cycle; and have multiple effects on other tissues and organs, such as bone, blood, kidney, and adipose.

Another steroid hormone made by the ovaries is **progesterone** (proh-JES-ter-ohn). Progesterone production peaks after ovulation and during pregnancy (*pro-* = "supporting," *ges-* = "a carrying; pregnancy"). Progesterone has multiple physiological effects, many of which help to prepare the body for pregnancy and support fetal development during pregnancy. It also affects smooth muscle tissue, body temperature, blood clotting, bone tissue, and metabolism.

We discuss the physiological effects of estrogens, progesterone, and testosterone more completely in the reproductive system chapter (see Chapter 26).

The Pineal Gland: Melatonin

« FLASHBACK

1. What is the epithalamus? (p. 433)

2. What is the reticular formation? (p. 436)

The **pineal gland** is part of the *epithalamus,* the diencephalon's posterior portion. Recall that the pineal gland secretes the neurohormone **melatonin** (mel-uh-TOH-nin; see Chapter 12). The trigger for melatonin secretion appears to be related to light and dark cycles. The pineal gland begins secreting melatonin when ambient light decreases in the evening, and it reaches its peak secretion over the nighttime hours. Melatonin's main target tissues are the sleep-regulating centers in the reticular formation of the brainstem, where it appears to adjust the sleep phase of the sleep/wake cycle in some people. Synthetic melatonin is often sold as a dietary supplement, to be used to promote sleep and prevent jet lag. However, its effectiveness for these uses is questionable.

Quick Check

☐ 1. What are the target tissues and effects of melatonin?

☐ 2. What are the target tissues and main effects of testosterone and estrogens?

Adipose Tissue: Leptin

Adipocytes produce a protein hormone known as **leptin.** Although it is a protein, leptin manages to cross the blood brain barrier to interact with its main target cells: neurons in the hypothalamus that control feeding. Leptin acts on these neurons to induce satiety, a feeling of fullness, which prevents overfeeding. Interestingly, the amount of leptin secreted appears to be related to the amount of adipose tissue present. In theory, this should prevent humans with excess adipose tissue from overeating. However, the mechanisms that regulate feeding are quite complex, involving multiple neural and hormonal factors. See *A&P in the Real World, Pseudoscience Exposed: Leptin and Obesity* for one of the limits to this hormone's effects.

The Heart: Atrial Natriuretic Peptide

Certain cardiac muscle cells contain stretch-sensitive ion channels. These channels open more widely when the blood volume inside the heart increases, and this stimulates the cardiac muscle cells to secrete the hormone **atrial natriuretic peptide,** or **ANP** (nay′-tree-yoo-RET-ik; *natri-* = "sodium"). ANP has two main target tissues: smooth muscle cells lining blood vessels, and the tubules of the kidneys. ANP triggers relaxation of smooth muscle

Pseudoscience Exposed: Leptin and Obesity

Leptin has been heavily researched for its possible pharmaceutical application in weight loss. The hypothesis was that leptin supplements would suppress the appetite, which would lead to weight loss. Unfortunately, this research demonstrated that administered leptin was not effective at reducing appetite in humans. However, this hasn't stopped many manufacturers of over-the-counter dietary supplements from marketing their leptin-containing supplements as "miracle weight loss" solutions.

cells in blood vessels, an effect that increases the vessel diameter, which is known as *vasodilation*. In the kidneys, it enhances excretion of sodium ions, an effect called *natriuresis* (nay′-tree-yoo-REE-sis). The process of natriuresis also enhances water excretion because it makes the fluid in the kidney tubules more concentrated. This causes less water to be retained by osmosis, more water to be excreted in the urine, and a lower blood volume. Both vasodilation and the decreased blood volume that accompanies natriuresis decrease blood pressure.

Quick Check

☐ 3. What are the hormones of the thymus gland, adipose tissue, and cardiac muscle cells? What are the target tissues and effects of each of these hormones?

The Kidneys: Erythropoietin

« FLASHBACK

1. What are erythrocytes, and where are they formed? (p. 145 and p. 186)

The kidneys play several endocrine roles, including the following:

- **Erythropoietin production.** The hormone **erythropoietin** (eh-rith′-roh-POY-eh-tin), or **EPO,** is secreted mainly by certain cells of the kidney in response to a decreased level of oxygen in the blood. EPO acts on the red bone marrow, where it stimulates the development of erythrocytes, which increases the oxygen-carrying capacity of the blood. We discuss the role of EPO in erythrocyte production more fully in the chapter on blood (see Chapter 19).

- **Renin secretion.** The kidney cells also secrete **renin** (REE-nin), an enzyme in blood that converts the plasma protein angiotensinogen to angiotensin-I. It is an important part of the renin-angiotensin-aldosterone system, which maintains blood pressure (see Chapters 18 and 24).

- **Conversion of vitamin D to its active form.** As mentioned in Module 16.3, it is not until vitamin D arrives in the kidneys that it is converted to its active form, calcitriol. Recall that this change is stimulated by parathyroid hormone.

The hormones, target tissues, and effects of the pancreas, the endocrine glands covered in this module, and the secondary endocrine organs and tissues are summarized in **Table 16.4**.

Quick Check

☐ 4. What are the endocrine roles of the kidneys?

Apply What You Learned

☐ 1. Individuals who live at extreme northern and southern latitudes experience winter with little to no periods of daylight and summer with little to no periods of darkness. Predict the effect these seasonal conditions will have on melatonin secretion and the sleep/wake cycle.

Table 16.4 Hormones Produced by Other Endocrine Organs and Hormone-Secreting Tissues

Organ/Tissue	Hormone(s)	Target Tissue(s)	Effects
Other Endocrine Organs			
Pancreas	• Insulin • Glucagon	Insulin: most cells, including those of the liver, skeletal and cardiac muscle, and adipocytes Glucagon: liver, muscle, adipose	Insulin: stimulates glucose uptake by most target cells, glycogenesis, and lipogenesis Glucagon: stimulates gluconeogenesis, glycogenolysis, ketogenesis, and fat and protein breakdown
Thymus	• Thymosin • Thymopoietin	T lymphocytes	Promote T lymphocyte maturation
Testes	Testosterone	• Male reproductive organs • Multiple other target tissues	• Androgenic effects • Anabolic effects
Ovaries	• Estrogens	Female reproductive organs	• Development of female secondary sex characteristics • Regulate the menstrual cycle • Multiple other effects
	• Progesterone	Multiple other target tissues	• Prepare the body for pregnancy • Support fetal development • Multiple other effects
Hormone-Secreting Tissues			
Pineal gland	Melatonin	Reticular formation of the brainstem	• Regulates the sleep/wake cycle • Promotes sleep
Adipose tissue	Leptin	Brain	Promotes satiety
Heart	Atrial natriuretic peptide (ANP)	• Smooth muscle cells in blood vessel walls • Kidney tubules	• Relaxes smooth muscle cells in blood vessels, causing vasodilation • Promotes natriuresis and fluid loss in the kidneys, decreasing blood volume • Lowers blood pressure as a result of above effects
Kidneys	• Erythropoietin	Red bone marrow	Increases rate at which erythrocytes are formed
	• Renin	Enzyme that activates angiotensinogen	Part of the renin-angiotensin-aldosterone system
	• Vitamin D (calcitriol)	Small intestines	Vitamin D activation

☐ 2. A genetic mutation in mice leads to a dysfunctional leptin receptor in the hypothalamus that does not bind leptin. Predict the consequences of this mutation.

☐ 3. Renal failure is generally characterized by an inability of the kidney cells to perform their many physiological functions. Predict those consequences of renal failure that extend beyond the urinary system.

See answers in Appendix A.

MODULE **16.7**
Three Examples of Endocrine Control of Physiological Variables

Learning Outcomes

1. Provide specific examples to demonstrate how hormones maintain homeostasis in the body.
2. Describe how hormones work together with the nervous system to mediate the stress response.

As we saw in previous modules, many regulated physiological variables are controlled by the endocrine system. It can be challenging to form a "big picture" view of how the endocrine system regulates these variables, because they are generally controlled by multiple hormones from multiple endocrine organs. In this module, we examine some important examples of hormone "teamwork" by looking at fluid balance, metabolism, and the stress response. This discussion will help you put together all the pieces of the endocrine system to build a greater understanding of how hormones work together to maintain homeostasis.

Hormonal Control of Fluid Homeostasis

Fluid homeostasis refers to maintenance of the necessary volume of water in the extracellular fluid. Although several hormones indirectly contribute to fluid homeostasis, the three main water-regulating hormones are ADH, aldosterone, and ANP. Let's summarize their interactions.

- **Normal conditions: Low levels of ADH and aldosterone are secreted continually and lead to average urine output** (**Figure 16.25a**). The hypothalamus and adrenal cortex continually secrete small amounts of ADH and aldosterone, respectively. This is necessary because huge volumes of blood are filtered through the kidneys every minute. In fact, the equivalent of the entire volume of blood plasma flows through the tubules of the kidneys about 60 times per day. Therefore, ADH and aldosterone must be continually secreted to ensure that most of the plasma volume is not lost to the urine. Aldosterone's important effects on electrolyte and acid-base homeostasis provide another reason why it must be continually secreted.

- **Decreased plasma volume and increased plasma solute concentration: ADH and aldosterone secretions rise, leading to decreased urine output** (**Figure 16.25b**). Two conditions signal the cells of the endocrine system that there is inadequate water in the body: a decreased plasma volume, which is sensed by specialized cells in the kidneys, and an increased solute concentration of the plasma, which is sensed by receptors in the hypothalamus. These conditions trigger increased secretions of aldosterone and ADH, respectively, which lead to greater water retention and lower urine output.

- **Increased plasma volume and decreased plasma solute concentration: ADH and aldosterone secretions decline and ANP secretion increases, leading to increased urine output** (**Figure 16.25c**). Through negative feedback loops, the secretion of ADH and aldosterone falls when the plasma volume rises and/or the plasma solute concentration decreases. If plasma volume continues to rise, ANP secretion from the heart is triggered. Together, these changes in hormone secretion result in water loss from the blood and an increased urine output.

Quick Check

☐ 1. Which hormones primarily control fluid homeostasis?

☐ 2. What is the role of each of these hormones with respect to fluid homeostasis?

Hormonal Control of Metabolic Homeostasis

Endocrine control of metabolic homeostasis is perhaps our most complex example because it involves several factors: controlling the metabolic rate of different cell populations, ensuring fuels are available during times of activity, storing metabolic fuels during times of rest, regulating growth, and feeding behaviors. Let's simplify things by looking at the hormones that maintain metabolic homeostasis during three periods: at rest while fasting, at rest while feeding, and during exercise.

- **At rest, while fasting: Thyroid hormones determine the basal metabolic rate.** As you read in Module 16.3, thyroid hormones set the basal metabolic rate by triggering processes that cause the body cells to consume ATP. Thyroid hormones also stimulate gluconeogenesis, which leads to an increase in the blood glucose concentration. During fasting, this increase helps to maintain a stable blood glucose level. Fasting also triggers the release of glucagon and an elevated level of growth hormone, which further raise the blood glucose level.

- **At rest, while feeding: Insulin secretion increases.** When feeding begins, the increase in blood glucose concentration causes insulin release. Insulin triggers glucose uptake by its target cells and the storage of excess glucose as glycogen

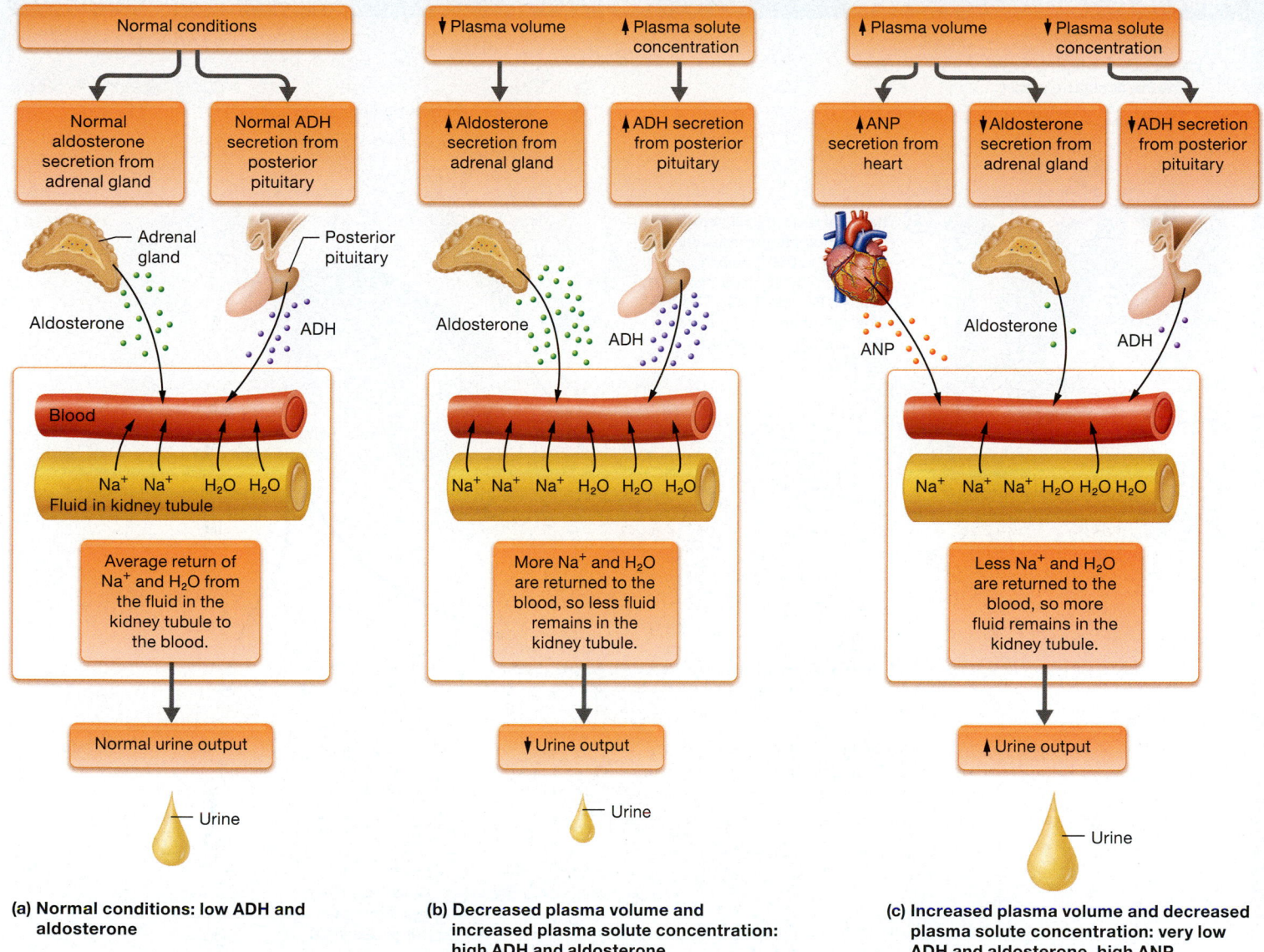

(a) **Normal conditions: low ADH and aldosterone**

(b) **Decreased plasma volume and increased plasma solute concentration: high ADH and aldosterone**

(c) **Increased plasma volume and decreased plasma solute concentration: very low ADH and aldosterone, high ANP**

Figure 16.25 **Summary of endocrine control of fluid homeostasis.**

and fat in adipocytes. The presence of protein in a meal leads to GH and IGF-1 release, which stimulates protein synthesis with the ingested amino acids. Protein in a meal also stimulates glucagon release to prevent hypoglycemia in the event of a carbohydrate-poor meal. Feeding ends with the help of both insulin and leptin, which promote satiety in the hypothalamus.

- **During exercise: Catecholamines control the metabolic rate and glucagon secretion increases.** When exercise begins, the sympathetic nervous system increases catecholamine secretion from the adrenal medulla. Catecholamines in the blood cause the metabolic rate of all cells to increase to a level higher than the basal rate set by thyroid hormones. This increased metabolic rate necessitates more metabolic fuels, and for this reason the α cells of the pancreas increase glucagon secretion. These actions are further discussed in our upcoming examination of the stress response.

Quick Check

☐ 3. Which hormones control metabolic homeostasis?

☐ 4. What is the role of each of these hormones with respect to metabolic homeostasis?

Putting It All Together: The Big Picture of the Hormonal Response to Stress

« FLASHBACK

1. What is the role of the hypothalamus in the sympathetic response? (p. 532)

2. What is the role of the parasympathetic nervous system? (p. 528)

The Big Picture of the Hormonal Response to Stress

Figure 16.26

Play Animation @ Mastering Anatomy & Physiology

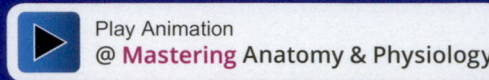

1 The body is subjected to a stressor such as exercise.
2 The initial response is two-fold:
• The hypothalamus releases GHRH and CRH, causing the anterior pituitary to release GH and ACTH (see Figure 16.10).
• The hypothalamus activates the sympathetic nervous system.
Hypothalamus
GHRH
CRH
Sympathetic nervous system
Somatotrophs
Corticotrophs
Anterior pituitary
ACTH
GH
3a ACTH and the sympathetic nervous system trigger the release of cortisol, aldosterone, and catecholamines (see Figure 16.21).
3b Sympathetic nervous system triggers the release of glucagon (see Figure 16.24).
Pancreas
Adrenal gland
Aldosterone
Cortisol
Catecholamines (epinephrine and norepinephrine)
Glucagon
4a Aldosterone increases Na^+ and H_2O retention in the kidneys, increasing blood pressure (see Figure 16.20).
Liver
Adipose tissue
↑Glucose
↑Fatty acids
4b GH, cortisol, catecholamines, and glucagon trigger an increased release of metabolic fuels from the liver and adipose tissue.
GHRH = Growth hormone–releasing hormone
CRH = Corticotropin-releasing hormone
GH = Growth hormone
ACTH = Adrenocorticotropic hormone

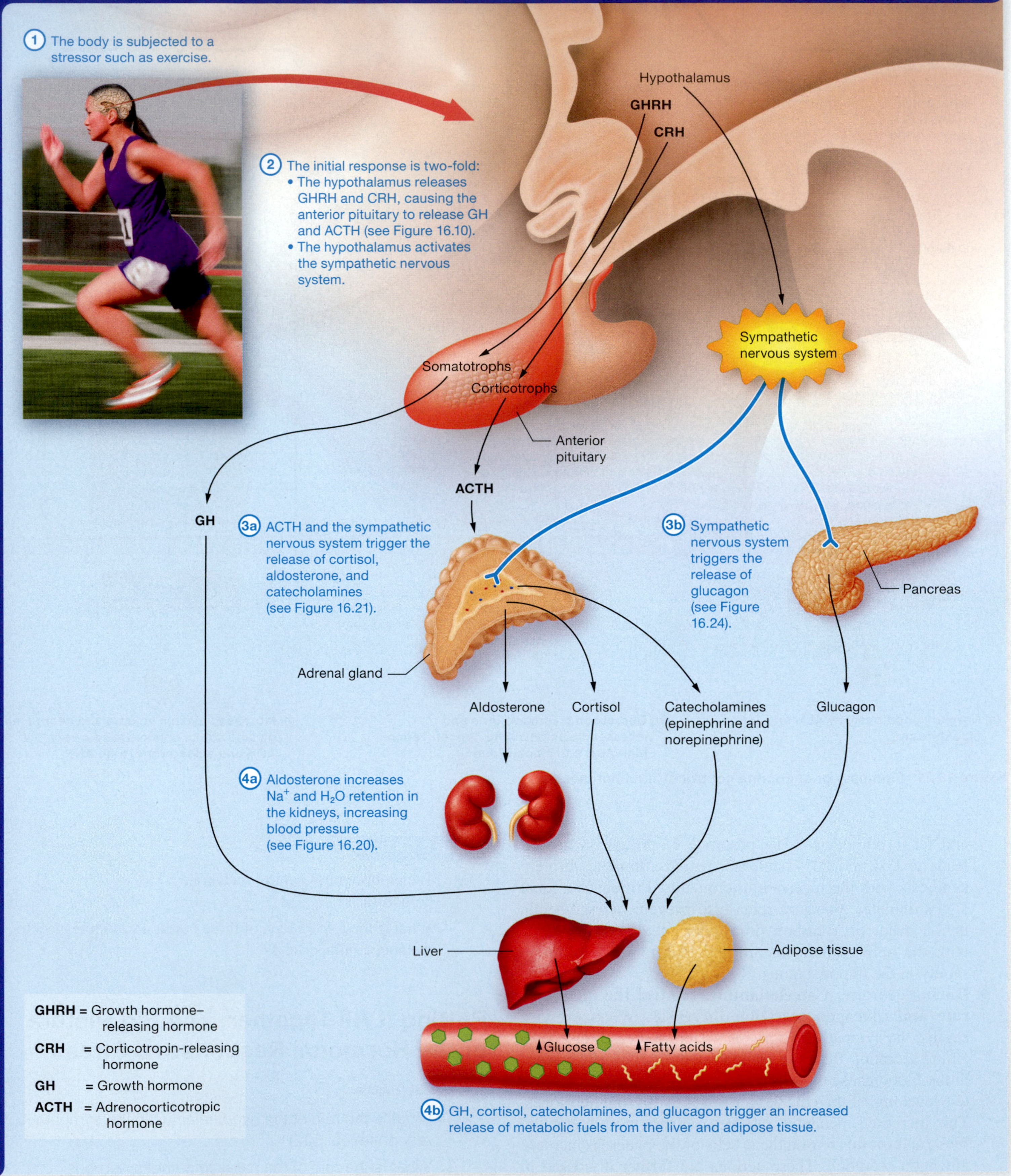

We first discussed regulation of the stress response with the sympathetic nervous system (see Chapter 14). There we pointed out that although it's common to think of a stressor as being something inherently *stressful,* a stressor can be anything from excitement to mild exercise. In this chapter, we have discussed endocrine regulation of the stress response with multiple hormones, including CRH, ACTH, and adrenal steroid hormones from the HPA axis, GH from the anterior pituitary, catecholamines from the adrenal medulla, and glucagon from the pancreas. Now let's put all the pieces of the nervous system and endocrine system together and examine how they work jointly to achieve the physiological changes witnessed in the body during the stress response.

The steps in the stress response are outlined in **Figure 16.26**. When the stressor is resolved, the hypothalamus decreases its stimulation of the sympathetic neurons, CRH secretion, and GHRH secretion. The decline in sympathetic activity and CRH secretion, along with an increase in parasympathetic nervous system activity, lowers ACTH, GH, and adrenal hormone levels and restores resting conditions.

Quick Check

- ☐ 5. Which hormones are involved in the stress response?
- ☐ 6. What is the role of each hormone in the stress response? How do these hormonal effects tie in with the response of the sympathetic nervous system?

Apply What You Learned

- ☐ 1. Mr. Dent is discovered to have an ADH-secreting tumor. Predict what will happen to the amount of aldosterone secreted by his adrenal cortex. What do you think will happen to the amount of ANP secreted by Mr. Dent's heart? Explain.
- ☐ 2. Ms. Prefect has a tumor of the hypothalamus that reduces its secretion of CRH. Will her body continue to secrete adrenal steroids and catecholamines as part of the stress response? Explain.
- ☐ 3. Your friend is on a no-carbohydrate, all-fat diet, leading to essentially no intake of dietary glucose. Predict which hormones might be secreted in excess to compensate for the decreased level of blood glucose. How will these hormones compensate?

See answers in Appendix A.

Chapter Summary

For everything you need to succeed in this course, go to Mastering Anatomy & Physiology. There you will find:

- Practice Tests
- Author Podcasts
- Big Picture Animations
- Concept Boost Video Tutors
- **iP2** Interactive Physiology 2.0 Tutorials
- *A&PFlix* A&P Flix 3D Animations
- Active-Learning Workbook

MODULE 16.1
Overview of the Endocrine System 584–592

- Cells of the **endocrine system** secrete **hormones** into the bloodstream that affect distant targets, in contrast to the neurons of the nervous system, which come close to their target cells and work through the release of neurotransmitters into synapses.
- Hormones may also be released as **paracrine** and **autocrine** signals.
- The seven endocrine organs include the anterior pituitary, thyroid, and parathyroid glands, the adrenal cortices, the thymus, the endocrine pancreas, and the testes or ovaries. **Neuroendocrine organs** secrete *neurohormones,* including those of the hypothalamus, the adrenal medulla, and the pineal gland.
- The two types of hormones are **amino acid–based hormones** and **steroid hormones.**
- Hormones may travel through the blood freely or bound to proteins.
- Hormones exert their effects by binding to **receptors** on their **target cells.**
- Hydrophilic hormones bind to receptors in the target cells' plasma membranes and trigger the formation of a **second messenger,** generally with the help of a **G-protein.** The second messenger then initiates changes within the cell.
- Hydrophobic hormones can bind to intracellular receptors. The resulting receptor-steroid complex binds to and interacts with the cell's DNA to trigger a change within the cell.
- Hormones may affect gene transcription, enzymes, cell division, and the permeability of the plasma membrane, and may cause secretion from other glands.
- Hormone secretion can be stimulated by hormonal, humoral, and neural stimuli.
- The **half-life** of a hormone is the time it takes for its plasma concentration to be reduced by half.
- The production and secretion of most hormones are regulated through negative feedback loops.

MODULE 16.2
The Hypothalamus and the Pituitary Gland
593–601

- The **hypothalamus** is the anteroinferior portion of the diencephalon.

- The **posterior pituitary** is composed of axons and axon terminals that store certain neurohormones produced by the hypothalamus:
  - **Antidiuretic hormone (ADH)** acts on kidney cells to promote water retention and increase the amount of water in the body.
  - **Oxytocin** triggers uterine contractions during childbirth and milk ejection during lactation.

- The hypothalamus controls the anterior pituitary through **releasing** and **inhibiting hormones** that are secreted into the blood vessels of the **hypothalamic-hypophyseal portal system.**

- The **anterior pituitary** is an endocrine gland that secretes the following hormones:
  - **Thyroid-stimulating hormone** stimulates growth of and secretion from the thyroid gland.
  - **Adrenocorticotropic hormone** stimulates secretion from the adrenal cortex and the adrenal medulla.
  - **Prolactin** stimulates milk production in the mammary glands.
  - **Luteinizing hormone (LH)** and **follicle-stimulating hormone (FSH)** are **gonadotropins.** In the male, they trigger the production of both testosterone and chemicals that bind and concentrate testosterone. In the female, they trigger the production of estrogens and progesterone, ovulation, and the maturation of oocytes.

- The acute effects of **growth hormone** include lipogenesis and gluconeogenesis. Its long-term effects are mediated by **insulin-like growth factor (IGF),** which stimulates rapid protein synthesis and cell division.

- Growth hormone disorders include **pituitary dwarfism** and **gigantism.**

MODULE 16.3
The Thyroid and Parathyroid Glands
601–608

- The **thyroid gland** consists of hollow, spherical **thyroid follicles** that are filled with **colloid** and lined by cuboidal *follicle cells.* Follicle cells produce and secrete thyroid hormones. Between follicles are **parafollicular cells** that secrete **calcitonin.**

- Three to five **parathyroid glands** are located on the posterior thyroid gland. They consist of **chief cells** that secrete **parathyroid hormone,** or **PTH.**

- Thyroid hormones consist of tyrosine residues attached to iodine atoms. **Triiodothyronine,** or T_3, has three iodine atoms, and **thyroxine,** or T_4, has four iodine atoms.
  - Thyroid hormones' actions include regulation of the metabolic rate, thermoregulation, promotion of growth and development, and synergistic effects with the sympathetic nervous system.
  - Thyroid hormone synthesis begins with the attachment of iodine atoms to **thyroglobulin** in the colloid, which is cleaved to T_3 or T_4 in follicle cells.
  - The production of thyroid hormones is regulated by a multi-tiered feedback loop involving TRH from the hypothalamus and TSH from the anterior pituitary.
  - **Hyperthyroidism** is characterized by weight loss, heat intolerance, disruptions in blood pressure and heart rhythm, goiter, and exophthalmos.
  - **Hypothyroidism** is characterized by weight gain, cold intolerance, slow heart rate and low blood pressure, and goiter.

- PTH increases the blood calcium ion concentration by stimulating osteoclasts, calcium ion absorption from the small intestine, and calcium ion retention from the kidney tubules.

- Calcitonin inhibits osteoclasts and transiently decreases the blood calcium ion concentration.

MODULE 16.4
The Adrenal Glands 608–614

- Each **adrenal gland** consists of an outer **adrenal cortex** and an inner **adrenal medulla.**

- **Mineralocorticoids** regulate fluid, electrolyte, and acid-base homeostasis. They are secreted by the **zona glomerulosa.** The main mineralocorticoid is **aldosterone,** which triggers sodium ion retention and potassium ion and hydrogen ion loss to the urine.

- **Glucocorticoids** are secreted by the **zona fasciculata** and **zona reticularis.** The main glucocorticoid is **cortisol.** Cortisol is an important part of the **stress response,** and triggers gluconeogenesis and protein and fat breakdown, and inhibits the inflammatory response. It is regulated by the *hypothalamic-pituitary-adrenal (HPA) axis.*

- **Androgenic steroids** are secreted by the cells of the zona reticularis. They have effects on reproductive organs and many other tissues.

- The adrenal medulla consists of **chromaffin cells** that secrete *catecholamines,* particularly **epinephrine,** when stimulated by the sympathetic nervous system.

MODULE 16.5
The Endocrine Pancreas 614–618

- The **pancreas** consists of two groups of cells: *acinar cells,* which secrete exocrine products into the digestive tract; and **pancreatic islets,** which secrete hormones.

- The hormone **glucagon** is produced by **alpha (α) cells** of the pancreatic islets. Its stimulates glycogen breakdown, gluconeogenesis, protein and fat breakdown, and the formation of *ketone bodies.*

- The hormone **insulin** is produced by the **beta (β) cells** of the pancreatic islets. Insulin stimulates the uptake of glucose, fats, and amino acids by cells, and the formation of glycogen and fat.
  - ○ Abnormally high insulin levels result in **hypoglycemia,** which may lead to confusion, weakness, dizziness, rapid breathing, coma, and eventually death.
  - ○ When insulin secretion is inadequate or the released insulin is ineffective, **hyperglycemia** results. The two most common causes of hyperglycemia are **type 1** and **type 2 diabetes mellitus.**

MODULE 16.6
Other Endocrine Glands and Hormone-Secreting Tissues 619–622

- The **pineal gland** secretes the hormone **melatonin,** which regulates the sleep/wake cycle.
- The **thymus** produces **thymosin** and **thymopoietin,** which stimulate T lymphocyte maturation.
- The **gonads** include the **testes** in the male and the **ovaries** in the female.
  - ○ The testes produce and activate the hormone **testosterone,** which has both androgenic and anabolic effects.
  - ○ The ovaries produce **estrogens** and **progesterone.** Estrogens trigger the development of female sex characteristics, regulate the menstrual cycle, and affect multiple other organs and tissues. Progesterone prepares the body for pregnancy and supports fetal development.
- Adipose tissue produces **leptin,** which induces satiety in the hypothalamus.
- Cardiac muscle cells produce **atrial natriuretic peptide,** which promotes vasodilation and natriuresis and water loss in the kidneys.
- The kidneys produce **erythropoietin,** which stimulates erythrocyte development, and the enzyme **renin,** which is part of a system that activates aldosterone.

MODULE 16.7
Three Examples of Endocrine Control of Physiological Variables 622–625

- Fluid balance is maintained by ADH, aldosterone, and ANP. ADH and aldosterone rises when the plasma solute concentration increases and the plasma volume decreases. ANP secretion rises when the plasma volume rises.
- Metabolic homeostasis is maintained by thyroid hormones, GH, glucagon, insulin, and catecholamines.
- The stress response involves the hormones of the HPA axis, GH, catecholamines, angiotensin-II, and glucagon.

Assess What You Learned

Scan the QR Code for additional practice test questions

https://goo.gl/4fOtHo

LEVEL 1   Check Your Recall

1. Mark the following statements as true or false. If a statement is false, correct it to make a true statement.

   a. The cells of the nervous system communicate via action potentials, whereas the cells of the endocrine system communicate via hormones.
   b. Autocrine signals affect the same cells that secrete them.
   c. The pancreas, thyroid gland, and parathyroid glands secrete neurohormones.
   d. Steroid hormones are hydrophilic molecules that bind to plasma membrane proteins as part of a second-messenger system.
   e. The secretion of most hormones is regulated by a negative feedback system.

2. Which of the following is *not* a potential effect that a hormone could have on its target cell?

   a. Activating genes in the DNA
   b. Stimulating cellular division
   c. Altering the permeability of the plasma membrane
   d. All of the above are potential effects of a hormone on its target cell.

3. Which of the following hormones is/are produced by the posterior pituitary?

   a. Antidiuretic hormone   c. Both a and b
   b. Oxytocin   d. Neither a nor b

4. How does ADH affect the amount of water in the body, and how does it accomplish this? How does this affect the osmolarity of the blood?

5. Fill in the blanks: Hypothalamic releasing and inhibiting hormones are released into the _____ system and affect secretion from the _____ gland.

6. List the target tissues and effects of the following anterior pituitary gland hormones.

   a. Thyroid-stimulating hormone
   b. Adrenocorticotropic hormone
   c. Prolactin
   d. Gonadotropins
   e. Growth hormone

7. The thyroid gland consists of:

   a. follicle cells that secrete calcitonin.
   b. spherical thyroid follicles that contain iodine-containing colloid.
   c. parafollicular cells that produce thyroid hormones.
   d. spherical thyroid follicles that surround parathyroid hormone–secreting cells.

8. Which of the following is *not* an effect of thyroid hormones?

   a. Regulation of the metabolic rate
   b. Promotion of growth and development
   c. Thermoregulation
   d. Synergism with the parasympathetic nervous system

9. Mark the following statements as true or false. If a statement is false, correct it to make a true statement.

   a. About 90% of the thyroid hormone produced is triiodothyronine (T_3).
   b. Thyroxine (T_4) is the more active of the two thyroid hormones.
   c. Thyroid hormones are produced by follicle cells.
   d. Iodine atoms are a key component of parathyroid hormone.
   e. Excess secretion of thyroid hormones produces weight gain, cold intolerance, and slow heart rate.

10. Fill in the blanks: A rise in free T_3 and T_4 would be expected to produce a(n) _____ in TRH and TSH secretion. A decrease in free T_3 and T_4 would be expected to produce a(n) _____ in TRH and TSH secretion.

11. Which of the following statements correctly describes the role of parathyroid hormone?

   a. Parathyroid hormone regulates the metabolic rate.
   b. Parathyroid hormone increases the blood calcium ion concentration.
   c. Parathyroid hormone decreases the blood sodium ion concentration.
   d. Parathyroid hormone causes water retention by the kidneys.

12. Fill in the blanks: The outer part of the adrenal gland is the _____, which secretes _____. The inner part of the adrenal gland is the _____, which secretes _____.

13. Which of the following is *not* an effect of aldosterone?

   a. Increased excretion of hydrogen ions from the fluid in the kidneys
   b. Increased retention of sodium ions from the fluid in the kidneys
   c. Increased retention of potassium ions from the fluid in the kidneys
   d. Increased retention of water from the fluid in the kidneys

14. Cortisol is:

   a. a potent inhibitor of the inflammatory response.
   b. an important part of the stress response.
   c. Both a and b are correct.
   d. Neither a nor b is correct.

15. Describe the components of the hypothalamic-pituitary-adrenocortical axis, and explain how it regulates secretion from the adrenal gland.

16. Which of the following hormones is *not* an integral part of the stress response?

   a. Epinephrine   c. Insulin
   b. Cortisol   d. Glucagon

17. Which of the following statements about the adrenal medulla is *false?*

   a. Secretion from the adrenal medulla is triggered by ACTH and the sympathetic nervous system.
   b. The adrenal medulla consists of glandular epithelial cells.
   c. The adrenal medulla is a modified postsynaptic sympathetic ganglion.
   d. The products of the adrenal medulla prolong the effects of the sympathetic response.

18. Explain how insulin and glucagon are antagonists.

19. Type 1 diabetes mellitus is characterized by _____, and type 2 diabetes mellitus is characterized by _____.

   a. hypoglycemia; destruction of the pancreatic β cells
   b. destruction of the pancreatic β cells; destruction of the pancreatic α cells
   c. insulin resistance; destruction of the pancreatic β cells
   d. destruction of the pancreatic β cells; insulin resistance

20. Match the following hormones with their correct descriptions.

   _____ Leptin
   _____ Atrial natriuretic peptide
   _____ Melatonin
   _____ Estrogens
   _____ Erythropoietin
   _____ Testosterone

   a. Produced by the pineal gland; regulates the sleep/wake cycle
   b. Produced by the kidneys; regulates red blood cell production
   c. Produced by the heart; promotes sodium ion loss in the kidneys and vasodilation
   d. Produced by the testes; promotes androgenic and anabolic actions
   e. Produced by adipose tissue; promotes satiety
   f. Produced by the ovaries; regulate the menstrual cycle and the development of secondary sex characteristics

21. Mark the following statements as true or false. If a statement is false, correct it to make a true statement.

   a. Both ADH and aldosterone increase the amount of water in the body and decrease the solute concentration of the blood.
   b. Thyroid hormones and insulin maintain blood glucose concentration during fasting.
   c. During exercise, the parasympathetic nervous system increases the metabolic rate.
   d. Insulin secretion rises during feeding; growth hormone and glucagon are secreted during feeding if protein is present in the meal.

LEVEL 2   Check Your Understanding

1. Predict the effects of a pancreatic tumor that secretes insulin. How would the effects change if the tumor secreted glucagon instead?

2. Females with hormone-secreting tumors of the adrenal cortex occasionally develop male secondary sex characteristics. Explain why this may happen.

3. A patient has a brain tumor that necessitates removal of his pituitary gland. Will its removal affect production of ADH and oxytocin? Explain.

LEVEL 3   Apply Your Knowledge

PART A: Application and Analysis

1. Ms. Reczkiewicz has her thyroid gland removed to treat hyperthyroidism. Her condition is stable after surgery, but one day later she develops symptoms of severe hypocalcemia. What has happened? How would you correct this problem?

2. A new diet guru claims hypersecretion of cortisol is the reason why so many people are obese. He is marketing a dietary supplement that is supposedly able to block cortisol secretion and lead to weight loss. How would cortisol hypersecretion cause weight gain? Would blocking normal cortisol secretion lead to weight loss? Why or why not?

3. Let's say that the dietary supplement in question 2 actually works and successfully blocks cortisol secretion. Could this drug potentially be harmful? Predict its effects on overall homeostasis.

4. Mr. Montez is a patient with type I diabetes mellitus. He presents with dizziness, rapid breathing, confusion, and weakness. You find out that he forgot to inject his normal dose of insulin this morning. Will his blood glucose concentration be normal? Explain. Your colleague suggests that Mr. Montez needs to ingest some sugar. Is this going to help him? Why or why not?

PART B: Make the Connection

5. What has likely happened to the pH of Mr. Montez's blood? What does this mean about the hydrogen ion concentration in his blood? How will his buffer systems respond to this change in pH? *(Connects to Chapter 2)*

6. You have read that aldosterone causes sodium ion retention from the kidneys. How would blocking aldosterone secretion decrease the amount of water retained from the fluid in the kidneys? *(Connects to Chapter 3)*

See answers in Appendix A.

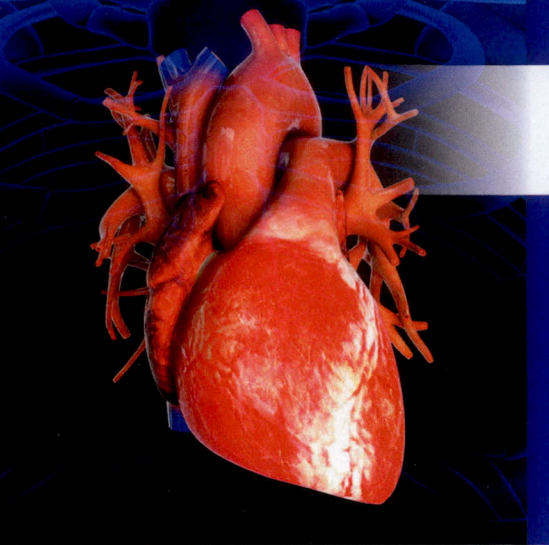

17

The Cardiovascular System I: The Heart

*For practice applying concepts to a clinical scenario, check out the **Running Case Study** for this chapter @ Mastering Anatomy & Physiology*

Computer-generated image: The heart is a muscular organ that pumps blood through the body.

Humans have long had a fascination with the heart, and we still equate it with our most treasured emotion: love. Through centuries of studying this amazing organ, we have learned about its many functions and how it contributes to the maintenance of homeostasis. In this chapter, we begin our tour of the **cardiovascular system,** which consists of the heart, blood vessels, and blood. The *heart* pumps **blood,** the tissue carrying oxygen and nutrients, into the **blood vessels,** a system of conduits that distributes it throughout the body. We'll discuss blood vessels and blood in subsequent chapters (see Chapters 18 and 19, respectively); for now let's explore this system's most prominent organ—the heart.

 **17.1**

Overview of the Heart

Learning Outcomes

1. Describe the position of the heart in the thoracic cavity.
2. Describe the basic surface anatomy of the chambers of the heart.
3. Explain how the heart functions as a double pump and why this is significant.

As you know, the muscular organ called the **heart** is a pump that contracts rhythmically to deliver blood to the body. But the heart actually consists of two pumps that propel blood to two different destinations, or *circuits*. In this module, we introduce the basic structure and location of the heart, the two locations to which the heart delivers blood, and other functions of this remarkable organ.

Location and Basic Structure of the Heart

« FLASHBACK

1. Where is the mediastinum located? (p. 17)

As you can see in **Figure 17.1a**, the heart is situated in the chest slightly to the left side in a subdivision of the thoracic cavity known as the **mediastinum** (mee´-dee-uh-STY-num). Recall from the introductory chapter that the mediastinum, located between the two lungs, houses not just the heart but also the great blood vessels, trachea, and esophagus (see Chapter 1). As we discuss shortly, the heart itself lies within a subcavity of the mediastinum called the *pericardial cavity* (**Figure 17.1b**).

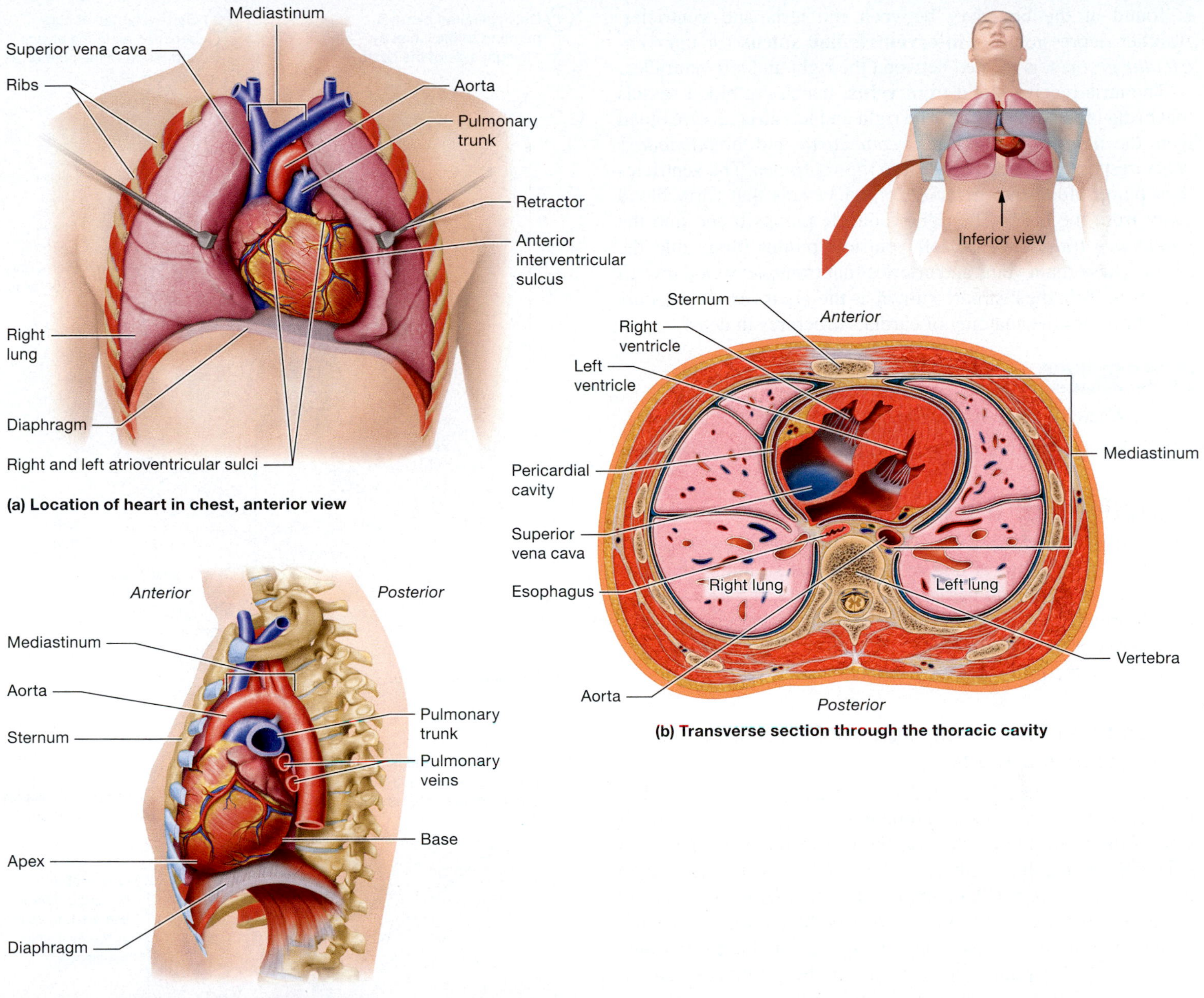

Mediastinum

Superior vena cava

Ribs

Aorta

Pulmonary trunk

Retractor

Anterior interventricular sulcus

Right lung

Diaphragm

Right and left atrioventricular sulci

(a) Location of heart in chest, anterior view

Inferior view

Sternum

Anterior

Right ventricle

Left ventricle

Pericardial cavity

Superior vena cava

Esophagus

Aorta

Right lung

Left lung

Mediastinum

Vertebra

Posterior

(b) Transverse section through the thoracic cavity

Anterior

Posterior

Mediastinum

Aorta

Sternum

Pulmonary trunk

Pulmonary veins

Apex

Base

Diaphragm

(c) Location of heart in chest, left lateral view

Figure 17.1 Location and basic anatomy of the heart in the thoracic cavity.

The heart doesn't really resemble the shape we generally think of as a heart—instead, it is somewhat cone-shaped. Its **apex,** which is the point of the cone, points toward the left hip, and its flattened **base** is its posterior side (not the inferior) facing the posterior rib cage (**Figure 17.1c**). In spite of the heart's incredible power and endurance, it is a relatively small organ, only about the size of your fist, and it generally weighs from 250 to 350 grams (slightly less than 1 pound).

The heart consists of four hollow chambers: the superior **right** and **left atria** (AY-tree-uh; singular, *atrium*) and the inferior **right** and **left ventricles** (VEN-trih-kulz; **Figure 17.2**). Externally, an indentation known as the **atrioventricular sulcus** (ay'-tree-oh-ven-TRIK-yoo-lur SUL-kuss; *sulcus* = "groove")

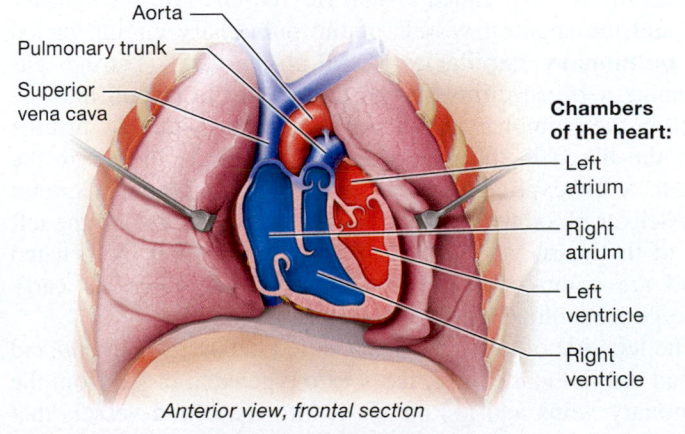

Aorta

Pulmonary trunk

Superior vena cava

Chambers of the heart:

Left atrium

Right atrium

Left ventricle

Right ventricle

Anterior view, frontal section

Figure 17.2 The chambers of the heart.

is found at the boundary between the atria and ventricles. Another depression, the **interventricular sulcus** (or *interventricular groove*), is located between the right and left ventricles.

The atria receive blood from **veins,** which are blood vessels that bring blood to the heart. The right and left atria receive blood from the *superior* and *inferior venae cavae* and the *pulmonary veins,* respectively, and deliver it to the ventricles. The ventricles then pump blood into **arteries,** blood vessels that carry blood away from the heart. The right ventricle pumps blood into the *pulmonary trunk,* and the left ventricle pumps blood into the *aorta.* These main veins and arteries that transport blood toward and away from the heart are known as the *great vessels.* Module 17.2 discusses the anatomy of cardiac structures in detail.

> **Quick Check**
>
> ☐ 1. Where is the heart located, and how large is it?
> ☐ 2. What are the heart's upper and lower chambers called?

Functions of the Heart

The primary function of the heart—to pump blood—seems readily apparent. But of course this isn't as straightforward as it sounds. As you read earlier in this module, the heart pumps blood through two different sets of vessels, or circuits. We take a closer look at these circuits and some of the heart's other functions in the following subsections.

Circulation of Blood through the Pulmonary and Systemic Circuits

The heart is divided functionally into right and left sides. The right side of the heart pumps blood to the lungs, whereas the left side of the heart pumps blood to the rest of the body. The right side of the heart, shaded blue in **Figure 17.3a**, is sometimes called the *pulmonary pump* (PULL-munn-ehr-ee; *pulmon-* = "lungs") because it pumps blood into a series of blood vessels leading to and within the lungs, collectively called the **pulmonary circuit.** ① The pulmonary arteries of the pulmonary circuit deliver blood that is oxygen-poor and carbon dioxide–rich, or *deoxygenated* (dee-AWK-suh-jeh-nay′-t'd), to the lungs. ② The process of *gas exchange* takes place between the tiny air sacs in the lung, called *alveoli* (al-vee-OH-lye; see Chapter 21), and the smallest vessels of the pulmonary circuit, called the **pulmonary capillaries** (KAP-uh-lehr-eez). During gas exchange, oxygen diffuses from the air in the alveoli into the blood in the pulmonary capillaries, and carbon dioxide diffuses from the blood in the pulmonary capillaries to the air in the alveoli, to be expired. ③ The veins of the pulmonary circuit then deliver blood that is oxygen-rich, or *oxygenated,* to the left side of the heart. Vessels and organs that transport oxygenated blood are color-coded red in this text, and those that carry deoxygenated blood are shown as blue.

The left side of the heart, often called the *systemic pump* and shaded red in **Figure 17.3b**, receives oxygenated blood from the pulmonary veins and ① pumps it into the blood vessels that serve the body, collectively known as the **systemic circuit.**

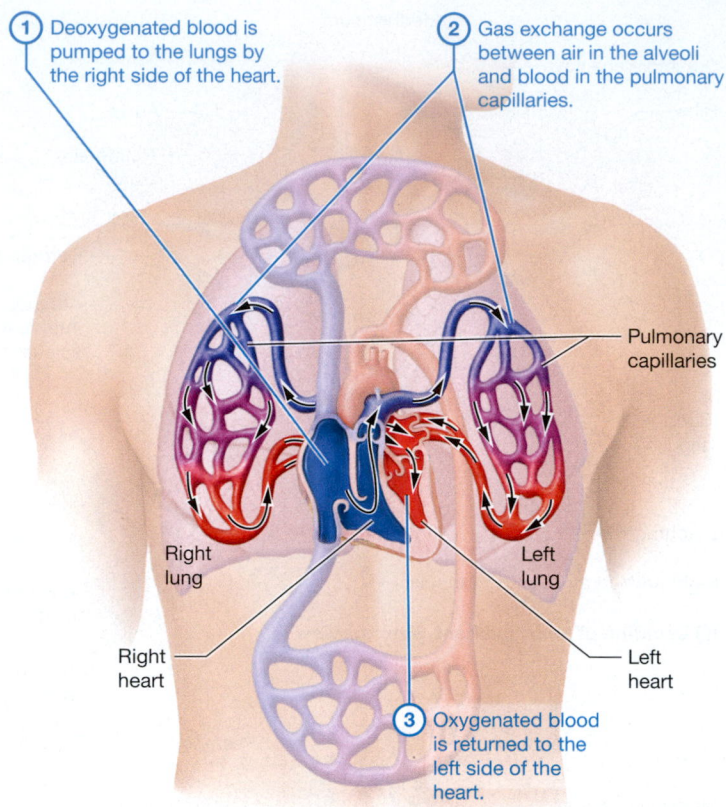

① Deoxygenated blood is pumped to the lungs by the right side of the heart.

② Gas exchange occurs between air in the alveoli and blood in the pulmonary capillaries.

Pulmonary capillaries

Right lung

Left lung

Right heart

Left heart

③ Oxygenated blood is returned to the left side of the heart.

(a) The pulmonary circuit

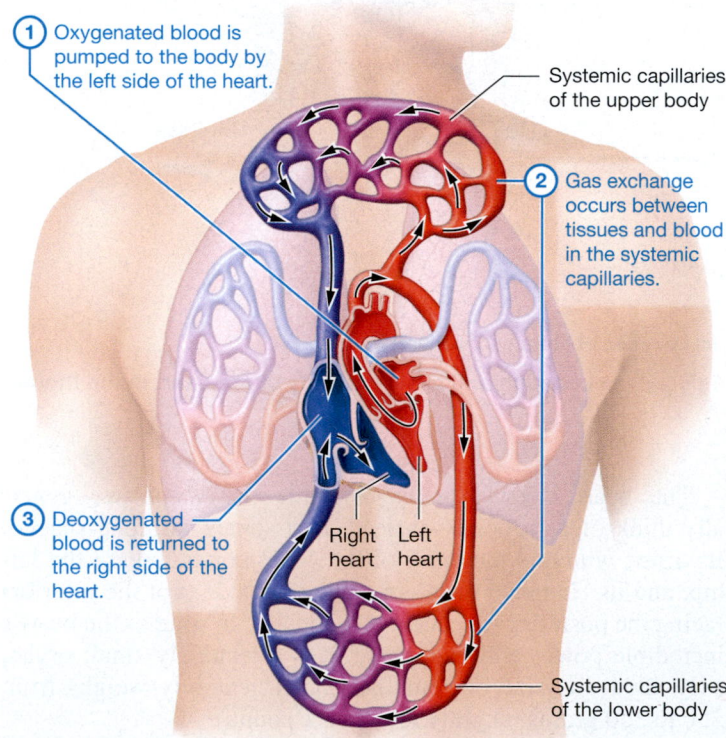

① Oxygenated blood is pumped to the body by the left side of the heart.

Systemic capillaries of the upper body

② Gas exchange occurs between tissues and blood in the systemic capillaries.

③ Deoxygenated blood is returned to the right side of the heart.

Right heart

Left heart

Systemic capillaries of the lower body

(b) The systemic circuit

Figure 17.3 The pulmonary and systemic circuits.

Play Animation
@ **Mastering** Anatomy & Physiology

In the systemic circuit, arteries deliver oxygenated blood to the smallest blood vessels, the systemic capillaries. Here ② gas exchange occurs again, except in reverse: Oxygen diffuses from the blood into the tissues, and carbon dioxide diffuses from the tissues into the blood. In addition to delivering oxygen, blood also delivers nutrients, picks up wastes to be excreted, and distributes hormones to their target cells throughout the body. As a result of gas exchange in the tissues, the blood becomes deoxygenated. ③ The veins of the systemic circuit must then deliver the blood back to the right side of the heart, which pumps it into the pulmonary circuit to become oxygenated again.

The pulmonary circuit is a low-pressure circuit. It delivers blood only a short distance to the lungs, so the forces resisting blood flow are low. In contrast, the systemic circuit is a high-pressure circuit. It must deliver blood a far greater distance to the entire rest of the body, and so the forces resisting blood flow are much higher. Keep in mind that, in spite of these pressure differences, both sides of the heart contract at the same time and eject the same volume of blood. We'll see in the next module that this has implications for the structure of the heart's chambers.

Other Functions of the Heart

The heart has other functions besides pumping blood. One of the most important is to help maintain homeostasis of the pressure that blood exerts on the blood vessels, also known as *blood pressure*. As we discuss in the next chapter, the rate and force of the heart's contraction are major factors that influence the blood pressure and the blood flow to organs. Recall that the heart (specifically the atria) also acts as an endocrine organ and produces a hormone called *atrial natriuretic peptide* (nay'-tree-yoor-ET-ik), or *ANP* (see Chapter 16). ANP lowers blood pressure by decreasing sodium ion retention in the kidneys. This reduces osmotic water retention and so the volume and pressure of blood in the blood vessels.

Heart Anatomy and Blood Flow Pathway

Learning Outcomes

1. Describe the layers of the pericardium and heart wall.
2. Describe the location and function of the coronary circulation and great vessels.
3. Describe the structure and function of the chambers, septa, valves, and other structural features of the heart.
4. Trace the pathway of blood flow through the heart, and explain how structures of the heart ensure that blood flows in a single direction.

As you read about in the previous module, the heart is a muscular organ that pumps blood through the body. The blood is contained within the heart's hollow spaces, its *lumen*. Like all organs, the heart is composed of different tissue layers that work together to ensure its proper functioning.

In this module we start with the body cavity in which the heart is found and the tissue layers of the heart wall. We then look at the detailed structure of the chambers of the heart, the great vessels, and the structures that ensure blood flow in a single direction—the heart's *valves*. The module concludes with an examination of the *coronary circulation,* the set of arteries and veins that supply and drain the heart muscle.

The Pericardium, Heart Wall, and Heart Skeleton

« FLASHBACK

1. What is a serous membrane? What does a serous membrane produce and what is its function? (p. 153)
2. What are the components of cardiac muscle tissue? (p. 148)
3. What is simple squamous epithelium? (p. 129)

You may have seen a surgeon in a movie splitting open a patient's thoracic cavity and dramatically exposing the beating heart. This makes for a great movie scene, but it's not very accurate. The heart isn't actually exposed in the thoracic cavity—instead, what you see when the thoracic cavity is first opened looks more like what is shown in **Figure 17.4a** (to read more about the surgical procedure in which the thoracic cavity is opened, see *A&P in the Real World: Thoracotomy* on p. 634). The membranous structure surrounding the heart is called the **pericardium** (pehr'-ih-KAR-dee-um; *peri-* = "around," *cardi-* = "heart"). The pericardium is a sac with two components: the **fibrous pericardium,** a tough outer layer that attaches the heart to surrounding structures, and the **serous pericardium,** a thin inner serous membrane that produces serous fluid (**Figure 17.4b**).

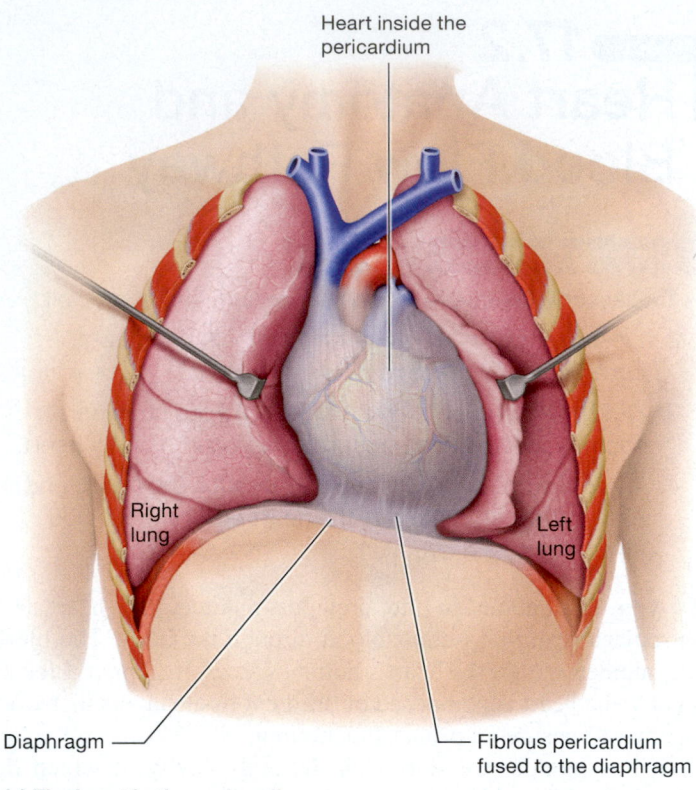

Heart inside the pericardium

Right lung

Left lung

Diaphragm

Fibrous pericardium fused to the diaphragm

(a) The heart in the pericardium

Thoracotomy

The thoracic cavity is opened in the surgical procedure known as a **thoracotomy** (thohr-uh-KAWT-uh-mee; -*otomy* = cutting into). A thoracotomy is performed when a surgeon must gain access to the thoracic organs, the surrounding blood vessels, or the anterior side of the thoracic vertebral column. It generally involves an incision in the chest wall and cutting through either the sternum or the ribs. The sternum or ribs are separated and held apart during the procedure with a surgical instrument called a *retractor* (or "rib spreader"), which creates a "window" into the thoracic cavity. When the surgical procedure is completed, the chest wall is closed, and a *chest tube* must be inserted to prevent air from leaking in to the thoracic cavity and potentially causing collapse of the lung.

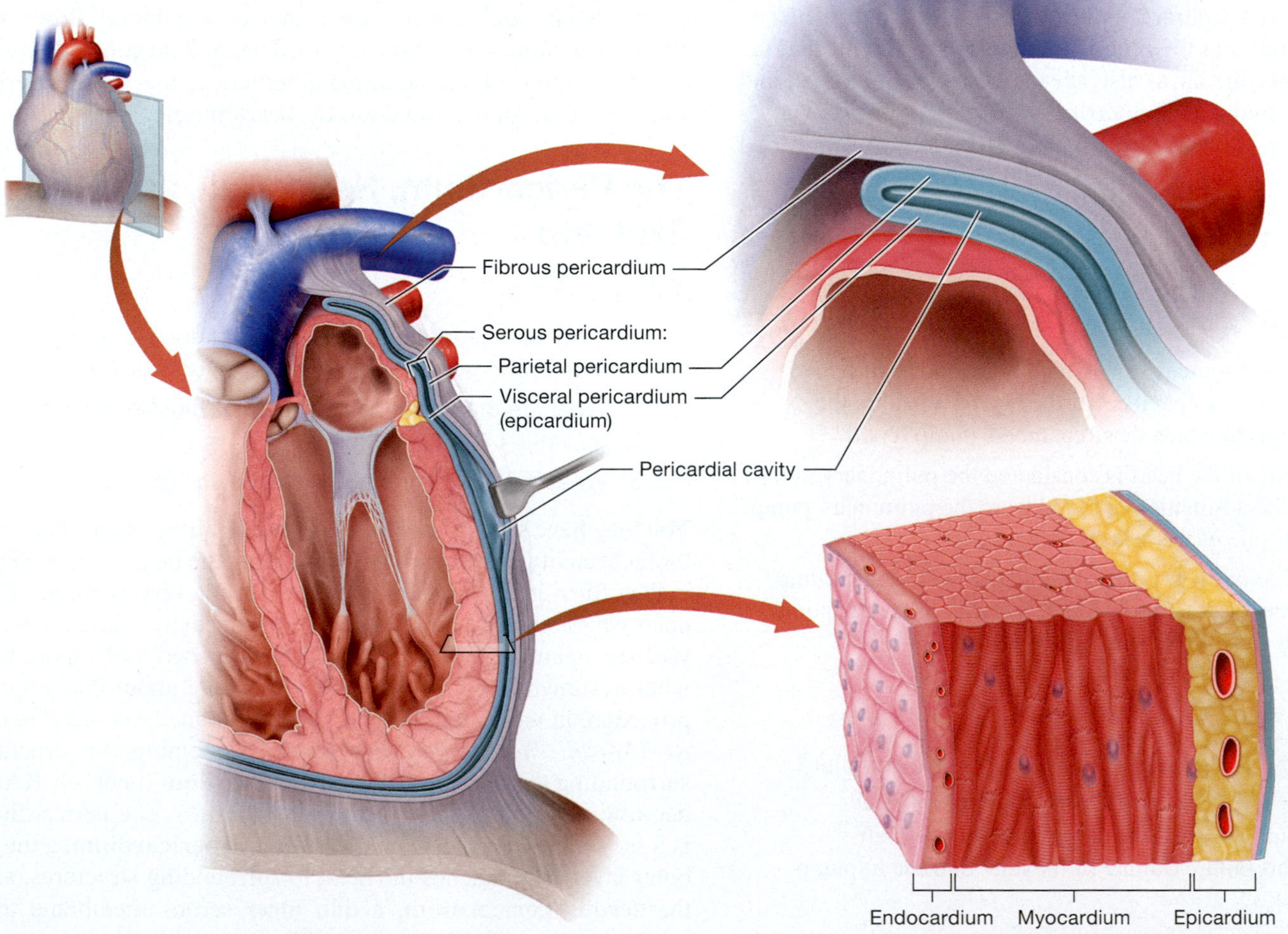

Fibrous pericardium

Serous pericardium:

Parietal pericardium

Visceral pericardium (epicardium)

Pericardial cavity

Endocardium   Myocardium   Epicardium

(b) Frontal dissection of the heart and pericardial cavity

(c) Layers of the heart wall

Figure 17.4 The pericardium and the layers of the heart wall.

The fibrous pericardium is composed of collagen bundles that make it tough and enable it to anchor the heart to structures such as the diaphragm and the great vessels. The collagen bundles also give the fibrous pericardium low distensibility—it doesn't change shape or size considerably when stretching forces are applied. This helps to prevent the chambers of the heart from overfilling with blood. (However, see *A&P in the Real World: Cardiac Tamponade* for a discussion of when this property becomes a disadvantage.)

Turning to the serous pericardium, notice in the figure that its outer layer, the **parietal pericardium,** is fused to the inner surface of the fibrous pericardium. The parietal pericardium encases the heart like a sac, but when it reaches the great vessels, it folds under itself and forms another layer that adheres directly to the heart. This inner layer is the **visceral pericardium,** also known as the **epicardium** (ep′-ih-KAR-dee-um; *epi-* = "on top"). The visceral pericardium is considered the most superficial layer of the heart wall. Although it might seem as if the serous pericardium is two separate structures—the parietal and visceral pericardia—remember that they are actually one continuous structure, as shown here:

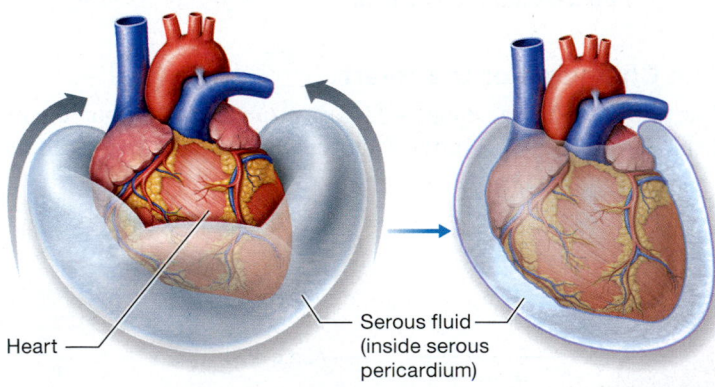

Heart — / Serous fluid —
                (inside serous
                pericardium)

Between the parietal and visceral pericardia we find the **pericardial cavity,** which is filled with a very thin layer of serous fluid, called **pericardial fluid.** This fluid acts as a lubricant, decreasing friction as the heart moves and preventing it from rubbing against the sternum.

The visceral pericardium rests on top of a thin layer of areolar connective tissue that contains large fat deposits (**Figure 17.4c**). These deposits tend to collect in the atrioventricular and interventricular sulci (which you could see in Figure 17.1a). Deep to this connective tissue we find the second and thickest heart wall layer, the **myocardium** (my′-oh-KAR-dee-um; *myo-* = "muscle"). The myocardium has two components: cardiac muscle tissue and a fibrous skeleton. Cardiac muscle tissue consists of **cardiac muscle cells,** or **myocytes** (MY-oh-sytz), and their surrounding extracellular matrix. Cardiac muscle cells are arranged in a spiral manner around the heart chambers, a structure that allows them to undergo a "wringing" action when they contract.

The **fibrous skeleton** is composed of dense irregular collagenous connective tissue. It is located in regions that need additional support, including within the interventricular and

interatrial septa (the muscular walls between the ventricles and atria, respectively) and around the heart valves (look ahead to Figure 17.7). The fibrous skeleton has multiple functions, which include giving the cardiac muscle cells something on which to pull when they contract, providing structural support, and acting as an insulator for the heart's electrical activity. We explore the structure of cardiac muscle cells in Module 17.3.

The lumen of the heart is lined by the third and deepest layer of the heart wall, the **endocardium** (en′-doh-KAR-dee-um; *endo-* = "within"). The endocardium is composed of a special type of simple squamous epithelium called **endothelium** (en′-doh-THEE-lee-um) and several layers of connective tissue with elastic and collagen fibers. The endothelial cells of the endocardium are continuous with the endothelial cells that line blood vessels and so share many of the same tasks. For example, the endothelial cells that make up the heart endocardium play a role similar to those found in the blood brain barrier—they form a sort of "blood heart barrier" that helps to regulate the concentration of electrolytes and other chemicals in the extracellular fluid of the myocardium.

Cardiac Tamponade

If the pericardial cavity becomes filled with excess fluid, **cardiac tamponade** (tam-poh-NAHD) may result. This condition has many potential causes, including trauma, certain cancers, kidney failure, recent thoracic surgery, and HIV. Regardless of the cause, the result is the same—the excess fluid in the pericardial cavity expands toward the heart because the fibrous pericardium is strong but not very flexible. This applies pressure to the outside of the heart, which squeezes the heart and reduces the capacity of the ventricles to fill with blood, compromising the amount of blood pumped with each beat. The treatment for cardiac tamponade may include a procedure in which the excess fluid is removed via a needle inserted into the pericardial cavity.

Quick Check

☐ 1. How do the fibrous pericardium and the serous pericardium differ?

☐ 2. Where is the pericardial cavity located?

☐ 3. What are the three layers of the heart wall, and of what types of tissue are they composed?

The Great Vessels, Chambers, and Valves of the Heart

As we discussed in Module 17.1, the heart consists of four chambers: two atria and two ventricles. The atria receive blood from veins, and pump blood into the ventricles through structures called **valves.** These valves have flaps that close when the ventricles contract, keeping the blood from moving backward. The contracting ventricles then eject blood into arteries, which carry the blood through either the systemic or the pulmonary circuit. We now turn to the structure of these chambers, the valves, and the vessels that enter and exit the heart.

The Great Vessels

The vessels that bring blood to and away from the heart, the **great vessels,** are the largest ones in the body. There are four main great vessels (**Figure 17.5**):

- **Major systemic veins: superior vena cava and inferior vena cava.** The two veins that drain the majority of the systemic circuit are the **superior** and **inferior venae cavae** (VEE-nee KAY-vee; singular, *vena cava*). The superior vena cava (SVC) drains deoxygenated blood from most veins superior to the diaphragm, and the inferior vena cava (IVC) drains deoxygenated blood from most veins inferior to the diaphragm. Note in Figure 17.5 that both veins have large openings into the posterior aspect of the right atrium, the chamber into which they drain blood.
- **Pulmonary trunk.** The large **pulmonary trunk** receives deoxygenated blood pumped from the right ventricle as

the first and widest vessel of the pulmonary circuit. The pulmonary trunk is located on the anterior side of the heart, nearly along the midline, where it originates from the right ventricle. After a short distance the trunk splits into the *right* and *left pulmonary arteries,* which bring deoxygenated blood to the right and left lungs, respectively. The pulmonary arteries branch extensively inside the lungs to become the tiny pulmonary capillaries where gases are exchanged.

- **Pulmonary veins.** After the blood is oxygenated in the pulmonary capillaries, it returns to the heart through a set of **pulmonary veins.** Most people have four pulmonary veins—two from each lung—that drain oxygenated blood into the posterior part of the left atrium.
- **Aorta.** The **aorta** (ay-OHR-tah) supplies the entire systemic circuit with oxygenated blood. The aorta is the largest and thickest artery in the systemic circuit—and, in fact, in the entire body. The aorta stems from the left ventricle as the *ascending aorta,* after which it curves to the left and makes a U-turn as the *aortic arch.* (Notice that the pulmonary trunk splits into the two pulmonary arteries just underneath the aortic arch.)

The Chambers of the Heart

We now turn to the heart's largest structures: its four chambers. As you have seen, the ventricles are the larger chambers.

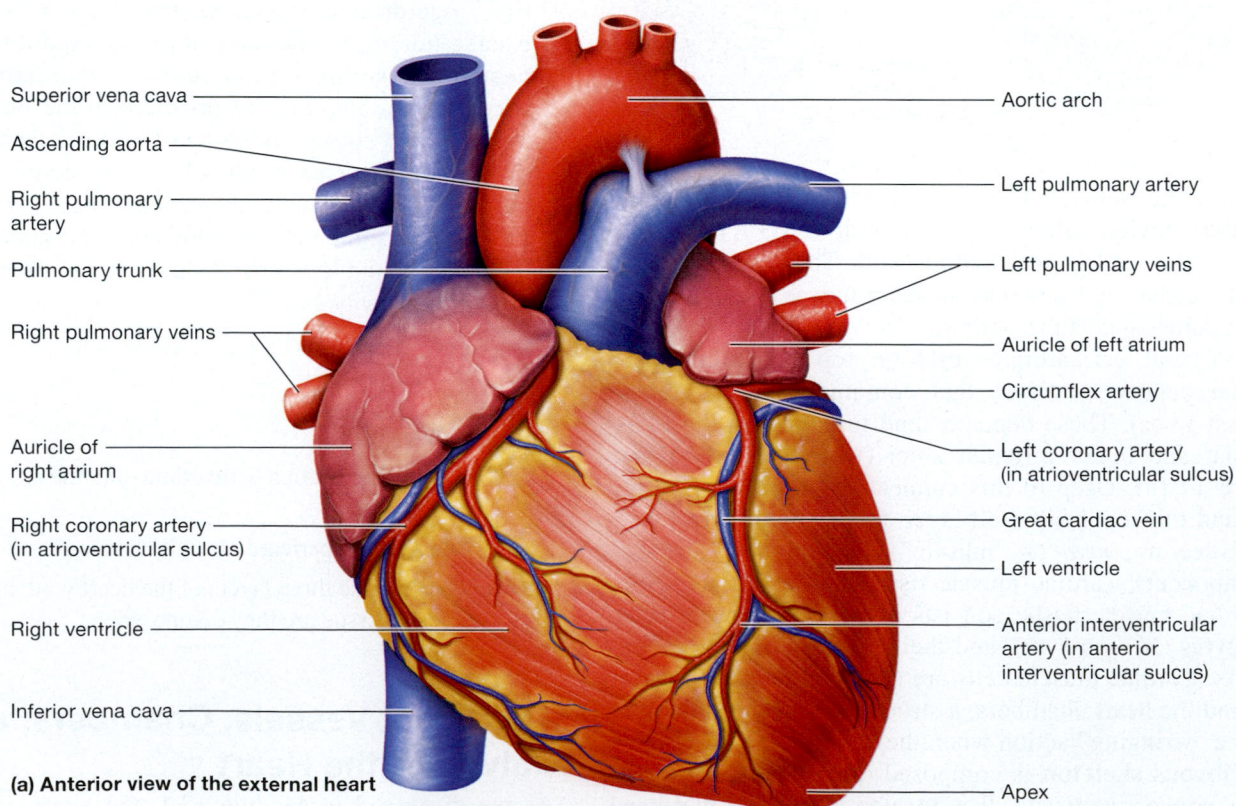

(a) Anterior view of the external heart

Labels (left, top to bottom): Superior vena cava; Ascending aorta; Right pulmonary artery; Pulmonary trunk; Right pulmonary veins; Auricle of right atrium; Right coronary artery (in atrioventricular sulcus); Right ventricle; Inferior vena cava

Labels (right, top to bottom): Aortic arch; Left pulmonary artery; Left pulmonary veins; Auricle of left atrium; Circumflex artery; Left coronary artery (in atrioventricular sulcus); Great cardiac vein; Left ventricle; Anterior interventricular artery (in anterior interventricular sulcus); Apex

Figure 17.5 The external anatomy of the heart.

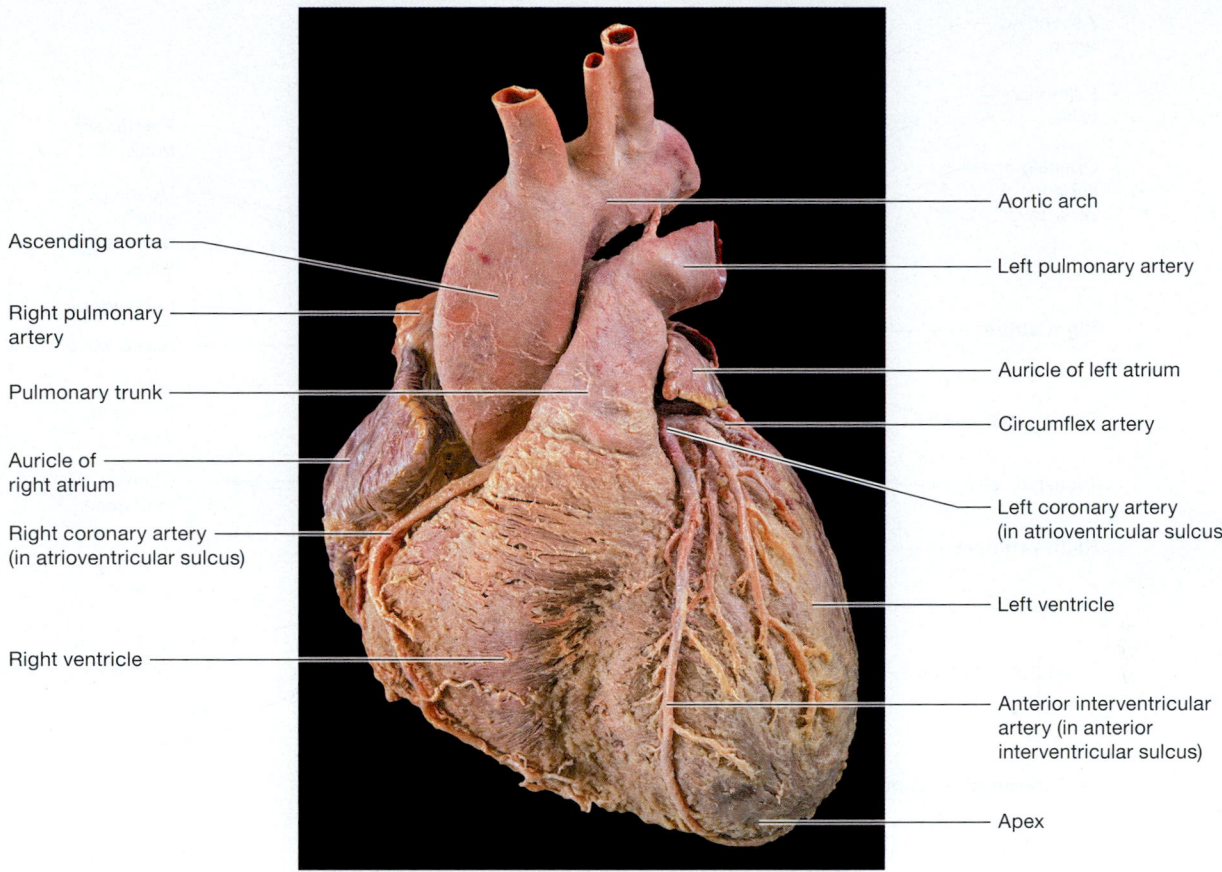

Aortic arch

Left pulmonary artery

Auricle of left atrium

Circumflex artery

Left coronary artery
(in atrioventricular sulcus)

Left ventricle

Anterior interventricular
artery (in anterior
interventricular sulcus)

Apex

Ascending aorta

Right pulmonary
artery

Pulmonary trunk

Auricle of
right atrium

Right coronary artery
(in atrioventricular sulcus)

Right ventricle

(b) Cadaver photo, anterior view of the external heart

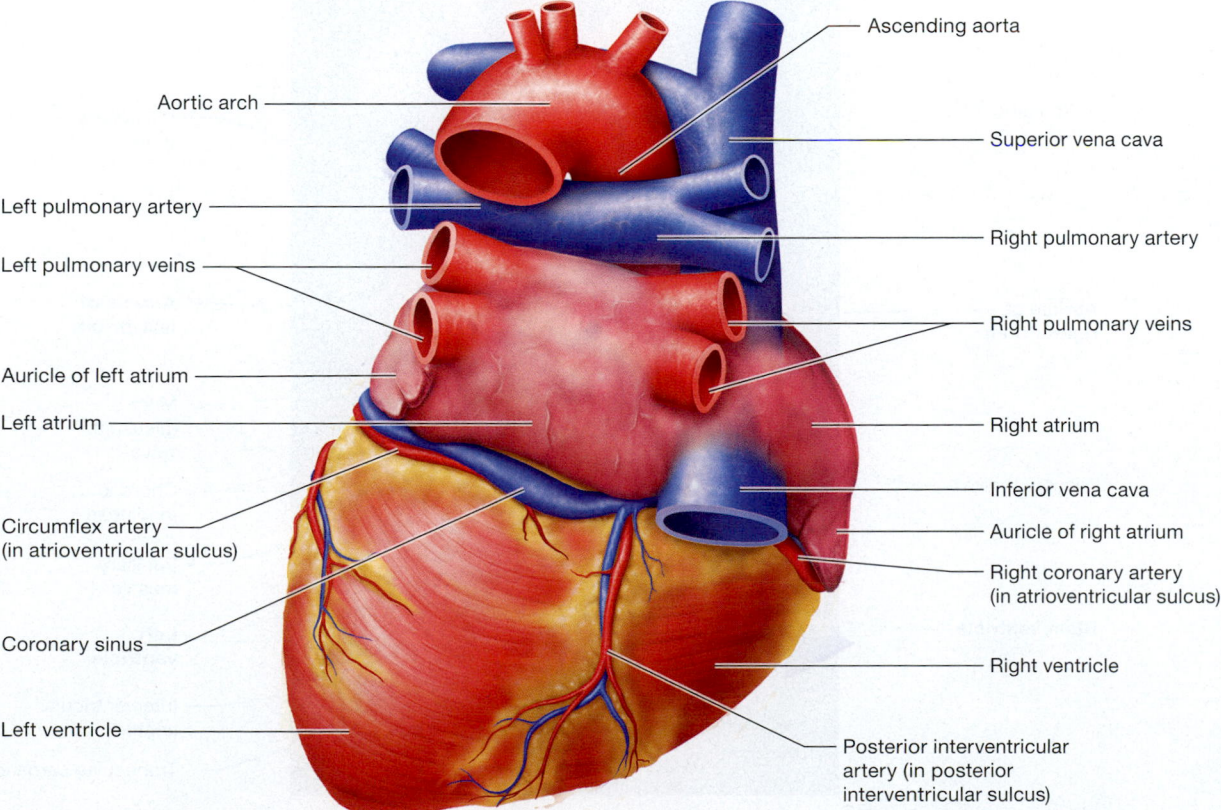

Ascending aorta

Superior vena cava

Right pulmonary artery

Right pulmonary veins

Right atrium

Inferior vena cava

Auricle of right atrium

Right coronary artery
(in atrioventricular sulcus)

Right ventricle

Posterior interventricular
artery (in posterior
interventricular sulcus)

Aortic arch

Left pulmonary artery

Left pulmonary veins

Auricle of left atrium

Left atrium

Circumflex artery
(in atrioventricular sulcus)

Coronary sinus

Left ventricle

(c) Posterior view of the external heart

Figure 17.5 The external anatomy of the heart. *(continued)*

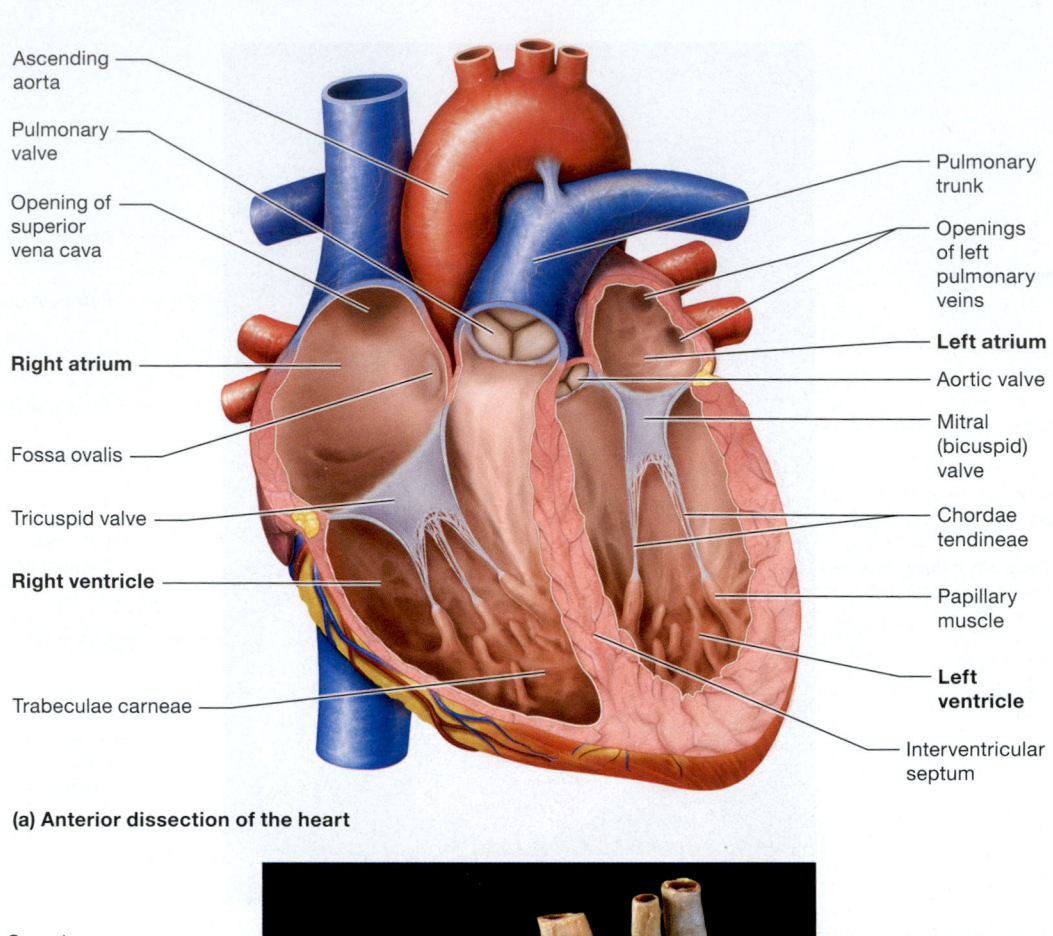

Ascending aorta

Pulmonary valve

Opening of superior vena cava

Right atrium

Fossa ovalis

Tricuspid valve

Right ventricle

Trabeculae carneae

Pulmonary trunk

Openings of left pulmonary veins

Left atrium

Aortic valve

Mitral (bicuspid) valve

Chordae tendineae

Papillary muscle

Left ventricle

Interventricular septum

(a) Anterior dissection of the heart

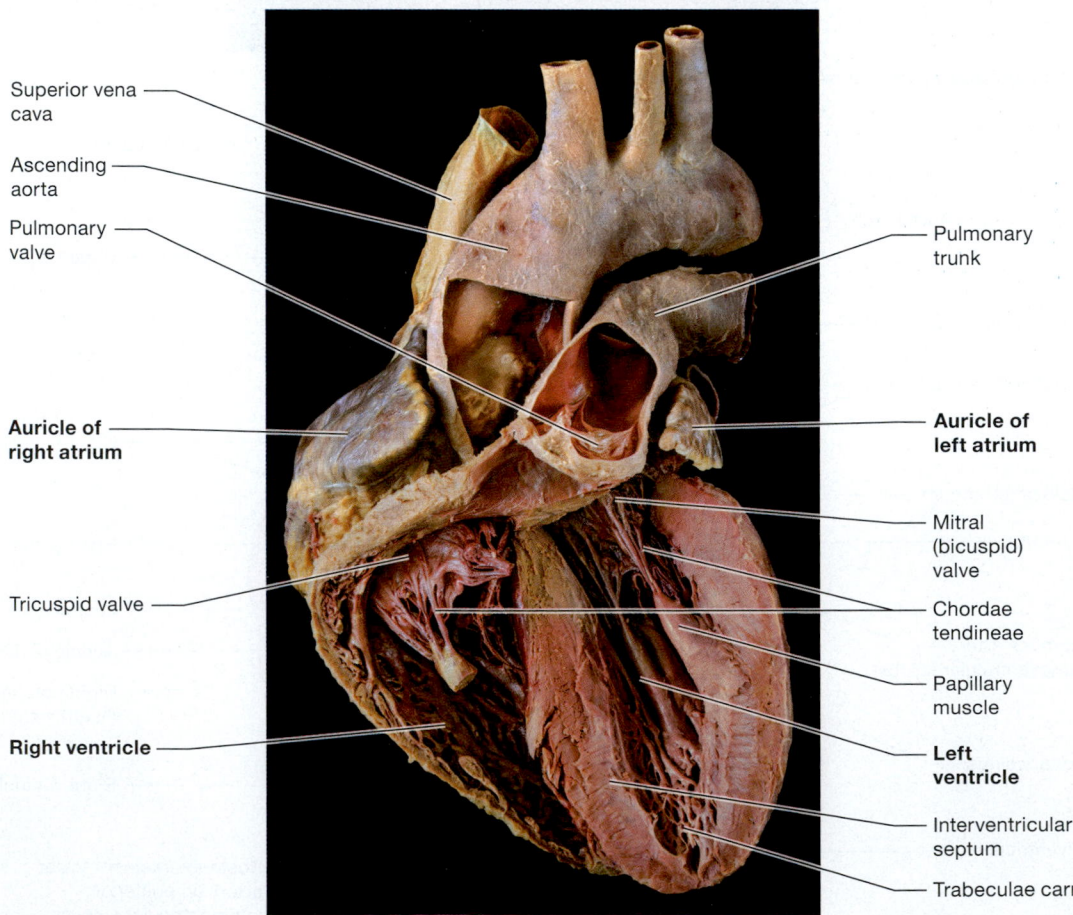

Superior vena cava

Ascending aorta

Pulmonary valve

Auricle of right atrium

Tricuspid valve

Right ventricle

Pulmonary trunk

Auricle of left atrium

Mitral (bicuspid) valve

Chordae tendineae

Papillary muscle

Left ventricle

Interventricular septum

Trabeculae carneae

(b) Cadaver photo, anterior dissection of the heart

Figure 17.6 The internal anatomy of the heart.

In addition, they have much thicker walls, which you can see in **Figure 17.6**. This makes the ventricles much stronger pumps, which is needed to generate the pressure that pumps blood through the pulmonary and systemic circuits.

The Atria The atria are not symmetrical in size, shape, or location. The right atrium is large, thin-walled, and located anteriorly. In contrast, the left atrium has a thicker wall, is somewhat smaller, and is located mostly on the posterior side of the heart, where it makes up much of the heart's base and posterior surface. Externally, each atrium has a muscular pouch called the **auricle** (OHR-ih-kul). The auricles, named for their resemblance to the external ear, expand to give the atria more space in which to hold blood. The auricle of the right atrium is much larger than that of the left atrium.

The internal surface of the right atrium has muscular ridges on its anterior side, called **pectinate muscles** (PEK-tin-et; *pectinate* = "combed"). Its posterior surface is smooth, and here we find the openings of the superior and inferior venae cavae and the coronary sinus. The left atrium is mostly formed by the pulmonary veins, and so internally its walls are smooth.

The two atria are separated by a thin wall, the **interatrial septum.** Notice in Figure 17.6a that there is a small indentation in the interatrial septum called the **fossa ovalis.** This indentation is the remnant of a hole known as the **foramen ovale** (fohr-AY-men oh-VAEH-lee), which is present in the interatrial septum of the fetal heart. The foramen ovale directs blood from the right atrium directly to the left atrium, bypassing the right ventricle and the pulmonary circulation (because fetal lungs are not yet functional). Normally the foramen ovale closes shortly before or after birth, leaving behind the fossa ovalis (the foramen ovale is discussed in the development chapter—see Chapter 27).

The Ventricles Like the atria, the right and left ventricles are asymmetrical, as you can see here:

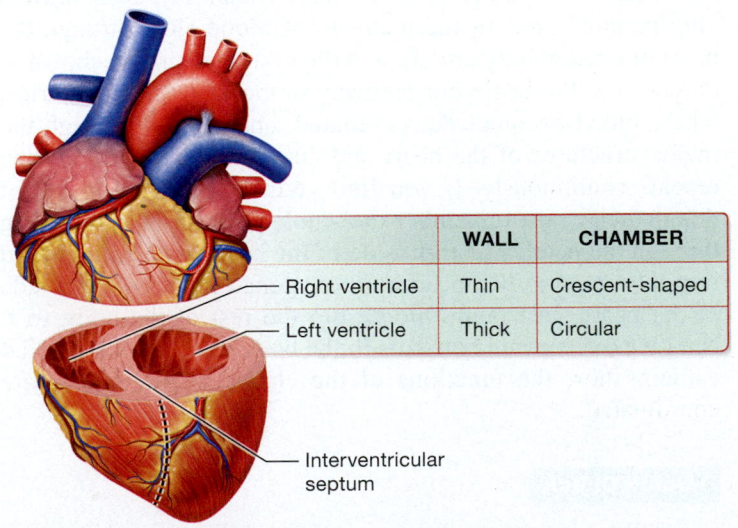

	WALL	CHAMBER
Right ventricle	Thin	Crescent-shaped
Left ventricle	Thick	Circular
Interventricular septum		

Notice that the right ventricle is wider and has thinner walls than the left ventricle. This is an example of the Structure-Function

Core Principle (p. 28)—because of the pressure differences in the pulmonary and systemic circuits, the right ventricle has little resistance against which to pump, whereas the left ventricle pumps against much greater resistance. This means that the left ventricle has to work harder than the right ventricle and, as a result, has greater muscle mass. In fact, the walls of the left ventricle are about three times thicker than those of the right ventricle. You can also see that the left ventricle is longer and circular in cross section, whereas the right ventricle is shorter and crescent-shaped in cross section.

Internally, both ventricles have a ridged surface created by irregular protrusions of cardiac muscle tissue (see Figure 17.6). These ridges are collectively referred to as **trabeculae carneae** (trah-BEK-yoo-lee kar-NEE-ee; "beams of flesh"). Each ventricular wall also contains finger-like projections, the **papillary muscles** (PAP-ih-lehr-ee; *papill-* = "nipple"), which attach by tendon-like cords called *chordae tendineae* (KOHR-dee ten-din-EE-ee) to the valves located between the atria and the ventricles. The trabeculae carneae and the papillary muscles ensure that the valves work properly, discussed in Module 17.4. A thick, muscular wall, the **interventricular septum,** separates the right and left ventricles. This septum contracts with the rest of the ventricular muscle and helps to expel blood into the pulmonary trunk and aorta.

The Valves of the Heart

Blood flow through the heart must occur in only one direction so that deoxygenated blood goes to the pulmonary circuit and oxygenated blood goes to the systemic circuit. Backflow of blood generally doesn't occur in the veins draining into the atria, as the atria are under very low pressure and blood mostly flows into the atria with the help of gravity and the pressure in the veins. However, pressure in the ventricles when they contract is very high, which could push blood back into the atria. Similarly, blood flowing out of the ventricles moves against gravity into high-pressure arteries, which could cause it to flow backward into the ventricles. To prevent this backflow, two sets of valves are present in the heart, which we examine next.

The Atrioventricular Valves The valves between the atria and the ventricles are called the right and left **atrioventricular (AV) valves** (**Figure 17.7a**). You can think of these valves as "inflow" valves, as blood must flow through them to enter the ventricles. The two AV valves consist of flaps, called *cusps,* that are composed of endocardium overlying thin extensions of the fibrous skeleton. Each valve is named for the number of cusps it contains: The **tricuspid valve** between the right atrium and right ventricle contains three cusps, and the **bicuspid valve** between the left atrium and left ventricle has two cusps. Note that the bicuspid valve is more commonly called by its clinical name, **mitral valve** (MY-trul), and this is the term used in this book. Figure 17.7a illustrates how the fibrous skeleton

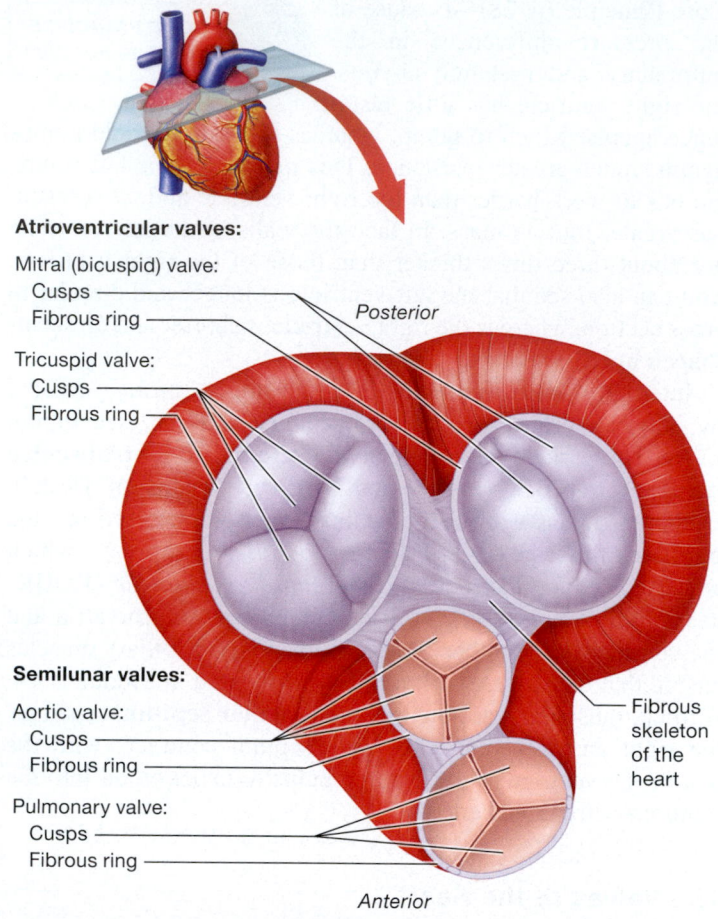

Atrioventricular valves:

Mitral (bicuspid) valve:
  Cusps
  Fibrous ring

Posterior

Tricuspid valve:
  Cusps
  Fibrous ring

Semilunar valves:

Aortic valve:
  Cusps
  Fibrous ring

Fibrous
skeleton
of the
heart

Pulmonary valve:
  Cusps
  Fibrous ring

Anterior

(a) The atrioventricular and semilunar valves

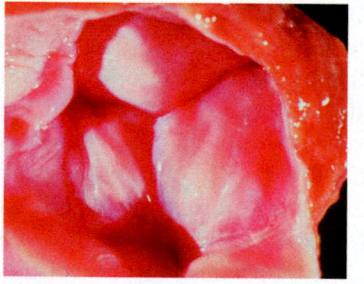 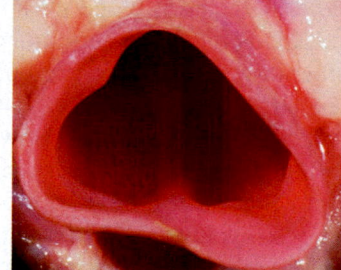

(b) The pulmonary valve, closed (left) and open (right)

Figure 17.7 Anatomy of the atrioventricular and semilunar valves.

supports these atrioventricular valves—two rings extend from a central core, and each ring surrounds the opening of a valve.

Extending from the inferior end of each cusp we find the **chordae tendineae** ("tendon strings"). The thin chordae tendineae attach to papillary muscles that contract just before the ventricles begin ejecting blood. This creates tension on the chordae tendineae, which keeps the valves closed. These actions prevent the cusps from *everting,* or ballooning back into the atria, when the pressure in the ventricles increases.

The Semilunar Valves

The blood in the pulmonary trunk and aorta is prevented from completely flowing back into the

ventricles by two valves called the **semilunar (SL) valves** (seh-mee-LOO-nur). You can think of these valves as "outflow" valves, since blood passes through them as it flows out of the heart. The word semilunar refers to the half-moon shape of the valves' three cusps, which are also composed of endocardium and a thin layer of the fibrous skeleton. The two SL valves are named according to the artery in which they reside: The **pulmonary valve** is located between the right ventricle and pulmonary trunk, and the **aortic valve** is posterior to it between the left ventricle and the aorta. The fibrous skeleton supports the aortic and pulmonary valves in much the same way it supports the AV valves. You can see what the pulmonary valve looks like closed and open in **Figure 17.7b.** No chordae tendineae or papillary muscles attach to the SL valves.

As you can imagine, dysfunctional valves can lead to many problems. *A&P in the Real World: Valvular Heart Disease* on p. 642 describes what happens when valves are not working correctly. Often, dysfunctional valves must be replaced surgically with either a mechanical valve or a real valve from a human cadaver or a pig.

Quick Check

☐ 4. What are the four main great vessels? From which structure(s) does each vessel receive blood? To which structure(s) does each vessel deliver blood?

☐ 5. How do the right and left ventricles differ in size and shape?

☐ 6. Why do you think it is important to ensure via heart valves that blood flows in only one direction in the heart?

Putting It All Together: The Big Picture of Blood Flow through the Heart

Now that we've discussed the heart's anatomy, let's draw a "big picture" view of the pathway of blood flow through the heart, the pulmonary circuit, and the systemic circuit, shown in **Figure 17.8.** We begin our pathway in the systemic capillaries, where blood becomes deoxygenated, and trace it through the major structures of the heart and circulation. This cycle then repeats continuously. If you find yourself forgetting parts of this pathway, just remember the whole point of blood flowing through the heart: The right side of the heart must pump blood into the lungs so it can become oxygenated, and the left side of the heart must pump blood into the rest of the body so it can give oxygen and nutrients to the body's cells. Module 17.4 explains how the functions of the chambers and valves are coordinated.

Quick Check

☐ 7. What is the overall pathway of blood flow through the heart, beginning and ending with the systemic capillaries, and including the valves and great vessels?

The Big Picture of Blood Flow through the Heart

Figure 17.8

Play Animation @ Mastering Anatomy & Physiology

(1) Blood in systemic capillaries delivers oxygen to body cells.

(2) Systemic veins return deoxygenated blood to the right atrium (see Figure 17.6).

(3) Blood passes from the right atrium through the tricuspid valve to the right ventricle (see Figures 17.6 and 17.7).

(4) The right ventricle pumps blood through the pulmonary valve to the pulmonary trunk (see Figures 17.3a and 17.7).

(5) The pulmonary trunk delivers blood to the pulmonary capillaries of the left and right lungs.

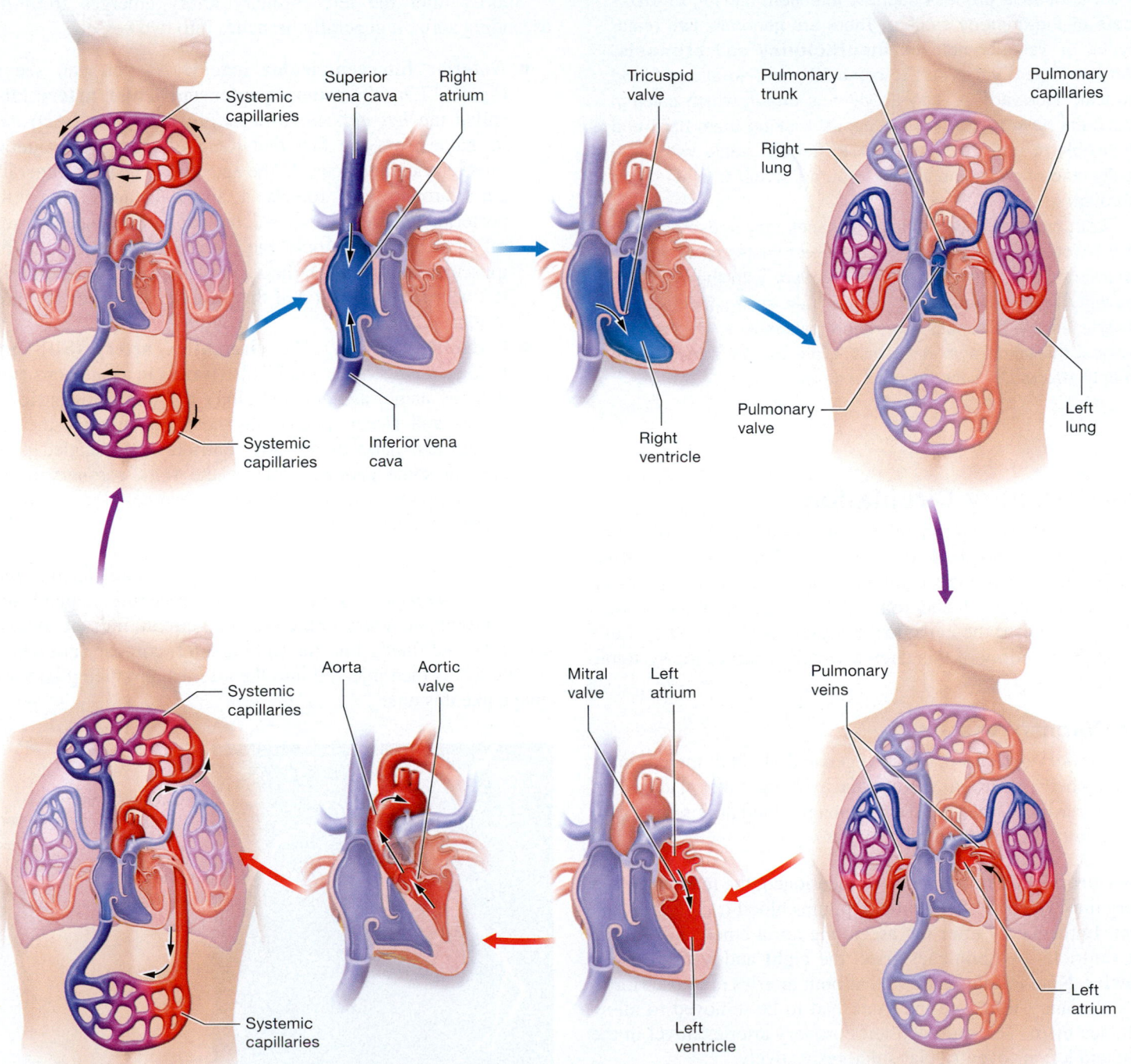

Systemic capillaries

Superior vena cava

Right atrium

Tricuspid valve

Pulmonary trunk

Right lung

Pulmonary capillaries

Systemic capillaries

Inferior vena cava

Pulmonary valve

Right ventricle

Left lung

Systemic capillaries

Aorta

Aortic valve

Mitral valve

Left atrium

Pulmonary veins

Systemic capillaries

Left ventricle

Left atrium

(9) The aorta delivers blood to the systemic capillaries (see Figure 17.3b), and the cycle continues.

(8) The left ventricle pumps blood through the aortic valve to the aorta.

(7) Blood passes from the left atrium through the mitral valve to the left ventricle.

(6) The blood becomes oxygenated in the pulmonary capillaries and the pulmonary veins return oxygenated blood to the left atrium.

Valvular Heart Diseases

Valvular heart diseases impair the function of one or more of the valves. Such diseases may be congenital (present at birth) or acquired from a disease process such as infection, cancer, or disorders of the immune system. There are generally two major types of valvular defects: **insufficiency** and **stenosis.** An insufficient valve fails to close fully and so allows blood to leak backward. A stenotic valve is one in which calcium deposits have built up in the cusps, making them hard and inflexible. Blood flows through a stenotic valve with difficulty, and often the heart has to pump harder to eject blood through it.

Both types of valvular heart diseases may cause an audible "swooshing" of blood when the heart beats, called a *heart murmur*. Other signs and symptoms vary with the type and severity of the disease but may include enlargement of the heart, fatigue, dizziness, and heart palpitations. Of the four valves, the mitral and aortic valves are the ones most commonly affected by valvular heart disease.

The Coronary Circulation

Although the heart's chambers are filled with blood, the myocardium is too thick for oxygen and nutrients to diffuse from inside the chambers to all the organ's cells. For this reason, the heart is supplied and drained by a set of blood vessels collectively called the **coronary circulation** (**Figure 17.9**). Let's look at the anatomy of this circuit, which is part of the systemic circulation.

The Coronary Vessels

The *coronary arteries* deliver oxygenated blood to the coronary capillary beds, where gas and nutrient exchange takes place within the myocardium. Then the deoxygenated blood drains from capillaries into a series of *coronary veins.*

The Coronary Arteries   As we mentioned, the main systemic artery into which the left ventricle pumps blood is the ascending aorta. Immediately after the ascending aorta emerges from the left ventricle, two branches arise: the **right** and **left coronary arteries.** Note in Figure 17.9a that both arteries are posterior to the pulmonary trunk, so the trunk had to be removed to allow us to see them. The right and left coronary arteries travel in the right and left atrioventricular sulci, respectively.

The right coronary artery continues to travel inferiorly and medially along the right atrioventricular sulcus, where it gives off several branches that supply the right atrium and ventricle (orange area in Figure 17.9a):

- **Marginal artery.** The largest branch is the **marginal artery,** so named because it typically arises near the inferior margin, or border, of the heart.

- **Posterior interventricular artery.** After the marginal artery branches off, the right coronary artery curls around to the posterior heart and travels in the posterior interventricular sulcus as the appropriately named **posterior interventricular artery.**

Shortly after the left coronary artery emerges from the ascending aorta, it generally branches into two vessels:

- **Anterior interventricular artery.** As you can see in Figure 17.9a, the **anterior interventricular artery** (also called the *left anterior descending artery,* or *LAD*) gets its name from the fact that it travels along the anterior interventricular sulcus. At the apex of the heart, it generally curls around and travels a short distance along the posterior interventricular sulcus. Along its course it gives off branches that supply the interventricular septum (the muscular wall between the ventricles), most of the anterior left ventricle, and some of the posterior left ventricle (area shaded yellow).

- **Circumflex artery.** The **circumflex artery** (SIR-kum-fleks; *circum-* = "around") also travels in a path consistent with its name, as it curves along the left atrioventricular sulcus and *flexes around* the heart. It supplies the left atrium and parts of the left ventricle (area shaded pink) and, in some people, replaces the right coronary artery in supplying the branch that becomes the posterior interventricular artery.

These arteries can be imaged with a procedure called *percutaneous coronary angiography.* In this procedure, a small tube is fed through an artery in the systemic circuit, into the ascending aorta, and finally into the right and left coronary arteries. A special dye is then injected into the arteries, producing an x-ray image like this one:

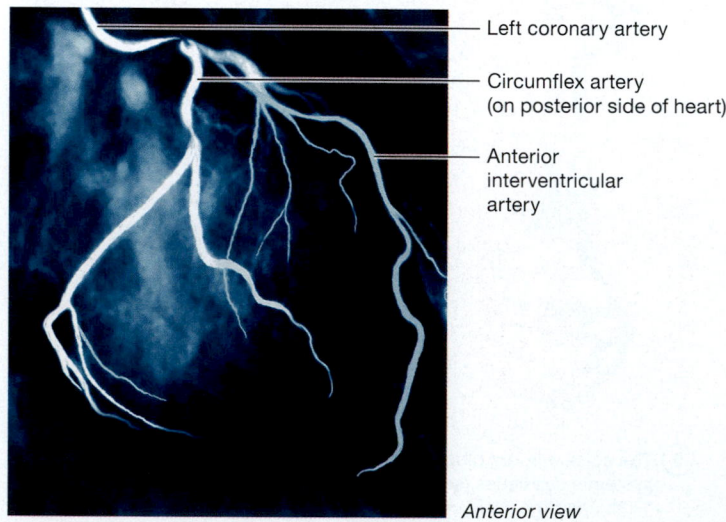

Left coronary artery

Circumflex artery (on posterior side of heart)

Anterior interventricular artery

Anterior view

Although the arterial supply of the heart is shown fairly simply in Figure 17.9a, it's not always quite this straightforward. The distribution of these blood vessels is inconsistent—some people

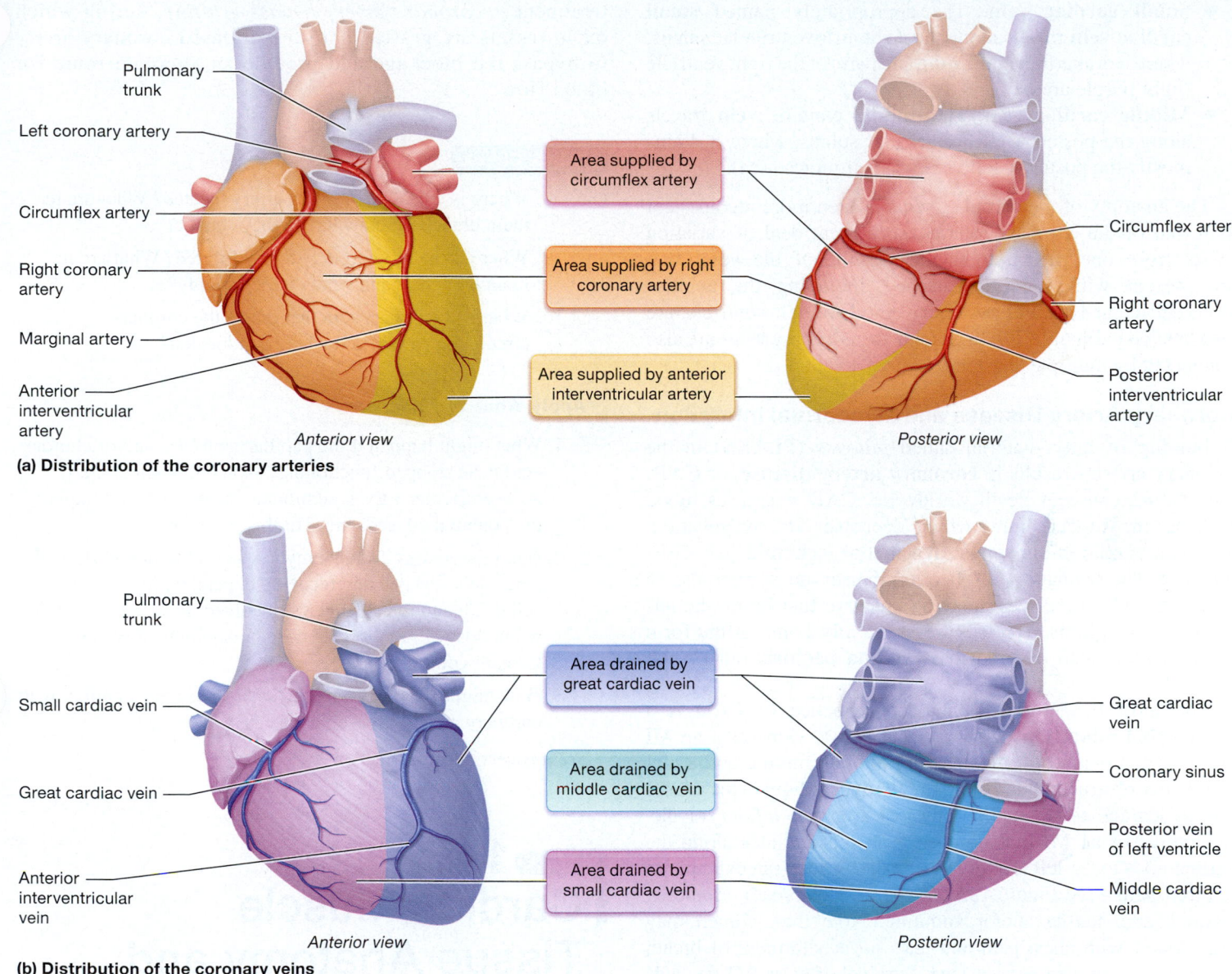

Pulmonary trunk

Left coronary artery

Circumflex artery

Right coronary artery

Marginal artery

Anterior interventricular artery

Area supplied by circumflex artery

Area supplied by right coronary artery

Area supplied by anterior interventricular artery

Circumflex artery

Right coronary artery

Posterior interventricular artery

Anterior view

Posterior view

(a) Distribution of the coronary arteries

Pulmonary trunk

Small cardiac vein

Great cardiac vein

Anterior interventricular vein

Area drained by great cardiac vein

Area drained by middle cardiac vein

Area drained by small cardiac vein

Great cardiac vein

Coronary sinus

Posterior vein of left ventricle

Middle cardiac vein

Anterior view

Posterior view

(b) Distribution of the coronary veins

Figure 17.9 **The coronary circulation.**

have certain arterial branches that others lack. Coronary arterial supply is further complicated by the formation of **anastomoses** (an′-ass-toh-MOH-seez; "coming together"), which are systems of channels formed between blood vessels. The coronary arteries may form anastomoses with one another, with branches from the pericardium, or even with arteries from outside the coronary circulation entirely. When blood flow to the myocardium is insufficient, occasionally new anastomoses will form to provide alternate routes of blood flow, or **collateral circulation,** to the myocardium. Collaterals formed in this manner help to protect the muscle cells from damage that could result from blocked vessels.

The Coronary Veins The most common pattern of venous flow in the coronary veins (also called the *cardiac veins*) is illustrated in Figure 17.9b. Generally, the majority of the heart's veins empty into a large venous structure on the posterior heart, called the **coronary sinus,** which drains into the posterior right atrium. The coronary sinus receives blood from three major veins:

● **Great cardiac vein.** The large **great cardiac vein** ascends along the anterior interventricular sulcus and then travels to the posterior side of the heart along the left atrioventricular sulcus. It drains the left atrium and much of both ventricles (blue area).

- **Small cardiac vein.** The appropriately named **small cardiac vein** travels along the right atrioventricular sulcus, where it drains the right atrium and parts of the right ventricle (light purple area).
- **Middle cardiac vein.** The **middle cardiac vein** travels along the posterior interventricular sulcus, where it drains mostly the posterior left ventricle (turquoise area).

The anatomy of the coronary veins is even more inconsistent than that of the coronary arteries, so a great deal of variation exists from one person to another. Some of the veins that are present with some consistency, including the anterior interventricular vein and posterior vein of the left ventricle, are also labeled in Figure 17.9. In general, the coronary veins are also connected by multiple anastomoses.

Coronary Artery Disease and Myocardial Infarction

A buildup of fatty material called *plaques* (PLAKS) in the coronary arteries results in **coronary artery disease,** or **CAD,** the leading cause of death worldwide. CAD decreases blood flow to the myocardium, which decreases its oxygenation. This is a condition known as **myocardial ischemia** (iss-KEE-mee-ah). Unfortunately, CAD often causes no symptoms, so a person with this disease may be unaware that he or she has it. When symptoms do arise, they generally come in the form of chest pain, also referred to as **angina pectoris** (an-JY-nuh pek-TOH-riss).

The most dangerous potential consequence of CAD is a **myocardial infarction (MI),** or *heart attack.* Generally, an MI occurs when plaques in the coronary arteries rupture and a clot forms that obstructs blood flow to the myocardium. This causes the myocardial tissue supplied by that artery to *infarct,* or die. Symptoms of an MI include chest pain that radiates along the dermatomes to the left arm or left side of the neck (an example of referred pain—see Figure 13.15), shortness of breath, sweating, anxiety, and nausea and/or vomiting. Note that women may not present with chest pain and may suffer shortness of breath or back, jaw, or arm pain instead. Survival after an MI depends on the extent and location of the damage. Cardiac muscle cells generally do not undergo mitosis, and so after an MI the dead cells are replaced with fibrous, noncontractile scar tissue. Death of part of the myocardium increases the workload of the remaining heart muscle.

The risk factors for CAD and MI include smoking, high blood pressure, poorly controlled diabetes, high levels of certain lipids in the blood, obesity, age over 40 for males and over 50 for females, genetics, and male sex. CAD is definitively diagnosed via angiography.

Treatments for CAD include lifestyle modifications and appropriate medications. If these two approaches fail, then invasive treatments are considered. A commonly performed invasive procedure is **coronary angioplasty** (AN-jee-oh-plas′-tee), during which a balloon is inflated in the blocked artery and a piece of wire-mesh tubing called a *stent* may be inserted into the artery to keep it open. A more invasive treatment is *coronary artery bypass grafting,* during which other vessels are grafted onto the diseased coronary artery to bypass the blockage and provide an alternate route for blood flow.

Quick Check

☐ 8. Where is the right coronary artery located? What are its main branches, and what do they supply?

☐ 9. Where is the left coronary artery located? What are its main branches, and what do they supply?

☐ 10. What is the largest vein that drains the coronary circulation called, and where is it located?

Apply What You Learned

☐ 1. What might happen if the papillary muscles and/or chordae tendineae stopped functioning? Would this affect the atrioventricular valves, semilunar valves, or both? Explain the potential consequences of this problem.

☐ 2. Heart tissue dies during a myocardial infarction (MI), and a person's survival and recovery depend on the extent of cell death and the chamber(s) involved. In which chamber would loss of function be most damaging to survival? Explain your response.

☐ 3. How might coronary artery anastomoses help a patient with coronary artery disease?

See answers in Appendix A.

MODULE **17.3**

Cardiac Muscle Tissue Anatomy and Electrophysiology

Learning Outcomes

1. Describe the histology of cardiac muscle tissue, and differentiate it from that of skeletal muscle.

2. Describe the phases of the cardiac muscle cell action potential, including the ion movements that occur in each phase, and explain the importance of the plateau phase.

3. Contrast the way action potentials are generated in cardiac pacemaker cells, cardiac contractile cells, and skeletal muscle cells.

4. Describe the parts of the cardiac conduction system, and explain how the system functions.

5. Identify the waveforms in a normal electrocardiogram (ECG), and relate the ECG waveforms to electrical activity in the heart.

As you are no doubt aware, your heart contracts without any conscious effort on your part. Like all muscle cells, cardiac muscle cells contract in response to electrical excitation in the form of **action potentials.** However, unlike skeletal muscle fibers and many smooth muscle cells, cardiac muscle cells do not require stimulation from the nervous system to generate action potentials. This is because their electrical activity is coordinated by a very small, unique population of modified cardiac muscle cells called **pacemaker cells.** These cells rhythmically and spontaneously generate action potentials that trigger the other type of cardiac muscle cell, known as **contractile cells,** to also have action potentials. This gives cardiac muscle the property of **autorhythmicity** (aw′-toh-rith-MISS-ih-tee)—it sets its own rhythm without a need for input from the nervous system.

We begin this module with an overview of the histology of cardiac muscle cells and their similarities to, and differences from, skeletal muscle fibers. The discussion then shifts to the cardiac action potentials that are generated by pacemaker cells and the group of pacemakers in the heart collectively called the *cardiac conduction system,* after which we compare pacemaker action potentials with contractile cell action potentials. The module concludes with an overview of *electrocardiograms,* one of the clinical topics pertaining to the heart's *electrophysiology.*

Histology of Cardiac Muscle Tissue and Cells

◄◄ FLASHBACK

1. What are gap junctions and desmosomes? (p. 127)

2. What is an electrical synapse? (p. 405)

3. What is excitation-contraction coupling? (p. 354)

Many components of the cardiac muscle cells illustrated in **Figure 17.10** should look familiar to you. Compare the diagram of cardiac muscle cells in this figure to the diagram of skeletal muscle fibers in Figure 10.4. Cardiac muscle cells, like skeletal muscle fibers, have *striations,* which means that the cells have alternating light and dark bands when viewed under the microscope. As with skeletal muscle fibers, these striations are due to the arrangement of the contractile proteins within the cardiac muscle cell. Both types of muscle tissue also have transverse tubules and a sarcoplasmic reticulum. These structural similarities reflect physiological similarities—both skeletal and cardiac muscle tissues have the same general function, to generate tension, and both generate tension through the sliding-filament mechanism of contraction

(see Module 10.2 for a review of this mechanism). This is a nice example of the Structure-Function Core Principle (p. 28).

CORE PRINCIPLE
Structure-Function

However, in spite of these similarities, there are notable differences between cardiac muscle cells and skeletal muscle fibers. Unlike skeletal muscle cells, cardiac muscle cells do not result from the fusion of multiple immature myoblasts and so do not form long fibers. Instead, they are individual, typically branched cells, usually with a single nucleus, and they are shorter and wider than skeletal muscle fibers. In addition, cardiac muscle cells contain abundant *myoglobin,* a protein that carries oxygen, and nearly half of their cytoplasmic volume is composed of mitochondria. Both of these facts reflect their high energy demands.

Figure 17.10 reveals another difference between cardiac and skeletal muscle cells: Structures called **intercalated discs** (in-TER-kuh-lay′-t'd; "discs inserted between") join adjacent cardiac muscle cells. Intercalated discs join pacemaker cells to contractile cells, and contractile cells to one another. Notice in the figure that the membranes of the cardiac muscle cells are highly folded where they meet at the discs. At the ends of these folds we find *desmosomes* that hold the cardiac muscle cells together.
Where the flat parts of the membranes touch are *gap junctions* that allow ions to

CORE PRINCIPLE
Cell-Cell Communication

rapidly pass from one cell to another, permitting communication among the cardiac muscle cells (an example of the Cell-Cell Communication Core Principle, p. 28).

These gap junctions are simply electrical synapses, which we discussed in the nervous system chapter (look back at Figure 11.19). Gap junctions in these discs allow the electrical activity generated by the pacemaker cells to rapidly spread to all cardiac muscle cells via electrical synapses. This permits the heart to contract as a unit and produce a coordinated *heartbeat.* For this reason, the cells of the heart are sometimes referred to as a *functional syncytium* (sin-SISH-um), the term for a large, multinucleated cell.

Like skeletal muscle fibers and other excitable cells, a cardiac muscle cell contains selective gated ion channels in its sarcolemma, including the following:

- **Voltage-gated sodium ion channels.** The voltage-gated sodium ion channels in the cardiac muscle cell sarcolemma open in response to voltage changes across the membrane. These channels are found in all cardiac muscle cells except certain pacemaker cells.
- **Calcium ion channels.** Calcium ion channels are present in all cardiac muscle cells. These channels exhibit voltage-gated opening but time-gated closing, meaning that they close after a certain period regardless of the voltage.
- **Potassium ion channels.** All types of cardiac muscle cells have one or more types of potassium ion channels. Some of these channels are ligand-gated, whereas others are voltage-gated.

● **Nonselective cation channels.** Unique to certain pacemaker cells is a type of nonselective cation channel known as the *HCN channel*. This voltage-gated channel is unusual in that it is activated by hyperpolarization. When opened, it allows sodium ions to enter the cell while potassium ions simultaneously exit.

The opening and closing action of these ion channels is responsible for both the pacemaker and contractile cardiac action

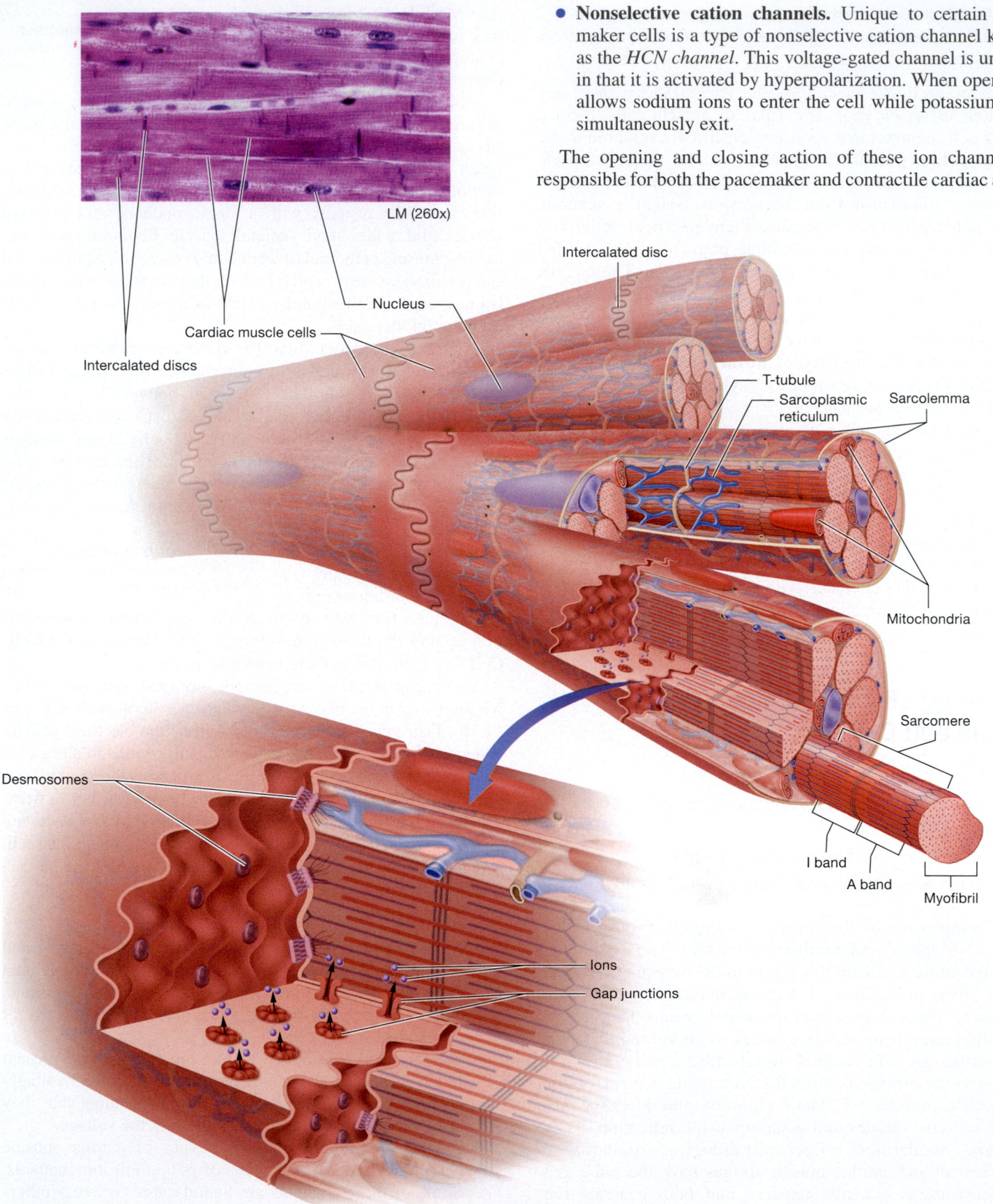

LM (260x)

Intercalated disc

Nucleus

Cardiac muscle cells

Intercalated discs

T-tubule

Sarcoplasmic reticulum

Sarcolemma

Mitochondria

Sarcomere

Desmosomes

I band

A band

Myofibril

Ions

Gap junctions

Figure 17.10 Cardiac muscle cells.

Practice art labeling
@ **Mastering** Anatomy & Physiology

potentials. Before moving on to these topics, let's review some electrophysiology terms.

StudyBOOST ⬆

Revisiting Electrophysiology

Yes, it's time to go back again to everyone's favorite topic in A&P—electrophysiology. Before you delve into cardiac electrophysiology, first bear in mind that the majority of this material isn't new to you. In fact, this is the third time that we have explored this topic in this book. Second, remember that although cardiac muscle cells may look exotic, they are simply excitable cells that behave in a similar way to the other excitable cells you have studied, including skeletal muscle fibers, smooth muscle cells, and neurons. Following are some simplified "refresher" definitions that you may find helpful.

- **Voltage:** A difference in electrical potential between two points.
- **Membrane potential:** The voltage or charge difference that exists across the membranes of all cells, including excitable cells.
- **Resting membrane potential:** The membrane potential of an excitable cell at rest (that is not being stimulated), which averages between –60 and –90 mV.
- **Ion gradient:** A difference in the concentration of ions across a plasma membrane.
- **Current:** The flow of ions or electrons with a chemical or electrical gradient.
- **Depolarization:** A change in resting membrane potential to a *less negative* value. This change occurs when positive charges (generally, sodium and/or calcium ions) rush into the cell.
- **Repolarization:** The return of a cell to its negative resting membrane potential. This occurs when positive charges (potassium ions) leave the cell.
- **Hyperpolarization:** A change in resting membrane potential to a *more negative* value. This occurs when positive charges (potassium ions) exit the cell (or when negative charges enter the cell) and cause its potential to become more negative than the resting membrane potential.

Last but not least, let's review those all-important ion gradients and the Gradients Core Principle (p. 28) in a bit more depth. Remember that the concentration of sodium 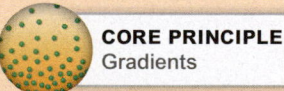 **CORE PRINCIPLE** Gradients and calcium ions in the extracellular fluid (ECF) is *higher* than that in the cytosol, and the concentration of potassium ions in the ECF is *lower* than that in the cytosol. For this reason, sodium and calcium ions tend to follow their concentration gradients to enter a cell when their channels open, whereas potassium ions follow their concentration gradient and leave the cell when their channels open. The sodium and potassium ion gradients are maintained by the Na⁺/K⁺ pump and the calcium ion gradient by a separate active transport pump. ∎

Quick Check

☐ 1. How do pacemaker and contractile cells differ? What is autorhythmicity?

☐ 2. What are intercalated discs? What is their function?

Cardiac Electrophysiology: Pacemaker Cells and the Cardiac Conduction System

⟪ FLASHBACK

1. How does an action potential occur in a skeletal muscle fiber? (p. 350)

Pacemaker cells make up only about 1% of the total number of cardiac muscle cells. There are three populations of these cells in the heart that are capable of spontaneously generating action potentials and setting the pace of the heart. These three cell populations are collectively called the **cardiac conduction system.** After looking at how pacemaker cells generate action potentials, we'll cover the components of the cardiac conduction system.

Pacemaker Cell Action Potentials

As with skeletal muscle fibers, an action potential in a pacemaker cell results from a reversal in membrane potential—the inside of the plasma membrane swings from negative (about –60 mV) to momentarily positive (about +10 mV). These changes happen because of voltage-gated ion channels in the sarcolemma, and because of unequal concentrations of calcium, sodium, and potassium ions on either side of the membrane that drive those ions in or out of the cell through the channels.

Figure 17.11 shows the electrical tracing produced by an action potential of a pacemaker cell, called a **pacemaker potential.** Notice that it's much different in appearance from those of a skeletal muscle cell (see Chapter 10). For one, the depolarization in a pacemaker cell occurs much more slowly. This is due in part to the lack of voltage-gated sodium ion channels in the sarcolemma of many pacemaker cells. In addition, a pacemaker cell action potential oscillates—that is, it never remains at a resting level and instead occurs in a cycle, with the last event triggering the first. This is possible because of the HCN channels—nonspecific cation channels that are unique to pacemaker cells. Recall that these channels open when the membrane hyperpolarizes, and their opening starts a slow depolarization. Figure 17.11 shows how this works, as described next.

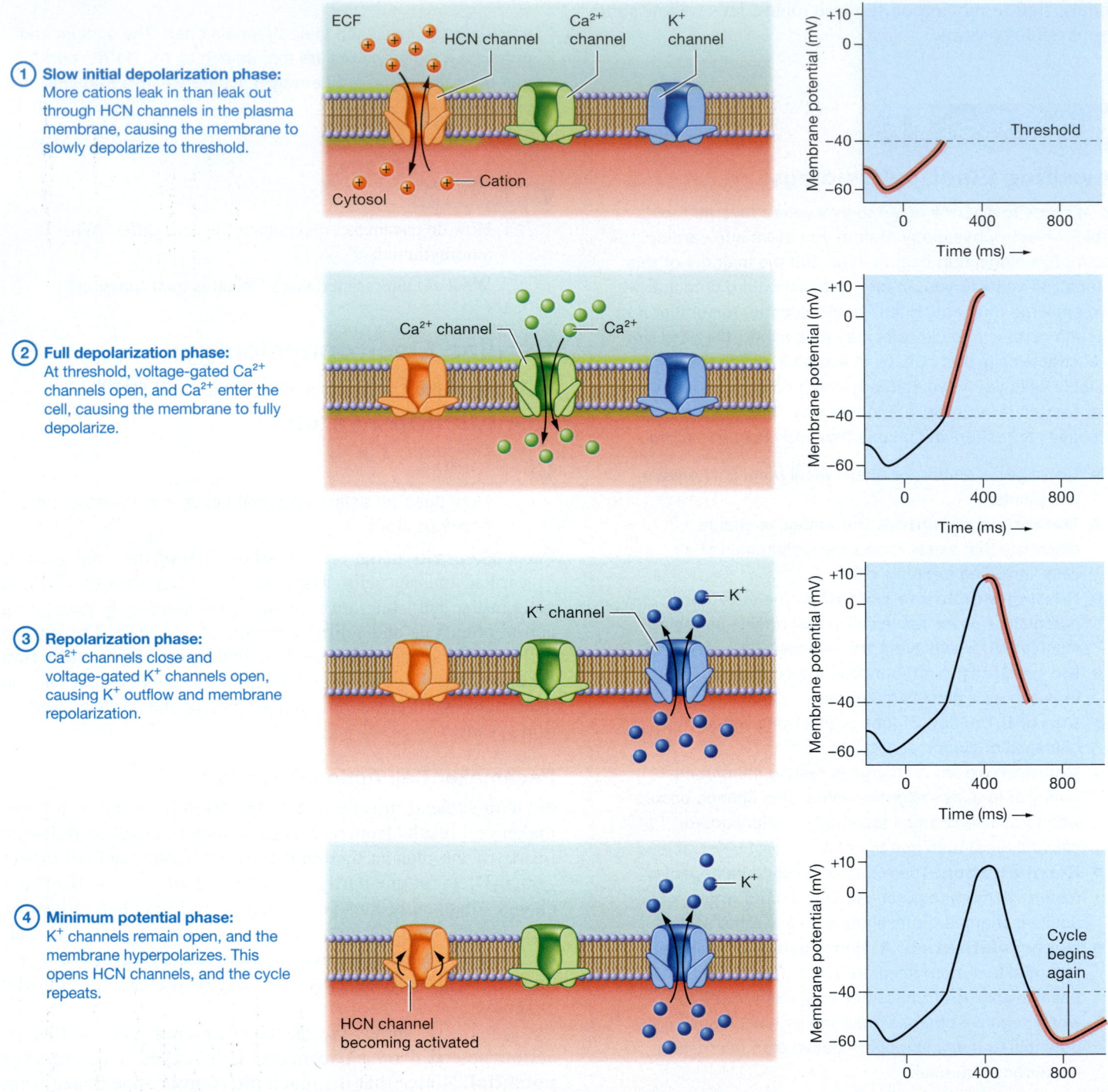

Figure 17.11 A pacemaker cell action potential.

① **Slow initial depolarization phase.** We start a pacemaker potential with the plasma membrane in a *hyperpolarized* state—it is at its minimum membrane potential of about –60 mV. The hyperpolarized membrane triggers the voltage-gated HCN channels in the membrane to begin to open. These channels allow more sodium ions to leak into the cell than potassium ions to leak out, which results in an overall slow depolarization to threshold.

② **Full depolarization phase.** When the membrane reaches a threshold value of –40 mV, voltage-gated calcium ion channels open, allowing calcium ions to enter the cell. This causes the membrane to fully depolarize, and is responsible for the slow upstroke of the tracing that you see in Figure 17.11.

③ **Repolarization phase.** Calcium ion channels are time-gated for closing, so after a certain time period (about

100–150 msec), they begin to close. At the same time, voltage-gated potassium ion channels begin to open. This allows potassium ions to exit the cell, and the membrane begins to repolarize.

④ **Minimum potential phase.** Potassium ion channels remain open until the membrane reaches its minimum potential. When this happens, the membrane is hyperpolarized, which triggers the potassium ion channels to begin closing, the HCN channels to begin opening, and the cycle begins again.

The events of a pacemaker cell action potential are such that the end of step ④ causes step ① to begin. It is this self-perpetuating, cyclic nature that allows pacemaker cells to pace the heart.

Anatomy of the Cardiac Conduction System

Pacemaker action potentials are spread quickly through the heart by the cardiac conduction system—a group of interconnected pacemaker cells. The cardiac conduction system includes the following three populations of pacemaker cells (left side of **Figure 17.12**):

● **Sinoatrial node.** The **sinoatrial node** (sy′-noh-AY-tree-ul), or **SA node,** is located in the right atrium slightly inferior and lateral to the opening of the superior vena cava. Under normal conditions, the cells of the SA node have the fastest intrinsic rate of depolarization—about 60–70 or more times per minute. However, this rate is subject to influence from the sympathetic and parasympathetic nervous systems.

● **Atrioventricular node.** The **atrioventricular node,** or **AV node,** is a cluster of pacemaker cells located posterior and medial to the tricuspid valve. These cells have a slower intrinsic rate of depolarization than those of the SA node—only about 40–50 action potentials per minute.

● **Purkinje fiber system.** The slowest group of pacemaker cells is collectively called the **Purkinje fiber system** (per-KIN-jee); its cells depolarize only about 20 times per minute. The cells of this system are sometimes called *atypical pacemakers,* because their action potentials rely on different ion channels and they function in a slightly different way. Note in Figure 17.12 that there are three components to the Purkinje fiber system: the AV bundle, the right and left bundle branches, and the terminal branches of the Purkinje fibers. The **atrioventricular bundle (AV bundle)** penetrates the heart's fibrous skeleton in the inferior interatrial septum and the superior interventricular septum. The **right** and **left bundle branches** course along the right and left sides of the interventricular septum, respectively. Then the **terminal branches** penetrate the ventricles and finally come into contact with the contractile cardiac muscle cells.

We show the cardiac conduction system as a discrete, two-dimensional structure for simplicity. Bear in mind, however, that these groups of cells are not actually visible on dissection of the heart. In addition, the system is far from flat—the Purkinje fiber system is a complex three-dimensional network of connected pacemaker cells that spreads throughout both ventricles.

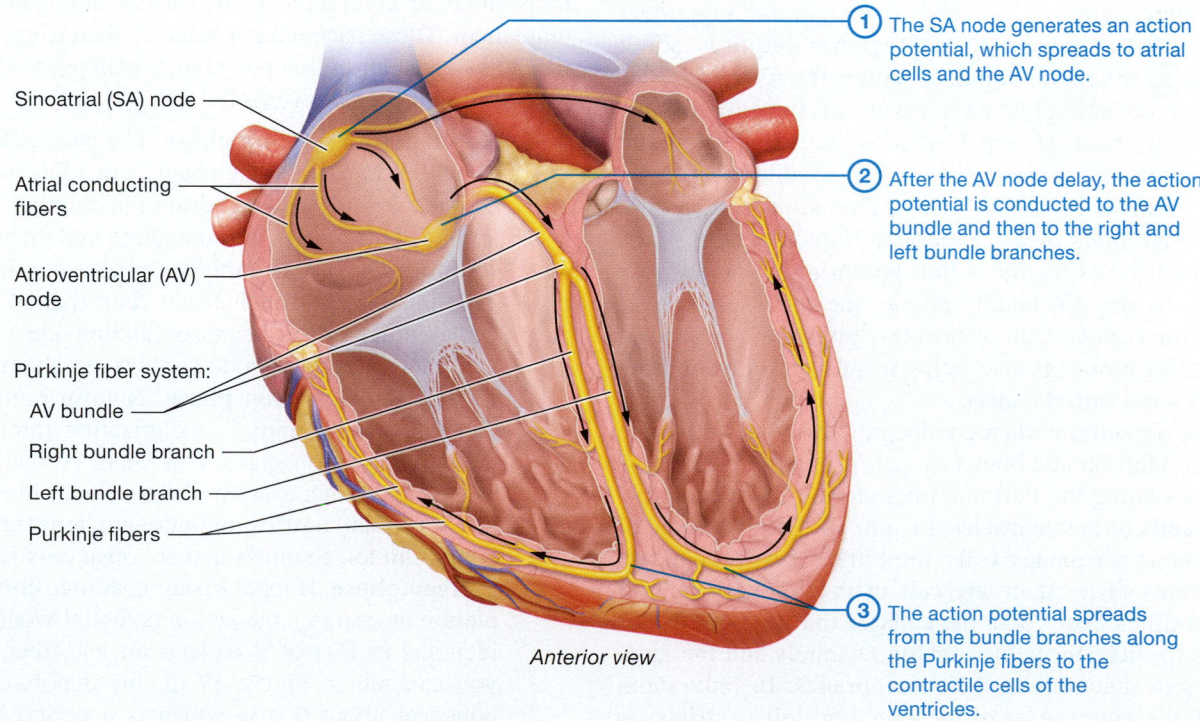

① The SA node generates an action potential, which spreads to atrial cells and the AV node.

② After the AV node delay, the action potential is conducted to the AV bundle and then to the right and left bundle branches.

③ The action potential spreads from the bundle branches along the Purkinje fibers to the contractile cells of the ventricles.

Sinoatrial (SA) node

Atrial conducting fibers

Atrioventricular (AV) node

Purkinje fiber system:

AV bundle

Right bundle branch

Left bundle branch

Purkinje fibers

Anterior view

Figure 17.12 The cardiac conduction system.

Pacing the Heart: Sinus Rhythm

Each of these populations of pacemaker cells can potentially pace the heart, or make it beat at a certain rate, but the one that depolarizes fastest is the one that "wins" and sets the heart rate. The other pacemakers of the cardiac conduction system will pace the heart only if the fastest pacemaker ceases to function.

The SA node is the normal pacemaker of the entire heart; electrical rhythms generated and maintained by the SA node are known as **sinus rhythms.** The AV node and Purkinje fiber system normally only conduct action potentials generated by the SA node. However, if the SA node ceases to function, the AV node can successfully pace the heart, albeit somewhat slowly. Note that the AV bundle of the Purkinje system is the only connection between the AV node and the ventricles. If this system is blocked, the SA node cannot pace the ventricles even if it is functioning normally. The Purkinje fiber system is capable of pacing the heart, but its slow rate of depolarization is not adequate to sustain life beyond a short period of time. Occasionally, a group of regular contractile cells or pacemaker cells other than the SA node will attempt to pace the heart at the same time as the SA node. This "extra" pacemaker is called an **ectopic pacemaker,** and can result in irregular heart rhythms.

Conduction Pathway through the Heart

Now that we know the components, let's focus on the sequence of electrical events in the heart for one heartbeat (see right side of Figure 17.12). Under normal conditions, ① the SA node generates an action potential, which spreads rapidly via gap junctions to the surrounding atrial cells. The impulses are then conducted by specialized *atrial conducting fibers* to the AV node, a process that altogether requires about 0.03 second. ② When the impulse reaches the AV node, conduction slows considerably as a result of two conditions: first, the low number of gap junctions between AV nodal cells, and second, the presence of the nonconducting fibrous skeleton that surrounds the AV node. This slow conduction, known as the *AV node delay,* generally lasts about 0.13 second. The time it takes for the action potential to spread from the SA node to the AV bundle allows the atria to depolarize (and contract) *before* the ventricles, giving the ventricles time to fill with blood. It also helps to prevent current from flowing backward into the atria.

The action potential is then conducted from the AV bundle to the right and left bundle branches. ③ At this point, depolarization spreads along the Purkinje fibers to the contractile cardiac muscle cells of the ventricles. In spite of their slow rate of depolarization as pacemaker cells, the Purkinje fibers conduct action potentials faster than any cell in the myocardium (as much as 150 times faster than the cells of the AV node). This speed is due to differences in their ion channels and the large numbers of gap junctions in their membranes. In only about 0.06 second, the entire mass of the right and left ventricles is depolarized, for a total action potential duration of about 0.22 second.

Quick Check

☐ 3. What are the three populations of cells in the heart that can act as pacemakers? How do they differ?

☐ 4. What is the importance of the AV node delay?

Cardiac Electrophysiology: Contractile Cells

《 FLASHBACK

1. How does an action potential occur in a skeletal muscle fiber? (p. 350)

2. What are tetanus and the refractory period? (p. 365)

3. What is the sliding-filament mechanism of contraction, and what role do calcium ions play in muscle contraction? (p. 355)

Contractile cells make up the great majority (99%) of cardiac muscle cells. We now turn to contractile cell action potentials, which are more similar to the action potentials you learned about in the muscle and nervous tissue chapters. Then we go on to see how excitation-contraction coupling works in contractile cells.

Contractile Cell Action Potentials

Action potentials generated by pacemaker cells are rapidly transmitted to contractile cells via the intercalated discs that unite them. These pacemaker potentials then trigger the contractile cells to have an action potential, which proceeds through the following sequence of events (**Figure 17.13**):

① **Rapid depolarization phase.** The pacemaker cell action potentials cause voltage changes in adjacent cells, which activate voltage-gated sodium ion channels in the sarcolemma. This causes an immediate and massive influx of sodium ions, which rapidly depolarizes the membrane. This upstroke is much faster than that of a pacemaker cell depolarization because calcium ion channels activate more slowly than do sodium ion channels.

② **Initial repolarization phase.** Note in Figure 17.13 that there is a small, initial repolarization immediately after the depolarization spike. This small repolarization is due to the abrupt inactivation of the sodium ion channels and to a very small outflow of potassium ions through selected potassium ion channels that are open only briefly.

③ **Plateau phase.** If repolarization were to continue to completion at step ②, the action potential would look nearly identical to that of a skeletal muscle fiber. However, as you can see in Figure 17.13, the depolarization is sustained at about 0 mV, which is a period known as the

plateau phase. This critical phase is mostly due to the slow opening of calcium ion channels and the resulting influx of calcium ions (certain sodium ion channels also remain open). As the calcium ions enter, potassium ions leave (through different channels), so there is very little net change in membrane potential. The calcium ion channels are also slow to close, which allows the plateau phase to last much longer than the initial depolarization phase.

Note that these are the same types of calcium ion channels as those in pacemaker cells; they simply play a different role in the cell's action potential.

④ **Repolarization phase.** The final phase of the action potential occurs when both the sodium and calcium ion channels return to their resting states and most of the potassium ion channels open. This allows positively charged potassium ions to exit the cardiac muscle cell,

① **Rapid depolarization phase:**
Voltage-gated Na⁺ channels activate and Na⁺ enter, rapidly depolarizing the membrane.

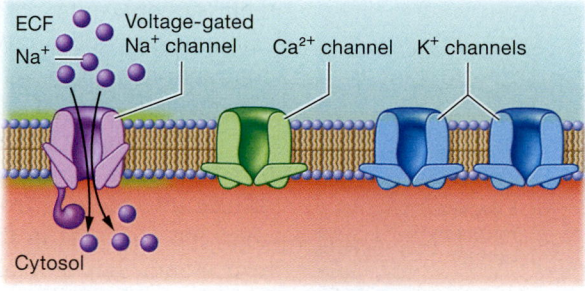
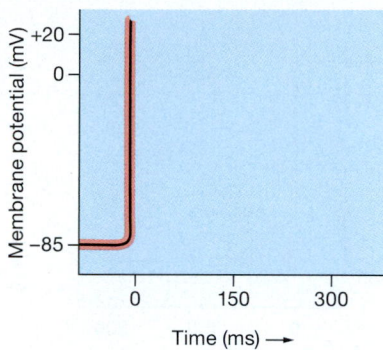

② **Initial repolarization phase:**
Na⁺ channels are inactivated and some K⁺ channels open; K⁺ leak out, causing a small initial repolarization.

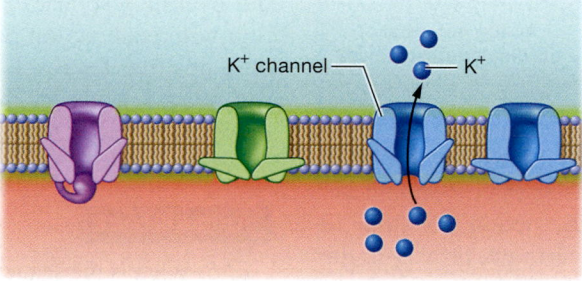
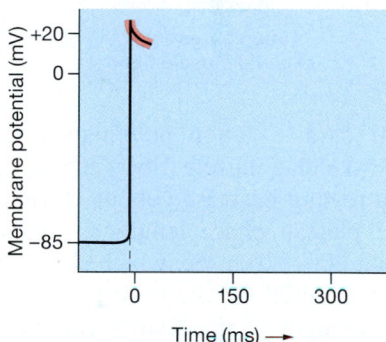

③ **Plateau phase:**
Ca²⁺ channels open and Ca²⁺ enter as K⁺ exit, prolonging the depolarization.

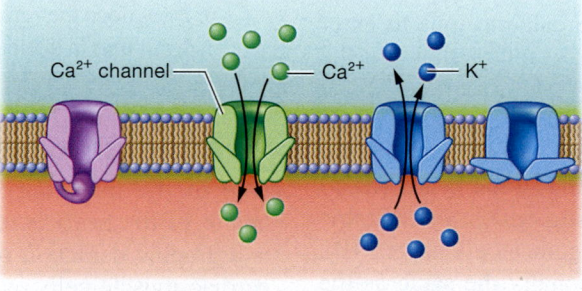
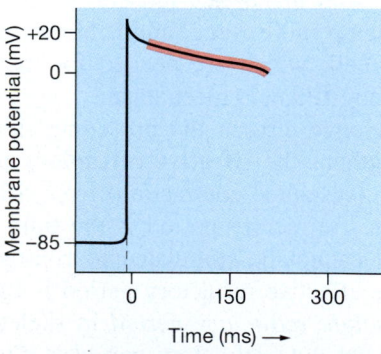

④ **Repolarization phase:**
Na⁺ and Ca²⁺ channels close as K⁺ continue to exit, causing repolarization.

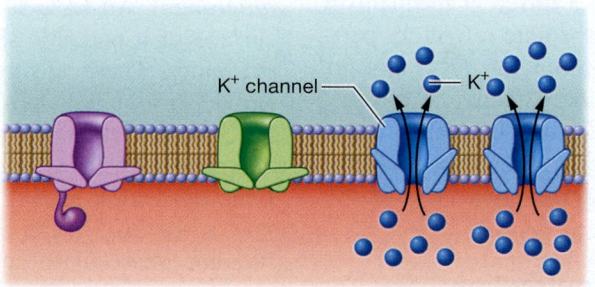
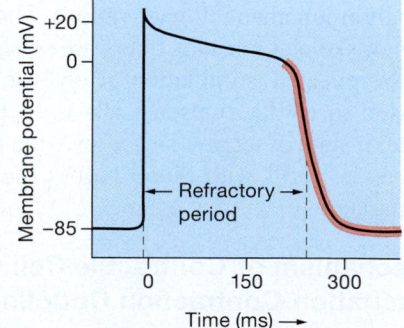

Figure 17.13 A contractile cell action potential.

Play Interactive Physiology 2.0
@ Mastering Anatomy & Physiology

and the membrane potential returns to its resting value of about −85 mV.

The sequence of events of a contractile cell action potential resembles that of a skeletal muscle fiber action potential (see Module 10.3 for a review) with one important exception: the plateau phase. We cannot overemphasize the importance of this phase. Compare the timing of action potentials in skeletal muscle fibers and cardiac contractile cells:

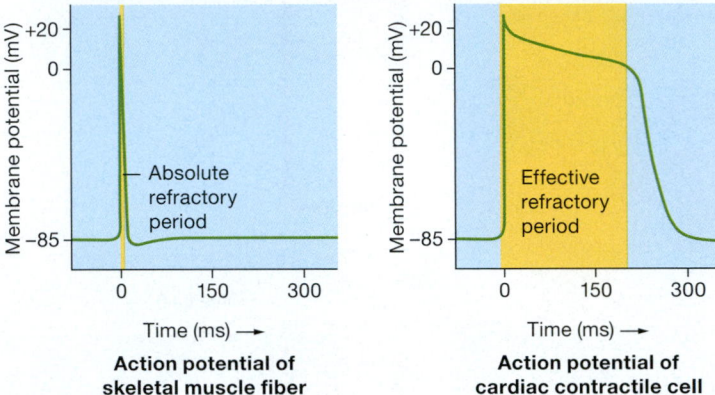

Action potential of skeletal muscle fiber

Action potential of cardiac contractile cell

If cardiac action potentials lasted only about 1–5 msec, like skeletal muscle fiber action potentials, this would lead to a resting heart rate about 15 times faster than it should be. The plateau phase lengthens the cardiac action potential to about 200–300 msec, which slows the heart rate, providing the time required for the heart to fill with blood. It also increases the strength of the heart's contraction. The prolonged action potential makes the muscle twitch last a long time, so it can develop more force, and it allows more calcium ions to enter the cell, which are needed for the cell's contraction via the sliding-filament mechanism.

Notice also in the preceding figure that the plateau phase lengthens the **effective refractory period,** which prevents *tetany* (sustained contraction) from occurring in the heart. Recall that a refractory period is the time during which an excitable cell cannot be stimulated to contract again (see Chapter 10). The effective refractory period is functionally the same as the *absolute refractory period* in skeletal muscle fibers, although the absolute refractory period is due to a different mechanism (sodium ion channel inactivation). The absolute refractory periods of skeletal muscles fibers are so short that tetanus can occur with repeated stimulation. However, the effective refractory period in cardiac muscle cells is so long that the cells cannot enter a state of tetany. This allows the heart to relax and the ventricles to refill with blood before the cardiac muscle cells are stimulated to contract again.

Mechanism of Contractile Cell Contraction: Excitation-Contraction Coupling

The mechanism for cardiac muscle cell contraction is very similar to that of skeletal muscle fiber contraction, and it occurs by the sliding-filament mechanism. In a stimulated cardiac muscle cell, the depolarization propagates through the sarcolemma and dives into the cell along the T-tubules, which causes the sarcoplasmic reticulum to release calcium ions. These ions bind to troponin, which allows actin and myosin to bind and a crossbridge cycle to begin.

However, there is an important difference between the contraction of the two cell types. Skeletal muscle fibers release nearly all the calcium ions required for contraction from their sarcoplasmic reticulum. For this reason, extracellular calcium ion concentration has little effect on the strength of contraction of a skeletal muscle fiber. However, the sarcoplasmic reticulum of cardiac muscle cells is much less extensive and not as well connected to the T-tubule system. For this reason, many of the calcium ions needed for contraction diffuse into the cell from the extracellular fluid through calcium ion channels in the T-tubules. These calcium ions bind to troponin, but they also stimulate the sarcoplasmic reticulum to release more calcium ions, in a phenomenon known as *calcium ion–induced calcium ion release*. This amplifies the overall response, ensuring a sufficiently strong contraction with each action potential. Notice that this is an example of a positive feedback loop—the initial stimulus of calcium ion release leads to an escalating response of progressively more calcium ion release.

Quick Check

☐ 5. What is the sequence of events of a contractile cell action potential? What is the significance of the plateau phase of this action potential?

☐ 6. How does the refractory period of cardiac muscle cells differ from that in skeletal muscle fibers? What is its purpose?

The Electrocardiogram

One of the most important clinical tools for examining the health of the heart is the **electrocardiogram,** or **ECG,** which is a graphic depiction of the electrical activity occurring in all cardiac muscle cells over a period of time (note that it is also sometimes called an *EKG* from the German term *Elektrokardiogramm*). An ECG is recorded by placing *electrodes* on the surface of a patient's skin: six on the chest and two on each extremity. These electrodes record the changes in electrical activity from unique positions in the heart, which register on the ECG as deflections, or **waves.** Keep in mind that a wave shows *changes* in electrical activity—if there is no *net* difference, there is no deflection shown on the ECG. However, even when the line on the ECG is flat between waves, the cells of the heart are in some phase of an action potential. Also note that the ECG can show only electrical changes that occur in contractile cells, because there are simply not enough pacemaker cells to measure with a standard ECG. For this reason, the ECG appears flat when pacemaker cells have their action potentials.

As you can see in **Figure 17.14**, an ECG recording generally consists of five waves, each of which represents the active depolarization or repolarization of different parts of the heart. Note that some of the waves point upward and others downward. Be careful not to equate these upward and downward deflections with depolarization and repolarization. Remember: The ECG tells you only that a net change in electrical activity has taken place. It does *not* tell you if that change is negative or positive across the membrane.

The waves of an ECG include the following:

- **P wave.** The small, initial **P wave** represents the depolarization of all cells within the atria except the SA node. The P wave nearly always registers as an upward deflection on the ECG. Figure 17.14 shows a flat segment immediately preceding the P wave that represents the time period during which the SA node depolarizes.
- **QRS complex.** The large **QRS complex,** which represents ventricular depolarization, is actually three separate waves. The Q wave is the first downward deflection, the R is the large upward deflection, and the S is the second downward deflection. The QRS is usually much larger in magnitude than the P wave because of the size differences between the atria and ventricles. Note that atrial repolarization occurs as the ventricles depolarize. This repolarization does produce a small electrical signal, but it is generally obscured by the much larger QRS complex and therefore not visible on the tracing.
- **T wave.** The small **T wave** occurs after the S wave of the QRS complex and represents ventricular repolarization. The T wave is an upward deflection under normal conditions. However, certain pathological states, such as myocardial ischemia, may cause the T waves to become inverted as a result of functional changes in the cells' electrical activity.

The periods between the waves represent important action potential phases, as well as the spread of electrical activity through the heart. *Intervals* include a component of at least one wave, whereas *segments* do not include any wave components. Three intervals are of note. The first is the **R-R interval,** the time between two successive R waves. This interval represents the entire duration of the generation and spread of an action potential through the heart, and can be measured to determine the heart rate. The second is the **P-R interval,** or the period from the beginning of the P wave to the beginning of the QRS complex. The P-R interval represents the time it takes for the depolarization from the SA node to spread through the atria to the ventricles; it includes the AV node delay. The third is the **Q-T interval,** the time from the beginning of the QRS complex to the end of the T wave. During the Q-T interval, the ventricular cells are undergoing action potentials.

Between the end of the S wave and the beginning of the T wave is the **S-T segment.** The S-T segment is flat because it is recorded during the plateau phase of the ventricles, and no net changes occur in electrical activity. Elevation or depression of the S-T segment is seen with many clinical conditions, most notably myocardial ischemia and myocardial injury and infarction.

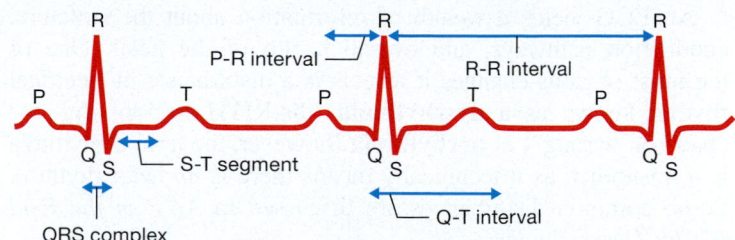

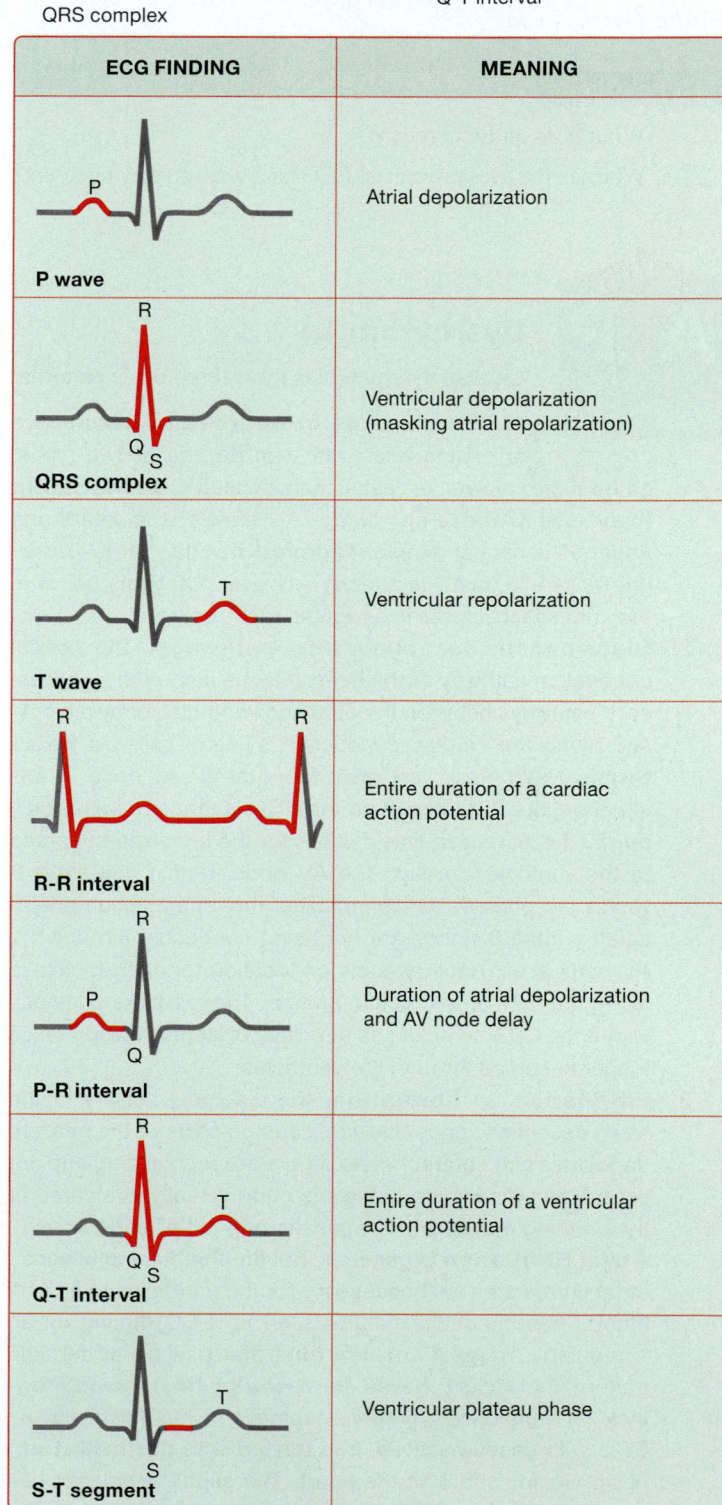

ECG FINDING	MEANING
P wave	Atrial depolarization
QRS complex	Ventricular depolarization (masking atrial repolarization)
T wave	Ventricular repolarization
R-R interval	Entire duration of a cardiac action potential
P-R interval	Duration of atrial depolarization and AV node delay
Q-T interval	Entire duration of a ventricular action potential
S-T segment	Ventricular plateau phase

Figure 17.14 A normal electrocardiogram (ECG) tracing.

An ECG yields a wealth of information about the structure, conduction pathways, and overall health of the heart. One of the most obvious changes it reveals is a disturbance in electrical rhythm known as a **dysrhythmia** (dis-RITH-mee-ah; *dys-* = "bad" or "wrong") or **arrhythmia** (however, the term arrhythmia is a misnomer, as it technically means there is *no* heart rhythm). Some common disturbances are discussed in *A&P in the Real World: Dysrhythmias*.

Quick Check

☐ 7. What does an ECG record?

☐ 8. What are the five waves in an ECG, and what do they represent?

Apply What You Learned

☐ 1. As their name suggests, calcium ion channel blockers block calcium ion channels in the sarcolemma of pacemaker and contractile cardiac muscle cells and slow calcium ion entry into the cell during an action potential. Predict the effects these drugs would have on the length and rate of generation of action potentials of a pacemaker cell. How would these drugs affect the strength of contraction of a contractile cell?

☐ 2. Certain drugs used to treat dysrhythmias, called *local anesthetics,* work by blocking sodium ion channels and slowing the entry of sodium ions into the cell during an action potential. Predict the effect of these drugs on the action potential of contractile cells.

Dysrhythmias

Cardiac dysrhythmias have three basic patterns:

1. **Disturbances in heart rate.** Disturbance in resting heart rate can be one of two types: going either slower or faster than expected. **Bradycardia** (bray´-dee-KAR-dee-uh; *brady-* = "slow") is a heart rate under 60 beats per minute. **Tachycardia** (tak´-ih-KAR-dee-uh; *tachy-* = "fast") is a heart rate over 100 beats per minute; sinus tachycardia is a regular, fast rhythm.

2. **Disturbances in conduction pathways.** The normal conduction pathway of the heart may be disrupted by accessory pathways between the atria and ventricles or by a blockage along the cardiac conduction system, called a **heart block.** Heart blocks are often found at the AV node. In this situation, the P-R interval on the ECG is longer than normal, due to the increased time it takes for the impulses to spread to the ventricles through the AV node. Notice that extra P waves are present, which indicates that some action potentials from the SA node are not being conducted through the AV node at all. Another common location for heart blocks is along the right or left bundle branch. These blocks generally widen the QRS complex, as the wave of depolarization takes longer to spread through the ventricles.

3. **Fibrillation.** In **fibrillation,** the electrical activity in the heart essentially goes haywire, causing parts of the heart to depolarize and contract while others are repolarizing and not contracting (fibrillating muscle is often visually compared to the writhing movement of a plastic bag full of earthworms). **Atrial fibrillation** is generally not life-threatening because atrial contraction isn't necessary for the ventricles to fill with blood. This condition manifests on an ECG tracing as an "irregularly irregular" rhythm (one that has no discernible pattern) that lacks P waves. **Ventricular fibrillation,** however, is immediately life-threatening and manifests on the ECG with chaotic activity. It is treated with **defibrillation,** or an electric shock to the heart. The shock depolarizes all ventricular muscle cells simultaneously and throws the cells into their refractory periods. Ideally, the SA node will resume pacing the heart after the shock is delivered. Ventricular

fibrillation is not the same as "flat-lining," which is a condition called **asystole** (ay-SIS-toh-lee). Defibrillation is not used for asystole because the heart is not fibrillating, so there is no electrical activity to reset. Instead, asystole is treated with CPR and pharmacological agents that stimulate the heart, such as atropine and epinephrine.

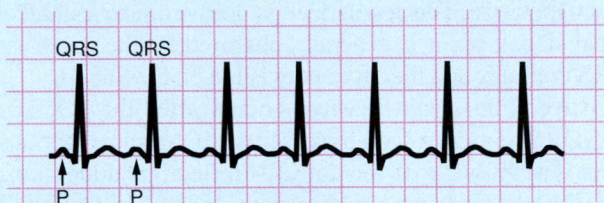

Sinus tachycardia

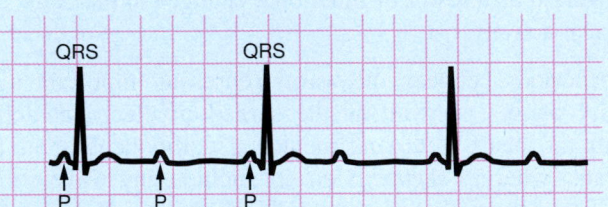

Heart block at AV node

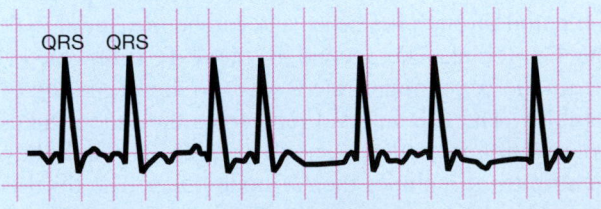

Atrial fibrillation

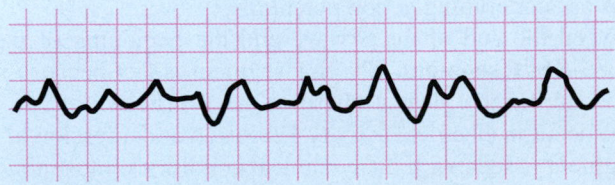

Ventricular fibrillation

Would local anesthetic drugs have a significant effect on the cells of the SA or AV node? Why or why not?

See answers in Appendix A.

MODULE 17.4
Mechanical Physiology of the Heart: The Cardiac Cycle

Learning Outcomes

1. Describe the phases of the cardiac cycle.
2. Relate the opening and closing of specific heart valves in each phase of the cardiac cycle to pressure changes in the heart chambers.
3. Relate the heart sounds and ECG waveforms to the normal mechanical events of the cardiac cycle.
4. Compare and contrast pressure and volume changes of the left and right ventricles and the aorta during one cardiac cycle.

So far in this chapter we've focused on the heart's electrophysiology. Now we turn to the heart's **mechanical physiology,** which refers to the actual processes by which blood fills the cardiac chambers and is pumped out of them. As we discussed earlier, cardiac muscle cells contract as a unit to produce one coordinated contraction, called a **heartbeat.** The muscle cells are arranged in a spiral pattern, producing a "wringing" action in the heart when it contracts. Pressure changes caused by contractions drive blood flow through the heart, with valves preventing backflow. The sequence of events that takes place within the heart from one heartbeat to the next is known as the **cardiac cycle.**

This module explains the events of the cardiac cycle and the sounds the heart produces as it beats. The module concludes with a big picture view that correlates cardiac electrophysiology with mechanical physiology.

The Relationship between Pressure Changes, Blood Flow, and Valve Function

You have read that the heart's contraction drives blood flow through its chambers and into the pulmonary trunk and aorta. However, its contraction does not open the valves through which the blood must pass. Instead, blood flows in response to pressure gradients. As the ventricles contract and relax, the pressure in the chambers changes,  **CORE PRINCIPLE** Gradients causing blood to push on the valves and either open or close them (an example of the Gradients Core Principle, p. 28).

Let's examine how this happens, using the left ventricle as our example. When the left ventricle contracts, its pressure rises above the pressures in the left atrium and the aorta, as

shown in **Figure 17.15a**. This causes blood to flow from the ventricle to the aorta and produces two changes in the valves, which occur at the same time: (1) The mitral valve is forced shut by blood pushing up on it; and (2) the aortic valve is forced open by the pressure of the outgoing blood. In **Figure 17.15b** you can see that when the ventricle relaxes, the opposite occurs: The pressure in the ventricle falls below those in the left atrium and in the aorta. The higher pressure from the blood in the left atrium forces the mitral valve open, allowing

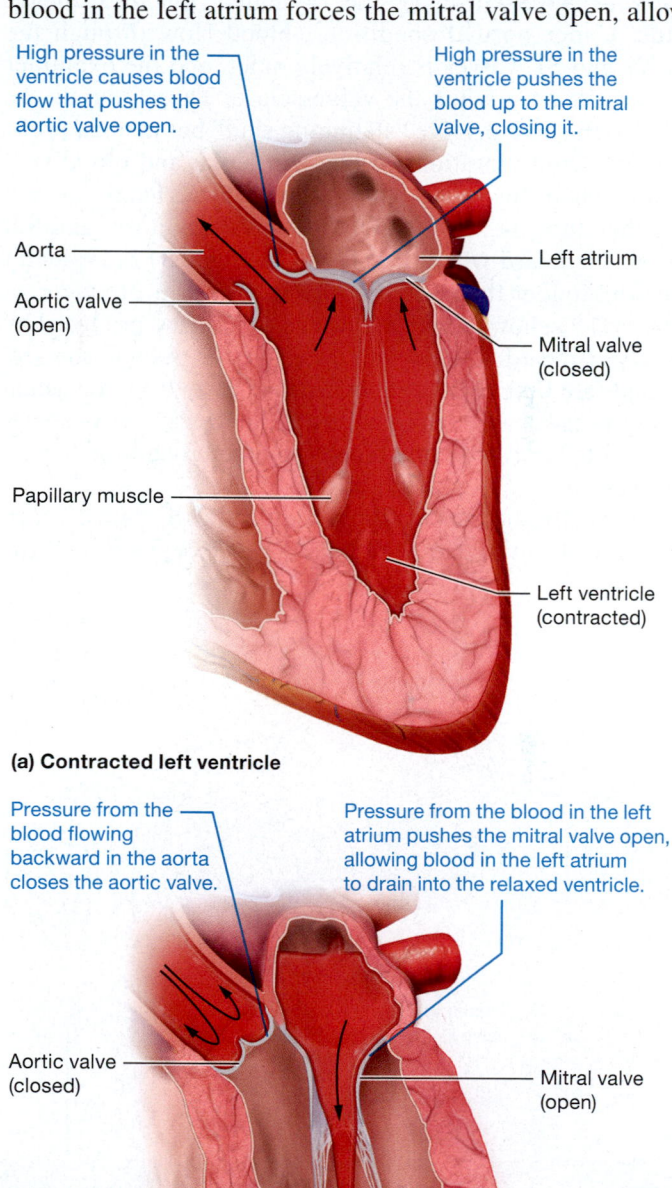

(a) Contracted left ventricle

(b) Relaxed left ventricle

Figure 17.15 Pressure changes, blood flow, and valve function.

the blood to drain into the relaxed ventricle. At the same time, the higher pressure from the blood in the aorta pushes the cusps of the aortic valve closed.

Heart Sounds

One of the most basic tools available to a health care provider is a *stethoscope,* which can be used to listen to, or *auscultate* (AWS-kul-tayt), the rhythmic "lub dub" **heart sounds.** Under normal conditions, blood flow through the open AV and SL valves is relatively quiet, and the examiner hears sounds only when the valves close. These sounds are not due to the actual valve "slamming shut" but instead probably result from vibrations of the ventricular and blood vessel walls when the valves shut. There are two heart sounds: **S1,** or the "lub," is heard when the AV valves close, and **S2,** the "dub," is heard when the SL valves close. S1 is typically longer and louder than S2, although it's lower in frequency.

Figure 17.16 shows where on the chest the sound produced by each valve is heard clearest with a stethoscope. As you can see, the sounds are best heard a slight distance away from the actual valves. This is because the sound waves are carried away somewhat from the valves by the blood moving through the ventricles or vessels.

Occasionally, heart sounds in addition to S1 and S2 can be heard with auscultation. *A&P in the Real World: Heart*

Heart Murmurs and Extra Heart Sounds

One of the more common findings on chest auscultation is an audible sound called a **heart murmur,** which occurs when blood flow through the heart is turbulent. Heart murmurs are generally caused by defective valves, although they may also result from defective chordae tendineae or holes in the interatrial or interventricular septum. Children, however, often have heart murmurs that do not represent defects.

Chest auscultation may also reveal extra heart sounds. The first, called **S3,** can occur just as blood begins to flow into the ventricles, right after S2. It results from the recoil of the ventricular walls as they are stretched and filled. Another heart sound is **S4,** which is heard when most of the blood has finished draining from the atria to the ventricles, just before S1. It typically results from blood being forced into a stiff or enlarged ventricle. Both S3 and S4 may represent pathology, but they can also occasionally be heard in a healthy heart.

Murmurs and Extra Heart Sounds discusses some of these other heart sounds and the problems that may cause them.

Quick Check

☐ 1. What causes the heart sounds S1 and S2?

Events of the Cardiac Cycle

Each cardiac cycle consists of one period of relaxation called **diastole** (dy-ASS-toh-lee) and one period of contraction called **systole** (SIS-toh-lee) for each chamber of the heart. The atrial and ventricular diastoles and systoles occur at different times as a result of the AV node delay we discussed earlier. However, both sides of the heart are working to pump blood into their respective circuits simultaneously.

We can divide the cardiac cycle into four main phases defined by the actions of the ventricles and the positions of the valves: filling, contraction, ejection, and relaxation. These four phases, illustrated in **Figure 17.17**, are as follows:

① **Ventricular filling phase.** The **ventricular filling phase** is the period during which blood drains from the atria into the ventricles. During ventricular filling, pressures in the left and right ventricles are lower than those in the atria, the pulmonary trunk, and the aorta. The higher pressures in the pulmonary trunk and aorta cause the semilunar valves to be closed, which prevents the flow of blood from the pulmonary trunk and aorta back into the ventricles. The atrioventricular valves, however, are open because of the higher atrial pressure, and blood flows down its pressure

Aortic valve
Pulmonary valve
1
2
3
4
5
Tricuspid valve
6
7
Mitral valve

AREA	Aortic	Tricuspid	Mitral	Pulmonary
LOCATION OF SOUND	Second intercostal space, right sternal border	Fifth intercostal space, left sternal border	Fifth intercostal space, mid-clavicular line	Second intercostal space, left sternal border
TIMING OF SOUND	Aortic valve is heard here during S2.	Tricuspid valve is heard here during S1.	Mitral valve is heard here during S1.	Pulmonary valve is heard here during S2.

Figure 17.16 Heart sounds.

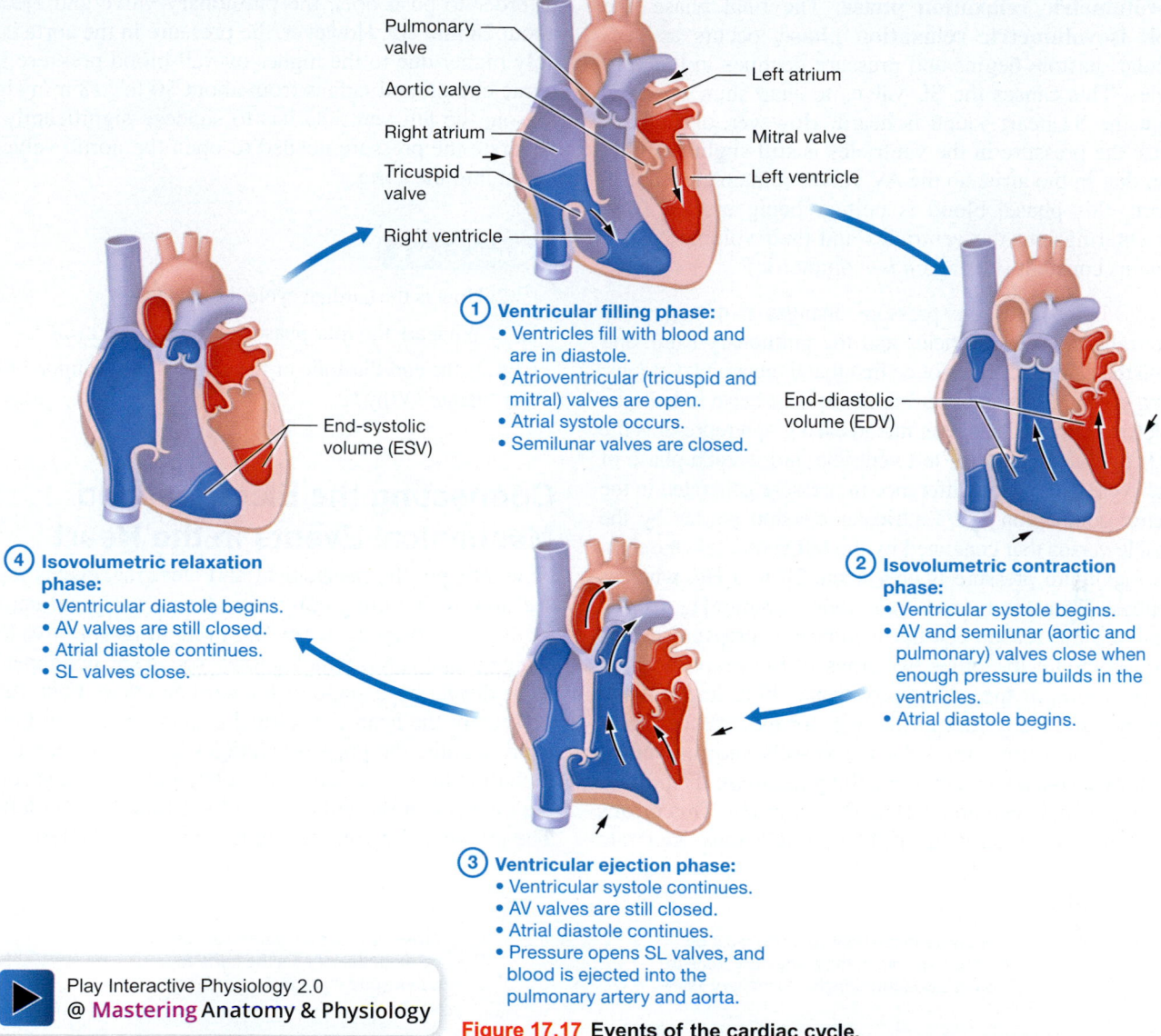

Pulmonary valve
Aortic valve
Right atrium
Tricuspid valve
Right ventricle
Left atrium
Mitral valve
Left ventricle

① Ventricular filling phase:
- Ventricles fill with blood and are in diastole.
- Atrioventricular (tricuspid and mitral) valves are open.
- Atrial systole occurs.
- Semilunar valves are closed.

End-diastolic volume (EDV)

End-systolic volume (ESV)

④ Isovolumetric relaxation phase:
- Ventricular diastole begins.
- AV valves are still closed.
- Atrial diastole continues.
- SL valves close.

② Isovolumetric contraction phase:
- Ventricular systole begins.
- AV and semilunar (aortic and pulmonary) valves close when enough pressure builds in the ventricles.
- Atrial diastole begins.

③ Ventricular ejection phase:
- Ventricular systole continues.
- AV valves are still closed.
- Atrial diastole continues.
- Pressure opens SL valves, and blood is ejected into the pulmonary artery and aorta.

▶ Play Interactive Physiology 2.0
@ **Mastering** Anatomy & Physiology

Figure 17.17 Events of the cardiac cycle.

gradient from the atria into the ventricles. Nearly 80% of the total blood volume of the atria drains passively in this manner into the ventricles. At the beginning of the ventricular filling phase, the atria are in diastole, but as blood continues to drain into the ventricles, the pressure gradient becomes smaller and filling slows. At this point, atrial systole takes place and the contracting atria eject a variable volume of blood into the ventricles—as much as the remaining 20% of blood volume or as little as just a few percent. At the end of atrial systole, each ventricle contains about 120 ml of blood. This value is known as the **end-diastolic volume (EDV)** because it is the ventricular volume at the end of ventricular diastole.

② **Isovolumetric contraction phase.** The beginning of ventricular systole occurs during the shortest phase of the cardiac cycle, called the **isovolumetric contraction phase** (aye′-soh-vawl-yoo-MET-rik; *iso-* = "same"). During this phase, the pressure in the ventricles rises

rapidly as the ventricles begin to contract. This high pressure closes the AV valves and causes the S1 heart sound to be heard. However, ventricular pressure is not yet high enough to push open the semilunar valves, so both sets of valves are closed and the ventricular volume does not change (which is why it's named *iso*volumetric).

③ **Ventricular ejection phase.** At the beginning of the **ventricular ejection phase,** the pressure in the ventricles becomes higher than that in the pulmonary trunk and aorta, and this pushes the SL valves open. The beginning of the ejection phase is marked by rapid outflow of blood from the ventricles. However, as this phase continues, the pressure in the pulmonary trunk and aorta approaches that in the ventricles. At this point, ejection of blood into the vessels decreases considerably. The ventricular ejection phase sees approximately 70 ml of blood pumped from each ventricle. This means that about 50 ml of blood remains in each ventricle, a volume known as the **end-systolic volume (ESV).**

④ **Isovolumetric relaxation phase.** The final phase, the short **isovolumetric relaxation phase,** occurs as ventricular diastole begins and pressure declines in the ventricles. This causes the SL valves to snap shut, at which point the S2 heart sound is heard. However, during this phase the pressure in the ventricles is still slightly higher than that in the atria, so the AV valves remain closed. So, during this phase, blood is neither being ejected from nor entering into the ventricles and their volume briefly remains constant (it is again *iso*volumetric).

Figure 17.18 illustrates the pressure changes that take place within the right and left ventricles and the pulmonary trunk and aorta during the cardiac cycle. Notice first that there are two graphs, which show you what happens during the cardiac cycle in both the right and left ventricles. The green line in each graph represents the pressure change in the right or left ventricle during each phase of the cardiac cycle. Notice the difference in pressure generated in the isovolumetric contraction and ventricular ejection phases by the right ventricle versus that generated by the left ventricle—the right ventricle's maximum pressure is only about 28 mm Hg, whereas the left ventricle's maximum pressure is about 118 mm Hg.

You can see the reason for this ventricular pressure difference if you compare the other two lines in the graphs, which show the pressures in the pulmonary trunk (blue line; for the right ventricle) and the aorta (purple line; for the left ventricle). Notice that the pressure curves for the vessels and the ventricles eventually cross. As you can see, the pressure in the pulmonary trunk varies between about 10 and 28 mm Hg. This means that the maximum pressure the right ventricle must generate

in order to push open the pulmonary valve and eject blood is about 28 mm Hg. However, the pressure in the aorta is considerably higher due to the higher overall blood pressure in the systemic circuit, and ranges from about 80 to 118 mm Hg. For this reason, the left ventricle has to squeeze significantly harder to generate the pressure needed to open the aortic valve and eject blood into the aorta.

Quick Check

☐ 2. How is the cardiac cycle defined?

☐ 3. What are the four phases of the cardiac cycle?

☐ 4. Is the end-diastolic or the end-systolic volume of blood larger? Why?

Connecting the Electrical and Mechanical Events in the Heart

Now let's put the mechanical and electrical events together and see how their timing correlates during a normal cardiac cycle. **Figure 17.19** presents a *cardiac cycle diagram* (also known as a *Wiggers diagram*), which shows you what is happening in the heart during each stage of the cardiac cycle. Four variables are presented: the heart's electrical events, visible on the ECG; the heart sounds; the pressure changes in the left side of the heart, including the aortic, left ventricular, and left atrial pressures; and the changes in the volume of blood found in the left ventricle. The pressure and volume graphs show only the left ventricle, for

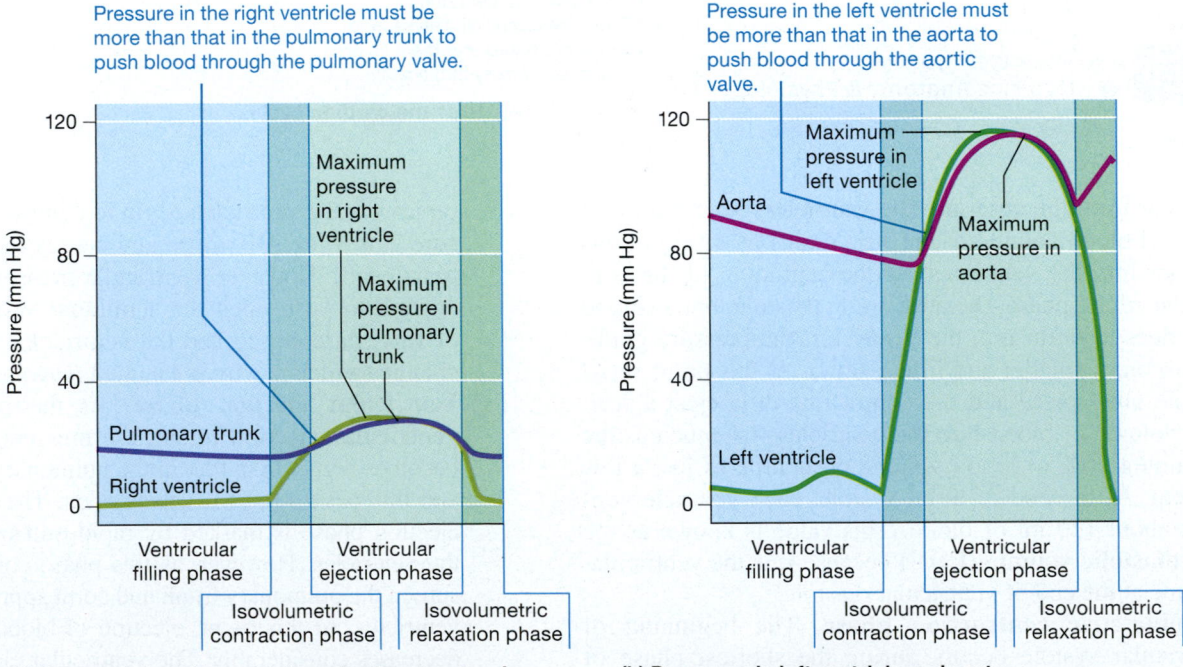

(a) Pressures in right ventricle and pulmonary trunk   (b) Pressures in left ventricle and aorta

Figure 17.18 Comparison of pressure changes in left and right ventricles during the cardiac cycle.

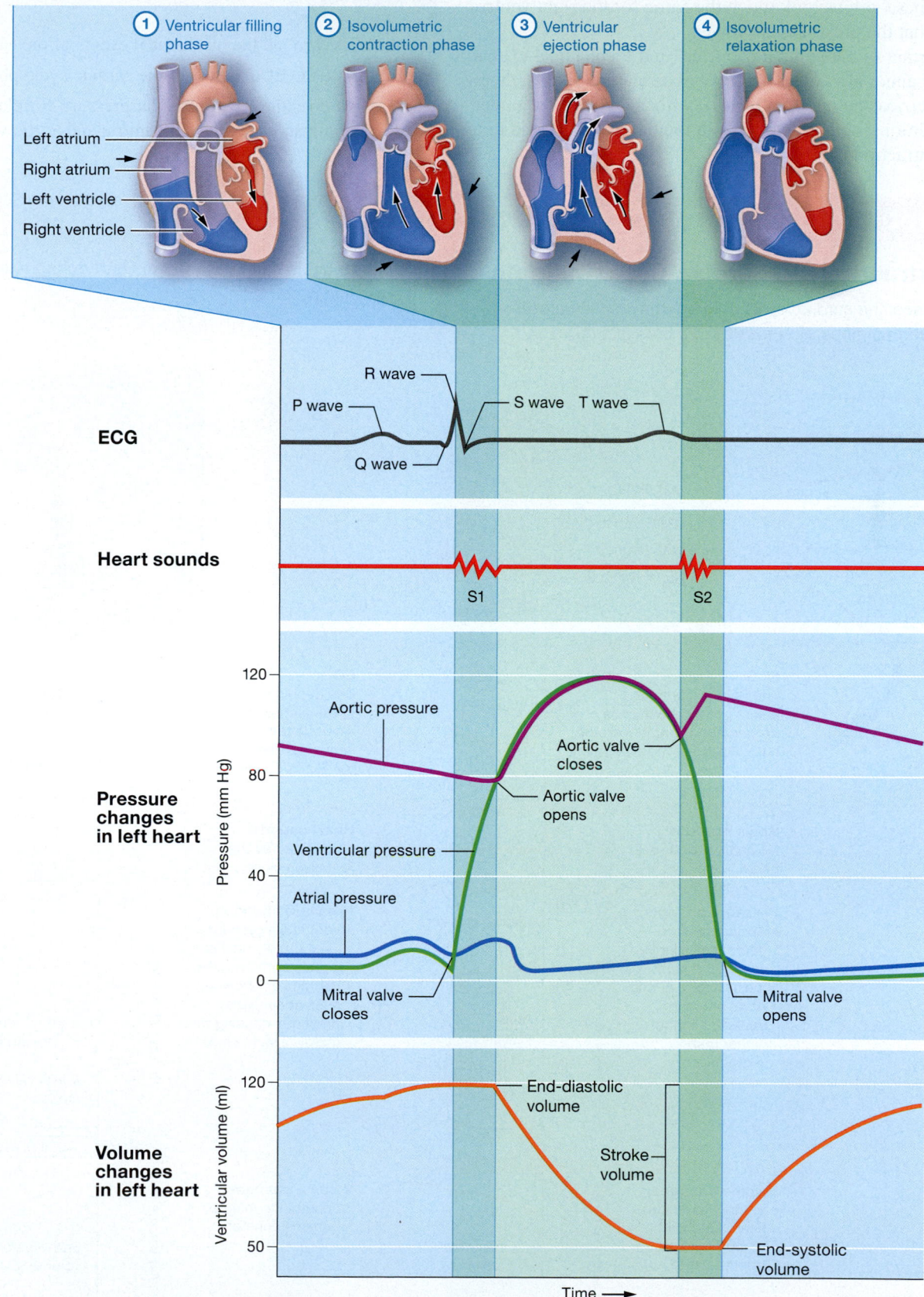

Figure 17.19 Cardiac cycle diagram showing an overview of electrical and mechanical events in the heart during the cardiac cycle. Note that this figure shows the cardiac cycle in only the left side of the heart.

simplicity; these graphs look much the same for the right ventricle, except that the pressures are lower.

It's important to realize that electrical and mechanical events in the heart cannot always be positively correlated. That is, there may be electrical activity with no heartbeat. However, under normal conditions, action potentials in the heart should lead to a physical contraction.

Quick Check

- ☐ 5. Walk through the mechanical events of the cardiac cycle.
- ☐ 6. Relate the ECG waves to the cardiac cycle phases.
- ☐ 7. How does the left ventricular pressure correlate with the opening and closing of the mitral and aortic valves?

ConceptBOOST

Deconstructing the Cardiac Cycle Diagram

At first glance, the cardiac cycle diagram may look like a jumbled collection of lines. So Figure 17.20 breaks it down and clarifies what is happening in the heart during each phase of the cardiac cycle. ■

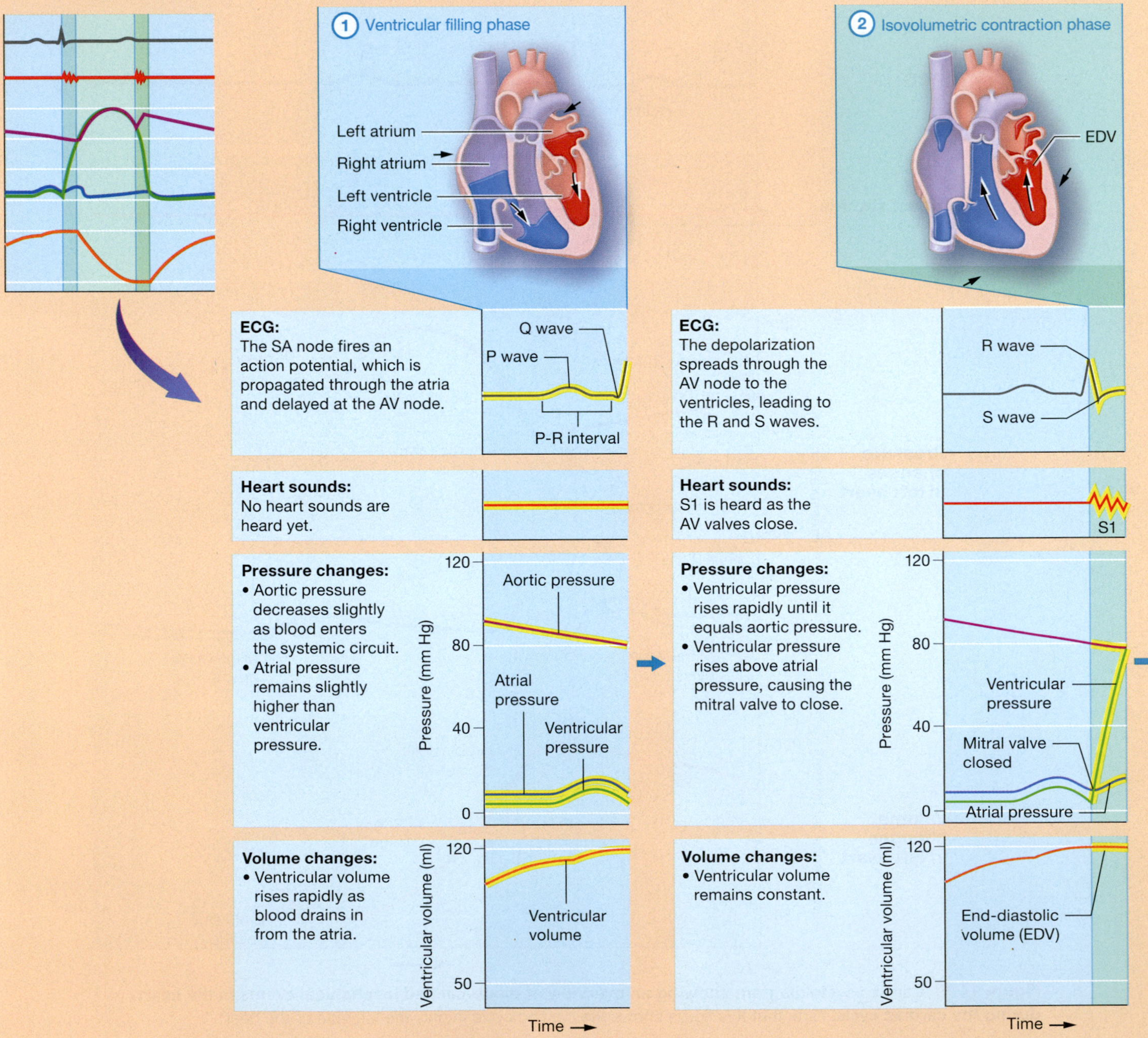

1 Ventricular filling phase
Left atrium
Right atrium
Left ventricle
Right ventricle
ECG:
The SA node fires an action potential, which is propagated through the atria and delayed at the AV node.
Q wave
P wave
P-R interval
Heart sounds:
No heart sounds are heard yet.
Pressure changes:
• Aortic pressure decreases slightly as blood enters the systemic circuit.
• Atrial pressure remains slightly higher than ventricular pressure.
Pressure (mm Hg)
120
80
40
0
Aortic pressure
Atrial pressure
Ventricular pressure
Volume changes:
• Ventricular volume rises rapidly as blood drains in from the atria.
Ventricular volume (ml)
120
50
Ventricular volume
Time
2 Isovolumetric contraction phase
EDV
ECG:
The depolarization spreads through the AV node to the ventricles, leading to the R and S waves.
R wave
S wave
Heart sounds:
S1 is heard as the AV valves close.
S1
Pressure changes:
• Ventricular pressure rises rapidly until it equals aortic pressure.
• Ventricular pressure rises above atrial pressure, causing the mitral valve to close.
Pressure (mm Hg)
120
80
40
0
Ventricular pressure
Mitral valve closed
Atrial pressure
Volume changes:
• Ventricular volume remains constant.
Ventricular volume (ml)
120
50
End-diastolic volume (EDV)
Time

Figure 17.20 **Electrical and mechanical events in the left side of the heart during each phase of the cardiac cycle.**

☐ 1. Sometimes health care providers will elect not to treat conditions such as atrial fibrillation in which there is no functional atrial contraction. Explain the logic behind this decision.

☐ 2. Predict what would happen to the end-systolic volume if the ventricles were diseased and failing to pump adequately.

Would this cause the end-diastolic volume to increase, decrease, or stay the same? Explain.

☐ 3. What happens in the heart electrically if the Q-T interval is prolonged? Which part of the cardiac cycle would this affect? What would this do to the heart rate?

See answers in Appendix A.

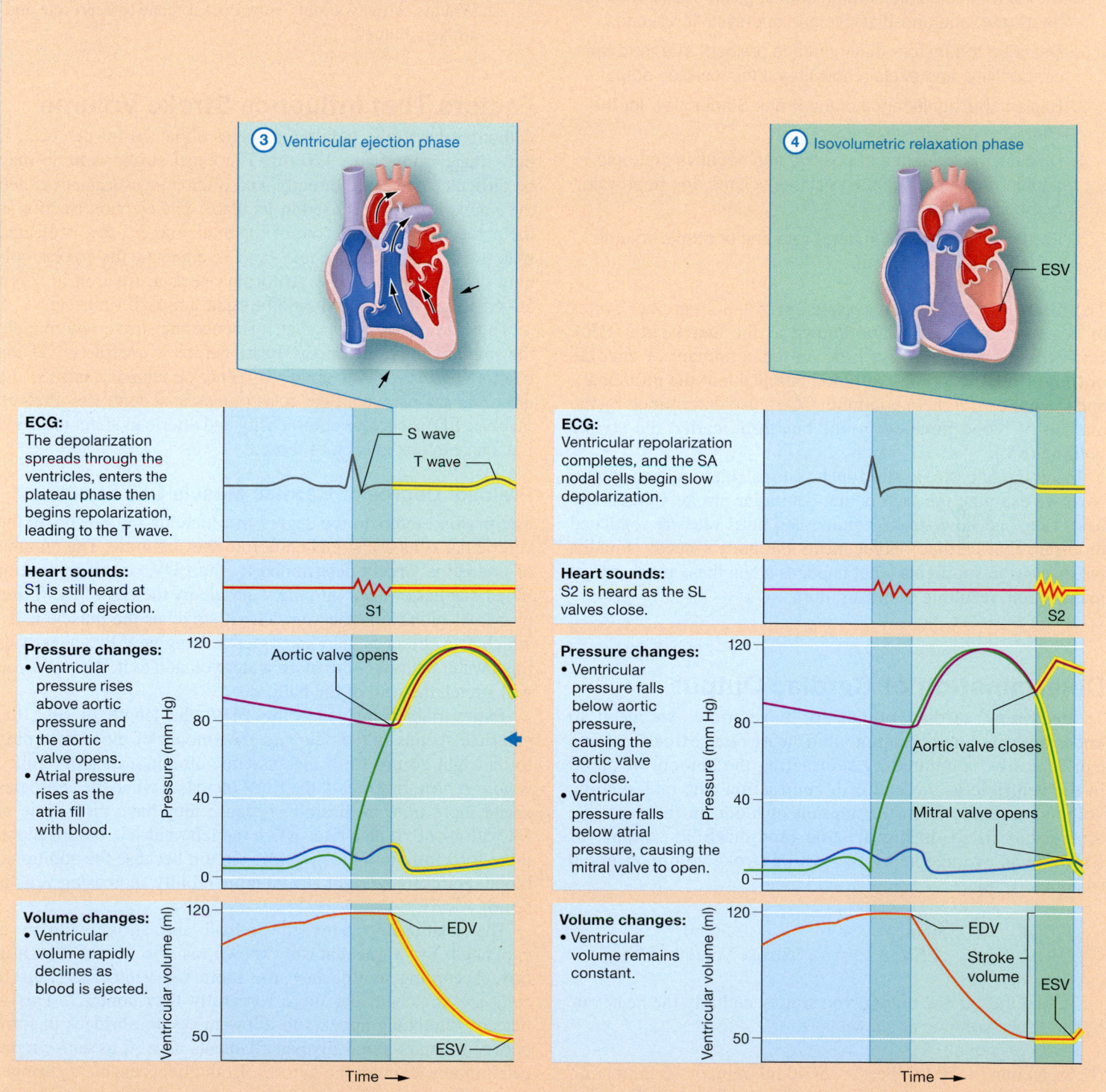

③ Ventricular ejection phase

④ Isovolumetric relaxation phase

— ESV

ECG:
The depolarization spreads through the ventricles, enters the plateau phase then begins repolarization, leading to the T wave.

S wave
T wave

ECG:
Ventricular repolarization completes, and the SA nodal cells begin slow depolarization.

Heart sounds:
S1 is still heard at the end of ejection.

S1

Heart sounds:
S2 is heard as the SL valves close.

S2

Pressure changes:
• Ventricular pressure rises above aortic pressure and the aortic valve opens.
• Atrial pressure rises as the atria fill with blood.

Pressure (mm Hg)
120
80
40
0

Aortic valve opens

Pressure changes:
• Ventricular pressure falls below aortic pressure, causing the aortic valve to close.
• Ventricular pressure falls below atrial pressure, causing the mitral valve to open.

Pressure (mm Hg)
120
80
40
0

Aortic valve closes

Mitral valve opens

Volume changes:
• Ventricular volume rapidly declines as blood is ejected.

Ventricular volume (ml)
120
50

EDV

ESV

Time →

Volume changes:
• Ventricular volume remains constant.

Ventricular volume (ml)
120
50

EDV

Stroke volume

ESV

Time →

Figure 17.20 Electrical and mechanical events in the left side of the heart during each phase of the cardiac cycle. (*continued*)

Cardiac Output and Regulation

Learning Outcomes

1. Define and calculate cardiac output, given stroke volume, heart rate, and end-diastolic and end-systolic volumes.

2. Describe the factors that influence preload, afterload, and contractility, and explain how they affect cardiac output.

3. Explain the significance of the Frank-Starling law for the heart.

4. Discuss the influence of positive and negative inotropic and chronotropic agents on stroke volume and heart rate, respectively.

5. Predict how changes in heart rate and/or stroke volume will affect cardiac output.

The heart undergoes an average of 60–80 cardiac cycles or beats per minute, a value known as the **heart rate (HR).** However, the heart rate isn't the only determinant of **cardiac output (CO),** the amount of blood pumped into the pulmonary and systemic circuits in 1 minute. CO is also determined by the amount of blood pumped in one heartbeat, called the **stroke volume (SV).**

In this module, we first find out how to calculate cardiac output. Next we examine the factors that determine stroke volume and heart rate, and how stroke volume and heart rate are regulated to ensure that cardiac output meets the body's needs. Finally, the discussion spotlights what happens when these mechanisms malfunction, resulting in heart failure.

Determination of Cardiac Output

To determine cardiac output for a ventricle, we need to know both its stroke volume and heart rate. Stroke volume can be easily calculated by subtracting the amount of blood in the ventricle at the end of a contraction (the end-systolic volume, or ESV) from the amount of blood in the ventricle after it has filled during diastole (end-diastolic volume, or EDV). In an average heart, the resting stroke volume is equal to about 70 ml:

$$
\begin{array}{ccccc}
120 \text{ ml} & - & 50 \text{ ml} & = & 70 \text{ ml} \\
\text{EDV} & - & \text{ESV} & = & \text{Stroke Volume (SV)}
\end{array}
$$

To find the cardiac output, you simply multiply the heart rate by the stroke volume, as shown here:

$$
\begin{array}{ccccc}
72 \text{ beats/min} & \times & 70 \text{ ml/beat} & = & 5040 \text{ ml/min, or } \sim 5 \text{ liters/min} \\
\text{HR} & \times & \text{SV} & = & \text{Cardiac Output (CO)}
\end{array}
$$

So, as you can see, resting cardiac output averages about 5 liters/min, which means that the right ventricle pumps about 5 liters into the pulmonary circuit and the left ventricle pumps the same amount into the systemic circuit in 1 minute. Normal adult blood volume is about 5 liters, so your entire supply of blood passes through your heart every minute.

Quick Check

☐ 1. What is stroke volume? How is stroke volume calculated?

☐ 2. What is cardiac output? How does it relate to heart rate and stroke volume?

Factors That Influence Stroke Volume

Although the stroke volume averages about 70 ml per beat, it may range from 50 to 120 ml. The exact stroke volume may be difficult to measure directly, and often a measurement called the *ejection fraction* is used in its place. The ejection fraction is the percentage of blood (out of the total amount) that is ejected with each ventricular systole, and is equal to the stroke volume divided by the EDV. A normal ejection fraction is about 50–65%, and this value should be equal for each ventricle.

There are three factors that influence the stroke volume: (1) the *preload* imposed on the heart before it contracts; (2) the heart's *contractility,* or ability to generate tension; and (3) the *afterload* against which the heart pumps as it contracts. Preload involves EDV, whereas contractility and afterload affect the ESV. Let's take a look at each of these.

Preload: Degree of Cardiac Muscle Cell Stretch

The **preload** refers to the degree to which the sarcomeres in the ventricular cells are stretched before they contract. The amount of preload is largely determined by the EDV, or the amount of blood that has drained into the ventricle by the end of the filling phase. As blood fills the ventricle, it stretches the muscle cells, which also stretches their sarcomeres. Imagine it like water filling a water balloon—the more water you add to it, the more you will stretch the wall of the balloon.

Two variables influence the EDV: the length of time the ventricle spends in diastole and the amount of blood returning to the right atrium from the systemic circuit, a quantity called *venous return.* In general, the EDV increases when the ventricles spend more time in diastole, because they have more time to fill with blood. It also rises when the left ventricle pumps blood more forcefully into the systemic circuit, because the additional blood returns to the right atrium more rapidly, increasing venous return.

The relationship between preload and stroke volume is explained by a mechanism known as the **Frank-Starling law.** According to this law, the more the ventricular muscle cells are stretched, the more forcefully they contract. This is because stretching appears to allow more crossbridges to form between the actin and myosin filaments as well as cause more calcium ions to enter the cytosol. Both effects enable a stronger contraction and a higher stroke volume. This relationship is particularly important during exercise, when cardiac output must increase to meet the body's needs.

Contractility

The heart's **contractility** is its intrinsic pumping ability, or ability to generate tension, in the absence of any external influences such as stretch. Contractility is difficult to measure directly; however, it can be estimated clinically by examining the velocity of blood being ejected from the ventricles. The relationship between contractility and stroke volume is pretty straightforward: Increasing contractility will increase the stroke volume and so decrease the ESV (because more blood is ejected during each cardiac cycle). Decreasing contractility will do the opposite: Decrease the stroke volume and increase the ESV (assuming that preload and afterload remain constant). Agents that affect contractility are known as *inotropic agents* (AYE-noh-troh-pik; *ino-* = "a muscle fiber," *trop-* = "change").

Physiological processes that increase heart rate, such as sympathetic nervous stimulation, often also affect contractility and so increase the force of contraction. This explains why stroke volume and heart rate generally increase together. Note, though, that this is true only up to a certain point—when the heart rate is too high, contractility decreases, as does preload. This happens because the heart is beating too rapidly to develop significant tension during each contraction. The result is a decrease in both stroke volume and cardiac output.

Afterload: Amount of Back Pressure

Afterload refers to the force that the right and left ventricles must overcome in order to eject blood into their respective arteries. It is largely determined by the blood pressure in the arteries of both the pulmonary and systemic circuits. As afterload increases, ventricular pressure must be greater to exceed the pressure in arterial pulmonary and systemic vessels and open the semilunar valves. An increase in afterload therefore generally causes a decrease in stroke volume, which in turn leads to a rise in the ESV of the ventricles. On the other hand, a decrease in afterload generally corresponds to a higher stroke volume and a lower ESV. Read *A&P in the Real World: Ventricular Hypertrophy* on p. 664 to discover what happens when preload and afterload are increased over time.

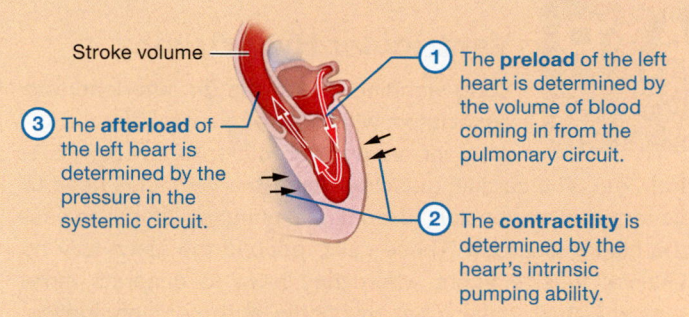

Let's put these three factors together to see how they interact to determine stroke volume. In our first scenario, we have a high preload and contractility and a low afterload, and all three together raise stroke volume:

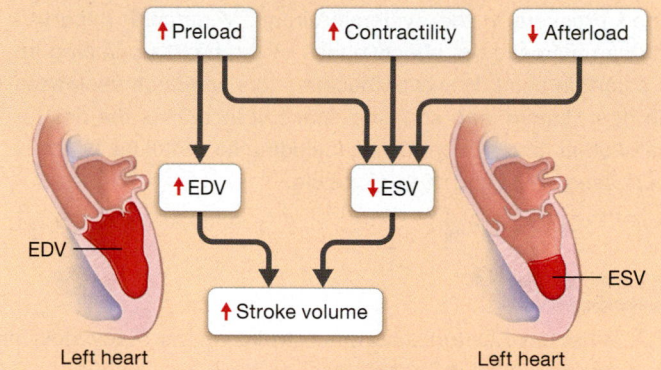

The high preload leads to a high EDV, and this volume of blood stretches the cardiac muscle cells. This, combined with increased contractility, leads to a more forceful contraction. Also, the afterload is low, so this forceful contraction isn't pumping against a great deal of resistance. A forceful contraction against low resistance leads to a high stroke volume and a low ESV.

In this second scenario, we have a low preload and contractility and a high afterload, which all decrease stroke volume:

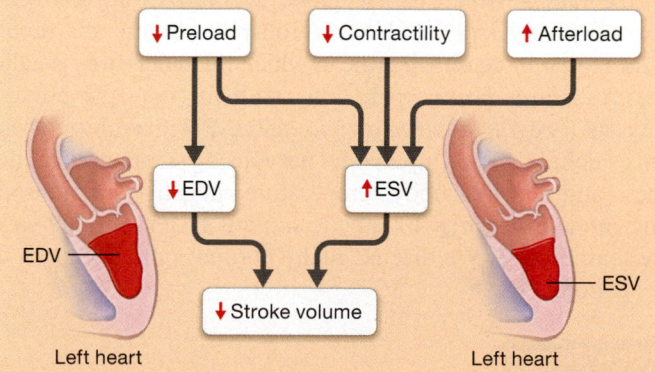

Notice here that the low preload leads to a low EDV, so now the cardiac muscle cells are much less stretched. When this is combined with a diminished contractility, the heart contracts weakly. In addition, the afterload is now high, so this weak contraction is pumping against a great deal of resistance. A weak contraction against high resistance leads to a low stroke volume, and more blood left in the ventricle after the contraction (a high ESV). ■

ConceptBOOST 》》》

Understanding How Changes in Preload, Contractility, and Afterload Affect Stroke Volume

You've just read about the factors that determine stroke volume—preload, contractility, and afterload. The following diagram illustrates these factors, using only the left ventricle for simplicity.

Ventricular Hypertrophy

Long-standing increases in afterload are associated with enlargement of the ventricles, or *ventricular hypertrophy* (hy-PER-troh-fee). The cardiac muscle cells of the ventricles need to generate more tension to continue pumping blood against the higher afterload. These cells respond the same way as skeletal muscle fibers when they have to generate more tension—they make more myofibrils and more organelles, and as a result they get bigger.

Right ventricular hypertrophy most often results from respiratory disease or high blood pressure in the pulmonary circuit, and left ventricular hypertrophy generally results from high blood pressure in the systemic circuit. Ventricular hypertrophy can increase the effectiveness of the heart's pumping up to a certain point. However, because this condition decreases the heart lumen, and so filling space, it increases the risk for many other cardiac conditions, including heart failure, which is discussed at the end of this module.

Quick Check

☐ 3. What three factors determine stroke volume? How does the stroke volume differ with each factor?

☐ 4. What is the Frank-Starling law, and how does it relate to preload?

Factors That Influence Heart Rate

The other determinant of cardiac output is the heart rate. Under normal conditions, the rate at which the SA node generates action potentials determines the heart rate. Factors that influence the rate at which the SA node depolarizes are known as *chronotropic agents* (KROHN-oh-troh-pik; *chrono-* = "time"). Anything that increases the rate at which this node fires is called a *positive chronotropic agent*; one with the opposite effect is known as a *negative chronotropic agent*. Positive chronotropic agents include the sympathetic nervous system, certain hormones, and elevated body temperature; negative chronotropic agents include the parasympathetic nervous system and decreased body temperature. We discuss many chronotropic agents in the next section.

Quick Check

☐ 5. What is a chronotropic agent?

Regulation of Cardiac Output

 **FLASHBACK**

1. What effects does the sympathetic nervous system have on the adrenal gland and the heart? (pp. 524–525)

2. What effects does the parasympathetic nervous system have on the heart? (p. 528)

Although the heart is autorhythmic, it still requires regulation to ensure that cardiac output meets the body's needs at all times. Cardiac output is regulated primarily by the nervous and endocrine systems, which influence both heart rate and stroke volume. Next we look at the effects of the nervous system, endocrine system, and other variables on the heart.

Cardiac Innervation and Regulation by the Nervous System

Recall that the two branches of the autonomic nervous system, or ANS, regulate our automatic functions (see Chapter 14). The *sympathetic nervous system* innervates the heart via a set of **sympathetic nerves** that stem from ganglia located along the spinal cord (**Figure 17.21**). The neurons of these nerves release the neurotransmitter norepinephrine, which increases cardiac output with both positive chronotropic and inotropic effects.

First, they increase the heart rate by raising the rate at which the SA node fires, up to 180–200 or more times per minute. Second, they cause more calcium ions to enter the cardiac muscle cell through calcium ion channels. This increases contractility, which raises stroke volume. Together, these two effects can dramatically increase cardiac output.

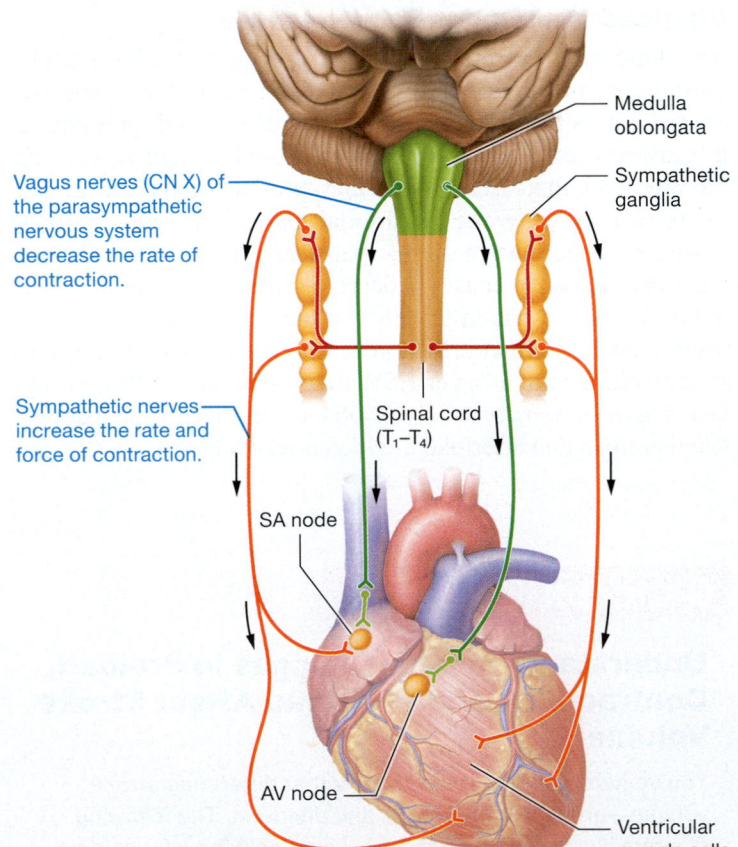

Figure 17.21 Innervation and nervous regulation of the heart.

Medulla oblongata

Sympathetic ganglia

Vagus nerves (CN X) of the parasympathetic nervous system decrease the rate of contraction.

Sympathetic nerves increase the rate and force of contraction.

Spinal cord (T₁–T₄)

SA node

AV node

Ventricular muscle cells

Note the major difference in how the nervous system controls the strength of cardiac and skeletal muscle contractions. The somatic nervous system stimulates more motor units, and so more skeletal muscle fibers, to elicit a stronger skeletal muscle contraction, a phenomenon called *recruitment.* However, all cardiac muscle cells work together for each contraction of the heart, regardless of the strength of the contraction. So the sympathetic nervous system cannot increase the number of cardiac muscle cells that are activated, and recruitment is not possible. For this reason, the sympathetic nervous system must control the strength of contraction instead by the number of calcium ions that enter the cardiac muscle cell.

The parasympathetic nervous system exerts essentially the opposite effects on the heart. As you can see in Figure 17.21, it innervates the heart by the left and right *vagus nerves* (CN X). Both nerves release acetylcholine when stimulated. Acetylcholine primarily affects the SA node, decreasing its rate of action potential generation, although it also slows the rate of conduction through the AV node. This negative chronotropic effect slows the heart rate and can even stop the heart temporarily if the parasympathetic stimulation is strong enough. The vagus nerves also weakly reduce ventricular contractility due to their mild negative inotropic effects.

Cardiac Regulation by the Endocrine System

Hormonal regulation of cardiac output occurs in various forms. For one, recall that the adrenal medulla is activated by the sympathetic nervous system, and in response it secretes the hormones epinephrine and norepinephrine into the blood (see Chapter 14). These hormones have the same effects as the sympathetic nervous system neurotransmitters—they are positive inotropic and chronotropic agents—but their effects are longer-lasting. Other hormones that have these same effects include thyroid hormone and glucagon produced by the pancreas.

Some hormones regulate cardiac output through their control of water balance. The amount of water in the blood (the blood volume) plays a significant role in determining the heart's preload and so its strength of contraction. Hormones such as aldosterone and antidiuretic hormone increase blood volume and preload, and so raise cardiac output. Others, such as atrial natriuretic peptide, decrease blood volume and preload, and therefore reduce cardiac output.

Other Factors That Influence Cardiac Output

Although the nervous and endocrine systems are the primary regulators of cardiac output, other variables influence cardiac output as well. For example, the concentration of certain electrolytes in the extracellular fluid has an important part in determining the length and magnitude of an action potential and cardiac output. Another factor that influences cardiac output is body temperature—the SA node fires more rapidly at higher body temperatures and more slowly at lower body temperatures. In addition, conditions such as age and physical fitness influence heart rate and cardiac output; younger children and the elderly often have a higher resting heart rate, whereas trained athletes often have a much lower resting heart rate. This is because exercise increases the stroke volume, so for the body to maintain a constant cardiac output, the heart rate must decrease. Factors that affect cardiac output are summarized in **Figure 17.22**.

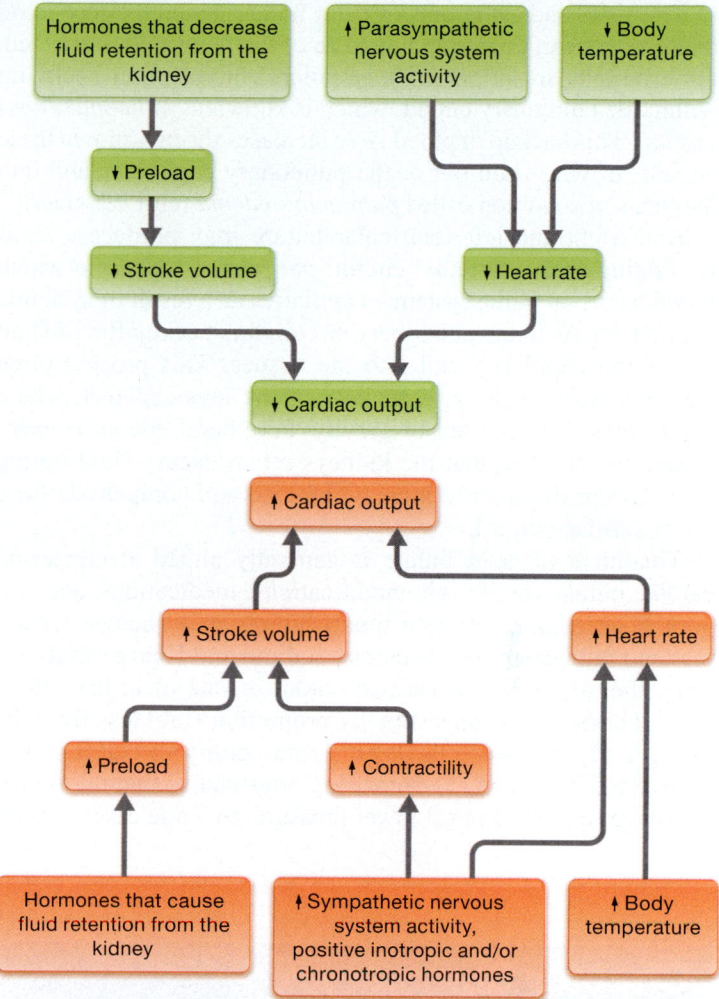

Figure 17.22 Regulation of cardiac output.

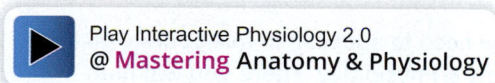

Play Interactive Physiology 2.0
@ **Mastering** Anatomy & Physiology

Quick Check

☐ 6. What effects does the sympathetic nervous system have on cardiac output?

☐ 7. What effects does the parasympathetic nervous system have on cardiac output?

☐ 8. How would a hormone that decreases the amount of water in the body affect cardiac output?

Heart Failure

Heart failure, formerly known as *congestive heart failure,* is defined as any condition that reduces the heart's ability to function effectively as a pump. Causes of heart failure include reduced contractility due to myocardial ischemia and/or myocardial infarction, valvular heart diseases, any disease of the heart muscle itself (known as *cardiomyopathy*), and electrolyte imbalances. Heart failure generally results in decreased stroke volume, which in turn reduces cardiac output.

The signs and symptoms of heart failure generally depend on the type of heart failure and the side of the heart that is affected. For example, in left ventricular failure, blood often backs up within the pulmonary circuit, which is known as *pulmonary congestion*. This backup of blood flow increases the pressure in these vessels, driving fluid out of the pulmonary capillaries and into the lungs, a condition called *pulmonary edema* (eh-DEE-mah).

Both right and left ventricular failure may produce a similar finding in the systemic circuit: *peripheral edema*, in which blood backs up in the systemic capillaries as a result of systemic congestion. As in the pulmonary circuit, this backup forces fluid out of the capillaries and into the tissues. This process often causes visible swelling, especially in the legs and feet, where fluid collects as a result of gravity. Peripheral edema is exacerbated by the fact that the kidneys retain excess fluid during heart failure (in order to increase preload and compensate for a lower cardiac output).

Treatment of heart failure is generally aimed at increasing cardiac output via lifestyle modifications, medications, and surgery if necessary. Lifestyle modifications may include weight loss and mild exercise plus dietary sodium and fluid restrictions. Drug therapy increases cardiac output in one of at least three ways: (1) reducing congestion by promoting fluid loss from the kidneys, (2) increasing the heart's contractility so that it pumps more effectively, and (3) decreasing afterload so that the ventricles have to pump against lower pressure. In some cases, a heart transplant and/or a surgically implanted pacemaker that electrically stimulates and paces the heart may be necessary.

Quick Check

☐ 9. How is heart failure defined?

Apply What You Learned

☐ 1. The drug digoxin (dih-JOK-sin) increases the concentration of intracellular calcium ions in cardiac muscle cells. Predict the specific effect this will have on stroke volume and cardiac output.

☐ 2. Predict what would happen to cardiac output if sympathetic nervous system stimulation of the heart had positive chronotropic effects only.

☐ 3. A patient has a heart rate of 70 beats per minute, an EDV of 110 ml, and an ESV of 70 ml.

   a. What is this patient's stroke volume?

   b. What is her ejection fraction?

   c. What is her cardiac output?

   d. Are the values that you calculated within normal ranges? Explain.

See answers in Appendix A.

Chapter Summary

For everything you need to succeed in this course, go to Mastering Anatomy & Physiology. There you will find:

- Practice Tests
- Author Podcasts
- Big Picture Animations
- Concept Boost Video Tutors
- iP2™ Interactive Physiology 2.0 Tutorials
- A&PFlix A&P Flix 3D Animations
- Active-Learning Workbook

MODULE 17.1
Overview of the Heart 630–633

- The **heart** is a two-sided pump that drives blood into blood vessels.

- The heart has two superior **atria** that receive blood from **veins** and two inferior **ventricles** that pump blood into **arteries.**

- The right side of the heart is the *pulmonary pump,* which delivers deoxygenated blood to the **pulmonary circuit.** The left side of the heart is the *systemic pump,* which delivers oxygenated blood to the **systemic circuit.**

- The heart also secretes the hormone *atrial natriuretic peptide.*

MODULE 17.2
Heart Anatomy and Blood Flow Pathway 633–644

- The heart wall has three layers: the outer **fibrous** and **serous pericardium,** the middle **myocardium,** and the inner **endocardium.** The serous pericardium is composed of the outer **parietal pericardium** and the inner **visceral pericardium,** between which is the serous fluid–containing **pericardial cavity.**

- The **coronary circulation** consists of the *coronary arteries,* which supply the myocardium with oxygenated blood, and the *coronary veins,* which drain deoxygenated blood from the myocardium. The two main coronary arteries are the **right coronary artery** and the **left coronary artery.** The **coronary sinus** is the large vessel that receives blood from the coronary veins and empties it into the right atrium.

- The four main great vessels are the **venae cavae (superior** and **inferior),** the **pulmonary trunk,** the **pulmonary veins,** and the **aorta.**

- Each atrium has an external **auricle.** The two atria are separated by the **interatrial septum.**

- The two ventricles are separated by the **interventricular septum.** The left ventricle has a thicker wall than the right because it pumps against higher pressure.

- Blood flowing into the ventricles passes through the **atrioventricular (AV) valves,** including the **tricuspid** and **mitral valves.** The **cusps** of each valve are attached to **papillary muscles** by **chordae tendineae** that prevent the cusps from everting when the ventricles contract.

- Blood flowing out of the ventricles passes through the **semilunar (SL) valves,** including the **pulmonary** and **aortic valves.**

MODULE 17.3
Cardiac Muscle Tissue Anatomy and Electrophysiology 644–655

- **Cardiac muscle cells** contract in response to **action potentials** via the sliding-filament mechanism. **Pacemaker cells** are **autorhythmic** and spontaneously generate action potentials. These action potentials trigger cardiac **contractile cells** to have action potentials.

- Cardiac muscle cells are short, branched, striated cells joined by **intercalated discs.**

- A contractile cell action potential results from a reversal in membrane potential. The stages of the action potential are as follows: rapid depolarization due to sodium ion influx; brief initial repolarization due to potassium ion efflux; a **plateau phase** due to calcium ion influx and simultaneous potassium ion efflux; and repolarization due to continued potassium ion efflux.

- Pacemaker cells depolarize rhythmically. They reach threshold through nonselective cation influx; depolarization is due to the inflow of calcium ions. Their repolarization and hyperpolarization are due to the outflow of potassium ions.

- Three populations of pacemaker cells exist in the heart, collectively called the **cardiac conduction system: the sinoatrial (SA) node,** which is the main pacemaker of the heart; the **atrioventricular (AV) node,** the heart's backup pacemaker; and the components of the **Purkinje fiber system.**

- An action potential in the heart normally takes the following path: SA node → atrial contractile cells → AV node where it is delayed → AV bundle → right and left bundle branches → Purkinje fibers → ventricular contractile cells.

- The **electrocardiogram** measures electrical changes occurring in the heart, which appear as **waves** on the recording.

MODULE 17.4
Mechanical Physiology of the Heart: The Cardiac Cycle 655–661

- **Mechanical physiology** describes the physiology of cardiac pumping.

- Pressure changes generated by ventricular contraction and relaxation cause the blood to flow between chambers and the valves to open and close.

- The **S1** heart sound is caused by the closing of the AV valves at the beginning of isovolumetric contraction. The **S2** heart sound is caused by the closing of the SL valves at the beginning of isovolumetric relaxation.

- The **cardiac cycle** is the sequence of events that takes place within the heart from one heartbeat to the next, during which each chamber has a relaxation period (**diastole**) and a contraction period (**systole**).

- There are four stages of the cardiac cycle: **ventricular filling, isovolumetric contraction, ventricular ejection,** and **isovolumetric relaxation.** The volume of blood in the ventricles at the end of ventricular filling is the **end-diastolic volume (EDV);** the volume of ventricular blood after ventricular ejection is the **end-systolic volume (ESV).**

- Blood is pumped out of the ventricles when the pressure in the ventricles is higher than the pressure in the vessels into which they pump the blood.

MODULE 17.5
Cardiac Output and Regulation 662–666

- **Cardiac output** is the amount of blood pumped out by each ventricle in 1 minute. It is equal to the **heart rate** multiplied by the **stroke volume.**

- Stroke volume is the amount of blood ejected from the right or left ventricle with each beat, and is equal to the EDV minus the ESV. Stroke volume is determined by the **preload,** the degree of stretch imposed on the cardiac muscle cells; **contractility,** the effectiveness with which the heart pumps; and **afterload,** the force against which the ventricles must pump.

- Heart rate is influenced by chronotropic agents such as the ANS and endocrine system.

- Cardiac output is regulated primarily by the endocrine and nervous systems.
  - Epinephrine and norepinephrine are positive *chronotropic* and *inotropic* agents.
  - Acetylcholine is primarily a negative chronotropic agent.
  - The endocrine system releases chronotropic and inotropic hormones, and other hormones that regulate water balance and preload.

Assess What You Learned

LEVEL 1 Check Your Recall

1. Mark the following statements as true or false. If a statement is false, correct it to make a true statement.

   a. The heart is located in the mediastinum slightly to the left of the midline.
   b. The heart consists of two superior ventricles and two inferior atria.
   c. Arteries always carry oxygenated blood away from the heart, and veins always carry deoxygenated blood toward the heart.
   d. The pulmonary circuit delivers blood from the right side of the heart to the lungs to become oxygenated.
   e. The heart plays a role in the regulation of blood pressure and secretes the hormone atrial natriuretic peptide.

2. The pericardial cavity is located between:

   a. the parietal pericardium and the fibrous pericardium.
   b. the fibrous pericardium and the myocardium.
   c. the parietal pericardium and the visceral pericardium.
   d. the epicardium and the endocardium.

3. Which of the following statements is *true?*

   a. The tricuspid valve is located between the right atrium and the right ventricle.
   b. The mitral valve is located between the pulmonary veins and the left atrium.
   c. The pulmonary valve is located between the pulmonary artery and the pulmonary veins.
   d. The aortic valve is located between the right ventricle and the aorta.

4. Match the following terms with the correct definition.

   _____ Auricle
   _____ Aorta
   _____ Coronary sinus
   _____ Papillary muscle
   _____ Fossa ovalis
   _____ Pectinate muscle
   _____ Venae cavae
   _____ Pulmonary trunk
   _____ Chordae tendineae
   _____ Pulmonary veins

   a. Drainage point for the coronary veins
   b. Extensions that attach papillary muscles to valves
   c. Remnant of a hole present in the fetal interatrial septum
   d. Two largest veins of the systemic circuit
   e. Flaplike extension from the right or left atrium
   f. Finger-like projections of ventricular muscle
   g. Main artery of the pulmonary circuit
   h. Veins that drain the pulmonary circuit
   i. Largest artery of the systemic circuit
   j. Ridges of muscle in the atria

5. Fill in the blanks: The coronary arteries are the first branches off the _____. The right coronary artery becomes the _____ on the posterior side of the heart. The left coronary artery branches into the _____ and the _____.

6. How do pacemaker cardiac muscle cells differ from contractile cardiac muscle cells? What is autorhythmicity?

7. Cardiac muscle cells are joined by structures called:

   a. T-tubules.
   b. tight junctions.
   c. sarcoplasmic reticulum.
   d. intercalated discs.

8. Mark the following statements as true or false. If a statement is false, correct it to make a true statement.

   a. The rapid depolarization phase of the contractile cell action potential is due to the opening of voltage-gated potassium ion channels.
   b. Pacemaker cells lack a distinct plateau phase.
   c. The plateau phase in contractile cells is due to the influx of calcium ions through calcium ion channels.
   d. The repolarization phase of the contractile cell is due to the potassium ions rushing into the cell through potassium ion channels.
   e. Open sodium ion channels cause hyperpolarization in pacemaker cells, which triggers HCN channels to open and begins a new action potential.

9. What are the effects of the plateau phase of the contractile cell action potential?

10. The _____ is the primary pacemaker of the heart.

    a. atrioventricular node
    b. sinoatrial node
    c. Purkinje fiber system
    d. atrioventricular bundle

11. The AV node delay:

    a. allows the atria and ventricles to depolarize and contract as a unit.
    b. allows the two ventricles to depolarize and contract separately.
    c. allows the atria and ventricles to depolarize and contract separately.
    d. speeds up the impulse transmission from the atria to the ventricles.

12. Explain what each of the following terms represents on an electrocardiogram (ECG).

    a. P wave
    b. QRS complex
    c. T wave
    d. P-R interval
    e. S-T segment

13. Mark the following statements as true or false. If a statement is false, correct it to make a true statement.

    a. Systole is the contraction portion of the cardiac cycle and diastole is the relaxation portion.
    b. Atrial systole is responsible for ejecting most of the blood into the ventricles during the ventricular filling phase of the cardiac cycle.
    c. The amount of blood in the ventricles at the end of the ventricular filling phase is the end-systolic volume.
    d. The ventricular ejection phase generally correlates with the S-T segment and the T wave on the ECG.

14. Which chamber generates the highest pressure during systole?

a. Right atrium

b. Right ventricle

c. Left atrium

d. Left ventricle

15. Fill in the blanks: The first heart sound is called _____ and it is caused by the closing of the _____ valves. It occurs at the beginning of the _____ phase of the cardiac cycle. The second heart sound is called _____ and it is caused by the closing of the _____ valves. It occurs at the beginning of the _____ phase of the cardiac cycle.

16. Cardiac output is equal to:

a. end-diastolic volume minus end-systolic volume.

b. heart rate multiplied by stroke volume.

c. stroke volume divided by end-diastolic volume.

d. heart rate multiplied by preload.

17. Fill in the blanks: An increase in preload causes a/an _____ in stroke volume in accordance with the _____ law. An increase in afterload causes a/an _____ in stroke volume. An increase in contractility causes a/an _____ in stroke volume.

18. Which of the following statements is *false?*

a. The sympathetic nervous system releases epinephrine and norepinephrine, which are positive chronotropic and inotropic agents.

b. The endocrine system regulates cardiac output through chronotropic and inotropic hormones and through hormones that regulate water balance.

c. The parasympathetic nervous system releases acetylcholine and epinephrine, which are strongly negative inotropic agents.

d. Factors such as electrolyte concentrations, body temperature, and age all affect cardiac output.

LEVEL 2 Check Your Understanding

1. A birth defect called *transposition of great vessels* results in the pulmonary trunk emanating from the left ventricle and the aorta stemming from the right ventricle.

a. Which ventricle is thicker-walled, and why?

b. Considering your answer to part (a), predict the potential effects of this birth defect.

2. Predict which would be more damaging to long-term survival: a blood clot lodged in the right coronary artery or one in the left coronary artery. Explain.

3. When the SA node doesn't function properly, the AV node takes over pacing the heart and produces what is known as a junctional rhythm. Explain why we don't see P waves on the ECG of an individual with such a rhythm.

4. Common findings in heart failure are fluid retention by the kidneys and stimulation of the heart by the sympathetic nervous system. How would both of these findings help the body to compensate for the failing heart?

LEVEL 3 Apply Your Knowledge

PART A: Application and Analysis

1. You are an athletic trainer who is working with someone planning to run a marathon. Your trainee tells you to give him a workout that will make his heart "beat faster than ever before." What do you tell him about the effects of too rapid a heart rate?

2. A newer drug, *ivabradine,* lowers the heart rate by blocking the nonselective HCN cation channels. Why would this action decrease the heart rate? Would this drug have an effect on pacemaker cells, contractile cells, or both? Explain.

3. Mr. Watson has been diagnosed with mitral insufficiency, or a malfunctioning mitral valve, which causes the valve to not close properly. Predict the signs and symptoms you might expect from a disease of this valve. What would happen to the patient's stroke volume and cardiac output? Explain. What might help improve his cardiac output?

PART B: Make the Connection

4. An experimental toxin makes the refractory period of cardiac muscle cells equal in length to that of skeletal muscle fibers. Predict the consequences of this toxin. *(Connects to Chapter 10)*

See answers in Appendix A.

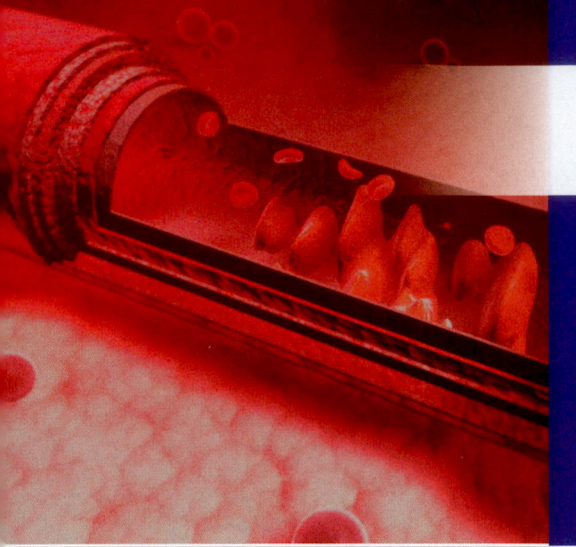

18

The Cardiovascular System II: The Blood Vessels

For practice applying concepts to a clinical scenario, check out the **Running Case Study** for this chapter @ Mastering Anatomy & Physiology

Computer-generated image: Blood vessels are tubular organs that distribute blood throughout the body.

We now turn our attention to the second portion of the cardiovascular system: the blood vessels, collectively called the **vasculature** (VASS-kyoo-lah-chur; *vascul-* = "vessel"). The vasculature of one individual consists of billions of blood vessels that transport blood to the tissues, where gases, nutrients, and wastes are exchanged, and then transport it back to the heart. If these blood vessels were situated end to end, they would measure over 60,000 miles long. But blood vessels are not merely the "pipes" of the cardiovascular system. They also regulate blood flow to tissues, control blood pressure, and secrete a variety of chemicals.

In the heart chapter, we discussed the two circuits that carry blood through the body: the *pulmonary circuit,* which transports blood between the heart and the lungs; and the *systemic circuit,* which transports blood between the heart and the rest of the body (see Chapter 17). Here we cover primarily the systemic circuit, as the pulmonary circuit is discussed with the respiratory system (see Chapter 21). The first part of this chapter (Modules 18.1 through 18.5) explores the basic structure and function of the different types of blood vessels; the physiology of blood pressure and circulation; and the physiology of gas, nutrient, and waste exchange. In the second part (Modules 18.6 through 18.8), we examine the anatomy of the vessels that make up the systemic circuit.

MODULE 18.1
Overview of Arteries and Veins

Learning Outcomes

1. Compare and contrast the structures of arteries and veins, and of arterioles and venules.
2. Define vascular anastomosis, and explain the significance of anastomoses.

As you read about in the heart chapter, the pulmonary and systemic circuits are composed of three kinds of blood vessels—arteries, capillaries, and veins (revisit Figure 17.8):

- **Arteries are the *distribution system* of the vasculature. Arteries** are vessels that travel away from the heart, branching into vessels of progressively smaller diameter. Recall that arteries in the pulmonary circuit carry deoxygenated blood, whereas those in the systemic circuit carry oxygenated blood.

- **Capillaries are the *exchange system* of the vasculature. Capillaries** are vessels of very small diameter that form branching networks called **capillary beds.** Gases, nutrients, wastes, and other substances are quickly exchanged between cells and the blood through the capillary walls.
- **Veins function as the *collection system* of the vasculature. Veins** drain blood from capillary beds and return it to the heart; they follow the opposite pattern of arteries—small veins merge with other veins to become progressively larger as they get closer to the heart. In the pulmonary circuit, veins transport oxygenated blood, whereas in the systemic circuit, they transport deoxygenated blood.

In this module, we introduce the basic structure of the blood vessel wall and the general features of arteries and veins. The basic structure and function of capillaries and capillary beds are discussed in Module 18.4.

Structure and Function of Arteries and Veins

⟪ FLASHBACK

1. What is smooth muscle tissue, and what are its basic properties? (p. 148)
2. What is the inner lining of the heart called? Of what tissue type is it composed? (p. 633)

Although blood vessels differ in both diameter and the thickness of their walls, they still share the same basic pattern of organization. All blood vessels are tubular organs with several tissue layers, or **tunics,** that surround a central space—the **lumen.**

The three tunics of the blood vessel wall are the *tunica intima, tunica media,* and *tunica externa* (**Figure 18.1**):

- **Tunica intima.** The innermost **tunica intima** is composed of **endothelium,** which consists of a sheet of simple squamous epithelium and its basal lamina. Recall that this endothelium is continuous with the inner lining of the heart, the *endocardium,* which is also composed of endothelium (see Chapter 17). Endothelial cells provide a smooth surface over which blood can flow with a minimum of friction and turbulence. In addition, endothelial cells produce many chemicals and proteins, such as *nitric oxide,* collagen, and clotting factors, and they contain enzymes within their plasma membrane. Deep to the endothelium we find a thin layer of subendothelial connective tissue and a layer of elastic fibers called the *internal elastic lamina.* The elastic fibers give the vessel the properties of *distensibility,* or the ability to stretch when subjected to increased pressure, and *elasticity,* or the ability to recoil back to the original size when the stretching force is removed.
- **Tunica media.** The middle layer of the blood vessel wall is the **tunica media** (*medi-* = "middle"). As you can see in Figure 18.1, it has two components: a layer of smooth muscle cells arranged in a circular manner around the

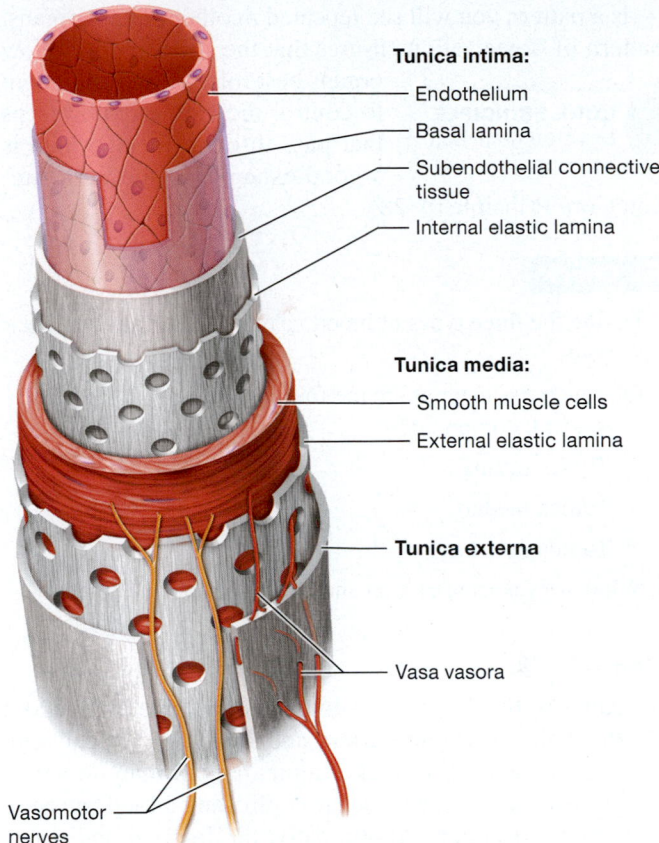

Tunica intima:
- Endothelium
- Basal lamina
- Subendothelial connective tissue
- Internal elastic lamina

Tunica media:
- Smooth muscle cells
- External elastic lamina

Tunica externa

Vasa vasora

Vasomotor nerves

Figure 18.1 The tunics (layers) of the blood vessel wall.

lumen, and another layer of elastic fibers called the *external elastic lamina.* The smooth muscle cells of the tunica media control the diameter of the blood vessel and so the amount of blood that flows to organs. Notice that these muscle cells are innervated by nerves from the sympathetic nervous system, called *vasomotor nerves.* These nerves stimulate the smooth muscle cells of the tunica media to contract, an action known as **vasoconstriction,** which narrows the diameter of the vessel. When sympathetic stimulation of the smooth muscle cells decreases, these cells relax and the vessel's diameter increases, a change called **vasodilation.** As discussed later, vasoconstriction and vasodilation are important determinants of blood pressure and the blood flow to organs and tissues.
- **Tunica externa.** The outermost **tunica externa** (also known as the *tunica adventitia*) is composed of dense irregular collagenous connective tissue that supports the blood vessel and prevents it from overstretching.

In Figure 18.1, you can see tiny vessels supplying the tunica media and tunica externa. These vessels are the **vasa vasora** (VAY-zuh vay-ZOH-ruh), or "vessels to the vessels." The vasa vasora supply oxygen and nutrients to the outer layers of the larger blood vessels, whose cells are too far away from the lumen to receive oxygen and nutrients by diffusion alone.

The structure of the blood vessel wall—an inner epithelial tissue lining, a layer of connective tissue, a layer of smooth muscle tissue, and an additional outer layer of tough connective tissue—is a pattern you will see repeated in other hollow organs. This pattern of organization ensures that the structure of hollow organs best follows their function: to control the flow of substances that pass through them, which is a good example of the Structure-Function Core Principle (p. 28).

 **CORE PRINCIPLE** Structure-Function

Quick Check

☐ 1. Define the three types of blood vessels in the cardiovascular system.

☐ 2. Of which tissue types are the following layers of the blood vessel wall composed?

    a. Tunica intima

    b. Tunica media

    c. Tunica externa

☐ 3. What are vasoconstriction and vasodilation?

Arteries

As you can see in **Figure 18.2**, there are notable differences between the walls of a typical artery and those of a typical vein. Most arteries have a much thicker tunica media than do veins, which signifies the arteries' role in controlling blood pressure and blood flow to organs. Additionally, the internal and external elastic laminae are much more extensive in arteries than in veins, which reflects the fact that arteries are under much higher pressure than are veins.

The thickness of each component of the arterial wall depends on the artery's size and function. We can generally divide arteries into three classes based on similarity in structure and function—*elastic arteries, muscular arteries,* and *arterioles.*

- **Elastic arteries.** The largest-diameter arteries are the **elastic arteries,** also known as the **conducting arteries.** As their name implies, elastic arteries have a very extensive elastic lamina. The tunica media of an elastic artery contains 40–70 sheets of elastic fibers arranged between thin layers of smooth muscle cells. These arteries, including the aorta and its immediate branches, are nearest the heart and therefore under the highest pressure of any vessels in the cardiovascular system. For this reason, they need to be extremely distensible, as they are greatly stretched with each ventricular systole. The relatively small amount of smooth muscle tissue in these vessels means that their diameter does not change significantly with stimulation from the vasomotor nerves.

- **Muscular arteries. Muscular arteries,** also called **distributing arteries,** are generally intermediate in diameter. They contain a well-developed tunica media composed primarily of smooth muscle cells. Most smaller branches off the aorta are considered muscular arteries, including most named arteries that supply organs, discussed in Module 18.6. Due to the predominance of smooth muscle cells, the diameter of muscular arteries does change significantly with vasoconstriction and vasodilation. This allows the nervous and endocrine systems to adjust local blood flow to different organs by changing the vessel diameter. These are the vessels most likely to become blocked, as you can discover in *A&P in the Real World: Atherosclerosis* on p. 674.

- **Arterioles.** The smallest arteries are the **arterioles** (ahr-TEER-ee-ohlz), which range in size from 0.3 mm (the size of the smallest pencil lead) to less than the diameter of a human hair (about 70–120 µm). Arterioles contain each of the three layers of the blood vessel wall, but the layers are extremely thin, and the tunica media contains only one to

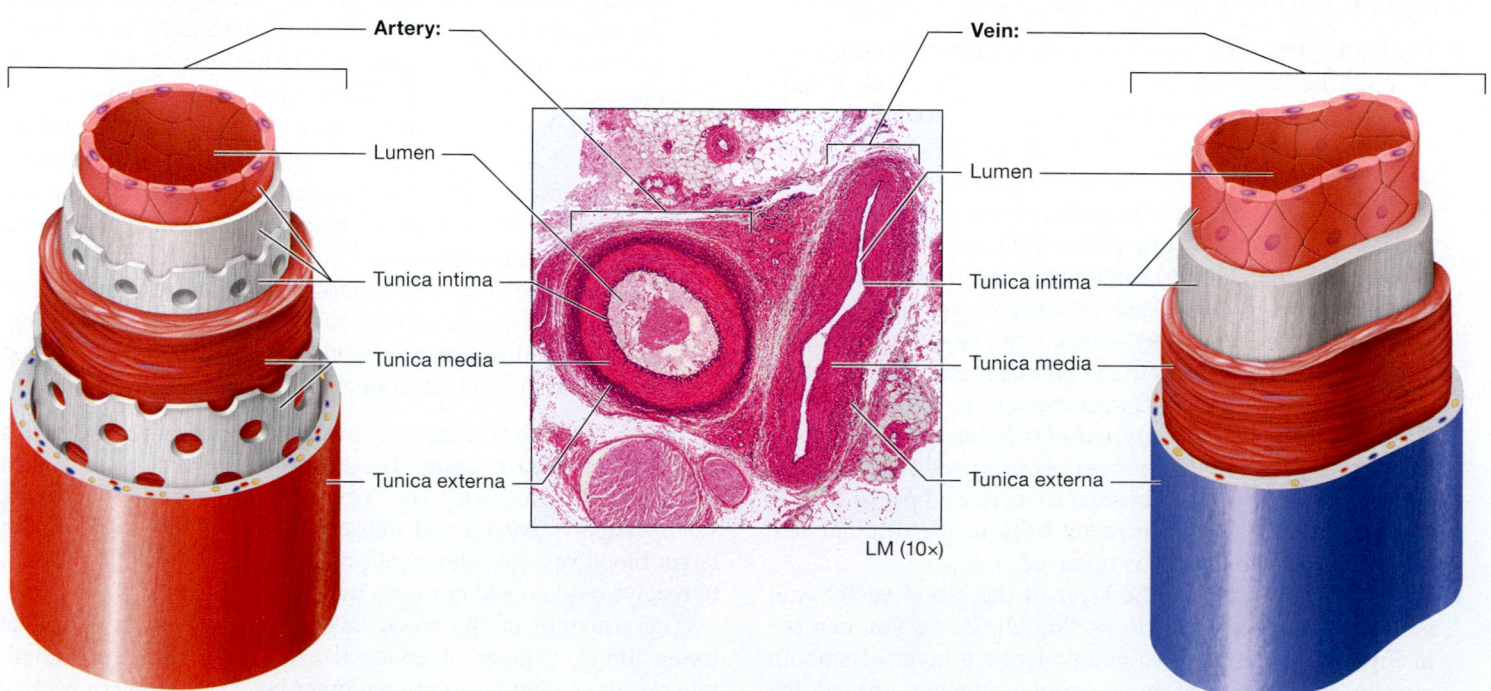

Figure 18.2 A comparison of the walls of arteries and veins. The photomicrograph shows an artery and a vein in cross section.

Artery: — Lumen — Tunica intima — Tunica media — Tunica externa

Vein: — Lumen — Tunica intima — Tunica media — Tunica externa

LM (10×)

three layers of smooth muscle cells. The smallest arterioles, called **metarterioles** (*meta-* = "boundary"), directly feed capillary beds in most tissues. The smooth muscle cells of metarterioles are confined mostly to a **precapillary sphincter** that encircles the metarteriole-capillary junction. Both vasomotor nerves and hormones in the blood can affect the diameter of arterioles; precapillary sphincters are mostly responsive to hormones and local tissue conditions. Vasoconstriction and vasodilation of these structures have a profound impact on the blood flow to individual tissues; in fact, arteriolar and/or precapillary sphincter constriction can completely cut off blood flow to tissues. We revisit the structure of arterioles and metarterioles in Module 18.4, where we explore capillary beds in detail.

In addition to the functions just discussed, certain arteries also play a role in monitoring blood pressure and detecting the concentration of certain chemicals in the blood. Pressure receptors, or **baroreceptors,** are found in the aorta, as well as in the *common carotid artery,* located in the neck. Also in the carotid artery and the aorta are groups of *chemoreceptors* that detect blood oxygen, carbon dioxide, and hydrogen ion concentrations. These pH receptors, covered in Module 18.3, send signals to the CNS.

Quick Check

☐ 4. Describe the basic properties of elastic arteries, muscular arteries, and arterioles.

☐ 5. Which type of artery controls blood flow to organs? Which controls blood flow to tissues?

Veins

Veins typically outnumber arteries and their lumens have a larger average diameter. For these reasons, up to 70% of the total blood in the body is located in the veins at any given moment (this includes both systemic and pulmonary veins). This feature allows veins to function as **blood reservoirs; Figure 18.3**

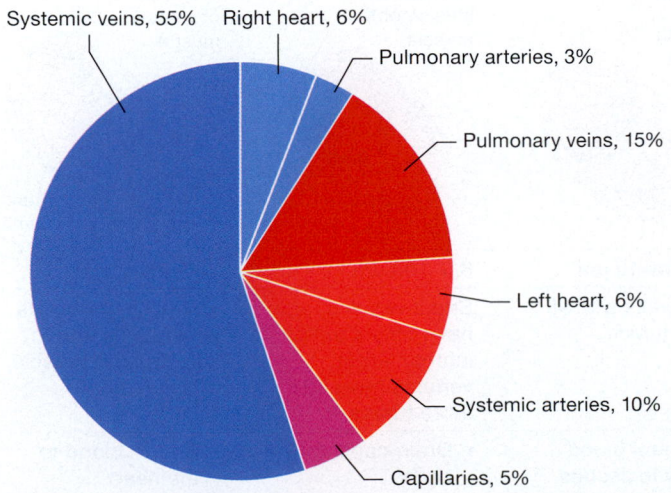

Systemic veins, 55%   Right heart, 6%

Pulmonary arteries, 3%

Pulmonary veins, 15%

Left heart, 6%

Systemic arteries, 10%

Capillaries, 5%

Figure 18.3 Blood distribution in the cardiovascular system.

shows how much of the blood is typically found in veins and how much in other parts of the cardiovascular system. When necessary, blood can be diverted from the veins to other parts of the body. In the histological section in Figure 18.2, the vein appears collapsed, whereas the artery is circular—this is because veins typically have much thinner walls, fewer elastic fibers, less smooth muscle, and larger lumens than arteries.

The smallest veins are the **venules,** which drain blood from capillary beds. The tiny **postcapillary venules** consist of little more than endothelium and some surrounding connective tissue. This structure enables them to exchange material with the surrounding interstitial fluid in much the same way that materials are exchanged in capillaries. The three tunics become more distinct as venules merge to become larger venules and then veins. Most veins have a thin tunica media with few smooth muscle cells, and their diameter changes only slightly with vasodilation and vasoconstriction.

Many veins contain **venous valves.** These valves are extensions of the tunica intima that overlap and prevent blood from flowing backward in the venous circuit. They are especially numerous in the veins of the legs, where blood flow toward the heart is strongly opposed by gravity.

Table 18.1 on p. 674 summarizes information on the types of arteries and veins.

Quick Check

☐ 6. How do veins differ structurally and functionally from arteries?

☐ 7. What are venules, and how do they differ from veins?

☐ 8. What are venous valves, and what are their functions?

Vascular Anastomoses

❮❮ FLASHBACK

1. What is an anastomosis? What purpose does this structure serve in the coronary circulation? (p. 643)

In the heart chapter you read about the **vascular anastomoses** (an-ass´-toh-MOH-seez) that are present within the coronary circulation (see Chapter 17). Recall that anastomoses are locations where vessels connect via pathways called **collateral vessels.** *Arterial anastomoses* exist in many organs such as the heart and the brain, as well as around joints. Interestingly, new arterial anastomoses can be formed when blood flow through an artery is insufficient to meet the tissue's metabolic needs. Tissues deprived of oxygen appear to secrete chemicals that trigger a process called *angiogenesis,* or the formation of new blood vessels. As these new arteries grow, they often connect parts of the circulation via collaterals and increase blood flow to the tissue.

The most common type of anastomosis is the *venous anastomosis,* in which neighboring veins are connected by small collaterals. Smaller veins are often so interconnected by collaterals that they form complex, weblike patterns that are

Atherosclerosis

The leading cause of death in the developed world is **atherosclerosis** (aeh-ther'-oh-skleh-ROH-sis), an arterial disease that affects large- and medium-sized muscular arteries. The disease is characterized by the formation of atherosclerotic **plaques** (PLAKS), which are buildups of lipids, cholesterol, calcium salts, and cellular debris within the arterial tunica intima, as shown here:

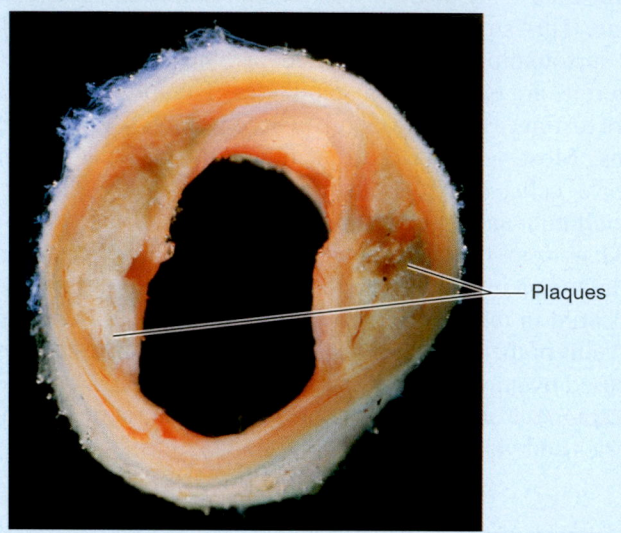

Plaques

Atherosclerotic plaques tend to form in regions where blood undergoes sudden changes in velocity and direction of flow, such as branching points or where the vessels curve.

Current evidence indicates that these plaques are generated in response to some sort of injury to the endothelium. The endothelium may be injured by high blood pressure, certain types of cholesterol, previous infections, the toxins in cigarette smoke, a high and prolonged blood glucose level, and a high blood concentration of a chemical called *homocysteine* (hoh'-moh-SIS-tuh-een). When the endothelium is injured, the vessel wall becomes inflamed, which attracts phagocytes that attempt to "clean up" the area. However, this inflammation actually results in damage to the blood vessel, and eventually the damaged area turns into a plaque.

Although the plaque is located in the tunica intima, changes also occur in the tunica media. Smooth muscle cells proliferate and secrete an extracellular matrix with protein fibers and ground substance that surrounds the plaque. If the contents of the plaque are exposed to blood, a clot may form in the vessel and obstruct blood flow entirely, as occurs in a myocardial infarction (heart attack) or stroke.

An estimated 10% of the world population has atherosclerosis, but its true prevalence is difficult to determine because most patients don't display symptoms in the early stages. The treatment of atherosclerosis primarily focuses on reducing factors that injure the endothelium. This includes dietary modification, physical activity, agents to lower cholesterol, control of blood glucose level, smoking cessation, and management of high blood pressure. In severe disease, surgery or other invasive procedures may be necessary to open or bypass occluded vessels.

Table 18.1 Types of Arteries and Veins

	Elastic arteries	Muscular arteries	Arterioles	Venules	Veins
Diameter	2.5–1.0 cm	1.0 cm–0.3 mm	0.3 mm–10 μm	8.0–100 μm	100 μm–1.5 cm
Structure	Large arteries with well-developed elastic laminae	Thick-walled arteries with a well-developed tunica media	Thin walls with all three tunics	Small venules have only a tunica intima; larger venules have all three tunics	Thin-walled vessels with a large lumen, little smooth muscle, and valves
Function(s)	• Conduct blood under high pressure to organs	• Control blood flow to organs • Regulate blood pressure	• Control blood flow to tissues • Feed capillary beds	• Drain capillary beds	• Return blood to the heart

sometimes visible beneath the skin, particularly in individuals with pale skin. A third type of anastomosis is the *arteriovenous anastomosis,* in which an artery empties directly into a vein without passing through a capillary bed. Examples of arteriovenous anastomoses are found in the skin and in the fetal circulation, where blood needs to be shunted between the arterial and venous systems to bypass certain organs.

Quick Check

☐ 9. What is an anastomosis?

Apply What You Learned

☐ 1. Predict the possible effects of defective elastic fibers in arteries.

☐ 2. Would you expect the same effects if venous elastic fibers were defective? Why or why not?

☐ 3. How does atherosclerosis change the structure of an artery? How does this impair the artery's ability to perform its functions?

See answers in Appendix A.

MODULE **18.2**
Physiology of Blood Flow

Learning Outcomes

1. Describe the factors that influence blood flow, blood pressure, and peripheral resistance.

2. Explain the relationships between vessel diameter, cross-sectional area, blood pressure, and blood velocity.

3. Explain how blood pressure varies in different parts of the systemic and pulmonary circuits.

4. Describe how blood pressure changes in the arteries, capillaries, and veins.

5. Explain how mean arterial pressure is calculated.

6. Describe the mechanisms that assist in the return of venous blood to the heart.

Public awareness of blood pressure has increased dramatically over the past two decades. Mountains of research have shown the dramatic effects that high blood pressure, or *hypertension,* has on heart disease, vascular disease, kidney disease, stroke risk, and more. But what exactly is blood pressure, and why does it affect the health of so many body systems? To answer these questions, we now turn our attention to *hemodynamics* (hee′-moh-dy-NAM-iks; "blood power"), or the physiology of blood flow in the cardiovascular system. In this module we first explore the basic concepts of circulation and define terms such

as blood pressure and blood flow. The discussion then shifts to those factors that determine blood pressure, to the blood pressure in different circuits, and to types of vessels in the cardiovascular system.

Introduction to Hemodynamics

◀◀ FLASHBACK

1. What is a gradient? (p. 26)

2. What is cardiac output? What is the average value for cardiac output? (p. 662)

Before we venture into the physiology of circulation, we need to address some basic concepts related to it, such as gradients. Recall that gradients—whether concentration gradients, pressure gradients, or electrical gradients—drive most processes in the body. The physiology of the cardiovascular system is no exception. As you learned in the last chapter, the pumping heart provides the force that drives blood through the vasculature. In an example of the Gradients Core Principle (p. 28), the heart drives 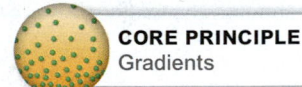 **CORE PRINCIPLE** Gradients blood through these vessels by creating a *pressure gradient.* The pressure is highest near the heart and decreases as we move away from it. Blood flows with this pressure gradient from the area of higher pressure near the heart to the area of lower pressure in the peripheral vasculature.

But what exactly is "pressure" with respect to blood? Defined simply, **blood pressure** is the outward force that the blood exerts on the walls of the blood vessels. Blood pressure is expressed in the units *millimeters of mercury,* or *mm Hg,* which is the force exerted by a column of mercury one millimeter in height. So, if the blood pressure is 40 mm Hg, the pressure in the blood vessel is equal to that generated by a column of mercury 40 mm in height. The blood pressure varies dramatically in different parts of the vasculature—it is highest in the large systemic arteries and lowest in the large systemic veins.

The magnitude of the blood pressure gradient is one main factor that determines the **blood flow,** or the volume of blood that flows per minute. Generally, blood flow matches cardiac output, which averages about 5–6 liters/min. Blood flow is *directly proportional* to the pressure gradient, meaning that blood flow increases when the pressure gradient increases and vice versa. The other factor that determines blood flow is **resistance,** defined as any impedance to blood flow. Blood flow is *inversely proportional* to resistance; that is, as resistance increases, blood flow decreases.

The velocity with which blood flows is largely determined by the cross-sectional area of the blood vessel. As the arterial system branches into more, progressively smaller vessels, the total cross-sectional area increases. This increase in area causes the velocity of blood flow to decrease. For this reason, the velocity of blood flow is fastest in the aorta and slowest in the capillaries.

ConceptBOOST))))

A Closer Look at Cross-Sectional Area and Velocity

The idea of a vessel's cross-sectional area may sound complex, but it's actually a combination of two fairly simple things. First, it refers to something from the first chapter of this book—a cross section, or transverse section, through a blood vessel. The "area" (A) is simply the calculation of the area of the circle made by the cross section. It is equal to pi (π) times the radius of the circle (r) squared (or $A = \pi r^2$; the radius is one-half the circle's diameter).

The radius of an arteriole or capillary is quite small, so the cross-sectional area of such a vessel will also be very small. However, these small vessels are far more numerous than large vessels with a large cross-sectional area. For this reason, the *total* cross-sectional area of the small vessels is much greater than that of the large vessels. You can see this in the very simple example below, in which our one large vessel has a cross-sectional area of 5 cm², and each of our five smaller vessels has an area of 2 cm²:

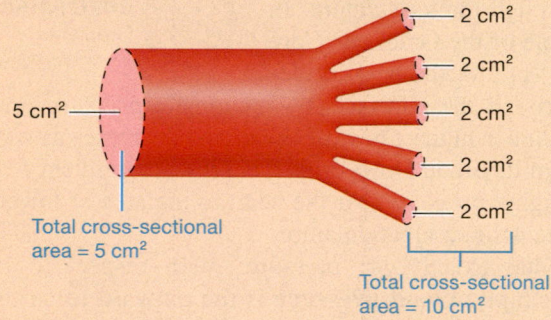

As you can see, the total cross-sectional area of the small vessels (10 cm²) is twice that of our large vessel. So, it's easy to see why cross-sectional area increases as arteries branch into millions of smaller vessels.

Let's apply this to velocity, which decreases as cross-sectional area increases.* This is because the same volume of blood is now filling the equivalent of a much larger container:

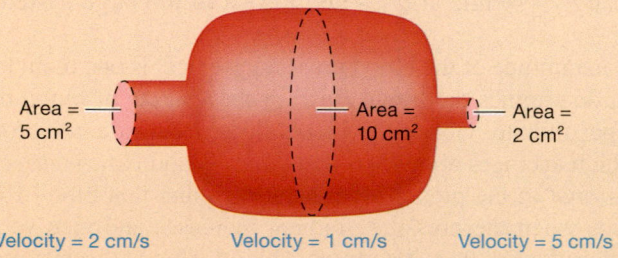

*Note that we are assuming a flow rate of 10 ml/s in determining velocity.

You can see that the blood flows fastest in the small container and slowest in the large one. This happens because blood flowing through a larger container is spread out over a wider area. So when cross-sectional area increases as vessels branch, the blood is more spread out, causing it to flow more slowly. This is important because a slower velocity of blood flow allows efficient gas and nutrient exchange in capillaries. ■

Quick Check

☐ 1. What is hemodynamics?

☐ 2. How are blood flow and the pressure gradient related? How are blood flow and resistance related?

☐ 3. How does cross-sectional area influence the velocity of blood flow?

Factors That Determine Blood Pressure

⟪ FLASHBACK

1. What are stroke volume and heart rate? (p. 662)
2. What is contractility, and how do changes in contractility affect cardiac output? (p. 663)

Multiple variables interact to determine the blood pressure in various parts of the vasculature. The three main factors that influence blood pressure are *resistance, cardiac output,* and *blood volume.* As we discuss in the next module, the body maintains blood pressure at a relatively constant set point by adjusting one or more of these three factors. Each variable is explored next; **Figure 18.4** provides an overview.

Peripheral Resistance

Anything that hinders blood flow through the vasculature contributes to the overall resistance of the circuit. The vessels near the heart are generally large, elastic vessels that contribute little to overall resistance. Instead, most resistance is encountered away from the heart, in the body's *periphery,* which has given rise to the term **peripheral resistance.** Peripheral resistance and blood pressure are directly related: *As peripheral resistance increases, blood pressure increases.*

Resistance is mainly determined by three variables: blood vessel radius, blood viscosity, and blood vessel length. Of these variables, vessel radius is the quickest to change, whereas vessel length is, of course, the slowest. Let's examine each more closely.

- **Blood vessel radius.** Vessel radius dramatically affects resistance. Resistance varies *inversely* with the vessel's radius (remember from basic math that radius is just one-half of the diameter of a circle, in this case a transverse section of the vessel). So, as a vessel's radius increases (i.e., the vessel dilates), the resistance to blood flow decreases, and vice versa.

- **Blood viscosity. Viscosity** is often described as the "thickness" of a liquid, but its technical definition is the inherent resistance that all liquids have to flow. The more viscous a liquid is, the more its molecules resist being put into motion and staying in motion. Blood has a relatively high viscosity due to the number of proteins and cells it contains. Generally, blood viscosity remains relatively constant, but it can be altered by states that change either the number of cells or proteins in the blood or the amount of water in the blood, such as dehydration. Peripheral resistance is raised by conditions that increase blood viscosity and is lowered by conditions that decrease blood viscosity. This makes

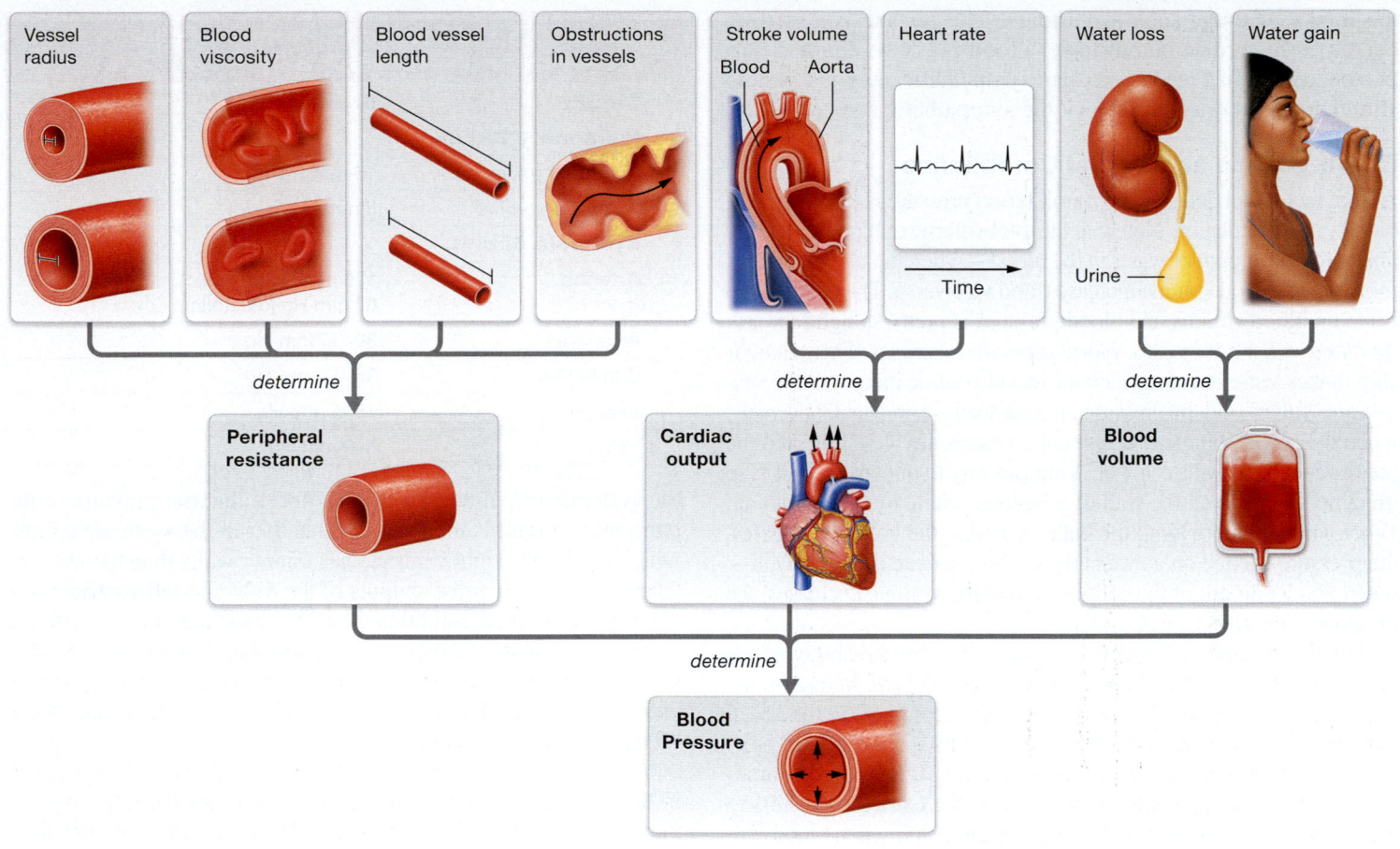

Figure 18.4 Factors that determine blood pressure.

▶ Play Interactive Physiology 2.0
@ **Mastering** Anatomy & Physiology

sense—the flow in a pipe filled with maple syrup (higher viscosity) is going to be much slower than the flow in a pipe of the same diameter filled with water (lower viscosity).

- **Blood vessel length.** The length of a blood vessel also influences resistance. Stated simply, the longer the blood vessel, the greater the resistance. Just as more pressure is required to propel water through a long hose than a short one, more pressure is needed to propel blood through a long vessel than a short one. This is one reason why resistance in the pulmonary circuit is so much lower than in the systemic circuit: The vessels in the pulmonary circuit are simply shorter and therefore offer less resistance to blood flow. This also explains in part why blood pressure rises with obesity—vessel length increases as body size increases.

A fourth factor that can influence peripheral resistance is the presence of obstructions within the blood vessels that are caused by certain disease states. Obstructions in the vessel, such as tumors, the fatty plaques seen with atherosclerosis, or blood clots, affect the way in which blood flows through the vessel. Ideally, blood flow through a vessel should be *laminar* (*laminar* = "layered"), in which the layer of blood nearest the blood vessel wall adheres to the wall due to friction (resistance) but the blood in the center of the vessel flows more freely. The flow becomes *turbulent* when there are obstructions within the vessel. In turbulent flow, blood doesn't move forward readily and requires more

force to move through the vessels, which increases resistance. Although laminar blood flow produces no audible sound, turbulent blood flow often does as the blood bounces along the vessel walls. These sounds are heard in the heart as murmurs and in the vessels as *bruits* (BROO-eez; *bruit* = "noise").

Cardiac Output

Recall that cardiac output (CO) is the product of *stroke volume* (the amount of blood pumped with each beat) times *heart rate* (the number of beats per minute) (look at Figure 18.4 again). Cardiac output and peripheral resistance are the two factors that determine the pressure gradient driving circulation. The relationship is expressed by the simple mathematical equation:

$$\underset{\substack{\textit{Pressure} \\ \textit{gradient}}}{\Delta P} = \underset{\substack{\textit{Cardiac} \\ \textit{output}}}{CO} \times \underset{\substack{\textit{Peripheral} \\ \textit{resistance}}}{PR}$$

Note here that the symbol Δ (Greek delta) signifies change, meaning that a pressure change is caused by altering cardiac output and/or peripheral resistance. This relationship indicates that anything changing cardiac output is also going to alter the pressure gradient that drives blood flow and therefore the blood pressure. So, generally speaking, *when cardiac output increases, blood pressure increases.* Such an increase occurs, for example, in response to sympathetic nervous system stimulation and the intake of drugs such as caffeine. The reverse is also true; decreases in cardiac output lower blood pressure. Cardiac output decreases in response to parasympathetic nervous system stimulation and drugs that block the sympathetic response.

Blood Volume and Vessel Compliance

A final factor that determines overall blood pressure is the volume of blood in the circulation. Note that the total volume of blood is directly linked to the amount of water in the blood—when the blood contains more water, blood volume increases, and vice versa. The relationship between blood volume and blood pressure is pretty straightforward: *As blood volume increases, blood pressure increases.* This relationship makes sense because when the blood volume increases, the vessels are "fuller" and the pressure on their walls is greater. Conversely, when the blood volume decreases, the vessels are "less full" and the pressure on their walls is lower. A simple way to understand this is to imagine the stress on the wall of a balloon when you fill it with air. When you add more air to the balloon so that the volume increases, the pressure exerted on its walls rises. The opposite is true as well—when you let air out of the balloon so that the volume decreases, the pressure exerted on its walls falls.

Small increases in blood volume are offset by the ability of the vessels to stretch, a property known as *compliance.* Veins are the most compliant vessels, and they stretch to accommodate the added fluid when blood volume increases, with only a small rise in pressure. However, when the veins cannot stretch further to accommodate additional increases in blood volume, the extra blood shifts to the arteries. Arteries are much less compliant, and when blood volume increases in arteries, overall blood pressure rises. Anything that decreases the compliance of arteries or veins, such as normal changes that occur with aging, makes the vasculature less able to adapt to increases in blood volume. So, when compliance decreases, even small increases in blood volume can raise the blood pressure.

Quick Check

☐ 4. What three main factors determine blood pressure? How does each factor influence it?

☐ 5. How does a change in vessel diameter affect peripheral resistance?

☐ 6. What is vessel compliance, and which vessels are the most compliant?

Blood Pressure in Different Portions of the Circulation

 **FLASHBACK**

1. Which ventricle is stronger and thicker? Why? (p. 639)

2. Define systole and diastole. (p. 656)

The term blood pressure is fairly generic and can refer to the pressure anywhere within

Table 18.2 Pressures in Pulmonary and Systemic Circuits	
Circuit	**Pressure**
Pulmonary Circuit	
Pulmonary arteries	15 mm Hg
Pulmonary veins	5 mm Hg
Systemic Circuit	
Arteries	120 mm Hg (systolic), 80 mm Hg (diastolic)
Arterioles	80–35 mm Hg
Capillaries	35–15 mm Hg
Venules	15–5 mm Hg
Veins	5–0 mm Hg

the systemic or pulmonary circuit. Recall that the pressure in the pulmonary circuit is much lower than that in the systemic circuit, which is why the right ventricle has thinner walls than the left (see Chapter 17). The cardiac outputs of the right and left ventricles are equal under normal conditions, but the resistance in the systemic circuit is far greater than that in the pulmonary circuit. For this reason, the pressures in the two circuits are dramatically different, averaging about 15 mm Hg in the pulmonary circuit and about 95 mm Hg in the systemic circuit.

As you can see in Table 18.2, the pressure doesn't change much in the pulmonary circuit—it remains fairly low from the pulmonary artery to the pulmonary veins. However, the pressure does change significantly as blood travels through the systemic circuit (**Figure 18.5**). We examine the pressure changes in the vessels of the systemic circuit more closely in the next subsections.

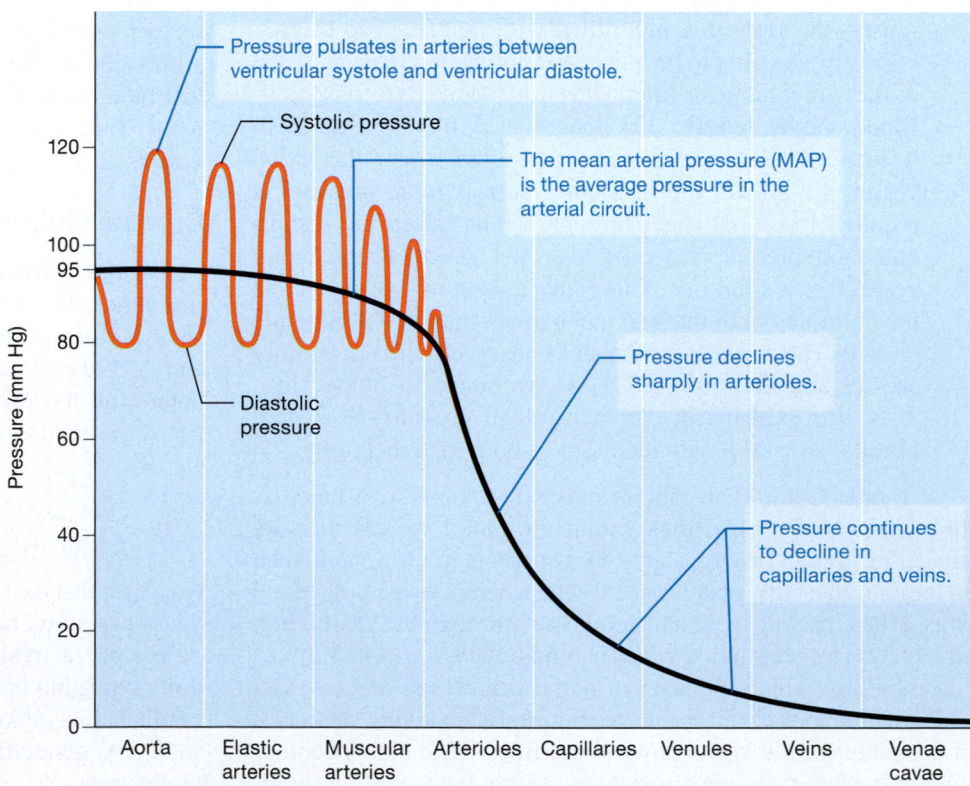

Figure 18.5 Pressure profile of the systemic circuit.

Systemic Arterial Pressure

Generally, when someone reports his or her "blood pressure," the person is actually reporting the pressure in the systemic arterial circuit. Figure 18.5 shows the pressure profile of the entire systemic circuit. Notice that it is highest in the aorta and the elastic arteries and that it declines slightly as it spreads throughout the muscular arteries. The black line in the graph shows the **mean arterial pressure,** or **MAP,** which is the average pressure in the systemic arteries during an entire cardiac cycle. The MAP generally measures about 95 mm Hg.

However, because the heart has both contraction and relaxation periods, the pressure gradient generated by the heart *pulsates:* It rises during ventricular systole and declines during ventricular diastole. This leads to two separate pressures in the arteries: a **systolic pressure,** which averages about 110–120 mm Hg, and a **diastolic pressure,** which averages about 70–80 mm Hg when the person is at rest. These pulsations are represented in Figure 18.5 by the orange wavy line that overlies the MAP line. The difference between the systolic and diastolic pressures—about 40 mm Hg—is known as the **pulse pressure.**

ConceptBOOST 》))

Taking a Closer Look at Systolic and Diastolic Pressures

Before moving on, let's slow down and take a minute to think about what the term blood pressure really means. Say we have a very distensible artery that contains no blood—it's completely empty, as in part (a) of the figure shown here. That artery will take on a certain "resting" size due to the smooth muscle in its walls.

But now say we add a really large volume of blood into the vessel, as in part (b). See what happens? The artery walls stretch and it changes size. Why does this happen? Because the *blood* is putting *pressure* on the artery walls. Next take the same volume of blood and force it into the artery with a pump, as in part (c). What does the artery look like now? It has stretched more, because the blood is putting even more pressure on the artery.

Pressure from pump

(a) Vessel at relaxed size, without blood **(b) Vessel filled with blood** **(c) Vessel filled with blood with pressure source driving blood into the vessel**

Let's apply this to the circulation. When the left ventricle of the heart is in diastole, the arteries resemble the artery in part (b). The ventricle isn't forcing blood into them, so the blood puts less pressure on their walls. This is why the diastolic pressure in a blood pressure reading is the smaller number. Conversely, when the left ventricle is in systole, the arteries resemble the artery in part (c). The ventricle is forcing more blood into them, and so the blood puts more pressure on their walls. This is why the systolic pressure in a blood pressure reading is the larger number.

Notice something else about the illustration: See how the vessel changes size between parts (b) and (c)? The artery is smaller during diastole, then larger during systole. This size change repeats with each cardiac cycle. We said that the pressure gradient generated by the heart pulsates, but the artery itself also pulsates. You can, of course, feel this in certain arteries close to the skin. The pulsation of the vessel explains why we call these points *pulses.*

Let's think about one final thing. When we consider blood pressure in terms of blood actually pushing against vessel walls, it becomes easier to understand the factors that determine blood pressure:

- *Cardiac output.* If cardiac output rises, the ventricles are going to force more blood into the arteries, and that will make the blood press more forcefully against the vessel walls.
- *Blood volume.* If we increase blood volume, it's like trying to add more air to a balloon. The vessel wall will stretch to an extent to accommodate the extra blood, but eventually the vessel won't be able to expand anymore and the blood will simply put pressure on its wall.
- *Resistance.* If we decrease the size of the vessels, but have the same volume of blood, it's like trying to fit a certain volume of air into a too-small balloon. The blood doesn't have as much "room" to move, and it presses against the vessel walls with greater force. ■

Pressure in the systemic circuit declines most sharply in the arterioles. As you can see in Figure 18.5, MAP decreases from about 80 mm Hg in the large arterioles to about 30 mm Hg in the smallest arterioles. Pulsations are generally measurable in the large arterioles but decline in smaller arterioles until they are imperceptible. This sharp decrease in pressure is due to the sharp increase in peripheral resistance in the arterioles. This may sound counterintuitive because you read earlier that increases in peripheral resistance raise blood pressure. But think about what happens when you kink a garden hose, which increases its resistance to flow. Pressure increases upstream from the kink (closer to the spigot), but pressure decreases downstream from the kink, which is evident from the fact that water flow slows or even stops downstream.

Measuring the Arterial Blood Pressure The arterial blood pressure is usually measured in the arm with an instrument called a *sphygmomanometer* (sfig′-moh-mah-NAW-muh-ter) and a stethoscope. A cuff is wrapped around the arm and inflated to a pressure higher than the expected systolic pressure. This cuts off blood flow through the main artery in the arm, the *brachial* artery (BRAY-kee-ul). The pressure in the cuff is then slowly decreased and blood flow resumes through the artery when the cuff pressure reaches the systolic pressure. Normally, flow through the artery is laminar and silent. However, when blood flow through the brachial artery resumes at the systolic pressure, it becomes turbulent. This produces audible *sounds of Korotkoff* that may be auscultated (listened to) with a stethoscope. The pressure at which the sounds are first auscultated is recorded as the systolic pressure. The sounds continue to be audible until the diastolic pressure is reached, at which point they become muffled and then disappear. The pressure at which the sounds are very soft is considered the most accurate estimate of diastolic pressure, although diastolic pressure is often recorded at the point when the sounds disappear.

Calculating the Mean Arterial Pressure MAP is difficult to measure directly, but we can use the measured systolic and diastolic pressures to estimate the MAP. You might expect the MAP to simply equal the average of the diastolic and systolic pressures. However, the heart spends more time in diastole than in systole, and this affects the MAP. For this reason, we calculate MAP by using the following equation:

$$MAP = \text{diastolic pressure} + \frac{1}{3}(\text{systolic pressure} - \text{diastolic pressure})$$

So we can say that the MAP is approximately equal to the diastolic pressure plus one-third of the pulse pressure. For a person with a blood pressure of 120/80, this gives us the following calculation:

$$MAP = 80\,mm\,Hg + \tfrac{1}{3}(120\,mm\,Hg - 80\,mm\,Hg) = 93\,mm\,Hg$$

Notice that this is close to the average value for MAP of about 95 mm Hg, mentioned earlier.

Systemic Capillary Pressure

Pressure continues to decline throughout the remainder of the systemic circuit. You can see in Figure 18.5 how the pressure changes through a capillary bed. At the arteriolar end, pressure is approximately the same as that in the small arterioles—about 35 mm Hg. However, at the venular end of the capillary bed, the pressure has decreased to about 15 mm Hg. This pressure decrease is largely due to the reduction in blood volume that takes place in capillaries (see Module 18.5).

Systemic Venous Pressure and Mechanisms of Venous Return

Pressure declines even further in venules and veins, dropping to only about 4 mm Hg in the inferior vena cava and as low as 0 mm Hg in the right atrium. The low pressure is largely due to the high compliance of veins and the declining resistance as these vessels merge and become larger.

Venous blood must be returned to the heart at the same rate it is pumped into the arteries. But the venous circuit is under such low pressure that there isn't much of a driving force to propel venous blood back to the heart. Veins have a higher cross-sectional area than arteries, which makes the flow of venous blood slower than that of arterial blood. Indeed, in some locations blood could even potentially flow backward, particularly in those veins that flow

against gravity. See *A&P in the Real World: Varicose Veins* for a common problem related to decreased venous return.

You have already read about two of the mechanisms that help return venous blood to the heart: (1) venous valves that prevent backward flow in the veins, and (2) smooth muscle in the walls of veins that may contract under sympathetic nervous system stimulation to increase the rate of venous return. Another mechanism that assists venous return is the *skeletal muscle pump* (**Figure 18.6**). Skeletal muscles surrounding the deeper veins of the upper and lower limbs squeeze the blood in the veins and propel it proximally (toward the heart) as they contract and relax.

There is a mechanism similar to the skeletal muscle pump for veins of the thoracic and abdominopelvic cavities that are not surrounded by skeletal muscles. This mechanism, known as the *respiratory pump,* is driven by the rhythmic changes in pressure in the thoracic and abdominopelvic cavities that occur with ventilation. During inspiration, high pressure in the abdominopelvic cavity creates a pressure gradient that pushes blood in the abdominal veins upward, and low pressure in the thoracic cavity allows thoracic veins to expand. The reverse occurs during expiration—abdominal veins expand and fill with blood while thoracic veins are squeezed.

Quick Check

☐ 7. How does mean arterial pressure differ from systolic pressure and diastolic pressure?

☐ 8. How does the pressure change in arterioles, capillaries, and veins? Why?

☐ 9. What are the mechanisms that assist in returning venous blood to the heart?

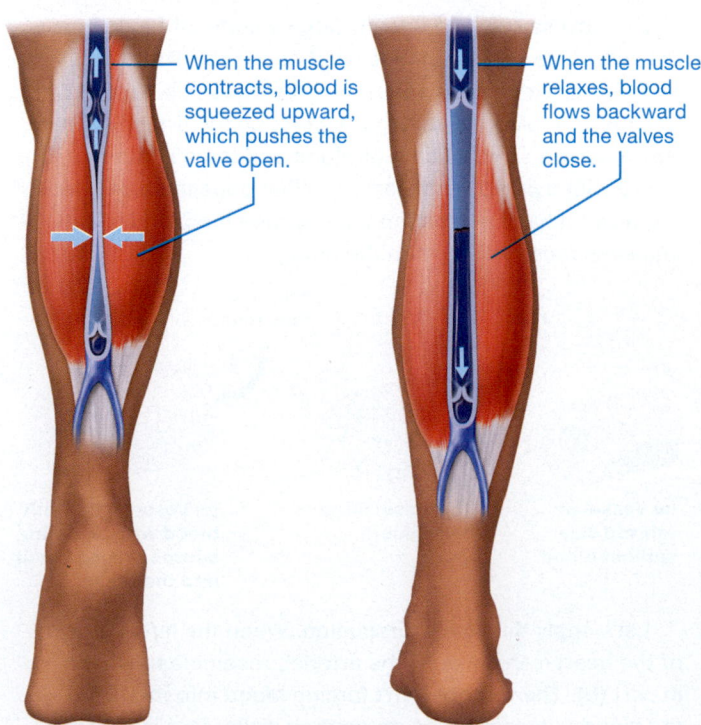

When the muscle contracts, blood is squeezed upward, which pushes the valve open.

When the muscle relaxes, blood flows backward and the valves close.

Figure 18.6 The skeletal muscle pump.

Varicose Veins

A&P in the Real World

The common condition of **varicose veins** is characterized by dilated, bulging, and often hardened veins. Varicose veins are frequently located in the superficial veins of the lower limb. Conditions that decrease the rate of venous return, such as pregnancy, standing upright for prolonged periods, and abdominal obesity, cause blood to pool in the veins of the lower limb. There also appears to be a genetic predisposition to the development of varicose veins. The lower limb's superficial veins are not supported by skeletal muscle pumps, so the extra blood volume stretches them, causing enlarged regions that are visible beneath the skin.

Another common location for varicose veins is around the anus, where they are called **hemorrhoids** (HEM-uh-roydz). High pressure in the abdominopelvic cavity during defecation or childbirth decreases the return of venous blood from the anal veins. These veins are also superficial and not well supported by surrounding tissues, and thus may weaken and dilate because of the high pressure.

Apply What You Learned

☐ 1. What effect would nicotine, which is a vasoconstrictor, have on peripheral resistance? Considering this effect, explain why cigarette smoking is a risk factor for developing high blood pressure (hypertension).

☐ 2. Diseases such as *arteriosclerosis* that harden the arteries cause a loss of arterial compliance. Predict the effect this would have on blood pressure.

☐ 3. Predict what would happen to a person's blood pressure in the case of severe blood loss.

See answers in Appendix A.

MODULE 18.3
Maintenance of Blood Pressure

Learning Outcomes

1. Describe the role of arterioles in regulating tissue blood flow and systemic arterial blood pressure.

2. Describe the local, hormonal, and neural factors that affect and regulate blood pressure.

3. Explain the main effects and importance of the baroreceptor reflex.

4. Explain how the respiratory and cardiovascular systems maintain blood flow to tissues via the chemoreceptor reflex.

5. Describe common causes of and treatments for hypertension.

A sufficient pressure gradient must be present at all times for blood flow to meet the body's needs. The mean arterial pressure must be constantly maintained at about 95 mm Hg, and any deviation from this set point triggers nervous, endocrine, and renal mechanisms that act to restore blood pressure to this level. We discuss these mechanisms in this module.

Short-Term Maintenance of Blood Pressure

« FLASHBACK

1. What effects does the sympathetic nervous system have on blood pressure? (p. 524)

2. How does the parasympathetic nervous system affect blood pressure? (p. 594)

Short-term control of blood pressure, such as that needed when you stand up from a sitting position, is primarily accomplished by the nervous system and certain hormones of the endocrine system. Such short-term effects are generally achieved by adjustment of resistance and cardiac output.

Nervous System Maintenance of Blood Pressure

Recall that the main branch of the nervous system in control of maintaining homeostasis is the autonomic nervous system (ANS) via its two divisions: the sympathetic and parasympathetic nervous systems (see Chapter 14). Both divisions have immediate effects on blood pressure. The upcoming subsections discuss these effects, as well as a reflex mediated by the ANS, called the *baroreceptor reflex,* and the effects mediated by vessel chemoreceptors.

Sympathetic and Parasympathetic Effects on Blood Pressure As we discussed in the ANS chapter, the sympathetic nervous system is the "fight or flight" division of the ANS that prepares the body for stressors such as exercise, even mild exercise like position changes (see Chapter 14). In **Figure 18.7a**, axons release the neurotransmitters **norepinephrine** and **epinephrine** onto cardiac muscle cells and the smooth muscle cells of blood vessels, to produce two immediate changes: (1) an increase in heart rate and contractility, which increases cardiac output; and (2) vasoconstriction of all types of vessels, but especially arterioles, which increases peripheral resistance. Both changes increase blood pressure. Note that primarily blood vessels serving and draining the digestive and urinary organs and the skin are constricted. The vessels serving the active skeletal muscles and heart actually dilate. This is critical during exercise, as it allows increased delivery of oxygen and nutrients to these tissues.

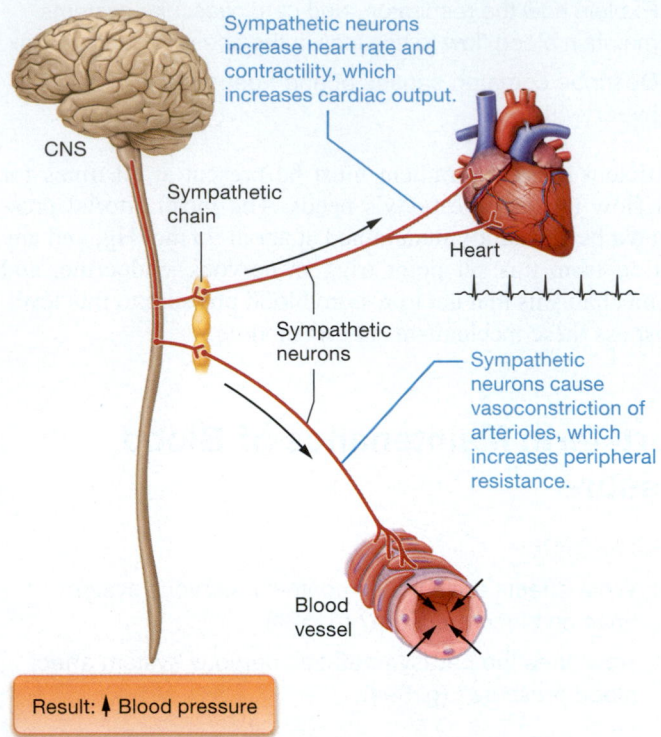

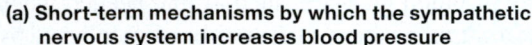

Result: ↑ Blood pressure

(a) Short-term mechanisms by which the sympathetic nervous system increases blood pressure

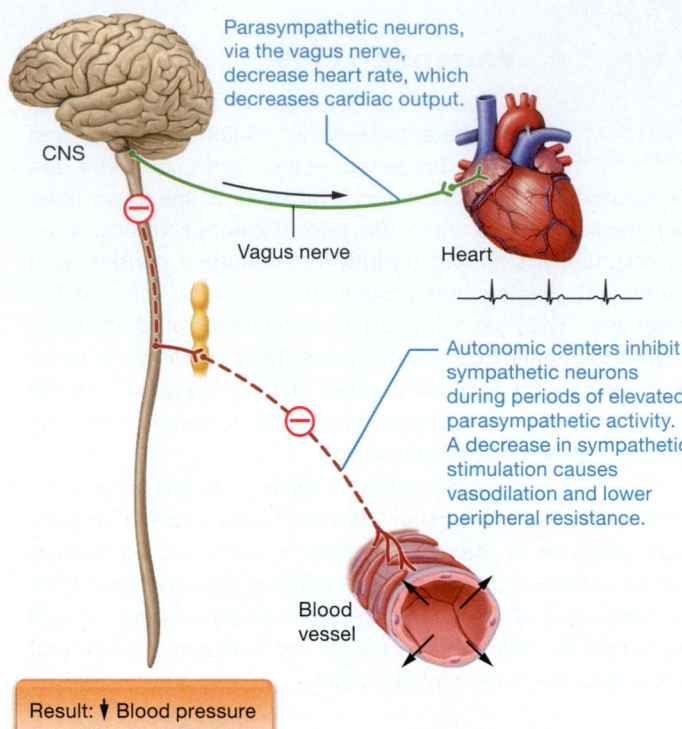

Result: ↓ Blood pressure

(b) Short-term mechanisms by which the parasympathetic nervous system decreases blood pressure

Figure 18.7 Effects of the autonomic nervous system on blood pressure.

The effects of the parasympathetic nervous system on blood pressure are opposite those of the sympathetic nervous system (**Figure 18.7b**). The axons of the parasympathetic system, via the vagus nerve, release **acetylcholine** primarily onto certain cardiac pacemaker cells (particularly the sinoatrial [SA] and atrioventricular [AV] nodes) and cardiac muscle cells. This slows the heart rate and has a mild effect on contractility, which decreases cardiac output and so blood pressure. Most axons of the parasympathetic nervous system do not innervate the smooth muscle cells of blood vessels. However, an increase in parasympathetic activity does allow vasodilation and a decrease in peripheral resistance. This effect occurs because autonomic centers in the brainstem increase the firing of parasympathetic neurons while inhibiting the firing of sympathetic neurons. When the sympathetic neurons are inhibited, the vessels dilate.

The Baroreceptor Reflex Blood pressure is monitored by the ANS with specialized mechanoreceptors in the vessels, called *baroreceptors.* Earlier in the chapter we described the baroreceptors located in areas of the common carotid artery and the aortic arch, the **carotid sinus** and **aortic sinus,** respectively. These receptors depolarize and fire rapid action potentials in response to pressure exerted on the arterial wall; this triggers the baroreceptor reflex.

The **baroreceptor reflex** arc, another example of the Feedback Loops Core Principle (p. 28), is a negative feedback loop

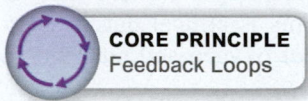

CORE PRINCIPLE
Feedback Loops

that responds to an increase in blood pressure. The reflex arc, shown in **Figure 18.8a**, proceeds by the following steps.

- **Stimulus: Blood pressure increases above normal range.**
- **Receptor: Baroreceptors in the carotid sinus (and aortic sinus) detect the increased pressure and fire action potentials at a faster rate.** An increase in blood pressure stretches the arteries. This stretch is detected by the baroreceptors, and they depolarize at an increased rate. This produces increasingly frequent action potentials.
- **Control center: The impulses travel to the medulla of the brainstem for integration.** The carotid sinus is innervated by cranial nerve IX, the glossopharyngeal nerve. Impulses travel via its parasympathetic afferent neurons to the control center for this negative feedback loop: the cardiovascular center in the medulla oblongata, where the impulses are integrated.
- **Effector/response: Autonomic centers in the medulla inhibit sympathetic activity, inducing vasodilation and decreased heart rate, lowering cardiac output.** Autonomic centers in the medulla inhibit sympathetic neurons from firing. This decrease in sympathetic activity causes vasodilation (note that parasympathetic neurons have no direct effect on most blood vessels). In addition, they stimulate vagal parasympathetic neurons that decrease cardiac output.
- **Homeostatic range and negative feedback: Blood pressure decreases, and negative feedback decreases response from the medulla.** These effects decrease both cardiac output and peripheral resistance, so blood pressure decreases to the normal range. When the baroreceptors are strongly stimulated, the decrease in cardiac output and blood pressure can be dramatic—the heart can nearly come

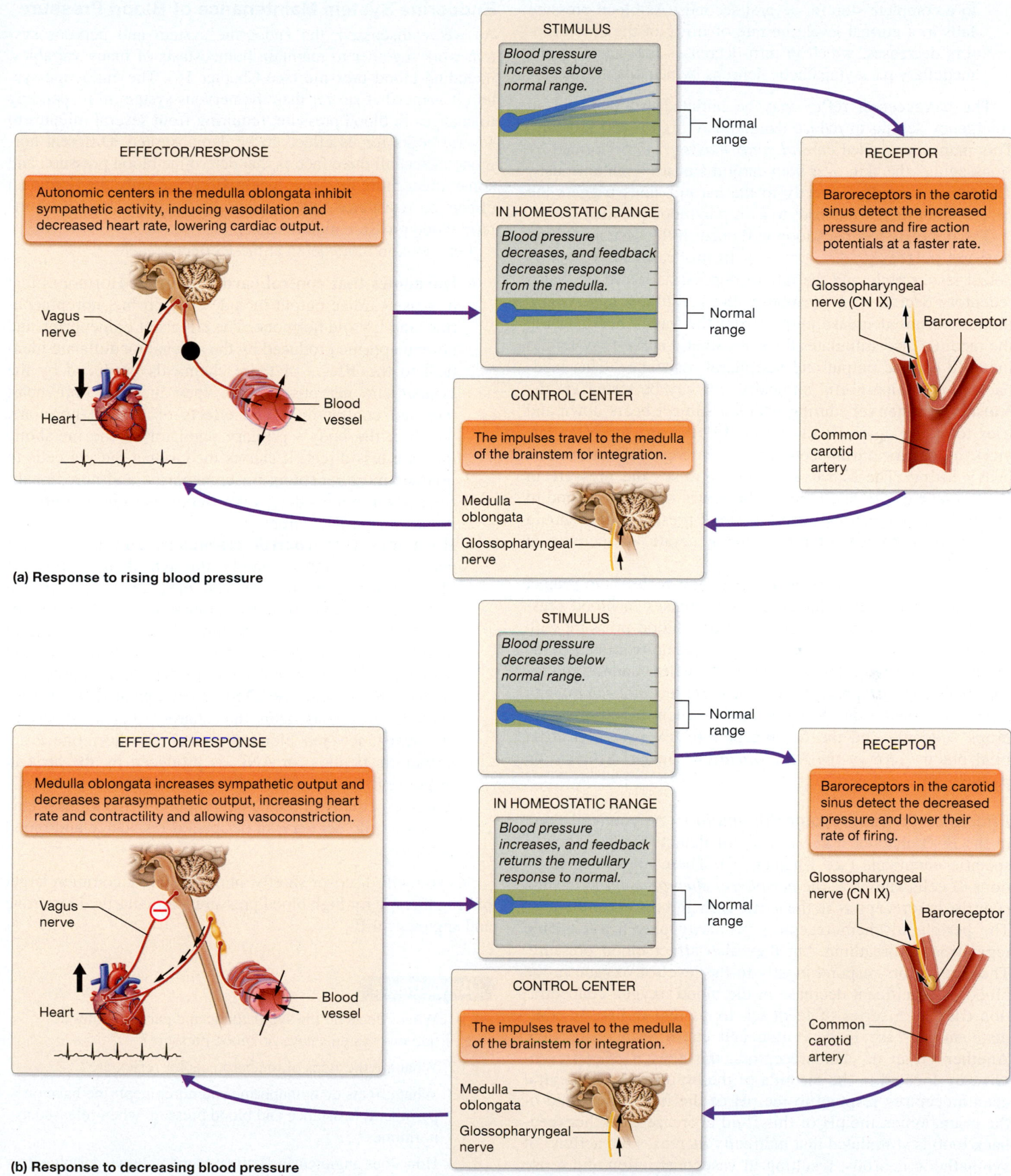

STIMULUS

Blood pressure increases above normal range.

— Normal range

RECEPTOR

Baroreceptors in the carotid sinus detect the increased pressure and fire action potentials at a faster rate.

Glossopharyngeal nerve (CN IX)

Baroreceptor

Common carotid artery

EFFECTOR/RESPONSE

Autonomic centers in the medulla oblongata inhibit sympathetic activity, inducing vasodilation and decreased heart rate, lowering cardiac output.

Vagus nerve

Heart

Blood vessel

IN HOMEOSTATIC RANGE

Blood pressure decreases, and feedback decreases response from the medulla.

— Normal range

CONTROL CENTER

The impulses travel to the medulla of the brainstem for integration.

Medulla oblongata

Glossopharyngeal nerve

(a) Response to rising blood pressure

STIMULUS

Blood pressure decreases below normal range.

— Normal range

RECEPTOR

Baroreceptors in the carotid sinus detect the decreased pressure and lower their rate of firing.

Glossopharyngeal nerve (CN IX)

Baroreceptor

Common carotid artery

EFFECTOR/RESPONSE

Medulla oblongata increases sympathetic output and decreases parasympathetic output, increasing heart rate and contractility and allowing vasoconstriction.

Vagus nerve

Heart

Blood vessel

IN HOMEOSTATIC RANGE

Blood pressure increases, and feedback returns the medullary response to normal.

— Normal range

CONTROL CENTER

The impulses travel to the medulla of the brainstem for integration.

Medulla oblongata

Glossopharyngeal nerve

(b) Response to decreasing blood pressure

Figure 18.8 **Maintaining homeostasis: regulation of blood pressure by the baroreceptor reflex.**

to a complete stop for several seconds. As blood pressure falls to a normal level, the rate of firing of the baroreceptors decreases, which in turn decreases the output of the medullary parasympathetic neurons by negative feedback.

The baroreceptor reflex may be initiated artificially in an emergency setting to reduce dangerously high blood pressure. This maneuver, called *carotid sinus massage,* is performed by "massaging" the skin over both carotid sinuses simultaneously. The pressure applied directly to the carotid sinus triggers the baroreceptor reflex and leads to a drop in blood pressure.

The opposite feedback loop will occur if the systemic blood pressure decreases (**Figure 18.8b**). In this case, the receptors detect less stretch, and the rate of depolarization of the baroreceptors decreases. In response, the medullary parasympathetic neurons decrease output from the vagus nerve, while the medulla also stimulates the sympathetic nervous system to increase cardiac output and peripheral resistance. This feedback loop is also useful clinically—it may be tested via the **Valsalva maneuver,** during which a subject bears down and tries to expire against a closed glottis (the airway in the larynx), as occurs during coughing, sneezing, defecation, and heavy lifting. The Valsalva maneuver raises the pressure in the thoracic cavity and reduces the return of venous blood to the heart, which causes a drop in blood pressure. This should trigger the baroreceptor reflex and generate an increase in heart rate.

The overall purpose of the baroreceptor reflex is to protect the body from sudden increases or decreases in blood pressure from moment to moment. It adjusts blood pressure with changes in position, such as when you get up from bed in the morning. However, the baroreceptor reflex cannot reduce chronic high blood pressure because these receptors adapt with repeated stimulation. What happens when blood pressure drops suddenly and the baroreceptor reflex isn't activated? Find out in *A&P in the Real World: Vasovagal Syncope* on p. 686.

Effects of Chemoreceptor Stimulation

As you read about in the PNS chapter, a *chemoreceptor* detects the presence of specific chemicals (see Chapter 13). There are two populations of cells known as the *peripheral chemoreceptors* located near the baroreceptors in the aortic arch and the carotid artery. The peripheral chemoreceptors primarily play a role in the regulation of breathing, but they also affect blood pressure. These receptors respond mostly to the level of oxygen in the blood. A significant decrease in the blood oxygen concentration triggers a series of feedback loops that indirectly stimulate an increase in heart rate and cause vasoconstriction. Another group of chemoreceptors, the *central chemoreceptors,* are located in the medulla of the brainstem. The central chemoreceptors respond to the pH of the interstitial fluid of the brain. When the pH of this fluid decreases, another feedback loop is stimulated that indirectly increases the activity of sympathetic neurons, resulting in vasoconstriction and a rise in blood pressure.

Endocrine System Maintenance of Blood Pressure

As we've discussed, the endocrine system and nervous system work together to maintain homeostasis of many variables, including blood pressure (see Chapter 16). The endocrine system is somewhat slower than the nervous system in responding to changes in blood pressure, requiring from several minutes to several hours for its effects to be demonstrated. Different hormones affect all three factors that determine blood pressure, and some affect more than one factor. However, only the hormonal effects on resistance and cardiac output are involved in short-term blood pressure maintenance.

Let's look in more detail at these effects:

- **Hormones that control cardiac output.** Hormones that increase cardiac output include epinephrine, norepinephrine, and thyroid hormone. The effects of epinephrine and norepinephrine produced by the adrenal medulla are identical to the effects of these chemicals produced by the sympathetic nervous system—they increase both heart rate and contractility. The effects of thyroid hormone, which is the body's primary regulator of the metabolic rate, are less direct. It causes the cardiac muscle cells to produce more receptors for epinephrine and norepinephrine, which allows these two chemicals to have a greater impact on cardiac output.

- **Hormones that control resistance.** Hormones affect peripheral resistance mainly through their control of blood vessel diameter. As you have read, epinephrine and norepinephrine from the adrenal medulla cause vasoconstriction and increase peripheral resistance, elevating blood pressure. Another hormone that affects resistance is angiotensin-II, produced as part of the renin-angiotensin-aldosterone system (RAAS). Angiotensin-II is a very powerful vasoconstrictor that sharply increases peripheral resistance and blood pressure. The hormone atrial natriuretic peptide, or ANP, is produced by the atria in response to increased blood volume. ANP causes vasodilation, especially of the vessels supplying the kidney. This causes a mild decrease in peripheral resistance and blood pressure.

As you will discover shortly, hormones are a common target of drug therapy for high blood pressure, particularly aldosterone and angiotensin-II.

Quick Check

☐ 1. What effects do the sympathetic and parasympathetic nervous systems have on blood pressure?

☐ 2. What are the steps of the baroreceptor reflex arc?

☐ 3. What effects do epinephrine and norepinephrine have on peripheral resistance and blood pressure when released as hormones?

☐ 4. How does angiotensin-II affect blood volume, peripheral resistance, and blood pressure?

Long-Term Maintenance of Blood Pressure by the Endocrine and Urinary Systems

FLASHBACK

1. How does antidiuretic hormone affect blood volume? (p. 594)
2. What are the overall effects of the renin-angiotensin-aldosterone system? (p. 610)

The long-term maintenance of blood pressure falls to the urinary system and certain hormones of the endocrine system that affect the kidneys. These systems control blood pressure by increasing or decreasing the amount of body water lost as urine, which affects blood volume.

The endocrine system regulates blood volume through the release of hormones, including ANP and angiotensin-II, as well as antidiuretic hormone (ADH) and aldosterone. When blood pressure increases, atrial cells secrete ANP, which causes the kidneys to excrete more water and sodium ions to decrease blood volume, and so blood pressure. Conversely, when blood pressure decreases, ADH secretion from the posterior pituitary increases. ADH triggers thirst and increases the amount of water retained by the kidneys. These effects raise blood volume and blood pressure. Another effect triggered when blood pressure decreases is secretion of the enzyme *renin* (REE-nin) from the kidneys as part of the previously mentioned RAAS. Renin begins the process that activates angiotensin-II, which induces thirst, causes sodium ion retention, and as a result increases blood volume. Angiotensin-II also triggers the secretion of aldosterone from the adrenal gland, which causes retention of sodium ions and water from the kidneys, increasing blood volume.

The urinary system's control over blood volume is fairly simple: If blood pressure increases, more water flows through the tiny filtering tubes of the kidneys, called tubules, than these cells can return to the blood. This water is then lost from the body as urine, and blood volume and blood pressure decrease. Conversely, if blood pressure decreases, less water flows through the kidneys' tubules and these cells have more time to reclaim the water and return it to the blood. This results in decreased urine production and a slight increase in blood volume and blood pressure.

Quick Check

☐ 5. Which hormones affect blood volume, and how do they affect it?

☐ 6. How do the kidneys influence blood pressure maintenance?

Summary of Blood Pressure Maintenance

The maintenance of blood pressure by the nervous, endocrine, and urinary systems is summarized in **Figure 18.9**. Each panel examines the effects of the systems on one of the factors that influence blood pressure—peripheral resistance, cardiac output, and blood volume.

MECHANISMS AFFECTING PERIPHERAL RESISTANCE	EFFECT
Endocrine Epinephrine Norepinephrine Angiotensin-II Renin secretion **Neural** Increased sympathetic activity	**Increased peripheral resistance** 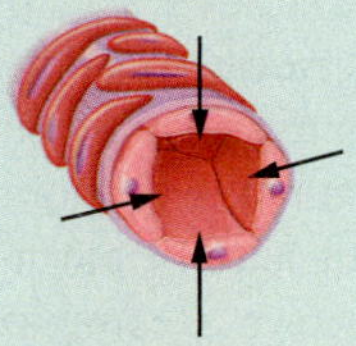
Endocrine Decreased renin secretion **Neural** Increased parasympathetic activity (vasodilation due to brainstem-mediated decrease in sympathetic activity)	**Decreased peripheral resistance** 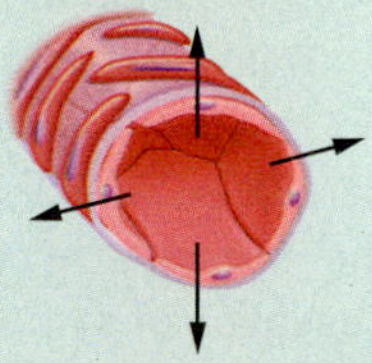

MECHANISMS AFFECTING CARDIAC OUTPUT	EFFECT
Endocrine Thyroid hormone Epinephrine Norepinephrine **Neural** Increased sympathetic activity Decreased parasympathetic activity	**Increased rate and force of contraction** 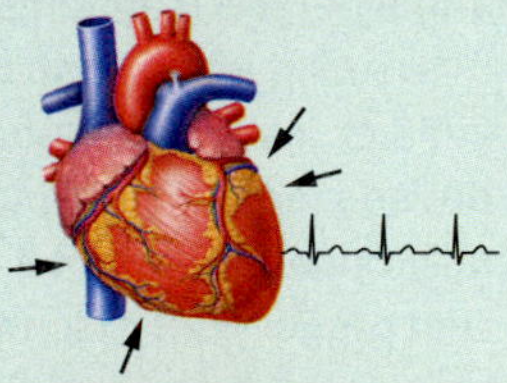
Endocrine Atrial natriuretic peptide **Neural** Increased parasympathetic activity Decreased sympathetic activity	**Decreased rate** 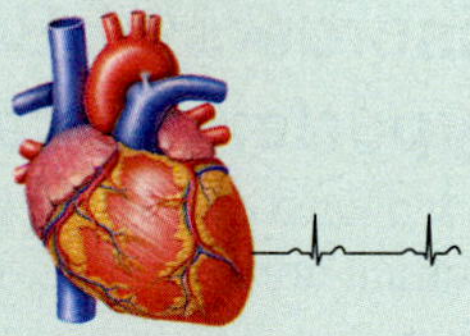

MECHANISMS AFFECTING BLOOD VOLUME	EFFECT
Endocrine Angiotensin-II ADH Aldosterone **Renal** Increased H_2O retention	**Increased blood volume** 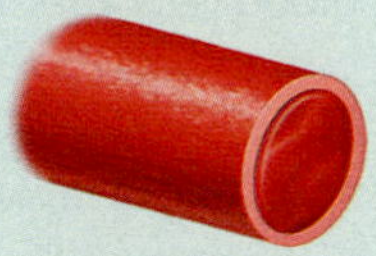
Endocrine Atrial natriuretic peptide **Renal** Decreased H_2O retention	**Decreased blood volume** 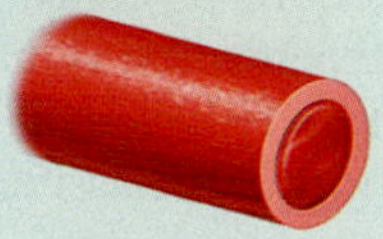

Figure 18.9 Blood pressure maintenance.

Vasovagal Syncope

A common scene in movies or TV is a person fainting from shock or fear. This is a legitimate physiological response called **vasovagal syncope** (VVS; vay-zoh-VAY-gul SING-kuh-pee), which can be triggered by stimuli such as pain, extreme emotional stress, trauma, and fright. The exact mechanism that initiates VVS is not known, but it seems to begin in the cerebral cortex and involve a complex series of neural events. Regardless of the initiating mechanism, it results in the brainstem autonomic centers tremendously increasing parasympathetic output and virtually eliminating sympathetic output.

This sudden autonomic imbalance has profound effects on blood pressure. First, the vagus nerve directly decreases cardiac output by slowing the heart rate and mildly decreasing the strength of the heart's contraction. Second, the near cessation of sympathetic output removes the sympathetic tone from blood vessels, which causes massive vasodilation and a significant decrease in peripheral resistance. The combined decrease in cardiac output and peripheral resistance leads to a dramatic and sudden fall in blood pressure. When blood pressure falls in this manner, blood flow to the brain is reduced, and a person may temporarily lose consciousness, or "faint."

Usually when blood pressure falls, the baroreceptor reflex initiates events to return blood pressure to normal. So why doesn't the baroreceptor reflex prevent VVS? The answer isn't fully understood, but it appears that the complex neural events that trigger VVS also inhibit the baroreceptor reflex.

Disorders of Blood Pressure: Hypertension and Hypotension

As we have discussed, blood pressure must be maintained within a certain normal range. If the blood pressure rises too high (a condition known as *hypertension*) or falls too low (a condition known as *hypotension*), severe disturbances in homeostasis may result.

Hypertension

An estimated 20% of the world's population has **hypertension,** or high blood pressure. Untreated or undertreated hypertension has a high cost to society and is associated with coronary artery disease (the leading cause of death in North America), stroke (the third leading cause of death), heart failure, certain types of dementia, kidney disease, and vascular disease. Since hypertension is the number one modifiable risk factor for each of these potentially life-threatening conditions, the healthcare field has focused intensive efforts on educating the public about this "silent killer."

Hypertension is diagnosed based on the average of two or more readings recorded on two or more successive visits to a healthcare provider. It is classified according to the following system:

- Normal (or *normotensive*): systolic less than 120 mm Hg and diastolic less than 80 mm Hg
- Pre-hypertension: systolic 120–139 mm Hg, diastolic 80–89 mm Hg

- Stage 1 hypertension: systolic 140–159 mm Hg, diastolic 90–99 mm Hg
- Stage 2 hypertension: systolic greater than or equal to 160 mm Hg, diastolic greater than or equal to 100 mm Hg

Note that pre-hypertension is a relatively new category that is intended to determine those individuals at risk of developing hypertension.

There are two types of hypertension: **essential** (or **primary**) **hypertension,** for which no cause is identifiable, and **secondary hypertension,** for which a cause is identifiable, such as narrowing of the arteries serving the kidneys (*renal artery stenosis*). Essential hypertension is far more common than secondary hypertension, accounting for about 95% of cases. The pathogenesis of essential hypertension is complex, and much about it remains uncertain. Current evidence indicates its development is based on multiple factors, including genetics, ethnic heritage (it is over twice as common in African Americans than in Americans of European descent), advancing age, dietary factors such as excessive salt intake, excess sympathetic tone, and abnormal activity in the RAAS.

Treatment of hypertension depends on the stage at which it is diagnosed. Two areas of key treatment in all stages are (1) lifestyle modifications, including smoking cessation, weight loss if needed, limited alcohol intake, and increased physical activity; and (2) dietary modifications, including decreased salt, cholesterol, and saturated fat intake. Drug therapy is aimed at modifying the three factors that influence blood pressure: cardiac output, blood volume, and peripheral resistance. Common targets for drug therapies include the RAAS, the tubules in the kidneys, calcium ion channels in vascular smooth muscle, and sympathetic receptors on the heart.

Hypotension

Hypotension is considered to be any abnormally low blood pressure. Technically, hypotension is defined by a systolic pressure lower than 90 mm Hg and/or a diastolic pressure lower than 60 mm Hg, but is generally diagnosed as such only if the individual shows symptoms. When symptoms do arise, they vary greatly with the severity of the hypotension. Mild hypotension may cause mild dizziness and lightheadedness. Severe hypotension, or **circulatory shock,** results in much more dramatic symptoms, including loss of consciousness and organ failure. In this condition, there is insufficient blood pressure to deliver oxygen and nutrients to the cells, which can be rapidly fatal.

There are many potential causes of hypotension, all of which relate to one or more of the three factors that determine blood pressure. Let's look at these more closely.

- **Reduced blood volume.** The most common cause of hypotension is reduced blood volume, or *hypovolemia*. This condition can result from blood loss; fluid losses from diarrhea, vomiting, or overuse of diuretics; or insufficient fluid intake. Severe blood loss can lead to *hypovolemic shock,* which is fatal unless blood volume is restored.
- **Decreased cardiac output.** A decrease in either stroke volume or heart rate can result in hypotension. Decreases in heart rate are generally due to medications prescribed to treat hypertension. A decrease in stroke volume is most

commonly caused by heart failure, or the inability of the heart to function efficiently as a pump. Severe heart failure can dramatically reduce cardiac output, and may result in a condition called *cardiogenic shock*.

● **Vasodilation.** Excessive vasodilation can produce profound hypotension. Many conditions can cause excessive vasodilation, including anti-hypertension medications, acidosis, and ANS dysfunction. Another potential cause of excessive vasodilation is the release of histamine into the blood during a severe allergic reaction, a condition called *anaphylactic shock*. Similar excessive vasodilation occurs with certain bacterial infections of the blood; this is called *septic shock*.

The treatment of hypotension depends largely on its severity and cause. Mild hypotension may be managed simply by increasing fluid intake or changing a person's medication. Severe hypotension and circulatory shock, however, require aggressive management aimed at the underlying cause of the shock. A patient may be given fluids and/or blood to increase fluid volume, medications to raise cardiac output, and systemic vasoconstrictors.

Quick Check

☐ 7. What is hypertension?

☐ 8. What is circulatory shock, and why is it dangerous?

Apply What You Learned

☐ 1. Explain how each of the following treatments for hypertension lowers blood pressure:

    a. Diuretics (drugs that increase the production of urine)

    b. Angiotensin-receptor blockers (drugs that block angiotensin receptors on smooth muscle cells)

    c. Beta blockers (drugs that block the effects of the sympathetic nervous system on the heart)

☐ 2. Urine production essentially stops in cases of extremely low blood pressure. Explain why this happens. (*Hint:* Consider that urine production affects blood volume.)

See answers in Appendix A.

MODULE
18.4
Capillaries and Tissue Perfusion

Learning Outcomes

1. Describe the different types of capillaries, and explain how their structure relates to their function.

2. Explain the roles of diffusion, filtration, and osmosis in capillary exchange.

3. Describe how autoregulation controls blood flow to tissues.

We now turn our attention to the anatomy and physiology of the smallest blood vessels: the capillaries. As you learned in the first module, capillaries generally are found in clusters called *capillary beds* that wind their way through the cells of most tissues in the body—exceptions include cartilage, the sclera and cornea of the eye, and epithelial tissue. The blood flow to a tissue through a capillary bed is known as **tissue perfusion**. As you might expect, tissue perfusion is tightly regulated to ensure that the metabolic needs of all tissues are met at all times.

In this module, we explore the structure and function of the three types of capillaries and capillary beds. We then examine tissue perfusion and the factors that control this process. The module concludes with a look at the perfusion in "special" circulations such as those of the heart and the brain.

Capillary Structure and Function

◀◀ FLASHBACK

1. What are lipid-soluble (nonpolar) substances? What are water-soluble (ionic and polar) substances? (p. 39)

2. What is transcytosis? (p. 85)

As you can see in **Figure 18.10**, capillaries are extremely thin vessels, with walls that are only about 0.2 μm in thickness. Each

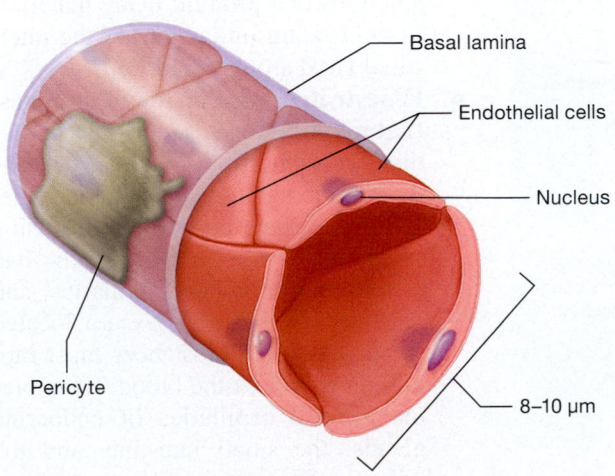

(a) A generalized capillary

Figure 18.10 **The structure of a generalized capillary.**

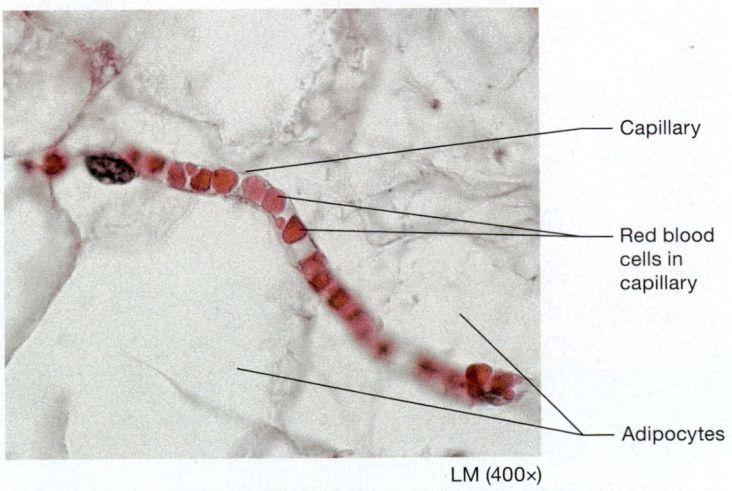

LM (400×)

(b) Photomicrograph of a capillary

capillary consists only of an endothelium rolled into a tube and a small amount of basal lamina secreted by the endothelial cells (see Figure 18.10a). Around some capillaries are cells called **pericytes** (PEHR-ih-sytz; *peri-* = "around") that have contractile filaments and appear to control blood flow through the capillary.

Capillaries measure, on average, about 50 μm in length and about 8–10 μm in diameter. The walls of most capillaries consist of one to three endothelial cells joined by tight junctions that curl around the capillary's entire circumference. In such capillaries, blood cells have to pass through in single file (see Figure 18.10b).

Capillary Exchange

The thinness of the capillary wall makes sense if you consider capillaries' primary function: Nutrients, gases, ions, and wastes must be able to cross the wall and travel between the blood in the capillary and the tissue cells. This movement of materials is known as **capillary exchange.** Materials are generally exchanged via three mechanisms that you have already read about (**Figure 18.11**):

- **Diffusion and osmosis through gaps and fenestrations.** The tight junctions between many endothelial cells are incomplete and leave small gaps between endothelial cells. In addition, some capillaries have small pores within their endothelial cells, called **fenestrations** (fen-eh-STRAY-shunz; *fenestr-* = "window"). Water is able to move freely through these pores by osmosis, and small solutes such as monosaccharides and amino acids can move if a concentration gradient is present.

- **Diffusion through the membranes of endothelial cells.** Lipid-soluble substances such as oxygen, carbon dioxide, and certain lipids can generally enter and exit the capillary by diffusing across the membrane of one side of the endothelial cell and out the membrane on the other side. These substances then enter either the interstitial fluid or the blood.

- **Transcytosis.** Larger substances must cross the endothelial cells by transcytosis. Recall that during transcytosis, substances are taken into the cell by endocytosis and then leave the other side of the cell by exocytosis (see Chapter 3).

Both diffusion and transcytosis involve a substance crossing the entire width of the endothelial cell. If capillaries contained thicker cells or multiple tissue layers, diffusion and transcytosis would occur far too slowly to meet the body's needs. This is another example of the Structure-Function Core Principle (p. 28).

CORE PRINCIPLE
Structure-Function

Types of Capillaries

Capillaries in different parts of the body have slightly different functions and, accordingly, they have slightly different structures. The three types of capillaries are *continuous, fenestrated,* and *sinusoidal* (**Table 18.3**); their characteristics are as follows:

- **Continuous capillaries.** The majority of capillaries in the body are **continuous capillaries,** located in the muscles, skin, and most nervous and connective tissues. Continuous capillaries are the least "leaky"—they permit the fewest substances to enter or exit the blood by the paracellular route because their endothelial cells are joined together by tight junctions. They contain small gaps between endothelial cells through which some small substances diffuse, but the majority move into or out of the capillaries by diffusion through the endothelial cells or by transcytosis. The capillaries in the brain that form the blood brain barrier are a modified type of continuous capillary, with specialized tight junctions that prohibit many materials from crossing into or out of the interstitial fluid around brain cells.

- **Fenestrated capillaries.** The **fenestrated capillaries** are much leakier than continuous capillaries because their endothelial cells contain fenestrations. These pores allow diffusion to take place much more quickly than it does in continuous capillaries, and so fenestrated capillaries are located in places where substances must rapidly enter or exit the blood. Examples include the capillaries of endocrine glands, the small intestine, and the kidneys.

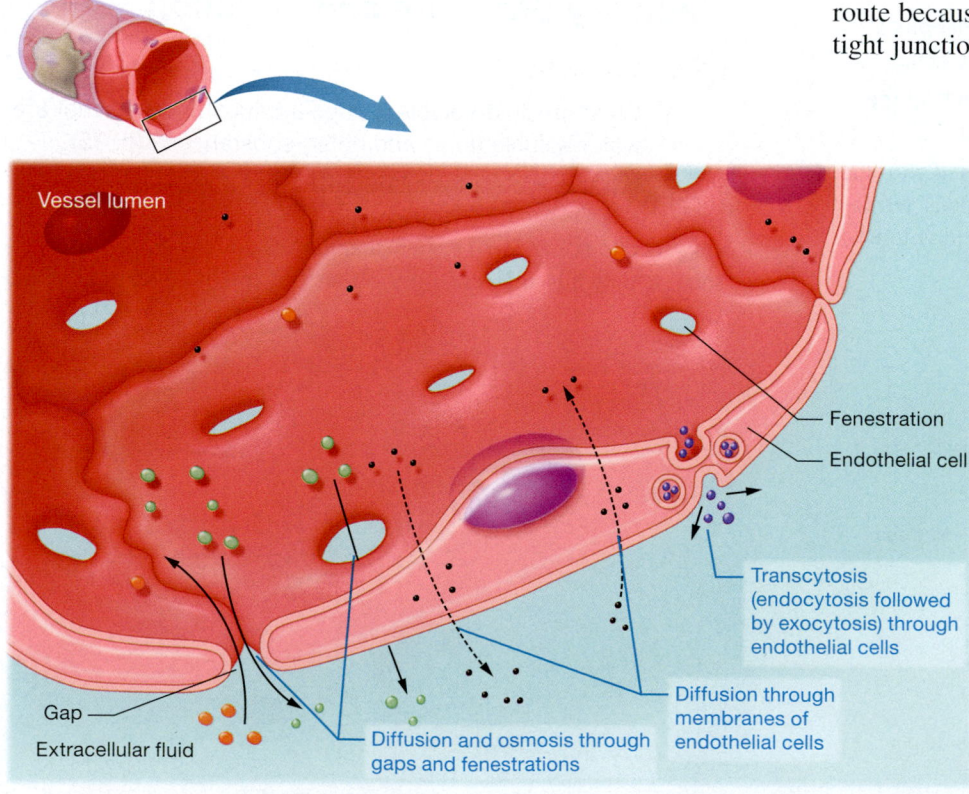

Vessel lumen

Fenestration
Endothelial cell

Transcytosis (endocytosis followed by exocytosis) through endothelial cells

Diffusion through membranes of endothelial cells

Gap
Extracellular fluid

Diffusion and osmosis through gaps and fenestrations

Figure 18.11 Capillary exchange mechanisms.

Table 18.3 Types of Capillaries

	Continuous capillaries	Fenestrated capillaries	Sinusoidal capillaries
	Tight junctions between cells	Fenestrations	Irregular basal lamina, Large pores, Spaces between endothelial cells
Structure	Endothelial cells joined by tight junctions	Contain fenestrations in the endothelial cells	Discontinuous sheet of endothelium, irregular basal lamina, very large pores
Location	• Skin • Most nervous and connective tissue • Muscle tissue	• Kidneys • Endocrine glands • Small intestine	• Liver • Lymphoid organs • Bone marrow • Spleen
Function	Least "leaky"—permit a narrow range of substances to cross the capillary walls	Moderately leaky—allow large volumes of fluid and larger substances to cross capillary walls	Leakiest—allow large substances such as cells to cross the capillary walls

• **Sinusoidal capillaries.** The leakiest capillaries are the **sinusoidal capillaries,** often referred to simply as **sinusoids** (SYN-yuh-soydz; *sinu-* = "fold" or "hollow"). As you can see in Table 18.3, sinusoids have a discontinuous sheet of endothelium, an irregular basal lamina, and very large pores in their endothelial cells. Sinusoids are typically three to four times larger in diameter than are other capillaries, and often have an irregular shape because their boundaries are determined by the organ in which they reside. They are located in organs and tissues such as the liver, spleen, lymphoid organs, and bone marrow. Their size, shape, and sluggish blood flow allow them to facilitate the transfer of large substances such as blood cells and large proteins between the interstitial fluid and the blood.

> **Quick Check**
>
> ☐ 1. Describe the structure and size of a typical capillary.
>
> ☐ 2. List three ways in which substances may cross a capillary to enter or exit the blood.
>
> ☐ 3. Describe the properties of the three types of capillaries, and explain how they differ.

Blood Flow through Capillary Beds

The flow of blood occurring within the body's capillary beds is collectively called the **microcirculation.** The microcirculation consists of two types of vessels: (1) the true capillaries, where materials are exchanged; and (2) a small, central vessel. Notice in **Figure 18.12a** how true capillaries form interweaving networks with multiple anastomoses. These true capillaries are fed by the proximal end of the central vessel, which is formed by either a small terminal arteriole or a metarteriole. Recall from Module 18.1 that each capillary-metarteriole junction contains a precapillary sphincter that controls the amount of blood flowing into the capillaries. At rest, only about 25% of the body's capillary beds are fully open.

True capillaries are drained at the distal end of the capillary bed by a portion of the central vessel called a **thoroughfare channel.** The thoroughfare channel lacks precapillary sphincters, which reflects the fact that this is the drainage end of the capillary bed. When the precapillary sphincters close, as in **Figure 18.12b,** blood flows straight from the metarteriole to the thoroughfare channel and into a venule without passing through the true capillaries. Blood does not flow backward into the true capillaries from the thoroughfare channels because the pressure gradient favors the movement of blood in one direction only.

Local Regulation of Tissue Perfusion

Tissue perfusion is critical for the survival of our cells. As you might expect, tissue perfusion is largely a function of arterial blood pressure—if the arterial pressure is too low, blood will not be adequately delivered to the tissues and they will be poorly perfused. In addition, tissue perfusion is also regulated by local factors within each individual tissue, a phenomenon known as **autoregulation,** or self-regulation. Autoregulation ensures that the correct amount of blood is delivered to match a tissue's level of activity. There are two main types of local autoregulatory controls: the *myogenic mechanism* and *metabolic controls*.

The Myogenic Mechanism

The **myogenic mechanism** relies on properties inherent in the vascular smooth muscle cells in the arterioles supplying capillary beds. Increases in arteriolar pressure open stretch-sensitive channels in the arteriolar smooth muscle cells. This initiates a depolarization in the membranes of these cells so they can contract without nervous system stimulation. For this reason,

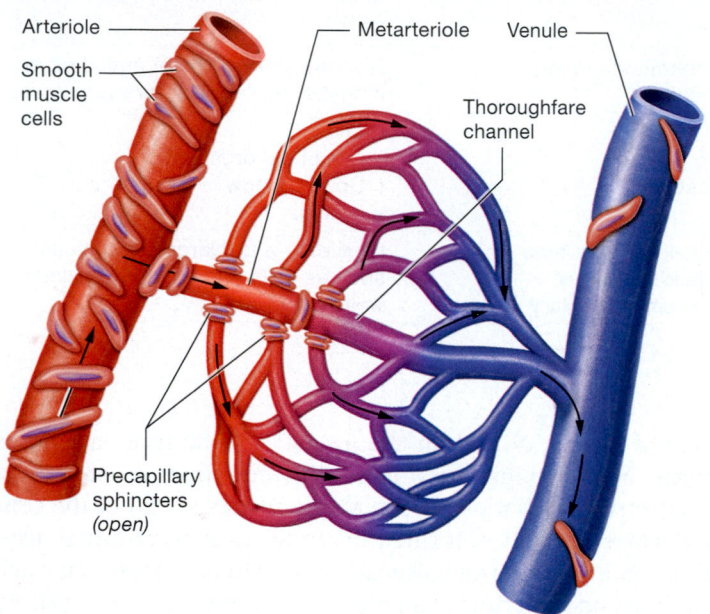

(a) Blood flow through a capillary bed when precapillary sphincters are open

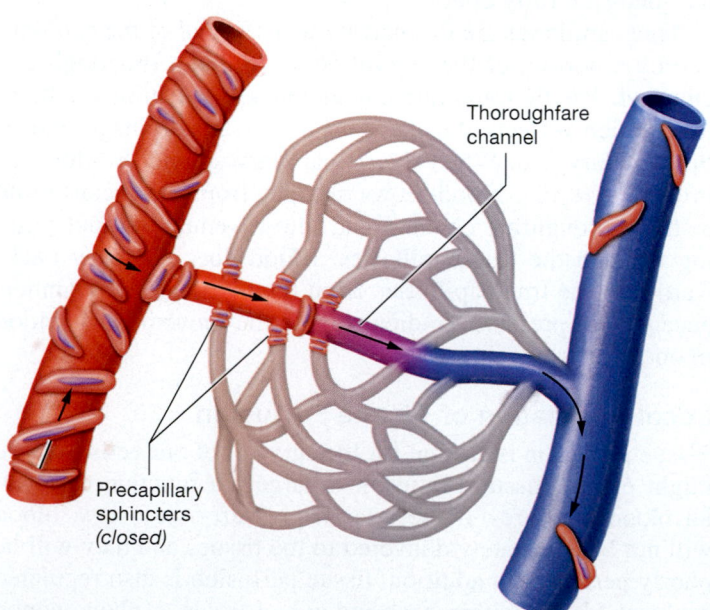

(b) Blood flow through a capillary bed when precapillary sphincters are closed

Figure 18.12 Structure of and blood flow through a capillary bed.

increases in arteriolar pressure lead to arteriolar vasoconstriction. The reverse response also occurs—arteriolar smooth muscle relaxes when arteriolar pressure decreases.

To understand how the myogenic response affects tissue perfusion, think about the relationships between the velocity of blood flow, the pressure gradient, and resistance. Recall that the velocity of blood flow is proportional to the pressure gradient. Normally, blood is only present in capillaries for less than 1 second, which means that all gases, nutrients, ions, and wastes must be exchanged between the blood and the tissues in this short time. However, if the velocity of blood flow increases, less capillary exchange takes place. Conversely, if velocity decreases, flow is inadequate and blood that has already undergone capillary exchange remains in the capillary.

The myogenic mechanism counters a change in blood flow by altering arteriolar resistance. The velocity of blood flow is related inversely to resistance, so if resistance increases, velocity decreases, and vice versa. So the myogenic mechanism slows blood flow by increasing resistance when arteriolar pressure rises. Conversely, it speeds up blood flow by decreasing resistance when arteriolar pressure lowers. Both changes maintain local tissue perfusion at a constant level and ensure that gases and other substances continue to be exchanged in adequate amounts even in the face of fluctuating blood pressure.

Metabolic Controls

A second type of autoregulatory mechanism is mediated by chemicals present in the interstitial fluid surrounding capillaries. These chemicals are the result of cellular metabolic (ATP-generating) activities, and so the mechanisms are called *metabolic controls.* Recall that cells consume oxygen to turn ADP and inorganic phosphate into ATP while generating carbon dioxide as a waste product (see Chapter 3). The carbon dioxide diffuses into the interstitial fluid and reacts with water to yield an acid called *carbonic acid.* So, the interstitial fluid of an actively metabolizing cell, such as a contracting skeletal muscle fiber, will contain a low concentration of oxygen and high concentrations of carbon dioxide and hydrogen ions. All three of these conditions cause the smooth muscle cells of local arterioles to relax, dilating the arterioles. This increases perfusion and ensures adequate oxygen and nutrient delivery to the actively metabolizing cells.

Metabolic controls also work in the opposite direction. Tissues whose cells are producing ATP slowly will have a high concentration of oxygen and low concentrations of carbon dioxide and hydrogen ions. These conditions cause constriction of local arterioles and a decrease in tissue perfusion.

Quick Check

☐ 4. What is tissue perfusion?

☐ 5. Explain what happens to arterioles and tissue perfusion via the myogenic mechanism when blood pressure increases or decreases.

☐ 6. Which chemicals produce changes in tissue perfusion? How do they affect tissue perfusion?

Tissue Perfusion in Special Circuits

« FLASHBACK

1. What are the two layers of the skin? Where is the blood supply of the skin located? (p. 161)
2. What are neurotransmitters? (p. 413)
3. What is the coronary circulation? What are the two main vessels of the coronary circuit? (p. 642)

Each organ has its own unique requirements in terms of blood flow and necessary nutrients. Although the autoregulatory mechanisms that we just discussed control blood flow, many organs also have their own specific regulatory mechanisms. In this section we address the regulatory mechanisms of the heart, brain, skeletal muscles, and skin. The special circulations of other organs are discussed in their respective chapters.

Heart

The heart is a fairly small organ—it is only about the size of a fist and weighs under 1 pound, which is less than 0.5% of total body mass (see Chapter 17). In spite of its small size, however, the heart receives about 5% of the total cardiac output via the coronary circulation. Interestingly, the perfusion pattern through the coronary circulation is the opposite of the rest of the systemic circuit, which generally increases during systole and decreases during diastole. In the heart this is reversed because it actually squeezes its own arteries during ventricular systole, so heart tissue perfusion *decreases* during systole. During diastole, the ventricles relax, blood resumes flowing through the coronary arteries, and tissue perfusion to the cardiac muscle cells increases. This is one reason why extreme increases in heart rate and/or force of contraction can be dangerous: The heart does not spend enough time in diastole and the myocardium is not adequately perfused. For this reason, people are advised to see a physician before starting an exercise program. Too intense a workout by an individual who is not adequately fit could lead to dangerous elevations in heart rate.

The main local autoregulatory mechanism of cardiac muscle tissue appears to be metabolic controls, in particular the concentration of oxygen in the cardiac interstitial fluid. Under normal resting conditions, blood flow to the heart is about 60–70 ml/min per 100 g of tissue. However, during strenuous exercise or other sympathetic activity, blood flow can increase to more than 250 ml/min per 100 g of tissue. This appears to happen because a low interstitial fluid oxygen level triggers the production of chemicals called vasodilators; these chemicals directly dilate the arterioles serving the myocardium and so greatly increase perfusion.

Brain

The brain is extremely intolerant of ischemia, more so than any other tissue in the body. A sudden decrease in tissue perfusion to the brain will result in loss of consciousness within seconds

(see *A&P in the Real World: Vasovagal Syncope* on p. 686 for an example). The brain accounts for only about 2% of total body mass, but receives about 15% of the total cardiac output. Given the high, constant metabolic demands of the brain, it will probably come as no surprise that its blood flow is tightly regulated. Autoregulatory mechanisms, including myogenic and metabolic controls, maintain cerebral blood flow at a nearly constant rate of about 750 ml/min.

Despite this constant blood supply, blood flow within the brain to different areas varies considerably with neuronal activity. This is in part due to standard metabolic controls; areas of the brain that have higher activity consume more oxygen and produce more carbon dioxide. However, it also appears to be caused by neurotransmitters produced by active neurons. Certain neurotransmitters are direct vasodilators, whereas others act on astrocytes (neuroglial cells) that are in contact with cerebral arterioles. When stimulated, the astrocytes induce production of vasodilators that open arterioles and so increase perfusion to active regions of the brain.

Skeletal Muscle

Blood flow changes dramatically in skeletal muscle tissue during exercise, increasing as much as 50-fold, which is known as *hyperemia* (hy′-per-EE-mee-uh; "excessive blood"). The mechanism behind skeletal muscle hyperemia lies in the structure of the arteries supplying skeletal muscle tissue. An artery that enters a skeletal muscle is known as a *feed artery*. Feed arteries branch into multiple arterioles and end in **terminal arterioles** that supply a capillary bed. During resting conditions, resistance in the feed arteries is relatively high and many of the terminal arterioles are constricted, which limits tissue perfusion. When exercise begins, as shown below, ① metabolic conditions around the skeletal muscle fiber dilate the terminal arterioles. This produces a small increase in tissue perfusion as more capillary beds open, but not enough of an increase to meet the demands of vigorously working skeletal muscle.

However, as exercise continues, the vessels proximal to the terminal arterioles also dilate. ② Next, the arterioles dilate, which significantly boosts tissue perfusion. ③ Finally, continuing exercise causes the feed arteries to dilate, which dramatically increases tissue perfusion:

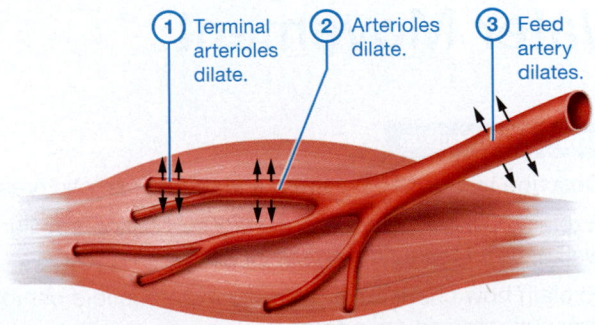

① Terminal arterioles dilate.   ② Arterioles dilate.   ③ Feed artery dilates.

This progressive, stepwise vasodilation allows the skeletal muscles to precisely match perfusion to the degree of exercise.

Skin

As we've explained, the skin is the largest organ in the body (see Chapter 5). Its blood supply is located in the dermis, and oxygen and nutrients must diffuse from capillaries in the dermis to the epidermis. Local autoregulation of the skin's blood flow does take place, especially in response to temperature—placing something warm directly on the skin will cause vasodilation, whereas something cold will produce vasoconstriction. However, the most important control over the skin's blood flow is neural via the sympathetic nervous system. This is because sympathetic neurons regulate blood flow in the skin as part of its temperature-regulation physiology. The sympathetic nervous system alters perfusion through the skin via vasoconstriction to conserve heat or vasodilation to release heat.

Quick Check

☐ 7. How is the pattern of blood flow in the coronary circulation different from that in the rest of the systemic circuit?

☐ 8. How is perfusion in the brain increased for areas with increased activity?

☐ 9. What is hyperemia? How is skeletal muscle tissue able to develop hyperemia?

Apply What You Learned

☐ 1. Predict what might happen if capillaries were composed of thicker epithelium.

☐ 2. A hypothetical poison destroys the tight junctions of the continuous capillaries of the brain. Predict the effects of such a poison.

☐ 3. In the initial stages of ventricular fibrillation (during which the heart produces no functional contraction), tissue perfusion in the heart actually *increases*. Explain.

See answers in Appendix A.

MODULE 18.5

Capillary Pressures and Water Movement

Learning Outcomes

1. Describe hydrostatic pressure and colloid osmotic pressure.

2. Explain how net filtration pressure across the capillary wall determines fluid movement across that wall.

3. Explain how changes in hydrostatic and colloid osmotic pressure may cause edema.

So far in this chapter we have presented the gases, electrolytes, nutrients, and wastes that are exchanged within the capillaries by diffusion or transcytosis. However, we have not yet addressed the

other substance that is exchanged in large volumes between the capillaries and the interstitial fluid: water. Water also moves into and out of the blood by passing through either the small pores of the capillaries or the fenestrations of the endothelial cells. Water doesn't cross a capillary wall by diffusion or transcytosis like other substances. Instead, the movement of water across a capillary is driven by a process called *filtration,* which is the movement of a fluid by a force such as pressure or gravity. In this module we examine this process and determine the overall amount of water lost or gained from a capillary.

Pressures at Work in a Capillary

⟪ FLASHBACK

1. During osmosis, do water molecules move toward or away from a solution with a higher solute concentration? (p. 76)

2. What is osmotic pressure? (p. 77)

Two basic pressures that drive water movement are at work within a capillary: *hydrostatic pressure* and *osmotic pressure.* These forces promote movement in opposite directions—hydrostatic pressure drives water out of the capillary, whereas osmotic pressure generally draws fluid into the capillary. Let's examine each in greater detail.

Hydrostatic Pressure

Hydrostatic pressure, or **HP,** is simply the force that a fluid exerts on the wall of its container. This definition should sound familiar, as we gave essentially the same one for blood pressure earlier. This is because blood is the fluid in a vessel that is creating the hydrostatic pressure. For this reason, blood pressure is equal to hydrostatic pressure.

You can see what happens here when we have a hydrostatic pressure *gradient* (i.e., the pressure is higher in one compartment than in another compartment):

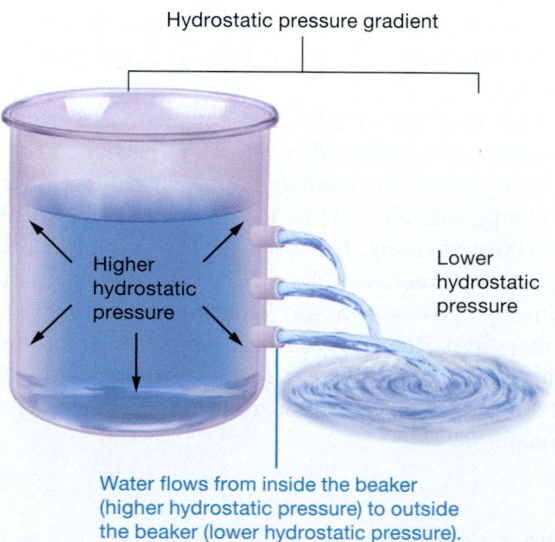

Hydrostatic pressure gradient

Higher hydrostatic pressure

Lower hydrostatic pressure

Water flows from inside the beaker (higher hydrostatic pressure) to outside the beaker (lower hydrostatic pressure).

Notice that fluid flows through the holes from the area of higher hydrostatic pressure to the area of lower hydrostatic

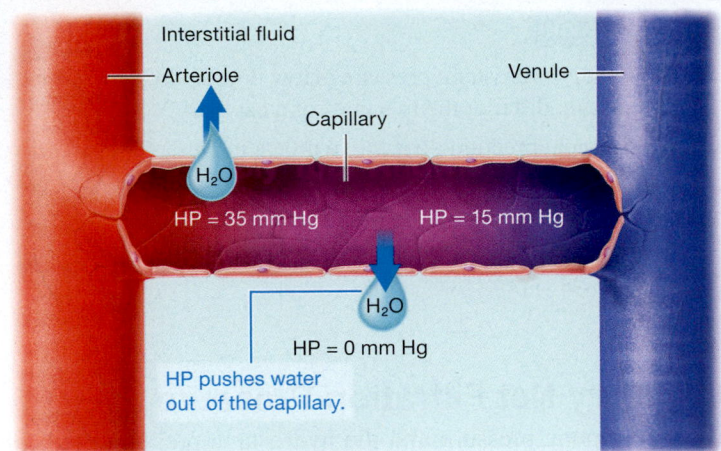

(a) Hydrostatic pressure (HP)

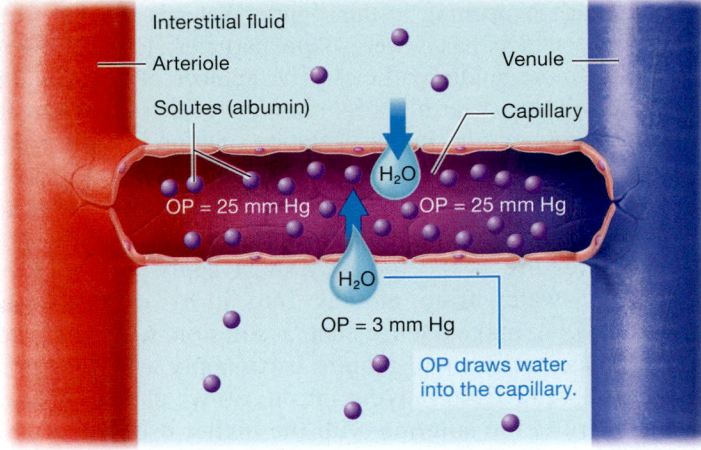

(b) Osmotic pressure (OP)

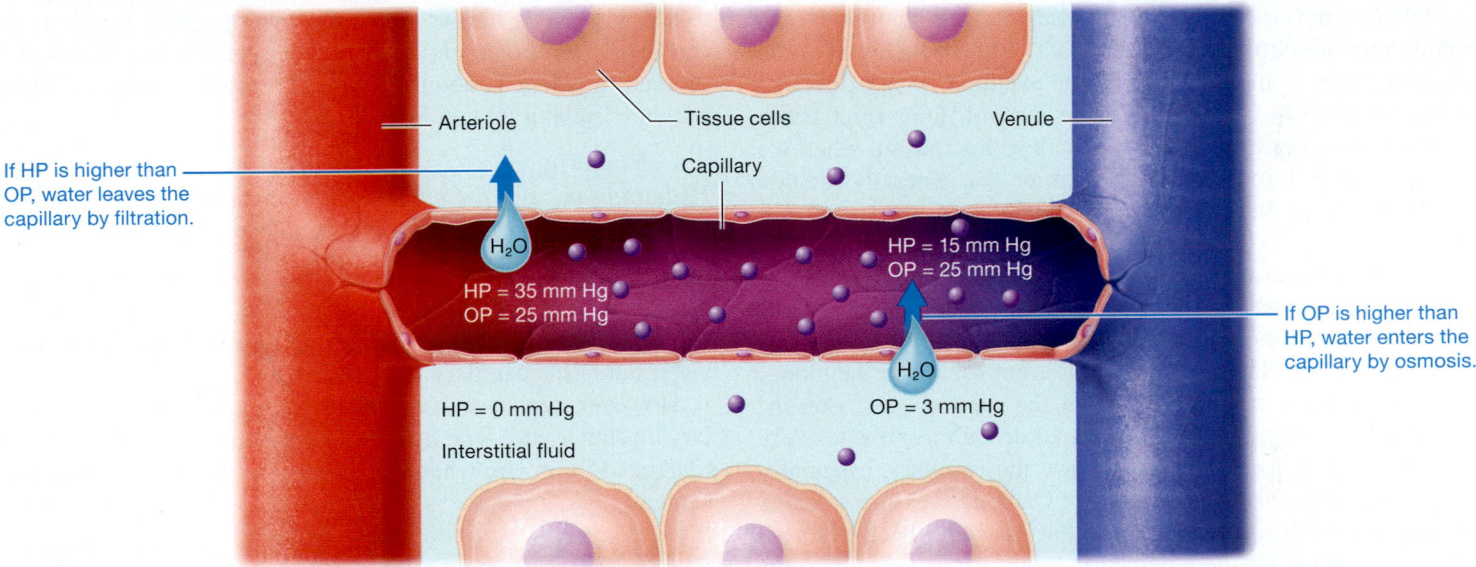

(c) How hydrostatic and osmotic pressures work together

Figure 18.13 **Hydrostatic and osmotic pressures in capillary blood and interstitial fluid.**

pressure until the gradient is extinguished. This fluid flow is a passive process known as **filtration,** which is something you witness any time you make a pot of coffee with a coffee filter.

How this process applies to water movement across a capillary is illustrated in **Figure 18.13a.** Notice first that the hydrostatic pressure in a capillary changes as we move from its arteriolar end, where it measures about 35 mm Hg, to its venular end, where it measures about 15 mm Hg. Second, note that the hydrostatic pressure of the interstitial fluid is so low that it is functionally 0 mm Hg. So, we have steep hydrostatic pressure gradients on both ends of the capillary that drive water *out* of the capillary and into the interstitial fluid.

Osmotic Pressure

In the cell chapter you read about one of the most essential passive transport processes in the body: osmosis (see Chapter 3). Recall that osmosis involves the movement of solvent from a

solution with a lower solute concentration to one with a higher solute concentration, in this way:

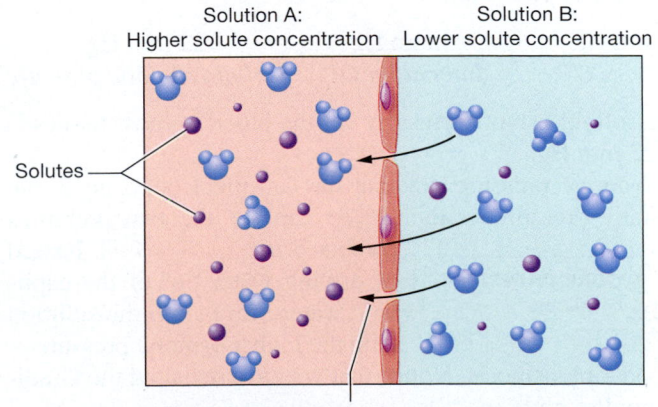

See what's happening in this figure: Water is moving from solution B, which has fewer solute particles, to solution A, which has more solute particles. The number of solute particles in a solution determines its *osmolarity,* also known as its *osmotic concentration.* In our example, solution B has a lower osmolarity than solution A.

One way to compare the osmolarity of two solutions is to compare their **osmotic pressures,** or **OP.** Recall that osmotic pressure is the force we must apply to a solution to prevent water from moving into it by osmosis (see Chapter 3). It makes sense that a solution with a higher solute concentration will require us to apply a bigger force to prevent water from moving into it. So we can say that as a general rule, the solution with the higher osmolarity also has the higher osmotic pressure. In our example, solution A has a higher osmolarity, so it is the solution with the higher osmotic pressure.

Note that in biological systems such as the human body, the membranes are permeable to more than water, which makes the situation more complicated than our example. Many different solutes can easily pass between the cytosol, interstitial fluid, blood, and other extracellular fluids. For this reason, when we consider the solution's osmotic pressure, we generally include only the solutes that do not cross membranes in significant amounts.

We apply this principle to the calculation of osmotic pressure in the blood. If you turn your attention to **Figure 18.13b,** you'll notice that the osmotic pressure of the blood in the capillary is about 25 mm Hg. There are many different solutes in the blood, but most of them can enter and exit the capillary. As a result, the blood's osmotic pressure is generated almost exclusively by proteins that are too large to leave the capillary, especially the protein *albumin* (al-BYOO-min). Osmotic pressure remains about the same throughout the capillary's length because these proteins do not leave the blood.

The interstitial fluid has far fewer "resident solutes" than the blood and so has a very low osmotic pressure of about 3 mm Hg. The difference in osmotic pressure between the blood and interstitial fluid creates an osmotic pressure gradient known as the **colloid osmotic pressure (COP),** or the *oncotic pressure.* We determine colloid osmotic pressure with simple subtraction:

25 mm Hg	−	3 mm Hg	=	22 mm Hg
Capillary OP		*Interstitial OP*		*Colloid osmotic pressure*

So, the colloid osmotic pressure for the blood in most tissues is about 22 mm Hg.

An osmotic pressure gradient has an effect opposite to the hydrostatic pressure gradient (you can see the two pressures combined in **Figure 18.13c).** Instead of driving water out of the capillary, water moves *into* the solution with the higher osmotic pressure—the blood—by osmosis. Notice that we have revisited the Gradients Core Principle (p. 28).

 **CORE PRINCIPLE** Gradients

Quick Check

☐ 1. What is hydrostatic pressure? How does hydrostatic pressure differ at the two ends of a capillary?

☐ 2. In which direction does the hydrostatic pressure gradient drive the movement of water in a capillary?

☐ 3. What is colloid osmotic pressure? In which direction does water move when an osmotic gradient is present?

Capillary Net Filtration Pressure

Colloid osmotic pressure and the hydrostatic pressure gradient cause water to move in opposite directions, which leads us to a question: Which one "wins" the water battle? Well, in a sense, they both do, but in different locations—look at **Figure 18.14** to see why. Let's first compare the two opposing forces at the capillary's arteriolar end, where we see that the hydrostatic pressure gradient is about 35 mm Hg and the colloid osmotic pressure is about 22 mm Hg:

35 mm Hg	−	22 mm Hg	=	13 mm Hg
Hydrostatic pressure driving water out		*Colloid osmotic pressure drawing water in*		*Overall force driving water out*

So, at the capillary's arteriolar end, the hydrostatic pressure gradient is larger and therefore "wins." The overall pressure, the **net filtration pressure (NFP),** is 13 mm Hg; this force drives water out of the capillary by filtration.

However, see what happens as we move toward the capillary's venular end—the hydrostatic pressure gradient decreases to about 15 mm Hg, whereas colloid osmotic pressure stays the same:

15 mm Hg	−	22 mm Hg	=	− 7 Hg
Hydrostatic pressure driving water out		*Colloid osmotic pressure drawing water in*		*Net filtration pressure drawing water in*

Here the net filtration pressure is a *negative* number, which means that water is drawn *into* the capillary, a process called *absorption.* So, at the venular end, colloid osmotic pressure "wins" and water is absorbed into the capillary by osmosis.

Now we can put all this together and examine the entire capillary to determine whether there is net water loss or gain in capillaries.

13 mm Hg	−	7 mm Hg	=	6 mm Hg
Arteriolar net filtration pressure		*Venular net filtration pressure*		*Overall net filtration driving water out*

So an overall net filtration pressure of about 6 mm Hg drives water *out* of the capillary by filtration. This doesn't sound like much, but it translates to about 2–4 liters of water lost from the blood to the interstitial fluid per day. Considering the entire blood volume averages only 5 liters, that is a significant amount of water. This lost volume of water is eventually returned to

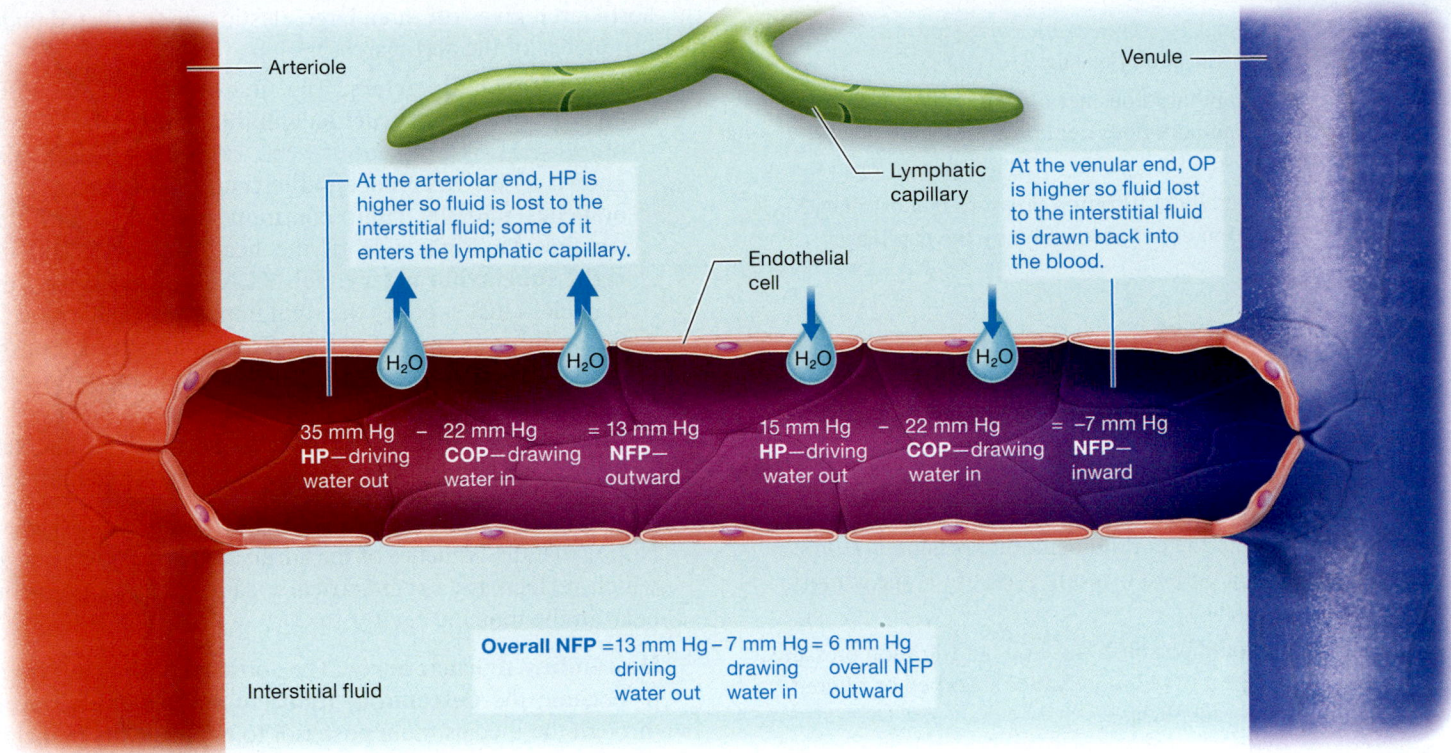

Figure 18.14 Net filtration pressure (NFP) in capillaries.

the blood by vessels of the *lymphatic system* via a mechanism discussed in the lymphatic and immune systems chapter (see Chapter 20).

Another Way to Think about Hydrostatic and Osmotic Pressures

Here's a simple visual analogy for hydrostatic and osmotic pressures. Imagine hydrostatic pressure as a faucet within the capillary that is always turned on, so water is always flowing out of it. Next, imagine osmotic pressure as a sponge within the capillary, constantly drawing water into it. At the arteriolar end, the "faucet" is turned on more strongly and spouting out more water than the "sponge" can reabsorb, so the capillary loses more water. At the venular end, the "faucet" is turned down and putting out less water, and so the "sponge" is able to reabsorb this water plus additional water in the interstitial fluid. So, at the venular end, the capillary gains more water. Over the entire length of the capillary, the faucet spouts out more water than the sponge can reabsorb, so there is a net loss of water from the capillary. ■

Edema

Edema (eh-DEE-mah; "swelling") is a condition characterized by an excessive amount of water in the interstitial fluid. Edema may be caused by either an increase in the capillary hydrostatic pressure gradient or a decrease in the colloid osmotic pressure (and sometimes a combination of the two). Several factors may increase the capillary hydrostatic pressure gradient, the most common of which is hypertension. Patients with hypertension have a greater capillary hydrostatic pressure at both the arteriolar and the venular ends of capillaries. This raises the overall net filtration pressure, which causes more water to be lost from the blood. This effect is especially pronounced in the legs and feet, a condition called *peripheral edema,* where the hydrostatic pressure gradient is already slightly higher due to the effects of gravity.

Even more profound edema may result from decreases in colloid osmotic pressure. This occurs in several conditions; one of the most common is liver failure due to chronic alcohol abuse, viral infection, or cancer. The liver makes the majority of the plasma proteins (such as albumin) that are responsible for osmotic pressure. When the liver fails, production of these proteins decreases significantly and colloid osmotic pressure drops. The excess water tends to accumulate in the interstitial fluid of the abdomen, a condition known as **ascites** (uh-SY-teez). Ascites may also result from severe protein malnutrition, in which case it is called *kwashiorkor* (kwah-shee-OHR-kor).

☐ 4. What is the net filtration pressure?

☐ 5. Where in the capillary does net filtration take place? Why? Where in the capillary does net absorption take place? Why?

☐ 6. What is the overall net filtration pressure in a capillary? Is there net filtration or net absorption by the capillary?

Apply What You Learned

☐ 1. What effect would hypotension have on the capillary hydrostatic pressure gradient and net filtration pressure?

☐ 2. Poison X destroys the tight junctions between endothelial cells and allows blood proteins to leak out of the capillary and enter the interstitial fluid. What would this do to the colloid osmotic pressure? (*Hint:* Remember that colloid osmotic pressure is the osmotic pressure gradient.)

☐ 3. What effect would poison X have on net filtration pressure? Would you expect to see more water absorbed or filtered from the capillaries? Why?

See answers in Appendix A.

MODULE 18.6

Anatomy of the Systemic Arteries

Learning Outcomes

1. Describe the patterns of arterial blood flow for the head and neck, the thoracic cavity, the abdominopelvic cavity, and the upper and lower limbs.

2. Identify major arteries of the systemic circuit.

3. Identify the major pulse points.

We now turn our attention to the anatomy of the systemic circuit arteries—the anatomy of the pulmonary circuit is covered in the chapters that discuss the heart and respiratory systems (see Chapters 17 and 21, respectively). We mentioned previously that the largest artery in the body is the **aorta,** which begins at the left ventricle. Notice in **Figure 18.15** that the aorta has four divisions (listed under "Arteries of the trunk"):

1. **Ascending aorta.** The initial segment of the aorta is the **ascending aorta,** which travels superiorly. Recall that the right and left coronary arteries supplying the myocardium branch from the ascending aorta (see Chapter 17).

2. **Aortic arch.** The ascending aorta curves to the left to become the **aortic arch.** As you can see in Figure 18.15, the

aortic arch gives off three large elastic arteries (listed under "Branches of the aortic arch"):

- **Brachiocephalic artery.** The first branch off the aortic arch is the short **brachiocephalic artery** (bray'-kee-oh-seh-FAL-ik; *brachi-* = arm, *cephal-* = head), also known as the **brachiocephalic trunk.** This short artery branches into the **right common carotid artery** that supplies the right side of the head and neck and the **right subclavian artery** (sub-KLAY-vee-in; "below the clavicle") that supplies the right upper limb and thorax.
- **Left common carotid artery.** The middle branch of the aortic arch is the **left common carotid artery,** which supplies the left side of the head and neck.
- **Left subclavian artery.** The third branch is the **left subclavian artery** that supplies the left upper limb and thorax.

Notice that the branches of the aortic arch are asymmetrical, which reflects the asymmetrical location and shape of the heart in the thoracic cavity.

3. **Descending thoracic aorta.** The aortic arch turns inferiorly to become the **descending thoracic aorta,** which travels through the mediastinum posterior to the heart. Its branches supply thoracic structures.

4. **Descending abdominal aorta.** The descending thoracic aorta pierces the diaphragm and enters the abdominopelvic cavity to become the **descending abdominal aorta,** whose branches supply the abdominal viscera. At about the level of the fourth lumbar vertebra, it splits into its terminal branches: the **right** and **left common iliac arteries** (EE-lee-ak). These arteries then split into the **internal iliac arteries,** which supply pelvic structures, and the **external iliac arteries** (listed under "Arteries of the lower limb"), which supply the lower limbs.

Now let's explore the anatomy and distribution of the major named branches of the aorta's divisions. As you read and study the accompanying figures, notice that many arteries are named for the organ or region of the body they supply. For example, the *brachial artery* supplies the arm and the *renal artery* supplies the kidney. Remember this tip as you progress through this module; it will make it easier to remember the names of these arteries.

Arteries of the Head and Neck

◀◀ FLASHBACK

1. Where are the transverse foramina located? What is their function? (p. 233)

2. Where is the carotid canal located? What structure passes through it? (p. 217)

Most of the arterial supply to the head and neck comes from the right and left common carotid arteries, although the subclavian

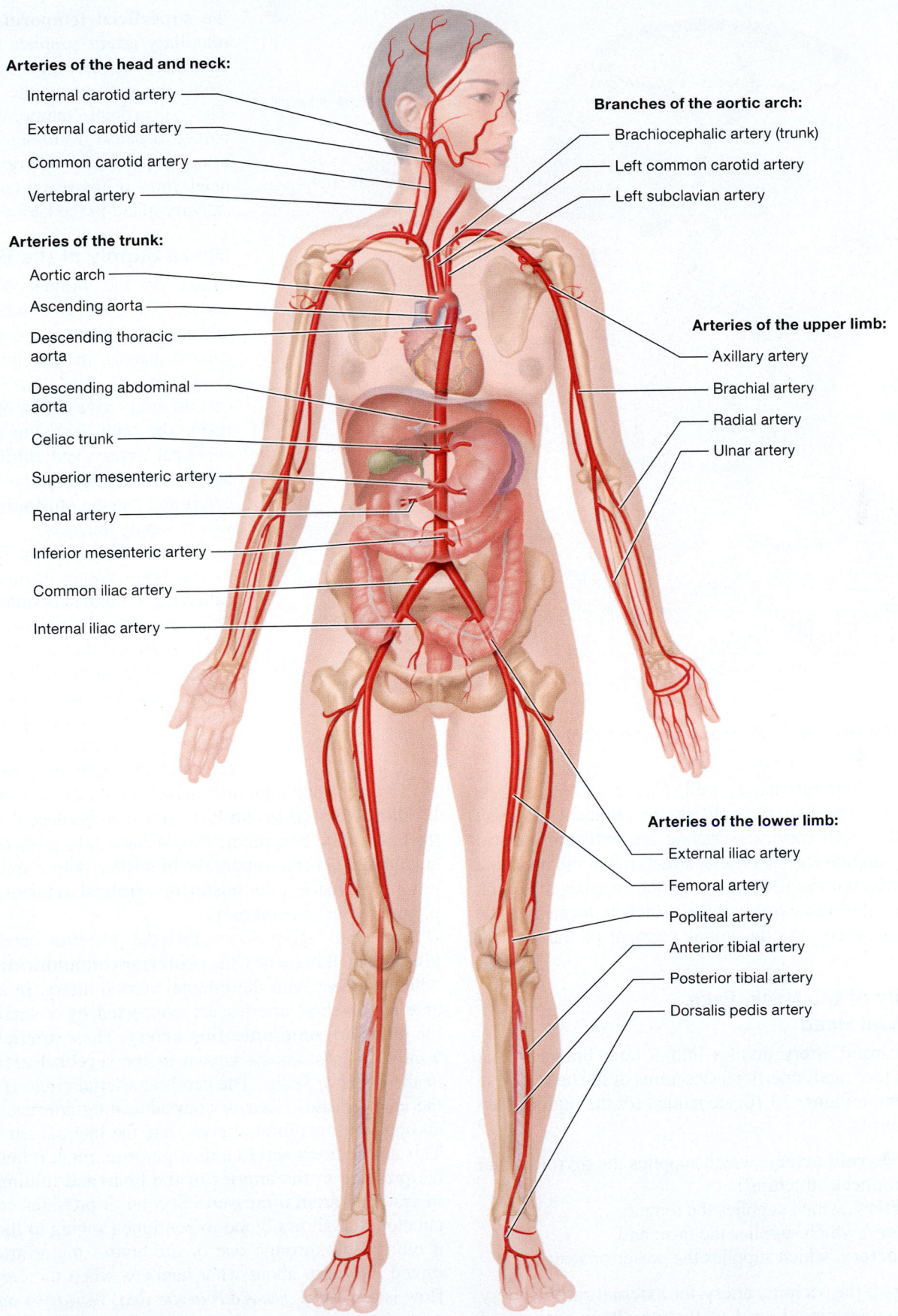

Arteries of the head and neck:

Internal carotid artery

External carotid artery

Common carotid artery

Vertebral artery

Arteries of the trunk:

Aortic arch

Ascending aorta

Descending thoracic aorta

Descending abdominal aorta

Celiac trunk

Superior mesenteric artery

Renal artery

Inferior mesenteric artery

Common iliac artery

Internal iliac artery

Branches of the aortic arch:

Brachiocephalic artery (trunk)

Left common carotid artery

Left subclavian artery

Arteries of the upper limb:

Axillary artery

Brachial artery

Radial artery

Ulnar artery

Arteries of the lower limb:

External iliac artery

Femoral artery

Popliteal artery

Anterior tibial artery

Posterior tibial artery

Dorsalis pedis artery

Figure 18.15 **The major systemic arteries.**

Practice art labeling
@ **Mastering** Anatomy & Physiology

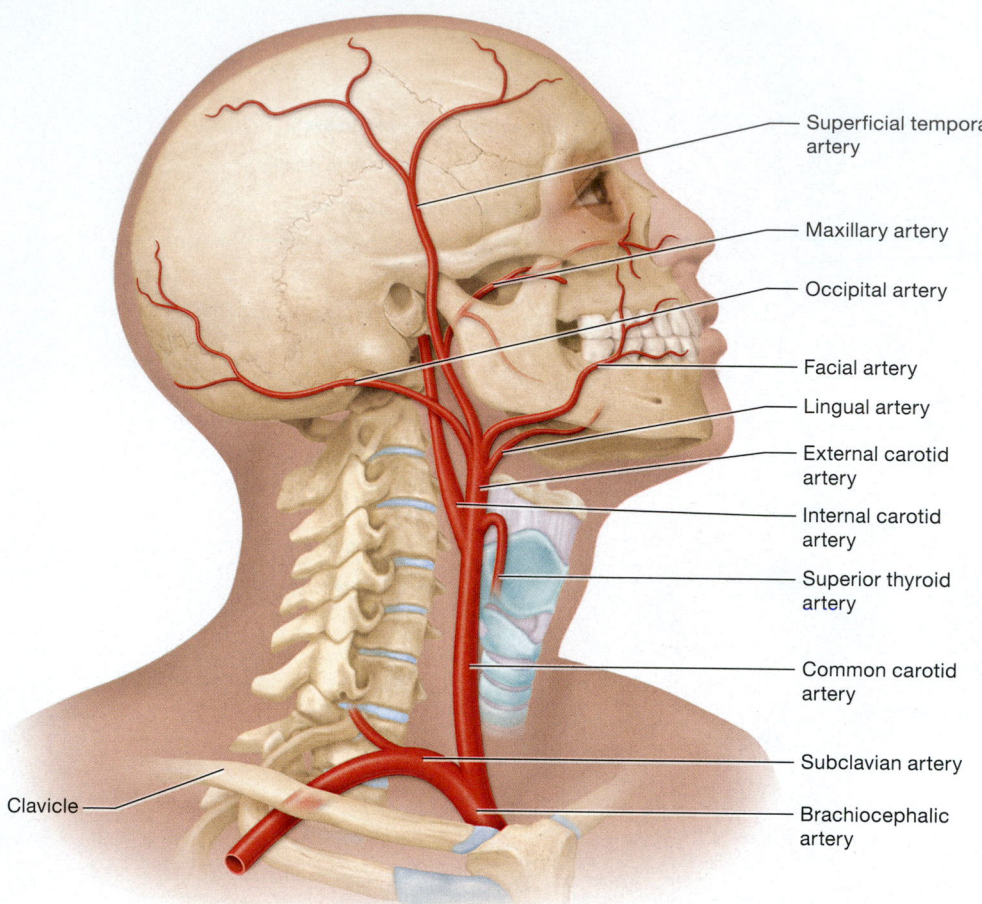

Superficial temporal artery

Maxillary artery

Occipital artery

Facial artery

Lingual artery

External carotid artery

Internal carotid artery

Superior thyroid artery

Common carotid artery

Subclavian artery

Brachiocephalic artery

Clavicle

Figure 18.16 Arteries of the neck, face, and superficial head.

arteries provide contributions, as well. Notice in **Figure 18.16** that the common carotid arteries split into two branches at about the level of the fourth cervical vertebra: the **external carotid artery,** which supplies the superficial structures of the head and face, and the **internal carotid artery,** which supplies the brain. At this split we find the *carotid sinus,* which is located in the common carotid artery and the initial segment of the internal carotid artery.

Blood Supply of the Neck, Face, and Superficial Head

The external carotid artery divides into several branches that serve the neck, face, and superficial structures of the head. These branches, shown in Figure 18.16, are named for the regions they supply, as follows:

- **superior thyroid artery,** which supplies the thyroid gland and anterior neck structures;
- **lingual artery,** which supplies the tongue;
- **facial artery,** which supplies the face; and
- **occipital artery,** which supplies the posterior scalp.

After giving off the occipital artery, the external carotid artery splits into its two terminal branches: the **maxillary artery** and

the **superficial temporal artery.** The maxillary artery supplies many deeper structures of the face, including the teeth, the gums, and the nasal cavity. The superficial temporal artery, so named because it crosses the temporal bone, supplies the lateral side of the head, most of the scalp, and the parotid salivary gland in the lateral cheek.

Blood Supply of the Brain

Much of the brain's blood supply comes from the two internal carotid arteries, which enter the skull via the carotid canals in the temporal bone (**Figure 18.17a** and **b**). The internal carotid artery gives off several branches inside the skull, including the **anterior cerebral artery** and **middle cerebral artery,** which supply the lobes of the cerebrum, and the **ophthalmic arteries,** which supply the eyes.

The remainder of the blood supply to the brain comes from two vessels called the **vertebral arteries,** which are branches of the right and left subclavian arteries. Recall that the vertebral arteries ascend through the neck by passing through the *transverse foramina* in the transverse processes of the cervical vertebrae (see Chapter 7). The two arteries enter the skull by passing through the foramen magnum, after which they fuse to form the single **basilar artery** (BAY-zih-lur), so named because it travels along the base of the brainstem. The basilar artery gives off numerous small branches that supply the brainstem before splitting into its terminal divisions: the **posterior cerebral arteries,** which supply the posterior cerebrum.

Notice in **Figure 18.17c** that the posterior cerebral arteries give off small branches, the **posterior communicating arteries,** which connect with the internal carotid artery. In addition, the anterior cerebral arteries are connected by a small collateral, the **anterior communicating artery.** These arteries are part of a circular anastomosis known as the **cerebral arterial circle,** or the *circle of Willis.* The cerebral arterial circle is made up of the anterior and posterior communicating arteries, the anterior and posterior cerebral arteries, and the internal carotid arteries. This anastomosis serves a dual purpose. First, it helps to equalize pressure in the arteries of the brain and minimize changes in systemic arterial pressure. Second, it provides collateral circulation that allows blood to continue flowing to the brain even if blood flow through one of the brain's major arteries is disrupted. To learn about what happens when the cerebral blood flow is disrupted, see *A&P in the Real World: Cerebrovascular Accident.*

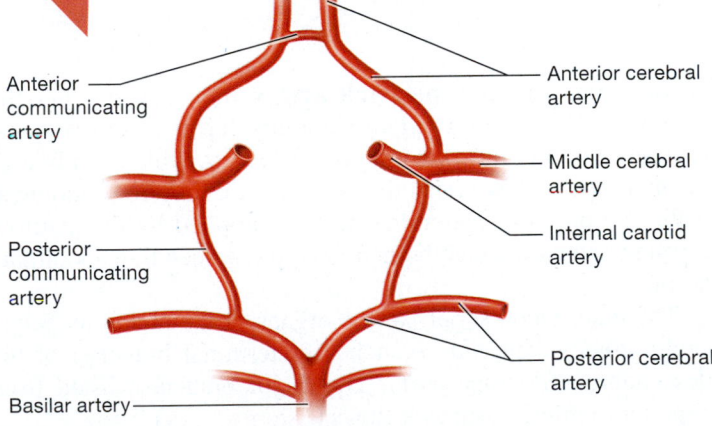

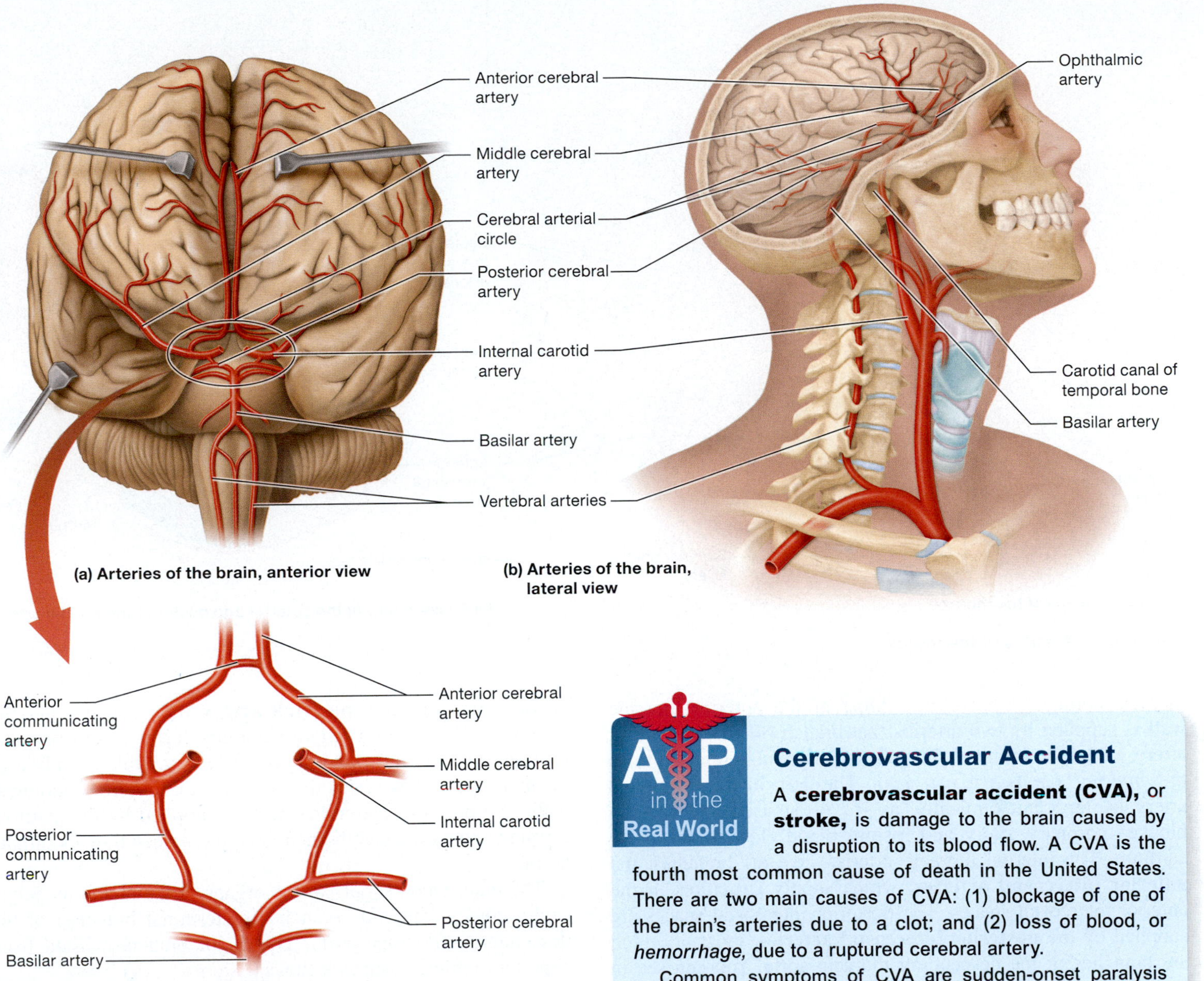

(a) Arteries of the brain, anterior view

Anterior cerebral artery

Middle cerebral artery

Cerebral arterial circle

Posterior cerebral artery

Internal carotid artery

Basilar artery

Vertebral arteries

(b) Arteries of the brain, lateral view

Ophthalmic artery

Carotid canal of temporal bone

Basilar artery

Anterior communicating artery

Anterior cerebral artery

Middle cerebral artery

Internal carotid artery

Posterior communicating artery

Posterior cerebral artery

Basilar artery

(c) Detail of cerebral arterial circle, anterior view

Figure 18.17 Arteries of the brain.

Quick Check

- ☐ 1. List the three branches of the aortic arch.
- ☐ 2. What are the two branches of the common carotid artery, and what do they supply?
- ☐ 3. What is the cerebral arterial circle, and what is its purpose? Which arteries contribute to it?

Arteries of the Thorax

The blood supply to the thorax and the thoracic organs is derived from branches of the subclavian arteries and the descending

Cerebrovascular Accident

A **cerebrovascular accident (CVA),** or **stroke,** is damage to the brain caused by a disruption to its blood flow. A CVA is the fourth most common cause of death in the United States. There are two main causes of CVA: (1) blockage of one of the brain's arteries due to a clot; and (2) loss of blood, or *hemorrhage,* due to a ruptured cerebral artery.

Common symptoms of CVA are sudden-onset paralysis or paresis (weakness), loss of vision, difficulty speaking or understanding speech, and headache. Symptoms often affect only one side of the body; however, they can occasionally occur on both sides of the body when multiple clots break off and block several vessels downstream.

Risk factors for CVA include hypertension, atherosclerosis (particularly in the carotid arteries), diabetes mellitus, cigarette smoking, hypercholesterolemia, and a cardiac dysrhythmia called atrial fibrillation. In addition, the risk of CVA increases with age, and women have a higher lifetime risk of CVA than do men.

The treatment of a CVA due to ischemia generally includes medications to dissolve the clot and thin the blood. A hemorrhagic stroke usually requires surgery to repair the damaged vessel. Regardless of its cause, a CVA requires prompt treatment, as any delay can result in permanent neurological damage or even death.

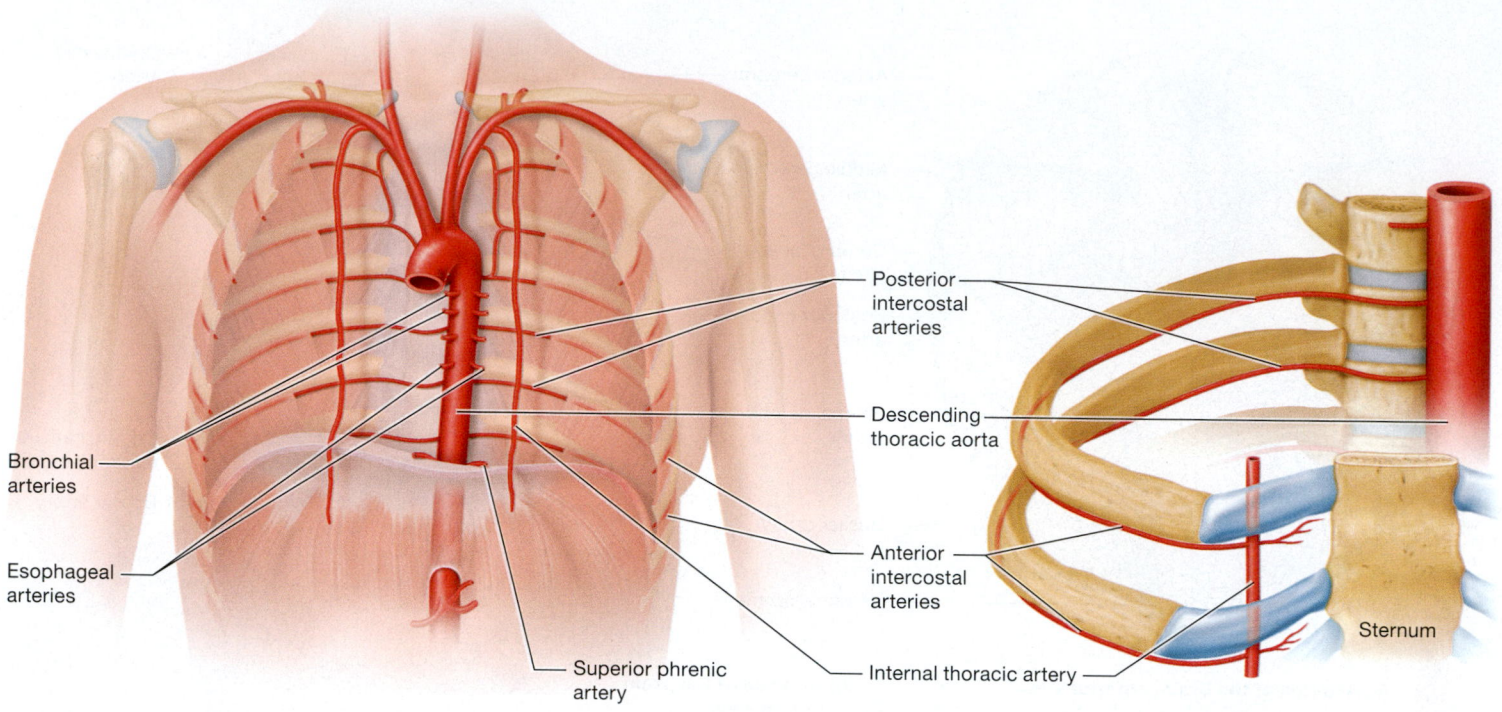

(a) Arterial supply of the thorax

(b) Anastomosis of the anterior and posterior intercostal arteries

Figure 18.18 Arteries of the thorax.

thoracic aorta (**Figure 18.18a**). Most of the anterior thoracic wall is supplied by two arteries that branch off the subclavian artery, called the **internal thoracic arteries** (also referred to as the *internal mammary arteries*). The left internal thoracic artery is important clinically, as it's often used to bypass a blocked coronary artery and restore blood flow to the myocardium. The internal thoracic arteries give off branches, the **anterior intercostal arteries,** which supply structures in the anterior intercostal spaces. The posterior intercostal spaces are supplied by the **posterior intercostal arteries,** most of which are branches of the descending thoracic aorta. You can see in **Figure 18.18b** that the anterior and posterior intercostal arteries meet and form anastomoses.

The organs of the thoracic cavity are supplied by arteries that primarily branch from the descending thoracic aorta. These arteries include the following:

- **bronchial arteries,** which supply the lungs (don't confuse these with the vessels of the pulmonary circuit—the bronchial arteries are separate and part of the systemic circuit);
- **esophageal arteries,** which supply the esophagus; and
- **superior phrenic arteries,** which supply portions of the diaphragm.

Certain thoracic arteries also supply structures of the abdomen (just as some abdominal arteries supply structures of the thorax), which we discuss next.

Arteries of the Abdominal Organs

The superior portion of the anterior abdominal wall is supplied by the termination of the internal thoracic artery, which

becomes the **superior epigastric artery** after it crosses into the abdomen. The superior epigastric artery forms extensive anastomoses with the **inferior epigastric artery,** which is a branch of the external iliac artery that supplies the inferior abdominal wall. The posterior abdominal wall is supplied by five pairs of **lumbar arteries,** which branch from the descending abdominal aorta.

The main blood supply to the organs of the abdominopelvic cavity comes from side branches or terminal branches of the descending abdominal aorta. The major branches, listed from superior to inferior, are as follows (**Figure 18.19a**):

1. **Inferior phrenic arteries.** With the superior phrenic arteries, the small **inferior phrenic arteries** supply blood to the diaphragm.
2. **Celiac trunk.** The short, unpaired **celiac trunk** (SEE-lee-ak; *celi-* = "abdominal cavity") is the next branch (**Figure 18.19b**). It splits immediately into three separate arteries that supply abdominal organs:
   - The **common hepatic artery** travels off to the right side of the body in the direction of the liver. It gives off two branches, the **right gastric artery,** which supplies the stomach, and the **gastroduodenal artery,** which supplies the stomach, pancreas, and duodenum (doo-AW-den-um; the first part of the small intestine). After it gives off the gastroduodenal artery, it becomes the **hepatic artery proper,** which supplies the liver and gallbladder.
   - The **splenic artery** travels toward the spleen and branches to supply the stomach and pancreas. Its ultimate destination is, of course, the spleen.

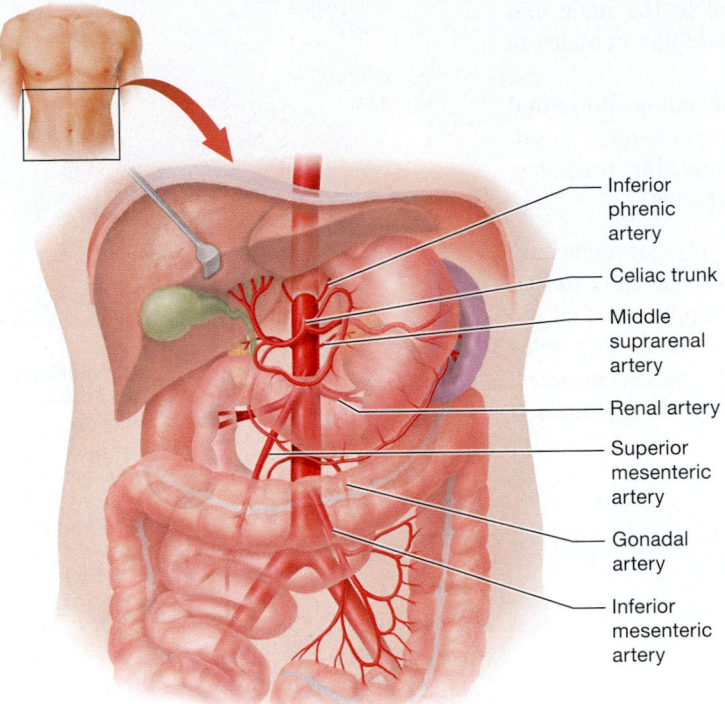

(a) Overview of the abdominal arteries

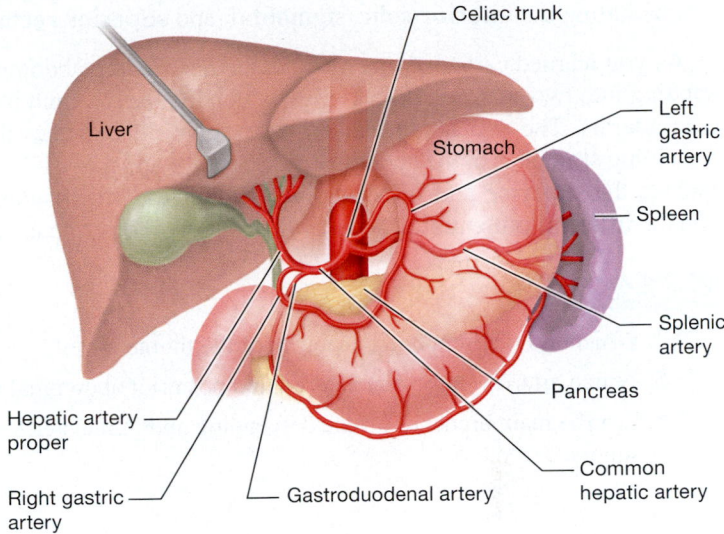

(b) Distribution of the celiac trunk

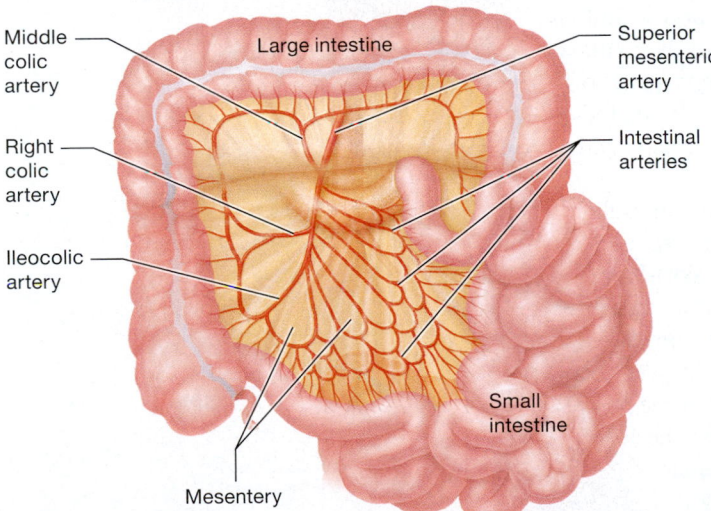

(c) Distribution of the superior mesenteric artery

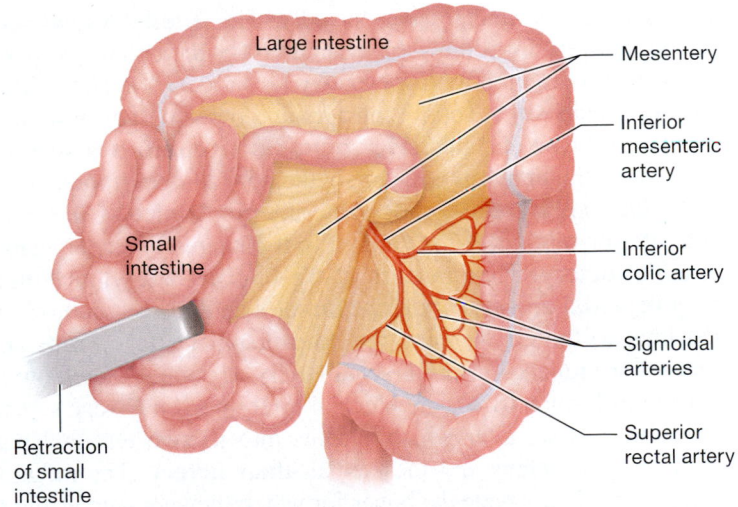

(d) Distribution of the inferior mesenteric artery

Figure 18.19 Arteries of the abdomen.

- The **left gastric artery** is the smallest branch of the celiac trunk. It supplies the stomach and forms an anastomosis with the right gastric artery.

3. **Middle suprarenal arteries.** The small, paired **middle suprarenal arteries** supply the adrenal glands that rest atop the kidneys (see Figure 18.19a).

4. **Superior mesenteric artery.** The small intestine is the site of the majority of nutrient absorption, and it therefore requires an extensive blood supply. This blood supply comes from the large, single **superior mesenteric artery** (**Figure 18.19c**). Its branches to the small intestine consist of numerous anastomosing **intestinal arteries** that travel through the

mesentery, a double fold of serous membrane surrounding the small intestine. Occlusion of the branches of the superior mesenteric artery can result in *acute mesenteric ischemia,* which may result in the death of intestinal tissue, or a *bowel infarction.* The superior mesenteric artery also supplies much of the large intestine via its branches the **ileocolic** (ee´-lee-oh-KOHL-ik), the **middle colic,** and the **right colic arteries.**

5. **Renal arteries.** The paired **renal arteries** branch from the descending abdominal aorta at about the level of the first and second lumbar vertebrae. As their name implies, they deliver blood to the kidneys.

6. **Gonadal arteries.** The small **gonadal arteries** deliver blood to the male and female gonads, or reproductive organs. They are called the **testicular arteries** in males and the **ovarian arteries** in females.

7. **Inferior mesenteric artery.** The final major branch off the descending abdominal aorta is the unpaired **inferior mesenteric artery.** As you can see in **Figure 18.19d**, it supplies the remainder of the large intestine via its anastomosing branches, including the **inferior colic, sigmoidal,** and **superior rectal arteries.**

As you learned earlier in this module, the descending abdominal aorta terminates by splitting into two common iliac arteries, which themselves split into internal and external iliac arteries. The two internal iliac arteries give off branches that supply most of the remaining structures in the abdominopelvic cavity and the pelvic musculature. Examples include the **superior** and **inferior gluteal arteries,** which supply the gluteal muscles, and the **internal pudendal arteries,** which supply the perineum and external genitalia.

> ### Quick Check
>
> ☐ 4. Which arteries supply the majority of the thoracic wall?
>
> ☐ 5. Which arteries supply the anterior and posterior abdominal wall?
>
> ☐ 6. List the main branches off the descending abdominal aorta and the structures they supply.

Arteries of the Upper Limb

The main arteries that supply the upper and lower limbs are like the streets in a city—their names can change unexpectedly. The arterial supply of the upper limb is derived entirely from the subclavian artery, which changes its name in the axillary region to become the **axillary artery** (**Figure 18.20**). The axillary artery gives off several branches that supply muscles and other tissues around the axilla, including the **anterior** and **posterior humeral circumflex arteries,** which surround the head of the humerus and supply the bone and the shoulder joint.

As the axillary artery passes the inferior border of the teres major muscle (the inferior portion of the "armpit"), it changes its name again to become the **brachial artery.** The branches of the brachial artery, including the posterior **deep brachial artery,** supply the muscles and other structures of the arm. Just proximal to the elbow joint, the deep brachial artery forms an anastomosis with other branches of the brachial artery to supply the elbow joint. As you can see in Figure 18.20, the brachial artery becomes very superficial in the antecubital fossa, after which it splits into its two terminal branches: the lateral **radial artery** and the medial **ulnar artery.** The radial and ulnar arteries travel alongside the bones for which they are named and supply structures of the forearm. They terminate in the palm, in arterial anastomoses called the **superficial** and **deep palmar arches.**

> ### Quick Check
>
> ☐ 7. Which artery supplies the upper limb?
>
> ☐ 8. Trace the arterial supply of the upper limb from the axilla to the hand.

Arteries of the Lower Limb

The pattern of blood flow through the lower limb resembles the pattern of blood flow through the upper limb. We start in each with a main vessel that changes its name in different regions and bifurcates into terminal branches. In the lower limb, the main vessel is the **femoral artery.** As you can see in **Figure 18.21**, the external

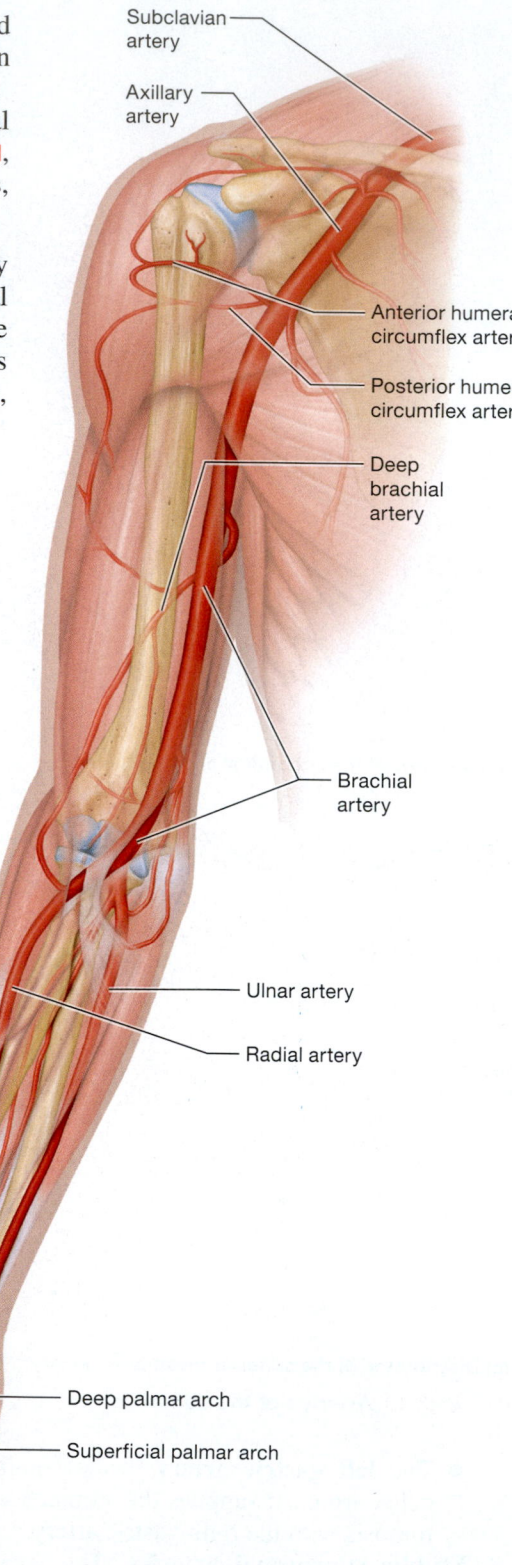

Figure 18.20 Arteries of the upper limb.

Subclavian artery

Axillary artery

Anterior humeral circumflex artery

Posterior humeral circumflex artery

Deep brachial artery

Brachial artery

Ulnar artery

Radial artery

Deep palmar arch

Superficial palmar arch

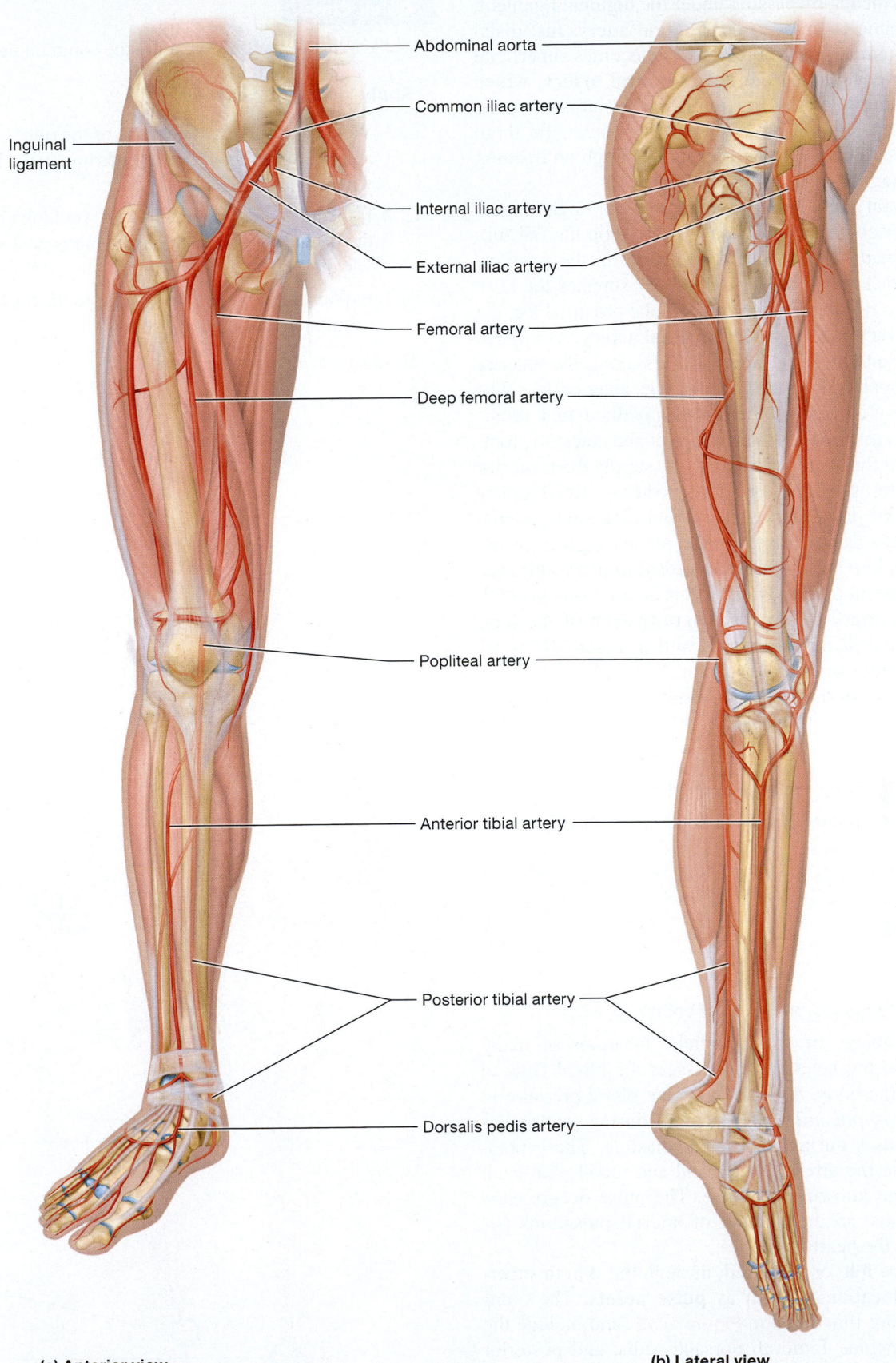

Abdominal aorta

Common iliac artery

Inguinal ligament

Internal iliac artery

External iliac artery

Femoral artery

Deep femoral artery

Popliteal artery

Anterior tibial artery

Posterior tibial artery

Dorsalis pedis artery

(a) Anterior view

(b) Lateral view

Figure 18.21 Arteries of the lower limb.

iliac artery enters the leg by passing under the inguinal ligament, after which its name changes to the femoral artery. Just distal to the inguinal ligament, the femoral artery becomes superficial and gives off a branch called the **deep femoral artery,** which supplies the structures of the hip joint, the femur, and many thigh muscles. The femoral artery then dives under the thigh muscles and passes into the posterior thigh through an opening in the adductor magnus muscle.

When the femoral artery emerges in the posterior thigh, it takes a new name, to become the **popliteal artery** (pop-lih-TEE-ul). This vessel is named for the fact that it resides in the *popliteal fossa* (the posterior knee). The popliteal artery supplies the knee joint and splits into two terminal branches in the proximal leg: the **anterior tibial artery** and the **posterior tibial artery.** As implied by its name, the anterior tibial artery travels along the anterior tibia, where it supplies the structures of the anterior leg. The anterior tibial artery is renamed the **dorsalis pedis artery** (dohr-SAL-iss PEE-diss) as it crosses the ankle joint and enters the foot.

The branches of the posterior tibial artery supply the posterior and lateral structures of the leg. The posterior tibial artery becomes superficial just distal to the medial malleolus (medial ankle bone), after which it travels to the plantar surface of the foot. Here it splits into the **medial** and **lateral plantar arteries.** A branch of the lateral plantar artery forms an anastomosis with the dorsalis pedis artery to form the **plantar arch** of the foot. (You can't see these plantar structures well in Figure 18.21, as they are on the plantar surface, or sole, of the foot.)

The major arteries of the body are summarized in **Table 18.4.**

Quick Check

☐ 9. Which artery supplies the lower limb?

☐ 10. Trace the arterial supply of the lower limb from the femur to the foot.

Pulse Points

◀◀ FLASHBACK

1. Why do arteries pulsate, but not veins? (p. 679)

We can take advantage of the superficial locations of many arteries to measure the heart rate and assess the blood flow to different parts of the body. As you know, the blood pressure in systemic arteries is *pulsatile*—it increases during ventricular systole and decreases during ventricular diastole. These pressure changes cause the arteries to expand and recoil with each heartbeat, a change known as a **pulse.** The pulse occurs each time the heart beats, so the number of arterial pulsations per minute is equal to the heart rate.

The pulse can be felt, or **palpated,** through the skin in superficial arteries at locations known as **pulse points.** The common pulse points are illustrated in **Figure 18.22**, and include the carotid, radial, brachial, femoral, dorsalis pedis, and posterior tibial arteries. Pulses may also be palpated in the ulnar, superficial temporal, and popliteal arteries; however, these are less common sites in the clinical setting.

Quick Check

☐ 11. What is a pulse? What are the common pulse points?

Apply What You Learned

☐ 1. Would a nearly total blockage of the right internal carotid artery completely cut off blood flow to the brain? Why or why not?

☐ 2. If the ulnar artery were severed, would the blood supply to the medial side of the hand and digits be disrupted? Why or why not?

☐ 3. Explain why irreparable damage to the femoral artery may result in loss of the lower limb.

See answers in Appendix A.

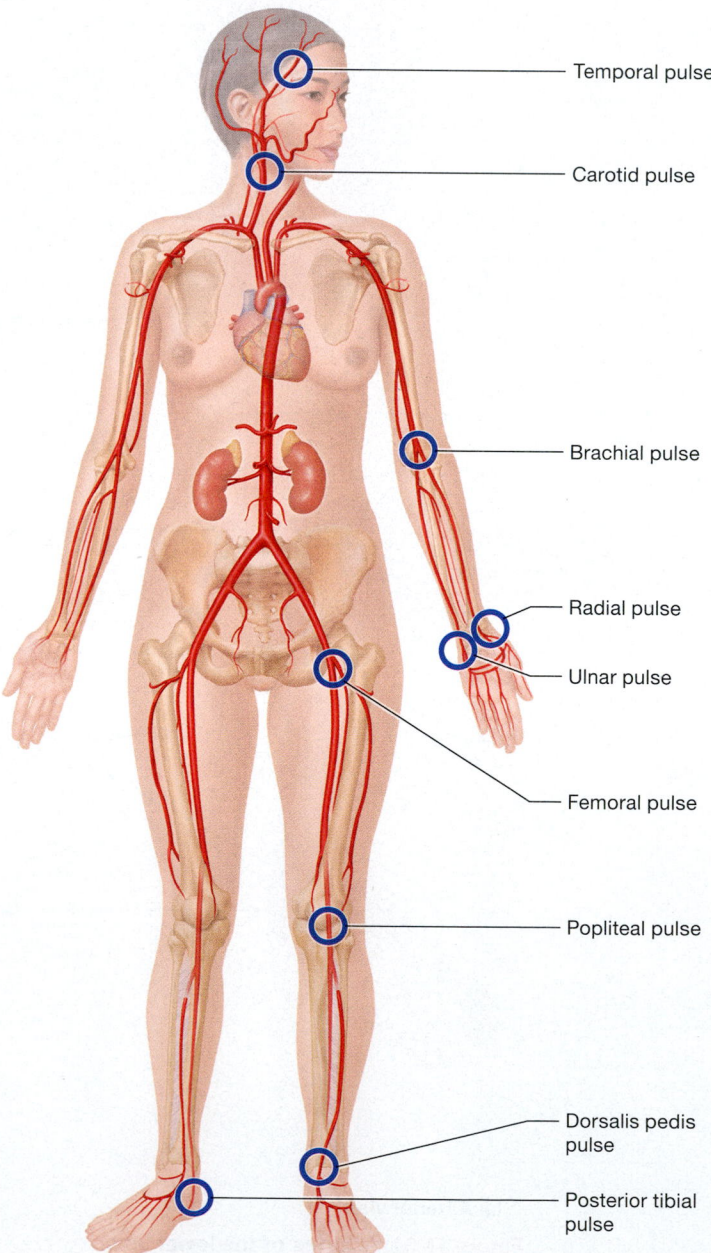

Figure 18.22 Common pulse points.

Table 18.4 Major Systemic Arteries

Artery	Description	Location	Structure(s) Supplied
Aorta	Largest artery in the body with four divisions: ascending aorta, aortic arch, descending thoracic aorta, descending abdominal aorta	Mediastinum and posterior abdominal wall	• Branches supply all structures in the body.
Brachiocephalic artery	First branch off the aortic arch	Right side of the mediastinum	• Branches supply the right side of the upper body.
Common carotid arteries	Right is a branch of the brachiocephalic artery; left is the second branch off the aortic arch.	Lateral neck	• Head • Neck • Brain
Subclavian arteries	Right is a branch of the brachiocephalic artery; left is the third branch off the aortic arch.	Deep to the clavicle	• Thorax • Upper limb
External carotid artery	Superficial branch of the common carotid artery	Lateral neck	• Neck • Face • Superficial head
Internal carotid artery	Deep branch of the common carotid artery	Deep neck, passes through the carotid canals into the brain	• Brain • Orbit
Vertebral arteries	Branches of the subclavian that fuse in the brain to form the basilar artery	Pass through transverse foramina of cervical vertebrae and enter the foramen magnum	• Brain
Celiac trunk	Large unpaired artery, second branch off the descending abdominal aorta	Abdominal cavity, just inferior to the diaphragm	• Structures of the abdominal cavity
Common hepatic artery	Branch off the celiac trunk; splits into the right gastric artery, the gastroduodenal artery, and the hepatic artery proper	Right side of the celiac trunk	• Liver • Stomach • Duodenum • Pancreas
Left gastric artery	Branch off the celiac trunk	Along the medial side of the stomach	• Stomach
Splenic artery	Third branch of the celiac trunk	Posterior to the stomach	• Spleen • Stomach • Pancreas
Superior mesenteric artery	Large unpaired artery; fourth branch off the descending abdominal aorta	Branches fan throughout small and large intestines.	• Small intestine • Most of the large intestine
Renal arteries	Approximately the fifth and sixth branches off the descending abdominal aorta	Travel to the kidneys	• Kidneys
Inferior mesenteric artery	Final branch off the descending abdominal aorta	Branches fan throughout terminal part of the large intestine.	• Remainder of the large intestine
Axillary artery	Continuation of the subclavian artery	Axilla	• Structures around the axilla
Brachial artery	Continuation of the axillary artery	Arm	• Arm
Radial artery	Terminal branch of the brachial artery	Lateral forearm	• Forearm and hand
Ulnar artery	Terminal branch of the brachial artery	Medial forearm	• Forearm and hand
Common iliac arteries	Terminal branches of the descending abdominal aorta	Posterior pelvic wall	• Structures of the pelvis and lower limb
Internal iliac artery	Terminal branch of the common iliac artery	Posterior pelvis	• Pelvic structures
External iliac artery	Terminal branch of the common iliac artery	Anterior pelvis	• Anterior abdominal wall
Femoral artery	Continuation of the external iliac artery	Anterior thigh	• Thigh
Popliteal artery	Posterior continuation of the femoral artery	Popliteal fossa	• Knee joint • Posterior leg • Posterior thigh
Anterior tibial artery	Terminal branch of the popliteal artery	Anterior tibia	• Anterior leg • Anterior foot
Dorsalis pedis artery	Continuation of the anterior tibial artery	Ankle and foot	• Ankle joint • Foot
Posterior tibial artery	Terminal branch of the popliteal artery	Posterior leg, becomes superficial around the medial malleolus	• Posterior leg • Plantar foot

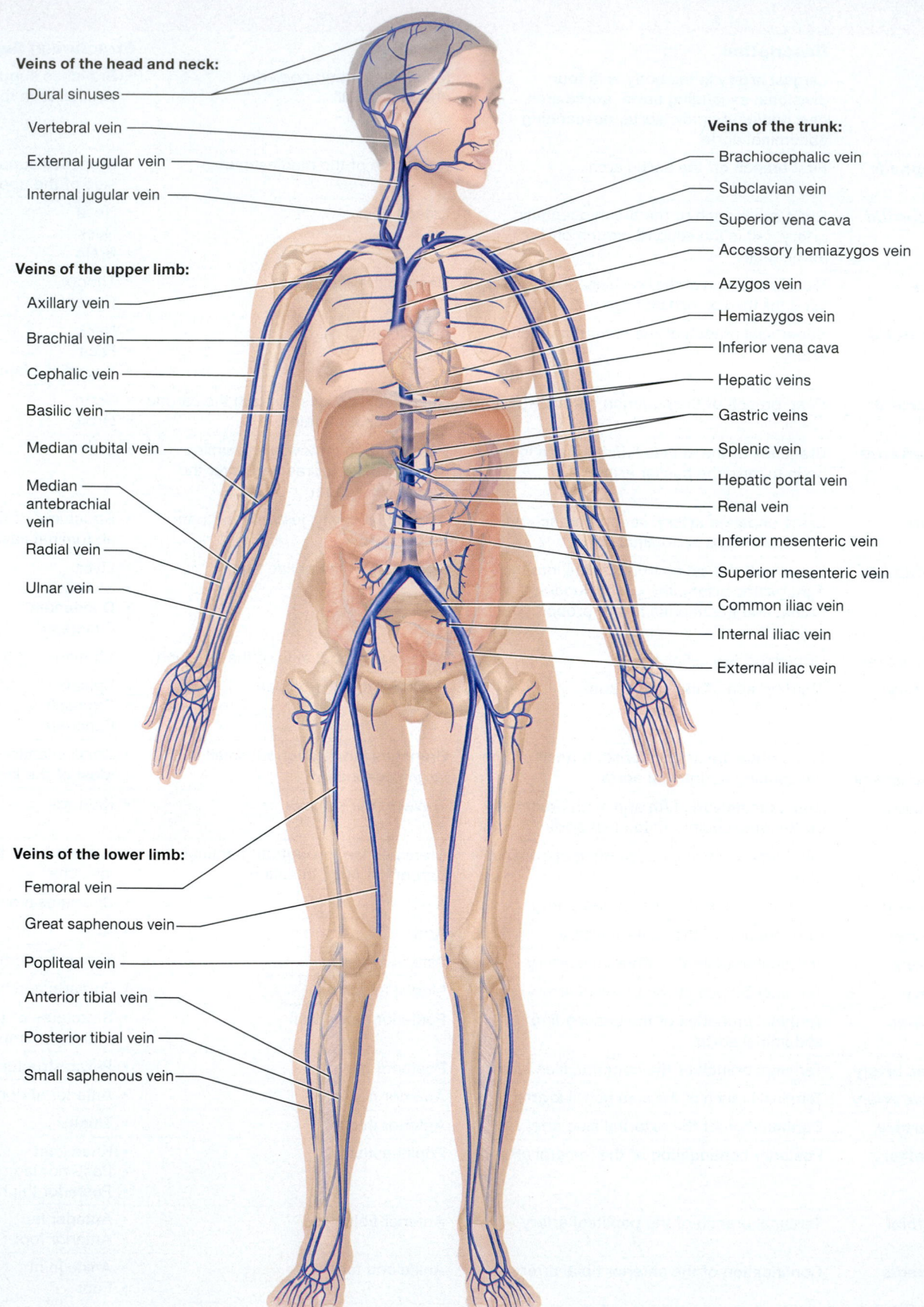

Veins of the head and neck:

Dural sinuses

Vertebral vein

External jugular vein

Internal jugular vein

Veins of the upper limb:

Axillary vein

Brachial vein

Cephalic vein

Basilic vein

Median cubital vein

Median antebrachial vein

Radial vein

Ulnar vein

Veins of the trunk:

Brachiocephalic vein

Subclavian vein

Superior vena cava

Accessory hemiazygos vein

Azygos vein

Hemiazygos vein

Inferior vena cava

Hepatic veins

Gastric veins

Splenic vein

Hepatic portal vein

Renal vein

Inferior mesenteric vein

Superior mesenteric vein

Common iliac vein

Internal iliac vein

External iliac vein

Veins of the lower limb:

Femoral vein

Great saphenous vein

Popliteal vein

Anterior tibial vein

Posterior tibial vein

Small saphenous vein

Figure 18.23 The major systemic veins.

Anatomy of the Systemic Veins

Learning Outcomes

1. Describe the patterns of venous blood drainage for the head and neck, the thoracic cavity, the abdominopelvic cavity, and the upper and lower limbs.

2. Identify major veins of the systemic circuit.

3. Describe the structure and function of the hepatic portal system.

Deoxygenated blood that has traveled through capillary beds drains into systemic veins to be returned to the heart. Most veins superior to the diaphragm, including those of the head, neck, thorax, and upper limbs, drain into the two **brachiocephalic veins** (listed under "Veins of the trunk"; **Figure 18.23**). Notice that two brachiocephalic veins are present, whereas only a single brachiocephalic artery is present (on the right side of the body). The two brachiocephalic veins in turn merge to form the **superior vena cava,** which empties into the right atrium.

Venous blood from the lower limbs and pelvis drains into the **external** and **internal iliac veins,** respectively. These two vessels merge to form the **common iliac veins,** which themselves merge to form the **inferior vena cava.** Most veins inferior to the diaphragm drain into the inferior vena cava as it ascends through the abdominal cavity. The inferior vena cava pierces the diaphragm and empties directly into the right atrium.

In this module we discuss the pathways taken by venous blood as it returns to the heart through the systemic circuit. You will notice that venous return in many ways parallels the arterial circuit, and many veins share the same names with the corresponding arteries, which makes their names easier to remember.

Veins of the Head and Neck

◀◀ FLASHBACK

1. What is the dura mater? What is the falx cerebri? (p. 451)

2. What are the dural sinuses, and where are they located? (p. 451)

3. What are arachnoid granulations, and what is their main function? (p. 452)

We begin with the veins that drain the structures of the head and neck. Although much of the venous drainage of the head and neck parallels its arterial supply, you'll see that the circuit of the brain is somewhat more complicated. These subsections examine the anatomy of those veins that drain the superficial structures of the head and neck and the venous system that drains the brain.

Venous Drainage of the Neck, Face, and Scalp

As you can see in **Figure 18.24**, three main veins drain the head and neck: the deep **internal jugular vein** and **vertebral vein** and the superficial **external jugular vein** (listed under "Veins of the neck"). Note that there is no common jugular vein. The external jugular and vertebral veins empty into the **subclavian vein,** which merges with the internal jugular vein to become the brachiocephalic vein. The vertebral vein primarily drains structures of the posterior neck and travels with the vertebral artery in the transverse foramina of cervical vertebrae. The veins that drain the face share the same names with many of the arteries that supply this region. The **superficial temporal, facial,** and **maxillary veins** (listed under "Veins of the head and face") drain into both the internal and external jugular veins.

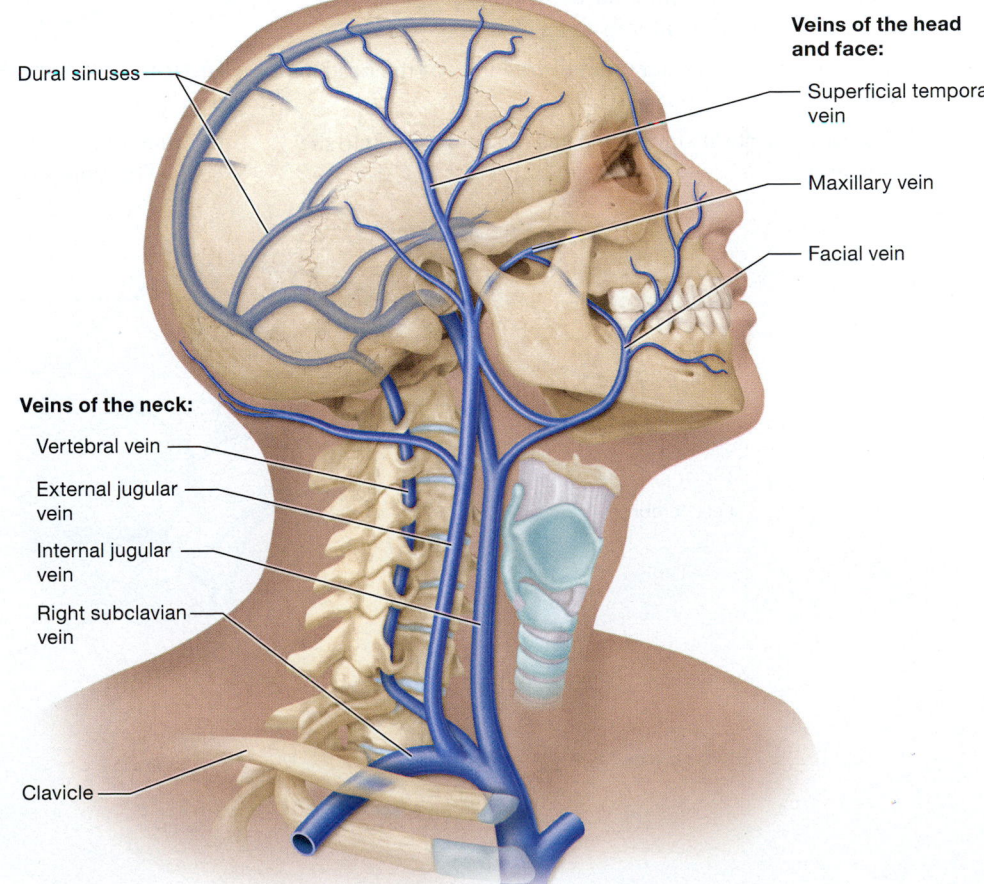

Figure 18.24 Veins of the neck and superficial head.

Venous Drainage of the Brain

Recall that the brain is surrounded by a tough connective tissue membrane called the dura mater (see Chapter 12). The dura mater consists of two layers of tissue that are fused over much of the brain. However, these layers separate in certain locations, and the spaces between these two layers form a system of venous channels known as the **dural sinuses** (Figure 18.25). Blood from the brain capillaries drains into **cerebral veins,** which then drain into the dural sinuses. In addition, the dural sinuses receive cerebrospinal fluid from the arachnoid granulations that project into these venous channels, which allows this fluid to return to the circulation.

The dural sinus system begins with the **superior sagittal sinus** in the longitudinal fissure of the brain. In the inferior falx cerebri is the **inferior sagittal sinus,** which drains posteriorly into the **straight sinus.** Notice in the figure how the straight sinus merges with the superior sagittal sinus to form the two **transverse sinuses** on the floor of the occipital bone. The transverse sinuses travel anteriorly and become the S-shaped **sigmoid sinuses,** which drain into the internal jugular vein. The sigmoid sinuses also receive blood from the anastomosing **cavernous sinuses,** which lie in the anterior and inferior cranial cavity and drain the ophthalmic veins.

Quick Check

☐ 1. Where do most veins superior to the diaphragm drain? Where do most veins inferior to the diaphragm drain?

☐ 2. Which veins drain the superficial structures of the head and face?

☐ 3. Where are the dural sinuses located? What drains into the dural sinuses?

Veins of the Thorax and Abdomen

The venous drainage of the thorax and abdomen is more complicated than their arterial supply, although many vessels do share the same names. The following subsections examine these vessels and the drainage systems of the thoracic wall, that of the abdomen and pelvis, and a special circulatory route called the *hepatic portal system.*

Venous Drainage of the Thoracic and Abdominal Walls

The drainage of the anterior thorax and abdomen is fairly straightforward—the anterior thorax is drained primarily by the **internal thoracic veins,** which empty into the brachiocephalic veins, and the anteroinferior abdomen is drained by the **inferior epigastric veins,** which empty into the external iliac veins. However, notice in Figure 18.26 that the posterior thoracic and abdominal walls are drained by an entirely different system collectively called the **azygos system** (ay-ZY-gus; *azygos* = "unpaired"). This system consists of three vessels: the **azygos vein** on the right and the **hemiazygos** (hem′-ee-ay-ZY-gus) and **accessory hemiazygos veins** on the left. The left **lumbar veins** (which is not shown in Figure 18.26) and most of the left **posterior intercostal veins** drain into the hemiazygos or accessory hemiazygos veins, which themselves drain into the azygos vein. The same veins on the right, along with the **bronchial** and **esophageal veins,** drain into the azygos vein. The azygos vein merges directly with the superior vena cava.

Venous Drainage of the Pelvic and Abdominal Organs

The pelvic organs are drained by veins that merge into the internal iliac vein, which then empties into the common iliac vein. The common iliac veins, as well as several veins of the

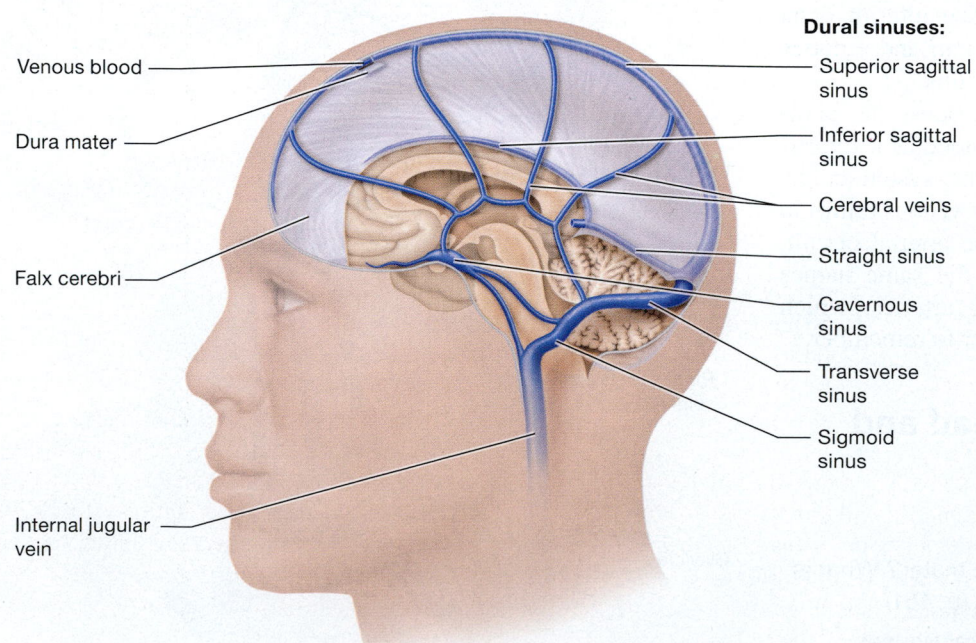

Dural sinuses:
- Superior sagittal sinus
- Inferior sagittal sinus
- Cerebral veins
- Straight sinus
- Cavernous sinus
- Transverse sinus
- Sigmoid sinus

Venous blood

Dura mater

Falx cerebri

Internal jugular vein

Figure 18.25 Dural sinuses and other veins of the brain.

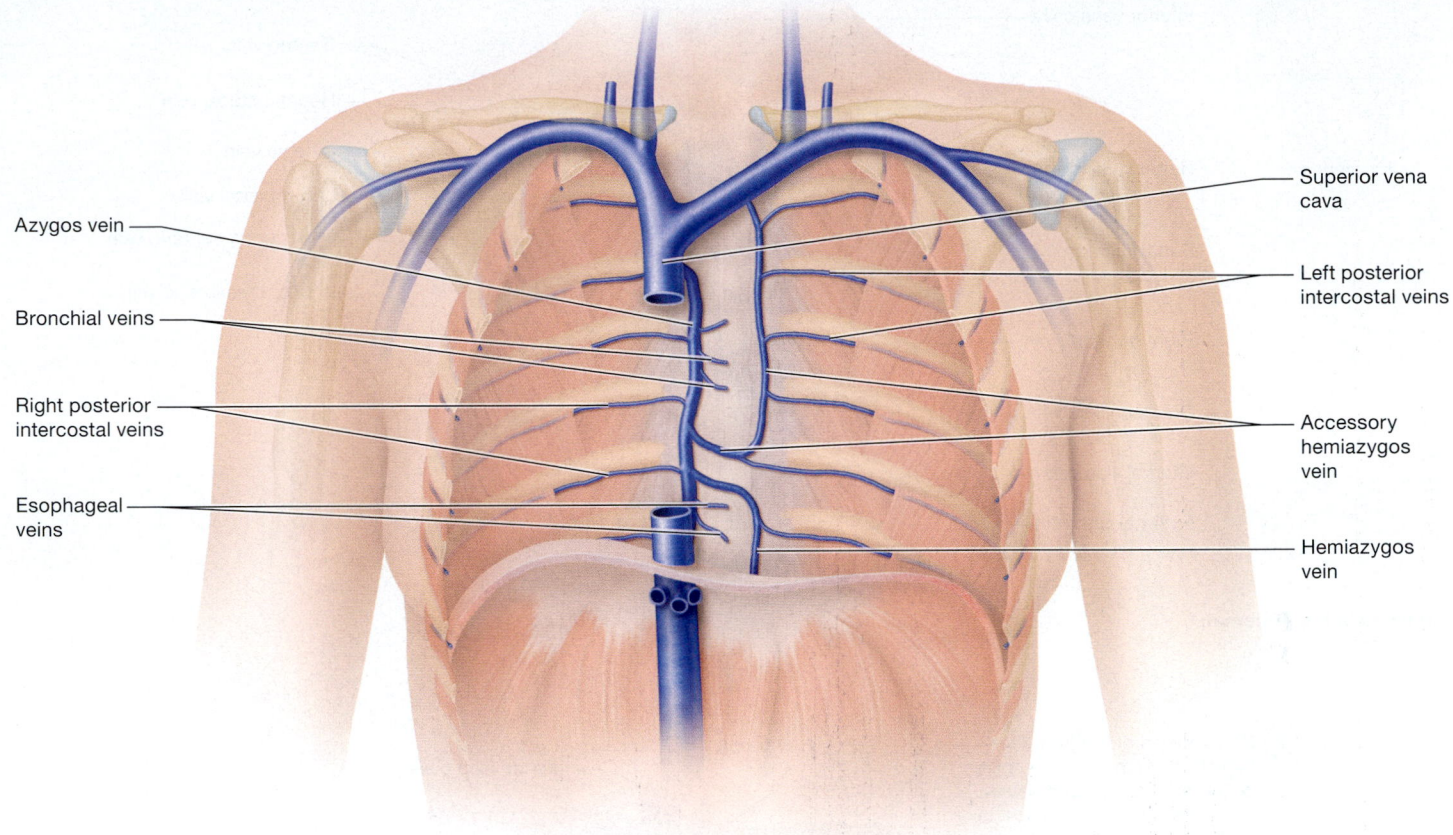

Figure 18.26 Veins of the posterior thorax.

abdominopelvic cavity, merge to form the inferior vena cava. These vessels are illustrated in **Figure 18.27a** and include the following:

- **right gonadal vein,** which drains the testes in males and the ovaries in females (the left gonadal vein typically drains into the left renal vein);
- **renal veins,** which drain the kidneys; and
- **hepatic veins,** which drain the liver.

Other veins of the abdominal cavity include the **splenic vein,** which drains the spleen, and the **gastric veins,** which drain the stomach. In addition, the **superior mesenteric vein** drains the small intestine and much of the large intestine, and the **inferior mesenteric vein** drains the remainder of the large intestine. These veins are part of a special circulation that is discussed next.

The Hepatic Portal System

Take note of something interesting about the splenic, gastric, and superior and inferior mesenteric veins in **Figure 18.27b**—they do not drain into the inferior vena cava. Instead, ① they merge with and drain into a large vein that enters the liver, called the **hepatic portal vein.** The hepatic portal vein branches extensively in the liver to form another set of capillary beds. This special type of circuit in which veins feed a capillary bed is known as a **portal system,** and this one is called the **hepatic portal system.**

The word portal means "gate," and in essence the hepatic portal system is the "gatekeeper" of the abdominal circulation. To understand this role, we must first address some of the varied functions of the liver. Many of the most basic metabolic reactions, such as the storage of glucose, the production of proteins from dietary amino acids, and the formation of new glucose, occur in the cells of the liver. In addition, the cells of the liver, called *hepatocytes* (heh-PAT-oh-sytz), contain an extensive network of smooth endoplasmic reticulum that allows them to carry out *detoxification reactions:* The enzymes interact with chemicals we ingest, catalyzing reactions that alter them to render them less harmful to the body. Finally, hepatocytes also contain enzymes that catalyze reactions to break down and process the products of red blood cell destruction.

Venous blood draining from the stomach and the small and large intestines contains both absorbed nutrients and ingested chemicals such as drugs, preservatives, alcohol, and more, and venous blood from the spleen contains products of red blood cell breakdown. The hepatic portal system in essence allows the liver to have "first dibs" on this venous blood and the nutrients and chemicals it contains. (See *A&P in the Real World: Drugs and the Hepatic Portal System* on p. 713 to read about how this fact influences the administration of medications.) ② This blood is delivered to the liver by the hepatic portal vein, which branches extensively to form a network of capillary beds that consist of

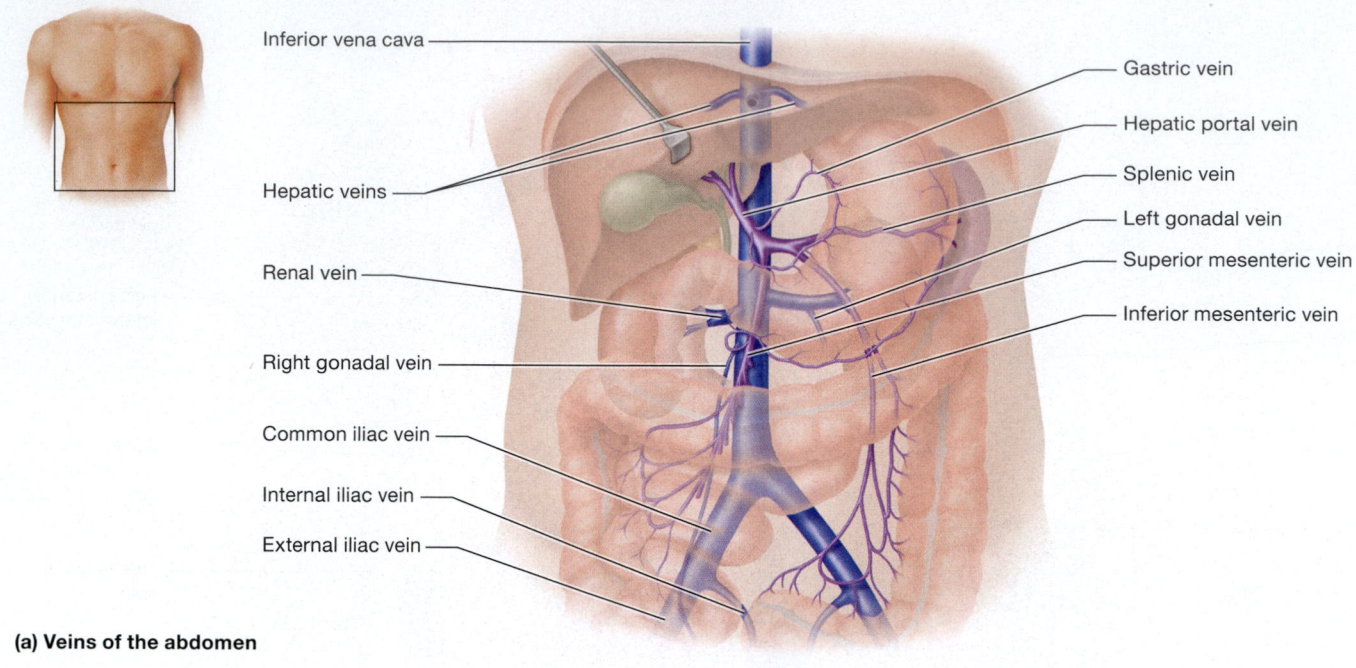

Inferior vena cava

Hepatic veins

Renal vein

Right gonadal vein

Common iliac vein

Internal iliac vein

External iliac vein

Gastric vein

Hepatic portal vein

Splenic vein

Left gonadal vein

Superior mesenteric vein

Inferior mesenteric vein

(a) Veins of the abdomen

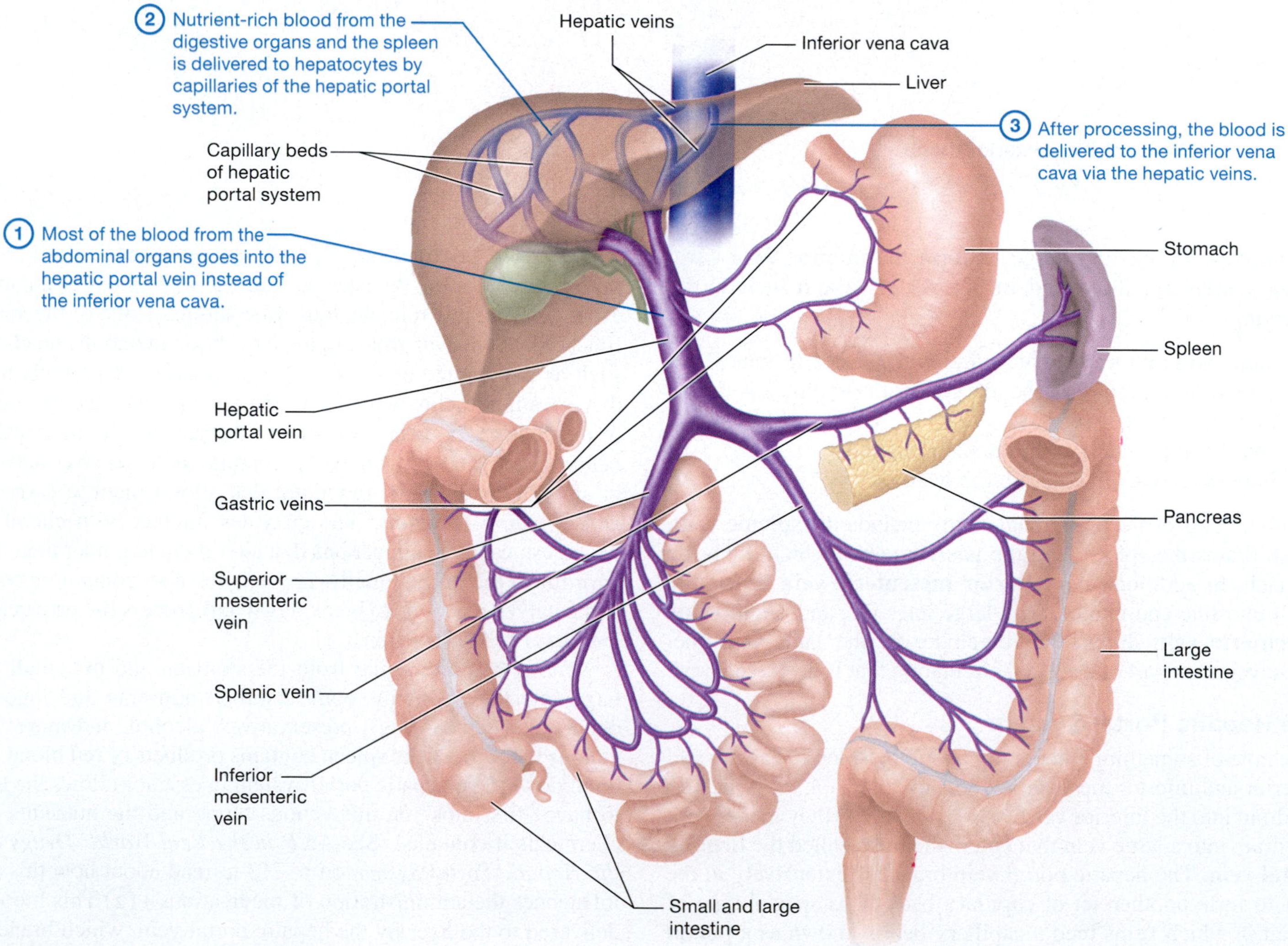

② Nutrient-rich blood from the digestive organs and the spleen is delivered to hepatocytes by capillaries of the hepatic portal system.

Capillary beds of hepatic portal system

① Most of the blood from the abdominal organs goes into the hepatic portal vein instead of the inferior vena cava.

Hepatic portal vein

Gastric veins

Superior mesenteric vein

Splenic vein

Inferior mesenteric vein

Small and large intestine

Hepatic veins

Inferior vena cava

Liver

③ After processing, the blood is delivered to the inferior vena cava via the hepatic veins.

Stomach

Spleen

Pancreas

Large intestine

(b) Organization of the hepatic portal system

Figure 18.27 Veins of the abdomen.

extremely leaky sinusoidal capillaries. Here, the nutrients, toxins, and products of red blood cell degradation exit the sinusoids to be processed by hepatocytes. In this way, the liver can process and detoxify this blood before it ever reaches the rest of the systemic circulation. ③ The processed blood then exits the liver via hepatic veins and joins the rest of the blood in the systemic circuit in the inferior vena cava.

Quick Check

☐ 4. How does drainage of the posterior body wall differ on the right and left sides?

☐ 5. Which abdominal vessels drain straight into the inferior vena cava?

☐ 6. How does blood circulate through the hepatic portal system? What is the function of this system?

Veins of the Upper Limb

The venous drainage of the upper limb consists of a superficial system and a deep system. The superficial venous system begins in the hand with the **dorsal venous arch,** which drains into a set

of highly anastomosing veins in the hand and wrist whose anatomy often varies from person to person (**Figure 18.28a**). Then the most common venous pattern in the forearm includes the lateral **cephalic vein,** the middle **median antebrachial vein,** and the medial **basilic vein** (listed under "Superficial veins" in **Figure 18.28b**). In the antecubital fossa, a small oblique vein called the **median cubital vein** often connects the cephalic and basilic veins. The median cubital vein is a common site for drawing blood.

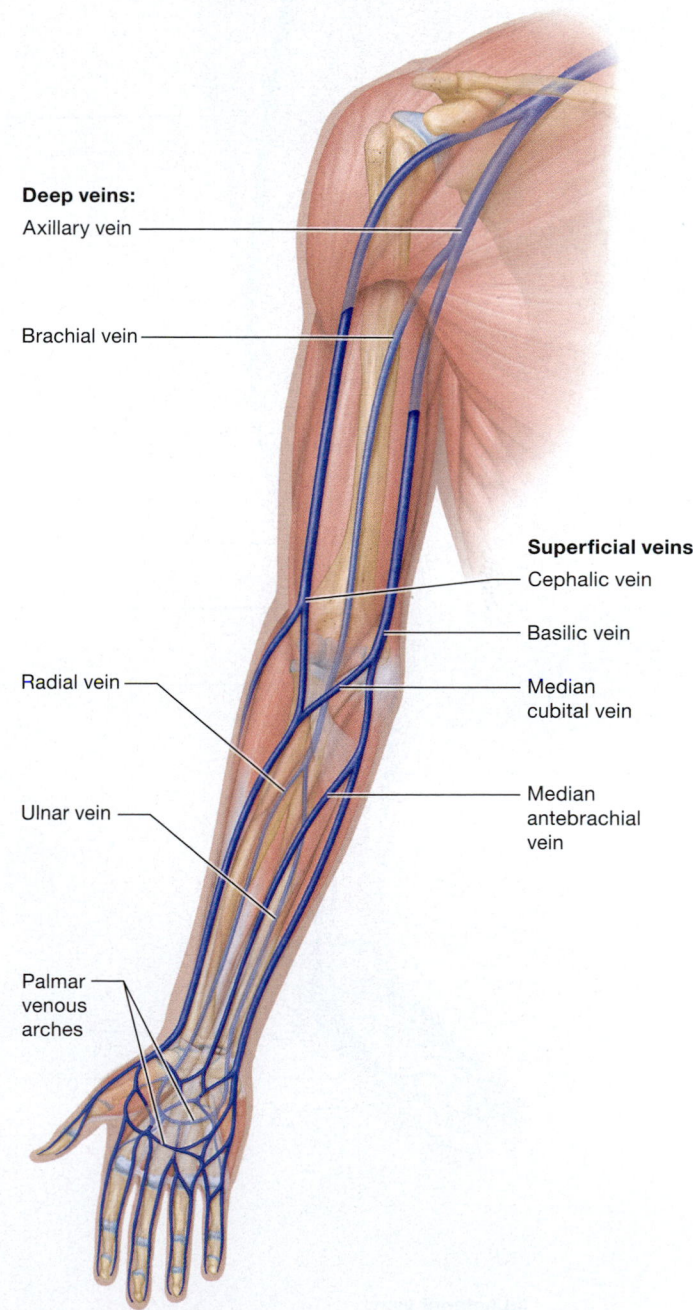

Deep veins:
Axillary vein

Brachial vein

Superficial veins:
Cephalic vein

Basilic vein

Median cubital vein

Radial vein

Median antebrachial vein

Ulnar vein

Palmar venous arches

Basilic vein

Cephalic vein

Dorsal venous arch

(a) Posterior aspect of the hand

(b) Anterior view of the upper limb

Figure 18.28 Veins of the upper limb.

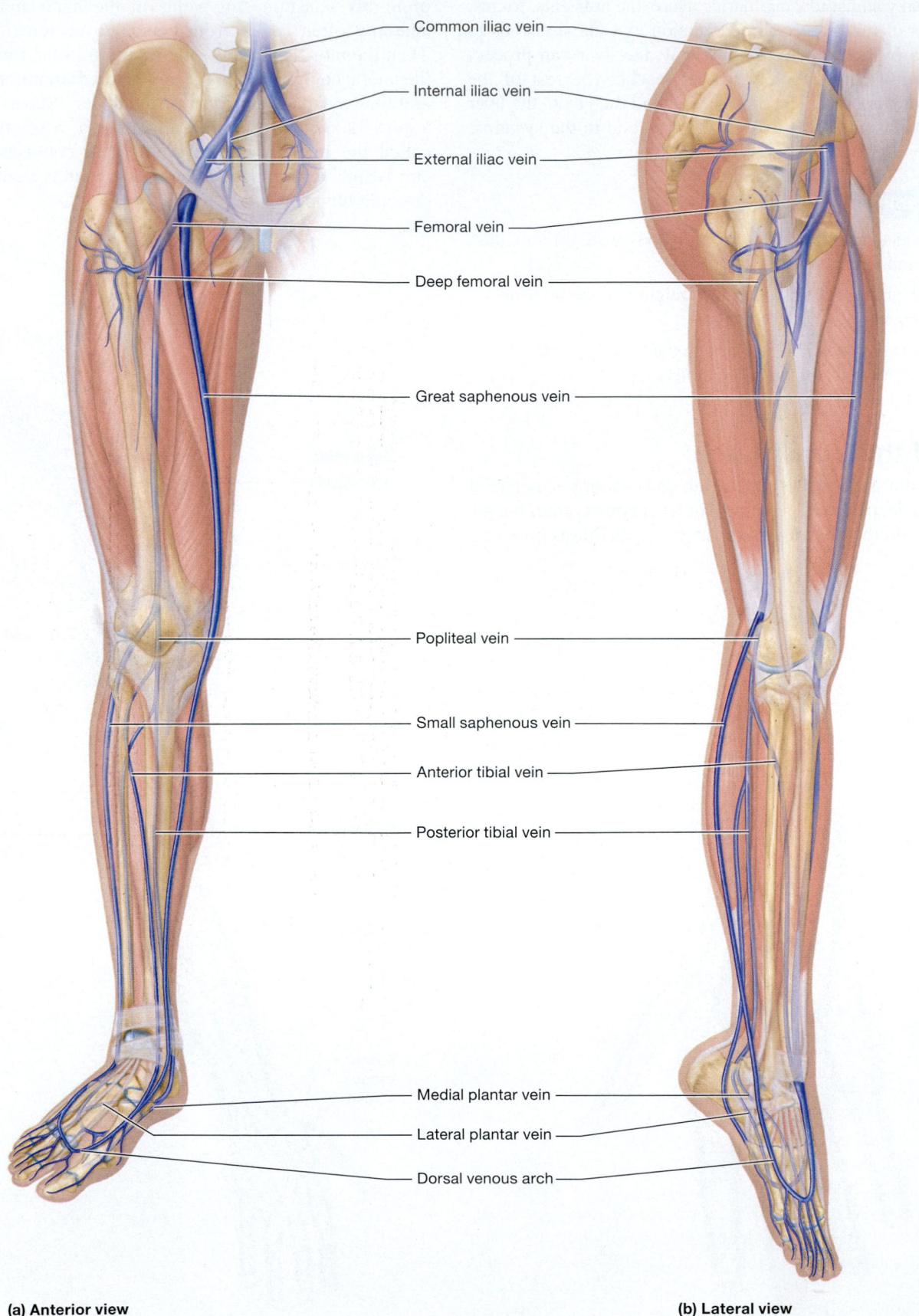

Common iliac vein

Internal iliac vein

External iliac vein

Femoral vein

Deep femoral vein

Great saphenous vein

Popliteal vein

Small saphenous vein

Anterior tibial vein

Posterior tibial vein

Medial plantar vein

Lateral plantar vein

Dorsal venous arch

(a) Anterior view

(b) Lateral view

Figure 18.29 Veins of the lower limb.

Drugs and the Hepatic Portal System

The hepatic portal system serves numerous vital functions, including protecting us from some of the nastier toxins we ingest. However, the liver doesn't have a way to discern "good chemicals" from "bad chemicals," and treats most medications exactly as it would most lethal toxins—its enzymes catalyze reactions that destroy them. This poses a problem because generally drugs need to reach the systemic capillary beds to have an effect, and they can't get there unless they can pass the gatekeeper that is the hepatic portal system.

The exact extent of hepatic metabolism varies for each drug, and this serves as a basis for drug dosage calculation. For example, if 100 mg of a drug is needed to produce a therapeutic effect, but 90% of the dose is destroyed by hepatic metabolism, the drug is administered in a 1000-mg dose. However, certain drugs are so thoroughly destroyed in the hepatic portal system that effectively none of the drug reaches the systemic circulation. These drugs must be administered by routes such as injection that allow them access to the systemic capillaries without first having to pass through the hepatic gatekeeper.

Vein Grafting

When the blood flow through an artery is blocked due to atherosclerosis or another disease, it is often necessary to bypass the artery to restore blood flow to a body part. This is done via a *bypass graft,* a procedure in which a vein is removed and grafted onto an artery in two places to give blood an alternate route of blood flow. A vein commonly used for this procedure is the great saphenous vein, because it is superficial and thus readily accessible. In addition, blood in the superficial structures that it drains can usually take alternate venous paths.

The vein is removed by a surgical procedure in which two small incisions are made in the medial leg. The vein is then isolated and tied off, and removed with a tug. The surgeon sews the harvested vein into place on the affected artery, allowing the blood to bypass the blockage.

The deep venous system also begins in the hand with the **palmar venous arches** (listed under "Deep veins" in Figure 18.28b). These anastomosing vessels drain into the **radial** and **ulnar veins** in the wrist and forearm, which largely parallel the radius and ulna. The radial and ulnar veins merge around the antecubital fossa to form the **brachial vein,** which unites with the basilic vein to form the **axillary vein.** The cephalic vein empties into the axillary vein in the proximal arm. The axillary vein becomes the subclavian vein after it passes deep to the clavicle.

Quick Check

☐ 7. What are the superficial veins of the upper limb?

☐ 8. What are the deep veins of the upper limb?

Veins of the Lower Limb

Like the veins of the upper limb, those of the lower limb are divided into a superficial system and a deep system (**Figure 18.29**). The superficial system begins with the **dorsal venous arch** in the foot; this drains into the anastomosing **great** and **small saphenous veins** (SAEH-feh-nus) in the ankle and leg. The great saphenous vein is located in the superficial medial leg and thigh and is a frequent location for varicose veins. This vein is also commonly harvested for bypass grafts, as described in *A&P in the Real World: Vein Grafting.* It empties into the **femoral vein** in the proximal thigh. The small saphenous vein is slightly lateral to the great saphenous vein and empties into the **popliteal**

vein in the popliteal fossa. Note that the superficial veins have collateral branches that connect them with the deep veins.

The system of deep veins of the lower limb mirrors the lower limb's arterial supply. The deep veins of the foot drain into the **medial** and **lateral plantar veins,** which drain into the **anterior** and **posterior tibial veins** in the ankle and leg. Near the popliteal fossa, these veins merge to form the popliteal vein. The popliteal vein crosses into the anterior thigh through the adductor magnus muscle, at which point it becomes the femoral vein. After the femoral vein passes deep to the inguinal ligament, it becomes the external iliac vein.

The major veins of the body are summarized in **Table 18.5**.

Quick Check

☐ 9. What are the two major superficial veins of the lower limb?

☐ 10. What is the main deep vein of the lower limb?

Apply What You Learned

☐ 1. Most veins in the systemic circuit have high compliance and can stretch to accommodate increases in blood volume. Do you think this is the case for the dural sinuses? Why or why not?

☐ 2. The antibiotic medication ticarcillin is not effective if given by mouth, because nearly the entire dose of the drug is destroyed by the hepatic portal system. However, this drug is effective when injected into the bloodstream. Explain.

☐ 3. The great saphenous vein is commonly harvested and used in coronary artery bypass grafts. Do you think removal of this vein would significantly affect venous drainage of the leg? Why or why not?

See answers in Appendix A.

Table 18.5 Major Systemic Veins

Vein	Description	Location	Structure(s) Drained
Brachiocephalic veins	Merge to form the superior vena cava	Mediastinum	• Most structures superior to the diaphragm
Internal iliac veins	Merge with the external iliac veins to form the common iliac vein	Posterior pelvis	• Pelvis
External iliac veins	Merge with the internal iliac veins to form the common iliac vein	Anterior pelvis	• Lower limb
Common iliac veins	Merge to form the inferior vena cava	Posterior pelvic wall	• Pelvis • Lower limb
Dural sinuses	Venous channels in the brain that receive blood and cerebrospinal fluid	Between the two layers of the dura in the brain	• Brain
Internal jugular veins	Merge with the subclavian veins to form the brachiocephalic veins	Deep neck, alongside the common carotid artery	• Brain • Superficial head • Neck
External jugular veins	Drain into the subclavian veins	Superficial neck	• Superficial face • Neck
Vertebral veins	Drain into the subclavian veins	Pass through the vertebral foramina of cervical vertebrae	• Posterior neck
Subclavian veins	Merge with the internal jugular veins to form the brachiocephalic veins	Deep to the clavicle	• Upper limb • Upper thorax
Azygos vein	Large unpaired vein	Right side of the posterior thoracic wall	• Posterior thoracic and abdominal wall on the right side • Esophagus • Lungs
Hemiazygos and accessory hemiazygos veins	Drain into the azygos vein	Posterior body wall on the left side	• Posterior body wall
Splenic vein	Drains into the hepatic portal vein	Posterior to the stomach	• Spleen
Gastric veins	Drain into the hepatic portal vein	Around the stomach	• Stomach
Superior mesenteric vein	Drains into the hepatic portal vein	Multiple branches around the small and large intestines	• Small intestine • Much of the large intestine
Inferior mesenteric vein	Drains into the hepatic portal vein	Multiple branches around the large intestine	• Remainder of the large intestine
Hepatic portal vein	Receives blood from the digestive organs and the spleen	Large vein that enters the liver	• Processes blood from the spleen and digestive organs
Hepatic veins	Drain into the inferior vena cava	Posterior liver	• Arterial blood from the liver • Venous blood from the hepatic portal system
Renal veins	Drain into the inferior vena cava	Posterior abdomen	• Kidneys
Cephalic veins	Superficial forearm veins, drain into the axillary vein	Lateral arm and forearm	• Superficial structures of the upper limb
Basilic veins	Superficial forearm veins, merge with the brachial veins to form the axillary veins	Medial arm and forearm	• Superficial structures of the upper limb
Radial and ulnar veins	Deep forearm veins, merge to form the brachial vein	Lateral forearm (radial), medial forearm (ulnar)	• Forearm
Brachial veins	Form from the union of the radial and ulnar veins	Anterior arm	• Arm
Axillary veins	Continuation of the brachial veins	Axilla	• Arm • Axilla
Great saphenous veins	Superficial veins of the lower limb	Medial lower limb	• Superficial structures of the lower limb
Anterior and posterior tibial veins	Merge to form the popliteal vein	Anterior and posterior leg	• Leg • Foot
Popliteal veins	Form from the union of the anterior and posterior tibial veins	Popliteal fossa	• Knee joint • Leg • Thigh
Femoral veins	Continuation of the popliteal veins	Anterior thigh	• Thigh

MODULE **18.8**
Putting It All Together: The Big Picture of Blood Vessel Anatomy

Learning Outcomes

1. Describe the general pathway of blood flow through the body.
2. Identify the arteries and veins of the body.

Up to this point in the chapter you've seen illustrations of the vessels in isolation. However, humans obviously don't have just arteries or just veins—we have both, which is what you will see in a laboratory or clinical setting. So, we have included a big picture view of the blood vessels to help you better appreciate the relationships these vessels have to one another and to the other organs of the body. In these figures, we continue to use the convention of showing arteries in red and veins in blue.

The blood vessels of the head and neck are illustrated in **Figure 18.30a** and **b**. In part b, the dissection on the left is superficial, and that on the right is deep. Surrounding structures have been faded back to more clearly show the vessels.

Figure 18.31 depicts the blood vessels of the abdomen. The superficial structures of the abdomen and parts of the abdominal organs have been removed to allow a clearer view of the vessels. For the same reason, the remaining structures have also been faded back.

The blood vessels of the upper and lower limbs are shown in **Figure 18.32**. As with other figures, the musculature of both the upper and lower limbs has been faded back to more clearly show the vessels deep to them.

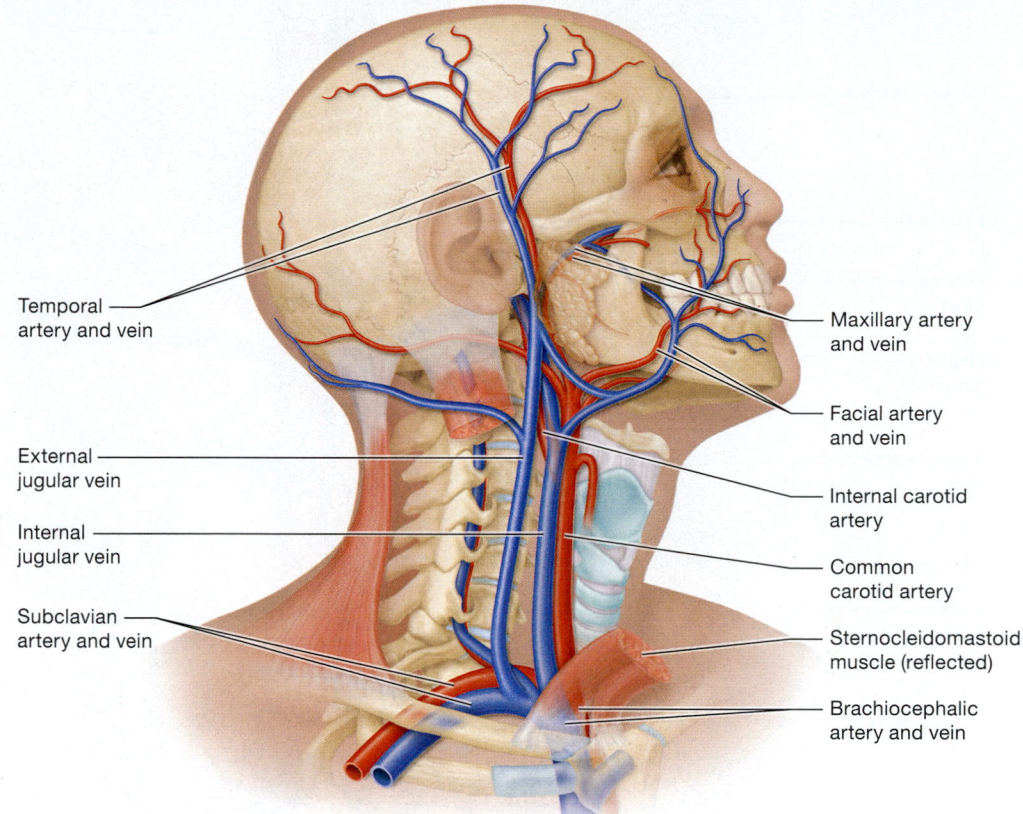

Temporal artery and vein

Maxillary artery and vein

Facial artery and vein

External jugular vein

Internal carotid artery

Internal jugular vein

Common carotid artery

Subclavian artery and vein

Sternocleidomastoid muscle (reflected)

Brachiocephalic artery and vein

(a) Vessels of the head

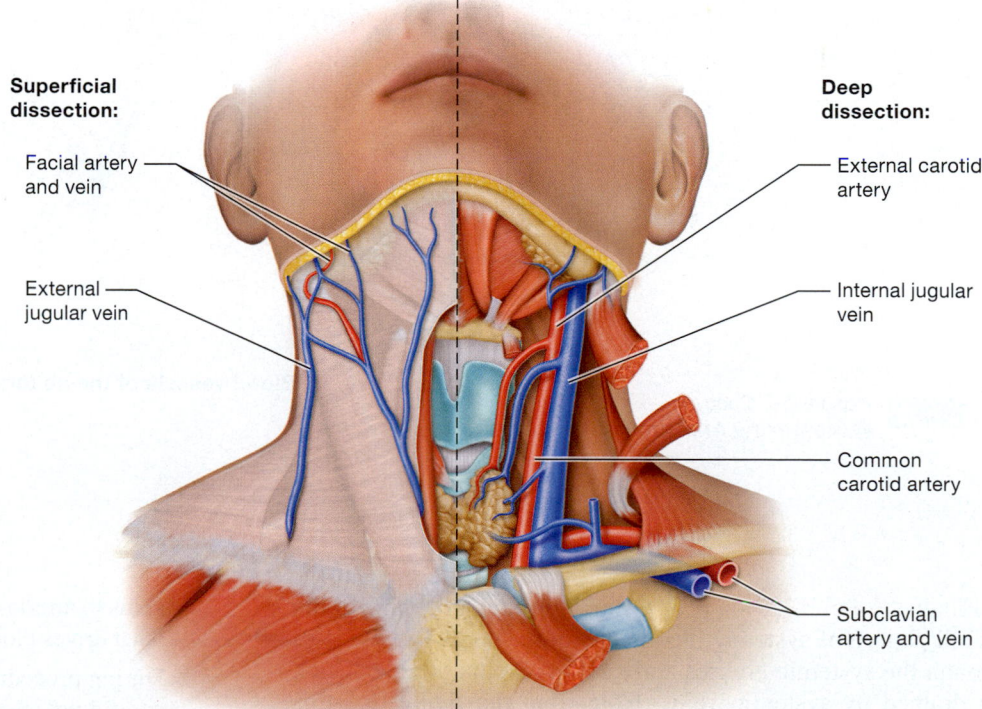

Superficial dissection:

Deep dissection:

Facial artery and vein

External carotid artery

External jugular vein

Internal jugular vein

Common carotid artery

Subclavian artery and vein

(b) Vessels of the neck

Figure 18.30 Blood vessels of the head and neck.

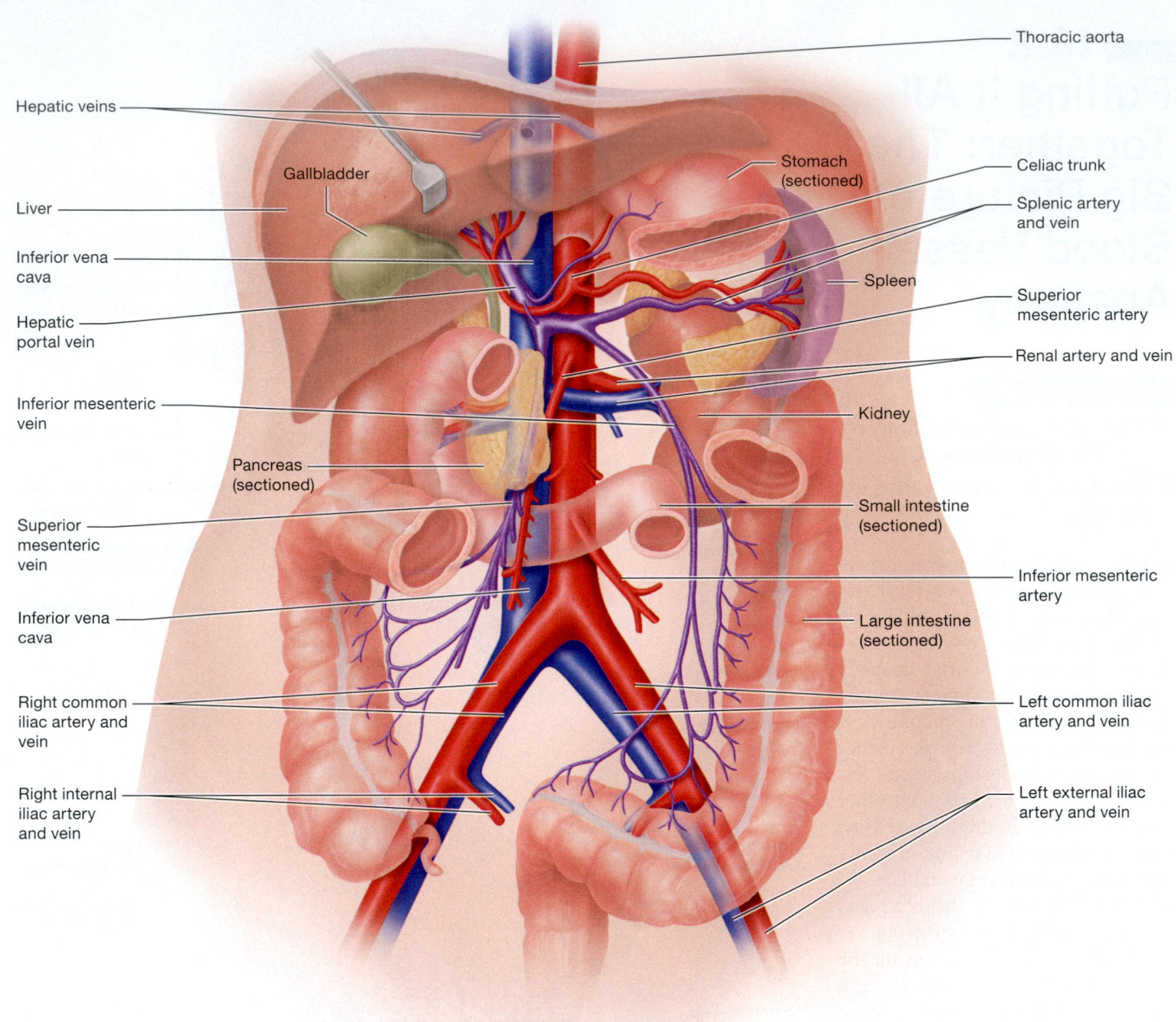

Figure 18.31 Blood vessels of the abdomen.

Explore PAL™ Cadaver
@ Mastering Anatomy & Physiology

Finally, **Figure 18.33** on p. 718 shows a schematic view of the overall pattern of systemic blood flow in the body. Blood flows through the systemic arteries and through capillary beds, which are drained by systemic veins. Note from the figure that this blood flow occurs only in one direction, from arteries to veins.

Apply What You Learned

☐ 1. As Figure 18.33 shows, blood can flow only one way through the blood vessels—it cannot go backward from veins to arteries. Explain why this is true. (*Hint:* Consider what drives blood flow.)

☐ 2. During procedures on the carotid artery, one must be careful not to enter or injure the internal jugular vein. Explain why this is so, considering their anatomical arrangement. Why is there minimal risk of hitting the external jugular vein?

See answers in Appendix A.

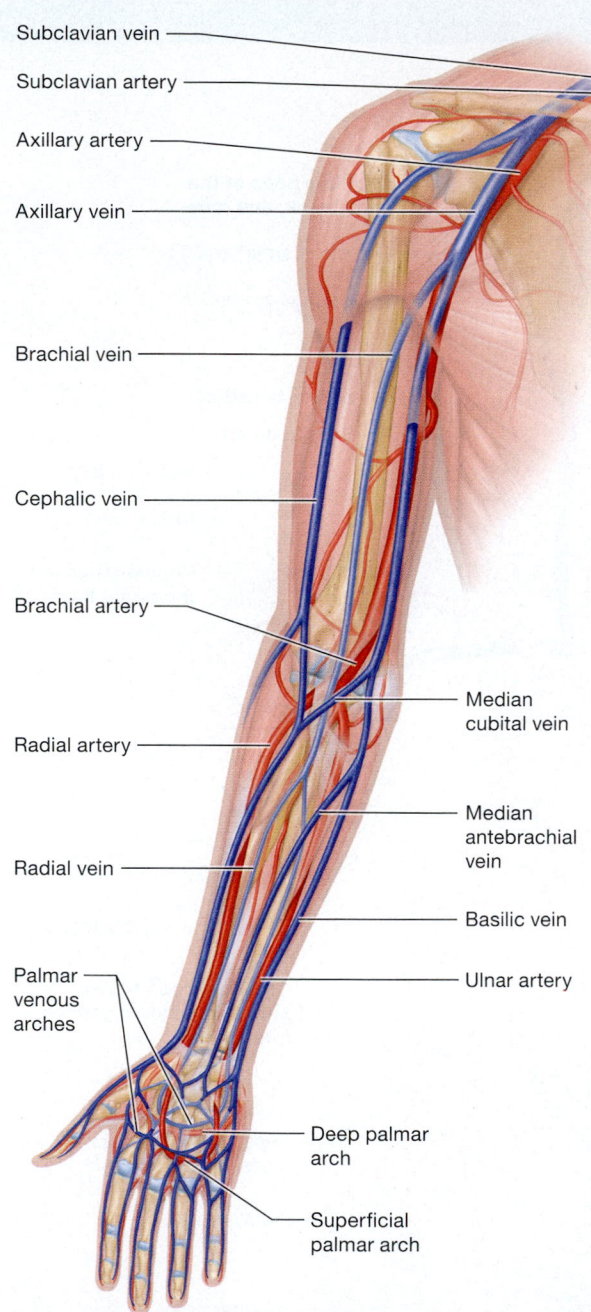

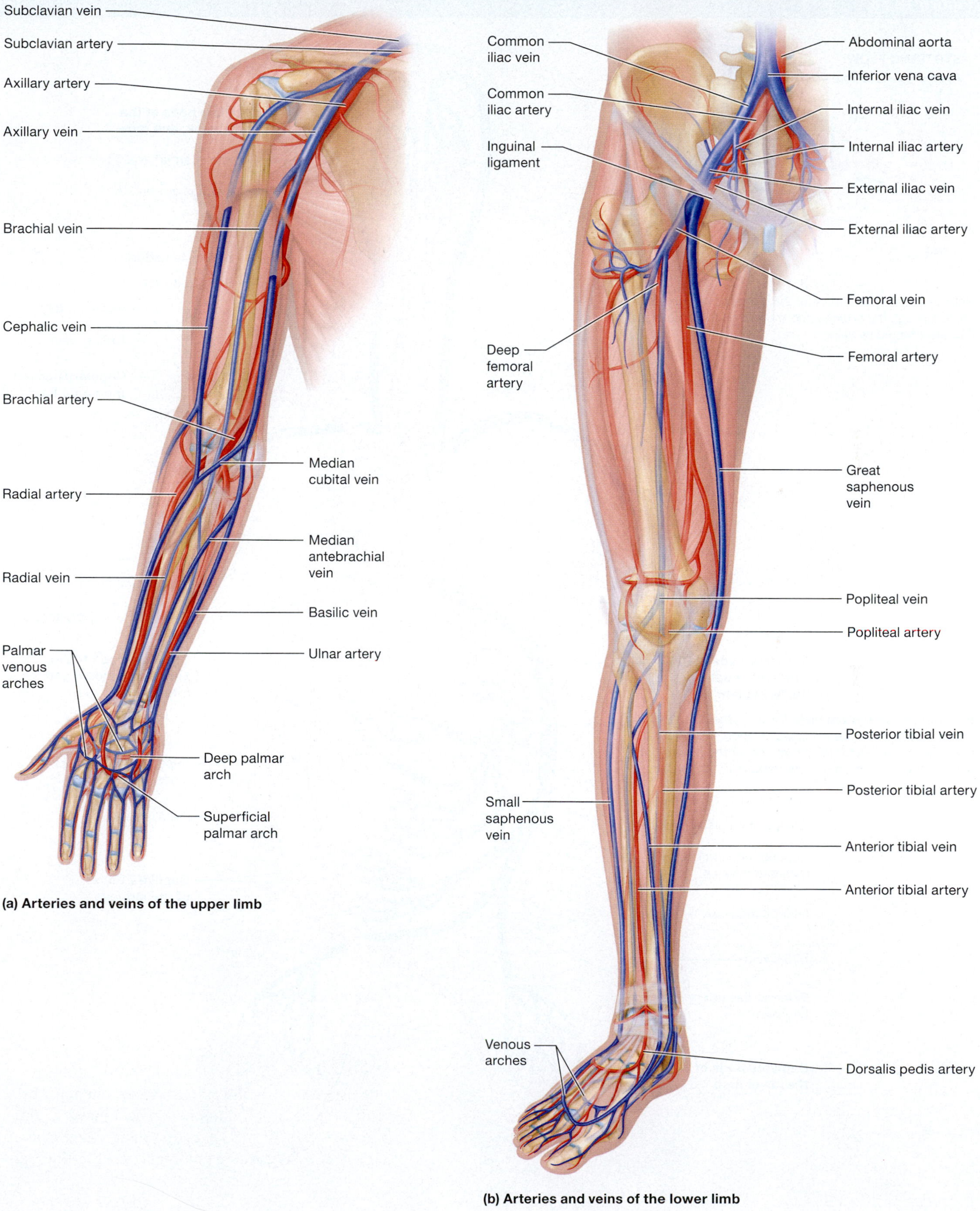

Subclavian vein

Subclavian artery

Axillary artery

Axillary vein

Brachial vein

Cephalic vein

Brachial artery

Radial artery

Radial vein

Palmar venous arches

Median cubital vein

Median antebrachial vein

Basilic vein

Ulnar artery

Deep palmar arch

Superficial palmar arch

(a) Arteries and veins of the upper limb

Common iliac vein

Common iliac artery

Inguinal ligament

Deep femoral artery

Small saphenous vein

Venous arches

Abdominal aorta

Inferior vena cava

Internal iliac vein

Internal iliac artery

External iliac vein

External iliac artery

Femoral vein

Femoral artery

Great saphenous vein

Popliteal vein

Popliteal artery

Posterior tibial vein

Posterior tibial artery

Anterior tibial vein

Anterior tibial artery

Dorsalis pedis artery

(b) Arteries and veins of the lower limb

Figure 18.32 Blood vessels of the upper and lower limbs.

The Big Picture of Systemic Blood Flow in the Body

Figure 18.33

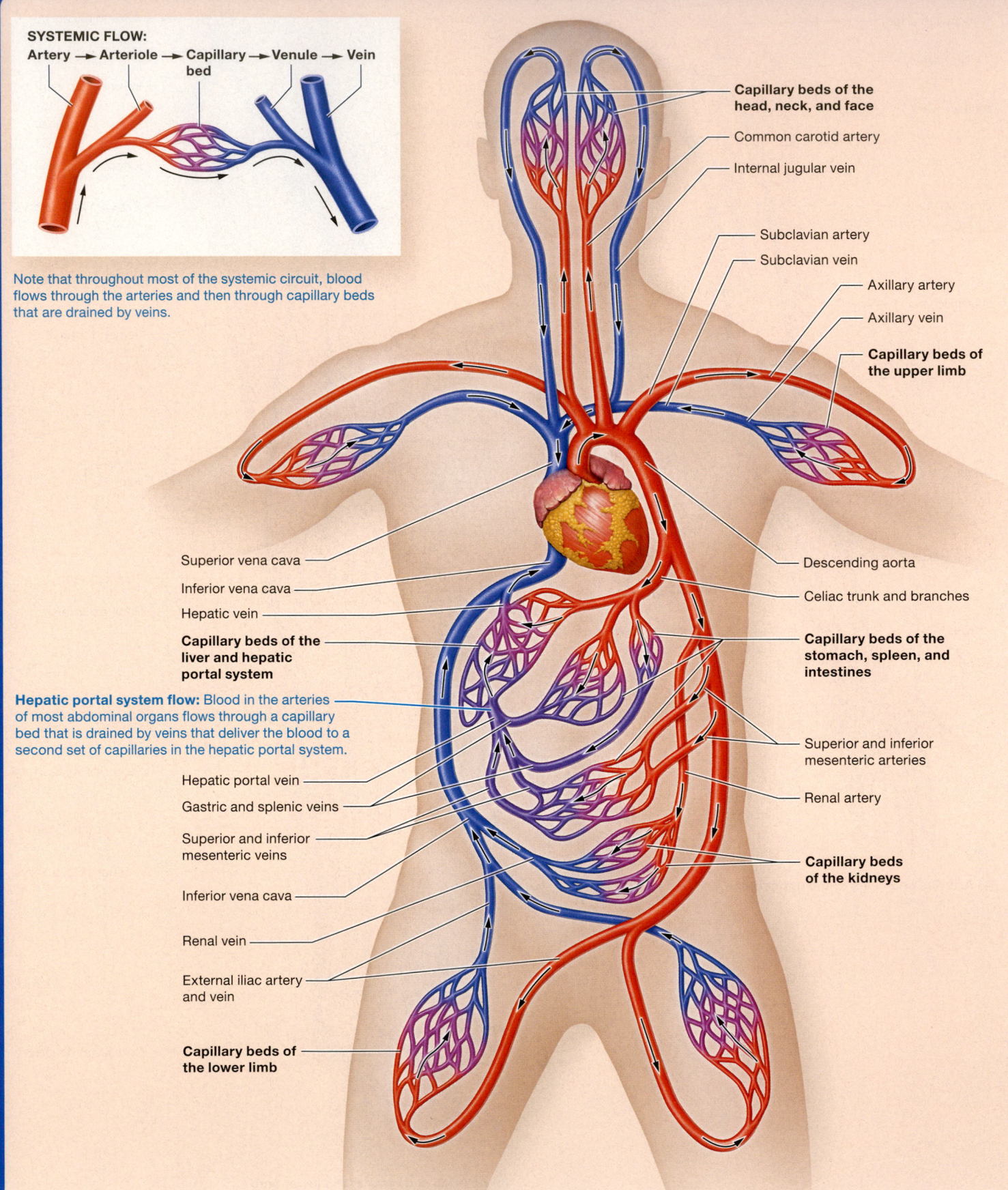

SYSTEMIC FLOW:

Artery → Arteriole → Capillary bed → Venule → Vein

Note that throughout most of the systemic circuit, blood flows through the arteries and then through capillary beds that are drained by veins.

Capillary beds of the head, neck, and face

Common carotid artery

Internal jugular vein

Subclavian artery

Subclavian vein

Axillary artery

Axillary vein

Capillary beds of the upper limb

Superior vena cava

Inferior vena cava

Hepatic vein

Capillary beds of the liver and hepatic portal system

Hepatic portal system flow: Blood in the arteries of most abdominal organs flows through a capillary bed that is drained by veins that deliver the blood to a second set of capillaries in the hepatic portal system.

Hepatic portal vein

Gastric and splenic veins

Superior and inferior mesenteric veins

Inferior vena cava

Renal vein

External iliac artery and vein

Capillary beds of the lower limb

Descending aorta

Celiac trunk and branches

Capillary beds of the stomach, spleen, and intestines

Superior and inferior mesenteric arteries

Renal artery

Capillary beds of the kidneys

Chapter Summary

For everything you need to succeed in this course, go to Mastering Anatomy & Physiology. There you will find:

 Practice Tests

 Author Podcasts

 Big Picture Animations

 Concept Boost Video Tutors

 iP2™ Interactive Physiology 2.0 Tutorials

 A&PFlix A&P Flix 3D Animations

 Active-Learning Workbook

MODULE 18.1
Overview of Arteries and Veins 670–675

- The pulmonary and systemic circuits consist of three types of vessels: **arteries,** the vasculature's distribution system; **capillaries,** the vasculature's exchange system; and **veins,** the vasculature's collection system.

- Blood vessels are organs with three tissue layers that surround a hollow **lumen:** the innermost **tunica intima,** the middle **tunica media,** and the outermost **tunica externa.**

- Arteries typically have a thicker wall, a thicker tunica media, and more extensive elastic laminae than do veins.

- There are three types of arteries: **elastic arteries,** the largest arteries; **muscular arteries,** which deliver blood to organs; and **arterioles,** which control blood flow to tissues.

- Certain arteries contain **baroreceptors** that monitor blood pressure and *chemoreceptors* that detect the oxygen, carbon dioxide, and hydrogen ion concentrations of the blood.

- **Venules** are the smallest veins that drain capillary beds. Veins function as **blood reservoirs** and many contain **venous valves** that prevent blood from flowing backward.

- Many vessels in the body are connected by **anastomoses.**

MODULE 18.2
Physiology of Blood Flow 675–681

- *Hemodynamics* is the study of the physiology of blood flow.

- The heart creates a *pressure gradient*—pressure is highest near the heart and decreases going away from the heart.

- **Blood pressure** is the force that blood exerts on the walls of the blood vessels.

- **Blood flow** is directly proportional to the pressure gradient but inversely proportional to **resistance,** which is defined as any impedance to blood flow.

- The velocity of blood flow decreases as cross-sectional area increases through the branching arterial system.

- Three factors determine blood pressure: *resistance, cardiac output,* and *blood volume.*
  - Decreased vessel radius, increased vessel length, increased blood viscosity, and the presence of an obstruction increase resistance and increase blood pressure.
  - As cardiac output increases, blood pressure also generally increases.
  - Increases in blood volume may lead to increases in blood pressure.

- The pressure in the pulmonary circuit averages about 15 mm Hg.

- The **mean arterial pressure (MAP)** in the systemic circuit is about 95 mm Hg. The **systolic pressure** is about 110–120 mm Hg, and the **diastolic pressure** is about 70–80 mm Hg. The difference between the systolic and diastolic pressures is the **pulse pressure.** The pressure in the arterioles is about 35 mm Hg.

- The MAP is calculated by adding one-third of the pulse pressure to the diastolic pressure.

- Blood pressure in the capillaries is about 35 mm Hg at the arteriolar end and about 15 mm Hg at the venular end. Pressure is about 4 mm Hg in the inferior vena cava.

- Venous valves, venoconstriction, *skeletal muscle pumps,* and the *respiratory pump* help return venous blood to the heart.

MODULE 18.3
Maintenance of Blood Pressure 681–687

- Short-term control of blood pressure is mediated by the nervous system and the endocrine system.
  - The sympathetic nervous system increases blood pressure via vasoconstriction and an increase in cardiac output. The parasympathetic nervous system decreases blood pressure through lowering cardiac output. Neural changes in blood pressure are mediated by the **baroreceptor reflex** arc and chemoreceptors. When the oxygen level decreases sharply, peripheral chemoreceptors trigger a rise in blood pressure. When the pH decreases, the central chemoreceptors trigger a rise in heart rate and blood pressure.
  - Thyroid hormone, **epinephrine,** and **norepinephrine** increase cardiac output. Epinephrine, norepinephrine, and angiotensin-II increase peripheral resistance.

- Long-term control of blood pressure is mediated by the urinary system and the endocrine system.
  - Aldosterone, antidiuretic hormone, and angiotensin-II increase blood volume. *Atrial natriuretic peptide* decreases blood volume.
  - The urinary system directly controls blood pressure by increasing or decreasing the amount of water lost from the body as urine.

- **Hypertension,** or high blood pressure, is defined as a systolic pressure over 140 mm Hg and/or a diastolic pressure over 90 mm Hg. **Hypotension** is a systolic pressure lower than 90 mm Hg and/or a diastolic pressure lower than 60 mm Hg.

MODULE 18.4
Capillaries and Tissue Perfusion 687–692

- **Tissue perfusion** is the blood flow through a capillary bed to a tissue.

- Substances may be exchanged across capillary walls by osmosis, diffusion through the endothelial cells, diffusion through small pores, and transcytosis.

- There are three types of capillaries, ordered from least to most leaky: **continuous capillaries, fenestrated capillaries,** and **sinusoidal capillaries.**

- The **microcirculation** is the total blood flow that takes place within the capillary beds.

- Tissue perfusion in most tissues is largely **autoregulated** to ensure that blood flow meets the cells' needs.
  - The **myogenic mechanism** involves reflexive vasoconstriction and vasodilation in response to stretching of smooth muscle in blood vessels.
  - Arterioles dilate when the concentration of O_2 decreases and the concentrations of CO_2 and H^+ increase. They constrict in the opposite conditions.

- Tissue perfusion in the heart decreases during ventricular systole and increases during ventricular diastole.

- The brain, skeletal muscle, and skin have their own autoregulatory systems for tissue perfusion.

MODULE 18.5
Capillary Pressures and Water Movement 692–696

- Water crosses a capillary by **filtration,** or the movement of a fluid by a force such as pressure or gravity.

- **Hydrostatic pressure** is the force that the blood exerts on the wall of the vessel. The hydrostatic pressure gradient forces water to leave the capillary and enter the interstitial fluid.

- **Osmotic pressure** is the force that must be applied to a solution in order to prevent osmosis—it is higher in solutions with higher osmolarity. Osmotic pressure is higher in blood than in interstitial fluid, which creates an osmotic pressure gradient—the **colloid osmotic pressure**—that draws water into the capillary.

- The **net filtration pressure** in a capillary is the difference between the hydrostatic pressure gradient and colloid osmotic pressure. The overall net filtration pressure in a capillary bed is about 6 mm Hg in favor of filtration.

- **Edema** is an excessive amount of water in the interstitial fluid.

MODULE 18.6
Anatomy of the Systemic Arteries 696–706

- The **ascending aorta** becomes the **aortic arch,** which has three branches: the **brachiocephalic artery,** the **left common carotid artery,** and the **left subclavian artery.**

- The **descending abdominal aorta** terminates by splitting into the two **common iliac arteries,** which in turn split into the **internal** and **external iliac arteries.**

- Most of the blood supply to the superficial head and face comes from branches of the **external carotid arteries.**

- The **internal carotid arteries** and the two **vertebral arteries** supply the brain. The two arterial systems communicate via the **cerebral arterial circle.**

- Branches of the **descending thoracic aorta** supply the lungs, esophagus, and diaphragm.

- The descending abdominal aorta has the following major branches: the **celiac trunk, superior mesenteric artery, renal arteries,** and **inferior mesenteric artery.**

- The upper limb is supplied by the subclavian artery, which becomes the **axillary artery** and then the **brachial artery.** The brachial artery splits into the **radial** and **ulnar arteries.**

- The lower limb is supplied by the external iliac artery, which becomes the **femoral artery.** This then becomes the **popliteal artery,** which splits into the **posterior** and **anterior tibial arteries** in the leg.

- The pulsations produced by changes in systolic and diastolic pressure may be palpated through the skin in **pulse points.**

MODULE 18.7
Anatomy of the Systemic Veins 707–714

- Most veins superior to the diaphragm drain into the **brachiocephalic veins,** which merge to form the **superior vena cava.** Venous blood from the lower limbs and pelvis drains into the **common iliac veins,** which merge to form the **inferior vena cava.**

- Blood from the head, face, and neck drains into the **vertebral, external jugular,** and **internal jugular veins.**

- Blood from the brain drains into the **dural sinuses,** which drain into the internal jugular vein.

- The anterior thoracic and abdominal walls drain into the **internal thoracic veins** and the external iliac veins. The posterior thoracic and abdominal walls drain into the **azygos system.**

- Abdominal veins that drain directly into the inferior vena cava include the **gonadal, renal,** and **hepatic veins.**

- The **splenic, gastric, superior mesenteric,** and **inferior mesenteric veins** drain into the **hepatic portal vein,** which delivers blood to hepatocytes that process nutrients, toxins, and the products of red blood cell destruction. Blood drains from the hepatic portal system via the hepatic veins into the inferior vena cava.

- The upper limb is drained by superficial and deep veins. The superficial veins include the lateral **cephalic vein,** the middle median antebrachial vein, and the medial **basilic vein.** The deep veins include the **brachial vein,** which becomes the **axillary vein.**

- The superficial veins of the lower limb drain into the **great** and **small saphenous veins.** The deep veins include the **femoral vein** in the anterior thigh.

MODULE **18.8**
Putting It All Together: The Big Picture of Blood Vessel Anatomy 715–718

- Figures 18.30–18.33 summarize the big picture of the circulation and of the systemic arteries and veins.

Assess What You Learned

Scan the QR Code for additional practice test questions

https://goo.gl/oFbDck

LEVEL 1 Check Your Recall

1. Mark the following statements as true or false. If a statement is false, correct it to make a true statement.

   a. Arteries are the exchange vessels of the cardiovascular system.
   b. The tunica intima is composed of a sheet of endothelium, connective tissue, and the internal elastic lamina.
   c. The tunica media consists of smooth muscle cells innervated by vasomotor nerves.
   d. Muscular arteries control blood flow at the tissue level.
   e. Veins have smaller lumens, more elastic fibers, and more smooth muscle than arteries.

2. Locations where vessels connect via collateral vessels are known as:

   a. thoroughfare channels.
   b. metarterioles.
   c. anastomoses.
   d. venules.

3. The carotid sinus contains:

   a. baroreceptors.
   b. chemoreceptors.
   c. metabolic controls.
   d. smooth muscle cells.

4. Which of the following factors would increase peripheral resistance?

   a. Increased blood viscosity
   b. Shorter vessel
   c. Vasodilation
   d. An increase in vessel radius

5. Which of the following would produce a decrease in blood pressure?

   a. Increased cardiac output
   b. Vasodilation
   c. Vasoconstriction
   d. Increased blood volume

6. Fill in the blanks: The two pressures within the systemic arterial circuit are the _____ pressure and the _____ pressure. The difference between these two pressures is the _____ pressure.

7. The lowest pressure in the systemic circuit occurs in the:

   a. arteries.
   b. arterioles.
   c. capillaries.
   d. venules.
   e. veins.

8. Explain the mechanisms that assist in the return of venous blood to the heart.

9. Mark the following statements as true or false. If a statement is false, correct it to make a true statement.

   a. The sympathetic nervous system increases blood pressure in the short term by increasing cardiac output and peripheral resistance.
   b. A sudden increase or decrease in blood pressure triggers the baroreceptor reflex, which is mediated by the medulla and the autonomic nervous system.
   c. Hormones that decrease blood volume and blood pressure include antidiuretic hormone and aldosterone.
   d. Angiotensin-II is a chemical that produces profound vasoconstriction.
   e. The kidneys control blood pressure directly through adjustment of blood volume.

10. Define each term:

    a. Pressure gradient
    b. Blood pressure
    c. Blood flow
    d. Resistance

11. Capillaries consist of:

    a. three thin tunics.
    b. only a thin tunica intima with a well-developed internal elastic lamina.
    c. a thin sheet of endothelium with a basal lamina.
    d. stratified epithelium.

12. List three ways in which substances can cross the capillary wall.

13. Which of the following structures is the "leakiest"?

    a. Continuous capillary
    b. Sinusoidal capillary
    c. Precapillary sphincter
    d. Fenestrated capillary

14. Mark the following statements as true or false. If a statement is false, correct it to make a true statement.

    a. Arterioles reflexively dilate when blood pressure increases.
    b. Increased concentrations of carbon dioxide and hydrogen ions and a decreased concentration of oxygen in the interstitial fluid cause local arteriolar constriction.
    c. Tissue perfusion to the heart decreases during systole and increases during diastole.
    d. The increase in tissue perfusion in skeletal muscle that occurs during exercise is known as hyperemia.
    e. The sympathetic nervous system causes vasodilation in the skin when body temperature decreases.

15. The hydrostatic pressure gradient drives water _____ capillaries, and the colloid osmotic pressure draws water _____ capillaries.

    a. out of; out of
    b. out of; into
    c. into; out of
    d. into; into

16. Net filtration pressure in most capillary beds favors:

    a. absorption.
    b. osmosis.
    c. filtration.
    d. secretion.

17. Match the following arteries with the correct description.

    _____ Radial artery
    _____ Celiac trunk
    _____ Basilar artery
    _____ Superior mesenteric artery
    _____ Dorsalis pedis artery
    _____ Femoral artery
    _____ Internal iliac artery
    _____ Renal artery
    _____ Internal carotid artery
    _____ Subclavian artery

    a. Supplies the small intestine and most of the large intestine
    b. Supplies the pelvis
    c. Supplies the upper limb
    d. Located in the lateral forearm
    e. Provides the blood supply to the lower limb
    f. Supplies the brain via the anterior and middle cerebral arteries
    g. Large single branch off the aorta that supplies the liver, stomach, duodenum, and spleen
    h. Supplies the foot and ankle
    i. Supplies the kidney
    j. Forms from the fusion of the two vertebral arteries

18. Which of the following is not a common pulse point?

    a. Femoral artery
    b. Subclavian artery
    c. Common carotid artery
    d. Brachial artery

19. Which of the following vessels does not drain into the hepatic portal vein?

    a. Splenic vein
    b. Inferior mesenteric vein
    c. Gastric vein
    d. Renal vein
    e. Superior mesenteric vein

20. Match the following veins with the correct description.

    _____ Cephalic vein
    _____ Great saphenous vein
    _____ Dural sinus
    _____ Azygos vein
    _____ Hepatic portal vein
    _____ Splenic vein
    _____ Internal jugular vein
    _____ Brachiocephalic veins
    _____ Brachial vein
    _____ Inferior mesenteric vein

    a. Drains the posterior abdominal and thoracic walls
    b. Two veins merge to form the superior vena cava
    c. Receives blood from the spleen and digestive organs
    d. Superficial vein in the medial leg
    e. Drains the brain and face
    f. Drains part of the large intestine
    g. Superficial vein in the lateral upper limb
    h. Deep vein of the arm
    i. Drains the spleen
    j. Drain the brain capillaries and cerebrospinal fluid

LEVEL 2 Check Your Understanding

1. Explain why a severed artery spurts blood, whereas a severed vein merely leaks blood.

2. Explain why a person who is 7 feet tall is likely to have higher blood pressure than a person whose height is 4 feet.

3. Collagen vascular diseases weaken the collagen in the tunica externa of blood vessels. Predict the effects of such a disease.

LEVEL 3 Apply Your Knowledge

PART A: Application and Analysis

1. You are babysitting two children who are having a contest to see who can hang upside-down for the longest time. At the end of the contest, both children feel dizzy and are worried they are sick. Explain to them why they feel this way and why the effect is only temporary. (*Hint:* Consider how the baroreceptor reflex responds when a person hangs upside-down.)

2. Mr. Gupta has been diagnosed with *nephrotic syndrome,* which is characterized by the loss of plasma proteins such as albumin into the urine. What impact will this have on his colloid osmotic pressure? How will this affect the net filtration pressure in his capillary beds? What effect will this likely cause?

3. Predict the effects of each of the following on systemic arterial blood pressure:

   a. The practice of "blood doping," which increases the number of red blood cells in the blood
   b. Caffeine consumption, which increases heart rate and causes vasoconstriction
   c. Blood loss from a bleeding ulcer

PART B: Make the Connection

4. Explain why the skin and lips of fair-skinned individuals may "turn blue" in very cold weather. (*Connects to Chapter 5*)

5. Ms. Rodgers has been diagnosed with *secretion of inappropriate ADH syndrome* (SIADH). What effect will SIADH have on the amount of water in her body? What symptoms would you expect from this condition? How would it affect her blood pressure and net filtration pressure? (*Connects to Chapter 16*)

See answers in Appendix A.

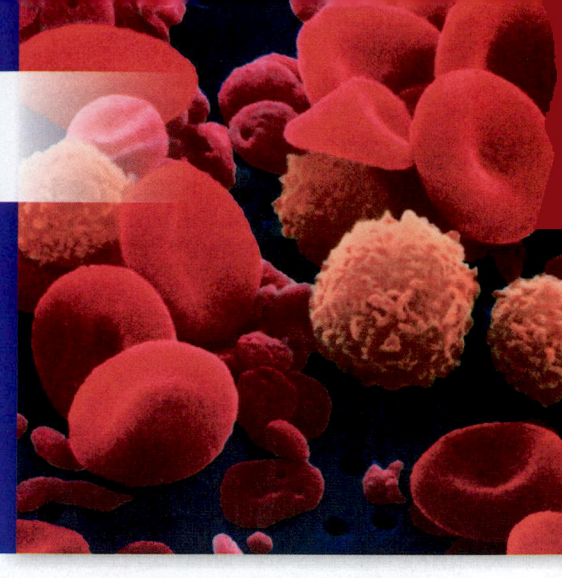

19
Blood

Blood has fascinated humans for millennia. The practice of "bloodletting"—removing large quantities of blood—was the most commonly performed medical procedure from ancient Egypt to 19th-century America. Although bloodletting has fallen out of favor for treating most conditions, we still draw blood to diagnose illness and monitor health. Indeed, blood figures prominently in the medical field—just think how many times you and the people around you have had "blood tests" of some sort.

For the past several chapters we have concentrated on the organ and system levels of organization. Now, we shift our focus to the tissue level as we explore the structure and function of blood. Many topics covered here will tie in closely with our next chapter, the immune system.

MODULE **19.1**
Overview of Blood

Learning Outcomes

1. Describe the major components of blood.
2. Describe the basic functions of blood.
3. Describe the overall composition of plasma, including the major types of plasma proteins, their functions, and where in the body they are produced.

About 5 liters of **blood,** a fluid connective tissue, circulate through the heart and blood vessels at all times. Blood makes up about 8% of total body weight. When a sample of blood is removed and examined under a microscope, its major components are visible: **plasma,** the liquid extracellular matrix; and **formed elements,** the cells and cell fragments that are suspended in plasma. Blood contains three types of formed elements: (1) *erythrocytes,* or *red blood cells (RBCs);* (2) *leukocytes,* or *white blood cells (WBCs);* and (3) tiny cellular fragments called *platelets.*

When a blood sample is placed in a centrifuge and spun rapidly, these components separate into three distinct layers. The top layer is plasma, which constitutes about 55% of the total blood volume. The middle layer, the **buffy coat,** makes up only about 1% of the total blood volume. It consists of leukocytes and platelets. The bottom layer, which accounts for the remaining 44% of blood, is a dark red precipitate that

For practice applying concepts to a clinical scenario, check out the **Running Case Study** *for this chapter* @ Mastering Anatomy & Physiology

Photo: Colored scanning electron micrograph (SEM) showing red and white blood cells and platelets.

723

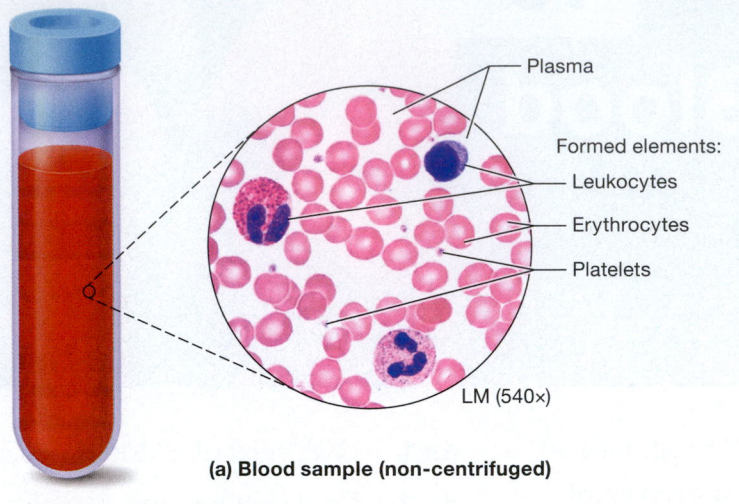

(a) Blood sample (non-centrifuged)

Plasma

Formed elements:

Leukocytes

Erythrocytes

Platelets

LM (540×)

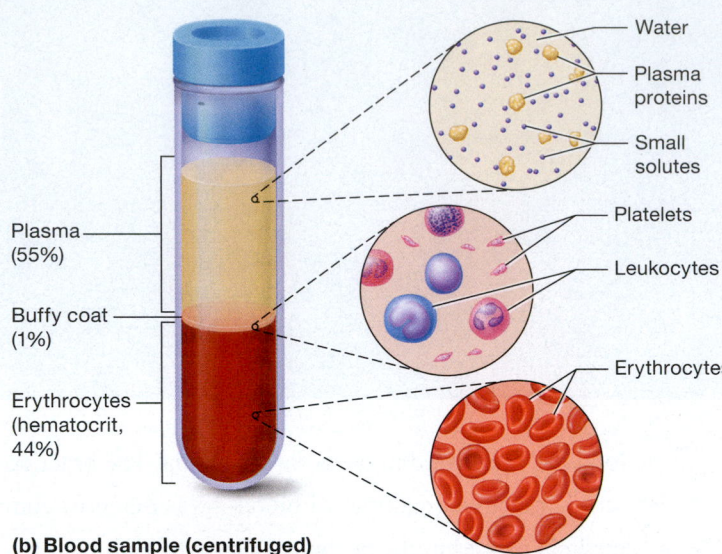

(b) Blood sample (centrifuged)

Water

Plasma proteins

Small solutes

Plasma (55%)

Buffy coat (1%)

Erythrocytes (hematocrit, 44%)

Platelets

Leukocytes

Erythrocytes

Figure 19.1 **The three visible layers of blood.**

Practice art labeling
@ Mastering Anatomy & Physiology

consists of erythrocytes. The percentage of blood composed of erythrocytes is called the **hematocrit** (heh-MAEH-toh-krit; *hemat-* = "blood") (**Figure 19.1**).

Overview of Blood Functions

« FLASHBACK

1. What is a buffer? (p. 49)
2. What are metabolic reactions? (p. 68)

Blood performs multiple functions, many of which revolve around its ability to access nearly every tissue in the body. These functions include the following:

- **Exchanging gases.** Oxygen is transported from the lungs to the tissues primarily by erythrocytes. Similarly, both erythrocytes and plasma transport carbon dioxide away from the tissues to the lungs.
- **Distributing solutes.** Plasma transports many solutes, including nutrients, hormones, and wastes. Additionally, blood transports ions and plays a role in regulating ion concentrations in the tissues.
- **Performing immune functions.** Both the cells (leukocytes) and proteins of the immune system use blood as a transport vehicle to reach almost any tissue in the body.
- **Maintaining body temperature.** Heat is a byproduct of many chemical reactions in the body, particularly metabolic reactions. Blood carries away the heat produced by an actively metabolizing tissue, which helps to maintain a constant temperature in that tissue.
- **Sealing damaged vessels by forming blood clots.** When a blood vessel is broken, platelets and certain proteins form a *blood clot* that seals the damaged vessel, preventing excessive blood loss.
- **Preserving acid-base homeostasis.** The pH of blood is generally maintained within the narrow range of 7.35–7.45.

The pH remains fairly constant because blood composition controls many of the body's most important buffer systems. Also, many plasma proteins act as buffers.

- **Stabilizing blood pressure.** One of the primary factors that determine blood pressure is the volume of blood in the circulation (see Chapter 18). Blood volume is vital to maintaining blood pressure at a constant level.

Our discussion, as it unfolds, will cover several of these functions, including gas transport, immune functions, and blood clotting. However, many specifics involving these roles of blood are addressed in later chapters.

Quick Check

☐ 1. Which substances does blood transport through the body?
☐ 2. What role does blood play in the immune system and in blood clotting?
☐ 3. How does blood regulate temperature and acid-base homeostasis?

Plasma

« FLASHBACK

1. Which types of molecules and/or compounds are hydrophilic? Which are hydrophobic? (p. 47)
2. What is colloid osmotic pressure? (p. 693)

Plasma is a pale yellow liquid consisting primarily of water, which accounts for about 90% of its volume. The amount of water in plasma is one factor that determines the thickness, or *viscosity,* of blood. If plasma contains less water, its viscosity increases and blood flow becomes sluggish.

Plasma proteins make up about 9% of plasma volume. These proteins—most made by the liver—are too large to fully dissolve in the water portion of plasma and so form a colloid. Important plasma proteins include:

- **Albumin.** As you read in Chapter 18, **albumin** (al-BYOO-min) is a relatively large protein produced by the liver that is responsible for blood's colloid osmotic pressure, or the pressure that draws water into the blood via osmosis, an example of the Gradients Core Principle (p. 28). To discover what happens when the liver cannot produce enough albumin, see *A&P in the Real World: Cirrhosis.*

 **CORE PRINCIPLE**
Gradients

- **Immune proteins.** The **γ-globulins,** also known as **antibodies,** are plasma proteins of the immune system. Unlike most plasma proteins, antibodies (see later in this chapter) are produced by leukocytes. The chapter on immunity will discuss antibodies in detail (see Chapter 20).
- **Transport proteins.** Recall that lipid-based compounds such as fats and steroids are hydrophobic (see Chapter 2). This feature makes their transport through blood problematic, because hydrophobic compounds tend to associate with one another and form clumps rather than associate with water molecules. These compounds may be transported safely through the blood by binding to **transport proteins** that are hydrophilic and can associate with water molecules. Examples of such transport proteins include globular proteins called **α-** and **β-globulins** and **lipoproteins.**
- **Clotting proteins.** A blood clot, which is a collection of platelets and *clotting proteins,* stops bleeding from an injured blood vessel. Module 19.5 gives details on clotting proteins.

The remaining 1% of plasma volume consists of a variety of small solutes that generally dissolve in the water portion of plasma and form a solution. The solutes include nutrients such as glucose and amino acids, nitrogenous wastes, ions, and small amounts of dissolved gases such as oxygen and carbon dioxide. These small substances are readily exchanged between the blood and the interstitial fluid in most capillary beds.

Table 19.1 summarizes the components of plasma.

Table 19.1 Components of Plasma

Plasma Component	Function
Water	90% of plasma volume; solvent that dissolves and transports many solutes through the body
Plasma Proteins	9% of plasma volume; multiple functions (see below)
Albumin	Maintains osmotic pressure
Immune proteins	Produced by leukocytes; function in immunity
Transport proteins	Bind and transport hydrophobic compounds through the blood
Clotting proteins	Stop blood loss from damaged vessels
Other Solutes	1% of plasma volume; multiple functions (see below)
Glucose, amino acids	Nutrition; used for protein synthesis
Ions	Electrolyte and acid-base homeostasis
Dissolved gases (small amounts of oxygen and carbon dioxide)	Oxygen delivered to the tissues; carbon dioxide delivered to the lungs to be exhaled
Wastes	Delivered to the appropriate organ for excretion

Quick Check

☐ 4. What are the four main categories of plasma proteins?

☐ 5. What are the other components of blood plasma?

Apply What You Learned

☐ 1. Predict the effect of a decreased albumin level in the plasma.

☐ 2. Explain why proteins can readily travel through plasma, but lipids cannot.

☐ 3. Predict the homeostatic consequences of extensive blood loss from an injury. (*Hint:* How does blood normally help maintain homeostasis?)

See answers in Appendix A.

Cirrhosis

A&P in the Real World

Liver disease, or *cirrhosis* (sih-ROH-sus), has many causes, including cancer, alcoholism, and viral hepatitis. A common disease in the United States, cirrhosis is the 10th leading cause of death for men and the 12th for women. One consequence of cirrhosis is a progressive decrease in the production of plasma proteins, particularly albumin and clotting proteins. The decrease in albumin leads to a reduction in osmotic pressure, which causes the blood to lose water to the extracellular spaces. This produces severe edema (swelling) in the abdominal area, a condition known as *ascites* (ah-SY-teez). The progressive decline in clotting proteins also causes easy bruising and delays clotting, which can be fatal.

MODULE 19.2
Erythrocytes and Oxygen Transport

Learning Outcomes

1. Describe the structure and functions of erythrocytes.
2. Discuss the structure and function of hemoglobin, as well as its breakdown products.
3. Explain the basic process of erythropoiesis and its regulation through erythropoietin.
4. Describe the causes and symptoms of anemia.

As you learned in the first module, erythrocytes average about 44% of the total blood volume, a value known as the hematocrit. The hematocrit is generally higher in males (40–50%) than in females (36–44%) due to males' typically larger body size and greater muscle and bone mass. The primary function of erythrocytes is oxygen and carbon dioxide transport, at which they are extremely effective.

The structure of erythrocytes enables them to carry out their vital tasks of gas transport. Even so, certain conditions can impair the oxygen-carrying capacity of blood, a topic covered later in this module.

Erythrocyte Structure

« FLASHBACK

1. What is a polypeptide? (p. 57)
2. What is an oxidation reaction? (p. 43)

A typical **erythrocyte** (eh-RITH-roh-syt; *erythr-* = "red"), or **red blood cell (RBC),** is a *biconcave disc:* a flattened, donut-shaped cell that is concave on both sides (**Figure 19.2a** and **b**). This shape gives erythrocytes a large surface-to-volume ratio, which is critical to their role in gas exchange. Erythrocytes are small cells, measuring only about 7.5 μm in diameter and 2.5 μm in width. At this size, you could fit about 70,000 erythrocytes on the head of a pin.

Notice in the figure that mature erythrocytes are anucleate (have no nucleus) and also lack most other organelles. This means that mature erythrocytes are not capable of carrying out oxidative catabolism or protein synthesis. In fact, an erythrocyte consists of little more than a plasma membrane surrounding cytosol filled with enzymes and about one billion molecules of the oxygen-binding protein **hemoglobin,** or **Hb** (HEE-muh-gloh-bin) (**Figure 19.2c**). The shape and composition of an erythrocyte

facilitate its transport of oxygen through the blood, an example of the Structure-Function Core Principle (p. 28).

CORE PRINCIPLE
Structure-Function

As you can see in **Figure 19.3,** hemoglobin is a large protein that consists of four polypeptide subunits: two alpha (α) chains and two beta (β) chains. Each polypeptide is bound to an iron-containing compound called a **heme group.** The heme binds to oxygen (O_2) in parts of the body where the oxygen level is high (such as the lungs), forming a structure called **oxyhemoglobin (HbO_2).** Where the oxygen level is low, as in the tissues surrounding systemic capillaries, hemoglobin releases oxygen to become **deoxyhemoglobin.**

Hemoglobin's interaction with oxygen produces a red color, and the shade of red is determined by how much oxygen is bound to it. When all four heme groups are bound to oxygen, the protein assumes a bright red color. For this reason, in systemic arteries, where nearly 100% of the hemoglobin is oxyhemoglobin, the blood is bright red. Conversely, when fewer of the heme groups are bound to oxygen, hemoglobin takes on a darker red-purple color. So, in systemic veins, where more of the hemoglobin is deoxyhemoglobin, the blood is generally a much darker shade of red. (Contrary to popular perception, systemic venous blood is not blue. Rather, veins simply *look* blue because of an optical illusion related to light refraction.)

In venous blood where the oxygen level is low, hemoglobin binds to carbon dioxide (CO_2), to form **carbaminohemoglobin.** Approximately 23% of the carbon dioxide in blood is transported in this manner. Hemoglobin also binds to carbon monoxide (CO), which forms **carboxyhemoglobin.** Unfortunately, carbon monoxide binds to hemoglobin more strongly than does oxygen and has an affinity for hemoglobin about 200 times greater than that of oxygen. Furthermore, CO changes the shape of hemoglobin so that the oxygen already bound to hemoglobin cannot be released into the tissues. This makes exposure to a

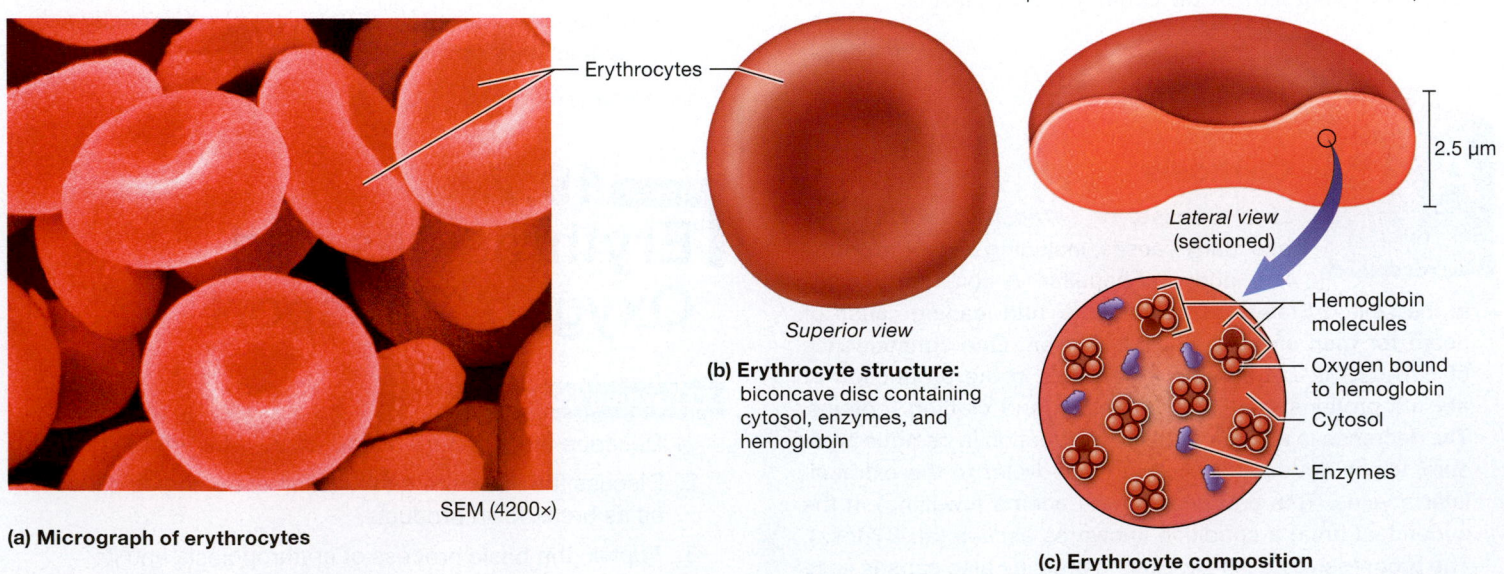

(a) **Micrograph of erythrocytes**

SEM (4200×)

Erythrocytes

7.5 μm

2.5 μm

Superior view

(b) **Erythrocyte structure:** biconcave disc containing cytosol, enzymes, and hemoglobin

Lateral view (sectioned)

Hemoglobin molecules

Oxygen bound to hemoglobin

Cytosol

Enzymes

(c) **Erythrocyte composition**

Figure 19.2 Erythrocyte structure and composition.

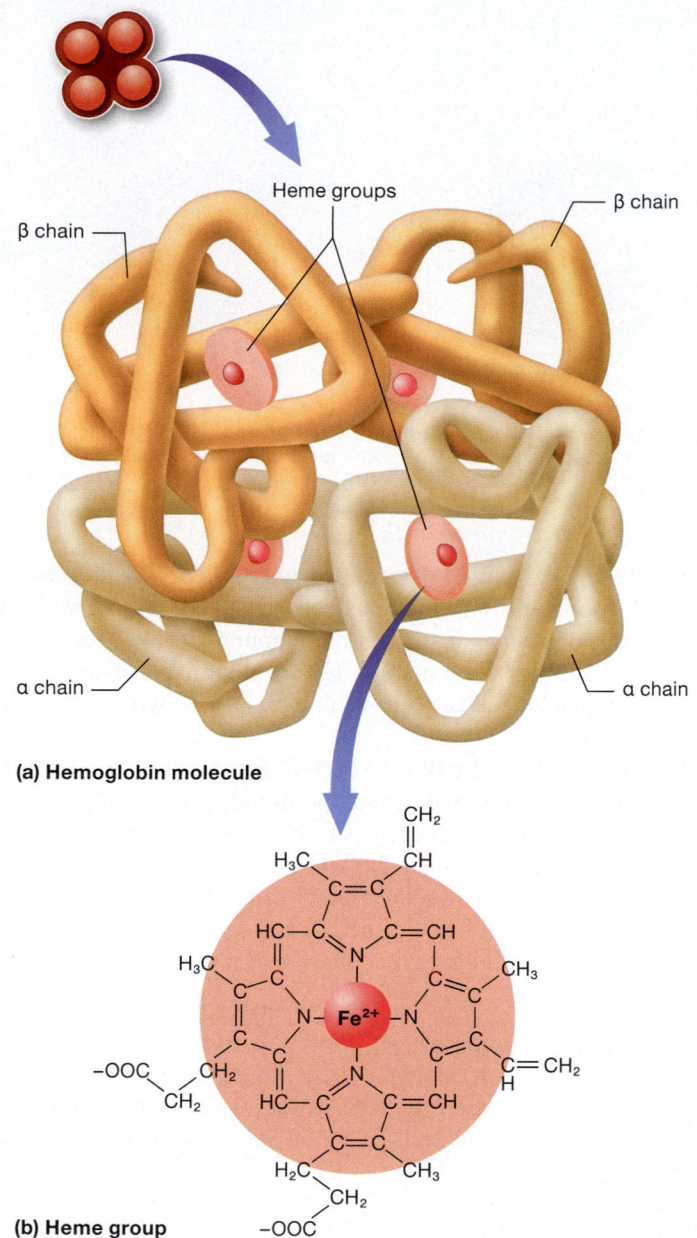

(a) Hemoglobin molecule

(b) Heme group

Figure 19.3 Hemoglobin structure. When the heme group in each hemoglobin binds oxygen, it becomes oxyhemoglobin.

high level of carbon monoxide lethal—it severely decreases the oxygenation of body tissues.

☐ 1. Describe the structure of a typical erythrocyte.

☐ 2. Differentiate between oxyhemoglobin, deoxyhemoglobin, carbaminohemoglobin, and carboxyhemoglobin.

Lifespan of an Erythrocyte

《 FLASHBACK

1. What is erythropoietin? (p. 620)

2. What is a sinusoid? Where are sinusoids located? (p. 689)

Erythrocytes lead rough lives. They tumble and squeeze through more than 100,000 blood vessels under high pressure and enormous shear forces, which leads to damage that accumulates over time. Erythrocytes lack the cellular machinery to repair this damage, so their average lifespan is only 100–120 days, and the body must continuously make new erythrocytes.

Erythrocyte Formation: Erythropoiesis

Recall that **hematopoiesis** (heh′-mah-toh-poy-EE-sis; *poiesis* = "make, produce") is the process that produces the formed elements in blood (see Chapter 6). Hematopoiesis occurs in **red bone marrow,** which houses the cells from which all formed elements arise: the **hematopoietic stem cells,** or **HSCs.**

Erythropoiesis (ee-rith′-roh-poy-EE-sis), the formation of erythrocytes, is part of the larger process of hematopoiesis (**Figure 19.4**). Erythropoiesis begins when HSCs differentiate into progenitor cells called *erythrocyte colony-forming units (CFUs)*. At this point they are *committed,* meaning they are able to become only a single cell type. Erythrocytes form at the incredible rate of about 250 *billion* cells per day.

Erythrocyte CFUs next differentiate into **proerythroblasts** (-*blast* = "immature cell"), a process requiring the presence of the hormone **erythropoietin** (ee-rith′-roh-POY-eh-tin), or **EPO,** secreted by the kidneys. Proerythroblasts develop into **erythroblasts,** which rapidly synthesize hemoglobin and other proteins. Notice that as erythroblasts mature, their nuclei shrink and are eventually ejected from the cells, at which point they are called **reticulocytes** (reh-TIK-yoo-loh-sytz′). These cells retain some organelles, particularly ribosomes. When they eject these remaining organelles by exocytosis, they enter the bloodstream by squeezing through the large pores in the sinusoidal capillaries of the bone marrow and are considered erythrocytes. The entire process takes 5–7 days.

Keep in mind that reticulocytes are normally found in the blood in small numbers, but they are able to bind and transport oxygen on their hemoglobin. For this reason, reticulocytes are sometimes released into the circulation if the number of circulating erythrocytes is inadequate. Indeed, an elevated reticulocyte count is a sign of slow blood loss, as occurs with a stomach ulcer.

Regulation of Erythropoiesis

Erythropoiesis is regulated by a negative feedback loop to maintain the hematocrit within normal limits (**Figure 19.5**). This is critical, because if the hematocrit drops too low, there will not be enough erythrocytes available to transport oxygen to the tissues. Conversely, if the hematocrit climbs too high, the blood becomes thicker (its viscosity increases), making blood flow sluggish.

This negative feedback loop primarily involves the hormone erythropoietin, with the concentration of oxygen in the blood influencing the production of this hormone. The synthesis of erythropoietin demonstrates the Feedback Loops Core Principle (p. 28).

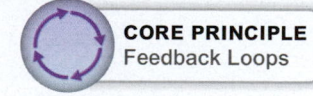

CORE PRINCIPLE
Feedback Loops

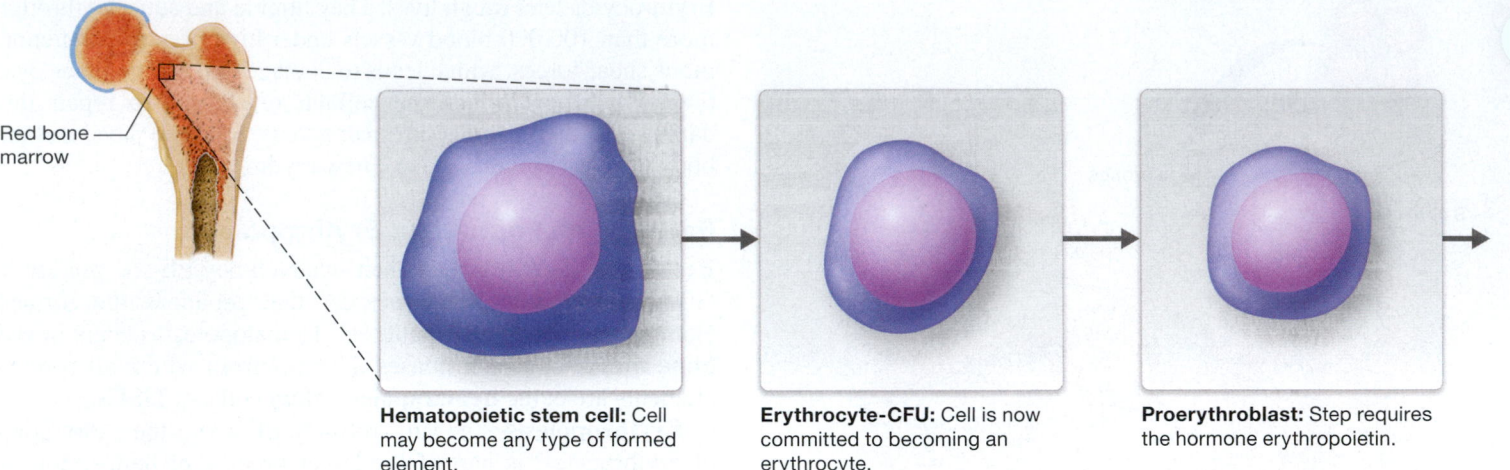

Hematopoietic stem cell: Cell may become any type of formed element.

Erythrocyte-CFU: Cell is now committed to becoming an erythrocyte.

Proerythroblast: Step requires the hormone erythropoietin.

Figure 19.4 Erythropoiesis: formation of erythrocytes.

The negative feedback loop that regulates erythropoiesis proceeds as follows:

- **Stimulus: The blood oxygen level decreases below normal range.** A decrease in the oxygen content of blood is usually due to respiratory problems, but it may also result from heart conditions or reduced availability of oxygen (e.g., at high altitudes).
- **Receptor: Kidney cells detect a low oxygen level.** Specialized cells in the kidneys act as chemoreceptors and monitor the blood's oxygen content. They are stimulated when the oxygen content falls below a certain level.
- **Control center: Kidneys produce more erythropoietin and release it into the blood.** The kidneys produce erythropoietin, which communicates with HSCs in the bone marrow. Here we see an example of the Cell-Cell Communication Core Principle (p. 28).

CORE PRINCIPLE
Cell-Cell Communication

- **Effector/response: Production of erythrocytes increases.** Erythropoietin has many effects. It speeds up the rate of erythropoiesis and reduces the amount of time needed for new erythrocytes to mature. Erythropoietin can also trigger the replacement of yellow bone marrow with red bone marrow.
- **Homeostasis: The blood level of oxygen rises to normal.** As more erythrocytes enter the blood, the hematocrit and oxygen-carrying capacity of the blood increase. Under normal circumstances, the oxygen content of blood will rise as a result. Chemoreceptors in the kidneys detect that the oxygen level has returned to normal, and erythropoietin production and release decline to normal.

In addition to erythropoietin, chemicals known as *growth factors* can increase the production of erythrocytes. Growth factors are produced by cell types such as endothelial cells in blood vessels and fibroblasts in

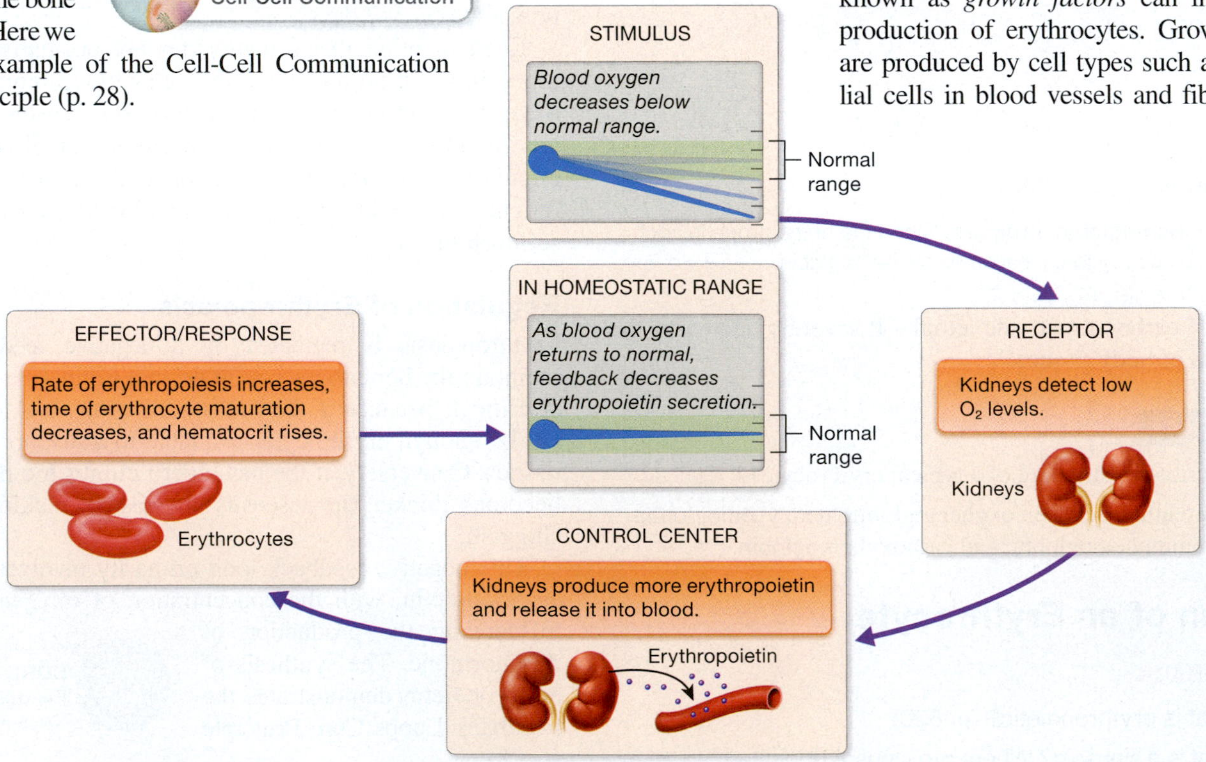

STIMULUS

Blood oxygen decreases below normal range.

Normal range

IN HOMEOSTATIC RANGE

As blood oxygen returns to normal, feedback decreases erythropoietin secretion.

Normal range

RECEPTOR

Kidneys detect low O_2 levels.

Kidneys

EFFECTOR/RESPONSE

Rate of erythropoiesis increases, time of erythrocyte maturation decreases, and hematocrit rises.

Erythrocytes

CONTROL CENTER

Kidneys produce more erythropoietin and release it into blood.

Erythropoietin

Figure 19.5 Regulation of erythropoiesis.

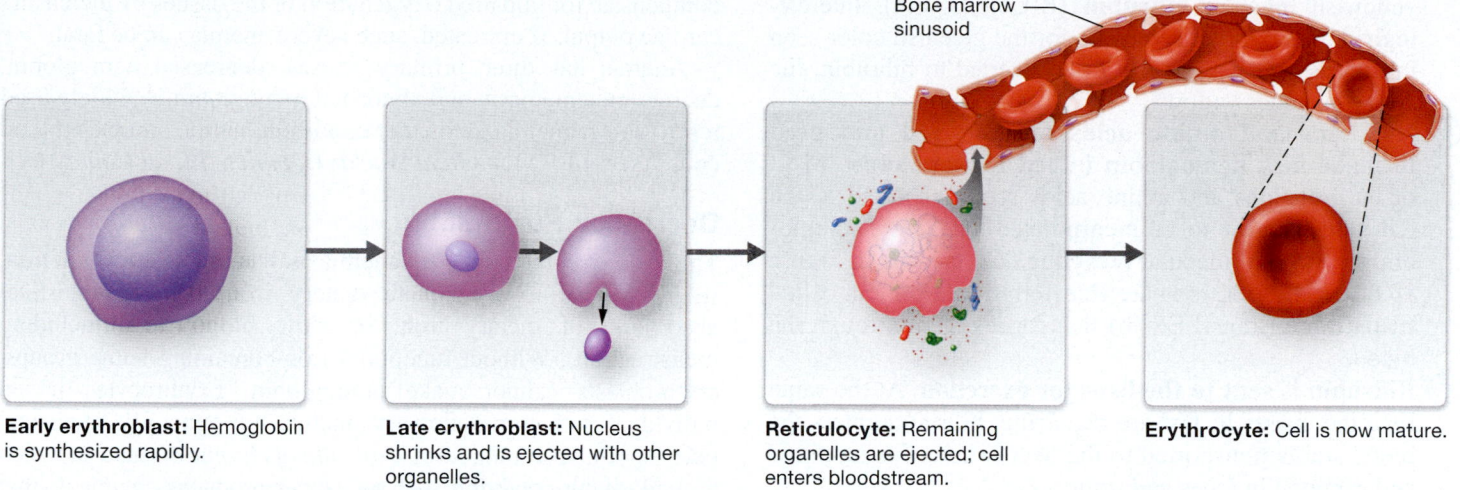

Early erythroblast: Hemoglobin is synthesized rapidly.

Late erythroblast: Nucleus shrinks and is ejected with other organelles.

Reticulocyte: Remaining organelles are ejected; cell enters bloodstream.

Erythrocyte: Cell is now mature.

Bone marrow sinusoid

Figure 19.4 Erythropoiesis: formation of erythrocytes. (*continued*)

connective tissue. They aid in the transition from yellow bone marrow to red bone marrow, and also trigger mitosis of hematopoietic stem cells.

Erythrocyte Death

As erythrocytes age, their plasma membranes become less flexible, making their passage through tiny capillaries more difficult. This is particularly the case in the sinusoids of the **spleen,** an organ located in the superior left abdominal cavity. Erythrocyte destruction proceeds by the following steps (**Figure 19.6**):

① **Erythrocytes become trapped in the sinusoids of the spleen.** Older erythrocytes are not flexible enough to exit the tortuous sinusoids.

② **Spleen macrophages digest erythrocytes.** In the sinusoids, erythrocytes encounter leukocytes called *macrophages* (discussed in Module 19.3). Macrophages are phagocytes that ingest and destroy older erythrocytes and other cells.

③ **Hemoglobin is broken down into amino acids, iron ions, and bilirubin.** The macrophages break down the hemoglobin of erythrocytes. The polypeptide chains of hemoglobin are broken down into amino acids, the iron ions are removed from the heme groups, and the remainder of the heme is converted first into the waste product **biliverdin,** a greenish pigment (*verd* = "green"). Generally, biliverdin is then converted into a second waste product, the

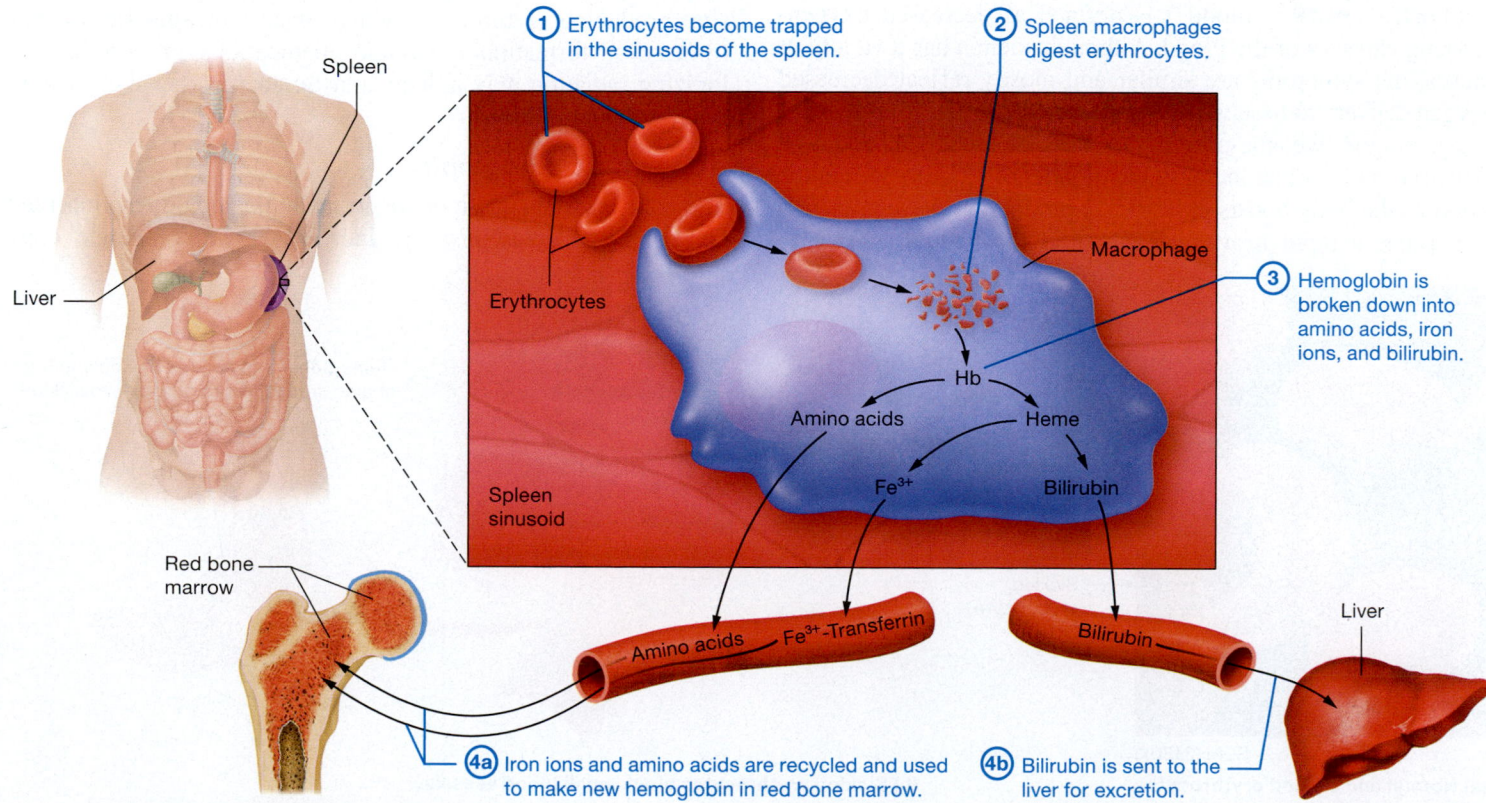

① Erythrocytes become trapped in the sinusoids of the spleen.

② Spleen macrophages digest erythrocytes.

Spleen

Liver

Erythrocytes

Macrophage

③ Hemoglobin is broken down into amino acids, iron ions, and bilirubin.

Hb

Amino acids

Heme

Fe^{3+}

Bilirubin

Spleen sinusoid

Red bone marrow

Amino acids

Fe^{3+}-Transferrin

Bilirubin

Liver

④a Iron ions and amino acids are recycled and used to make new hemoglobin in red bone marrow.

④b Bilirubin is sent to the liver for excretion.

Figure 19.6 Erythrocyte death.

yellowish pigment **bilirubin** (BIL-ih-roo-bin). Interestingly, biliverdin is responsible for the greenish color seen in bruises. As the biliverdin is converted to bilirubin, the bruise becomes yellow.

(4a) Iron ions and amino acids are recycled and used to make new hemoglobin in red bone marrow. Most of the iron ions and amino acids are transported to the red bone marrow to be incorporated into new hemoglobin, a form of molecular recycling. You can see in Figure 19.6 that the iron ions are transferred to a protein called **transferrin** (trans-FER-in) that carries them through the blood.

(4b) Bilirubin is sent to the liver for excretion. At the same time the events in **(4a)** are occurring, bilirubin enters the blood and is transported to the liver, where it is modified and excreted in feces and urine.

Occasionally, trauma or disease requires removal of the spleen. If the spleen is no longer present, older erythrocytes are destroyed by macrophages in the liver and bone marrow sinusoids.

> **Quick Check**
>
> ☐ 3. Walk through the basic steps of erythropoiesis.
> ☐ 4. What is the lifespan of an erythrocyte?
> ☐ 5. How are erythrocytes and hemoglobin recycled?

Anemia

The common condition **anemia** (uh-NEE-mee-uh; *a/an-* = "without," *-emia* = "blood") is defined as decreased oxygen-carrying capacity of the blood. Although anemia has a variety of causes, the symptoms are similar and mostly reflect decreased oxygen delivery to tissues. They include pallor (pale skin and nail beds), fatigue, weakness, and shortness of breath. In addition, many types of anemia increase the number of reticulocytes in the blood as the body boosts erythrocyte production. Severe anemia may cause a rapid heart rate, which is the body's attempt to compensate for impaired oxygenation of the tissues by increasing cardiac output. If untreated, such severe anemia can be fatal.

Anemia has three primary causes: decreased hemoglobin, decreased hematocrit, and abnormal hemoglobin. A widely used test to detect anemia (and other conditions) is the complete blood count (see *A&P in the Real World: Complete Blood Count*).

Decreased Hemoglobin

The most common type of anemia is **iron-deficiency anemia,** which is due to inadequate dietary iron, reduced intestinal absorption of dietary iron, or slow blood loss (including menstruation). Without functional iron-containing heme groups, erythroblasts cannot make hemoglobin. Erythrocytes in an individual with iron-deficiency anemia are generally small and pale. Another common form is *anemia of chronic disease,* in which an underlying condition such as cancer produces chemicals that interfere with the transport of iron from the liver to the red bone marrow. Decreased hemoglobin may also be due to malnutrition, vitamin B6 deficiency, certain drugs, poisoning with heavy metals such as lead, or even pregnancy.

Decreased Hematocrit

Many factors can reduce the number of erythrocytes in the blood, which lowers the hematocrit. One is blood loss, which may be acute, as in an injury, or chronic, as in a bleeding stomach ulcer. **Pernicious anemia** results from vitamin B12 deficiency, which interferes with DNA synthesis of rapidly dividing cells, including hematopoietic cells in bone marrow. Erythrocyte destruction can lead to **hemolytic anemia.** Causes of hemolytic anemia include bacterial infections, diseases of the immune system or liver, and lead poisoning. Finally, the red bone marrow may stop producing erythrocytes, which results in **aplastic anemia** (*-plasis* = "formation"). Certain medications or exposure to ionizing radiation may induce aplastic anemia, but the cause in many cases is unknown.

Abnormal Hemoglobin

The most common cause of abnormal hemoglobin is the inherited condition *sickle-cell disease*. Individuals with a single copy

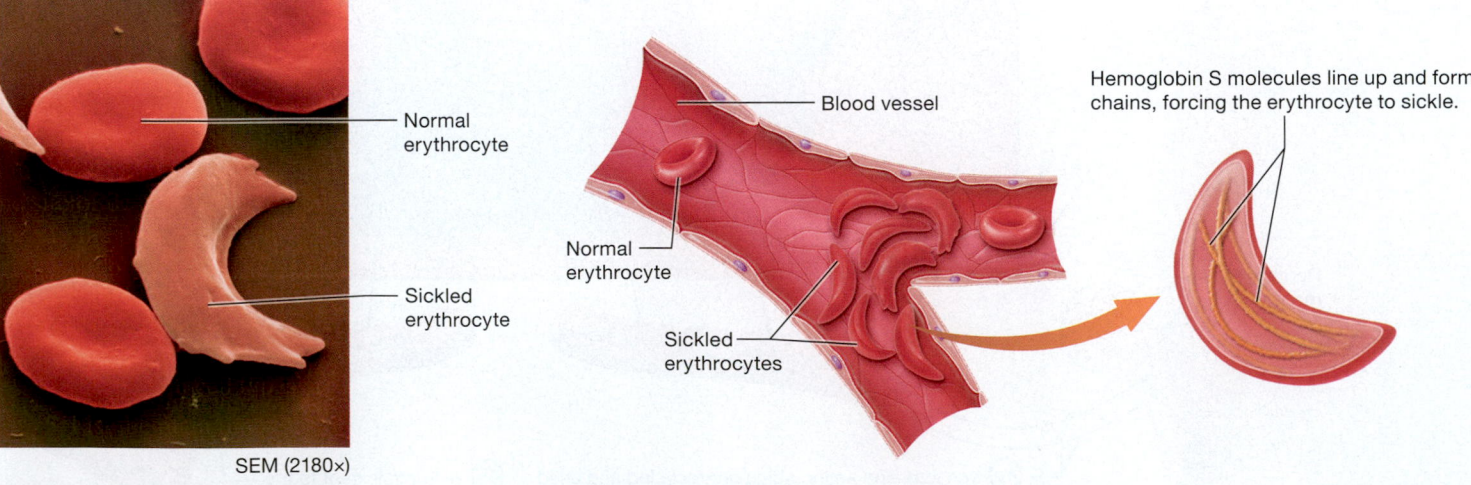

(a) **Normal and sickled erythrocytes**

SEM (2180×)

(b) **Sickled erythrocytes block small blood vessels.**

Normal erythrocyte

Sickled erythrocyte

Blood vessel

Normal erythrocyte

Sickled erythrocytes

Hemoglobin S molecules line up and form chains, forcing the erythrocyte to sickle.

Figure 19.7 Erythrocytes in sickle-cell disease.

Complete Blood Count

An important test for anemia and many other conditions is a **complete blood count,** or **CBC.** Blood is drawn and examined under a microscope or by an automated analyzer to evaluate the number and characteristics of blood cells. The factors examined include:

- erythrocyte count, the number of erythrocytes reported as number of cells per milliliter, which is used to calculate the hematocrit;
- concentration of hemoglobin in the sample;
- erythrocyte characteristics such as size, volume, and concentration of hemoglobin in their cytosol; and
- number and volume of platelets.

Most CBCs also examine the numbers and types of leukocytes. The immune system chapter provides a closer look at this portion of the CBC (see Chapter 20).

of a defective gene have *sickle-cell trait* and are generally asymptomatic. Such individuals are resistant to the mosquito-borne disease *malaria,* which is endemic to Africa and other tropical locations. Individuals with two copies of the defective gene have **sickle-cell disease,** in which they produce an abnormal hemoglobin called **hemoglobin S,** or **HbS.** When the oxygen level is low, HbS proteins line up in a row, forcing the erythrocytes into a curved "sickle" shape (**Figure 19.7**). The sickle-shaped cells get stuck in capillary beds, which leads to ischemia and tissue damage. The sickled cells are eventually destroyed; this lowers the number of circulating erythrocytes and causes anemia.

Quick Check

☐ 6. What are the common symptoms of anemia?

☐ 7. Summarize the three primary causes of anemia.

Apply What You Learned

☐ 1. Explain how an erythrocyte's structure relates to its functions.

☐ 2. Athletes have been known to abuse erythropoietin. Explain what potential benefit an athlete might hope to derive from erythropoietin. What are the risks of abusing erythropoietin?

See answers in Appendix A.

MODULE
19.3
Leukocytes and Immune Function

Learning Outcomes

1. Compare and contrast the relative prevalence and morphological features of the five types of leukocytes.

2. Describe functions for each of the five major types of leukocytes.

3. Discuss the difference in leukopoiesis of granulocytes and agranulocytes.

Leukocytes (LEWK-oh-sytz; *leuk-* = "white"), or **white blood cells (WBCs),** are larger than erythrocytes and have prominent nuclei. Also unlike erythrocytes, leukocytes don't generally function within the blood, but instead use it as a transport vehicle to reach nearly all tissues in the body. When they reach their destination, they adhere to the walls of capillaries or venules and exit the blood by squeezing between the endothelial cells of the vessels.

Leukocytes come in two basic varieties, shown in **Figure 19.8**: (1) **granulocytes** (GRAN-yoo-loh-sytz), which contain cytoplasmic granules that the cells release when activated, and (2) **agranulocytes** (ay-GRAN-yoo-loh-sytz), which lack visible cytoplasmic granules. The following sections examine the properties of both granulocytes and agranulocytes, as well as the process of leukocyte formation. We'll focus on leukocyte functions in detail in the lymph and immune chapter (see Chapter 20).

Granulocytes

≪ FLASHBACK

1. What is phagocytosis? (p. 83)

2. What are the main functions of lysosomes and lysosomal enzymes? (p. 93)

Granulocytes are easy to distinguish from other cell types by their unusual nuclei (see Figure 19.8). Although many granulocytes may appear to have multiple nuclei, each cell actually has a single nucleus with multiple lobes that are connected by thin bands of nuclear material. All granulocytes contain general lysosomal granules as well as specific granules unique to each type of granulocyte.

Cells must be treated with specific dyes or stains so we can see their features under a microscope (see Chapter 4). Cells in a blood smear are treated with two dyes: the red, acidic dye *eosin* and the dark purple, basic (alkaline) dye *methylene blue.* We classify granulocytes into three categories based on the color of their granules after being stained: *neutrophils, eosinophils,* and *basophils.*

Neutrophils

Neutrophils (noo-TROH-fills; "neutral liking") are the most common type of leukocyte, making up about 60% of total leukocytes in the blood. The neutrophils' granules take up both the acidic and basic dyes (hence the name *neutro*phil), staining their cytoplasm a light lilac color (see Figure 19.8). Their nuclei typically have three to five variably shaped lobes, which is the reason for their other name: **polymorphonucleocytes** (pah´-lee-mohr-foh-NOO-klee-oh-sytz; *poly-* = "many," *morpho-* = "shape"), sometimes shortened to **PMNs** or **polys.**

Neutrophils are attracted to injured cells by chemicals released by the damaged cells, a process called **chemotaxis** (kee-moh-TAK-sis). This process occurs with any type of cellular injury, from trauma caused by stepping on a nail to infection by bacteria.

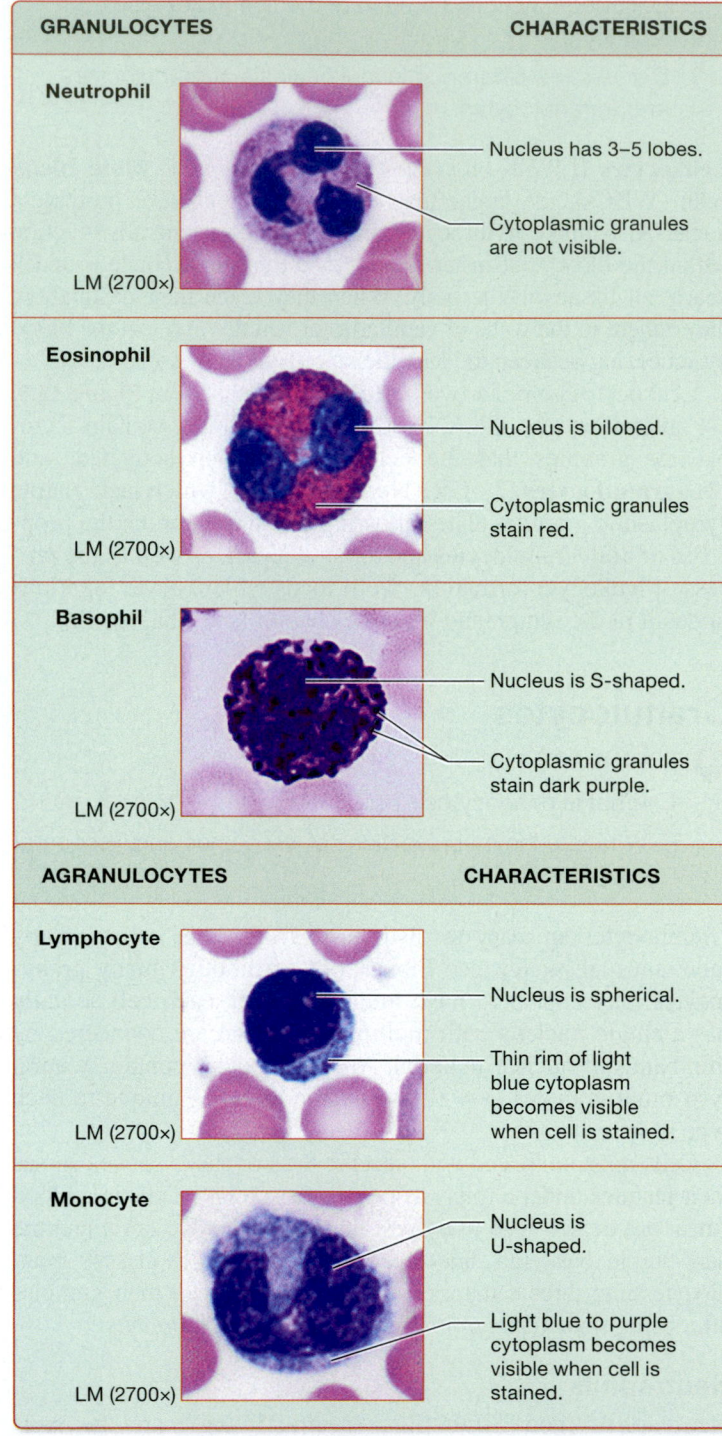

GRANULOCYTES	CHARACTERISTICS
Neutrophil LM (2700×)	Nucleus has 3–5 lobes. Cytoplasmic granules are not visible.
Eosinophil LM (2700×)	Nucleus is bilobed. Cytoplasmic granules stain red.
Basophil LM (2700×)	Nucleus is S-shaped. Cytoplasmic granules stain dark purple.

AGRANULOCYTES	CHARACTERISTICS
Lymphocyte LM (2700×)	Nucleus is spherical. Thin rim of light blue cytoplasm becomes visible when cell is stained.
Monocyte LM (2700×)	Nucleus is U-shaped. Light blue to purple cytoplasm becomes visible when cell is stained.

Figure 19.8 Classes of leukocytes.

View PAL™ Histology
@ Mastering Anatomy & Physiology

When neutrophils reach the damaged tissue, they exit the blood and release the contents of their granules. These contents directly kill bacterial cells, enhance inflammation, and attract more neutrophils and leukocytes to the area. Neutrophils are also very active phagocytes that ingest and destroy bacterial cells.

Eosinophils

As you can infer from their name, the granules of **eosinophils** (ee′-oh-SIN-oh-filz; "eosin liking") take up the dye eosin and so appear red. Eosinophils are relatively rare leukocytes that account for fewer than 3% of total leukocytes in the blood. Their nuclei are bilobed, resembling a barbell, with two circular lobes connected by a thin strand of nuclear material.

Eosinophils are involved in the body's response to infection with parasitic worms and in allergic reactions. Their granules contain substances that assist in these functions, including enzymes and toxins that are specific to parasites, as well as chemicals that mediate (bring about) inflammation. In addition, eosinophils are phagocytes and ingest foreign compounds that have been bound by proteins of the immune system.

Basophils

Basophils (BAY-zoh-filz; "basic liking"), the least common leukocyte, make up less than 1% of total leukocytes in the blood. As implied by their name, basophils contain granules that take up the basic dye methylene blue, which stains them dark purple-blue. Note that the granules almost completely obscure the typically S-shaped nuclei in these cells.

Basophils release chemicals from their granules that mediate inflammation. We discuss inflammation in the chapter on immunity (see Chapter 20).

Quick Check

☐ 1. How do granulocytes and agranulocytes differ?

☐ 2. Describe the basic structure and function of the three types of granulocytes.

Agranulocytes

Lymphocytes (LIM-foh-syts) and monocytes (MAH-noh-syts) are the two kinds of agranulocytes. Both cell types lack visible cytoplasmic granules, although like granulocytes, they do contain lysosomes.

Lymphocytes

The second most numerous type of leukocyte, **lymphocytes,** make up about 30–34% of total leukocytes in the blood. They contain large, spherical nuclei and generally a thin rim of light blue cytoplasm that is visible when stained (see Figure 19.8).

There are two basic types of lymphocytes: **B lymphocytes** (or **B cells**) and **T lymphocytes** (or **T cells**). Both B and T lymphocytes are activated by cellular markers called **antigens** (AN-tih-jenz). Antigens, which are generally glycoproteins, are present on all cells and most biological compounds.

Although B and T lymphocytes are structurally similar, they differ in their functions. Activated B lymphocytes produce proteins called **antibodies** that bind to antigens and remove them from tissues. Each population of B lymphocytes secretes antibodies with a specific structure that allows them to bind to only one unique antigen. Populations of T lymphocytes also show specificity for individual antigens. However, activated T lymphocytes do not produce antibodies. Instead, populations of T lymphocytes have specific receptors for individual antigens. When the receptors are bound, T lymphocytes activate other components of the immune system and directly destroy abnormal body cells, such as cancer cells and those that are virally infected. Antibodies and T lymphocyte receptors, then, are structured in such a way that they bind to only one unique antigen, an example of the Structure-Function Core Principle (p. 28).

CORE PRINCIPLE
Structure-Function

Monocytes

The largest leukocytes are the **monocytes,** which account for 4–8% of the total leukocyte population. Monocytes are easily distinguished from other leukocytes by their large U-shaped nuclei and light blue or purple cytoplasm that becomes visible when the cells are stained.

Monocytes remain in the blood for only a few days before they exit the capillaries and enter the tissues, where some mature into very active phagocytes called **macrophages** (MAK-roh-fehj-uhz; "big eater"). Macrophages ingest dead and dying cells (such as old erythrocytes or those damaged from trauma), bacteria, antigens, and other cellular debris. In addition, they activate other parts of the immune system by displaying phagocytosed antigens to other leukocytes.

Quick Check

☐ 3. Compare the two types of agranulocytes.

☐ 4. How do B and T lymphocytes differ?

☐ 5. What does a monocyte become in the tissues? What is its primary function?

Leukocyte Formation: Leukopoiesis

Leukocytes form in the bone marrow by the process of **leukopoiesis** (loo'-koh-poy-EE-sis). Like erythrocytes, all leukocytes arise from hematopoietic stem cells (HSCs). The HSCs divide and split into two cell lines:

- the **myeloid cell line,** which produces most of the formed elements, including erythrocytes and platelets; and
- the **lymphoid cell line,** which forms lymphocytes.

The myeloid cell line differentiates early into blast cells that are committed to becoming monocytes or granulocytes. Monocytes develop from committed cells called **monoblasts** and precursor cells called **promonocytes.** All granulocytes derive from **myeloblasts** that differentiate into precursor cells known as **promyelocytes.** In the final precursor stage, the cells are referred to as **band cells** or **stab cells,** which develop into mature granulocytes that enter the bloodstream. If someone has an active infection, it is common to find band cells in the blood as the immune system attempts to increase the numbers of circulating leukocytes.

The lymphoid cell line differentiates first into **lymphoblasts,** at which point the cells are committed to becoming lymphocytes. Lymphoblasts develop into precursor cells called **prolymphocytes,**

which then become mature B and T lymphocytes. These cells mature in different locations: B lymphocytes remain in the bone marrow, whereas T lymphocytes migrate to the thymus gland in the mediastinum to complete their maturation (remember "B" for "bone marrow" and "T" for "thymus").

Figure 19.9 summarizes the process of leukopoiesis. Sometimes myeloid or lymphoid cells undergo abnormal changes, resulting in the uncontrolled cell division of cancer (see *A&P in the Real World: Leukemias* on p. 738).

Quick Check

☐ 6. From which cell type do all leukocytes derive?

☐ 7. How do the lymphoid and myeloid cell lines differ?

☐ 8. Where do T and B lymphocytes mature?

Apply What You Learned

☐ 1. Predict the type of leukocyte that would respond to the following conditions.

    a. Bacterial infection

    b. Viral infection

    c. Parasitic infection

    d. Tissue injury due to trauma

See answers in Appendix A.

MODULE 19.4
Platelets

Learning Outcomes

1. Explain how platelets differ structurally from the other formed elements of blood.

2. Discuss the role of the megakaryocyte in the formation of platelets.

Platelets, the smallest of the formed elements, are involved in *hemostasis,* a process that stops blood loss from an injured blood vessel. We now examine the characteristics of platelets and how they form in bone marrow. Module 19.5 explores their role in hemostasis.

Platelet Characteristics

« FLASHBACK

1. What normally happens during telophase of mitosis? (p. 115)

2. What are the functions of actin and myosin proteins? (p. 342)

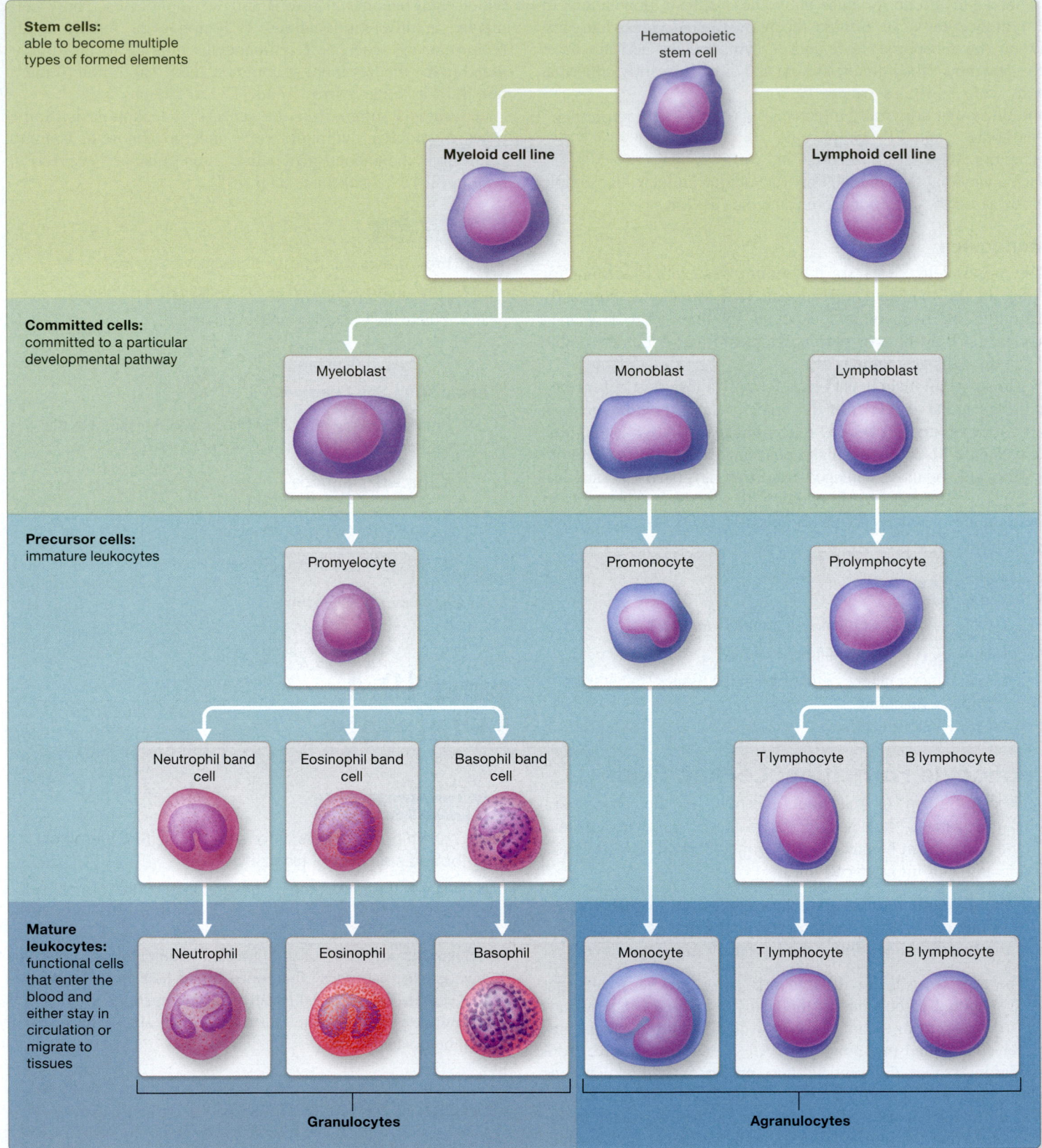

Figure 19.9 Leukopoiesis: formation of leukocytes.

Platelets (PLAYT-letz) are not true cells, but instead small fragments of cells surrounded by a plasma membrane. As you can see in **Figure 19.10a**, platelets lack nuclei and most other organelles. However, they do contain multiple types of granules that house clotting factors and enzymes, small numbers of mitochondria, and glycogen deposits that enable them to carry out oxidative catabolism. These tiny structures also contain cytoskeletal components, including a system of microtubules that are associated with actin and myosin filaments.

Platelet Formation

Like all formed elements, platelets begin as hematopoietic stem cells (**Figure 19.10b**). Their precursor cells, called **megakaryoblasts** (meg′-uh-KEHR-ee-yoh-blasts; "immature big nucleus cells"), derive from the myeloid cell line. During development, megakaryoblasts become **megakaryocytes,** which undergo repeated rounds of mitosis. However, unlike other cells, megakaryocytes do not undergo cytokinesis during anaphase and telophase of mitosis. The end result is a massive cell with multiple copies of DNA located in a single nucleus (you can see the size difference between a megakaryocyte and an erythrocyte in Figure 19.10b).

Under the influence of hormones such as *thrombopoietin,* mature megakaryocytes extend ribbon-like "arms," or extensions, filled with cytosol, granules, and several organelles. As you can see in Figure 19.10b, these arms extend through the clefts in the bone marrow sinusoids and into the bloodstream. The force of the blood coursing past the arms lops off small pieces that become platelets. Each arm can give rise to thousands of platelets. Most remain in the general circulation, where they have a lifespan of only 7–10 days. When platelets have reached old age, the liver and spleen remove them from circulation.

Quick Check

☐ 1. Describe the structure of a platelet.

☐ 2. How are platelets formed?

Putting It All Together: The Big Picture of Formed Elements

Let's now take what we just discussed about platelets and compare it with what you've read about the other formed elements. **Figure 19.11**, The Big Picture of Formed Elements, presents a summary of hematopoiesis followed by the properties of each of the blood's formed elements for easy comparison and reference.

Apply What You Learned

☐ 1. Many toxins and poisons interfere with a cell's ability to undergo protein synthesis. Will this affect platelets that are already in the bloodstream? Why or why not? (*Hint:* Which cellular structure is required to undergo protein synthesis? What is the structure of a platelet?)

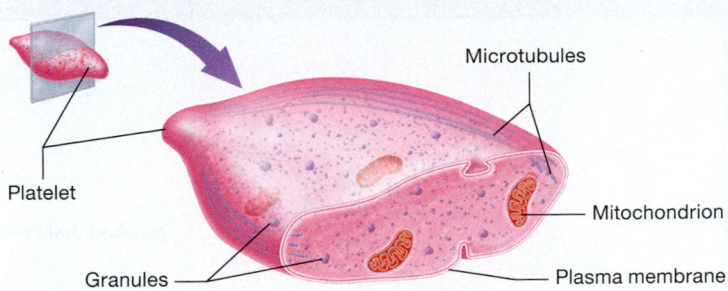

(a) Platelet structure

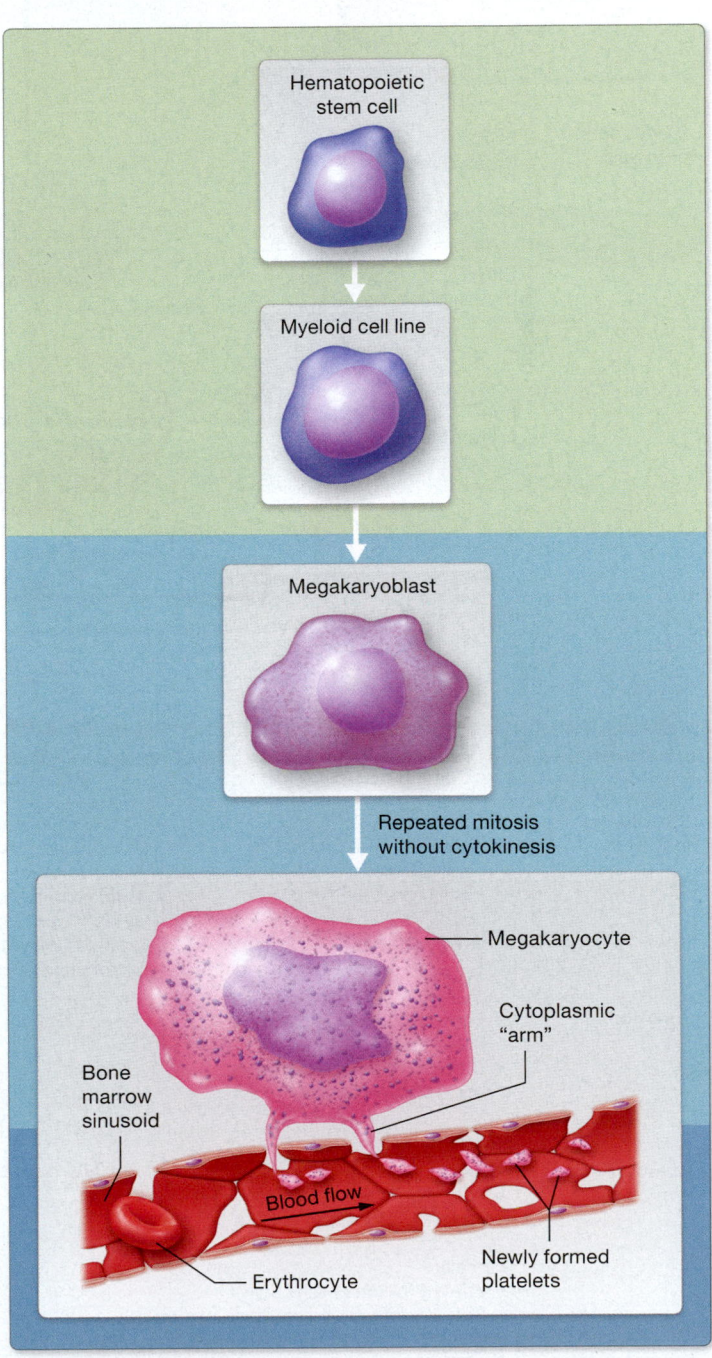

(b) Formation of platelets

Figure 19.10 Structure and formation of platelets.

The Big Picture of Formed Elements

Figure 19.11

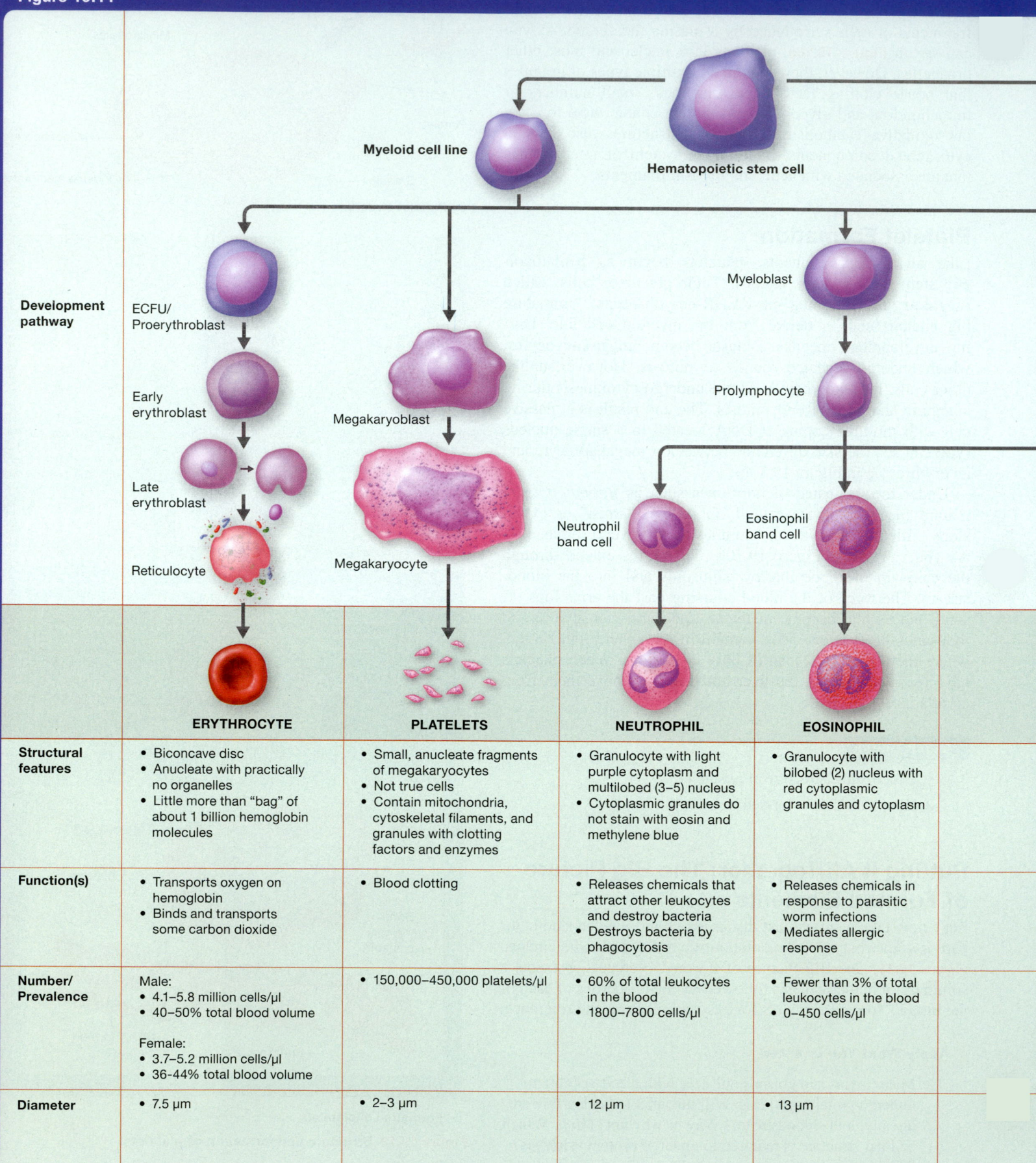

	ERYTHROCYTE	PLATELETS	NEUTROPHIL	EOSINOPHIL	
Structural features	• Biconcave disc • Anucleate with practically no organelles • Little more than "bag" of about 1 billion hemoglobin molecules	• Small, anucleate fragments of megakaryocytes • Not true cells • Contain mitochondria, cytoskeletal filaments, and granules with clotting factors and enzymes	• Granulocyte with light purple cytoplasm and multilobed (3–5) nucleus • Cytoplasmic granules do not stain with eosin and methylene blue	• Granulocyte with bilobed (2) nucleus with red cytoplasmic granules and cytoplasm	
Function(s)	• Transports oxygen on hemoglobin • Binds and transports some carbon dioxide	• Blood clotting	• Releases chemicals that attract other leukocytes and destroy bacteria • Destroys bacteria by phagocytosis	• Releases chemicals in response to parasitic worm infections • Mediates allergic response	
Number/ Prevalence	Male: • 4.1–5.8 million cells/µl • 40–50% total blood volume Female: • 3.7–5.2 million cells/µl • 36-44% total blood volume	• 150,000–450,000 platelets/µl	• 60% of total leukocytes in the blood • 1800–7800 cells/µl	• Fewer than 3% of total leukocytes in the blood • 0–450 cells/µl	
Diameter	• 7.5 µm	• 2–3 µm	• 12 µm	• 13 µm	

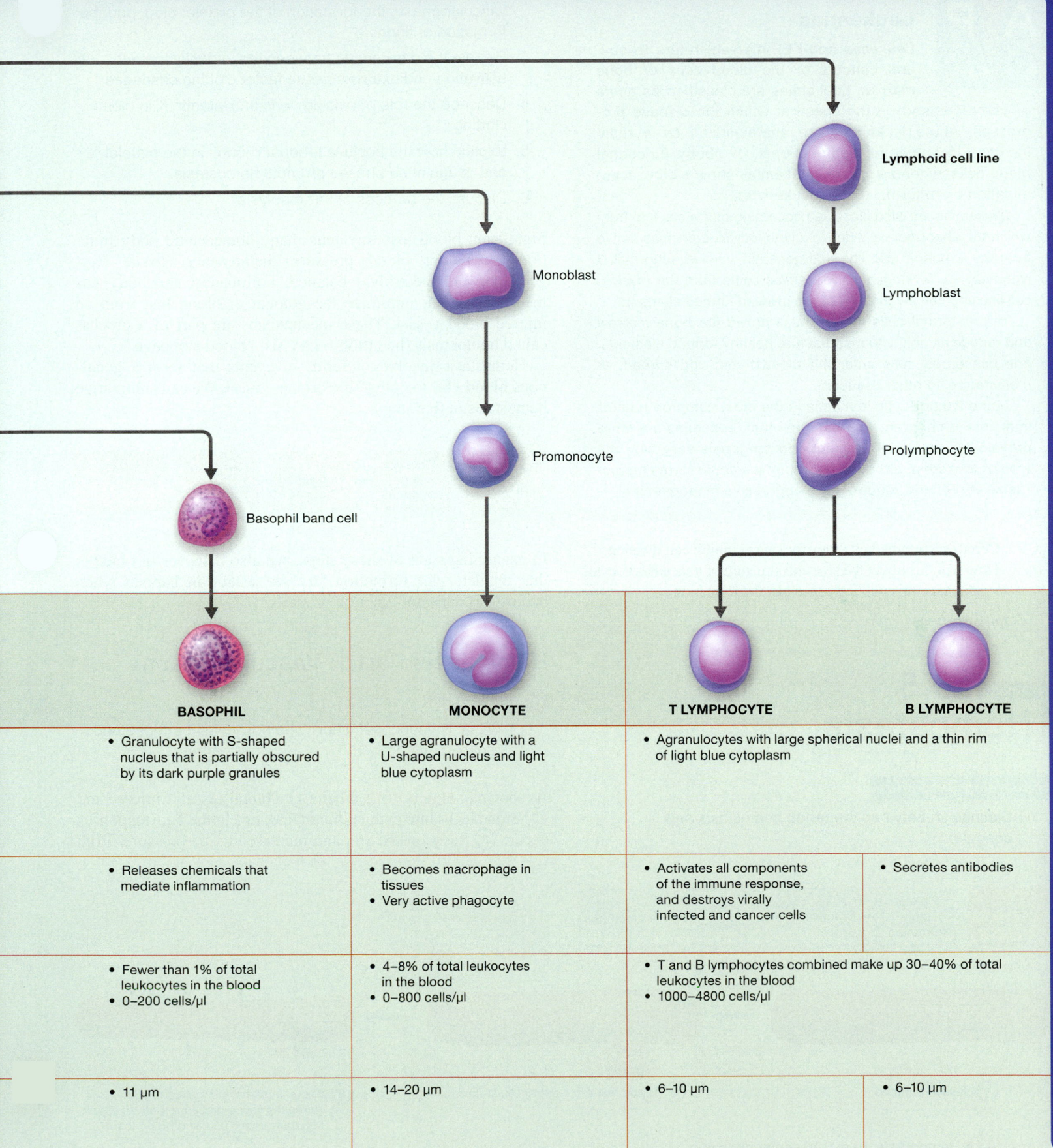

Basophil band cell	Monoblast	Lymphoid cell line	
	Promonocyte	Lymphoblast	
		Prolymphocyte	
BASOPHIL	**MONOCYTE**	**T LYMPHOCYTE**	**B LYMPHOCYTE**
• Granulocyte with S-shaped nucleus that is partially obscured by its dark purple granules	• Large agranulocyte with a U-shaped nucleus and light blue cytoplasm	• Agranulocytes with large spherical nuclei and a thin rim of light blue cytoplasm	
• Releases chemicals that mediate inflammation	• Becomes macrophage in tissues • Very active phagocyte	• Activates all components of the immune response, and destroys virally infected and cancer cells	• Secretes antibodies
• Fewer than 1% of total leukocytes in the blood • 0–200 cells/μl	• 4–8% of total leukocytes in the blood • 0–800 cells/μl	• T and B lymphocytes combined make up 30–40% of total leukocytes in the blood • 1000–4800 cells/μl	
• 11 μm	• 14–20 μm	• 6–10 μm	• 6–10 μm

Leukemias

AP in the Real World

Leukemia (loo-KEE-mee-ah) refers to several cancers of the blood cells or bone marrow. Leukemias are classified as *acute* or *chronic* based on the speed at which the disease progresses. Acute leukemias are characterized by a rapid increase in immature, nonfunctional, or poorly functional blood cells, whereas chronic leukemias show a slow accumulation of abnormal mature leukocytes.

Leukemias are also classified according to the cell line from which the abnormal cells derive. *Lymphocytic leukemias* derive from the lymphoid cell line and generally involve abnormal B lymphocytes. *Myelogenous leukemias* come from the myeloid cell line and can involve any of the myeloid formed elements.

The abnormal cells in leukemias crowd the bone marrow and reduce its ability to manufacture healthy formed elements. The cancerous cells enter the bloodstream and spread, or *metastasize*, to other tissues.

Acute lymphocytic leukemia is the most common type of leukemia in children, whereas the other leukemias are more prevalent in adults. Treatment and prognosis vary with the type of leukemia, although generally the acute forms metastasize earlier and require more aggressive management.

☐ 2. Certain chemotherapy drugs for cancer inhibit cell division. How will this affect the size and structure of a megakaryocyte? Predict how this will influence platelet formation.

See answers in Appendix A.

MODULE **19.5**
Hemostasis

Learning Outcomes

1. Distinguish between the terms hemostasis and coagulation.

2. Describe the process of hemostasis, including the vascular phase, the formation of the platelet plug, and the formation of fibrin.

3. Explain the differences between the intrinsic/contact activation and extrinsic/tissue factor clotting cascades.

4. Describe the role of calcium ions and vitamin K in blood clotting.

5. Explain how the positive feedback loops in the platelet and coagulation phases promote hemostasis.

6. Discuss the process of thrombolysis.

Significant blood loss threatens many homeostatic body functions, including blood pressure maintenance, tissue oxygenation, and electrolyte balance. Fortunately, the body has mechanisms that minimize the amount of blood lost from an injured blood vessel. These mechanisms are part of a process called **hemostasis** (hee-moh-STAY-sis; "blood stoppage").

Hemostasis involves a series of events that form a gelatinous **blood clot** to "plug" the broken vessel. We can summarize hemostasis in five steps:

In examining each of these steps, we also discover the factors that regulate clot formation and see what can happen when hemostasis does not function properly.

Hemostasis Part 1: Vascular Spasm

« FLASHBACK

1. What is vasoconstriction, and how does it affect blood flow? (p. 671)

As shown in **Figure 19.12**, when ① a blood vessel is injured and ② blood leaks into extracellular fluid, two immediate responses occur: ③ vasoconstriction and increased tissue pressure. Both

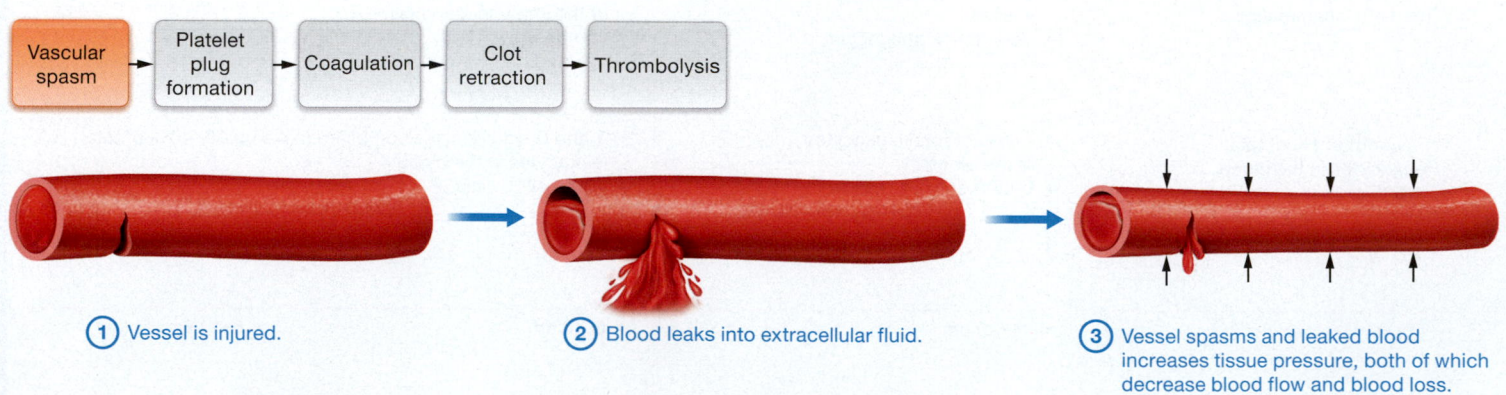

① Vessel is injured.

② Blood leaks into extracellular fluid.

③ Vessel spasms and leaked blood increases tissue pressure, both of which decrease blood flow and blood loss.

Figure 19.12 Hemostasis, part 1: Vascular spasm.

responses decrease the blood vessel diameter. Recall that such a decrease leads to a local reduction in blood pressure and blood flow, which minimizes blood loss from the injured vessel (see Chapter 17). This activity has the same effect as applying compression or a tourniquet to a bleeding wound. Surgeons also put this principle into practice by clamping off blood vessels during surgery—the external pressure collapses the vessel and reduces blood loss.

Quick Check

☐ 1. What purpose does vascular spasm serve in the process of hemostasis?

Hemostasis Part 2: Platelet Plug Formation

⟪ FLASHBACK

1. What are the layers of the blood vessel wall? Where are collagen fibers located in the vessel wall? (p. 671)

Although vasoconstriction and increased tissue pressure can minimize blood loss from an injured vessel, they cannot stop bleeding entirely. For this, we require a "plug" that acts like a cork. This plug consists primarily of platelets, and so is known as the **platelet plug.**

Normally, platelets aren't "sticky"—a good thing, because otherwise they would adhere to tissues and block blood flow. However, **Figure 19.13** shows that platelets adhere to the injury site. Why? When a vessel is injured, ① collagen fibers and chemicals in the vessel's tunica adventitia are exposed. The injured endothelial cells release a glycoprotein called *von Willebrand factor* (fahn VIL-uh-brant), or vWF, which ② binds to

CORE PRINCIPLE
Cell-Cell Communication

receptors on the surface of the platelets' plasma membranes. Together, the exposed collagen and vWF make the platelets sticky so they adhere to one another and the vessel wall. Part of this process demonstrates the Cell-Cell Communication Core Principle (p. 28): Endothelial cells produce vWF, which communicates with platelets.

The binding of vWF and collagen to platelets triggers a series of events within platelets known collectively as **platelet activation.** Activated platelets release the contents of their granules—ATP, ADP, serotonin, calcium, clotting factors, and other chemicals—by exocytosis. ③ Many of these factors attract and activate nearby platelets and cause them to clump together, or **aggregate**. Platelet aggregation forms the platelet plug. This plug seals the injured vessel temporarily, but the repair is short-lived without a molecular "glue" that can hold the platelets together.

Quick Check

☐ 2. What is the role of platelets in hemostasis?
☐ 3. How are platelets triggered to aggregate?

Hemostasis Part 3: Coagulation

The glue that binds platelets, endothelial cells, and other formed elements together is a sticky, threadlike protein called **fibrin** (FY-brin). Fibrin converts the soft, liquid platelet plug into a more solid mass by the process of **coagulation** (koh-aeh′-gyoo-LAY-shun; *coag-* = "bring together"). We normally find fibrin in plasma and in platelets in its inactive form, **fibrinogen** (fy-BRIN-oh-jen). Fibrinogen is converted into fibrin by the **coagulation cascade,** a series of reactions that occur at

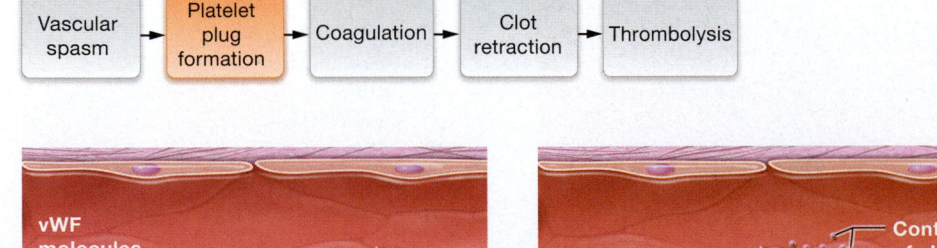

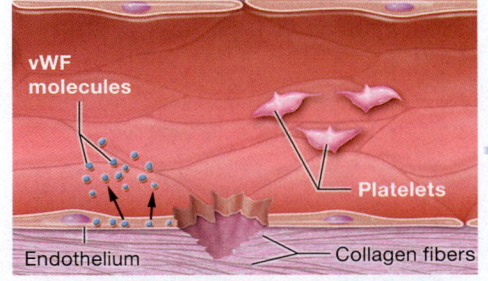

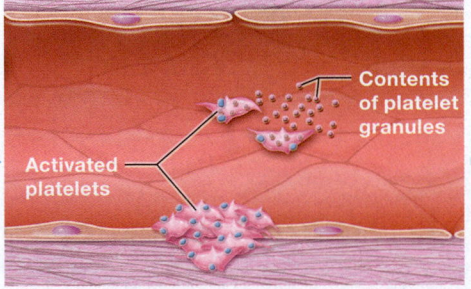

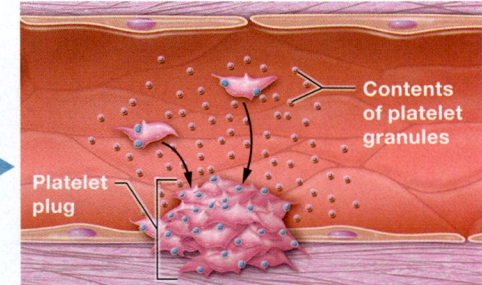

① Vessel injury exposes collagen fibers and causes endothelial cells to produce von Willebrand factor (vWF).

② vWF binds platelets, which activates them and causes them to stick to the exposed collagen and to one another. Activated platelets release the contents of their granules.

③ Contents of platelet granules activate and attract more platelets, which aggregate to form the platelet plug.

Figure 19.13 Hemostasis, part 2: Platelet plug formation.

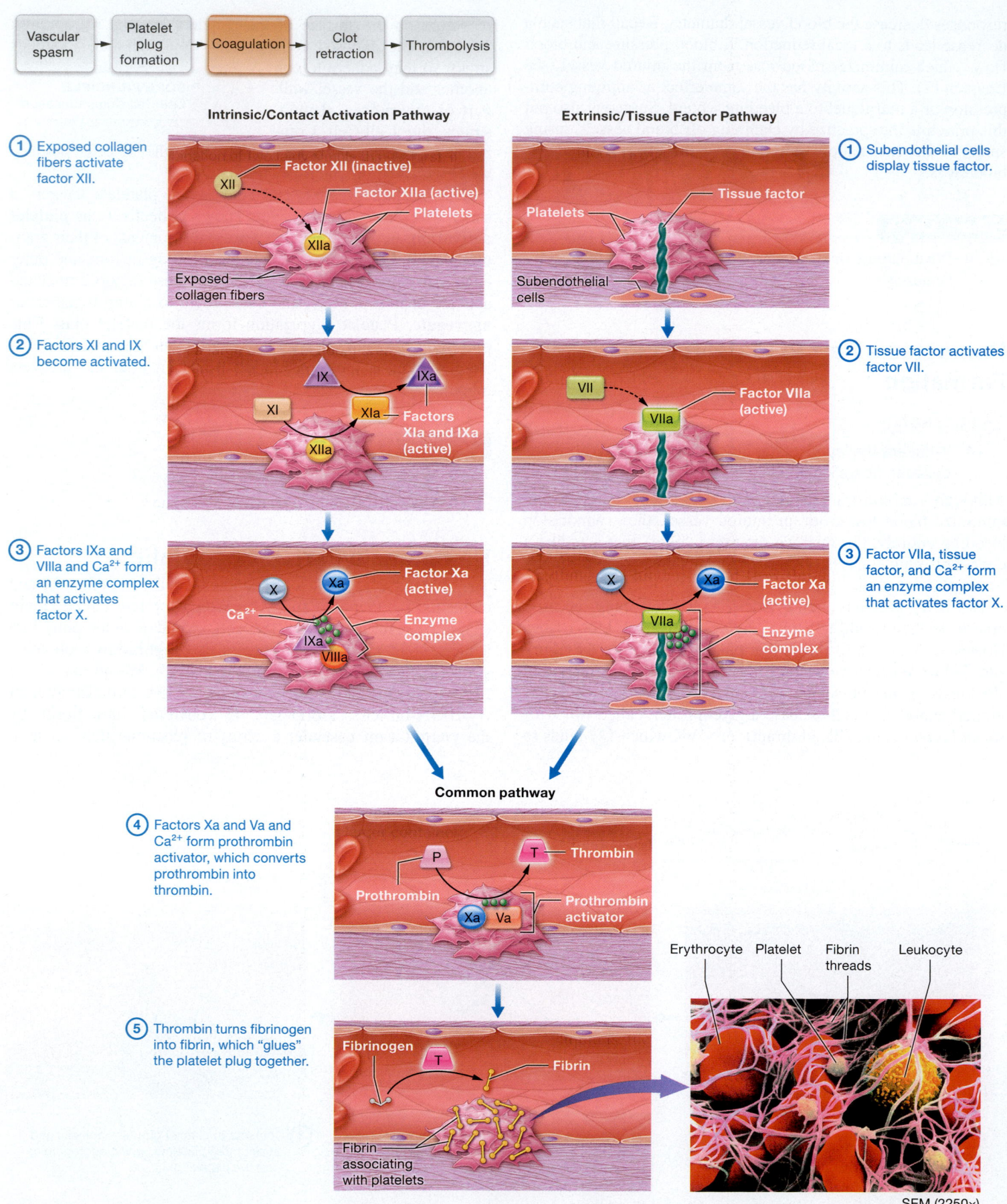

Intrinsic/Contact Activation Pathway

① Exposed collagen fibers activate factor XII.

Factor XII (inactive)
Factor XIIa (active)
Platelets
Exposed collagen fibers

② Factors XI and IX become activated.

Factors XIa and IXa (active)

③ Factors IXa and VIIIa and Ca²⁺ form an enzyme complex that activates factor X.

Factor Xa (active)
Enzyme complex

Extrinsic/Tissue Factor Pathway

① Subendothelial cells display tissue factor.

Tissue factor
Platelets
Subendothelial cells

② Tissue factor activates factor VII.

Factor VIIa (active)

③ Factor VIIa, tissue factor, and Ca²⁺ form an enzyme complex that activates factor X.

Factor Xa (active)
Enzyme complex

Common pathway

④ Factors Xa and Va and Ca²⁺ form prothrombin activator, which converts prothrombin into thrombin.

Thrombin
Prothrombin
Prothrombin activator

⑤ Thrombin turns fibrinogen into fibrin, which "glues" the platelet plug together.

Fibrinogen
Fibrin
Fibrin associating with platelets

Erythrocyte Platelet Fibrin threads Leukocyte

SEM (2250×)

Figure 19.14 Hemostasis, part 3: Coagulation cascade.

the surface of the platelets and/or damaged endothelial cells (**Figure 19.14**).

The coagulation cascade relies on **clotting factors.** Most clotting factors are enzymes produced by the liver that circulate in the blood in their inactive forms. Each is named with a Roman numeral (I–XIII), although many factors have a descriptive name as well (**Table 19.2** on p. 742). Unfortunately, the factors were assigned the Roman numeral in the order of their discovery, not in the order in which they operate in the coagulation cascade. This is why the cascade does not proceed in numerical sequence.

Keep in mind that the synthesis of four clotting factors— factors II, VII, IX, and X—depends on the presence of vitamin K. These vitamin K–dependent clotting factors are important targets of drug therapy that blocks coagulation.

The coagulation cascade proceeds by two pathways: the *intrinsic pathway*, also known as the *contact activation pathway*, and the *extrinsic pathway*, also known as the *tissue factor pathway*. Both converge to a common pathway that ends with fibrin formation.

Intrinsic/Contact Activation Pathway

The **intrinsic pathway** is so named because all the factors required for it to proceed are located in ("intrinsic to") the blood. The events are initiated when the inactive protein clotting factor XII comes into contact with exposed collagen fibers from damaged endothelium. This is the source of its other name—the **contact activation pathway.** Note that this term is now often preferred, because it is considered more descriptive than the name "intrinsic."

The steps of the intrinsic/contact activation pathway proceed as follows (see Figure 19.14, left side):

1. **Exposed collagen fibers activate factor XII.** The negative charges on the surface of the exposed collagen proteins attract and bind a clotting protein in the plasma called **factor XII.** Activated factor XII is known as **factor XIIa** ("a" stands for "activated").

2. **Factors XI and IX become activated.** Factor XIIa is an enzyme that activates another clotting protein in the plasma, **factor XI.** Factor XIa, in turn, enzymatically activates another clotting protein, **factor IX.**

3. **Factors IXa and VIIIa along with calcium ions form an enzyme complex that activates factor X.** The intrinsic/ contact activation pathway ends by forming a large enzyme complex that consists of factor IXa, another protein called **factor VIIIa,** and calcium ions. This enzyme complex activates **factor X,** which becomes factor Xa. Note that calcium ions are required for the factors to associate with platelets in the platelet plug.

The intrinsic/contact activation pathway isn't just activated by exposed collagen fibers—activation occurs any time blood contacts a surface with negative charges, including glass test tubes. For this reason, blood in a test tube will coagulate unless treated with a chemical to prevent clotting.

Extrinsic/Tissue Factor Pathway

The **extrinsic pathway,** which occurs simultaneously with the intrinsic/contact activation pathway, has this name because it is initiated by a factor outside ("extrinsic to") the blood. This factor is a protein known as **tissue factor,** which is the source of its alternate name, the **tissue factor pathway.** Again, this newer name is now often preferred, as it is more descriptive than the term "extrinsic." (Note that another advantage to using the newer names is that "intrinsic" pathway and "extrinsic" pathway sound very similar and are easier to get confused. "Contact activation" pathway and "tissue factor" pathway are much less ambiguous.)

The extrinsic/tissue factor pathway proceeds as shown in Figure 19.14 (right side):

1. **Subendothelial cells display tissue factor.** Beneath the endothelium is a layer of loose connective tissue called the *subendothelium.* When an injury to the blood vessel penetrates the endothelium, it damages the subendothelium beneath it. Damaged *subendothelial cells,* mostly fibroblasts, display tissue factor in their plasma membranes.

2. **Tissue factor activates factor VII.** The clotting protein **factor VII** circulates in the blood in its inactive form. When it comes into contact with tissue factor, it changes in structure and becomes the active form, **factor VIIa.**

3. **Factor VIIa, tissue factor, and calcium ions form an enzyme complex that activates factor X.** Factor VIIa forms a complex with tissue factor and calcium ions that enzymatically cleaves factor X into factor Xa. The calcium ions play the same role here as they did in the intrinsic/ contact activation pathway: They are required for the interaction of factors with platelets in the platelet plug.

The end result of both pathways is the same: activation of factor X to the active enzyme factor Xa. Although the extrinsic/ tissue factor pathway is thought to be the more important path to factor Xa production, both pathways interact and rely heavily on each other to move to completion. A deficiency in any clotting protein can disrupt the entire coagulation cascade.

At this point, the two pathways converge to the common pathway.

Common Pathway

In the common pathway, fibrin is produced (see Figure 19.14):

4. **Factors Xa and Va, along with calcium ions, form prothrombin activator, which converts prothrombin into thrombin.** The common pathway begins with the formation of *prothrombin activator,* a large enzyme complex that consists of factor Xa, another protein called **factor Va,** and calcium ions. Prothrombin activator converts the inactive protein **prothrombin** to the active enzyme **thrombin.**

5. **Thrombin turns fibrinogen into fibrin, which "glues" the platelet plug together.** In the final step of the coagulation cascade, thrombin acts on the inactive protein fibrinogen, which is both present in plasma and released by platelets. Thrombin converts fibrinogen into fibrin, its active form. The fibrin threads form a mesh that glues together the platelet plug, other formed elements, and damaged endothelial cells, sealing the damaged blood vessel.

Table 19.2 Clotting Factors

Factor	Alternate Name	Description	Function
I	Fibrinogen	Plasma protein made by the liver and platelets	• Becomes fibrin, sticky protein threads that "glue" the platelet plug together in the common pathway
II	Prothrombin	Plasma protein made by the liver; synthesis requires vitamin K	• Becomes thrombin, which converts fibrinogen to fibrin
III	Tissue factor	Glycoprotein in the plasma membranes of cells external to the endothelium	• Binds factor VIIa • Activates the extrinsic/tissue factor clotting pathway
IV	Calcium ion	Ion in the plasma	• Multiple roles, including activation of factor X and prothrombin
V	Labile factor	Plasma protein made by the liver; stored in platelets	• Part of the complex that activates prothrombin in the common pathway
VII*	Stable factor	Plasma protein made by the liver; synthesis requires vitamin K	• Part of the complex that activates factor X in the extrinsic/tissue factor pathway
VIII	Antihemophilic factor A	Plasma protein made by the liver	• Part of the complex that activates factor X in the intrinsic/contact activation pathway
IX	Christmas factor	Plasma protein made by the liver; synthesis requires vitamin K	• Part of the complex that activates factor X in the intrinsic/contact activation pathway
X	Stuart factor	Plasma glycoprotein made by the liver; synthesis requires vitamin K	• Part of the complex that activates prothrombin in the common pathway
XI	Plasma thromboplastin antecedent	Plasma protein made by the liver and stored in platelets	• Activates factor IX in the intrinsic/contact activation pathway
XII	Hageman factor	Plasma glycoprotein made by the liver	• Activates factor XI in the intrinsic/contact activation pathway
XIII	Fibrin stabilizing factor	Plasma protein made by the liver and stored in platelets	• Activated by thrombin • Stabilizes the fibrin mesh

*Note that there is no factor VI.

Here's an interesting fact about thrombin: It catalyzes the activation of factors V and VIII. However, thrombin is produced *after* these factors are activated. How is this possible?

The answer is that the enzymes have a small amount of activity before thrombin is produced. As this activity leads to thrombin production, thrombin acts on the factors to convert them to their active forms, which accelerates thrombin production. So, as more thrombin is produced, the response is progressively amplified, which is an example of a *positive feedback loop*. The positive feedback loop is self-limiting for two reasons. First, clotting factors are rapidly inactivated by enzymes in plasma, so their action is fairly brief and limited to the site of injury. Second, fibrin inhibits the production of thrombin via a negative feedback loop, so thrombin activity decreases as the fibrin level increases. Notice that here we have another example of the Feedback Loops Core Principle (p. 28): Coagulation proceeds by a positive feedback loop, but it is ultimately regulated by negative feedback.

 **CORE PRINCIPLE** Feedback Loops

ConceptBOOST »))

Making Sense of the Coagulation Cascade

What's the best way to approach the coagulation cascade? Remember that the entire process has three simple goals:

- **Goal 1: Produce factor Xa.** The goal of both pathways is to produce the active enzyme factor Xa. Although the reactions in each pathway are different, the end result is the same.
- **Goal 2: Produce thrombin.** The product of both pathways, factor Xa, accomplishes the second goal of coagulation by producing the enzyme thrombin. This happens in the common pathway.
- **Goal 3: Produce fibrin.** Thrombin, in turn, accomplishes the third goal of coagulation: producing fibrin to hold the platelet plug together and seal the wound. This is the final result of the common pathway.

Develop a global understanding of these three goals, and the details will be much easier to remember. ■

Quick Check

☐ 4. What is the coagulation cascade?

☐ 5. How do the intrinsic/contact activation and extrinsic/tissue factor pathways differ?

☐ 6. What are the overall goals of the intrinsic/contact activation, extrinsic/tissue factor, and common pathways?

Hemostasis Part 4: Clot Retraction

As the coagulation cascade completes, the actin and myosin fibers in platelets contract via a mechanism similar to that of the actin and myosin fibers in a skeletal muscle fiber. This action, known as **clot retraction,** brings the edges of the wounded vessel closer together, much as sutures (or "stitches") do with the edges of a skin wound (**Figure 19.15**). Clot retraction also squeezes **serum** (SEER-um)—a fluid consisting of plasma minus the clotting proteins—out of the clot. (Think of it as squeezing water out of a wet rag.)

Quick Check

☐ 7. What is the purpose of clot retraction?

Hemostasis Part 5: Thrombolysis

After a wound has healed, the blood clot is no longer necessary and it dissolves in a process called **thrombolysis** (thrahm-BAH-luh-sis; -*lysis* = "split"). Thrombolysis begins with **fibrinolysis** (fy´-brih-nuh-LY-sis)—the breakdown of the fibrin glue that was produced in the coagulation cascade. Compared to the formation of fibrin, its dissolution is relatively straightforward (**Figure 19.16**):

① **Endothelial cells release tissue plasminogen activator (tPA).** Thrombolysis begins when healed endothelial cells produce and release the enzyme **tissue plasminogen**

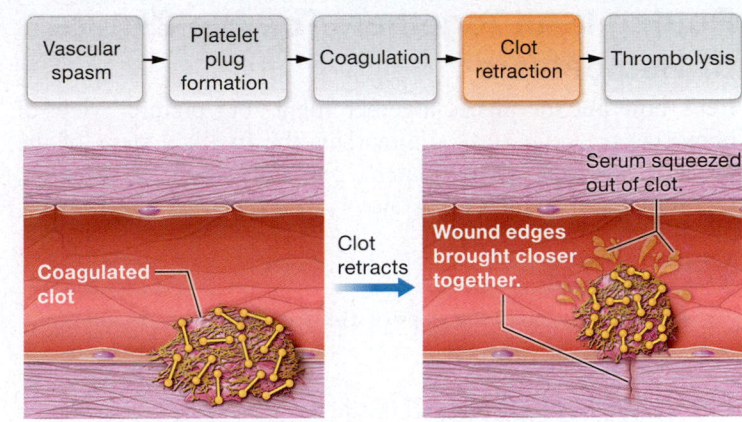

Figure 19.15 Hemostasis, part 4: Clot retraction.

(plaz-MIN-oh-jen) **activator,** or **tPA.** Thrombolysis can also be initiated by a similar enzyme called **urokinase** (yoo-roh-KY-nayz), which is produced by cells of the kidney and found in the plasma and interstitial fluid.

② **tPA activates plasminogen.** The inactive enzyme **plasminogen,** which is normally found in plasma, has a high affinity for fibrin proteins and binds them as they are incorporated into the blood clot. As a result, every blood clot contains a significant amount of plasminogen. When tPA contacts plasminogen, it catalyzes the reaction that converts it to the active enzyme **plasmin.**

③ **Plasmin degrades fibrin, and the clot dissolves.** Plasmin catalyzes the reaction that degrades both fibrin and fibrinogen. This causes the remaining components of the clot to dissociate from the endothelium.

Quick Check

☐ 8. What is thrombolysis?

☐ 9. Explain the roles of tPA and plasmin in thrombolysis.

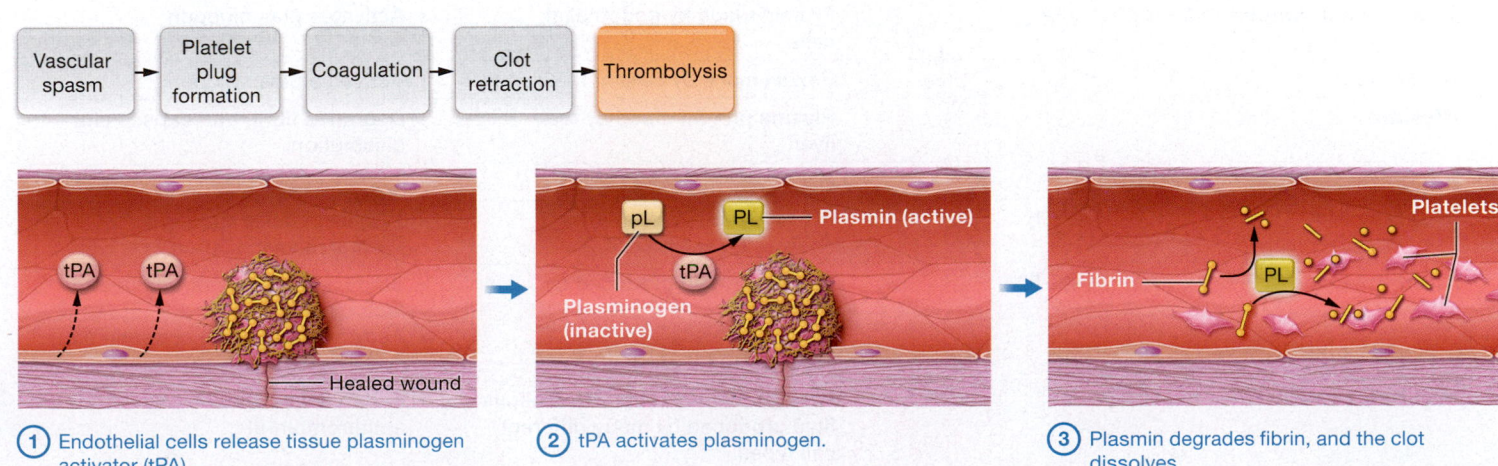

① Endothelial cells release tissue plasminogen activator (tPA).

② tPA activates plasminogen.

③ Plasmin degrades fibrin, and the clot dissolves.

Figure 19.16 Hemostasis, part 5: Thrombolysis.

Putting It All Together: The Big Picture of Hemostasis

Let's now put the pieces together for a "big picture" view of hemostasis. Figure 19.17 summarizes the five key steps of the process: vascular spasm, platelet plug formation, coagulation, clot retraction, and thrombolysis.

Quick Check

☐ 10. Walk through the steps of clot formation and dissolution.

Regulation of Clotting

Blood clotting is tightly regulated to keep its positive feedback cycle in check. Endothelial cells produce chemicals that regulate the first and second stages of clot formation. Two of these chemicals are **prostacyclin** (PRAHS-tuh-sy-klin), which inhibits platelet aggregation, and nitric oxide, which causes vasodilation.

Prostacyclin is a prostaglandin, a group of chemicals with many functions, including triggering inflammation. Anti-inflammatory medications such as ibuprofen inhibit prostaglandin formation. Some of them also inhibit prostacyclin formation, a fact that accounts for the increased risk of blood clots with these drugs.

Endothelial cells and hepatocytes produce substances called **anticoagulants** (an′-tee-koh-AG-yoo-lunts) that inhibit coagulation. Some examples include:

- **Antithrombin-III.** One of the most important anticoagulants, the protein **antithrombin-III (AT-III),** binds and inhibits the activity of both factor Xa and thrombin. It inhibits thrombin that has already formed and prevents the formation of new thrombin.
- **Heparan sulfate.** Heparan sulfate is a polysaccharide that enhances the activity of AT-III. Heparan sulfate is similar to another anticoagulant called *heparin,* which is often used clinically. We discuss its applications in *A&P in the Real World: Anticlot Medications* on p. 746.
- **Protein C.** Active **protein C** catalyzes the reactions that degrade factors Va and VIIIa. To become active, protein C requires another protein, called *protein S.*

Table 19.3 summarizes thrombolytic (clot-destroying) agents and factors that regulate clotting.

Quick Check

☐ 11. How does prostacyclin regulate hemostasis?

☐ 12. What are the functions of antithrombin-III, heparan sulfate, and protein C? How do they affect hemostasis?

Disorders of Clotting

A clotting disorder—a condition in which clotting is not regulated properly—can have drastic consequences for homeostasis. There are two types of clotting disorders: (1) *bleeding disorders,* when the blood is unable to clot, and (2) *hypercoagulable conditions,* when clots form at improper times and tissue locations.

Bleeding disorders often result from clotting protein deficiencies; for example, **hemophilia A** is caused by a shortage of factor VIII, and **hemophilia B** by inadequate factor IX. Many bleeding disorders can be treated by replacement of the missing clotting factor or protein with periodic infusions.

The opposite problem, a hypercoagulable condition, can result in the formation of an inappropriate clot, a condition called **thrombosis.** The clot, or **thrombus** (plural, *thrombi*), is dangerous because it can obstruct blood flow through a vessel. In addition, a piece of the thrombus, called a **thromboembolus** (plural, *thromboemboli,* often shortened to *emboli*), may break off and lodge in smaller vessels downstream from the thrombus.

Table 19.3 Thrombolytic Agents and Regulators of Clot Formation			
Factor	**Alternate Name**	**Description**	**Function**
Thrombolytic Agents			
Tissue plasminogen activator	tPA	Protein made by endothelial cells	• Activates plasminogen
Urokinase		Protein made by cells of the kidney	• Activates plasminogen
Plasmin		Plasma protein made by the liver	• Degrades fibrin and causes clot dissolution
Regulators of Clot Formation			
Prostacyclin	PGI$_2$	Prostaglandin made by endothelial cells	• Inhibits platelet aggregation • Promotes vasodilation
Antithrombin-III	AT-III	Plasma protein produced by the liver	• Binds and inhibits thrombin and factor Xa • Activity enhanced by heparin and heparan sulfate
Heparan sulfate		Polysaccharide in the extracellular fluid produced by many different cell types	• Enhances activity of antithrombin-III
Protein C		Plasma protein produced by the liver; synthesis requires vitamin K	• Degrades factors Va and VIIIa

The Big Picture of Hemostasis

Figure 19.17

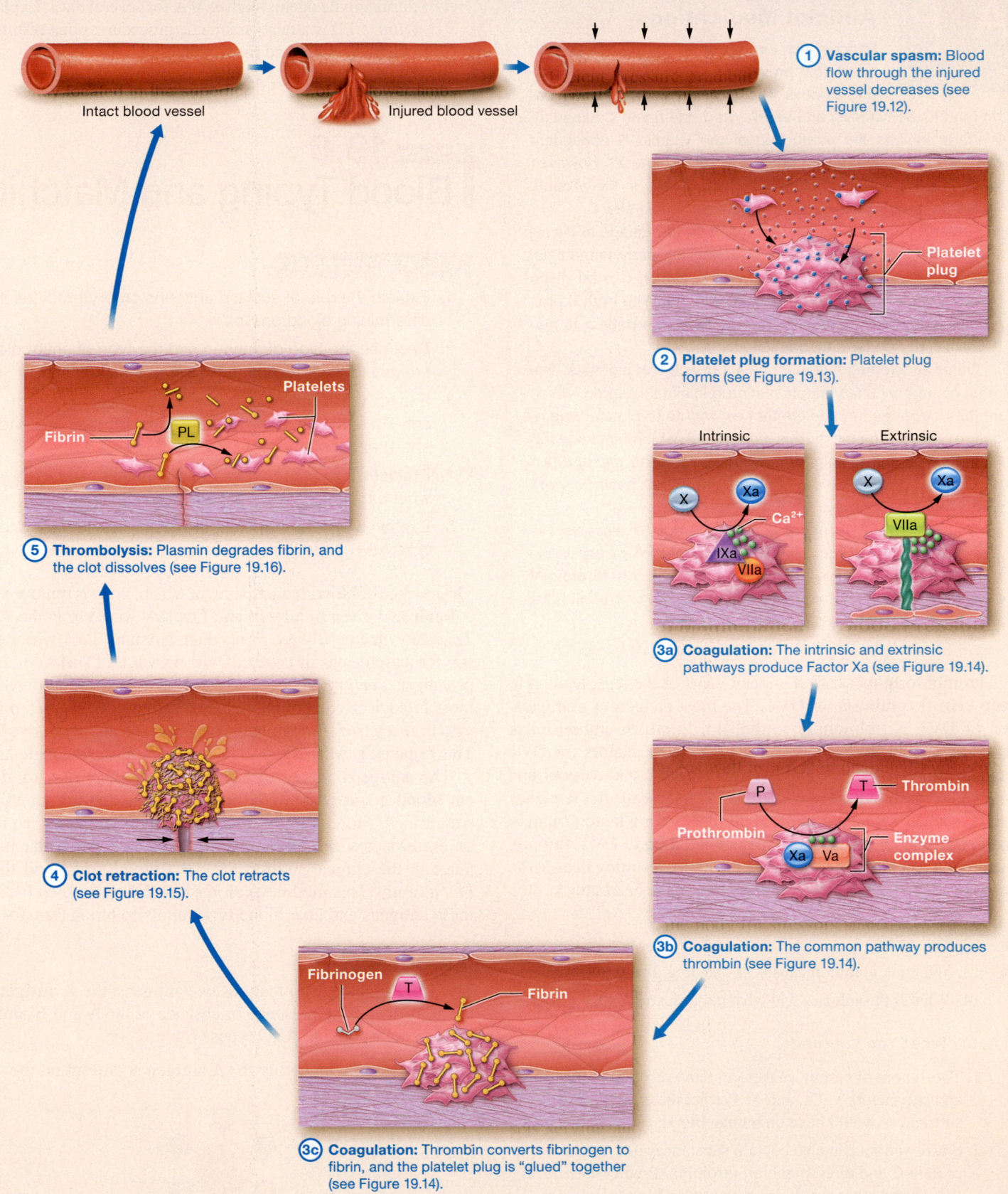

Intact blood vessel

Injured blood vessel

① **Vascular spasm:** Blood flow through the injured vessel decreases (see Figure 19.12).

Platelet plug

② **Platelet plug formation:** Platelet plug forms (see Figure 19.13).

Intrinsic

Extrinsic

③a **Coagulation:** The intrinsic and extrinsic pathways produce Factor Xa (see Figure 19.14).

Prothrombin

Thrombin

Enzyme complex

③b **Coagulation:** The common pathway produces thrombin (see Figure 19.14).

Fibrinogen

Fibrin

③c **Coagulation:** Thrombin converts fibrinogen to fibrin, and the platelet plug is "glued" together (see Figure 19.14).

④ **Clot retraction:** The clot retracts (see Figure 19.15).

⑤ **Thrombolysis:** Plasmin degrades fibrin, and the clot dissolves (see Figure 19.16).

Fibrin

Platelets

Anticlot Medications

Patients with thrombi or emboli are treated with drugs that prevent one step in the blood clotting process. The anticoagulants include heparin (and its derivatives) and **warfarin** (trade name Coumadin). Warfarin inhibits the production of the vitamin K–dependent clotting factors by the liver (factors II, VII, IX, and X). However, warfarin does not affect clotting factors already in the plasma, so the drug requires 2–3 days to show noticeable effect.

Heparin is generally used in hospitals to manage and prevent emboli because its onset of action is nearly immediate. However, it is injected, and so warfarin, which can be given by mouth, is preferred in outpatient settings. With both drugs, careful monitoring is required to ensure that a patient is neither over- nor under-anticoagulated.

Other anticlotting medications act directly on platelets. The most commonly used antiplatelet drug is simple aspirin, which inhibits enzymes in platelets that contribute to platelet aggregation. Another medication is *clopidogrel*, which blocks platelet receptors and inhibits platelet activation. The *glycoprotein IIb/IIIa inhibitors*, used commonly in heart attack patients, work by a similar mechanism.

Occasionally, it is necessary to attempt to dissolve the clot with a thrombolytic agent. The enzyme tPA or urokinase is used when blood flow must be restored rapidly to prevent tissue damage, as, for example, when thrombi or emboli have caused a stroke or heart attack.

Thrombi form most often in deep veins of the legs, leading to **deep vein thrombosis,** or **DVT.** The most dangerous complication of DVT is **pulmonary embolism,** in which emboli break off and lodge in the vessels of the lungs. Risk factors for DVT include prolonged immobility, which causes blood to pool and potentially clot, and other conditions that increase the tendency to clot, such as surgery, trauma, abnormal clotting factors, anticoagulant deficiencies, and pregnancy. In addition to DVT, thrombi and emboli can block the vessels of the brain, resulting in a *stroke,* or the heart, causing a *myocardial infarction,* or heart attack.

Quick Check

☐ 13. What are common causes of bleeding disorders?

☐ 14. What is thrombosis, and what may cause it?

Apply What You Learned

☐ 1. The venom of certain snakes contains substances that activate factors V, IX, and X. Predict the effects these substances would have on human blood.

☐ 2. Snake venom can also contain substances that activate protein C and directly inhibit thrombin. Predict the effects these substances would have on human blood. How do these effects differ from those in question 1?

☐ 3. It is generally recommended that patients who have received cardiopulmonary resuscitation (CPR) not be given thrombolytic agents such as tPA for several days. Explain. (*Hint:* CPR is a fairly traumatic procedure, often resulting in broken bones and tissue damage.)

See answers in Appendix A.

MODULE 19.6
Blood Typing and Matching

Learning Outcomes

1. Explain the role of surface antigens on erythrocytes in determining blood groups.

2. Describe the type of antigen and the type of antibodies present in each ABO and Rh blood type.

3. Explain the differences between the development of anti-Rh antibodies and the development of anti-A and anti-B antibodies.

4. Predict which blood types are compatible, and explain what happens during a transfusion reaction.

5. Explain why blood type O− is the universal donor and type AB+ is the universal recipient.

The practice of **blood transfusion,** in which blood is removed from a **donor** and given to a **recipient,** first saw wide use in the 1800s. Tragically, many early recipients died. Not until 1901 was the reason for these deaths discovered—the presence of surface markers, or antigens, on erythrocytes. As Module 19.3 noted, all cells and most biological compounds have unique antigens. The immune system recognizes foreign antigens and responds to remove them. This response is what killed so many early transfusion patients.

The antigens on erythrocytes, which give rise to different **blood groups,** are carbohydrate chains that are genetically determined. More than 30 different types of antigens are found on erythrocytes, but two antigen groups have particular importance in the clinical setting: the *ABO blood group* and the *Rh blood group.* This module explores how blood is typed based on these antigens and how blood types apply to blood transfusions.

Blood Typing

The **ABO blood group** features two antigens: the **A antigen** and the **B antigen.** The presence or absence of the A and B antigens gives us four possible ABO types (**Table 19.4**):

- **type A,** in which only the A antigen is present on the erythrocytes;

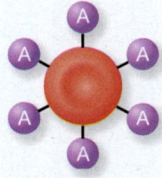

Table 19.4 The Eight Major Blood Types

Blood Type	Prevalence in U.S. Population	Antigens Present on Erythrocyte Surface	Antibodies Present in Plasma*	May Receive From	May Donate To
AB+	3%	A, B, Rh	None	Universal recipient	AB+
AB−	1%	A, B	Anti-Rh	AB−, A−, B−, O−	AB+, AB−
A+	34%	A, Rh	Anti-B	A+, A−, O+, O−	AB+, A+
A−	6%	A	Anti-B, anti-Rh	A−, O−	AB+, AB−, A+, A−
B+	9%	B, Rh	Anti-A	B+, B−, O+, O−	AB+, B+
B−	2%	B	Anti-A, anti-Rh	B−, O−	AB+, AB−, B+, B−
O+	38%	Rh	Anti-A, anti-B	O+, O−	AB+, A+, B+, O+
O−	7%	None	Anti-A, anti-B, anti-Rh	O−	Universal donor

*Assuming prior exposure to Rh antigens for Rh-negative individuals.

- **type B,** in which only the B antigen is present;

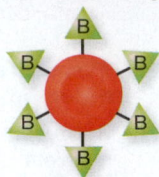

- **type AB,** in which both the A and B antigens are present; and

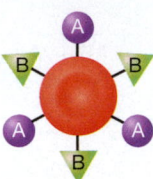

- **type O,** in which neither the A nor B antigens are present.

Note that there is no "O" antigen. The "O" simply means that both the A and B antigens are absent.

The **Rh blood group** (so named because it was first discovered in the rhesus monkey) features the **Rh antigen,** also known as the *D antigen.* Individuals with an Rh antigen on their erythrocytes are **Rh-positive** (abbreviated "Rh+"), and those lacking an Rh antigen are **Rh-negative** (abbreviated "Rh−").

The ABO and Rh blood groups combine to give the eight common blood types. For example, the erythrocytes in a type AB− individual feature the A and B antigens, but no Rh antigen, whereas the erythrocytes in a type O+ individual feature only the Rh antigen, without A or B:

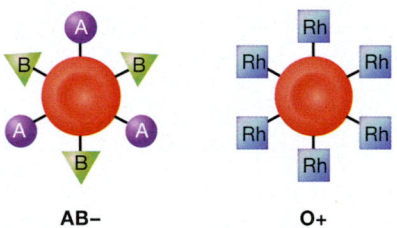

AB−   O+

Type O+ is the most common blood type, present in about 38% of the U.S. population. AB− is the least common, found in only about 1% of the U.S. population.

We can determine blood type in the laboratory, using antibodies that recognize and bind to individual antigens on erythrocytes. Antibodies are produced by cells derived from activated B lymphocytes, and they bind to foreign antigens. Each unique antibody has a unique antigen to which it binds. As you can see in **Figure 19.18,** antibodies cause bound antigens to clump together, or **agglutinate.** For this reason, antibodies are sometimes called **agglutinins** (uh-GLOO-tuh-ninz). Ultimately, agglutination promotes destruction of the erythrocytes, a reaction known as **hemolysis** (hee-MAW-luh-sis; *lysis* = "to separate or split").

To determine blood type, a blood sample is treated with three antibodies:

- **Anti-A antibodies** bind and agglutinate A antigens.

- **Anti-B antibodies** bind and agglutinate B antigens.
- **Anti-Rh antibodies** bind and agglutinate Rh antigens.

As you can see in **Figure 19.19,** if agglutination occurs in response to an antibody, that antigen is present on the erythrocyte. If agglutination does not occur, that antigen is absent.

Quick Check

☐ 1. What determines the eight common blood types?

☐ 2. A sample reacts with anti-A and anti-B antibodies, but does not react with anti-Rh antibodies. What is the blood type?

Blood Transfusions

Your immune system recognizes the antigens on your erythrocytes as "self" antigens, meaning they belong to you. Under normal circumstances, your immune system does not produce antibodies to self antigens, because if it did, your antibodies would bind your own antigens. However, your immune system does produce antibodies to *foreign* antigens. This means that antibodies are present in your plasma only if the antigens are normally *absent* on your erythrocytes.

Note that anti-A and anti-B antibodies are *pre-formed,* meaning they are present in the plasma even if the individual has never been exposed to those antigens. Anti-Rh antibodies, however, are produced only if a person has been exposed to blood containing Rh antigens. For this reason, an Rh− individual generally has no anti-Rh antibodies unless he or she has been exposed (sensitized) to Rh+ erythrocytes. This has particular relevance for pregnant Rh− women, which is discussed in *A&P in the Real World: Hemolytic Disease of the Newborn, or Erythroblastosis Fetalis.*

Here's an example: Sue has type B− blood, which means that her erythrocytes have B antigens only. As you can see here, her immune system cannot make anti-B antibodies, because if it does, she would agglutinate her own erythrocytes. However, Sue's erythrocytes lack the A and Rh antigens, so she can safely produce anti-A and anti-Rh antibodies.

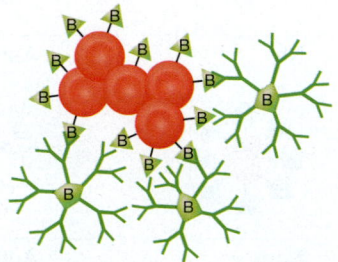

Anti-B antibodies would agglutinate Sue's erythrocytes.

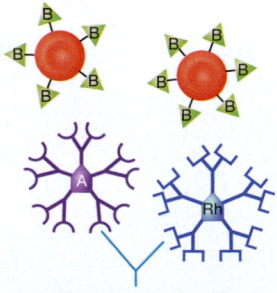

Anti-A and anti-Rh antibodies do not cause agglutination.

Let's apply this to blood transfusions: A patient cannot receive any blood containing antigens that his or her immune system would recognize as foreign. So, in our example, Sue cannot receive blood with A or Rh antigens, which includes types A+,

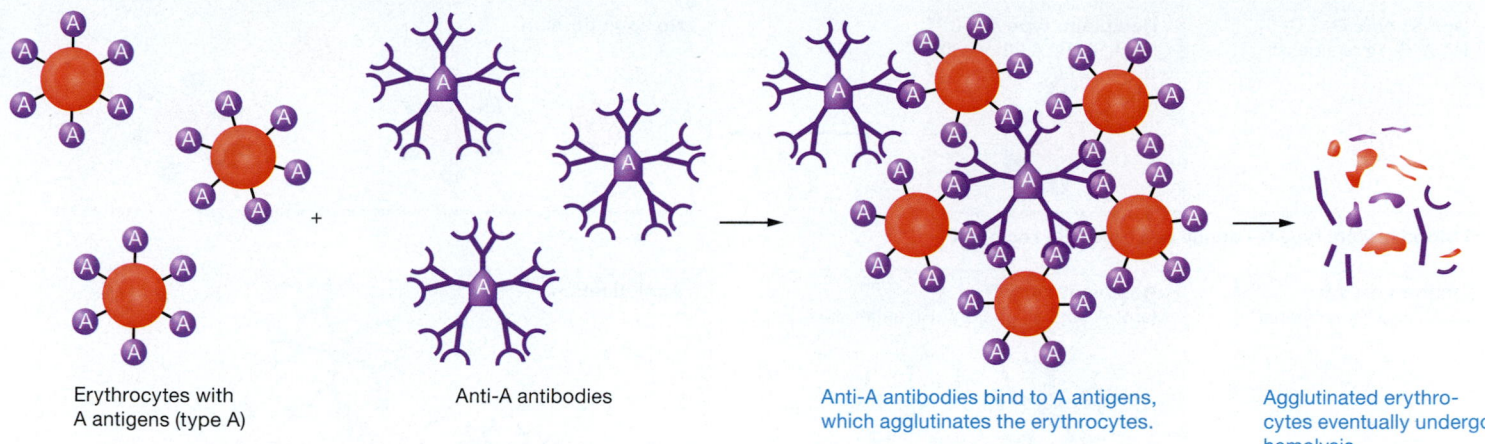

Erythrocytes with A antigens (type A)

Anti-A antibodies

Anti-A antibodies bind to A antigens, which agglutinates the erythrocytes.

Agglutinated erythrocytes eventually undergo hemolysis.

Figure 19.18 How antibodies agglutinate erythrocytes.

A−, B+, AB+, AB−, and O+. Sue can safely receive blood only from the blood types that do not have A and Rh antigens: B− and O−. These two blood types are a **match** for Sue's blood type.

What if Sue wants to donate blood? We have already established that she has B antigens on her erythrocytes. This means her blood can be given only to recipients who lack the anti-B antibody. Refer to Table 19.4, and you'll see that the blood types that lack anti-B antibodies, and so are a match for Sue's blood, are types B+, B−, AB+, and AB−. All the other blood types have anti-B antibodies that will agglutinate her donated erythrocytes. This agglutination, known as a **transfusion reaction,** destroys the donor erythrocytes, possibly leading to kidney failure and death.

Figure 19.20a shows another example:

- Donor Ed has O− blood, which has none of the antigens of the ABO or Rh groups.

- Recipient Oscar has AB+ blood, which has none of the anti-A, anti-B, or anti-Rh antibodies.

To determine if we have a match, we simply check to see if Oscar makes any antibodies that might bind to Ed's antigens. Obviously, Ed has no antigens of the ABO or Rh groups and Oscar has no anti-A, anti-B, or anti-Rh antibodies, so nothing could possibly interact. In fact, Ed can safely donate to a recipient with any blood type because Ed's O− erythrocytes lack these antigens. For this reason, blood type O− is considered the **universal donor.** Similarly, Oscar's AB+ blood lacks the major antibodies, so he can safely receive any blood type. For this reason, blood type AB+ is considered the **universal recipient.**

AP in the Real World

Hemolytic Disease of the Newborn, or Erythroblastosis Fetalis

When an Rh-negative mother carries and gives birth to an Rh-positive fetus, **hemolytic disease of the newborn,** also known as **erythroblastosis fetalis,** results. During birth, fetal erythrocytes enter the mother's blood, which stimulates her immune system to produce anti-Rh antibodies. This carries little risk for her first such pregnancy. However, in subsequent pregnancies, maternal anti-Rh antibodies can cross the placenta into the blood of the fetus, hemolyzing (rupturing) fetal erythrocytes if they are Rh+ and potentially killing the fetus. A characteristic finding is the presence of immature erythroblasts in fetal blood (giving the condition its other name, erythroblastosis fetalis).

Fortunately, this disease can be effectively prevented with blood type screening. If a pregnant woman is Rh-negative, she is given an injection of Rho(D) *immune globulin*, which contains anti-Rh antibodies. These antibodies bind any fetal cells that have entered the maternal circulation, which prevents maternal leukocytes from producing and secreting their own anti-Rh antibodies.

BLOOD TYPE	**Anti-A** antibodies agglutinate A antigens.	**Anti-B** antibodies agglutinate B antigens.
A	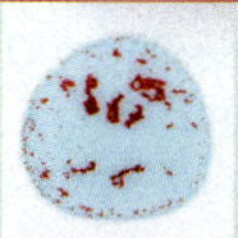	
B	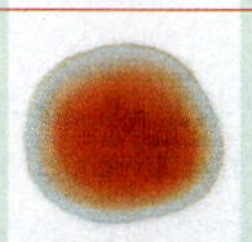	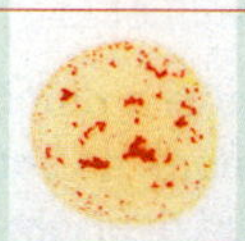
AB	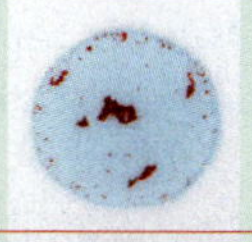	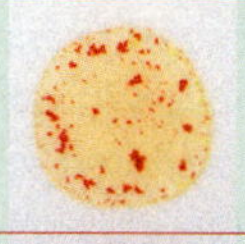
O	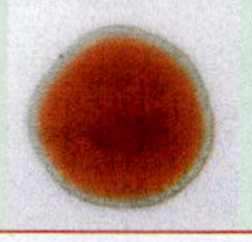	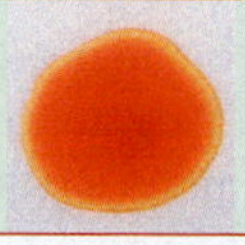

Figure 19.19 Blood type testing. Blood samples from four patients are combined with antibodies. Agglutination indicates that a specific antigen is present on that patient's erythrocytes.

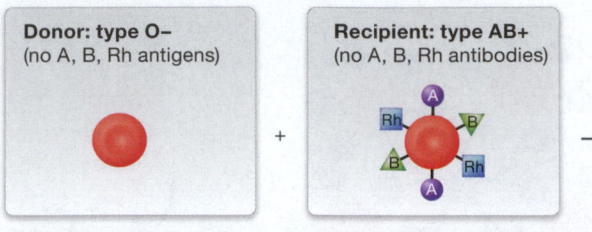

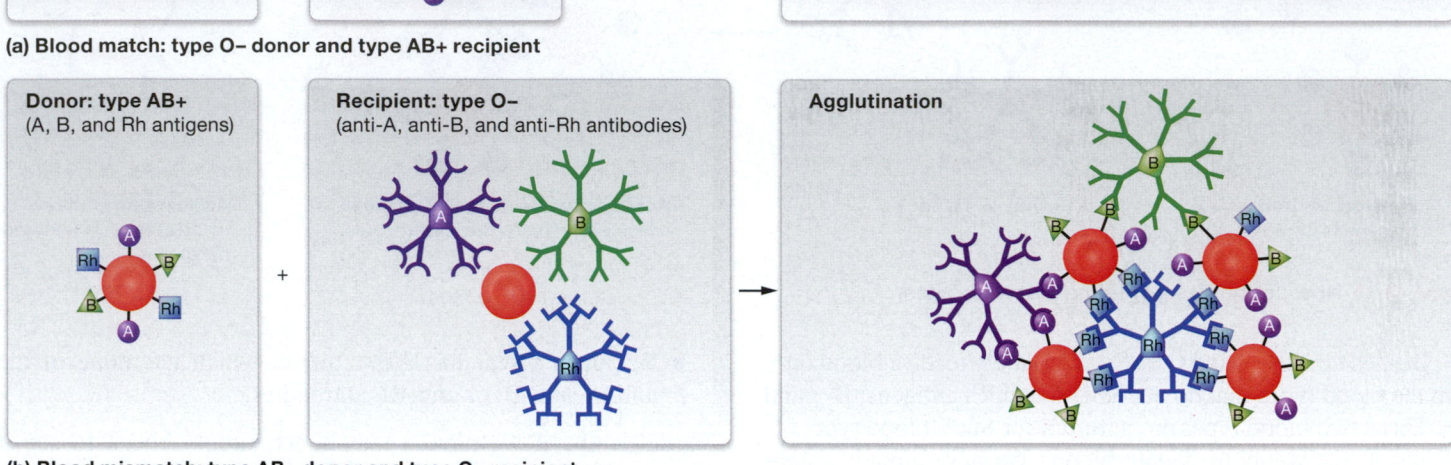

(a) Blood match: type O– donor and type AB+ recipient

(b) Blood mismatch: type AB+ donor and type O– recipient

Figure 19.20 Matching blood types for blood transfusions.

Let's flip the donors and recipients around now (**Figure 19.20b**):

- Donor Oscar has AB+ blood, which has A, B, and Rh antigens.
- Recipient Ed has O– blood, which has anti-A, anti-B, and anti-Rh antibodies.

Ed has three types of antibodies that would bind Oscar's antigens, so infusing Oscar's erythrocytes into Ed would produce a transfusion reaction. Indeed, donor erythrocytes with any antigens at all will produce a transfusion reaction in Ed. So, individuals with blood type O– may receive blood only from other type O– individuals. Additionally, because Oscar's erythrocytes have all three types of antigens, he can donate blood only to recipients with no anti-A, anti-B, and/or anti-Rh antibodies in their blood—in other words, only type AB+ individuals.

ConceptBOOST))))

What about the Donor's Antibodies?

At this point you may be wondering, "But can't the donor's antibodies bind and destroy the recipient's antigens?" The answer is yes, donor antibodies can bind to a recipient's antigens, and unless the blood types are exactly matched, some donor antibodies might destroy a few of the recipient's erythrocytes. The key word is *few*—only a *few* of the recipient's erythrocytes are likely to be harmed.

Take a closer look at our example in Figure 19.20a. Donor Ed's erythrocytes have no antigens of the ABO or Rh groups, but his blood does have anti-A, anti-B, and anti-Rh antibodies. For the sake of simplicity, let's say the sample of Ed's donated blood contains 100 antibodies. Ed's blood

is given to recipient Oscar, who has type AB+ blood. Now, in Oscar's bloodstream we find:

- donated O– erythrocytes from Ed;
- 100 anti-A, anti-B, and anti-Rh antibodies from Ed's blood; and
- Oscar's own AB+ erythrocytes.

Ed's 100 antibodies might destroy 100 of Oscar's erythrocytes. But, Oscar has received millions of new erythrocytes from Ed, so he won't really miss the 100 that he lost.

Now reverse the donors and recipients (see Figure 19.20b): AB+ Oscar becomes the donor, and O– Ed the recipient. Ed receives a sample of Oscar's blood, and in Ed's bloodstream we find:

- donated AB+ erythrocytes from Oscar;
- no major erythrocyte antibodies from Oscar;
- Ed's own O– erythrocytes; and
- thousands to millions of Ed's own anti-A, anti-B, and anti-Rh antibodies.

That's quite a difference in the number of antibodies. And what's more, as soon as Ed's leukocytes come into contact with Oscar's erythrocytes, Ed's immune system makes millions more of these antibodies. And the result? A transfusion reaction in which Ed's antibodies destroy every one of Oscar's donated erythrocytes.

See the difference? The number of antibodies in the donor's blood is relatively low, which limits the amount of damage they do to the recipient's erythrocytes. But the recipient's immune system can make millions more antibodies, more than enough to destroy all the donated erythrocytes. So remember—we think about the donor's *antigens* and the recipient's *antibodies*. ■

Quick Check

☐ 3. How are blood types matched for blood transfusions?

☐ 4. Which blood type is the universal donor? Why?

☐ 5. Which blood type is the universal recipient? Why?

Apply What You Learned

☐ 1. Mr. Jones, whose blood type is A−, has been injured and requires a blood transfusion. His friend, Mr. Ramirez,

wants to donate blood to Mr. Jones. Mr. Ramirez's blood type is AB+. Can he donate to Mr. Jones? Explain.

☐ 2. Ms. Wright also wants to donate blood to Mr. Jones. Her blood type is O+. Can she donate blood to Mr. Jones? Explain.

☐ 3. Explain why a person with blood type O− can donate to someone with any blood type, but can receive blood only from other O− donors.

See answers in Appendix A.

Chapter Summary

For everything you need to succeed in this course, go to **Mastering** Anatomy & Physiology. There you will find:

 Practice Tests

Author Podcasts

▶ Big Picture Animations

))) Concept Boost Video Tutors

iP2™ Interactive Physiology 2.0 Tutorials

A&PFlix A&P Flix 3D Animations

Active-Learning Workbook

MODULE 19.1
Overview of Blood 723–725

- Three types of formed elements are found in blood: **erythrocytes, leukocytes,** and **platelets.**
- Plasma makes up 55% of the total blood volume. It consists of water, **plasma proteins,** and dissolved solutes.
- Functions of blood include gas exchange; solute distribution; immunity; temperature, acid-base, and blood pressure homeostasis; and sealing damaged vessels.

MODULE 19.2
Erythrocytes and Oxygen Transport
725–731

- Erythrocytes consist mainly of a plasma membrane around cytosol with enzymes and the protein **hemoglobin.**
- Hemoglobin is a protein with four polypeptide subunits, each with an iron-containing **heme group.** The heme groups bind to oxygen to form **oxyhemoglobin.**

- **Erythropoiesis** occurs in red bone marrow and begins with **hematopoietic stem cells.** Erythropoiesis requires the hormone **erythropoietin.**
- Old and damaged erythrocytes are destroyed in the **spleen.** The iron and polypeptide chains of hemoglobin are recycled, and the heme group is converted into **bilirubin.**
- **Anemia** is the decreased oxygen-carrying capacity of the blood.

MODULE 19.3
Leukocytes and Immune Function 731–733

- **Granulocytes** are leukocytes with visible cytoplasmic granules and lobed nuclei.
  - ○ **Neutrophils,** the most common leukocyte, kill bacteria by phagocytosis.
  - ○ **Eosinophils** respond to parasitic worm infections and are involved in allergic responses.
  - ○ The granules of **basophils** mediate inflammation.
- **Agranulocytes** lack visible cytoplasmic granules.
  - ○ **Lymphocytes** are the second most numerous leukocyte. The two types of lymphocytes are **B lymphocytes,** which secrete **antibodies,** and **T lymphocytes,** which destroy cancer or virally infected cells.
  - ○ **Monocytes** mature into **macrophages.**
- **Leukopoiesis** is the process of leukocyte formation. All leukocytes are derived from hematopoietic stem cells, which divide into the **lymphoid cell line** and the **myeloid cell line.**

MODULE 19.4
Platelets 733–738

- Platelets are anucleate fragments of massive **megakaryocytes,** which are derived from the myeloid cell line.

MODULE 19.5
Hemostasis 738–746

- Hemostasis—stopping blood loss from an injured vessel—is carried out by platelets and **clotting factors.** A **blood clot** is a collection of platelets, clotting proteins, and other formed elements. Hemostasis has five phases: (1) vascular spasm, (2) formation of a **platelet plug,** (3) **coagulation,** (4) clot retraction, and (5) **thrombolysis.**
- Clotting is regulated by **anticoagulants,** including **antithrombin-III** and **protein C.**

MODULE 19.6
Blood Typing and Matching 746–751

- Two major groups of erythrocyte **antigens** are the **ABO blood group** and the **Rh blood group.**
- Blood type is based on the presence or absence of antigens from the ABO and Rh groups.
- **Antibodies** bind and **agglutinate** antigens. Three types of antibodies may be present in blood: **anti-A, anti-B,** and **anti-Rh antibodies.**
- Antigens and antibodies are the basis for blood matching in **blood transfusions,** in which blood is transferred from a compatible **donor** to a **recipient.** Type O− is the **universal donor,** and type AB+ is the **universal recipient.**

Assess What You Learned

Scan the QR Code for additional practice test questions

https://goo.gl/cEVFaS

LEVEL 1 Check Your Recall

1. Which of the following is *not* a formed element of blood?

   a. Erythrocyte
   b. Leukocyte
   c. Mast cell
   d. Platelet

2. Which of the following plasma proteins is responsible for osmotic pressure?

   a. γ-Globulins
   b. Albumin
   c. α-Globulins
   d. Clotting proteins

3. List the seven major functions of blood.

4. Mark the following statements as true or false. If a statement is false, correct it to make a true statement.

   a. Erythrocytes are biconcave discs with prominent nuclei.
   b. The main function of erythrocytes is to transport oxygen on the protein hemoglobin.

   c. Hemoglobin forms oxyhemoglobin when it binds to oxygen.
   d. Hemoglobin consists of two polypeptide chains bound to a heme group.

5. Erythropoiesis requires stimulation from the hormone:

   a. thrombin.
   b. thrombopoietin.
   c. thymosin.
   d. erythropoietin.

6. Fill in the blanks: Erythrocytes are derived from stem cells called _____, circulate in the blood for approximately _____ days, and are destroyed by an organ called the _____.

7. Anemia is defined as:

   a. a decreased oxygen-carrying capacity of the blood.
   b. a decreased iron content of the blood.
   c. decreased bone marrow function.
   d. abnormalities in hemoglobin.

8. Leukocytes are:

   a. nucleated cells that function in blood clotting.
   b. nucleated cells that function in immunity.
   c. anucleate cells that function in blood clotting.
   d. anucleate cells that function in immunity.

9. Match the following leukocytes with the correct definition.

   _____ Basophil
   _____ B lymphocyte
   _____ Neutrophil
   _____ Monocyte
   _____ T lymphocyte
   _____ Eosinophil

   a. Destroys bacteria; directly phagocytoses bacteria
   b. Responds to parasitic worm infection and mediates the allergic response
   c. Activates all parts of the immune response; directly kills cancer or virally infected cells
   d. Secretes inflammatory mediators
   e. Agranulocyte that matures into macrophage
   f. Agranulocyte that secretes antibodies

10. Fill in the blanks: Lymphocytes are derived from the _____ cell line, whereas the other leukocytes are derived from the _____ cell line.

11. Platelets are derived from cells called:

    a. thromboblasts.
    b. leukoblasts.
    c. megakaryocytes.
    d. thrombokaryocytes.

12. Number the steps of hemostasis in order, putting 1 by the first event, 2 by the second, and so on.

    _____ The intrinsic/contact activation and extrinsic/tissue factor pathways produce factor Xa.
    _____ The clot retracts.
    _____ Thrombin converts fibrinogen to fibrin, and fibrin glues the plug together.
    _____ Platelets are activated, and the platelet plug forms.
    _____ Vasoconstriction and increased tissue pressure decrease blood flow through the vessel.
    _____ Tissue plasminogen activator activates plasmin, which degrades fibrin.
    _____ The common pathway produces thrombin.

13. How do the intrinsic/contact activation and extrinsic/tissue factor coagulation pathways differ? How are they similar?

14. What are the overall goals of the common pathway of coagulation?

15. Which of the following is *not* an anticlotting agent produced by endothelial cells?

a. Prostacyclin   c. Antithrombin-III
b. Protein C   d. Warfarin

16. Tissue plasminogen activator, urokinase, and plasmin are important components of:

a. coagulation.   c. platelet plug formation.
b. fibrinolysis.   d. hemostasis.

17. Fill in the blanks: The two most clinically important groups of antigens on erythrocytes are the _____ and _____ blood groups.

18. Which antigens does a person with blood type A− have on the surface of his or her erythrocytes?

a. A antigens   d. Both a and c are correct.
b. B antigens   e. All of the above
c. Rh antigens

19. Which of the following antibodies does a person with type O+ blood have in his or her plasma?

a. Anti-A antibodies
b. Anti-B antibodies
c. Anti-Rh antibodies
d. Both a and b are correct.
e. All of the above

20. Mr. Reczkiewicz has blood type AB−. Which of the following blood types could be safely donated to Mr. Reczkiewicz, assuming he has had prior exposure to Rh+ blood? (Circle all that apply.)

a. Type O+   d. Type AB+
b. Type A−   e. Type O−
c. Type B−

LEVEL 2 Check Your Understanding

1. Explain how blood, being a liquid, enables all its components (formed elements and plasma) to perform their functions.

2. Predict how abnormal hemoglobin proteins that contain only two iron ions, instead of four, would affect homeostasis.

3. The anticoagulant drug warfarin primarily disrupts the extrinsic/tissue factor coagulation pathway. Explain why disrupting only this pathway disrupts the entire coagulation cascade.

4. Cirrhosis of the liver often reduces production of many types of plasma proteins, including albumin and clotting factors. Predict the effects on the body of decreased numbers of these proteins in the plasma. Would this also affect the number of γ-globulins in the plasma? Why or why not?

LEVEL 3 Apply Your Knowledge

PART A: Application and Analysis

1. A blood sample from your patient shows that she has decreased numbers of neutrophils. Predict the effects of this condition. How would it differ if numbers of T lymphocytes were decreased instead?

2. Mr. Jackson presents to the emergency room with a minor wound that has bled for several days. An examination of his medical history reveals that Mr. Jackson has hemophilia A, which is caused by a deficiency of factor VIII. Your co-worker suggests that you give Mr. Jackson some platelets to stop the bleeding. Will this help your patient? Why or why not?

3. Ms. Wu, whose blood type is O−, requires a blood transfusion. Her family members volunteer to donate blood. Their blood types are as follows: her son, type B−; her husband, type B+; her daughter, type O+. Which family members could safely donate blood to Ms. Wu? Who could not? Explain.

PART B: Make the Connection

4. Elise is a 36-year-old woman who has volunteered to donate red bone marrow to a patient in need of a bone marrow transplant for aplastic anemia. The physician performing the bone marrow extraction is an intern, and it is her first time doing the procedure. She asks that you prepare a site on Elise's tibia for the procedure. Is this a good place for red bone marrow to be extracted, considering Elise's age? Why or why not? Can you suggest an alternative location? Explain. *(Connects to Chapter 6)*

See answers in Appendix A.

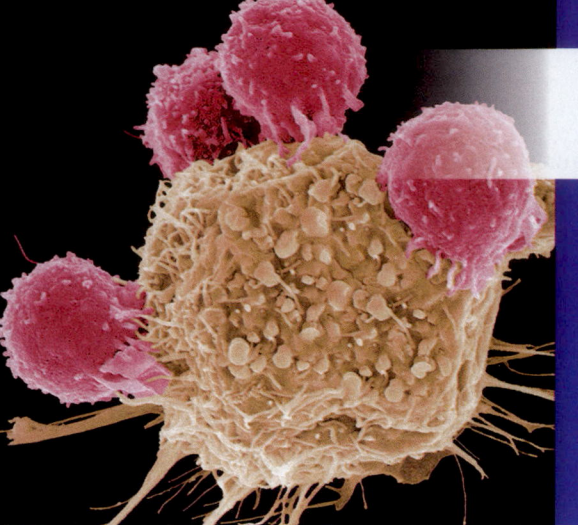

20

The Lymphatic System and Immunity

*For practice applying concepts to a clinical scenario, check out the **Running Case Study** for this chapter @ Mastering Anatomy & Physiology*

Photo: This scanning electron micrograph (SEM) shows cytotoxic T cells surrounding and attacking a cancer cell.

We come into contact with literally trillions of microbes every day, so why aren't we constantly sick? The short answer is that our bodies protect us from infection by many of these microbes. How do they do this? Well, that requires a longer answer. Remember in our exploration of the structure and function of blood, we introduced formed elements known as *leukocytes,* or *white blood cells* (*WBCs*), and immune proteins located in the plasma (see Chapter 19). Now we focus our attention on these cells and proteins, which make up part of a larger system collectively called the **immune system,** which works to defend the body against internal and external threats. Note that the immune system is not technically a system according to the definition you have learned. This system contains no organs or tissues of its own, and instead consists of cells and proteins located in the blood and the tissues of other systems, including the lymphatic system (lim-FAEH-tik). As we'll see, the **lymphatic system** is a group of organs and tissues that not only works with the immune system but also participates in a number of functions such as fluid homeostasis.

The immune and lymphatic systems function together in what is broadly called **immunity:** the set of diverse processes that protect the body from both cellular injury and disease-causing cells and substances known as *pathogens* (PATH-uh-jenz; *patho-* = "disease"). These functions are crucial to our homeostasis, which is evidenced by the devastating consequences of diseases such as *AIDS* (*acquired immunodeficiency syndrome*) that result from the loss of certain components of the immune system. Although the lymphatic system is technically its own system and has functions independent of immunity, the two systems have such close structural and functional relationships that it makes sense to cover them together in a single chapter.

We begin by introducing the structures and functions of the lymphatic system. Next, we tie the two systems together as we examine the cells and proteins of the immune system. The chapter concludes with a focus on how immune disorders affect the body's homeostasis.

MODULE 20.1
Structure and Function of the Lymphatic System

here will be revisited several times throughout this chapter, which reflects the close level of integration between the lymphatic and immune systems.

Functions of the Lymphatic System

◀◀ FLASHBACK

1. What is the net filtration pressure in blood capillaries? (p. 692)
2. How much water is lost from blood capillaries to the interstitial fluid every day? (p. 694)

The lymphatic system has three basic functions that are carried out by its vessels and organs. These functions include the following:

- **Regulation of interstitial fluid volume.** As you read about in the blood vessels chapter, the net filtration pressure in blood capillaries favors filtration, meaning that water is lost from the plasma in the blood to the interstitial fluid (see Chapter 18). The amount of fluid that is lost from the plasma is significant, measuring about 2–4 liters per day. This fluid must be returned to the circulation, or both the blood volume and blood pressure will drop too low to maintain homeostasis. The vessels that perform this function are the lymphatic vessels, which pick up excess interstitial fluid, transport it through the body, and deliver it back to the cardiovascular system. When the interstitial fluid exits the extracellular space and enters the lymphatic vessels, it is known as **lymph** (LIMF). This means that interstitial fluid and lymph are very similar in composition.
- **Absorption of dietary fats.** The breakdown products of fats in the diet are too large to pass through the tiny spaces between the endothelial cells of blood capillaries. However, they are able to enter small lymphatic vessels in the small intestine. The dietary fats travel through the lymphatic vessels and are delivered to the blood with lymph.
- **Immune functions.** The lymphatic system has an important role in the immune system. As we discuss shortly, the lymphoid organs filter pathogens from the lymph and blood. They also house several types of leukocytes, and play a part in their maturation.

Learning Outcomes

1. Describe the major functions of the lymphatic system.
2. Compare and contrast lymphatic vessels and blood vessels in terms of structure and function.
3. Explain the mechanisms of lymph formation, and trace the pathway of lymph circulation through the body.
4. Describe the basic structure and cellular composition of lymphatic tissue, and relate them to the overall functions of the lymphatic system.
5. Describe the structures and functions of the lymphoid organs.

The lymphatic system consists of two main components, both of which are illustrated in **Figure 20.1**: (1) a system of blind-ended tubes called **lymphatic vessels,** and (2) **lymphatic tissue** and **lymphoid organs,** including clusters of *lymphoid follicles* such as the *tonsils,* as well as the *lymph nodes, spleen,* and *thymus.* In this module, we introduce the basic functions of these vessels and organs and then turn to their structure. Much of what we discuss

Quick Check

☐ 1. What are the main functions of the lymphatic system?

Lymphatic Vessels and Lymph Circulation

◀◀ FLASHBACK

1. What are endothelial cells? (p. 687)
2. How are continuous capillaries structured? (p. 688)
3. What are the mechanisms that assist in the return of venous blood to the heart? (p. 675)

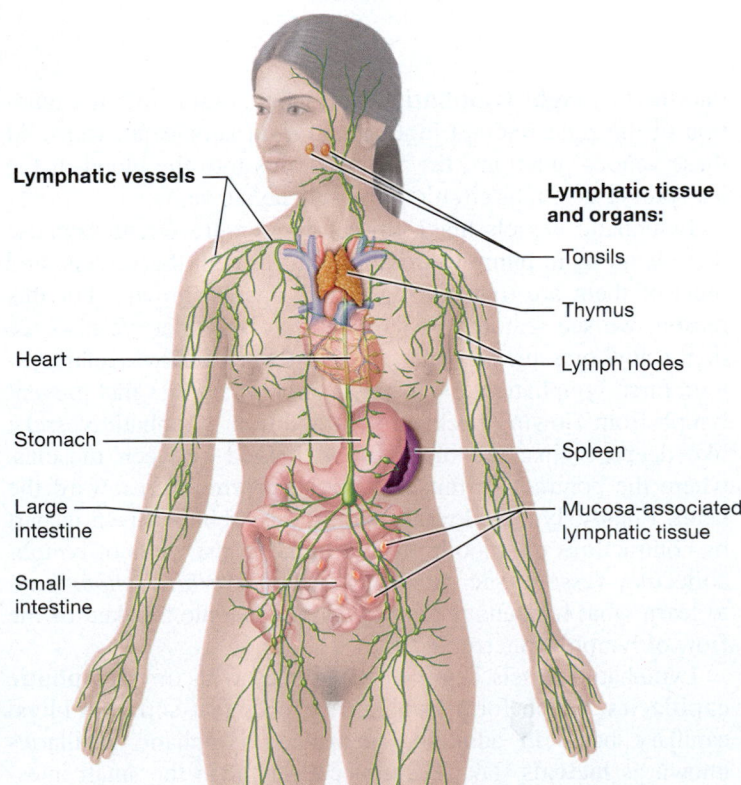

Lymphatic vessels

Lymphatic tissue and organs:
— Tonsils
— Thymus
— Lymph nodes
— Spleen
— Mucosa-associated lymphatic tissue

Heart
Stomach
Large intestine
Small intestine

Figure 20.1   Overview of the lymphatic system.

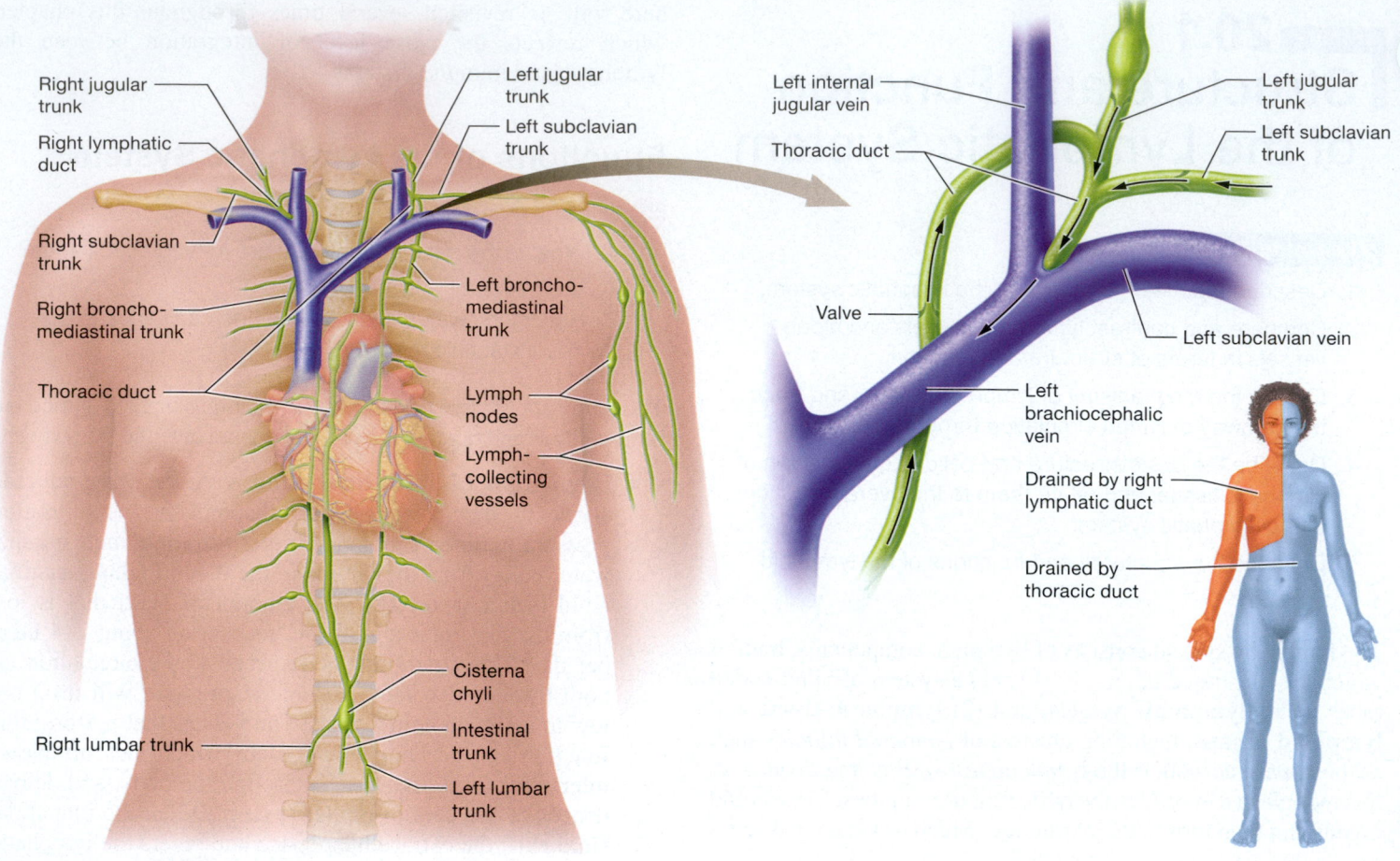

(a) The main lymph trunks and ducts

(b) Drainage point of the thoracic duct

Figure 20.2 Main lymphatic trunks and ducts.

Lymph is collected in vessels appropriately called **lymph-collecting vessels,** which merge to form larger vessels, the **lymph trunks.** There are nine lymphatic trunks that drain the lymph from body regions, illustrated in **Figure 20.2a**:

- *lumbar trunks,* which receive lymph from the lower limbs and pelvic area;
- *the intestinal trunk,* which receives fat-containing lymph from the small lymphatic vessels in the small intestine;
- *jugular trunks,* which receive lymph from the head and neck;
- *bronchomediastinal trunks* (brong′-koh-meh-dee-ah-STY-nul), which receive lymph from the thoracic cavity; and
- *subclavian trunks,* which receive lymph from the upper limbs.

Note in Figure 20.2a that the intestinal and lumbar trunks all drain into a large, swollen-looking vessel called the **cisterna chyli** (sis-TER-nuh KY-lee; "reservoir of digestive juice"). The cisterna chyli and the other lymphatic trunks drain into one of two **lymphatic ducts.** The cisterna chyli and the trunks from the left side of the body drain into the **thoracic duct,** the largest lymphatic duct, which runs along the anterior vertebral column and drains into the junction of the left internal jugular and left subclavian veins (**Figure 20.2b**). In fact, the thoracic duct drains all of the lower body as well as the left side of the upper body. The remaining trunks from the right upper side of the body drain

into the tiny **right lymphatic duct,** which drains into the junction of the right internal jugular and right subclavian veins. At these venous junctions, the lymph drains into the blood of the low-pressure venous circuit at the subclavian veins.

Lymphatic vessels make up a low-pressure circuit because there is no main pump to drive lymph through the vessels, and most of them are transporting lymph against gravity. For this reason, we see features in lymphatic vessels that we also see in the low-pressure venous circuit of the cardiovascular system. First, lymphatic vessels have *lymphatic valves* that prevent lymph from flowing backward. In addition, lymphatic vessels, like deeper veins, are often found lodged between muscles, where the contracting muscles massage lymph up toward the heart. Finally, lymph flow through the vessels is driven in part by contractions of smooth muscle found in the walls of lymph-collecting vessels. See *A&P in the Real World: Lymphedema* to learn what happens if lymph vessels become blocked or the flow of lymph is interrupted.

Lymphatic vessels begin in the tissues with tiny **lymphatic capillaries,** which form weblike networks that surround blood capillary beds. In addition, specialized lymphatic capillaries known as **lacteals** (lak-TEE-ulz) collect fat in the small intestine. As you can see in **Figure 20.3a**, lymphatic capillaries differ both structurally and functionally from blood capillaries. For

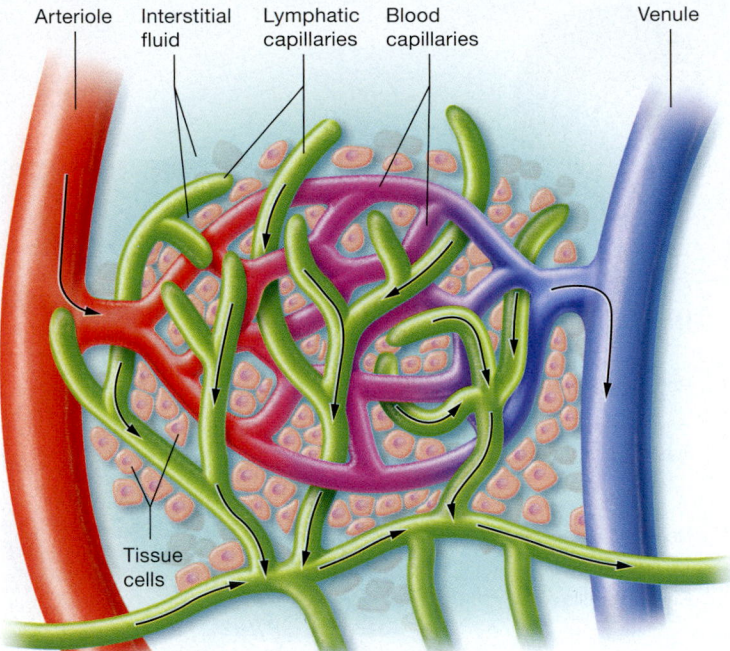

(a) **Lymphatic capillaries surrounding a capillary bed**

(b) **Lymphatic capillary structure and function**

Figure 20.3 **Structure and function of lymphatic capillaries.**

one, lymphatic capillaries are blind-ended, which makes the lymphatic vasculature a one-way system that only moves lymph away from the tissues. Blood capillaries, of course, form a two-way system that moves blood both toward and away from the tissues.

Notice in **Figure 20.3b** that unlike the endothelial cells that form the walls of blood capillaries, the cells of lymphatic capillary walls are not tightly joined, and instead are able to flap open and closed. Fluid that leaks from the blood capillaries increases the pressure in the interstitial fluid, which forces the lymphatic endothelial cells apart and allows large volumes of fluid to enter the lymphatic capillaries. When the pressure in the interstitial fluid decreases, the endothelial cells flap shut. In this way, the lymphatic system plays the crucial role of precisely controlling the amount of fluid between our cells. Note in Figure 20.3b that these vessels are also leaky enough to allow cells such as *macrophages* and other immune cells to enter the lymph.

Lymphedema

The role of the lymphatic system in regulating the volume of interstitial fluid becomes readily apparent in **lymphedema** (limf-eh-DEE-muh). **Edema,** commonly known as *swelling,* is defined as an accumulation of excess interstitial fluid. Many conditions can cause mild to moderate edema, including trauma, vascular disease, and heart failure. However, the edema seen with lymphedema is typically severe and can be disfiguring.

Lymphedema is generally due to removal of lymphatic vessels during surgery or blockage of the vessels from pathogens such as parasites. Both conditions prevent the lymphatic vessels from transporting excess interstitial fluid back to the cardiovascular system. The fluid therefore accumulates in the tissues of the affected body part, causing it to enlarge. The photo shows a case of lymphedema in the arm of a breast cancer patient, resulting from the surgical removal of lymph nodes.

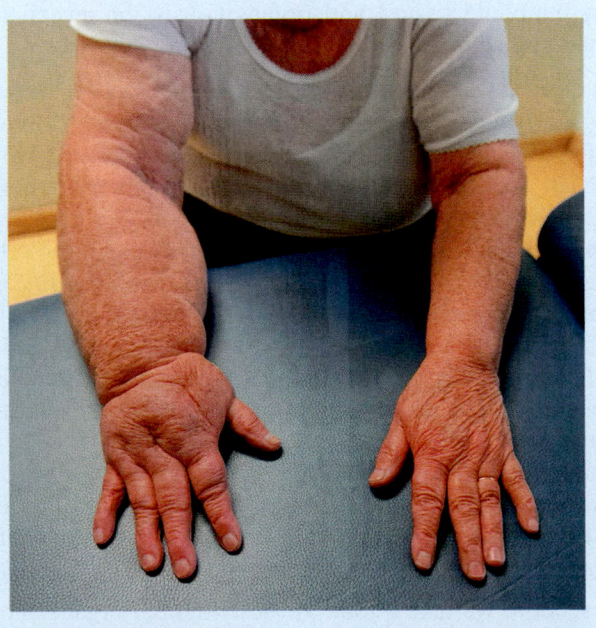

The extreme permeability of the lymphatic capillaries allows pathogens such as bacteria and cancer cells in the interstitial fluid to have an easier time entering the lymphatic capillaries than the blood capillaries. So, these pathogens can spread to other parts of the body through the lymphatic system. However, along the pathway of lymphatic vessels, we find clusters of the lymphoid organs called lymph nodes. Lymph nodes limit the spread of pathogens through the body by acting as filters, trapping pathogens and preventing them from traveling elsewhere.

Quick Check

☐ 2. How do lymphatic capillaries differ from blood capillaries in structure and function?

☐ 3. What is the pathway of lymph flow from the right foot to the point where it enters the blood circulation?

Lymphoid Tissues and Organs

《 FLASHBACK

1. What is reticular tissue? (p. 139)

2. What are splenic sinusoids, and where are they located? (p. 689)

3. Where do T lymphocytes migrate to mature? (p. 733)

The predominant tissue type of the lymphatic system is a type of loose connective tissue called *reticular tissue* that contains specialized cells and thin *reticular fibers* (look back at Figure 4.15). These reticular fibers interweave to form "nets" that trap disease-causing pathogens, which is an example of the Structure-Function Core Principle (p. 28). In the lymphatic system, reticular tissue is often simply called **lymphoid tissue.** Lymphoid tissue is found in lymphoid organs and also in independent clusters that have varying degrees of organization.

 **CORE PRINCIPLE** Structure-Function

Most lymphoid tissues and organs consist of several main cell types, shown in **Figure 20.4**:

- **Lymphocytes.** Lymphoid organs house lymphocytes, including *B lymphocytes* and *T lymphocytes* (more commonly referred to as *B cells* and *T cells,* respectively). As we discussed previously, B lymphocytes and T lymphocytes are agranulocytes with diverse immune functions (see Chapter 19).
- **Phagocytes.** There are two main types of phagocytic cells in lymphoid tissues and organs: *macrophages* and *dendritic cells.* Recall that **macrophages** (MAK-roh-fay-jez; "big eaters") are mature monocytes that are very active phagocytes. **Dendritic cells** are leukocytes with spiny processes that resemble the dendrites of neurons. They play an important role in activating lymphocytes, which we discuss in a later module.
- **Reticular cells.** Reticular fibers, composed of a specialized, thin type of collagen protein, are produced by **reticular cells.** Reticular cells are particularly abundant in organs such as the spleen and lymph nodes.

In this section, we take a look at the lymphoid organs and tissues and see how these cells function within them.

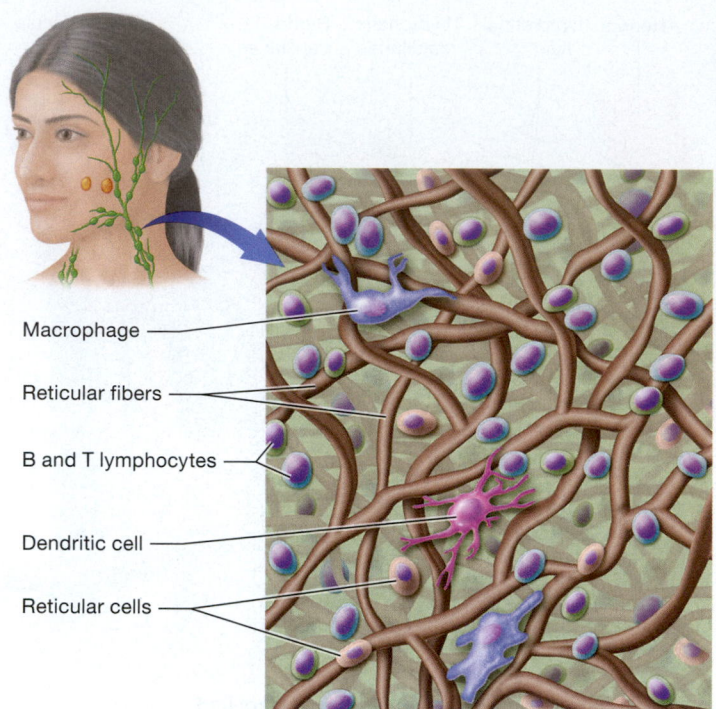

Figure 20.4 Microscopic structure of lymphoid organs.

Macrophage
Reticular fibers
B and T lymphocytes
Dendritic cell
Reticular cells

Mucosa-Associated Lymphatic Tissue (MALT)

The mucous membranes, such as those of the oral and nasal cavities, are all vulnerable parts of the body because they are exposed to large numbers of pathogens. These areas are protected by loosely organized clusters of lymphoid tissue collectively called **mucosa-associated lymphatic tissue (MALT).** MALT is found scattered throughout the gastrointestinal tract, the respiratory passages, and, to a limited extent, the genitourinary tract.

Much of the MALT in the body consists of loosely organized clusters of B and T cells. Most MALT lacks any kind of connective tissue capsule. However, certain types of specialized MALT are partially encapsulated by connective tissue.

Specialized MALT consists of roughly spherical clusters called **lymphoid follicles,** or **lymphoid nodules,** that consist primarily of B cells. When lymphoid follicles have been exposed to pathogens, they develop light-staining regions, the **germinal centers,** where the B cells are actively dividing. In the gastrointestinal tract, we find these specialized clusters of MALT in three locations: the **tonsils,** located around the oral and nasal cavities; **Peyer's patches** (PY-erz), or *aggregated lymphoid nodules,* located in the last portion of the small intestine (called the *ileum*); and the **appendix,** located off the large intestine.

The three main tonsils are illustrated in **Figure 20.5**. The first is the **pharyngeal tonsil** (fuh-RIN-jee-uhl; also known as the *adenoid* [AD-eh-noyd]), which is located in the posterior nasal cavity (the nasopharynx). The **palatine tonsils** are found in the posterolateral oral cavity, whereas the **lingual tonsil** is found at the base of the tongue. The epithelium lining the tonsils indents deeply in several locations, forming *tonsillar crypts* that trap bacteria and

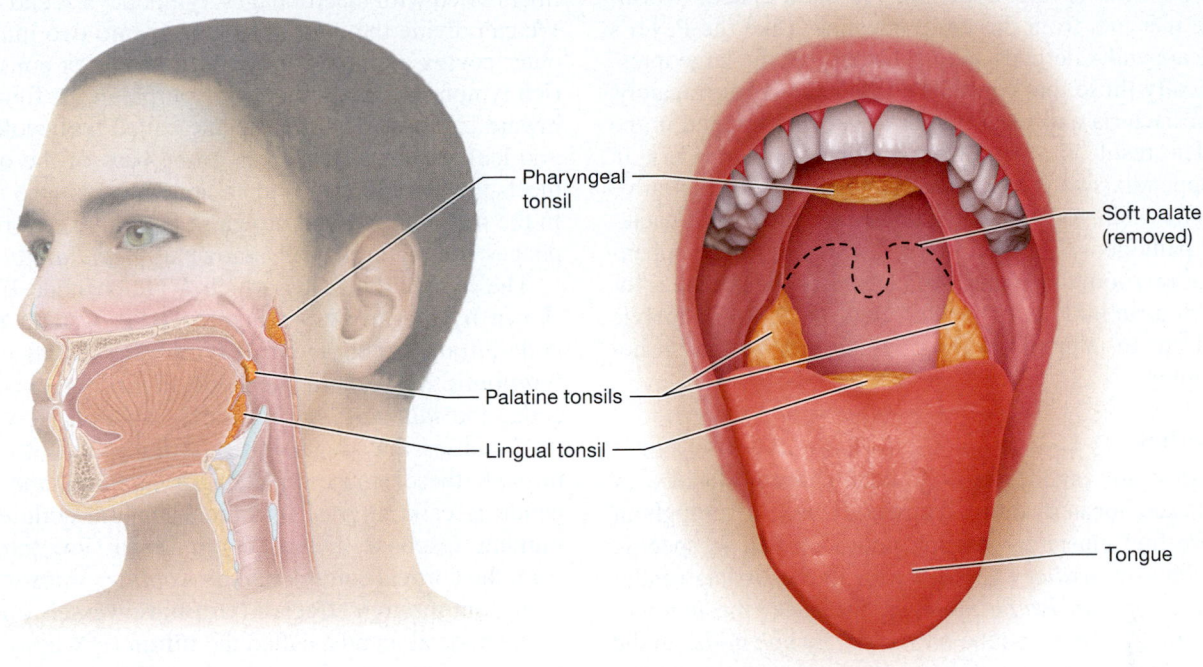

(a) Location of tonsils in the oral and nasal cavities

(b) Anterior view of the oral cavity

Figure 20.5   Location of the tonsils.

debris. The tonsils' location puts them into contact with a large number of potential pathogens. As a result, they commonly become inflamed, a condition known as *tonsillitis*.

Peyer's patches are also exposed to a large number of potential pathogens. The last portion of the small intestine joins to the large

intestine, which houses a tremendous number of bacteria. These bacteria generally are not pathogenic in the large intestine, but if they gain entry to the small intestine, they could cause disease. The position of the Peyer's patches allows them to defend against any bacteria that have escaped from the large intestine (**Figure 20.6**).

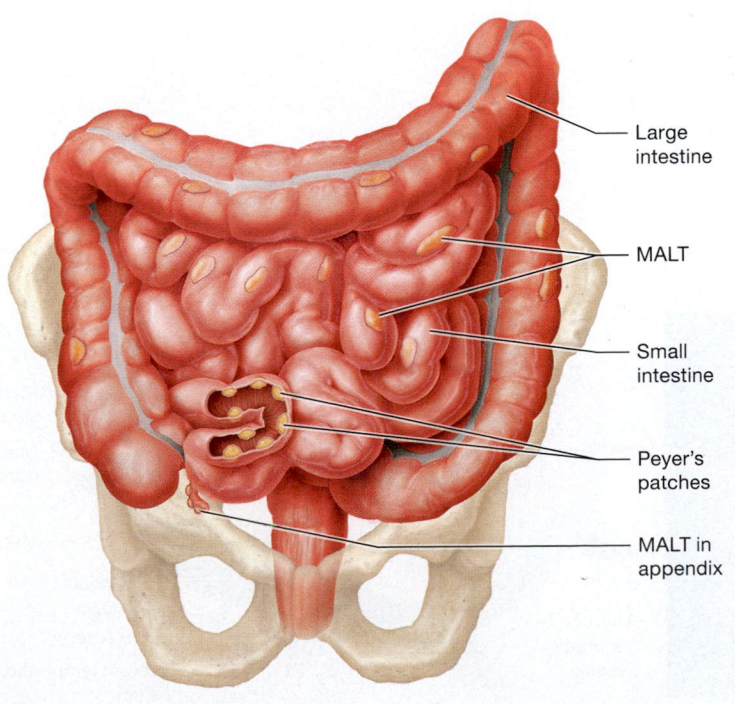

(a) Location of MALT in intestines

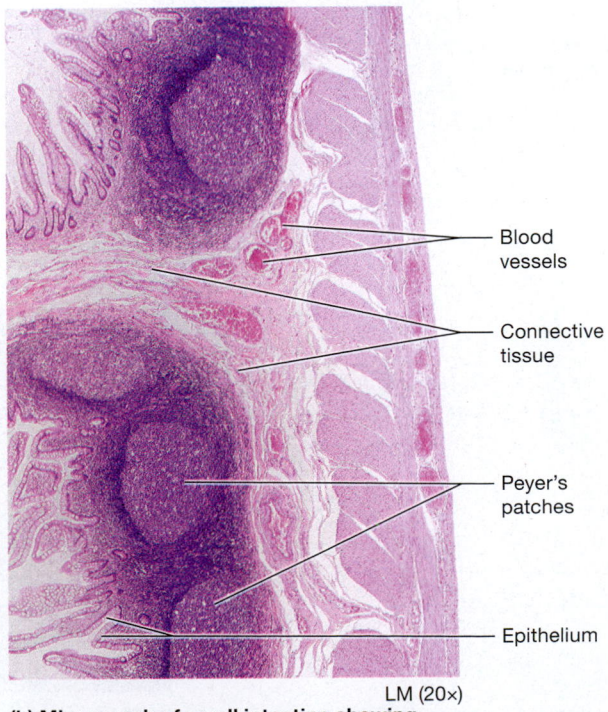

LM (20×)

(b) Micrograph of small intestine showing Peyer's patches

Figure 20.6   MALT of the intestines.

Finally, we turn to the appendix—a blind-ended, worm-shaped tube that juts from the large intestine. Like the Peyer's patches, the appendix defends against bacteria in the large intestine, specifically those that could be pathogenic. Unfortunately, it is easy for bacteria and fecal matter to become trapped in the appendix. The result is a condition known as *appendicitis,* in which the appendix swells much like lymph nodes do (as you've likely noticed) when you are sick and they are filled with bacteria or other pathogens. The blind-ended structure of the appendix gives it little room to expand, which creates the potential for rupture. Such a rupture spills bacteria into the abdominopelvic cavity and leads to internal bleeding, which may be fatal if not treated promptly.

Lymph Nodes

Lymph nodes are small, vaguely bean-shaped clusters of lymphatic tissue located along lymphatic vessels throughout the body. We find clusters of lymph nodes located in specific areas, including the *axillary lymph nodes* in the axillae (under the arms), the *cervical lymph nodes* in the neck, the *inguinal lymph nodes* in the groin, and the *mesenteric lymph nodes* in the abdominal cavity around the abdominal organs (**Figure 20.7a**). **Figure 20.7b** shows an inguinal lymph node in a cadaver.

Note in **Figure 20.7c** that lymph nodes feature an external connective tissue capsule that surrounds a network of reticular

fibers filled with macrophages, lymphocytes, and dendritic cells. We can divide the interior of a node into two main regions: the outer **cortex** and inner **medulla.** The cortex consists of B cell–rich lymphoid follicles (colored purple in the figure) divided by inward extensions of the capsule called **trabeculae** (trah-BEK-yoo-lee). In the darker areas at the base of the cortex where it meets the medulla, we find an area that contains mostly T cells. In the medulla (colored brown in the figure) are rows of macrophages and mature B cells known as *medullary cords.*

The basic process by which a lymph node filters lymph is shown by the arrows in Figure 20.7c. Lymph flows into the node through multiple small lymphatic vessels called **afferent lymphatic vessels.** In the node, it first enters a hollow area called the *subcapsular sinus,* after which it flows through *cortical sinuses* that run next to the trabeculae. As lymph passes through these sinuses, pathogens and immune cells such as dendritic cells migrate into the lymphoid nodule to initiate an immune response. Lymph continues to flow through the node from the cortical sinuses to the *medullary sinuses,* after which it drains out through **efferent lymphatic vessels** on the other side of the node at an area called the **hilum** (HY-lum). Lymph nodes are remarkably effective at their job and trap approximately 90% of the pathogens in lymph. This prevents these pathogens from being delivered to the blood, where they could easily spread to other tissues and organs.

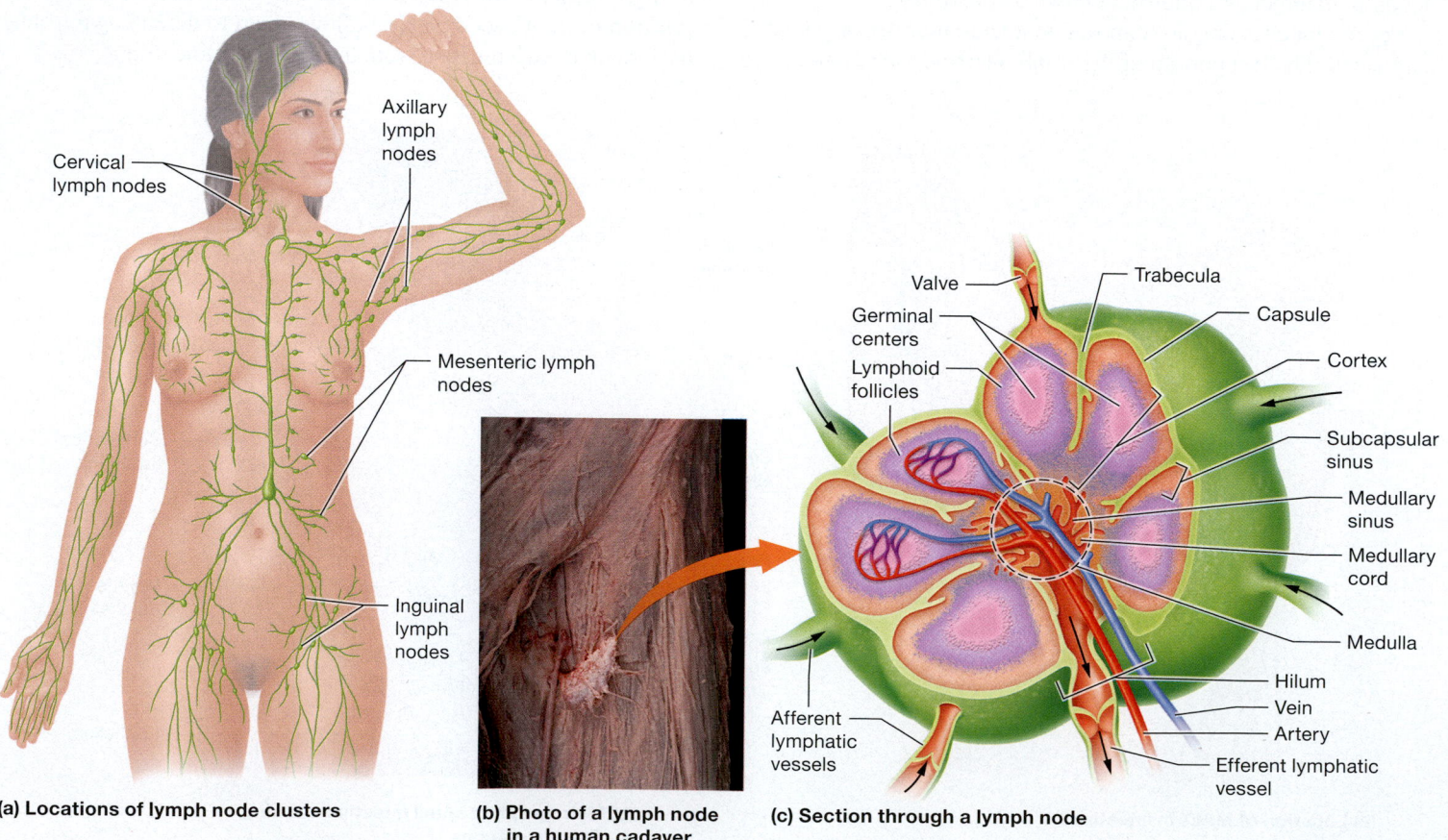

(a) Locations of lymph node clusters

(b) Photo of a lymph node in a human cadaver

(c) Section through a lymph node

Figure 20.7 Location, structure, and function of lymph nodes.

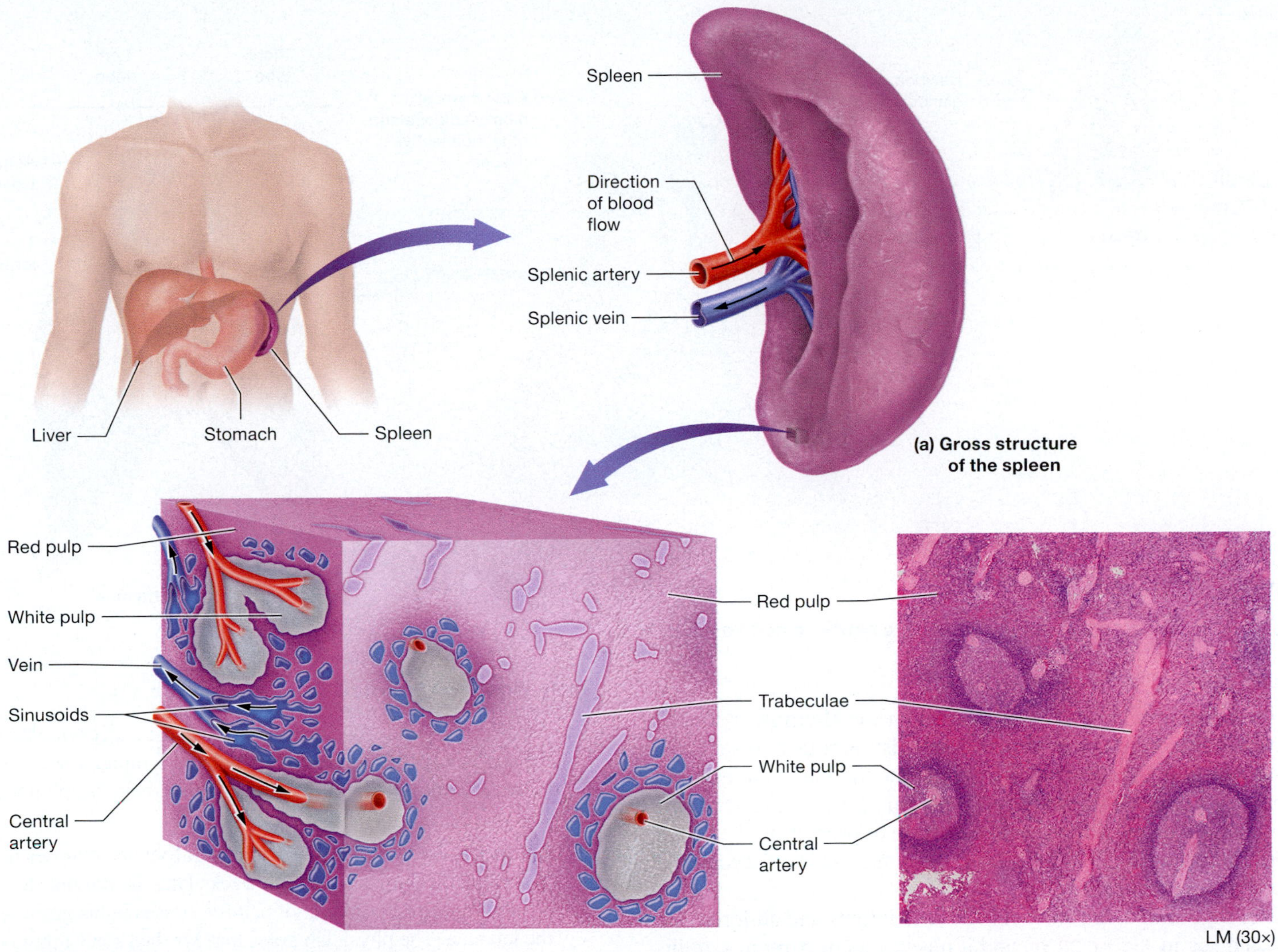

Spleen

Direction
of blood
flow

Splenic artery

Splenic vein

**(a) Gross structure
of the spleen**

Liver   Stomach   Spleen

Red pulp

White pulp

Vein

Sinusoids

Central
artery

Red pulp

Trabeculae

White pulp

Central
artery

LM (30×)

(b) Microscopic structure of the spleen

Figure 20.8 **Structure of the spleen.**

View **PAL** Histology
@ **Mastering** Anatomy & Physiology

When pathogens enter a lymph node and B cells are activated, the germinal centers of the lymphoid follicles enlarge. This causes the whole lymph node to swell, a condition commonly known as "swollen glands." Note, however, that this term is incorrect—lymph nodes are not glands and they secrete no products. Certain conditions such as bacterial and viral infections cause the swollen lymph nodes to become painful. Other conditions, such as allergies or the spread of cancerous cells through lymphatic vessels, generally cause the nodes to enlarge with little pain.

The Spleen

Having covered the organs and tissues that protect us from pathogens on our mucous membranes and in the lymph, we now turn to the organ that protects us from pathogens in the blood: the spleen. The **spleen,** which is the largest lymphoid organ in the body, is located on the lateral side of the left upper quadrant of the abdominopelvic cavity. It is a purplish-brown organ about the size of a large bar of soap (**Figure 20.8a**). Recall that, like all lymphoid organs, the spleen's internal structure consists of a network of reticular fibers made by reticular cells. Within this reticular tissue framework we find two distinct histological regions: **red pulp,** which contains macrophages that destroy old erythrocytes; and **white pulp,** which filters pathogens from the blood and contains leukocytes (**Figure 20.8b**).

Let's take a closer look at the white pulp. Figure 20.8b shows that the white pulp surrounds branches of the splenic artery called the *central arteries.* Immediately surrounding the arteries in the center of the white pulp we find a zone that contains mostly T cells. Just outside this zone are lymphoid follicles composed of a central core of B cells, and on the outer rim of the white pulp are macrophages, dendritic cells, and some B and T cells.

When blood enters the spleen, it flows through a series of arterioles until it comes to the sinusoids—the spleen's large, exceptionally leaky capillaries. Here pathogens encounter macrophages and dendritic cells, which process portions of the pathogens.

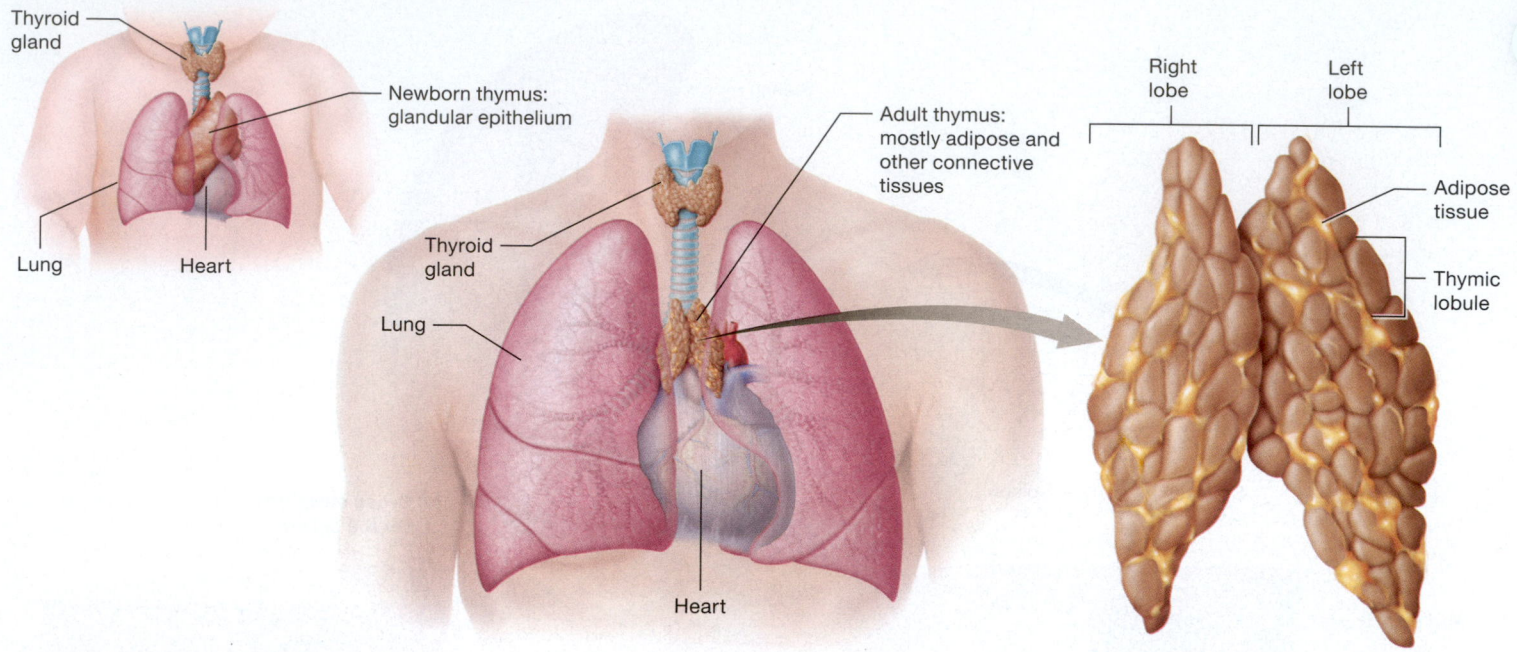

(a) Thymus in newborn and adult

(b) Structure of adult thymus

Figure 20.9 Comparison of the thymus in a newborn and an adult.

The Thymus

In the endocrine chapter, you read about the **thymus,** a small, encapsulated organ in the superior mediastinum that consists of two lobes (see Chapter 16). Recall that the thymus secretes hormones that enable it to carry out its primary function: generating a population of functional T cells capable of protecting the body from pathogens. Unlike the spleen and lymph nodes, the thymus doesn't trap pathogens.

The thymus is large and very active in infants and children, as you can see in **Figure 20.9a**, and it reaches its maximum size in individuals about 12–14 years old. After this point, it begins to atrophy, and the thymic tissue is gradually replaced with fat. As the thymus atrophies, its production of T cells decreases as well. Indeed, by age 65, the rate of T cell production falls to about 2% of the rate at which an infant produces T cells. This drop in production does not normally lead to a decline in immune function, however, because of the enormous number of T cells produced during infancy and childhood.

The adult thymus, shown in **Figure 20.9b**, consists of subunits called *thymic lobules* (or thymic corpuscles), which are visible externally as the "lumps" on the surface of the thymus. Each lobule contains two regions: an outer cortex and an inner medulla. The cortex contains densely packed T cells. The medulla contains fewer of these cells, and is thought instead to be mostly the site of destruction of certain populations of T cells that could react to the body's own cells (see Module 20.4). Notice that there are no lymphoid follicles in the thymus because it lacks B cells.

☐ 4. What main cell types are located in lymphoid organs?

☐ 5. List the main functions of MALT, lymph nodes, the spleen, and the thymus.

☐ 1. A patient has undergone surgery for breast cancer in which the axillary lymph nodes and associated lymphatic vessels were removed. Predict the effects of removing lymph nodes and vessels.

☐ 2. Mr. Jackson steps on a nail with his left foot and chooses not to seek medical treatment. Two weeks later, he presents to his physician with a complaint of painful masses in his groin on the left side. The physician notes that Mr. Jackson's inguinal lymph nodes are swollen. What has likely happened?

☐ 3. Predict the potential consequences of removing each of the following lymphoid organs:

   a. Spleen

   b. Thymus (in an adult)

   c. Thymus (in a newborn)

   d. Tonsils

See answers in Appendix A.

MODULE 20.2
Overview of the Immune System

1. Identify the differences between innate and adaptive immunity, and explain how the two types of immunity work together.

2. Describe the basic differences between antibody-mediated and cell-mediated immunity.

3. Describe the roles that surface barriers play in immunity.
4. Describe the cells and proteins that make up the immune system.
5. Explain how the immune and lymphatic systems are connected structurally and functionally.

As we have discussed, the immune system consists primarily of a group of many different types of leukocytes and proteins that defend the body from anything that causes cellular injury. This includes injury resulting from trauma and pathogens. We commonly think of pathogens as microorganisms that cause *infection,* meaning they invade a tissue, increase in number, and cause cellular damage. Such microorganisms include bacterial cells, viruses, fungi, and parasites. However, many other cells and substances are pathogenic, including cancer cells, bacterial toxins, and toxins found in animal venoms.

Together, the components of the immune system offer three lines of defense against pathogens:

- The *first line of defense* involves cutaneous and mucous membranes that act as surface barriers to block the entry of pathogens into the body.
- The *second line of defense* consists of the responses of the cells and proteins that make up innate immunity.
- The *third line of defense* consists of the responses of the cells and proteins of adaptive immunity.

Before we consider the inner workings of the immune system, let's survey its parts and processes. In this module, we examine the two basic types of immunity, the surface barriers that protect us, the components of the immune system, and the connections between the lymphatic and immune systems. The following modules explore the functional aspects and components of the immune system more closely.

Types of Immunity

Immunity is classified according to the way it responds to different pathogens or forms of cellular injury. As implied by its name, **innate,** or **nonspecific, immunity** responds to all pathogens or classes of pathogen in the same way. The innate immune system consists of *antimicrobial proteins* and certain cells. Note that innate immunity responds quickly and is the dominant response to pathogens for the first 12 hours after exposure. This occurs because the cells and proteins exist in the bloodstream, even in the absence of a stimulus (hence the term "innate").

The other type of immunity is **adaptive immunity,** the components of which respond to unique markers called **antigens** (AN-tih-jenz). Antigens are present on all cells, including our own cells, and most biological molecules, and they identify a cell or molecule as belonging to a specific group. Adaptive immunity is also known as **specific immunity** because its cells and proteins respond individually to only a specific, individual antigen. This is in contrast to innate immunity, which reacts in the same manner to every potential pathogen.

There are two "arms" of the adaptive immune system. The first, **cell-mediated immunity,** is brought about by two types of T cells. The second, **antibody-mediated immunity** (also known as

humoral immunity), is carried out by B cells and proteins they produce, called *antibodies.* Adaptive immunity responds more slowly than innate immunity because one must be exposed to a specific antigen for the response to be initiated. It is therefore also known as **acquired immunity.** Adaptive immunity takes 3–5 days to mount a response, but after this point, it is the dominant response.

An important difference between the innate and adaptive immunities is that adaptive immunity has the capacity for *immunological memory,* in which exposure to an antigen is "remembered" by specific lymphocytes and antibodies. This allows a more rapid and efficient response on subsequent exposures. Innate immunity lacks the capacity for immunological memory and will respond in the same way with repeat exposure to a pathogen.

Before we move on, keep in mind that adaptive immunity and innate immunity are not independently functioning arms of the immune system. Each type of immunity relies on the other, and the response to a pathogen involves a highly integrated series of events within both parts of the immune system. Indeed, dysfunction within a component of either adaptive or innate immunity may lead to catastrophic failure of the entire system (see Module 20.7).

Quick Check

☐ 1. How do innate immunity and adaptive immunity differ?
☐ 2. What are the three lines of defense?

Surface Barriers

« FLASHBACK

1. Where are mucous membranes located? What do they secrete? (p. 153)
2. What substances does the skin secrete? What are the functions of these secretions? (p. 162)

The first line of defense against any potential threat to the body is the coverings that line body surfaces: the skin and the mucous membranes, and certain products they secrete. Both the skin and mucosae are **surface barriers,** meaning that they provide a continuous physical barrier to block the entry of potential pathogens into the body.

We've previously discussed the skin's numerous protective functions (see Chapter 5). Besides being continuous and covering nearly every external surface of the body, skin is also relatively resistant to mechanical stresses, thanks to its several layers of epithelial cells filled with the hard protein *keratin.* The body's other type of surface barrier is the mucous membrane. Recall that mucous membranes line all passageways in the body that open to the outside, including the respiratory, gastrointestinal, and genitourinary tracts (see Chapter 4). Mucous membrane epithelia lack keratin and are generally thinner than skin epithelia, so they are less resistant to mechanical abrasion.

The importance of surface barriers is evident any time the surface barrier is compromised. Think about what happens when you receive a wound to your skin—pathogens likely entered the body when the wound occurred, so you must wash it thoroughly to remove them. You must then cover the wound with a bandage to prevent future invasion from pathogens. In this case, the bandage acts as the new surface barrier until the skin is healed and again able to prevent pathogen invasion.

The skin and mucosae provide more than a simple barrier. They also produce numerous protective secretions, including the following:

- Sebaceous glands in the skin secrete *sebum*, an oily mixture with a slightly acidic pH that deters the growth of most pathogenic organisms.
- Mucous membranes produce *mucus*, a thick, sticky substance that traps debris and many pathogens and protects the underlying cells from chemical and mechanical trauma. In addition, many mucous membranes, particularly those in the respiratory tract, have cilia projecting from their surface. Cilia in these mucosae beat in unison to propel substances trapped in mucus out of the respiratory tract.
- The mucosa of the stomach secretes acid, which kills ingested pathogens.
- Cells of the skin and mucosae produce *defensins*, antimicrobial peptides that damage the membranes of pathogens.

Finally, the skin and several mucosae in the body house microorganisms, mostly bacterial species, known as *normal flora*. These bacteria are generally not disease-causing in individuals with functional immune systems. They perform important functions on our surface barriers in protecting us from disease, including competing with pathogenic bacteria for space and resources, which limits their growth; secreting substances that kill pathogenic bacteria; and secreting substances that acidify surface barriers to deter growth of pathogens.

Of course, surface barriers, the products they produce, and the organisms they host can't protect us from every pathogen we encounter, in part because the pathogens have evolved mechanisms to circumvent them. To read about some of these mechanisms, see *A&P in the Real World: How Pathogens Can Evade Surface Barriers.*

Quick Check

☐ 3. What are the body's main surface barriers, and how do they function in its defense?

Overview of Cells and Proteins of the Innate and Adaptive Immune Systems

《 FLASHBACK

1. Where are most leukocytes produced? (p. 733)
2. What are phagocytes? (p. 731)

The cells and proteins of the innate and adaptive immune systems produce the responses of the second and third lines of defense. The main cells of the immune system are the different types of leukocytes, which we've already discussed: the *agranulocytes* (B and T lymphocytes and monocytes), which lack cytoplasmic granules; and the *granulocytes* (neutrophils, eosinophils, and basophils), which contain cytoplasmic granules (look back at Figure 19.8). Many cells of innate immunity can function as **phagocytes**—cells that "eat" foreign or damaged

How Pathogens Can Evade Surface Barriers

Pathogens have evolved multiple mechanisms to evade the defenses provided by surface barriers and facilitate their invasion of our bodies. The following are three examples of such pathogens:

- *Collagenase-producing bacteria.* Certain bacterial species, notably *Clostridium perfringens* (klaws-TRID-ee-um per-FRIN-genz), produce enzymes called *collagenases* (kah-LAJ-eh-nay´-zez), which catalyze reactions that degrade the protein collagen. When a wound is infected with *C. perfringens,* the bacteria begin to produce collagenases that destroy collagen in the basement membrane and dermis. This allows them access to deeper tissues such as skeletal muscle, eventually resulting in massive tissue death. As the bacteria destroy tissues, they produce large quantities of gas, leading to the common name of *gas gangrene* (GANG-green). To make matters worse, *C. perfringens* also produces enzymes that catalyze the destruction of neutrophils, making the immune system much less able to respond to the invasion.

- *Phagocytosis-resistant fungi.* Pathogenic organisms that come into contact with mucosae are generally ingested by local macrophages, minimizing the risk of infection. However, some fungi are quite resistant to phagocytosis. For example, the fungal yeast cells that cause the disease *blastomycosis* (blas´-toh-my-KOH-sis) have very thick walls that are difficult for phagocytes to ingest. Still other fungal species, such as those producing the respiratory diseases *cryptococcosis* (krip´-toh-kok-KOH-sis) and *histoplasmosis* (hiss´-toh-plaz-MOH-sis), actually survive inside macrophages. This makes the fungi invisible to the immune system, and the macrophages spread the disease as they travel through the blood and lymph to different tissues.

- *Acid-tolerant pathogens.* Secretions produced by surface barriers are generally slightly to very acidic, which deters or kills most pathogens. But certain pathogens tolerate or even prefer an acidic pH. For example, the poliovirus is able to survive the highly acidic environment of the stomach when it is ingested, allowing it to replicate in the digestive tract and potentially invade the rest of the body. In addition, many species of fungus and the bacterium *Helicobacter pylori* (hel´-ih-koh-BAK-ter py-LOH-rye) prefer an acidic pH, allowing them to infect the skin and stomach, respectively.

cells—although neutrophils and macrophages are the most active phagocytes. As you read earlier in this module, lymphocytes are the primary cells of adaptive immunity.

Other types of leukocytes also function in immunity. One is the **natural killer (NK) cell,** a type of lymphocyte that acts primarily in innate immunity. NK cells are generally located in the blood and spleen, and are formed in the bone marrow from the same cell line as lymphocytes. Another cell type is one you encountered in the previous module: dendritic cells, immune cells that are located in many lymphoid organs. Dendritic cells are part of the innate immune response, but their main role is to activate the T cells of adaptive immunity.

The other main components of the immune system are groups of different types of proteins. You have already been introduced to one group of proteins—antibodies, the proteins produced by B lymphocytes that function in adaptive immunity. Another group of proteins, collectively called the *complement system,* functions in innate immunity. The final type is the **cytokines** (SY-toh-kynz), a diverse group of proteins secreted by cells of both innate and adaptive immunity. Cytokines have a variety of effects, including regulating the development and activity of immune cells; we will discuss their many other effects in later modules.

The upcoming modules examine the roles of each of these cells and proteins in more detail. Then we look into how they interact to produce a functional immune response.

Quick Check

☐ 4. List the main types of cells and proteins of the immune system.

How the Lymphatic and Immune Systems Work Together

As we mentioned in the previous module, the lymphatic and immune systems are closely connected both structurally and functionally. Some of the ways in which they are connected include the following:

- **Lymphoid organs and tissues provide a residence for cells of the immune system.** B cells, T cells, and macrophages frequently take up residence in lymphoid organs such as the lymph nodes, MALT, and the spleen. In addition, they provide a place for lymphocyte development.
- **Lymphoid organs and tissues trap pathogens for the immune system.** The fine networks of reticular fibers in lymphoid tissues form "nets" that trap pathogens so that leukocytes may interact with them more easily.
- **Lymphoid organs activate cells of the immune system.** Lymphoid organs house cells such as dendritic cells and macrophages, which play a crucial role in activating B and T cells. In addition, the thymus is required for the selection of a functional population of T cells.

The lymphatic system plays a greater role in adaptive immunity than in innate immunity, although lymphoid organs do house cells of innate immunity, such as macrophages. In the upcoming modules, we explore more closely how these two systems interact.

Quick Check

☐ 5. How are the immune system and lymphatic system connected structurally and functionally?

Apply What You Learned

☐ 1. *Interstitial cystitis* is characterized by erosion of the mucous membrane lining the urinary bladder. Predict the potential consequences of this condition.

☐ 2. Your friend argues that the third line of defense of the immune system is the most important and that the first and second are relatively unimportant. What do you tell him?

See answers in Appendix A.

MODULE 20.3
Innate Immunity: Internal Defenses

Learning Outcomes

1. Describe the roles that phagocytic and nonphagocytic cells and plasma proteins such as complement and interferon play in innate immunity.

2. Walk through the stages of the inflammatory response, and describe its purpose.

3. Describe the process by which fever is generated, and explain its purpose.

As you learned in the previous module, innate immunity is present in the body without needing exposure to a specific antigen. For this reason, innate immunity is typically quick to respond to a pathogen, generally within seconds to a few hours. Innate immunity consists of two main components: (1) a group of antimicrobial proteins, including the antimicrobial complement proteins and a variety of cytokines; and (2) several types of cells, including neutrophils, macrophages, and NK cells. In this module, we explore these components and finish with an examination of how they interact to produce the *inflammatory response.*

Cells of Innate Immunity

⟪ FLASHBACK

1. How do granulocytes and agranulocytes differ? (p. 731)

2. What are the three types of granulocytes, and what are their properties? (p. 731)

3. What is phagocytosis? Walk through its basic steps. (p. 83)

Pathogens that have moved past our surface barriers next meet our bodies' second line of defense: the cells and proteins of innate immunity, which were introduced in the previous module. We can divide the cells of innate immunity into two types, phagocytic cells and nonphagocytic cells. The phagocytic cells include macrophages, *neutrophils* (NOO-troh-filz), *eosinophils* (ee′-oh-SIN-uh-filz), and dendritic cells. You have read about the process by which cells ingest particles and other cells, called **phagocytosis** (fay′-goh-sy-TOH-sis)—you may wish to review this process by looking back at Figure 3.12 before continuing. The nonphagocytic cells include NK cells and basophils.

Phagocytes

Certain granulocytes and agranulocytes can function as phagocytes. For example, the agranulocytes known as monocytes exit the bloodstream and take up residence in various tissues. After they do so, they enlarge and undergo a variety of other changes to become macrophages, which are very active phagocytes. Some macrophages remain fixed in certain tissues, whereas others roam freely in the body.

Macrophages are activated by a variety of stimuli, including certain substances present on pathogens, chemicals secreted by damaged cells, and signals from the cells of adaptive immunity. Activated local macrophages are generally the first cells to respond to a cellular injury, where they ingest other cells and cellular debris, as shown here:

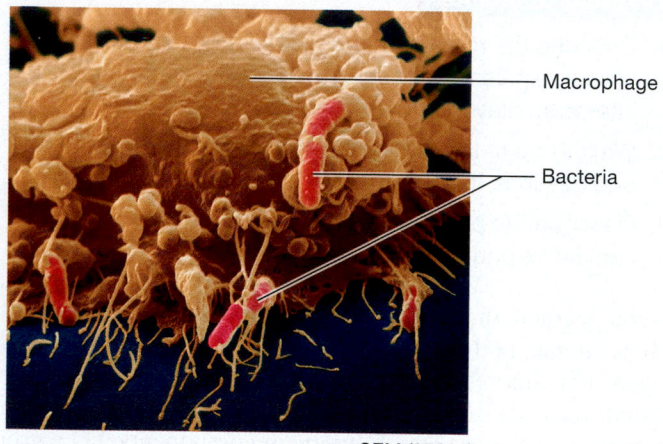

SEM (2700×)

They kill pathogens they have ingested with chemicals, including hydrogen peroxide and hypochlorous acid (the active component in bleach). Macrophages also have *cytotoxic* effects, meaning that they can secrete these substances onto pathogens that are too large to ingest. In addition, they can function as **antigen-presenting cells (APCs),** cells that display portions of the pathogens (antigens) they ingest on their plasma membranes. T cells respond to these antigens by becoming activated. Activated T cells in turn secrete substances that increase the activity of the macrophages, in an example of a positive feedback loop. This is one of many instances of innate and adaptive immunity working together.

The most predominant granulocyte is the **neutrophil.** Like macrophages, neutrophils are highly effective phagocytes. When they are activated, their oxygen consumption increases, a phenomenon called the *respiratory burst*. This allows them to create chemicals such as hydrogen peroxide, hypochlorous acid, and hydroxide anions, which kill ingested pathogens. Neutrophils can ingest many types of cells, but are particularly effective at destroying bacterial pathogens. Neutrophils generally reside in the blood and must be recruited to damaged tissues by chemical signals.

The granulocytes known as **eosinophils** are phagocytes that can migrate from the blood to the tissues where they are needed. However, their phagocytic functions appear to be less critical than those of neutrophils, and they are primarily involved in responses to parasitic pathogens. When a person is infected with a parasite, eosinophils cover it and release the contents of their granules. The chemicals from the granules damage the parasite and either destroy it or make it easier for other immune cells to destroy.

The final group of phagocytic cells is the dendritic cells. Like macrophages, these cells function as antigen-presenting cells; indeed, dendritic cells are the most important APCs in the immune response. The substances they ingest are presented to T cells (and, to a lesser extent, B cells), which are then activated.

Other Cells of Innate Immunity

The first group of nonphagocytic cells we'll cover is the natural killer (NK) cells, introduced in the previous module. NK cells are lymphocytes with the remarkable ability to recognize cancerous cells and cells infected with certain viruses in spite of the fact they cannot recognize antigens. They appear to do this by scanning the cells for irregularities and by reacting to cells that have bound to antibodies (another example of innate and adaptive immunity working together). NK cells are also cytotoxic, releasing substances that destroy their target cells. In addition, they secrete an antimicrobial cytokine that activates macrophages and enhances phagocytosis.

A relatively rare cell that is generally nonphagocytic is the **basophil.** Basophils are granulocytes whose granules contain chemicals that mediate inflammation (*inflammatory mediators*). Basophils are located primarily in the blood, although a related type of cell called the **mast cell** is located in mucous membranes. Like regular basophils, mast cells contain granules with chemicals that trigger inflammation, particularly that involved in allergic responses.

Quick Check

☐ 1. Match the following cell types with their correct definition.

____ Macrophage	a. Granulocyte that attacks parasites
____ NK cell	b. Granulocyte that secretes inflammatory mediators
____ Eosinophil	
____ Dendritic cell	c. Highly active phagocytic agranulocyte
____ Neutrophil	d. Highly active phagocytic granulocyte
____ Basophil	e. Cell that is cytotoxic to cancerous cells and to cells infected with certain viruses
	f. Cell found in lymphoid organs that has long processes and acts as an antigen-presenting cell

☐ 2. Which of the cells in question 1 do not directly destroy a pathogen or abnormal cell?

Antimicrobial Proteins

In addition to cells, the innate immune response is mediated by a variety of plasma **antimicrobial proteins,** including complement proteins and several types of cytokines. In this section we examine the roles of these proteins and how they are produced and activated. As with most of the topics we have discussed in this module, note that complement and cytokines have roles in adaptive immunity, as well, and are revisited later.

Complement

The group of proteins collectively known as the **complement system** consists of around 30 plasma proteins that are produced primarily by the liver. Note that complement proteins are designated with a "C" and a number.

Complement proteins circulate primarily in their inactive forms and must be activated by a complex cascade of events mediated by enzymes. Note in **Figure 20.10** that there are three main series of enzymatic reactions, or pathways, by which complement proteins may be activated: (1) the *classical pathway*, which begins when inactive complement proteins bind to antibodies bound to antigen; (2) the *lectin pathway*, which is initiated when proteins called *lectins* bind to carbohydrates on the surface of microbes; and (3) the *alternative pathway*, which begins at the cleavage of an inactive complement protein called C3 into its active form C3b. This can also occur when inactive complement proteins encounter foreign cells such as bacteria. The three pathways converge when C3b is activated, which in turn cleaves the inactive protein C5 into its active component C5b.

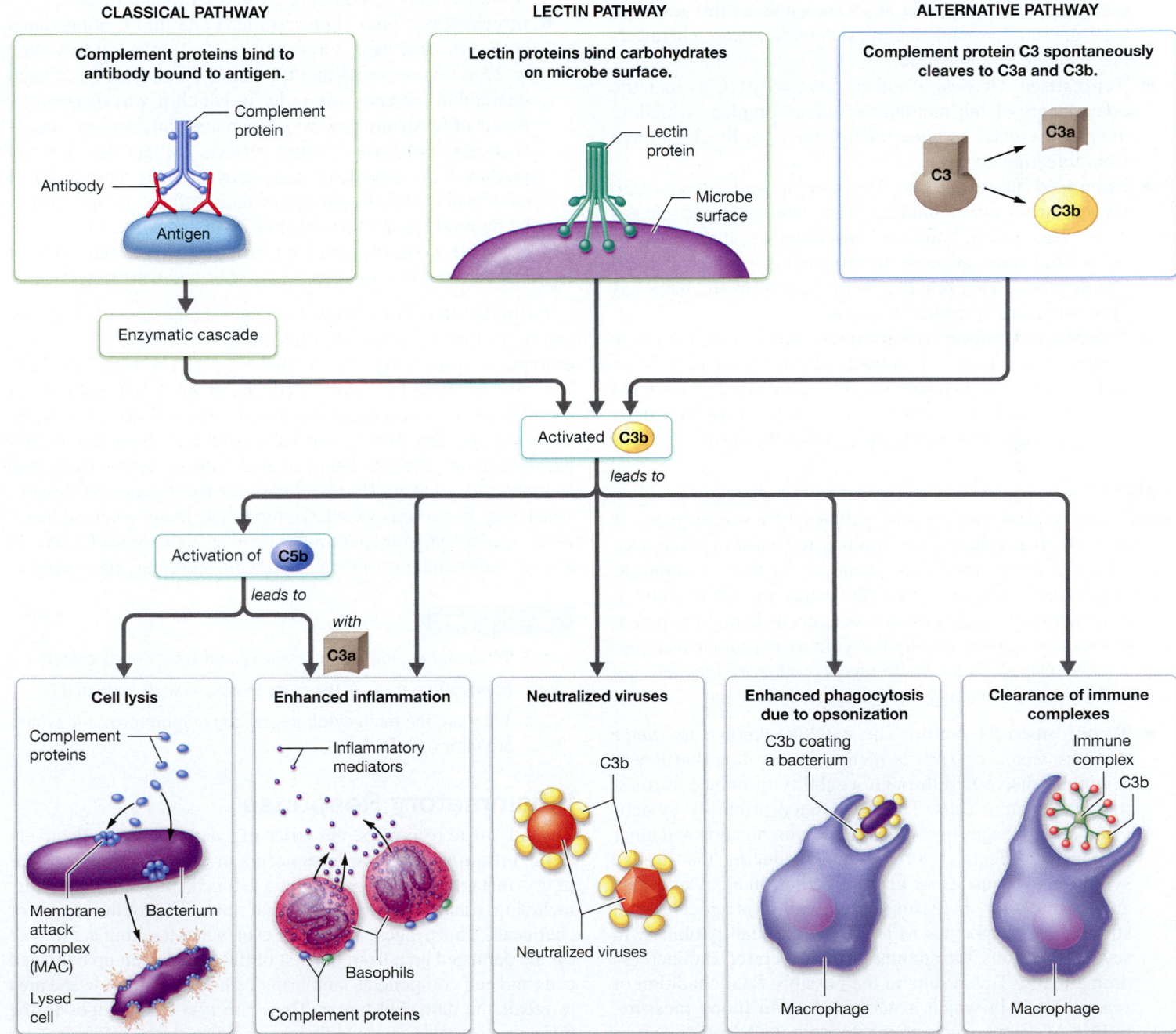

Figure 20.10 **Pathways for activation of the complement system.**

Activated complement proteins lead to the following main effects, illustrated in Figure 20.10:

- **Cell lysis.** Some complement proteins are able to *lyse,* or "pop," the plasma membranes of pathogens, leading to their destruction. This is mediated by C5b, which binds to the surface of a pathogen and provides a docking site for several other activated complement proteins. Together these complement proteins form a structure collectively known as the **membrane attack complex,** or **MAC,** which inserts itself into the plasma membrane of the target cell, creating a pore that causes it to lyse.

- **Enhanced inflammation.** The *inflammatory response* is a nonspecific response to cellular injury. Several complement proteins, particularly C5a and C3a, enhance this response by triggering basophils and mast cells to release chemicals that mediate inflammation.

- **Neutralized viruses.** Another function of C3b and the components of the membrane attack complex is binding to certain viruses and *neutralizing* them, or blocking them from infecting host cells.

- **Enhanced phagocytosis.** The complement protein C3b acts as an *opsonin,* binding both pathogens and phagocytes. This action, known as **opsonization,** makes phagocytes bind more strongly to the pathogen and enhances phagocytosis. This function is evident from the name, as opsonin means "prepare for eating."

- **Clearance of immune complexes.** C3b also binds to *immune complexes*—clusters of antigens bound to antibodies—and triggers their phagocytosis. This clears these complexes from the circulation, which is critical to preventing them from lodging in different tissues around the body.

Cytokines

Recall that cytokines are proteins produced by several types of immune cells that enhance the immune response in some way. Interestingly, many cytokines induce "flu-like" symptoms, including fever, chills, and aches (the aches are due to stimulation of inflammation). Although one tends to blame the pathogen for these symptoms, usually it is your own immune response that actually causes them. There are several cytokines that are involved in innate immunity, including the following:

- **Tumor necrosis factor.** The cytokine known as *tumor necrosis factor,* or *TNF,* is named for the fact that it was originally discovered through its ability to induce necrosis (death) of tumor cells. TNF is secreted primarily by activated macrophages in response to certain bacteria and other pathogens. Effects of TNF include inducing the flu-like symptoms we just described, attracting phagocytes to the area of infection, increasing the activity of phagocytes, and stimulating phagocytes to release additional cytokines. In severe infections, the amount of TNF secreted can increase dramatically. This results in the possibly fatal condition of *septic shock,* in which a marked drop in blood pressure, failure of organs such as the kidneys and liver, inappropriate blood clotting, and a severe drop in blood glucose concentration seriously disrupt homeostasis.

- **Interferons.** The cytokines known as **interferons** (in-ter-FEER-onz), abbreviated **IFN,** are produced by a variety of cells, including macrophages, dendritic cells, NK cells, and cells of adaptive immunity. Interferons are generally produced in response to infection with intracellular agents such as viruses or intracellular bacteria, and are named for their capacity to "interfere" with the ability of the pathogens to infect other cells. One of their primary actions is to inhibit viral replication inside host cells. In addition, they activate various components of both innate and adaptive immunity, such as stimulating nearby cells to produce antimicrobial proteins. Interferons are actually partly responsible for producing some of the symptoms associated with flu-like illnesses, including muscle aches and fever.

- **Interleukins.** *Interleukins* (in-ter-LOO-kinz), abbreviated as *IL* followed by a number (e.g., IL-2), constitute a class of 29 cytokines (note that the number in the name of each interleukin indicates the order in which it was discovered, not its order in any process). The name "interleukin" means "between leukocytes," which reflects the fact that they are produced by leukocytes and many of their actions affect other leukocytes. Examples of these effects include stimulating production of neutrophils by the bone marrow, stimulating NK cells, triggering the production of certain types of interferons from different leukocytes, and activating T cells.

Cytokines have been studied for their therapeutic uses in conditions such as persistent viral infections and certain cancers. For example, a specific type of interferon is used to treat infections like hepatitis B and C (viral infections of the liver) and cancers like melanoma (cancer of the melanocytes in the skin). The adverse effects of cytokine therapy revolve around their induction of flu-like symptoms. Patients being treated with cytokines often feel as though they have a flu-like illness for the duration of therapy, which may last several months or more. The many potential therapeutic uses of cytokines have made them an active area of research, and our understanding of them is growing and changing rapidly.

Quick Check

☐ 3. What is the complement system, and what are its roles?

☐ 4. In what two ways is the complement system activated?

☐ 5. What are the main cytokines of innate immunity, and what are their roles?

Inflammatory Response

Like all innate responses, the series of events known collectively as the **inflammatory response** occurs in reaction to any cellular injury. Inflammation begins when a cell is damaged by anything, including trauma, bacterial or viral invasion, toxins, heat, or chemicals. This triggers a series of events that together act to wall off the damaged area from the rest of the body, clean up damaged cells and cell components, and bring cells and proteins to the area to repair the damaged tissue. There are two basic stages to the inflammatory response: (1) Damaged cells release inflammatory mediators that cause local changes in the damaged tissue; and (2) phagocytes arrive at the area and clean up the damaged tissue.

Part 1: Release of Inflammatory Mediators and the Cardinal Signs of Inflammation

As shown in **Figure 20.11**, after tissue damage occurs (such as that caused by the splinter in step ①), the inflammatory response begins when the damaged cells and nearby mast cells release inflammatory mediators (step ②). These mediators can include histamine, serotonin, cytokines, a peptide called *bradykinin* (brayd-ee-KY-nin), and a group of related lipids known as *prostaglandins* (prahs-tuh-GLAN-dinz) and *leukotrienes* (loo-koh-TRY-eenz). Prostaglandins and leukotrienes are important targets for medications that block the inflammatory response, discussed in *A&P in the Real World: Anti-inflammatory Medications* on p. 770. In addition, certain activated complement proteins trigger the release of inflammatory mediators from cells such as basophils and mast cells, and act to a limited degree as inflammatory mediators themselves.

Inflammatory mediators trigger several effects, which you have almost certainly seen first-hand on occasion (shown in step ③): The injured area becomes red and swollen, feels warm to the touch, and hurts. These four signs—redness, heat, swelling (or *edema;* eh-DEE-muh), and pain—are known as the *cardinal signs of inflammation,* and they are due to the following actions of inflammatory mediators (see Figure 20.11):

● **Vasodilation of arterioles.** Inflammatory mediators such as histamine and bradykinin are potent *vasodilators*—they relax the smooth muscle in arterioles supplying the damaged tissue, causing the vessels to "open up" and blood to flow through them more freely. This results in increased blood flow to the damaged tissue, and the area becomes congested with blood, a condition known as *hyperemia* (hyper- = "above" or "elevated," -*emia* = "blood"). Hyperemia accounts for the redness and heat that accompany inflammation (because blood is warmer than the body's surface).

● **Increased capillary permeability.** The structure of most capillaries prevents them from leaking all of their contents to the interstitial fluid. Inflammatory mediators increase the permeability, or "leakiness," of local capillary beds. This allows protein-rich fluid to leak from the blood vessels into the tissue spaces, and leads to the cardinal sign of

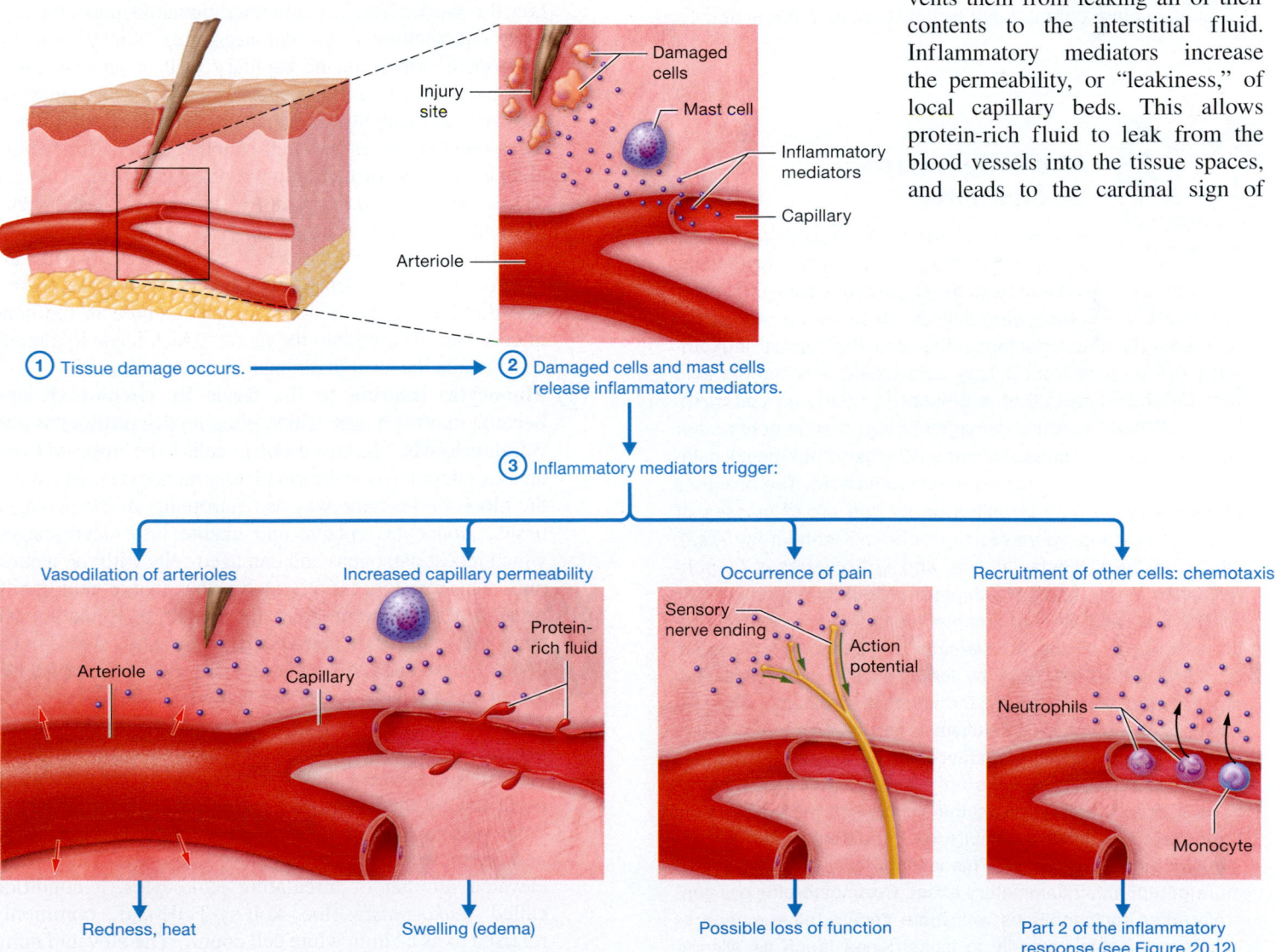

Figure 20.11 The inflammatory response, part 1: effects of inflammatory mediators.

swelling. The proteins in the fluid include clotting proteins such as fibrinogen, complement proteins, and proteins needed for tissue repair.

- **Occurrence of pain.** Many of the inflammatory mediators, particularly bradykinin and prostaglandins, trigger action potentials in the peripheral processes of sensory neurons. This leads to the extremely important and underappreciated effect of *pain,* which serves numerous protective functions. Pain lets us know when our tissues are being damaged so that we can try to avoid further damage. In some cases, it also causes a temporary loss of function—you're unlikely to run on your broken ankle when it's throbbing—which allows the body to repair the damage.
- **Recruitment of other cells: chemotaxis.** The final effect of inflammatory mediators is the recruitment of leukocytes to the damaged area through a process known as **chemotaxis** (kee-moh-TAK-sis; *-taxis* = "movement"). Inflammatory mediators attract and activate a number of leukocytes, particularly macrophages and neutrophils. Note that as this is occurring, complement proteins are also being activated.

Anti-inflammatory Medications

Pain is one of the more unpleasant effects of the inflammatory response. Fortunately, many medications have been developed to reduce inflammation and the accompanying pain by blocking the production of prostaglandins. Prostaglandins and the related leukotrienes are derived from a fatty acid called arachidonic acid (ah-rak′-ih-DAHN-ik) that is present in nearly all cell membranes. When the cell is damaged or triggered in some other way, the enzyme *phospholipase A2* (fos-foh-LY-payz) catalyzes a reaction that cleaves arachidonic acid. The products of this reaction may be acted on by two broad classes of enzymes: *cyclooxygenases* (sy-kloh-AWK-seh-jen-ay′-zez), which produce prostaglandins, and *lipoxygenases* (ly-poh-AWK-seh-jen-ay′-zez), which produce leukotrienes.

The first, and largest, group of anti-inflammatory medications consists of the *nonsteroidal anti-inflammatory drugs,* or *NSAIDs.* NSAIDs work by inhibiting the cyclooxygenase enzyme and preventing the formation of prostaglandins. An example of a common anti-inflammatory medication available over the counter is *ibuprofen* (aye-byoo-PROH-fen).

The *corticosteroids* make up a second group of medications. Corticosteroids mimic the actions of the body's own hormone cortisol, which inhibits the formation of both prostaglandins and leukotrienes. This leads to a wider-ranging and more potent anti-inflammatory effect. Corticosteroids are generally used for conditions with more severe inflammation or inflammation due primarily to leukotrienes (such as allergy-related inflammation). Examples of common corticosteroids include *cortisone* and *prednisone.*

Part 2: Phagocyte Response

The second component of the inflammatory response deals with the phagocytes that ingest pathogens and cellular debris. We can divide the arrival and activation of phagocytes into stages, based on which phagocytes enter the area and the processes occurring there. The stages, shown in **Figure 20.12**, proceed as follows:

① **Local macrophages are activated.** Within minutes of a cellular injury, macrophages already present in the tissue enlarge and begin to phagocytize pathogens and damaged cells. Though few in number, these "first responders," the only phagocytes present within the first hour or so of the inflammatory response, perform the critical function of containing invading pathogens.

② **Neutrophils migrate by chemotaxis to the damaged tissue and phagocytize bacteria and cellular debris.** After several minutes, neutrophils from the blood migrate to the damaged tissue. Inflammatory mediators and activated complement proteins attract neutrophils and enable them to leave the blood and enter the tissue. Notice in Figure 20.12 how this works: The inflammatory mediators make the capillary endothelium in the damaged area "sticky," and the neutrophils adhere to the capillary wall, a process called *margination.* The inflammatory mediators also increase capillary permeability, as we discussed earlier, which provides enough spaces between endothelial cells for neutrophils to squeeze through into the damaged tissue. This is a process known as *diapedesis* (dy′-uh-peh-DEE-sis). Once the neutrophils are in the tissue, they then begin to destroy bacteria and other cellular debris. Note that neutrophils are able to ingest only a single microbe before they themselves also die. During this time, neutrophils stored in the bone marrow are released into the blood, which leads to a rapid, acute rise in the level of circulating neutrophils.

③ **Monocytes migrate to the tissue by chemotaxis and become macrophages, which phagocytize pathogens and cellular debris.** The next group of cells to be attracted to the area by chemotaxis is the circulating monocytes, which exit the blood in the same way as neutrophils. In the damaged tissue, monocytes enlarge and mature into macrophages, which ingest pathogens and damaged cells with the neutrophils. This response is much slower than that of neutrophils, because it takes several hours for monocytes to mature and there are far fewer monocytes than neutrophils stored in the bone marrow and the circulation. However, monocytes live longer and are more aggressive than neutrophils.

④ **The bone marrow increases production of leukocytes, leading to leukocytosis.** The final response occurs in the bone marrow, where leukocytes are produced. Cytokines produced by the activated phagocytes act on cells in the bone marrow to increase the production of neutrophils and monocytes over the next 3–4 days. This leads to an elevated number of circulating leukocytes, a condition called *leukocytosis* (loo′-koh-sy-TOH-sis), commonly referred to as a "high white cell count." The elevated numbers of leukocytes allow the damaged area to be cleared and any pathogens removed so that cells such as fibroblasts can begin the process of healing.

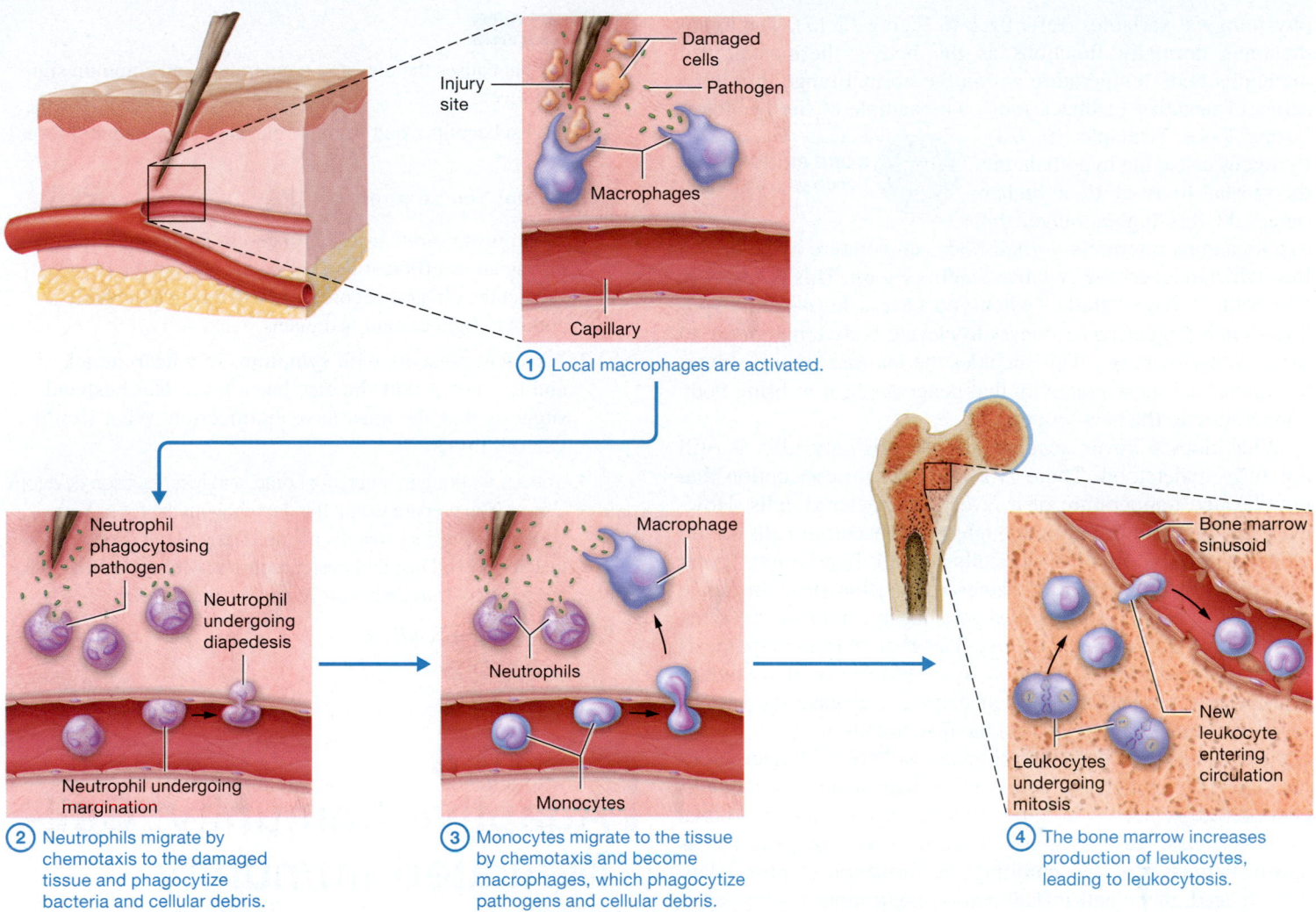

Figure 20.12 **The inflammatory response, part 2: phagocyte response.** This response involves the processes of margination, diapedesis, chemotaxis, and phagocytosis.

The neutrophils and macrophages in the damaged tissue go to work digesting pathogens and damaged cells. Eventually, all of the neutrophils and many of the macrophages die themselves. The accumulation of dead leukocytes, dead tissue cells, and fluid leads to a whitish mixture known as **pus** (a wound filled with pus is said to be *purulent* [PYOOR-uh-lent]). It is common to associate pus with infection; however, pus indicates only that an inflammatory response took place. Even so, much more pus is typically found with inflammation due to certain bacterial infections than would be produced by other types of cellular injury. Indeed, certain bacterial species, known as *pyogenic bacteria*, are known for producing quite purulent wounds. The pus is generally absorbed into the surrounding tissues and lymphatic vessels when healing has completed.

Quick Check

☐ 6. What are the four cardinal signs of inflammation?

☐ 7. What are inflammatory mediators, and what do they do?

☐ 8. Explain the role of phagocytes in the inflammatory response.

Fever

« FLASHBACK

1. Which part of the brain is responsible for temperature homeostasis? (p. 441)

2. What effects are triggered when this part of the brain senses that body temperature is too low or too high? (p. 441)

One of the most recognizable, and unpleasant, signs of illness is fever. **Fever** is defined simply as a body temperature above the normal range, which is generally between 36 and 38° C (or 97 and 99° F); an individual with a fever is referred to as *febrile* (FEB-ry'l). Many conditions can cause fever, but in this section we primarily focus on those related to the immune system.

Like inflammation, fever is an innate response to cellular injury. Fever is initiated when chemicals called **pyrogens** (PY-roh-jenz; *pyro-* = "fire," *-gen* = "causing") are released from damaged cells or certain bacteria. Pyrogens act on the hypothalamus, the small component of the diencephalon (the central core of the brain) that regulates the homeostasis of many regulated physiological variables (refer back to Figure 12.17). The hypothalamus normally functions as the body's thermostat, and maintains body temperature within the normal range through a series of negative feedback loops, an example of the Feedback Loops Core Principle (p. 28). Pyrogens cause the hypothalamic thermostat to reset to a higher range. At this higher range, the

CORE PRINCIPLE
Feedback Loops

hypothalamus interprets *normal* body temperature as being too low, which triggers the negative feedback loop. This is why you feel cold, or have "chills," when you have a fever—the hypothalamus is triggering responses to elevate body temperature to the new, higher range. This includes the familiar sign of shivering, increased muscle activity that generates heat to bring body temperature to the new set point.

What does a fever accomplish? Surprisingly, this is still not fully understood. There is a common misconception that the elevated temperature of a fever kills bacterial cells. However, any temperature high enough to kill bacterial cells would almost certainly kill all of our cells as well. In addition, fevers can occur in the absence of bacterial infection (it is an innate response, after all), so this idea just doesn't hold water. What good, then, does a fever do us? The best current hypothesis is that our phagocytes function more efficiently at a slightly higher body temperature, which helps to increase the rate of recovery from whatever caused the inflammation.

Fevers may be alleviated, or made to "break," either spontaneously or through the action of certain medications. Medications that reduce an inflammation-mediated fever are known as *antipyretics*. Most antipyretics work in the same manner as anti-inflammatories, by inhibiting the formation of prostaglandins. Indeed, many anti-inflammatory medications such as aspirin and ibuprofen are also antipyretics. Regardless of what makes the fever break, it results from the hypothalamus being reset back to the normal temperature range. This causes the hypothalamus to sense the febrile temperature as being too high, and it triggers negative feedback mechanisms to lower the body's temperature. These include sweating and dilation of the blood vessels serving the skin, which makes the skin appear red or flushed. Both conditions make you feel hot, although your body temperature is actually decreasing.

From a clinical perspective, fever is a critical warning sign of inflammatory processes occurring somewhere in the body. Although we tend to associate fevers with infection, remember that fever is a nonspecific response and can tell you only that widespread inflammation is present. This is evidenced by the fact that fevers may be seen in the absence of any infectious pathogens following severe trauma, burns, or even childbirth. It is also important to remember that occasionally fevers are not related to inflammation. Conditions such as heat stroke, certain brain diseases, and reactions to medications can cause fevers, and these fevers will not reduce with administration of anti-inflammatory medications. We discuss these and other causes of elevated body temperature in the metabolism chapter (see Chapter 23).

Table 20.1 summarizes the components of the first and second lines of defense.

Quick Check

☐ 9. What causes the elevated temperature and common signs of fever?

☐ 10. What benefits result from the elevated temperature of fever?

Apply What You Learned

☐ 1. *Neutropenia* (noo'-troh-PEE-nee-uh) is the condition of having an insufficient level of neutrophils in the blood. Predict the effect this condition will have on the body's ability to fight certain pathogens.

☐ 2. A patient presents with symptoms of a heart attack, and it is noted that she also has a fever. Her husband suggests that she must have an infection. What would you tell him?

☐ 3. You are working in a nursing home, and Mrs. Jackson develops a fever. You prepare to call her doctor about the fever. Your co-worker suggests that it's not necessary to call the doctor, and that you should just administer aspirin to Mrs. Jackson to make her fever go away. What do you tell your co-worker?

See answers in Appendix A.

MODULE 20.4
Adaptive Immunity: Cell-Mediated Immunity

Learning Outcomes

1. Explain the differences between antigens, haptens, antigenic determinants, and self antigens.

2. Describe the processes of T lymphocyte activation, differentiation, and proliferation, including the roles of antigen-presenting cells.

3. Compare and contrast the classes of T lymphocytes.

4. Describe the two types of major histocompatibility complex antigens, and explain their functions.

5. Describe the purpose of immunological memory, and explain how it develops.

6. Describe the process by which a transplanted organ or tissue is rejected, and explain how this may be prevented.

The first arm of the adaptive immune system is cell-mediated immunity. As we discussed, cell-mediated immunity involves the different classes of T cells, including the **helper T (T_H) cells,** also known as **CD4 cells,** and the **cytotoxic T (T_C) cells,** also known as **CD8 cells.** ("CD" stands for "cluster of differentiation," which refers to specific molecules within the plasma membrane that differentiate one cell type from another.) These cells respond primarily to cells infected with intracellular pathogens (viruses and intracellular bacteria), cancer cells, and foreign cells such as those from a transplanted organ.

Table 20.1 Summary of the First and Second Lines of Defense

Component	Description	Function(s)
First Line of Defense: Surface Barriers		
Skin	Stratified squamous, keratinized epithelium with deeper connective tissue; contains sebaceous and sweat glands	Provides hard, continuous external layer that protects from pathogens and mechanical trauma; secretions deter microbial growth
Mucous membranes	Epithelial membranes that line all surfaces that open to the outside of the body; contain goblet cells that secrete mucus	Continuous surface lined with sticky mucus that traps pathogens and prevents their interaction with deeper cells
Secretions	Sebum produced by sebaceous glands; mucus produced by mucous membranes (mucosae); acid secreted by the stomach; defensins produced by the skin and mucosae	Kill or deter the growth of pathogens; trap pathogens and debris
Normal flora	Mostly bacterial species that live on skin and mucosae, generally without causing disease	Competing with pathogenic bacteria for space and resources; secreting substances that kill and/or deter the growth of other bacteria
Second Line of Defense: Internal Defenses		
Cells		
Phagocytes	Includes macrophages, neutrophils, eosinophils, and dendritic cells	Ingest pathogens, damaged cells, and cellular debris; release toxic substances onto pathogens; present antigens to T cells
NK (natural killer) cells	Nonphagocytic cells related to lymphocytes	Target and destroy certain cancer cells and certain virally infected cells
Basophils and mast cells	Nonphagocytic cells with granules that contain inflammatory mediators	Mediate the inflammatory response
Antimicrobial Proteins		
Complement	Group of around 30 plasma proteins produced by the liver	Mediate cell lysis; enhance inflammation; bind and neutralize viruses; enhance phagocytosis through opsonization; trigger clearance of immune complexes
Tumor necrosis factor	Cytokine secreted primarily by activated macrophages in response to certain bacteria and other pathogens	Induces death of tumor cells; causes flu-like symptoms; attracts and stimulates phagocytes
Interferons	Cytokine produced by several cell types, generally in response to infection with intracellular pathogens	Block viruses from infecting cells near already infected cells; induce flu-like symptoms
Interleukins	Large class of cytokines produced by several immune cells	Multiple effects including inducing flu-like symptoms, stimulating neutrophil production by the bone marrow, stimulating NK cells and the production of interferon, and activating T cells
Innate Responses to Cellular Injury		
Inflammation	Nonspecific response to cellular injury characterized by vasodilation, increased capillary permeability, pain, and phagocyte mobilization	Attracts immune cells to an area of tissue damage to clean up cellular debris, clear the area of any pathogens, and facilitate healing
Fever	Resetting of the hypothalamic set point for normal body temperature to a higher value by pyrogens	Uncertain purpose; possibly helps phagocytes function more efficiently

In this module, we trace the cell-mediated response, starting with T cell activation and finishing with the effects of T_H and T_C cells. As you read, note not only the differences between the different types of immunity but also the extensive interactions between the cell-mediated, antibody-mediated, and innate immune responses. The module concludes with an examination of how cell-mediated immunity is involved in rejection of transplanted tissues and organs, and what can be done to prevent this rejection.

Antigens

Recall from earlier in the chapter that an antigen is a substance that B or T cells recognize. The name "antigen" comes from the phrase "*anti*body *gen*erating," a reference to their ability to generate the production of antibodies from B cells. Antigens are usually peptides, but they may also be substances such as carbohydrates, lipids, and metals (nickel is a notoriously antigenic metal). These other substances must usually be coupled with a peptide to function fully as an antigen.

Only certain antigens generate a response from the immune system. Those with this capability are called *immunogens* (im-MYOO-noh-jenz; "immune generating"). Antigens present on your own cells, called **self antigens,** are not immunogens in your body. Some very small antigens, called *haptens,* are immunogenic only if they are attached to a protein carrier. An important example of a hapten is urushiol (oo-ROO-shee-ohl), the toxin in poison ivy. When urushiol enters the skin, it is oxidized and the product of this reaction binds to skin proteins, making it immunogenic.

Antigens may be classified as exogenous or endogenous. An *exogenous antigen* is one that originates outside the cell and must be taken into the cell by phagocytosis. An *endogenous antigen*, however, is one of two things: (1) a foreign antigen present on a pathogen that lives inside your cell, such as an intracellular bacterium, or (2) a foreign or self antigen encoded by your DNA. The second scenario includes normal self antigens, foreign cancer antigens made by mutated DNA, and foreign viral antigens (when a virus infects your cell, it inserts its genetic material into your DNA, triggering your cell to synthesize millions of copies of its proteins, including its antigens). It's easy to think of exogenous as meaning "foreign" and endogenous as meaning "self," but don't forget that endogenous antigens can be foreign or self.

T Cell Response to Antigen Exposure

⟪ FLASHBACK

1. What is the function of a lysosome? (p. 93)
2. What are the functions of the rough endoplasmic reticulum? (p. 91)

T cells are formed in the bone marrow, but they leave the bone marrow and migrate to the thymus to mature (**Figure 20.13**). As

T cells mature, they undergo gene rearrangements that lead to a huge variety of genetically distinct T cells. Every distinct T cell has a receptor on its surface, called (appropriately enough) the **T cell receptor** that binds a specific antigen. The unique portion of the antigen to which the receptor binds is known as its *antigenic determinant*. Each population of T cells that can respond to a specific antigen is known as a **clone.** There are millions of different clones in the immune system, but only a few cells of each clone exist in the body at any given time.

Some T cell clones are capable of recognizing and responding to pathogens, whereas others are not. The thymus "screens" these cells and mediates the destruction of those clones that cannot recognize antigens. This ensures that an individual is **immunocompetent,** or able to mount a normal response to foreign antigens. Other T cell clones, known as *self-reactive T cells,* recognize self antigens as foreign and would attack your cells if released into the circulation. These self-reactive T cells are also destroyed, ensuring **self tolerance** and preventing our T cells from attacking our own cells. Over 95% of all T cells die in the thymus in this manner. T cells that do survive this screening are released into the circulation when they mature. These cells, which are known as *naïve* T cells (ny-EEV) because they have not yet encountered their specific antigens, eventually reside in the blood, lymphoid organs, or other lymphatic tissue.

Naïve T cells must bind their specific antigens before they become activated. In these subsections, we see how activation occurs and discuss the molecules required for this process to take place: *major histocompatibility complex (MHC) molecules.*

MHC Molecules and Antigen Presentation

T cells cannot directly interact with an antigen. Instead, they can interact only with pieces of antigen bound to glycoproteins, called **major histocompatibility complex (MHC) molecules**

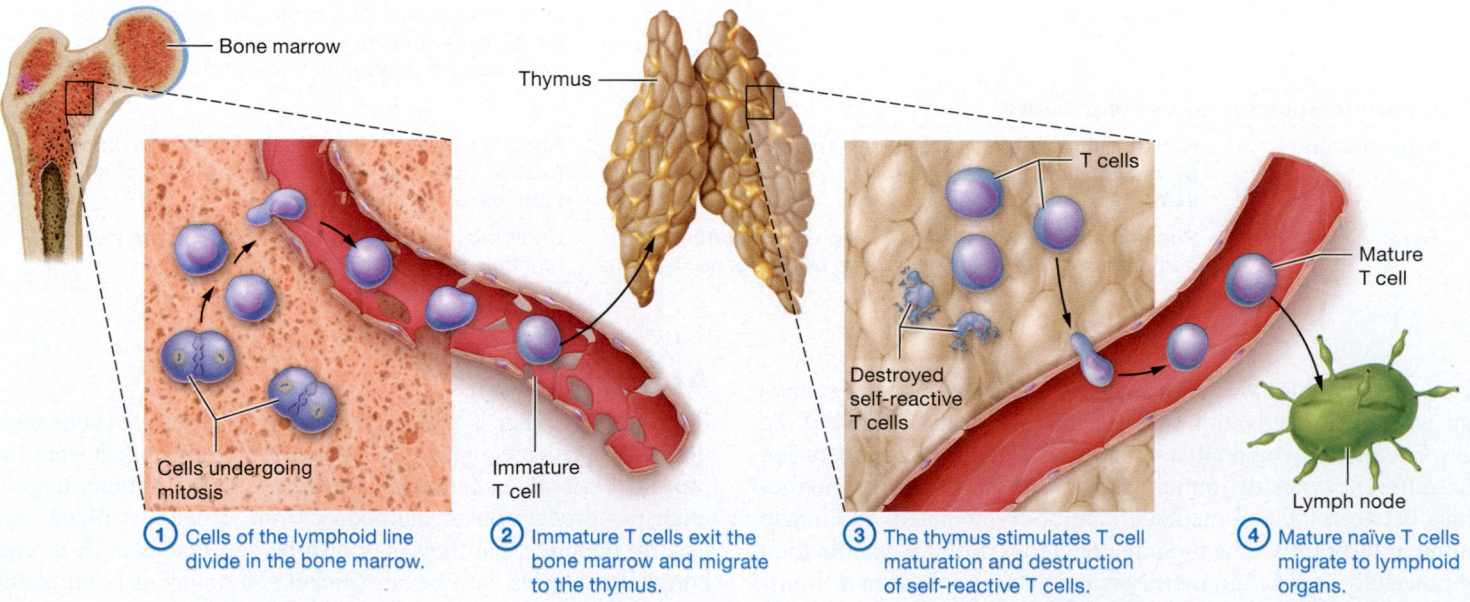

① Cells of the lymphoid line divide in the bone marrow.

② Immature T cells exit the bone marrow and migrate to the thymus.

③ The thymus stimulates T cell maturation and destruction of self-reactive T cells.

④ Mature naïve T cells migrate to lymphoid organs.

Bone marrow

Thymus

Cells undergoing mitosis

Immature T cell

T cells

Destroyed self-reactive T cells

Mature T cell

Lymph node

Figure 20.13 T cell maturation.

(hiss'-toh-kawm-pat-ih-BIL-ih-tee). The name comes from the fact that MHC molecules are major determinants of compatibility among tissue and organ donors and recipients (remember that *histo-* means "tissue"). MHC molecules are found on nearly all nucleated cells. Note, however, that they are not present on the surface of erythrocytes, which is why blood is considerably easier to donate than other organs and tissues.

MHC molecules in essence serve as "docking sites" for specific components of antigens that are then displayed to naïve T cells. There are two types of MHC molecules: **Class I MHC molecules** are found on the surface of the plasma membrane on nearly all nucleated cells, whereas **class II MHC molecules** are found only on the surfaces of antigen-presenting cells, including dendritic cells, macrophages, and B lymphocytes. Naïve cytotoxic T (T_C) cells generally interact only with class I MHC molecules, whereas naïve helper T (T_H) cells generally interact with class II MHC molecules. The two classes of molecules also differ in the types of antigens they display. Class I MHC molecules present endogenous antigens, whereas class II MHC molecules present exogenous antigens.

ConceptBOOST))))

Why Do We Need Both Class I and Class II MHC Molecules?

At this point you may be wondering why we need two types of MHC molecules. The answer lies in the differing functions of T_H and T_C cells. As you've discovered, naïve T_H cells are activated by antigen-presenting cells (APCs) bearing portions of antigen on their class II MHC molecules, as shown here:

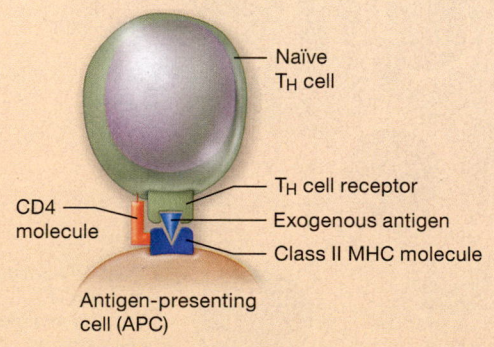

The APCs aren't diseased themselves; rather, they function as a warning system that alerts the T_H cells to a problem somewhere in the body. The T_H cells then stimulate other parts of the innate and adaptive defenses to combat the threat.

Compare this to what happens with a naïve T_C cell. A naïve T_C cell is activated by a *diseased* cell bearing an antigen on a class I MHC molecule, as shown here:

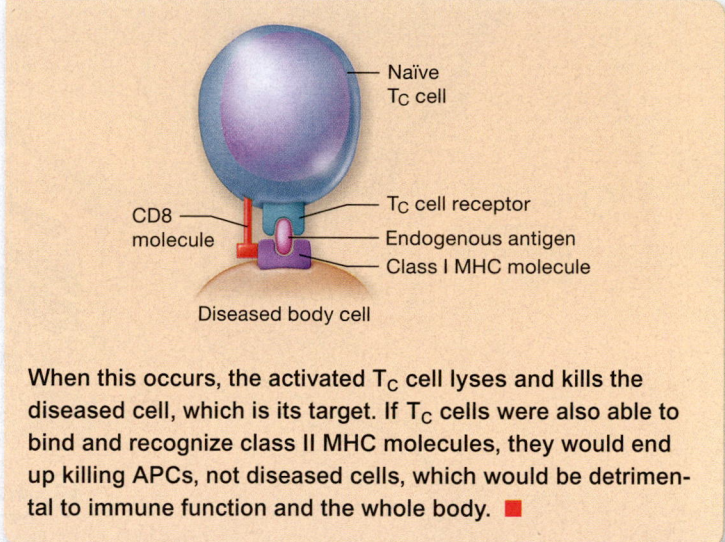

When this occurs, the activated T_C cell lyses and kills the diseased cell, which is its target. If T_C cells were also able to bind and recognize class II MHC molecules, they would end up killing APCs, not diseased cells, which would be detrimental to immune function and the whole body. ∎

The basic steps by which a class I MHC molecule processes and displays an endogenous antigen are as follows (**Figure 20.14a**):

1. The cell synthesizes either a self antigen or a foreign antigen. (Note that some foreign endogenous antigens, such as those from intracellular bacteria, are not synthesized by the cell.)
2. The antigen is broken down by enzymes in the cytosol.
3. An antigen fragment containing the antigenic determinant is transported into the rough endoplasmic reticulum (RER) and is coupled with a class I MHC molecule in the RER membrane.
4. The MHC-antigen complex leaves the RER by a vesicle and is inserted into the cell's plasma membrane.

Now look at **Figure 20.14b** to contrast how exogenous antigens are displayed by class II MHC molecules. The basic steps are as follows:

1. The cell ingests a pathogen by phagocytosis.
2. The phagocytic vesicle fuses with a lysosome; the pathogen is degraded and its antigens are fragmented.
3. The lysosome fuses with a vesicle from the RER that contains class II MHC molecules, and an antigen fragment binds to the MHC molecule.
4. The MHC-antigen complex is inserted into the cell's plasma membrane.

Note that both processes end up with the same result: portions of antigens displayed on the plasma membrane attached to MHC molecules. These MHC-antigen complexes then interact with and activate T cells.

T Cell Activation, Clonal Selection, and Differentiation

Now let's take a look at how naïve T cells interact with MHC molecules to become activated. The process of naïve T cell activation is very similar for T_H and T_C cells, the main difference being the type of MHC molecule involved. For this reason, we

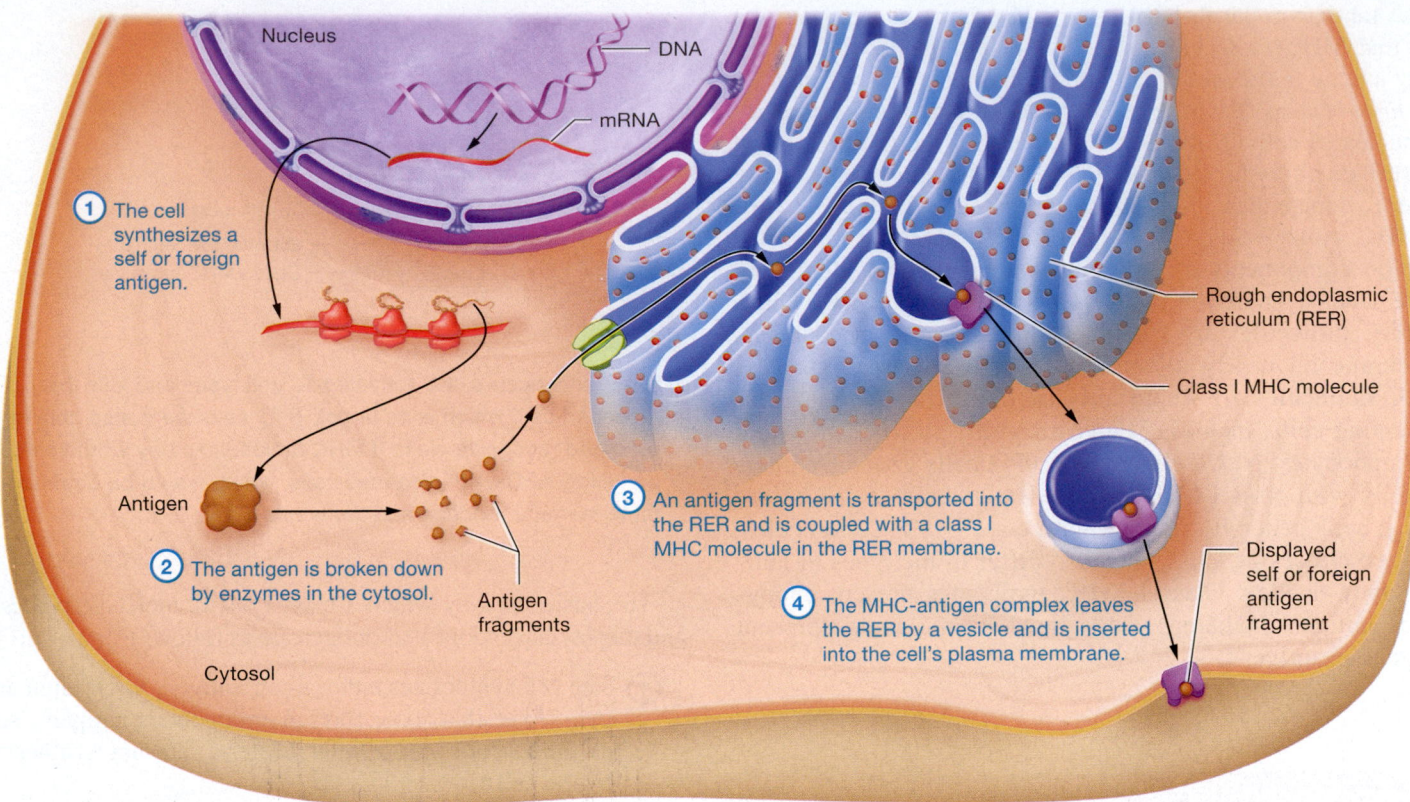

(a) Class I MHC molecules process and display endogenous antigens.

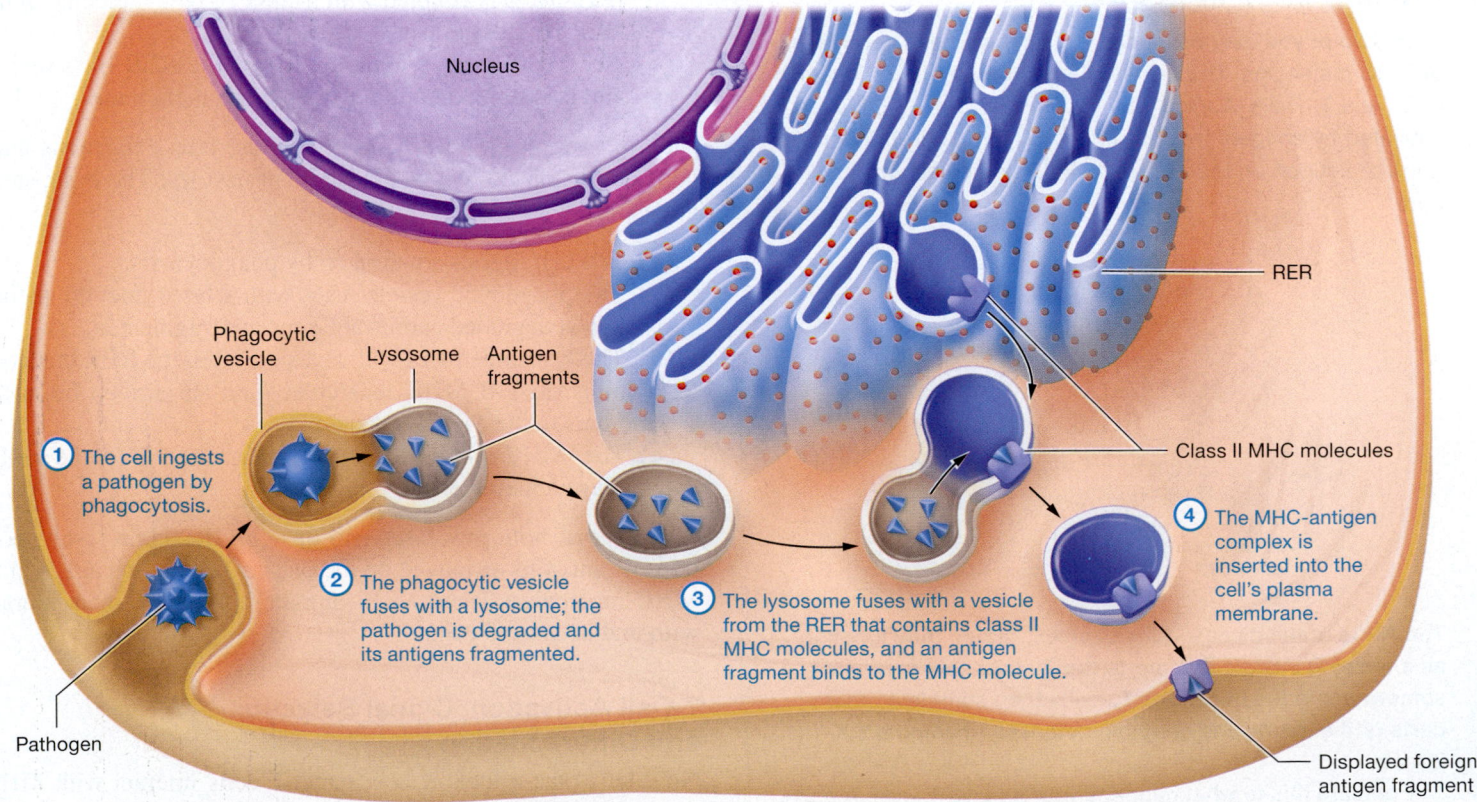

(b) Class II MHC molecules process and display exogenous antigens.

Figure 20.14 **Antigen processing and display by MHC molecules.**

present them both together here. The overall process, shown in **Figure 20.15**, consists of the following steps:

① **Dendritic cells display the antigen fragments on their MHC molecules, and the MHC-antigen complex binds to the receptor of a specific naïve T_H or T_C cell clone.** The process of T cell activation begins with a dendritic cell processing and displaying antigen fragments on its MHC molecules. It then presents the antigen to a particular T cell clone, which is specific for its individual

MHC-antigen complex. When the T cell receptor recognizes and binds this complex, multiple changes are triggered inside the T cell and the process of activation begins. This process is known as **clonal selection** because the antigen "selects" a particular T cell clone.

② **The T_H or T_C cell binds a co-stimulator and becomes activated.** T cell receptors normally have low affinity for their MHC-antigen complexes, which is a protective mechanism that prevents unnecessary naïve T cell activation. This is particularly true of T_C cells, because their

① Dendritic cells display the antigen fragments on their MHC molecules, and the MHC-antigen complex binds to the receptor of a specific naïve T_H or T_C cell clone.

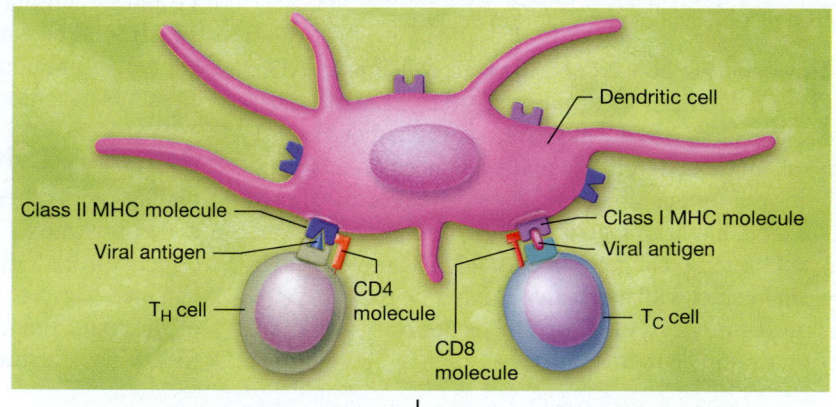

② The T_H or T_C cell binds a co-stimulator and becomes activated.

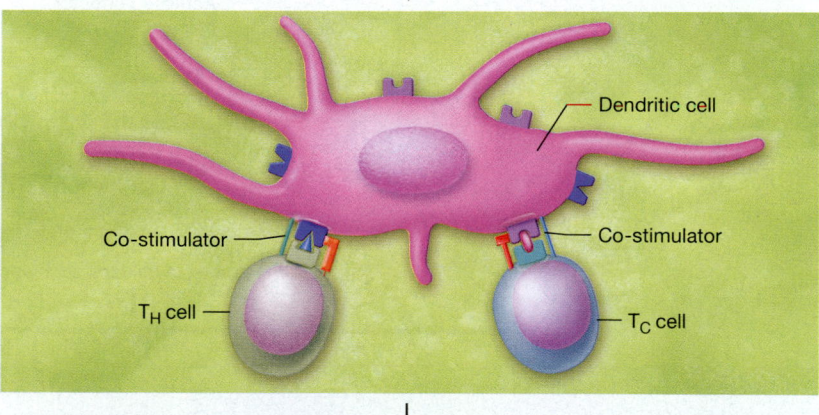

③ The activated T_H or T_C cell clone proliferates and differentiates into effector cells and memory T cells.

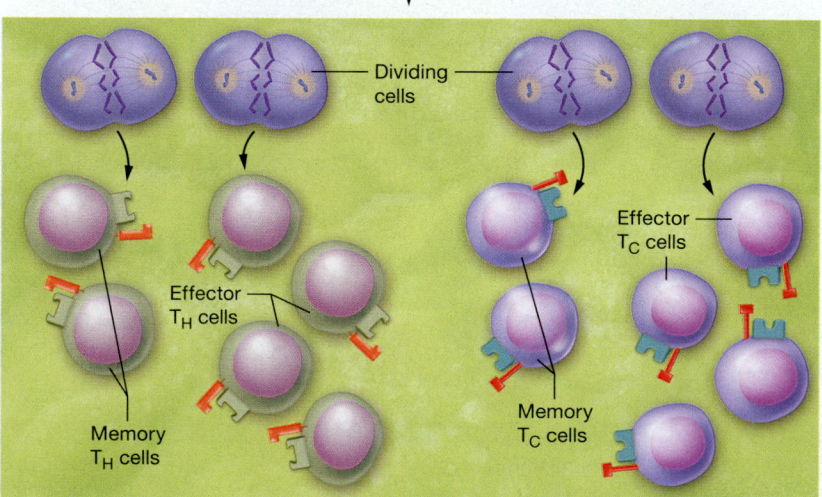

Figure 20.15   T cell activation, clonal selection, and differentiation. Note that T_H and T_C cells interact with different MHC molecules.

effects are so destructive when they are activated. Full T cell activation requires the interaction of the naïve T cell with other molecules on the cells, called *co-stimulators.*

③ **The activated T_H or T_C cell clone proliferates and differentiates into effector cells and memory T cells.** T_H or T_C cells increase in number when activated and differentiate into *effector cells,* or those that cause the immediate effects, and *memory T cells.* **Memory T cells** are responsible for cell-mediated **immunological memory,** in which these cells respond more quickly and efficiently to subsequent exposures to an antigen. One reason for this efficiency is that memory T cells have no need of a co-stimulator, which makes the response considerably more rapid.

You may be wondering how a dendritic cell, which is an APC, can activate a T_C cell. Don't APCs display class II MHC molecules? Well, remember that class I MHC molecules are found on the surface of nearly all nucleated cells, including dendritic cells. And dendritic cells have a unique ability to undergo a process known as *cross-presentation,* in which they are able to display some of the ingested viral and bacterial antigens on class I MHC molecules. In this way, they can interact with and activate naïve T_C cells.

At the end of this process, effector T cells are fully activated. The next section examines their functions.

Quick Check

☐ 1. What are MHC molecules?

☐ 2. How do class I and class II MHC molecules differ?

☐ 3. How are T cells activated?

Effects of T Cells

 FLASHBACK

1. What is apoptosis? (p. 117)

T_H and T_C cells have very different roles, although they do interact and depend on each other to function properly. We now examine the cells' functions, how they interact, and how they work with elements of antibody-mediated adaptive immunity.

Role of T_H Cells

Unlike most other immune cells, helper T cells have no phagocytic or cytotoxic abilities. Instead, T_H cells primarily exert their effects through the secretion of cytokines that then activate and enhance various components of the immune response. Some of the main functions of activated effector T_H cells include the following, illustrated in **Figure 20.16**.

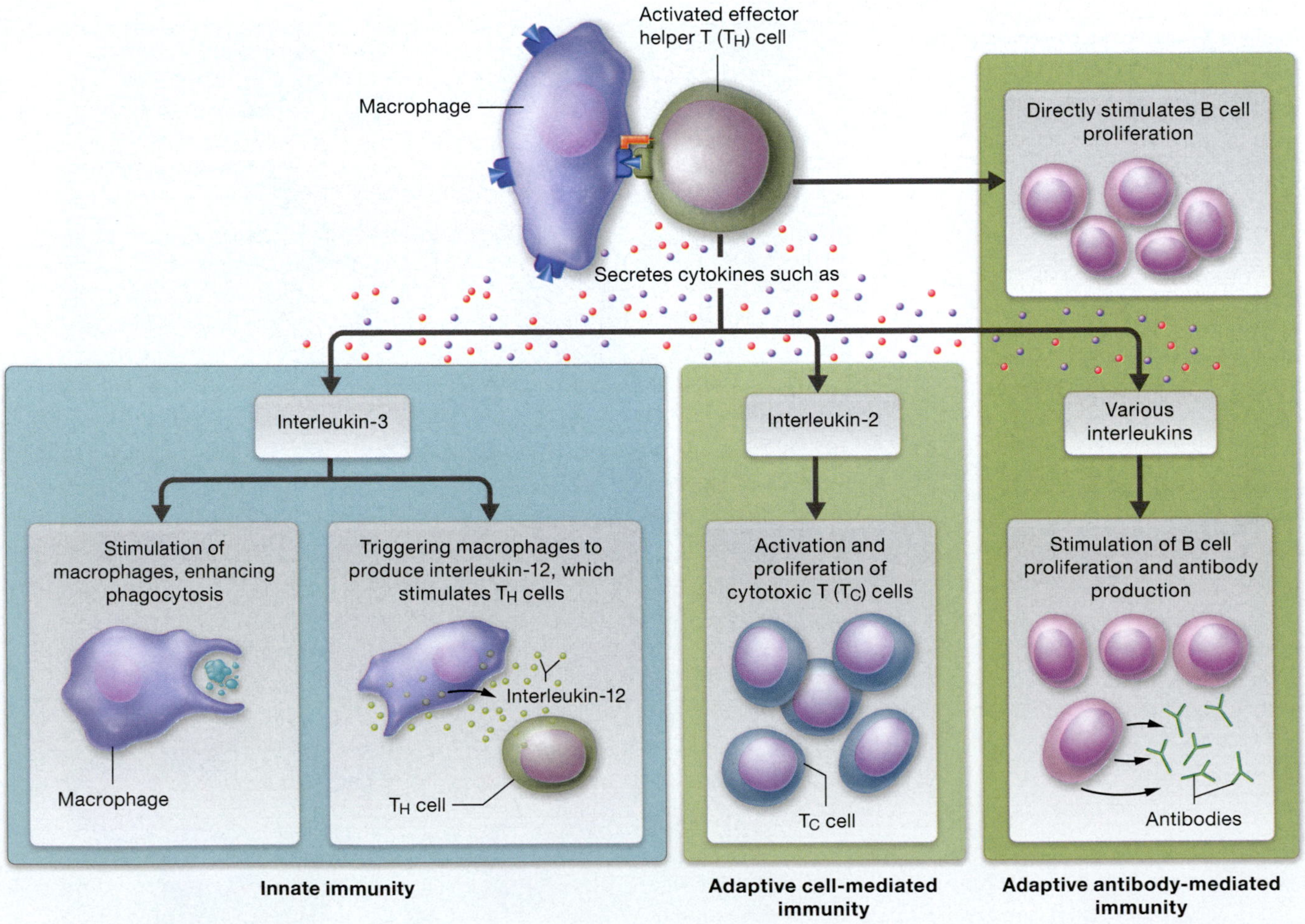

Activated effector
helper T (T_H) cell

Macrophage

Directly stimulates B cell
proliferation

Secretes cytokines such as

| Interleukin-3 | Interleukin-2 | Various interleukins |

Stimulation of macrophages, enhancing phagocytosis

Triggering macrophages to produce interleukin-12, which stimulates T_H cells

Interleukin-12

Macrophage

T_H cell

Activation and proliferation of cytotoxic T (T_C) cells

Stimulation of B cell proliferation and antibody production

T_C cell

Antibodies

Innate immunity

Adaptive cell-mediated immunity

Adaptive antibody-mediated immunity

Figure 20.16 Effects of T_H cells.

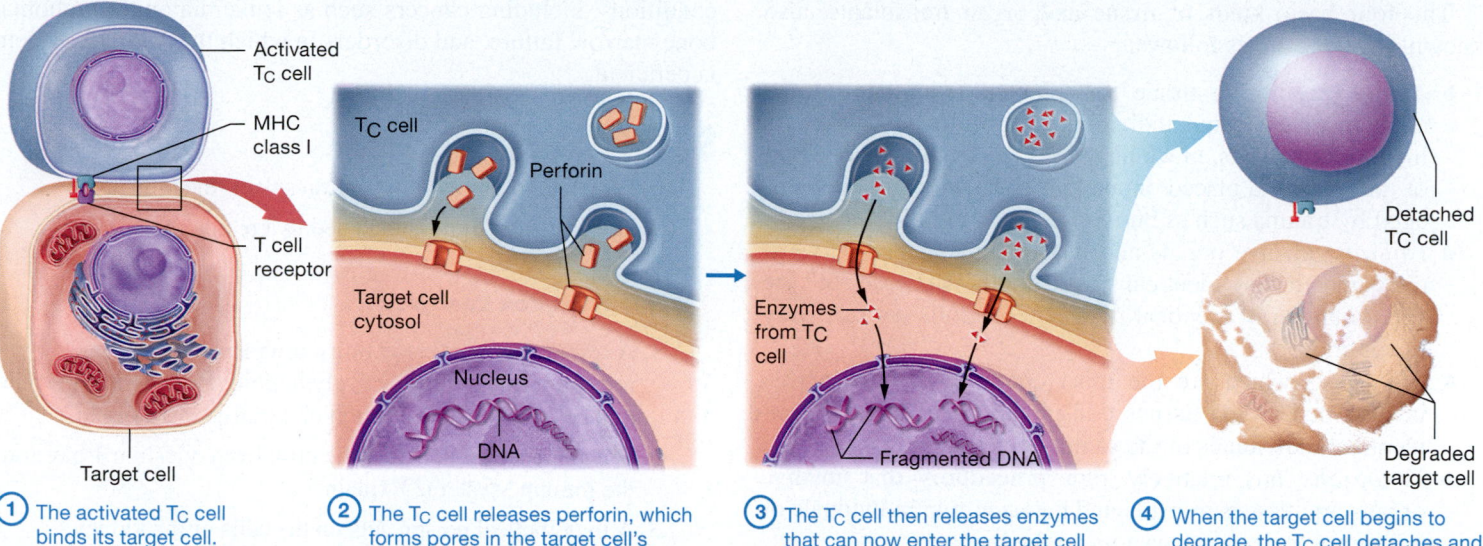

① The activated T$_C$ cell binds its target cell.

② The T$_C$ cell releases perforin, which forms pores in the target cell's plasma membrane.

③ The T$_C$ cell then releases enzymes that can now enter the target cell and fragment its DNA.

④ When the target cell begins to degrade, the T$_C$ cell detaches and searches for a new target cell.

Figure 20.17 Function of T$_C$ cells.

- **Innate immunity: stimulation of macrophages.** T$_H$ cells secrete the cytokine *interleukin-3 (IL-3)*, which stimulates macrophages to become more efficient phagocytes. It also causes macrophages to produce interleukin-12, which stimulates T$_H$ cells to generate more IL-3 in a positive example of the Feedback Loops Core Principle (p. 28).

CORE PRINCIPLE
Feedback Loops

- **Adaptive cell-mediated immunity: activation of T$_C$ cells.** T$_H$ cells secrete the cytokine *interleukin-2 (IL-2)*, which is required to activate T$_C$ cells. Indeed, in the absence of T$_H$ cells and IL-2, most T$_C$ cells fail to activate and become unresponsive to the antigen. IL-2 also stimulates the proliferation of T$_C$ cells.

- **Adaptive antibody-mediated immunity: stimulation of B cells.** T$_H$ cells directly bind to B cells and stimulate them to proliferate and differentiate. They also secrete various interleukins that stimulate B cell proliferation and increase antibody production. T$_H$ cells are so important to B cell activation that they derive their name from this role—they are required to "help" initiate B cell response.

T$_H$ cells are required for normal function of all components of the immune system, including innate, antibody-mediated, and cell-mediated immunity. For this reason, failure of T$_H$ cells to function, which we see with conditions such as acquired immunodeficiency syndrome (AIDS), can lead to failure of the entire immune response. We discuss AIDS and the virus that causes it in Module 20.7.

Role of T$_C$ Cells

The primary function of cytotoxic T cells is evident from their name—they kill other cells, specifically those with foreign antigens bound to class I MHC molecules. Their ability to interact with class I MHC molecules means they can detect abnormalities in any cell type with a nucleus, which is critical for the detection of cancer cells, foreign cells, and cells infected with intracellular pathogens such as viruses and bacteria. T$_C$ cells are activated in the same way as T$_H$ cells, with the addition that they require IL-2 from T$_H$ cells to activate fully. This protective mechanism prevents abnormal T$_C$ cell activation.

As you can see in **Figure 20.17,** ① an activated T$_C$ cell binds its target cell, after which ② it releases a protein called **perforin.** Perforin forms pores in, or perforates, the target cell's plasma membrane. ③ The T$_C$ cell then releases enzymes that can now enter the target cell's cytosol. These enzymes catalyze reactions that degrade target cell proteins and eventually lead to fragmentation of the target cell's DNA and its death. T$_C$ cells also bind to proteins on the plasma membrane of target cells that induce the process of *apoptosis,* or programmed cell death (see Chapter 3). ④ When the target cell begins to degrade, the T$_C$ cell detaches and searches for a new target cell.

Although T$_C$ cells perform critical functions in terms of maintaining an organism's health, they can also cause problems. Certain types of allergies and the rejection of transplanted organs are mediated by T$_C$ cells. We discuss allergies in Module 20.7, and organ transplantation next.

Quick Check

☐ 4. How do T$_H$ cells differ from other immune cells? What are their main functions?

☐ 5. What are the main functions of T$_C$ cells?

Organ and Tissue Transplantation and Rejection

The ability to transplant organs and tissues between nonidentical individuals was one of the great medical advances of the 20th century. Heart, lung, liver, kidney, pancreas, and bone marrow transplants save thousands of lives every year, with ever-increasing success rates.

The four basic kinds of tissue and organ transplants, also known as *grafts,* are as follows:

- *Autografts* involve tissue transplanted from one site to another in the same individual. Examples of autografts include skin grafts, in which skin is removed from one part of the body and placed on another part to repair skin damaged by trauma such as burns.
- *Isografts* involve organs and tissues transplanted between two genetically identical individuals. Because of the small number of identical twins, these grafts are relatively uncommon.
- *Allografts,* which are the most common type of grafts, involve organs and tissues transplanted between two non-identical individuals of the same species.
- *Xenografts* are relatively rare procedures that involve organs and tissues transplanted between two individuals of different species, such as a pig and a human.

Autografts and isografts result in no response from T_C cells because the antigens bound to class I MHC molecules are not recognized as foreign. However, both allografts and xenografts do contain antigens that the organ recipient's immune system recognizes as foreign. This leads to a reaction from the immune system that, if left untreated, results in *rejection* of the organ or tissue. A rejected organ or tissue first fails to function properly, and then its cells die as T_C cells destroy them, a condition known as *necrosis.* Rejection can lead to the death of the transplant recipient from organ failure and from blood clots and other complications of necrotic tissue.

Clearly, graft rejection is something that must be prevented, and this is done in two ways. The first way involves ensuring that allograft antigens are as similar as possible to the recipient's antigens. This is determined through a process that screens the antigens most likely to cause rejection—those associated with MHC molecules, which are encoded on genes called *human leukocyte antigen* (*HLA*) genes. A donor individual who has a high percentage of HLA similarity to the recipient is said to be a *match.* Exact HLA matches are rare and most likely to be found within families. For example, siblings have a 25% chance of being exact HLA matches. Fortunately, exact matches aren't necessary, and grafts can do well even with one or two HLA mismatches out of four.

The other means of preventing graft rejection is by suppressing the immune response with medications, which is known as *immunosuppressive therapy.* Of course, immunosuppressive therapy is not without serious risks, and dramatically increases a patient's risk of infection, even with agents that are normally nonpathogenic.

One of the most commonly performed transplant procedures is a bone marrow transplant, also referred to as a *hematopoietic stem cell transplant,* or *HSCT.* The procedure removes healthy hematopoietic stem cells from the blood or bone marrow of a donor and infuses them into a recipient. Typically, recipients are treated with immunosuppressive drugs and radiation to kill the diseased bone marrow prior to transplantation; this also minimizes the risk of rejection. HSCTs are used to treat a variety of conditions, including cancers such as leukemia and lymphoma, bone marrow failure, and disorders in which the immune system is deficient.

Quick Check

☐ 6. Which cells are involved in transplant rejection?

☐ 7. How are donor grafts matched to a recipient?

Apply What You Learned

☐ 1. A researcher is investigating a drug to prevent organ rejection; this drug selectively inhibits T_H cells. How would this help to prevent rejection of a transplanted organ?

☐ 2. What other effects would the drug from question 1 have on the immune system? Explain.

☐ 3. A hypothetical poison acts on the cells of the kidney and induces a change in the genes for the self antigens displayed on their class I MHC molecules. Predict what might happen if this poison were administered to someone.

See answers in Appendix A.

MODULE 20.5
Adaptive Immunity: Antibody-Mediated Immunity

Learning Outcomes

1. Describe the process of B cell activation and proliferation.
2. Describe the five major classes of antibodies, and explain their structure and functions.
3. Compare and contrast the primary and secondary immune responses.
4. Explain how vaccinations induce immunity.
5. Compare and contrast active immunity and passive immunity.

We now look to the second arm of adaptive immunity: antibody-mediated immunity, also called *humoral immunity.* (Humoral immunity refers to the fact that its components were first identified in the body's fluids, which were known historically as *humors.*) As we discussed earlier, antibody-mediated immunity involves B cells and proteins secreted by B cells, called **antibodies.** Just as T cells have T cell receptors, B cells have **B cell receptors** that bind to specific antigens, and a group of B cells that bind to a specific antigen is known as a clone. The antibodies secreted by a B cell clone bind to the same antigen as the B cell receptor. In fact, we'll see that the structure of a B cell receptor is almost identical to that of an antibody.

The antibody-mediated immune response has three basic phases. The first phase involves a B cell clone recognizing its specific antigen, which triggers it to undergo changes and start secreting antibodies. The second phase begins when the antibody level in the blood rises dramatically. Antibodies, also known as **immunoglobulins** or **gamma globulins,** are the "action" component of antibody-mediated immunity, in that they are directly responsible for its actions or effects in mediating the destruction of antigens to which they bind. The third and final phase of antibody-mediated immunity is the persistence of a population of B cells called *memory B cells* that react much more rapidly and efficiently if the antigen is encountered again.

This module examines the phases of the antibody-mediated response. It also considers how we can take advantage of the third phase of the response, the persistence of memory cells, to *vaccinate*, or *immunize*, ourselves against a variety of pathogens.

Phase 1: B Cell Activation, Clonal Selection, and Differentiation

B cells develop in the bone marrow from the lymphoid cell line, where billions of B cells are produced each day (**Figure 20.18**). They remain in the bone marrow to mature, but only about 10% of these cells finish their maturation process. This is largely because B cell clones that recognize self antigens (self-reactive B cells) are destroyed. This prevents the development of *autoimmunity*, in which B cells recognize self antigens as foreign and produce *autoantibodies* that bind self antigens. (We discuss autoimmune disorders in Module 20.7.)

Naïve B cell clones that complete maturation enter the circulation and eventually take up residence in lymphoid organs such as the spleen and lymph nodes. When antigens enter the body, they are captured in these lymphoid organs. However, if the B cells are not exposed to their specific antigens within a few days to a few weeks, the cells die.

Naïve B cells that do encounter their antigens become activated by the following process, illustrated in **Figure 20.19**:

1. **The B cell clone binds its antigen and is activated.** The antigen binds to a B cell receptor on the surface of a specific B cell clone. As with T cell activation, this process is called clonal selection. A B cell clone bound to an antigen is said to be *sensitized.*

2. **The sensitized B cell processes the antigen and presents it on its class II MHC molecules. It then binds to a T_H cell, which secretes cytokines that activate the B cell.** The B cell processes the antigen and displays it on its class II MHC molecules. A T_H cell then binds the class II MHC molecules on the sensitized B cell. This binding starts a series of events inside the cell that triggers transcription of antibody genes and multiple other changes that activate the B cell. 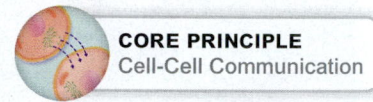 **CORE PRINCIPLE** Cell-Cell Communication
The two cells interacting and influencing one another is an example of the Cell-Cell Communication Core Principle (p. 28).

3. **The B cell divides repeatedly; the resulting cells differentiate into plasma cells and memory B cells.** Among the changes triggered in the B cells when they are activated is rapid cell division, which results in an explosion in the population of these cells. In addition to rapid proliferation, the resulting B cells differentiate into two populations of cells: (1) **plasma cells,** which secrete antibodies; and (2) **memory B cells,** which are long-lived cells that do not secrete antibodies but will respond to antigens upon a second exposure. Dendritic cells and other antigen-presenting cells continue to expose the plasma cells and other B cells to the antigen, and in response they keep proliferating and differentiating.

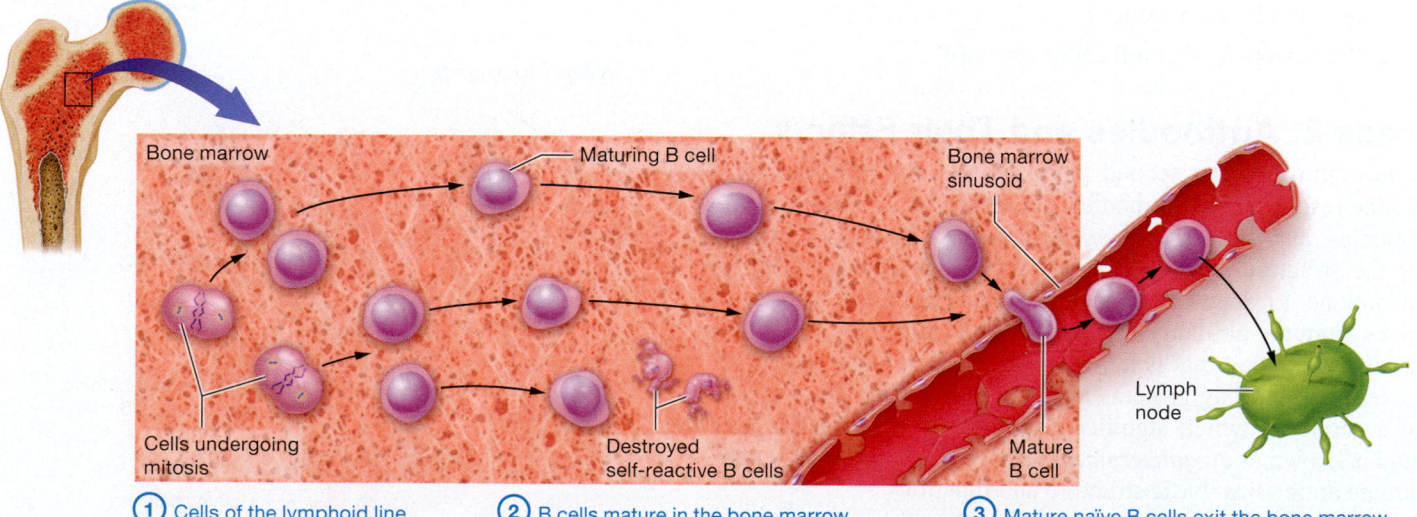

1. Cells of the lymphoid line divide in the bone marrow.

2. B cells mature in the bone marrow, where self-reactive B cells are destroyed.

3. Mature naïve B cells exit the bone marrow and take up residence in lymphoid organs.

Figure 20.18 **B cell maturation.**

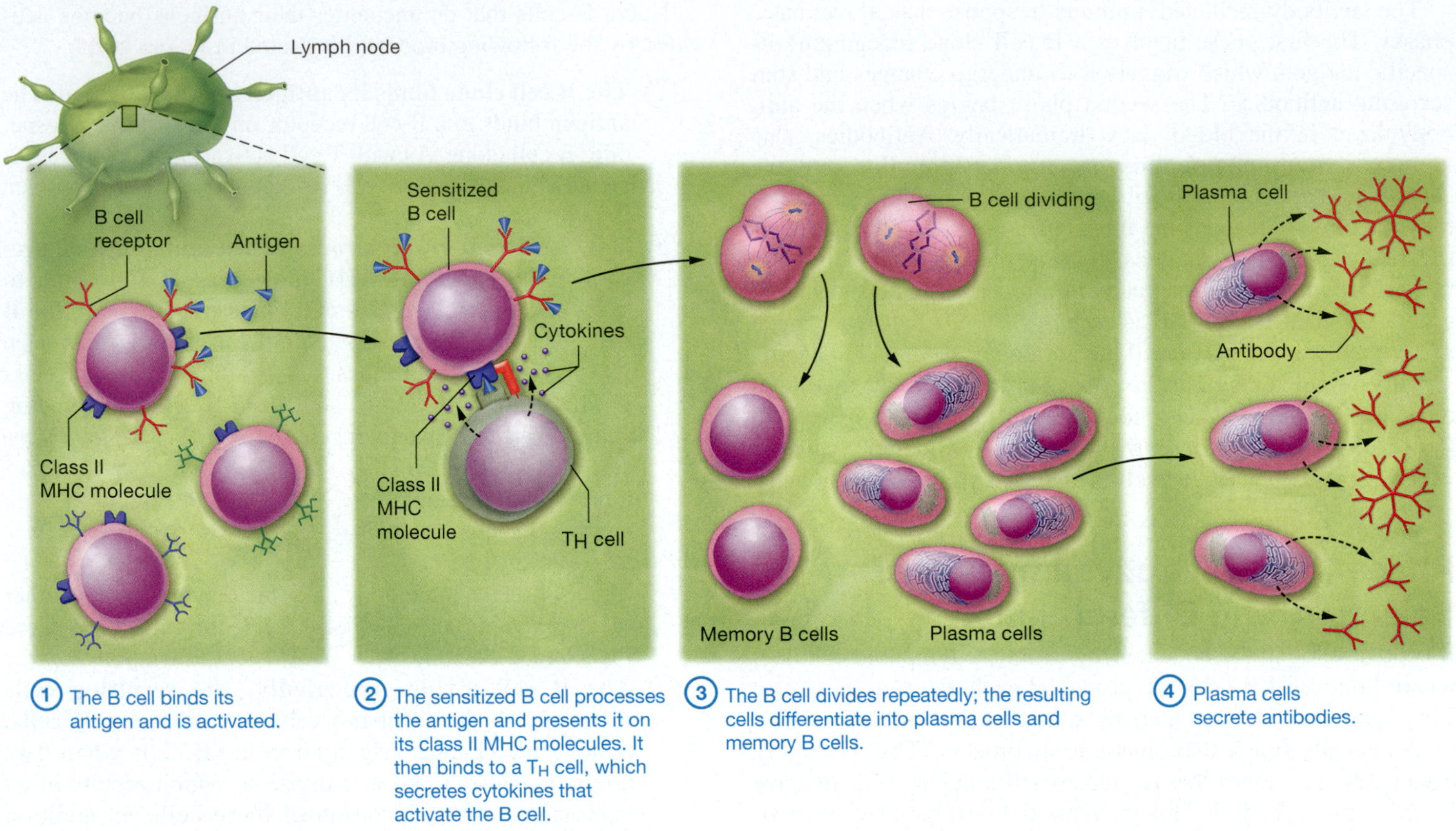

Figure 20.19

Lymph node

B cell receptor   **Antigen**

Class II MHC molecule

Sensitized B cell

Cytokines

Class II MHC molecule

TH cell

B cell dividing

Memory B cells   **Plasma cells**

Plasma cell

Antibody

① The B cell binds its antigen and is activated.

② The sensitized B cell processes the antigen and presents it on its class II MHC molecules. It then binds to a TH cell, which secretes cytokines that activate the B cell.

③ The B cell divides repeatedly; the resulting cells differentiate into plasma cells and memory B cells.

④ Plasma cells secrete antibodies.

Figure 20.19   B cell activation, clonal selection, and differentiation.

④ **Plasma cells secrete antibodies.** Plasma cells continue to divide and increase their numbers as they actively secrete antibodies. Note in Figure 20.19 that the rough endoplasmic reticulum is prominent in plasma cells, which reflects their active protein synthesis.

Quick Check

☐ 1. How is a B cell activated?

☐ 2. What happens after a B cell is activated?

Phase 2: Antibodies and Their Effects

We now move to the second phase of the antibody-mediated immune response: the antibodies and their effects. The study of antibodies, also known as *serology* (see-RAWL-uh-jee), examines the structure and functions of antibodies as well as their applications to medicine and research. The name serology derives from the antibody-containing fluid called *serum*, which is the part of blood that is left over when you remove the cells, platelets, and clotting proteins from the rest of the plasma (note that a serum in which significant quantities of antibodies are found is known as an *antiserum*). Let's now survey serology and examine antibodies' basic structure and functions.

Antibody Structure and Classes

The basic subunit of an antibody, shown in **Figure 20.20,** is a Y-shaped protein formed from four peptide chains, two *heavy*

(*H*) and two *light* (*L*) *chains*. Each of these chains has two types of regions: (1) **constant (C) regions,** which are relatively similar among antibodies and are responsible for many of the antibodies' effects; and (2) **variable (V) regions,** which are unique sequences of amino acids responsible for antigen recognition and binding. An antibody has V regions at the tips of the two arms of the protein, so the basic subunit of an antibody has two **antigen-binding sites,** one on each arm.

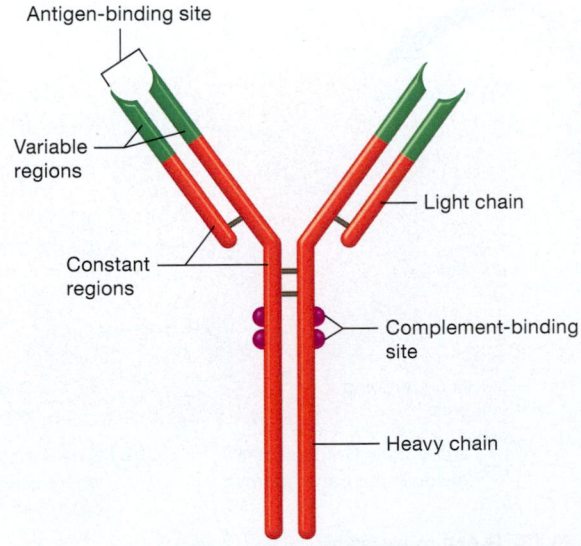

Antigen-binding site

Variable regions

Constant regions

Light chain

Complement-binding site

Heavy chain

Figure 20.20   The basic structure of an antibody monomer.

As you can see here, the single antibody subunit, called a *monomer,* can be combined with other subunits to form a larger structure such as a *dimer,* which consists of two subunits, or a *pentamer,* which consists of five subunits:

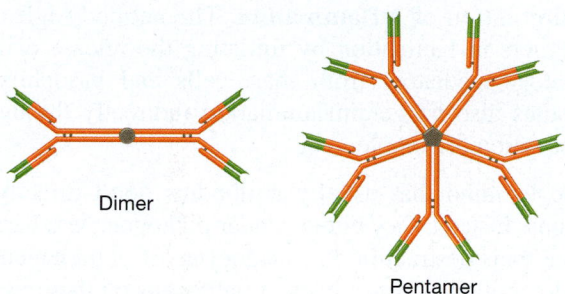

Dimer

Pentamer

There are five basic classes of antibody, which are grouped according to the structure of their C regions. Each antibody is named with the two-letter abbreviation "Ig," which stands for "immunoglobulin," followed by a letter that designates its class. This gives us the types IgG, IgA, IgM, IgE, and IgD, which you can remember with the mnemonic GAMED. Let's take a closer look at these classes:

- **IgG.** The most prevalent antibody in the body is **IgG,** which accounts for about 75–80% of the antibodies in serum. It consists of a single subunit. IgG is the only antibody able to cross from the blood of a pregnant woman to her developing fetus through a structure called the *placenta.*
- **IgA.** IgA is usually a dimer, consisting of two Y-shaped subunits, which gives this antibody four antigen-binding sites. IgA is present in secretions from the skin, mucous membranes, and exocrine glands (i.e., tears, saliva, sweat, and breast milk).
- **IgM.** The largest antibody is the pentamer **IgM** (think "M" for "massive"), which consists of five subunits in a starlike pentamer, for a total of 10 antigen-binding sites. IgM is generally the first antibody secreted by plasma cells when the body is invaded by a pathogen. Note that IgM also exists as a single subunit embedded in the B cell plasma membrane, where it functions as a B cell receptor.
- **IgE.** The single-subunit antibody **IgE** is generally present in very low amounts in the body's fluids. IgE binds to two types of antigen: (1) antigens associated with parasitic pathogens such as tapeworms, and (2) environmental antigens known as *allergens,* which are linked with inflammatory reactions called *allergies.* IgE molecules bind to mast cells in mucous membranes, and when they come into contact with their specific antigens, they trigger mast cells to release the contents of their granules, a process called *degranulation.* Mast cell granules contain inflammatory mediators such as histamine that initiate a localized inflammatory response. The ensuing inflammation is responsible for common allergy symptoms such as a runny nose and watery eyes. We discuss allergies in more detail in Module 20.7.
- **IgD.** The antibody **IgD** is unique because it is the only antibody not secreted by B cells in significant amounts. Rather,

its single subunit is located on the surface of B cells, where it acts as an antigen receptor that helps activate B cells in a similar manner to IgM.

Although all antibodies in a particular class feature similar C regions, each antibody has a unique V region that determines its ability to recognize and bind a certain antigen. There is remarkable variability in antibodies' V regions, and there are an estimated one million or more different antibodies in the body. This means that your immune system is capable of recognizing at least one million unique antigens, which is certainly a feat. This is made possible by a genetic mechanism found only in lymphocytes whereby the genes that code for antibodies undergo random recombination, producing a huge number of potential genes and antibody products.

Note that a B cell clone that starts out producing a certain class of antibody, such as IgM, has the ability to switch to production of a different class, such as IgG. The mechanism that allows for this process, known as *class switching,* involves genetic rearrangements to yield different antibody C regions, while maintaining the same antibody V regions. In this way, the antibody can have different effects while binding the same type of antigen.

Functions of Secreted Antibodies

The actions of antibodies are based on their ability to bind antigens, which leads to multiple effects on pathogens. Many of these effects represent an integration of adaptive and innate immunity—the antibodies of adaptive immunity often activate phagocytes and complement proteins of innate immunity. The basic effects of secreted antibodies are shown in **Figure 20.21** and include the following:

- **Agglutination and precipitation.** Antibodies can bind to antigens on more than one cell. This creates a clump of cells that are cross-linked by their attachment to antibodies. The clumping of cells is known as **agglutination.** Similar to agglutination is the process called **precipitation,** which involves soluble antigens (proteins and other biological molecules) instead of whole cells. Both agglutination and precipitation allow these antigen-antibody complexes to precipitate out of body fluids and therefore make the complexes easier for phagocytes to ingest. IgM is the most potent agglutinating and precipitating antibody as a result of its 10 antigen-binding sites.
- **Opsonization.** As you read about, opsonization involves proteins such as complement coating the pathogen and activating phagocytes. IgG antibodies are also opsonins able to coat pathogens and bind and activate phagocytes, which greatly enhances phagocytosis.
- **Neutralization.** Bacterial toxins, viral proteins, and animal venoms are molecules with specific components that are harmful. Antibodies bind to these components, as well as certain viruses and bacteria, and prevent them from interacting with our cells. This renders the toxin inactive, which is known as **neutralization.** Most neutralizing antibodies are of either the IgG or the IgA class.

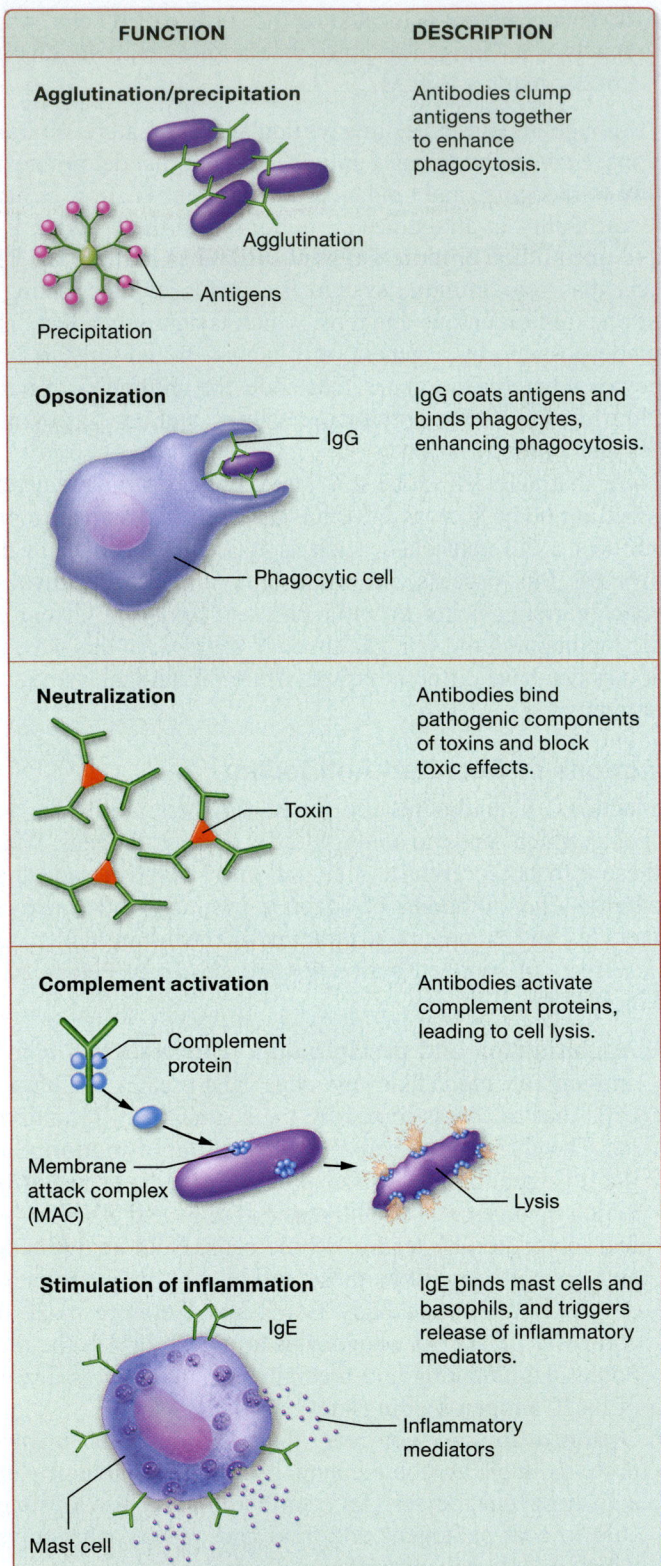

FUNCTION	DESCRIPTION
Agglutination/precipitation	Antibodies clump antigens together to enhance phagocytosis.
Opsonization	IgG coats antigens and binds phagocytes, enhancing phagocytosis.
Neutralization	Antibodies bind pathogenic components of toxins and block toxic effects.
Complement activation	Antibodies activate complement proteins, leading to cell lysis.
Stimulation of inflammation	IgE binds mast cells and basophils, and triggers release of inflammatory mediators.

Figure 20.21 Functions of antibodies.

- **Complement activation.** Several antibodies, particularly IgM and IgG, bind and activate the complement proteins of innate immunity (see the classical pathway in Figure 20.10). When several antibodies bind a single cell, their complement-binding sites are exposed. This allows complement to activate and lyse the foreign cell with its membrane attack complex. This effect is particularly important in defense against cellular pathogens such as bacteria and is partly responsible for our reaction to foreign cells such as donated erythrocytes.

- **Stimulation of inflammation.** The antibody IgE directly triggers inflammation by initiating the release of inflammatory mediators from mast cells and basophils. Antibodies also trigger inflammation indirectly through their activation of complement.

Keep in mind that usually antibodies don't directly "kill" pathogens. Instead, they either render pathogens less harmful or facilitate their destruction by phagocytes or complement. Only rarely do antibodies cause lysis of pathogens on their own.

The structures and functions of antibodies are summarized in **Figure 20.22**.

CLASS		FUNCTION
IgG	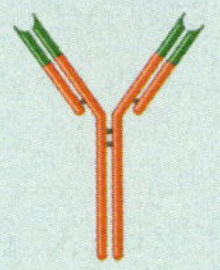	• Makes up the majority of antibodies in serum • The only antibody that can cross the placenta from mother to fetus • Functions in opsonization, neutralization, and complement fixation
IgA	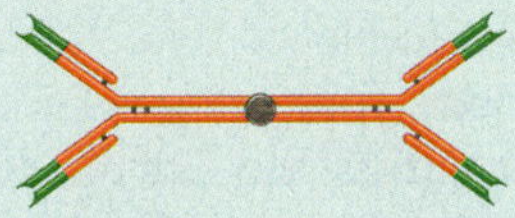	• Found in secretions such as breast milk and saliva • Functions in agglutination and neutralization
IgM	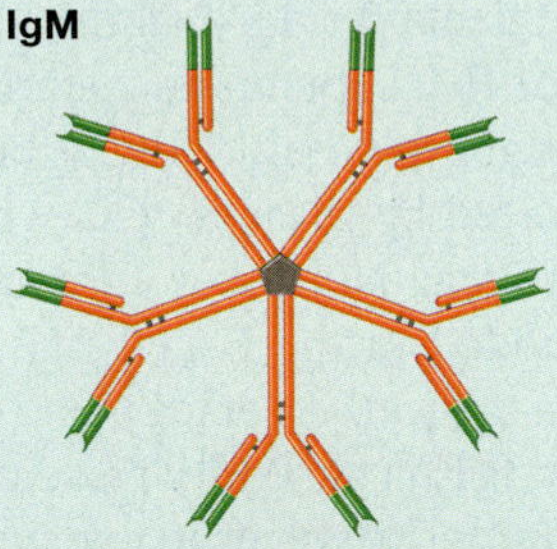	• The first antibody secreted on exposure to an antigen • Potent agglutinating and precipitating agent • Functions in complement fixation
IgE	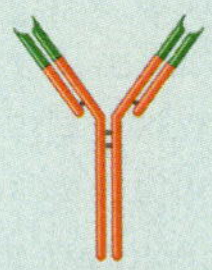	• Binds mast cells and basophils and triggers their degranulation, facilitating inflammation, particularly in the allergic response
IgD	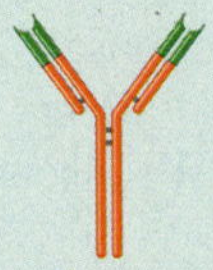	• Antibody found exclusively on the surface of B cells • Has a role in B cell sensitization and activation

Figure 20.22 Antibody classes.

Table 20.2 Comparison of the Primary and Secondary Immune Responses

Characteristic	Primary Immune Response	Secondary Immune Response
Lag phase	4–5 days	1–3 days
Time until antibody peak	7–14 days	3–5 days
Primary antibody	IgM	IgG
Duration of response	14–21 days	28 days and beyond

Quick Check

☐ 3. What are the five classes of antibody, and how do they differ?

☐ 4. List the five main functions of antibodies.

Phase 3: Immunological Memory

As we discussed, activated B cells divide into two populations: plasma cells and memory B cells. The memory B cells are responsible for antibody-mediated immunological memory. Like memory T cells, the memory B cells respond more efficiently when the antigen is encountered a second time. In this section we explore how the immune response differs after initial exposure to an antigen and how we apply the principles of immunological memory to prevent disease with vaccinations.

The Primary and Secondary Immune Responses

The first time you are exposed to an antigen, the response proceeds the way that was described in an earlier section: A B cell specific for that antigen recognizes it, the activated B cell proliferates and differentiates into plasma and memory B cells, and plasma cells begin to secrete antibodies. This response is called the **primary immune response.** The primary immune response is effective, but it's slow—there is an initial 4- to 5-day *lag phase* as the B cells proliferate, differentiate into plasma cells and memory cells, and begin to secrete antibodies. Antibody levels peak about 7–14 days after the antigen is encountered. It is during the lag phase that you generally feel "sick."

The next time you are exposed to the same antigen, the response is much different. The memory B cells formed during the primary immune response are long-lived, persisting in the

body from years to an entire lifetime. When these memory B cells encounter the antigen for which they are specific, the **secondary immune response** begins. There are several key differences between the two responses (summarized in **Table 20.2**). First, the secondary immune response has a shorter lag phase (about 1–3 days), and its antibody levels peak more rapidly (3–5 days) and reach a peak 100–1000 times larger. Also, the main antibody involved in the secondary response is IgG, thanks to the fact that many of the memory B cells have already undergone class switching, whereas it's IgM in the primary response. In addition, the secondary response lasts longer than the primary response. Finally, the antibodies secreted in the secondary immune response are more effective. IgG antibodies, produced in higher amounts in the secondary immune response, bind more tightly to and have a higher affinity for their antigens. For this reason, you may not feel "sick" when exposed to the same antigen or may experience only mild symptoms. We take advantage of this secondary immune response with vaccinations, our next topic.

Vaccinations

A **vaccination,** also known as an **immunization,** involves exposing an individual to an antigen to elicit a primary immune response and generate memory cells. Then if the individual is exposed to the antigen a second time, a secondary immune response will occur and symptoms will be minimal.

A vaccine may be any of several types, depending on its components, which can include the following:

- **Live, attenuated vaccines.** Certain pathogens can be *attenuated* (ah-TEN-yoo-ay′-tid), meaning that their ability to cause disease has been greatly reduced. When these pathogens are administered as a vaccine, they can still divide to a limited extent in the body and so induce a primary immune response. The ability of the pathogens to divide generates a strong primary immune response due to their extended presence in the body. Live, attenuated vaccines are completely safe for most individuals, although there is an extremely small risk that the organisms could revert to a pathogenic form. Therefore, these vaccines are often avoided in patients with compromised immune systems.

- **Inactivated vaccines.** Vaccines may also be prepared with inactivated, or "killed," pathogens. Inactivated pathogens are not capable of dividing in the body, so they generate a weaker primary immune response and generally require several repeat vaccinations, known as "boosters."

- **Subunit vaccines.** With certain pathogens, only the portion of the pathogen that causes disease is required to develop immunity. For example, the bacteria responsible for tetanus and diphtheria secrete disease-causing toxins. Vaccinations called *toxoids* contain inactivated toxins from these bacteria, which induce the immune system to produce antibodies to the toxins. Other subunits that may be used in a vaccine are the antigenic polysaccharides found in the outer component of certain bacterial walls. Like inactivated vaccines, subunit vaccines tend to induce a weaker primary immune response and typically require boosters to maintain immunity.

One of the biggest challenges to effective vaccination is posed by pathogens that have a high rate of *mutation*—random changes in their genetic makeup. Mutations often change the pathogens' antigens, which alters the B cells and antibodies needed to respond to those antigens. So the initial exposure confers immunity only to the original strain, not the mutated strains. This has been a major issue in the development of effective vaccines for *human immunodeficiency virus* (*HIV*) and tuberculosis (a vaccine for this disease exists, but it is unreliable). The pathogens mutate so rapidly that a vaccine cannot reliably protect an individual from infection.

The influenza viruses constitute another group of pathogens with a high mutation rate. Their rapid mutation rate means that a new vaccine must be prepared every year. The vaccine typically contains the dominant viral strain from the previous year, the dominant strain from the current year, and another strain that could become prevalent. This method of preparing the annual "flu shot" isn't perfect— sometimes the viruses mutate into new strains, or an undetected strain becomes the dominant one. Still, the vaccine does provide significant protection from influenza most years, and healthcare providers encourage their patients to be vaccinated every year. It should be noted that one cannot "get the flu" from an influenza vaccination, as it contains inactivated virus (note, though, that the nasal FluMist vaccine contains an attenuated strain). If an individual gets sick shortly after receiving the vaccine, the illness is likely due to a different strain of influenza, an inadequate amount of time for the immune system to mount a primary response, or another virus entirely.

Vaccinations are undeniably among the most important medical discoveries of the past 200 years. Before the discovery of vaccinations, the only way to develop memory cells for a pathogen was to be exposed to it, generally via infection. Unfortunately, many did not survive this initial exposure or suffered serious health consequences as a result. Vaccines have eliminated the virus that causes smallpox, once a major threat to human survival, and have nearly eliminated the virus that causes the paralytic disease polio. The only places where polio persists are those countries that have stopped vaccination programs or those in which members of the population have no access to vaccinations. Vaccination is, unfortunately, a practice about which many misconceptions exist; for a discussion of one of the most common misconceptions, see *A&P in the Real World, Pseudoscience Exposed: The Myth of Vaccines and Autism.*

A&P in the Real World: Pseudoscience Exposed: The Myth of Vaccines and Autism

If you do an Internet search for "vaccines and autism" you will receive hundreds of thousands of hits, many of which will claim that vaccines cause the developmental disorder autism. The concern about a possible link between vaccinations and autism arose when it was noticed that many children first started showing symptoms of autism around the time of their childhood vaccinations. The thought was that perhaps *thimerosal,* a mercury-containing preservative found in most vaccines, was the culprit. Certain forms of mercury, after all, are known to cause brain damage.

However, there was one major problem with this hypothesis: The form of mercury that causes brain damage is methylmercury, but thimerosal contains *ethylmercury.* Nonetheless, independent researchers across the globe have examined thimerosal, but no link between thimerosal and autism has ever been found. Indeed, thimerosal was removed from vaccines in Denmark in 1992, and the autism rate actually *increased* by a small amount after this was done. Although no evidence supports the idea that thimerosal is harmful, the United States removed thimerosal from most vaccines in 2001 as a result of public concerns.

Researchers have also explored other potential links between vaccinations and autism, but the results have been the same: No link can be established. Unfortunately, many people continue to staunchly advocate that vaccines are the root cause of autism, even with an absolute lack of evidence. This has generated concern among well-meaning parents and led to a decline in the number of individuals who receive vaccinations. The consequence of decreased vaccination rates is an increase in the number of cases of potentially lethal diseases, such as measles, whooping cough, and meningitis, which raises serious public health concerns.

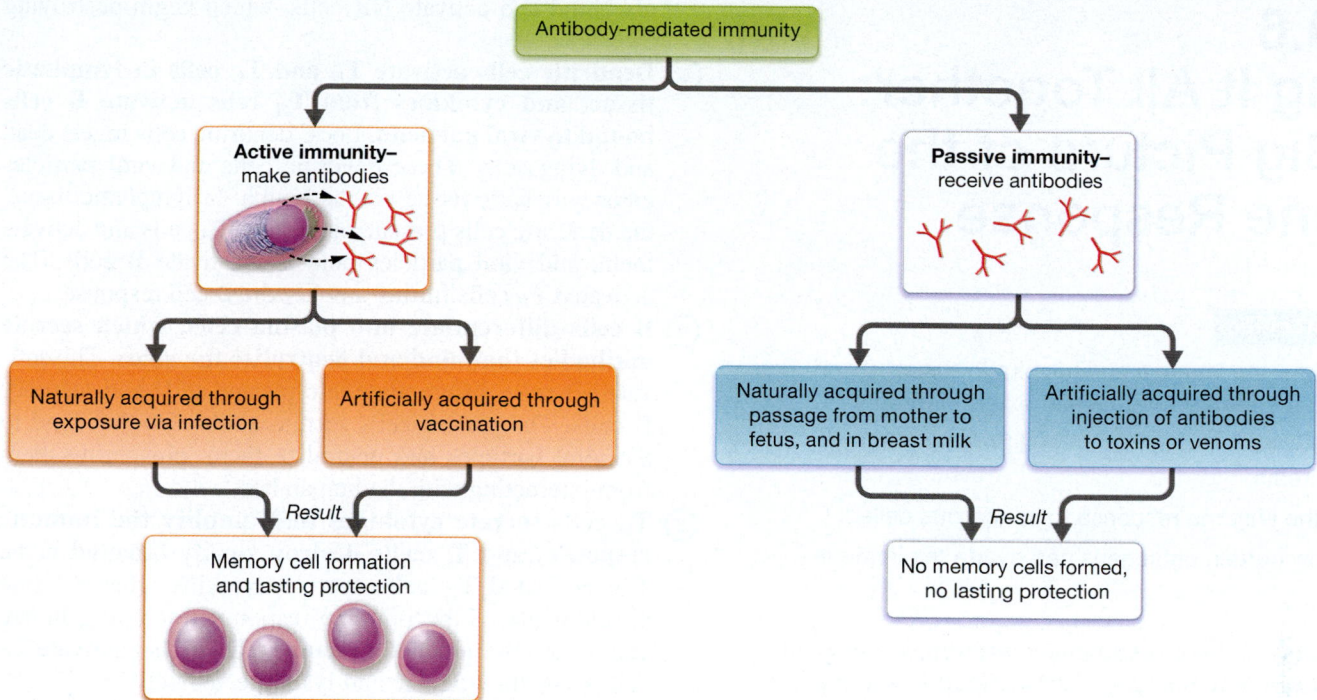

Figure 20.23   Active and passive antibody-mediated immunity.

Active and Passive Antibody-Mediated Immunity

There are two types of antibody-mediated immunity, active and passive (**Figure 20.23**). The type of immunity we have been discussing is **active immunity,** so named because the body's cells actively respond to an antigen. Active immunity may be received naturally through exposure to an antigen via infection or artificially via a vaccination. It results in the production of memory cells and large numbers of antibodies and is, therefore, relatively long-lasting, ranging from years to a lifetime. The length of time during which active immunity lasts depends on several factors, particularly the extent of exposure. For example, children exposed to the chickenpox virus through moderate infection tend to develop lifelong immunity. However, those children exposed to the virus via a very mild infection may develop immunity for only a few years and become infected again in later childhood or in adulthood. (A vaccination for the chickenpox virus is now available, so this scenario mostly applies to unvaccinated children and adults.)

Passive immunity is found when preformed antibodies are passed from one organism to another. Passive immunity may be naturally acquired, for example, when the IgG of a pregnant woman crosses from her blood into that of her fetus, or artificially acquired, from an injection with preformed antibodies. This form of immunity lasts only the amount of time that the antibodies stay in the bloodstream, which is about 3 months on average. Even though it is temporary, passive immunity is critical and life-saving in many instances. For example, passive immunity is the sole source of antibodies that an infant has at birth. In addition, preformed antibodies are administered when a person is exposed to animal venom (as an *antivenin*), certain toxin-producing bacteria (as an *antitoxin* or *toxoid*) to neutralize the toxins, and certain viruses such as hepatitis A.

Quick Check

☐ 5. How do the primary and secondary immune responses differ?

☐ 6. What is the purpose of a vaccination, and what are the three types of vaccines?

☐ 7. How do active immunity and passive immunity differ?

Apply What You Learned

☐ 1. A patient has a hypothetical disorder that prevents the development of memory cells. Predict the effects this disorder would have on the primary and secondary immune responses.

☐ 2. The disorder *hypogammaglobulinemia* is characterized by a decrease in the secretion of one or more classes of antibody. Predict the effect of decreased secretion of:

a. IgA.

b. IgG.

c. IgG in a pregnant female.

d. IgM.

☐ 3. You are injured in a farming accident, and in the emergency room you are administered injections of both *tetanus antitoxin,* which contains antibodies to the tetanus toxin, and tetanus antigens. Which injection provided active immunity, and which passive? Explain.

See answers in Appendix A.

Putting It All Together: The Big Picture of the Immune Response

Learning Outcomes

1. Describe how the immune and lymphatic systems work together to respond to internal and external threats.
2. Explain how the immune response differs for different types of threats.
3. Describe the immune response to cancerous cells.
4. Explain how certain pathogens can evade the immune response.

The immune response involves so many different components working simultaneously that it's often difficult to see the big picture of how the body responds to a threat. This module helps you see the "forest for the trees," as we walk through the basic immune response in three different scenarios: infection with the common cold virus, infection with a strain of toxin-producing bacteria, and a cancerous tumor. Along the way, we discuss how the parts of the immune system work with the lymphatic system to form an integrated response that protects the body from a variety of threats. We conclude the module with a look at how certain pathogens are able to evade the immune response and cause serious homeostatic disturbances.

Scenario 1: The Common Cold

Let's start with one of the most frequent infectious conditions in the world: the *common cold*. The common cold is due to a variety of viruses that typically have a high mutation rate, which is why researchers have yet to develop a vaccine for the pathogens. Although it might seem that you frequently get sick with a cold, in truth most pathogens, viruses included, are deterred by the body's surface barriers and so never even gain entry into the body. When a virus does gain entry, the response proceeds by the following steps, which are illustrated in **Figure 20.24**:

1. **Infected cells trigger an inflammatory response and secrete interferons.** The virus enters through the respiratory tract and infects local cells. This triggers the infected cells to secrete certain interferons and other inflammatory mediators that initiate the inflammatory response.

2. **Interferons prevent infection in neighboring cells and activate natural killer (NK) cells, which lyse infected cells.** The interferons secreted by the infected cells travel to nearby cells and prevent viral infection. Interferons

also bind and activate NK cells, which begin destroying infected cells.

3. **Dendritic cells activate T_H and T_C cells in lymphatic tissue, and cytokines from T_H cells activate B cells bound to viral particles.** Local dendritic cells ingest dead and dying cells. These dendritic cells and viral particles enter lymphatic tissue via the lymph. In lymphatic tissue, the dendritic cells present antigens to T_H cells and activate them, and viral particles bind and activate B cells. The activated T_H cells further amplify the B cell response.

4. **B cells differentiate into plasma cells, which secrete antibodies that bind and neutralize the virus.** The activated B cells differentiate into plasma cells and memory B cells. The plasma cells secrete antibodies that bind to the viral particles and neutralize them, preventing them from interacting with the human host cells.

5. **T_H cells secrete cytokines that amplify the immune response, and T_C cells destroy virally infected cells.** The activated T_H cells secrete cytokines that activate all elements of the immune response, including innate and antibody-mediated immunity. They also activate T_C cells, which recognize and lyse infected cells.

You may be wondering at this point why you still feel sick, when all of this is going on in your immune system. There are actually two reasons. For one, many unpleasant symptoms of colds are due to your own immune system. The inflammatory response triggers increased blood flow and capillary permeability in the upper respiratory tract, and the fluid that leaks out of capillaries tends to leak out of your nose. Interferons also induce many of the other symptoms that you experience, including fever and body aches. The other reason you feel sick is that the cold viruses mutate rapidly, and new strains for which you lack memory cells emerge every year. Your body must mount a primary immune response for each new type of mutated virus, which takes time to complete. During this time, you feel sick as the virus does local damage and your immune system works to combat it.

Quick Check

☐ 1. How do innate immunity and adaptive immunity work together to defeat a cold virus?

Scenario 2: Bacterial Infection

In this second scenario, we examine the immune response to the bacterium *Streptococcus pyogenes*. This bacterium causes a variety of infections, the most common of which is *streptococcal pharyngitis* (faer-en-JY-tis; commonly called "strep throat"—recall that the *pharynx* is the throat). The process proceeds by the following steps, illustrated in **Figure 20.25** on p. 790:

1. **Bacteria damage cells and induce an inflammatory response, as local macrophages ingest the bacteria.** The bacteria damage cells in and around the pharynx, and the damaged cells release inflammatory mediators

The Big Picture
of the Immune Response to the Common Cold

Figure 20.24

Play Animation
@ Mastering Anatomy & Physiology

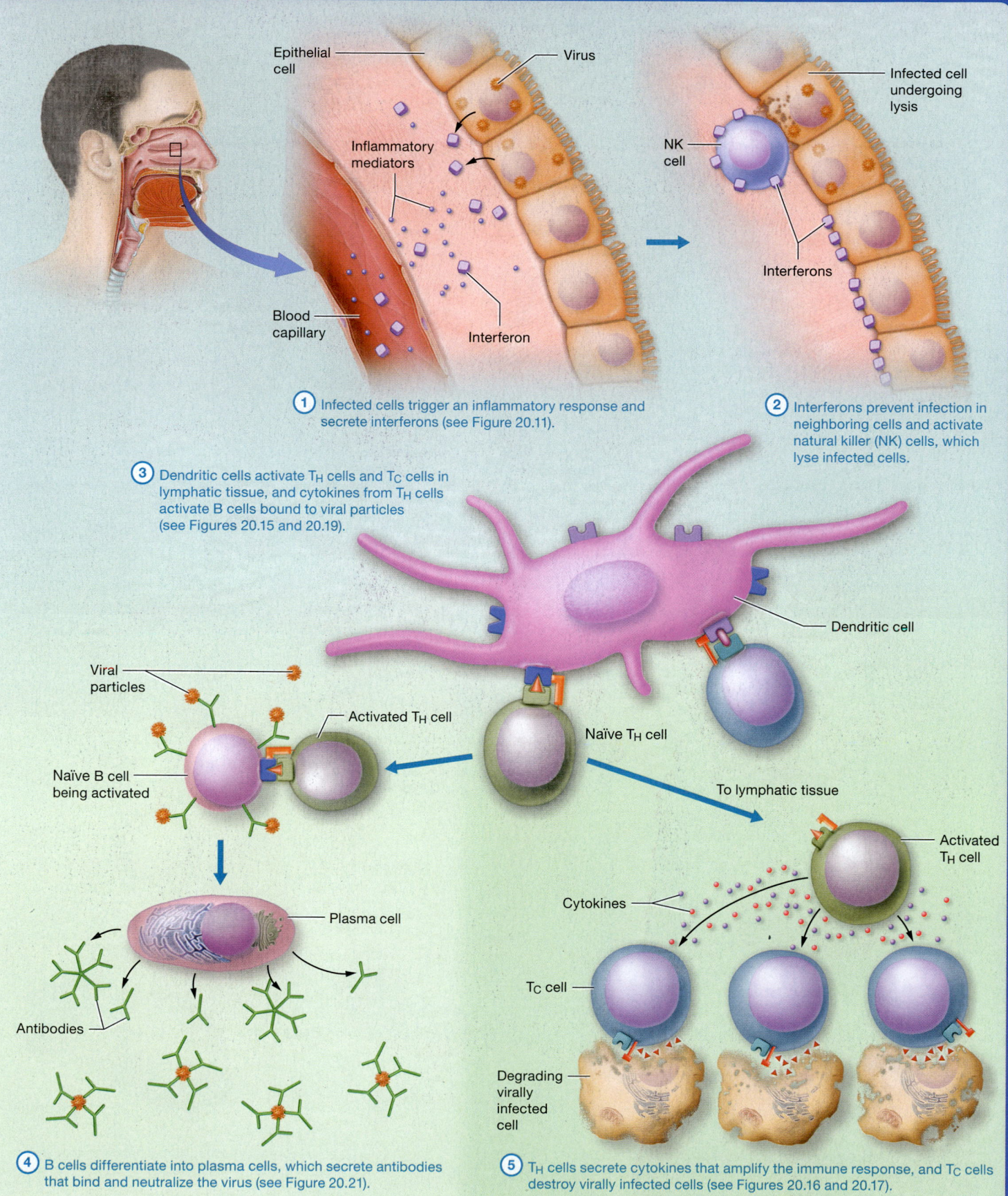

Epithelial cell

Virus

Inflammatory mediators

Infected cell undergoing lysis

NK cell

Interferons

Blood capillary

Interferon

① Infected cells trigger an inflammatory response and secrete interferons (see Figure 20.11).

② Interferons prevent infection in neighboring cells and activate natural killer (NK) cells, which lyse infected cells.

③ Dendritic cells activate T_H cells and T_C cells in lymphatic tissue, and cytokines from T_H cells activate B cells bound to viral particles (see Figures 20.15 and 20.19).

Dendritic cell

Viral particles

Activated T_H cell

Naïve T_H cell

Naïve B cell being activated

To lymphatic tissue

Activated T_H cell

Cytokines

Plasma cell

T_C cell

Antibodies

Degrading virally infected cell

④ B cells differentiate into plasma cells, which secrete antibodies that bind and neutralize the virus (see Figure 20.21).

⑤ T_H cells secrete cytokines that amplify the immune response, and T_C cells destroy virally infected cells (see Figures 20.16 and 20.17).

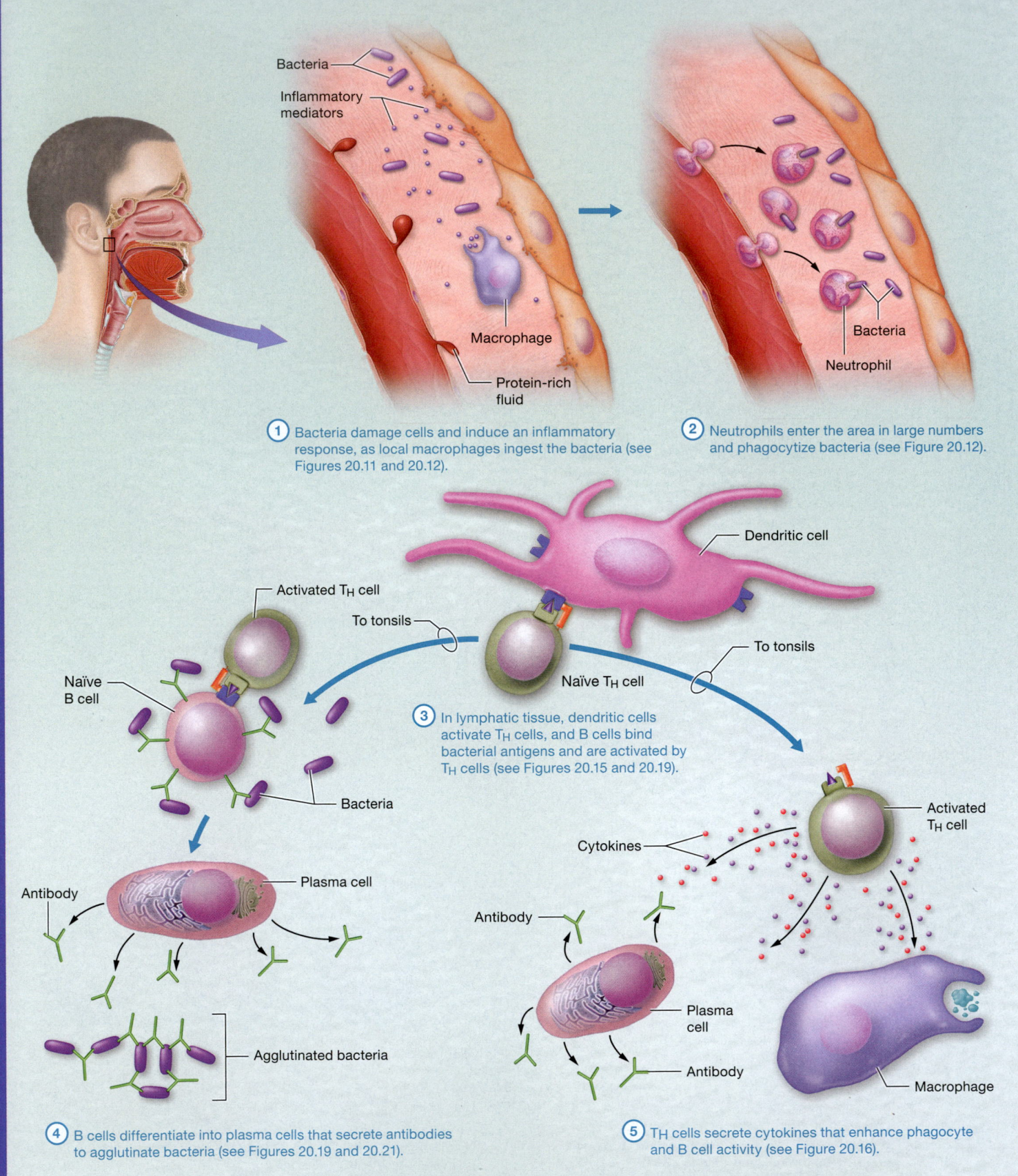

Bacteria

Inflammatory mediators

Macrophage

Protein-rich fluid

Bacteria

Neutrophil

1 Bacteria damage cells and induce an inflammatory response, as local macrophages ingest the bacteria (see Figures 20.11 and 20.12).

2 Neutrophils enter the area in large numbers and phagocytize bacteria (see Figure 20.12).

Dendritic cell

Activated T_H cell

To tonsils

To tonsils

Naïve B cell

Naïve T_H cell

Activated T_H cell

Bacteria

3 In lymphatic tissue, dendritic cells activate T_H cells, and B cells bind bacterial antigens and are activated by T_H cells (see Figures 20.15 and 20.19).

Cytokines

Antibody

Plasma cell

Antibody

Plasma cell

Antibody

Agglutinated bacteria

Macrophage

4 B cells differentiate into plasma cells that secrete antibodies to agglutinate bacteria (see Figures 20.19 and 20.21).

5 T_H cells secrete cytokines that enhance phagocyte and B cell activity (see Figure 20.16).

that induce an inflammatory response. This inflammation is responsible for one of the first signs of streptococcal pharyngitis: a sore throat. Local macrophages become activated and begin to phagocytize the bacteria.

② **Neutrophils enter the area in large numbers and phagocytize bacteria.** The inflammatory mediators attract circulating neutrophils to the area, which phagocytize bacteria. They also trigger the bone marrow to release its store of neutrophils and increase their production. In many cases, this is sufficient to contain and resolve the infection, and adaptive immunity is either only weakly activated or not activated at all.

③ **In lymphatic tissue, dendritic cells activate T_H cells, and B cells bind bacterial antigens and are activated by T_H cells.** If the infection is not resolved by the neutrophils and macrophages alone, dendritic cells and bacteria eventually enter the lymph and migrate to local lymphoid organs such as the tonsils. Here the bacterial antigens get processed by dendritic cells and presented to naïve T_H cells. Once activated, the T_H cells stimulate naïve B cells, which have already bound to bacterial antigens.

④ **B cells differentiate into plasma cells that secrete antibodies to agglutinate bacteria.** The activated B cells differentiate into plasma cells, which secrete antibodies that bind and agglutinate bacteria. The antibody-bacteria complexes are ingested by phagocytes such as macrophages and neutrophils.

⑤ **T_H cells secrete cytokines that enhance phagocyte and B cell activity.** The activated T_H cells secrete cytokines that increase the effectiveness of neutrophils and macrophages and also increase the activities of B cells.

Note that the immune response to this bacterium did not involve T_C cells. However, an intracellular bacterium, or one that lives inside its host cell, causes a response that involves T_C cells because its antigens are displayed on class I MHC molecules. The predominant cell type that responds can give you information about the likely infectious agent, which is discussed in *A&P in the Real World: Complete Blood Count with Differential.*

Quick Check

☐ 2. Walk through the basic steps of the immune response to a bacterial infection.

Scenario 3: Cancer

Our final scenario involves a disease that is responsible for over 8 million deaths per year worldwide: *cancer.* Cancer cells are formerly normal body cells that have undergone mutations causing them to de-differentiate (become less specialized), lose control of their cell cycles, and lose their attachments to the surrounding cells and extracellular matrix (see Chapter 3). These mutations result in a mass of unspecialized cells called a *malignant tumor,* whose cells are capable of indefinite growth and can *metastasize* (meh-TAS-tuh-syz), or spread, through the lymphatic system or blood to other parts of the body. The cancer

Complete Blood Count with Differential

One of the first laboratory tests ordered when a patient is admitted to the hospital is a *complete blood count* (*CBC*). A component of the CBC is the leukocyte count, which can indicate if inflammation is present somewhere in the body. Unfortunately, inflammation is nonspecific and an elevated leukocyte count tells you little about the cause of inflammation. For this reason, a *differential,* which measures the relative prevalence of the different types of leukocytes in the blood, is ordered along with the CBC.

A differential can give you a great deal of information. First, if the infection is caused by bacteria, neutrophils tend to be selectively elevated because of their extensive role in the eradication of bacteria. Second, viral pathogens tend to induce primarily a lymphocyte-dominant response, so the numbers of lymphocytes in the blood would likely be elevated with a viral infection. Finally, inflammation due to parasites or allergies usually causes an elevated level of eosinophils in the blood.

cells invade and destroy healthy tissues as they crowd out normal cells and compete for nutrients. (Note that *benign tumors* [beh-NYN] lack the ability to metastasize and do not divide indefinitely. Benign tumors are not considered cancerous.)

The immune system is quite effective at eliminating cancers before they ever get the chance to damage other tissues. This effectiveness is largely due to certain cells of the immune system, especially T cells and NK cells, that "scan" the cells in the body for tumor antigens, a function known as **immune surveillance.** Immune surveillance is absolutely critical to the prevention of tumor development.

When tumor cells are discovered via immune surveillance, they are eliminated by the following process, shown in **Figure 20.26** on p. 792:

① **Cancer cells damage surrounding cells and induce an inflammatory response.** Cancer cells invade surrounding, healthy cells, which secrete inflammatory mediators that induce an inflammatory response.

② **NK cells migrate to the area and begin to destroy the cancer cells and secrete interferons.** Inflammatory mediators attract NK cells to the area, and they begin to destroy the cancer cells and secrete certain interferons, which also induce cancer cell death. In addition, macrophages are activated, and they secrete tumor necrosis factor, which also induces cell death.

③ **Dendritic cells ingest cellular debris and migrate to lymph nodes, where they activate naïve T_H and T_C cells.** Cancer cell debris produced as a result of tumor death is then ingested by dendritic cells, which then migrate to the lymph and lymph nodes. In the lymph nodes, dendritic cells present antigens to naïve T_H cells and T_C cells.

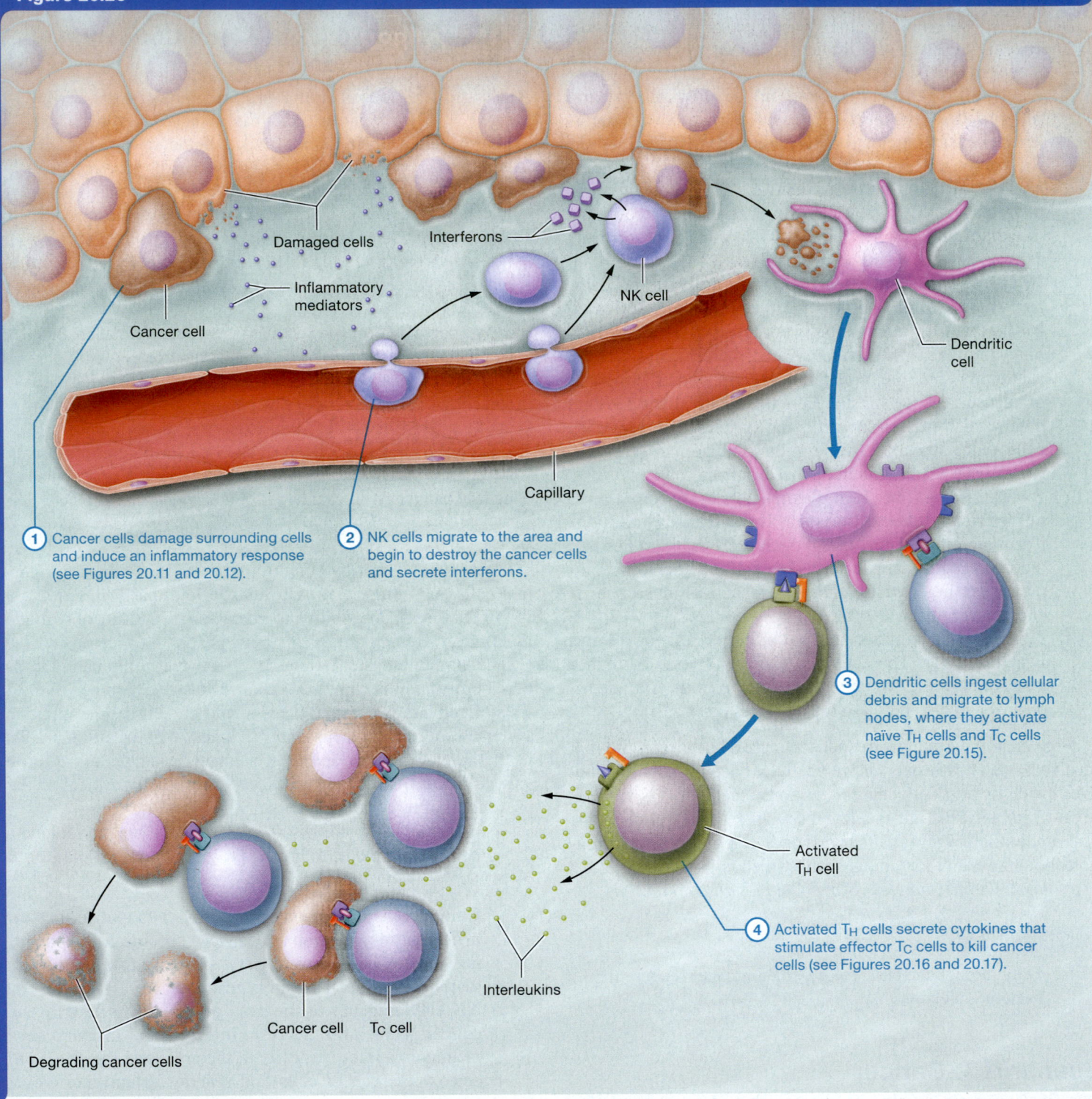

Damaged cells

Interferons

Inflammatory mediators

NK cell

Cancer cell

Dendritic cell

Capillary

① Cancer cells damage surrounding cells and induce an inflammatory response (see Figures 20.11 and 20.12).

② NK cells migrate to the area and begin to destroy the cancer cells and secrete interferons.

③ Dendritic cells ingest cellular debris and migrate to lymph nodes, where they activate naïve T$_H$ cells and T$_C$ cells (see Figure 20.15).

Activated T$_H$ cell

④ Activated T$_H$ cells secrete cytokines that stimulate effector T$_C$ cells to kill cancer cells (see Figures 20.16 and 20.17).

Interleukins

Cancer cell   T$_C$ cell

Degrading cancer cells

④ **Activated T$_H$ cells secrete cytokines that stimulate effector T$_C$ cells to kill cancer cells.** The activated T$_H$ cells secrete interleukin-2, which fully activates T$_C$ cells. These cells can then destroy any cancer cells they encounter.

As you are likely aware, this process isn't always 100% effective, and cancer cells are often able to escape the immune response. We address how this happens in the next section.

Quick Check

☐ 3. How does the immune system survey the body for cancer cells?

☐ 4. Walk through the steps of the immune response to cancerous cells.

Pathogens That Evade the Immune Response

Although the immune response works most of the time in preventing serious homeostatic disruptions, certain pathogens have evolved ways to evade elements of the innate and/or adaptive immune responses. For example, many viruses block or inhibit cytokines such as interferons. Other viruses are able to block display of viral antigens on class I MHC molecules, which prohibits T_C cell activation. Bacteria also have ways of evading the immune response, including adaptations that allow them to attach more strongly to host cells, secretion of substances that destroy antibodies, and inhibition of phagocytosis. Such microorganisms are especially pathogenic and difficult for the immune system to eradicate.

Cancer cells are also able to evade the immune response. Sometimes the cells simply grow too quickly and overwhelm the immune system. Other cancer cells have mutated so that they express low numbers of class I MHC molecules, which prevents the expression of adequate numbers of tumor antigens to fully activate T_C cells. In addition, tumor cells provide poor co-stimulatory signals, which prevents T_H cell activation and subsequent T_C cell activation. Finally, tumor cells secrete cytokines that suppress T cell activity. All of these factors combine to make many cancers able not only to evade the immune response but to cause the death of the patient as well.

Quick Check

☐ 5. Explain how microorganisms evade the immune response.

☐ 6. How do cancer cells escape the immune response and spread to other tissues?

Apply What You Learned

☐ 1. A virus develops a mutation that prevents the activation of T_C cells. Predict the consequences of this mutation. What other components of the immune response would need to compensate for the lack of activated T_C cells?

☐ 2. What would happen if cancer cells developed a mutation similar to that in question 1?

☐ 3. A patient presents to the emergency room with a fever. What does the fever indicate?

☐ 4. Blood work on the patient in question 3 shows an elevated level of neutrophils in her blood. What does this tell you about the possible cause of the fever? Explain.

See answers in Appendix A.

MODULE 20.7
Disorders of the Immune System

Learning Outcomes

1. Describe the characteristics of the different types of hypersensitivity disorders.

2. Describe the common immunodeficiency disorders.

3. Explain why HIV targets certain cell types, and describe the effects this virus has on the immune system.

4. Describe the characteristics of common autoimmune disorders.

Given the importance of the immune system, it is probably no surprise that disorders of this system can result in catastrophic homeostatic imbalances. Disorders take three forms: (1) The immune system may overreact and damage tissues, which results in a *hypersensitivity disorder;* (2) one or more components of the immune system may fail, which is an *immunodeficiency disorder;* and (3) the immune system may treat self antigens as foreign and attack the body's own tissues, which is an *autoimmune disorder.* This module explains each type of disorder, including the causes, symptoms, and treatments.

Hypersensitivity Disorders

You have probably experienced at least one type of the group of immune system dysfunctions known as the **hypersensitivity disorders,** in which the immune system's response causes tissue damage. There are four types of hypersensitivity disorder (numbered I–IV) that are classified according to the exact immune components causing the hypersensitivity.

Type I: Immediate Hypersensitivity

The most common type of hypersensitivity disorder is **type I,** or **immediate, hypersensitivity,** which affects a tremendous number of people—about 20% of the U.S. population. This type of hypersensitivity is more commonly known as **allergies,** and the disorders that accompany it are called *allergic disorders.* We touched on this form of hypersensitivity earlier with our discussion of IgE (the "allergy antibody") and mast cells.

Immediate hypersensitivity reactions occur when an individual reacts to a foreign antigen, an allergen. Common allergens include pollen, dust mites, pet dander, peanuts, shellfish, and bee venom. As shown in **Figure 20.27**, in the first exposure of an individual to an allergen, ① an allergen binds a B cell, and ② the B cell differentiates into plasma cells that begin to secrete antibodies, just as they normally would when exposed to a foreign antigen. But there is one key difference between this response and a normal response: The plasma cells secrete IgE instead of IgG or IgM. This first exposure generates a primary immune response, which occurs slowly and produces few symptoms, but does result in the formation of large numbers of IgE molecules that coat mast cells and basophils. Such mast cells and basophils are said to be *sensitized.*

Subsequent exposures to the allergen in a sensitized person result in rapid responses that occur within a few minutes. This happens because ③ when the allergen binds the sensitized mast cell or basophil, the IgE molecules on the cell form cross-links. These links cause the cell to release the inflammatory mediators in its granules, including histamine, leukotrienes, and prostaglandins. These substances trigger the vasodilation and increased capillary permeability of an inflammatory response.

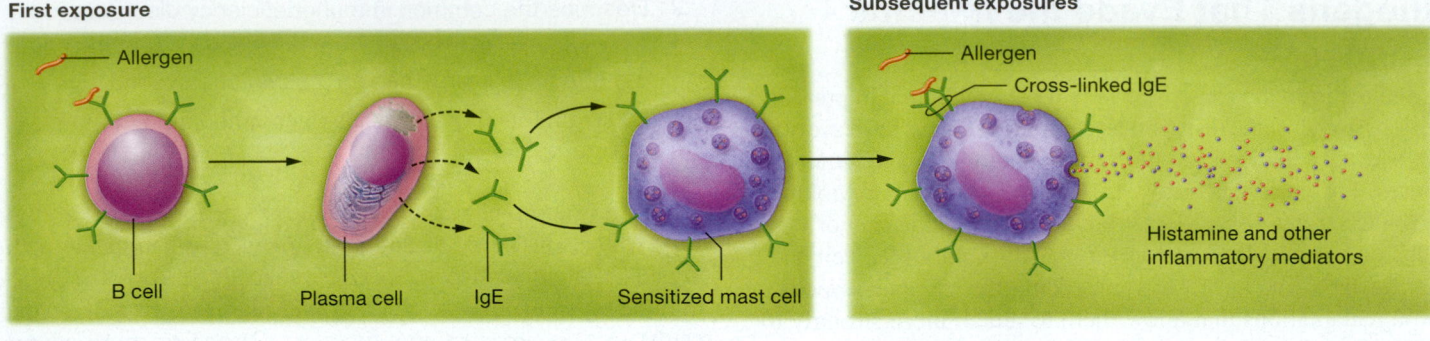

First exposure

① An allergen binds a B cell. ② The B cell differentiates into plasma cells that secrete IgE antibodies, which bind to a mast cell, sensitizing it.

B cell · Plasma cell · IgE · Sensitized mast cell

Subsequent exposures

③ The allergen binds the sensitized mast cell, and IgE molecules on the cell form cross-links that cause the cell to release inflammatory mediators from its granules, triggering an inflammatory response.

Allergen · Cross-linked IgE · Histamine and other inflammatory mediators

Figure 20.27 Type I hypersensitivity response.

The symptoms of immediate hypersensitivity depend on the scope of the reaction. Local reactions, such as those in the nasal cavity upon exposure to pollen, produce the common symptoms of runny nose and itchy eyes. Allergen exposure may also result in a skin rash in which small areas of the skin appear red and elevated, known as *hives,* or *urticaria* (er′-tih-KAER-ee-ah). Urticaria are often itchy but generally fade within 30 minutes. More potent reactions occur in patients with the allergic respiratory disease *asthma* (AZ-mah), in which allergen exposure severely limits the ability to breathe by triggering inflammation, smooth muscle spasm, and excess mucus secretion in the respiratory passages. These reactions can be treated by a variety of medications, as discussed in *A&P in the Real World: Treatments for Allergies.*

The most dramatic immediate hypersensitivity reaction, **anaphylactic shock** (aeh′-nah-fah-LAK-tik), involves a systemic release of histamine and other inflammatory mediators. Anaphylactic shock consists of several life-threatening events, including (1) severe spasm of the smooth muscle of the respiratory tract, particularly the smooth muscle in the larynx (the "voice box"); (2) systemic vasodilation, which causes blood pressure to drop and decreases blood flow to all organs, including the brain; and (3) increased capillary permeability in all of the body's capillaries, which further lowers blood pressure and causes body-wide swelling, as there is a massive loss of fluid to the tissue spaces and lungs. Anaphylactic shock is easily fatal if not treated immediately. Treatment generally consists of an injection of epinephrine, which causes relaxation of the smooth muscle of the airways and systemic vasoconstriction. Patients with a known risk for anaphylactic shock generally carry an epinephrine "pen" with them that allows self-administration of epinephrine in case of exposure to the allergen.

Type II: Antibody-Mediated Hypersensitivity

In **type II hypersensitivity,** also known as *antibody-mediated hypersensitivity,* the antibodies produced by the immune response that bind to foreign antigens also bind to self antigens. This

Treatments for Allergies

Given the enormous number of people who suffer from allergic disorders, the market for medications that treat allergies is huge. Most medications available over the counter are *antihistamines,* or medications that block cells' receptors for histamine. This prevents the pro-inflammatory effects of histamine and limits the symptoms experienced by allergy sufferers. Unfortunately, histamine is also involved in a variety of other processes, both as a hormone and as a neurotransmitter. This leads to the most common side effect of antihistamines: drowsiness. Indeed, some antihistamines are so sedating that they are used more as sleep aids than as allergy treatments. Newer antihistamines have been modified so they are less able to cross the blood brain barrier, and are therefore somewhat less sedating.

Other classes of anti-allergy medications and treatments include the following:

- *Antileukotriene agents* block the enzyme that produces leukotrienes, which inhibits many aspects of the allergic inflammatory response.
- *Corticosteroids* block the synthesis of leukotrienes and prostaglandins and are potent inhibitors of allergic inflammation. They are commonly used for related allergic disorders such as asthma.
- *Allergen immunotherapy,* commonly known as "allergy shots," involves administration of the allergen in increasing doses, with the aim of inducing tolerance to the allergen and a diminished IgE response.

reaction occurs in three situations: (1) Foreign antigens bind to normal self antigens; (2) donor erythrocytes infused into an individual are mismatched in the ABO/Rh antigen groups; and (3) self-reactive B cells are not destroyed in the bone marrow, which leads to autoimmunity (discussed shortly).

An example of the first situation is the reaction to antibiotic medications such as penicillin. These drugs are able to bind to erythrocytes in certain patients, which alters their antigens and causes them to be recognized as foreign. Activated B cells secrete antibodies that lead to complement activation and complement-mediated lysis of erythrocytes.

Type III: Immune Complex–Mediated Hypersensitivity

The reactions of **type III hypersensitivity** are mediated by *immune complexes,* or clusters of soluble antigens (those not attached to the cell surface) bound to antibodies. Immune complexes are generally cleared by phagocytes, but certain complexes are difficult for macrophages to ingest. These complexes deposit in various places in the body, including the capillary beds in the kidneys, blood vessel walls, the synovial membrane of joints, and the choroid plexus in the brain. When they deposit in these organs and tissues, they initiate an inflammatory reaction that attracts neutrophils and causes damage to the affected areas.

Type IV: Delayed-Type Hypersensitivity

The final type of hypersensitivity is **type IV,** or **delayed-type, hypersensitivity (DTH).** DTH is unique in that it is mediated by T cells rather than antibodies. In these cases, T_H cells recognize antigens bound to MHC molecules as foreign and mediate their destruction by activating and recruiting macrophages and in some cases T_C cells. This reaction generally takes 2–3 days to manifest, hence its name. As with type I hypersensitivity, T_H cells must be sensitized by an initial exposure, and the reaction occurs with subsequent exposures.

One of the most common types of DTH reactions is *contact dermatitis,* in which skin comes into contact with an allergen such as the oils in poison ivy or poison oak, certain metals, or other chemicals that can form complexes with skin proteins. The allergens are taken in and displayed by dendritic cells in the skin, which activate T_H cells. The T_H cells then secrete cytokines that recruit macrophages. The macrophages arrive in large numbers and destroy the self cells with the foreign antigens through phagocytosis and the release of enzymes and other chemicals that also damage surrounding cells. This results in a rash that is itchy and occasionally painful.

Other forms of DTH are caused by intracellular pathogens that are not easily cleared by the immune response. One such pathogen is the bacterium *Mycobacterium tuberculosis,* which causes the respiratory infection tuberculosis. In this case, the DTH reaction begins as a normal immune response to a pathogen. However, this pathogen is difficult to eradicate, and the response often persists, with macrophages continually secreting enzymes and chemicals that nonspecifically damage surrounding cells. See *A&P in the Real World: The Tuberculin Skin Test* to find out how the DTH reaction can be used to identify tubercular infection.

The Tuberculin Skin Test

We can take advantage of the DTH reaction to determine whether a person has been exposed to the bacterium that causes tuberculosis. The **Mantoux test** (mahn-TOO), also known as the **PPD test** (for purified protein derivative), involves injection of a small amount of a protein derived from the cell wall of the *Mycobacterium tuberculosis* bacterium just underneath the skin. If the individual has been exposed to tuberculosis, he or she will have sensitized T cells and will mount a DTH response. This leads to an area of *induration*—a hard, swollen area caused by the infiltration of macrophages and subsequent cellular destruction. A person who has not been previously exposed to the tuberculosis antigens will have no area of induration because he or she will lack sensitized T cells.

Quick Check

☐ 1. Define the four types of hypersensitivity reactions.

☐ 2. Explain how type IV hypersensitivity differs from the other three types.

Immunodeficiency Disorders

A decrease in the function of one or more components of the immune system results in an **immunodeficiency disorder.** In the preceding modules, you have seen how the various parts of the immune response are interconnected. For this reason, when one component of the immune response fails, it generally leads to failure of other components, and the results can be catastrophic. There are two basic types of immunodeficiency disorders: *primary immunodeficiencies,* which are genetic or developmental in nature, and *secondary immunodeficiencies,* which are acquired through infection, trauma, cancer, or certain medications. In the following sections, we focus on some of the more common primary and secondary immunodeficiency disorders, and examine their effects on the immune response as a whole.

Primary Immunodeficiency Disorders

Primary immunodeficiency disorders may impair either innate or adaptive immunity. The most common dysfunctions of innate immunity involve deficient complement proteins or abnormalities in phagocytes (which may be present in too low a number, demonstrate deficient bacterial killing, or be unable to properly adhere to pathogens). Patients with defective phagocytes and complement proteins are at much higher risk for bacterial infections and parasitic infections.

Common dysfunctions of adaptive immunity include the *hypogammaglobulinemias* (hy´-poh-gaeh´-muh-glob´-yoo-lin-EE-mee-uhz), which are characterized by a decrease in one or more types of antibodies. Another common form of primary immunodeficiency involving adaptive immunity is a cluster of disorders referred to as **severe combined immunodeficiency,** or **SCID.** The different forms of SCID are caused by failures of lymphoid cell lines in the bone marrow, which affect B cells, NK cells, and T cells to varying degrees. Regardless of the underlying cellular involvement, the forms of SCID share common features, including low circulating levels of lymphocytes, a condition known as **leukopenia** (loo′-koh-PEE-nee-uh; -*penia* = deficiency), failure of the thymus to develop, and failure of cell-mediated immunity. This results in recurrent infections that typically cause death early in life unless the patient lives in a sterile environment (a "bubble"). However, it is extraordinarily difficult to maintain a completely sterile environment, and such measures only temporarily prolong a patient's life.

Secondary Immunodeficiency Disorders and AIDS

Secondary immunodeficiency disorders have multiple forms. Many of these disorders are induced artificially to combat cancers originating in the bone marrow or to prevent transplant rejection. Others are a result of cancers of immune cells and lymphoid organs that depress the immune response in some way. However, the most common cause of secondary immunodeficiency by far is the virally induced disease **acquired immunodeficiency syndrome,** or **AIDS.**

AIDS is caused by **human immunodeficiency virus 1 (HIV-1),** which is spread through contact with infected blood, semen, vaginal fluid, or breast milk. HIV-1 preferentially binds and interacts with cells displaying CD4 molecules. The affinity of the virus for CD4 molecules is due to a glycoprotein on its surface that fits into a CD4 molecule the way a key fits into a lock. This makes cells that display large numbers of CD4 molecules, particularly T_H cells, specifically at risk for infection. Keep in mind that certain strains of HIV-1 can also interact with CD4 molecules on other immune cells, such as monocytes, macrophages, and dendritic cells.

After an HIV-1 virion (VEE-ree-ahn), or viral particle, has bound to a CD4 molecule, it interacts with other cell surface molecules that allow it to gain entry into the host cell. The events that occur inside the cell differ from many other viruses because HIV-1 is a *retrovirus,* which means that it contains an RNA genome and reproduces with the help of an enzyme called *reverse transcriptase.* As implied by its name, reverse transcriptase catalyzes the reverse process of normal transcription—instead of transcribing DNA into RNA, this enzyme catalyzes the transcription of viral RNA into DNA. The DNA copy of the viral genome is then inserted into the host DNA, where it triggers the production of viral RNA and proteins. The infected cell eventually lyses, and the new virions are released to infect new cells.

The progression of HIV-1 infection can be divided into three phases, although the course may vary from one individual to another. The first phase, the *acute phase,* lasts about 3 months and is characterized by a sharp decline in T_H cells and a sharp rise in HIV-1 virions. Patients may exhibit flu-like symptoms during this phase, although the condition often goes unnoticed. The second

phase, the *chronic phase,* begins with the production of antibodies to HIV-1 virions, an initial slight recovery in the number of T_H cells, and a decline in the number of HIV-1 virions. This is due to the actions of B cells, T_C cells, various cytokines, and the remaining T_H cells. The chronic phase may last 8 or more years in untreated individuals, during which many patients show few to no signs of HIV-1 infection. It is during the final phase when an individual is said to have AIDS, which is characterized by progressively declining numbers of T_H cells and progressively increasing numbers of HIV-1 virions. Without treatment, the duration of the final phase is generally no more than 3 years.

The signs and symptoms of AIDS are due largely to the destruction of T_H cells. As you have read, T_H cells are required for almost all parts of the innate and adaptive immune responses to function properly. For this reason, the loss of T_H cells causes the entire adaptive immune response to fail (some innate responses remain). This failure leads to recurrent infections, particularly with agents that are *opportunistic,* or not generally pathogenic in immunocompetent patients. For example, respiratory infection with a species of the fungus *Pneumocystis* (noo-moh-SIS-tus) is extremely common in AIDS patients but very rare in individuals with a healthy immune system. Other consequences of AIDS include cancers such as *Kaposi's sarcoma* (kap-OH-seez), which affects the blood vessels and leads to purple-red skin lesions, shown in **Figure 20.28**; damage to the central and peripheral nervous systems; and severe muscle wasting. Patients generally succumb to complications from infection or cancer.

AIDS was discovered in 1981 primarily in homosexual males and intravenous drug abusers. Since that time, it has spread across the globe to now-epidemic proportions. Over one million individuals in the United States are infected with HIV-1, and although the disease remains more prevalent in homosexual males, it is being found with increasing frequency in heterosexual males and females. Worldwide, over 36 million were living with HIV-1 as of the end of 2016. Note that a staggering percentage of these individuals with HIV-1 live in sub-Saharan Africa—about 66%—and the majority of these patients are females.

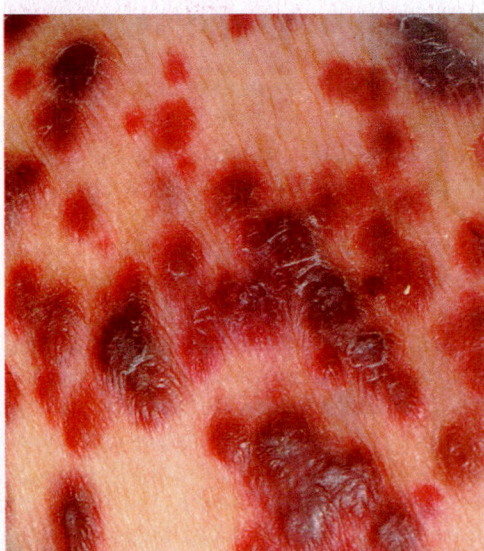

Figure 20.28 **The skin of an AIDS patient with Kaposi's sarcoma.**

Clearly, the development of treatments for HIV and AIDS is a global priority. Currently, a number of drug therapies are available for HIV-1. There are three main mechanisms by which most of these drugs work: (1) inhibiting reverse transcriptase, (2) inhibiting viral enzymes needed to synthesize mature virions, and (3) blocking the entry of HIV-1 into its target cells. The drugs are typically administered in combination as a "cocktail" to inhibit as many aspects of the viral replication cycle as possible.

Although drug cocktails do prolong patients' lives, none actually cure the disease. In addition, the newer drug cocktails are expensive, require following a strict regimen, and cause unpleasant adverse effects. Furthermore, due to the error-prone nature of reverse transcriptase, HIV-1 is prone to mutations, some of which may render certain drugs ineffective. For these reasons, the best way to effectively curb the spread of HIV-1 and put a halt to the epidemic is to develop a vaccine for the virus. Unfortunately, multiple issues make this a considerably more difficult task than with other pathogens. For one, the virus's high mutation rate selects for mutants that are resistant to immunity. An additional factor is that most vaccines stimulate the development of antibodies, but antibodies to HIV-1 in most individuals provide little protection from the disease because the virus has a high mutation rate. Research into how to best combat these and other issues is ongoing, but progress has unfortunately been slow. Indeed, the eradication of this epidemic is proving to be one of the greatest public health challenges of modern times.

Quick Check

☐ 3. How do primary and secondary immunodeficiency disorders differ?

☐ 4. Why is AIDS so devastating to the immune system?

Autoimmune Disorders

Normally, several processes ensure immune self tolerance—that the immune system does not react to self antigens and damage the body's own tissues and cells. However, sometimes these processes fail, and the result is populations of self-reactive T cells or of B cells that secrete antibodies that bind to self antigens, called *autoantibodies*. This produces an **autoimmune disorder.** Autoimmune disorders may be localized and affect only one organ or tissue, or they may be systemic and affect multiple parts of the body. Autoimmunity results in a type II, III, or IV hypersensitivity reaction, which is what produces the organ and tissue damage characteristic of the disease.

Many situations could lead to the development of autoimmunity, including the following:

- **Release of self antigens not previously encountered by T cells.** T cells are generally exposed to self antigens during development, and those self-reactive T cells are destroyed. However, some antigens are *sequestered,* meaning that they are not exposed to the developing T cells. For example, a protein in myelin, the substance surrounding the axons of many neurons, is normally sequestered by the blood brain barrier and T cells are not exposed to it. But, infection or trauma might release this protein and its antigens into the circulation, activating T cells specific for this antigen. This is believed to be the mechanism behind the development of the autoimmune disease *multiple sclerosis,* in which the myelin sheath is attacked.

- **Foreign antigens mimic self antigens.** Certain viral and bacterial antigens closely resemble normal self antigens. Normally, T cells specific for these antigens do not attack self cells because of the lack of co-stimulatory signals. However, if an individual comes into contact with these pathogens, co-stimulatory signals activate these T cells, which then attack self cells. This occurs with the disease *rheumatic fever* (roo-MAT-ik), which occurs after infection with a specific species of streptococcal bacterium whose antigens resemble those on the heart muscle.

- **Cells may inappropriately express class II MHC molecules.** As we discussed earlier, only B cells and antigen-presenting cells normally display class II MHC molecules. For poorly understood reasons, other cell types may also inappropriately express class II MHC molecules. This activates T cells and triggers an immune response to these normal self antigens. This appears to be the case with type 1 diabetes mellitus, in which the immune system destroys the insulin-producing cells of the pancreas.

- **Certain pathogens nonspecifically activate B cells.** Many pathogens can induce the production of cytokines that nonspecifically activate B cells, resulting in the production of autoantibodies. This is the proposed mechanism behind the disease *systemic lupus erythematosus* (eh′-rih-them-ah-TOH-sis), in which infection with a certain virus is followed by production of antibodies to proteins in DNA, erythrocytes, platelets, and leukocytes.

Each of these situations also relies on multiple genetic factors that increase an individual's tendency to develop autoimmunity.

Quick Check

☐ 5. What is an autoimmune disorder? What produces the tissue damage characteristic of these disorders?

☐ 6. What are the potential causes of autoimmune disorders?

Apply What You Learned

☐ 1. Certain drugs have been designed to inhibit the release of the contents of mast cell granules. Predict the effect of these drugs. Which type of hypersensitivity would they most affect?

☐ 2. A hypothetical treatment for HIV-1 infection increases antibody production by B cells. Is this likely to help treat this infection? Why or why not?

☐ 3. Autoimmune disorders are often treated with agents that suppress the immune system. How would this alleviate some of the symptoms of these disorders? Predict a potential adverse effect of these agents.

See answers in Appendix A.

Chapter Summary

For everything you need to succeed in this course, go to **Mastering** Anatomy & Physiology. There you will find:

 Practice Tests

Author Podcasts

Big Picture Animations

Concept Boost Video Tutors

iP2™ Interactive Physiology 2.0 Tutorials

A&PFlix A&P Flix 3D Animations

Active-Learning Workbook

MODULE 20.1
Structure and Function of the Lymphatic System 755–762

- The lymphatic system consists of **lymphatic vessels** and **lymphoid organs** that regulate the interstitial fluid volume, absorb dietary fats, and assist in immune functions.

- Lymphatic vessels are a series of blind-ended vessels that return lymph to the blood via the **right lymphatic duct** or the **thoracic duct.**

- Lymphoid organs contain **macrophages,** lymphocytes, granulocytes, **dendritic cells,** and **reticular cells.**

- **Mucosa-associated lymphatic tissue,** or **MALT,** is located along the mucous membranes. Specialized MALT includes the **tonsils, Peyer's patches,** and **appendix.**

- **Lymph nodes** are encapsulated clusters of lymphatic tissue located along lymphatic vessels. Lymph nodes filter lymph as it travels back to the cardiovascular system and trap pathogens.

- The **spleen** filters the blood. The **white pulp** removes pathogens, and the **red pulp** destroys old erythrocytes.

- The **thymus** is the site of maturation of T cells. The thymus is active in infants and children, and begins to atrophy after the early teenage years.

MODULE 20.2
Overview of the Immune System 762–765

- **Innate immunity** does not require exposure to **antigens** to produce a response, and it responds to pathogens or classes of pathogens in the same way.

- The first line of defense involves surface barriers; the second line of defense involves the cells and proteins of innate immunity;

and the third line of defense involves the cells and proteins of adaptive immunity.

- **Adaptive immunity** requires exposure to a specific antigen to mount a response. The two arms of adaptive immunity are **antibody-mediated immunity** and **cell-mediated immunity.**

- **Surface barriers** consist of the skin that lines the body's external surfaces and the mucous membranes.

MODULE 20.3
Innate Immunity: Internal Defenses 765–772

- Phagocytes of innate immunity include **macrophages, neutrophils, eosinophils,** and **dendritic cells.** Dendritic cells and macrophages function as **antigen-presenting cells.**

- Nonphagocytic cells of innate immunity include **NK cells, basophils,** and **mast cells.**

- The **complement system** is a group of 30 or more plasma proteins produced primarily by the liver that are involved in cell lysis, enhancing phagocytosis, enhancing inflammation, and neutralizing certain viruses.

- The cytokines involved in innate immunity include *tumor necrosis factor* (*TNF*), **interferons,** and *interleukins.*

- The **inflammatory response** occurs in response to any cellular injury.
  ○ Injured cells secrete *inflammatory mediators* that trigger vasodilation, increased capillary permeability, and **chemotaxis.**
  ○ In the second part of the inflammatory response, local macrophages are activated, and blood-borne neutrophils and monocytes are attracted to the area.
  ○ **Fever** is a body temperature above the normal range of 36–38° C. Fever is initiated when **pyrogens** reset the hypothalamic temperature to a higher range.

MODULE 20.4
Adaptive Immunity: Cell-Mediated Immunity 772–780

- T cells form in the bone marrow but migrate to the thymus to mature.

- Cell-mediated immunity involves T_H, or **CD4, cells** and T_C, or **CD8, cells.** These cells respond to cancer cells, cells infected with intracellular pathogens, and foreign cells.

- T cells interact with pieces of antigen bound to **major histocompatibility,** or **MHC, molecules. Class I MHC molecules,** which are located on all nucleated cells, display endogenous antigens and interact with T_C cells. **Class II MHC molecules,** which are located on antigen-presenting cells and other immune cells, display exogenous antigens and interact with T_H cells.

- Naïve T cells are activated when the T cell receptor binds to an MHC–antigen complex in the presence of a *co-stimulator*. Activated T cells proliferate and differentiate into *effector cells* and **memory T cells.**

- T_H cells activate T_C cells, stimulate macrophages and B cells, and enhance the cells and proteins of innate immunity.

- T_C cells release **perforin,** which perforates the plasma membrane so that enzymes can enter the cell and fragment its DNA.

- The four basic kinds of transplants are *autografts, isografts, allografts,* and *xenografts.*

MODULE 20.5
Adaptive Immunity: Antibody-Mediated Immunity 780–787

- B cells develop and mature in the bone marrow.

- Antibody-mediated immunity is mediated by B cells that secrete **antibodies,** which bind to specific antigens.

- When a naïve B cell encounters its antigen, the antigen binds and triggers the B cell to proliferate and differentiate into antibody-secreting **plasma cells** and **memory B cells.**

- Antibodies are Y-shaped proteins with antigen-binding sites on the tips of their arms. There are five classes of antibody: the monomers **IgG, IgE,** and **IgD;** the dimer **IgA;** and the pentamer **IgM.**

- The functions of antibodies include **agglutination, precipitation,** and opsonization, all of which enhance phagocytosis; **neutralization;** activation of complement; and stimulation of inflammation.

- The first exposure to an antigen generates the slower **primary immune response.** Subsequent exposures to antigens generate the faster, more efficient **secondary immune response.**

- **Vaccinations** expose an individual to an antigen to generate memory cells so that subsequent antigen exposures will trigger a secondary immune response.

- **Active immunity** is generated by exposure to an antigen and results in the production of memory cells. **Passive immunity** is generated by receiving preformed antibodies from another source.

MODULE 20.6
Putting It All Together: The Big Picture of the Immune Response 788–793

- The immune response to a viral infection involves a series of events that culminates in the destruction of virally infected cells by T_C cells.

- The immune response to a bacterial infection involves phagocytosis of bacteria by neutrophils and macrophages, and the activation of antibody-secreting plasma cells.

- The immune response to cancer culminates in the activation of T_C cells, which kill tumor cells.

- Pathogens evade the immune response through evading phagocytosis, mutation, and blocking secretion of certain cytokines.

MODULE 20.7
Disorders of the Immune System 793–797

- A **hypersensitivity disorder** occurs when the immune system's response causes tissue damage. There are four classes: **types I, II, III,** and **IV.**

- A decrease in the function of one or more components of the immune system results in an **immunodeficiency disorder.** A common example is **acquired immunodeficiency syndrome,** or **AIDS,** which is caused by the destruction of T_H cells by the **human immunodeficiency virus (HIV-1).**

- **Autoimmune disorders** occur when the immune system recognizes normal cells as foreign due to self-reactive T and B cells and the production of *autoantibodies.*

Assess What You Learned

Scan the QR Code for additional practice test questions

https://goo.gl/BKDijC

LEVEL 1 Check Your Recall

1. Which of the following is not a function of the lymphatic system?

   a. Absorption of dietary fats
   b. Secretion of digestive enzymes
   c. Regulation of interstitial fluid volume
   d. Housing and maturation of leukocytes

2. Mark the following statements as true or false. If a statement is false, correct it to make a true statement.

   a. Lymphatic vessels constitute a one-way system that delivers interstitial fluid from the blood vessels to the extracellular space.
   b. Lymph from the lower limbs drains into the right lymphatic duct.
   c. Fat-containing lymph from the intestines drains into the cisterna chyli.
   d. The thymus is the site of maturation of B cells.
   e. Clusters of MALT located around the oral and nasal cavities are known as Peyer's patches.

3. Fill in the blanks: The lymphoid organ that filters the blood is the _____, and the lymphoid organ that filters the lymph is the _____.

4. Fill in the blanks: Nonspecific immunity is also known as
   _____. Specific immunity is also known as _____.
   Specific immunity has the capacity for immunological
   _____.

5. Which of the following make up the body's first line of defense?

   a. Surface barriers
   b. Cells and proteins of adaptive immunity
   c. Cells and proteins of innate immunity
   d. All of the above

6. Which of the following does *not* describe the relationship
   between the lymphatic and immune systems?

   a. Lymphoid organs and tissues house cells of the immune system.
   b. Lymphoid organs and tissues trap pathogens for the immune
      system.
   c. Lymphoid organs activate cells of the immune system.
   d. Lymphoid organs and tissues create the surface barriers of
      the immune system.

7. Mark the following statements as true or false. If a statement is
   false, correct it to make a true statement.

   a. Surface barriers contain substances in their secretions that
      kill pathogens and deter their growth.
   b. Phagocytic cells of innate immunity include NK cells and
      basophils.
   c. NK cells are cytotoxic cells that lyse and kill cancer cells
      and cells infected with certain viruses.
   d. Interleukin-1 is a cytokine that prevents viral replication in
      infected cells.
   e. Fever is generated by pyrogens that reset the temperature set
      point of the hypothalamus to a lower value.

8. Which of the following functions is/are performed by
   complement proteins?

   a. Cell lysis
   b. Opsonization
   c. Enhancing inflammation
   d. Only a and b
   e. All of the above

9. Fill in the blanks: Injured tissue releases chemicals
   called _____ that mediate the four cardinal signs
   of inflammation, which are _____, _____,
   _____, and _____.

10. When naïve B cells are activated, they differentiate into:

    a. plasma cells.
    b. memory cells.
    c. both a and b.
    d. none of the above.

11. Match the following antibodies with the correct definition.

    _____ IgD   a. Antibody found in secretions
    _____ IgM   b. Most common antibody; crosses the placenta
    _____ IgG   c. Antibody involved in allergies and parasitic
    _____ IgA   infections
    _____ IgE   d. Pentamer and potent agglutinating agent
                e. Antibody bound to the B cell plasma
                   membrane

12. Mark the following statements as true or false. If a statement is
    false, correct is to make a true statement.

    a. In the primary immune response, the lag phase lasts about
       5 days, during which time B cells proliferate and
       differentiate.
    b. The secondary immune response is mediated by plasma cells.
    c. Subunit vaccines consist of pathogens that are alive but
       unable to cause disease.
    d. Vaccinations are given to induce the production of the
       inflammatory response.

13. Explain how active immunity and passive immunity differ.

14. On which type(s) of cells are class I MHC molecules located?

    a. Antigen-presenting cells only
    b. B cells only
    c. All nucleated body cells
    d. Antigen-presenting cells and B cells

15. Fill in the blanks: Class I MHC molecules display _____
    antigens and activate _____ cells. Class II MHC
    molecules display _____ antigens and activate
    _____ cells.

16. Which of the following is *not* a role of T_H cells?

    a. Stimulation of macrophages
    b. Activation of naïve T_C cells
    c. Stimulation of naïve B cells
    d. Stimulation of clonal selection

17. Mark the following statements as true or false. If a statement is
    false, correct it to make a true statement.

    a. The cells involved in organ and transplant rejection are
       primarily B cells.
    b. The immune response to a viral infection involves NK cells
       and different kinds of lymphocytes.
    c. Neutrophils are a critical component of the response to a
       parasitic infection.
    d. Cancer cells are destroyed by NK cells, T_C cells, and
       macrophages.
    e. Dendritic cells may activate both T_H and T_C cells.

18. Type I hypersensitivity is due to release of inflammatory
    mediators from _____, and type IV hypersensitivity is
    due to the actions of _____.

    a. T_H cells; macrophages
    b. neutrophils; T_H cells
    c. B cells; T_C cells
    d. mast cells; T_H cells

19. Explain why a patient with AIDS due to HIV is at an increased
    risk of infection and of developing certain cancers.

20. Which of the following conditions may lead to the
    development of autoimmunity?

    a. Self antigens not previously encountered by T cells are
       released into the circulation.
    b. Foreign antigens mimic self antigens.
    c. Cells may inappropriately display class II MHC molecules.
    d. Certain pathogens nonspecifically activate B cells.
    e. All of the above are correct.

LEVEL 2 Check Your Understanding

1. Eileen presents with flu-like symptoms, including fever, chills, and body aches.

   a. What part of the immune response is likely producing these symptoms?

   b. You order blood work and find that the number of neutrophils in the blood is greatly elevated. What does this tell you about the likely type of pathogen involved? Explain.

   c. What might it mean if lymphocytes instead of neutrophils were elevated? Explain your reasoning.

2. Complement proteins are crucial for stimulating phagocytes to clear immune complexes. Predict what type of hypersensitivity disorder might result from complement deficiencies. Explain what other consequences might arise from a complement deficiency.

3. Terrence has severe asthma and allergies, and is placed on a medication that blocks the functioning of IgE. How would this medication alleviate his symptoms?

LEVEL 3 Apply Your Knowledge

PART A: Application and Analysis

1. Your friend tests her snake-wrangling skills on an eastern diamondback rattlesnake, and is bitten and envenomated (injected with the snake's venom). In the emergency department, doctors administer the antivenin *CroFab,* which contains antibodies to the toxins in snake venom. Is this an example of active or passive immunity? Will this confer any lasting protection to your friend?

2. Carla presents to her physician with a complaint of swollen axillary lymph nodes. Her physician performs a breast exam and finds a tumor on the same side as the swollen nodes.

   a. Explain what process has likely caused her nodes to swell.

   b. Carla undergoes surgery to have the tumor and affected lymph nodes and surrounding lymphatic vessels removed. Predict any complications Carla might have after the surgery that could arise from removal of the lymph nodes and vessels.

3. Mr. White has been diagnosed with *febrile neutropenia,* a condition characterized by the presence of a fever with an extremely low number of neutrophils in the blood. Blood work demonstrates the presence of bacteria in his blood. Explain why neutropenia might lead to a bacterial infection. Why is a bacterial infection particularly dangerous for Mr. White?

PART B: Make the Connection

4. Mr. White developed neutropenia as a consequence of cancer chemotherapy, which destroyed much of his bone marrow. What other components of the immune system would be harmed by bone marrow destruction? Would you expect his hematocrit to be elevated or decreased? What effects would you expect to see from this change in hematocrit? *(Connects to Chapter 19)*

5. The chemotherapeutic drugs Mr. White is taking affect all cells that undergo rapid mitosis, such as those of the skin. What effect would this have on the functions of the skin? How could this affect his immunity? *(Connects to Chapter 5)*

See answers in Appendix A.

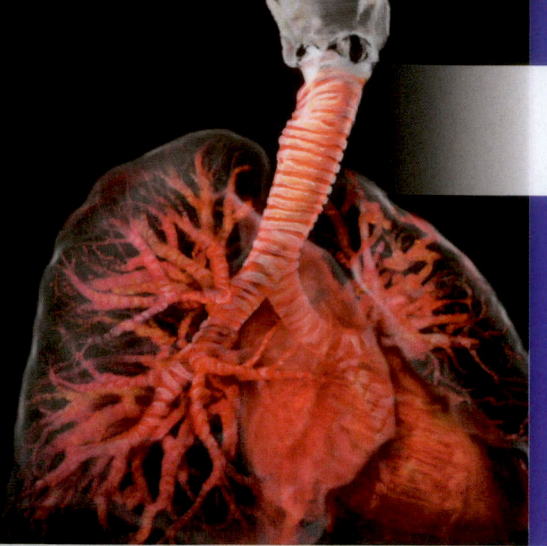

21

The Respiratory System

*For practice applying concepts to a clinical scenario, check out the **Running Case Study** for this chapter @ Mastering Anatomy & Physiology*

Computer-generated image: The human respiratory tract, highlighting the trachea and bronchial tree.

As the American Lung Association says, "You can go three weeks without food, and three days without water, but only three minutes without air." Under most conditions, cells can survive only a few minutes if a person's respiratory functions cease. The organs of the **respiratory system** work ceaselessly to bring our cells the oxygen needed for survival. They also coordinate other functions, such as maintaining acid-base homeostasis and producing sound. In this chapter we explore the structures of the respiratory system, and link these structures to how breathing and gas exchange occur, how homeostasis of the respiratory system is maintained, and what happens when this homeostasis is disrupted.

MODULE **21.1**
Overview of the Respiratory System

Learning Outcomes

1. Describe and distinguish between the upper and lower respiratory tracts.
2. Describe and distinguish between the conducting and respiratory zones of the respiratory tract.
3. Describe the major functions of the respiratory system.
4. Define and describe the four respiratory processes—pulmonary ventilation, pulmonary gas exchange, gas transport, and tissue gas exchange.

In this module, we introduce some basic structures and functions of the respiratory system; we will revisit many of them in greater depth later in the chapter. Let's begin with a brief introductory tour of the structures that support the respiratory system's many functions.

Anatomy of the Respiratory System: An Overview

« FLASHBACK

1. What is elastic connective tissue, and what is its function? (p. 139)
2. What is a serous membrane? What does it secrete? (p. 19)

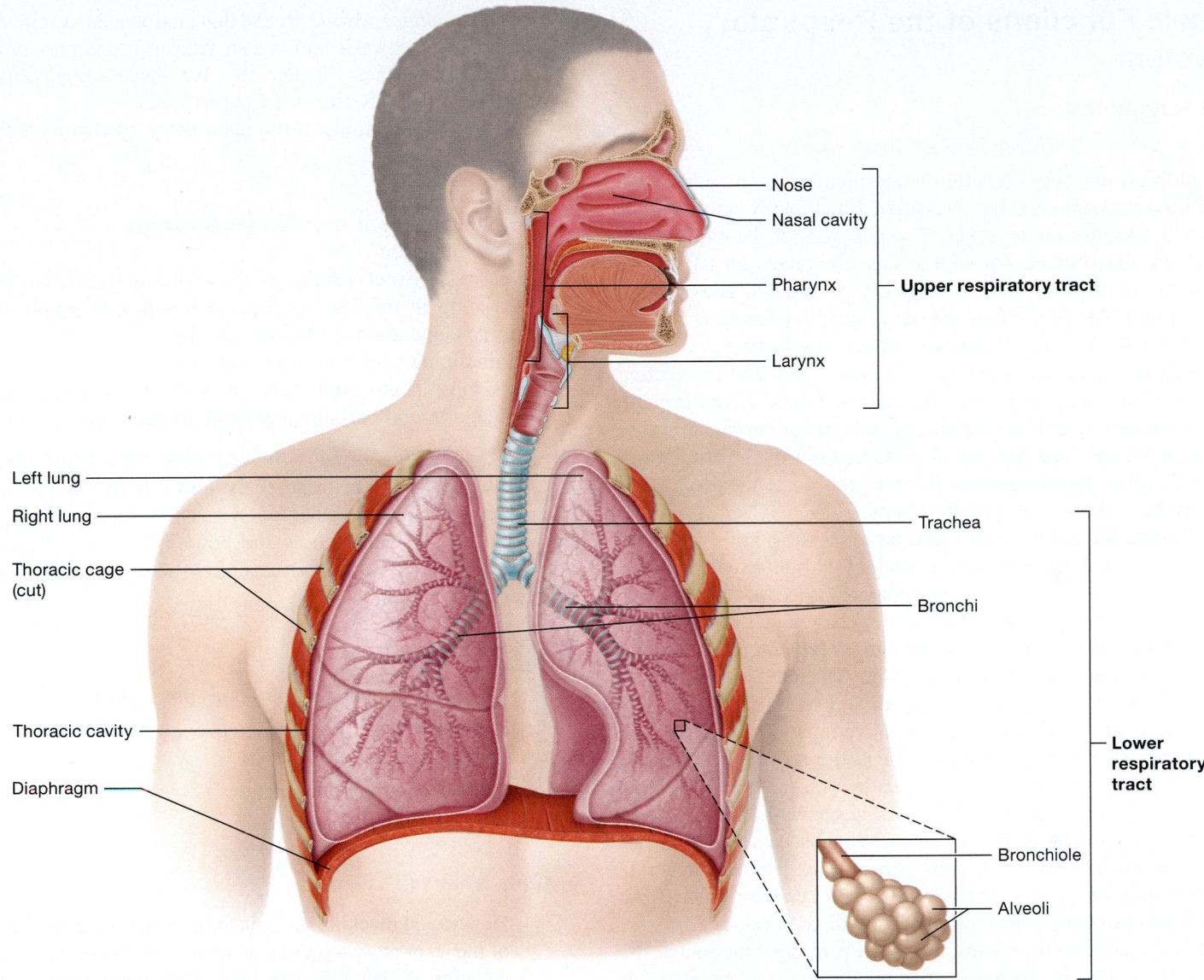

Figure 21.1 Organs of the respiratory system.

The organs of the respiratory system, shown in **Figure 21.1,** reside in the head, the neck, and the thoracic cavity. Notice that these organs include the blood vessels of the pulmonary circuit; the structures of the thoracic (rib) cage, including the respiratory muscles; the paired lungs; and the **respiratory tract,** which consists of the hollow passages that collectively transport gases. Each component of the respiratory tract has a unique gross and histological structure. These components include the following:

- the *nose* and the *nasal cavity,* which is encased by cranial and facial bones;
- the *pharynx,* also called the *throat;*
- the *larynx* in the anterior neck;
- the *trachea* in the mediastinum; and
- the *bronchial tree,* branching passages that begin as *bronchi* (BRONG-kye; singular, *bronchus*) and divide successively until they become tiny *bronchioles.*

The structures of the respiratory tract are classified anatomically into upper and lower tracts. The **upper respiratory tract** includes the passageways from the nasal cavity to the larynx. The **lower respiratory tract** includes the passageways from the trachea to the respiratory tract's terminal structures, the alveoli. **Alveoli** (al-vee-OHL-aye; singular, *alveolus*) are tiny air sacs, arranged in grapelike clusters, through which gases are exchanged. The most conspicuous structures of the respiratory system are the *lungs,* paired spongy organs in the thoracic cavity that are enclosed by the diaphragm and thoracic cage. Each lung is actually just a collection of about 150 million alveoli with the surrounding blood vessels and elastic connective tissue in which the alveoli and branches of the respiratory tract are embedded. We explore in detail the structure of the respiratory tract components and lungs in Module 21.2.

Quick Check

- ☐ 1. What are the main structures of the respiratory system?
- ☐ 2. Is the larynx part of the upper or lower respiratory tract?
- ☐ 3. Where are alveoli? What is their basic function?

Basic Functions of the Respiratory System

« FLASHBACK

1. What is the function of a buffer? (p. 49)

In addition to being classified anatomically, the organs of the respiratory system are also classified functionally into the conducting and respiratory zones. The passages of the **conducting zone** are the conduits through which air travels on its way in and out of the body as it is *inspired,* or inhaled, and *expired,* or exhaled. Air is filtered, warmed, and moistened as it travels through the many branches of the conducting zone. The **respiratory zone** is where gases are exchanged. The conducting zone includes structures from the nose and nasal cavity through the bronchioles, and the respiratory zone includes structures that contain alveoli. The functional classification of the respiratory tract is more precise and useful than the anatomical one, and this is the system we use in this chapter.

Respiration is the process that provides the body's cells with oxygen and removes the waste product carbon dioxide. Respiration actually consists of four separate processes: (1) pulmonary ventilation, (2) pulmonary gas exchange, (3) gas transport in the blood, and (4) tissue gas exchange. *Pulmonary ventilation* (PULL-muh-nehr-ee; *pulmon-* = "lung"), often shortened to *ventilation,* is the movement of air in and out of the lungs. The remaining processes refer to the movement of gases through the body: *Pulmonary gas exchange* is the movement of gases between the lungs and the blood, *gas transport* is the movement of gases through the blood, and *tissue gas exchange* is the movement of gases between the blood and the tissues. (Note that *cellular respiration,* in which cells consume oxygen and generate carbon dioxide as they produce ATP, is a separate process not involved in oxygen delivery and carbon dioxide removal.)

It's no surprise that respiration is a primary function of the respiratory system. However, the respiratory system also has several other vital roles. The mechanism for speech and sound production is found in the respiratory system, as we will discuss in Module 21.2. In addition, recall that the neurons for the sense of smell are located in the olfactory epithelium, a patch of modified epithelial tissue in the roof of the nasal cavity (see Chapter 15).

Other functions of the respiratory system have to do with pressure changes in the thoracic cavity. When you hold your breath and bear down, the pressure increases in the abdominopelvic cavity. This pressure helps to expel the cavity's contents during urination, defecation, and childbirth. The pressure changes generated also help to propel lymph and venous blood in the thoracic and abdominopelvic cavities, which are under very low pressure.

However, you may be less familiar with another important task of the respiratory system that has to do with maintaining homeostasis. Proper functioning of the respiratory system is critical in maintaining acid-base balance in the extracellular fluid, as this system controls one of the primary buffer systems in the body. In addition, it synthesizes an enzyme involved in the production of *angiotensin-II*. Recall that angiotensin-II is involved in acid-base homeostasis and is also vital in blood pressure and fluid homeostasis (see Chapter 16). We discuss angiotensin-II again in the urinary chapter (see Chapter 24).

To recap, the functions of the respiratory system include the following:

- Respiration
- Producing speech and other vocalizations
- Detecting odors
- Helping to expel contents of the abdominopelvic cavity
- Assisting in the flow of venous blood and lymph in the thoracic and abdominopelvic cavities
- Maintaining acid-base homeostasis
- Assisting in the production of angiotensin-II for maintenance of blood pressure and fluid homeostasis

Given all these functions, it makes sense that when there is an imbalance in the respiratory system's homeostasis, tissue oxygenation and multiple other functions are disrupted. Throughout this chapter, we look at how the body maintains respiratory homeostasis and the various effects of homeostatic imbalances of the respiratory system.

Quick Check

☐ 4. List and define the four processes that make up respiration.

☐ 5. How does the respiratory system contribute to the maintenance of homeostasis?

☐ 6. List and describe four functions of the respiratory system besides respiration.

Apply What You Learned

☐ 1. How would blocking the structures of the conducting zone of the respiratory tract impair ventilation? How would the situation change if the structures of the respiratory zone were blocked? Explain.

☐ 2. What homeostatic disturbances would you expect to see with diseases of the respiratory system? Explain.

See answers in Appendix A.

MODULE 21.2

Anatomy of the Respiratory System

Learning Outcomes

1. Trace the pathway through which air passes during inspiration.

2. Describe the gross anatomical features and function of each region of the respiratory tract, the pleural and

thoracic cavities, and the pulmonary blood vessels and nerves.

3. Describe the histology of the different regions of the respiratory tract, the types of cells present in alveoli, and the structure of the respiratory membrane.

4. Explain how the changes in epithelial and connective tissue in air passageways relate to their function.

5. Describe the structure of the lungs and pleural cavities.

In this module, we explore the structure of each component of the conducting and respiratory zones of the respiratory tract. As you read, note how the subtle differences in structure at the gross and microscopic levels enable each component of the respiratory tract to carry out its function.

The Nose and Nasal Cavity

≪ FLASHBACK

1. What are goblet cells, and what are their functions? (p. 134)

2. Which bones form the anterior, superior, and lateral walls of the nasal cavity? (p. 214)

3. In which bones are the paranasal sinuses located? What are the functions of the paranasal sinuses? (p. 215)

The respiratory tract begins with the **nose** and the **nasal cavity,** which is the cavity posterior to the nose. (The mouth and oral cavity are part of the gastrointestinal tract—see Chapter 22.)

The nose and nasal cavity perform the following important functions:

- warm and humidify the inhaled air;
- filter out debris from inhaled air and secrete antibacterial substances;
- house olfactory receptors; and
- enhance the resonance of the voice (which is why your voice sounds funny when you have a "stuffed-up" nose).

Let's take a closer look at the external structure of the nose and the internal structure of the nasal cavity.

External Nasal Anatomy

The nose is covered with skin and supported internally by muscle, bone, and cartilage. Both bone and hyaline cartilage form the framework of the nose, including the paired **nasal bones** superiorly, followed by the **lateral cartilages** and the **alar cartilages** (AY-lahr) inferiorly (**Figure 21.2**). This framework gives rise to several surface features, including the *root* and *bridge* of the nose (the areas between the eyebrows and eyes, respectively), the *dorsum nasi* (NAYZ-aye; the anterior margin of the nose), and the *apex,* or tip, of the nose. Lateral to the apex are the flared **alae** (AYL-ee; "wings"), which surround the paired openings called the **nostrils,** or **anterior nares** (NEHR-eez), anteriorly and laterally. Individual variations in nose size and shape are typically due to differences in the structure of its cartilage framework.

Internal Nasal Anatomy and Histology

When you inhale, air goes through the nostrils and enters the nasal cavity, a hollow space framed by bone and cartilage.

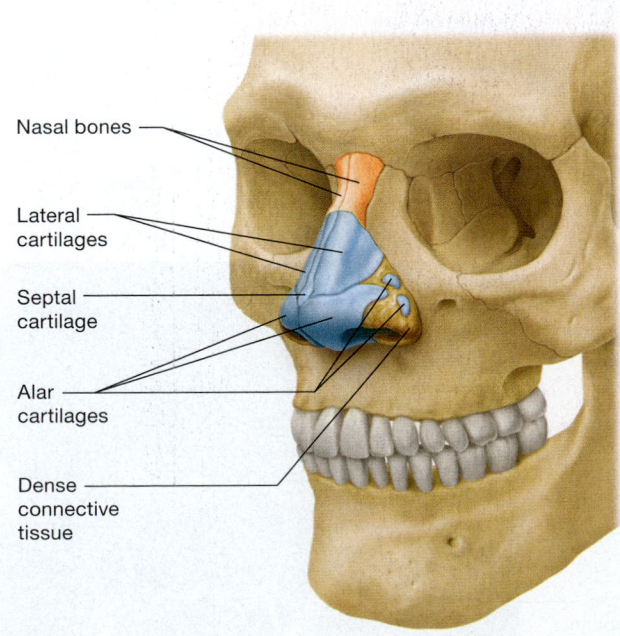

(a) Internal structures

Nasal bones

Lateral cartilages

Septal cartilage

Alar cartilages

Dense connective tissue

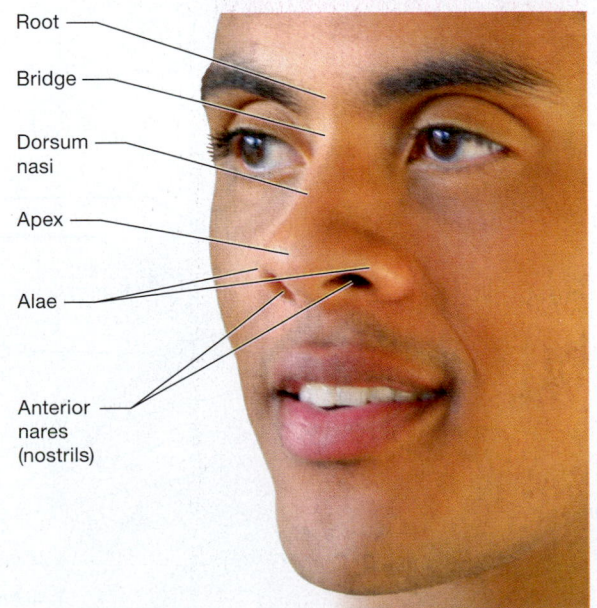

(b) External structures

Root

Bridge

Dorsum nasi

Apex

Alae

Anterior nares (nostrils)

Figure 21.2 Anatomy of the nose.

Notice in **Figure 21.3** that the nasal cavity is larger than you might have thought. It extends from its two anterior openings, the nostrils or anterior nares, to its two posterior openings, called the **posterior nares** (singular, *naris*). The nasal cavity is divided into right and left portions by the *nasal septum,* which, like the framework of the nose, is composed of both bone and hyaline cartilage. The anteriormost part of the nasal cavity, just inside the nostril, is the **vestibule.** The vestibule contains bristle-like hairs that prevent objects such as insects and airborne particles from entering the nasal cavity.

Although the nasal cavity is a hollow structure, it contains very little empty space. Rather, it is filled with bony projections known as **nasal conchae** (KAHN-kee; singular, *concha*) (refer back to Figure 7.12). There are three sets of nasal conchae: the *superior* and *middle nasal conchae,* both part of the ethmoid bone; and the *inferior nasal conchae,* which are independent bones. The nasal conchae curl around three narrow passages called the **superior, middle,** and **inferior nasal meatuses** (mee-AY-tuss-ez). This arrangement causes air flow through the meatuses to be turbulent. The turbulent air flow extracts dust and other debris from the air, much in the same way that the turbulent air flow of a clothes dryer helps to remove lint from your clothes. The foreign debris sticks to the mucus lining the nasal conchae, so it does not reach the deeper structures of the respiratory tract. The turbulence also slightly delays the movement of the air, during which time it picks up moisture from the mucous membranes.

Figure 21.3 also shows hollow cavities called the **paranasal sinuses,** or simply *sinuses,* which are connected to the nasal cavity via small passageways located between the nasal conchae. The paranasal sinuses are located within the frontal, ethmoid, sphenoid, and maxillary bones, and so are named the **frontal, ethmoid, sphenoid,** and **maxillary sinuses,** respectively (see

Chapter 7). Like the nasal cavity, they warm, humidify, and filter the air. They also lighten the skull and enhance voice resonance. Their epithelium is continuous with that of the nasal cavity, which allows the movement of air and mucus between the two areas. Unfortunately, this arrangement also allows infections to spread from the nasal cavity into the paranasal sinuses.

Now let's look a little closer at how the type of cells in each of these areas allows them to perform their functions. The vestibule is lined with stratified squamous epithelium, which enables it to be more resistant to mechanical stresses such as abrasion (scratching). Posterior to the vestibule, the epithelium changes to become two types of mucous membrane: the **olfactory mucosa** and the **respiratory mucosa** (also called *respiratory epithelium*). The olfactory mucosa is located on the roof of the nasal cavity, where it houses olfactory receptors (see Figure 21.3a). These bipolar neurons project through the holes in the cribriform plate called *olfactory foramina* to contact the overlying olfactory bulbs in the brain (see Chapter 15).

The remainder of the nasal cavity is lined by the respiratory mucosa, which is composed of pseudostratified ciliated columnar epithelium (see Figure 4.5d). Interspersed throughout the ciliated columnar cells are specialized unicellular glands called **goblet cells,** which secrete mucus. As one of the main functions of the nasal cavity is air filtration, this ciliated, goblet cell–rich epithelium is an example of how form follows function in the nasal cavity. Foreign particles get trapped in the mucus made by the goblet cells, and the cilia beat in unison to propel the debris and mucus toward the posterior nasal cavity, where the mixture travels down the pharynx and is eventually swallowed or expelled by sneezing or blowing your nose. (Almost all microorganisms that make it to the stomach are destroyed by the acid found there.)

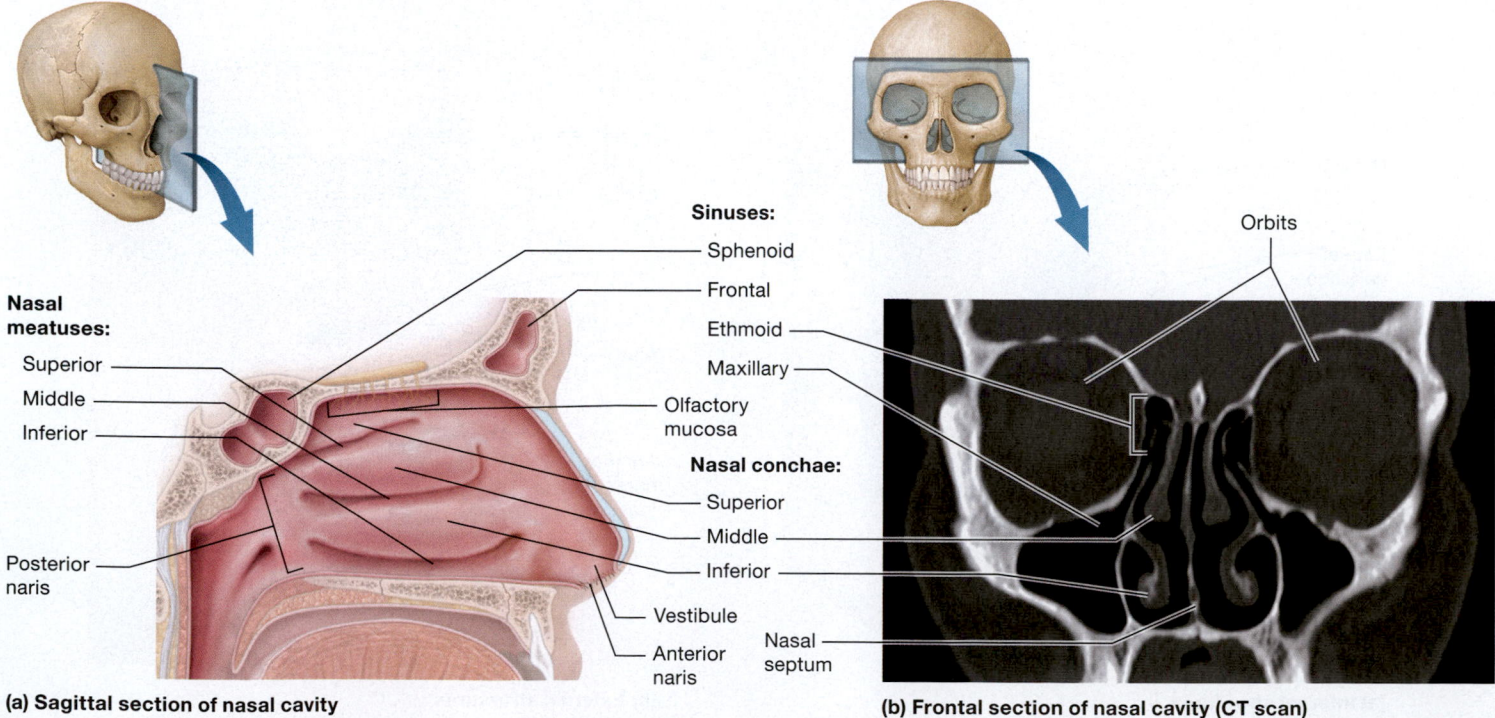

Nasal meatuses:
Superior
Middle
Inferior

Posterior naris

Sinuses:
Sphenoid
Frontal
Ethmoid
Maxillary

Olfactory mucosa

Nasal conchae:
Superior
Middle
Inferior

Vestibule

Anterior naris

Orbits

Nasal septum

(a) Sagittal section of nasal cavity

(b) Frontal section of nasal cavity (CT scan)

Figure 21.3 Anatomy of the nasal cavity.

Deep to the respiratory mucosa is a lamina propria richly supplied with blood vessels. Interestingly, the lamina propria contains a number of large venous sinusoids (see Chapter 18) that can become engorged with blood, an anatomical feature similar to the erectile tissue of the penis. The venous sinusoids fill with blood on one side of the nasal cavity, partially obstructing the flow of air through that side, which helps to prevent the mucosa from drying out. About every 30 minutes, the side filled with blood switches, and air flows through the previously obstructed side. This partial blockage is not usually noticeable, but it may become apparent when the nasal mucosa is inflamed due to allergy or a viral infection such as a cold.

The Pharynx

After inspired air has passed through the posterior nares, it enters the next segment of the respiratory tract, the **pharynx** (FEHR-inks), commonly called the throat. Anatomically, the pharynx has three divisions: the *nasopharynx,* the *oropharynx,* and the *laryngopharynx* (**Figure 21.4**).

The **nasopharynx** (nayz-oh-FEHR-inks) sits posterior to the nasal cavity, beginning at the posterior nares. Like this cavity, it is lined with pseudostratified ciliated columnar epithelium;

the two structures therefore perform the same functions of warming, humidifying, and filtering the inspired air. Within the nasopharynx we find the opening of the *pharyngotympanic tube,* a structure that connects the middle ear with the pharynx (see Chapter 15). The nasopharynx also houses the *pharyngeal tonsil,* which is composed of specialized lymphatic tissue that traps pathogens entering the nasal cavity.

The inferior border of the nasopharynx is a part of the soft palate called the *uvula* (YOOV-yuh-luh). The uvula and the soft palate move posteriorly during swallowing to prevent food or liquid from entering the nasopharynx and nasal cavity. However, sometimes this mechanism fails, especially when you laugh, and the unfortunate result is that food or liquid may come out of your nose.

The next segment of the pharynx is the **oropharynx** (ohr-oh-FEHR-inks), located posterior to the oral cavity. Both food and air pass through the oropharynx, and so it is lined with a protective nonkeratinized stratified squamous epithelium rather than pseudostratified ciliated columnar epithelium, which is too thin to be protective. It extends from the uvula to the tip of the larynx, called the *epiglottis.* It houses three tonsils: a pair of *palatine tonsils* on either side of the uvula, and a single *lingual tonsil* at the base of the tongue.

The last portion of the pharynx, the **laryngopharynx** (luh-ring´-goh-FEHR-inks), extends from the hyoid bone to the

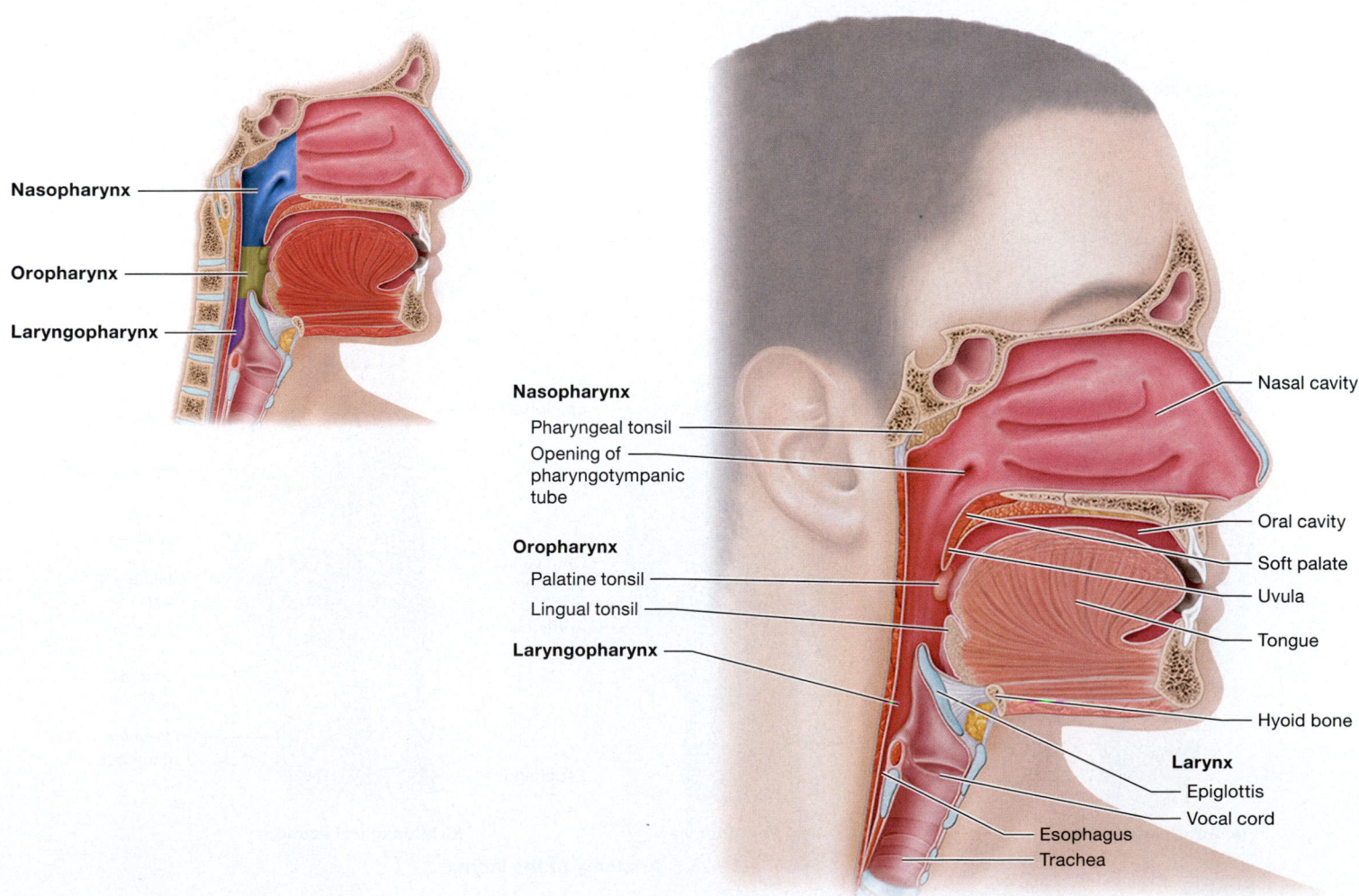

Figure 21.4 Anatomy of the pharynx.

esophagus, the passage that conveys food to the stomach. Anteriorly, it opens into the larynx, and posteriorly it continues as the esophagus. Like the oropharynx, it is a passageway for both food and air, and so is lined with a nonkeratinized stratified squamous epithelium that protects it from abrasion.

Quick Check

☐ 1. Match the following terms with the correct definition or description.

___Respiratory mucosa

___Nasopharynx

___Laryngopharynx

___Oropharynx

___Paranasal sinuses

___Goblet cell

   a. Posterior to the oral cavity, lined with stratified squamous epithelium

   b. Extends from the posterior nares to the uvula

   c. Unicellular gland that secretes mucus

   d. Extends from the hyoid bone to the esophagus

   e. Pseudostratified ciliated columnar epithelium

   f. Hollow cavities that are connected to the nasal cavity

The Larynx

« FLASHBACK

1. What are the basic properties of hyaline cartilage? (p. 142)

2. What are the basic properties of elastic cartilage? (p. 142)

Inspired air moves from the laryngopharynx into the next part of the respiratory tract: the **larynx** (LEHR-inks) (**Figure 21.5**). This short passage lies anterior to the esophagus and extends from about the third to the sixth cervical vertebra (see Figure 21.4). The larynx keeps food and liquids out of the rest of the respiratory tract and, as its common name implies, it houses the vocal cords, which are involved in sound production.

Superior to the vocal cords, the larynx is lined with a stratified squamous nonkeratinized epithelium that is continuous with that of the laryngopharynx. This prevents the larynx surface from abrasion due to contact with food. Inferior to the vocal cords, the epithelium changes to a pseudostratified ciliated columnar epithelium, which we saw in the nasal cavity and the nasopharynx. The cilia in the larynx propel mucus and debris upward and out; when we "clear our throats," we are expelling this mucus from the larynx.

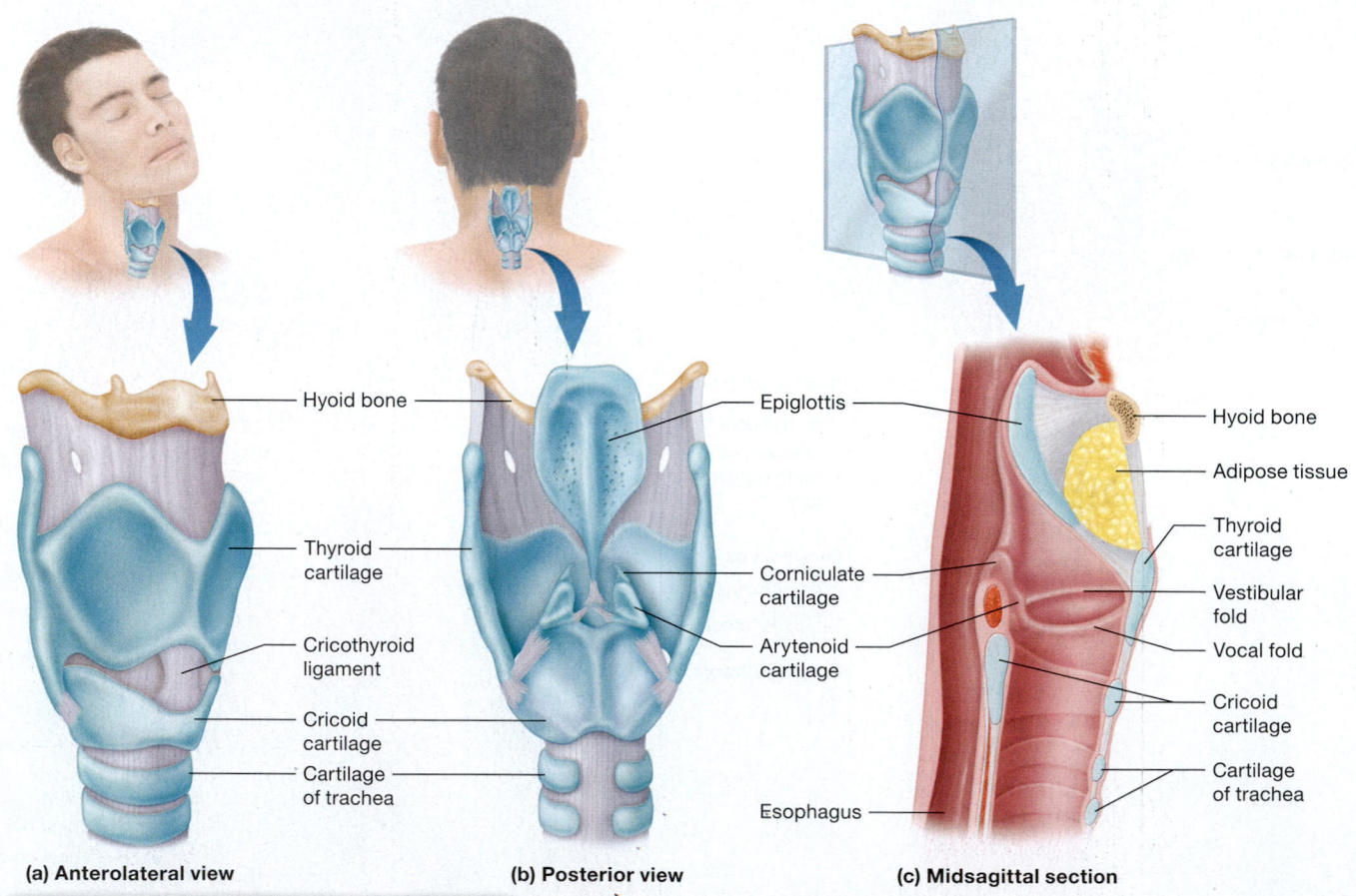

(a) Anterolateral view   (b) Posterior view   (c) Midsagittal section

Figure 21.5 Anatomy of the larynx.

Explore PAL™ Cadaver @ Mastering Anatomy & Physiology

Cartilage Framework of the Larynx

The flexible framework for the larynx is made of nine pieces of cartilage (see Figure 21.5). This cartilage framework is supported by muscles that attach the larynx to other structures of the neck and by muscles within the larynx itself (see Chapter 9 for a review). The three largest cartilages are unpaired, whereas the other six occur as three pairs of smaller cartilages.

The three unpaired cartilages are the thyroid cartilage, the epiglottis, and the cricoid cartilage. The shieldlike **thyroid cartilage** is the largest piece of the larynx's framework, forming its anterior and superior walls (see Figure 21.5). It is attached superiorly to the hyoid bone and inferiorly to the cricoid cartilage by fibrous membranes. The thyroid cartilage is prominent anteriorly, and is known as the "Adam's apple." This common name refers to the fact that the larynx is typically more prominent in males than in females, due to the influence of male sex hormones on its shape and size during puberty.

Posterior to the thyroid cartilage is the flap of elastic cartilage called the **epiglottis.** Notice in Figure 21.5b that the base of the epiglottis is attached to the posterior side of the thyroid cartilage, and its superior edge is free-standing. Normally, this superior edge stands upright to allow air to enter the larynx at its opening, called the **glottis** (look ahead to Figure 21.6). During swallowing, however, the larynx is elevated by its surrounding muscles so that the epiglottis covers the glottis, preventing food and liquids from entering the larynx. The tongue also pushes the epiglottis down during swallowing, helping to keep the larynx sealed off. Occasionally this mechanism fails, usually when you are laughing or talking while eating or drinking, and food or liquid "goes down the wrong pipe" (or is aspirated) into the larynx instead of the esophagus. This triggers a cough reflex that removes the offending bit of food or liquid.

The third piece of unpaired cartilage is the **cricoid cartilage** (KRY-koyd) (see Figure 21.5). The cricoid cartilage is inferior to the thyroid cartilage, and is attached to it by a thin membrane called the *cricothyroid ligament*. This ligament is the site for a procedure sometimes performed on a choking victim, called a *cricothyroidotomy*. During this procedure, the cricothyroid ligament is cut and a sterile tube is inserted into the larynx, which restores the airway.

The remaining six laryngeal cartilages occur in three pairs, and make up part of the posterior and lateral walls of the larynx. These small cartilages include the following:

- *Arytenoid cartilages.* The arytenoid (uh-RIH-tin-oyd) cartilages are triangular pieces of cartilage that are involved in sound production (see Figure 21.5b and c). They attach to the *vocal folds* and the intrinsic muscles of the larynx, and we discuss them in the next subsection.
- *Corniculate cartilages.* Notice in Figure 21.5b and c that the arytenoid cartilages are capped by small pieces of cartilage—these are the tiny corniculate (kohr-NIK-yoo-layt) cartilages. Like the arytenoids, the corniculate cartilages function in sound production.
- *Cuneiform cartilages.* The cuneiform (kyoo-NEE-ih-form) cartilages are found in the lateral wall of the larynx, where they help to support the epiglottis.

The thyroid, cricoid, and the bulk of the arytenoid cartilages are hyaline cartilage, whereas the rest are elastic fibrocartilage. Within this cartilaginous framework of the larynx, we find the elastic structures involved in sound production, which we address next.

Mucosal Folds and Sound Production

As shown in **Figure 21.6**, the inner surface of the larynx isn't smooth. Rather, folds of its mucosa project into its lumen. The superior pair of folds extends from the arytenoid cartilages to the thyroid cartilage. These are the **vestibular folds,** or the *false vocal cords*. As their common name implies, the vestibular folds play no role in sound production. However, they do have the important function of closing off the glottis during swallowing.

Inferior to the vestibular folds we find another set of mucosal folds called the **vocal folds,** or the *true vocal cords*. Like the vestibular folds, the vocal cords have a posterior attachment to the arytenoid cartilages and an anterior attachment to the thyroid cartilage. At the core of the vocal cords are elastic bands called the *vocal ligaments*. Notice in Figure 21.6 that the vocal cords appear whitish. They have this appearance because the mucosa that overlies the vocal ligaments is thin, making the white vocal ligaments visible.

The muscles of the larynx control the length and tension of the vocal cords. When these muscles contract, they pull on the arytenoid and corniculate cartilages and cause them to pivot. Figure 21.6 shows the effect of this pivoting motion—when the cartilages rotate inward, they *adduct* the vocal cords, and the glottis narrows (see Figure 21.6a). When the cartilages rotate outward, they *abduct* the vocal cords, and the glottis opens (see Figure 21.6b). This figure shows two superior views of each position, one as it would appear in a living throat and one of the cartilage structure.

Sound is produced as expired air passes over the vocal ligaments. The loudness of the sound is determined by the force of the airstream—the more forceful the expiration, the louder the sound. The sound's pitch is largely determined by the tension of the vocal ligaments and the speed of their vibration, just like plucking the strings of a guitar. A higher pitch is produced when the vocal ligaments are tightly adducted, because they are tense and vibrate more rapidly (see Figure 21.6a). A lower pitch results when the vocal ligaments are more abducted, because they are looser and vibrate more slowly (see Figure 21.6b). Adult males typically have a deeper (lower-pitched) voice than females because their vocal ligaments are longer and thicker as a result of their wider larynx, and thus vibrate more slowly.

Note that the movement of air across the vocal cords produces only a buzzing sound, not speech. Speech requires the coordinated efforts of the structures superior to the glottis, including the muscles of the pharynx, the soft palate, the tongue, and the lips.

The Trachea

The inspired air passes from the larynx into the **trachea** (TRAY-kee-uh), which delivers this air to the lower structures

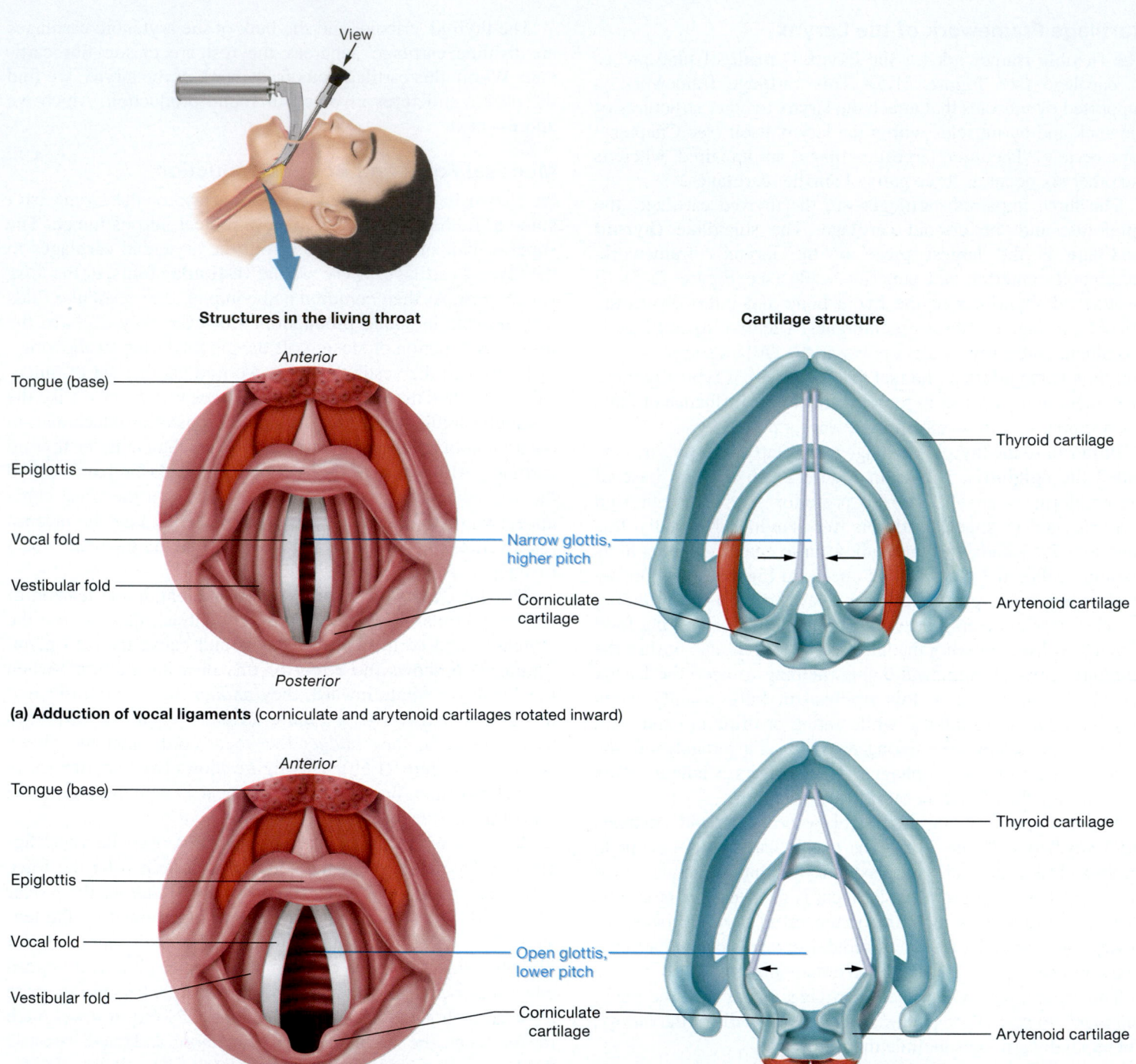

Structures in the living throat

Cartilage structure

Anterior

Tongue (base)

Epiglottis

Vocal fold

Vestibular fold

Narrow glottis, higher pitch

Corniculate cartilage

Thyroid cartilage

Arytenoid cartilage

Posterior

(a) Adduction of vocal ligaments (corniculate and arytenoid cartilages rotated inward)

Anterior

Tongue (base)

Epiglottis

Vocal fold

Vestibular fold

Open glottis, lower pitch

Corniculate cartilage

Thyroid cartilage

Arytenoid cartilage

Posterior

(b) Abduction of vocal ligaments (corniculate and arytenoid cartilages rotated outward)

Figure 21.6 Changes in the vocal ligaments during speech.

of the respiratory tract (**Figure 21.7a**). The trachea is a hollow structure about 2 cm in diameter and about 10–12 cm long. It begins in the inferior part of the neck and extends to the mediastinum.

As you can see in **Figure 21.7b**, rings of hyaline cartilage cover the anterior and lateral surfaces of the trachea, but not its posterior surface, giving these rings a C shape. The cartilage rings are rigid enough to support the trachea and keep it open, or

patent (PAY-tent), but they are also flexible enough to allow the trachea to change diameter during pulmonary ventilation.

On its posterior surface, where the cartilage is absent, the trachea is covered with elastic connective tissue and a band of smooth muscle. For this reason, the posterior surface of the trachea is soft, an arrangement that allows the esophagus to expand during swallowing. When you swallow a very large mouthful of food without adequately chewing it first, the esophagus

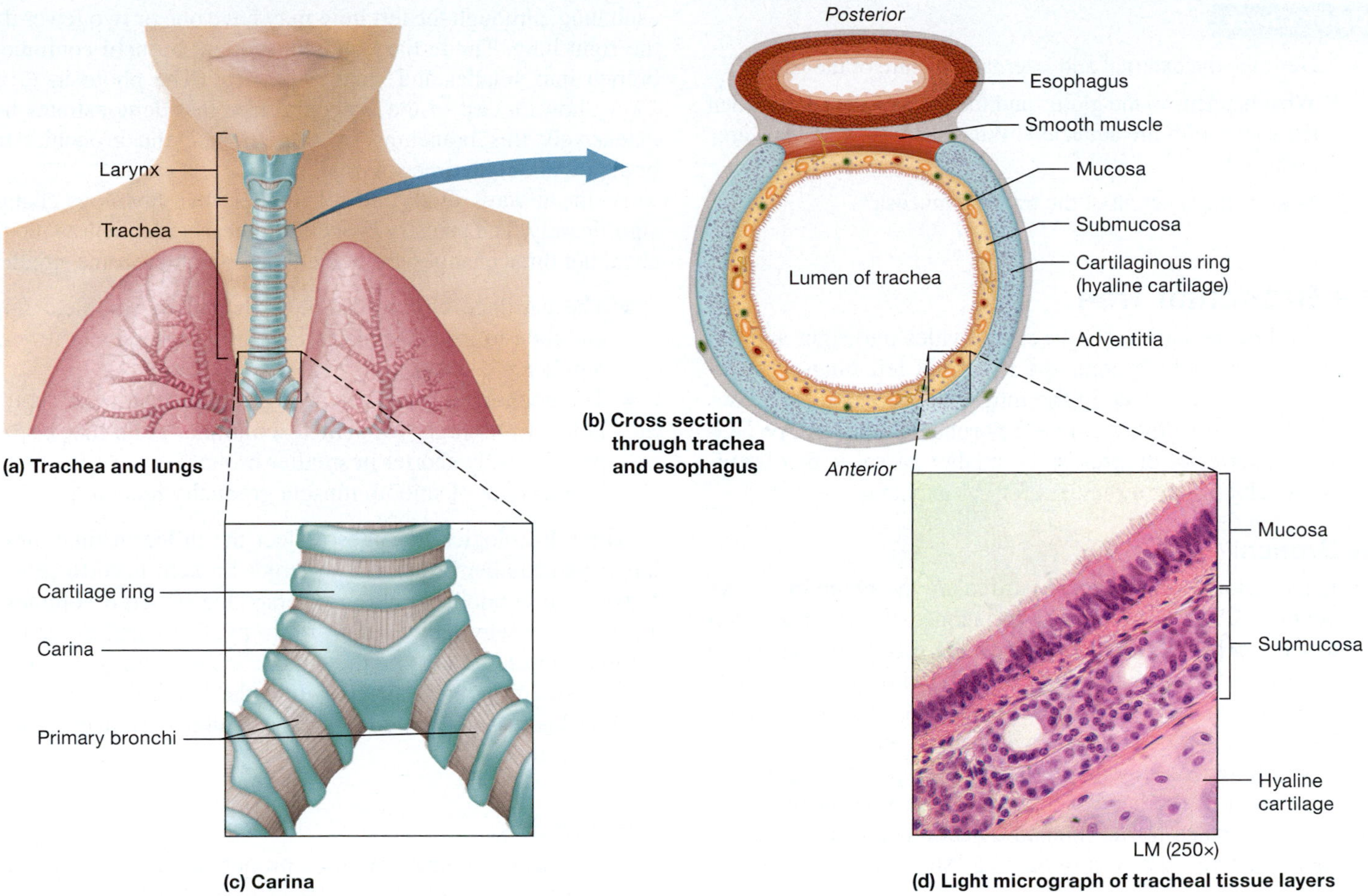

Posterior

Esophagus

Smooth muscle

Mucosa

Submucosa

Cartilaginous ring
(hyaline cartilage)

Lumen of trachea

Adventitia

**(b) Cross section
through trachea
and esophagus**

Anterior

Larynx

Trachea

(a) Trachea and lungs

Cartilage ring

Carina

Primary bronchi

(c) Carina

Mucosa

Submucosa

Hyaline
cartilage

LM (250×)

(d) Light micrograph of tracheal tissue layers

Figure 21.7 **Anatomy of the trachea.**

expands into the trachea a significant amount, causing a temporary feeling of discomfort until the food has passed through the esophagus.

The last tracheal cartilage ring is called the **carina** (kuh-RY-nuh). The carina is different in appearance from the other rings, forming a "hook" of cartilage that curves down and back (**Figure 21.7c**). From the anterior view it forms partial rings that surround the first branches of the bronchial tree. The mucosa of the carina contains sensory receptors that trigger a violent cough reflex if any foreign material comes into contact with them.

Note in **Figure 21.7b** and **d** that the trachea follows the same histological pattern we have seen for most hollow organs in the body, with an inner *mucosa,* a middle *submucosa,* and an outer *adventitia.* The **mucosa** (myoo-KOH-suh) of the trachea is continuous with that of the inferior larynx and consists of respiratory epithelium (pseudostratified ciliated columnar) with numerous goblet cells that secrete mucus. The mucus lines the trachea and traps foreign debris; the cilia beat in a coordinated fashion that propels the mucus up to the pharynx to be either swallowed or expelled. The *submucosa* is primarily loose connective tissue, deep to which are the trachea's cartilage rings. The outermost *adventitia* (ad-ven-TISH-uh) is dense irregular connective tissue that anchors the trachea to the surrounding structures.

Cigarette smoking alters the structure, as well as the function, of the epithelial lining of the respiratory tract. You can read more about these effects and one of their common consequences in *A&P in the Real World: Smoker's Cough.*

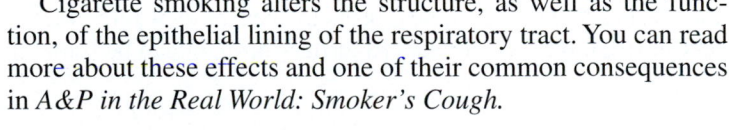

Smoker's Cough

The deep, rattling sound of "smoker's cough" is fairly easy for most of us to recognize. This cough is linked directly to the numerous adverse effects of cigarette smoke on the respiratory system. First, the chemicals in cigarette smoke act as irritants, increasing mucus secretion from the goblet cells. Second, these chemicals partially paralyze, and eventually destroy, the cilia lining the respiratory tract. As a result, more mucus is present, but the cilia are less able to sweep it out of the airways. A cough develops, as this is the only way the body can prevent mucus buildup. The good news for smokers is that cilia regrow, and so smoker's cough usually diminishes or disappears within a few months after they stop smoking. It's never too late to quit!

- ☐ 2. Describe the external and internal structure of the larynx.
- ☐ 3. What happens to the glottis and the pitch of the sound when the vocal cords are adducted? What happens when they are abducted?
- ☐ 4. What is the function of the tracheal mucosa?

The Bronchial Tree

At the level of the carina, the trachea divides into right and left *primary bronchi,* which enter the right and left lungs, respectively (**Figure 21.8a**). Once in the lung, each bronchus branches to form what is collectively called the bronchial tree (BRAWNG-kee-uhl)—a series of progressively smaller passages that terminate in tiny alveoli, the structures for gas exchange.

The Bronchi

The first structures of the bronchial tree are the **bronchi.** As we just mentioned, the first bronchi are named the right and left **primary bronchi,** which branch from the distal portion of the trachea. Figure 21.8a shows that the two primary bronchi differ in appearance: The right primary bronchus is wider, shorter, and straighter, and the left is narrower, longer, and more horizontal. These differences are due primarily to the position of the left lung in relation to the heart. The right primary bronchus is the more common place for an inhaled object (such as a peanut) to become lodged because it is more of a "straight shot" from the trachea.

Inside the lungs, the primary bronchi branch into **secondary bronchi.** Notice that there are three secondary bronchi in the right lung, and only two in the left lung due to the space needed for the heart. The secondary bronchi then branch into smaller **tertiary bronchi.** Generally, 10 tertiary bronchi can be found in

each lung, although the left lung may have one or two fewer than the right lung. The tertiary and subsequent bronchi continue to branch into smaller and smaller bronchi. The photo in **Figure 21.8b** shows a cast of the bronchial tree that demonstrates how extensively this branching occurs. In fact, the bronchial tree branches about 26 times in total.

As the airways divide and get smaller, their histology changes significantly. The primary bronchi are nearly identical to the trachea, but three changes are evident as bronchi become smaller:

- The cartilage changes from C-shaped to complete rings, and then to irregular plates that are progressively fewer in number.
- The epithelium gradually changes from respiratory epithelium in the larger bronchi to columnar cells that become progressively shorter in smaller bronchi.
- The amount of smooth muscle gradually increases.

These histological changes reflect the different functions of larger and smaller bronchi. The larger bronchi need to serve as fairly rigid conduits to allow the passage of large volumes of air, which is why they contain more hyaline cartilage and less smooth muscle. However, the smaller airways must be able to change size to control air flow into the bronchioles and alveoli. For this reason, they contain less hyaline cartilage and more smooth muscle.

Bronchioles

The smallest airways of the bronchial tree are its tiny **bronchioles** (BRAWNG-kee-ohlz) (**Figure 21.9**). By definition, bronchioles are less than 1 mm in diameter and lack cartilage. Bronchioles also feature a thicker ring of smooth muscle and simple cuboidal epithelium with few cilia and few, if any, goblet cells. This is because most dust and debris have been removed by the time the air reaches the bronchioles. Bronchioles continue

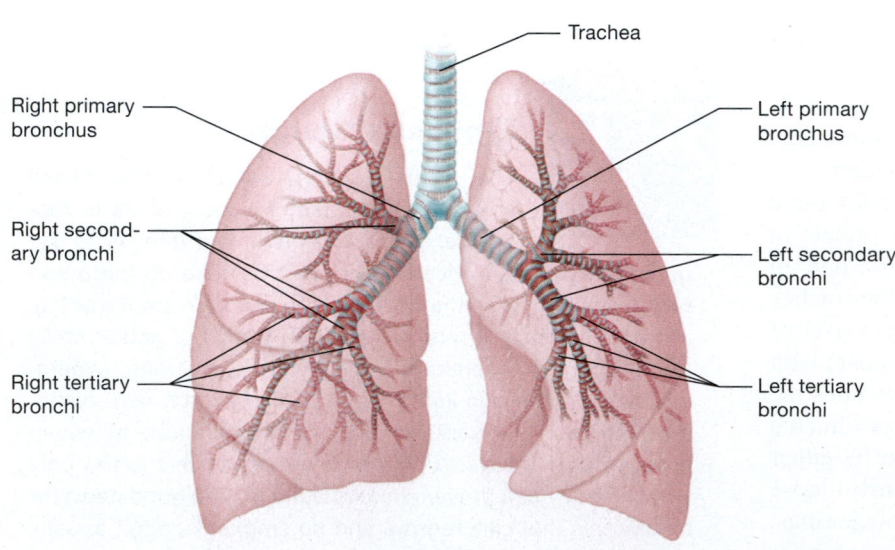

(a) Conducting zone passages and bronchial tree

Trachea

Right primary bronchus

Right secondary bronchi

Right tertiary bronchi

Left primary bronchus

Left secondary bronchi

Left tertiary bronchi

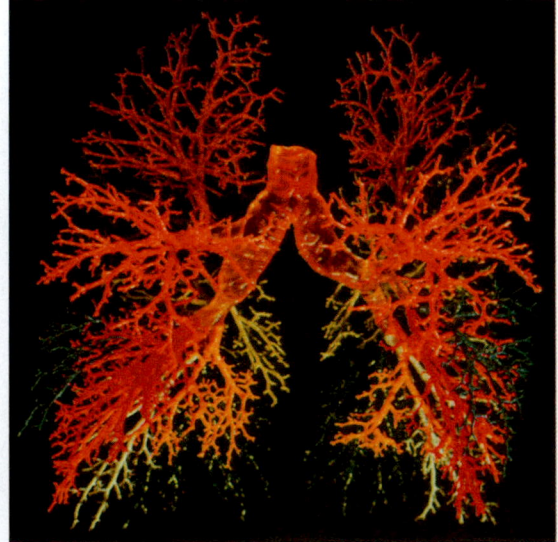

(b) Cast of bronchial tree

Figure 21.8 Branching pattern of the bronchial tree.

to branch until they become tiny *terminal bronchioles,* which are the final part of the conducting airways. To give you a sense of perspective, your lungs contain about 65,000 terminal bronchioles.

Each terminal bronchiole gives rise to two or more smaller **respiratory bronchioles,** which are surrounded by very thin bands of smooth muscle. These bronchioles have alveoli budding off their walls and so are the beginning of the respiratory tract's respiratory zone. Each respiratory bronchiole then branches into two or more smaller **alveolar ducts,** short passages that also contain alveoli along their walls. The alveolar ducts finally terminate in **alveolar sacs,** which are grapelike clusters of alveoli.

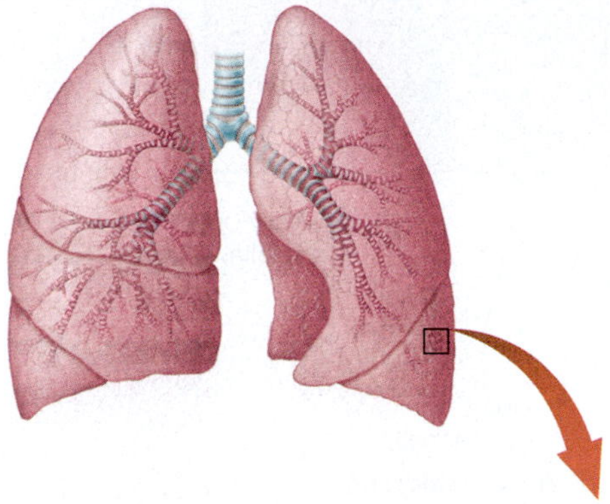

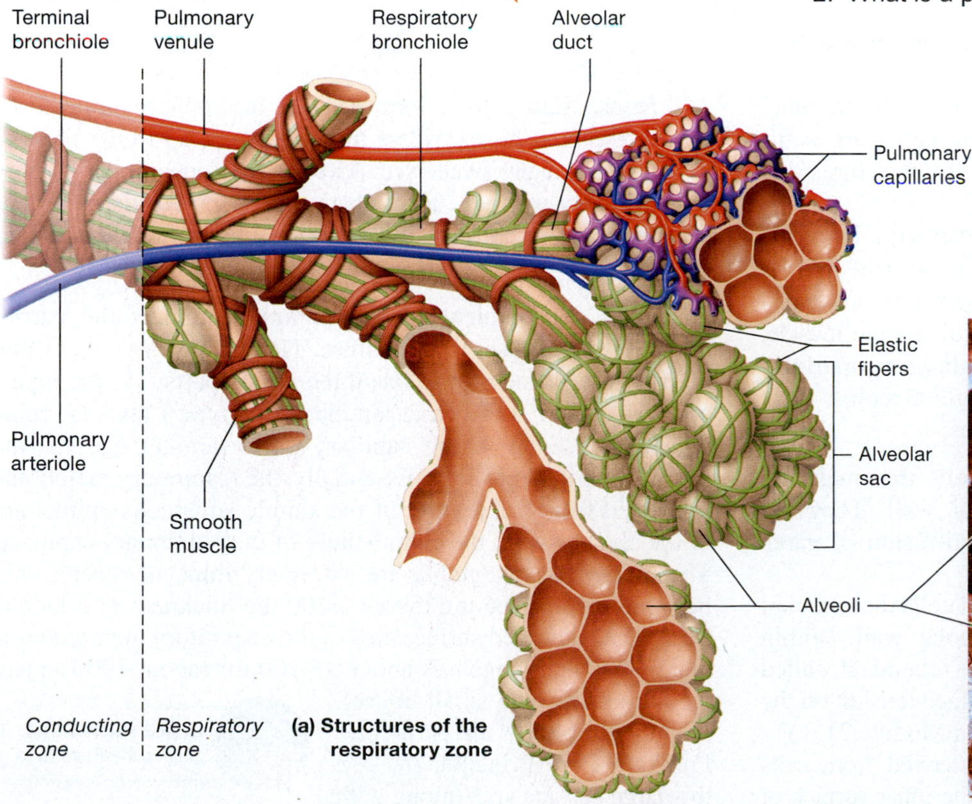

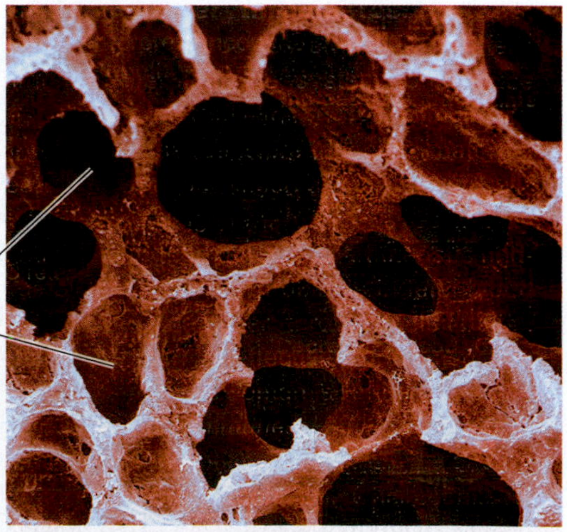

At the alveoli, we have finally arrived at the site for gas exchange. As a review, to get to the alveoli, our inhaled air had to pass through the following:

Nares → Nasal cavity → Nasopharynx → Oropharynx → Laryngopharynx → Larynx → Trachea → Primary bronchi → Secondary bronchi → Tertiary bronchi → Multiple branches of bronchi → Bronchioles → Terminal bronchioles → Respiratory bronchioles → Alveolar ducts → Alveolar sacs

Alveoli and the Respiratory Membrane

« FLASHBACK

1. What is the function of simple squamous epithelium? (p. 129)

2. What is a phagocyte? What is its function? (p. 83)

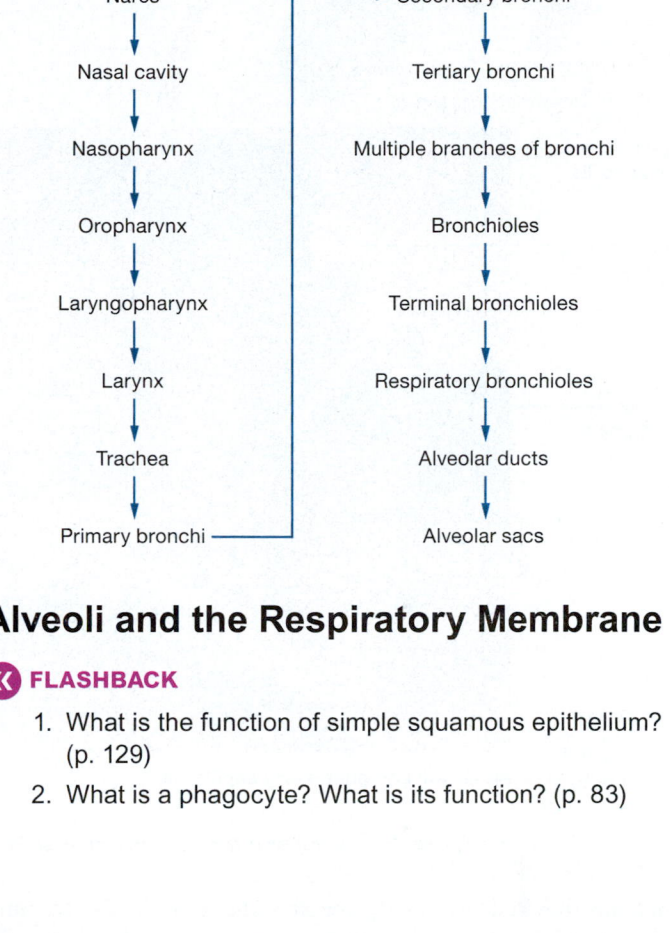

(a) **Structures of the respiratory zone**

Terminal bronchiole — Pulmonary venule — Respiratory bronchiole — Alveolar duct — Pulmonary capillaries — Elastic fibers — Alveolar sac — Alveoli

Pulmonary arteriole — Smooth muscle

Conducting zone | Respiratory zone

(b) **Electron micrograph of alveoli**   SEM (95×)

Figure 21.9 Anatomy of the respiratory zone.

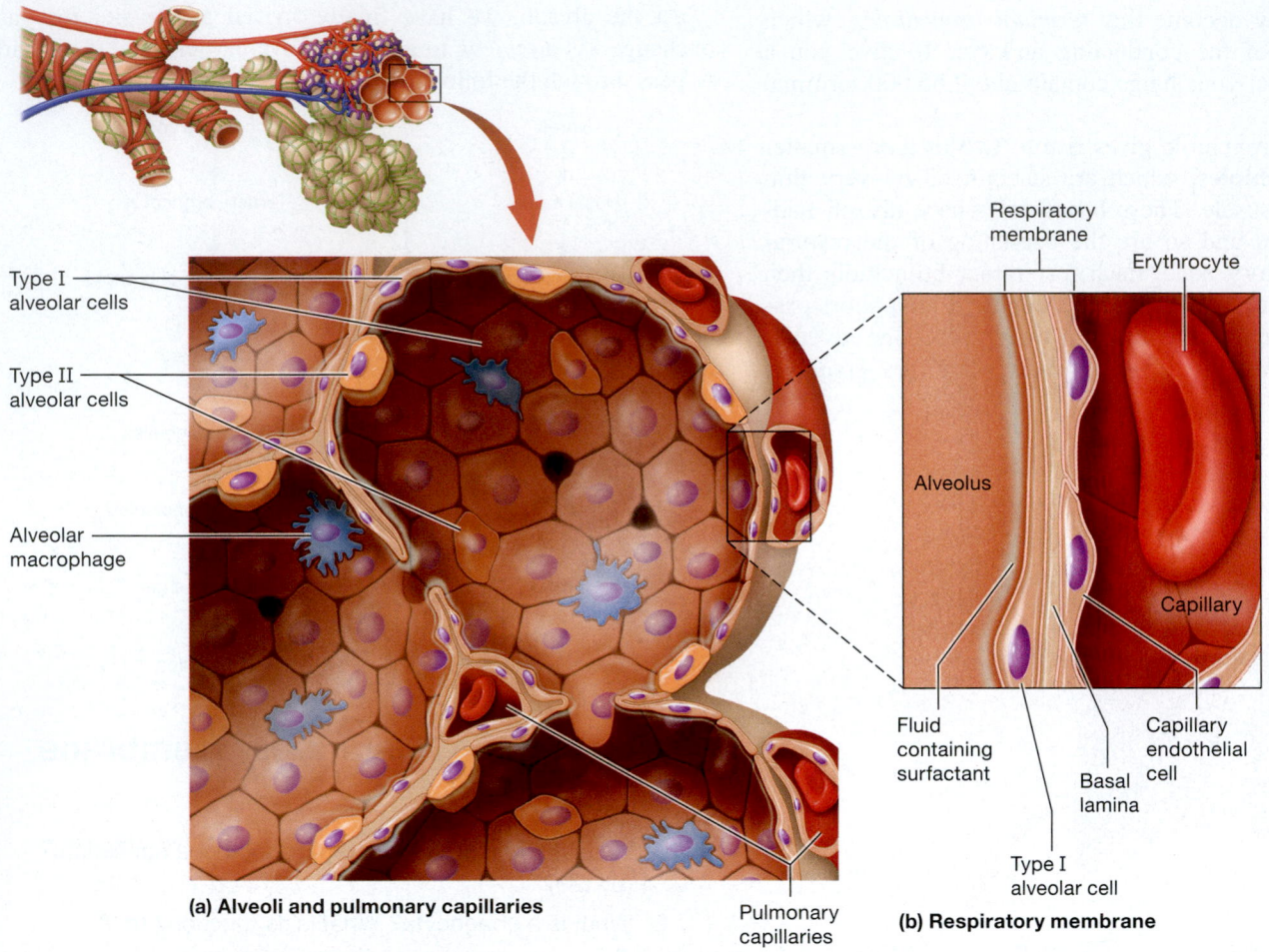

Type I
alveolar cells

Type II
alveolar cells

Alveolar
macrophage

(a) Alveoli and pulmonary capillaries

Pulmonary
capillaries

Respiratory
membrane

Erythrocyte

Alveolus

Capillary

Fluid
containing
surfactant

Basal
lamina

Capillary
endothelial
cell

Type I
alveolar cell

(b) Respiratory membrane

Figure 21.10 Structures of the alveoli and the respiratory membrane.

When the inhaled air finally reaches the alveoli, the terminal structures of the respiratory tract, the gases in the air are available to diffuse into the blood. Let's focus on the structures that accomplish this gas exchange.

We find most alveoli in groups called alveolar sacs, but some do protrude from the walls of respiratory bronchioles and alveolar ducts. Notice in Figure 21.9a that externally alveoli are surrounded by a network of elastic fibers, very thin bands of smooth muscle, and pulmonary capillaries that arise from the pulmonary arteriole.

Figure 21.10a shows the structure of a single alveolus. Each round, thin-walled alveolus has three cell types:

- **Type I alveolar cells** are squamous cells that make up about 90% of the cells in the alveolar wall. They are exceedingly thin, which permits rapid diffusion of gases across their plasma membranes.
- **Type II alveolar cells** are small cuboidal cells that account for about 10% of the cells in the alveolar wall. Within their cytoplasm are the precursors to a chemical called *surfactant,* which helps to reduce the surface tension on the alveoli. (We discuss surfactant in detail in Module 21.3.)
- **Alveolar macrophages** are phagocytes derived from cells formed in the bone marrow. They roam the inner surface of the alveoli, cleaning up and digesting any stray debris that was not filtered out in the bronchial tree (hence their alternate

name, "dust cells"). Most alveolar macrophages migrate up to the bronchioles, where they are swept up to the pharynx and eventually swallowed. Alveolar macrophages are directly involved with an agent that causes a potentially serious respiratory infection; see *A&P in the Real World: Tuberculosis.*

The very thin type I alveolar cells help make up what is known as the **respiratory membrane,** which is the barrier through which gases must diffuse. Note in **Figure 21.10b** that the respiratory membrane has three major parts: (1) the type I alveolar cells; (2) the basal lamina of the type I alveolar cells, which is fused with the capillary basal lamina; and (3) the capillary endothelial cells. Essentially, the respiratory membrane is formed from the joining of the simple squamous epithelium of the alveolus with the endothelium of the pulmonary capillary. These three parts together are extremely thin, measuring only between 0.2 and 0.6 μm (about 1/100 the thickness of a human hair). The combined surface area of the respiratory membranes is huge—for both lungs it is about 80–100 m² (about 1000 square feet—the size of a small house). In an example of the Structure-Function Core Principle (p. 28), this large surface area (along with

CORE PRINCIPLE
Structure-Function

the thinness of the respiratory membrane) makes pulmonary gas exchange a very efficient process.

Tuberculosis

The respiratory infection **tuberculosis,** or **TB,** is caused by the bacterium *Mycobacterium tuberculosis.* The bacterium is spread easily via the air, generally when an infected individual coughs or sneezes. About 90% of tuberculosis infections are *latent,* or asymptomatic. The remaining 10% of infections are *active,* meaning that the infected individual shows symptoms of the disease. Symptoms include a persistent cough with blood-tinged sputum, fever, night sweats, and weight loss. Tuberculosis may also involve organs outside the respiratory system, such as the bones or the central nervous system, resulting in additional, organ-specific symptoms.

Active tuberculosis infection begins when the bacteria reach the alveoli. Here they enter alveolar macrophages and trigger an inflammatory response that attracts lymphocytes and other immune cells. Eventually, the lymphocytes attempt to surround and "wall off" the infected area, which results in the formation of a structure known as a *granuloma.* The granuloma can protect neighboring tissues and, in some cases, can lead to the infection becoming latent. However, it can also lead to death of the normal cells within the granuloma, a condition called *necrosis.*

Diagnosis of active tuberculosis is based on the finding of granulomas on chest x-ray; a positive tuberculin skin test, which indicates the presence of antibodies to the bacterium; and the presence of the bacteria in the sputum. Tuberculosis is notoriously difficult to treat, as the structure of *M. tuberculosis* renders many common antibiotics ineffective. In addition, most strains of the bacterium are resistant to multiple drugs that were previously effective, further limiting the choice of antibiotic therapy. Treatment of active infections usually consists of a cocktail of several antibiotics taken for a period of 6 months. Individuals with latent infections are treated with a single antibiotic to prevent the infection from becoming active. Untreated active infections have about a 50% mortality rate, and untreated latent infections have a greater chance of becoming active.

The Lungs and Pleurae

« FLASHBACK

1. Where are the pleural cavities located? Where is the mediastinum? (p. 17)

The left and right **lungs** are separated by the heart and the mediastinum. A lung's **base** rests on the diaphragm muscle, and its superior **apex** sits just above the clavicle (**Figure 21.11a**). The anterior, posterior, and lateral surfaces of the lung come into contact with the rib cage, and for this reason are called the *costal surfaces* of the lung (*cost-* = "rib"). The lung's medial surface comes into contact with the structures of the mediastinum, and so is named the *mediastinal surface* (mee´-dee-uh-STY-nul) (**Figure 21.11b**). The mediastinal surface of both lungs

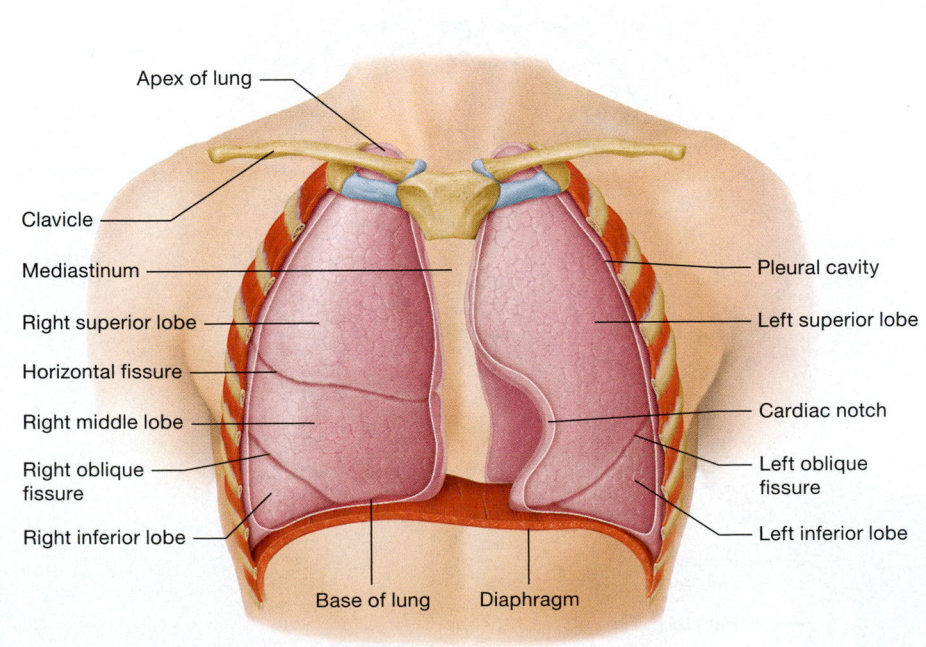

(a) Anterior view of right and left lungs

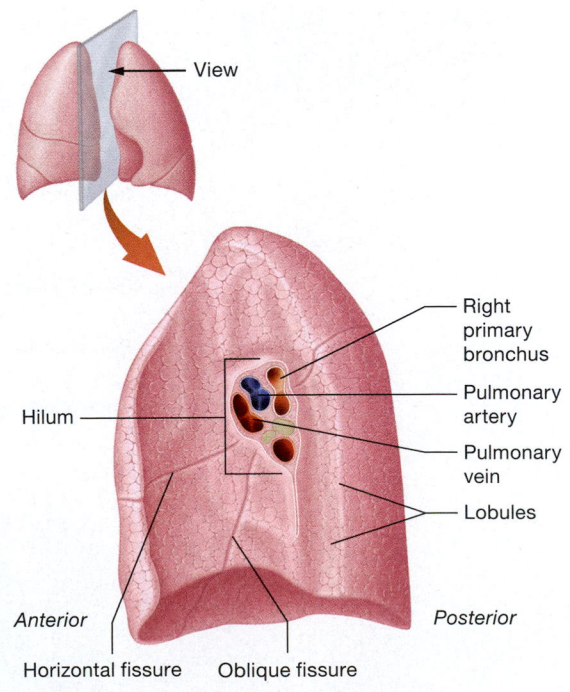

(b) Mediastinal surface of right lung

Figure 21.11 Anatomy of the lungs and associated structures.

contains a triangular depression called the **hilum** (HY-lum), where the primary bronchi, blood vessels, lymphatic vessels, and nerves enter and exit the lung. Additionally, the mediastinal surface of the left lung comes into contact with the heart, resulting in a groove called the **cardiac notch.**

As you can see in Figure 21.11a, each lung is divided into **lobes.** The left lung has two lobes because of the space necessary for the heart: the left superior lobe and the left inferior lobe, separated from each other by an *oblique fissure.* The right lung has three lobes: the right superior, the right middle, and the right inferior lobes, separated from one another by the *horizontal* and *oblique fissures.* Each secondary bronchus supplies one lobe of a lung, which explains why the secondary bronchi are often called *lobar bronchi.*

The lobes are divided by thin walls of connective tissue into **bronchopulmonary segments** (brong´-koh-PULL-muh-nehr-ee). Each tertiary bronchus serves one bronchopulmonary segment (for this reason they are often called *segmental bronchi*). As with tertiary bronchi, there are usually 10 bronchopulmonary segments. Each bronchopulmonary segment is further subdivided into **lobules,** hexagonal structures about the size of a dime.

Each lung is served by a large **pulmonary artery,** which brings deoxygenated blood from the right ventricle of the heart to the lungs (see Figure 21.11b and Chapter 17). Oxygenated

blood leaves the lungs and returns to the left atrium of the heart by smaller **pulmonary veins.** However, neither the pulmonary artery nor the pulmonary veins supply the tissues of the lungs with oxygenated blood and nutrients. Instead, this task is accomplished by the *bronchial arteries,* which are part of the systemic circuit, not the pulmonary circuit.

Each lung is enclosed by a subcavity of the thoracic cavity called the **pleural cavity** (PLOOR-ul; **Figure 21.12**; see Chapter 1). The pleural cavities are found between two layers of a serous membrane; this arrangement is very similar to that of the pericardial cavity around the heart (refer back to Figure 17.4). The outer pleural membrane, the **parietal pleura** (puh-RY-eh-tul), is fused to the structures surrounding the lungs, including the rib cage and the diaphragm. At the hilum, the parietal pleura turns over on itself to form the inner pleural membrane, the **visceral pleura** (VISS-er-ul). The visceral pleura is continuous with the surface of the lobes of the lungs and dives into each of the lung's fissures.

These pleural membranes secrete a thin layer of serous fluid, often called **pleural fluid,** that fills the pleural cavities. Just as serous fluid in the pericardial cavities lubricates the beating heart, pleural fluid lubricates the lungs as they expand and contract, reducing friction. As you will see in Module 21.3, the pressure within the pleural cavity also helps to prevent the lungs from collapsing when you take a breath, because the pleural fluid helps the visceral pleura adhere to the parietal

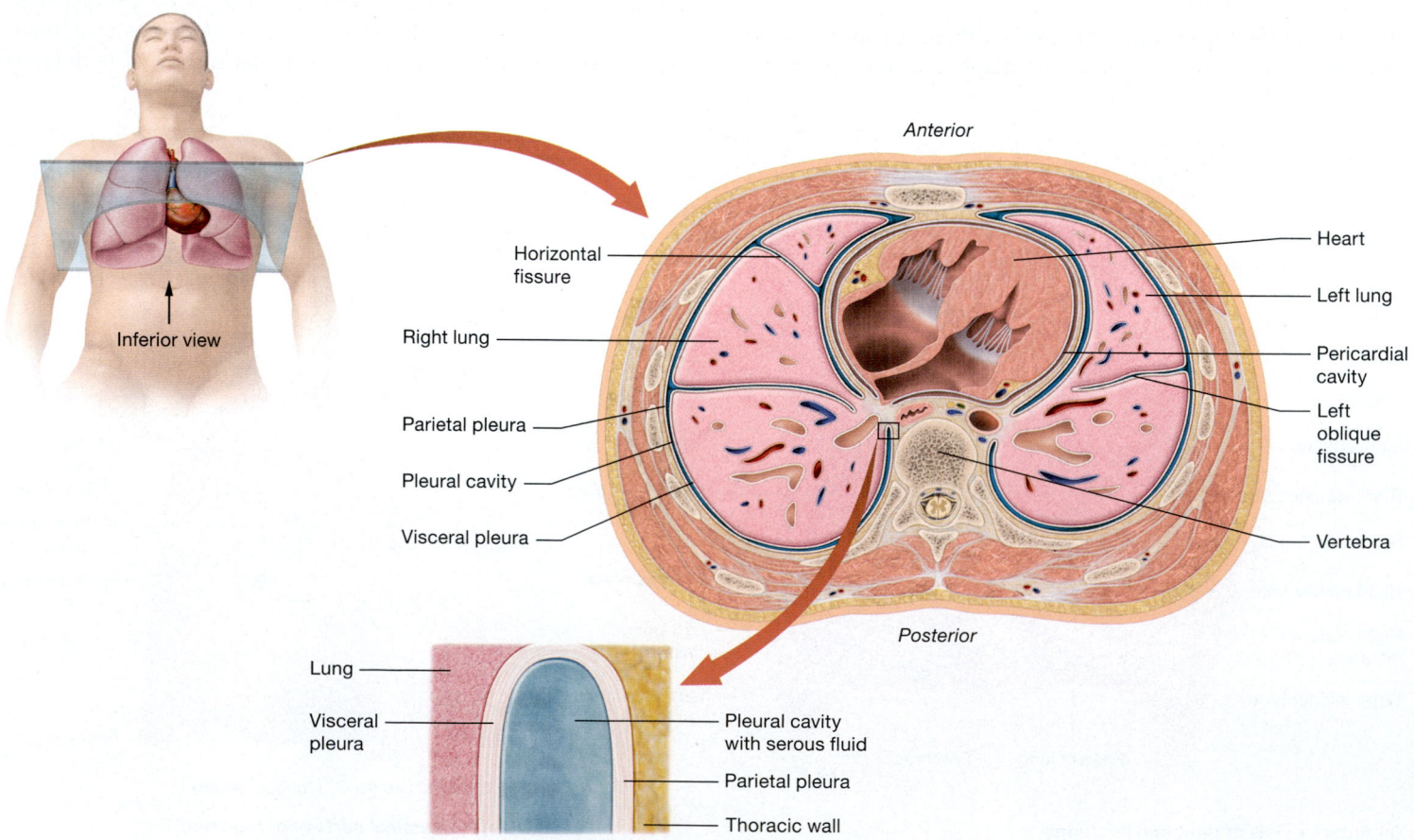

Figure 21.12 The pleurae and pleural cavities. You are looking at an inferior view of this transverse section of the trunk.

pleura. See *A&P in the Real World: Pleuritis and Pleural Friction Rub* to read about what happens when the pleural membranes are inflamed.

The changes we have seen in the histology of the different portions of the respiratory tract are summarized in **Table 21.1**. Now

let's move on to how the air gets into these structures through the process of breathing.

Quick Check

- ☐ 5. How does the epithelium of the bronchial tree change as it divides into smaller branches?
- ☐ 6. Trace the pathway from the primary bronchi to the alveolar sacs.
- ☐ 7. What structures make up the respiratory membrane?
- ☐ 8. Explain the structure of the pleural cavities.

Apply What You Learned

- ☐ 1. How does the form of the respiratory epithelium follow its function?
- ☐ 2. Explain why a person with laryngitis (inflammation of the larynx) "loses" his or her voice.
- ☐ 3. Predict what would happen if the tracheal hyaline cartilage rings were absent. What might happen if they were made of bone rather than cartilage?
- ☐ 4. How does form follow function in type I alveolar cells?

See answers in Appendix A.

Table 21.1 Function and Composition of Regions of the Conducting and Respiratory Zones

Structure	Function	Type(s) of Epithelia	Other Histological Features
Nasal cavity	• Conduit for air • Detects odors • Warms, moistens, filters air • Sound resonance	Olfactory, pseudostratified ciliated columnar	Richly supplied with blood vessels for maximal warming and humidification
Nasopharynx	• Conduit for air • Warm, moisten, filter air	Pseudostratified ciliated columnar	—
Oropharynx and laryngopharynx	Conduit for food and air	Stratified squamous nonkeratinized	—
Larynx	• Proper routing of food and air • Sound production • Protection of the airways • Warming and moistening of air	Superior to the vocal cords: stratified squamous; inferior to the vocal cords: pseudostratified ciliated columnar	Framework is hyaline and elastic cartilage; contains the vestibular and vocal folds
Trachea	• Conduit for air • Warms, moistens, filters air	Pseudostratified ciliated columnar	Has C-shaped rings of cartilage for flexible support
Bronchi	• Conduit for air • Warm, moisten, filter air	Pseudostratified columnar to simple columnar	Large bronchi contain plates of hyaline cartilage; amount of cartilage decreases as the diameter of the bronchi decreases
Bronchioles	Smooth muscle controls air flow, some filtering of air	Simple columnar to simple cuboidal	Lack cartilage; contain large amounts of smooth muscle
Respiratory bronchioles and alveolar ducts	Gas transport, gas exchange	Simple cuboidal to simple squamous	Very little smooth muscle; have alveoli budding from their walls
Alveoli	Gas exchange	Extremely thin simple squamous (type I alveolar cells)	Also contain cuboidal type II alveolar cells and alveolar macrophages

Pulmonary Ventilation

Learning Outcomes

1. Describe how pressure and volume are related, and explain how this relationship applies to pulmonary ventilation.

2. Explain how the inspiratory muscles, accessory muscles of inspiration, and accessory muscles of expiration change the volume of the thoracic cavity.

3. Explain how the values for atmospheric pressure, intrapulmonary pressure, and intrapleural pressure change with inspiration and expiration.

4. Explain how each of the following factors affects pulmonary ventilation: airway resistance, pulmonary compliance, and alveolar surface tension.

5. Describe and identify the values for the respiratory volumes and the respiratory capacities.

The first process of respiration is **pulmonary ventilation,** or breathing. It consists of two phases: **inspiration,** or inhaling, which brings air into the lungs, and **expiration,** or exhaling, which moves air out of the lungs. In this module, we first examine the relationship between pressure and volume and then explore how this relationship provides the driving force for pulmonary ventilation. We also consider physical factors that affect the ability of the lungs to do their job, such as the diameter of the airways. Finally, we describe the volumes of air that can be measured when a person breathes and explain the use of these measured volumes to determine how well the lungs are functioning.

The Pressure-Volume Relationship

◀◀ FLASHBACK

1. What is a pressure gradient? (p. 26)

What causes air to move into and out of the body when we breathe? To answer this question, first remember that any time molecules are moved in the body without the use of additional energy, a *gradient* of some sort is involved. Air is a mixture of gas molecules, and the movement of gas molecules depends on **pressure gradients,** that is, differences in pressure between one area and another. Gas molecules move from areas of higher pressure to areas of lower pressure. In an example of the Gradients  **CORE PRINCIPLE** Gradients

Core Principle (p. 28), these pressure gradients are what drive the movement of air molecules during inspiration and expiration.

One of the simplest ways to change the pressure in a container is to change its volume. This relationship between pressure and volume is described by **Boyle's law,** which states that at a constant temperature and constant number of gas molecules, the pressure and volume of a gas are inversely related—as the volume of a container increases, the pressure the gas exerts

on the container decreases. The opposite is also true—as the volume of a container decreases, the pressure the gas exerts on the container increases. These changes are easy to visualize with a syringe, like the ones you've seen in a doctor's office. When you depress the plunger, you decrease the syringe's volume and increase the chamber's pressure. When you pull the plunger out, you increase the syringe's volume and decrease the chamber's pressure, as you can see here:

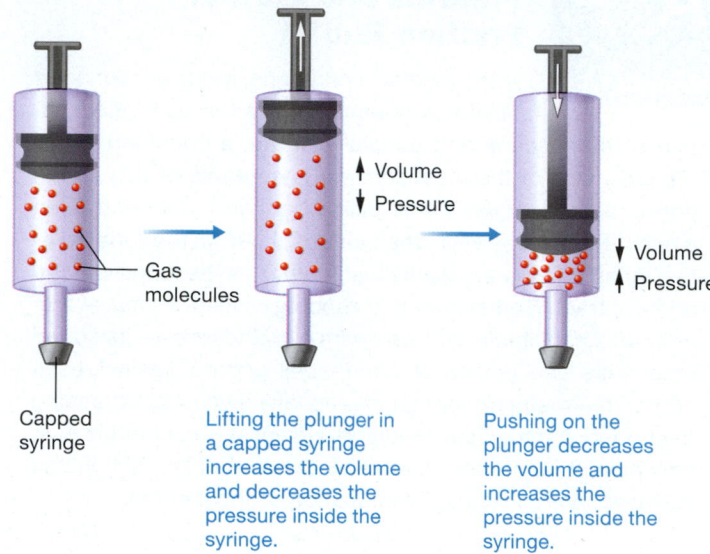

Capped syringe

Gas molecules

↑ Volume
↓ Pressure

↓ Volume
↑ Pressure

Lifting the plunger in a capped syringe increases the volume and decreases the pressure inside the syringe.

Pushing on the plunger decreases the volume and increases the pressure inside the syringe.

This example used a capped syringe, so gas molecules were trapped in the chamber. But when the syringe is opened, you can see how the gas molecules move in relation to the pressure gradient. Before the syringe's plunger is moved, the pressures inside and outside the chamber are equal and the gas molecules are at equilibrium. However, when we move the plunger, the volume and pressure of the chamber change, which causes gas molecules to flow along the pressure gradients:

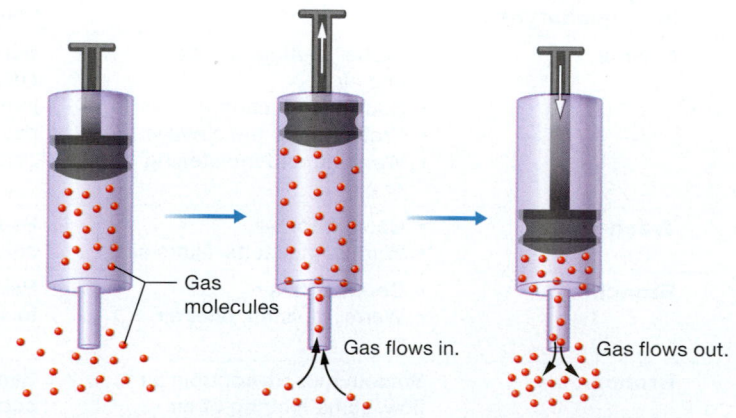

Gas molecules

Gas flows in.

Gas flows out.

In an open syringe, gas pressure and volume are at equilibrium.

Lifting the plunger decreases the pressure in the syringe chamber, creating a pressure gradient. Gas molecules move to the area of lower pressure inside the syringe.

Pushing on the plunger increases the pressure in the syringe chamber, reversing the gradient. Gas molecules again move to the area of lower pressure outside the syringe.

Now let's see how pressure and volume changes in the lungs lead to pulmonary ventilation.

☐ 1. What drives the movement of gases?

☐ 2. What happens to the pressure of a container if its volume increases? What happens to its pressure if its volume decreases?

The Process of Pulmonary Ventilation

◀◀ FLASHBACK

1. What is the action of the diaphragm muscle? (p. 307)

2. What are the actions of the internal and external intercostal muscles? (p. 307)

In a similar way to our syringe example, pulmonary ventilation involves volume changes in the thoracic cavity and lungs that lead to the creation of a pressure gradient. This gradient, in turn, causes air to move into or out of the lungs. We'll discuss each of these processes and then see how they combine to drive ventilation.

Pressure Gradients of Ventilation

Generally speaking, three pressure gradients influence pulmonary ventilation:

- **Atmospheric pressure.** The molecules that make up air are every bit as subject to the force of gravity as we are. The pull of gravity on the air around us creates what is known as *atmospheric pressure*. At sea level, atmospheric pressure is about 760 mm Hg. It increases as you go below sea level, and it decreases as you rise above sea level.

- **Intrapulmonary pressure.** The air pressure within the alveoli is called the *intrapulmonary pressure*. It rises and falls with inspiration and expiration, but it always eventually equalizes with atmospheric pressure due to pressure gradients reaching equilibrium.

- **Intrapleural pressure.** *Intrapleural pressure* is the pressure found within the pleural cavity. Like intrapulmonary pressure, intrapleural pressure also rises and falls with inspiration and expiration. However, intrapleural pressure does not equalize with atmospheric pressure. Instead, it is normally about 4 mm Hg *less* than intrapulmonary pressure. This difference is due to the slight suction effect created by the tendency of the lung to collapse and the chest wall to expand. Additionally, the pleural fluid is constantly pumped out of the pleural cavity and into the lymphatic vessels, which creates a slight vacuum that in effect "glues" the two membranes together.

Figure 21.13 on p. 820 shows how these pressures change during quiet ventilation. In step ① of this figure, the lungs are at rest between breaths. The intrapulmonary pressure in the alveoli is the same as the atmospheric pressure at 760 mm Hg, so no air flows in or out of the lungs. The intrapleural pressure is 4 mm Hg below atmospheric, at 756 mm Hg.

For inspiration to occur, the intrapulmonary pressure must be less than atmospheric pressure. This is accomplished by increasing the volume of the lungs, which decreases the intrapulmonary pressure to about 758 mm Hg (see Figure 21.13, step ②). As intrapulmonary pressure decreases, air moves into the lungs. Notice that during normal inspiration, the intrapleural pressure remains *lower* than the intrapulmonary pressure.

Air will continue to move into the lungs until the gradient no longer exists, which happens when intrapulmonary pressure equals atmospheric pressure, as in step ③ of Figure 21.13. At this point between inspiration and expiration, no air movement occurs, as there is no pressure gradient to drive its movement.

Air moves out of the lungs when a pressure gradient is created during expiration and intrapulmonary pressure rises above atmospheric pressure to about 762 mm Hg (see Figure 21.13, step ④). The lungs do not fully deflate during expiration because of the slight suction effect from the intrapleural pressure and the outward recoil of the chest wall. Both forces oppose the lungs' elastic recoil, keeping the lungs inflated and most of the alveoli open at all times. Expiration stops when intrapulmonary pressure equals atmospheric pressure (returning to step ① of the figure). Throughout these changes, the intrapleural pressure stays about 4 mm Hg less than the intrapulmonary pressure.

What happens if the intrapleural pressure increases to a level at or above atmospheric pressure? Under these conditions, the intrapleural pressure no longer exerts a suction effect that prevents the lungs from collapsing. Indeed, the added pressure actually enhances the lungs' elastic recoil, and the lungs immediately collapse, as shown here:

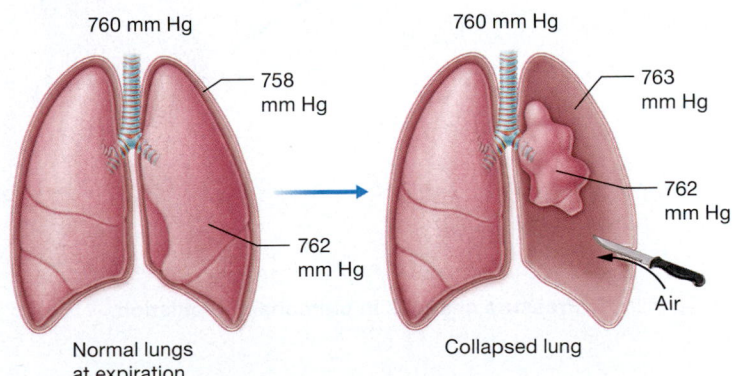

760 mm Hg — 758 mm Hg — 762 mm Hg

Normal lungs at expiration

760 mm Hg — 763 mm Hg — 762 mm Hg — Air

Collapsed lung

When the lung is fully collapsed, intrapulmonary pressure is so high that the body is unable to bring it below atmospheric pressure, and inspiration cannot occur. Many things can increase intrapleural pressure above atmospheric, including excess pleural fluid (a *pleural effusion*), air in the pleural cavity (a *pneumothorax;* noo-moh-THOR-aks), or blood in the pleural cavity (a *hemothorax*). The preceding illustration shows a pneumothorax caused by the trauma of a knife wound, but note that trauma is only one potential cause of lung collapse.

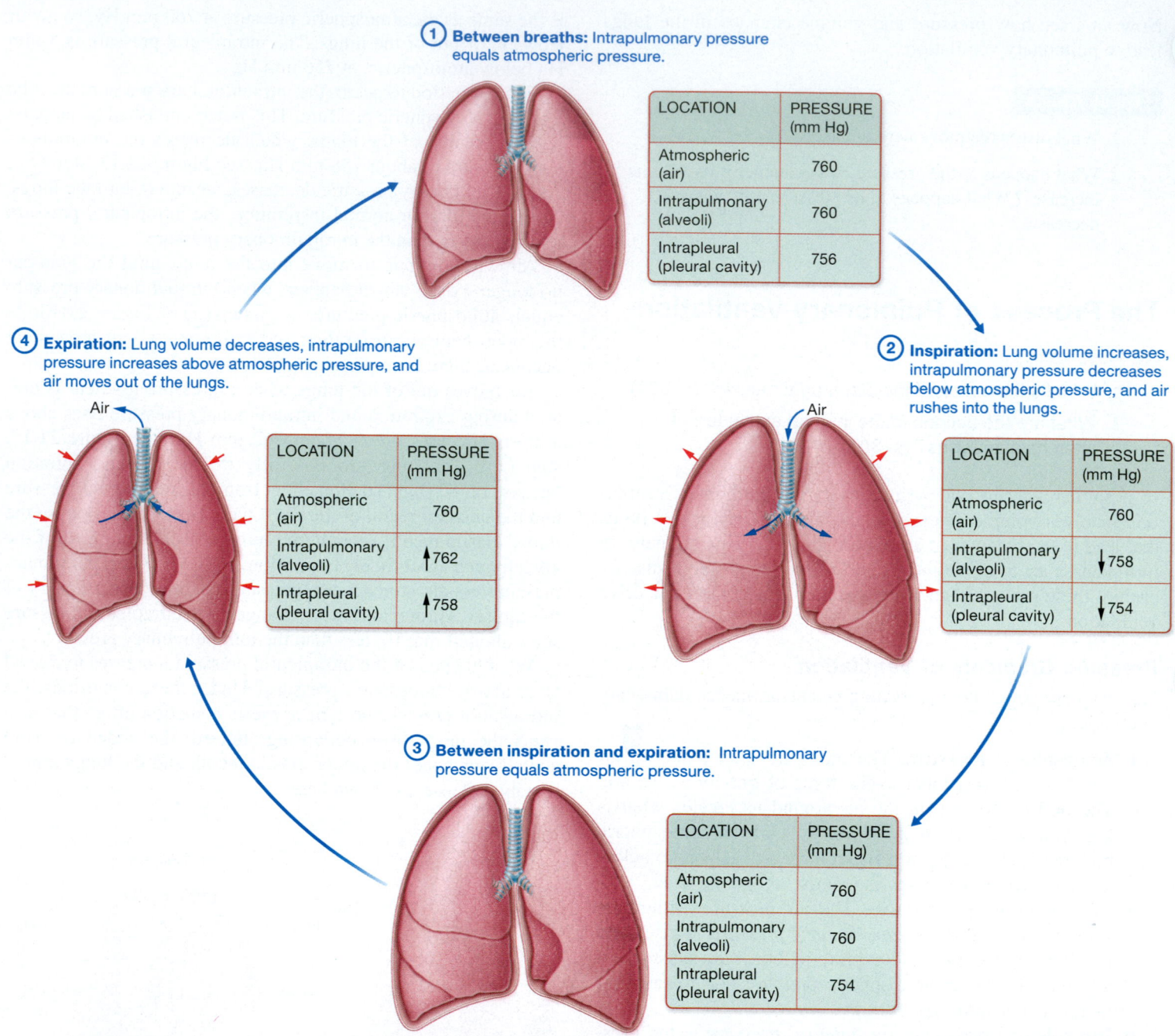

① Between breaths: Intrapulmonary pressure equals atmospheric pressure.

LOCATION	PRESSURE (mm Hg)
Atmospheric (air)	760
Intrapulmonary (alveoli)	760
Intrapleural (pleural cavity)	756

④ Expiration: Lung volume decreases, intrapulmonary pressure increases above atmospheric pressure, and air moves out of the lungs.

Air

LOCATION	PRESSURE (mm Hg)
Atmospheric (air)	760
Intrapulmonary (alveoli)	↑762
Intrapleural (pleural cavity)	↑758

② Inspiration: Lung volume increases, intrapulmonary pressure decreases below atmospheric pressure, and air rushes into the lungs.

Air

LOCATION	PRESSURE (mm Hg)
Atmospheric (air)	760
Intrapulmonary (alveoli)	↓758
Intrapleural (pleural cavity)	↓754

③ Between inspiration and expiration: Intrapulmonary pressure equals atmospheric pressure.

LOCATION	PRESSURE (mm Hg)
Atmospheric (air)	760
Intrapulmonary (alveoli)	760
Intrapleural (pleural cavity)	754

Figure 21.13 Pressure changes in pulmonary ventilation.

Mechanics of Inspiration and Expiration

The lungs themselves have no skeletal muscle tissue, so they cannot change their volume on their own. Instead, they rely on skeletal muscles of the thoracic cavity, called **inspiratory muscles,** to increase their volume (**Figure 21.14**). The inspiratory muscles don't increase the volume of the lungs directly. Rather, they increase the size of the thoracic cavity, which indirectly increases the volume of the lungs. This functional relationship is due to the pleural membranes. Recall that the parietal pleura

is attached to the inner surface of the thoracic cavity, so when the thoracic cavity increases in height and diameter, the parietal pleura is pulled with it. The parietal pleura, in turn, pulls on the visceral pleura, which pulls the lungs outward, increasing their volume.

The main inspiratory muscle is the **diaphragm muscle** (DY-ah-fram), which is located between the thoracic and abdomino-pelvic cavities. The other inspiratory muscles are the **external intercostal muscles,** which are located between the ribs. As

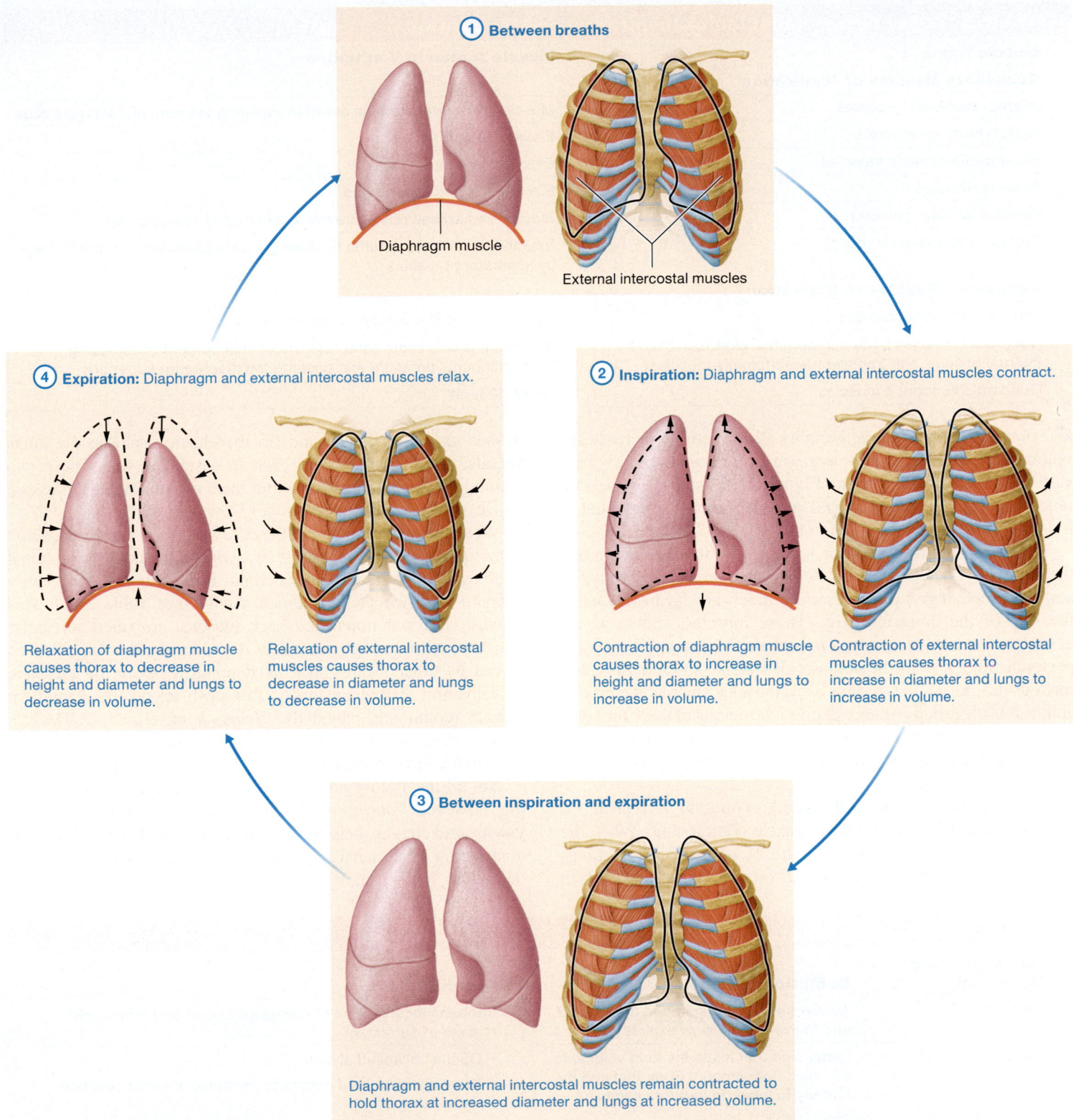

① Between breaths

Diaphragm muscle

External intercostal muscles

④ Expiration: Diaphragm and external intercostal muscles relax.

Relaxation of diaphragm muscle causes thorax to decrease in height and diameter and lungs to decrease in volume.

Relaxation of external intercostal muscles causes thorax to decrease in diameter and lungs to decrease in volume.

② Inspiration: Diaphragm and external intercostal muscles contract.

Contraction of diaphragm muscle causes thorax to increase in height and diameter and lungs to increase in volume.

Contraction of external intercostal muscles causes thorax to increase in diameter and lungs to increase in volume.

③ Between inspiration and expiration

Diaphragm and external intercostal muscles remain contracted to hold thorax at increased diameter and lungs at increased volume.

Figure 21.14 Volume changes in pulmonary ventilation: structure and function of the inspiratory muscles in quiet breathing.

shown in Figure 21.14 step ①, when relaxed between breaths, the diaphragm muscle is dome-shaped and bulges up into the thoracic cavity. This keeps the volume of the lungs low. Similarly, the relaxed external intercostal muscles cause the rib cage, and therefore the lungs, to be at a reduced size.

As shown in Figure 21.14 step ②, when the diaphragm muscle contracts, it moves from its relaxed dome shape and pulls down to become flat. This action increases the height of the thoracic cavity. When the external intercostal muscles contract, they pull the rib cage superiorly and anteriorly, which increases the height

Table 21.2 Accessory Muscles of Inspiration and Expiration

Muscle Name	Muscle Action in Ventilation
Accessory Muscles of Inspiration	
Internal intercostal muscles	Muscle fibers near sternum assist in elevating sternum and thoracic cage
Pectoralis minor muscles	Elevate superior five ribs
Sternocleidomastoid muscles	Elevate sternum
Scalene muscles	Elevate first and second ribs
Serratus anterior muscles	Elevate and spread ribs, increasing diameter of thoracic cage
Erector spinae muscle group	Extend the vertebral column to allow greater expansion of thoracic cage by inspiratory muscles
Accessory Muscles of Expiration	
Internal intercostal muscles	Lateral muscle fibers depress thoracic cage
Abdominal muscles (rectus abdominis, external oblique, internal oblique, and transversus abdominis muscles)	Depress thoracic cage; compress abdominal contents, increasing intra-abdominal pressure and pushing the diaphragm superiorly
Quadratus lumborum muscles	Fixate 12th rib

and diameter of the thoracic cavity. Both actions increase the volume of the lungs, which decreases the intrapulmonary pressure. Between breaths, as in Figure 21.14 step ③, the diaphragm and external intercostal muscles remain contracted to hold the lungs at the increased volume.

When inspiration is deep or forced, both the diaphragm and the external intercostal muscles contract more forcefully. In addition, another group of muscles contracts and further increases the size of the thoracic cavity. These muscles, known as the *accessory muscles of inspiration,* include the internal intercostal, pectoralis minor, sternocleidomastoid, scalene, and serratus anterior muscles, as well as certain back muscles such as the erector spinae muscle group (**Table 21.2** gives the details of these muscles' actions). The resulting additional increase in lung volume further decreases intrapulmonary pressure, moving air into the lungs.

Unlike inspiration, normal (quiet) expiration is typically a passive process that requires no muscle contractions. When the inspiratory muscles relax, as in Figure 21.14 step ④, two things happen: (1) the diaphragm resumes its original dome shape, which

pushes up on the lungs; and (2) the elastic tissue in the lungs **recoils,** and the lungs snap back to a smaller size. These two actions decrease lung volume and raise intrapulmonary pressure above atmospheric pressure, so air flows out of the lungs.

When expiration is forced, the *accessory muscles of expiration* come into play. These muscles, which include the internal intercostal, abdominal, and certain back muscles, forcefully decrease the size of the thoracic cavity (see Table 21.2). This is why your abdominal and back muscles are often sore after you have had a cough for a few days. Expiration may also be forced by other means, such as slapping a person on the back or delivering abdominal thrusts that push up on the diaphragm muscle (commonly called the *Heimlich maneuver;* HYM-lik). Both of these actions are used on people who are choking, in the hope that a forceful expiration will dislodge the obstruction.

Not all inspiratory and expiratory movements are for breathing, however. *Nonrespiratory movements,* which include sighs, yawns, and coughs, help to keep the alveoli inflated and prevent obstruction of the airways. The nonrespiratory movements are defined in **Table 21.3**.

Table 21.3 Nonrespiratory Movements

Nonrespiratory Movement	Definition	Function
Sigh	A slow and deep inspiration that is held and followed by a slow expiration	Reopens local groups of collapsed alveoli and stimulates release of surfactant
Yawn	Large sigh that takes the lung volume to the maximum amount of air that can be forcibly inhaled (*inspiratory capacity*)	• Opens collapsed alveoli • Yawning when tired occurs to minimize alveolar collapse during sleep • Yawning after sleep opens alveoli that have collapsed during sleep
Sneeze	Deep inspiration followed by a large, forceful expiration through the nose at a velocity of about 100 miles per hour	Clears foreign or irritating substances from the nasal cavity
Cough	Similar to a sneeze except the initial inspiration is small or absent; velocity can approach 500 miles per hour	Clears the larynx, trachea, or lower airways

The Big Picture of Pulmonary Ventilation

Figure 21.15

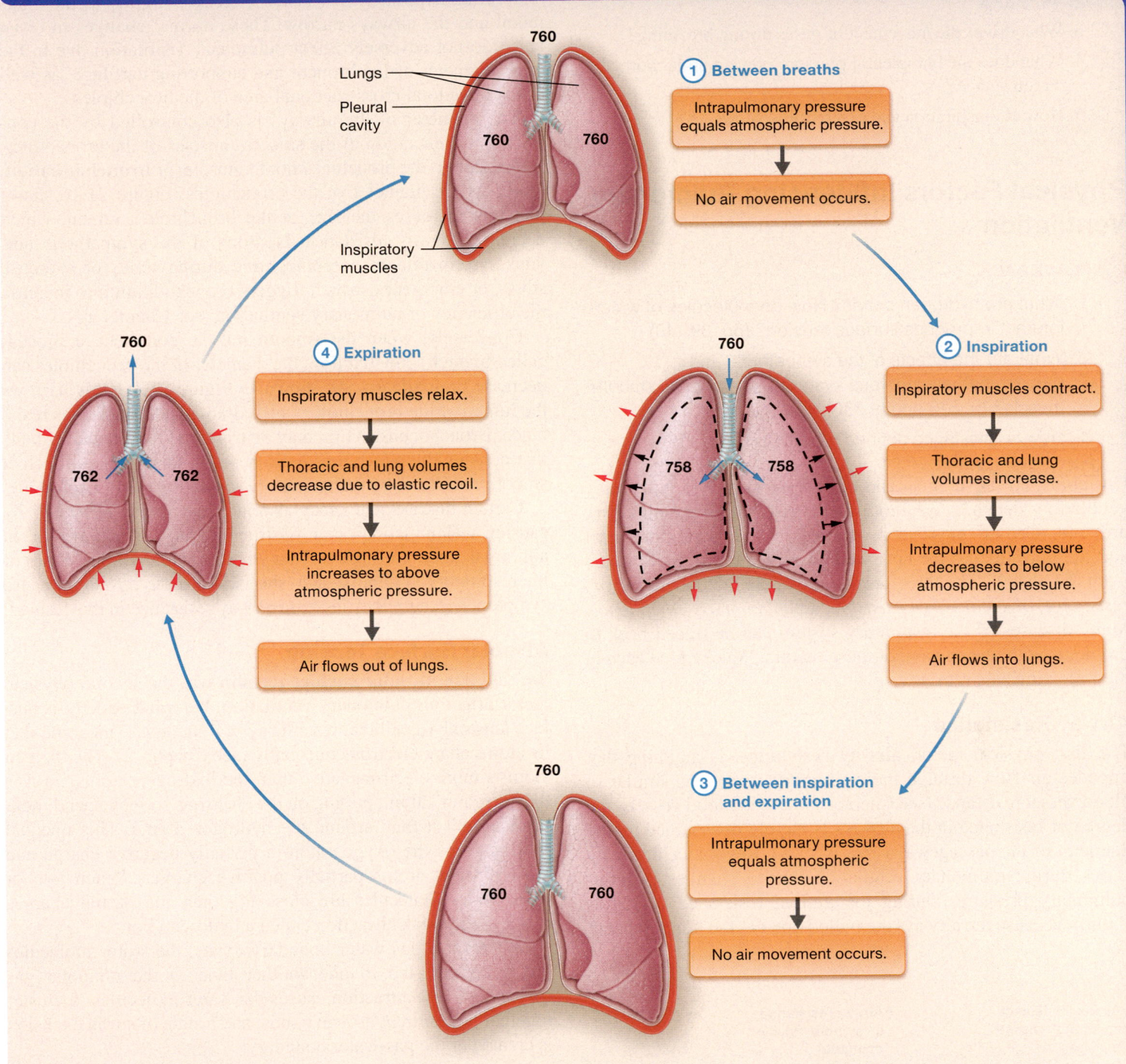

760

Lungs
Pleural cavity

Inspiratory muscles

760
760   760

① Between breaths

Intrapulmonary pressure equals atmospheric pressure.

No air movement occurs.

④ Expiration

Inspiratory muscles relax.

Thoracic and lung volumes decrease due to elastic recoil.

Intrapulmonary pressure increases to above atmospheric pressure.

Air flows out of lungs.

760

762   762

760

758   758

② Inspiration

Inspiratory muscles contract.

Thoracic and lung volumes increase.

Intrapulmonary pressure decreases to below atmospheric pressure.

Air flows into lungs.

760

760   760

③ Between inspiration and expiration

Intrapulmonary pressure equals atmospheric pressure.

No air movement occurs.

Putting It All Together: The Big Picture of Pulmonary Ventilation

Let's put all this together now as we review the process of pulmonary ventilation (**Figure 21.15**). ① Between breaths, the intrapulmonary and atmospheric pressures are equal, so no air moves between them. ② During inspiration, the inspiratory muscles contract, which increases the volume of the lungs. This decreases intrapulmonary pressure below atmospheric pressure, and air flows into the lungs. ③ Between inspiration and expiration, intrapulmonary pressure is again equal to atmospheric pressure and no air movement occurs. ④ During expiration, the lungs' volume decreases because of relaxation of the inspiratory muscles and the elastic recoil of the lungs. This increases intrapulmonary pressure above atmospheric pressure, and air flows out of the lungs.

823

☐ 3. What drives the movement of gases during breathing?

☐ 4. What does the intrapleural pressure prevent under normal circumstances?

☐ 5. How are inspiration and expiration achieved?

Physical Factors Influencing Pulmonary Ventilation

《 FLASHBACK

1. What are hydrogen bonds? How do molecules of water interact to produce surface tension? (pp. 39–40)
2. What are the effects of the sympathetic and parasympathetic nervous systems on bronchial smooth muscle? (p. 523 and p. 530)
3. How is resistance defined? (p. 676)

Our ability to move air in and out of our lungs via pulmonary ventilation depends on more than just the intrapulmonary pressure. Certain physical properties of the respiratory tract and the lungs determine how well gases are transported to and from the alveoli. The three primary physical factors are *airway resistance, alveolar surface tension,* and *pulmonary compliance.* Many diseases of the respiratory system can be traced back to problems with one or more of these factors. We examine each in the following subsections.

Airway Resistance

The first physical factor, **airway resistance,** is anything that impedes air flow through the respiratory tract; it is similar to blood vessel resistance (see Chapter 18). As in blood vessels, the degree of resistance in the airways is largely determined by the diameter of the passageway (**Figure 21.16**). Resistance normally varies during pulmonary ventilation due to changes in intrapulmonary pressure. During inspiration, resistance decreases slightly because the airways are pulled open as the lungs expand,

and during expiration, resistance increases slightly as the lungs recoil and the airways narrow. These normal changes in resistance do not adversely affect pulmonary ventilation due to the large diameter of the bronchi and supporting cartilage, as well as the huge total cross-sectional area of the bronchioles.

The diameter of the airways is also controlled by the contraction or relaxation of the smooth muscles of the bronchioles. Relaxation of the bronchial smooth muscle, or **bronchodilation,** increases the diameter of the bronchioles. Figure 21.16 shows that this increases the size of the bronchiole's lumen, which decreases airway resistance. Neurons of the sympathetic nervous system release norepinephrine during times of exercise, stress, or emergency, which triggers bronchodilation to increase the efficiency of pulmonary ventilation (see Chapter 14).

Conversely, when the smooth muscle contracts, a process called **bronchoconstriction,** the diameter of the bronchioles can decrease dramatically. As shown in Figure 21.16, this narrows the lumen of the bronchiole, which dramatically increases resistance. Bronchoconstriction may be triggered by inhaled irritants that activate the parasympathetic neurons serving the bronchial smooth muscle or by inflammatory mediators (see Chapter 20).

Certain disease states may also increase airway resistance by causing inflammation or an obstruction such as a tumor or an abnormal amount of mucus. Many bacterial and viral diseases of the respiratory system, such as the viral disease *influenza,* cause symptoms because they increase airway resistance in this fashion.

Alveolar Surface Tension

We said that **alveolar surface tension** was the second physical factor affecting pulmonary ventilation, but what exactly is surface tension? Recall that water is a *polar molecule,* meaning that its atoms share electrons unequally (see Chapter 2). The oxygen atom is more electronegative than the hydrogen atoms, so it is the "greedier" atom, pulling on the electrons so they spend more time around it than around the hydrogen atoms. This unequal sharing gives the oxygen atom a partially negative charge and the hydrogen atoms partially positive charges. When two or more water molecules are close together, the partial charges result in the weak attraction called a hydrogen bond.

Wherever a gas-water boundary exists, the water molecules are more attracted to one another than to the nonpolar gas molecules. This attraction causes the water molecules to cluster together and form hydrogen bonds, and a state of surface tension is created at the gas-water boundary:

↓Airway resistance as smooth muscle relaxes

↑Airway resistance as smooth muscle contracts

Smooth muscle

↑Diameter of airway

↓Diameter of airway

Figure 21.16 Relationship between airway resistance and airway diameter.

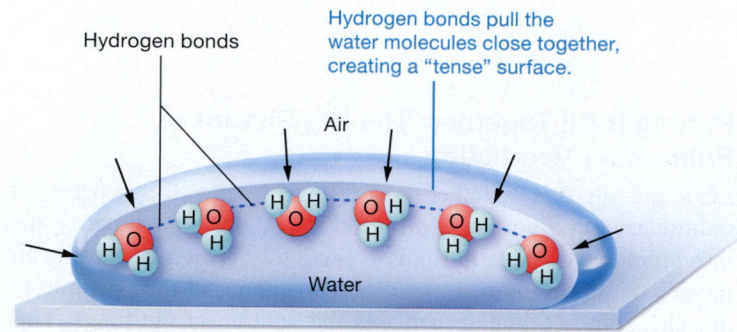

Hydrogen bonds

Hydrogen bonds pull the water molecules close together, creating a "tense" surface.

Air

Water

The cells of the alveoli, like all cells in the body, are coated in a liquid film composed mainly of water. These water molecules attract one another, and their attraction is greatest when the alveoli are smallest during expiration. Left unopposed, this attraction would create high surface tension that would collapse the alveoli during expiration (a condition known as **atelectasis;** at'-eh-LEK-tuh-sis), and make it very difficult to re-inflate them during inspiration (**Figure 21.17a**).

Fortunately, the attraction isn't left unopposed. The liquid film coating the alveoli isn't pure water, but rather it contains a chemical called pulmonary **surfactant.** The function of this surfactant, which is produced by type II alveolar cells, is to reduce alveolar surface tension. Surfactant is similar in structure to a detergent—it has one end that is polar and another end that is nonpolar. When surfactant is added to water, its polar ends interact with water molecules while its nonpolar ends will not interact with water molecules. The net effect of this action is that the hydrogen bonds between water molecules are disrupted, which reduces alveolar surface tension. The reduced surface tension allows the alveoli to remain partially open, even during expiration (**Figure 21.17b**).

Pulmonary Compliance

The third physical factor that influences the effectiveness of gas exchange is called pulmonary compliance. **Pulmonary compliance** refers to the ability of the lungs and the chest wall to stretch, a property known as *distensibility*. Note that compliance does not refer to *elasticity*, which is the ability of a tissue to return to its original state when stretched. Rather, it is the ability of the tissue to stretch in the first place. Pulmonary compliance is primarily determined by three factors:

- **Degree of alveolar surface tension.** As you just learned, surface tension on the alveoli resists their inflation, but surfactant counters this tension.
- **Distensibility of elastic tissue in the lungs.** The elastic tissue in the lungs gives them not only their ability to recoil but also the ability to stretch during inflation.
- **Ability of the chest wall to move.** The chest wall must also stretch during inspiration, as the lungs and chest wall move together.

Factors that decrease pulmonary compliance make pulmonary ventilation less efficient and increase the work of breathing. For example, diseases that decrease surfactant production can result in alveolar surface tension that is too high, which makes the alveoli difficult to inflate during inspiration. (See *A&P in the Real World: Infant Respiratory Distress Syndrome* to learn about how this problem can affect premature infants.) In addition, anything that damages the lung, including infections such as tuberculosis or foreign particles such as coal dust, can destroy the lungs' elastic tissue. In these situations, the destroyed elastic tissue is replaced with dense irregular collagenous connective tissue, producing a state called *fibrosis* (see Chapter 4). This stiffens the lungs, decreasing their compliance, and makes them difficult to inflate.

Quick Check

- ☐ 6. What is airway resistance? What is the main factor that determines the amount of airway resistance?
- ☐ 7. How does surfactant decrease surface tension?
- ☐ 8. What is pulmonary compliance? What three factors influence pulmonary compliance?

Pulmonary Volumes and Capacities

Measuring the volumes of air that a person exchanges with each breath is a useful way to assess the state of his or her pulmonary function. Such a measurement can be obtained with an instrument called a **spirometer** (spy-RAHM-eh-ter; *spiro-* = "breathing"). The graph produced by a spirometer shows a record of a

Infant Respiratory Distress Syndrome

When surfactant is inadequate, it takes tremendous effort to keep the alveoli inflated between breaths because alveolar surface tension increases and lung compliance decreases. Many premature newborns face this difficulty because surfactant is not produced in significant amounts until the last 10–12 weeks of gestation (as it's not needed during the fetal period). These infants are said to have **infant respiratory distress syndrome (RDS)**. Risk factors for RDS include prematurity, male sex, maternal history of diabetes, family history of RDS, and caesarean delivery.

Treatment is aimed at delivering natural or synthetic surfactant by inhalation. Another treatment is *continuous positive airway pressure (CPAP)*, which delivers slightly pressurized air that prevents the alveoli from collapsing during expiration. In cases of severe RDS, mechanical ventilation may be necessary. However, the main goal of the medical community is to prevent RDS from happening at all, by recognizing at-risk patients and preventing premature births.

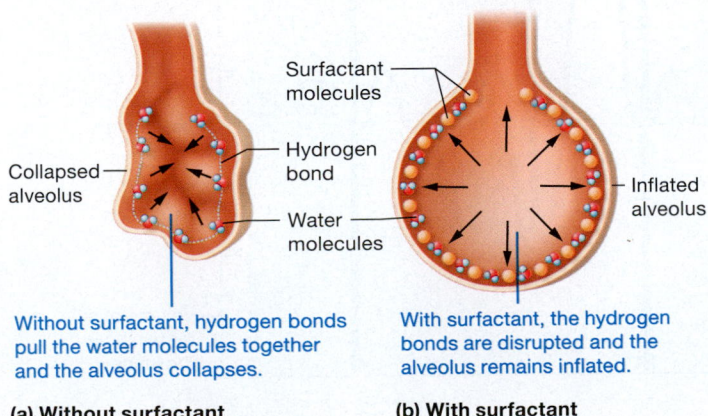

Without surfactant, hydrogen bonds pull the water molecules together and the alveolus collapses.

With surfactant, the hydrogen bonds are disrupted and the alveolus remains inflated.

(a) Without surfactant   **(b) With surfactant**

Figure 21.17 Effect of surfactant on alveolar surface tension.

person's normal and forceful inspirations and expirations. Three volumes can be measured—the tidal volume and the inspiratory and expiratory reserve volumes (see the green arrows in **Figure 21.18**):

- **Tidal volume (TV)** is the amount of air inspired or expired during normal, quiet ventilation. In the average healthy adult, the TV is about 500 ml of air. If we multiply the TV by the number of breaths per minute, we get the **minute volume,** the total volume of air that moves in and out of the lungs each minute. For the average adult, who takes about 12 breaths per minute, the minute volume is about 6 liters per minute. Note, however, that only about 350 ml of the air in a tidal inspiration actually makes it into the respiratory zone. The air that remains in the conducting zone airways is said to be in the **anatomical dead space.** For this reason, some health professionals prefer to think in terms of the **alveolar ventilation rate (AVR),** as opposed to the minute volume. The AVR is the volume of air that reaches the alveoli multiplied by the breaths per minute, and it averages about 4.2 liters per minute.
- **Inspiratory reserve volume (IRV)** refers to the volume of air that can be forcibly inspired after a normal tidal inspiration. The IRV averages 2100–3300 ml of air, depending on a person's sex and size.
- **Expiratory reserve volume (ERV)** is essentially the IRV in reverse. It refers to the amount of air that can be forcibly expired after a normal tidal expiration. The ERV averages about 700–1200 ml of air.

You may have noticed that the IRV is larger than the ERV by about 1400–2100 ml of air. This difference exists because even after the most forceful expiration, some air remains in the lungs, a volume known as the **residual volume (RV)**. The RV remains in the lungs due mostly to the intrapleural pressure and outward recoil of the chest wall, which keep the lungs slightly inflated. Unlike the TV, IRV, and ERV, the RV is measured not by spirometry but by other techniques.

To get a better picture of a person's pulmonary function, two or more of the pulmonary volumes can be combined to create **pulmonary capacities** (see the blue arrows in Figure 21.18). The following four pulmonary capacities are commonly considered:

- **Inspiratory capacity** is the total amount of air that a person can inspire after a tidal expiration. It is equal to the TV plus the IRV.
- **Functional residual capacity** is the amount of air that is normally left in the lungs after a tidal expiration. It is the sum of the ERV and the RV.
- **Vital capacity** represents the total amount of exchangeable air, or the total amount of air that can move in and out of the lungs. It is equal to the sum of the TV, IRV, and the ERV. The vital capacity is measured in the laboratory with a spirometer as the **forced vital capacity (FVC)**, in which the subject follows a maximal inspiration with a maximal exhalation. This measurement can give important information about the presence and type of pulmonary disease, which we discuss in Module 21.8.
- **Total lung capacity** represents the total amount of exchangeable and nonexchangeable air in the lungs. It is the total of all four pulmonary volumes: IRV, TV, ERV, and RV.

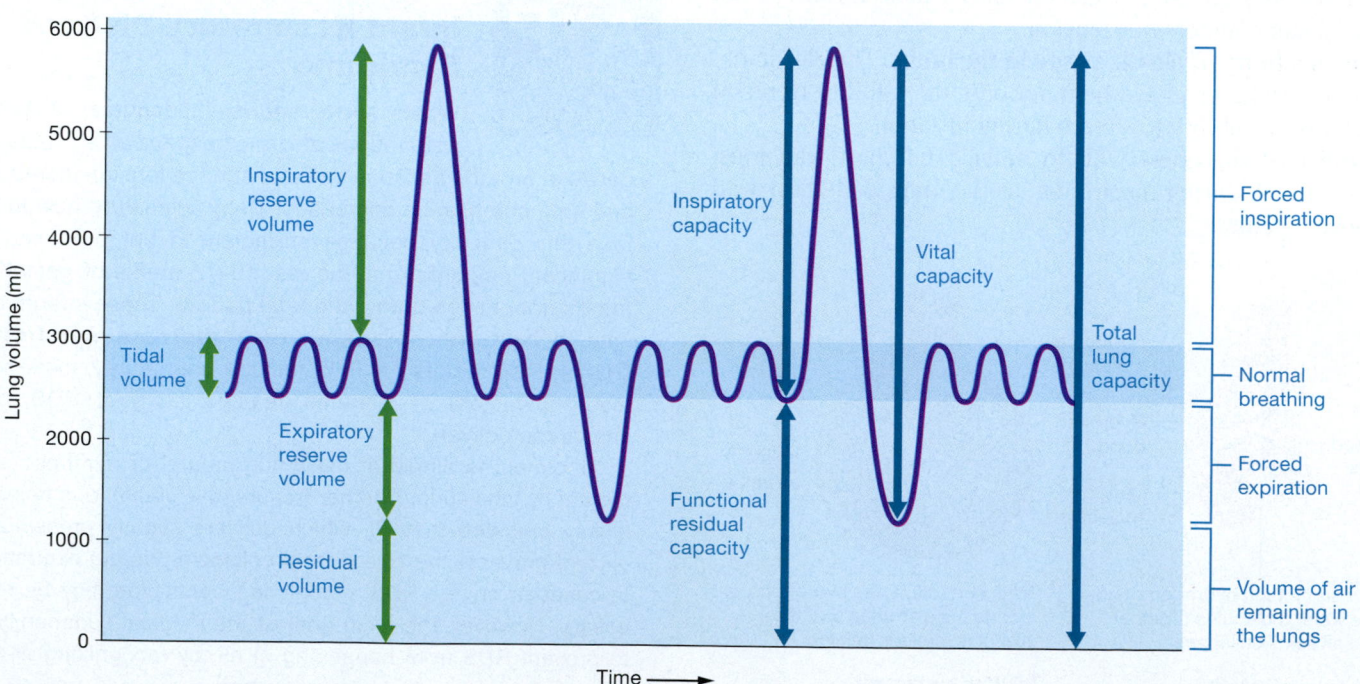

Figure 21.18 Graph of pulmonary volumes and capacities.

Practice art labeling
@ Mastering Anatomy & Physiology

Table 21.4 Pulmonary Volumes and Capacities

Measurement	Average Value (Female, Male)	Definition
Pulmonary Volumes		
Tidal volume (TV)	500 ml, 500 ml	Volume of air exchanged with normal, quiet breathing
Inspiratory reserve volume (IRV)	1900 ml, 3100 ml	Maximum volume of air that can be forcibly inspired after a tidal inspiration
Expiratory reserve volume (ERV)	700 ml, 1200 ml	Maximum volume of air that can be forcibly expired after a tidal expiration
Residual volume (RV)	1100 ml, 1200 ml	Volume of air that remains in the lungs after a forced expiration
Pulmonary Capacities		
Inspiratory capacity	2400 ml, 3600 ml	Total amount of air that can be inspired; equal to tidal volume plus inspiratory reserve volume
Functional residual capacity	1800 ml, 2400 ml	Total amount of air that normally remains in the lungs after a tidal expiration; equal to the residual volume plus expiratory reserve volume
Vital capacity	3100 ml, 4800 ml	Total amount of exchangeable air; equal to sum of the tidal volume, expiratory reserve volume, and inspiratory reserve volume
Total lung capacity	4200 ml, 6000 ml	Total amount of exchangeable and nonexchangeable air; equal to the sum of all the pulmonary volumes

Table 21.4 summarizes the pulmonary volumes and capacities and gives average values for each in adults. Note that except for tidal volume, the values for all volumes and capacities are lower in females than in males. These values also vary with a person's height, age, and health.

Quick Check

☐ 9. What are three measurable pulmonary volumes?

☐ 10. What is the vital capacity?

Apply What You Learned

☐ 1. A person is injured in a car accident, and the emergency room physician determines that the patient's left pleural cavity has filled with blood. What effect will this have on the intrapleural pressure in that pleural cavity? What will this do to the left lung? Explain.

☐ 2. Melvin Gartner has been admitted to the hospital with a diagnosis of pneumonia.

   a. You note that the passages of his respiratory tract can become filled with mucus. What impact will this have on the effectiveness of his pulmonary ventilation? Explain.

   b. He has also developed *pulmonary edema*, a condition in which fluid accumulates in and around the alveoli and increases surface tension. What impact will this have on the effectiveness of his pulmonary ventilation? Explain.

   c. Mr. Gartner is given pulmonary function tests, and the following values are obtained: tidal volume, 400 ml; inspiratory reserve volume, 3050 ml; expiratory reserve volume, 650 ml. Calculate his inspiratory capacity and vital capacity. Are his values normal?

See answers in Appendix A.

MODULE 21.4
Gas Exchange

Learning Outcomes

1. Describe the relationship of Dalton's law and Henry's law to pulmonary and tissue gas exchange and to the amounts of oxygen and carbon dioxide dissolved in plasma.

2. Describe oxygen and carbon dioxide pressure gradients and net gas movements in pulmonary and tissue gas exchange.

3. Explain how oxygen and carbon dioxide movements are affected by changes in partial pressure gradients.

4. Describe the mechanisms of ventilation-perfusion matching.

5. Explain the factors that maintain oxygen and carbon dioxide gradients between blood and tissue cells.

Pulmonary ventilation only brings new air into the alveoli and removes oxygen-poor air from the alveoli. It doesn't actually bring oxygen into the blood or remove carbon dioxide from the blood, nor does it deliver oxygen to our cells. For this to occur, we rely on *gas exchange*, which consists of two processes: *pulmonary gas exchange* and *tissue gas exchange*. As we noted in Module 21.1, **pulmonary gas exchange** involves the exchange of gases between the alveoli and the blood, and **tissue gas exchange** involves the exchange of gases between the blood in systemic capillaries and the body's cells.

In this module, we look at what drives pulmonary and tissue gas exchange, including the pressure that a gas exerts and its ability to dissolve in water. We also consider various characteristics of the body that make these processes more or less efficient, such as the surface area and blood supply available for gas exchange. Let's begin with two gas laws that apply to gas exchange.

The Behavior of Gases

The first gas law that applies to gas exchange is called **Dalton's law of partial pressures.** Dalton's law states that each gas in a mixture exerts its own pressure, called its *partial pressure,* so the total pressure of a gas mixture is the sum of the partial pressures of its gases. The partial pressure of a gas is symbolized by P_{gas} and is equal to its percent of the gases times the total pressure of all gases.

Let's turn our attention to atmospheric pressure first. We know that air is a mixture of gases, made up of about 78% nitrogen and 21% oxygen, with the remaining 1% composed of carbon dioxide, argon, and other gases. The total atmospheric pressure is equal to the partial pressures of all these gases:

$$P_{N_2} + P_{O_2} + P_{Ar} + P_{others} = \text{Atmospheric pressure}$$
$$(760 \text{ mm Hg at sea level})$$
$$P_{N_2} = 0.78 \times 760 = 593 \text{ mm Hg}$$
$$P_{O_2} = 0.21 \times 760 = 160 \text{ mm Hg}$$
$$P_{Ar} = [0.0093] \times 760 = 7 \text{ mm Hg}$$

The partial pressures of a gas determines where it moves by diffusion. As we discussed previously, a gas follows a pressure gradient and moves from an area of higher pressure to an area of lower pressure. The greater the pressure difference, the more rapidly the gas diffuses. This is the rationale for giving 100% oxygen to a person in respiratory distress—it increases the partial pressure of oxygen in the lungs, causing oxygen to diffuse into the person's blood at a faster rate.

A related gas law that applies to gas exchange deals with the ability of a gas to dissolve in water, termed its *solubility* in water. Just as a pinch of salt dissolves in water, some gases also dissolve to a certain extent when mixed with water. **Henry's law** states that the degree to which a gas dissolves in a liquid is proportional to both its partial pressure and its solubility in the liquid. The simplest example of this is the carbon dioxide dissolved in carbonated drinks. The carbon dioxide is added to the solution under high pressure, which causes the molecules to go into solution. When you open the can or bottle containing the drink, the pressure immediately decreases as the carbon dioxide molecules leave the solution and move from the area of higher pressure (the drink) to the area of lower pressure (the surrounding atmosphere).

Henry's law also explains the behavior of gases we find in air when they come in contact with water in the body:

- Although nitrogen gas has a high partial pressure in air (making up about 78% of its volume), very little nitrogen is present in the plasma. The reason is its low solubility in water, too low for any of it to dissolve in appreciable amounts.
- Even though oxygen gas has a lower partial pressure than nitrogen (as we said, oxygen makes up about 21% of the gas in air), a small amount of oxygen is found dissolved in plasma. Why? Its solubility in water, although fairly low, is still higher than that of nitrogen.
- Carbon dioxide has a very low partial pressure (making up less than 1% of the volume of air), yet it is found dissolved in plasma. Why? For a gas, it has a relatively high solubility in water. In fact, it is about 20 times more soluble in water than is oxygen.

Henry's law forms the basis for a treatment called *hyperbaric oxygen therapy.* See *A&P in the Real World: Hyperbaric Oxygen Therapy* for more information.

Quick Check

☐ 1. How does the pressure gradient between two gas mixtures influence the diffusion of those gases?

☐ 2. Why do we find more carbon dioxide dissolved in the plasma than oxygen?

Pulmonary Gas Exchange

As we have already discussed, pulmonary gas exchange (also called *external respiration;* not to be confused with cellular respiration) is the diffusion of gases between the alveoli and the blood. During pulmonary gas exchange, oxygen diffuses from the air in the alveoli into the blood in the pulmonary capillaries, and carbon dioxide flows in the opposite direction. This exchange converts the oxygen-poor, carbon dioxide–rich blood delivered by the pulmonary artery into oxygen-rich blood that contains less carbon dioxide. This oxygenated blood is then delivered via the pulmonary veins to the left side of the heart to be distributed to all the tissues of the body (see Chapter 17).

Hyperbaric Oxygen Therapy

In hyperbaric oxygen therapy, a person is placed in a chamber and exposed to a higher-than-normal partial pressure of oxygen. This higher partial pressure increases the amount of oxygen that is dissolved in the plasma, which in turn increases overall oxygen delivery to the tissues.

Hyperbaric oxygen is useful in treating conditions that benefit from having a greater percentage of oxygen dissolved in the plasma, such as severe blood loss, crush injuries, anemia (a decreased oxygen-carrying capacity of the blood), chronic wounds, certain infections, and burns. Hyperbaric oxygen therapy is also used for decompression sickness (also known as "the bends"), a condition generally seen in divers who have ascended from a dive too rapidly. Decompression sickness is caused by dissolved gases in the blood coming out of solution and forming gas bubbles in the bloodstream. Hyperbaric oxygen forces the gases back into solution, eliminating the bubbles.

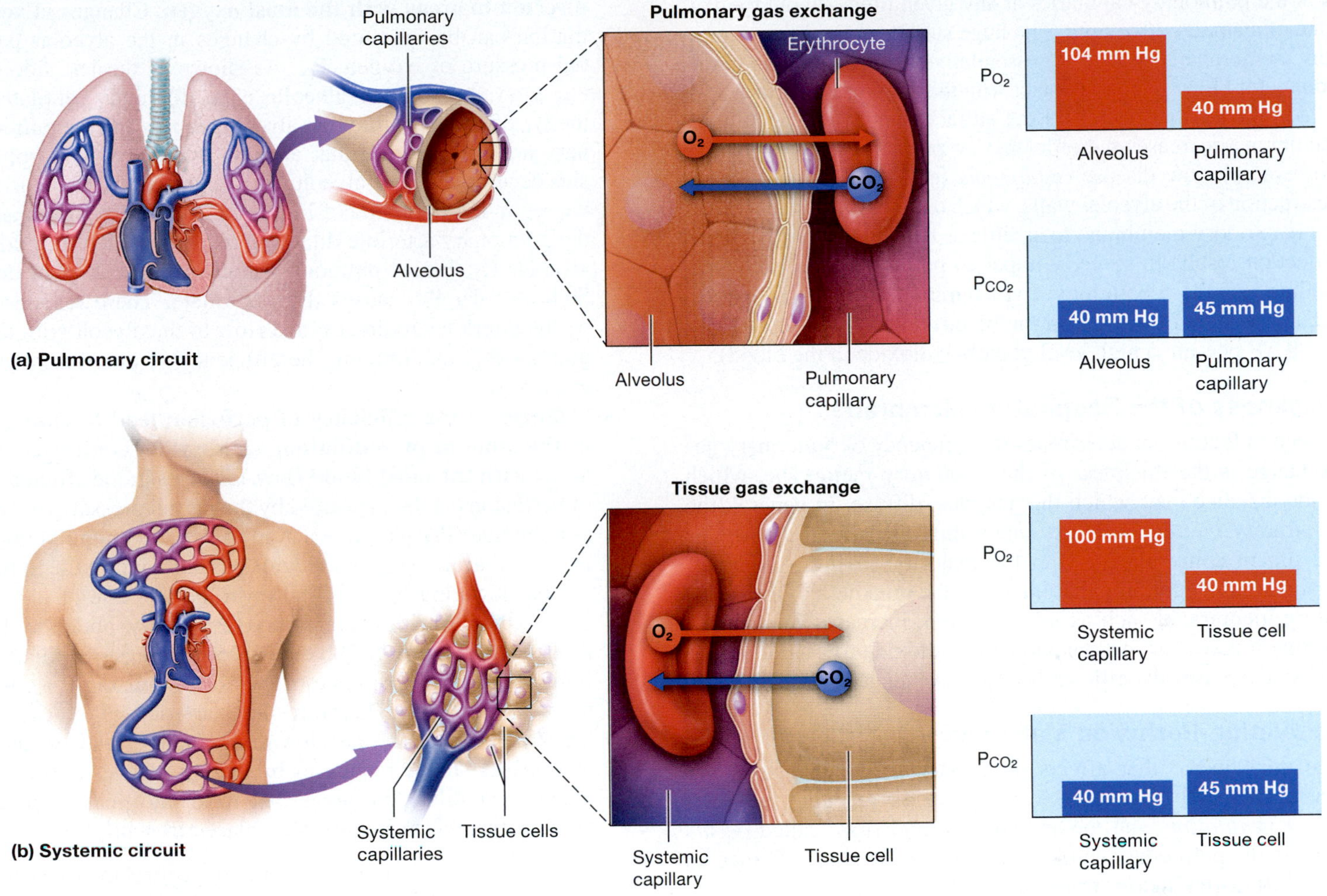

Pulmonary capillaries

Pulmonary gas exchange

Erythrocyte

O_2

CO_2

Alveolus

Pulmonary capillary

P_{O_2}

104 mm Hg	40 mm Hg
Alveolus	Pulmonary capillary

P_{CO_2}

40 mm Hg	45 mm Hg
Alveolus	Pulmonary capillary

Alveolus

(a) Pulmonary circuit

Tissue gas exchange

O_2

CO_2

Systemic capillary

Tissue cell

P_{O_2}

100 mm Hg	40 mm Hg
Systemic capillary	Tissue cell

P_{CO_2}

40 mm Hg	45 mm Hg
Systemic capillary	Tissue cell

(b) Systemic circuit

Systemic capillaries   Tissue cells

Figure 21.19 Pulmonary and tissue gas exchange.

Figure 21.19 shows the processes of pulmonary and tissue gas exchange (note that we cover tissue gas exchange after the entire pulmonary process).

Like the movement of all gases, pulmonary gas exchange is driven by pressure gradients (an example of the Gradients Core Principle, p. 28); in this case we are dealing with the differences in partial pressure between the oxygen and carbon dioxide in the alveoli and in the blood.

CORE PRINCIPLE
Gradients

As you can see in Figure 21.19a, the deoxygenated blood entering the pulmonary capillaries has a very low partial pressure of oxygen (P_{O_2}) of about 40 mm Hg, compared with the air in the alveoli, which has a P_{O_2} of about 104 mm Hg. This steep gradient drives the movement of oxygen from the alveoli to the blood very efficiently. The movement of oxygen is so efficient that it takes only about one-quarter of a second for equilibrium to occur (i.e., to achieve the same P_{O_2} in both the pulmonary capillary blood and the alveoli).

Notice, however, that the pressure gradient for carbon dioxide isn't as steep as the gradient for oxygen. The partial pressure of carbon dioxide (P_{CO_2}) in the blood of pulmonary capillaries is about 45 mm Hg, and in the alveoli it is about 40 mm Hg.

Yet even with this smaller gradient, approximately equal amounts of carbon dioxide and oxygen are exchanged. This is due again to Henry's law. Remember that it is not just partial pressure that determines gas movement, but also solubility. As you read, carbon dioxide is about 20 times more soluble than oxygen in water, so it moves extremely rapidly across the respiratory membrane and doesn't require a big pressure gradient.

Factors Affecting Efficiency of Pulmonary Gas Exchange

Aside from the partial pressures of oxygen and carbon dioxide, three additional factors impact the efficiency of pulmonary gas exchange: (1) the surface area of the respiratory membrane, (2) the thickness of the respiratory membrane, and (3) the degree to which the amount of air flow matches the amount of blood flow. Let's examine each of these factors.

Surface Area of the Respiratory Membrane

We have already discussed the extremely large surface area of the respiratory membrane in Module 21.2. Recall that for the average person, it is about 80–100 m² for both lungs (roughly 1000 square feet). Consider for a moment that only about 75–100 ml of blood is in the pulmonary capillaries at any given time—about the volume of a small coffee mug. The huge surface area of the respiratory membrane is enough for essentially every erythrocyte in this volume of blood to be in contact with the respiratory surface.

Any factor that decreases the surface area of the respiratory membrane decreases the efficiency of pulmonary gas exchange. For example, the disease *emphysema* (em-fih-SEE-mah) causes destruction of the alveolar walls, which reduces the surface area of the respiratory membrane to as little as one-fifth of normal. This reduction results in severely impaired pulmonary gas exchange, leading to the conditions **hypoxemia** (hy'-pawk-SEE-mee-uh; a low level of oxygen in the blood) and **hypercapnia** (hy'-pur-KAP-nee-uh; a high level of carbon dioxide in the blood).

Thickness of the Respiratory Membrane

A second factor that determines the efficiency of pulmonary gas exchange is the thickness of the respiratory membrane, which is the distance over which the gas must diffuse. In general, the respiratory membrane is extremely thin, measuring as little as 0.2 μm in some places, with the majority having a width of about 0.6 μm. Anything that increases the thickness of the respiratory membrane, such as inflammation, increases the amount of time it takes for oxygen and carbon dioxide to be exchanged and so decreases the efficiency of gas exchange.

Ventilation-Perfusion Matching

The third factor that affects pulmonary gas exchange is the degree of match between the amount of air reaching the alveoli, or *ventilation*, and the amount of blood flow, called *perfusion*, in the pulmonary capillaries. This relationship is known as **ventilation-perfusion matching** (also called *ventilation-perfusion coupling*, because the two processes are coupled together).

Two phenomena keep ventilation and perfusion closely matched:

- **Changes in alveolar ventilation lead to changes in perfusion, so blood flow is directed to areas with the most oxygen.** Changes in ventilation can be measured by changes in the alveolar partial pressure of oxygen, P_{O_2}. As shown in the left side of **Figure 21.20a**, when an alveolus is insufficiently ventilated, the P_{O_2} of the air within that alveolus drops, and the pulmonary arteriole serving that alveolus constricts. The opposite occurs when an alveolus is adequately ventilated, as we see in the right side of Figure 21.20a. The P_{O_2} rises and the pulmonary arteriole dilates in response. Therefore, low alveolar P_{O_2} causes pulmonary arterioles to constrict, and high alveolar P_{O_2} causes them to dilate. These responses by the arterioles redirect blood flow to the alveoli with the greatest P_{O_2}, maximizing the efficiency of pulmonary gas exchange.

- **Changes in the efficiency of perfusion lead to changes in the amount of ventilation, so air flow is directed to areas with the most blood flow.** Changes in the efficiency of perfusion can be measured by the partial pressure of carbon dioxide (P_{CO_2}) in alveolar arterioles. When the pulmonary capillaries bring blood with less carbon dioxide to the alveoli, lowering P_{CO_2}, as shown in the left side of **Figure 21.20b**, the bronchiole constricts. Conversely, notice in the right side of Figure 21.20b that when pulmonary capillaries bring more carbon dioxide to the alveoli, raising P_{CO_2}, the bronchiole dilates. Therefore, high arteriolar P_{CO_2} causes bronchioles to dilate, and low arteriolar P_{CO_2} causes them to constrict. These responses by the bronchioles allow the lungs to ventilate the areas that have the greatest blood flow, so that gas exchange takes place efficiently.

These two processes combine to ensure that pulmonary ventilation and pulmonary capillary perfusion match each other as closely as possible. A value called the **ventilation/perfusion (V/Q) ratio** describes this match. Hypothetically, the V/Q ratio should be about 1.0, but in the human body it averages around 0.8–0.9. The V/Q ratio is a bit lower than expected because the lungs have slightly higher

Normal P_{O_2} and P_{CO_2} in alveolus and arteriole
(ventilation and perfusion matched)

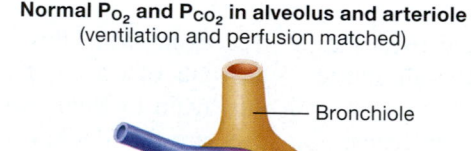
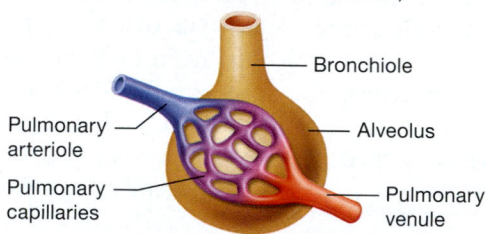

- Bronchiole
Pulmonary arteriole
- Alveolus
Pulmonary capillaries
- Pulmonary venule

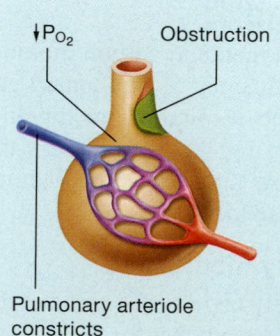

Low P_{O_2} in alveolus

↓P_{O_2} Obstruction

Pulmonary arteriole constricts

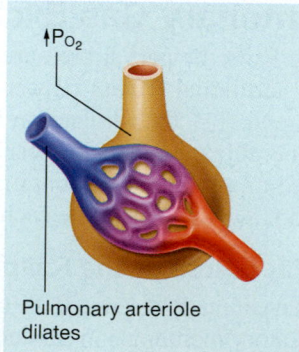

High P_{O_2} in alveolus

↑P_{O_2}

Pulmonary arteriole dilates

(a) Changes in ventilation (alveolar P_{O_2}) lead to changes in perfusion.

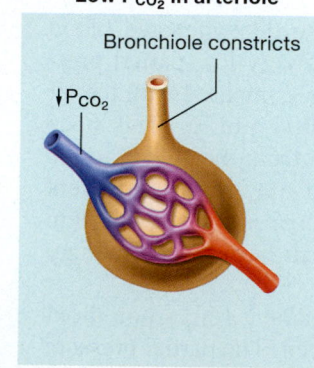

Low P_{CO_2} in arteriole

Bronchiole constricts

↓P_{CO_2}

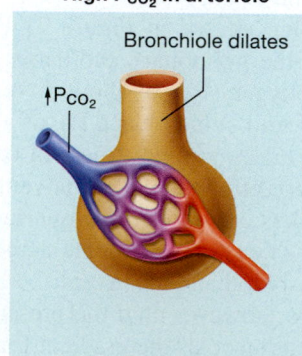

High P_{CO_2} in arteriole

Bronchiole dilates

↑P_{CO_2}

(b) Changes in perfusion (arteriolar P_{CO_2}) lead to changes in ventilation.

Figure 21.20 Ventilation-perfusion coupling.

perfusion than ventilation at their bases. This is due to the effects of gravity on the blood in the pulmonary arteries. Read *A&P in the Real World: V/Q Mismatch* for information on what happens when these processes are out of sync.

V/Q Mismatch

When ventilation does not match perfusion to an area of the lung, the result is called a *V/Q mismatch*. Common causes of a V/Q mismatch include pneumonia, asthma, pulmonary edema, or any other condition that results in less air getting to the alveoli. In cases like these, a right-to-left *shunt* is said to exist, meaning that blood is "shunted" from the right side of the heart to the left side without receiving adequate oxygenation in the pulmonary circulation.

Another potential cause of a V/Q mismatch is a reduction in blood flow to the lungs. The most common scenario in this case is a blockage in a pulmonary artery or arteriole, called a *pulmonary embolus.* Many substances can act as pulmonary emboli, including pieces of a blood clot, gas bubbles, fat globules, and amniotic fluid (the fluid surrounding a fetus in a pregnant woman). Such a condition creates what is known as *alveolar dead space*—the alveoli in a segment of the lung are ventilated but are not perfused, so no pulmonary gas exchange takes place.

Quick Check

☐ 3. What takes place during pulmonary gas exchange? Which pressure gradients drive this process?

☐ 4. How do the surface area and thickness of the respiratory membrane influence the efficiency of pulmonary gas exchange?

☐ 5. What is the ventilation/perfusion ratio, and what two mechanisms keep the ratio close to 1?

Tissue Gas Exchange

◀◀ FLASHBACK

1. What is stratified epithelial tissue? (p. 132)
2. What are capillary beds, and how are they structured? (p. 687)

The exchange of oxygen and carbon dioxide between the blood and the tissues is called tissue gas exchange (or *internal respiration*). Tissue gas exchange works via essentially the same mechanisms as pulmonary gas exchange (look back at Figure 21.19b). In tissue gas exchange, oxygen moves from the blood in the systemic capillaries into the cells of the tissues.

As with pulmonary gas exchange, the main driving force for tissue gas exchange is a pressure gradient—the difference in the P_{O_2} between blood and tissues. The cells are constantly using oxygen for cellular respiration (see Chapter 3), and for this reason the P_{O_2} in the tissues is low, averaging about 40 mm Hg. In contrast, the P_{O_2} in the systemic capillaries is high, measuring about 100 mm Hg. This steep pressure gradient facilitates rapid exchange of oxygen between the capillaries and the tissues.

The pressure gradient functions in the reverse direction for carbon dioxide. The cells of the tissues produce large quantities of carbon dioxide as a waste product of cellular respiration, so the P_{CO_2} in the tissues is relatively high (about 45 mm Hg). The P_{CO_2} in the blood is lower (about 40 mm Hg), and this pressure gradient, combined with carbon dioxide's high solubility in water, facilitates its rapid transport into the systemic capillaries.

Factors Affecting Efficiency of Tissue Gas Exchange

As with pulmonary gas exchange, physical factors play a role in the efficiency of tissue gas exchange. These factors include the following:

- **The surface area available for gas exchange.** Most capillaries exist in branching, weblike capillary beds (see Chapter 18). This branching increases the surface area available for tissue gas exchange and makes it a more efficient process.
- **The distance over which diffusion must occur.** Stratified epithelial tissues can have only so many layers of living cells because oxygen can diffuse only so far (see Chapter 4). Cells that are too distant from capillaries do not receive adequate oxygen and die.
- **The perfusion of the tissue.** Tissue gas exchange is most efficient when tissues are adequately supplied with blood via the capillaries. Inadequate perfusion explains why diseases that affect the small blood vessels can lead to damage and death of a tissue. If a tissue is not adequately perfused, oxygen is not delivered to it and carbon dioxide is not removed.

Quick Check

☐ 6. What are three factors that influence the efficiency of tissue gas exchange?

Apply What You Learned

☐ 1. The condition *acute respiratory distress syndrome* (*ARDS*) is characterized by damage to the respiratory membrane that causes it to thicken. What effect will this thickening have on pulmonary gas exchange? Explain.

☐ 2. What effect will low-oxygen conditions have on pulmonary arterioles? What effect will such conditions have on blood pressure in the pulmonary circuit? (*Hint:* You may wish to review blood vessel resistance in Chapter 18.)

☐ 3. The toxins in cigarette smoke damage the cells of the tissue capillaries and also cause arterioles supplying capillary beds to constrict. What effect will this have on tissue gas exchange?

See answers in Appendix A.

MODULE
21.5
Gas Transport through the Blood

Learning Outcomes

1. Describe the ways in which oxygen is transported in blood, including the reversible reaction for oxygen binding to hemoglobin.

2. Interpret the oxygen-hemoglobin dissociation curve, and describe the factors that affect the curve.

3. Describe the ways in which carbon dioxide is transported in blood, including the reversible reaction that converts carbon dioxide and water to carbonic acid.

4. Predict how changing the partial pressure of carbon dioxide will affect the pH of plasma.

5. Describe the conditions hyperventilation and hypoventilation.

We discussed pulmonary and tissue gas exchange together because of their similarities, but in doing so we skipped an important step: **gas transport** in the blood. Remember that oxygen and carbon dioxide have limited solubility in water. Therefore, in order for pulmonary and tissue gas exchange to occur, both of these gases must undergo chemical reactions that enable their safe transport through the bloodstream and exchange in the lungs and tissues. We explore these reactions in this module. First, we look at the importance of the protein *hemoglobin* in transporting oxygen, and then at three ways of transporting

carbon dioxide in the blood, including the role of a substance known as *bicarbonate*. We also describe how ventilation and carbon dioxide can affect the body's acid-base homeostasis.

Oxygen Transport

« FLASHBACK

1. What is hemoglobin's main function? (p. 726)
2. Describe the structure of hemoglobin. (p. 727)

As you learned from our discussion of Henry's law, oxygen does not dissolve in great amounts in the plasma because its solubility in water is rather low. In fact, only about 1.5% of the total oxygen transported by the blood travels unbound. Most of the oxygen is transported on the protein **hemoglobin (Hb)**. In this section, we examine how oxygen is carried in this manner.

Transport of O_2 on Hemoglobin

Hemoglobin is a protein found in erythrocytes that consists of four polypeptide chains, each including a **heme group** (see Figure 19.3 for review). Each of these heme groups contains one iron ion that can bind one molecule of oxygen, so each Hb protein can transport up to four molecules of oxygen. About 280 million Hb proteins are present in each erythrocyte, giving each one the ability to carry over a billion oxygen molecules.

Hb binds and releases O_2 via two reactions known as *loading* and *unloading reactions,* respectively (**Figure 21.21**). These reactions can be summarized as follows:

$$HHb \quad + \quad O_2 \quad \rightleftharpoons \quad HbO_2$$

Deoxyhemoglobin   *Oxyhemoglobin*

① During loading, oxygen from alveoli binds to hemoglobin (Hb) in the pulmonary capillaries, converting it to oxyhemoglobin (HbO₂).

② Oxygen-rich blood travels to the heart, which pumps it to the systemic circulation.

③ During unloading, Hb in the systemic capillaries releases oxygen to the tissue cells.

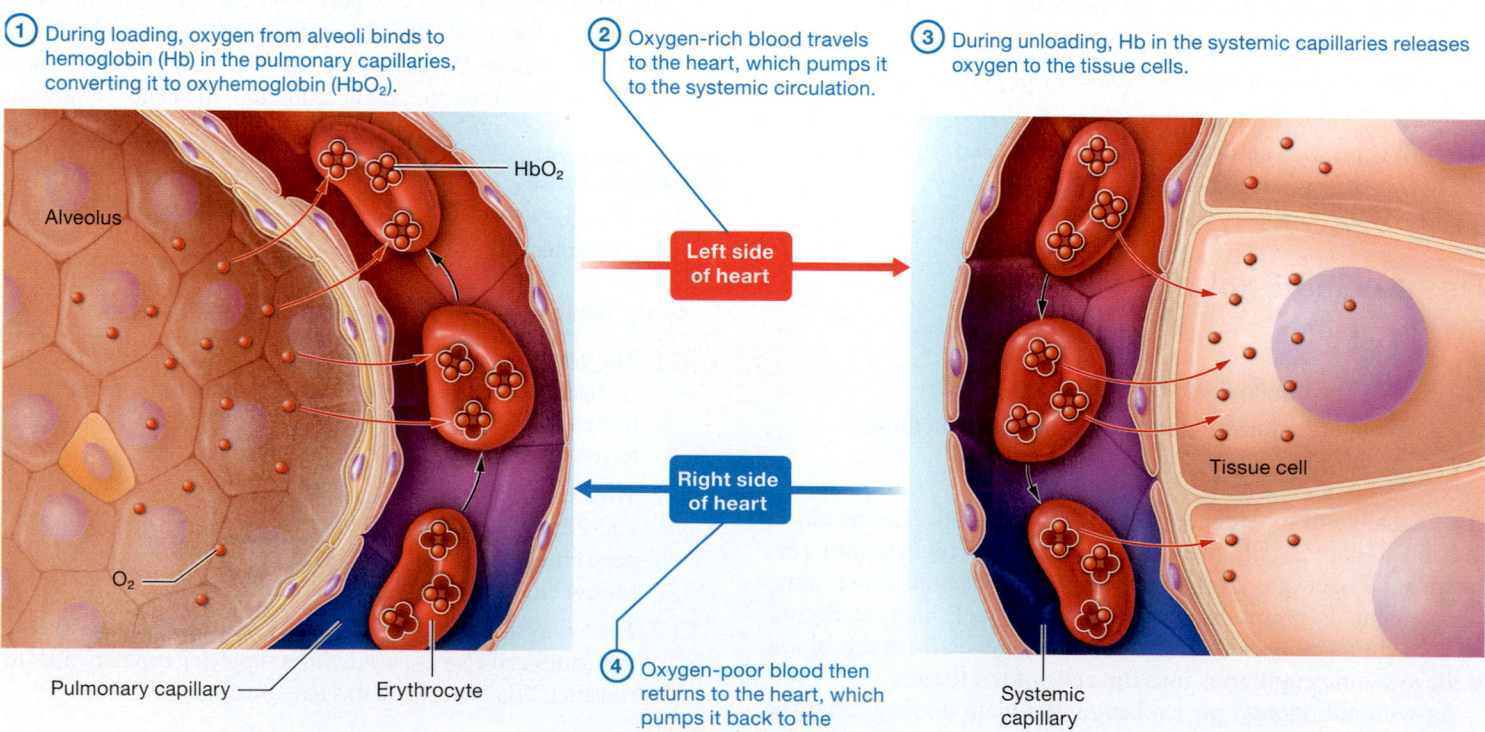

Alveolus

HbO₂

Left side of heart

Right side of heart

O₂

Pulmonary capillary — Erythrocyte

④ Oxygen-poor blood then returns to the heart, which pumps it back to the pulmonary circulation.

Tissue cell

Systemic capillary

Figure 21.21 Transport of oxygen: loading and unloading of oxygen on hemoglobin in erythrocytes.

1. **During loading, oxygen from alveoli binds to hemoglobin in the pulmonary capillaries.** This loading reaction converts hemoglobin from **deoxyhemoglobin (HHb)** to **oxyhemoglobin (HbO$_2$).** An Hb with one to three oxygen molecules attached is said to be *partially saturated.* When an Hb is fully loaded with four oxygen molecules, it is said to be *fully saturated.*

2. **With the Hb fully saturated, the oxygen-rich blood travels to the left side of the heart, which pumps it to the systemic circulation.**

3. **During unloading, Hb in the systemic capillaries releases oxygen to the tissue cells.** Without the unloading reaction, Hb would never give any of its oxygen to the tissues that need it.

4. **Oxygen-poor blood then returns to the right side of the heart, which pumps it back to the pulmonary circulation.**

Let's look at the unloading reaction more closely. The amount of oxygen unloaded from Hb must strictly match the tissues' metabolic needs. Cells that are actively carrying out metabolic reactions require large amounts of oxygen, and if Hb does not unload adequate oxygen, the cells will not be able to produce adequate ATP. At the same time, cells with a slower metabolic rate do not require large amounts of oxygen. If Hb were to give these cells too much oxygen, toxicity and cellular damage could result.

Hb's ability to load or unload oxygen, and therefore its saturation, depends on two factors:

- the P$_{O_2}$ in the lungs or tissues, and
- the tightness with which Hb binds oxygen, also called the *affinity,* or the *bond strength,* of Hb.

Let's examine how each of these factors ensures appropriate tissue oxygenation.

Effect of P$_{O_2}$ on Hemoglobin Saturation

The percentage of Hb bound to oxygen is called the **percent saturation of Hb.** One of the main determinants of the percent saturation of Hb is the P$_{O_2}$ of the blood and tissues. The higher the P$_{O_2}$ of the blood, the more the loading reaction is favored, because more O$_2$ molecules are available to bind to Hb. The opposite is also true.

The relationship between P$_{O_2}$ and percent saturation of Hb is shown in the **oxygen-hemoglobin dissociation curve,** illustrated in **Figure 21.22.** Let's see what this graph tells us about the relationship between these two factors:

- **Percent O$_2$ saturation of Hb.** These values are on the left-hand side of the graph. Notice that the Hb in arterial blood at point A normally has a percent saturation of about 97–100%. This blood has passed through the lungs and heart but has not yet reached the tissues to unload. After unloading O$_2$ in the tissues, the Hb in venous blood at point B has a percent saturation of about 75%. This means that the venous blood returning to the heart and lungs of a person at rest still has about 75% of the oxygen it can hold bound to its Hb. But note how this situation can change when oxygen demands by the tissues change. Point C

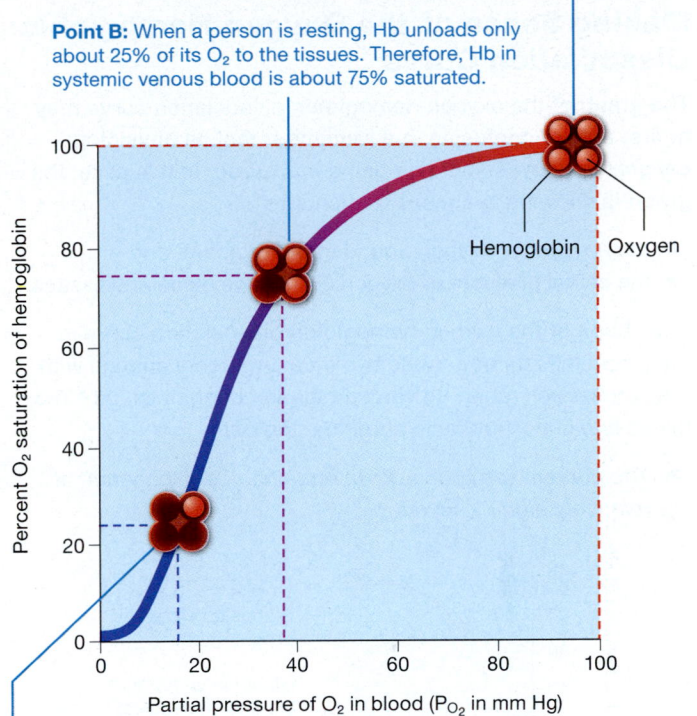

Point A: Hemoglobin (Hb) is almost 100% saturated in systemic arterial blood, as no O$_2$ has been unloaded to the tissues.

Point B: When a person is resting, Hb unloads only about 25% of its O$_2$ to the tissues. Therefore, Hb in systemic venous blood is about 75% saturated.

Hemoglobin | Oxygen

Partial pressure of O$_2$ in blood (P$_{O_2}$ in mm Hg)

Point C: When a person exercises vigorously, Hb unloads most of its oxygen in the tissues. In such cases, the Hb entering venous blood is only about 25% saturated.

Figure 21.22 **The oxygen-hemoglobin dissociation curve.**

represents the percent saturation of Hb in venous blood during vigorous exercise, when the tissues require much more oxygen. In this situation, the percent saturation of Hb can drop significantly, to 25% or even less.

- **Blood P$_{O_2}$** Along the bottom of the graph is the partial pressure of oxygen in blood, the P$_{O_2}$. Notice that the P$_{O_2}$ in arterial blood is close to 100 mm Hg (at point A), and the P$_{O_2}$ in venous blood of a person at rest is about 40 mm Hg (at point B). Then notice at point C that during vigorous exercise the P$_{O_2}$ of venous blood drops significantly as the tissues consume more oxygen.

You can see that this curve has an S shape, which results from Hb's changing affinity for oxygen as the P$_{O_2}$ changes. What's happening? When Hb is fully saturated, it hangs on to the oxygen molecules tightly, which matches the red-shaded part of the curve in Figure 21.22. In this region, even a big change in P$_{O_2}$ doesn't alter the percent saturation of Hb significantly. However, when Hb loses its first oxygen molecule, it changes shape, and it becomes easier for the second and third oxygen molecules to unload from Hb. The purple-shaded curve on the graph shows the resulting sharp decrease in the percent saturation of Hb. Finally, the fourth and last oxygen molecule is bound least tightly to Hb. Notice in the blue-shaded part of the curve that it takes a drastically low P$_{O_2}$ for Hb to unload the fourth oxygen molecule.

ConceptBOOST))))

Making Sense of the Oxygen-Hemoglobin Dissociation Curve

The graph of the oxygen-hemoglobin dissociation curve may at first appear confusing, but remember that all physiological graphs show something being *measured*. In this case, the graph is showing two measurements:

- the amount of oxygen bound to hemoglobin, and
- the partial pressure of oxygen (P_{O_2}) in the blood and tissues.

The shape of the oxygen-hemoglobin dissociation curve in the graph tells us how these two measurements change with respect to each other. So what do these changes tell us? They tell us two important facts about hemoglobin:

- *The percent saturation of Hb remains nearly constant at relatively high P_{O_2} levels.*

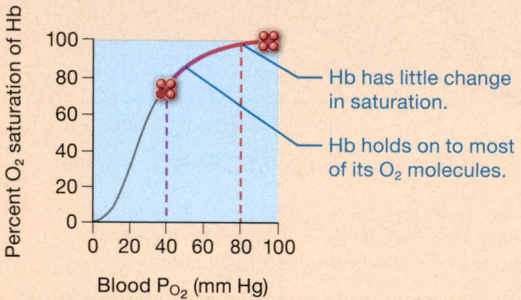

Notice that the percent saturation of Hb changes very little when the P_{O_2} drops from 100 mm Hg to 80 mm Hg. This means that even when oxygen levels in arterial blood are lower than normal, hemoglobin holds on to most of its oxygen molecules. This is important because arterial P_{O_2} never gets above about 80–85 mm Hg in some individuals, such as those at high altitudes or those with pulmonary disease. So this property of Hb is crucial because it allows Hb to hang on to its oxygen molecules until it reaches the much lower P_{O_2} levels found in the capillaries.

- *The percent saturation of Hb drops rapidly when P_{O_2} levels are low.*

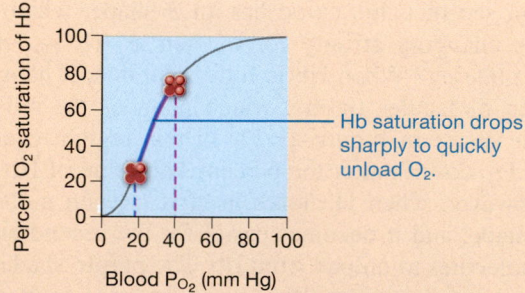

- Notice that when the P_{O_2} drops to about 40 mm Hg, which is generally the P_{O_2} in the capillary beds of a person at rest, the percent saturation of Hb drops sharply and rapidly.

Remember that blood is present in the capillaries for an extremely short time—only about 0.75 second. For this reason, Hb must be able to unload its oxygen molecules rapidly. This property of Hb is therefore important because it allows the Hb not only to unload its oxygen where it is needed in the capillary beds, but also to do so at a fast enough pace to ensure that the cells receive sufficient oxygen.

Take note of one final aspect of this graph: From the shape of this curve you can see why arterial P_{O_2} levels of 40–50 mm Hg, which occur when one stops breathing, are not compatible with life. At such a low P_{O_2}, Hb unloads all its oxygen in the arterial blood instead of the capillaries. Because free oxygen molecules in arterial blood cannot diffuse in appreciable amounts through the arterial wall to reach the cells, the cells are starved of oxygen and, as a result, they die. ■

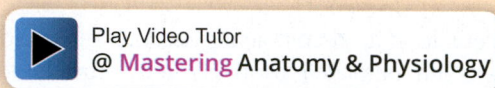
Play Video Tutor
@ Mastering Anatomy & Physiology

Effect of Affinity on Hemoglobin Saturation

Four main factors can change the affinity of Hb for oxygen by altering Hb's shape. The first three factors are (1) the temperature, (2) the pH of the blood, and (3) the P_{CO_2} (**Figure 21.23**) (we'll get to the fourth shortly). In general, increases in temperature, hydrogen ion concentration, and P_{CO_2} all decrease Hb's affinity for oxygen, which results in very small shifts in Hb structure that favor the unloading of more oxygen into the tissues. This scenario makes sense if you remember that a cell actively producing ATP needs more oxygen and also produces heat, hydrogen ions, and CO_2 as byproducts. The opposite is also true—as the temperature, acidity, and P_{CO_2} decrease, as occurs in tissues with a slower metabolic rate, Hb binds oxygen more strongly, and less oxygen is unloaded. This phenomenon is known as the **Bohr effect.**

The fourth factor that influences Hb's affinity for oxygen is the level of a chemical called **BPG (2,3-bisphosphoglycerate).** Erythrocytes produce BPG during a side reaction of glycolysis (look ahead to Chapter 23). BPG binds with Hb, reducing its affinity for oxygen, which increases the unloading of oxygen into the tissues. Erythrocytes produce greater amounts of BPG when Hb is less saturated with oxygen. (See *A&P in the Real World: Carbon Monoxide Poisoning* to read about the effects of increasing Hb's affinity for oxygen.)

BPG is a key factor that allows people to live at higher altitudes or with poor pulmonary function, as both situations result in lower P_{O_2} levels and less saturation of Hb with oxygen. In an example of the Cell-Cell Communication Core Principle (p. 28), the BPG level also increases in response to certain hormones, such as norepinephrine, epinephrine, thyroxine, testosterone, and human growth hormone. Each of these hormones promotes cellular metabolism and for this reason increases the cellular demand for oxygen.

CORE PRINCIPLE
Cell-Cell Communication

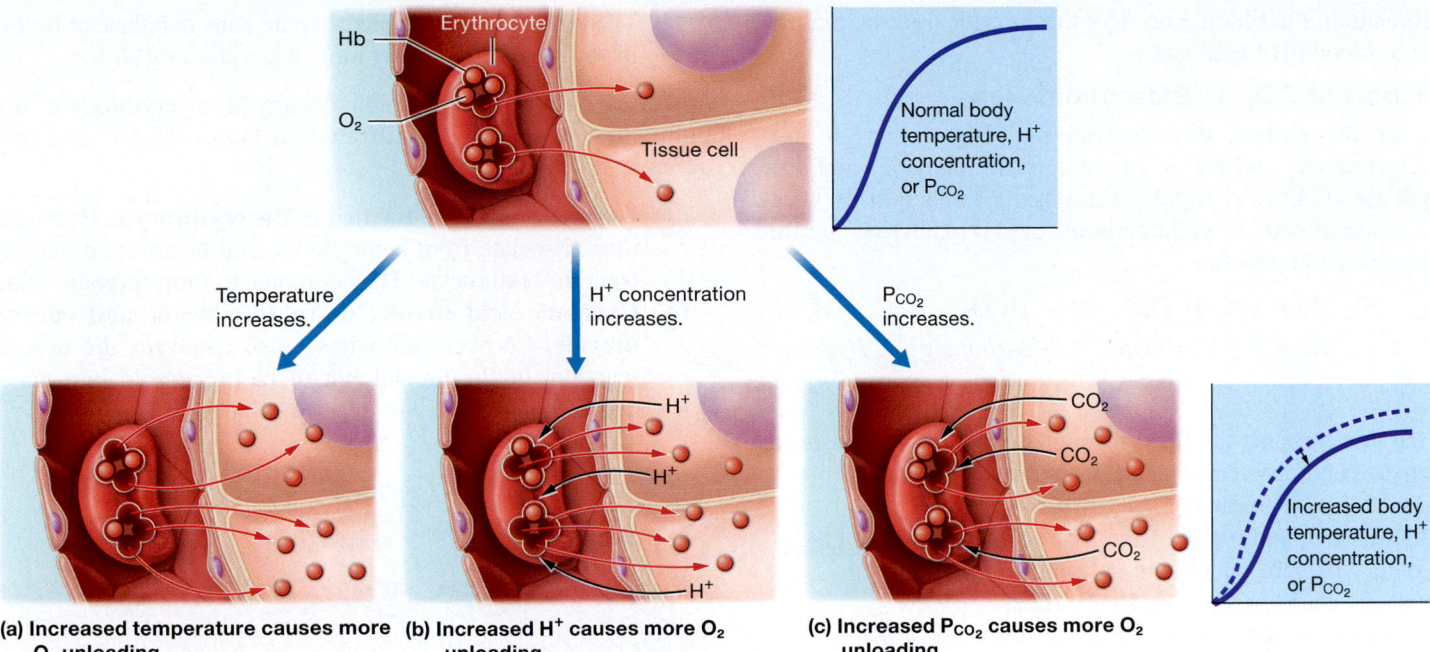

(a) **Increased temperature causes more O_2 unloading.**

(b) **Increased H^+ causes more O_2 unloading.**

(c) **Increased P_{CO_2} causes more O_2 unloading.**

Figure 21.23 **Effect of temperature, hydrogen ion concentration, and P_{CO_2} on oxygen unloading from hemoglobin.**

Carbon Monoxide Poisoning

Carbon monoxide (CO) poisoning is familiar to most people from a common scene in television and movies: a suicide or homicide that occurs when a car sits running in a closed garage. This is indeed one way to get CO poisoning, but it is certainly not the only way. Carbon monoxide is produced from burning organic compounds. The colorless, odorless, and tasteless gas is found in smoke from fires and cigarettes, and in exhaust fumes from portable heaters, stoves, and engines.

CO produces its toxic effects by binding reversibly with Hb, producing *carboxyhemoglobin*. Not only does this binding cause CO to occupy some of Hb's oxygen-carrying spots, it also causes a change in Hb's shape that increases its affinity for oxygen. This change means that Hb binds oxygen much more strongly and will not adequately release it to the body's tissues.

CO has an affinity for Hb about 200–230 times that of oxygen, so even small concentrations of this gas can cause serious problems. Symptoms of CO poisoning include confusion, dizziness, and nausea. In severe cases, it can cause seizures, coma, and death. The main treatment is 100% oxygen, delivered at either atmospheric or hyperbaric pressure.

Quick Check

☐ 1. How is the majority of oxygen transported through the bloodstream?

☐ 2. How do temperature, pH, P_{CO_2}, and BPG affect Hb's affinity for oxygen?

☐ 3. Why is the S shape of the oxygen-hemoglobin dissociation curve important?

Carbon Dioxide Transport

« FLASHBACK

1. What is an enzyme? What is its main function? (p. 43)

2. What is the function of a buffer? (p. 49)

3. What is the normal pH range for the blood? (p. 49)

The carbon dioxide produced by our cells' metabolic reactions is transported through the blood from tissue cells to the alveoli. Carbon dioxide is transported in the blood in three ways: dissolved in plasma, bound to Hb, and as bicarbonate ions. Let's take a look at each of these methods.

- **Dissolved in the plasma.** Recall from our earlier discussion that carbon dioxide has a higher solubility in water than does oxygen. For this reason, more carbon dioxide than oxygen dissolves in plasma, and about 7–10% of the total carbon dioxide is transported in this manner.

- **Bound to Hb within erythrocytes.** About 20% of carbon dioxide is transported by Hb. Unlike oxygen, carbon dioxide doesn't bind to the heme group of Hb. Instead, carbon dioxide binds to Hb's polypeptide chains. Carbon dioxide binds easily and reversibly with Hb, forming **carbaminohemoglobin** (not to be confused with carboxyhemoglobin, which is the combination of carbon monoxide with Hb).

- **As bicarbonate ions in blood.** The majority of carbon dioxide in the blood (about 70%) travels as a totally different compound, **bicarbonate ion** (HCO_3^-).

The third method is the major way of transporting carbon dioxide, so the upcoming subsections discuss how carbon dioxide is converted to bicarbonate ions in erythrocytes. We then examine how ventilation controls the amount of carbon dioxide and bicarbonate in the blood, and why this is critical to the maintenance of blood pH homeostasis.

Transport of CO_2 as Bicarbonate Ions

Most of the carbon dioxide entering the plasma diffuses into erythrocytes, where it encounters the enzyme **carbonic anhydrase (CA)**. CA rapidly catalyzes the reaction of water with carbon dioxide to yield carbonic acid (H_2CO_3) via the following reversible reaction:

$$CO_2 \;+\; H_2O \;\rightleftharpoons\; H_2CO_3 \;\rightleftharpoons\; HCO_3^- \;+\; H^+$$

| Carbon dioxide | Water | Carbonic acid | Bicarbonate ion | Hydrogen ion |

Note that carbonic acid doesn't linger for long, instead dissociating to become bicarbonate and hydrogen ions. Most bicarbonate ions then diffuse into the plasma, and many of the hydrogen ions bind with Hb, which acts as a buffer to resist pH change (see Chapter 2). Bicarbonate ions have a negative charge, so this influx of negative charges into the plasma could be problematic. Fortunately, the outward movement of bicarbonate ions is balanced by the movement of chloride ions into the erythrocyte, a phenomenon known as the **chloride shift.**

In **Figure 21.24a**, the forward reaction of carbon dioxide and water to carbonic acid and then to bicarbonate and hydrogen ions is shown in an erythrocyte in a systemic capillary. The process is as follows:

1. **Carbon dioxide diffuses from the cells into the erythrocyte.** Carbon dioxide is formed in tissue cells as a byproduct of the cells' catabolic reactions, after which it diffuses into the systemic capillary and erythrocyte.

2. **Carbon dioxide is converted to carbonic acid.** The enzyme carbonic anhydrase catalyzes the reaction of carbon dioxide and water to form carbonic acid (H_2CO_3).

3. **Carbonic acid dissociates into bicarbonate (HCO_3^-) and hydrogen ions.** At the pH in the cytosol, much of the carbonic acid dissociates into bicarbonate and hydrogen ions.

4. **Hydrogen ions bind to hemoglobin.** The newly formed hydrogen ions bind to hemoglobin, which serves as a buffer and helps to minimize pH changes from the hydrogen ions.

5. **Bicarbonate ions enter the plasma (as chloride ions enter the erythrocyte).** Bicarbonate ions leave the erythrocyte and enter the plasma.

The outward flow of bicarbonate ions is balanced by the inward flow of chloride ions in the chloride shift.

The reverse reaction, which occurs in an erythrocyte in a pulmonary capillary, is illustrated in **Figure 21.24b** and proceeds as follows:

1. **Carbonic acid is re-formed in the erythrocyte.** Hydrogen ions dissociate from hemoglobin, and bicarbonate ions re-enter the erythrocyte. They combine to form carbonic acid.

2. **Carbonic acid breaks down into water and carbon dioxide.** Carbonic anhydrase also catalyzes the reverse reaction, or the breakdown of carbonic acid into water and carbon dioxide.

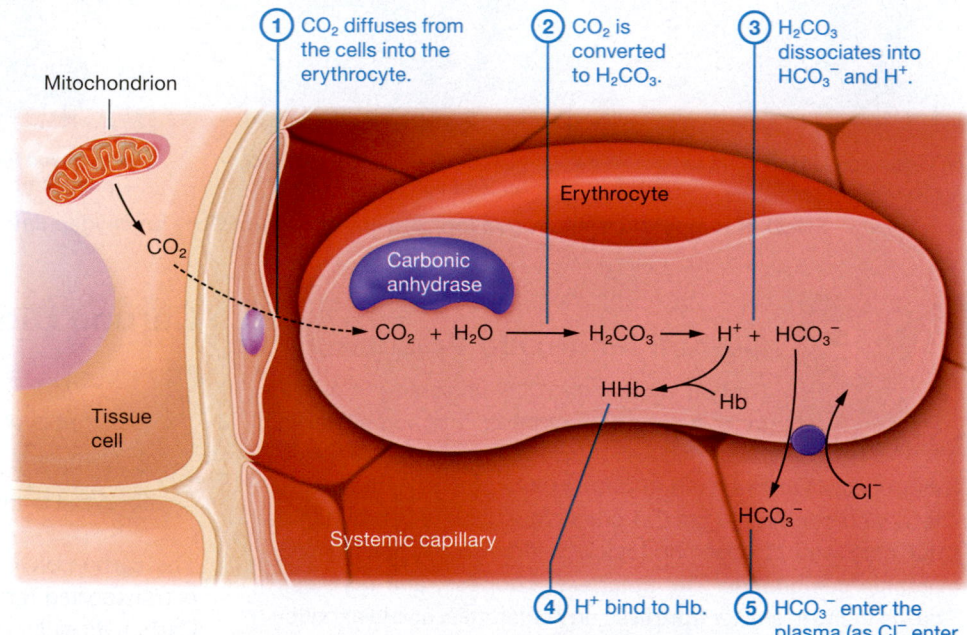

(a) **Bicarbonate ion formation in an erythrocyte in a systemic capillary**

1. CO_2 diffuses from the cells into the erythrocyte.
2. CO_2 is converted to H_2CO_3.
3. H_2CO_3 dissociates into HCO_3^- and H^+.
4. H^+ bind to Hb.
5. HCO_3^- enter the plasma (as Cl^- enter the erythrocyte).

(b) **Carbon dioxide formation in an erythrocyte in a pulmonary capillary**

1. H_2CO_3 is re-formed in the erythrocyte.
2. H_2CO_3 breaks down into H_2O and CO_2.
3. CO_2 diffuses into the alveolus.

Figure 21.24 Transport of carbon dioxide: the conversion of carbon dioxide and water into carbonic acid in erythrocytes.

Communication Core Principle (p. 28), the VRG's inspiratory neurons stimulate motor neurons to

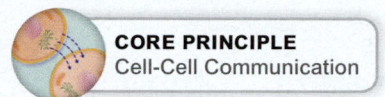

CORE PRINCIPLE
Cell-Cell Communication

trigger action potentials in the neurons of the **phrenic nerve** supplying the diaphragm and in the **intercostal nerves** supplying the external intercostal muscles. VRG neurons also stimulate neurons that innervate many accessory muscles of inspiration and expiration.

- **Dorsal respiratory group.** The neurons of the **dorsal respiratory group (DRG)** nuclei are found in the posterior medulla. Their primary role is to integrate sensory information from the blood and the lungs and relay it to other respiratory nuclei, although some DRG neurons are involved in inspiration as well.

Many other neurons related to respiratory functions are located in the cranial nerve nuclei surrounding the DRG and VRG. These neurons mostly innervate the accessory muscles of inspiration and/or expiration, and some bring sensory information from the blood and lungs back to the DRG. We have so many redundant and overlapping respiratory centers because we must continue to breathe while we sleep.

Quick Check

☐ 1. Which collection of neurons generates the basic rhythm for breathing?

☐ 2. What are the functions of the dorsal and ventral respiratory groups?

Control of the Rate and Depth of Ventilation

Although the RPG seems to set the basic rate and pattern of ventilation, clearly your ventilation rate changes as the needs of your body change. For example, why does your rate and depth of ventilation increase during and after exercise? And why do you gasp for air after holding your breath? The answer to these questions lies in negative feedback loops that rely on information

CORE PRINCIPLE
Feedback Loops

gathered by chemoreceptors, an example of the Feedback Loops Core Principle (p. 28).

Recall that a *chemoreceptor* is a specialized cell that responds to a change in the concentration of a certain chemical (see Chapter 13). Two types of chemoreceptors are involved in the control of ventilation: central and peripheral. Let's examine what each of these chemoreceptors does, as well as our ability to voluntarily control ventilation.

Function of Central Chemoreceptors

We discussed with Figure 21.25 how a person's pattern of ventilation can change his or her blood pH. Other mechanisms can also affect blood pH, such as an increased metabolic rate. But how does the body deal with these changes?

The answer is a negative feedback cycle started by the **central chemoreceptors,** which are neurons in the medullary reticular formation. These chemoreceptors detect changes in the P_{CO_2} and pH of the blood by monitoring the hydrogen ion concentration of the extracellular fluid of the brain and cerebrospinal fluid (CSF).

Alterations in arterial P_{CO_2} are the most powerful stimuli that induce changes in ventilation. In fact, the urge to breathe after holding your breath comes largely from the slight accumulation of carbon dioxide during the period of breath holding. Carbon dioxide in the blood diffuses through the blood brain barrier into the extracellular fluid of the brain and through the choroid plexus into the CSF, where it combines with water to form bicarbonate and hydrogen ions. Both these hydrogen ions and carbon dioxide molecules interact directly with the central chemoreceptors. (Note, however, that the blood brain barrier is relatively impermeable to hydrogen ions, so changes in *arterial* hydrogen ion concentration provide a relatively weak stimulus for the central chemoreceptors to induce changes in ventilation.)

When subjected to changes in P_{CO_2} and hydrogen ion concentration, the central chemoreceptors trigger one of the following responses via classic negative feedback loops:

- **High P_{CO_2} triggers hyperventilation.** As you can see in **Figure 21.28a**, in the presence of high arterial P_{CO_2} and/or high hydrogen ion concentration, the central chemoreceptors trigger the RPG to set an increased rate and depth of ventilation, or hyperventilation. Recall that during hyperventilation, carbon dioxide leaves the blood at a rapid pace. Its departure causes a decrease in blood P_{CO_2} and hydrogen ions—and therefore CSF P_{CO_2} and hydrogen ion concentration—and restores homeostasis. This is why your respiratory rate increases during and after exercise: More carbon dioxide is produced by cells during exercise due to the rapid rate of ATP synthesis. The higher P_{CO_2} is detected by the central chemoreceptors, and they stimulate hyperventilation to rid the body of excess carbon dioxide.

- **Low P_{CO_2} triggers hypoventilation.** In the presence of low arterial P_{CO_2} and/or low hydrogen ion concentration, as shown in **Figure 21.28b**, the central chemoreceptors elicit the opposite response—they trigger the RPG to slow the rate and depth of ventilation, or hypoventilation, so that the body retains carbon dioxide. This response causes an increase in blood P_{CO_2} and hydrogen ions and restores homeostasis.

A sudden change in arterial P_{CO_2} can alter ventilation within seconds to minutes. This response is initially strong but diminishes over the next day or two because the central chemoreceptors become less sensitive over time. In addition, bicarbonate ions move from the blood (through the blood brain barrier) into the brain extracellular fluid to counteract the pH imbalance. This is partly why some patients with pulmonary disease and chronically elevated arterial P_{CO_2} have a nearly normal ventilation rate. Certain drugs such as opiates and alcohol also blunt the response of the central chemoreceptors to an increased P_{CO_2}, inhibiting one of the main feedback loops the body uses to detect and rectify impaired ventilation. Such drugs can produce *respiratory depression,* especially in overdoses.

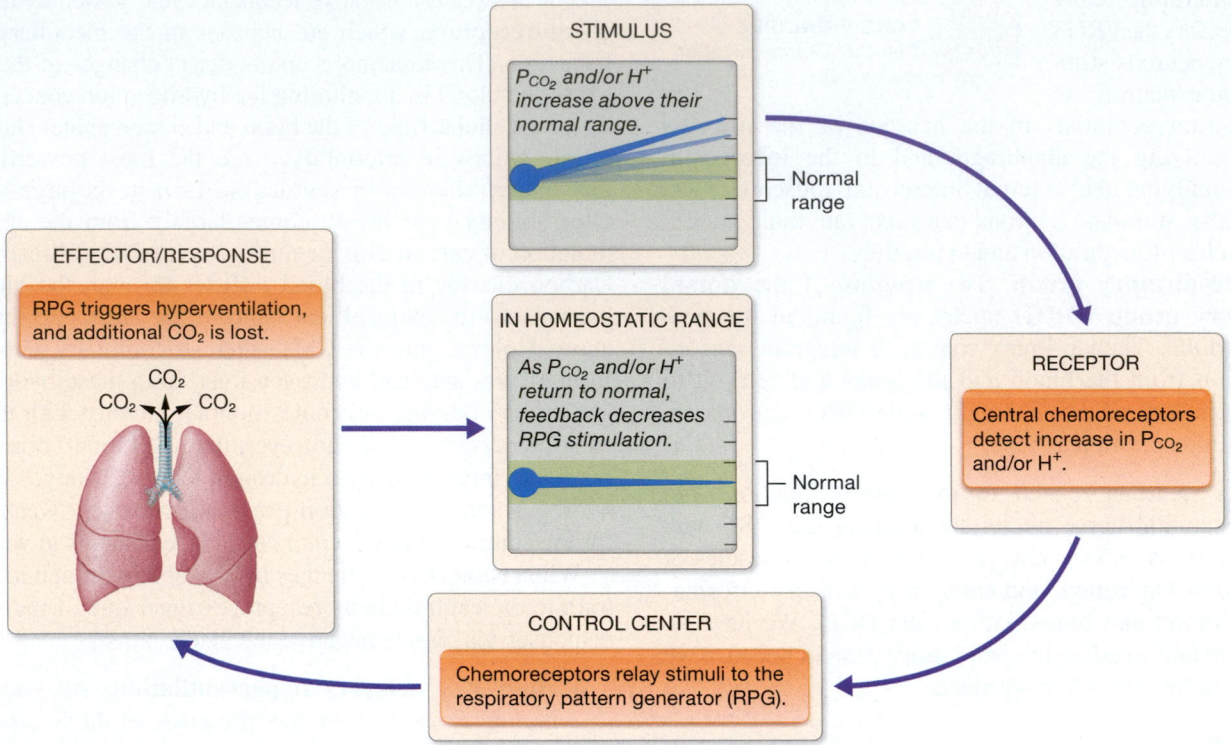

(a) **Response to increased arterial P$_{CO_2}$ and/or H$^+$ concentration by a negative feedback loop**

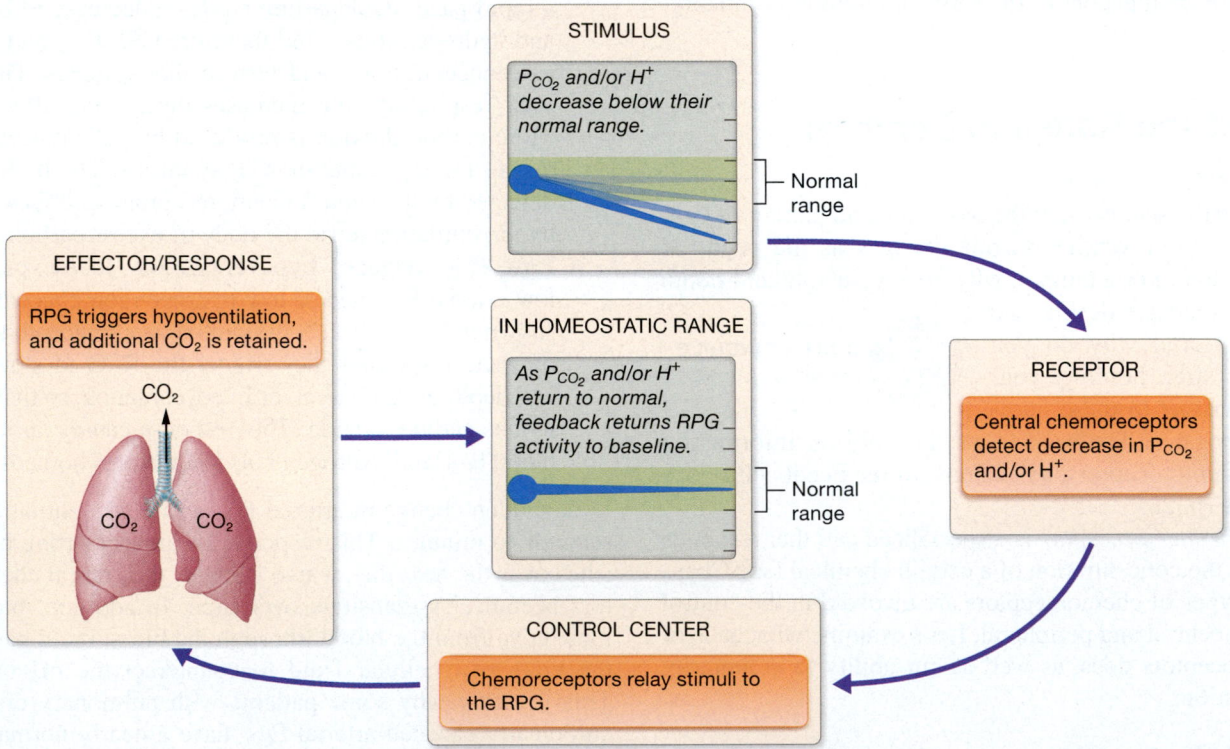

(b) **Response to decreased arterial P$_{CO_2}$ and/or H$^+$ concentration by a negative feedback loop**

Figure 21.28 **Role of the central chemoreceptors in regulation of blood pH via regulation of blood pH via the rate of ventilation.**

Influence of Peripheral Chemoreceptors

The P_{CO_2} and hydrogen ion concentration of the brain extracellular fluid are not the only factors that influence the RPG. The **peripheral chemoreceptors** are specialized groups of cells located in the carotid arteries and in the aorta, called the **carotid bodies** and **aortic bodies,** respectively. These chemoreceptors detect a variety of stimuli, including the P_{CO_2} and hydrogen ion concentration of the arterial blood, but are most sensitive to the P_{O_2} of the arterial blood. When the arterial P_{O_2} falls below about 70 mm Hg (recall that normal arterial P_{O_2} is about 100 mm Hg), the carotid and aortic bodies send signals to the DRG via neurons of the glossopharyngeal nerve (CN IX) and the vagus nerve (CN X). These signals are relayed to multiple places in the brainstem, including the VRG, the central chemoreceptors, and the RPG, which triggers an increase in the rate and depth of ventilation. See how this process works at high altitudes in *A&P in the Real World: High-Altitude Acclimatization* on p. 844.

Stretch Receptors and Voluntary Control

As we've established, the rate and depth of ventilation increase in response to a decrease in blood pH or P_{O_2}, and an increase in P_{CO_2}. But as you've surely noticed, you can only increase the depth of ventilation so much—there comes a point at which you cannot further inspire and the inspiratory muscles reflexively relax. This response is due to a negative feedback loop initiated by *pulmonary stretch receptors* in the walls of the trachea and bronchi. These receptors are activated when one takes a deeper breath than normal, and action potentials are sent via neurons that travel with the glossopharyngeal and vagus nerves (look back at Figure 21.27). The ultimate effect of stimulating pulmonary stretch receptors is stimulation of the DRG, which inhibits the inspiratory muscles and so protects the lungs from overinflation and helps to maintain eupnea.

The final type of control of ventilation is voluntary control. Although we discuss the control of ventilation as if it were purely involuntary, we are able to exert some voluntary control over ventilation. This voluntary control is mediated by the cerebral cortex and bypasses the respiratory centers. These signals travel instead through the spinal cord to neurons of the appropriate spinal nerves.

The control mechanisms of ventilation are summarized in **Figure 21.29**.

Quick Check

☐ 3. Where are the central chemoreceptors located? What do they detect?

☐ 4. What do the central chemoreceptors trigger if arterial P_{CO_2} or hydrogen ion concentration, or both, increase? What do they do if either or both decrease?

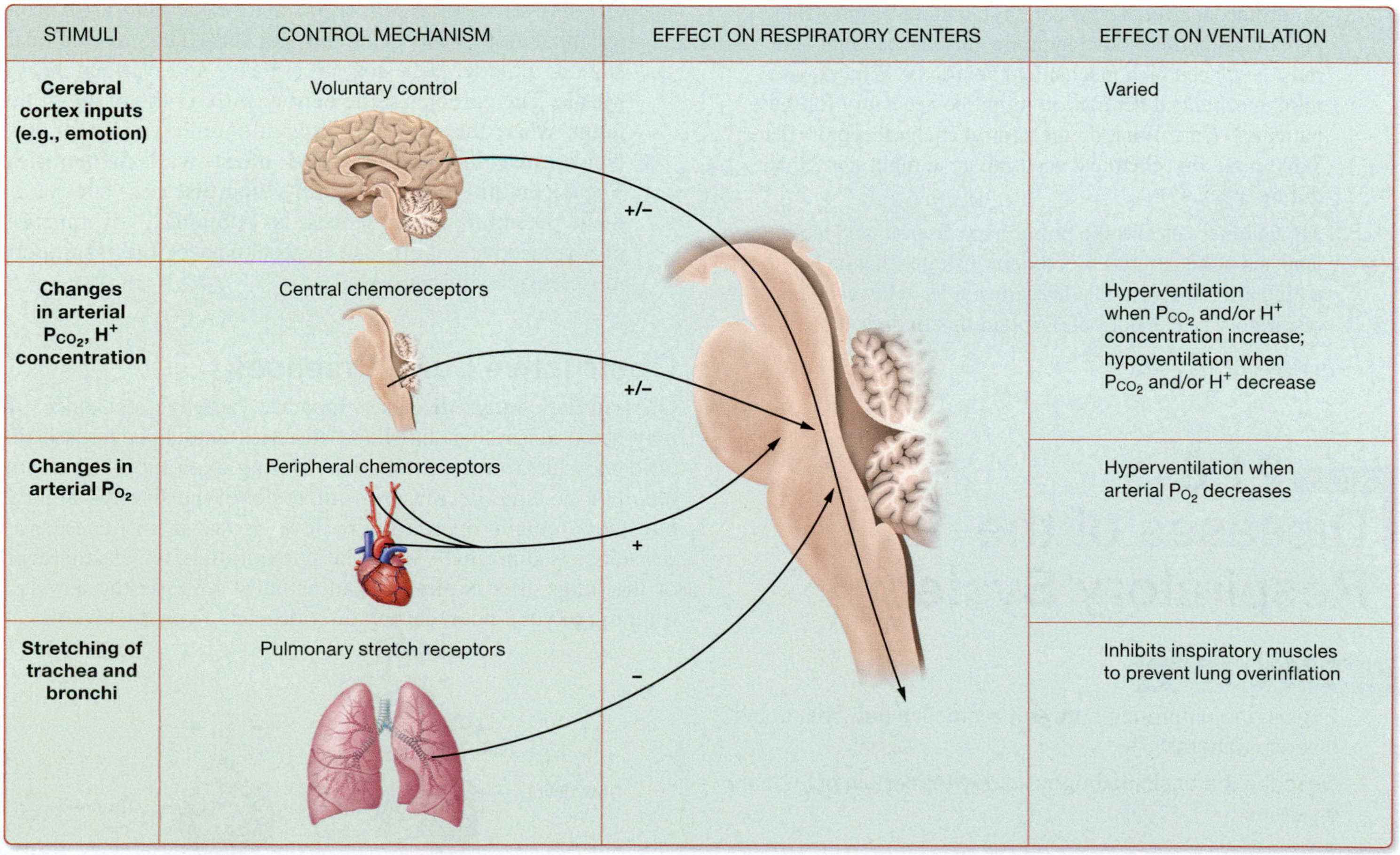

STIMULI	CONTROL MECHANISM	EFFECT ON RESPIRATORY CENTERS	EFFECT ON VENTILATION
Cerebral cortex inputs (e.g., emotion)	Voluntary control	+/−	Varied
Changes in arterial P_{CO_2}, H^+ concentration	Central chemoreceptors	+/−	Hyperventilation when P_{CO_2} and/or H^+ concentration increase; hypoventilation when P_{CO_2} and/or H^+ decrease
Changes in arterial P_{O_2}	Peripheral chemoreceptors	+	Hyperventilation when arterial P_{O_2} decreases
Stretching of trachea and bronchi	Pulmonary stretch receptors	−	Inhibits inspiratory muscles to prevent lung overinflation

Figure 21.29 Control mechanisms of ventilation.

High-Altitude Acclimatization

A&P in the Real World

Experienced mountain climbers know that when they gradually increase their elevation over a period of days rather than hours, they are able to tolerate much lower atmospheric oxygen levels. This phenomenon, called *high-altitude acclimatization,* helps the peripheral chemoreceptors increase the rate and depth of breathing. This hyperventilatory response allows the body to maintain an acceptable blood P_{O_2}.

The reason that a slow ascent is necessary lies in the sensitivity of the peripheral chemoreceptors. The longer they are exposed to a low P_{O_2} the *more* they stimulate an increase in ventilation. A gradual ascent up a mountain allows more time for the peripheral chemoreceptors to be exposed to a low P_{O_2} and therefore a greater increase in ventilation. This practice permits experienced mountain climbers to reach great elevations without supplemental oxygen.

☐ 5. What is the role of the peripheral chemoreceptors in the control of ventilation?

Apply What You Learned

☐ 1. Individuals with congenital central hypoventilation syndrome lack functional central chemoreceptors. These patients sometimes have hypoxic episodes at night and must be placed on a mechanical ventilator. Why do you think mechanical ventilation might be necessary for these patients? (*Hint:* What do the central chemoreceptors detect? Why might this chemical accumulate at night more than during the day?)

☐ 2. Mr. Gardner has chronic pulmonary disease, and laboratory analysis of his arterial blood demonstrates that he has a high arterial P_{CO_2} and a low arterial P_{O_2}. However, his respiratory rate is normal. Explain this finding.

See answers in Appendix A.

MODULE 21.8

Diseases of the Respiratory System

Learning Outcomes

1. Explain the difference between restrictive and obstructive disease patterns.
2. Describe the basic pathophysiology for certain pulmonary diseases.

Lung diseases are the third leading cause of death in the United States. Every year, they take the lives of nearly 350,000 people and cost the American economy more than 150 billion dollars. A history of smoking is implicated in many of these diseases. Two generally recognized patterns of lung disease can be identified—*restrictive* and *obstructive*—although many diseases have characteristics of both. In this module we describe these patterns and the reasons for them.

Restrictive Lung Diseases

Restrictive lung diseases decrease pulmonary compliance and reduce the effectiveness of *inspiration*. The most common causes of decreased pulmonary compliance are increased alveolar surface tension and destruction of the elastic fibers of the lung. Restrictive lung diseases decrease the inspiratory capacity, vital capacity, and total lung capacity, and make effective pulmonary ventilation difficult.

Individually, the restrictive lung diseases are not as prevalent as some of the more common obstructive diseases, but collectively their number is large. Some of the more common restrictive diseases are the following:

- **Idiopathic pulmonary fibrosis.** The primary problem in idiopathic pulmonary fibrosis is inflammation of the lung tissue, followed by destruction of the lungs' elastic tissue and its replacement with thick bundles of collagen fibers. The cause of the disease is unknown, but about 66–75% of patients have a history of heavy smoking.
- **Pneumoconiosis.** Pneumoconiosis (noo′-moh-koh-nee-OH-sis) refers to a collection of diseases that arise from the inhalation of inorganic dust particles, such as coal dust, silicon dioxide, asbestos, fiberglass, and various heavy metals. The particles settle permanently in the tissue of the lungs, where they cause inflammation followed by fibrosis.
- **Neuromuscular diseases and chest wall deformities.** These conditions are not purely lung diseases, but one of their potential consequences is pulmonary dysfunction in a restrictive pattern due to weakness of the inspiratory muscles or a stiff chest wall.

Obstructive Lung Diseases

Obstructive lung diseases increase airway resistance. It may seem surprising, but these diseases actually decrease the efficiency of *expiration*. Normally during expiration, the elastic recoil of the lungs decreases the diameter of the airways slightly, but they remain open to air flow. However, when airway resistance is abnormally high before expiration, the elastic recoil of the lungs after expiration can actually collapse the airways, trapping oxygen-poor, carbon dioxide–rich air in the alveoli:

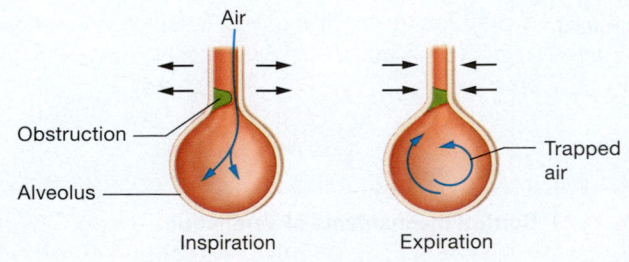

Inspiration   Expiration

A patient with an obstructive lung disease will often purse his or her lips and breathe out slowly to minimize the compression of the airways and prevent air trapping. Patients also often hyperinflate their lungs to decrease air trapping by enlarging their airways. Typically, with an obstructive disease pattern, the residual volume increases, which decreases the vital capacity. Let's look at some common obstructive lung diseases, including emphysema, asthma, and lung cancer.

Chronic Obstructive Pulmonary Disease

Over 16 million Americans are affected by **chronic obstructive pulmonary disease,** or **COPD,** and the disorder is the sixth leading cause of death worldwide. COPD is defined as persistent airway obstruction that is not fully reversible. Nearly all patients with COPD have a history of cigarette smoking, although genetic and environmental factors may be involved as well. There are three subtypes of COPD: emphysema, small airway disease, and chronic bronchitis.

- **Emphysema.** As we mentioned, **emphysema** is characterized by destruction of the structures of the respiratory zone and a loss of alveolar surface area. Most cases of emphysema are due to cigarette smoke, as the resins in the smoke destroy the lungs' elastic fibers. About 1–2% of cases are due to a genetic mutation causing deficiency of an enzyme that prevents the destruction of lung tissue by other enzymes. When this protective enzyme is deficient, the elastic tissue of the lungs is destroyed at a rapid pace.
- **Small airway disease.** Some degree of disease of the bronchioles usually accompanies emphysema. In small airway disease, the bronchioles narrow and are typically plugged with mucus.
- **Chronic bronchitis.** Patients with **chronic bronchitis** have excessive mucus in the airways that must be cleared by coughing. This condition is almost exclusively due to cigarette smoke, which increases the number and size of mucous glands and goblet cells and paralyzes the cilia of the respiratory epithelium.

Most patients with COPD eventually exhibit a mixture of all three subtypes. COPD symptoms generally develop over a period of years. They include shortness of breath, cough, and excessive mucus production. Patients may exhibit a prolonged expiratory phase, an elevated arterial P_{CO_2}, and a decreased arterial P_{O_2}, especially as the disease worsens. With severe disease, patients may exhibit **cyanosis,** a bluish color to the skin and mucous membranes caused by a higher amount of deoxyhemoglobin in arterial blood (deoxyhemoglobin has a dark red-purple color). Those with severe disease may also develop a "barrel chest" as a result of lung hyperinflation.

One of the first and most important steps in the treatment of COPD is smoking cessation. Other treatments are aimed at increasing oxygenation through supplemental oxygen therapy and reducing airway resistance. In severe cases, lung transplant surgery may be an option.

Asthma

Asthma (AZ-muh) is a condition that affects about 7–10% of the worldwide population. In this obstructive disease, the airways are hyperresponsive to a trigger, which is most often dust mites, mold, pollen, or animal dander (see Chapter 20). Other triggers include exercise, certain drugs such as aspirin, infections, and air pollutants. When people with asthma are exposed to a trigger, three responses occur in their airways: (1) bronchoconstriction; (2) inflammation; and (3) increased production of excessively thick mucus. Each of these three responses causes a significant increase in airway resistance. Even when not exposed to a trigger, the airways of many asthma patients are mildly inflamed, a state that contributes to their hyperresponsiveness.

Asthma manifests episodically in *attacks* of varying frequency, during which patients suffer shortness of breath, wheezing (whistling during inspiration or expiration), and sometimes coughing. Most attacks are short-lived, and patients typically recover their normal pulmonary function after the attack has ended. However, in some patients, a persistent state of airway obstruction—called *status asthmaticus*—may develop and last for days or weeks.

Treatment of acute asthma attacks is aimed at reducing airway resistance through bronchodilators that bind beta-2 receptors on bronchiolar smooth muscle cells and trigger their relaxation. Longer-term treatments are aimed at preventing or decreasing the frequency of attacks by reducing the inflammation of the airways with anti-inflammatory steroids and decreasing their responsiveness to allergens.

Lung Cancer

What is the number one cancer killer of women? Most people would answer "breast cancer," but in fact it is lung cancer. This disease is the highest cancer killer of both women and men. The term **lung cancer** refers to tumors arising from the epithelium lining the bronchi, bronchioles, and alveoli (**Figure 21.30**). There are many

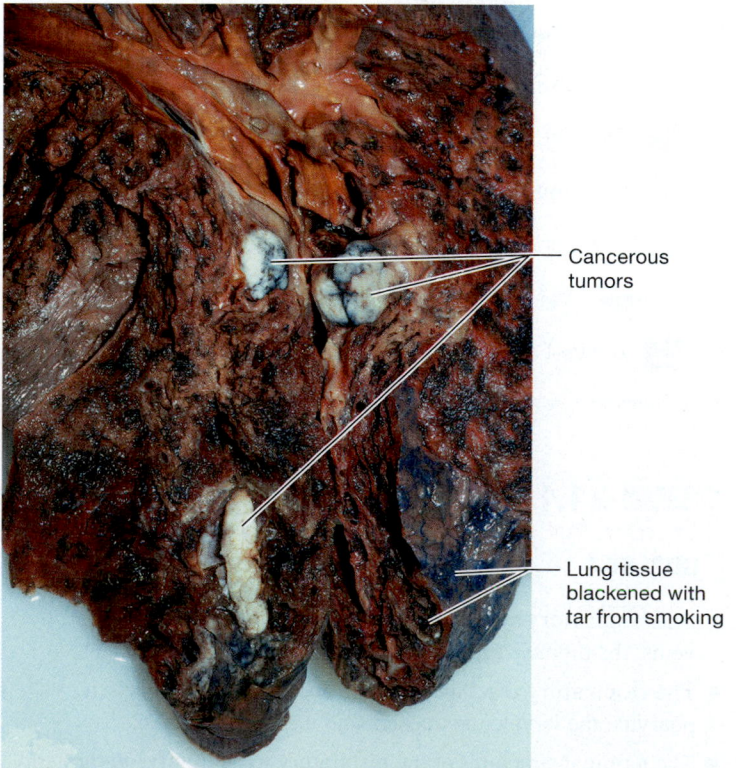

Cancerous tumors

Lung tissue blackened with tar from smoking

Figure 21.30 Cancerous tumor of the lung.

types of lung tumors, each with a different predominant cell type, clinical course, and rate of spread, or *metastasis* (see Chapter 3).

The number one risk factor for lung cancer is cigarette smoking. Smoking raises the risk 13-fold, and heavy smoking as much as 60- or 70-fold. Even passive, or "second-hand," smoke increases the risk of developing lung cancer. Research estimates this risk to be about one and a half times that of a nonsmoker. Other factors that increase the risk for lung cancer include genetics and exposure to certain chemicals or radiation.

The symptoms of lung cancer are fairly nonspecific. They include cough, blood-streaked sputum, chest pain, and recurrent infections. Initially, the symptoms have an obstructive pattern as the airways narrow due to tumor growth. However, some classes of lung cancer can result in the accumulation of fluid in the pleural cavity, which may lead to collapse of the alveoli and a decrease in lung compliance. This, in turn, causes a restrictive pattern of symptoms. Additionally, as lung tumors enlarge, the disease may take on a restrictive pattern as the normal lung tissue is infiltrated by the expanding tumor.

Lung cancer is difficult to detect early because of the nonspecific symptoms. Also, since most people with lung cancer have a history of smoking, the condition may be confused with COPD. Treatment varies depending on the class of tumor, but it may include surgery to remove the tumor and affected lymph nodes, radiation therapy, and chemotherapy.

Quick Check

☐ 1. What are the differences between obstructive and restrictive lung diseases?

☐ 2. What are the three subtypes of COPD? What is the main risk factor for developing COPD or lung cancer?

☐ 3. What are the three airway responses to asthma?

Apply What You Learned

☐ 1. Obstructive diseases increase the residual volume. Explain how this change decreases the vital capacity.

☐ 2. Cigarette smoke destroys the cilia of the respiratory epithelium. Why would this create the cough seen with chronic bronchitis?

☐ 3. Why does a person with asthma hyperventilate during an asthma attack? (*Hint:* Think about the level of carbon dioxide in the patient's blood during an attack.)

See answers in Appendix A.

Chapter Summary

For everything you need to succeed in this course, go to Mastering Anatomy & Physiology. There you will find:

- Practice Tests
- Author Podcasts
- Big Picture Animations
- Concept Boost Video Tutors
- **iP2** Interactive Physiology 2.0 Tutorials
- **A&PFlix** A&P Flix 3D Animations
- Active-Learning Workbook

- **Lungs** are paired spongy organs in the thoracic cavity that consist of millions of alveoli.

- The organs of the respiratory tract may be classified anatomically and functionally:
  - Anatomically, they are classified as either the **upper respiratory tract** or the **lower respiratory tract.**
  - Functionally, they are classified as part of the **conducting zone** or the **respiratory zone.**

- **Respiration** consists of four separate processes: pulmonary ventilation, pulmonary gas exchange, gas transport, and tissue gas exchange.

- The respiratory system also helps to maintain acid-base, fluid, and blood pressure homeostasis; assists in the flow of venous blood and lymph; produces speech; detects odors; and helps expel the abdominal contents.

MODULE 21.1
Overview of the Respiratory System
802–804

- The **respiratory system** includes the pulmonary arteries and veins, the thoracic cage, the lungs, and the respiratory tract.
- The **respiratory tract** includes the nose and nasal cavity, the pharynx, the larynx, the trachea, and the bronchial tree.
- The terminal structures of the respiratory tract are **alveoli** through which gases are exchanged.

MODULE 21.2
Anatomy of the Respiratory System
804–817

- The **nose** and **nasal cavity** moisten, warm, and filter the inspired air; house olfactory receptors; and enhance voice resonance.
  - The nose has a framework of hyaline cartilage and bone.
  - The nasal cavity extends from the **anterior nares** to the **posterior nares.** It is divided into right and left portions by the *nasal septum.*

- The nasal cavity is lined mostly with **respiratory mucosa.** A portion is lined with **olfactory mucosa.**

- The **pharynx** has three divisions: the **nasopharynx,** the **oropharynx,** and the **laryngopharynx.**

- The **larynx** protects the airway and produces sound.
  - It is framed by nine pieces of cartilage: the **epiglottis,** the **thyroid cartilage,** the **cricoid cartilage,** and the paired *arytenoid, corniculate,* and *cuneiform* cartilages.
  - The epiglottis seals off the larynx during swallowing to prevent food or liquid from entering the respiratory tract.
  - The larynx contains two sets of mucosal folds: the **vestibular folds** and the **vocal folds.** The vocal folds vibrate to produce sound.

- The **trachea** extends from the inferior neck to the mediastinum.

- The first part of the **bronchial tree** is the conducting zone (**primary bronchi** to *terminal bronchioles*), which delivers air from the trachea to its second part, the respiratory zone (**respiratory bronchioles** to **alveolar sacs**), where gas exchange takes place.

- Alveoli are where gas exchange takes place.
  - Alveoli have three cell types: squamous **type I alveolar cells;** surfactant-secreting **type II alveolar cells;** and **alveolar macrophages.**
  - The **respiratory membrane** consists of the type I alveolar cells, a shared basal lamina, and the endothelial cells of the pulmonary capillaries.

- The lungs consist of the alveoli and the surrounding elastic connective tissue. Each lung is divided into **lobes** and is enclosed by its own **pleural cavity,** which is the space between the **parietal** and **visceral pleurae.**

MODULE 21.3
Pulmonary Ventilation 818–827

- **Pulmonary ventilation** depends on pressure gradients that drive **inspiration** and **expiration.**

- **Boyle's law** states that the pressure and volume of a gas are inversely related.

- Three pressures are at work during the process of ventilation: *atmospheric pressure, intrapulmonary pressure,* and *intrapleural pressure.*

- During inspiration, the volume of the lungs increases, decreasing intrapulmonary pressure below atmospheric pressure. During expiration, the volume of the lungs decreases, increasing intrapulmonary pressure above atmospheric pressure.

- Inspiration is brought about by the **inspiratory muscles,** which include the **diaphragm** and **external intercostal muscles.** The inspiratory muscles increase the volume of the lungs.

- Expiration during normal quiet breathing is passive due to the elastic **recoil** of the lungs. As the lungs recoil, their volume decreases.

- The respiratory system can perform several nonrespiratory movements.

- Three physical factors determine the effectiveness of pulmonary ventilation:
  - **Airway resistance** is the impedance to air flow, and it decreases the effectiveness of pulmonary ventilation.
  - **Alveolar surface tension** is the attraction created by hydrogen bonding between water molecules that tends to collapse the alveoli. **Surfactant** reduces alveolar surface tension.
  - **Pulmonary compliance** refers to the ability of the lungs to stretch. If pulmonary compliance decreases, the effectiveness of pulmonary ventilation decreases.

- Lung function may be assessed with a **spirometer.** The pulmonary volumes measured are the **tidal volume (TV),** the **inspiratory reserve volume (IRV),** and the **expiratory reserve volume (ERV).** A fourth pulmonary volume not assessed by spirometry is the **residual volume (RV).**

- Pulmonary volumes may be combined to yield **pulmonary capacities,** including the **inspiratory capacity, vital capacity, functional residual capacity,** and **total lung capacity.**

MODULE 21.4
Gas Exchange 827–831

- During **pulmonary gas exchange,** oxygen diffuses from the alveoli to the blood, and carbon dioxide diffuses from the blood to the alveoli.

- During **tissue gas exchange,** oxygen diffuses from the blood to the tissues, and carbon dioxide diffuses from the tissues into the blood.

MODULE 21.5
Gas Transport through the Blood 832–838

- About 98.5% of oxygen in blood is transported on **hemoglobin (Hb).**

- The ability of Hb to load or unload oxygen depends on the P_{O_2} in the lungs or tissues and the *affinity* of Hb for oxygen.

- The **oxygen-hemoglobin dissociation curve** shows the relationship between the percent saturation of Hb and the P_{O_2}.

- Increases in temperature, hydrogen ion concentration, P_{CO_2}, and **BPG** facilitate the unloading of oxygen from Hb.

- Some carbon dioxide dissolves in plasma (7–10%) and binds to Hb (20%); most is converted to **bicarbonate ions** (70%) in a reaction catalyzed by the enzyme **carbonic anhydrase.**

- One of the primary buffer systems in the body is the **carbonic acid–bicarbonate buffer system.** Increasing P_{CO_2}, as occurs with **hypoventilation,** causes respiratory acidosis. Decreasing P_{CO_2}, as occurs with **hyperventilation,** causes respiratory alkalosis.

MODULE 21.6
Putting It All Together: The Big Picture of Respiration 838–840

- The big picture of respiration is shown in Figure 21.15.

MODULE **21.7**
Neural Control of Ventilation 840–844

- Groups of neurons in the medulla of the brainstem maintain **eupnea**: The **respiratory pattern generator** (**RPG**) in the brainstem sets the basic pattern of ventilation. These neurons are assisted by neurons in the **ventral respiratory group** (**VRG**) and the **dorsal respiratory group** (**DRG**) of the pons and medulla.

- **Central chemoreceptors** of the medulla respond to changes in hydrogen ion concentration and P_{CO_2} of the brain extracellular fluid. If the P_{CO_2} increases, they trigger hyperventilation. If the P_{CO_2} decreases, they trigger hypoventilation.

- The **peripheral chemoreceptors** in the **aortic** and **carotid bodies** respond to the P_{O_2} of arterial blood.

MODULE **21.8**
Diseases of the Respiratory System
844–846

- **Restrictive lung diseases** decrease pulmonary compliance (ability to stretch) and the effectiveness of inspiration.

- **Obstructive lung diseases** increase airway resistance and decrease the effectiveness of expiration.
  - **Chronic obstructive pulmonary disease** (**COPD**) is persistent airway obstruction that is not fully reversible.
  - **Asthma** is an obstructive disease in which the airways are hyperresponsive to a trigger of some sort.
  - **Lung cancer** refers to tumors that arise from the epithelium of the lung tissue.

Assess What You Learned

Scan the QR Code for additional practice test questions

https://goo.gl/zllqtj

LEVEL 1 Check Your Recall

1. Which of the following are functions of the respiratory system? Circle all that apply.
   a. Providing for speech
   b. Regulating the autonomic nervous system
   c. Maintaining the acid-base balance of the blood
   d. Temperature homeostasis
   e. Raising the pressure in the abdominopelvic cavity
   f. Assisting in blood pressure regulation

2. Fill in the blanks: Air enters the lungs through the hollow passages known collectively as the _____, which terminate in grapelike clusters called _____. The lungs are encased in the _____ membranes.

3. Mark the following statements as true or false. If a statement is false, correct it to make a true statement.
   a. The trachea contains O-shaped rings of hyaline cartilage.
   b. Goblet cells secrete serous fluid into the respiratory tract.
   c. The framework of the larynx is formed by nine pieces of cartilage, the largest of which is the thyroid cartilage.
   d. The pseudostratified ciliated columnar epithelium of the respiratory tract warms, filters, and humidifies the inspired air.
   e. The epithelium of the oropharynx changes from stratified columnar epithelium to simple squamous epithelium to enable it to resist abrasion from food.

4. The function of the epiglottis is to:
   a. contract muscularly to cover the laryngopharynx.
   b. vibrate to produce sound.
   c. trigger a cough reflex.
   d. cover the glottis during swallowing.

5. Fill in the blanks: The structures that vibrate to produce sound are called the_____. A higher-pitched sound is produced when they are _____; a lower-pitched sound is produced when they are _____.

6. Which of the following structural changes does *not* take place as we progress down the bronchial tree?
   a. More smooth muscle tissue
   b. More goblet cells
   c. Less cartilage
   d. Decrease in the height of the epithelium

7. All the following statements about the alveoli and respiratory membrane are false. Correct each to make a true statement.
   a. Type I alveolar cells secrete a chemical called surfactant.
   b. Alveolar macrophages are squamous epithelial cells that constitute 90% of the total cells in the alveoli.
   c. The respiratory membrane consists of the type II alveolar cells, the pulmonary capillaries' endothelial cells, and their shared basal lamina.
   d. The respiratory membrane must be thick to function effectively in gas exchange.

8. Which of the following statements about pulmonary ventilation is *false?*
   a. Normal expiration requires the use of the expiratory muscles to decrease lung volume.
   b. The inspiratory muscles increase lung volume, which decreases intrapulmonary pressure.
   c. For inspiration to occur, intrapulmonary pressure must decrease below atmospheric pressure.
   d. The intrapleural pressure is less than the intrapulmonary pressure; this prevents the lungs from collapsing during expiration.

9. Match each term with the correct definition.

_____ Airway resistance
_____ Surface tension
_____ Surfactant
_____ Pulmonary compliance
_____ V/Q ratio

a. A detergent-like chemical secreted by bronchial smooth muscle that reduces surface tension
b. The matching of ventilation to perfusion
c. Largely determined by the diameter of the airways
d. Caused by the formation of hydrogen bonds between water molecules
e. Determined by the surface tension of the alveoli, the elastic tissue of the lungs, and the condition of the chest wall

10. Mark the following statements as true or false. If a statement is false, correct it to make a true statement.

a. The functional residual capacity is the volume of air normally left in the lungs after a tidal expiration.
b. The tidal volume is the amount of air left in the lungs after maximal expiration.
c. The inspiratory capacity is equal to the vital capacity plus the tidal volume.
d. The vital capacity is the total amount of exchangeable air.

11. Which of the following does *not* affect the efficiency of pulmonary gas exchange?

a. The surface area of the respiratory membrane
b. The degree of match of ventilation to perfusion
c. The percent saturation of hemoglobin
d. The thickness of the respiratory membrane

12. Henry's law states that the degree to which a gas dissolves in a liquid is determined by its:

a. partial pressure.
b. solubility.
c. surface tension.
d. Both a and b are correct.
e. All of the above are correct.

13. Fill in the blanks: When the alveolar P_{O_2} decreases, the pulmonary arterioles _____. When the arteriolar P_{CO_2} increases, the bronchioles _____.

14. Mark the following statements as true or false. If a statement is false, correct it to make a true statement.

a. Oxygen is transported primarily by dissolving in the plasma.
b. Each Hb protein can carry up to four oxygen molecules.
c. Hb in arterial blood is normally almost fully saturated.
d. When the P_{O_2} of arterial blood drops slightly, the percent saturation of Hb drops dramatically.
e. Increased P_{CO_2}, temperature, and hydrogen ion concentration all increase Hb's affinity for oxygen.

15. Match the following terms with the correct description.

_____ Bicarbonate ion
_____ Hemoglobin
_____ BPG
_____ Carbonic anhydrase
_____ Chloride shift
_____ Carbaminohemoglobin

a. The movement of anions into the erythrocyte to balance the outward movement of bicarbonate
b. Hemoglobin bound to carbon dioxide
c. Enzyme that catalyzes the conversion of carbon dioxide and water into carbonic acid
d. The main way carbon dioxide is transported in the blood
e. Protein that transports oxygen in the blood
f. Substance that decreases hemoglobin's affinity for oxygen

16. Fill in the blanks: Hyperventilation causes a/an _____ in P_{CO_2}, which causes a/an _____ in blood pH. Hypoventilation causes a/an _____ in P_{CO_2}, which causes a/an _____ in blood pH.

17. The basic rhythm for breathing is maintained by the:

a. respiratory pattern generator of the pons.
b. dorsal respiratory group of the medulla.
c. ventral respiratory group of the pons.
d. respiratory pattern generator of the medulla.

18. Which of the following groups of neurons stimulate the diaphragm and intercostal muscles?

a. The dorsal respiratory group
b. The ventral respiratory group
c. The respiratory pattern generator
d. The central chemoreceptors

19. Fill in the blanks: The main chemical factor(s) to which the central chemoreceptors of the medulla respond is/are _____, whereas the main chemical factor to which the peripheral chemoreceptors respond is _____.

20. Which of the following statements is *false?*

COPD has three subclasses: emphysema, small airway disease, and chronic bronchitis.

a. An obstructive disease pattern is characterized by increased airway resistance, which makes expiration difficult.
b. Asthma is characterized by bronchoconstriction, airway inflammation, and excessive mucus secretion.
c. Research has demonstrated that cigarette smoking only marginally increases the risk of lung cancer.

LEVEL 2 Check Your Understanding

1. Explain what would happen to the pressure inside a cylinder if you decreased its volume, according to Boyle's law.

2. If you swallow a large bite of food without properly chewing it first, you will feel discomfort during ventilation. Explain this, considering the arrangement of the trachea and the esophagus.

3. Predict what would happen to the tidal volume and inspiratory reserve volume if the phrenic nerves were severed. Which muscles would contract to try to compensate for this?

4. Inflammation of the epiglottis, called *epiglottitis,* is considered an emergency. Explain why this situation is so serious.

LEVEL 3 Apply Your Knowledge

PART A: Application and Analysis

1. When a person hyperventilates, what happens to his or her blood pH? A person who is hyperventilating is often told to breathe into a paper bag. Why might this help to correct the pH imbalance?

2. Some athletes train at higher altitudes. Over time, this results in higher levels of erythrocytes and the chemical BPG. What would be the advantages to these changes?

3. You and a friend are having a contest to see who can hold his or her breath the longest. Your friend hyperventilates before holding his breath, and subsequently wins the contest. Why did hyperventilation give him an advantage?

4. Prolonged vomiting can result in significant loss of hydrogen ions from the blood, and a subsequent increase in the blood's pH. Would you expect a person suffering from this condition to hypoventilate or hyperventilate? Explain.

PART B: Make the Connection

5. Mrs. Jordan is brought to the emergency room by paramedics, who report that she lost consciousness while at the grocery store. You notice that her respiratory rate is elevated and her breath has a "fruity" odor. You suspect that she is suffering from diabetic ketoacidosis.

   a. What is diabetic ketoacidosis? *(Connects to Chapter 16)*

   b. Does Mrs. Jordan likely have type I or type II diabetes mellitus? Explain. *(Connects to Chapter 16)*

   c. What has likely happened to the pH of her blood, and why? *(Connects to Chapters 16 and 19)*

   d. Explain why she is hyperventilating.

6. What happens to the metabolic rate of skeletal muscle tissue during exercise? What waste products are produced from metabolic reactions? How and why does this affect your rate of ventilation during exercise? *(Connects to Chapter 10)*

See answers in Appendix A.

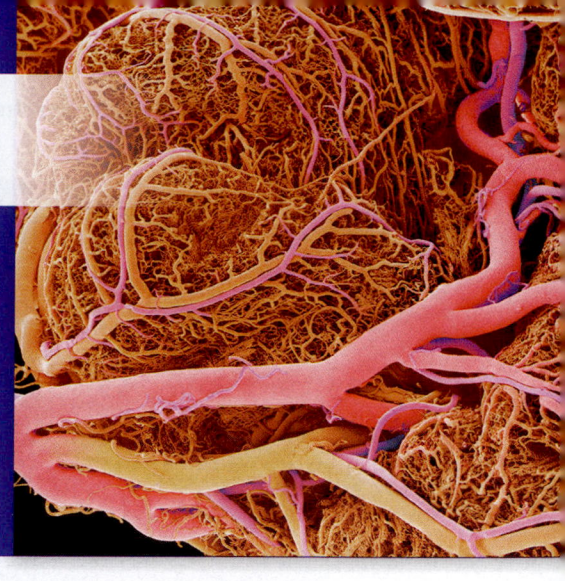

22

The Digestive System

We all must eat to survive, but most of us don't think about eating except when we're hungry. We generally understand *why* we eat, as we know that our bodies require nutrients—carbohydrates, lipids, and proteins—to maintain homeostasis. But do you ever think about what happens *when* we eat?

Nutrients are available to us in the form of food, but this comes with a small problem: The food isn't in a form that we can immediately use. If you have a peanut butter and jelly sandwich, you can't just put that sandwich straight into your bloodstream. Our bodies therefore require an organ system to take the sandwich into the body and break it down so that its nutrients can be delivered to cells via the blood. The set of organs that accomplishes this task is the **digestive system.**

This chapter explores the many structures and functions of the digestive system. We look at the anatomy of its organs, and how they break down food and deliver nutrients to the blood. We also examine how hormones and the autonomic nervous system regulate these organs and their processes. The chapter concludes with a "big picture" look at the processes carried out by the digestive system. But first, let's start with an overview of its basic structures and functions.

MODULE **22.1**

Overview of the Digestive System

Learning Outcomes

1. Describe the major functions of the digestive system.
2. Describe the histological structure and function of each of the four layers of the alimentary canal wall.
3. Explain the basic anatomy, organization, and regulation of the digestive system.

The organs of the digestive system are located from the head to the abdominopelvic cavity (**Figure 22.1**). The digestive system consists of two types of organs: the organs of the *alimentary canal,* also known as the **gastrointestinal (GI) tract,** or **digestive tract;** and the accessory organs. The **alimentary canal** is a continuous passage through which food moves. It consists of the *oral cavity* (mouth), *pharynx, esophagus, stomach, small intestine,* and *large intestine.* The **accessory organs** are not part of the alimentary canal

For practice applying concepts to a clinical scenario, check out the **Running Case Study** *for this chapter* @ Mastering Anatomy & Physiology

Photo: SEM showing the extensive blood supply of the small intestine, which allows it to rapidly absorb digested nutrients.

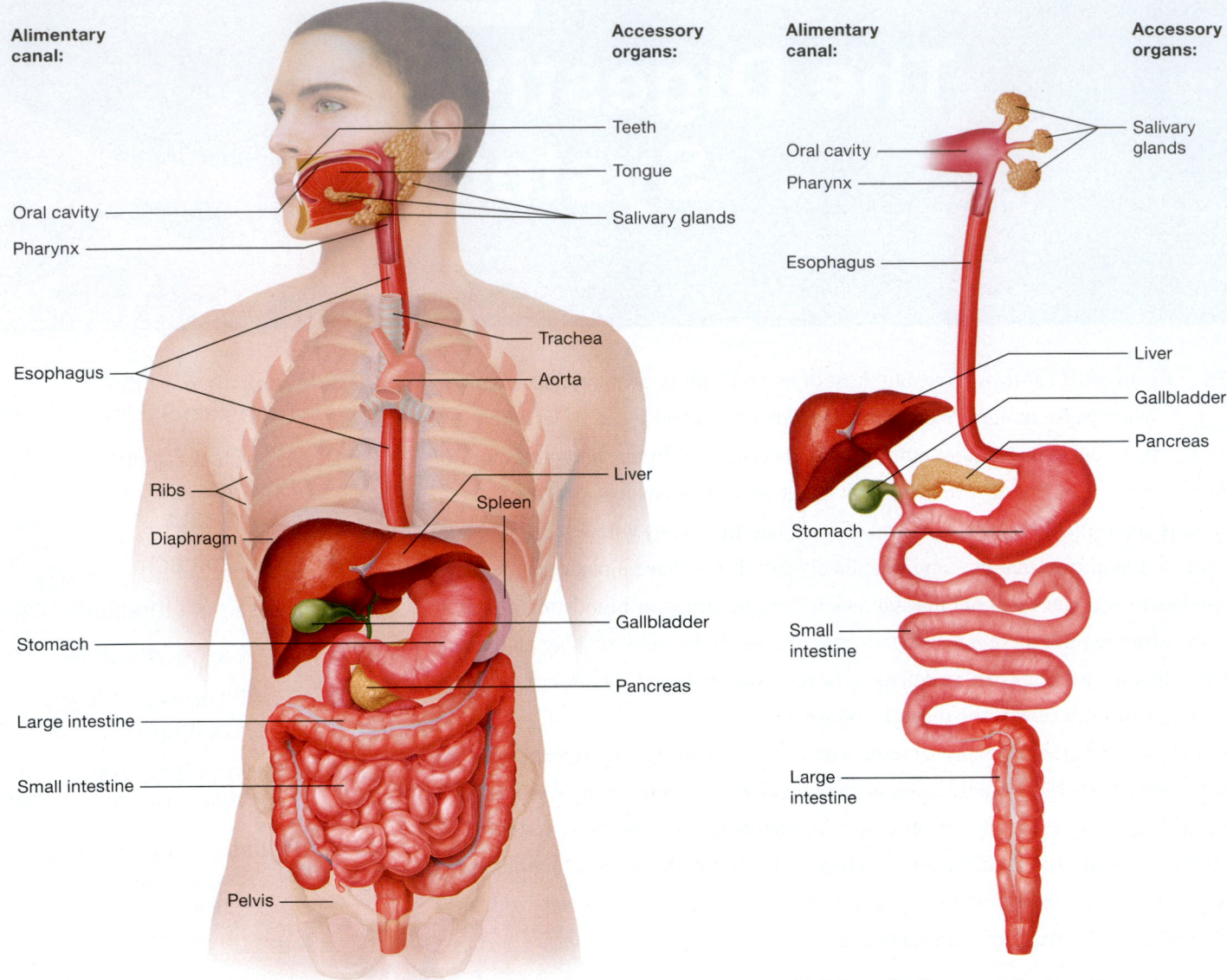

Alimentary canal:
- Oral cavity
- Pharynx
- Esophagus
- Ribs
- Diaphragm
- Stomach
- Large intestine
- Small intestine
- Pelvis

Accessory organs:
- Teeth
- Tongue
- Salivary glands
- Trachea
- Aorta
- Liver
- Spleen
- Gallbladder
- Pancreas

(a) Digestive system anatomy

Alimentary canal:
- Oral cavity
- Pharynx
- Esophagus
- Stomach
- Small intestine
- Large intestine

Accessory organs:
- Salivary glands
- Liver
- Gallbladder
- Pancreas

(b) Schematic diagram of the digestive system

Figure 22.1 Overview of the digestive system.

but assist in digestion in some way. They are located around the alimentary canal and include the *teeth, tongue, salivary glands, liver, gallbladder,* and *pancreas.* Food generally does not come into direct contact with the accessory organs, although the teeth and tongue are exceptions.

Let's survey all these organs to see just what part they play in digestion and to understand their basic structure and organization. We examine each in greater detail in subsequent modules.

Basic Digestive Functions and Processes

The most fundamental function of the digestive system is to take food into the body and break it down into its component nutrients so that they can be used by body cells. But the digestive system also plays a critical role in fluid and electrolyte homeostasis—it takes in water and electrolytes such as sodium and potassium

ions via the diet and delivers them to the blood. Other functions of the digestive system include ingesting vitamins and minerals, producing hormones, and excreting metabolic wastes.

The digestive system must perform the following six basic processes to carry out these functions:

- **Ingestion.** Food and water are brought into the digestive system by **ingestion,** which occurs via the mouth under normal conditions.
- **Secretion.** Digestive organs contain both endocrine and exocrine glands that **secrete** a variety of substances—such as mucus, enzymes, acid, and hormones—to aid other digestive processes.
- **Propulsion.** Ingested food and liquids pass from one digestive organ to the next by the process of **propulsion.** Propulsion is accomplished largely by rhythmic contractions of the smooth muscle of the alimentary canal called **peristalsis** (pehr-uh-STAL-sis) and is aided by mucus secreted by multiple organs.

- **Digestion.** Food breakdown occurs by the process of **digestion.** There are two kinds of digestion. The first is **mechanical digestion,** in which digestive organs physically break food down into smaller pieces via processes such as chewing and mixing food by movements by the muscles of the alimentary canal. In the second, **chemical digestion,** enzymes secreted by digestive organs catalyze reactions that break the chemical bonds within food particles until only small compounds remain.
- **Absorption.** Once food particles are mechanically and chemically digested, nutrients move through the wall of the alimentary canal into blood or lymphatic vessels by a process called **absorption.** Water, electrolytes, and vitamins are also absorbed into the blood in the same manner.
- **Defecation.** Certain ingested materials are not digestible or usable by the body. Such materials continue their transit through the alimentary canal until they exit the body as *feces* through **defecation.** Defecation also provides the body with a way to eliminate certain metabolic wastes. Note that defecation is simply a specialized form of propulsion.

The organs of the alimentary canal carry out all these basic processes. Food is ingested at the mouth and propelled through the canal from one organ to another as it is digested mechanically and chemically. The nutrients are absorbed into the bloodstream, and the indigestible substances are removed from the body by defecation. Substances secreted by various digestive organs aid each of these processes. Most of the accessory organs do not come into direct contact with food, and so they are involved in secretion, mechanical digestion, and chemical digestion. The digestion and absorption of each nutrient are covered together in a later module after we have discussed the anatomy of the digestive system.

Quick Check

☐ 1. What are the functions of the digestive system?

☐ 2. What are the six basic processes of the digestive system?

Organization of the Digestive System

« FLASHBACK

1. What is a serous membrane, and what is its function? (p. 153)
2. Which blood vessels supply and drain the digestive organs? (p. 701 and p. 710)
3. What is a mucous membrane, and what are goblet cells? (p. 134 and p. 153)

Most of the digestive organs—with the exceptions of the oral cavity, pharynx, and esophagus—reside in the abdominopelvic cavity. Here they share a common set of serous membranes, blood vessels, and nerves, which the following subsections examine. We also look at the general pattern in which the tissue layers of the alimentary canal are organized, which is a pattern that will look familiar—inner mucosa, middle loose connective tissue and smooth muscle, and outer dense connective tissue.

The Peritoneal Membranes

The abdominopelvic cavity houses the largest serous membrane in the body, the **peritoneal membrane** (pehr´-ih-toh-NEE-uhl), or **peritoneum** (**Figure 22.2a**). Like the serous membranes that surround the heart and lungs, the peritoneal membrane consists of two layers. The outer layer is the **parietal peritoneum** (puh-RY-eh-tul), which lines the inner surface of the body wall. Where the parietal peritoneum meets the abdominal organs, it folds in on itself to become the inner **visceral peritoneum,** or serosa, which forms the outer tissue layer of such organs. Between these two peritoneal layers we find the **peritoneal cavity,** which contains serous fluid. This fluid lubricates organs as they slide past one another. Note that the peritoneal cavity is much larger than the pleural or pericardial cavity. Abdominal trauma or infection can inflame the peritoneum, as you can read about in *A&P in the Real World: Peritonitis.*

Organs located entirely within the peritoneal cavity are **intraperitoneal.** Some abdominal organs are partly or completely outside the peritoneal cavity; such organs are said to be **retroperitoneal** (reh´-troh-pehr-ih-toh-NEE-uhl; *retro-* = "behind"). Some organs, though, are partly intraperitoneal and partly retroperitoneal. For example, in Figure 22.2a, the pancreas is shown as retroperitoneal, but if we took a more lateral section, we would find that part of it is actually intraperitoneal.

Notice in Figure 22.2a how the visceral peritoneum folds over on itself around certain organs, particularly the small intestine, to form structures called **mesenteries** (MEZ-en-tehr-eez). The mesenteries support and bind these organs together and keep the small intestine in a particular shape that fits within the abdominopelvic cavity. This function is key, as we don't want loops of intestine wandering around our abdominopelvic cavities. The mesenteries also house blood vessels, nerves, and

Peritonitis

Peritonitis is inflammation of the peritoneum (pehr´-ih-tuh-NY-tis; *-itis* = "inflammation of"). Peritonitis results when substances such as blood or the contents of an abdominal organ leak into the peritoneal cavity. Usually this is due to abdominal trauma that ruptures a blood vessel or abdominal organ, and often involves a bacterial infection.

Abdominal pain is the most common symptom of peritonitis. Most patients experience *rebound tenderness,* in which they feel little pain when a hand is placed on their abdomen, but significant pain when the hand is removed. This is caused by the inflamed peritoneal membranes snapping back into place when the hand leaves the abdomen.

The treatment for peritonitis may involve antibiotic therapy and surgery to correct its underlying cause. Most patients recover with appropriate treatment, but it can prove fatal if left untreated.

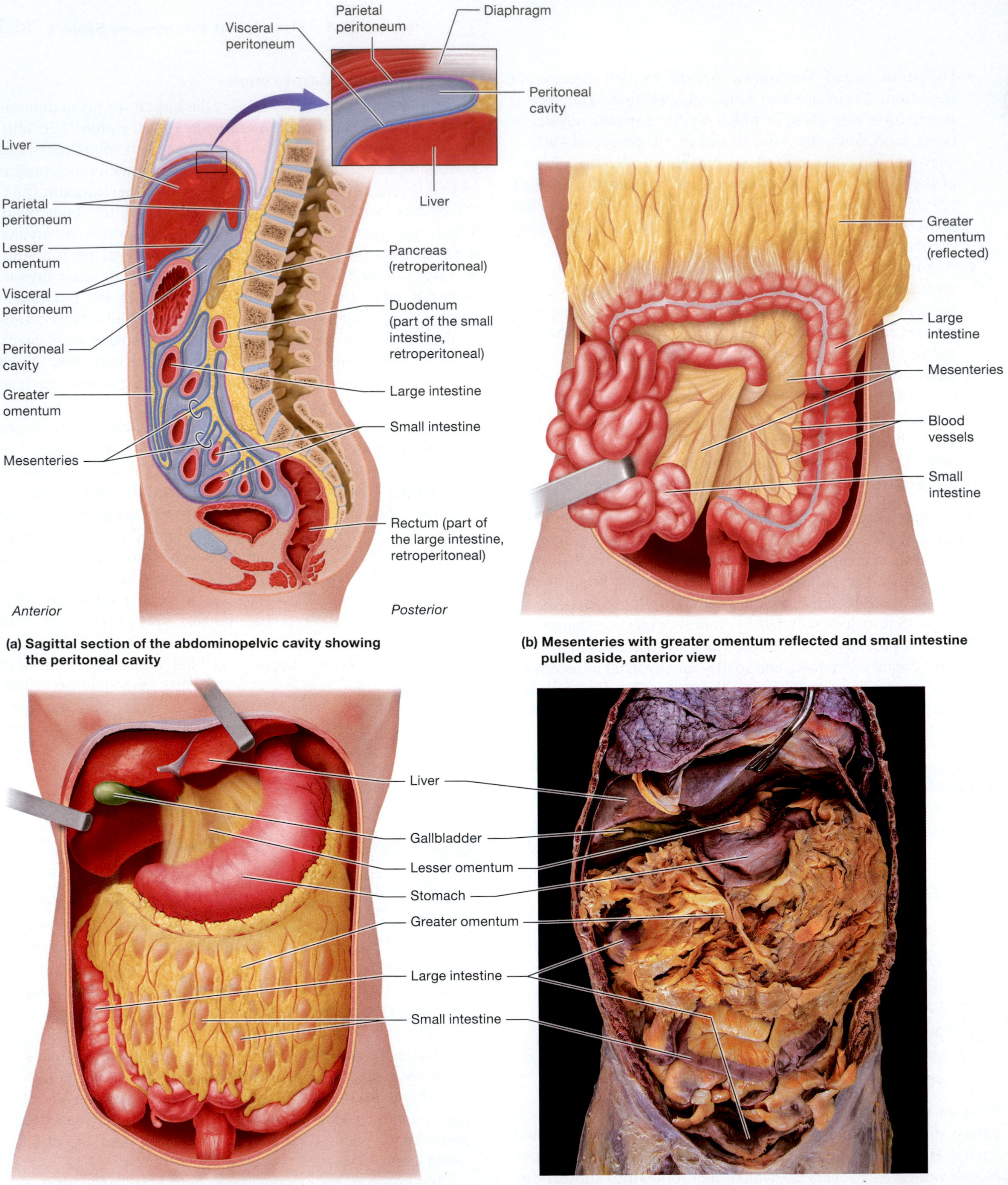

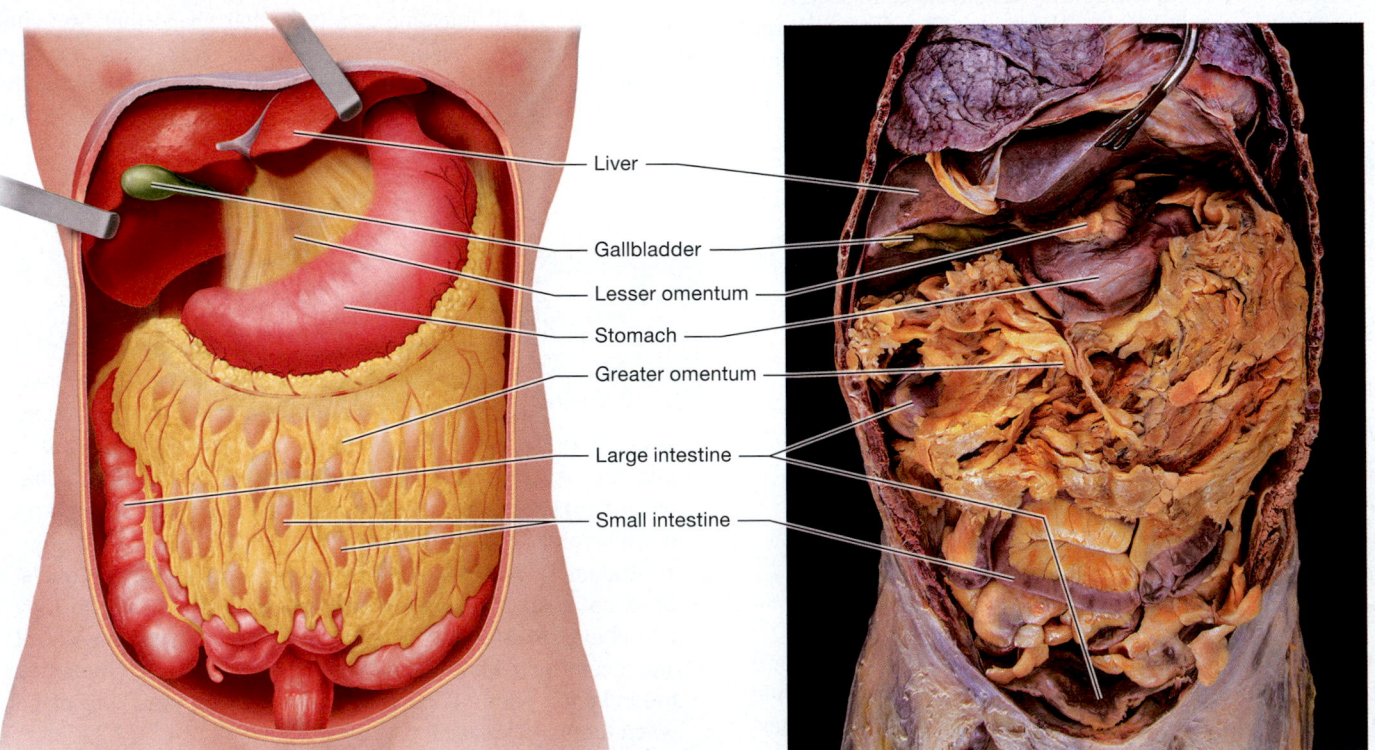

Liver

Parietal peritoneum

Lesser omentum

Visceral peritoneum

Peritoneal cavity

Greater omentum

Mesenteries

Visceral peritoneum

Parietal peritoneum

Diaphragm

Peritoneal cavity

Liver

Pancreas (retroperitoneal)

Duodenum (part of the small intestine, retroperitoneal)

Large intestine

Small intestine

Rectum (part of the large intestine, retroperitoneal)

Anterior

Posterior

(a) Sagittal section of the abdominopelvic cavity showing the peritoneal cavity

Greater omentum (reflected)

Large intestine

Mesenteries

Blood vessels

Small intestine

(b) Mesenteries with greater omentum reflected and small intestine pulled aside, anterior view

Liver

Gallbladder

Lesser omentum

Stomach

Greater omentum

Large intestine

Small intestine

(c) Anterior view of the abdominal organs and omenta, illustration (left), and cadaver photo (right)

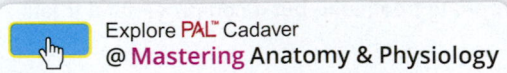

Explore **PAL**™ Cadaver
@ **Mastering** Anatomy & Physiology

Figure 22.2 The abdominopelvic cavity houses the peritoneum, the largest serous membrane in the body.

lymphatic vessels, anchoring them in place (**Figure 22.2b**). The mesentery attached to much of the large intestine is often called the *mesocolon.*

There are two mesenteries—the greater omentum and lesser omentum—that are especially prominent. The **greater omentum** (oh-MEN-tum; *omentum* = "apron"), which is unique among the mesenteries in that it consists of four layers of folded visceral peritoneum, is named for the fact that it covers the abdominal organs like an apron (**Figure 22.2c**). The greater omentum, which is the first structure visible when the abdominal cavity is opened, extends from the base of the stomach down into the pelvis. The **lesser omentum** is a smaller mesentery that extends from the medial surface of the stomach to the liver. Both the greater and lesser omenta generally accumulate adipose tissue between their folds.

Blood and Nerve Supply to the Abdominal Digestive Organs

The organs of the digestive system are extensively supplied with blood vessels and nerves. You read previously about the blood vessels that supply and drain the abdominal digestive organs, which are collectively called the **splanchnic circulation** (SPLANK-nik; *splan-* = "organ"; see Chapter 18). The arterial supply of the digestive organs consists of branches from the abdominal aorta, including the *celiac trunk, superior mesenteric artery, inferior mesenteric artery,* and branches from each of these arteries. The digestive organs are drained by a set of veins that drain into the *hepatic portal vein.* The hepatic portal vein then delivers the blood to the liver for processing. Blood drains

from the liver by a set of *hepatic veins,* which in turn deliver blood to the inferior vena cava.

Recall that there are two branches of the *autonomic nervous system,* or *ANS*: the *sympathetic nervous system* and the *parasympathetic nervous system* (see Chapter 14). As we introduced in the PNS chapter (see Chapter 13), a cluster of nerves is called a *plexus.* The nerves of the sympathetic and parasympathetic divisions that serve the abdominal digestive organs are located in three main clusters: the *celiac plexus,* the *superior mesenteric plexus,* and the *inferior mesenteric plexus.* Notice that these clusters have the same names as the arteries that supply the digestive organs. In addition, a self-contained branch of the autonomic nervous system known collectively as the **enteric nervous system,** or **ENS** (en-TEHR-ik; *enter-* = "intestine"), supplies the alimentary canal from the esophagus to the anus (the terminal portion of the large intestine). Nerve plexuses of the ENS work with the sympathetic and parasympathetic nervous systems to control secretion from and motility of the alimentary canal.

Histology of the Alimentary Canal

The organs of the alimentary canal follow the same general tissue pattern of other hollow organs we have studied: concentric layers of tissue surround a space called the **lumen** (LOO-men). Like other hollow organs, those of the alimentary canal contain an inner epithelium, a layer of connective tissue, a layer of smooth muscle, and an outer layer of connective tissue. Most regions of the alimentary canal have four named tissue layers (**Figure 22.3**): *mucosa, submucosa, muscularis externa,* and either the *serosa* or *adventitia.*

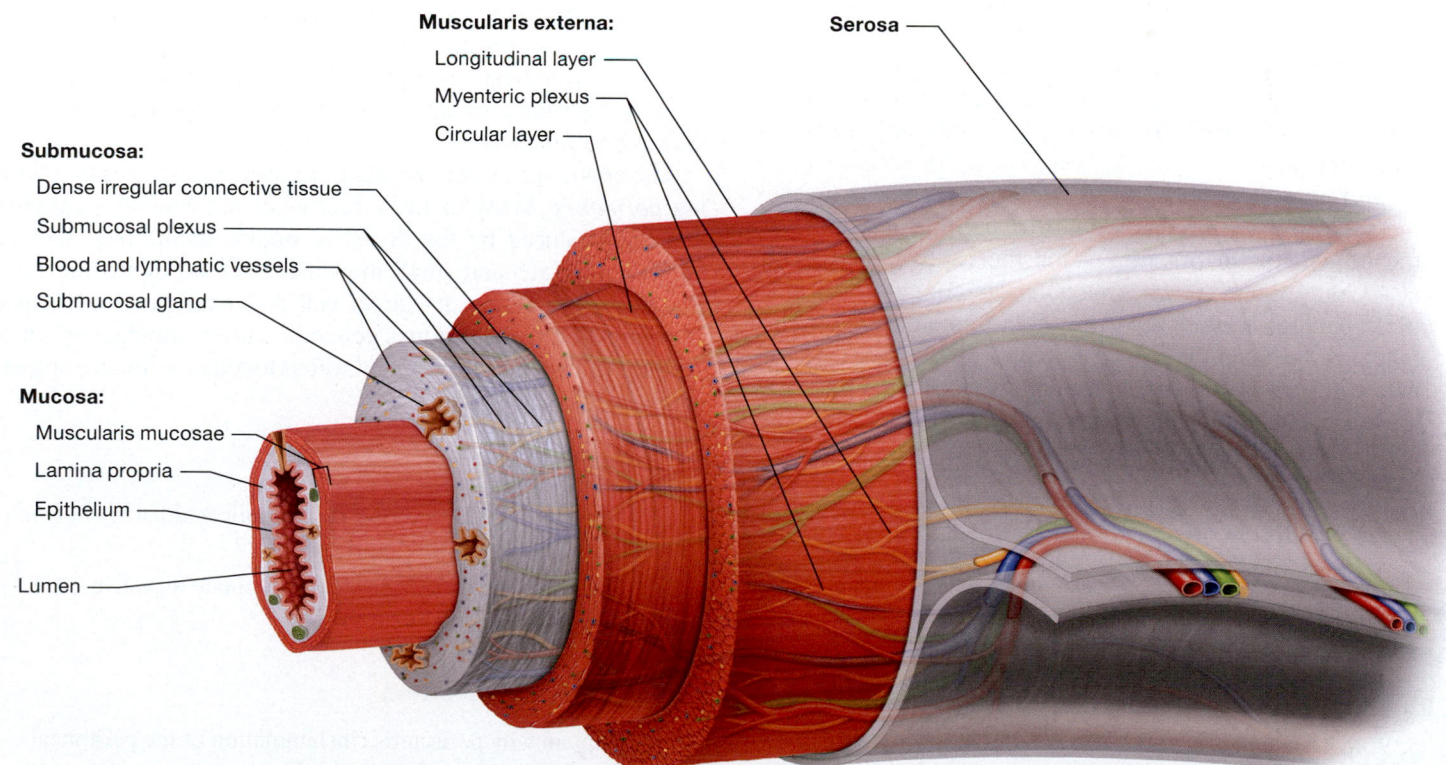

Muscularis externa:
Longitudinal layer
Myenteric plexus
Circular layer

Serosa

Submucosa:
Dense irregular connective tissue
Submucosal plexus
Blood and lymphatic vessels
Submucosal gland

Mucosa:
Muscularis mucosae
Lamina propria
Epithelium

Lumen

Figure 22.3 **The basic tissue organization of most of the alimentary canal.**

- **Mucosa.** The innermost **mucosa** consists of three components: First is a layer of epithelium facing the lumen, followed by a thin layer of loose connective tissue called the **lamina propria,** and finally two thin layers of smooth muscle together known as the **muscularis mucosae.** The epithelium from the stomach to the end of the large intestine is simple columnar epithelium with copious mucus-secreting cells such as goblet cells. The mucus coats the epithelium, protecting it and the underlying tissues from ingested food and chemicals secreted by digestive organs. The mucosa also houses *regenerative epithelial cells* that have a high rate of mitosis. These cells allow the mucosa to replace epithelial cells as they are damaged or sloughed off in the alimentary canal. The lamina propria houses blood and lymphatic vessels, glands, and mucosa-associated lymphatic tissue (MALT; see Chapter 20). The two layers of the muscularis mucosae are arranged in different directions—the inner layer is circular and the outer layer is longitudinal. As discussed later, this arrangement allows propulsion as the two layers contract alternately.
- **Submucosa.** The **submucosa** is composed of dense irregular connective tissue with blood and lymphatic vessels and submucosal glands. Here we find nerve clusters of the enteric nervous system, referred to as the **submucosal plexus.** Each plexus regulates secretion from and blood flow to its area of the alimentary canal.
- **Muscularis externa.** The **muscularis externa** is a thick muscular layer composed of smooth muscle in most of the alimentary canal. We generally find two layers of smooth muscle that are arranged in the same manner as in the muscularis mucosae, with inner circular and outer longitudinal layers. The motility of the muscularis externa is regulated by groups of nerves of the enteric nervous system, called the **myenteric plexus** (*my-* = "muscle," *enter-* = "intestine").
- **Serosa or adventitia.** The outer connective tissue layer is the **serosa** in the organs within the peritoneal cavity and the **adventitia** in organs outside the cavity. The serosa, also called the *visceral peritoneum,* is composed of simple squamous epithelial tissue and loose connective tissue, whereas the adventitia is composed of dense irregular connective tissue. Both structures support digestive organs and anchor them to surrounding structures.

Each organ of the alimentary canal has a slightly different variation of this basic plan. These variations ensure that the form of each organ best follows its function.

Quick Check

☐ 3. Where is the peritoneal cavity located?

☐ 4. Where does blood from the abdominal digestive organs drain?

☐ 5. Which branches of the nervous system supply the digestive organs?

☐ 6. What are the four main tissue layers of the alimentary canal?

Regulation of Motility by the Nervous and Endocrine Systems

《 FLASHBACK

1. What are the two branches of the autonomic nervous system, and how do they differ with respect to regulating digestion? (p. 518)

2. What are hormones and paracrines? (pp. 586–587)

As you will see throughout this chapter, movement, or **motility,** of the alimentary canal is a key process in every region of the canal. In the oral cavity, the pharynx, the superior portion of the esophagus, and the last portion of the large intestine, motility is due to skeletal muscle. In the rest of the alimentary canal, motility is the work of smooth muscle. Motility takes several forms, including *swallowing, churning, peristalsis,* and *defecation.* Some of these movements propel food through the canal, whereas others mix it as part of mechanical digestion.

Each type of motility is regulated by the nervous system and/or the endocrine system. Nervous system regulation of motility is accomplished by the nerves of the ANS. The sympathetic and parasympathetic branches of the ANS generally have opposite effects on gastrointestinal motility—sympathetic activity inhibits digestive processes, and parasympathetic activity stimulates them.

In addition, motility is regulated by the ENS. The functions stimulated by the ENS are often called *short reflexes* because the reflex pathways are confined to local neurons. In contrast, functions stimulated by the ANS are known as *long reflexes* because they must travel outside the local digestive neurons to the CNS to function.

The endocrine system regulates digestive processes by secreting *hormones.* Many of these hormones are actually paracrines that are produced by the digestive organs themselves, particularly the stomach and small intestine. Each hormone binds to a receptor on or within its target cell and regulates some aspect of its functioning. Hormones generally either stimulate or inhibit motility of the alimentary canal, often together with one or more neural mechanisms.

Quick Check

☐ 7. How do the effects of the sympathetic and parasympathetic nervous systems on digestion differ?

☐ 8. How does the endocrine system regulate digestive processes?

Apply What You Learned

☐ 1. Explain why peritonitis (inflammation of the peritoneal membranes) can affect multiple digestive organs.

☐ 2. Disorders of the digestive system often disrupt the homeostasis of many regulated variables. Explain why.

☐ 3. Many forms of chemotherapy for cancer treatment kill rapidly dividing cells. Predict the effect these drugs will have on the epithelium of the alimentary canal.

See answers in Appendix A.

MODULE
22.2
The Oral Cavity, Pharynx, and Esophagus

Learning Outcomes

1. Discuss the structure and basic functions of the oral cavity, the different types of teeth, and the tongue.

2. Describe the structure and function of the salivary glands, their respective ducts, and the products secreted by their cells.

3. Describe and classify the regions of the pharynx with respect to passage of food and/or air through them.

4. Describe the structure and function of the esophagus, including the locations of skeletal and smooth muscle within its wall.

5. Explain the process of deglutition, including the changes in position of the glottis and larynx.

The digestive system begins at the *oral cavity,* or *mouth.* Four digestive processes take place here: ingestion, secretion, chemical and mechanical digestion, and propulsion. Although the oral cavity is technically part of the alimentary canal, it houses two accessory organs: the *teeth* and the *tongue.* In addition, three pairs of accessory organs, the *salivary glands,* are located in and around the oral cavity. Together, these organs turn ingested food into a moist, chewed mass called a **bolus** (BOH-lus; "ball"). The posterior oral cavity, pharynx, and esophagus then deliver the bolus to the stomach through the process of *swallowing.* In this module, we look at the structure and function of the oral cavity and its associated accessory organs, the pharynx, and the esophagus. Then we examine the process of swallowing.

Structure of the Oral Cavity

◀◀ FLASHBACK

1. Which bones form the hard palate? (p. 226)

2. What is the main function of the soft palate and uvula? (p. 807)

The oral cavity, illustrated in **Figure 22.4**, is the area posterior to the teeth and bounded by the palate and tongue. The lateral walls of the oral cavity are formed by the *cheeks,* which are composed largely of the buccinator muscles and lined internally

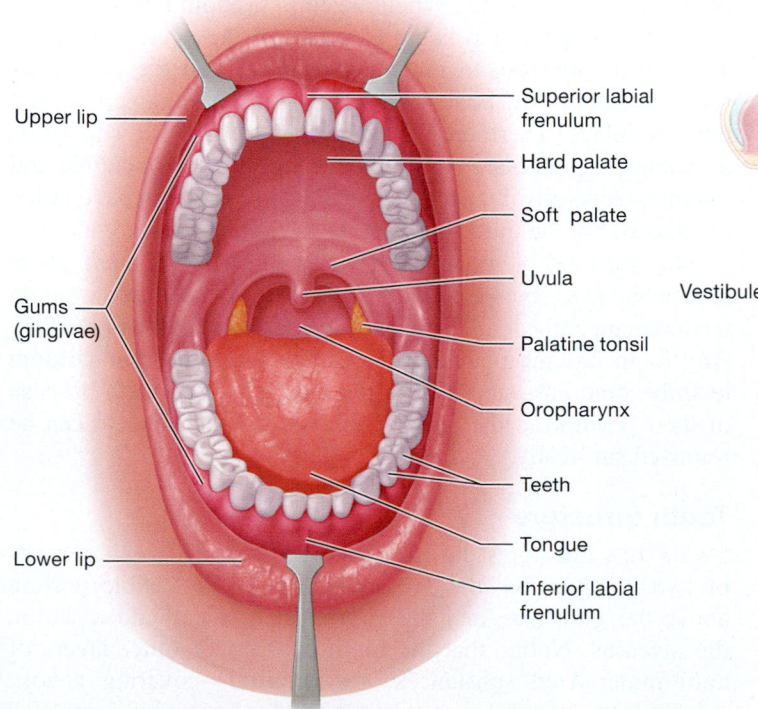

(a) Anterior view of the oral cavity

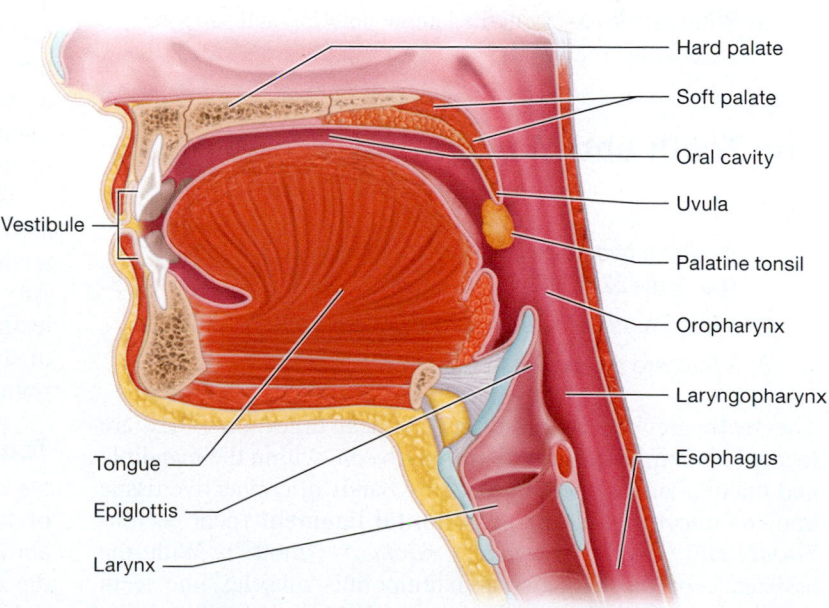

(b) Oral cavity and pharynx, sagittal section

Figure 22.4 **Oral cavity and pharynx.**

by stratified squamous nonkeratinized epithelium. The cheeks terminate anteriorly as the *lips,* which are formed by the orbicularis oris muscle and covered with stratified squamous keratinized epithelium. The integument covers most of the lips, but the portion near the mouth, called the *red margin,* contains less keratin than the surrounding skin. As a result, the red margin is fairly translucent and the blood in the vessels of the dermis is more visible, giving the lips a slight reddish tint. The mouth's inferior wall is composed of muscles of the tongue and muscles that attach to the hyoid bone.

Posterior to the lips and cheeks we find the **gums,** or *gingivae* (JIN-jih-vay). The gums are covered with stratified squamous nonkeratinized epithelium overlying connective tissue and the maxilla and mandible, where the teeth are housed. A narrow band of mucosa called the **labial frenulum** (FREN-yoo-lum) attaches the internal surfaces of the upper and lower lips to the gums on the midline. The narrow space between the teeth and gums and the internal surfaces of the lips and cheeks is the **vestibule.** The space posterior to the teeth and gums is the **oral cavity proper.**

The superior boundary, or "roof," of the mouth is the **palate** (PAL-it), which consists of two portions: the anterior two-thirds is the **hard palate,** and the posterior one-third is the **soft palate.** The hard palate consists of stratified squamous epithelium and connective tissue covering the palatine processes of the maxillary bones and the palatine bones. The surface of the hard palate is slightly rough, which assists in mechanical digestion. The arch-shaped soft palate consists of stratified squamous epithelium overlying skeletal muscle. Extending inferiorly from the soft palate is a projection called the **uvula** (YOO-vyoo-luh). When we swallow, the soft palate and uvula move posteriorly to prevent food from entering the nasal cavity.

Quick Check

☐ 1. Which structures form the lateral, anterior, and superior walls of the oral cavity?

The Teeth and Mastication

《 FLASHBACK

1. In which two bones are the teeth located? (pp. 219–220)
2. Which muscles are involved in mastication? (p. 299)
3. What are the components of bone tissue? (p. 189)

The **teeth** are key organs of mechanical digestion. They are located in their bony sockets, called *alveoli,* within the mandible and maxilla and are held in place by bands of connective tissue known collectively as the **periodontal ligament** (pehr´-ee-oh-DAHN-tuhl; *peri-* = "around," *odont-* = "tooth"). With the assistance of the masseter and temporalis muscles, the teeth **masticate,** or chew, ingested food, grinding it into smaller pieces. Mastication aids digestion by increasing the overall surface area of the food, giving digestive enzymes more places to catalyze the reactions of chemical digestion.

There are three types of teeth, which are classified according to their shape (**Figure 22.5a**):

- **Incisors.** The **incisors** are the central teeth that are broad and flat with a narrow crown. They are specialized for cutting off pieces of food. The middle two incisors are the *central incisors,* and those to either side are the *lateral incisors.*
- **Canines.** The **canines** (KAY-nynz), also known as **cuspids,** are on either side of the incisors. Their pointed crowns are specialized for ripping and tearing.
- **Molars.** The teeth posterior and lateral to the canines are the **premolars** and the **molars.** Both types of molars have broad crowns with rounded projections called **cusps** that are specialized for grinding.

Humans develop two sets of teeth: one set of "baby teeth" and one set of "permanent teeth." In the next subsections, we examine the differences between these two sets of teeth and the general structure of a tooth.

Primary and Secondary Dentition

The "baby teeth" are the **primary dentition,** or **deciduous teeth** (dih-SIJ-oo-us; *decid-* = "falling off"). There are 20 deciduous teeth, with 4 incisors, 2 canines, and 4 molars in both the mandible and maxilla (**Figure 22.5b**). The first deciduous teeth to erupt are generally the lower central incisors at about 6 months of age. Deciduous teeth continue to erupt at a rate of about one pair of teeth per month until the age of 24 months, at which time all 20 teeth are usually present.

The **secondary dentition** is the set of **permanent teeth,** which are situated above the primary dentition in the maxilla and below it in the mandible. When the child is about 6 years of age, these teeth enlarge and begin to press on the deciduous teeth. This causes the root to gradually dissolve and the deciduous tooth falls out of the bone. The permanent tooth then erupts and takes its place. There are 32 permanent teeth, with 4 incisors, 2 canines, 4 premolars, and 6 molars in both the mandible and maxilla. Generally, by age 12 all the deciduous teeth have fallen out and all but the third set of secondary molars have erupted.

The third set of molars, known as the *wisdom teeth,* erupt somewhat later, between ages 17 and 21. Sometimes wisdom teeth remain embedded in the bone, a condition called *impaction.* Any tooth has the potential to become impacted, but wisdom teeth become impacted more often than any other teeth because of their position at the back of the jaw. Impacted teeth can be removed surgically.

Tooth Structure

Figure 22.6 illustrates the structure of a tooth. A tooth consists of two components: the **crown,** which is the visible portion above the gum line, and the **root,** which is embedded within the alveolus. Notice that the tooth consists of outer layers of hard mineralized substances—the **enamel**—covering a soft, inner gelatinous substance called the **pulp.** Enamel is composed almost fully of secreted calcium hydroxyapatite crystals with only a small amount of organic material. This makes enamel

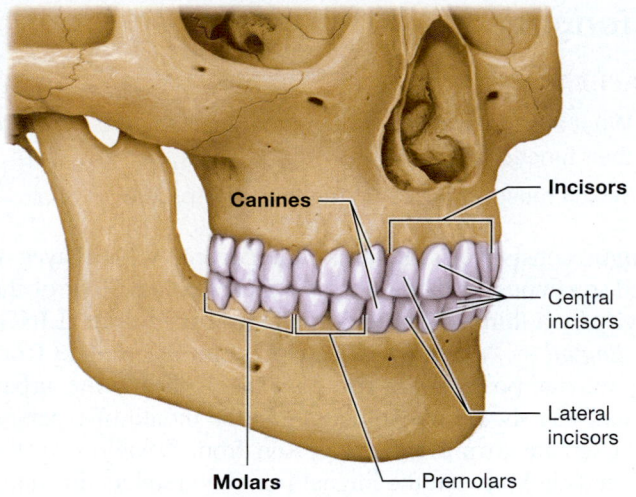

	INCISORS		CANINES	MOLARS			
	Central incisors	Lateral incisors		Premolars	First	Second	Third
UPPER JAW							
LOWER JAW							

Canines — **Incisors**

— Central incisors

— Lateral incisors

Molars — Premolars

(a) The three main classes of teeth

Primary dentition:

— Central incisors (7.5 mo)

— Lateral incisors (9 mo)

— Canines (18 mo)

— Molars (14–24 mo)

Maxillary teeth

Mandibular teeth

— Molars (12–20 mo)

— Canines (16 mo)

— Lateral incisors (7 mo)

— Central incisors (6 mo)

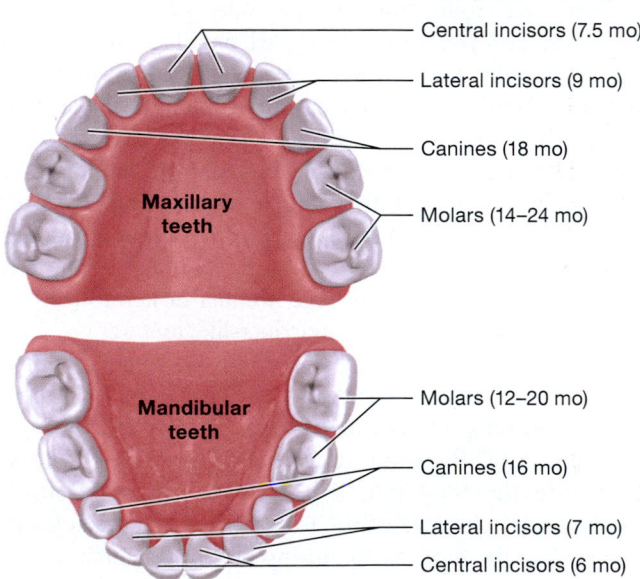

Secondary dentition:

— Central incisors (7–8 yr)

— Lateral incisors (8–9 yr)

— Canines (11–12 yr)

— Premolars (10–12 yr)

— Molars (6–13 yr)

Maxillary teeth

Mandibular teeth

— Third molars, or "wisdom teeth" (17–21 yr)

— Molars (6–13 yr)

— Premolars (10–12 yr)

— Canines (9–10 yr)

— Lateral incisors (7–8 yr)

— Central incisors (6–7 yr)

(b) Primary and secondary dentition, with approximate age of eruption

Figure 22.5 Types of teeth and the primary and secondary dentition.

the hardest substance in the body, allowing it to endure the forces that accompany chewing. The cells that secrete enamel deteriorate after the tooth erupts, so the body cannot repair damaged enamel.

The oral cavity houses hundreds of different bacterial species, collectively called the *normal flora.* In individuals with a healthy immune system and good oral hygiene, the normal

flora generally cause few problems, and indeed may protect us from more pathogenic bacteria. However, damaged enamel sets the stage for tooth decay, which is explored in *A&P in the Real World: Dental Caries* on p. 866.

On the outer portion of the root of the tooth, we find a different kind of mineralized bonelike tissue known as **cementum.** Cementum is composed of about half calcium hydroxyapatite

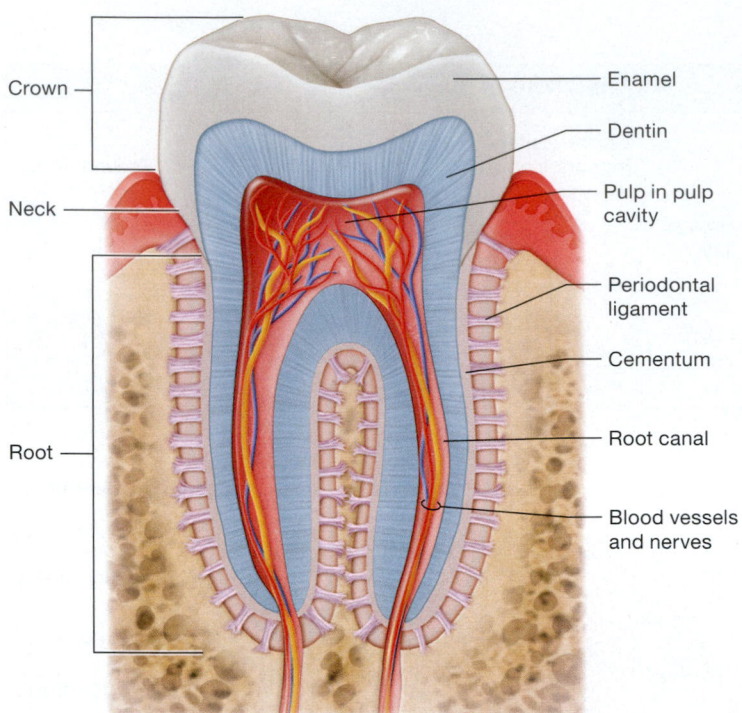

Crown

Neck

Root

Enamel

Dentin

Pulp in pulp cavity

Periodontal ligament

Cementum

Root canal

Blood vessels and nerves

Figure 22.6 Structure of a tooth.

crystals and half organic compounds such as collagen fibers. This is approximately the same composition that bone has, and for this reason cementum is about as hard as bone. The periodontal ligament extends collagen fibers into the cementum, which helps this substance "cement" the tooth in place.

The inner layer of mineralized tissue in both the root and crown is **dentin,** which is composed of about 70% calcium hydroxyapatite crystals. This makes dentin the second hardest material in the body. However, dentin also has some degree of elasticity, which is critical for preventing the overlying enamel from fracturing when chewing hard substances. Unlike enamel, dentin is formed throughout life by cells called *odontoblasts* that line its inner surface.

The final component of a tooth is the inner pulp, which is composed of loose connective tissue and housed within the central **pulp cavity.** Pulp contains blood vessels and nerves that supply the other tissues of the tooth with nutrients and innervation. The pulp cavity extends into the root via the thin **root canal.** Occasionally, the pulp becomes infected, which results in inflammation and generally a great deal of pain. When the infection risks the health of the tooth, a *root canal procedure* may be performed to remove the pulp from the pulp cavity and root canal, and fill the newly hollow space with an inert material.

Quick Check

☐ 2. What are the three types of teeth?

☐ 3. How do the primary dentition and secondary dentition differ?

☐ 4. What are the mineral and nonmineral components of a tooth?

The Tongue

‹‹ FLASHBACK

1. What are the four types of tongue papillae, and what are their functions? (p. 541)

2. Which muscles make up the tongue? (p. 300)

The **tongue** consists of skeletal muscle covered with a layer of stratified squamous epithelium. It is attached to the floor of the oral cavity by a thin band called the **lingual frenulum** (LING-gwuhl; *lingual* = "tongue"), which prevents the tongue from moving too far posteriorly. You have likely heard the urban legend that you should put something in the mouth of a person having a seizure to prevent the person from "swallowing the tongue" and choking, but the lingual frenulum renders this quite impossible. (Note, however, that a person having a seizure can easily choke on objects placed in the mouth, so you should never attempt this.)

Recall that the epithelium of the tongue is arranged into small projections called **papillae** (puh-PIL-ee; refer to Figure 15.4). There are four kinds of papillae: *filiform, fungiform, circumvallate,* and *foliate papillae.* All papillae except filiform contain epithelium with sensory receptors called *taste buds,* which detect chemicals associated with different taste sensations. Filiform papillae play no role in taste, and are instead covered with stratified squamous keratinized epithelium. The keratinized cells make the surface of the tongue rough, which assists in mechanical digestion. Human tongues have a limited number of keratinized cells, so our tongues do not feel rough to the touch. Cats' tongues, however, have a large number of keratinized cells, which are responsible for their sandpaper-like texture.

Two groups of skeletal muscles control tongue movement: extrinsic and intrinsic. The **extrinsic muscles** control the position of the tongue. Extrinsic muscles, involved during the ingestion phase of digestion, move the tongue during chewing and help turn the food into a bolus. The **intrinsic muscles** control the shape and size of the tongue. Intrinsic muscles push the food against the hard palate during chewing, which assists in mechanical digestion, and also push the bolus posteriorly during swallowing.

Quick Check

☐ 5. What are the functions of the tongue pertaining to digestion?

The Salivary Glands

‹‹ FLASHBACK

1. Which antibody class is found in salivary gland secretions? (p. 783)

2. How do the parasympathetic and sympathetic nervous systems affect secretion from salivary glands? (p. 523 and p. 530)

The other accessory organs associated with the mouth are the **salivary glands** (SAL-uh-vehr-ee). There are three pairs of salivary glands, all of which secrete **saliva** (suh-LY-vuh), a fluid containing water, enzymes, mucus, and other solutes, through a duct into the oral cavity. The upcoming subsections explore the anatomy and histology of these glands, as well as the components and functions of saliva.

Anatomy and Histology of the Salivary Glands

Figure 22.7 illustrates the three pairs of salivary glands:

- **Parotid glands.** The **parotid glands** (puh-RAWT-id; *para-* = "beside," *ot-* = "ear") are large glands located over the masseter muscle just anterior to the ear. These glands secrete saliva through the **parotid duct,** which passes over the masseter muscle and pierces the buccinator muscle to open into the oral cavity near the second molar. The parotid glands secrete 25–30% of total saliva.
- **Submandibular glands.** The smaller **submandibular glands** are located just medial to the inferior portion of the body of the mandible. They secrete saliva through the **submandibular duct,** which empties into the floor of the oral cavity. In spite of their smaller size, the submandibular glands are very active, and secrete 65–70% of total saliva.
- **Sublingual glands.** As implied by their name, the **sublingual glands** (sub-LING-gwuhl) are situated inferior to the tongue. The sublingual glands secrete saliva

through several small **sublingual ducts** that empty into the oral cavity just under the tongue. These are the smallest salivary glands and secrete only about 5% of total saliva.

The basic secretory cell of the salivary glands is the **acinar cell** (ASS-uh-nahr). There are two main types of acinar cells in salivary glands: **serous cells,** which secrete a water-based fluid with enzymes and other solutes; and **mucous cells,** which secrete mucus. The secretions from serous cells are involved in digestive processes, and are generally released just before or during eating. However, secretions from mucous cells are primarily involved in keeping the oral mucosa moist, and so are released continually.

The three main types of salivary glands differ in the proportion of mucous and serous cells they contain. The parotid glands have only serous cells, the submandibular glands have mostly serous cells and a small number of mucous cells, and the sublingual glands contain mostly mucous cells. For this reason, the parotid glands secrete mainly water and enzymes, the submandibular glands secrete enzymes mixed with some mucus, and the sublingual glands secrete mainly mucus with a small amount of enzymes.

Saliva

Saliva consists primarily of water; electrolytes such as sodium, chloride, and potassium ions; and variable amounts of mucus, depending on the type of salivary gland. It also contains the following components:

- **Salivary amylase.** The first digestive enzyme that ingested food encounters is **salivary amylase** (AM-uh-layz). It catalyzes the beginning of carbohydrate digestion, breaking down large polysaccharides into smaller polysaccharides.

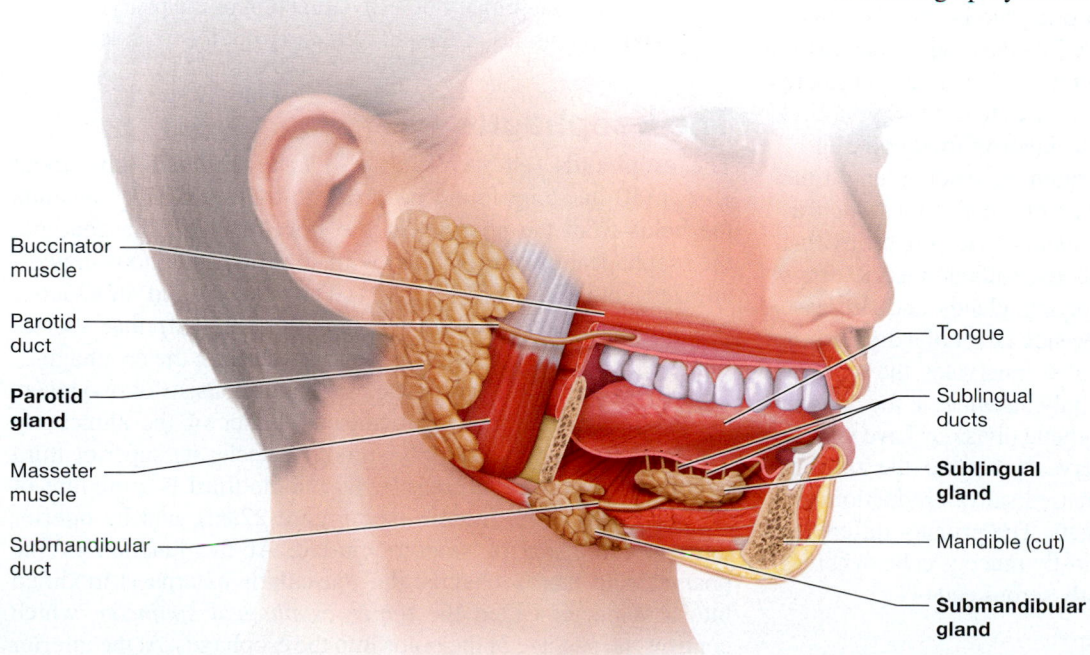

Figure 22.7 Anatomy of the salivary glands.

- **Lysozyme.** The chemical **lysozyme** (LY-soh-zy'm) catalyzes the perforation of bacterial plasma membranes. This allows bacteria-killing substances in the saliva to enter and kill the bacteria.
- **Secretory IgA.** Recall that the antibody immunoglobulin A, or IgA, is found in the body's secretions, including saliva (see Chapter 20). IgA binds specific antigens on pathogens and mediates their destruction.
- **Bicarbonate ions.** During eating, when flow rates of saliva are high, basic bicarbonate ions (HCO_3^-) are added to saliva. Their primary function is to neutralize any acid from the stomach that regurgitates into the esophagus.

Saliva performs several functions in the oral cavity, including moistening, lubricating, and cleansing the oral mucosa. In addition, its lysozyme and IgA deter the growth of pathogenic bacteria in the oral cavity. Saliva also functions in (1) mechanical digestion, by moistening and helping to mix ingested food into a bolus so it can be swallowed, and (2) chemical digestion, by the actions of salivary amylase. Finally, many food polymers dissolve in the water of saliva, and the resulting monomers then stimulate taste receptors on the tongue.

Salivation

Salivation refers to the process of secretion from the three pairs of salivary glands. Salivation is primarily controlled by the parasympathetic nervous system in a reflex arc. The arc begins with sensory stimuli such as the smell or taste of food, which communicate this information to the *salivatory nucleus* in the brainstem. Parasympathetic fibers from the brainstem exit via the facial nerve (cranial nerve VII) to innervate the submandibular and sublingual glands, and via the glossopharyngeal nerve (cranial nerve IX) to innervate the parotid glands. Neurons from these nerves release acetylcholine (ACh) onto the acinar cells, which triggers the acinar cells to secrete saliva. Even the mere smell of food can trigger this reflex, which is why you find yourself drooling when you smell your favorite food cooking.

The importance of ACh in salivation is evident any time ACh receptors are blocked. A number of medications such as antihistamines for allergies have *anticholinergic* properties that block the effects of ACh around the body. Anticholinergic drugs bind to the ACh receptors on the salivary glands and decrease saliva secretion, leading to the common side effect of a dry mouth.

The sympathetic nervous system also innervates the salivary glands. Interestingly, this is one of the only instances in the body in which the sympathetic and parasympathetic divisions have similar effects on an organ. Sympathetic nerves innervate the salivary glands and their ducts, triggering increased saliva production and facilitating saliva transport through ducts. The primary difference is that sympathetic nerves stimulate mostly mucous cells, whereas parasympathetic nerves stimulate mostly serous cells.

Quick Check

☐ 6. What are the three types of paired salivary glands, and where are they located?

☐ 7. What are the components of saliva, and what are its functions?

☐ 8. Walk through the steps of the parasympathetic reflex arc that stimulates salivation.

The Pharynx

« FLASHBACK

1. What are the three divisions of the pharynx? (p. 807)
2. What are pharyngeal constrictor muscles? (p. 301)

As noted previously, the **pharynx** (FEHR-inks), or throat, sits posterior to the nasal and oral cavities (see Chapter 21). It consists of three divisions: the *nasopharynx, oropharynx,* and *laryngopharynx.* Of these three divisions, only the oropharynx and laryngopharynx are part of the alimentary canal. Like the oral cavity, both are lined with stratified squamous epithelium to protect them from abrasion by food. The oropharynx houses two sets of tonsils—the *palatine tonsils* and the *lingual tonsils.* The tonsils perform defensive functions and help protect the remainder of the alimentary canal from any pathogens that enter the body via the oral and nasal cavities.

The primary function of the pharynx is propulsion in the form of swallowing, during which the bolus passes through the pharynx and into the esophagus. Recall that the pharynx is surrounded by three pairs of skeletal muscles: the *upper, middle,* and *lower pharyngeal constrictor muscles* (see Chapter 9). These muscles contract sequentially during swallowing and propel the bolus inferiorly.

Quick Check

☐ 9. What is the primary digestive function of the pharynx? Which muscles of the pharynx carry out this function?

The Esophagus

The **esophagus** (eh-SAWF-uh-gus) is a muscular tube about 25 cm (10 in.) long found posterior to the trachea. It transmits the bolus from the pharynx to the stomach. Like the pharynx, the esophageal mucosa is lined with stratified squamous nonkeratinized epithelium (**Figure 22.8**). The mucosa and submucosa contain *esophageal glands* (eh-sah'-fuh-JEE-uhl) that secrete mucus to lubricate the bolus as it passes through the esophagus.

The muscularis externa of the esophagus consists of two layers of muscle, but it differs from the remainder of the alimentary canal. Rather than being only smooth muscle, its superior third is composed of skeletal muscle, its middle third is a mixture of skeletal and smooth muscle (see Figure 22.8a), and its inferior third is composed of smooth muscle. At the junction of the pharynx and the esophagus, the muscularis externa is modified into a sphincter called the *upper esophageal sphincter,* which controls the passage of the bolus into the esophagus. At the inferior end of the esophagus is another sphincter, the **gastroesophageal sphincter** (gas'-troh-eh-sah-fuh-JEE-uhl), also known as the *lower esophageal sphincter,* that regulates the passage of the bolus

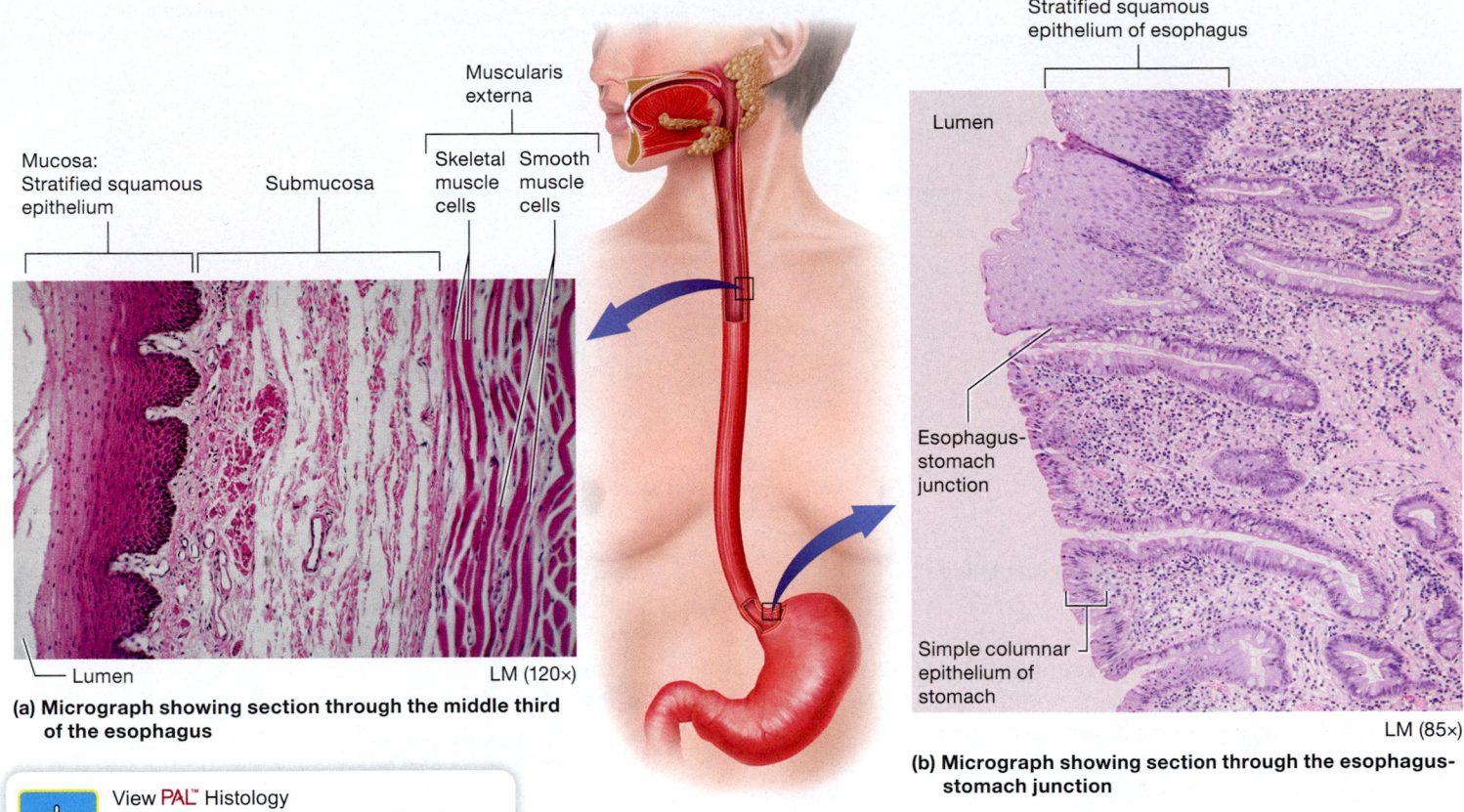

Mucosa:
Stratified squamous
epithelium

Submucosa

Muscularis
externa

Skeletal
muscle
cells

Smooth
muscle
cells

Lumen

LM (120×)

**(a) Micrograph showing section through the middle third
of the esophagus**

Stratified squamous
epithelium of esophagus

Lumen

Esophagus-
stomach
junction

Simple columnar
epithelium of
stomach

LM (85×)

**(b) Micrograph showing section through the esophagus-
stomach junction**

View PAL™ Histology
@ **Mastering** Anatomy & Physiology

Figure 22.8 Histology of the esophagus.

into the stomach. This sphincter also prevents the contents of the stomach from re-entering the esophagus.

The primary functions of the esophagus are propulsion and a small amount of secretion of mostly mucus. During swallowing, the skeletal muscle and smooth muscle of the muscularis undergo peristalsis, which pushes the bolus inferiorly. Although the esophagus' thick epithelium prevents any significant absorption from taking place, it does protect the esophagus from abrasion by food. Note that this is another example of the Structure-Function Core Principle (p. 28).

CORE PRINCIPLE
Structure-Function

Quick Check

☐ 10. Where is the esophagus located? What are its major functions?

☐ 11. How do the muscularis externa and mucosa in the esophagus differ from these tissues in most of the alimentary canal?

Swallowing

The process of **swallowing,** or **deglutition** (dee-gloo-TISH-un; *degluto-* = "to swallow"), is a specialized type of propulsion that pushes a bolus of food from the oral cavity through the pharynx and esophagus to the stomach. Swallowing relies on the coordinated action of the upper alimentary canal, including

the soft palate, pharynx, and esophagus (**Figure 22.9**). The tongue also plays a role in this process, and is the only accessory organ to directly participate in motility. Swallowing consists of three phases: voluntary, pharyngeal, and esophageal.

① **Voluntary phase: The tongue pushes the bolus posteriorly toward the oropharynx.** During the **voluntary phase,** the tongue pushes the bolus superiorly against the hard palate and posteriorly toward the oropharynx. The voluntary phase is the only stage of swallowing under conscious control.

② **Pharyngeal phase: The bolus enters the oropharynx; the soft palate and epiglottis seal off the nasopharynx and larynx, respectively.** The bolus passes from the pharynx to the esophagus during the **pharyngeal phase.** This phase involves the involuntary contraction of skeletal muscles such as the pharyngeal constrictor muscles, and is controlled by the *swallowing reflex,* a reflex initiated by the medulla oblongata. The reflex arc is initiated when the bolus contacts sensory receptors in the oropharynx. This triggers the elevation of the uvula and soft palate, which prevents the bolus from going into the nasopharynx. It also triggers the elevation of the larynx. Notice in Figure 22.9 that the bolus pushes the epiglottis down. This, together with the elevation of the larynx, prevents *aspiration,* or the entry of food into the larynx.

③ **Esophageal phase: Peristaltic waves move the bolus down the esophagus to the stomach.** During the final

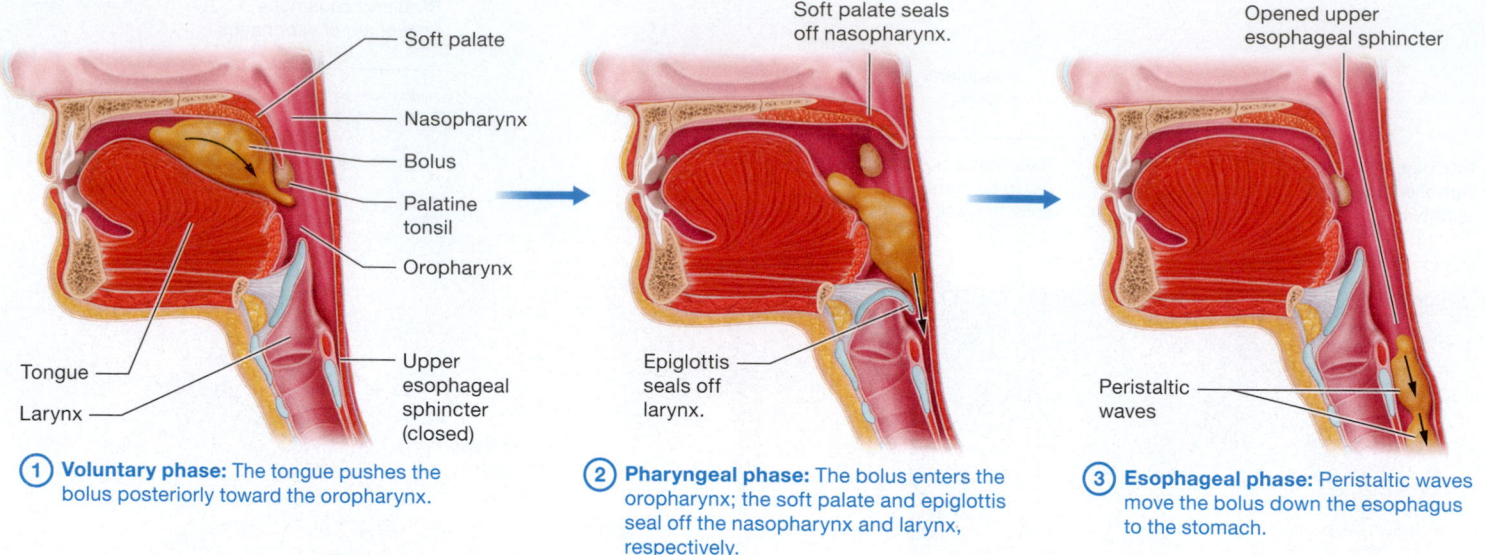

① **Voluntary phase:** The tongue pushes the bolus posteriorly toward the oropharynx.

② **Pharyngeal phase:** The bolus enters the oropharynx; the soft palate and epiglottis seal off the nasopharynx and larynx, respectively.

③ **Esophageal phase:** Peristaltic waves move the bolus down the esophagus to the stomach.

Figure 22.9 The process of swallowing (deglutition).

phase of swallowing, the **esophageal phase,** the bolus passes through the esophagus to the stomach. This phase begins as the upper esophageal sphincter relaxes. Neurons of the enteric nervous system then stimulate the inner circular and outer longitudinal layers of the muscularis externa to undergo peristalsis and "massage" the bolus inferiorly toward the stomach.

Notice that control of swallowing is almost entirely neural. The voluntary phase is under control of the cerebral cortex, and the remaining two phases are regulated by the medulla and the enteric nervous system. The ANS does not directly control any phase of swallowing, but it does influence the esophageal phase—the parasympathetic division stimulates peristalsis, and the sympathetic division inhibits it.

Quick Check

☐ 12. Walk through the steps of deglutition.

☐ 13. What controls the process?

Apply What You Learned

☐ 1. Inadequate production of saliva is associated with tooth decay. Why do you think saliva and dental health are linked?

☐ 2. Chemical warfare agents called *nerve gases* block the enzyme acetylcholinesterase, which prevents the breakdown of acetylcholine and so increases the amount of time it is active at cholinergic synapses. Explain why drooling (excessive salivation) is an effect of these agents.

☐ 3. Mr. Kekoa suffers from Parkinson's disease, which diminishes his swallowing reflex and decreases the activity of the muscularis externa of his esophagus. Predict what complications may result.

See answers in Appendix A.

MODULE **22.3**

The Stomach

1. Describe the structure and function of the different regions of the stomach.

2. Describe the structure of the gastric glands and the functions of the types of cells they contain.

3. Explain how hormones, nervous system stimulation, and the volume, chemical composition, and concentration of the chyme affect motility in the stomach.

4. Discuss the function, production, and regulation of the secretion of hydrochloric acid.

5. Explain the effects of the cephalic phase, gastric phase, and intestinal phase on the functions of the stomach and small intestine, and give examples for each phase.

The esophagus pierces the diaphragm muscle and passes through the *esophageal hiatus* (hy-AY-tus) to empty into the J-shaped **stomach.** In the sections that follow, we examine the gross and microscopic anatomy of the stomach. The discussion then turns to secretion from and motility of the stomach, and how the endocrine and nervous systems regulate these activities.

Gross Anatomy of the Stomach

As you saw in Figure 22.1, the stomach sits primarily in the left upper quadrant just inferior to the diaphragm muscle. The stomach's convex left side is known as its *greater curvature,* and its concave right side is its *lesser curvature.* **Figure 22.10** shows that the stomach has five anatomical regions, discussed next.

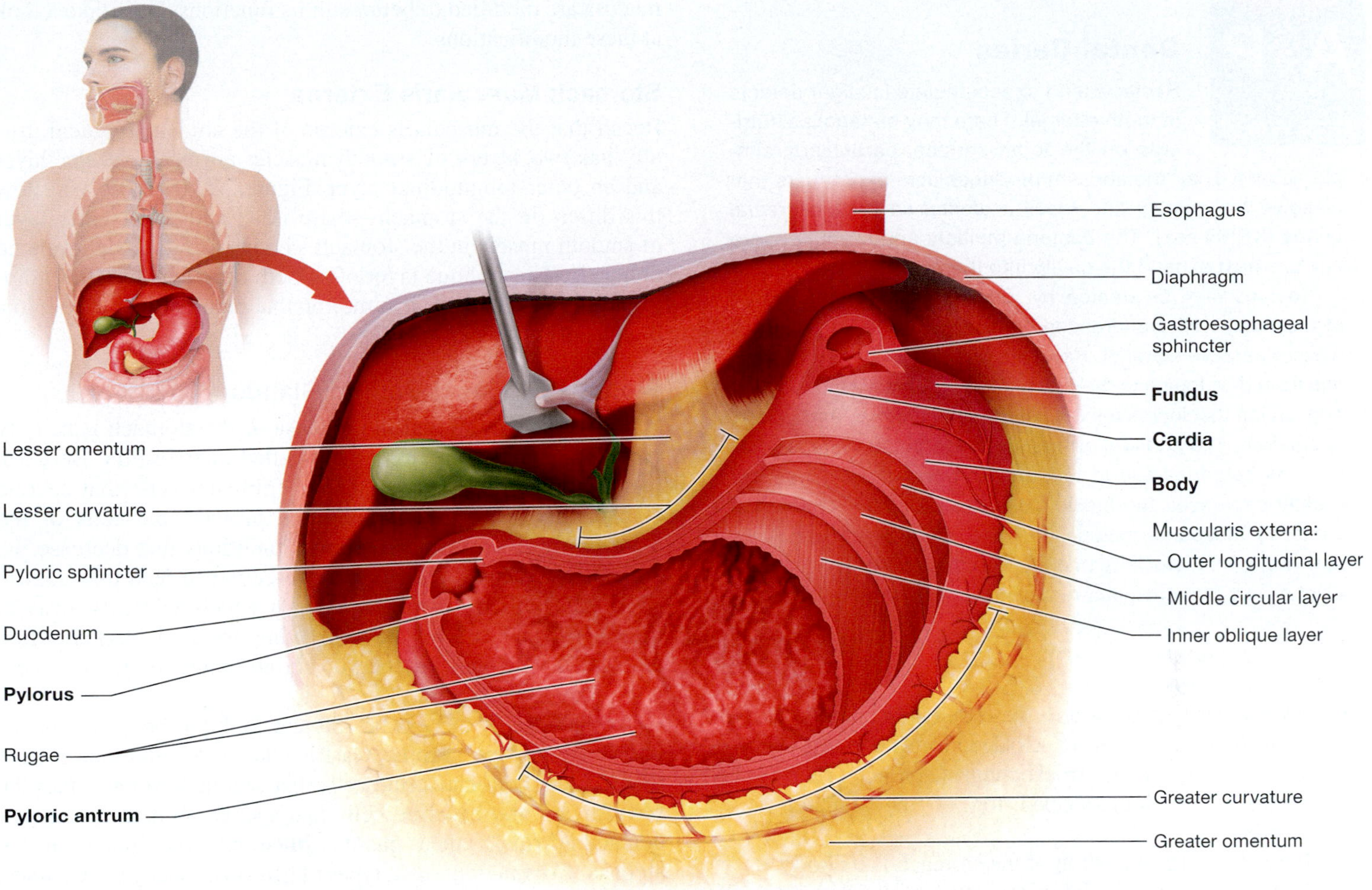

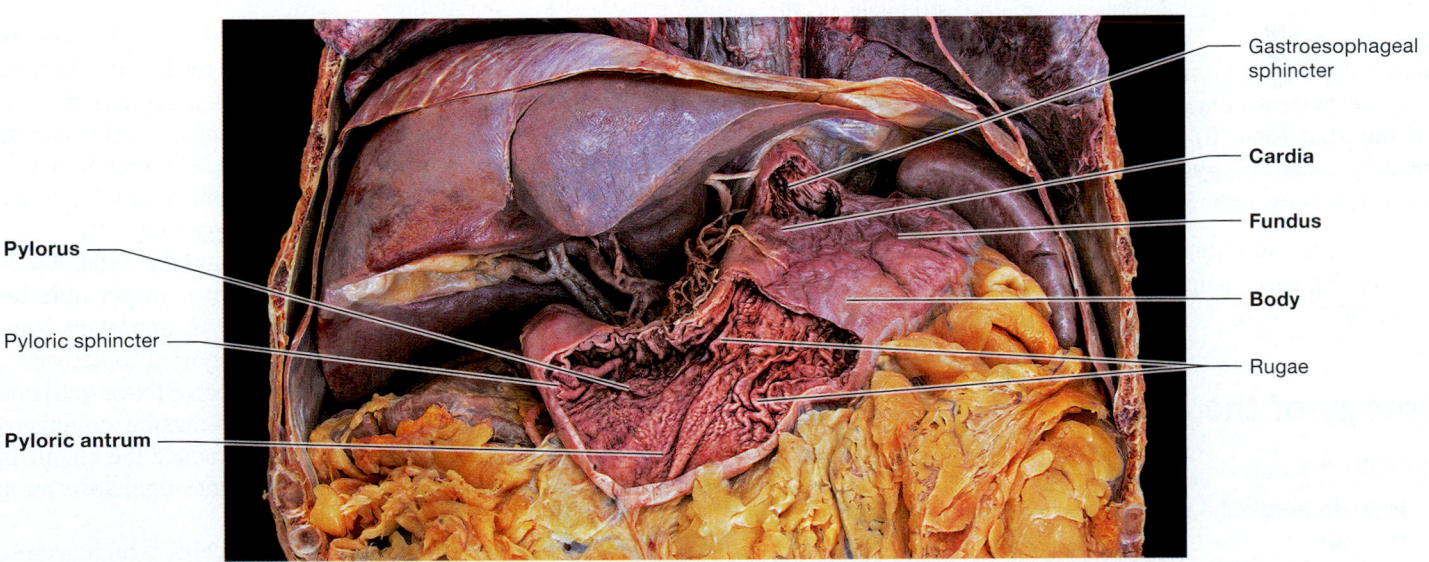

Esophagus

Diaphragm

Gastroesophageal
sphincter

Fundus

Cardia

Body

Muscularis externa:

Outer longitudinal layer

Middle circular layer

Inner oblique layer

Greater curvature

Greater omentum

Lesser omentum

Lesser curvature

Pyloric sphincter

Duodenum

Pylorus

Rugae

Pyloric antrum

(a) Regions and tissue layers of the stomach

Gastroesophageal
sphincter

Cardia

Fundus

Body

Rugae

Pylorus

Pyloric sphincter

Pyloric antrum

(b) Stomach and surrounding organs, cadaver photo

Figure 22.10 Gross anatomy of the stomach.

Explore PAL™ Cadaver
@Mastering Anatomy & Physiology

Dental Caries

Bacteria tend to accumulate in slight defects in tooth enamel. There they metabolize nutrients on the tooth surface, particularly simple sugars. Their metabolism produces acid byproducts that dissolve the enamel and result in *dental cavities,* or *dental caries* (KEHR-eez). The bacteria multiply and produce more acids, which expand the cavity into the underlying dentin.

Cavities may be treated by removing the damaged area of the tooth (with the dreaded dental drill) and filling it with a hard restorative material. However, it is better to prevent cavities from developing in the first place. One key weapon is fluoride, an ion that increases the hardness of enamel, particularly in children. The addition of fluoride to the public water supply in many countries and to toothpastes and mouthwashes has greatly decreased the incidence of dental caries. Other preventative measures include brushing and flossing regularly, applying dental sealants that fill defects in the enamel, taking calcium supplements, and reducing sugar intake.

- **Cardia.** The region where the esophagus empties into the stomach is the **cardia** (KAR-dee-ah). The cardia receives the bolus when the gastroesophageal sphincter relaxes.
- **Fundus.** The dome-shaped top of the stomach is its **fundus.**
- **Body.** The largest portion of the stomach is its **body.**
- **Pyloric antrum.** The inferior portion of the stomach is the **pyloric antrum** (py-LOHR-ik AN-trum).
- **Pylorus.** The terminal portion of the stomach is the **pylorus** (*pylor-* = "gatekeeper"), which abuts the first portion of the small intestine, the *duodenum.* Like the cardia, the pylorus contains a sphincter that controls the flow of ingested food. In the pylorus, a thick ring of smooth muscle called the **pyloric sphincter** regulates the flow of materials between the stomach and the small intestine.

Notice that the interior of the stomach contains folds called **rugae** (ROO-ghee). Rugae allow the stomach to expand considerably.

Histology of the Stomach

 **FLASHBACK**

1. How do endocrine glands and exocrine glands differ? (p. 134)
2. What is an acid? How does an acid affect the pH of a solution? (p. 48)

The stomach has the same four tissue layers as the rest of the alimentary canal, with a mucosa, submucosa, muscularis externa, and serosa. However, the stomach's muscularis externa and mucosa are modified to better suit its functions. Let's take a look at these modifications.

Stomach Muscularis Externa

Recall that the muscularis externa of the alimentary canal usually has two layers of smooth muscle: an inner circular layer and an outer longitudinal layer. Figure 22.10 shows you how this differs in the stomach—there is an additional inner layer of smooth muscle in the stomach's body with its fibers oriented obliquely. This oblique layer of smooth muscle allows the stomach to perform **churning,** a motion that pummels the food into a liquid called **chyme** (KY'M).

Stomach Mucosa: Gastric Glands

Figure 22.11a shows that the mucosa of the stomach is heavily indented to form deep structures called **gastric pits.** Between the gastric pits we find columnar epithelial cells that secrete a thick mucus layer that lines and protects the cells of the stomach from its own secretions. Conditions that decrease the amount of mucus secreted by these cells can lead to a *gastric ulcer* (also called a *peptic ulcer*), in which acid eats away at the mucosa and exposes the underlying tissues. Ulcers are often associated with the presence of certain bacteria and/or excessive production of acid.

At the base of the gastric pits are multiple branched glands known as **gastric glands.** Gastric glands are unusual in that they contain both endocrine cells that secrete hormones into the bloodstream and exocrine cells that secrete an acidic, enzyme-containing fluid called **gastric juice** into the lumen of the stomach. We find four main types of cells in or near gastric glands, each of which secretes a different product. From superficial to deep, these cell types are as shown in **Figure 22.11b** and **c.**

- **Mucous neck cells.** The cells located near the top, or "neck," of the gland are called **mucous neck cells.** As their name implies, these cells secrete mucus much like the surface epithelial cells. However, these surface cells secrete alkaline mucus, whereas mucous neck cells secrete acidic mucus. This prevents their mucus from neutralizing the acid produced by other cells known as parietal cells.
- **Parietal cells.** The next cells are the **parietal cells,** which secrete the hydrochloric acid (HCl) that is responsible for the acidic pH of gastric juice. Acid is an important component of gastric juice because it is required to activate a precursor enzyme called **pepsinogen** (pep-SIN-oh-jen) and also because it destroys most disease-causing organisms we ingest. In addition, parietal cells produce the chemical **intrinsic factor,** which is required for intestinal absorption of vitamin B12, found in various foods.
- **Chief cells.** Next down are the **chief cells,** which secrete the inactive precursor enzyme pepsinogen. When pepsinogen encounters an acidic pH, it becomes the active enzyme **pepsin,** which begins protein digestion in the stomach.
- **Diffuse neuroendocrine system cells.** Located at the very bottom of the gland are the **diffuse neuroendocrine system**

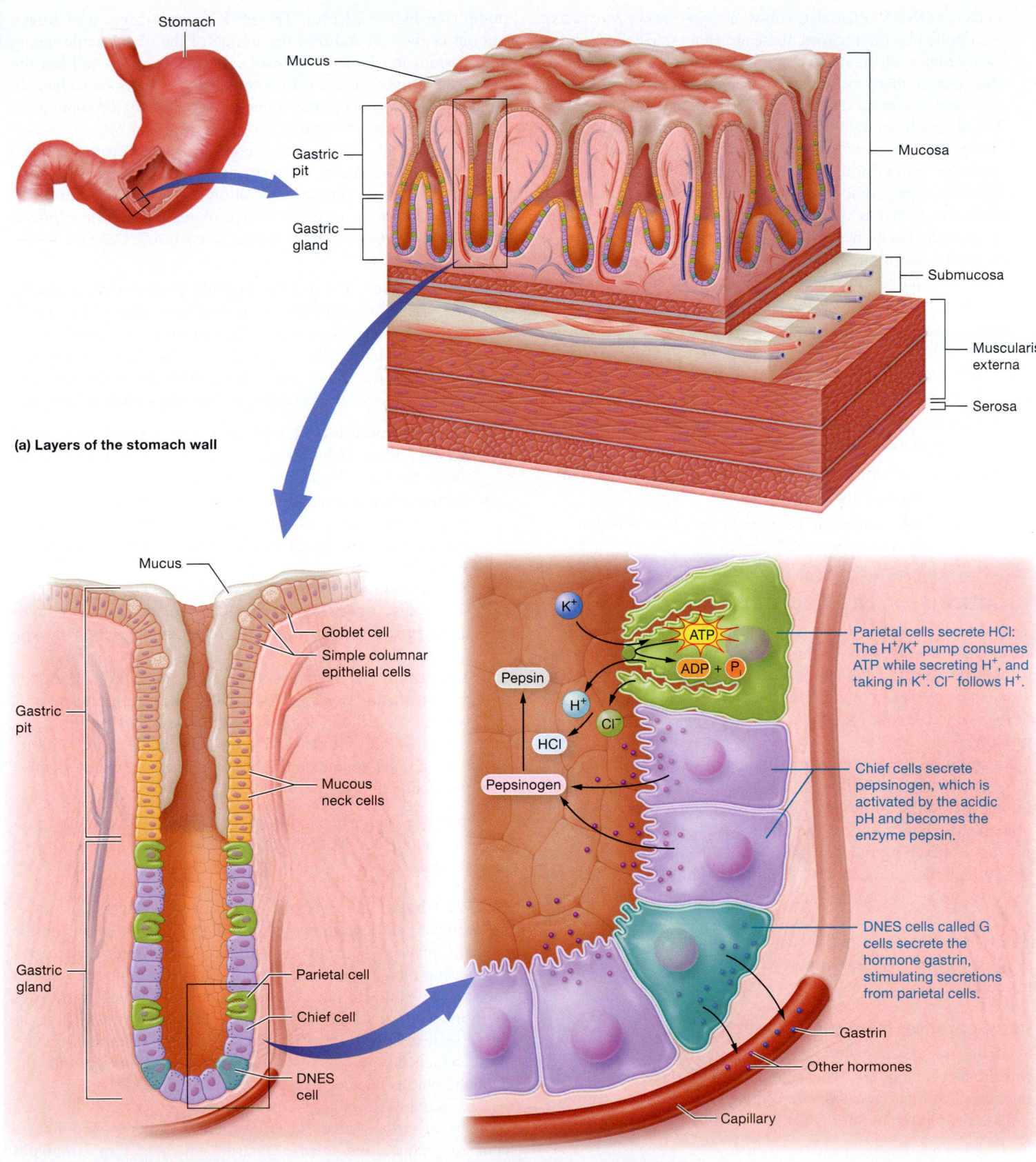

Stomach

Mucus

Mucosa

Submucosa

Muscularis externa

Serosa

Gastric pit

Gastric gland

(a) Layers of the stomach wall

Mucus

Goblet cell

Simple columnar epithelial cells

Gastric pit

Mucous neck cells

Gastric gland

Parietal cell

Chief cell

DNES cell

(b) Section through a gastric pit and gastric gland

K+

ATP

ADP + P_i

Pepsin

H+

Cl−

HCl

Pepsinogen

Parietal cells secrete HCl: The H+/K+ pump consumes ATP while secreting H+, and taking in K+. Cl− follows H+.

Chief cells secrete pepsinogen, which is activated by the acidic pH and becomes the enzyme pepsin.

DNES cells called G cells secrete the hormone gastrin, stimulating secretions from parietal cells.

Gastrin

Other hormones

Capillary

(c) Functions of gastric gland cells

Figure 22.11 Structure and functions of the stomach mucosa and gastric glands.

cells (or **DNES cells;** also known as *enteroendocrine cells*). As implied by their names, these are endocrine cells found in many places through the body—hence the term "diffuse"—that secrete many of the same products as neurons. As you can see in Figure 22.11c, DNES cells are closest to the blood vessels in the underlying submucosa, which gives the hormones they produce ready access to the blood. DNES cells are found throughout the digestive system, and several different types are located within gastric glands. The DNES cells illustrated in the figure* are called **G cells,** which secrete the hormone **gastrin.** This hormone, along with the hormone *histamine* produced by a different type of DNES cell, stimulates acid secretion from parietal cells.

Quick Check

☐ 1. What are the five regions of the stomach, and where are they located?

☐ 2. Why does the stomach have an extra layer of smooth muscle in its muscularis externa?

☐ 3. What are the primary cell types of the gastric glands, and what does each type of cell produce?

☐ 4. What is the relationship between hydrochloric acid and pepsiongen in the stomach?

Functions of the Stomach

The stomach performs three primary functions: secretion, propulsion, and digestion. Note that absorption of digested nutrients is not a main role of the stomach, but small amounts of chemicals such as alcohol may be absorbed here. This is why drinking alcohol without eating food first can lead to rapid intoxication and vomiting.

As you have read, the gastric glands of the stomach secrete many chemicals, including hormones, acid, mucus, and pepsinogen, which becomes the digestive enzyme pepsin. Pepsin catalyzes the beginning of the chemical digestion of proteins, and the stomach's churning actions begin mechanical digestion. When the stomach has completed churning, it propels the food, now chyme, through the pyloric sphincter into the duodenum. Secretion from the stomach, particularly of acid, and its motility are tightly regulated by the endocrine and nervous systems. We examine this regulation in the next two subsections.

Acid Secretion from the Stomach

Recall that the gastric glands of the stomach secrete multiple exocrine products, including hydrochloric acid (HCl) released by parietal cells. The hydrogen ions are secreted by an ATP-consuming pump in the plasma membrane, called the H^+/K^+ pump, or proton

*Choi E, Roland JT, Barlow BJ, et al. 2014. Cell lineage distribution atlas of the human stomach reveals heterogeneous gland populations in the gastric antrum. *Gut* 63:1711–1720.

pump (see Figure 22.11c). The H^+/K^+ pump drives a hydrogen ion out of the cell and into the lumen of the gland while taking a potassium ion into the cytosol. The chloride ion follows the hydrogen ion via passive diffusion through a membrane channel. Acid secretion occurs in this manner continuously throughout the day between meals, at what is known as the *basal rate*.

During eating, gastric acid secretion changes from the basal rate under the influence of the parasympathetic nervous system and multiple hormones. Secretion can be divided into three phases based on the primary source of regulation: the *cephalic phase, gastric phase,* and *intestinal phase* (**Figure 22.12**).

Cephalic Phase The initial **cephalic phase** is mediated by the sight, smell, taste, or even thought of food. This phase, which is directed by the CNS, prepares the stomach to receive food by increasing the release of hydrogen ions into it. These stimuli trigger output from the *vagus nerve,* the main nerve of the parasympathetic nervous system, resulting in four physiological effects:

- **Direct stimulation of hydrogen ion release.** The vagus nerves release ACh onto parietal cells, which directly stimulates them to release hydrogen ions.
- **Stimulation of gastrin secretion.** Vagal nerve stimulation triggers the release of a peptide that stimulates G cells to release the hormone gastrin. Gastrin triggers hydrogen ion secretion.
- **Stimulation of histamine secretion.** The vagus nerve stimulates a specific type of DNES cell to release the hormone **histamine.** Like gastrin, histamine triggers hydrogen ion secretion.
- **Inhibition of somatostatin secretion.** The hormone **somatostatin** (soh-mah´-toh-STAT-in) is produced by a type of DNES cell located in the antrum of the stomach. Somatostatin inhibits acid secretion. Vagal nerve stimulation inhibits somatostatin release, which has the effect of increasing hydrogen ion secretion.

Together, the four changes during the cephalic phase are responsible for about 30–40% of total hydrogen ion secretion from parietal cells.

Gastric Phase The **gastric phase** begins when food enters the stomach, and continues the stimulation provided during the cephalic phase. There are two stimuli that trigger acid secretion during the gastric phase. The first is the simple presence of food in the stomach. Distention of the stomach wall stimulates neurons of the ENS and sensory receptors involved with vagus nerve reflexes. Both the vagus nerve and ENS neurons release ACh, which has the same four effects we discussed with the cephalic phase. The net effect is to increase acid secretion.

The second stimulus is the presence of partially digested proteins in gastric juice. These protein fragments stimulate G cells to produce and release gastrin, which in turn stimulates acid secretion. The acid activates pepsin, which catalyzes protein digestion. Digested proteins stimulate more gastrin release, which triggers further acid secretion and more protein digestion. Note that this is a positive feedback loop and we have again visited

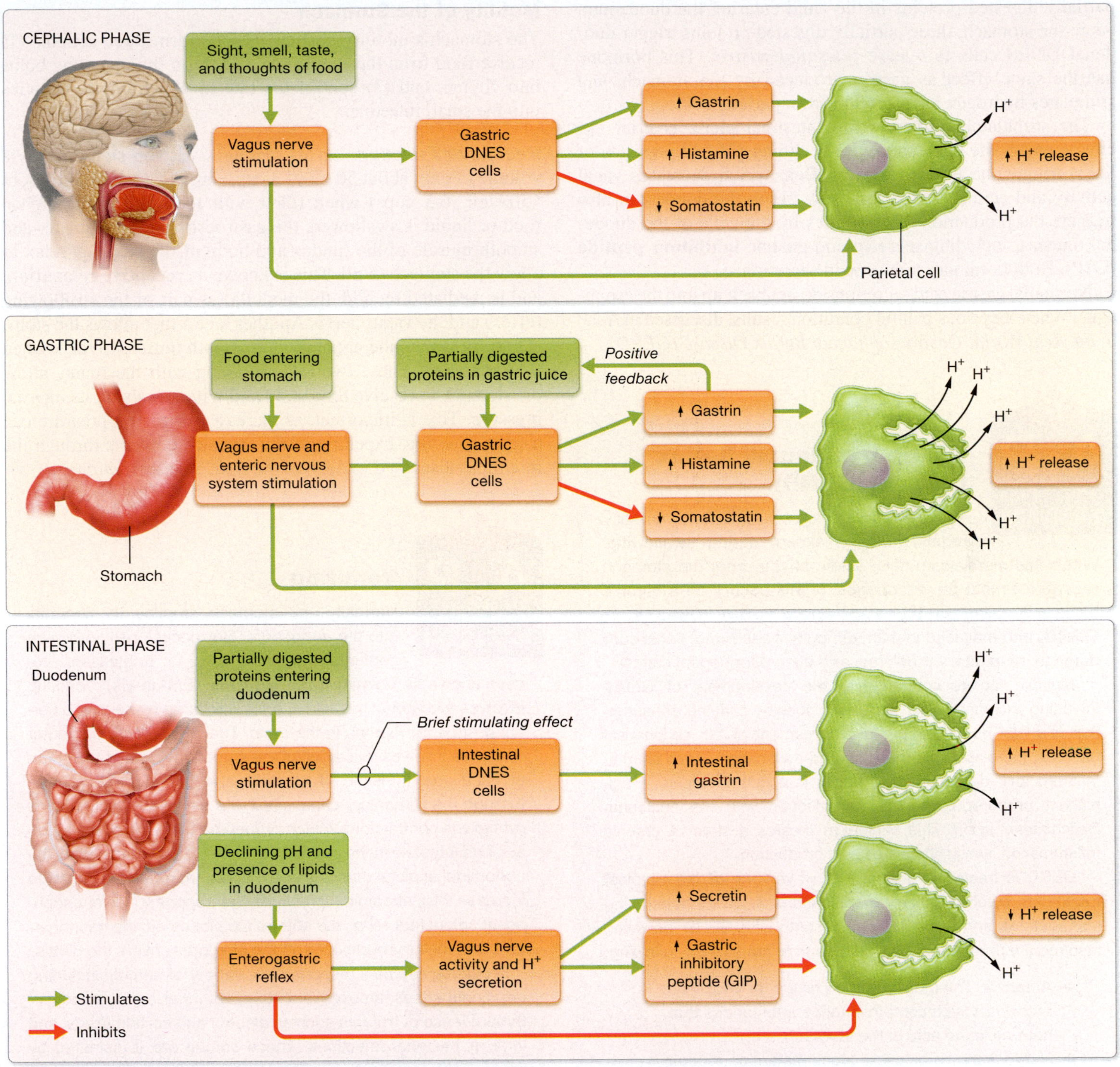

CEPHALIC PHASE

Sight, smell, taste, and thoughts of food → Vagus nerve stimulation → Gastric DNES cells → ↑ Gastrin → / ↑ Histamine → / ↓ Somatostatin → Parietal cell → ↑ H⁺ release (H^+, H^+)

GASTRIC PHASE

Food entering stomach / Partially digested proteins in gastric juice → *Positive feedback* → Vagus nerve and enteric nervous system stimulation → Gastric DNES cells → ↑ Gastrin / ↑ Histamine / ↓ Somatostatin → ↑ H⁺ release (H^+ ×6)

Stomach

INTESTINAL PHASE

Duodenum

Partially digested proteins entering duodenum → Vagus nerve stimulation — *Brief stimulating effect* → Intestinal DNES cells → ↑ Intestinal gastrin → ↑ H⁺ release (H^+)

Declining pH and presence of lipids in duodenum → Enterogastric reflex → Vagus nerve activity and H⁺ secretion → ↑ Secretin / ↑ Gastric inhibitory peptide (GIP) → ↓ H⁺ release (H^+)

→ Stimulates
→ Inhibits

Figure 22.12 The three phases of acid secretion from the stomach.

the Feedback Loops Core Principle (p. 28). Other stimuli trigger gastrin secretion as well, including caffeine and alcohol. Together, gastrin and the neural stimulation of parietal cells during the gastric phase account for about 50–60% of total acid secretion.

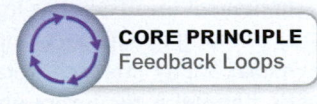

CORE PRINCIPLE
Feedback Loops

The positive feedback loop in the stomach does not continue indefinitely. Once the acid concentration reaches a certain

level—a pH of about 2.0—a negative feedback mechanism is initiated. Extremely low pH levels trigger somatostatin release, which in turn decreases acid secretion.

Intestinal Phase The final phase of gastric acid secretion, the **intestinal phase,** is responsible for the remaining approximately 10% of acid secretion, after which further acid secretion is inhibited. The intestinal phase is triggered by the presence of partially digested proteins in the fluid entering the duodenum. As in the stomach, these partially digested proteins trigger duodenal DNES cells to release *intestinal gastrin.* This hormone has the same effect as gastrin produced by the stomach, and stimulates hydrogen ion secretion from parietal cells.

The stimulatory effect of the intestinal phase is brief. As chyme enters the duodenum, the declining pH and presence of lipids trigger the **enterogastric reflex,** which decreases vagal activity and acid secretion. The low pH in the duodenum also triggers the production of hormones by the cells of the duodenal mucosa, including **secretin** and **gastric inhibitory peptide (GIP).** Both hormones reduce acid secretion.

Normally, gastric acid secretions do not back up into the esophagus. When they do, a painful condition results, discussed in *A&P in the Real World: Gastroesophageal Reflux Disease (GERD).*

Gastroesophageal Reflux Disease (GERD)

The gastroesophageal sphincter normally remains closed except during swallowing. When this mechanism fails, acidic chyme from the stomach *regurgitates* into the esophagus. If this occurs on a chronic basis, it is called **gastroesophageal reflux disease,** or **GERD,** and may lead to pain, difficulty swallowing, vocal cord damage, respiratory problems, and even esophageal cancer.

Multiple factors contribute to the development of GERD, including increased acid secretion, obesity, caffeine consumption, a *hiatal hernia* (abnormal enlargement of the esophageal hiatus in the diaphragm muscle), and medications such as anti-inflammatory drugs. Another factor that contributes to a significant proportion of cases is infection with the bacterium *Helicobacter pylori.* This bacterium creates a state of chronic inflammation, which stimulates acid production.

GERD is treated in a variety of ways. One of the simplest treatments involves dietary modifications, such as consuming less caffeine and alcohol. Most patients with GERD, however, also require drug therapy, including one or more of the following:

- *Antacids.* The most common drugs for GERD are *antacids,* which contain alkaline substances that neutralize the acid in the stomach.
- *H_2 blockers.* H_2 blockers block histamine receptors on parietal cells. This lowers the release not only of histamine-mediated acid but also of gastrin- and ACh-mediated acid.
- *Proton pump inhibitors.* Proton pump inhibitors are powerful inhibitors of the H^+/K^+ pump. These drugs generally result in the most drastic decrease in acid secretion.

Note that if *H. pylori* infection is present, antibiotics must be given in addition to acid-reducing medications to eradicate the bacteria.

Motility of the Stomach

The stomach's motility enables it to perform three actions: (1) receive food from the esophagus, (2) churn the incoming bolus into chyme, and (3) control the rate at which chyme empties into the small intestine.

Receptive Function In its resting state, the stomach has a volume of only about 50 ml (0.2 cup), but it can expand to about 1500 ml (6.3 cups) when filled with food and liquid. When food or liquid is swallowed, the gastroesophageal sphincter and smooth muscle of the fundus and body of the stomach relax to allow the stomach to fill. This is known as **receptive relaxation,** and is mediated by both the medulla (as part of the swallowing reflex) and the vagus nerve. Another factor that allows the stomach to fill is the inherent ability of smooth muscle to relax when it is stretched. These two factors, along with the rugae, allow the stomach to receive food and fluid without raising its internal pressure. This is important, as increases in gastric pressure can trigger *vomiting,* expelling of the stomach contents through the mouth, as discussed in *A&P in the Real World: Vomiting.*

Vomiting

Normally, the stomach empties its contents into the duodenum. But occasionally stomach contents move backward, an unpleasant process known as **vomiting,** or **emesis** (EM-uh-sis). Vomiting involves a complex motor response mediated by several different groups of neurons in the brain. The first part of this motor response involves relaxation of the smooth muscle of the stomach and small intestine, which slows and then stops propulsion through the alimentary canal. Next, the smooth muscle begins *retrograde* contractions, which propel the contents of the stomach and intestine in the opposite direction. At the same time, the abdominal skeletal muscles and diaphragm muscle contract to increase intra-abdominal pressure; the upper and lower esophageal sphincters relax; the soft palate closes off the nasopharynx; and the larynx elevates so the epiglottis covers the glottis.

Vomiting can be a response to a variety of stimuli, including emotional stress, an overfilled stomach, inner ear dysfunction, the presence of irritants such as alcohol and certain drugs, and viral or bacterial infections. These stimuli are transmitted by several different sensory afferent neurons, including those of the vagus nerve, mostly to a group of neurons in the medulla known as the *solitary nucleus.* These neurons integrate the incoming stimuli and then mediate the motor responses of vomiting. Drugs that treat vomiting are known as *anti-emetics.* Many act by blocking neurotransmitter receptors in the medulla to prevent these neurons from initiating vomiting.

Although vomiting is unpleasant, it can sometimes be beneficial. Many stomach irritants that stimulate vomiting could be harmful if they were absorbed into the bloodstream, and expelling them from the body prevents this from happening.

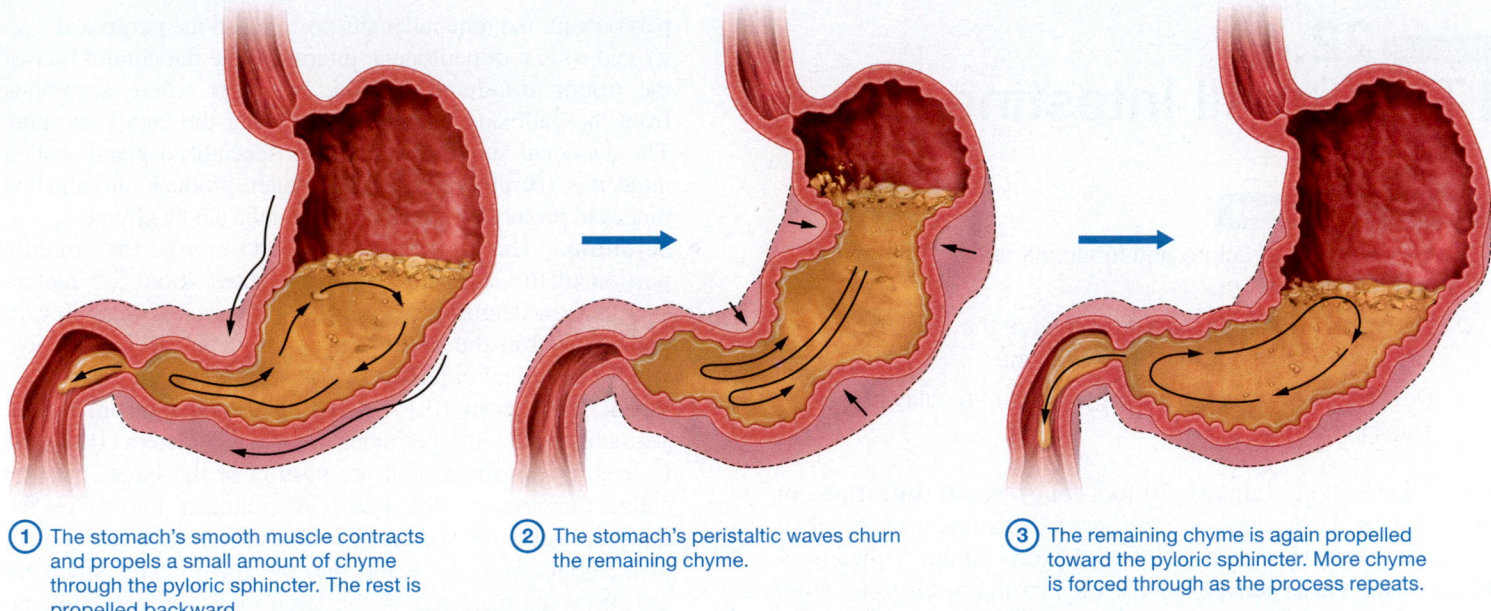

① The stomach's smooth muscle contracts and propels a small amount of chyme through the pyloric sphincter. The rest is propelled backward.

② The stomach's peristaltic waves churn the remaining chyme.

③ The remaining chyme is again propelled toward the pyloric sphincter. More chyme is forced through as the process repeats.

Figure 22.13 The process of churning in the stomach.

Churning Function   After a meal, the smooth muscle layers of the stomach begin to produce waves of peristalsis for churning (**Figure 22.13**). In this process, ① the stomach's smooth muscle produces peristaltic contractions that propel a small amount of chyme through the pyloric sphincter into the duodenum. The rest of the chyme is pushed backward into the stomach. ② The peristaltic waves then churn and mix the remaining chyme, and ③ the process repeats.

The peristaltic waves are initiated and controlled by a group of specialized cells collectively called the *gastric pacemaker.* The pacemaker controls the rate of the waves, which remains relatively constant at about three per minute. Their strength, however, is subject to regulation by several factors, including stimulation from the vagus nerve and secretion of certain hormones. For example, certain DNES cells of the stomach produce the hormone **serotonin,** which stimulates gastric motility. Another hormone with the same effect is **intestinal gastrin,** which is produced by duodenal DNES cells.

Emptying Function   The final function of gastric motility is to control the movement of chyme into the duodenum. Different materials pass through the pyloric sphincter at different rates. Liquids such as water and saline move rapidly from the stomach to the duodenum with essentially no delay. Solids must be converted to a nearly liquid state before they are able to enter the small intestine. This occurs relatively easily with carbohydrates and proteins, which mix with gastric juice with few problems. It may take much longer with lipids, because they require extensive churning due to their nonpolar structure.

A key factor that determines the rate of gastric emptying is the amount and composition of chyme in the duodenum. When sensory receptors in the wall of the duodenum detect a high degree of stretch, a low pH, a high lipid composition, and/or a high solute concentration in the chyme, they trigger a negative feedback loop that delays gastric emptying. This feedback loop is mediated by both the vagus nerve and hormones secreted by the duodenum, including *secretin, cholecystokinin,* and *gastric inhibitory peptide.* Each of these hormones is also involved in secretion from the stomach and other digestive organs, and will be discussed shortly.

Control of gastric emptying is critical because the duodenum must mix the incoming chyme thoroughly before it moves to the rest of the small intestine. There are two reasons for this: (1) Chyme is acidic, and the duodenum must mix it with bicarbonate ions to avoid damaging the intestinal mucosa; and (2) chyme is generally very concentrated and must be diluted with water from pancreatic juice to prevent the chyme from drawing water into the intestinal lumen by osmosis. The duodenum can process incoming chyme only so quickly, and therefore must receive small amounts at a time.

Quick Check

☐ 5. What are the three phases of acid secretion in the stomach? Which hormones and neural stimuli trigger each phase?

☐ 6. What are the basic functions of gastric motility?

☐ 7. Why is control of gastric emptying into the duodenum so critical?

Apply What You Learned

☐ 1. *Pyloric stenosis* (a hardened pyloric sphincter) causes the stomach contents to be emptied slowly and erratically. What symptoms might accompany this disorder? (*Hint:* What effect would pyloric stenosis have on gastric pressure?)

☐ 2. A procedure called a *vagotomy,* in which the vagus nerve is severed, is occasionally performed to decrease acid secretion by the stomach. How would such a procedure lower acid secretion? What other digestive processes would it affect and in what way?

See answers in Appendix A.

MODULE 22.4
The Small Intestine

Learning Outcomes

1. Describe the structure and functions of the duodenum, jejunum, and ileum.

2. Discuss the histology and functions of the circular folds, villi, and microvilli of the small intestine.

3. Describe the functions and regulation of motility in the small intestine.

The 6-meter-long (almost 20-foot-long) **small intestine,** or *small bowel,* is the longest portion of the alimentary canal (it is considerably shorter in a living person—about 3 meters or 10 feet—than in a cadaver because of smooth muscle tone). Four main processes occur in the small intestine: secretion, digestion, absorption, and propulsion.

Cells of the small intestine, known as *enterocytes,* produce multiple digestive enzymes, hormones, and mucus. These enzymes, along with those released by the pancreas, are responsible for the bulk of chemical digestion. Indeed, no further nutrient digestion takes place by human enzymes once food leaves the small intestine (bacteria in the large intestine do digest certain nutrients). After the nutrients are digested chemically, they are absorbed across the enterocytes into either the blood or the lymph, along with water, vitamins, and other substances. Like the stomach, the small intestine also mixes and propels its contents along its length and into the large intestine. Let's look at the structure and functions of the small intestine.

Divisions of the Small Intestine

The small intestine consists of three divisions—the *duodenum, jejunum,* and *ileum* (**Figure 22.14**). These divisions are distinguishable by the following slight histological differences:

- **Duodenum.** The **duodenum** (doo-AH-den-um), the initial segment of the small intestine, begins at the pylorus. The shortest of the three divisions, it is only about 25 cm (10 in.) long. The duodenum arches into a "C" shape as it curves around the pancreas. Only the proximal portion of the duodenum is within the

peritoneum; the remainder sits posterior to the peritoneal cavity and so is retroperitoneal. Internally, the duodenum houses the **major duodenal papilla,** which is where secretions from the gallbladder and pancreas enter the small intestine. The duodenal submucosa contains specialized glands called *duodenal (Brunner's) glands,* which produce an alkaline mucus to protect the duodenum from the acidic chyme.

- **Jejunum.** The **jejunum** (jeh-JOO-num), the middle portion of the small intestine, measures about 2.5 meters (7.5 feet) in length. It begins at the *duodenojejunal flexure* and sits within the peritoneal cavity. It is the most active site for chemical digestion and absorption.

- **Ileum.** The **ileum** (ILL-ee-um), the small intestine's final segment, is also intraperitoneal. About 3.6 meters (10.8 feet) in length, it terminates at the portion of the large intestine called the *cecum* (SEE-kum). A sphincter known as the **ileocecal valve** (ill'-ee-oh-SEE-kuhl) controls the movement of materials from the ileum into the cecum. This sphincter also prevents materials in the large intestine from flowing backward into the ileum. This function is key because, as we discuss shortly, the large intestine houses a great number of bacteria that could cause illness if they entered the ileum.

The small intestine has the same four tissue layers as the rest of the alimentary canal. But, as always, the form of the small intestine has specializations that better enable it to perform its functions, as we'll see next.

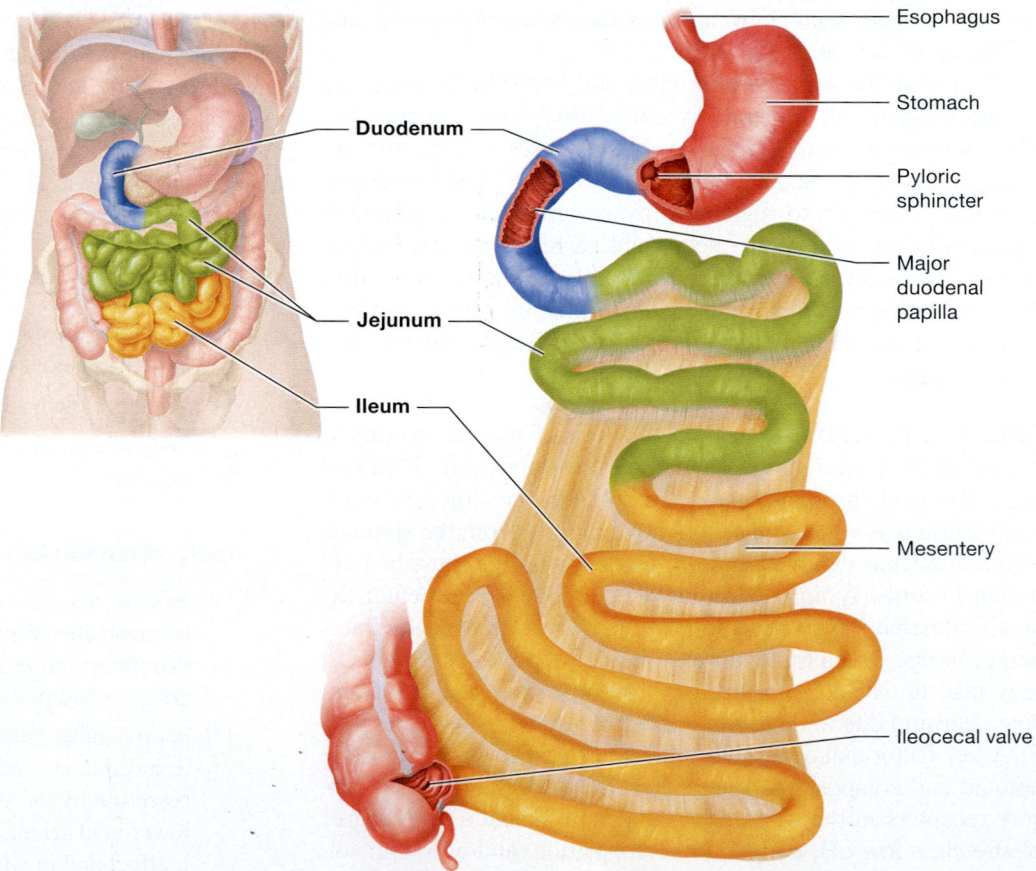

Figure 22.14 Gross anatomy of the small intestine.

☐ 1. What are the three divisions of the small intestine?

Structure and Functions of the Small Intestine

« FLASHBACK

1. What are microvilli? What is the function of microvilli? (p. 98)

The internal surface of the small intestine contains three progressively smaller types of folds. Most of the absorption that occurs within the alimentary canal takes place in the small intestine, and these folds increase the surface area available for absorption

about 400 to 600 times—another example of the Structure-Function Core Principle (p. 28).

CORE PRINCIPLE
Structure-Function

The largest type of folds, which are visible with the naked eye as ridges in the wall, are called **circular folds** (or *plicae circulares;* **Figure 22.15a**). Circular folds involve both the mucosa and submucosa of the small intestine. These folds not only increase surface area but also slow down the transit of chyme through the small intestine, which gives the nutrients more time to be digested, and the small intestine cells, called **enterocytes** (EN-tehr-oh-sytz), more time to absorb nutrients.

The smaller two types of folds are not visible to the naked eye. The mucosa folds into projections called **villi** (VILL-aye; singular, *villus;* **Figure 22.15b**). Notice that each villus consists of a layer of enterocytes and occasional mucus-secreting goblet

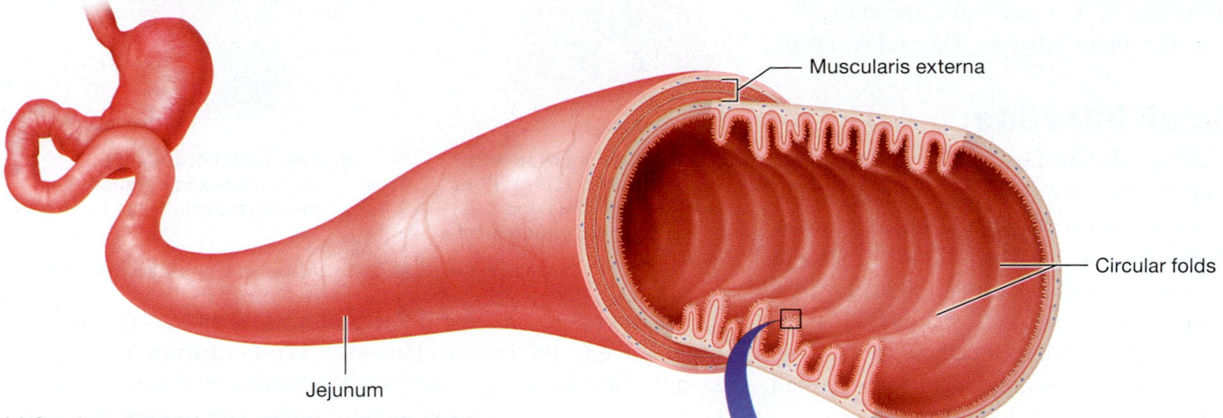

Muscularis externa

Circular folds

Jejunum

(a) Section of the jejunum showing circular folds

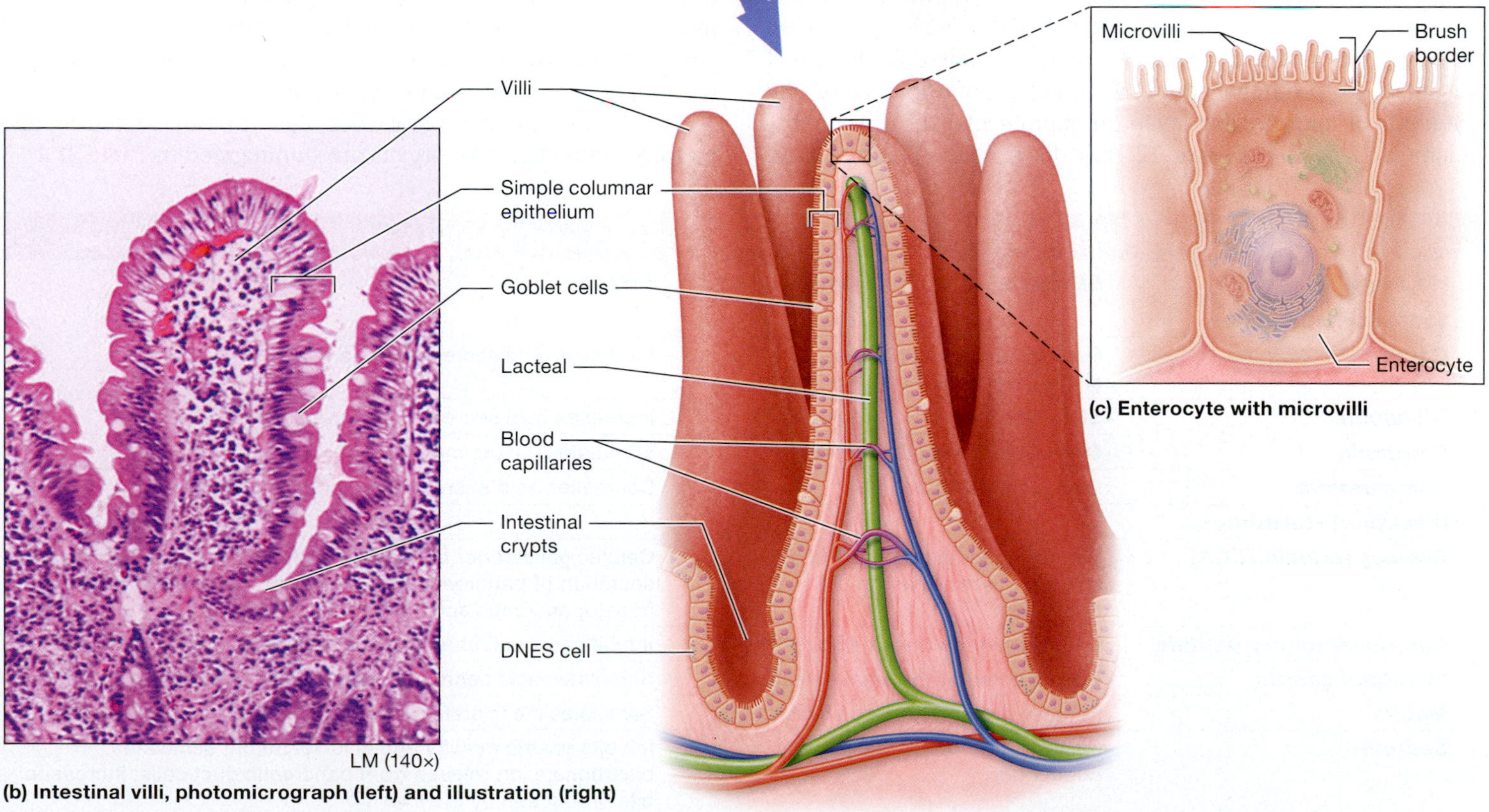

Villi

Simple columnar epithelium

Goblet cells

Lacteal

Blood capillaries

Intestinal crypts

DNES cell

LM (140×)

(b) Intestinal villi, photomicrograph (left) and illustration (right)

Microvilli

Brush border

Enterocyte

(c) Enterocyte with microvilli

Figure 22.15 Folds of the small intestine.

cells surrounding a central core of blood capillaries and a lymphatic vessel called a *lacteal*. Between villi, the mucosa indents to form **intestinal crypts,** which house glands that also contain hormone-secreting DNES cells.

The smallest folds, the **microvilli,** are found in the plasma membrane of the enterocytes (**Figure 22.15c**). Each enterocyte has as many as 3000 microvilli, which gives the cell the appearance of a bristle brush, or **brush border,** on microscopic examination. Associated with the brush border are numerous digestive enzymes produced and secreted by enterocytes, such as sucrase, maltase, and lactase, which catalyze reactions that break down disaccharides, and peptidases, which catalyze reactions that break down peptides.

Quick Check

☐ 2. Why is the internal surface of the small intestine arranged into progressively smaller folds? What are these folds called?

Motility of the Small Intestine

Small intestinal motility differs during fasting and eating. During fasting, the small intestine exhibits slow, rhythmic contractions along its length in a pattern called the **migrating motor complex.** These contractions clear any remaining material from the small intestine, including leftover food and secretions. The migrating motor complex requires about 2 hours to push digesting food from the duodenum to the ileocecal valve. This movement is controlled by both the ENS and a hormone called *motilin,* which is produced by cells in the duodenal mucosa.

The small intestine undergoes two types of movement during eating: peristalsis and segmentation (**Figure 22.16**). As with other regions of the alimentary canal, peristalsis in the small intestine is accomplished by alternating contractions of the longitudinal and circular layers of smooth muscle in the muscularis externa (see Figure 22.16a). Its primary function is to

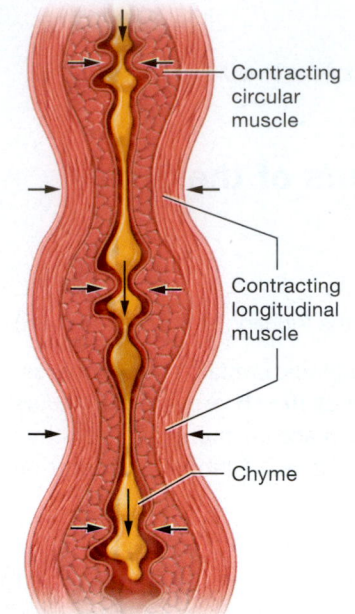
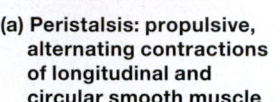
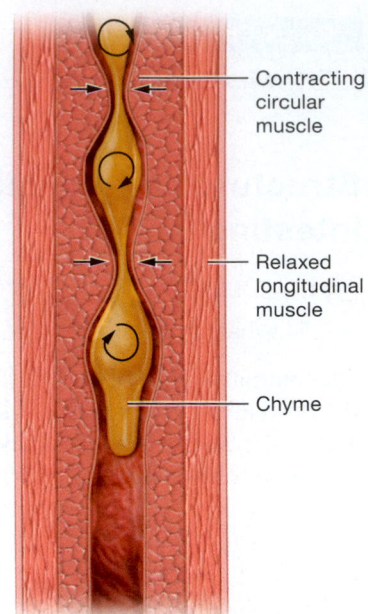

(a) **Peristalsis: propulsive, alternating contractions of longitudinal and circular smooth muscle**

(b) **Segmentation: mixing contractions of circular smooth muscle only**

Figure 22.16 Peristalsis and segmentation in the small intestine.

propel chyme toward the ileum and ultimately through the ileocecal valve to the cecum. However, **segmentation,** also known as intestinal churning, involves contractions of only the circular layer of smooth muscle, which produces a squeezing motion (see Figure 22.16b). The primary functions of segmentation are mechanical digestion and mixing the chyme with intestinal and pancreatic enzymes as well as bile. The vagus nerve appears to regulate both peristalsis and segmentation.

The hormones and paracrines that control secretion and motility of the digestive organs are summarized in **Table 22.1**.

Table 22.1 Hormones and Paracrines Involved in Secretion and Motility of the Digestive Organs		
Hormone	**Stimulus for Production**	**Effects**
Stomach Hormones		
Gastrin	Partially digested proteins; stimulation from the vagus nerve	Increases acid secretion by parietal cells
Histamine	Stimulation from the vagus nerve	Increases acid secretion by parietal cells
Serotonin	Distention of the stomach	Stimulates gastric motility
Somatostatin	Decreasing stomach pH	Decreases acid secretion by parietal cells
Intestinal Hormones		
Cholecystokinin (CCK)	Partially digested proteins and lipids in chyme entering the duodenum	Causes gallbladder to contract and release bile; stimulates secretion of pancreatic enzymes from acinar cells; relaxes hepatopancreatic sphincter
Gastric inhibitory peptide	Chyme entering the small intestine	Inhibits acid secretion from parietal cells
Intestinal gastrin	Chyme entering the small intestine	Stimulates acid secretion from parietal cells
Motilin	Released regularly during fasting	Stimulates the migrating motor complex of the small intestine
Secretin	Partially digested proteins in the duodenum	Inhibits gastric motility and acid secretion; stimulates bicarbonate ion release from pancreatic duct cells; increases bile production by the liver
Vasoactive intestinal peptide	Partially digested proteins in the duodenum	Inhibits acid secretion by parietal cells; stimulates pancreatic secretion; increases intestinal blood flow

☐ 3. How does motility of the small intestine differ during fasting and during eating?

☐ 4. What are segmentation and peristalsis, and how do they differ?

☐ 1. The condition known as *celiac disease* (SEE-lee-ak), or *sprue,* is caused by an allergy to gluten, a protein found naturally in certain grains. Celiac disease leads to the destruction of the microvilli and even the villi of the small intestine. Predict the effects of this disease.

☐ 2. Certain poisons stimulate the parasympathetic nervous system, which increases the activity of the vagus nerve. Predict the effect this would have on the motility of the stomach and small intestine. Are there processes that would be unaffected by altered vagus nerve activity? Explain.

See answers in Appendix A.

MODULE **22.5**
The Large Intestine

1. Describe the gross and microscopic anatomy of the divisions of the large intestine.

2. Describe the defecation reflex and the functions of the internal and external anal sphincters.

3. Discuss conscious control of the defecation reflex.

The **large intestine,** or *large bowel,* runs along the border of the abdominal cavity, surrounding the small intestine and other abdominal organs like a frame (**Figure 22.17a**). About 1.5 meters (5 feet) long, it is so named because it has a larger diameter than the small intestine. The large intestine receives material from the small intestine that was not digested or absorbed. You might think this means the small intestine does the "important" work and the large intestine is simply a passageway for the leftovers, known as **feces,** or *fecal matter.* However, the large intestine is very active in absorbing water and electrolytes. Absorption in the large intestine is critical for maintaining fluid, electrolyte, and acid-base homeostasis. When this absorption is disrupted, as with bacterial or viral gut infections,

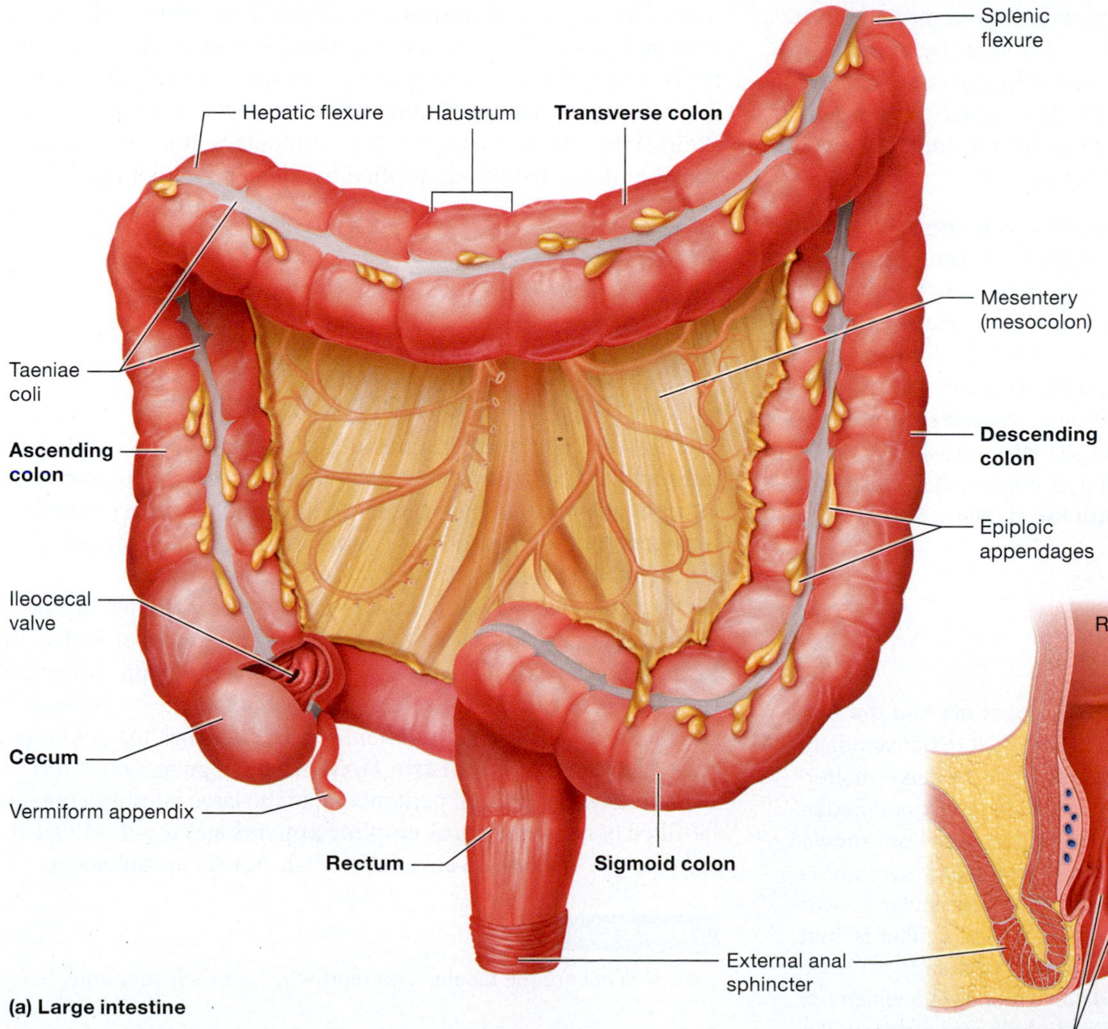

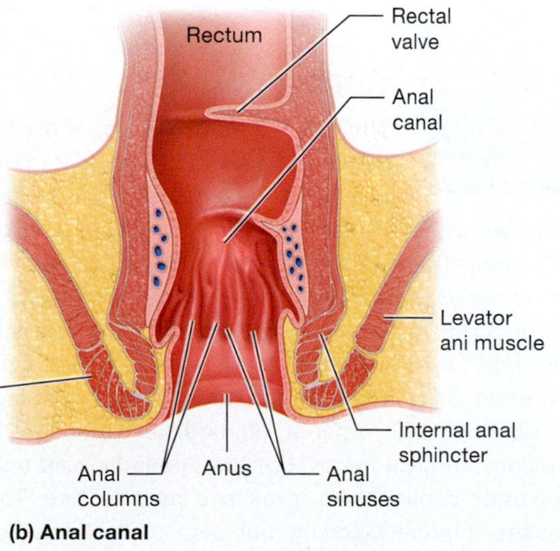

(a) Large intestine

(b) Anal canal

Figure 22.17 Gross anatomy of the large intestine.

dehydration as well as electrolyte and acid-base disturbances may result.

The tasks of the large intestine also include secretion (primarily in the form of mucus), propulsion, and defecation. In addition, it houses numerous bacteria that perform important functions such as synthesizing vitamins. The upcoming sections examine the gross and microscopic anatomy of the large intestine, the bacteria that inhabit it, and its basic functions.

Gross Anatomy of the Large Intestine

The large intestine is made up of three segments: the cecum, the colon, and the rectum. As you read in the last module, the first portion of the large intestine is the **cecum,** a blind pouch (*cecum-* = "blind") that is intraperitoneal and located in the right lower quadrant of the abdomen. The cecum features a smaller blind-ended pouch extended from its posteroinferior end, called the **vermiform appendix** (VER-muh-form uh-PEN-diks; "wormlike appendage"), which is generally shortened to simply *appendix* (see Figure 22.17a). The appendix was long considered a vestigial organ, or one left over from evolution that served no function. However, it is now understood that it houses multiple lymphatic nodules and plays a role in the immune system. Despite its small size, an inflamed appendix can lead to big problems, as discussed in *A&P in the Real World: Appendicitis.*

The next and longest portion of the large intestine is the **colon** (KOH-lun), which is divided into four portions:

- **Ascending colon.** The retroperitoneal **ascending colon** travels superiorly along the right side of the abdomen from the right lower quadrant to the right upper quadrant. When it reaches the liver, it makes a sharp left-hand turn at a junction called the **hepatic flexure,** also known as the *right colic flexure.*
- **Transverse colon.** At the hepatic flexure, the ascending colon becomes the intraperitoneal **transverse colon,** so named because it passes transversely across the superior abdominal cavity. At the spleen, it takes a sharp turn inferiorly at a junction called the **splenic flexure,** also known as the *left colic flexure.*

Appendicitis

The small size of the appendix and the fact that it is blind-ended cause it to occasionally become blocked, generally by fecal matter. Bacteria within the feces multiply in the appendix and cause infection. This results in inflammation, a condition known as *appendicitis* (ah-pen´-dih-SY-tus). Signs and symptoms of appendicitis include abdominal pain, particularly over the right lower quadrant; rebound tenderness similar to that experienced with peritonitis; nausea; and vomiting.

Appendicitis requires immediate treatment, which is typically surgical removal of the appendix. Left untreated, the appendix continues to swell and may rupture. This not only causes internal bleeding but also spills bacteria-filled fecal material into the peritoneal cavity, leading to peritonitis. These two conditions can be fatal if not properly managed.

- **Descending colon.** The splenic flexure gives rise to the retroperitoneal **descending colon,** which passes along the left side of the abdominal cavity.
- **Sigmoid colon.** In the left lower quadrant, the descending colon becomes the S-shaped **sigmoid colon** (*sigmoid* = "S-shaped"), which is intraperitoneal and passes toward the sacrum.

After the sigmoid colon, the large intestine continues as the **rectum** (REK-tum). This portion of the large intestine runs anterior to the sacrum and is retroperitoneal. The walls of the rectum feature horizontal folds called **rectal valves,** which allow the passage of *flatus* (gas) without risking the simultaneous passage of feces.

The rectum ends at the **anal canal,** the last portion of the large intestine (**Figure 22.17b**). Recall from the muscular system chapter that the anal canal passes through the levator ani muscle in the floor of the pelvic cavity (see Figure 9.16). The walls of the anal canal feature longitudinal grooves called **anal columns.** Between the anal columns are **anal sinuses,** which secrete mucus when feces pass through the anal canal during defecation. The terminal portion of the anal canal has two sphincters. The first is the involuntary **internal anal sphincter,** which is simply the thickened circular layer of the muscularis externa. The internal anal sphincter is supplied by parasympathetic motor neurons. The second is the voluntary **external anal sphincter,** which is composed of skeletal muscle. As this sphincter is voluntary, it is innervated by somatic motor neurons controlled by the cerebral cortex.

Quick Check

- ☐ 1. What are the regions of the large intestine?
- ☐ 2. How do the internal and external anal sphincters differ?

Histology of the Large Intestine

The mucosa of the large intestine lacks villi and its cells lack microvilli. These structural adaptations reflect the fact that nutrient absorption is not the large intestine's primary function. Like much of the alimentary canal, its mucosa is rich with goblet cells that secrete protective and lubricating mucus.

The muscularis externa of the large intestine is unique in that its longitudinal layer is not continuous throughout most of its length. Instead, this layer is gathered into three bands or ribbons of muscle called **taeniae coli** (TEE-nee-ee KOHL-ahy; *taeniae* = "ribbons"; see Figure 22.17a). Their constant tension bunches the colon into pockets referred to as **haustra** (HAW-struh; singular, *haustrum*).

The serosa, or visceral peritoneum, of the large intestine contains fat-filled pouches known as **epiploic appendages** (ep-ih-PLOH-ik; *epiploic* = "membrane-covered"). Their function is unknown.

Quick Check

- ☐ 3. What are the taeniae coli, and what are their functions?

Bacteria in the Large Intestine

The large intestine hosts a staggering number of bacteria; indeed, there are about 10 times more bacteria in the large intestine than cells in the human body, and bacteria make up as much as 60% of the dry mass of feces. These bacteria, the

normal flora or **gut flora,** consist of about 500 different bacterial species that coexist with humans in a *symbiotic* (mutually beneficial) relationship. Humans provide the bacteria with the environment they need to survive, and the bacteria perform a number of useful functions for humans, including the following:

- **Produce vitamins.** Bacteria produce vitamins such as vitamin K, which is necessary for blood clotting.
- **Metabolize undigested materials.** Bacteria metabolize carbohydrates such as soluble fibers that the small intestine is unable to digest, converting them into fatty acids and other compounds the body can absorb and use. This also aids in the absorption of certain vitamins and electrolytes. A somewhat unfortunate byproduct of this metabolism is the production of gas within the intestine that is released as flatus.
- **Deter the growth of harmful bacteria.** The normal flora prevent the growth of pathogenic, or disease-causing, microorganisms by competing for nutrients and producing chemicals that kill certain harmful bacterial species.
- **Stimulate the immune system.** During infancy, the normal flora induce immune tolerance to their own antigens. At the same time, they stimulate the development of mucosa-associated lymphatic tissue (MALT) and the production of antibodies to pathogens. This creates a favorable environment for the normal flora while also protecting the host from pathogenic bacteria.

Long-term antibiotic therapy, particularly antibiotics taken by mouth, can disrupt these functions by killing large numbers of normal flora. This can result in multiple problems, particularly infection with antibiotic-resistant pathogenic organisms such as the bacterium *Clostridium difficile*. Note that this situation may also occur in individuals who unintentionally consume antibiotics by eating the meat of animals fed antibiotics.

Quick Check

☐ 4. What functions are performed by the large intestine's normal flora?

Motility of the Large Intestine and Defecation

The large intestine has two functional segments: proximal and distal. The *proximal large intestine* consists of the ascending and transverse colon, and the *distal large intestine* consists of the descending and sigmoid colon, rectum, and anal canal.

The proximal large intestine is the primary site of water and electrolyte absorption and bacterial activity, and exhibits two main types of motility. The first is a type of segmentation, or churning, similar to what we saw in the small intestine, in which the circular muscle of each haustrum contracts repeatedly. This swirls the material around in the haustrum, which aids in water and electrolyte absorption. These contractions are controlled primarily by local neurons of the ENS and are triggered by stretch.

The other type of motility in the proximal large intestine is a propulsive motion known as a **mass movement,** or **mass peristalsis.** During a mass movement, multiple haustra undergo peristalsis, which propels their contents toward the distal large intestine. Mass movements occur three to four times per day, and appear to be triggered by food consumption, which initiates reflexes controlled by the ENS.

A small amount of absorption takes place in the distal large intestine, mostly of water. However, its main role is to store fecal material until it is ready to be expelled during defecation. For this reason, the distal large intestine is much less motile than the proximal large intestine. When mass movements force fecal material into the normally empty rectum, it initiates the parasympathetic-mediated **defecation reflex** (**Figure 22.18**):

① **Stretch receptors transmit the sensation of rectal distention to the spinal cord.** The walls of the rectum house stretch receptors that are activated by the presence of fecal material stretching the wall. Sensory neurons transmit these impulses to the spinal cord.

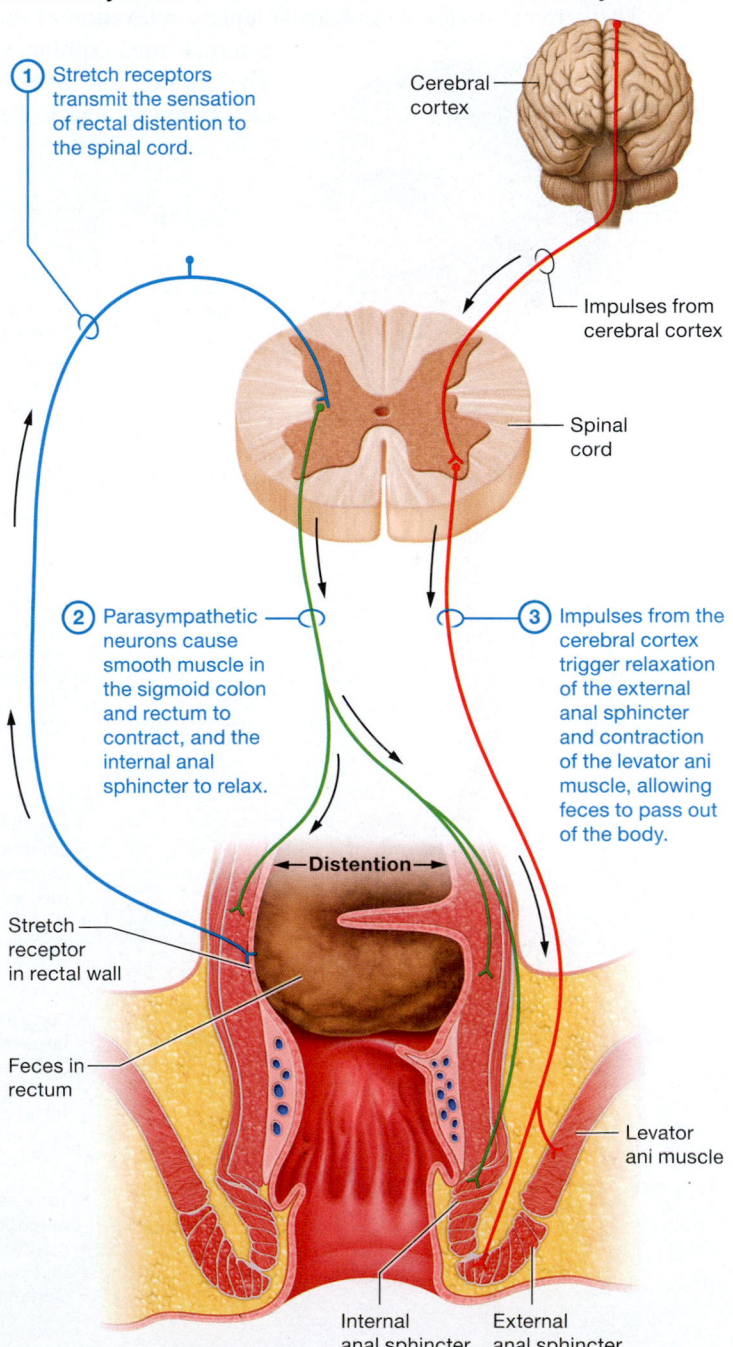

Involuntary reflex:

① Stretch receptors transmit the sensation of rectal distention to the spinal cord.

② Parasympathetic neurons cause smooth muscle in the sigmoid colon and rectum to contract, and the internal anal sphincter to relax.

Stretch receptor in rectal wall

Feces in rectum

Voluntary control:

Cerebral cortex

Impulses from cerebral cortex

Spinal cord

③ Impulses from the cerebral cortex trigger relaxation of the external anal sphincter and contraction of the levator ani muscle, allowing feces to pass out of the body.

←Distention→

Levator ani muscle

Internal anal sphincter

External anal sphincter

Figure 22.18 Defecation reflex.

(2) **Parasympathetic neurons cause smooth muscle in the sigmoid colon and rectum to contract, and the internal anal sphincter to relax.** In an infant or young child, relaxation of the internal anal sphincter results in release of fecal matter from the anal canal. In older children and adults, the defecation reflex ends here and there is conscious contraction of the external anal sphincter if it is not an appropriate time to defecate. The reflex is initiated again with the next mass movement, and steps (1) and (2) repeat. However, if the time is appropriate for defecation to occur, then we proceed to the next step:

(3) **Impulses from the cerebral cortex trigger relaxation of the external anal sphincter and contraction of the levator ani muscle, allowing feces to pass out of the body.** The cerebral cortex stimulates voluntary relaxation of the external anal sphincter. The levator ani muscle also contracts, which ele-

vates the anus. These three movements combined allow the feces to pass. Note that this activity is assisted by contraction of the abdominal muscles against a closed glottis, a procedure called the *Valsalva maneuver,* which increases intra-abdominal pressure.

Anything that increases or decreases the motility of the large intestine affects the amount of water present in feces. When motility increases, the large intestine does not have enough time to absorb water from fecal material. This produces watery feces, a condition known as *diarrhea* (dy-ah-REE-ah). Factors that may increase motility include irritation of the colon due to bacterial or viral infections or drugs that stimulate the parasympathetic nervous system. Conversely, when motility decreases, the large intestine absorbs too much water and the fecal material becomes hard, a condition called *constipation.* The large intestine's motility may be slowed by drugs such as opiate narcotics or those that block the effects of acetylcholine (ACh).

Figure 22.19 summarizes the structural and functional properties of the large intestine and other organs of the alimentary canal.

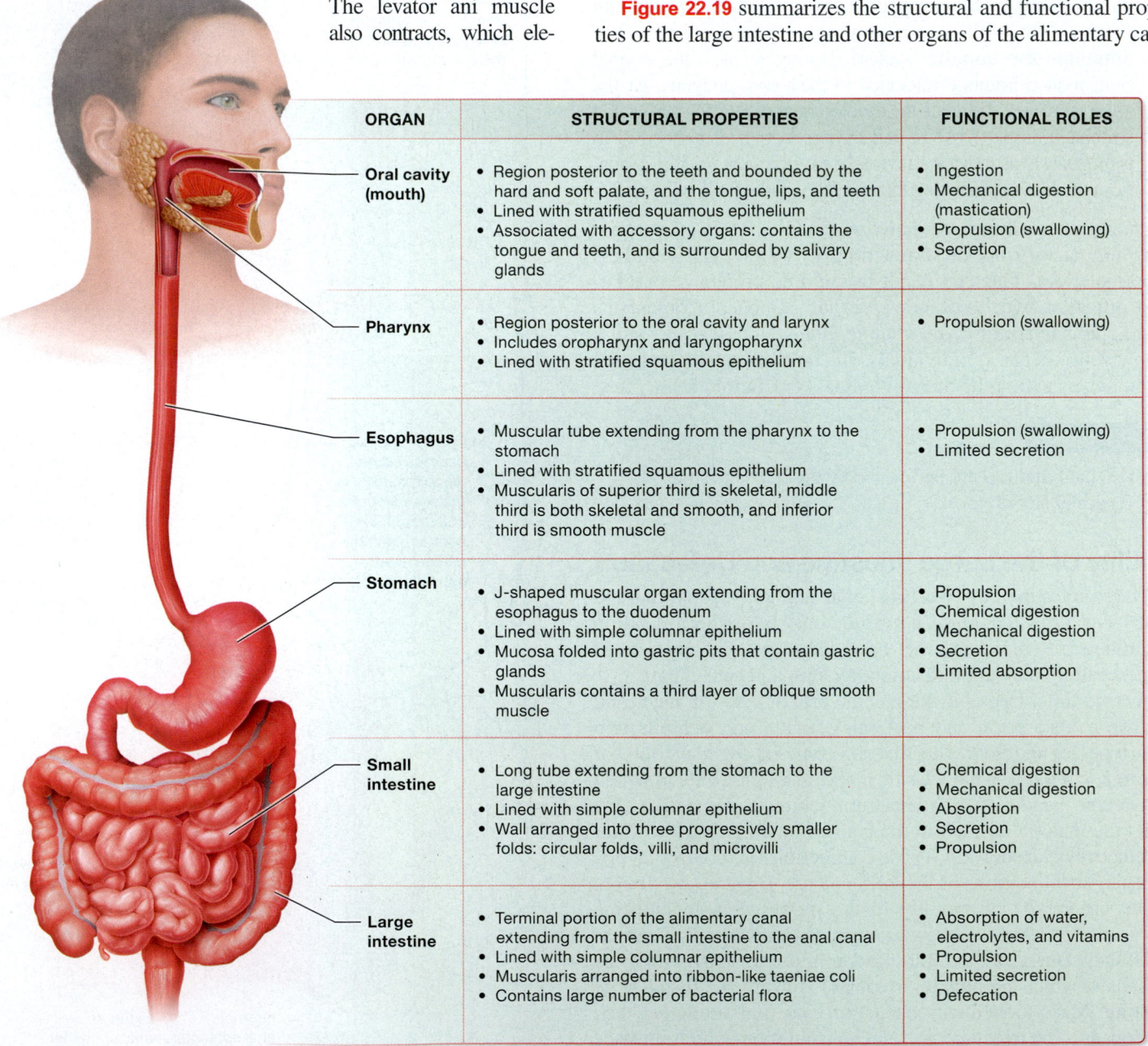

ORGAN	STRUCTURAL PROPERTIES	FUNCTIONAL ROLES
Oral cavity (mouth)	• Region posterior to the teeth and bounded by the hard and soft palate, and the tongue, lips, and teeth • Lined with stratified squamous epithelium • Associated with accessory organs: contains the tongue and teeth, and is surrounded by salivary glands	• Ingestion • Mechanical digestion (mastication) • Propulsion (swallowing) • Secretion
Pharynx	• Region posterior to the oral cavity and larynx • Includes oropharynx and laryngopharynx • Lined with stratified squamous epithelium	• Propulsion (swallowing)
Esophagus	• Muscular tube extending from the pharynx to the stomach • Lined with stratified squamous epithelium • Muscularis of superior third is skeletal, middle third is both skeletal and smooth, and inferior third is smooth muscle	• Propulsion (swallowing) • Limited secretion
Stomach	• J-shaped muscular organ extending from the esophagus to the duodenum • Lined with simple columnar epithelium • Mucosa folded into gastric pits that contain gastric glands • Muscularis contains a third layer of oblique smooth muscle	• Propulsion • Chemical digestion • Mechanical digestion • Secretion • Limited absorption
Small intestine	• Long tube extending from the stomach to the large intestine • Lined with simple columnar epithelium • Wall arranged into three progressively smaller folds: circular folds, villi, and microvilli	• Chemical digestion • Mechanical digestion • Absorption • Secretion • Propulsion
Large intestine	• Terminal portion of the alimentary canal extending from the small intestine to the anal canal • Lined with simple columnar epithelium • Muscularis arranged into ribbon-like taeniae coli • Contains large number of bacterial flora	• Absorption of water, electrolytes, and vitamins • Propulsion • Limited secretion • Defecation

Figure 22.19 Summary of the structure and function of the organs of the alimentary canal.

22.7 examines the secretory products of each organ and their roles.

The Pancreas

« FLASHBACK

1. What are the endocrine functions of the pancreas? (p. 615)

Recall from the endocrine system chapter that the **pancreas** is a gland with both endocrine and exocrine functions (see Chapter 16). Its endocrine secretions, the hormones *insulin* and *glucagon,* are released from pancreatic islets into the blood and affect most cells in the body. Its exocrine secretions are enzymes secreted primarily by clusters of acinar cells. These enzymes are released into ducts of the pancreas that empty into the alimentary canal, where they help with digestion. We examine the structure of the pancreas and the acinar cells, as well as their secretions, in the next subsections.

Anatomy and Histology of the Pancreas

The pancreas is mostly located in the left upper quadrant of the abdomen, where it extends from the duodenum to the spleen; most of it is retroperitoneal. It has three portions (**Figure 22.20a**): (1) a wide **head** that contacts the duodenum, (2) a middle **body,** and (3) a thinner **tail** that tapers off toward the spleen. Running down the middle of the pancreas is the **main pancreatic duct,** which receives secretions from acinar cells. Near the duodenum, the main pancreatic duct merges with the duct from the liver and gallbladder, after which it empties into the duodenum.

Acinar cells are modified simple cuboidal epithelial cells. They are found in clusters known as **acini** (ASS-uh-nye; singular, *acinus;* **Figure 22.20b**). Each acinus surrounds a small duct into which its cells secrete their products. The duct cells also secrete products, discussed shortly. Most of these small ducts merge and drain into the main pancreatic duct, although some secrete into the smaller *accessory pancreatic duct.*

Pancreatic Juice

The collective secretions of the pancreatic acinar and duct cells are known as **pancreatic juice.** It consists of water and multiple digestive enzymes and other proteins. In addition, the duct cells secrete bicarbonate ions, a base, which make pancreatic juice alkaline. This helps to neutralize the acidic chyme that enters the duodenum from the stomach and protects the duodenum from damage by the acid. The digestive enzymes, secreted by acinar cells, are crucial in chemical digestion and catalyze the reactions that digest carbohydrates, proteins, lipids, and nucleic acids. Module 22.7 discusses the individual enzymes.

Like gastric secretion, pancreatic secretion occurs at a basal rate between meals. During eating, pancreatic secretion rises due to parasympathetic and hormonal stimulation

---

Quick Check

☐ 5. Why is control of motility in the large intestine important to fluid balance?

☐ 6. Which types of movements occur in the proximal large intestine?

☐ 7. Summarize the steps of the defecation reflex.

Apply What You Learned

☐ 1. Patients taking oral antibiotics for a bacterial infection are often advised to take supplements containing bacteria present in the normal flora. Why are patients counseled to do this?

☐ 2. What might happen to step 3 of the defecation reflex (see Figure 22.18) if the spinal cord were cut? Explain.

See answers in Appendix A.

MODULE 22.6
The Pancreas, Liver, and Gallbladder

Learning Outcomes

1. Describe the gross and microscopic structure of the pancreas and its digestive functions.

2. Describe the gross and microscopic structures of the liver and gallbladder.

3. Describe the functions of the liver pertaining to digestion.

4. Explain the structural and functional relationship between the liver and the gallbladder.

5. Explain how pancreatic and biliary secretions are regulated.

We turn now to the structure and function of the final three organs of the digestive system: the pancreas, liver, and gallbladder. Like the salivary glands, these accessory organs are *exocrine glands* that secrete a product through a duct to the outside of the body (see Chapter 4). You might find this strange, because the accessory organs secrete their products into the alimentary canal, and the alimentary canal seems to be inside the body. However, the lining of the alimentary canal is actually considered an external body surface because it is open to the outside on both ends. Think of it like a paper towel roll—its outer surface is akin to the body's skin, and its inner surface is akin to the alimentary canal. Both its surfaces are in contact with the outside environment and thus are considered "external."

This module explores the structures and functions of the pancreas, liver, and gallbladder. The neural and hormonal signals that regulate their secretion are also discussed. Module

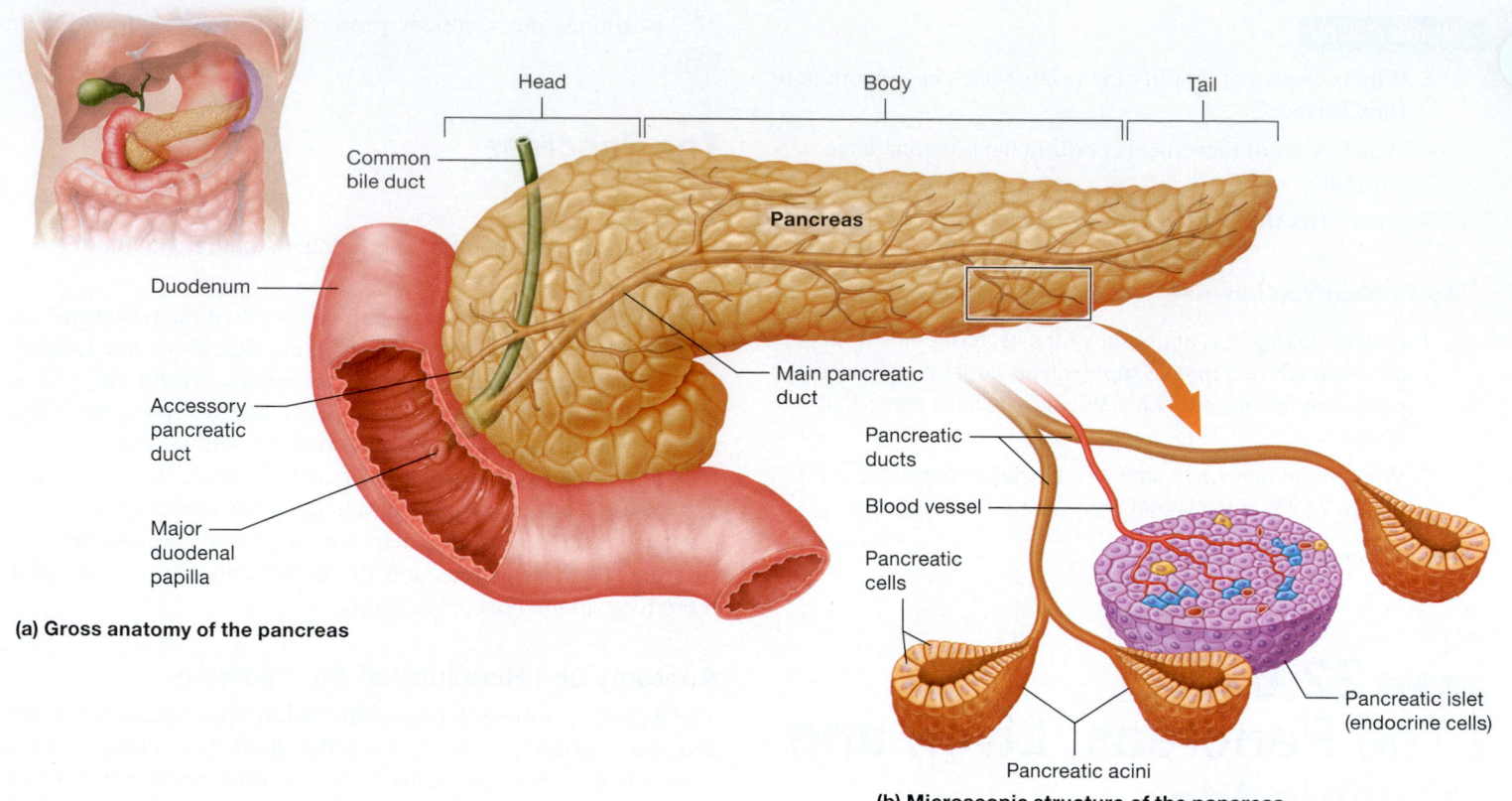

Head   Body   Tail

Common bile duct

Pancreas

Duodenum

Accessory pancreatic duct

Major duodenal papilla

Main pancreatic duct

(a) Gross anatomy of the pancreas

Pancreatic ducts

Blood vessel

Pancreatic cells

Pancreatic islet (endocrine cells)

Pancreatic acini

(b) Microscopic structure of the pancreas

Figure 22.20 Anatomy and histology of the pancreas.

(**Figure 22.21**). One hormonal mediator of pancreatic secretion is **cholecystokinin** (kohl´-eh-sih-stoh-KY-nin), or **CCK.** CCK is produced by duodenal DNES cells in response to the presence of lipids and partially digested proteins in the duodenum. It acts on acinar cells to trigger the secretion of digestive enzymes and other proteins. Another stimulatory hormone is *secretin,* which is released by duodenal cells in response to acid and lipids in the duodenum. Secretin primarily triggers duct cells to secrete bicarbonate ions.

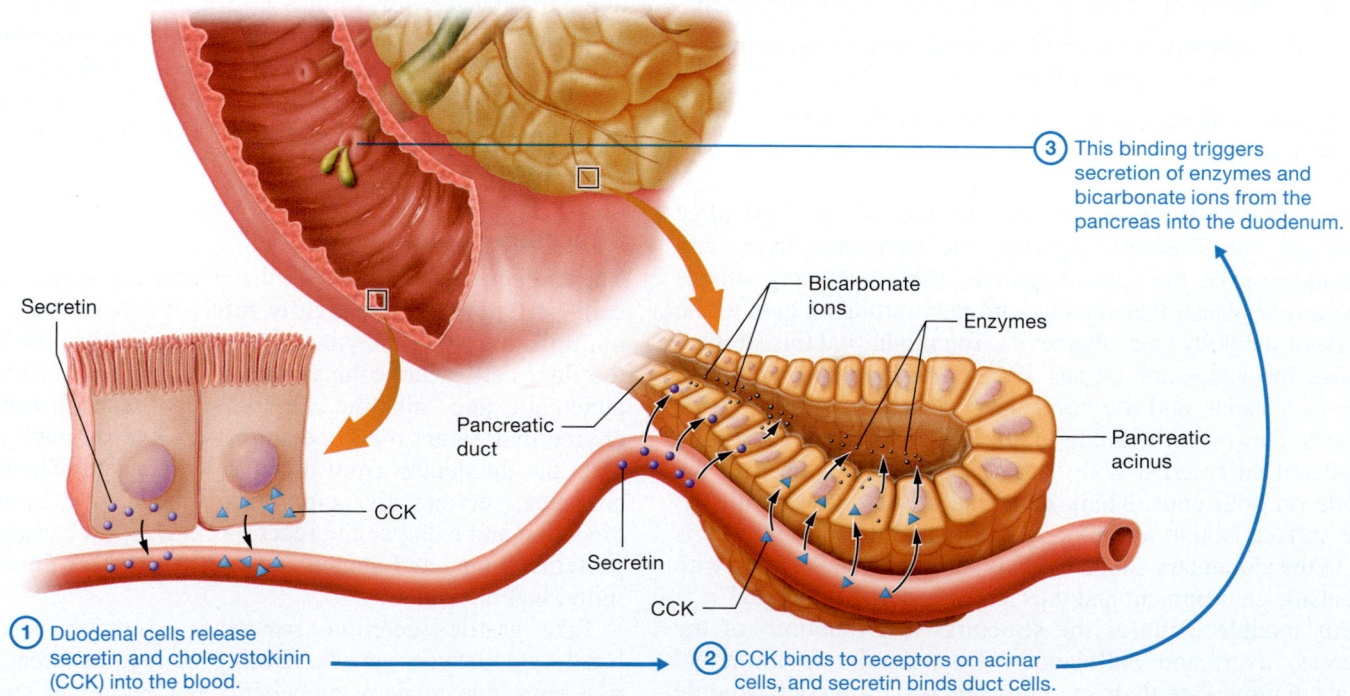

③ This binding triggers secretion of enzymes and bicarbonate ions from the pancreas into the duodenum.

Secretin

Bicarbonate ions

Enzymes

Pancreatic duct

Pancreatic acinus

CCK

Secretin

CCK

① Duodenal cells release secretin and cholecystokinin (CCK) into the blood.

② CCK binds to receptors on acinar cells, and secretin binds duct cells.

Figure 22.21 Secretion of pancreatic juice.

☐ 1. Which cells secrete pancreatic juice, and what are its components?

☐ 2. How do CCK and secretin affect pancreatic secretion?

The Liver and Gallbladder

FLASHBACK

1. What are sinusoids? (p. 689)
2. Which veins drain into the hepatic portal vein? (p. 710)
3. What is the hepatic portal system? (p. 709)

The pyramid-shaped **liver,** one of the largest organs in the body, is located in the right upper quadrant nestled against the inferior surface of the diaphragm muscle. On the liver's posterior side, we find the small sac known as the **gallbladder.** The liver and gallbladder have close anatomical and functional relationships, so we discuss them together. In the upcoming subsections, we look at the anatomy and histology of the liver and the gallbladder, their basic functions, and their main digestive secretion, *bile.*

Anatomy of the Liver

The liver is wrapped in a thin connective tissue capsule, and most of it is covered by the visceral peritoneum. It is composed of four lobes: the large **right lobe** and **left lobe** (**Figure 22.22a**)—located

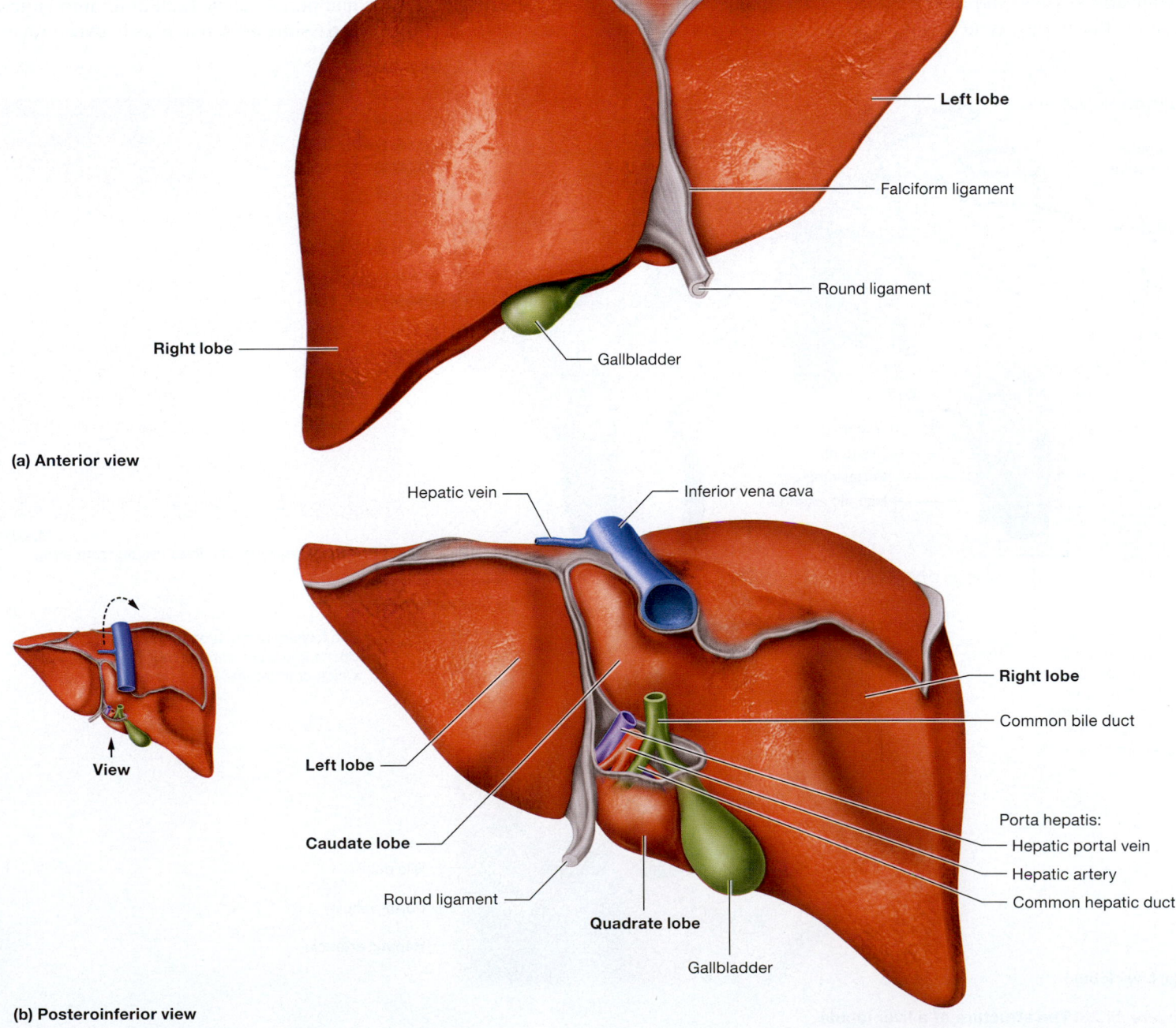

(a) Anterior view

(b) Posteroinferior view

Figure 22.22 Gross anatomy of the liver.

on the right and left sides of the liver, respectively—and the smaller **caudate lobe** (KAW-dayt) and **quadrate lobe,** located on the posterior side of the liver's right lobe (**Figure 22.22b**). The right and left lobes are separated by a fold of visceral peritoneum called the **falciform ligament** (FALL-suh-form), which also anchors the liver to the anterior abdominal wall. On its inferior surface we find the **round ligament,** a remnant of the *umbilical vein* that was present in the fetus.

On the liver's posterior side we find an indentation called the **porta hepatis** (POHR-tuh heh-PAEH-tis; *port-* = "gate," *hepat-* = "liver"). Numerous blood vessels enter and exit the liver at the porta hepatis, including the *hepatic artery,* which brings oxygen-rich blood to the liver; and the *hepatic portal vein,* which brings nutrient-rich, deoxygenated blood to the liver from multiple abdominal organs. In addition to blood vessels, we find nerves, lymphatic vessels, and the *common hepatic duct* entering and exiting the porta hepatis. Note that the hepatic veins, which

drain blood from the sinusoids, empty into the inferior vena cava on the liver's superior surface.

Histology of the Liver

The basic unit of the liver is the **liver lobule** (LAWB-yool; **Figure 22.23**). Liver lobules are separated from one another by septa that branch in from the connective tissue capsule of the liver. A liver lobule is composed of flattened plates of cells, called **hepatocytes** (heh-PAT-oh-sytz), arranged in the shape of a hexagon and stacked on each other. In the center of the lobule is a small **central vein.** At each of the six corners of the lobule, we find three structures collectively referred to as a **portal triad:** (1) a branch of the hepatic artery called a *hepatic arteriole,* (2) a branch of the portal vein called a *portal venule,* and (3) a small **bile duct** that carries bile.

The hepatic arteriole and portal venule both drain into large, leaky capillaries, the **hepatic sinusoids,** that pass between rows

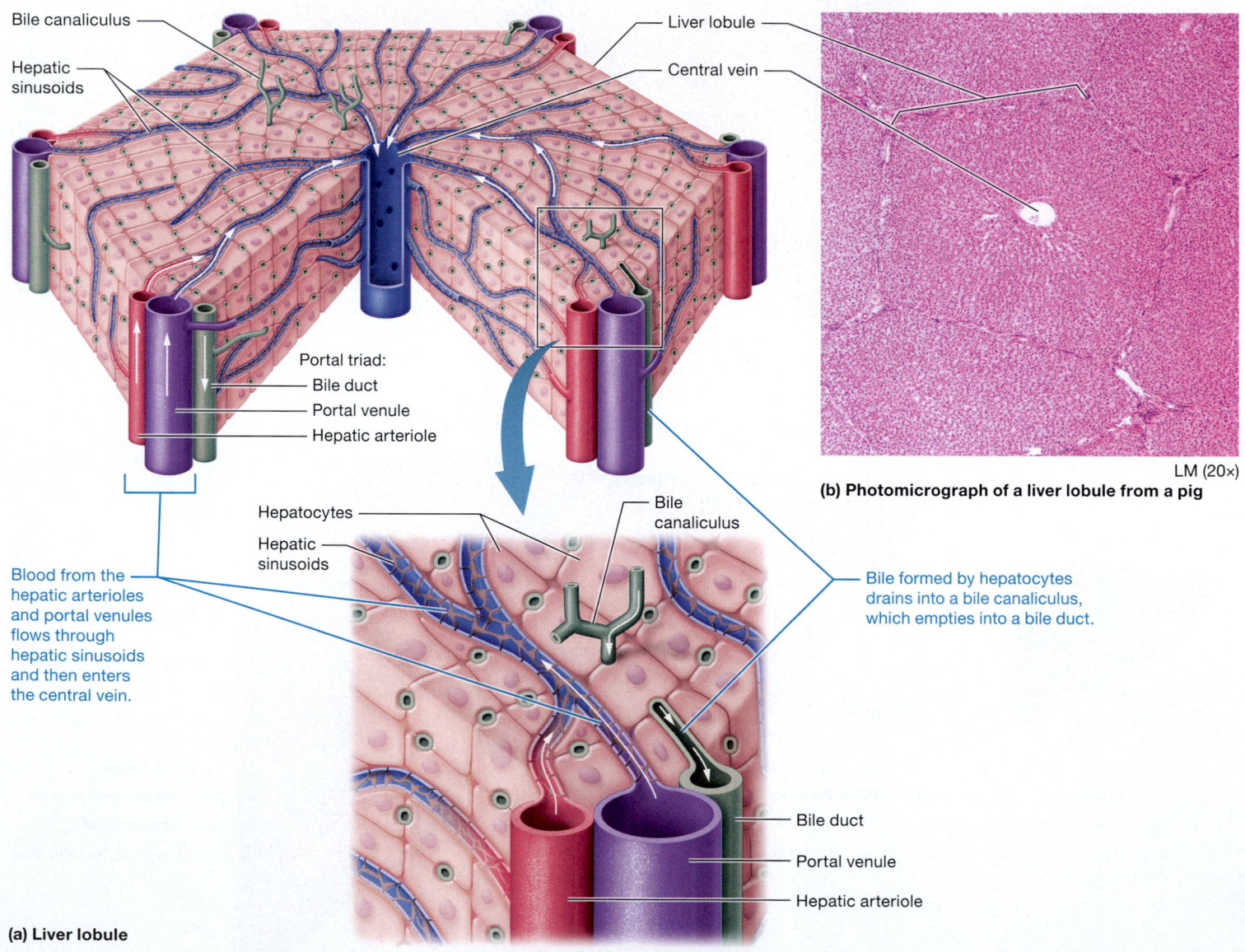

Bile canaliculus

Hepatic sinusoids

Liver lobule

Central vein

Portal triad:
- Bile duct
- Portal venule
- Hepatic arteriole

Blood from the hepatic arterioles and portal venules flows through hepatic sinusoids and then enters the central vein.

Hepatocytes

Hepatic sinusoids

Bile canaliculus

Bile formed by hepatocytes drains into a bile canaliculus, which empties into a bile duct.

Bile duct

Portal venule

Hepatic arteriole

LM (20×)

(b) Photomicrograph of a liver lobule from a pig

(a) Liver lobule

Figure 22.23 The structure of a liver lobule.

of hepatocytes. Recall from the blood vessels chapter that sinusoids have walls that are not continuous, which allows large substances to enter and exit (see Chapter 18). Blood flows slowly through the sinusoids as materials are exchanged between the blood and hepatocytes, eventually draining into the central vein. The central veins merge and drain into the hepatic veins, which in turn feed into the inferior vena cava. Notice that bile flows through the liver lobule in the opposite direction, from the hepatocytes into tiny ducts called *bile canaliculi* (kah′-nuh-LIK-yoo-lye; singular, *canaliculus*), which eventually drain into a bile duct.

Functions of the Liver

The liver has one of the most diverse sets of functions of any organ in the body. Like the pancreas, it releases both endocrine and exocrine secretions, and it converts harmful chemicals into nontoxic substances that the body can eliminate. We take a closer look at the functions of the liver in the next two subsections.

Bile Production   The liver's main digestive function is to produce **bile,** a liquid that contains multiple components, including water, electrolytes, and organic compounds. Bile serves two critical functions: (1) It is required for the digestion and absorption of lipids; and (2) it is the mechanism by which the liver excretes wastes and other substances that the kidneys cannot excrete. One of its main organic compounds is **bile salts,** which are derived from cholesterol. Bile salts are *amphiphilic,* meaning they have both polar and nonpolar parts. This allows them to interact with both lipids and the watery environment of the small intestine. When bile is released into the duodenum, the bile salts coat the lipids and physically break them apart into smaller pieces, a process known as *emulsification* (ee-mul′-sih-fih-KAY-shun). Although emulsification is mechanical digestion, it is necessary for the chemical digestion and absorption of lipids, which we further explore in Module 22.7.

Bile also contains varying amounts of materials that the liver excretes, including cholesterol, waste products, and harmful chemicals (toxins) such as heavy metals. Many of these products are not reabsorbed by the small or large intestine and so pass into the feces. An important example of the liver's excretory function involves *bilirubin* (BIL-ih-roo-bin), a waste product that results from the breakdown of hemoglobin by the spleen. Hepatocytes secrete bilirubin into bile, and the normal flora in the large intestine convert it to *urobilinogen* (yoor′-oh-bih-LIN-oh-jen). Bacteria convert most of the urobilinogen to *stercobilin,* which is brown and is responsible for the brown color of feces. A small portion of the urobilinogen is reabsorbed by the large intestine and ends up back in the hepatic portal system. The majority of this urobilinogen is simply re-secreted into the bile, although a small portion remains in the blood and is excreted by the kidneys (note that yellow urobilinogen is largely responsible for the color of urine).

Other Functions of the Liver   The liver performs a host of other functions, including the following (note that some of these functions are only peripherally related to the digestive system, so they are not discussed in detail here):

- **Nutrient metabolism.** The liver processes nutrients obtained from the diet. Carbohydrates and proteins absorbed from the alimentary canal are delivered to the liver by the hepatic portal vein, and lipids by the hepatic arteries. Within the liver, some of these nutrients are stored for later use, modified into another form, or used to synthesize other compounds. For example, some of the glucose taken in by the liver is stored in the form of glycogen, and many of the dietary amino acids are used to synthesize plasma proteins such as albumin and clotting proteins. We discuss the liver's role in multiple metabolic processes in the metabolism chapter (see Chapter 23).
- **Detoxification.** The liver detoxifies substances produced by the body, such as the previously mentioned bilirubin. In addition, the liver processes substances that we eat or drink, some of which are toxins harmful to the body (such as alcohol). These substances are generally delivered to the liver first via the hepatic portal vein, where the liver converts them into less harmful materials that can be excreted in bile or in urine. The liver also metabolizes many drugs, such as antibiotics. Individuals with impaired liver function metabolize drugs more slowly than those with normal liver function, and the drugs remain in their systems much longer. Does the liver ever need help with detoxification? For more information, see *A&P in the Real World, Pseudoscience Exposed: Do We Really Need to "Detox"?*
- **Excretion.** Recall that the liver directly excretes bilirubin in bile. Several other substances the liver processes are excreted in bile, particularly drugs such as certain antibiotics. The liver also modifies substances so that they can be excreted by the kidneys.

Pseudoscience Exposed: Do We Really Need to "Detox"?

The manufacturers of dietary supplements and alternative health products tell us that our bodies are toxic wastelands filled with disease-causing toxins that only their products can remove (for a price, of course). These "detoxification" products are big business, accounting for a large portion of the profits for the multibillion dollar alternative health industry. Many people ask if these products work, but a better question is whether we actually need them in the first place. So, are our bodies swimming with toxins? Do we all need to "detox"?

The simple answer is no, and we have our livers to thank for it. The liver is a magnificent organ that works tirelessly to convert harmful chemicals, or toxins, into nonharmful substances that can be excreted in bile or urine. There is simply no herbal supplement, foot pad, or magic bracelet that can even come close to doing the job our livers do every day. And what's more, the liver doesn't charge us for its services.

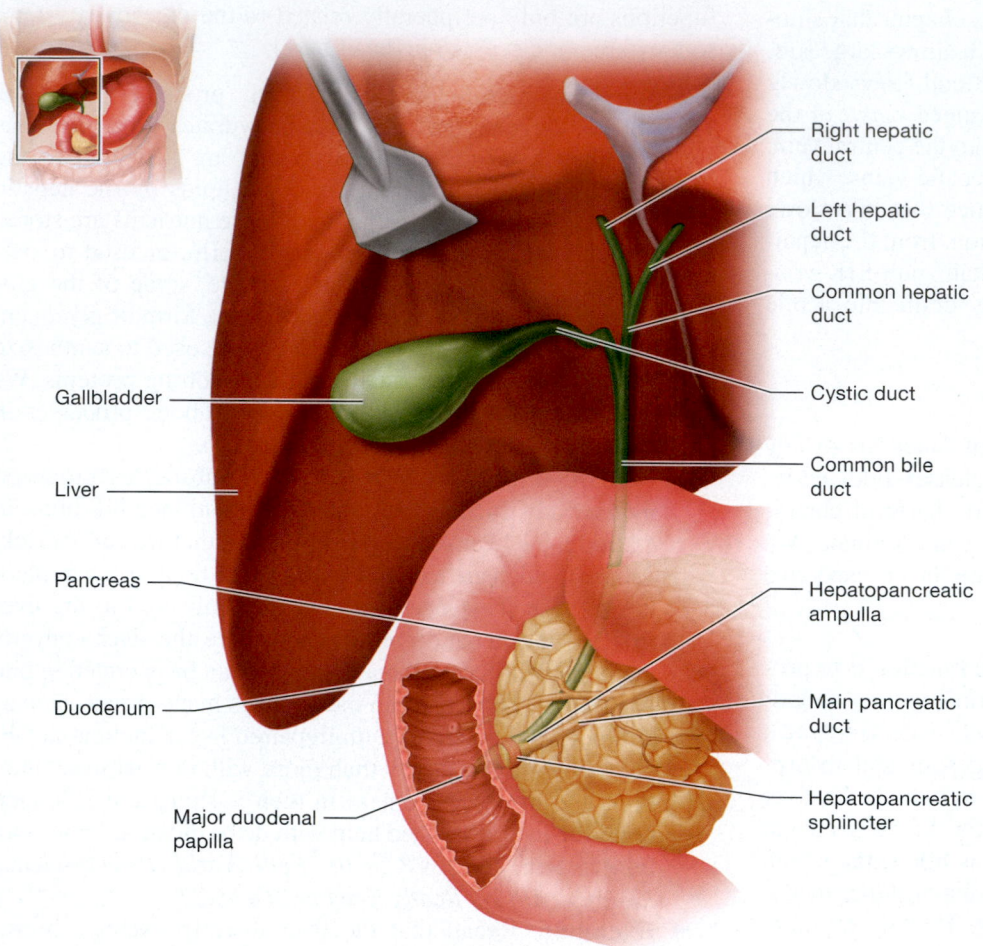

Right hepatic duct

Left hepatic duct

Common hepatic duct

Gallbladder

Cystic duct

Liver

Common bile duct

Pancreas

Hepatopancreatic ampulla

Duodenum

Main pancreatic duct

Major duodenal papilla

Hepatopancreatic sphincter

Figure 22.24 Structure of the gallbladder and its ducts.

The Gallbladder and Its Relationship to the Liver

The gallbladder, a small sac that sits on the posterior liver, receives most of the bile from the common hepatic duct (**Figure 22.24**). The gallbladder stores bile, concentrates it (removing water), and releases it when stimulated. Bile release is stimulated by the hormone CCK, which triggers contraction of the smooth muscle in the wall of the gallbladder. This causes the gallbladder to release bile into the **cystic duct** (SIS-tik).

As you can see in Figure 22.24, the cystic duct joins the common hepatic duct to form the **common bile duct.** The common bile duct joins the main pancreatic duct near the duodenum to form the **hepatopancreatic ampulla** (heh-PAT-oh-payn′-kree-at-ik am-POOL-uh). The ampulla is surrounded by a ring of smooth muscle called the **hepatopancreatic sphincter,** which controls the emptying of bile and pancreatic fluids into the duodenum. Recall that the contents of the hepatopancreatic ampulla then empty into the duodenum at the major duodenal papilla.

Bile contains calcium salts and cholesterol, and both can precipitate and form hard lumps called **gallstones** (**Figure 22.25**). Gallstones are generally asymptomatic, but occasionally they become lodged in the cystic duct or the common bile duct and block the outflow of bile. This results in abdominal pain,

particularly when eating a fatty meal, and the presence of undigested fats in the feces. The feces also take on a clay color due to the lack of bilirubin excretion. Painful gallstones are generally treated by surgically removing the gallbladder, a procedure known as a *cholecystectomy* (koh′-leh-sis-TEK-toh-mee). The absence of a gallbladder may temporarily reduce the ability of the small intestine to digest fats. However, over time the common hepatic duct enlarges and takes over some of the gallbladder's functions.

Bile Secretion

The liver produces bile continually but generally does not secrete it at a basal rate, the way pancreatic and gastric juices are released. Instead, bile secretion does not occur until the gallbladder contracts and the sphincter at the hepatopancreatic ampulla relaxes. This activity is mostly accomplished by cholecystokinin (CCK) and to a small extent by the vagus nerve (**Figure 22.26**). Other factors influence bile secretion, including the hormone secretin, which stimulates bile production and release by hepatocytes.

However, the most potent stimulus for bile production and release is bile itself, specifically bile salts. Recall from our earlier discussion of the liver that bile salts are reabsorbed in the last section of the ileum and transported back to the liver through the hepatic portal vein. As bile salts re-enter the liver, bile secretion rises dramatically in a positive feedback loop. Bile secretion continues into the duodenum until the duodenum empties, at which point CCK and secretin levels decline.

Figure 22.27 summarizes the structural and functional properties of the accessory digestive organs.

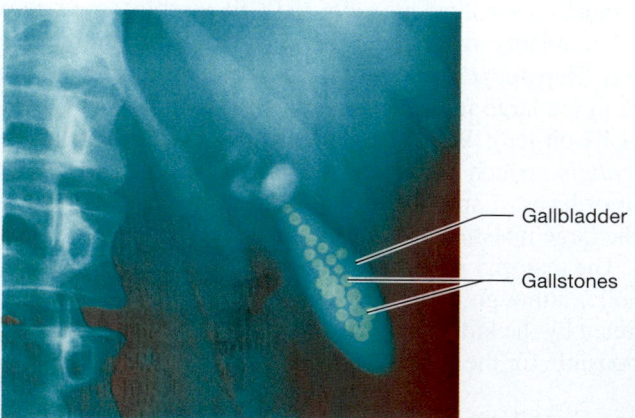

Gallbladder

Gallstones

Figure 22.25 Radiograph of a gallbladder showing gallstones.

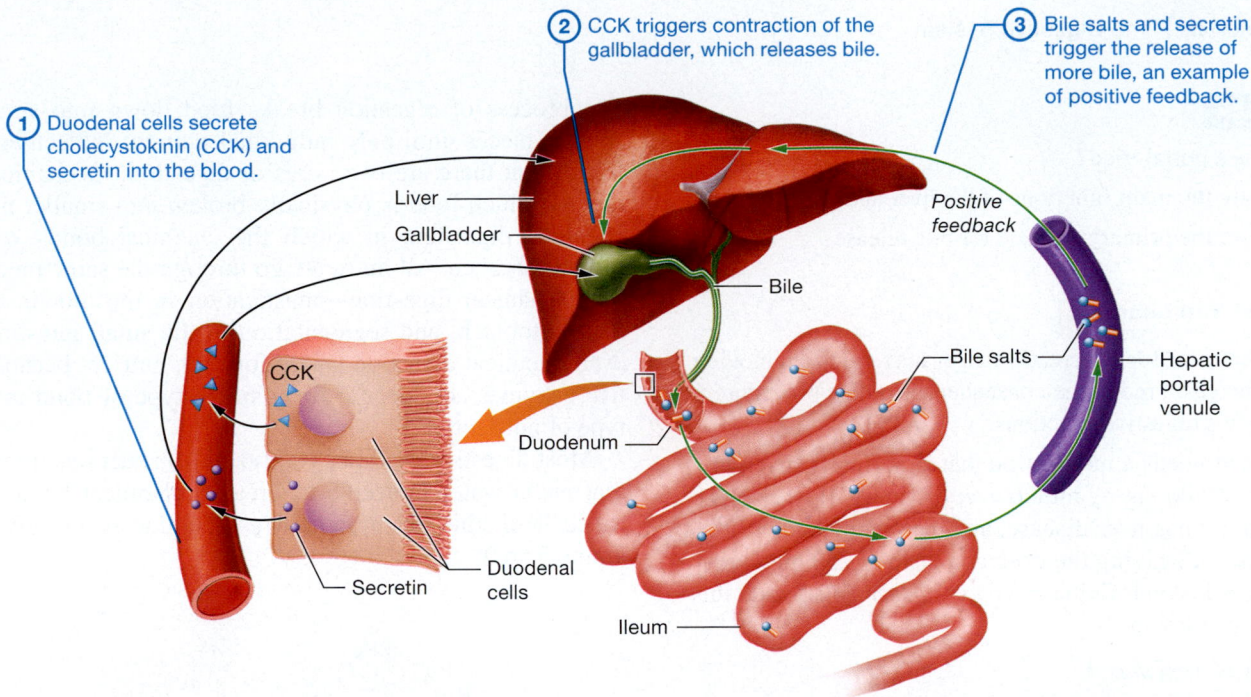

Figure 22.26 Secretion of bile.

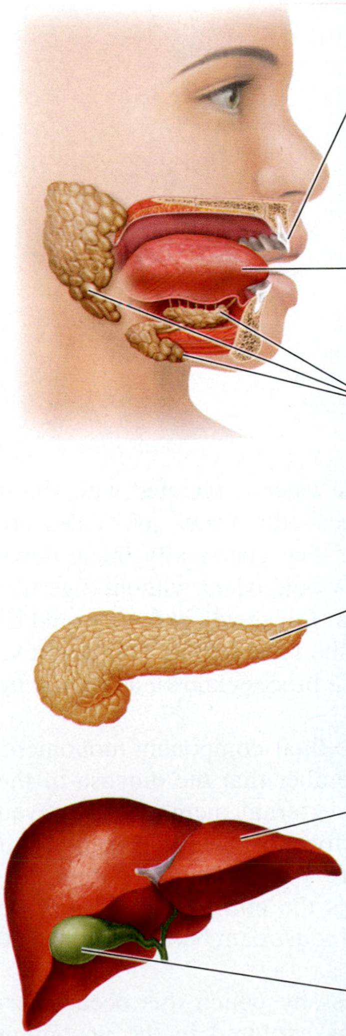

ORGAN	STRUCTURAL PROPERTIES	FUNCTIONAL ROLES
Teeth	• Two sets: primary and secondary dentition • Consist of a crown above the gum and a root embedded in bone • Composed of inner pulp cavity surrounded by dentin, which in turn is surrounded by enamel or cementum	• Mechanical digestion (mastication)
Tongue	• Consists of skeletal muscle with overlaying stratified squamous epithelium • Surface contains papillae	• Mechanical digestion • Propulsion (swallowing) • Sense of taste
Salivary glands	• Three sets: parotid glands, submandibular glands, and sublingual glands • Consist of mucous cells and serous cells	• Secrete saliva, which assists in chemical digestion, deters the growth of harmful microorganisms, and moistens food to assist in swallowing and mechanical digestion • Chemical digestion of carbohydrates
Pancreas	• Consists of pancreatic acini, composed of acinar cells surrounding a duct	• Secretes enzymes that catalyze chemical digestion of lipids, carbohydrates, proteins, and nucleic acids • Secretes bicarbonate ions to neutralize acidic chyme
Liver	• Consists of hexagonal liver lobules surrounding a central vein • Liver lobules contain plates of hepatocytes	• Mechanical digestion (via bile production) • Excretion (excretes wastes in bile)
Gallbladder	• Muscular sac on the posteroinferior liver	• Mechanical digestion (stores, concentrates, and releases bile)

Figure 22.27 Summary of the structure and function of the accessory digestive organs.

Quick Check

☐ 3. What is a portal triad?

☐ 4. What are the main functions of the liver and gallbladder?

☐ 5. What are the primary triggers for bile release?

Apply What You Learned

☐ 1. The treatment for pancreatic cancer may require removing the pancreas. Predict the consequences of this procedure for the body's digestive functions.

☐ 2. A patient needs a medication that is normally given at a dose of 500 mg by mouth every 8 hours. However, the patient has liver disease, and the pharmacist recommends giving the drug at a dose of 250 mg every 12 hours instead. Explain why the pharmacist made this recommendation.

See answers in Appendix A.

MODULE **22.7**
Nutrient Digestion and Absorption

Learning Outcomes

1. Explain the process of enzymatic hydrolysis reactions of nutrients.

2. Describe the enzymes involved in chemical digestion, including their activation, substrates, and end products.

3. Describe the process of emulsification, explain its importance, and discuss how bile salts are recycled.

4. Explain the processes involved in absorption of each type of nutrient, fat-soluble and water-soluble vitamins, and vitamin B12.

We covered the anatomy and basic functions of the digestive organs, and so we can now look more closely at two of the key digestive processes: digestion and absorption. This module explores the physical and chemical breakdown and absorption of dietary carbohydrates, proteins, lipids, and nucleic acids. The discussion then shifts to the absorption of water, electrolytes, and vitamins, which occurs in both the small and large intestines. But first, let's introduce some basic principles of digestion and absorption.

Overview of Digestion and Absorption

 **FLASHBACK**

1. What is a hydrolysis reaction? (p. 51)

The process of digestion breaks food down into smaller and smaller pieces until only individual nutrient molecules remain. Recall that there are two types of digestion: mechanical digestion, in which food is physically broken into smaller parts, and chemical digestion, in which the chemical bonds within the food are broken. All nutrients go through the same mechanisms of mechanical digestion—mastication in the mouth, churning in the stomach, and segmentation in the small intestine. However, chemical digestion varies for each nutrient because digestive enzymes are specific for a single type of bond in a single type of nutrient.

Most digestive enzymes catalyze *hydrolysis* reactions—those that use a water molecule to break a chemical bond, as illustrated here (this example shows a peptide bond between two amino acids):

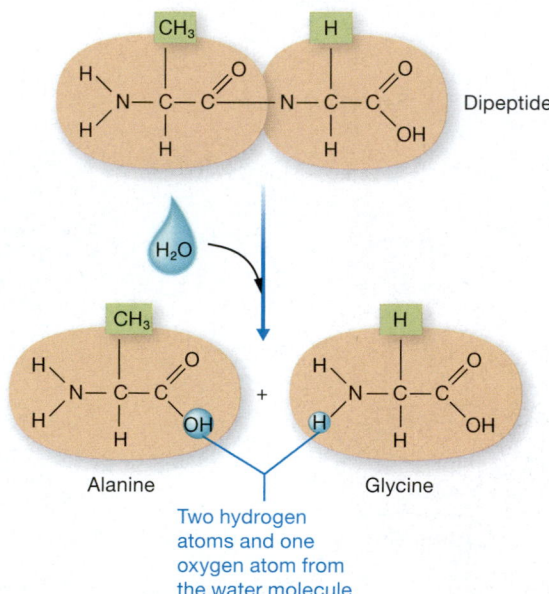

This is one reason why so much water is secreted with fluids like gastric and pancreatic juices—the water molecules are key components of the reactions that chemically break down food. These hydrolysis reactions would occur without digestive enzymes, but they would occur far too slowly and we would be unable to extract nutrients from the food we eat. Our digestive enzymes speed up the reactions, a process known as *enzymatic hydrolysis.*

Once nutrients are digested to their component monomers, they must enter the body. Remember that the mucosa of the alimentary canal is actually an external membrane, so even though it might seem that the food is "inside" the body, it technically is not until it is absorbed. Nutrients and other substances must be absorbed across the epithelial lining of the alimentary canal and enter the bloodstream before they can be delivered to body cells.

As with digestion, the process by which this occurs varies slightly for each nutrient, as explored in the upcoming sections.

Understanding Absorption in the Alimentary Canal

To travel from the lumen of the alimentary canal to the blood, a substance must cross barriers. In Figure 22.28, you can see the path a typical substance takes through an enterocyte of the small intestine. Note that sometimes small molecules such as water can cross between enterocytes instead of through them, but most substances we discuss take the route in Figure 22.28.

The apical side of the plasma membrane—see step ① in Figure 22.28—is the most significant barrier to the absorption of digested nutrients. Most substances require a transport protein or channel to cross the membrane, and if no such protein exists, the substance will remain in the lumen and eventually be excreted in the feces. This can be beneficial and help to prevent absorption of harmful chemicals. One example is elemental mercury (the liquid mercury found in certain thermometers). In this form, mercury has great difficulty crossing a plasma membrane by diffusion, and no transporter or channel protein exists for it in the plasma membrane. So, if elemental mercury is accidentally ingested, most of it passes straight through the alimentary canal, unabsorbed. ■

Fluid in lumen of small intestine

Apical side of plasma membrane

Channel or transport protein

① A molecule crosses the apical side of the enterocyte plasma membrane via a channel or transport protein.

Cytosol of enterocyte

② The molecule travels through the cytosol.

Basal side of plasma membrane

③ The molecule exits from the basal side of the plasma membrane and enters the interstitial fluid.

Interstitial fluid in core of villus

④ The molecule passes through or between endothelial cells of an intestinal capillary and enters the blood.

Blood in capillary

Figure 22.28 Path taken by a substance being absorbed into the blood through an enterocyte.

Digestion and Absorption of Carbohydrates

« FLASHBACK

1. How do polysaccharides, disaccharides, and monosaccharides differ? (p. 62)
2. How does secondary active transport allow substances to cross the plasma membrane? (p. 81)
3. What is facilitated diffusion? (p. 76)

Carbohydrate digestion begins in the mouth with the help of salivary amylase from the salivary glands, which catalyzes the reactions that break long polysaccharides into shorter *oligosaccharides.* Salivary amylase generally accomplishes little actual chemical digestion because the food simply is not in the mouth long enough for it to have much effect. Some minor digestion by salivary amylase continues in the stomach, but stomach acid generally inactivates this enzyme before it can complete the task.

Chemical digestion of carbohydrates resumes in the small intestine (Figure 22.29). Here polysaccharides and oligosaccharides encounter a pancreatic enzyme called **pancreatic amylase.** This enzyme is very similar in structure and function to salivary amylase, and ① catalyzes the reactions that break the remaining polysaccharides into oligosaccharides.

Oligosaccharide digestion is completed by reactions catalyzed by enzymes that are products of enterocytes on the brush border of the small intestine (*brush border enzymes*), including the enzymes **lactase, maltase,** and **sucrase.** Lactase catalyzes only one reaction: the digestion of the sugar lactose (found in milk and milk-based products) into glucose and galactose (see *A&P in the Real World: Lactose Intolerance*). ② Both maltase and sucrase

Lactose Intolerance

Many adults lack the enzyme lactase and as a result cannot digest the milk sugar lactose, a condition known as *lactose intolerance.* Most people produce lactase as infants, but production of the enzyme declines as we age, with up to 75% of adults showing decreased lactase activity. The percentage of lactose intolerance varies with ethnic heritage—it is found in only about 5% of the population of northern Europe, but about 90% of the population in certain regions of Asia.

Without lactase, lactose cannot be digested, and the undigested lactose remains in the small intestine and is eventually delivered to the large intestine. Lactose in the large intestine acts as an osmotic particle that can draw water into the feces. This can lead to abdominal cramping and diarrhea when lactose is consumed. Lactose intolerance can be managed by avoiding lactose-containing foods or by taking over-the-counter medications that contain supplemental lactase.

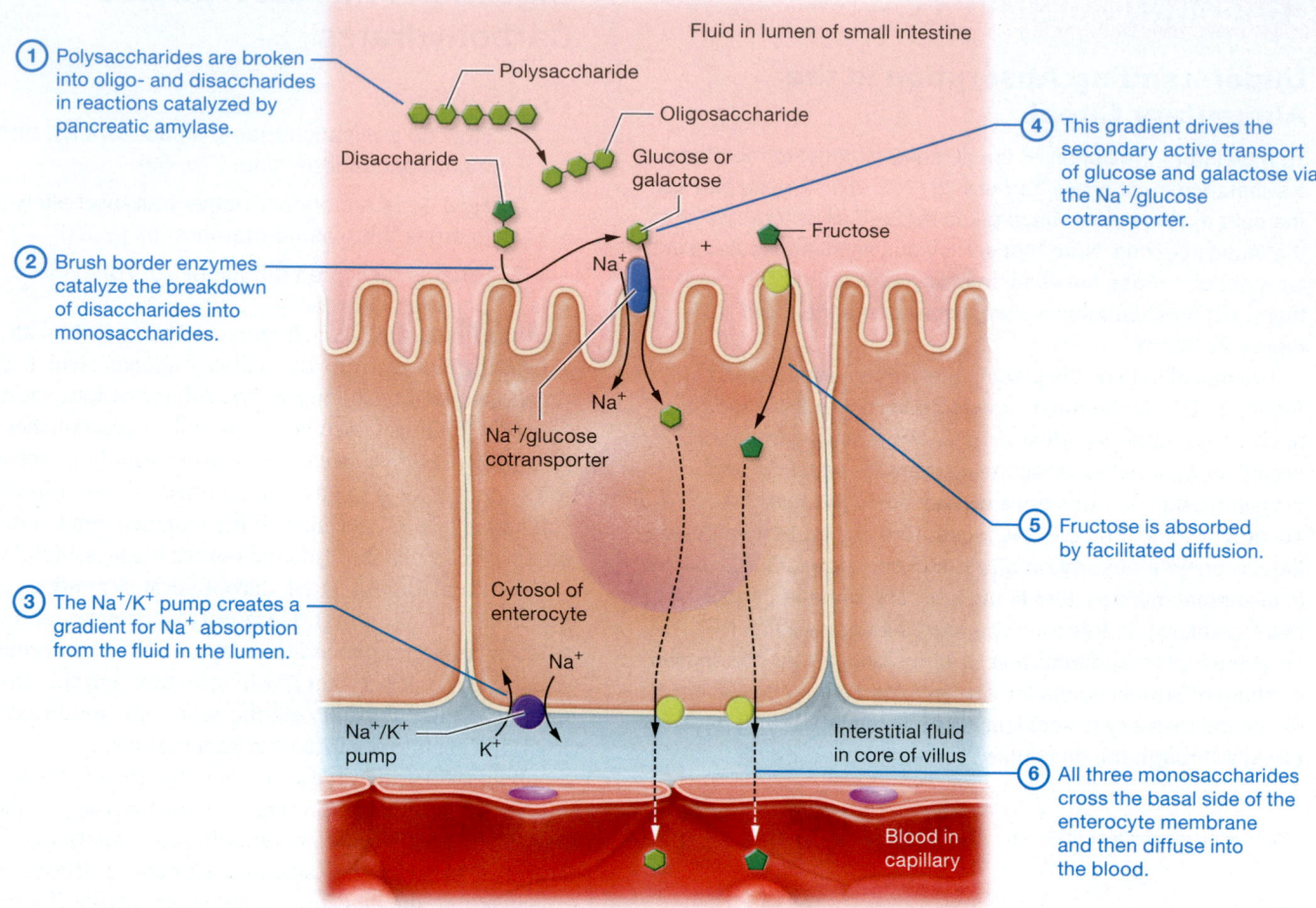

1 Polysaccharides are broken into oligo- and disaccharides in reactions catalyzed by pancreatic amylase.

Fluid in lumen of small intestine

Polysaccharide

Oligosaccharide

Disaccharide

Glucose or galactose

4 This gradient drives the secondary active transport of glucose and galactose via the Na⁺/glucose cotransporter.

2 Brush border enzymes catalyze the breakdown of disaccharides into monosaccharides.

Fructose

Na^+

Na^+

Na⁺/glucose cotransporter

5 Fructose is absorbed by facilitated diffusion.

Cytosol of enterocyte

3 The Na⁺/K⁺ pump creates a gradient for Na⁺ absorption from the fluid in the lumen.

Na^+

Na⁺/K⁺ pump

K^+

Interstitial fluid in core of villus

6 All three monosaccharides cross the basal side of the enterocyte membrane and then diffuse into the blood.

Blood in capillary

Figure 22.29 Carbohydrate digestion and absorption in the small intestine.

catalyze reactions that break oligosaccharides into disaccharides and monosaccharides. The end result of all this activity is the production of the monosaccharides glucose, fructose, and galactose.

After carbohydrates have been digested into monosaccharides, they are ready to be absorbed. Notice in Figure 22.29 that both glucose and galactose are transported across the enterocyte's apical membrane by the same mechanism. Note also that the concentrations of glucose and galactose are higher in the enterocyte's cytosol than they are in the lumen of the small intestine. This means that absorption of glucose and galactose requires these molecules to move *against* their concentration gradients.

In the cell chapter, we discussed the fact that anytime we move a substance against its concentration gradient, energy must be expended (see Chapter 3). The energy in this case comes via a secondary active transport mechanism known as the **Na⁺/glucose cotransporter.** Recall that secondary active transport is a form of membrane transport in which ATP is used indirectly to move a chemical across the membrane. In this case, the Na⁺/K⁺ pump consumes ATP to drive sodium ions out of the enterocyte and into the lumen of the small intestine. This creates a concentration gradient that favors the movement of sodium ions back into the enterocyte, another example of the

Gradients Core Principle (p. 28) and of the importance of gradients in anatomy and physiology.

CORE PRINCIPLE
Gradients

Remember that a concentration gradient is actually a form of potential energy. The cell harnesses this energy by 4 allowing sodium ions to diffuse back into the enterocyte, and this energy drives the transport of glucose and galactose into the enterocyte.

Fructose is unable to bind the Na⁺/glucose cotransporter, so it crosses the apical enterocyte membrane by a separate mechanism (see Figure 22.29). 5 Fructose binds a channel that mediates its facilitated diffusion across the membrane. Facilitated diffusion is a passive process and requires no net input of energy, and so a concentration gradient must be present or fructose will not be absorbed.

All three monosaccharides 6 cross the basal enterocyte membrane by the same facilitated diffusion mechanism. This process involves a membrane protein that is very similar to the one that helps fructose cross the apical enterocyte membrane. After the monosaccharides cross the basal membrane, they diffuse through the extracellular fluid and into the capillaries in the villus. Once in the blood, they are delivered to the liver via the hepatic portal vein for processing.

☐ 1. Where does carbohydrate digestion take place?

☐ 2. Which enzymes are involved in digesting carbohydrates?

☐ 3. How do glucose, galactose, and fructose cross the apical enterocyte membrane?

Digestion and Absorption of Proteins

« FLASHBACK

1. What is endocytosis? (p. 83)

2. What are Peyer's patches? (p. 758)

Chemical digestion of proteins does not begin until they reach the stomach, where they encounter the enzyme pepsin. Recall that the chief cells of the gastric glands produce the inactive precursor pepsinogen. Pepsinogen requires a pH of about 2 to become pepsin, and pepsin is inactivated completely at a pH of 7. Activated pepsin catalyzes reactions that digest proteins into smaller polypeptides, oligopeptides, and some free amino acids. However, the reactions catalyzed by pepsin digest only about 10–15% of the proteins in ingested food. Individuals who do not produce pepsin do not appear to have any less ability to digest proteins.

The remainder of protein digestion takes place in the small intestine with the help of pancreatic enzymes and brush border enzymes (step ① of **Figure 22.30**). There are five pancreatic enzymes that digest proteins, all of which are released as inactive precursors. This protects the pancreas from *autodigestion,* or digestion of its cells by its own enzymes.

The first pancreatic enzyme to become activated is the precursor *trypsinogen* (trip-SIN-oh-jen), which becomes the active enzyme **trypsin** (TRIP-sin) when it encounters enzymes on the intestinal brush border. Trypsin in turn catalyzes the reactions that convert the other pancreatic enzymes to their active forms, and activates additional trypsinogen as it is secreted by the pancreas. These enzymes catalyze reactions that digest proteins and polypeptides into oligopeptides and some free amino acids.

The final enzymes to act on proteins are associated with the enterocytes. There are multiple brush border enzymes that catalyze the digestion of oligopeptides into free amino acids. These enzymes are limited in the reactions they are able to catalyze, so some small oligopeptides (two to three amino acids long) remain undigested. These small oligopeptides are taken up by enterocytes, and in the enterocyte cytoplasm they encounter more enzymes that catalyze their breakdown into free amino acids.

To be absorbed in the small intestine, proteins consumed for nutrition must generally be broken down into small oligopeptides

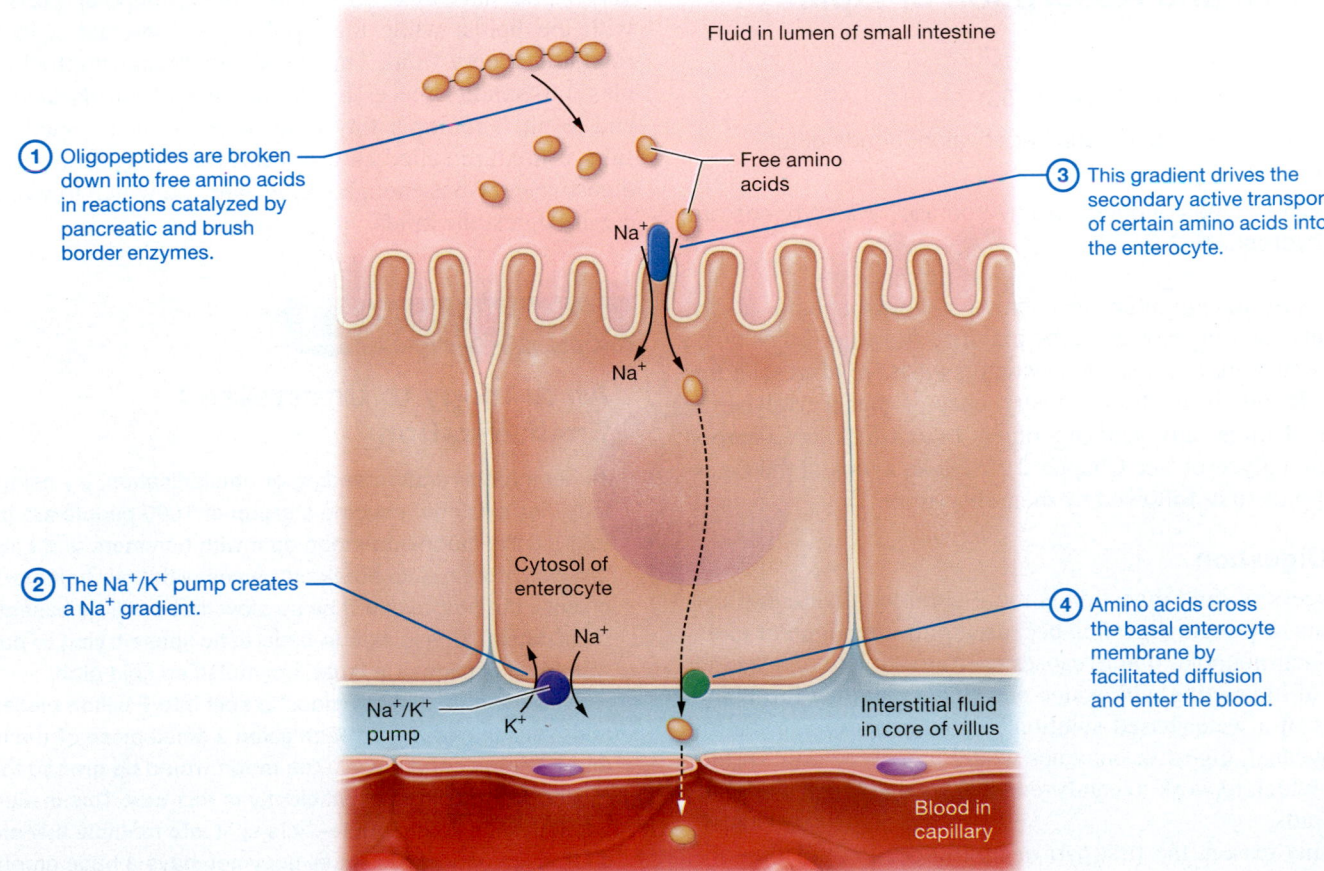

Figure 22.30 Amino acid digestion and absorption in the small intestine.

and free amino acids. However, humans do have the ability to absorb small amounts of whole, undigested proteins involved in immunity by the process of endocytosis. Specialized cells overlying the Peyer's patches in the small intestine ingest these proteins and deliver them to the lymphatic tissue in the patches. This is particularly important in the development of the immune system in newborn infants.

Oligopeptides and free amino acids cross the enterocyte apical membrane primarily by secondary active transport membrane proteins that use a sodium ion gradient established by the Na^+/K^+ pump steps ② and ③ of Figure 22.30. Within the enterocyte, oligopeptides are broken down into free amino acids. The free amino acids then exit the basal enterocyte membrane by facilitated diffusion (step ④), after which they enter the capillaries in the villus. Like carbohydrates, the amino acids are then delivered to the liver for processing via the hepatic portal vein.

Quick Check

☐ 4. Where does protein digestion begin, and with which enzyme?

☐ 5. How does protein digestion occur in the small intestine?

☐ 6. How are amino acids absorbed across the apical and basal enterocyte membranes?

Digestion and Absorption of Lipids

⟪ FLASHBACK

1. What are triglycerides? (p. 53)

2. How do polar and nonpolar covalent bonds differ? (p. 39)

3. Which types of bonds are hydrophilic? Which bonds are hydrophobic? (p. 47)

The class of organic molecules known as lipids includes a variety of fats and oils. However, the majority (90%) of lipids taken in by the diet are *triglycerides,* which is what we discuss in this section. Recall from the chemistry chapter that a triglyceride consists of three fatty acid monomers bound to a three-carbon core called glycerol (see Chapter 2). We take a look at the digestion of lipids first, followed by their absorption.

Lipid Digestion

The process of lipid digestion is more complex than either carbohydrate or protein digestion because lipids are nonpolar molecules. Their nonpolar nature causes lipids to stick to one another instead of interacting with water, and as a result they form large globules in a water-based solution. If this happened in the alimentary canal, digestive enzymes would have very little surface area on which to work to catalyze reactions that break the bonds in the lipids.

For this reason, the first part of lipid digestion must involve physically breaking up large lipid globules into smaller globules by mechanical digestion to give digestive enzymes more

surface area on which to work. This is accomplished first by physical processes such as mastication in the mouth, churning in the stomach, and segmentation in the small intestine. However, even after these processes, the lipid globules would still be too large for efficient chemical digestion to take place. To mechanically break lipids into even smaller globules, we need the help of bile salts.

Recall that bile salts are amphiphilic molecules with polar and nonpolar parts. We use the following schematic structure to represent a bile salt:

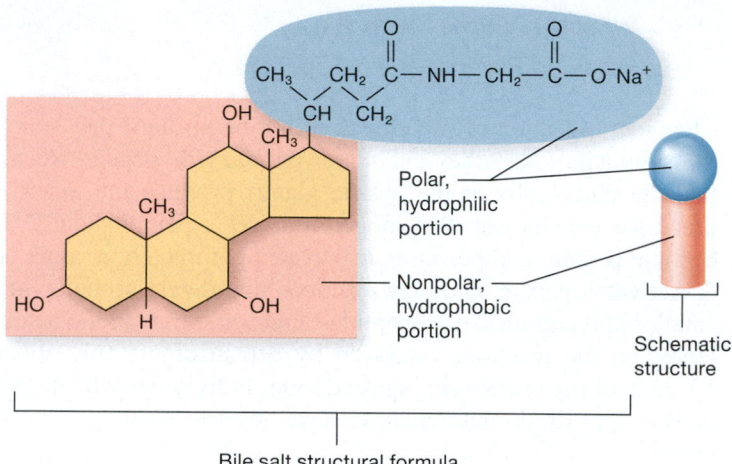

Bile salt structural formula

When bile salts mix with lipids, their nonpolar parts interact with the lipids, while their polar parts interact with the surrounding watery fluid. This physically breaks up the lipid globules into smaller pieces by the process of **emulsification.** The end result is multiple tiny lipid droplets each coated with bile salts, a mixture called an **emulsion** (ee-MUL-shun). This gives digestive enzymes enough surface area on which to work to efficiently digest the lipids.

Study BOOST ⬆

An Analogy to Understand Emulsification

To get a better understanding of emulsification, try using the following analogy: Imagine a group of 1000 people are trying to grind the moon into moon dust with hammers and chisels. They will continue to break off pieces with their hammers and chisels, but progress will be so slow that it will be essentially nonexistent. This is akin to pancreatic lipase trying to catalyze the digestion of a large, unemulsified lipid glob.

Now imagine that the moon is split into 7 billion pieces, and 7 billion people are each given a small piece of the moon and a hammer and chisel. The moon would be ground to dust much more quickly and efficiently in this way. This is akin to an emulsified lipid globule—it is split into multiple tiny pieces so that the pancreatic lipase enzymes have a huge amount of area on which to catalyze their reactions. ■

The process of lipid digestion proceeds as follows (**Figure 22.31**):

(1) **Lipids are broken apart by stomach churning and broken down in reactions catalyzed by gastric lipase.** Lipid digestion begins in the stomach with the help of the enzyme **gastric lipase.** Gastric lipase catalyzes the reactions that remove one fatty acid from triglycerides, leaving some free fatty acids and diglycerides. About 15% of total dietary fats are digested this way.

(2) **Lipids enter the small intestine and are emulsified by bile salts.** The undigested and partially digested triglycerides enter the small intestine, where they interact with bile salts. Bile salts coat the lipids and physically break them apart into smaller pieces. This is purely a mechanical process—no bonds are broken during emulsification.

(3) **Pancreatic lipase catalyzes reactions that digest the lipids into free fatty acids and monoglycerides.** The pancreas releases an enzyme, **pancreatic lipase,** that catalyzes lipid breakdown. In this process, triglycerides are digested into monoglycerides and free fatty acids.

(4) **Bile salts remain associated with the digested lipids to form micelles.** After chemical digestion by lipase is complete, the bile salts and digested lipids stay together in structures known as **micelles** (my-SELZ). If the digested lipids did not remain associated with bile salts, they would re-form into large globules and absorption would not be possible.

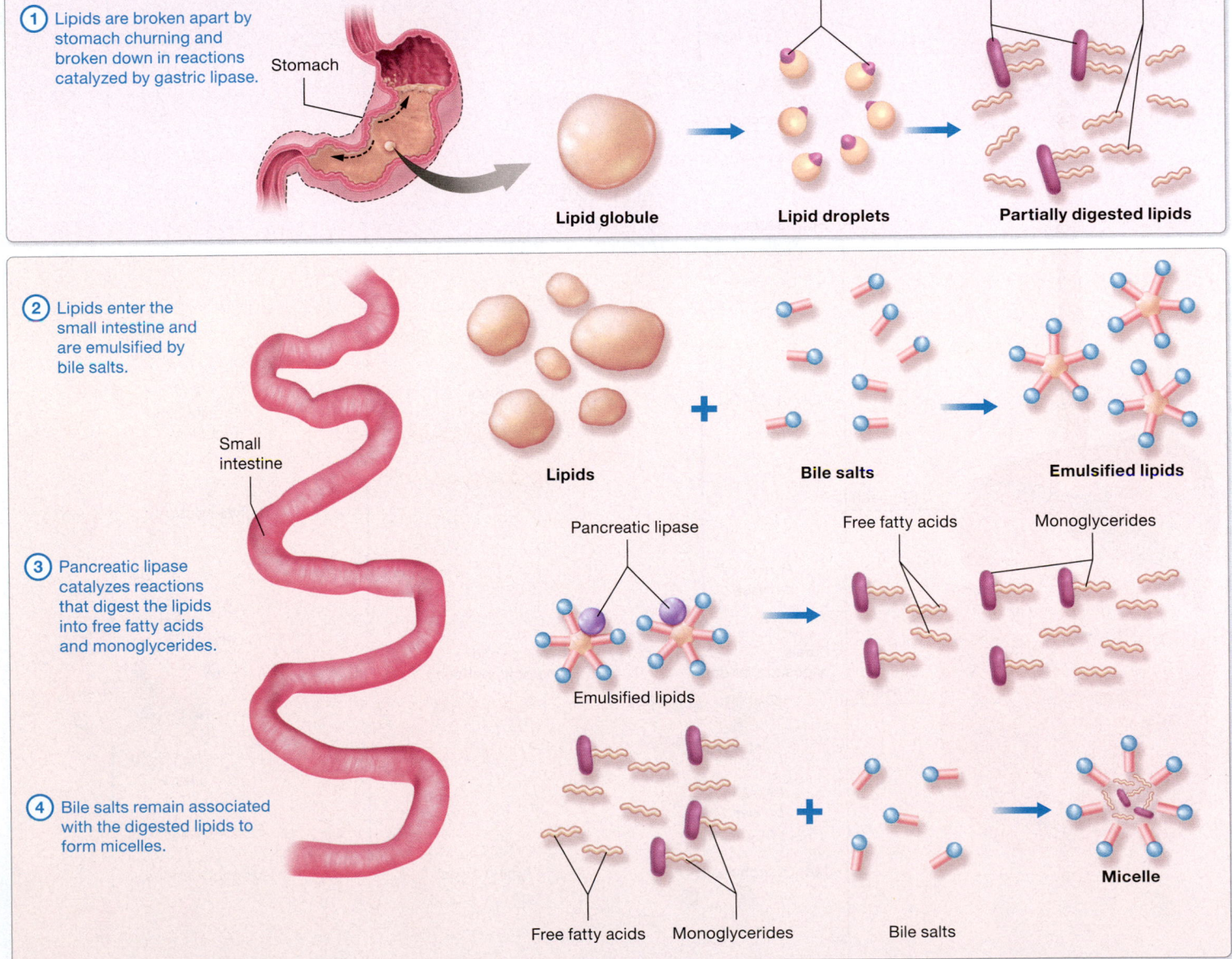

Figure 22.31 Digestion of lipids.

Figure 22.32 summarizes the processes of digestion for carbohydrates, proteins, and lipids.

Lipid Absorption

Just as lipid digestion is more complicated than that of carbohydrates and proteins, lipid absorption is also more complex. Again, this is due to lipids' nonpolar nature. Lipids face several water-based barriers that deter them from passing into the cytosol of the enterocytes, including the mucus lining of the small intestine and the polar phosphate heads of the enterocytes'

plasma membranes. For this reason, lipids require assistance to approach the enterocyte membranes and eventually cross them.

Let's walk through the steps of lipid absorption (**Figure 22.33**):

① **Micelles escort lipids to the enterocyte plasma membrane.** Digested triglycerides and other lipids such as cholesterol and *fat-soluble vitamins* (such as vitamin D) remain associated with bile salts in micelles. Notice that these bile salts have their polar portions on the outside,

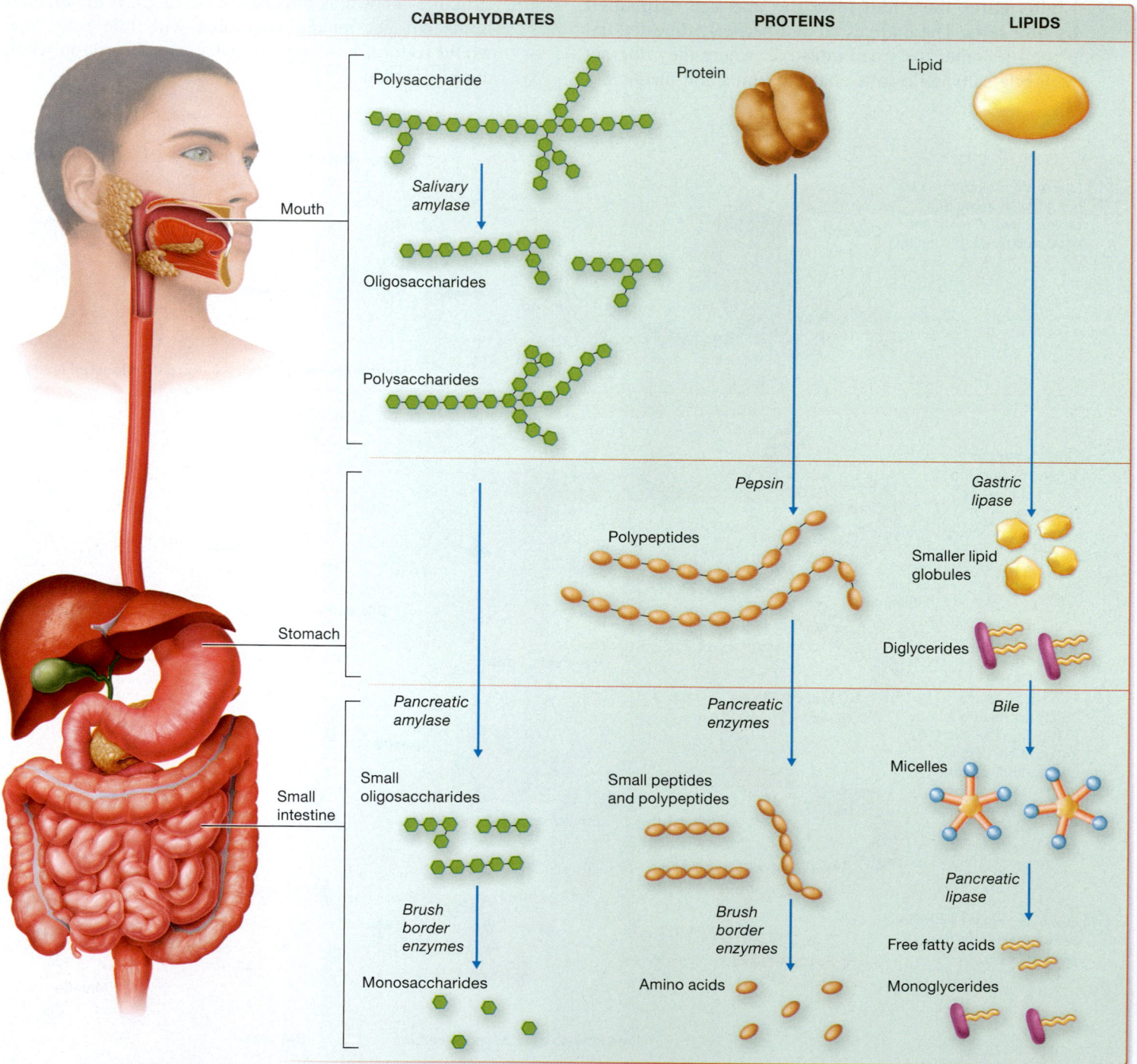

Figure 22.32 Summary of the digestion of carbohydrates, proteins, and lipids.

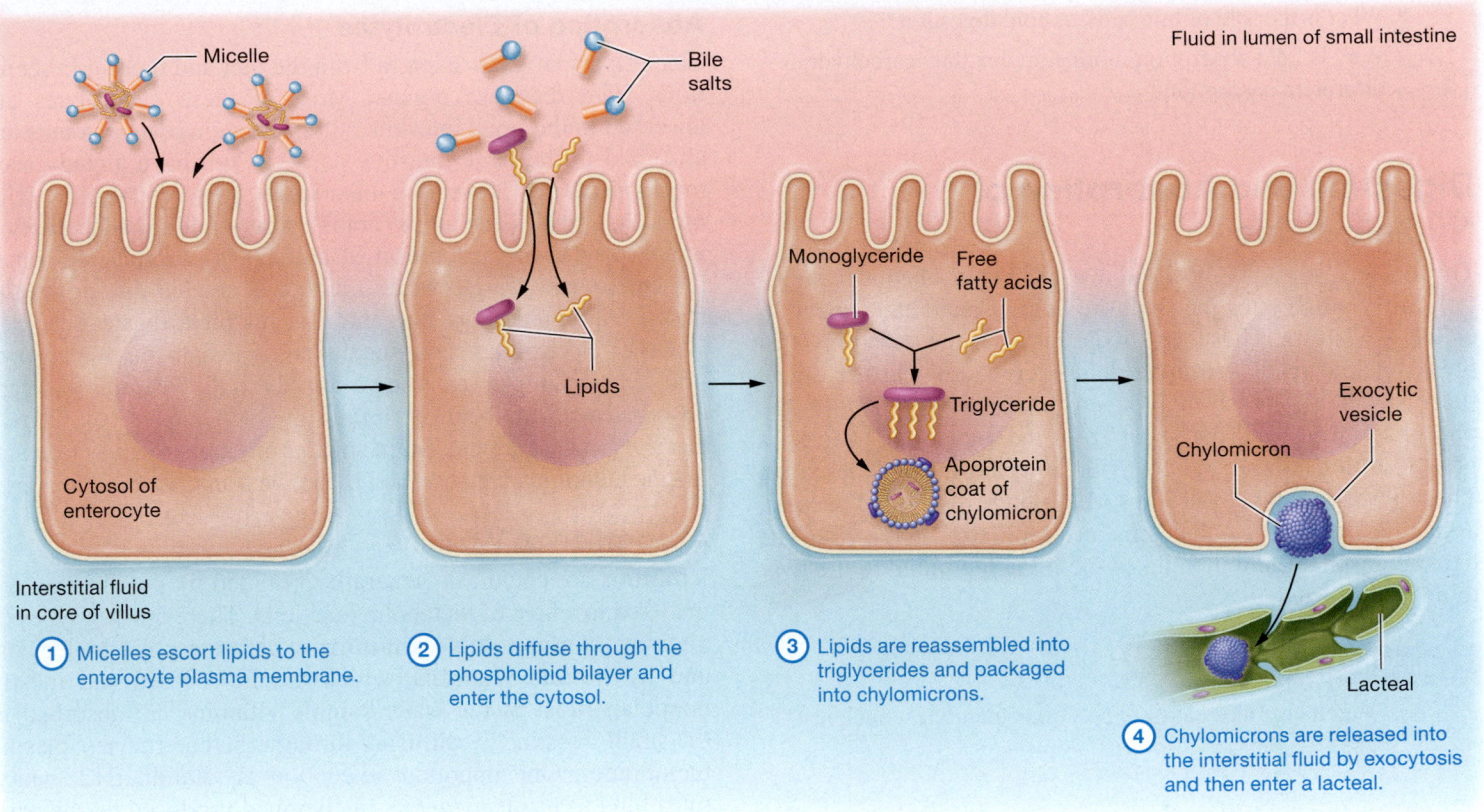

Figure 22.33 Lipid absorption in the small intestine.

facing the water-based environment, and their nonpolar portions on the inside, facing the digested lipids. The outer polar portion of the micelle allows it to approach the polar mucus layer and phosphate heads of the plasma membrane.

② **Lipids diffuse through the phospholipid bilayer and enter the cytosol.** Once the lipids in the micelles are at the plasma membrane, the attraction of the fatty acid tails of the phospholipid bilayer draws the lipids into the cytosol. This generally occurs by simple diffusion, although some lipids use facilitated diffusion by carrier-mediated transport. Note that the leftover bile salts in the micelles remain in the lumen of the small intestine and are reabsorbed by active transport mechanisms in the ileum.

③ **Lipids are reassembled into triglycerides and packaged into chylomicrons.** Within the enterocyte, enzymes catalyze reactions that turn the free fatty acids and monoglycerides back into triglycerides. The triglycerides are packaged with cholesterol, other dietary lipids, phospholipids, and lipid-binding proteins known as *apoproteins* into structures called **chylomicrons** (ky-loh-MY-krahnz). Chylomicrons are similar to micelles in that their nonpolar lipids face the inside and their polar portions face the outside. This allows chylomicrons to travel with the polar water molecules in blood.

④ **Chylomicrons are released into the interstitial fluid by exocytosis and then enter a lacteal.** The newly formed

chylomicrons are packaged into vesicles by the Golgi apparatus and released by exocytosis from the basal enterocyte membrane into the interstitial fluid in the core of the villus. Chylomicrons vary in size but are generally large—in fact, too large to enter the capillaries in the villus. However, lacteals, which are also found in the core of the villus, have valves in their walls that allow large substances to enter and exit. Chylomicrons therefore enter a lacteal, where they join the lymph.

Unlike proteins and carbohydrates, lipids are not delivered directly to the liver via the hepatic portal vein after absorption. However, the hepatic portal vein delivers the leftover bile salts to the liver, where they are used to make new bile. The lipids travel within chylomicrons through the lymphatic vessels and eventually to the thoracic duct, where they join the blood with the rest of the lymph as the thoracic duct meets the junction of the left internal jugular and left subclavian veins. As chylomicrons travel through blood capillaries, lipids are progressively removed and enter cells. Lipids removed from chylomicrons in the liver are processed by a number of metabolic processes. Chylomicrons that have been depleted of all their lipids are taken in by hepatocytes and dismantled.

Quick Check

☐ 7. Where does lipid digestion begin and with which enzyme?

8. What is the role of bile salts in lipid digestion?

9. Why is lipid absorption more complex than carbohydrate or protein absorption?

Digestion and Absorption of Nucleic Acids

The nucleic acids in the food we eat begin chemical digestion in the small intestine with the help of pancreatic enzymes called *nucleases*. These enzymes catalyze the reactions that break nucleic acids into individual nucleotides. Further digestion occurs via brush border enzymes, which remove the phosphate group and the sugar from the nucleotide to leave a phosphate ion, ribose or deoxyribose, and a nitrogenous base. These three substances are absorbed via primary and secondary active transport mechanisms into capillaries in the villi. The hepatic portal system then takes them via the blood to the liver to be metabolized.

Quick Check

10. Which enzymes catalyze reactions that digest nucleic acids?

Absorption of Water, Electrolytes, and Vitamins

In addition to nutrients, the alimentary canal absorbs large quantities of water, electrolytes (such as sodium, potassium, and calcium ions), and vitamins. Much of this absorption occurs in the small intestine, but a significant quantity occurs in the large intestine as well. The next subsections examine the absorption of each of these substances.

Absorption of Water

On average, over 9 liters of water enter the small intestine each day. About 2 liters of water are ingested, and the remaining liters are secreted into the alimentary canal by the alimentary canal itself and accessory organs. Of these 9 liters, about 8 liters are absorbed into the enterocytes of the small intestine, and most of the remaining water is absorbed into the enterocytes of the large intestine. This leaves only about 0.1 liter of water to be excreted in feces.

Water absorption occurs exclusively by osmosis. This requires both the cytosol of the enterocytes and the extracellular fluid in the villi to be *more* concentrated than the fluid in the lumen of the intestine to drive water movement. For this reason, anything that causes excess solutes to remain in the lumen of the intestines will hold onto water and decrease its reabsorption. This fact can be exploited in the use of drugs known as *osmotic laxatives,* which soften the feces. These drugs consist of solutes that either are not absorbed or are given in amounts in excess of what can be absorbed. As a result, water is retained in the feces, softening them and making them easier to eliminate.

Absorption of Electrolytes

Electrolytes are both taken in from the diet and present in secretions from digestive organs. Most of these electrolytes are absorbed in the small intestine, although significant amounts are absorbed in the large intestine as well. You have already seen mechanisms for absorption of sodium ions (cotransport with monosaccharides and amino acids). This sodium ion absorption is also key in the absorption of anions—as sodium ions enter the enterocytes, an electrical gradient is created that drives the absorption of anions such as chloride and bicarbonate ions.

Other electrolytes have specialized mechanisms for absorption. For example, as you've discovered, calcium ions are absorbed with the help of vitamin D (see Chapter 16). Certain electrolytes that are present in smaller amounts, such as iron and magnesium ions, are absorbed by active transport mechanisms.

Absorption of Vitamins

Vitamins are chemicals generally provided by the diet that are involved in a host of metabolic reactions. There are two types of vitamins: **water-soluble vitamins,** which are polar molecules; and **fat-soluble vitamins,** which are lipid-based and mostly nonpolar. Most of the water-soluble vitamins are absorbed in the small intestine by diffusing through the enterocytes' plasma membranes. One important exception is vitamin B12, which must bind to *intrinsic factor*—a chemical produced by the parietal cells of the stomach—to be absorbed in the ileum. A shortage of intrinsic factor can impair vitamin B12 absorption, as discussed in *A&P in the Real World: Intrinsic Factor and Vitamin B12 Deficiency.* Fat-soluble vitamins are packaged into micelles with fats and other lipids and are absorbed with them. The fat-soluble vitamins include A, D, E, and K. Vitamins are covered in more depth in the metabolism and nutrition chapter (see Chapter 23).

Intrinsic Factor and Vitamin B12 Deficiency

Vitamin B12 plays a role in the metabolic functions of every cell in the body; it is particularly important to the formation of erythrocytes and the normal functioning of the nervous system. Its absorption requires the presence of intrinsic factor (IF). An IF-B12 complex forms and then binds to the enterocyte plasma membrane, which enables its absorption.

Many conditions can impair production of intrinsic factor, including an autoimmune condition that attacks the stomach, atrophy of the gastric mucosa from ulcers or infections, and surgical removal of the stomach for weight loss or cancer treatment. Without intrinsic factor, a deficiency of vitamin B12 results, even if adequate B12 is present in the diet. This deficiency can lead to pernicious anemia and nervous system problems and can be treated only with B12 injections.

Table 22.2 Digestive Enzymes

Enzyme(s)	Source	Reaction Catalyzed
Carbohydrates		
Salivary amylase	Salivary glands	Polysaccharides into smaller polysaccharides and oligosaccharides
Pancreatic amylase	Pancreatic juice	Polysaccharides into oligosaccharides
Maltase, sucrase, lactase	Intestinal brush border	Oligosaccharides into monosaccharides
Proteins		
Pepsin	Chief cells of gastric glands (secreted as precursor pepsinogen)	Proteins into polypeptides and oligopeptides
Trypsin	Pancreatic juice	Oligopeptides into small peptides; activates itself and other pancreatic enzymes
Chymotrypsin	Pancreatic juice	Oligopeptides into small peptides
Carboxypeptidase	Pancreatic juice	Oligopeptides into small peptides
Dipeptidase and tripeptidase	Intestinal brush border	Dipeptides and tripeptides into amino acids
Lipids		
Gastric lipase	Gastric glands	Triglycerides into free fatty acids and diglycerides
Pancreatic lipase	Pancreatic juice	Triglycerides into free fatty acids and monoglycerides
Nucleic Acids		
Nucleases	Pancreatic juice	Nucleic acids into nitrogenous bases and simple sugars

To help you study what's been covered in this module, look to **Table 22.2**, which summarizes the digestive enzymes.

Quick Check

☐ 11. How is water absorbed in the small and large intestines?

☐ 12. How is sodium ion absorption achieved? How does it affect the absorption of other electrolytes?

Apply What You Learned

☐ 1. The weight loss drug orlistat, available over the counter under the name Alli, blocks the activity of pancreatic lipase. Predict the effects of this drug. Why might it help a person lose weight?

☐ 2. Many dietary supplements contain supplemental digestive enzymes that manufacturers claim are needed to digest food properly.

    a. What do you think will happen to these enzymes in the stomach after they are ingested? (*Hint:* Remember that enzymes are proteins.)

    b. Do you think that these supplements are likely to be useful? Why or why not?

☐ 3. Would bile salts be effective if they were polar molecules instead of amphiphilic molecules? What if they were nonpolar molecules? Why or why not?

See answers in Appendix A.

MODULE **22.8**
Putting It All Together: The Big Picture of Digestion

Learning Outcomes

1. Describe the overall big picture of digestion and digestive processes.

Now let's put together the information we covered in previous modules to create a big picture of digestion. **Figure 22.34** traces the digestion pathway, starting with the ingestion of food and ending with the defecation of the remaining, indigestible portion.

Quick Check

☐ 1. In which parts of the alimentary canal does chemical digestion take place?

☐ 2. How do the accessory organs assist digestion?

Apply What You Learned

☐ 1. Patients who are unable to eat by mouth often have a feeding tube placed into the stomach. Which digestive processes are missed with this type of feeding?

See answers in Appendix A.

The Big Picture of Digestion

Figure 22.34

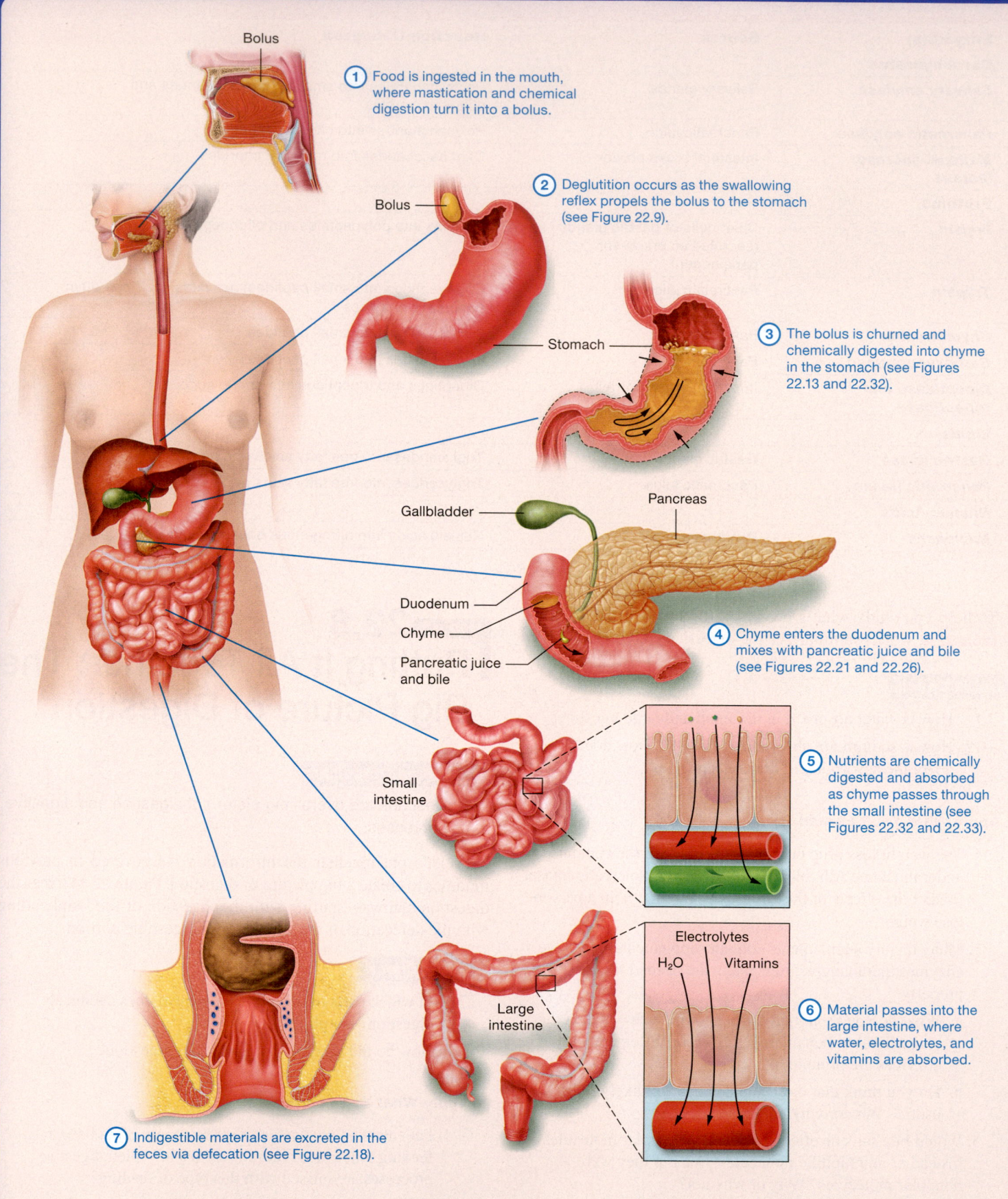

Bolus

1 Food is ingested in the mouth, where mastication and chemical digestion turn it into a bolus.

Bolus

2 Deglutition occurs as the swallowing reflex propels the bolus to the stomach (see Figure 22.9).

Stomach

3 The bolus is churned and chemically digested into chyme in the stomach (see Figures 22.13 and 22.32).

Gallbladder

Pancreas

Duodenum

Chyme

Pancreatic juice and bile

4 Chyme enters the duodenum and mixes with pancreatic juice and bile (see Figures 22.21 and 22.26).

Small intestine

5 Nutrients are chemically digested and absorbed as chyme passes through the small intestine (see Figures 22.32 and 22.33).

Large intestine

Electrolytes

H_2O

Vitamins

6 Material passes into the large intestine, where water, electrolytes, and vitamins are absorbed.

7 Indigestible materials are excreted in the feces via defecation (see Figure 22.18).

Chapter Summary

For everything you need to succeed in this course, go to Mastering Anatomy & Physiology. **There you will find:**

- 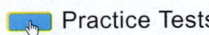 Practice Tests
- Author Podcasts
- Big Picture Animations
- Concept Boost Video Tutors
- **iP2** Interactive Physiology 2.0 Tutorials
- **A&PFlix** A&P Flix 3D Animations
- Active-Learning Workbook

- The **tongue** epithelium is arranged into **papillae.** The tongue is involved in **mechanical digestion,** and it pushes the bolus posteriorly during swallowing.

- The **salivary glands** are located around the oral cavity, into which they secrete **saliva.** The three pairs of salivary glands are the **parotid glands, submandibular glands,** and **sublingual glands.** Saliva consists primarily of water, with mucus, **salivary amylase, lysozyme,** and secretory IgA. **Salivation** is controlled by a parasympathetic nervous system reflex arc mediated by the facial and glossopharyngeal nerves.

- Food passes from the oral cavity to the **pharynx,** and then into the **esophagus,** where waves of peristalsis massage it to the stomach.

- **Swallowing,** or **deglutition,** is a type of propulsion partly under conscious control and partly mediated by the *swallowing reflex.*

MODULE 22.1
Overview of the Digestive System 851–857

- The digestive system consists of the **alimentary canal** and the **accessory organs.**

- The functions of the digestive system include food, water, electrolyte, and vitamin intake; hormone production; and waste excretion.

- There are six basic digestive processes: **ingestion, secretion, propulsion, digestion, absorption,** and **defecation.**

- The **peritoneum** consists of the **parietal peritoneum** and the **visceral peritoneum.** Between these two layers is the **peritoneal cavity.**

- The abdominal digestive organs are supplied and drained by the **splanchnic circulation.**

- The organs of the alimentary canal have four tissue layers: the **mucosa, submucosa, muscularis externa,** and either the **serosa** or the **adventitia.**

- The motility of the alimentary canal is controlled by the **enteric nervous system (ENS),** the parasympathetic nervous system, and hormones.

MODULE 22.2
The Oral Cavity, Pharynx, and Esophagus 857–864

- The **teeth masticate** and grind food into a **bolus.**
  - The three classes of teeth are **incisors, canines,** and **molars.** There are 20 **primary,** or **deciduous, teeth** and 32 **secondary,** or **permanent, teeth.**
  - A tooth consists of a **crown,** the portion above the gum line, and the **root,** the portion embedded in the alveolus.

MODULE 22.3
The Stomach 864–871

- The **stomach** has five regions: the **cardia, fundus, body, pyloric antrum,** and **pylorus.** It controls emptying into the duodenum via the **pyloric sphincter.**
  - The muscularis externa of the stomach has a third oblique layer that performs **churning.**
  - The stomach's mucosa houses **gastric glands** with four main cell types: **diffuse neuroendocrine (DNES) cells,** which produce hormones; **chief cells,** which secrete **pepsinogen; parietal cells,** which secrete hydrochloric acid and **intrinsic factor;** and **mucous neck cells,** which secrete acidic mucus.

- Gastric acid is secreted by H^+/K^+ pumps in the membranes of parietal cells. Secretion is stimulated by the vagus nerve, **histamine,** and **gastrin.** It is inhibited by **somatostatin** and the **enterogastric reflex.**

- The stomach's motility has three functions: (1) receptive function, (2) churning function, and (3) emptying function. The rate of gastric emptying is controlled by the composition of chyme in the duodenum and hormones such as **secretin, cholecystokinin,** and **gastric inhibitory peptide.**

MODULE 22.4
The Small Intestine 872–875

- The **small intestine** has three regions: the **duodenum, jejunum,** and **ileum.**
  - To increase surface area available for nutrient absorption, the internal surface of the small intestine is arranged into progressively smaller folds: **circular folds, villi,** and **microvilli.**

○ Villi surround a central core of capillaries and a *lacteal.* Between villi are **intestinal crypts** that house glands.

- The small intestine's motility between meals is characterized by slow, rhythmic contractions of the **migrating motor complex.** During feeding, the small intestine exhibits **segmentation** and peristalsis.

- The **large intestine** has four components: the **cecum, colon, rectum,** and **anal canal.**
- The large intestine houses a great number of bacteria that produce vitamins, metabolize undigested materials, deter the growth of harmful bacteria, and stimulate the immune system.
- The proximal large intestine undergoes propulsive **mass movements.** When mass movements propel fecal material into the distal colon and rectum, the **defecation reflex** is initiated, which is mediated by the parasympathetic nervous system and the cerebral cortex.

- The **pancreas** is an endocrine and exocrine gland. Its exocrine portion consists of **acinar cells** that secrete **pancreatic juice,** which consists of water, bicarbonate ions, and multiple digestive enzymes.
- The liver is in the right upper quadrant and consists of four lobes.
  ○ The liver's main digestive function is the production of **bile,** which contains **bile salts,** water, and metabolic wastes such as bilirubin.
  ○ Other functions of the liver include nutrient metabolism, detoxification, excretion, production of plasma proteins, and storage of nutrients.
- The **gallbladder** receives bile from the common hepatic duct and stores it for later release.
- The main trigger for bile release from the gallbladder and liver is cholecystokinin.

- Carbohydrate digestion begins in the mouth with salivary amylase. It continues in the small intestine with **pancreatic amylase** and the *brush border enzymes* **lactase, maltase,** and **sucrase.** Glucose and galactose are absorbed across the apical enterocyte membrane with the secondary active transport **Na$^+$/glucose cotransporter.** Fructose crosses the apical membrane by

facilitated diffusion. All three monomers cross the enterocyte's basal membrane by facilitated diffusion.

- Protein digestion begins in the stomach with the enzyme **pepsin.** It continues in the small intestine with pancreatic enzymes such as **trypsin** and brush border peptidases. Amino acids are absorbed across the enterocyte's apical and basal membranes primarily by facilitated diffusion.
- Absorbed amino acids and carbohydrates enter the capillaries in the villus, and are delivered to the hepatic portal system.
- Lipid digestion is more complicated because lipids are nonpolar.
  ○ Lipid globules are broken apart physically by chewing, by churning, and by **emulsification** by bile salts.
  ○ Emulsified lipids are broken down in reactions catalyzed by **pancreatic lipase** into two free fatty acids and a monoglyceride.
  ○ Digested lipids remain associated with other lipids and bile salts to form **micelles.**
- Digested lipids are absorbed into the enterocyte by diffusion. In the cytosol, fatty acids are reassembled into triglycerides and packaged into **chylomicrons.** Chylomicrons exit the enterocyte by exocytosis and enter a lacteal, where they travel through the lymph before being delivered to the blood.
- Water is absorbed across the small and large intestines by osmosis. Electrolytes are absorbed across the small and large intestines as well by passive and active transport mechanisms.

- The big picture of digestion is shown in Figure 22.34.

Assess What You Learned

Scan the QR Code for additional practice test questions

https://goo.gl/bTl2lg

LEVEL 1 Check Your Recall

1. Which of the following is not one of the six basic processes carried out by the digestive system?

   a. Propulsion
   b. Gas exchange
   c. Secretion
   d. Absorption

2. The peritoneal cavity is located:

   a. around each of the digestive organs.
   b. between layers of mesentery.
   c. superior to the greater omentum.
   d. between the parietal and visceral peritoneum.

3. Mark the following statements as true or false. If a statement is false, correct it to make a true statement.

   a. The mucosa from the stomach to the anus consists of an inner layer of stratified columnar epithelium.
   b. The muscularis externa of most of the alimentary canal consists of inner circular and outer longitudinal layers of smooth muscle.
   c. The soft palate and uvula move posteriorly during swallowing to prevent food from entering the laryngopharynx.
   d. The mucosa of the esophagus, pharynx, and oral cavity contains simple squamous epithelium to protect it from abrasion.

4. Match the following terms with the correct definition.

   _____ Chief cells
   _____ Parietal cells
   _____ Gastrin
   _____ Pyloric sphincter
   _____ Diffuse neuroendocrine (DNES) cells
   _____ Gastroesophageal sphincter
   _____ Pepsin
   _____ Chyme

   a. Hormone that stimulates multiple digestive processes
   b. Enzyme that begins protein digestion
   c. Produce acid and intrinsic factor
   d. Liquid produced as a result of stomach churning
   e. Produce pepsinogen
   f. Cells in gastric glands that produce hormones
   g. Controls passage of bolus from esophagus to stomach
   h. Controls passage of stomach contents to duodenum

5. The common hepatic duct and main pancreatic duct enter into the:

   a. pyloric sphincter.
   b. duodenum.
   c. ileum.
   d. cecum.

6. What are the three folds of the small intestine called? What is their purpose?

7. Which of the following is *not* one of the functions of the bacterial flora of the large intestine?

   a. Metabolism of undigested carbohydrates
   b. Deterring the growth of harmful bacteria
   c. Chemical digestion of dietary proteins
   d. Stimulation of the immune system
   e. Vitamin production

8. Trace the pathway that food takes through the entire alimentary canal, from the oral cavity to the anal canal.

9. Mark the following statements as true or false. If a statement is false, correct it to make a true statement.

   a. The crown of a tooth is surrounded by an outer layer of pulp, which is the hardest substance in the body.
   b. The filiform papillae of the tongue are keratinized, which makes them rough and able to assist in mechanical digestion.
   c. The three pairs of salivary glands are the parotid, submandibular, and sublingual glands.
   d. The exocrine cells of the pancreas and salivary glands are islet cells.

10. Which of the following statements about accessory organ secretions is *not* true?

    a. Hepatocytes produce bile, which drains out of the liver via the common hepatic ducts.
    b. Saliva contains secretory IgA and lysozyme, which play an important role in preventing the growth of pathogenic bacteria in the oral cavity.
    c. Pancreatic juice contains digestive enzymes and bicarbonate ions to neutralize the acidic chyme.
    d. The gallbladder produces bile, which drains out of the gallbladder via the cystic duct.

11. Which of the following best describes the microscopic structure of the liver?

    a. Hexagonal plates of hepatocytes surrounding a central vein with portal triads at each corner
    b. Octagonal plates of hepatocytes surrounding a central artery with portal triads at each corner
    c. Irregular groups of hepatocytes surrounding a central vein with bile ducts at each corner
    d. Hexagonal plates of hepatocytes surrounding a portal vein with bile ducts at each corner

12. Match the following digestive enzymes with the digestive reaction they catalyze.

    _____ Salivary amylase
    _____ Sucrase
    _____ Trypsin
    _____ Pancreatic lipase
    _____ Pancreatic amylase
    _____ Pepsin

    a. Breaks down sucrose into glucose and fructose in small intestine
    b. Digests lipids in small intestine
    c. Breaks down proteins into polypeptides in stomach
    d. Breaks down polysaccharides into oligosaccharides in small intestine
    e. Assists in protein digestion in small intestine and activates other protein-digesting enzymes
    f. Breaks down polysaccharides into oligosaccharides in the mouth

13. Which of the following best describes the role of bile salts in lipid digestion?

    a. Bile salts begin chemical digestion of lipids.
    b. Bile salts bind to lipids and physically break them apart into monoglycerides and free fatty acids.
    c. Bile salts bind to lipids and physically break them apart into smaller pieces.
    d. Bile salts interact with pancreatic lipase and catalyze lipid breakdown.

14. Which of the following is/are absorbed in the small intestine by the Na^+/glucose cotransporter?

    a. Glucose
    b. Galactose
    c. Fructose
    d. Both a and b are correct.
    e. All of the above are correct.

15. The primary hormone that triggers the secretion of pancreatic juice and bile is:

    a. gastric inhibitory peptide.
    b. cholecystokinin.
    c. motilin.
    d. somatostatin.

16. Fill in the blanks: Lipids associate with bile salts in the small intestine to form _____. In the enterocyte, lipids are packaged with other lipids into _____.

17. Which of the following statements about absorption in the small and large intestines is *false?*

    a. Absorption of water occurs by secondary active transport.
    b. Absorption of sodium ions creates an electrical gradient that drives absorption of chloride and bicarbonate anions.
    c. Fat-soluble vitamins are absorbed with other lipids.
    d. Vitamin B12 requires intrinsic factor to be absorbed in the small intestine.

18. Mark the following statements as true or false. If a statement is false, correct it to make a true statement.

    a. Motility of the alimentary canal is regulated by the enteric nervous system, the parasympathetic nervous system, and hormones.
    b. The esophageal and pharyngeal phases of swallowing are mediated by the swallowing reflex of the brainstem and the enteric nervous system.
    c. Smooth muscle of the stomach contracts when food enters from the esophagus.
    d. Control of gastric emptying into the duodenum is important to allow the chyme to be neutralized and diluted by pancreatic secretions.
    e. During eating, the small intestine exhibits motion in the migrating motor complex pattern.

19. Defecation is stimulated by:

    a. stretch receptors in the rectum, initiating the defecation reflex in the spinal cord.
    b. segmentation of haustra in the proximal large intestine.
    c. contraction of the external anal sphincter.
    d. activation of the defecation reflex by the presence of fecal material in the proximal large intestine.

20. Which of the following stimuli mediates salivation?

    a. Neurons of the ENS
    b. ACh from parasympathetic neurons
    c. The hormone secretin
    d. Feedback from the cerebral cortex

21. Which of the following occurs/occur during the cephalic phase of gastric acid secretion?

    a. Stimulation of gastrin and histamine release
    b. Direct stimulation of parietal cells
    c. Inhibition of somatostatin release
    d. Both a and b are correct.
    e. All of the above are correct.

22. How does absorption of lipids differ from absorption of carbohydrates and proteins in the small intestine?

    a. Lipids are absorbed into a capillary; carbohydrates and proteins are absorbed into a lacteal.
    b. Lipids are not absorbed in the small intestine.
    c. Lipids are absorbed into a lacteal; carbohydrates and proteins are absorbed into a capillary.
    d. They are all absorbed into the same structure.

LEVEL 2 Check Your Understanding

1. You and your friend are having dinner and you tell a funny joke. Your friend laughs at the same time he tries to swallow some water, and the water comes out his nose. Why has this happened?

2. General anesthesia eliminates the swallowing reflex but does not eliminate the feedback loops that cause vomiting. Considering this, explain why patients are advised not to eat or drink anything for at least 12 hours prior to undergoing surgery.

3. Drugs with anticholinergic side effects block ACh receptors in the peripheral nervous system, including those on digestive organs. Predict the effects such drugs would have on motility and secretion on the specific organs of the digestive system.

4. Your friend insists that absorption is the most important process carried out by the digestive system. What do you think of this claim? How would you respond?

LEVEL 3 Apply Your Knowledge

PART A: Application and Analysis

1. Mr. Williams presents to your clinic with a complaint of abdominal pain in the right upper quadrant. He says that the pain worsens when he eats, particularly when he eats fatty meals. He has noticed that his stool has been an unusual clay color recently. You perform an ultrasound of his abdomen and find that gallstones are blocking his common bile duct, preventing bile from entering the duodenum.

    a. Why are his symptoms worse when he consumes a high-fat meal? Would you expect his symptoms to worsen or lessen when he consumes a meal consisting only of carbohydrates? Explain.
    b. You test Mr. Williams's stool and find high amounts of undigested fats. Explain this finding.
    c. Explain why his feces have become a clay color instead of a normal brown color.

2. A surgical procedure known as *gastric bypass* involves removing all or part of the stomach.

    a. Ms. Anthony has undergone gastric bypass and her stomach has been removed entirely. She goes out to dinner and eats an extremely large meal. Explain what will happen in her small and large intestine following this meal.
    b. How would stomach removal affect absorption of vitamin B12? Explain.

PART B: Make the Connection

3. Individuals experiencing prolonged vomiting lose a great deal of hydrochloric acid with the vomitus. Predict the effect this loss of acid will have on the pH of the blood. How will the respiratory system respond to the change in pH? *(Connects to Chapters 2 and 21)*

4. Irritant laxatives increase the amount of water in the feces by triggering an inflammatory response in the intestinal mucosa.

Explain why an inflammatory response causes this effect. *(Connects to Chapter 20)*

5. Predict what types of nervous system damage might lead to a loss of the defecation reflex. *(Connects to Chapters 12 and 13)*

See answers in Appendix A.

23
Metabolism and Nutrition

*For practice applying concepts to a clinical scenario, check out the **Running Case Study** for this chapter @ Mastering Anatomy & Physiology*

Computer-generated image: The adenosine triphosphate (ATP) molecule shown here is the cell's main source of chemical energy; the cell uses ATP to fuel its processes.

Have you ever noticed that when you're watching television, reading a magazine, or surfing the Internet, hardly a moment passes that you aren't faced with advertisements for products and diets promising to "rev your metabolism," "stimulate fat metabolism," or "cut calories"? Many of these ads make outrageous promises along the lines of, "Eat all you want and still lose 30 pounds in 2 weeks!" Although it is easy to spot exaggerated claims like these, others are more difficult to see. Most ads use terms that sound scientific and seem to be based on good physiological reasoning, which often makes it difficult to separate fact from fiction.

In our everyday lives, we generally use the term "calorie" to refer to the energy content of the food we eat. But *calorie* actually refers to a measure of heat: One calorie is the amount of heat required to raise the temperature of one gram of water by one degree Celsius. In terms of the human diet, the unit is actually the **Calorie** (with a capital C), which is really a **kilocalorie.** The Calorie is the amount of heat required to raise the temperature of one *kilogram* of water by one degree Celsius. So, we use heat units to measure the amount of energy in food. The references to heat production and energy don't end with the Calorie; you've likely heard that certain activities or dietary supplements will help you to "burn calories" or "burn fat" or even "turn your body into a fat-melting furnace." But why are these heat-related terms associated with food, diets, and metabolism? And can you really "burn calories" and "rev" your metabolism? This chapter explores these topics as part of a discussion of human metabolism and nutrition.

MODULE 23.1
Overview of Metabolism and Nutrition

Learning Outcomes

1. Define the terms metabolism, catabolism, and anabolism, and identify the nutrients the body is able to use for fuel.

2. Compare and contrast endergonic and exergonic reactions.

3. Describe the process of phosphorylation.

4. Describe the hydrolysis of ATP, and explain why this reaction is exergonic.

5. Explain what happens in an oxidation-reduction reaction and how electrons are transferred between reactants, including NADH and FADH$_2$.

Metabolism is defined as the sum of the body's chemical reactions. Metabolism includes four basic processes: (1) harnessing the energy in the chemical bonds of nutrients obtained from the diet that may be used to make **adenosine triphosphate,** or **ATP;** (2) converting one type of chemical into another for the cell's synthesis reactions; (3) carrying out synthesis reactions and assembling macromolecules such as proteins, polysaccharides, nucleic acids, and lipids; and (4) breaking down macromolecules into their monomers or other smaller compounds. Metabolism occurs through many series of enzyme-catalyzed reactions that are referred to as *metabolic pathways.* Each reaction in a series is characterized by a small chemical change in a reactant as the product is produced, and generally involves the addition or removal of one or more atoms or electron pairs.

In this module we explore the introductory concepts of metabolism, beginning with the phases of metabolism and different types of nutrients. Then we examine how the cell uses energy to carry out reactions, and the sources of energy for cellular metabolism, including ATP.

Phases of Metabolism: Catabolism and Anabolism

❰❰ FLASHBACK

1. What are catabolic and anabolic reactions? (p. 43)

2. Define the monomers of carbohydrates, proteins, and lipids, and describe their characteristic features. (p. 50)

Metabolism is generally considered to have two phases. The first is a series of reactions collectively known as **catabolism** (kah-TAEH-bohl-izm; *cata-* = "down"). Recall that a *catabolic reaction* is one in which a substance is broken down into smaller parts (see Chapter 2). Although certain reactions in a catabolic pathway may require the input of energy, the process as a whole releases energy that the cell can harness to drive reactions such as ATP production. To provide materials for this process, macromolecules that we ingest are broken down into nutrient monomers by our digestive system. The body uses three types of nutrient monomers for its catabolism to generate ATP:

- **Glucose.** Both dietary carbohydrates and those stored by the body are generally degraded by catabolic pathways to the monosaccharide glucose. Glucose catabolism is carried out by every cell in the body, and it is the preferred fuel for many cells, including those of the brain and the liver.

- **Fatty acids.** Catabolic pathways degrade lipids (mostly triglycerides) into fatty acids and glycerol. These compounds then enter their own catabolic pathways and are used by the cell.

- **Amino acids.** Proteins are degraded by catabolic pathways into amino acids. Like lipids, amino acids then enter their own catabolic pathways, during which they are converted into other compounds for use by the cell.

The second phase of metabolism is known as **anabolism** (aeh-NAEH-bohl-izm; *ana-* = "up"). As we discussed previously, an *anabolic reaction* is one in which smaller compounds are combined to make a larger compound (see Chapter 2). Via anabolism, cells build proteins, nucleic acids, carbohydrates, lipids, and all other macromolecules. Anabolic reactions require energy, and anabolism is driven by the energy liberated from catabolism.

Quick Check

☐ 1. How do catabolism and anabolism differ?

☐ 2. What three types of nutrients can the cell use to generate ATP?

Energy Requirements of Metabolic Reactions

❰❰ FLASHBACK

1. How do endergonic and exergonic reactions differ? (p. 42)

In the preceding section, we spoke of energy being released from catabolic reactions and required by anabolic reactions. In other words, catabolic reactions are generally *exergonic,* and anabolic reactions are generally *endergonic.* Recall from the chemistry chapter (see Chapter 2) that we define these terms as follows:

- **Exergonic reactions.** Reactions that release energy are known as **exergonic reactions.** The products of these reactions have *less* energy than their reactants, and the remaining energy is released during the reaction. To understand why energy is released, we need to discuss the *law of conservation of energy,* which states that energy cannot be created or destroyed; it can only change forms. This means that the energy present before the reaction must equal the energy present after the reaction. Since the products of an exergonic reaction have less energy than the reactants, energy must be released so that the total energy is the same before and after the reaction. Most catabolic reactions are exergonic.

- **Endergonic reactions.** An **endergonic reaction** is one that requires the input of energy to proceed. This is because the products of the reaction have *more* energy than the reactants. To be in line with the law of conservation of energy, energy must be added to the reactant side so that the energy on both sides of the equation is equal. Anabolic reactions tend to be endergonic.

Endergonic and exergonic reactions in the cell are *coupled*—essentially, the cell "pays for" an endergonic reaction with the energy released by an exergonic reaction. To do so, the cell needs

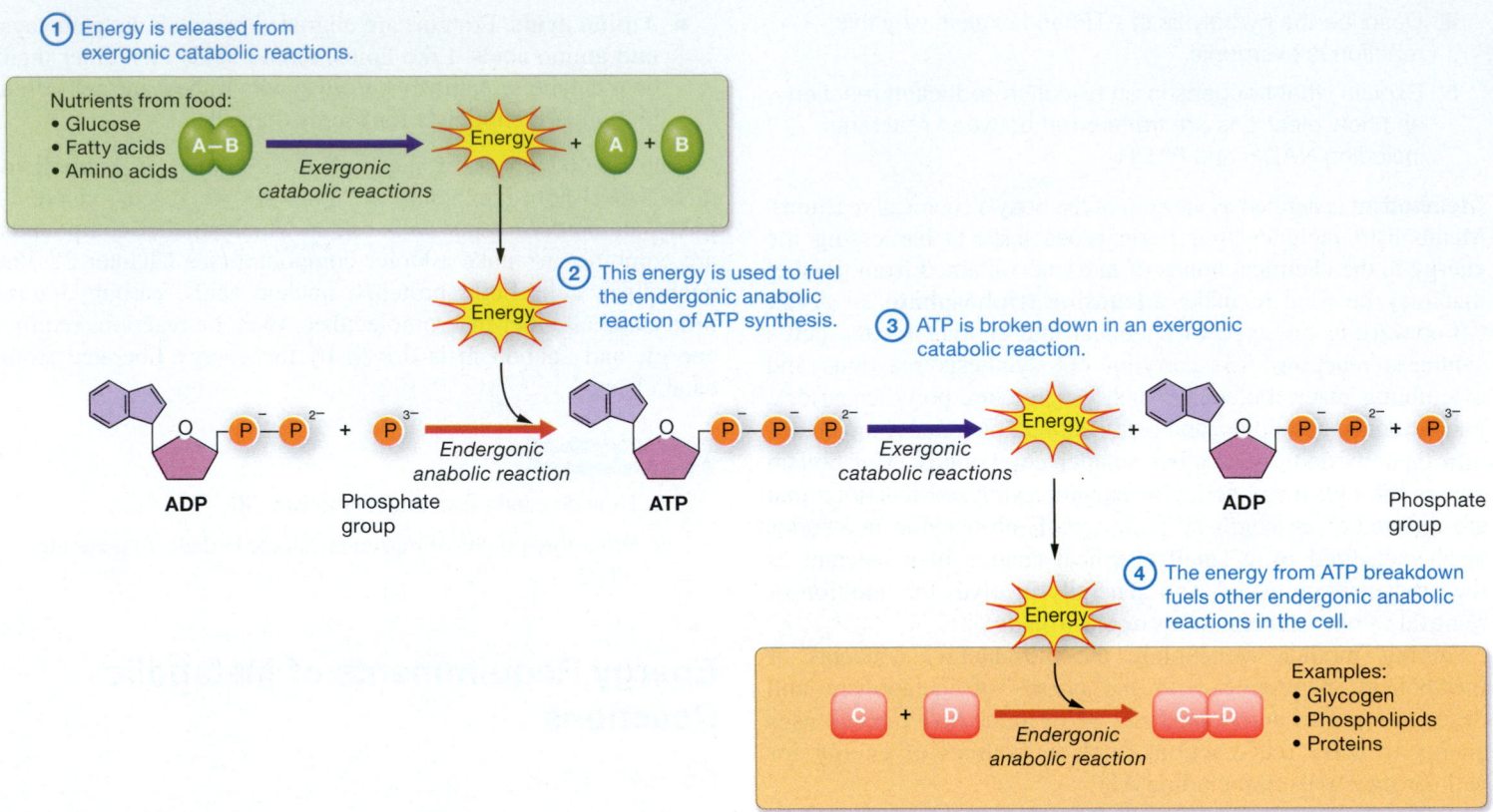

Figure 23.1 Coupling of endergonic and exergonic reactions.

some sort of "money" with which to pay—that is, some way to store and transfer the released energy. This comes in the form of compounds such as ATP. An overview of how this process works is shown in **Figure 23.1**:

① **Energy is released from exergonic catabolic reactions.** Nutrients such as glucose, fatty acids, and amino acids are broken down via catabolic reactions that eventually release energy.

② **This energy is used to fuel the endergonic anabolic reaction of ATP synthesis.** Synthesis of compounds such as ATP is an endergonic process. The energy for this reaction comes from the exergonic reactions in the previous step.

③ **ATP is broken down in an exergonic catabolic reaction.** In another exergonic catabolic reaction, the third phosphate group from ATP is removed and energy is released.

④ **The energy from ATP breakdown fuels other endergonic anabolic reactions in the cell.** ATP breakdown is a highly exergonic process that can fuel a number of endergonic reactions in the cell.

ATP is a critical component of these reactions, so we look more closely at it and its reactions next.

Quick Check

☐ 3. How do endergonic and exergonic reactions differ?
☐ 4. Why are endergonic and exergonic reactions coupled?

Adenosine Triphosphate (ATP) and Phosphorylation

《 FLASHBACK

1. What is chemical energy? (p. 42)
2. What happens during a hydrolysis reaction? (p. 51)

ATP is the cell's main source of energy "money," but how can such a simple compound drive nearly all cellular processes? To begin, recall what happened when we combined two unstable atoms, such as sodium and chlorine—they engaged in an exergonic reaction and formed products that have less energy and were therefore more stable (look back at Figure 2.4). In the process, they released their excess energy in a small explosion. The same principle is at work with ATP, because ATP, like lone sodium and chlorine atoms, is unstable. The phosphate groups of ATP have negative charges, and as a result they repel one another strongly. The repelling charges put a strain on the compound, making these bonds highly unstable and the compound reactive.

The bond between the second and third phosphate groups of ATP is particularly unstable because these groups repel each other strongly, and a great deal of energy is released when this bond is broken. Some sources call this bond a "high-energy bond." This term is not correct, as there is no more energy contained in this bond than in any other chemical bond. This reaction is highly exergonic only because the bond is the most unstable one in ATP due to the repelling forces of the second and third phosphate groups.

The cell can harness energy from ATP by removing its third phosphate group in a *hydrolysis* reaction. In this reaction, the bond between the second and third phosphate groups is broken by a water molecule, resulting in free phosphate (P_i) and *adenosine diphosphate,* or *ADP.* As we have discussed previously, these reactions are catalyzed by enzymes called *ATPases.*

Although ATP hydrolysis is highly exergonic, the cell is able to harness only about 40% of this energy to perform work. The low number reflects the fact that exergonic reactions are not completely efficient, and much of the released energy is lost as heat. In the cell, about 60% of the energy from ATP hydrolysis is lost as heat. This may seem wasteful, but bear in mind that most processes that burn fuel, such as a wooden log on a fire, lose about 70–90% of their energy as heat—which explains why fires and running engines feel hot to the touch. So ATP hydrolysis is pretty efficient, relatively speaking.

The energy that the cell harnesses from ATP drives cellular processes in two ways. First, it can be used by the cell to directly "pay for" certain reactions. ATP is used in this manner to fuel processes such as muscle contraction or the movement of a ribosome along a strand of messenger RNA.

The second mechanism by which ATP "pays for" most cellular processes is slightly more complex. In this process, called **phosphorylation** (faws-fohr′-uh-LAY-shun), ATP donates a phosphate group to a reactant, with the help of an enzyme. Phosphorylation energizes the reactant in much the same way it energizes ADP—by making the resulting product unstable. This instability favors the conversion of the reactant to the product.

A simple example of this is the conversion of the amino acid glutamate to a different amino acid, glutamine. This reaction requires the addition of ammonia to glutamate. ATP assists this reaction as follows: (1) ATP is hydrolyzed, yielding ADP and P_i. (2) The energy released in the first step is used in an endergonic reaction to add the P_i to glutamate, which produces an unstable compound:

$$\text{Glutamate} + P_i + \text{Energy} \rightarrow \text{Glutamate-P}$$

A glutamate-P molecule is unstable for the same reason that ATP is unstable—the glutamate and P repel each other, which places strain on its structure. (3) The P is removed from glutamate and replaced with an ammonia group, forming glutamine, the product:

$$\text{Glutamate-P} + NH_3 \rightarrow \text{Glutamine} + P_i + \text{Energy}$$

This reaction is exergonic because the product, glutamine, has less energy than the original glutamate-P.

Quick Check

☐ 5. Why is the removal of the third phosphate group from ATP highly exergonic?

☐ 6. How does ATP fuel cellular processes?

Nutrients and ATP Generation

« FLASHBACK

1. What is an oxidation-reduction reaction? (p. 43)

Before we discuss ATP production, we need to take a look at how a cell uses the chemical energy in nutrients to drive ATP synthesis. This involves a type of reaction discussed in the chemistry chapter called an *oxidation-reduction reaction.* In this reaction, electrons are transferred from one reactant to another (see Chapter 2). The next subsections examine oxidation-reduction reactions and the compounds to which electrons are transferred: *electron carriers.*

Oxidation-Reduction Reactions

People often speak of "burning fat" when they exercise, and you may assume that this type of burning differs from burning wood in a fire. Surprisingly, however, this is not really the case. Although we obviously don't have mini fires raging inside our cells, the same basic reaction that takes place when wood burns in a fire also occurs within our cells. When something burns, some of its electrons are transferred from one substance to another; this is an exergonic process that generates a great deal of heat. You can see this with the explosive sodium-chlorine reaction shown in **Figure 23.2**, in which heat is generated as the elements burn. As we discussed, such reactions are called **oxidation-reduction reactions** (see Chapter 2). The substance that loses electrons is **oxidized,** and the substance that gains electrons is **reduced** (a useful mnemonic is "OIL RIG," which stands for "**O**xidation **I**s **L**oss; **R**eduction **I**s **G**ain").

When a wooden log burns, the compounds in the wood are oxidized—they lose electrons to the oxygen in the air, which is reduced. As electrons are stripped from the wood, energy is released as heat, which keeps the fire hot and burning. Fires will not burn in the absence of oxygen because the oxygen must be present to accept electrons from the wood. Similarly, fires will not burn when the wood has been oxidized to ashes, because electrons have been removed and none are available to give to oxygen.

Essentially the same process takes place when our cells "burn" nutrients—only the oxidation-reduction reactions in our cells occur over a series of steps rather than all at once, as with a log in a fire. During catabolism, nutrients such as glucose are oxidized and their electrons are passed to other compounds, which are reduced. The flow of electrons from one compound to another is then used to perform cellular work, including the synthesis of ATP.

Figure 23.2 Energy release during an oxidation-reduction reaction.

Electron Carriers

Nutrient oxidation releases a lot of chemical energy, but it produces very little ATP directly. Instead, it is the electrons stripped off the nutrients that generate most of the body's ATP, as their flow produces an **electromotive force** that can do work.

During nutrient catabolism, nutrients are oxidized, and their electrons are removed (like the electrons of zinc in our battery example) and passed to compounds with higher electron affinities, called **electron carriers** (which would be analogous to the copper in the battery). This electron transfer generates an electromotive force that is used by the cell to perform work, the same way electricity from a battery can power a light bulb or a motor. The majority of the cell's electron carriers are integral and peripheral proteins found within the mitochondrial membranes. However, two electron carriers, **nicotinamide adenine dinucleotide (NAD^+)** and **flavin adenine dinucleotide (FAD),** are located within the cytosol and the mitochondrial matrix. These two electron carriers are the first to accept electrons from nutrients during catabolism. NAD^+ does this as follows:

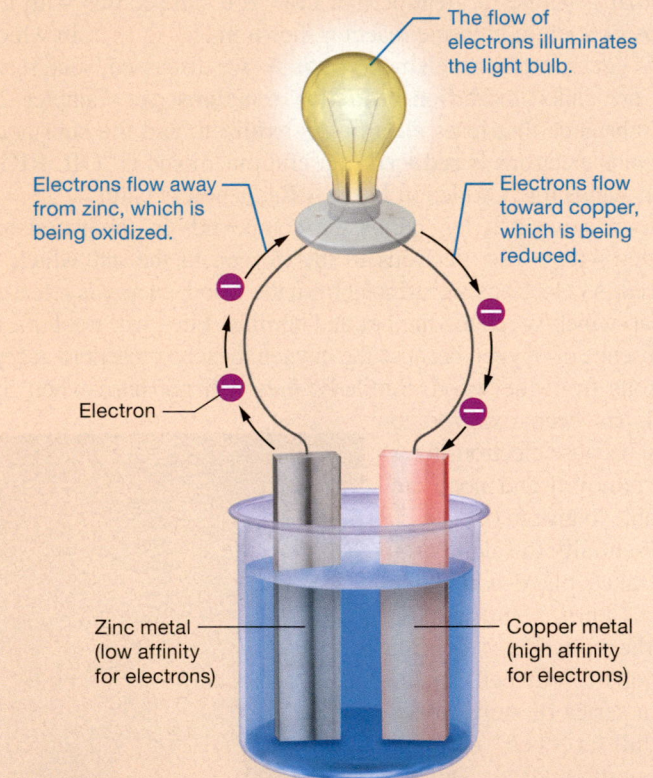

$$NAD^+ + H^+ + 2e^- \longrightarrow NADH$$

Note that the electrons (purple dots) are carried by and added with hydrogen atoms (not hydrogen ions, which have no electrons). Therefore, when reduced, NAD^+ becomes NADH (and FAD becomes $FADH_2$). These electrons will eventually be passed to other electron carriers to generate an electromotive force linked to ATP production. The specifics of these processes are discussed in the next two modules.

ConceptBOOST))

How Electron Movement Can Be Harnessed to Do Work

Using the movement of electrons to power the work of producing ATP is done in a way very similar to the function of a simple battery. A battery contains separate compartments with two metals (common examples are zinc and copper) that have differing *electron affinities*—that is, they attract electrons with unequal strength. As you can see here, when the two compartments are connected via a metal wire, electrons leave the metal with the lower electron affinity (zinc) and flow toward the metal with the higher electron affinity (copper):

The flow of electrons illuminates the light bulb.

Electrons flow away from zinc, which is being oxidized.

Electrons flow toward copper, which is being reduced.

Electron

Zinc metal (low affinity for electrons)

Copper metal (high affinity for electrons)

In the process, zinc is oxidized and copper is reduced. If this metal wire is placed within a device such as a light bulb or a motor, the spontaneous flow of electrons through the motor provides a force, the electromotive force, which can perform work. A battery "dies" when the metal with low electron affinity is completely oxidized and so can no longer supply electrons to the metal in the other compartment. Our cells contain "biological batteries" that work by this same principle. ■

Quick Check

☐ 7. In an oxidation-reduction reaction, what happens to the chemical that is oxidized? What happens to the chemical that is reduced?

☐ 8. What is electron affinity? Do electrons flow toward a chemical with a higher or lower electron affinity?

☐ 9. What is the electromotive force?

Apply What You Learned

☐ 1. Would the reaction involving two stable chemicals likely be endergonic or exergonic? Explain. How would this change if the two chemicals were unstable? Why?

☐ 2. Explain how a fire that burns coal is similar to the reactions that take place in your cells.

☐ 3. Riboflavin (ry-boh-FLAY-vin), also called vitamin B2, is required for synthesis of FAD. Predict the effects of a diet that is deficient in riboflavin.

See answers in Appendix A.

23.2
Glucose Catabolism and ATP Synthesis

Learning Outcomes

1. Explain the overall reaction for glucose catabolism.
2. Describe the processes of glycolysis, formation of acetyl CoA, and the citric acid cycle.
3. Describe the process of the electron transport chain.
4. Discuss the process of chemiosmosis and its role in ATP production.
5. Give the energy yield of each part of glucose catabolism and the overall energy yield of glucose catabolism.

Glucose catabolism refers to the reactions that involve the breakdown of the monosaccharide glucose and the use of the chemical energy in its bonds to drive ATP synthesis. These reactions are critical, as glucose is the preferred carbohydrate fuel for essentially every cell in the body. Many cells can use the monosaccharides fructose and galactose (which are isomers of glucose), as well, but they are almost always converted to glucose before they are catabolized.

Glucose catabolism has two main parts, each of which generates ATP: (1) *glycolysis* (gly-KAWL-uh-sis), a series of reactions in the cytosol that split glucose; and (2) the *citric acid cycle,* a series of reactions in the mitochondrial matrix that breaks down glucose further. Then, to use the energy liberated by glucose catabolism, a series of oxidation-reduction reactions known as *oxidative phosphorylation* takes place in the inner mitochondrial membrane. This series of reactions involves the transfer of electrons between electron carriers known collectively as the *electron transport chain* (*ETC*), and this leads to ATP synthesis. You can see how these processes are related in **Figure 23.3**. The upcoming sections present each in greater depth.

Overview of Glucose Catabolism and ATP Synthesis

« FLASHBACK

1. How does creatine phosphate work in a skeletal muscle fiber? (p. 360)
2. How do oxidative catabolism and glycolytic catabolism differ? (pp. 360–363)

Glucose contains a great deal of potential energy in its bonds, and that energy is released when it is catabolized. The energy released from glucose catabolism can be used to produce ATP in two ways:

- **Substrate-level phosphorylation. Substrate-level phosphorylation** involves the transfer of a phosphate group

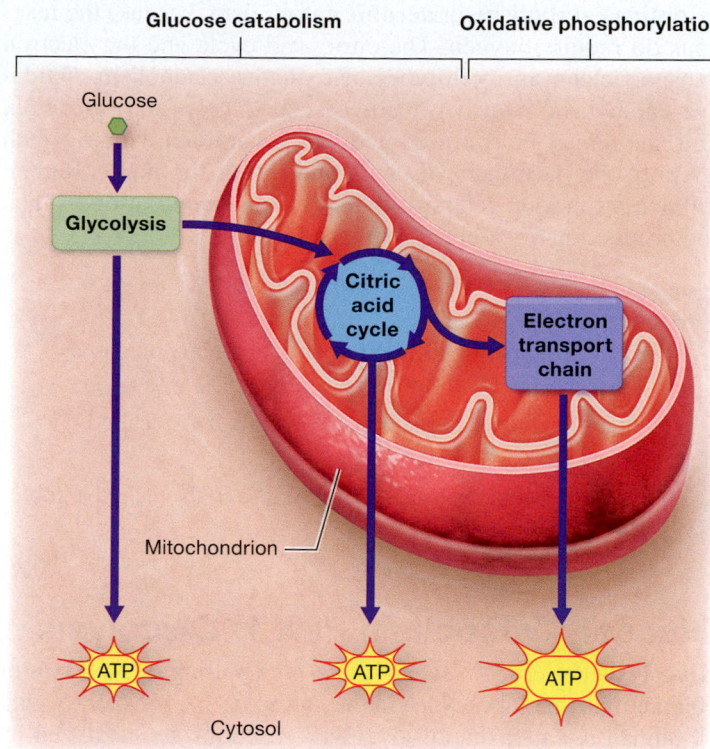

Figure 23.3 Overview of glucose catabolism and oxidative phosphorylation.

directly from a phosphate-containing chemical (the substrate) to ADP, to form ATP. A simple example is one that you saw in the muscle tissue chapter—that of creatine phosphate (see Chapter 10). Recall that creatine phosphate donates its phosphate group to ADP, producing ATP:

Creatine phosphate + ADP → ATP + Creatine

In this reaction, creatine phosphate is the substrate that is phosphorylating ADP to yield ATP. Similar reactions occur during glycolysis and the citric acid cycle, as energy released from glucose catabolism drives the transfer of a phosphate group from a substrate to ADP.

- **Oxidative phosphorylation.** During **oxidative phosphorylation,** the energy from the flow of electrons is harnessed in a process that generates ATP (recall our earlier battery example). Oxidative phosphorylation produces a great deal more ATP for the cell than does substrate-level phosphorylation. Note that oxidative phosphorylation occurs in the mitochondria as part of the ETC. The reactions of glucose catabolism in glycolysis and the citric acid cycle reduce electron carriers, which are then fed into the ETC and participate in oxidative phosphorylation.

The reactions of glucose catabolism can be divided broadly into two classes. The first is known as **glycolytic catabolism,** or **anaerobic catabolism,** because the reactions can take place in the absence of oxygen (an-aehr-OH-bik; *an-* = "without," *aero-* = "air"). As implied by the name, the process of glycolysis involves glycolytic catabolism. The second class is called

oxidative catabolism, or **aerobic catabolism,** because the reactions do require oxygen. The citric acid cycle and the electron transport chain are both types of oxidative catabolism. Oxidative catabolism is also sometimes called *cellular respiration.* This should not be confused with pulmonary ventilation, which refers to the movement of air in and out of the lungs. Rather, cellular respiration refers to the cell's consumption of oxygen and production of carbon dioxide and water.

Quick Check

☐ 1. How do substrate-level phosphorylation and oxidative phosphorylation differ?

☐ 2. What is glycolytic catabolism? Why is it also called anaerobic catabolism?

☐ 3. What is oxidative catabolism? Why is it also called aerobic catabolism?

Glucose Catabolism Part 1: Glycolysis

Glycolysis (*glyco-* = "sugar," *lysis* = "splitting") is a series of 10 reactions that takes place in the cytosol (**Figure 23.4**). As with all human metabolism, all glycolytic reactions are catalyzed by enzymes. Its name suggests what happens during glycolysis: Six-carbon glucose is split into two three-carbon sugars called **pyruvate** (py-ROO-vayt; also referred to as *pyruvic acid*). Glycolysis occurs in every cell in the body, even cells that lack mitochondria and other organelles, such as erythrocytes. The reactions of glycolysis can be broadly separated into two phases: the *energy investment phase,* during which a small amount of ATP is actually consumed, and the *energy payoff phase,* during which NADH and ATP are made. Although glycolysis consists of 10 separate reactions (5 in each phase), we have combined the reactions into five steps for simplicity. Let's look at what happens in each phase.

Energy Investment Phase

The energy investment phase of glycolysis consists of five reactions that have been condensed here into three major steps. They proceed as follows:

①　**First phosphorylation: Glucose is phosphorylated by ATP, yielding glucose-6-phosphate and ADP.** The six-carbon glucose is phosphorylated in the first reaction of this phase, "spending" an ATP molecule. As you can see in Figure 23.4, a phosphate group is removed from ATP and attached to glucose, producing *glucose-6-phosphate* (and ADP). This reaction is critical to glucose catabolism because it effectively traps the glucose in the cytosol and prevents it from leaving the cell.

②　**Second phosphorylation: The carbon atoms in glucose-6-phosphate are rearranged, which is then phosphorylated by another ATP, yielding fructose-1,6-bisphosphate and ADP.** In reactions two and three, the carbons of glucose-6-phosphate are rearranged and a second phosphorylation reaction takes place, which consumes another molecule

of ATP. This reaction yields the six-carbon sugar fructose-1,6-bisphosphate (and ADP).

③　**Cleavage: The six-carbon fructose-1,6-bisphosphate is split, and two three-carbon compounds are formed.** During reactions four and five, the six-carbon sugar fructose is split, yielding two three-carbon compounds: glyceraldehyde-3-phosphate and dihydroxyacetone-3-phosphate. The latter compound undergoes another reaction and converts into a molecule of glyceraldehyde-3-phosphate.

At the end of the energy investment phase of glycolysis, we have "spent" two molecules of ATP and are left with two three-carbon compounds, each of which contains a phosphate group. These compounds are now ready to enter the energy payoff phase.

Energy Payoff Phase

The energy payoff phase of glycolysis consists of five reactions that we have condensed into two major steps. During these reactions, the phosphate groups of the three-carbon sugars are transferred to ADP, to yield ATP, and the compounds are oxidized to produce NADH. Remember that each three-carbon compound produced in the energy investment phase goes through these reactions, so there are two rounds of energy payoff reactions for each glucose. The reactions proceed as follows:

④　**Oxidation: Glyceraldehyde-3-phosphate is phosphorylated and oxidized by NAD^+ to yield NADH and 1,3-bisphosphoglycerate, which then donates a P_i to ADP, yielding ATP.** During reaction six, each of the glyceraldehyde-3-phosphates is given a second phosphate group, after which they are oxidized and their electrons are transferred to NAD^+. This produces two molecules of the reduced NADH. In reaction seven, one of the phosphate groups of each 1,3-bisphosphoglycerate is transferred to ADP, yielding two molecules of 3-phosphoglycerate and two molecules of ATP by substrate-level phosphorylation.

⑤　**ATP synthesis: The carbon atoms in the two molecules of 3-phosphoglycerate are rearranged to form phosphoenolpyruvate, which donates a P_i to ADP, yielding ATP and pyruvate.** After rearranging the carbon atoms in the two molecules of 3-phosphoglycerate to produce two molecules of phosphoenolpyruvate in reactions eight and nine, the remaining phosphate groups on the phosphoenolpyruvate are transferred to ADP via substrate-level phosphorylation in reaction ten. This produces two ATP and two three-carbon pyruvates.

Overall Yield of Glycolysis

So, in a nutshell, what have we accomplished during glycolysis? At the conclusion of the energy payoff phase, we have:

- spent two ATP,
- synthesized four ATP,
- synthesized two NADH, and
- split glucose into two three-carbon pyruvates.

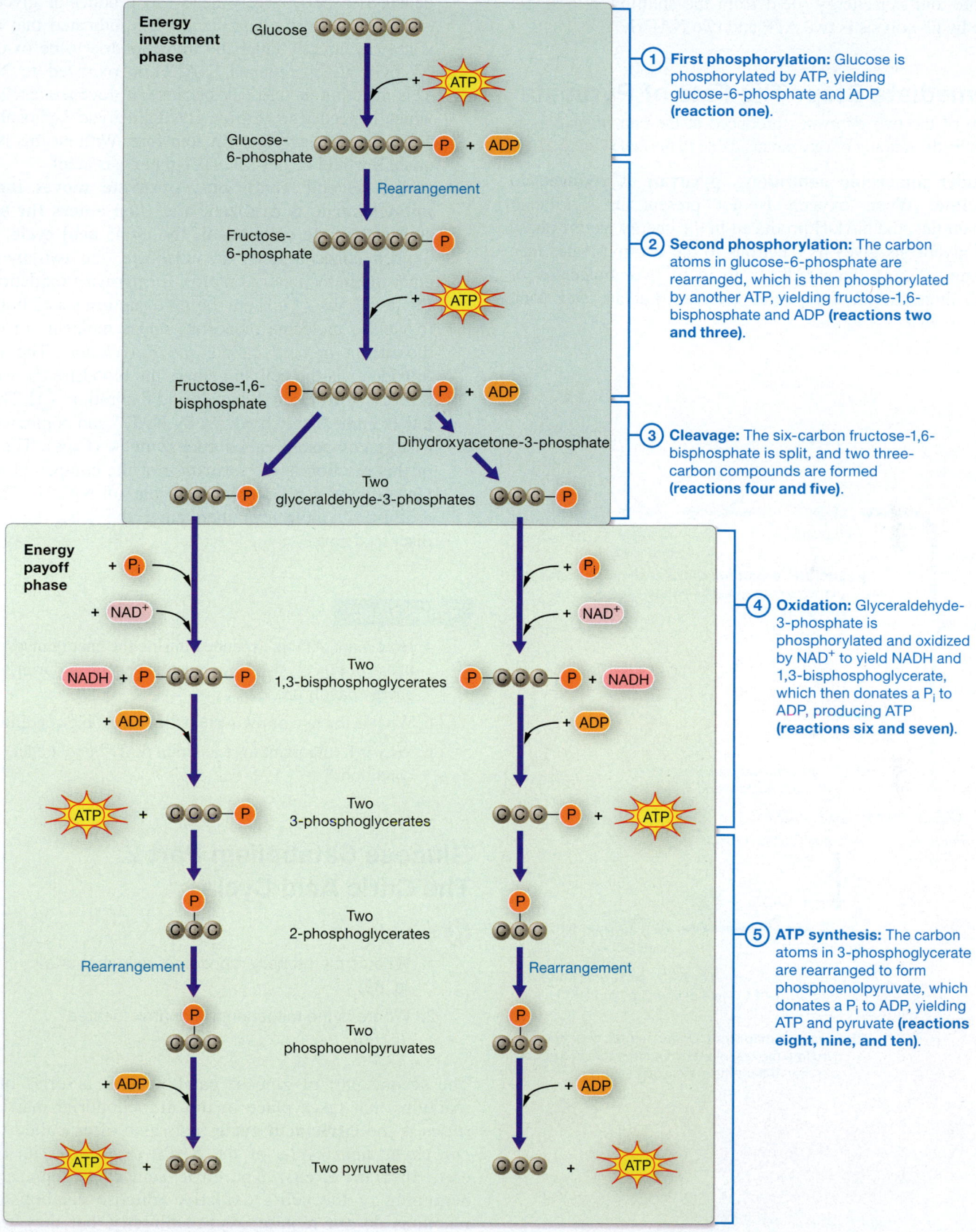

Figure 23.4 The reactions of glycolysis.

Therefore, our *net* energy yield from the splitting of a single glucose by glycolysis is two ATP and two NADH.

Intermediate Step: The Fate of Pyruvate

The fate of the two pyruvates produced at the end of glycolysis depends on the amount of oxygen available to the cell (**Figure 23.5**):

- **Under anaerobic conditions, pyruvate is reduced to lactate.** When oxygen is not present in significant quantities, the NADH produced in the energy payoff phase of glycolysis gives its electrons to the two pyruvates (see Figure 23.5a). This reduces them to two molecules of the three-carbon compound **lactate.** (Lactate was long

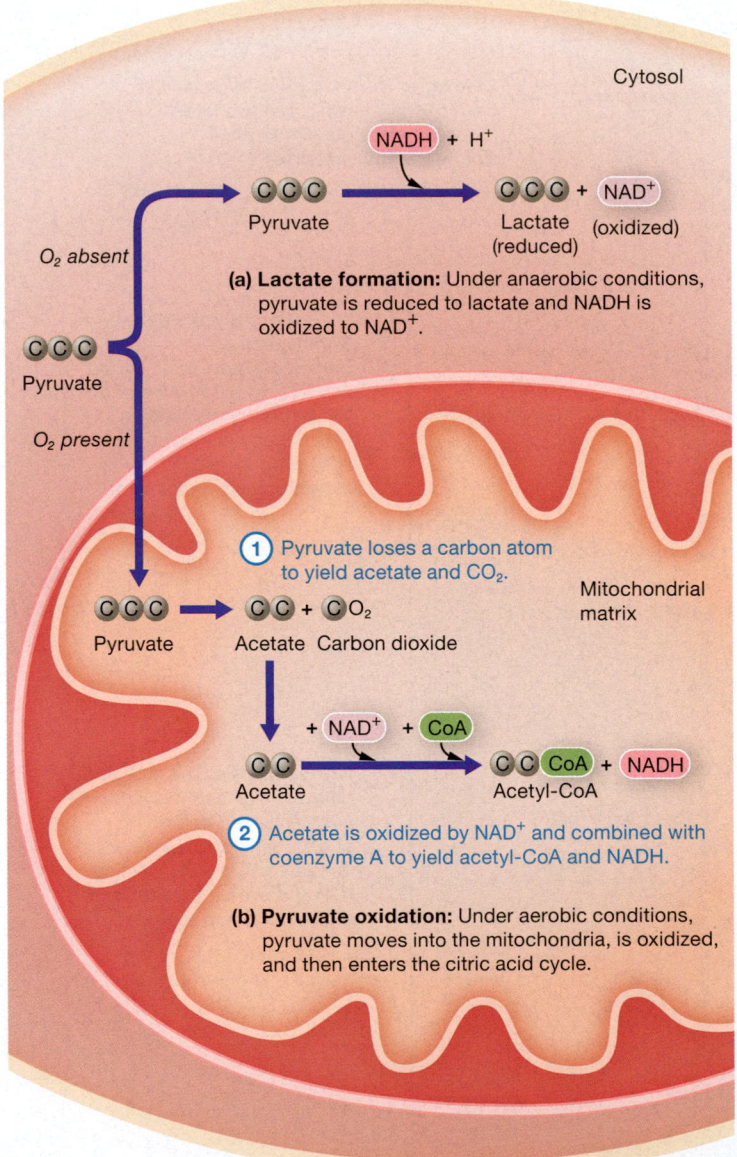

(a) Lactate formation: Under anaerobic conditions, pyruvate is reduced to lactate and NADH is oxidized to NAD^+.

(b) Pyruvate oxidation: Under aerobic conditions, pyruvate moves into the mitochondria, is oxidized, and then enters the citric acid cycle.

① Pyruvate loses a carbon atom to yield acetate and CO_2.

② Acetate is oxidized by NAD^+ and combined with coenzyme A to yield acetyl-CoA and NADH.

Figure 23.5 Intermediate step: the fate of pyruvate after glycolysis. Note that two molecules of pyruvate are produced for every molecule of glucose broken down.

believed to be a largely dead-end product of glycolysis, but research in the past decade has indicated that lactate is able to directly enter the mitochondria to be oxidized.) As pyruvate is reduced, NADH is oxidized to NAD^+. This reaction is critical because it regenerates NAD^+, which is needed to oxidize glyceraldehyde-3-phosphate to 1,3-bisphosphoglycerate in step four. Without this NAD^+, no ATP would be generated in glycolysis at all.

- **Under aerobic conditions, pyruvate moves into the mitochondria, is oxidized, and then enters the second step of glucose catabolism, the citric acid cycle.** When oxygen concentrations are adequate, the two pyruvates enter the mitochondrial matrix for pyruvate oxidation (see Figure 23.5b). ① There the pyruvate loses a carbon atom to yield a molecule of acetate and a molecule of carbon dioxide, a process called *decarboxylation.* The carbon dioxide exits the cell and enters the bloodstream, where it is released from the lungs during expiration. ② The acetate is immediately oxidized by NAD^+ and combined with a large compound called **coenzyme A (CoA).** The result of this reaction is the formation of the compound **acetyl-CoA** (uh-SEE-til) and a molecule of NADH. The two acetyl-CoA molecules then enter the citric acid cycle, discussed next.

Quick Check

☐ 4. How much ATP is expended during the energy investment phase of glycolysis? How much is generated during the energy payoff phase?

☐ 5. What is the fate of pyruvate under anaerobic conditions?

☐ 6. Why is it important to regenerate NAD^+ under anaerobic conditions?

Glucose Catabolism Part 2: The Citric Acid Cycle

⟪ FLASHBACK

1. What is the primary function of mitochondria? (p. 89)

2. Where is the mitochondrial matrix located? (p. 89)

The second part of glucose catabolism is a series of eight reactions that takes place in the mitochondrial matrix; this series is the **citric acid cycle** (it is also often called the *tricarboxylic acid cycle,* or the *Krebs cycle* after its discoverer, Sir Hans Krebs). The citric acid cycle represents the beginning of the cell's oxidative glucose catabolism. The reactions do not require oxygen directly, but they will not proceed unless sufficient oxygen is present. These reactions produce a relatively small amount of ATP by substrate-level phosphorylation.

When acetyl-CoA from pyruvate oxidation enters the citric acid cycle, it proceeds as follows (**Figure 23.6**; note that we have condensed the eight reactions into four steps):

① **Citrate synthesis: Acetyl-CoA combines with oxaloacetate to form citrate and CoA.** In reaction one, the acetyl-CoA "donates" its two-carbon acetyl group to a four-carbon compound called oxaloacetate, forming the six-carbon compound **citrate** (for which the cycle was named) and leaving CoA. Acetyl-CoA in this reaction acts a little like an ATP—acetyl-CoA has a great deal of potential energy, as its bond is unstable. When the bond is broken, the resulting CoA and acetate have less energy, and the released energy helps to unite acetate and oxaloacetate.

② **First oxidation: Citrate is rearranged, then oxidized by NAD⁺, generating CO_2 and NADH.** In reactions two through four, the atoms in the six-carbon citrate are rearranged and oxidized, eventually forming the four-carbon compound *succinyl-CoA* (containing the same CoA that was part of acetyl-CoA). The two removed carbons are lost as two molecules of carbon dioxide, and the removed electrons are accepted by two NAD^+, producing two NADH.

③ **ATP synthesis: Succinyl-CoA is converted to succinate and CoA, while forming ATP.** In the fifth reaction, the CoA is removed from succinyl-CoA. In this reaction we again see an unstable bond between CoA and its partner. The energy released when CoA is removed from succinate is used to drive the formation of ATP by substrate-level phosphorylation (sometimes the very similar compound *GTP*, or *guanosine triphosphate*, is produced instead).

④ **Second oxidation: Succinate is oxidized by FAD and NAD⁺, generating FADH₂ and NADH, and is converted back to oxaloacetate.** Reactions six through eight of the citric acid cycle see succinate oxidized to other four-carbon compounds until it is converted back into oxaloacetate. As the compounds are oxidized, their electrons are transferred to FAD and NAD^+, generating one each of $FADH_2$ and NADH.

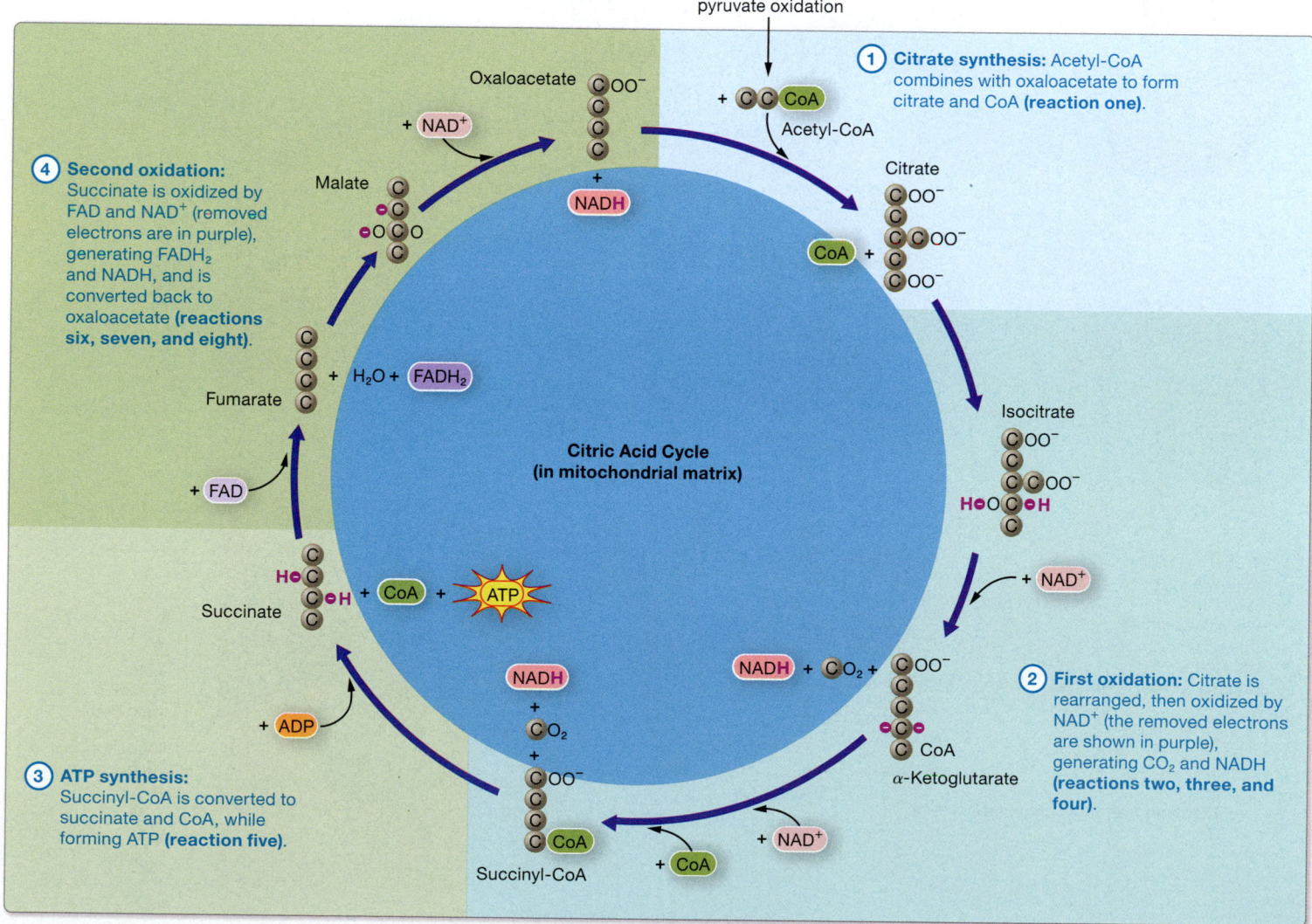

Figure 23.6 The citric acid cycle. For each glucose broken down, this cycle "turns" twice.

By the end of the citric acid cycle, we have regenerated oxaloacetate, the compound with which we started in step one. This is why this process is called a *cycle*—it begins and ends with the same compound.

Note that each glucose yields two acetyl-CoA molecules, so for each glucose, this cycle "turns" twice. Keeping this in mind, we can tally up the overall yield per glucose from the start of glucose catabolism:

- ten NADH (two from glycolysis, two from pyruvate oxidation, and six from the citric acid cycle),
- two $FADH_2$, and
- four ATP (two from glycolysis and two from the citric acid cycle).

As you can see, we have generated very little energy in the form of ATP from these reactions. However, we have conserved the chemical energy in the original glucose by passing off its electrons to NADH and $FADH_2$. Both of these electron carriers will now move on to the process of ATP synthesis with the electron transport chain.

ATP Synthesis: The Electron Transport Chain and Oxidative Phosphorylation

« FLASHBACK

1. How is potential energy present in a concentration gradient? (p. 75)

We are finally at the stage where most of the potential energy of glucose is used to make ATP: oxidative phosphorylation. Oxidative phosphorylation involves three processes: (1) the transfer of electrons between electron carriers, (2) the generation and maintenance of a hydrogen ion concentration gradient, and (3) the use of this gradient to drive the release of ATP. This section examines these processes.

Electron Transfer and Proton Pumping

The energy stored in the electrons generated in glucose catabolism and accepted by NADH and $FADH_2$ is released by a group of more than 15 different electron carriers collectively called the **electron transport chain (ETC).** As you can see in **Figure 23.7a**, the electron carriers of the ETC are contained

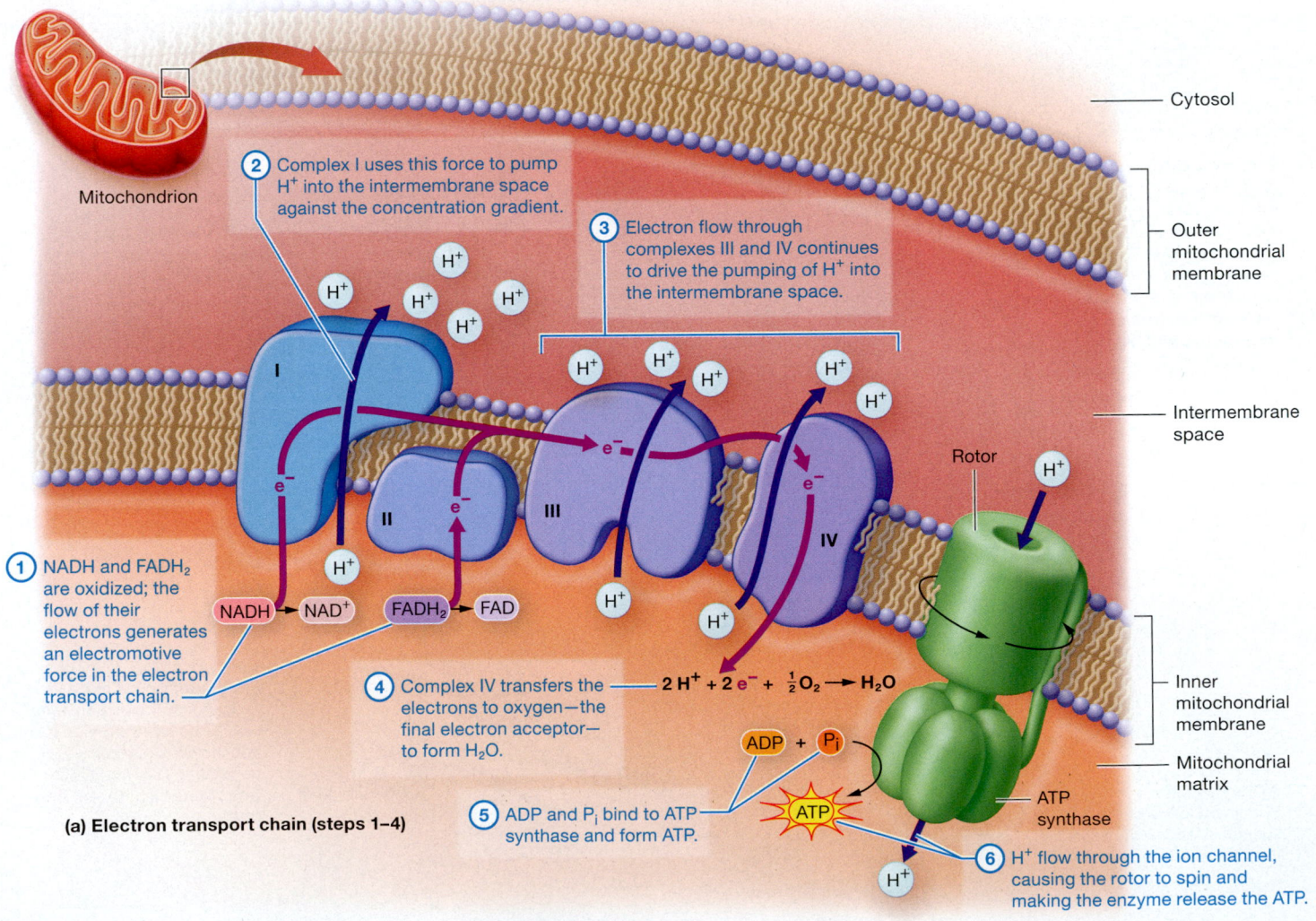

(a) Electron transport chain (steps 1–4)

(b) Oxidative phosphorylation (steps 5 and 6)

Figure 23.7 The electron transport chain and oxidative phosphorylation.

within four large enzyme complexes (numbered I–IV) embedded in the inner mitochondrial membrane.

These carriers are also sometimes called the *electron transfer chain,* in recognition of the fact that electrons are transferred between them. Remember that these electrons are usually passed between carriers as part of hydrogen atoms. The events of oxidative phosphorylation unfold as follows (see Figure 23.7a):

① **NADH and FADH$_2$ are oxidized; the flow of their electrons generates an electromotive force in the electron transport chain.** NADH is oxidized to NAD$^+$, its electrons are passed to the first electron carrier in enzyme complex I, and the electron carrier is reduced. As this process continues, the flow of electrons acts as a "cellular battery" and generates an electromotive force in the ETC. (FADH$_2$ is oxidized in a similar manner using a different electron carrier in complex II.)

② **Complex I uses this force to pump hydrogen ions into the intermembrane space against the concentration gradient.** Complex I harnesses the energy of this electromotive force, just as we saw with our earlier battery example, to pump hydrogen ions against the concentration gradient. Hydrogen ions are pumped from the mitochondrial matrix, where their concentration is relatively low, into the intermembrane space, where their concentration is higher. This first establishes and then maintains a steep hydrogen ion concentration gradient. (Hydrogen ions are also often called *protons* because that is their sole subatomic particle, so the movement of hydrogen ions is sometimes referred to as *proton pumping.*) Note that some of the electrons may also be transferred through complex II, which does not contribute to hydrogen ion pumping.

③ **Electron flow through complexes III and IV continues to drive the pumping of hydrogen ions into the intermembrane space.** As you can see in the figure, electrons continue to be transferred, first to complex III and then complex IV. As they gain electrons, the electron carriers in these complexes are reduced and hydrogen ions continue to be pumped into the intermembrane space.

④ **Complex IV transfers the electrons to oxygen—the final electron acceptor—to form H$_2$O.** Complex IV transfers electrons to the final electron acceptor, oxygen. In this reaction, oxygen is reduced and accepts the electrons while combining with two hydrogen ions. The end result of this reduction is the following reaction:

$$2\,H^+ + 2\,e^- + \frac{1}{2}\,O_2 \rightarrow H_2O$$

In other words, one oxygen atom forms a covalent bond with two hydrogen atoms to produce water. This step is absolutely critical—without oxygen acting as the final electron acceptor, glucose catabolism stops at glycolysis. In fact, this is the main reason why we breathe oxygen, and why we cannot live without it. For information on what happens when this step is blocked, see *A&P in the Real World: Cyanide and the ETC* on p. 916.

At this point, you may be wondering why electrons have to pass through so many electron carriers before they get to oxygen. Although it would be simpler to transfer the electrons directly from NADH and FADH$_2$ to oxygen, doing so would cause most of the energy of the reaction to be lost as heat. The fact that this process proceeds by several small steps allows more of the released energy to be harnessed to do work, pumping hydrogen ions into the intermembrane space against the concentration gradient.

ConceptBOOST)))

Why Do We Breathe?

Breathing, during which we inspire oxygen-rich air and expire carbon dioxide-rich air, is one of our most fundamental processes necessary to sustain life. But why, exactly do we need to inspire oxygen and expire carbon dioxide?

To answer this question, first let's dispel a common misconception that oxygen is somehow converted into carbon dioxide in the body's cells. Although it is true that both carbon dioxide and oxygen have to do with catabolism, the two do not directly interact in any way, nor does one ever become the other.

Next, let's clarify the roles and fates of oxygen and carbon dioxide:

- **Oxygen accepts the final electrons in the ETC to become a molecule of water.** The oxygen that we inhale is picked up by the bloodstream and delivered to our cells. In the cell, it ends up in the inner mitochondrial membrane, where it accepts the final electrons in the ETC. When oxygen accepts the electrons, it becomes part of a molecule of water. So, the oxidation-reduction reactions in the mitochondria generate water as a final end product (sometimes called metabolic water).

- **Carbon dioxide comes from carbons lost during carbohydrate catabolism.** Before pyruvate enters the citric acid cycle, it loses one of its carbons as a molecule of carbon dioxide. Similarly, carbons are removed during the citric acid cycle and are also lost as carbon dioxide. As we've mentioned, these carbon dioxide molecules diffuse through the cytosol, exit the cell, and enter the bloodstream; when they reach the lungs, they are exhaled.

Now we can answer the question of why we breathe, which is really two questions. First, why do we inhale oxygen? Acting as the final electron acceptor might seem to be such a small task that oxygen performs, given all the processes that take place in the human body. But if oxygen doesn't do this job for about 4 minutes, we can't sustain life. This is because depriving a cell of oxygen is like hitting the "off" switch of the ETC, which then shuts off the vast majority of the cell's ATP production. So, clearly we must take in enough oxygen to act as the final electron acceptor so that we can produce enough ATP to meet the body's needs.

The second question is why do we have to exhale the carbon dioxide we produce? You know from the respiratory chapter that if carbon dioxide accumulates in the blood, it affects the blood pH, making it more acidic. The human body has very little tolerance for pH swings, and a drop in pH of just 0.4 from the normal range can cause coma and death. For this reason, it's important that we balance carbon dioxide production with carbon dioxide removal by expiration.

Finally, let's think about this in big picture terms. We know that the human body generates a huge amount of ATP per day—about 132 *pounds*. So, consider our tissues, organs, and organ systems, and how many of them are dedicated, directly or indirectly, to ATP production. The heart pumps oxygen and carbon dioxide to and from the lungs through blood vessels. Hemoglobin transports oxygen to cells. The digestive organs digest and absorb nutrients for cells to oxidatively catabolize, and multiple endocrine glands produce hormones that ensure nutrient fuels are available to cells. And last, the respiratory system, obviously, brings in oxygen and removes carbon dioxide. All these processes enable continuous ATP production to support our constant, enormous need for ATP. ∎

ATP Synthesis

At this point, the energy generated by the "cellular battery" has pumped hydrogen ions against the concentration gradient from the mitochondrial matrix into the intermembrane space. The result of this pumping is that we now have a steep concentration gradient, with more hydrogen ions in the intermembrane space than in the mitochondrial matrix. We also have an electrical gradient, with more positive charges in the intermembrane space than in the

 **CORE PRINCIPLE** Gradients

matrix, bringing us back to the Gradients Core Principle (p. 28).

The two gradients combined make up an electrochemical gradient, which is a source of potential energy (see Chapter 11). Overall, the potential energy in the bonds of the glucose has been transformed into the potential energy of this electrochemical gradient. This potential energy becomes kinetic energy when hydrogen ions flow through an ion channel across the membrane. The kinetic energy is then used to perform work.

Using an electrochemical gradient to do work is a type of mechanism known as *chemiosmosis*. You can visualize how chemiosmosis works with the simple analogy of a hydroelectric dam—when the dam is opened, the potential energy of the water becomes kinetic energy as the water flows, and the flow of water generates a force that can be used to perform work, such as turning a turbine to produce electricity.

The flow of hydrogen ions back into the matrix down the electrochemical gradient drives the work of ATP synthesis by a mechanism surprisingly similar to our dam example. In this case, the "turbine" is part of a large protein in the inner mitochondrial membrane called **ATP synthase.** This protein

functions as both an ion channel through the membrane and an enzyme that catalyzes the production of ATP, as you can see in **Figure 23.7b**. The process continues as follows:

⑤ **ADP and P$_i$ bind to ATP synthase and form ATP.** ATP synthesis begins when ADP and P$_i$ bind to the enzyme portion of ATP synthase, uniting them and forming ATP. Since this enzyme is most stable when bound to ATP, it holds on to it tightly. ATP cannot be used by the cell until it is released from this enzyme.

⑥ **Hydrogen ions flow through the ion channel, causing the rotor to spin and making the enzyme release the ATP.** Now the energy from glucose catabolism finally comes into play. The electrochemical gradient causes hydrogen ions to flow into the matrix through the ion channel of ATP synthase, which generates a force called the *proton motive force*. The electrochemical energy of the proton motive force causes a portion of the enzyme called the *rotor* to start spinning just like a turbine. The spinning rotor causes the enzyme to release its ATP. This process is repeated, with a new ADP and P$_i$ binding and joining into ATP, and the spinning rotor causing ATP to be released. So, the majority of the energy harnessed from the oxidation of glucose is used to create a hydrogen ion gradient. The energy of the hydrogen ion gradient then causes ATP synthase to release the newly formed ATP for the cell to use.

Quick Check

☐ 7. What products are generated by the citric acid cycle for each glucose?

☐ 8. What is the role of oxygen in glucose catabolism?

☐ 9. What is the potential energy in the bonds of glucose used to fuel *directly*?

Putting It All Together: The Big Picture of Glucose Catabolism and ATP Synthesis

Now that we have progressed through all the parts of glucose catabolism and ATP synthesis, we can summarize the events and figure out how much ATP we have formed from glucose (**Figure 23.8**). The overall reaction of glucose catabolism is as follows:

$$C_6H_{12}O_6 + 6\,O_2 \rightarrow 6\,H_2O + 6\,CO_2 + 38\,ATP + heat$$

Here we are assuming that adequate oxygen is available and oxidative catabolism can take place.

ConceptBOOST 📣

ATP Yield from Oxidative Catabolism

It can be difficult to keep track of all the ATP that is generated from the different processes of oxidative catabolism, so let's

The Big Picture of Glucose Catabolism and Oxidative Phosphorylation

Figure 23.8

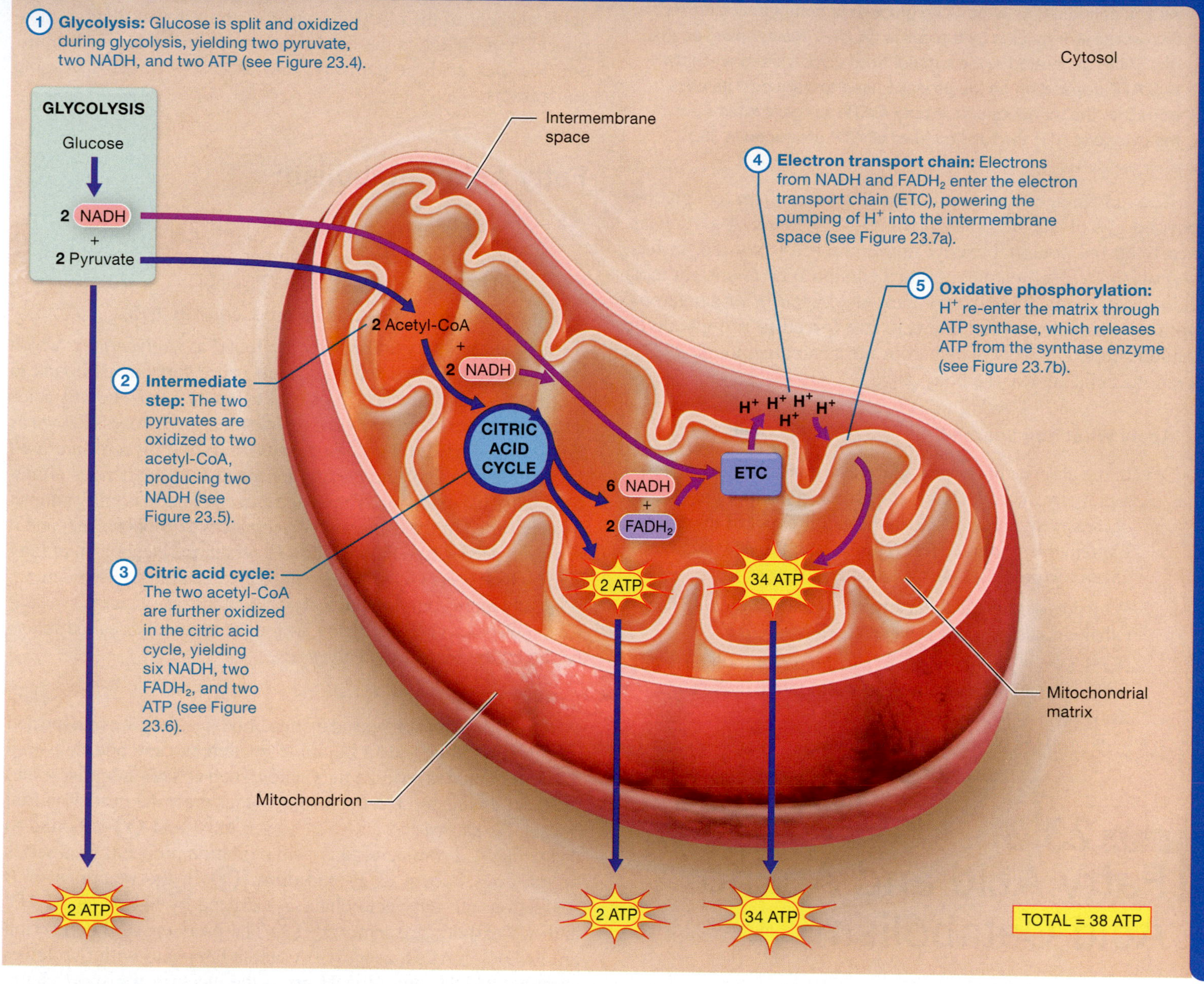

① **Glycolysis:** Glucose is split and oxidized during glycolysis, yielding two pyruvate, two NADH, and two ATP (see Figure 23.4).

Cytosol

GLYCOLYSIS

Glucose

2 NADH
+
2 Pyruvate

Intermembrane space

④ **Electron transport chain:** Electrons from NADH and FADH$_2$ enter the electron transport chain (ETC), powering the pumping of H$^+$ into the intermembrane space (see Figure 23.7a).

⑤ **Oxidative phosphorylation:** H$^+$ re-enter the matrix through ATP synthase, which releases ATP from the synthase enzyme (see Figure 23.7b).

② **Intermediate step:** The two pyruvates are oxidized to two acetyl-CoA, producing two NADH (see Figure 23.5).

2 Acetyl-CoA
+
2 NADH

CITRIC ACID CYCLE

6 NADH
+
2 FADH$_2$

H$^+$ H$^+$ H$^+$ H$^+$
H$^+$

ETC

③ **Citric acid cycle:** The two acetyl-CoA are further oxidized in the citric acid cycle, yielding six NADH, two FADH$_2$, and two ATP (see Figure 23.6).

2 ATP

34 ATP

Mitochondrial matrix

Mitochondrion

2 ATP

2 ATP

34 ATP

TOTAL = 38 ATP

calculate the total produced per glucose. It's easy enough to count the ATP produced from glycolysis per glucose—two—and from the citric acid cycle per glucose—also two. But what about the NADH and FADH$_2$? The oxidation-reduction reactions of one NADH release enough energy to pump enough hydrogen ions into the intermembrane space to drive the release of three ATP molecules from ATP synthase (we're rounding up a bit to use whole numbers here). The energy released from the oxidation-reduction reactions of FADH$_2$ is slightly less, and yields enough of a gradient to drive the release of about two ATP molecules. So, we can say that NADH correlates with the production of roughly three ATP molecules, and FADH$_2$ with roughly two.

This brings us to the total shown in the table in the next column.

Process	ATP from glucose catabolism	Electron carriers from glucose catabolism	Electron carriers going through ETC and oxidative phosphorylation	Final yield of ATP
Glycolysis	2 ATP			2
		2 NADH	2 × 3 ATP	6
Pyruvate oxidation		2 NADH	2 × 3 ATP	6
Citric acid cycle	2 ATP			2
		6 NADH	6 × 3 ATP	18
		2 FADH$_2$	2 × 2 ATP	4
				TOTAL = 38 ATP

As you can see then, when one glucose is completely oxidized, the overall energy gain by the cell is 38 ATP. (Note, however, that the NADH generated in the cytosol from glycolysis may yield 4 or 6 ATP depending on the process used to transport its electrons into the mitochondria. If only 4 ATP are generated from the cytosolic ATP, the total number of ATP produced is 36.) ■

The advantages of continuing the catabolism of glucose beyond glycolysis are pretty obvious—if we stop at glycolysis, nearly all the potential energy in the glucose is wasted. In fact, glycolysis by itself cannot produce enough ATP to sustain most cells in the body. This is why lack of oxygen is so dangerous; without adequate oxygen, the cells are forced to use glycolytic catabolism exclusively, which does not generate enough ATP to keep the cells, and so the body, alive.

Apply What You Learned

☐ 1. In the absence of oxygen, glucose is consumed extremely rapidly by the cell. However, glucose consumption decreases significantly when the cell is exposed to oxygen. Explain this finding. (*Hint:* Think about the ATP yield of anaerobic processes versus aerobic processes.)

☐ 2. Why do you think that the amount of carbon dioxide you exhale increases when you exercise?

☐ 3. Predict what would happen to a cell if a mutation resulted in the formation of insufficient oxaloacetate in the citric acid cycle.

See answers in Appendix A.

MODULE 23.3
Fatty Acid and Amino Acid Catabolism

Learning Outcomes

1. Describe lipolysis, deamination, and transamination.
2. Summarize the β-oxidation of fatty acids, and explain how it leads to ATP production and relates to ketogenesis.
3. Explain how amino acid catabolism leads to ATP production.
4. Describe the effect of amino acid catabolism on ammonia and urea production.

The cell is able to catabolize fatty acids and amino acids in addition to glucose. As we progress through this module, you will notice that fatty acid and amino acid catabolism shares many of the same products and pathways seen in glucose catabolism. This is because the same basic principles are at work with the catabolism of all three nutrients—each is converted into chemicals that the cell can oxidize, and its electrons are removed and passed to electron carriers such as NADH and $FADH_2$. Just as

with glucose, the electrons removed from fatty acids and amino acids are sent to the electron transport chain for oxidative phosphorylation, which is the final common pathway for each nutrient. We look at these pathways and processes in the upcoming subsections.

Fatty Acid Catabolism

« FLASHBACK

1. What are the chemical components and properties of fatty acids and triglycerides? (p. 53)
2. When are ketone bodies produced? (p. 615)

Recall that most fats in the body are found as *triglycerides,* which contain three long hydrocarbon chains called **fatty acids** bound to the modified sugar **glycerol** (look back at Figure 2.18). The fatty acids and glycerol are liberated by an enzyme-catalyzed process known as *lipolysis* (ly-PAWL-uh-sis). Both fatty acids and glycerol are used for fuel, but 95% of the energy generated by triglycerides comes from fatty acid oxidation. The catabolism of glycerol is fairly simple: It is converted to the compound glyceraldehyde-3-phosphate, and it enters glycolysis (see Figure 23.4). However, the catabolism of fatty acids, which occurs by a process called *beta-oxidation (β-oxidation),* is slightly more complicated. Fatty acid catabolism generates acetyl-CoA molecules via β-oxidation, as well as compounds called *ketone bodies* by the process of *ketogenesis.*

β-oxidation

When fatty acids are released from adipocytes, they are transported through the bloodstream bound to protein carriers and delivered to tissues that can oxidize fatty acids, such as skeletal muscle and the heart. Inside these cells, the fatty acids enter the mitochondrial matrix. As shown in **Figure 23.9**, each fatty acid ① is bound to coenzyme A. From here, fatty acid oxidation proceeds by a series of reactions known as β-**oxidation.** ② In the reactions of β-oxidation, the fatty acid chain is oxidized by NAD^+ and FAD, and two carbons are removed. ③ The products of this reaction are $FADH_2$, NADH, acetyl-CoA, and a fatty acid with two fewer carbons. $FADH_2$ and NADH go to the electron transport chain, and acetyl-CoA goes to the citric acid cycle. ④ Another CoA binds to the shorter fatty acid, and rounds of β-oxidation are

Cyanide and the ETC

The poison *cyanide* (SY-uh-ny'd) binds to a key enzyme of the ETC and inactivates it. This makes the cells unable to produce adequate ATP, even when the blood oxygen level is normal or elevated. Cyanide exposure is fatal only seconds after inhalation or injection; however, it may take several minutes to several hours for the poison to be fatal after ingestion or skin exposure. Several agents can be given as antidotes to cyanide poisoning, most of which either reverse its actions or bind to it and render it less harmful.

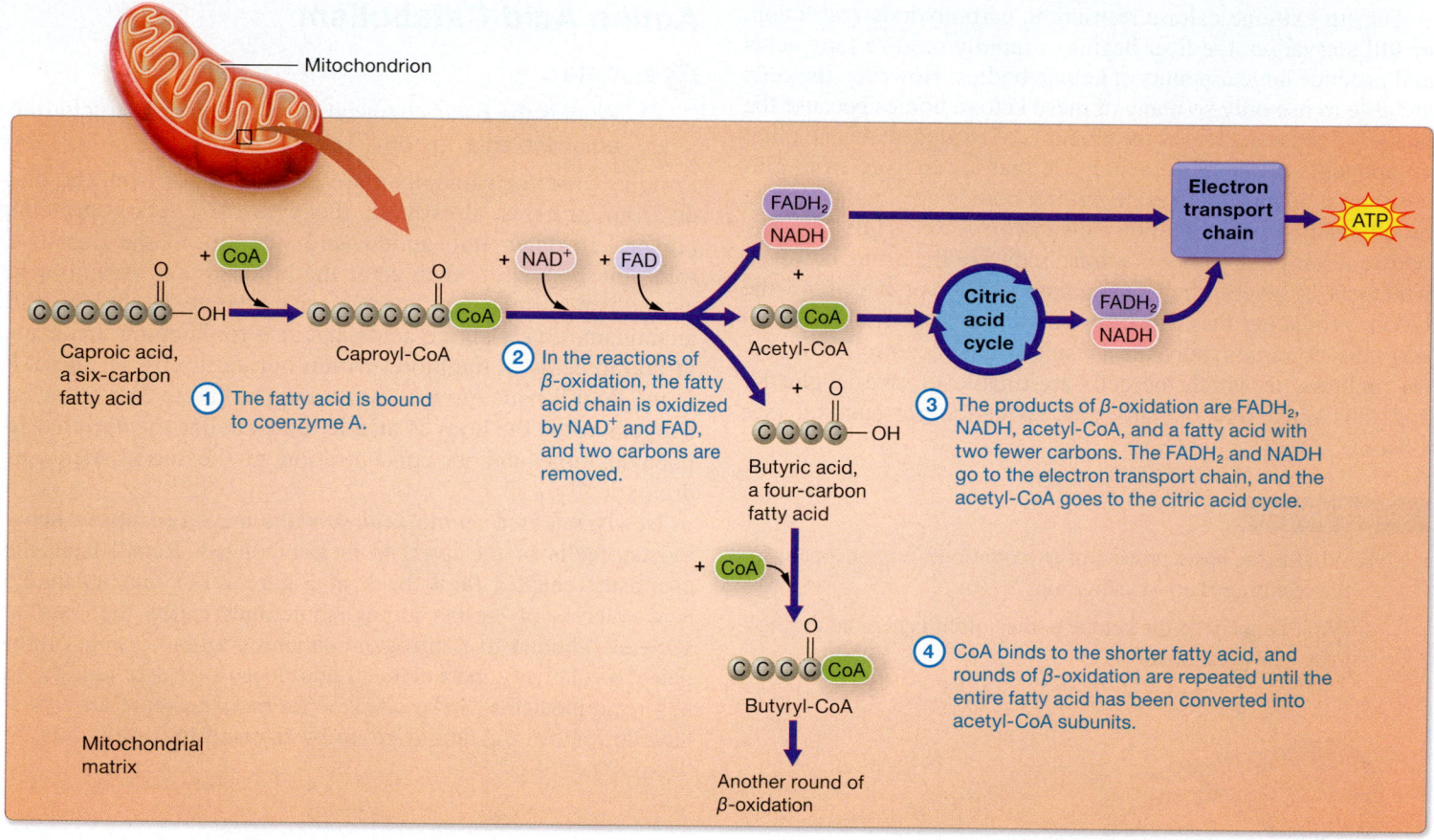

Figure 23.9 Fatty acid catabolism and β-oxidation.

repeated until the entire fatty acid has been converted into two-carbon acetyl-CoA subunits—fatty acids that have six carbons yield three acetyl-CoA molecules, fatty acids with four carbons yield two acetyl-CoA molecules, and so forth. The released acetyl-CoA molecules are then fed into the citric acid cycle and further oxidized by the same pathway that we discussed with glucose catabolism. The NADH and FADH₂ from β-oxidation and from the citric acid cycle donate their electrons to the electron transport chain, and ATP is generated.

You may have heard that fat has more than twice the "calories" of carbohydrate or protein. This means that the oxidation of fats releases at least twice as much energy (or heat) as the other two classes of nutrients. Why, though, does fatty acid catabolism generate so much more energy? Let's examine this using the 16-carbon fatty acid palmitic acid as an example: With each round of β-oxidation, one NADH and one FADH₂ are produced as the oxidation-reduction reactions progress. For palmitic acid, β-oxidation will repeat seven times to break the whole fatty acid into two-carbon units. When β-oxidation of palmitic acid is complete, we have generated seven NADH and FADH₂ each, which will then enter the electron transport chain to generate 35 ATP.

But we haven't finished with our ATP-generating potential, as we have also produced eight acetyl-CoA molecules that will be fed into the citric acid cycle and generate a total

of 96 more ATP. Clearly, then, the energy yield from fatty acids is much higher than that from a molecule of glucose, whose oxidation generates a maximum of 38 ATP. Note, though, that the energy yield from fatty acid oxidation will vary depending on the length of the fatty acid chain; those with longer chains yield more ATP, whereas those with shorter chains yield fewer.

Ketogenesis

Much of the β-oxidation in the body takes place in the liver, and under certain conditions the acetyl-CoA produced by β-oxidation can be converted into compounds called **ketone bodies** (KEE-tohn). This process, **ketogenesis,** occurs in the mitochondrial matrix and involves the combination of two acetyl-CoA molecules into a four-carbon ketone body. The liver cannot metabolize ketone bodies in any appreciable amount under normal conditions, so they are released into the bloodstream and delivered to other cells. In the cells, ketone bodies are split into two molecules of acetyl-CoA, after which they enter the citric acid cycle and are oxidized. Some tissues, such as those of the heart and skeletal muscle, take up and oxidize ketone bodies quite readily. Other tissues, such as the brain, are not as able to oxidize ketone bodies except under starvation conditions.

During extreme caloric restriction, carbohydrate restriction, or full starvation, the liver begins to rapidly oxidize fatty acids and produce large amounts of ketone bodies. However, the cells are able to use only so many of these ketone bodies because the lack of glucose decreases the amounts of oxaloacetate and other components of the citric acid cycle that are derived from glucose breakdown. As a result, ketone bodies accumulate in the blood, leading to a condition called *ketosis* (kee-TOH-sis). The body excretes many excess ketone bodies in the urine; however, if ketosis is severe, they accumulate and lower the pH of the blood, producing the dangerous condition *ketoacidosis*. Recall that this condition may also result from uncontrolled diabetes mellitus, in which the cells are unable to take in glucose and thus must rely on fatty acid oxidation for survival (see Chapter 16).

Quick Check

☐ 1. What is the basic process of β-oxidation? What happens to the resulting acetyl-CoA groups?

☐ 2. What happens to the ketone bodies often generated during fatty acid catabolism?

Amino Acid Catabolism

◀◀ FLASHBACK

1. What is the basic chemical composition and structure of an amino acid? (p. 56)

Proteins used for catabolism are derived either from the diet or from proteins already in the cytosol. Dietary proteins are broken down into amino acid subunits in the digestive tract; these subunits then enter the bloodstream. Amino acid catabolism is important not just for oxidative catabolism and anabolism of new amino acids, but also for keeping the levels of amino acids in the blood within normal limits. Read *A&P in the Real World: Phenylketonuria* on p. 920 to find out what happens when the body is unable to catabolize the amino acid phenylalanine and its concentration in the blood increases dramatically.

Newly released amino acid subunits are taken up by hepatocytes (cells of the liver) to be catabolized. Recall from the chemistry chapter (look back at Figure 2.21) that an amino acid consists of carbon atoms, sometimes called the "carbon skeleton," bound to a nitrogen-containing amino group. Nitrogen is a relatively inert element and doesn't oxidize readily, so before amino acids can be oxidized for fuel, the hepatocytes first have to remove the amino group by **transamination** (*trans-* = "across").

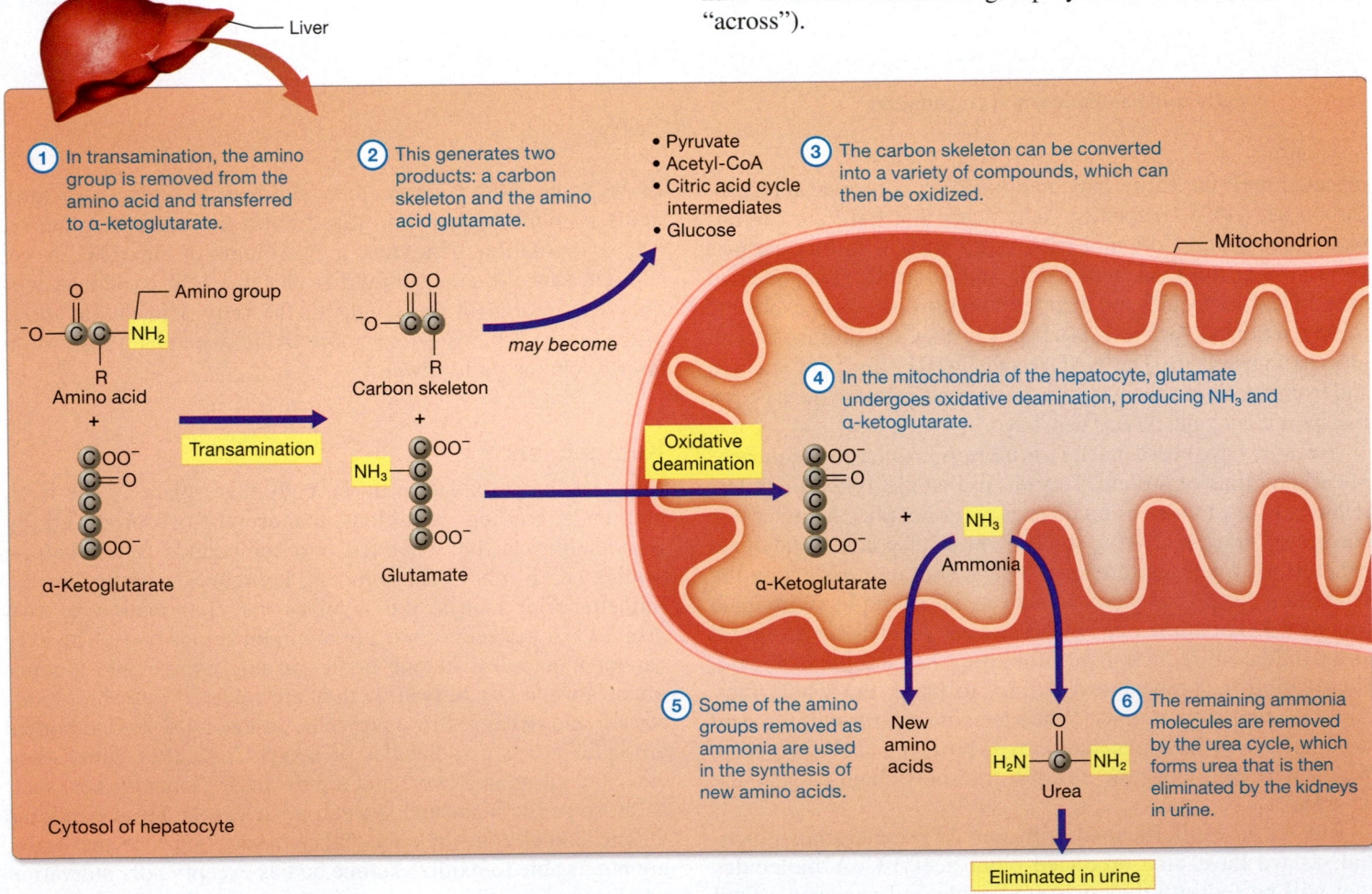

Figure 23.10 Amino acid catabolism.

The Big Picture of Nutrient Catabolism

Figure 23.11

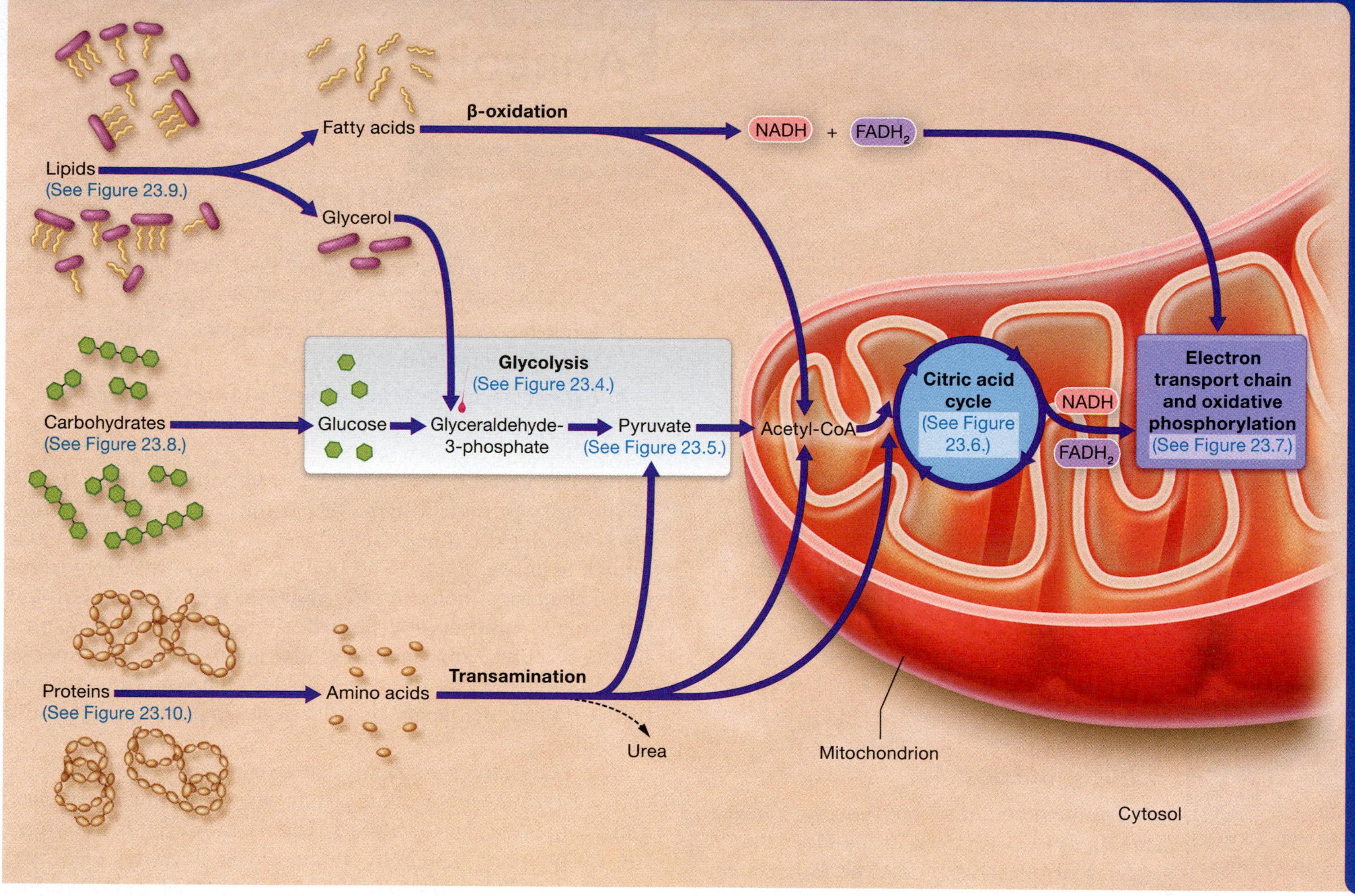

The events of amino acid catabolism occur by the following steps, shown in **Figure 23.10**:

1. **In transamination, the amino group is removed from the amino acid and transferred to α-ketoglutarate.** The removed amino group is transferred to the five-carbon compound α-ketoglutarate, one of the components of the citric acid cycle.

2. **This generates two products: a carbon skeleton and the amino acid glutamate.**

3. **The carbon skeleton can be converted into a variety of compounds, which can then be oxidized.** The carbon skeleton itself may become certain compounds, including pyruvate, acetyl-CoA, other citric acid cycle compounds, and even glucose. These compounds can then be oxidized via the reactions of glycolysis or the citric acid cycle.

4. **In the mitochondria of the hepatocyte, glutamate undergoes oxidative deamination, producing NH_3 and α-ketoglutarate.** Glutamate "carries" the amino group to the mitochondria of the hepatocyte. In the mitochondria, glutamate undergoes a process called *oxidative deamination,* during which the amino group

is removed, forming ammonia (NH_3) and re-forming α-ketoglutarate.

5. **Some of the amino groups removed as ammonia are used in the synthesis of new amino acids.**

6. **The remaining ammonia molecules are removed by the urea cycle, which forms urea that is then eliminated by the kidneys in urine.** The remaining ammonia must be removed from the body, because it is toxic to cells. This occurs via the **urea cycle** (yoo-REE-uh), which takes place in hepatocytes. During the urea cycle, two ammonia groups are combined with carbon dioxide to form the compound **urea.** Most of the urea resulting from this cycle is then eliminated in the urine by the kidneys.

Putting It All Together: The Big Picture of Nutrient Catabolism

The large numbers of catabolic reactions we have discussed can seem overwhelming. To help you study them, the pathways of fatty acid, glucose, and amino acid catabolism are summarized in **Figure 23.11**.

919

☐ 3. Why is it necessary to remove the amino group from amino acids during their catabolism?

☐ 4. What is the fate of the carbon skeletons of amino acids?

☐ 5. Why must the body eliminate the ammonia that results from transamination?

☐ 1. Explain how the pathways of protein, lipid, and carbohydrate catabolism are similar. How are they different?

☐ 2. Explain why kidney failure could result in the accumulation of nitrogen-containing waste products in the blood.

☐ 3. What does it mean when we say that fats have more "calories" than carbohydrates or proteins? Why is this true? Would this depend on the specific fatty acids consumed? Explain.

See answers in Appendix A.

Phenylketonuria

Phenylketonuria (fee´-nul-kee-tohn-YOOR-ee-uh), or PKU, was one of the first genetic defects of metabolism detected in humans. It is a relatively rare disorder, affecting between 1 in 10,000 and 1 in 14,000 newborns. PKU results from a deficiency in one of the enzymes involved in the catabolism of the amino acid phenylalanine. Normally, phenylalanine is degraded to acetyl-CoA; however, when the necessary enzyme is deficient, phenylalanine accumulates in the blood and is excreted in the urine (hence the name of the disease). A high level of phenylalanine interferes with normal development of the nervous system, and severe intellectual disability (formerly known as mental retardation) may result.

In the 1960s, it was discovered that the effects of PKU can be prevented by early detection and strict dietary modifications. For this reason, at this time states began to require that newborns be screened for the disease shortly after birth. Today, screening is required by law in every state before newborns leave the hospital. Many other countries mandate testing, including Canada, Japan, and most of western and eastern Europe.

The primary treatment of PKU is a specialized diet. Those diagnosed with PKU must limit their intake of protein-rich foods and foods with added phenylalanine, such as the artificial sweetener aspartame. Foods with aspartame have a warning label to alert those with PKU to the phenylalanine content.

Anabolic Pathways

1. Describe the processes of glycogenesis, glycogenolysis, and gluconeogenesis.

2. Describe the process by which fatty acids are synthesized and stored as triglycerides in adipose tissue.

3. Explain how nutrients may be converted to amino acids and lipids if needed.

4. Explain the metabolic fate of excess dietary proteins and carbohydrates.

As we discussed in Module 23.1, catabolism tells only part of the story of metabolism. To get the full picture, we must also consider the other side of metabolism: anabolism, also known as *biosynthesis*. The body's anabolic reactions serve many purposes, including nutrient storage; synthesis of new nutrients; synthesis of structural elements such as phospholipids, proteins, and nucleic acids; and synthesis of special compounds such as FAD and NAD^+. This module focuses primarily on the anabolic reactions that store and build nutrients.

The body stores nutrients for two reasons: (1) When we eat, we take in more energy than we can immediately use; and (2) the body needs a ready supply of nutrients to oxidize for maintaining metabolic homeostasis between the times we eat. The two main nutrient energy storage forms we find in the body are *glycogen* (GLY-koh-jen), which consists of linked glucose units, and adipose (fat), which is made up mostly of triglycerides. Interestingly, often the glucose in glycogen is derived from amino acids or glycerol; similarly, often the fatty acids of triglycerides are derived from glucose and amino acids. Let's examine how the body stores and manufactures nutrients.

Glucose Anabolism

« FLASHBACK

1. What is a polysaccharide? (p. 52)

2. How does the body store glucose? (p. 52)

There are two anabolic processes involving glucose. The first is *glycogenesis,* which is the storage of excess glucose taken in from the diet. The second is *gluconeogenesis,* which is the formation of new glucose from noncarbohydrate precursors. The next two subsections explore these processes.

Glycogenesis

Glucose is stored in the form of **glycogen,** a large, branching compound composed of thousands of glucose units. The process of glycogen synthesis, which is called **glycogenesis,** involves a series of enzyme-catalyzed reactions that add glucose units to the growing glycogen molecule, as shown in **Figure 23.12**. Although nearly all cells are able to store glycogen in cytoplasmic inclusions the majority of the body's glycogen is stored in the liver and skeletal muscle tissue.

When the glucose concentration in the bloodstream decreases, hormones such as glucagon and epinephrine trigger the breakdown of glycogen, a catabolic process called **glycogenolysis** (gly′-koh-jen-AWL-ih-sis). This releases glucose molecules into the bloodstream, which raises the concentration of blood glucose. Glycogen stores do not last long, however, and are depleted after about half a day (much faster when a person is engaged in strenuous activity).

Gluconeogenesis

When glycogen supplies are exhausted, the body needs to turn to alternative methods to keep up with the needs of cells that require glucose. The body's main method of doing so is **gluconeogenesis** (gloo′-koh-nee-oh-JEN-eh-sis; literally, "making new glucose"), or synthesis of glucose from noncarbohydrate precursors (**Figure 23.13**). During gluconeogenesis, the hepatocytes and certain cells of the kidney perform reactions that convert three- and four-carbon compounds into glucose. Sources of these compounds include the following:

- glycerol from triglyceride catabolism,
- pyruvate and lactate from glycolysis,
- intermediate compounds from the citric acid cycle, and
- certain amino acids called *glucogenic* amino acids.

The one major nutrient that cannot be converted into glucose is a fatty acid. This is because fatty acid oxidation can produce

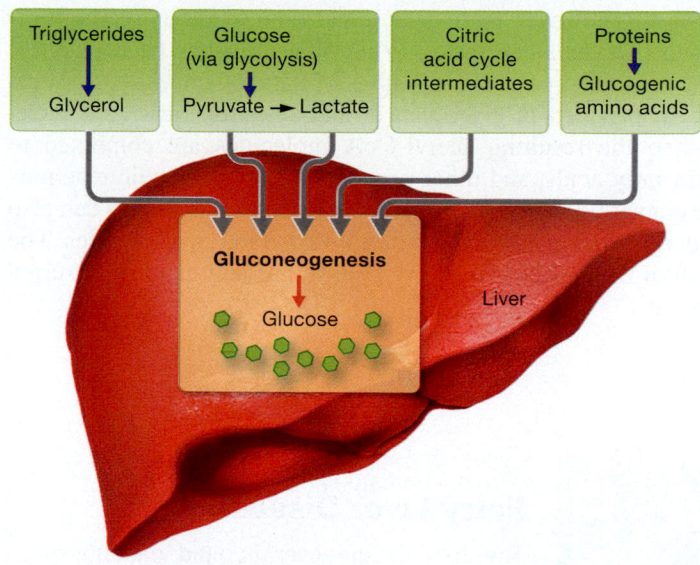

Figure 23.13 Gluconeogenesis.

only acetyl-CoA, and acetyl-CoA cannot be used by mammals to generate new glucose. The glucose synthesized by gluconeogenesis is either released into the bloodstream and delivered to cells or made into glycogen.

Quick Check

☐ 1. How is glucose stored in the body?

☐ 2. Which precursors can be used by the body for gluconeogenesis?

☐ 3. Why can fatty acids not be used as precursors for gluconeogenesis?

Fatty Acid Anabolism

The synthesis of fatty acids somewhat resembles β-oxidation in reverse; however, they are still separate pathways that require different enzymes and take place in different locations. Fatty acid synthesis occurs by a process called **lipogenesis** (ly′-poh-JEN-eh-sis). Lipogenesis takes place in the cytosol. It involves a large enzyme complex called *fatty acid synthase,* which catalyzes reactions that progressively lengthen the fatty acid chain by two-carbon units. Most of the fatty acids are attached to glycerol and assembled into triglycerides in the endoplasmic reticulum, after which they are stored in adipocytes. The majority of the body's energy is stored as triglycerides in adipose tissue, although some triglycerides are also stored in the liver. See *A&P in the Real World: Fatty Liver Disease* on p. 922 to discover what happens when too many fatty acids are stored in the liver.

Not all fatty acids and glycerol in triglycerides begin as dietary fats; often they start out as glucose or amino acids. After eating, the body takes the glucose obtained from food and uses it for its immediate needs. Excess glucose is stored as

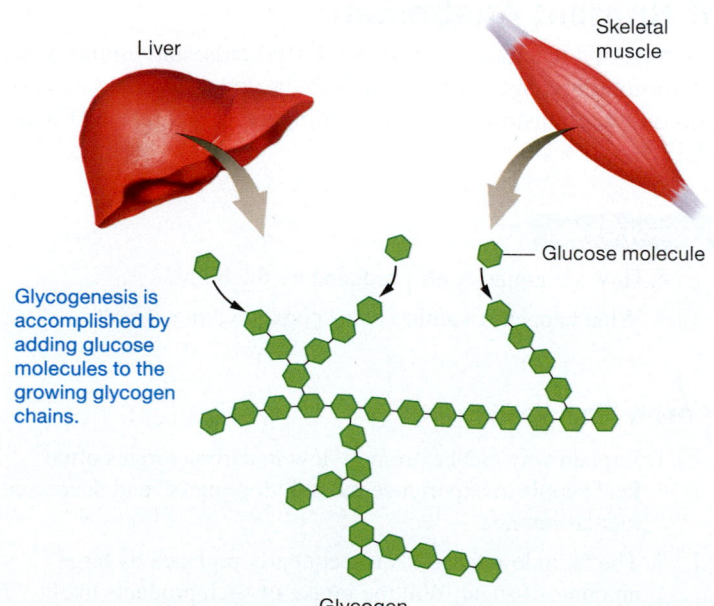

Glycogenesis is accomplished by adding glucose molecules to the growing glycogen chains.

Figure 23.12 Glycogenesis.

glycogen; however, there is an upper limit to the amount of glycogen that cells can make. When this limit has been reached, glucose undergoes the beginning of catabolism via glycolysis, but the products of glycolysis do not enter the citric acid cycle. Rather, the resulting acetyl-CoA molecules are combined to form fatty acids, and other products of the glycolytic reactions are converted into glycerol. Excess dietary amino acids can also undergo similar reactions to be converted into triglycerides. The truth of the matter is that any type of nutrient can be converted into fat.

Fatty Liver Disease

A&P in the Real World

The role of the liver in lipid metabolism becomes obvious in the condition called *steatohepatitis* (stee´-aeh-toh-hep´-uh-TY-tis; *steat-* = "fat"), commonly known as "fatty liver disease" (FLD). FLD has many causes, including alcohol abuse, obesity, diabetes, and many drugs and toxins. It is characterized by the accumulation of triglycerides and other fats in hepatocytes, as shown in the micrograph on the right here, next to one of a normal liver:

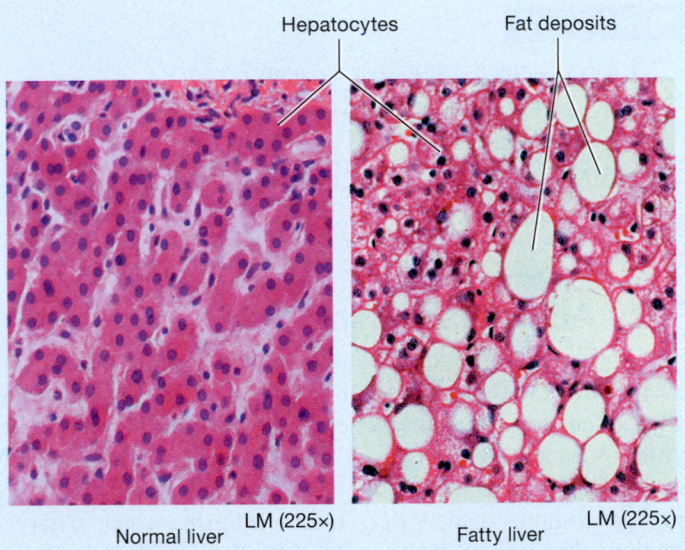

Hepatocytes   Fat deposits

LM (225×)   LM (225×)

Normal liver   Fatty liver

The exact mechanism by which FLD develops isn't precisely known, but may include decreased capacity for mitochondrial β-oxidation, increased fatty acid synthesis, increased delivery of fatty acids to the liver, and deficient export of triglycerides from the liver to other cells.

Untreated, FLD may progress to cirrhosis and hepatic failure. However, FLD is generally reversible if the underlying cause is addressed. Treatments include dietary modifications with weight loss, abstinence from alcohol, and certain medications.

Quick Check

☐ 4. How are fatty acids synthesized?

☐ 5. How are nonlipids converted into fats?

Amino Acid Anabolism

We have already discussed protein synthesis and seen how strings of amino acids are joined by peptide bonds in a sequence determined by the DNA of one gene (see Chapter 3). The body can synthesize 11 of the 20 amino acids found in human proteins. These amino acids are primarily synthesized by reactions that add an amino group to carbon skeletons such as α-ketoglutarate, pyruvate, and oxaloacetate. The remaining amino acids (known as *essential amino acids*) must be supplied by the diet, which we discuss in Module 23.7.

Unlike carbohydrates and lipids, amino acids are not stored to any significant extent in the body's proteins. A cell needs to synthesize only so many proteins; once this limit has been reached, any excess amino acids obtained by the diet are converted to other compounds for storage. All nucleated cells produce proteins, although some cells, such as skeletal muscle cells, undergo higher than average rates of protein synthesis due to the composition of their cells.

Some amino acids are converted to glucose by gluconeogenesis and either used immediately or incorporated into glycogen. Others are converted to fatty acids and stored in the body's adipose tissue. You may have heard the claim by individuals promoting high-protein diets that proteins cannot be stored as fat. This is, of course, untrue.

Putting It All Together: The Big Picture of Nutrient Anabolism

The anabolic reactions of glucose, fatty acids, and amino acids are summarized in **Figure 23.14**. To help you study both catabolism and anabolism of nutrients, compare this figure to Figure 23.11.

Quick Check

☐ 6. How are amino acids produced by the body?

☐ 7. What happens to amino acids consumed in excess?

Apply What You Learned

☐ 1. Explain why diets extremely low in carbohydrates often lead people to experience mental "fogginess" and decreased mental stamina.

☐ 2. The fat in low-fat foods is frequently replaced by large amounts of sugar. Will the intake of such products likely lead to fat loss? Explain.

See answers in Appendix A.

The Big Picture of Nutrient Anabolism

Figure 23.14

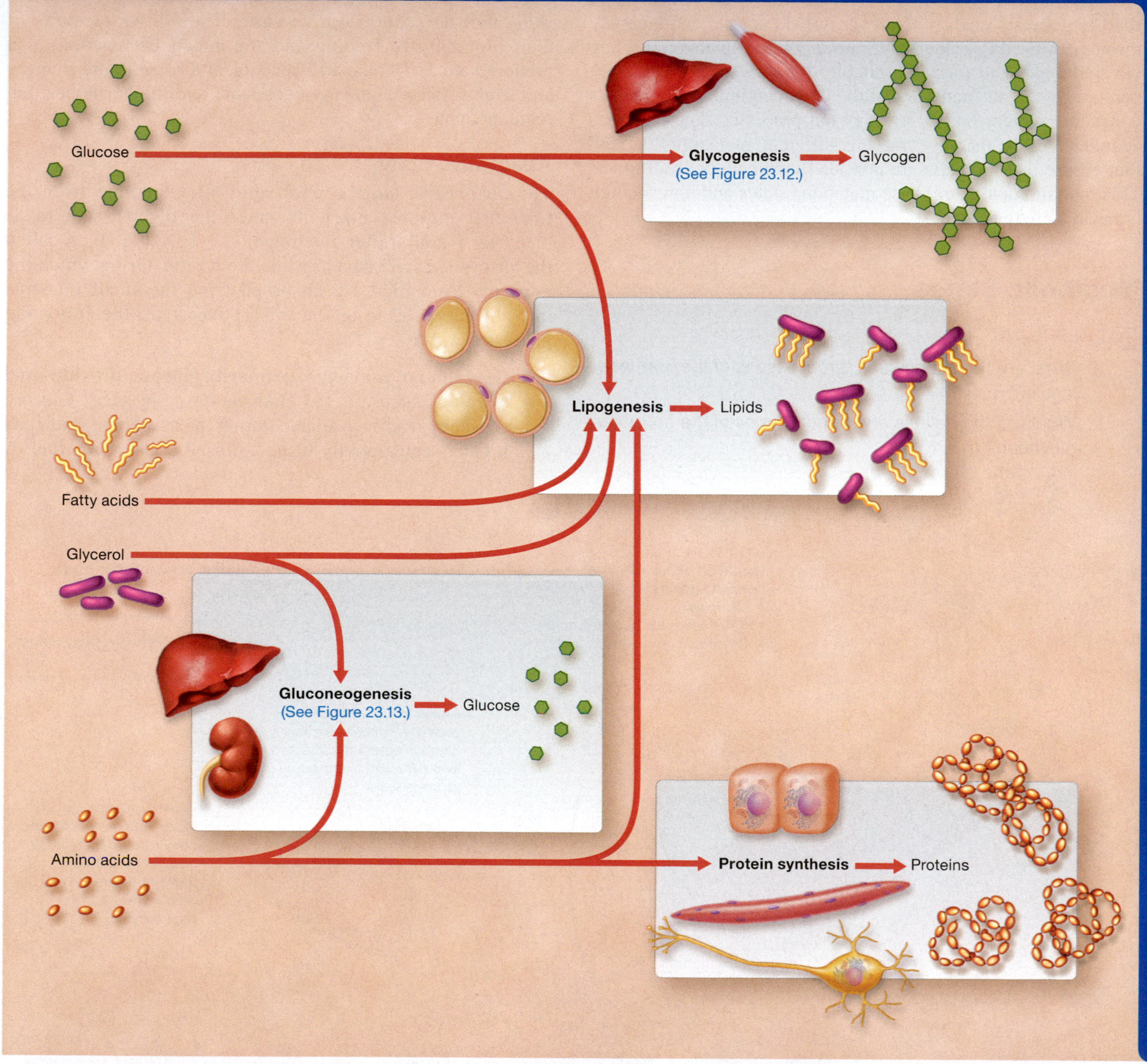

Glucose
Glycogenesis
(See Figure 23.12.)
Glycogen
Lipogenesis
Lipids
Fatty acids
Glycerol
Gluconeogenesis
(See Figure 23.13.)
Glucose
Amino acids
Protein synthesis
Proteins

MODULE **23.5**

Metabolic States and Regulation of Feeding

Learning Outcomes

1. Compare and contrast the processes that occur in the absorptive and postabsorptive states.
2. Explain the roles of insulin and glucagon in the absorptive and postabsorptive states.
3. Describe the role of the hormones insulin and glucagon in regulating glucose and amino acid catabolism and anabolism.
4. Explain the significance of glucose sparing for neural tissue in the postabsorptive state.
5. Describe how feeding behaviors are regulated.

We have established that the body oxidizes the nutrients in food to generate energy. However, we do not eat continuously throughout the day; rather, nutrients are generally taken in when we are hungry and food is available. Yet even when food is not being ingested, the activities of the body continue, and the cells constantly require energy to carry out processes. The body has adapted to this fact by entering different metabolic "states" that ensure that the cells are provided with energy at all times. This module focuses on these metabolic states and how feeding behaviors are regulated.

Metabolic States

« FLASHBACK

1. What are the actions and target tissues of the hormone insulin? (p. 616)
2. What are the actions and target tissues of the hormone glucagon? (p. 615)

The body has two basic metabolic states: the *absorptive state* and the *postabsorptive state*. They differ in terms of their proximity to feeding and the metabolic reactions that predominate. Let's take a closer look at the metabolic changes and events that accompany each state and the hormones that regulate them.

Absorptive State

The **absorptive state** occurs right after feeding, from the time that ingested nutrients enter the bloodstream to as many as 4 hours after feeding (**Figure 23.15a**). It is called the absorptive state because it is the period during which the nutrients from food are absorbed from the small intestine into the blood. During the absorptive state, the following processes occur:

- **Oxidation of nutrients, primarily glucose, provides fuel to cells.** Glucose is the main fuel used by cells. Glucose is usually readily available during the absorptive state, and it enters cells rapidly to be oxidized as the main fuel. A

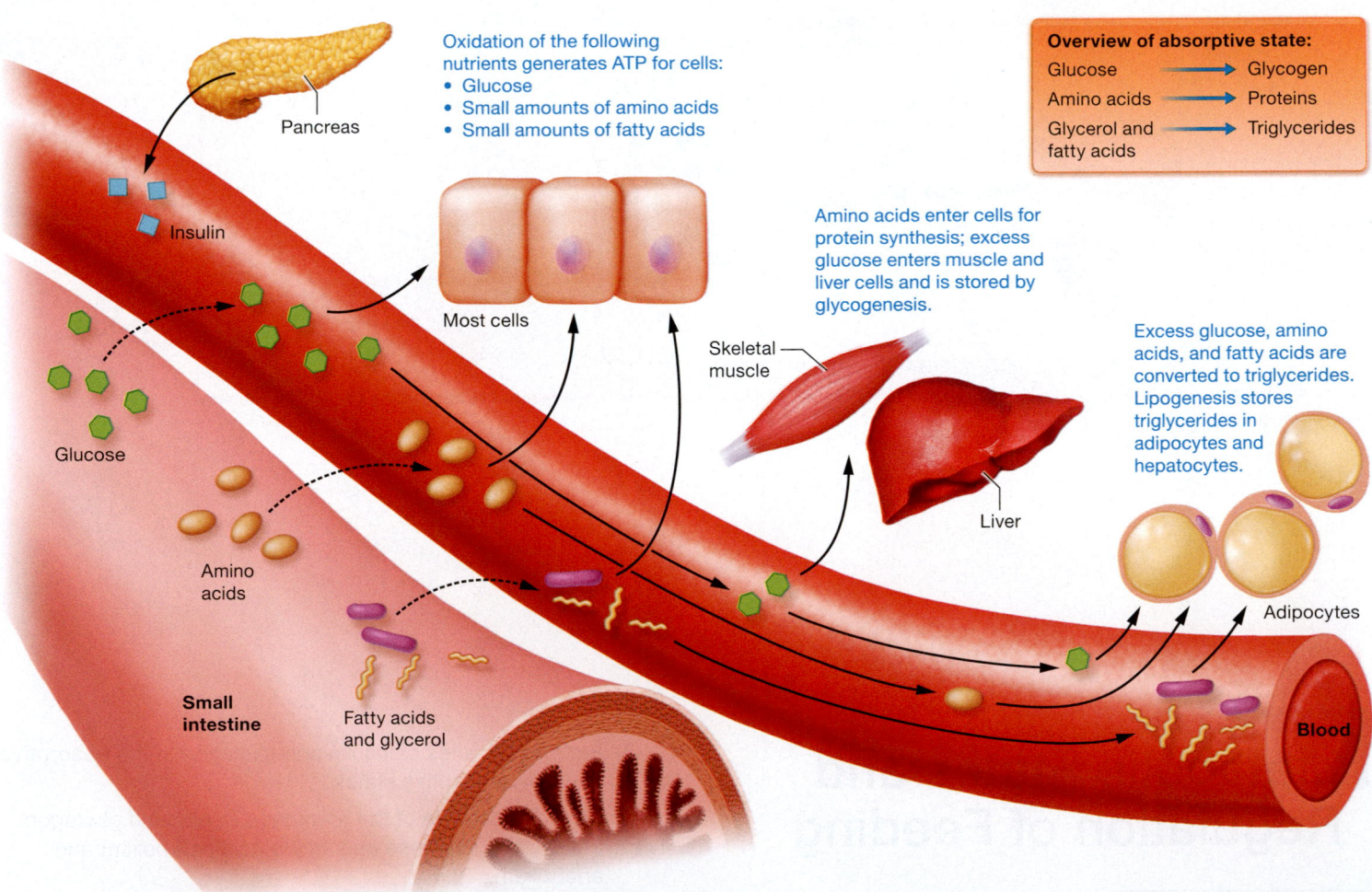

(a) Changes in the absorptive state triggered by insulin.

Figure 23.15 Comparison of the absorptive and postabsorptive states.

small fraction of the absorbed fatty acids and glycerol are oxidized by different tissues for fuel; similarly, a small percentage of amino acids are deaminated and oxidized for fuel.

- **Synthesis of amino acids into proteins provides structural materials to cells.** Many ingested amino acids enter cells, where they are used for protein synthesis.
- **Glycogenesis stores excess glucose in skeletal muscle cells and hepatocytes.** Glucose that is not oxidized for energy is stored as glycogen.
- **Lipogenesis stores triglycerides in adipocytes and hepatocytes.** The majority of triglycerides are simply stored in adipose tissue or the liver. Excess glucose is also converted into fatty acids or glycerol and stored as fat, and some amino acids are converted to fatty acids and stored as well.

These changes are largely controlled by the pancreatic hormone *insulin*. As we discussed in the endocrine chapter, insulin release is triggered by a rising level of glucose in the blood (see Chapter 16). This hormone targets most body cells and stimulates the uptake of glucose by cells, lowering the concentration of glucose in the blood. It also acts on the liver

and several other tissues, initiating the anabolic processes we just covered. In the absorptive state, then, cells take in fuels and, under the control of insulin, use or store them.

Postabsorptive State

Once nutrient absorption is complete, the body enters the **postabsorptive state** (**Figure 23.15b**). The postabsorptive state generally begins about 4 hours after feeding, although this varies depending on the amount and types of nutrients ingested. For most individuals, the body is generally in the postabsorptive state in late morning, late afternoon, and most of the night. During this state, the body slows and then ceases anabolic processes such as glycogenesis, lipogenesis, and protein synthesis because the steady supply of nutrients ends during the postabsorptive period. In their place, the following processes occur:

- **Breakdown of proteins in muscle cells releases glucogenic amino acids into the blood.** If necessary, muscle proteins are broken down to release glucogenic amino acids that can be taken up by cells and converted to glucose.

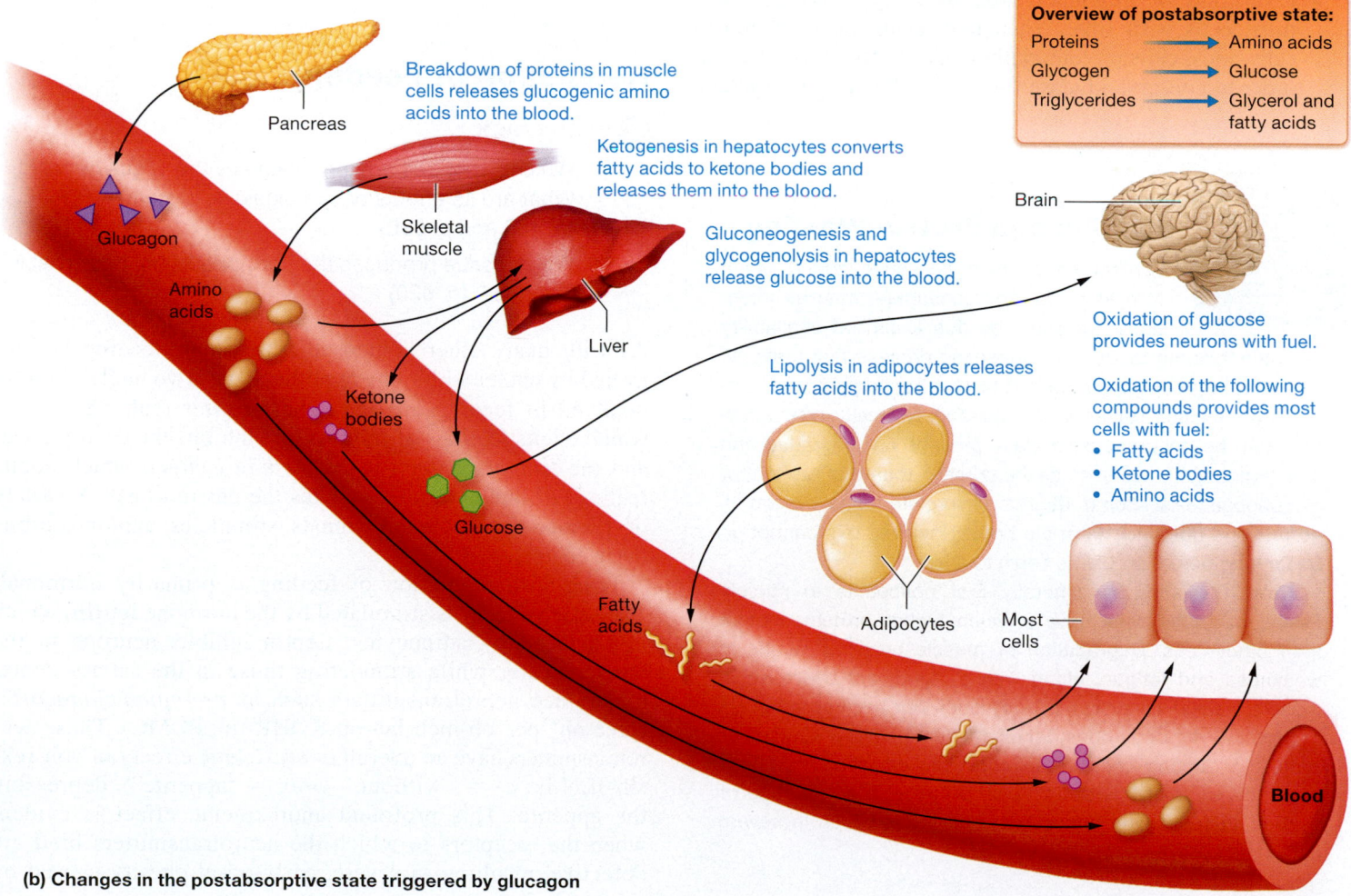

(b) **Changes in the postabsorptive state triggered by glucagon**

Figure 23.15 Comparison of the absorptive and postabsorptive states. (*continued*)

See *A&P in the Real World: Fasting and Protein Wasting* to find out more about protein breakdown during the postabsorptive state.

- **Ketogenesis in hepatocytes converts fatty acids to ketone bodies and releases them into the blood.** Many of the fatty acids oxidized in the liver are converted to ketone bodies, which are released into the blood. The advantage to releasing ketones is that cells of the nervous system are able to oxidize them for fuel when glucose is unavailable, whereas these cells are not capable of oxidizing fatty acids to any significant extent.
- **Gluconeogenesis and glycogenolysis in hepatocytes release glucose into the blood.** The concentration of blood glucose begins to decline during the postabsorptive state. To ensure that cells requiring glucose, such as those of the brain, are not deprived of their primary fuel, gluconeogenesis and glycogenolysis in hepatocytes begin to raise the concentration of blood glucose.
- **Lipolysis in adipocytes releases fatty acids into the blood.** The limited concentration of blood glucose means that cells will need another source of fuel, namely, the fuel in fats. For this reason, lipolysis begins and fatty acids are released into the bloodstream from adipocytes.
- **Oxidation of nutrients such as fatty acids provides most cells with fuel.** When fatty acids are delivered to cells, most cells preferentially begin to catabolize them in order to conserve glucose for cells of the nervous system,

a phenomenon called **glucose sparing.** Cells outside the nervous system can also use ketone bodies and amino acids for fuel.

Several hormones help the body adapt during the postabsorptive state. One of these is the pancreatic hormone *glucagon,* which is released in response to a decreasing blood glucose concentration. Its main target tissue is the liver, where it triggers glycogenolysis and gluconeogenesis (see Chapter 16 for a review of how blood glucose homeostasis is maintained). Other hormones that play a role during the postabsorptive state include the catecholamines epinephrine and norepinephrine, which stimulate lipolysis in adipose tissue and glycogenolysis in skeletal muscle. In addition, cortisol stimulates gluconeogenesis and processes that release glucogenic precursors into the blood.

Quick Check

☐ 1. Why does anabolism dominate in the absorptive state?

☐ 2. Would you expect to find a high level of ketone bodies in the blood during the absorptive state?

☐ 3. Which metabolic activities are prevalent during the postabsorptive state?

Regulation of Feeding

《 FLASHBACK

1. Which structures are innervated by the vagus nerve? What are its effects on the digestive system? (p. 486 and p. 530)

2. Which tissue produces the hormone leptin, and what is its effect? (p. 620)

As with many other homeostatic variables, feeding is controlled by nuclei within the hypothalamus. Two nuclei directly involved in feeding are the *satiety center* (suh-TY-eh-tee), which elicits feelings of fullness and inhibits the desire to eat, and the *hunger center* (or the *feeding center*), which elicits feelings of hunger and stimulates the desire to eat. A variety of hormonal and neural signals stimulates and/or inhibits these centers.

Long-term regulation of feeding is primarily hormonal. The satiety center is stimulated by the hormone **leptin,** which is produced by adipocytes. Leptin inhibits neurons in the hunger center while stimulating those in the satiety center to produce neurotransmitters such as *pre-opiomelanocortin* (pree-oh'-pee-oh-meh-lan-oh-KOHR-tin; POMC). These neurotransmitters have an overall *anorexigenic* effect (an'-oh-rek-sih-JEN-ik; *a-* = "without," *-orex* = "appetite"), depressing the appetite. This profound anorexigenic effect is evident when the receptors to which the neurotransmitters bind are defective, resulting in severe, early-onset obesity caused by

A&P in the Real World

Fasting and Protein Wasting

Fasting, or a prolonged period during which little to no food is consumed, may be intentional, or could be due to limited availability of nutrient-dense foods or to a disease process that prevents ingestion or processing of nutrients. One consequence of fasting is a condition called *protein wasting.* Normally, after a few hours in the postabsorptive state, proteins are degraded and their amino acids are used as the primary source of carbon for gluconeogenesis. Some degree of protein degeneration is normal and does not harm the body. However, this cannot go on for long before problems begin to arise.

Protein wasting is generally first noticeable in muscle tissue, as those cells have a much higher protein content than other cells. Degradation of muscle proteins results in weakness and fatigue, along with a significant decrease in muscle mass. However, the body does not just rob muscles of protein—the body does not discriminate among sources of protein or among cell populations. This can lead to a loss of critical structural and functional proteins in a multitude of tissues, including the heart, liver, and brain, which in severe cases may prove fatal.

uncontrolled overeating. The discovery of leptin generated a great deal of excitement due to its potential as a novel agent to suppress appetite and cause weight loss. Unfortunately, studies of leptin have shown that administering excess leptin does not result in decreased appetite or weight loss in humans.

The hunger center is stimulated by a decreasing level of leptin secretion and a rising level of secretion of the hormone **ghrelin** (GRAY-lin) from the cells of the stomach mucosa. Ghrelin stimulates neurons in the hunger center to produce neurotransmitters such as *neuropeptide Y* and *orexins,* both of which are *orexigenic,* promoting hunger. Interestingly, ghrelin levels decrease in obese individuals after gastric bypass surgery (removal of part of the stomach), which is hypothesized to partially account for their reduced food intake and subsequent weight loss.

A variety of short-term signals can also inhibit or stimulate feeding. Feeding triggers the release of several hormones, such as insulin, that have actions similar to those of leptin, which in turn decrease food intake. In addition, feeding stretches the stomach walls and initiates the release of gastrointestinal hormones such as cholecystokinin. Both stimuli cause the vagus nerve to indirectly suppress the hunger center and decrease the release of orexins and neuropeptide Y. The concentration of certain solutes in the blood can also directly stimulate or inhibit centers in the hypothalamus. For example, a low level of glucose in the blood (hypoglycemia) stimulates the hunger center and the release of orexins.

Of course, feeding behaviors are more complex than the signals we have discussed. A number of psychological and environmental factors affect feeding, as well, which has hampered efforts to find a "cure-all" medication that suppresses appetite.

Quick Check

☐ 4. Which centers in the hypothalamus regulate feeding?

☐ 5. What are the actions of leptin?

☐ 6. How do the actions of leptin differ from those of ghrelin?

Apply What You Learned

☐ 1. How would the absorptive state differ in a person who has completely eliminated carbohydrates from his or her diet?

☐ 2. Type 1 diabetes mellitus results from the inability to produce the hormone insulin. Predict how this condition changes the normal events of the absorptive state.

☐ 3. Certain antidepressants trigger the production of neurotransmitters by the neurons of the hypothalamic feeding center. Predict the effects these medications have on appetite and feeding.

See answers in Appendix A.

MODULE
23.6
The Metabolic Rate and Thermoregulation

Learning Outcomes

1. Define metabolic rate and basal metabolic rate.
2. Describe factors that affect metabolic rate.
3. Differentiate between radiation, conduction, convection, and evaporation.
4. Explain the importance of thermoregulation in the body.

The metabolic modifications that occur during the absorptive and postabsorptive states ensure that the body has a ready supply of fuels that can be oxidized to drive its processes. The total amount of energy expended by the body to power all its processes is called the **metabolic rate.** A consequence of this energy expenditure is a great deal of heat production. Indeed, metabolism is the primary source of body heat, and metabolic processes are an important part of body temperature regulation, or *thermoregulation*. In this module, we examine the metabolic rate and how it is measured. Then our attention shifts to heat production, the process of thermoregulation, and what happens when thermoregulation is impaired.

Metabolic Rate

The metabolic rate is expressed in a unit we encountered in the chapter opener: the Calorie or, more accurately, the kilocalorie. Recall that a kilocalorie is defined as the amount of heat required to raise the temperature of one kilogram of water by one degree Celsius. Why would we use a measure of heat to express the body's metabolic rate? The answer is because most of the body's energy is eventually lost to the environment as heat. Consider the following facts:

- The liberation of the chemical energy in foods is inefficient, and much of the energy liberated by catabolism is lost as heat.
- The use of ATP to drive cellular processes is similarly inefficient, and much of the chemical energy of ATP use is lost as heat.
- Much of the energy from the processes driven by ATP, such as muscle contraction, is also lost as heat.

Because of this conversion of energy to heat, a process called *thermogenesis,* heat production is a good way to measure energy expenditure. Heat production is determined by **calorimetry** (kal′-ohr-IM-eh-tree), in which we measure the actual amount of heat produced by an individual's body. The value obtained from calorimetry is then used to directly determine the person's metabolic

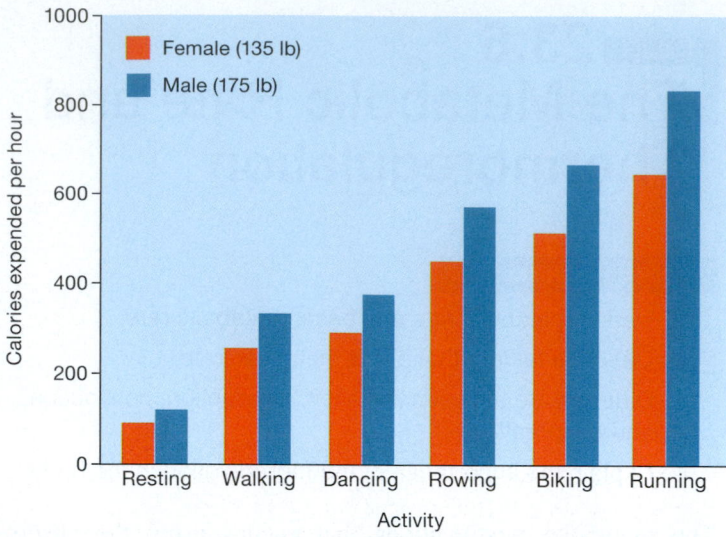

Figure 23.16 Approximate energy expenditures in adults during various activities.

rate. However, because this process is physically difficult to perform, most often the metabolic rate is determined indirectly by measuring the amount of oxygen a person consumes, as this is directly related to the amount of energy generated and thus heat produced.

One of the values that calorimetry helps to measure is the **basal metabolic rate,** or **BMR,** which is defined as the minimal rate of metabolism for an awake individual, or the "energy cost of living." Even when the body is completely at rest, it is still using a considerable amount of energy to perform the processes necessary to maintain life, such as keeping the heart beating and filtering the blood through the kidneys. A person is said to be functioning at his or her BMR when the following conditions are met: (1) The person has not eaten for the past 12 hours and has had a restful night of sleep; (2) the person is not performing physical activity and has not performed strenuous physical activity for at least 1 hour; (3) the person is not under physical or emotional stress; and (4) the temperature surrounding the person is constant and comfortable.

Of course, the body's energy expenditure is not going to precisely equal the BMR most of the time. Rather, it will vary at different times of the day and with different activities (**Figure 23.16**). A variety of factors influence the metabolic rate, including hormones such as thyroid hormone and growth hormone, fever, nutritional status, physical activity, and certain drugs. For example, during exercise the metabolic rate can increase 10-fold or more above the BMR, and during sleep it decreases to about 10–15% below the BMR. Additionally, the BMR varies from person to person, depending on factors such as sex and muscle mass (as skeletal muscle is a very metabolically active tissue). Is it possible to change your BMR? Find out in *A&P in the Real World, Pseudoscience Exposed: "Rev" Your Metabolism.*

Quick Check

☐ 1. What is a Calorie?

☐ 2. What is the basal metabolic rate (BMR)? When is a person functioning at his or her BMR?

Heat Exchange between the Body and the Environment

« FLASHBACK

1. How does water take heat with it when it evaporates? (p. 46)

Heat, like solutes, water, and pressure, will move with a gradient, again reminding us of the Gradients Core Principle (p. 28). Specifically, heat moves from an area of higher temperature to one of lower temperature. The heat produced by the body's metabolic reactions is exchanged with the environment in this manner, mostly through the skin. Heat is exchanged by four mechanisms: *radiation, conduction, convection,* and *evaporation* (**Figure 23.17**).

CORE PRINCIPLE
Gradients

Radiation

The transfer of heat from one object to another through electromagnetic waves is known as **radiation.** The body loses heat to the environment when the body temperature is higher than the surrounding air, which is generally the case. Radiation is the reason why you feel warmth from the sun. It's also why you might feel cold when you initially lie under a blanket but gradually feel warmer—your body heat is transferred to the air, and the heated air is trapped by the blanket. Note that this also works in reverse when the air temperature is higher than body temperature—the body will gain heat. This is why exposure to extreme heat for a long time can be dangerous. Radiation is responsible for about 50% of the body's total heat exchange with the environment.

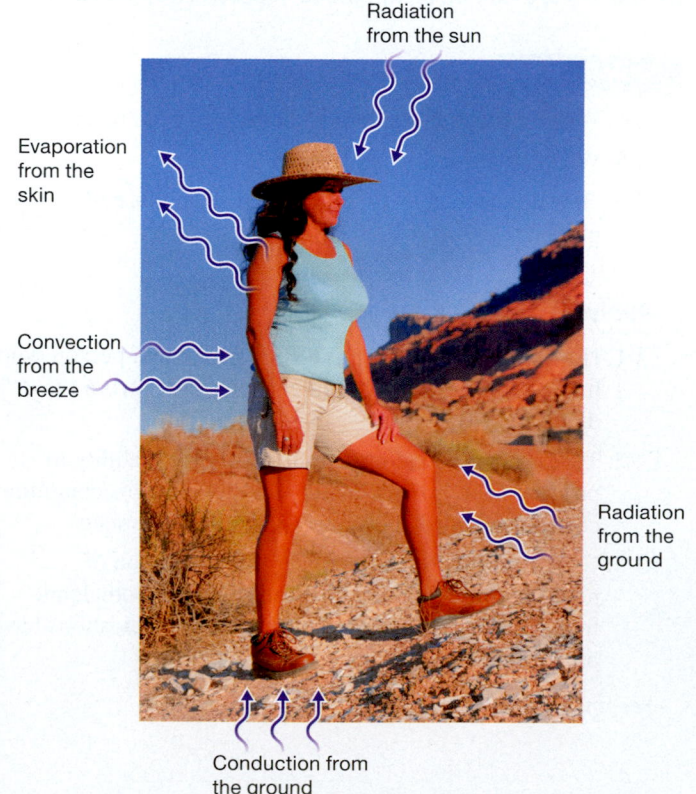

Figure 23.17 Mechanisms of heat transfer.

Pseudoscience Exposed: "Rev" Your Metabolism

You have probably heard about a lot of things that claim to speed up, or "rev," your metabolism, or increase your BMR. It generally is possible to increase your BMR. For example, you've already seen that exercise can increase the metabolic rate by up to 10 times. But this is just a temporary effect—after exercise stops, the metabolic rate gradually returns to its basal level. However, consistent exercise, particularly weight training, can affect the BMR. Adipose tissue is very energy-efficient, and so has a low metabolic rate. By contrast, skeletal muscle tissue is energy-inefficient and has a much higher metabolic rate. For this reason, an individual who loses adipose tissue and gains muscle tissue will generally increase his or her BMR.

But what about those over-the-counter diet pills that claim to "rev" your metabolism? The first thing you need to understand about dietary supplements in the United States is that manufacturers are not required to demonstrate safety *or* efficacy in order to sell their products. Also, quality testing is not required, so these products might not even contain the ingredients as claimed. Most of these dietary supplements claim to contain varying amounts of stimulants, such as the following:

- *Caffeine.* Caffeine increases the metabolic rate temporarily, but over the long term, body cells become insensitive to its effects.
- *Green tea.* The active components in green tea do appear to increase the BMR. However, a certain dosage and a specific part of the plant are required to achieve effectiveness. There is no way to be certain that any given dietary supplement contains adequate amounts of green tea or the correct components.
- *Capsaicin.* A component of peppers, capsaicin (kap-SAY-sin), has been shown to increase the metabolic rate. However, its use is limited by adverse effects and stringent dosage requirements.

When dealing with over-the-counter dietary supplements, it is best to be cautious. They are often quite expensive and are not required to be proven safe or effective in order to be marketed and sold. Besides, the best way to "rev" your metabolism can be free: a regular exercise program.

Conduction and Convection

Similar to radiation is **conduction,** in which heat is transferred from one object to another through direct contact. This can work through solids, liquids, and gases such as air, although it occurs much faster through contact with solids and liquids than with air. For example, conduction is the reason why you feel cold when you jump in a cold pool—your body heat is directly transferred to the water, lowering your body temperature—or why your feet feel hot when you walk on hot pavement.

Conduction works hand in hand with another mechanism of heat exchange: **convection,** in which heat is transferred through a liquid or gaseous medium. The simplest example of convection can be seen in many houses. Have you ever noticed that heating vents in a room are located near the floor and air conditioning vents are located near the ceiling? The reason for this is because warm air expands and rises from the floor, whereas cool air contracts, making it denser and causing it to fall into the room. Convection works the same way in body heat transfer; body heat rises away from the body and is replaced by falling cooler, denser air. Convection is also responsible for the cooling effect of a fan or breeze; the cooler, denser air in the breeze contracts and displaces the warmer air surrounding you.

Convection aids conduction because the cooler air around the body provides a steeper temperature gradient for heat transfer between the body and the surrounding air. Together these two processes account for about 15–20% of the body's heat transfer with the environment.

Evaporation

The final heat transfer mechanism is **evaporation,** during which water changes from a liquid to a gas. Water in liquid form absorbs a great deal of heat from its surroundings, and this heat enables liquid water to change into a gas. Water in and on the body absorbs body heat, removing the heat as it evaporates.

Evaporation occurs from the skin, lungs, and oral and nasal mucosae continuously without stimulation from the nervous system, producing what's known as *insensible heat loss.* This accounts for about 10% of the body's total heat transfer under resting conditions. When body temperature increases, the hypothalamus triggers *sweating,* which can result in the remaining 20–25% of heat loss, in addition to water loss. We discuss sweating in the next section.

Quick Check

- ☐ 3. What is the main heat transfer mechanism between the environment and the body? How does this mechanism transfer heat?
- ☐ 4. How do convection and conduction work together to transfer heat?
- ☐ 5. Why does evaporation have a cooling effect?

Thermoregulation: Body Temperature Regulation

« FLASHBACK

- 1. How does the integumentary system contribute to thermoregulation? (p. 162)
- 2. How does thyroid hormone assist in thermoregulation? (p. 602)
- 3. What is the mechanism behind a fever? (p. 771)

As you can tell from the preceding Flashback questions, we have discussed the regulation of body temperature, or thermoregulation, several times in this book. The amount of heat in the body is a key regulated variable that must be maintained within a narrow range so that all systems function properly. You have already learned about how metabolism produces heat, so let's find out how the body handles this heat and regulates its internal temperature.

Body Temperature

In spite of often significant swings in air temperature, body temperature remains relatively constant in the range of 35.8–38.2° C (96–101° F), averaging about 37.5° C (99.5° F). Note that this refers to the **core body temperature**, which is that of deep structures in the body, such as the liver and brain. It might strike you that this measurement differs from what is called "normal" on most oral and skin thermometers. This difference exists because measurements in these locations do not accurately reflect the core temperature due to variations with activity and air temperature, averaging instead about 35.6–37° C (97.7–98.6° F). Historically, the most accurate way to determine core body temperature is via a thermometer inserted into the rectum, but the inherent discomfort of this method now precludes its regular use. Certain newer thermometers have the ability to measure the temperature of the blood in the temporal artery near the surface of the skin, and this provides a noninvasive and generally accurate reading of core temperature.

Core body temperature fluctuates throughout the day. You may have noticed that you feel colder in the morning than in the evening. This reflects the fact that core body temperature is highest in the evening and decreases by about 1° C in the morning. In women, core body temperature also fluctuates with the ovarian cycle, which proves useful for fertility planning.

Regulation of Body Temperature

Like most other homeostatic variables, body temperature is regulated by a negative feedback loop, another example of the Feedback Loops Core Principle (p. 28). The feedback loop that responds to a rising body temperature, illustrated in **Figure 23.18a**, proceeds as follows:

CORE PRINCIPLE
Feedback Loops

- **Stimulus: Body temperature increases above the normal range.**
- **Receptor: Thermoreceptors in the skin and the hypothalamus detect the increased body temperature.** Many factors can cause body temperature to rise, including stimulation from the sympathetic nervous system, muscle activity, inflammation, and elevated environmental temperature. Temperature-sensitive *thermoreceptors* are located throughout the body, particularly in the skin and hypothalamus. These receptors detect the current higher temperature, and then send this information to the hypothalamus.
- **Control center: The heat-loss center in the hypothalamus is activated.** The hypothalamic thermoregulatory centers, which act as the control center, compare the current temperature with the temperature set point, and find that they are too far apart. This stimulates a hypothalamic nucleus, the **heat-loss center,** which sends signals to effectors.
- **Effector/response: Blood vessels in the skin dilate, and sweat glands release sweat.** Several effectors then produce responses that reduce body temperature. One response is dilation of blood vessels in the skin. Recall that this increases blood flow through the skin and promotes heat loss (see Chapter 5). Another response is increased sweating from the skin's sweat glands. However, sweating only produces significant heat loss if the humidity in the air isn't too high. In high-humidity conditions, the rate of evaporation decreases, which is why humid heat feels hotter than "dry heat."
- **Homeostatic range and negative feedback: As body temperature returns to the homeostatic range, the hypothalamus stops stimulating the effectors, which decreases the responses.** When the thermoreceptors detect a return to the homeostatic body temperature range, they stop stimulating the hypothalamus through negative feedback, which in turn ends stimulation of the effector organs, decreasing or shutting down the responses.

A similar feedback loop is triggered by a falling body temperature, commonly caused by a low environmental air or water temperature, and proceeds as follows (**Figure 23.18b**):

- **Stimulus: Body temperature decreases below the normal range.**
- **Receptor: Thermoreceptors in the skin detect the decreased body temperature.** In this case, it is primarily thermoreceptors in the skin that detect the lower temperature. This is because the external environment generally reaches temperatures low enough to activate the receptors well before the body's internal environment can become that cold. The skin's thermoreceptors then send this information to the hypothalamus.
- **Control center: The heat-promoting center in the hypothalamus is activated.** Again the thermoregulatory centers of the hypothalamus act as the control center. These centers compare the current temperature with the temperature set point, and find that they are too far apart. They then activate the **heat-promoting center,** which sends signals to effectors.
- **Effector/response: Blood vessels in the skin constrict, shivering in skeletal muscles is triggered, and the metabolic rate increases.** These signals trigger several effectors to produce responses that raise the body's temperature. For one, it causes constriction of blood vessels in the skin. This diverts blood from the skin to deeper tissues and minimizes heat loss through radiation, conduction, and convection. It also increases heat production through promotion of involuntary muscle contractions known as **shivering.** Finally, it stimulates the release of thyroid hormones from the thyroid gland and the release of the catecholamines epinephrine and norepinephrine from the adrenal medulla. Both effects increase the metabolic rate of their target cells (an example of the Cell-Cell Communication Core Principle, p. 28), and therefore heat production.

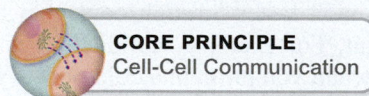

CORE PRINCIPLE
Cell-Cell Communication

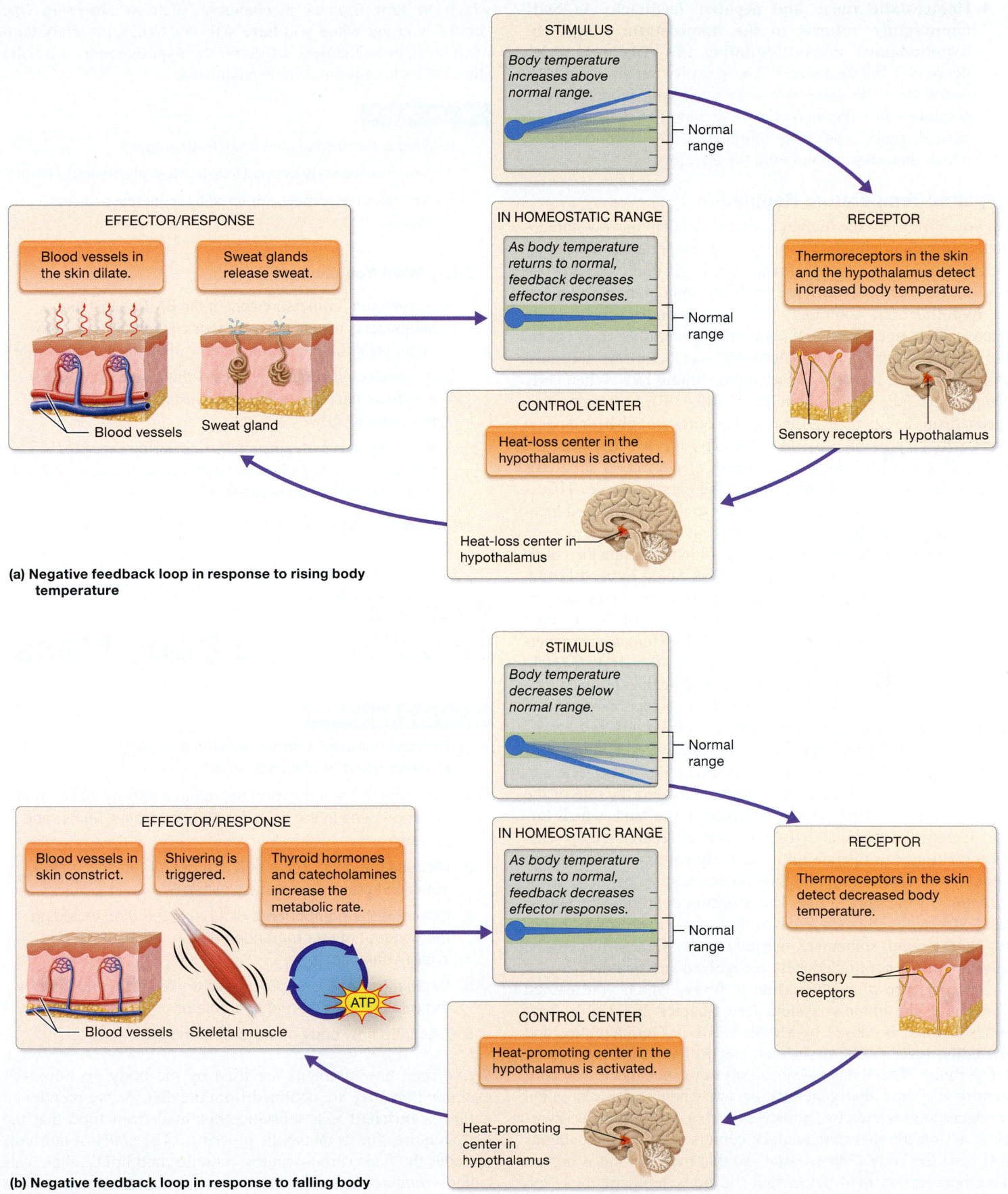

STIMULUS

Body temperature increases above normal range.

— Normal range

RECEPTOR

Thermoreceptors in the skin and the hypothalamus detect increased body temperature.

Sensory receptors   Hypothalamus

IN HOMEOSTATIC RANGE

As body temperature returns to normal, feedback decreases effector responses.

— Normal range

CONTROL CENTER

Heat-loss center in the hypothalamus is activated.

Heat-loss center in hypothalamus

EFFECTOR/RESPONSE

Blood vessels in the skin dilate.

Sweat glands release sweat.

Blood vessels   Sweat gland

(a) Negative feedback loop in response to rising body temperature

STIMULUS

Body temperature decreases below normal range.

— Normal range

RECEPTOR

Thermoreceptors in the skin detect decreased body temperature.

Sensory receptors

IN HOMEOSTATIC RANGE

As body temperature returns to normal, feedback decreases effector responses.

— Normal range

CONTROL CENTER

Heat-promoting center in the hypothalamus is activated.

Heat-promoting center in hypothalamus

EFFECTOR/RESPONSE

Blood vessels in skin constrict.

Shivering is triggered.

Thyroid hormones and catecholamines increase the metabolic rate.

Blood vessels   Skeletal muscle   ATP

(b) Negative feedback loop in response to falling body temperature

Figure 23.18 Maintaining homeostasis: regulation of core body temperature by negative feedback loops.

- **Homeostatic range and negative feedback: As body temperature returns to the homeostatic range, the hypothalamus stops stimulating the effectors, which decreases the responses.** The negative feedback mechanisms work the same way as they did with increased temperature—the thermoreceptors detect the return to the normal range and stop stimulating the hypothalamus, which then stops stimulating the effectors.

Impaired Temperature Regulation

As you have seen, the human body has a fairly narrow homeostatic temperature range. When the core temperature moves outside this range, significant disruptions in all body systems can result. An abnormal decrease in body temperature is known as **hypothermia,** commonly due to prolonged exposure to decreased temperatures. Hypothermia generally slows down the body's enzymes, which slows the metabolic rate and most of its processes, including heart and breathing rate. When body temperature falls below about 32° C (89.6° F), heat-promoting mechanisms cease, mental function becomes sluggish, and coma and death are possible. An individual who survives hypothermia may suffer from *frostbite,* or the death of peripheral structures such as the fingers, toes, and tips of the ears and nose. This is due to the constriction of blood vessels to these peripheral locations as part of the body's heat-conserving mechanisms.

Humans are more tolerant in general of hypothermia than of an abnormally elevated temperature, which is called **hyperthermia.** Technically, any temperature higher than the normal core temperature range could be considered hyperthermia, but the defining temperature is usually above 38.3° C (101° F). Humans can generally survive internal temperatures up to about 40.6–41.1° C (105–106° F), depending on their age and general health. Temperatures at or above 42.8° C (109° F) are generally incompatible with life.

Some cases of hyperthermia, known collectively as *heat illness,* are caused by prolonged exposure to elevated temperatures in which the body's ability to cool itself is insufficient. The elevated temperatures increase enzyme activity and the rate of the body's reactions. This actually produces more heat, which further increases enzyme activity in a positive feedback loop. This loop continues to operate until the body's temperature rises to about 41.1° C (106° F), at which point enzymes begin to denature and organs start to fail. The resulting condition, known as **heat stroke,** is characterized by hot, dry skin; rapid breathing; and a rapid, sometimes irregular pulse. Heat stroke can be quickly fatal if not immediately recognized and treated.

Another type of hyperthermia is **fever,** which you learned about with the immune system (see Chapter 20). Unlike heat illness, which is caused by elevated external temperatures and humidity, fever is due to an increase in the set point for body temperature. Recall that a fever is part of the nonspecific inflammatory response that generally occurs when chemicals called *pyrogens* are secreted by leukocytes such as macrophages. Pyrogens act on the thermoregulatory centers of the hypothalamus and reset the body's "thermostat" so that the set point is higher. Thermoreceptors then detect that the body temperature is too low and trigger the heat-promoting center in the hypothalamus,

which in turn triggers mechanisms such as shivering (the "chills" you get when you have a fever). When the body temperature rises to the new set point, the hypothalamus normally shuts off its heat-promoting mechanisms.

Quick Check

☐ 6. What is the normal core body temperature?

☐ 7. How does the body respond to a decrease in core temperature?

☐ 8. How does the response differ with an increase in core temperature?

Apply What You Learned

☐ 1. A condition for measurement of the BMR is that the individual is not under physical or emotional stress. How might physical or emotional stress affect the metabolic rate?

☐ 2. If a product increased lipolysis without increasing the metabolic rate, would this likely result in weight loss? Explain why or why not.

☐ 3. What happens to the metabolic rate during exercise? Explain why you feel warm when you exercise, even if the environmental temperature is cold.

See answers in Appendix A.

MODULE **23.7**
Nutrition and Body Mass

Learning Outcomes

1. Define the terms nutrient, essential nutrient, macronutrient, and micronutrient.

2. Describe the dietary sources, relative energy yields, and common uses in the body for carbohydrates, lipids, and proteins.

3. Discuss the major roles and sources of fat- and water-soluble vitamins and dietary minerals.

4. Describe the components of a balanced diet, including the concept of recommended dietary allowances and energy balance.

5. Explain how the liver processes cholesterol, and identify the roles of the different lipoproteins.

6. Describe how body mass is determined.

We've seen how nutrients are used by the body, so now let's examine how they are obtained from the diet. As we mentioned earlier, a **nutrient** is a substance obtained from food that the body requires for its metabolic processes. The study of nutrients includes the fuels carbohydrates, proteins, and lipids; chemicals called *vitamins* and *minerals;* and other structural compounds taken in from the diet, such as cholesterol.

The upcoming sections take a look at the dietary sources of these nutrients and the amounts of each nutrient required for optimal health. The end of the module concentrates on how nutrient intake contributes to body mass and the problems that may arise when excessive intake leads to obesity. First, however, let's get acquainted with some basics of nutrition science.

Overview of Nutrients

There are five main categories of nutrients. The first three—carbohydrates, proteins, and lipids—are known as **macronutrients** because they make up the bulk of our diet and are required in relatively large amounts. The other two categories, vitamins and minerals, are called **micronutrients** because they are required from the diet in much smaller amounts. Note that water is often considered a sixth class of nutrient, but we discuss its many roles in the body in the fluids and electrolytes chapter (see Chapter 25).

For most types of nutrients, the amount required from the diet to maintain health has been calculated by the U.S. Department of Agriculture (USDA) and is referred to as the *recommended dietary allowance,* or *RDA.* The RDA for each nutrient depends on several factors, including a person's age and sex, the number and types of processes in which the nutrient is involved, and whether or not the body can produce the nutrient. Many compounds, such as glucose, several types of amino acids and fatty acids, as well as some vitamins, can be made by the body. However, the body is unable to produce certain compounds called **essential nutrients** (so named because they are an essential part of the diet). There are as many as 50 essential nutrients, including certain amino acids and fatty acids, most vitamins, and all minerals.

Quick Check

☐ 1. How do macronutrients and micronutrients differ?

☐ 2. What is an essential nutrient?

Macronutrients

 **FLASHBACK**

1. How do saturated and unsaturated fats differ? (p. 53)

We take in a wide variety of carbohydrates, lipids, and proteins through the diet, some of which are used for fuel, and others for structural compounds; still others cannot be digested and so simply pass through the body unaltered. The following sections examine the dietary sources, the recommended dietary allowances, and the functions for each of these macronutrients (**Table 23.1**).

Carbohydrates

Dietary carbohydrates consist of monosaccharides such as glucose, disaccharides such as sucrose, and polysaccharides such as starch. Most of these forms of carbohydrates come from plant sources, although there are animal sources as well, notably the disaccharide lactose in milk. We find monosaccharides and disaccharides (nutritionally simple sugars) naturally in fruits, honey, milk, and sugar cane, as well as added to many foods and drinks, including breakfast cereals, candy, juices, and sodas. Polysaccharides, sometimes called *complex carbohydrates,* are found in foods such as grains and vegetables.

Certain polysaccharides known collectively as **fiber** are not fully digestible by humans because we lack the digestive enzymes for their catabolic reactions. *Insoluble fiber,* which is found in whole grains, many fruit skins, bran, and some vegetables, is not digestible at all and passes through the digestive tract and into the feces virtually unaltered. *Soluble fiber,* however, is partially digestible by bacteria in the colon into compounds that can be absorbed into the bloodstream. This type of fiber is found in the pulp of many fruits, grains such as oats, and many vegetables. Soluble fiber has been the topic of much research, as some of the absorbed compounds have been found to have beneficial health effects, such as lowering the level of cholesterol in the blood.

Table 23.1 Summary of Macronutrients

Macronutrient Class	Dietary Sources	Daily Requirement (for Adults)	Functions
Carbohydrates • Monosaccharides and disaccharides (simple sugars) • Polysaccharides (complex carbohydrates)	• Fruits, dairy products, sugar cane, sugar beets, candy, additives to many foods and drinks • Rice, pasta, bread, potatoes, vegetables, glycogen in meat	45–65% of total caloric intake	• Oxidized for energy • Used in production of other substances • Incorporated into multiple structural molecules
Lipids • Saturated • Unsaturated	Meats, eggs, dairy products, plant oils	< 30% of total caloric intake	• Oxidized for fuel • Used to form structural molecules • Glycerol used for gluconeogenesis
Proteins	Meats, legumes, nuts, seeds, dairy products, certain grains, eggs	10–35% of total caloric intake	• Used to form structural molecules • Used to manufacture enzymes, antibodies, and hormones • Oxidized for fuel • Used for gluconeogenesis

Nutritionists generally recommend that an individual's daily intake of Calories consist of about 45–65% carbohydrates. These carbohydrates should be mostly polysaccharides, rather than monosaccharides or disaccharides, because polysaccharides tend to be more nutrient-dense. Many foods with primarily simple sugars are highly processed and nutrient-poor, and for this reason are sometimes said to have "empty calories."

In the past decade, carbohydrates have gotten a lot of negative attention in the media. A number of diet plans say that carbohydrates are the main reason for most people's expanding waistlines. Nevertheless, carbohydrates remain an essential part of the diet. Glucose is the preferred fuel source for the brain and other tissues such as the liver, and the body will go to great lengths to ensure that the glucose level in blood remains constant. Diet plans that severely restrict or even eliminate carbohydrates may lead to some fat loss as the body initiates processes like β-oxidation and ketogenesis, but they also cause protein breakdown to release amino acids for gluconeogenesis. This can lead to wasting of skeletal muscle tissue and other vital organs such as the heart. In addition, ketone bodies can accumulate in the blood and potentially affect the blood's pH, as you discovered earlier.

Lipids

Several types of lipids are ingested in the diet, including triglycerides, cholesterol, and certain vitamins. This discussion is restricted to those lipids used as a fuel source—namely, the triglycerides or neutral fats (cholesterol, a structural lipid, is discussed later in this module), which consist of three fatty acids bound to glycerol. As we discussed previously, **saturated fatty acids** have hydrocarbon chains with no double bonds between carbon atoms (see Figure 2.17). Saturated fats are found primarily in animal fats, including butter, cream, cheese, fatty meats, and lard. Certain plant products have saturated fats as well, including coconut oil, cottonseed oil, and cacao (the main ingredient in chocolate).

Unsaturated fatty acids have hydrocarbon chains with one or more double bonds between carbon atoms. An unsaturated fatty acid with a single double bond is a *monounsaturated fat,* and one with multiple double bonds is a *polyunsaturated fat.* Unsaturated fats can be classified according to the location of their final double bond in the fatty acid chain. This is indicated by the Greek letter omega, followed by a number indicating the location of the carbon (e.g., omega-3 fatty acid). They may also be classified according to the orientation of hydrogen atoms around the carbon atoms. Those with hydrogen atoms on opposite sides are known as *trans* fats; those with hydrogen atoms on the same side are *cis* fats. *Trans* fats have gotten a great deal of publicity in recent years, as they have demonstrated many negative health consequences. (For more information, look back at *A&P in the Real World: The Good, the Bad, and the Ugly of Fatty Acids,* in Chapter 2.)

Fat is an important nutrient; in fact, there are two **essential fatty acids,** those that the human body can't make and must be supplied in the diet. The first, the polyunsaturated omega-6 fatty acid *linoleic acid,* is an important building block for many lipids in the body. The second is the polyunsaturated omega-3 fatty acid *linolenic acid.* Both are present in most plant oils, and linolenic acid is found in abundance in fish oils.

Present USDA guidelines recommend that fats make up no more than 30% of an individual's total caloric intake. The majority of these fats should be unsaturated fats, some of which may have potential health benefits. For example, studies have demonstrated possible positive effects on cardiovascular health from ingestion of omega-3 fatty acids. Saturated fat intake should be no more than 10% of the total caloric intake, as many correlations have been found between the consumption of large amounts of saturated fats and certain diseases, particularly coronary artery disease and certain cancers.

Proteins

As we have discussed throughout this text, proteins and amino acids are important molecular fuels, structural molecules, and enzymes. They are also used in the synthesis of other compounds, such as nucleic acids, glycoproteins, and glucose. Dietary sources of proteins include animal proteins from meats, dairy products, and eggs, as well as plant proteins from legumes, nuts, seeds, certain grains, and soy. Current guidelines recommend that about 10–35% of your total caloric intake come from proteins. This number varies with each individual; for example, children and athletes typically have higher protein requirements than do older, sedentary individuals.

Twenty amino acids make up the body's proteins, 11 of which, called **nonessential amino acids,** can be synthesized from carbon skeletons. The other nine **essential amino acids** must be supplied by the diet (**Table 23.2**). A dietary protein is considered a **complete protein** when it supplies all the essential amino acids; an **incomplete protein** lacks one or more of them. Most animal proteins are complete proteins, but of the plant proteins, only soy is complete. For this reason, there is a common misconception that individuals eating vegetarian diets cannot obtain all the essential amino acids. However, vegetarians can readily obtain them by combining incomplete proteins. The classic example of a good combination is rice and beans.

A dietary protein that has been getting a great deal of media attention in recent years is *gluten.* Gluten is actually a protein composite, a combination of two proteins occurring naturally in wheat and certain other grass-based grains. It is found in foods prepared directly from wheat as well as in foods in which it is

Table 23.2 Essential and Nonessential Amino Acids

Essential Amino Acids	Nonessential Amino Acids
Histidine	Alanine
Isoleucine	Arginine
Leucine	Asparagine
Lysine	Aspartate
Methionine	Cysteine
Phenylalanine	Glutamate
Threonine	Glutamine
Tryptophan	Glycine
Valine	Proline
	Serine
	Tyrosine

used as an additive to improve the food's texture and increase its protein content. An indeterminate percentage of the population, variably given as between 4% and 7%, has a sensitivity to gluten. The symptoms range from mild digestive irritation to small intestine damage in the 1% who have an autoimmune disease known as *celiac disease* (SEE-lee-ak). Although the percentage of individuals with gluten sensitivity is relatively small, gluten has been incorrectly reported by the media to be potentially harmful to nearly everyone and the cause of a host of diseases. However, the only people who truly need to avoid gluten are those with gluten sensitivity; there is no evidence that it is unhealthy for the remaining 93% of the population.

Quick Check

- ☐ 3. What are dietary sources of monosaccharides, disaccharides, and polysaccharides?
- ☐ 4. Which fatty acids are essential? What are the main sources of these fatty acids?
- ☐ 5. How does a complete protein differ from an incomplete protein?

Micronutrients

《 FLASHBACK

1. How do hydrophobic and hydrophilic substances differ? (p. 47)
2. What are some minerals important to human physiology? (p. 33)

The micronutrients—vitamins and minerals—are chemicals that are not used for fuel but that play critical roles in nearly all our physiological processes. The vitamins and minerals required in our diets, their sources, and their roles are examined next.

Vitamins

Some media reports tend to associate the term vitamin with anything that's deemed "good for you." But what are vitamins, and what do they do in the body? A **vitamin** (from its historical name, *vital amine*) is an organic compound required for the body's functions. There are 13 known vitamins, and their letter names indicate the order in which they were discovered. Some vitamins, such as vitamins A and D, act as hormones and influence the functions of other cells. Others, particularly the B vitamins, act as enzyme *cofactors,* or nonprotein compounds required for the enzyme to function properly. Still others, including vitamins E and C, are important antioxidants that detoxify free radicals formed as a normal part of the body's metabolism.

Vitamins are classified by their structure, which determines their water or lipid solubility. The **fat-soluble vitamins** have a structure similar to that of cholesterol, with one or more hydrocarbon rings and long hydrocarbon chains. For this reason, they are nonpolar and are therefore hydrophobic and soluble in other lipids. The four fat-soluble vitamins are vitamins A, D, E, and K. As we have seen, fat-soluble vitamins bind bile salts in the small intestine and are absorbed via micelles with other lipids (see Chapter 22). In the body, excess fat-soluble vitamins can be stored in adipose tissue. This means that excess stores of fat-soluble vitamins have the potential to reach toxic levels.

Vitamin C and the B vitamins are **water-soluble vitamins,** meaning that they have predominantly polar bonds and are therefore hydrophilic and soluble in water. With the exception of vitamin B12, they are absorbed from the gastrointestinal tract with water and polar solutes. Recall that vitamin B12 must bind to the compound intrinsic factor, produced by the stomach, to be absorbed (see Chapter 22). Water-soluble vitamins are not stored in the body in any appreciable amount, and any ingested in excess are generally excreted by the kidneys into the urine.

The 13 vitamins are summarized in **Table 23.3**. It covers their roles, their sources, and the effects of their deficiency and excess.

Table 23.3 Summary of Vitamins

Vitamin	Dietary Sources and RDA	Functions	Symptoms of Deficiency	Symptoms of Excess
Fat-Soluble				
Vitamin A (retinol)	Leafy green vegetables, yellow vegetables; 0.7–0.9 mg	• Required for low-light and color vision • Functions as growth factor for epithelium • Needed for growth and development	• Delayed growth • Night blindness and potential total blindness • Immune dysfunction	• Liver damage • Nausea • Decreased appetite • Osteoporosis • CNS dysfunction • Death from extreme doses
Vitamin D	Synthesized in skin when exposed to certain wavelengths of UV light, supplemented in milk and other foods; 15–20 µg (unless exposure to sunlight is inadequate)	• Required for calcium ion homeostasis and bone growth	• Rickets (in children) • Osteoporosis	• Multiple organ dysfunction due to hypercalcemia

(continued)

Table 23.3 Summary of Vitamins *(continued)*

Vitamin	Dietary Sources and RDA	Functions	Symptoms of Deficiency	Symptoms of Excess
Vitamin E (tocopherol)	Vegetable oils, nuts, seeds, dairy products; 15 mg	• Antioxidant	• Anemia • CNS dysfunction	• Nausea • Fatigue • Vision disturbances
Vitamin K	Leafy green vegetables, produced by intestinal bacteria; 90–120 µg	• Required for synthesis of clotting factors II, VII, IX, and X	• Bleeding disorders	• Liver dysfunction • Excessive clotting in individuals taking the drug warfarin
Water-Soluble				
Vitamin B1 (thiamine)	Legumes, whole grains, certain meats; 1.1–1.2 mg	• Coenzyme in many catabolic pathways	• Disease beriberi—peripheral nerve dysfunction, heart disease	• Decreased blood pressure • Nausea
Vitamin B2 (riboflavin)	Dairy products, eggs, certain vegetables, added to many grains; 0.9–1.1 mg	• Component of FAD and so critical component of catabolic pathways	• Skin disorders • Dysfunction of other epithelial membranes	• Skin dryness and itching
Vitamin B3 (niacin)	Nuts, meats, grains, potatoes; 11–12 mg	• Component of NAD and so critical component of catabolic pathways	• Disease pellagra—CNS, gastrointestinal, and skin dysfunction	• Vasodilation, leading to skin flushing and burning • Death from extreme doses
Vitamin B5 (pantothenic acid)	Meats, dairy products, grains; 5 mg	• Required to synthesize coenzyme A	• Delayed growth • CNS dysfunction	None known
Vitamin B6 (pyridoxine)	Meat, whole grains, citrus fruits, vegetables; 1.3–1.7 mg	• Coenzyme in many metabolic pathways	• Peripheral nerve dysfunction • CNS dysfunction • Anemia • Delayed growth	• CNS dysfunction, including coma • Possible death from extreme doses
Vitamin B7 (biotin)	Red meat, organ meats, eggs, legumes, nuts, dairy products; 25–30 µg	• Coenzyme in many anabolic pathways	• Skin disorders • Neuromuscular dysfunction • Fatigue	None known
Vitamin B9 (folic acid)	Leafy vegetables, fruits, grains; 400 µg, 600 µg in pregnancy	• Coenzyme in many metabolic pathways	• Delayed growth • Anemia • Birth defects if deficient during pregnancy	• Can mask symptoms of vitamin B12 deficiency
Vitamin B12	Dairy products, meat, added to many other foods; 2.4 µg	• Coenzyme in nucleic acid metabolism • Required for development of erythrocytes	• Disease pernicious anemia	• Polycythemia
Vitamin C	Citrus fruits, red and green peppers; 75–90 mg	• Antioxidant • Coenzyme in collagen synthesis	• Disease scurvy—deterioration of skin and epithelial membranes	• Nausea and vomiting • Kidney stones

Minerals

As we discussed, a **mineral** is any element other than carbon, hydrogen, oxygen, and nitrogen that is required by living organisms (look back at Chapter 2). Many minerals in the human body exist in their ionized forms, as this is the form in which they are most stable. Like vitamins, minerals are not used for fuel, but they are an important component of many physiological processes. There are seven minerals called *major minerals* that are required in moderate amounts for optimal health: calcium ions, chloride, magnesium ions, phosphorus, potassium ions, sodium ions, and sulfur. In addition to these seven, very small amounts of many other minerals, the *trace elements,* are necessary, including iodine, iron, selenium, and zinc. The biochemical roles of 10 trace elements have been established; many others are hypothesized, but research hasn't yet defined their roles.

Mineral-rich foods include fruits, vegetables, dairy products, and certain meats. Foods that are highly refined are typically very mineral-poor, as are high-fat foods. Many very low-carbohydrate diets prohibit intake of foods rich in minerals, which could lead to mineral deficiencies. Are minerals consumed in excess quantities beneficial? What about vitamins? Find out in *A&P in the Real World, Pseudoscience Exposed: Vitamin and Mineral Megadoses* on p. 940.

The functions, sources, and symptoms of mineral deficiencies and excesses are summarized in **Table 23.4**.

Quick Check

☐ 6. What roles do vitamins play in the body?

☐ 7. What are the major minerals required from the diet?

Table 23.4 Summary of Minerals

Mineral	Dietary Sources	Functions	Symptoms of Deficiency	Symptoms of Excess
Major Minerals				
Calcium ions	Dairy products, legumes, dark green vegetables	• Required for bone and tooth synthesis and maintenance • Important for nerve function • Important in muscle contraction	• Osteoporosis • Muscle spasms • Nerve dysfunction	• Kidney stones • Disturbances in heart rhythm • Muscle weakness
Chloride	Table salt	• Osmotic particle • Required for synthesis of stomach acid • Important in acid-base balance	• Muscle cramps	• Fluid retention
Magnesium ions	Green vegetables, whole grains, dairy products	• Cofactor in many enzymes • Important in bone formation • Extracellular cation	• Muscle cramps	• Diarrhea • Disturbances in heart rhythm
Phosphorus	Meat, dairy, grains, legumes	• Important component of nucleotides • Required for bone and tooth synthesis • Component of buffer system	• Osteoporosis • Fatigue • Muscle weakness	None known
Potassium ions	Sea salt, many fruits such as bananas, meats, dairy products, vegetables	• Important cation in electrophysiological processes • Osmotic particle	• Disturbances in heart rhythm • Heart failure • Muscle weakness	• Disturbances in heart rhythm • Death if severe excess
Sodium ions	Table salt	• Important cation in electrophysiological processes • Osmotic particle	• Disturbances in heart rhythm • Muscle weakness	• Disturbances in heart rhythm • Water retention • Hypertension
Sulfur	Many different proteins	• Component of certain amino acids	• Inability to produce certain amino acids and the proteins in which they occur	None known
Trace Elements				
Chromium	Yeast, meats, some vegetables	• Required for insulin activity	• Decreased effectiveness of insulin	None known
Cobalt	Meats, dairy products	• Component of vitamin B12	• Same symptoms as vitamin B12 deficiency	None known
Copper	Sea animals, nuts, legumes	• Component of coenzymes of the electron transport chain • Required for hemoglobin synthesis	• Anemia	• Liver failure • CNS dysfunction

(continued)

Table 23.4 Summary of Minerals *(continued)*

Mineral	Dietary Sources	Functions	Symptoms of Deficiency	Symptoms of Excess
Fluorine	Sea animals, added to most drinking water	• Prevention of tooth decay	• Higher incidence of tooth decay	None known
Iodine	Sea animals, dairy products, added to salt	• Required for synthesis of thyroid hormone	• Hypothyroidism	None known
Iron	Meats, eggs, legumes, certain grains, certain vegetables	• Required for synthesis of hemoglobin	• Iron-deficiency anemia	• Nausea and vomiting • Liver damage
Manganese	Nuts, grains, fruits, vegetables	• Cofactor in several enzymes • Potential multiple roles in physiology	• Abnormalities in bone and cartilage • Possible anemia	None known
Molybdenum	Legumes, grains	• Cofactor in many enzymes • Involved in urea processing	• Decreased ability to process amino acids	None known
Selenium	Meats, whole grains, sea animals	• Antioxidant • Enzyme cofactor	• Muscle cramps	• Nausea and vomiting • Fatigue
Zinc	Meats, grains	• Component of digestive enzymes • Component of carbonic anhydrase • Required for normal sperm production in males	• Impaired immunity • Low sperm count in males • Diminished reproductive drive	• CNS dysfunction • Difficulty walking

Structural Lipid: Cholesterol

 **FLASHBACK**

1. What is the function of cholesterol in the body? (p. 56)
2. What are some functions of the liver? (p. 881)

A discussion of nutrition would be incomplete without mention of the steroid **cholesterol.** Although cholesterol isn't oxidized as a fuel source, it is important in many of the body's anabolic reactions. As you've read throughout this book, cholesterol is modified to produce a variety of substances, including steroid hormones, vitamin D, and bile salts. In addition, cholesterol is an important structural molecule, as it forms part of plasma membranes. This section discusses how cholesterol is processed by the body, particularly the liver, and what can happen when the level of cholesterol in the blood is elevated.

Cholesterol Processing by the Liver

Cholesterol is found in the diet in animal-based foods such as meats and dairy products. The body generally has little to no need for dietary cholesterol, as it can be synthesized by the liver. Normally, about 15% of the body's cholesterol comes from the diet; the remaining 85% is made in the liver. Recall that cholesterol is packaged in the small intestine with other lipids as micelles, which escort these nonpolar compounds to the enterocyte membrane, where they are absorbed into the cell. In the enterocytes, cholesterol and other lipids are packaged into chylomicrons, which are then absorbed into a lacteal and eventually delivered to the liver for processing (see Chapter 22).

The liver also plays a key role in distributing cholesterol throughout the body. As we noted previously, cholesterol is a steroid and so is nonpolar and hydrophobic. For this reason, it must be packaged along with other lipids onto carrier proteins called **lipoproteins** for safe transport through the blood. The structure of a typical lipoprotein and its lipids is shown here:

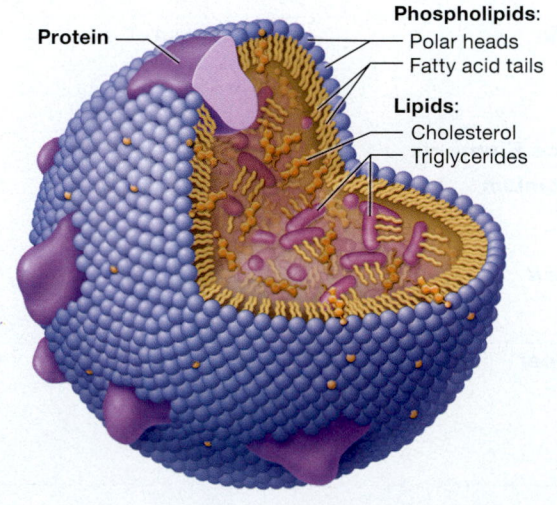

There are several types of lipoproteins that differ in their ratios of lipid to protein, which affects their density. Those carrying more lipids than proteins are larger and have a lower density, whereas those with fewer lipids and more proteins are smaller and have a higher density.

Let's look at these types in more detail (**Figure 23.19**):

● **Very low-density lipoproteins.** The **very low-density lipoproteins (VLDLs)** contain a higher percentage of lipids than proteins, and so have a relatively low density. VLDLs are triglyceride-rich particles that deliver triglycerides from the liver to other tissues, primarily adipose tissue for storage and muscle tissue for immediate use. Triglycerides

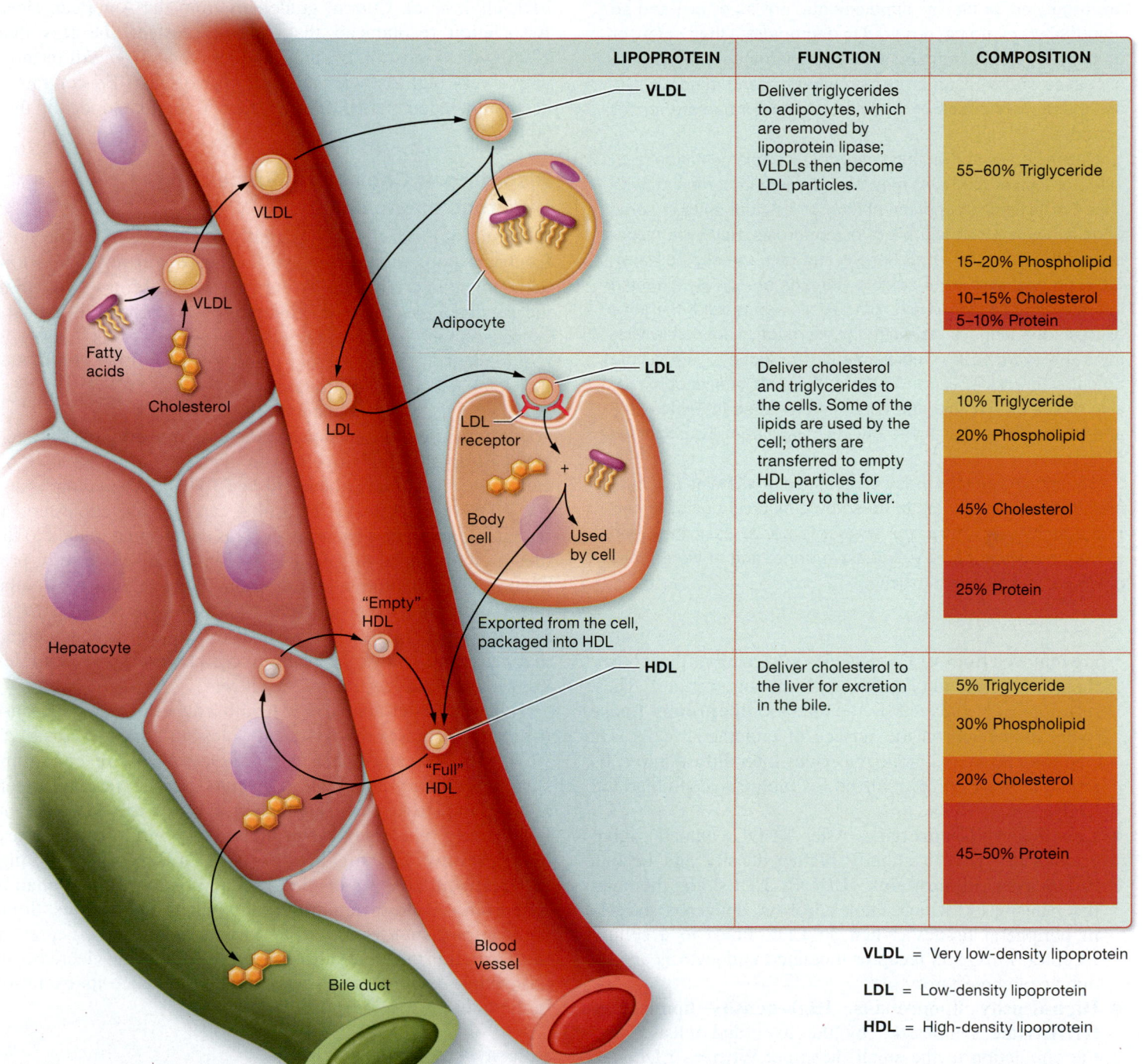

LIPOPROTEIN	FUNCTION	COMPOSITION
VLDL	Deliver triglycerides to adipocytes, which are removed by lipoprotein lipase; VLDLs then become LDL particles.	55–60% Triglyceride 15–20% Phospholipid 10–15% Cholesterol 5–10% Protein
LDL	Deliver cholesterol and triglycerides to the cells. Some of the lipids are used by the cell; others are transferred to empty HDL particles for delivery to the liver.	10% Triglyceride 20% Phospholipid 45% Cholesterol 25% Protein
HDL	Deliver cholesterol to the liver for excretion in the bile.	5% Triglyceride 30% Phospholipid 20% Cholesterol 45–50% Protein

VLDL = Very low-density lipoprotein

LDL = Low-density lipoprotein

HDL = High-density lipoprotein

Figure 23.19 Cholesterol distribution by lipoproteins.

Pseudoscience Exposed: Vitamin and Mineral Megadoses

If something is good for you in small doses, it must be even better for you in large doses, right? That certainly seems to be the thinking of many dietary supplement manufacturers. In the United States, vitamins and minerals are regulated as dietary supplements, not as drugs, and so manufacturers are not required to demonstrate their safety or efficacy. It's not uncommon to see supplements with extreme excesses of the RDAs of certain vitamins and minerals. So, are these megadoses of vitamins and minerals safe and/or effective?

Research has shown that *small* dose supplementation (e.g., 100% or less of the RDA) may be of benefit in certain individuals. For example, supplementation of calcium salts in some people may reduce their risk of osteoporosis. However, megadoses are generally of no benefit and can actually be harmful. High doses of water-soluble vitamins are simply excreted into the urine by the kidneys, but toxicity can result if the dose exceeds the kidneys' capacity for excretion. Fat-soluble vitamins are excreted less efficiently by the kidneys and so the potential for toxicity is much higher. The toxic effects of excessive doses of vitamins and minerals are listed in Tables 23.3 and 23.4—notice that these effects range from liver damage to death.

So, should you take those megadoses of vitamins and minerals? Unless you're told to do so by a medical doctor to correct a significant deficit, the answer is no. At best, they lead to the production of very expensive urine, and at worst, they could cause significant toxicity.

are removed from VLDL particles via reactions that convert them into free fatty acids and monoglycerides. These reactions are catalyzed by the enzyme **lipoprotein lipase,** which is located on the surface of endothelial cells. The free fatty acids and monoglycerides then diffuse into cells, where they are either oxidized for fuel or reassembled into triglycerides.

- **Low-density lipoproteins.** After VLDLs unload triglycerides, they have a slightly higher density and become **low-density lipoproteins (LDLs).** LDLs are the main mechanism by which cholesterol is delivered to cells in peripheral tissues. When an LDL binds to a cellular receptor, it triggers receptor-mediated endocytosis and the uptake of cholesterol.

- **High-density lipoproteins. High-density lipoproteins (HDLs)** are synthesized by the liver and released into the circulation or the small intestine. Whereas LDLs and VLDLs transfer lipids to the body's cells, HDLs transfer lipids from the cells in peripheral tissues to the liver. In the liver, cholesterol is taken up by hepatocytes and excreted from the body via the bile.

You may have heard VLDL and LDL particles described as "bad" cholesterol. This refers to the fact that these particles deliver cholesterol to tissues and can contribute to disease processes such as atherosclerosis (the formation of fatty plaques in arteries). In contrast, HDL particles are described as "good" cholesterol because a higher level of HDL particles will lead to less total cholesterol deposition in peripheral tissues.

The ideal value for total cholesterol is less than 200 mg/dl, but just as important as total cholesterol is the ratio of HDL to LDL cholesterol. Current guidelines from the American Heart Association recommend that LDL cholesterol be less than 129 mg/dl in most individuals, and less than 70 mg/dl in those with a very high risk for cardiovascular disease. By contrast, it is recommended that HDL be *at least* 50 mg/dl and ideally over 60 mg/dl.

Disorders of Cholesterol Processing: Hypercholesterolemia

Cholesterol production is partially regulated by a negative feedback loop, so that increases in intake will result in a limited decline in production by the liver, which brings us again to the Feedback Loops Core Principle (p. 28). However, if intake exceeds about 15% of the total requirement for cholesterol, then the total blood concentration of cholesterol is likely to increase, a condition called **hypercholesterolemia** (hy'-per-koh-less'-ter-ah-LEE-me-uh). Individuals with hypercholesterolemia typically have high total cholesterol and LDL levels, along with a low HDL level. Another factor that can increase cholesterol is increased production by the liver. This can also be related to the diet—a high intake of saturated fats increases cholesterol production while inhibiting its excretion. *Trans* fats have a similar effect, increasing the production of LDLs and decreasing that of HDLs. Hypercholesterolemia can also be due to genetic defects that cause the liver to produce excess cholesterol (a condition known as *familial hypercholesterolemia*).

CORE PRINCIPLE
Feedback Loops

Hypercholesterolemia is a health concern largely because it greatly increases the risk of developing atherosclerosis. As we mentioned, atherosclerosis is associated with conditions such as coronary artery disease, myocardial infarction (heart attack), and stroke (see Chapter 18). For this reason, treatment is generally recommended in individuals with a total cholesterol level over 200 mg/dl and/or an LDL level over 130 mg/dl. Treatment consists of dietary modification in combination with exercise and several different types of medications. Some medications, such as the group known as the *statins,* prevent the formation of cholesterol by the liver. Others decrease its absorption from the small intestine and enhance its excretion in the bile.

Quick Check

☐ 8. Why is HDL considered "good" cholesterol, whereas LDL and VLDL are considered "bad" cholesterol?

☐ 9. What role does the liver play in cholesterol metabolism?

Diet and Body Mass

As mentioned in the introduction to this chapter, diet products, supplements, and books are everywhere, and they form the basis of a billion-dollar industry. Clearly, food issues and weight loss are on the minds of many, many people in the United States. And there is a good reason for this—a significant portion of the American population weighs more than what is generally considered healthy in relation to height.

The term *body mass* refers to the amount of matter in the body; the more commonly used term *body weight* refers to the resulting force exerted on this mass by gravity. An individual's weight in and of itself often doesn't give a clinician enough information, as it varies with several factors, particularly an individual's height. For this reason, many healthcare providers and nutritionists instead calculate a number called the **body mass index (BMI),** which is obtained with the following equation:

$$BMI = \frac{Weight\ (kg)}{[Height\ (m)]^2}$$

As you can see from the equation, BMI takes height into account, and so provides a somewhat more reliable index from which to determine relative body mass. From the BMI, a person can be classified as underweight (BMI < 18.5), normal weight (BMI = 18.5–24.9), overweight (BMI = 25.0–29.9), or obese (BMI ≥ 30.0). Note that these numbers are only reference ranges, not absolutes, because the BMI does not consider factors such as muscle mass, body type, or ethnicity. For example, many highly trained athletes have BMIs in the overweight range due to their larger than average muscle mass.

This section first looks at factors that affect an individual's body mass. Then we turn to how this applies to a diet that meets all nutritional needs and is generally considered healthy. Finally, we examine obesity, the problems that accompany it, and the many ways in which people try to address it.

Energy Balance

Multiple factors determine an individual's recommended body mass, including genetics, age, sex, body type, and degree of physical training. However, the rate at which someone gains or loses body mass is largely dependent on a principle known as **energy balance,** or **nitrogen balance,** the difference between energy intake and energy expenditure. The total amount of energy intake an individual requires to meet his or her daily needs depends on a number of factors, including age, sex, muscle mass, activity level, and genetic factors. The precise number of Calories an individual requires can be estimated by calculating the BMR and factoring in the activity level. For most adults, this ranges from 2000 to 2500 Calories.

So, how does someone gain, lose, or maintain body mass? The answer is pretty simple:

- If the total number of Calories taken in equals the total number expended, body mass remains stable.

- If the total number of Calories taken in is greater than the total number expended, a condition known as *positive energy balance,* body mass increases.
- If the total number of Calories taken in is less than the total number expended, a condition known as *negative energy balance,* body mass decreases.

Essentially, if you eat more Calories than you burn, you will gain weight. This principle is the basic idea behind all varieties of diets—those intended to help an individual gain weight, lose weight, or maintain a given weight.

Healthy Diet

An individual's diet is considered healthy when it meets his or her nutritional needs in terms of energy requirement; types and amounts of macronutrients, vitamins, minerals, and essential amino acids and fatty acids; and no more than the recommended amounts of other substances such as cholesterol. Healthy diets are known to promote health, as their name implies, and reduce the risk for illnesses such as heart disease, stroke, cancer, and diabetes.

But what is a healthy diet? In 2011, the USDA released guidelines called **MyPlate,** seen in **Figure 23.20.** MyPlate shows the proportions of the types of foods that should make up meals. As you can see, the guidelines recommend filling your plate half with vegetables and fruits, and half with grains and proteins. In addition, the small circle represents a serving of a low-fat dairy product such as milk or yogurt.

Since the release of MyPlate, other institutions have issued similar guides that include healthy oils and replace the dairy recommendation with water. These different nutritional recommendations from so many sources can make it confusing to figure out exactly what is healthy for you. The best thing to do in the face of multiple, ever-changing guidelines is use good sense—you really can't go wrong with a diet rich in vegetables, fruits, whole grains, and lean protein in amounts that meet, but don't exceed, your body's energy requirements.

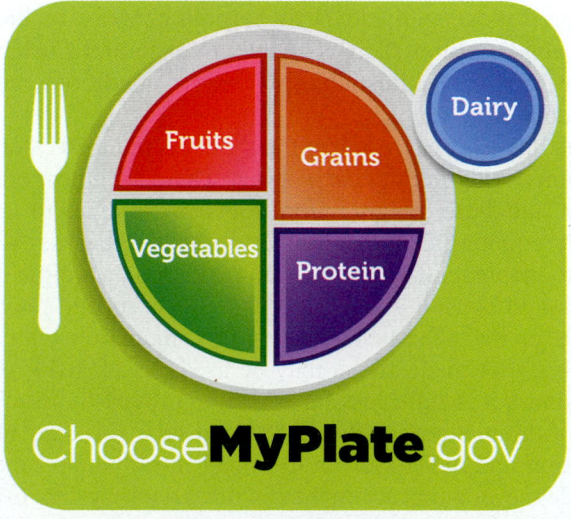

Figure 23.20 Current USDA healthy diet guidelines.

Obesity

The prevalence of adults who are overweight and **obese** in the United States is alarming—according to the Centers for Disease Control (CDC), approximately 34% are overweight, 33.8% obese (have a BMI > 30), and 5.7% morbidly obese (BMI > 40). Even more alarming is the trend that shows a relatively rapid increase in obesity rates; in 1960, only 13.4% of adults were considered obese, and 0.9% morbidly obese. Despite this increase in obesity, however, the percentage of adults meeting the BMI guidelines for overweight has remained nearly unchanged over the past 50 years. Similarly alarming is the obesity rate of 16.9% for children aged 2–19 years. This is an increase from only 5.0% in 1971.

Why is obesity such a concern? The answer has many factors. Chief among them are the health concerns that disproportionately face obese individuals. Obesity is one of the leading preventable causes of death worldwide, and decreases a person's lifespan by an average of 6–7 years. It increases the risk of type 2 diabetes, coronary artery disease, osteoarthritis, certain types of cancer, hypertension, and obstructive sleep apnea (periods in which breathing stops during sleep). Another concern with obesity is social in nature. Obese individuals, particularly children, can face social stigmatization. Finally, obesity has an economic cost, as well, both to the individual and to society. Obesity and its related conditions cost the U.S. government an estimated 100 billion dollars per year. In addition, an obese person often faces employment discrimination and pays more for health insurance premiums.

Obesity has many potential causes, but the number one cause is positive energy balance—that is, a caloric intake that exceeds the number of Calories that are burned. In the United States, this is partly due to the widespread availability of food, particularly high-Calorie, high-fat, processed foods with few nutrients (e.g., fast food). This fact, combined with a sedentary lifestyle, can lead to an energy imbalance. In addition, a genetic connection to obesity exists in many individuals. For example, abnormalities in genes controlling appetite and the metabolic rate can contribute to the development of obesity. Other, less common factors that predispose individuals to becoming obese include endocrine abnormalities such as hypothyroidism and Cushing's syndrome, certain medications, psychiatric conditions, and lower socioeconomic status.

The mainstay of treatment for obesity is reduced caloric intake combined with an exercise program. There are a host of diet programs out there, ranging from low-carbohydrate to low-fat to simply counting Calories. Each type of plan has its advantages and disadvantages, and anyone embarking on a weight loss journey should consult a dietician or physician to find the best plan for his or her needs. Sometimes, in conjunction with diet and exercise, diet medications may be used. As of 2014, two types of agents for weight loss have been approved by the Food and Drug Administration (FDA). The first type (orlistat) blocks fat digestion and absorption in the small intestine, and the second includes several that act on the central nervous system to decrease appetite. In cases of morbid obesity, surgery may be considered.

Surgical procedures, such as the "lap band," which constricts the stomach and prevents it from overfilling, or gastric bypass, which removes a portion of the stomach, can yield significant weight loss results, but only when combined with reduced caloric intake and exercise.

In addition, an ever-increasing number of diet plans and over-the-counter dietary supplements claim to produce miraculous weight loss results. Here we definitely have a case of buyer beware—many are the modern-day equivalents of snake oil, as they promise something they cannot possibly deliver. Most of these plans and products use scientific-sounding terms in their advertisements (a classic tactic of pseudoscience), which can make it difficult to determine if they are legitimate or a waste of money. Some warning signs to look for include promises of unreasonable results ("Eat all you want and still lose weight!"), diet plans that nearly or fully omit entire food groups or severely restrict caloric intake, and any product that uses fear or claims of "conspiracies" to induce you to buy. Remember also that manufacturers of weight loss dietary supplements are not required to test their ingredients for safety or efficacy, which allows them to make any claims about their product and weight loss without having to actually prove that the claims are true.

The best defense for your health (and your bank account) is knowledge and good sense. If something sounds too good to be true, trust your instincts. There are no miraculous instant solutions for weight loss. If you want to lose weight, you have to burn more energy than you take in; in other words, eat less and exercise more—simple, but usually not easy.

Quick Check

□ 10. What conditions create a positive or negative energy balance? What effect will each have on body mass?

□ 11. What are the causes and consequences of obesity?

Apply What You Learned

□ 1. Some low-carbohydrate diets prohibit almost all consumption of carbohydrates, including fruits and vegetables. Predict the potential consequences of these diets.

□ 2. Mr. Alcozer comes into a clinic for a follow-up appointment after having been diagnosed with high cholesterol. The nurse notes that his total cholesterol has decreased from 250 mg/dl to 230 mg/dl, his LDL cholesterol has remained the same at 200 mg/dl, and his HDL cholesterol has decreased from 50 mg/dl to 30 mg/dl. Do these numbers represent an overall improvement in his condition? Why or why not?

□ 3. A friend has embarked on a diet plan whose advertisements say that he can eat as much of anything as he wants and still lose weight. Explain to your friend, in physiological terms, whether or not this claim is likely to be true and why.

See answers in Appendix A.

Chapter Summary

> **For everything you need to succeed in this course, go to** Mastering **Anatomy & Physiology. There you will find:**
>
> 🔲 Practice Tests
>
> 🔘 Author Podcasts
>
> ▶️ Big Picture Animations
>
> 🔊 Concept Boost Video Tutors
>
> **iP2** Interactive Physiology 2.0 Tutorials
>
> *A&PFlix* A&P Flix 3D Animations
>
> ✏️ Active-Learning Workbook

MODULE 23.1
Overview of Metabolism and Nutrition
902–906

- **Metabolism** is the sum of the body's chemical reactions, and includes harnessing the chemical energy in *nutrients* to make **ATP,** converting one type of substance into another, carrying out synthesis reactions **(anabolism),** and degrading macromolecules **(catabolism).**

- **Exergonic reactions** release energy; **endergonic reactions** require energy.

- ATP is an unstable compound. When a phosphate group is removed in a hydrolysis reaction catalyzed by an *ATPase,* the reaction is highly exergonic.

- ATP drives reactions either directly by providing energy from its hydrolysis or by **phosphorylating** a reactant.

- During an **oxidation-reduction reaction,** electrons are transferred from one reactant to another.

- Electrons in the cell are passed from nutrients to different **electron carriers,** including **FAD** and **NAD⁺,** which generates an **electromotive force** that provides the energy to generate ATP.

MODULE 23.2
Glucose Catabolism and ATP Synthesis
907–916

- Glucose catabolism proceeds by **glycolytic catabolism** and the **citric acid cycle,** which generate ATP, NADH, and $FADH_2$. The electron carriers are fed into the **electron transport chain (ETC),** where more ATP is generated.

- The energy released from glucose catabolism generates ATP via either **substrate-level phosphorylation** or **oxidative phosphorylation.**

- Oxygen is neither used nor required for **anaerobic catabolism;** however, oxygen is required for **aerobic catabolism.**

- **Glycolysis** is a series of 10 anaerobic reactions that occur in the cytosol, during which glucose is split into two three-carbon **pyruvate** molecules. Glycolysis yields, overall, two molecules of ATP and one molecule of NADH.

- Under anaerobic conditions, pyruvate is reduced to **lactate** and NADH is oxidized back into NAD⁺. Under aerobic conditions, pyruvate moves into the mitochondria and is oxidized to yield **acetyl-CoA** and carbon dioxide.

- The citric acid cycle is a series of eight aerobic reactions that take place in the mitochondrial matrix during which oxidation-reduction reactions generate NADH and $FADH_2$.

- Oxidative phosphorylation consists of three processes:
  ○ Electrons are transferred from NADH and $FADH_2$ to the electron transport chain (ETC). The final electron acceptor is oxygen, which is reduced to become water.
  ○ The electromotive force drives the pumping of hydrogen ions into the mitochondrial intermembrane space.
  ○ Hydrogen ions flow back into the matrix through **ATP synthase,** which causes it to release its newly formed ATP.

- The overall yield from all parts of glucose catabolism and ATP synthesis is about 38 ATP per glucose molecule.

MODULE 23.3
Fatty Acid and Amino Acid Catabolism
916–920

- Triglycerides are catabolized in the following way: **Glycerol** is converted into glyceraldehyde-3-phosphate and fed into glycolysis. **Fatty acids** are oxidized in two-carbon units by the process **β-oxidation,** which yields seven each of NADH and $FADH_2$ and a molecule of acetyl-CoA.

- Under certain conditions β-oxidation can produce **ketone bodies.**

- Amino acid catabolism begins in the liver with **transamination.** This process yields a carbon skeleton that is fed into oxidative catabolism, and the amino acid glutamate, which is eliminated from the body via the **urea cycle.**

MODULE 23.4
Anabolic Pathways 920–923

- Anabolism is responsible for nutrient synthesis and storage, as well as synthesis of proteins, nucleic acids, NAD⁺, and FAD.

- Glucose is stored as **glycogen,** primarily in the liver and in skeletal muscle tissue. Glycogen is synthesized by the process **glycogenesis** and broken down by **glycogenolysis.**

- **Gluconeogenesis** is the synthesis of glucose from noncarbohydrate precursors.
- Most triglycerides are stored in adipose tissue and in the liver. **Lipogenesis** takes place in the cytosol, where fatty acids are built in two-carbon units, then combined with glycerol to form triglycerides.
- Excess glucose can be converted into fatty acids and glycerol.
- Cells can synthesize 11 of 20 amino acids from carbon skeletons. Excess amino acids are converted to glucose or fatty acids.

MODULE 23.5
Metabolic States and Regulation of Feeding 923–927

- The **absorptive state** takes place from the time food is ingested to the time it is fully absorbed. In this state, glucose is the main fuel, glycogenesis and lipogenesis increase, and protein synthesis takes place.
- The **postabsorptive state** begins after ingested nutrients are fully absorbed. In this state, glycogenesis, lipogenesis, and protein synthesis decrease, and gluconeogenesis, glycogenolysis, lipolysis, and ketogenesis increase.
- Feeding is regulated by two hypothalamic nuclei: the *satiety center* and the *hunger center.*
- Long-term regulation of feeding is done by **leptin,** which is *anorexigenic,* and **ghrelin,** which is *orexigenic.*
- Short-term regulation of feeding involves insulin, digestive hormones, and the vagus nerve, which stimulate the satiety center. Hypoglycemia stimulates the hunger center.

MODULE 23.6
The Metabolic Rate and Thermoregulation 927–932

- The total amount of energy expended by the body to drive all its processes is the **metabolic rate.** The **basal metabolic rate (BMR)** is the "energy cost of living."
- Heat may be exchanged between the body and the environment through **radiation, conduction, convection,** and **evaporation.**
- **Thermoregulation** is the mechanism by which the body maintains a stable internal temperature. Normal **core body temperature** averages about 37.5° C (99.5° F).
  ○ When body temperature increases, the hypothalamic **heat-loss center** promotes vasodilation of skin blood vessels and sweating.
  ○ When body temperature decreases, the hypothalamic **heat-promoting center** triggers constriction of skin blood vessels, shivering, and an increased metabolic rate.
- Impaired thermoregulation can lead to **hypothermia** or **hyperthermia.**

MODULE 23.7
Nutrition and Body Mass 932–942

- Carbohydrates, proteins, and lipids are **macronutrients,** and vitamins and minerals are **micronutrients.** Nutrients that cannot be made by the body are **essential nutrients.**
- Monosaccharides and disaccharides are found in foods such as fruits, milk, honey, and sugar cane. Polysaccharides are found in grains and vegetables. **Fiber** is a polysaccharide not digestible by humans.
- Triglycerides are the main dietary lipids used as a fuel source. Fatty acids bound to glycerol in triglycerides may be **saturated** or **unsaturated.**
- There are 20 amino acids in the body's proteins. Eleven amino acids can be produced by the body; the other nine are **essential.**
- There are 13 known **vitamins,** organic compounds that act as hormones, cofactors for enzymes, and antioxidants. **Fat-soluble vitamins** include vitamins A, D, E, and K. **Water-soluble vitamins** include vitamin C and the B vitamins.
- A **mineral** is any element other than carbon, hydrogen, oxygen, and nitrogen that is required by living organisms.
- **Cholesterol** is a steroid molecule obtained from animal products and produced by the liver. Cholesterol is attached to **lipoproteins** in the liver. **Very low-density lipoproteins (VLDLs)** deliver triglycerides to peripheral tissues; **low-density lipoproteins (LDLs)** deliver cholesterol to peripheral tissues; and **high-density lipoproteins (HDLs)** deliver cholesterol to the liver for excretion in the bile.
- The **body mass index (BMI)** is a measurement of body weight relative to height that provides a range for normal *body mass.*
- The difference between energy intake and energy expenditure is known as **energy balance.**
- Individuals who have a body mass index over 30.0 are defined medically as **obese.** Obesity increases the risk for many diseases.

Assess What You Learned

Scan the QR Code for additional practice test questions

https://goo.gl/gLQZg0

LEVEL 1 Check Your Recall

1. Which of the following statements is false?
   a. Metabolism is the sum of all the reactions in the body.
   b. Anabolic pathways are exergonic and release energy.
   c. Catabolic pathways break larger substances into smaller substances.
   d. The body "pays for" anabolic reactions with the energy from catabolic reactions.

2. Which nutrient is the preferred fuel for essentially every cell in the body?

   a. Glucose   c. Glycerol
   b. Fatty acids   d. Amino acids

3. Fill in the blanks: A/an _____ reaction releases energy because the products have less energy than the reactants, whereas a/an _____ reaction consumes energy because the products have more energy than the reactants.

4. Mark the following statements as true or false. If a statement is false, correct it to make a true statement.

   a. During an oxidation-reduction reaction, the compound with the greatest electron affinity accepts electrons and is oxidized.
   b. The electromotive force is generated by the flow of electrons from one substance that is oxidized to another substance that is reduced.
   c. NAD^+ and FAD are electron carriers in the cell.
   d. ATP is the "money" that the cell uses to "pay for" the cell's exergonic reactions.

5. Does ATP possess a "high-energy bond"? Explain. How does ATP provide energy to drive a reaction or process?

6. Place the following events of glucose catabolism in the correct order, placing a 1 by the first event, a 2 by the second, and so forth.

   _____ Electrons are passed between electron carriers in the inner mitochondrial membrane, and hydrogen ions are pumped into the intermembrane space.
   _____ ATP is generated in the cytosol by substrate-level phosphorylation; NADH is generated as well.
   _____ Hydrogen ions pass through the channel of ATP synthase and ATP is released from the enzyme.
   _____ Acetyl-CoA is combined with oxaloacetate to form citrate.
   _____ ATP is consumed to split glucose into two three-carbon compounds.
   _____ Pyruvate loses a carbon, forming acetyl-CoA, carbon dioxide, and NADH.
   _____ Citrate undergoes a series of oxidation-reduction reactions that generate ATP, NADH, and $FADH_2$.

7. Which of the following statements is false?

   a. Oxaloacetate is regenerated at the completion of the citric acid cycle.
   b. Glycolysis takes place in the cytosol.
   c. The energy yield from oxidative phosphorylation is much greater than the yield from substrate-level phosphorylation.
   d. The electromotive force directly generates ATP.

8. What is the role of oxygen in oxidative phosphorylation?

9. The amino groups from amino acid catabolism are eliminated as which of the following compounds?

   a. Uric acid
   b. Ammonium ions
   c. Ammonia
   d. Urea

10. Mark the following as true or false. If a statement is false, correct it to make a true statement.

   a. The reactions of β-oxidation crop fatty acids into three-carbon pyruvate units.
   b. Most amino acids are converted into products that are oxidized by β-oxidation.
   c. The final common pathway for carbohydrate, protein, and lipid catabolism is the same.
   d. Fat oxidation generates more ATP than protein or carbohydrate catabolism.
   e. Fat and protein catabolism take place primarily in the cytosol.

11. When fats are catabolized by the liver, some of the acetyl-CoA is converted into:

   a. amino acids.   c. glucose.
   b. ketone bodies.   d. glycerol.

12. Fill in the blanks: Glucose is stored by the process of _____ and new glucose is generated by the process of _____.

13. Which of the following can be used to synthesize fatty acids?

   a. Glucose   c. Both a and b
   b. Amino acids   d. Neither a nor b

14. Identify the following characteristics as belonging to the absorptive state or the postabsorptive state.

   a. Ketogenesis occurs. _____
   b. Gluconeogenesis and glucose sparing take place. _____
   c. Glycogenesis and lipogenesis take place. _____
   d. Protein synthesis decreases or stops. _____
   e. Glucose is readily available and is used as the primary fuel for most cells. _____
   f. Lipolysis releases fatty acids into the blood. _____

15. The hormone that directly stimulates the hunger center is:

   a. ghrelin.
   b. cholecystokinin.
   c. insulin.
   d. leptin.

16. Why is heat production a good way to measure the body's metabolic rate?

17. Which of the following best defines the basal metabolic rate?

   a. The rate of heat production during exercise
   b. The rate at which glucose is catabolized
   c. The amount of energy needed for the body's anabolic reactions
   d. The energy cost of living

18. Mark the following statements as true or false. If a statement is false, correct it to make a true statement.

   a. Conduction is energy transfer between the body and the environment via electromagnetic waves.
   b. Thermoreceptors send temperature information to centers in the hypothalamus.
   c. Heat-loss mechanisms include sweating and vasodilation of the blood vessels in the skin.
   d. The body temperature measured on the skin or in the mouth is generally higher than the core body temperature.

19. Fill in the blanks: A/an_____ is a nutrient that the body requires in large quantities, whereas a/an_____ is one that the body requires in much smaller quantities. A/an _____ is a nutrient that the body cannot produce and so must be supplied from the diet.

20. Which of the following statements is false?

    a. Fiber is a polysaccharide that is completely or partially indigestible by the human body.
    b. A complete protein contains all the essential amino acids.
    c. Saturated fats contain hydrogen atoms arranged on opposite sides of the carbon-carbon double bond.
    d. Linoleic and linolenic acid are essential fatty acids.

21. How do fat-soluble and water-soluble vitamins differ? Which vitamins are fat-soluble?

22. Mark the following statements as true or false. If a statement is false, correct it to make a true statement.

    a. Vitamins and minerals are used as fuel sources for cells.
    b. High-density lipoproteins deliver cholesterol from the liver to peripheral tissues.
    c. Negative energy balance refers to the condition in which fewer Calories are taken in than are expended.
    d. Obesity is defined as a body mass index of 30.0 or greater.

LEVEL 2 Check Your Understanding

1. Exercise is said to "burn" fat. In what ways is this analogy true?

2. Certain diet plans claim that a person may eat all the protein he or she wants, as protein is not stored by the body. Is this claim completely accurate? Explain.

3. Erythrocytes lack organelles, including mitochondria. What sort of carbohydrate catabolism can erythrocytes undergo? Explain.

4. You are performing a lab experiment to measure your lab partner's metabolic rate. To perform the experiment, your lab partner is hooked up to a machine that measures her oxygen consumption.

    a. Why is the amount of oxygen that your lab partner consumes a good way to measure her metabolic rate?

    b. Your lab partner just ran up three flights of stairs to get to lab on time. Will the measured metabolic rate be her basal metabolic rate? Why or why not?

5. People with extremely restricted caloric intake often complain of being cold. What does this signify about their metabolic rate?

LEVEL 3 Apply Your Knowledge

PART A: Application and Analysis

1. A patient presents to a clinic with acute ingestion of the poison cyanide, which blocks the functioning of the electron transport chain. He is exhibiting many of the symptoms seen with suffocation (lack of breathing). Explain this finding. Would the level of oxygen in the body be normal, elevated, or decreased in cyanide poisoning? Explain.

2. Your friend hears about the latest fad diet that consists entirely of fat. Predict the initial effects such a diet might have on your friend's concentration, memory, and other nervous system functions. Would you recommend this diet to anyone? Why or why not?

3. Certain dietary supplements for weight loss contain drugs that are claimed to block the absorption of polysaccharides such as starch. If a "starch blocker" actually worked and fully prevented the absorption of carbohydrates into the bloodstream, what effects would this have on protein and lipid metabolism?

PART B: Make the Connection

4. The condition hyperthyroidism is characterized by an increase in the level of thyroid hormone. People with untreated hyperthyroidism often complain of feeling hot. Explain this finding. *(Connects to Chapter 16)*

5. When you are exercising, your respiratory rate increases and you exhale more carbon dioxide. Explain this. What will happen to the pH of the blood if the respiratory rate does not increase? Why? *(Connects to Chapter 21)*

See answers in Appendix A.

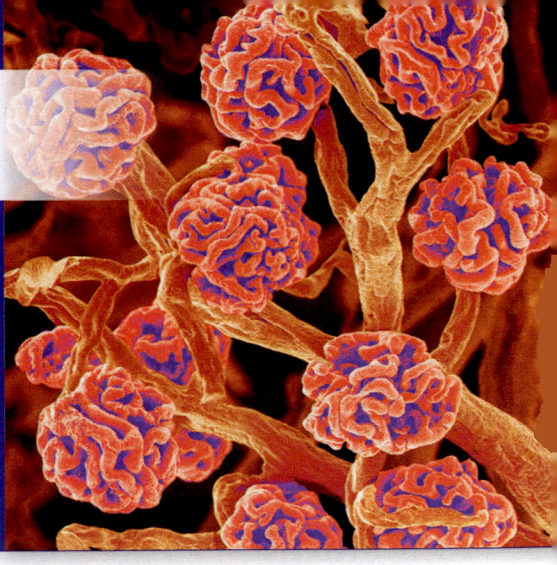

The Urinary System

Animals living in an aquatic environment face little risk of becoming dehydrated. However, animals that started to spend more time on dry land millions of years ago needed mechanisms to conserve water and prevent dehydration. The organ system that performs this function in humans—the **urinary system**—is the topic of this chapter. The organs of the urinary system are organs of **excretion;** that is, they remove wastes and water from the body. Specifically, the urinary system "cleans the blood" of *metabolic wastes,* which are substances produced by the body that it cannot use for any purpose. However, as you will learn in this chapter, the urinary system does far more: This system is also essential for removing toxins, maintaining homeostasis of many variables (including blood pH and blood pressure), and producing erythrocytes. Read on to discover how the urinary system is vital to your body's homeostasis.

Photo: This colorized scanning electron micrograph shows glomeruli, the filtering units of the kidneys.

24.1
Overview of the Urinary System

Learning Outcomes

1. List and describe the organs of the urinary system.
2. Describe the major functions of the kidneys.

The urinary system is composed of the paired *kidneys* and the *urinary tract*. The **kidneys** filter the blood to remove metabolic wastes and then modify the resulting fluid, which allows these organs to maintain fluid, electrolyte, acid-base, and blood pressure homeostasis. This process produces **urine,** a fluid that consists of water, electrolytes, and metabolic wastes. Then the remaining organs of the urinary system—those of the urinary tract—transport, store, and eventually eliminate urine from the body. In this module, we first examine the basic structures of the urinary system, and then turn to the functional roles of the kidneys.

Overview of Urinary System Structures

As you can see in **Figure 24.1**, the kidneys resemble their namesake, the kidney bean, in both shape and color. The kidneys are situated against the posterior abdominal wall and are *retroperitoneal* (reh′-troh-pehr′-ih-tohn-EE-ul; *retro-* = "behind") organs, meaning they are located posterior to the peritoneal membranes. Note, however, that the two kidneys differ slightly in position—the left kidney extends from about T12 to L3, whereas the right kidney sits slightly lower on the abdominal wall because of the position

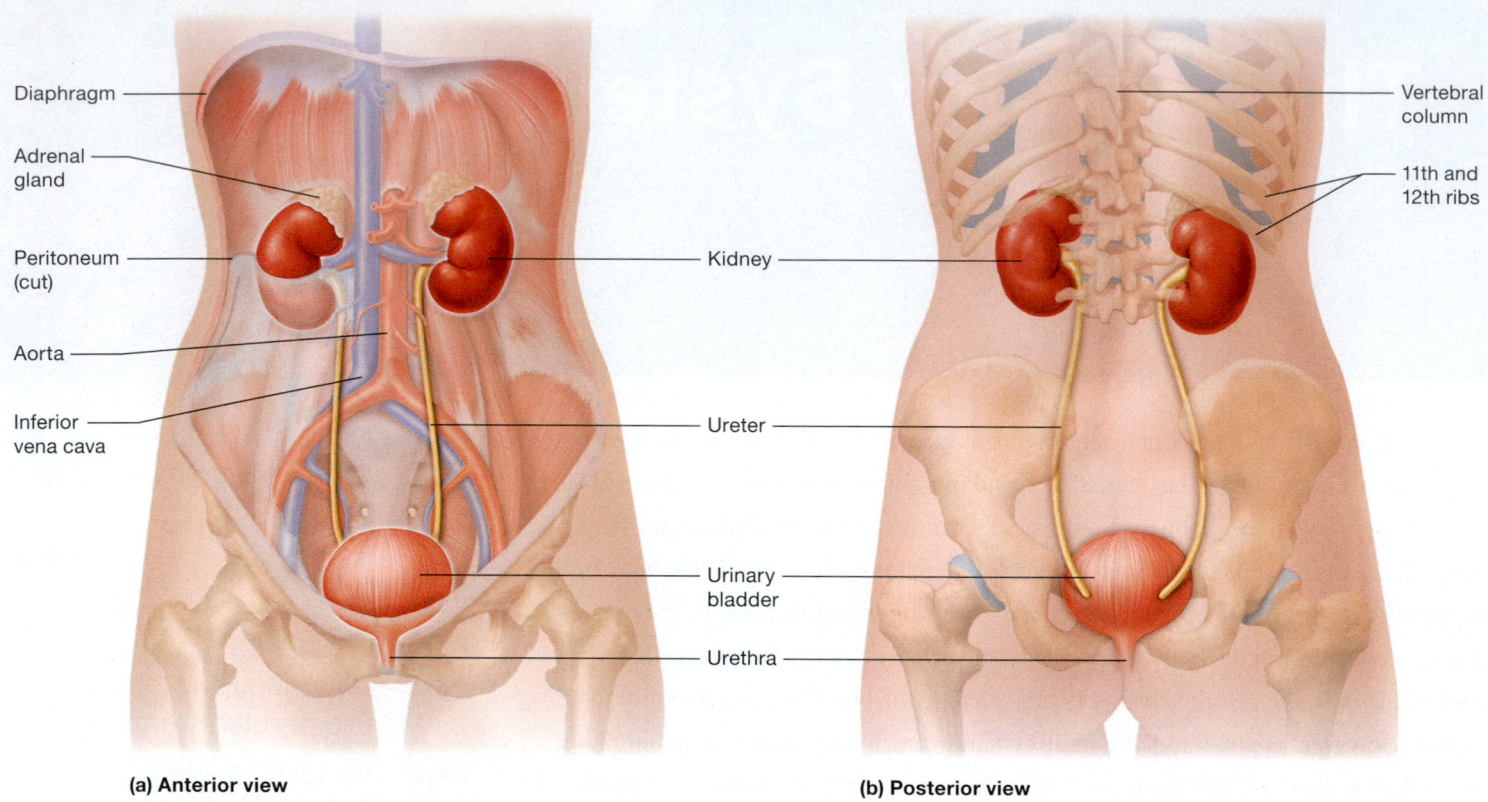

Figure 24.1 Organs of the urinary system in a female.

of the liver. The superior portions of both kidneys are partially protected by the 11th and 12th pairs of ribs. Each kidney is capped by an adrenal gland (*ad-* = "near," *ren-* = "kidney"); these glands perform endocrine functions and secrete a variety of hormones (see Chapter 16).

The **urinary tract** is composed of the paired *ureters* (YOOR-eh-terz), the *urinary bladder,* and the *urethra* (yoo-REE-thruh). Urine leaves each kidney through one of the two ureters, tubes that run along the posterior body wall, connecting the kidneys with the hollow urinary bladder. The bladder, which stores urine, sits on the floor of the pelvic cavity. Urine is expelled from the body through the passage called the urethra, which connects the urinary bladder with the outside of the body.

Quick Check

☐ 1. What are the organs of the urinary system?

Overview of Kidney Function

◀◀ FLASHBACK

1. What are three factors that determine blood pressure? (p. 676)
2. What is erythropoietin, and what is its main function? (p. 727)

The regulation of homeostasis by the urinary system takes place in the kidneys, so let's take a quick look at what the kidneys do and how they do it. The kidneys perform the following functions:

- **Removal of metabolic wastes.** As we have discussed, the kidneys filter the blood, removing metabolic wastes. These wastes are eliminated from the body via the urine.
- **Maintenance of fluid and electrolyte balance.** The kidneys maintain the homeostasis of blood solute concentration, or osmolarity, by conserving or eliminating water and electrolytes such as sodium, potassium, and calcium ions.
- **Maintenance of acid-base balance.** The kidneys assist in the long-term maintenance of blood pH homeostasis by conserving or eliminating hydrogen (H^+) and bicarbonate (HCO_3^-) ions.
- **Maintenance of blood pressure.** The kidneys directly influence systemic blood pressure through their control of blood volume. Additionally, they secrete an enzyme that influences both blood volume and peripheral resistance.
- **Regulation of erythropoiesis.** The kidneys regulate erythrocyte production in the bone marrow by releasing the hormone *erythropoietin* (eh-rith′-roh-POY-eh-tin; see Chapter 19).
- **Performing other metabolic functions.** The kidneys play many important metabolic roles, including detoxifying substances in the blood, activating vitamin D, and making new glucose through the process of *gluconeogenesis* (gloo′-koh-nee′-oh-JEN-eh-sis).

☐ 2. What are the basic functions of the kidneys?

Apply What You Learned

☐ 1. Inflammation of the peritoneal membranes, or *peritonitis,* can cause dysfunction of multiple organs in the abdominal cavity. Would you expect peritonitis to affect the kidneys? Explain.

☐ 2. Explain why a patient with long-term renal failure might have a decreased number of erythrocytes in his or her blood.

See answers in Appendix A.

MODULE 24.2
Anatomy of the Kidneys

Learning Outcomes

1. Describe the external structure of the kidney, including its location, support structures, and coverings.
2. Trace the path of blood flow through the kidneys.
3. Describe the major structures, subdivisions, and histology of the renal corpuscle, renal tubule, and collecting system.
4. Trace the pathway of filtrate flow through the nephron and collecting system.
5. Compare and contrast cortical and juxtamedullary nephrons.

Let's now take a closer look at the anatomy of the kidneys. In this module, we explore first the external and internal anatomy of the kidneys. We then turn our attention to the structure and basic roles of the kidneys' functional units: the *nephrons* (NEF-rawnz; *nephro-* = "kidney"). We conclude the module by examining the two main types of nephrons.

External Anatomy of the Kidneys

The kidneys are held in place on the posterior body wall and protected by three external layers of connective tissue (**Figure 24.2**). From superficial to deep, these layers are as follows:

- **Renal fascia.** The **renal fascia** (FASH-ee-uh) is a layer of dense irregular collagenous connective tissue that anchors each kidney to the peritoneum and to the fascia covering the muscles of the posterior abdominal wall.
- **Adipose capsule.** The middle and thickest layer, called the **adipose capsule,** consists of adipose tissue that wedges each kidney in place and shields it from physical shock. During prolonged starvation, the body uses the fatty acids in the adipose capsule of the kidney for fuel. This causes the kidney to droop, a condition called *nephroptosis* (nef-rawp-TOH-sis; *-ptosis* = "drooping").
- **Renal capsule.** The **renal capsule** is an extremely thin layer of dense irregular collagenous connective tissue that covers the exterior of each kidney like plastic wrap. It protects the kidney from infection and physical trauma.

Without its connective tissue coverings, a typical adult kidney is about the size of a large bar of soap (11 cm long, 6 cm wide, and 3 cm thick) and weighs about 150 grams. On the medial surface we find an opening called the **hilum** (HY-lum), through which

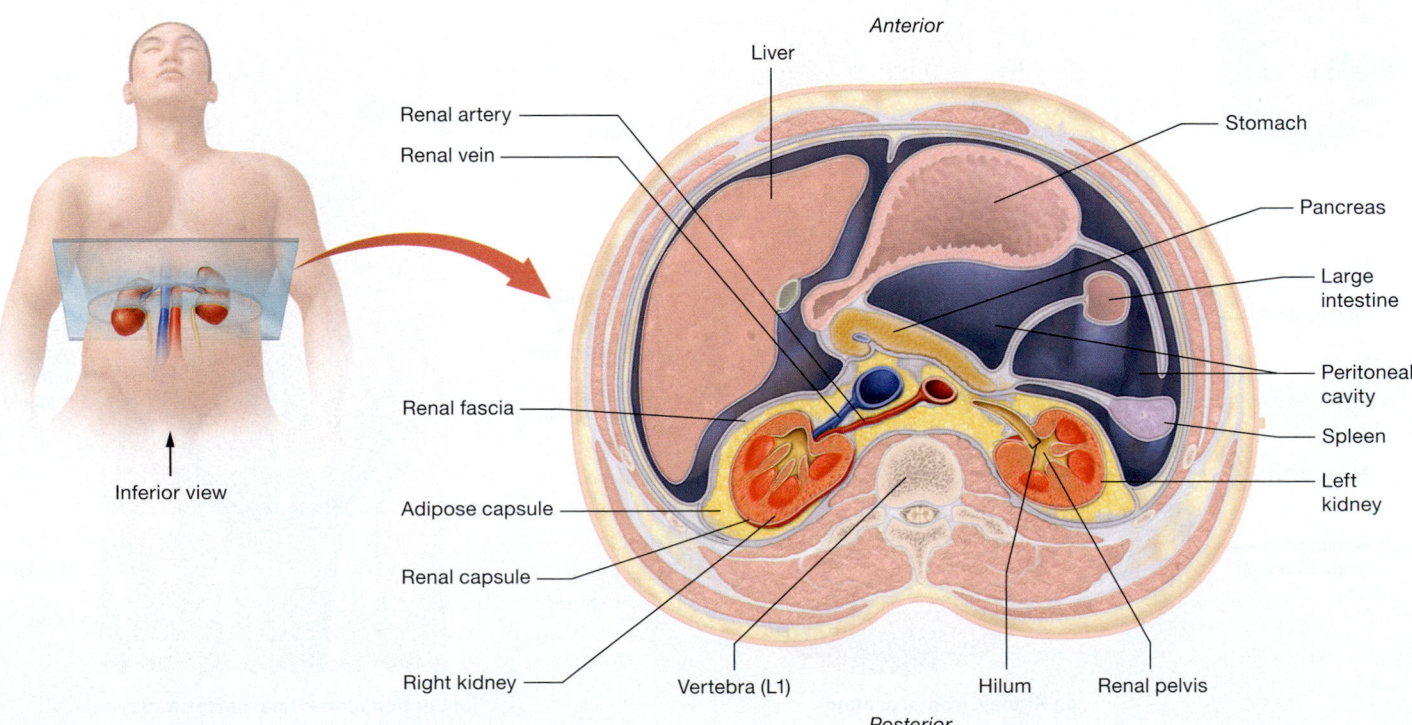

Figure 24.2 Position and external structure of the kidneys. Note that you are looking at an inferior view of this transverse section of the trunk.

the *renal artery, renal vein,* renal nerves, and ureter enter and exit the kidney. The hilum opens to a central cavity called the **renal sinus,** which is lined by the renal capsule and filled with urine-draining structures and adipose tissue. This connective tissue anchors the ureter, blood vessels, and nerves in place.

Quick Check

☐ 1. What are the three connective tissue coverings of the kidney?

Internal Anatomy of the Kidneys

A frontal section of the kidney reveals the three distinct regions of this organ: the outermost **renal cortex,** the middle **renal medulla,** and the inner **renal pelvis** (**Figure 24.3a**). Together, the renal cortex and the renal medulla make up the urine-forming portion of the kidney. The renal pelvis and associated structures drain urine that the cortex and medulla have formed.

Notice in Figure 24.3a that the renal cortex is reddish-brown. This is due to its rich blood supply—it houses 90–95% of the kidney's blood vessels. At specific points, extensions of the renal cortex called **renal columns** pass through the renal medulla toward the renal pelvis. The renal columns house blood vessels that branch from the renal artery as they travel to the outer portion of the cortex.

Within the renal medulla we also find cone-shaped **renal pyramids** (or *medullary pyramids*), which are separated from one another by a renal column on each side. Notice that the renal pyramids are darker in color and appear striped, reflecting that they are made up of parallel bundles of small tubes, with fewer blood vessels than in the renal cortex.

The renal cortex and renal medulla of each kidney contain over one million microscopic filtering structures called **nephrons.** Nephrons are the functional units of the kidney—each one is capable of filtering the blood and producing urine. **Figure 24.3b** shows the basic structure of a nephron, which consists of two main components: the globe-shaped *renal corpuscle,* and a long, snaking tube of epithelium called the *renal tubule.* Notice how these components are arranged in the kidney. The renal corpuscle and the majority of the renal tubule reside in the renal cortex, whereas varying amounts of the renal tubule dip into the renal medulla.

The tip of each renal pyramid tapers into a slender **papilla** (pah-PIL-uh), which borders on the first urine-draining structure, a cup-shaped tube called a **minor calyx** (KAY-liks; plural *calyces,* KAL-ih-seez). Urine from three to four minor calyces drains into a larger **major calyx** (shown in Figure 24.3a). Two to three major calyces, in turn, drain urine into the large collecting chamber that is the renal pelvis, which leads into the ureter. Smooth muscle tissue in the walls of the calyces and renal pelvis contracts to help propel urine toward the ureter. Both the calyces and the renal pelvis reside in the renal sinus.

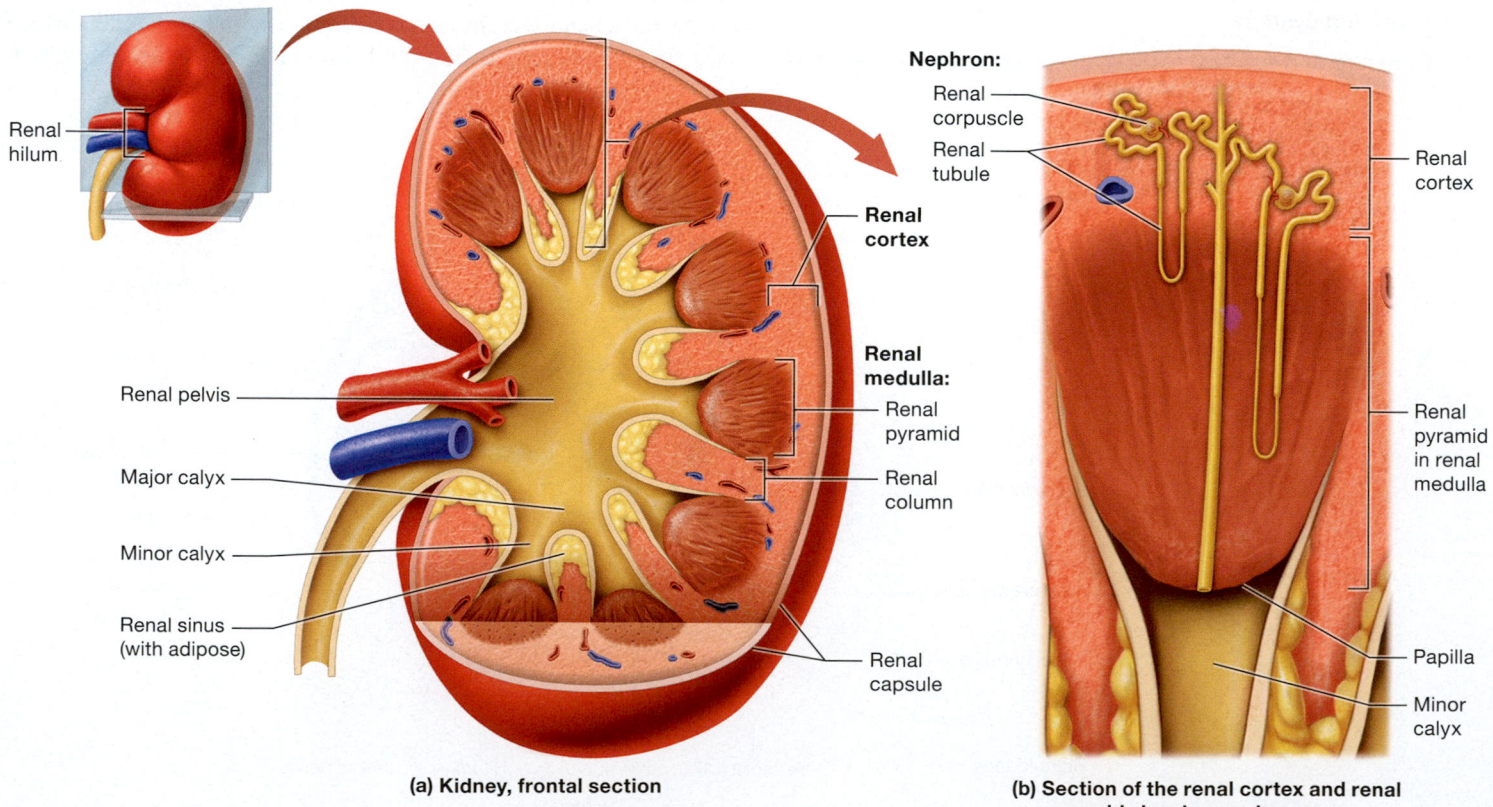

(a) Kidney, frontal section

(b) Section of the renal cortex and renal pyramid showing nephrons

Figure 24.3 Internal anatomy of the kidney, including the nephron.

Quick Check

☐ 2. What are the three regions of the kidney, and how do they differ structurally?

☐ 3. What is the functional unit of the kidney?

Blood Supply of the Kidneys

◀◀ FLASHBACK

1. How is blood delivered to and drained from a typical capillary bed? (p. 671)

2. What are the major arteries and veins that supply and drain the kidneys? (p. 716)

The kidneys receive approximately one-fourth of the total cardiac output—about 1200 ml per minute—from the right and left **renal arteries,** which branch from the abdominal aorta. The renal arteries fan out into ever-smaller vessels as they pass through the renal sinus to the renal columns and cortex. **Figure 24.4** traces the path of blood flow through the kidney. From largest to smallest, the arteries shown in Figure 24.4 are as follows: **①** Renal artery → **②** Segmental artery → **③** Interlobar artery → **④** Arcuate artery → **⑤** Interlobular (cortical radiate) artery.

You learned in the blood vessels chapter that systemic capillary beds are fed by arterioles and drained by venules (see Chapter 18). But in the kidneys, we find an unusual capillary bed that is both fed and drained by arterioles (see Figure 24.4b). In the kidney's renal cortex, the interlobular arteries branch into tiny **afferent arterioles** (*affer-* = "to carry toward"), which feed a ball-shaped capillary bed called the **glomerulus** (gloh-MEHR-yoo-lus; *glom-* = "ball"). The glomerulus, which consists of glomerular capillaries and supporting cells, is part of the renal corpuscle of the nephron. The glomerular capillaries are then drained by a second arteriole called the **efferent arteriole** (*effer-* = "to carry away"). The efferent arteriole feeds a second capillary bed known as the **peritubular capillaries,** which together form a plexus, or network, around the renal tubule of each nephron. The order of this part of the system is as follows: **⑥** Afferent arteriole → **⑦** Glomerulus → **⑧** Efferent arteriole → **⑨** Peritubular capillaries.

The venous route of blood out of the kidney parallels the arterial path. Groups of peritubular venules unite into a series of progressively larger venous vessels. The sequence, from smallest to largest (see Figure 24.4), is as follows: **⑩** Interlobular vein → **⑪** Arcuate vein → **⑫** Interlobar vein → **⑬** Renal vein.

Note that no segmental veins are present in the kidneys; the interlobar veins merge in the renal sinus to form the large **renal vein.** The renal vein then exits the kidney through the hilum and empties into the inferior vena cava.

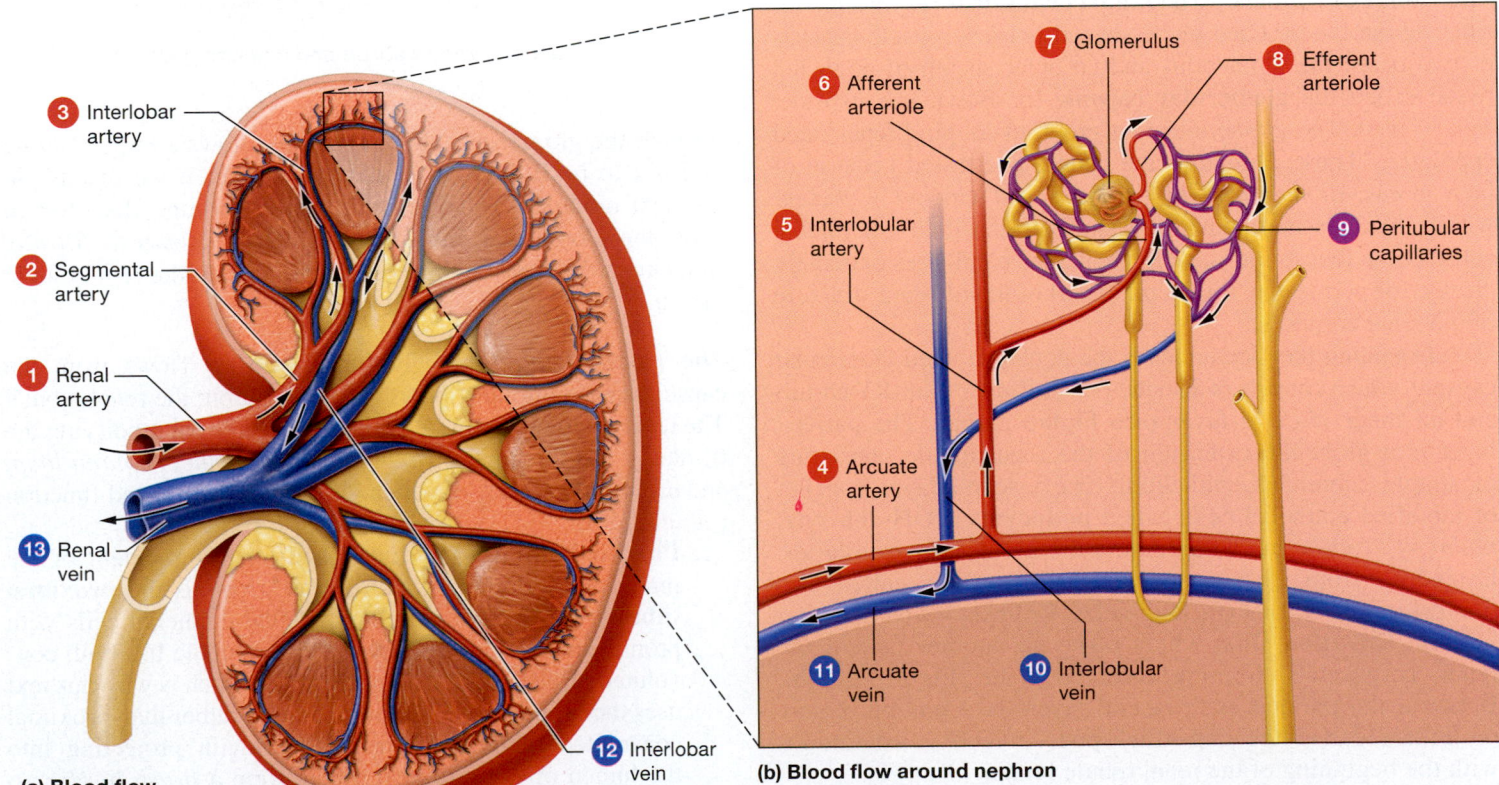

③ Interlobar artery

② Segmental artery

① Renal artery

⑬ Renal vein

⑫ Interlobar vein

(a) Blood flow through kidney

⑦ Glomerulus

⑥ Afferent arteriole

⑧ Efferent arteriole

⑤ Interlobular artery

⑨ Peritubular capillaries

④ Arcuate artery

⑪ Arcuate vein

⑩ Interlobular vein

(b) Blood flow around nephron

Figure 24.4 **Blood flow through the kidney.**

Quick Check

☐ 4. Trace the sequence of blood flow through the kidneys from the renal artery to the renal vein.

☐ 5. How does the arrangement of capillary beds in the kidneys differ from those elsewhere in the body?

Microanatomy of the Kidney: The Nephron and Collecting System

《 FLASHBACK

1. What are fenestrated capillaries? (p. 689)

As you learned in this module, most functions carried out by the kidneys occur in their tiny nephrons (**Figure 24.5**). Nephrons filter the blood and modify the filtered fluid as it passes through the renal tubules. This fluid then leaves the nephron and drains into the tubules of the *collecting system* (which are not considered part of the nephron), where it is further modified until it finally becomes urine. Let's look more closely at each of these components.

The Nephron

As we introduced earlier, the nephron has two main divisions: the renal corpuscle and the renal tubule. Both structures are composed of multiple parts.

The Renal Corpuscle   The **renal corpuscle** is responsible for filtering the blood. Each globe-shaped renal corpuscle consists of two parts: the glomerulus and an outer sheath of epithelial tissue called the **glomerular capsule** (or Bowman's capsule; **Figure 24.6**). The glomerulus is a group of looping fenestrated capillaries. These capillaries are called fenestrated because of large pores, or *fenestrations* (fenestre- = "window"), present within their plasma membranes and between their endothelial cells. These fenestrations, which make the capillaries extremely "leaky," or permeable, form a main part of the filtering structure of the renal corpuscle.

Surrounding the glomerulus is the double-layered glomerular capsule, which consists of an outer *parietal layer* (puh-RY-eh-tul) and an inner *visceral layer* (see Figure 24.6a). The parietal layer is a globelike extension of the renal tubule consisting of simple squamous epithelium. The visceral layer consists of modified epithelial cells called **podocytes** (POH-doh-sytz; *podo-* = "foot") that wrap around the glomerular capillaries. Extending from each podocyte are extensions called *foot processes,* or *pedicels* (PED-ih-selz). Pedicels weave together to form **filtration slits,** which make up another part of the renal corpuscle's filtering structure (see Figure 24.6b). Between the parietal and visceral layers is a hollow region, or lumen, called the **capsular space,** which is continuous with the beginning of the renal tubule lumen.

The podocytes and fenestrated glomerular capillaries form part of a complex membrane that filters blood flowing

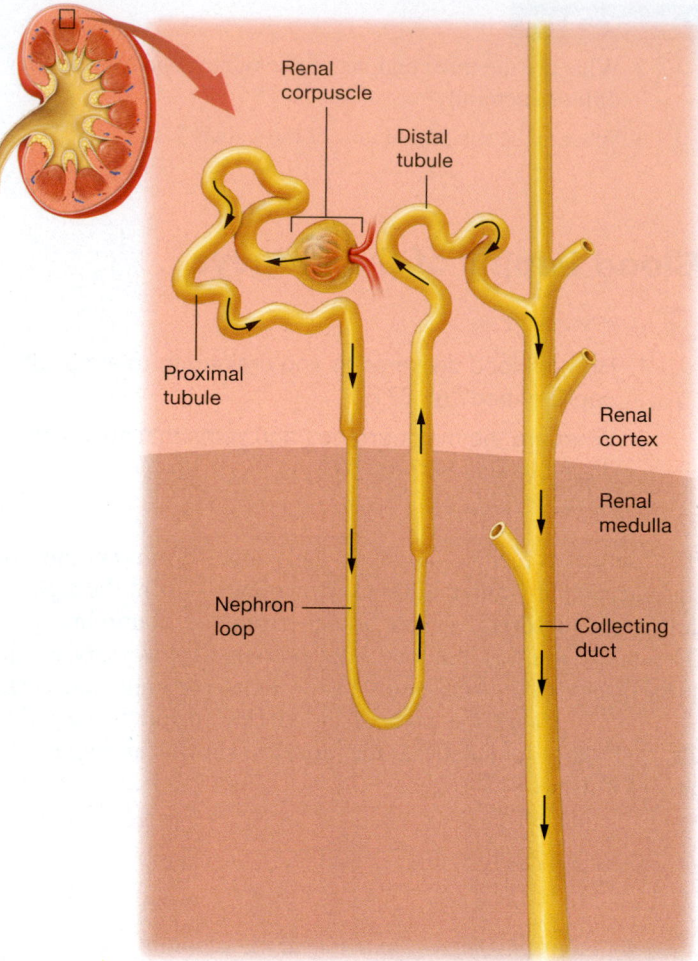

Figure 24.5   A generalized nephron and collecting system.

through the glomerulus. This structure allows a large volume of fluid to be filtered from the blood (which we discuss in the next module). The fluid that passes through the filter to leave the glomerular capillaries, which is known as *filtrate,* first enters the capsular space, then flows into the renal tubule lumen.

The Renal Tubule   Newly formed filtrate flows from the capsular space into the "pipes" of the nephron: the renal tubule. The renal tubule is a winding tube responsible for modifying the filtrate. It has three regions: the *proximal tubule, nephron loop,* and *distal tubule,* each of which differs in structure and function (**Figure 24.7**).

1. **Proximal tubule.** The initial and longest segment of the renal tubule through which filtrate flows is the **proximal tubule;** it consists of simple cuboidal epithelial cells with prominent microvilli. This part of the tubule has both convoluted (coiled) and straight sections, which is why this text uses the broader term proximal tubule (rather than proximal convoluted tubule). The many microvilli projecting into the lumen of the proximal tubule form a *brush border,* so named because the fine projections resemble the bristles on a brush. This border greatly increases surface area.

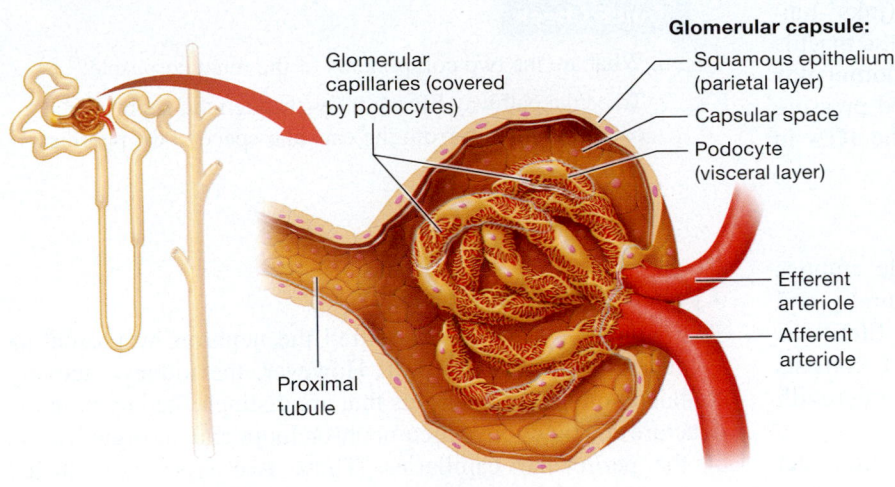

Glomerular capsule:
Glomerular capillaries (covered by podocytes)
Squamous epithelium (parietal layer)
Capsular space
Podocyte (visceral layer)
Efferent arteriole
Afferent arteriole
Proximal tubule

(a) The renal corpuscle

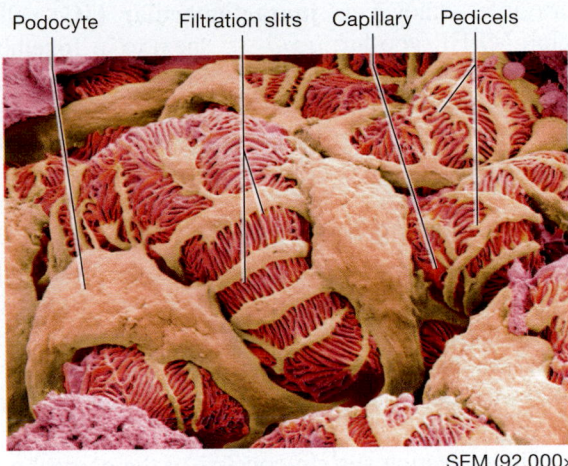

Podocyte Filtration slits Capillary Pedicels

SEM (92,000×)

(b) SEM of capillary surrounded by podocytes

Figure 24.6 **The renal corpuscle.**

2. **Nephron loop.** The remaining filtrate flows on to the **nephron loop,** also known as the *loop of Henle,* which is the only part of the renal tubule that dips into the renal medulla. The nephron loop has two limbs: The **descending limb** travels toward the renal medulla, turns 180°, and becomes the **ascending limb,** which climbs back toward the renal cortex. Most of the descending limb is composed of simple squamous epithelium, and so is often called the **thin descending limb.** In some nephrons this thin segment also forms the bend region and part of the ascending limb, so there is also a **thin ascending limb.** However, the majority of the ascending limb is composed of thicker simple cuboidal epithelium, and is referred to as the **thick ascending limb.**

3. **Distal tubule.** The final segment through which the filtrate passes in the renal tubule is the **distal tubule.** (Again, this text uses the broader term, as this segment of the tubule has both convoluted and straight sections.) Like the proximal tubule, the distal tubule is composed of simple cuboidal epithelium. However, the distal tubule has sparse microvilli and so lacks a brush border.

The Juxtaglomerular Apparatus

At the transition point between the ascending limb of the nephron loop and the distal tubule, we find a tightly packed group of cells called the **macula densa** (MAK-yoo-luh DEN-suh; "dense spot"). The macula densa comes into contact with modified smooth muscle cells in the afferent and efferent

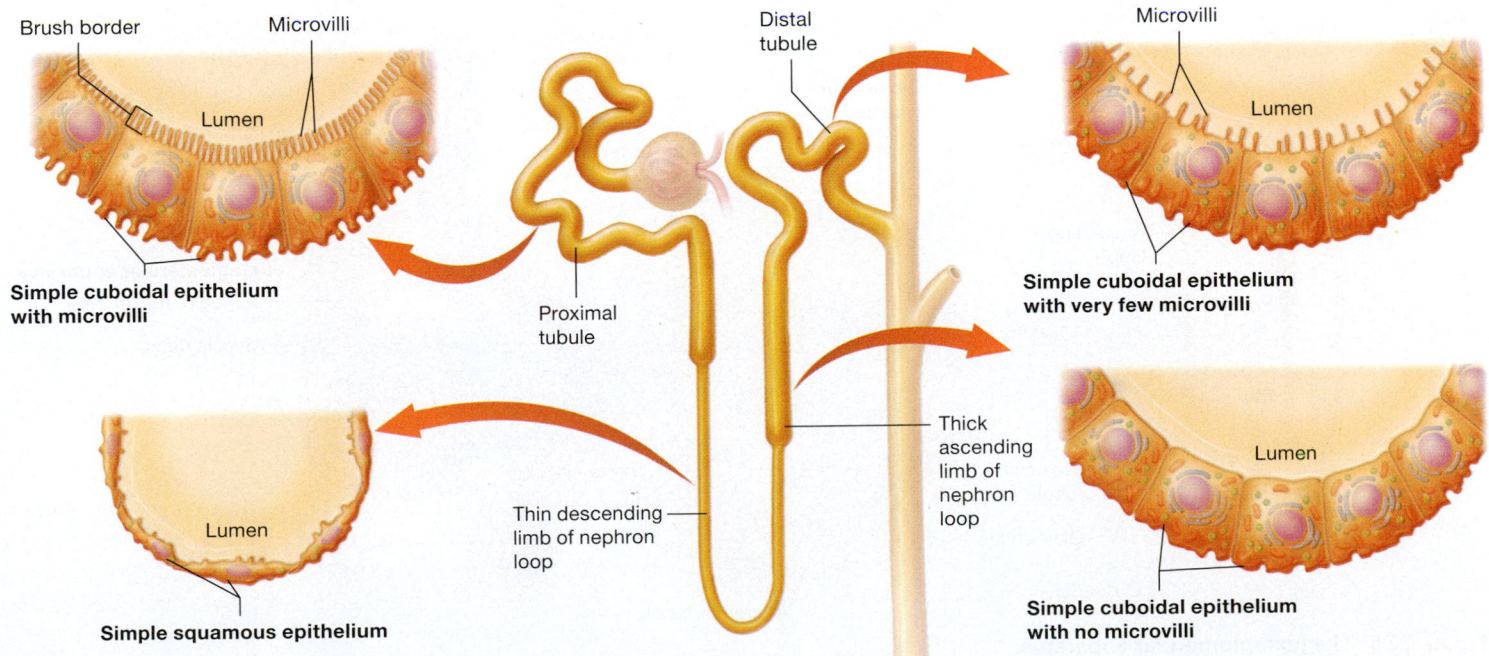

Brush border Microvilli
Lumen
Simple cuboidal epithelium with microvilli

Lumen
Simple squamous epithelium

Distal tubule
Proximal tubule
Thin descending limb of nephron loop
Thick ascending limb of nephron loop

Microvilli
Lumen
Simple cuboidal epithelium with very few microvilli

Lumen
Simple cuboidal epithelium with no microvilli

Figure 24.7 **Structural characteristics of the renal tubule.**

arterioles, known as **juxtaglomerular (JG) cells** (juks'-tuh-gloh-MEHR-yoo-lur; *juxta-* = "next to"). Together, the macula densa and JG cells form a structure called the **juxtaglomerular apparatus** (JGA; **Figure 24.8**), which regulates blood pressure and the glomerular filtration rate. We will revisit the JGA in Module 24.4.

The Collecting System

The nephron ends where filtrate in the distal tubule empties into the **collecting system,** another series of structurally and functionally distinct tubules that further modify the filtrate as it passes through them. Most of the collecting system consists of simple cuboidal or columnar epithelium with few microvilli (**Figure 24.9**).

The collecting system consists of the cortical collecting duct and the medullary collecting system. The distal tubule empties filtrate into the first part of the collecting system, the *cortical collecting duct,* which is found within the renal cortex. The cortical collecting duct is made up of simple cuboidal epithelial cells. Note that each cortical collecting duct drains several distal tubules. As the cortical collecting duct passes into the renal medulla, it becomes the *medullary collecting duct.* Deep within the renal medulla, several medullary collecting ducts empty filtrate into a larger **papillary duct.** The papillary duct contains low columnar epithelial cells. The *medullary collecting system* includes the medullary collecting ducts and the papillary ducts. When the filtrate reaches the end of the papillary duct, it is urine. The urine exits at the papilla of the renal pyramid into a minor calyx. The formation of crystals within the tubules of the collecting system can block the flow of filtrate and lead to intense pain. You can find out more about this condition in *A&P in the Real World: Nephrolithiasis.*

Quick Check

☐ 6. What are the two components of the renal corpuscle?

☐ 7. Trace the pathway filtrate takes through the nephron and collecting system from the capsular space to the papillary duct.

Types of Nephrons

The previous discussion simplified the nephron by presenting a "generalized" version of it. However, the kidneys actually contain two types of nephrons that are distinguished by both the structure and function of their nephron loops and the organization of the peritubular capillaries. These two types are labeled *cortical* and *juxtamedullary* (juk'-stah-MED-yoo-lehr-ee).

About 80% of nephrons are **cortical nephrons,** so named because they are located primarily in the renal cortex (**Figure 24.10a** on p. 956). The renal corpuscles of cortical nephrons are situated in the *outer* renal cortex, and they have very short nephron loops that either just dip into the superficial part of the renal medulla or never leave the cortex. Peritubular capillaries supply blood to the nephron loops of cortical nephrons.

The other, less numerous, type of nephron in the kidney is the **juxtamedullary nephron.** Notice in **Figure 24.10b** that the renal corpuscle of the juxtamedullary nephron sits close to the boundary between the renal cortex and the renal medulla. In addition, it has a long nephron loop that burrows deeply into the renal medulla. Here the loop is surrounded by a ladder-like network of capillaries called the **vasa recta** (VAY-zuh REK-tuh; "straight vessels") arising from the efferent arteriole. Like the peritubular capillaries, the vasa recta empty into interlobular

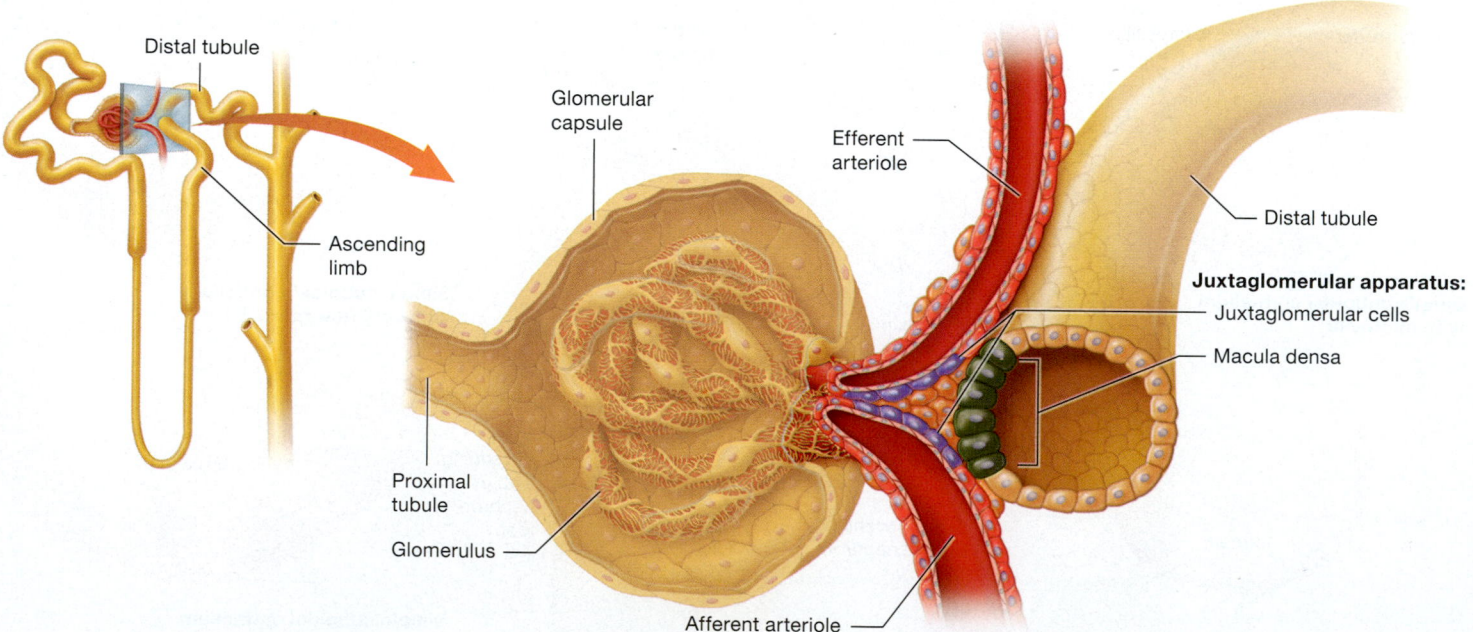

Figure 24.8 The juxtaglomerular apparatus.

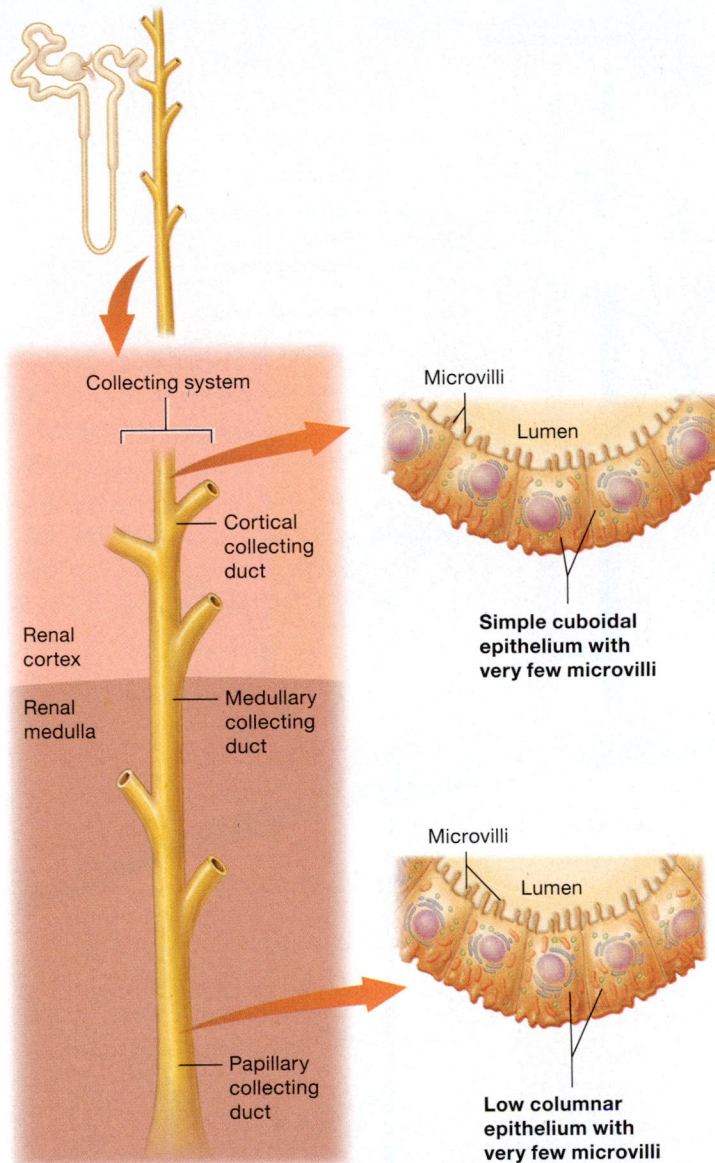

Figure 24.9 **Structural characteristics of the collecting system.**

veins. Notice that the portion of the nephron that lies in the cortex is surrounded by branches of regular peritubular capillaries. This unique structure allows the juxtamedullary nephrons to function as part of a system that controls the volume and concentration of urine, concepts we discuss in Module 24.6.

Apply What You Learned

☐ 1. Predict the effects of a condition that results in gradual loss of microvilli from the proximal tubule.

☐ 2. Predict the potential effects of abnormally narrow renal arteries, a condition called *renal artery stenosis,* on the ability of the kidneys to carry out their functions.

See answers in Appendix A.

Nephrolithiasis

Nephrolithiasis (nef´-roh-lith-AYE-uh-sis; *lith-* = "stone") is characterized by the formation of **renal calculi** (KAL-kyoo-lye), commonly known as *kidney stones.* Renal calculi are crystalline structures composed most commonly of calcium oxalate salts. They form when the concentrations of these ions, as well as solutes such as sodium ions, hydrogen ions, and uric acid, are present in the filtrate in higher than normal amounts. This condition is known as *supersaturation,* and supersaturated ions are more likely to come out of solution and crystallize. Risk factors for supersaturation include dehydration; a diet high in fat, animal protein, and salt; and obesity.

Typically, the crystals form in the nephron loop, distal tubule, and/or collecting system. Most crystals simply pass unnoticed into the urine. However, sometimes the crystals adhere to the epithelium of the tubules, particularly in the collecting system, and form *seed crystals* that lead to the formation of stones. The stones may remain in the collecting system or may break off and lodge in the calyces, renal pelvis, and ureter. Stones lodged within the urinary system cause the most common symptom of nephrolithiasis: severe pain, known as *renal colic,* that radiates from the lumbar region to the pubic region. Other symptoms include blood in the urine, sweating, nausea, and vomiting.

Nephrolithiasis can be diagnosed in several ways, including computed tomography scans and an **intravenous pyelogram** (PY-loh-gram; *pyelo-* = "pelvis"), or **IVP.** An IVP is a radiograph of the urinary system that uses a contrast medium such as iodine to reveal the structure of the renal pelvis, the major and minor calyces, and the ureters and urinary bladder. An example of an IVP with a renal calculus blocking the left ureter is shown here:

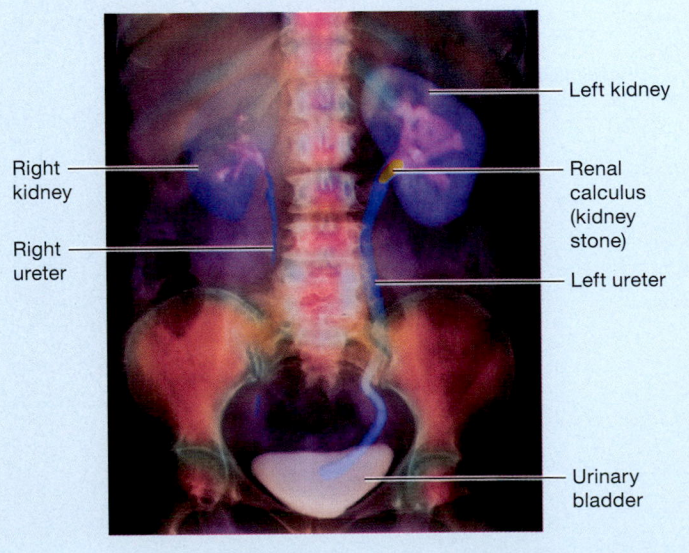

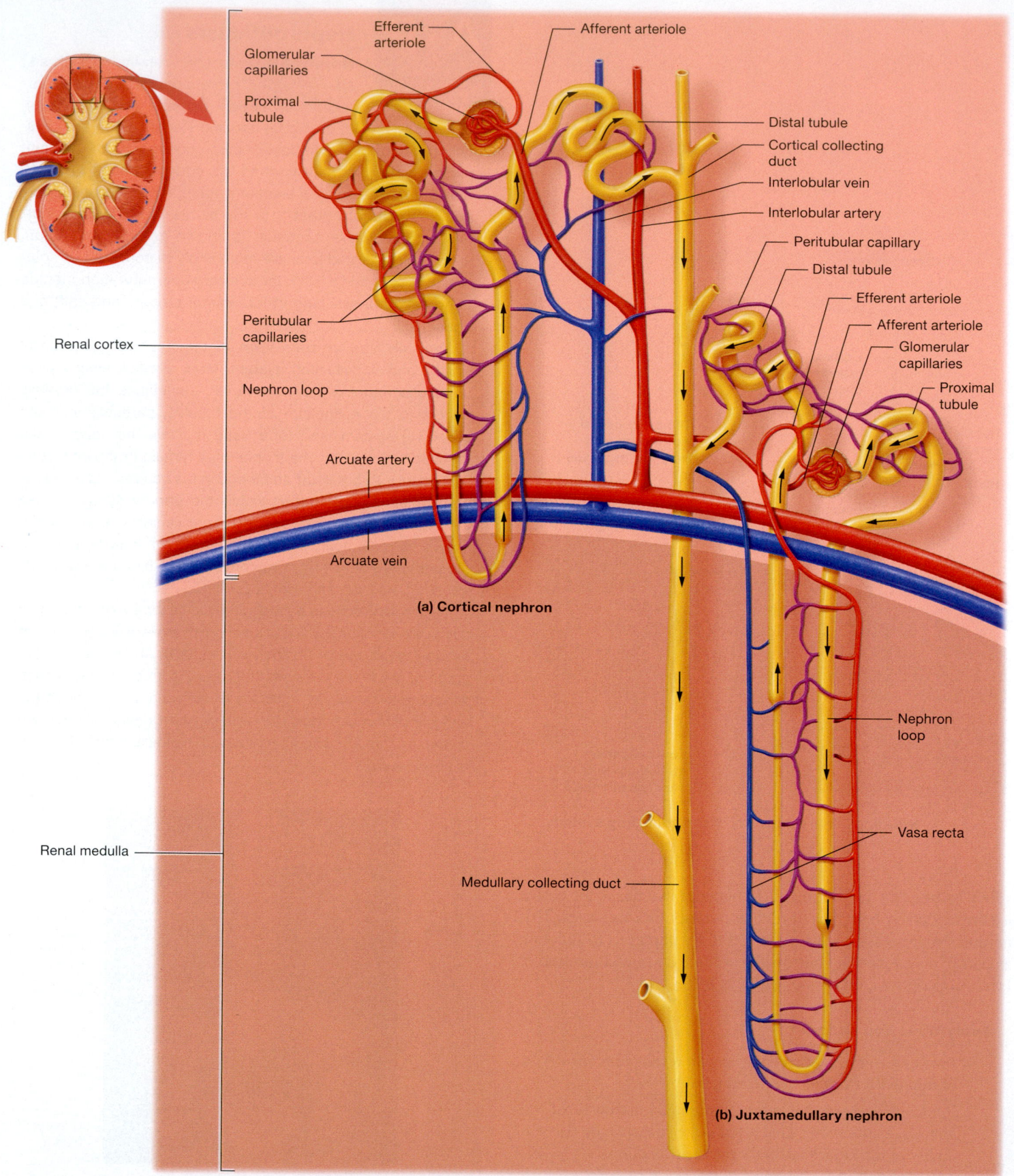

Efferent arteriole

Afferent arteriole

Glomerular capillaries

Proximal tubule

Distal tubule

Cortical collecting duct

Interlobular vein

Interlobular artery

Peritubular capillary

Distal tubule

Efferent arteriole

Afferent arteriole

Glomerular capillaries

Proximal tubule

Renal cortex

Peritubular capillaries

Nephron loop

Arcuate artery

Arcuate vein

(a) Cortical nephron

Renal medulla

Nephron loop

Vasa recta

Medullary collecting duct

(b) Juxtamedullary nephron

Figure 24.10 **Cortical and juxtamedullary nephrons.**

Overview of Renal Physiology

Learning Outcomes

1. Describe the three major processes in urine formation.

Nephrons and the collecting system carry out three basic physiological processes that allow the kidneys to perform their functions: *filtration, reabsorption,* and *secretion.* These functions, illustrated in **Figure 24.11**, are as follows:

- **Glomerular filtration.** The first process performed by the nephron is to filter the blood, a process known as **glomerular filtration.** This takes place as blood passes through the glomerular capillaries and some of the plasma is filtered into the surrounding capsular space. The endothelial cells of the glomerulus form part of a filter that is selective based on size. This enables the filter to hold back cells and most proteins, which remain in the blood, but permit some of the smaller substances—including water, electrolytes (such as sodium and potassium ions), acids and bases (such as hydrogen and bicarbonate ions), organic compounds, and metabolic wastes—to exit the blood and enter the glomerular capsule. The fluid and solutes that enter the capsular space of the nephron form the **filtrate,** also known as *tubular fluid.*

- **Tubular reabsorption.** The next process the nephron and collecting system perform is to modify the filtrate as it flows through the tubules. Much of this modification involves reclaiming substances from the filtrate, such as water, glucose, amino acids, and electrolytes, and returning them to the blood. This process is known as **tubular reabsorption.** The majority of the water and solutes filtered at the glomerulus are reabsorbed. Most reabsorption takes place in the proximal tubule and nephron loop. However, some more precisely controlled reabsorption also occurs in the distal tubule and collecting system; this process allows the kidneys to vary the amounts of different substances reabsorbed to maintain homeostasis as the needs of the body change.

- **Tubular secretion.** The filtrate is also modified by **tubular secretion,** which is essentially tubular reabsorption in the reverse direction. Notice in Figure 24.11 that during tubular secretion substances are moved from the peritubular capillary blood into the filtrate to be eventually excreted. Secretion happens in most parts of the nephron and collecting system, although different substances may be secreted more in one area than another. Tubular secretion helps maintain electrolyte and acid-base homeostasis and removes toxins from the blood that did not enter the filtrate via filtration.

In the next two modules, Modules 24.4 and 24.5, we examine each of these processes and how they contribute to the overall function of the kidneys and urinary system. In Module 24.6 we see how these processes allow the kidneys to control the volume and concentration of urine produced.

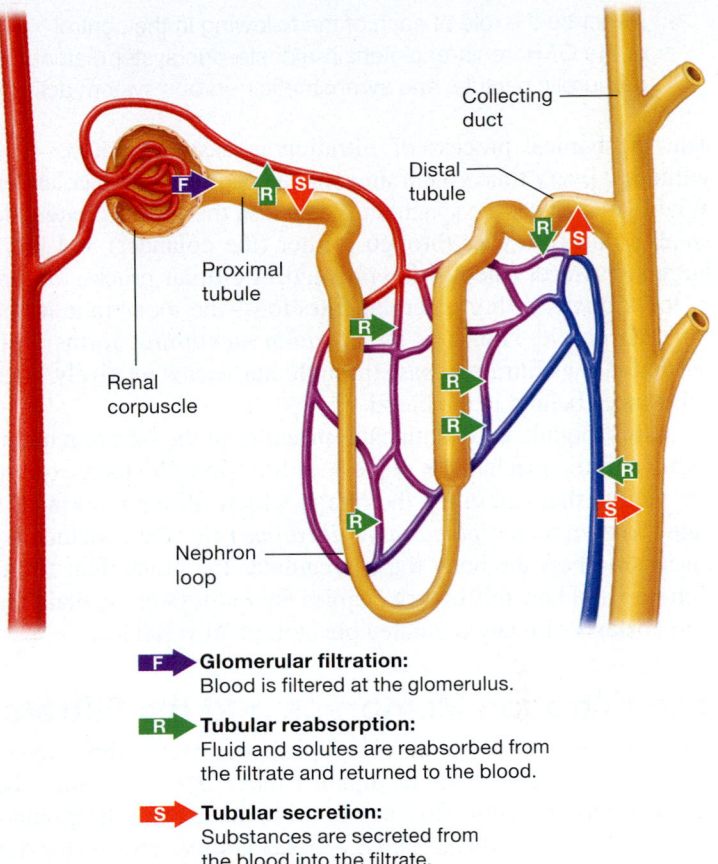

F ▸ **Glomerular filtration:**
Blood is filtered at the glomerulus.

R ▸ **Tubular reabsorption:**
Fluid and solutes are reabsorbed from the filtrate and returned to the blood.

S ▸ **Tubular secretion:**
Substances are secreted from the blood into the filtrate.

Figure 24.11 Three physiological processes carried out by the kidneys.

 Play Interactive Physiology 2.0
@ **Mastering** Anatomy & Physiology

Quick Check

☐ 1. What happens during glomerular filtration?

☐ 2. How do tubular reabsorption and tubular secretion differ?

Renal Physiology I: Glomerular Filtration

Learning Outcomes

1. Describe the structure of the filtration membrane.

2. Define the glomerular filtration rate (GFR) and its average value.

3. Explain how the hydrostatic and colloid osmotic pressures combine to yield the net filtration pressure in the glomerulus.

4. Predict specific factors that will increase or decrease the GFR.

5. Explain how the myogenic and tubuloglomerular feedback mechanisms affect the GFR.

6. Describe the role of each of the following in the control of the GFR: renin-angiotensin-aldosterone system, atrial natriuretic peptide, and sympathetic nervous system activity.

The mechanical process of filtration is easy to picture—you witness it every time you drain cooked spaghetti into a colander. When you "filter" the spaghetti, you cause the filtrate (water and small solutes) to pass through a filter (the colander) and leave large substances (the noodles) behind. A similar process occurs in the kidneys with glomerular filtration—the membrane in the renal corpuscle, known as the *filtration membrane,* forms a filter that allows filtrate to pass through, but leaves relatively large substances behind in the blood.

In this module we examine the structure of the filtration membrane and the mechanical process of filtration. We then explore the factors that determine the rate at which filtrate is formed, a value known as the *glomerular filtration rate.* We conclude by discussing how the body tightly regulates the glomerular filtration rate and how this in turn enables the kidneys to maintain the homeostasis of many regulated physiological variables.

The Filtration Membrane and the Filtrate

The inner part of the glomerular capsule consists of three layers that act as barriers, or filters; together these layers are called the **filtration membrane.** This membrane is made up of the glomerular capillary endothelial cells, a basal lamina, and podocytes (the visceral layer of the glomerular capsule) (**Figure 24.12**). Let's look at each of these layers, from deep to superficial (the direction of filtrate flow), more closely.

1. **Fenestrated glomerular capillary endothelial cells.** Like all capillaries, the glomeruli are composed of endothelial cells. Recall, however, that the glomerular endothelial cells are **fenestrated:** They have pores that make them leakier than most capillaries. The gaps between the glomerular endothelial cells are relatively large (about 70–100 nm) but are still small enough to prevent blood cells and platelets from exiting the capillaries.

2. **Basal lamina.** This thin layer of extracellular matrix gel separates the glomerular endothelial cells from the podocytes. Collagen fibers within the basal lamina form a meshwork that acts like a colander, preventing substances with a diameter greater than 8 nm from entering the capsular space. This effectively blocks the passage of most plasma proteins. In addition, the collagen fibers have negative charges that repel negatively charged plasma proteins, even those smaller than 8 nm in diameter.

3. **Podocytes (visceral layer of the glomerular capsule).** The podocytes composing the visceral layer of the glomerular capsule make up the third and finest filter in the filtration membrane. Note in Figure 24.12 that their finger-like pedicels wrap around the glomerular capillaries and interlace to form filtration slits. These narrow slits allow only substances with a diameter less than 6–7 nm to enter the capsular space. Albumin, the most prevalent plasma protein, has a diameter of 7.1 nm, so it is normally prevented from entering the filtrate.

The fluid and solutes that pass through the filtration membrane and enter the capsular space make up the filtrate. The size of the

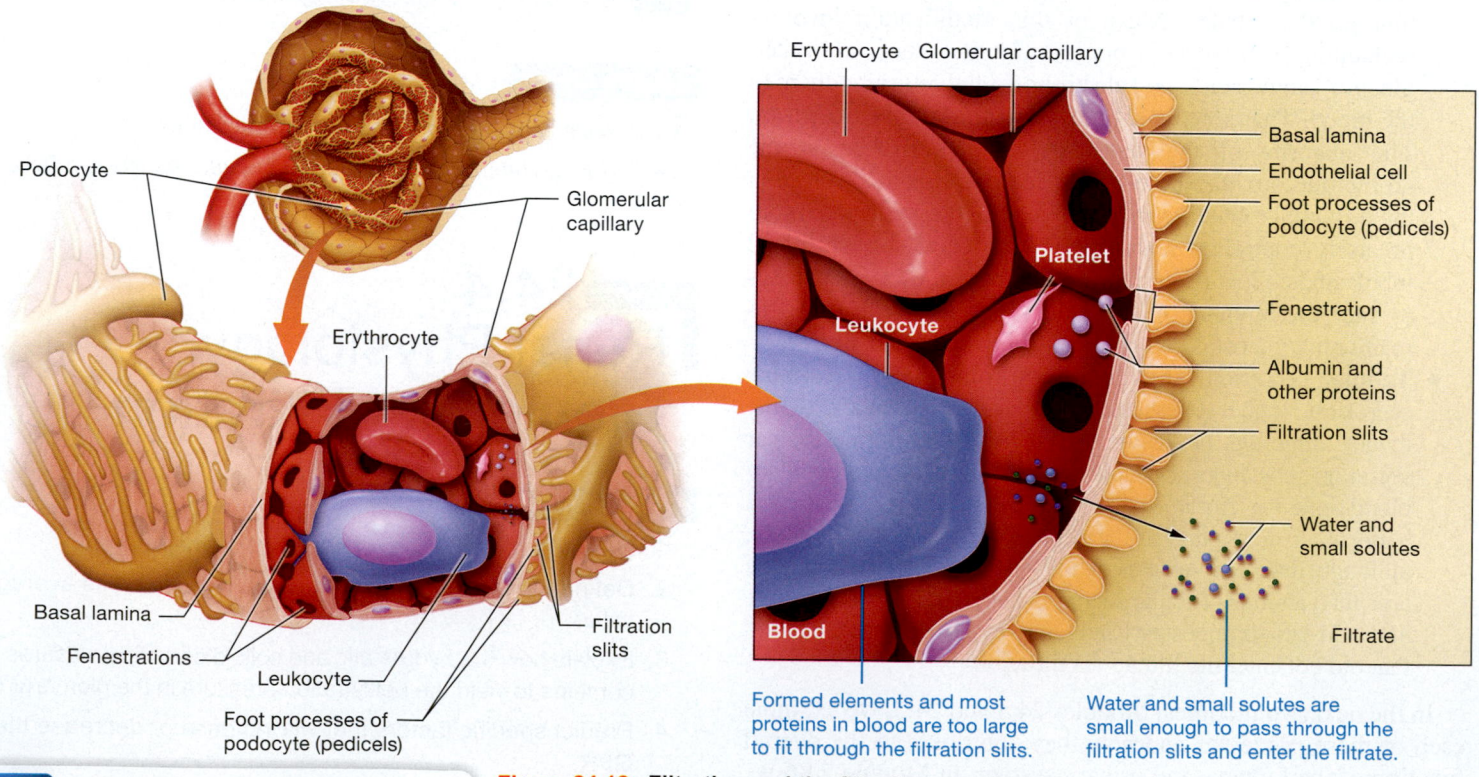

Podocyte

Glomerular capillary

Erythrocyte

Basal lamina

Fenestrations

Leukocyte

Foot processes of podocyte (pedicels)

Filtration slits

Erythrocyte   Glomerular capillary

Basal lamina

Endothelial cell

Foot processes of podocyte (pedicels)

Platelet

Fenestration

Leukocyte

Albumin and other proteins

Filtration slits

Blood

Water and small solutes

Filtrate

Formed elements and most proteins in blood are too large to fit through the filtration slits.

Water and small solutes are small enough to pass through the filtration slits and enter the filtrate.

Figure 24.12 Filtration and the filtration membrane.

Play Animation
@ Mastering Anatomy & Physiology

pores in this membrane determines the filtrate composition. Notice in Figure 24.12 that the filtration membrane allows water and small dissolved solutes, including glucose, electrolytes, very small proteins, and amino acids, to leave the blood and enter the capsular space, but prevents the exit of formed elements, such as cells and platelets, and most proteins. Another important substance filtered from the blood is a group of compounds known as *nitrogenous wastes.* These compounds include *urea* and *ammonium ions* (NH_4^+), which are waste products of protein metabolism; *creatinine,* a waste product of the creatine kinase reaction that occurs in muscle cells and other body cells; and *uric acid,* a waste product of nucleic acid metabolism. All are relatively small substances that pass easily through the pores of the filtration membrane.

The percentage of plasma that passes through the filtration membrane to enter the capsular space and become filtrate is called the *filtration fraction.* An average filtration fraction is 20%, which means that about one-fifth of the plasma that flows through the glomerulus exits the blood and enters the capsular space. The filtration fraction is so large because of the looping arrangement of the glomerular capillaries. This looping increases their surface area dramatically—if you were to stitch all the surfaces of the filtration membranes of one kidney together, their combined  area would be about 6 m² (about the size of a small bedroom). In an example of the Structure-Function Core Principle (p. 28), this large surface area makes filtration through this stack of ever-finer filters a very efficient process.

CORE PRINCIPLE
Structure-Function

Quick Check

☐ 1. What are the three components of the filtration membrane?

☐ 2. What is the function of the filtration membrane?

☐ 3. Which substances are found in the filtrate?

The Glomerular Filtration Rate (GFR)

« FLASHBACK

1. Define osmosis. (p. 76)

2. What is colloid osmotic pressure? (p. 693)

3. What is vascular resistance? (p. 676)

Filtrate is formed at the remarkably rapid rate of about 125 ml/min. This value, the amount of filtrate formed by both kidneys in 1 minute, is known as the **glomerular filtration rate (GFR).** Over the course of a day, the kidneys produce about 180 liters of filtrate (to put this into perspective, imagine filling your gastrointestinal tract with a similar volume by drinking 90 two-liter bottles of soda in 1 day). This is an impressive feat, considering that we have only about 3 liters of plasma. This means that your entire plasma volume is filtered by your kidneys about 60 times per day. (Read about what happens to the GFR when there is inflammation of the filtration membrane in *A&P in the Real World: Glomerulonephritis.*)

The kidneys are able to filter blood so efficiently in part because the glomerular capillaries are remarkably permeable. However, even with fenestrated capillaries, filtration will happen only if a pressure gradient is present to push water and solutes through the filtration membrane (an example of the Gradients Core Principle, p. 28). In this section we discuss the forces that allow this process to occur.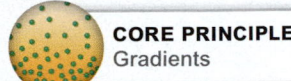

CORE PRINCIPLE
Gradients

Filtration Pressures

First let's review the two forces that drive fluid movement in a typical capillary bed:

- **Hydrostatic pressure.** *Hydrostatic pressure* is the force of a fluid on the wall of its container. In the case of blood capillaries, hydrostatic pressure is equal to the blood pressure, and it tends to push water *out of* the capillary and into the interstitial space.
- **Colloid osmotic pressure.** *Colloid osmotic pressure* (*COP*) is the pressure created by proteins (primarily albumin) in the plasma. The osmotic gradient created by these proteins pulls water *into* the capillaries by osmosis.

Recall that these two forces work together in a capillary bed to determine the *net filtration pressure* (*NFP*) of the bed (see Chapter 18). The net filtration pressure determines the direction of water movement between the capillaries and the interstitial fluid. Simply stated, water moves out of the capillary if hydrostatic pressure is higher than COP, or into the capillary if COP is higher than hydrostatic pressure.

Net Filtration Pressure at the Glomerulus

We can apply these same two principles to the glomerular capillaries. However, the situation is slightly more complex in the glomerulus, because we have a third force to factor in: the

A P in the Real World

Glomerulonephritis

Glomerulonephritis (gloh-mehr´-yoo-loh-nef-RY-tiss) is a common condition that involves damage to and destruction of the glomeruli, with resulting inflammation of the glomerular capillaries and the filtration membrane. Recall that the major hallmarks of inflammation are both increased blood flow and increased capillary permeability. This causes the filtration membrane to become excessively leaky, leading to the loss of blood cells and proteins to the urine. As nephrons are further damaged and destroyed by inflammation, the GFR decreases, which triggers compensatory mechanisms to maintain the GFR. This causes additional loss of cells and proteins to the urine. Eventually, these mechanisms are insufficient to restore the GFR, and toxic substances can accumulate in the blood. This condition may eventually lead to complete renal failure, in which the GFR is insufficient for the kidneys to perform their functions.

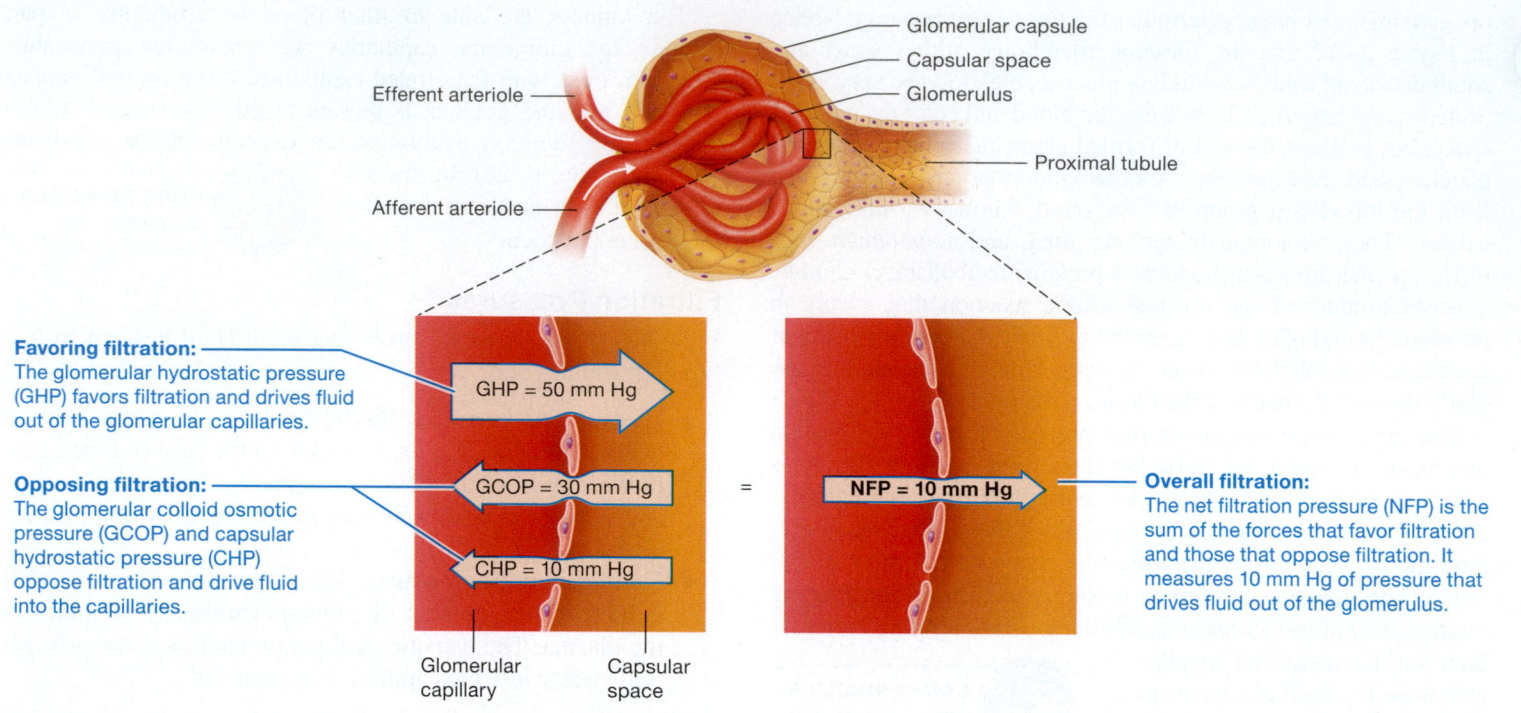

Favoring filtration:
The glomerular hydrostatic pressure (GHP) favors filtration and drives fluid out of the glomerular capillaries.

GHP = 50 mm Hg

Opposing filtration:
The glomerular colloid osmotic pressure (GCOP) and capsular hydrostatic pressure (CHP) oppose filtration and drive fluid into the capillaries.

GCOP = 30 mm Hg

CHP = 10 mm Hg

=

NFP = 10 mm Hg

Overall filtration:
The net filtration pressure (NFP) is the sum of the forces that favor filtration and those that oppose filtration. It measures 10 mm Hg of pressure that drives fluid out of the glomerulus.

Glomerular capillary

Capsular space

Efferent arteriole

Glomerular capsule
Capsular space
Glomerulus

Proximal tubule

Afferent arteriole

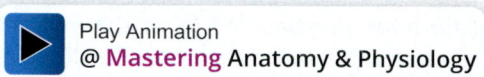

Play Animation
@ **Mastering** Anatomy & Physiology

Figure 24.13 Net filtration pressure in the glomerular capillaries.

hydrostatic pressure of the fluid in the capsular space. Let's examine each of these forces, shown in **Figure 24.13**:

- **Glomerular hydrostatic pressure.** The **glomerular hydrostatic pressure (GHP)**, which is largely determined by the systemic blood pressure, measures about 50 mm Hg. This pressure is considerably higher than that of a typical capillary bed (which ranges from 17 to 35 mm Hg). This is because blood leaving such a capillary bed enters a low-resistance venule, whereas blood leaving the glomerulus enters a high-resistance efferent arteriole. This arteriole's high resistance is due to its smooth muscle and its small diameter—notice in Figure 24.13 that the diameter of the efferent arteriole is smaller than that of the afferent arteriole. This causes fluid to back up and push against the walls of the glomerular capillaries, which *favors* its movement through the filtration membrane.
- **Glomerular colloid osmotic pressure.** Like the COP in typical capillaries, the **glomerular colloid osmotic pressure (GCOP)** is created by the presence of proteins such as albumin in the plasma. The GCOP averages about 30 mm Hg, slightly higher than the COP in a typical capillary bed, because the blood in the glomerulus is a bit more concentrated. The reason for this is that water leaves the glomerular blood rapidly through the filtration membrane, which causes any solutes left in the blood to increase in concentration. The higher the blood's solute concentration, the higher its osmotic pressure. The GCOP *opposes* filtration, drawing water into the capillaries by osmosis.
- **Capsular hydrostatic pressure.** Like an emptying kitchen sink with the faucet turned on, the fluid in the capsular

space can drain into the renal tubule only so quickly. The rapidly accumulating filtrate inside the capsular space of a nephron builds up a hydrostatic pressure of its own, called the **capsular hydrostatic pressure (CHP).** This pressure (about 10 mm Hg) tries to push water into the glomerular capillaries and so *opposes* filtration.

We can combine these three forces to yield the glomerular **net filtration pressure (NFP),** the total pressure gradient available to drive water across the filtration membrane and into the capsular space. To find the glomerular NFP, we subtract the two forces that oppose filtration (GCOP and CHP) from the one that favors filtration (GHP):

$$\begin{aligned} \text{NFP} &= \text{GHP} - (\text{GCOP} + \text{CHP}) \\ &= 50 \text{ mm Hg} - (30 \text{ mm Hg} + 10 \text{ mm Hg}) \\ &= 10 \text{ mm Hg} \end{aligned}$$

So, we find a net filtration pressure of about 10 mm Hg in the glomerular capillaries. This relatively high pressure, combined with the leakiness of the glomerular capillaries and their large surface area, yields an average GFR of about 125 ml/min. To gain perspective on this, compare the GFR to the rate at which systemic capillary beds lose water—about 1.5 ml/min, quite a significant difference.

Quick Check

☐ 4. What is the GFR?

☐ 5. Which three pressures combine to determine the net filtration pressure? Which pressure(s) promote filtration? Which pressure(s) oppose filtration?

Factors That Affect the Glomerular Filtration Rate

≪ FLASHBACK

1. What are the primary functions of the hormones angiotensin-II and atrial natriuretic peptide? (p. 610)

2. What happens to a blood vessel when blood pressure increases and the vessel is stretched? (p. 671)

Glomerular filtration is essentially the "gatekeeper" of renal physiology because it begins the process of waste removal. The GFR determines how rapidly the blood is cleansed of metabolic wastes, how effectively the kidneys can carry out both tubular reabsorption and tubular secretion, and how well the kidneys are able to maintain fluid, electrolyte, and acid-base homeostasis in the body. For these reasons, anything that affects the GFR influences all functions of the kidney.

Several factors regulate and affect the GFR. These factors may originate within the kidney itself or may be due to the function of systems outside the urinary system. The internal mechanisms are called *autoregulation,* and have to do with maintaining the GFR. The external factors, including neural and hormonal factors, are part of broader systems that work to maintain systemic blood pressure and affect the GFR in doing so. We discuss each of these factors in the upcoming sections. But first let's focus on the physical changes that are used by all these mechanisms to control the GFR.

ConceptBOOST))))

How Changes in Arteriolar Diameter Influence the GFR

As we discussed earlier, filtration will occur only when a net pressure gradient in the glomerulus drives fluid out of the blood and into the capsular space. The size of this gradient determines how much filtration takes place—a small gradient will lead to only minimal filtration, whereas a large gradient leads to heavy filtration. There are several ways to adjust the size of the pressure gradient in the glomerulus, but one of the easiest ways is to change the diameter of the afferent (entering) and efferent (leaving) arterioles. When either arteriole constricts or dilates, this changes the glomerular hydrostatic pressure (GHP), which also changes the entire net filtration pressure gradient.

You can think of blood flowing in and out of the glomerulus as being similar to water flowing in and out of a sink, where the afferent arteriole is the faucet, the basin is the glomerulus, and the efferent arteriole is the drainpipe. Keep this analogy in mind as we explore how this mechanism works:

- Vasoconstriction of the afferent arteriole "turns down the faucet." As you can see in the figure at the top of the next column, this allows less blood to flow into the glomerulus, which decreases the GHP and the GFR.

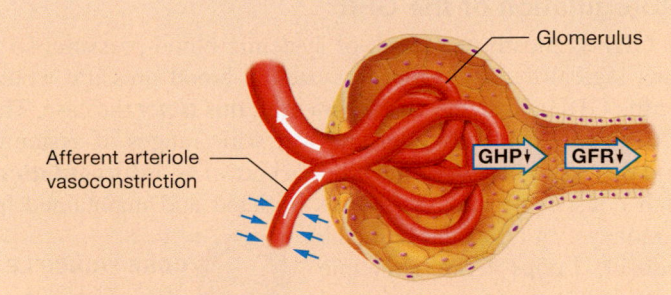

- Vasoconstriction of the efferent arteriole "clogs the drain." This causes blood to back up within the glomerulus, which increases the GHP and the GFR:

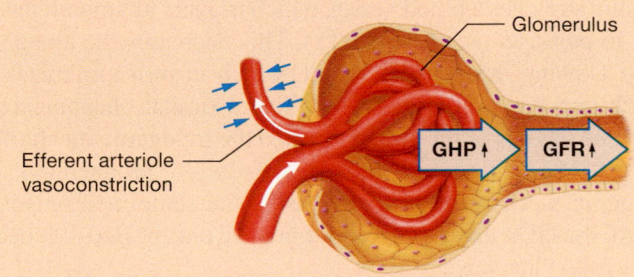

Vasodilation has the opposite effects:

- Vasodilation of the afferent arteriole "turns up the faucet." This increases the GHP and the GFR:

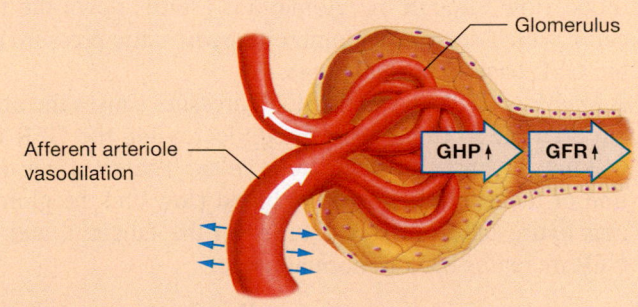

- Vasodilation of the efferent arteriole "unclogs the drain," allowing increased flow out of the glomerulus. This decreases the GHP and the GFR:

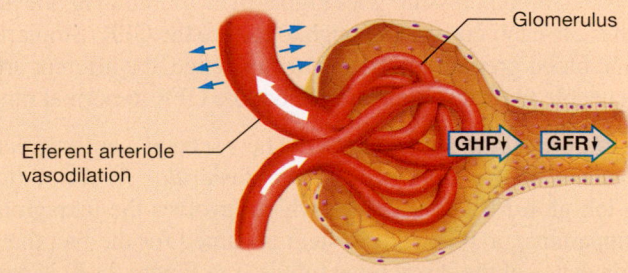

In the upcoming subsections, you'll see that the main mechanisms the body uses to control or maintain the GFR work by causing one or more of these physical changes. ■

Autoregulation of the GFR

The GHP is mostly the result of systemic blood pressure, so it seems logical that a change in systemic blood pressure would alter the GHP and so the GFR. However, this is *not* the case. The GFR remains relatively constant over wide ranges of systemic blood pressure because of the process known as **autoregulation,** which consists of local responses initiated and maintained by the kidneys. In an example of the Feedback Loops Core Principle (p. 28), autoregulation consists of two negative feedback processes: the *myogenic mechanism* and *tubuloglomerular feedback.*

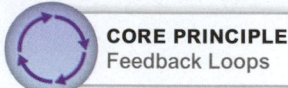

CORE PRINCIPLE
Feedback Loops

The Myogenic Mechanism Recall that an increase in blood pressure stretches a blood vessel, which triggers its smooth muscle cells to constrict (see Chapter 18). This decreases the amount of blood flowing through the vessel, which in turn minimizes the stretch imposed on the vessel. We find a similar phenomenon in the blood vessels of the kidney, termed the **myogenic mechanism** (my-oh-JEN-ik; *myo-* = "muscle," *-genic* = "producing"). The myogenic mechanism works by using the same mechanisms to control the GFR that we discussed in the Concept Boost. It occurs in the following way:

- An *increase* in systemic blood pressure stretches the afferent arteriole and leads to an increase in the GFR. This causes the muscle cells to contract, constricting the arteriole. Constriction of the afferent arteriole decreases the blood flow through the glomerulus ("turns down the faucet"), which decreases glomerular hydrostatic pressure and the GFR back to normal range.
- A *decrease* in the systemic blood pressure causes the afferent arteriole to be less stretched, decreasing the GFR and making its smooth muscle cells relax. The resulting dilation of the arteriole raises the blood flow and the glomerular hydrostatic pressure ("turns up the faucet"), and the GFR increases back to normal range.

The myogenic mechanism acts rapidly—it can restore the normal GFR within seconds of even a significant change in blood pressure. Note, however, that these mechanisms can restore the GFR over a systolic blood pressure range of about 80–180 mm Hg (a normal range for systolic pressure is 100–120 mm Hg). If the systolic blood pressure goes higher than 180 mm Hg, the afferent arteriole can't constrict enough to restore the GFR. Similarly, if systolic blood pressure drops below this range, the afferent arteriole is unable to dilate enough to restore the GFR to normal range.

Tubuloglomerular Feedback The second autoregulatory mechanism of the kidneys is *tubuloglomerular feedback* (too´-byoo-loh-gloh-MEHR-yoo-lur), which involves the juxtaglomerular apparatus, or JGA. The system is named for the fact that the macula densa of the distal renal tubule (see Figure 24.8) is part of a negative feedback loop that controls pressure in the glomerulus. The basic feedback loop proceeds by the following steps:

1. If the GFR increases, the volume of filtrate flowing through the renal tubule increases.

2. The increased filtrate volume leads to an increased delivery of sodium and chloride ions to the macula densa cells in the distal tubule, causing them to absorb more of these ions from the filtrate.

3. In an example of the Cell-Cell Communication Core Principle (p. 28), the macula densa cells release ATP into the interstitial fluid. Some ATP is converted to the nucleoside adenosine, which leads to constriction of the afferent arteriole.

CORE PRINCIPLE
Cell-Cell Communication

4. The GFR decreases back toward normal range.

If the high levels of sodium and chloride ions in the filtrate persist for more than a few minutes, the macula densa cells signal JG cells of the afferent arterioles to reduce their release of the enzyme *renin* (REE-nin). The overall effect of this action is to increase the GFR and promote urine formation. This allows the JGA to play a part in maintaining fluid and sodium ion homeostasis.

With both the myogenic mechanism and tubuloglomerular feedback at work, the GFR remains remarkably consistent through changes in blood pressure, which happen with some frequency. For example, every time you sneeze or get up off the couch, your blood pressure changes momentarily. However, these mechanisms are insufficient to maintain the GFR when dealing with very large changes in blood pressure. If the systolic blood pressure rises above 180 mm Hg, the GFR and urine output will increase dramatically. Conversely, if the systolic blood pressure drops below about 70 mm Hg, glomerular filtration will cease because the compensatory mechanisms cannot create an NFP that favors filtration. At this point, no urine will be produced, a life-threatening condition called *anuria* (an-YOO-ree-uh; "absence of urine formation").

Quick Check

☐ 6. What happens to the GFR when the afferent arteriole constricts? What happens to the GFR when it dilates?

☐ 7. How does tubuloglomerular feedback affect the GFR?

Hormonal Effects on the GFR

Recall that *hormones* are chemical messengers released into the bloodstream that regulate the functions of other cells (see Chapter 16). Like the two autoregulatory mechanisms, hormones that affect the GFR do so by adjusting the glomerular hydrostatic pressure. However, hormones affect the GFR as part of a larger system that regulates systemic blood pressure. One hormone that regulates the GFR is *angiotensin-II* (an´-jee-oh-TEN-sin), a component of the *renin-angiotensin-aldosterone system*. Another set of regulating hormones consists of the *natriuretic peptides* (nay´-tree-yer-ET-ik; *natri-* = "sodium") produced by the heart. First let's examine the former hormone system.

The Renin-Angiotensin-Aldosterone System The **renin-angiotensin-aldosterone system (RAAS)** is a complex system with a primary function of maintaining systemic blood pressure; it preserves the GFR as a secondary effect. For this reason, it acts on much more than just the afferent and/or efferent arterioles of the glomeruli. The RAAS also significantly impacts tubular reabsorption in the nephron and collecting system in order to influence electrolyte balance and blood volume, in addition to causing changes within the body as a whole.

This system may be triggered into action by three conditions: (1) stimulation from neurons of the sympathetic nervous system, (2) low blood pressure, or (3) stimulation from the macula densa cells in response to low sodium and chloride ion concentration

in the filtrate. Note that often the system is "turned on" by a combination of these three factors.

The RAAS proceeds by the following steps, illustrated in **Figure 24.14**:

① **Systemic blood pressure decreases, causing a decrease in the GFR.**

② **JG cells release renin.** Decreased blood flow through the afferent arteriole triggers the JG cells to release the enzyme **renin** into the bloodstream.

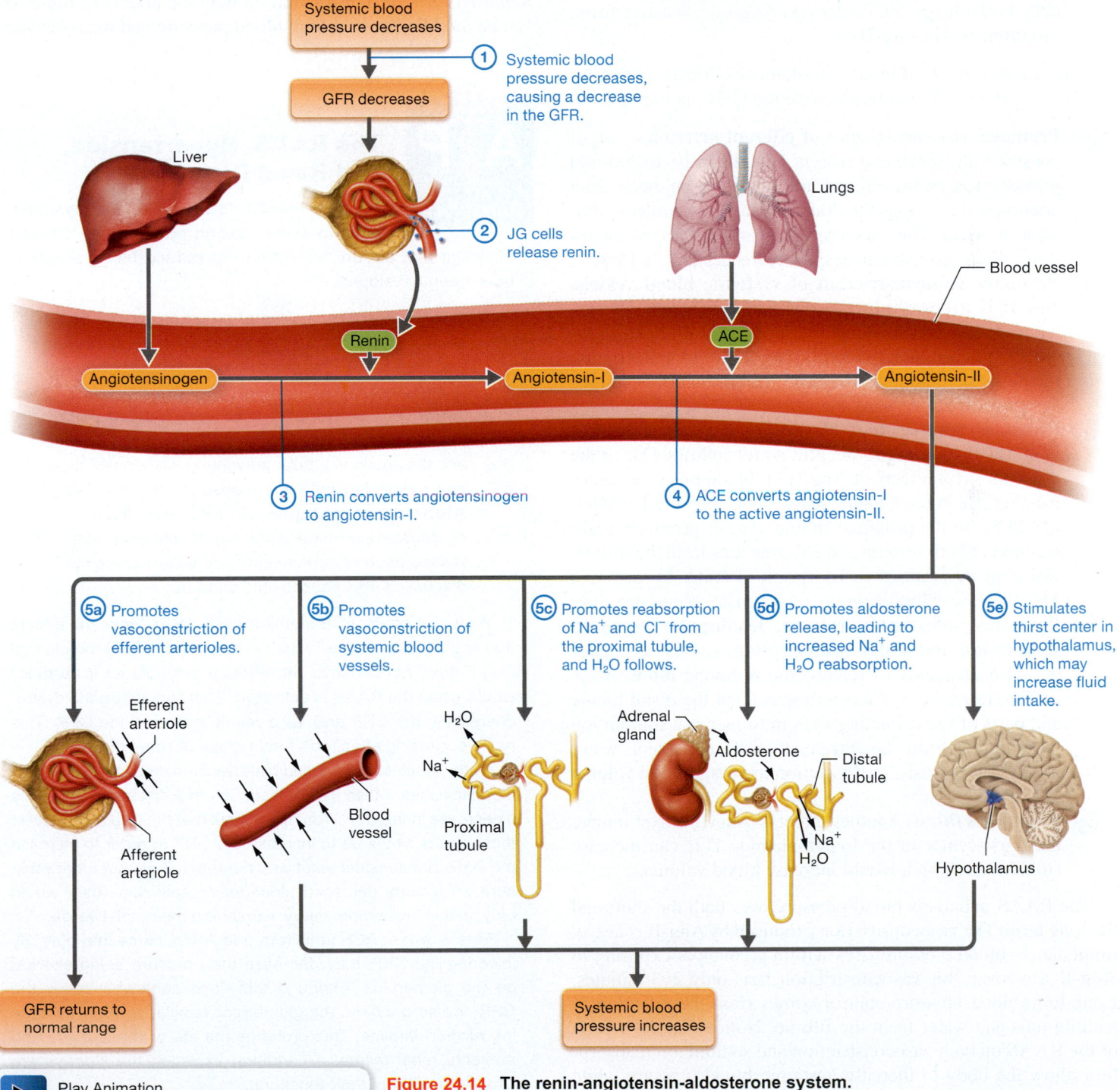

Figure 24.14 The renin-angiotensin-aldosterone system.

Play Animation
@ Mastering Anatomy & Physiology

③ **Renin converts angiotensinogen to angiotensin-I.** Renin circulates until it encounters an inactive protein produced by the liver, called **angiotensinogen** (an′-jee-oh-ten-SIN-oh-jen). Renin catalyzes the conversion of angiotensinogen to a product with minimal activity, **angiotensin-I (Ang-I).**

④ **ACE converts angiotensin-I to the active angiotensin-II.** Ang-I circulates through the blood until it encounters an enzyme called **angiotensin-converting enzyme (ACE),** which is made by cells such as the endothelial cells in the lungs. ACE converts Ang-I to its active form, **angiotensin-II (Ang-II).**

Ang-II has several different, simultaneous effects that influence both systemic blood pressure and the GFR, including:

⑤ₐ **Promotes vasoconstriction of efferent arterioles.** Ang-II constricts all renal blood vessels, but it has a 10- to 100-fold greater effect on the efferent arterioles than on the afferent arterioles. This "clogs the drain," raising glomerular hydrostatic pressure. This increased pressure, in turn, maintains the GFR at a normal rate despite the reduced blood flow.

⑤ᵦ **Promotes vasoconstriction of systemic blood vessels.** Ang-II is a powerful vasoconstrictor of nearly all systemic vessels (it causes effects more than 40 times stronger than those of norepinephrine). This vasoconstriction increases peripheral resistance, which in turn increases systemic blood pressure.

⑤𝒸 **Promotes reabsorption of sodium and chloride ions from the proximal tubule, and water follows.** One of the most powerful effects of Ang-II on blood pressure comes from its role in renal tubule reabsorption (covered in Module 24.5). In the proximal tubule, Ang-II promotes reabsorption of both sodium and chloride ions from the filtrate, which in turn causes reabsorption of water by osmosis. This increases blood volume, which raises blood pressure.

⑤𝒹 **Promotes aldosterone release, leading to increased sodium ion and water reabsorption.** Ang-II stimulates the adrenal glands to release the hormone **aldosterone** (al-DAHS-ter-ohn). Aldosterone acts on the distal tubule and parts of the collecting system to increase sodium ion reabsorption from the filtrate. If ADH is present, water follows by osmosis, so this action increases blood volume and blood pressure.

⑤ₑ **Stimulates thirst.** Another effect of Ang-II is to stimulate the thirst center in the hypothalamus. This can increase fluid intake, which would increase blood volume.

The RAAS maintains blood pressure over both the short and the long term. The vasoconstriction produced by Ang-II is nearly immediate—blood pressure rises within seconds of exposure to Ang-II. However, this vasoconstriction lasts only 2–3 minutes. Long-term blood pressure control comes from the retention of sodium ions and water from the filtrate. Notice that the effects of the RAAS on both vasoconstriction and sodium ion reabsorption allow the body to increase systemic blood pressure while preserving the GFR.

In addition to its effects on blood pressure, the RAAS is also critical to maintaining sodium ion balance. Renin secretion increases when the plasma sodium ion concentration decreases so that more sodium ions are reabsorbed. Conversely, renin secretion decreases when the plasma sodium ion concentration increases, so the kidneys excrete more sodium ions. The RAAS is so effective at this task that sodium ion intake can increase up to 50 times and the level of sodium ions in the plasma will remain relatively unchanged.

See *A&P in the Real World: The RAAS, Hypertension, and Renal Disease* for information on how the effects of the RAAS can be modified to treat high blood pressure and renal disease.

The RAAS, Hypertension, and Renal Disease

Because the effects of the RAAS on systemic blood pressure are so potent, three classes of drugs that act on this system to reduce blood pressure have been developed:

- **ACE inhibitors,** a class of drugs developed from snake venom, block angiotensin-converting enzyme and therefore inhibit the conversion of angiotensin-I to angiotensin-II.
- **Angiotensin-receptor blockers (ARBs)** block the angiotensin receptors on the cells of blood vessels and the proximal tubule, preventing vasoconstriction and sodium ion and water reabsorption, respectively.
- **Aldosterone antagonists** block the effects of aldosterone on the distal tubule and decrease reabsorption of sodium ions and water; as we've discussed, this causes mild diuresis.

As a side effect, ACE inhibitors and ARBs block the effects that angiotensin-II (Ang-II) has on the kidneys. This means that Ang-II does *not* constrict the efferent arteriole as it normally would when the RAAS is activated. This "opens up the drain," decreasing the GHP and, as a result, reducing the GFR. This has a surprising effect on certain types of renal disease.

Both diabetes mellitus and hypertension damage the glomerular capillaries, which in turn leads to renal disease and causes a decrease in the GFR. You might think that the treatment goal in these cases would be to increase the GHP in order to increase the GFR. But consider what would happen if we put more pressure on already damaged glomerular capillaries—they would only sustain even more injury, worsening the renal disease.

This is where ACE inhibitors and ARBs come into play. By lowering the GHP, they decrease the pressure being exerted on the glomerulus. Although this does somewhat lower the GFR, it also prevents the glomerular capillaries from sustaining further damage. This prolongs the life of the kidneys and prevents *renal failure*, a condition in which the kidneys are unable to perform their functions.

Atrial Natriuretic Peptide Recall that certain cells of the atria in the heart produce the hormone **atrial natriuretic peptide (ANP)** (see Chapter 16). ANP, which is released when the volume of blood in the atria increases, lowers blood volume and blood pressure and thereby reduces the workload of the heart. One way that ANP accomplishes this task is by dilating the afferent arterioles and constricting the efferent arterioles of the glomeruli, a combination that "turns up the faucet" and "clogs the drain." This increases the glomerular hydrostatic pressure, raising the GFR. A high GFR leads to more fluid loss, which decreases blood volume and effectively lowers blood pressure. Note that ANP is part of a system that works to decrease the systemic blood pressure; it simply happens to do so by affecting the GFR. We consider the other effects of ANP on the kidney in Module 24.5.

Quick Check

☐ 8. What are the basic steps of the renin-angiotensin-aldosterone system?

☐ 9. What are the effects of the RAAS on the GFR?

☐ 10. What are the effects of ANP on the GFR?

Neural Regulation of the GFR

Neural control of the GFR is chiefly mediated by the sympathetic division of the autonomic nervous system. As with the hormonal systems just described, the sympathetic nervous system affects the GFR as part of a larger system to control systemic blood pressure. Sympathetic neurons release norepinephrine during times of increased sympathetic activity, which causes constriction of most systemic blood vessels, including the afferent arterioles, and so elevates systemic blood pressure. You might think this should decrease glomerular hydrostatic pressure and the GFR, but the sympathetic effect on the GFR isn't quite that simple. Instead, its overall effect depends on how much it is stimulated:

- If the level of sympathetic stimulation is low (e.g., during mild exercise such as walking or jogging), sympathetic neurons trigger the JG cells to release renin. This leads to the formation of a low level of Ang-II, which raises systemic blood pressure and maintains the GFR. This makes sense—your body has no need to decrease the GFR if you're going for a jog or nervous about a test.

- If the level of sympathetic stimulation is high (e.g., during severe blood loss), a large amount of renin is secreted, and the blood concentration of Ang-II increases dramatically. A high level of Ang-II will actually constrict both the afferent and efferent arterioles, decreasing the GFR. This is especially important in cases of severe hypotension and dehydration, as it helps the body to minimize fluid loss and preserve blood volume and blood pressure. In these cases, perfusion of vital organs (brain, heart, etc.) takes precedence over filtration in the kidneys.

Table 24.1 summarizes the autoregulatory, hormonal, and neural mechanisms that control the GFR.

Table 24.1 Summary of Control of the Glomerular Filtration Rate

Stimulus	Responding Mechanism	Main Effect(s)	Effect on the GFR
Autoregulation			
Increased stretching of the afferent arteriole (due to increased blood pressure)	Myogenic mechanism	• Vasoconstriction of the afferent arteriole	Decrease
Increased sodium ion delivery to the macula densa cells (due to an increased GFR)	Tubuloglomerular feedback	• Vasoconstriction of the afferent arteriole • Vasodilation of the efferent arteriole	Decrease
Hormonal Mechanisms			
Sympathetic nervous system activated; a decreased GFR; decreased systemic blood pressure	Renin-angiotensin-aldosterone system	• Vasoconstriction of the efferent arteriole • Systemic vasoconstriction • Reabsorption of sodium ions from the filtrate in the proximal tubule • Release of aldosterone • Reabsorption of water	Maintain at normal level (decrease at high levels of activation)
Increased systemic blood pressure	Atrial natriuretic peptide	• Vasodilation of the afferent arteriole • Vasoconstriction of the efferent arteriole • Decreased reabsorption of sodium ions • Increased water loss	Increase
Neural Mechanisms			
Multiple	Sympathetic nervous system	• Constriction of all vessels, including the afferent and efferent arterioles • Stimulates RAAS	Maintain at normal level (decrease at high levels of activation)

☐ 11. How does the sympathetic nervous system affect the GFR at both low and high levels of stimulation?

Renal Failure

If a person's GFR decreases, he or she may enter **renal failure,** a condition in which the kidneys are unable to carry out their vital functions. Many conditions may lead to renal failure, including factors that decrease blood flow to the kidneys, diseases of the kidneys themselves, and anything that obstructs urine outflow from the kidneys. Short-term renal failure, known as *acute renal failure* or *acute kidney injury,* is common among hospitalized patients and may resolve completely with treatment of the underlying cause. Some people develop *chronic* (long-term) *renal failure,* which is defined as a decrease in the GFR lasting 3 months or longer.

The biggest risk factors for developing chronic renal failure are diabetes mellitus and hypertension. Symptoms depend on the severity of the renal failure, and those with mild renal failure might not notice any symptoms. As renal function declines, patients experience fatigue, edema, nausea, and loss of appetite. Severe renal failure, in which the GFR is less than 50% of normal, results in a condition known as **uremia** (yoo-REEM-ee-uh). Uremia is characterized by a buildup of waste products and by fluid, electrolyte, and acid-base imbalances. Untreated, it can lead to coma, seizures, and death.

When a patient develops the signs and symptoms of uremia, **dialysis** (dy-AL-uh-sis) treatment may be initiated. There are two types of dialysis. The first is **hemodialysis,** which temporarily removes an individual's blood and passes it through a filter that removes metabolic wastes and extra fluid, and normalizes electrolyte and acid-base balance. Hemodialysis must be performed three times per week at a dialysis clinic. The second type is **peritoneal dialysis,** in which dialysis fluid is placed into the peritoneal cavity, allowed to circulate for several hours, and then drained. A patient is able to undergo peritoneal dialysis nightly at home, so this is often the preferred treatment for individuals needing long-term dialysis.

☐ 1. Ms. Douglas has advanced liver disease; because her liver is no longer able to produce plasma proteins, her colloid osmotic pressure has decreased. Predict the effects that this loss of pressure will have on the net filtration pressure and the GFR in her nephrons.

☐ 2. Certain drugs that treat high blood pressure cause vasodilation of systemic arteries and arterioles, including those in the kidneys. What effect would these drugs have on the GFR? How would the myogenic mechanism and tubuloglomerular feedback respond to this change in the GFR?

☐ 3. Mr. Adams is taking an ACE inhibitor and an angiotensin-receptor blocker, both drugs that block the RAAS, for his high blood pressure. He complains that when he tries to engage in physical activity, he feels faint. He is asked to exercise on a treadmill, and his blood pressure remains very low when he exercises, rather than rising with his level of physical activity. Explain how his medications could be causing his current problem.

See answers in Appendix A.

24.5
Renal Physiology II: Tubular Reabsorption and Secretion

1. Describe how and where water, organic compounds, and ions are reabsorbed in the nephron by both passive and active processes.

2. Describe the location(s) in the nephron where tubular secretion occurs.

3. Describe how the renin-angiotensin-aldosterone system, antidiuretic hormone, and atrial natriuretic peptide each work to regulate reabsorption and secretion.

As we've discussed, the rate of filtrate formation is very high, so the kidneys have to work to reclaim most of the filtrate and return it to the bloodstream. Recall that this occurs by the process of *tubular reabsorption* in the nephrons and collecting system. Of the 180 liters of water filtered by the kidneys every day, only about 1.8 liters (1%) ultimately leave the body as urine, which means that the kidneys are very efficient at reclaiming water. In addition, about 99% of the solutes filtered are subsequently reabsorbed. Given the magnitudes of these quantities, it's not surprising that the kidneys use approximately 20% of the body's ATP at rest.

Since our kidneys reabsorb so much of what they filter, you may wonder why our bodies need to filter blood through the kidneys at all. First of all, certain waste products are not reabsorbed. In addition, in spite of the high GFR, some waste substances do not pass through the filtration membranes of the glomeruli and therefore don't enter the filtrate. This necessitates the process of *tubular secretion,* which occurs in the renal tubules and collecting ducts at the same time as tubular reabsorption. These nonreabsorbed and secreted substances are often toxic when present in high concentrations, and the high GFR and process of secretion ensure their rapid clearance from the body. In addition, the presence of a large volume of filtrate in the nephrons gives the kidneys precise control over fluid, electrolyte, and acid-base balance. If substances are not needed, they are excreted with the urine. If they are needed, they will be reabsorbed into the blood.

In the upcoming sections, we examine the processes of tubular reabsorption and secretion more closely. We also trace the filtrate as it flows through the different parts of the nephron and collecting system, and discuss how it changes in both composition and concentration to finally become urine.

Principles of Tubular Reabsorption and Secretion

◀◀ FLASHBACK

1. What are tight junctions, and what are their functions? (p. 127)
2. What are facilitated diffusion, primary active transport, and secondary active transport? (p. 76 and p. 80)

This section deals with how substances move between the inside of the renal tubule and the surrounding peritubular capillaries. We also examine how solute and water movement are driven by either active or passive transport processes. Let's start by reviewing some of the principles we've learned that we can now apply to tubular reabsorption and secretion.

Paracellular and Transcellular Transport Routes

When discussing the movement of solutes and water, we first need to discuss exactly where and how the substances are moved. As you can see in **Figure 24.15**, in tubular reabsorption, substances must pass from the filtrate in the lumen (inside) of the tubule, across or between the tubule cells, into the interstitial fluid, and finally across or between the endothelial cells of the peritubular capillaries to re-enter the blood. In tubular secretion, substances move in the opposite direction. But exactly how do these substances cross the tubule cells?

As you learned in the histology chapter, substances can be transported by two different routes across epithelia: paracellular or transcellular (look back at Figure 4.6). For a quick review, these processes are described as follows:

- **Paracellular route.** On the paracellular route (*para-* = "beside"), substances pass *between* adjacent tubule cells. The tight junctions between the tubule cells are just leaky enough to allow some substances such as small ions and water to move passively between them, particularly in the proximal tubule.
- **Transcellular route.** On the transcellular route (*trans-* = "across"), substances such as glucose and amino acids must move *through* the tubule cells. A reabsorbed substance first crosses the *apical* membrane of the tubule cell (the membrane facing the tubule lumen), then travels through the cytosol, and finally exits the cell through the *basolateral* membrane (the side of the membrane facing the interstitial fluid).

Transport along the paracellular route is passive, requiring no energy in the form of ATP, because it consists of either water moving by osmosis or solutes moving with their concentration gradients. The same is true for certain solutes taking the transcellular route. However, other solutes moving along the transcellular route travel against their concentration gradients and so require energy, either directly or indirectly, from ATP. Secretion is an active process, so it must occur via the transcellular route across the tubule cell membrane.

Reabsorbed substances that have entered the interstitial fluid may then cross the endothelial cells of the blood vessel and enter the blood. These substances can follow the same routes into the

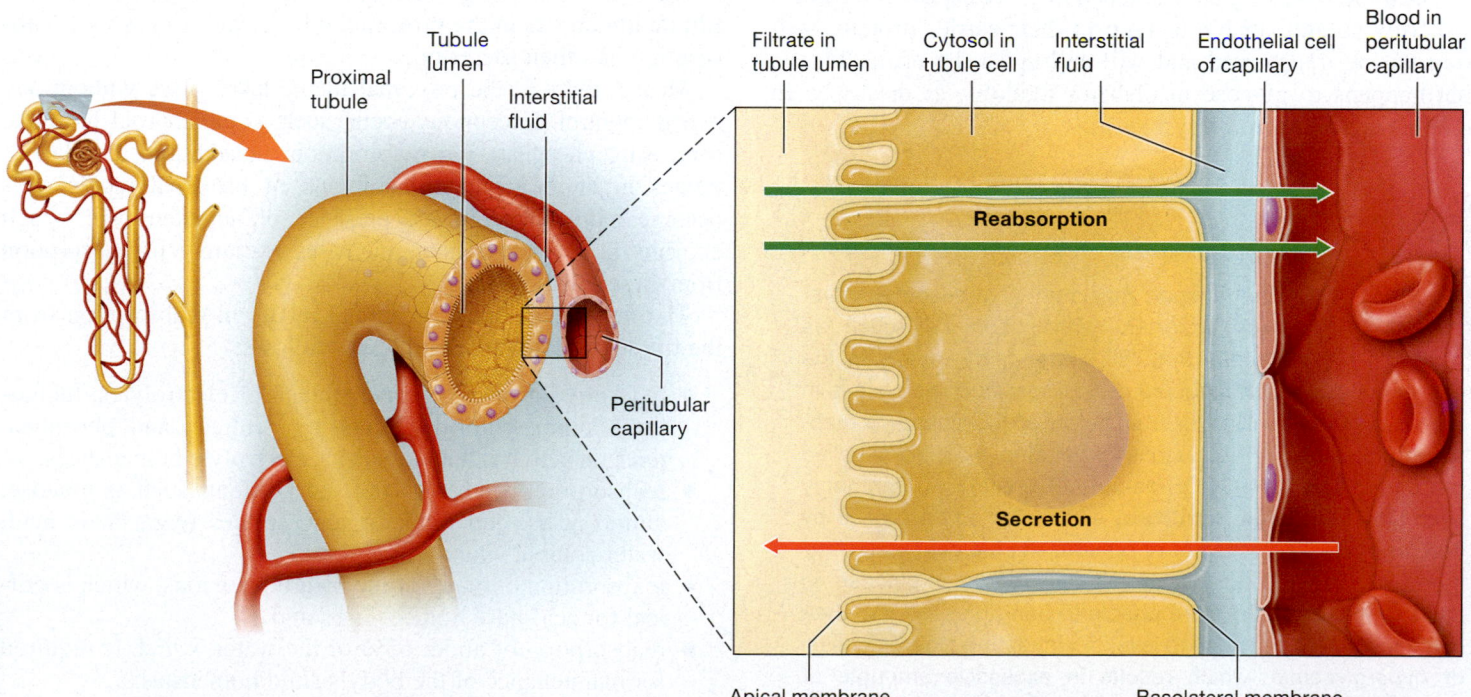

Figure 24.15 Barriers to tubular reabsorption and secretion.

capillary that they followed to exit the tubule—they may take the paracellular route or the transcellular route. Generally, these processes are passive, and solutes move by diffusion and water by osmosis.

Carrier-Mediated Transport and the Transport Maximum

Most of the solutes that are reabsorbed and secreted via the transcellular route require the use of a carrier protein in the tubule cell plasma membrane. Recall that there are three ways in which cells use carrier proteins to transport solutes (see Chapter 3):

- *facilitated diffusion,* in which a carrier protein passively transports a solute with its concentration gradient, without using energy from ATP (an example of the Gradients Core Principle, p. 28);  **CORE PRINCIPLE** Gradients
- *primary active transport,* in which a carrier protein "pump" directly uses ATP to move a solute against its concentration gradient; and
- *secondary active transport,* in which a concentration gradient set up by a primary active transport pump is used to drive the transport of a second solute against its concentration gradient via another carrier protein.

There are two types of active transport carrier proteins: *Antiport pumps* (or *antiporters*) move two or more solutes in opposite directions, and *symport pumps* (or *symporters*) move two or more solutes in the same direction (see Chapter 3). Both have a limited number of sites on which they can transport solutes, much as a train has only a certain number of seats for passengers. If all their sites become filled, the carrier proteins are said to be **saturated,** as they have reached their **transport maximum (T_M).** Any solutes unable to bind to their carrier proteins will likely not be transported and will end up in the urine. This is what happens to glucose in diabetes mellitus, as discussed in *A&P in the Real World: Glycosuria.*

Glycosuria

The transport maximum (T_M) becomes especially important with substances such as glucose that "fill seats" fairly rapidly in the carriers of the proximal tubule cells. If too much glucose is in the filtrate, the T_M will be reached before all the glucose is reabsorbed, so there will be glucose in the urine, a condition called *glycosuria* (gly´-koh-SOOR-ee-ah). Glycosuria is commonly seen with **diabetes mellitus,** a disorder characterized by defects in the production of or response to the pancreatic hormone *insulin.* Insulin causes most cells to take in glucose; in its absence, these cells are unable to bring glucose into their cytosol. This leads to a high level of circulating blood glucose, or *hyperglycemia,* which results in excessive amounts of glucose in the filtrate and ultimately in the urine.

Quick Check

☐ 1. What are the two routes of reabsorption and secretion?

☐ 2. What are the three types of transport processes that involve carrier proteins in the renal tubule and collecting system?

Reabsorption and Secretion in the Proximal Tubule

« FLASHBACK

1. What is the function of the Na⁺/K⁺ pump? In which direction(s) does it move sodium and potassium ions? (p. 80)

2. What is the function of the carbonic acid–bicarbonate ion buffer system? (p. 49)

The remainder of this module follows the filtrate from the capsular space through the nephron and collecting system as it is modified by tubular reabsorption and secretion. We start, of course, with the proximal tubule. Recall that the cells of the proximal tubule have prominent microvilli that provide these cells with a large surface area. This facilitates the remarkably rapid reabsorption that occurs in this very active segment of the renal tubule. In fact, the proximal tubule is the most metabolically active part of the nephron, as most of the filtrate is reabsorbed here—its Na⁺/K⁺ pumps alone consume about 6% of the body's ATP at rest. In addition to all of this reabsorption, a great deal of secretion takes place in the proximal tubule as well. The following sections examine the changes that the filtrate undergoes in the proximal tubule; we discuss first reabsorption and then secretion.

Most activity in the proximal tubule takes place without any outside control. Exceptions occur, such as parathyroid hormone decreasing phosphate ion reabsorption. In addition, the Na⁺/K⁺ pumps in the basolateral membrane of proximal tubule cells become more active in the presence of angiotensin-II, which explains how this hormone increases sodium ion reabsorption from these cells.

The main roles of the proximal tubule in reabsorption from the filtrate back to the blood are as follows:

- reabsorption of a large percentage of electrolytes, including sodium, chloride, potassium, sulfate, and phosphate ions, an activity that is vital for electrolyte homeostasis;
- reabsorption of nearly 100% of nutrients such as glucose, amino acids, and other organic solutes (e.g., lactic acid, water-soluble vitamins);
- reabsorption of many of the bicarbonate ions, which is critical for acid-base homeostasis; and
- reabsorption of about 65% of the water, which is required for maintenance of the body's fluid homeostasis.

Let's look more closely at each of these roles.

Sodium Ion Reabsorption

We begin with the reabsorption of sodium ions, because this process turns out to be the key to reabsorbing many other substances in the proximal tubule. Much of the sodium ion reabsorption occurs by the transcellular route through sodium ion leak channels on the apical surface of the proximal tubule cell. Another mechanism for transcellular sodium ion reabsorption involves three types of carrier proteins in the apical membranes of proximal tubule cells:

- carrier proteins specific for sodium ions that enable facilitated diffusion of sodium ions from the filtrate into the tubule cells,
- Na^+ symporters that bring sodium ions from the filtrate into the cells with other solutes (such as glucose), and
- Na^+/H^+ antiporters that bring sodium ions into the cells while secreting hydrogen ions into the filtrate.

Both the leak channels and the carrier proteins rely on a concentration gradient created by Na^+/K^+ pumps in the basolateral membrane. These pumps continually drive sodium ions out of the tubule cells and into the interstitial fluid. This creates a relatively low sodium ion concentration in the cytosol of the proximal tubule cells (about 12 milliosmoles, or mOsm). The sodium ion concentration in the filtrate is significantly higher (about 142 mOsm), which is a steep concentration gradient that is critical for the secondary active transport of many other solutes.

Reabsorption of Organic Solutes and Ions

The cells of the first half of the proximal tubule contain **Na^+/glucose symporters.** The symporters use the sodium ion gradient created by Na^+/K^+ pumps to carry both glucose and sodium ions from the filtrate into the tubule cell, an example of secondary active transport (**Figure 24.16**). Once in the cell, glucose is transported via facilitated diffusion into the interstitial fluid, where it diffuses into the peritubular capillaries. Other symporters in the apical membrane of the proximal tubule cells function in a similar fashion, allowing the secondary active transport of sodium ions and another solute, such as an ion (e.g., SO_4^{2-} or HPO_4^{2-}) or an organic solute (e.g., amino acids or lactic acid).

The reabsorption of sodium ions also leads to the reabsorption of anions such as chloride ions in another way, too. As sodium ions are passively transported out of the tubule lumen, the lumen accumulates a net negative charge. This creates an electrical gradient that pushes the negatively charged chloride ions across the epithelium through the paracellular route. These

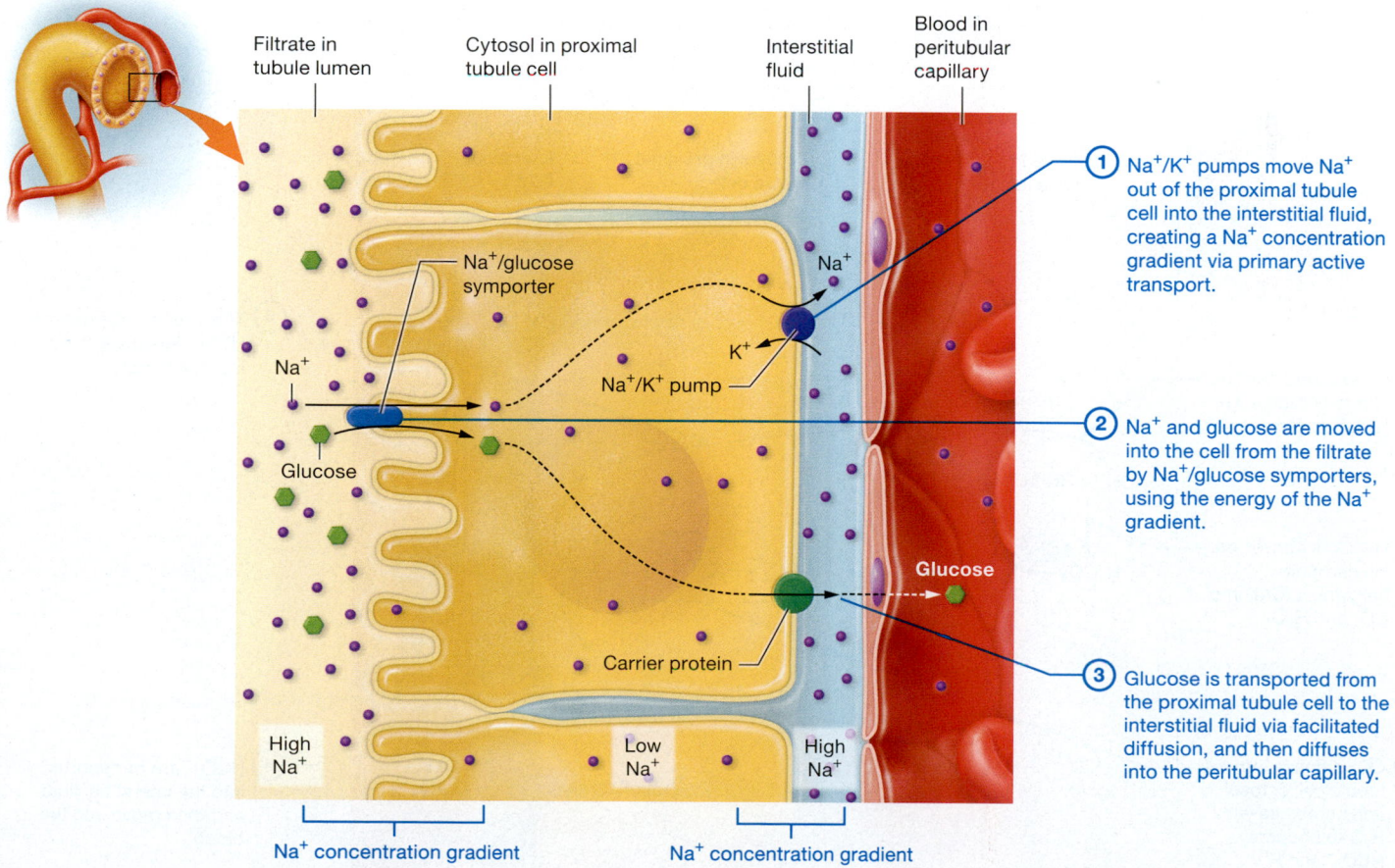

Figure 24.16 **Glucose reabsorption in the proximal tubule.**

chloride ions similarly follow sodium ions into the interstitial fluid and into the plasma, as well.

Bicarbonate Ion Reabsorption

Bicarbonate ion reabsorption from the proximal tubule involves principles that you learned in the chemistry chapter about the carbonic acid–bicarbonate buffer system (see Chapter 2). Recall that carbon dioxide (CO_2) in the blood reacts with water (H_2O) to produce carbonic acid (H_2CO_3); this reaction is catalyzed by the enzyme *carbonic anhydrase (CA)*. Newly formed carbonic acid spontaneously dissociates into bicarbonate (HCO_3^-) and hydrogen (H^+) ions. The complete reaction is as follows:

$$CO_2 + H_2O \underset{\text{anhydrase}}{\overset{\text{Carbonic}}{\rightleftharpoons}} H_2CO_3 \rightleftharpoons H^+ + HCO_3^-$$

Remember that carbonic anhydrase also catalyzes the reverse reaction, turning carbonic acid into carbon dioxide and water.

Bicarbonate reabsorption from the proximal tubule occurs in a roundabout way (**Figure 24.17**). The process somewhat resembles a game of "fetch": The cell "tosses" hydrogen ions into the filtrate, and the hydrogen ion "fetches" a bicarbonate ion, which is then brought back into the cell. This toss is accomplished with the help of another carrier protein: the **Na⁺/H⁺ antiporter.** This carrier protein transports sodium ions into the cell while

secreting hydrogen ions from the cell into the filtrate. Here's how the process unfolds:

1. **Hydrogen ions secreted into the filtrate combine with bicarbonate ions to form carbonic acid.** Hydrogen ions are secreted from the cytosol of the proximal tubule cell into the filtrate by the Na^+/H^+ antiporter. The hydrogen ions react with bicarbonate ions in the filtrate to form carbonic acid.

2. **Carbonic acid is converted, via carbonic anhydrase, to carbon dioxide and water.** Carbonic anhydrase on the apical plasma membrane of the tubule cell catalyzes the conversion of carbonic acid into carbon dioxide and water.

3. **Carbon dioxide diffuses into the tubule cell cytosol and combines with water to become bicarbonate and hydrogen ions.** Carbon dioxide diffuses into the cytosol of the tubule cell with its gradient. Carbonic anhydrase within the cell catalyzes the conversion of carbon dioxide and water into carbonic acid, which then dissociates into a bicarbonate ion and a hydrogen ion.

4. **Bicarbonate ions are transported into the interstitial fluid and then move into the blood.** The bicarbonate ion is transported across the basolateral membrane to enter the interstitial fluid and then diffuses into the peritubular capillaries.

5. **The process repeats as hydrogen ions are again secreted into the filtrate.** The hydrogen ions that were released when carbonic acid dissociated are recycled,

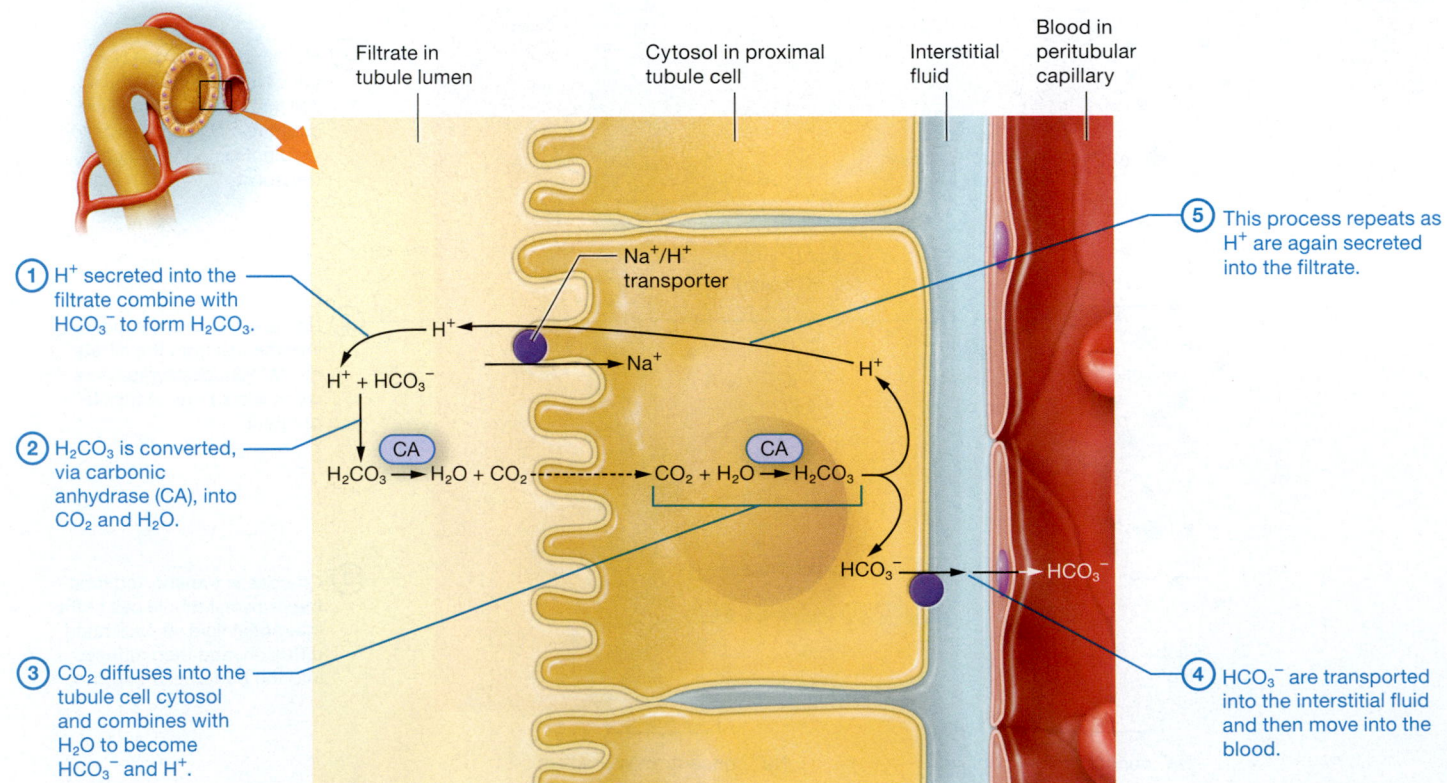

Figure 24.17 Bicarbonate ion reabsorption in the proximal tubule.

as they are again secreted into the filtrate by the Na^+/H^+ antiporter, and the process is repeated.

Though this process may seem complicated, it is effective and allows the cells of the proximal tubule to reabsorb about 90% of the bicarbonate ions from the filtrate. This is a key component of the body's ability to maintain the pH of the blood within a very specific range, 7.35–7.45 (see Chapter 25).

Obligatory Water Reabsorption and Its Effect on Other Electrolytes

By the time the filtrate has reached the second half of the proximal tubule, many of the sodium ions as well as glucose and other organic compounds have been reabsorbed. The accumulation of solutes in the cytosol and interstitial fluid draws water out of the filtrate by osmosis along both paracellular and transcellular routes. Remember that in osmosis, water moves to the solution with a higher solute concentration.

This type of water reabsorption is called **obligatory water reabsorption,** because water is "obliged" to follow solute movement (**Figure 24.18**). A kind of water channel in the plasma membrane called an **aquaporin** (ah-kwah-POHR-in) greatly enhances rapid water reabsorption. These channels, which are located in both the apical and basolateral membranes of proximal tubule cells, allow water to move through these cells via the transcellular route.

As obligatory water reabsorption continues, the concentration of solutes, such as potassium, calcium, and magnesium ions, rises in the filtrate. This creates a concentration gradient that favors their diffusion into or between the proximal tubule cells. Notice that in this process, active reabsorption of solutes stimulates further reabsorption of water by osmosis. This in turn stimulates passive reabsorption of other solutes.

Secretion in the Proximal Tubule

In addition to the hydrogen ion secretion we discussed earlier, other substances are secreted into the filtrate by the proximal tubule cells, including many nitrogenous waste products and drugs. In the first half of the proximal tubule, most of the uric acid in the filtrate is reabsorbed, but nearly all of it is secreted back into the filtrate in the second half of the tubule. Additionally, ammonium ions (NH_4^+), creatinine, and small amounts of urea are also secreted. Drugs such as penicillin and morphine have significant renal secretion. These drugs must be taken often (typically 3–5 times per day), because medicine lost through renal secretion must be replaced in order to maintain relatively consistent blood levels.

Quick Check

☐ 3. Which substances are reabsorbed from the proximal tubule? Which of these substances are reabsorbed using the sodium ion gradient?

☐ 4. What is obligatory water reabsorption?

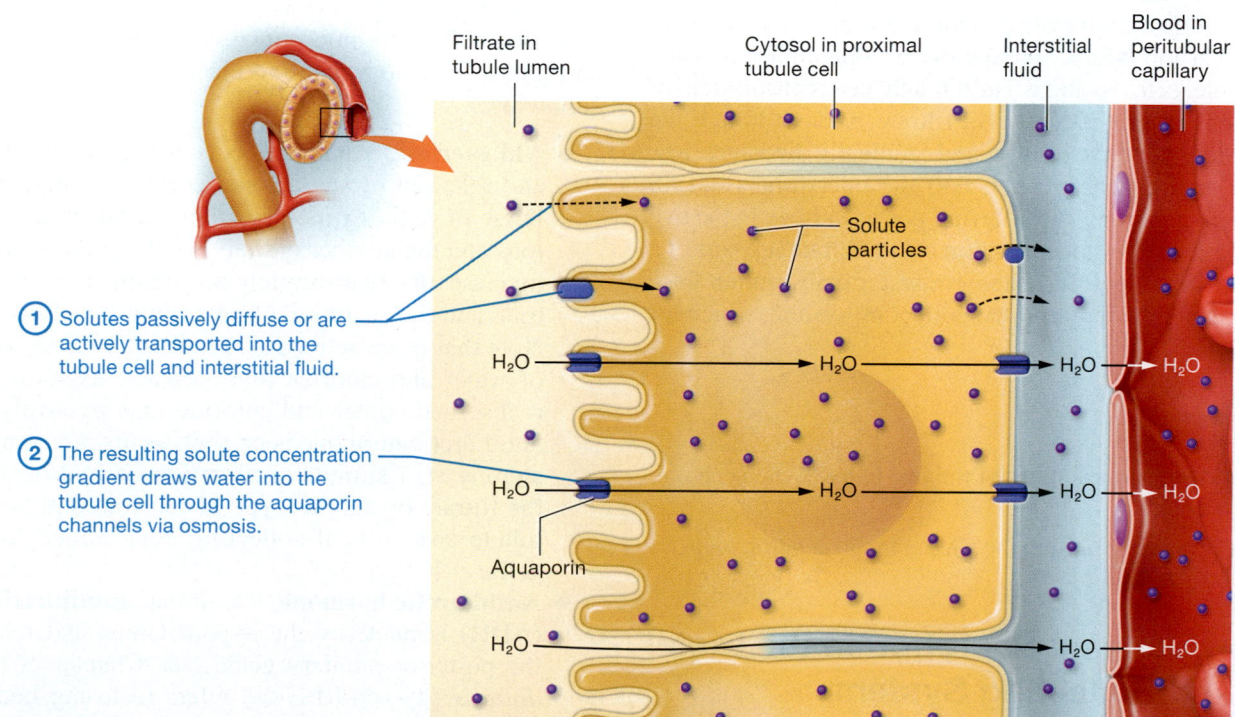

1. Solutes passively diffuse or are actively transported into the tubule cell and interstitial fluid.

2. The resulting solute concentration gradient draws water into the tubule cell through the aquaporin channels via osmosis.

Filtrate in tubule lumen
Cytosol in proximal tubule cell
Interstitial fluid
Blood in peritubular capillary
Solute particles
Aquaporin

Figure 24.18 **Obligatory water reabsorption in the proximal tubule.**

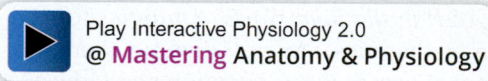

Play Interactive Physiology 2.0
@ **Mastering** Anatomy & Physiology

Reabsorption in the Nephron Loop

When the filtrate reaches the nephron loop, it barely resembles the original filtrate—about 60–70% of the electrolytes and water have been reabsorbed from the filtrate and returned to the blood, in addition to most of the organic solutes, such as glucose and amino acids. As the filtrate flows through the nephron loop, it undergoes further losses: Approximately 20% of the total water, 25% of the total sodium and chloride ions, and a significant portion of the other remaining ions are reabsorbed and returned to the blood.

In the proximal tubule, you saw that water reabsorption is proportional to solute reabsorption. For this reason, the filtrate in the proximal tubule has the same concentration, or *osmolarity,* as the interstitial fluid, about 300 mOsm. In the nephron loop, however, the filtrate's osmolarity changes as it flows through the loop. This is due to the differing permeabilities in the ascending and descending limbs of the nephron loop.

The thin descending limb of the nephron loop is freely permeable to water, but much less permeable to solutes such as sodium and chloride ions. So, water can move out of the thin descending limb cells by osmosis, but few solutes follow. This causes the osmolarity of the filtrate to increase as it passes down the descending limb and rounds the bend of the loop.

The cells of the thick ascending limb are impermeable to water, but they transport NaCl into the tubule cells with the use of **$Na^+/K^+/2Cl^-$ symporters.** This secondary active transport system brings one sodium, one potassium, and two chloride ions into the tubule cell, relying on the Na^+/K^+ pump on the basolateral membrane to create a favorable sodium ion gradient. Remember that the Na^+/K^+ pumps drive potassium ions back *into* the tubule cell, so there isn't much net reabsorption of potassium ions in this process. As filtrate passes through the ascending limb, it loses solutes and gradually becomes less concentrated as ions are pumped into the interstitial fluid. In juxtamedullary nephrons, the differing permeabilities in the two limbs and the changing concentration of the filtrate are part of a larger system that allows for extensive water reabsorption from the filtrate in both the loop and the collecting system, which we cover in Module 24.6.

Quick Check

☐ 5. How do the permeabilities of the two limbs of the nephron loop differ?

☐ 6. What is reabsorbed from the filtrate in the nephron loop?

Reabsorption and Secretion in the Distal Tubule and Collecting System

By the time the filtrate enters the first part of the distal tubule, about 85% of the water and 90% of the sodium ions have been reabsorbed. For this reason, the rate of filtrate flow in this part of the tubule is significantly slower (about 20 ml/min) than

it was in the early proximal tubule (about 120 ml/min). As you learned earlier, the cells of the distal tubule lack significant microvilli. This reflects the Structure-Function Core Principle (p. 28)—most of the water reabsorption has already taken place,

CORE PRINCIPLE
Structure-Function

so the cells do not require extensive microvilli to reabsorb much of the remaining water and solutes. However, reabsorption of the remaining water and sodium ions is still critical. If we excreted the remaining water and sodium ions in the urine, we would lose about 29 liters of water and a significant portion of our sodium ions every day. This situation would be incompatible with life.

The early distal tubule is structurally and functionally similar to the ascending limb of the nephron loop. However, the latter portion of the distal tubule is very similar to the cortical collecting duct, so we discuss them together here. Then we examine the medullary collecting system, which differs structurally and functionally from these other areas.

The Late Distal Tubule and Cortical Collecting Duct: Hormone Regulation

The primary cell type of the late distal tubule and cortical collecting duct is known as the **principal cell.** These cells have hormone receptors that determine their function, and so the majority of their activity is regulated by hormones in order to fine-tune water, electrolyte, and acid-base balance. For this reason, the water reabsorption here is called **facultative water reabsorption,** because water is reabsorbed in accordance with the body's needs (*facultative* means "able to adapt to a need"). The hormones involved in facultative water reabsorption as well as water and electrolyte balance include the following:

- **Aldosterone.** Aldosterone is a steroid hormone made and released by the adrenal cortex. It interacts with the DNA of cells to increase their permeability to sodium ions and the number of their Na^+/K^+ pumps. Both actions increase the reabsorption of sodium ions from the filtrate and the secretion of potassium ions into the filtrate. Note that these actions also indirectly cause reabsorption of water and chloride ions, because as sodium ions are reabsorbed, water and chloride ions passively follow (if antidiuretic hormone is present, as discussed next). Aldosterone also stimulates secretion of hydrogen ions into the filtrate by another type of cell found in the late distal tubule and cortical collecting duct, called **intercalated cells.**

- **Antidiuretic hormone.** Recall that **antidiuretic hormone (ADH)** is made by the hypothalamus and released from the posterior pituitary gland (see Chapter 16). Note that *diuresis* (dy-yoo-REE-sis) refers to losing body water to the urine, and a **diuretic** is an agent that promotes diuresis (see *A&P in the Real World: Diuretics* to read about some common diuretics). An *anti*-diuretic, therefore, refers to an agent that causes water retention and reduces urine output.

ADH exerts these effects by causing aquaporins to be inserted into the apical membranes of principal cells, permitting rapid water reabsorption. In the absence of ADH, principal cells are barely permeable to water, and a larger volume of water is lost in the urine.

- **Atrial natriuretic peptide (ANP).** ANP triggers *natriuresis* (nay′-tree-yoo-REE-sis), or urinary excretion of sodium ions. It also appears to inhibit release of ADH and aldosterone, causing fewer sodium ions (and also less water) to be reabsorbed, and so more sodium ions and water to appear in the urine.

The Medullary Collecting System

As filtrate flows through the medullary collecting ducts and the papillary ducts, this is the kidney's last chance to regulate fluid, electrolyte, and acid-base balance before the filtrate becomes urine. The principal cells of the medullary collecting system have the following properties:

- They are impermeable to water in the absence of ADH. However, in the presence of ADH, they reabsorb large volumes of water.
- They are permeable to urea, which allows some urea to move down its concentration gradient into the interstitial fluid.
- They continue to reabsorb ions such as sodium, chloride, and bicarbonate from the filtrate.

In addition, both principal and intercalated cells in this region actively secrete hydrogen ions from the interstitial fluid into the filtrate against a very high concentration gradient; they can increase the concentration of hydrogen ions in the filtrate about 900 times. We revisit the medullary collecting system in Module 24.6, as it is part of a mechanism that allows the kidneys to concentrate urine and conserve water.

Quick Check

☐ 7. How does aldosterone impact reabsorption and secretion in the late distal tubule and cortical collecting duct?

☐ 8. How does ADH influence water reabsorption? In which parts of the nephron and collecting system does ADH act?

How Tubular Reabsorption and Secretion Maintain Acid-Base Balance

The kidneys can adjust reabsorption and secretion of hydrogen and bicarbonate ions, which helps maintain pH homeostasis of the extracellular fluids, including blood. Thus far, you have seen two examples of acid-base regulation throughout the renal tubule and collecting system: The cells of the proximal tubule secrete hydrogen ions as a way to reabsorb bicarbonate ions, and the principal and intercalated cells of the late distal tubule and collecting system actively secrete hydrogen ions.

Another mechanism in renal tubule cells is stimulated when the pH of the blood becomes abnormal. If the pH of the blood *decreases,* making it too acidic, enzymes in the

tubule cells will remove the amino group ($-NH_2$) from the amino acid glutamine in the cytosol. In doing this, the cells generate two ammonia molecules (NH_3) and two bicarbonate ions. The ammonia is then secreted, and the bicarbonate ions

A&P in the Real World

Diuretics

Diuretics are drugs that work on different sections of the renal tubule to block solute reabsorption. This decreases osmotic water reabsorption, increasing urine output. For this reason, diuretics are often called "water pills." The most commonly used classes of diuretics include the following:

- **Loop diuretics.** As their name implies, the *loop diuretics* work on the nephron loop. They inhibit the $Na^+/K^+/2Cl^-$ channels in the loop's ascending limb, which in turn decreases water reabsorption from the loop's descending limb. These drugs also disrupt the countercurrent mechanism in juxtamedullary nephrons (see p. 977), which makes them some of our most powerful diuretics. Note that a consequence of loop diuretics is that they also block the reabsorption of potassium ions. This leads to excessive potassium ions lost to the urine, causing loop diuretics to sometimes be called *potassium-wasting* diuretics.

- **Thiazide diuretics.** *Thiazide diuretics* target the sodium ion transporter in the distal tubule, blocking reabsorption of both sodium and chloride ions. This creates an osmotic gradient that favors less reabsorption of water, which increases urine production. Thiazide diuretics are also potassium-wasting. Their mechanism of action leads to a higher concentration of sodium ions in the filtrate when it reaches the collecting system. This, in turn, stimulates increased sodium ion reabsorption through channels in collecting duct cells, which indirectly leads to potassium ion secretion.

- **Potassium-sparing diuretics.** A low potassium ion level, known as *hypokalemia,* is a very real concern with patients taking potassium-wasting diuretics. For this reason, these drugs are sometimes paired with *potassium-sparing diuretics,* or those that cause diuresis while preventing potassium ion loss. One example of a potassium-sparing diuretic is *spironolactone,* which blocks aldosterone's effects in the distal tubule. This causes mild diuresis but also prevents aldosterone-stimulated potassium ion secretion. Another example is *triamterene,* which blocks the sodium ion channels in the cells of the collecting system, preventing sodium ion reabsorption and potassium ion secretion. These are the channels that cause potassium-wasting with thiazide diuretics, and so triamterene is sometimes paired with a thiazide.

The Big Picture of Tubular Reabsorption and Secretion

Figure 24.19

▶ Play Animation @ Mastering Anatomy & Physiology

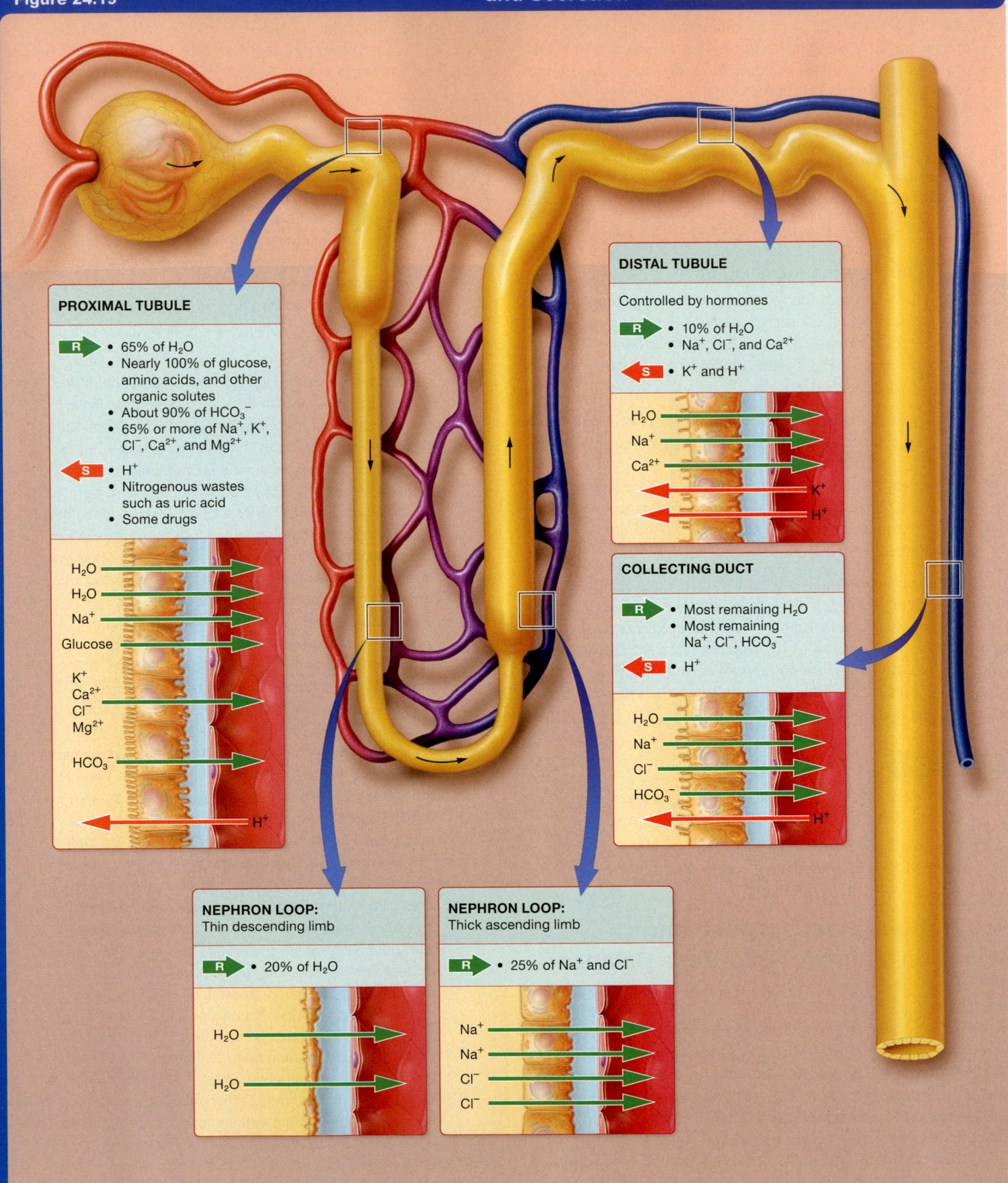

PROXIMAL TUBULE

R
- 65% of H_2O
- Nearly 100% of glucose, amino acids, and other organic solutes
- About 90% of HCO_3^-
- 65% or more of Na^+, K^+, Cl^-, Ca^{2+}, and Mg^{2+}

S
- H^+
- Nitrogenous wastes such as uric acid
- Some drugs

H_2O
H_2O
Na^+
Glucose
K^+
Ca^{2+}
Cl^-
Mg^{2+}
HCO_3^-
H^+

DISTAL TUBULE

Controlled by hormones

R
- 10% of H_2O
- Na^+, Cl^-, and Ca^{2+}

S
- K^+ and H^+

H_2O
Na^+
Ca^{2+}
K^+
H^+

COLLECTING DUCT

R
- Most remaining H_2O
- Most remaining Na^+, Cl^-, HCO_3^-

S
- H^+

H_2O
Na^+
Cl^-
HCO_3^-
H^+

NEPHRON LOOP: Thin descending limb

R
- 20% of H_2O

H_2O
H_2O

NEPHRON LOOP: Thick ascending limb

R
- 25% of Na^+ and Cl^-

Na^+
Na^+
Cl^-
Cl^-

are reabsorbed. In the filtrate, the ammonia molecules bind and buffer hydrogen ions to form the ammonium ion (NH_4^+). Both ammonia buffering and bicarbonate ion reabsorption help to raise the pH of the blood back to normal. If the pH of the blood *increases,* making it too alkaline, the tubule cells will reabsorb fewer bicarbonate ions from the filtrate, excreting them in the urine and lowering the pH of the blood.

Putting It All Together: The Big Picture of Tubular Reabsorption and Secretion

As you've seen, both tubular reabsorption and secretion are vital processes, and enormous quantities of water and solutes cross the tubule cells every day. Filtrate entering the proximal tubule contains water; ions such as sodium, potassium, chloride, calcium, and bicarbonate; and organic solutes such as glucose, amino acids, and metabolic wastes. By the time the filtrate leaves the papillary ducts to become urine, most of the water and solutes have been reclaimed. In **Figure 24.19** we summarize the substances that are filtered, reabsorbed, and secreted by the nephrons and collecting system of the kidneys.

The final product that exits the papillary ducts is urine; the properties of this fluid are discussed in Module 24.7.

Apply What You Learned

□ 1. You discover a new toxin that blocks the reabsorption of all sodium ions from the proximal tubule. What effect would this drug have on the reabsorption of water and other electrolytes from this tubule?

□ 2. What effect would this toxin have on the reabsorption of glucose and bicarbonate ions?

□ 3. Respiratory conditions can cause chronic *hypoventilation* that leads to a decreased blood pH. Predict how the kidneys will respond to this change in pH.

See answers in Appendix A.

MODULE 24.6
Renal Physiology III: Regulation of Urine Concentration and Volume

Learning Outcomes

1. Explain why the differential permeability of specific sections of the renal tubule is necessary to produce concentrated urine.

2. Predict specific conditions that cause the kidneys to produce dilute versus concentrated urine.

3. Explain the role of the nephron loop, the vasa recta, and the countercurrent mechanism in the concentration of urine.

In the previous two modules on renal physiology, we followed the path of the filtrate from its formation via glomerular filtration to its modification in the renal tubules and collecting system by reabsorption and secretion. Now we are moving on to discuss the variable water reabsorption in the late distal tubule and collecting system, and the effect of this on urine concentration and volume.

As you learned in the previous module, about 85% of water reabsorption in the kidney is obligatory—water is "obliged" by osmosis to follow solutes that have been reabsorbed. The last 15% of water reabsorption is facultative water reabsorption, which is adjusted by hormones to meet the body's needs and maintain fluid homeostasis. Facultative water reabsorption is what determines final urine concentration and volume. In this module, we examine how the kidneys adjust facultative water reabsorption to produce either dilute or concentrated urine, and the role that hormones, particularly ADH, play in this process.

Osmolarity of the Filtrate

Let's first trace the filtrate through the nephron and follow how its osmolarity can change on its way to becoming urine. The filtrate that exits the blood and enters the renal tubule initially is *iso-osmotic,* or equally osmotic, to the plasma at 300 mOsm. Recall that in the nephron loop, however, the filtrate's osmolarity changes because of the differing permeabilities of its ascending and descending limbs.

The thin descending limb of the nephron loop is permeable to water but not solutes, as you learned, so water flows from the filtrate to the interstitial fluid by osmosis, but very few solutes follow. This causes the filtrate to become progressively more concentrated as it travels down the loop. By the time it reaches the bottom of the loop, it averages approximately 900 mOsm, about 3 times more concentrated than plasma (the exact concentration depends on the length of the nephron loop—longer loops contain more concentrated filtrate at the bottom).

As the filtrate enters the thick ascending limb, sodium and other ions are pumped out of the filtrate and into the interstitial fluid. However, because this part of the limb is virtually impermeable to water, water can't follow these solutes. So the concentration of the filtrate decreases as it moves up the ascending limb of the loop. By the time the filtrate exits the thick ascending limb and moves into the distal tubule, its osmolarity is generally less than that of filtrate at the same level in the thin descending limb. In the early distal tubule, ions continue to leave the filtrate while water stays behind, and the concentration of the filtrate decreases even further, to about 100 mOsm, or 3 times *less* concentrated than plasma.

Once the filtrate enters the late distal tubule and collecting system, principal cells may begin facultative water reabsorption,

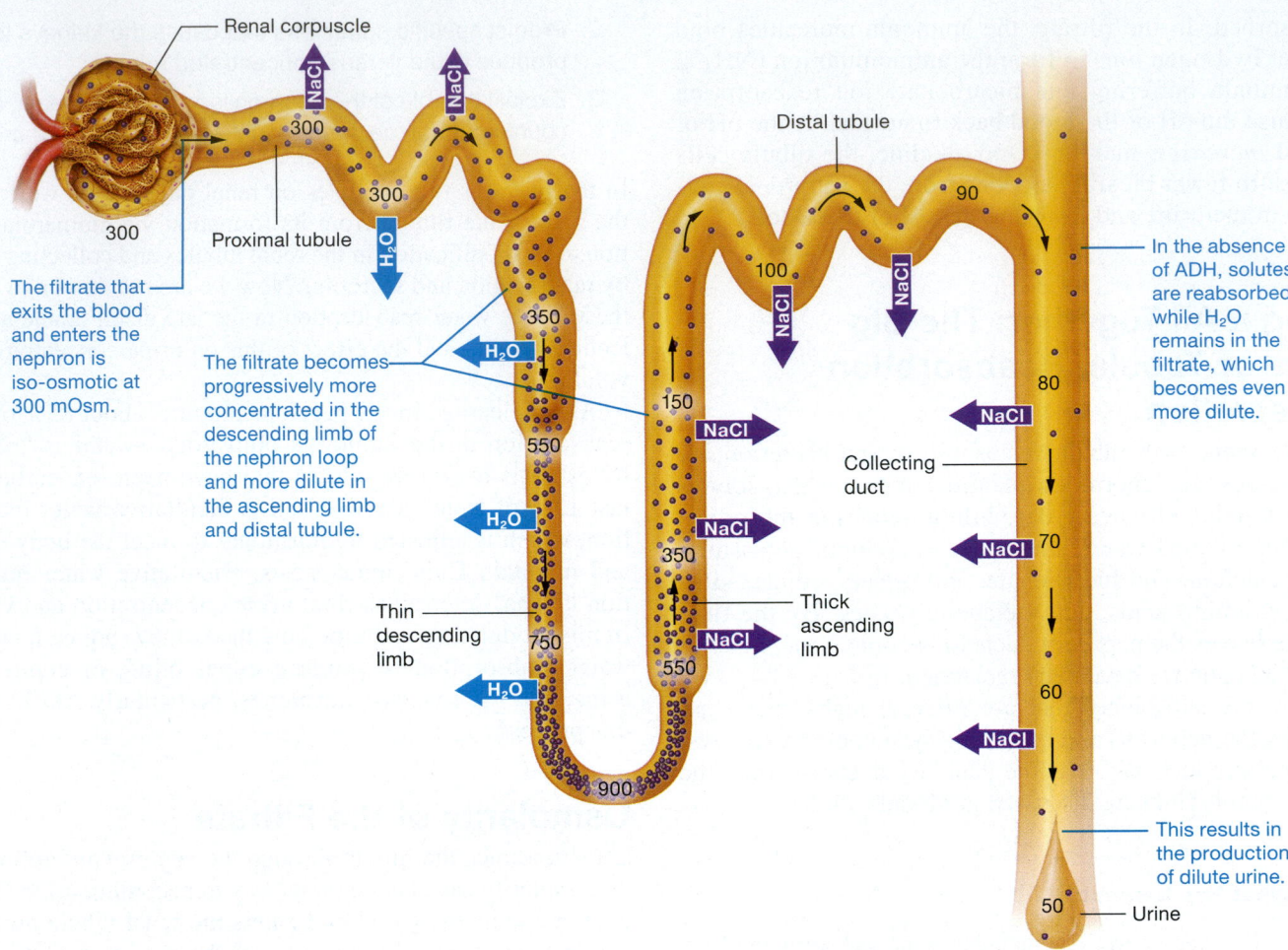

Renal corpuscle

Distal tubule

The filtrate that exits the blood and enters the nephron is iso-osmotic at 300 mOsm.

Proximal tubule

The filtrate becomes progressively more concentrated in the descending limb of the nephron loop and more dilute in the ascending limb and distal tubule.

Thin descending limb

Thick ascending limb

Collecting duct

In the absence of ADH, solutes are reabsorbed while H_2O remains in the filtrate, which becomes even more dilute.

This results in the production of dilute urine.

Urine

Figure 24.20   Formation of dilute urine.

so the concentration of the filtrate varies with the amount of water reabsorbed. If less water is reabsorbed, the concentration of the filtrate remains low as it passes through the late distal tubule and collecting system. The end result is the production of *dilute urine,* or urine with a concentration less than 300 mOsm. If, however, more water is reabsorbed, the concentration of the filtrate progressively increases as it passes through the late distal tubule and collecting system. The end result is the production of *concentrated urine,* or urine with a concentration greater than 300 mOsm. The upcoming sections examine the means by which both dilute urine and concentrated urine are produced.

☐ 1. How does the concentration of the filtrate change as it passes through the renal tubule and collecting system?

☐ 2. What are dilute urine and concentrated urine?

Production of Dilute Urine

The kidneys produce dilute urine when the solute concentration of the body's extracellular fluid is too low (which means the extracellular fluid contains excess water) (**Figure 24.20**). The

filtrate entering the late distal tubule is already less concentrated than the surrounding interstitial fluid. For this reason, the kidneys simply have to "turn off" facultative water reabsorption, to produce dilute urine. This is accomplished through a reduction in ADH release, which renders the principal cells of the late distal tubule and collecting system essentially impermeable to water.

Note that the principal cells continue to reabsorb sodium and chloride ions from the filtrate, so the number of solutes in the filtrate decreases while the amount of water remains the same. This causes the concentration of the filtrate to progressively decrease as it passes through the collecting system. When there is an excessive amount of water in the extracellular fluid, the osmolarity of the urine can fall as low as 50 mOsm, which is only one-sixth the osmolarity of plasma.

Typically, the volume of urine produced also increases when the urine is very dilute. As you learned, the normal volume of urine produced is about 1.8 liters per day. However, this value can increase dramatically when ADH secretion is low. Note that urine volume is also influenced by many factors, including fluid intake, general health, diet, and blood pressure. It is also affected by diuretics, which act on various parts of the renal tubule to block reabsorption of water or solutes and promote diuresis.

Quick Check

☐ 3. Under what condition do the kidneys produce dilute urine?

☐ 4. How do the kidneys produce dilute urine?

The Countercurrent Mechanism and the Production of Concentrated Urine

Our kidneys are quite effective at conserving water and can produce urine with a concentration up to about 1200 mOsm. (This is impressive, but the kidneys of the Australian desert hopping mouse can produce urine with a concentration as high as 10,000 mOsm.) Concentrated urine results from "turning on" facultative water reabsorption in the principal cells of the late distal tubule and collecting system by the release of ADH. However, facultative water reabsorption requires more than simply the presence of ADH. Recall that water reabsorption happens only by osmosis, and osmosis will occur only if an osmotic gradient is present to drive it. This means that facultative water reabsorption takes place only if the interstitial fluid surrounding the nephron is *more* concentrated than the filtrate (an example of the Gradients Core Principle, p. 28).

CORE PRINCIPLE
Gradients

The interstitial fluid within the renal cortex has about the same osmolarity as the interstitial fluid elsewhere in the body, approximately 300 mOsm. The filtrate entering the late distal tubule and cortical collecting duct has an osmolarity of about 100 mOsm. So the interstitial fluid around the late distal tubule and collecting duct is more concentrated than the filtrate, which creates a small but significant gradient to drive water reabsorption. However, by the time the filtrate enters the medullary collecting duct, enough water has been reabsorbed that its osmolarity is about equal to that of interstitial fluid—300 mOsm—and the gradient disappears.

If the osmolarity of the interstitial fluid of the whole renal medulla remained 300 mOsm, no further osmosis would occur and water reabsorption in the renal tubules would be finished. So, to continue to reabsorb water and produce concentrated urine, the nephrons must work to create an osmotic gradient within the renal medulla. This gradient, known as the **medullary osmotic gradient,** starts with the medullary interstitial fluid at a concentration of 300–400 mOsm at the cortex/medulla border, and increases to about 1200 mOsm, 4 times more concentrated than plasma, at the deepest regions of the medulla.

The medullary osmotic gradient is established and maintained by a system called the **countercurrent mechanism,** which is a mechanism that involves fluids flowing in opposite directions that exchange materials or heat. The countercurrent mechanism has three components: (1) a *countercurrent multiplier* system in the nephron loops of juxtamedullary nephrons, (2) the recycling of urea in the medullary collecting ducts, and (3) a *countercurrent exchanger* in the vasa recta.

Establishing the Medullary Osmotic Gradient: The Nephron Loop and the Countercurrent Multiplier

Recall that there are two types of nephrons: cortical and juxtamedullary. Up to this point in the chapter, we have been discussing the physiology of both types of nephron. Now, however, we shift our attention specifically to the physiology of juxtamedullary nephrons—those with long nephron loops that descend deeply into the renal medulla. Within these long nephron loops we find a system called the **countercurrent multiplier,** which helps to establish the medullary osmotic gradient.

In this system, the term *countercurrent* refers to the fact that the filtrate in the two limbs of the nephron loop flows in opposite directions—the filtrate in the descending limb flows toward the renal pelvis, and the filtrate in the ascending limb flows back up toward the renal cortex. It is called a *multiplier* because the two limbs of the loop are close enough that a small osmotic difference at the top of the loop is multiplied as we move farther down the loop.

As depicted in **Figure 24.21**, the system works like this:

① **NaCl is actively transported from the filtrate in the thick ascending limb into the interstitial fluid, raising its NaCl concentration.** $Na^+/K^+/2Cl^-$ symporters pump NaCl from the cells of the thick ascending limb into the interstitial fluid. Recall that this segment of the nephron loop is impermeable to water, so water can't follow the NaCl out of the tubule. The accumulation of NaCl in the interstitial fluid increases its osmolarity.

② **The NaCl pumped into the interstitial fluid draws water out of the filtrate in the thin descending limb into the interstitial fluid by osmosis.** The concentrated interstitial fluid creates an osmotic gradient that draws water from the filtrate in the thin descending limb into the interstitial fluid.

③ **Due to the continuing loss of water, the NaCl concentration of the filtrate increases as it approaches the bottom of the loop.** Remember that the nephron loop's thin descending limb is impermeable to NaCl, so it remains in the filtrate as water leaves. For this reason, the filtrate becomes progressively more concentrated as it reaches the bottom of the nephron loop.

④ **The high NaCl concentration of the filtrate that reaches the thick ascending limb allows the NaCl reabsorption to continue.** The filtrate reaches the thick ascending limb with a very high NaCl concentration. This is key because the symporters in the thick ascending limb function in proportion to the NaCl concentration: The higher the NaCl concentration in the filtrate, the more that is pumped into the interstitial fluid. This is how the countercurrent multiplier is able to achieve such a high NaCl concentration in the interstitial fluid in the deepest part of the medulla. Notice that it reaches a concentration of about 1200 mOsm, a significantly higher value than is found in the filtrate in most nephrons, which reaches a concentration of about 900 mOsm. However, as the filtrate moves up the thick ascending limb, NaCl is pumped out and the NaCl concentration in the filtrate decreases. As this occurs, the pumps will progressively drive out less NaCl into the interstitial fluid, and so its NaCl concentration will progressively decrease as we move higher up in the medulla.

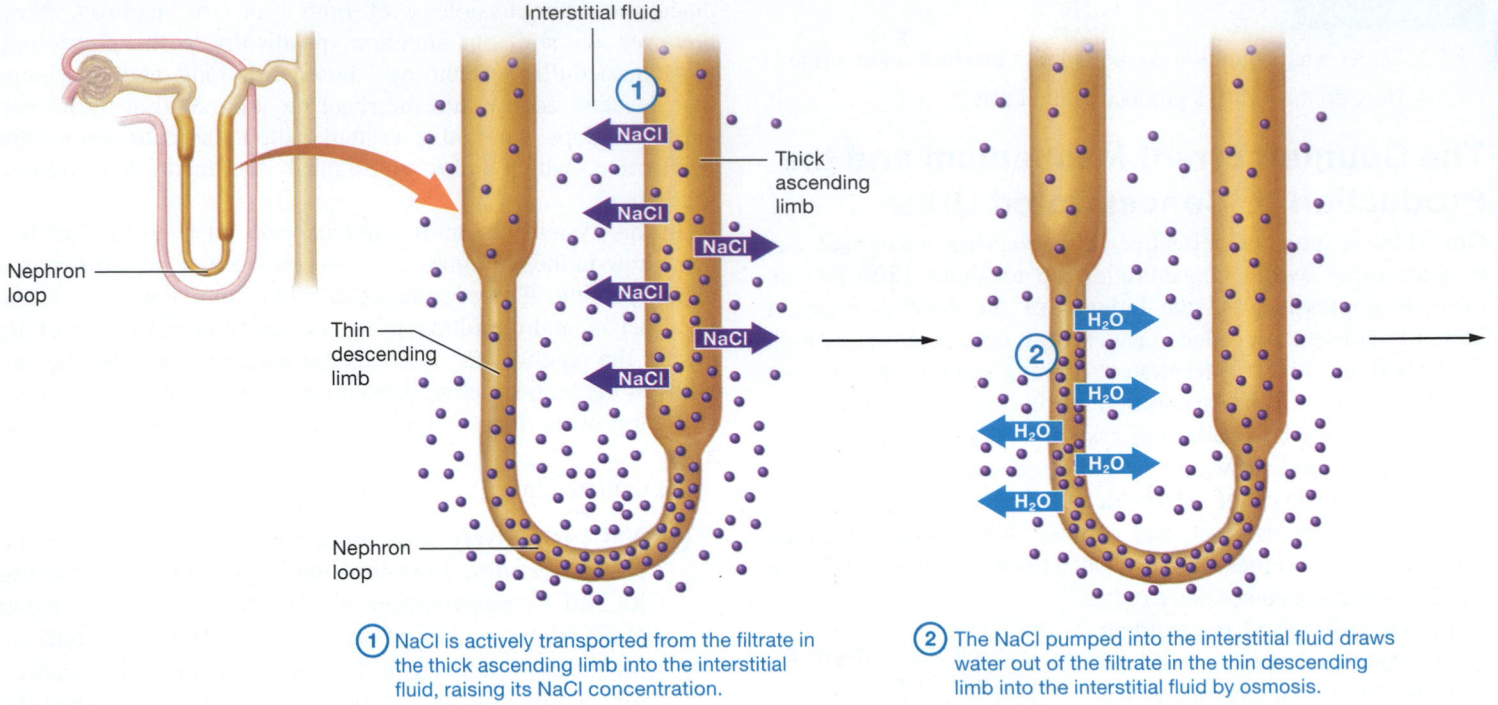

① NaCl is actively transported from the filtrate in the thick ascending limb into the interstitial fluid, raising its NaCl concentration.

② The NaCl pumped into the interstitial fluid draws water out of the filtrate in the thin descending limb into the interstitial fluid by osmosis.

Figure 24.21 **The countercurrent multiplier in the nephron loop.**

Play Animation
@ **Mastering** Anatomy & Physiology

As the symporters begin pumping NaCl into the interstitial fluid in step ④, we return to step ①. But note that all of these steps are actually occurring at the same time; we separated them here for simplicity.

ConceptBOOST))))

Demystifying the Countercurrent Multiplier

Let's be honest: The countercurrent multiplier is a difficult concept for everyone to understand at first. To demystify it, first consider the "why" of the system: to help the kidneys conserve water and produce concentrated urine. Next, we can consider the "what" of the system (as in what it does): It pumps large amounts of NaCl into the interstitial fluid of the renal medulla to establish a concentration gradient. Finally, we can consider the "how" of the system (as in how it does this). In the descending limb, water leaves so that NaCl is left behind in the filtrate. This NaCl can then be pumped out into the interstitial fluid, fulfilling the role of the countercurrent multiplier.

The second thing to think about is that the countercurrent multiplier isn't the typical stepwise process that we discuss in A&P. Even when we talk about cycles, they usually have a distinct beginning point and ending point that repeats. But the countercurrent multiplier is more like a perpetual motion machine—each step relies on the previous step, and there really isn't a distinct starting and stopping point. So don't try to understand it in terms of steps. Instead, think of it as being

like a pendulum continually swinging around in a circle or an audio track that's forever stuck on repeat:

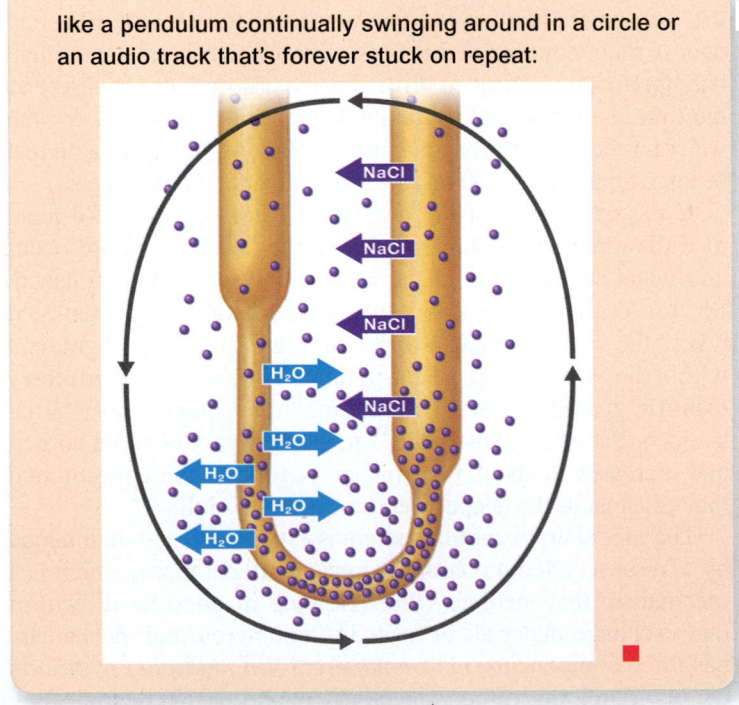

Contributing to the Medullary Osmotic Gradient: The Medullary Collecting System and Urea Recycling

NaCl isn't the only solute in the interstitial fluid of the medulla. Another solute helps to establish the medullary osmotic gradient: urea, which diffuses into the interstitial fluid from the medullary

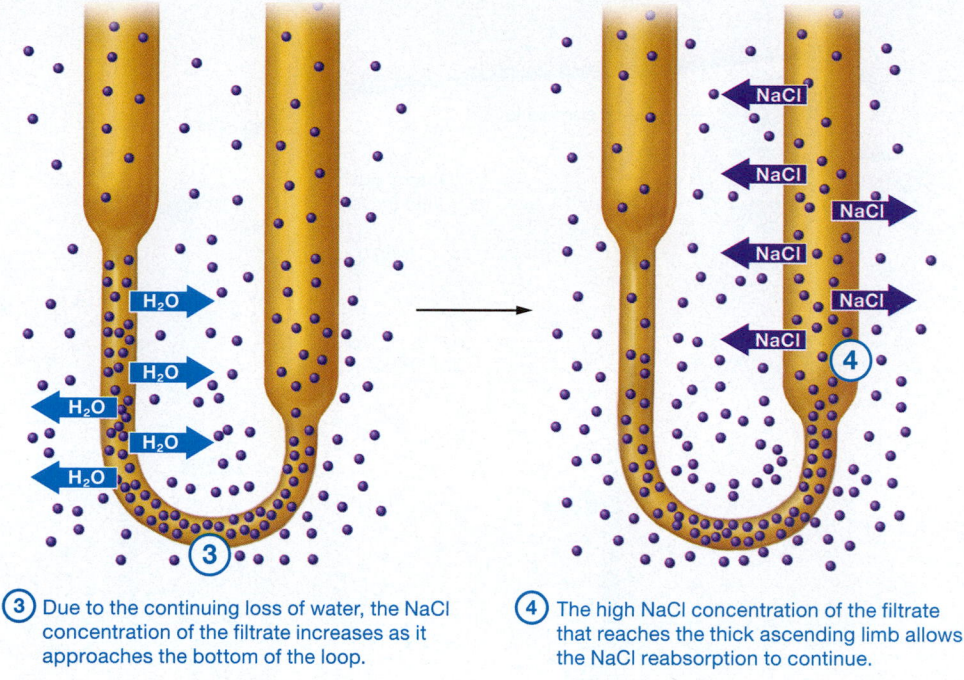

③ Due to the continuing loss of water, the NaCl concentration of the filtrate increases as it approaches the bottom of the loop.

④ The high NaCl concentration of the filtrate that reaches the thick ascending limb allows the NaCl reabsorption to continue.

Figure 24.21 *(continued)*

collecting duct by a process called *urea recycling.* Here's how the process works: As water is reabsorbed from the filtrate, urea becomes more concentrated in the remaining fluid. In the medullary collecting ducts and papillary ducts, urea follows its concentration gradient and passively diffuses out of the filtrate and into the interstitial fluid, where it contributes to the gradient. However, some of the urea enters the thin descending limb of the nephron loop to rejoin the filtrate. The urea remains in the filtrate until it reaches the medullary collecting duct, where it can again diffuse into the interstitial fluid. Note, though, that the urea diffusing out of the medullary collecting duct constitutes only a small amount of the total urea; much of the urea remains in the filtrate and is excreted in the urine.

Maintaining the Medullary Osmotic Gradient: The Vasa Recta and the Countercurrent Exchanger

Normally when solutes are reabsorbed into the interstitial fluid, they simply enter the blood in the peritubular capillaries and are carried away from the nephron. But this would present a problem in the renal medulla, because we can't produce concentrated urine without that osmotic gradient in the interstitial fluid created by the countercurrent multiplier and urea recycling. This means that the reabsorbed sodium and chloride ions can't be immediately swept away by the blood. However, the renal medulla has to have blood vessels, as its cells need oxygen and nutrients. For this reason, we have a special system of blood flow around the nephron loops of juxtamedullary nephrons to maintain the medullary osmotic gradient: the vasa recta.

Like the limbs of the nephron loop, the vasa recta descend into the renal medulla, and then, following a hairpin turn, ascend toward the renal cortex. But notice in **Figure 24.22** that

they are arranged so that their flow is opposite to that of the nephron loop—the descending vasa recta parallel the ascending limb of the nephron loop, and the ascending vasa recta parallel the descending limb. This arrangement of countercurrent flow allows the vasa recta to act as a **countercurrent exchanger** that can supply blood to the renal medulla cells while maintaining the medullary osmotic gradient.

Let's break down how countercurrent exchange works. First, notice in the figure that the blood within the descending vasa recta has a concentration of about 300 mOsm. This means that ① as the blood descends into the medulla it is hypo-osmotic to the interstitial fluid. As we saw earlier, this situation causes water to leave the blood and enter the interstitial fluid by osmosis. In addition, more NaCl is present in the interstitial fluid than in the blood, so NaCl diffuses from the interstitial fluid into the blood. The blood in the vasa recta continues to pick up NaCl and lose water as it descends deeper into the renal medulla. By the time the blood reaches the deepest part of the medulla, it has a concentration of about 1200 mOsm.

You might think this would cause all of the NaCl in the interstitial fluid to be "swept away" and the gradient to disappear. However, notice what happens in step ② as the vasa recta ascend through the medulla: The gradient is now reversed, and the blood is hyperosmotic to the interstitial fluid. The NaCl that was picked up by the descending vasa recta diffuses out of the blood of the ascending vasa recta and back into the interstitial fluid, and water moves by osmosis from the interstitial fluid into the blood. By the time the ascending vasa recta exit the renal medulla, the blood has approximately the same concentration (about 300 mOsm) it had when it entered the renal medulla.

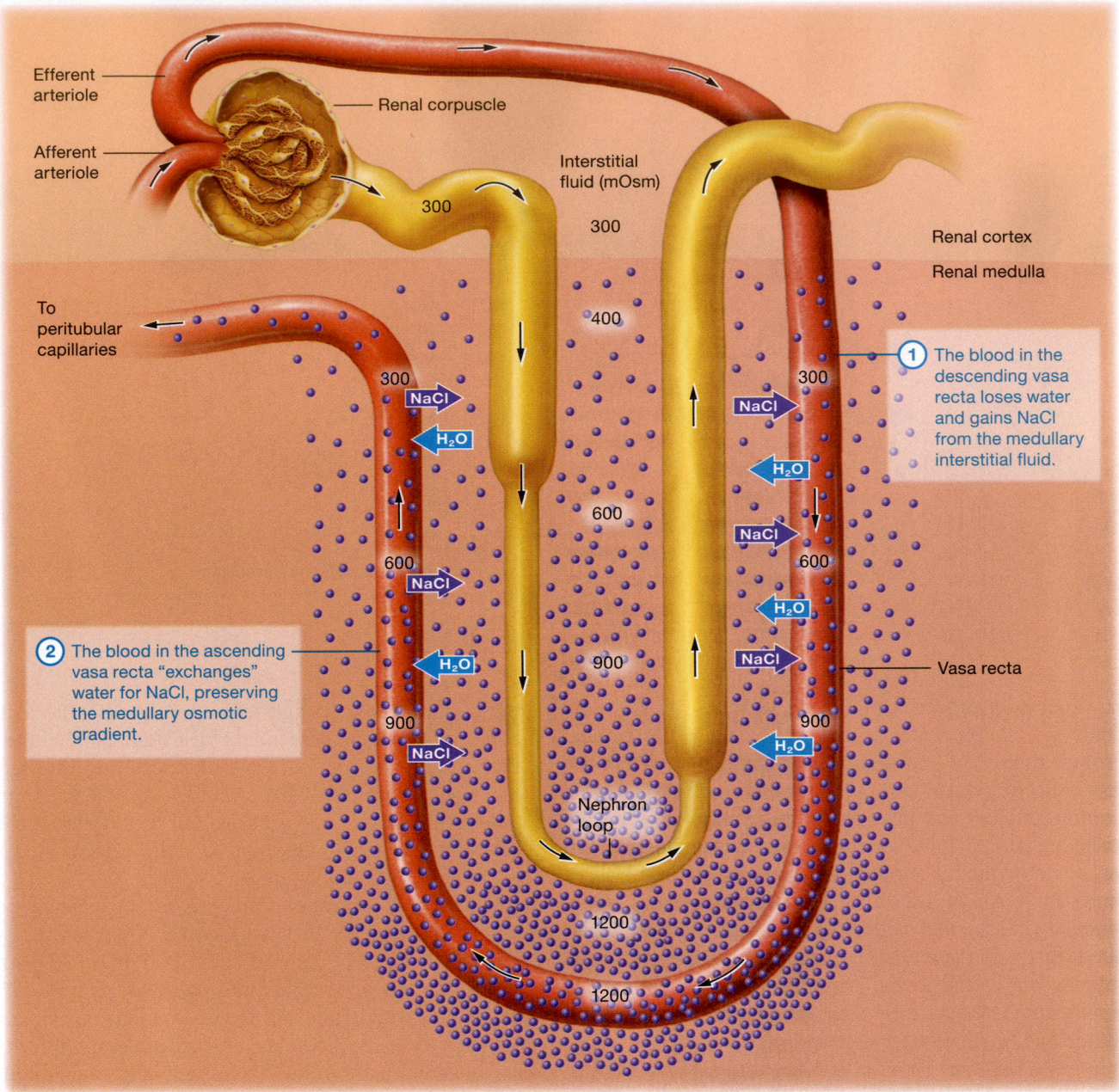

Figure 24.22 Maintenance of the medullary osmotic gradient by the vasa recta and the countercurrent exchanger.

Take note of the "big picture" of what happens in the vasa recta: The descending vasa recta pick up NaCl and lose water, and then the ascending vasa recta "exchange" these two substances by picking up water and losing NaCl. In the end, the cells of the renal medulla obtain oxygen and nutrients but we maintain the medullary osmotic gradient established by the countercurrent mechanism and urea recycling. Note that Figure 24.22 is a simplified model, as numerous vasa recta and nephrons are actually in proximity to one another in the kidney.

How the Countercurrent Mechanism Produces Concentrated Urine

The countercurrent system is complicated, so let's take time to summarize its function. First, keep in mind that the entire function of this system is to conserve water for the body when needed. Water can be reabsorbed from collecting ducts only if ADH is present and an osmotic gradient is there to drive osmosis. This medullary osmotic gradient doesn't exist on its own, so the kidneys must create and maintain it in the following ways:

- The countercurrent multiplier of the thick ascending limb establishes the medullary interstitial gradient by pumping NaCl into the interstitial fluid.
- Continued solute reabsorption, including urea recycling, from the filtrate in the medullary collecting duct adds to the gradient.
- The countercurrent exchanger of the vasa recta allows perfusion of the inner medulla while maintaining the medullary interstitial gradient.

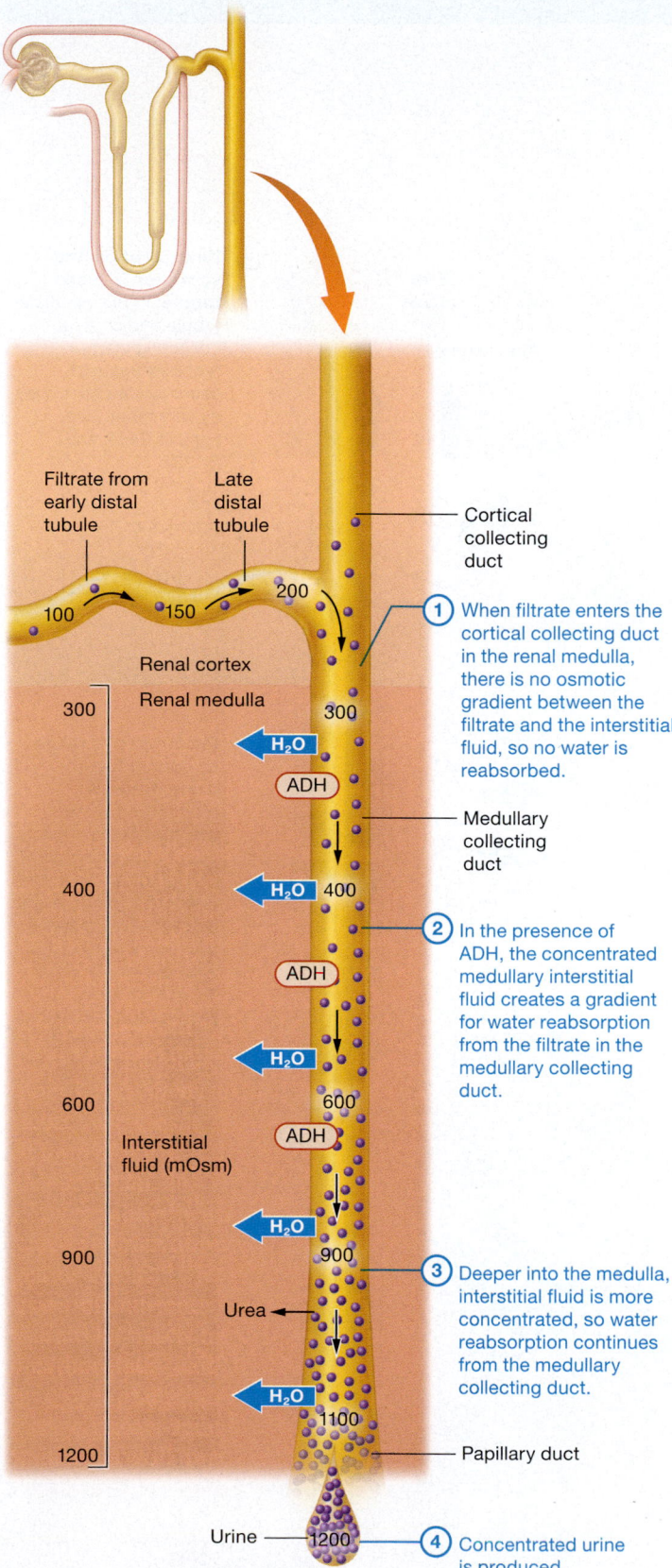

Figure 24.23   Formation of concentrated urine.

Play Animation
@ Mastering Anatomy & Physiology

Figure 24.23 shows how these combine to produce concentrated urine:

(1) **When filtrate enters the cortical collecting duct in the renal medulla, there is no osmotic gradient between the filtrate and the interstitial fluid, so no water is reabsorbed.** Filtrate entering the renal medulla has the same concentration as the interstitial fluid, so no osmotic gradient is present to drive water reabsorption.

(2) **In the presence of ADH, the concentrated medullary interstitial fluid creates a gradient for water reabsorption from the filtrate in the medullary collecting duct.** The interstitial fluid in the renal medulla is more concentrated than the filtrate. So as the filtrate passes deeper into the renal medulla through the medullary collecting duct, if ADH is present, water is drawn into the interstitial fluid by osmosis, thanks to the aquaporins in the principal cells and the medullary osmotic gradient. Water is reabsorbed until the filtrate and interstitial fluid are iso-osmotic.

(3) **Deeper into the medulla, interstitial fluid is more concentrated, so water reabsorption continues from the medullary collecting duct.** The process continues because the interstitial fluid becomes progressively more concentrated in the deep renal medulla, allowing continued water reabsorption.

(4) **Concentrated urine is produced.** This process produces urine with a concentration up to 1200 mOsm. We cannot make more concentrated urine, because after this point there is no longer a gradient to drive osmosis.

Find out what happens to urine concentration when too much ADH is secreted in *A&P in the Real World: SIADH* on p. 983.

Quick Check

☐ 5. What three factors allow the kidney to produce and maintain the medullary osmotic gradient?

☐ 6. How is concentrated urine produced?

Apply What You Learned

☐ 1. Alcohol inhibits the release of ADH. Predict how this inhibition will influence urine volume and concentration.

☐ 2. Drugs called *loop diuretics* block the $Na^+/K^+/2Cl^-$ symporters in the thick ascending limb of the nephron loop.

   a. Predict the effect these drugs will have on the gradient in the renal medulla, and the resulting effect on water reabsorption from the medullary collecting system. How will this impact urine volume?

   b. Predict what might happen if the drug instead had the opposite effect—it caused *more* sodium and chloride ions to be reabsorbed into the medullary interstitial fluid.

See answers in Appendix A.

The Big Picture of Renal Physiology

Figure 24.24

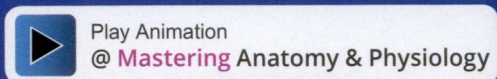

Play Animation @ **Mastering** Anatomy & Physiology

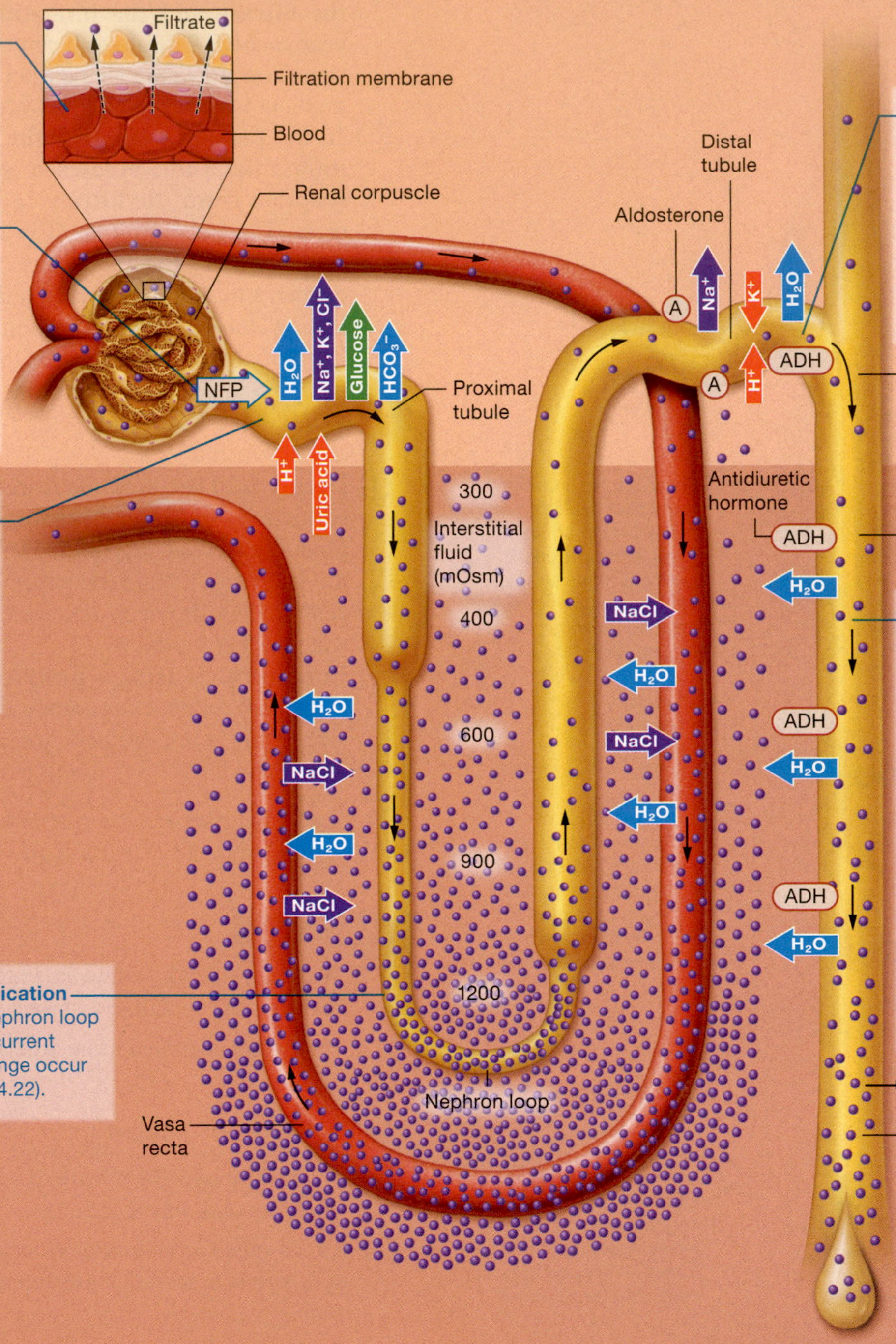

① Glomerular filtration: In the renal corpuscle, filtrate is formed as blood is filtered through the filtration membrane (see Figure 24.12).

Filtrate

Filtration membrane

Blood

Renal corpuscle

② GFR and its regulation: The GFR is determined by the net filtration pressure in the renal corpuscle, which is influenced by many factors, such as angiotensin-II (see Figures 24.13 and 24.14).

NFP

H₂O · Na⁺, K⁺, Cl⁻ · Glucose · HCO₃

H⁺ · Uric acid

Proximal tubule

③ Reabsorption and secretion in proximal tubule: The proximal tubule is the site of extensive tubular reabsorption and select secretion (see Figure 24.19).

300

Interstitial fluid (mOsm)

400

600

900

1200

H₂O

NaCl

H₂O

NaCl

④ Countercurrent multiplication and exchange: In the nephron loop and vasa recta, countercurrent multiplication and exchange occur (see Figures 24.21 and 24.22).

Vasa recta

Nephron loop

NaCl

H₂O

NaCl

H₂O

NaCl

H₂O

Distal tubule

Aldosterone

A · Na⁺ · K⁺ · H₂O

A · H⁺ · ADH

Antidiuretic hormone

ADH

H₂O

ADH

H₂O

ADH

H₂O

ADH

H₂O

⑤ Reabsorption and secretion in distal tubule: In the late distal tubule and cortical collecting duct, reabsorption and secretion are controlled by hormones (see Figures 24.20 and 24.23).

Cortical collecting duct

Renal cortex
Renal medulla

Medullary collecting duct

⑥ Production of dilute or concentrated urine: Water is reabsorbed in the medullary collecting duct in the presence of ADH and the medullary concentration gradient (shown here). Water is not reabsorbed in the absence of ADH. The amount of water reabsorbed determines whether dilute or concentrated urine will be produced (see Figure 24.23).

Urea

Urine

Syndrome of inappropriate ADH secretion, or **SIADH,** is characterized by the secretion of excess antidiuretic hormone. SIADH has multiple causes, including ADH-secreting tumors (certain lung tumors are especially prone to this). Regardless of the cause, the result is the same: excessive fluid retention leading to a decreased plasma osmolarity and a high urine osmolarity.

MODULE 24.7
Putting It All Together: The Big Picture of Renal Physiology

Learning Outcomes

1. Describe the overall process by which blood is filtered and filtrate is modified to produce urine.

Let's now take a big picture look at all the processes involved in renal physiology. We can break these processes down into six broad steps outlined in **Figure 24.24**. Note that Figure 24.24 shows the production of concentrated urine only.

Apply What You Learned

☐ 1. Glomerulonephritis, or inflammation of the glomerulus, results in excessively leaky glomerular capillaries and damaged glomeruli. The damaged and destroyed glomeruli cause the GFR to decrease. Which compensatory mechanisms would you expect to be triggered, and what effects would they have?

☐ 2. Certain diuretics block the effects of carbonic anhydrase in the proximal tubule. Predict the effects these drugs would have on the pH of the blood. How might the kidneys compensate for this?

See answers in Appendix A.

MODULE 24.8
Urine and Renal Clearance

Learning Outcomes

1. Explain how the physical and chemical properties of a urine sample are determined and relate these properties to normal urine composition.

2. Explain how filtration, reabsorption, and secretion determine the rate of excretion of any solute.

3. Explain how renal clearance rate can be used to measure the GFR.

In the previous modules, we discussed how blood is filtered and how filtrate is modified as it flows through the nephron and collecting system. Now we look at the final product that exits the papillary ducts and is excreted by the body: the urine. In this module, we examine the components of normal urine and how abnormalities in the urine can signify problems in the kidneys and elsewhere in the body. We also explore another way in which the health of the kidneys may be assessed, through a value known as *renal clearance.*

Urine Composition and Urinalysis

As you learned in earlier modules, urine is the fluid that remains after tubular reabsorption and secretion have taken place. It normally contains water; sodium, potassium, chloride, and hydrogen ions; phosphates; sulfates; and metabolic waste products such as urea, creatinine, ammonia, and uric acid. It may also contain trace amounts of bicarbonate, calcium, and magnesium ions.

Certain conditions cause abnormalities in the urine, so an analysis of the urine, or **urinalysis,** can be a valuable tool in the diagnosis of disease. The variables examined in a typical urinalysis are as follows:

- **Color.** Urine is colored by a yellow pigment called *urochrome* (YOOR-oh-krohm), a breakdown product of hemoglobin. The more concentrated the urine, the less water is present, so the same amount of urochrome makes the urine darker yellow. The color of urine may also be altered by certain foods, vitamins, drugs, and food dyes, or by the presence of blood.

- **Translucency.** Regardless of color, urine should always be translucent (you should be able to see light through it). Cloudy urine typically indicates an infection but may also indicate that the urine contains large quantities of protein.

- **Odor.** Recently voided urine should have a mild odor. If urine is allowed to sit out, however, bacteria metabolize the urea in the urine to produce ammonia, giving it a stronger odor. The odor of urine may also be altered by certain disease states, such as diabetes mellitus or infection, or by eating certain foods, such as asparagus.

- **pH.** The pH of urine is normally around 6.0—slightly acidic—but it can range from 4.5 to 8.0. The reason for the minimum pH of 4.5 is that the hydrogen ion transport pumps in the distal tubule and collecting system cannot pump against a higher hydrogen ion gradient than this.

- **Specific gravity.** Specific gravity compares the amount of solutes in a solution to the amount in deionized water. Deionized water has no solutes and is assigned a specific gravity of 1.0; urine has solutes, so its specific gravity will be higher than that of water. This value typically ranges from 1.001 (very dilute urine) to 1.035 (very concentrated urine).

Other properties of urine, such as the presence and relative levels of certain solutes, are analyzed with urinalysis test strips. These strips are dipped in the urine, and their chemical indicators change color in the presence of specific substances. We commonly test for blood, protein, leukocytes, glucose, and more. If the kidneys are functioning properly, these substances should not be present in urine in significant amounts.

Quick Check

☐ 1. What is the normal composition of urine?

☐ 2. What are its normal characteristics?

Renal Clearance

Evaluation of renal function is very important in clinical settings to monitor the health of the kidneys. Although urinalysis may provide valuable clues about renal function, a more complete assessment is provided by measuring the rate at which the kidneys remove a substance from the blood, a process known as **renal clearance.** The renal clearance of the chemical is then used to estimate the GFR. Renal clearance and the GFR both are measured in the same units: milliliters of plasma per minute.

For a substance to provide an accurate measure of renal clearance and the GFR, the substance should be completely filtered and neither reabsorbed nor secreted. Substances secreted by renal tubules have a renal clearance greater than their GFR, whereas those that are reabsorbed have a renal clearance less than their GFR. To measure renal clearance, we therefore have a limited group of substances from which to choose. Two such commonly used substances are creatinine and inulin.

Creatinine, as you learned earlier, is a waste product of the metabolism of muscle and other cells. Nearly all of the creatinine produced is excreted by the kidneys. When the kidneys are impaired, the level of creatinine in the blood tends to rise. Generally, a plasma creatinine level above 1.2 mg/dl (milligrams per deciliter) is considered abnormal, but this varies with age, sex, and body mass. Creatinine excretion may be used to estimate the GFR by comparing the amount of creatinine excreted in the urine to the plasma concentration of creatinine. The main difficulty with using creatinine as an indicator of glomerular filtration is that between 15% and 50% of creatinine in the urine arrived there via secretion, not filtration. So a patient with a very low GFR may still have a nearly normal result from this test.

A more accurate assessment of the GFR can be obtained using the substance *inulin* (IN-yoo-lin). Inulin (not to be confused with the hormone insulin) is a complex carbohydrate found in plants such as garlic and artichokes that is filtered by the glomerulus, but is neither reabsorbed nor secreted by the renal tubule or collecting system. The GFR may be measured by injecting inulin and comparing its excretion in the urine to its plasma concentration.

Quick Check

☐ 3. What is renal clearance, and what is it used to estimate?

Apply What You Learned

☐ 1. Metabolic acidosis is characterized by a decreased blood pH from the accumulation of metabolic acids. Predict the effects this condition will have on the pH of the urine. What effect would you expect the opposite condition, metabolic alkalosis, to have on the urine pH?

☐ 2. Dietary supplementation with creatine phosphate is popular among athletes for its supposed performance-enhancing effects. What effect would creatine phosphate supplementation have on the amount of creatinine in the blood, and therefore the amount of creatinine that the kidneys must excrete?

See answers in Appendix A.

MODULE **24.9**

Urine Transport, Storage, and Elimination

Learning Outcomes

1. Describe the structure and functions of the ureters, urinary bladder, and urethra.

2. Relate the anatomy and histology of the bladder to its function.

3. Compare and contrast the male and female urinary tracts.

4. Describe the micturition reflex.

5. Describe voluntary control of micturition.

Newly formed urine drains from the papillary ducts to minor calyces, then to major calyces, and finally to a renal pelvis. The renal pelves then drain urine into the organs of the urinary tract. In this module, we examine the structure and function of the organs of the urinary tract. We conclude with a look at the process by which urine is expelled from the body, *micturition* (mik-choo-RISH-un).

Anatomy of the Urinary Tract

« FLASHBACK

1. What are the properties of transitional epithelium? (p. 135)

The urinary tract consists of the two ureters, the urinary bladder, and the urethra (**Figure 24.25a**). The upcoming subsections examine their anatomy and histology and the differences between the male and female urinary tracts.

Ureters

The **ureters** transport urine from the kidneys to the urinary bladder. The ureters are generally about 25–30 cm long and 3–4 mm

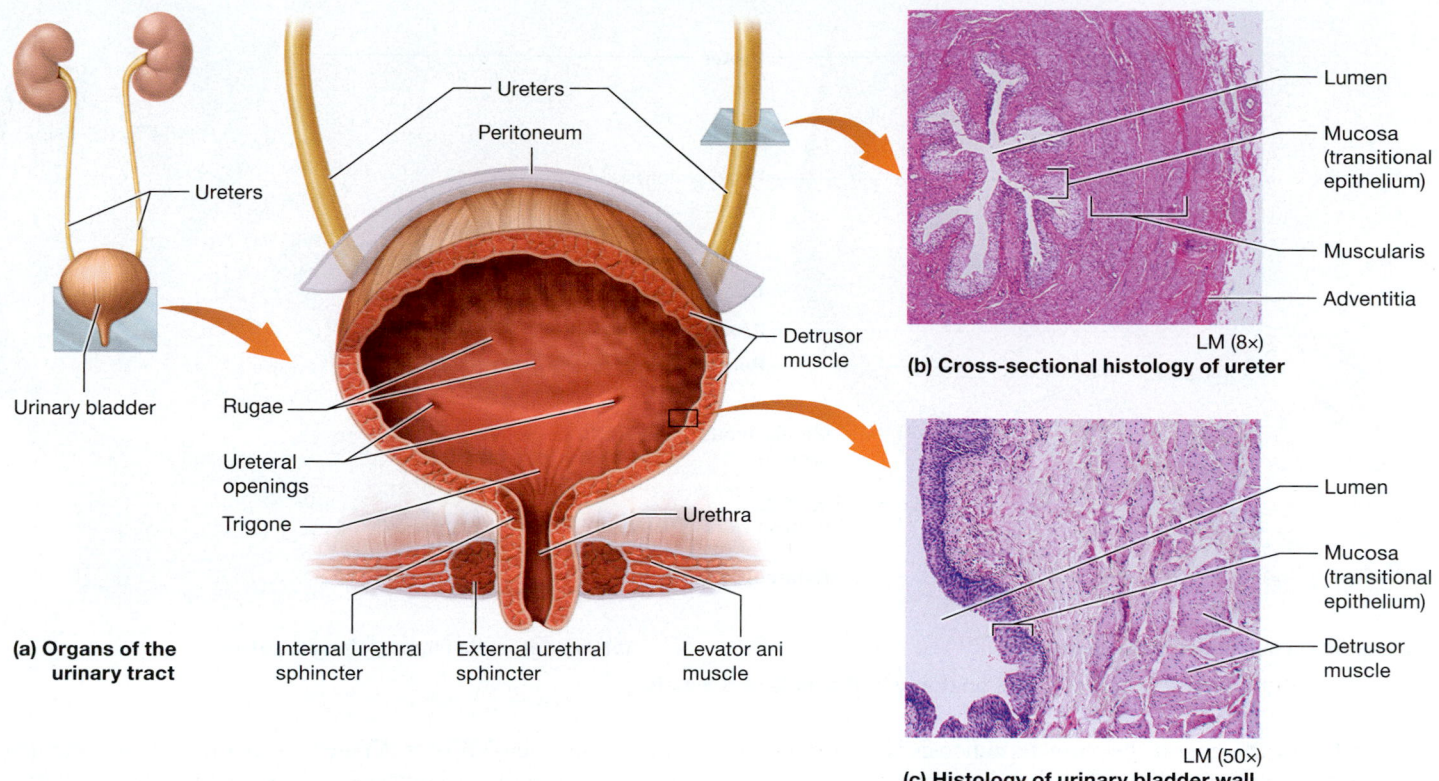

Figure 24.25 **Anatomy of the urinary tract.**

in diameter in an adult. They begin at roughly the level of the second lumbar vertebra, travel behind the peritoneum, and empty into the urinary bladder.

Like any hollow organ, the ureters have a multilayered wall. The three layers are named as follows, from superficial to deep (**Figure 24.25b**):

1. **Adventitia.** This superficial layer, known as the **adventitia** (ad-ven-TISH-uh), is fibrous connective tissue that supports the ureters.
2. **Muscularis.** The smooth muscle–containing middle layer is the **muscularis.** Like the smooth muscle of the alimentary canal, the smooth muscle in the muscularis contracts rhythmically via *peristalsis.* Waves of peristalsis course through the muscularis as often as five times per minute, depending on the rate of urine production, to propel urine toward the urinary bladder.
3. **Mucosa.** The innermost layer is the **mucosa,** a mucous membrane composed of transitional epithelium and its underlying basal lamina. As you learned in Chapter 4, transitional epithelium is stratified with cells that can change from having a dome shape to being squamous. This property allows the epithelium to expand and recoil.

The ureters drain into the posterior, inferior urinary bladder. In this region a mechanism prevents urine from flowing backward through the ureter. As each ureter passes along the posterior urinary bladder, it travels obliquely through a "tunnel" in the bladder wall. As urine collects in the bladder, the pressure rises and compresses this tunnel, pinching the ureter closed and preventing backflow of urine.

Urinary Bladder

The **urinary bladder** is a hollow, distensible organ that sits on the floor of the pelvic cavity, suspended by a fold of parietal peritoneum. It collapses when empty, but when distended, it becomes pear-shaped, and can hold up to about 700–800 ml of urine in both sexes (though it doesn't necessarily expand to full capacity in females because of the position of the uterus).

Like the ureters, the wall of the urinary bladder has three tissue layers (**Figure 24.25c**). From superficial to deep, these layers are as follows:

1. **Adventitia.** The adventitia, the most superficial layer, is composed of areolar connective tissue. On the superior surface of the urinary bladder is an additional serosa, which is a fold of the parietal peritoneum.
2. **Detrusor muscle.** The middle tissue layer is composed of smooth muscle known as the **detrusor muscle** (dee-TROO-sohr; "to push down"). The muscle fibers are arranged into inner longitudinal, middle circular, and outer longitudinal layers. The detrusor muscle forms a circular band around the opening of the urethra, called the **internal urethral sphincter,** shown in Figure 24.25a.
3. **Mucosa.** The mucosa is composed of transitional epithelium with an underlying basal lamina. It is a mucous membrane that produces mucus, protecting the bladder epithelium from urine. When the bladder is not full, folds of mucosa called *rugae* (ROO-ghee) are visible. What would happen if this mucus were not present? See *A&P in the Real World: Interstitial Cystitis* on p. 986 to find out.

Notice that the floor of the urinary bladder contains a triangular area called the **trigone** (TRY-gohn; "triangle"). The trigone lacks

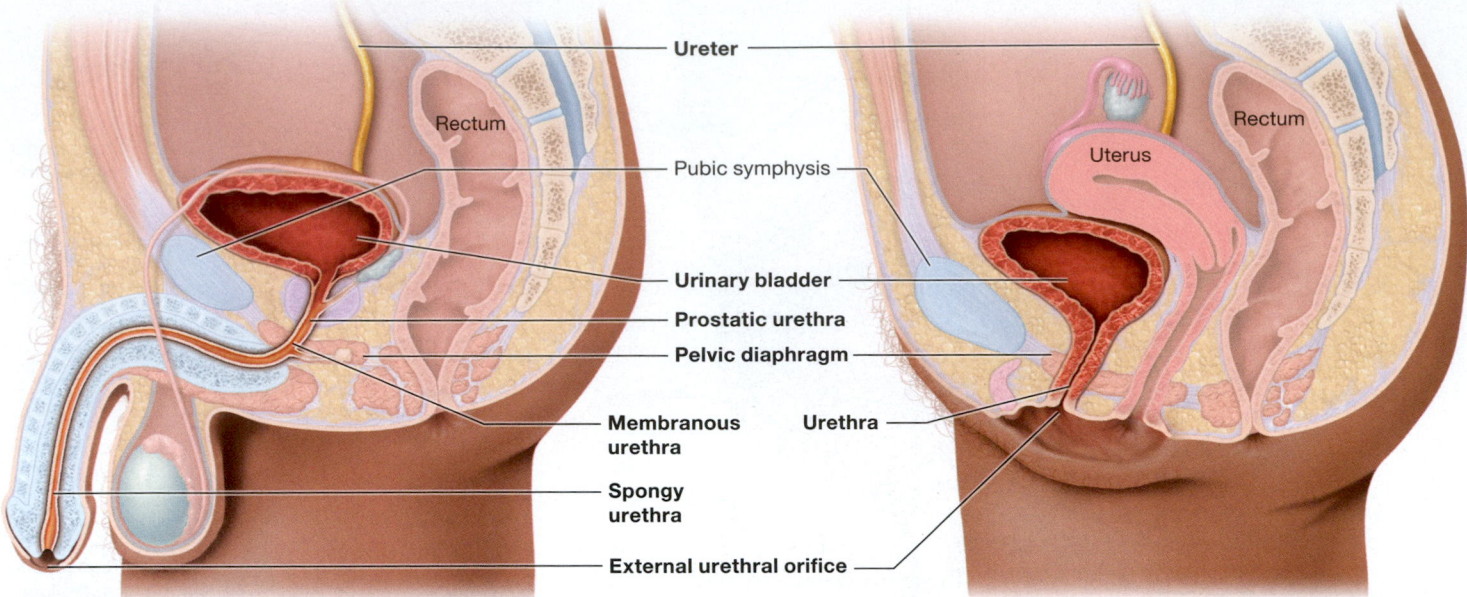

(a) Sagittal section through male pelvis

(b) Sagittal section through female pelvis

Figure 24.26 Comparison of urinary tract anatomy in the male and female.

rugae and appears smooth because its mucosa is tightly bound to the underlying muscularis. The two posterior corners of the trigone are formed by the two ureteral orifices (openings). These orifices have mucosal flaps that act as valves to prevent backflow of urine during elimination. The apex of the trigone is formed by the opening to the urethra, the **internal urethral orifice.**

Notice in **Figure 24.26** that the position of the urinary bladder differs in males and females. In males, it is anterior to the rectum; in females, it is anterior to the vagina and inferior to the uterus.

Urethra

The **urethra** is the terminal portion of the urinary tract; it drains urine from the urinary bladder to the outside of the body. Like the rest of the urinary tract, the urethra has an outer adventitia, middle muscularis, and inner mucosa. It begins at the internal urethral orifice in the urinary bladder, which is surrounded by the internal urethral sphincter. This sphincter remains closed unless urine is being eliminated. A second urethral sphincter, the **external urethral sphincter,** is formed from the *levator*

ani muscle (leh-VAY-ter AY-nee; sometimes called the *urogenital diaphragm*), the muscular floor of the pelvic cavity. This sphincter is composed of skeletal muscle and is under voluntary control. There is no sphincter at the external orifice to the urethra in either males or females.

The urethra differs in males and females structurally, as shown in Figure 24.26, which means that functional differences must exist as well. The female urethra is shorter (about 4 cm in length) and opens at the **external urethral orifice** between the vagina and the clitoris. It serves exclusively as a passage for urine.

The male urethra is considerably longer (about 20 cm in length) and consists of three regions:

1. **Prostatic urethra.** As the urethra exits the urinary bladder, it passes through the *prostate gland,* which sits inferior to the urinary bladder. This section of the urethra is called the **prostatic urethra.**
2. **Membranous urethra.** The shortest segment, known as the **membranous urethra,** passes through the levator ani muscle.
3. **Spongy urethra.** The longest segment of the male urethra, the **spongy urethra,** passes through the penis to the external urethral orifice. It is called the spongy urethra because it passes through an erectile body of the penis called the *corpus spongiosum.*

The location of the male urethra allows it to serve a dual purpose: It transports both urine and semen (see Chapter 26).

Quick Check

☐ 1. What are the three tissue layers of the organs of the urinary tract?

☐ 2. What are the functions of the ureters and urinary bladder?

☐ 3. How does the urethra differ structurally and functionally in males and females?

Interstitial Cystitis

Inadequate mucus production by the mucosa of the urinary bladder can cause the disease *interstitial cystitis* (sis-TY-tis; *IC*). This condition allows the acid and other toxic substances in urine to damage the underlying mucosa and other tissues. IC is characterized by frequent urination and pelvic pain resulting from mucosal ulcerations. The underlying cause of IC is unknown, and unfortunately it responds poorly to treatment. Over the long term, IC can lead to bladder scarring and contraction.

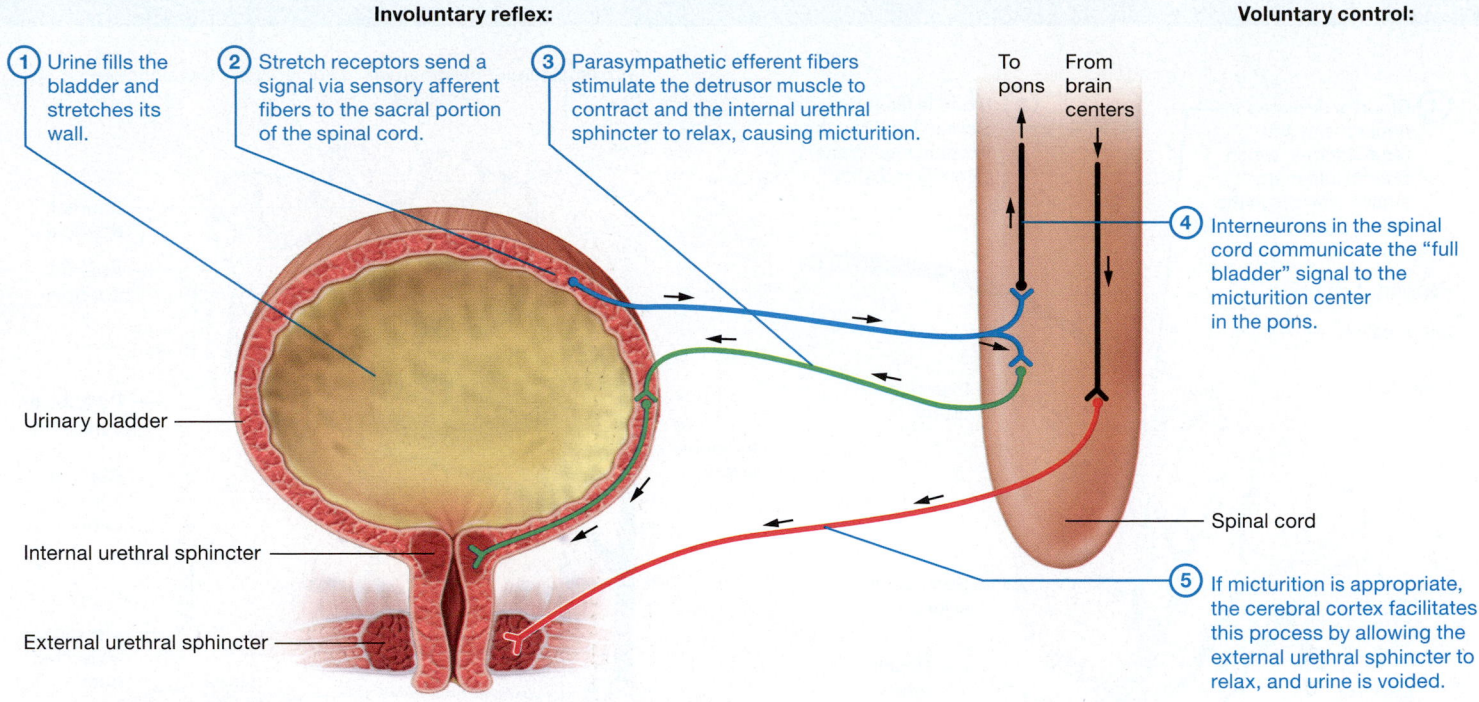

Involuntary reflex:

(1) Urine fills the bladder and stretches its wall.

(2) Stretch receptors send a signal via sensory afferent fibers to the sacral portion of the spinal cord.

(3) Parasympathetic efferent fibers stimulate the detrusor muscle to contract and the internal urethral sphincter to relax, causing micturition.

Voluntary control:

To pons   From brain centers

(4) Interneurons in the spinal cord communicate the "full bladder" signal to the micturition center in the pons.

Urinary bladder

Internal urethral sphincter

External urethral sphincter

Spinal cord

(5) If micturition is appropriate, the cerebral cortex facilitates this process by allowing the external urethral sphincter to relax, and urine is voided.

Figure 24.27 Micturition.

Micturition

《 FLASHBACK

1. What are afferent and efferent neurons? (p. 476)
2. What is a reflex arc? (p. 506)

Micturition, also called urination or voiding, is the discharge of urine from the urinary bladder to the outside of the body. It is activated by a reflex arc called the **micturition reflex,** which is mediated by the parasympathetic nervous system. This reflex arc is carried out by three components: (1) stretch receptors in the wall of the urinary bladder, (2) sensory afferent nerve fibers that convey this information to the sacral portion of the spinal cord (S2 and S3), and (3) parasympathetic efferent fibers that travel to the detrusor muscle and internal urethral sphincter.

Micturition is shown in **Figure 24.27.** When urine fills the bladder and stretches its walls (1), the micturition reflex occurs. In infants and young children, this involuntary reflex is the primary mechanism by which the urinary bladder is emptied. This reflex arc then proceeds as follows:

(2) Stretch receptors send a signal via sensory afferent fibers to the sacral portion of the spinal cord.

(3) Parasympathetic efferent fibers stimulate the detrusor muscle to contract and the internal urethral sphincter to relax, causing micturition.

As children mature, however, pathways develop between these parasympathetic neurons and the brain that allow control over the external urethral sphincter. At this point, micturition is predominantly controlled by the **micturition center** in the pons, making the process voluntary. When the bladder is full, two

additional steps then occur simultaneously with the involuntary process (see Figure 24.27):

(4) Interneurons in the spinal cord communicate the "full bladder" signal to the micturition center in the pons.

(5) If micturition is appropriate, the cerebral cortex facilitates this process by allowing the external urethral sphincter to relax, and urine is voided.

If micturition is not appropriate, then the detrusor muscle relaxes, the internal and external urethral sphincters remain closed, and the urge to urinate passes. The reflex generally initiates again within about an hour. This cycle repeats until the sensation of having to urinate becomes more acute. By the time about 500–600 ml has accumulated in the urinary bladder, the urge to urinate becomes too strong, voluntary control over the external urethral sphincter is lost, and micturition occurs. After micturition, the bladder contains only about 10 ml of urine.

Quick Check

☐ 4. What are the steps of the micturition reflex?

☐ 5. How is micturition consciously controlled?

Apply What You Learned

☐ 1. Predict what would happen if the epithelium of the urinary tract were made of simple squamous epithelium instead of transitional epithelium.

☐ 2. How would a spinal cord injury above the level of S1–S2 affect micturition in that patient? How would the situation change if the injury were below S1–S2?

See answers in Appendix A.

The Big Picture of Urine Formation, Storage, and Elimination

Figure 24.28

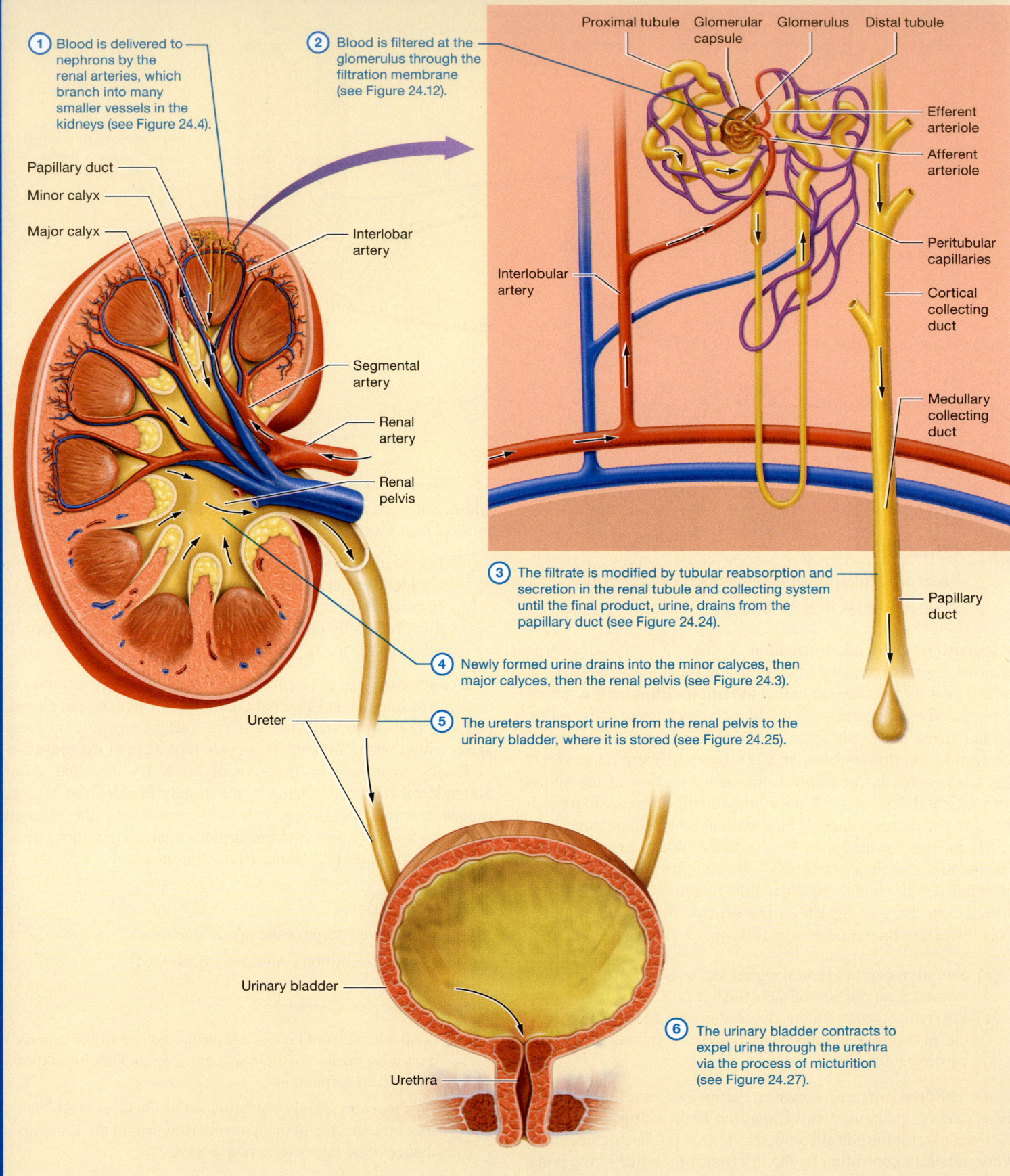

1. Blood is delivered to nephrons by the renal arteries, which branch into many smaller vessels in the kidneys (see Figure 24.4).

2. Blood is filtered at the glomerulus through the filtration membrane (see Figure 24.12).

Proximal tubule
Glomerular capsule
Glomerulus
Distal tubule

Papillary duct
Minor calyx
Major calyx

Interlobar artery

Efferent arteriole

Afferent arteriole

Interlobular artery

Segmental artery

Peritubular capillaries

Cortical collecting duct

Renal artery

Renal pelvis

Medullary collecting duct

3. The filtrate is modified by tubular reabsorption and secretion in the renal tubule and collecting system until the final product, urine, drains from the papillary duct (see Figure 24.24).

Papillary duct

4. Newly formed urine drains into the minor calyces, then major calyces, then the renal pelvis (see Figure 24.3).

5. The ureters transport urine from the renal pelvis to the urinary bladder, where it is stored (see Figure 24.25).

Ureter

Urinary bladder

6. The urinary bladder contracts to expel urine through the urethra via the process of micturition (see Figure 24.27).

Urethra

MODULE **24.10**

Putting It All Together: The Big Picture of Urine Formation, Storage, and Elimination

Learning Outcomes

1. Describe the overall process by which urine is formed, stored, and eliminated.

Now that we have examined all of the processes and organs of the urinary system, let's look at one final big picture view so that we can trace the process of urine formation, storage, and elimination. The overall process is shown in **Figure 24.28**, with references back to individual figures throughout the chapter that you can consult for more details.

Apply What You Learned

☐ 1. Rory presents with a *staghorn calculus*, a huge kidney stone that forms in the renal pelvis and extends into the major and minor calyces. Rory's staghorn calculus is preventing urine from draining from his renal pelvis into his ureter. How do you think this is affecting the GFR of the affected kidney? Why? (Hint: How do you think a staghorn calculus would affect the capsular hydrostatic pressure? What would that do to net filtration pressure and thus the GFR?)

See answers in Appendix A.

Chapter Summary

For everything you need to succeed in this course, go to Mastering Anatomy & Physiology. There you will find:

- Practice Tests
- Author Podcasts
- Big Picture Animations
- Concept Boost Video Tutors
- **iP2™** Interactive Physiology 2.0 Tutorials
- **A&PFlix** A&P Flix 3D Animations
- Active-Learning Workbook

MODULE **24.1**

Overview of the Urinary System 947–949

- The organs of the **urinary system** are the **kidneys** and the ureters, urinary bladder, and urethra.
- The kidneys remove waste products from the blood and produce **urine;** maintain fluid, electrolyte, and acid-base balance, as well as blood pressure; regulate erythrocyte production; and perform other metabolic functions.

MODULE **24.2**

Anatomy of the Kidneys 949–956

- The kidneys are paired retroperitoneal organs with three layers of protective coverings.
- Each kidney consists of over one million **nephrons.**

- Internally, the kidney consists of three regions: the superficial **renal cortex,** the middle **renal medulla,** and the innermost **renal pelvis.**
- The kidney is supplied by the large **renal artery.** The artery branches multiple times, and in the renal cortex, branches give off tiny **afferent arterioles,** each of which feeds a **glomerulus,** which is then drained by an **efferent arteriole.** The efferent arteriole feeds a second capillary bed, the **peritubular capillaries.** These vessels lead into a series of veins that drain into the **renal vein.**
- Nephrons have two main parts:
  ○ The **renal corpuscle,** which consists of the **glomerulus** and the surrounding **glomerular capsule;** and
  ○ The renal tubule, which consists of the **proximal tubule,** the **nephron loop,** and the **distal tubule.**
- The **juxtaglomerular apparatus** is a specialized region of the nephron where the **macula densa** comes into contact with the **JG cells** of the afferent arteriole.
- Nephrons drain into the **collecting system,** which consists of the *cortical collecting ducts* and the *medullary collecting system.*
- There are two types of nephrons: **cortical nephrons,** which have short nephron loops, and **juxtamedullary nephrons,** which have long nephron loops.

MODULE **24.3**

Overview of Renal Physiology 957

- Nephrons carry out three basic physiological processes:
  ○ **Glomerular filtration,** resulting in the production of **filtrate;**
  ○ **Tubular reabsorption,** reclaiming solutes and water from the filtrate; and
  ○ **Tubular secretion,** transporting substances from the blood into the filtrate.

MODULE 24.4
Renal Physiology I: Glomerular Filtration
957–966

- The **filtration membrane** includes the glomerular capillary endothelial cells, the basal lamina of the glomerulus, and the podocytes (visceral layer of the glomerular capsule). It permits only water and small solutes to pass into the filtrate.

- The rate at which filtrate is formed is called the **glomerular filtration rate (GFR);** it averages about 125 ml/min. The GFR is determined by the net filtration pressure (NFP) in the glomerulus, which averages about 10 mm Hg.

- Maintenance of the GFR is crucial to homeostasis of fluid, electrolyte, and acid-base balance.
  - The kidneys can autoregulate the GFR by the **myogenic mechanism** and *tubuloglomerular feedback*, which modify the diameter of the glomerular afferent arterioles.
  - The **renin-angiotensin-aldosterone system** (**RAAS**) works to maintain the GFR as part of the larger process of maintaining systemic blood pressure. The system culminates in the formation of the hormone **angiotensin-II.**
  - **Atrial natriuretic peptide** increases the GFR by dilating the afferent arteriole.
  - The sympathetic nervous system may maintain or decrease the GFR, depending on the level of stimulation.

MODULE 24.5
Renal Physiology II: Tubular Reabsorption and Secretion 966–975

- Substances may be reabsorbed by passing *between* the tubule cells (the **paracellular route**) or *through* them (the **transcellular route**). Substances are secreted only through the transcellular route.

- Reabsorption and secretion occur in the renal tubule and collecting system:
  - In the proximal tubule, sodium ion reabsorption drives **obligatory water reabsorption,** reabsorption of many solutes, and hydrogen ion secretion. About 65% of the water, much of the HCO_3^-, nearly all the amino acids and glucose, and variable amounts of electrolytes are reabsorbed here.
  - In the nephron loop, another 15–20% of the water and 20–25% of the NaCl are reabsorbed.
  - Reabsorption and secretion in the late distal tubule and cortical collecting ducts are controlled by hormones:
  - **Aldosterone** causes reabsorption of sodium ions and secretion of potassium and hydrogen ions.
  - **Antidiuretic hormone** (**ADH**) stimulates the reabsorption of water through the formation of **aquaporin** channels.
  - The medullary collecting system secretes hydrogen ions, reabsorbs many electrolytes, and reabsorbs most of the remaining water from the filtrate under the influence of ADH.
  - The pH of the blood is regulated by adjusting the secretion of hydrogen ions and the reabsorption of bicarbonate ions.

MODULE 24.6
Renal Physiology III: Regulation of Urine Concentration and Volume 975–983

- **Facultative water reabsorption** in the late distal tubule and collecting system determines the concentration and volume of urine that the kidneys produce.

- Dilute urine is produced by decreasing the release of ADH, which causes less water to be reabsorbed.

- The ability to produce concentrated urine relies on ADH and a **medullary osmotic gradient** to drive the reabsorption of water by osmosis.
  - The medullary osmotic gradient is established by the **countercurrent multiplier** in the nephron loops of juxtamedullary nephrons.
  - Urea diffuses into the interstitial fluid from the medullary collecting system, which contributes to the medullary osmotic gradient.
  - The vasa recta maintain the medullary osmotic gradient by acting as a **countercurrent exchanger.**
  - The result of the medullary osmotic gradient in the presence of ADH is continued water reabsorption and a low volume of concentrated urine.

MODULE 24.7
Putting It All Together: The Big Picture of Renal Physiology 983

- The big picture of renal physiology is shown in Figure 24.24.

MODULE 24.8
Urine and Renal Clearance 983–984

- Urine is composed of mostly water with solutes such as electrolytes and metabolic wastes.

- Renal function may be assessed by measuring the rate at which certain substances are removed from the blood by the kidney, known as **renal clearance,** as a way to estimate the GFR.

MODULE 24.9
Urine Transport, Storage, and Elimination 984–987

- The ureters, urinary bladder, and urethra comprise the **urinary tract.**
- The **ureters** are muscular tubes that undergo peristalsis to propel urine toward the urinary bladder.
- The wall of the **urinary bladder** consists of the inner **mucosa,** the middle **detrusor muscle,** and the outer **adventitia.**
- The **urethra** contains two sphincters: the involuntary **internal urethral sphincter** and the voluntary **external urethral sphincter.**
- **Micturition** is the process whereby urine is voided from the urinary bladder through the urethra. It occurs via the **micturition reflex,** mediated by the parasympathetic nervous system.

MODULE **24.10**
Putting It All Together: The Big Picture of Urine Production, Storage, and Elimination 988–989

- The big picture of urine formation, storage, and elimination is shown in Figure 24.28.

Assess What You Learned

Scan the QR Code for additional practice test questions

https://goo.gl/rOQzfM

LEVEL 1   Check Your Recall

1. What are the four main organs of the urinary system?

2. Which of the following is not a physiological process carried out by the kidneys?

   a. Blood pressure regulation
   b. Tubular reabsorption
   c. Tubular secretion
   d. Glomerular filtration
   e. All of the above are physiological processes carried out by the kidneys.

3. Mark the following statements as true or false. If a statement is false, correct it to make a true statement.

   a. The kidneys are retroperitoneal and covered by three layers of connective tissue.
   b. Internally, the kidneys consist of an outer renal medulla, a middle renal pelvis, and an inner renal cortex.
   c. The first capillary bed of the kidneys is the peritubular capillaries, which are fed by the afferent arteriole and drained by the efferent arteriole.
   d. Filtrate flows from the renal corpuscle to the distal tubule, the nephron loop, the proximal tubule, and into the collecting system.

4. Cortical and juxtamedullary nephrons differ in the:

   a. lengths of their nephron loops.
   b. structure of the capillaries surrounding them.
   c. structure of their renal corpuscles.
   d. Both a and b are correct.
   e. Both b and c are correct.

5. Describe the structure of the filtration membrane.

6. Which of the following substances would pass through the filtration membrane to become part of the filtrate under normal circumstances? (Circle all that apply.)

   a. Sodium ions
   b. Albumin
   c. Glucose
   d. Erythrocytes
   e. Leukocytes
   f. Amino acids
   g. Urea

7. Mark the following statements as true or false. If a statement is false, correct it to make a true statement.

   a. Sodium ions and glucose are cotransported into the proximal tubule cell by secondary active transport.
   b. The distal tubule reabsorbs sodium ions and secretes potassium and hydrogen ions in response to ADH.
   c. Sodium ion reabsorption creates a gradient that helps drive the reabsorption of water and many other solutes from the proximal tubule.
   d. ADH triggers water reabsorption from the nephron loop.
   e. Obligatory water reabsorption occurs in the distal tubule and collecting system.

8. Fill in the blanks for the following statements:

   a. When the GFR decreases, the macula densa releases chemicals to _____ the afferent arteriole.
   b. The sympathetic nervous system _____ the blood vessels supplying the kidney to _____ the glomerular filtration rate.
   c. The enzyme _____ is released by JG cells in response to a decrease in the GFR.
   d. The enzyme _____ converts angiotensin-I to angiotensin-II.
   e. Generally, angiotensin-II _____ systemic blood pressure while _____ the GFR.

9. Which of the following is *false* about the GFR?

   a. The GFR averages about 120 ml/min.
   b. The GFR increases when the afferent arteriole dilates.
   c. The GFR decreases when the efferent arteriole constricts.
   d. The GFR decreases when the afferent arteriole constricts.

10. The route by which substances are reabsorbed by crossing through the cells of the renal tubule and collecting system is known as the:

    a. paracellular route.
    b. transcellular route.
    c. primary active transport route.
    d. facultative route.

11. Fill in the blanks: Glomerular hydrostatic pressure _____ filtration; colloid osmotic pressure and capsular hydrostatic pressure _____ filtration.

    a. favors; favor
    b. opposes; oppose
    c. favors; oppose
    d. opposes; favor

12. Dilute urine is produced when decreased levels of _____ are secreted:

    a. aldosterone
    b. atrial natriuretic peptide
    c. ADH
    d. none of the above

13. Which of the following conditions does not contribute to the creation and/or maintenance of the medullary osmotic gradient?

    a. The countercurrent exchanger of the vasa recta
    b. The countercurrent multiplier of the nephron loops of cortical nephrons
    c. The countercurrent multiplier of the nephron loops of juxtamedullary nephrons
    d. The permeability of the medullary collecting system to urea and other ions

14. Fill in the blanks: The kidneys produce _____ urine when the osmolarity of the body's fluids increases. They produce _____ urine when the osmolarity of the body's fluids decreases.

15. Normal urine should have which of the following properties? Circle all that apply.

   a. Translucency   c. Cloudy appearance
   b. Yellowish pigment   d. pH less than 4.5

16. The GFR may be estimated by measuring the rate at which certain substances are removed from the blood, which is known as:

   a. renal clearance.
   b. plasma creatinine.
   c. glomerular hydrostatic pressure.
   d. inulin estimation.

17. Fill in the blanks for each of the following statements:

   a. The process by which urine is eliminated is called _____, and it is mediated by reflexes involving the _____ nervous system.
   b. The mucosa of the organs of the urinary tract is lined with _____ epithelium.
   c. The three layers of smooth muscle in the urinary bladder are known as the _____ muscle.
   d. The female urethra provides a passageway for _____, whereas the male urethra provides a passageway for _____ and _____.

LEVEL 2 Check Your Understanding

1. Predict the effects the following scenarios would have on glomerular filtration:

   a. Having excess proteins in the blood, increasing colloid osmotic pressure
   b. Having low arterial blood pressure (hypotension)
   c. Having high arterial blood pressure (hypertension)

2. Trace the pathway taken by a molecule of urea through the kidney from the glomerulus to the renal pelvis if the urea is recycled.

3. Explain why urinary tract infections, which involve the urethra and urinary bladder, are much more common in females than males.

4. Why must the kidneys establish a concentration gradient in the interstitial fluid of the renal medulla in order to produce concentrated urine?

LEVEL 3 Apply Your Knowledge

PART A: Application and Analysis

1. Drugs that treat hypertension, or high blood pressure, have the following actions. Discuss the specific effect that each drug will have on the kidneys.

   a. Blocking the action of aldosterone on the kidneys
   b. Blocking the receptor for angiotensin-II on blood vessels and in the renal tubule cells
   c. Blocking the $Na^+/Cl^-/2K^+$ transport pumps in the thick ascending limb of the nephron loop

2. Mr. Wu is a patient with kidney disease who presents to your clinic for monitoring. You notice on his chart that his GFR was estimated through inulin administration to be about 35 ml/min. What does this tell you about the health of his kidneys? Mr. Wu is taking a medication that is normally excreted from the body in the urine. You order blood work and find that the concentration of this medication in his plasma is much higher than normal. How does his decreased GFR explain the elevated level of medication in his plasma?

3. Deana is a 4-year-old girl with a rare genetic defect that causes the Na^+/glucose symporters in the proximal tubule to reabsorb fewer glucose and sodium ions than normal. Predict the effects this defect will have on the composition and volume of Deana's urine. Explain why you would expect to see increased activity of the tubuloglomerular feedback and the renin-angiotensin-aldosterone system in Deana's kidneys.

PART B: Make the Connection

4. Explain how each of the drugs in question 1 from this section would lower blood pressure. (*Connects to Chapter 17*)

5. What might it mean if you found a high concentration of urobilinogen in your patient's urine? (*Hint:* Consider the source of urobilinogen.) (*Connects to Chapter 19*)

See answers in Appendix A.

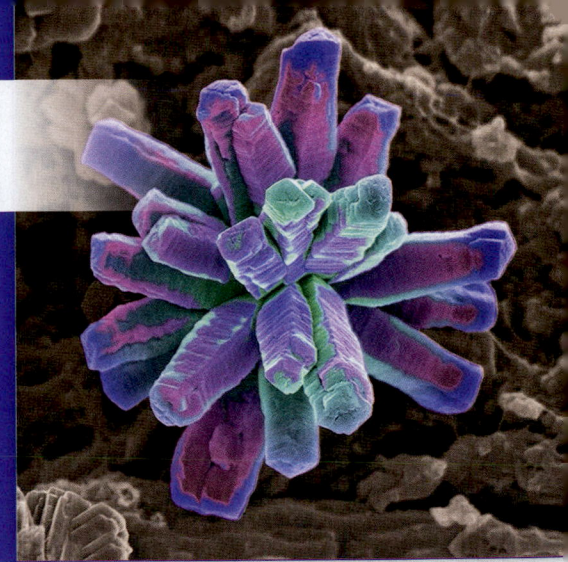

25

Fluid, Electrolyte, and Acid-Base Homeostasis

Y ou have read about the topics of this chapter—body fluids, electrolytes, and acids and bases—multiple times already in this book. We looked at the roles of water in the body in the chemistry chapter (see Chapter 2), the roles of electrolytes with muscle and nervous tissue (see Chapters 10, 11, and 17), acid-base homeostasis with the respiratory system (see Chapter 21), and all three with the digestive and urinary systems (see Chapters 22 and 24). Now it's time to examine these three critical physiological variables on their own so that you can see the overall picture of how they affect the body and how the body works to maintain them within their normal ranges. The upcoming modules survey body fluids, electrolytes, and acid-base balance in detail, including their effects on homeostasis in the presence of an imbalance. The chapter concludes with an example of how all three variables are regulated together.

MODULE **25.1**

Overview of Fluid, Electrolyte, and Acid-Base Homeostasis

Learning Outcomes

1. Explain the concept of balance with respect to fluids and electrolytes and acids and bases.
2. Define the terms body fluid, electrolyte, acid, base, pH scale, and buffer.

Fluids, electrolytes, and acids and bases are not exactly new to you at this point in your study of anatomy and physiology. Still, it's helpful to quickly review the basic principles of these three variables. This module provides such a review before we delve into the details of their regulation. You may also wish to refer back to the chemistry chapter for more background information about these concepts (see Chapter 2).

Body Fluids

The term body fluids is used to encompass all the body's water-based liquids, including blood plasma, interstitial fluid, cytosol, cerebrospinal fluid, lymph, exocrine secretions, and other specialized fluids. Each of these fluids has a slightly different composition, but the main component of each is water. As water is their main component, the study of **fluid balance**—maintaining the appropriate volume and concentration of the body's intracellular and extracellular fluids—is largely the study of water balance.

*For practice applying concepts to a clinical scenario, check out the **Running Case Study** for this chapter @ Mastering Anatomy & Physiology*

Photo: Colored scanning electron micrograph (SEM) showing a calcium phosphate crystal, which can form in the body when phosphate ions are present in a high concentration.

The functions of water in the body's fluids are diverse. As we've described, water is a polar solvent, meaning that compounds with ionic and polar covalent bonds can dissolve in it. This allows water to transport and distribute many different solutes throughout the body. Water also distributes body heat, cushions organs and tissues, and lubricates organs and tissues as they move.

Water obeys the *principle of mass balance*—in other words, what is gained by the body must equal what is lost by the body. Multiple factors impact fluid balance, such as the amount of water ingested, the amount of physical activity, kidney function, medications, and digestive activities. When an imbalance results in inadequate or excess water in the body, the body generally compensates, but if it cannot, serious disturbances to homeostasis can occur. As you will see in the modules that follow, such disturbances affect more than just body fluids—they affect electrolyte and acid-base balance, too, so their effects are wide-ranging.

Quick Check

☐ 1. What is a body fluid?

☐ 2. What is balance with respect to body fluids?

Electrolytes

《 FLASHBACK

1. How is an electrolyte defined? (p. 50)

2. What are the main electrolytes in the body? (p. 936)

As we've discussed, an **electrolyte** is a substance that dissociates into ions when placed in water, whereas a **nonelectrolyte** is a substance that generally has covalent bonds and does not dissociate into charged particles in water. Electrolytes are named for the fact that the resulting charged particles conduct electricity in a solution of water (**Figure 25.1**). This is, of course, why you are warned to never allow electrical devices to come into contact with tap water when you're in it. Tap water contains electrolytes, and so will conduct an electric current from the device. Your body also contains multiple electrolytes, and so this electric current will flow from the device to the tap water and finally into your body, possibly resulting in death.

Throughout this book, you have seen the importance of several electrolytes, particularly sodium, potassium, and calcium ions. Note that we have to call them "ions" because if we didn't include that term, we would be referring to the atomic, metal form. The difference is significant—in their atomic forms, sodium and potassium metals both explode when placed in water, and calcium metal causes a reaction that emits hydrogen gas. In their ionic forms, they simply dissolve into the water of the body's fluids, where they have many important roles. Other examples of electrolytes include chloride ions, acids, bases, and some proteins.

Like water balance, *electrolyte balance* in the body is maintained through the principle of mass balance, by which the amount of electrolytes gained through the diet is equal to that lost from the body via various routes. There are numerous mechanisms, most of which are hormonal, that act to maintain a state

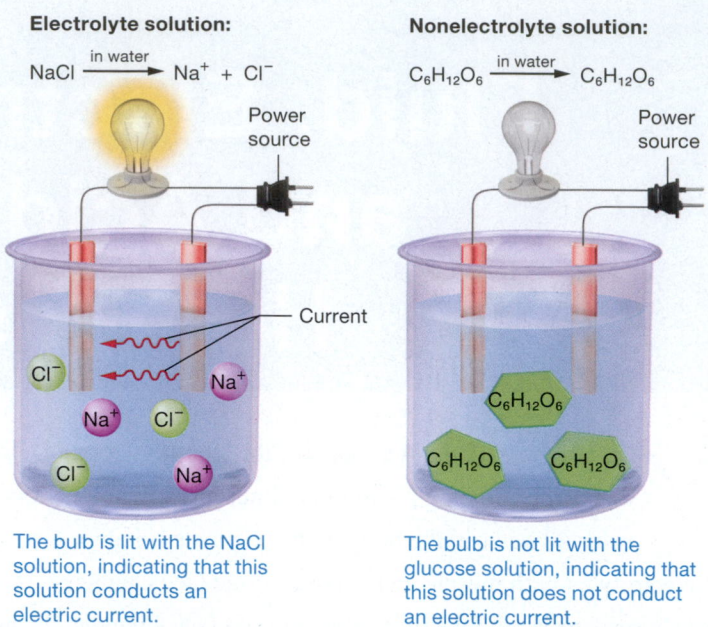

Figure 25.1 Comparison of electrical conduction in electrolytes and nonelectrolytes.

of electrolyte balance. Consequently, the numbers of these ions in the blood stay relatively stable. But equally important to the total number of ions in the body is their concentration, which depends not only on the number of ions but also on the amount of water in the body. For this reason, fluid balance is a critical factor that helps to determine electrolyte balance. We examine how this works, and the effects of imbalances, in Module 25.3.

Quick Check

☐ 3. How does an electrolyte differ from a nonelectrolyte?

☐ 4. What is electrolyte balance?

Acids, Bases, and pH

《 FLASHBACK

1. How are acids and bases defined? (p. 48)

2. What is the pH scale? (p. 48)

As we've covered, an **acid** is a chemical that dissociates in water to release a hydrogen ion (H^+), as in the following reaction that uses HCl as an example:

$$HCl \rightarrow H^+ + Cl^-$$
Hydrochloric acid   *Hydrogen ion*   *Chloride ion*

Common acids in the human body include hydrochloric acid (HCl) in the stomach and carbonic acid (H_2CO_3) in the blood. Hydrogen ions released from acids play many important physiological roles, such as digestion of ingested food, inactivation of bacteria and other pathogens, and intracellular digestion by lysosomes. In addition, recall that creation of a hydrogen ion gradient is a critical component of ATP synthesis by mitochondria (see Chapter 23).

A **base,** or **alkali,** by contrast, is a chemical that accepts a hydrogen ion in a solution, such as the following:

$$NaOH \quad + \quad HCl \quad \rightarrow \quad NaCl \quad + \quad H_2O$$

Sodium hydroxide (base)   Hydrochloric acid   Salt   Water

The result of this reaction is generally a salt and a molecule of water. Most bases that you encounter in the laboratory have an electrolyte bound to OH^-, the *hydroxide ion.* However, in the human body, the most common base is the *bicarbonate ion,* or HCO_3^-. Bicarbonate ions and certain other bases are part of *buffer systems,* meaning that they bind hydrogen ions released from acids to resist dramatic swings in the hydrogen ion concentration of body fluids.

Recall that the hydrogen ion concentration of a solution is measured by the **pH scale.** As you've read, an increased hydrogen ion concentration results in a solution with a lower pH. Conversely, a solution with a lower hydrogen ion concentration has a higher pH. Solutions with a pH less than 7 are acidic, those with a pH greater than 7 are basic, and those with a pH of exactly 7 are neutral (see Chapter 2).

Quick Check

☐ 5. How do acids and bases differ?

☐ 6. Which pH values are acidic, basic, and neutral?

Apply What You Learned

☐ 1. A patient suffers from renal failure and so produces less urine. If this patient increased his water intake, what state of fluid balance or imbalance would exist? Why?

☐ 2. Would you expect this same patient's electrolytes to be balanced or imbalanced? Explain.

See answers in Appendix A.

MODULE 25.2
Fluid Homeostasis

Learning Outcomes

1. Describe the fluid compartments, and explain how each contributes to the total body water.

2. Compare and contrast the relative concentrations of major electrolytes in intracellular and extracellular fluids.

3. Explain how osmotic pressure is generated, and compare and contrast the roles that hydrostatic and osmotic pressures play in the movement of water between fluid compartments.

4. Describe the routes of water gain and loss from the body.

5. Describe the mechanisms that regulate water intake and output, and explain how dehydration and overhydration develop.

You've probably heard the oft-repeated advice that the average person should consume at least eight glasses of water (at 8 fluid ounces each) per day. This is widely reported as fact by many media outlets, websites, and diet and health books. However, this value equates to nearly 2 liters of pure water per day, which is quite a lot of water. Other claims about water and hydration have been made, such as its purported health benefits, its ability to promote weight loss, and the effects of caffeine on hydration. But are any of these claims true? And just how much water does a person really need to consume every day?

This module answers these questions as it examines the most abundant chemical compound in the human body. We first discuss the amount of water in the body and where it is located. Next, we revisit a basic principle we have seen many times in this book: water movement due to hydrostatic pressure and osmosis. Finally, we address water gains and losses and the mechanisms by which fluid balance is regulated.

Total Body Water

You may have heard or read that the human body is composed of about 60% water. However, this value refers to the amount of water in the "standard man," a reference point that considers the standard values for a 20- to 30-year-old 70-kg (154-lb) male of European descent. For this standard man, the amount of water in the body is about 42 kg (92.5 lb). One kilogram of water is equal to one liter, so this equates to 42 liters (11 gal) of water. This volume represents a value known as the **total body water.** To visualize how much water this is, gather 21 two-liter bottles of soda and place them side by side. The amount of liquid in those bottles is equal to the total body water for this hypothetical individual.

Of course, most people don't fit the profile of the standard man, and so total body water differs from this value. An individual's body mass, age, sex, and amount of adipose tissue all affect total body water. For example, women tend to have a lower percentage of total body water than men because they have a higher percentage of body fat, which contains less water than other tissues (and because men generally have more muscle, a tissue that contains a lot of water). In addition, infants of both sexes have a higher total body water, averaging about 65%, whereas individuals over 60 years of age average a total body water of about 52% for men and 46% for women.

Quick Check

☐ 1. What is total body water?

☐ 2. What factors affect total body water?

Fluid Compartments: Intracellular and Extracellular Fluids

There are two basic places, or *compartments,* where we find body fluids: inside the cells, the **intracellular compartment,** and the area outside the cells, the **extracellular compartment.** The intracellular compartment isn't a true single compartment, but is instead trillions of microscopic compartments inside the plasma membranes of our cells. The fluid within the intracellular compartment is, of course, **cytosol** (or *intracellular fluid [ICF]*), which accounts for about 60% of the body's fluids, or about 26 liters. The extracellular compartment is filled with **extracellular fluid (ECF),** made up of a variety of fluids that you have studied throughout this book. Two of the fluids that we have discussed at length are *plasma,* the fluid portion of the blood, and *interstitial fluid,* the fluid in the spaces between the cells. Plasma accounts for approximately 8% (3 liters) of the total body water, and interstitial fluid for about 32% (13 liters) (**Figure 25.2**).

The solute composition of plasma and interstitial fluid is similar—both contain various types of ions in similar amounts. The main difference in the composition of the two fluids is in protein content: Plasma has a much higher protein content than interstitial fluid, which has virtually no proteins. As you can see in **Figure 25.3**, however, the solute composition of the ECF and the cytosol varies starkly. The concentrations of sodium, chloride, calcium, and bicarbonate ions are much higher in the ECF than in the cytosol. In contrast, proteins and potassium, magnesium, sulfate, and monohydrogen phosphate (HPO_4^{2-}) ions have a much higher concentration in the cytosol than in the ECF. The concentration gradients of these solutes are due to factors such as particle size and the reliance of certain solutes on transport proteins or pumps to cross the membrane. Many of these concentration gradients, particularly those of sodium, potassium, and calcium ions, play important physiological roles.

Quick Check

☐ 3. Where are the intracellular and extracellular compartments located?

☐ 4. How do the compositions of cytosol, interstitial fluid, and plasma differ?

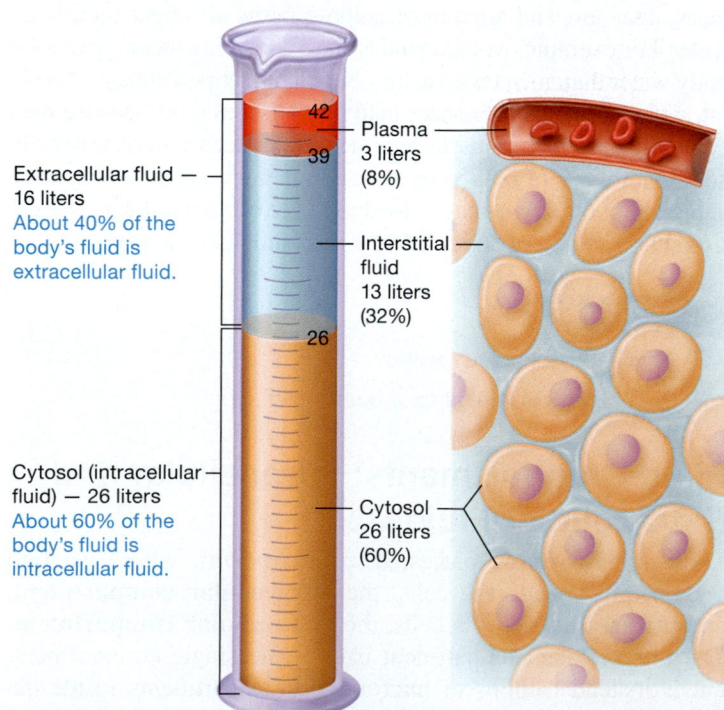

Figure 25.2 The distribution of water in the body.

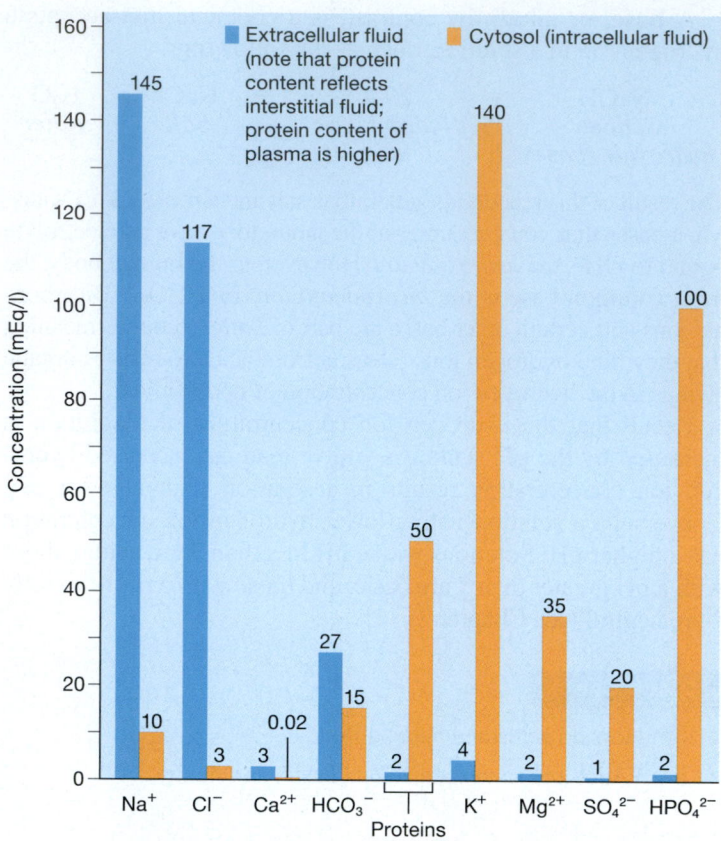

Figure 25.3 The solute composition of extracellular fluid and cytosol.

Movement of Water between Compartments

🜲 FLASHBACK

1. In which direction does water move during osmosis? (p. 76)

2. What are osmotic and hydrostatic pressures? In which directions do they drive water movement? (p. 77)

3. What is tonicity? (p. 78)

Although solute movement between fluid compartments is restricted, water generally moves freely between them. We have discussed this before in this book, so let's just quickly review. Recall from the blood vessels chapter (see Chapter 18) that the direction in which water moves between compartments is influenced by two pressure gradients: hydrostatic pressure and osmotic pressure, another example of the Gradients Core Principle (p. 28). The hydrostatic pressure gradient, or the force that the fluid exerts on cells, tends to result in water moving from an area of higher hydrostatic pressure to one with lower hydrostatic pressure.

CORE PRINCIPLE
Gradients

The osmotic pressure gradient, or the difference between the osmotic pressures of two solutions, tends to result in water moving by osmosis toward the solution with the higher osmotic pressure. A solution's osmotic pressure is determined by its *osmolarity,* or the number of solute particles present in the solution. This means that water moves toward the solution with the higher osmolarity, or the higher number of solute particles.

In the next three subsections, we look more closely at how changes in hydrostatic pressure and osmotic pressure of the plasma influence water movement. Then, we see how these two pressures combine to affect the volume of the interstitial fluid and the cytosol.

Hydrostatic Pressure

The effect of hydrostatic pressure on water movement with respect to the plasma is relatively simple. Hydrostatic pressure in this context is just the force of the water on the capillary wall. The hydrostatic pressure of the plasma is higher than that of the interstitial fluid, so normally water is forced out of the capillary. When the hydrostatic pressure of the plasma increases due to a higher water volume, more water is pushed out of the capillary into the interstitial fluid. This in turn increases the hydrostatic pressure of the interstitial fluid above that of the cytosol, creating a gradient that drives water into the cell.

When the hydrostatic pressure of the plasma decreases, we don't normally see water move from the interstitial fluid into the capillary, because the plasma's hydrostatic pressure is still significantly higher than that of the interstitial fluid. Instead, we simply find that less water moves out of the capillary and into the interstitial fluid. We generally only see the gradient reversed—that is, water moving from the interstitial fluid into the plasma—in extreme cases, such as during severe blood loss (known as *hypovolemic shock*).

Osmotic Pressure and Tonicity

Now we'll consider osmotic pressure. In the cell chapter, you read about tonicity—a comparison of the ability of two solutions to cause water movement by osmosis (look back at Figure 3.9). As we just described, the ECF and cytosol generally have the same osmotic pressure. Such solutions, which are referred to as *isotonic,* have the same ability to cause water movement by osmosis. So, there is usually no net osmotic movement of water between the two solutions. However, when the ECF's solute concentration is altered, its osmotic pressure changes, and so its ability to cause water movement by osmosis also changes. There are two possible types of ECF solute variations, hypotonic and hypertonic, and they cause water movement as follows:

- **Hypotonic ECF causes a cell to gain water.** The ECF is *hypotonic* to the cytosol when it has a lower osmotic pressure than the cytosol and so a lesser ability to cause water movement by osmosis. In this case, it's the cytosol that has a greater osmotic pressure, which gives it a greater ability to drive osmosis. For this reason, a cell in a hypotonic ECF gains water and will swell.
- **Hypertonic ECF causes a cell to lose water.** The ECF is *hypertonic* to the cytosol when it has a higher osmotic pressure than that of the cytosol and so has a greater ability to cause water movement by osmosis. So, a cell in

a hypertonic ECF loses water because it leaves the cytosol and enters the ECF. This causes the cell to shrivel, or *crenate.*

How Hydrostatic Pressure and Osmotic Pressure Influence Water Movement

Now we can examine how these two pressures combine to affect fluid movement between compartments. First, consider water movement between the two types of ECF: the plasma and interstitial fluid. In most blood capillaries, the plasma has a high hydrostatic pressure at the capillary's arteriolar end—much higher than that of the interstitial fluid. This forces water out of the capillary and into the interstitial fluid. At the capillary's venular end, the plasma has a high osmotic pressure but a lower hydrostatic pressure. As a result, most of the water lost from the plasma at the arteriolar end is drawn back into the capillary by osmosis. The capillary does lose a small amount of water, and this is returned to the plasma via the lymphatic system (look again at Figure 20.3 for a review).

Figure 25.4 illustrates the effect that this water movement has on the cytosol of a cell. Notice that under normal conditions, the interstitial fluid's hydrostatic and osmotic pressures are very close to those of the cytosol, so not much of a gradient exists between the two compartments and there is no significant net movement of water between them. However, if water is gained or lost from the plasma, its hydrostatic and osmotic pressures change. This in turn influences how much water leaves the plasma and enters the interstitial fluid, and so the hydrostatic and osmotic pressures of interstitial fluid also change. The differences in interstitial fluid pressures then directly influence the volume of the cytosol. Let's see how this works.

In Figure 25.4a, the plasma has gained water. This has increased its hydrostatic pressure but also made the plasma hypotonic by diluting it, which has decreased its osmotic pressure. Both situations—a higher hydrostatic pressure and a lower osmotic pressure—will cause water to move out of the plasma and into the interstitial fluid. This then increases the hydrostatic pressure and decreases the osmotic pressure of the interstitial fluid, both of which drive water into the cytosol. As a result, the cell swells.

In Figure 25.4b, we have the opposite situation of a water deficit due to dehydration. This has caused the hydrostatic pressure to decrease, leading to less water leaving the plasma, and it has also made the plasma hypertonic, meaning it now has a higher osmotic pressure. The hypertonic plasma will draw water out of the cytosol and interstitial fluid and into the plasma by osmosis. For this reason, the cell has shrunk.

Quick Check

- ☐ 5. In which directions do hydrostatic pressure and osmotic pressure move water?
- ☐ 6. If a cell is in interstitial fluid that is hypertonic to its cytosol, how does this affect the cell? Why?
- ☐ 7. What happens if the interstitial fluid is hypotonic to the cytosol of the cell?

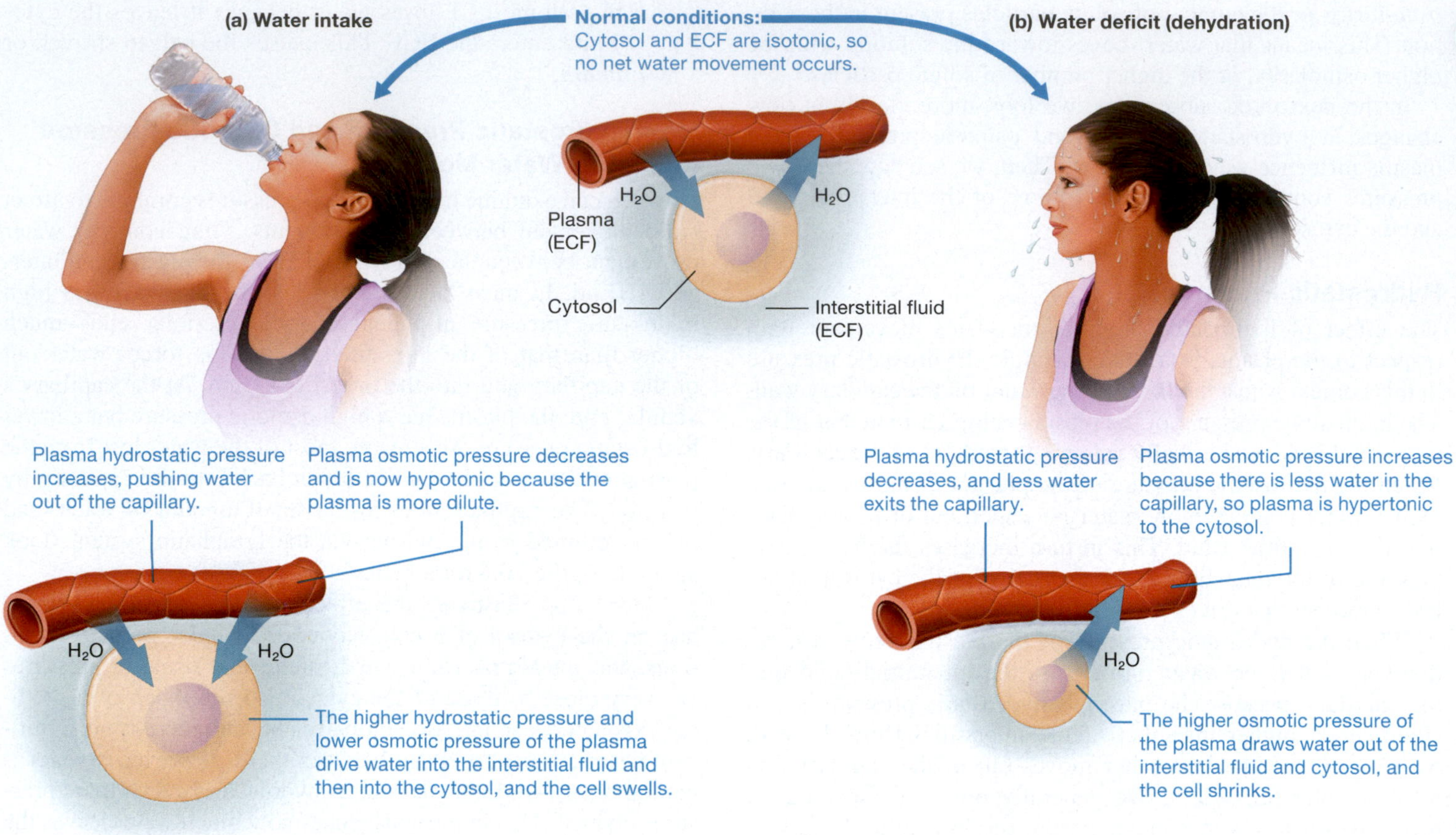

(a) Water intake

Normal conditions:
Cytosol and ECF are isotonic, so no net water movement occurs.

(b) Water deficit (dehydration)

H_2O H_2O

Plasma (ECF)

Cytosol —   — Interstitial fluid (ECF)

Plasma hydrostatic pressure increases, pushing water out of the capillary.

Plasma osmotic pressure decreases and is now hypotonic because the plasma is more dilute.

Plasma hydrostatic pressure decreases, and less water exits the capillary.

Plasma osmotic pressure increases because there is less water in the capillary, so plasma is hypertonic to the cytosol.

H_2O H_2O

H_2O

The higher hydrostatic pressure and lower osmotic pressure of the plasma drive water into the interstitial fluid and then into the cytosol, and the cell swells.

The higher osmotic pressure of the plasma draws water out of the interstitial fluid and cytosol, and the cell shrinks.

Figure 25.4 Fluid movements between compartments.

Water Losses and Gains

⟪ FLASHBACK

1. What does oxygen become after it accepts the final electrons in the electron transport chain? (p. 913)

2. What effect does angiotensin-II have on thirst and the amount of water in the body? (p. 962)

3. What are baroreceptors? (p. 682)

In the module introduction, we posed a common question: How much water do you need to consume every day? To answer this, we must first find out how water is lost and gained from the body. The next two subsections focus on these topics, and the third answers the question.

Factors That Influence Water Loss

Water is lost from the body in several ways, but most is lost via the kidneys in urine (**Figure 25.5a**). A certain amount of urine, called the **obligatory water loss,** must be produced each day irrespective of fluid intake. This is because the body must excrete a minimum amount of solutes to prevent buildup of toxic substances and electrolyte imbalances. The kidneys are not capable of producing urine with an osmolarity greater than 1200 mOsm, so a certain amount of water must be excreted with the solutes. The obligatory water loss for most people is about

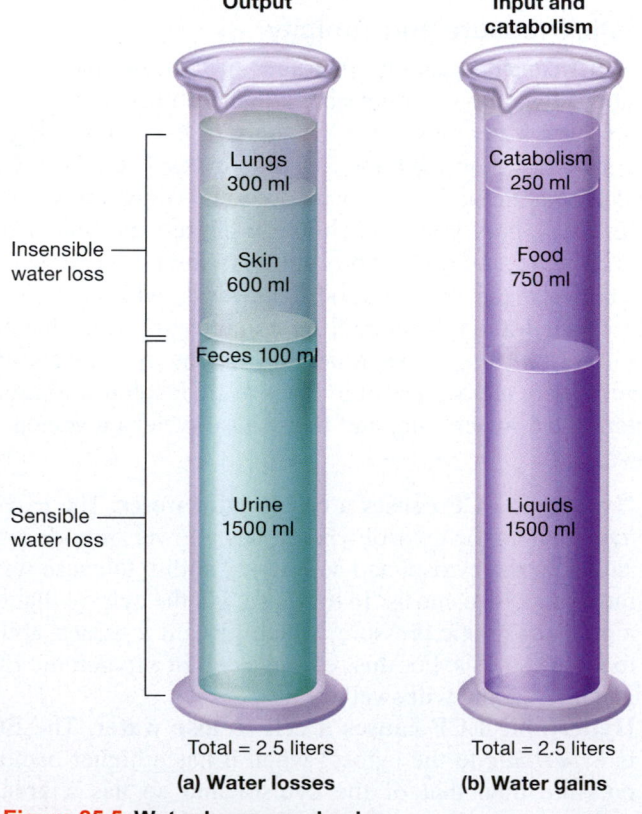

Output

Input and catabolism

Insensible water loss
- Lungs 300 ml
- Skin 600 ml
- Feces 100 ml

Sensible water loss
- Urine 1500 ml

- Catabolism 250 ml
- Food 750 ml
- Liquids 1500 ml

Total = 2.5 liters

Total = 2.5 liters

(a) Water losses

(b) Water gains

Figure 25.5 Water losses and gains.

500 ml, even if a person has not taken in any water at all. This water loss is why a human stranded without water is in trouble—the kidneys must continue to produce this volume of urine, and without water to replace the loss, a human can succumb to dehydration in a matter of days.

Normally, of course, we do consume liquids throughout the day, and so urine output is about 1500 ml, greater than obligatory water loss. In addition to urine, a small amount of water, about 100 ml, is also lost in the feces. Together, the water loss from these two sources is known as **sensible water loss** because we normally notice or "sense" it happening. About 600 ml is also lost from the skin due to sweating and evaporation, and another 300 ml is lost during expiration of humidified air. We normally don't notice when this water loss occurs, and for this reason it is called **insensible water loss** (the degree to which sweating is insensible, however, depends on the relative humidity—it can be considered sensible at higher humidity levels). Between the sensible and insensible water losses, most people will lose about 2.5 liters of water per day. This value fluctuates with water intake, physical activity, food intake, and other factors.

Factors That Influence Water Gain

As you can see in **Figure 25.5b**, the body gains water from three main sources. The first source is our own catabolic reactions. Recall that when oxygen molecules accept the final pair of electrons in the electron transport chain, a molecule of water is formed. Water is also formed as a byproduct of several other catabolic reactions. Together, water from all these reactions is called **metabolic water** (or *water of oxidation*), and it amounts to a total gain of about 250 ml per day. The other two sources are the intake of water via food and liquid in the diet. Water from food totals about 750 ml per day, and water from liquids about 1500 ml per day. All three sources combined generally equal 2.5 liters, or about the same amount of water lost every day.

The mechanism that drives water intake from liquids is known as the **thirst mechanism,** which is controlled by the hypothalamus. Many stimuli can trigger thirst, but by far the most potent is an increase in the osmolarity of the plasma. As you can see in **Figure 25.6a**, changes in plasma osmolarity are detected by neurons called **osmoreceptors,** located in the *thirst center* of the hypothalamus. An increase in the solute concentration of the ECF causes water to move out of the osmoreceptors, which triggers the sensation of thirst. This mechanism is very sensitive, as an increase in plasma osmolarity of even 2–3% is enough to initiate thirst.

When plasma osmolarity decreases, water moves back into the osmoreceptors and thirst is inhibited. Interestingly, however, we feel relieved of thirst as soon as we ingest liquids, well before they are absorbed and have the chance to affect osmoreceptors. This is believed to be due to receptors in the pharynx that detect the ingestion of liquid. Such a response is critical to preventing excessive liquid intake. If the liquid were not detected until ECF osmolarity decreased to the normal range—which takes about 30–60 minutes—one could continue drinking and overhydration could result.

Other stimuli that can trigger thirst include a decreased plasma volume and the hormone angiotensin-II (**Figure 25.6b**). A decrease in plasma volume, which is generally accompanied by a reduction in blood pressure, is detected by baroreceptors in several arteries. It is a much less potent stimulus for thirst, as a decrease of 10–15 mm Hg is required. Decreased plasma volume also stimulates the juxtaglomerular cells in the kidneys and initiates the *renin-angiotensin-aldosterone system* (*RAAS*), which culminates in the formation of angiotensin-II. Both mechanisms further enhance the thirst sensation.

There is a common claim that by the time the thirst mechanism is activated, you are already dehydrated. Proponents therefore advise that an individual consume water nearly continuously throughout the day in order to prevent dehydration. This is, of course, not scientifically accurate; we have said that it takes only a 2–3% change in the osmolarity of the ECF to stimulate thirst. However, there are occasions when we need to continuously administer fluids. In such cases in the medical field, it is common to use an *intravenous line,* discussed in *A&P in the Real World: Intravenous Fluids.*

How Much Water Do We Need to Drink?

Now we have enough information to answer the question of how much water a person should drink. Simply put, water intake should balance water loss. Since on average a person loses about 2.5 liters per day, then water gains should equal about 2.5 liters per day. On the surface, this might seem to support the "eight 8-oz glasses of water" advice. However, recall from Figure 25.5b that we usually obtain about 750 ml of water from the foods that we eat, and we produce about 250 ml of water from oxidative catabolism. So this leaves only about 1.5 liters of water that we need to drink. Of course, this amount will vary with activity. A person engaged in strenuous physical activity

Intravenous Fluids

Sometimes it's necessary to bypass the gastrointestinal system to get fluids into an individual rapidly. In these cases, a medical professional generally starts an **intravenous line,** often just called an **IV.** With this line, fluids can be administered directly into a vein, allowing them to have a nearly immediate effect on the body.

An IV consists of a flexible catheter that is inserted into a patient's vein, with an external port onto which bags of different fluids can be attached. Fluid from the bag drips into a chamber and then into a line at a specified rate, allowing the rate of fluid administration to be controlled. Most often, an IV is placed in a vein in the upper limb, which is known as a *peripheral line.* However, when intravenous access is required over the long term, a larger port may be inserted in a vein such as the subclavian or internal jugular vein. These lines are known as *central lines.*

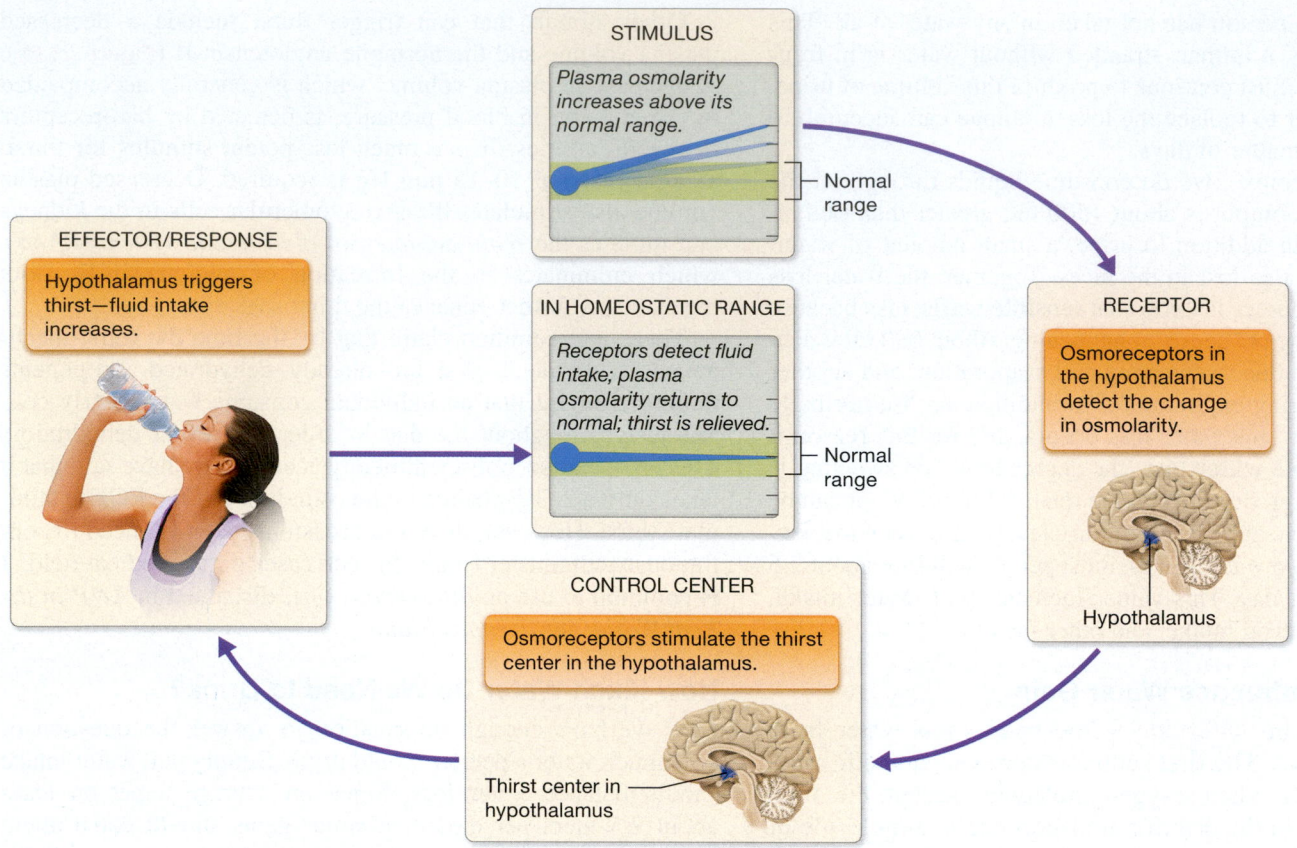

STIMULUS

Plasma osmolarity increases above its normal range.

— Normal range

RECEPTOR

Osmoreceptors in the hypothalamus detect the change in osmolarity.

Hypothalamus

IN HOMEOSTATIC RANGE

Receptors detect fluid intake; plasma osmolarity returns to normal; thirst is relieved.

— Normal range

EFFECTOR/RESPONSE

Hypothalamus triggers thirst—fluid intake increases.

CONTROL CENTER

Osmoreceptors stimulate the thirst center in the hypothalamus.

Thirst center in hypothalamus

(a) Control of thirst response due to increased plasma osmolarity by a negative feedback loop

STIMULUS

Plasma volume decreases below its normal range.

— Normal range

RECEPTOR

Baroreceptors in the arteries detect the change in blood pressure due to decreased plasma volume.

Juxtaglomerular (JG) cells in nephrons detect the change as well, activating the renin-angiotensin-aldosterone system (RAAS).

Baroreceptors

Heart

JG cells

IN HOMEOSTATIC RANGE

Receptors detect fluid intake; plasma volume returns to normal; thirst is relieved.

— Normal range

EFFECTOR/RESPONSE

Hypothalamus triggers thirst—fluid intake increases.

CONTROL CENTER

Baroreceptors and angiotensin-II from RAAS stimulate the thirst center in the hypothalamus.

Thirst center in hypothalamus

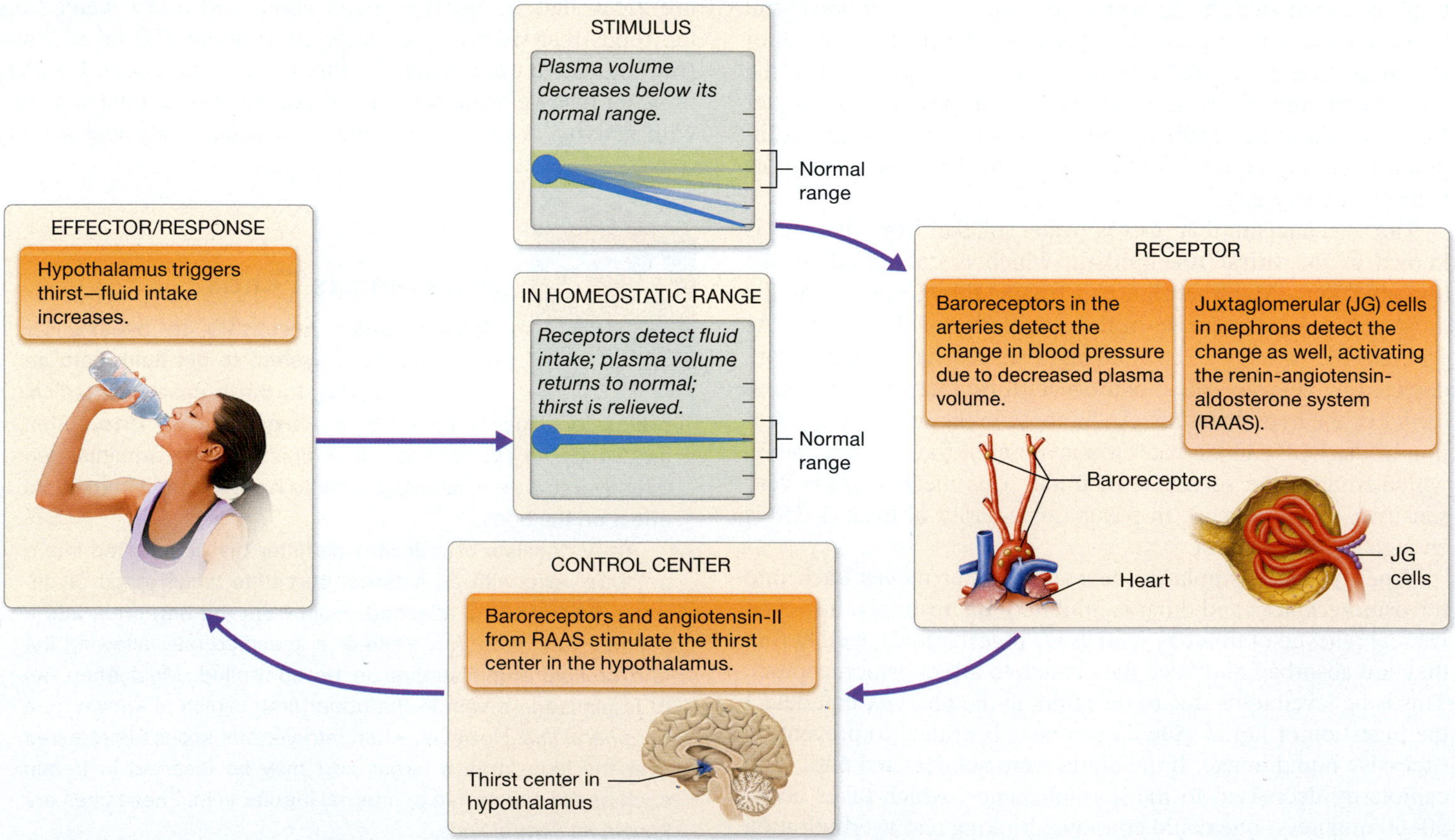

(b) Control of the thirst response due to decreased plasma volume by a negative feedback loop

Figure 25.6 Maintaining homeostasis: regulation of thirst by negative feedback loops.

can lose 4–5 liters of fluid through sweating, and naturally this volume needs to be replaced to prevent serious imbalances.

Many claim that there are health benefits to consuming more water than we need. The claims seem logical on the surface, but are they true? Several studies have examined this question and found that there were no noticeable benefits in individuals who consume more water than needed to replace normal losses. The individuals weren't harmed by the increased water intake but were not benefited by it, either. Many also claim that consuming excess water will help a person lose weight. This is also unfortunately not scientifically accurate—water intake does stretch the stomach, but it does not trigger the signals that make us feel full.

Quick Check

☐ 8. How is thirst stimulated?

☐ 9. How are fluids lost from the body?

☐ 10. What are the water requirements for an average person?

Hormonal Regulation of Fluid Balance

≪ FLASHBACK

1. What is antidiuretic hormone, and what are its actions on the kidneys? (p. 594)

The mechanisms that lead to water loss and gain are imprecise. You can't always take a drink when you're thirsty, and sometimes you might drink liquid when you're not thirsty. For these reasons, the body needs ways to ensure that water input matches water output. There are four main hormones that perform this task. By far the most important of these hormones is **antidiuretic hormone (ADH).** We have discussed ADH and the other three hormones that affect fluid balance—angiotensin-II, aldosterone, and atrial natriuretic peptide (ANP)—several times in this book (see Chapters 16 and 24).

ADH is produced by the hypothalamus and released in the posterior pituitary gland. Its main target tissues are the cells of the distal tubule and collecting system of the kidney. It triggers the insertion of water channels called *aquaporins* into their plasma membranes, which allows for the reabsorption of water from the kidneys back into the ECF by osmosis. When the ADH level rises, more water is reabsorbed, the ECF volume increases, and urine production decreases. When ADH is at its maximal secretion, the only water lost to the urine is that due to obligatory water loss. Conversely, when the ADH level falls, less water is reabsorbed, the ECF volume decreases, and urine production increases.

Of course, ADH secretion by itself isn't the whole story of water reabsorption in the collecting system. Recall from the urinary system chapter that even with large numbers of aquaporins in the plasma membranes of tubule cells, osmosis will not occur unless an osmotic gradient is present—the interstitial fluid in the medulla must have more solutes than the filtrate. This gradient is set up and maintained by the countercurrent system within the nephron loops of juxtamedullary nephrons (look back at Figure 24.21 for a review of this concept).

Quick Check

☐ 11. What is the role of ADH in fluid balance?

☐ 12. How is ADH secretion stimulated?

Imbalances of Fluid Homeostasis

The amount of water in your body may increase and decrease throughout the day. Generally, the body compensates for these changes by altering urine output and there is no noticeable disturbance in fluid balance. However, some conditions can either overwhelm the compensatory mechanisms or prevent the body from compensating at all. Certain imbalances can lead to a change in the osmolarity of the ECF, whereas others produce an *isosmotic* state, meaning that the ECF has the same tonicity as the cytosol. In this section, we look at both types of imbalances more closely.

Dehydration: Decreased Volume and Increased Osmolarity

Dehydration is defined as decreased volume and increased concentration of the ECF. Common causes for dehydration include profuse sweating, prolonged diarrhea and/or vomiting, certain endocrine conditions, and overuse of diuretics. In such cases, the water loss decreases the plasma volume, increasing its solute concentration and osmotic pressure. The changes in solute concentration lead to electrolyte imbalances in addition to water imbalances, which are covered in the next module.

Initially, a person becoming dehydrated will experience a dry mouth and thirst. This situation progresses to dryness of the skin and decreased urine production (called *oliguria;* ohl′-ih-GYU-ree-uh); eventually, fever may develop and organ function is impaired. The condition, and its associated electrolyte imbalances, is fatal if hydration is not restored.

Dehydration may be treated by consuming pure water, but this should be approached with caution. If water is consumed in moderate amounts, it will mildly dilute the ECF, making it slightly hypotonic to the cytosol. This will trigger a shift of water back into the cells, rehydrating them. However, if large amounts of water are consumed in a short period of time, the ECF can become overly dilute, causing too much water to enter the cells and resulting in cellular swelling. For this reason, treating dehydration with sports drinks is preferable in many cases, as sports drinks contain glucose and electrolytes, making them hypertonic to pure water.

Overhydration: Increased Volume and Decreased Osmolarity

Overhydration, also known as **hypotonic hydration,** results when the volume of the ECF increases and its osmotic pressure falls. Cells exposed to the hypotonic ECF swell as water enters the cells by osmosis. Normally, when water is consumed, hypotonic hydration is prevented by a decline in ADH production, which leads to loss of the excess water in the urine. However, this mechanism can fail when renal function is impaired, ADH is secreted abnormally, or extreme amounts of water are consumed in a short time period (a condition called *water toxicity*).

Hypotonic hydration results in electrolyte imbalances due to dilution of the ECF. Particularly affected is the sodium ion concentration, which can fall dramatically, leading to *hyponatremia*. Together, the hyponatremia and cellular swelling can cause severe disturbances to homeostasis, particularly to the central nervous system. The swelling of neurons, or *cerebral edema*, affects their ability to function and can lead to mental status changes, seizures, coma, and even death. Treatment consists of administering hypertonic fluids to draw the water back out of the cell and into the interstitial fluid.

Isosmotic Fluid Imbalances

Disruptions in homeostasis can result even when the ECF remains isotonic to the cytosol. For example, fluid depletion occurs in cases of blood loss, but the amount of solutes and water lost are the same. For this reason, the plasma's osmotic pressure is unchanged and there is no osmotic gradient to pull water into or out of a cell. This type of fluid imbalance is known as *hypovolemia*, and, if untreated, it can lead to shock, organ failure, and death due to extremely low blood pressure. Generally, if the blood products cannot be immediately replaced, isotonic fluids are administered to help restore ECF volume.

The same situation can exist with fluid excess, or *hypervolemia*—the ECF volume can increase without a significant change in osmotic pressure. A common example is **edema** (eh-DEE-muh), in which fluid accumulates in the interstitial fluid, causing swelling. As with hypovolemia, the fluid is isotonic to the cytosol, so the volume of the cells doesn't change. Edema can have many causes, including high hydrostatic pressure, as seen with hypertension, a decrease in the return of venous blood to the heart, and/or excess sodium ion retention; low colloid osmotic pressure; removal of lymphatic vessels; or release of inflammatory mediators. The excess interstitial fluid can make basic movements such as walking difficult and can also cause organ dysfunction. Edema may be managed by treating the underlying cause, using compression garments to prevent fluid accumulation, and by removing excess fluid through drugs such as diuretics.

Quick Check

☐ 13. How does dehydration affect the volume of the cytosol?

☐ 14. What are some causes of overhydration, and how does it affect cells?

☐ 15. How do dehydration and overhydration differ from isosmotic fluid imbalances?

Apply What You Learned

☐ 1. Would you expect an organ such as the liver, whose cells are protein-rich, to contain more or less water than lipid-rich adipose tissue? Explain your answer.

☐ 2. A nurse has been working for 11 hours on an overnight shift. He is asked to start an intravenous line and hang a bag of isotonic 0.9% sodium chloride. He accidentally grabs a bag of hypertonic 9% sodium chloride instead. What is likely to happen to his patient? Why?

☐ 3. Under what condition might it be beneficial to administer hypertonic sodium chloride to a patient?

See answers in Appendix A.

MODULE **25.3**
Electrolyte Homeostasis

Learning Outcomes

1. Describe the function of the most prevalent electrolytes found in body fluids.

2. Describe the hormonal regulation of electrolyte levels in the plasma.

3. Explain how calcium ion regulation is related to phosphate ions.

Technically, an electrolyte is any compound that ionizes in solution, but in the body we generally refer to electrolytes as salts that dissociate into ions such as sodium, potassium, calcium, magnesium, and chloride ions. The most critical systems for maintenance of electrolyte balance are the endocrine and urinary systems. The importance of preserving electrolyte balance cannot be overstated, as imbalances can very easily be fatal. In this module, we examine the roles and regulation of sodium, potassium, calcium, phosphate, chloride, and magnesium ions. We also look at the most commonly encountered electrolyte imbalances—those of sodium, potassium, and calcium ions.

Sodium Ions

« FLASHBACK

1. What is the function of the Na^+/K^+ ATPase pump? (p. 80)

2. What role do sodium ions play in the action potential? (p. 350)

Sodium ions are the most abundant extracellular cation, measuring about 135–145 mEq/l (milliequivalents per liter, a common way to express electrolyte concentration); in contrast, the levels in the cytosol are very low. This is due in large part to two factors: (1) the Na^+/K^+ ATPase pump, which uses ATP to pump sodium ions out of the cytosol against their concentration gradient; and (2) the relatively low permeability of the plasma membrane to sodium ions, which prevents these ions from leaking back into the cytosol in large numbers. This high extracellular concentration of sodium ions makes them one of the most important types of osmotic particles in the ECF.

The steep sodium ion gradient between the cytosol and ECF also makes this ion critical for all our electrophysiological processes. Recall that cells have a negative resting membrane potential—the inside of the cell is negative compared to the outside. When sodium ion channels open, the sodium ions follow their concentration gradient and rush into the cell. This causes the cell to gain positive charges and *depolarize:*

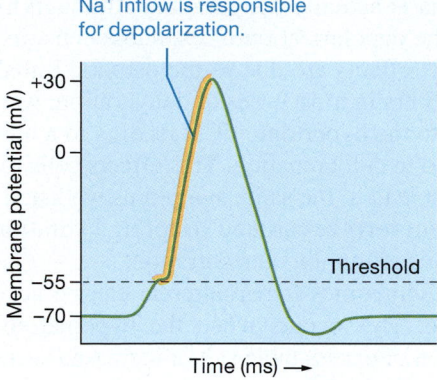

Depolarization is a key event in the functioning of all excitable cells, including neurons and the three different types of muscle cells.

Due to the critical roles of sodium ions in many processes, an imbalance can have a significant impact on overall homeostasis. This section looks at how sodium ion balance is maintained and then what happens when the sodium ion concentration increases or decreases abnormally.

Regulation of Sodium Ion Concentration

You might be surprised to learn how little salt a person would need to ingest in one sitting to prove fatal. As little as 2 teaspoons can raise plasma sodium ion concentration by 10 mEq/l, and 2 tablespoons by 30 mEq/l, which can easily cause death. In the typical American diet, we don't usually ingest that much salt at one time, but we tend to consume

a lot of it over the course of a day. Most people consume about 9 grams of salt in a day, which equals about 2 teaspoons. Yet in spite of this significant sodium ion intake, the concentration of sodium ions in the ECF remains stable, thanks to several mechanisms. This is critical not only to sodium ion balance but also to fluid balance, as water reabsorption in the kidneys depends on a gradient consisting largely of sodium ions in the interstitial fluid (see Chapter 24).

The two main hormones that increase sodium ion retention are *angiotensin-II* and *aldosterone* (**Figure 25.7**). A low extracellular sodium ion concentration is one of the major stimuli that triggers the RAAS. The low sodium ion concentration is detected by receptors in the macula densa of the nephron, which leads to the release of renin and the eventual formation of angiotensin-II. As we've covered, angiotensin-II increases the reabsorption of sodium ions from the proximal tubule. Angiotensin-II also stimulates aldosterone secretion, which causes sodium ion reabsorption from the distal tubule. Note that both actions also indirectly increase water reabsorption, as water follows sodium ions by osmosis. When secretion of both hormones decreases, the number of sodium ions (as well as the amount of water) reabsorbed decreases. Another hormone that influences sodium ion balance is *atrial natriuretic peptide* (*ANP*), which decreases sodium ion (and water) reabsorption. Together, these three hormones maintain a relatively stable plasma sodium ion concentration in the face of sometimes significant swings in sodium ion intake.

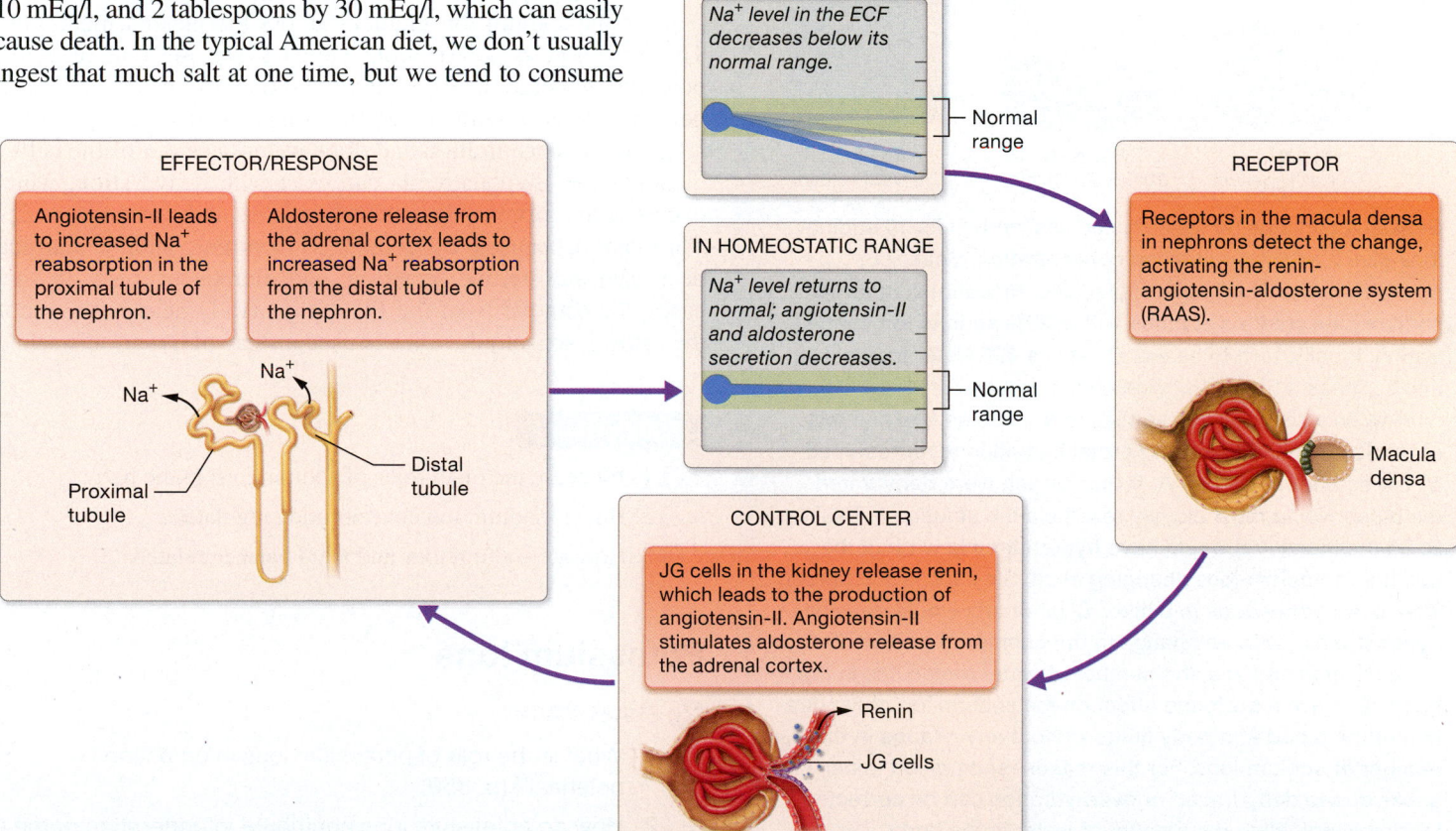

Figure 25.7 **Maintaining homeostasis: regulation of the extracellular concentration of sodium ions by a negative feedback loop.**

Sodium Ion Imbalances

The concentration of sodium ions in the body's fluids is determined by both the absolute number of sodium ions in the body fluids and the number of water molecules. So anything that abnormally increases or decreases the number of sodium ions and/or the number of water molecules can cause a sodium ion imbalance. This potential imbalance comes in two forms: *hypernatremia,* an elevated sodium ion concentration, and *hyponatremia,* a decreased sodium ion concentration.

ConceptBOOST))))

Why Does the Amount of Water in the Body Affect the Sodium Ion Concentration?

You might be wondering how the amount of water in the body can affect the number of sodium ions in the ECF. The answer is that it actually doesn't have any effect on the absolute number of sodium ions. Instead, it changes the *concentration* of sodium ions in body fluids, meaning the percentage of sodium ions relative to water molecules. Let's look at a simplified example here. Say the following three beakers each contain solutions of NaCl with 10 sodium ions (for simplicity we'll omit the chloride ions) and a different number of water molecules:

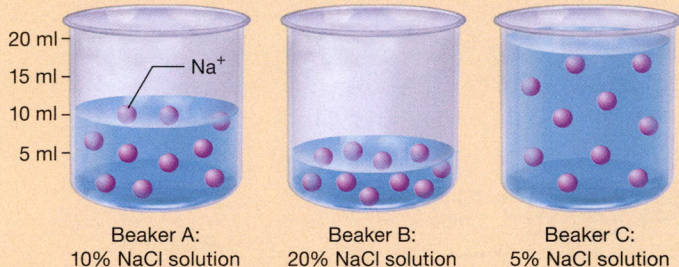

| Beaker A: | Beaker B: | Beaker C: |
| 10% NaCl solution | 20% NaCl solution | 5% NaCl solution |

Beaker A contains 90 water molecules, and so its 10 sodium ions give it a 10% sodium ion concentration. Beaker B, however, has only 40 water molecules. Now those same 10 sodium ions create a solution with a 20% sodium ion concentration. Finally, turn to beaker C—it has 190 water molecules, and so it has a 5% sodium ion concentration.

Now let's see how this applies to the human body. In this example, an individual with normal hydration is represented by the solution in beaker A. If that person were dehydrated, the blood would most closely resemble the solution in beaker B. So the individual would have hypernatremia without the number of sodium ions changing at all. Similarly, if the person were overhydrated, as in beaker C, he or she would develop hyponatremia, with no change in the number of sodium ions.

So, as you can see, the number of water molecules in the body can have a profound effect on the sodium ion concentration of the blood and body fluids without any change in the number of sodium ions. For this reason, sodium ion imbalances due to dehydration or overhydration can be corrected by simply adjusting the amount of water in the body. ■

Hypernatremia **Hypernatremia** (hy'-per-nah-TREE-mee-ah; *natri* = "sodium") is defined as a sodium ion concentration in the blood above 145 mEq/l. The most common cause of hypernatremia is actually dehydration. Although hypernatremia can impact the depolarization of excitable cells, its most immediate, obvious effects are due to the osmotic imbalances that it causes. As occurs in most cases of dehydration, water moves out of the cell into the hypertonic ECF, leading to a reduced volume of cytosol and to cell crenation. This affects cells throughout the body, and can lead to the same symptoms we listed for dehydration. Treatment revolves around restoring hydration, which will normalize the sodium ion concentration.

Note that sometimes hypernatremia can be accompanied by overhydration. This occurs when the hypernatremia is due to rapid ingestion of excess table salt or increased secretion of aldosterone, both of which increase water retention by the kidneys. Rehydration will be little help in these cases; for this reason, it's critical for a clinician to determine the primary cause of the hypernatremia so the proper course of treatment can be initiated.

Hyponatremia **Hyponatremia** is defined as a plasma sodium ion concentration less than 135 mEq/l. Like hypernatremia, hyponatremia is often not related to the absolute number of sodium ions but, rather, is due to overhydration. One cause of hyponatremia with overhydration is hypersecretion of ADH, a condition known as *syndrome of inappropriate ADH secretion* (*SIADH*). As with overhydration, hyponatremia can result in cellular swelling due to movement of water into the cytosol. It can also lead to problems with electrophysiology; specifically, it reduces the sodium ion gradient across the plasma membrane. This can slow depolarization, which slows the rate of action potential generation. Hyponatremia due to overhydration is generally treated with hypertonic saline to restore the plasma sodium ion concentration and draw water back out of the cells.

Sometimes hyponatremia can exist with dehydration, which occurs with a primary loss of sodium ions and water molecules. This can be caused by prolonged vomiting and/or diarrhea, decreased aldosterone secretion, and diuretic overuse. In such cases, the osmolarity of the ECF is generally nearly isotonic to the cytosol, so isotonic saline is used instead of hypertonic saline.

Quick Check

☐ 1. What are the main roles of sodium ions in the body?
☐ 2. How is sodium ion concentration regulated?
☐ 3. How are sodium ions and fluid balance related?

Potassium Ions

 **FLASHBACK**

1. What is the role of potassium ions in an action potential? (p. 350)
2. How do potassium ions contribute to generation of the resting membrane potential? (p. 349)

Whereas sodium ions are the most abundant extracellular cation, potassium ions are the most abundant intracellular cation. Indeed, the extracellular potassium ion concentration averages from 3.9 to 4.5 mEq/l, far lower than the average of 139 mEq/l in the cytosol. The concentration of potassium ions is higher in the cytosol of cells than in any other body fluids due primarily to the actions of the Na⁺/K⁺ pump. This steep concentration gradient, like that of sodium ions, is critical to the functioning of neurons and the three types of muscle cells in the body. The movement of potassium ions out of the cell down their concentration gradient through potassium ion channels is responsible for the repolarization phase of the action potential, shown here:

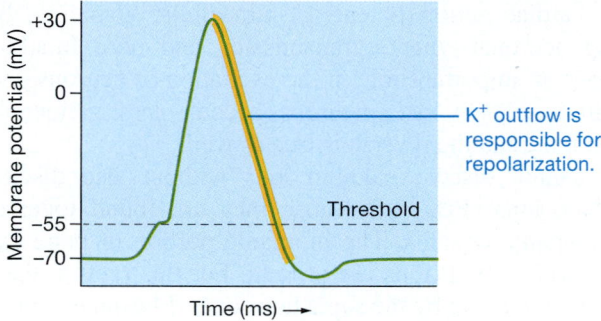

The repolarization phase is critical to the functioning of excitable cells because it returns cells to their resting membrane potential. A cell cannot be re-excited until it returns to this resting value. This means that unless repolarization occurs, a neuron cannot fire another action potential and muscle cells cannot contract. In short, without repolarization, the nervous system shuts down, the heart stops, and the muscles for breathing cease functioning.

Not only are potassium ions critical for action potentials but they are also the main factor responsible for generating the resting membrane potential. Recall that small numbers of potassium ions are constantly being lost from the cell to the ECF through leak channels. This loss of positive charges from the cytosol to the ECF helps to generate the negative resting membrane potential. As you've seen throughout this book, the negative resting membrane potential is the basis for essentially all the body's electrophysiology.

We briefly discussed the effects of potassium ion imbalances on the action potentials of the heart (see Chapter 17). Let's take a more detailed look at how the potassium ion concentration of the body fluids is regulated and at the wide-ranging effects that can result from imbalances.

Regulation of Potassium Ion Concentration

Significant numbers of potassium ions can be added to the body fluid from the diet, particularly when you eat a meal rich in fruits and vegetables. Such foods could raise the potassium ion concentration to a lethal level if all the ions remained in the ECF. Fortunately, several mechanisms within the endocrine and urinary systems help to prevent this from happening. For one, hormones such as insulin, aldosterone, and epinephrine stimulate the uptake of potassium ions by cells. This shifts the newly ingested potassium ions into the intracellular compartment.

The remaining excess potassium ions must be excreted, which is mostly handled by the kidneys via tubular secretion. Recall that potassium ions in the filtrate are reabsorbed, not secreted, in the proximal tubule and the nephron loop (see Chapter 24). This makes potassium ion secretion possible only in the distal tubule and collecting system. As you have learned, this is under the control of aldosterone, which triggers secretion of potassium ions (and hydrogen ions) and reabsorption of sodium ions. One of the more important stimuli for aldosterone release is a rising extracellular concentration of potassium ions. Conversely, when the extracellular potassium ion concentration falls, the secretion of aldosterone also generally falls, and more potassium ions are conserved. Note that aldosterone secretion is stimulated by several other factors that are unrelated to potassium ions, including decreased blood volume and blood pressure (and any other stimulus that triggers secretion of angiotensin-II) and a decreased blood pH. For this reason, potassium ion imbalances can result when these conditions arise.

Potassium Ion Imbalances

When the mechanisms we just described fail or when excess potassium ions are taken in or lost, potassium ion imbalances can result. Keep in mind that water imbalances can impact potassium ion concentration as well, but the effect is not as noticeable as with sodium ion concentration because fewer potassium ions are present in the ECF. The effects of excess potassium ions, or *hyperkalemia,* and insufficient potassium ions, or *hypokalemia,* are primarily electrophysiological in nature. Let's turn our attention to these imbalances.

Hyperkalemia We define **hyperkalemia** (hy′-per-kay-LEE-mee-uh; *kalem-* = "potassium") as a plasma potassium ion concentration above 4.5 mEq/l. Hyperkalemia is arguably one of the most dangerous electrolyte imbalances; indeed, a high dose of potassium ions can cause death almost immediately. Mild hyperkalemia alters the resting membrane potential—the excess of potassium ions in the ECF causes fewer potassium ions to leave the cell through leak channels. This means that the cell retains more positive charges, and as a result the resting membrane potential is more positive at rest than is normally the case (i.e., the membrane is slightly depolarized). For this reason, an action potential is more easily generated and the cells are more excitable with mild to moderate hyperkalemia. However, with severe hyperkalemia, the membrane can become so depolarized at rest that the cells are no longer excitable. In effect, this shuts down every excitable cell in the body, including neurons and muscle cells, and the result can be nearly instantaneous death.

Mild hyperkalemia is generally the result of renal failure or aldosterone insufficiency. It may also be due to widespread tissue damage that occurs with severe burns or trauma, as the damaged cells rupture and release their potassium ions into the ECF. Other potential causes include drugs such as aldosterone blockers and one commonly used for heart failure (find out more about this in *A&P in the Real World: Digoxin Toxicity and Hyperkalemia*). Severe hyperkalemia is almost always due to ingestion or administration of excess potassium ions. This can be intentional, as with a lethal injection, or accidental, as occurs when a medical professional mistakenly gives the wrong dose of potassium salts.

Hypokalemia   A plasma potassium ion concentration below 3.9 mEq/l is known as **hypokalemia.** The most common cause of hypokalemia is the use of diuretics, many of which lead to excess potassium ion loss in the urine. As with hyperkalemia, the effects of hypokalemia are primarily electrophysiological. However, whereas hyperkalemia makes the resting membrane potential more positive, hypokalemia makes it more negative, meaning that the cells become hyperpolarized. This change happens because more potassium ions exit the cell through leak channels, and the loss of positive charges makes the cell more negative. As a result, the cell becomes more difficult to stimulate, which leads to muscle weakness, mental status changes, and a slowed heart rate.

Quick Check

☐ 4. What is the main role of potassium ions in the body?

☐ 5. How is the concentration of potassium ions in the ECF regulated?

☐ 6. What happens to the resting membrane potential of excitable cells and their generation of action potentials in hyperkalemia and hypokalemia?

Digoxin Toxicity and Hyperkalemia

Digoxin (dih-JAWKS-in) is a drug derived from the foxglove plant that is used to treat heart failure. Its effectiveness as a heart failure agent comes from the fact that it blocks the functioning of the Na^+/K^+ pump in cardiac muscle cells. This blocking activity effectively lengthens the action potentials of the cardiac muscle cells, enabling a longer, stronger contraction.

Unfortunately, as with many drugs, digoxin comes with potential adverse effects, including abdominal pain and dizziness. In addition, the range between a therapeutic blood concentration of the drug and a toxic concentration is very small. As a result, it is relatively easy to develop *digoxin toxicity.* One of the main symptoms of digoxin toxicity is hyperkalemia. This is not true hyperkalemia in that the total number of potassium ions in the body is not elevated. Instead, these potassium ions are concentrated in the wrong compartment—the Na^+/K^+ pump is unable to drive them into the cytosol, and so they are not shifted from the interstitial fluid to the cytosol. Instead, they remain in the interstitial fluid around cardiac muscle cells. From here they diffuse into the blood, leading to the elevated blood concentration of potassium ions.

Calcium and Phosphate Ions

 **FLASHBACK**

1. What are the effects of parathyroid hormone? (p. 606)

2. How does vitamin D influence calcium ion homeostasis? (p. 203)

Calcium ions play many critical roles in the body. As you have learned, they are one of the main components of osseous tissue, giving bone its hardness and resistance to compression. In fact, about 99% of the total calcium ions in the body are located in bone as part of hydroxyapatite salts (see Chapter 6). Calcium ions are also required for muscle contraction, the plateau phase of the cardiac action potential, intracellular signaling, blood clotting, neuronal synaptic transmission, and more. In addition, they have an important role in the excitation of neurons—when calcium ion levels rise, neurons become less permeable to sodium ions, and the reverse is true as well.

We cannot discuss calcium ions without also discussing phosphate ions (PO_4^{3-}) because they are found together in hydroxyapatite crystals. The inorganic portion of bone cannot be built unless both ions are present. For this reason, they are regulated in tandem by the same hormones. Phosphate ions have other important functions in the body, for example, as an integral part of the structure of ATP. This section first examines the hormones and mechanisms that control phosphate and calcium ion levels in the ECF. The discussion then focuses on calcium ions, exploring what happens when these mechanisms are inadequate and imbalances result.

Regulation of Calcium and Phosphate Ion Homeostasis

The calcium ion concentration of the ECF measures about 8.7–10.4 mg/dl and rarely varies from this range by more than a few percentage points in either direction. This strict control is thanks to compensatory mechanisms that occur at three main sites: bone tissue, the distal tubules and collecting systems of the kidneys, and the small intestine. Simply put, when the level of calcium ions in the ECF falls, calcium ions are released from the bone by osteoclasts, more are reabsorbed from the filtrate in the kidneys, and more are absorbed from ingested food and liquids in the small intestine. Conversely, when the calcium ion level rises, calcium ions are deposited into bone by osteoblasts, fewer are reabsorbed from the filtrate, and fewer are absorbed from the small intestine.

The level of phosphate ions, however, is not nearly as tightly regulated, and the extracellular phosphate ion concentration varies throughout the day. Generally, the body maintains a slightly lower phosphate ion concentration than calcium ion concentration. This is because cells require a certain number of free calcium ions in their cytosol. When phosphate ions are present in large numbers, they bind calcium ions and form solid calcium phosphate salts, reducing the number of free calcium ions. Therefore, the body must excrete more phosphate ions and retain more calcium ions.

The two main hormones involved in calcium and phosphate ion regulation are *parathyroid hormone (PTH)* and *vitamin D3 (calcitriol)*. However, many other hormones influence this balance as well, including calcitonin, cortisol, and estrogens. Recall that PTH is released from the parathyroid glands in response to a decrease in the calcium ion concentration of the blood, and it triggers osteoclast activity and calcium ion reabsorption in the kidneys (see Chapter 16). In addition, it decreases the reabsorption of phosphate ions.

Another important effect of PTH is the activation of vitamin D3, which is a potent stimulator of calcium ion absorption by the small intestine. Vitamin D3 also assists PTH in increasing osteoclast activity and calcium ion reabsorption from the kidneys. With regard to phosphate ions, vitamin D3 triggers their absorption from the small intestine. This increases the concentration of phosphate ions in the ECF, in contrast to the effects of PTH, which decreases their level. This helps to ensure that the level of phosphate ions in the ECF does not fall too low.

Calcium Ion Imbalances

Unlike changes in phosphate ion concentration, small changes in calcium ion concentration can have immediate, dramatic impacts on many different body functions. Let's look at the effects of an abnormal increase and decrease in blood calcium ion concentration, known as *hypercalcemia* and *hypocalcemia*, respectively.

Hypercalcemia **Hypercalcemia** is defined as a plasma calcium ion concentration above 10.5 mg/dl. This condition is usually due to hyperparathyroidism but may also be caused by certain cancers, excess vitamin D, certain bone disorders, and renal failure. Hypercalcemia has wide-ranging effects on multiple organ systems. The most immediately noticeable effects are on the nervous system. As we discussed earlier, an elevated calcium ion concentration in the ECF makes neurons less permeable to sodium ions, which diminishes the ability of the neurons to depolarize. This in turn decreases the neurons' activity, leading to mental sluggishness, reduced reflex activity, and weak muscle activity (due to lack of stimulation from motor neurons).

Hypercalcemia affects other tissues, as well. Recall that calcium ions are responsible for the plateau phase of the cardiac action potential, which lengthens and strengthens the heart's contraction. When excess calcium ions are present, the plateau phase is shortened, making the contraction shorter and weaker. Hypercalcemia can cause decreased appetite and constipation due to decreased activity of gastrointestinal smooth muscle. It can also cause kidney stones, bone pain, and frequent urination.

Hypocalcemia A plasma calcium ion concentration lower than 8.7 mg/dl is considered to be **hypocalcemia.** The condition is generally due to a decrease in PTH, ineffective PTH (as occurs with renal failure), and vitamin D deficiency, although many other factors can contribute to it. Hypocalcemia has the opposite effect on neurons from hypercalcemia—the neurons become hyperexcitable, as more sodium ions enter and depolarize the neurons, stimulating action potentials. When hypocalcemia is

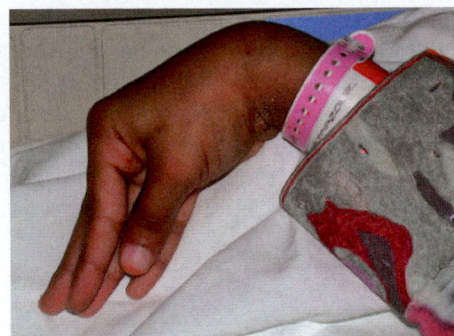

Figure 25.8 Carpopedal spasm of hypocalcemia.

severe, neurons begin firing spontaneously, leading to repetitive stimulation of skeletal muscles and tetanic (sustained) contractions. For example, the characteristic *carpopedal spasm* of hypocalcemia is shown in **Figure 25.8**. In addition, the plateau phase of the cardiac muscle cell action potential is lengthened, leading to excessively long and strong contractions. A calcium ion concentration below about 4 mg/dl is generally lethal because of these and other cardiac effects and tetanic spasm of the respiratory muscles.

Quick Check

- ☐ 7. What are the roles of calcium ions in normal physiology?
- ☐ 8. How are the concentrations of calcium and phosphate ions regulated?
- ☐ 9. What are the effects of hypercalcemia and hypocalcemia on the nervous system and the heart?

Other Ions Critical to Human Physiology

Many other ions play a role in human physiology, including the anion chloride and the cation magnesium. Let's briefly discuss their roles here.

Chloride Ions

Chloride ions have a variety of functions in the body. Their high concentration in the ECF makes them an important osmotic particle, along with sodium ions. Chloride ions are also a critical component of some other processes, including the production of hydrochloric acid secreted by the stomach. In addition, chloride ions are involved in the secretion of newly formed bicarbonate ions from erythrocytes. The concentration of chloride ions in the ECF is mostly determined by the kidneys, where the reabsorption of these ions is largely coupled to that of sodium ions. Chloride ions follow sodium ions passively from the filtrate to the ECF due to the electrical gradient created by sodium ion reabsorption. In addition, several transporters move the two ions together.

Magnesium Ions

Magnesium ions are critical to multiple cellular processes, as they are involved in the activation of many enzymes. In addition, magnesium salts are an important component of bone tissue. In fact, about half of the body's magnesium ions are located in bone; most of the rest reside in the cytosol, with only about 1% in the ECF. For this reason, the kidneys must excrete the majority of the magnesium ions that are ingested. Maintaining this low extracellular concentration of magnesium ions is critical because if the level in the ECF rises too high, it blocks the entry of calcium ions into cells, which can in turn lead to hypercalcemia.

Quick Check

☐ 10. What are the roles of chloride and magnesium ions in human physiology?

☐ 11. How is chloride ion reabsorption in the kidneys determined?

☐ 12. How is the concentration of magnesium ions in the ECF regulated?

Apply What You Learned

☐ 1. Addison's disease results in decreased secretion of aldosterone. What effect would this have on the levels of sodium, potassium, and chloride ions in the ECF?

☐ 2. How would the electrolyte imbalance in question 1 affect fluid balance?

☐ 3. Explain how the situation examined in questions 1 and 2 would change if, instead, excess aldosterone were secreted.

See answers in Appendix A.

MODULE 25.4
Acid-Base Homeostasis

Learning Outcomes

1. Explain the factors that determine the pH of blood, and describe how it is maintained within its normal range.

2. Describe the buffer systems that help to keep the pH of the body's fluids stable.

3. Describe the relationship of P_{CO_2} and bicarbonate ions to blood pH.

4. Describe the role of the respiratory system in regulating blood pH, and predict how hypo- and hyperventilation will affect blood pH.

5. Explain the mechanisms by which the kidneys secrete or retain hydrogen and bicarbonate ions, and describe how these processes affect blood pH.

6. Discuss the concept of compensation to correct respiratory and metabolic acidosis and alkalosis.

We now turn our attention toward ions with a different role in the body: hydrogen ions and various bases such as bicarbonate

ions. The normal hydrogen ion concentration of body fluids is equal to a pH range of about 7.35–7.45. The body must maintain the pH within this narrow, slightly alkaline, range because even a slight deviation can have disastrous consequences for our cells. Blood pH is maintained through a number of mechanisms, the most important of which are the respiratory system, the urinary system, and the body's **buffer systems.** There are two major types of buffer systems: *chemical buffer systems* and *physiological buffer systems.* Chemical buffers are, as the name suggests, chemical systems that work to buffer fluids in the body, whereas physiological systems are functions of organ systems that work to buffer fluids. In this module, we examine how the respiratory, urinary, and buffer systems maintain the pH of the blood and what happens when the pH moves out of its normal range. First, let's look at the sources of acids and bases in body fluids.

Sources of Acids and Bases in the Body

« FLASHBACK

1. How is carbon dioxide converted into carbonic acid? (p. 836)

Acids and bases in the body come from two main sources: (1) those that are formed as a normal part of metabolic processes, and (2) those that are ingested as a part of the diet. Throughout this book, you have seen how the body produces multiple metabolic acids. The biggest source of metabolic acids is carbon dioxide, which is a byproduct of glucose catabolism (see Chapter 23). Recall that carbon dioxide isn't an acid, but when it reacts with water, it forms carbonic acid (H_2CO_3), which can then dissociate to a hydrogen ion and a bicarbonate ion (HCO_3^-). Carbon dioxide is known as a **volatile acid** because in solution it is able to change states from carbonic acid back to carbon dioxide gas. This allows carbon dioxide to be eliminated from the body through both the lungs and the kidneys.

The body's reactions produce many other metabolic acids. For example, lactic acid is a byproduct of glycolysis, uric acid is produced from the breakdown of nucleic acids, the intermediates of the citric acid cycle are acids, and acidic ketone bodies are produced from fat breakdown. Such acids are known as **fixed acids** (or *nonvolatile acids*) because they cannot change to a gaseous state and so must be eliminated by the kidneys.

The diet is also a source of both acids and bases. Protein-rich foods contain amino acids, fat-rich foods contain fatty acids, and some fruits contain acids such as citric acid. Some foods, such as certain fruits and vegetables, also contribute small numbers of base ions, particularly bicarbonate. Note that although these foods do contribute some amounts of acids and bases to the body, they represent an exceedingly small minority that have essentially no effect on overall pH—the vast majority of acids and bases come from the body's own metabolic reactions, not from outside sources. In spite of this some in the alternative health industry promote a diet high in alkaline foods. See *A&P in the Real World, Pseudoscience Exposed: Alkaline Diets* for more information.

You may have noticed that there are a lot of metabolic sources of hydrogen ions, but relatively few for base ions. For this reason, the body's acid-base balance requires mechanisms that deal

Pseudoscience Exposed: Alkaline Diets

A popular fad in the alternative medicine business is the "alkaline diet," in which a person is advised to eat only foods with a basic pH (certain fresh fruits and vegetables) and avoid all acidic foods (all animal products, beans, seeds, and whole grains). What is the purpose of such a diet? Well, according to proponents, we all have blood that is too acidic due to our diets, and acid, they say, causes inflammation, which causes disease. An alkaline diet is therefore the literal cure for every disease, up to and including cancer. As you might expect, these assertions are generally made by people trying to sell you something.

Before we open our wallets, we need to examine these ideas more closely. For the first claim, that dietary foods and liquids affect the pH of the blood, remember that the pH of stomach acid is extremely low—about 1–3—and this is true no matter what you eat. Also, remember that the pancreas secretes bicarbonate ions to neutralize the acidic stomach contents entering the duodenum. So, regardless of the alkalinity or acidity of your diet, the food you eat is going to leave your stomach at a pH of about 2, and travel through your small intestine at a pH of about 7.

The second claim, that all diseases are the result of an acidic body fluid pH, is distorting real physiology. It's true that during an inflammatory response, the pH in the inflamed area falls due to phagocyte activity. But this pH decline is the *result* of the inflammation, not the *cause* of the inflammation. Saying the reverse is a little like saying a dented bumper caused a car accident.

So should you open your wallet for this diet? Well, you be the judge. Proponents do make one true claim, which is that an alkaline environment kills cancer cells. They're right: Cancer cells can't survive in a very alkaline environment, but neither can healthy cells. If the pH is high enough to kill cancer cells, all cells die. This is one of those times when it's a good thing that the diet can't possibly physiologically work as advertised.

with buffering hydrogen ions, excreting hydrogen ions, and conserving base ions. These mechanisms are examined next.

☐ 1. What are the major sources of acids for the body?
☐ 2. Why are acids the main focus of the body's acid-base balance mechanisms?

Chemical Buffer Systems

❮❮ FLASHBACK

1. How is a base related to an acid? (p. 48)
2. Which body system controls the carbonic acid–bicarbonate ion buffer system? How does this system affect the number of hydrogen ions in the body? (pp. 837–838)

Before discussing buffer systems, let's review a few definitions from chemistry (see Chapter 2). Recall that a *strong acid* is one that releases most of its hydrogen ions when placed in water, which lowers the pH of the solution. A *weak acid,* however, releases relatively few hydrogen ions in solution, and so has a much smaller impact on pH. Conversely, a *strong base* is one that binds and removes a great number of hydrogen ions from the solution, increasing the solution's pH. A *weak base* binds relatively few hydrogen ions in a solution and so has less impact on its pH.

The body's **chemical buffer systems,** which consist of a weak acid and its conjugate weak base, function to resist large swings in pH. When a strong acid is added to a solution, the weak base of the buffer system binds the released hydrogen ions and removes them from solution. When a strong base is added to a solution, the buffer system's weak acid releases its hydrogen ions to bind the base ions. Both situations minimize pH changes and help maintain acid-base homeostasis. The buffering capacity of the body's fluids is known as the *alkaline reserve.* When the alkaline reserve is depleted, acid-base imbalances can result.

There are three main chemical buffer systems in the body: the *carbonic acid–bicarbonate ion buffer system*, the *phosphate buffer system*, and the *protein buffer system*. These three systems are explored in the next subsections.

Carbonic Acid–Bicarbonate Ion Buffer System

The most important chemical buffer system in the blood is the **carbonic acid–bicarbonate ion buffer system,** which consists of the weak acid carbonic acid and its weak conjugate base, the bicarbonate ion. One source of carbonic acid is the reaction that we described earlier in which carbon dioxide reacts with water in erythrocytes, producing carbonic acid. When carbonic acid dissociates, it yields a hydrogen ion and a bicarbonate ion:

$$CO_2 + H_2O \leftrightarrow H_2CO_3 \leftrightarrow HCO_3^- + H^+$$

$$\text{Carbon dioxide} \quad \text{Water} \quad \text{Carbonic acid} \quad \text{Bicarbonate ion} \quad \text{Hydrogen ion}$$

Bicarbonate ions are also found as salts, primarily sodium bicarbonate ($NaHCO_3$). Normally, sodium bicarbonate completely dissociates in solution to form sodium and bicarbonate ions:

$$NaHCO_3 \leftrightarrow Na^+ + HCO_3^-$$

$$\text{Sodium bicarbonate} \quad \text{Sodium ion} \quad \text{Bicarbonate ion}$$

How do carbonic acid and bicarbonate ion buffer a solution? First, remember that a beaker of pure water contains an equal number of hydrogen and hydroxide ions, as shown at the top of **Figure 25.9**. Next, let's look at this process in an acidic solution. The upper left illustration in Figure 25.9a shows a strong acid, such as hydrochloric acid, being added to a beaker of water. Notice that the acid immediately donates its hydrogen ions to the solution, which increases its hydrogen ion concentration and therefore reduces its pH.

Now see what happens when we add the same acid to a solution buffered by sodium bicarbonate, shown at the bottom left in Figure 25.9a. Notice that here (after sodium bicarbonate dissociates), the hydrogen ion from the strong acid is accepted by the bicarbonate ion, forming carbonic acid. Since carbonic acid is a

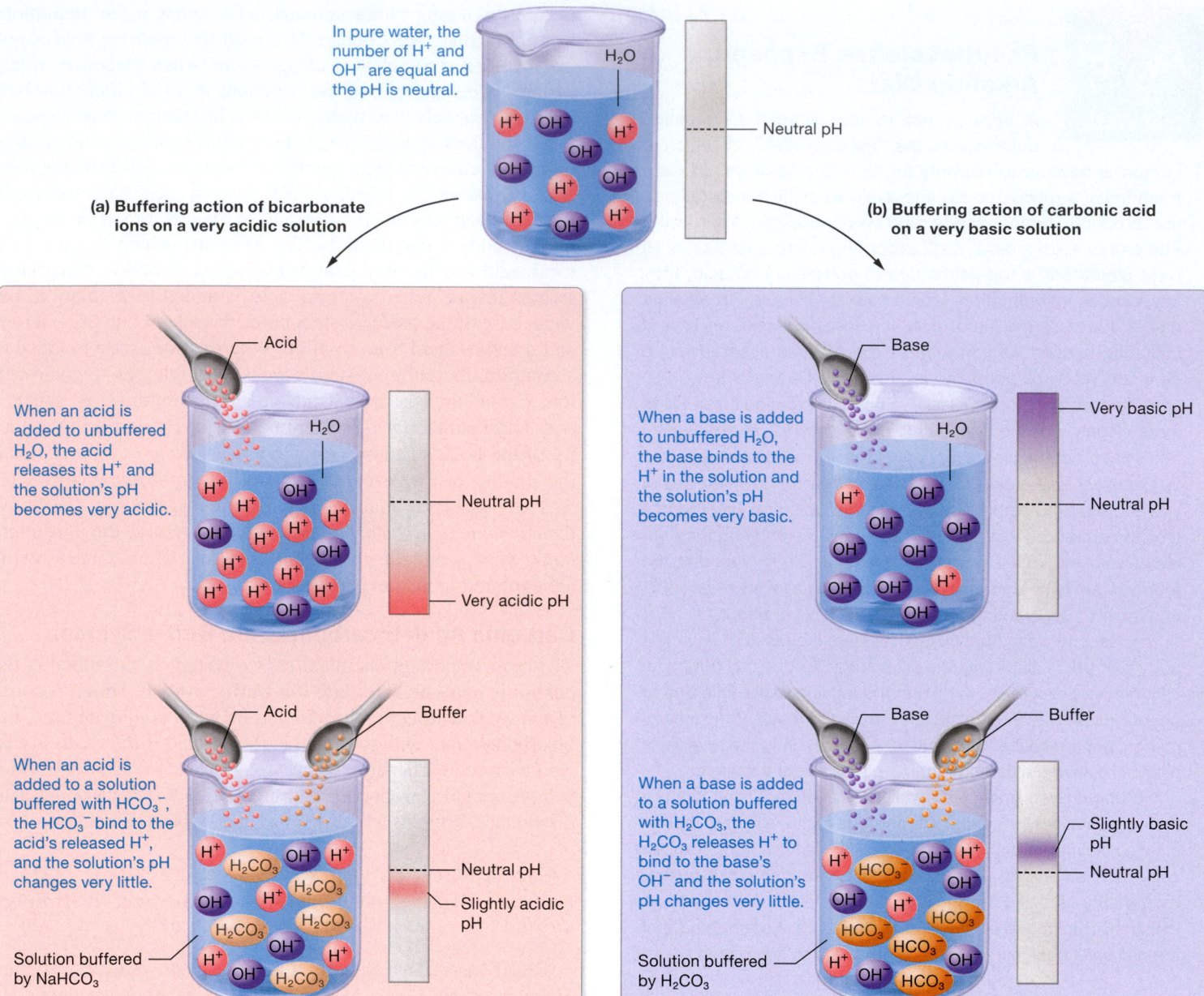

Figure 25.9 The effects of buffers.

weak acid, it doesn't release its hydrogen ions as readily, minimizing any pH changes in the solution.

Carbonic acid can similarly buffer a very basic solution. For example, see what happens when a strong base such as sodium hydroxide (NaOH) is added to water, as shown in the upper right illustration in Figure 25.9b. In this unbuffered solution, the released hydroxide ions remove many of the hydrogen ions from the solution and raise its pH, making it very basic. When carbonic acid is present, however, as shown at the bottom right in Figure 25.9b, it releases bicarbonate and hydrogen ions into the solution. The hydrogen ion binds to the hydroxide ion released from sodium hydroxide, forming water and a molecule of sodium bicarbonate. The bicarbonate ion is a weak base and will not remove significant numbers of hydrogen ions from the solution.

Phosphate Buffer System

The **phosphate buffer system** consists of the weak acid dihydrogen phosphate ($H_2PO_4^-$) and its conjugate base, hydrogen phosphate (HPO_4^{2-}). As in the carbonic acid–bicarbonate ion buffer system, the acid and conjugate base balance excess hydroxide and hydrogen ions, respectively, keeping the pH of a solution stable. The concentration of hydrogen phosphate is too low in the ECF—only about 8% of the bicarbonate ion concentration—to make it an important buffer here. However, it has a higher concentration in the cytosol and in the filtrate within the kidney tubules. For this reason, it is an important buffer in these locations.

Protein Buffer System

A huge variety of proteins is found in the body, particularly in the cytosol of our cells and in plasma. As you've discovered, proteins are composed of strings of amino acids, each containing a carboxylic acid group (−COOH). In solution, some of these carboxylic acid groups ionize to their weak conjugate base form, allowing them to act as a **protein buffer system.** When hydrogen ions are added to the cytosol, they can bind to these bases and be removed from the cytosol, preventing large swings in pH. There are many important intracellular protein buffers; in fact, about 60–70% of the body's total buffering capacity takes the form of protein buffers in the cytosol. One example of an intracellular protein buffer is the hemoglobin inside erythrocytes, which acts as a buffer for hydrogen ions formed from the dissociation of carbonic acid.

Quick Check

☐ 3. How can carbon dioxide become an acid?

☐ 4. How do carbonic acid and bicarbonate ions buffer the plasma?

☐ 5. What are the components of the phosphate and protein buffer systems?

Physiological Buffer Systems: Respiratory and Renal Regulation of Blood pH

« FLASHBACK

1. What happens to the pH of the blood when a person hypoventilates or hyperventilates? (pp. 837–838)

2. How is bicarbonate ion absorption coupled with hydrogen ion secretion? (p. 970)

3. What effect does aldosterone have on hydrogen ion secretion in the kidneys? (p. 972)

The chemical buffers that we just discussed respond to *changes* in the hydrogen ion concentration of the body fluids in order to maintain it within its normal range. However, for these systems to work, the body fluids must contain adequate numbers of base ions, particularly bicarbonate ions. The body must also have a way to rid itself of hydrogen ions in order to maintain the normal, slightly alkaline pH. Two systems perform these tasks: the respiratory system, which controls the amount of volatile acids in the ECF, and the urinary system, which controls the amount of fixed acids and bicarbonate ions in the ECF. These systems can also work to correct hydrogen ion imbalances. For this reason, these systems are sometimes referred to as **physiological buffer systems.** You have read about these mechanisms previously in this book (see Chapters 21 and 24), but let's review them now.

Respiratory System Effects on Blood pH

In the previous section, we saw that the carbonic acid in body fluids is formed from the reaction of carbon dioxide with water. The lungs directly control the amount of carbon dioxide in the blood, so it follows that the lungs also control, to a significant extent, the levels of carbonic acid and hydrogen ions. Under normal conditions, the amount of carbon dioxide expired by the lungs matches the amount of carbon dioxide produced by metabolic reactions, which is why your respiratory rate doesn't remain constant. When you are physically active, glycolytic catabolism and oxidative catabolism occur more rapidly and more carbon dioxide is generated. As a result, your respiratory rate increases to match the higher level of carbon dioxide production. The opposite is also true—when you are sedentary, you generate less carbon dioxide and so your respiratory rate decreases.

The respiratory system also influences the number of bicarbonate ions in the plasma. Remember that the reaction between carbon dioxide and water occurs rapidly in erythrocytes due to the presence of the enzyme *carbonic anhydrase* (refer back to Figure 21.24). When carbonic acid is formed in erythrocytes, much of it dissociates into bicarbonate and hydrogen ions. The hydrogen ions that form bind to hemoglobin, which buffers the cytosol of erythrocytes. The bicarbonate ions, however, are transported into the plasma, where they buffer fixed acids.

Urinary System Effects on Blood pH

The kidneys work with the lungs in two ways to determine the number of hydrogen ions and base ions in the blood. First, the kidneys can excrete fixed acids that the lungs cannot excrete, such as lactic acid, ketones, phosphoric acid, uric acid, and ammonium ions (NH_4^+). About 25% of the body's hydrogen ions are eliminated in this manner.

Second, the kidneys contribute to acid-base homeostasis by controlling the concentration of bicarbonate ions in the blood. Generally, the kidneys reabsorb all the bicarbonate ions from the filtrate. Recall, though, that there are no true transport mechanisms to move bicarbonate ions from the filtrate into the tubule cells. For this reason, other, more roundabout methods for reabsorption are needed. In addition, the kidneys can actually manufacture new bicarbonate ions when the pH of the blood falls, and secrete bicarbonate ions when the pH of the blood rises. We examine these mechanisms next.

Hydrogen Ion Secretion and Bicarbonate Ion Reabsorption in the Proximal Tubule

The proximal tubule cells secrete hydrogen ions from fixed acids by a secondary active transport process. Recall that this process is coupled to bicarbonate ion reabsorption—the secreted hydrogen ions bind to bicarbonate ions in the filtrate to form carbonic acid. Carbonic anhydrase then catalyzes the conversion of the carbonic acid to carbon dioxide and water, both of which enter the kidney tubule cell. Inside the cell, they react again with carbonic anhydrase to form carbonic acid, which dissociates into bicarbonate and hydrogen ions. The bicarbonate ion leaves the cell and enters the blood, whereas the hydrogen ion is recycled, as it is re-secreted to bind another bicarbonate ion in the filtrate (Figure 24.17 provides a diagram of this process). This process accounts for reabsorption of nearly 100% of the bicarbonate ions in the filtrate and secretion of large numbers of hydrogen ions.

Another process that occurs in the proximal tubule secretes hydrogen ions while actually forming *new* bicarbonate ions. As you can see in **Figure 25.10**:

① **Glutamine enters a proximal tubule cell and undergoes reactions that yield two ammonia molecules and two bicarbonate ions.**

② **The two molecules of ammonia combine with hydrogen ions, producing two acidic ammonium ions. The two ammonium ions are secreted into the filtrate for excretion in urine.**

③ **Bicarbonate ions are transported into the interstitial fluid and then into the blood.**

Although a huge number of hydrogen ions are secreted in the proximal tubule, the pH of the filtrate doesn't change all that much. The explanation for this is that many of the secreted hydrogen ions are recycled during the reabsorption of bicarbonate ions, and so the concentration of hydrogen ions in the filtrate increases only about 3 or 4 times. In addition, the hydrogen ions that are not recycled are either buffered as ammonium ions or bind to other buffers in the filtrate, particularly phosphate buffers. These two facts combined mean that the filtrate pH reaches a minimum value of only about 6.7 in the proximal tubule.

Hydrogen Ion Secretion in the Distal Tubule and Collecting System The major acidification of the urine takes place in the late distal tubule and collecting system by cells called *intercalated cells* (in-TER-kuh-lay′-ted). These cells have primary active transport pumps in their plasma membranes that pump hydrogen ions into the filtrate. Notice in **Figure 25.11** that the source of the hydrogen ions is actually carbon dioxide. The process proceeds by the following steps:

① **Carbon dioxide diffuses from the blood into the distal tubule cell.** Carbon dioxide from the blood and interstitial fluid diffuses freely into the tubule cells when a concentration gradient is present.

② **Carbon dioxide reacts with water to form carbonic acid, which then dissociates into hydrogen and**

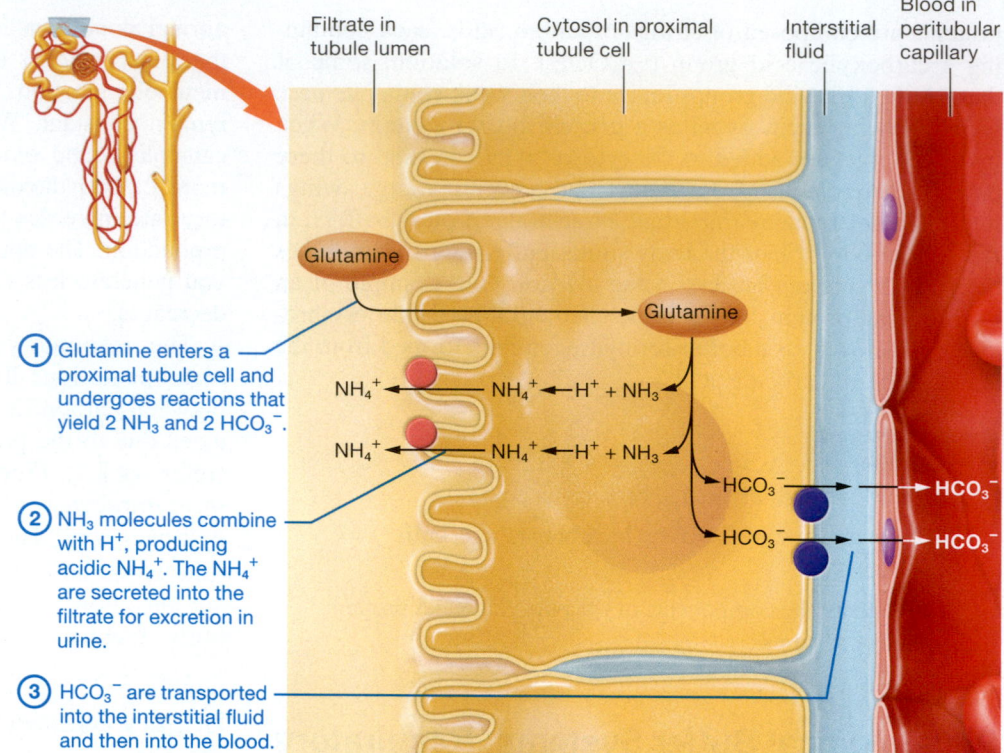

① Glutamine enters a proximal tubule cell and undergoes reactions that yield 2 NH_3 and 2 HCO_3^-.

② NH_3 molecules combine with H^+, producing acidic NH_4^+. The NH_4^+ are secreted into the filtrate for excretion in urine.

③ HCO_3^- are transported into the interstitial fluid and then into the blood.

Figure 25.10 Production of bicarbonate ions from glutamine in the proximal tubule.

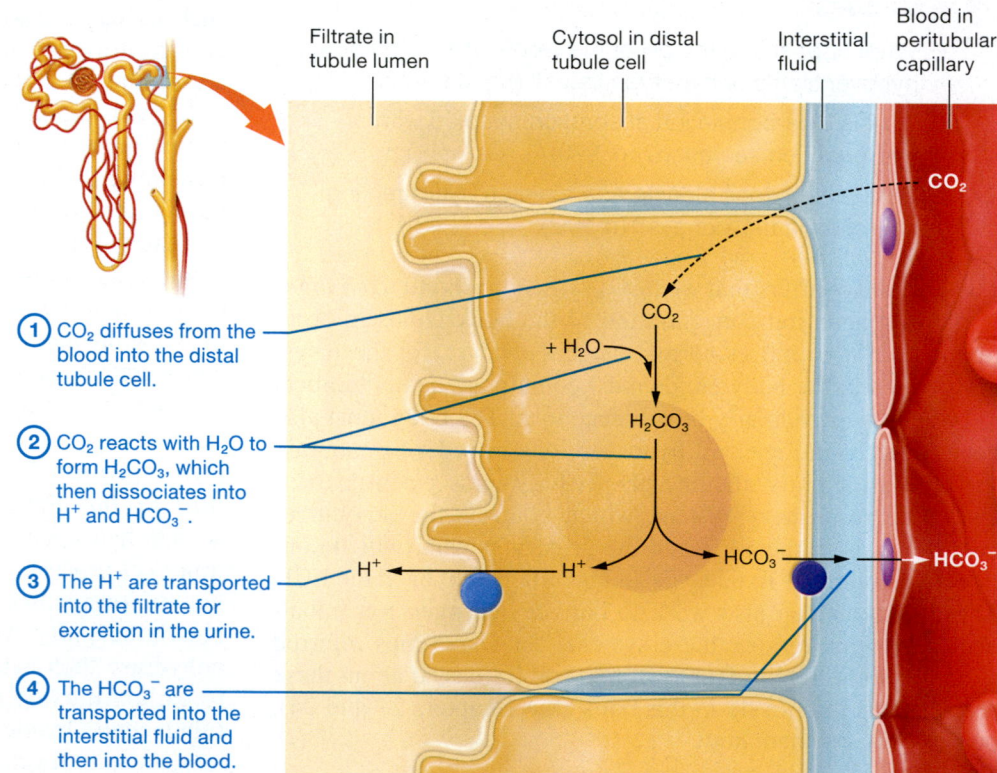

① CO_2 diffuses from the blood into the distal tubule cell.

② CO_2 reacts with H_2O to form H_2CO_3, which then dissociates into H^+ and HCO_3^-.

③ The H^+ are transported into the filtrate for excretion in the urine.

④ The HCO_3^- are transported into the interstitial fluid and then into the blood.

Figure 25.11 Secretion of hydrogen ions in the distal tubule and collecting system.

bicarbonate ions. In the distal tubule cell, carbon dioxide and water undergo a reaction catalyzed by carbonic anhydrase to form carbonic acid. The carbonic acid dissociates into hydrogen and bicarbonate ions.

③ **The hydrogen ions are transported into the filtrate for excretion in the urine.** The newly formed hydrogen ions are transported into the filtrate by active transport pumps, including one that exchanges hydrogen ions for potassium ions. Note that these hydrogen ions, unlike the ones in the proximal tubule, are not recycled.

④ **The bicarbonate ions are transported into the interstitial fluid and then into the blood.** The newly formed bicarbonate ions are transported across the tubule cell plasma membrane into the interstitial fluid and blood.

Although the body needs to be able to eliminate excess hydrogen ions in the urine to maintain acid-base homeostasis, a pH in the filtrate that is too acidic inhibits hydrogen ion pumps. If the pH of the filtrate reaches about 4.5, the pumps are unable to secrete any more hydrogen ions. To avoid this, the secreted hydrogen ions in the distal tubule filtrate bind to hydrogen phosphate ions (HPO_4^{2-}) of the phosphate buffer system, forming dihydrogen phosphate ($H_2PO_4^-$). This allows the hydrogen ions to be excreted in urine without contributing to its acidity.

The process in the distal tubule and collecting system accounts for only about 5% of the total hydrogen ion secretion, but it's the major factor that determines the pH of the urine. This is largely because the hydrogen ions secreted from the distal tubule and collecting system are not recycled. As a result, the concentration of hydrogen ions in the filtrate can increase up to 900 times that of the interstitial fluid. Note that this process is influenced by aldosterone. When the aldosterone level rises, the intercalated cells secrete more hydrogen ions. Conversely, when the aldosterone level falls, fewer hydrogen ions are secreted.

> ### Quick Check
>
> ☐ 6. How do the rate and depth of ventilation correlate to the pH of the ECF?
>
> ☐ 7. Explain how hydrogen ion secretion is coupled with bicarbonate ion reabsorption in the nephron.
>
> ☐ 8. How are new bicarbonate ions formed in the cells of the nephron?

Acid-Base Imbalances

◀◀ FLASHBACK

1. What is the cause of diabetic ketoacidosis? (p. 616)
2. Which type of acid is produced by cells of the gastric glands? (p. 865)

Normally, the body's buffer systems ensure that the pH of body fluids remains within a narrow range, even when acids or bases are added or lost. But when these buffer systems are overwhelmed by a significant loss or gain of acids or bases, they can fail, resulting in a pH imbalance. The two basic types of pH imbalance are *acidosis* and *alkalosis*. The body attempts to correct both types of imbalances through a process called *compensation,* which involves **respiratory compensation** and/or **renal compensation.**

Although compensation can bring the pH closer to the normal range, it is not a permanent solution to the imbalance. Rather, the permanent solution lies in correcting the root cause of the imbalance. Most pH imbalances are secondary to other causes, such as respiratory disorders, kidney failure, or gastrointestinal problems. For this reason, treatment is generally aimed at the primary cause, and when this is resolved, the pH returns to the normal range. The upcoming subsections explore the types and causes of pH imbalances and how compensation occurs.

Acidosis

Acidosis is defined as a body fluid pH of less than 7.35 (note that *acidemia* refers to low blood pH specifically). The pH of a patient with acidosis generally remains in the alkaline range (it is rare to see a pH less than 7.0), but it is still more acidic than the normal range. Acidosis develops when more hydrogen ions are added than the body's buffers can bind or when the number of buffers such as bicarbonate ions decreases. This causes neurons to become less excitable, leading to signs and symptoms of nervous system depression, such as slurred speech, drowsiness, and unresponsiveness. Left untreated, acidosis can lead to coma and, in severe cases, death.

There are many different causes of acidosis, which can be broadly divided into two classes: *respiratory acidosis* and *metabolic acidosis.* Irrespective of the cause of acidosis, the body's response is generally the same: decrease the number of hydrogen ions and increase the number of base ions, particularly bicarbonate.

Respiratory Acidosis and Compensation We define the condition **respiratory acidosis** as a decrease in the pH of body fluids due to excess carbon dioxide. The carbon dioxide excess happens when ventilation decreases, which is known as *hypoventilation* (see Chapter 21). Hypoventilation leads to the accumulation of carbon dioxide because less of it is being exhaled. This interferes with the carbonic acid–bicarbonate ion buffer system and causes a shift toward excessive carbonic acid, lowering the pH of the blood.

There are three general causes of respiratory acidosis: suppressed ventilation from brainstem dysfunction, blockage of air passages in the lungs, and decreased gas exchange in the alveoli. Whatever the cause, the respiratory and urinary systems attempt to compensate for the acidosis.

Respiratory compensation begins within minutes of the pH disturbance, as excess hydrogen ions in the ECF stimulate chemoreceptors in the brain. These neurons then trigger an increase in the rate and depth of ventilation in an attempt to rid the body of excess carbon dioxide (and indirectly hydrogen ions). This is why an individual in respiratory distress, which initially causes hypoventilation, will generally have an increased respiratory rate. It might seem that the individual is breathing rapidly because of an inability to "catch" his or her breath, but the elevated respiratory rate is generally due to elevated carbon dioxide, not low oxygen.

Respiratory compensation for respiratory acidosis is often not terribly effective because disordered ventilation was the initial cause of the problem. Such cases therefore rely more on renal compensation: The kidney tubule cells absorb all the available bicarbonate ions from the filtrate, secrete hydrogen ions, and produce new bicarbonate ions from glutamine. Renal compensation takes several hours to days to have a significant effect on overall blood pH, so these actions by the kidneys won't be immediately noticeable.

Metabolic Acidosis and Compensation **Metabolic acidosis** is defined as the addition of hydrogen ions to the ECF (from acids other than carbon dioxide) or loss of bicarbonate ions. The term refers to the fact that the condition can result from the addition of metabolic acids such as lactic acid, uric acid, and ketones to the blood. There are many different causes of metabolic acidosis, including prolonged diarrhea (which results in the loss of bicarbonate ions to the feces), the production of excess metabolic acids (such as occurs with diabetic ketoacidosis), failure of the kidneys to reabsorb bicarbonate ions or secrete hydrogen ions, and the ingestion of acidic drugs and toxins such as methanol.

The first compensatory response to metabolic acidosis is respiratory compensation, which occurs in the form of hyperventilation. In fact, hyperventilation in the absence of overt respiratory distress is one of the main signs that can point a clinician toward a diagnosis of metabolic acidosis.

The extent to which renal compensation occurs depends on the health of the kidneys. Obviously, if renal failure is the cause of acidosis, the kidneys aren't going to be able to contribute to the compensation. However, if the cause is unrelated to the kidneys, they will work overtime to eliminate hydrogen ions and retain bicarbonate ions, as we saw with respiratory acidosis.

ConceptBOOST

How Can Respiratory Changes Compensate for Metabolic Acidosis?

We have just established that the respiratory rate increases in metabolic acidosis as part of respiratory compensation. But how does changing the level of carbon dioxide in the blood help when the problem is metabolic in nature? The answer is that respiratory compensation cannot correct the actual cause of the metabolic imbalance. It can, however, correct the resulting pH imbalance. Remember that the pH of a solution depends only on the *number* of hydrogen ions present, not on the source of the hydrogen ions. Take a look at the following simple graphic, which shows two hypothetical samples of blood, one from a person with normal pH (on the left) and one from a person suffering from metabolic acidosis (on the right). The hydrogen ions from carbonic acid are represented in purple, and those from the metabolic acids are represented in blue:

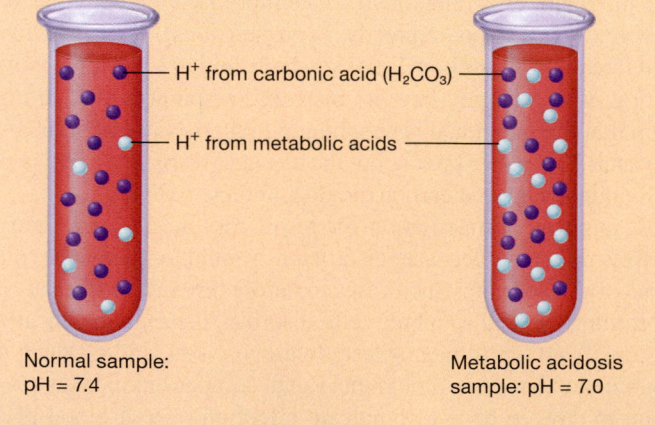

Normal sample: pH = 7.4

H⁺ from carbonic acid (H_2CO_3)

H⁺ from metabolic acids

Metabolic acidosis sample: pH = 7.0

Notice that the normal sample of blood has a total of 20 hydrogen ions (15 from carbonic acid and 5 from metabolic acids) and a pH of 7.4. The metabolic acidosis sample, however, has a total of 30 hydrogen ions as a result of the excess metabolic acids and a pH of 7.0.

Now let's see what happens when the person with metabolic acidosis starts hyperventilating and exhaling more carbon dioxide. As carbon dioxide molecules are exhaled, the number of hydrogen ions in the blood decreases. You can see this in the blood sample on the right.

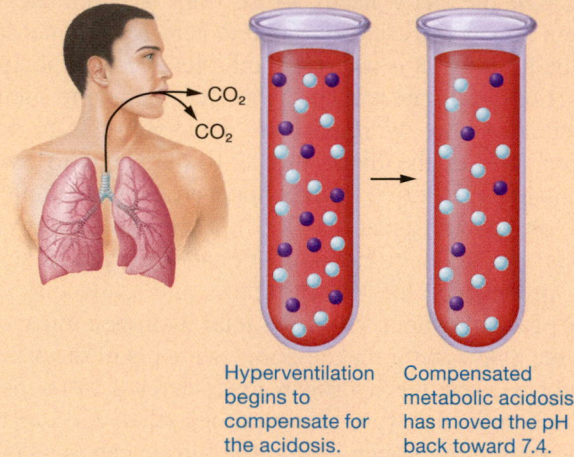

CO_2

CO_2

Hyperventilation begins to compensate for the acidosis.

Compensated metabolic acidosis has moved the pH back toward 7.4.

Notice that this sample of blood has 20 hydrogen ions (15 from metabolic acids, 5 from carbonic acid), the same number as in the initial tube, with a normal pH of 7.4. At this point, we have a compensated metabolic acidosis. The condition that caused the acidosis in the first place has not been resolved, as shown by the fact that there is still an elevated number of metabolic hydrogen ions. But the total number of hydrogen ions has gone down, thanks to respiratory compensation.

Of course, our diagrams here are very simplified and the pH values in the figures aren't realistic given the numbers of hydrogen ions illustrated. But the basic principle still stands—respiratory compensation for metabolic acidosis works by decreasing the total number of hydrogen ions in the blood. As you will see in the next section, the same situation exists for metabolic alkalosis but in reverse—carbon dioxide is retained to *increase* the total number of hydrogen ions in the body and return the pH to normal. ■

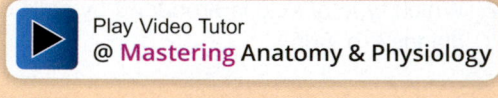
Play Video Tutor
@ Mastering Anatomy & Physiology

Alkalosis

Alkalosis is defined as a body fluid pH greater than 7.45 (*alkalemia* refers specifically to an elevated blood pH). This occurs when more base ions are added than the buffers can handle or when the number of hydrogen ions in the body fluids decreases. Alkalosis increases the excitability of neurons, and so they fire action potentials inappropriately. This results in sensory symptoms such as numbness and tingling and motor symptoms such as muscle twitches or tetanic contractions. Untreated, alkalosis can lead to seizures and death due to sustained contractions of respiratory muscles.

As with acidosis, there are two types of alkalosis: *respiratory alkalosis* and *metabolic alkalosis.* The compensatory response for both involves excretion of bicarbonate ions and retention of hydrogen ions.

Respiratory Alkalosis and Compensation

The condition **respiratory alkalosis** is caused by a loss of carbon dioxide through the lungs due to hyperventilation. Generally, this results from psychological states that increase the rate of breathing. The condition may also occur when an individual is at extremely high altitudes and the low P_{O_2} of the atmosphere triggers hyperventilation.

If hyperventilation is psychological in nature, there generally isn't much of a metabolic compensatory response. Usually, if the pH of the blood becomes too high, the individual quickly loses consciousness and the respiratory centers in the brainstem slow the respiratory rate and retain carbon dioxide to try to return the fluid pH to the normal range. If the hyperventilation is physiological in nature, however, the compensatory response primarily involves the kidneys excreting bicarbonate ions and retaining hydrogen ions.

Metabolic Alkalosis and Compensation

Metabolic alkalosis results from a loss of hydrogen ions or an excess of bicarbonate ions. Hydrogen ion loss is commonly the result of prolonged vomiting and the loss of acidic stomach contents. It may also be produced by certain diuretics or excess aldosterone, both of which lead to increased sodium ion reabsorption with concurrent hydrogen ion secretion. An increase in the bicarbonate ion level can happen when excessive bicarbonate salts have been ingested; the "culprit" is usually oral antacids taken for stomach ulcers or severe acid reflux.

Compensation for metabolic alkalosis is both respiratory and renal. The respiratory response to a decreased hydrogen ion concentration is hypoventilation. This retains carbon dioxide and increases the number of hydrogen ions in the body. The renal response to metabolic alkalosis is the same as for respiratory alkalosis—the kidneys retain hydrogen ions and secrete bicarbonate ions into the urine.

Arterial Blood Gases

Acid-base imbalances can be difficult to distinguish from one another with standard blood tests, which examine venous blood. To get an accurate picture of a patient's acid-base balance, we need instead to look at the arterial blood. In a procedure to determine *arterial blood gases* (*ABGs*), blood is drawn from the radial or brachial artery, and the pH, P_{CO_2}, and concentration of bicarbonate ions in the blood are measured. From these values, it is possible in many cases to surmise the type of acid-base imbalance and the degree of compensation that has occurred. If little to no compensation has occurred, the

imbalance is called *uncompensated;* if some compensation has occurred, but the pH of the blood is still abnormal, it is called *partially compensated;* and if the pH of the blood is normal, but the P_{CO_2} and the bicarbonate ion level are abnormal, it is called *fully compensated.*

Each type of acid-base disturbance has a characteristic set of ABG findings, which are as follows:

- *Respiratory acidosis.* The primary abnormality seen with respiratory acidosis is an elevated arterial P_{CO_2} and lower than normal blood pH. If the bicarbonate ion level is elevated, then renal compensation has begun and the acidosis is partially compensated.
- *Metabolic acidosis.* The values for metabolic acidosis are slightly more complex, but the general rule of thumb is that you will have a decreased pH and lower bicarbonate ion level in the blood, although this might not always be the case. The P_{CO_2} is also generally lower than normal because of respiratory compensation. Due to the rapid effects of respiratory compensation, most cases of metabolic acidosis are partially compensated.
- *Respiratory alkalosis.* The main findings with respiratory alkalosis are fairly straightforward—an elevated blood pH and a low P_{CO_2}. If the bicarbonate ion level is also low, this signifies partial compensation by the kidneys.
- *Metabolic alkalosis.* Metabolic alkalosis shows an elevated blood pH and a high bicarbonate ion level. If respiratory compensation has occurred, the P_{CO_2} will be elevated.

Keep in mind that these are just general guidelines; many cases of acid-base imbalance are quite complex and do not fit these profiles.

Table 25.1 will help you study and remember the causes and compensations of acidosis and alkalosis.

Quick Check

☐ 9. How do metabolic acidosis and respiratory acidosis differ?

☐ 10. How does the body compensate for metabolic acidosis and respiratory alkalosis?

☐ 11. What are the main differences between metabolic alkalosis and respiratory alkalosis?

Apply What You Learned

☐ 1. Mr. Wong presents with pneumonia and asthma that has progressively worsened over several days, and has led to obstruction of his airways. The physician in the emergency department orders arterial blood gases. Predict what you are likely to find in the results of this test.

☐ 2. You note that Mr. Wong's respiratory rate is elevated. Explain why he is hyperventilating. Will this likely help his pH imbalance?

☐ 3. A urinalysis shows that the pH of Mr. Wong's urine is 4.5. Are his kidneys going to be able to compensate any further for his acidosis? Why or why not?

See answers in Appendix A.

Table 25.1 Summary of Acid-Base Disorders

Type of Disorder	Characteristic Blood Gas Findings (without Compensation)	Causes	Main Methods of Compensation
Respiratory acidosis	• Decreased pH • Increased P_{CO_2}	• Blockage of air passages in the lungs • Decreased gas exchange in the alveoli • Decreased ventilation from brainstem dysfunction	• Hyperventilation to remove CO_2 • Retention of HCO_3^- by the kidneys • Excretion of H^+ by the kidneys
Respiratory alkalosis	• Increased pH • Decreased P_{CO_2}	• Psychological states that cause hyperventilation • Brain tumor or other damage to respiratory centers • Hyperventilation due to high altitudes and low atmospheric P_{O_2}	• Slowed respiratory rate • Excretion of HCO_3^- and retention of H^+ by the kidneys (if alkalosis is physiological rather than psychological)
Metabolic acidosis	• Decreased pH • Decreased HCO_3^-	• Prolonged diarrhea • Excessive production of metabolic acids • Renal failure • Ingestion of acidic drugs or toxins	• Hyperventilation • Retention of HCO_3^- from the kidneys • Excretion of H^+ by the kidneys * (The second and third responses are only possible if renal failure is not the cause of metabolic acidosis.)
Metabolic alkalosis	• Increased pH • Increased HCO_3^-	• Prolonged vomiting • Ingestion of certain diuretics or large doses of oral antacids • Excess aldosterone	• Slowed respiratory rate • Retention of H^+ by the kidneys • Excretion of HCO_3^- by the kidneys

MODULE 25.5
An Example of Fluid, Electrolyte, and Acid-Base Homeostasis

Learning Outcomes

1. Provide specific examples to demonstrate how the cardiovascular, endocrine, and urinary systems respond to maintain homeostasis of fluid volume, electrolyte concentration, and pH in the body.

You've seen in this chapter that the maintenance of fluid, electrolyte, and acid-base homeostasis requires the input of the two systems in the body whose primary responsibilities are the maintenance of homeostasis: the endocrine and nervous systems. You've also discovered that control of all three variables is intertwined. All three must be balanced together, as imbalances of one variable cause imbalances of the others. To illustrate this, we examine the effects of and response to a common imbalance: dehydration from inadequate fluid intake.

The process shown in **Figure 25.12** first deals with the homeostatic imbalances that occur with dehydration, and then with how all these imbalances are corrected together. The general sequence of events is another example of the Feedback Loops Core Principle (p. 28), as it follows a multistep feedback loop. The steps are as follows:

CORE PRINCIPLE
Feedback Loops

① **Total body water decreases, leading to the following imbalances of electrolytes and extracellular fluid pH:**
  - decreased blood volume and possibly blood pressure;
  - increased extracellular electrolyte concentrations, particularly that of sodium ions;
  - increased osmolarity of the ECF, which draws water out of the cytosol of cells; and
  - increased concentrations of metabolic acids in the ECF (due to fewer water molecules)—metabolic acidosis may develop.

② **Juxtaglomerular cells release renin, which leads to formation of angiotensin-II.** The decreased blood pressure and blood volume trigger renin release from the juxtaglomerular cells of the kidneys, leading to the production of more angiotensin-II.

③ **Angiotensin-II has multiple effects:**
③a **It causes vasoconstriction, increases sodium ion reabsorption (and water follows by osmosis), and stimulates thirst.** One of the most immediate effects of angiotensin-II is to increase blood pressure through vasoconstriction. It also acts on the proximal tubule of

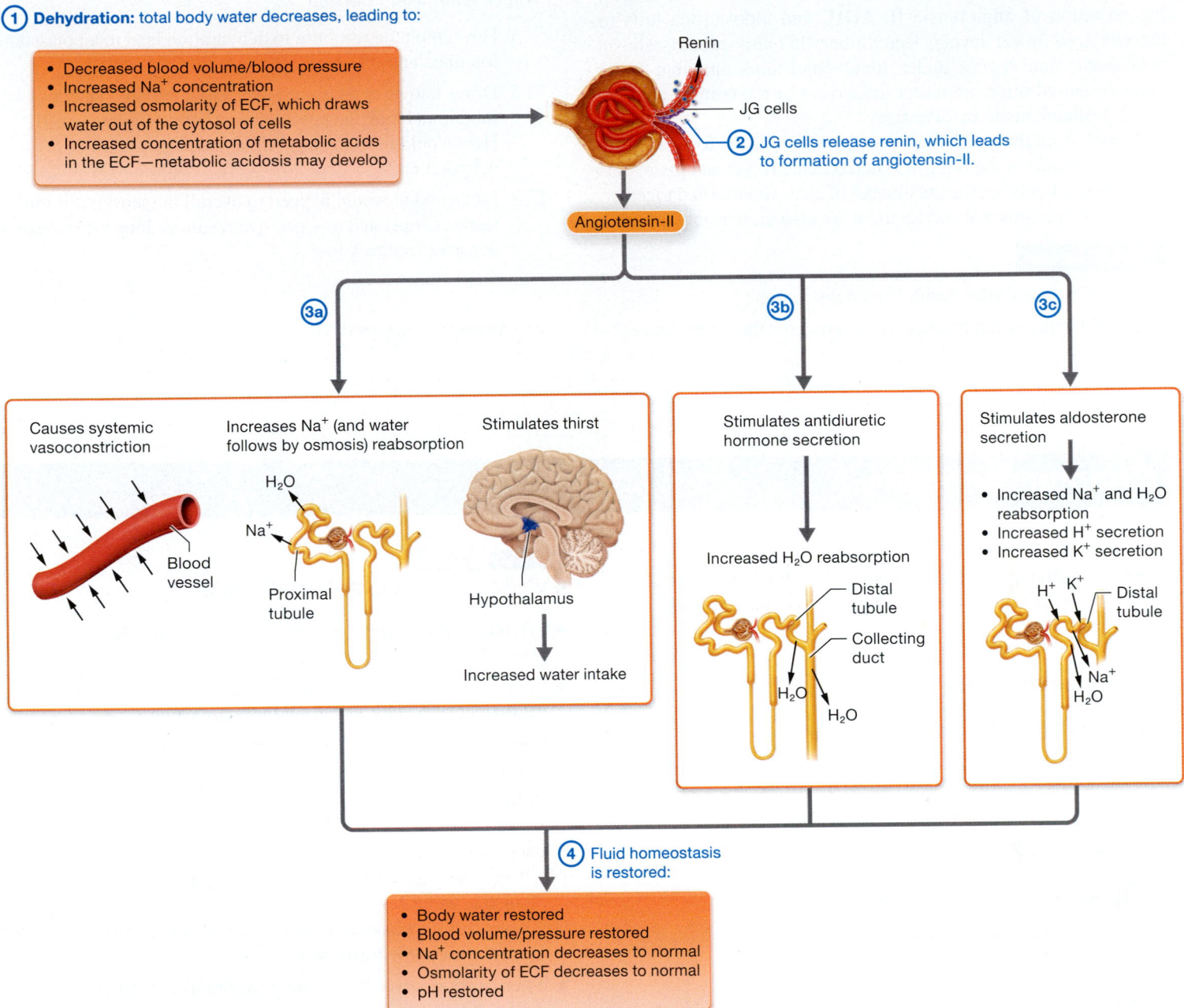

① **Dehydration:** total body water decreases, leading to:

- Decreased blood volume/blood pressure
- Increased Na^+ concentration
- Increased osmolarity of ECF, which draws water out of the cytosol of cells
- Increased concentration of metabolic acids in the ECF—metabolic acidosis may develop

Renin

JG cells

② JG cells release renin, which leads to formation of angiotensin-II.

Angiotensin-II

③a

Causes systemic vasoconstriction

Blood vessel

Increases Na^+ (and water follows by osmosis) reabsorption

H_2O

Na^+

Proximal tubule

Stimulates thirst

Hypothalamus

Increased water intake

③b

Stimulates antidiuretic hormone secretion

Increased H_2O reabsorption

Distal tubule

Collecting duct

H_2O

H_2O

③c

Stimulates aldosterone secretion

- Increased Na^+ and H_2O reabsorption
- Increased H^+ secretion
- Increased K^+ secretion

H^+ K^+

Distal tubule

Na^+

H_2O

④ Fluid homeostasis is restored:

- Body water restored
- Blood volume/pressure restored
- Na^+ concentration decreases to normal
- Osmolarity of ECF decreases to normal
- pH restored

Figure 25.12 Physiological responses to dehydration.

the nephron to increase fluid retention from the kidneys through stimulation of sodium ion reabsorption. Finally, it also stimulates thirst, which will lead to increased water intake if conditions permit.

③b It stimulates antidiuretic hormone (ADH) secretion. Several stimuli, primarily angiotensin-II and the increased osmolarity of the ECF, cause more ADH to be released. This dramatically increases water reabsorption from the distal tubule and collecting system.

③c It stimulates aldosterone secretion. Angiotensin-II also stimulates aldosterone release. Aldosterone triggers sodium ion reabsorption from the distal tubule and collecting system, and water follows by osmosis. It also

stimulates potassium and hydrogen ion secretion. This hydrogen ion secretion helps to restore the normal pH of the blood.

④ **Fluid homeostasis is restored.** All the actions of angiotensin-II, ADH, and aldosterone combined increase the amount of water in the body. This restores the total body water, blood volume, and blood pressure, and decreases the osmolarity of the ECF back to normal. The increased amount of water also restores the normal sodium ion concentration. Note that although these mechanisms caused sodium ion reabsorption, more water molecules are reabsorbed than sodium ions, so the concentration still returns to normal. Finally, the excretion of hydrogen ions restores the normal pH of the body fluids.

When the imbalanced variables return to their normal ranges, the secretion of angiotensin-II, ADH, and aldosterone falls to the previous lower levels. Remember that this process shows how correction occurs under ideal conditions and that some cases of dehydration are either too severe or too complicated for these mechanisms to resolve fully.

Notice from this discussion that the mechanisms for correcting water imbalance in the body also correct electrolyte and pH imbalances. So, not only are the imbalances of each variable tied together, but the mechanisms that correct them are also inextricably linked.

Quick Check

☐ 1. What imbalances result from dehydration?

☐ 2. What hormonal mechanisms correct for these imbalances?

Apply What You Learned

☐ 1. How could the response to dehydration lead to a potassium ion imbalance? Explain.

☐ 2. Drugs known as ACE (angiotensin-converting enzyme) inhibitors block the production of active angiotensin-II. How would this blockage affect the response to dehydration?

☐ 3. Predict what would happen to overall homeostasis if this response operated as a positive feedback loop rather than a negative feedback loop.

See Answers in Appendix A.

Chapter Summary

For everything you need to succeed in this course, go to Mastering Anatomy & Physiology. There you will find:

 Practice Tests

Author Podcasts

▶ Big Picture Animations

))) Concept Boost Video Tutors

iP2 Interactive Physiology 2.0 Tutorials

A&PFlix A&P Flix 3D Animations

Active-Learning Workbook

MODULE 25.1
Overview of Fluid, Electrolyte, and Acid-Base Homeostasis 993–995

- Body fluids include all the body's water-based liquids. They function in transport, regulation of body temperature, cushioning, and lubrication.

- Water obeys the *principle of mass balance*—what is gained by the body must equal what is lost by the body.

- An **electrolyte** is a substance that dissociates into ions in water; the resulting charged particles conduct an electric current. Electrolytes in the body obey the principle of mass balance.

- An **acid** is a chemical that dissociates in water to release a hydrogen ion. A **base** is a chemical that dissociates in water to form an ion that can accept a hydrogen ion.

MODULE 25.2
Fluid Homeostasis 995–1002

- The **total body water** of the standard man is about 42 kg, or about 60% of the total body weight.

- Body fluids are in two general compartments: the **intracellular compartment** and the **extracellular compartment.**

- Water movement between the plasma and cytosol is determined by osmotic pressure and hydrostatic pressure.

- **Obligatory water loss** is the amount of water, about 500 ml, that must be lost each day in the urine in order to excrete a minimum amount of solutes.
  ○ Water loss from the kidneys and feces make up the body's **sensible water loss.**
  ○ Water loss via evaporation and during expiration make up the body's **insensible water loss.**

- Water intake is driven by the **thirst mechanism.** Changes in plasma osmolarity are detected by **osmoreceptors** in the thirst center of the hypothalamus.

- On average, a person needs to consume about 1.5 liters of water per day, although this amount varies with activity.

- **Antidiuretic hormone** (ADH) increases water reabsorption from the distal tubules and collecting system of the kidneys. ADH secretion is stimulated by osmoreceptors, angiotensin-II, decreased blood volume, and decreased blood pressure.

- **Dehydration** is characterized by decreased volume and increased osmolarity of the ECF, and tends to cause cellular crenation. **Overhydration** features an increased volume and decreased osmolarity of the ECF, and tends to cause cellular swelling. Isosmotic fluid imbalances include *hypovolemia* and **edema.**

MODULE 25.3
Electrolyte Homeostasis 1002–1008

- Sodium ions are more concentrated in the ECF. This concentration gradient for sodium ions is critical for depolarization of excitable cells and for the maintenance of ECF osmolarity.
  - **Hypernatremia** is an abnormal increase in the plasma sodium ion concentration. The most common cause of hypernatremia is dehydration.
  - **Hyponatremia** is an abnormal decrease in the plasma sodium ion concentration. The most common cause is overhydration.
- Potassium ions are more concentrated in the cytosol than the ECF. They are largely responsible for generating the negative resting membrane potential, and they cause the repolarization phase of the action potential.
  - When potassium ions are ingested, aldosterone, insulin, and epinephrine trigger their uptake by cells.
  - **Hyperkalemia** has many sources and causes the resting membrane potential to become more positive than normal, and so cells are more easily excitable.
  - **Hypokalemia** is most commonly caused by diuretics. It causes the resting membrane to hyperpolarize, and so cells are more difficult to stimulate.
- Calcium ions are required for bone tissue, synaptic transmission, contraction of all three types of muscle tissue, blood clotting, and as intracellular messengers. They are regulated in tandem with phosphate ions.
  - PTH increases calcium ion reabsorption from the kidneys, inhibits phosphate reabsorption, stimulates osteoclasts, and activates vitamin D3.
  - **Hypercalcemia** is generally caused by hyperparathyroidism. It decreases the excitability of neurons and shortens the plateau phase of the cardiac action potential.
  - **Hypocalcemia** is generally due to decreased or ineffective PTH or inadequate vitamin D. Neurons become more excitable, which can lead to tetanic muscle contractions.
- Chloride ions are important osmotic particles in the ECF. They, along with sodium ions, are generally reabsorbed from the kidneys.
- Magnesium ions are found in bone and are important in many biochemical reactions in cells. Their low concentration in the ECF is maintained by renal excretion.

MODULE 25.4
Acid-Base Homeostasis 1008–1016

- The pH of the blood is maintained between 7.35 and 7.45 by the urinary system, respiratory system, and **buffer systems.**
- Sources of acids include those formed during metabolism, including the **volatile acids,** formed from carbonic acid, and **fixed acids,** such as lactic acid, amino acids, and fatty acids.
- **Chemical buffer systems** consist of a weak acid and its conjugate weak base. Their function is to resist large swings in pH. The *alkaline reserve* is the buffering capacity of the body's fluids.
  - The **carbonic acid–bicarbonate ion buffer system** is the most important buffer system for the blood.
  - The **phosphate buffer system** is an important buffer system for the cytosol and the filtrate within the kidney tubules.
  - In **protein buffer systems,** many proteins such as hemoglobin can bind hydrogen ions and act as buffers.
- The respiratory system and urinary system are the body's **physiological buffer systems.** The respiratory system controls the amount of carbon dioxide in the body. The kidneys control the amount of hydrogen ions and bicarbonate ions reabsorbed and/or secreted.
- Acid-base imbalances include **acidosis** and **alkalosis.** They are corrected by the respiratory and urinary systems, a process called **compensation.**
  - **Respiratory acidosis** is caused by an abnormal decrease in ventilation and retention of carbon dioxide. **Metabolic acidosis** is caused by the addition of hydrogen ions to the ECF from a source other than carbon dioxide or by the loss of bicarbonate ions.
  - **Respiratory alkalosis** is due to abnormal hyperventilation. **Metabolic alkalosis** results from a loss of hydrogen ions or an excess of bicarbonate ions.
  - *Arterial blood gases* measure the pH, P_{CO_2}, and bicarbonate ion content of the arterial blood.

MODULE 25.5
An Example of Fluid, Electrolyte, and Acid-Base Homeostasis 1016–1018

- An imbalance in one variable will lead to imbalances in other variables, as evidenced by dehydration. In dehydration, the decreased amount of water in the body leads to increased levels of sodium ions (hypernatremia) and acidosis.
- Mechanisms that correct dehydration also correct the accompanying hypernatremia and acidosis.

Assess What You Learned

Scan the QR Code for additional practice test questions

https://goo.gl/HfmCm6

LEVEL 1 Check Your Recall

1. Which of the following statements best describes the principle of mass balance?
   a. The amount of a variable that is gained by the body through ingestion equals the amount that is lost from the body.
   b. The body maintains a stable mass at all times.
   c. The amount of each variable brought into the body must be balanced by all other variables.
   d. The amount of a variable ingested is regulated by a positive feedback loop.

2. How does an electrolyte differ from a nonelectrolyte?

3. What happens to the pH of a solution when hydrogen ions are added?

a. The pH increases.

b. The pH decreases.

c. The pH does not change.

d. The pH does not measure hydrogen ion concentration.

4. As a percentage of body weight, the total body water tends to be higher in _____ and lower in _____.

a. infants; men   c. men; infants

b. women; men   d. infants; women

5. Fill in the blanks. A cell in a/an _____ fluid will lose water, and a cell in a/an _____ fluid will gain water.

6. Mark the following statements as true or false. If a statement is false, correct it to make a true statement.

a. The thirst mechanism is mediated by osmoreceptors located in the cerebral cortex.

b. The sensible water loss includes the water lost from the body via the skin and the respiratory system.

c. The main hormone that regulates fluid balance is antidiuretic hormone (ADH).

d. Atrial natriuretic peptide promotes water retention in the kidneys and increases the amount of water in the body.

e. Dehydration is characterized by a decreased volume and increased osmolarity of the ECF.

7. Which of the following is *false* with respect to sodium ions in human physiology?

a. Sodium ions are the most abundant extracellular cation.

b. Sodium ions are an important osmotic particle in the ECF.

c. The entry of sodium ions into a cell causes depolarization.

d. Sodium ions are more concentrated in the cytosol than in the ECF.

8. List the effects of each of the following hormones on electrolyte balance. Note that some hormones affect more than one electrolyte.

a. Angiotensin-II   d. Vitamin D

b. Aldosterone   e. Atrial natriuretic peptide

c. Parathyroid hormone

9. Which of the following effects tend to be caused by hypernatremia?

a. Inhibition of ADH secretion

b. Cellular crenation

c. Cellular swelling

d. Increased urine production

10. Mark the following statements as true or false. If a statement is false, correct it to make a true statement.

a. Potassium ions are responsible for the repolarization phase of the action potential.

b. Insulin, aldosterone, and epinephrine stimulate the uptake of potassium ions into cells.

c. Hyperkalemia tends to decrease the resting membrane potential and hyperpolarize excitable cells.

d. Hypocalcemia causes neurons to become hyperexcitable, leading to potential tetanic contractions.

e. Chloride ions are generally reabsorbed from the kidneys, along with bicarbonate ions.

11. The biggest source of metabolic acids in the body is:

a. lactic acid.

b. ketone bodies.

c. carbon dioxide.

d. uric acid.

12. What is the main buffer system of the ECF?

a. Protein buffer system

b. Carbonic acid–bicarbonate ion buffer system

c. Phosphate buffer system

d. None of the above

13. Explain what happens to the pH of a buffered solution when hydrogen ions are added. Why does this happen?

14. Fill in the blanks: An increase in ventilation _____ the pH of the blood due to a/an _____ of carbon dioxide in the blood. A decrease in ventilation _____ the pH of the blood due to a/an _____ of carbon dioxide in the blood.

15. Which of the following mechanisms is/are used by the kidneys to regulate the pH of the blood? Circle all that apply.

a. Hydrogen ions are secreted from the proximal and distal tubules and the collecting system.

b. Hydrogen ions are reabsorbed from the nephron loop.

c. New bicarbonate ions are formed from glutamine and carbon dioxide in the interstitial fluid that enters proximal tubule cells.

d. Bicarbonate ions can be secreted.

e. Bicarbonate ions are reabsorbed directly from the filtrate.

16. Mark the following statements as true or false. If a statement is false, correct it to make a true statement.

a. Respiratory acidosis is caused by hypoventilation.

b. Renal compensation for acid-base disturbances begins within minutes, whereas respiratory compensation begins within several hours.

c. Respiratory compensation for metabolic acidosis consists of hypoventilation.

d. Renal compensation for metabolic alkalosis consists of reabsorption of hydrogen ions and secretion of bicarbonate ions.

e. You would expect to find an elevated P_{CO_2} in arterial blood gas analysis of a patient with respiratory acidosis.

17. Which of the following statements correctly describe(s) the role of aldosterone?

a. Aldosterone triggers sodium ion and so water reabsorption.

b. Aldosterone triggers potassium ion secretion.

c. Aldosterone triggers hydrogen ion secretion.

d. Both a and b are correct.

e. All of the above are correct.

18. How does angiotensin-II help to restore fluid balance when a person is dehydrated?

LEVEL 2 Check Your Understanding

1. Your friend argues that *all* water conducts electricity, regardless of what it contains. You prepare three liquids to test this hypothesis: one with deionized water (with no solutes, only water molecules), one with 5% glucose in water, and one with 5% sodium chloride in water. Which of these solutions, if any, will conduct an electric current? Explain.

2. A woman begins a diet and exercise regimen and loses 30 lb. Will her total body water increase, decrease, or stay the same? Why?

3. Explain how the amount of water in the body affects the concentration of ions and solutes in the ECF.

4. Diabetic ketoacidosis is characterized by an increased level of ketone bodies, which causes metabolic acidosis. A patient in diabetic ketoacidosis will have an altered rate of ventilation. Will the patient be hyperventilating or hypoventilating? How will a change in the rate and depth of ventilation compensate for an acidosis that is metabolic in nature?

LEVEL 3 Apply Your Knowledge

PART A: Application and Analysis

1. Elise Anderson is a 6-year-old girl who presents to the emergency department with a history of vomiting for the past 3 days. The nurse notices that her respiratory rate is abnormally low. What is the likely reason for this change in ventilation? Predict what Elise's arterial blood gas values would show.

2. What do you think has happened to Elise's ECF volume and osmolarity over the past 3 days? Will this lead to a change in the volume of water in the cytosol of her cells? Explain.

3. Ms. Johanssen is a patient in the hospital. The nurse examines her laboratory reports and notices that she has developed hyperkalemia and acidosis over the past several days. On closer examination of her medical chart, the nurse also sees that her physician recently doubled her dose of *spironolactone*, an aldosterone-blocking diuretic. How does this explain her laboratory findings?

4. A laboratory printout of arterial blood gases indicates that a patient has an increased P_{CO_2}, decreased pH, and normal bicarbonate ion concentration. Is this patient in acidosis or alkalosis? Is the pH disturbance respiratory or metabolic in nature? Explain your reasoning. How long do you think the patient has had this pH disturbance? (*Hint:* Look at the bicarbonate ion concentration. What system controls the concentration of bicarbonate ions, and how quickly does it compensate for pH disturbances?)

PART B: Make the Connection

5. What happens to the concentration of sodium ions in the ECF if you consume a large amount of salt without consuming any water? How will this affect the osmotic pressure of the ECF? Why could this lead to an elevation in blood pressure? (*Connects to Chapter 19*)

See answers in Appendix A.

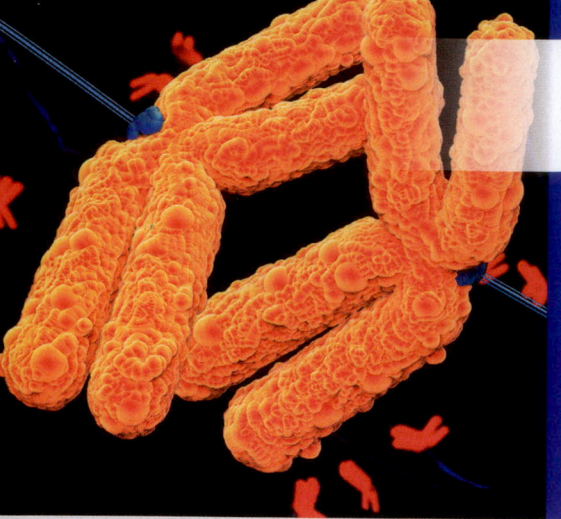

26

The Reproductive System

*For practice applying concepts to a clinical scenario, check out the **Running Case Study** for this chapter @ Mastering Anatomy & Physiology*

Computer-generated image: Meiosis occurs only in the sex cells of testes and ovaries. This image shows pairs of chromosomes (orange) being pulled to opposite ends of a cell by spindle fibers (blue).

So far, we have discussed organ systems that are absolutely required for survival. For example, how long could we live without a functional heart or brain? However, there is one body system that is not technically necessary for survival: the **reproductive system** (although it does produce important hormones that affect other systems). In fact, this system appears to almost hibernate throughout most of an individual's childhood until **puberty**—the period of life when reproductive organs mature and become functional.

This chapter explores the male and female reproductive organs. We also look at how the reproductive systems are regulated and how the sex hormones influence other body systems. (We will save discussion of pregnancy and the functions of the female reproductive tract in this process for the development and heredity chapter—see Chapter 27.) Let's begin by looking at the main parts of the reproductive system and the process of cell division that produces the *gametes*, or sex cells.

MODULE 26.1
Overview of the Reproductive System and Meiosis

Learning Outcomes

1. Compare and contrast the basic structure and function of the male and female reproductive systems.
2. Contrast the overall processes of mitosis and meiosis.
3. Describe the stages of meiosis.

In all animals, two forms of cell division take place: *mitosis* and *meiosis*. **Mitosis** (my-TOH-sis) is the cell division undergone by all *somatic,* or body, cells that are capable of division (see Chapter 3). (Recall that some cells are *amitotic,* or unable to divide once they have differentiated.) Mitosis produces "daughter" cells that are genetically identical to the "mother" cell. *Meiosis* (my-OH-sis), however, is a special type of cell division that occurs only in cells destined to become gametes. This module explores the process of meiosis and compares it to mitosis. Before we begin, however, let's introduce some basic principles and terms associated with the reproductive system.

Introduction to the Male and Female Reproductive Systems

《 FLASHBACK

1. What are the male and female gonads? (p. 619)
2. What hormones do the gonads produce? (p. 619)

As you are likely aware, the male and female reproductive systems differ in many ways. However, they also share a number of similarities, including the following:

- In both males and females, the most important organs in the reproductive system are the **gonads,** or **primary sex organs.** These are the *testes* in males and the *ovaries* in females.
- The gonads secrete **sex hormones,** including *testosterone* and *estrogens.*
- The gonads also produce **gametes,** or sex cells, through the process of meiosis. In males the gametes are *sperm;* in females the gametes are *ova.*
- Both males and females have additional organs and structures that contribute to the functioning of the reproductive system. These are known as **accessory reproductive organs.**

The similarities between the male and female reproductive systems reflect their similar function. Keeping this in mind will help you as you proceed through this chapter.

Quick Check

☐ 1. What are the male and female gonads? What are their functions?

Overview of Meiosis

All human somatic cells with a nucleus have 46 chromosomes, and these chromosomes are paired so that there are 23 pairs of chromosomes in each nucleus (see Chapter 3). Somatic cells are called **diploid** (DIP-loyd; *di* = "two," *ploid* = "having chromosome sets"), or **2n,** because the chromosomes are paired. One set, or 23, of these chromosomes consists of the **paternal chromosomes** that came from the father of the individual. The other set, the **maternal chromosomes,** came from the individual's mother.

Each member of a pair of chromosomes has the same **genes,** which are segments of the DNA that code for a specific protein. For this reason, the pairs are called **homologous chromosomes** (huh-MAH-luh-gus; "sharing a similar structure"), shown here:

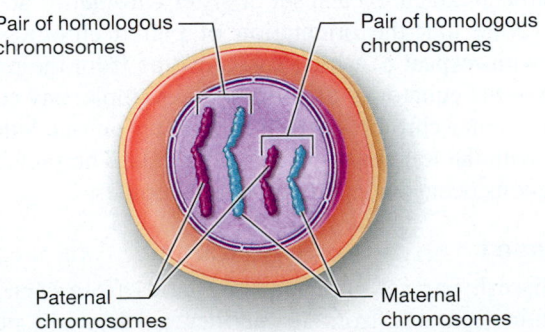

Pair of homologous chromosomes — Pair of homologous chromosomes

Paternal chromosomes — Maternal chromosomes

However, the matching genes on the homologous chromosomes may have different variants, or **alleles** (uh-LEELZ). For example, let's say that one homologous pair of chromosomes includes the gene for hair color. The paternal chromosome of this homologous pair might have an allele for brown hair, whereas the maternal chromosome of this pair might have an allele for blond hair. (For an in-depth discussion of this topic, see Chapter 27.)

During **fertilization,** the process by which a sperm and ovum fuse, the single-celled offspring called a **zygote** is formed. The zygote is the cell that will eventually divide to produce all the cells in a new individual. For this reason, it must contain the correct number of chromosomes—46. This means that a sperm and ovum must each contribute 23 chromosomes—otherwise, the zygote would end up with 92. This is made possible by the process of **meiosis,** during which a cell divides to form daughter cells with half the number of chromosomes.

As with mitosis, just before meiosis begins, all the cell's DNA is replicated so that each new cell will get a complete copy. The result of DNA replication is a cell with two identical copies of its chromosomes, which are called **sister chromatids** (KROH-muh-tidz). Each pair of sister chromatids is connected in the center by a region known as a *centromere,* which you can see here:

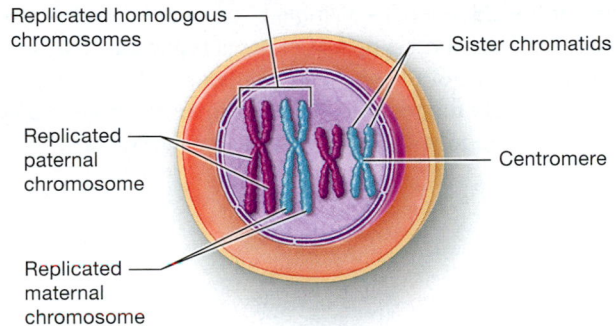

Replicated homologous chromosomes — Sister chromatids

Replicated paternal chromosome — Centromere

Replicated maternal chromosome

Recall from the cell chapter (see Chapter 3) that when a cell is not dividing, its DNA is in the form of **chromatin,** which is one very long piece of DNA with its associated proteins. After replication, when cell division begins, chromatin threads coil tightly and condense into the barlike chromosomes.

Also like mitosis, meiosis has four basic phases: prophase, metaphase, anaphase, and telophase. However, in meiosis each of these phases occurs twice. Meiosis includes two successive divisions, **meiosis I,** or the *first meiotic division,* and **meiosis II,** or the *second meiotic division.*

Meiosis I (First Meiotic Division)

《 FLASHBACK

1. What are spindle fibers? (p. 112)
2. What are centrioles? (p. 97)

The first meiotic division separates the homologous pairs to produce **haploid** (or **1n**) cells, which have only half the number of chromosomes of the original cell. For this reason, this process is sometimes referred to as *reduction division.*

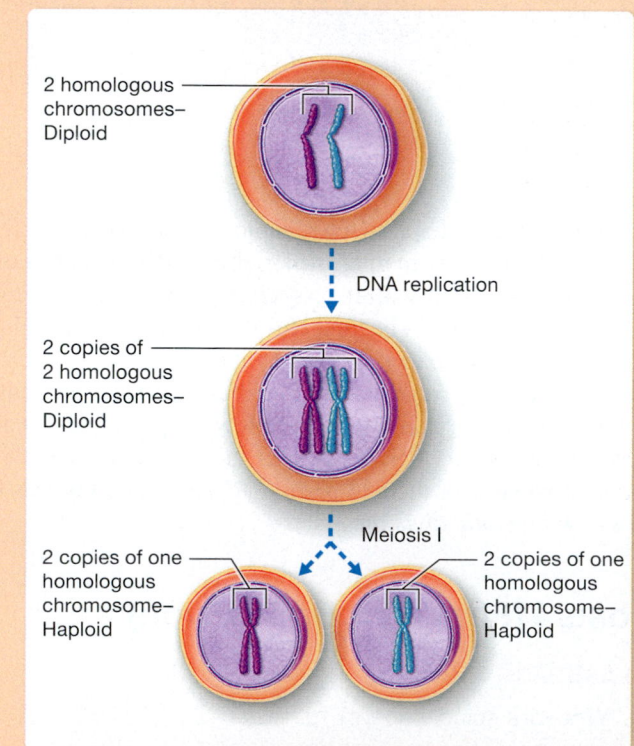

ConceptBOOST))))

Understanding Diploid and Haploid Cells

There is often confusion as to why the daughter cells at the end of meiosis I are haploid, because they might *appear* to have the same absolute number of chromosomes as the starting mother cell (four in our simplified example in Figure 26.1). However, remember that you are not seeing four individual chromosomes—you are seeing two sets of sister chromatids. A set of sister chromatids is simply two identical copies of a *single* homologous chromosome.

Here's a simple analogy. Say your A&P class requires two textbooks: one class textbook and one lab manual. Each book contains similar but slightly different information. The two books represent a pair of homologous chromosomes found in a diploid cell—each homologous chromosome contains similar but slightly different information in its DNA. Now imagine that the bookstore accidentally ships you two copies of the lab manual that got stuck together instead of a separate lab manual and class textbook. You now have two identical copies of the lab manual—a single book—not two individual books. This represents a pair of sister chromatids—they are simply two copies of the exact same genetic information. You can see a simple progression here, using an example of a cell with two homologous chromosomes in the mother cell:

2 homologous chromosomes– Diploid

DNA replication

2 copies of 2 homologous chromosomes– Diploid

Meiosis I

2 copies of one homologous chromosome– Haploid

2 copies of one homologous chromosome– Haploid

So, the diploid mother cell here has two homologous chromosomes, or the lab manual and the class textbook. But by the end of meiosis I, the resulting cells have only one

replicated homologous chromosome, or two joined copies each, of either the lab manual or the textbook. The number of distinct chromosomes (or books) has been reduced by half, from two to one, making the cell now haploid. Note that we have simplified this somewhat because, as you will see, sister chromatids are not necessarily identical at the end of meiosis I. However, the basic principle remains: A set of sister chromatids represents a single homologous chromosome. ■

Even though the DNA is in its noncondensed form (chromatin) while undergoing replication, the first cell diagram we see in **Figure 26.1** shows what the chromosomes would look like if they had already assumed their characteristic shape. Notice that one chromosome in each pair is labeled "paternal" and the other is labeled "maternal."

Early Prophase I

At the beginning of prophase I, the DNA of the chromosomes has already been replicated. The chromatin then condenses so that individual sister chromatids become visible through a microscope as thin lines. At the same time, nucleoli disappear, the nuclear membrane temporarily breaks down, and spindle fibers and centrioles form and align at opposite poles.

Middle to Late Prophase I

During middle to late prophase I, the homologous chromosomes move together and pair up tightly side by side in a process called **synapsis** (sih-NAP-sis). When the two homologous chromosomes are touching, they are called **tetrads** (TEH-tradz). While touching, some sister chromatids break in one or more places and homologous chromosomes exchange segments of DNA, forming chromatids with new combinations of genetic information in a step referred to as **crossing over.** This is the first process that shuffles the alleles, and changes the chromosomes so that the sister chromatids are *no longer identical.* This will result in genetically unique cells at the end of meiosis.

Metaphase I

At the time of metaphase I, the sister chromatids of the homologous pairs line up on either side of the equator, or midline of the cell, forming a *double line* of chromosomes. Spindle fibers then branch out from the centrioles at opposite ends of the poles of the cell and attach to each set of sister chromatids at the centromere. Note that the orientation of paired chromosomes is random with respect to which chromosome from the pair is on one side of the equator or the other. For example, any combination of maternal chromosomes may line up on one side of the equator, with the remainder on the other side. The same random combinations occur with paternal chromosomes.

Anaphase I

In anaphase I, one set of sister chromatids from each homologous pair separates from its partner as they are pulled to opposite poles of the cell by the spindle fibers. Due to the

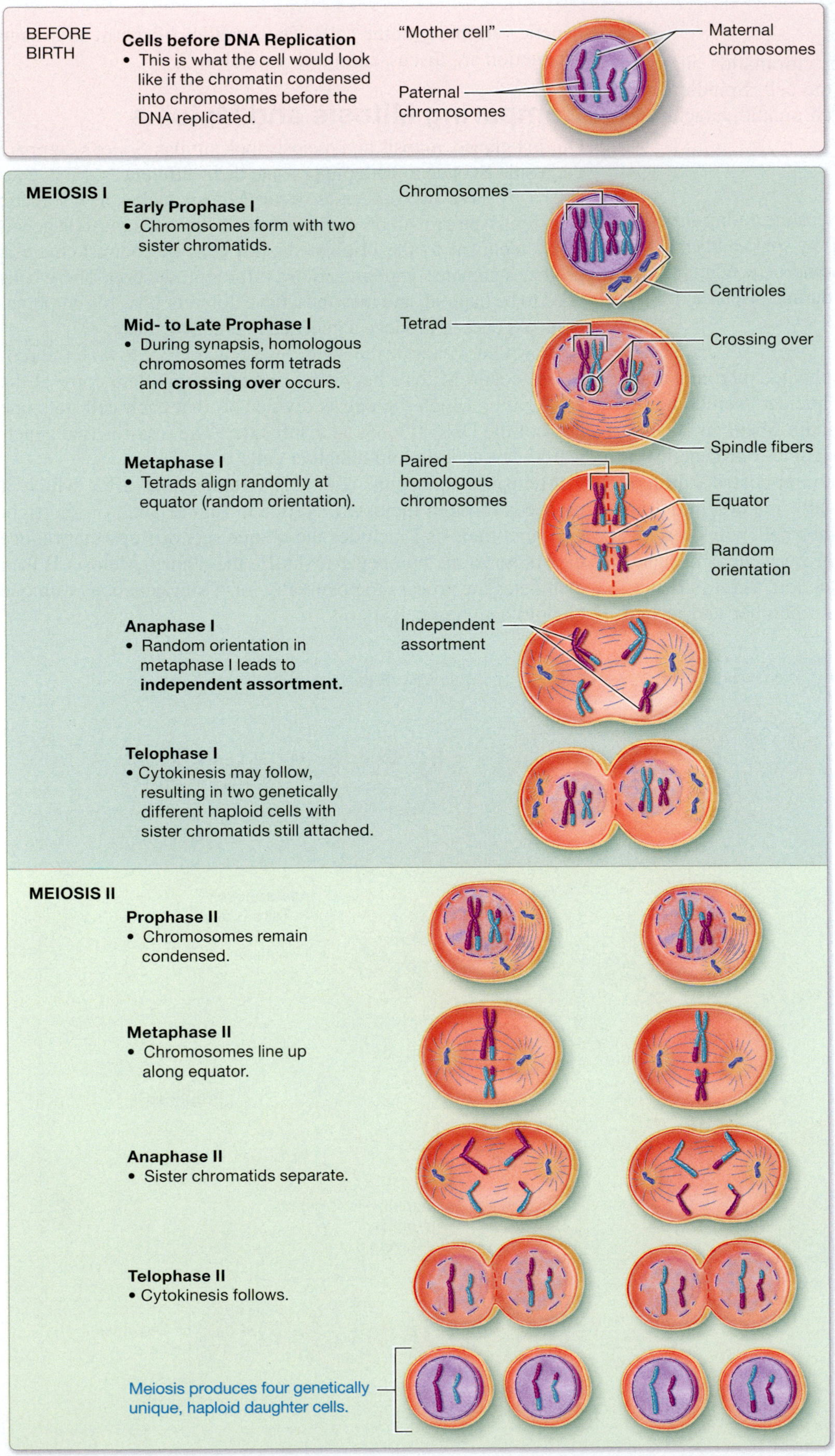

BEFORE BIRTH	**Cells before DNA Replication** • This is what the cell would look like if the chromatin condensed into chromosomes before the DNA replicated.	"Mother cell" — Maternal chromosomes Paternal chromosomes

MEIOSIS I

Early Prophase I
• Chromosomes form with two sister chromatids.

Chromosomes — Centrioles

Mid- to Late Prophase I
• During synapsis, homologous chromosomes form tetrads and **crossing over** occurs.

Tetrad — Crossing over — Spindle fibers

Metaphase I
• Tetrads align randomly at equator (random orientation).

Paired homologous chromosomes — Equator — Random orientation

Anaphase I
• Random orientation in metaphase I leads to **independent assortment.**

Independent assortment

Telophase I
• Cytokinesis may follow, resulting in two genetically different haploid cells with sister chromatids still attached.

MEIOSIS II

Prophase II
• Chromosomes remain condensed.

Metaphase II
• Chromosomes line up along equator.

Anaphase II
• Sister chromatids separate.

Telophase II
• Cytokinesis follows.

Meiosis produces four genetically unique, haploid daughter cells.

Figure 26.1 **The stages of meiosis.** For simplicity, the cell is shown with only two pairs of homologous chromosomes.

random orientation of homologous pairs that occurred during metaphase I, separation of the maternal and paternal chromosomes is random as well. This process is called **independent assortment** and is the second shuffling of the alleles, further increasing genetic variability of the daughter cells.

Telophase I

By the time we reach telophase I, the sets of sister chromatids arrive at opposite ends of the cell. A nuclear membrane may again form around them, but this does not always occur. **Cytokinesis** (sy′-toh-kuh-NEE-sis), or division of the cytoplasm, then takes place, separating the cells.

At the end of this phase, we have two haploid cells, each containing 23 individual homologous chromosomes, with each chromosome having two sister chromatids. Also, these two cells are genetically different from each other, due to crossing over and independent assortment having taken place.

Quick Check

☐ 2. Which mechanisms increase the genetic variability of daughter cells?

☐ 3. Why are cells haploid at the end of meiosis I?

Meiosis II (Second Meiotic Division)

The second meiotic division separates the chromatids of each chromosome. The resulting cells will still be haploid, but they will have only one copy of each chromatid rather than two copies of each chromatid. As before, there are four stages: prophase II, metaphase II, anaphase II, and telophase II.

Prophase II

The second prophase event of meiosis is the same as prophase of mitosis, and it occurs in both cells produced from meiosis I. The nuclear membrane, if it re-formed, breaks down, and the sister chromatids remain condensed.

Metaphase II

During metaphase II, each pair of sister chromatids aligns to form a single line along the equator of the cell. Spindle fibers extend from the centrioles to each sister chromatid, attaching at the centromere.

Anaphase II

The sister chromatids of each chromosome separate during anaphase II and are pulled to opposite ends of the cell by spindle fibers. Each sister chromatid is now called a chromosome or a nonreplicated chromosome. Cytokinesis generally begins during anaphase II.

Telophase II

Finally, during telophase II, the nonreplicated chromosomes arrive at opposite ends of the cell. The nuclear membrane reforms, cytokinesis continues, and the spindle fibers and centrioles disassemble. One original cell has become four genetically unique daughter cells, each with 23 nonreplicated chromosomes. In males, these four daughter cells mature into four sperm cells. In females, only one daughter cell will go on to potentially become an ovum. However, the basic steps of meiosis are the same for sperm and ovum production, as they are necessary to achieve the correct chromosome number and ensure genetic variability in the cells. The specifics of sperm and ovum production are discussed in later modules.

Comparing Mitosis and Meiosis

On first glance, mitosis and meiosis look similar. So let's compare them side-by-side to better appreciate their differences. First, consider *why* these two processes occur: Mitosis makes new somatic cells for tissue growth or repair, and these new cells must be genetically identical to the original diploid cell. In contrast, meiosis produces gametes—sperm and ova—for reproduction. These cells need to be haploid, and they must have different possible combinations of genes to produce a genetically unique gamete.

Now that we know the "why," we can apply it to *how* the processes occur. Mitosis begins by making a complete copy of the DNA, after which the cell is divided so that each cell gets one copy of the DNA (**Figure 26.2**, left side). The result is two genetically identical, diploid daughter cells.

Meiosis also begins with DNA replication, after which it proceeds through meiosis I. Notice in Figure 26.2, on the right side, that meiosis I includes the unique sets of steps that reduce the chromosome number *and* shuffle the alleles. Meiosis II then completes the process to potentially form four genetically unique haploid daughter cells.

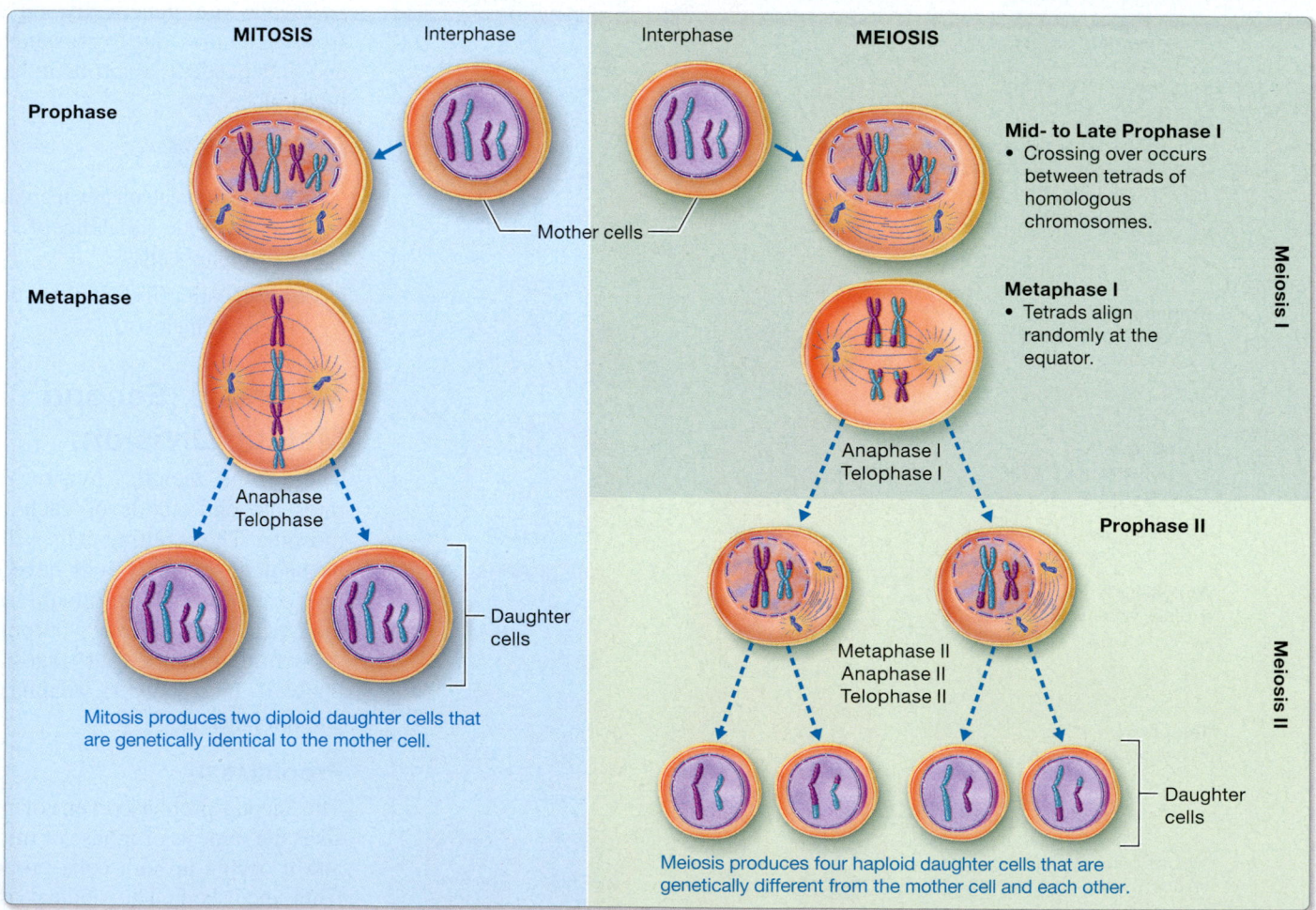

Figure 26.2 A comparison of the phases of mitosis and meiosis.

☐ 4. How many daughter cells can potentially be produced from one mother cell in meiosis?

☐ 5. How do daughter cells differ from the mother cell?

☐ 6. How do mitosis and meiosis differ?

Apply What You Learned

☐ 1. What do you think would happen if crossing over and independent assortment no longer occurred during meiosis? (*Hint:* Compare the final cells of mitosis to those of meiosis.)

☐ 2. Suppose that a homologous chromosome pair fails to separate during the first meiotic division, a problem known as nondisjunction. Predict the effects on potential offspring.

See answers in Appendix A.

MODULE **26.2**

Anatomy of the Male Reproductive System

Learning Outcomes

1. Describe the structure and functions of the male reproductive system.

2. Trace the pathway that sperm travel from the seminiferous tubules to the external urethral orifice of the penis.

3. Describe the organs involved in semen production.

4. Discuss the composition of semen and its role in sperm function.

We turn now to the anatomy of male reproductive structures (**Figure 26.3**). In this module, we first examine the paired *testes,* or male gonads. The testes produce sperm, which travel through a series of passages known collectively as the *duct system,* including the ducts through the *penis,* which are the next topics we discuss. We then move on to the *accessory glands,* which secrete fluid into the ducts that becomes part of *semen,* the fluid that accompanies sperm. Finally, we conclude with external structures, including the *scrotum* and *spermatic cord.*

Testes

« FLASHBACK

1. Where is the abdominopelvic cavity? (p. 17)

2. What is the peritoneum? (p. 20)

The paired **testes** (or *testicles;* singular, *testis*) are ovoid structures approximately 4 cm (1.6 in.) long and 2.5 cm (1 in.) wide. They are located outside the abdominopelvic cavity in a saclike structure, the **scrotum,** that is composed of skin, connective tissue, and smooth muscle. Connective tissue surrounds each testis and divides it internally into about 250 internal segments known as **lobules** (**Figure 26.4a**), which contain one to four tightly coiled, looped **seminiferous tubules** (seh´-men-IF-er-us). These tiny tubules are the sites of sperm production.

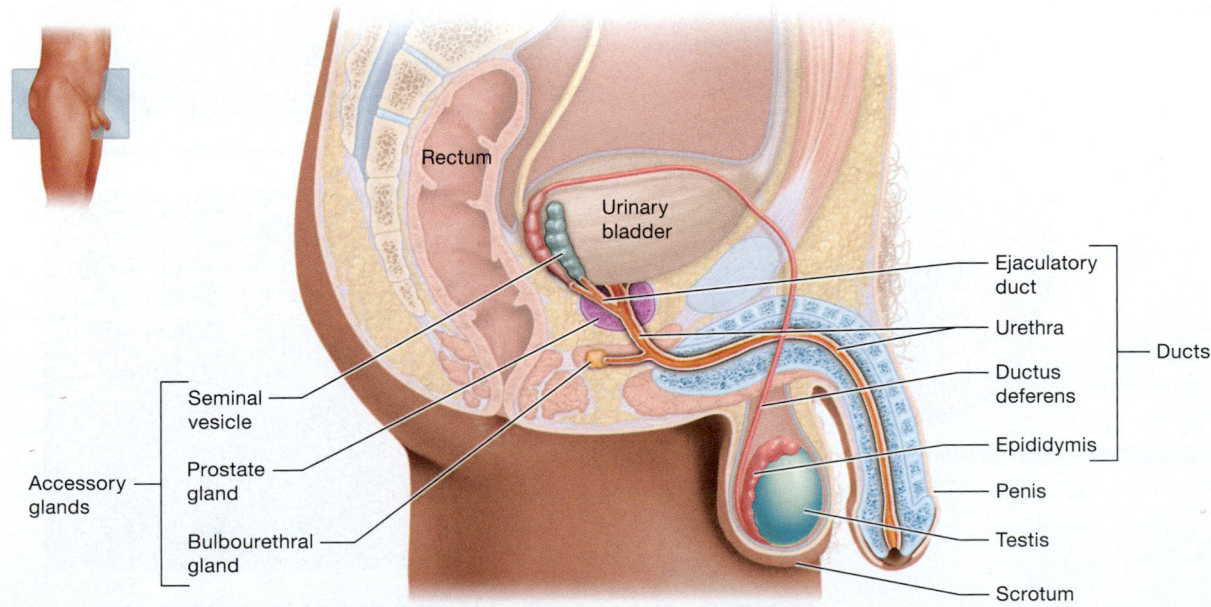

Rectum

Urinary bladder

Seminal vesicle

Prostate gland

Accessory glands

Bulbourethral gland

Ejaculatory duct

Urethra

Ductus deferens

Ducts

Epididymis

Penis

Testis

Scrotum

Figure 26.3 **Internal organs of the male reproductive system, midsagittal section.**

The testes perform two very important functions. They (1) produce sperm cells and (2) secrete androgen hormones, most importantly *testosterone.* The seminiferous tubules contain two types of cells: **spermatogenic cells,** which form sperm, and **sustentacular cells,** which have several functions that support the production of sperm. One such function includes the production of *testicular fluid,* which contains nutrients for developing sperm cells. Between the seminiferous tubules lie **interstitial cells** (**Figure 26.4b** and **c**), which produce and secrete androgens, mostly testosterone, into the surrounding interstitial fluid (hence their name—interstitial cells are in the interstitial space and secrete into the interstitial fluid).

Seminiferous tubules are surrounded by muscle-like **myoid cells,** which contract to push sperm and testicular fluids through the tubules and out of the testes. The seminiferous tubules of each lobule merge into a single **straight tubule** that moves sperm into the **rete testis** (REE-tee TES-tus), a network of tubules on the posterior side of the testis. Sperm then pass through **efferent ductules** and enter the first portion of the duct system, the *epididymis.*

Each testis receives blood from a **testicular artery** that branches from the abdominal aorta superior to the pelvis. A network of veins called the **pampiniform venous plexus** drains blood from the testes into *testicular veins.* The testes are innervated by both the sympathetic and parasympathetic divisions of the autonomic nervous system. In addition, a high number of nociceptors (pain receptors) and thermoreceptors are concentrated in the testes and surrounding connective tissues. These nociceptors explain in part why testicular injuries are often so painful.

Quick Check

☐ 1. Which cell type in the testes produces sperm? Which cell type produces testosterone?

☐ 2. What is the specific site of sperm production in the testes?

Duct System

« FLASHBACK

1. What are microvilli? (p. 98)

2. What other system includes the urethra as an important structure? (p. 948)

After leaving the testis, the sperm travel through a system of ducts: the *epididymis, ductus deferens, ejaculatory duct,* and *urethra.* The duct system transports sperm and is also involved in **ejaculation** (ee-jak′-yoo-LAY-shun), the process by which semen is expelled from the penis. As sperm move through these ducts, they are supported and nourished by secretions, collectively known as *semen,* from accessory structures (discussed in the next section).

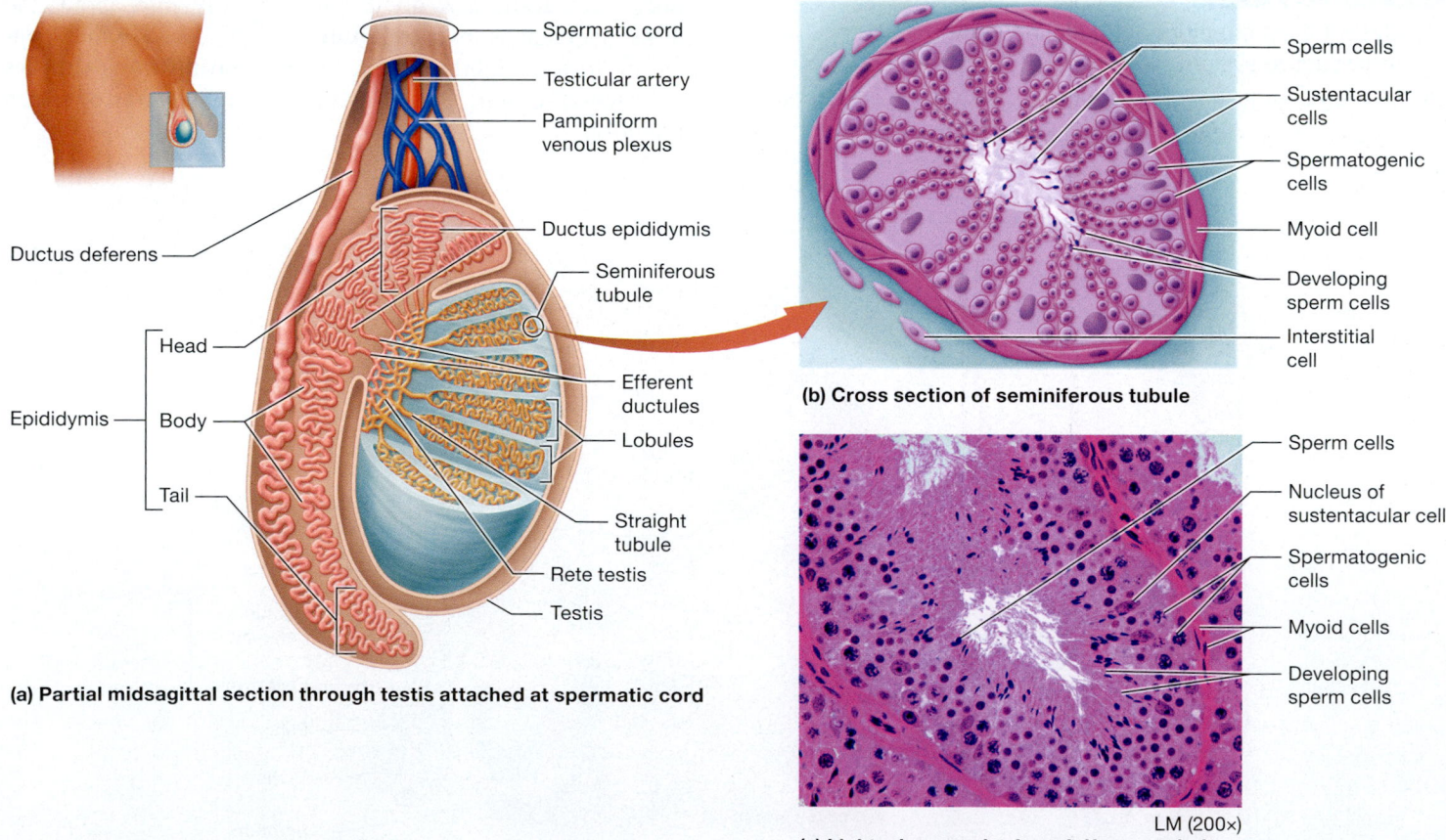

(a) Partial midsagittal section through testis attached at spermatic cord

(b) Cross section of seminiferous tubule

(c) Light micrograph of seminiferous tubule

LM (200×)

Figure 26.4 Internal structures of the testis and epididymis.

Epididymis

The **epididymis** (ep´-ih-DID-uh-miss) is the site of sperm maturation and storage and is filled with many ductules. It consists of a single tube called the **ductus epididymis** that passes through all regions of the epididymis. The ductus epididymis is very long—about 6 m (19.7 ft)—but is so tightly coiled and twisted that it fits inside the 3.8 cm (1.5 in.) length of the epididymis. As you can see in Figure 26.4a, each epididymis sits on the superior and posterior surface of a testis and includes head, body, and tail regions. The *head* of the epididymis sits on its superior surface and contains the efferent ductules of the testis. Sperm move from the head to the *body* of the epididymis, which is its narrow midportion, and finally into the *tail,* its smaller, inferior portion.

The mucosa lining the hollow lumen of the ductus epididymis consists of pseudostratified epithelial cells with long, nonmotile microvilli (or *stereocilia*) that absorb excess testicular fluid, pass nutrients to the sperm in the lumen, and complete the maturation process of the sperm. The long length of the ductus epididymis allows ample time for sperm to mature, and also permits sperm to be stored for several months if ejaculation does not occur. Here we see an example of the Structure-Function Core Principle (p. 28). Sperm that are not ejaculated are resorbed.

CORE PRINCIPLE
Structure-Function

Ductus Deferens

The ductus epididymis widens within the tail of the epididymis and becomes the **ductus deferens** (or *vas deferens*). The ductus deferens is long (40–45 cm, or 15.7–17.7 in.) and thin (about 2.5 mm, or 0.1 in., in diameter). Notice in Figure 26.4a that it extends along the posterior border of the epididymis inside the *spermatic cord* with the testicular arteries, veins, and nerves. It enters the pelvic cavity through a mostly fibrous tunnel called the *inguinal canal* (look ahead to Figure 26.6), after which it passes along the lateral side of the bladder, over the ureter, and then along the posterior bladder. Here it ends in a saclike wider area called the **ampulla** (**Figure 26.5a**).

The organization of the tissue layers of the ductus deferens is the same as you have seen for other hollow organs: an inner mucosa, a middle layer of smooth muscle, and an outer adventitia layer. The mucosa of the ductus deferens consists of pseudostratified columnar epithelium resting on loose connective tissue. The muscularis consists of three layers of smooth muscle; the inner and outer layers are longitudinal, and the middle layer is circular. During ejaculation, the smooth muscle creates strong contractions that rapidly squeeze the sperm forward along the tract. Like the epididymis, the ductus deferens can store sperm for several months and resorb any sperm not ejaculated by that time.

Ejaculatory Duct

Sperm next move into the short **ejaculatory duct,** which is approximately 2 cm (0.8 in.) long. It is located at the junction of the ampulla of the ductus deferens and the duct of an accessory gland called the *seminal vesicle.* Each ejaculatory duct then travels through another accessory gland, the *prostate gland,* and empties into the urethra (see Figure 26.5a).

Urethra

Remember from your study of the urinary system that the **urethra,** which is about 18–20 cm (7.1–7.9 in.) long, connects the urinary bladder to the external body surface. In males, it moves both urine and semen (not at the same time, however), so it belongs to both urinary and reproductive systems. You can see in Figure 26.5a that the male urethra has three regions:

- the *prostatic urethra* is surrounded by the prostate gland;
- the *membranous urethra* passes through the external urethral sphincter; and
- the *spongy urethra* extends through the penis and terminates at the *external urethral orifice.*

As with the ductus deferens, smooth muscle in the wall of the urethra contracts during ejaculation to help expel semen through the external urethral orifice. The mucosa of the urethra contains several small, mucus-secreting *urethral glands* that empty into the urethra and contribute to semen.

Quick Check

☐ 3. What is the function of the epididymis? How does it perform this function?

☐ 4. Trace the pathway that sperm take from the seminiferous tubules to the outside of the body.

Penis

« FLASHBACK

1. Where are the iliac arteries and veins? (p. 716)
2. What muscles are found in the floor of the perineum? (p. 310)

The **penis** delivers sperm into the female reproductive tract and is therefore called the male *copulatory organ.* At the base of the penis is the **root.** The **body,** or *shaft,* enlarges into the end of the penis, called the **glans penis,** which is where the external urethral orifice is located (see Figure 26.5a). The skin covering the penis is loose and forms a circular fold called the **prepuce** (PREE-pyoos), or *foreskin,* around the glans. The prepuce may be removed shortly after birth in a procedure called *circumcision.* There is a great deal of controversy about the medical necessity of circumcision, with many studies showing no measurable benefit.

Internally, the penis contains three cylindrical erectile bodies, or *corpora* (singular, *corpus*), covered by sheaths of dense fibrous connective tissue, which you can see in the transverse section in **Figure 26.5b.** Each erectile body is a spongy network of connective tissue and smooth muscle filled with vascular spaces. The first two erectile bodies are the paired dorsal **corpora cavernosa** (KOHR-pohr-uh kaeh-ver-NOH-suh). At the base of the penis, the two corpora cavernosa split to form the two **crura** (KROO-ruh; singular, *crus*) of the penis that attach to the ischial rami. The other erectile body is the ventral **corpus spongiosum**

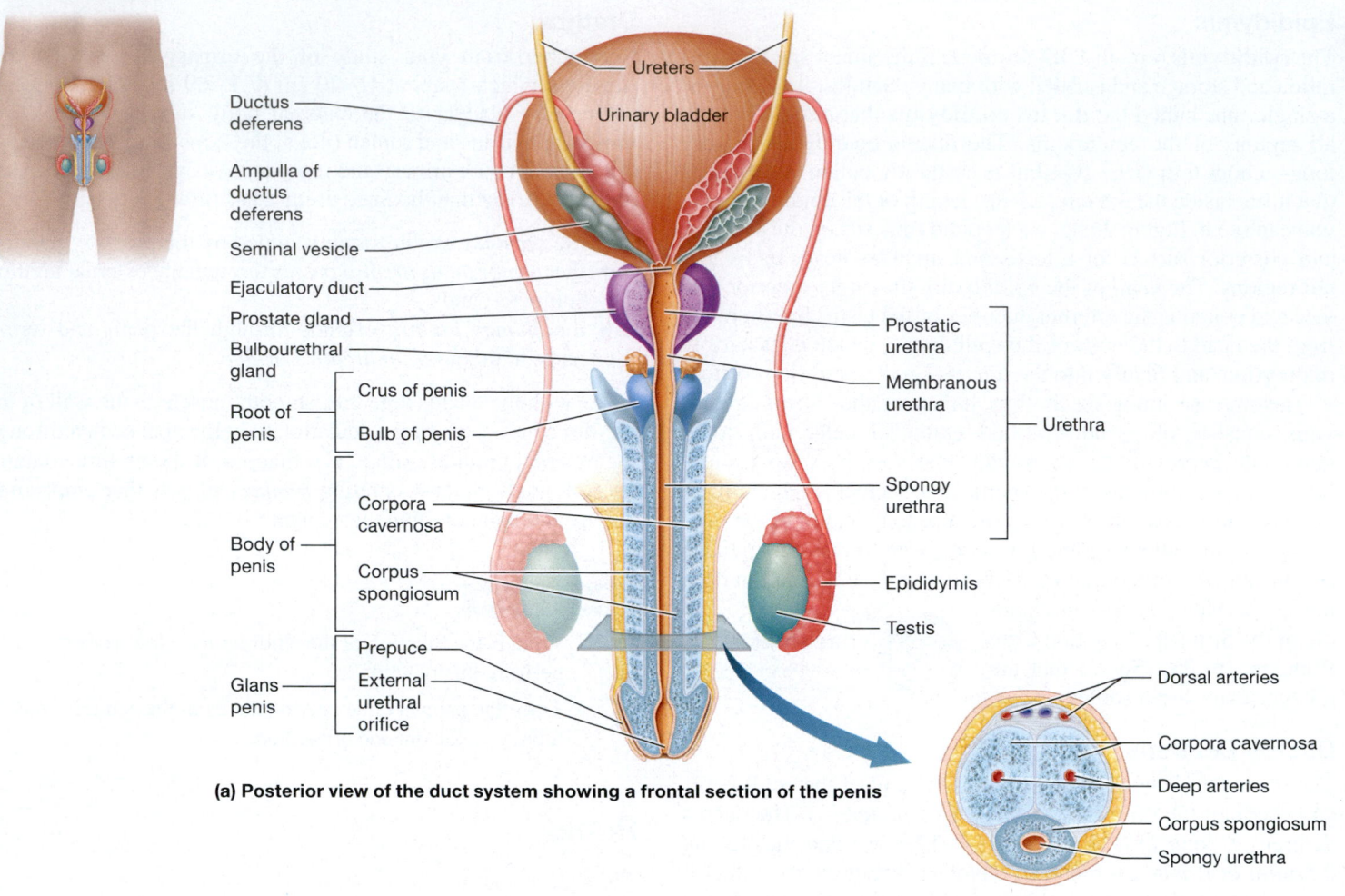

Ureters

Urinary bladder

Ductus deferens

Ampulla of ductus deferens

Seminal vesicle

Ejaculatory duct

Prostate gland

Bulbourethral gland

Root of penis
- Crus of penis
- Bulb of penis

Body of penis
- Corpora cavernosa
- Corpus spongiosum

Glans penis
- Prepuce
- External urethral orifice

Prostatic urethra

Membranous urethra

Urethra

Spongy urethra

Epididymis

Testis

(a) Posterior view of the duct system showing a frontal section of the penis

Dorsal arteries

Corpora cavernosa

Deep arteries

Corpus spongiosum

Spongy urethra

(b) Cross section of the penis with dorsal surface on top

Figure 26.5 Male reproductive duct system and penis.

(KOR-pus spun-jee-OH-sum), which expands to form the **bulb** of the penis at the base. You can see in Figure 26.5a how the bulb sits between the two crura; together these structures make up the root of the penis and connect the penis to the pelvic bones. The corpus spongiosum also expands at its distal end to form the glans. During sexual excitement, the vascular spaces fill with blood, which causes the penis to enlarge and become rigid. This event, called *erection,* is discussed in Module 26.3.

The blood supply to the penis comes from the internal iliac arteries, which give rise to the internal penile arteries; these divide to form the *dorsal arteries* and the *deep arteries*. The **dorsal arteries** supply the penile skin, fascia, and corpus spongiosum, and the **deep arteries** feed the corpora cavernosa.

The penis is innervated by many sensory and motor fibers, including both sympathetic and parasympathetic nerves. Tactile, pressure, and temperature receptors are abundant here. These are important for sexual stimulation as well as control of muscles that regulate the distance of the scrotum from the body wall.

Together, the penis and scrotum make up the **external male genitalia.** They are part of the male **perineum,** a diamond-shaped area between the thighs bordered by the pubic symphysis anteriorly, the ischial tuberosities laterally, and the coccyx posteriorly. The perineum is separated into the **urogenital triangle** (yoo′-roh-JEN-uh-tul), which contains the base of the penis and the scrotum, and the **anal triangle,** which contains the anus. The floor of the perineum is formed by muscles (described in Chapter 9).

Quick Check

☐ 5. What happens to the corpora cavernosa and the corpus spongiosum during sexual excitement?

☐ 6. Why are sensory receptors abundant in penile tissue?

Accessory Sex Glands

《 FLASHBACK

1. What kinds of tissues make up an exocrine gland? (p. 134)

2. How do exocrine glands secrete their products? (p. 137)

The accessory sex glands of the male reproductive system are exocrine glands. They produce most of the liquid portion of semen, the mixture of secretions and sperm expelled during ejaculation, and help lubricate the outside of the penis during intercourse. The accessory glands include the *seminal vesicles, prostate,* and *bulbourethral glands* (see Figure 26.5a).

Seminal Vesicles

The paired **seminal vesicles** (SEH-min-ul) are found on the posterior surface of the urinary bladder near the ampullae of the ductus deferens. Each seminal vesicle is about 15 cm (5.9 in.) long, but is coiled and folded until it is about 5 cm (2.0 in.) long. The duct of each seminal vesicle converges with the ductus deferens to form the ejaculatory ducts. The exterior of each gland has a fibrous capsule that covers a thick layer of smooth muscle. Deep to this is the mucosa layer, which is made of a pseudostratified columnar epithelium that secretes *seminal fluid.*

   Seminal fluid, a yellowish secretion that makes up approximately 60–70% of total semen volume, contains:

- fructose, a sugar that sperm use as a nutrient to make ATP;
- prostaglandins, which stimulate smooth muscle contractions in both the male and female reproductive tracts and increase sperm viability; and
- coagulating proteins and enzymes (a coagulating protein, together with an enzyme from the prostate, forms a temporary clot of semen in the female reproductive tract).

In addition, the pH of the seminal fluid is alkaline (basic) to help neutralize acidic fluids from the male urethra and the female reproductive tract.

Prostate

The **prostate gland** (PRAHS-tayt; be careful not to call it the "pros*trate*") is an egg-sized gland that sits just inferior to the urinary bladder and surrounds the urethra and ejaculatory ducts. Covered by an outer fibrous capsule, it is made up of 20–30 tubular glands embedded in a mass of smooth muscle and dense connective tissue. During ejaculation, prostatic smooth muscle contracts and squeezes the *prostatic secretions* into the prostatic urethra via several ducts. The prostate gland tends to enlarge with age; you can find out more about this in *A&P in the Real World: Benign Prostatic Hyperplasia and Prostate Cancer.*

   Prostatic secretions (prah-STAT-ik) are milky in appearance and make up about 20–30% of semen volume. They contain:

- citrate, a sugar that sperm use as a nutrient to make ATP;
- prostate-specific antigen (PSA) and other enzymes, which dissolve the clot of semen that initially forms in the female reproductive tract so that the sperm can move deeper into the tract; and
- antimicrobial chemicals, which inhibit the growth of some bacteria to decrease the risk of infection in the female reproductive system.

The pH of the prostatic fluid is also alkaline to help neutralize acids from the male urethra and the female reproductive tract.

A&P in the Real World

Benign Prostatic Hyperplasia and Prostate Cancer

The prostate slowly enlarges from birth to puberty and then expands rapidly until about age 30. Further enlargement often occurs after age 45. If the prostate is noncancerous but expands to the point of pushing on the urethra, the condition is called **benign prostatic hyperplasia,** or **BPH.**

   Sometimes an enlarged prostate is secondary to prostate cancer, which is the second most common cancer in U.S. men (after non-melanoma skin cancer). Prostate cancer is one of the leading causes of cancer death among men of all races. Risk factors for the development of prostate cancer include genetic predisposition, African ethnic heritage, age over 50 years, obesity, and poor diet. Screening for prostate cancer usually includes a digital rectal examination—in which a medical professional palpates the prostate through the anterior rectal wall—and assessment of the blood prostate-specific antigen (PSA) level.

   A PSA blood level below 2.5 ng/ml is considered normal. When the number of cells in the prostate increases, as in either BPH or prostate cancer, the PSA rises. If there are any suspicious areas, such as lumps, or if the PSA level is over 4.0 ng/ml, tissue biopsy specimens from different sites in the prostate are taken to look for cancerous cells.

   Prostate cancer may be treated with surgery, radiation therapy, chemotherapy, or hormone therapy. In some men, the tumors grow very slowly and have a low risk of metastasis. In these cases, regular monitoring of the disease may be all that is required. However, some tumors are quite aggressive and metastasize to other tissues rapidly. For this reason, screening and early detection are critically important.

Bulbourethral Glands

The paired **bulbourethral glands** (bul'-boh-yoo-REETH-ruhl), or *Cowper's glands,* are located at the base of the penis, on either side of the membranous urethra. The size of a small marble, each round gland is about 1 cm (0.4 in.) in diameter with a short duct that connects to the urethra.

   In response to sexual stimulation, the bulbourethral glands secrete a thick alkaline, mucus-like fluid that helps neutralize any acidic urine remaining in the urethra prior to ejaculation. It also lubricates the urethra for the passage of ejaculate. Unlike the prostate gland, bulbourethral glands grow smaller with age and are very small in older men.

Semen

 FLASHBACK

1. What is coagulation? What are clotting proteins? (p. 725 and p. 739)
2. What is a flagellum? (p. 99)

Semen is a milky white, somewhat sticky mixture of sperm and fluids from the testes, seminal vesicles, prostate, bulbourethral glands, and urethral glands. Sperm contribute only about 5% to semen's total volume. Each typical **ejaculate,** which is the amount of semen expelled during one ejaculation, measures 2.5–5 ml in volume and contains 40–750 million sperm cells. This may seem like a large number of sperm; however, only a small fraction will ever reach the ovum.

About 5 minutes after semen is ejaculated, it coagulates because of the combination of a clotting protein from the seminal vesicles and a clotting enzyme from the prostate. This coagulation prevents the sperm from leaking back out of the female reproductive tract. Fifteen to 30 minutes later, the clot breaks down because of PSA and other anticoagulant enzymes from the prostate. As it dissolves, the sperm are then gradually released from the clot, are activated, and begin their ascent up the female reproductive tract.

The alkaline pH of semen helps make the sperm fully motile, and the sperm begin the process of **capacitation** (kuh-paeh′-suh-TAY-shun). By this process, sperm undergo changes that enable them to penetrate and fertilize an immature female gamete. They will finish capacitation in the female reproductive tract, where secretions will prepare the sperm cell's plasma membrane to fuse with that of the female gamete.

Prostaglandins from the prostate are thought to thin the mucus in the female reproductive tract and may stimulate contractions, which pull the semen farther into the tract, facilitating fertilization. Antimicrobial chemicals from the prostate kill some bacteria, including *Escherichia coli,* which live in the large intestine and may cause illness if allowed to colonize the reproductive tract.

Quick Check

□ 7. How many sperm are in each milliliter of semen?

□ 8. What are the functions of semen?

Support Structures: Scrotum and Spermatic Cord

❮❮ FLASHBACK

1. What is fascia? (p. 161)

2. What are thermoreceptors? (p. 162)

The *scrotum,* the sac that contains the testes, is located between the thighs (**Figure 26.6**). A midline **septum** divides the scrotum into two internal compartments, one for each testis. The location of the septum is marked externally by a ridgelike seam at the midline, called the **raphe** (RAY-fee), which continues anteriorly along the ventral side of the penis and posteriorly to the anus. The wall of the scrotum includes an outer layer of skin, a layer of superficial fascia, and a layer of smooth muscle called the **dartos muscle.**

Extending from the scrotum is the **spermatic cord,** a tube that is made up largely of layers of fascia and contains the ductus deferens, which passes into the pelvic cavity. The spermatic cord also contains blood vessels, nerves, and lymphatic vessels, as well as smooth muscle tissue known collectively as the **cremaster muscle** (kree-MASS-ter), which controls the height of the testes. It enters the pelvic cavity by passing through the *external inguinal ring,* and ends at the **inguinal canal,** a passageway into the abdomen. Note that sometimes the abdominal contents can protrude through the inguinal canal, leading to an *inguinal hernia.* This condition is about 25 times more common in men than in women, due mostly to the large size of the inguinal canal and external ring in males, necessary to accommodate the spermatic cord.

The scrotum and spermatic cord together help to support sperm production through temperature regulation. The core body temperature of 37° C is too warm for producing large numbers of viable sperm. Under normal circumstances, the temperature of the scrotum is about 3° C lower, which is the optimal temperature for developing sperm. This temperature difference is controlled in two ways.

First, remember that the spermatic cord contains the testicular artery and an extensive network of veins, the pampiniform venous plexus, which surrounds the testicular artery before draining into the testicular veins. In another example of the Structure-Function Core Principle (p. 28), this unique anatomical arrangement of blood vessels results in a *countercurrent heat exchange.* When warm arterial blood passes into the testis, the cooler blood from the pampiniform plexus passes close by, moving in the opposite direction. Some of the heat from the arterial blood transfers to the cooler blood, making the blood that enters the testis a few degrees cooler.

CORE PRINCIPLE
Structure-Function

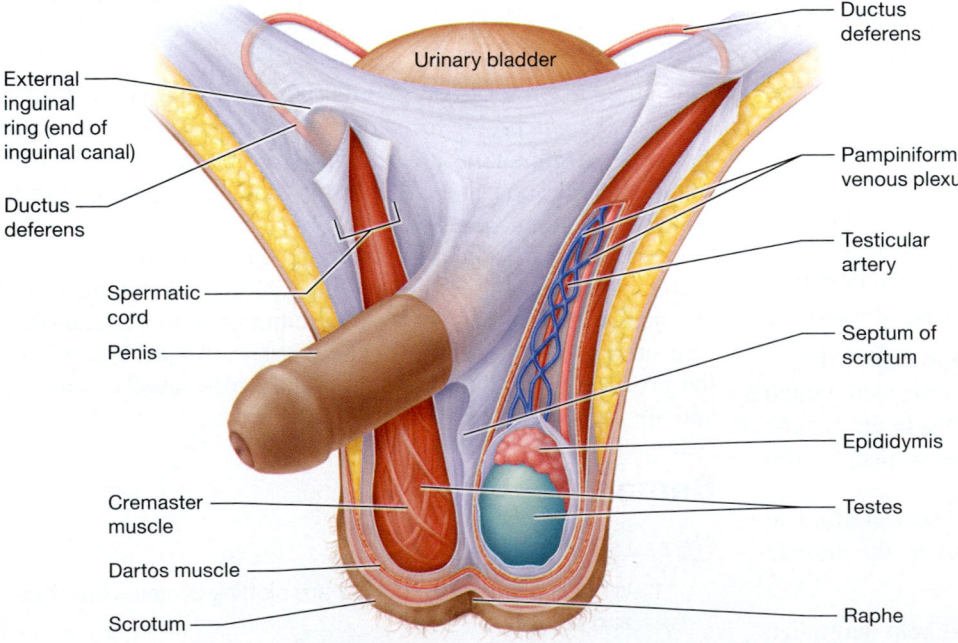

External inguinal ring (end of inguinal canal)

Ductus deferens

Spermatic cord

Penis

Cremaster muscle

Dartos muscle

Scrotum

Urinary bladder

Ductus deferens

Pampiniform venous plexus

Testicular artery

Septum of scrotum

Epididymis

Testes

Raphe

Figure 26.6 The scrotum and spermatic cord.

Second, the scrotum can be cooled or heated slightly by controlling its proximity to the body cavity. For example, when the external temperature is cold, the testes are pulled closer to the pelvic floor and the warmth of the body wall, and the scrotum becomes shorter and heavily wrinkled. This decreases the surface area of the scrotum and reduces heat loss. When the external temperature is hot, the scrotal sac relaxes and becomes *flaccid,* or limp. This increases the surface area for cooling, which may involve sweating, and moves the testes farther away from the warm body trunk. These changes in scrotal surface area are controlled by the dartos muscle, which contracts to wrinkle the scrotal skin, and the cremaster muscle, which contracts to elevate the testes.

The scrotum is innervated by many sensory and motor fibers and includes both sympathetic and parasympathetic innervation. In addition to the thermoreceptors listed previously, the scrotum has a high number of tactile receptors important in sexual stimulation. The cremaster muscle also often contracts when a man is sexually aroused to protect the area during sexual intercourse.

Table 26.1 lists functions of the male reproductive organs.

Table 26.1 Functions of the Male Reproductive Structures

Structure	Function
Internal Genitalia	
Testis	Produces sperm cells as well as testosterone and inhibin
Duct System	
Epididymis	Promotes sperm cell maturation; stores sperm; moves sperm cells to ductus deferens
Ductus deferens	Stores sperm; moves sperm to ejaculatory duct
Ejaculatory duct	Transports sperm from ductus deferens into urethra
Urethra	Transports semen out of penis
Accessory Glands	
Seminal vesicle	Secretes alkaline fluid with nutrients, prostaglandins, and a coagulating enzyme that support sperm, enhance their motility, and maintain semen in the female reproductive tract after ejaculation
Prostate gland	Secretes slightly alkaline fluid with nutrients and several components that support sperm, help activate sperm for fertilization, and function as an anticoagulant; also has immunological functions
Bulbourethral gland	Secretes mucus to lubricate glans penis and neutralize acidic traces of urine in urethra
External Genitalia	
Scrotum	Encloses, protects, and regulates temperature of testes
Penis	Moves urine and semen out of body; inserts into vagina during intercourse to deposit semen

Quick Check

☐ 9. What is the benefit of having the testes located in the scrotum rather than the abdominopelvic cavity?

☐ 10. Which part of the duct system passes through the inguinal canal inside the spermatic cord?

Apply What You Learned

☐ 1. A patient has a blockage that prevents secretions from the seminal vesicles. How might this affect the amount of semen produced?

☐ 2. A man is taking a medication that triggers involuntary muscle contractions as a side effect. In addition, the weather is very hot. Predict how this combination of factors might affect his production of viable sperm. (*Hint:* What is one function of the muscles in the scrotum?)

See answers in Appendix A.

MODULE 26.3
Physiology of the Male Reproductive System

Learning Outcomes

1. Relate the general stages of meiosis to the specific process of spermatogenesis.
2. Discuss endocrine regulation of spermatogenesis.
3. Discuss the events and endocrine regulation of male puberty.
4. Describe the male sexual response.
5. Describe male secondary sex characteristics and their role in reproductive system function.
6. Explain the effects of aging on reproductive function in males.

Now that we have examined the anatomy of the male reproductive system, we turn our attention to its physiology. In this module, we first discuss how sperm cells develop via a process known as *spermatogenesis*. Next we focus on how the hypothalamus and anterior pituitary regulate the production of hormones responsible for reproduction in males. The module closes with a look at the male sexual response, the changes that take place in males at puberty, and the changes in male reproductive function that occur with age.

Spermatogenesis

◀◀ FLASHBACK

1. What is the function of mitochondria? What is ATP? (p. 60 and p. 89)
2. What is peristalsis? (p. 852)

Spermatogenesis (sper-mat′-oh-JEN-uh-sis) is the process of male gamete, or **sperm cell,** development. It begins at puberty and continues throughout life. Every day, a healthy young adult male makes about 400 million sperm.

Spermatogenesis occurs in the seminiferous tubules of the testes through the division of *spermatogenic,* or sperm-forming, cells. **Spermatogonia** (sper-mat′-oh-GOHN-ee-ah; singular, *spermatogonium*), the stem cells that begin the process, are diploid cells with 46 chromosomes. They are found along the basement membrane of the seminiferous tubules, as you can see in **Figure 26.7**.

Before puberty, spermatogonia divide by mitosis, but only to produce more spermatogonia. When puberty begins, they become functional stem cells and divide, again by mitosis, to maintain the stem cell line and create more spermatogonia. However, some of the cells resulting from mitosis differentiate into diploid **primary spermatocytes** (sper-MAT-oh-sytz; step ①). As spermatogonia divide into these two cell types, the cell destined to become a primary spermatocyte is pushed toward the lumen of the seminiferous tubule while the spermatogonium

takes its place along the basement membrane to replace its parent cell. The primary spermatocyte then undergoes meiosis I, forming two smaller, joined haploid cells called **secondary spermatocytes** (step ②). The two secondary spermatocytes rapidly undergo meiosis II and produce four small haploid **spermatids** (step ③). Notice in Figure 26.7 that the spermatids are located at the lumen of the tubule.

Quick Check

☐ 1. What are the steps of spermatogenesis?

Sustentacular Cells

《 FLASHBACK

1. What is the function of tight junctions? (p. 127)
2. What are microtubules? (p. 97)

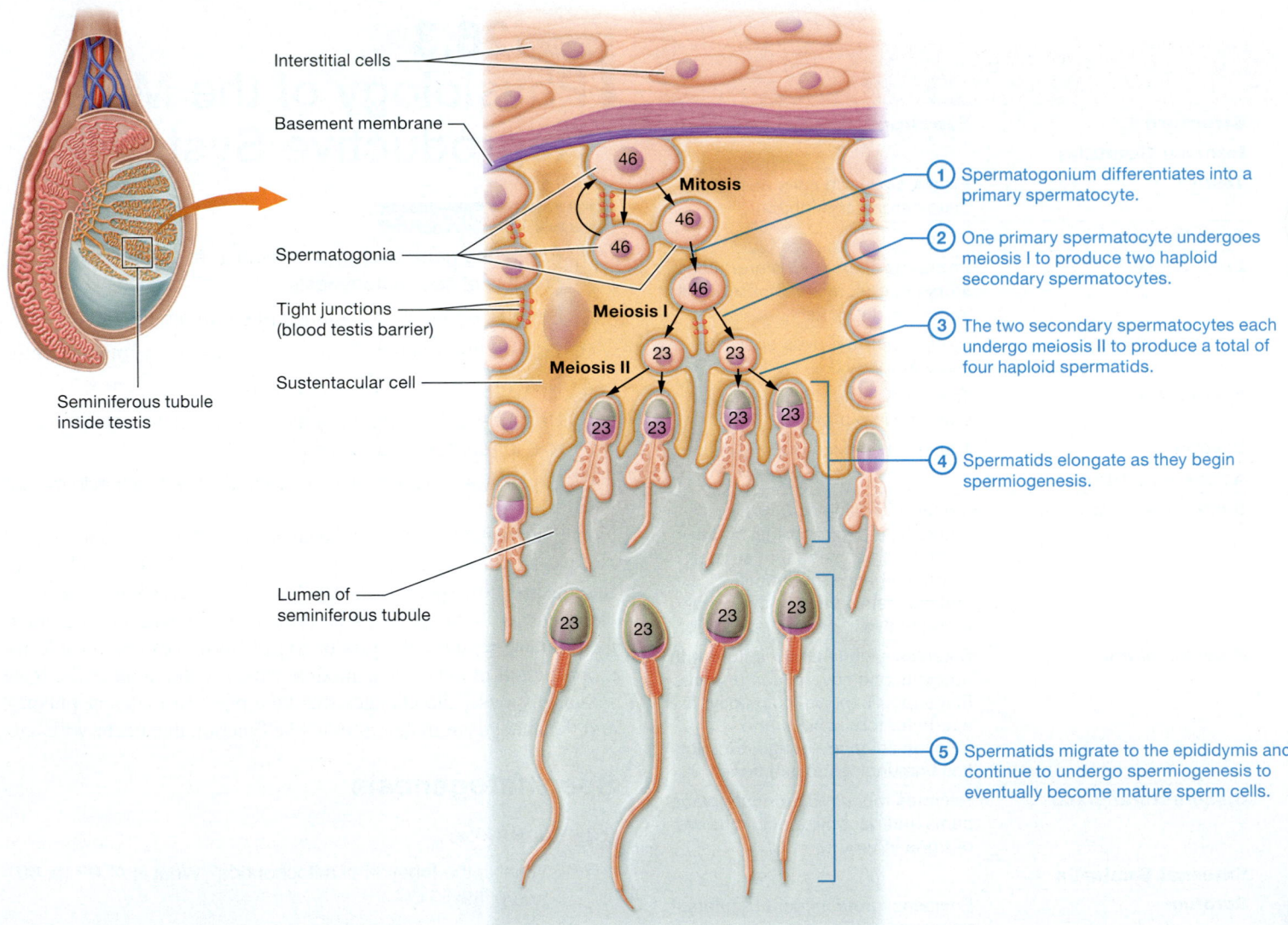

Interstitial cells

Basement membrane

46
Mitosis
46

Spermatogonia
46
46

Tight junctions
(blood testis barrier)
Meiosis I
46

Sustentacular cell
Meiosis II
23   23

Seminiferous tubule
inside testis
23   23   23   23

Lumen of
seminiferous tubule
23   23   23   23

① Spermatogonium differentiates into a primary spermatocyte.

② One primary spermatocyte undergoes meiosis I to produce two haploid secondary spermatocytes.

③ The two secondary spermatocytes each undergo meiosis II to produce a total of four haploid spermatids.

④ Spermatids elongate as they begin spermiogenesis.

⑤ Spermatids migrate to the epididymis and continue to undergo spermiogenesis to eventually become mature sperm cells.

Figure 26.7 The stages of spermatogenesis in the seminiferous tubules. Spermatogenesis begins with a spermatogonium that has 46 chromosomes, and produces 4 spermatids that have 23 chromosomes each.

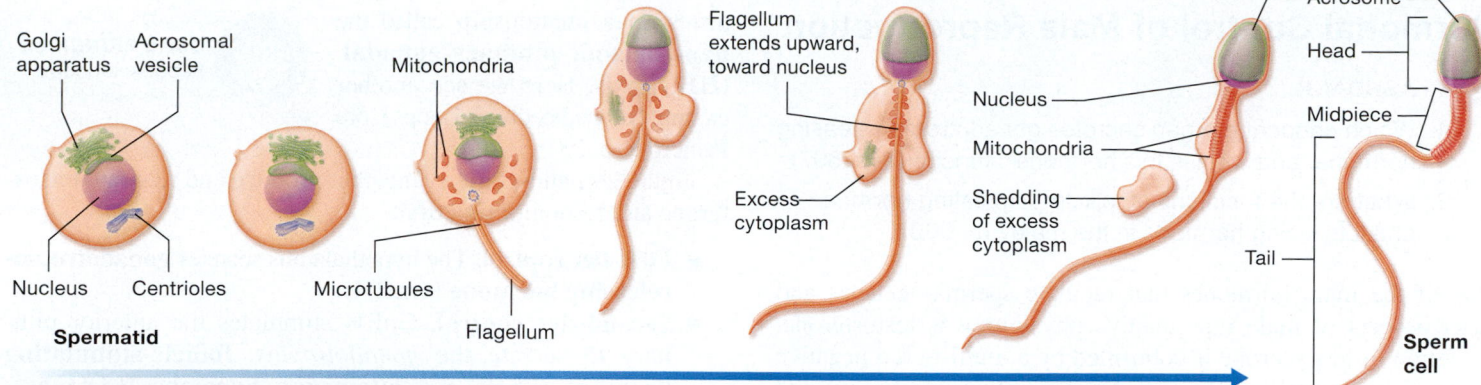

(a) Maturation from a spermatid to a sperm cell

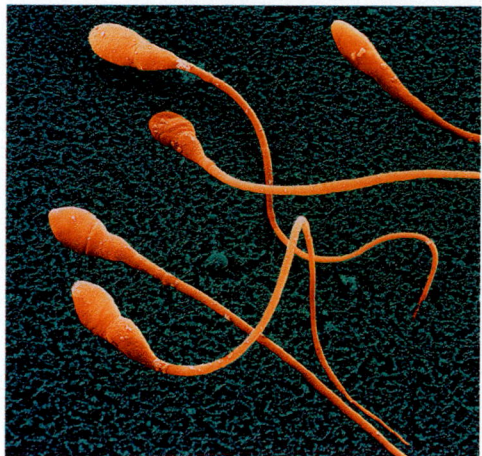

(b) SEM of mature sperm cells   SEM (1650×)

Figure 26.8 Spermiogenesis.

The yellow cells in Figure 26.7 that you can see surrounding the spermatogenic cells are supporting cells called **sustentacular cells** (sus'-ten-TAK-yoo-luhr), also referred to as *nurse cells*. These large cells extend from the basement membrane to the lumen of the tubule and are bound to one another by tight junctions. These tight junctions form a **blood testis barrier** that separates the forming sperm cells from the immune system. Immune cells in the bloodstream must be prevented from encountering the new antigens on the genetically unique sperm cells, as they would recognize these antigens as foreign and mount an immune response against them. Notice, however, that the spermatogonia are outside the blood testis barrier. This is possible because they are genetically identical to other body cells, so their antigens are not recognized by immune cells as foreign. Indeed, they need to be outside the blood testis barrier for hormones and paracrines to reach them and stimulate spermatogenesis.

In addition to forming the blood testis barrier, sustentacular cells perform many other functions important to spermatogenesis, which explains why they are nicknamed nurse cells. These functions include the following:

- providing structural support for spermatogonia development by maintaining the environment around the cells;
- secreting substances that stimulate mitosis of spermatogonia and initiation of meiosis in response to testosterone and *follicle-stimulating hormone;*

- providing nutrients to the dividing cells;
- phagocytizing damaged spermatogenic cells and excess cytoplasm released from maturing spermatids; and
- producing *androgen-binding protein (ABP)* and *inhibin* (see the next module), which help regulate spermatogenesis.

Quick Check

☐ 2. How do sustentacular cells support developing sperm?

Spermiogenesis

The maturation process of sperm is called **spermiogenesis** (sper-mee'-oh-JEN-uh-sis) and involves changing the size and shape of the cell. Spermiogenesis begins in the testes as the spermatid elongates and sheds excess cytoplasm (step ④ in Figure 26.7). As it continues, the following occur (step ⑤): A structure known as the **acrosome** (AK-roh-sohm), which contains digestive enzymes and helps fertilize the female gamete, forms over the nucleus; more mitochondria are produced and move to their positions around the beginning part of the flagellum; microtubules in the cell form the flagellum, or tail; and cells lose their connections to sustentacular cells and are released into the lumen of the seminiferous tubule.

The changing appearance of the cell as it matures from a spermatid to a sperm cell is shown in **Figure 26.8**. Each sperm cell has head, midpiece, and tail regions and is about 55–60 μm (0.0021–0.0024 in.) long and about 3 μm (0.0001 in.) wide (see Figure 26.8a). The **head** includes the nucleus and the acrosome and very little cytoplasm; the **midpiece** contains the mitochondria; and the **tail** is a flagellum surrounded by the plasma membrane. The tail moves in a mature sperm cell because of ATP provided by the mitochondria of the midpiece.

Even though the sperm now has a flagellum, it is still nonmotile and will migrate to the epididymis to finish maturing. Sperm move slowly, taking about 12 days to reach the tail of the epididymis, where they can remain viable for several months. If the sperm are not ejaculated by this time, the body resorbs them. Together, the processes of spermatogenesis and spermiogenesis take about 60–70 days to complete.

Quick Check

☐ 3. How do spermatogenesis and spermiogenesis differ?
☐ 4. At what point is a sperm cell mature?

Hormonal Control of Male Reproduction

« FLASHBACK

1. Which endocrine organ secretes gonadotropin-releasing hormone, and what is this hormone's function? (p. 597)
2. What are the functions of follicle-stimulating hormone and luteinizing hormone in the male? (p. 600)

One of the main hormones that regulate spermatogenesis and other aspects of male reproductive physiology is testosterone. Secretion of testosterone is controlled by a multi-tiered negative feedback loop involving the hypothalamus, anterior pituitary, and testes, in a relationship called the **hypothalamic-pituitary-gonadal (HPG) axis.** Here we see another example of the Feedback Loops Core Principle (p. 28).

CORE PRINCIPLE
Feedback Loops

Figure 26.9 shows the events that produce and regulate testosterone and testicular function:

- **First-tier control.** The hypothalamus secretes **gonadotropin-releasing hormone (GnRH).**
- **Second-tier control.** GnRH stimulates the anterior pituitary to secrete the *gonadotropins*: **follicle-stimulating hormone (FSH)** and **luteinizing hormone** (LOO-tee´-

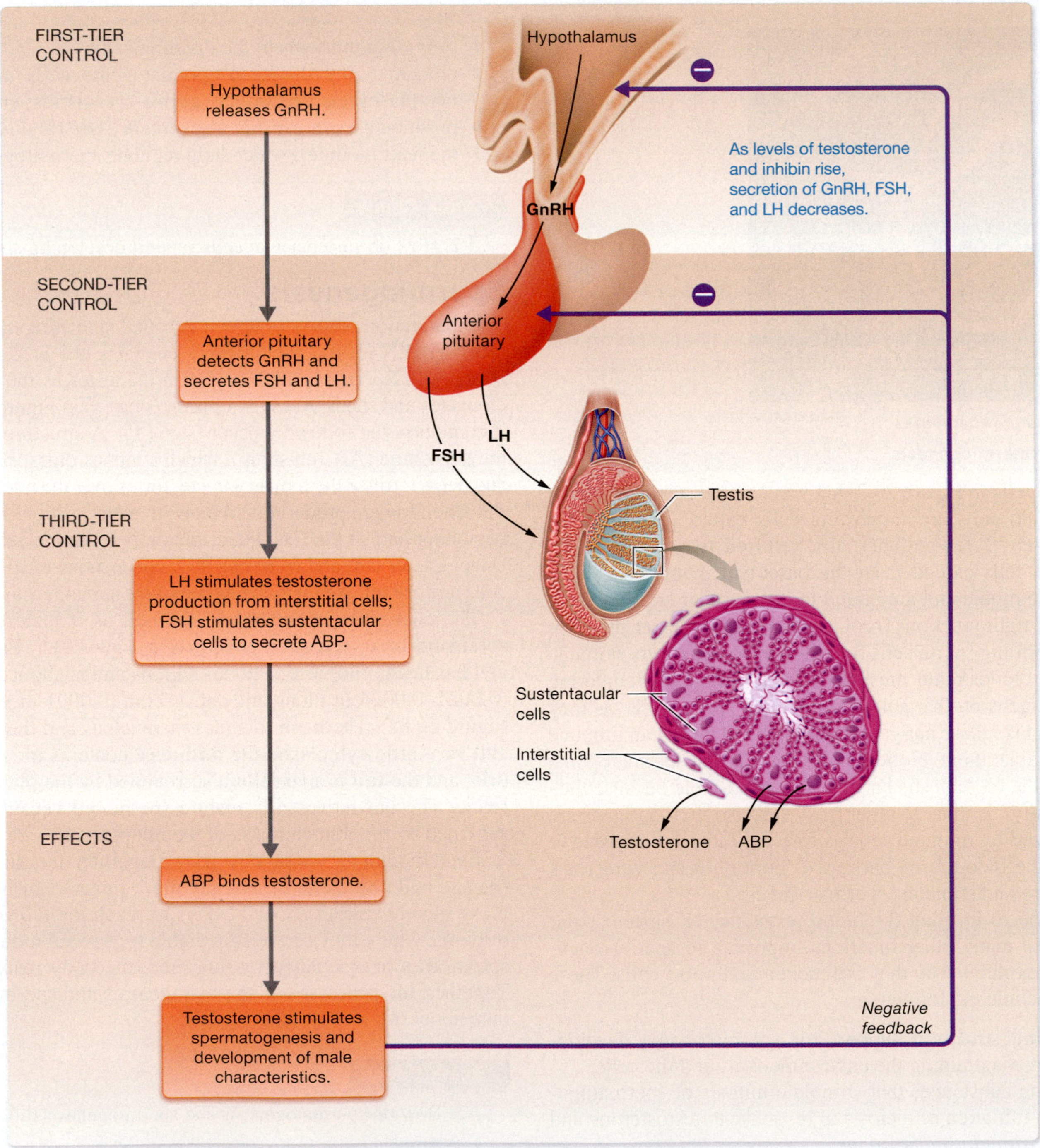

FIRST-TIER CONTROL

Hypothalamus

Hypothalamus releases GnRH.

GnRH

As levels of testosterone and inhibin rise, secretion of GnRH, FSH, and LH decreases.

SECOND-TIER CONTROL

Anterior pituitary

Anterior pituitary detects GnRH and secretes FSH and LH.

FSH   LH

Testis

THIRD-TIER CONTROL

LH stimulates testosterone production from interstitial cells; FSH stimulates sustentacular cells to secrete ABP.

Sustentacular cells

Interstitial cells

EFFECTS

Testosterone   ABP

ABP binds testosterone.

Testosterone stimulates spermatogenesis and development of male characteristics.

Negative feedback

Figure 26.9 Hormonal regulation of testicular function via the hypothalamic-pituitary-gonadal (HPG) axis.

uh-nyz-ing) (LH), also called *interstitial cell-stimulating hormone* (*ICSH*) in males.

- **Third-tier control.** The testes are the target organs of FSH and LH. LH stimulates the interstitial cells to produce testosterone. FSH stimulates the sustentacular cells to secrete **androgen-binding protein (ABP),** and the hormone *inhibin*.

Testosterone is the main hormone that directly stimulates spermatogenesis. However, recall that testosterone-producing interstitial cells are located between the seminiferous tubules, yet the target cells for spermatogenesis—spermatogonia and sustentacular cells—are *within* the seminiferous tubules. For this reason, sustentacular cells secrete androgen-binding protein in response to FSH, which binds and concentrates testosterone in the seminiferous tubules around its target cells. Much of the testosterone also enters the blood, where it affects other tissues and leads to the development of male characteristics (discussed shortly).

When the level of testosterone rises to a certain value, GnRH secretion falls by negative feedback, which in turn causes a decline in the levels of FSH and LH secretion. In addition, the hormone inhibin, secreted by sustentacular cells, decreases the secretion of FSH. Sustentacular cells also convert some testosterone to estrogens, which directly inhibits testosterone production by interstitial cells.

It usually takes a few years to establish the HPG axis, but once established, the testosterone level and spermatogenesis remain fairly stable until about age 70. At this age, it is relatively common to see gradual decreases in both. However, sometimes conditions can disrupt testicular function, leading to infertility, which is explored in *A&P in the Real World: Male Infertility.*

You can look ahead to Table 26.3 on p. 1056 to see a summary of the functions of the major reproductive hormones in both males and females.

Quick Check

- ☐ 5. What is the relationship between the hypothalamus, anterior pituitary gland, and testes in the HPG axis?
- ☐ 6. On what type of cell do FSH and LH act in males, and what do they stimulate in their target cells?
- ☐ 7. What are the reproductive functions of testosterone and androgen-binding protein?

Male Sexual Response

« FLASHBACK

1. What is the basic function of the parasympathetic division of the autonomic nervous system? (p. 526)
2. What is the basic function of the sympathetic division of the autonomic nervous system? (p. 519)

For fertilization to occur, sperm cells must reach female oocytes deep inside the female reproductive tract. To optimize the chances that fertilization will occur, sperm should enter into the female reproductive tract as far as possible. This is usually accomplished by **sexual intercourse,** also called *copulation* or *coitus* (KOY-tus), most commonly defined as the insertion of the male penis into the copulatory organ of the female, the *vagina*.

Male Infertility

Infertility is the inability to produce a pregnancy after 1 year of unprotected intercourse. Approximately 40 percent of all infertility cases result from male infertility, which is usually due to a low sperm count. A normal level is 40–750 million sperm per ejaculate; a count of less than 15 million sperm cells per milliliter of semen usually indicates infertility.

A low sperm count can result from any sort of damage to the testes, such as physical trauma, exposure to radiation, or disease. It could also be due to developmental defects. During normal development, the testes begin forming inside the abdominopelvic cavity and then descend into the scrotum. If a testis does not descend into the scrotum, a disorder called cryptorchidism, sperm cells will not be produced. In addition, inadequate secretion of GnRH, FSH, LH, or testosterone for any reason will also lower sperm count.

The basic phases of the male sexual response include *erection* and *ejaculation*. **Erection** is the enlargement and stiffening of the penis, which allows the penis to enter the vagina. An erection results when the erectile tissue in the penis becomes engorged with blood. Normally, blood vessels in the penis that supply erectile tissue are constricted, and the penis is in a **flaccid,** or relaxed, state. During sexual arousal, a parasympathetic reflex triggers the release of nitric oxide (NO) from blood vessel endothelial cells. This relaxes smooth muscle in the walls of the arterioles of the penis, dilating them, and within the blood sinuses in erectile tissue, which allows large amounts of blood to enter the tissue. The combination of increased blood flow and widening blood sinuses results in an erection. The erection is maintained, at least partially, by the compression of the veins that drain the penis. Note that the corpus spongiosum does not become as rigid as the corpora cavernosa, which allows the urethra to remain open during ejaculation so that sperm and semen can pass.

Erection can occur because of a variety of sexual stimuli, including touch; mechanical stimulation of the penis; and erotic sights, sounds, and smells. Emotions and thoughts, controlled by the higher thought centers of the brain, may cause or prevent erections. **Impotence,** or *erectile dysfunction* (*ED*), is the inability to maintain an erection long enough to participate in sexual intercourse (you can read more about this in *A&P in the Real World: Erectile Dysfunction* on p. 1038).

Ejaculation is the process by which semen is expelled from the penis. Although erection is under parasympathetic control, ejaculation is under *sympathetic* control. It actually occurs in two stages, *emission* and *expulsion*. **Emission** is the movement of sperm and testicular fluid, as well as secretions from the prostate gland and seminal vesicle, into the urethra. As semen accumulates in the urethra with the increase in sexual tension, sensory impulses are stimulated and pass into the sacral portion of the spinal cord. Motor neurons from the spinal cord stimulate skeletal muscles at the base of the erectile columns of the penis to rhythmically contract, resulting in expulsion of the semen from the urethra. **Orgasm** refers to

Erectile Dysfunction

A&P in the Real World

Various psychological and physical factors may cause erectile dysfunction (ED). Psychological influences include stress, depression, and anxiety, while common physical causes include cardiovascular disease and diabetes mellitus. Obesity, tobacco and alcohol use, and certain prescription medications can also play a role. Older men have a greater risk of erectile dysfunction because they are more likely to have physical conditions associated with it. Also, the amount of connective tissue in the erectile tissue of the penis increases with age, reducing blood flow to the penis.

Treatment for ED includes medications that permit blood vessels supplying the penis to remain dilated, thereby maintaining an erection. Other treatments may involve surgery to repair damaged blood vessels or to insert penile implants.

the time period during which feelings of pleasure are experienced. In males, this coincides with ejaculation.

Orgasm and ejaculation are quickly followed by **resolution,** or *relaxation.* Blood vessels in the erectile tissues and blood sinuses constrict, forcing blood out of the penis, which soon becomes flaccid again. After ejaculation, a *latent period,* or **refractory period,** follows; during this time, ranging from minutes to hours, a man is unable to achieve another orgasm. The refractory period tends to get longer as men age.

Quick Check

☐ 8. What is the role of the parasympathetic nervous system in the male sexual response? What is the role of the sympathetic nervous system?

☐ 9. What are the two stages of ejaculation? What occurs during these stages?

Effects of Testosterone on Other Body Systems

 **FLASHBACK**

1. What are the androgenic and anabolic effects of testosterone? (p. 619)

2. Which endocrine organs secrete androgens? (p. 612)

Testosterone is known as the male steroid hormone for more reasons than its importance in stimulating spermatogenesis. Without testosterone, many characteristics we identify as "male," or "masculine," would not be distinguishable. The level of this hormone increases dramatically at puberty and is responsible for the growth, maturation, and maintenance of male reproductive organs, including the testes, scrotum, penis, and duct system. Testosterone also triggers the development of **male secondary sex characteristics,** which are features induced in nonreproductive organs that are associated with adult males.

Puberty

Throughout a boy's childhood, testosterone from the testes and androgens from the adrenal cortex are secreted in low, but sufficient, amounts to inhibit GnRH release from the hypothalamus. This continues until puberty, when the hypothalamus becomes much less sensitive to inhibition by the androgens, causing the GnRH level to increase. GnRH stimulates the HPG axis, and the blood testosterone level begins to rise, triggering spermatogenesis and the appearance of male secondary sex characteristics.

In most boys, puberty begins between 12 and 14 years of age, but this age varies. Usually within a few years after the onset of puberty, the hormone levels stabilize. As soon as spermatogenesis begins, males are capable of sexual reproduction, even if hormone levels are still fluctuating.

Secondary Sex Characteristics

Male secondary sex characteristics develop primarily at puberty, but may continue to develop into early adulthood in some males. They include the following:

- growth of pubic, axillary, and facial hair;
- enhanced hair growth on the chest or other body areas in some men;
- an enlarged larynx and thicker vocal cords, resulting in a noticeable laryngeal prominence ("Adam's apple") and deepening voice; and
- thickened skin and increased secretions of sebaceous glands (the oilier skin often results in acne).

Testosterone also causes **somatic (body) effects,** many of which are anabolic. Under the influence of testosterone, more calcium salts are deposited in bone, making it denser. In addition, skeletal muscle mass increases, accounting for the higher muscle mass in males compared to females, and all organs in general are larger. This accounts for males' larger cardiac output and total lung capacity. Testosterone also has a direct effect on the production of erythropoietin, causing males to have a higher number of erythrocytes than females.

Finally, testosterone has an influence on behavior. For example, it is the basis for the male **libido,** which is the desire for sexual activity. However, the long-held idea that testosterone is responsible for aggression is highly disputed, and much evidence suggests that it is not true.

Quick Check

☐ 10. Why does the level of GnRH increase at the beginning of puberty?

☐ 11. What are the most noticeable secondary sex characteristics caused by testosterone?

☐ 12. What are the anabolic effects of testosterone?

Effects of Aging: Male Climacteric

Male climacteric (kly-MAK-ter-ik), or *andropause,* is the period in which reproductive functions begin to decline in men. These changes are usually gradual and typically begin in the early 50s, but may or may not produce noticeable changes in all men. The size and weight of the testes may decrease and the number of sustentacular and interstitial cells may begin to decline. When sustentacular cells produce less inhibin and interstitial cells produce less testosterone, GnRH increases the production of FSH and LH because negative feedback decreases. If FSH and LH levels remain high, it indicates that the testes are not producing enough testosterone to suppress their production. A reduced level of testosterone results in the gradual production of fewer sperm cells, although this is not absolute, as many men in their 70s or 80s have fathered children. In addition, a low level of testosterone is associated with depressed mood and fatigue in some men and leads to a decrease in muscle strength and some decrease in bone density. However, bone density typically declines more slowly in aging men than in women, because the loss of testosterone is gradual.

Probably the most noticeable change for most men during climacteric is enlargement of the prostate gland. This compresses the prostatic urethra, which often causes difficulty with urination.

Quick Check

☐ 13. Why does sperm cell production decrease as men age?

☐ 14. Which organs not directly associated with the reproductive system are affected by a decreased testosterone level?

Apply What You Learned

☐ 1. If a man's anterior pituitary did not secrete LH, how would this affect spermatogenesis?

☐ 2. Predict how a drug that inhibits nitric oxide production might affect a man's sexual response.

☐ 3. Explain why a man experiencing emotional or physical stress is often unable to develop or maintain an erection. (*Hint:* Which branch of the autonomic nervous system is involved in the stress response?)

See answers in Appendix A.

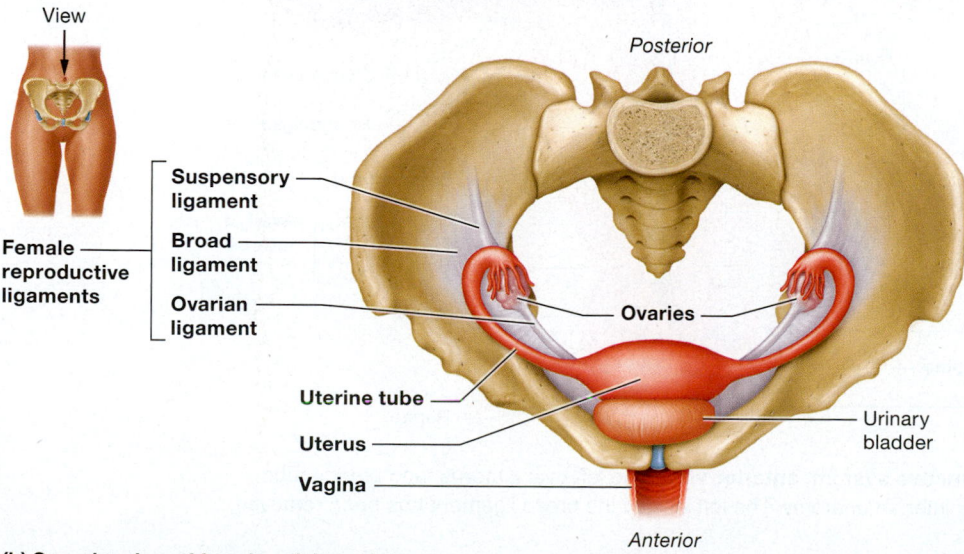

(a) **Midsagittal section of female pelvic cavity**

Posterior — Peritoneum, Rectum, Rectouterine pouch

Anterior — Suspensory ligament, Uterine tube, Ovary, Round ligament, Uterus, Vesicouterine pouch, Vagina, Urethra, Urinary bladder

View

Female reproductive ligaments — Suspensory ligament, Broad ligament, Ovarian ligament

Posterior — *Anterior*

Ovaries, Uterine tube, Uterus, Vagina, Urinary bladder

(b) **Superior view of female pelvic cavity**

Figure 26.10 Internal organs of the female reproductive system.

MODULE **26.4**

Anatomy of the Female Reproductive System

Learning Outcomes

1. Describe the structure and functions of the female reproductive organs.

2. Describe the histology of the uterine wall.

3. Trace the pathway of the female gamete from the ovary to the uterus.

The male and female reproductive systems share a number of similarities. Both systems produce sex hormones and gametes, and both have a series of ducts to move the gametes. However, the female reproductive system must also support a developing **conceptus**—the offspring from fertilization through birth—and nourish an infant, and these functional differences lead to structural differences.

The internal female genitalia include the *ovaries, uterine tubes, uterus,* and *vagina.* See **Figure 26.10a** for a midsagittal view of these organs and some of the ligaments that hold them in place. In this view you can also see how these organs are related to the urinary bladder and the rectum. Notice that the peritoneum dips down on either side of the uterus. This creates two "pouches" between the uterus and the bladder and rectum—the *vesicouterine pouch* (vez′-uh-koh-YOO-ter-in) and the *rectouterine pouch* (rek′-toh-YOO-ter-in), respectively.

Unlike the male reproductive organs, the female reproductive organs are located mostly in the pelvic cavity. This can be appreciated by looking at **Figure 26.10b**, which shows how the organs are arranged from a superior view (imagine you are holding a pelvis and looking down into it). We examine these organs in this module before turning to the *external female genitalia* and *mammary glands*.

Ovaries

« FLASHBACK

1. What are the main functions of estrogens and progesterone? (p. 619)

The paired **ovaries** are the main female gonads and are found on the lateral walls of the pelvic cavity (see Figure 26.10b). Each ovary weighs only about 6–8 g and is about the size of a walnut in women who are actively releasing ova on a monthly cycle. These organs appear relatively smooth before puberty, but as a woman matures, they take on a bumpy appearance due to scars left behind when ova are released approximately once each month.

The ovaries perform two main functions: They (1) produce ova and (2) secrete several hormones. Among these hormones are the *estrogens,* which include estradiol, estrone, and estriol, with estradiol being the most abundant. We will use the term estrogens to refer to all these hormones collectively. The ovaries also produce *progesterone* (proh-JES-ter-ohn), inhibin, and relaxin. (The function of relaxin is discussed in Chapter 27.)

As in many other organs, the superficial portion of the ovary is the cortex and the deeper portion is the medulla (**Figure 26.11**). The **ovarian cortex** is the site of *oogenesis,* the production of female gametes. In females, gametes are produced inside saclike structures called *follicles,* which develop and mature along with the gametes. The **ovarian medulla** houses blood vessels, lymphatic vessels, and nerves.

Each ovary is held in place by three ligaments. First is the **broad ligament,** a large, flat ligament that attaches to the ovaries, uterine tubes, and uterus and connects them to the bony pelvis (you can see this attachment to the pelvis in Figure 26.10b). Next is the **ovarian ligament,** which connects the medial surface of the ovary to the uterus, near the attachment of the uterine tube. Finally, the **suspensory ligament** (or *infundibulopelvic ligament;* in-fun-DIB-yoo-loh-pel′-vik) connects the lateral surface of the ovary to the pelvic wall. Notice in Figure 26.11 that this ligament carries the **ovarian artery** and **ovarian vein.**

Quick Check

☐ 1. What are the main functions of the ovaries?
☐ 2. Which three ligaments support the ovary, and to which structures are they attached?

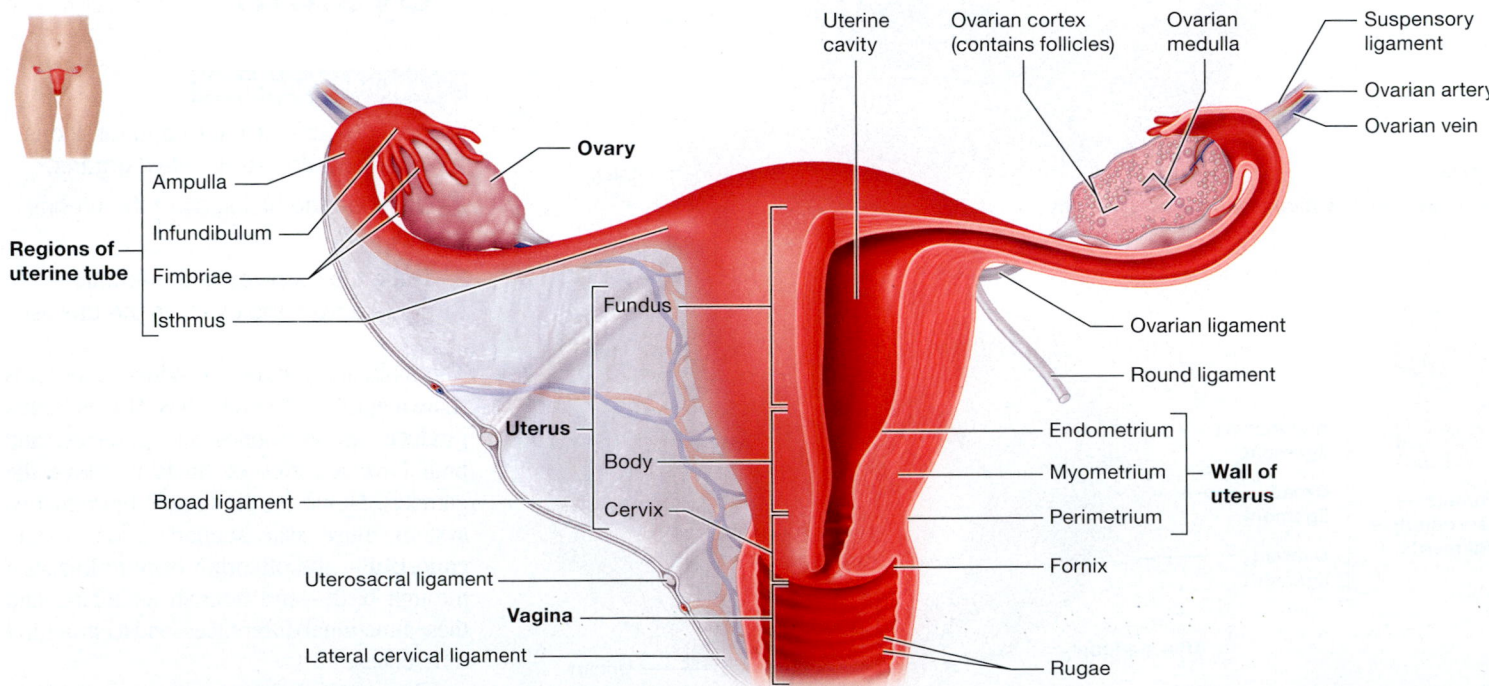

Figure 26.11 Internal organs of the female reproductive system, anterior view. The left ovary, uterus, and uterine tube and the vagina are cut in a frontal section to show the internal anatomy. The left side of the broad ligament has been removed.

Uterine Tubes

« FLASHBACK

1. What is the function of cilia? (p. 99)

The **uterine tubes** (YOO-ter-ihn), also called *fallopian tubes* or *oviducts,* form the first part of the duct system of the female reproductive system (see Figure 26.11). Externally, they are covered by the peritoneum and are supported along their length by part of the broad ligament. Internally, they are lined by smooth muscle and simple columnar epithelium, with cells that are both ciliated and nonciliated.

Each uterine tube is approximately 10 cm (3.9 in.) long and extends medially from the ovary to the superior and lateral region of the uterus. At its proximal end, the uterine tube connects to the uterus in a constricted region called the **isthmus.** The distal end of each uterine tube expands as it curves around the ovary, forming the **ampulla,** which is where fertilization of an oocyte normally occurs. The ampulla connects to the **infundibulum,** an open funnel-shaped structure. At the end of the infundibulum are ciliated, finger-like projections known as **fimbriae** (FIM-bree-aye) that drape over the ovary and "catch" an oocyte that is released from an ovary during a process known as *ovulation.*

Quick Check

☐ 3. What structures catch an ovulated oocyte and move it toward the uterus?

☐ 4. What is the usual site of fertilization?

Uterus

The **uterus** (YOO-ter-uhs), known commonly as the womb, is part of the pathway through which sperm travel to reach the uterine tubes. It is also the site where a fertilized ovum implants, and it provides protection and nutritional support for a developing conceptus. In addition, contractions of the uterus are responsible for ejecting the fully developed conceptus (known as a fetus) during childbirth. When pregnancy does not occur, the uterus is the source of menstrual flow.

The uterus is located in the pelvis anterior to the rectum and posterior to the urinary bladder. A hollow, thick-walled, muscular organ, it is similar in size and shape to an inverted pear. It weighs 30–40 g, but it is usually a little larger in women who have borne children. Like the ovaries, the uterus is supported by several ligaments: inferiorly by the **lateral cervical ligaments,** which extend from the cervix and superior vagina to the lateral walls of the pelvis; posteriorly by the paired **uterosacral ligaments** (yoo′-ter-oh-SAYK-ruhl), which connect the uterus to the sacrum; anteriorly by the **round ligaments,** which connect the uterus to the anterior body wall (you can see this attachment in Figure 26.10); and laterally by the broad ligaments. In addition to these ligaments, the uterus also receives a great deal of support from the muscles of the pelvic floor.

There are three regions of the uterus: the main portion, or **body;** the rounded region superior to the entrance of the uterine tubes, called the **fundus;** and the **cervix,** the narrow neck that projects into the vagina inferiorly. The most inferior portion of the cervix, which serves as the entrance into the uterus, is known as the *external os.* It is this portion of the cervix that is generally observed during a pelvic examination. Glands in the mucosa of the cervical canal secrete mucus that covers the external os, acting as a protective plug.

The wall of the uterus is composed of three layers, which are labeled in Figure 26.11. The **perimetrium** (pehr-ee-MEE-tree-um) is the outermost serous layer and is an extension of the parietal peritoneum. The **myometrium** is the thick middle layer, composed of bundles of smooth muscle. It contracts rhythmically during orgasm and during childbirth to expel the fetus from the mother's body and in some women, to a lesser degree, during the menstrual cycle. The innermost **endometrium** lines the uterine cavity and is a mucous membrane composed of simple columnar epithelium on a layer of connective tissue called the lamina propria.

Quick Check

☐ 5. What are the ligaments that support the uterus?

☐ 6. From deep to superficial, what are the layers of the uterine wall?

Vagina

« FLASHBACK

1. What are rugae? (p. 865)

2. What is glycogen? (p. 52)

The **vagina** (vuh-JY-nuh) is the female organ of copulation because it receives the penis (and semen) during sexual intercourse. It also provides a passageway for delivery of a fetus and for menstrual flow. The vagina is a thick-walled tube, approximately 8–10 cm (3.1–3.9 in.) long that extends from the cervix to the exterior of the body (see Figure 26.10). It parallels the anterior urethra, lying between the urinary bladder and the rectum. The superior end of the vaginal canal surrounds the external os of the cervix and produces a recess called the vaginal **fornix** (shown in Figure 26.11). The wall of the vagina is covered with transverse ridges, or **rugae,** which stimulate the penis during intercourse. Generally, the lumen of the vagina is quite small and, except where it is held open by the cervix, its walls touch one another. However, the vagina stretches considerably during intercourse and childbirth.

The vaginal mucosa is a stratified squamous epithelium, allowing it to withstand friction. It has no glands, but is lubricated by cervical mucus and secretions from the epithelial cells, which increase during sexual excitement. The epithelial cells also secrete *glycogen,* which resident bacteria break down to produce ATP and lactic acid. This helps maintain the acidic pH that keeps the vagina healthy and free of infection. Earlier in the chapter, we mentioned that the acidic pH of the vagina is also hostile to sperm. This is one reason why the alkaline semen is so important for maintaining sperm health in the female reproductive tract.

In most women, the mucosa near the distal vaginal orifice forms an incomplete partition called the *hymen.* The hymen is very vascular and may bleed when it ruptures, which happens in about 50% of women during their first experience of sexual intercourse. However, it may also rupture from an injury or during activities such as sports.

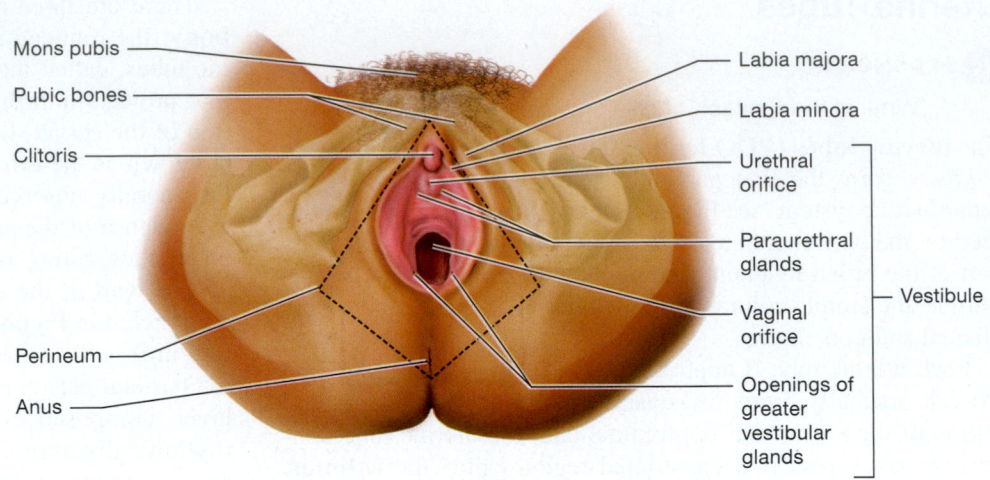

Figure 26.12 The female perineum.

Labels: Mons pubis, Pubic bones, Clitoris, Perineum, Anus, Labia majora, Labia minora, Urethral orifice, Paraurethral glands, Vestibule, Vaginal orifice, Openings of greater vestibular glands

Quick Check

☐ 7. What are the functions of the vagina?

☐ 8. Which cells produce the lactic acid that gives the vagina an acidic pH? What is the importance of maintaining this acidic pH?

Female External Genitalia

The **external female genitalia,** or **vulva,** are the female reproductive structures that lie external to the vagina (**Figure 26.12**). The external genitalia include the *mons pubis, labia majora, labia minora, clitoris,* and structures associated with the *vestibule.* Although the external genitalia are often referred to as the "vagina," the vagina is located internally and is not part of the external genitalia.

The **mons pubis** (mahnz PYOO-bis) is a fatty, rounded area overlying the pubic symphysis that is covered with hair after puberty. Posterior to the mons pubis are two elongated, fatty protective skin folds, the **labia majora** (LAY-bee-uh), which enclose two thinner skin folds called the **labia minora.** Enclosed within the labia minora is a recess known as the **vestibule.** The vestibule contains the **urethral orifice** and **vaginal orifice;** the **paraurethral glands** (pehr-uh-yoo-REETH-ruhl), which discharge mucus into the urethra; and the openings of the **greater vestibular glands** (*Bartholin's glands*), which release a small amount of mucus into the vestibule, helping to lubricate the opening of the vagina during intercourse.

Anterior to the vestibule is the **clitoris** (KLIT-ur-us), a small, protruding structure composed of erectile tissue. It corresponds to the corpora cavernosa of the penis and is important in the female sexual response. Just as in the penis, erectile tissue of the clitoris becomes engorged with blood and is innervated by many sensory, motor and autonomic fibers. Tactile, pressure,

and temperature receptors are abundant here as well, and are also important for sexual stimulation.

The female **perineum** is a diamond-shaped region located between the pubic arch anteriorly, the coccyx posteriorly, and the ischial tuberosities laterally (see Figure 26.12). The tissues of the perineum overlie the muscles of the pelvic outlet, and the posterior ends of the labia majora overlie the central tendon, which is the insertion of most muscles supporting the pelvic floor.

The region between the vagina and anus is the *clinical perineum.* During childbirth some doctors will cut a straight incision, called an **episiotomy** (eh-pee-zee-AWT-uh-mee), through the skin and muscle of this area if a tear appears likely. Episiotomies are done because straight incisions usually heal more quickly than jagged tears.

Quick Check

☐ 9. What are two functions of the female external genitalia?

☐ 10. How are the external genitalia of the female similar to those of the male? How are they different?

Mammary Glands

≪ FLASHBACK

1. What are the layers of the skin? (p. 165)
2. What is the basic structure of a sweat gland? (p. 177)

The **mammary glands** are part of the integumentary system and are actually modified sweat glands. Their major biological role is to produce milk to nourish a newborn infant, so they are important only when reproduction has already been accomplished. They are present in both sexes but normally function only in females.

Each mammary gland is contained within a rounded skin-covered breast in the hypodermis region, superficial to the

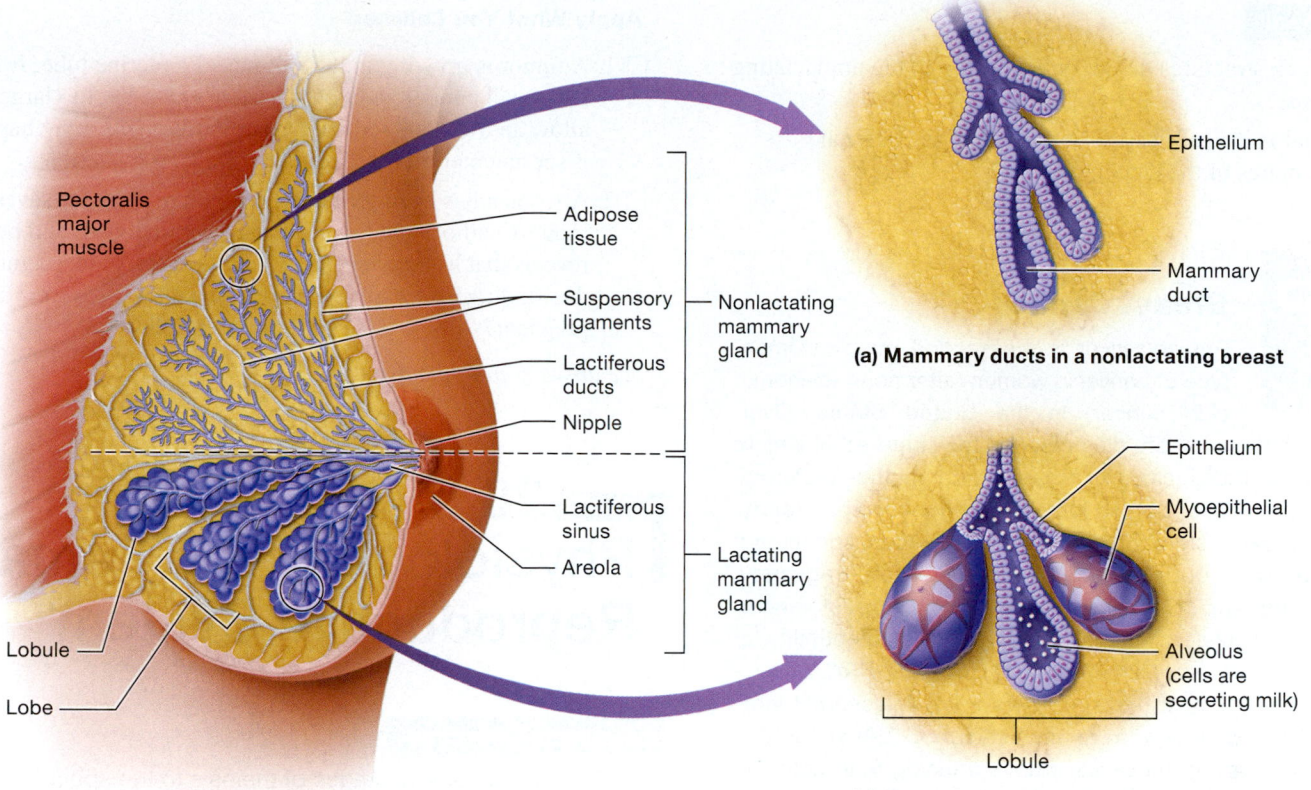

(a) Mammary ducts in a nonlactating breast

(b) Ducts and alveoli of a lactating breast

Figure 26.13 **Internal anatomy of the female breast.**

pectoralis major muscle of the chest (**Figure 26.13**). Just below the center of each breast is an area of pigmented skin, the **areola,** which surrounds a central protruding **nipple** through which milk exits the gland.

Internally, each mammary gland consists of 15–25 **lobes** that radiate around and open at the nipple. The lobes are padded and separated from one another by fibrous connective tissue and fat, and they receive support from **suspensory ligaments** that attach the breast to the underlying muscle fascia and the overlying dermis. Within each lobe are smaller units called **lobules,** which contain glandular **alveoli** that produce milk when a woman is lactating (compare Figures 26.13a and b). Contraction of **myoepithelial cells** that surround the alveoli helps propel milk toward the nipples.

When milk is being produced, it passes from the alveoli into a series of **mammary ducts,** which flow into **lactiferous ducts** that open to the outside at the nipple. Just deep to the areola, each lactiferous duct has a dilated region called a **lactiferous sinus,** where milk accumulates during nursing. (Chapter 27 discusses the process and regulation of lactation in more detail.)

In nonpregnant women, the glandular structure of the breast is mostly undeveloped and alveoli are not present. For this reason, breast size is largely due to the amount of *adipose tissue* in the breast. One of the most common sites of cancer in women is the breast (see *A&P in the Real World: Breast Cancer* on p. 1044).

Table 26.2 summarizes the functions of the female reproductive organs.

Table 26.2 Functions of the Female Reproductive Structures

Structure	Function
Internal Genitalia	
Ovary	Produces oocytes and estrogens, progesterone, and inhibin
Uterine tube	Moves oocyte or fertilized ovum toward uterus; site of fertilization and early stages of development
Uterus	Protects and sustains conceptus during pregnancy; cyclic shedding results in menstrual flow
Vagina	Receives penis during sexual intercourse; provides passageway for menstrual flow, for sperm, and for infant during birth
External Genitalia	
Labia majora	Enclose and protect other external reproductive structures
Labia minora	Enclose vestibule and protect openings of vagina and urethra
Clitoris	Erectile tissue innervated with sensory nerve endings
Vestibule	Recess that contains external openings of vagina and urethra
Greater vestibular gland	Secretes mucus, which helps lubricate the opening of the vagina

Quick Check

☐ 11. Which structures do not fully develop in the nonlactating breast?

☐ 12. What is the pathway of milk through the internal structures of the lactating breast?

A&P in the Real World

Breast Cancer

Breast cancer is the second most common type of cancer in women (after non-melanoma skin cancer) in the United States. Over 200,000 cases a year are diagnosed, a few hundred of which occur in men. Risk factors for breast cancer include maternal relatives with breast cancer, longer reproductive span (early first menstrual cycle coupled with menstruation continuing until a later age), obesity, no pregnancies or first pregnancy at or after the age of 35, and the presence of breast cancer genes. In particular, two genes that increase susceptibility to breast cancer have been identified: *BRCA1* and *BRCA2*.

Unfortunately, most breast cancers are painless until they become advanced and have metastasized. Therefore, the best way to prevent the cancer from spreading is to detect it early. Monthly self-examinations can locate lumps that may be malignant. Mammograms (x-rays of the breast) can detect small areas of increased tissue density:

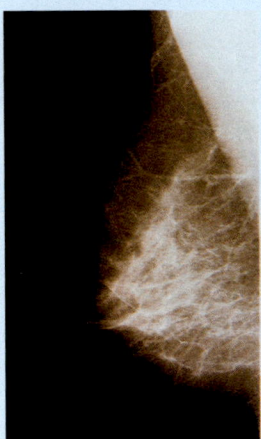
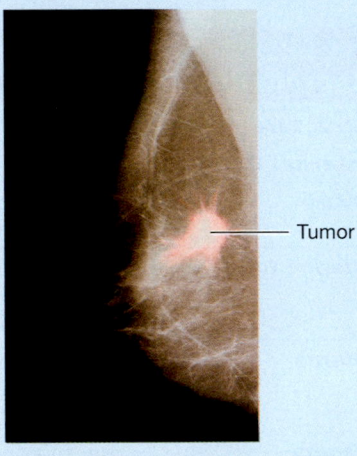

Tumor

| (a) Normal breast tissue | (b) Breast tissue with cancerous tumor |

A mammogram can detect many small malignancies that are not yet palpable in a self-examination. Most physicians recommend that women over age 50 have a mammogram every 2 years and women with a family history of breast cancer have regular mammograms before age 40.

Treatment for breast cancer may involve chemotherapy, radiation therapy, lumpectomy (surgical removal of the tumor and immediate surrounding tissue), or mastectomy (surgical removal of the breast and possibly the underlying pectoral muscles and axillary lymph nodes). Several chemotherapeutic drugs may decrease the risk of recurrence or disease progression.

Apply What You Learned

☐ 1. A tumor is growing in the isthmus of a uterine tube. It is still small enough to allow sperm to enter, but too large to allow an ovulated oocyte to pass. Predict what may happen if sperm enter this uterine tube following ovulation.

☐ 2. A woman has been trying to become pregnant for several months without success. A scan of her abdomen and pelvis reveals that her uterine tubes lack fimbriae. How might this finding help explain her lack of success in becoming pregnant?

See answers in Appendix A.

MODULE **26.5**
Physiology of the Female Reproductive System

Learning Outcomes

1. Relate the general stages of meiosis to the specific process of oogenesis.
2. Describe the events of the ovarian cycle and the uterine cycle.
3. Discuss endocrine regulation of oogenesis.
4. Analyze graphs depicting the typical female monthly sexual cycle and correlate ovarian activity, hormonal changes, and uterine events.
5. Describe female secondary sex characteristics and their role in reproductive system function.
6. Describe the physiological changes associated with menopause, and explain the fertility changes that precede menopause.

Female gametes are produced by the process of *oogenesis.* This module examines oogenesis and how it differs from spermatogenesis. We then shift our attention to the female's *ovarian* and *uterine cycles* and how the cyclic fluctuation of hormones controls these cycles. The module concludes with an examination of the female sexual response and the changes that occur at puberty, as well as how the reproductive cycle changes as a woman ages.

Oogenesis

In females the process of gamete development is **oogenesis** (oh′-oh-JEN-uh-sis), which produces **ova** (singular, *ovum*). Oogenesis differs from spermatogenesis in many ways. For one, oogenesis actually begins before a female infant is born but is then arrested until puberty. At that time, it resumes and remains active as part of the ovarian cycle until ending at *menopause,* which usually occurs when a woman is 45–55 years old. The left side of **Figure 26.14** summarizes the stages of oogenesis:

- **Before birth.** The stem cells of females, called **oogonia** (oh′-oh-GOH-nee-ah; singular, *oogonium*), undergo their mitotic divisions before birth, during months 2–7 of the

OOGENESIS (development of an ovum)

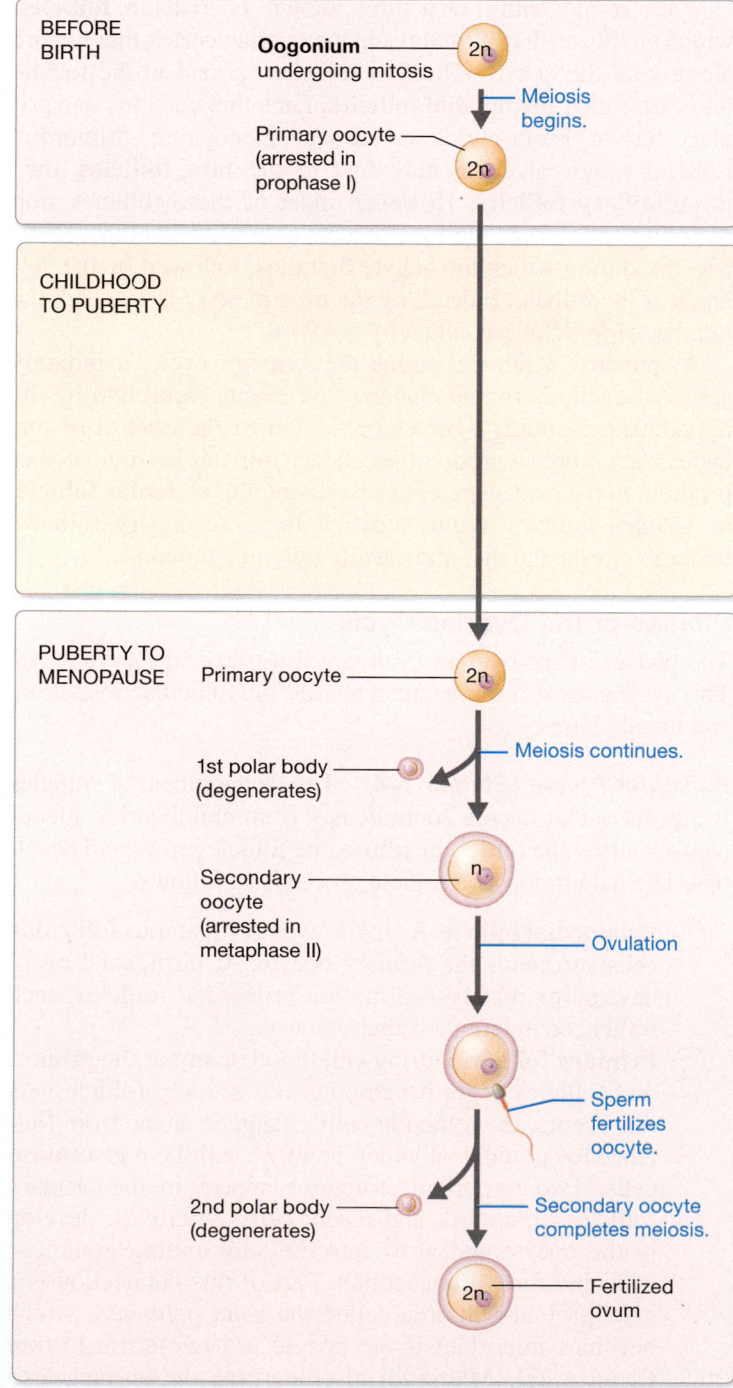

Figure 26.14 The stages of oogenesis. Oogenesis begins with an oogonium that has 46 chromosomes, and produces 1 secondary oocyte that has 23 chromosomes; when fertilized, this secondary oocyte becomes an ovum with 46 chromosomes.

fetal period. This means that a female produces all the oogonia she will ever make before she has even been born. At about seven months of fetal development, the total number of oogonia is about 7 million; however, this number drops sharply as many of the oogonia begin to die. The remaining 1–2 million oogonia begin to undergo meiosis I, at which point they are called **primary oocytes.** The primary oocytes proceed as far as prophase I, after which their development is arrested.

- **Childhood.** The primary oocytes remain in prophase I and do not develop further by meiosis until a girl reaches puberty. At that time, rising hormone levels trigger the start of the ovarian cycle. Note that during childhood, primary oocytes continue to degenerate—by the time a girl reaches puberty, the number has dropped to around 300,000.
- **Puberty to Menopause.** About once a month after puberty, 20–30 primary oocytes are stimulated to continue development. Usually one primary oocyte (or more in some women) will complete meiosis I to produce two haploid cells that are very different in size. The smaller cell, the first **polar body,** contains DNA but very little cytoplasm and often degenerates. The larger cell, the **secondary oocyte,** contains DNA and most of the cytoplasm plus absorbed extracellular fluid. The extra cytoplasm in the secondary oocyte ensures that if it is fertilized, it will have adequate nutrients for its 4- to 7-day journey toward the uterus. The large amount of cytoplasm is also the reason why only the secondary oocyte has the potential to become an ovum.

As the ovarian cycle progresses, the secondary oocyte begins meiosis II, but its development is again arrested, this time in metaphase II. If the secondary oocyte is fertilized, it will complete meiosis II and divide to form an ovum and the second polar body. If fertilization does not occur, the secondary oocyte will remain in metaphase II, and will eventually be shed with the menses. Of a woman's 300,000 primary oocytes, only about 400–500 will become mature secondary oocytes during her reproductive lifetime.

ConceptBOOST)))

Spermatogenesis versus Oogenesis

Both spermatogenesis and oogenesis produce haploid gametes with unique combinations of alleles. However, there are notable differences between the two processes. Let's compare them for a moment to see why these functional differences exist.

One key distinction is the number of cells produced. Sperm are produced continuously and in high numbers after the onset of puberty. This high sperm production occurs for two important reasons. First, sperm must reach the secondary oocyte to fertilize it, and to do so they have to travel through the inhospitable female reproductive tract. Most won't survive, and so the more sperm that enter the reproductive tract, the better the chances of fertilization. Second, sperm cells must penetrate the secondary oocyte by secreting enzymes. However, a single sperm cell by itself cannot produce enough of this enzyme—instead, hundreds of sperm are needed.

In contrast, oocyte production is complete before the female is even born, and from puberty to menopause she generally releases only one secondary oocyte per month. This makes sense—there is no reason for the female reproductive system to produce multiple secondary oocytes at one time. In fact, it would actually be dangerous to produce many secondary oocytes simultaneously, as the female system has a difficult time supporting more than two or three conceptuses during one pregnancy.

Another difference is in the structure of the gametes. As you just read, when a secondary oocyte forms, it receives most of the nutrient-rich cytoplasm, and the extra set of DNA is discarded in the polar body. The nutrients in the cytoplasm help support the developing conceptus until it begins obtaining nutrients from the mother. The situation differs for a sperm cell: It only needs to deliver genetic material to the developing oocyte, so excess cytoplasm is removed during spermiogenesis to keep the sperm cell small.

The following table provides you with a summary of the key differences between spermatogenesis and oogenesis for easy reference. ■

Characteristic	Spermatogenesis	Oogenesis
Time of onset	Begins at puberty	Begins before birth; primary oocytes arrested at prophase I
What happens at puberty	Begins and continuously produces sperm cells	Meiosis resumes; a primary oocyte is stimulated to mature to a secondary oocyte each month
Number of cells produced	Produces millions of small, motile sperm each day	Produces one large secondary oocyte (arrested in metaphase II) each month
Result of meiosis	Four haploid spermatids	One haploid ovum and two haploid polar bodies that generally degenerate
When process ends	Continues until death	Continues until menopause at age 45–55
Duration of meiosis	Continuous throughout life	Pauses twice: at prophase I from prenatal development to puberty, and at metaphase II from ovulation to fertilization

Quick Check

☐ 1. When in the life cycle of a female does oogenesis begin?

☐ 2. When is development of an oocyte arrested, and why?

☐ 3. How many ova are produced at the end of oogenesis? Why?

Ovarian Follicles and the Ovarian Cycle

Oocytes reside within structures known as **ovarian follicles,** which are blister-like structures in the ovarian cortex that mature along with the oocyte. The first follicles formed in the female fetus are called **primordial follicles.** Each one contains one primary oocyte. From childhood through menopause, primordial follicles progressively mature first to **primary follicles,** then to **secondary follicles.** However, most of these follicles stop maturing and instead die by a process called *atresia* (ah-TREE-zee-ah), during which the oocyte first dies, followed by the collapse of the follicle. Indeed, by the time puberty is reached, the number of follicles has fallen by 50–90%.

At puberty, a female begins the **ovarian cycle,** a monthly series of cyclic hormone changes and events controlled by the hypothalamic-pituitary-gonadal axis. Under the control of this cycle, each month a group of secondary follicles is stimulated to progress to the next stage of development, the **vesicular follicle,** or *tertiary follicle.* Again, most of these secondary follicles undergo atresia and die, and usually only one remains.

Phases of the Ovarian Cycle

The phases of the ovarian cycle are illustrated in **Figure 26.15**. This cycle consists of three main phases: the follicular, ovulation, and luteal phases.

Follicular Phase (Stages 1–4)   The development of follicles is a process that occurs continuously from childhood to menopause. During the **follicular phase,** the follicle grows and develops. The maturation of a follicle proceeds as follows:

① **Primordial follicle.** A single layer of squamous **follicular cells** surrounds the primary oocyte. At birth, each ovary has approximately 1–2 million primordial follicles, each follicle containing a primary oocyte.

② **Primary follicle.** During childhood, many of the primordial follicles begin developing into primary follicles. As this occurs, the follicular cells change in shape from flattened to cuboidal, at which point we call them **granulosa cells.** Two important structures appear in the primary follicles: microvilli and *thecal cells.* Microvilli develop in the oocyte and grow into the surrounding granulosa cells and form a connection. Part of this connection is a glycoprotein-rich area called the *zona pellucida,* which becomes important if the oocyte is later fertilized (see Chapter 27). Microvilli also increase the surface area available for the granulosa cells to transfer materials to the growing primary oocyte through gap junctions. Only a small number of the primary follicles will continue development.

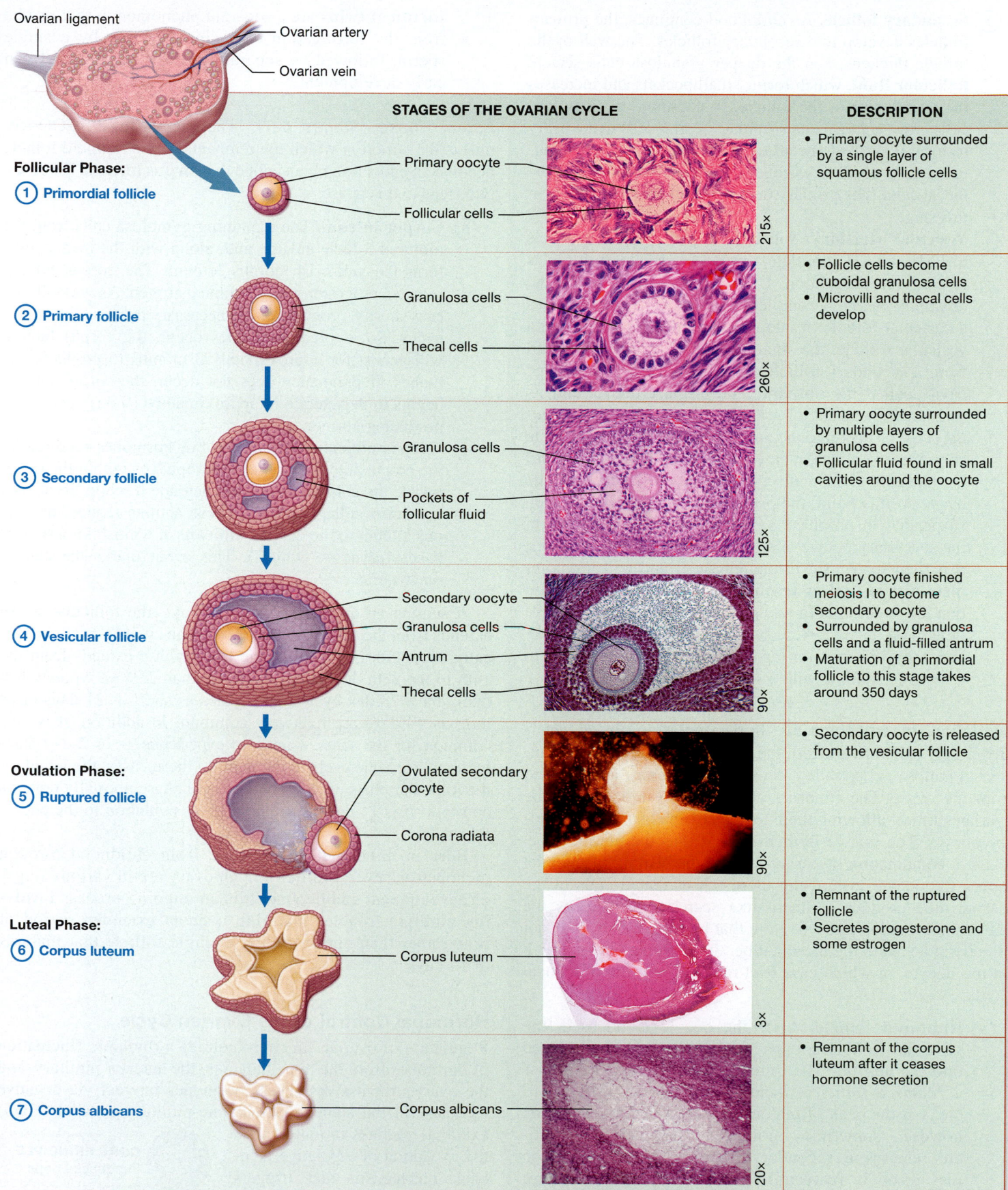

Ovarian ligament			STAGES OF THE OVARIAN CYCLE		DESCRIPTION

Ovarian ligament
Ovarian artery
Ovarian vein

Follicular Phase:

① Primordial follicle
- Primary oocyte
- Follicular cells
215x

- Primary oocyte surrounded by a single layer of squamous follicle cells

② Primary follicle
- Granulosa cells
- Thecal cells
260x

- Follicle cells become cuboidal granulosa cells
- Microvilli and thecal cells develop

③ Secondary follicle
- Granulosa cells
- Pockets of follicular fluid
125x

- Primary oocyte surrounded by multiple layers of granulosa cells
- Follicular fluid found in small cavities around the oocyte

④ Vesicular follicle
- Secondary oocyte
- Granulosa cells
- Antrum
- Thecal cells
90x

- Primary oocyte finished meiosis I to become secondary oocyte
- Surrounded by granulosa cells and a fluid-filled antrum
- Maturation of a primordial follicle to this stage takes around 350 days

Ovulation Phase:

⑤ Ruptured follicle
- Ovulated secondary oocyte
- Corona radiata
90x

- Secondary oocyte is released from the vesicular follicle

Luteal Phase:

⑥ Corpus luteum
- Corpus luteum
3x

- Remnant of the ruptured follicle
- Secretes progesterone and some estrogen

⑦ Corpus albicans
- Corpus albicans
20x

- Remnant of the corpus luteum after it ceases hormone secretion

Figure 26.15 The ovarian cycle.

③ **Secondary follicle.** As childhood continues, the primary follicles develop into secondary follicles. The wall of the follicle thickens, and the deeper granulosa cells secrete **follicular fluid,** which forms small pockets and increases the overall size of the follicle. In addition, the granulosa cells grow, enlarge, and stimulate nearby cells in the ovary to form a layer of **thecal cells** around the follicle. Granulosa cells produce estrogens from the secretions of the thecal cells. The primary oocyte is still growing slowly at this stage.

④ **Vesicular (tertiary) follicle.** At puberty, rising levels of FSH and LH stimulate a group of secondary follicles to continue maturation. Usually only one follicle—the *dominant follicle*—completes this process and becomes a vesicular follicle. As you can see in Figure 26.15, in a vesicular follicle, the smaller pockets of follicular fluid from a secondary follicle merge to form a single large cavity called the **antrum.** The primary oocyte and its surrounding capsule of granulosa cells, known as *cumulus cells,* project into the antrum. As the vesicular follicle continues to enlarge, it may reach 15–25 mm (0.6–1.0 in.) in diameter and create a bulge in the surface of the ovary. At this point, the primary oocyte, which has been suspended in prophase I, completes meiosis I to form the secondary oocyte and the first polar body. However, it is then suspended in metaphase II and will not complete meiosis unless fertilization occurs. Keep in mind that the maturation of any *one* individual oocyte and follicle likely takes 3–4 cycles to complete. So it will take about 90–120 days for the dominant follicle to mature to a vesicular follicle with a secondary oocyte.

Ovulation Phase (Stage 5)   In the **ovulation phase,** generally simply called **ovulation,** the secondary oocyte and associated granulosa cells, called the **corona radiata,** are released from the ovary. The fimbriae of the uterine tube sweep the ovarian surface, allowing them to "catch" the ovulated secondary oocyte. The oocyte is then carried into the uterine tube, where a combination of peristalsis and the beating of the cilia moves the oocyte toward the uterus. Nonciliated cells in the uterine tube produce a mucus-like secretion that keeps the oocyte moist and nourished. Note that because the ovary is not directly attached to the uterine tube, the fimbriae may not pick up the oocyte, in which case it is released into the peritoneal cavity.

⑤ **Ruptured follicle.** After the oocyte is released, the vesicular follicle collapses and ruptured vessels bleed into the antrum. The remaining structure is known as a *ruptured follicle.* Usually only one vesicular follicle is at the peak of maturation when ovulation occurs; however, sometimes more than one vesicular follicle releases a secondary oocyte. This phenomenon may result in **fraternal twins,** or *nonidentical twins.* It occurs more commonly in some families and also increases in frequency with advancing maternal age.

Identical twins are a separate phenomenon that results from the fertilization of a single oocyte by a single sperm, followed by separation of the dividing cells in early development.

Luteal Phase (Stages 6–7)   The **luteal phase** (LOO-tee-uhl) is the period in which the remnants of the ruptured follicle become an endocrine organ called the **corpus luteum.** There are two steps to this phase:

⑥ **Corpus luteum.** The remaining granulosa cells from the ruptured follicle enlarge and, along with the thecal cells, form the yellowish corpus luteum. The corpus luteum secretes progesterone and some estrogen. As we will discuss shortly, progesterone is necessary to maintain a pregnancy. So, if pregnancy does occur, the corpus luteum will persist for approximately 3 months to produce hormones. If pregnancy does not occur, the corpus luteum begins to degenerate in approximately 10 days and stops producing hormones.

⑦ **Corpus albicans.** When levels of hormones secreted by the corpus luteum decline, macrophages within the ovary invade the corpus luteum and degrade it while fibroblasts lay down collagen fibers. These actions reduce the corpus luteum to the **corpus albicans,** a whitish knot of scar tissue (*albus* = "white"). This event marks the end of one ovarian cycle.

Assuming an ovarian cycle of 28 days, the follicular phase extends from the 1st to the 14th day of the cycle, ovulation typically occurs on day 14, and the luteal phase extends from the 14th to the 28th day. However, fewer than 25% of women normally have 28-day cycles, and cycles as short as 21 days or as long as 40 days are relatively common. In addition, it is also common for the same woman to experience 1- or 2-day fluctuations from one cycle to the next. In these cases, the length of the follicular phase varies, but the luteal phase normally remains constant: It is 14 days from the time of ovulation to the end of the cycle.

Bear in mind that at any time from childhood through menopause, several follicles in an ovary are in various stages of development and decline. It is an ongoing process involving multiple follicles maturing over an extended period of time rather than one involving a single follicle that matures in 14 days.

Hormonal Control of the Ovarian Cycle

Regulation of ovarian function requires a rhythmic fluctuation of hormones from the hypothalamus, the anterior pituitary, and the ovaries themselves. These hormones interact via negative feedback in the same hypothalamic-pituitary-gonadal (HPG) axis that operates in males, with the exception of a positive feedback mechanism that triggers ovulation. Here we see another example of the Feedback Loops Core Principle (p. 28). During

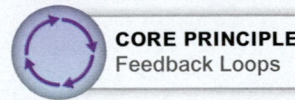

CORE PRINCIPLE
Feedback Loops

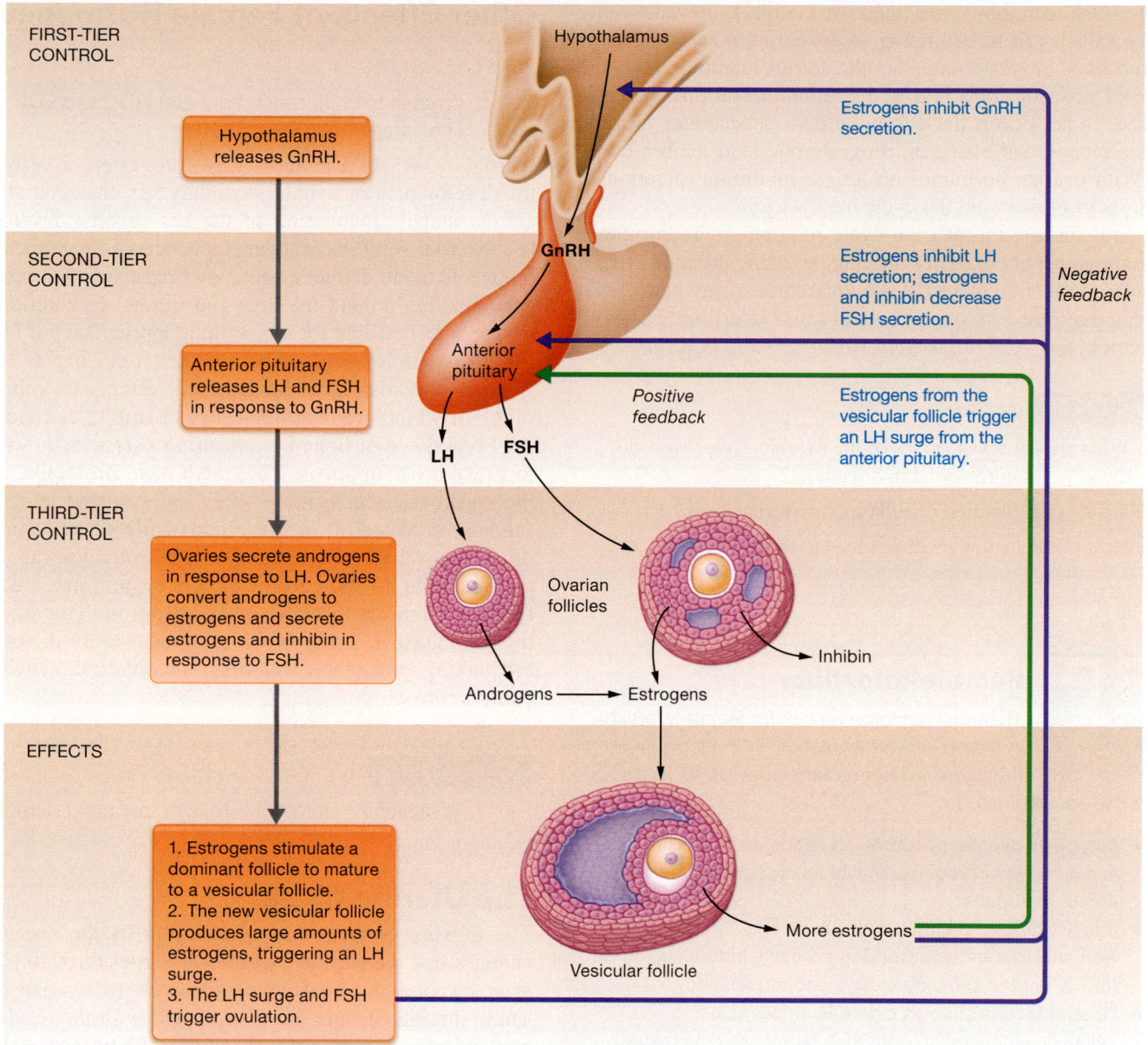

FIRST-TIER CONTROL

Hypothalamus

Hypothalamus releases GnRH.

Estrogens inhibit GnRH secretion.

GnRH

SECOND-TIER CONTROL

Anterior pituitary releases LH and FSH in response to GnRH.

Estrogens inhibit LH secretion; estrogens and inhibin decrease FSH secretion.

Negative feedback

Anterior pituitary

LH   FSH

Positive feedback

Estrogens from the vesicular follicle trigger an LH surge from the anterior pituitary.

THIRD-TIER CONTROL

Ovaries secrete androgens in response to LH. Ovaries convert androgens to estrogens and secrete estrogens and inhibin in response to FSH.

Ovarian follicles

Inhibin

Androgens → Estrogens

EFFECTS

1. Estrogens stimulate a dominant follicle to mature to a vesicular follicle.
2. The new vesicular follicle produces large amounts of estrogens, triggering an LH surge.
3. The LH surge and FSH trigger ovulation.

More estrogens

Vesicular follicle

Figure 26.16 **Hormonal regulation of ovarian function via the hypothalamic-pituitary-gonadal (HPG) axis.**

a 28-day cycle, levels of hypothalamic, anterior pituitary, and ovarian hormones fluctuate as follows (**Figure 26.16**):

- **First-tier control.** The hypothalamus releases gonadotropin-releasing hormone (GnRH).
- **Second-tier control.** GnRH stimulates the anterior pituitary to release FSH and LH.
- **Third-tier control.** The ovaries are the target organ of FSH and LH. LH stimulates thecal cells in the follicle to secrete androgens, which diffuse through the basement membrane to the granulosa cells. FSH stimulates granulosa cells in follicles to convert these androgens to estrogens. FSH also stimulates granulosa cells to secrete inhibin.

Effects The estrogens and inhibin produced have various effects throughout the reproductive system. Estrogens stimulate the dominant follicle to continue developing into a vesic-

ular follicle. The granulosa cells of the new vesicular follicle continue to produce estrogens in increasing amounts. In most cases, a rising concentration of the target hormone would trigger a negative feedback loop; that is, you might expect the rising levels of estrogens to cause the anterior pituitary to decrease LH release. However, the opposite occurs: When the granulosa cells eventually produce enough estrogens, they exert *positive feedback* on the anterior pituitary, and trigger a large increase in LH release. This increase is known as the *LH surge*. The LH surge, accompanied by a rise in FSH secretion, triggers ovulation. The resulting corpus luteum produces additional progesterone, estrogens, and inhibin.

Negative Feedback FSH stimulates the granulosa cells to secrete inhibin. This, along with increasing levels of estrogens, now exerts negative feedback on the hypothalamus and pituitary. Estrogens inhibit GnRH and LH secretion, and inhibin inhibits FSH secretion. This negative feedback prevents other primary follicles from beginning to develop too early. If fertilization occurs, progesterone from the corpus luteum inhibits the release of gonadotropins and will continue to do this. If fertilization does not occur, the corpus luteum degenerates into the corpus albicans, and estrogen, progesterone, and inhibin levels drop. With ovarian hormones no longer inhibiting secretion of GnRH, its level rises and the cycle begins again.

As you can see, the ovarian cycle requires many hormones working together at the right times, and exerting the right type of feedback, to function properly. An imbalance in any part of this cycle can interfere with ovulation and cause infertility. For more information, see *A&P in the Real World: Female Infertility.*

Quick Check

☐ 4. What are the seven stages of the ovarian cycle? How do these stages correspond to oogenesis?

☐ 5. What is the function of FSH in the ovarian cycle?

☐ 6. How does positive feedback lead to the LH surge?

Female Infertility

The most common reason for female infertility is a problem with ovulation. Without ovulation, there are no secondary oocytes to fertilize. Possible causes include:

- polycystic ovarian syndrome, a condition that results in excessive androgens, which decreases the production of estrogens;
- primary ovarian insufficiency, which occurs when a woman's ovaries stop working normally before age 40 (this is not the same as early menopause); and
- lifestyle factors such as cigarette smoking, being severely underweight, poor diet, stress, and excessive athletic training, all of which can decrease the production of estrogens; and obesity, which increases the production of estrogens to the point that they block LH and FSH secretion by negative feedback.

Infertility may be related to causes outside the ovary, as well. For example, blockage of uterine tubes due to pelvic inflammatory disease (discussed in Module 26.7), endometriosis, or uterine fibroids (noncancerous clumps of tissue and muscle on the internal walls of the uterus) can all lead to difficulty conceiving.

Treatment for infertility in women depends on the cause. Medications, including synthetic gonadotropins, may stimulate ovulation. Abdominal surgery may be necessary to clear blocked uterine tubes or remove endometriosis or fibroids. If no other treatments are successful, in vitro fertilization (IVF) is an option. In this procedure, oocytes and sperm, from the mother and father or from donors, are fertilized in a laboratory and then implanted in the mother's uterus.

Other Effects of Female Hormones

« FLASHBACK

1. What is the difference between HDL and LDL cholesterol? (p. 939)

Estrogens and, to a lesser degree, progesterone also stimulate the development of female secondary sex characteristics. These effects include maturation of the sex organs, development of the external genitalia, and maintenance of anatomical features unique to adult females, such as breast development and fat accumulation around the hips and thighs. In addition, progesterone is responsible for maintaining a pregnancy if fertilization occurs, which is discussed in more detail in Chapter 27.

Estrogens also have far-reaching effects on other tissues, many of which are beneficial. For example, estrogens reduce the breakdown of bone by inhibiting osteoclasts, which helps to prevent osteoporosis. They also have protective effects on the cardiovascular system—they increase the level of HDLs ("good" cholesterol) in the blood while decreasing the level of LDLs ("bad" cholesterol). However, not all of estrogens' effects on the cardiovascular system are always beneficial, as they promote blood coagulation. This accounts for the increased risk of developing blood clots in women who are taking estrogen-containing medications such as oral contraceptives.

Quick Check

☐ 7. What are the effects of estrogens and progesterone?

The Uterine Cycle

The **uterine cycle,** or *menstrual cycle,* is the series of cyclic changes that the uterine endometrium goes through each month as it responds to the fluctuating levels of ovarian hormones. These uterine changes are coordinated with the levels of estrogen and progesterone released during the phases of the ovarian cycle, which are controlled by FSH and LH released from the anterior pituitary. Let's first look at the structures of the endometrium that are affected by the changing hormone levels.

The endometrium has two main strata, or layers. The **stratum functionalis,** or *functional layer,* undergoes cyclic changes in response to ovarian hormones. It detaches from the uterine wall and is shed as a discharge of roughly 35–50 ml of blood and other materials from the vagina approximately once per month during **menstruation.** The thinner, deeper **stratum basalis,** or *basal layer,* does not thicken in response to ovarian hormones, but forms a new stratum functionalis after menstruation ends.

Numerous **endometrial (uterine) glands** change in length as endometrial thickness changes. The main vascular supply of the uterus comes from the *uterine arteries,* which arise from the internal iliac arteries in the pelvis. After branching several times, they eventually form the **spiral (coiled) arteries** of the stratum functionalis.

Phases of the Uterine Cycle

Now let's examine the three phases of the uterine cycle. For simplicity, we will assume a 28-day cycle (**Figure 26.17**):

① **Menstrual phase, days 1–5.** During the menstrual phase, the uterus sheds the stratum functionalis, resulting in menstruation.

② **Proliferative (preovulatory) phase, days 6–14.** During the proliferative phase, a new stratum functionalis develops with endometrial glands and spiral arteries and veins. As this new layer thickens (proliferates), its endometrial glands enlarge and its spiral arteries and veins increase in number. Cervical mucus is thick and sticky until just before ovulation, when it thins to facilitate the passage of sperm into the uterus. At the end of the proliferative phase around day 14, ovulation occurs.

③ **Secretory (postovulatory) phase, days 15–28.** Regardless of the normal length of a woman's cycle, this phase usually lasts 14 days. During the secretory phase, the spiral arteries convert the stratum functionalis to secretory mucosa and endometrial glands secrete a nutritious glycogen-rich fluid, also known as *uterine milk,* into the uterine cavity to sustain the conceptus.

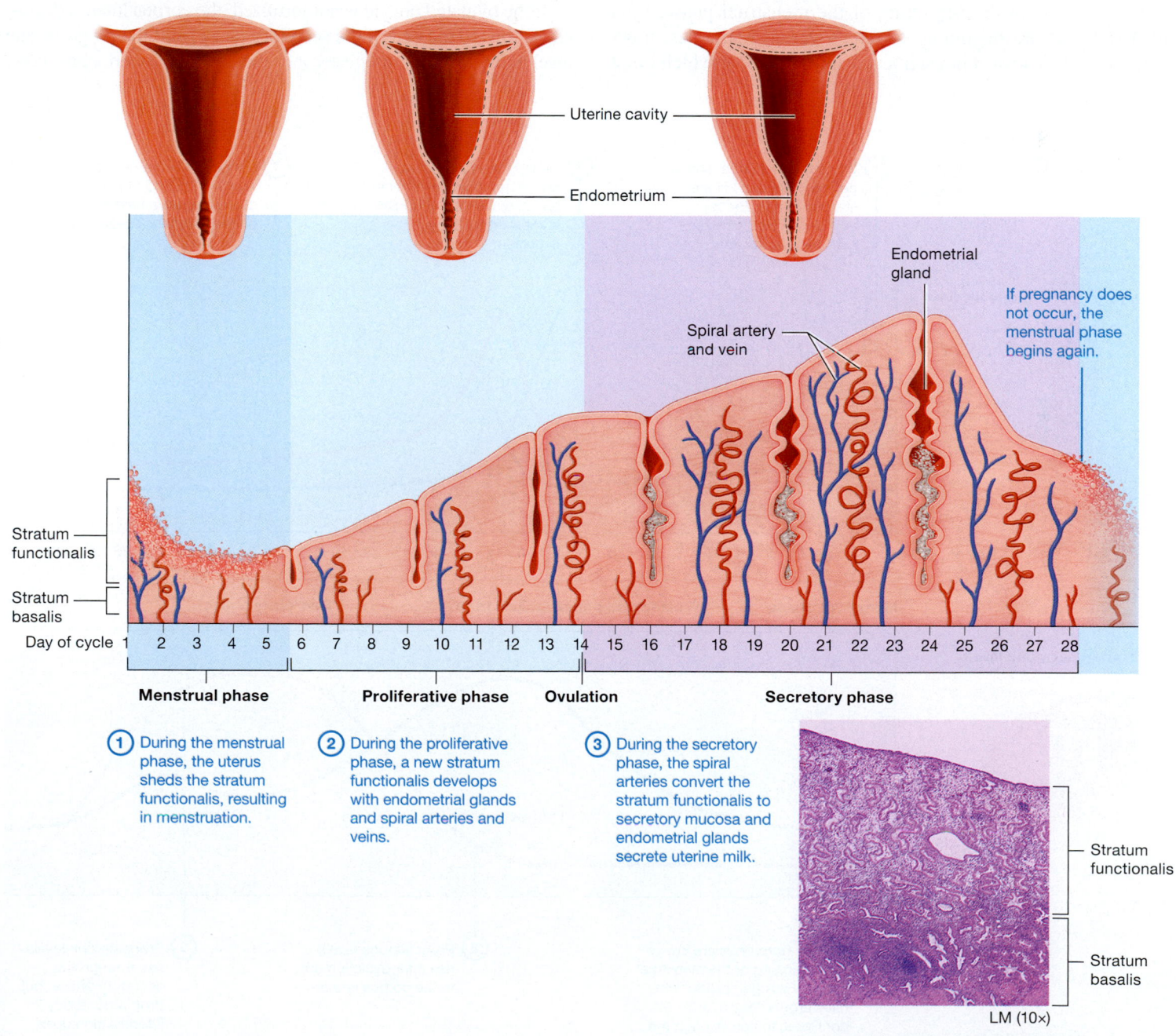

Uterine cavity

Endometrium

Endometrial gland

Spiral artery and vein

If pregnancy does not occur, the menstrual phase begins again.

Stratum functionalis

Stratum basalis

Day of cycle 1 2 3 4 5 6 7 8 9 10 11 12 13 14 15 16 17 18 19 20 21 22 23 24 25 26 27 28

Menstrual phase   **Proliferative phase**   **Ovulation**   **Secretory phase**

① During the menstrual phase, the uterus sheds the stratum functionalis, resulting in menstruation.

② During the proliferative phase, a new stratum functionalis develops with endometrial glands and spiral arteries and veins.

③ During the secretory phase, the spiral arteries convert the stratum functionalis to secretory mucosa and endometrial glands secrete uterine milk.

Stratum functionalis

Stratum basalis

LM (10×)

Endometrium during secretory phase

Figure 26.17 Endometrial changes during the uterine cycle.

If pregnancy does not occur, the cells of the stratum functionalis die and, on day 28, the menstrual phase begins again. If pregnancy does occur, the secretory phase will continue and the uterus will continue to develop.

Hormonal Control of the Uterine Cycle

Regulation of uterine function also requires a rhythmic fluctuation of hormones from the hypothalamus, pituitary, and ovaries. Again, for simplicity we will assume a 28-day cycle as we examine the changes that occur during the uterine cycle.

Figure 26.18 presents two graphs: one showing the changing levels of gonadotropins during the phases of the uterine cycle, and another showing the changing levels of ovarian hormones. As you can see, at the beginning of the menstrual phase, FSH and LH levels are beginning to rise. Ovarian hormones, however, are at their lowest normal levels until day 5, at which point

the growing ovarian follicles start to produce more estrogens. As the proliferative phase begins, the rising levels of estrogens cause a new stratum functionalis to generate and trigger endometrial cells to synthesize progesterone receptors. The high levels of estrogens trigger release of more FSH and LH, and their levels peak around day 14, which causes ovulation. At the start of the secretory phase, the level of progesterone from the corpus luteum begins to rise, while the amount of estrogen secreted decreases. The high progesterone level and falling levels of estrogens inhibit the anterior pituitary, which, as you can see in the graph, leads to low amounts of LH and FSH. This progesterone also stimulates the continuing development of the stratum functionalis during the secretory phase.

If the ovulated oocyte is not fertilized, the corpus luteum degenerates toward the end of the secretory phase, and the progesterone level falls. Without progesterone, the stratum functionalis layer

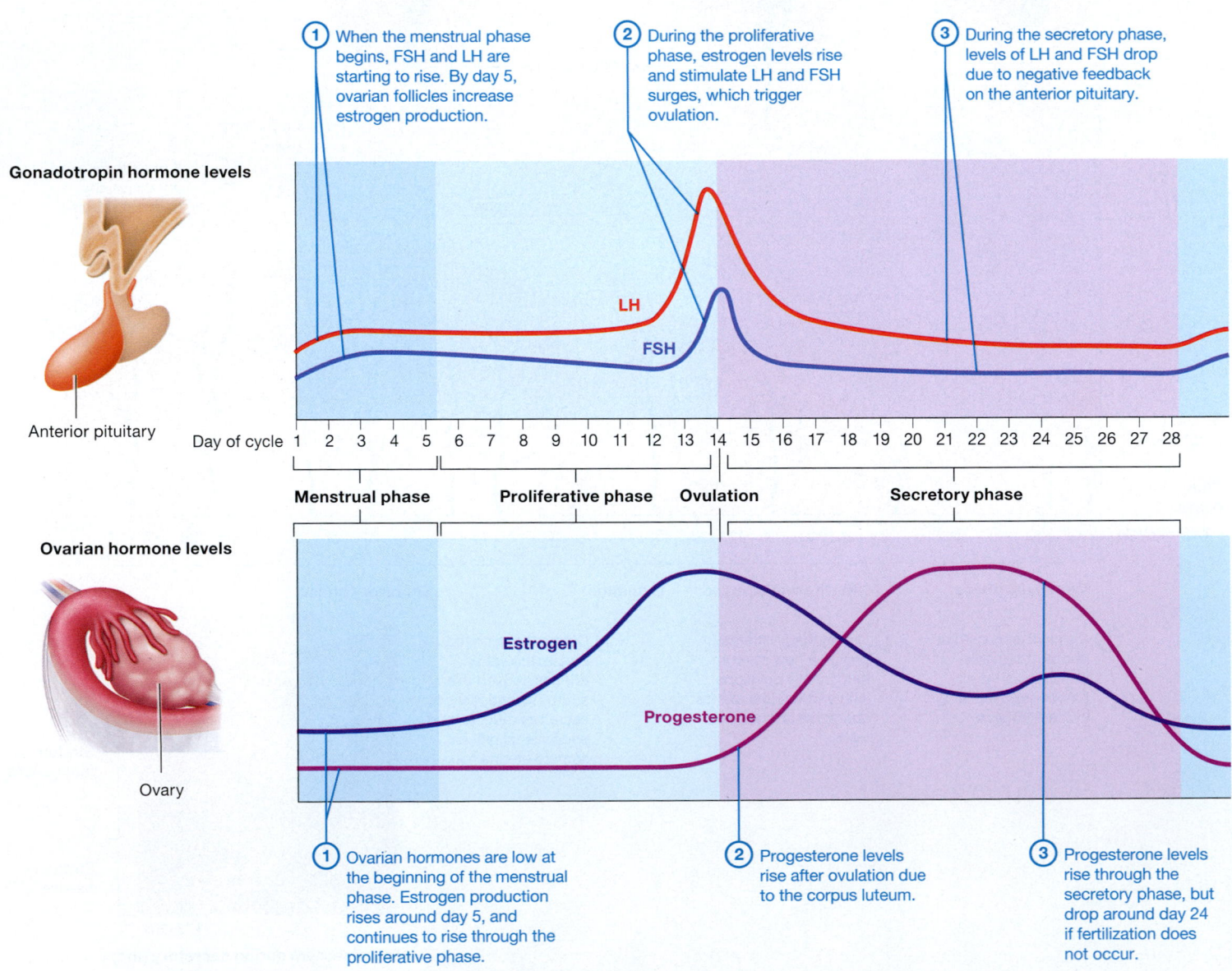

Gonadotropin hormone levels

① When the menstrual phase begins, FSH and LH are starting to rise. By day 5, ovarian follicles increase estrogen production.

② During the proliferative phase, estrogen levels rise and stimulate LH and FSH surges, which trigger ovulation.

③ During the secretory phase, levels of LH and FSH drop due to negative feedback on the anterior pituitary.

Anterior pituitary

LH

FSH

Day of cycle 1 2 3 4 5 6 7 8 9 10 11 12 13 14 15 16 17 18 19 20 21 22 23 24 25 26 27 28

Menstrual phase   Proliferative phase   Ovulation   Secretory phase

Ovarian hormone levels

Ovary

Estrogen

Progesterone

① Ovarian hormones are low at the beginning of the menstrual phase. Estrogen production rises around day 5, and continues to rise through the proliferative phase.

② Progesterone levels rise after ovulation due to the corpus luteum.

③ Progesterone levels rise through the secretory phase, but drop around day 24 if fertilization does not occur.

Figure 26.18 Pituitary and ovarian hormone interactions during the uterine cycle.

loses its blood supply and breaks down. This triggers the beginning of a new menstrual phase. If fertilization does take place, the outer layer of the developing conceptus begins to secrete the LH-like hormone **human chorionic gonadotropin (hCG),** which keeps the corpus luteum from degenerating and the progesterone level high. hCG is unique to the developing conceptus, which makes it useful as a target compound for home pregnancy tests.

Quick Check

☐ 8. What is the function of estrogen and progesterone in the uterine cycle?

☐ 9. How do levels of ovarian hormones and gonadotropins influence one another?

Female Sexual Response

The **female sexual response** is similar to that of males in most respects. During sexual excitement, the vaginal mucosa, vestibule, and breasts engorge with blood; the clitoris and nipples become erect; and increased activity of the vaginal mucosa and vestibular glands lubricates the vestibule. For the most part, these events are similar to the erection phase in men. Tactile and psychological stimuli, mediated along autonomic pathways, promote sexual excitement.

As in males, orgasm refers to the period during which women experience intense feelings of pleasure. However, unlike in males, no ejaculation takes place. The uterus does exhibit peristaltic waves of contraction, and the cervix pushes down somewhat into the vagina. If semen is present, these actions may draw semen farther into the reproductive tract, but female orgasm is not required for conception. Females do not experience a refractory period, so they may have multiple orgasms during a single sexual experience.

The female libido, or sex drive, may be related to the androgen *dehydroepiandrosterone* (*DHEA*), produced by the adrenal cortex in women. Androgens and estrogens both appear to influence sexual behavior because of their interactions in the hypothalamus. In addition, many women experience variations in libido in response to monthly fluctuations of estrogens and progesterone.

Quick Check

☐ 10. What are the similarities between the male and female sexual responses?

☐ 11. What are the differences between the male and female sexual responses?

Puberty and Menopause

❮❮ FLASHBACK

1. What is adipose tissue? (p. 139)

2. What is leptin? (p. 620)

As we have discussed previously, levels of estrogens and progesterone increase dramatically at puberty and begin to decline during menopause. Let's take a look at these two periods in a woman's life and the physiological changes that accompany them.

Puberty and Menarche

For most girls, puberty begins between ages 9 and 11, when secretions of estrogens and progesterone increase from the ovaries. Before puberty, the low level of GnRH from the hypothalamus keeps LH and FSH levels low, which in turn results in low levels of estrogens and progesterone. At puberty, the hypothalamus becomes much less sensitive to inhibition by estrogens and progesterone, and the GnRH level increases. GnRH stimulates the HPG axis, and blood estrogens and progesterone levels begin to rise, resulting in the appearance of female secondary sex characteristics.

The first sign of puberty in girls is budding breasts. This is followed by the development of **female secondary sex characteristics,** those features that are associated with identifying someone as female and are induced in nonreproductive organs by the female hormones, mostly estrogens. These secondary characteristics include:

- the appearance of pubic and axillary hair;
- an increase in the overall amount of adipose tissue in the subcutaneous layer, with additional deposits in the breasts and around the hips and thighs;
- increased secretions of sebaceous glands (the oilier skin often results in acne); and
- skeletal changes, including increased height and widening pelvis.

Approximately 2 years after the onset of puberty, females experience their first episode of menstrual bleeding, called **menarche** (MEN-ahr-kee). Perhaps because of the physical toll exerted by pregnancy on the female body, menarche (and the accompanying cyclic events) will not occur unless a girl has at least 15–17% body fat, and adult menstruation will cease if body fat drops too low. One hormone in particular, *leptin,* secreted by adipocytes, stimulates gonadotropin secretion. For this reason, if the leptin level declines because body fat is too low, gonadotropin levels decrease as well and directly affect the HPG axis.

Quick Check

☐ 12. What is menarche, and when does it usually occur in females?

☐ 13. What are the female secondary sex characteristics?

Effects of Aging: Menopause

After puberty, reproductive cycles usually continue at regular intervals into the later 40s or early 50s, when they become increasingly irregular. Eventually, the cycles stop completely. The time from the onset of irregular menstrual cycles to their complete cessation, which may be up to 5 years in some women, is called the **female climacteric.**

The Big Picture of Hormonal Regulation of the Ovarian and Uterine Cycles

Figure 26.19

▶ Play Animation
@ Mastering Anatomy & Physiology

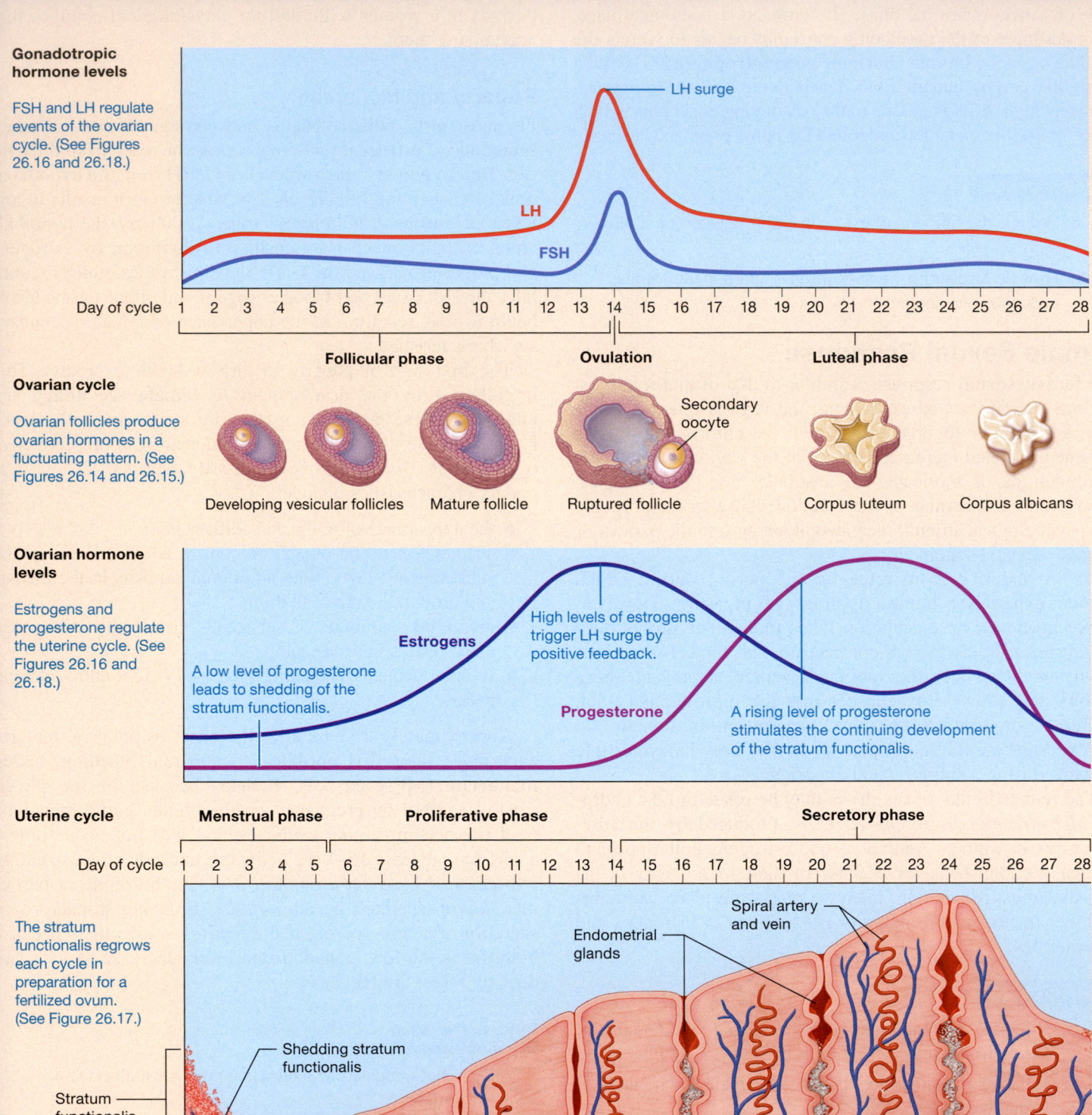

Gonadotropic hormone levels

FSH and LH regulate events of the ovarian cycle. (See Figures 26.16 and 26.18.)

LH surge

LH

FSH

Day of cycle 1 2 3 4 5 6 7 8 9 10 11 12 13 14 15 16 17 18 19 20 21 22 23 24 25 26 27 28

Follicular phase   Ovulation   Luteal phase

Ovarian cycle

Ovarian follicles produce ovarian hormones in a fluctuating pattern. (See Figures 26.14 and 26.15.)

Developing vesicular follicles   Mature follicle

Secondary oocyte

Ruptured follicle   Corpus luteum   Corpus albicans

Ovarian hormone levels

Estrogens and progesterone regulate the uterine cycle. (See Figures 26.16 and 26.18.)

Estrogens

High levels of estrogens trigger LH surge by positive feedback.

A low level of progesterone leads to shedding of the stratum functionalis.

Progesterone

A rising level of progesterone stimulates the continuing development of the stratum functionalis.

Uterine cycle

The stratum functionalis regrows each cycle in preparation for a fertilized ovum. (See Figure 26.17.)

Menstrual phase   Proliferative phase   Secretory phase

Day of cycle 1 2 3 4 5 6 7 8 9 10 11 12 13 14 15 16 17 18 19 20 21 22 23 24 25 26 27 28

Spiral artery and vein

Endometrial glands

Shedding stratum functionalis

Stratum functionalis

Stratum basalis

When menstruation has not taken place for at least 1 year, **menopause** (MEN-ah-pawz) has occurred. It takes place because the number of primary follicles left to respond to FSH and LH is quite low after approximately three decades of ovarian cycles and also because of **atresia**—a process in which immature follicles degenerate and are resorbed during the follicular phase of the menstrual cycle. Those follicles that remain in the ovaries are less responsive to gonadotropins, and so less estrogen and progesterone are secreted. As a result, ovulation does not occur and the blood concentration of estrogens declines sharply, although many women continue to synthesize some estrogen from adrenal androgens. However, secretions of FSH and LH from the pituitary are no longer inhibited, so these hormones may be continuously released for some time.

The lower concentration of estrogens and progesterone may change the female secondary sex characteristics. The breasts, vagina, uterus, and uterine tubes may shrink, and the pubic and axillary hair may thin. Often bone density decreases, which may lead to osteoporosis, and the skin becomes thinner.

Some women experience unpleasant vasomotor signs, including sensations of heat called "hot flashes." A hot flash may last up to 5 minutes and may be accompanied by chills and/or sweating. These vasomotor symptoms may result from changes in the rhythmic secretion of GnRH by the hypothalamus in response to declining concentrations of sex hormones. Women may also experience migraine headaches, backaches, fatigue, and mood swings during menopause.

Some women experience menopause before age 45, which is called **early menopause.** If it occurs before age 40, it is **premature menopause.** Early or premature menopause can happen for several reasons, including chemotherapy or radiation treatments for cancer, surgery to remove the ovaries or uterus, and autoimmune diseases. In some cases, the reason may simply be genetics: Women with a family history of early menopause are more likely to experience early menopause.

Quick Check

☐ 14. What is menopause?

☐ 15. Why does menopause occur?

Putting It All Together: The Big Picture of Hormonal Regulation and Female Reproductive Cycles

The ovarian and uterine cycles are intimately tied together. **Figure 26.19** shows the big picture of how the events of the two cycles correlate, and how they relate to the levels of anterior pituitary hormones.

Table 26.3 summarizes the functions of the major reproductive hormones in both males and females, identifying the source, the target tissue(s), and the body's response to each hormone.

Apply What You Learned

☐ 1. A new drug is able to stop the atresia of primary oocytes *without* increasing the number of secondary oocytes ovulated each cycle. Predict the effect this drug would have on the ability of a woman to become pregnant.

☐ 2. A young healthy woman begins to take a daily herbal supplement that mimics the effects of progesterone. Predict how this might affect her ovarian and uterine cycles.

☐ 3. A 38-year-old woman has decided not to have children. She is considering a form of birth control called tubal ligation, which involves closing off the uterine tubes. However, she is concerned that this procedure will push her into early menopause. How would you address her concerns?

See answers in Appendix A.

MODULE 26.6
Methods of Birth Control

Learning Outcomes

1. Explain why changes in cervical mucus can predict a woman's monthly fertility.

2. Provide examples of how birth control methods relate to normal reproductive function.

Birth control, or contraception, refers to any procedure or device that is intended to prevent pregnancy. In this module, we explore common methods of contraception.

Behavioral Methods

Abstinence—the avoidance of intercourse—is the only 100% reliable method of avoiding pregnancy. Periodic abstinence, also called the *rhythm method,* is based on avoiding intercourse near the time of ovulation. The rhythm method has a high failure rate partly because it is often difficult to predict the exact date of ovulation. Also, intercourse must be avoided at least 7 days before ovulation and at least 2 days after ovulation, so there are no surviving sperm and fertile oocytes in the reproductive tract at the same time.

Withdrawal, also called *coitus interruptus,* involves withdrawing the penis before ejaculation. This method also has a high failure rate because it is difficult to time withdrawal accurately, and because sperm may be present in pre-ejaculatory fluid. In addition, sperm ejaculated anywhere on the vulva can potentially enter the female reproductive tract.

Quick Check

☐ 1. Why do most behavioral methods of birth control have high failure rates?

Table 26.3 Major Reproductive Hormones in Males and Females

Hormone	Source	Target Tissue(s)	Response
Males			
Gonadotropin-releasing hormone (GnRH)	Hypothalamus	Anterior pituitary	Stimulates secretion of LH and FSH
Follicle-stimulating hormone (FSH)	Anterior pituitary	Testes (sustentacular cells)	Stimulates production of androgen-binding protein (ABP), inhibin; necessary for spermatogenesis
Luteinizing hormone (LH)	Anterior pituitary	Testes (interstitial cells)	Stimulates production and secretion of testosterone
Testosterone	Testes (interstitial cells)	Testes	Stimulates spermatogenesis
		Other body tissues	Stimulates development and maintenance of reproductive organs; responsible for development of secondary sex characteristics
Inhibin	Testes (sustentacular cells)	Anterior pituitary	Inhibits FSH secretion
Females*			
Gonadotropin-releasing hormone (GnRH)	Hypothalamus	Anterior pituitary	Stimulates secretion of LH and FSH
Follicle-stimulating hormone (FSH)	Anterior pituitary	Ovarian follicles and granulosa cells	Stimulates follicles to begin development; stimulates granulosa cells to convert androgens to estrogens
Luteinizing hormone (LH)	Anterior pituitary	Ovarian follicles and thecal cells	Stimulates production of gonadal hormones (e.g., estrogens); stimulates follicles to finish maturation; stimulates ovulation; contributes to development of corpus luteum
Estrogens	Ovaries (follicles) and granulosa cells	Uterus	Stimulate proliferation of endometrial cells; stimulate uterine contractions
		Mammary glands	Stimulate development of ducts of mammary glands
		Other body tissues	Stimulate development of secondary sex characteristics
		Hypothalamus and anterior pituitary	Before ovulation, cause positive feedback to increase LH and FSH secretion
		Ovaries	Inhibit adjacent follicles
Progesterone	Ovaries (corpus luteum)	Uterus	Causes hypertrophy of endometrial cells; stimulates secretion of mucus from uterine glands; helps maintain pregnancy and prevents uterine contractions
		Mammary glands	Stimulates alveoli of mammary glands to develop during pregnancy
		Other body tissues	Stimulates development of secondary sex characteristics
Inhibin	Ovaries (follicles) and granulosa cells	Anterior pituitary	Inhibits FSH secretion

*Other hormones secreted in females, primarily during pregnancy, include relaxin, prolactin, oxytocin, and human chorionic gonadotropin. (These hormones are covered in Chapter 27.)

Barrier Methods

Barrier methods of birth control prevent sperm from moving beyond the vagina. These methods include condoms, diaphragms, cervical caps, and the sponge (**Figure 26.20**).

Most *condoms* (see Figure 26.20a) are made of synthetic materials or latex, although some are made from animal membranes (lamb intestine). Male condoms, worn on the penis, are the most common type of condom in use. Female condoms are inserted into the vagina and include a flexible ring that covers the cervix and an outer ring that covers the external genitalia. In addition to reducing the risk of pregnancy, synthetic and latex condoms lower the risk of sexually transmitted infections (discussed in Module 26.7).

Diaphragms or *cervical caps* are placed inside the vagina and cover the cervix to block sperm (see Figure 26.20b). Both of these barriers require a prescription and a physical exam by a

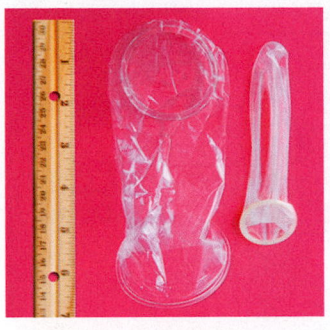
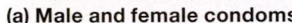

(a) Male and female condoms

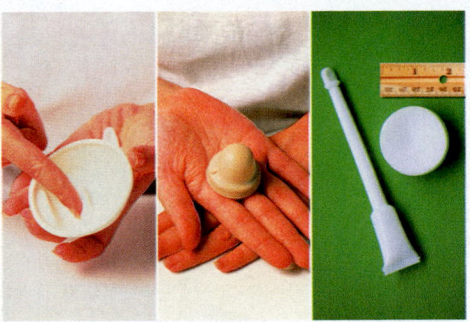

(b) Diaphragm, cervical cap, and sponge

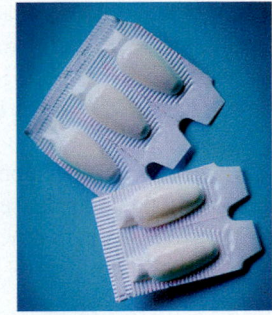

(c) Spermicide

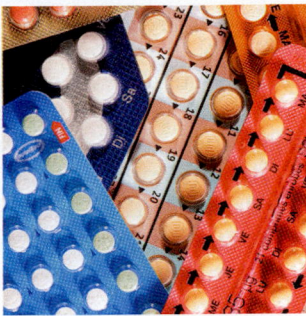

(d) Oral contraceptives

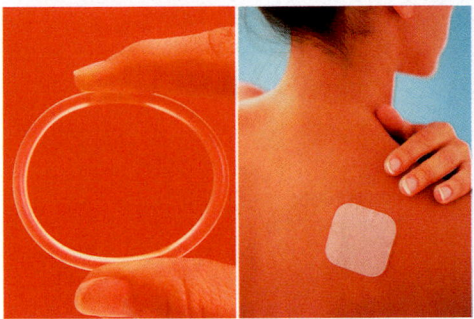

(e) Vaginal ring and skin patch

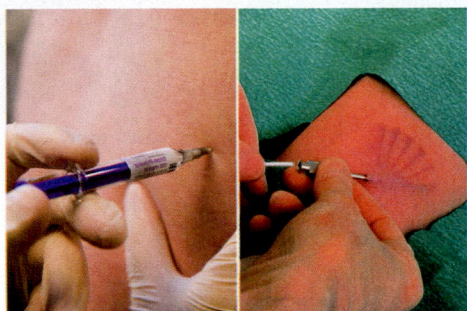

(f) Progesterone injections and implants

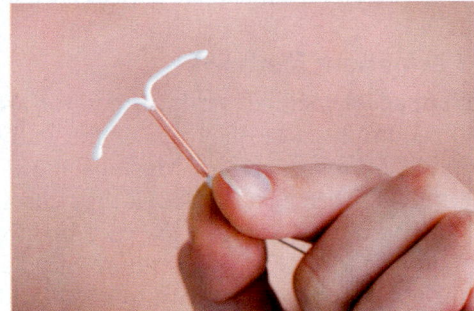

(g) IUD

Figure 26.20 **Temporary methods of birth control.**

healthcare provider for proper fit. A similar form of birth control that requires no fitting is the *sponge,* a foam disk that is inserted before intercourse to cover the cervix.

Many barrier methods, particularly the diaphragm, cervical cap, and sponge, are made more effective by using a chemical **spermicide,** which kills or damages sperm cells it contacts. Spermicides are available without a prescription and may be purchased as foams, creams, or jellies (see Figure 26.20c).

Quick Check

☐ 2. What are some common barrier methods of contraception?

Hormonal Methods

Oral contraceptives, or *birth control pills,* contain synthetic estrogens and progesterone (see Figure 26.20d). The artificially high levels of estrogens and progesterone have a negative feedback effect on the release of FSH and LH. Low levels of FSH and LH prevent follicle maturation and, subsequently, ovulation. In addition, they interfere with the buildup of the endometrial lining necessary for implantation of a conceptus. A side benefit is that oral contraceptives often reduce the amount of monthly bleeding during menstruation. In fact, some reduce the number of menstrual cycles to only about four per year.

Oral contraceptives are very effective at preventing pregnancy if taken as prescribed. However, they may cause side effects. Some of these are relatively minor, such as nausea, fluid retention, and breast tenderness, whereas others are potentially more severe. The risk of serious side effects, such as

cardiovascular or liver problems, increases in women over 35, particularly those who smoke or have a history of hypertension or bleeding disorders.

One alternative to the combined estrogen–progesterone pill is the *mini-pill,* which contains only synthetic progesterone. A high level of progesterone thickens the cervical mucus and makes it difficult for sperm to reach the egg. The mini-pill is slightly less effective than the combined pill.

Hormonal alternatives to a daily pill include:

- vaginal rings containing estrogen and progesterone, which are worn for the 3 weeks or so between menstrual cycles (see Figure 26.20e);
- skin patches containing estrogen and progesterone that are replaced weekly (see Figure 26.20e);
- intramuscular injections of progesterone every 3 months (see Figure 26.20f); and
- thin rods containing progesterone, which are implanted under the skin of the arm and remain in the body for approximately 3 years (see Figure 26.20f)

Quick Check

☐ 3. How do oral contraceptive pills prevent pregnancy?

☐ 4. Why is a progesterone-only contraceptive less effective than one containing estrogen and progesterone?

Intrauterine Methods

An **intrauterine device (IUD)** is a small T-shaped device made of plastic and copper that is placed into the uterus by a health provider, where it can remain for approximately 10 years. It prevents pregnancy by not allowing a fertilized ovum to implant into the uterine wall. In addition, copper is released inside the womb and spreads to the uterine tubes, where it poisons the sperm and oocyte (see Figure 26.20g).

An **intrauterine system (IUS)** is a small T-shaped device similar to the IUD, but it is made of plastic and releases a small amount of progesterone each day to prevent pregnancy. The IUS can remain in the uterus for approximately five years.

Quick Check

☐ 5. How do intrauterine devices prevent pregnancy?

Permanent Methods

Permanent methods of birth control are referred to as "sterilization" because they render the individual sterile, or incapable of reproduction. Most of these techniques require a surgical procedure. Although they can sometimes be reversed with additional surgery, the second surgery is successful less than 45% of the time. Therefore, permanent methods are best for people who do not wish to have children at any point in the future.

Male sterilization is accomplished by a surgical procedure called a **vasectomy** (vaeh-SEK-tuh-mee) in which the physician cuts through and then either *ligates* (ties off) or *cauterizes* (burns) the ends of the ductus deferens (**Figure 26.21a**). This prevents sperm from leaving the man's body with the semen during ejaculation. After a vasectomy, sperm are still produced, but they degenerate and are destroyed by phagocytosis. Semen is also still produced because the vas deferens is cut below the area where the accessory structures secrete the other components of semen. Also, because the blood vessels are unaffected, the blood testosterone level remains normal. For these reasons, a vasectomy has no effect on sexual desire or performance. Since some sperm remain in the vas deferens, unprotected intercourse should be postponed until all sperm have been phagocytized, which may take up to 12 weeks.

A similar surgical procedure, called **tubal ligation** (lye-GAY-shun), is performed in women (**Figure 26.21b**). During this procedure, a surgeon makes an insertion into the wall of the body cavity, cuts through the uterine tubes, and either ligates or cauterizes their ends. This prevents the oocyte from passing through the uterine tubes and the sperm from reaching the oocyte.

Nonsurgical permanent methods of birth control include *tubal implants*. These small metal springs are placed in each uterine tube and do not require incisions in the abdomen. Over time, scar tissue grows around each implant and permanently blocks the tubes. Three months after the procedure, an x-ray is taken of the uterine tubes, which have been filled with contrast dye, to determine if they are completely blocked.

Unfortunately, no perfect form of birth control exists. **Table 26.4** shows the failure rates of the available contraceptive methods.

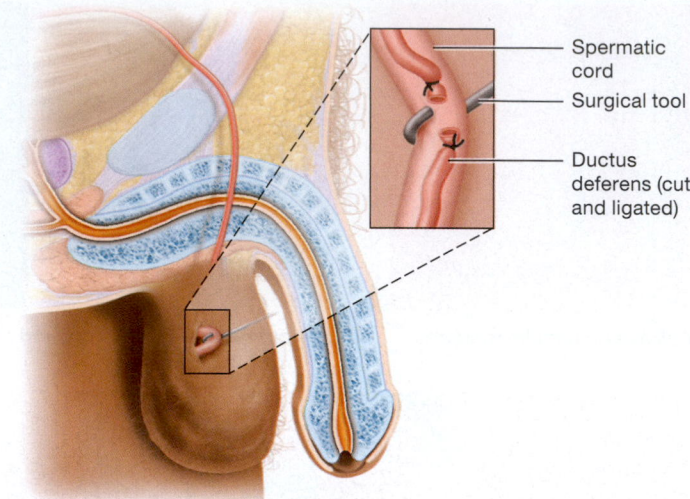

(a) **Vasectomy**

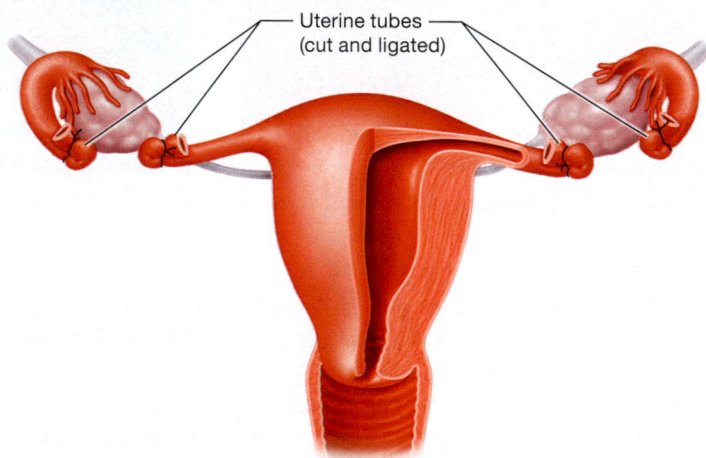

(b) **Tubal ligation**

Figure 26.21 Permanent methods of birth control.

Quick Check

☐ 6. Which methods of birth control are also called sterilization? Why?

Apply What You Learned

☐ 1. Choose an appropriate method of contraception for each of the following individuals and briefly explain your reasoning:

   a. A woman is 25 years old. She is married and is in a committed relationship. She currently has no children but wants children "someday." She has experienced blood clots due to a coagulation disorder in the past.

   b. A man is 25 years old. He is single and has multiple sexual partners. He may want children at some point in his life, but he is not sure.

See answers in Appendix A.

Table 26.4 Failure Rates for Common Methods of Birth Control

Method (most successful to least successful within each category)	Rate of Failure (pregnancies per 100 users)	
	Perfect Use (method used correctly and consistently)	Typical Use (method may be used incorrectly or inconsistently)
Behavioral Methods		
Complete abstinence	0	Unknown
Withdrawal method	4	27
Rhythm method	3–5	25
Permanent Methods		
Vasectomy	0.10	0.15
Tubal ligation and implants	0.5	0.5
Hormonal Methods		
Skin patch	0.1	1–2
Vaginal contraceptive ring	0.1	1–2
Birth control pill	0.3	1–2
Hormone injections	0.3	1–2
Implants	0.3	1–2
Mini-pill	0.5	2
Intrauterine Devices and Systems	0.2–0.6	0.2–0.8
Barrier Methods		
Male and female condoms	2–5	15–21
Diaphragm, cervical cap, sponge (with spermicide)	6–9	16–32
Spermicides (foams, creams, jellies) used alone	18	29
No Birth Control	85	85

MODULE 26.7
Sexually Transmitted Infections (STIs)

Learning Outcomes

1. Predict factors or situations affecting the reproductive system that could disrupt homeostasis.
2. Describe the causes, symptoms, and potential complications of common sexually transmitted infections.

A **sexually transmitted infection (STI),** also called a *sexually transmitted disease* (*STD*) or *venereal disease,* is an infectious disease spread through sexual contact. The most important cause of reproductive disorders, STIs often cause infertility in adults and deformities or even death in a fetus.

We have already discussed the most notorious STI, HIV and AIDS, in the chapter on the lymphatic system and immunity (see Chapter 20). This module presents several of the most common STIs, and distinguishes between those caused by bacteria and parasites and those caused by viruses.

Bacterial and Parasitic STIs

FLASHBACK

1. What is a pathogen? (p. 754)
2. What is the conjunctiva? (p. 545)

The first group of STIs we discuss are those caused by pathogenic, or disease-causing, bacteria and parasites. Both types of pathogens are foreign cells that invade tissues and create an inflammatory reaction. The common bacterial and parasitic STIs we discuss in this section are chlamydia, gonorrhea, syphilis, and trichomoniasis.

Chlamydia

Chlamydia (klah-MID-ee-uh) is a common STI caused by the bacterium *Chlamydia trachomatis*. Even though it is easily treated with antibiotics, an estimated 3 million chlamydial infections occur annually in the United States. This may be because chlamydia produces no symptoms in most people. If symptoms do occur, they may not appear until several weeks after exposure and most commonly include painful urination, pain in the pelvic region and testes, and vaginal or penile discharge. Newborns may become infected in the birth canal, and as a result develop respiratory tract infections and a form of conjunctivitis (infection of the conjunctiva) called *trachoma,* which can lead to blindness.

In women, untreated chlamydia can spread into the uterus and uterine tubes and cause *pelvic inflammatory disease* (*PID*), which scars the uterine tubes. Scar tissue may prevent fertilization by completely blocking sperm, resulting in infertility, or it may partially block the uterine tube, preventing a fertilized ovum from moving into the uterus and causing it to implant in the uterine tube (see Chapter 27).

Gonorrhea

Gonorrhea (gawn-oh-REE-uh), commonly called "the clap," is caused by the bacterium *Neisseria gonorrhoeae.* Gonorrhea infects an estimated 800,000 to 1 million people each year in the United States. It is treatable with antibiotics, but antibiotic-resistant strains of the bacteria are becoming more prevalent. Although some men have no symptoms, most exhibit clinical signs that include painful urination and penile discharge, which ranges in color from white to yellow to green. Most women are asymptomatic, but painful urination and vaginal bleeding may occur. As with chlamydia, untreated gonorrhea can lead to PID in women and blindness in newborns. Gonorrhea and chlamydia often occur together.

Syphilis

Syphilis (SIFF-ih-liss) is caused by the spiral-shaped bacterium *Treponema pallidum.* Approximately 50,000 individuals are infected with syphilis each year in the United States. The bacterium incubates in the body for up to 6 weeks and then produces a small painless lesion called a *chancre* (SHAN-kur), at which point the disease is transmissible to other people via contact with the chancre. In males, the chancre usually forms on the penis, but in females, the chancre may form inside the vagina and go undetected. The chancre generally heals within 6 weeks, and the disease seems to have run its course. However, the second stage of the disease appears a few weeks later, when a pink skin rash appears all over the body. Fever and joint pain are common, and hair loss sometimes occurs. Symptoms again subside, this time within about 3 months, but the disease has not been cured—it has entered the latent stage.

Symptoms of syphilis can come and go for years, although the immune system may eventually kill the bacteria. It is also possible for the disease to progress to *tertiary syphilis,* which produces destructive lesions in the brain and damages the cardiovascular organs, bones, and joints, damage that can be fatal.

Antibiotics can treat the infection but cannot repair damage done by the bacterium. Infected newborns may exhibit no symptoms at first, but can have many health problems, including impaired vision and hearing, if not treated with antibiotics.

Trichomoniasis

Trichomoniasis (trik′-uh-moh-NY-ah-sis) is an STI caused by the protozoan parasite *Trichomonas vaginalis.* An estimated 3–4 million people have the infection, which can be easily cured with antibiotics. Trichomoniasis may cause itching, burning, or redness of the genitals and, in women, may result in a white to greenish vaginal discharge with a strong odor. However, as with several other STIs, many individuals exhibit no symptoms. Infection during pregnancy can lead to early birth of the baby and low birth weight.

☐ 1. Which of the four STIs discussed in this section are bacterial? Which are parasitic?

☐ 2. What are the stages of untreated syphilis?

Viral STIs

◀◀ FLASHBACK

1. What is interferon? (p. 768)

Unlike bacterial and parasitic infections, viral infections are caused by small nonliving viruses taking over the cellular organelles responsible for protein synthesis and using them to make copies of themselves. The new viral particles are then released from the cell and go on to infect other cells. Common viral STIs discussed in this section are human papillomavirus and herpes simplex virus 2.

Genital Warts: Human Papillomavirus (HPV)

Genital warts, caused by over 40 types of the **human papillomavirus (HPV),** are the most common STI in the United States. Nearly 80 million people are currently infected with HPV, and about 14 million new infections occur each year.

Genital warts associated with HPV occur in just under 400,000 people a year. Warts may appear on the penis, perineum, or anus in males and on the cervix, vaginal wall, perineum, or anus in females. The most common risk to newborns is warts in the throat, called *recurrent respiratory papillomatosis,* a rare condition that may result from the mother having genital warts during late pregnancy. Genital warts can be treated with cryosurgery, laser surgery, or *interferon,* a cytokine released by infected cells that helps combat intracellular pathogens.

Many cases of cervical cancer are caused by HPV (see *A&P in the Real World: Cervical Cancer*), although most of the strains that cause genital warts do not lead to cancer. A vaccine against HPV can protect against some of the most common disease-causing strains, including those that cause 70% of cervical cancers; most cases of HPV-induced penile, vulvar, vaginal, and anal cancers; and 90% of genital warts. No medication is available to treat HPV after it has been contracted, which underlies the importance of preventing it.

Cervical Cancer

Cervical cancer occurs most often in women between the ages of 30 and 50, particularly those with a history of sexually transmitted infections, cervical inflammation, or multiple pregnancies. It is frequently caused by the human papillomavirus (HPV), which is transmitted sexually.

Cervical cancer used to be a leading cause of cancer death for women in the United States. However, in the past few decades, both the number of cases and the number of deaths from cervical cancer have decreased significantly, and are projected to decrease further as the HPV vaccine becomes more widespread. This decline is due in large part to the Pap (Papanicolaou) smear test, which detects precancerous cells and early-stage cancers before symptoms are noticeable. The Pap test involves scraping loose cells from the external os of the cervix and examining them microscopically. Cells showing signs of abnormal development, called *dysplasia*, warrant further investigation, including visual examination of the cervix or a biopsy to determine if cancerous cells are present. If cancerous cells are detected, surgical removal of the tumor or even a hysterectomy (removal of the uterus) may be required. In addition, radiation therapy or chemotherapy may be needed.

Genital Herpes

Genital herpes is usually caused by the herpes simplex virus type 2. Approximately 750,000 people in the United States are infected each year, but many are unaware that they harbor the virus. The disease causes blisters that are often extremely painful on the external genitalia of both sexes and sometimes on the thighs and buttocks as well. The virus is most easily transmitted when such blisters occur, if there is direct skin-to-skin contact. However, a person is contagious even during the dormant stage, which can last for weeks to years. Herpes infections can be passed to a fetus, resulting in neonatal herpes, a potentially fatal infection. Genital herpes may also increase the risk of HIV infections and some cancers.

The only treatment for genital herpes is antiviral medication, which shortens or sometimes prevents outbreaks. Daily use of antiviral medication may reduce the risk of transmission to sexual partners.

Quick Check

☐ 3. What is the relationship between human papillomavirus and certain cancers?

☐ 4. When is an individual with genital herpes contagious?

Apply What You Learned

☐ 1. Jim is presenting for a college sports physical. He says he has multiple sexual partners and practices safe sex occasionally. However, he states that he is sure he does not have any STIs because he has never had any lesions or other symptoms. What do you tell him, and why?

☐ 2. Why does the Centers for Disease Control (CDC) recommend that girls as young as 11 or 12 be vaccinated against the human papillomavirus? Why does the CDC recommend that boys also receive the vaccination?

See answers in Appendix A.

Chapter Summary

For everything you need to succeed in this course, go to Mastering Anatomy & Physiology. There you will find:

 Practice Tests

 Author Podcasts

 Big Picture Animations

 Concept Boost Video Tutors

iP2™ Interactive Physiology 2.0 Tutorials

A&PFlix A&P Flix 3D Animations

 Active-Learning Workbook

MODULE 26.1
Overview of the Reproductive System and Meiosis 1022–1027

- The **gonads** produce **gametes** and **sex hormones.**

- Gametes, or **sperm** and **ova,** are produced by meiosis. During **meiosis I,** pairs of **homologous chromosomes,** each consisting of two **sister chromatids,** separate to produce **haploid (1n)** cells with 23 chromosomes. Also, crossing over and independent assortment take place. During **meiosis II,** homologous chromosomes separate into individual chromatids, potentially producing four haploid cells.

- Mitosis occurs in almost every cell type in the body, but meiosis occurs only in the testes and ovaries, to produce gametes. In addition, mitosis produces two **diploid (2n)** cells that are genetically identical to the original cell, but meiosis produces up to four haploid cells that are genetically unique.

Anatomy of the Male Reproductive System
1027–1033

- The paired **testes,** or male gonads, produce sperm cells and secrete the male hormone *testosterone.*

- **Seminiferous tubules** in the testes contain *spermatogenic cells* that form sperm and **sustentacular cells,** which support sperm cells and form the **blood testis barrier.** The tubules are surrounded by **interstitial cells,** which produce testosterone.

- The male duct system includes the **epididymis,** where sperm maturation occurs; the **ductus deferens,** which transports sperm into the abdominopelvic cavity; the **ejaculatory duct,** which extends through the prostate gland; and the **urethra,** which transports sperm out of the body through the penis.

- Accessory structures that produce semen include the **seminal vesicles,** which are located on the posterior bladder and produce about 30% of the volume of semen; the **prostate gland,** which sits inferior to the bladder and produces about 60% of the volume of semen; and the **bulbourethral glands,** which are found at the root of the penis and produce mucus.

- The external genitalia of the male include the **penis,** the male copulatory organ, and the **scrotum,** which contains the testes and epididymis.

Physiology of the Male Reproductive System 1033–1039

- **Spermatogenesis** is the process of sperm cell development that occurs in the seminiferous tubules of the testes. The end result is four total haploid **spermatids. Spermiogenesis** is the maturation of spermatids into motile sperm cells. It occurs in the epididymis. Both processes begin at puberty and continue throughout life.

- Testosterone secretion and spermatogenesis are controlled by the **hypothalamic-pituitary-gonadal (HPG) axis.**
  - The hypothalamus releases **gonadotropin-releasing hormone (GnRH),** which stimulates the anterior pituitary to secrete **follicle-stimulating hormone (FSH)** and **luteinizing hormone (LH).**
  - LH stimulates interstitial cells to produce testosterone, which stimulates spermatogenesis; maturation of sex organs; development of secondary sex characteristics; and anabolic effects, including bone growth and increased muscle mass.
  - FSH stimulates sustentacular cells to secrete **androgen-binding protein (ABP)** and inhibin.

- The male sexual response begins with **erection** of the penis. This is followed by **ejaculation,** which is the expulsion of semen.

- **Puberty** in boys usually begins between the ages of 12 and 14, when GnRH and so the testosterone levels begin to rise.

- **Male secondary sex characteristics** include pubic, axillary, and facial hair; deeper voice; thicker skin; increased bone density;

larger skeletal muscles; increased basal metabolic rate and red blood cell count; and increased volume of body fluids.

Anatomy of the Female Reproductive System 1039–1044

- The paired **ovaries,** or female gonads, produce ova and secrete the hormones *estrogens, progesterone,* inhibin, and relaxin.

- Oocytes are located within **follicles,** where they develop and mature. After an oocyte is ovulated, it is picked up by the **fimbriae** of the **uterine tube,** which is the site of fertilization.

- The **uterus** sits in the pelvis anterior to the rectum and posterosuperior to the urinary bladder. The **endometrium** lines the uterus and is the site of implantation of the fertilized ovum; the other layers are the **myometrium** and **perimetrium.**

- The **vagina** is the female copulatory organ and provides a passageway for delivery of an infant and for menstrual flow.

- The **vulva** of the female includes the **mons pubis, labia majora** and **labia minora, clitoris,** and **vestibule.** The vestibule contains the external openings of the urethra and the vagina.

- The **mammary glands** are modified sweat glands composed of adipose and fibrous connective tissue. They are filled with glandular **lobes** that contain lobules with **alveoli.** If pregnancy occurs, the alveoli produce milk.

Physiology of the Female Reproductive System 1044–1055

- **Oogenesis** is the process by which an **ovum** develops.
  - Before a female is born, **oogonia** divide and differentiate to produce **primary oocytes** arrested in prophase I until puberty.
  - About once per month after puberty, a primary oocyte completes meiosis I to produce one large **secondary oocyte** and one small **polar body.** The secondary oocyte proceeds to metaphase II, where its development is arrested unless it is fertilized. If fertilized, it completes meiosis II to yield the ovum and second polar body.

- The **ovarian cycle** is the monthly series of events associated with maturation of oocytes and their follicles. It includes the following phases:
  - In the **follicular phase,** several **primary follicles** mature into **secondary follicles.** Usually one secondary follicle will develop into a **vesicular follicle,** and the primary oocyte inside will become a secondary oocyte.
  - **Ovulation** releases the secondary oocyte into the peritoneal cavity.
  - In the **luteal phase,** the ruptured follicle forms the **corpus luteum,** which secretes progesterone and estrogens for the rest of the cycle.

- The HPG axis controls the ovarian cycle.
  - The hypothalamus releases **GnRH**, which stimulates the anterior pituitary to secrete **FSH** and **LH.**
  - FSH and LH stimulate follicles to grow and secrete estrogens and inhibin, which inhibit the release of more FSH and LH.
  - One follicle eventually produces enough estrogens to exert positive feedback on the HPG axis, resulting in an LH surge and a rise in FSH. This triggers ovulation and transforms the ruptured follicle into the corpus luteum.
  - The corpus luteum produces progesterone and estrogens and will continue to do so if fertilization occurs. If fertilization does not occur, the progesterone level drops, and the cycle repeats.
- Estrogens promote maturation of sex organs, development of external genitalia, development of breast tissue, preservation of bone mass, and maintenance of vascular health.
- The **uterine cycle,** the series of cyclic changes in the endometrium that occur each month in response to fluctuating ovarian hormones, includes the following phases, regulated by the HPG axis:
  - In the **menstrual phase,** the uterus sheds the **stratum functionalis** of the endometrium through the vagina. This phase results from rising FSH and LH levels and normal/low levels of estrogens and progesterone.
  - In the **proliferative phase,** a new stratum functionalis grows and thickens as estrogen levels rise. Ovulation occurs in the ovary at the end of this phase.
  - In the **secretory phase,** the endometrium prepares for implantation of a conceptus by continuing to grow. If fertilization does not occur, the menstrual phase will follow.
- The **female sexual response** is similar to that of males, and erectile tissues also engorge with blood. However, orgasm in females is not accompanied by ejaculation.
- Puberty usually begins when girls are 9–11 years old, when the GnRH level increases. GnRH stimulates the HPG axis, and estrogens and progesterone levels rise, which stimulates oogenesis to resume and female sex characteristics to develop.
- **Female secondary sex characteristics** include changes in bone structure, such as widening of the pelvis; an increase in adipose tissue in general and particularly in the breasts and around the hips and thighs; and hair in the pubic and axillary areas.

MODULE 26.6
Methods of Birth Control 1055–1059

- Birth control, or contraception, refers to any procedure or device intended to prevent pregnancy.
- Methods of birth control include behavioral modification such as *abstinence;* **barrier methods,** such as condoms; hormonal methods, such as **oral contraceptives;** intrauterine methods; and permanent methods, such as **vasectomy, tubal ligation,** and tubal implants.

MODULE 26.7
Sexually Transmitted Infections (STIs)
1059–1061

- **Sexually transmitted infections (STIs)** are the most important cause of reproductive disorders and often result in infertility, fetal deformities, and even fetal death.
- Common STIs include HIV and AIDS, **chlamydia, gonorrhea, syphilis, genital warts (HPV), genital herpes,** and **trichomoniasis.**

Assess What You Learned

Scan the QR Code for additional practice test questions
https://goo.gl/OQbC1y

LEVEL 1 Check Your Recall

1. Fill in the blanks: At the end of meiosis, each cell has _____ chromosomes and they are genetically _____ from the original cell.

2. Match the specific phase of meiosis with the correct description.

  _____ Prophase I
  _____ Metaphase I
  _____ Anaphase I
  _____ Telophase I
  _____ Prophase II
  _____ Metaphase II
  _____ Anaphase II
  _____ Telophase II

  a. Homologous chromosomes arrive at opposite poles.
  b. Homologous chromosomes line up.
  c. Chromosomes line up at the equator.
  d. Independent assortment occurs.
  e. Separated sister chromatids arrive at opposite poles.
  f. Sister chromatids remain condensed.
  g. Crossing over occurs.
  h. Sister chromatids pull apart.

3. Mark the following statements about spermatogenesis as true or false. If a statement is false, correct it to make a true statement.

  a. Spermatogenesis is the process of sperm cell development. It begins at puberty and ends when a man reaches approximately 45–55 years of age.
  b. Spermatogonia are the stem cells that begin the process of spermatogenesis and divide by mitosis.
  c. Primary spermatocytes undergo the first meiotic division and form two smaller diploid cells called secondary spermatocytes.
  d. Spermiogenesis is the maturation process of sperm cells and involves changing the size and shape of the cell.

4. Which of the following structures is the site of sperm maturation?

   a. Epididymis   d. Urethra
   b. Ductus deferens   e. Scrotum
   c. Ejaculatory duct

5. Which of the following occurs in the testes?

   a. Production of gametes
   b. Secretion of testosterone
   c. Production of semen
   d. Both a and b
   e. Both b and c

6. What are the three regions of the male urethra?

7. Match the component of the glandular secretions with the correct accessory sex gland. A gland may match with more than one component.

   _____ Seminal vesicle   a. Citrate
   _____ Prostate   b. Fructose
   _____ Bulbourethral gland   c. Prostaglandins
                               d. Coagulating enzyme
                               e. Mucus
                               f. Proteolytic enzymes

8. Fill in the blanks: Internally, the penis contains three cylindrical erectile bodies, the paired _____ and the _____.

9. Which of the following is the main function of the scrotum?

   a. The scrotum contains the testes and provides additional blood supply to them because of the location outside the abdominopelvic cavity.
   b. The scrotum contains the testes and maintains the temperature of the testes about 3° C lower than body temperature, which is the optimal temperature for developing sperm.
   c. The scrotum contains the testes and maintains the temperature of the testes about 3° C higher than body temperature, which is the optimal temperature for developing sperm.
   d. The scrotum contains the testes and contains the blood testis barrier to separate the seminiferous tubules from the bloodstream.

10. Number the sequence of events in the hormonal regulation of male reproduction, placing a 1 by the first event, a 2 by the second, and so forth.

    _____ Luteinizing hormone (LH) stimulates interstitial cells to produce testosterone and follicle-stimulating hormone (FSH) stimulates sustentacular cells to secrete inhibin and androgen-binding protein, which keeps some of the testosterone near the spermatogenic cells.
    _____ Testosterone and inhibin both exert negative feedback controls on the hypothalamus and anterior pituitary.
    _____ Androgen-binding protein binds testosterone.
    _____ Gonadotropin-releasing hormone (GnRH) stimulates the anterior pituitary to secrete FSH and LH.
    _____ The hypothalamus releases GnRH.

11. Fill in the blanks: _____ is the enlargement and stiffening of the penis, which results when the tissue in the penis becomes engorged with blood, and _____ is the process by which semen is expelled from the penis.

12. Which of the following hormones is/are *not* produced by the ovaries?

    a. Relaxin
    b. Inhibin
    c. Estrogens
    d. Oxytocin
    e. Progesterone

13. Which of the following ligaments does *not* attach to the ovaries?

    a. Lateral cervical ligaments
    b. Broad ligament
    c. Ovarian ligament
    d. Suspensory ligament
    e. All of these attach to the ovaries.

14. Fill in the blanks: The _____ is the site of fertilization in the female reproductive system, and the _____ is the site of implantation of the fertilized ovum.

15. Fill in the blanks: Within the lobes of the mammary glands are smaller units called _____, which contain glandular _____ that produce milk when a woman is lactating.

16. Mark the following statements about oogenesis as true or false. If a statement is false, correct it to make a true statement.

    a. Oogenesis begins before a female is born and continues for the remainder of her life.
    b. About once a month after puberty begins, some primary oocytes are stimulated to continue development. Usually one will complete the first meiotic division to produce two haploid cells that are the same size.
    c. The fate of all polar bodies is the same; they all degenerate and are reabsorbed by the body.
    d. Only if a secondary oocyte is fertilized will it complete the second meiotic division and expel the last polar body.

17. Match the follicle stage with the correct description.

    _____ Primordial follicle   a. Several layers of granulosa
    _____ Primary follicle   cells with pockets of
    _____ Secondary follicle   follicular fluid surround the
    _____ Vesicular follicle   primary oocyte.
    _____ Corpus luteum   b. Knot of whitish scar tissue
    _____ Corpus albicans   c. Yellow glandular-looking
                                      endocrine structure
                                  d. A single layer of squamous
                                      follicular cells surrounds the
                                      primary oocyte.
                                  e. One or two layers of
                                      cuboidal granulosa cells
                                      surround the primary oocyte.
                                  f. Several layers of granulosa
                                      cells surround the secondary
                                      oocyte and project into the
                                      antrum.

18. Number the sequence of events in the hormonal regulation of female reproduction, placing a 1 by the first event, and so forth.

_____ Increasing levels of estrogens and inhibin exert negative feedback controls on the hypothalamus and pituitary to inhibit the release of more FSH and LH, leaving only one follicle to survive.

_____ LH and FSH surge stimulates the primary oocyte of one follicle to complete meiosis to metaphase II, which triggers ovulation.

_____ FSH and LH stimulate follicles to grow and secrete estrogens and inhibin.

_____ Progesterone from the corpus luteum inhibits the release of gonadotropins and will continue to do this if pregnancy is achieved.

_____ The hypothalamus releases GnRH.

_____ A single follicle produces enough estrogens to exert positive feedback on the hypothalamus and anterior pituitary to trigger an LH surge.

_____ The GnRH level rises to begin the cycle over.

_____ LH transforms the ruptured follicle into the corpus luteum, which produces progesterone and some estrogens.

_____ If pregnancy does not occur, the corpus luteum degenerates into the corpus albicans and the progesterone level drops.

_____ GnRH stimulates the anterior pituitary to secrete FSH and LH.

19. Mark the following statements about the uterine cycle as true or false. If a statement is false, correct it to make a true statement.

a. During the proliferative phase, the endometrium prepares for the implantation of the conceptus by enlarging the endometrial glands and secreting glycogen into the uterine cavity.

b. During the secretory phase, the endometrium generates a new stratum functionalis.

c. During the menstrual phase, the uterus sheds the stratum basalis of the endometrium, which results in the discharge of blood and tissue from the vagina.

20. Which of the following hormones stimulates the generation of a new stratum functionalis in the endometrium?

a. Estrogens
b. Progesterone
c. Inhibin
d. LH
e. FSH

21. Which of the following methods of birth control is classified as "sterilization"?

a. Oral contraceptive pill
b. Tubal ligation
c. Intrauterine device
d. The patch

22. Which of the following sexually transmitted infections is/are caused by a virus? Mark all that apply.

a. Chlamydia
b. Genital warts
c. Genital herpes
d. Syphilis
e. Gonorrhea

LEVEL 2 Check Your Understanding

1. The second meiotic division is very similar to mitosis. How would you explain the differences between the first meiotic division and mitosis?

2. Sustentacular cells in the testes are nicknamed "nurse cells." Explain why the nickname is appropriate.

3. How might the female orgasm increase the chance for fertilization to occur?

4. Explain why oral contraceptives, which artificially raise levels of estrogens and progesterone, prevent pregnancy.

LEVEL 3 Apply Your Knowledge

PART A: Application and Analysis

1. Mr. Hassan has recently started getting up during the night to urinate and has also noticed that he is having trouble emptying his bladder. He has made a doctor's appointment, but his wife has already told him that the doctor will want to check his prostate. Do you think his wife is correct? Explain your answer.

2. A professional male athlete has been taking a synthetic testosterone supplement for several months. He and his wife have been trying to have a baby, with no success, and have decided to begin treatment for infertility. Before testing the man's sperm count, the doctor discovers that the man's LH level is very low. Why do you think his LH level is low? Would the low LH level affect sperm production?

3. Fabiola, a 17-year-old college student, suffers from an eating disorder. Her percentage of body fat has dropped to 14%. Her menstrual cycles were irregular, and now they have stopped altogether. Why do you think this has happened? Could the lack of menstrual periods be beneficial in this situation? Explain.

PART B: Make the Connection

4. A 60-year-old man has been taking an herbal medication for impotence. Recently, he has been having periods of dizziness when he is exercising and has started to check his blood pressure, which drops after he takes the medication. Why do you think a medication that prevents impotence also affects blood pressure? (*Connects to Chapter 18*)

See answers in Appendix A.

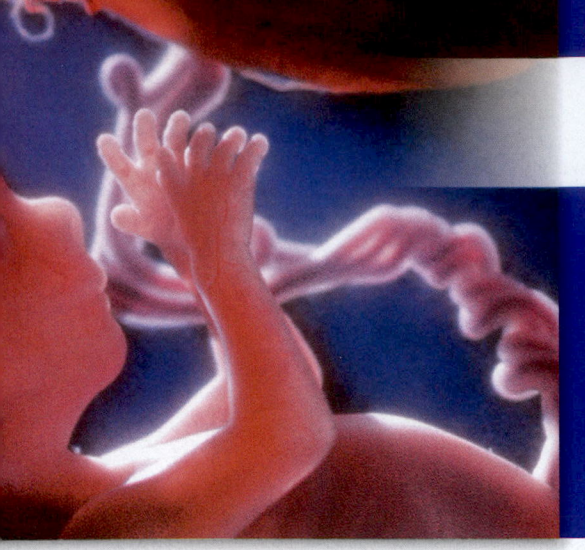

27

Development and Heredity

For practice applying concepts to a clinical scenario, check out the **Running Case Study** *for this chapter* @ Mastering Anatomy & Physiology

Photo: Image of a human fetus during the fourth month of development.

Every day, the population of the earth increases by over 200,000 people. Perhaps because births are so common, it's easy to forget that the development of a new human life is an incredible event. That a single, fertilized oocyte can grow into a complex, independent individual made of trillions of cells is truly astounding.

The reproductive system chapter examined the anatomy and physiology of the male and female reproductive systems and described the formation of *gametes*— the *sperm* and *oocytes* (see Chapter 26). Let's now explore the developmental processes that occur when a fertilized oocyte develops into a new individual. This chapter covers the major stages of development from fertilization through birth, and then the process of birth itself. Next, our focus shifts to what happens to both the mother and newborn in the first few months after delivery. The chapter ends with a brief description of how genetic characteristics are transmitted from parent to offspring—a mechanism called *heredity*.

MODULE 27.1
Overview of Human Development

Learning Outcomes

1. Explain the difference between prenatal and postnatal development.
2. Describe the periods of prenatal development.

Human development begins at *fertilization,* with the union of gametes, and continues throughout life. **Developmental biology** is the science that studies the changes in form and function that take place during these processes. It encompasses **embryology** (em′-bree-AH-luh-jee), the study of the events that occur during the **prenatal period** (pree-NAY-tuhl; *nata-* = "birth")—the approximately 38 weeks of development inside the mother's uterus. It also includes the **postnatal period,** which includes the period from birth through all life stages (although "postnatal" may also refer to the short period following birth). Let's look a bit more closely at these periods.

The Process of Prenatal Development

⟪ FLASHBACK

1. What is the function of mitosis? (p. 111)

2. What layers make up the wall of the uterus? (p. 1041)

The prenatal period is what people usually refer to as pregnancy. The term **pregnancy** embraces all the events from **conception**—when an oocyte is fertilized and begins to develop—until birth, and includes the anatomical and physiological changes to the mother during this time. The growing offspring is generally referred to as the **conceptus** (kun-SEP-tuss). It develops during the **gestation period** (jes-TAY-shun) of pregnancy, which extends from the mother's last menstrual period until birth, or for approximately 280 days, or 40 weeks. The gestation period is longer than the prenatal period because it includes the 2 weeks after the mother's last menstrual cycle—a date that women can usually recall. We will focus on the approximately 38 weeks of prenatal development, which is divided into three stages (**Figure 27.1**):

	PRE-EMBRYONIC PERIOD	EMBRYONIC PERIOD	FETAL PERIOD
Developmental stage	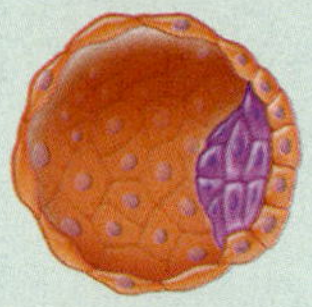 Blastocyst	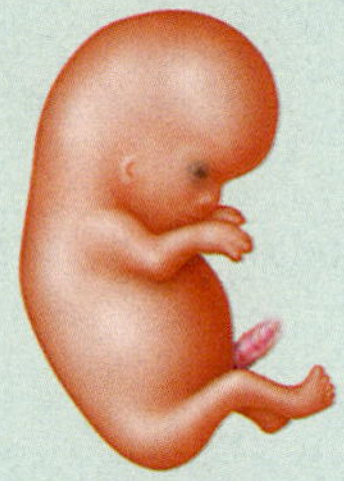 Embryo	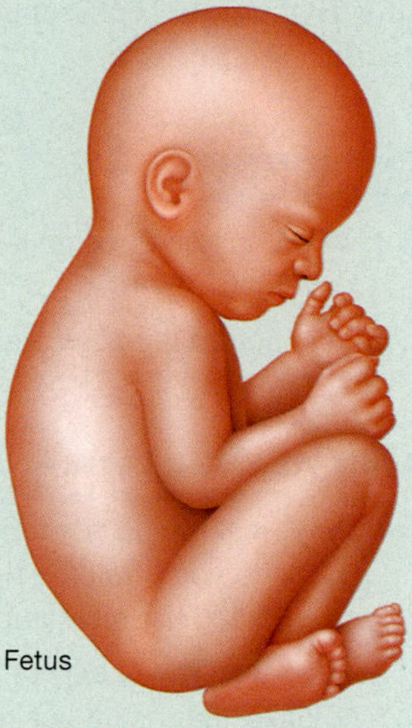 Fetus
Events	**Weeks 1 and 2:** • Zygote divides mitotically many times to produce a multicellular blastocyst that implants in the uterus.	**Weeks 3 through 8:** • Blastocyst grows, folds, and forms rudimentary organ systems. • Conceptus is now called an embryo.	**Weeks 9 through 38 (until birth):** • Conceptus is now called a fetus. • It grows larger and develops until its organ systems can function without assistance from the mother.

Figure 27.1 Summary of developmental changes during the prenatal period. Note that these illustrations are not to scale.

- **Pre-embryonic.** The **pre-embryonic period** lasts for the first 2 weeks after fertilization. Recall from the reproductive system chapter that the female gamete is the **secondary oocyte** (see Chapter 26). When a secondary oocyte is fertilized by a **sperm cell,** the male gamete, the secondary oocyte undergoes meiosis II to become an ovum. The male and female nuclei then fuse to become a single-celled *zygote.* During the pre-embryonic stage, the zygote divides mitotically many times to produce a multicellular structure, the **blastocyst** (BLAS-toh-sist), which *implants* in, or attaches to, the endometrium of the uterus.
- **Embryonic.** The **embryonic period** extends from week 3 through week 8 of gestation. During this stage, the conceptus grows, folds, and forms rudimentary organ systems. As you might expect from the name of the period, it is now called an **embryo** (EM-bree-oh).
- **Fetal.** The final stage, the **fetal period,** lasts from week 9 through week 38 of gestation, or until birth. In this stage, the conceptus is now called a **fetus** (FEE-tus); it grows larger and continues to develop until its organ systems can function without assistance from the mother.

Several intermediate stages, particularly in the embryonic period, are responsible for the progressive tissue and organ development

that occurs during the prenatal period. We discuss these intermediate stages in more detail in the next three modules. After birth, the fetus becomes a **newborn,** and the postnatal period of life begins.

If any part of this process is interrupted or defective, reproduction will not be successful. **Infertility** is the inability to produce a pregnancy after 1 year of unprotected sexual intercourse. Individuals who are infertile can benefit from techniques that increase their chances of having children (see *A&P in the Real World: Assisted Reproductive Technology* on p. 1069).

The Postnatal Period

In the postnatal period, rapid developmental changes continue to take place, particularly in the first few months, although they are not as dramatic as in the prenatal period. This period can be broadly divided into five stages: the *neonatal period* (nee-oh-NAYT-uhl), *infancy, childhood, adolescence,* and *adulthood* (*maturity*). The neonatal period extends from birth to 1 month of age and will be the focus of our study of the postnatal period in Module 27.6. Infancy picks up after the neonatal period, lasting from 1 month through 2 years of age, and childhood lasts until the period of sexual maturation, or puberty, begins. At this point, adolescence begins, and finishes when the individual is capable of sexual reproduction.

Adulthood extends from the end of adolescence until death, and is the stage at which development ends and the degeneration of tissues and organs, called **senescence** (seh-NESS-ens), begins. Eventually senescence will lead to death, even without a specific disease state.

Quick Check

☐ 1. What are the three major periods of prenatal development?

☐ 2. When does the postnatal period of life begin?

Apply What You Learned

☐ 1. A typical pregnancy lasts 38 weeks from fertilization. Why do you think the first 3 months of a pregnancy are considered the most important developmentally?

☐ 2. Why would you expect a baby who is born 10 weeks early to have significant health problems?

See answers in Appendix A.

MODULE **27.2**

Pre-embryonic Period: Fertilization through Implantation (Weeks 1 and 2)

Learning Outcomes

1. Describe the process of fertilization.

2. Describe the events of sperm capacitation, acrosomal reaction, sperm penetration, cortical reaction, and fusion of pronuclei.

3. Explain the formation and function of the extraembryonic membranes.

The pre-embryonic period begins with fertilization of the secondary oocyte by a sperm cell, and continues until the blastocyst moves into the uterus and *implants* in the endometrium of the uterine wall. During this period, some cells of the blastocyst begin to form *extraembryonic membranes* (ek-strah-em′-bree-AHN-ik; "outside the embryo"), which help support the developing conceptus.

The upcoming sections examine fertilization, the formation of the blastocyst, and implantation. We then look at the structure and function of the extraembryonic membranes that develop during these processes.

Fertilization

◀◀ FLASHBACK

1. What are gametes? (p. 1023)

2. How do spermatogenesis and oogenesis differ? (pp. 1045–1046)

3. What are the components of semen? (p. 1032)

As discussed in the reproductive system chapter, the process of **meiosis** (my-OH-sis) produces **gametes,** or sex cells (see Figure 26.1). Meiosis results in genetically unique *haploid* cells (*n*), with half the number of chromosomes normally found in body cells. Each gamete contains 23 chromosomes; when gametes combine, they restore the correct *diploid* number (2*n*) of 46 to future cells of the new offspring.

Sperm cells are ejaculated in **semen** (SEE-men; *semen* = "seed"), which is a mixture of sperm and fluid secretions from the male reproductive glands. The fluid secretions in semen contain chemicals that make the female reproductive tract more hospitable to sperm, which gives more sperm an opportunity to compete in fertilizing the secondary oocyte.

Each ejaculate contains 40–750 million sperm cells. However, only a small fraction of this number ever reaches the ovulated secondary oocyte. Many sperm are destroyed by acidic vaginal secretions. Still other sperm fail to penetrate the mucus of the cervical canal or, if they do, are destroyed by immune cells in the uterus. Sperm motility is also influenced by a woman's stage in her reproductive cycle. When a woman is in the secretory phase of her reproductive cycle, the level of estrogen is high and the cervical mucus is thin and watery, which promotes sperm motility. However, if the female is in the proliferative phase of her reproductive cycle, the progesterone level is high and the cervical mucus is viscous, which inhibits sperm motility. Also, many sperm survive these obstacles only to go up the wrong uterine tube (the one that currently has no ovulated oocyte).

In fact, no oocyte may be available at all when the sperm enter the female reproductive tract. An ovulated secondary oocyte is viable for only 24 hours after it is released from the ovary. Because oocytes travel slowly through the uterine tubes, they are usually in the distal end of the uterine tube, called the **ampulla** (am-POOL-uh), when they are fertilized. Pregnancy most frequently occurs when sperm enter the female reproductive tract during a 3-day window—from 2 days before ovulation to 1 day after—but sperm can remain viable in the female's reproductive tract for up to 5 days.

Fertilization is the fusion of a sperm cell and secondary oocyte to form a zygote. Remember that a secondary oocyte is released during ovulation because meiosis II has been suspended in metaphase (see Figure 26.14). The secondary oocyte divides to form an *ovum* only if it is fertilized.

Both gametes contribute equally to the chromosome number and the genetic makeup of the new offspring, but other functional roles of the gametes are unique. For example, a sperm cell must travel a long distance to reach the oocyte, so it needs to be small and motile. In contrast, a secondary oocyte is relatively large because it must provide the cellular machinery necessary for the first week of development. Cilia lining the uterine tubes and gentle peristaltic contractions propel the oocyte toward the uterus, so the oocyte is not motile. These structural differences between male and female gametes demonstrate the Structure-Function Core Principle (p. 28).

CORE PRINCIPLE
Structure-Function

Several events must happen before fertilization can occur between the sperm cell and the oocyte (**Figure 27.2**):

(1) **Sperm undergo capacitation as they migrate to the oocyte.** A sperm cell moves by a whiplike motion of its *flagellum,* or "tail." Prostaglandins in semen are also thought to stimulate uterine contractions, which assist with moving the sperm farther into the female reproductive tract. Although female orgasm is not required for fertilization, it also results in uterine contractions and, again, may help in moving sperm. All these processes together are called **sperm migration.**

Assisted Reproductive Technology

According to a report compiled in 2014 by the Centers for Disease Control, one in eight couples in the United States had trouble getting pregnant or sustaining a pregnancy. In males, the most common cause of infertility is a low sperm count. Infertility in females may be caused by hormonal problems associated with ovulation, diseases of the ovary, obstruction of the uterine tubes, or conditions in which the uterus is not able to receive or maintain a fertilized ovum. A common treatment for infertility involves direct manipulation of the reproductive process using **assisted reproductive technology,** or **ART.**

One type of ART is *intrauterine insemination* (IUI), also called *artificial insemination,* in which sperm that have been washed and concentrated are placed directly into the woman's uterus at the time of ovulation. Sperm then swim up the uterine tube and fertilize the oocyte. A similar procedure is *gamete intrafallopian transfer* (GIFT), in which sperm and oocytes are placed into the woman's uterine tube. Fertilization and all developmental steps must occur naturally in both procedures.

For other procedures, oocytes are fertilized outside the female's reproductive tract. For example, in *zygote intrafallopian transfer* (ZIFT), a new zygote is immediately placed into the woman's uterine tube. The zygote will then need to develop into a blastocyst and implant. With *in vitro fertilization* (IVF), oocytes are incubated with sperm cells to allow fertilization and most or all of the steps of the pre-embryonic period to complete. At this point, the blastocyst or embryo is transferred into the woman's uterus, where it must implant naturally. In some cases, the oocytes are directly injected with a sperm cell or a sperm cell nucleus to fertilize the oocyte.

Before it can fertilize an oocyte, a sperm must undergo **capacitation** (kuh-pah'-sih-TAY-shun), a series of functional changes that make it fully motile and modify its plasma membrane so it can fuse with the plasma membrane of the oocyte. Capacitation occurs as components of the seminal fluid weaken the sperm's plasma membrane; fluids in the female reproductive tract may contribute to this process, as well. In addition, the oocyte itself attracts the sperm to its location, using chemical signals (a process called *chemotaxis*). Researchers have identified a receptor on human sperm cells that responds to a chemical secreted by oocytes, which appears to be progesterone.

(2) **The acrosomal reaction releases enzymes from the head of the sperm.** An ovulated secondary oocyte is surrounded by a layer of *granulosa* cells called the **corona radiata** (kuh-ROH-nuh ray-dee-AH-tuh). To reach the secondary oocyte, sperm must penetrate this cell layer. This is accomplished through a combination of sperm movements and *hyaluronidase*, an enzyme released from the membrane-bound sac on the head of the sperm called the **acrosome** (AK-roh-sohm). This enzyme catalyzes reactions that break down the protein layer that serves as a barrier between the granulosa cells.

Immediately deep to the oocyte's corona radiata is the **zona pellucida** (ZOH-nuh peh-LOO-sih-duh), a layer of extracellular matrix secreted by the oocyte that contains **sperm-binding receptors.** These receptors are an important site for communication between cells, an example of the Cell-Cell Communication Core Principle (p. 28). When a sperm cell binds

CORE PRINCIPLE
Cell-Cell Communication

to a sperm-binding receptor in the zona pellucida, calcium ion channels in the sperm's plasma membrane open, allowing calcium ions to flow into the cell. The resulting increase in the intracellular calcium ion level triggers the **acrosomal reaction,** which releases acrosomal enzymes, including additional hyaluronidase and *acrosin*. These enzymes catalyze reactions that digest holes in the zona pellucida. Numerous sperm cells are needed to release enough hyaluronidase and acrosin to break through the corona radiata and zona pellucida, so a single sperm cell is not capable of fertilizing an oocyte alone. This is why males with sperm counts below 20 million have an impaired ability to produce offspring.

(3) **The sperm binds to the plasma membrane of the oocyte.** When the first sperm cell penetrates the zona pellucida and binds to the oocyte's plasma membrane, changes in the oocyte and zona pellucida occur so that no more sperm can enter.

(4) **Sperm entry stimulates the cortical reaction, which destroys sperm-binding receptors.** Specialized secretory vesicles called *cortical granules* that are present in the oocyte contain enzymes that catalyze reactions to

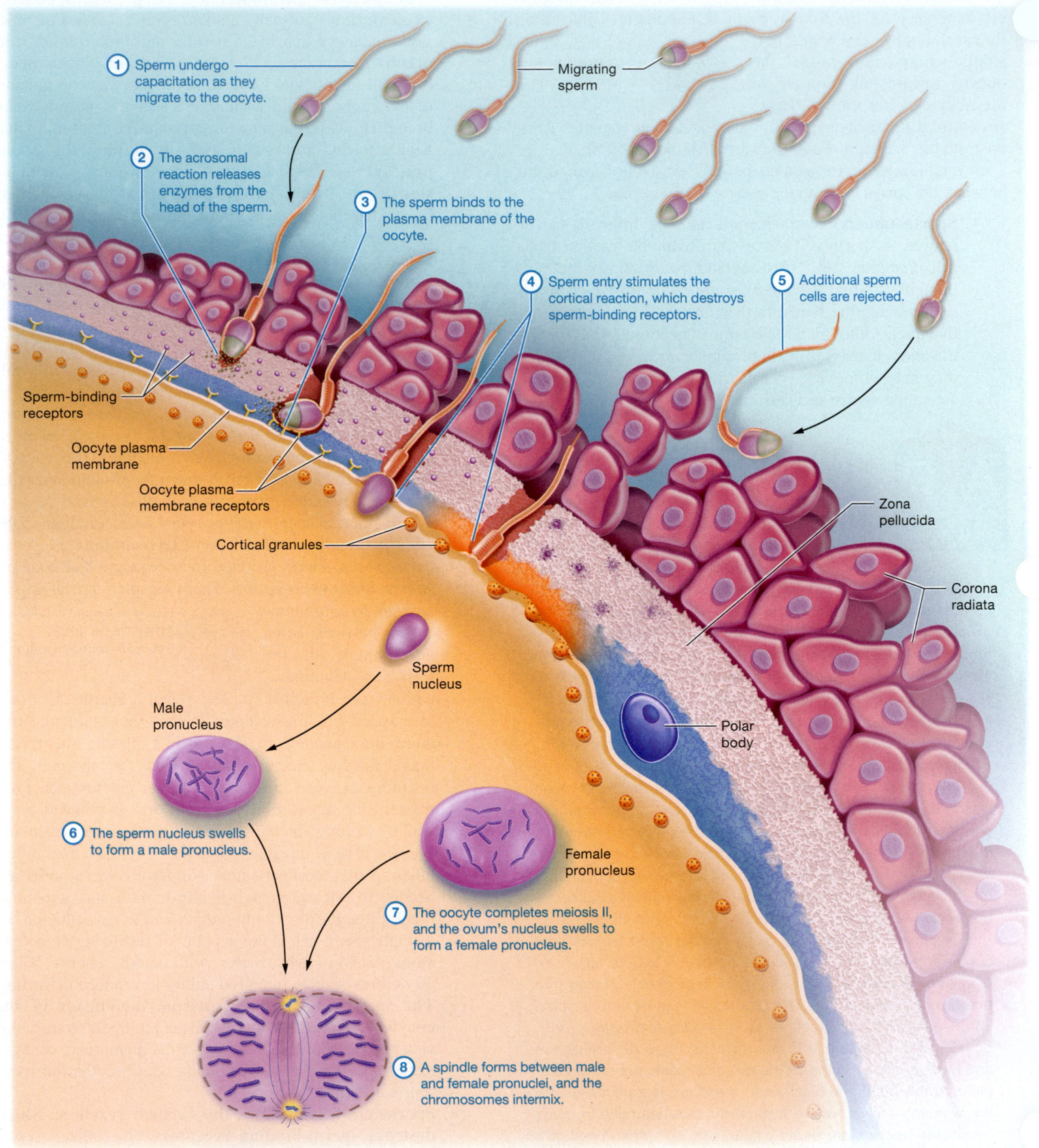

Figure 27.2 **Events leading to fertilization.**

destroy the sperm-binding receptors. Sperm entry triggers the release of the enzymes from the cortical granules, a process termed the **cortical reaction.**

⑤ **Additional sperm cells are rejected.** The cortical reaction prevents additional sperm cells from binding. On rare occasions, more than one sperm cell enters the secondary oocyte, a phenomenon known as **polyspermy.** In this case development cannot proceed because the fertilized oocyte is triploid ($3n$)—has three sets of chromosomes—which is fatal to the zygote.

⑥ **The sperm nucleus swells to form a male pronucleus.** The plasma membranes of the sperm and oocyte bind and fuse together; the cytoplasmic contents of the sperm cell, including the nucleus, are then released into the cytoplasm of the oocyte. The sperm nucleus swells to form the **male pronucleus.**

⑦ **The oocyte completes meiosis II, and the ovum's nucleus swells to form a female pronucleus.** At the same time, the fusion of the plasma membranes signals the secondary oocyte to complete meiosis II and become an ovum. The released polar body is expelled but sometimes remains visible at the edge of the ovum. The ovum's nucleus also swells and becomes the **female pronucleus.**

⑧ **A spindle forms between male and female pronuclei, and the chromosomes intermix.** Spindle fibers form between the male and female pronuclei, and the pronuclear membranes break down, allowing the chromosomes to mingle.

Quick Check

☐ 1. What is capacitation, and where does this process occur?

☐ 2. What are the steps of fertilization?

Cleavage and Blastocyst Formation

« FLASHBACK

1. What are the areas of the uterine tube? (p. 1041)

2. What happens to the oocyte during ovulation? (p. 1048)

The next steps of development are shown in **Figure 27.3**. When the secondary oocyte is fertilized, chromosomes from the male and female pronuclei combine in the process of *amphimixis*, and the diploid zygote is formed. Approximately 30 hours after fertilization occurs, the zygote begins **cleavage**—a series of rapid mitotic divisions that produce small, genetically identical cells called **blastomeres** (BLAS-toh-meerz). The divisions occur too quickly for cells to grow; for this reason, the cell number increases, but the cell size becomes progressively smaller. Each subsequent division takes less time than the previous one, so that by the second day there are 4 cells and by the end of the third day there are

16 cells. At this stage, the cells start to differentiate and will be the building blocks of all future tissues of the developing conceptus. At this point, the conceptus is known as a **morula** (MOHR-yoo-luh; "little mulberry") and remains covered by the zona pellucida.

Notice in the figure that while these mitotic divisions are taking place, the conceptus has moved through the uterine tube and into the uterine cavity. Around day 4, a glycogen-rich secretion from endometrial glands in the lumen of the uterus, called *uterine milk,* nourishes the conceptus. The conceptus, now called a blastocyst, hatches from the zona pellucida and the blastomeres reorganize to surround an internal fluid-filled cavity.

By the time the blastocyst is ready for implantation, it has two distinct cell populations: an outer layer of large, flattened cells, called **trophoblast cells** (TROHF-oh-blast; *troph-* = "nourish"), that surround the fluid-filled cavity; and an inner cluster of rounded cells, the **inner cell mass,** or *embryoblasts.* The trophoblast cells will participate in forming the *placenta,* the temporary organ that provides nutrients and oxygen to the conceptus. They also suppress the mother's immune system, preventing it from attacking the conceptus. The inner cell mass will form the *embryo proper*—the developing body.

Sometimes during these early cell divisions, within the first 8 days after fertilization, the cells separate into two groups and develop into two individuals, called **monozygotic** (mahn′-oh-zy-GAWT-ik; *mono-* = "single"), or **identical, twins.** Monozygotic twins are genetically identical, or practically so, and have a nearly identical appearance. They usually share the same placenta. Very rarely, identical triplets or even quadruplets result from separations during these early cell divisions. A more common type of twinning results from the ovulation of two secondary oocytes at the same time, which are then fertilized by two separate sperm cells. These individuals are **dizygotic** (dy-zy-GAWT-ik; *di-* = "double"), or **fraternal, twins,** and may be different sexes and as dissimilar in appearance as any siblings. Multiple ovulations can result in triplets, quadruplets, or even greater numbers of offspring. However, this is extremely rare, except with the use of assisted reproductive technology that artificially stimulates the release of multiple oocytes.

Quick Check

☐ 3. What is the function of cleavage of the zygote?

☐ 4. How does a morula differ from a blastocyst?

Implantation

« FLASHBACK

1. What is a corpus luteum, and what does it secrete? (p. 1048)

2. Which hormones are responsible for regulating the uterine cycle? (pp. 1048–1050)

Approximately 4–7 days after fertilization, the blastocyst begins to attach to the endometrium of the uterus in the process of *implantation.* The trophoblast invades the **stratum functionalis**

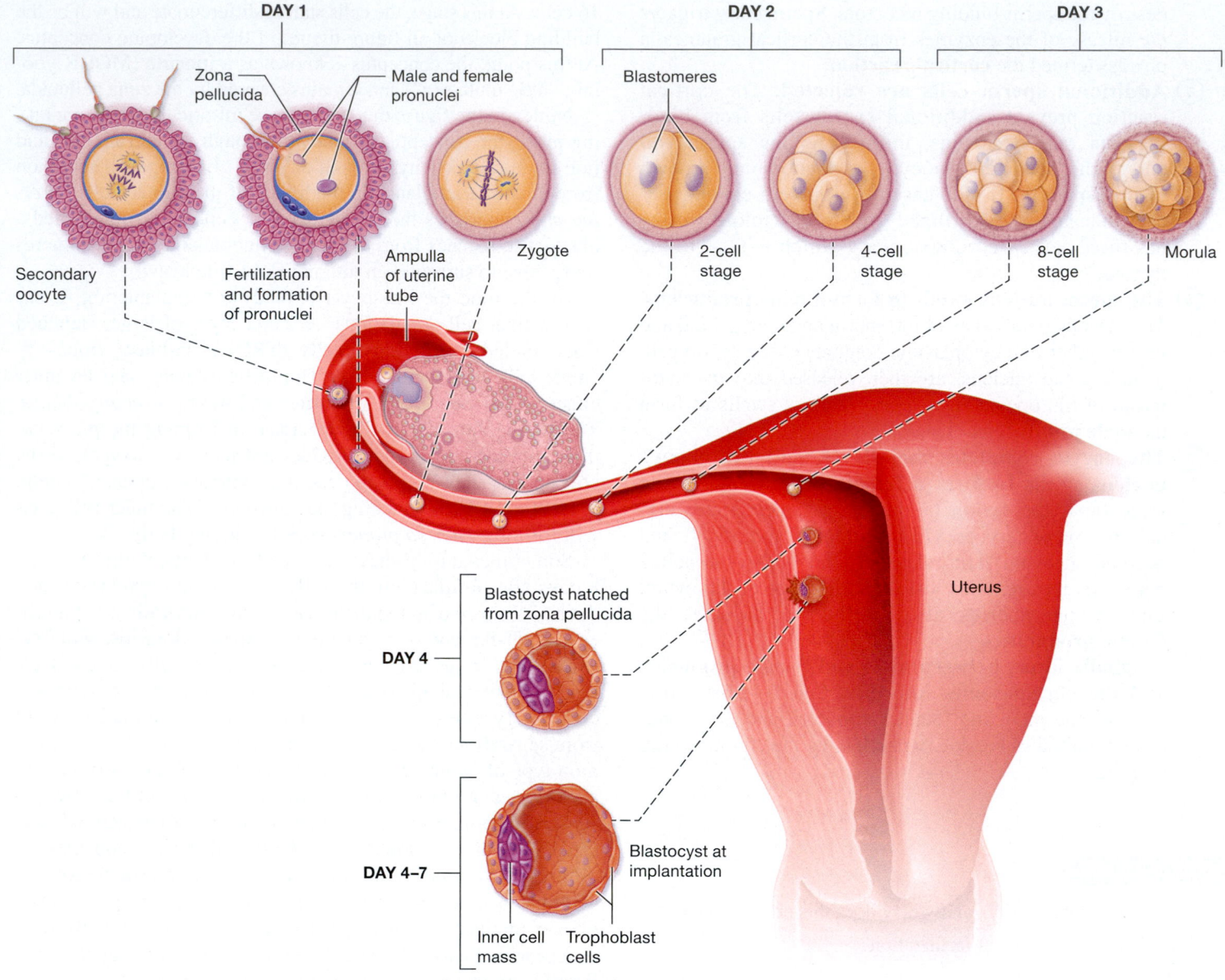

Figure 27.3 Steps of the pre-embryonic period: fertilization through implantation.

(funk-shun-AL-iss) layer of the endometrium by secreting digestive enzymes that catalyze reactions to degrade the endometrial lining (**Figure 27.4**).

Trophoblast cells divide into two layers: (1) the inner **cytotrophoblast** (sy′-toh-TROHF-oh-blast), or **cellular trophoblast,** where cells remain intact; and (2) the outer **syncytiotrophoblast** (sin-SISH-ee-oh-trohf′-oh-blast; *syncyt-* = "cells together"), or *syncytial trophoblast* (sin-SISH-ee-al), where the cells' plasma membranes disappear and their cytoplasmic contents merge. This makes the syncytiotrophoblast functionally a large, single, multinucleated cell. Almost immediately after formation, the syncytiotrophoblast begins to secrete enzymes to digest uterine cells. Interestingly, it is nonantigenic, which means that it does not trigger an immune reaction from the maternal tissue it is invading.

The syncytiotrophoblast progressively invades the uterine lining over the next several days. It reaches maternal blood vessels on or before day 12, at which point it surrounds them and digests the walls of the vessels. This causes the maternal blood to pool and fill cavities called **lacunae,** or *intervillous spaces.* The syncytiotrophoblast also makes contact with nutrients in the uterine glands. By day 16, the blastocyst moves farther into the lining and becomes covered by regrowth of the maternal epithelial cells on the surface. When the blastocyst is embedded in the uterine wall and completely covered by uterine epithelial cells, it has fully implanted.

The syncytiotrophoblast is also responsible for secreting the hormone **human chorionic gonadotropin (hCG)** (koh-ree-AHN-ik goh-nad′-oh-TROHP-in), which stimulates the corpus

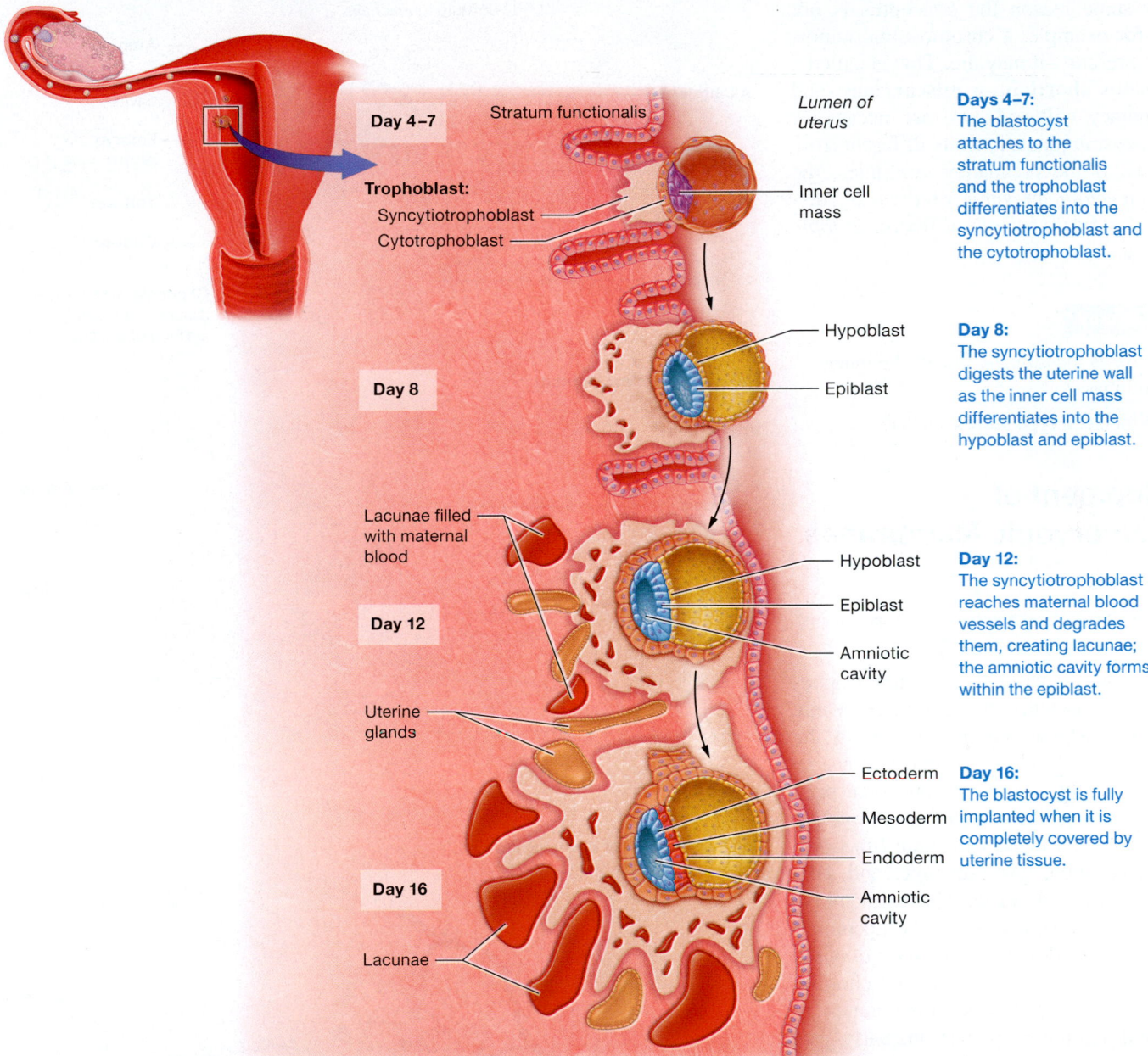

Day 4–7

Stratum functionalis

Lumen of
uterus

Trophoblast:

Syncytiotrophoblast

Cytotrophoblast

Inner cell
mass

Days 4–7:
The blastocyst
attaches to the
stratum functionalis
and the trophoblast
differentiates into the
syncytiotrophoblast and
the cytotrophoblast.

Day 8

Hypoblast

Epiblast

Day 8:
The syncytiotrophoblast
digests the uterine wall
as the inner cell mass
differentiates into the
hypoblast and epiblast.

Lacunae filled
with maternal
blood

Day 12

Uterine
glands

Hypoblast

Epiblast

Amniotic
cavity

Day 12:
The syncytiotrophoblast
reaches maternal blood
vessels and degrades
them, creating lacunae;
the amniotic cavity forms
within the epiblast.

Ectoderm

Mesoderm

Endoderm

Amniotic
cavity

Day 16

Lacunae

Day 16:
The blastocyst is fully
implanted when it is
completely covered by
uterine tissue.

Figure 27.4 Implantation of the blastocyst in the uterine lining.

luteum in the ovary (a structure formed after an ovarian follicle ruptures) to secrete the hormones *estrogen* and *progesterone* and helps promote placental development. Progesterone suppresses menstruation by maintaining the endometrium of the uterus and prolonging the secretory phase of the uterine cycle. Some hCG is secreted into the urine; because hCG is produced only during pregnancy, it is the basis for pregnancy tests. The hCG level rises in the mother's blood until the end of the second month of development, when a membrane called the *chorion* (KOHR-ee-ahn) takes over the secretion of estrogen and progesterone. At that point, hCG is no longer needed, and its level declines and remains low for the remainder of the pregnancy.

Notice in Figure 27.4 that during the same time period, the *inner cell mass,* or developing offspring, separates from the trophoblast and differentiates into two layers: the superior **epiblast** and the inferior **hypoblast.** Together, these layers form a flat bilaminar embryonic disc (*bilaminar* = "having two layers"). A small cavity appears within the epiblast and will enlarge to become the *amniotic cavity* (am-nee-AWT-ik), which will surround the conceptus and fill with fluid. The bilaminar embryonic disc will eventually become the three primary germ layers that differentiate to produce all tissues in the body: *ectoderm, mesoderm,* and *endoderm*. These three layers are the blueprint for future organ development and are discussed in Module 27.3.

If for some reason the conceptus is not viable—for example, a chromosomal abnormality is present—it may die. This is called a **spontaneous abortion,** or **miscarriage,** and the pregnancy will end. On rare occasions, the blastocyst implants in a site different from the uterus. If the pregnancy continues, the result is an *ectopic pregnancy* (*ectop-* = "displaced"; see *A&P in the Real World: Ectopic Pregnancy*).

Quick Check

☐ 5. What is the ultimate fate of the inner cell mass?

☐ 6. Which cell type secretes hCG?

Development of Extraembryonic Membranes

❮❮ FLASHBACK

1. What are the structures of the urinary system? (pp. 947–948)

The bilaminar embryonic disc and trophoblast together produce the **extraembryonic membranes,** which first appear during the second week of development but continue to develop during the embryonic and fetal periods. Collectively, these membranes protect the conceptus and assist with vital functions, including nutrition, gas exchange, and storage and removal of waste. The extraembryonic membranes we discuss in the upcoming subsections include the *yolk sac, amnion, allantois,* and the already mentioned chorion. **Figure 27.5** illustrates these membranes in week 3 of development (part a), and early (part b) and late (part c) in week 4.

Yolk Sac

The **yolk sac** arises from hypoblast cells and is the first extraembryonic membrane to develop. In some animals such as birds, the yolk sac is the main nutritional source for the developing offspring, but in humans, cells from the yolk sac form part of the digestive tract and are the source of the first blood cells and blood vessels. In addition, they produce the first *germ cells,* the precursors of the gametes, that will eventually migrate into the developing ovaries and testes.

Amnion

The **amnion** (AM-nee-ahn) is a transparent membrane that develops from the

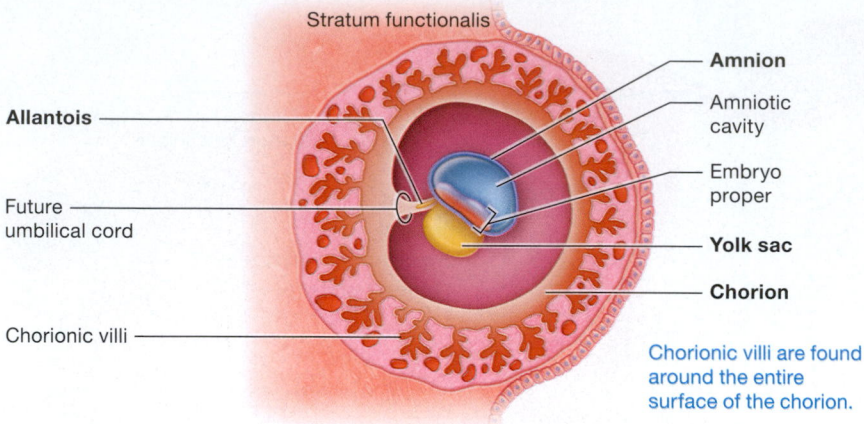

Chorionic villi are found around the entire surface of the chorion.

(a) Week 3 of development

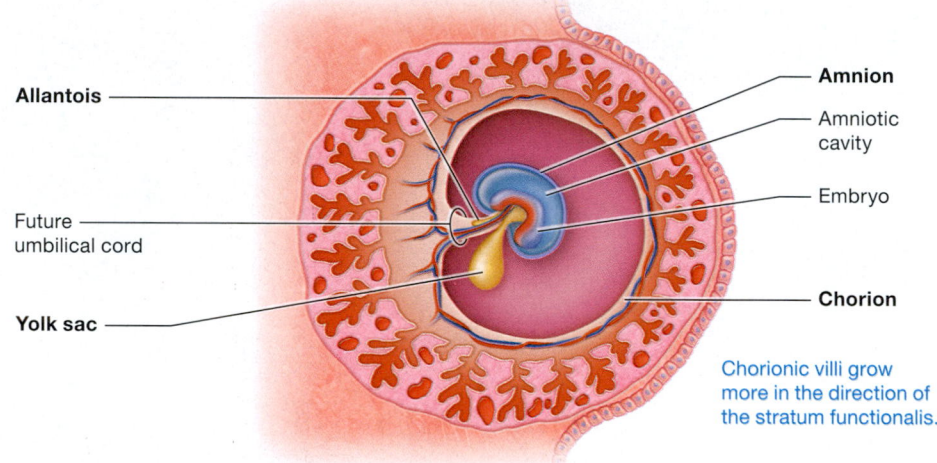

Chorionic villi grow more in the direction of the stratum functionalis.

(b) Early week 4 of development

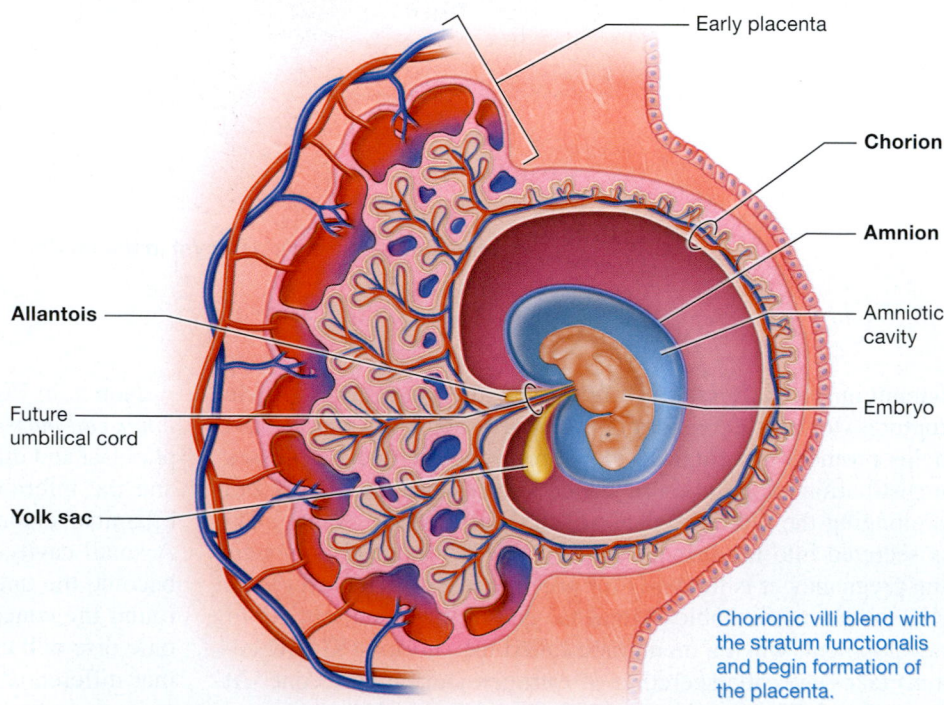

Chorionic villi blend with the stratum functionalis and begin formation of the placenta.

(c) Late in week 4 of development

Figure 27.5 Formation of the extraembryonic membranes.

Ectopic Pregnancy

In an *ectopic pregnancy*, a conceptus implants and grows in any location other than the endometrium of the uterus. In the United States, 1–2% of all pregnancies are ectopic. Almost all these are "tubal pregnancies," meaning that they are in the uterine tubes; however, they can occur in other locations, including the abdominal cavity, ovary, or cervix. An ectopic pregnancy presents a large risk to the mother, as only the uterus is able to expand and sustain a growing conceptus. If an ectopic pregnancy does not terminate on its own in a spontaneous abortion, medical intervention is necessary to remove the conceptus. Otherwise, the region in which it implants can eventually rupture and result in serious bleeding and possibly even death of the mother.

Table 27.1 Summary of the Four Extraembryonic Membranes

Extraembryonic Membrane	Primary Functions
Yolk sac	• Contributes to the formation of the digestive tract • Source of the first blood cells, blood vessels, and germ cells
Amnion and amniotic cavity	Amnion: • Surrounds the conceptus • Produces amniotic fluid Amniotic cavity contains amniotic fluid, which: • Protects the conceptus from trauma • Helps maintain a constant temperature • Allows symmetrical muscle development • Prevents drying out and adhesion of body parts during growth
Allantois	• Forms the base for the umbilical cord that links the conceptus to the placenta • Ultimately becomes part of the urinary bladder
Chorion	• Encloses all other extraembryonic membranes • Forms the chorionic villi • Forms the main embryonic part of the placenta

epiblast. It grows to completely enclose the conceptus in the fluid-filled **amniotic cavity,** which is penetrated only by the *umbilical cord* (um-BIL-ih-kuhl), whose vessels will connect the growing conceptus to the mother's blood supply (see Figure 27.5c).

The amnion secretes **amniotic fluid** into the cavity. Amniotic fluid protects the conceptus from trauma and helps maintain a constant temperature. It also allows freedom of movement, which is important for symmetrical muscle development, and prevents body parts from drying out and adhering to one another during growth. At first, all the fluid forms by filtering the mother's blood, but as the embryo's kidneys become functional toward the end of the eighth or ninth week, its own urine contributes to amniotic fluid volume.

Allantois

The **allantois** (uh-LAN-toh-iss) is a small outpocketing of the caudal (posterior) wall of the yolk sac that develops during the third week. In the allantois early blood vessel formation begins, and these become the umbilical vessels that will connect the conceptus to the placenta. The tissue around the allantois becomes the connecting stalk, which forms the remainder of the umbilical cord. Inside the conceptus, a portion of the allantois known as the *allantoic duct* runs from the umbilicus to the urinary bladder, which it drains.

Chorion

The **chorion,** the outermost extraembryonic membrane that develops from the cytotrophoblast and syncytiotrophoblast, encloses all the other membranes and the conceptus (see Figure 27.5). As the chorion develops, it forms outgrowths called the **chorionic villi** (singular, *villus*), which are initially found around the entire surface of the chorion. However, as the pregnancy progresses, chorionic villi grow only in the direction of the stratum functionalis layer of the endometrium and do not continue to surround the other membranes and the conceptus.

By late in week 4 of development, the chorionic villi are blended with the stratum functionalis layer and begin to form the principal embryonic part of the placenta.

Table 27.1 summarizes the functions of the four extraembryonic membranes.

Quick Check

☐ 7. What are the functions of amniotic fluid?

☐ 8. What are the main functions of each of the extraembryonic membranes?

Apply What You Learned

☐ 1. What would happen if the mother's immune system recognized the syncytiotrophoblast as foreign tissue? (*Hint:* Remember that the syncytiotrophoblast dissolves the walls of maternal blood vessels.)

☐ 2. What would happen to progesterone and estrogen levels if the trophoblast in an implanted blastocyst failed to secrete hCG?

See answers in Appendix A.

MODULE 27.3
Embryonic Period: Week 3 through Week 8

At the end of the pre-embryonic period, the blastocyst is fully implanted, the inner cell mass has differentiated into the bilaminar embryonic disc, and the extraembryonic membranes have developed. We then enter the **embryonic period,** which begins at the third week of development and continues through the eighth week. During this stage, the conceptus is referred to as an embryo.

The embryonic period starts with a process called **gastrulation** (gas-troo-LAY-shun), which is the rearrangement and migration of the cells of the bilaminar embryonic disc to form the **trilaminar embryonic disc.** Trilaminar means "having three layers," so this is the stage at which the three germ layers develop. The three germ layers will become all the major organ systems in the process of *organogenesis* by the end of week 8, although for the most part the organs are not functional until later in development. Also during the embryonic period, the placenta forms and begins to provide nutrition and oxygen to the embryo and remove wastes. Let's now walk through the processes of gastrulation and organogenesis.

Gastrulation and Formation of Germ Layers

In week 3, we start with a bilaminar embryonic disc containing two cell layers: the epiblast and hypoblast. At this time, a thin groove on the dorsal surface of the epiblast develops called the **primitive streak.** The primitive streak elongates along the cephalic-caudal (head-tail) line of the embryo (**Figure 27.6a**), establishing the head and tail regions, the right and left sides, and the dorsal and ventral (posterior and anterior) surfaces of the embryo.

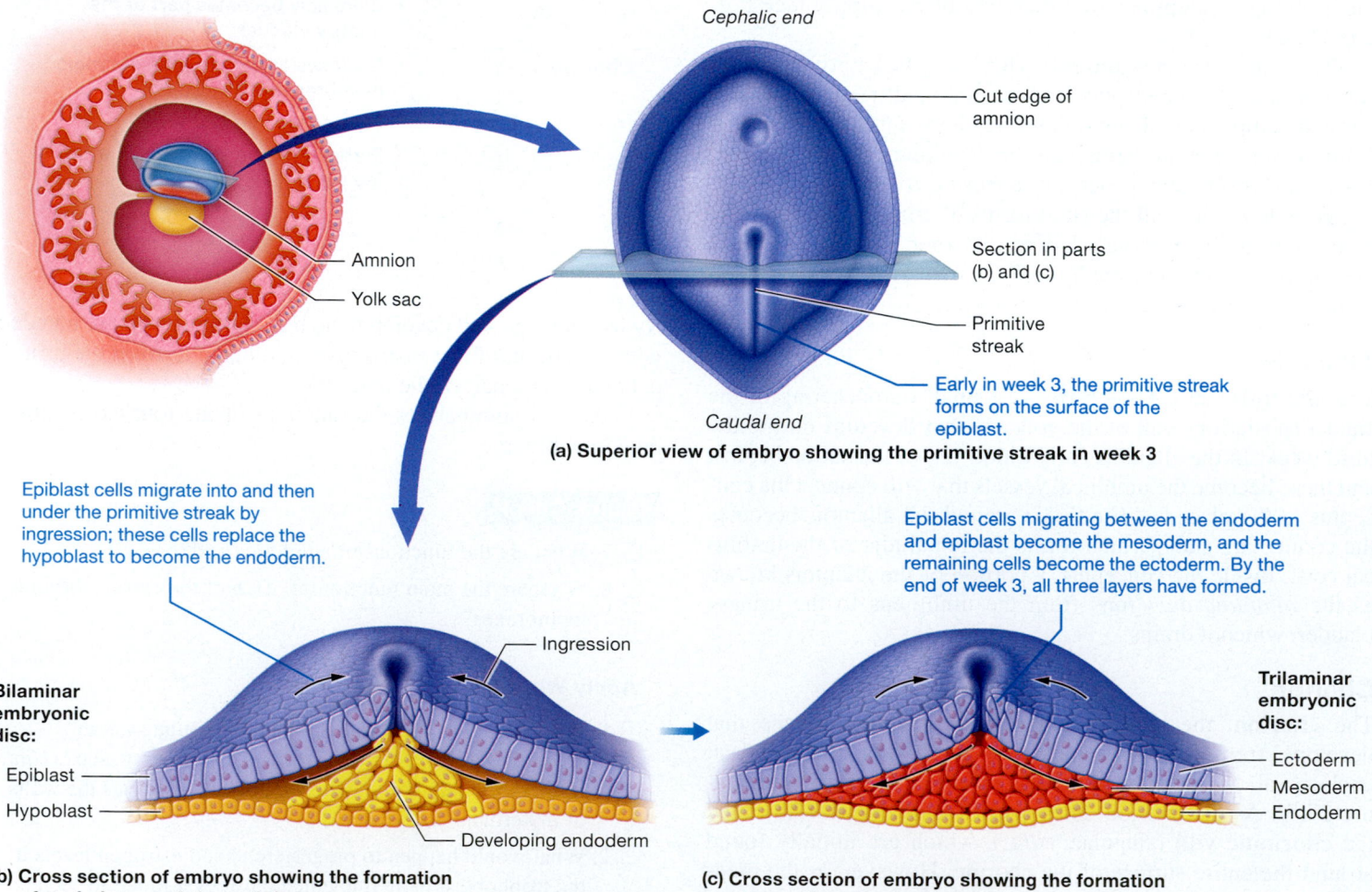

Figure 27.6 Gastrulation and formation of the three primary germ layers.

When the primitive streak has developed, gastrulation begins. It starts as some cells detach from the epiblast layer and move into and then underneath the primitive streak in a process called *ingression*. The first cells that migrate in this way become the inner germ layer, or **endoderm,** which replaces the hypoblast (**Figure 27.6b**). The next cells migrate to a position between the epiblast and endoderm, and this becomes the middle layer, the **mesoderm.** The remaining cells of the epiblast form the outer layer, the **ectoderm** (**Figure 27.6c**). By the end of week 3, all three primary germ layers have formed, making the embryo a trilaminar embryonic disc. At this point, the ectoderm and endoderm are tightly packed and surround the loosely organized mesoderm.

As you can see in Figure 27.6, the trilaminar embryonic disc is fairly flat. This begins to change during the fourth week of development, when two types of folding cause the flat embryonic disc to become cylindrical. The first, **cephalocaudal folding** (sef´-ah-loh-KAW-dul; *cephalo-* = "head," *caud-* = "tail"), occurs in the head and tail regions of the embryo, as its name implies (**Figure 27.7a**). The second type of folding, **transverse,** or **lateral, folding,** occurs when the left and right sides of the embryo curve and fold toward the midline (**Figure 27.7b**). As the sides move closer together, the yolk sac is pinched off, leaving only a small area of its endoderm called the **vitelline duct** (vy-TEL-een).

Cephalocaudal folding creates the future head and buttocks regions of the embryo. Transverse folding creates the future trunk region and almost immediately forms the **primitive gut,** which later becomes the digestive tract inside the abdominal cavity.

☐ 1. What is the primitive streak, and what does it establish?

☐ 2. Which cells form the three germ layers?

☐ 3. What are the two types of embryonic folding?

Organogenesis

« FLASHBACK

1. What are the layers of the skin? (p. 161)

2. What are the divisions of the nervous system? (p. 382)

In the process of **organogenesis** (ohr-gan´-oh-JEN-ih-sis), the three primary germ layers differentiate into organs and organ systems. When organogenesis begins, the primary germ layers have undergone cephalocaudal and transverse folding and are beginning to differentiate, and the embryo is approximately 1.5 mm (0.6 in.) long. When the embryonic period ends at the end of week 8, the embryo will be approximately 25 mm (1.0 in.) long and will have recognizable organ systems. Although not all the organs will be functional, some will begin to work.

Let's consider each of the major structures that will form by examining the differentiation of each primary germ layer. We'll start with the structures formed by the most superficial layer and end with those of the deepest.

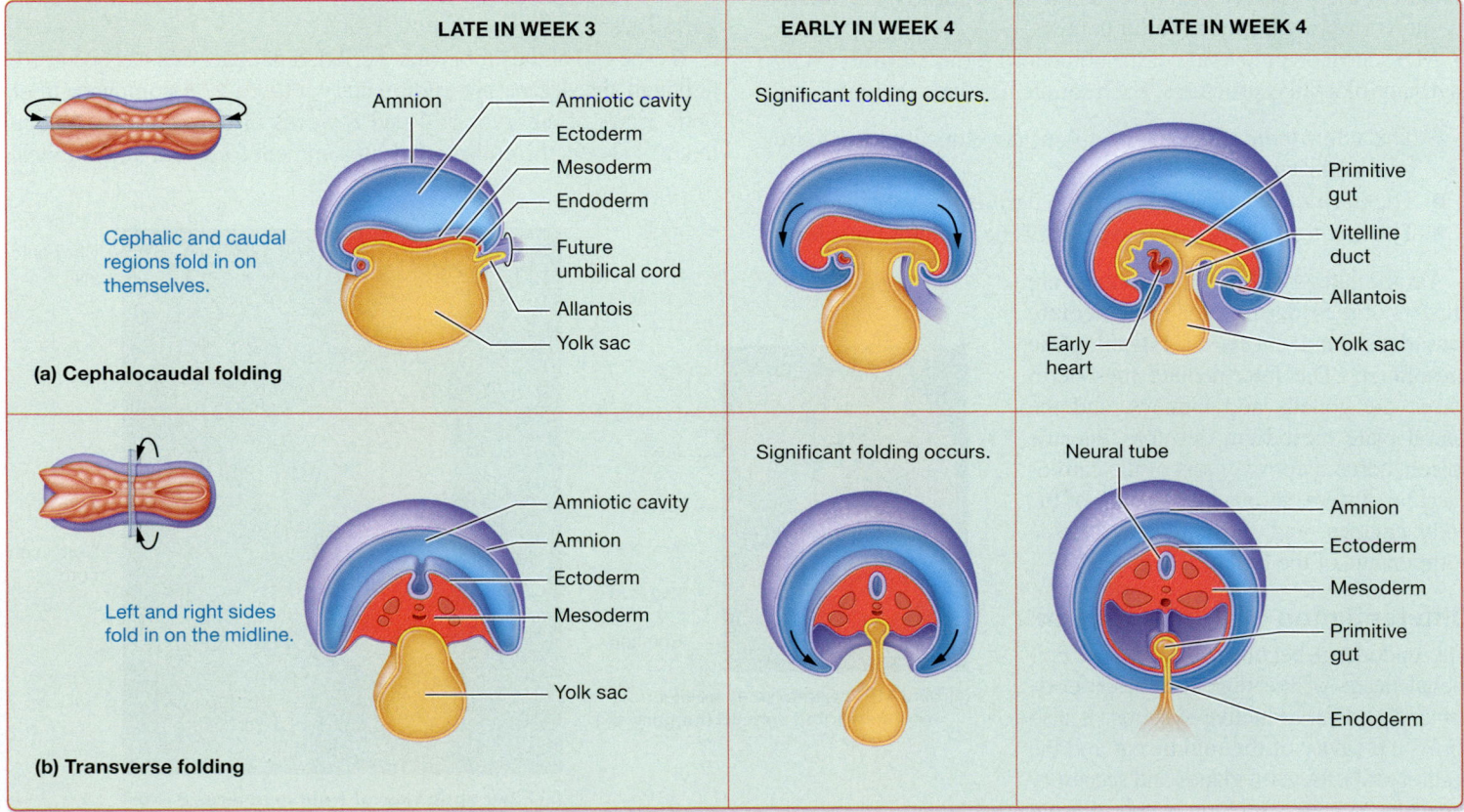

Figure 27.7 Embryonic folding and developmental changes during weeks 3 and 4 of development.

Differentiation of the Ectoderm

After the embryo undergoes cephalocaudal and transverse folding, the ectoderm is on the external surface of the cylindrical embryo. This arrangement is appropriate because the ectoderm will form the epidermis of the skin. The ectoderm is also responsible for forming the majority of the nervous system and sense organs. In fact, the first major event of organogenesis is **neurulation** (NOOR-uh-lay-shun), which produces the brain and spinal cord. Cells of the ectoderm thicken and form the *neural plate,* which folds inward and deepens. The edges fuse to form a *neural tube,* which will pinch off from the ectodermal layer. The anterior end of the neural tube will become the brain, and the remainder will become the spinal cord. By the end of the fourth week, the anterior end of the neural tube will develop into three enlarged areas called the **primary brain vesicles:** the *forebrain, midbrain,* and *hindbrain* (refer back to Figure 12.3).

Between the layers of the ectoderm that form the epidermis and the neural tube are special cells called **neural crest cells.** These cells will migrate throughout the embryo and give rise to various structures, including cranial, spinal, and sympathetic ganglia and associated nerves; pigment cells of the skin; cells of the adrenal medulla; and connective tissue cells.

Differentiation of the Mesoderm

Immediately deep to the primitive streak, some mesodermal cells at the midline form a structure called the **notochord** (NOH-toh-kohrd), which serves to support and organize the embryo around a central axis. Remnants of the notochord will remain in the intervertebral discs as the notochord is surrounded by the vertebrae during development.

Mesoderm on either side of the notochord forms **somites** (SOH-mytz), or blocklike structures. Each somite has three regions:

- The *sclerotome* (SKLER-oh-tohm) develops into the vertebrae and ribs.
- The *dermatome* develops into the dermis layer of the skin.
- The *myotome* develops into most of the skeletal muscles.

On the lateral sides of the somites are clusters of mesoderm called **intermediate mesoderm** and sheets of **lateral plate mesoderm.** The intermediate mesoderm forms the gonads and kidneys, and the lateral plate mesoderm develops into the spleen, adrenal cortex, most of the cardiovascular system, serous membranes of the body cavities, and the connective tissue components of the limbs.

Differentiation of the Endoderm

The endoderm becomes the internal epithelial layers of the digestive, respiratory, urinary, and reproductive systems. It also forms the cavity of the middle ear and the auditory tube. Several glands and accessory digestive organs—including the thyroid gland; parathyroid glands; thymus; parts of

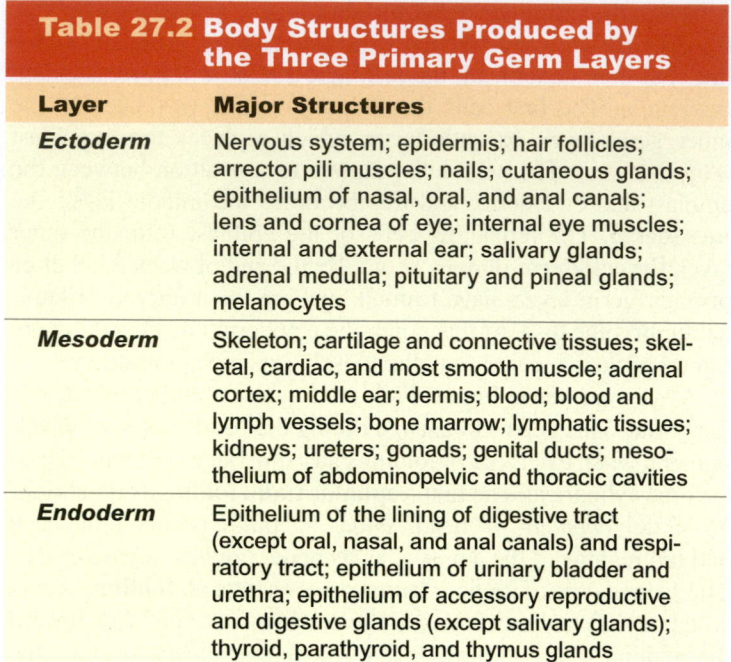

Table 27.2 Body Structures Produced by the Three Primary Germ Layers

Layer	Major Structures
Ectoderm	Nervous system; epidermis; hair follicles; arrector pili muscles; nails; cutaneous glands; epithelium of nasal, oral, and anal canals; lens and cornea of eye; internal eye muscles; internal and external ear; salivary glands; adrenal medulla; pituitary and pineal glands; melanocytes
Mesoderm	Skeleton; cartilage and connective tissues; skeletal, cardiac, and most smooth muscle; adrenal cortex; middle ear; dermis; blood; blood and lymph vessels; bone marrow; lymphatic tissues; kidneys; ureters; gonads; genital ducts; mesothelium of abdominopelvic and thoracic cavities
Endoderm	Epithelium of the lining of digestive tract (except oral, nasal, and anal canals) and respiratory tract; epithelium of urinary bladder and urethra; epithelium of accessory reproductive and digestive glands (except salivary glands); thyroid, parathyroid, and thymus glands

the palatine tonsils; and the majority of the liver, gallbladder, and pancreas—also form from the endoderm.

Summary of Organogenesis

All body structures derive from the three primary germ layers. **Table 27.2** summarizes additional structures produced by each germ layer.

By the end of the embryonic period, organogenesis has occurred, although the organs are rudimentary. **Figure 27.8** compares fiber optic views of embryos at 4 and 8 weeks of development. Notice that at 4 weeks the embryo has a prominent forebrain and somites,

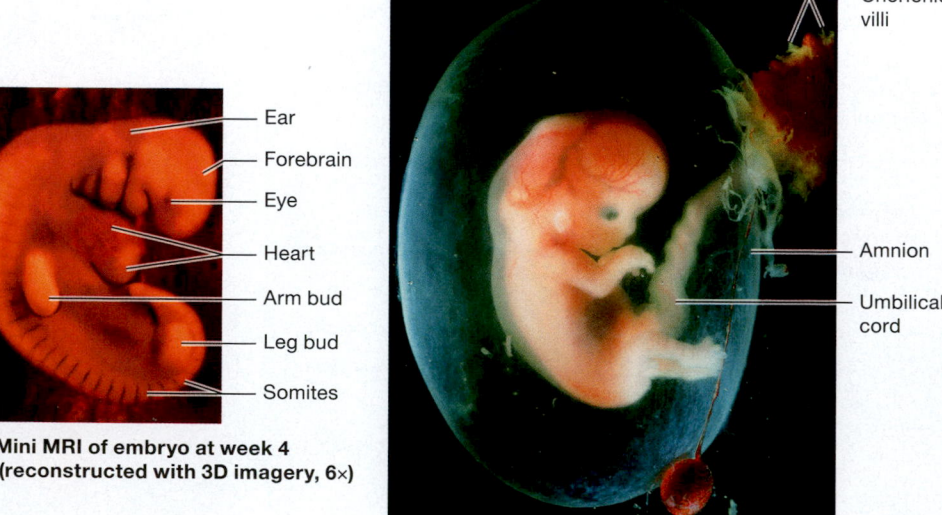

Ear
Forebrain
Eye
Heart
Arm bud
Leg bud
Somites

(a) Mini MRI of embryo at week 4 (reconstructed with 3D imagery, 6×)

Chorionic villi
Amnion
Umbilical cord

(b) Fiber optic view of embryo at week 8 (1.5×)

Figure 27.8 Development during the embryonic period.

while at 8 weeks the embryo looks essentially human, albeit with different body proportions.

During this period (weeks 3–8), the embryo is especially sensitive to **teratogens** (teh-RAT-oh-jenz; *terato-* = "monster"), or substances that can cause birth defects or even death. Some infectious agents that produce maternal illness, particularly rubella, are teratogenic. Other teratogens include alcohol, nicotine, and many over-the-counter and prescription drugs, such as aspirin, anticoagulants, and some antibiotics. For example, the acne medication Accutane® has resulted in life-threatening birth defects. Possibly the most notorious teratogen is the sedative thalidomide, which was prescribed to thousands of pregnant women in the 1960s to alleviate *morning sickness* (nausea), but resulted in many infants with severely deformed, short, flipper-like legs and arms.

Quick Check

☐ 4. What is the first major event that occurs during organogenesis?

☐ 5. List two structures formed by each of the primary germ layers.

Apply What You Learned

☐ 1. Why does exposure to teratogens during organogenesis increase the risk of birth defects?

☐ 2. Why do you think the first major event in organogenesis is neurulation? (*Hint:* Think about the function of the brain and spinal cord.)

See answers in Appendix A.

MODULE **27.4**

Fetal Period: Week 9 until Birth (about Week 38)

Learning Outcomes

1. Describe the formation and function of the placenta.
2. Identify the major events of fetal development.
3. Describe the specific structures unique to the fetal circulation.

The **fetal period** of development extends from the beginning of week 9 until birth, which usually occurs near week 38. By the 12th week, the placenta is fully functional, providing nutrition to the fetus for the rest of the gestation period. Let's look at the development of both the placenta and the fetus during this period, and finish with a big picture look at prenatal development.

Placentation

⟪ FLASHBACK

1. What is diffusion? (p. 75)
2. How does hemoglobin transport oxygen? (p. 832)

During the pre-embryonic period, the developing conceptus receives nutrients from two sources: uterine milk secreted from endometrial glands, and digested endometrial cells from the uterine wall. The conceptus will continue to receive nutrition from digested endometrial cells until the placenta is fully formed at about week 12, during the fetal period.

The **placenta,** a temporary organ that is shed after the infant is born, is the site of exchange of oxygen, nutrients, and wastes between the mother and fetus. In addition, the placenta produces hormones to support the pregnancy. **Placentation** (plah-sen-TAY-shun) is the process of forming the disc-shaped placenta, which attaches to the uterine wall and to the embryo or fetus through the umbilical cord (**Figure 27.9a**). Placentation begins during implantation, but most placental growth occurs in the fetal period. If the placenta grows in the wrong place, it may block the cervix of the mother's uterus, resulting in a potentially serious condition during birth (see *A&P in the Real World: Placenta Previa*).

The **umbilical cord** connects the center of the placenta to the fetus' *umbilicus* (belly button). The cord normally contains two **umbilical arteries** that carry deoxygenated fetal blood away from the fetal heart to the placenta, and one **umbilical vein** that carries oxygen and nutrients toward the fetal heart. The umbilical cord also contains a soft connective tissue with a jelly-like consistency, called *Wharton's jelly,* that insulates and protects the umbilical arteries and vein; its external surface is covered by the amniotic membrane.

Recall that when the embryonic period begins, the blastocyst has already implanted in the uterus and chorionic villi have penetrated uterine blood vessels to form lacunae filled with maternal blood. Eventually, these lacunae merge into a single blood-filled cavity called the **placental sinus** (**Figure 27.9b**). Once placentation has begun, the stratum functionalis of the

Placenta Previa

In **placenta previa** (PREE-vee-uh), a part of or all the placenta attaches to the inferior portion of the uterus, near or even covering the cervix. Although most cases result in live births, placenta previa may cause spontaneous abortion or premature birth. The condition is also dangerous to the mother. As the uterus stretches to accommodate fetal and placental growth, the placenta may hemorrhage or become infected, which increases the risk of maternal mortality. Typically, caesarean section—surgical removal of the fetus (described in Module 27.5)—is the preferred method of delivery in cases of placenta previa.

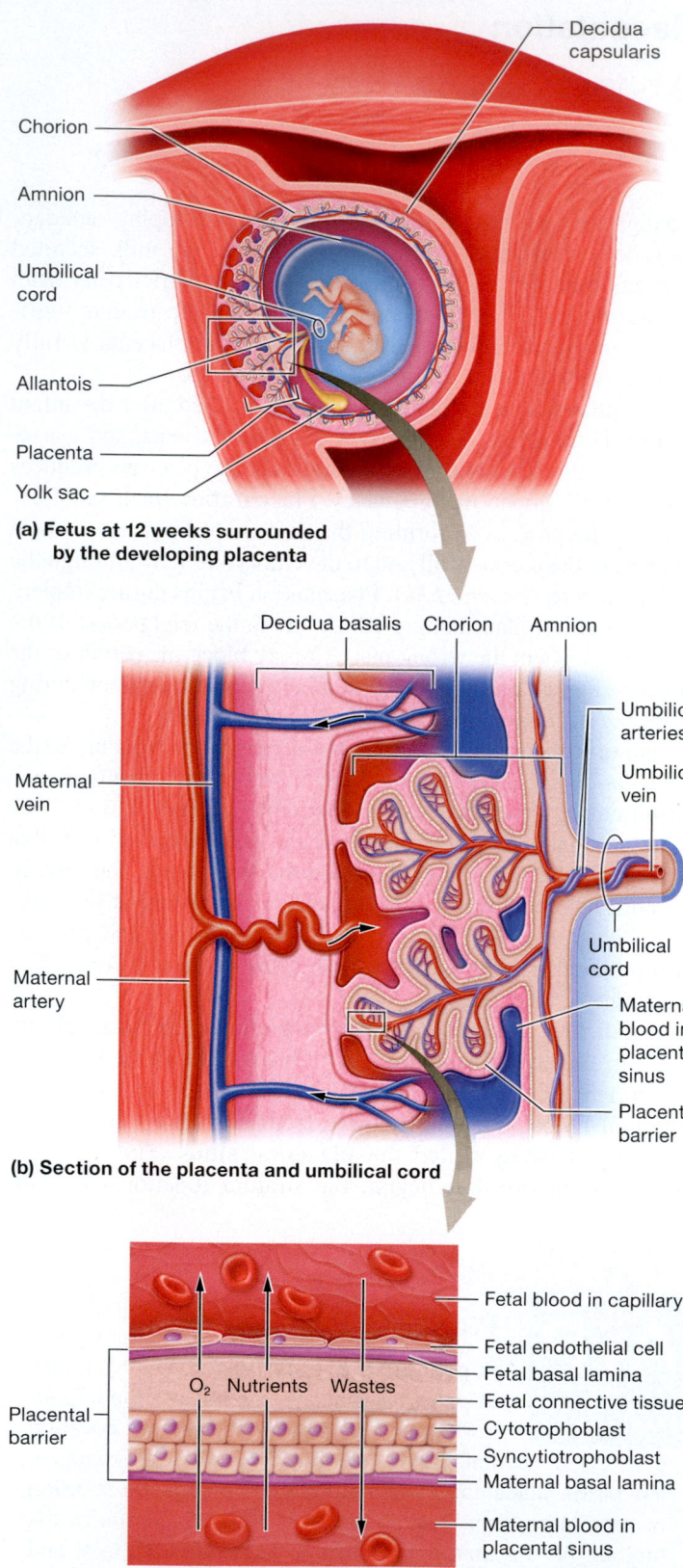

(a) Fetus at 12 weeks surrounded by the developing placenta

Decidua capsularis

Chorion

Amnion

Umbilical cord

Allantois

Placenta

Yolk sac

(b) Section of the placenta and umbilical cord

Decidua basalis Chorion Amnion

Maternal vein

Maternal artery

Umbilical arteries

Umbilical vein

Umbilical cord

Maternal blood in placental sinus

Placental barrier

(c) Schematic showing movement of substances across the placental barrier between maternal and fetal blood

Placental barrier

O_2 Nutrients Wastes

Fetal blood in capillary

Fetal endothelial cell

Fetal basal lamina

Fetal connective tissue

Cytotrophoblast

Syncytiotrophoblast

Maternal basal lamina

Maternal blood in placental sinus

Figure 27.9 Development and nutritive functions of the placenta.

uterus is known as the **decidua** (deh-SID-yoo-uh). The region of the endometrium that lies beneath the fetus becomes the **decidua basalis,** and the region that surrounds the uterine cavity forms the **decidua capsularis** (see Figure 27.9a).

The placenta is a unique organ because it develops from both fetal and maternal structures. The fetal portion is formed by the chorionic villi, and the maternal portion by the decidua basalis. The chorionic villi contain fetal blood vessels that are surrounded by maternal blood in the placental sinus; however, maternal blood and fetal blood do not mix because they are separated by the **placental barrier.** As you can see in **Figure 27.9c**, the barrier consists of the maternal and fetal basal laminae, fetal connective tissue, the cytotrophoblast, and the syncytiotrophoblast.

Some substances are exchanged across the placental barrier between the fetal and maternal blood supplies. This movement of substances occurs through diffusion—an important example of the Gradients Core Principle (p. 28). Since the maternal and fetal

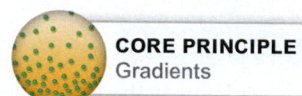

CORE PRINCIPLE
Gradients

blood supplies are so close, oxygen and nutrients diffuse from the maternal blood into the fetus for transport to tissues, and waste diffuses from the fetus to the maternal blood for removal (see Figure 27.9c). Let's focus in detail on the maternal and fetal blood flow through the placenta.

Maternal blood flows to the decidua basalis layer of the uterus and then through maternal arteries into the placental sinus. Maternal blood touches the placental barrier, which allows diffusion of substances, but not passage of blood cells. Think of the placental barrier as a "screen" that allows certain gases, nutrients, and wastes to get through because they are small enough to fit through the "holes." Once diffusion has taken place, the maternal blood, now deoxygenated and carrying wastes, flows into the maternal veins and back into the mother's cardiovascular system.

Fetal blood leaves the fetus and enters the two umbilical arteries (shown as blue in Figure 27.9 because the blood is deoxygenated). Blood flows into the placenta and moves into fetal capillaries in the chorionic villi. These capillaries are on the fetal side of the placental barrier. Fetal blood then picks up oxygen and nutrients and delivers wastes by diffusion, but again, no blood cells move through the barrier. Fetal hemoglobin has a greater affinity for oxygen than does maternal hemoglobin, so it picks up oxygen from the mother's blood at the placental barrier. Next, the oxygenated blood leaves the placenta and flows back to the fetus through the single umbilical vein (notice it is red because the blood is now oxygenated).

If fetal blood and maternal blood were to mix, it could be very dangerous for both the mother and fetus. For example, if the mother has blood type A and the fetus has blood type B, both could die because of the agglutination of erythrocytes caused by the incompatibilities between their blood types (see Chapter 19).

In addition to its nutritive functions, the placenta also functions as an endocrine organ. The placenta assumes the production of hCG as it grows and results in the subsequent reduction of the syncytiotrophoblast. In addition, the corpus luteum relinquishes production of progesterone and estrogens to the placenta by the end of the third month of gestation. The placenta also produces *human placental lactogen* and *placental prolactin,* which help prepare the mammary glands for milk production, and *relaxin,* which relaxes the body's muscles, joints, and ligaments, presumably to facilitate stretching during delivery of the newborn.

Quick Check

☐ 1. What are the functions of the placenta?

☐ 2. Which structure connects the conceptus to the placenta?

Fetal Development

« FLASHBACK

1. What is surfactant? (p. 825)
2. What are the four chambers of the heart? (p. 631)

During fetal development, the growth of the fetus is characterized by the maturation of tissues and organs. In addition, fetal growth is rapid and fetal size increases dramatically. Measurements of length are usually taken as crown-rump length (CRL) or crown-heel length (CHL). An embryo 25 mm (1.0 in.) long can grow to be as large as 50 cm (19.7 in.; CHL) during the fetal period. Weight increases as well, most obviously in the last 8 weeks of development. The average weight of a full-term fetus ranges from 3.2 to 3.5 kg (7 to 8 lb).

One way that a fetus and newborn infant differ noticeably is in their body proportions. At the beginning of the fetal stage, the head is disproportionately large and the lower limbs are short. During the fetal period leading up to birth, the body proportions become more like those of an infant. After birth, body proportions will continue to change until adulthood, when the head is the smallest proportionally to the rest of the body.

Major Events in Fetal Development

Let's summarize the most important changes that occur during the 6 months of the fetal period (months 3–9 of the pregnancy).

- **Month 3:** The fetal body lengthens as its head growth slows, while the upper limbs grow to their birth length. Ossification begins in most bones. In the fetus' face, the eyes are well developed, the eyelids are fused, the bridge of the nose develops, and the external ears are present but not prominent. During this period, the genitals are distinguishable as male or female. By the end of the third month, the approximate CRL is 9 cm (3.5 in.).
- **Month 4:** Body growth is rapid during the fourth month. The lower limbs lengthen and joints are forming as the skeleton continues to ossify. The kidneys are well formed, the digestive glands are forming, and the heartbeat can be heard with a stethoscope. The fetus begins to develop reflexes, and will startle and turn away from loud noises and bright lights. The CRL is about 14 cm (5.5 in.) by the end of the fourth month.

- **Month 5:** Growth slows during the fifth month and the lower limbs achieve their correct proportions relative to the body. Hair grows on the head, and the skin is covered by a downy hair called **lanugo** (lah-NOO-goh) and a white-to-gray secretion composed of shed epithelial cells and sebum known as **vernix caseosa** (kay-see-OH-suh). Both lanugo and vernix caseosa cover the skin to protect it from amniotic fluid. Brown adipose tissue forms and helps with heat production after the fetus is born. The mother can feel movements—a phenomenon referred to as *quickening*—as skeletal muscles begin to contract. The approximate CRL is 19 cm (7.5 in.) by the end of the fifth month.
- **Month 6:** The fetus gains significant weight during the sixth month. The eyebrows and eyelashes appear, and the eyelids, which have been fused since the third month, are partially open. The fetus' skin is wrinkled and translucent. Importantly, the fetal lungs begin to produce surfactant, which is key to survival if the infant is born prematurely. The approximate CRL is 23 cm (9.1 in.) by the end of the sixth month.
- **Month 7:** During the seventh month, the eyelids open completely, and fat is deposited in subcutaneous tissue so the skin is slightly smoother, but still wrinkled and red. The fetus usually turns upside down, assuming the **vertex** position. In males, the testes begin to descend through the inguinal canal and into the scrotum. By the end of the seventh month, the approximate CRL is 28 cm (11 in.).
- **Months 8 and 9:** The two final months see the fetal neurons forming networks, and organs growing and developing. In addition, blood cells form in bone marrow, and the digestive and respiratory systems complete development during the ninth month. The fetal skin is less wrinkled, and lanugo is shed. In males, the testes complete their descent into the scrotum. The approximate CRL is 30 cm (12 in.) by the end of the eighth month and 36 cm (14 in.) by the end of the ninth month.

Figure 27.10 compares fiber optic views of fetuses at months 3 and 5 of development. As you can see, at 3 months the limbs are short and the head is very large. At 6 months the limbs are much longer and the head is closer to the postnatal proportion.

Approximately 266 days after fertilization, a fetus is considered full-term and ready to be born. The scalp usually has hair, and the fingers and toes have well-developed nails. If, however, a fetus is born before that time, its tissues and organs may not have developed completely. Read about the major possible complications of being born prematurely in *A&P in the Real World: Prematurity* on p. 1082.

Fetal Circulation

All organ systems grow and develop after birth, but one system must make rapid changes soon after birth to adjust to life without a placenta—the fetal cardiovascular system. The greatest anatomical changes between the prenatal and postnatal states occur in this system. The fetus depends on the

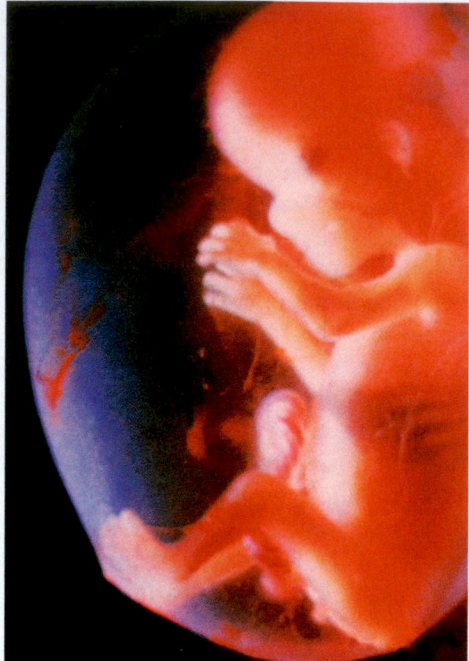
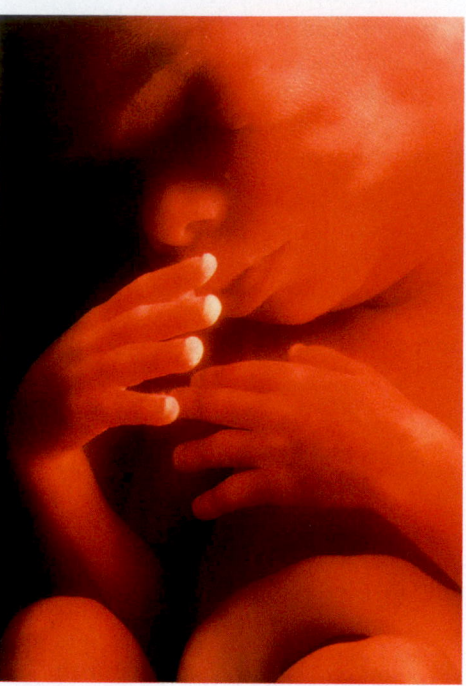

(a) Fiber optic view of fetus at month 3

(b) Fiber optic view of fetus at month 5

Figure 27.10 Development during the fetal period (images are approximately actual size).

placenta and umbilical cord for its blood supply. Cardiovascular structures seen only during prenatal development include the umbilical arteries and vein, and three circulatory "shortcuts" called **vascular shunts:** the *ductus venosus, foramen ovale,* and *ductus arteriosus* (**Figure 27.11a**).

As we know from our discussion of the placenta, the umbilical arteries take fetal blood to the placenta for exchange with maternal blood, and the umbilical vein returns that blood to the fetus. Blood that enters the fetus from the umbilical vein travels to the liver first. However, the immature liver cannot perform many of its filtering or storage functions (a job done by the maternal liver), so complete perfusion is not necessary before birth. For this reason, most blood from the umbilical vein bypasses the liver via the **ductus venosus** (veh-NOH-suhs), which connects to the inferior vena cava and flows to the right atrium of the heart.

The fetus depends on maternal blood for its oxygen, so there is little need for all the blood that enters the right side of the heart to travel to the lungs, which are not fully functional until just after birth. For this reason, two vascular shunts bypass the fetal pulmonary circulation:

Prematurity

An infant is considered premature if it is born more than 3 weeks before full-term, which is 38 weeks. More than 12% of the babies born in the United States each year are premature. The earlier the birth, the more complications the infant is likely to experience. Most commonly, premature infants suffer from respiratory, digestive, and thermoregulatory difficulties.

The most common respiratory problem is **infant respiratory distress syndrome (IRDS).** IRDS develops in infants born before 24 weeks because surfactant is not produced in the alveoli of the lungs until this time. Without surfactant, alveoli collapse during exhalation and are difficult to re-inflate due to alveolar surface tension. Synthetic surfactants can help premature infants breathe until their lungs develop sufficiently.

Premature infants face many other difficulties. For example, they often have minimal sucking and swallowing reflexes and require feeding tubes. In addition, the underdeveloped liver cannot produce enough clotting factors or albumin, so they bleed easily and may develop swelling, or edema. Also, their hypothalamus has not completely developed, so they cannot maintain a constant body temperature and must be kept warm in an incubator.

Premature birth is the most common cause of infant death in the United States. However, due to advances in neonatal care, survival rates have increased to almost 30% for infants born at 23 weeks of pregnancy, about 50–60% of those born at 24 weeks, about 75% born at 25 weeks, and more than 90% born at 27–28 weeks. Note, though, that infants who survive often face lifelong issues such as cerebral palsy, learning disabilities, and hearing and vision loss.

- The **foramen ovale** (foh-RAY-men oh-VAH-lay) is a hole in the interatrial septum that directly connects the right and left atria. The foramen ovale allows blood to "skip" ahead and avoid the pulmonary circulation by moving directly from the right to the left side of the heart and the systemic circulation.
- The **ductus arteriosus** (ahr-TEER-ee-oh'-sis) is a short passage that connects the pulmonary trunk to the aorta. The ductus arteriosus allows blood to move from the pulmonary trunk directly into the aorta.

After birth, these shunts and the umbilical vessels close, and the regular circulatory pattern begins. Within a year in most infants, the flaps in the interatrial septum seal and the foramen ovale becomes a depression called the **fossa ovalis** (**Figure 27.11b**). In some cases, however, the flap does not fuse shut, but remains closed anyway because of the high pressure in the left atrium. Resistance and pressure changes in the pulmonary trunk and aorta cause the ductus arteriosus to close approximately 3 months after birth. The remnant of the ductus arteriosus is a permanent cord, the **ligamentum arteriosum.** The ductus venosus similarly degenerates and becomes the fibrous **ligamentum venosum.** The umbilical arteries become the **medial umbilical ligaments,** and the umbilical vein becomes the **ligamentum teres** (TEHR-eez), also called the *round ligament of the liver.*

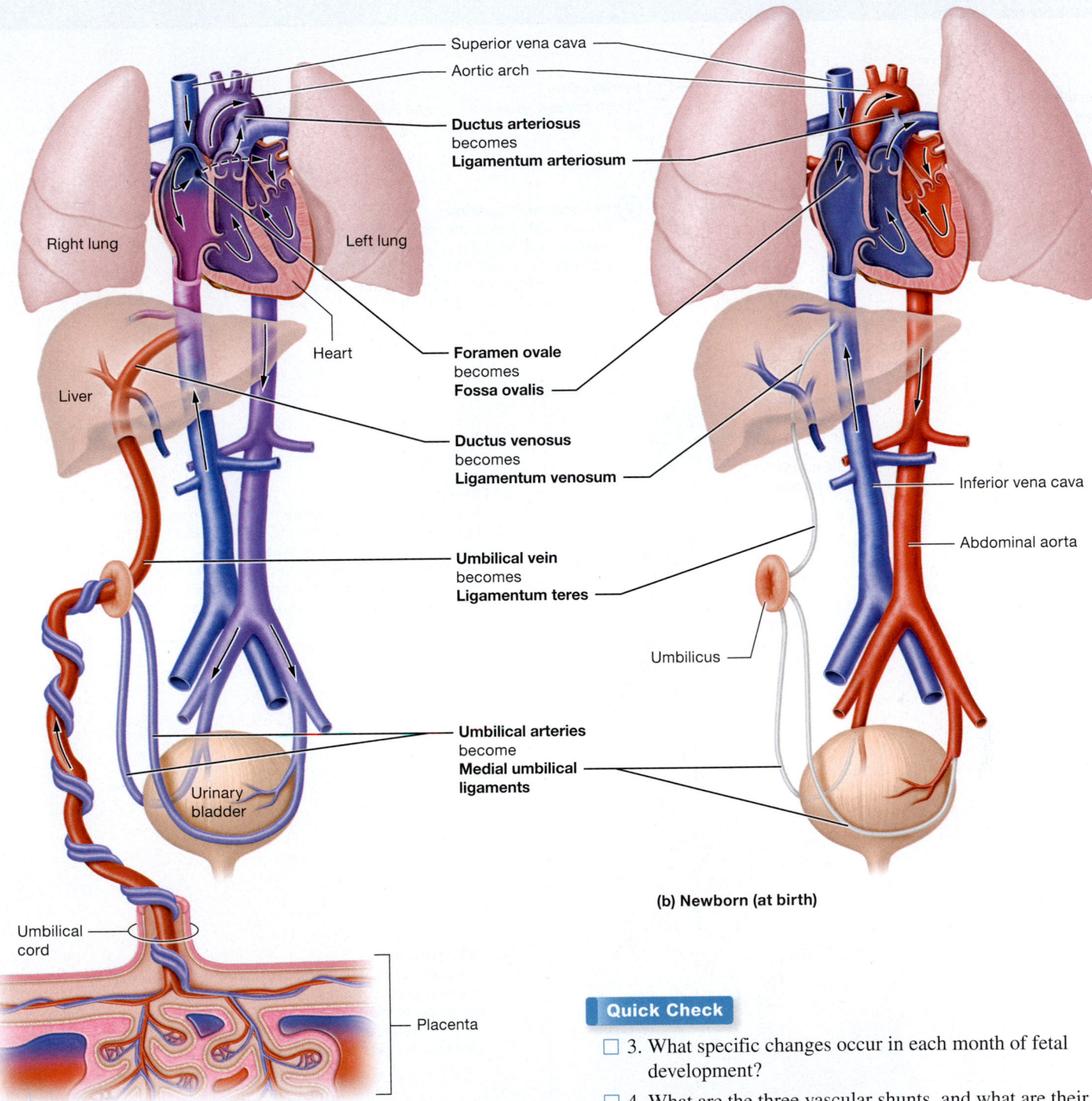

Superior vena cava

Aortic arch

Ductus arteriosus
becomes
Ligamentum arteriosum

Right lung

Left lung

Heart

Liver

Foramen ovale
becomes
Fossa ovalis

Ductus venosus
becomes
Ligamentum venosum

Inferior vena cava

Abdominal aorta

Umbilical vein
becomes
Ligamentum teres

Umbilicus

Umbilical
arteries
become
**Medial umbilical
ligaments**

Urinary
bladder

Umbilical
cord

Placenta

(a) Fetus (38 weeks)

(b) Newborn (at birth)

Figure 27.11 Comparison of fetal and newborn cardiovascular systems.

Putting It All Together: The Big Picture of Prenatal Development

Figure 27.12 summarizes key events of prenatal development. This figure starts with fertilization and ends with week 38, just before birth usually takes place.

Quick Check

☐ 3. What specific changes occur in each month of fetal development?

☐ 4. What are the three vascular shunts, and what are their functions?

Apply What You Learned

☐ 1. If a pregnant woman wants to know the sex of her baby as early as possible, when should she schedule an ultrasound exam? (*Hint:* Ultrasound allows visualization of the fetus by applying a probe to the skin of the mother's abdomen.)

☐ 2. If the ductus venosus of an infant does not close after birth, what consequences might this have?

See answers in Appendix A.

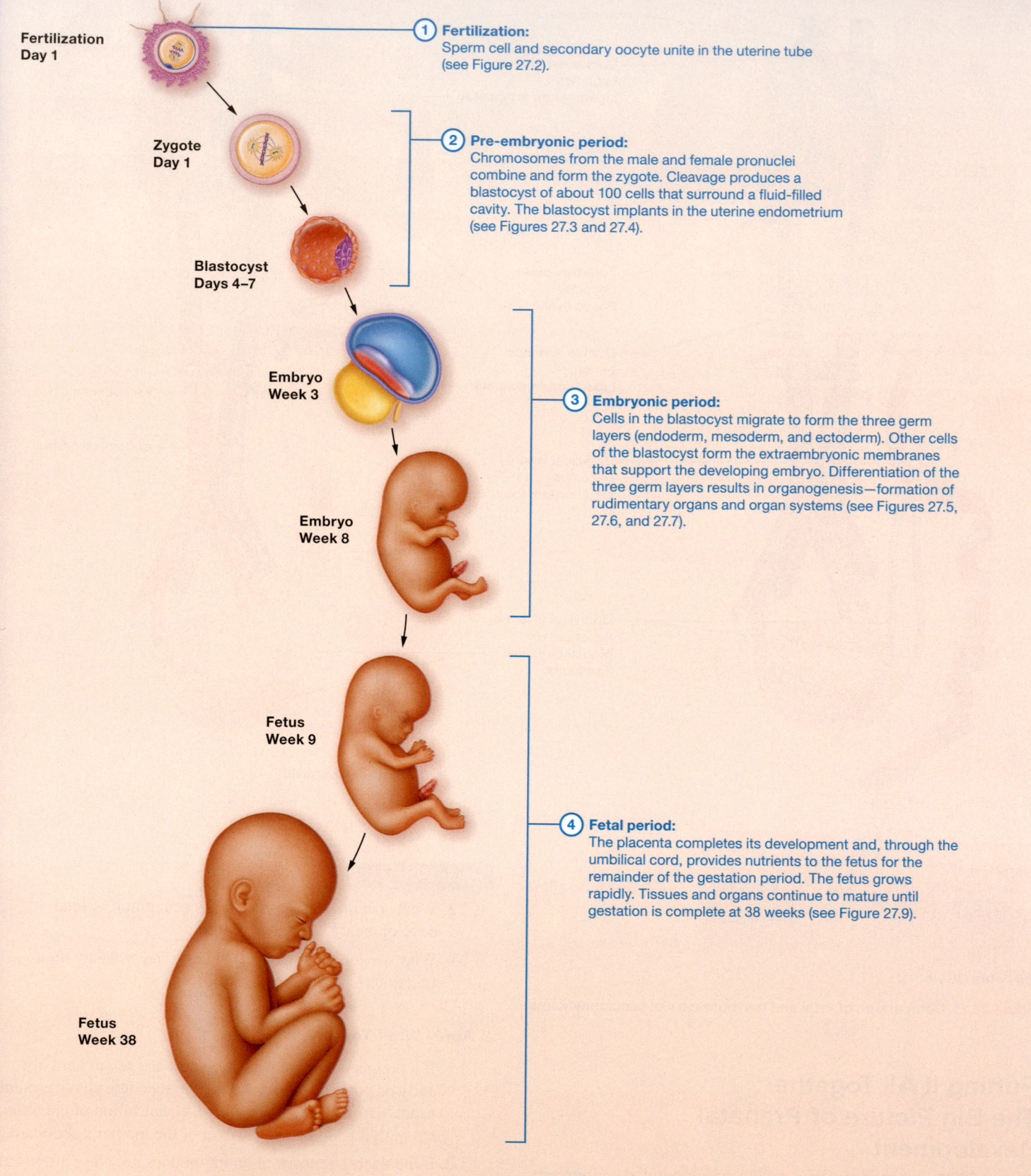

**Fertilization
Day 1**

**Zygote
Day 1**

**Blastocyst
Days 4–7**

**Embryo
Week 3**

**Embryo
Week 8**

**Fetus
Week 9**

**Fetus
Week 38**

1 Fertilization:
Sperm cell and secondary oocyte unite in the uterine tube (see Figure 27.2).

2 Pre-embryonic period:
Chromosomes from the male and female pronuclei combine and form the zygote. Cleavage produces a blastocyst of about 100 cells that surround a fluid-filled cavity. The blastocyst implants in the uterine endometrium (see Figures 27.3 and 27.4).

3 Embryonic period:
Cells in the blastocyst migrate to form the three germ layers (endoderm, mesoderm, and ectoderm). Other cells of the blastocyst form the extraembryonic membranes that support the developing embryo. Differentiation of the three germ layers results in organogenesis—formation of rudimentary organs and organ systems (see Figures 27.5, 27.6, and 27.7).

4 Fetal period:
The placenta completes its development and, through the umbilical cord, provides nutrients to the fetus for the remainder of the gestation period. The fetus grows rapidly. Tissues and organs continue to mature until gestation is complete at 38 weeks (see Figure 27.9).

MODULE 27.5
Pregnancy and Childbirth

Learning Outcomes

1. Describe hormonal changes during pregnancy and the effects of the hormones.

2. Describe the functional changes in the maternal reproductive, endocrine, cardiovascular, respiratory, digestive, urinary, and integumentary systems during pregnancy.

3. Explain the hormonal events that initiate and regulate labor.

4. Describe the three stages of labor.

So far, we have examined the changes that occur in the developing conceptus. Now, let's turn our attention to the dramatic anatomical and physiological changes that occur in the mother's body during this 9-month process of pregnancy. It is important to remember that each woman, and in fact each pregnancy, is different. Some pregnancies are uneventful, whereas others result in significant medical complications. Some women feel ill, while others feel wonderful. As it is not possible to cover all the potential complications and changes that might happen during pregnancy, we concentrate on the most important events. We conclude this module with the most dramatic event in this process—childbirth.

Maternal Changes during Pregnancy

« FLASHBACK

1. How do hormone levels fluctuate during a female's ovarian and menstrual cycles? (p. 1054)

2. What are the functions of cortisol and aldosterone? (pp. 609–610)

3. What are the normal curvatures of the spine? (pp. 230–231)

From a clinical standpoint, pregnancy is subdivided into three **trimesters**—each lasting 3 months—which differ slightly from the subdivisions of embryonic development.

- **First trimester: months 1–3 of pregnancy.** Recall that within the first 12 weeks of development, all pre-embryonic and embryonic development is completed and fetal development begins. So, by the end of the first trimester, the rudiments of all the major organ systems are present. This fact makes the first trimester the most critical stage of development. This is also the reason why most spontaneous abortions occur during the first trimester, because of chromosomal abnormalities that prevent the organ systems from developing correctly. Some women are unaware that they are pregnant until a good portion of this trimester is over, while others realize they are pregnant more quickly. Many women find the first trimester to be plagued with morning sickness, fatigue, and other symptoms such as breast tenderness. There are some women who do not suffer from the typical symptoms of early pregnancy and still others who experience increased energy levels.

- **Second trimester: months 4–6 of pregnancy.** During the second trimester, the fetus continues to grow and develop, and the pregnancy usually becomes obvious as the uterus and abdomen expand. Many women feel relief from morning sickness during this trimester and may feel more energetic, although some women continue to experience these or other symptoms. Reaching the second trimester makes many women begin to feel comfortable that the pregnancy is "safe" because spontaneous abortions are less common during the fetal period of development. Additionally, this is the time period in which many women feel the movements of the growing fetus for the first time.

- **Third trimester: months 7–9 of pregnancy.** The fetus grows rapidly during the third trimester and gains a significant amount of weight. The woman's uterus and abdomen enlarge further, and many women experience new symptoms related to the size of the fetus, such as backaches and pressure on internal organs, which are now competing for space in the abdominal cavity. In addition, the weight of the fetus rests on the urinary bladder, increasing the urge to urinate. As the delivery nears, many women grow more uncomfortable as the fetus settles into a lower position in the abdomen.

The rest of this section takes a look at the hormonal, anatomical, and physiological changes that occur during these three trimesters.

Hormonal Changes

During a typical ovarian cycle, the corpus luteum degenerates about 2 weeks after ovulation. Consequently, the concentrations of estrogen and progesterone decline rapidly, the uterine lining breaks down, and the endometrium sloughs off as menstrual flow. If this were to occur following implantation, the blastocyst would be lost in a spontaneous abortion.

Remember that when the blastocyst implants in the uterine lining, the syncytiotrophoblast begins secreting human chorionic gonadotropin, or hCG, which maintains the corpus luteum so that it will continue secreting estrogen and progesterone. For this reason, the inner lining of the uterine wall, the stratum functionalis, continues to grow and develop. At the same time, the hormones stimulated by hCG inhibit the release of *follicle-stimulating hormone* (*FSH*) and *luteinizing hormone* (*LH*) by the anterior pituitary gland, which stops the normal ovarian cycle. Secretion of hCG continues at a high level for about 2 months, then declines by the end of the fourth month, but by this time, the corpus luteum is no longer needed, because the placenta secretes sufficient estrogen and progesterone. For the remainder of the pregnancy, placental estrogen and placental progesterone maintain the uterine lining.

In addition to estrogen and progesterone, several other hormones have important roles during pregnancy. Along with progesterone, **relaxin** from the placenta helps suppress uterine contractions until the birth process begins. When labor nears, relaxin loosens the connective tissue of the pubic symphysis and sacroiliac joints, which allows greater movement of the joints to help the fetus pass through the birth canal.

The placenta also secretes a variety of other hormones. For example, the hormone called **human placental lactogen (hPL),** or *human chorionic somatomammotropin* (hCS; soh-maeh'-toh-MAM-oh-troh-pin), helps stimulate breast development and prepares the mammary glands to secrete milk. It may also affect a pregnant woman's metabolism of glucose by inhibiting the release of insulin, which in turn raises the glucose level in the blood supply, allowing more to go to the fetus. A decreased maternal blood glucose level results in increased fatty acid metabolism by the mother. Another placental hormone is **melanocyte-stimulating hormone (MSH),** which is partially responsible for darkening of the areolae and nipples of the breast as well as changes in skin tone in some women. Finally, the placenta also produces **corticotropin-releasing hormone (CRH),** which current evidence indicates is important in determining the length of pregnancy and the timing of childbirth. In addition, CRH increases secretion of *cortisol* from the adrenal cortex, which is needed for maturation of the fetal lungs and production of surfactant. The placenta's secretion of CRH begins about the 12th week of development and rises significantly toward the end of pregnancy.

Other endocrine glands increase secretion during pregnancy as well. Aldosterone from the adrenal cortex promotes reabsorption of sodium ions from kidney tubules, which causes water to follow by osmosis. This leads to an overall increase in blood volume during pregnancy. Parathyroid hormone from the parathyroid glands helps to maintain a high concentration of maternal blood calcium ions because fetal demand for calcium ions is high to support bone development. **Prolactin** secreted from the anterior pituitary stimulates milk production by the mammary glands. Finally, **oxytocin** from the fetal and maternal

hypothalamus is secreted during the second and third trimesters and peaks during labor to stimulate uterine contractions and allow milk release from the mammary glands.

Anatomical and Physiological Changes

Many anatomical and physiological changes take place to support pregnancy and prepare a woman's body for *childbirth* and *lactation* (lak-TAY-shun), or breastfeeding. These changes affect multiple organ systems.

Reproductive System The uterus enlarges greatly over the 9-month pregnancy and eventually extends upward from its normal location in the pelvic cavity to reach the level of the xiphoid process of the sternum. Before pregnancy, the uterus is approximately the size of a fist, but by the fourth month, the uterus fills the pelvic cavity and has already begun to push into the abdominal cavity (**Figure 27.13a** and **b**). The bulk of uterine enlargement is not because of the developing fetus—which is only about 90 mm (3.5 in.) long from crown to rump at this time—but because of hypertrophy of the myometrium, placental growth, and accumulation of amniotic fluid. In the last few weeks of pregnancy, estrogens promote the formation of gap junctions between the uterine muscle cells to facilitate coordinated contractions during childbirth (see Chapter 10).

By about 28 weeks, or 7 months, the uterus reaches above the umbilicus (**Figure 27.13c**). As pregnancy progresses, the uterus continues to expand, displacing abdominal organs upward, compressing the diaphragm, and pressing on the urinary bladder. As a result, a pregnant woman may be unable to eat large meals, may develop heartburn, and may have to urinate frequently. The increased size of the anterior abdomen changes the center of gravity and some women develop *lordosis,* which is accentuated

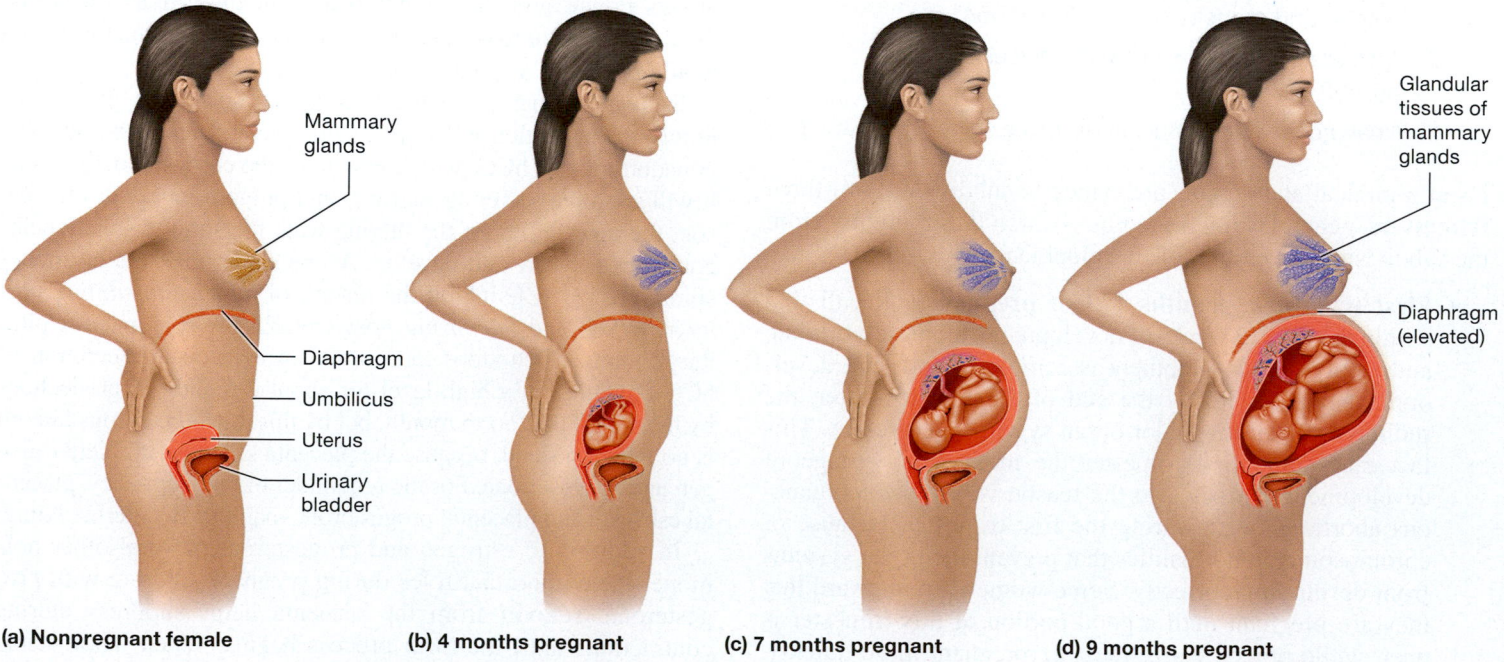

(a) Nonpregnant female (b) 4 months pregnant (c) 7 months pregnant (d) 9 months pregnant

Figure 27.13 Changes in a woman's body during pregnancy.

lumbar curvature, and backaches during the last few months of pregnancy (**Figure 27.13d**).

Notice in the figure that the mammary glands also undergo significant changes in preparation for lactation after birth. They are typically sore during the first trimester because of enlargement due to increased levels of estrogen and progesterone. Glandular tissue grows, and additional **alveoli**—the structures of the breast that produce milk—develop under the influence of prolactin. Typically, the breasts enlarge because of the increase in glandular tissue and additional adipose tissue that accumulates during the pregnancy (refer back to Figure 26.13).

Cardiovascular System   The growing placenta requires increased blood volume, and as the offspring grows, it needs more oxygen and produces more waste that must be removed. These changes cause the woman's blood volume and, subsequently, cardiac output, to increase. Blood pressure may rise to some extent, but typically, pregnant women have a lower hematocrit (number of erythrocytes) than nonpregnant women. This makes the blood slightly thinner, which decreases resistance and tends to prevent a significant rise in blood pressure. As the uterus grows and presses on pelvic blood vessels, venous return from the lower limbs may be impaired and sometimes results in *varicose veins* and leg swelling, or *edema.*

Perhaps because of the close association between maternal and fetal circulations, consequences arise when these systems do not work normally. See *A&P in the Real World: Preeclampsia* to read about a potentially serious condition that highlights this association.

Respiratory System   As previously mentioned, more oxygen is needed during pregnancy, and more carbon dioxide is produced to be exhaled. Progesterone makes the medullary respiratory center more sensitive to carbon dioxide, which stimulates removal of more carbon dioxide from the blood by increasing the rate and depth of ventilation. However, during the latter part of pregnancy, the uterus may prevent the diaphragm from contracting completely and the lungs from expanding completely (see Figure 27.13d). Other effects on the respiratory system are mediated by estrogen, which may cause the nasal mucosa to swell, resulting in nasal congestion and occasional *epistaxes*, or nosebleeds. These changes can contribute to shortness of breath in some women.

Digestive System   For normal development of the fetus, good nutrition is necessary all during pregnancy. A pregnant woman typically needs about 300 extra calories each day to supply both herself and the fetus. Specifically, protein, calcium salts, iron salts, and folic acid requirements for increase from the recommended daily levels required for nonpregnant women. The increased intake of food, water retention, and the growing fetus lead to weight gain. The average weight gain is about 13 kg (28.7 lb), but there are significant differences among women.

During the first trimester, many pregnant women suffer from *morning sickness*, nausea and sometimes vomiting that occur in the morning, as well as nausea and vomiting that may occur throughout the day. These symptoms are thought to be related to elevated hCG, estrogen, and progesterone levels, but no definitive cause has been identified. Some women suffer from an extreme form of morning sickness called **hyperemesis gravidarum** (hy'-per-EM-uh-sis grav-ih-DEHR-em). This results in excessive vomiting leading to dehydration, electrolyte imbalances, and weight loss in the mother. Severe cases require hospitalization and administration of fluids intravenously.

Some women experience food cravings and/or food aversions for reasons we have yet to completely identify. It could relate to the fact that some foods, such as bland carbohydrates, might contain fewer toxins and so be "safer" for the conceptus. For example, some women crave fruits and vegetables and tend to temporarily lose interest in caffeinated or spicy foods.

Another digestive system effect is linked to progesterone. This hormone relaxes the smooth muscle of the gastrointestinal tract and slows peristalsis, resulting in constipation.

Urinary System   The fetus produces metabolic wastes that the mother's kidneys must excrete. These wastes, combined with the mother's higher blood volume, cause the glomerular filtration rate to rise by up to 50% during pregnancy. However, even as more urine is being produced, progesterone signals relaxation of the smooth muscle of the ureters and may slow the movement of urine out of the urinary tract. The uterus tends to compress the urinary bladder as it expands (see Figure 27.13), sometimes leading to *stress incontinence,* or unintentional loss of urine.

Integumentary System   Changes in the skin during pregnancy are more apparent in some women than in others. Some experience increased pigmentation around the eyes and cheekbones in a masklike pattern called **chloasma** (kloh-AZ-muh), in the areolae of the breasts, and along the midline of the lower abdomen. *Stretch marks,* or *striae* (STREE-ay), over the abdomen can occur as the uterus enlarges rapidly and the elastic fibers in the dermis break. In addition, progesterone and estrogen decrease the normal hair loss of some women, who subsequently experience

Preeclampsia

Approximately 10% of pregnant women in the United States experience a dangerous complication known as *preeclampsia* (pree'-eh-KLAMP-see-uh). The causes are not known, but may involve insufficient blood flow to the uterus or a problem with the mother's immune system. Its occurrence seems to be related to the number of fetal cells that enter the maternal circulation. It leads to the fetus being deprived of oxygen and the mother developing edema, hypertension, and high protein levels in urine (proteinuria).

Undiagnosed preeclampsia can lead to *eclampsia,* a potentially fatal condition that can result in seizures and put both mother and fetus at risk. Unfortunately, the only treatment for preeclampsia is to deliver the baby, but once that is done the condition usually resolves within a few weeks.

thicker hair during pregnancy. However, nails may become stronger and thicker or more brittle and fragile during pregnancy. A woman's hair and nails typically return to normal within 6 months of childbirth.

Quick Check

☐ 1. What is the first hormone secreted that prevents a menstrual cycle, thereby blocking removal of an implanted blastocyst?

☐ 2. How do a woman's nutritional requirements change during pregnancy?

Parturition

FLASHBACK

1. What is the "lesser pelvis"? (p. 247)
2. What is the function of oxytocin? (p. 596)

Pregnancy ends with childbirth, or **parturition** (par-too-RIH-shun), when the fetus is expelled from the uterus through the vagina. The series of events that expel the fetus are collectively called **labor**. Although birth is estimated to take place during week 38 of pregnancy, many occur 1–2 weeks earlier or later.

The series of events involved in initiating and regulating childbirth are as follows (**Figure 27.14**):

① **The fetal adrenal cortex produces cortisol, which stimulates the placenta to secrete a high level of estrogen.** The initial stimulus for labor comes from the fetus. As the time of childbirth nears, the fetus begins to secrete a high level of cortisol. The cortisol binds to receptors on the placenta and stimulates it to produce large amounts of estrogen.

② **The high estrogen level stimulates the uterus to form oxytocin receptors.** The rising estrogen level stimulates the formation of additional oxytocin receptors on the myometrium. It also overrides the quieting influence of progesterone on the myometrium. The myometrium becomes "irritable," leading to irregular contractions. These **Braxton-Hicks contractions** are referred to as *false labor* because they do not result in the three stages of labor. Some women experience strong Braxton-Hicks contractions and because of this may think they have begun the process of labor. However, true labor results in uterine contractions that increase in intensity and regularity and cause changes in the cervix.

③ **Both the fetal hypothalamus and maternal hypothalamus secrete oxytocin, which stimulates the placenta**

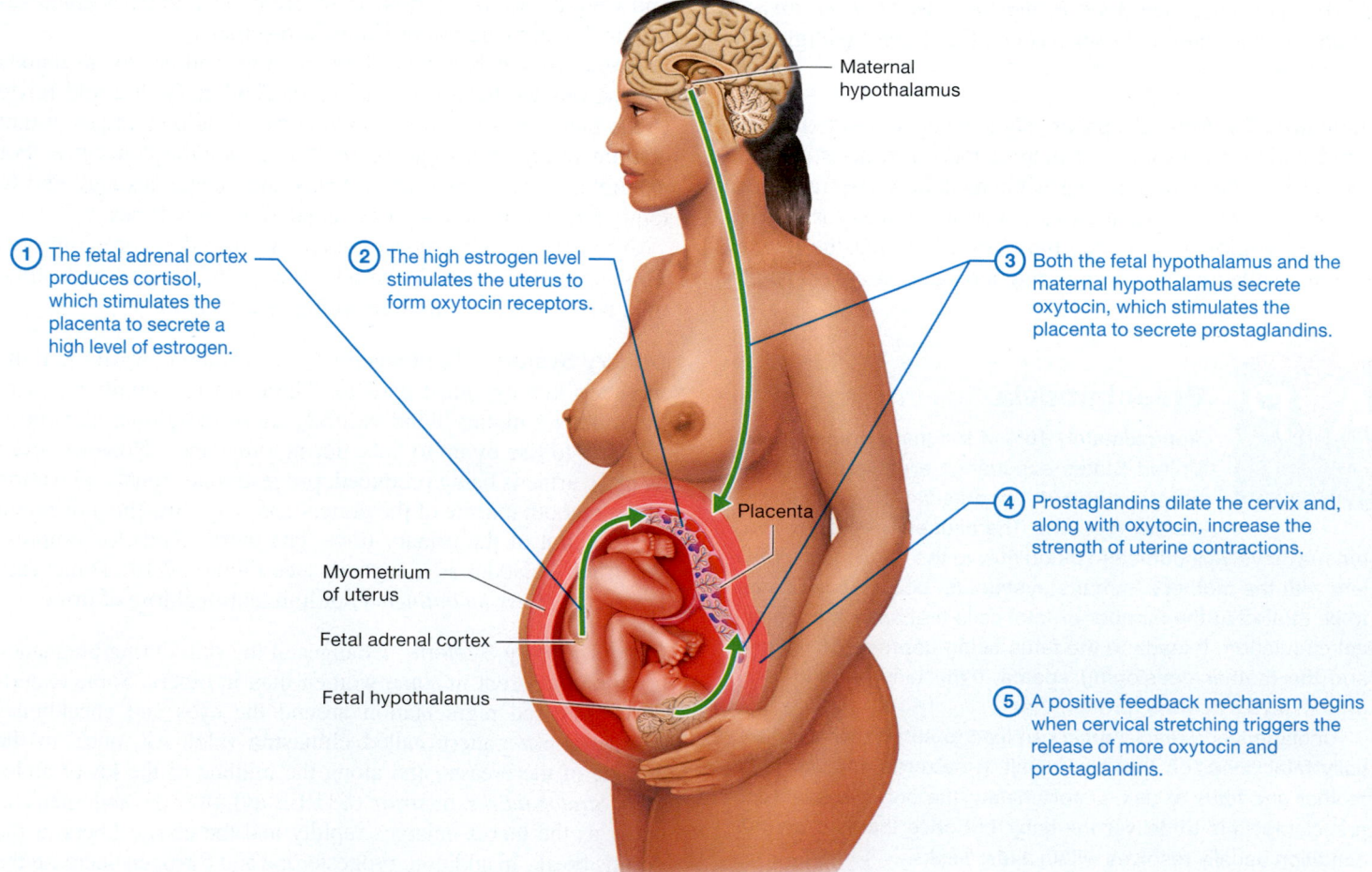

Maternal hypothalamus

① The fetal adrenal cortex produces cortisol, which stimulates the placenta to secrete a high level of estrogen.

② The high estrogen level stimulates the uterus to form oxytocin receptors.

③ Both the fetal hypothalamus and the maternal hypothalamus secrete oxytocin, which stimulates the placenta to secrete prostaglandins.

Placenta

④ Prostaglandins dilate the cervix and, along with oxytocin, increase the strength of uterine contractions.

Myometrium of uterus

Fetal adrenal cortex

Fetal hypothalamus

⑤ A positive feedback mechanism begins when cervical stretching triggers the release of more oxytocin and prostaglandins.

Figure 27.14 Initiation and regulation of childbirth by a positive feedback mechanism.

to secrete prostaglandins. Now that the uterus has more receptors for oxytocin, the fetus and mother both increase oxytocin production. This causes the placenta to produce chemicals called *prostaglandins*. It also leads to early, usually weak, contractions of the myometrium.

④ **Prostaglandins dilate the cervix and, along with oxytocin, increase the strength of uterine contractions.** These prostaglandins cause the cervix to open or *dilate*, which is necessary for the fetus to exit the uterus. Both prostaglandins and oxytocin cause the uterine contractions to progressively increase in strength and frequency.

⑤ **A positive feedback mechanism begins when cervical stretching triggers the release of more oxytocin and prostaglandins.** As the head of the fetus pushes on and stretches the cervix, more oxytocin is released. As more oxytocin is released, the myometrium contracts more forcefully and the placenta secretes more prostaglandins. Both effects cause the cervix to stretch more, which stimulates the release of more oxytocin in a positive feedback loop. This brings us back to the Feedback Loops Core Principle (p. 28).

CORE PRINCIPLE
Feedback Loops

During pregnancy, the rising level of progesterone leads to the formation of a *mucous plug* at the cervix. When the cervix begins to dilate slightly, the mucus (which may be clear or bloody) is discharged into the vagina. Labor may begin within hours after the mucous plug is discharged, or it may not occur for up to 2 weeks.

In some pregnancies, labor does not begin as it should or, for some reason, the mother wants to choose the time she gives birth. In these situations, labor can be initiated, or **induced,** in a clinical setting. Prostaglandin gels may be applied to the cervix to stimulate dilation and an intravenous solution of synthetic oxytocin, called *Pitocin,* is administered.

Labor can be subdivided into three stages—the *dilation, expulsion,* and *placental stages.* Let's explore what happens in each of these stages.

Dilation Stage

The **dilation stage** is the time from the onset of labor until the cervix is fully dilated to 10 cm (3.9 in.) in diameter (**Figure 27.15a** and **b**). When labor begins, contractions originate in the upper part of the uterus and move toward the vagina. As labor progresses, contractions become stronger and the lower part of the uterus becomes involved.

Dilation is the longest stage of labor, but its length varies. The range is usually longer (8–24 hours) in women who are having their first child and shorter (4–12 hours) in those who have previously given birth. As long as the fetus is in the *vertex,* or head-first, presentation, the fetus' head is forced against the cervix, helping it to thin and dilate. At some point during this stage, the amnion ruptures and releases the amniotic fluid, which is the event known as the "water breaking" or "rupture of membranes." If the amnion does not rupture on its own by the end of this stage, it may need to be ruptured manually. Toward the end of the dilation stage, the head of the fetus enters the lesser pelvis and the fetus rotates so that its head can pass through the pelvic outlet.

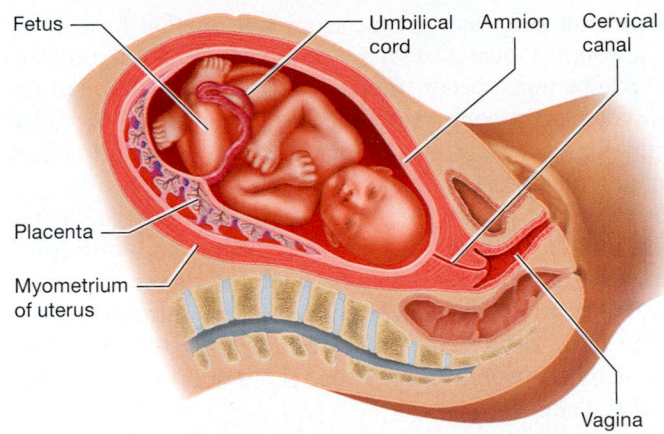

(a) **Normal position of fully developed fetus before labor begins**

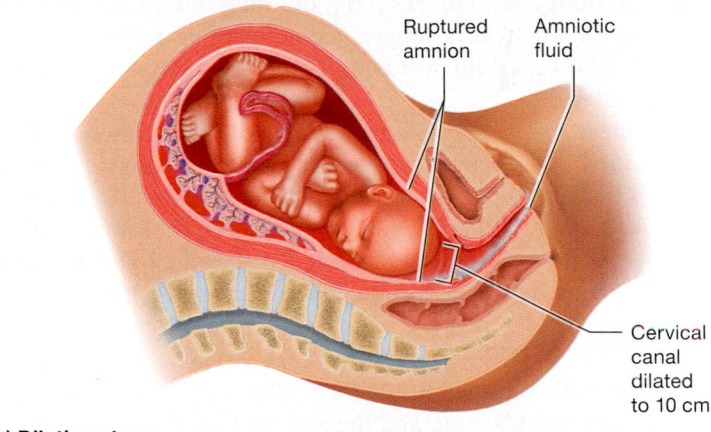

(b) **Dilation stage**

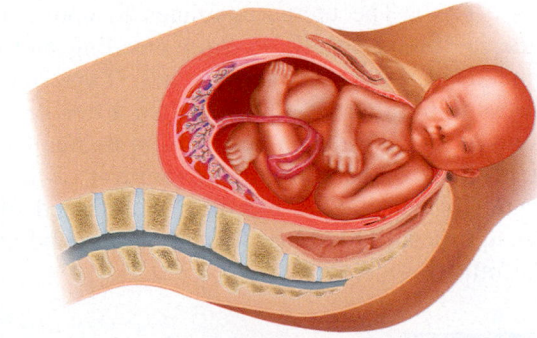

(c) **Expulsion stage**

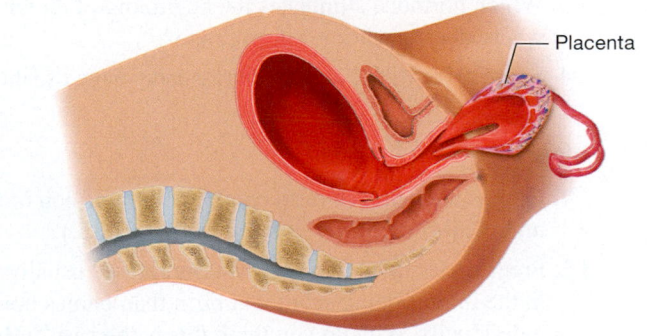

(d) **Placental stage**

Figure 27.15 The stages of labor.

Expulsion Stage

The **expulsion stage** is the time from full dilation to delivery of the newborn (**Figure 27.15c**). Once the cervix is fully dilated, strong contractions occur frequently, often assisted by the mother voluntarily contracting her abdominal muscles, called "pushing" or "bearing down." This stage may take up to a few hours, but typically ranges from 30 minutes to 1 hour.

When the first part of the fetus's head distends the vagina, **crowning** has occurred. An incision, called an **episiotomy** (eh-pih'-zee-AWT-uh-mee), may be done at this time to widen the vaginal orifice and reduce tearing of the surrounding tissues. The fetus' head is then born, followed by the rest of the body, which is delivered more easily. The umbilical cord is then clamped and cut.

Under certain circumstances, a vaginal delivery is not possible or is unsafe for the mother and fetus. For instance, in a **breech birth,** or *buttocks-first,* or other *nonvertex* presentation, labor often lasts longer. Complications may arise, such as the umbilical cord wrapping around the fetus or becoming compressed.

In these cases, a **caesarean section,** or **C-section,** may be performed. A C-section is a surgical procedure in which an incision is made through the abdominal and uterine walls in order to remove the fetus and placenta. In most cases, delivery in subsequent pregnancies will also require C-section, although vaginal delivery after a prior C-section is sometimes an option.

Placental Stage

The **placental stage** is the time after delivery of the newborn when the placenta and the attached fetal extraembryonic membranes, collectively called the **afterbirth,** are delivered (**Figure 27.15d**). The uterus continues to contract, compressing uterine blood vessels to limit bleeding and pulling the placenta away from the uterine wall. This stage typically takes about 30 minutes, although often a healthcare provider will simply tug on the umbilical cord to manually remove the afterbirth more quickly. Complete removal of the placenta is important to stop uterine bleeding. A retained placenta can lead to significant blood loss for the mother and an increased risk for infection.

MODULE **27.6**
Postnatal Changes in the Newborn and Mother

Learning Outcomes

1. Describe the major changes that occur to the newborn during the neonatal period.

2. Describe the major anatomical and physiological changes to the mother after parturition.

3. Describe the structure and function of the mammary glands.

4. Explain the hormonal regulation of lactation.

Recall that the *postnatal period* of development lasts from birth until death. In this section we focus on the neonatal period and the immediate changes the newborn and mother must make to regain homeostasis after birth.

Changes in the Newborn

《 FLASHBACK

1. What is brown adipose tissue? (p. 141)

2. What are the functions of IgG and IgA antibodies? (p. 783)

The first 4 weeks after birth constitute the **neonatal period** for the newborn, or **neonate.** Immediately after birth, the newborn must make significant physiological adjustments to adapt to life outside the uterus. It must now do for itself all the physiological functions that the mother had been doing, including breathing, obtaining nutrients, excreting waste, and maintaining body temperature.

To determine whether a newborn is adjusting appropriately, its physical status is evaluated at 1 minute and 5 minutes after birth, based on five criteria: skin color, pulse rate, respiratory level, muscle activity, and grimace (which assesses reflexes). Each criterion is given a score of 0–2, and the total is called the **Apgar score** (**Table 27.3**) An Apgar score of 8–10 indicates a healthy baby. Lower scores signify a problem in one or more of the physiological functions assessed.

The most dramatic event after the birth is the first breath. Remember that before birth, most blood bypassed the lungs, so in addition to inflating the alveoli of the lungs for the first time, the first breath also causes the blood pressure in the lungs to drop and decreases pressure in the right side of the heart. For this reason, blood flows from the left side of the heart to push flaps over the foramen ovale, which closes the opening. Due to this initial change in respiration, fetal circulatory routes are converted to the newborn circulatory system (look back at Figure 27.11). The neonatal respiratory rate (30–60 breaths per minute) and heart rate (120–140 beats per minute) are significantly higher than those of adults. These higher rates result because compared to adults, neonates have smaller lung capacity and the physical size of their heart in relation to their body is reduced.

Table 27.3 Apgar Scoring for Newborns

Physiological Criteria	Score 0	Score 1	Score 2
Skin color	Pale to blue	Limbs blue, body pink	Pink; in dark-skinned babies, the color inside the mouth, the lips, and the palms of the hands and soles of the feet is examined.
Pulse rate	No pulse	100 beats per minute or lower	Over 100 beats per minute
Respiratory level	No breathing	Slow and/or irregular breathing; weak cry	Normal regular breathing; strong cry
Muscle activity	No movement; muscles flaccid	Some movement; poor muscle tone	Active movement; good muscle tone
Grimace (reflex involving face and body, initiated by mechanically stimulating plantar surface of foot)	No response	Facial grimace; some body movement	Facial grimace; significant body movement

Newborns lose heat more rapidly than adults because they have a larger ratio of surface area to volume. However, keep in mind that newborns are born with brown adipose tissue which is a special tissue deposited during the fifth month of fetal development that produces heat rather than storing energy. Eventually, most of this brown adipose tissue is metabolized and is not replaced. However, as an infant grows, its metabolic rate increases, and it accumulates more subcutaneous white (or adult) adipose tissue which helps insulate against heat loss.

Although the kidneys are almost fully developed at birth, they are less capable than adult kidneys of concentrating urine. Consequently, newborns have a higher rate of water loss and require more fluid intake, relative to body weight, than adults do. It helps that their food source is liquid during the first few months of life.

Babies are born with a near-adult level of IgG antibodies acquired passively from the mother through the placenta. However, maternal IgG breaks down rapidly and is virtually gone by 1 year after birth. Depending on the disease, maternal IgG may be sufficient to prevent infection after exposure until the infant's immune system is fully developed. In addition, breastfed infants receive added protection from gastroenteritis thanks to IgA antibodies in the *first milk,* or *colostrum.*

Other notable changes occur in the nervous and skeletal systems. During the first year of life, the nervous system grows and many neurons develop a myelin sheath. As for the skeleton, various areas of it are cartilaginous rather than ossified at birth; these will be converted to bone over time.

In some cases, infants are born with anatomical malformations; these are called *congenital disorders* because the deformity exists from birth. Two congenital disorders involve a newborn's vascular shunts (see *A&P in the Real World: Patent Ductus Arteriosus and Patent Foramen Ovale*).

Quick Check

☐ 1. What are the criteria used to determine a newborn's Apgar score?

☐ 2. What is the benefit to the newborn of acquiring IgG antibodies from the mother?

☐ 3. Explain what happens to each structure in the fetal circulation after the infant is born.

Changes in the Mother

⮜ FLASHBACK

1. What is the function of endorphins? (p. 415)

2. What are the functions of complement proteins and interferon? (pp. 767–768)

Now let's shift our focus to the mother. For her, the first 6 weeks following birth make up the **postpartum period.** The woman's body undergoes several changes to return to pre-pregnancy homeostasis and to begin and then maintain the production of breast milk—the process of lactation. Most of these changes are completed within 6 weeks, but lactation can continue up to a year or more.

Anatomical and Physiological Changes

Just as the mother's body changed during pregnancy, many adjustments occur after birth to reverse those changes and restore

Patent Ductus Arteriosus and Patent Foramen Ovale

Failure of an infant's ductus arteriosus to close after birth results in **patent ductus arteriosus** (PAY-tent). This condition causes difficulties because it allows oxygen-poor blood being carried by the pulmonary artery to mix with oxygen-rich blood being carried by the aorta. In addition, this condition causes an increase in pulmonary blood pressure that can damage the heart and lungs. If not corrected, patent ductus arteriosus leads to cardiac dysrhythmias and possibly heart failure, because the heart must work harder to supply blood to body tissues.

Similarly, if an infant's foramen ovale does not close because the septal flaps fail to seal after birth, **patent foramen ovale** results. As the heart grows larger, the patent foramen ovale also enlarges. In most cases, there are no symptoms and no treatment is prescribed. However, evidence increasingly shows that in adults, patent foramen ovale is associated with strokes because it increases the formation of inappropriate clots. Patent foramen ovale may also increase the risk of migraine headaches.

homeostasis to the mother's system. Immediately after birth, endorphins released by the pituitary gland allow the mother to ignore fatigue and pain to take care of the new infant. However, endorphin production is short-lived, and these positive feelings soon dissipate. In addition, within a few days after birth, the levels of estrogen and progesterone fall significantly because they are no longer needed to maintain the uterine lining. The level of corticotropin-releasing hormone (CRH) also declines because the placenta is no longer present. Sudden reductions in the levels of estrogen, progesterone, and CRH have been linked to a severe form of depression called **postpartum depression.** Women who suffer from this condition may require medical treatment.

For most women, the respiratory and digestive systems resume normal function soon after birth because the need for oxygen and nutrients returns to pre-pregnancy levels and the respiratory and digestive organs are no longer compressed. The digestive symptoms associated with pregnancy—such as nausea, vomiting, heartburn, and constipation—usually stop after childbirth.

Some of the excess fluid volume retained in the mother during pregnancy is removed during birth through blood loss and expulsion of the amniotic fluid. More of the fluid is removed as **lochia** (LOH-kee-uh), a bloody, yellowish discharge that is expelled from the uterus for up to 6 weeks after birth. Lochia is usually heavy for the first week or so and then lightens for the remainder of the time. The urinary system removes still more excess fluid. When the CRH level declines, the level of aldosterone also declines and urine production increases. When the fluid volume returns to normal, cardiac output also returns to the pre-pregnancy level.

Other changes, such as altered pigmentation in the skin, usually return to the pre-pregnancy state. The thickened hair is also usually temporary, as many women experience significant hair loss after pregnancy. However, some changes, such as stretch marks and varicose veins, are often irreversible.

Lactation

Lactation refers to the production and release of **breast milk** from the mammary glands. In the last month of pregnancy, increasing levels of placental estrogen, progesterone, and human placental lactogen stimulate the production of prolactin from the anterior pituitary. Estrogen and prolactin stimulate alveoli in the mammary gland to grow and lactiferous ducts to branch. However, high levels of estrogen and progesterone inhibit the actual production of milk.

For the first few days after birth, the mammary glands secrete a thick yellowish fluid called **colostrum** (kohl-AWS-trum).

Colostrum has very little fat and less lactose than milk, but it contains protein and is rich in IgA antibodies. These antibodies from the mother provide passive immunity and protect against certain infections, such as bacterial ear infections, respiratory syncytial virus, and gastrointestinal viruses that cause diarrhea.

A few days after birth, true breast milk starts to be produced, because with the placenta gone, the levels of estrogen and progesterone decline and prolactin is not inhibited. Breast milk has a high fat content and contains substances such as essential fatty acids, growth factors, enzymes to help digest the milk, IgA, and immune chemicals such as complement proteins and interferon. Infants can easily digest and absorb it. Breast milk also acts as a mild laxative and helps cleanse **meconium** (mih-KOH-nee-um), a tarry, green fecal material made up of bile and other wastes that is first excreted by the newborn. Removing meconium quickly is important because fetal red blood cells break down rapidly after birth and, as a result, bilirubin can accumulate in the infant's tissues, causing jaundice.

Oxytocin is responsible for the actual *ejection* of milk from the alveoli of the mammary glands, a process called the **let-down reflex** (**Figure 27.16**). Let-down occurs when oxytocin binds to *myoepithelial cells* surrounding the glands and causes

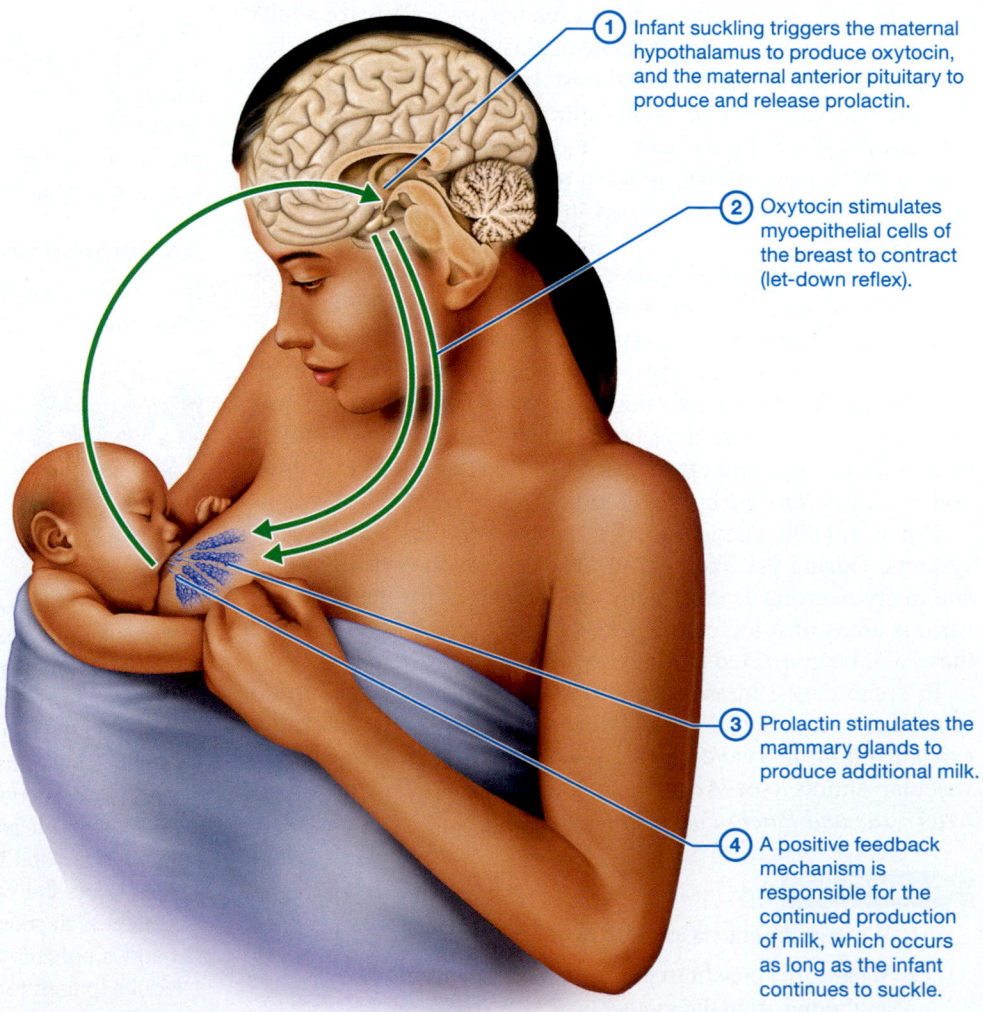

1. Infant suckling triggers the maternal hypothalamus to produce oxytocin, and the maternal anterior pituitary to produce and release prolactin.

2. Oxytocin stimulates myoepithelial cells of the breast to contract (let-down reflex).

3. Prolactin stimulates the mammary glands to produce additional milk.

4. A positive feedback mechanism is responsible for the continued production of milk, which occurs as long as the infant continues to suckle.

Figure 27.16 Hormonal regulation of lactation (the let-down reflex) by a positive feedback mechanism.

them to contract and eject the milk. During nursing, oxytocin also stimulates the recently emptied uterus to contract, which helps return it to pre-pregnancy size.

Continual milk production is an example of a positive feedback mechanism. In an example of the Feedback Loops Core Principle (p. 28), continuous milk production depends on mechanical stimulation of the nipples, which is provided by the infant's suck-

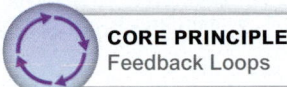

CORE PRINCIPLE
Feedback Loops

ling. Mechanoreceptors in the nipple trigger secretion of bursts of prolactin, which stimulate the production of milk needed for the next feeding. Interestingly, suckling also inhibits the release of gonadotropin-releasing hormone (GnRH) from the hypothalamus, which in turn decreases the release of follicle-stimulating hormone (FSH) and luteinizing hormone (LH) from the anterior pituitary. This inhibits both the ovarian and menstrual cycles, and for a certain period after birth (about 6–24 months), a breastfeeding woman will not ovulate or menstruate.

Milk production recedes when the infant no longer suckles. At this point, the mammary glands return to their pre-pregnancy state.

Quick Check

☐ 4. Describe the benefits of colostrum for an infant.

☐ 5. What role do hormones play in the let-down reflex and continual milk production?

Apply What You Learned

☐ 1. Why do infants begin receiving vaccinations for certain diseases immediately after birth?

☐ 2. As a newborn grows, his or her milk requirements increase. How does breast milk production increase to keep up with the demand?

See answers in Appendix A.

MODULE 27.7
Heredity

Learning Outcomes

1. Define chromosome, gene, allele, homologous, homozygous, heterozygous, genotype, and phenotype.

2. Analyze genetics problems involving dominant and recessive alleles, incomplete dominance, codominance, and multiple alleles.

3. Explain how polygenic inheritance differs from inheritance that is controlled by only one gene.

4. Explain how environmental factors can modify gene expression.

5. Discuss the role of sex chromosomes in sex determination and sex-linked inheritance.

How does a single-celled zygote become a human made of trillions of cells? And how does it "know" whether that individual should have brown or blue eyes, or brown or red hair? The answers to these questions are the same—it's in the genes. Recall from the cell chapter that **genes** are segments of the DNA that code for a specific protein (see Chapter 3). Proteins make up the bulk of body structures, so genes are basically the blueprints for tissues and organs. The genes are already present in that first small cell, and they are in every nucleated cell that is produced thereafter by mitotic divisions.

Although all healthy individuals have the same tissues and organs, the exact blueprint for each person is unique. Genes have different variants, or *alleles,* that allow for a variety of combinations. Most inherited traits are actually determined by multiple alleles or the interaction of more than one gene. In this module, we discuss the principles of heredity and examine the interactions of various genes and their alleles.

Introduction to Heredity

◀◀ FLASHBACK

1. What are homologous chromosomes? (p. 1023)

Heredity is the transmission of genetic characteristics from parent to offspring through the genes. The study of heredity and how it is transmitted is the field of **genetics.**

Collectively, all of an individual's DNA is referred to as the **genome;** each genome contains around 20,000 genes with about 3.2 billion nucleotides. These nucleotides are folded and compressed to form the 46 **chromosomes** that are found in all somatic cells with a nucleus (*chromo-* = "colored," *soma-* = "body," so named because they stain darkly). These chromosomes are paired so that there are 23 pairs of **homologous chromosomes** in each nucleus (see Figure 3.27). Each pair of homologous chromosomes contains one from each parent, so in total our cells have 23 maternal chromosomes and 23 paternal chromosomes. For this reason, the cells are called *diploid* (*di-* = "two"), meaning that they have two copies of each chromosome.

Chromosomes 1 through 22 are known as **autosomes** (AW-toh-sohmz). Each member of a pair of autosomes carries the same genes. The twenty-third pair are the **sex chromosomes,** the X and Y chromosomes, which determine the sex of the individual. In humans, a female normally has two X chromosomes, whereas a male has one X and one Y chromosome (in this case, the X and Y chromosomes are not truly homologous, as they contain different genes). A full set of the 23 chromosome pairs can be arranged and displayed in what is called a **karyotype** (KEHR-ee-oh-ty'p). To produce a karyotype, cells are chemically induced to divide, and then arrested in metaphase. A computer then arranges the chromosomes into pairs according to size and structure (look back at Figure 3.27).

The presence of abnormalities in the genome results in a **genetic disorder.** There are many possible abnormalities in an individual's DNA, ranging from a single incorrect nucleotide in a gene to the addition or subtraction of an entire chromosome or set of chromosomes. Let's now take a closer look at genes and examine how the genetic makeup, called the *genotype,* relates to how it manifests physically as a trait, which is known as the *phenotype.*

Genes and Alleles

Recall that homologous chromosomes have pairs of genes that code for the same trait. These genes are found at the same location, or **locus,** of that chromosome (see Chapter 26). However, these genes are not necessarily identical—they may have slightly different sequences of nucleotides. These variants of a gene are called **alleles** (uh-LEELZ). For example, both maternal and paternal chromosomes have genes that code for eye color. However, one gene may have an allele that codes for brown eyes, whereas the other gene may have an allele coding for blue eyes. If two alleles code for the same trait (such as blue eyes), they are said to be **homozygous** (hoh-moh-ZY-gus; *homo-* = "same") for that trait. If the two alleles are different (one codes for blue eyes, whereas the other codes for brown), they are **heterozygous** (het′-er-oh-ZY-gus; *hetero-* = "different").

Some alleles mask or suppress the expression of another allele for the same gene. An allele that can mask another is called **dominant,** and the allele that is masked is known as **recessive.** Alleles are generally represented with letters: Dominant alleles are shown as capital letters and recessive alleles as lowercase letters. Let's use the example of freckles to further explain these concepts. Freckles are the result of a dominant trait, which we represent with an "*F*." A lack of freckles is the result of a recessive trait, so we represent it with an "*f.*" Individuals with the genetic makeup *FF* are said to be homozygous dominant, those with *Ff* are heterozygous, and those with *ff* are homozygous recessive. Those who are homozygous dominant (*FF*) will have freckles. In addition, those who are heterozygous (*Ff*) will also have freckles, because the dominant allele masks the recessive one. Only individuals who are homozygous recessive, or *ff,* will have no freckles.

Note that not all alleles are dominant or recessive, and many factors often influence the inheritance of a trait. Also, most genes have more than two different alleles in the population (such as multiple choices for eye color), but each individual receives only two alleles—one from each parent.

Genotype versus Phenotype

An individual's genetic makeup is called the **genotype** (JEEN-oh-ty′p). The physical expression of an individual's genotype in the form of a trait is referred to as the **phenotype** (FEE-noh-ty′p). In the freckles example just described, *FF* and *Ff* are different genotypes, but they produce the same phenotype: the appearance of freckles. The genotype *ff* would yield the phenotype of no freckles. Although a person with a heterozygous genotype displays only the dominant phenotype, the recessive allele remains in his or her genotype. If this individual reproduces, he or she can pass on the recessive trait, which may appear as the phenotype in the offspring or future generations.

Quick Check

☐ 1. What is the genome? How many chromosomes are in the human genome?

☐ 2. What is the difference between homozygous and heterozygous alleles?

☐ 3. How does a genotype differ from a phenotype?

Patterns of Inheritance

« FLASHBACK

1. What is sickle cell anemia? (pp. 730–731)
2. What are the ABO blood groups? (pp. 746–747)
3. Which photoreceptors produce high-acuity color vision? (p. 553)

Some human phenotypes can be linked to a single gene pair, but most inherited traits are determined by multiple alleles or by the interaction of several genes. Environmental influences may also affect gene expression and ultimately a person's phenotype. The upcoming subsections look at some of the different patterns of inheritance, and the resulting genotypes and phenotypes.

Dominant and Recessive Traits

The pattern of inheritance known as **autosomal dominant-recessive inheritance** refers to the interaction of alleles that are strictly dominant or recessive. We already discussed this type of inheritance with the example of freckles in the previous section. The dominant allele (*F*) is always expressed in the phenotype, whether the individual is homozygous (*FF*) or heterozygous (*Ff*). Recessive alleles are expressed only if two of them are present (*ff*), because there is no dominant allele to mask the recessive one.

Other examples of autosomal dominant traits include dimples, astigmatism (irregularly shaped cornea or lens of the eye that causes blurred vision), and free (unattached) earlobes. The corresponding recessive traits are lack of dimples, nondistorted vision, and attached earlobes.

A **Punnett square** is a simple diagram used to predict the possible offspring genotypes that might occur if a man and woman have a child. **Figure 27.17** shows a Punnett square for the possible genotypes of offspring from a man and woman who are both heterozygous for dimples, or *Dd*. Remember that when sperm cells and oocytes form during meiosis, each receives only one allele. For this reason, one oocyte might have the dominant "*D*"

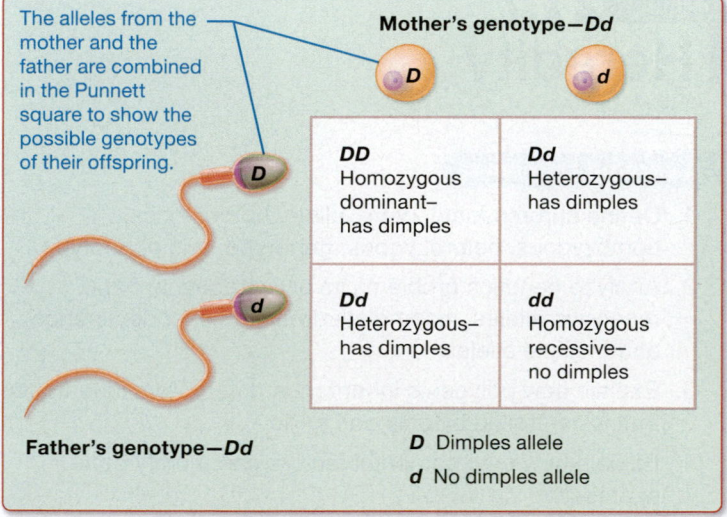

Figure 27.17 Possible offspring with a dominant-recessive trait: dimples.

allele, whereas the other has the recessive "*d*" allele, and the same options exist for the sperm cell. We can see in the figure that when both parents are heterozygous for dimples, the probabilities of offspring with each possible genotype are the following: *DD* (1/4, or 25%), *Dd* (1/2, or 50%), and *dd* (1/4, or 25%). The phenotypes of these offspring are as follows: dimples (*DD* or *Dd*, totaling 3/4, or 75%), and no dimples (*dd*, 1/4, or 25%).

Each time fertilization occurs, a new sperm and oocyte are involved, so the probabilities remain the same for each child. In other words, if the first child has no dimples, this does not mean the second child will have dimples. Each child has a 25% chance of lacking dimples.

Traits with Incomplete Dominance

Relatively few human traits follow a strict dominant-recessive inheritance pattern. In the pattern known as **incomplete dominance,** the heterozygous genotype results in a phenotype that is "between" the phenotypes of individuals who are homozygous dominant and homozygous recessive.

An example of incomplete dominance in humans involves the sickling gene that causes sickle-cell anemia. Individuals who are homozygous for the recessive sickling gene (*ss*) have sickle-cell anemia and will produce an abnormal hemoglobin, called hemoglobin S, which causes erythrocytes to bend into a sickle shape. The sickled cells block small blood vessels, causing pain, organ damage, and even death. Heterozygous individuals (*Ss*) have *sickle-cell trait* and make both hemoglobin S and normal hemoglobin. Most individuals with sickle-cell trait are healthy, but they can experience sickle-cell crises, particularly if they experience a low blood oxygen level. Individuals who are homozygous for normal hemoglobin (*SS*) produce no hemoglobin S.

Figure 27.18 shows a Punnett square with a mother and father who are both heterozygous for the sickling gene (*Ss*) and so have sickle-cell trait. The probabilities of each possible genotype are: *SS* (1/4, or 25%), *Ss* (1/2, or 50%), and *ss* (1/4, or 25%). The potential phenotypes of the offspring are as follows: normal hemoglobin (*SS;* 25%), sickle-cell trait (*Ss;* 50%), and sickle-cell anemia (*ss;* 25%).

Multiple-Allele Traits

Many human traits involve the interaction of multiple alleles. A good example of one such trait is the ABO blood group. Recall from the blood chapter that your blood type is determined by the presence or absence of the A and B antigens on erythrocytes (see Chapter 19). The allele for antigen A is represented as I^A, while the allele for antigen B is I^B. The allele that codes for a lack of either antigen (resulting in blood type O) is represented with an *i*.

ABO blood types exhibit the phenomenon of **codominance,** meaning that some alleles are equally dominant and so are equally expressed. I^A and I^B are codominant alleles, while *i* is recessive. Possible genotypes and phenotypes of the ABO blood group are shown here:

Genotype	Phenotype
$I^A I^A$	Blood type A
$I^A i$	Blood type A
$I^B I^B$	Blood type B
$I^B i$	Blood type B
$I^A I^B$	Blood type AB
ii	Blood type O

Figure 27.19a shows a Punnett square with a mother who has type A blood and is heterozygous for the trait ($I^A i$), and a father who has type B blood and is also heterozygous for the trait ($I^B i$). There are four possible genotypes for their offspring, with equal probabilities: $I^A I^B$ (1/4, or 25%), $I^A i$ (1/4, or 25%), $I^B i$ (1/4, or 25%), and *ii* (1/4, or 25%). The phenotypes of these individuals would be AB, A, B, and O, respectively.

Sex-Linked Traits

Sex-linked traits are those that are specifically expressed on the X chromosome or Y chromosome. The Y chromosome contains genes that determine male phenotypic traits, which explains why females have the genotype *XX* and males have the genotype *XY*. All oocytes carry an X chromosome, so female gametes do not determine the sex of an offspring. Sperm cells, however, carry either an X chromosome or a Y chromosome, so the male gamete determines the offspring's sex.

As you can see from this scanning electron micrograph, the X chromosome is significantly larger than the Y chromosome:

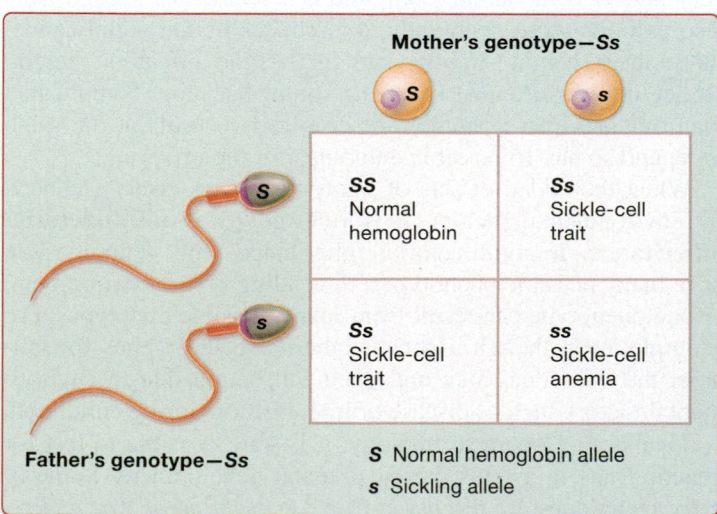

Figure 27.18 Possible offspring with an incomplete dominant trait: sickle-cell anemia.

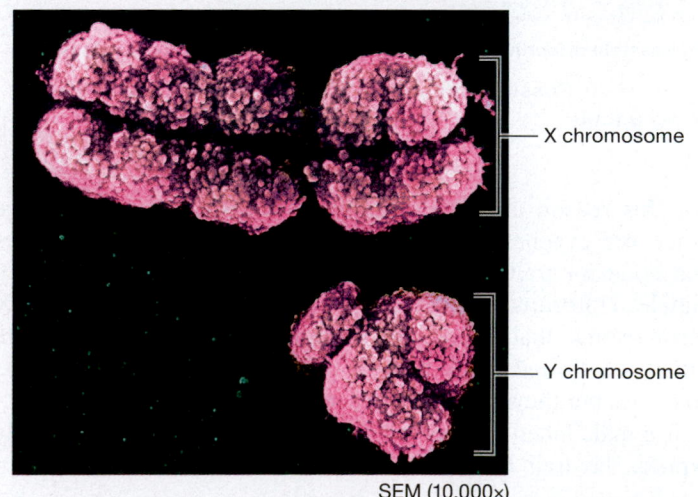

SEM (10,000×)

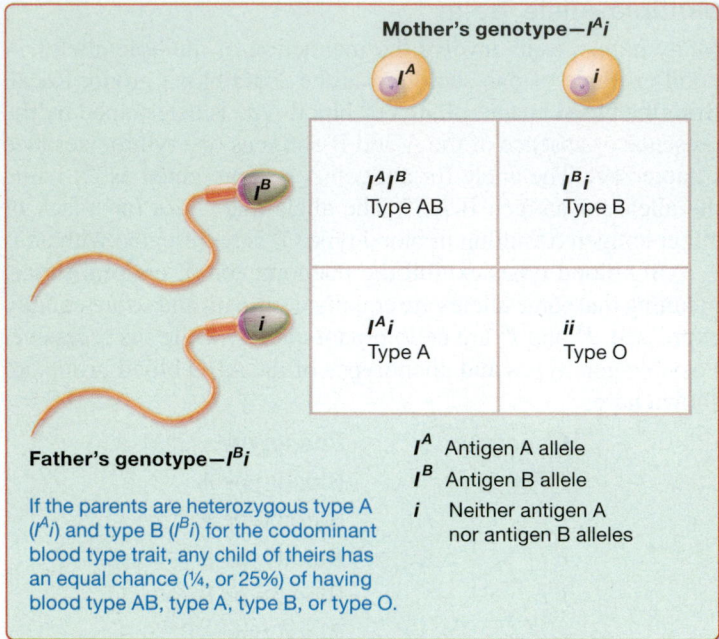

Mother's genotype—$I^A i$

	I^A	i
I^B	$I^A I^B$ Type AB	$I^B i$ Type B
i	$I^A i$ Type A	ii Type O

Father's genotype—$I^B i$

I^A Antigen A allele
I^B Antigen B allele
i Neither antigen A nor antigen B alleles

If the parents are heterozygous type A ($I^A i$) and type B ($I^B i$) for the codominant blood type trait, any child of theirs has an equal chance (¼, or 25%) of having blood type AB, type A, type B, or type O.

(a) Possible offspring with a multiple-allele trait: blood type

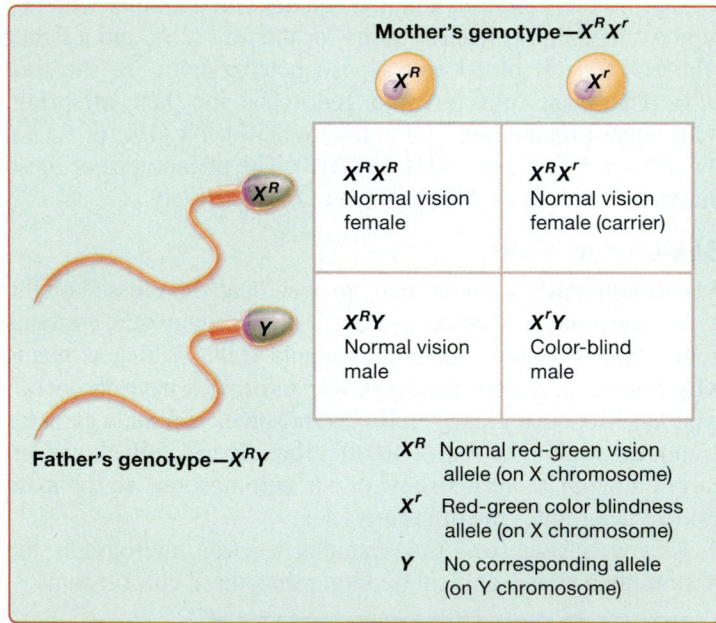

Mother's genotype—$X^R X^r$

	X^R	X^r
X^R	$X^R X^R$ Normal vision female	$X^R X^r$ Normal vision female (carrier)
Y	$X^R Y$ Normal vision male	$X^r Y$ Color-blind male

Father's genotype—$X^R Y$

X^R Normal red-green vision allele (on X chromosome)
X^r Red-green color blindness allele (on X chromosome)
Y No corresponding allele (on Y chromosome)

(b) Possible offspring with X-linked trait: red-green color blindness

Figure 27.19 Possible offspring with multiple-allele traits and X-linked traits.

For this reason, the X chromosome holds approximately five times more genes than the Y chromosome, including genes that code for traits other than those that make an individual a female. Unfortunately, there can be alleles for genes on the X chromosome that encode harmful traits, producing what are called **sex-linked, or X-linked, disorders.** (*Y-linked disorders do occur, but they are extremely rare.*)

If a male inherits a trait for an X-linked disorder, he will likely express the trait because the Y chromosome does not have a corresponding allele to influence its expression. However, a female will express the trait for an X-linked disorder only if she is homozygous for the recessive allele. A female who is heterozygous for a recessive allele does not show the phenotype of the trait, but she is a *carrier* of that allele. This means that she can pass the allele on to her offspring. If she passes this on to a female child, her daughter will also be a carrier. If she passes it on to a male child, her son will express the trait. Note that a recessive allele is never passed from a father to a son because the son receives only the father's Y chromosome. A father can, however, pass on a recessive allele to a daughter, because he contributes one of her X chromosomes.

Examples of X-linked disorders include Duchenne muscular dystrophy (a muscle-destroying disease; see Chapter 10), certain forms of baldness, color blindness, and hemophilia (a disorder of blood clotting; see Chapter 19). Let's look further into what happens with color blindness. In red-green color blindness, the recessive allele results in defective cones that do not allow the affected individual to distinguish between red and green. **Figure 27.19b** shows a Punnett square with a mother who is heterozygous for red-green color blindness ($X^R X^r$), and a father with one allele for normal red-green vision ($X^R Y$). The four possible genotypes of offspring are as follows, with equal probabilities: $X^R X^R$ (1/4, or 25%), $X^R X^r$ (1/4, or 25%), $X^R Y$ (1/4, or 25%), and $X^r Y$ (1/4, or 25%). The phenotypes of these individuals are, respectively: female with normal vision, female carrier with normal vision, male with normal vision, male with red-green color blindness. So while a female child of these parents has no risk of red-green color blindness, a male child has a 50% chance of being red-green color blind.

Polygenic Inheritance Patterns

Most inherited traits are controlled by the combined effects of two or more genes, a condition called **polygenic inheritance** (pawl-ee-JEN-ik). Examples of traits with polygenic inheritance include height, skin color, and eye color. **Figure 27.20** shows a simplified example of polygenic inheritance in which two genes regulate height. The first gene exhibits three genotypes: *TT* is very tall, *Tt* is moderately tall, and *tt* is short. The second gene also exhibits three genotypes: *AA* reduces height significantly, *Aa* reduces height slightly, and *aa* has no effect on height. Notice that this Punnett square is somewhat more complicated than our previous examples—it crosses two traits at the same time, and so has 16 possible outcomes for the offspring.

When the added effects of many genes are also influenced by environmental factors, it is referred to as **multifactorial inheritance.** In multifactorial inheritance, one genotype can have many possible phenotypes, depending on the environment, or one phenotype can result from many possible genotypes. For example, even though a person inherits several genes for tallness, the individual may not reach full height due to environmental factors such as disease or malnutrition during childhood.

Thanks to modern technology, it is now possible to test for genetic traits in a fetus during prenatal development. Some of these techniques are discussed in *A&P in the Real World: Prenatal and Newborn Genetic Screening.*

Figure 27.20 Possible offspring with polygenic inheritance: height.

Prenatal and Newborn Genetic Screening

Cells and amniotic fluid may be withdrawn and analyzed to test for chromosomal abnormalities. Many genetic disorders can be detected by such procedures, including Down syndrome, sickle-cell anemia, and hemophilia. These tests are usually recommended for women who will be 35 years or older at delivery, as their oocytes are older, which increases the risk of chromosomal abnormalities. They are also recommended for women who know they or the father are carriers of inherited diseases, or when possible fetal abnormalities are discovered on ultrasound.

In **amniocentesis** (am′-nee-oh-sen-TEE-sis), which is done between 14 and 20 weeks of pregnancy, amniotic fluid is withdrawn using a needle inserted into the amniotic cavity, as shown to the right.

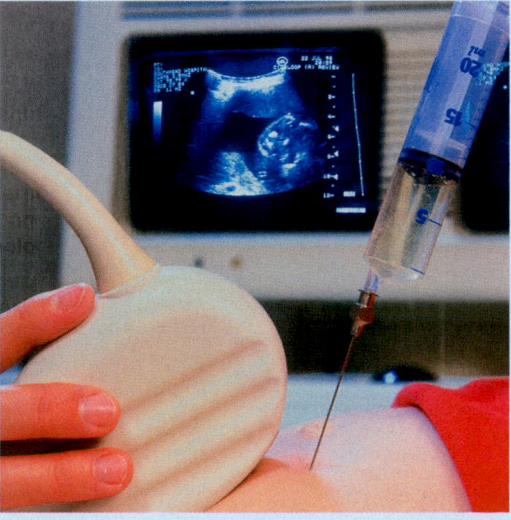

Similarly, **chorionic villi sampling,** which is usually performed between 10 and 12 weeks of pregnancy, withdraws chorionic villi tissue using a suction catheter guided through the vagina and into the uterus. It is also possible to remove chorionic villi by inserting a needle through the abdominal cavity. Both procedures require an ultrasound transducer, which uses sound waves to monitor the location and position of the fetus and the movement of the needle or catheter. Although the risk of amniocentesis or chorionic villi sampling is low, there is a possibility that the test could harm the fetus.

After birth, newborn screening procedures involve the analysis of blood samples to determine if the baby has a genetic or metabolic disorder. Prompt identification of any disorders can allow for early treatment and prevent long-term complications.

Quick Check

☐ 4. If a trait is autosomal dominant, how many alleles are necessary for the trait to be seen in the phenotype?

☐ 5. What is codominance?

☐ 6. What is an X-linked disorder? How can a female be a carrier of such a disorder?

Apply What You Learned

☐ 1. Draw a Punnett square to determine if it is possible for a man and a woman who both have attached earlobes to have a child with free earlobes. (*Hint:* Free earlobes

are autosomal dominant, and the trait follows the strict dominant-recessive inheritance pattern.)

☐ 2. A couple who are expecting a baby come to you for genetic counseling. The father is red-green color blind. The mother has normal color vision, but her father had red-green color blindness. What are the chances that the baby will be color blind? Are the probabilities different for male and female offspring? If so, why?

☐ 3. Determine the possible genotypes and phenotypes of offspring from a mother with the blood type AB and a father with the blood type O.

See answers in Appendix A.

Chapter Summary

For everything you need to succeed in this course, go to Mastering Anatomy & Physiology. There you will find:

- 📺 Practice Tests
- 🎙️ Author Podcasts
- ▶️ Big Picture Animations
-))) Concept Boost Video Tutors
- **iP2™** Interactive Physiology 2.0 Tutorials
- **A&PFlix** A&P Flix 3D Animations
- ✏️ Active-Learning Workbook

MODULE 27.1
Overview of Human Development
1066–1068

- A pregnancy consists of three **trimesters,** each lasting 3 months. The **pre-embryonic period** includes the first 2 weeks, the **embryonic period** includes weeks 3–8, and the **fetal period** includes weeks 9–38, or until birth.

- The **postnatal period** includes the changes from newborn to toddler, childhood, and adolescence into adulthood.

MODULE 27.2
Pre-embryonic Period: Fertilization through Implantation (Weeks 1 and 2)
1068–1075

- **Fertilization** is the fusion of a **sperm cell** and **secondary oocyte** to form a zygote. Sperm must migrate to the oocyte, undergo

capacitation, and penetrate the **corona radiata** and **zona pellucida** to successfully fertilize an oocyte.

- In the oocyte, the sperm cell swells to form the **male pronucleus.** The oocyte completes meiosis and forms the **female pronucleus,** and *amphimixis* fuses the two pronuclei to form the diploid zygote.

- The zygote undergoes **cleavage** to become a **blastocyst,** which differentiates into the **trophoblast** and an **inner cell mass.** The layers of the trophoblast produce **human chorionic gonadotropin (hCG)** and digest the uterine lining into which the blastocyst implants. hCG stimulates the corpus luteum to secrete estrogen and progesterone and promotes placental development.

- The inner cell mass forms the embryo and the **extraembryonic membranes,** including the **amnion, yolk sac, allantois,** and **chorion.**

MODULE 27.3
Embryonic Period: Week 3 through Week 8
1076–1079

- At the start of the embryonic period, **cephalocaudal folding** and **lateral folding** transform the embryo to its cylindrical shape. **Gastrulation** leads to the development of the three primary germ layers.

- **Organogenesis** involves differentiation of the germ layers into rudimentary organ systems. The three germ layers differentiate into the following:
  - The **ectoderm** forms the epidermis of the skin, the nervous system, and the sense organs.
  - The **mesoderm** forms the skeletal structures, muscles, most other organs (except the epithelial linings), mesenteries, and dermis.
  - The **endoderm** forms the linings and glands of the digestive and respiratory tracts and several endocrine glands.

MODULE **27.4**
Fetal Period: Week 9 until Birth (about Week 38) 1079–1084

- The **placenta** forms when the syncytiotrophoblast produces **chorionic villi** that extend into the **stratum functionalis** of the endometrium. The placenta provides oxygen and nutrients to the conceptus and secretes estrogen and progesterone to maintain the uterine lining.

- The **fetal period** is a time of rapid growth and maturation of tissues and organs.

- The fetal cardiovascular system includes **shunts** that bypass the pulmonary circuit and the liver. In the newborn, these shunts close.

MODULE **27.5**
Pregnancy and Childbirth 1085–1090

- Estrogen, progesterone, **relaxin, placental lactogen,** and **prolactin** maintain the pregnancy and prepare for delivery and lactation.

- During pregnancy, the mother's uterus eventually fills most of her abdominopelvic cavity and compresses her abdominal organs. Glandular tissue in her mammary glands grows in preparation for **lactation.**

- Physiological changes in the mother during pregnancy include an increase in blood volume and cardiac output, weight gain, and an increase in the glomerular filtration rate.

- Pregnancy terminates with **parturition,** or **childbirth,** when the fetus is expelled from the uterus through the vagina.

- Labor is initiated by a positive feedback mechanism that requires the secretion of **prostaglandins** and **oxytocin** to dilate the cervix and contract the uterus.

- The stages of labor include the **dilation, expulsion,** and **placental stages.**

MODULE **27.6**
Postnatal Changes in the Newborn and Mother 1090–1093

- The first 4 weeks after birth for the neonate constitute the **neonatal period.** The time period after birth for the mother is called the **postpartum period.**

- At 1 minute and 5 minutes after birth, an **Apgar score** is recorded to evaluate a newborn's physical status.

- The newborn's first breath changes the blood pressure in the pulmonary circulation, closing fetal vascular shunts.

- The mother's system returns to homeostasis after birth as estrogen and progesterone levels decline, the uterus returns to its pre-pregnancy size and secretes **lochia** for up to 6 weeks, and organ system functions return to pre-pregnancy levels.

- **Lactation** refers to the production and release of breast milk from the mammary glands. A positive feedback mechanism involving mechanical stimulation of the mammary glands is responsible for continued milk production.

MODULE **27.7**
Heredity 1093–1098

- The transmission of genetic characteristics from parent to offspring through the **genes** is called **heredity; genetics** is the science that studies heredity.

- Genes are segments of DNA that code for a specific protein, and **alleles** are variants of genes. The genetic makeup of an individual is referred to as the **genotype,** and the physical expression of an individual's genotype is the **phenotype.**

- **Autosomal dominant-recessive inheritance** refers to the interaction of alleles that are strictly **dominant** or **recessive,** which means that a dominant trait masks a recessive trait.

- Many traits involve **incomplete dominance,** or the interaction of multiple alleles.

- **Sex-linked traits** are those that are specifically expressed on the X or Y chromosome, called the **sex chromosomes.**

- **Polygenic inheritance** occurs when the combined effects of two or more genes produce a trait.

- **Multifactorial inheritance** involves the influence of environmental factors in addition to the combined effects of two or more genes to produce a trait.

Assess What You Learned

Scan the QR Code for additional practice test questions

https://goo.gl/iWcUyc

LEVEL 1 Check Your Recall

1. Match the correct time period of gestation with the prenatal period of development:

_____ Pre-embryonic period	a. Weeks 3–8
_____ Fetal period	b. Weeks 1–2
_____ Embryonic period	c. Weeks 9–38

2. Fill in the blanks: During the pre-embryonic period, the conceptus divides into hundreds of cells by a process called _____ and becomes a/an _____, which implants in the uterine wall.

3. Mark the following statements about fertilization as true or false. If a statement is false, correct it to make a true statement.

    a. Fertilization is the fusion of a sperm and a secondary oocyte to form a new cell called a zygote.

    b. Prostaglandins in semen may assist with the movement of sperm farther into the female reproductive tract by stimulating uterine contractions.

    c. Secondary oocytes are covered with a layer of cells called the zona pellucida.

    d. The cortical reaction allows additional sperm cells to bind to the oocyte.

4. Which of the following areas in the female reproductive tract is the site of fertilization?

   a. Uterus
   b. Ampulla of the uterine tube
   c. Vagina
   d. Cervix of the uterus
   e. Any of these areas are possible sites of fertilization.

5. Number the sequence of events in fertilization.

   _____ Formation of male and female pronuclei
   _____ Acrosomal reaction
   _____ Rejection of sperm
   _____ Binding of sperm cells
   _____ Cortical reaction
   _____ Fusion of pronuclei

6. Fill in the blanks. The blastocyst has two distinct cell populations: large, flat cells called the _____, and a cluster of round cells called the _____.

7. Match the extraembryonic membrane with the correct statement:

   _____ Amnion
   _____ Yolk sac
   _____ Allantois
   _____ Chorion

   a. Forms part of the digestive tract and is the first extraembryonic membrane to form
   b. Forms the base for the umbilical cord and later becomes part of the urinary bladder
   c. Encloses the conceptus and secretes fluid to fill the sac
   d. Forms outgrowths that blend with the stratum functionalis layer of the uterus

8. Fill in the blanks: _____ is the process of forming the placenta, which is a disc-shaped organ that attaches to the fetus via a/an _____.

9. Match the structures of the embryo with the primary germ layer from which each developed. Each germ layer will have more than one structure.

   _____ Endoderm
   _____ Mesoderm
   _____ Ectoderm

   a. Nervous system
   b. Linings of the digestive tract
   c. Skeletal structures
   d. Epidermis of the skin
   e. Thyroid and parathyroid glands
   f. Skeletal muscles

10. What are the two types of embryonic folding?

11. Match the fetal developments to the correct month of gestation.

    _____ Month 3
    _____ Month 4
    _____ Month 5
    _____ Month 6
    _____ Month 7
    _____ Months 8 and 9

    a. A downy hair called lanugo develops.
    b. Organs specialize and grow.
    c. Lungs begin to produce surfactant.
    d. The heartbeat can be heard with a stethoscope.
    e. Fetus turns to the vertex position.
    f. Ossification begins in most bones.

12. Match the structures found in the fetal circulation with what each fetal structure becomes after birth:

    _____ Foramen ovale
    _____ Ductus arteriosus
    _____ Ductus venosus

    a. Ligamentum venosum
    b. Fossa ovalis
    c. Ligamentum arteriosum

13. The _____ is responsible for secreting the hormone human chorionic gonadotropin (hCG).

    a. Syncytiotrophoblast
    b. Ovary
    c. Endometrium
    d. Morula

14. Which of the following hormones secreted during pregnancy assists with increasing the mother's blood volume?

    a. Estrogen
    b. Progesterone
    c. Cortisol
    d. Aldosterone
    e. Parathyroid hormone

15. Number the sequence of events that are responsible for initiating labor.

    _____ Hypothalamus of the mother and fetus stimulate the placenta to secrete prostaglandins.
    _____ Fetal cortisol levels increase and stimulate the placenta to release estrogen.
    _____ Prostaglandins and oxytocin stimulate the uterus to contract.
    _____ Estrogen stimulates the myometrium to form oxytocin receptors.

16. Mark the following statements about the stages of labor as true or false. If a statement is false, correct it to make a true statement.

    a. The dilation stage is the time from the onset of labor until the cervix is fully expanded.
    b. The placental stage is the time after delivery of the infant when the placenta and the attached fetal extraembryonic membranes are delivered.
    c. The expulsion stage of labor is usually longer than the dilation or placental stages.

17. Fill in the blanks: The first _____ (time frame) after birth constitute(s) the _____ period of life for the newborn.

18. Number the sequence of events that initiate lactation.

    _____ Estrogen and progesterone levels decrease when the placenta is expelled and prolactin stimulates milk production.
    _____ In the last month of pregnancy, placental estrogen, progesterone, and placental lactogen stimulate the production of prolactin.
    _____ Estrogen and prolactin stimulate acini in the mammary gland to grow and lactiferous ducts to branch.
    _____ Oxytocin binds to myoepithelial cells in the mammary glands and causes them to contract and eject milk.

19. Fill in the blanks: _____ are segments of DNA that code for a specific protein, and the human _____ includes over 20,000 of these segments.

20. Which pattern of inheritance includes examples of codominance?

  a. Autosomal dominant and recessive traits
  b. Multiple-allele traits
  c. Sex-linked traits
  d. Polygenic inheritance

LEVEL 2 Check Your Understanding

1. The prenatal period typically lasts 38 weeks. From the clinical viewpoint, a pregnancy typically lasts 40 weeks. What is the reason for this difference?

2. Why is the single-celled zygote approximately the same size as the multicellular blastocyst before it implants in the uterine wall?

3. Why do most spontaneous abortions occur in the first trimester?

4. Explain why many pregnant women frequently feel the urge to urinate during their last trimester.

LEVEL 3 Apply Your Knowledge

PART A: Application and Analysis

1. Ms. Noble is in the first trimester of her pregnancy when she suddenly feels severe pain in her lower abdomen. She rushes to the hospital where it is discovered that she has an ectopic pregnancy. The physician wants to do immediate surgery, but Ms. Noble argues that the embryo could simply move to the uterus if they wait. What would you tell this patient?

2. A new allergy medicine has just been approved by the Food and Drug Administration (FDA); however, it is not approved for pregnant women. What might be the reason for this exclusion?

3. Determine the probability of a couple having a child with a widow's peak if the conditions are as follows: The gene for widow's peak follows strict autosomal dominant-recessive inheritance. Widow's peak (W) is dominant, and no widow's peak (w) is recessive. The genotype of the mother is Ww, and the genotype of the father is ww.

4. Elise is a freshman college student taking Introduction to Biology. In lab, she is asked her parents' blood types to complete a Punnett square. She knows that she has type A blood and that her mother has type B, but she does not know her father's blood type. Elise's lab partner assures her that she can figure out her father's blood type just by knowing her own and her mother's. Is her lab partner correct? Explain why or why not.

PART B: Make the Connection

5. Laura has diabetes and is 6 months pregnant. Lately, her blood pressure and blood glucose levels have been higher than normal. What factors related to the pregnancy might cause Laura's blood pressure and blood glucose to rise? *(Connects to Chapter 18)*

See answers in Appendix A.

APPENDIX A   ANSWERS

CHAPTER 1

Apply What You Learned

MODULE 1.1

1. and **2.** Answers will be unique for each student, so none are provided here.

MODULE 1.2

1. It involves the *chemical* thyroid hormone made by *cells* in the *tissues* of the *organ* the thyroid gland. This gland is part of the *endocrine system*. The entire *organism* can be affected by this condition. **2.** As living organisms, a human and a rose plant share the features common to all life forms: composed of cells, metabolism, responsiveness, movement (the movement of the rose plant is internal), growth, excretion, and reproduction.

MODULE 1.3

1. The answer to this question will, of course, be different for each student. Here is one example: The mole is located in the left anterior buccal region, 1 centimeter lateral to the oral region, and 4 centimeters superior to the mental region. **2.** Either a frontal section or a transverse section would allow you to view both lungs at the same time.

MODULE 1.4

1. The fragment entered the pelvic cavity, then passed into the abdominal cavity, where it likely encountered the peritoneal cavity (and the parietal and visceral peritoneal membranes). The fragment then pierced the diaphragm to enter the thoracic cavity and finally the pleural cavity (passing through the parietal and visceral pleural membranes). **2.** Inadequate serous fluid impairs an organ's ability to move within the cavity without friction. This could lead to pain as the two layers of the serous membrane rub together. **3.** This condition will impair an organ's ability to expand, as the excess serous fluid will compress the organ.

MODULE 1.5

1. Such cells will likely be unable to perform most of their original functions because they will have lost the cellular structures required to perform them. **2.** With a positive feedback loop, sweating would trigger more sweating, which would in turn trigger even more sweating, and the response would continue to amplify. This would lead to a lower than normal body temperature and dehydration. **3.** A gradient is present any time there is more of a substance in one area than another and the two areas are connected. In this case, there are a large number of molecules of air freshener producing the scent near the outlet, and the number of molecules decreases as you move farther away from the outlet.

Assess What You Learned

Level 1

1. anatomy; physiology
2. a. False: You should study using *various learning modalities*.
   b. False: You should budget your time for studying at least *one week* before an exam.
   c. True
   d. False: It is better to *get enough sleep* before an exam.
3. It is easier to navigate through familiar material; your professor prepares lectures with the assumption that you have read the assignment; your brain is best able to perceive concepts, ideas, and facts that it expects to see and that are familiar.
4. d   5. b

6. a. False: Histology is the division of microscopic anatomy that studies the *tissue* level of organization.
   b. True   c. True
   d. False: The anatomical position features the person facing forward, feet shoulder width apart, and palms facing *anteriorly*.

7. _d_ Anterior   _h_ Inferior
   _g_ Lateral   _c_ Distal
   _e_ Proximal   _b_ Medial
   _a_ Posterior   _f_ Superior

8. a   9. c   10. d   11. cranial, vertebral

12. thoracic, abdominopelvic   13. b, c, f   14. b   15. c

16. a. False: Structure and function are closely related at *all levels of organization*.
    b. True
    c. False: *Negative* feedback loops are triggered by a deviation from a homeostatic set point and are shut down when conditions return to the set point.
    d. False: The effects of *positive* feedback loops are amplified to create an escalating response.
    e. True

17. d

Level 2

1. The skull is hard and able to withstand tension (pulling) and compression (pushing) forces, so it is well suited to providing protection to underlying structures. The skull also has a number of cavities for housing delicate structures and a number of prominences (bulges) to which muscles can attach.

2. a. The esophagus is located in the mediastinum of the thoracic cavity, along the body's midline. It begins inferior to the oral cavity in the cervical region and ends at the stomach in the epigastric region. It is located posterior to the trachea and great blood vessels and anterior to the spinal column.
   b. The brain is located within the cranial cavity, deep to the skull. It is superior to the spinal cord and extends from the frontal region to the occipital region.
   c. The urinary bladder in a female is located in the inferior pelvic cavity, inferior to the uterus, and along the midline. It is located deep to the skin, muscle, and bone of the pelvis.

Level 3

1. The organs that could be causing the pain in the right upper quadrant include parts of the large intestine, the liver, the gallbladder, the right kidney, and portions of the small intestine. The cause of the pain could also involve the peritoneal cavity.

2. The surgeon will first have to cut through the skin and muscles of the anterior abdomen. Next, the surgeon will enter the peritoneal cavity by cutting through the parietal peritoneum. Finally, the pancreas is deep to the stomach, so the surgeon will have to move the stomach out of the way in order to reach the pancreas.

3. The incisions will be unlikely to pass through the parietal peritoneum or enter the peritoneal cavity because the kidneys are retroperitoneal—behind the peritoneum. If the incision were made on the anterior lumbar region, then both sides of the parietal peritoneum would need to be incised to reach the kidneys. First, the parietal peritoneum on the anterior body wall would be cut, and then the parietal peritoneum on the posterior body wall would be cut.

4. A negative feedback loop would restore the blood pressure to normal levels and then shut off; a positive feedback loop would restore the blood pressure to normal levels and then cause it to continue to drop below its set point. The baroreceptor reflex is therefore likely to be a negative feedback loop, as a positive feedback loop would cause another homeostatic imbalance: low blood pressure.

CHAPTER 2
Apply What You Learned

MODULE 2.1

1. One sample is likely a different isotope of carbon, meaning that it has a different number of neutrons than the other sample. The added neutrons gave one of the samples a greater mass. **2.** Lithium has three protons, indicated by its atomic number, and three electrons. Its mass number of 7 tells us that it has four neutrons.

MODULE 2.2

1. Fluorine is much more reactive than neon. This is due to the valence electrons—fluorine has seven in its outer shell, and neon has eight. Therefore, neon obeys the octet rule and is relatively stable, and fluorine does not obey the octet rule and is reactive. **2.** No, it would not. Hydrogen bonds only form between two molecules with partial charges. A molecule of H_2 has a nonpolar covalent bond in which electrons are shared equally and neither hydrogen atom has a partial charge. **3.** Sodium atoms have a single valence electron, and so each of these atoms has only one electron to share. If two of them were to form a nonpolar covalent bond, each atom would still have only two valence electrons and would not obey the octet rule. For this reason, the Na_2 molecule would be very unstable and would degrade very quickly.

MODULE 2.3

1. You feel hot during and after exercise due to the production of heat, which is a byproduct of the conversion of chemical energy in the cells to mechanical energy of movement. **2.** An unstable molecule is one that tends to react with other unstable molecules in order to form stable ones. If most of our biological molecules were unstable, they would not maintain their structure or function in the body. **3.** Without enzymes, reactions in our cells would proceed at an extremely slow pace—too slow a pace to maintain life.

MODULE 2.4

1. The water around a fetus keeps it buoyant within the uterus and cushions it from mechanical trauma. The water also helps to maintain a stable temperature around the fetus. **2.** Heat produced by the body is normally absorbed by the water in and around the cells, which changes very little in temperature. When the amount of water in the body is depleted, the heat is absorbed less efficiently, which can lead to an overall increase in body temperature.

MODULE 2.5

1. Glucose would be the most soluble. A triglyceride would not be soluble at all due to its nonpolar structure. A large protein might have polar or even ionic bonds, but is likely too large to be fully soluble in water. **2.** The body's reactions are catalyzed by enzymes, most of which are proteins. At high temperatures, enzymes lose their structure or are denatured. A denatured enzyme is no longer able to function, and so the rates of the body's reactions would fall precipitously. **3.** Most enzymes are proteins. The structure of every protein in the body is coded for by the DNA. If the code of DNA for an enzyme is changed via a gene defect, then the structure of the enzyme will change. An enzyme's structure is very specific for its function, so the change in structure could render it nonfunctional.

Assess What You Learned

Level 1

1. a. True
  b. False: Protons have a positive charge, *neutrons have no charge,* and electrons have a negative charge.
  c. True
  d. False: Every element has a characteristic number of protons, which is called the element's *atomic* number.

2. atomic number; mass number   **3.** d

4. Electrons are transferred from a metal to a nonmetal in an ionic bond. In a covalent bond, however, electrons are shared between two nonmetals.

5. a. polar covalent   d. nonpolar covalent
  b. ionic   e. ionic
  c. nonpolar covalent   f. polar covalent

6. Hydrogen bonds are weak attractions between partially negative atoms and partially positive hydrogen atoms in one or more molecules with polar covalent bonds.

7. During a chemical reaction, electrons may be transferred from one atom to another, or chemical bonds may be broken, rearranged, and/or formed.

8. __c__ Endergonic reaction   __h__ Oxidation-reduction reaction
  __f__ Potential energy   __b__ Chemical energy
  __e__ Electrical energy   __d__ Catabolic reaction
  __g__ Anabolic reaction   __a__ Kinetic energy

9. b   **10.** d   **11.** b   **12.** hydrophobic; hydrophilic

13. a. True
  b. False: As a solution's pH rises, it becomes more *basic.*
  c. True
  d. False: A solution with a pH of 10 has *fewer* hydrogen ions than a solution with a pH of 2.

14. d

15. Triglycerides consist of three fatty acids bound to glycerol. They are highly nonpolar molecules. Phospholipids consist of two fatty acids and a phosphate group bound to glycerol. The two fatty acids are highly nonpolar, but the phosphate group is highly polar, making the overall molecule amphiphilic.

16. a. True
  b. False: The main storage form of glucose in the human body is *glycogen.*
  c. False: Lipids contain *hydrocarbon chains* and are therefore *nonpolar* covalent molecules.
  d. False: Proteins are composed of strings of *amino* acids linked by *peptide* bonds.

17. The structure of a protein is closely related to its function. If a protein loses its structure, which is known as denaturation, then the protein is unable to function.

18. a

19. a. RNA   d. DNA
  b. DNA   e. DNA
  c. RNA   f. RNA

20. d

Level 2

1. The product of the radioactive decay will not be the same element. An element is identified by its atomic number, or number of protons. If the number of protons changes, then the atomic number and the identity of the element also change.

2. Lipids are nonpolar and therefore hydrophobic, which causes them to clump together in the water-based environment of the digestive tract. Carbohydrates and proteins are generally polar, hydrophilic molecules, and so they mix and interact with the contents of the digestive tract more easily.

3. The difference in polarity is due to the ratio of oxygen atoms to carbon and hydrogen atoms in the two molecules. Monosaccharides have proportionally more oxygen atoms, and they are arranged in such a way that the overall molecule is polar covalent. Fatty acids have fewer oxygen molecules and long hydrocarbon chains that make the overall molecule nonpolar.

Level 3

1. Cellulose is composed of glucose molecules linked by glycosidic bonds. If humans were to break down the cellulose, it would lead to the release of glucose molecules that would be absorbed into the bloodstream. This would actually increase the calories consumed and not lead to weight loss.

2. Food is very unlikely to affect the pH of the blood. However, if it did, the extra hydrogen ions would simply bind the conjugate bases of buffers, which would minimize the pH change.

3. Enzymes are necessary for essentially all of our chemical reactions, including the production of ATP and synthesis of proteins. With these processes inhibited, the cell ceases functioning and dies.

4. When the pH of a solution increases, a buffer releases hydrogen ions to decrease the solution's pH. When the pH of the solution returns to its normal value, the buffer stops releasing hydrogen ions. This is an example of a negative feedback loop. If this were a positive feedback loop, the buffer would bind progressively more hydrogen ions as the pH increased.

CHAPTER 3
Apply What You Learned

MODULE 3.1

1. Bacteria must have different mechanisms for modifying proteins, since they lack much of the cellular machinery for protein modification (Golgi and ER). They may also be limited in the amount of ATP they can produce because they lack mitochondria, and they have to carry out their cellular reactions in the cytosol, as they lack membrane-enclosed organelles. 2. Cells are specialized so that their structure enables them to carry out their functions. If a cell loses its specialized structure and becomes more generalized, it will also lose its ability to carry out these specific functions.

MODULE 3.2

1. The nonpolar parts of the detergent would interact with the fatty acid tails of the phospholipids, and the polar parts of the detergent would interact with the phosphate heads. These interactions would pull the phospholipid bilayer apart and rupture the cell. 2. Structural proteins of the plasma membrane help to maintain the membrane's stability, and when this is disrupted, the plasma membrane falls apart. When the plasma membrane falls apart, the cell dies.

MODULE 3.3

1. a. The hypotonic solution will cause an individual's red blood cells to swell, as water will move into the more concentrated cytosol by osmosis.
   b. Under normal conditions, this solution will cause no net change in cell volume by osmosis.

c. In a hypertonic solution, the red blood cells will crenate because water will move out of the cells into the more concentrated extracellular solution by osmosis.

2. The energy for secondary active transport relies on concentration gradients created by primary active transport pumps, especially the Na^+ gradient produced by the Na^+/K^+ pump. When this pump is disrupted, the cell is not able to establish this gradient, preventing normal secondary active transport.

MODULE 3.4

1. It would block production of the bulk of the cell's ATP, resulting in insufficient ATP to fuel the cell's many energy-requiring processes. 2. Phagocytes are cells that ingest large particles, which are then broken down inside the cells by products released from organelles such as lysosomes. Defective lysosomal enzymes would inhibit the phagocytes' ability to adequately digest these ingested particles, particularly bacteria. 3. This poison inhibits protein synthesis. Proteins in the cell may act as enzymes, structural molecules, transport molecules, and more. Without the ongoing synthesis of these proteins, a cell will rapidly die.

MODULE 3.5

1. Microtubules are assembled and disassembled as needed by a cell—when a microtubule is no longer needed, it is disassembled and then a new one is reassembled to suit a different need. If microtubules could not be assembled, then essentially all cellular functions and elements that require them would be hindered. The cell's shape would become abnormal; organelles would not be held in position; the endoplasmic reticulum would collapse; and the chromosomes would not separate into two sets during mitosis. 2. Both cilia and flagella have a core of microtubules that are critical to their structure and movement. Blocking microtubule assembly would inhibit their motility and therefore their functions. 3. Intermediate filaments consist of protein filaments wrapped around one another. This creates a ropelike filament that is quite strong and much harder to disassemble than either microtubules or actin filaments, both of which are assembled from globular proteins.

MODULE 3.6

1. Ribosomes are made in the nucleolus, and they include rRNA encoded by genes in that structure. The nucleolus is a component of the nucleus. Anucleate cells would therefore be unable to produce new ribosomes. 2. The DNA in a cell is packaged into long chromatin threads consisting of the DNA molecule itself wrapped around various proteins such as histones. To isolate the DNA, the proteins must first be degraded. 3. Cells that are very active in protein synthesis need more ribosomes and larger nucleoli due to the prominent role ribosomes play in this process.

MODULE 3.7

1. The TGA triplet corresponds to the ACU codon, which codes for the amino acid threonine. The mutated triplet corresponds to the ACG codon, which also codes for the amino acid threonine. Therefore, this particular mutation will not change the structure of the protein. The second mutation corresponds to the codons CGC mutating to CCC. This will result in the change of amino acid from arginine to proline, which will change the protein's structure. 2. In inhibiting ribosomes, the poison ricin specifically inhibits translation. Protein transcription will not be impacted, but the ribosomes will be unable to bind the mRNA transcript and begin translation.

MODULE 3.8

1. The process of cytokinesis divides the cytoplasm and other organelles between the two daughter cells. If cytokinesis did not take place,

the result would be a huge cell with several times the normal cytoplasmic volume. **2.** Periwinkle will inhibit mitosis by preventing the formation of the mitotic spindle from microtubules, which prevents the separation of chromosomes. Yew will inhibit it by preventing the disassembly of the spindle, which will block the cell's ability to re-form the nuclear membrane and chromatin.

Assess What You Learned

Level 1

1. d **2.** plasma membrane; cytoplasm; nucleus

3. The cytosol (intracellular fluid) is the fluid compartment in the cells, and the extracellular fluid is the fluid outside the cells. The two are kept separate by the cells' plasma membranes.

4. b

5. a. True
  b. False: Cholesterol provides the plasma membrane with stability in the face of changing *temperatures.*
  c. False: Membrane *glycoproteins and glycolipids* are vital for cell-cell recognition.
  d. True
  e. False: The overall structure of the plasma membrane is a mosaic with the components *moving fluidly within the membrane.*

6. Passive transport processes require no expenditure of energy by the cell. Active transport processes require energy expenditure in the form of ATP.

7. _d_ Osmosis   _c_ Simple diffusion
   _h_ Secondary active transport _b_ Primary active transport
   _g_ Exocytosis   _f_ Pinocytosis
   _a_ Phagocytosis   _e_ Facilitated diffusion

8. into, swell; out of, crenate

9. _f_ Peroxisome   _b_ Golgi apparatus
   _d_ Ribosome   _c_ Lysosome
   _h_ Smooth endoplasmic _a_ Rough endoplasmic reticulum
      reticulum
   _e_ Mitochondrion   _g_ Vesicle

10. The smooth and rough endoplasmic reticulum completes the synthesis of lipids and proteins. They are then sent by vesicles to the Golgi apparatus, which modifies and sorts the products, and then packages them into another vesicle. The vesicles are sent to lysosomes, other organelles, or the plasma membrane.

11. a. False: Actin filaments combine with myosin motor proteins to provide the cell with *movement.*
  b. False: *The flagella* found on sperm cells propel the cells through a liquid medium.
  c. True
  d. False: Intermediate filaments are ropelike structures composed of *fibrous* proteins.
  e. True
  f. False: Endocytosis, exocytosis, muscle cell contraction, and cellular "crawling" are all mediated by *actin* filaments.

12. d

13. Chromatin is condensed during cell division by being coiled repeatedly around histone proteins. This enables precise division of the chromosomes into two complete sets when the cell divides.

14. a

15. a. In a gene, each *triplet* specifies one amino acid in a protein sequence.
  b. A transcription factor must bind to the promoter region of a gene before the enzyme *RNA polymerase* is able to bind and begin transcription.

  c. The enzyme RNA polymerase builds a strand of *messenger* RNA, whose codons are complementary to DNA's triplets.
  d. Proteins destined for secretion from the cell enter the *rough ER* after translation, to be folded and modified.
  e. During translation, amino acids are delivered by the *tRNA molecule.*

16. a. _8_   f. _2_
   b. _6_   g. _5_
   c. _3_   h. _7_
   d. _1_   i. _4_
   e. _9_

17. a

18. The cell cycle must be tightly regulated in order to balance cell division with cell death. Improperly regulated cell division may lead to a progressive deterioration of the tissue (if cells do not divide enough) or an increase in the number of cells in a tissue (if the cells divide too often).

19. a. False: During the G_0 phase of the cell cycle, the cell stalls until conditions for division are more favorable.
  b. False: During *anaphase,* the sister chromatids are pulled apart and the chromosomes move to the opposite poles of the cell.
  c. False: The main enzyme that builds the new DNA strands during DNA synthesis is *DNA polymerase.*
  d. True

20. _g_ G_1 _h_ Telophase
   _d_ Metaphase   _c_ M phase
   _e_ S phase   _b_ G_2
   _a_ Cytokinesis   _f_ Anaphase

Level 2

1. a. Diffusion is movement of solute molecules from a high concentration to a low concentration.
  b. Osmosis is movement of a solvent from a solution with a lower solute concentration to one with a higher solute concentration.
  c. Primary active transport is transport of a substance across the plasma membrane against its concentration gradient with the use of energy from ATP.
  d. Secondary active transport moves a substance across the plasma membrane against its concentration gradient using the energy from another substance's concentration gradient.
  e. Transcription means copying the code of a gene by creating a complementary strand of mRNA.
  f. Translation means translating the code in the mRNA into a protein with the help of tRNA and ribosomes.
  g. DNA synthesis is the replication of a cell's DNA in preparation for cell division.
  h. Mitosis is division of a cell's nuclear material between two daughter cells.

2. Cells with a very high demand for ATP will be disproportionately affected by mitochondrial diseases. Examples include those of muscles, the nervous system, the liver, and the kidneys.

3. a. Cilia are short, motile extensions from the cell. Their structure enables them to beat rhythmically to sweep substances past the cell.
  b. Microvilli are folds of the plasma membrane that increase the surface area for absorption, which increases the efficiency of absorption in cells for which absorption is a primary function.
  c. The ropelike structure of intermediate filaments allows them to be flexible but very strong and therefore provide the cell with structural support.
  d. Lysosomes are composed of a phospholipid bilayer surrounding digestive enzymes in an acidic environment. This structure

allows them to digest particles within the cell while preventing the enzymes in acid from damaging the rest of the cell.

e. The nuclear envelope is composed of a double phospholipid bilayer with large nuclear pores. It contains the chromatin and other nuclear structures within the nucleus while allowing fairly large molecules such as mRNA to move between the nucleoplasm and cytosol.

4. Healthy cells produce only specific proteins in specific amounts due to mechanisms that regulate gene expression. In cancer cells, these mechanisms may be lost due to changes in the DNA, causing them to produce inappropriate proteins.

5. Some cell types are subject to a lot of wear and tear because of their location and/or function, including the cells of the skin and the gastrointestinal tract. Such cells must have a rapid rate of mitosis to replace dead or damaged cells. Cells that have slow or absent cell division include those that have complex structure and/or do not have the cellular machinery for cell divisions. These cells include neurons and muscle cells.

Level 3

1. The fluids were likely hypertonic instead of isotonic. The hypertonic fluids caused the water to leave the patient's cells by osmosis, which led to his cells becoming dehydrated.

2. Introns are noncoding DNA, meaning that they do not code for any protein in the human body. Even if a virus were to "activate" them, it wouldn't result in the production of a protein, as they do not code for anything. In addition, introns are removed from mRNA transcripts before they are translated.

3. This poison would prevent the synthesis of tubulin proteins and therefore microtubules. This would inhibit the development of the mitotic spindle, and mitosis would be unable to proceed.

4. Microtubules are responsible for maintaining cell shape and the positions of organelles. Without microtubules, the organelles would disperse, particularly the members of the endomembrane system. In addition, vesicles and other cell structures would not be transported from place to place, disrupting vesicular transport within the cell and between the cell and the environment. Finally, if the cell has a flagellum or cilia, they would be unable to function. These conditions would result in the death of the cell.

5. An enzyme increases the rate of biological reactions by lowering their activation energy. Without enzymes, reactions occur so slowly that they functionally don't occur at all. Methotrexate, in stopping the synthesis of folic acid, also stops the synthesis of nucleotides. Without nucleotides, a cell cannot build new mRNA strands during transcription, make ATP molecules, make ribosomes, or replicate DNA. If present in a high enough concentration, this will kill the cell.

CHAPTER 4
Apply What You Learned

MODULE **4.1**

1. This would cause the tissues to be unable to resume their original shape and size once the stretching forces were removed. 2. Tight junctions make the spaces between the cells relatively impermeable to water and many solutes. Nonfunctional tight junctions in blood vessels would allow much of the fluid and solutes in blood to leak out of the vessel and into the extracellular fluid. 3. Desmosomes help to increase a tissue's mechanical strength. This is particularly important in tissues subject to high amounts of stress, like the skin. If the skin's desmosomes were not working properly, the skin would not be as resistant to mechanical stresses and could break down more easily.

MODULE **4.2**

1. The thinness of the epithelium of the lungs allows gases to diffuse across this barrier rapidly and efficiently. If the epithelium thickens, the distance over which the gases must diffuse will increase, making the process slower and less efficient. 2. The thickness of the stratified squamous epithelium in the skin helps to resist mechanical trauma and other stresses. As the epithelium gets progressively thinner, it is less able to resist these mechanical stresses, resulting in more skin lacerations. 3. Like the epithelium in the lung, the simple columnar epithelium in the digestive tract allows substances to diffuse across it. If the digestive tract had stratified columnar epithelium, it would be far too thick to allow efficient diffusion of water and nutrients into the bloodstream.

MODULE **4.3**

1. Collagen fibers are very tough but not very elastic or extensible. Replacement of elastic fibers with collagen fibers would therefore reduce a tissue's elasticity and extensibility and instead make it quite tough. 2. Hyaline cartilage has far more ground substance than fibrocartilage, which has far more protein fibers. This makes fibrocartilage much less smooth as a surface for articulation. 3. Osteoblasts are bone-building cells, and osteoclasts are bone-destroying cells. If osteoclasts were more active, there would be excessive destruction of bone and as a result the bone would be much weaker.

MODULE **4.4**

1. a. Skeletal muscle tissue is composed of contractile muscle fibers. Dense irregular connective tissue is composed of fibroblasts, bundles of collagen fibers, and ground substance.
   b. Skeletal muscle fibers generate tension, whereas the components of dense irregular connective tissue provide an organ with strength and resistance to mechanical stresses.
   c. None of the components of dense irregular connective tissue are contractile, and so the functional abilities of the tissue would be completely altered.

2. The intercalated discs allow the cardiac muscle fibers to contract in unison, producing a uniform contraction of the heart. If they were not functional, the cardiac muscle cells would not contract in unison and the heart would not have uniform, functional contractions.

MODULE **4.5**

1. Cells of brain tumors are generally neuroglial cells, as they are capable of mitosis. 2. An axon sends messages to target cells, including other neurons, muscle cells, and glands. If the axon is damaged, a neuron can still receive messages from other neurons, but can no longer send messages to target cells.

MODULE **4.7**

1. Serous fluid is thin and watery, whereas mucus is thick and sticky. Serous fluid in the respiratory tract would not be sticky enough to trap inhaled debris, and serous fluid in the digestive tract would not be thick enough to protect the epithelium from damaging substances like stomach acid. 2. Synovial fluid lubricates a joint so that two articulating bones can move smoothly against each other. A lack of synovial fluid could lead to a "dry joint" that doesn't move smoothly and causes pain on movement.

MODULE **4.8**

1. Most of the damaged cardiac muscle tissue will not regenerate; instead, it will undergo fibrosis. This results in the formation of collagenous connective tissue, which lacks the contractile properties of cardiac muscle tissue. For this reason, function of the heart muscle rarely returns to normal. 2. Nutrition is a key factor in wound healing by both fibrosis and regeneration. Individuals who are homeless generally have poor nutritional status, and their diets often lack factors that are essential for healing a wound.

Assess What You Learned

Level 1

1. Connective tissues have ECM as their main component, which allows them to serve their functions of binding and supporting. Epithelial tissue is composed of sheets of cells with little ECM, which enables it to form a barrier between the body and the external environment and between the organs and fluid-filled cavities.

2. a. connective
   b. muscle
   c. epithelial
   d. nervous
   e. muscle
   f. connective

3. a. Collagen fibers are very tough fibers that make a tissue resistant to tension and pressure.
   b. Glycosaminoglycans create a concentration gradient to draw water into the ECM by osmosis and trap it there; this makes a tissue more resistant to compression.
   c. Reticular fibers are smaller fibers that interweave to form supportive networks around the cells of many tissues and also form "webs" in certain organs to trap foreign cells.
   d. Proteoglycans are large molecules that make a tissue firmer and more resistant to compression and also act as a barrier to substances diffusing through the ECM.
   e. Glycoproteins hold cells in their places in a tissue and connect cells to one another and to components of the ECM.
   f. Elastic fibers have the ability to be stretched and return to their original length, and so give a tissue the properties of distensibility and elasticity.

4. impermeable; mechanical strength

5. a. True   b. True
   c. False: Epithelial tissue is *avascular*.   d. True
   e. False: Pseudostratified epithelium appears to be *stratified* but is actually *simple epithelium*.
   f. False: *Simple* epithelia are specialized to allow substances to cross their cells rapidly.

6. __c__ Simple squamous   __f__ Simple columnar
   __e__ Pseudostratified columnar   __b__ Transitional
   __a__ Keratinized stratified squamous   __d__ Simple cuboidal

7. a. Endocrine glands release hormones into the bloodstream and contain no ducts. Exocrine glands release substances into a body cavity or the outside of the body through a duct.
   b. A unicellular gland is a single cell that produces and secretes its product. A multicellular gland is a group of cells that function together to produce and release a product.
   c. During holocrine secretion, product accumulates until the cell ruptures and releases its product. During merocrine secretion, the cell secretes its product by exocytosis.

8. b

9. a. False: *Adipocytes* store lipids in a large inclusion in their cytoplasm.
   b. False: Loose connective tissue features *ground substance* as its primary component.
   c. True
   d. False: *Hyaline cartilage* provides a smooth surface on which bones may articulate with little friction.
   e. False: The ECM of bone tissue consists of calcium phosphate crystals *and an organic component*.

10. __f__ Hyaline cartilage   __g__ Fibrocartilage
    __d__ Elastic cartilage   __h__ Cardiac muscle
    __b__ Smooth muscle   __c__ Dense regular elastic tissue
    __a__ Dense irregular   __e__ Reticular connective tissue
    connective tissue

11. c   12. Neurons; neuroglia; dendrite; axon

13. a. Mucous membranes are composed of *epithelium* and the underlying loose connective tissue.
    b. *Goblet* cells secrete mucus.
    c. Synoviocytes are the secretory cells of *synovial membranes*.
    d. The cutaneous membrane is composed of *stratified* squamous epithelium, loose connective tissue, and dense irregular connective tissue.
    e. Serous membranes line all *body cavities*.

14. Epithelial tissues, many connective tissues, and smooth muscle tissue often heal by regeneration because their cells retain the capacity to divide by mitosis. Nervous tissue and other types of muscle tissue generally repair by fibrosis because their cells usually are unable to undergo mitosis.

Level 2

1. You would not expect bleeding to occur, because epithelial tissue is avascular, or lacking blood vessels.

2. Collagen fibers are an important component of all tissues listed, giving them strength and the ability to resist tension. In vitamin C deficiency, collagen synthesis is impaired, which leads to bone pain and weakness, skin fragility, tooth loss, joint pain, and myriad other symptoms.

3. Mucus protects many tissues, including the airways and the gastrointestinal tract. Inadequate mucus would allow bacteria and other irritants to have easier access to these tissues and would put the digestive tract at risk from its own acid and enzymes. Too much mucus could interfere with both breathing and digestion, processes that depend on transporting gases or nutrients across the epithelium.

Level 3

1. Desmosomes hold the epithelial cells of the skin together so that the skin functions as a unit. When the desmosomes are attacked and rendered nonfunctional, the cells will not be held to one another or to the underlying basement membrane. This leads to skin fragility and separation of the skin layers, which manifests as large blisters.

2. Each of these components of the ECM makes a tissue firm. When the ECM of all tissues has more collagen fibers, GAGs, and glycoproteins, the tissues will become firmer than normal, eventually reaching a point where they are stiff and immoveable.

3. The lungs contain elastic fibers, which enable the lungs to be both distensible and elastic. When elastic fibers are replaced with non-distensible and nonelastic collagen fibers, the lungs progressively lose their ability to stretch and then return to their original size after the stretching force is removed. This can severely limit one's ability to breathe.

4. The simple epithelia in these organs enable substances to rapidly cross the cells and move from one place to another. If the simple epithelia were replaced with keratinized stratified squamous epithelia, substances would either cross the epithelia too slowly or be unable to cross them at all.

5. a. The epithelium is likely to be simple, as the single layer of cells in simple epithelia allow substances such as ions to cross more efficiently and rapidly.
   b. When sodium ions move from one compartment to another, an osmotic gradient is created that causes water to follow by osmosis.

6. When a cell undergoes mitosis, it must replicate its DNA and then divide the DNA and the cytoplasm equally between the two daughter cells. For a cell such as a skeletal muscle fiber

that is multinucleated, it would be extremely difficult to accurately replicate and divide the DNA from many separate nuclei into two daughter cells.

7. Cells such as fibroblasts should have large numbers of protein-making organelles, including ribosomes, rough ER, and the Golgi apparatus. High rates of protein synthesis also require a great deal of ATP, so one would expect to find large numbers of mitochondria in these cells as well.

CHAPTER 5
Apply What You Learned

MODULE 5.1

1. The cells of the epidermis receive oxygen and nutrients from blood vessels located in the dermis. If these vessels were constricted for several hours, very little oxygen and very few nutrients would reach the epidermal cells, which could damage them and lead to their subsequent death. 2. The skin would feel cold, as the heat from the blood would be directed to deeper tissues. 3. Complications experienced by burn victims include difficulty regulating body temperature and fluid homeostasis, increased susceptibility to infection and harm from the environment, and decreased or abnormal sensation due to damaged nerves.

MODULE 5.2

1. There are touch receptors in the epidermis (Merkel cells), which is why you are able to feel the cut. However, there are no blood vessels in the epidermis, which is why the cut does not bleed. 2. It could potentially affect these superficial cells because they would have increased access to blood vessels. In addition, the thickness of the skin might increase, as the cells in the stratum granulosum might continue dividing if they gained access to high amounts of oxygen and nutrients.

MODULE 5.3

1. Her wound likely extends down to the hypodermis, which is where we find adipose tissue. 2. The wound runs perpendicular to the tension lines in the anterior forearm. Wounds that run perpendicular to skin tension lines are more likely to scar, as they tend to gape.

MODULE 5.4

1. The skin of an individual with vitiligo tends to appear mottled, with alternating patches of darker-pigmented and lighter-pigmented skin. 2. Breanna might want to tell Max that he should take an anatomy and physiology course! Melanin does naturally shield the nuclei of keratinocytes from the effects of UV rays, but only to a certain point. Excess exposure to this radiation will mutate the DNA and potentially burn the skin of anyone, irrespective of skin pigmentation. 3. The reddish hue imparted to fair skin by hemoglobin depends on the amount of oxygen to which the hemoglobin is bound. If the blood is well oxygenated, it will be bright red and give the skin a pink hue. If the blood is poorly oxygenated, it will be much less red and could even develop a faint purplish hue.

MODULE 5.5

1. The amount that a hair grows depends on the rate of cell division and the amount of time the cells spend in the growth phase. Increasing the rate of cell division and/or increasing the amount of time that hair is in the growth phase could increase the length of the eyelashes. 2. It is true that hair and nails are composed largely of the protein keratin. However, protein and amino acids are needed by dividing cells in the nail and hair matrices, not by the cells in the hair shaft and nail plate, which are dead. The products could potentially affect future hair and nail growth, but they will not change the present state of hair or nails.

MODULE 5.6

1. She has suffered a second-degree burn, which has involved all or part of the dermis. She is unlikely to develop significant scarring from the burn, but the UV radiation could have damaged her DNA. 2. This carcinogen in all likelihood will not lead to any type of cancer unless the skin is damaged or the exposure is repeated. Because hydrophobic lipid-like substances coat the keratinocytes of the epidermis, the hydrophilic carcinogen will have difficulty penetrating the skin to the deeper layers where dividing cells are located.

Assess What You Learned

Level 1

1. An organ by definition is two or more tissues that work together to achieve common functions. The skin consists of keratinized stratified squamous epithelium, loose connective tissue, and dense irregular connective tissue. These three tissues form a barrier between the environment and the internal body and perform a host of other functions.

2. b   3. c

4. Dermal blood vessels constrict when in a cold environment, to direct heat to deeper tissues and minimize heat loss to the environment.

5. _2_ Stratum spinosum   _4_ Stratum lucidum
   _5_ Stratum corneum   _3_ Stratum granulosum
   _1_ Stratum basale

6. a

7. a. False: *Keratinocytes* account for the bulk of the epidermis.
   b. False: Keratinocytes begin life in the stratum *basale* and gradually are pushed into the stratum *corneum*.
   c. True
   d. False: *Melanocytes* produce the pigment melanin.

8. c

9. Dermal papillae house capillary loops that provide oxygenated blood to the epidermal cells. They also house sensory receptors that detect changes in the environment.

10. b

11. a. True
    b. False: Melanin is produced by melanocytes and covers the nuclei of neighboring *keratinocytes*.
    c. False: Carotene is a *yellow-orange* pigment that accumulates in the stratum corneum.
    d. False: *Decreased* amounts of blood flowing through the dermis lead to pallor in the skin.
    e. True

12. d   13. hair shaft, hair root, hair follicle, matrix

14. b   15. holocrine, merocrine

16. _c_ Eccrine sweat gland   _e_ Mammary gland
    _a_ Sebaceous gland   _b_ Apocrine sweat gland
    _d_ Ceruminous gland

17. Sweat is a water-based secretion that functions primarily in thermoregulation and is secreted via merocrine secretion. Sebum is a lipid-based secretion that provides a hydrophobic barrier and is secreted via holocrine secretion.

18. b   19. b

Level 2

1. Both hair and nails are derived from the epidermis, which is avascular. Since these structures have no blood vessels, there is no bleeding when they are damaged unless the injury involves the underlying dermis.

2. Hair shafts consist of dead keratinocytes. Dead cells do not carry out chemical reactions, and so they have no need of vitamins. For shampoos and conditioners to have any effect on the hair, the vitamins would have to reach the cells in the hair matrix, and this is unlikely to happen. The claims are therefore very likely to be invalid.

3. An organ consists of two or more tissues. The hair and nails consist only of epidermal cells, and are therefore not organs. However, if the surrounding connective tissue is considered, they could be called organs.

Level 3

1. The patient has likely suffered heat stroke. The bright red color of the skin is due to the body's attempt to release heat via dilation of the vessels in the dermis.

2. The lipid-based toxins in poison ivy easily penetrate the epidermis and reach the cells of the deep epidermis and the dermis. However, the water-soluble toxins in venom cannot pass through the lipid-based substance that coats keratinocytes in significant amounts, and so the venom does not reach the deep epidermis and the dermis.

3. a. The immediate concern is malignant melanoma, because her lesion has the characteristics of this particular cancer.
b. Exposure to UV radiation mutates the DNA of cells in the skin. This DNA mutation is what triggers the production of more melanin by melanocytes. For this reason, any tanning, irrespective of the source, comes from the initial DNA damage and is therefore not completely safe.

4. Sebum helps to repel water from the skin, along with the lipid-based substance produced by the keratinocytes. If sebum were water-soluble, the skin would be less water-resistant and more prone to gain or lose water to the external environment.

5. a. Both chemicals are components of the ECM. Collagen is a protein fiber that gives the skin tensile strength, and hyaluronic acid is a glycosaminoglycan that traps water in the ECM and increases its firmness.
b. Most proteins are large polar compounds.
c. Since these compounds are large and polar, they are unlikely to penetrate the water-resistant lipid coating of the epidermis in significant amounts.
d. The collagen and hyaluronic acid would need to reach the dermis in order to have any effect. Since they are unlikely to penetrate the epidermis in any significant amounts, they are not likely to be effective.

6. A second-degree burn involves epithelium and connective tissue, which generally heal by regeneration unless the damage is extensive. However, muscle tissue generally heals by fibrosis, so healing of a severe third-degree burn will involve some degree of fibrosis.

CHAPTER 6
Apply What You Learned

MODULE **6.1**

1. If the spongy bone were on the outside, the thin trabeculae would be subjected to more mechanical trauma than usual, which would damage their structure. In addition, the bone marrow housed within the trabeculae could also sustain damage without the outer shell of compact bone to protect it. **2.** The epiphyseal plate is the structure from which a bone grows in length. If this structure is damaged, longitudinal growth could be impacted, and the bone on one side of the body could be shorter than the same bone on the other side of the body. **3.** The nutrient artery supplies a good portion of the blood flow to the bone. If this artery is damaged, portions of the bone could be cut off from the normal blood supply, which could lead to further bone injury. A bone with a damaged nutrient artery will take longer to heal because it will lack sufficient blood flow to deliver adequate oxygen, nutrients, and other components vital to healing.

MODULE **6.2**

1. Bones affected by osteogenesis imperfecta are extremely brittle, as they lack the collagen fibers to give them tensile strength. This makes them much more likely to fracture when subjected to even normal physical forces. **2.** Excess osteoclast activity would mean that more ECM would be broken down than built, which would lead to progressive weakening of the bone. **3.** The function of compact bone is to provide a hard outer shell that resists all types of forces and protects the inner spongy bone and marrow. The concentric rings of bone matrix that characterize compact bone make it extremely strong and hard and enable it to perform these functions.

MODULE **6.3**

1. Osteoclasts function in both types of ossification to resorb the initial primary bone so that it can be replaced with secondary bone. If osteoclasts failed to function, the weaker, less organized primary bone would remain instead of being replaced. In addition, nonfunctional osteoclasts would result in the failure of the medullary cavity to form and enlarge during endochondral ossification. **2.** Primary bone is much weaker than secondary bone because it has little inorganic matrix, which makes it much less able to resist compressive forces. In addition, the irregularly arranged collagen bundles make primary bone less resistant to tensile forces, which further weakens it.

MODULE **6.4**

1. The decreased rate of mitosis would reduce the number of chondroblasts at the epiphyseal plate, and therefore slow the production of hyaline cartilage and the longitudinal growth of the bones. In addition, malnutrition would lead to an insufficient intake of calcium salts, vitamin D, vitamin C, and other nutrients needed for ossification, which could weaken the bones and further delay their growth. **2.** Growth hormone also promotes appositional bone growth, so if it is secreted in excess, the bones will grow in width, even if they can no longer grow in length. **3.** A young girl with excessive estrogen secretion is likely to have a significant increase in longitudinal bone growth, but is also likely to suffer from premature closure of the epiphyseal plates. Her height would increase in the short term, but her adult height might be less than average.

MODULE **6.5**

1. Load-bearing exercise exerts compression and tension on bones, which increases osteoblast activity. Astronauts in space for a long period are not able to engage in load-bearing activities, and so their osteoblasts are less stimulated than normal, leading to reduced bone mass. **2.** A tumor that secretes parathyroid hormone would trigger excessive osteoclast activity, and the resulting breakdown of bone would lead to more frequent bone fractures. In addition, it would cause the concentration of calcium ions in the blood to be much higher than normal because the normal negative feedback loop would be ineffective.

Assess What You Learned

Level 1

1. a

2.

e	Long bone	c	Short bone
g	Epiphysis	b	Periosteum
a	Nutrient artery	d	Diaphysis
h	Flat bone	f	Medullary cavity

3. Yellow bone marrow consists largely of adipose tissue, whereas red bone marrow contains the hematopoietic stem cells that form all formed elements of the blood.

4. a. True
   b. False: *Both the organic and inorganic matrices of bone are* responsible for the strength of bone tissue.
   c. False: Collagen fibers are one of the predominant parts of the *organic* matrix.
   d. True
   e. False: *Osteoclasts* are responsible for bone resorption, and *osteoblasts* are responsible for bone deposition.
   f. True

5. osteon; lamellae, central canal; perforating canals; lacunae, canaliculi

6. d   7. b

8. a. Endochondral   d. Intramembranous
   b. Both   e. Both
   c. Intramembranous   f. Endochondral

9. The primary ossification center is generally in the diaphysis of a long bone, whereas the secondary ossification centers are generally in the epiphyses.

10. a   11. c

12. $\dfrac{4}{\dfrac{1}{\dfrac{3}{2}}}$

13. a. Growth hormone increases both longitudinal and appositional growth by increasing the rate of mitosis of chondrocytes in the epiphyseal plate, increasing the activity of osteogenic cells, and directly stimulating the osteoblasts in the periosteum.
   b. Testosterone increases the rate of appositional bone growth, the rate of calcium ion deposition, and the rate of mitosis at the epiphyseal plate. It also accelerates closure of the epiphyseal plate.
   c. Estrogen increases the rate of mitosis of chondrocytes at the epiphyseal plate, inhibits osteoclasts, and accelerates closure of the epiphyseal plate.

14. osteoblasts, organic matrix; inorganic matrix, vesicles; osteoclasts, hydrogen ions, enzymes

15. a. True   b. True
   c. False: The greater the load the bone must carry, the *less* bone that is resorbed by osteoclasts.
   d. True
   e. False: Parathyroid hormone increases the blood calcium ion concentration by increasing the activity of *osteoclasts*.
   f. True

16. d

17. $\dfrac{3}{\dfrac{1}{\dfrac{4}{2}}}$

Level 2

1. An individual who is not bearing weight regularly is not placing an adequate load on his or her bones. This leads to insufficient stimulation of bone deposition, and so bone resorption predominates. This leads to progressive loss of bone mass.

2. The skin is involved in the production of vitamin D, and the kidneys and intestines must be able to absorb calcium ions into

the blood. If any of these organs are diseased and unable to perform their functions, levels of vitamin D and/or calcium ions in the blood will be inadequate to maintain bone mass.

3. The epiphyseal plate is composed of hyaline cartilage. Any disease that affects hyaline cartilage will impair the function of the epiphyseal plate, and therefore the structure and function of the bone as it grows.

Level 3

1. The periosteum is where the majority of the osteoblasts reside. Osteoblasts are needed in fracture repair to rebuild the bone and bridge the gap between bone fragments. If the periosteum is stripped, bone healing will occur much more slowly because the population of osteoblasts will be dramatically reduced.

2. Testosterone and other anabolic steroids accelerate the closure of the epiphyseal plates. If anabolic steroids are taken during adolescence, the elevated levels of androgens will cause the plates to close even more rapidly, which will lead to shorter stature.

3. Lucy lives in an environment in northern Canada where she likely receives inadequate amounts of ultraviolet radiation. This has decreased the amount of vitamin D synthesis taking place in her skin. This decrease, combined with the lack of vitamin D supplementation, has resulted in a severe lack of vitamin D overall and the consequent leaching of calcium ions out of her bones. Such a condition has led to her stunted growth and weakened bones.

4. Lysosomes are organelles that generally perform digestive functions. Their main enzymes for catalyzing these breakdown reactions are functional only at an acidic pH. The lysosomes of osteoclasts contain enzymes responsible for catalyzing reactions that break down the organic component of bone. If the lysosomes are unable to maintain an acidic pH, these enzymes cannot function. This will lead to an imbalance in which bone deposition abnormally dominates bone resorption.

CHAPTER 7
Apply What You Learned

MODULE 7.1

1. The bones of the axial skeleton protect delicate vital structures, such as the spinal cord, brain, heart, and lungs. The bones of the appendicular skeleton are more important for movement than protection, and so damage to these bones is likely to be less damaging to overall homeostasis. 2. Openings typically allow the passage of important structures, such as nerves, blood vessels, and the spinal cord. Openings that are too small could compress these important structures and impair their functions. Those that are too large could lead to instability and inadequate support of the structures.

MODULE 7.2

1. a. The ethmoid bone has likely been fractured. The superior part of the ethmoid forms both the roof of the nasal cavity and part of the floor of the cranial cavity. A fracture of this bone could connect the two cavities, allowing the fluid to leak out from around the brain through the nose.
   b. This portion of the sphenoid bone forms the posterior orbit through which the nerve for vision passes. A fracture in this location could damage the nerve, leading to vision problems.

2. Ms. Midna likely has an infection in her frontal and maxillary sinuses, leading to her pain. The paranasal sinuses are connected to the nasal cavity via small bony canals, which allow infectious agents to pass from one location to the other.

MODULE **7.3**

1. Ms. Cho has also likely fractured her fourth rib, which articulates with the transverse process. The ribs are attachment points for muscles of breathing, and so each breath pulls on the fractured rib, causing pain.

2. a. The abnormal curvature is scoliosis.

  b. The muscle weakness has allowed the curvature to develop. The curvature, in turn, compressed the discs in the lower thoracic region, leading to their herniation.

MODULE **7.4**

1. The trochlear notch of the ulna forms a joint with the distal humerus. Normally, the notch is fairly deep, so the two bones fit together well and are stable. However, if the notch were abnormally shallow, the bones would not fit together as snugly and the stability of the joint would be compromised. **2.** Mr. Heller has likely fractured his clavicle. The clavicle normally braces the upper limb in its proper position. When it is fractured, this function is disrupted, and the upper limb hangs in a more medial position than normal.

MODULE **7.5**

1. Many characteristics can be examined, including the overall shape, the angle of the pubic arch, the shape of the pelvic inlet, the orientation of the ischial tuberosities, and the shape and size of the sacrum. **2.** The head of the femur fits into the acetabulum, which is normally fairly deep. This gives the hip joint a fair amount of stability, which would be lacking were the acetabulum too shallow. The lack of stability is likely putting abnormal strain on her pelvic bones, femur, and surrounding structures, leading to her pain.

Assess What You Learned

Level 1

  1. c, d, and f   **2.** b   **3.** sagittal; coronal, squamous, lambdoid

  4. a. False: The four paranasal sinuses include the frontal, *ethmoidal*, sphenoidal, and *maxillary* sinuses.
    b. True   **c.** True
    d. False: The *mastoid* process of the temporal bone is a thick, posterior projection.
    e. False: The most conspicuous feature of the *occipital* bone is the foramen magnum.

  5. c   **6.** e   **7.** a

  8. a. False: The thoracic and sacral curvatures are the vertebral column's *convex* curvatures.
    b. False: A vertebral disc is composed of an *outer* anulus fibrosus and an *inner* nucleus pulposus.
    c. True
    d. False: The *sacral and coccygeal* vertebrae are fused in an adult.

  9. d   **10.** xiphoid process; manubrium, clavicle

  11. True ribs (1–7) attach to the sternum by their own costal cartilage; false ribs (8–12) do not attach directly to the sternum. Ribs 8–10 attach to the costal cartilage of true ribs; floating ribs (11–12) do not attach to the sternum at all.

  12. b   **13.** humerus; ulna, radius   **14.** c   **15.** a

  16. a. False: The *acetabulum* articulates with the head of the femur at the hip joint.
    b. True   **c.** True
    d. False: The two pubic bones articulate at the *pubic symphysis*.
    e. True

  17. c   **18.** tibia, fibula; patella   **19.** d   **20.** e

Level 2

  1. The atlas lacks a body and has large superior articular facets that articulate with the occipital bone. The axis has a superior projection called the dens that fits into the large vertebral foramen of the atlas. This allows the two bones to move around each other at the atlantoaxial joint.

  2. Bones serve as levers for muscles, and so of course abnormal bone structure can cause muscles to function improperly. But even minor bone abnormalities can disrupt muscle function, because muscles attach to bone prominences such as tuberosities, trochanters, and tubercles. If these projections are abnormal in shape or size, then the muscle attachment may be affected.

  3. The knee joint is formed by the medial and lateral femoral and tibial condyles; the elbow joint is formed by the capitulum and trochlea of the humerus, the head of the radius, and the trochlear notch of the ulna. The elbow joint is more stable because of the shapes of the articulating bones, which allow a tighter fit than in the knee joint.

Level 3

  1. Generally, the bones involved are the perpendicular plate of the ethmoid bone and/or the vomer. When the nasal septum shifts to one side, it obstructs part of the nasal cavity, which is the first portion of the respiratory tract. This can make breathing through the nose difficult and necessitate breathing through the mouth instead.

  2. Kyphosis is an abnormal curvature of the spine in the anterior direction. This places stress on the anterior portion of the vertebrae, specifically the vertebral bodies, which are the part of the vertebrae most likely fractured. (Such a fracture is called a wedge fracture.)

  3. The person who performed CPR likely performed it incorrectly, doing compressions over the xiphoid process or over the ribs instead of over the body of the sternum.

  4. a. Anterior portion of the head (frontal region of the head)
    b. Superior, anterior cervical region
    c. Medial side of the forearm (medial antebrachial region)
    d. Medial leg (medial crural region)
    e. Located between the ribs
    f. Anterior pelvic region

CHAPTER 8

Apply What You Learned

MODULE **8.1**

1. The bones of the mature skull are joined by joints functionally classified as synarthroses. The sutures between skull bones are interlocked and become fused once an individual reaches skeletal maturity. This is important because these joints have evolved for protection of the underlying brain. If these joints were freely moveable or even slightly moveable, this protective function would be compromised. **2.** The joint resulting from this surgical procedure resembles a synarthrosis. The fused joint would no longer have a synovial cavity or hyaline articular cartilage and would be more stable than mobile.

MODULE **8.2**

1. Such diseases can completely disrupt the structure of the gomphoses because the periodontal ligament that locks a tooth into the alveolus of the mandible or maxilla would not be strong enough to withstand the demands of chewing. The teeth would be at risk for falling out or being damaged, as they would no longer be structurally sound. When the structure is compromised, the function may be lost. **2.** If the sutures in the skull were fully fused at birth, the brain tissues could not grow within the skull. The sutures do not fuse until later in life to allow for brain growth. **3.** Synchondroses are classified functionally as immoveable, and immobility is vital to their function. Usually these joints are involved in protection of and stability for the surrounding anatomical structures. Any motion at these joints would compromise these vital functions.

MODULE 8.3

1. Ligaments are important in supporting synovial joints. If the ligaments are too loose, they will not fulfill their supportive role, and the synovial joint will be more likely to allow abnormal motion, leading to injury such as dislocation. On the other hand, cartilaginous and fibrous joints do not rely on ligaments for stability, as these two joint types are inherently more stable. **2.** Tendons generally pass over or around synovial joints on their way to their insertions on bones. When the muscle associated with an injured tendon is contracted, the tendon is pulled taut over the joint and this causes pain. As the joint is moved over and over again, the injured tendon will constantly be a source of pain.

MODULE 8.4

1. The condition drop foot leaves the foot in a plantarflexed position. Individuals with this condition drag their toes as they walk because they are unable to flatten their feet. As you might expect, this makes the knee and hip work harder, as they have to pull the foot up farther so that the pointed toes clear the ground with each step. **2.** Like any disorder or disease that affects the function of ligaments, Marfan syndrome can cause hypermobility of joints. In a person with this syndrome, a biaxial joint could have a significantly greater range of motion, allowing some hyperextension and possibly circumduction. The joint could be at risk for dislocation, as well, because the increased range of motion reduces its stability.

MODULE 8.5

1. a. The anterior cruciate ligament runs from an anterior insertion site on the tibia to the posterior aspect of the femur. When the knee is extended, this ligament is pulled taut to prevent hyperextension. The injured ligament would permit an increase in the amount of extension or hyperextension, allowing the tibia to move too far anteriorly on the femur.

b. The tibial collateral ligament supports the medial aspect of the knee joint and prevents the lower leg from moving too far laterally relative to the thigh at the knee joint. The injured ligament would allow too much lateral movement of the lower leg relative to the thigh at the knee joint.

c. The posterior cruciate ligament is found just posterior to the anterior cruciate ligament and runs from the anteroinferior femur to the posterior side of the tibia. This ligament is pulled tightly when the knee is flexed to prevent hyperflexion and dislocation of the tibia from the femur. The injured ligament would allow an increased amount of flexion at the knee, and the tibia would be at risk for being dislocated from the femur posteriorly.

2. This change to the structure of the thumb joint would completely alter the functional capability of the hand. The special movement of opposition would be lost, as a uniaxial hinge joint would not allow that type of movement. The thumb would move around only a single axis, allowing for flexion and extension similar to that of the other digits of the hand, but not allowing abduction and adduction. Grasping and holding objects, abilities that are taken for granted, would be severely limited if the thumb joint were uniaxial.

Assess What You Learned

Level 1

   1. b   **2.** c

3. a. cartilaginous   d. fibrous
   b. synovial   e. fibrous
   c. cartilaginous

4. d

5. a. True   b. False: A syndesmosis is a type of *fibrous* joint.
   c. False: Cartilaginous joints are synarthroses *or amphiarthroses*.
   d. True

6. articular cartilage; synovial membrane; articular capsule

7. Synovial fluid lubricates a joint to reduce friction, absorbs shock, and helps deliver nutrients to cartilage cells.

8. c   **9.** b

10. __c__ Plane joint   __e__ Condylar joint
   __f__ Saddle joint   __a__ Pivot joint
   __b__ Ball-and-socket joint   __d__ Hinge joint

11. a. Flexion decreases the angle between two bones, and extension increases the angle between two bones.
   b. Adduction moves a body part toward the body's midline, and abduction moves it away from the midline.
   c. Rotation is the turning of a bone along its own longitudinal axis.
   d. Circumduction is movement of a bone in a cone-shaped radius.
   e. Dorsiflexion is the movement of the foot and ankle toward the leg, and plantarflexion is the movement of the foot and ankle away from the leg.
   f. Elevation is movement of a body part in a superior direction, and depression is movement in an inferior direction.

12. a. False: The knee and the elbow are *uniaxial* joints.
   b. True   c. True
   d. False: The *knee* joint is stabilized by the medial and lateral menisci.
   e. False: The *shoulder* joint is less stable than the *hip* joint, but it allows more motion.

13. c   **14.** a

15. The acetabulum is relatively deep, which allows the femoral head to fit inside it snugly. The fit is enhanced by the acetabular labrum, the articular capsule, and multiple ligaments.

Level 2

1. a. The primary function of fibrous joints is stability, or maintaining bones in certain positions relative to other bones. These joints consist of bones united by collagen fibers, which prevent them from moving freely and give them stability.
   b. Most cartilaginous joints (except the epiphyseal plate) must both be stable and allow for motion. The fibrocartilage that joins the bones in these joints is flexible, which allows some motion, but it is also tough, which provides stability.
   c. Synovial joints must allow for a significant amount of motion between two bones. This is made possible by the fact that the two articulating bones are separated by a fluid-filled joint cavity.

2. The tendon of the biceps brachii muscle passes through the articular capsule of the shoulder joint as it attaches to its origin point. For this reason, inflammation of the tendon in this area leads to inflammation in the articular capsule of the shoulder joint. This causes pain on movement of the joint.

3. The head of the humerus doesn't fit very snugly into the glenoid cavity, as the cavity is too shallow. The glenoid labrum helps improve this fit and provides the shoulder with extra stability. If the glenoid labrum is smaller than normal, the degree of fit decreases, leading to instability and possibly chronic dislocations of the shoulder joint.

Level 3

1. The sutures are fused fibrous joints that are synarthroses. No motion is allowed at these joints, and therefore the claims are unlikely to be true.

2. Lauren has likely injured her anterior cruciate ligament, which normally prevents hyperextension of the knee and anterior displacement of the tibia. She has also likely injured her tibial collateral ligament, as it normally prevents the abnormal lateral tibial motion seen in the exam.

3. Fibrocartilage is very tough and functions as a good shock absorber in a joint. However, it contains large numbers of protein fibers, particularly collagen, and so it isn't as smooth as hyaline cartilage and is not as effective at reducing friction. For this reason, a synovial joint with fibrocartilage will not function as well as one with hyaline cartilage.

CHAPTER 9
Apply What You Learned

MODULE 9.1

1. The name would be sternohyoid muscle. 2. The temporal bone serves as the origin, as it is the more fixed end of the muscle attachment. The insertion is on the mandible, which is moved toward the origin during contraction of the temporalis. 3. A first-class lever system is used to dislodge the large rock, because the fulcrum is in the middle. The work is accomplished at a mechanical advantage because the fulcrum is close to the load.

MODULE 9.2

1. He will be unable to raise his eyebrows (and wrinkle the skin of his forehead) and he will have difficulty squinting. 2. This will interfere with his ability to pull his cheeks inward as in performing sucking motions and whistling. It will also significantly decrease his ability to elevate his mandible and hinder his ability to masticate food. 3. Contracture of the erector spinae muscle group on the left side would tend to pull the vertebral column to that side, leading to the development of a curved vertebral column (scoliosis).

MODULE 9.3

1. The external intercostals muscles raise and spread the ribs during inspiration, which will tug on the bruised ribs, causing pain. 2. Quiet expiration doesn't require any muscle contractions, as relaxation of the diaphragm muscle and the elastic recoil are generally adequate to cause expiration. This doesn't put significant stress on the ribs, and so wouldn't cause significant pain. However, forced expiration engages muscles such as the internal intercostal muscles, which would pull on the bruised ribs, causing pain.

MODULE 9.4

1. Exercises involving abduction of the arm against resistance would help to re-strengthen the deltoid muscle, as it is the prime agonist of arm abduction. 2. The first part of the movement, reaching up to grab the next segment of rope, involves the triceps brachii, deltoid, trapezius (superior and inferior portion), levator scapulae, and serratus anterior muscles. The second part of the movement, where you pull yourself up the rope, involves the biceps brachii, brachialis, latissimus dorsi, teres major, trapezius (middle and inferior sections), and the rhomboid major and minor muscles. (Note that this list is not exhaustive, as there are many muscles involved in such a complex motion.)

MODULE 9.5

1. Both muscles abduct and medially rotate the thigh, so she will be unable to perform these motions on the affected limb. 2. Both abducting and medially rotating the thigh are necessary movements for normal walking. In addition, both muscles stabilize the pelvis during walking. With these muscles paralyzed, Ms. Sadler's gait will be less steady and she will have difficulty clearing the ground as she walks. She will need to compensate for the loss of these muscles by using other muscles that abduct the thigh, such as the sartorius and tensor fascia lata muscles. 3. The muscles involved are likely the fibularis longus and fibularis brevis muscles. Several muscles plantarflex the foot, but only these two muscles plantarflex as well as evert the foot.

Assess What You Learned

Level 1

1. c   2. d

3. _c_ Brevis   _b_ Rectus
   _a_ Digitorum   _d_ Flexor
   _f_ Hallucis   _e_ Pronator

4. c   5. d

6. a. True
   b. False: The orbicularis *oculi* muscle is located around the eye.
   c. True   d. True

7. d   8. a   9. b   10. a

11. _e_ Sternocleidomastoid muscle
    _c_ Transversus abdominis muscle
    _a_ Internal oblique muscle
    _b_ Rectus abdominis muscle
    _f_ Splenius capitis muscle
    _d_ Quadratus lumborum muscle

12. a

13. a. False: The pectoralis major and *deltoid* muscles are antagonists.
    b. False: The latissimus dorsi muscle is a major *adductor* of the arm.
    c. True   d. True

14. _b_ Biceps brachii muscle
    _e_ Pronator teres muscle
    _a_ Extensor digitorum muscle
    _f_ Brachialis muscle
    _d_ Triceps brachii muscle
    _c_ Flexor carpi radialis muscle

15. b

16. _f_ Vastus medius muscle
    _a_ Gracilis muscle
    _d_ Gluteus medius muscle
    _c_ Gastrocnemius muscle
    _b_ Biceps femoris muscle
    _e_ Tibialis anterior muscle

Level 2

1. A flexor will flex its body part, which is likely the finger or digits, based on the digitorum part of the name. Profundus means deep, so this is likely a muscle that is deep to other muscles. Based on this information, this is a muscle deep in the anterior forearm that flexes the digits.

2. The diaphragm muscle inserts into its own central tendon, and so when it contracts, it essentially pulls on itself, causing it to move down and flatten. This movement also pulls on the structures to which it is attached, in particular the lungs, which increases the height of the thoracic cavity.

3. These are postural muscles, and their continual contraction is necessary to keep us in an upright position.

4. Each of the three parts of the trapezius muscle has a different origin and insertion, and so each has a different action.

5. The levator ani muscle surrounds the external urethral and anal sphincters. Strengthening the levator ani muscle can help to improve muscle tone around these sphincters, which may help to better control them.

6. The rectus femoris muscle crosses two joints—the hip joint and the knee joint—so it must have at least two actions. Looking at the way it crosses the hip joint, we can infer that it flexes the hip. However, when we look at how it crosses the knee joint, we can infer that it extends the leg.

Level 3

1. The muscles affected include the orbicularis oris, buccinator, risorius, and levator labii superiorus muscles.

2. Ms. Cho was likely using her accessory muscles of inspiration while she had the respiratory infection. This led to muscle soreness because these muscles generally don't get used for this purpose and they got a 2-week-long "workout."

3. Elise has potentially strained the hamstring muscle group (the biceps femoris, semitendinosus, and semimembranosus muscles), which are causing her pain when she extends her leg, and her tibialis anterior and tibialis posterior muscles, which are causing her pain when she inverts her foot.

4. Throwing a punch involves several motions, including flexion and extension of the arm and forearm, pronation of the forearm, and protraction of the scapula. Muscles that Chris can strengthen to help with these movements include the pectoralis major and deltoid muscles for flexion and extension of the arm; the biceps brachii, brachialis, and triceps brachii muscles for flexion and extension of the forearm; the pronator quadratus and pronator teres muscles for pronation of the forearm; and the serratus anterior and pectoralis minor muscles for protraction of the scapula.

5. a. The bone involved in a hip fracture is the femur; it was likely the femoral neck that was fractured.
   b. Muscles that attach to the proximal femur include the gluteus medius muscle (the greater trochanter), the iliopsoas muscle (the lesser trochanter), the vastus muscles of the quadriceps femoris muscle group (the proximal femoral shaft), and the adductor muscle group (the proximal femoral shaft).
   c. Strengthening muscle actions involve thigh abduction for the gluteus medius muscle, thigh extension for the gluteus maximus muscle, thigh adduction for the adductor muscles, and leg extension for the vastus muscle group.

CHAPTER 10
Apply What You Learned

MODULE 10.1

1. We would expect the new type of muscle tissue to have structural elements similar to those of the other three types of muscle tissue. This is due to the structure-function relationship—the new muscle tissue type has the same function as the other types, so it would likely have a similar structure. 2. A muscle cell that is distensible but not elastic would be able to stretch but would not be able to return to its original resting length. A cell that is not distensible would not be able to stretch at all, irrespective of its elasticity.

MODULE 10.2

1. The spring shape of titin allows it to perform its primary functions, which are to hold thick filaments in place, help return the thick filaments to their resting positions after a contraction, and provide the muscle fiber with elasticity. Were titin to lose this structure,

it would not be able to adequately perform its function. The thick filaments would be less likely to return to their normal resting positions after a contraction, and the sarcomere might not return to its normal resting length. 2. The basic mechanism of contraction involves the sliding of thick and thin filaments past one another. If they were stationary in a myofibril, no contraction would occur and no tension would be generated.

MODULE 10.3

1. Blocking the Na^+/K^+ pump would disrupt the resting membrane potential and eventually deplete the Na^+ and K^+ gradients. This would diminish the amount of Na^+ in the extracellular fluid available for depolarization and the amount of K^+ in the cytosol available for repolarization. Eventually, action potentials would cease. 2. Tetrodotoxin will prevent Na^+ from entering the cytosol when the voltage-gated channels are triggered to open by the end-plate potential. This will inhibit the depolarization phase of the action potential and prevent an action potential from propagating along the sarcolemma. 3. If the voltage-gated potassium ion channels in the sarcolemma were blocked, it would prevent the cell from repolarizing. A cell that cannot repolarize will never be able to depolarize again, and so no further action potentials will be generated or propagated.

MODULE 10.4

1. The progressive destruction of ACh receptors leads to a progressive decline in the strength of end-plate potentials. This will make it harder for a neuron to elicit action potentials, resulting in muscle weakness and partial or full paralysis. 2. The troponin-tropomyosin interaction is necessary to regulate contraction and relaxation. Were troponin unable to bind tropomyosin, it would not be able to hold tropomyosin in place when the muscle is not contracting, and it would not be able to move it away from the active sites on actin to initiate a contraction. 3. The action potential would be unable to reach and open calcium ion channels in the terminal cisternae without T-tubules. This would prevent calcium ions from being released from the SR, which would in turn prevent contraction.

MODULE 10.5

1. Creatine phosphate, even when provided in excess, yields only enough energy to fuel activities lasting a few seconds. Activities lasting longer than this will be fueled by glycolytic and oxidative catabolism, and extra creatine phosphate will be of no benefit. 2. Long-lasting activities typically use all the available glucose and require the use of other fuels such as fatty acids. This is why aerobics are said to "burn fat." Short sprints will not produce this same effect because they use primarily glycolytic catabolism and glucose as a fuel.

MODULE 10.6

1. When you lift a heavy box, the muscle fibers will likely be stimulated at a very high rate, about 80–100 times per second. This will result in wave summation and fused tetanus, which are needed to produce a strong enough contraction. 2. The difference in strength has to do with the difference in starting length of the muscle fibers. When your friend's arm was flexed at the elbow, her muscle fibers were shortened and therefore able to produce less tension due to increased overlap of the thin and thick filaments. However, when she picked it up from a relaxed or neutral position, the overlap of the filaments was ideal for maximal tension production. 3. This new fiber type would likely use primarily oxidative catabolism, as it has a ready supply of oxygen from myoglobin and blood vessels and a large number of mitochondria in which to carry out oxidative catabolism. Such fibers tend to be fatigue-resistant because they continue to produce adequate quantities of ATP for as long as chemical fuels are available.

MODULE **10.7**

1. When more tension is needed, fast motor units are recruited in addition to slow motor units. These fast motor units are less fatigue-resistant than slow motor units, which is why you are likely to feel fatigued more quickly. **2.** Lowering the weight is accomplished by an isotonic eccentric contraction, which involves lengthening of the muscle fibers. Such contractions can cause injury due to overstretching if they are not done in a controlled manner.

MODULE **10.8**

1. An athlete who runs marathons likely undergoes endurance training. This individual probably has a higher number of type I fibers, and the training would likely increase the number of mitochondria, the amount of oxidative enzymes, and the blood flow. Each of these adaptations would make the athlete's muscle fibers more resistant to fatigue. **2.** If your friend had truly "used up all of his oxygen," it would be because he was no longer breathing. Instead, the rapid rate of breathing, or excess postexercise oxygen consumption, is due to the body's attempt to correct homeostatic imbalances that result from exercise.

MODULE **10.9**

1. No, you would not find recruitment in single-unit smooth muscle tissue and cardiac muscle tissue, as both types contain muscle cells that are connected by gap junctions and so function as a unit. When the cells contract, they all contract together, so there are no cells left to recruit. **2.** The released calcium ions from the SR, combined with an increase in calcium ion binding to calmodulin, would trigger the activation of myosin light-chain kinase. This would lead to contraction of the muscle cells, and constriction (narrowing) of the airways. **3.** Inactivating myosin light-chain kinase is a key part of smooth muscle cell relaxation. Taking this medication will hopefully cause the smooth muscle cells of the airways to relax, allowing them to open and restore normal breathing.

Assess What You Learned

Level 1

1. a. False: a property of all muscle cells is *distensibility,* which means that the tissue is able to stretch.
   b. True
   c. False: a muscle cell's plasma membrane is called the *sarcolemma.*
   d. True

2. Skeletal muscle fibers are typically long, thin cells. Cylindrical myofibrils make up about 80% of their sarcoplasm, along with a large number of mitochondria and multiple nuclei. The myofibrils are surrounded by sarcoplasmic reticulum.

3. a

4. __c__ Z-disc   __e__ H zone
   __d__ Sarcomere   __f__ I band
   __a__ A band   __b__ M line

5. The thick and thin filaments slide past one another, which generates tension, as described by the sliding-filament mechanism of contraction. As multiple sarcomeres generate tension, the entire muscle fiber contracts.

6. a. True
   b. False: The concentration of Na⁺ is highest in the *extracellular fluid,* and the concentration of K⁺ is highest in the *cytosol.*
   c. True   **d.** True

7. The three components of the neuromuscular junction are (1) the axon terminal of the motor neuron, which contains synaptic vesicles with the neurotransmitter acetylcholine; (2) the narrow synaptic cleft; and (3) the motor end plate of the muscle fiber, which contains acetylcholine receptors.

8. __2__
   __4__
   __1__
   __5__
   __3__

9. c   **10** a   **11.** c

12. a. False: muscle fibers generate more tension if the resting length of their sarcomeres is *100–120% of their optimal length.*
   b. True
   c. False: muscles that require a great deal of precise control will have *small* motor units.
   d. True

13. a   **14.** b   **15.** glycolytic; oxidative

16. c   **17.** a, c, d, e

18. Excessive postexercise oxygen consumption (EPOC) is due to the body's efforts to correct the disturbances to homeostasis that were brought on by exercise, including dissipating heat, restoring ion concentrations, and correcting blood pH.

19. Functions of smooth muscle include peristalsis, regulating the flow through organs such as blood vessels, and the formation of sphincters.

20. c

21. a. smooth   **e.** cardiac, some smooth
   b. cardiac   **f.** skeletal, cardiac
   c. skeletal, cardiac   **g.** smooth
   d. smooth   **h.** skeletal

Level 2

1. You would expect to find larger motor units in the back. The hand requires precise motor control, which necessitates multiple small motor units.

2. This would cause the resting membrane potential to become progressively less negative (more positive), as potassium ions wouldn't leak out of the cell but sodium ions would continue to leak into the cell.

3. Neostigmine prolongs the activity of ACh in the synaptic cleft by preventing its nearly immediate breakdown by AChE. This prolongs the skeletal muscle contraction.

4. Cardiac muscle cells and some smooth muscle cells are stimulated by pacemaker cells that stimulate the cells to have action potentials. Such cells do not require stimulation from an external nerve supply.

Level 3

1. Curare would prevent acetylcholine from binding to the motor end plate and initiating contraction, effectively paralyzing the recipient. This could be useful in the treatment of muscle spasms and in cases in which a patient must be paralyzed, such as during surgery. An overdose would be lethal because it would paralyze the respiratory muscles and prevent breathing.

2. Glycogen stores are one of the main sources of energy for short, powerful bursts of activity. Therefore, protein loading in an athlete engaging in these types of activities would actually decrease his or her performance. An athlete engaging in light endurance-type activities would not likely suffer a great decrease in performance, as glycogen is not the primary fuel for these activities.

3. Damage to the nerves has impaired their ability to stimulate the muscle fibers, which means that the fibers will not receive the signal from the neuron to have an action potential or contract. However, the muscle fibers are still able to contract when stimulated by an electrode because the electrode triggers the initiation of an action potential along their sarcolemmas, which then leads to a contraction.

4. Blocking the release of calcium ions from the SR will prevent calcium ions from flooding the cytosol and binding to troponin. This will prevent actin and myosin from binding and the initiation of contraction.

5. The segment of DNA that codes for a single protein is a gene. Each gene is a unique sequence of nucleic acids in the DNA that specifies a unique sequence of amino acids. When this sequence is modified, the resulting protein and its amino acid sequence are modified, often leading to a nonfunctional protein. Jesse's genetic defect prevents normal muscle relaxation by preventing the removal of calcium ions from the cytosol after a contraction.

6. ATP is required to fuel multiple aspects of muscle contraction, including cocking of the myosin head, running the pumps in the SR, and running the Na^+/K^+ pumps in the sarcolemma. A decrease in ATP production would therefore interfere with all aspects of muscle tissue function. Other organs and tissues with high metabolic rates would also be affected, including the kidneys, the heart, the brain, and the liver.

CHAPTER 11

Apply What You Learned

MODULE 11.1

1. PNS neurons of the somatic sensory division deliver temperature, smell, and sight stimuli pertaining to the coffee to the CNS. In the CNS, different neurons integrate these stimuli. Then, PNS neurons of the somatic motor division coordinate lifting the cup of coffee to your mouth and taking a sip. 2. An injury to the visceral motor division would likely be the most threatening to survival. This division innervates the viscera, including the heart, the smooth muscle of hollow organs, and the glands. Damage to this division would therefore produce disturbances in variables such as heart rate and blood pressure, which could prove fatal.

MODULE 11.2

1. Microglia are phagocytes that ingest dead or damaged cells in the brain, including neurons and other neuroglia. A brain injury results in damaged cells, and as a result, more microglia will be present. 2. In a patient with this syndrome, conduction will slow significantly in PNS neurons due to destruction of myelin. This will decrease the delivery speed of motor impulses to muscles, leading to partial or full paralysis. It will also slow the delivery of sensory impulses to the CNS, leading to decreased sensation. 3. Ms. Karabekian might regain some sensation, but she is unlikely to recover fully. Ganglia contain the cell bodies of PNS neurons. These neurons have some limited ability to regenerate if the cell body is intact. However, since her cell bodies were damaged in the injury, her neurons are unlikely to regenerate.

MODULE 11.3

1. The gradients for Na^+ and K^+ change very little with a single action potential. However, the gradients will start to dissipate with repeated action potentials, and the poison will prevent the pumps from restoring them, reducing the ability of the neuron to have

an action potential. 2. In this hypothetical situation, the opening of sodium ion channels would lead to Na^+ leaving the cytosol and entering the extracellular fluid. This would cause the cell to lose positive charges, and the cell would hyperpolarize instead of depolarize. 3. The impulses are conducted relatively slowly because temperature sensations are carried on type C fibers, which are small and unmyelinated.

MODULE 11.4

1. The influx of calcium ions in the axon terminal is what triggers exocytosis of the synaptic vesicles. If these calcium ion channels are blocked, exocytosis will not be triggered, and synaptic transmission will effectively be halted. 2. Calcium ions are more concentrated in the extracellular fluid, so when calcium ion channels are opened, calcium ions follow their gradient and flow into the cytosol of the neuron. The influx of positive charges produces an EPSP. Repeated EPSPs make an action potential more likely to occur. 3. To make an action potential more likely, the number of EPSPs needs to be increased. This can occur via temporal summation, in which one or more neurons fire repeated action potentials and trigger multiple EPSPs in rapid succession. It can also occur via spatial summation, in which multiple neurons fire action potentials and trigger EPSPs all at the same time.

MODULE 11.5

1. Glutamate binds ionotropic postsynaptic receptors that trigger EPSPs in the postsynaptic neurons. Blocking glutamate receptors will prevent the generation of these EPSPs and reduce the number of action potentials fired in the CNS. 2. Glycine typically triggers IPSPs and so has an inhibitory effect in the CNS. Blocking glycine receptors will increase the amount of inappropriate excitation in the CNS, leading to abnormal motor, sensory, and mental processes. 3. Such blockage would prolong the actions of the neurotransmitters in the synaptic cleft, increasing excitation of the postsynaptic neurons. If you prolonged the actions of inhibitory neurotransmitters, you would decrease the excitation of postsynaptic neurons.

MODULE 11.6

1. Such circuits are required for certain types of muscle contractions, as these often involve both agonist and antagonist muscle groups, of which one group must be stimulated while the other must be inhibited. 2. Prolonged abuse of the drug leads to synaptic fatigue; the number of dopamine receptors declines over time, and the remaining receptors become less sensitive to dopamine.

Assess What You Learned

Level 1

1. c 2. d

3. _c_ Schwann cells   _e_ Oligodendrocytes
   _f_ Ependymal cells   _b_ Satellite cells
   _a_ Microglial cells   _d_ Astrocytes

4. a. False: Aggregates of *rough endoplasmic reticulum* form dark-staining Nissl bodies within the cell body.
   b. False: The *cell body* contains a high density of ribosomes, rough endoplasmic reticulum, and Golgi apparatus.
   c. True   d. True

5. b   6. internode; nodes of Ranvier

7. absolute refractory period; relative refractory period

8. d

9. a. ES   d. CS
   b. CS   e. ES
   c. B

10. a

11. _d_ GABA   _c_ Glutamate

    _g_ Dopamine   _e_ Endorphins

    _a_ Substance P   _f_ Norepinephrine

    _b_ Acetylcholine

12. c   **13.** d

14. a. True

b. False: The concentration of Na^+ is highest in the *extracellular fluid*, and the concentration of K^+ is highest in the *cytosol*.

c. True   d. True

e. False: An *action* potential is a change in membrane potential that conducts the long-distance signals of the nervous system.

15. a. 2   d. 3

b. 5   e. 4

c. 1

16. a. False: An *inhibitory* postsynaptic potential is caused by K^+ or Cl^- channels opening in the membrane of the postsynaptic neuron.

b. True

c. False: An *excitatory* postsynaptic potential causes the membrane potential of the postsynaptic neuron to approach threshold.

d. True

Level 2

1. Such a drug would not only prevent the conduction of action potentials but also block the generation of local potentials. This would prevent the depolarization of the axon hillock to threshold, and thus the generation of action potentials.

2. In this case, the drug would prevent the neuron from repolarizing. The neuron would remain depolarized and unable to generate another action potential.

3. The cell body must be intact because this is where the nucleus and all cellular "machinery" for protein synthesis are housed. Without an intact cell body, the neuron won't be able to produce the proteins needed for regeneration.

4. In continuous conduction, the initial segment of the axolemma depolarizes to threshold and has an action potential, which depolarizes the next segment of the axolemma to threshold. This continues down the length of the axon. Saltatory conduction is faster than continuous conduction as only the nodes of Ranvier must be depolarized to threshold, allowing action potentials to "leap" down the length of the axon.

Level 3

1. Initially, synaptic transmission is increased as the actions of acetylcholine are prolonged. However, eventually the postsynaptic cell becomes refractory to stimulation because its membrane is unable to repolarize, and so it cannot be depolarized again or have another action potential. This will decrease the excitation at synapses that use acetylcholine, including the neuromuscular junction.

2. Opening Cl^- channels will cause the neurons to hyperpolarize, making it very difficult to stimulate them to have an action potential. This slows neuronal activity in the brain, which puts the patient "to sleep" for the procedure.

3. Albert's nervous system will partially or completely shut down, as action potentials will be unable to propagate along his axons. This will result in paralysis, loss of autonomic functions, and mental impairment. The severity of the symptoms will depend on the amount of poison he ingested, but could include a

decrease in all functions of the brain and nervous system. Death results from moderate to severe poisoning.

4. Lithium would simply make the tetrodotoxin poisoning worse, as it would further decrease the ability of the neurons to generate and conduct action potentials. The drug should not be administered while the patient recovers from the poison.

5. The poison would prevent the spread of action potentials along the sarcolemma of muscle fibers. The end-plate potentials at the motor end plate would not be directly affected, though, as they are generated by ligand-gated ion channels. However, because of the effect on action potentials, the neurons supplying the muscle fibers would be unable to stimulate those fibers.

6. If depolarization triggered a negative feedback loop, the initial depolarization to threshold would not lead to further depolarization. Instead, it would lead to repolarization and the cell would return to its resting membrane potential.

CHAPTER 12
Apply What You Learned

MODULE **12.1**

1. An injury to the brainstem would be most damaging in terms of survival because the brainstem is responsible for maintaining basic aspects of homeostasis, such as breathing and blood pressure. **2.** Damage to the cerebrum, specifically the frontal lobe, could lead to changes in personality. This is because of association areas such as the prefrontal cortex, which house neurons involved in higher functions like personality and conscience.

MODULE **12.2**

1. The basic functions that enable life are controlled primarily by the hypothalamus and brainstem. For this reason, such functions can continue even in the absence of cortical activity. However, an individual in a vegetative state does not have any true sensory awareness of pain or other stimuli and lacks the ability to initiate voluntary movement, since that requires upper motor neurons of the motor cortices. **2.** Damage to the basal nuclei will cause a patient to have difficulty initiating movement. Damage to the cerebellum will cause a patient to have difficulty controlling ongoing movement. **3.** The gyri and sulci increase the surface area of the cerebral cortex, which in essence allows more neurons to "fit" in a small area. A lack of gyri and sulci severely reduces the surface area of the cerebral cortex and thus the number of neurons located within it. **4.** Effects of damage to the thalamus depend on which portions are injured. Because the thalamus controls the inflow of information to the cerebral cortex, thalamic injury can lead to deficits in perception and awareness of sensory stimuli. Such injury can also affect an individual's ability to initiate voluntary movement. Severe damage to the thalamus will essentially cut off the cerebral cortex from the rest of the nervous system, leading to permanent coma or a persistent vegetative state.

MODULE **12.3**

1. The structures mainly responsible for these changes are the hypothalamus and the reticular formation. The cerebral cortex also plays a role, as it is the cortical mediation of "nervous" feelings that trigger the hypothalamus to stimulate the reticular formation nuclei. **2.** Orexins stimulate appetite and wakefulness. Therefore, an individual with such a tumor is likely to experience an increase in appetite and difficulty falling asleep at night. **3.** Under most conditions, muscle groups are paralyzed during REM sleep, which prevents sleeping individuals from acting out their dreams. These muscle groups are not fully paralyzed during non-REM sleep, and thus a sleeper can potentially engage in sleeptalking or sleepwalking.

MODULE 12.4

1. The lateralization of certain functions means that some functions are performed primarily by the right or left side of the cerebrum. Mr. Jacobs can expect difficulties with attention, facial recognition, "negative" emotions such as anger, and interpretation of the emotional aspects of language. Furthermore, damage to sensory and motor regions of the right hemisphere will impair his ability to sense stimuli from the left or command muscles on the left side of his body. 2. Ms. Marcos will have deficits in the formation of new declarative memories, as immediate and short-term declarative memories are processed by the hippocampus before becoming long-term memories. Existing long-term memories and new and old nondeclarative memories will be largely unaffected by the surgery. 3. The amygdala is a key player (although not the only one) in the highly subjective feelings that we experience. Thus, Ms. Marcos can expect to have some emotional abnormalities, although the degree to which this will be a problem for her is difficult to predict.

MODULE 12.5

1. CSF makes the brain buoyant, preventing it from crushing itself under its own weight. A decreased amount of CSF will put pressure on the brain, leading to headaches and possible impairment of various sensory and motor functions, depending on the degree of CSF loss. 2. Normally, the capillaries of the brain are highly impermeable to most large molecules due to the blood brain barrier. Disrupting the tight junctions of the blood-brain barrier with mannitol will increase the permeability of the capillaries and allow more molecules of a drug to exit the blood and enter the ECF.

MODULE 12.6

1. The dural sinuses are located between the periosteal and meningeal layers of the cranial dura. The spinal cord has no periosteal dura; instead, the dura there is composed of a single layer. For this reason, there is no place for a dural sinus in the spinal cord. 2. The resident has likely not advanced the needle far enough and so has withdrawn material from the epidural space rather than the subarachnoid space. The epidural space contains adipose tissue and some blood vessels but no CSF. The CSF is located in the subarachnoid space.

MODULE 12.7

1. Areas of the body that have a higher density of sensory receptors have a larger proportional area of the primary somatosensory cortex dedicated to them. The fingertips have a higher density of sensory receptors than the back does, so their area of the primary somatosensory cortex is larger than that for the back. 2. The pathways of the anterolateral system, which carry pain stimuli from the left side of the body, decussate in the posterior horn and travel up the right side of the spinal cord. The pathways of the right posterior columns, which carry tactile stimuli, travel in the right side of the spinal cord and do not decussate until the brainstem. Therefore, the injury is on the right side of the spinal cord. 3. These drugs bind to receptors in the brainstem and spinal cord, and block the transmission and perception of pain stimuli. They do not treat the cause of the pain, so when their effects wear off, the perception of pain will return.

MODULE 12.8

1. The globus pallidus sends inhibitory impulses to the thalamus, blocking the initiation of inappropriate movement. Degeneration of the globus pallidus would remove this inhibition, and the individual would experience inappropriate, involuntary movements. 2. The caudate nucleus, putamen, and substantia nigra are instrumental in initiating movement by inhibiting the globus pallidus. With their degeneration, an individual would have difficulty in initiating movement, as the globus pallidus would inhibit movement continuously. 3. Cerebellar damage will cause an individual to appear clumsy and uncoordinated, as the cerebellum will be unable to detect motor error and/or unable to send feedback to the primary motor cortex to correct for motor error. 4. The stroke likely occurred on the left side of Ms. Nazari's brain. Motor pathways initiate in one cerebral hemisphere, and then most of the fibers decussate in the medulla to serve the opposite side of the body. Therefore, damage to the left hemisphere will mostly affect the right side of the body.

Assess What You Learned

Level 1

1. a

2. a. False: Humans use *100%* of their brains.
   b. True
   c. False: The right and left lateral ventricles are the largest of the ventricles in the brain and are located in the *cerebrum.*
   d. False: The *brainstem* is responsible for our basic, involuntary homeostatic functions and reflexes.

3. c   4. b

5. a. True
   b. False: The cerebral cortex is composed of *gray* matter.
   c. True
   d. False: The prefrontal cortex is located in the frontal lobe and is concerned with *behavior, personality, learning, and memory.*

6. d

7. _c_ Amygdala   _b_ Medulla oblongata
   _b_ Thalamus   _d_ Hypothalamus
   _f_ Hippocampus   _e_ Cerebellum
   _h_ Midbrain   _g_ Pons

8. a

9. Cerebrospinal fluid is formed by the choroid plexuses. The capillaries of the choroid plexuses are fenestrated, and so fluid and electrolytes leak out of them and into the space under ependymal cells, which then secrete the CSF into the brain's ventricles.

10. The blood brain barrier consists of the endothelial cells of the capillaries in the brain, which are joined by more tight junctions than most other capillaries and have limited capacity for endocytosis and exocytosis, as well as the ependymal cells that surround the capillaries of the choroid plexuses. This barrier keeps many chemicals out of the ECF of the brain and the CSF, protecting the brain from many different types of toxins.

11. a. True   b. True
    c. False: The epidural space around the spinal cord is a *true* space.
    d. False: The posterior horn of spinal gray matter contains the cell bodies of *sensory* neurons.
    e. False: The corticospinal tracts are the main *motor* tracts in the spinal cord.

12. medulla, posterior horn of the spinal cord

13. The fingertips, lips, and hands have the greatest representation within the primary somatosensory cortex because of the importance of manual dexterity and facial expression to human activities.

14. d

15. cerebral cortex, plan and initiate voluntary movement; anterior horn of the spinal cord, directly innervate skeletal muscles

16. a. _4_   d. _1_
    b. _6_   e. _2_
    c. _3_   f. _5_

17. a. True
  b. False: The cerebellum *does not* monitor the initiation of movement but *does* monitor ongoing movements.
  c. True   d. True
18. b   **19.** hypothalamus, brainstem reticular formation
20. b
21. _f_ REM sleep   _c_ Beta waves
  _d_ Stage I sleep   _e_ Theta waves
  _a_ Delta waves   _b_ Stage IV sleep
22. d
23. long-term potentiation, hippocampus, cerebral cortex
24. c

Level 2

1. The basal nuclei inhibit abnormal or involuntary movements. In a person with Huntington's disease, you will therefore see involuntary movements and vocalizations due to loss of inhibition.

2. The basal nuclei are responsible for the initiation of movement, and the cerebellum is responsible for monitoring ongoing movement. An individual with damage to the basal nuclei will have difficulty in initiating a movement. An individual with damage to the cerebellum, however, will have difficulty in continuing a movement in a smooth, fluid, balanced fashion after it has been initiated.

3. The hippocampus is involved in the formation of declarative memories, but not their storage (they are stored in the cerebral cortex). Nondeclarative memories are not formed by the hippocampus, but rather by the motor cortices, the cerebellum, and the basal nuclei, so damage to the hippocampus will not affect these memories.

Level 3

1. Damage to the posterior of the head is most likely to affect the occipital lobe and perhaps parts of the parietal and temporal lobes. Occipital lobe damage could interfere with interpretation of visual stimuli. Temporal lobe damage could interfere with many different functions, such as memory, hearing, and language. Parietal lobe damage could disrupt interpretation of various types of somatic sensory stimuli.

2. Ms. Norris likely has sustained damage to her reticular formation, which is the part of the brainstem responsible for mediating the sleep/wake cycle. Damage to the reticular formation can prevent the normal arousal of the cerebral cortex seen in an awake individual, and the patient therefore remains unarousable.

3. Arlene will lose tactile sensation and muscle control on the left side of her body. These stimuli are carried on the posterior columns and corticospinal tracts, respectively, and these tracts do not decussate until they reach the medulla. Therefore, at the level of the thoracic spinal cord, they carry information from the same side. Pain perception, however, could be lost from the right side of her body. Pain and other touch stimuli are carried by the anterolateral system, which decussates in the spinal cord. Damage to the left side of the spinal cord will therefore affect the second-order neurons serving the right side of the body.

4. Orexins are neurotransmitters that induce feeding. Blocking orexin release could potentially decrease appetite and lead to weight loss. However, orexins are also involved in the sleep/wake cycle. Normally, they are released on waking and help to maintain a state of wakefulness throughout the day via stimulation of the reticular formation. Blocking orexin release could therefore potentially lead to daytime drowsiness.

5. Endorphin release does not stop the transmission of pain impulses—they are detected normally by peripheral nerves. However, endorphins block the perception of pain by inhibiting the delivery of these impulses to the cerebral cortex. A decrease in pain perception allows an athlete to continue exercise with less discomfort and will temporarily minimize fatigue.

CHAPTER 13
Apply What You Learned

MODULE **13.1**

1. A nerve is a group of axons and small blood vessels surrounded by connective tissue. Damage to a single motor or sensory neuron would produce far less functional impairment than damage to an entire nerve, which contains multiple sensory and motor neurons.
2. Like a muscle, a nerve consists of multiple cells bound together in fascicles by connective tissue sheaths. The fascicles, in turn, are surrounded and held together by another connective tissue sheath.
3. Damage to the visceral sensory or motor division would be the most life-threatening, because these divisions innervate vital organs such as the heart, blood vessels, and lungs and maintain homeostasis.

MODULE **13.2**

1. a. CN VII, CN IX, and CN X detect taste.
  b. CN XII provides motor innervation to most muscles of the tongue.
  c. CN III, CN IV, and CN VI provide motor innervation to the extrinsic eye muscles that move the eyes.
  d. CN VIII, specifically the vestibular portion, is responsible for balance and equilibrium.
  e. CN V innervates the muscles of mastication that close the jaw.

2. The cranial nerves that innervate the salivary glands are CN VII and CN IX. If the activity of these nerves is inhibited, less saliva will be produced, and the individual will experience a dry mouth.

MODULE **13.3**

1. The phrenic nerve, which innervates the diaphragm and so is the main nerve that drives breathing, is formed of axons from C_3, C_4, and C_5. Damage at or above the level of C_4 would render this nerve unable to function.

  2. a. Inability to flex the wrist and fingers; numbness or pain on the anterior thumb, the anterior second and third digits, and the lateral side of the fourth digit
  b. Inability to dorsiflex and evert the ankle; numbness or pain around the knee, the anterior and distal leg, and the dorsum of the foot
  c. Inability to flex the hip and extend the knee; numbness or pain along the anterior thigh and the medial thigh, leg, and foot
  d. Significant impairment of most muscles of the lower limb except the anterior thigh muscles; numbness or pain on the posterior lower limb and part of the anterior leg
  e. Reduced ability to flex the forearm and difficulty moving the hand; numbness or pain along the fifth digit and the medial side of the fourth digit

MODULE **13.4**

1. Loss of their myelin sheaths would cause these neurons to conduct action potentials much more slowly than normal. This would delay the arrival at the cerebellum of critical feedback on motor error. 2. Reading Braille relies largely on Merkel cell fibers, as they have the finest spatial resolution and are numerous in the fingertips. 3. Extremes in temperature and painful stimuli can cause serious homeostatic imbalances. If our thermoreceptors and nociceptors were rapidly adapting, we would not be adequately sensitive to these ongoing stimuli.

MODULE **13.5**

1. a. Upper motor neurons of the premotor cortex are involved in planning movement and selecting a motor program, and so damage to them would lead to a loss of specific motor skills.

b. Damaged upper motor neurons of the primary motor cortex cannot stimulate lower motor neurons, leading to uncontrolled muscle contractions mediated by those lower motor neurons and the spinal cord.

c. A lower motor neuron is the only path for stimulating the muscle fibers of its motor unit, so damage to lower motor neurons produces paralysis or paresis.

2. Information from proprioceptors allows the cerebellum to detect and correct motor error. Damaged proprioceptors (or their sensory neurons) would not effectively communicate motor error to the cerebellum, leading to dyskinesia (difficulty executing smooth movements).

MODULE **13.6**

1. The feedback from muscle spindles helps ensure that movement is fluid and prevents damage from excessive stretching of a muscle. Inhibition of muscle spindles would result in jerky movements, difficulty maintaining proper posture and balance, and potential muscle damage from excessive stretching. **2.** The damage to the facial nerve would impair the motor arm of the corneal blink reflex, so the eyelid wouldn't close even though the stimulus would be detected. **3.** Lower motor neurons directly trigger muscle contraction, so if they are not functioning, muscles will not contract after the stretch stimulus. Upper motor neurons do not directly contact muscle, so lower motor neurons may continue to stimulate muscles to contract when the stretch receptors are triggered.

Assess What You Learned

Level 1

1. a. False: The *visceral* sensory division of the PNS detects sensory information from the organs in the thoracic and abdominopelvic cavities.

b. True   c. True

d. False: The term nerve is *not* the equivalent of the term neuron. (*Nerves are composed of the axons of many neurons.*)

e. True

2. anterior ramus, posterior ramus

3. a. Peripheral nerve: a collection of axons, connective tissue sheaths, and blood vessels

b. Nerve plexus: a network of nerves that innervates a specific body region

c. Posterior root ganglion: a collection of cell bodies of sensory neurons located in the posterior root outside the spinal cord

4. a. CN VIII   Vestibulocochlear nerve __e__
  b. CN V   Trigeminal nerve __k__
  c. CN XII   Hypoglossal nerve __g__
  d. CN VI   Abducens nerve __a__
  e. CN X   Vagus nerve __j__
  f. CN I   Olfactory nerve __c__
  g. CN XI   Accessory nerve __h__
  h. CN III   Oculomotor nerve __l__
  i. CN VII   Facial nerve __b__
  j. CN II   Optic nerve __i__
  k. CN IX   Glossopharyngeal nerve __d__
  l. CN IV   Trochlear nerve __f__

5. CN I, II, and VIII are sensory; CN III, IV, VI, XI, and XII are motor; CN V, VII, IX, and X are mixed.

6. __d__ Phrenic nerve   __a__ Radial nerve
  __h__ Median nerve   __g__ Intercostal nerves

__e__ Femoral nerve   __c__ Common fibular nerve
__f__ Tibial nerve   __b__ Musculocutaneous nerve

7. b   **8.** d

9. Many spinal nerves carry both somatic and visceral neurons, so visceral sensations travel along the same pathways as somatic sensations. This causes referred pain.

10. b

11. a. __4__   d. __5__
  b. __3__   e. __2__
  c. __1__

12. Upper motor neurons are located in the motor areas of the cerebral cortex; they do not directly contact skeletal muscle fibers. They are involved in making the "decision" to move, as well as planning and monitoring movement. Lower motor neurons directly stimulate skeletal muscle fibers to contract, but they are not involved in the planning of movement.

13. Upper motor neurons select a motor program. They stimulate the caudate nucleus and putamen, which inhibit the globus pallidus. This allows the thalamus to stimulate the upper motor neurons, which in turn stimulate lower motor neurons. The lower motor neurons then stimulate muscle fibers to contract.

14. a   **15.** Muscle spindles, Golgi tendon organs   **16.** c

17. a. True

b. False: *Muscle spindles* detect stretch in a simple stretch reflex.

c. False: A *simple stretch* reflex is a monosynaptic reflex with only one synapse in the spinal cord.

d. False: The crossed-extension reflex occurs simultaneously with the *flexion* reflex.

18. The sensory arm is mediated by the glossopharyngeal nerve, and the motor arm is mediated by the vagus nerve.

Level 2

1. CN I: identify smells; CN II: test vision; CN III, IV, and VI: test ability to follow a moving finger with the eyes; CN V: test ability to open and close the jaw or test sensation on the face; CN VII: test ability to make different facial expressions or test ability to taste; CN VIII: test hearing and have patient stand on one foot to assess balance; CN IX and X: test ability to swallow and/or taste; CN XI: test ability to shrug shoulders and rotate neck; CN XII: test ability to move the tongue.

2. The change would be potentially harmful. If the nociceptors were rapidly adapting, we would be unable to detect the continuing presence of painful stimuli, which would likely lead to tissue damage.

3. All spinal nerves result from the fusion of the anterior and posterior roots, and thus all carry both sensory and motor signals.

Level 3

1. One of the simplest tests is to assess Delia's reflexes. Reflexes are diminished to absent with lower motor neuron disorders and are normal to exaggerated with upper motor neuron disorders. It would also help to look at her muscle tone, which is decreased with lower motor neuron disorders and generally increased with upper motor neuron disorders.

2. Jason is likely to experience weakness or paralysis of the muscles of his arm, forearm, wrist, hand, and digits. He will also experience sensory deficits of varying severity in the arm, depending on how severely damaged the nerves were. He may also have pain associated with peripheral neuropathy caused by the injury.

3. The pain could be related to the heart, but is more likely to be referred pain from the stomach or the pancreas, which tend to refer to the area along the diaphragm because of their dermatomal distribution.

4. CIPA affects all somatic and visceral sensory neurons, and patients are unable to feel pain or distinguish temperatures or textures. This condition is dangerous because Maria will not be able to feel pain that leads to tissue damage, and may suffer burns or severe injuries.

5. Sweating is an important aspect of thermoregulation, as it helps to cool the body. Maria's inability to produce sweat will make her prone to overheating from normal physical activities, leading to hyperthermia, which can prove fatal.

CHAPTER 14
Apply What You Learned

MODULE 14.1

1. An individual with autonomic neuropathy would have difficulty maintaining homeostasis of many vital functions, including blood pressure, digestion, and urine production. An individual with somatic motor neuron neuropathy would have difficulty controlling voluntary muscle movements, and could eventually exhibit paralysis of affected muscles. 2. The sympathetic nervous system is the "fight or flight" division, meaning that it prepares the body for emergencies. Blocking many of the effects of this system would make Mrs. Williams less able to regulate her blood pressure and make other necessary adjustments during an emergency.

MODULE 14.2

1. Activating the receptors will dampen the sympathetic response, as these receptors inhibit preganglionic sympathetic neurons. Such a drug could be used to lower blood pressure and heart rate. 2. When preganglionic sympathetic neurons function normally, they stimulate postganglionic cell bodies by releasing ACh onto nicotinic receptors. Blocking these receptors would prevent stimulation of postganglionic sympathetic neurons, which is the same as stimulating receptors on the preganglionic neurons. 3. Epinephrine and norepinephrine from the adrenal medulla will still stimulate this patient's heart and blood vessels to increase his heart rate and blood pressure during exercise.

MODULE 14.3

1. Atropine will block the muscarinic ACh receptors on all parasympathetic target cells (and on some sympathetic target cells, too). This leaves the sympathetic nervous system unopposed, and all of its effects are amplified, including elevated heart rate, increased blood pressure, and pupil dilation. 2. Stimulation of ACh release will trigger many effects that the parasympathetic nervous system has on its target cells. These include increasing saliva release (drooling), pupil constriction and adjustment of the lens for near vision (poor distance vision), and excess digestive activity (diarrhea).

MODULE 14.4

1. Autonomic centers and the cell bodies of many parasympathetic neurons are located in the brainstem, and so damage to it causes autonomic dysfunction. 2. Dominance of parasympathetic tone would lead to bronchoconstriction because parasympathetic input would outweigh sympathetic input to the smooth muscle of the airways.

Assess What You Learned

Level 1

1. b
2. thoracolumbar, thoracic and, lumbar regions of the spinal cord
3. c

4. Adrenergic receptors bind to norepinephrine and epinephrine and are located on many target cells of the sympathetic nervous system. Cholinergic receptors bind acetylcholine. They are found on sympathetic ganglia neurons and certain sympathetic target cells, as well as on all parasympathetic ganglia neurons and target cells.

5. e

6. Sympathetic neurons stimulate the adrenal medulla to secrete epinephrine and norepinephrine, which function as hormones, into the bloodstream. This allows epinephrine and norepinephrine to reach target cells that are not innervated by sympathetic neurons and prolongs the effects of sympathetic activation.

7. d

8. a. sympathetic
   b. parasympathetic
   c. parasympathetic
   d. sympathetic
   e. sympathetic
   f. sympathetic
   g. parasympathetic

9. a. False: The parasympathetic nervous system generally *increases* the secretion from digestive glands.
   b. True   c. True
   d. False: *Parasympathetic* tone controls the resting rate of the heart.

10. Nicotinic, muscarinic   11. b   12. d

Level 2

1. a. Sympathetic nervous system: the system that maintains homeostasis in times of exercise, emergency, and emotion
   b. Parasympathetic nervous system: the system that maintains homeostasis in times of rest; promotes digestion, diuresis, and defecation

2. The sympathetic nervous system increases the rate and force of heart contraction and constricts blood vessels that supply the digestive, integumentary, and urinary systems while dilating blood vessels that supply skeletal muscles. This increases blood pressure and routes blood to the skeletal muscles so that they can continue to function as you run. Sympathetic neurons also dilate the bronchioles so that more oxygen is available to be delivered to skeletal muscles and the heart, and they increase the metabolic rate so that there is more ATP for the heart and skeletal muscles to consume. These neurons also trigger the release of metabolic fuels so that the cells can continue to produce ATP at a higher rate. Finally, they increase sweat secretion to remove heat from the body.

3. Once the race is over, the parasympathetic nervous system restores the heart rate and blood pressure to resting levels, which ensures that all organs receive the proper amount of blood flow. The absence of sympathetic stimulation causes the blood vessels serving the digestive, integumentary, and urinary systems to dilate, which promotes their functions. Parasympathetic neurons trigger secretions by digestive glands so that digestive processes may take place. They also lower the metabolic rate and promote the storage of metabolic fuels to save them for the next time the sympathetic nervous system is activated.

Level 3

1. Cranial nerves III, VII, IX, and X would be affected by this drug. Potential adverse effects include excessive salivation, constriction of the pupils, increased blood flow to the skin (flushing), blurry distance vision, low heart rate and blood pressure, bronchoconstriction, and excessive digestive activity.

2. Yes, the pairing of these two drugs will cause problems for the patient. The albuterol will be unable to bind and activate the β_2 receptors because the propranolol will block these receptors.

The albuterol is therefore unlikely to be effective, and her asthma symptoms will not be alleviated.

3. The symptoms of Horner syndrome mimic those of excessive parasympathetic activity because there is no sympathetic activity to balance it. Mr. Chevalier is likely to experience constricted pupils; blurry distance vision; decreased sweating on the face; flushing of the skin due to vasodilation; and droopy eyelids due to reduced sympathetic stimulation of the superior tarsal muscle, which opens the eye.

4. This poison results in excessive stimulation of skeletal muscles as a result of prolonged action at the motor end plate. Eventually, the skeletal muscle fibers become so overstimulated that they are unable to respond to further stimulation, which leads to paralysis. The ANS effects mimic those of cholinergic drugs, causing a low heart rate and blood pressure, drooling, increased blood flow to the skin (flushing), bronchoconstriction, and constriction of the pupils.

CHAPTER 15
Apply What You Learned

MODULE 15.1

1. Stimuli for the special senses (with the exception of olfaction) are detected by specialized receptor cells instead of neurons. For example, taste chemicals are detected by gustatory cells, which then transmit the stimuli to a sensory neuron. The receptor cells must transduce, or convert, the stimuli from a chemical signal to an electrical signal.

MODULE 15.2

1. The burning sensation is caused by the chemicals binding to sensory receptors of the trigeminal nerve rather than olfactory neurons. It therefore isn't actually caused by a "smell" at all. 2. The olfactory bulb and tracts lie just inferior to the frontal lobe and would be compressed by such a tumor. This would interfere with the delivery of olfactory stimuli to the inferomedial temporal lobe, decreasing the ability to identify and perceive odorants.

MODULE 15.3

1. Such a tumor will prevent production of saliva. Without saliva, the taste chemicals will not dissolve and bind to receptors on the gustatory cells, so taste sensation will be impaired. 2. If the facial, glossopharyngeal, and vagus nerves are intact, taste sensation is preserved, and the individual will be able to distinguish the tastes of hot peppers and French fries. However, he or she will not detect a difference in the spiciness, or "heat," of the two foods, because this is detected by the trigeminal nerve. 3. The limbic system stores emotional reactions to food, and it created an association between eating coleslaw and severe vomiting. For this reason, coleslaw remains something you have no desire to eat.

MODULE 15.4

1. Pathogens may travel from the nasal cavity through the nasolacrimal duct, lacrimal sac, and canaliculi to reach the conjunctiva and cause infection and inflammation. 2. The lateral rectus muscle in the left eye is not functioning properly. 3. If rods were damaged, one would expect a loss of night vision and peripheral vision rather than a loss of high-acuity color vision that is seen with macular degeneration. This is because rods are located in the periphery of the retina and are responsible for vision in very low light. They cannot, however, produce high-acuity color vision, as is seen with the cones that are lost with macular degeneration.

MODULE 15.5

1. Lenses that correct hyperopia increase the convergence of rays. Therefore, the focal point will fall in front of your retina. 2. cGMP wouldn't be converted to GMP and would continue to keep sodium ion channels open, depolarizing the rod. It would remain depolarized and would be unable to respond to light. 3. Blindness would occur in the right half of the visual field (lateral side of the left eye and medial side of the right eye) because all the visual signals from that half are carried in the left optic tract.

MODULE 15.6

1. This person would experience dryness and a buildup of debris in the external auditory canal. 2. Contraction of the tensor tympani muscle reduces vibration of the tympanic membrane, and the energy of sound waves is less likely to reach the inner ear. A persistent contraction of the muscle would reduce the amount of sound energy reaching the inner ear, and hearing would be impaired.

MODULE 15.7

1. The sound is loud and has a high pitch. High-pitched sounds produce vibration of the basilar membrane at the base of the cochlea, and loud sounds produce greater vibration. 2. Such a hearing aid is unlikely to help Eva very much, as the problem is in the receptors or the vestibulocochlear nerve. Devices that utilize bone conduction help individuals with conduction deafness, not sensorineural deafness.

MODULE 15.8

1. The utricle detects tilting of the head, and dysfunction of it could lead to dizziness and loss of balance. 2. The vestibulo-ocular reflex produces eye movement in the direction opposite to that of head movement, so your eyes move to the left. The oculomotor, trochlear, and abducens nuclei receive input from the vestibular nuclei and produce a motor response to this input. 3. The loss of balance could be due to dysfunction of one of the components of the vestibular system in the inner ear (the semicircular ducts and ampulla and/or the utricle and saccule). An ear, nose, and throat specialist will be able to test the structures of the inner ear to determine if any of these are indeed the source of her symptoms.

MODULE 15.9

1. Any of the structures in the visual pathway may be nonfunctional: the cells in the retina, the axons of the optic nerve or tract, the lateral geniculate nucleus of the thalamus, the optic radiations, or the primary visual cortex. 2. Olfaction will not be affected because the olfactory pathway does not travel through the thalamus. The pathways for gustation, vision, hearing, and vestibular sensation pass through the thalamus to reach various areas of the cortex where conscious awareness occurs. Damage to the thalamus prevents this information from reaching the cortex, and the patient will lack awareness of these sensations.

Assess What You Learned

Level 1

1. __c__ Rod or cone   __b__ Olfactory neuron
    __e__ Hair cell in cochlea   __a__ Hair cell in vestibule
    __d__ Gustatory cell

2. b   3. odorant, action, olfactory   4. d   5. a

6. __b__ Sweet   __a__ Bitter
    __e__ Sour   __c__ Umami
    __d__ Salty

7. b   8. d   9. a

10. If you look directly at an object, its image will be focused on the fovea of the retina. There are only cones in the fovea, which do not function in dim light. To see the object, you must look to the side of it, so that the image falls adjacent to the fovea, where rods are located.

11. a. Photons are absorbed by rhodopsin in *rods*.

b. *Only cones* are found in the fovea centralis.

c. In the dark, rods and cones *do not* produce action potentials. (*Rods and cones produce only graded potentials.*)

12. a

13. a. False: The *malleus* is connected to the tympanic membrane.
   b. True
   c. False: The auditory canal is separated from the middle ear by the *tympanic membrane*.
   d. False: The cochlear duct is filled with *endolymph*.
   e. True
   f. False: The spiral organ is located in the scala *media* (*cochlear duct*).

14. High-frequency sounds cause the basilar membrane to vibrate near the base, where it is narrow and stiff; low-frequency sounds cause it to vibrate near the tip, where it is wide and flexible. Different populations of neurons are activated in the cochlear nerve, which the brain interprets as different frequencies.

15. Turning the head causes the endolymph of the semicircular duct to lag behind and push the cupula to one side. This bends the stereocilia and kinocilium on each hair cell, causing it to depolarize or hyperpolarize, which changes the amount of neurotransmitters released onto dendrites in the neurons of the vestibular nerve.

16. c   **17.** basilar, tectorial   **18.** True   **19.** b

20. crista ampullaris, cupula   **21.** e

Level 2

1. You would gradually lose your sense of smell because there would be no replacement of the olfactory neurons as they died off.

2. You would be unable to move your right eye to the right, since the lateral rectus muscle would be paralyzed.

3. The structures after the chiasma carry stimuli from both eyes, so damage there would cause impairment in both. If only one eye is affected, the damage must be in the retina or the optic nerve.

4. The following structures on the right side may be damaged: optic tract, lateral geniculate nucleus, optic radiations, and/or primary visual cortex.

5. The endolymph in the semicircular canals keeps moving after you stop, due to inertia. This bends the stereocilia and kinocilia on the hair cells, which the brain interprets as movement.

Level 3

1. There could be damage to the nuclei of the facial, glossopharyngeal, and/or vagus nerves, or the primary gustatory cortex in the parietal lobe may be impaired. More commonly, there is some type of deficit in olfaction, since it is highly involved in our sense of taste.

2. Myopia is caused by an abnormally long eyeball. While staring at a computer screen might give her eye strain and a headache, it is not going to change the shape of her eyeballs and give her myopia.

3. If the pharyngotympanic tube didn't open, the pressure on each side of the tympanic membrane would not be able to equalize. If the pressure difference increased, the membrane could rupture and hearing would be impaired.

4. If the round window couldn't move, there would be no fluid movement in the scala vestibuli or the scala tympani, since fluid cannot be compressed.

5. The axons of the olfactory nerve run through the cribriform plate of the ethmoid and may be torn if the plate is fractured. Much of what we perceive as taste is actually stimulation of the olfactory receptors, so impairment of olfaction leads to a reduction in the sense of taste.

6. The orbicularis oculi muscle may be paralyzed, preventing the eyelid from closing completely. If the eye cannot close, tears cannot be spread over the sclera and cornea, leading to dryness of the eye and potential damage.

CHAPTER 16
Apply What You Learned

MODULE 16.1

1. Its structural similarity to cholesterol indicates that it is likely hydrophobic, which means it may bind to receptors in the plasma membrane, in the cytosol, or in the nucleus.

2. a. Secretion of hormone X is likely to increase.
   b. Hormone X should increase the level of chemical Y back to its normal range.
   c. The secretion of hormone X should decrease by negative feedback.

MODULE 16.2

1. With low TRH, both TSH secretion and thyroid hormone secretion will decrease. **2.** The extra ADH will abnormally increase the amount of water in the body and decrease the osmolarity of the extracellular fluid. **3.** Such a tumor would increase the secretion of GH and IGF, raising their levels in the blood. In a child, this could lead to pituitary gigantism; in an adult, it could lead to acromegaly.

MODULE 16.3

1. Thyroid hormone increases the metabolic rate, which could promote weight loss. Unfortunately, such a supplement could mimic the signs and symptoms of hyperthyroidism. It could also decrease the secretion of a person's own thyroid hormone through negative feedback, resulting, at best, in no effect. **2.** The patient's levels of T_3 and T_4 are likely low. The low levels are causing the anterior pituitary to secrete extra TSH in an attempt to stimulate T_3 and T_4 production and secretion from the thyroid gland. However, the extra TSH is not effective, because of either iodine deficiency or Hashimoto thyroiditis. **3.** The result would likely be hypercalcemia and weakened bones due to excessive osteoclast activity.

MODULE 16.4

1. An ACTH-secreting tumor will increase the cortisol level in the blood, but the elevated cortisol level will decrease CRH secretion through negative feedback.

2. a. The sodium ion concentration could potentially decrease, and the potassium ion concentration could increase.
   b. Blocking aldosterone could increase the hydrogen ion concentration of the blood, decreasing the pH.
   c. By decreasing water reabsorption, these drugs could cause the blood concentration of solutes other than sodium to increase slightly.

3. Large amounts of epinephrine and norepinephrine will mimic a prolonged sympathetic response, leading to elevated heart rate and blood pressure, decreased digestive and urinary activity, elevated metabolic rate, and more.

MODULE 16.5

1. A glucagon-secreting tumor would cause hyperglycemia and ketosis, and thus resemble diabetes mellitus. **2.** The signs and symptoms would be almost the exact opposite of those in question 1: Hypoglycemia would result from excess glucose uptake by target cells. **3.** The sympathetic nervous system triggers the release of glucagon to increase the release of metabolic fuels. Drugs that block sympathetic neurons could hypothetically decrease the release of glucagon, although the effect would likely be minor due to other stimuli that trigger glucagon release.

MODULE 16.6

1. Melatonin secretion is stimulated by a decreasing light level. So when the light level is low all day, secretion will increase and individuals might feel more tired than normal. When the light level is high all day, secretion will decrease and individuals might feel less able to fall asleep. **2.** Leptin promotes satiety, and so mice with this mutation are likely to always feel hungry and therefore to overeat. **3.** The consequences will be wide-ranging and include a decrease in erythrocyte production due to decreased erythropoietin, a loss of bone mass due to lower levels of vitamin D (and so a decreased blood calcium ion concentration), and disrupted blood pressure regulation due to abnormal renin secretion.

MODULE 16.7

1. The ADH-secreting tumor will increase the blood volume and decrease the blood osmolarity. This will reduce secretion of aldosterone and stimulate secretion of ANP in order to lower the blood volume (and blood pressure). **2.** The amounts of adrenal steroids and catecholamines might decrease slightly, but many other stimuli trigger secretion of these hormones. For this reason, the amounts secreted will not likely decrease significantly. **3.** Glucagon secretion will increase in order to raise the concentration of glucose in the blood.

Assess What You Learned

Level 1

1. a. True b. True
   c. False: The pancreas, thyroid gland, and parathyroid glands secrete *endocrine hormones*.
   d. False: *Most amino acid–based hormones* are hydrophilic molecules that bind to plasma membrane proteins as part of a second-messenger system.
   e. True

2. d **3.** d

4. ADH increases the amount of water in the body by increasing water retention from the fluid in the kidneys and so decreasing the amount of water lost from the body as urine. This reduces the osmolarity of the blood.

5. hypothalamic-hypophyseal portal, anterior pituitary

6. a. Thyroid-stimulating hormone: target tissue = thyroid gland; stimulates production and release of thyroid hormones, growth and development of the thyroid gland
   b. Adrenocorticotropic hormone: target tissue = adrenal cortex (and to a small extent medulla); stimulation of secretion from the adrenal gland, primarily glucocorticoids
   c. Prolactin: target tissue = mammary gland; stimulation of milk production
   d. Gonadotropins: target tissues = ovaries and testes; FSH stimulates production of chemicals that bind and concentrate testosterone by the testes in males and triggers the production of estrogens and maturation of ovarian follicles in females; LH promotes the production of testosterone in males and the production of estrogens and progesterone and the release of an oocyte in females.
   e. Growth hormone: target tissues = bone, muscle, adipose, liver, and cartilage; promotes fat breakdown, gluconeogenesis, protein synthesis, and cell division

7. b **8.** d

9. a. False: About 90% of the thyroid hormone produced is *thyroxine (T₄)*.
   b. False: *Triiodothyronine (T₃)* is the more active of the two thyroid hormones.
   c. True

d. False: Iodine molecules are a key component of *thyroid hormone*.
   e. False: *Decreased* secretion of thyroid hormones produces weight gain, cold intolerance, and slow heart rate.

10. decrease; increase **11.** b

12. adrenal cortex, steroid hormones; adrenal medulla, catecholamines

13. c **14.** c

15. The hypothalamus produces CRH, which stimulates secretion of ACTH from the anterior pituitary. ACTH, in turn, stimulates secretion from the adrenal cortex. When levels of adrenal hormones rise, production of CRH and ACTH decreases through negative feedback.

16. d **17.** b

18. Glucagon triggers changes that increase the concentration of glucose in the blood, such as glycogen breakdown and gluconeogenesis. Insulin triggers changes that decrease the concentration of glucose in the blood, including the formation of glycogen; uptake of glucose, fatty acids, and amino acids; and storage of these molecules.

19. d

20. _e_ Leptin   _f_ Estrogens
      _c_ Atrial natriuretic peptide   _b_ Erythropoietin
      _a_ Melatonin   _d_ Testosterone

21. a. True
   b. False: Thyroid hormones and *glucagon* maintain blood glucose concentration during fasting.
   c. False: During exercise, the *sympathetic* nervous system increases the metabolic rate.
   d. True

Level 2

1. An insulin-secreting tumor would result in extreme hypoglycemia, as the excess insulin would trigger glucose uptake and storage and removal from the blood. A glucagon-secreting tumor would have the opposite effect, and lead to hyperglycemia due to excess glucose release and formation.

2. These tumors may produce androgenic steroids, which, when present in excess, promote the development of male secondary sex characteristics in females.

3. The hormones of the anterior pituitary will need to be replaced, because most of them are required for the function of other glands in the body, including the adrenal and thyroid glands. ADH and oxytocin will also need to be replaced, because although they are produced in the hypothalamus, they are released in the posterior pituitary.

Level 3

1. Her parathyroid glands were removed with her thyroid gland, and her symptoms are the result of a parathyroid hormone deficiency. Simply replacing her parathyroid hormone should be enough to restore calcium ion homeostasis.

2. Abnormal hypersecretion of cortisol does cause weight gain by increasing appetite and fluid retention and decreasing the metabolic rate. However, blocking *normal* cortisol secretion will not lead to weight loss because normal levels of cortisol do not cause weight gain.

3. If the supplement actually blocked normal cortisol secretion, it could lead to significant problems with glucose and other aspects of homeostasis, resulting in symptoms similar to those of Addison disease.

4. Mr. Montez's blood glucose level will be elevated (hyperglycemia) because glucose is not entering the cells. Ingesting sugar will not help because his cells cannot take in the glucose, and will simply lead to even more profound hyperglycemia.

5. He is likely in diabetic ketoacidosis, meaning that his blood will have a high concentration of ketone bodies, increasing hydrogen ion concentration and lowering the pH of the blood. His buffer systems will respond by binding and removing the excess hydrogen ions from the blood, restoring the pH to normal.

6. Blocking aldosterone will decrease the retention of sodium ions from the fluid in the kidneys. This will increase the solute concentration of this fluid, which will cause water to remain in the fluid due to osmosis. Overall, the amount of water retained by the kidneys will decrease.

CHAPTER 17
Apply What You Learned

MODULE **17.1**

1. The right side of the heart pumps deoxygenated blood into the pulmonary circuit. Failure of the right ventricle to pump adequately would reduce the oxygenation of the blood that normally occurs in the lungs. **2.** In left-sided heart failure, oxygenated blood is not delivered to tissues adequately, so cells potentially won't receive enough oxygen and nutrients.

MODULE **17.2**

1. Dysfunction of the papillary muscles and/or chordae tendineae would affect only the atrioventicular valves, as these muscles and cords do not attach to the semilunar valves. Such dysfunction could allow the AV valves to evert, letting blood leak back into the atria from the ventricles. **2.** Death of cells in the left ventricle leads to the worst prognosis because it must pump blood to the entire systemic circuit. If the left ventricle fails, none of the body's cells are supplied with adequate oxygenated blood. The right ventricle does not have to contract as strongly to pump blood through the lower-resistance pulmonary circuit. **3.** Coronary artery anastomoses provide alternate routes of blood flow when the flow through a vessel is blocked.

MODULE **17.3**

1. Slowing calcium ion entry into the cell lengthens the plateau phase and therefore lengthens the contraction. These drugs also reduce the calcium ion concentration in the sarcoplasm, which leads to weaker contractions. **2.** Blocking sodium ion channels in contractile cardiac cells would interfere with their ability to depolarize and have an action potential. These types of sodium ion channels are not present in the SA or AV nodal cells, so a local anesthetic would not directly affect their action potentials.

MODULE **17.4**

1. Most of the blood drains from the atria to the ventricles passively, without any need for the atria to contract. For this reason, atrial fibrillation may not significantly affect end-diastolic volume in certain patients. **2.** Heart failure generally causes the end-systolic volume to increase because the heart is unable to pump as much of the blood with each beat. The end-diastolic volume could potentially decrease if venous return decreases due to blood backing up in the pulmonary and systemic circuits. **3.** If the Q-T interval is prolonged, the ventricles take longer than normal to complete an action potential. Under some circumstances, a long Q-T interval can lower the heart rate and can lengthen and strengthen the ventricular contraction, or ventricular ejection.

MODULE **17.5**

1. Digoxin increases contractility, which will raise stroke volume and thus cardiac output. **2.** In this case, the heart would beat faster but not contract more strongly. This could actually lower cardiac output by lowering stroke volume.

3. a. The stroke volume is 40 ml.
   b. Her ejection fraction is about 36%.
   c. Her cardiac output is about 2.8 liters/min.
   d. A normal resting cardiac output is about 5 liters/min, so this patient's is lower than normal.

Assess What You Learned

Level 1

1. a. True
   b. False: The heart consists of two *inferior* ventricles and two *superior* atria.
   c. False: Arteries *generally* carry oxygenated blood away from the heart, and veins *generally* carry deoxygenated blood toward the heart.
   d. True   e. True

2. c   3. a

4. <u>e</u> Auricle   <u>j</u> Pectinate muscle
   <u>i</u> Aorta   <u>d</u> Venae cavae
   <u>a</u> Coronary sinus   <u>g</u> Pulmonary trunk
   <u>f</u> Papillary muscle   <u>b</u> Chordae tendineae
   <u>c</u> Fossa ovalis   <u>h</u> Pulmonary veins

5. ascending aorta; posterior interventricular artery; anterior interventricular artery, circumflex artery

6. Pacemaker cells are autorhythmic, meaning they spontaneously generate rhythmic action potentials. They contain fewer contractile elements and lack Na^+ channels.

7. d

8. a. False: The rapid depolarization phase of the contractile cell action potential is due to the opening of voltage-gated *sodium* ion channels.
   b. True   c. True
   d. False: The repolarization phase of the contractile cell is due to the potassium ions rushing *out of* the cell through potassium channels.
   e. False: Open *potassium* ion channels cause hyperpolarization in pacemaker cells, which triggers HCN channels to open and begins a new action potential.

9. The plateau lengthens and strengthens the contraction, lengthens the action potential, and lengthens the cardiac refractory period, thus preventing tetanus.

10. b   11. c

12. a. P wave: depolarization of the atria
    b. QRS complex: depolarization of the ventricles
    c. T wave: repolarization of the ventricles
    d. P-R interval: the delay at the AV node
    e. S-T segment: the plateau phase of ventricular action potentials

13. a. True
    b. False: Atrial systole is responsible for ejecting *about 20%* of the blood into the ventricles during the inflow phase of the cardiac cycle.
    c. False: The amount of blood in the ventricles at the end of the inflow phase is the end-*diastolic* volume.
    d. True

14. d

15. S1, atrioventricular; isovolumetric contraction; S2, semilunar; isovolumetric relaxation

16. b   17. increase, Frank-Starling; decrease; increase

18. c

Level 2

1. a. The left ventricle is thicker-walled because it pumps against a greater resistance and therefore needs to be stronger and have greater muscle mass.

   b. This defect will cause the right ventricle to grow larger, as it has to pump against much greater resistance in the systemic circuit. Also, the left ventricle will atrophy because it is pumping against the lower resistance of the pulmonary circuit.

2. Both would cause significant problems and potential death of cardiac muscle and of the individual. However, blockage of the left coronary artery would be more damaging to long-term survival because that artery supplies the left ventricle. The left ventricle pumps blood into the systemic circuit, and so if its functions are compromised, body cells are not supplied with oxygen and nutrients.

3. A junctional rhythm is bradycardic, meaning that the heart rate is below 60 beats per minute, averaging about 40 beats per minute. The decrease in heart rate also lowers the cardiac output. We do not see P waves on the ECG because the atria are not depolarizing.

4. Heart failure results in a decrease in cardiac output. To compensate and increase the cardiac output, both stroke volume and heart rate increase. Fluid retention raises preload, which increases stroke volume. Sympathetic nervous system stimulation increases contractility through positive inotropic effects, which raises stroke volume and also elevates heart rate through positive chronotropic effects.

Level 3

1. The trainee should be informed that increasing his heart rate up to a certain point will increase his cardiac output. However, too much of an increase in heart rate will actually decrease cardiac output because the heart will have inadequate time to fill with and eject blood.

2. The HCN channels slowly allow cations to enter the pacemaker cell so that it may reach threshold and fire an action potential. If the HCN channel is blocked, cations will enter more slowly and so the rate of depolarization will slow. This will decrease the heart rate. HCN channels are found only in SA and AV nodal cells, so this drug would affect only those cells.

3. The signs and symptoms of this condition include easy fatigability and weakness. This is because the condition allows blood to flow backward into the left atrium instead of the aorta, which reduces stroke volume and cardiac output. Mr. Watson's cardiac output might be improved by first fixing the defective valve. He might also be helped by drugs that increase contractility or decrease afterload, both of which increase stroke volume.

4. The refractory period of skeletal muscle fibers is quite short, which allows the fibers to be repeatedly stimulated and generate a sustained contraction. If the cardiac muscle cells had a similarly short refractory period, they would be able to maintain a sustained contraction if repeatedly stimulated. This could potentially result in death, as the heart chambers would not be able to go through diastole, as is usual in the cardiac cycle.

CHAPTER 18
Apply What You Learned

MODULE 18.1

1. The elastic fibers in an artery allow the vessel to stretch and recoil to its original size after being stretched. Defective elastic fibers would make a vessel either noncompliant (not able to stretch) or unable to recoil to its original size after stretching. 2. Veins would also be affected, as they are very compliant. However, they are under lower pressure than arteries and so the effects would be less obvious. 3. Atherosclerosis makes an artery narrow and hard through damage to the endothelium. This interferes with its function of delivering blood to capillary beds by decreasing the amount of blood that can flow through the artery.

MODULE 18.2

1. Vasoconstriction increases resistance, which raises blood pressure. For this reason, regular consumption of a vasoconstrictor such as nicotine will increase the risk of developing hypertension. 2. Arteriosclerosis increases blood pressure because the vessels are no longer able to stretch to accommodate small increases in blood volume. 3. Severe blood loss causes a significant decrease in blood volume, which decreases blood pressure.

MODULE 18.3

1. a. Diuretics decrease blood volume, which decreases blood pressure.

   b. Angiotensin-receptor blockers block the effects of angiotensin-II on blood vessels, decreasing resistance and therefore blood pressure.

   c. Beta blockers decrease the rate and force of heart contraction, decreasing cardiac output and blood pressure.

2. Extremely low blood pressure decreases blood flow to the kidneys, which in turn decreases filtration of fluid into the tubules of the kidneys. In addition, hormones such as ADH will trigger increased retention of water from the kidneys, further decreasing urine output. Reduced urine output decreases loss of blood volume.

MODULE 18.4

1. Thicker epithelium would slow transport of substances across the epithelium, which would considerably decrease the efficiency of nutrient and gas exchange between capillaries and tissues. 2. This poison would partially destroy the blood brain barrier, allowing more potentially toxic substances and pathogens into the brain's interstitial fluid than would normally occur. 3. Perfusion to the heart muscle decreases during systole, so if the heart momentarily isn't contracting, perfusion will actually increase (this will stop, though, once the pressure gradient driving blood flow has ceased).

MODULE 18.5

1. Hypotension would reduce the force driving water out of the capillaries, potentially causing the net filtration pressure to favor absorption instead of filtration. 2. Osmotic pressure is normally higher in the blood than in the interstitial fluid because of blood proteins. However, if blood proteins entered the interstitial fluid, the osmotic pressure there would rise. As the concentrations of the blood and interstitial fluid get closer in value, there is less force to drive the movement of water back into the capillary by osmosis. 3. Poison X would cause the overall net filtration pressure to strongly favor filtration, resulting in potentially massive fluid loss from the blood to the interstitial fluid.

MODULE 18.6

1. Although blockage of the right internal carotid artery would decrease blood flow to the brain, it wouldn't cut it off entirely due to the structure of the cerebral arterial circle, which provides alternate routes of blood flow to the brain. 2. The blood supply to the hand is provided by the superficial and deep palmar arches, which form from an anastomosis of the ulnar and radial arteries. So, although blood flow to the hand and digits would be reduced if the ulnar artery were severed, it would not be completely disrupted. 3. The femoral artery is the only major artery that supplies the entire lower limb. If it is irreparably damaged, little can be done to restore blood flow to the lower limb, resulting in loss of the limb.

MODULE **18.7**

1. The dural sinuses are not formed by regular veins, but rather are venous channels between layers of the dura mater. The dura mater is not very distensible. Therefore, compliance of the dural sinuses is low compared to that of other veins. **2.** A drug such as ticarcillin is ineffective when administered by mouth, because most of it never enters the capillary beds of the regular systemic circuit. However, if such a drug is injected into a vein, it will go to the heart, then the arteries, and then the capillary beds to be delivered to the tissues, where it will have an effect. **3.** Removal of the great saphenous vein does decrease venous drainage slightly. However, the venous system is characterized by a large number of anastomoses, so generally the venous blood is simply diverted to another vein.

MODULE **18.8**

1. Blood always flows down a pressure gradient. The pressure in the arteries is higher than the pressure in the veins, so blood can flow only from arteries to veins. **2.** The internal jugular vein is adjacent to the carotid artery, whereas the external jugular vein is more superficial and is separated from the carotid artery by muscles in the neck.

Assess What You Learned

Level 1

1. a. False: *Capillaries* are the exchange vessels of the cardiovascular system.
   b. True   c. True
   d. False: Muscular arteries control blood flow at the *organ* level.
   e. False: *Arteries* have smaller lumens, more elastic fibers, and more smooth muscle than *veins*.

2. c   **3.** a.   **4.** a   **5.** b

6. systolic, diastolic; pulse   **7.** e

8. Venous valves prevent blood from flowing backward; smooth muscle in the walls of veins can contract to reduce the volume; the skeletal muscle pump and the respiratory pump move blood toward the heart.

9. a. True   b. True
   c. False: Hormones that *increase* blood volume and blood pressure include antidiuretic hormone and aldosterone.
   d. True   e. True

10. a. A pressure gradient is a situation in which pressure is higher in one area than in another.
   b. Blood pressure is the force exerted on a blood vessel wall by the blood.
   c. Blood flow is the volume of blood that flows per minute.
   d. Resistance is any impedance to blood flow.

11. c

12. Substances may cross the wall (1) through gaps or pores in the endothelial layer, (2) through the membranes of endothelial cells by diffusion, or (3) through the endothelial cells by transcytosis.

13. b

14. a. False: Arterioles reflexively *constrict* when blood pressure increases.
   b. False: Increased concentrations of carbon dioxide and hydrogen ions and a decreased concentration of oxygen in the interstitial fluid cause local arteriolar *dilation*.
   c. True   d. True
   e. False: The sympathetic nervous system causes *vasoconstriction* in the skin when body temperature decreases.

15. b   **16.** c

17. _d_ Radial artery   _e_ Femoral artery
    _g_ Celiac trunk   _b_ Internal iliac artery
    _j_ Basilar artery   _i_ Renal artery
    _a_ Superior mesenteric artery   _f_ Internal carotid artery
    _h_ Dorsalis pedis artery   _c_ Subclavian artery

18. b   **19.** d

20. _g_ Cephalic vein   _i_ Splenic vein
    _d_ Great saphenous vein   _e_ Internal jugular vein
    _j_ Dural sinus   _b_ Brachiocephalic veins
    _a_ Azygos vein   _h_ Brachial vein
    _c_ Hepatic portal vein   _f_ Inferior mesenteric vein

Level 2

1. The blood in an artery is under much higher pressure than the blood in a vein. For this reason, blood is propelled out of a severed artery with much greater force than it is out of a severed vein.

2. A very tall person simply has more blood vessels that are longer. This increases the resistance of the circuit, and the increase in resistance leads to an increase in blood pressure.

3. The collagen in the tunica externa prevents blood vessels from overstretching when the blood pressure increases during systole. Dysfunctional collagen fibers weaken the vessels, and they could abnormally expand, potentially causing an aneurysm.

Level 3

1. A person hanging upside-down increases the blood flow to the head, which increases the pressure detected by the baroreceptors in the carotid sinus. This will trigger the baroreceptor reflex, which will lead to an overall decrease in systemic blood pressure. When a person in this situation returns to an upright position, his or her blood pressure will be quite low, leading to dizziness and temporary feelings of being ill.

2. The loss of albumin will decrease colloid osmotic pressure and cause the net filtration pressure to further favor filtration. Eventually, this will cause an excessive loss of fluid from the blood to the interstitial fluid, leading to edema.

3. a. An increase in the number of erythrocytes will increase the viscosity of the blood, so resistance and blood pressure will increase.
   b. After ingestion of caffeine, both vasoconstriction and increased cardiac output will increase the blood pressure.
   c. A bleeding ulcer will decrease blood volume, which will in turn decrease blood pressure.

4. The precapillary sphincters in the skin cut off blood flow to the skin to divert blood to deeper tissues as a heat conservation mechanism. The lack of oxygenated blood in the capillaries of the skin causes the skin to take on a bluish hue, particularly in the lips, where the epidermis is thin.

5. SIADH will increase the amount of water retained from the kidneys and therefore the blood volume. Increased blood volume will elevate blood pressure and raise the hydrostatic pressure in capillary beds, causing the net filtration pressure to further favor filtration.

CHAPTER 19
Apply What You Learned

MODULE **19.1**

1. A decreased amount of albumin in the plasma will decrease its colloid osmotic pressure. As a result, less water will be drawn into capillaries at the venous end of a capillary bed, leading to a net loss of more water than normal from the plasma. The water lost to the interstitial fluid will accumulate there, resulting in edema. **2.**

Proteins are composed of largely polar amino acids (although some have nonpolar R groups) and therefore mix with polar water. Lipids are composed largely of nonpolar hydrocarbons and therefore do not mix with polar water. Instead, lipids would congregate, forming a clump in the blood, which could block blood flow. For this reason, they must be transported through the blood on plasma proteins. **3.** Extensive blood loss impairs the ability of blood to carry out its homeostatic functions. This will disrupt gas exchange, solute distribution, the volume of extracellular fluid and cytosol, the body's acid-base balance, and blood pressure.

MODULE 19.2

1. An erythrocyte is essentially a "bag" of cytosol with hemoglobin molecules. This structure allows an erythrocyte to carry a great number of oxygen molecules (and some carbon dioxide), and its disc shape gives it a large surface area for rapid diffusion of gases. In addition, the lack of most organelles and the disc shape make the erythrocyte flexible so that it can fit through small vessels without sustaining damage. **2.** Erythropoietin increases a person's hematocrit and therefore the oxygen-carrying capacity of the blood. An athlete who can carry more oxygen in his or her blood will likely have greater endurance and resistance to muscular fatigue. However, an abnormally elevated hematocrit increases the viscosity of the blood and makes blood flow sluggish and could potentially block the flow through blood vessels.

MODULE 19.3

1. a. The primary leukocyte that responds to bacterial infections is the neutrophil, so one would expect to see elevated levels of neutrophils in this case.
   b. Virally infected cells are destroyed by T lymphocytes, so we see elevated levels of T lymphocytes with viral infections.
   c. The main leukocyte that responds to parasitic infections is the eosinophil, so elevated eosinophil levels are seen with such infections.
   d. Both macrophages and neutrophils ingest dead and dying cells after tissue injury, so one may see elevated levels of both monocytes and neutrophils in the blood after trauma.

MODULE 19.4

1. This will not affect platelets that have already formed, which lack nuclei and other organelles required for protein synthesis. It will, however, affect the formation of new platelets via thrombopoiesis. **2.** Megakaryocytes undergo repeated rounds of mitosis to become cells with a large cytoplasmic mass. If their mitosis is inhibited, they will be much smaller cells (and fewer in number), resulting in decreased levels of platelets.

MODULE 19.5

1. The activation of these three factors will stimulate coagulation, causing the blood to begin to clot abnormally. **2.** These substances will have the opposite effects of those in question 1—they will inhibit coagulation and lead to abnormal bleeding. **3.** tPA activates plasmin, which catalyzes the reactions that degrade fibrin and break up clots. After CPR, there is generally some amount of internal injury to blood vessels, and clots prevent blood loss. If tPA is administered, these clots will dissolve and the patient will bleed internally.

MODULE 19.6

1. Mr. Ramirez cannot donate blood safely to Mr. Jones. With type A− blood, Mr. Jones has anti-B antibodies and possibly anti-Rh antibodies in his plasma (if he has been exposed to Rh+ blood previously). These antibodies will bind to the B and Rh antigens of Mr. Ramirez's blood. **2.** Ms. Wright can donate blood to Mr. Jones safely only if he has

not been exposed to Rh+ blood previously. If he has had such prior exposure, his anti-Rh antibodies will bind to her Rh antigens. **3.** A person with type O− blood has no A, B, or Rh antigens on the surface of the erythrocytes, and so there can be no reaction with a recipient's anti-A, anti-B, or anti-Rh antibodies. However, this person's plasma contains anti-A, anti-B, and possibly anti-Rh antibodies, which would react with any blood type other than O−.

Assess What You Learned

Level 1

1. c   **2.** b

3. Gas exchange, solute distribution, immune functions, maintaining body temperature, blood clotting, maintaining acid-base homeostasis, stabilizing blood pressure

4. a. False: Erythrocytes are biconcave discs with *no nuclei as mature cells.*
   b. True   c. True
   d. False: Hemoglobin consists of *four* polypeptide chains *each* bound to a heme group.

5. d   **6.** hematopoietic stem cells, 100–120, spleen

7. a   **8.** b

9. _d_ Basophil   _e_ Monocyte
   _f_ B lymphocyte   _c_ T lymphocyte
   _a_ Neutrophil   _b_ Eosinophil

10. lymphoid, myeloid   **11.** c

12.
```
 3
 6
 5
 2
 1
 7
 4
```

13. The intrinsic/contact activation pathway begins when inactive factor XII comes into contact with exposed collagen, whereas the extrinsic/tissue factor pathway begins when subendothelial cells display tissue factor, which activates factor VII. In addition, the extrinsic/tissue factor pathway involves fewer steps to the activation of factor X (thrombin) than does the intrinsic/contact activation pathway. The pathways are similar in that they both culminate in the activation of factor X, which in turn leads to the common pathway and the activation of thrombin and the production of fibrin.

14. The common pathway converts fibrinogen to fibrin. Fibrin then "glues" the platelet plug and other formed elements together to form a blood clot.

15. d   **16.** b   **17.** ABO, Rh   **18.** a   **19.** d

20. b, c, and e

Level 2

1. The formed elements of blood must be distributed throughout the entire body to perform their functions—deliver oxygen to the tissues, perform immune functions, and clot damaged vessels—which is possible only in a liquid connective tissue. In addition, the plasma proteins and dissolved solutes must also be distributed throughout the body, which again is possible only in a liquid medium.

2. Oxygen molecules bind to the iron-containing heme groups, of which there are normally four. If only two iron ions are present, there are half as many binding sites for oxygen molecules, which means the blood is capable of carrying half as much oxygen. This will lead to a reduction in oxygenation of the tissues, which will trigger erythropoietin production and increase the number of erythrocytes in the blood. The excess erythrocytes will increase the oxygen-carrying capacity of the blood but will also raise the risk for blood clots, as an increased number of erythrocytes will make blood flow sluggish.

3. Disrupting either the intrinsic/contact activation or extrinsic/tissue factor pathway leads to disruption of the entire process because both pathways rely heavily on each other to proceed to completion. In addition, the extrinsic/tissue factor pathway is the most important path to the production of factor Xa, which is required for the common pathway.

4. A decrease in albumin will result in a decrease in osmotic pressure and a resulting loss of water from blood vessels into the interstitial fluid. A decrease in clotting proteins will interfere with the blood's ability to clot and may prolong bleeding time when an injury happens. Cirrhosis will not affect the production of γ-globulins, as they are produced by B lymphocytes, not the liver.

Level 3

1. Neutrophils are very active phagocytes, particularly in response to a bacterial infection. A deficiency in neutrophils will make a patient especially at risk for bacterial infections. T lymphocytes are involved in directly killing virally infected cells and cancer cells, but they are also responsible for activating other cells and components of the immune system. With a decrease in T lymphocytes, a patient is at risk for all types of infectious agents and certain cancers.

2. Platelets will not help Mr. Jackson, as he is able to form a platelet plug. He has a deficiency in a clotting factor, which prevents coagulation. This means that fibrin will not be produced and the platelet plug will not be "glued" together. Giving him platelets will only increase the size of the ineffectual platelet plug.

3. None of her family members could safely donate blood to Ms. Wu. Her blood type is O⁻, meaning she lacks A, B, and Rh antigens and has anti-A, anti-B, and anti-Rh antibodies (we know that she has anti-Rh antibodies because her daughter is type O⁺, so she has been exposed to the Rh antigen). All of the individuals who wish to donate to Ms. Wu have B and/or Rh antigens, so they cannot donate.

4. Since Elise is an adult, she is unlikely to have red bone marrow in her tibia, as much of the red bone marrow in an adult is replaced with fat-filled yellow bone marrow. A better place to withdraw red bone marrow from Elise would be the iliac crest, which does contain red bone marrow in an adult and is relatively superficial and easily accessed surgically.

CHAPTER 20
Apply What You Learned

MODULE 20.1

1. Removal of the lymph nodes and vessels will increase the patient's risk of infection and also decrease the return of excess interstitial fluid back to the cardiovascular system, leading to edema. **2.** Bacteria or other pathogens were likely introduced into the wound, after which they traveled through the lymphatic vessels to the inguinal lymph nodes, where they were trapped.

3. a. Removal of the spleen will increase the risk of infection slightly. In addition, destruction of old erythrocytes will have to take place elsewhere, such as in the plasma and liver.

b. Removal of the thymus in an adult will likely have few consequences, as T cell maturation has already taken place.

c. Removal of the thymus in a newborn will stop T cell maturation, leading to severe dysfunction in the entire immune system.

d. Removal of the tonsils will increase the risk of infection slightly, particularly around the oral and nasal cavities.

MODULE 20.2

1. Erosion of this mucous membrane compromises the first line of defense against pathogens and tissue damage. As a result, the urinary bladder is at higher risk of infection and damage from chemicals in urine. **2.** All three lines of defense rely on one another to function properly. If one line fails, the other two have great difficulty in protecting the body.

MODULE 20.3

1. A deficiency in neutrophils puts one at great risk of bacterial infection, as neutrophils are one of the main types of cells responsible for killing bacteria. **2.** Although it is possible that the patient has an infection, the fever could also be caused by inflammation due to tissue damage from a heart attack. **3.** A fever is a sign that inflammation is occurring in the body. Although aspirin might make the fever decrease, it will not treat the underlying cause of the inflammation. Thus, further examination is warranted any time a patient in such a setting develops an unexpected fever.

MODULE 20.4

1. T_H cells secrete interleukin-2, which is necessary for T_C cells to become fully activated. If T_H cells are inhibited, T_C cells will do less damage to the cells of the donated organ. **2.** The drug would decrease many other components of the immune response, including innate and antibody-mediated immunity. **3.** The change in the self antigens would likely cause the cells to be recognized by T_C cells as foreign, and T_C cells would attack and destroy the kidney cells.

MODULE 20.5

1. The lack of memory cells would effectively prevent the development of immunological memory, and a secondary immune response would be impossible. For this reason, every encounter with an antigen would initiate a primary immune response.

2. a. Decreased secretion of IgA will decrease the effectiveness of surface barriers that release IgA with their secretions.

b. IgG is the main antibody secreted in the secondary immune response. For this reason, a deficiency in IgG will prolong the duration and increase the severity of infection with many different types of pathogens.

c. A deficiency of IgG in a pregnant female puts both the mother and fetus at higher risk for developing prolonged and severe infections. It is particularly dangerous for the fetus, as maternal IgG is the only source of immunity before and immediately after birth.

d. A decrease in IgM will prolong the primary immune response and potentially increase the severity of an infection, as IgM is generally the first antibody to arrive and contain an infection.

3. The tetanus antitoxin provided passive immunity, as it contains preformed antibodies that will last only about 3 months. The tetanus antigen stimulates your immune system to produce its own antibodies, and so provides active immunity.

MODULE 20.6

1. An inability to activate T_C cells would decrease the ability of the body to fight off the viral infection, as T_C cells would not lyse and kill infected cells. Essentially every other component of the immune response, particularly T_H, NK, and B cells, would need to compensate for the lack of T_C cells. **2.** The result would be similar—T_C cells would be unable to detect and destroy the cancer cells. **3.** The fever indicates that an inflammatory response is occurring in the patient's body. **4.** The elevated level of neutrophils points to a bacterial infection as the possible cause of the fever, but this can't be determined definitively without more information.

MODULE 20.7

1. Such a drug will decrease allergen-mediated inflammation by preventing the release of mast cell granules. It would primarily affect type I hypersensitivity. **2.** The antibodies secreted by B cells would bind to a specific HIV-1 antigen, which could temporarily slow the infection. However, as the virus mutates, the antibodies would no longer bind the viral antigens, and a new B cell clone would be required. Note that the B cell response also depends on the levels of T_H cells—if there is a very low number of T_H cells, B cells won't activate fully or produce the higher levels of antibodies. **3.** Suppressing the immune system will decrease the production of autoantibodies, reducing some of the symptoms of an autoimmune disorder. However, this suppression will also raise the risk of infection.

Assess What You Learned

Level 1

1. b

2. a. False: Lymphatic vessels constitute a one-way system that delivers fluid from the *extracellular space to the blood vessels*.
   b. False: Lymph from the lower limbs drains into the *thoracic* duct.
   c. True
   d. False: The thymus is the site of maturation of *T* cells.
   e. False: Clusters of MALT located around the oral and nasal cavities are known as *tonsils*.

3. spleen, lymph node

4. innate immunity; acquired immunity; memory

5. a   **6.** d

7. a. True
   b. False: Phagocytic cells of innate immunity include *macrophages* and *neutrophils*.
   c. True
   d. False: *Interferon* is a cytokine that prevents viral replication in infected cells.
   e. False: Fever is generated by pyrogens that reset the temperature set point of the hypothalamus to a *higher* value.

8. e

9. inflammatory mediators, edema, redness, pain, heat

10. c

11. _e_ IgD   _a_ IgA
     _d_ IgM   _c_ IgE
     _b_ IgG

12. a. True
   b. False: The secondary immune response is mediated by *memory* cells.
   c. False: *Attenuated* vaccines consist of pathogens that are alive but unable to cause disease.
   d. False: Vaccinations are given to induce the production of *memory cells*.

13. Active immunity involves exposure to an antigen and a subsequent primary immune response in which antibodies and memory cells are generated. Passive immunity involves the transfer of preformed antibodies into an individual. Passive immunity does not result in the production of memory cells or a secondary immune response.

14. c   **15.** endogenous, cytotoxic T; exogenous, helper T

16. d

17. a. False: The cells involved in organ and transplant rejection are primarily T_C cells.
   b. True
   c. False: Neutrophils are a critical component of the response to a *bacterial* infection.
   d. True   e. True

18. d

19. HIV mediates the destruction of T_H cells, and these cells are responsible for activating most elements of the specific and nonspecific immune responses. Without these cells, a patient is at risk for bacterial, viral, and parasitic infections. In addition, immune surveillance is compromised, leading to an increased risk for certain cancers.

20. e

Level 2

1. a. The body aches are a result of inflammation, and the fever and chills are a result of pyrogens resetting the hypothalamic temperature set point. Both inflammation and fever are part of the innate immune response.
   b. As neutrophils are the body's main response to bacteria, the infection is likely bacterial in nature. The elevated number of neutrophils means that the body is producing more neutrophils to deal with the increased number of bacteria.
   c. If lymphocytes were elevated instead, this would indicate a possible viral infection, as lymphocytes are the main immune system response to viruses. It could also indicate other generalized causes of systemic inflammation, such as cancer or a severe injury.

2. A complement deficiency could result in a type III hypersensitivity disorder, as the uncleared immune complexes could deposit in tissues and create inflammatory reactions. It could also depress other processes that complement proteins support, including lysing bacteria, enhancing inflammation, neutralizing viruses, and enhancing phagocytosis through opsonization.

3. IgE is the main antibody that responds to allergens. Individuals with type I hypersensitivity disorders tend to produce excess IgE that responds abnormally to allergens. Blocking the functioning of IgE would prevent it from activating basophils and mast cells, and therefore reduce Terrence's symptoms.

Level 3

1. This is an example of passive immunity, as the antibodies are preformed and not made by your friend's plasma cells. The antivenin will not confer lasting immunity; however, your friend will produce her own antibodies after exposure to the snake venom via an active response, which will have a longer-lasting effect.

2. a. Cancer cells likely broke off from the tumor at the start of metastasis. Once in the interstitial fluid, they entered a lymph vessel and were delivered to a lymph node. Here, they encountered T cells, dendritic cells, and other immune cells that mounted an inflammatory response, which resulted in the swelling of the nodes.
   b. Removal of lymphatic vessels will inhibit the drainage of excess interstitial fluid, causing it to accumulate in the extracellular space. This results in sometimes significant edema. Removal of the lymph nodes reduces the body's ability to respond to pathogens, although it is likely that the pathogens would simply be caught by other, neighboring lymph nodes.

3. Neutrophils are the body's main defense against bacteria, so when the number of neutrophils is low, the body is unable to respond as it normally would to bacterial pathogens. This is particularly dangerous because the normal response to a bacterial infection is an *increase* in neutrophil numbers; the fact that the neutrophil count is already too low, combined with the inability to increase it, could make the bacterial infection rapidly lethal.

4. One would expect to see a decrease in the numbers of all types of leukocytes, as they are all produced by the bone marrow. The change would also cause a decrease in the hematocrit, as erythrocytes are also produced by the bone marrow. The decreased hematocrit would result in symptoms of anemia: fatigue, weakness, and pallor.

5. The cells of the epidermis normally undergo rapid mitosis to replace cells that die and are sloughed. If the epidermis is unable to undergo rapid mitosis, the skin will thin and be more prone to injury. This reduces its ability to function as the first line of defense of the immune system.

CHAPTER 21
Apply What You Learned

MODULE **21.1**

1. Blocking the structures of the conducting zone would prevent delivery of air to and from the structures where gas exchange occurs. Blocking structures of the respiratory zone would directly impair gas exchange, as this is where it takes place. **2.** The number one homeostatic disturbance would be a lack of oxygenation of the cells. Respiratory diseases also lead to imbalances in the pH of the body's fluids and potential abnormalities in blood pressure and fluid balance.

MODULE **21.2**

1. The pseudostratified columnar epithelium with goblet cells is a component of the mucous membrane lining much of the respiratory tract. It secretes mucus, which traps the inhaled debris, and the cilia projecting from the apical surfaces of the epithelial cells sweep the debris out of the respiratory tract. **2.** Inflammation of the larynx and vocal folds makes the folds much less elastic. As a result, they don't vibrate when air passes over them, and thus the person "loses" his or her voice. **3.** The tracheal rings keep the trachea patent, and if they were absent, the trachea could potentially collapse and block ventilation. The flexible nature of the cartilage rings allows the trachea to remain firmly open yet able to move with the body and surrounding organs. Bone is not flexible, and if the rings were composed of bone, the trachea would be stiff and inflexible. **4.** The type I alveolar cells are simple squamous epithelial cells. They are extremely thin, a feature that allows gases to cross their apical and basal membranes very rapidly.

MODULE **21.3**

1. The blood in the left pleural cavity will increase the intrapleural pressure. If the intrapleural pressure rises above the intrapulmonary pressure, the patient's left lung will collapse.

2. a. The increased amount of mucus in the respiratory tract will increase airway resistance, which will decrease the effectiveness of pulmonary ventilation.

b. Pulmonary edema increases surface tension on the cells of the alveoli. This causes atelectasis, which also decreases the effectiveness of pulmonary ventilation.

c. Inspiratory capacity = 3450 ml; vital capacity = 4100 ml. Inspiratory capacity is slightly lower than normal, and vital capacity is significantly lower than normal.

MODULE **21.4**

1. An increase in the thickness of the respiratory membrane increases the distance over which gases must diffuse to enter and exit the blood. This will decrease the effectiveness of pulmonary gas exchange. **2.** Low partial pressures of oxygen in the alveoli cause the pulmonary arterioles to constrict. If this happens over the entire

pulmonary circuit, blood pressure in this circuit will increase. **3.** This will decrease the amount of oxygen and nutrients delivered to the tissues. Not only will blood flow be reduced by the vasoconstriction, but also the damage to capillaries will cause them to thicken, which will decrease the efficiency of gas exchange.

MODULE **21.5**

1. Anything that increases hemoglobin's affinity for oxygen will decrease its ability to release oxygen to the tissues. As a result, the amount of oxygen in the venous blood will remain nearly as high as in the arterial circuit, and the tissue cells will be starved of oxygen.

2. a. A fever increases the temperature of the tissues, which will increase the unloading of oxygen from hemoglobin.

b. A lower number of hydrogen ions in the blood will decrease the unloading of oxygen from hemoglobin.

c. An elevated level of carbon dioxide in the blood will increase the unloading of oxygen from hemoglobin.

3. The forward reaction must take place in the capillary beds of the systemic circuit in order to convert carbon dioxide to a bicarbonate ion. In the pulmonary circuit, the reverse reaction must take place to convert the bicarbonate ion back into carbon dioxide so that it may be exhaled.

MODULE **21.7**

1. At night during sleep, the rate and depth of breathing decrease, which causes the P_{CO_2} to increase. The central chemoreceptors, which detect P_{CO_2}, are needed to ensure that the rate and depth of ventilation do not fall so low that pH, P_{CO_2}, and P_{O_2} homeostasis is disrupted. Without functional central chemoreceptors, mechanical ventilation may be needed to ensure that the rate and depth of ventilation are kept at a relatively constant level. **2.** The central and peripheral chemoreceptors lose their sensitivity over time when exposed to continually high partial pressures of carbon dioxide. This patient has chronic pulmonary disease, and the lab results indicate that this is what has happened.

MODULE **21.8**

1. The residual volume is the amount of air that remains in the lungs after expiration, or the air that is not exchangeable. The vital capacity includes all of the exchangeable volumes of air. If the residual volume increases, the volumes of exchangeable air all decrease, which in turn decreases vital capacity. **2.** Without functional cilia, mucus and inhaled debris are not cleared from the respiratory tract in the normal way, and the only way to clear the tract is with a deep cough. **3.** During an asthma attack, a person hypoventilates due to airway obstruction. This increases the retention of carbon dioxide, which in turn increases the amount of hydrogen ions in the blood. The elevated level of hydrogen ions stimulates the central chemoreceptors, which then increase the rate of ventilation to blow off the excess carbon dioxide.

Assess What You Learned

Level 1

1. a, c, e, f   **2.** respiratory tract, alveoli; pleural

3. a. False: The trachea contains *C*-shaped rings of hyaline cartilage.
b. False: Goblet cells secrete *mucus* into the respiratory tract.
c. True   d. True
e. False: The epithelium of the oropharynx changes from *pseudostratified ciliated* columnar epithelium to *stratified* squamous epithelium to enable it to resist abrasion from food.

4. d   **5.** true vocal cords (vocal folds); adducted, abducted

6. b

7. a. Type *II* alveolar cells secrete a chemical called surfactant.

b. *Type I alveolar cells* are squamous epithelial cells that constitute 90% of the total cells in the alveoli.

c. The respiratory membrane consists of the type *I* alveolar cells, the pulmonary capillaries' endothelial cells, and their shared basal lamina.

d. The respiratory membrane must be *thin* to function effectively in gas exchange.

8. a

9. __c__ Airway resistance   __e__ Pulmonary compliance
    __d__ Surface tension   __b__ V/Q ratio
    __a__ Surfactant

10. a. True

b. False: The *residual* volume is the amount of air left in the lungs after maximal expiration.

c. False: The inspiratory capacity is equal to the *inspiratory reserve volume* plus the tidal volume.

d. True

11. c   **12.** d   **13.** constrict; dilate

14. a. False: Oxygen is transported primarily by *binding to hemoglobin*.

b. True   c. True

d. False: When the P_{O_2} of arterial blood drops slightly, the percent saturation of Hb drops *only slightly*.

e. False: Increased P_{CO_2}, temperature, and hydrogen ion concentration all *decrease* Hb's affinity for oxygen.

15. __d__ Bicarbonate ion   __c__ Carbonic anhydrase
    __e__ Hemoglobin   __a__ Chloride shift
    __f__ BPG   __b__ Carbaminohemoglobin

16. decrease, increase; increase, decrease

17. d   **18.** b

19. arterial P_{CO_2} and hydrogen ions within the extracellular fluid of the brain; arterial P_{O_2}.

20. d

Level 2

1. Boyle's law states that pressure and volume are inversely related. This means that as the volume of a container increases, the pressure decreases, and as the volume of a container decreases, the pressure increases. The pressure inside the cylinder in this case would therefore increase as its volume decreased.

2. The esophagus is located on the posterior side of the trachea, which lacks cartilage rings of support. When the esophagus expands, it pushes into the trachea's soft posterior side, giving the feeling of being less able to breathe in spite of the fact that the food is in the esophagus, not the respiratory tract.

3. Both volumes would decrease dramatically, as the phrenic nerve is responsible for innervating the diaphragm and stimulating it to contract. Without the phrenic nerve, the main muscle of inspiration would be unable to contract, and inspiration would be severely decreased. The accessory muscles of inspiration and the external intercostals would try to increase the volume of the thoracic cavity in an effort to compensate.

4. The epiglottis covers the larynx during swallowing to prevent food and liquid from entering the respiratory tract. If the epiglottis is swollen, it can potentially seal off the larynx during ventilation as well, and cause the person to suffocate.

Level 3

1. A person loses carbon dioxide during hyperventilation, which increases the pH of his or her blood. Breathing into a paper bag can help because the bag contains exhaled carbon dioxide, which the person will rebreathe. This returns the partial pressure of carbon dioxide in the blood to normal and reduces the pH.

2. An individual with more erythrocytes is able to carry more oxygen in the blood, which increases the capacity for endurance. Elevated levels of BPG decrease hemoglobin's affinity for oxygen, making it easier to unload in the tissues. This also increases the capacity for endurance.

3. When your friend hyperventilated, he decreased the amount of carbon dioxide in his blood. This made the accumulation of carbon dioxide as he held his breath have less effect on the overall pH of his blood, resulting in less stimulation of the central chemoreceptors. This enabled him to hold his breath for a longer time.

4. Such a person would hypoventilate. The decrease in hydrogen ions in his or her blood would reduce the conversion of bicarbonate ions to carbonic acid, which could dissociate into CO_2 and water. Central chemoreceptors would detect the resulting decline in P_{CO_2} and trigger hypoventilation. That in turn would retain carbon dioxide and decrease the blood pH to a more normal range.

5. a. Diabetic ketoacidosis is the accumulation of ketones in the blood as a result of excessive catabolism of fatty acids. It results from insufficient insulin, which makes the individual unable to take glucose into the cells.

b. Mrs. Jordan likely has type I diabetes mellitus, the type that is more prone to lead to ketoacidosis.

c. As the name implies, ketoacidosis results in a decrease in blood pH. This is due to the accumulation of ketones, which are organic acids.

d. Her hyperventilation was triggered to correct the pH of her blood. As she hyperventilates, she blows off carbon dioxide, which decreases the production of carbonic acid and therefore reduces the total number of hydrogen ions in her blood.

6. The rate of ATP production increases during exercise to meet the demands of the contracting muscle fibers. A waste product of this ATP production is carbon dioxide. The accumulated carbon dioxide triggers the central chemoreceptors to increase the rate of ventilation in order to rid the body of it.

CHAPTER 22
Apply What You Learned

MODULE **22.1**

1. The peritoneal membranes surround a number of organs within the abdominal cavity. For this reason, it is quite easy for an infection to spread rapidly from one organ to another. **2.** The digestive system plays a role in many physiological processes that contribute to the maintenance of homeostasis, including regulation of fluid and electrolyte balance, acid-base balance, and nutrient balance. A disorder of the digestive system can therefore interfere with any of these processes. **3.** The epithelium of the alimentary canal contains regenerative cells that have a rapid rate of mitosis to replace the cells that are sloughed off or damaged. Chemotherapy drugs that cause the death of rapidly dividing cells will kill many of these cells, reducing the ability of the mucosa of the alimentary canal to maintain its epithelial lining.

MODULE **22.2**

1. Saliva washes away many of the acids and other material produced by the normal flora in the mouth that cause tooth decay. In addition, it contains substances such as lysozyme and IgA that kill pathogenic bacteria. Any condition that decreases the production of saliva will create favorable conditions for bacterial activity and tooth decay. **2.** Secretion of saliva is stimulated by parasympathetic neurons that release acetylcholine (ACh) onto target cells. Anything that increases the amount of ACh will also increase the production and release of saliva, which results in drooling. **3.** Mr. Kekoa's difficulty in swallowing can lead to aspiration of food because the swallowing reflex doesn't properly trigger elevation of the larynx. In addition, the decrease in activity of the muscularis externa of the esophagus will decrease peristaltic activity, potentially allowing the bolus to become lodged in the esophagus.

MODULE **22.3**

1. When the stomach cannot empty normally, the pressure inside it rises. This distends the stomach, which can trigger a vomiting reflex. For this reason, individuals with pyloric stenosis can suffer from frequent vomiting, resulting in malnutrition and fluid, electrolyte, and acid-base imbalances. **2.** The vagus nerve stimulates acid secretion from parietal cells, so severing this nerve will decrease acid secretion. However, it will also decrease the activation of pepsinogen, which affects protein digestion in the stomach.

MODULE **22.4**

1. Celiac disease can severely decrease the surface area available for digestion and absorption of nutrients in the small intestine. This leads to the presence of undigested nutrients and excess unabsorbed water and electrolytes in the feces, which will likely cause chronic diarrhea. In addition, it can lead to malnutrition and fluid, electrolyte, and acid-base imbalances. **2.** Such a poison would increase the motility of the stomach by increasing the activity of the vagus nerve, which could lead to vomiting. However, much of the churning motion in the stomach is controlled by ENS neurons and hormones such as serotonin and intestinal gastrin, and these control mechanisms would be unaffected by the poison. In the small intestine, peristalsis and segmentation would rise due to increased vagal activity, which would speed up the transit of the ingested food through the alimentary canal, possibly leading to diarrhea. The migrating motor complex would be largely unaffected, as it is under the influence of the ENS and motilin.

MODULE **22.5**

1. Oral antibiotics often kill off the normal flora in the large intestine. The main danger from this is that pathogenic bacteria that are not killed by the antibiotic may take over and cause infection. Ingesting supplements containing the normal flora helps to replace flora killed by the antibiotic, preventing pathogenic bacteria from causing infection. **2.** Cutting of the spinal cord would prevent conscious control of defecation, as the spinal cord would no longer be able to relay impulses from the cerebral cortex to voluntarily control the external anal sphincter and levator ani muscle.

MODULE **22.6**

1. Pancreatic acini produce pancreatic juice, which contains bicarbonate ions, water, and multiple digestive enzymes needed to digest all types of nutrients. Removal of the pancreas would therefore severely impact chemical digestion, and supplementary digestive enzymes would need to be administered. **2.** Under normal conditions, the liver will metabolize the drug and steadily decrease its concentration in the blood over a period of several hours. However, in a patient with liver disease, this function will be performed much less efficiently, prolonging the life of the drug in the blood. As a result, a lower dose is needed.

MODULE **22.7**

1. By preventing the activity of pancreatic lipase, orlistat blocks chemical digestion of lipids, and therefore also their absorption. This decreases the amount of dietary fat that the body absorbs, which could potentially lead to weight loss. Unfortunately, it can also lead to a deficiency of fat-soluble vitamins.

2. a. Being proteins, the supplemental digestive enzymes are likely to be damaged and possibly denatured by the acid in the stomach, and pepsin will begin digesting them into smaller polypeptides.

b. Such supplements are fairly useless, as the enzymes will be rendered nonfunctional in the stomach unless they are treated with a special coating to protect them from acid and pepsin. A healthy person probably doesn't need supplemental digestive enzymes unless a specific deficiency is noted.

3. If bile salts were polar molecules, they would be unable to interact with lipids. If they were nonpolar, they would be unable to interact with water molecules.

MODULE **22.8**

1. The parts of the digestive process that will be missed when a patient is fed directly into the stomach are mechanical digestion by mastication and a small amount of chemical digestion by salivary amylase.

Assess What You Learned

Level 1

1. b   **2.** d

3. a. False: The mucosa from the stomach to the anus consists of an inner layer of *simple* columnar epithelium.
   b. True   c. True
   d. False: The mucosa of the esophagus, pharynx, and oral cavity contains *stratified* squamous epithelium to protect it from abrasion.

4. __e__ Chief cells   __f__ DNES cells
   __c__ Parietal cells   __g__ Gastroesophageal sphincter
   __a__ Gastrin   __b__ Pepsin
   __h__ Pyloric sphincter   __d__ Chyme

5. b

6. The three folds of the small intestine are microvilli, folds of the enterocyte plasma membrane; villi, folds of the mucosa; and circular folds, folds of the mucosa and submucosa. These folds increase the surface area of the small intestine for absorption and chemical digestion.

7. c

8. Oral cavity → oropharynx → laryngopharynx → esophagus → stomach → duodenum → jejunum → ileum → cecum → ascending colon → transverse colon → descending colon → sigmoid colon → rectum → anal canal

9. a. False: The crown of a tooth is surrounded by an outer layer of *enamel,* which is the hardest substance in the body.
   b. True   c. True
   d. False: The secretory cells of the pancreas and salivary glands are *acinar* cells.

10. d   **11.** a

12. __f__ Salivary amylase   __b__ Pancreatic lipase
    __a__ Sucrase   __d__ Pancreatic amylase
    __e__ Trypsin   __c__ Pepsin

13. c   **14.** d   **15.** b

16. micelles; chylomicrons   **17.** a

18. a. True   b. True
   c. False: Smooth muscle of the stomach *relaxes* when food enters from the esophagus.
   d. True
   e. False: During *fasting*, the small intestine exhibits motion in the migrating motor complex pattern.

19. a   **20.** b   **21.** e   **22.** c

Level 2

1. Normally during swallowing, the uvula and soft palate move posteriorly to close off the nasopharynx and direct food down into the laryngopharynx. However, when your friend laughs, he exhales through his nose and this mechanism fails, causing the water to go up into the nasopharynx and nasal cavity instead of down into the laryngopharynx.

2. A patient who has food or liquid in his or her gastrointestinal tract may vomit during surgery as a result of the effects of the general anesthesia. With no reflex for swallowing in place (and no protection for the respiratory tract), the vomit can potentially go into the larynx, causing a patient to choke.

3. In general, such drugs inhibit the motility and secretion of most organs of the digestive system. They decrease salivation and secretion from the stomach, and they slow the movement of the stomach, small intestine, and large intestine. This can lead to dry mouth and constipation.

4. Absorption is indeed an important process of the digestive system, but each process of the system is required for the whole to work. For example, nutrients cannot be absorbed unless they are physically and chemically digested, they cannot be digested if they are not propelled through the alimentary canal, and they cannot be propelled if they are not first ingested. All the processes are connected, and one cannot be deemed "more important" than any other.

Level 3

1. a. Bile release increases when fatty foods are ingested, and the contraction of the gallbladder onto the stones causes Mr. Williams's pain. If he ate a carbohydrate-only meal, his symptoms would likely be much less severe, as bile is not required for carbohydrate digestion.
   b. His bile duct is likely blocked, preventing the release of bile into his small intestine. Bile is needed to emulsify fats so that they can be chemically digested by pancreatic lipase. In the absence of bile, pancreatic lipase will be unable to digest significant amounts of fats, leading to the presence of these fats in the feces.
   c. Bile contains the pigment stercobilin, which is responsible for the brown color of feces. Since bile is not being released into the small intestine, stercobilin isn't either, resulting in the production of nonpigmented feces.

2. a. Total removal of the stomach means that it is not present to control emptying into the duodenum. Instead, the bolus will be "dumped" straight into the duodenum, which will be unable to mix the bolus with pancreatic juice. As a result, nutrients will not be absorbed fully and the concentration of chyme in the small intestine will be too high. This will cause water to be drawn out of the large and small intestines, resulting in diarrhea.
   b. Absorption of vitamin B12 is dependent on production of intrinsic factor by the stomach. With the stomach gone, no intrinsic factor will be produced, preventing the absorption of vitamin B12 in the small intestine.

3. When acid is lost through vomiting, more acid must be secreted from the parietal cells of the stomach. This decreases the hydrogen ion concentration of the blood, leading to an increase in its pH. This increase in pH will cause a decrease in respiratory rate as the body attempts to retain carbon dioxide to lower the pH to a normal level.

4. The inflammatory response leads to vasodilation, along with increased capillary permeability. Both of these factors result in water loss, which is the desired effect of the laxative.

5. Three types of nervous system damage can lead to a loss of the defecation reflex (or a loss of conscious control over the reflex): Damage to the peripheral nerves that innervate stretch receptors, the internal anal sphincter, and/or the external anal sphincter can lead to loss of the reflex. Damage to the spinal cord can impair the ability of these neurons to communicate with their target cells. Finally, damage to the cerebral cortex can impair a person's ability to determine the appropriate time for defecation.

CHAPTER 23
Apply What You Learned

MODULE 23.1

1. The reaction of two stable chemicals would likely be endergonic, as it requires energy to break bonds in a stable chemical. If the two substances were unstable, the reaction could be exergonic, as they would release energy as they became more stable. 2. A fire that burns coal involves oxidation-reduction reactions, in which the coal is oxidized, and the fire results from the energy released by the oxidation. In your body cells, fuels such as glucose are oxidized, and the released energy is harnessed in the synthesis of ATP. 3. A riboflavin deficiency will affect almost all cells and decrease their ability to make ATP. The first cells affected will be those that are highly active metabolically, such as neurons, muscle fibers, and liver cells.

MODULE 23.2

1. In the absence of oxygen, glucose is catabolized by glycolysis, which yields only 2 ATP. For this reason, large amounts of glucose must be catabolized in order to meet the needs of the cell. When oxygen is available, glucose is catabolized by oxidative catabolism, which generates more ATP (38 total). So, less glucose must be broken down to produce the same amount of ATP. 2. The rate at which oxidative catabolism takes place must increase during exercise to meet the body's higher needs. Because carbon dioxide is produced as a waste product from these processes, a higher rate of oxidative catabolism also leads to a higher rate of carbon dioxide production. 3. Oxaloacetate is required in the first step of the citric acid cycle, where it combines with acetyl-CoA. This mutation would cause less oxaloacetate to be available to react with acetyl-CoA, and therefore fewer electron carriers would be reduced and fewer ATP produced.

MODULE 23.3

1. The pathways of catabolism for all three nutrients finish by a common pathway in the citric acid cycle, and all three generate reduced electron carriers for the electron transport chain. They differ in the way in which they are oxidized—lipids are oxidized via β-oxidation, amino acids must be deaminated before they are converted to citric acid cycle intermediates or glucose, and glucose is split into two pyruvate. 2. Urea, a nitrogen-containing waste product, formed in the deamination of amino acids, is eliminated by the kidneys. Kidney failure decreases the rate at which urea can be excreted, causing it to build up to toxic levels. 3. Lipids generally have more carbons than glucose or amino acids. For this reason, the catabolism of lipids generates more reduced electron carriers, and so more ATP, than does the catabolism of glucose or amino acids.

MODULE 23.4

1. Glucose is the fuel that is preferred by neurons, and depriving neurons of glucose can decrease the efficiency of their activity. 2. If the total caloric content of the food is unchanged, then eating it will not lead to fat loss. The extra sugar will simply be converted to lipids and stored in adipose tissues.

MODULE 23.5

1. Normally, glucose is the main fuel used by cells during the absorptive state, but if carbohydrates are unavailable, cells will instead oxidize lipids and amino acids. In addition, glycerol and glucogenic amino acids will be converted to glucose for those cells that require it as their main fuel source. 2. In the absence of insulin, most

body cells will be unable to take up glucose, and the processes that normally take place during the absorptive state will decrease or stop altogether. Blood glucose levels will climb, the body will catabolize lipids and amino acids for fuel, and ketone production will be high. **3.** Stimulating the neurons of the feeding center in the hypothalamus in this way will cause hunger and likely trigger feeding.

MODULE 23.6

1. Stressors activate the sympathetic nervous system and trigger the release of epinephrine and norepinephrine from sympathetic neurons and the adrenal medulla. This will elevate the metabolic rate above its basal level. **2.** Stimulating lipolysis without increasing the metabolic rate will not cause weight loss because the released fatty acids will not be catabolized. Instead, they will be stored in adipocytes in a different location. **3.** The metabolic rate increases during exercise to provide more ATP for contracting skeletal muscles and the heart. The reactions involved in ATP synthesis are not completely efficient, and much of the energy is lost as heat. This heat makes you feel warm, regardless of the environmental temperature.

MODULE 23.7

1. There are many potential consequences to such diets, depending on which low-carbohydrate foods are consumed. Decreased intake of fruits and vegetables will lower the amount of certain vitamins and minerals, as well as fiber, that is consumed. This could lead to changes in digestive habits, such as constipation. In addition, the lack of glucose in the diet will increase the rate at which gluconeogenesis takes place, which could cause protein breakdown to liberate glucogenic amino acids. **2.** Mr. Alcozer's condition has not improved—if anything, it is actually worse. Ideally, his HDL should have increased, as HDL particles take cholesterol to the liver for excretion. Instead, his HDL decreased. In addition, his LDL should have decreased, as the LDL particles deposit cholesterol in tissues, but his LDL remained unchanged. **3.** It's a matter of energy balance—if he takes in more than he oxidizes for fuel, then those excess nutrients will be stored. It doesn't matter what type of nutrient is consumed—consuming it in excess will not allow for weight loss unless energy expenditure is also increased.

Assess What You Learned

Level 1

1. b   **2.** a   **3.** exergonic, endergonic

4. a. False: During an oxidation-reduction reaction, the compound with the greatest electron affinity accepts electrons and is *reduced*.
b. True
c. True
d. False: ATP is the "money" that the cell uses to "pay for" the cell's *endergonic* reactions.

5. ATP does not contain a "high-energy bond"; rather, its terminal bond is very unstable. When this bond is broken, the released energy can either directly fuel a process or trigger the phosphorylation of another chemical to make a reaction more energetically favorable.

6. 6   2   7   4   1   3   5

7. d

8. Oxygen is the final electron acceptor in the electron transport chain.

9. d

10. a. False: The reactions of β-oxidation crop fatty acids into *two-carbon acetyl-CoA* units.
b. False: Most amino acids are converted into products that are oxidized by *the citric acid cycle*.

c. True   **d.** True
e. False: Fat and protein catabolism takes place primarily in the *mitochondria*.

11. b   **12.** glycogenesis, gluconeogenesis   **13.** c

14. a. postabsorptive   **d.** postabsorptive
b. postabsorptive   **e.** absorptive
c. absorptive   **f.** postabsorptive

15. a

16. Heat is a byproduct of every metabolic reaction, and nearly all of the energy in the human body is eventually lost to the environment as heat.

17. d

18. a. False: *Radiation* is energy transfer between the body and the environment via electromagnetic waves.
b. True   **c.** True
d. False: The body temperature measured on the skin or in the mouth is generally *lower* than the core body temperature.

19. macronutrient, micronutrient; essential nutrient

20. c

21. Fat-soluble vitamins have a cholesterol-based structure and are hydrophobic, so the body can store these compounds in adipose tissue when present in excess. Water-soluble vitamins have a polar structure and are hydrophilic, and so are excreted by the kidneys in the urine when present in excess. The fat-soluble vitamins include vitamins A, D, E, and K.

22. a. False: *Carbohydrates, lipids, and proteins* are used as fuel sources for cells.
b. False: *Low-density* lipoproteins deliver cholesterol from the liver to peripheral tissues.
c. True   **d.** True

Level 2

1. When the body has exhausted its supply of glycogen during exercise, it begins to catabolize fatty acids. The fatty acids are broken down by β-oxidation, and the resulting acetyl-CoAs are sent to the citric acid cycle. Here they are oxidized, which is the same process that occurs when wood is burned in a fire.

2. Amino acids from proteins may be converted to glucose via gluconeogenesis, and any excess glucose could be stored as fat.

3. Erythrocytes are only able to undergo glycolysis, and so all NADH is reoxidized to NAD^+ and lactic acid is produced instead of pyruvate.

4. a. Oxygen is the final electron acceptor in the final common pathway (the electron transport chain) for carbohydrates, lipids, and proteins. The amount of oxygen consumed is therefore a good indicator of the rate at which the body's many metabolic reactions are occurring.
b. The criteria for measuring the basal metabolic rate include no engagement in strenuous physical activity within the past hour. For this reason, the metabolic rate measured will not be your lab partner's basal rate.

5. Heat is a byproduct of all metabolic reactions. A person with extremely restricted caloric intake feels cold because his or her metabolic rate has slowed and less heat is being generated.

Level 3

1. A person who is suffocating is unable to produce adequate ATP via oxidative phosphorylation to sustain life. The same situation exists with cyanide poisoning—even though the oxygen level is normal to elevated, electrons cannot be transferred to the oxygen molecules, resulting in insufficient ATP production.

2. A diet that consists of nothing but fat would, of course, contain inadequate protein and carbohydrates. Glucose is the preferred fuel of the brain under all but near-starvation conditions, and so this diet would have a negative effect on concentration, memory, and other nervous system functions. In addition, it would lack the essential amino acids needed to make body proteins, which would accelerate muscle wasting and other forms of protein breakdown. A diet like this would not be advisable for anyone.

3. If starches were not absorbed, this would likely increase the body's reliance on fats and proteins for fuel and to make new glucose. This would lead to protein breakdown to release amino acids for gluconeogenesis and accelerated lipolysis, resulting in potential ketosis.

4. Thyroid hormone is a major determinant of the metabolic rate. Excess thyroid hormone, as seen with hyperthyroidism, will elevate the metabolic rate above normal levels, which means that more heat is produced as a byproduct.

5. As you exercise, the rate at which you catabolize metabolic fuels increases, which results in the production of more carbon dioxide. This elevated CO_2 level triggers the increased respiratory rate to remove the extra CO_2 from the blood. If the respiratory rate does not increase, the CO_2 can accumulate and decrease the pH of the blood.

CHAPTER 24
Apply What You Learned

MODULE 24.1

1. Peritonitis would be unlikely to impact the kidneys, as they are retroperitoneal, or located posterior to the peritoneal membranes. 2. The kidneys produce the hormone erythropoietin, which stimulates erythrocyte production. In renal failure, all aspects of kidney function decrease, including production of erythropoietin. The lower levels of the hormone cause decreased rates of erythropoiesis.

MODULE 24.2

1. The microvilli of the proximal tubule increase the surface area available for absorption of water and other solutes. If the number of microvilli decreases, the cells will be unable to reabsorb the same amount of water and solutes. As a result, urine volume would increase and solutes such as glucose and amino acids would be found in the urine. 2. Anything that decreases the blood flow to the kidneys will lower the rate of filtration of fluid from the blood into the nephrons and therefore decrease the regulation of fluid, electrolyte, and acid-base balance performed by the kidneys.

MODULE 24.4

1. A decrease in colloid osmotic pressure will decrease the force that retains water in the blood in the glomerulus. As a result, the net filtration pressure will increase in favor of filtration, and more filtrate will be produced. 2. Such drugs would increase the GFR temporarily by increasing the glomerular hydrostatic pressure due to the dilation of renal arteries and afferent arterioles. However, these changes would be countered very rapidly by the myogenic mechanism and tubuloglomerular feedback, both of which would trigger constriction of the afferent arterioles to decrease the GFR. 3. Drugs that block the RAAS will inhibit overall vasoconstriction, and this can reduce the vasoconstriction that normally occurs with physical activity. Without this normal vasoconstriction, the blood pressure remains low, even when it would normally rise.

MODULE 24.5

1. The drug would significantly decrease the amount of water and certain electrolytes reabsorbed from the proximal tubule. If fewer sodium ions are reabsorbed, there will be a smaller electrical gradient for reabsorbing chloride, a smaller osmotic gradient for reabsorbing water, and therefore a smaller concentration gradient for reabsorption of other electrolytes. 2. Glucose molecules are reabsorbed with sodium ions as a result of secondary active transport. Bicarbonate ions are reabsorbed as a result of the Na^+/H^+ antiporter in the proximal tubule. If all sodium ion reabsorption were blocked, the glucose/sodium ion cotransporter and Na^+/H^+ antiporter would not function, and high amounts of glucose and bicarbonate ions would appear in the urine. 3. Hypoventilation will lead to acidosis, which causes the kidneys to increase their secretion of hydrogen ions and increase the reabsorption of bicarbonate ions.

MODULE 24.6

1. With lower levels of ADH, less water will be reabsorbed from the distal tubule and collecting system. This will increase urine volume and decrease urine concentration.

2. a. Since less sodium chloride is pumped into the interstitial fluid, the gradient for water reabsorption from the medullary collecting system will be decreased. This will reduce the volume of water reabsorbed and thus increase urine volume.
   b. If the drug instead stimulated the $Na^+/K^+/2Cl^-$ pumps, a steeper gradient in the renal medulla would be created for water reabsorption. This would cause a person to produce more concentrated urine and retain more water, possibly leading to overhydration.

MODULE 24.7

1. The decreased GFR would trigger the RAAS and tubuloglomerular feedback, both of which would lead to increased glomerular hydrostatic pressure and a slight increase in the GFR. 2. Blocking carbonic anhydrase will result in the reabsorption of fewer sodium and bicarbonate ions. This will lead to a decreased pH of the blood, as fewer bicarbonate ions will be present to buffer the blood. The resulting acidosis will cause the kidneys to increase secretion of hydrogen ions and retain bicarbonate ions from the distal tubule.

MODULE 24.8

1. With metabolic acidosis, you would expect the pH of the urine to be low; however, it will not likely drop below 4.5 irrespective of the degree of acidosis. With metabolic alkalosis, the urine pH will rise as more hydrogen ions are retained and bicarbonate ions are excreted. 2. Creatine phosphate is metabolized to creatinine, and so an increase in consumption of creatine will lead to an increased amount of creatinine in the blood. This will increase the workload of the kidneys, as they must excrete the excess creatinine.

MODULE 24.9

1. Simple squamous epithelium is very thin and would not protect the underlying connective and muscle tissue of the ureters and urinary bladder. In addition, the simple squamous cells would not be as able to change shape with the changing size of these structures, and would be more prone to tears and other structural disruptions. 2. A spinal cord injury above S1–S2 would prevent voluntary control of micturition, as the impulses would be unable to reach the micturition center in the pons. An injury below S1–S2 could damage the neurons in the involuntary pathway of the micturition reflex. This could lead to urinary retention, because the signals that trigger the emptying of the urinary bladder would be interrupted.

MODULE 24.10

1. The staghorn calculus is causing a "backup" in the kidney by preventing urine from draining from the major and minor calyces into the renal pelvis. The backup is increasing the amount of filtrate in the collecting system and nephron, which in turn increases the amount of filtrate that remains in the capsular space. This increases the capsular hydrostatic pressure, a force that opposes filtration. An increased capsular hydrostatic pressure causes the net filtration pressure to fall, which leads to a drop in the GFR.

Assess What You Learned

Level 1

1. The four main organs of the urinary system are the kidneys, ureters, urinary bladder, and urethra.

2. e

3. a. True
b. False: Internally, the kidneys consist of an outer renal *cortex,* a middle renal *medulla,* and an inner renal *pelvis.*
c. False: The kidneys' first capillary bed is the *glomerulus,* which is fed by the afferent arteriole and drained by the efferent arteriole.
d. False: Filtrate flows from the renal corpuscle to the *proximal* tubule, the nephron loop, the *distal* tubule, and into the collecting system.

4. d

5. The filtration membrane consists of a series of three progressively finer filters, including the glomerular endothelial cells, the basal lamina, and the filtration slits of the podocytes.

6. a, c, f, g

7. a. True
b. False: The distal tubule reabsorbs sodium ions and secretes potassium and hydrogen ions in response to *aldosterone.*
c. True
d. False: ADH triggers water reabsorption from the *late distal tubule and collecting system.*
e. False: *Facultative* water reabsorption occurs in the distal tubule and collecting system.

8. a. dilate
b. constricts; reduce
c. renin
d. angiotensin-converting enzyme
e. increases; maintaining

9. c   **10.** b   **11.** c   **12.** c   **13.** b

14. concentrated; dilute   **15.** a, b   **16.** a

17. a. micturition; parasympathetic
b. transitional
c. detrusor
d. urine; urine and semen

Level 2

1. a. An increase in colloid osmotic pressure will increase the forces opposing filtration in the glomerulus, so the GFR will decrease.
b. This condition will decrease the glomerular hydrostatic pressure, and therefore decrease the force favoring filtration, leading to a reduced GFR. Note that the renin-angiotensin-aldosterone system would compensate somewhat for this, as would tubuloglomerular feedback.
c. Hypertension will increase the glomerular hydrostatic pressure and increase the force favoring filtration, leading to an increase in the GFR. Note that tubuloglomerular feedback and the myogenic mechanism will compensate for this.

2. Glomerulus → proximal tubule → nephron loop → distal tubule → cortical collecting duct → medullary collecting duct → reabsorbed into the interstitial fluid → enters thin descending limb of nephron loop → distal tubule → cortical collecting duct → medullary collecting duct → papillary duct → minor calyx → major calyx → renal pelvis.

3. The urethra in females is much shorter and only a short distance from the anus and vagina, where large numbers of bacteria normally reside. The male urethra is much longer and not anatomically close to the anus, making urinary tract infections less common.

4. Water reabsorption, which is needed to produce concentrated urine, occurs by osmosis. Osmosis is the movement of water from a solution with a lower solute concentration to one with a higher solute concentration. So, for osmosis to occur across the medullary collecting ducts, the interstitial fluid must be more concentrated than the filtrate.

Level 3

1. a. Aldosterone promotes the retention of Na^+ and water in the late distal tubule and cortical collecting duct. Adverse effects of blocking aldosterone include high plasma K^+ and acidosis, as aldosterone normally promotes the secretion of K^+ and H^+.
b. Ang-II normally causes constriction of the efferent arteriole and systemic blood vessels. If you block the receptors for Ang-II on the vessels of the kidney, you could potentially dilate the efferent arteriole and lower the GFR.
c. These pumps are largely responsible for creating the gradient needed for countercurrent multiplication and the production of concentrated urine. Blocking these pumps leads to a large volume of dilute urine. However, these drugs also block the reabsorption of K^+, which can lead to low plasma K^+ concentrations.

2. Mr. Wu's GFR, at 35 ml/min, is much lower than normal, which means he is in renal failure and inadequate amounts of blood are being filtered each minute. The plasma concentrations of the drug are higher than expected, therefore, because the drug is not being eliminated as it should by the kidneys.

3. Sodium ion reabsorption in the proximal tubule drives the reabsorption of many solutes, so the concentration of these solutes, including glucose and many electrolytes, would increase in the urine. Sodium ions also drive the reabsorption of water by osmosis, and therefore the volume of urine would increase. The presence of excess sodium ions in the filtrate would trigger the macula densa to release chemicals that cause constriction of the afferent arteriole and a decrease in the GFR. The macula densa will also trigger the release of renin from JG cells, which activates the RAAS and culminates in systemic vasoconstriction and other effects of Ang-II.

4. If you block aldosterone, you also block sodium ion and water reabsorption, which reduces blood volume and blood pressure. Drugs that block angiotensin-II receptors on blood vessels will lower blood pressure by allowing for systemic vasodilation. The result of blocking pumps in the nephron loop and reducing the medullary osmotic gradient will be the production of a large volume of dilute urine, which decreases blood volume and blood pressure.

5. Urobilinogen is a breakdown product of hemoglobin that is excreted by the kidneys. Excess urobilinogen in the urine could indicate an increased plasma concentration of urobilinogen, which could be caused by an elevated rate of erythrocyte destruction.

CHAPTER 25

Apply What You Learned

MODULE 25.1

1. The kidneys produce urine, which aids in fluid balance by excreting excess water. If the kidneys cannot do this and fluid intake increases, water will not obey the principle of mass balance—a water imbalance will exist. **2.** The excess water in the patient's body will dilute the body fluids and change the concentration of extracellular electrolytes, leading to an imbalance.

MODULE 25.2

1. Lipids repel water and are hydrophobic, so lipid-rich cells contain less water than cells such as hepatocytes that are rich in other substances. **2.** A 9% sodium chloride solution is very hypertonic to the cytosol and will draw water out of the patient's cells by osmosis. As a result, the patient's cells will shrink, or crenate. This solution could also affect the patient's sodium ion balance. **3.** If a patient is overhydrated, meaning that too much water is in the plasma and cytosol, hypertonic saline could be used to draw the excess water out of the cells.

MODULE 25.3

1. The levels of sodium and chloride ions in the ECF will decrease because their reabsorption isn't stimulated by aldosterone. In addition, the level of potassium ions will increase because their secretion isn't stimulated by aldosterone. **2.** Water normally follows reabsorbed sodium and chloride ions by osmosis. With the reabsorption of these ions decreased, water reabsorption will decrease as well. **3.** Excess aldosterone would lead to reabsorption of large numbers of sodium and chloride ions, and also water by osmosis. It would also lead to a low blood potassium ion level as a result of excessive secretion of potassium ions.

MODULE 25.4

1. Mr. Wong is likely to have developed respiratory acidosis. Due to the long period over which the acidosis developed, it will likely be partially compensated. The arterial blood gases will probably show a low pH, high P_{CO_2}, and high HCO_3^-. **2.** Hyperventilation is a response to an acidotic state, as it rids the body of excess carbon dioxide. However, it is unlikely to be effective in this case due to Mr. Wong's state of impaired gas exchange. **3.** The hydrogen ion pumps of the kidneys are unable to pump against a gradient greater than a pH of 4.5, so they will be unable to secrete further hydrogen ions (unless some of the hydrogen ions in the filtrate are buffered).

MODULE 25.5

1. Dehydration triggers aldosterone secretion, which stimulates potassium ion secretion. This could lead to hypokalemia if large amounts of aldosterone are released. **2.** The blockage would hamper the ability of the body to adjust to dehydration, as the response begins with angiotensin-II. Note that hormones such as ADH would still be secreted due to stimulation of osmoreceptors, so the body would be able to partially respond to dehydration. **3.** As a positive feedback loop, the response would continue even after fluid balance was restored, leading to overhydration and other electrolyte and pH imbalances.

Assess What You Learned

Level 1

   1. a

   2. An electrolyte dissociates into ions in a solution, and these ions can conduct an electric current. A nonelectrolyte does not dissociate into ions in a solution, and the solution cannot conduct an electric current.

   3. b   **4.** d   **5.** hypertonic, hypotonic

   6. a. False: The thirst mechanism is mediated by osmoreceptors located in the *hypothalamus*.
   b. False: The *insensible* water loss includes the water lost from the body via the skin and the respiratory system.
   c. True

   d. False: Atrial natriuretic peptide promotes water *loss from* the kidneys and *decreases* the amount of water in the body.
   e. True

7. d

8. a. Angiotensin-II: promotes sodium ion reabsorption from the proximal tubule
   b. Aldosterone: promotes sodium ion reabsorption and potassium ion secretion from the distal tubule and collecting system
   c. Parathyroid hormone: promotes calcium ion reabsorption and inhibits phosphate ion reabsorption from the kidneys, as well as stimulating osteoclast activity
   d. Vitamin D: promotes calcium ion and phosphate ion absorption from the small intestine
   e. Atrial natriuretic peptide: inhibits sodium ion reabsorption in the distal tubule and collecting system

9. b

10. a. True   b. True
   c. False: *Hypokalemia* tends to decrease the resting membrane potential and hyperpolarize excitable cells.
   d. True
   e. False: Chloride ions are generally reabsorbed from the kidneys, along with *sodium* ions.

11. c   **12.** b

13. The pH changes very little when hydrogen ions are added to a buffered solution because the buffer's weak conjugate base binds the hydrogen ions, taking them out of solution.

14. increases, decrease; decreases, increase

15. a, c, d

16. a. True
   b. False: *Respiratory* compensation for acid-base disturbances begins within minutes, whereas *renal* compensation begins within several hours.
   c. False: Respiratory compensation for metabolic acidosis consists of *hyperventilation*.
   d. True   e. True

17. e

18. Angiotensin-II has several effects that result in water reabsorption: It promotes sodium ion reabsorption, which creates an osmotic gradient for water reabsorption; it stimulates thirst; it stimulates release of ADH; and it triggers the release of aldosterone.

Level 2

   1. Only the solution with 5% sodium chloride will conduct electricity. Deionized water contains no electrolytes, and glucose does not dissociate into ions in water and is therefore a nonelectrolyte.

   2. Her body water will increase with the loss of adipose tissue, as adipose tissue contains very little water.

   3. The amount of water does not affect the absolute number of ions and other solutes, but it does affect the percentage of ions and solutes relative to water. So if the total amount of water decreases, the relative proportion of solutes increases, and their concentration rises.

   4. The patient is likely to be hyperventilating. This removes carbon dioxide from the ECF, which reduces the hydrogen ion content. The source of the hydrogen ions doesn't matter, only that their number decreases.

Level 3

   1. Elise is hypoventilating because she likely has metabolic alkalosis due to the loss of hydrogen ions from vomiting. The hypoventilation will help her retain carbon dioxide and reduce the pH of her ECF. Her arterial blood gases will likely show an elevated pH, high bicarbonate ion concentration, and elevated P_{CO_2}.

2. The volume of her ECF has likely decreased due to fluid loss from the vomiting, and the osmolarity has likely increased. This could draw fluid out of her cells and into the ECF, resulting in cell crenation.

3. By blocking aldosterone, the drug has inhibited the secretion of both potassium and hydrogen ions from the kidneys. This has caused retention of excess potassium and hydrogen ions, leading to hyperkalemia and acidosis.

4. The patient likely has respiratory acidosis. Renal compensation has not yet occurred, as indicated by the normal bicarbonate ion concentration, so the acidosis is only a few hours old.

5. The sodium ion concentration will increase, resulting in an increase in the osmolarity and osmotic pressure of the ECF. This will draw water out of the cells and into the ECF, which leads to an increased blood volume. The higher blood volume will tend to elevate the blood pressure.

CHAPTER 26

Apply What You Learned

MODULE 26.1

1. Without crossing over and independent assortment, chromosomes would remain intact and would separate consistently, resulting in ova or sperm with the same alleles. 2. One kind of gamete would get both homologous chromosomes, and the other kind of gamete would get neither one. Therefore, the offspring would either receive an extra chromosome or have a missing chromosome. The effects on the offspring would vary, depending on the chromosome involved.

MODULE 26.2

1. The seminal vesicles contribute approximately 60–70% of the volume of semen, so the amount of semen produced would decrease. 2. If the dartos and cremaster muscles of the scrotum contract, the scrotum will be pulled close to the abdominopelvic cavity. If the weather is hot, this will raise the temperature of the scrotum to body temperature and will decrease the production of sperm.

MODULE 26.3

1. LH stimulates interstitial cells in the connective tissue around the seminiferous tubules to produce testosterone, which is necessary for sperm production. 2. Nitric oxide relaxes and dilates smooth muscle in the walls of the arterioles of the penis and within the blood sinuses in erectile tissue. This allows blood to enter the tissue and results in an erection. Without nitric oxide, an erection would not occur. 3. Stress activates the sympathetic nervous system, which suppresses the parasympathetic response needed for an erection.

MODULE 26.4

1. If sperm enter the tube and fertilize an oocyte, the fertilized ovum will not be able to pass through the tube to the uterus, causing it to remain in the much smaller uterine tube. 2. Fimbriae "catch" an ovulated oocyte. A lack of fimbriae would cause the oocyte to be released into the peritoneal cavity instead, where sperm likely cannot reach it.

MODULE 26.5

1. The drug would be unlikely to have much of an effect on an adult female's ability to become pregnant unless she suffered from excessive atresia of primary oocytes. A woman normally has about 300,000 primary oocytes at puberty. These oocytes do undergo atresia as she ages, but she still normally has far more than she will ever ovulate during her lifetime. Note, however, that if the drug stops the primary oocytes from undergoing atresia after menopause, it could allow her to become pregnant at a more advanced age. 2. Taking a supplement that produces a continuously high level of progesterone in a woman's body will inhibit the release of gonadotropins, which will prevent the ovarian cycle from beginning again. Also, mimicking the effects of progesterone will maintain the endometrial lining of the uterus, preventing menstruation. 3. The uterine tubes are not involved in the production of hormones responsible for the ovarian and uterine cycles. Therefore, the surgery will have no effect on the onset of menopause.

MODULE 26.6

1. a. She could use any of the barrier methods (condoms, cervical cap, or sponges) and spermicides. In addition, she could use an IUD, but should not consider an IUS or any hormonal contraceptives because of her bleeding disorder. Tubal ligation is not an option because of her age and unknown future plans.
   b. The man should be using condoms for each sexual encounter and should include spermicides if possible. A vasectomy is not an option because of his age and unknown future plans.

MODULE 26.7

1. Many cases of STIs are asymptomatic, and some individuals with an infection never develop symptoms or lesions. However, even an asymptomatic individual can pass an STI on to other partners. Furthermore, some diseases enter a latent period when they are inactive, and then manifest at a later date. 2. Girls should be vaccinated for HPV before they become sexually active to avoid contracting the virus in the future. This will significantly minimize their risk of developing cervical and other cancers and genital warts. Boys should receive the vaccine to prevent them from spreading the virus to sexual partners and also to reduce their risk of contracting certain cancers and genital warts.

Assess What You Learned

Level 1

1. 23, different

2.
g Prophase I		f Prophase II
b Metaphase I		c Metaphase II
d Anaphase I		h Anaphase II
a Telophase I		e Telophase II

3. a. False: Spermatogenesis is the process of sperm cell development. It begins at puberty and *continues throughout life*.
   b. True
   c. False: Primary spermatocytes undergo the first meiotic division and form two smaller *haploid* cells called secondary spermatocytes.
   d. True

4. a   5. d

6. The prostatic urethra, the membranous urethra, and the spongy urethra

7.
b,c,d Seminal vesicle
a,f Prostate
e Bulbourethral gland

8. corpora cavernosa, corpus spongiosum   9. b

10.
3
5
4
2
1

11. Erection, ejaculation   12. d   13. a

14. uterine tube, uterus   15. lobules, alveoli

16. a. False: Oogenesis begins before a female is born and continues *until she reaches menopause* (usually between 45 and 55 years of age).

    b. False: About once a month after puberty begins, some primary oocytes are stimulated to continue development. Usually one will complete the first meiotic division to produce two haploid cells, *one of which is large and the other small.*

    c. True   d. True

17. _d_ Primordial follicle   _f_ Vesicular follicle

     _e_ Primary follicle   _c_ Corpus luteum

     _a_ Secondary follicle   _b_ Corpus albicans

18.   _4_   _5_

       _6_   _10_

       _3_   _7_

       _8_   _9_

       _1_   _2_

19. a. False: During the *secretory* phase, the endometrium prepares for the implantation of the embryo by enlarging the endometrial glands and secreting glycogen into the uterine cavity.

    b. False: During the *proliferative* phase, the endometrium generates a new stratum functionalis.

    c. False: During the menstrual phase, the uterus sheds the stratum *functionalis* of the endometrium, which results in the discharge of blood and tissue from the vagina.

20. a   21. b   22. b and c

Level 2

1. The main purpose of the first meiotic division is to create gametes with half the number of chromosomes that were in the original cell. To accomplish this, homologous pairs of chromosomes are separated, rather than the sister chromatids of each chromosome. Also, during the first meiotic division, genetic variability is increased by two processes—crossing over and independent assortment—which randomly "shuffle" the paternal and maternal genes. The purpose of mitosis is to create two cells that are genetically identical to the original cell.

2. In addition to producing androgen-binding protein and inhibin, sustentacular cells create the blood-testis barrier by forming tight junctions with each other to separate the forming sperm cells from the immune system. This prevents the immune system from attacking the genetically unique sperm cells. Also, these cells secrete testicular fluid, which helps transport sperm through the lumen of the seminiferous tubules, and they provide nutrients to the dividing cells. They phagocytize damaged spermatogenic cells and the excess cytoplasm released from maturing spermatids.

3. During the female orgasm, the uterus exhibits peristaltic waves of contraction and the cervix pushes down somewhat into the vagina. If semen is present, it is possible that these actions may draw the semen farther into the female reproductive tract, which could increase the chances that fertilization will occur.

4. The artificially high levels of estrogens and progesterone have a negative feedback effect on the release of FSH and LH. Low levels of FSH and LH prevent follicle maturation and subsequently ovulation. Low levels also interfere with the buildup of the endometrial lining of the uterus necessary for implantation of an embryo.

Level 3

1. His wife is correct. The prostate gland surrounds the prostatic urethra and enlarges as men age. As it enlarges, it pushes in on the urethra and frequently affects urination.

2. High blood levels of testosterone will inhibit secretion of FSH and LH from the anterior pituitary, which is needed for testosterone production in the testes. LH stimulates interstitial cells to produce testosterone, and this local production of testosterone is required for spermatogenesis.

3. Body fat is essential to support a pregnancy. The brain monitors body fat using hormones such as leptin. Leptin stimulates gonadotropin secretion, so if leptin levels drop due to low body fat, gonadotropin secretion will decrease as well. This will affect the HPG axis and cause amenorrhea. For a female with such a low percentage of body fat, a pregnancy would be particularly stressful, so the fact that she is not having her menstrual cycle is actually beneficial to her at this point.

4. Erections occur when the corpora of the penis are engorged with blood. Most herbal medications for impotence increase the effects of the nitric oxide that is already present, which results in vasodilation of blood vessels to the penis. However, other blood vessels are often also affected, which can lower vascular resistance and decrease systemic blood pressure.

CHAPTER 27
Apply What You Learned

MODULE 27.1

1. The most fundamental events, such as formation of the organs, occur during the first trimester (first 3 months). **2.** The last 10 weeks of fetal development are important to complete the growth of the organs.

MODULE 27.2

1. The syncytiotrophoblast has to touch maternal cells and be able to dissolve maternal blood vessels without being attacked by the maternal immune system. If this cell were antigenic, the pregnancy would probably not progress beyond implantation because the mother's immune system would destroy the conceptus. **2.** Without secretion of hCG at the beginning of a pregnancy, the corpus luteum would degenerate and maternal levels of progesterone and estrogen would decrease.

MODULE 27.3

1. Teratogens are chemicals or biological agents, such as bacteria, that can cross the placenta and harm the conceptus. During the pre-embryonic and embryonic periods of development, tissues and organs are forming and are most susceptible to these agents. **2.** The nervous system integrates and allows communication between all of the developing tissues.

MODULE 27.4

1. She should schedule an ultrasound exam by the end of the third month of development. **2.** The ductus venosus acts as a shunt between the hepatic portal vein and the inferior vena cava. If it remains open after birth, some of the blood from the gastrointestinal tract will bypass the liver and not be filtered. Toxins will be allowed to pass to other organs, and nutrients that should be stored in the liver will also bypass the organ.

MODULE 27.5

1. The umbilical cord is usually not below the fetus's head in the vertex birth, so it stays behind the fetus. If the legs are touching the cervix, the umbilical cord can collect under the fetus, and during labor, the fetus may press on the cord. **2.** Even with one fetus, the uterus takes up a large space in the abdominal cavity when the fetus is full term. If there are two or more fetuses, the space is even more limited, and each fetus has less space to grow. Also, the fetuses may be born earlier because crowding inside the uterus might trigger contractions.

MODULE 27.6

1. Newborns receive some antibody protection from their mothers, but they need protection from diseases that might make them seriously ill or even be fatal. **2.** Milk production is controlled by a positive feedback mechanism. Prolactin stimulates the production of milk, and suckling by the infant increases prolactin production. The more the infant suckles, the more prolactin is produced, and, subsequently, the more milk is produced.

MODULE 27.7

1. Because both parents have attached earlobes, they must each be homozygous for the recessive allele. Using E for the dominant allele and e for the recessive allele, the Punnett square is as follows:

	e	e
e	ee	ee
e	ee	ee

Because none of the offspring genotypes include a dominant E allele, it is not possible for these parents to produce a child with free earlobes.

2. The mother's father had only one X chromosome to pass on, so she must have inherited his X^r allele. Because she is not color blind, she must have inherited X^R from her mother. Therefore, her genotype is $X^R X^r$ and she is a carrier. Her partner's genotype is $X^r Y$ because he is color blind. This is the Punnett square:

	X^r	Y
X^R	$X^R X^r$	$X^R Y$
X^r	$X^r X^r$	$X^r Y$

The offspring has a 50% chance of being color blind, regardless of sex.

3. The parents' genotypes produce the following offspring genotypes: $I^A i$ (1/2, or 50%) and $I^B i$ (1/2, or 50%). The phenotypes are blood type A (1/2, or 50%) and blood type B (1/2, or 50%).

Assess What You Learned

Level 1

1. _b_ Pre-embryonic period   _a_ Embryonic period
   _c_ Fetal period

2. cleavage, blastocyst

3. a. True   b. True

   c. False: Secondary oocytes are covered with a layer of cells called the *corona radiata*.
   d. False: The cortical reaction *prevents* additional sperm cells *from binding* to the oocyte.

4. b

5. _5_ a.   _2_ d.
   _1_ b.   _3_ e.
   _4_ c.   _6_ f.

6. trophoblast, inner cell mass

7. _c_ Amnion   _b_ Allantois
   _a_ Yolk sac   _d_ Chorion

8. Placentation, umbilical cord

9. _b, e_ Endoderm   _a, d_ Ectoderm
   _c, f_ Mesoderm

10. Cephalocaudal and lateral

11. _f_ Month 3   _c_ Month 6
   _d_ Month 4   _e_ Month 7
   _a_ Month 5   _b_ Months 8 and 9

12. _b_ Foramen ovale   _a_ Ductus venosus
   _c_ Ductus arteriosus

13. a   **14.** d

15. _3_ a.   _4_ c.
   _1_ b.   _2_ d.

16. a. True   b. True
   c. False; The *dilation stage* of labor is usually longer than the *expulsion* or placental stages.

17. 4 weeks, neonatal

18. _3_ a.   _2_ c.
   _1_ b.   _4_ d.

19. Genes, genome   **20.** b

Level 2

1. From the mother's perspective, a pregnancy begins at the last menstrual cycle rather than at fertilization. This is typically 2 weeks longer because ovulation and fertilization occur approximately 14 days after menstruation.

2. The zygote is a very large cell because the ovum is very large. Cleavage occurs quickly, and cells produced from these mitotic divisions do not have time to grow before they divide again.

3. If there are significant chromosomal abnormalities, with a few exceptions, the embryo cannot survive past the embryonic period.

4. The weight of the rapidly growing fetus rests on the urinary bladder, increasing the urge to urinate.

Level 3

1. Once the blastocyst implants, it embeds in the lining and is covered by endometrial cells. Detachment and movement would kill the embryo, so the only choice is to perform surgery to physically remove the embryo and try to prevent damage to the uterine tube.

2. The drug may cross the placenta and injure the fetus. The embryonic stage is particularly vulnerable to such effects because the tissues and organs are still forming.

3. The Punnett square is as follows:

	w	w
W	Ww	Ww
w	ww	ww

The probability of a child having a widow's peak is 50%.

4. Her lab partner is wrong. Elise can only narrow her father's blood type to two possibilities. Her mother has type B, so her genotype could be $I^B I^B$ or $I^B i$. However, Elise doesn't have type B or AB, so she must have inherited an i allele from her mother. Therefore, Elise's genotype must be $I^A i$, and her mother's must be $I^B i$. Elise must have inherited the I^A allele from her father. Therefore, his genotype could be $I^A I^A$, $I^A i$, or $I^A I^B$, and his blood type could be A or AB.

5. The growing placenta requires increased blood volume, and as the fetus grows, it needs more oxygen and produces more wastes that must be removed. Because of these changes, Laura's blood volume and, subsequently, her cardiac output, have increased. Blood pressure may rise because of increased blood volume. Also, some women develop preeclampsia, which results in high blood pressure. The placenta secretes human placental lactogen, which inhibits maternal insulin and increases blood glucose. The mother's cortisol levels also increase to promote maturation of fetal lungs, and this too can increase blood glucose.

Appendix B
The Metric System

Measurements, Units, and Equivalents

Measurement	Metric Unit	Metric-to-English Equivalent	English-to-Metric Equivalent
Length *SI unit* = *meter*	Kilometer (km)	1 km = 0.62 mile	1 mile = 1.61 km
	Meter (m)	1 m = 1.09 yards or 3.28 feet	1 yard = 0.914 m 1 foot = 0.305 m
	Centimeter (cm)	1 cm = 0.394 inch	1 inch = 2.54 cm
	Millimeter (mm)	1 mm = 0.039 inch	1 inch = 25.4 mm
	Micrometer or micron (μ or μm)	1 μm = 0.0039 inch	1 inch = 2540 μm
	Nanometer (nm)	1 nm = 0.0000039 inch	1 inch = 2,540,000 nm
Mass *SI unit* = *kilogram*	Kilogram (kg)	1 kg = 2.205 pounds	1 pound = 0.4536 kg
	Gram (g)	1 g = 0.0353 ounce	1 ounce = 28.35 g
	Milligram (mg)	1 mg = 0.015 grain	1 grain = 64.799 mg
	Microgram (μg)	1 μg = 0.000015 grain	1 grain = 64,799 μg
Area *SI unit* = *square meter*	Square meter (m^2)	1 m^2 = 1.1960 square yards or 10.764 square feet	1 square yard = 0.8361 m^2 1 square foot = 0.0929 m^2
	Square centimeter (cm^2)	1 cm^2 = 0.155 square inch	1 square inch = 6.4516 cm^2
Volume: solids *SI unit* = *cubic meter*	Cubic meter (m^3)	1 m^3 = 1.3080 cubic yards or 35.315 cubic feet	1 cubic yard = 0.7646 m^3 1 cubic foot = 0.0283 m^3
	Cubic centimeter (cm^3)	1 cm^3 = 0.0610 cubic inch	1 cubic inch = 16.387 cm^3
	Cubic millimeter (mm^3)	1 mm^3 = 0.000610 cubic inch	1 cubic inch = 1638.7 mm^3
Volume: liquids and gases *SI unit* = *liter*	Kiloliter (kl)	1 kl = 264.17 gallons	1 gallon = 0.003785 kl or 3.785 l
	Liter (l)	1 l = 0.264 gallon or 1.057 quarts	1 quart = 0.946 l 1 pint = 0.473 ml
	Milliliter (ml)	1 ml = 0.034 fluid ounce	1 fluid ounce = 29.57 ml
	Microliter (μl)	1 μl = 0.000034 fluid ounce	1 fluid ounce = 29,570 μl
Temperature *SI unit* = *Celsius degree or Kelvin*	Degrees Celsius (°C) Degrees Kelvin (K = °C + 273)	°F = 9/5(°C) + 32	°C = 5/9(°F − 32)

*SI units are the internationally recognized standard metric system measurements that are used in science and commerce. The abbreviation SI is derived from the French *Système International* (international system). Note that the liter is not an official SI unit, but its use is accepted.

Appendix C
Laboratory Reference Values

Blood Chemistry (Comprehensive Metabolic Panel)

Substance	Normal Range: Adult*	Potential Causes of Abnormal Values
Albumin	3.9–5.0 g/dl	• Low albumin is seen with malnutrition, renal failure, hepatic disease. • Elevated albumin is most commonly seen with dehydration.
Alkaline phosphatase	44–147 IU/l	Elevated level is seen with certain cancers; diseases of the bone or liver; bile duct obstruction.
Alanine aminotransferase (ALT)	8–37 IU/l	Elevated level is seen with hepatic disease or failure.
Aspartate aminotransferase (AST)	10–34 IU/l	Elevated level is seen with hepatic disease or failure, severe muscle injury.
Blood-urea-nitrogen (BUN)	7–20 mg/dl	• Low BUN typically indicates overhydration or malnutrition. • Elevated BUN typically indicates dehydration or renal, heart, or hepatic failure.
Calcium ions	8.5–10.9 mg/dl	• Low calcium ion concentration is seen with malnutrition; nutritional deficiency of vitamin D, vitamin K, or magnesium; pancreatitis; low parathyroid hormone levels. • Elevated calcium ion concentration is seen with hyperparathyroidism, renal disease, certain cancers.
Carbon dioxide	20–29 mmol/l	• Low carbon dioxide level may indicate compensation for metabolic acidosis, renal failure. • Elevated level generally indicates respiratory disease; may also indicate compensation for metabolic alkalosis.
Chloride ions	96–106 mmol/l	• Level tends to be decreased with respiratory diseases and overhydration. • Level tends to be increased with dehydration, excess cortisol secretion, renal failure.
Creatinine	0.8–1.4 mg/dl	• Level is elevated in renal failure, severe dehydration.
Glucose	70–100 mg/dl (fasting) 70–125 mg/dl (nonfasting)	• Low blood glucose may indicate excess insulin, hypoglycemia, hepatic disease. • High blood glucose may indicate diabetes mellitus, pancreatitis, hyperthyroidism.
Potassium ions	3.7–5.2 mEq/l	• Low level of potassium ions may indicate overuse of certain diuretics, excess aldosterone secretion. • Elevated level of potassium ions may indicate renal failure, dehydration, adrenal insufficiency.
Sodium ions	136–144 mEq/l	• Low level of sodium ions may indicate overhydration, use of certain diuretics, diarrhea, adrenal insufficiency. • Elevated level of sodium ions may indicate dehydration, renal failure, elevated cortisol levels.
Total bilirubin	0.2–1.9 mg/dl	• Elevated level may indicate liver disease, bile duct obstruction, hemolysis.
Total protein	6.3–7.9 g/dl	• Low protein is seen with malnutrition, renal failure, hepatic disease. • Elevated protein is seen with certain cancers such as multiple myeloma, renal or hepatic failure.

*Source: National Institutes of Health.

Arterial Blood Gases

Measurement	Normal Range: Adult at Sea Level*	Potential Causes of Abnormal Values
Partial pressure of oxygen (PaO2)	75–100 mmHg	Low PaO_2 usually indicates respiratory dysfunction; may be seen at high altitudes.
Partial pressure of carbon dioxide (PaCO2)	38–42 mmHg	• Low $PaCO_2$ indicates compensated metabolic acidosis or respiratory alkalosis. • Elevated $PaCO_2$ indicates compensated metabolic alkalosis or respiratory acidosis.
Blood pH	7.38–7.42	• Low pH means acidemia due to metabolic or respiratory causes. • Elevated pH means alkalemia due to metabolic or respiratory causes.
Oxygen saturation (SaO2)	94–100%	Low oxygen saturation usually indicates respiratory dysfunction; may be seen at high altitudes.
Bicarbonate (HCO3⁻)	22–28 mEq/l	• Low HCO_3^- indicates metabolic acidosis or compensated respiratory alkalosis. • Elevated HCO_3^- indicates metabolic alkalosis or compensated respiratory acidosis.

*Source: National Institutes of Health.

Complete Blood Count with Differential

Component	Normal Range: Adult Male*	Normal Range: Adult Female*	Potential Causes of Abnormal Values
Red blood cell (erythrocyte) count	4.1–5.8 million cells/μl	3.7–5.2 million cells/μl	• Low total erythrocyte count may indicate bleeding; destruction of erythrocytes; vitamin or mineral deficiency (iron, vitamin B12, folate); bone marrow disorder or damage; renal failure. • High total erythrocyte count may indicate lung disease, high altitude, dehydration, or excess erythropoietin.
Hemoglobin	12.7–17.1 g/dl	10.4–15.2 g/dl	Causes of elevated or depressed hemoglobin levels generally parallel the causes of abnormal erythrocyte count.
Hematocrit (%)	38.0–50.3%	32.0–45.0%	Causes of elevated or depressed hematocrit generally parallel the causes of abnormal erythrocyte count.
White blood cell (leukocyte) count	3.9–12.1 thousand cells/μl	3.9–12.5 thousand cells/μl	• Low total white cell count may indicate cancer or diseases of the bone marrow, immunosuppression due to drugs or infection (HIV). • High total white cell count may indicate infection, inflammation, allergic disorders, trauma, leukemia.
• **Neutrophils (%)**	39.7–77.3%	39.6–77.8%	Elevated neutrophil count most commonly indicates bacterial infection or severe systemic inflammation.
• **Lymphocytes (%)**	17.8–51.8%	17.8–52.8%	Elevated lymphocytes may indicate viral infection, certain bacterial infections, leukemia.
• **Monocytes (%)**	0–12%	0–12%	Elevated monocytes may indicate chronic infection or inflammation, certain types of leukemia.
• **Eosinophils (%)**	0–8%	0–8%	High eosinophil level most commonly indicate allergic disease or parasitic infection.
• **Basophils (%)**	0–2%	0–2%	High number of basophils may indicate certain types of allergic reactions or inflammation.
Platelets	157–441 thousand platelets/μl	172–453 thousand platelets/μl	• Low platelet level is associated with certain infections; certain drugs; hepatic disease; bone marrow disease or destruction. • Elevated platelet count is associated with certain cancers, autoimmune diseases, certain types of anemia.

*Source: Centers for Disease Control and Prevention.

Urinalysis

Characteristic or Component	Normal Range or Value: Adult*	Potential Causes of Abnormal Values
Color	Light to dark yellow, slight amber hue	• Hydration status: urine is darker in dehydration, lighter in overhydration. • Certain drugs and foods may change urine color to red, blue, brown, or another color. • Blood in the urine from bacterial infections, kidney stones, renal cancer, etc., may give it a red hue.
Clarity	Clear	Urine appears cloudy during infection, renal failure, and with kidney or bladder stones.
pH	4.5–8.0	pH affected by diet (high-protein diets lead to acidic urine; high-fat diets lead to alkaline urine), kidney stones, bacterial infection, certain drugs, and renal failure.
Specific gravity	1.005–1.025	Low specific gravity indicates overhydration, renal disease, diabetes insipidus; high specific gravity generally indicates dehydration.
Urobilinogen	0.5–1.0 mg/dl	Elevated urobilinogen may indicate hemolysis, hepatic disease, and constipation. Low urobilinogen may indicate obstructive biliary disease.
Bilirubin	Negative	Presence often indicates hepatic failure.
Blood (hemoglobin)	< 3 erythrocytes/ml (read as negative on urinalysis test strip)	Blood in the urine may indicate renal failure, infection, kidney stones.
Protein	< 150 mg/dl (read as negative on urinalysis test strip)	Proteinuria may indicate heart failure, fever, muscle damage, renal failure, glomerulonephritis, diabetic nephropathy.
Glucose	< 130 mg/dl	Elevated urine glucose is seen with diabetes mellitus, pregnancy.
Ketones	Negative	Ketones in the urine may be present with uncontrolled diabetes mellitus, low-carbohydrate diets, starvation, severe exercise, vomiting, and pregnancy.
Nitrite	Negative	Nitrites generally indicate a urinary tract infection.
Leukocyte esterase	Negative	Positive leukocyte esterase generally indicates a urinary tract infection.

*Source: Medscape.

Appendix D
Scientific Method

Science Is Both a Body of Knowledge and a Process

Science is two things: *knowledge* (organized, reliable information) about the natural world and the *process* we use to get that knowledge. The process of science, or the way scientific knowledge is acquired, is generally called the **scientific method,** although in practice this term encompasses a variety of methods. The scientific knowledge presented in this book was obtained slowly over time by the scientific method.

Scientific knowledge enables us to describe and predict the natural world. Through the scientific method, scientists strive to accumulate information that is as free as possible of bias, embellishment, or interpretation.

The Scientific Method Is a Process for Testing Ideas

Although there is more than one way to gather information about the natural world, the scientific method is a systematic process for developing and testing predictions (**Figure D.1**). You probably already use the scientific method, or at least elements of it, in your own everyday problem solving.

1. **Observe and generalize.** When we observe the world around us and make generalizations from what we learn, we are employing *inductive reasoning* (drawing conclusions from the specific to the general case). Usually, we don't even think about it and don't bother to put our observations and generalizations into any kind of formal language, but we do it just the same. For example, you are probably convinced that it will *always* be colder in winter than in summer (a generalization) because you have observed that every winter in the past was colder than the preceding summer (specific observation). The difference between common experience and good science is that science uses generalization to make a prediction that can be tested. Consider an example from clinical research:

Observation 1: Patients given a particular drug (call it Drug X) have lower blood pressures than patients not given the drug.

Observation 2: Independently, researchers in Canada showed that Drug X lowers blood pressures in two separate patient populations.

Generalization: Drug X lowers blood pressure in all humans.

2. **Formulate a hypothesis.** Observations and generalizations are used to develop a *hypothesis*. A **hypothesis** is a tentative statement about the natural world. Importantly, it is a statement that can lead to testable deductions.

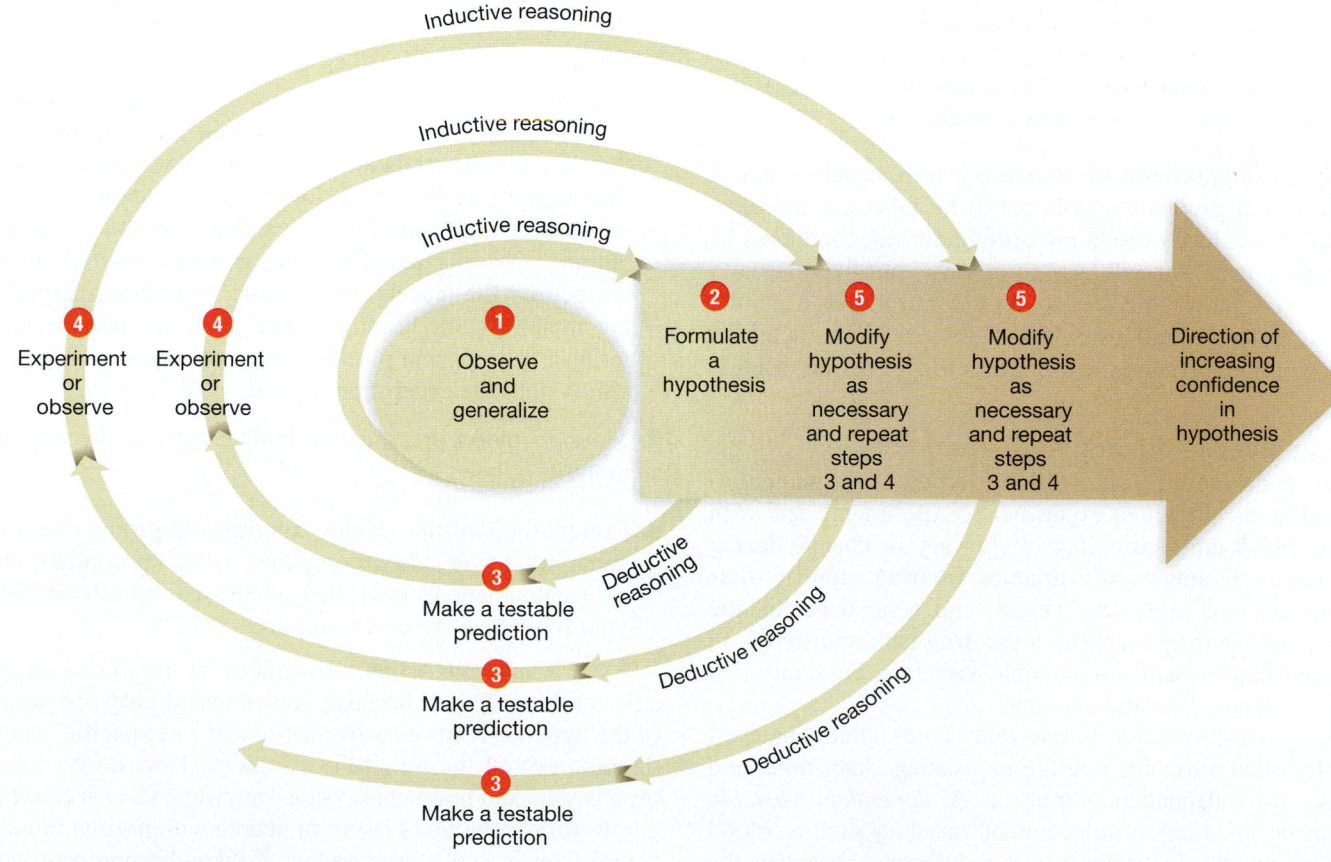

Figure D.1 **The scientific method.**

Hypothesis: Drug X is an effective treatment for high blood pressure in humans.

3. **Make a testable prediction.** Hypotheses that cannot be tested are idle speculation, or essentially hot air. But many hypotheses are so sweeping and comprehensive that ways must be found to test them under a variety of conditions. For example, you probably would not be convinced that Drug X is effective for all people under all conditions until you had at least tested it in quite a few people under many different conditions. To have confidence in your hypothesis, you must make testable predictions (also called *working hypotheses*) based on the hypothesis and then test them one at a time. Predictions employ *deductive reasoning* (applying the general case to the specific). Often they are put in the form of an *if . . . then* statement, in which the *if* part of the statement is the hypothesis. For example,

Prediction: If Drug X is an effective treatment for high blood pressure in humans, *then* 10 mg/day of Drug X will lower blood pressure in people with high blood pressure within one month.

Notice that the prediction is very specific. In this example, the prediction specifies the dose of drug, the medical condition of the persons on whom it will be tested, the expected effect of the drug if the prediction is correct, and a specified time period for the test. Its specificity makes it testable—yes or no, true or false.

4. **Experiment or observe.** The truth or falsehood of your prediction is determined by observation or by experimentation. An **experiment** is a carefully planned and executed manipulation of the natural world to test your prediction. The experiment that you conduct (or the observations you make) will depend on the specific nature of the prediction.

5. **Modify the hypothesis as necessary and repeat steps 3 and 4.** If your prediction turns out to be false, for instance, the drug didn't lower blood pressure under the conditions of your experiment, you will have to modify your hypothesis to fit the new findings and repeat steps 3 and 4. For example, perhaps the drug would lower blood pressure if you increased the dose or administered the drug for a longer period of time.

Designing and Conducting the Experiment

After you've developed your working hypothesis (prediction) you'll need to test it with an experiment. Experiments deal with **variables,** which are factors that might vary or change during an experiment. Examples of variables in drug clinical trials include the subjects' age, sex, weight, and general health; the drug dose; and the time over which the drug is administered. In a **controlled experiment,** all possible variables are controlled so that they cannot affect the outcome.

An *independent variable* is one that stands alone and isn't changed by other variables you are measuring. Age, time, and drug dose are independent variables. A *dependent variable* is one that depends on an independent variable, such as blood pressure or weight. To remember the difference between the two, think about it like this: An independent variable can cause a change in a dependent variable, but a dependent variable can't cause a change in an independent variable. For example, a certain dose of a drug (independent) can cause a change in blood pressure (dependent), but a change in blood pressure (dependent) can't cause a change in the dose of the drug (independent).

Wherever possible, scientists design experiments so that just one independent variable, called the *controlled variable,* is manipulated between groups. In our example, the controlled variable is drug treatment. So, with Drug X, you could follow these steps in your controlled experiment (**Figure D.2**):

1. Select a large group of human subjects with high blood pressure.

2. *Randomly* divide the larger pool of subjects into two groups. Designate one the **experimental group** and the other the **control group.** The importance of random assignment to the two groups is that all other independent variables that might affect the outcome (such as age, sex, or previous health problems) are automatically equalized between the two groups. In effect, the control group accounts for all unknown factors.

3. Treat subjects in the two groups *exactly* the same except that only the experimental group gets the drug. Treat the experimental group with 10 mg/day of Drug X—our controlled variable—for one month and the control group with a **placebo** (pluh-SEE-boh), or "false treatment." If you deliver Drug X in a pill, the placebo should be an identical-looking pill with no drug in it. If you administer Drug X as an injection in a saline solution, the placebo should be an injection of the same volume of saline.

When working with human subjects, scientists must take the power of suggestion into account. For example, if subjects in a study of the effectiveness of a pain medication were told they were in the medication group and not the placebo group, they might rate their pain lower. To eliminate the power of suggestion as a variable, researchers conduct experiments "blind," meaning that the subjects aren't told whether they are getting the placebo or the drug. Sometimes experiments are done "double-blind," so that even the person administering the drugs and placebos does not know which group is which until the experiment is over.

4. Measure blood pressures in both groups at the end of one month.

5. Compare the results. If the experimental group's blood pressures are lower when compared using appropriate statistical (mathematical) tests, then your prediction is verified and your hypothesis receives support.

Even if your prediction turns out to be true, however, you're still not done. This is because you've tested only one small part of the hypothesis (its effectiveness under one specific set of conditions), not all the infinite possibilities. How do we know the drug is safe, and how safe is "safe," anyway? Does it cause a dangerous drop in blood pressure in people with normal blood pressures? Does it cause birth defects if taken by pregnant women? Does long-term use lead to kidney failure? Does the drug's

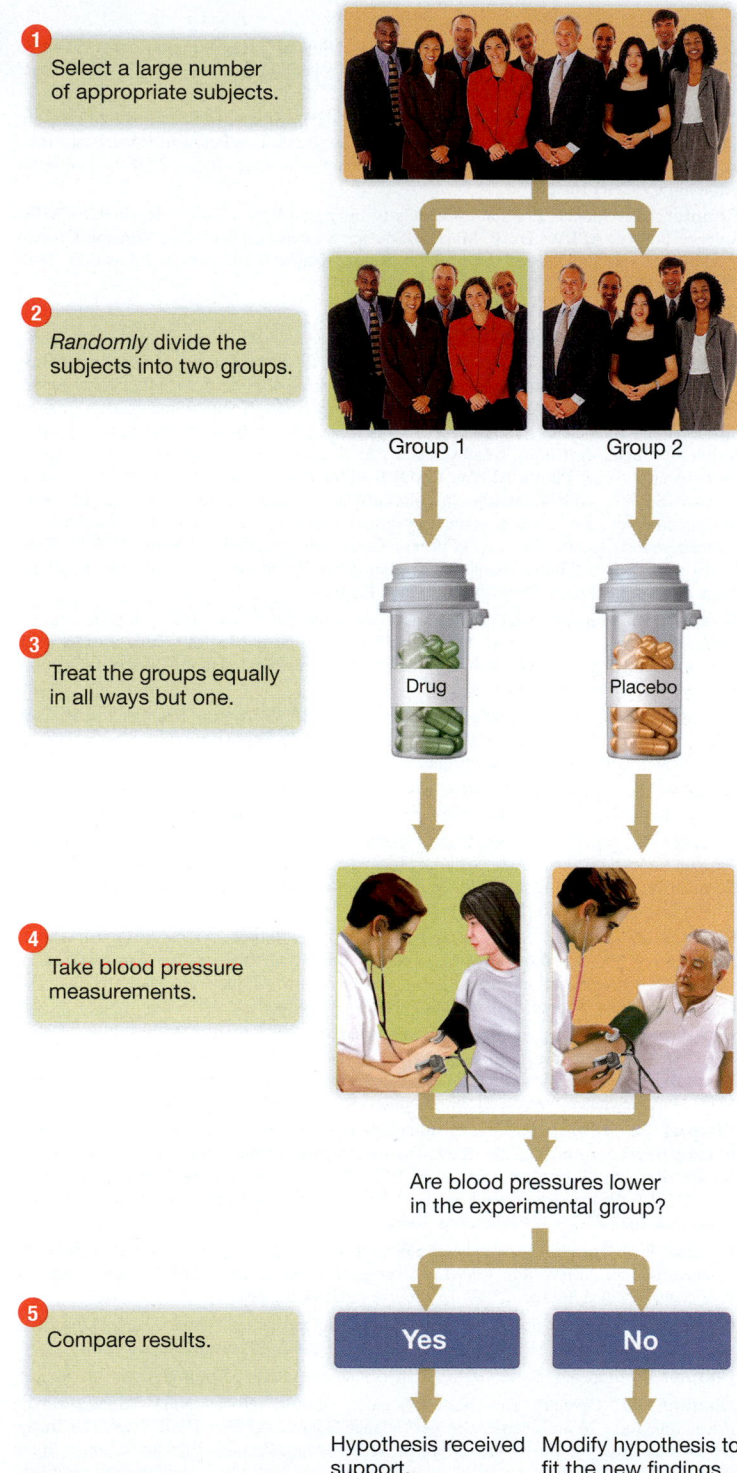

① Select a large number of appropriate subjects.

② *Randomly* divide the subjects into two groups.

Group 1   Group 2

③ Treat the groups equally in all ways but one.

Drug   Placebo

④ Take blood pressure measurements.

Are blood pressures lower in the experimental group?

⑤ Compare results.

Yes   No

Hypothesis received support.   Modify hypothesis to fit the new findings.

Figure D.2 The steps in a controlled experiment.

effectiveness diminish over time? Specific predictions may have to be stated and tested for each of these questions before the drug is declared effective and safe and allowed on the market.

Repeating the Results Is Key
A single study in support of a hypothesis is a good first step, but it's only a first step. For any hypothesis to gain credibility, the results must be repeated by many other researchers. Often results that initially seem promising fail the test of repeatability, and the hypothesis is eventually discarded. Only after many scientists have tried repeatedly (and failed) to disprove a hypothesis do they begin to have more confidence in it. A hypothesis cannot be proved true; it can only be supported or disproved.

As this example shows, the scientific method is a process of elimination. It's a very good method, though not perfect. It may be limited by our approach, level of technology, and even by our preconceived notions and cultural biases. We move toward the best explanation for the moment, with the understanding that it may change in the future.

Making the Findings Known
New information is not of much use if hardly anybody knows about it. For that reason, scientists need to let others know of their findings. Often, they publish the details in scientific journals to announce their findings to the world. Articles in *peer-reviewed* journals are subjected to the scrutiny of several experts (the scientists' peers) who must approve the article before it can be published. Peer-reviewed journals often contain the most current scientific information.

An unspoken assumption in any conclusion is that the results are valid only for the conditions under which the experiment was done. This is why scientific articles go into such detail about exactly how the experiment was performed. Complete documentation allows other scientists to repeat the experiments themselves or to develop and test their own predictions based on the findings of others.

Try to apply the scientific method to a hypothesis dealing with some aspect of evolution or to a global problem such as cancer or AIDS, and you begin to appreciate how scientists can spend a lifetime of discovery in science (and enjoy every minute). At times, the process seems like three steps forward and two steps back. You can bet that at least some of what you learn in this book will not be considered accurate 10 years from now. Nevertheless, even through our mistakes we make important new observations, some of which may lead to rapid advances in science and technology. Just in the past 100 years, we've developed antibiotics, sent people to the moon, and put computers on millions of desks.

Many people associate science with certainty, whereas in reality scientists are constantly dealing with uncertainty. That is why we find scientists who don't agree or who change their minds. Building and testing hypotheses is slow, messy work, requiring that scientists constantly question and verify each other.

A Well-Tested Hypothesis Becomes a Theory
How many times have you heard the phrase, "It's just a theory!", said in an attempt to discredit a scientific idea? Many people think that a *theory* is a form of idle speculation or a guess. Scientists use the word quite differently. To scientists, a **theory** is a broad hypothesis that has been extensively tested and supported over time, and that explains a broad range of scientific facts with a high degree of reliability. A theory is the highest status that any hypothesis can achieve. Even theories, however, may be modified over time as new and better information emerges.

Adapted from *Human Biology: Concepts and Current Issues* by Michael D. Johnson

APPENDIX D   SCIENTIFIC METHOD

Credits

Glossary

A band The component of the sarcomere that appears darker because it contains thick filaments and areas in which the thick and thin filaments overlap.

Abdominal cavity The portion of the abdominopelvic cavity located between the bony pelvis and the diaphragm.

Abdominopelvic cavity (ab-dom′-ih-noh-PEL-vik) A ventral body cavity located within the area inferior to the diaphragm that encompasses the area of the abdomen and the pelvis.

Abduction (ab-DUHK-shuhn) An angular movement whereby the body part moves away from the midline of the body or another point.

ABO blood group A blood group that classifies blood type based on the presence or absence of two antigens: the A and B antigens.

Absorption The transport of a substance across an epithelial lining into the bloodstream.

Absorptive state The time period from the point when ingested nutrients enter the bloodstream to about 4 hours after feeding.

Accommodation The rounding of the lens to focus light from near objects on the retina.

Acetabulum (ah′-suh-TAB-yoo-lum) The bony socket formed by the ilium, ischium, and pubis into which the femur fits to form the hip joint.

Acetylcholine (ACh) (ah-SEE-til-koh′-leen) A neurotransmitter involved in a wide variety of processes, including those of the autonomic nervous system and muscle contraction.

Acetylcholinesterase (AChE) (ah-SEET-′l-kohl-in-ess′-teh-rayz) An enzyme located in the synaptic cleft that degrades acetylcholine.

Acid A compound that releases one or more hydrogen ions when placed in water; has a pH less than 7.0.

Acromioclavicular joint (AC joint) The articulation between the acromion of the scapula and the lateral portion of the clavicle.

Acromion (ah-KROH-mee-ahn) The anterosuperior projection of the scapula that articulates with the clavicle.

Acrosome (AK-roh-sohm) A membrane-enclosed sac on the head of the sperm cell that contains digestive enzymes.

Actin (AK-tin) A bead-shaped contractile protein found in muscle fibers and motile cells.

Actin filaments (microfilaments) Thin protein filaments of the cytoskeleton composed of two intertwining actin strands that provide a cell with structural support, bear tension, help to maintain its shape, and function in cellular motion.

Action potential A quick, temporary change in the membrane potential of a cell in a single region of the plasma membrane.

Activation energy (E_a) The energy required for an adequately strong collision to occur between atoms for a reaction to take place.

Active immunity A type of immunity mediated by an individual's own immune system after exposure to an antigen.

Active transport A type of membrane transport against a gradient that requires energy expenditure in the form of ATP.

Adaptive immunity (specific immunity) The arm of the immune system that consists of cells and proteins that must bind to specific antigens to become activated.

Adduction (ad-DUK-shuhn) An angular movement whereby the body part moves toward the midline of the body or another reference point.

Adenine A purine nitrogenous base found in nucleic acids and compounds such as ATP.

Adenosine triphosphate (ATP) A nucleotide consisting of adenine, ribose, and three phosphate groups; the main energy currency of human cells.

Adipocyte (AD-ih-poh-sy′t) The main cell type in adipose tissue; each cell contains a large lipid droplet in its cytoplasm.

Adipose tissue A type of connective tissue proper that contains adipocytes; found in locations such as the hypodermis and around organs.

Adrenal cortex The outer region of the adrenal gland that produces steroid hormones.

Adrenal gland An endocrine gland located at the superior pole of each kidney; consists of an outer cortex and an inner medulla.

Adrenal medulla The inner region of the adrenal gland; consists of chromaffin cells that release epinephrine and norepinephrine when stimulated by the sympathetic nervous system.

Adrenergic receptors (ad-ren-ER-jik) Receptors in the plasma membranes of the target cells of sympathetic neurons that bind to catecholamines.

Adrenocorticotropic hormone (ACTH) A hormone released by the adrenal cortex that stimulates the production of mostly glucocorticoids from the adrenal cortex.

Afferent arterioles Arterioles that deliver blood to the glomeruli in the kidney.

Afterload The force that the right and left ventricles must overcome in order to eject blood into their respective arteries.

Agonist A muscle that provides the principal force required in a movement; also known as the *prime mover*.

Agranulocyte (ay-GRAN-yoo-loh-sy′t) A class of leukocyte without visible cytosolic granules; includes lymphocytes and monocytes.

Airway resistance The impedance to air flow within the respiratory tract.

Albumin (al-BYOO-mun) The most abundant plasma protein; largely responsible for the plasma's colloid osmotic pressure.

Aldosterone (al-DAHS-ter-ohn) A hormone produced by the adrenal cortex that increases reabsorption of sodium ions (and indirectly) water from the distal tubule, as well as the secretion of potassium and hydrogen ions.

Alimentary canal (gastrointestinal, or GI, tract) A series of hollow digestive organs through which food passes as it is ingested, digested, and absorbed; also carries out defecation.

Alleles (uh-LEELZ) Different variants of the same gene.

Alpha cells (α cells) Cells of the pancreatic islets that secrete the hormone glucagon.

Alpha motor neurons (α motor neurons) Lower motor neurons that innervate extrafusal muscle fibers.

Alveoli (al-vee-OHL-aye; singular, *alveolus*) Thin-walled structures in the lungs across which gases are exchanged.

Amino acid A monomer consisting of carbon, hydrogen, oxygen, and nitrogen atoms.

Amino acid–based hormone A chemical messenger consisting of one or more amino acids; most amino acid–based hormones are hydrophilic and so bind plasma membrane receptors.

Amnion (AM-nee-ahn) A membrane that develops from embryonic tissues; completely encloses the embryo in the fluid-filled amniotic cavity.

Amniotic fluid The fluid in the amniotic cavity that protects the developing conceptus.

Amphiarthrosis (am′-fee-ahr-THROH-sis) A functional classification of articulations in which the joint allows for only a small amount of motion between articulating bones.

Ampulla (am-PEWL-uh) An enlarged area at the base of each semicircular duct in the ear that contains receptor cells that detect rotation of the head.

Amygdala (uh-MIG-duh-luh) A component of the limbic system in the brain that functions in the behavioral expression of emotion.

Anabolic reaction (an-uh-BAWL-ik) A reaction in which smaller substances bond to form larger substances.

Anal canal The terminal portion of the large intestine that contains sphincters to regulate the passage of feces during defecation.

Anaphase The third phase of mitosis during which the spindle fibers begin to pull chromatids to opposite poles of the cell.

Anatomical position The standard position in which an anatomical specimen is presented, in which a subject is facing forward, arms at the side with the palms facing outward, and the feet are shoulder width apart.

Androgenic steroids Hormones produced by the adrenal cortex that have androgenic effects on a variety of tissues.

Anemia (ah-NEE-mee-uh) A condition in which the oxygen-carrying capacity of the blood is reduced due to a decreased amount of hemoglobin, a decreased number of erythrocytes, and/or abnormal hemoglobin.

Angiotensin-converting enzyme (ACE) An enzyme produced by the lungs and other tissues that converts angiotensin-I to the more active angiotensin-II.

Angiotensin-II (Ang-II) A hormone that has widespread effects on the body, which include causing vasoconstriction of all vessels, restoring the glomerular filtration rate, promoting ADH and aldosterone release, and increasing sodium ion and water reabsorption from the kidneys.

Anion (AN-aye-awn) A negative ion with more electrons than protons.

Antagonist A muscle generally located on the opposite side of a joint from its agonist that opposes or slows the action of the agonist.

Anterior (an-TEER-ee-ur) Toward the front; also known as *ventral*.

Anterior body cavity The major body cavity located mostly on the anterior side of the body.

Anterior cruciate ligament (ACL) A ligament found within the synovial cavity of the knee that, when the knee joint is extended, tightens to prevent hyperextension.

Anterior horn The anterior region of gray matter in the spinal cord that contains the cell bodies of motor neurons.

Anterior pituitary gland The anterior portion of the pituitary gland that is derived from glandular epithelium; produces many tropic hormones and growth hormone.

Anterior ramus The division of a spinal nerve that travels to the anterior side of the body.

Anterior root An extension from the anterior horn of the spinal cord containing the axons of lower motor neurons; some also contain the axons of visceral motor neurons.

Anterolateral system A set of ascending tracts in the spinal cord that convey sensations relating to pain, temperature, and non-discriminative touch.

Antibody A protein produced by activated B lymphocytes that binds to a specific antigen and facilitates its removal from a tissue.

Antibody-mediated immunity (humoral immunity) The arm of adaptive immunity that is mediated by B lymphocytes and antibodies.

Anticoagulant (an′-tee-koh-AG-yoo-lunt) A compound that inhibits some portion of the coagulation cascade.

Anticodon A three-nucleotide portion of transfer RNA that bonds to a specific messenger RNA codon.

Antidiuretic hormone (ADH) A hormone produced by the hypothalamus and stored in the posterior pituitary that causes the insertion of aquaporin channels in the cells of the distal tubule and collecting system to allow water reabsorption.

Antigen (AN-tih-jen) A unique glycoprotein marker found on the surface of almost all cells and biological chemical substances.

Antigen-presenting cells Immune cells that process and display antigen fragments on their class II major histocompatibility molecules to activate other immune cells such as helper T cells.

Antithrombin-III (AT-III) An anticoagulant that inhibits the activity of factor Xa and thrombin.

Aorta (ay-OHR-tuh) The largest artery in the body; arises from the left ventricle and delivers blood to the systemic circuit.

Aortic sinus An area within the aortic arch that houses baroreceptors.

Aortic valve The semilunar valve located between the left ventricle and the aorta.

Apocrine glands (AP-oh-krin) Sweat glands found in the axillae, the anal area, and the areola that secrete a thick sweat rich in proteins.

Appendicular region (ap′-en-DIK-yoo-lur) The portion of the body consisting of the upper and lower limbs.

Appositional growth The process by which bones grow in width.

Aquaporin (ah-kwah-POHR-in) A protein channel in the plasma membrane of renal tubule and collecting duct cells that allows the passage of water from the filtrate to the interstitial fluid.

Aqueous humor (AY-kwee-us) The watery fluid that fills the anterior cavity of the eyeball.

Arachnoid mater (ah-RAK-noyd) The middle meninx that surrounds the brain and the spinal cord.

Arbor vitae (AHR-bohr VEE-tay) Branching tracts of white matter located within the cerebellum.

Arm The proximal portion of the upper limb from the elbow to the shoulder that consists of the humerus.

Arrector pili A small band of smooth muscle cells that causes a hair to stand erect when contracted.

Arteries The vasculature's distribution system; pulmonary arteries carry deoxygenated blood to the lungs, and systemic arteries carry oxygenated blood to the body's cells.

Arterioles (ahr-TEER-ee-olz) The smallest arteries that control blood flow to individual tissues.

Articular capsule (ahr-TIK-yoo-luhr) A double-layered structure surrounding synovial joints that holds the articulating bones together.

Articulation The location where two or more bones come together; also known as a *joint*.

Association areas Areas of the cerebral cortex that integrate different types of information.

Association fibers White matter fibers in the brain that connect the cerebral cortex in adjacent gyri.

Astigmatism (ah-STIG-mah-tizm) An irregular curvature of the lens or cornea that produces uneven refraction and blurred vision.

Astrocyte (ASS-troh-sy't) A neuroglial cell of the central nervous system that facilitates the formation of the blood brain barrier, regulates the extracellular environment of the brain, anchors neurons and blood vessels in place, and repairs damaged brain tissue.

Asystole (ay-SIS-toh-lee) A condition in which there is no electrical activity in the heart, resulting in a flattened line on the ECG.

Atlas The first cervical vertebra.

Atmospheric pressure The pressure exerted by the air of the surrounding atmosphere.

Atom The smallest unit of matter that still retains its original properties.

Atomic nucleus (NOO-klee-us) The central core of an atom where the protons and neutrons reside.

Atomic number The number of protons found in an atom; the atomic number is unique to each element.

ATP synthase A large protein in the inner mitochondrial membrane that functions as an ion channel and an enzyme that catalyzes the production of ATP.

Atria (AY-tree-uh) The superior chambers of the heart that receive blood from veins and pump blood into the ventricles.

Atrial natriuretic peptide (ANP) A hormone produced by cells of the heart that triggers the excretion of sodium ions (and indirectly water) from the renal tubule of the kidneys.

Atrioventricular bundle (AV bundle) The first component of the heart's Purkinje fiber system that propagates action potentials from the atria to the ventricles.

Atrioventricular node (AV node) A cluster of pacemaker cells in the inferior right atrium that delays the propagation of the cardiac action potential from the atria to the ventricles.

Atrioventricular valves (AV valves) Valves located between the atria and ventricles of the heart.

Atrophy (AEH-troh-fee) A decrease in the size of the cell.

Auditory ossicles (AWSS-ih-kulz) A chain of three bones (malleus, incus, and stapes) that connect the tympanic membrane to the oval window.

Auricle (heart) (OHR-ih-kul) A muscular pouch extending from an atrium of the heart that allows the atrium to hold more blood.

Autocrine (AW-toh-krin) A chemical messenger that influences the function of the same cell or cell type that produced and secreted it.

Autoimmune disorder A condition characterized by the production of autoantibodies, or antibodies that bind and react to self antigens.

Autonomic ganglion A collection of cell bodies of postganglionic autonomic neurons.

Autonomic nervous system (ANS) (aw-toh-NAHM-ik) The division of the peripheral nervous system that controls homeostatic responses of the organs of many other systems; also known as the *visceral motor division of the PNS.*

Autonomic tone The constant amount of activity exhibited by each branch of the ANS on specific target cells.

Autoregulation A process in which a tissue regulates aspects of its own structural and functional properties.

Autorhythmicity (aw′-toh-rith-MISS-ih-tee) A property whereby a tissue sets its own rhythm of action potential generation without external influences.

Autosomal dominant-recessive inheritance Pattern of inheritance involving autosomal alleles that are strictly dominant or recessive.

Axial region (AKS-ee-ul) The portion of the body consisting of the head, neck, and trunk.

Axis The second cervical vertebra.

Axolemma (aks-oh-LEM-uh) The plasma membrane of an axon.

Axon A single extension of a neuron that can generate action potentials; generally carries information away from the cell body.

Axon hillock The initial portion of an axon where an action potential is generated.

Axon terminal A knoblike structure at the end of an axon that contains synaptic vesicles with neurotransmitters.

Azygos system (ay-ZY-gus) The group of veins—including the azygos vein, hemiazygos vein, and accessory hemiazygos vein—that drain structures of the posterior thorax.

B lymphocyte The class of immune cell responsible for the antibody-mediated arm of adaptive immunity.

Ball-and-socket joint A multiaxial joint in which the articulating surface of one bone is ball-shaped and fits into a cup-shaped socket in the other bone, allowing for movement around three axes.

Baroreceptor reflex A negative feedback loop that responds to an increase in blood pressure by triggering the parasympathetic nervous system to decrease blood pressure.

Baroreceptors Internal sensory receptors that detect pressure.

Basal metabolic rate (BMR) The minimal rate of caloric expenditure for an awake individual.

Basal nuclei Nuclei within the cerebral hemispheres that function in the initiation

of movement and the prevention of inappropriate movements.

Base (alkali) A compound that accepts a hydrogen ion from an acid; has a pH greater than 7.0.

Basement membrane Two layers of extracellular matrix—the basal lamia and the reticular lamina—that anchor an epithelial tissue to the underlying connective tissue.

Basophil (BEH-zuh-fill) A class of granulocyte that is involved in mediating inflammation.

Beta cells (β cells) Cells of the pancreatic islets that secrete the hormone insulin.

Beta oxidation (β oxidation) The process of fatty acid oxidation, during which two-carbon compounds are removed as acetyl-CoA and $FADH_2$ and NADH are generated.

Biaxial joints A functional classification of synovial joints in which motion between articulating bones occurs around two axes.

Bile A liquid that contains water, electrolytes, and organic compounds such as bile salts; it is required for the digestion and absorption of lipids, and it is a mechanism by which the liver excretes wastes.

Biochemistry The study of the chemistry of living beings.

Bipolar neuron A neuron with one axon and one dendrite.

Blastocyst (BLAS-toh-sist) The early stage of development during which implantation occurs.

Blood The fluid connective tissue consisting of formed elements and plasma.

Blood brain barrier The collective term for the tight junctions between the endothelial cells of the brain capillaries, their basal laminae, and astrocytes; keeps cerebrospinal fluid and brain extracellular fluid separate from the blood, protecting the brain from certain substances in the blood.

Blood clot A collection of platelets, erythrocytes, fibrin, and other clotting proteins that prevents further loss of blood from an injured blood vessel.

Blood flow The volume of blood that flows through the systemic or pulmonary circuit in one minute.

Blood pressure The hydrostatic pressure exerted on a blood vessel wall by the blood.

Blood testis barrier A barrier formed by sustentacular cells that prevents immune cells in the bloodstream from encountering the new antigens formed on the genetically unique sperm cells.

Body mass index (BMI) A calculation of body mass that takes into account an individual's height; allows the classification of an individual as underweight, normal weight, overweight, or obese.

Bolus (BOH-lus) A moistened, partially digested ball of ingested food that is swal-

lowed and transported to the stomach via the pharynx and esophagus.

Bone deposition The formation of new bone by osteoblasts.

Bone remodeling The continual process of bone growth and bone loss that takes place within healthy bone.

Bone resorption The destruction of bone tissue by osteoclasts.

Boyle's law A gas law that states that pressure and volume are inversely proportional.

Brachial plexus A group of nerves consisting of the anterior rami from C5 to T1 that innervates structures within and around the upper limb.

Brain The organ of the central nervous system that performs multiple integrative functions.

Brain waves Characteristic patterns of electrical activity in the brain that can be measured on an electroencephalogram (EEG).

Brainstem The most inferior part of the brain; responsible for automatic, subconscious functions.

Broca's area A premotor multimodal association area for speech located in the anterolateral frontal lobe; responsible for the ability to produce the motor actions necessary for language.

Bronchi (BRONG-kye; singular, *bronchus*) Larger passages of the bronchial tree.

Bronchial tree (BRONG-kee-uhl) Branching series of passageways that conducts air to and from the respiratory zone of the respiratory tract.

Bronchioles (BRONG-kee-olz) Passages of the bronchial tree that are less than one millimeter in diameter.

Bronchoconstriction The process by which the diameter of the bronchioles decreases, generally due to smooth muscle contraction.

Bronchodilation The process by which the diameter of the bronchioles increases, generally due to smooth muscle relaxation.

Buffer system A chemical system that resists changes in pH and prevents large swings in the pH when acid or base is added to the solution.

Bulbourethral gland (bul′-boh-yoo-REETH-ruhl) An accessory gland in the male reproductive system that secretes thick alkaline fluid that helps neutralize acidic urine remaining in the urethra prior to ejaculation.

Bundle branches The second component of the heart's Purkinje fiber system that propagates action potentials along the right and left sides of the interventricular septum.

Bursa (BER-suh) A synovial fluid–filled structure located between bones, tendons, and muscles of certain synovial joints; minimizes friction between moving parts.

Calcaneus (kal-KEHN-ee-us) The posterior tarsal bone; "heel bone."

Calcitonin (kal-sih-TOH-nin) A hormone produced by the parafollicular cells of the

GLOSSARY

thyroid gland; decreases blood calcium ion concentration by stimulating osteoblasts to build bone.

Calcitriol (kal-sih-TRY-ohl) The active form of vitamin D; increases the number of calcium ions absorbed from the small intestine.

Calmodulin (Cam) (kal-MAWD-yoo-lin) A protein in the cytosol of smooth muscle cells that binds to calcium ions during a contraction.

Calorie The amount of heat required to raise the temperature of one kilogram of water by one degree Celsius; also known as the *kilocalorie*.

Calorimetry (kal′-ohr-IM-eh-tree) The process by which the amount of body heat an individual produces is measured.

Calvarium The superior portion of the cranial cavity; also called the *cranial vault*.

Capillaries The vasculature's exchange system through which oxygen, nutrients, water, and wastes are exchanged between the blood and the cells; tiny vessels with only an endothelium and basal lamina.

Carbohydrate An organic compound composed of carbon, hydrogen, and oxygen atoms, often in the ratio of 1:2:1.

Carbonic acid–bicarbonate buffer system The major buffer system in the blood; consists of the weak acid carbonic acid and its conjugate base, the bicarbonate ion.

Carbonic anhydrase (CA) An enzyme that catalyzes the reversible reaction in which carbon dioxide and water are combined to become carbonic acid.

Cardiac conduction system The three populations of pacemaker cells in the heart: the sinoatrial node, the atrioventricular node, and the Purkinje fiber system.

Cardiac cycle The sequence of mechanical events in the heart that takes place from one heartbeat to the next.

Cardiac muscle A tissue type composed of wide, branching cardiac muscle cells and the surrounding endomysium.

Cardiac muscle cells (myocytes) Striated, branched muscle cells joined by intercalated discs; make up the majority of cardiac muscle tissue.

Cardiac output (CO) The volume of blood pumped into the pulmonary and systemic circuits in one minute.

Cardiovascular system The organ system made up of the heart and the blood vessels; delivers and drains blood to and from tissues.

Carina (kar-AYE-nuh) The final cartilage ring of the trachea where cough receptors are housed.

Carotid sinus Area in the common carotid artery near its bifurcation into the internal and external carotid arteries that houses baroreceptors.

Carpals The short bones of the wrist.

Cartilaginous joints (kar′-tih-LAJ-ih-nuhs) A structural classification of articulations in which the joints are held together by cartilage and have neither a synovial cavity nor an articular capsule.

Catabolic reaction (kat-uh-BAWL-ik) A reaction in which larger substances are broken down into smaller ones.

Cataract (KAT-ah-rakt) A condition in which clouding of the lens interferes with vision.

Catecholamines (kat′-eh-KOHL-ah-meenz) A group of neurotransmitters derived from the amino acid tyrosine; includes norepinephrine, epinephrine, and dopamine.

Cation (KAT-aye-on) A positive ion with more protons than electrons.

Cauda equina (KOW-duh eh-KWY-nuh) A large bundle of nerve roots given off at the terminal end of the spinal cord.

Caudate nuclei (KOW-dayt) Basal nuclei consisting of two C-shaped rings of gray matter; work with the putamen to inhibit the globus pallidus and initiate movement.

Cecum (SEE-kum) A blind pouch that is the first portion of the large intestine.

Celiac trunk (SEE-lee-ak) The single large vessel that branches off the superior descending abdominal aorta; branches into vessels that supply many abdominal organs.

Cell The smallest unit capable of carrying out the functions of life.

Cell-adhesion molecules (CAMs) Glycoproteins in the extracellular matrix that hold cells in a tissue together.

Cell body The central portion of a neuron that contains the nucleus and the bulk of the organelles.

Cell cycle The ordered series of events from cell formation to cell division.

Cell-mediated immunity The arm of adaptive immunity that is mediated by T lymphocytes.

Cell metabolism The collection of reactions that a cell carries out to maintain life.

Central canal A small, cerebrospinal fluid–filled canal located in the center of the spinal cord.

Central nervous system (CNS) The division of the nervous system consisting of the brain and the spinal cord.

Central sulcus A sulcus of each cerebral hemisphere that separates the frontal lobes from the parietal lobes.

Centrioles A pair of organelles within the centrosome that consist of modified microtubules and other proteins; play an important role in cell division.

Centrosome The central area of the cell from which microtubules emanate.

Cerebellum (sehr-uh-BEL-um) The posteroinferior portion of the brain; responsible for regulating ongoing movement and the correction of motor error.

Cerebral aqueduct A hollow, cerebrospinal fluid–filled channel that connects the

third ventricle in the diencephalon with the fourth ventricle in the brainstem.

Cerebral arterial circle A vascular anastomosis on the inferior surface of the brain that provides alternate routes of blood flow to the regions of the brain.

Cerebral cortex The outer two millimeters of gray matter of the cerebrum that function in conscious processes.

Cerebral hemispheres (cerebrum) The two large, superior structures of the brain that are responsible for higher mental functions, control of movement, and interpretation of sensations.

Cerebrospinal fluid (CSF) A fluid similar in composition to blood plasma (but with fewer proteins) that bathes and surrounds the brain and spinal cord.

Cerumen (seh-ROO-min) A modified sweat secreted by ceruminous glands in the external ear.

Ceruminous glands (seh-ROO-min-us) Modified sweat glands located in the external ear.

Cervical plexus A group of nerves consisting of the anterior rami from C1 to C4 that innervates primarily structures around the head and the neck.

Cervical vertebrae Seven bones of the vertebral column located in the neck.

Cervix The narrow neck, or outlet, of the uterus that projects into the vagina inferiorly.

Chemical A substance with a unique molecular composition that is used in or produced by chemical processes.

Chemical bond An energy relationship between atoms or ions caused by the sharing or transfer of valence electrons.

Chemical buffer system A system that resists changes in the pH of the body's fluids; consists of a weak acid and its conjugate base.

Chemical energy Energy stored in the chemical bonds of all molecules and compounds.

Chemical reaction The process that occurs when chemical bonds are formed, broken, or rearranged, or electrons are transferred between reactants.

Chemical synapse A type of synapse in which a presynaptic neuron releases neurotransmitters to trigger a change in a postsynaptic neuron.

Chemistry The study of matter and its interactions.

Chemoreceptor (KEE-moh-reh-sep′-ter) A sensory receptor that responds to a change in the concentration of a specific chemical in the air or the body's fluids.

Chemotaxis (kee-moh-TAK-sis) The process by which cells are attracted to a particular location in the body via chemicals released by other cells.

Cholecystokinin (CCK) (kohl′-eh-sih-stoh-KY-nin) A hormone produced by duodenal enteroendocrine cells; triggers the secretion of digestive enzymes and other proteins from the pancreas.

Cholesterol A lipid that is used as a precursor for many substances in the body, including bile salts and steroid hormones.

Cholinergic receptors (kohl-in-ER-jik) Receptors in the plasma membranes of the target cells of parasympathetic neurons and certain sympathetic neurons that bind acetylcholine.

Chondrocyte (KAHN-droh-sy't) The main cell type in cartilage that maintains the extracellular matrix of the tissue.

Chordae tendineae (KOHR-dee ten-din-EE-ee) Fibrous cords that attach papillary muscles to cusps of the atrioventricular valves.

Chorion (KOHR-ee-ahn) The outermost extraembryonic membrane that forms from maternal and embryonic tissues.

Chorionic villi (kohr-ee-AWN-ik VIL-aye; singular, *villus*) Outgrowths of the chorion that grow in the direction of the stratum functionalis layer of the endometrium.

Choroid (KOHR-oyd) The pigmented portion of the vascular layer of the eyeball that contains many blood vessels to nourish the retina.

Choroid plexuses Groups of brain capillaries that come into contact with ependymal cells and produce cerebrospinal fluid.

Chromaffin cells (KROH-maf-in) Cells found in the adrenal medulla that secrete epinephrine and norepinephrine when stimulated by the sympathetic nervous system.

Chromatin (KROH-muh-tin) A long, thin strand of DNA and its associated proteins.

Chromosome (KROH-moh-sohm) The tightly coiled and condensed form of chromatin found in a cell preparing for cell division.

Chylomicron (ky-loh-MY-krahn) A large particle composed of triglycerides packaged with cholesterol, other dietary lipids, phospholipids, and lipid-binding apoproteins that delivers dietary lipids to cells.

Chyme (KYME) A liquid mixture that results from the stomach churning a bolus and mixing it with gastric juice.

Cilia (SILL-ee-uh) Microtubule-containing extensions from the plasma membrane that sweep substances past the cell.

Ciliary body (SILL-ee-ehr-ee) A ring of smooth muscle that alters the shape of the lens to focus light on the retina.

Circadian rhythm (sir-KAY-dee-an) A biological process that follows a set pattern over the 24-hour day.

Circumduction (sir-kuhm-DUHK-shuhn) An angular movement that is the sum total of flexion-extension and abduction-adduction.

Citric acid cycle The second part of glucose catabolism; a series of eight reactions that takes place in the mitochondrial matrix during which sugars are oxidized and ATP and reduced electron carriers are generated.

Clavicle The anterosuperior bone of the pectoral girdle that spans from the scapula to the sternum.

Clitoris (KLIT-ur-us) A small, protruding structure of the female vulva composed of erectile tissue; important in the female sexual response.

Clonal selection The process by which an antigen selects and activates a specific T or B lymphocyte clone.

Clotting factors Proteins found in the blood in their inactive forms that are involved in the coagulation cascade.

Coagulation (koh-aeh-gyoo-LAY-shun) The process of formation of fibrin, which "glues" the components of a blood clot together.

Coagulation cascade A series of enzyme-catalyzed reactions involving clotting factors that culminates in the activation of fibrinogen to fibrin.

Coccygeal vertebrae Three to five small, fused vertebrae of the inferior vertebral column.

Cochlea (KOHK-lee-uh) The bony spiral-shaped portion of the inner ear important in hearing.

Codon (KOH-dawn) A three-nucleotide portion of messenger RNA that specifies a single amino acid.

Cognition The mental functions that include processing and responding to complex external stimuli, recognizing related stimuli, processing internal stimuli, and planning appropriate responses to stimuli.

Collagen fibers Protein fibers in the extracellular matrix of a tissue that are composed of the protein collagen and give a tissue tensile strength.

Collateral circulation Alternate routes of blood flow within a tissue.

Collateral ganglia Autonomic sympathetic ganglia located near the neurons' target organs.

Collecting system The series of tubules that receive filtrate from the distal tubules in the kidneys; responsible for fine-tuning fluid, electrolyte, and acid-base balance.

Colloid (KAWL-oyd) A mixture in which the substances are in different phases but the particles will not settle out when left undisturbed; the gelatinous material in thyroid follicles in which iodine and thyroid hormones are stored.

Colloid osmotic pressure (COP) An osmotic pressure gradient created by proteins within the plasma; water moves into the plasma from the interstitial fluid due to plasma's higher colloid osmotic pressure.

Colon (KOH-lun) The longest portion of the small intestine that consists of the ascending, transverse, descending, and sigmoid colon.

Columnar cell An epithelial cell with a column shape that is taller than it is wide.

Commissural fibers (kawm-ih-SHUR-ul) White matter fibers that connect the right and left cerebral hemispheres.

Compact bone The hard, dense bone tissue located on the exterior of a bone; composed of repeating units called osteons.

Complement system A group of proteins that function in innate immunity.

Complementary base pairing The formation of hydrogen bonds between the nucleic acid bases adenine and thymine (or uracil) and between guanine and cytosine.

Compound A molecule composed of two or more atoms of different elements.

Concentration gradient A situation in which a differential concentration of a substance exists, meaning that a high concentration of the substance is present in one location and a low concentration in a neighboring area.

Conducting zone The portion of the respiratory tract that conducts air toward and away from the alveoli.

Conduction The transfer of heat from one object to another through direct contact.

Condylar joint (KAHN-duh-luhr) A biaxial joint in which the oval-convex surface of one bone fits into a shallow concave depression on the other bone, allowing for movement around two axes.

Cones Photoreceptors that have a cone-shaped outer segment and respond to wavelengths of light that we perceive as color; responsible for high-acuity color vision.

Conjunctiva (kawn-junk-TY-vuh) A transparent membrane that covers the inside of the eyelids and the anterior sclera.

Connective tissue A type of tissue characterized by extensive extracellular matrix; functions in support, protection, and transport.

Connective tissue proper A type of connective tissue that generally functions to connect and support other tissues.

Continuous capillaries Capillaries whose endothelial cells are joined by tight junctions; the least leaky type of capillary.

Continuous conduction A relatively slow type of neuronal action potential conduction that occurs in unmyelinated axons in which every segment of the axolemma is depolarized.

Contractility The intrinsic pumping ability of the heart.

Contraction The generation of tension by a muscle cell.

Control center The component of a feedback loop that receives a stimulus from a receptor and regulates the output of an effector.

Conus medullaris (KOHN-us med-yoo-LEHR-us) The cone-shaped terminal portion of the spinal cord between L_1 and L_2.

Convection The transfer of heat through a liquid or gaseous medium.

Convergence The medial movement of the eyes to focus on a near object.

GLOSSARY

Converging circuit A type of neural circuit in which several input neurons converge on a single output neuron.

Core body temperature The temperature of deeper structures in the body such as the liver; averages 35.8–38.2° C (96–101° F).

Cornea (KOHR-nee-uh) A transparent structure on the anterior eye that refracts light as it enters the eye.

Coronary circulation The set of arteries, capillaries, and veins that supply blood to and drain blood from the myocardium of the heart.

Corpora cavernosa (KOHR-pohr-uh kae-ver-NOH-suh) Paired dorsal erectile bodies of spongy connective tissue and smooth muscle filled with vascular spaces in the penis.

Corpus albicans A whitish knot of scar tissue that forms as the corpus luteum degrades.

Corpus callosum The largest group of commissural fibers in the brain; connects the right and left cerebral hemispheres.

Corpus luteum An endocrine gland that forms from a ruptured ovarian follicle; secretes progesterone and some estrogens.

Corpus spongiosum (KOHR-pus spun-jee-OH-sum) Ventral erectile body of spongy connective tissue and smooth muscle filled with vascular spaces in the penis.

Cortical nephrons Nephrons with short nephron loops that are either confined to the renal cortex or extend only into the superficial renal medulla.

Cortical reaction The process by which cortical granules of the secondary oocyte release enzymes to destroy the sperm-binding receptors on the oocyte plasma membrane.

Corticonuclear tracts Motor tracts that control the muscles of the head and neck via the lower motor neurons of the cranial nerves.

Corticospinal tracts The largest descending spinal tracts; consist of the axons of upper motor neurons that control movement of muscles below the head and neck via stimulation of lower motor neurons of spinal nerves.

Cortisol (KOHR-tih-zohl) The main glucocorticoid produced by the adrenal cortex; stimulates gluconeogenesis as well as fat and protein breakdown and inhibits the inflammatory response.

Costal cartilage The cartilage that attaches the ribs to the sternum.

Countercurrent mechanism The mechanism in the nephron loops of juxtamedullary nephrons of the kidneys that establishes a concentration gradient in the medullary interstitial fluid in order to produce concentrated urine.

Covalent bonding Chemical bonding resulting from the sharing of electrons between two nonmetals or a nonmetal and hydrogen.

Cranial base The inferior portion of the cranial cavity that cradles the brain.

Cranial bones The eight bones of the skull that encase the brain.

Cranial cavity A dorsal body cavity located within the skull that houses the brain.

Cranial nerve A nerve containing axons whose origin or destination is the brain.

Creatine phosphate (CP) (KREE-uh-teen FAWS-fayt) An immediate energy source for certain cell types that donates a phosphate group to adenosine diphosphate (ADP).

Creatinine The metabolic waste product of creatine phosphate catabolism.

Crista ampullaris (KRIS-tah am-pyoo-LEHR-is) A cluster of receptor cells and supporting cells in the ampulla that detect head rotation.

Crossbridge cycle A series of events in a muscle fiber during which a myosin head grabs onto a series of actin subunits in the thin filament and pulls the thin filament progressively closer to the M line of the sarcomere.

Crossed-extension reflex A polysynaptic reflex that occurs concurrently with the flexion reflex; triggers extension of the limb on the opposite side of the body from a painful stimulus.

Crossing over An exchange of segments of DNA between homologous chromosomes.

Cuboidal cell An epithelial cell with a shape that is about as wide as it is tall.

Cutaneous membrane (kyoo-TAYN-ee-us) Another name for the skin or integument.

Cyclic adenosine monophosphate (cAMP) A common second messenger derived from ATP.

Cytokines (SY-toh-kynz) Proteins produced by several types of leukocytes that enhance the immune response in some way.

Cytokinesis (sy′-toh-kin-EE-sis) The division of the cytoplasm between two daughter cells that occurs near the end of cell division.

Cytology (sy-TAWL-uh-jee) The scientific study of cells.

Cytoplasm The part of a cell located inside the plasma membrane that consists of the cytosol and the organelles.

Cytosine A pyrimidine nitrogenous base found in nucleic acids.

Cytoskeleton A network of protein filaments within a cell that supports the cell, maintains its shape, holds its organelles in place, and functions in cellular motion.

Cytosol (intracellular fluid) (SY-toh-sawl) The fluid located inside a cell that contains water and a variety of solutes.

Dalton's law of partial pressures A gas law stating that the total pressure of a container is equal to the sum of the partial pressures of the gases in the mixture.

Declarative memory (fact memory) The memory of things that are readily available to consciousness that could in principle be expressed aloud.

Decussation The crossing over of tracts of white matter from one side of the brain or spinal cord to the other side.

Deep A position that is farther away from the surface of the body or more within the body's interior.

Defecation reflex Reflex arc triggered by the presence of feces in the anus; mediated by the parasympathetic nervous system and controlled by the cerebral cortex.

Deglutition (swallowing) (dee-gloo-TISH-un) A specialized type of propulsion that pushes a bolus of food from the oral cavity through the pharynx and esophagus to the stomach.

Dehydration A condition characterized by a decreased volume and increased concentration of the extracellular fluid.

Dehydration synthesis A reaction in which two monomers are linked by a covalent bond and that results in the formation of a water molecule.

Denaturation The destruction of the three-dimensional structure of a protein and the subsequent loss of its function.

Dendrite (DEN-dryt) A process extending from a neuron that receives input from other neurons.

Dendritic cells A class of leukocytes that often function as phagocytes and antigen-presenting cells and are located in places such as the epidermis of the skin.

Dens The superior projection from the axis that fits inside the atlas to form the atlanto-axial joint.

Dense irregular connective tissue A type of connective tissue proper that contains collagen fibers arranged in irregular bundles; found in locations such as joint and organ capsules and the dermis.

Dense regular collagenous connective tissue A type of connective tissue proper that contains collagen fibers arranged in regular bundles; found in locations such as tendons.

Dense regular elastic connective tissue A type of connective tissue proper that contains elastic fibers arranged in regular bundles; found in locations such as large blood vessels.

Dentin The inner mineralized material of a tooth composed largely of calcium hydroxyapatite crystals.

Deoxyribonucleic acid (DNA) A double-stranded nucleic acid that is found in the cell nucleus and contains the code for every protein in the body.

Depolarization A temporary increase in a cell's membrane potential.

Depression The movement of a joint in which the corresponding body part moves in an inferior direction.

Dermal papillae (DER-mul pah-PILL-ee) Projections of the papillary layer of the dermis that indent the overlying epidermis and house capillary loops and sensory receptors.

Dermatome (DER-muh-tohm) A region of the skin supplied by a specific spinal nerve.

Dermis The deeper layer of the skin consisting of loose and dense irregular collagenous connective tissue.

Desmosomes (DEZ-moh-sohmz) Intercellular junctions that hold cells together in a tissue to increase the tissue's resistance to mechanical stresses.

Diaphragm muscle (DY-uh-fram) The dome-shaped muscle located between the thoracic and abdominal cavities; the main muscle for breathing.

Diaphysis (dy-AF-eh-sis) The shaft of a long bone.

Diarthrosis (dy-ahr-THROH-sis) A functional classification of articulations in which the joint allows for a wide range of motion between articulating bones.

Diastole (dy-ASS-toh-lee) The period during which the heart is relaxing.

Diastolic pressure The blood pressure in the arteries when the ventricles are in diastole.

Diencephalon (dy′-en-SEF-ah-lahn) The central core of the brain; made up of the thalamus, hypothalamus, epithalamus, and subthalamus.

Diffusion A type of passive transport in which solutes move with their concentration gradients from an area of high solute concentration to one of low solute concentration.

Digestion The process of breaking down a substance into smaller subunits; may be physical or chemical.

Digestive system The collection of organs concerned with ingestion and digestion of food; absorption of nutrients; propulsion of ingested food through the digestive tract; and defecation of indigestible substances.

Diploë (dip-LOH-ee) The spongy bone tissue of flat bones.

Diploid (2n) A cell with 23 pairs of homologous chromosomes.

Dipole (DY-pohl) A partially positive and partially negative pole that forms as a result of a polar covalent bond.

Disaccharide (dy-SAK-uh-ryd) A carbohydrate with two monosaccharides joined by dehydration synthesis.

Distal A position that is farther away from the point of origin, usually the trunk.

Distal tubule The final segment of the renal tubule in which filtrate modification is controlled by hormones to fine-tune fluid, electrolyte, and acid-base balance.

Distensibility A property of a cell by which it can be stretched without sustaining damage.

Diverging circuit A type of neural circuit in which a single input neuron contacts several output neurons.

DNA synthesis (replication) The process during which a cell's DNA is duplicated prior to cell division.

Dominant trait A trait for which one allele is expressed in the phenotype over another allele.

Dorsiflexion (dohr-see-FLEK-shuhn) An angular movement of the foot at the ankle in which the angle between the foot and the leg decreases.

Ductus deferens (vas deferens) A tube that carries semen from the epididymis through the inguinal canal into the pelvic cavity.

Duodenum (doo-AH-den-um) The initial 25-cm-long segment of the small intestine into which bile and pancreatic juice are secreted.

Dura mater (DOO-rah MAH-ter) The tough outermost meninx that surrounds the brain and the spinal cord.

Dural sinuses The set of venous channels that are located between two layers of dura mater and drain the cerebral veins of the brain.

Dynamic equilibrium The ability to maintain balance during linear or rotational acceleration of the head or body.

Dyspnea (DISP-nee-uh) A condition of abnormal breathing.

Dysrhythmia (DIS-rith-mee-uh) Any disturbance in the electrical rhythm of the heart.

Eccrine sweat glands (EK-rin) Glands in the skin that produce a watery sweat as part of the body's thermoregulatory mechanisms.

Ectoderm The outermost germ layer of an embryo.

Edema (eh-DEE-mah) The collection of excess water in the interstitial fluid that results in swelling.

Effector Cell or organ that causes a physiological response.

Efferent arterioles Arterioles that drain blood from the glomeruli in the kidney.

Ejaculate The semen expelled during one ejaculation; typically measures 2.5–5 ml in volume and contains 40–750 million sperm cells.

Elastic arteries The largest arteries that are near the heart and so under the highest pressure; have a great deal of elastic fibers.

Elastic cartilage A type of cartilage with elastic fibers as the predominant element; found in locations such as the external ear.

Elastic fibers Protein fibers in the extracellular matrix of a tissue that give a tissue distensibility and elasticity.

Elastic filament A myofilament that consists of the structural protein titin.

Elasticity A property of a cell by which it will return to its resting length after it has been stretched.

Electrical synapse A type of synapse in which two cells are joined by gap junctions that allow the bidirectional flow of ions.

Electrocardiogram (ECG) A graphic representation of all of the action potentials occurring in the heart.

Electrochemical gradient The sum of the electrical gradient and chemical (concentration) gradient for an ion.

Electroencephalogram (EEG) A recording of the electrical activity of the brain.

Electrolyte (ee-LEK-troh-lyt) Cations and anions that result when ionic compounds are placed in a solution; these ions will conduct an electric current.

Electromotive force A force generated by the flow of electrons or ions that can perform work.

Electron A negatively charged subatomic particle located in electron shells around the atomic nucleus.

Electron transport chain (ETC) A group of electron carriers in the inner mitochondrial membrane through which electrons are passed as part of oxidative phosphorylation.

Electrophysiology The study of electrical changes in the body's cells and the physiological processes that accompany these changes.

Element A substance that cannot be broken into simpler substances by chemical means; consists of one or more atoms with the same atomic number.

Elevation The movement of a joint in which the corresponding body part moves in a superior direction.

Embryo (EM-bree-oh) The stage of development from the first to the eighth week of gestation during which the rudimentary organ systems are formed.

Emulsification (ee-mul′-sih-fih-KAY-shun) A process in which an amphiphilic compound physically breaks lipids into smaller clusters through polar and nonpolar interactions.

End-diastolic volume (EDV) The volume of blood present in the ventricles at the end of ventricular filling.

End-plate potential A depolarization at the motor end plate of a skeletal muscle fiber that can trigger an action potential.

End-systolic volume (ESV) The volume of blood present in the ventricles at the end of the ventricular ejection phase.

Endergonic reaction (en-der-GAHN-ik) A reaction in which the products have more energy than the reactants; requires the input of energy to proceed.

Endocardium (en′-doh-KAR-dee-um) The inner simple squamous epithelial layer of the heart wall; continuous with the endothelium of the blood vessels.

Endochondral ossification (en-doh-KAHN-drul) The process by which most bones form from a hyaline cartilage model.

Endocrine gland (EN-doh-krin) A gland that secretes a hormone or hormones directly into the bloodstream to influence the functions of distant target cells.

Endocrine system The group of organs that secretes hormones, which influence the functions of other cells and tissues.

Endocytosis (en′-doh-sy-TOH-sis) A type of vesicular transport in which substances are taken into the cell via an infolding of the plasma membrane.

Endoderm The innermost germ layer of an embryo.

Endomembrane system A group of organelles consisting of the endoplasmic reticulum, Golgi apparatus, and lysosomes that together work to process substances imported to and exported from the cell.

Endometrium The innermost tissue layer of the uterus composed of simple columnar epithelium.

Endoneurium (en′-doh-NOOR-ee-um) The connective tissue that surrounds an axon within a peripheral nerve.

Endosome (EN-doh-sohm) The vesicle into which substances are ingested in endocytosis; it forms from a pinched-off portion of the plasma membrane.

Endosteum (en-DAHS-tee-um) The connective tissue membrane lining the internal surfaces of a bone.

Endothelium A modified simple squamous epithelium that lines the inner surface of blood vessels and the heart.

Energy The capacity to do work.

Energy balance The difference between energy intake and energy expenditure; also known as *nitrogen balance*.

Enteric nervous system (ENS) (en-TEHR-ik) A self-contained branch of the autonomic nervous system that extends from the esophagus to the anus and regulates secretion and motility of the digestive organs.

Enzyme A biological catalyst, nearly always a protein, speeding up a reaction without changing its products or reactants.

Eosinophil (ee′-oh-SIN-uh-fill) A class of granulocyte that is involved in the immune response to parasites and in allergic reactions.

Ependymal cell (eh-PEN-dih-mul) A ciliated neuroglial cell of the central nervous system that lines the hollow cavities of the brain and spinal cord; cilia beat to circulate cerebrospinal fluid.

Epidermal ridges Folded patterns of the epidermis in thick skin that enhance gripping ability.

Epidermis (eh-puh-DER-mis) The superficial layer of the skin consisting of stratified squamous keratinized epithelium.

Epididymis (ep′-uh-DID-uh-miss) Portion of the male duct system that is the site of sperm maturation and storage.

Epidural space The fat-filled space between the spinal dura mater and the internal surface of the vertebral cavity.

Epiglottis A flap of elastic and hyaline cartilage that covers the larynx during swallowing to prevent food and liquid from entering the lungs.

Epinephrine A catecholamine that functions largely as a hormone released by the adrenal medulla and that is also released as a

neurotransmitter by certain neurons in the sympathetic nervous system.

Epineurium (ep′-ih-NOOR-ee-um) The connective tissue sheath that surrounds a peripheral nerve.

Epiphyseal line (eh-PIF-ih-see-uhl) A calcified remnant of the epiphyseal plate.

Epiphyseal plate A structure composed of hyaline cartilage from which long bones grow in length.

Epiphysis (eh-PIF-ih-sis) The end of a long bone.

Epithelial tissue (ep′-ih-THEE-lee-ul) A tissue type composed of epithelial cells and a thin basal layer of extracellular matrix; functions to cover and line all body surfaces and hollow organs.

Eponychium (ep-oh-NIK-ee-um) The most distal part of the proximal nail fold; consists of stratum corneum and covers the proximal part of the nail body.

Erythrocyte (red blood cell, or RBC) The primary formed element of the blood; biconcave disc-shaped cell that transports oxygen and carbon dioxide on its hemoglobin.

Erythropoiesis (ee-rith′-roh-poy-EE-sis) The process of differentiation and maturation of erythrocytes.

Erythropoietin (ee-rith′-roh-POY-eh-tin) The hormone produced by the kidneys that increases the rate of erythrocyte production and decreases the time it takes for them to mature.

Esophagus A tubular organ in the posterior thoracic cavity that transmits food from the pharynx to the stomach.

Essential nutrient A nutrient that cannot be manufactured by the body and so must be obtained from the diet.

Eupnea (YOOP-nee-uh) The condition of having a normal rate and pattern of ventilation.

Evaporation The transformation of water from a liquid to a gas; in the process, water must absorb a great deal of heat from its surroundings.

Eversion (ee-VER-zhuhn) A rotational movement of the foot in which the plantar surface rotates laterally away from the midline of the body.

Excess postexercise oxygen consumption (EPOC) The persisting increased rate of breathing during the recovery period after completing exercise.

Exchange reaction A reaction in which one or more atoms or electrons from the reactants are exchanged.

Excitability A property of a cell in which it is responsive in the presence of various stimuli.

Excitation-contraction coupling The linking of muscle fiber excitation via an action potential, with contraction via the release

of calcium ions from the sarcoplasmic reticulum.

Excitatory postsynaptic potential (EPSP) A change in the membrane potential of a neuron in which it becomes less negative due to the influx of positive ions.

Excretion The process by which waste products are removed from the body.

Exergonic reaction (eks-er-GAWN-ik) A reaction in which the products have less energy than the reactants; releases energy once completed.

Exocrine gland (EKS-oh-krin) A gland that secretes a product through a duct to the external surface of the body or into the respiratory, gastrointestinal, and/or genitourinary tract.

Exocytosis (eks′-oh-sy-TOH-sis) A type of vesicular transport in which substances are released from the cell via the fusion of a vesicle with the plasma membrane.

Expiration The movement of air out of the lungs.

Extension (eks-TEN-shuhn) An angular movement whereby the angle increases between articulating bones, moving the bones away from each other.

Extensor A muscle that contracts to increase the angle of the joint on which it operates.

Exteroceptor (EK-ster-oh-sep′-ter) A sensory receptor that responds to a stimulus exterior to the body.

Extracellular compartment The fluid compartment located outside cells that contains extracellular fluid.

Extracellular fluid (ECF) The fluid located in the spaces between cells.

Extracellular matrix The substance in a tissue outside of the cells that consists of extracellular fluid, ground substance, and protein fibers.

Extrafusal muscle fibers (ek-strah-FYOO-zul) Regular contractile skeletal muscle fibers.

Facial bones The 14 bones of the skull that form the framework of the face.

Facilitated diffusion A type of diffusion in which substances move through a membrane using a protein channel or carrier.

Facultative water reabsorption Water reabsorption that is controlled by hormones to maintain a constant extracellular fluid osmolarity.

False rib A rib that attaches to the costal cartilage of another rib rather than attaching directly to the sternum.

Falx cerebri (seh-REE-bry) A double fold of meningeal dura that dives into the longitudinal fissure and separates the two cerebral hemispheres.

Fascicle (FASS-ih-kul) A bundle of axons in a nerve surrounded by the perineurium.

Fast-twitch fibers Skeletal muscle fibers with high myosin ATPase activity that proceed more rapidly through their crossbridge cycles; generate rapid but generally short-duration contractions.

Fatty acid A lipid with a hydrocarbon chain bound to a carboxylic acid group.

Feces The collection of indigestible or undigested food, bacteria, and water in the large intestine that is eliminated via defecation.

Feedback loop A homeostatic control mechanism in which a change in a regulated variable causes effects that feed back and affect that same variable.

Femur The only bone of the thigh.

Fenestrated capillaries Capillaries whose cells have fenestrations to allow large volumes of fluids and solutes to be exchanged.

Fenestrations (fen-eh-STRAY-shunz) Small pores within endothelial cells of capillaries.

Fertilization The union of a sperm and ovum to form a zygote.

Fetus (FEE-tus) The stage of development from weeks 9 through 38 of gestation.

Fever A body temperature above the normal range of 36–38 °C (97–99 °F).

Fiber Polysaccharides that are not fully digestible by humans due to a lack of the enzymes required to catabolize them.

Fibrillation A condition in which muscle cells are contracting individually due to chaotic electrical activity.

Fibrin (FY-brin) A sticky, threadlike protein that "glues" the formed elements and clotting factors together in a blood clot.

Fibrinolysis (fy-brin′-uh-LY-sis) The process by which fibrin is degraded during thrombolysis.

Fibroblast A cell type within connective tissue proper that produces components of the extracellular matrix.

Fibrocartilage A type of cartilage with collagen fibers as the predominant element; found in locations such as the intervertebral discs.

Fibrosis (fy-BROH-sis) The process by which a damaged tissue is replaced with fibrous connective tissue during healing.

Fibrous joints A structural classification of articulations in which the joints are held together by dense regular collagenous connective tissue and have neither a synovial cavity nor an articular capsule.

Fibula (FIB-yoo-luh) The lateral leg bone.

Filtrate The fluid resulting from glomerular filtration that flows through the renal tubule and collecting system.

Filtration The movement of a fluid by a force such as hydrostatic pressure or gravity.

Filtration membrane A structure consisting of the glomerular endothelial cells, basal lamina, and podocytes through which blood is filtered to produce filtrate in the kidneys.

Filtration slits Tiny spaces between the foot processes of podocytes surrounding the glomerular capillaries.

Filum terminale (FY-lum ter-mee-NAL-ay) The inferior continuation of the spinal pia mater; continues after the end of the spinal cord through the vertebral cavity and anchors into the first coccygeal vertebra.

First-class lever A lever system in which the fulcrum is located between the applied force and the load to be moved.

Fissure A deep groove between major brain structures or lobes of the cerebrum.

Fixator A muscle that holds a bone in place, allowing other muscles to move the bone and joint more effectively.

Fixed acid An acid that cannot be converted to a gaseous form and so must be excreted by the kidneys.

Flagellum (flah-JEL-um) A single motile extension from a cell that propels the cell forward; found in humans only on sperm cells.

Flat bone A bone that is flat in shape.

Flavin adenine dinucleotide (FAD) One of the main electron carriers involved in nutrient catabolism; when reduced, it becomes $FADH_2$.

Flexion (FLEK-shuhn) An angular movement whereby the angle decreases between articulating bones, bringing the bones closer together.

Flexion reflex (withdrawal reflex) A polysynaptic reflex initiated by painful stimuli that triggers withdrawal of the affected body part.

Flexor A muscle that contracts to decrease the angle of the joint on which it operates.

Floating rib A rib that lacks an attachment to the sternum.

Fluid mosaic model The model that describes the plasma membrane as a dynamic structure consisting of a phospholipid bilayer with multiple components interspersed throughout.

Follicle-stimulating hormone (FSH) Hormone produced by the anterior pituitary gland that stimulates reproductive functions in males and females.

Fontanel A "soft spot" in the skull of a fetus and infant in which the cranial sutures have not yet fused.

Foramen ovale (fohr-AY-men oh-VAH-lay) An opening in the interatrial septum in the fetal heart that allows blood to bypass the pulmonary circuit.

Forearm The distal portion of the upper limb that consists of the radius and the ulna.

Fossa ovalis An indentation in the interatrial septum that marks the former location of the foramen ovale in the fetal heart.

Fourth ventricle A hollow, cerebrospinal fluid–filled cavity located within the brainstem.

Fovea centralis (FOH-vee-uh sen-TRAH-liss) An area of the retina with the highest density of cones that produces high-acuity vision.

Frank-Starling law A mechanism by which a ventricular muscle cell of the heart contracts more forcefully when it is stretched.

Frontal lobes The two anterior lobes of the cerebral hemispheres; responsible for planning and executing movement and complex mental functions.

Frontal plane A plane of section that divides the body or body part into anterior and posterior portions; also known as the *coronal plane*.

Fulcrum The pivot or hinge point around which movement is produced when a muscle or group of muscles contracts.

Funiculus (foo-NIK-yoo-luss) A general region of white matter in the spinal cord that is organized into tracts.

Fused tetanus (complete tetanus) A type of wave summation in which a muscle fiber is stimulated rapidly and the muscle fiber is not allowed to relax between contractions.

G-proteins Intracellular enzyme complexes involved in a variety of second-messenger systems.

Gallbladder The small, hollow organ on the posterior side of the liver that stores and releases bile.

Gametes Sex cells (sperm and ova) that are produced in the gonads.

Gamma motor neurons (γ motor neurons) Lower motor neurons that innervate intrafusal muscle fibers.

Ganglion A cluster of neuronal cell bodies in the peripheral nervous system.

Gap junctions Intercellular junctions that connect the cytosol of neighboring cells and allow water and solutes to pass between them.

Gastrin A hormone secreted by G cells of gastric glands that triggers acid secretion from parietal cells.

Gastrulation (gas-troo-LAY-shun) The rearrangement and migration of the cells of the bilaminar embryonic disc to form the three germ layers.

Gene A segment of DNA that specifies the amino acid sequence for a single protein.

Genetic code A list of the amino acids coded for by a specific DNA triplet or mRNA codon.

Genome (JEE-nohm) Collectively, all of an individual's DNA.

Genotype (JEEN-oh-ty′p) The genetic makeup of an individual.

Gestation period (jeh-STAY-shun) The period that extends from the mother's last menstrual period until birth, which is approximately 280 days, or 40 weeks.

Glenoid cavity (GLEN-oyd) The lateral depression of the scapula that articulates with the humerus at the shoulder joint.

Globus pallidus Two basal nuclei that sit medial to the putamen; inhibit the initiation of movement by inhibiting the thalamus.

Glomerular capsule The double-layered epithelium that surrounds the glomeruli of the kidneys; encloses the capsular space.

GLOSSARY

Glomerular filtration The process by which blood is filtered in the glomeruli of the kidneys.

Glomerular filtration rate (GFR) The rate at which filtrate is produced in the glomeruli of the kidneys.

Glomerular hydrostatic pressure (GHP) The blood pressure within the glomerular capillaries; favors filtration into the capsular space.

Glomerulus (gloh-MEHR-yoo-lus) Ball-shaped capillary bed where the blood is filtered in the kidney.

Glottis The opening between the vocal cords of the larynx through which air passes.

Glucagon A hormone produced by the pancreas that triggers changes to raise the concentration of glucose in the blood.

Glucocorticoids (gloo′-koh-KOHR-tih-koydz) A class of hormones produced by the adrenal cortex that are part of the body's response to a stressor.

Gluconeogenesis (gloo′-koh-nee-oh-JEN-eh-sis) The formation of new glucose from noncarbohydrate precursors, including glucogenic amino acids and glycerol.

Glucose (GLOO-kohs) The monosaccharide used as the primary fuel by most cells in the body.

Glycerol A modified sugar alcohol that is found in triglycerides and can be used as a glucogenic precursor.

Glycogen (GLY-koh-jen) A polysaccharide that is the storage form of glucose in animals.

Glycogenesis The formation of glycogen from glucose monomers.

Glycogenolysis (gly′-koh-jen-AWL-ih-sis) The breakdown of glycogen to release glucose monomers into the blood.

Glycolysis (gly-KAWL-uh-sis) A series of 10 reactions that takes place in the cytosol during which glucose is split and small amounts of ATP and NADH are generated.

Glycolytic catabolism (anaerobic catabolism) A series of ATP-producing reactions that occur in the cytosol of cells in which glucose is broken down into two molecules of pyruvate; these reactions do not require oxygen to proceed.

Goblet cell A unicellular exocrine gland that secretes mucus.

Golgi apparatus (GOHL-jee) A group of membrane-enclosed sacs that receive and modify products from the endoplasmic reticulum for export or use in the cell.

Golgi tendon organs Mechanoreceptors located within tendons near the muscle-tendon junction; monitor the tension generated by a muscle contraction.

Golgi tendon reflex A reflex mediated by Golgi tendon organs that causes relaxation in response to increased tension within a muscle.

Gomphosis (gahm-FOH-sis) A fibrous joint structurally and a synarthrosis functionally that is found between each tooth and the bony socket in either the mandible or the maxilla.

Gonads The primary organs of the reproductive system that produce gametes and hormones; the male gonads are testes and the female gonads are ovaries.

Gradient A condition in which more of something exists in one area than in another, and the two areas are connected.

Granulocyte (GRAN-yoo-loh-sy't) A class of leukocyte with visible cytosolic granules; includes neutrophils, eosinophils, and basophils.

Gray matter Collections of unmyelinated axons, dendrites, and cell bodies that appear gray.

Great saphenous veins (SAF-en-us) Large superficial veins of the medial lower limb.

Greater pelvis The portion of the pelvis superior to the pelvic brim.

Gross anatomy The study of structures of the human body that can be seen with the unaided eye.

Ground substance An amorphous component of the extracellular matrix containing water, proteoglycans, glycosaminoglycans, and other substances.

Growth An increase in the size of an individual cell or an increase in the number of cells; occurs when anabolic processes outweigh catabolic processes.

Growth hormone (GH) A hormone produced by the anterior pituitary gland that promotes fat breakdown and gluconeogenesis; also stimulates the release of insulin-like growth factor, which promotes growth and protein synthesis.

Guanine A purine nitrogenous base found in nucleic acids.

Gustation (gus-TAY-shun) The sense of taste.

Gyrus An elevated ridge of the superficial cerebrum.

Hair Small, filamentous structures that project from the surface of the skin and are composed of dead, keratinized cells.

Hair follicle An infolding of the epidermis in which a hair is embedded in the skin.

Hair follicle receptors Free nerve endings wrapped around the base of a hair follicle in the dermis or hypodermis.

Hallux The anatomical term for the big toe.

Haploid (1n) A cell with 23 chromosomes.

Heart The muscular organ that pumps blood to the pulmonary and systemic circuits.

Heart rate (HR) The number of heartbeats that occurs in a set period of time.

Heart sounds Noises heard during auscultation of the heart that are caused by vibrations of the ventricular wall when the valves close.

Heartbeat A single, coordinated contraction of cardiac muscle cells.

Hematocrit (heh-MAEH-toh-krit) The portion of the blood that consists of erythrocytes.

Hematopoiesis (heh′-mah-toh-poy-EE-sis) The process of differentiation and maturation of the formed elements of blood.

Hematopoietic stem cells Cells located in red bone marrow that can become any type of formed element of the blood.

Hemoglobin The iron-containing protein in erythrocytes that binds and carries oxygen throughout the blood.

Hemolysis (hee-MAH-luh-sis) The rupture of erythrocytes.

Hemostasis (hee-moh-STAY-sis) The process by which blood loss is stopped from a damaged blood vessel.

Henry's law A gas law stating that the degree to which a gas dissolves in a liquid is proportional to both its partial pressure and its solubility in the liquid.

Hepatic portal system A system that delivers blood from the digestive organs and the spleen to hepatocytes via the hepatic portal vein for processing of nutrients, wastes, and toxins.

Heredity (heh-RED-ih-tee) The transmission of genetic characteristics from parent to offspring through the genes.

Heterozygous (het′-er-oh-ZY-gus) Two alleles of a gene that code for different traits.

High-density lipoproteins (HDLs) Lipoproteins that transfer lipids from cells in peripheral tissues to the liver for excretion.

Hinge joint A uniaxial joint in which the convex articular surface of one bone fits into the concave depression of the other bone, creating a hinge that allows for motion around one axis.

Hippocampus (hip-poh-KAM-pus) A component of the limbic system in the brain that functions in memory and learning.

Histology (hiss-TAWL-uh-jee) The study of tissues.

Holocrine secretion (HOH-loh-krin) A type of exocrine secretion in which the product accumulates in the cell until the cell ruptures and dies.

Homeostasis (hoh′-mee-oh-STAY-sis) The maintenance of the body's stable internal environment.

Homologous chromosomes (huh-MAH-luh-gus) A set of maternal and paternal chromosomes with the same genes.

Homozygous (hoh-moh-ZY-gus) Two alleles of a gene that code for the same trait.

Hormone A chemical messenger secreted into the blood that triggers changes within its target cells.

Human anatomy The study of the structure of the human body.

Human chorionic gonadotropin (kohr-ee-AWN-ik goh-NAEH-doh-troh′-pin; hCG) A hormone produced by the outer layer of a developing embryo that keeps the corpus luteum from degenerating.

Human physiology (fiz'-ee-AWL-oh-gee) The study of the functions of the human body.

Humerus The only bone of the upper arm.

Hyaline cartilage (HY-ah-lin) A type of cartilage with ground substance as the predominant element; found in locations such as the articulating ends of bones, between the sternum and ribs, and in the nose.

Hydrogen bonds Weak attractions between partially positive hydrogen atoms of one compound or functional group and partially negative atoms of another compound or functional group.

Hydrolysis (hy-DRAWL-uh-sis) A reaction in which a molecule of water is added to a polymer, splitting it into component monomers.

Hydrophilic (hy-droh-FIL-ik) A property of substances with partially or fully charged ends (i.e., polar covalent and ionic compounds) that will dissolve when placed in water.

Hydrophobic (hy-droh-FOH-bik) A property of substances with no dipole (nonpolar covalent bonds) that will not dissolve when placed in water.

Hydrostatic pressure (HP) The pressure that a fluid exerts on the wall of its container.

Hydroxyapatite A calcium phosphate–based mineral that makes up the inorganic portion of the osseous tissue extracellular matrix.

Hyoid bone A bone in the superior neck to which muscles and ligaments attach.

Hyperextension (hy'-per-eks-TEN-shuhn) An angular movement whereby the joint allows for motion beyond the anatomical position of the articulation.

Hyperopia (hy'-per-OH-pee-ah) A condition in which the eye is too short or the cornea is too flat and the lens cannot round up enough for near vision.

Hyperpolarization A change in the membrane potential of an excitable cell to a value more negative than its resting membrane potential.

Hypersensitivity disorder A condition characterized by an exaggerated response of the immune system to a foreign antigen.

Hypertension High blood pressure; generally defined as a systolic pressure of 140 mm Hg or higher and/or a diastolic pressure of 90 mm Hg or higher.

Hypertonic solution A solution that has a greater ability to cause water movement by osmosis than the reference solution due to its higher number of solute particles.

Hypertrophy (hy-PER-troh-fee) An increase in cell size.

Hyperventilation An increased rate of ventilation.

Hypodermis (hy-poh-DER-mis) The layer of adipose tissue deep to the skin; not considered part of the integument.

Hyponychium The fold of stratum corneum that anchors the distal end of the nail plate to the underlying epithelium.

Hypotension Any abnormally low blood pressure, defined by a systolic pressure lower than 90 mm Hg and/or a diastolic pressure lower than 60 mm Hg.

Hypothalamic-pituitary-gonadal (HPG) axis The multi-tiered feedback loops of the hormones of the hypothalamus, anterior pituitary, and gonads.

Hypothalamus The small anterior and inferior component of the diencephalon of the brain; responsible for many homeostatic functions, producing the hormones oxytocin and antidiuretic hormone, and producing releasing and inhibiting hormones that affect the anterior pituitary gland.

Hypotonic solution A solution that has a lesser ability to cause water movement by osmosis than the reference solution due to its smaller number of solute particles.

Hypoventilation A decreased rate of pulmonary ventilation.

I band The component of the sarcomere that appears lighter because it contains only thin filaments.

Ileum (ILL-ee-um) The final and longest segment of the small intestine, measuring about 3.6 meters in length.

Ilium (ILL-ee-um; plural, *ilia*) The largest bone of the pelvis that forms its superior, lateral, and posterior walls.

Immune system A group of cells and proteins that works to defend the body against internal and external threats.

Immunodeficiency disorder A condition characterized by a deficit in one or more components of the immune response.

Immunological memory The phenomenon whereby cells of adaptive immunity respond more quickly and efficiently to subsequent exposures to an antigen.

Inclusions Clusters of identical storage materials in the cytosol of certain cell types.

Inferior Away from the head or toward the tail; also known as *caudal* (KAW-d'l).

Inferior vena cava The large vein that drains mostly structures inferior to the diaphragm; empties into the right atrium.

Inflammatory response The series of events that occurs in reaction to any cellular injury; involves vasodilation, increased capillary permeability, pain, swelling, and decreased function.

Inhibin A hormone secreted by the testes and ovaries that inhibits FSH secretion.

Inhibitory postsynaptic potential (IPSP) A change in the membrane potential of a neuron in which it becomes more negative due to the outflow of positive ions or the influx of negative ions.

Innate immunity (nonspecific immunity) The arm of the immune system that consists of cells and proteins that are not specific for individual antigens and that always respond to cell injury and pathogens in the same manner.

Inorganic compounds Compounds that do not contain carbon and hydrogen atoms bound to one another.

Insensible water loss A water loss that is not "sensed" or noticed; includes water lost to sweat and ventilation.

Insertion (muscle) The end of a muscle attached to the structure that will be moved when the muscle contracts.

Inspiration The movement of air into the lungs.

Inspiratory muscles Muscles that increase the volume of the lungs, which decreases the intrapulmonary pressure below that of atmospheric pressure, causing inspiration.

Insulas (IN-syoo-lahz) Deep lobes of the cerebrum; have functions relating to taste and to viscera.

Insulin A hormone produced by the pancreas that triggers changes to lower the concentration of glucose in the blood.

Insulin-like growth factor (IGF) A hormone produced largely by the liver in response to growth hormone; stimulates protein synthesis and cell division.

Integral protein A protein within a membrane that spans the width of the membrane.

Integument (in-TEG-you-ment) The skin and its accessory structures.

Integumentary system (in-teg'-yoo-MEN-tuh-ree) The organ system consisting of the skin and its accessory structures.

Intercalated disc (in-TER-kuh-lay'-t'd) Specialized structure that connects adjacent cardiac muscle cells and contains gap junctions and desmosomes.

Intercostal space The space located between two ribs.

Interferon (IFN) (in-ter-FEER-awn) A cytokine family that is generally produced in response to infection by intracellular pathogens; they inhibit viral replication inside host cells and activate components of innate and adaptive immunity.

Intermediate filaments Ropelike protein filaments of the cytoskeleton that hold organelles in place, form much of the framework of the cell, and help a cell withstand mechanical stresses.

Internal carotid arteries Branches of the common carotid arteries that supply the brain.

Internal jugular veins Veins that drain the brain and parts of the head and neck.

Interneurons Neurons between sensory and motor neurons that perform integrative functions.

Internodes The myelinated segment of an axon.

Interoceptor (IN-ter-oh-sep'-ter) A sensory receptor that responds to a stimulus originating within the body.

GLOSSARY

Interphase   The phase of the cell cycle during which the cell grows, replicates its DNA, and prepares for cell division.

Interstitial cells   Cells of the testes that produce androgens, mostly testosterone.

Intervertebral disc   The fibrocartilage pad between two vertebrae that absorbs shock and supports the vertebral column.

Intracellular compartment   The fluid compartment located within cells that contains cytosol or intracellular fluid.

Intrafusal muscle fibers   Specialized muscle fibers that are contractile only at the poles and are found in muscle spindles; innervated by sensory neurons that detect stretch.

Intramembranous ossification   (in'-trah-MEM-brah-nus) The process by which certain flat bones form from a mesenchymal model.

Intrapleural pressure   The pressure within the pleural cavities.

Intrapulmonary pressure   The pressure exerted by the air within the lungs.

Inversion   (in-VUR-zhuhn) A rotational movement of the foot in which the plantar surface rotates medially toward the midline of the body.

Ion   (AYE-awn) A charged particle that has lost or gained one or more electrons.

Ionic bond   (aye-AWN-ik) A chemical bond resulting from the transfer of electrons and resulting attraction between a positive metal cation and negative nonmetal anion.

Iris   A pigmented ring of smooth muscle cells that surrounds the pupil and controls the amount of light that enters the eye.

Irregular bone   A bone that has an irregular shape and that does not fit into any other category.

Ischium   (ISS-kee-um) The posteroinferior bone of the pelvis.

Isomers   (AYE-soh-merz) Two or more compounds with the same molecular formula but different three-dimensional arrangement of their atoms.

Isometric contraction   (aye-soh-MET-rik) A type of muscle contraction in which the tension generated is equal to that of the external load, and so the muscle cell remains at a constant length.

Isotonic concentric contraction   A type of muscle contraction in which the tension generated is greater than that of the external load, and so the muscle cell shortens with the contraction.

Isotonic eccentric contraction   A type of muscle contraction in which the tension generated is less than that of the external load, and so the muscle cell lengthens with the contraction.

Isotonic solution   A solution that has the same ability to cause water movement by osmosis as the reference solution due to having the same number of permeating solute particles.

Isotope   (AYE-soh-tohp) An element that has the same atomic number as another element but a different mass number (due to a different number of neutrons).

Isovolumetric contraction   (aye'-soh-vol-yoo-MET-rik) The period of the cardiac cycle during which the ventricles are contracting but their volume does not change because the pressure is not high enough to open the semilunar valves.

Isovolumetric relaxation   The period of the cardiac cycle during which the ventricles are relaxing but their volume does not change because the pressure is not high enough to open the atrioventricular valves.

Jejunum   (jeh-JOO-num) The second segment of the small intestine, measuring about 2.5 meters in length.

Joint   The location where two or more bones come together; also known as an *articulation*.

Juxtaglomerular apparatus (JGA)   The location where the macula densa contacts the juxtaglomerular cells of the afferent arteriole; monitors and maintains the glomerular filtration rate.

Juxtamedullary nephrons   Nephrons with long nephron loops that dip deeply into the renal medulla; play a role in the production of concentrated urine.

Keratin   (KEHR-uh-tin) A hard, fibrous protein that increases a tissue's resistance to mechanical stresses.

Keratinocyte   (kehr'-uh-TIN-oh-sy't) The main cell type found in the epidermis.

Ketone bodies   (KEE-tohn) Four-carbon compounds formed during lipid catabolism that a cell can catabolize to form two molecules of acetyl-CoA.

Kidneys   Paired retroperitoneal organs that filter the blood to remove wastes and maintain fluid, electrolyte, and acid-base homeostasis.

Kinetic energy   Energy in motion or action.

Labia   (LAY-bee-uh) Two sets of fatty protective skin folds that are posterior to the mons pubis; includes the outer labia majora and smaller inner labia minora.

Lacrimal apparatus   (LAK-ruh-mul) A collection of structures involved in the production and drainage of tears.

Lactation   (lak-TAY-shun) The production and release of breast milk from the mammary glands.

Lacteal   (lak-TEE-ul) A specialized lymphatic capillary in an intestinal villus that receives absorbed fats in addition to lymph.

Lamellated corpuscle   A sensory receptor located in the reticular layer of the dermis that detects pressure and vibration stimuli.

Large intestine   A tubular organ that is the last part of the alimentary canal; responsible for the absorption of substances such as water and electrolytes and the propulsion of feces out of the body.

Laryngopharynx   (lah-ring'-goh-FEHR-inks) The portion of the pharynx that is posterior to the larynx; serves as a passageway for food, liquids, and air.

Larynx   (LEHR-inks) The segment of the respiratory tract in the anterior neck that houses the vocal cords.

Lateral   A position that is farther away from the midline of the body or a body part.

Lateral horn   The lateral region of gray matter in the thoracic and lumbar spinal cord that contains the cell bodies of autonomic neurons.

Lateral malleolus   The distal end of the fibula; lateral ankle bone.

Lateral ventricles   Hollow, cerebrospinal fluid–filled cavities located within the two cerebral hemispheres.

Leak channel   An ion channel that is always open to allow specific ions to cross the plasma membrane with their concentration gradients.

Leg   The distal portion of the lower limb that consists of the tibia and the fibula.

Length-tension relationship   The relationship between the length of the sarcomeres of a muscle fiber while at rest and the amount of tension that can be generated by a contraction.

Lens   A transparent structure posterior to the pupil of the eye that refracts light rays to focus them on the retina.

Leptin   A hormone produced by adipocytes that promotes satiety (feelings of fullness after eating).

Lesser pelvis   The portion of the pelvis inferior to the pelvic brim.

Leukocyte (white blood cell, or WBC)   (LEWK-oh-sy't) A class of formed elements of the blood that perform functions relating to the immune system.

Leukopoiesis   (loo'-koh-poy-EE-sis)   The process of differentiation and maturation of leukocytes.

Levator   (LEE-vay-tohr) A muscle whose function is to elevate a body part.

Ligament   A strand of dense regular collagenous connective tissue that connects one bone to another to strengthen and reinforce an articulation.

Ligand   (LY-gand) A chemical that binds to a receptor in the plasma membrane of a cell.

Ligand-gated channel   An ion channel in the plasma membrane of a cell that opens or closes in response to the binding of a specific ligand to a receptor or to the channel.

Limbic system   A functional brain system that participates in learning, memory, emotion, and behavior; consists of the cingulate gyri, the parahippocampal gyri, the hippocampi, and the amygdalae.

Linea alba   (LINN-ee-uh AHL-bah) The midline connective tissue seam that separates the right and left rectus abdominis muscles.

Lipid (LIP-id) A nonpolar organic compound that contains primarily carbon and hydrogen atoms; examples include fats and oils.

Lipogenesis (ly′-poh-JEN-eh-sis) The process by which fatty acid and triglyceride synthesis occurs.

Liver The large organ in the right upper quadrant of the abdominopelvic cavity; functions include diverse metabolic activities, filtering blood from most abdominal organs, and bile production.

Local potential A small change in the membrane potential in a specific region of a cell's plasma membrane.

Long bone A bone with a shape that is longer than it is wide.

Long-term potentiation (LTP) An increase in synaptic activity between associated neurons; the mechanism by which hippocampal neurons encode long-term declarative memories.

Longitudinal fissure The deep groove that separates the two cerebral hemispheres.

Longitudinal growth The process by which long bones grow in length.

Loose connective tissue A type of connective tissue proper in which ground substance is the dominant element; also known as *areolar connective tissue*.

Low-density lipoproteins (LDLs) Cholesterol-rich particles that deliver cholesterol to cells in peripheral tissues.

Lower limb The portion of the appendicular skeleton that consists of the thigh, leg, ankle, and foot.

Lower motor neuron Motor neurons of the PNS that directly innervate skeletal muscles.

Lumbar plexus A group of nerves consisting of the anterior rami from L1 to L4 that innervates structures within the pelvis and lower limb.

Lumbar vertebrae Five large, blocky bones of the vertebral column located inferior to the thoracic vertebrae and superior to the sacrum.

Lumen The space enclosed by a hollow organ.

Lungs Paired spongy organs of the respiratory system that consist of alveoli and surrounding elastic connective tissue.

Luteinizing hormone (LH) Hormone produced by the anterior pituitary gland that stimulates reproductive functions in males and females.

Lymph (LIMF) The watery liquid located within lymphatic vessels; composition is similar to that of interstitial fluid but with fewer proteins.

Lymph nodes Clusters of lymphatic tissue located along lymphatic vessels throughout the body; they filter lymph.

Lymphatic system (lim-FAEH-tik) A group of organs and tissues with functions including immunity and fluid homeostasis.

Lymphatic vessels Blind-ended vessels that transport lymph from the interstitial fluid to the blood.

Lymphocyte (LIM-foh-sy′t) A class of agranulocyte that has diverse functions in the immune system; includes B and T lymphocytes.

Lysosome (LY-soh-sohm) A spherical membrane-enclosed organelle that contains digestive enzymes.

Macronutrients Nutrients required in relatively large amounts in the diet; include carbohydrates, proteins, and lipids.

Macrophage (MAK-roh-fehj) A phagocyte that ingests damaged cells, bacteria, and cellular debris.

Macula (MAK-yoo-luh) A cluster of receptor cells (hair cells) in the lining of the utricle and saccule responsible for detection of head tilting and linear movement.

Macula densa (MAK-yoo-luh DEN-suh) A group of cells in the segment of the renal tubule between the ascending limb of the nephron loop and the distal tubule whose function is to monitor flow through the nephron.

Major histocompatibility complex (MHC) molecules (hiss′-toh-kom-pat′-ih-BIL-ih-tee) Glycoproteins found on the surface of cells that bind and display antigen fragments.

Mammary gland Exocrine gland contained within a rounded skin-covered breast; consists of lobules with glandular alveoli that produce milk.

Manubrium (muh-NOO-bree-um) The superior portion of the sternum.

Mass number The number of protons plus the number of neutrons found in an atom.

Mast cell A tissue-bound immune cell found in many types of connective tissue that secretes inflammatory mediators.

Mastication The process of chewing during which food is physically broken into smaller particles by the teeth, the tongue, and the hard palate.

Matrix The group of cells at the base of a hair that are actively dividing; the area from which a hair grows.

Matter The material of our universe; anything that has mass and occupies space.

Mean arterial pressure (MAP) The average pressure in the systemic circuit during a complete cardiac cycle.

Mechanically gated ion channel An ion channel that opens or closes in response to mechanical deformation.

Mechanoreceptor (mek′-ah-noh-ree-SEP-ter) A sensory receptor that responds to a mechanical deformation of the cell.

Medial A position that is closer to the midline of the body or a body part.

Medial malleolus (mal-lee-OH-lus; plural, *malleoli*) The distal end of the tibia; medial ankle bone.

Mediastinum (meh′-dee-ass-TY-num) The portion of the thoracic cavity located between the lungs.

Medulla oblongata The final portion of the brainstem.

Medullary cavity (MED-yoo-lehr-ee) The largely hollow interior portion of the diaphysis of a long bone.

Megakaryocyte Cell located in the bone marrow from which platelets are derived.

Meiosis (my-OH-sis) A type of cell division that produces gametes.

Melanin (MEL-uh-nin) A pigment produced by melanocytes in the epidermis that ranges in color from brown-black to orange-red; responsible for much of the pigmentation of the skin, hair, and irises of the eyes.

Melanocyte A cell located in the epidermis of the skin that produces the pigment melanin.

Melatonin (mel-uh-TOH-nin) A hormone produced by the pineal gland that regulates the sleep/wake cycle.

Membrane A sheet of one or more tissues that lines a body surface or body cavity.

Membrane potential The difference in voltage between the extracellular fluid and the cytosol in the area near the plasma membrane.

Membranous urethra The short middle portion of the male urethra.

Menopause (MEN-uh-pawz) The cessation of regular menstrual cycles that normally occurs between the ages of 45 and 55 years.

Menstruation The discharge of blood and other materials from the vagina that occurs when a secondary oocyte is not fertilized.

Merkel cell A sensory receptor located in the stratum basale of the epidermis.

Merocrine secretion (MEHR-oh-krin) A type of exocrine secretion in which the product is secreted by exocytosis.

Mesentery Double folds of visceral peritoneum located around certain abdominal organs, such as the small and large intestines.

Mesoderm The middle germ layer of an embryo.

Messenger RNA (mRNA) A strand of RNA that contains the transcript for a single gene.

Metabolic acidosis A condition characterized by a decrease in the pH of the body's fluids; due to an accumulation of acids from sources other than carbon dioxide (such as ketone bodies) or a loss of bicarbonate ions.

Metabolic alkalosis A condition characterized by an increase in the pH of the body's fluids; due to a loss of metabolic acids (such as stomach acid) or an accumulation of bicarbonate ions or other bases.

Metabolic water Water produced as a byproduct of metabolic reactions.

Metabolism The sum of the body's chemical reactions.

Metacarpals The long bones of the hand.

Metaphase The second phase of mitosis during which sister chromatids line up along the equator of the cell.

Metarterioles The smallest arterioles that directly feed capillaries in many tissues.

Metatarsals   The long bones of the foot.

Micelle   (my-SEL) A small vesicle composed of an outer layer of bile salts surrounding an inner core of chemically digested lipids.

Microcirculation   The total blood flow that takes place within the capillary beds.

Microglial cell   A neuroglial cell of the central nervous system that acts as a phagocyte.

Micronutrients   Nutrients required in relatively small amounts in the diet; include vitamins and minerals.

Microscopic anatomy   The study of structures of the human body that require the use of a microscope for observation.

Microtubules   Hollow rods of the cytoskeleton composed of tubulin subunits that help a cell maintain its size and shape, position organelles, and play a role in motion within the cell.

Microvilli   Highly folded extensions of the plasma membrane that increase its surface area for absorption.

Micturition   (mik-choo-RISH-un) The process of voiding urine from the urinary bladder through the urethra.

Midbrain   The first portion of the brainstem.

Midsagittal plane   A plane of section that divides the body or body part into equal right and left portions; also known as the *median plane*.

Mineral   Any element other than carbon, hydrogen, oxygen, and nitrogen that is required by living organisms.

Mineralocorticoids   (min′-er-ah-loh-KOHR-tih-koydz) A class of hormones produced by the adrenal cortex that regulates fluid, electrolyte, and acid-base balance.

Mitochondria   (my′-toh-KAWN-dree-uh; singular, *mitochondrion*) Organelles surrounded by a double membrane that produce the bulk of the cell's ATP by oxidative catabolism.

Mitosis   (my-TOH-sis) The division of somatic cells that results in two identical daughter cells.

Mitotic spindle   A structure that emanates from centrioles and consists of microtubules that attach to sister chromatids at the centromere; pulls chromatids to opposite poles of the cell during mitosis.

Mitral valve   (MY-trul) Atrioventricular valve located between the left atrium and left ventricle; also known as the bicuspid valve.

Mixture   Two or more substances that are physically combined such that their chemical properties remain unaltered.

Molecule   (MAWL-eh-kyul) Two or more atoms joined by a chemical bond.

Monocyte   (MAH-noh-sy′t) A class of agranulocyte that matures into macrophages, very active phagocytes.

Monomer   (MAHN-oh-mer) A single molecular subunit that can be combined with another monomer to build a larger compound.

Monosaccharide   (mahn′-oh-SAK-uh-ry′d) A carbohydrate monomer.

Monosynaptic reflex   A reflex arc consisting of a single synapse between a sensory and motor neuron within the spinal cord or brainstem.

Morula   (MOHR-yoo-luh) The early developing conceptus composed of 16 or more cells.

Motility   The movement of a cell or tissue.

Motor division (efferent division)   The division of the peripheral nervous system that relays motor information from the central nervous system to muscle and glandular tissues.

Motor end plate   The specialized region of the skeletal muscle fiber plasma membrane that contains receptors for acetylcholine.

Motor learning   The correction of motor error by the cerebellum.

Motor neuron (efferent neuron)   A neuron that transmits motor impulses from the central nervous system to a muscle or gland cell.

Motor program   The group of neurons selected to perform a given action.

Motor unit   The group of muscle fibers innervated by a single motor neuron.

Mucosa-associated lymphatic tissue (MALT)   Loosely organized clusters of lymphoid tissue located along mucous membranes.

Mucous membrane   A sheet of epithelium and the underlying basement membrane that produces mucus; lines internal hollow organs that open to the outside.

Mucus   A thick, sticky secretion from goblet cells and other mucous membranes; protects underlying epithelial cells and traps foreign debris.

Multiaxial joints (triaxial joints)   (muhl′-tee-AK-see-uhl; try-AK-see-uhl) A functional classification of synovial joints in which motion between articulating bones occurs around three axes.

Multipolar neuron   A neuron with one axon and two or more dendrites.

Multi-unit smooth muscle   Smooth muscle cells that are able to contract individually.

Murmur   An abnormal sound heard in the heart caused by the turbulent flow of blood through a valve or other heart opening.

Muscle fiber   An alternate name for a skeletal muscle cell.

Muscle relaxation   The return of a muscle cell to its resting length due to the decreasing concentration of calcium ions in the cytosol.

Muscle spindle   A specialized bundle of intrafusal muscle fibers innervated by sensory neurons that detect stretch.

Muscle tone   The small amount of tension produced by a muscle at rest due to the involuntary activation of motor units by the brain and spinal cord.

Muscle twitch   A single cycle of contraction and relaxation of a muscle fiber generated by a single action potential.

Muscular arteries   Medium-sized arteries that control blood flow to organs; have a thick tunica media with a large amount of smooth muscle.

Muscular fatigue   An inability to maintain a given level of intensity of a particular exercise.

Muscular system   The body system that consists of the skeletal muscles and the associated connective tissues.

Mutation   An alteration in the DNA.

Myelin   (MY-eh-lin) A fatty substance that envelops and insulates the axons of certain neurons, increasing the speed of action potential conduction; formed from the plasma membranes of oligodendrocytes and Schwann cells.

Myelination   (my′-eh-lin-AY-shun) The process of formation of the myelin sheath by Schwann cells or oligodendrocytes.

Myocardium   (my′-oh-KAR-dee-um) The layer of the heart wall that consists of cardiac muscle cells and the heart's fibrous skeleton.

Myofibrils   (my-oh-FY-brilz) Long, cylindrical organelles composed of muscle proteins in a muscle fiber.

Myofilaments   Muscle proteins that make up a myofibril in a muscle fiber.

Myogenic mechanism   (my-oh-JEN-ik) An autoregulatory mechanism in blood vessels by which the degree of stretch of the vessel wall triggers a reflex that maintains blood flow to a tissue.

Myoglobin   (MY-oh-glohb-in) An oxygen-binding protein in muscle cells that increases the amount of oxygen immediately available to the cell.

Myometrium   The thick middle layer of the uterus composed of bundles of smooth muscle.

Myopia   (my-OH-pee-uh) A condition in which the eye is too long or the cornea is too curved and the lens cannot flatten enough for far vision.

Myosin   (MY-oh-sin) A club-shaped contractile protein found in muscle fibers and cells that are motile.

Nail   A hard structure located at the distal end of a digit that consists of dead, keratinized epithelial cells.

Nail matrix   The proximal portion of a nail that consists of living, dividing cells from which a nail grows.

Nail plate   The portion of a nail that rests on the superficial epidermal nail bed.

Nasal cavity   The two-sided cavity within the anterior skull that houses the sensory receptors for olfaction and serves as the first portion of the respiratory tract.

Nasal conchae   (KAWN-kee; singular, *concha*) Ridges of bone projecting from the lateral walls of the nasal cavity that make the air flow within the nasal cavity turbulent.

Nasal septum   The bony and cartilaginous structure that separates the two sides of the nasal cavity.

Nasopharynx   (nayz-oh-FEHR-inks) The portion of the pharynx that is posterior to the nasal cavity; normally serves as a passageway for air only.

Negative feedback loop The response in which the change in a regulated variable in one direction results in actions that cause changes in the variable in the opposite direction.

Neocortex The most recent part of our brains to evolve; makes up the majority of the cerebral cortex.

Neonatal period (nee-oh-NAYT-uhl) The first 4 weeks after birth.

Nephron (NEF-rahn) The functional unit of the kidney where blood is filtered and the resulting filtrate is modified.

Nephron loop The second segment of the renal tubule in which water and electrolytes are reabsorbed; consists of a descending limb and an ascending limb.

Nerves Organs of the peripheral nervous system that consist of bundles of axons, connective tissue sheaths, and blood vessels.

Nervous system The body system consisting of the brain, spinal cord, and nerves; one of the chief homeostatic systems in the body.

Net filtration pressure (NFP) The difference between the opposing forces of the hydrostatic pressure gradient and colloid osmotic pressure.

Net filtration pressure of the glomerulus (NFP) The overall filtration pressure in the glomerulus that consists of the sum of the pressure that favors filtration (glomerular hydrostatic pressure) and those that oppose filtration (glomerular colloid osmotic pressure and capsular hydrostatic pressure).

Neural circuits Specific patterns of connections between neuronal pools.

Neural integration The process by which a neuron integrates all of the postsynaptic potentials from multiple presynaptic neurons.

Neural tube A structure formed by the fourth week of development from which nervous tissue, the brain, and the spinal cord arise.

Neuroendocrine organ An organ that consists of nervous tissue but also secretes hormones.

Neuroglial cell (noor-oh-GLEE-uhl) A supporting cell of nervous tissue.

Neuromuscular junction (NMJ) The location where a neuron communicates with a muscle fiber.

Neuron (NOOR-ahn) A nerve cell capable of sending and receiving messages in the form of local and action potentials.

Neuronal pools Groups of interneurons in the central nervous system that process specific types of stimuli.

Neuropeptides A group of peptide neurotransmitters produced by the cell body of a neuron.

Neurotransmitters Chemical messengers produced by neurons that communicate with target cells.

Neutral A solution with a pH of exactly 7.0.

Neutral fat (triglyceride) (try-GLISS-er-iyd) A lipid consisting of three fatty acids bound to glycerol.

Neutron (NOO-tron) A subatomic particle with no charge that resides in the atomic nucleus.

Neutrophil (noo-TROH-fill) A class of granulocyte and very active phagocyte that ingests damaged cells and bacteria.

Nicotinamide adenine dinucleotide (NAD$^+$) One of the main electron carriers involved in nutrient catabolism; when reduced, it becomes NADH.

Nociception (NOH-sih-sep-shun) The detection and perception of painful stimuli.

Nociceptor (NOH-sih-sep-ter) A sensory receptor that responds to painful stimuli.

Node of Ranvier (myelin sheath gap) (rahn-vee-AY) The unmyelinated segment of an axon between two internodes.

Nonaxial joints (nahn-AX-ee-uhl) A functional classification of synovial joints in which motion between articulating bones occurs in one or more planes but does not move around an axis.

Nondeclarative memory The memory of things that are largely procedural or skill-based.

Nondiscriminative touch The senses such as deep pressure that lack the fine spatial resolution of the tactile senses.

Nonpolar covalent bond A covalent bond in which electrons are shared equally among the atoms in the bond.

Non–rapid eye movement sleep (non-REM sleep) The name for stages I–IV of the sleep cycle.

Norepinephrine A catecholamine that functions largely as a neurotransmitter released by neurons of the sympathetic nervous system and that is also released in small amounts by the adrenal medulla.

Nostrils The paired anterior openings to the nasal cavity.

Nuclear envelope The double-layered phospholipid membrane surrounding the nucleus.

Nucleolus (noo-klee-OH-luss) The structure located in the nucleus where ribosomes are assembled.

Nucleotide (NOO-klee-oh-ty'd) An organic compound consisting of a nucleic acid base, a monosaccharide, and one or more phosphate groups.

Nucleus (cellular) The biosynthetic center of the cell that contains the cell's DNA and much of its RNA.

Nucleus (in CNS) A cluster of neuronal cell bodies in the central nervous system.

Nucleus pulposus (NOO-klee-us pull-POH-sis) The inner, gelatinous, shock-absorbing portion of an intervertebral disc.

Nutrient A substance obtained from food that the body requires for its metabolism.

Obligatory water loss The amount of urine that must be produced daily irrespective of fluid intake to excrete wastes and other solutes.

Obligatory water reabsorption Water reabsorption in the kidneys that occurs without the regulation of hormones and irrespective of the medullary concentration gradient.

Obstructive lung disease A type of pulmonary disease in which the airway resistance increases.

Obturator foramen (AHB-too-ray-tohr fohr-AY-mun) The large hole in the anterior pelvis formed by the ilium and pubis.

Occipital lobes The posterior lobes of the cerebrum; process all information relating to vision.

Octet rule The principle that states that an atom is most stable with eight electrons in its valence shell.

Olecranon (oh-LEK-ruh-nahn) The posterior, proximal projection from the ulna; commonly known as the "elbow."

Olfaction (ohl-FAK-shun) The sense of smell.

Olfactory mucosa A small patch of specialized epithelium on the superior surface of the nasal cavity that contains olfactory receptor cells.

Oligodendrocyte (oh-lig'-oh-DEN-droh-syt) A neuroglial cell of the central nervous system that myelinates certain axons.

Omentum (oh-MEN-tum) A specialized mesentery composed of four layers of folded visceral peritoneum that covers the digestive organs in the anterior abdominal cavity.

Oogenesis (oh'-oh-JEN-uh-sis) The process of female gamete, or ova, production in females.

Opioids (OH-pee-oydz) A group of neuropeptides that block pain perception and have various effects on behavior.

Opposition The special movement that occurs only at the first carpometacarpal joint, in which the thumb travels across the palmar surface of the hand.

Optic chiasma (KYE-az-muh) An X-shaped structure formed by the meeting of the optic nerves at the midline; location where some of the axons of the optic nerve cross to the other side of the brain.

Oral cavity The space enclosed by the teeth anteriorly and laterally, the palate superiorly, the tongue inferiorly, and the oropharynx posteriorly; the first portion of the alimentary canal.

Orbit The cavity in the skull that houses the eyeball and its associated structures.

Organ Two or more tissues combined to produce a structure that has a recognizable shape and that performs a specialized task.

Organ system Two or more organs that work together to carry out a broad function in the body.

Organelles (ohr-guh-NELZ) Structures inside the cell that carry out specific, compartmentalized functions.

Organic compounds Compounds that contain carbon atoms bonded to hydrogen atoms.

Orgasm The time period during which feelings of pleasure are experienced during sexual intercourse; in males, this coincides with ejaculation.

Origin (muscle) The less moveable attachment point of a muscle on a bone.

Oropharynx (ohr-oh-FEHR-inks) The portion of the pharynx that is posterior to the oral cavity; serves as a passageway for food, liquids, and air.

Osmoreceptors Neurons in the thirst center of the hypothalamus that detect the osmolarity of the extracellular fluid.

Osmosis (oz-MOH-sis) A type of passive transport in which solvent, usually water, moves through a selectively permeable membrane from an area of lower solute concentration to one of higher solute concentration.

Osmotic pressure (OP) The force that would need to be applied to a solution to stop water from moving into it by osmosis.

Osseous tissue (AWSS-ee-us) The hard, dense tissue that forms the majority of a bone.

Osteoblast (AWSS-tee-oh-blast) An immature bone cell that secretes osteoid and aids in the deposition of bone's inorganic matrix.

Osteoclast (AWSS-tee-oh-klast) A bone-destroying cell that secretes chemicals to dissolve the organic and inorganic matrices of bone tissue.

Osteocyte (AWSS-tee-oh-sy't) A mature bone cell that monitors and maintains the surrounding bone matrix.

Osteoid (AWSS-tee-oyd) The organic component of the extracellular matrix of osseous tissue; consists of collagen fibers, proteoglycans, glycosaminoglycans, glycoproteins, and bone-specific proteins.

Osteon (AWSS-tee-on) The functional unit of compact bone; consists of concentric rings of bone matrix called lamellae that surround a central canal.

Ova (singular, ovum) Female gametes.

Ovarian cycle The monthly series of events associated with the maturation of an oocyte and its follicle in an ovary.

Ovarian follicle A small structure inside the ovarian cortex that contains a developing oocyte.

Ovaries The female gonads; the site of ova production and hormone secretion.

Ovulation (ah-vyoo-LAY-shun) The process by which the ovary expels a secondary oocyte.

Oxidation-reduction reaction A type of exchange reaction in which electrons are exchanged among reactants.

Oxidative catabolism (aerobic catabolism) A series of reactions that occur in the mitochondria in the presence of oxygen during which electrons are removed from carbon-based compounds and the energy released is used to fuel the synthesis of ATP.

Oxidative phosphorylation The formation of ATP by harnessing the energy from the flow of electrons to create a hydrogen ion gradient in mitochondria.

Oxygen-hemoglobin dissociation curve A graphic representation of the relationship between the partial pressure of oxygen in the blood and the degree of saturation of hemoglobin.

P wave The first wave on an ECG; represents the depolarization of the atria.

Pacemaker cell A cell that depolarizes spontaneously and triggers action potentials in neighboring cells.

Pacemaker potential An action potential spontaneously generated by a pacemaker cell in cardiac or smooth muscle.

Palate (PAL-it) The roof of the oral cavity; consists of the bony hard palate anteriorly and the muscular soft palate posteriorly.

Pancreas An endocrine and exocrine organ located inferior and posterior to the stomach; secretes enzymes and other products for digestion and the hormones insulin, glucagon, and somatostatin.

Pancreatic islet A cluster of endocrine cells in the pancreas that produces and secretes the hormones insulin, glucagon, and somatostatin.

Pancreatic juice A liquid consisting of water, bicarbonate ions, and enzymes produced by pancreatic acinar cells and released into the small intestine during digestion.

Papillae (pah-PILL-ee) Finger-like protrusions on the surface of the tongue, some of which contain taste buds.

Papillary layer (PAP-ih-lehr-ee) The superficial layer of the dermis composed of loose connective tissue.

Papillary muscles Projections of muscle from the right and left ventricular walls that attach to the atrioventricular valves via chordae tendineae.

Paracrine (PAR-uh-krin) A chemical messenger secreted into the extracellular fluid to influence nearby target cells.

Paranasal sinuses A group of hollow cavities that are continuous with the nasal cavity; located within the frontal, maxillary, sphenoid, and ethmoid bones.

Parasagittal plane (pehr'-ah-SAJ-ih-tul) A plane of section that divides the body or body part into unequal right and left portions.

Parasympathetic nervous system "Rest and digest" division of the ANS that promotes maintenance functions such as defecation, diuresis, and digestion.

Parathyroid gland A group of three to five small endocrine glands located on the posterior thyroid gland; produces parathyroid hormone.

Parathyroid hormone (PTH) A hormone produced by the parathyroid gland that increases the concentration of calcium ions in the extracellular fluid.

Parietal lobes (pah-RY-eh-tal) Lobes of the cerebral hemispheres located posterior to the frontal lobes; responsible for processing and integrating sensory information, and also function in attention.

Parotid glands (puh-RAWT-id) Large salivary glands located over the lateral mandible that secrete watery saliva.

Parturition (childbirth) The end of pregnancy when the fetus is expelled from the uterus through the vagina.

Passive immunity A temporary type of immunity mediated by preformed antibodies that are passed from one organism to another.

Passive transport A type of membrane transport with a gradient that does not require energy expenditure in the form of ATP.

Patella The sesamoid bone located within the tendon of the quadriceps femoris muscle.

Pectinate muscles (PEK-tin-et) Muscular ridges on the interior surface of the anterior atria, particularly the right atrium.

Pectoral girdle Bones that support the upper limb and hold it in its proper position; consists of the clavicle and scapula.

Pelvic cavity (PEL-vik) The portion of the abdominopelvic cavity located within the bony pelvis.

Pelvic girdle The bony structure that attaches the lower limbs to the axial skeleton and that supports the weight of the upper body.

Pelvic inlet The superior opening of the bony pelvis.

Pelvic outlet The opening located at the inferior boundary of the lesser pelvis.

Pelvis The bowl-shaped bony structure formed by the two coxal bones and the sacrum.

Pelvic bones The two bones that, together with the sacrum, form the pelvis; consist of three fused bones, the ilium, ischium, and pubis.

Penis Male copulatory organ that delivers sperm into the female reproductive tract.

Peptide A polymer consisting of two or more amino acids joined by a peptide bond via dehydration synthesis.

Peptide bond A covalent bond between the amino group of one amino acid and the carboxyl group of a second amino acid.

Pericardial cavity The very narrow potential space between the parietal and visceral pericardial membranes; contains a thin layer of serous fluid.

Pericardium (pehr'-ih-KAR-dee-um) The outer layer of the heart wall; consists of the fibrous pericardium and the serous pericardium.

Perineum (payr-uh-NEE-um) The diamond-shaped area on the pelvic floor bordered by the pubic symphysis anteriorly, the ischial tuberosities laterally, and the coccyx posteriorly.

Perineurium (pehr′-ih-NOOR-ee-um) The connective tissue that surrounds a fascicle of a peripheral nerve.

Periosteum (pehr′-ee-AWSS-tee-um) A membrane surrounding bones; composed of dense irregular connective tissue with osteogenic cells and osteoblasts.

Peripheral nervous system (PNS) The division of the nervous system consisting of the cranial and spinal nerves.

Peripheral protein A protein associated with either the intracellular or extracellular face of the plasma membrane.

Peripheral resistance Any impedance to blood flow found in the periphery of the vasculature.

Peristalsis (pehr-uh-STAL-sis) Rhythmic contractions of layers of smooth muscle that move material through a hollow organ.

Peritoneal cavity The narrow space between the visceral and parietal peritoneal membranes that is filled with serous fluid.

Peritoneum A set of double serous membranes around several abdominal organs; consists of the inner visceral peritoneum and the outer parietal peritoneum.

Peritubular capillaries Capillary beds that are fed by the efferent arterioles and surround the renal tubule.

Peroxisome (puh-RAWKS-ih-sohm) An organelle with enzymes that detoxify certain substances, metabolize fatty acids, and produce certain phospholipids.

pH The negative logarithm of the hydrogen ion concentration of a solution.

Phagocyte (FAYG-oh-syt) A cell that ingests foreign cells, dead cells, and other cellular debris by phagocytosis.

Phagocytosis (fayg′-oh-sy-TOH-sis) "Cellular eating"; a type of endocytosis that involves the ingestion of large particles.

Phalanx (FAY-langks) A long bone of the fingers or toes.

Pharyngotympanic tube (fah-ring′-goh-tim-PAN-ik) The passageway that connects the middle ear and the nasopharynx.

Pharynx (FEHR-inks) The region of the respiratory tract that is located posterior to the nasal cavity, oral cavity, and larynx.

Phenotype (FEE-noh-ty′p) The way an individual's genotype is expressed physically.

Phosphate buffer system A buffer that consists of the weak acid dihydrogen phosphate ($H_2PO_4^-$) and its conjugate base, hydrogen phosphate (HPO_4^{2-}).

Phospholipid (FAWSS-foh-lip-id) A lipid consisting of two fatty acids and a phosphate group bound to glycerol.

Phospholipid bilayer A double layer of phospholipids in which the fatty acid tails face one another and the phosphate heads face the surrounding aqueous environment; forms the plasma membrane and the membranes around organelles.

Photoreceptor A sensory receptor that responds to a change in light stimuli.

Phrenic nerve (FREN-ik) The main nerve of the cervical plexus that innervates the diaphragm, the main muscle of breathing; consists of the anterior rami from C_3 to C_5.

Physiological buffer systems Actions taken by the respiratory and urinary systems to excrete hydrogen ions and/or retain bicarbonate ions in order to maintain homeostasis of the pH of the body's fluids.

Pia mater (PEE-ah) The inner, delicate meninx that surrounds the brain and spinal cord.

Pineal gland (PIN-ee-al) The part of the diencephalon that functions as a neuroendocrine organ by secreting melatonin.

Pinocytosis (peen′-oh-sy-TOH-sis) "Cellular drinking"; a type of endocytosis that involves the ingestion of smaller amounts of material through indentations in the plasma membrane coated with specific proteins.

Pivot joint A uniaxial joint in which the rounded surface of one bone fits into a groove on the surface of another bone, allowing for rotational movement around one axis.

Placenta (plah-SEN-tuh) A temporary organ in the lining of the uterus through which oxygen, nutrients, and wastes are exchanged between the mother and conceptus.

Plane joint A synovial joint in which the flat surfaces of each articulating bone allow for nonaxial gliding motions.

Plantarflexion (plant-uhr-FLEK-shuhn) An angular movement of the foot at the ankle in which the angle between the foot and the leg increases.

Plasma The fluid extracellular matrix of blood; consists of water, proteins, and dissolved solutes.

Plasma cells Activated B cells that secrete antibodies.

Plasma membrane The selectively permeable barrier surrounding a cell that is composed of a bilayer of phospholipids and other components such as proteins and cholesterol.

Plasmin An enzyme that catalyzes the reaction that degrades fibrin and so removes a blood clot.

Plateau phase The portion of the cardiac action potential during which calcium ions enter the cardiac muscle cell as potassium ions exit the cell; lengthens and strengthens the resulting contraction of the cell.

Platelet (PLAYT-let) The cellular fragment of a megakaryocyte that participates in blood clotting.

Platelet plug A group of aggregated platelets that collect over an injured blood vessel to prevent blood loss.

Pleural cavity (PLOOR-ul) A thin potential space between the parietal and visceral pleural membranes surrounding the lungs; contains a thin layer of serous fluid.

Podocytes (POH-doh-sytz) Cells that make up the inner visceral layer of the glomerular capsule and surround the glomerular capillaries.

Polar body The smaller cell that is produced during meiosis of oogenesis; contains DNA but very little cytoplasm and generally does not divide further.

Polar covalent bond A covalent bond in which electrons spend more time around the more electronegative atom(s), which results in the formation of a dipole.

Pollex The anatomical name of the thumb.

Polymer (PAWL-ih-mer) A compound formed by joining two or more monomers chemically.

Polysaccharide (pawl′-ee-SAK-ah-ryd) A large carbohydrate consisting of multiple monosaccharides joined by dehydration synthesis.

Polysynaptic reflex A reflex arc involving multiple synapses between the PNS and CNS.

Pons The middle portion of the brainstem.

Positive feedback loop A type of feedback loop in which the effector's activity increases, reinforcing the initial stimulus and amplifying the response of the effector.

Postabsorptive state The period of time from about 4 hours after feeding until the next meal.

Postcentral gyrus A gyrus located posterior to the central sulcus of each cerebral hemisphere in the parietal lobes; contains the primary somatosensory cortex.

Posterior Toward the back; also known as *dorsal*.

Posterior body cavity The major body cavity located mostly on the posterior side of the body.

Posterior columns/medial lemniscal system A set of ascending tracts in the spinal cord that convey sensations relating to discriminative touch and proprioception; the posterior columns are located in the spinal cord, and they become the medial lemniscus after they decussate in the medulla.

Posterior cruciate ligament (PCL) A ligament found within the synovial cavity of the knee joint that, when the knee is flexed, tightens to prevent the tibia from being displaced posteriorly on the femur.

Posterior horn The posterior region of gray matter in the spinal cord that contains the cell bodies of second-order sensory neurons.

Posterior pituitary gland The posterior portion of the pituitary gland that is derived from nervous tissue; stores ADH and oxytocin produced by the hypothalamus.

Posterior ramus (RAY-muss; plural, *rami*) The division of a spinal nerve that travels to the posterior side of the body.

Posterior root   The extension from the posterior horn of the spinal cord containing the axons of somatic sensory neurons.

Posterior root ganglion   The swollen area of the posterior root where the cell bodies of somatic sensory neurons are housed.

Postganglionic neuron   An autonomic neuron whose cell body is located in an autonomic ganglion in the PNS.

Postsynaptic neuron   A neuron in a synapse that receives a message from a presynaptic neuron.

Postsynaptic potential   A positive or negative change in the membrane potential of a neuron as a result of synaptic transmission.

Potential energy   Energy that is stored, ready to be released and used to do work.

Power stroke   A pivoting motion of a myosin head in which it moves from its cocked position to a relaxed position, pulling actin with it as it relaxes.

Precentral gyrus   A gyrus located anterior to the central sulcus of each cerebral hemisphere in the frontal lobes; contains the primary motor cortex.

Pre-embryonic period   The period of human development from fertilization through implantation.

Prefrontal cortex   A multimodal association area in the frontal lobe responsible for many higher mental functions.

Preganglionic neuron   (PREE-gang-glee-awn´-ik) An autonomic neuron whose cell body is located in the CNS in the spinal cord or brainstem.

Pregnancy   The time period that encompasses the events from conception until birth.

Preload   The length or degree of stretch of the sarcomeres in the ventricular cells of the heart before they contract.

Presbyopia   (prez´-bee-OH-pee-uh) A condition in which the lens stiffens with age and cannot focus on near objects.

Pressure gradient   A condition in which there are two connected areas with different gas, hydrostatic, or osmotic pressures.

Presynaptic neuron   A neuron in a synapse that delivers a message to a target cell.

Primary active transport   A type of active transport in which ATP is expended to move a substance against its concentration gradient using a protein pump.

Primary motor cortices   Two motor areas in the frontal lobes of the cerebrum (in the precentral gyri) that house upper motor neurons involved in the conscious planning and initiation of movement.

Principle of complementarity of structure and function   A core principle of anatomy and physiology; states that the structure of a chemical, cell, tissue, or organ is always such that it best suits its function.

Principle of myoplasticity   The principle stating that the structure of a muscle will change in accordance with its functional use.

Projection fibers   White matter fibers that connect two areas of the cerebral cortex with one another, with other parts of the brain, and/or with the spinal cord.

Propagation   The spread of an action potential along the length of a cell's plasma membrane.

Prophase   The first phase of mitosis during which the chromatin condenses into chromosomes and the nuclear envelope disperses.

Proprioceptor   A sensory receptor that responds to a change in position of a body part, particularly a ligament or tendon.

Propulsion   The movement of a substance through a hollow organ via peristaltic contractions of smooth muscle layers.

Prostate   (PRAHS-tayt) An accessory gland in the male reproductive system that produces prostatic secretions—a component of semen.

Prostatic urethra   The initial segment of the male urethra that passes through the prostate gland.

Protein   (PROH-teen) An organic compound consisting of one or more polypeptide chains folded into a specific structure.

Protein buffer system   A buffer that consists of proteins whose amino acids contain carboxylic acid groups that ionize to a weak base capable of binding released hydrogen ions.

Protein C   An anticoagulant that catalyzes the reactions that degrade clotting factors Va and VIIIa.

Protein synthesis   The process of making a protein from DNA to RNA to assembled amino acids.

Proton   A positively charged subatomic particle located in the atomic nucleus.

Protraction   (proh-TRAK-shuhn) The movement of a joint in which the corresponding body part moves in an anterior direction.

Proximal   (PRAWKS-ih-mul) A position that is closer to the point of origin, usually the trunk.

Proximal tubule   The first segment of the renal tubule in which water, electrolytes, and organic nutrients are reabsorbed.

Pseudostratified columnar epithelium   (soo´-doh-STRAT-ih-fy'd) A single layer of columnar epithelial cells and the underlying basal lamina; the nuclei of the cells are uneven, giving the tissue the appearance of being stratified; found in locations such as the lining of the nasal cavity and much of the respiratory tract.

Pseudounipolar neuron   (soo´-doh-yoo-nih-POH-lur) A neuron with two axons—a peripheral process that brings input to the cell body and a central process that brings input to a target cell; formerly referred to as *unipolar neurons*.

Puberty   The period of life when reproductive organs mature and become functional.

Pubic symphysis   The pad of fibrocartilage between the two pubic bones.

Pubis   (PYOO-biss) The anteroinferior bone of the pelvis.

Pulmonary compliance   The distensibility of the lungs and chest wall, or the ability of the lungs and chest wall to stretch.

Pulmonary gas exchange   The exchange of oxygen and carbon dioxide across the respiratory membrane between the air in the alveoli and the blood in the pulmonary capillaries.

Pulmonary valve   A semilunar valve located between the right ventricle and the pulmonary trunk.

Pulmonary veins   Veins that transport oxygenated blood from the pulmonary circuit to the left atrium of the heart.

Pulmonary ventilation   The physical movement of air in and out of the lungs.

Pulse pressure   The difference between the systolic and diastolic pressures.

Punnett square   A simple diagram used to predict the possible offspring genotypes that might occur, given the genotypes of the parents.

Pupil   An opening in the iris through which light enters the eye.

Purkinje fiber system   (per-KIN-jee) A group of atypical pacemaker cells located in the ventricles.

Putamen   (poo-TAY-men) Two basal nuclei associated with the caudate nuclei; the two structures work together to inhibit the globus pallidus and initiate movement.

Pyloric sphincter   A ring of smooth muscle between the stomach and duodenum that controls the flow of chyme between the two organs.

QRS complex   The second wave complex on an ECG; represents depolarization of the ventricles.

Radiation   The transfer of heat from one object to another through electromagnetic waves.

Radius   The lateral bone of the forearm.

Range of motion   A measurement of the amount of movement that a particular joint is capable of under normal circumstances.

Rapid eye movement sleep (REM sleep)   The stage of the sleep cycle characterized by rapid, back-and-forth eye movements during which most dreaming occurs.

Receptor   A protein within a membrane that binds to a ligand.

Receptor-mediated endocytosis   A type of endocytosis in which specific substances are taken into the cell after binding with a receptor on the plasma membrane surface.

Receptor potential   A temporary depolarization caused by stimulation of a sensory receptor.

Recessive trait   A trait for which an allele is masked in the phenotype by a dominant allele.

Recoil The return of an elastic tissue to its original shape and size.

Recruitment An increase in the number of motor units of a skeletal muscle that are stimulated in order to produce a contraction with greater tension.

Rectum (REK-tum) The portion of the large intestine between the sigmoid colon and the anal canal; located along the sacrum.

Red bone marrow Hematopoietic tissue located within certain bones that produces all of the formed elements of the blood.

Referred pain Painful stimuli that are perceived as cutaneous pain along a specific dermatome.

Reflex A programmed, automatic motor response to a sensory stimulus.

Refraction Bending of light rays when they pass through a translucent object.

Refractory period The period during which an excitable cell either cannot respond to another stimulus (the absolute refractory period) or requires a stronger stimulus to respond (the relative refractory period).

Regeneration The process by which a damaged tissue is replaced with the same tissue during healing.

Regional anatomy The study of the human body taken from the perspective of specific body regions.

Renal arteries Branches of the descending abdominal aorta that supply the kidneys.

Renal clearance The removal of a solute from the body by the kidneys.

Renal compensation An adjustment of the secretion or retention of hydrogen and/or bicarbonate ions to compensate for an increase or decrease in the pH of the body's fluids.

Renal corpuscle A structure in the kidneys consisting of the glomerulus and the surrounding glomerular capsule.

Renal cortex The outer region of the kidneys that consists of blood vessels and most components of a kidney's nephrons.

Renal medulla The middle region of the kidneys that consists of renal pyramids and renal columns.

Renal pelvis The inner region of the kidneys that receives urine drained from the major and minor calyces.

Renal pyramid The triangular component of the renal medulla that consists of parallel bundles of tubules.

Renal veins Veins that drain the kidneys into the inferior vena cava.

Renin (REE-nin) An enzyme produced by JG cells in response to a decline in systemic blood pressure or sympathetic nervous system stimulation; converts angiotensinogen to angiotensin-I.

Renin-angiotensin-aldosterone system (RAAS) A system that results in the formation of angiotensin-II and aldosterone; the system raises systemic blood pressure and preserves blood flow to the kidneys.

Repolarization Movement of a cell's membrane potential back toward resting level after a depolarization has taken place.

Reposition The movement of the first carpometacarpal joint back to anatomical position from an opposed position.

Reproduction The production of new cells within an organism or the production of offspring.

Reproductive system An organ system concerned with the production of gametes, hormones, and offspring.

Respiration The group of processes that combine to provide the body's cells with oxygen and remove the waste product carbon dioxide.

Respiratory acidosis A condition characterized by a decrease in the pH of the body's fluids due to hypoventilation.

Respiratory alkalosis A condition characterized by an increase in the pH of the body's fluids due to hyperventilation.

Respiratory compensation An adjustment of the respiratory rate to compensate for an increase or decrease in the pH of the body's fluids.

Respiratory membrane A thin structure across which gases are exchanged, consisting of the alveolar epithelium, basal lamina, and the endothelial cells of pulmonary capillaries.

Respiratory mucosa The pseudostratified ciliated columnar epithelium that makes up much of the respiratory tract.

Respiratory pattern generator The collection of neurons in the ventral respiratory column of the reticular formation that generates the basic rhythm for breathing.

Respiratory system The group of organs consisting of the lungs and the respiratory tract that is responsible for gas exchange and other functions.

Respiratory tract A series of hollow organs, beginning with the nasal cavity and ending with the alveolar sacs, through which air is conducted and exchanged in the lungs.

Respiratory zone The portion of the respiratory tract that contains alveoli and so participates in gas exchange.

Responsiveness The property of living organisms by which they sense and react to changes in their environment.

Resting membrane potential The voltage difference across the plasma membrane of a cell when it is not being stimulated.

Restrictive lung disease A type of pulmonary disease in which the compliance of the lungs is decreased.

Reticular fibers Thin protein fibers in the extracellular matrix of a tissue that form "nets" within organs such as the spleen and lymph nodes and form supportive networks around blood vessels and nerves.

Reticular formation A group of more than 100 nuclei in the brainstem that are involved in a variety of functions, such as the maintenance of homeostasis and the sleep/wake cycle.

Reticular layer The deeper layer of the dermis that consists of dense irregular collagenous connective tissue.

Reticular tissue A type of connective tissue proper that contains reticular fibers; found in locations such as the spleen and lymph nodes.

Reticulocyte (reh-TIH-kyu-loh-sy't) An immature erythrocyte released into the circulation that still has a nucleus and some organelles.

Retina (RET-ih-nuh) A thin membrane lining the inside of the posterior eye that contains cells specialized to detect light as well as neurons that give rise to the optic nerve.

Retraction (ree-TRAK-shuhn) The movement of a joint in which the corresponding body part moves in a posterior direction.

Rh blood group A blood group that classifies blood type based on the presence or absence of the Rh antigen.

Rib cage Twelve pairs of ribs that form much of the thoracic cage via posterior attachments to the thoracic vertebrae and anterior attachment to the sternum or to costal cartilage (excepting ribs 11 and 12).

Ribonucleic acid (RNA) A single-stranded nucleic acid that is involved in the transcription and translation of the code in the DNA; also forms the bulk of ribosomes.

Ribosomal RNA (rRNA) A type of RNA produced in the nucleolus that serves as a main component of ribosomes.

Ribosomes (RY-boh-sohmz) Granular, non–membrane-enclosed organelles that are the site of protein synthesis.

Rods Photoreceptors that have a rod-shaped outer segment and that are sensitive to low-light conditions; responsible for low-acuity, black and white vision in dim light.

Rotation A nonangular pivoting motion in which one bone rotates or turns on a longitudinal axis that runs down the shaft of the bone.

Rotator cuff Supportive structure around the shoulder joint formed by the tendons of four muscles: subscapularis, supraspinatus, infraspinatus, and teres minor; provides the shoulder with most of its structural support, strength, and stability.

Rough endoplasmic reticulum (RER) A series of winding membranes whose surfaces are studded with ribosomes; functions in protein synthesis and modification.

Round ligaments Fibrous cords that connect the uterus to the anterior body wall.

Ruffini endings Slowly adapting receptors that respond to stretch and movement.

S1 The first heart sound auscultated during the cardiac cycle; caused by the closing of the atrioventricular valves.

S2   The second heart sound auscultated during the cardiac cycle; caused by the closing of the semilunar valves.

Sacral plexus   A group of nerves consisting of the anterior rami from L_4 to S_4 that mostly innervates structures of the pelvis and lower limb.

Sacrum   Five fused vertebrae that articulate laterally with the pelvic bones and form the posterior boundary of the pelvis.

Saddle joint   A biaxial joint in which the concave surface of one bone fits into the convex surface of the other bone, allowing for movement around two axes.

Sagittal plane   (SAJ-ih-tul) A plane of section that divides the body or body part into right and left portions.

Saliva   (suh-LY-vuh) A fluid secreted by salivary glands into the oral cavity that consists of water, mucus, salivary amylase, lysozyme, secretory IgA, and other solutes.

Salivary glands   (SAL-uh-vehr-ee) A set of three pairs of glands around the oral cavity that secrete saliva into it.

Salt   An inorganic compound consisting of a nonmetal anion and a metal cation joined by an ionic bond.

Saltatory conduction   A rapid type of neuronal action potential conduction that occurs in myelinated axons in which only the nodes of Ranvier are depolarized.

Sarcolemma   (sar-koh-LEM-uh) The plasma membrane of a muscle fiber.

Sarcomere   (SAR-koh-meer) The functional unit of muscle contraction; consists of the area of the myofibril from one Z-disc to the next Z-disc.

Sarcoplasm   (SAR-koh-plazm) The cytoplasm of a muscle fiber.

Sarcoplasmic reticulum (SR)   (sar-koh-PLAZ-mik reh-TIK-yoo-lum) The specialized smooth endoplasmic reticulum of a muscle fiber that stores calcium ions.

Satellite cell   A neuroglial cell that surrounds cell bodies of neurons in the peripheral nervous system.

Saturated fatty acid   A fatty acid with no double bonds between any of its carbon atoms.

Scapula   The posterolateral bone of the pectoral girdle that articulates with the humerus and the clavicle.

Scar tissue   Fibrous connective tissue that forms during fibrosis.

Schwann cell   A neuroglial cell of the peripheral nervous system that myelinates the axons of certain neurons.

Sclera   (SKLER-uh) The "white" part of the eyeball; thick collagenous structure that helps to maintain the shape of the eye.

Scrotum   (SKROH-tum) A saclike structure located outside the abdominopelvic cavity of the male that contains the paired testes.

Sebaceous glands   (suh-BAY-shus) Glands associated with hair follicles that secrete sebum via holocrine secretion.

Sebum   (SEE-bum) An oily secretion from sebaceous glands that lubricates and moisturizes the surface of the skin.

Second-class lever   A lever system in which the fulcrum is located farther from the applied force, with the load moved in between.

Second messenger   A chemical formed inside a cell that triggers some sort of change within the cell.

Secondary active transport   A type of active transport in which a primary active transport pump establishes a concentration gradient; the potential energy of this gradient is then used to fuel the transport of a second substance against its concentration gradient.

Semen   (SEE-muhn) A mixture of sperm and fluids from the testes, seminal vesicles, prostate, and bulbourethral and urethral glands.

Semiconservative replication   A type of DNA replication in which each daughter cell receives one original strand and one newly replicated strand of DNA.

Semilunar valves   Valves located between the ventricles and the arteries into which they pump blood.

Seminal vesicles   (SEH-min-ul) Accessory glands in the male reproductive system that produce seminal fluid—a component of semen.

Seminiferous tubules   (seh′-men-IF-er-us) Tightly coiled tubes inside the testes that are the site of sperm production.

Senescence   (seh-NESS-ens) The period of life beginning in adulthood that involves degeneration of tissues and organs.

Sensible water loss   A water loss that is "sensed" or noticed; includes water lost to the urine and the feces.

Sensory division (afferent division)   The division of the peripheral nervous system that detects changes in the environment and transmits them to the central nervous system.

Sensory neuron (afferent neuron)   A neuron that detects changes in the environment and transmits this information to the central nervous system.

Sensory receptor   The specialized part of a neuron that detects changes in the internal and/or external environment.

Sensory transduction   The conversion of a sensory stimulus to an electrical signal.

Serous fluid   A thin, watery secretion from a serous membrane that lubricates an organ in a cavity within the serous membrane.

Serous membranes   (SEER-us) Thin sheets of tissue that envelop certain organs and produce serous fluid.

Serum   (SEER-um) A liquid consisting of plasma without clotting factors.

Sesamoid bone   (SEH-suh-moyd) A bone located within a tendon.

Set point   The normal range of values of a regulated variable.

Sex-linked traits   Traits coded for by alleles that are located on the X or Y chromosome.

Short bone   A bone that is about as long as it is wide.

Simple columnar epithelium   A single layer of columnar epithelial cells and the underlying basal lamina; found in locations such as lining the digestive tract and many other hollow organs.

Simple cuboidal epithelium   A single layer of cuboidal epithelial cells and the underlying basal lamina; found in locations such as the kidney tubules, many endocrine glands, and the ducts of many exocrine glands.

Simple diffusion   A type of diffusion in which solutes move through a membrane or fluid from an area of high solute concentration to an area of low solute concentration without the aid of a protein channel or carrier.

Simple epithelia   A type of epithelial tissue composed of a single layer of epithelial cells and the underlying basal lamina.

Simple squamous epithelium   A single layer of squamous epithelial cells and the underlying basal lamina; found in locations such as the alveoli of the lungs, the kidneys, and lining the inner surface of blood vessels.

Simple stretch reflex   A monosynaptic reflex triggered by muscle stretch, which produces an automatic contraction of the muscle to counter the stretch.

Single-unit smooth muscle (unitary smooth muscle)   Smooth muscle cells that contract together as a single unit.

Sinoatrial node (SA node)   (sy′-noh-AY-tree-ul) A cluster of pacemaker cells in the superior right atrium that normally sets the pace for the heart.

Sinus rhythm   The normal electrical rhythm of the heart generated by the sinoatrial node.

Sinusoidal capillaries   Capillaries whose cells have large pores and gaps between them to permit large volumes of fluids, solutes, and cells to be exchanged.

Sister chromatids   (KROH-mah-tidz) The two identical copies of a replicated chromosome.

Skeletal muscle   A tissue type composed of multinucleate skeletal muscle cells and the surrounding endomysium.

Skeletal system   The body system that consists of the bones and their associated skeletal cartilages.

Skeleton   The complete set of bones of the skeletal system; perform the common functions of protection, providing mechanical levers for movement, support, and housing hematopoietic tissue.

Skin   An organ consisting of the superficial epidermis and deep dermis.

Skull The superior portion of the axial skeleton that houses the brain and special sense organs; consists of cranial and facial bones.

Sleep A reversible and normal suspension of consciousness.

Sliding-filament mechanism The mechanism of contraction of a muscle cell in which the thin and thick filaments slide past one another while generating tension.

Slow-twitch fibers Skeletal muscle fibers with low myosin ATPase activity that proceed relatively slowly through their cross-bridge cycles; generate slower but generally longer-lasting contractions.

Small intestine A tubular organ in the abdominopelvic cavity in which ingested food is digested, absorbed, and propelled through the remainder of the alimentary canal.

Smooth endoplasmic reticulum (SER) A series of winding membranes whose surfaces are not studded with ribosomes; functions in detoxification reactions, calcium ion storage, and lipid synthesis.

Smooth muscle A tissue type composed of flattened, spindle-shaped, uninucleate smooth muscle cells and the surrounding endomysium.

Sodium-potassium pump (Na⁺/K⁺ pump) A primary active transport pump consisting of a protein ATPase enzyme in the plasma membrane that pumps three Na^+ from the cytosol into the extracellular fluid and two K^+ from the extracellular fluid into the cytosol.

Solute (SAWL-yoot) The component present in a solution in lower amounts that is dissolved in the solvent; often a solid substance.

Solution A mixture in which the substances are in the same phase and the particles will not settle out when left undisturbed.

Solvent The component present in a solution in higher amounts; generally a liquid that dissolves a solute.

Somatic motor division A subdivision of the PNS that provides motor innervation to the skeletal muscles.

Somatic sensory division A subdivision of the PNS that provides sensory innervation to the skin, muscles, and joints.

Somatostatin (soh′-mah-toh-STAH-tin) Hormone produced by the hypothalamus that inhibits the release of growth hormone from the anterior pituitary gland.

Somatotopy A term that refers to the organization of the primary somatosensory cortex in which each part of the body is represented by a specific region in the cortex.

Somites (SOH-mytz) Blocklike structures of the mesoderm that develop into specific structures.

Spatial summation The additive effect of excitatory postsynaptic potentials triggered by multiple presynaptic neurons firing action potentials simultaneously.

Sperm Male gametes.

Spermatic cord Supporting structure of the male reproductive system extending from the scrotum to the inguinal canal.

Spermatogenesis (sper-mat′-oh-JEN-uh-sis) The process of sperm cell development.

Spermiogenesis (sper-mee′-oh-JEN-uh-sis) The maturation process of sperm.

Sphincter muscle A muscle (either skeletal or smooth) whose circular arrangement around an opening controls passage of materials or protects structures found at that opening.

Spinal cord The organ of the central nervous system that connects the brain with the peripheral nervous system and performs certain integrative functions.

Spinal nerve A short nerve that results from the fusion of the anterior and posterior roots of the spinal cord.

Spinothalamic tracts The largest white matter tracts of the anterolateral system in the spinal cord.

Spinous process A posterior projection from a cervical, thoracic, or lumbar vertebra.

Spiral organ (organ of Corti) A cluster of receptor cells and supporting cells at the boundary between the scala tympani and the cochlear duct that transduce the energy from sound waves into action potentials.

Spirometer (spy-RAHM-eh-ter) A laboratory instrument that measures volumes of air exchanged with ventilation.

Splanchnic circulation (SPLANK-nik) The collection of blood vessels that supplies and drains the digestive organs in the abdominopelvic cavity.

Splanchnic nerves Sympathetic nerves that synapse on collateral ganglia in the abdominopelvic cavity.

Spleen The large lymphoid organ in the left upper quadrant of the abdominopelvic cavity that filters blood and processes old erythrocytes.

Spongy bone The bone tissue located inside a bone; composed of small spicules of bone called trabeculae.

Spongy urethra The portion of the male urethra that passes through the corpus spongiosum of the penis.

Squamous cell (SKWAY-muss) An epithelial cell with a flat shape.

Static equilibrium The ability to maintain balance when the head and body are not moving.

Stem cell A cell that is undifferentiated and may develop into other cell types.

Sternum The central bone of the thoracic cage to which ribs attach.

Steroid A lipid consisting of a core of four hydrocarbon rings attached to other chemical groups.

Steroid hormones Chemical messengers derived from cholesterol, with a core of hydrocarbon rings; steroid hormones are hydrophobic and so interact with plasma membrane or intracellular receptors.

Stomach The J-shaped organ in the abdominopelvic cavity that is responsible for chemical and mechanical digestion and propulsion of ingested food.

Stratified columnar epithelium Several layers of cuboidal epithelial cells and the underlying basal lamina; lines the ducts of certain exocrine glands.

Stratified cuboidal epithelium Several layers of cuboidal epithelial cells and the underlying basal lamina; lines the ducts of certain exocrine glands.

Stratified epithelia A type of epithelial tissue composed of two or more layers of epithelial cells and the underlying basal lamina.

Stratified squamous epithelium Several layers of squamous epithelial cells and the underlying basal lamina; located in the oral cavity and esophagus, the anus, and vagina.

Stress response The series of physiological changes that maintains homeostasis when the body is faced with a stressor; mediated largely by the sympathetic nervous system and cortisol.

Striations (stry-AY-shunz) Alternating light and dark bands seen in skeletal and cardiac muscle cells.

Stroke volume (SV) The volume of blood ejected by a ventricle in one heartbeat.

Subarachnoid space The space between the arachnoid mater and the pia mater.

Subcutaneous tissue The adipose and other tissue deep to the skin.

Subdural space The very narrow space between the dura mater and the arachnoid mater.

Sublingual glands (sub-LING-gwuhl) Salivary glands located under the tongue that secrete mucus-rich saliva.

Submandibular glands Small salivary glands that are located on the medial side of the mandible.

Substance P A neuropeptide involved in the perception of pain.

Substantia nigra (sub-STAN-chah NY-grah) A nucleus in the midbrain of the brainstem that works with the basal nuclei to control and initiate movement.

Substrate-level phosphorylation The formation of ATP by transferring a phosphate group directly from a phosphate-containing compound to ADP.

Sulcus A shallow groove on the surface of the brain that separates two gyri.

Superficial A position that is closer to the surface of the body.

Superior Toward the head; also known as *cranial.*

Superior vena cava The large vein that drains most structures superior to the diaphragm; empties into the right atrium.

Surface anatomy The study of the surface markings of the human body.

Surface tension A visible film on the top of a water-based solution resulting from the formation of hydrogen bonds between adjacent water molecules.

Surfactant An amphiphilic chemical produced by type II alveolar cells that disrupts hydrogen bonds between water molecules and thus reduces surface tension.

Suspension A mixture in which the substances are in different phases and the particles settle out when left undisturbed.

Sustentacular cells (sus′-ten-TAK-yoo-luhr) Supporting cells that separate the forming sperm cells from the immune system.

Suture (SOO-cher) A fibrous joint structurally and a synarthrosis functionally that is found between the bones of the skull.

Sweat glands (sudoriferous glands) (soo′-doh-RIF-er-us) Glands in the skin that produce the watery secretion known as sweat.

Sympathetic chain ganglia A collection of autonomic ganglia located in a chain along the vertebral column.

Sympathetic nervous system "Fight or flight" division of the ANS that prepares the body for exercise, emergency, and emotion.

Symphysis (SIM-fuh-sis) A cartilaginous joint structurally and an amphiarthrosis functionally in which the articulating bones are connected by a fibrocartilaginous pad.

Synapse (SIN-aps) The location where a presynaptic neuron communicates with its target cell.

Synapsis (sih-NAP-sis) The pairing up of homologous chromosomes to form tetrads during meiosis.

Synaptic cleft The small space between the axon terminal of a presynaptic neuron and its target cell.

Synaptic delay The short delay between the arrival of the action potential at the axon terminal of a presynaptic neuron and the postsynaptic potential of a postsynaptic neuron.

Synaptic transmission The process by which a presynaptic neuron communicates with a postsynaptic neuron through either chemical or electrical synapses.

Synaptic vesicle A membrane-enclosed structure in an axon terminal that contains neurotransmitters.

Synarthrosis (sin-ahr-THROH-sis) A functional classification of articulations in which the joint allows no motion between bones.

Synchondrosis (sin-kahn-DROH-sis) A cartilaginous joint structurally and a synarthrosis functionally in which the articulating bones are held together by hyaline cartilage.

Syndesmosis (sin-dez-MOH-sis) A fibrous joint structurally and an amphiarthrosis functionally in which the articulating bones are joined by an interosseous membrane or ligament.

Synergist A muscle that works together with the agonist to make the movement more efficient and smooth.

Synovial cavity A unique structural element of a synovial joint characterized by a fluid-filled space between the articulating bones of a synovial joint.

Synovial fluid The fluid secreted by the synovial membrane to provide lubrication, shock absorption, and nutrients to the articulating bones and articular cartilage as well as providing a medium in which metabolic functions can occur.

Synovial joints (sih-NOH-vee-uhl) A structural classification of articulations in which the joints have a fluid-filled synovial cavity between articulating bones that are enclosed within an articular capsule.

Synovial membrane The inner layer of the articular capsule that secretes synovial fluid into the synovial cavity.

Systemic anatomy The study of the human body taken from the perspective of individual organ systems.

Systole (SIS-toh-lee) The period during which the heart is contracting.

Systolic pressure The blood pressure in the arteries when the ventricles are in systole.

T lymphocyte The leukocyte responsible for the cell-mediated arm of adaptive immunity.

T wave The third wave on an ECG; represents repolarization of the ventricles.

Tactile corpuscle A sensory receptor located in the dermal papillae of the skin that detects light touch stimuli.

Tactile senses The senses that pertain to fine or discriminative touch, including vibration, two-point discrimination, and light touch.

Talus (TAYL-us) The tarsal bone that articulates with the tibia and fibula at the ankle joint.

Tarsals Short bones of the ankle and foot.

Taste buds Clusters of receptor cells and supporting cells on the tongue that detect taste sensations.

Telodendria (tee′-loh-DEN-dree-uh) Small branches at the end of an axon.

Telophase (TEEL-uh-fayz) The final phase of mitosis during which the cell is cleaved into two daughter cells and the nuclear membranes re-form.

Temporal lobes Cerebral lobes located on the lateral surface of the cerebrum; perform functions related to hearing, language, memory, and emotions.

Temporal summation The additive effect of excitatory postsynaptic potentials triggered by a single presynaptic neuron that fires action potentials in rapid succession.

Terminal cisternae (sis-TER-nee) Enlarged regions of the sarcoplasmic reticulum where it contacts T-tubules.

Testes Male gonads that are the site of sperm and testosterone production.

Testosterone Hormone that stimulates spermatogenesis and the development of male sex characteristics.

Thalamus (THAL-uh-muss) The central and largest component of the diencephalon of the brain, consisting of two egg-shaped masses of gray matter; edits and sorts information entering the cerebrum.

Thermoreceptor A sensory receptor that responds to thermal stimuli.

Thermoregulation The process by which the body maintains a relatively constant internal temperature using several negative feedback loops.

Thick filament A myofilament composed of many molecules of the protein myosin.

Thick skin Skin that has five epidermal layers and in which the epidermal layers, particularly the stratum corneum, are generally quite thick; located on the palms of the hands and the soles of the feet.

Thigh The proximal portion of the lower limb that consists of the femur.

Thin filament A myofilament composed of molecules of the proteins actin, troponin, and tropomyosin.

Thin skin Skin that lacks a stratum lucidum and has thinner epidermal layers; located everywhere on the body except the palms of the hands and the soles of the feet.

Third-class lever A lever system in which the applied force is closer to the fulcrum, with the load moved farther away from the fulcrum.

Third ventricle A hollow, cerebrospinal fluid–filled cavity located in the center of the diencephalon.

Thirst mechanism The main mechanism that drives fluid intake; controlled by the hypothalamus.

Thoracic cage A bony "cage" that encases the thoracic cavity; consists of the sternum, twelve pairs of ribs, and the thoracic vertebrae.

Thoracic cavity (thoh-RAH-sik) A ventral body cavity located within the area superior to the diaphragm that encompasses the area of the thorax.

Thoracic vertebrae Twelve bones of the vertebral column that articulate laterally with the ribs and are located inferior to the cervical vertebrae and superior to the lumbar vertebrae.

Thrombin The active form of an enzyme that catalyzes the reaction converting fibrinogen to its active form, fibrin.

Thrombolysis (thrahm-BAH-luh-sis) The process by which a blood clot or thrombus is degraded.

Thrombus A blood clot; a collection of platelets, erythrocytes, fibrin, and other clotting proteins that prevents further loss of blood from a blood vessel.

Thymine A pyrimidine nitrogenous base found in DNA.

Thymus An endocrine gland in the anterior mediastinum that houses developing T lymphocytes and secretes paracrines that stimulate their maturation.

Thyroid cartilage The broad, shield-shaped cartilage framing the anterosuperior portion of the larynx.

Thyroid follicle The spherical subunit of the thyroid gland in which thyroid hormones are produced and stored.

Thyroid gland An endocrine gland located in the anterior neck that produces thyroid hormones and calcitonin.

Thyroid-stimulating hormone (TSH) The anterior pituitary hormone that stimulates the production and release of thyroid hormone and growth of the thyroid gland.

Thyroxine (T_4) (thy-RAWKS-een) A hormone produced by the thyroid gland that contains four iodine atoms; less active form of the hormone, which can be converted into T_3 by cells when needed.

Tibia (TIB-ee-uh) The medial leg bone.

Tight junctions Intercellular junctions that hold cells tightly together in a tissue and make the space between them relatively impermeable.

Tissue A group of structurally and functionally related cells and their extracellular matrix.

Tissue gas exchange The exchange of oxygen and carbon dioxide between the blood in the systemic capillaries and the tissue cells.

Tissue perfusion The amount of blood flowing to a tissue through a capillary bed.

Tissue plasminogen activator (tPA) (plaz-MIN-noh-jen) An enzyme produced by healed endothelial cells that catalyzes the activation of plasminogen to plasmin.

Tissue repair The process by which tissue damage is repaired.

Tonicity (toh-NISS-ih-tee) A comparison between the ability of two solutions separated by a selectively permeable membrane to cause water movement by osmosis.

Tonsils Specialized clusters of lymphatic tissue located around the oral and nasal cavities.

Total body water The total amount of water in the intracellular and extracellular compartments of the body.

Trabeculae (trah-BEK-yoo-lee) Spicules of bone that make up spongy bone tissue.

Trachea (TRAY-kee-uh) The hollow organ that conducts air to and from the larynx and bronchial tree of the respiratory tract.

Tract A bundle of white matter (myelinated axons) in the central nervous system.

Transcription The process by which a specific gene is copied onto a complementary strand of messenger RNA during protein synthesis.

Transcytosis A type of vesicular transport in which a substance is taken into one side of the cell by endocytosis and released from the other side of the cell by exocytosis.

Transduction The process of converting energy in the form of light, sound, movement, or touch into a neural signal.

Transfer RNA (tRNA) RNAs that contain an amino acid and an anticodon that bonds to a specific mRNA codon.

Transitional epithelium Several layers of epithelial cells and the underlying basal lamina found in the organs of the urinary system; cells can change shape from dome-shaped to squamous when stretched.

Translation The process by which a messenger RNA transcript is read by a ribosome and an amino acid string is synthesized.

Transverse plane A plane of section that divides the body or body part into superior and inferior or proximal and distal portions; also known as the *horizontal plane* or *cross-section*.

Transverse tubules (T-tubules) Hollow inward extensions of the muscle fiber sarcolemma that surround myofibrils; filled with extracellular fluid.

Triad A T-tubule and two adjacent terminal cisternae in a muscle fiber.

Tricuspid valve Atrioventricular valve located between the right atrium and right ventricle.

Triiodothyronine (T_3) (try′-aye-oh-doh-THY-roh-neen) A hormone produced by the thyroid gland that contains three iodine atoms; form of the hormone that has higher physiological activity in promoting growth and development, regulating the metabolic rate and heat production, and promoting the effects of the sympathetic nervous system on its target cells.

Trophoblast cells (TROHF-oh-blast) Large flattened cells of the blastocyst that participate in forming the placenta.

Tropomyosin (trohp′-oh-MY-oh-sin) A filamentous regulatory protein that covers the active sites of actin subunits in a thin filament.

Troponin (TROH-poh-nin) A regulatory protein with three subunits that binds tropomyosin and calcium ions in a thin filament.

True rib A rib that attaches to the sternum by its own costal cartilage.

Tubular reabsorption The process by which water and solutes are reclaimed from the filtrate in the renal tubules and collecting systems and returned to the blood.

Tubular secretion The process by which solutes are removed from the blood and placed into the filtrate in the renal tubules for excretion from the body.

Tubuloglomerular feedback (too′-byoo-loh-gloh-MEHR-yoo-luhr) A negative feedback mechanism in which the macula densa and JG cells regulate the GFR.

Tunica externa The outer layer of a blood vessel's wall; composed of fibrous connective tissue to support the vessel.

Tunica intima The inner layer of a blood vessel's wall; consists of endothelium, a basal lamina, and a varying number of elastic fibers.

Tunica media The middle layer of a blood vessel's wall that consists largely of smooth muscle.

Tympanic membrane (tim-PAN-ik) The membrane that separates the external auditory canal from the middle ear; vibrates when struck by sound waves and transmits the vibrations to the auditory ossicles.

Ulna The medial bone of the forearm.

Umbilical cord Structure that connects the center of the placenta to the umbilicus of the embryo or fetus; contains two umbilical arteries and one umbilical vein.

Unfused tetanus (incomplete tetanus) A type of wave summation in which a muscle fiber is stimulated rapidly and only allowed to partially relax between contractions.

Uniaxial joints (yoo′-nee-AK-see-uhl) A functional classification of synovial joints in which motion between articulating bones occurs around only one axis.

Universal donor An individual with blood type O– who is able to donate blood to any blood type due to the absence of A, B, and Rh antigens on his or her erythrocytes.

Universal recipient An individual with blood type AB+ who is able to receive donated blood from any blood type due to the absence of anti-A, anti-B, and anti-Rh antibodies in his or her plasma.

Unsaturated fatty acid A fatty acid with one or more double bonds between its carbon atoms.

Upper limb The portion of the appendicular skeleton that consists of the arm, forearm, wrist, and hand.

Upper motor neurons Interneurons located in the frontal lobe of the cerebral cortex that are involved in the conscious planning of movement; initiate movement via lower motor neurons of the PNS.

Uracil A pyrimidine nitrogenous base found in RNA in place of thymine.

Urea cycle The process that combines two molecules of ammonia with carbon dioxide to produce urea, which is eliminated by the kidneys.

Ureters (YOOR-eh-terz) Muscular tubes that transport urine from the kidneys to the urinary bladder.

Urethra (yoo-REE-thrah) The final segment of the urinary tract through which urine is voided.

Urinalysis A test performed on the urine to analyze its volume, color, and chemical characteristics.

Urinary bladder The hollow organ of the urinary tract that stores urine and contracts to expel it from the body.

Urinary system The body system consisting of the kidneys, ureters, urinary bladder, and urethra; filters the blood of wastes and works to maintain fluid, electrolyte, and acid-base homeostasis.

Urinary tract The organs of the urinary system (ureters, urinary bladder, and urethra) that transport, store, and eliminate urine.

Urine A liquid produced by the kidneys that consists of water, electrolytes, and metabolic wastes.

Uterine cycle (menstrual cycle) The series of cyclic changes that the uterine endometrium goes through each month as it responds to the fluctuating levels of ovarian hormones.

Uterine tubes (fallopian tubes) Tubes that transmit an oocyte from the ovaries to the uterus.

Uterus (YOO-ter-uhs) The organ in which a conceptus implants and develops until birth.

Uvula (YOO-vyoo-luh) A portion of the soft palate suspended in the posterior oral cavity; seals off the nasopharynx during swallowing to prevent food and liquid from entering the nasal cavity.

Vaccination The administration of an antigen to elicit a primary immune response and generate memory cells.

Vagina (vuh-JY-nuh) The female organ of copulation that receives the penis and semen during sexual intercourse; it is also the passageway for delivery of an infant and for menstrual flow.

Valence electrons The electrons in an atom's outer shell; involved in chemical reactions and the formation of chemical bonds.

Vasa recta (VAY-zuh REK-tuh) Capillary beds that surround the nephron loops of juxtamedullary nephrons.

Vasa vasora (VAY-zuh vay-ZOH-ruh) Small blood vessels that supply the walls of larger blood vessels.

Vascular anastomoses (an-ass'-toh-MOH-seez) Locations where blood vessels connect by collateral vessels.

Vasculature (VASS-kyoo-luh-chur) The collective term for the body's blood vessels.

Vasoconstriction The narrowing of a blood vessel due to contraction of vascular smooth muscle.

Vasodilation The opening of a blood vessel due to relaxation of vascular smooth muscle.

Veins The vasculature's collection system; systemic veins collect deoxygenated blood from systemic capillaries and return it to the right side of the heart, and pulmonary veins collect oxygenated blood from pulmonary capillaries and transport it to the left side of the heart.

Venous valves Flaps of endothelium in certain veins that prevent blood from flowing backward.

Ventilation/perfusion (V/Q) ratio The ratio of pulmonary ventilation of a segment of the lung to its blood flow (perfusion).

Ventricles (of brain) (VEN-trih-kulz) Cerebrospinal fluid–filled cavities located within the brain.

Ventricles (of heart) The inferior chambers of the heart that receive blood from the atria and pump blood into arteries.

Ventricular ejection phase The period of the cardiac cycle during which blood is ejected from the ventricles into the pulmonary trunk and aorta.

Ventricular filling phase The period of the cardiac cycle during which blood drains from the atria into the ventricles.

Venules Small veins that drain blood from capillary beds.

Vermiform appendix (VER-muh-form uh-PEN-diks) A blind-ended extension from the cecum of the large intestine that contains lymphatic nodules.

Vermis (VER-miss) The structure that connects the two lobes of the cerebellum.

Vertebral arteries Arteries supplying the brain that travel through the transverse foramina of the cervical vertebrae.

Vertebral cavity A posterior body cavity located within the vertebral column that houses the spinal cord.

Vertebral column A skeletal structure consisting of 24 individual vertebrae, 5 fused vertebrae of the sacrum, and 3–5 fused vertebrae of the coccyx; together encase the spinal cord and nerves within the vertebral cavity.

Very low-density lipoproteins (VLDLs) Triglyceride-rich particles that deliver triglycerides from the liver to other tissues.

Vesicle (VESS-ih-kul) A small, membrane-enclosed body in the cytosol that generally functions in transport or storage.

Vesicular transport (vess-IK-yoo-lur) A type of active transport in which materials are moved into and out of the cell by vesicles.

Vestibule The bony portion of the inner ear involved in detecting head position and linear movement.

Villi (VILL-aye; singular, *villus*) Folds of the mucosa and submucosa of the small intestine that increase surface area for absorption.

Visceral reflex arc A series of events in which a sensory stimulus in an organ leads to a predictable visceral motor response mediated by the ANS.

Viscosity The thickness of a liquid or the inherent resistance all liquids have to flow.

Visual field The area one can observe with one or both eyes when focusing on a central point.

Vital capacity The total volume of exchangeable air within the lungs; equal to the tidal volume plus the expiratory and inspiratory reserve volumes.

Vitamin An organic compound required in small amounts for specific functions within the body.

Vitreous humor The thick fluid located in the posterior cavity of the eyeball.

Vocal folds A set of elastic mucosal folds in the larynx that vibrate to produce sound.

Volatile acid An acid that can be converted to a gaseous form that may be excreted through the lungs.

Voltage The difference in electrical potential between two points.

Voltage-gated channel An ion channel in the plasma membrane of a cell that opens or closes in response to a change in the membrane potential.

Wave summation A phenomenon in which repeated stimulation of a muscle fiber by a motor neuron results in muscle twitches with progressively greater tension.

Wernicke's area (VER-nik-eez) An integrative multimodal association area for speech located in the temporal and parietal lobes; responsible for the ability to understand and produce intelligible language.

White matter Collections of myelinated axons that appear white.

Wrist The portion of the upper limb that articulates proximally with the forearm and distally with the hand.

Xiphoid process (ZY-foyd) The inferior portion of the sternum.

Yellow bone marrow Adipose tissue located within most bones of an adult.

Yolk sac The first extraembryonic membrane; forms part of the digestive tract and is the source of the first blood cells, blood vessels, and the first germ cells.

Z-disc The component of the sarcomere that contains structural proteins to which thin and elastic filaments attach.

Zygote The earliest stage of human development; consists of a single cell formed from the fusion of an ovum and a sperm cell.

Index

Prefixes, Suffixes, Word Roots, and Combining Forms

Many of the words we use in everyday English have their roots in other languages, particularly Greek and Latin. This is especially true for anatomical terms. The lists that follow include some of the foreign prefixes, suffixes, word roots, and combining forms that are part of many of the biological and anatomical terms used in this text.

PREFIXES

a-, an-, without or lacking: *anesthesia* = without feeling, particularly pain

ab-, away from, departing: *abduct* = move away from midline

ad-, to or toward: *adduct* = move toward midline

af-, toward: *afferent neurons* = neurons that send stimuli toward the CNS

ante-, before: *antebrachium* = the forearm (or the region before the arm)

anti-, against: *antidiuretic hormone* = decreases urine output, working against diuresis

auto-, self: *autoimmune disease* = inappropriate immune response against the body's own tissues

bi-, two: *biceps* = a muscle with two heads or attachment sections

brady-, slow: *bradycardia* = slower than normal heart rate

circum-, around: *circumduction* = circular movement around a point

co-, con-, with, together: *coenzyme* = molecule that temporarily combines with an enzyme to help it function

contra-, against, opposite: *contralateral* = on the opposite side of the body from the point of reference

cyan-, blue: *cyanotic* = bluish color of skin due to hypoxia

de-, reversal, loss, undoing: *depolarization* = reversal of membrane polarity

di-, two: *disaccharide* = molecule made of two joined simple sugars

dia-, across, through: *dialysis* = diffusion across a membrane

dys-, difficult, painful: *dysmenorrhea* = painful menstruation

ecto-, out, away from: *ectoderm* = outer layer of embryonic tissue

ectop-, displaced: *ectopic pregnancy* = embryo implants outside the uterus

ef-, away from: *efferent neurons* = neurons that send stimuli away from the CNS

em-, en-, in, inside: *embolus* = abnormal object moving within blood vessels

endo-, within, inner: *endocytosis* = taking larger particles into the cell

epi-, over, upon: *epidermis* = outermost layer of the skin over the dermis

eryth-, red: *erythrocyte* = red blood cell

eu-, well, true: *eukaryotic cell* = cell with a true membrane-bounded nucleus

ex-, exo-, outer, outside: *exocytosis* = moving larger particles out of the cell

extra-, beyond, outside: *extracellular fluid* = fluid outside of most body cells

hemi-, half: *hemisphere* = left or right half of the brain

hetero-, different: *heterozygous* = having two different alleles for the same trait

homo-, same, equal: *homozygous* = having two identical alleles for the same trait

hyal-, clear: *hyaline cartilage* = type of cartilage with a fairly clear matrix

hyper-, over, excessive: *hypertension* = higher than normal blood pressure

hypo-, below, under: *hypotension* = lower than normal blood pressure

juxta-, next to, near: *juxtaglomerular* = near the glomerulus of the kidney

in-, im-, not, unable: *involuntary* = not under the conscious control of the brain

infra-, beneath: *infraorbital* = below the eye socket

inter-, between: *intercostal* = between the ribs

intra-, within: *intravenous* = within a vein

iso-, same: *isotonic* = having the same osmotic pressure

leuk-, white: *leukocyte* = white blood cell

macro-, large: *macrophage* = large phagocytic cell

mal-, bad: *malnutrition* = poor nutrition

medi-, middle: *mediastinum* = middle portion of the thoracic cavity

mega-, large: *megakaryocyte* = large cell that produces platelets

melan-, black: *melanocyte* = cell that produces dark pigment molecules

mes-, middle: *mesoderm* = middle layer of embryonic tissue

meta-, after, beyond: *metastasis* = spread of cancer beyond its point of origin

micro-, small: *microvilli* = small projections of the cell's plasma membrane

milli-, thousandth: *millimeter* = one thousandth of a meter

mono-, single: *monosaccharide* = a single simple sugar molecule

neo-, new: *neonatal* = referring to a newborn infant up to 4 weeks of age

non-, not: *nonpolar covalent bond* = bond with no poles or charged areas

oligo-, few: *oligosaccharide* = smaller carbohydrates made of three (or a few more) simple sugars

ortho-, straight, correct: *orthopedics* = medical field related to correcting musculoskeletal conditions

para-, near, at the side: *parathyroid glands* = glands on the back side of the thyroid

per-, through: *perforin* = protein that produces a small hole through the plasma membrane of a cell targeted for destruction

peri-, surrounding, around: *perimysium* = connective tissue wrapping around a fascicle (bundle) of muscle cells

poly-, many: *polysaccharide* = molecule made of many joined simple sugars

post-, after: *postpartum* = after childbirth

pre-, before: *presynaptic* = before the synapse of two neurons

presby-, old: *presbyopia* = age-related farsightedness due to loss of lens elasticity

pro-, first, promoting: *prolactin* = hormone that promotes mammary gland milk production

quad-, four: *quadriceps* = a muscle with four heads or attachment sections

re-, returning, doing again: *reflux* = return flow of fluid (e.g., esophageal)

retro-, behind: *retroperitoneal* = organs behind the abdominal cavity membranes

semi-, half: *semilunar valve* = valve with a half-moon shape

sub-, below, under: *submandibular* = below the mandible

super-, supra-, above, excessive: *supraspinatus* = muscle above the spine of the scapula

syn-, together: *synapse* = area where two neurons come together

tachy-, fast, rapid: *tachycardia* = faster than normal heart rate

trans-, through, across: *transversus abdominis* = muscle with fibers that run across the abdominal cavity

tri-, three: *triglyceride* = a lipid containing three fatty acid molecules

ultra-, beyond: *ultraviolet* = wavelength of light beyond the range of normal human vision

xanth-, yellow: *xanthosis* = yellowish discoloration of tissue (not the same as jaundice)

SUFFIXES

-al, -ac, pertaining to: *menopausal* = pertaining to menopause in women

-algia, pain: *neuralgia* = sharp pain caused by nerve damage

-aps, juncture: *synapse* = junction of two neurons

-arche, beginning: *menarche* = start of menstrual cycles in girls

-ary, associated with, referring to: *dietary* = associated with food and nutrition

-ase, suffix on names of most enzymes: *acetylcholinesterase* = enzyme that catalyzes the breakdown of acetylcholine

-asthenia, weakness: *myasthenia* = disease that produces muscle weakness

-blast, sprout, immature cell: *osteoblast* = bone cell that builds bone tissue

-centesis, draining: *amniocentesis* = removal of a sample of amniotic fluid

-cide, kill: *bactericide* = a chemical that kills bacteria

-clast, -clasia, breaking down: *osteoclast* = bone cell that breaks down bone tissue

-crine, secrete, release: *endocrine* = gland that releases hormones into the blood

-ectomy, -tomy, to excise surgically: *nephrectomy* = removal of the kidney

-ell, -elle, small: *organelles* = small structures in cells performing specific functions

-emia, condition of the blood: *leukemia* = abnormal development of leukocyte-producing bone marrow tissue

-fferent, carry: *efferent arteriole* = carries blood away from the glomerulus of the kidney

-form, -forma, shape: *lentiform nucleus* = lens-shaped area of gray matter in the brain

-gen, initiating or causing agent: *allergen* = substance capable of evoking an allergic response

-genesis, production, creation: *gluconeogenesis* = production of glucose molecules

-gram, recorded data: *electrocardiogram* = a graphic report of impulse transmission through the heart's conduction system

-graph, instrument that records data: *electrocardiograph* = machine that records impulse transmission through the heart's conduction system

-ia, -sia, condition of, process of: *hypoxia* = delivery of too little oxygen to the body or a tissue

-iasis, abnormal condition of: *mydriasis* = potentially abnormal dilation of the pupil

-ic, -ac, pertaining to: *pancreatic* = pertaining to the pancreas

-ism, condition of: *hypothyroidism* = condition in which one produces too little thyroid hormone

-itis, inflammation of: *arthritis* = inflammation of the joints

-lemma, husk, shell, covering: *neurilemma* = covering around axons of neurons

-lith, stone: *otoliths* = crystalline structures resembling microscopic pebbles in the inner ear

-lysis, splitting: *hemolysis* = rupture of erythrocytes

-malacia, softening: *osteomalacia* = a disorder such as rickets that produces abnormal bone softening

-megaly, growing, enlarged: *acromegaly* = excessive thickening of bones

-metric, -metry, measurement, length: *spirometry* = measuring lung volumes

-odyn, pain, painful: *ophthalmodynia* = pain centered around the eyes

-oid, like or resembling: *lipoid* = a chemical resembling a lipid in composition

-ology, study of: *histology* = study of tissues

-oma, tumor: *melanoma* = skin tumor originating in melanocytes

-opia, vision: *myopia* = nearsightedness

-ory, pertaining to: *inflammatory* = pertaining to inflammation

-osis, condition of: *osteoporosis* = condition in which bones lose density and weaken

-pathy, disease state: *cardiomyopathy* = disease of heart muscle

-penia, lack of: *leukopenia* = low white blood cell count

-phagy, eating: *geophagy* = compulsion to eat soil, possibly driven by mineral deficiencies

-phasia, speech: *aphasia* = loss of the ability to speak

-philic, liking, loving: *hydrophilic* = describes molecules that are attracted to and freely dissolve in water

-phob, fearing, disliking: *acrophobia* = fear of heights

-plasia, growth: *neoplasia* = new abnormal cell growth

-plasm, matter or substance: *cytoplasm* = liquid substance inside the cell

-plasty, reconstruction of a part: *mammoplasty* = breast reshaping or reconstruction after surgery

-plegia, paralysis: *paraplegia* = paralysis of both legs

-pnea, breathing: *apnea* = absence of respiration

-rrhagia, -rrhage, discharge, break out: *hemorrhage* = excessive bleeding

-rrhea, flow: *diarrhea* = frequent excess flow of watery loose stools

-some, -soma, body: *phagosome* = vesicle containing a particle that a cell has "eaten," or brought in via endocytosis

-spire, breathing: *inspire* = to breathe in

-stalsis, compression: *peristalsis* = wavelike squeezing of tubular organs

-stasis, to stop, arrest: *hemostasis* = process of stopping blood loss

-tensin, -tension, pressure: *hypertension* = high blood pressure

-tomy, to cut: *mastectomy* = surgical removal of the breast

-tonic, pressure, tension: *isotonic* = same amount of osmotic pressure on both sides of a membrane

-trophy, growth: *hypertrophy* = excess growth of a tissue

-ty, state or condition of: *obesity* = having an abnormal amount of body fat

-ule, tiny, small: *tubule* = small tube (e.g., microscopic collecting tubules in the kidneys)

-uria, urine: *hematuria* = blood in the urine

WORD ROOTS WITH COMMON COMBINING VOWELS

acr/o-, limb: *acromegaly* = enlargement of the limbs

aden/o-, gland: *adenoma* = tumor of a gland

alve/o-, small cavity: *alveoloplasty* = surgical reconstruction of a dental alveolus (tooth socket)

ang/i-, vessel: *angiogram* = examination of a vessel

arter/i-, artery: *polyarteritis* = inflammation of many arteries

arthr/o-, joint, articulation: *arthralgia* = joint pain

bar/o-, pressure: *baroreceptor* = nerve cell that detects pressure

bil/i-, bile: *endobiliary* = within the bile duct

brach/i-, arm: *antebrachial* = relating to the forearm

bronch/o-, bronchus (air passage): *intrabronchial* = within the bronchus

carcin/o-, cancer: *carcinogen* = chemical or condition that can cause cancer

card/i-, heart: *pericardiocentesis* = draining of the space around the heart

cephal/o-, relating to the head: *cephalodynia* = head pain or headache

cerv/i-, neck: *cervicalgia* = neck pain

chol/e-, bile: *cholecystectomy* = removal of the bile-storing gallbladder

chondr/o-, cartilage: *chondrosarcoma* = tumor of the cartilaginous soft tissue

chrom/o-, color: *hypochromasia* = the condition of decreased color

corn/e-, horn: *corneocyte* = hard cell in the outer layer of the skin or a horn in a nonhuman animal

corp/u-, body: *corpuscle* = small body

cortic/o-, bark (outer part of a structure): *corticosteroid* = steroid hormone produced by the outer part of the adrenal gland

cran/i-, head, skull: *intracranial* = within the head or skull

cusp/i-, point: *tricuspid* = having three points or cusps

cutane/o-, skin: *percutaneous* = through the skin

cyst/o-, bladder: *cystolith* = stone in the urinary bladder

cyt/o-, cell: *cytology* = study of cells

dactyl/o-, digits (fingers, toes): *polydactyly* = condition of having more than the normal number of fingers and/or toes

dendr/o-, branch: *dendritic cell* = cell with extensions resembling the branches of a tree

dent/o-, dont/o-, tooth: *endodontic* = pertaining to the inner structures of the tooth

derm/o-, dermat/o-, skin: *intradermal* = within the skin

diastol/i-, standing between: *diastolic pressure* = blood pressure between beats of the heart

dips/i-, thirst: *polydipsia* = condition of elevated level of thirst

encephal/o, brain: *encephalitis* = inflammation within the brain

enter/o-, intestine: *enterohepatic* = pertaining to the intestines and the liver

esth/e-, feeling, sensation: *hyperesthesia* = excessive feeling or sensitivity

gastr/o-, stomach: *gastroesophageal* = pertaining to the stomach and esophagus

gest/u-, carried: *gestational* = pertaining to the period during which a fetus is carried by a pregnant woman

gingiv/o-, gums: *supragingival* = pertaining to the area superior to the gums

glom/o, glomerul/o-, ball: *glomerulonephritis* = inflammation of the glomerular capillaries of the kidney

gloss/o-, tongue: *microglossia* = abnormal smallness of the tongue

gluc/o-, glyc/o-, sugar: *gluconeogenesis* = formation of glucose molecules

hepat/o-, liver: *perihepatic* = around the liver

hist/o-, tissue: *histopathology* = study of disease states of tissues

hydr/o-, water: *dehydrate* = to decrease the amount of water in a fluid

hyster/o-, uterus: *hysterectomy* = surgical removal of the uterus

iatr/o-, treatment: *iatrogenic* = caused by treatment

inguin/o, groin: *retroinguinal* = behind the inguinal ligament in the groin

jugul/o-, throat: *jugulodigastric* = location in the throat near the digastric muscles

kal/e-, potassium: *hyperkalemia* = condition of elevated potassium ions in the blood

kary/o-, nucleus: *eukaryote* = cell with a true nucleus

kerat/o-, horn: *hyperkeratosis* = condition in which the outermost layer of the skin develops horny deposits

kin/e-, movement: *kinesiology* = study of movement

lab/i-, lip: *supralabial* = above the lip

lact/o-, milk: *lactorrhea* = excess or abnormal flow of milk

lacun/o-, space: *lacunae* = small spaces within bone tissue

lamell/o-, plate or thin sheet: *circumferential lamellae* = thin, circular plates of bone tissue

lamin/o-, layer: *laminar flow* = condition in which fluid flows in parallel layers

lapar/o-, abdomen: *laparotomy* = incision into the abdomen

lig/a-, tie, binding: *ligament* = structure that binds two bones together

lip/o-, fat (adipose): *lipolysis* = breakdown of fat

men/u, menstr/u-, monthly: *dysmenorrhea* = painful monthly discharge

metr/i-, uterus: *endometrium* = inner lining of the uterus

mut/a-, change: *mutagen* = agent generating a change in DNA

my/o-, muscle: *cardiomyopathy* = disease of heart muscle cells

myel/o-, marrow: *myeloblast* = immature cell of the bone marrow

nat/a-, birth: *prenatal* = before birth

natr/i-, sodium: *hyponatremia* = condition of decreased sodium ions in the blood

necr/o-, death, dead: *necrosis* = condition of death

nephr/o-, kidney: *nephroptosis* = drooping of a kidney

neur/o-, nerve: *neurology* = study of the nervous system

noct/o, nyct/o-, night: *nocturia* = urination occurring at night

ocul/o-, eye: *monocular* = pertaining to one eye

onc/o-, cancer: *oncogene* = a gene involved in the pathogenesis of cancer

opt/i, ophthalm/o-, eye: *ophthalmoscope* = instrument for examining the eye

orchid/o-, testis: *orchidectomy* = removal of a testis

oste/o-, bone: *osteoblast* = immature bone cell

ot/o-, ear: *otoscope* = instrument for examining the ear

ov/u-, egg: *ovular* = pertaining to an egg, or ovum

path/o-, disease: *pathogenesis* = development of disease

ped/i-, child: *pediatrics* = medical specialty focusing on children

phag/o-, eat: *autophagy* = self eating; condition of a cell digesting itself

pharyng/o-, throat: *glossopharyngeal* = pertaining to the tongue and throat

phleb/o-, vein: *phlebotomy* = incision into a vein

phot/o-, light: *photophobia* = fear or dislike of light

phys/i-, function: *pathophysiology* = study of the altered physiology of disease states

pin/o-, drink: *pinocytosis* = condition of cell drinking

plex/u-, network, twisted: *nerve plexus* = network of nerves

pneum/o-, pneumat/o-, air, breath: *pneumothorax* = air in the thoracic cavity

pneumon/o-, lung: *pneumonitis* = inflammation of the lung

pod/o-, foot: *podiatrist* = medical specialist of the foot and ankle

proct/o-, rectum: *proctoscope* = instrument used to examine the rectum

pseud/o-, false: *pseudoanemia* = pallor of mucous membranes and skin without other signs of true anemia

psych/i-, mind: *psychopathology* = study of diseases of the mind

pulmon/o, lung: *intrapulmonary* = within the lung

pyel/o-, pelvis: *pyelonephritis* = inflammation of the renal pelvis

py/o-, pus: *pyoderma* = condition causing pus to form in the skin

pyr/o-, fire: *pyrolysis* = breakdown of a chemical by elevated temperature

ren/i-, kidney: *adrenal* = pertaining to an area next to the kidney

rhin/o-, nose: *rhinorrhea* = drainage from the nose

sarc/o-, flesh, soft tissue: *sarcolemmal* = pertaining to the outer covering of a muscle fiber (the plasma membrane)

scler/o-, hard: *scleroderma* = hardening of the skin

sigm/o-, S-shaped: *sigmoidoscopy* = examination of the S-shaped portion of the colon

sin/u-, cavity: *sinusoid* = resembling a sinus or cavity

son/o-, sound: *sonogram* = data recorded using sound waves

spir/o-, breathe: *spirometry* = measurement of breathing

stat/i-, to stop, standing still: *hydrostatic* = pertaining to fluids not in motion

systol/i-, contract, standing together: *systolic pressure* = blood pressure during the heart's contraction

therm/o-, temperature: *thermogenesis* = generation of heat

thromb/o-, clot: *thrombosis* = abnormal condition of a blood clot

tom/o-, cut, slice: *dermatome* = instrument that takes slices of skin

tox/o-, poison, toxin: *neurotoxin* = toxin causing damage to the nervous system

tympan/o-, drum: *tympanoplasty* = reconstruction of the eardrum

urin/o, urine: *oliguria* = production of little urine

vas/o-, vascul/o, vessel: *extravascular* = outside a blood vessel

vesic/o-, vesicul/o-, bladder, small sac: *vesiculotomy* = surgical incision of the seminal vesicle

viscer/o-, organ: *visceral* = pertaining to an organ

vit/a-, life: *vital* = pertaining to life